eText Reference

Algebra and Trigonometry

Third Edition

Kirk Trigsted
University of Idaho

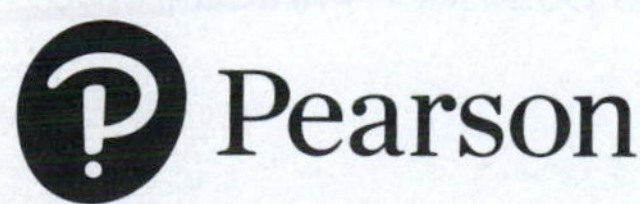

Director, Portfolio Management, Collegiate Math: Anne Kelly
Senior Editor: Dawn Murrin
Editorial Administrator: Joseph Colella
Manager, Courseware QA: Mary Durnwald
Vice President, Production and Digital Studio: Ruth Berry
Producer, Production and Digital Studio: Audra Walsh
Manager, Content Development: Kristina Evans
Managing Producer: Karen Wernholm
Content Producer: Ron Hampton
Product Marketing Director: Erin Kelly
Product Marketing Manager: Stacey Sveum
Product Marketing Assistant: Shannon McCormack
Field Marketing Manager: Peggy Lucas
Field Marketing Assistant: Adrianna Valencia
Cover Concept: Adrianna Valencia

For my wife, Wendy, and our children—
Benjamin, Emily, Gabrielle, and Isabelle.

1 18

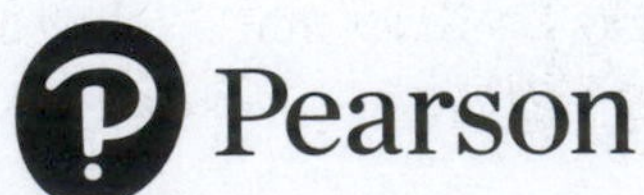

ISBN-13: 978-0-13-476800-7
ISBN-10: 0-13-476800-0

Contents

Preface

Introduction

This *eText Reference* contains the pages of Kirk Trigsted's *Algebra and Trigonometry* in a portable, bound format. Students can use the *eText Reference* to review the eText material anytime, anywhere.

A Note to Students

This eText was created for you and was specifically designed to be read online. Unlike a traditional textbook, I have created content that gives you, the reader, the ability to be an active participant in your learning. The eText pages have large, readable fonts and were designed to avoid scrolling. Throughout the material, I have carefully integrated hundreds of hyperlinks to definitions, previous chapters, interactive videos, animations, and other important content. Many of the videos and animations require you to actively participate. Take some time to "click around" and get comfortable with the navigation of the eText and explore its many features. To log into your MyLab™ Math course, you must first register at www.pearson.com/mylab/math. This *eText Reference* is a bound, printed version of the eText that provides a place for you to do practice work and summarize key concepts from the online videos and animations.

Before you attempt each homework assignment, read the appropriate section(s) of the eText. At the beginning of each section (starting in Chapter 1), you will encounter a feature called **Things to Know**. This feature includes all the prerequisite objectives from previous sections that you will need to successfully learn the material presented in the new section. If you do not think that you have a basic understanding of any of these objectives, click on any of the hyperlinks and rework those objectives, taking advantage of the videos or animations.

Try testing yourself by working through the corresponding You Try It exercises. Remember, you learn math by *doing* math! The more time you spend working through the videos, animations, and exercises, the more you will understand. If your instructor assigns homework in MyLab Math or MathXL, rework the exercises until you get them right.

At the end of each section, you will find a **Chapter Summary**. The chapter summaries highlight important concepts and definitions with side-by-side examples and videos to make it easy for you to study key concepts. Be sure to go back and read the eText at anytime while you are working on your homework. This eText is catered to your educational needs, and I hope you enjoy the experience.

A Note to Instructors

I have taught with MyLab Math for many years and have experienced first-hand how fewer and fewer students are using traditional textbooks. As the use of technology plays an ever-increasing role in how we are teaching our students, it is only natural to have a textbook that mirrors the way our students are learning. I am excited to have written an eText from the ground up to be used as an online, interactive tool for students to read while working in MyLab Math. I wrote this eText entirely from an online perspective, keeping MyLab Math and its existing functionality specifically in mind. Every hyperlink, video, and animation was strategically integrated within the context of each page to maximize the student learning experience. All of the interactive media was designed so students could actively participate while they learn math.

I am a proponent of students learning terms and definitions. Therefore, I have created hyperlinks throughout the text to the definitions of important mathematical terms. I have also inserted a significant amount of just-in-time review throughout the text by creating links to prerequisite topics. Students have the ability to reference these review materials with just a click of the mouse.

Each section has five **reading assessment questions** for those instructors who like to assign reading. These questions are conceptual in nature and were designed to test students on their reading comprehension. Each question was specifically designed to give specific feedback that directs the student back to the appropriate pages of the eText. Note that these questions are static. Students are given two chances to obtain the correct answer. Students will not be able to regenerate a similar exercise. The reading assessment questions can be identified in the Homework/Test Manager by the code "RA."

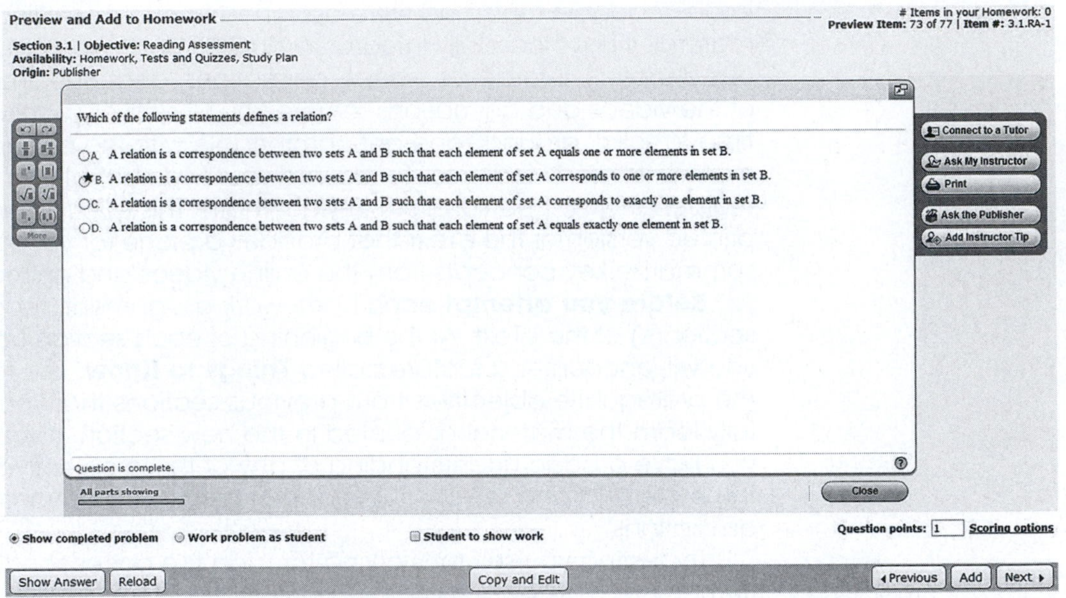

The first edition of *Algebra and Trigonometry* included several multipart exercises. I received feedback from many users who liked to use these multipart exercises for homework but often wished that they could assign only one of the parts for testing. This led to the creation of a new type of exercise called brief exercises. The multipart exercises are now called step-by-step exercises.

Step-by-Step Exercises

The step-by-step exercises were designed to use the power of MathXL to systematically walk the student through some of the more complex, conceptual topics. For example, instead of simply asking for the graph of quadratic function, the step-by-step exercise walks the student through the entire graphing process by asking for all of the important aspects of a parabola. The step-by-step exercises can be readily identified in the Homework/Test Manager by the code "SbS" that precedes the exercise number. An example of such an exercise follows.

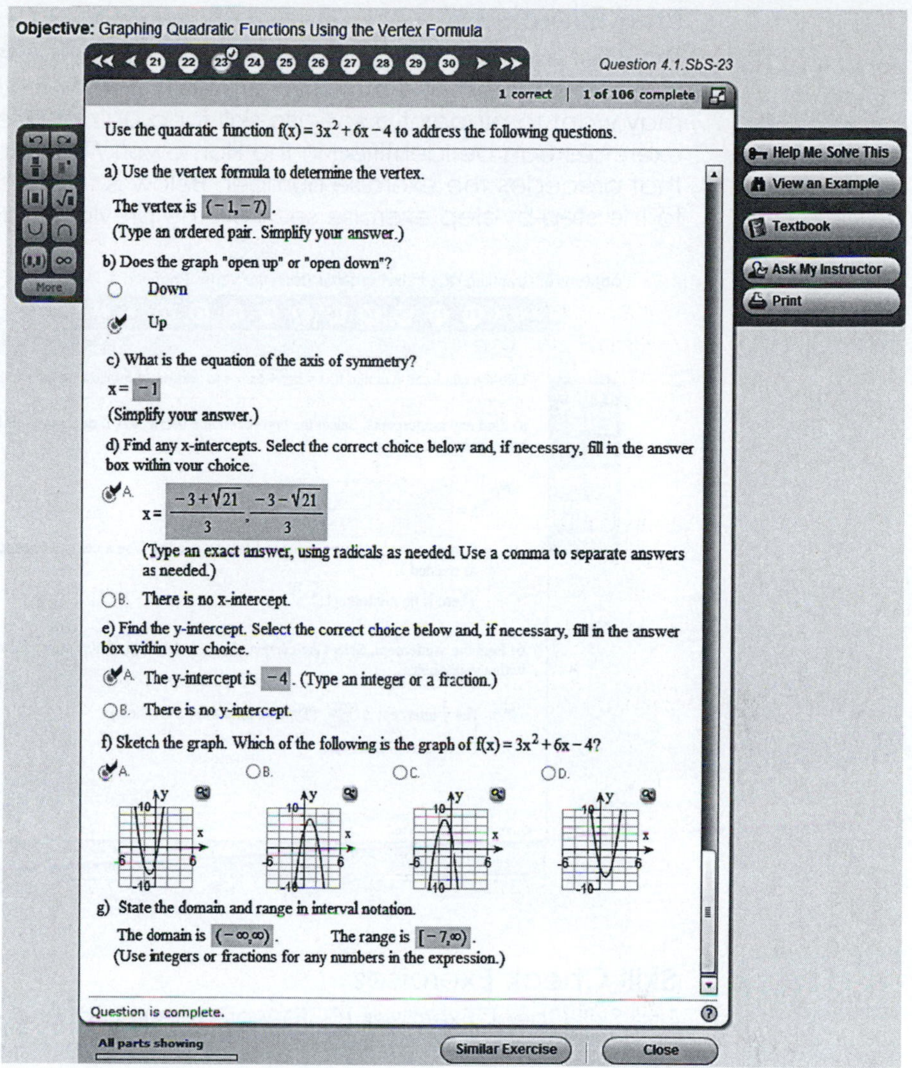

Objective: Graphing Quadratic Functions Using the Vertex Formula
Question 4.1.SbS-23
1 correct | 1 of 106 complete
Use the quadratic function f(x) = 3x² + 6x − 4 to address the following questions.
a) Use the vertex formula to determine the vertex.
The vertex is (−1, −7).
(Type an ordered pair. Simplify your answer.)
b) Does the graph "open up" or "open down"?
Down
Up
c) What is the equation of the axis of symmetry?
x = −1
(Simplify your answer.)
d) Find any x-intercepts. Select the correct choice below and, if necessary, fill in the answer box within your choice.
A. x = (−3 + √21)/3, (−3 − √21)/3
(Type an exact answer, using radicals as needed. Use a comma to separate answers as needed.)
B. There is no x-intercept.
e) Find the y-intercept. Select the correct choice below and, if necessary, fill in the answer box within your choice.
A. The y-intercept is −4. (Type an integer or a fraction.)
B. There is no y-intercept.
f) Sketch the graph. Which of the following is the graph of f(x) = 3x² + 6x − 4?
A. B. C. D.
g) State the domain and range in interval notation.
The domain is (−∞,∞). The range is [−7,∞).
(Use integers or fractions for any numbers in the expression.)
Question is complete.
All parts showing
Help Me Solve This
View an Example
Textbook
Ask My Instructor
Print
More
Similar Exercise
Close

In the exercise sets at the end of each section of this *eText Reference*, all step-by-step exercises are labelled with an **SbS** icon.

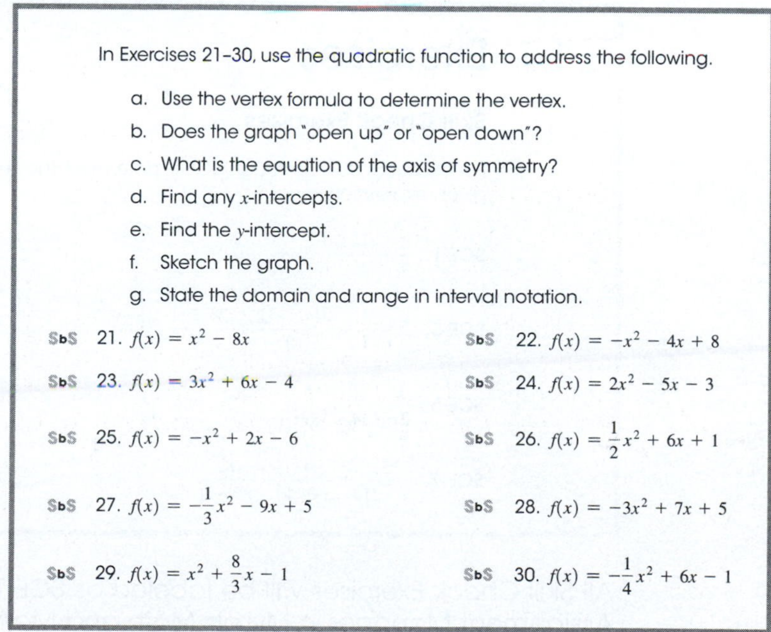

In Exercises 21–30, use the quadratic function to address the following.

a. Use the vertex formula to determine the vertex.
b. Does the graph "open up" or "open down"?
c. What is the equation of the axis of symmetry?
d. Find any x-intercepts.
e. Find the y-intercept.
f. Sketch the graph.
g. State the domain and range in interval notation.

SbS 21. $f(x) = x^2 - 8x$

SbS 22. $f(x) = -x^2 - 4x + 8$

SbS 23. $f(x) = 3x^2 + 6x - 4$

SbS 24. $f(x) = 2x^2 - 5x - 3$

SbS 25. $f(x) = -x^2 + 2x - 6$

SbS 26. $f(x) = \frac{1}{2}x^2 + 6x + 1$

SbS 27. $f(x) = -\frac{1}{3}x^2 - 9x + 5$

SbS 28. $f(x) = -3x^2 + 7x + 5$

SbS 29. $f(x) = x^2 + \frac{8}{3}x - 1$

SbS 30. $f(x) = -\frac{1}{4}x^2 + 6x - 1$

Brief Exercises

The brief exercises are copies of the step-by-step exercises and are designed to test one concept of a multistep problem and were designed for instructors who may want to pinpoint a specific skill for a quiz and testing purposes. The brief exercises can be identified in the Homework/Test Manager by the code "BE" that precedes the exercise number. Below is the brief exercise that corresponds to the step-by-step exercise seen on the previous page.

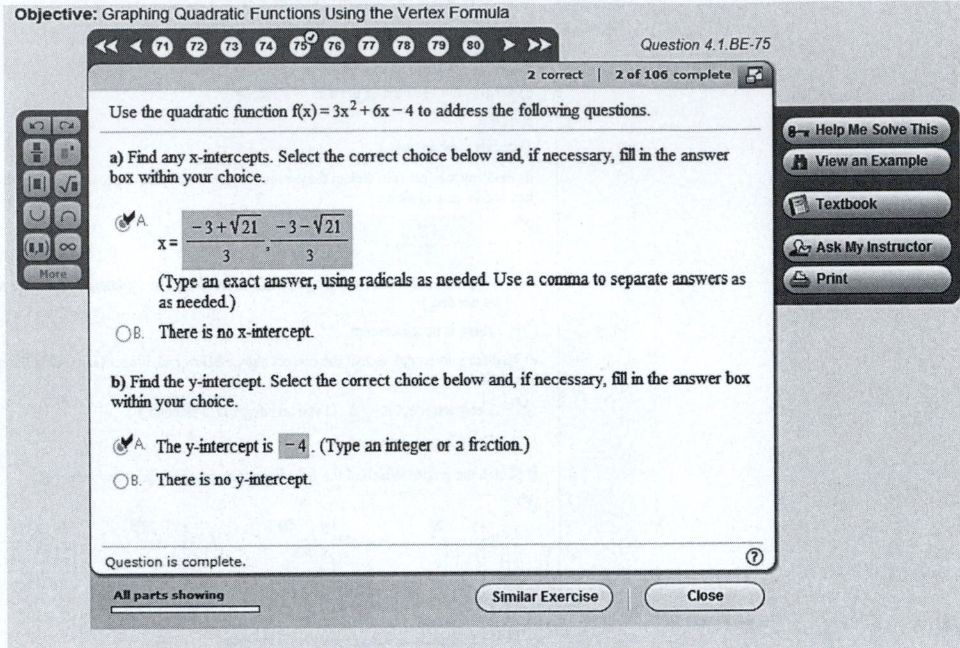

Skill Check Exercises

The Skill Check Exercises (SCE) were created based on comments submitted by users of the Trigsted precalculus series. Skill Check Exercises help students make algebra connections as they work through exercises that present new concepts. These exercises appear at the beginning of most exercises sets and are assignable in MyLab Math.

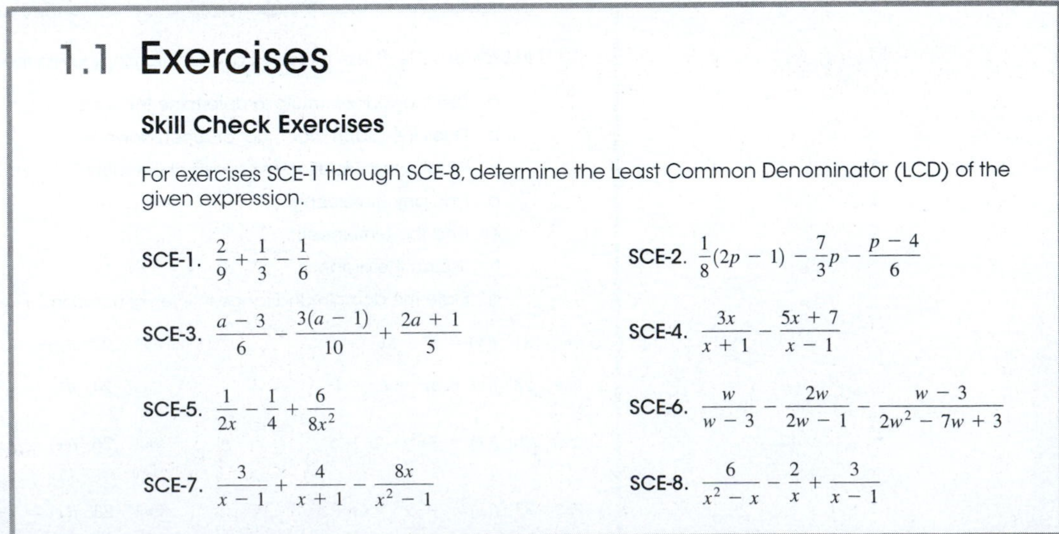

1.1 Exercises

Skill Check Exercises

For exercises SCE-1 through SCE-8, determine the Least Common Denominator (LCD) of the given expression.

SCE-1. $\frac{2}{9} + \frac{1}{3} - \frac{1}{6}$

SCE-2. $\frac{1}{8}(2p - 1) - \frac{7}{3}p - \frac{p - 4}{6}$

SCE-3. $\frac{a - 3}{6} - \frac{3(a - 1)}{10} + \frac{2a + 1}{5}$

SCE-4. $\frac{3x}{x + 1} - \frac{5x + 7}{x - 1}$

SCE-5. $\frac{1}{2x} - \frac{1}{4} + \frac{6}{8x^2}$

SCE-6. $\frac{w}{w - 3} - \frac{2w}{2w - 1} - \frac{w - 3}{2w^2 - 7w + 3}$

SCE-7. $\frac{3}{x - 1} + \frac{4}{x + 1} - \frac{8x}{x^2 - 1}$

SCE-8. $\frac{6}{x^2 - x} - \frac{2}{x} + \frac{3}{x - 1}$

All Skill Check Exercises will be labeled as SCE-1, SCE-2, SCE-3, and so forth in the Assignment Manager in MyLab Math and MathXL.

Resources for Success

MyLab Math Online Course (access code required)

Reach every student with MyLab Math

MyLab Math is the teaching and learning platform that empowers you to reach every student. By combining trusted author content with digital tools and a flexible platform, MyLab Math personalizes the learning experience and improves results for each student.

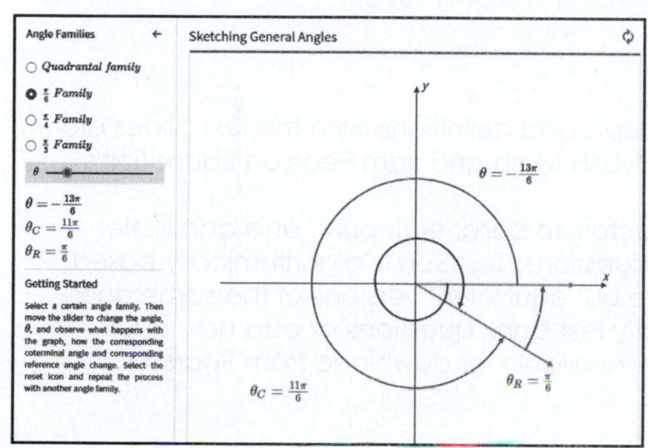

New! Guided Visualizations 

bring mathematical concepts to life, helping you visualize the concepts through directed explorations and purposeful manipulation. Guided Visualizations are integrated into the eText and can be assigned in MyLab Math to encourage active learning, critical thinking, and conceptual understanding.

New! Video Assessment Questions

are assignable MyLab Math exercises tied to the video program. These questions are designed to check students' understanding of the important math concepts covered in the video. Many video assessment questions utilize a new MyLab Math exercise type called **Drag & Drop Exercises.**

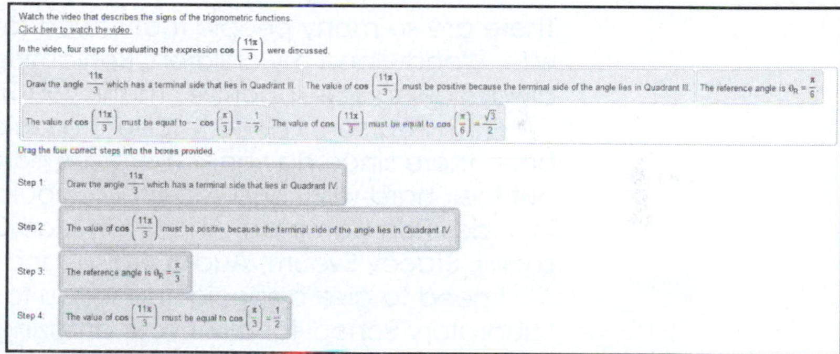

Objective: Use Descartes' rule of signs.

With each question you work, you are building your skills.

Watch the video and then solve the problem given below.

Click here to watch the video.

Determine the different possibilities for the numbers of positive, negative, and nonreal complex zeros for the following function.

$f(x) = -3x^3 + 2x^2 - 3x - 8$

What is the possible number of positive real zeros of this function?

(Use a comma to separate answers as needed.)

Enter your answer in the answer box and then click Check Answer.

2 parts remaining

Skill Builder

offers adaptive practice that is designed to increase students' ability to complete their assignments. By monitoring student performance on their homework, Skill Builder adapts to each student's needs and provides just-in-time, in-assignment practice to help them improve their proficiency of key learning objectives.

Guided Notebook

The *Guided Notebook* is a printed, interactive workbook created by the author to guide students through the eText. The workbook asks students to write down key definitions and work through important examples before they start their homework.

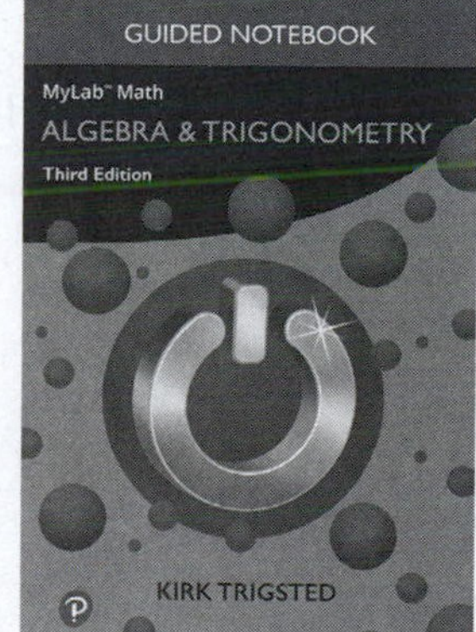

Support and Resources

MyLab Math for Algebra and Trigonometry Student Access Kit, by Kirk Trigsted
0-13-475159-0/978-0-13-475159-7

Guided Notebook

0-13-476799-3/978-0-13-476799-4

The *Guided Notebook* is a printed, interactive workbook created by the author to guide students through the eText. The *Guided Notebook* asks students to write down key definitions and work through important examples before they start their homework. This resource is available in a loose-leaf, three-hole-punched format to provide the foundation for a personalized course notebook. Instructors can also customize the *Guided Notebook* files found within MyLab Math.

Instructor Resources

PowerPoint® slides present key concepts and definitions from the text. Slides are available for download from within MyLab Math and from Pearson Education's online catalog.

TestGen® software enables instructors to build, edit, print, and administer tests using a computerized bank of questions. TestGen is algorithmically based, allowing instructors to create multiple but equivalent versions of the same question or test. Instructors can also modify test bank questions or add new questions. The software and test bank are available for download from Pearson Education's online catalog.

Acknowledgments

There are so many people that I need to thank. First, I would like to thank my wife, Wendy, and my children, Benjamin, Emily, Gabrielle, and Isabelle for always being so supportive. Thank you Trigsted team!

From Pearson, I need to give a special thank-you to Dawn Murrin who has been there since day one! There are countless others from Pearson that, without their hard work and expertise, I could never have finished this project. This includes Ron Hampton, Anne Kelly, Joe Colella, Kristina Evans, Ruth Berry, Peggy Lucas, Stacey Sveum, Audra Walsh, and Vicki Dreyfus.

I need to give a special thank-you to the instructors at LSU and the LSU Laboratory School for all of your amazing insight into improving this edition and for your wonderful suggestions for the many new updated exercises in this edition. Thank you all so much for all of your continued support. I am looking forward to working with all of you in the near future as we start looking ahead at the next edition. You have such an amazing team. This includes Phoebe Rouse, Debra Kopsco, Stephanie Kurtz, Selena Oswalt, Lindsay Waddell, Aimee Welch-James, and Amy Rouse. Also, thank you Selena for your diligence creating the answer section!

The list on the following pages includes our reviewers. Please accept my deepest apologies if I have omitted anyone. I am so very grateful for all of your help. I am truly touched by all of your continued support.

—Kirk Trigsted

Mohamed Baghzali, North Dakota State University
Teri Barnes, McLennan Community College
Linda Barton, Ball State University
Sam Bazzi, Henry Ford Community College
Molly Beauchman, Yavapai College
Brian Beaudrie, Northern Arizona University
Annette Benbow, Tarrant County College
Patricia Blus, National Louis University
Nina Bohrod, Anoka-Ramsey Community College
Barbara Boschmans, Northern Arizona University
David Bramlet, Jackson State University
Densie Brown, Collin County Community College
Connie Buller, Metropolitan Community College
Joe Castillo, Broward Community College
Mariana Coanda, Broward Community College
Alicia Collins, Mesa Community College
Earl W. Cook, Chattahoochee Valley Community College
April Church, East Carolina University
Kemba C. Countryman, Chattahoochee Valley Community College
Douglas Culler, Midlands Technical College
Momoyo Dahle, Las Positas Community College
Diane Daniels, Mississippi State University
Emmett Dennis, Southern Connecticut State University
Donna Densmore, Bossier Parish Community College
Debbie Detrick, Kansas City Kansas Community College
Timothy Doyle, University of Illinois at Chicago
Christina Dwyer, Manatee Community College
Stephanie Edgerton, Northern Arizona University
Jeanette Eggert, Concordia University
Brett Elliott, Southeastern Oklahoma State University
Amy Erickson, Georgia Gwinnett College
Nicki Feldman, Pulaski Technical College
Catherine Ferrer, Valencia Community College
Gerry Fitch, Louisiana State University
Cynthia Francisco, Oklahoma State University
Robert Frank, Westmoreland County Community College
Jim Frost, St. Louis Community College–Meramec
Angelito Garcia, Truman College
Dr. Sunshine Gibbons, Hillsborough Community College
Lee Gibson, University of Louisville
Charles B. Green, University of North Carolina–Chapel Hill
Jeffrey Hakim, American University
Mike Hall, Arkansas State University
Melissa Hardeman, University of Arkansas, Little Rock
Celeste Hernandez, Richland College
Pamela Howard, Boise State University
Jeffrey Hughes, Hinds Community College
Eric Hutchinson, College of Southern Nevada
Robert Indrihovic, Florence-Darlington Technical College
Philip Kaatz, Mesalands Community College
Cheryl Kane, University of Nebraska–Lincoln
Robert Keller, Loras College
Mike Kirby, Tidewater Community College
Susan Knights, Boise State University
Marie Kohrmann, North Lake College
Debra Kopcso, Louisiana State University
Stephanie Kurtz, Louisiana State University
Jennifer LaRose, Henry Ford Community College
Jeff Laub, Central Virginia Community College
Jennifer Legrand, St. Charles Community College
Kurt Lewandowski, Clackamas Community College
Oscar Macedo, University of Texas at El Paso
Shanna Manny, Northern Arizona University
Matthew Michaelson, Glendale Community College (AZ)
Pamela Spurlock Mills, University of Arkansas
Peggy L. Moch, Valdosta State University
Susan Moosai, Florida Atlantic University
Rebecca Morgan, Wayne State University
Rebecca Muller, Southeastern Louisiana University
Veronica Murphy, Saint Leo University
Charlie Naffziger, Central Oregon Community College
Prince Raphael A. Okojie, Fayetteville State University
Enyinda Onunwor, Stark State University
Selena Oswalt, Louisiana State University
Shahla Peterman, University of Missouri–St. Louis
Sandra Poinsett, College of Southern Maryland
Steve Proietti, Northern Essex Community College
Michael Puente, Richland College
Ray Purdom, University of North Carolina at Greensboro
Nancy Ressler, Oakton Community College
Mary Revels, Southeast Community College
Joe Rody, Arizona State University
Cheryl Roddick, San Jose State University
Amy Rouse, Louisiana State University Laboratory School
Phoebe Rouse, Louisiana State University
Patricia Rowe, Columbus State Community College
Amy Rushall, Northern Arizona University
Chie Sakabe, University of Idaho
Dr. Kristina Sampson, Lone Star College–CyFair
Jorge Sarmiento, County College of Morris
John Savage, Montana State University–College of Technology
Julie Sawyer, University of Idaho
Victoria Seals, Gwinnett Technical College
Susan P. Sherry, Northern Virginia Community College
Randell Simpson, Temple College
Rita Sowell, Volunteer State Community College
John Squires, Cleveland State Community College

Todd Stine, Harrisburg Area Community College
Pam Stogsdill, Bossier Parish Community College
Eleanor Storey, Front Range Community College
Robert Strozak, Old Dominion University
Lalitha Subramanian, Potomac State College of West Virginia University
Mary Ann Teel, University of North Texas
Gwen Terwilliger, University of Toledo
Jamie Thomas, University of Wisconsin–Manitowoc
Keith B. Thompson, Davidson County Community College
Terry Tiballi, State University of New York at Oswego
Suzanne Topp, Salt Lake Community College
Diann Torrence, Delgado Community College
Philip Veer, Johnson County Community College
Marcia Vergo, Metropolitan Community College
Lindsay Waddell, Louisiana State University
Kimberly Walters, Mississippi State University
Aimee Welch-James, Louisiana State University Laboratory School
Jacci White, Saint Leo University
Ralph L. Wildy Jr., Georgia Military College
Sherri M. Wilson, Fort Lewis College
Xuezheng Wu, Madison College
Janet Wyatt, Metropolitan Community College–Longview
Ghidei Zedingle, Normandale Community College
Brian Zimmerman, Phillips Community College of University of Arkansas

CHAPTER R

Review Chapter

CHAPTER R CONTENTS

R.1 Real Numbers

OBJECTIVES

1 Understanding the Real Number System

2 Writing Sets Using Set-Builder Notation and Interval Notation

3 Determining the Intersection and Union of Sets and Intervals

4 Understanding Absolute Value

SECTION R.1 EXERCISES

OBJECTIVE 1 UNDERSTANDING THE REAL NUMBER SYSTEM

A **set** is a collection of objects. Each object in the set is called an **element** or a **member** of the set. We typically use braces { } to enclose all elements of a set. Capital letters are sometimes used to name a set. For example, suppose we define sets A and B as

$$A = \{1, 2, 3, 4, 5, 6, 7, 8, 9, 10\} \quad \text{and} \quad B = \{2, 4, 6, 8, 10\}.$$

Notice that each element of set B is also an element of set A. We say that set B is a **subset** of set A and write $B \subset A$. The set of **real numbers** consists of several subsets of numbers. We now define these subsets.

> **Definition** Natural Numbers
>
> A **natural number**, or **counting number**, is an element of the set $\mathbb{N} = \{1, 2, 3, 4, 5, \ldots\}$.

Adding the element 0 to the set of natural numbers gives the set of *whole numbers*, denoted by the symbol $\mathbb{W}$.

Definition Whole Numbers

A **whole number** is an element of the set $\mathbb{W} = \{0, 1, 2, 3, 4, 5, \ldots\}$.

TIP Every natural number is also a whole number, so $\mathbb{N} \subset \mathbb{W}$.

Adding the **negative numbers** $-1, -2, -3, -4, \ldots$ to the set of whole numbers gives the set of *integers*, denoted by $\mathbb{Z}$.

Definition Integers

An **integer** is an element of the set $\mathbb{Z} = \{\ldots, -4, -3, -2, -1, 0, 1, 2, 3, 4, \ldots\}$.

TIP Every whole number is also an integer, so $\mathbb{N} \subset \mathbb{W} \subset \mathbb{Z}$.

The **integers** are still "whole" in the sense that they are not **fractions**. If fractions are included, the result is the set of *rational numbers*, denoted by $\mathbb{Q}$.

Definition Rational Numbers

A **rational number** is an element of the set $\mathbb{Q} = \left\{ \frac{p}{q} \middle| p \text{ and } q \text{ are integers, } q \neq 0 \right\}$.

A rational number can be written as the **quotient** of two integers p and q as long as $q \neq 0$. $\frac{p}{q}$ is called the **fraction form** for the rational number. p is the **numerator**, and q is the **denominator**. Examples of rational numbers include $\frac{3}{4}, -\frac{7}{8}$, and $\frac{13}{11}$.

TIP Every integer is also a rational number because the integer can be written with a denominator of 1. For example, $3 = \frac{3}{1}, 0 = \frac{0}{1}$, and $-8 = \frac{-8}{1}$. So, $\mathbb{N} \subset \mathbb{W} \subset \mathbb{Z} \subset \mathbb{Q}$.

Rational numbers are often written in **decimal form**. When a fraction is written as a **decimal**, the result is either a **terminating decimal**, such as $\frac{3}{4} = 0.75$ and $-\frac{7}{8} = -0.875$, or a **repeating decimal**, such as $\frac{5}{6} = 0.8333\ldots = 0.8\overline{3}$ and $\frac{13}{11} = 1.181818\ldots = 1.\overline{18}$.

TIP Every rational number can be expressed as a terminating or a repeating decimal. Conversely, every number that can be written as a terminating or a repeating decimal is a rational number.

Numbers that cannot be written as terminating or repeating decimals form the set of *irrational numbers*, denoted by $\mathbb{I}$.

Definition Irrational Numbers

An **irrational number** is an element of the set

$$\mathbb{I} = \{x \mid x \text{ is a number that is not rational}\}.$$

An irrational number cannot be written as the quotient of two integers, and its decimal form does not repeat or terminate. Square roots and other numbers expressed with radicals are often irrational numbers. For example, $\sqrt{3} = 1.7320508\ldots$ and $\sqrt[3]{5} = 1.7099759\ldots$ have decimal forms that continue indefinitely without a repeating pattern, so each is an irrational number. Another example of an irrational number is $\pi = 3.1415926\ldots.$

CAUTION Do not automatically assume that a number expressed with a radical is an irrational number. Some numbers expressed with radicals are rational. For example, $\sqrt{9} = 3$, $\sqrt{2.25} = 1.5$, and $\sqrt[3]{8} = 2$ are all rational numbers.

If the set of rational numbers is combined with the set of irrational numbers, the result is the set of *real numbers*, denoted by $\mathbb{R}$.

Definition Real Numbers

A **real number** is an element of the set

$$\mathbb{R} = \{x \mid x \text{ is a rational number or an irrational number}\}.$$

Real numbers are used in everyday activities such as computing salaries, making purchases, and measuring distances. As we progress through algebra, we will occasionally encounter numbers that are not real. For example, $\sqrt{-4}$ is not a real number because there is no real number that can be multiplied by itself to equal -4. We will define such *non-real complex numbers* later in this eText. Barring those, every number you can think of is a real number.

Figure 1 gives a visual of the set of real numbers that shows the following relationships:

- Every real number is either rational or irrational;
- All natural numbers are whole numbers;
- All whole numbers are integers; and
- All integers are rational numbers.

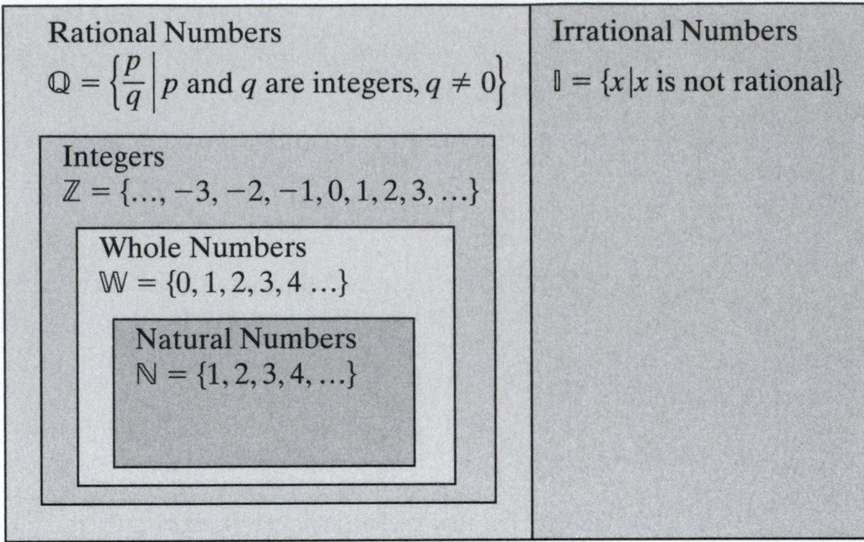

Figure 1 The real numbers

Real numbers can be represented by points located on a number line called the *real axis*. The point corresponding to 0 is called the *origin*. Figure 2 illustrates several real numbers plotted on a number line. In this text, we use a solid circle (•) to represent a point on a number line.

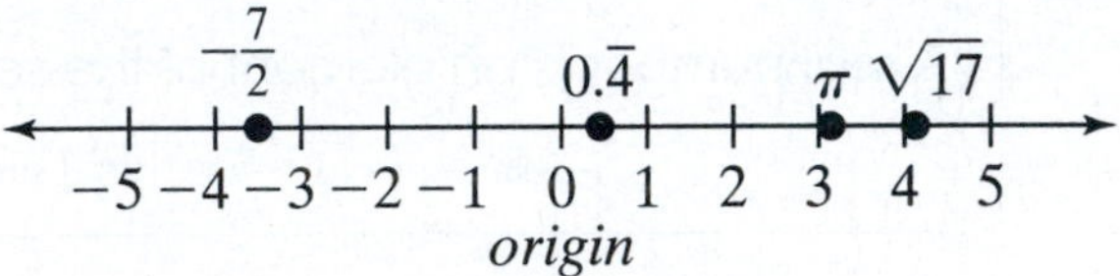

Figure 2

Example 1 Classify Real Numbers

Classify each number in the set $\left\{-5, -\frac{1}{3}, 0, \sqrt{3}, 1.\overline{4}, \sqrt{36}, 2\pi, 11\right\}$ as a natural number, whole number, integer, rational number, irrational number, and/or real number.

Solution

The number -5 is an integer, rational number, and real number.

The number $-\frac{1}{3}$ is a rational number and a real number.

The number 0 is a whole number, integer, rational number, and real number.
The number $\sqrt{3}$ is an irrational number and a real number.
The number $1.\overline{4}$ is a rational number and a real number.
The number $\sqrt{36}$ is a natural number, whole number, integer, rational number, and real number.
The number 2π is an irrational number and a real number.
The number 11 is a natural number, whole number, integer, rational number, and real number.

Watch this **video** to see the entire solution to this example. ●

You Try It Work through the following You Try It problem.

Work Exercises 1–6 in this textbook or in the MyLab Math Study Plan.

OBJECTIVE 2 WRITING SETS USING SET-BUILDER NOTATION AND INTERVAL NOTATION

Set-builder notation can be a convenient way to describe certain sets of numbers. For example, we used set-builder notation to describe the set of **rational numbers** and the set of **irrational numbers**. Suppose that we want to describe the set of all real numbers less than 10. Using set-builder notation, we write this set as $\{x|x < 10\}$. This set is read as "the set of all x such that x is less than 10." Figure 3 illustrates how we can represent this set on a number line.

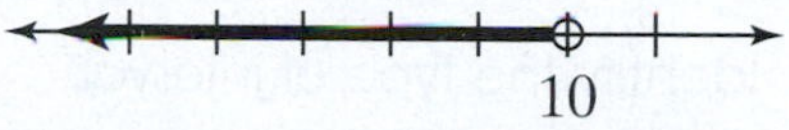

Figure 3 The set $\{x|x < 10\}$ represented on a number line.

The open circle (○) at the number 10 in Figure 3 represents that 10 is *not* included in the set. If we wanted to include the number 10, we would have used a solid circle (•). The arrow extending to the left indicates that the set goes on indefinitely toward negative infinity. In this text, we often use **interval notation** to describe sets of numbers such as the set displayed in Figure 3. The interval that describes this set is called an open infinite interval and is written as $(-\infty, 10)$. It is extremely important to be able to write intervals of numbers in both set-builder notation and interval notation. **Table 1** illustrates the different types of intervals and the corresponding set-builder notation.

Table 1

Type of Interval and Graph	Interval Notation	Set-Builder Notation
Open interval 	(a, b)	$\{x\|a < x < b\}$
Closed interval 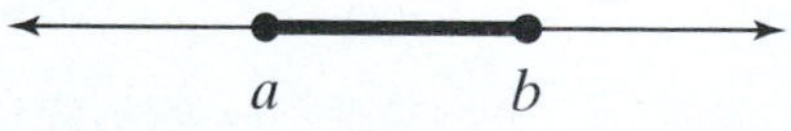	$[a, b]$	$\{x\|a \le x \le b\}$
Half-open intervals 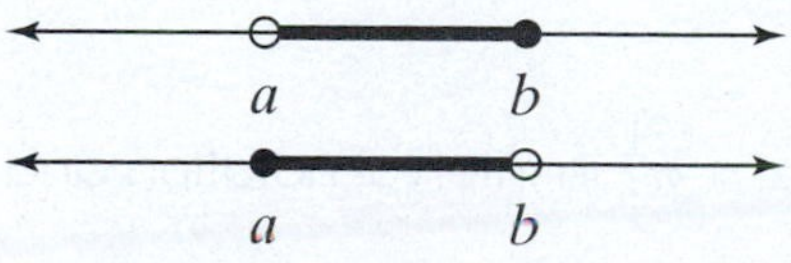	$(a, b]$ $[a, b)$	$\{x\|a < x \le b\}$ $\{x\|a \le x < b\}$
Open infinite intervals 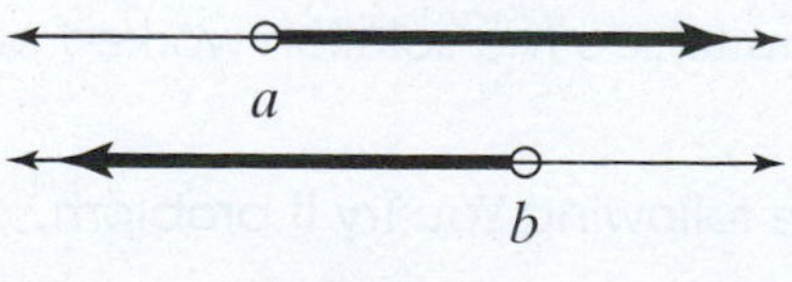	(a, ∞) $(-\infty, b)$	$\{x\|x > a\}$ $\{x\|x < b\}$

(*Continued*)

Table 1 *Continues*

Type of Interval and Graph	Interval Notation	Set-Builder Notation
Closed infinite intervals a b	$[a, \infty)$ $(-\infty, b]$	$\{x \mid x \geq a\}$ $\{x \mid x \leq b\}$

Example 2 Write a Set Using Set-Builder Notation and Interval Notation

Given the set sketched on the following number line,

a. Identify the type of interval.

b. Write the set using set-builder notation.

c. Write the set using interval notation.

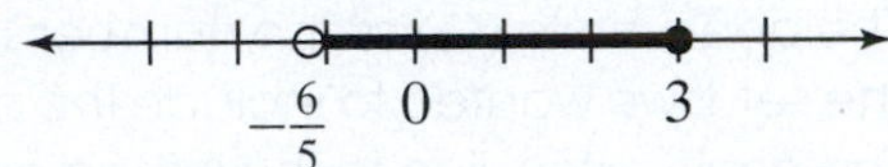

Solution

a. This interval is called a half-open interval. The solid circle at 3 indicates that the set includes the number 3. The open circle at $-\frac{6}{5}$ indicates that the set does not include $-\frac{6}{5}$.

b. This set can be described in set-builder notation as $\left\{x \middle| -\frac{6}{5} < x \leq 3\right\}$.

c. We write this set in interval notation as $\left(-\frac{6}{5}, 3\right]$.

Watch this **video** to see each part of the solution to this example.

Example 3 Write a Set Using Set-Builder Notation and Interval Notation

a. Write the set $\left[-\frac{1}{3}, \infty\right)$ in set-builder notation and graph the set on a number line.

b. Write the set $\left\{x \middle| -\frac{7}{2} < x \leq \pi\right\}$ in interval notation and graph the set on a number line.

Solution Watch the video to see the solution worked out in detail.

You Try It Work through the following You Try It problem.

Work Exercises 7–18 in this textbook or in the MyLab Math Study Plan.

OBJECTIVE 3 DETERMINING THE INTERSECTION AND UNION OF SETS AND INTERVALS

We are often interested in looking at two or more sets at a time.

INTERSECTION OF SETS

When two sets share a common **element**, we say that the common element is contained in the **intersection** of the two sets.

Intersection

For any two sets A and B, the **intersection** of A and B is given by $A \cap B$ and represents the **elements** that are in set A **and** in set B.

$$A \cap B = \{x \mid x \text{ is an element of } A \textbf{ and } \text{an element of } B\}$$

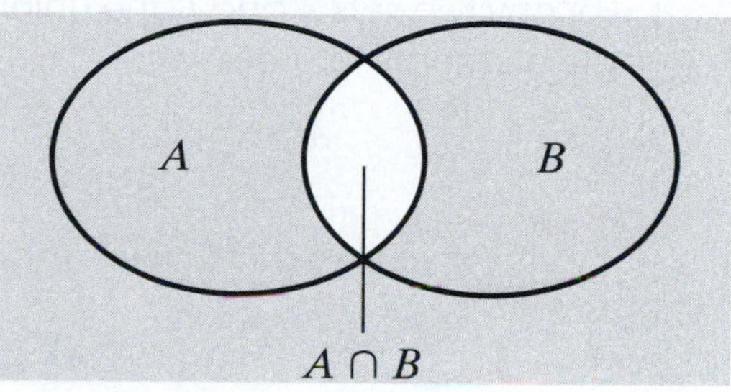

Notice that the intersection of two sets is the overlap of the two sets. Much like the intersection of two roads is the region common to both roads, the intersection of two sets is the set of elements that are common to both sets. If there is no overlap, the intersection is the **empty set**, and we write $A \cap B = \{\ \}$ or $A \cap B = \varnothing$.

Example 4 Find the Intersection of Sets

Let $A = \left\{-5, 0, \frac{1}{3}, 11, 17\right\}$, $B = \{-6, -5, 4, 17\}$, and $C = \left\{-4, 0, \frac{1}{4}\right\}$.

a. Find $A \cap B$. **b.** Find $B \cap C$.

Solution Watch the **video** to verify the following solutions.

a. The set $A \cap B$ is the set of elements that are in both A and B. Both sets contain the numbers -5 and 17. So,

$$A \cap B = \left\{-5, 0, \frac{1}{3}, 11, 17\right\} \cap \{-6, -5, 4, 17\} = \{-5, 17\}.$$

b. Set B and set C have **no** elements in common. Therefore, the intersection of sets B and C is the empty set. We write

$$B \cap C = \{-6, -5, 4, 17\} \cap \left\{-4, 0, \frac{1}{4}\right\} = \varnothing.$$

You Try It Work through the following You Try It problem.

Example 5 Finding the Intersection of Two Sets

Let $A = \{x \mid x > -2\}$ and $B = \{x \mid x \leq 5\}$. Find $A \cap B$, the intersection of the two sets.

Solution Check your answer, or watch the video for the solution.

You Try It Work through the following You Try It problem.

UNION OF SETS

When considering two sets, we are often interested in listing the elements that belong to either set. Elements that belong to either set A or set B is called the **union** of sets A and B.

> **Union**
>
> For any two sets A and B, the **union** of A and B is given by $A \cup B$ and represents the elements that are in set A **or** in set B.
>
> $$A \cap B = \{x \mid x \text{ is an element of } A \textbf{ and } \text{an element of } B\}$$
>
> 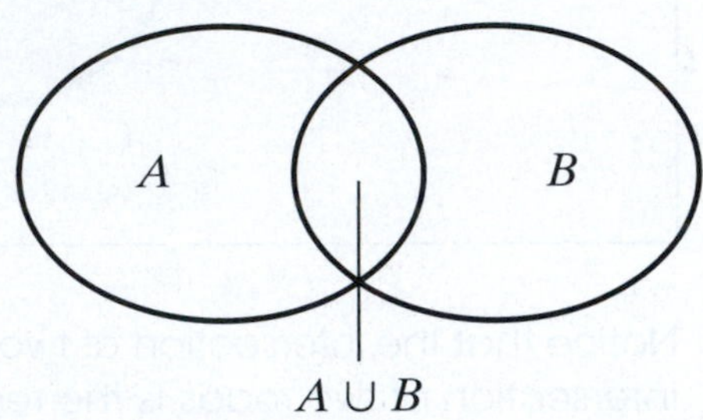
>

Notice that the union of two sets is the combination of the two sets. The union of two sets contains elements that are just in set A, just in set B, or in both A and B. Elements that appear in both sets A and B are only listed once when writing the union of the sets.

Example 6 Finding the Union of Two Sets

Let $A = \{1, 3, 4, 5, 7, 10, 12\}$ and $B = \{2, 4, 6, 8, 10, 12\}$. Find $A \cup B$, the **union** of the two sets.

Solution Read the following, or watch this video for the solution.

The union is the set of all unique values that are in either set A or in set B.

$$A \cup B = \{1, 2, 3, 4, 5, 6, 7, 8, 10, 12\}$$

Notice that while 4, 10, and 12 occurred in both sets, these numbers were listed only once in the union.

You Try It Work through this You Try It problem.

Example 7 Finding the Union of Two Sets

Let $A = \{x|x < -2\}$ and $B = \{x|x \geq 5\}$. Find $A \cup B$, the union of the two sets.

Solution Check your answer, or watch the video for the solution.

You Try It Work through this You Try It problem.

Work Exercises 19–30 in this eText or in the MyLab Math Study Plan.

Example 8 Find the Intersection of Sets

Find the intersection of the following sets and graph the set on a number line.

a. $[0, \infty) \cap (-\infty, 5]$ **b.** $((-\infty, -2) \cup (-2, \infty)) \cap [-4, \infty)$

Solution Watch the video to verify the following solutions.

a. $[0, \infty) \cap (-\infty, 5] = [0, 5]$

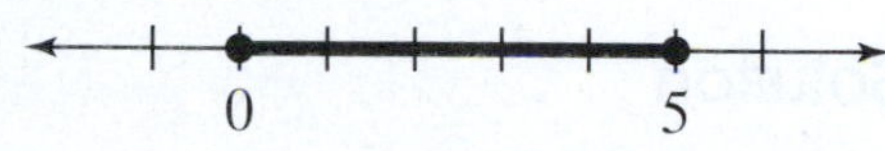

b. $((-\infty, -2) \cup (-2, \infty)) \cap [-4, \infty) = [-4, -2) \cup (-2, \infty)$

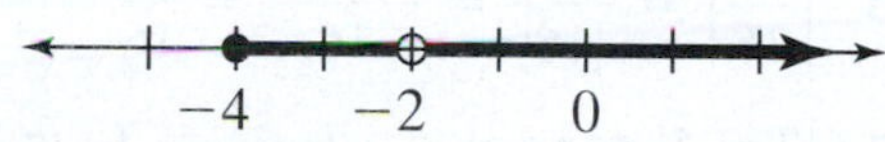

You Try It Work through the following You Try It problem.

Work Exercises 31–36 in this textbook or in the MyLab Math Study Plan.

OBJECTIVE 4 UNDERSTANDING ABSOLUTE VALUE

The absolute value of a real number a is defined as the distance between a and 0 on a number line and is denoted $|a|$. For example, $|-5| = 5$ because the number -5 is 5 units away from 0 on a number line. See Figure 4.

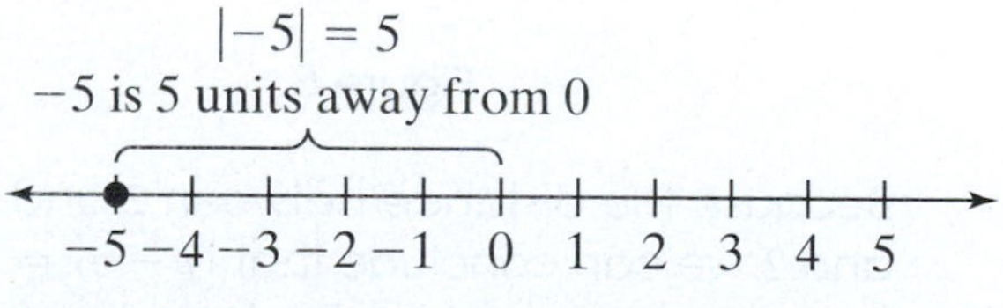

$|-5| = 5$

-5 is 5 units away from 0

−5 −4 −3 −2 −1 0 1 2 3 4 5

Figure 4 The absolute value of -5 is 5 because -5 is 5 units away from 0.

Definition Absolute Value

The absolute value of a real number a is defined by

$$|a| = \begin{cases} a & \text{if } a \geq 0 \\ -a & \text{if } a < 0 \end{cases}$$

and represents the distance between a and 0 on a number line.

Because the absolute value of a number a represents the *distance* from a to 0 on a number line, it follows that $|a| \geq 0$ for any real number a. This property and several other properties of absolute value are stated as follows.

Properties of Absolute Value

For all real numbers a and b,

1. $|a| \geq 0$ **2.** $|ab| = |a||b|$ **3.** $\left|\frac{a}{b}\right| = \frac{|a|}{|b|}, b \neq 0$

4. $|a|^2 = a^2$ **5.** $|a - b| = |b - a|$

Example 9 Evaluate Absolute Value Expressions

Evaluate the following expressions involving absolute value.

a. $|-\sqrt{2}|$ b. $\left|\frac{-4}{16}\right|$ c. $|2 - 5|$

Solution

a. $|-\sqrt{2}| = -(-\sqrt{2}) = \sqrt{2}$

b. $\left|\frac{-4}{16}\right| = \frac{|-4|}{|16|} = \frac{-(-4)}{16} = \frac{4}{16} = \frac{1}{4}$

c. $|2 - 5| = |-3| = -(-3) = 3$ or $|2 - 5| = |5 - 2| = |3| = 3$ ●

Let's take a closer look at the absolute value expression $|2 - 5|$ from Example 9c and interpret its meaning, geometrically. The expression $|2 - 5|$ represents the distance between the numbers 2 and 5 on a number line. Notice in Figure 5 that the distance between 2 and 5 on a number line is three units.

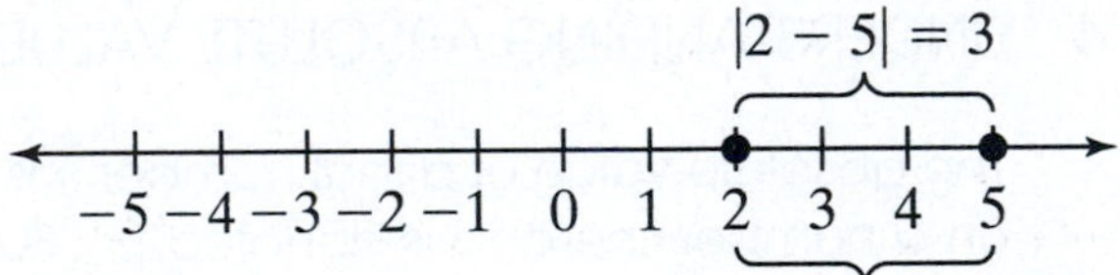

Figure 5

Because the distance between 2 and 5 is the same as the distance between 5 and 2, we can conclude that $|2 - 5| = |5 - 2|$. Thus, for any real numbers a and b, $|a - b| = |b - a|$, which is precisely property 5 of absolute values. We now restate property 5 as the distance between two real numbers on a number line.

Distance between Two Real Numbers on a Number Line

For any real numbers a and b, the distance between a and b on a number line is defined by $|a - b|$ or $|b - a|$.

Example 10 Find the Distance between Two Real Numbers

Find the distance between the numbers -5 and 3 using absolute value.

Solution Let $a = -5$ and $b = 3$. The distance between a and b is given by $|a - b|$ or $|b - a|$. $|a - b| = |-5 - 3| = |-8| = 8$. So, the distance between -5 and 3 is 8 units.

You Try It **Work through the following You Try It problem.**

Work Exercises 37–44 in this textbook or in the MyLab Math Study Plan.

R.1 Exercises

In Exercises 1–4, classify each number as a natural number, whole number, integer, rational number, irrational number, and/or real number. Each number may belong to more than one set.

1. 259
2. $-11{,}401$
3. $-\dfrac{43}{7}$
4. $-\sqrt{17}$

In Exercises 5 and 6, given each set of real numbers, list the numbers that are a) natural numbers, b) whole numbers, c) integers, d) rational numbers, and/or e) irrational numbers.

5. $\left\{-17, -\sqrt{5}, -\dfrac{25}{19}, 0, 0.331, 1, \dfrac{\pi}{2}\right\}$
6. $\left\{-11, -3.\overline{2135}, -\dfrac{3}{9}, \dfrac{\sqrt{7}}{2}, 21.1\right\}$

In Exercises 7–10, given the set sketched on the number line, a) identify the type of interval, b) write the set using set-builder notation, and c) write the set using interval notation.

7. 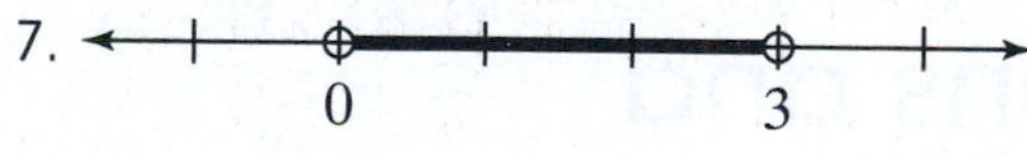

8. 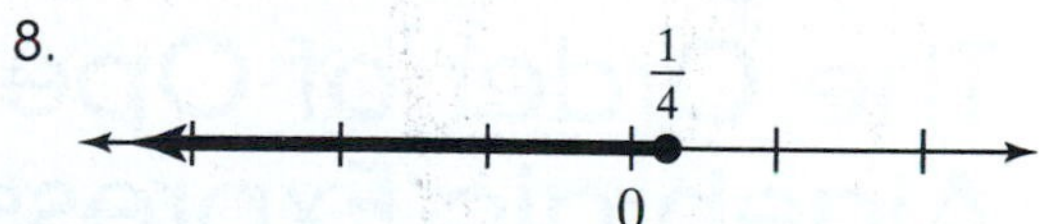

9. 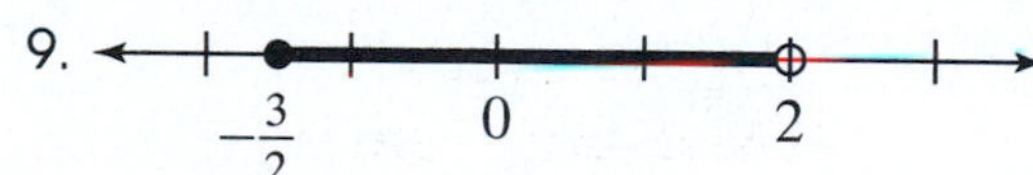

10. 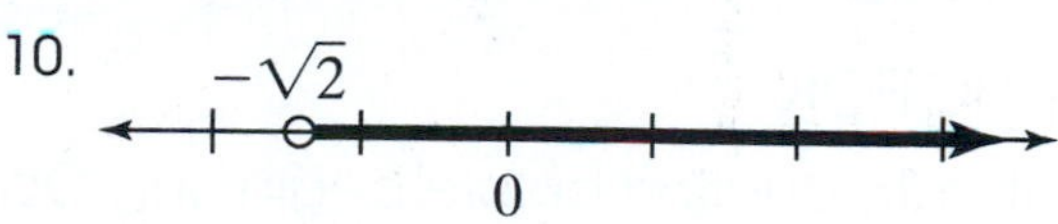

In Exercises 11–14, write each interval in set-builder notation and graph the set on a number line.

11. $\left[-\dfrac{1}{2}, 5\right]$
12. $\left(0, \dfrac{5}{2}\right)$
13. $(-\infty, 3]$
14. $(-1, \infty)$

In Exercises 15–18, write the given set in interval notation and graph the set on a number line.

15. $\left\{x \middle| -\dfrac{5}{2} \le x \le 1\right\}$
16. $\{x|0 \le x < 3\}$
17. $\left\{x \middle| x \ge \dfrac{3}{4}\right\}$
18. $\{x|x > -4\}$

In Exercises 19–22, find the indicated set given $A = \{-2, 0, 1, 5, 6, 9, 15\}$, $B = \{-6, -4, -2, 0, 2, 4, 6\}$, and $C = \{-1, 3, 7, 11, 20\}$.

19. $A \cap B$
20. $A \cap C$
21. $A \cup B$
22. $B \cup C$

In Exercises 23–26, find $A \cap B$.

23. $A = \{x|x < 9\}; B = \{x|x \geq 2\}$

24. $A = \{x|x > 5\}; B = \{x|x \geq -3\}$

25. $A = \{x|x \leq -2\}; B = \{x|x \leq 0\}$

26. $A = \{x|x \leq 14\}; B = \{x|0 < x < 18\}$

In Exercises 27–30, find $A \cup B$.

27. $A = \{x|x \leq -3\}; B = \{x|x \geq 5\}$

28. $A = \{x|x \leq -8\}; B = \{x|x < 6.4\}$

29. $A = \{x|x < 10\}; B = \{x|x > -1\}$

30. $A = \{x|x \geq 0\}; B = \{x|-7 \leq x < 0\}$

In Exercises 31–36, find the intersection of the given sets.

31. $(-\infty, \infty) \cap (-\infty, 3)$

32. $(-8, 5] \cap (-12, 3)$

33. $[0, \infty) \cap (-\infty, 4]$

34. $((-\infty, 1) \cup (1, \infty)) \cap (-\infty, \infty)$

35. $((-\infty, 0) \cup (0, \infty)) \cap [-3, \infty)$

36. $(-\infty, 10) \cap (-5, \infty) \cap (-\infty, 1]$

In Exercises 37–41, evaluate the absolute value expression.

37. $|286|$

38. $|-2.2|$

39. $|-2 \cdot 3|$

40. $-\left|\frac{-2 \cdot 6}{4}\right|$

41. $|2 - \pi|$

In Exercises 42–44, find the distance between the given two numbers using absolute value.

42. 3 and 14

43. -5 and 7

44. -24 and -1.5

R.2 The Order of Operations and Algebraic Expressions

INTRODUCTION

Read this introduction before beginning Objective 1.

OBJECTIVES

1 Understanding the Properties of Real Numbers

2 Using Exponential Notation

3 Using the Order of Operations to Simplify Numeric and Algebraic Expressions

SECTION R.2 EXERCISES

Introduction to Section R.2

In this section, we learn how to simplify **algebraic expressions**. An algebraic expression consists of one or more terms that include **variables**, constants, and operating symbols such as + or −. For example, $4x^3y^4 - 1$ is an algebraic expression consisting of two terms. The letters x and y used in this expression are called **variables** and are used to represent any number. Before we work with algebraic expressions, it is important to understand some basic properties of real numbers. Our understanding of these properties will enhance our ability to simplify numeric and algebraic expressions.

OBJECTIVE 1 UNDERSTANDING THE PROPERTIES OF REAL NUMBERS

We all know that when adding two real numbers or multiplying two numbers, the order in which we perform the addition or multiplication does not affect the result. For example, $3 + 5 = 8$ and $5 + 3 = 8$, so $3 + 5 = 5 + 3$. Similarly, $(-4)(6) = -24$ and $(6)(-4) = -24$, so $(-4)(6) = (6)(-4)$. The operations of addition and multiplication are said to be commutative. The commutative property of addition and multiplication are stated on the following page.

Commutative Property of Addition

If a and b are real numbers, then $a + b = b + a$.

Commutative Property of Multiplication

If a and b are real numbers, then $ab = ba$.

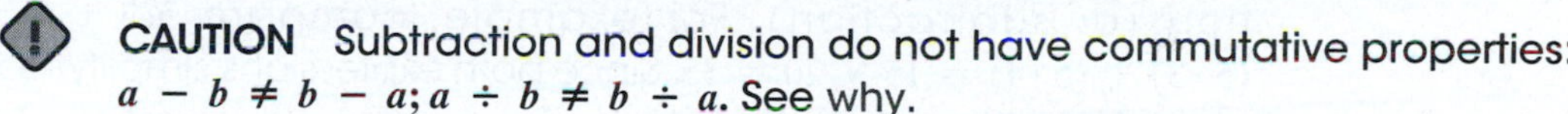

CAUTION **Subtraction and division do not have commutative properties:** $a - b \neq b - a; a \div b \neq b \div a$. See why.

Next, let's look at the *associative properties*. The **associative property of addition** states that regrouping terms does not affect the sum. For example, $(3 + 5) + 4 = 8 + 4 = 12$ and $3 + (5 + 4) = 3 + 9 = 12$, so $(3 + 5) + 4 = 3 + (5 + 4)$.

Associative Property of Addition

If a, b, and c are real numbers, then $(a + b) + c = a + (b + c)$.

The **associative property of multiplication** states that regrouping factors does not affect the product. For example, $(-2 \cdot 5) \cdot 3 = (-10) \cdot 3 = -30$ and $-2 \cdot (5 \cdot 3) = -2 \cdot (15) = -30$, so $(-2 \cdot 5) \cdot 3 = -2 \cdot (5 \cdot 3)$.

Associative Property of Multiplication

If a, b, and c are real numbers, then $(a \cdot b) \cdot c = a \cdot (b \cdot c)$.

CAUTION **Subtraction and division do not have associative properties:** $(a - b) - c \neq a - (b - c)$; $(a \div b) \div c \neq a \div (b \div c)$. See why.

Example 1 Use the Properties of Real Numbers to Rewrite Expressions

a. Use the commutative property of addition to rewrite $11 + y$ as an equivalent expression.

b. Use the commutative property of multiplication to rewrite $7(y + 4)$ as an equivalent expression.

c. Use the associative property of addition to rewrite $(2 + 3x) + 8$ as an equivalent expression.

d. Use the associative property of multiplication to rewrite $8(pq)$ as an equivalent expression.

Solution

a. By the commutative property of addition, the algebraic expression $11 + y$ is equivalent to $y + 11$.

b. By the commutative property of multiplication, the algebraic expression $7(y + 4)$ is equivalent to $(y + 4) \cdot 7$.

c. By the associative property of addition, the algebraic expression $(2 + 3x) + 8$ is equivalent to $2 + (3x + 8)$.

d. By the associative property of multiplication, the algebraic expression $8(pq)$ is equivalent to $(8p)q$.

You Try It **Work through the following You Try It problem.**

Work Exercises 1–11 in this textbook or in the MyLab Math Study Plan.

The **distributive property** states that multiplication distributes over addition (or subtraction). For example, compare $5(3 + 4) = 5(7) = 35$ with $(5 \cdot 3) + (5 \cdot 4) = 15 + 20 = 35$. Since both expressions simplify to 35, they are equivalent: $5(3 + 4) = (5 \cdot 3) + (5 \cdot 4)$.

Distributive Property

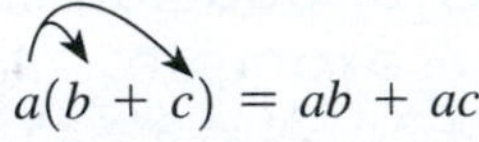

If a, b, and c are real numbers, then $a(b + c) = ab + ac$.

Because multiplication is **commutative**, we can also write the distributive property as

$$(b + c)a = ba + ca.$$

The distributive property also applies to subtraction.

$$a(b - c) = ab - ac$$

The distributive property extends to sums (or differences) involving more than two terms.

$$a(b + c + d) = ab + ac + ad$$

Example 2 Using the Distributive Property

Use the distributive property to multiply.

a. $4(x + 9)$ b. $(x - 4) \cdot 2$ c. $-5(x + y - 4)$

Solution

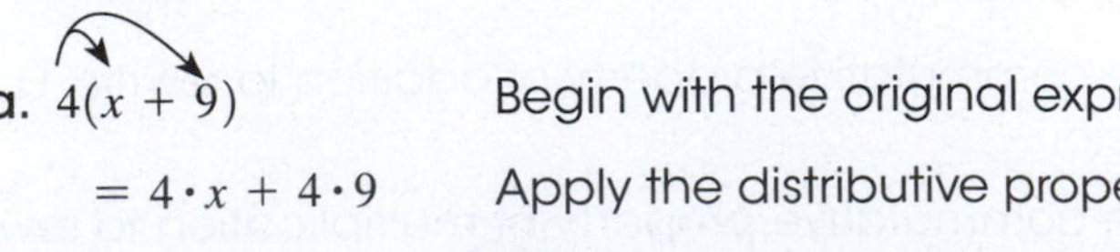

a. $4(x + 9)$ Begin with the original expression

$= 4 \cdot x + 4 \cdot 9$ Apply the distributive property

$= 4x + 36$ Find each product

b. and c. Try to complete these problems on your own. Check your answers to parts b and c, or watch this video for complete solutions to all three parts.

You Try It Work through this You Try It problem.

Work Exercises 12–15 in this eText or in the MyLab Math Study Plan.

Example 3 Using the Distributive Property

Use the distributive property to rewrite each sum or difference as a product.

a. $xz + yz$ **b.** $2xw - xy + 5xz$

Solution

a. The algebraic expression $xz + yz$ contains two terms, xz and yz. Both terms contain the same variable, z. We can therefore use the **distributive property** to rewrite the original expression as

$$xz + yz = z(x + y).$$

b. Watch this video for a complete solution to this example.

You Try It Work through this You Try It problem.

Work Exercises 16–18 in this eText or in the MyLab Math Study Plan.

OBJECTIVE 2 USING EXPONENTIAL NOTATION

Consider the **product** $2 \cdot 2 \cdot 2 \cdot 2 \cdot 2$. The **factor** 2 is repeated five times. We can write this product as the *exponential expression* 2^5. The superscript number 5 indicates that the factor 2 is repeated five times. The number 5 is called an *exponent*, or *power*, and the number 2 is called the *base*.

$$\underbrace{2 \cdot 2 \cdot 2 \cdot 2 \cdot 2}_{5 \text{ factors of } 2} = 2^5 \leftarrow \text{Exponent} \quad (2 \uparrow \text{Base})$$

We read 2^5 as "2 raised to the fifth power."

> **Definition** Exponential Expression
>
> If a is a **real number** and n is a **natural number**, then the **exponential expression** a^n represents the product of n factors of a.
>
> $$a^n = \underbrace{a \cdot a \cdot a \cdot \ldots \cdot a}_{n \text{ factors of } a}$$
>
> a is the **base** of the expression, and n is the **exponent** or **power**.

We read a^n as "a raised to the nth power" or "a to the nth." An exponent of 2 is usually read as "squared," whereas an exponent of 3 is read as "cubed." For example, 5^2 is "five squared" and 5^3 is "five cubed."

An **exponent of 1** is usually not written. For example, $2^1 = 2$.

We **simplify** numeric **exponential expressions** by multiplying the **factors**. For example, $2^5 = 2 \cdot 2 \cdot 2 \cdot 2 \cdot 2 = 32$. When a negative sign is involved, we must see if the negative sign is part of the **base**. For example, consider $(-5)^2$. The parentheses around -5 indicate that the negative sign is part of the base. So, $(-5)^2 = (-5)(-5) = 25$. In the expression -5^2, however, the negative sign is not part of the base. Instead, we need to find the **opposite** of 5^2. So, $-5^2 = -(5 \cdot 5) = -25$.

CAUTION **A common error that occurs when simplifying exponential expressions is to multiply the base by the exponent.**
For example, $\mathbf{10^2 = 10 \cdot 10 = 100}$, not $\mathbf{10 \cdot 2 = 20}$.

Example 4 Using Exponential Notation

a. Rewrite the expression $5 \cdot 5 \cdot 5$ using exponential notation. Then identify the base and the exponent.

b. Rewrite the expression $4x \cdot 4x \cdot 4x \cdot 4x \cdot 4x \cdot 4x$ using exponential notation. Then identify the base and the exponent.

c. Identify the base and the exponent of the expression $(-3)^4$, and then evaluate.

d. Identify the base and the exponent of the expression -2^5, and then evaluate.

Solution

a. We can rewrite $5 \cdot 5 \cdot 5$ as 5^3. The base is 5, and the exponent is 3.

b. The expression $4x \cdot 4x \cdot 4x \cdot 4x \cdot 4x \cdot 4x$ can be rewritten in exponential notation a $(4x)^6$. The base is $4x$ and the exponent is 6.

c. The base of the expression $(-3)^4$ is -3, and the exponent is 4. Therefore, $(-3)^4 = (-3) \cdot (-3) \cdot (-3) \cdot (-3) = 81$.

d. The expression -2^5 can be rewritten as $-1 \cdot 2^5$. The base of the exponential expression is 2, and the exponent is 5. Thus, $-2^5 = -1 \cdot 2^5 = -1 \cdot 2 \cdot 2 \cdot 2 \cdot 2 \cdot 2 = -1 \cdot 32 = -32$.

You Try It **Work through the following You Try It problem.**

Work Exercises 19–25 in this textbook or in the MyLab Math Study Plan.

OBJECTIVE 3 USING THE ORDER OF OPERATIONS TO SIMPLIFY NUMERIC AND ALGEBRAIC EXPRESSIONS

The **numeric expression** $2 + 3 \cdot 4$ contains two **operations**: addition and multiplication. Depending on which operation is **performed** first, a different answer will result.

Add first, then multiply: $2 + 3 \cdot 4 = 5 \cdot 4 = 20$
Multiply first, then add: $2 + 3 \cdot 4 = 2 + 12 = 14$
(Different results)

For this reason, mathematicians have agreed on a set **order of operations**, which determines the priority in which operations should be performed. For example, we perform multiplication before addition from left to right. So the correct **evaluation** for our numeric expression above is 14, not 20.

Order of Operations

1. **Parentheses (or other grouping symbols)** Evaluate operations within parentheses (or other grouping symbols) first, starting with the innermost set and working out.
2. **Exponents** Work from left to right evaluating any exponential expressions as they occur.
3. **Multiplication and Division** Work from left to right and perform any multiplication or division operations as they occur.
4. **Addition and Subtraction** Work from left to right and perform any addition or subtraction operations as they occur.

Example 5 Use the Order of Operations to Simplify a Numeric Expression

Simplify each expression.

a. $-2^3 + [3 - 5 \cdot (1 - 3)]$. b. $\dfrac{|2 - 3^3| + 5}{5^2 - 4^2}$

Solution Watch the interactive video to verify that

a. $-2^3 + [3 - 5 \cdot (1 - 3)] = 5$ and b. $\dfrac{|2 - 3^3| + 5}{5^2 - 4^2} = \dfrac{10}{3}$

You Try It Work through the following You Try It problem.

Work Exercises 26–33 in this textbook or in the MyLab Math Study Plan.

Example 6 Use the Order of Operations to Evaluate an Algebraic Expression

Evaluate the algebraic expression $-x^3 - 4x$ for $x = -2$.

Solution

$-x^3 - 4x$	Write the original algebraic expression.
$= -(-2)^3 - 4(-2)$	Substitute -2 for x.
$= -(-8) - 4(-2)$	$(-2)^3 = -8$.
$= 8 + 8$	Simplify.
$= 16$	

You Try It Work through the following You Try It problem.

Work Exercises 34–41 in this textbook or in the MyLab Math Study Plan.

When simplifying algebraic expressions, we must remove all grouping symbols and combine **like terms**. Like terms are two or more terms consisting of the same variable (or variables) raised to the same power. For example, $3x^2$ and $5x^2$ are like terms because each term contains the same variable, x, raised to the same power, 2.

Example 7 Simplify an Algebraic Expression

Simplify each algebraic expression.

a. $3x^2 - 2 + 5x^2$ b. $-3(4a - 7) + 5(3 - 5a)$

Solution

a. $3x^2 - 2 + 5x^2$ Write the original algebraic expression.

$= 3x^2 + 5x^2 - 2$ Use the commutative property of addition to rearrange the terms.

$= (3 + 5)x^2 - 2$ Use the distributive property $ab + ac = (b + c)a$ to combine like terms.

$= 8x^2 - 2$ Simplify.

b. $-3(4a - 7) + 5(3 - 5a)$ Write the original algebraic expression.

$= -3(4a) - 3(-7) + 5(3) + 5(-5a)$ Use the distributive property.

$= -12a + 21 + 15 - 25a$ Simplify.

$= -12a - 25a + 21 + 15$ Use the commutative property of addition to rearrange the terms.

$= (-12 - 25)a + 21 + 15$ Use the distributive property $ab + ac = (b + c)a$ to combine like terms.

$= -37a + 36$ Simplify.

You Try It Work through the following You Try It problem.

Work Exercises 42–51 in this textbook or in the MyLab Math Study Plan.

R.2 Exercises

In Exercises 1–3, use the commutative property of addition to rewrite each expression as an equivalent expression.

1. $13 + m$
2. $ab + c$
3. $11w + 10t$

In Exercises 4–6, use the commutative property of multiplication to rewrite each expression as an equivalent expression.

4. $3 \cdot z$
5. mn
6. $6(v + 4)$

In Exercises 7–9, use the associative property of addition to rewrite each expression as an equivalent expression.

7. $4 + (a + 11)$
8. $(d + a) + 30$
9. $(zw + a) + y$

In Exercises 10 and 11, use the associative property of multiplication to rewrite the expression as an equivalent expression.

10. $45 \cdot (y \cdot z)$
11. $8(2(y + z))$

In Exercises 12–15, use the distributive property to multiply the given expressions.

12. $12(11a + 4)$ 13. $(x + y)20$ 14. $15(a - 10 + 6b)$ 15. $(h + 2v - 6)5$

In Exercises 16–18, use the distributive property to rewrite each sum or difference as a product.

16. $wx + zx$ 17. $ab - bc$ 18. $qp - rq + qt$

In Exercises 19–21, write each expression using exponential notation. Then identify the base and the exponent.

19. $4 \cdot 4 \cdot 4$ 20. $w \cdot w \cdot w \cdot w$ 21. $(-6y) \cdot (-6y) \cdot (-6y) \cdot (-6y) \cdot (-6y)$

In Exercises 22–25, identify the base and exponent of each expression, and then evaluate.

22. 8^3 23. -5^4 24. $(-4)^3$ 25. $(2z)^5$

In Exercises 26–33, simplify each expression using the order of operations.

26. $7 - (4 \cdot 6 - 26)$ 27. $2 + 8^3 \div (-56) - 7$ 28. $21^3 \div \dfrac{3^2}{5 - 2} - (-1)$

29. $-\dfrac{7}{6}\left(\dfrac{1}{2}\right) + \dfrac{5}{9} \div \dfrac{7}{6}$ 30. $10^2 + 20^2 \div 5^2$ 31. $3 \cdot (18 + 3)^2 - 4 \cdot (8 - 3)^2$

32. $\dfrac{7 \cdot 3 - 2^2}{16 - 2^3}$ 33. $\dfrac{|-5^2 + (-3)^2| + 4 \cdot 5}{6 \div 2 - 3 \cdot 2^2}$

In Exercises 34–41, evaluate each algebraic expression for the indicated variable.

34. $4 - 3a$ for $a = 4$ 35. $30 \div 3x$ for $x = -2$

36. $-z^2 - 6z$ for $z = -3$ 37. $\dfrac{6b - 9b^2}{b^2 - 6}$ for $b = 3$

38. $54 \div 3^2 \cdot n(n - 4)$ for $n = 7$

39. $|8x^2 - 5y|$ for $x = -1$ and $y = 3$

40. $b^2 - 4ac$ for $a = -4, b = 3,$ and $c = -2$

41. $\dfrac{y_2 - y_1}{x_2 - x_1}$ for $x_1 = 4, x_2 = 6, y_1 = -2,$ and $y_2 = 7$

In Exercises 42–51, simplify each algebraic expression.

42. $-7x - 10x$ 43. $10z^2 - z^2$

44. $-7y^5 - 7y^5$ 45. $7a - 20 - 17a + 50$

46. $3k + 3k^2 + 7k + 7k^2$ 47. $5(3t - 5) - 6t$

48. $6 - 3[4 - (7r - 2)]$ 49. $-5(3w - 9) + 2(2w + 7)$

50. $56p^2 - 36 - [4(p^2 - 9) + 1]$ 51. $\dfrac{3}{4}x - \dfrac{4}{3} + \dfrac{7}{2}x + \dfrac{5}{6}$

R.3 The Laws of Exponents; Radicals

OBJECTIVES

1 Simplifying Exponential Expressions Involving Integer Exponents

2 Evaluating Radicals

3 Simplifying Expressions of the form $\sqrt[n]{a^n}$

4 Simplifying Exponential Expressions Involving Rational Exponents

5 Simplifying Radical Expressions Using the Product Rule

6 Simplifying Radical Expressions Using the Quotient Rule

SECTION R.3 EXERCISES

OBJECTIVE 1 SIMPLIFYING EXPONENTIAL EXPRESSIONS INVOLVING INTEGER EXPONENTS

In **Section R.2**, we introduced exponential notation. In this section, we learn how to simplify exponential expressions of the form $b^{m/n}$, where m/n is a **rational number**. We start by simplifying expressions involving *positive* integer exponents.

Properties of Positive Integer Exponents

Suppose m and n are positive integers and a and b are real numbers, then

1. $b^m b^n = b^{m+n}$
2. $\frac{b^m}{b^n} = b^{m-n}$, where $m > n$ and $b \neq 0$

   $\frac{b^m}{b^n} = 1$, where $m = n$

   $\frac{b^m}{b^n} = \frac{1}{b^{n-m}}$, where $m < n$ and $b \neq 0$
3. $(b^m)^n = b^{mn}$
4. $(ab)^n = a^n b^n$
5. $\left(\frac{a}{b}\right)^n = \frac{a^n}{b^n}$, where $b \neq 0$

Although these five properties are true for positive integer exponents, they are actually true for *all* integer exponents, including exponents that are negative integers or zero. By property (1), for positive integer exponents m and n, we know that $b^m b^n = b^{m+n}$. If we want this property to hold true for $n = 0$, then $b^m b^0 = b^{m+0} = b^m$ if and only if $b^0 = 1$. This suggests the following rule.

Zero Exponent Rule

If b is a real number such that $b \neq 0$, then $b^0 = 1$.

Similarly, if we let $n = -m$, then $b^m \cdot b^{-m} = b^{m+(-m)} = b^0 = 1$ if and only if b^{-m} is the reciprocal of b^m or $b^{-m} = \frac{1}{b^m}$. This is known as the reciprocal rule of exponents.

Reciprocal Rule of Exponents

If b is a real number such that $b \neq 0$ and m is a positive integer, then $b^{-m} = \frac{1}{b^m}$.

We now state the following laws of exponents, which hold true for *all* integer exponents.

Laws of Exponents

Suppose m and n are integers and a and b are real numbers, then

1. $b^m b^n = b^{m+n}$	Product rule of exponents
2. $\frac{b^m}{b^n} = b^{m-n}$, where $b \neq 0$	Quotient rule of exponents
3. $(b^m)^n = b^{mn}$	Power rule of exponents
4. $(ab)^n = a^n b^n$	Product to power rule of exponents
5. $\left(\frac{a}{b}\right)^n = \frac{a^n}{b^n}$, where $b \neq 0$	Quotient to power rule of exponents
6. $b^0 = 1$	Zero exponent rule
7. $b^{-m} = \frac{1}{b^m}$, where $b \neq 0$	Reciprocal rule of exponents

Example 1 Simplify Exponential Expressions Using the Laws of Exponents

Simplify each exponential expression. Write your answers using positive exponents. Assume all variables represent positive real numbers.

a. $-3^{-4} \cdot 3^4$ **b.** $5z^4 \cdot (-9z^{-5})$ **c.** $\left(\frac{a^{-3}b^4c^{-6}}{2a^5b^{-4}c}\right)^{-3}$

Solution

a. $-3^{-4} \cdot 3^4 = -3^{-4+4}$

$= -3^0$

$= -1$

b. $5z^4 \cdot (-9z^{-5}) = 5 \cdot (-9) \cdot z^4 \cdot z^{-5}$

$= -45 \cdot z^{4-5}$

$= -45 \cdot z^{-1}$

$= \frac{-45}{z}$

c. $\left(\frac{a^{-3}b^4c^{-6}}{2a^5b^{-4}c}\right)^{-3} = \left(\frac{b^4b^4}{2a^5a^3cc^6}\right)^{-3}$

$= \left(\frac{b^8}{2a^8c^7}\right)^{-3}$

$= \frac{b^{-24}}{2^{-3}a^{-24}c^{-21}}$

$= \frac{2^3a^{24}c^{21}}{b^{24}}$

$= \frac{8a^{24}c^{21}}{b^{24}}$

(eText Screens R.3-1–R.3-30)

You may want to watch the **interactive video** to see these three solutions worked out in detail.

You Try It **Work through the following You Try It problem.**

Work Exercises 1–30 in this textbook or in the MyLab Math Study Plan.

OBJECTIVE 2 EVALUATING RADICALS

Before we discuss exponential expressions with rational exponents, we must first define a radical.

A number a is said to be an nth root of b if $a^n = b$. For example, 4 is a square root (2nd root) of 16 because $(4)^2 = 16$. The number -4 is also a square root of 16 because $(-4)^2 = 16$. Similarly, 3 and -3 are 4th roots of 81 because $(3)^4 = 81$ and $(-3)^4 = 81$. It follows that every positive real number has two real nth roots when n is an even integer. This is not the case if n is an odd integer. Every real number has exactly one nth root when n is an odd integer. For example, the only real cube root (3rd root) of -8 is -2 because $(-2)^3 = -8$. We use the symbol $\sqrt[n]{b}$ to mean the *positive* nth root of b, also called the **principal** n**th root** of a number b. The symbol $\sqrt[n]{}$ is called a **radical sign**. The integer n is called the **index**. The complete expression $\sqrt[n]{b}$ is called a **radical**. We now define the principal nth root.

Definition Principal nth Root

For any integer $n \geq 2$, the **principal** n**th root** of a number b is defined as

$$\sqrt[n]{b} = a \quad \text{if and only if} \quad a^n = b.$$

If n is even and $b \geq 0$, then $a \geq 0$.

If n is even and $b < 0$, then a is not a real number.

If n is odd and $b \geq 0$, then $a \geq 0$.

If n is odd and $b < 0$, then $a < 0$.

For a real principal nth root to exist when n is a positive integer, the **radicand** must be greater than or equal to zero. Otherwise, the nth root is not a real number.

Example 2 Simplify Radical Expressions

Simplify each radical expression.

a. $\sqrt{144}$ b. $\sqrt[3]{\dfrac{1}{27}}$ c. $\sqrt[6]{-64}$

Solution

a. $\sqrt{144} = 12$ because $12^2 = 144$.

b. $\sqrt[3]{\dfrac{1}{27}} = \dfrac{1}{3}$ because $\left(\dfrac{1}{3}\right)^3 = \dfrac{1}{27}$.

c. The radical expression $\sqrt[6]{-64}$ has no real principal 6th root because there is no real number a such that $a^6 = -64$.

You Try It **Work through the following You Try It problem.**

Work Exercises 31–36 in this textbook or in the MyLab Math Study Plan.

OBJECTIVE 3 SIMPLIFYING EXPRESSIONS OF THE FORM $\sqrt[n]{a^n}$

Suppose that you were asked to evaluate the expression $\sqrt{a^2}$. At first glance, you might mistakenly think that $\sqrt{a^2} = a$. However, this is not necessarily correct. Consider the following examples:

$$\sqrt{5^2} = \sqrt{25} = 5 \text{ and}$$
$$\sqrt{(-5)^2} = \sqrt{25} = 5$$

As you can see from the two examples above, the result of both radicals was the positive number 5. We can rewrite these two radicals as

$$\text{w}\sqrt{5^2} = |5| = 5 \text{ and}$$
$$\sqrt{(-5)^2} = |-5| = 5.$$

The two results above lead to the following property for simplifying expressions of the form $\sqrt[n]{a^n}$, where n is a positive even integer.

> **Simplifying Expressions of the Form $\sqrt[n]{a^n}$, where n is a Positive Even Integer**
>
> If a is a real number and if n is an even positive integer, then $\sqrt[n]{a^n} = |a|$.

Now consider the following two examples of radicals of the form $\sqrt[n]{a^n}$, where the index n is the positive *odd* integer, 3:

$$\sqrt[3]{2^3} = \sqrt[3]{8} = 2 \text{ and}$$
$$\sqrt[3]{(-2)^3} = \sqrt[3]{-8} = -2$$

This result leads to the following property for simplifying expressions of the form $\sqrt[n]{a^n}$, where n is a positive *odd* integer.

> **Simplifying Expressions of the Form $\sqrt[n]{a^n}$, where n Is a Positive Odd Integer**
>
> If a is a real number and if n is an odd positive integer, then $\sqrt[n]{a^n} = a$.

Example 3 Simplify Radical Expressions of the Form $\sqrt[n]{a^n}$

Simplify each radical expression

a. $\sqrt[7]{z^7}$ b. $\sqrt[4]{x^4}$ c. $\sqrt[6]{(-3)^6}$

Solution

a. The radical expression $\sqrt[7]{z^7}$ is of the form $\sqrt[n]{a^n}$, where the index, n, is 7 which is a positive odd integer. Therefore, $\sqrt[7]{z^7} = z$.

b. The radical expression $\sqrt[4]{x^4}$ is of the form $\sqrt[n]{a^n}$, where the index, n, is 4 which is a positive even integer. Therefore, $\sqrt[4]{x^4} = |x|$.

(eText Screens R.3-1–R.3-30)

c. The radical expression $\sqrt[6]{(-3)^6}$ is of the form $\sqrt[n]{a^n}$, where the index, n, is 6 which is a positive even integer. Therefore, $\sqrt[6]{(-3)^6} = |-3| = 3$.

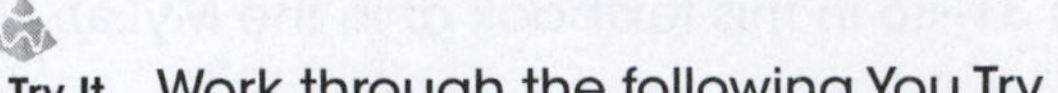

You Try It Work through the following You Try It problem.

Work Exercises 37–40 in this textbook or in the MyLab Math Study Plan.

TIP Sometimes when simplifying radical expressions involving variables we will encounter directions that say, "assume all variables represent positive real numbers." When this is the case then we do not have to use absolute value bars when simplifying radicals of the form $\sqrt[n]{a^n}$ when n is an even integer. If we assume the variable is a positive real number in Example 3b, then the expression $\sqrt[4]{x^4}$ simplifies as $\sqrt[4]{x^4} = |x| = x$ since x is positive.

OBJECTIVE 4 SIMPLIFYING EXPONENTIAL EXPRESSIONS INVOLVING RATIONAL EXPONENTS

We can use our knowledge of radicals to simplify expressions involving **rational** exponents. We start by defining the expression $b^{1/n}$.

> **Definition** $b^{1/n}$
>
> For any integer $n \geq 2$ and any real number b, we define the expression $b^{1/n}$ as
>
> $$b^{1/n} = \sqrt[n]{b}$$
>
> provided that the expression $\sqrt[n]{b}$ is a real number*.

Example 4 Evaluate Expressions Involving Rational Exponents

Simplify each expression.

a. $4^{1/2}$ **b.** $(-27)^{1/3}$ **c.** $\left(\dfrac{1}{64}\right)^{1/6}$

Solution

a. $4^{1/2} = \sqrt{4} = 2$ **b.** $(-27)^{1/3} = \sqrt[3]{-27} = -3$ **c.** $\left(\dfrac{1}{64}\right)^{1/6} = \sqrt[6]{\dfrac{1}{64}} = \dfrac{1}{2}$

Each exponent from Example 4 was a **rational number** of the form $\dfrac{1}{n}$. The following definition provides a way to deal with expressions involving rational exponents of the form $\dfrac{m}{n}$.

> **Definition** $b^{m/n}$
>
> If $\dfrac{m}{n}$ is a **rational number** with $n \geq 2$ and b is a real number, then we define the expression $b^{m/n}$ as
>
> $$b^{m/n} = \sqrt[n]{b^m} = (\sqrt[n]{b})^m$$
>
> provided that the expression $\sqrt[n]{b}$ is a **real number***.

The laws of exponents that were stated previously for integer exponents hold true for rational exponents as well and are worth stating again.

Laws of Exponents

Suppose s and t are rational numbers and a and b are real numbers, then

1. $b^s b^t = b^{s+t}$	Product rule of exponents
2. $\frac{b^s}{b^t} = b^{s-t}$, where $b \neq 0$	Quotient rule of exponents
3. $(b^s)^t = b^{st}$	Power rule of exponents
4. $(ab)^s = a^s b^s$	Product to power rule of exponents
5. $\left(\frac{a}{b}\right)^s = \frac{a^s}{b^s}$, where $b \neq 0$	Quotient to power rule of exponents
6. $b^0 = 1$	Zero exponent rule
7. $b^{-s} = \frac{1}{b^s}$, where $b \neq 0$	Reciprocal rule of exponents

When we simplify expressions involving rational exponents, we always simplify until all exponents are positive and until each base appears only once in the expression.

Example 5 Simplify Expressions Involving Rational Exponents

Simplify each expression using positive exponents. Assume all variables represent positive real numbers.

a. $(9)^{3/2}$ b. $(-32)^{-3/5}$ c. $\frac{(125x^4y^{-1/4})^{2/3}}{(x^2y)^{1/3}}$

Solution

a. $(9)^{3/2} = \sqrt{9^3} = (\sqrt{9})^3 = 3^3 = 27$

b. $(-32)^{-3/5} = (\sqrt[5]{-32})^{-3} = (-2)^{-3} = \left(\frac{1}{-2}\right)^3 = -\frac{1}{8}$

c. $\frac{(125x^4y^{-1/4})^{2/3}}{(x^2y)^{1/3}} = \frac{125^{2/3}x^{8/3}y^{-1/6}}{x^{2/3}y^{1/3}} = \frac{(\sqrt[3]{125})^2x^{8/3-2/3}}{y^{1/3+1/6}} = \frac{25x^2}{y^{1/2}}$

Watch the **interactive video** to see each solution worked out in detail.

You Try It **Work through the following You Try It problem.**

Work Exercises 41–53 in this textbook or in the MyLab Math Study Plan.

OBJECTIVE 5 SIMPLIFY RADICAL EXPRESSIONS USING THE PRODUCT RULE

We can develop a *product rule for radicals* that is similar to the **product-to-power rule for exponents**. Look at the radical expression $\sqrt[n]{ab}$.

$$\begin{aligned}\sqrt[n]{ab} &= (ab)^{\frac{1}{n}} && \text{Rewrite the radical as a rational exponent.}\\ &= a^{\frac{1}{n}}b^{\frac{1}{n}} && \text{Apply the product-to-power rule for exponents.}\\ &= \sqrt[n]{a}\sqrt[n]{b} && \text{Convert back to radical expressions.}\end{aligned}$$

This **product rule for radicals** works in both directions: $\sqrt[n]{ab} = \sqrt[n]{a}\sqrt[n]{b}$ and $\sqrt[n]{a}\sqrt[n]{b} = \sqrt[n]{ab}$. So, we can use the rule to multiply radicals and to simplify radicals.

Product Rule for Radicals

If $\sqrt[n]{a}$ and $\sqrt[n]{b}$ are real numbers, then $\sqrt[n]{a}\sqrt[n]{b} = \sqrt[n]{ab}$.

CAUTION The index on each radical must be the same in order to use the product rule for radicals.

Example 6 Using the Product Rule to Multiply Radicals

Multiply. Assume all variables represent positive real numbers.

a. $\sqrt{3}\cdot\sqrt{7}$ b. $\sqrt{5}\cdot\sqrt{20}$ c. $\sqrt[5]{9x^2}\cdot\sqrt[5]{14y^3}$ d. $\sqrt[3]{-2}\cdot\sqrt[3]{4}$

Solutions Watch this video, or read through the following solutions.

In each case, use the **product rule for radicals** and then simplify.

a. $\underbrace{\sqrt{3}\cdot\sqrt{7}}_{\text{Original expression}} = \underbrace{\sqrt{3\cdot 7}}_{\text{Product rule for radicals}} = \underbrace{\sqrt{21}}_{\text{Simplify the radicand}}$

b. $\underbrace{\sqrt{5}\cdot\sqrt{20}}_{\text{Original expression}} = \underbrace{\sqrt{5\cdot 20}}_{\text{Product rule for radicals}} = \underbrace{\sqrt{100}}_{\text{Simplify the radicand}} = \underbrace{10}_{\text{Evaluate}}$

c. $\underbrace{\sqrt[5]{9x^2}\cdot\sqrt[5]{14y^3}}_{\text{Original expression}} = \underbrace{\sqrt[5]{9x^2\cdot 14y^3}}_{\text{Product rule for radicals}} = \underbrace{\sqrt[5]{126x^2y^3}}_{\text{Simplify the radicand}}$

d. $\underbrace{\sqrt[3]{-2}\cdot\sqrt[3]{4}}_{\text{Original expression}} = \underbrace{\sqrt[3]{-2\cdot 4}}_{\text{Product rule for radicals}} = \underbrace{\sqrt[3]{-8}}_{\text{Simplify the radicand}} = \underbrace{-2}_{\text{Evaluate}}$

You Try It Work through this You Try It problem.

Work Exercises 54–57 in this eText or in the MyLab Math Study Plan.

We now learn how to simplify radicals in which the **radicand** is not a perfect nth power. We do this by rewriting the radicand as a product of perfect nth powers and any remaining factors.

Work through this **animation** to see two examples of how to simplify radicals that are not perfect nth powers.

Using the Product Rule to Simplify Radical Expressions of the Form $\sqrt[n]{a}$

Step 1 Write the radicand as a product of two factors, one being the largest possible perfect nth power.

Step 2 Use the product rule for radicals to take the nth root of each factor.

Step 3 Simplify the nth root of the perfect nth power.

Example 7 Using the Product Rule to Simplify Radicals

Use the **product rule** to simplify. Assume all variables represent positive real numbers.

a. $\sqrt{700}$ **b.** $\sqrt[3]{40}$ **c.** $\sqrt[4]{x^8y^5}$ **d.** $\sqrt{50x^4y^3}$

Solutions

Follow the **three-step process**.

a. The largest factor of 700 that is a **perfect square** is 100.

$\sqrt{700}$	Begin with the original expression.
$=\sqrt{100\cdot 7}$	Factor 700 into the product $100\cdot 7$.
$=\sqrt{100}\cdot\sqrt{7}$	Use the product rule for radicals.
$=10\sqrt{7}$	Evaluate $\sqrt{100}$.

b. The largest factor of 40 that is a **perfect cube** is 8.

$\sqrt[3]{40}$	Begin with the original expression.
$=\sqrt[3]{8\cdot 5}$	Factor 40 into the product $8\cdot 5$.
$=\sqrt[3]{8}\cdot\sqrt[3]{5}$	Use the product rule for radicals.
$=2\sqrt[3]{5}$	Evaluate $\sqrt[3]{8}$.

c. Note that x^8 is a **perfect 4th power** because $x^8=(x^2)^4$. Also, $y^5=y^4\cdot y$. So, the largest factor of x^8y^5 that is a perfect 4th power is x^8y^4.

$=\sqrt[4]{x^8y^5}$	Begin with the original expression.
$=\sqrt[4]{x^8y^4\cdot y}$	Factor x^8y^5 into the product $x^8y^4\cdot y$.
$=\sqrt[4]{x^8y^4}\cdot\sqrt[4]{y}$	Use the product rule for radicals.
$=x^2y\sqrt[4]{y}$	Simplify $\sqrt[4]{x^8y^4}$.

d. Try to simplify this **radical expression** on your own. Check your **answer**, or watch this **video** for a complete solution.

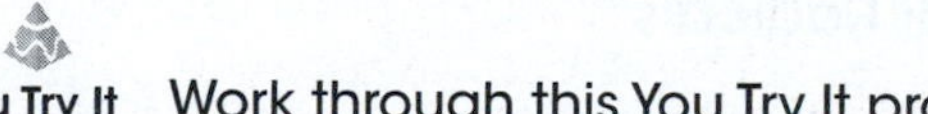

You Try It Work through this You Try It problem.

Work Exercises 58–67 in this eText or in the MyLab Math Study Plan.

Example 8 Using the Product Rule to Multiply and Simplify Radicals

Multiply and simplify. Assume all variables represent positive real numbers.

a. $3\sqrt{10}\cdot 7\sqrt{2}$ **b.** $2\sqrt[3]{4}\cdot 5\sqrt[3]{6}$ **c.** $\sqrt[4]{18x^3}\cdot\sqrt[4]{45x^2}$

Solutions First, use the **product rule** to multiply **radicals**. Then, simplify using the three-step process.

a. $3\sqrt{10}\cdot 7\sqrt{2}$	Begin with the original expression.
$= 3\cdot 7\cdot \sqrt{10}\cdot \sqrt{2}$	Rearrange factors to group radicals together.
$= 3\cdot 7\cdot \sqrt{10\cdot 2}$	Use the product rule for radicals.
$= 21\sqrt{20}$	Multiply $3\cdot 7$; multiply $10\cdot 2$.
$= 21\sqrt{4\cdot 5}$	4 is the largest factor of 20 that is a perfect square.
$= 21\sqrt{4}\cdot \sqrt{5}$	Use the product rule for radicals.
$= 21\cdot 2\sqrt{5}$	Simplify $\sqrt{4}$.
$= 42\sqrt{5}$	Multiply $21\cdot 2$.

b. and **c.** Try to work these problems on your own. Check your **answers**, or watch this **interactive video** for complete solutions to all three parts. ●

You Try It Work through this You Try It problem.

Work Exercises 68–73 in this eText or in the MyLab Math Study Plan.

Be careful not to confuse an **exponent** with the **index** of a **radical**, or vice versa. Watch this **animation** to explore this distinction in more detail.

OBJECTIVE 6 SIMPLIFY RADICAL EXPRESSIONS USING THE QUOTIENT RULE

We can now develop a *quotient rule for radicals*. Consider the **radical expression** $\sqrt[n]{\frac{a}{b}}$.

$\sqrt[n]{\dfrac{a}{b}} = \left(\dfrac{a}{b}\right)^{\frac{1}{n}}$	Rewrite the radical as a rational exponent.
$= \dfrac{a^{\frac{1}{n}}}{b^{\frac{1}{n}}}$	Apply the **quotient-to-power rule for exponents.**
$= \dfrac{\sqrt[n]{a}}{\sqrt[n]{b}}$	Convert back to radical expressions.

The **quotient rule for radicals** is used to simplify radical expressions involving fractions.

> **Quotient Rule for Radicals**
>
> If $\sqrt[n]{a}$ and $\sqrt[n]{b}$ are **real numbers** and $b \neq 0$, then $\sqrt[n]{\dfrac{a}{b}} = \dfrac{\sqrt[n]{a}}{\sqrt[n]{b}}$.

CAUTION The index on each radical must be the same in order to use the quotient rule for radicals.

Like the product rule, the quotient rule works in both directions: $\sqrt[n]{\dfrac{a}{b}} = \dfrac{\sqrt[n]{a}}{\sqrt[n]{b}}$ and $\dfrac{\sqrt[n]{a}}{\sqrt[n]{b}} = \sqrt[n]{\dfrac{a}{b}}$.

We now add more requirements for a radical expression to be **simplified**.

> **Simplified Radical Expression**
>
> For a **radical expression** to be **simplified**, it must meet the following three conditions:
>
> **Condition 1.** The **radicand** has no **factor** that is a **perfect power** of the **index** of the radical.
>
> **Condition 2.** The radicand contains no fractions or **negative exponents**.
>
> **Condition 3.** No **denominator** contains a radical.

As we saw in **Example 7**, the **product rule for radicals** can be used to help resolve issues with Condition 1. The **quotient rule for radicals** is used to resolve issues with Conditions 2 and 3.

Example 9 Using the Quotient Rule to Simplify Radicals

Use the quotient rule to simplify. Assume all variables represent positive real numbers.

a. $\sqrt{\dfrac{25}{64}}$ b. $\sqrt[3]{\dfrac{16x^5}{27}}$ c. $\sqrt[4]{\dfrac{7x^4}{625}}$ d. $\sqrt{\dfrac{3x^9}{48x^3}}$

Solutions In each case, we use the **quotient rule for radicals**.

a. $\underbrace{\sqrt{\dfrac{25}{64}}}_{\text{Original expression}} = \underbrace{\dfrac{\sqrt{25}}{\sqrt{64}}}_{\text{Quotient rule for radicals}} = \underbrace{\dfrac{5}{8}}_{\text{Evaluate each square root}}$

b. $\underbrace{\sqrt[3]{\dfrac{16x^5}{27}}}_{\text{Original expression}} = \underbrace{\dfrac{\sqrt[3]{16x^5}}{\sqrt[3]{27}}}_{\text{Quotient rule for radicals}} = \underbrace{\dfrac{\sqrt[3]{8x^3\cdot 2x^2}}{\sqrt[3]{27}}}_{\text{Write as product using perfect-cube factors}} = \underbrace{\dfrac{\sqrt[3]{8x^3}\cdot\sqrt[3]{2x^2}}{\sqrt[3]{27}}}_{\text{Product rule for radicals}} = \underbrace{\dfrac{2x\sqrt[3]{2x^2}}{3}}_{\text{Simplify the cube roots}}$

c. and **d.** Try using the quotient rule to simplify these expressions on your own. Check your **answers**, or watch this **video** for complete solutions. ●

You Try It **Work through this You Try It problem.**

Work Exercises 74–79 in this eText or in the MyLab Math Study Plan.

In Example 9, we used the quotient rule for radicals to remove fractions from the radicand. In Example 10, we will use the quotient rule to remove **radicals** from a denominator.

Example 10 Using the Quotient Rule to Simplify Radicals

Use the quotient rule to **simplify**. Assume all **variables** represent positive numbers.

a. $\dfrac{\sqrt{240x^3}}{\sqrt{15x}}$ b. $\dfrac{\sqrt[3]{-500z^2}}{\sqrt[3]{4z^{-1}}}$ c. $\dfrac{\sqrt{150m^9}}{\sqrt{3m}}$ d. $\dfrac{\sqrt{45x^5y^{-3}}}{\sqrt{20xy^{-1}}}$

Solutions In each case, we use the **quotient rule for radicals.**

a. $\dfrac{\sqrt{240x^3}}{\sqrt{15x}}$ Begin with the original expression.

$= \sqrt{\dfrac{240x^3}{15x}}$ Use the quotient rule for radicals.

$= \sqrt{16x^2}$ Simplify the radicand.

$= 4x$ Simplify.

b. $\dfrac{\sqrt[3]{-500z^2}}{\sqrt[3]{4z^{-1}}}$ Begin with the original expression.

$= \sqrt[3]{\dfrac{-500z^2}{4z^{-1}}}$ Use the quotient rule for radicals.

$= \sqrt[3]{\dfrac{-500}{4} \cdot z^{2-(-1)}}$ Divide factors and subtract exponents in radicand.

$= \sqrt[3]{-125z^3}$ Simplify radicand.

$= -5z$ Simplify the cube root.

c. and d. Try to simplify each radical expression on your own. Check your **answers**, or watch this **interactive video** for complete solutions to all four parts.

You Try It Work through this You Try It problem.

Work Exercises 80–85 in this eText or in the MyLab Math Study Plan.

R.3 Exercises

In Exercises 1–30, use the laws of exponents to simplify each expression using positive exponents only. Assume all variables represent nonzero real numbers.

1. $3^2 \cdot 3^3$
2. $k^4 \cdot k^5$
3. $x^5 \cdot x^9 \cdot x^{12}$
4. $(a^5b^4)(a^7b^2)$
5. $\dfrac{t^9}{t^2}$
6. $(r^5)^{-7}$
7. $(w^{-8})^{-6}$
8. $\dfrac{n}{n^{-7}}$
9. $\dfrac{m^{-6}}{m^{-9}}$
10. $(-5y)^0$
11. -6^0
12. 2^{-2}
13. -4^{-2}
14. n^{-3}
15. $5x^{-6}$
16. $-7y^{-2}$
17. $\dfrac{1}{t^{-2}}$
18. $\dfrac{5}{y^{-9}}$
19. $\dfrac{a^{-11}}{b^{-4}}$
20. $\dfrac{-5p^{-7}q^4}{w^5}$
21. $b^{11}b^{-4}$
22. $a^{-11}a^{-3}$
23. $3^{-4} \cdot 3$
24. $\dfrac{z^{14}}{z^{-4}}$
25. $(-5x^2y^5)^2$
26. $\dfrac{24x^3y^5}{32x^8y^{-7}}$
27. $\dfrac{(a^{-3}b)^{-2}}{(a^3b^{-1})^2}$
28. $(16a^{-3}bc^{-6})(2ab)^{-5}$

29. $\left(\dfrac{q^3p^4w^5}{q^{-3}p^{-4}w^{-5}}\right)^{-2}$ 30. $\dfrac{(3^{-1}x^{-1}y^{-1})^{-2}(3x^{-4}y^3)(9x^{-2}y^3)^0}{(3x^{-3}y^{-5})^2}$

In Exercises 31–36, evaluate the radical or state that it is not a real number.

31. $\sqrt{169}$ 32. $\sqrt{\dfrac{1}{25}}$ 33. $\sqrt[3]{216}$

34. $\sqrt[3]{-343}$ 35. $\sqrt[4]{-16}$ 36. $\sqrt[3]{-\dfrac{1}{1000}}$

In Exercises 37–40, simplify the radical expression.

37. $\sqrt[5]{y^5}$ 38. $\sqrt[4]{(-9)^4}$ 39. $\sqrt{k^2}$ 40. $\sqrt[5]{8^5}$

In Exercises 41–53, simplify each exponential expression. Write any negative exponents as positive exponents. Assume all variables represent positive real numbers.

41. $25^{1/2}$ 42. $\left(\dfrac{1}{16}\right)^{1/4}$ 43. $-81^{1/4}$ 44. $16^{3/4}$

45. $8^{-4/3}$ 46. $x^{-1/3}$ 47. $x^{-3/4} \cdot x^{7/4}$ 48. $x^{1/3} \cdot x^{1/4} \cdot x^{-1/2}$

49. $(x^5y^{10})^{1/5}$ 50. $\dfrac{(x^2y^4)^{4/7}(xy^2)^{1/7}}{x^{4/7}y^{4/7}}$ 51. $(9x^2y^{-5/2})^{1/2}$

52. $(32x^{-5}y^{10})^{-1/5}(xy^{1/5})$ 53. $\left(\dfrac{x^{-3/4}y^{1/4}}{x^{-1/4}}\right)^{-8}$

In Exercises 54–57, multiply. Assume all variables represent positive real numbers.

54. $\sqrt[4]{5} \cdot \sqrt[4]{7}$ 55. $\sqrt{3} \cdot \sqrt{12}$ 56. $\sqrt[3]{5x^2} \cdot \sqrt[3]{25x}$ 57. $\sqrt[4]{5x} \cdot \sqrt[4]{15y^3}$

In Exercises 58–67, use the product rule to simplify. Assume all variables represent positive real numbers.

58. $\sqrt{72}$ 59. $\sqrt[3]{56}$ 60. $-\sqrt{63}$ 61. $\sqrt{48x^3}$

62. $\sqrt[3]{-125x^6}$ 63. $3\sqrt{125y^9}$ 64. $7\sqrt[3]{128x^4y^5}$ 65. $\sqrt{a^{15}}$

66. $\sqrt[4]{x^5y^8z^9}$ 67. $x^2y\sqrt[4]{x^5y^{11}}$

In Exercises 68–73, multiply and simplify. Assume all variables represent positive real numbers

68. $\sqrt{6} \cdot \sqrt{30}$ 69. $4\sqrt{3x} \cdot 7\sqrt{6x}$ 70. $\sqrt{20x^3y} \cdot \sqrt{18xy^4}$

71. $\sqrt[3]{9x^2} \cdot \sqrt[3]{6x^2}$ 72. $2x\sqrt[3]{5y^2} \cdot 3x\sqrt[3]{25y}$ 73. $\sqrt[4]{8x^2} \cdot \sqrt[4]{6x^3}$

In Exercises 74–85, use the quotient rule to simplify. Assume all variables represent positive real numbers.

74. $\sqrt{\dfrac{49}{81}}$ 75. $\sqrt{\dfrac{12x^3}{25y^2}}$ 76. $\sqrt[3]{\dfrac{11x^3}{8}}$ 77. $\sqrt[3]{\dfrac{16x^9}{125y^6}}$

78. $\sqrt[4]{\dfrac{21x^5}{16}}$ 79. $\sqrt{\dfrac{6x^5}{54x}}$ 80. $\dfrac{\sqrt{45}}{\sqrt{5}}$ 81. $\dfrac{\sqrt{96x^6y^8}}{\sqrt{12x^2y^5}}$

82. $\dfrac{\sqrt{120x^9}}{\sqrt{3x^{-1}}}$ 83. $\dfrac{\sqrt[3]{5000}}{\sqrt[3]{5}}$ 84. $\dfrac{\sqrt[3]{-72m}}{\sqrt[3]{3m^{-2}}}$ 85. $\dfrac{\sqrt{3x^5y^3}}{\sqrt{48xy^5}}$

R.4 Polynomials

OBJECTIVES

1 Understanding the Definition of a Monomial and a Polynomial

2 Adding and Subtracting Polynomials

3 Multiplying Polynomials

4 Dividing Polynomials Using Long Division

SECTION R.4 EXERCISES

OBJECTIVE 1 UNDERSTANDING THE DEFINITION OF A MONOMIAL AND A POLYNOMIAL

A **term** is a **constant**, a **variable**, or the **product** of a constant and one or more variables raised to **powers**. If a term contains a single constant **factor** (possibly a simplified fraction) and if none of the variable factors can be combined using the **rules for exponents**, then it is a **simplified term**. Examples of simplified terms include

$$4x^3,\ -\frac{2}{5}x^2y^3,\ x^2,\ 35x^{-3},\ -3yz^4,\ 9y^{1/2},\ \frac{8}{x},\text{ and }10.$$

The terms $7x^3x^2$ and $\dfrac{4x^4}{6}$ are not simplified. Do you see why?

Some simplified terms are also *monomials*.

> **Definition** Monomial
>
> A **monomial** is a simplified term in which all variables are raised to non-negative **integer** powers and no variables appear in any denominator.

Every **monomial** has both a *degree* and a *coefficient*.

> **Definition** Degree of a Monomial
>
> The **degree of a monomial** is the sum of the **exponents** on the variables.

For example, the degree of $4x^3$ is 3, and the degree of $-\dfrac{2}{5}x^2y^3$ is $2 + 3 = 5$.

Definition Coefficient of a Monomial

The **coefficient of a monomial** is the **constant** factor. It is sometimes called the **numerical coefficient**.

For example, the coefficient of $4x^3$ is 4, and the coefficient $-\frac{2}{5}x^2y^3$ is $-\frac{2}{5}$.

Other examples of monomials include x^5, $-\sqrt{2}x^2y^7$, and πxyz. The degree and coefficient of each of these monomials is shown in the box at right.

Monomial	Degree	Coefficient
x^5	5	1
$-\sqrt{2}x^2y^7$	9	$-\sqrt{2}$
πxyz	3	π

You Try It Work through the following You Try It problem.

Work Exercises 1–5 in this textbook or in the MyLab Math Study Plan.

A **polynomial** is the sum or difference of one or more monomials. A polynomial with two terms is called a **binomial.** A polynomial with three terms is called a **trinomial.** The **degree** of a polynomial is the degree of the highest monomial term.

Example 1 Identify Polynomials

Determine whether each algebraic expression is a polynomial. If the expression is a polynomial, state the degree.

a. $8x^3y^2 - 2$ **b.** $5w^5 - 3y^2 - \frac{6}{w}$ **c.** $7p^8m^4 - 5p^5m^3 + 2pm$

Solution

a. The expression $8x^3y^2 - 2$ is a polynomial. Because this polynomial has two terms, we say that $8x^3y^2 - 2$ is a binomial. The degree of this binomial is 5.

b. The expression $5w^5 - 3y^2 - \frac{6}{w}$ is not a polynomial because the term $\frac{6}{w}$ is equivalent to $6w^{-1}$. For an expression to be a polynomial, all exponents must be nonnegative integers.

c. The expression $7p^8m^4 - 5p^5m^3 + 2pm$ is a polynomial. Because this polynomial has three terms, we say that $7p^8m^4 - 5p^5m^3 + 2pm$ is a trinomial. The degree of this trinomial is 12.

You Try It Work through the following You Try It problem.

Work Exercises 6–10 in this textbook or in the MyLab Math Study Plan.

OBJECTIVE 2 ADDING AND SUBTRACTING POLYNOMIALS

To add or subtract polynomial expressions, we combine **like terms**. Recall that like terms have the same variables raised to the same corresponding exponents. For example, the monomial expressions $3x^2y^3$ and $5x^2y^3$ are like terms.

We can add or subtract these terms using the reverse of the **distributive property**.

$$3x^2y^3 + 5x^2y^3 = (3 + 5)x^2y^3 = 8x^2y^3$$
$$3x^2y^3 - 5x^2y^3 = (3 - 5)x^2y^3 = -2x^2y^3$$

Example 2 Add and Subtract Polynomials

Add or subtract the polynomials as indicated.

a. $(6x^4y^2 - 2xy^3) + (7xy^3 - 8x^4y^2)$ b. $(6x^4y^2 - 2xy^3) - (7xy^3 - 8x^4y^2)$

Solution

a. To add polynomials, we remove the parentheses and then combine the like terms.

$(6x^4y^2 - 2xy^3) + (7xy^3 - 8x^4y^2)$	Write the original expression.
$= 6x^4y^2 - 2xy^3 + 7xy^3 - 8x^4y^2$	Remove the parentheses.
$= (6x^4y^2 - 8x^4y^2) + (7xy^3 - 2xy^3)$	Collect like terms.
$= (6 - 8)x^4y^2 + (7 - 2)xy^3$	Use the reverse of the **distributive property** to combine like terms.
$= -2x^4y^2 + 5xy^3$	Simplify.

b. To subtract polynomials, we remove the parentheses, making sure to change the sign of each term of the second polynomial and then combine like terms.

$(6x^4y^2 - 2xy^3) - (7xy^3 - 8x^4y^2)$	Write the original expression.
$= 6x^4y^2 - 2xy^3 - 7xy^3 + 8x^4y^2$	Remove the parentheses. Change the sign of each term of the second polynomial.
$= (6x^4y^2 + 8x^4y^2) + (-7xy^3 - 2xy^3)$	Collect like terms.
$= (6 + 8)x^4y^2 + (-7 - 2)xy^3$	Use the reverse of the **distributive property** to combine like terms.
$= 14x^4y^2 - 9xy^3$	Simplify.

You Try It Work through the following You Try It problem.

Work Exercises 11–18 in this textbook or in the MyLab Math Study Plan.

For the remainder of this section, we will focus our attention mainly on polynomial expressions in one variable.

> **Definition** Polynomial in One Variable
>
> The algebraic expression $a_nx^n + a_{n-1}x^{n-1} + a_{n-2}x^{n-2} + \cdots + a_1x + a_0$ is a polynomial in one variable of degree n, where n is a nonnegative integer. The numbers $a_0, a_1, a_2, \ldots, a_n$ are called the **coefficients** of the polynomial.

We typically write polynomials in descending order, meaning that we first list the monomial with the highest degree, then the monomial with the next highest degree, and so on. Polynomials that are written in descending order are said to be written in **standard form**.

Example 3 Add and Subtract Polynomials

Add or subtract as indicated and express your answer in standard form.

a. $(-x^5 + 3x^4 - 9x^2 + 7x + 9) + (x^4 - 5x^5 + x^3 - 3x^2 - 8)$

b. $(-x^5 + 3x^4 - 9x^2 + 7x + 9) - (x^4 - 5x^5 + x^3 - 3x^2 - 8)$

Solution

a. To add polynomials, we remove the parentheses then combine the like terms.

$(-x^5 + 3x^4 - 9x^2 + 7x + 9) + (x^4 - 5x^5 + x^3 - 3x^2 - 8)$	Write the original expression.
$= -x^5 + 3x^4 - 9x^2 + 7x + 9 + x^4 - 5x^5 + x^3 - 3x^2 - 8$	Remove the parentheses.
$= (-x^5 - 5x^5) + (3x^4 + x^4) + x^3 + (-9x^2 - 3x^2) + 7x + (9 - 8)$	Collect like terms.
$= -6x^5 + 4x^4 + x^3 - 12x^2 + 7x + 1$	Simplify and write in standard form.

b. To subtract polynomials, we remove the parentheses, making sure to change the sign of each term of the second polynomial and then combine the like terms.

$(-x^5 + 3x^4 - 9x^2 + 7x + 9) - (x^4 - 5x^5 + x^3 - 3x^2 - 8)$	Write the original expression.
$= -x^5 + 3x^4 - 9x^2 + 7x + 9 - x^4 + 5x^5 - x^3 + 3x^2 + 8$	Remove the parentheses and change the sign of each term of the second polynomial.
$= (-x^5 + 5x^5) + (3x^4 - x^4) - x^3 + (-9x^2 + 3x^2) + 7x + (9 + 8)$	Collect like terms.
$= 4x^5 + 2x^4 - x^3 - 6x^2 + 7x + 17$	Simplify and write in standard form.

Watch the **video** to see each solution worked out in detail.

You Try It Work through the following You Try It problem.

Work Exercises 19–25 in this textbook or in the MyLab Math Study Plan.

OBJECTIVE 3 MULTIPLYING POLYNOMIALS

To find the product of two monomials in one variable, we multiply the coefficients together and use the **product rule of exponents** to add any exponents. For example, $(8x^4)(-3x^7) = (8)(-3) \cdot x^4 \cdot x^7 = -24x^{11}$. To multiply a monomial by a polynomial with more than one term, we need to use the **distributive property** multiple times. Thus, if m is a monomial and $p_1, p_2, p_3, \ldots\ p_n$ are the terms of a polynomial, then

$$m(p_1 + p_2 + p_3 + \cdots + p_n) = mp_1 + mp_2 + mp_3 + \cdots + mp_n.$$

Example 4 Multiply a Polynomial by a Monomial

Multiply $(-2x^5)(3x^3 - 5x^2 + 1)$.

Solution

$$\begin{aligned}(-2x^5)(3x^3 - 5x^2 + 1) &= -2x^5(3x^3) - 2x^5(-5x^2) - 2x^5(1)\\ &= -6x^8 + 10x^7 - 2x^5\end{aligned}$$

When we find the product of any two polynomials, we use the **distributive property** to multiply each term of the first polynomial by each term of the second polynomial and then simplify.

Example 5 Multiply Polynomials

Find the product and express your answer in standard form.

$$(2x^2 - 3)(3x^3 - x^2 + 1)$$

Solution Watch the **video** to verify that $(2x^2 - 3)(3x^3 - x^2 + 1) = 6x^5 - 2x^4 - 9x^3 + 5x^2 - 3$.

You Try It Work through the following You Try It problem.

Work Exercises 26–31 in this textbook or in the MyLab Math Study Plan.

THE FOIL METHOD

When we find the product of two binomials, we can use a technique known as the FOIL method. FOIL is an acronym that stands for **F**irst, **O**utside, **I**nside, and **L**ast.

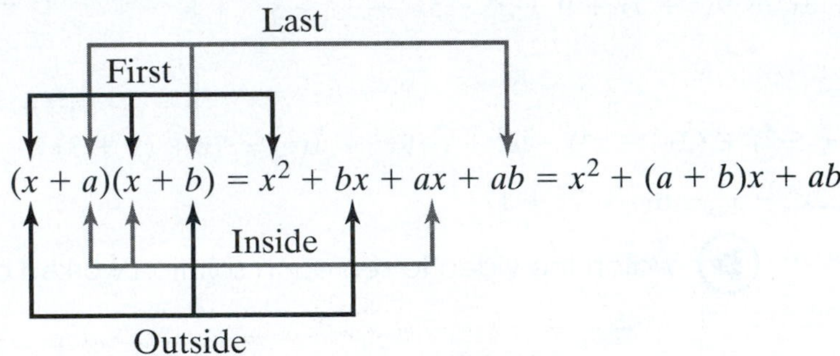

We illustrate the FOIL method by finding the product of the binomials $3x + 5$ and $2x + 7$.

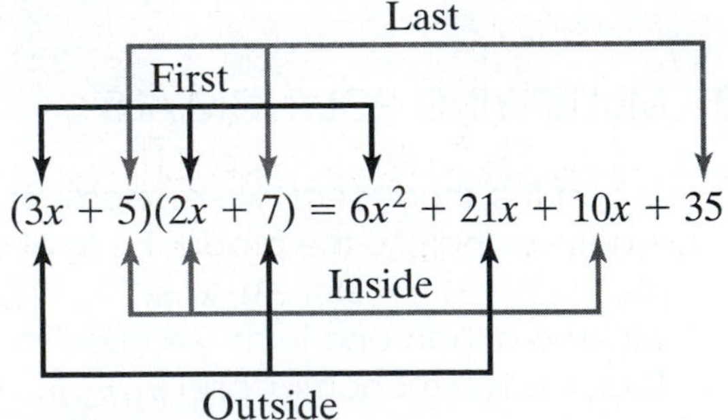

The product of the **F**irst terms:	$(3x)(2x) = 6x^2$
The product of the **O**utside terms:	$(3x)(7) = 21x$
The product of the **I**nside terms:	$(5)(2x) = 10x$
The product of the **L**ast terms:	$(5)(7) = 35$

We now simplify by combining like terms:

$$6x^2 + 21x + 10x + 35 = 6x^2 + 31x + 35$$

Therefore, $(3x + 5)(2x + 7) = 6x^2 + 31x + 35$.

Watch this **animation** to see how the FOIL method works.

You Try It Work through the following You Try It problem.

Work Exercises 32–40 in this textbook or in the MyLab Math Study Plan.

Example 6 Multiply Binomials Using the FOIL Method

Find the product: $(x + 4)(x - 4)$.

Solution

$(x + 4)(x - 4)$	Write the original expression.
$= x^2 - 4x + 4x - 16$	Use the FOIL method.
$= x^2 - 16$	Combine like terms.

Notice that the product of the two binomials in Example 6 is also a binomial. (There is no x-term.) The product in Example 6 is a special product known as the **difference of two squares** because the resulting binomial is the difference of two perfect square terms. In general, $(a + b)(a - b) = a^2 - b^2$. This formula is worth remembering. There are other special products that are worth remembering as well. These products include the **square of a binomial** and the **cube of a binomial**. We now list some special products.

Special Products

Let a and b represent any real number or algebraic expression, then

$(a + b)(a - b) = a^2 - b^2$	Difference of two squares
$(a + b)^2 = a^2 + 2ab + b^2$	Squaring a binomial
$(a - b)^2 = a^2 - 2ab + b^2$	
$(a + b)^3 = a^3 + 3a^2b + 3ab^2 + b^3$	Cubing a binomial
$(a - b)^3 = a^3 - 3a^2b + 3ab^2 - b^3$	

Determine whether you can use these special product formulas to find the products in Example 7.

Example 7 Use Special Product Formulas

Find each product using a special product formula.

a. $(2x + 3)(2x - 3)$ **b.** $(3x + 2y)^2$ **c.** $(5 - 3z)^3$

Solutions

a. We can use the difference of two squares formula $(a + b)(a - b) = a^2 - b^2$ with $a = 2x$ and $b = 3$ to get $(2x + 3)(2x - 3) = (2x)^2 - (3)^2 = 4x^2 - 9$.

b. We can use the squaring a binomial formula $(a + b)^2 = a^2 + 2ab + b^2$ with $a = 3x$ and $b = 2y$ to get $(3x + 2y)^2 = (3x)^2 + 2(3x)(2y) + (2y)^2 = 9x^2 + 12xy + 4y^2$.

c. Using the cubing a binomial formula $(a - b)^3 = a^3 - 3a^2b + 3ab^2 - b^3$ with $a = 5$ and $b = 3z$ we get

$$(5 - 3z)^3 = (5)^3 - 3(5)^2(3z) + 3(5)(3z)^2 - (3z)^3$$
$$= 125 - 225z + 135z^2 - 27z^3$$

(eText Screens R.4-1–R.4-21)

You Try It Work through the following You Try It problem.

Work Exercises 41–52 in this textbook or in the MyLab Math Study Plan.

OBJECTIVE 4 DIVIDING POLYNOMIALS USING LONG DIVISION

We can long divide two polynomials in much the same way as the long division of two integers.

For example, suppose we want to divide the polynomial $2x^2 - 7x + 4$ by the polynomial $x + 8$. The polynomial $2x^2 - 7x + 4$ is the **dividend** and $x + 8$ is the **divisor**. We set up this problem using the familiar symbol used for long division with the dividend on the inside of the symbol and the divisor on the outside.

$$x + 8\overline{)2x^2 - 7x + 4}$$

We now illustrate the long division process.

Step 1. Divide the first term of the dividend, $2x^2$, by the first term of the divisor, x. Write the result, $2x$, above the first term of the dividend.

$$\begin{array}{r} 2x \\ x + 8\overline{)2x^2 - 7x + 4} \end{array} \qquad \frac{2x^2}{x} = 2x, \text{ or equivalently, } 2x \cdot x = 2x^2.$$

Step 2. Multiply $2x$ by $x + 8$ and line up like terms vertically.

$$\begin{array}{r} 2x \\ x + 8\overline{)2x^2 - 7x + 4} \\ \underline{2x^2 + 16x} \end{array} \qquad 2x(x + 8) = 2x^2 + 16x.$$

Step 3. Subtract $2x^2 + 16x$ from $2x^2 - 7x$. We perform the subtraction by **changing the sign** of each term of the polynomial being subtracted and then combining like terms.

$$\begin{array}{r} 2x \\ x + 8\overline{)2x^2 - 7x + 4} \\ \underline{-2x^2 - 16x} \\ -23x \end{array} \qquad \longleftarrow \text{Change the signs: } -(2x^2 + 16x) = -2x^2 - 16x.$$

Step 4. Bring down the constant 4.

$$\begin{array}{r} 2x \\ x + 8\overline{)2x^2 - 7x + 4} \\ \underline{-2x^2 - 16x} \downarrow \\ -23x + 4 \end{array}$$

Step 5. Divide $-23x$ by the first term of the divisor, x. Write the result, -23, above the second term of the dividend.

$$\begin{array}{r} 2x - 23 \\ x + 8\overline{)2x^2 - 7x + 4} \\ \underline{-2x^2 - 16x} \\ -23x + 4 \end{array} \qquad \frac{-23x}{x} = -23, \qquad \text{or equivalently, } -23 \cdot x = -23x.$$

Step 6. Multiply -23 by $x + 8$ and line up like terms vertically.

$$\begin{array}{r} 2x - 23 \\ x + 8\overline{)2x^2 - 7x + 4} \\ \underline{-2x^2 - 16x} \\ -23x + 4 \\ \underline{-23x - 184} \end{array} \qquad -23(x + 8) = -23x - 184.$$

Step 7. Subtract $-23x - 184$ from $-23x + 4$.

$$\begin{array}{r} 2x - 23 \\ x + 8\overline{)2x^2 - 7x + 4} \\ \underline{-2x^2 - 16x} \\ -23x + 4 \\ \underline{+23x + 184} \\ 188 \end{array}$$

Change the signs. $-(23x - 184) = 23x + 184$

We know that we are done because 188 is of lower degree (degree zero) than the degree of the divisor (degree one). The polynomial $2x - 23$ is called the **quotient**, and the constant 188 is the **remainder**. We can always check our long division process by verifying that the dividend is equal to the product of the quotient and divisor plus the remainder.

$$\text{dividend} = (\text{divisor})(\text{quotient}) + \text{remainder}$$

$$2x^2 - 7x + 4 \stackrel{?}{=} (x + 8)(2x - 23) + 188$$

$$2x^2 - 7x + 4 \stackrel{?}{=} 2x^2 - 7x - 184 + 188$$

$$2x^2 - 7x + 4 = 2x^2 - 7x + 4 \checkmark$$

Example 8 Divide Polynomials

Find the quotient and remainder when $3x^4 + x^3 + 7x + 4$ is divided by $x^2 - 1$.

Solution Using the long division symbol, place the dividend on the inside of the symbol and the divisor on the outside of the symbol. We write both polynomials in standard form, inserting coefficients of 0 for any missing terms.

$$x^2 + 0x - 1\overline{)3x^4 + x^3 + 0x^2 + 7x + 4}$$

The long division process is shown here. Watch the video to see every step of this process.

$$\begin{array}{r} 3x^2 + x + 3 \\ x^2 + 0x - 1\overline{)3x^4 + x^3 + 0x^2 + 7x + 4} \\ \underline{-(3x^4 + 0x^3 - 3x^2)} \\ x^3 + 3x^2 + 7x \\ \underline{-(x^3 + 0x^2 - x)} \\ 3x^2 + 8x + 4 \\ \underline{-(3x^2 + 0x - 3)} \\ 8x + 7 \end{array}$$

The quotient is $3x^2 + x + 3$, and the remainder is $8x + 7$.

You Try It Work through the following You Try It problem.

Work Exercises 53–56 in this textbook or in the MyLab Math Study Plan.

R.4 Exercises

In Exercises 1–5, determine whether the algebraic expression is a monomial. If the expression is a monomial, state the degree and coefficient.

1. $5x^4$
2. $\dfrac{13}{x^2}$
3. $-x^4y^8$
4. $2x^2 - 3y^5$
5. 0

In Exercises 6–10, determine whether the algebraic expression is a polynomial. If the expression is a polynomial, state the degree.

6. $5x^3 - \dfrac{13}{x^2}$
7. π
8. $3wx^2 - \sqrt{5}$
9. $\dfrac{2x^2 - 3y^5}{x^5 - 1}$
10. $2p^3m^4 - 5p^2m^3 + 2p^4m - 4pm^3 + 2p^4m^3$

In Exercises 11–18, perform the indicated operations.

11. $(9x^2y - 6xy) + (8x^2y - xy)$
12. $(17p^2q + 21pq^2) - (10pq^2 - 13p^2q)$
13. $(a^3 + b^3) + (2a^3 - 3a^2b + 9b^3)$
14. $(5x^3 + xy - 7) + (-2x^3 - 4xy + 3)$
15. $(2m^2 + n) - (4m^2 - 16mn - 5n)$
16. $(3a^4b^2 + 7a^2b^2 - 5ab) - (2a^4b^2 - 8a^2b^2 + ab)$
17. $\left(\dfrac{2}{5}x^2 + \dfrac{5}{6}xy - \dfrac{3}{8}y^2\right) + \left(-\dfrac{4}{5}x^2 + \dfrac{1}{9}xy - \dfrac{1}{4}y^2\right)$
18. $\left(\dfrac{9}{10}a - \dfrac{1}{4}b\right) - \left(\dfrac{3}{10}a - \dfrac{5}{8}b\right)$

In Exercises 19–25, perform the indicated operations. Write each answer in standard form.

19. $(4x + 5) + (-5x + 9)$
20. $(x^2 + 11x + 7) + (3x - 13)$
21. $(9x^3 - 12x^2 + 11x + 5) - (3x^2 - 5x + 11)$
22. $(4x^5 + 8x^3 + 9x) + (7x^4 - 2x^3 + 8x^2)$
23. $(4x^2 - 4x + 4) + (3x^2 - 5x + 3) - (7x^2 + 8)$
24. $\left(\dfrac{1}{3}x^2 - \dfrac{1}{4}x + \dfrac{1}{5}\right) + \left(\dfrac{1}{6}x^2 + \dfrac{1}{3}x - \dfrac{1}{2}\right)$
25. $\left(\dfrac{2}{7}x^3 + \dfrac{3}{5}x^2 - \dfrac{4}{9}x - \dfrac{5}{6}\right) - \left(\dfrac{3}{8}x^3 - \dfrac{4}{3}x^2 + \dfrac{5}{6}x - \dfrac{2}{3}\right)$

In Exercises 26–31, find the indicated product. Write each answer in standard form.

26. $x(x^2 - x + 7)$
27. $-5x^4(7x^6 - 2)$
28. $2x^2(3x^5 - 2x^3 + 7x - 4)$
29. $(x + 3)(x^2 - 7x - 6)$
30. $-\dfrac{1}{2}x^2(2x^2 + 4x + 6)$
31. $\dfrac{1}{3}x^3(5x^3 + 6x^2 - 2x - 9)$

In Exercises 32–40, find the indicated product using the FOIL method. Write each answer in standard form.

32. $(x + 3)(x + 2)$
33. $(x + 9)(x - 4)$
34. $(x - 5)(x - 7)$
35. $(3x + 2)(x + 5)$
36. $(5x - 4)(3x + 2)$
37. $(2x - 7)(5x - 8)$
38. $(-x - 8)(-3x - 7)$
39. $\left(x + \frac{1}{3}\right)\left(x - \frac{1}{2}\right)$
40. $\left(\frac{2}{3}x + \frac{3}{4}\right)\left(\frac{3}{5}x + \frac{5}{6}\right)$

In Exercises 41–52, find each product using a special product formula.

41. $(x - 10)(x + 10)$
42. $(6x + 2)(6x - 2)$
43. $(x + 5)^2$
44. $(4x - 9)^2$
45. $(8x + y)(8x - y)$
46. $(x + 4)^3$
47. $(x - 6)^3$
48. $(7p - 2q)^2$
49. $(3x + 2)^3$
50. $(x^2y^3 - 4xy^2)(x^2y^3 + 4xy^2)$
51. $\left(\frac{2}{5}xy + 3\right)^2$
52. $(x - 4)(x + 4)(x^2 - 16)$

In Exercises 53–56, find the quotient and remainder.

53. $7x^3 - 5x^2 + x + 4$ is divided by $x + 3$
54. $9x^4 - 2x^2 + 2x + 9$ is divided by $x^2 + 5$
55. $6x^5 - 5x^2 + 7x + 6$ is divided by $2x^3 - 1$
56. $7 - x^2 + x^4$ is divided by $x^2 + x + 9$

R.5 Operations with Radicals

OBJECTIVES

1 Add and Subtract Radical Expressions

2 Multiply Radical Expressions

3 Rationalize Denominators of Radical Expressions

SECTION R.5 EXERCISES

OBJECTIVE 1 ADD AND SUBTRACT RADICAL EXPRESSIONS

In Section R.4 we learned how to simplify **polynomial expressions** by combining like terms. Recall that **like terms** involve two or more expressions that have the same variables raised to the same corresponding exponents. For example, $3x^2y^3$ and $5x^2y^3$ are like terms.

We can add or subtract these terms using the reverse of the **distributive property**.

$$3x^2y^3 + 5x^2y^3 = (3 + 5)x^2y^3 = 8x^2y^3$$

$$3x^2y^3 - 5x^2y^3 = (3 - 5)x^2y^3 = -2x^2y^3$$

$4xy^3$ and $2x^2y^2$ are not like terms, so we cannot add or subtract them. They have the same variables, but the variables in each term do not have the same corresponding exponents. This idea is true with **radicals** as well. We can only add and subtract **like radicals**.

> **Definition** Like Radicals
>
> Two radicals are **like radicals** if they have the same **index** and the same radicand.

Watch this **animation** to explore the concept of like radicals in more detail.

We add and subtract like radicals in the same way that we add and subtract **like terms**: reverse the **distributive property** to factor out the like radical and then simplify.

Example 1 Adding and Subtracting Radical Expressions

Add or subtract.

a. $\sqrt{15} + 7\sqrt{15}$ b. $10\sqrt{5} - 9\sqrt[4]{5}$ c. $\sqrt[3]{\frac{4}{27}} + 8\sqrt[3]{4}$

Solutions Watch this **video** for the complete solutions, or read through the following.

a. $\sqrt{15} + 7\sqrt{15}$ Begin with the original expression.

$= \sqrt{15} + 7\sqrt{15}$ Identify like radicals.

$= (1 + 7)\sqrt{15}$ Use the reverse distributive property.

$= 8\sqrt{15}$ Simplify.

b. The **indices** are different, so $\sqrt{5}$ and $\sqrt[4]{5}$ are not **like radicals**. We cannot simplify $10\sqrt{5} - 9\sqrt[4]{5}$ further since $\sqrt{5}$ and $\sqrt[4]{5}$ are not like radicals.

c. $\sqrt[3]{\frac{4}{27}} + 8\sqrt[3]{4}$ Begin with the original expression.

$= \frac{\sqrt[3]{4}}{\sqrt[3]{27}} + 8\sqrt[3]{4}$ Use the quotient rule.

$= \frac{\sqrt[3]{4}}{3} + 8\sqrt[3]{4}$ Simplify.

$= \frac{\sqrt[3]{4}}{3} + 8\sqrt[3]{4}$ Identify like radicals.

$= \left(\frac{1}{3} + 8\right)\sqrt[3]{4}$ Use the reverse distributive property.

$= \frac{25}{3}\sqrt[3]{4}$ or $\frac{25\sqrt[3]{4}}{3}$ Simplify.

You Try It Work through this You Try It problem.

Work Exercises 1–4 in this eText or in the MyLab Math Study Plan.

Example 2 Adding and Subtracting Radical Expressions

Add. Assume **variables** represent non-negative values.

$$12\sqrt[3]{7x^2} + 4\sqrt[3]{7x^2}$$

Solution Read through the following solution, or watch this **video**.

$12\sqrt[3]{7x^2} + 4\sqrt[3]{7x^2}$	Begin with the original expression.
$= 12\sqrt[3]{7x^2} + 4\sqrt[3]{7x^2}$	Identify like radicals.
$= (12 + 4)\sqrt[3]{7x^2}$	Use the reverse distributive property.
$= 16\sqrt[3]{7x^2}$	Simplify.

You Try It Work through this You Try It problem.

Work Exercises 5–13 in this eText or in the MyLab Math Study Plan.

CAUTION Sometimes it is necessary to simplify radicals first before adding or subtracting.

Example 3 Adding and Subtracting Radical Expressions

Add or subtract.

a. $\sqrt{54} + 6\sqrt{72} - 3\sqrt{24}$ b. $\sqrt[3]{24} - \sqrt[3]{192} + 4\sqrt[3]{250}$

Solutions

a. $\sqrt{54} + 6\sqrt{72} - 3\sqrt{24}$	Begin with the original expression.
$= \sqrt{9\cdot 6} + 6\sqrt{36\cdot 2} - 3\sqrt{4\cdot 6}$	Factor.
$= \sqrt{9}\cdot\sqrt{6} + 6\sqrt{36}\cdot\sqrt{2} - 3\sqrt{4}\cdot\sqrt{6}$	Use the product rule.
$= 3\sqrt{6} + 6\cdot 6\sqrt{2} - 3\cdot 2\sqrt{6}$	Simplify radicals.
$= 3\sqrt{6} + 36\sqrt{2} - 6\sqrt{6}$	Multiply factors.
$= (3 - 6)\sqrt{6} + 36\sqrt{2}$	Collect like radicals.
$= -3\sqrt{6} + 36\sqrt{2}$	Simplify.

b. Work the problem on your own. Check your **answer**, or watch this **video** for a detailed solution.

You Try It Work through this You Try It problem.

Work Exercises 14–24 in this eText or in the MyLab Math Study Plan.

Example 4 Adding and Subtracting Radical Expressions

Add or subtract. Assume **variables** represent non-negative values.

a. $\sqrt[3]{27m^5n^4} + 2mn\sqrt[3]{m^2n} - m\sqrt[3]{m^2n^4}$

b. $2a\sqrt{16ab^3} + 4\sqrt{9a^2b} - 5\sqrt{4a^3b^3}$

Solutions

a. $\sqrt[3]{27m^5n^4} + 2mn\sqrt[3]{m^2n} - m\sqrt[3]{m^2n^4}$ Begin with the original expression.

$= \sqrt[3]{27m^3n^3 \cdot m^2n} + 2mn\sqrt[3]{m^2n} - m\sqrt[3]{n^3 \cdot m^2n}$ Factor.

$= \sqrt[3]{27m^3n^3} \cdot \sqrt[3]{m^2n} + 2mn\sqrt[3]{m^2n} - m\sqrt[3]{n^3} \cdot \sqrt[3]{m^2n}$ Use the product rule.

$= 3mn\sqrt[3]{m^2n} + 2mn\sqrt[3]{m^2n} - mn\sqrt[3]{m^2n}$ Simplify radicals.

$= (3 + 2 - 1)mn\sqrt[3]{m^2n}$ Collect like radicals.

$= 4mn\sqrt[3]{m^2n}$ Simplify.

b. Work the problem on your own. Check your **answer**, or watch this **video** for a detailed solution.

You Try It Work through this You Try It problem.

Work Exercises 25–30 in this eText or in the MyLab Math Study Plan.

Example 5 Adding and Subtracting Radical Expressions

Add or subtract. Assume **variables** represent non-negative values.

a. $\dfrac{\sqrt{45}}{6x} - \dfrac{4\sqrt{20}}{5x}$ b. $\dfrac{\sqrt[4]{a^5}}{3} + \dfrac{a\sqrt[4]{a}}{12}$ c. $\dfrac{3x^3\sqrt{24x^3y^3}}{2x\sqrt{3x^2y}} - \dfrac{x^2\sqrt{10xy^4}}{\sqrt{5y^2}}$

Solutions Try to work the problems on your own. Remember to find a **common denominator** before adding or subtracting fractions. Check your **answers**, or watch this **interactive video** for detailed solutions to all three parts.

You Try It Work through this You Try It problem.

Work Exercises 31–34 in this eText or in the MyLab Math Study Plan.

OBJECTIVE 2 MULTIPLY RADICAL EXPRESSIONS

To multiply radical expressions, follow the same approach as when multiplying **polynomial expressions**. Use the **distributive property** to multiply each term in the first expression by each term in the second. Then simplify the resulting products and combine **like terms** and **like radicals**.

Example 6 Multiplying Radical Expressions

Multiply. Assume **variables** represent non-negative values.

a. $5\sqrt{2x}(3\sqrt{2x} - \sqrt{3})$

b. $\sqrt[3]{2n^2}(\sqrt[3]{4n} + \sqrt[3]{5n})$

Solutions

a. $5\sqrt{2x}(3\sqrt{2x} - \sqrt{3})$ Begin with the original expression.

$= 5\sqrt{2x} \cdot 3\sqrt{2x} - 5\sqrt{2x} \cdot \sqrt{3}$ Use the distributive property.

$= 5 \cdot 3 \cdot \sqrt{2x} \cdot \sqrt{2x} - 5 \cdot \sqrt{2x} \cdot \sqrt{3}$ Rearrange factors.

$= 15\sqrt{4x^2} - 5\sqrt{6x}$ Multiply.

$= 15 \cdot 2x - 5\sqrt{6x}$ Simplify the first radical.

$= 30x - 5\sqrt{6x}$ Simplify.

b. Try this problem on your own. Check your answer, or watch this video for a detailed solution.

You Try It Work through this You Try It problem.

Work Exercises 35–38 in this eText or in the MyLab Math Study Plan.

Example 7 Multiplying Radical Expressions

Multiply. Assume all variables represent non-negative values.

a. $(7\sqrt{2} - 2\sqrt{3})(\sqrt{2} - 5)$ b. $(\sqrt{m} - 4)(3\sqrt{m} + 7)$

Solutions

a. Use the FOIL method to multiply the radical expressions. The two first terms are $7\sqrt{2}$ and $\sqrt{2}$. The two outside terms are $7\sqrt{2}$ and -5. The two inside terms are $-2\sqrt{3}$ and $\sqrt{2}$. The two last terms are $-2\sqrt{3}$ and -5.

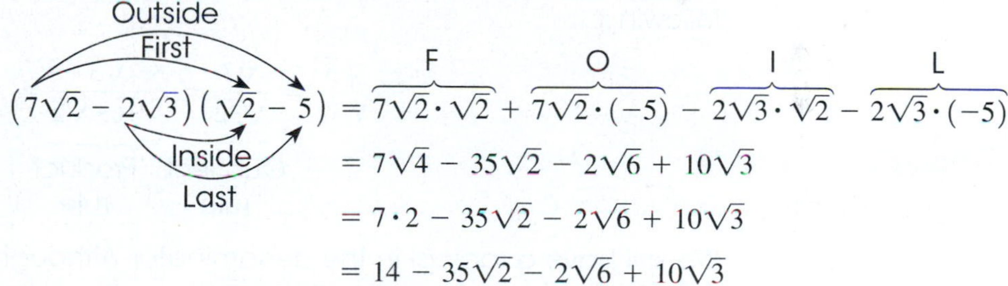

$$(7\sqrt{2} - 2\sqrt{3})(\sqrt{2} - 5) = \overbrace{7\sqrt{2} \cdot \sqrt{2}}^{F} + \overbrace{7\sqrt{2} \cdot (-5)}^{O} - \overbrace{2\sqrt{3} \cdot \sqrt{2}}^{I} - \overbrace{2\sqrt{3} \cdot (-5)}^{L}$$

$$= 7\sqrt{4} - 35\sqrt{2} - 2\sqrt{6} + 10\sqrt{3}$$

$$= 7 \cdot 2 - 35\sqrt{2} - 2\sqrt{6} + 10\sqrt{3}$$

$$= 14 - 35\sqrt{2} - 2\sqrt{6} + 10\sqrt{3}$$

b. Try this problem on your own. Check your answer, or watch this video for a detailed solution.

You Try It Work through this You Try It problem.

Work Exercises 39–44 in this eText or in the MyLab Math Study Plan.

Example 8 Using Special Products to Multiply Radical Expressions

Multiply. Assume all variables represent non-negative values.

a. $(\sqrt{y} + 3)(\sqrt{y} - 3)$

b. $(3\sqrt{x} - 2)^2$

Solutions

a. $(\sqrt{y}+3)(\sqrt{y}-3)$ is a **product of conjugates**. We use the rule for $(A+B)(A-B)$ with $A=\sqrt{y}$ and $B=\mathbf{3}$.

$$(A+B)(A-B)=A^2-B^2 \qquad \text{Write the product of conjugates rule.}$$

$$(\sqrt{y}+\mathbf{3})(\sqrt{y}-\mathbf{3})=(\sqrt{y})^2-(\mathbf{3})^2 \qquad \text{Substitute } \sqrt{y} \text{ for } A \text{ and } \mathbf{3} \text{ for } B.$$

$$=y-9$$

b. This **expression** has the form of the **square of a binomial difference** $(A-B)^2$. Use the **special product** rule $(A-B)^2=A^2-2AB+B^2$ with $A=3\sqrt{x}$ and $B=2$. Finish working this problem on your own. Check your **answer**, or watch this **video** for a detailed solution. ●

You Try It Work through this You Try It problem.

Work Exercises 45–50 in this eText or in the MyLab Math Study Plan.

In part (a) of Example 8, the product of **conjugates** involving **square roots** resulted in an expression without any **radicals**. This result is true in general and occurs because the product of conjugates equals the **difference of two squares**.

OBJECTIVE 3 RATIONALIZE DENOMINATORS OF RADICAL EXPRESSIONS

In Section R.3, we saw that a **simplified radical expression** has no **radicals** in the **denominator**. We used the **quotient rule** to eliminate radicals from the denominator of a radical expression. However, the quotient rule is not always enough. For example, simplifying $\sqrt{\frac{3}{32}}$ using the quotient rule and **product rule** yields the following:

$$\sqrt{\frac{3}{32}}=\underbrace{\frac{\sqrt{3}}{\sqrt{32}}}_{\text{Quotient rule}}=\underbrace{\frac{\sqrt{3}}{\sqrt{16}\cdot\sqrt{2}}}_{\text{Product rule}}=\frac{\sqrt{3}}{4\sqrt{2}}$$

We still have a radical in the denominator. Although there is nothing wrong with this, it is common practice to remove radicals from the denominator so that the denominator is a **rational number**. This process is called **rationalizing the denominator**.

> **Rationalizing a Denominator with One Term**
>
> To rationalize a denominator with a single radical of **index** n, multiply the **numerator** and denominator by a radical of index n so that the radicand in the denominator is a **perfect** n**th power**.

In the example from the previous page, we see that there is a radical in the denominator of the expression $\frac{\sqrt{3}}{4\sqrt{2}}$. Multiplying the numerator and denominator of this expression by the same radical is equivalent to multiplying by 1, which will not change the value of the expression. Therefore,

$$\underbrace{\frac{\sqrt{3}}{4\sqrt{2}}}_{\text{Original expression}} = \underbrace{\frac{\sqrt{3}}{4\sqrt{2}} \cdot \frac{\sqrt{2}}{\sqrt{2}}}_{\text{Multiply by 1}} = \underbrace{\frac{\sqrt{6}}{4\sqrt{4}}}_{\text{Product rule}} = \underbrace{\frac{\sqrt{6}}{4(2)} = \frac{\sqrt{6}}{8}}_{\text{Simplify}}.$$

Example 9 Rationalizing Denominators with Square Roots

Rationalize the denominator. Assume all variables represent non-negative values.

a. $\dfrac{\sqrt{7}}{\sqrt{5}}$ b. $\sqrt{\dfrac{3}{10x}}$

Solutions

Watch this video, or read through the following solutions.

a. Since the **denominator** contains a **square root**, multiply the **numerator** and denominator by a square root so that the **radicand** in the denominator is a **perfect square**. $5 \cdot 5 = 5^2 = 25$ is a perfect square, so multiply the numerator and denominator by $\sqrt{5}$.

$$\underbrace{\frac{\sqrt{7}}{\sqrt{5}}}_{\text{Original expression}} = \underbrace{\frac{\sqrt{7}}{\sqrt{5}} \cdot \frac{\sqrt{5}}{\sqrt{5}}}_{\text{Multiplying by 1}} = \underbrace{\frac{\sqrt{35}}{\sqrt{25}}}_{\text{Product rule}} = \underbrace{\frac{\sqrt{35}}{5}}_{\text{Simplify}}$$

b. Using the **quotient rule**, write $\sqrt{\dfrac{3}{10x}} = \dfrac{\sqrt{3}}{\sqrt{10x}}$. Since the denominator contains a square root, multiply the numerator and denominator by a **square root** so that the **radicand** in the denominator is a **perfect square**. $10x \cdot 10x = (10x)^2 = 100x^2$ is a perfect square, so multiply the numerator and denominator by $\sqrt{10x}$.

$$\underbrace{\sqrt{\frac{3}{10x}}}_{\text{Original expression}} = \underbrace{\frac{\sqrt{3}}{\sqrt{10x}}}_{\text{Quotient rule}} = \frac{\sqrt{3}}{\sqrt{10x}} \cdot \underbrace{\frac{\sqrt{10x}}{\sqrt{10x}}}_{\text{Multiplying by 1}} = \underbrace{\frac{\sqrt{30x}}{\sqrt{100x^2}}}_{\text{Product rule}} = \underbrace{\frac{\sqrt{30x}}{10x}}_{\text{Simplify}}$$

You Try It Work through this You Try It problem.

Work Exercises 51–54 in this eText or in the MyLab Math Study Plan.

Example 10 Rationalizing Denominators with Cube Roots or Fourth Roots

Rationalize the denominator. Assume all variables represent non-negative values.

a. $\sqrt[3]{\dfrac{11}{25x}}$

b. $\dfrac{\sqrt[4]{7x}}{\sqrt[4]{27y^2}}$

Solutions

a. Using the **quotient rule**, write $\sqrt[3]{\dfrac{11}{25x}} = \dfrac{\sqrt[3]{11}}{\sqrt[3]{25x}}$. For **roots greater than 2**, it is helpful to write the **radicand** of the denominator in **exponential form**.

$$\sqrt[3]{25x} = \sqrt[3]{5^2x^1}$$

Since the denominator contains a **cube root**, we multiply the numerator and denominator by a cube root so that the radicand in the denominator is a **perfect cube**. To do this, we need the **factors** in the radicand to have **exponents** of 3 or **multiples** of 3. We have two 5's and one x, so we need one more 5 and two more x's to get $5^2x^1 \cdot 5^1x^2 = 5^3x^3 = (5x)^3$, which is a perfect cube. Therefore, multiply the numerator and denominator by $\sqrt[3]{5^1x^2} = \sqrt[3]{5x^2}$.

$$\underbrace{\sqrt[3]{\frac{11}{25x}}}_{\text{Original expression}} = \underbrace{\frac{\sqrt[3]{11}}{\sqrt[3]{25x}}}_{\text{Quotient rule}} = \underbrace{\frac{\sqrt[3]{11}}{\sqrt[3]{25x}} \cdot \frac{\sqrt[3]{5x^2}}{\sqrt[3]{5x^2}}}_{\text{Multiplying by 1}} = \underbrace{\frac{\sqrt[3]{55x^2}}{\sqrt[3]{125x^3}}}_{\text{Product rule}} = \underbrace{\frac{\sqrt[3]{55x^2}}{5x}}_{\text{Simplify}}$$

b. Try this problem on your own. Check your **answer**, or watch this **video** for a detailed solution.

You Try It Work through this You Try It problem.

Work Exercises 55 and 56 in this eText or in the MyLab Math Study Plan.

CAUTION Though not required, it is best to rationalize the denominator last when simplifying radical expressions since the radicand in the denominator will then involve simpler expressions.

Example 11 Rationalizing Denominators

Simplify each expression first and then rationalize the denominator. Assume all variables represent non-negative values.

a. $\sqrt{\dfrac{3x}{50}}$ b. $\dfrac{\sqrt{18x}}{\sqrt{27xy}}$ c. $\sqrt[3]{\dfrac{-4x^5}{16y^5}}$

Solutions Try these problems on your own. Check your **answers**, or watch this **video** for detailed solutions to all three parts.

You Try It Work through this You Try It problem.

Work Exercises 57 and 58 in this eText or in the MyLab Math Study Plan.

The goal of **rationalizing a denominator** is to write an equivalent expression without **radicals** in the denominator. Earlier in this section, we noted that multiplying **conjugates** involving **square roots** results in an expression without any radicals. So, we can use the **product of conjugates** to rationalize the denominator of a radical expression whose **denominator** contains two **terms** involving one or more square roots.

Rationalizing a Denominator with Two Terms

To rationalize a denominator with two terms involving one or more square roots, multiply the numerator and denominator by the conjugate of the denominator.

Example 12 Rationalizing Denominators with Two Terms

Rationalize the denominator. Assume all variables represent non-negative values.

a. $\dfrac{2}{\sqrt{3}+5}$ b. $\dfrac{7}{3\sqrt{x}-4}$ c. $\dfrac{\sqrt{y}-3}{\sqrt{y}+2}$

Solutions

a. Since the **denominator** has two **terms** and involves a **square root**, multiply the **numerator** and denominator by $\sqrt{3}-5$, which is the **conjugate** of the denominator.

$$\frac{2}{\sqrt{3}+5}=\frac{2}{\sqrt{3}+5}\cdot\frac{\sqrt{3}-5}{\sqrt{3}-5}$$ Multiply numerator and denominator by $\sqrt{3}-5$.

$$=\frac{2(\sqrt{3}-5)}{(\sqrt{3})^2-(5)^2}$$ Multiply numerators and multiply denominators.

$$=\frac{2(\sqrt{3}-5)}{3-25}$$ Simplify the denominator.

$$=\frac{2(\sqrt{3}-5)}{-22}$$

$$=\frac{{}^{1}\cancel{2}(\sqrt{3}-5)}{-\cancel{22}_{11}}$$ Divide out a common factor of 2.

$$=-\frac{\sqrt{3}-5}{11}\text{ or }\frac{5-\sqrt{3}}{11}$$ Simplify.

b. Since the **denominator** has two **terms** and involves a **square root**, multiply the **numerator** and denominator by the **conjugate** of the denominator, $3\sqrt{x}+4$.

$$\frac{7}{3\sqrt{x}-4}=\underbrace{\frac{7}{3\sqrt{x}-4}\cdot\frac{3\sqrt{x}+4}{3\sqrt{x}+4}}_{\text{Multiply numerator and denominator by }3\sqrt{x}+4}=\underbrace{\frac{7(3\sqrt{x}+4)}{(3\sqrt{x})^2-(4)^2}}_{\text{Multiply numerators and multiply denominators}}=\underbrace{\frac{7(3\sqrt{x}+4)}{9x-16}}_{\text{Simplify}}\text{ or }\underbrace{\frac{21\sqrt{x}+28}{9x-16}}_{\text{Distribute}}$$

The result can be left in factored form, or we can distribute in the final step. Leaving the numerator factored until the end helps with dividing out **common factors**, if they are present.

c. Work this problem on your own. Check your **answer**, or watch this **video** for a detailed solution.

You Try It Work through this You Try It problem.

Work Exercises 59–64 in this eText or in the MyLab Math Study Plan.

R.5 Exercises

In Exercises 1–50, assume that all variables represent non-negative values and be sure to simplify your answer.

In Exercises 1–34, add or subtract, if possible.

1. $4\sqrt{7} + 9\sqrt{7}$
2. $\sqrt[3]{\frac{32}{27}} - \sqrt[3]{108}$
3. $3\sqrt{2} + 5\sqrt{6}$
4. $3\sqrt{7} - 4\sqrt[3]{6} + 2\sqrt{7} - 8\sqrt[3]{6}$
5. $\sqrt[3]{6x} - 7\sqrt[3]{6x}$
6. $2\sqrt[3]{4y} - 7\sqrt[3]{4y}$
7. $\sqrt{3x} + 3\sqrt{x} - 9\sqrt{3x}$
8. $6x\sqrt{11} - 13x\sqrt{11}$
9. $\frac{\sqrt{x}}{4} + \frac{3}{8}\sqrt{x}$
10. $\frac{4\sqrt{3x}}{9} + \frac{2\sqrt{3x}}{15} - \frac{7\sqrt{3x}}{6}$
11. $10\sqrt{5} - 3\sqrt{x} - 4\sqrt{5} + 9\sqrt{x}$
12. $6\sqrt{a} + 2\sqrt{b} - a\sqrt{3}$
13. $4xy\sqrt[3]{2} + 3xy\sqrt[3]{2} - 5x\sqrt[3]{2}$
14. $\sqrt{18} + \sqrt{98}$
15. $\sqrt{45} - \sqrt{80}$
16. $3\sqrt[3]{32} + \sqrt[3]{500}$
17. $3\sqrt{48} - \sqrt{24}$
18. $\sqrt{\frac{3}{16}} + \sqrt{\frac{27}{4}}$
19. $5\sqrt{12} - \sqrt{75} + 3\sqrt{20}$
20. $9\sqrt{7} - 6\sqrt{72} + 4\sqrt{18} - 5\sqrt{112}$
21. $\sqrt[3]{72} - \sqrt[3]{9}$
22. $6\sqrt[3]{54} + 2\sqrt{8} - 4\sqrt[3]{16}$
23. $7\sqrt[3]{54} + 3\sqrt[3]{8} - 5\sqrt[3]{16}$
24. $\sqrt[3]{\frac{3}{125}} + \sqrt[3]{81}$
25. $2\sqrt[4]{81x^5} - 3\sqrt[4]{x^5}$
26. $6\sqrt{20x^3} + 4x\sqrt{45x}$
27. $\sqrt{100t^3} + 6\sqrt{t^2} - \sqrt{25t^3}$
28. $9\sqrt[3]{y^4} - 3y\sqrt[3]{125y} + \sqrt[3]{64y^4}$
29. $3a^2\sqrt{ab^3} + \sqrt{16a^5b^3} - 5b\sqrt{a^5b}$
30. $\sqrt[3]{81m^3n} - 4\sqrt[3]{3m^3n} + 3m\sqrt[3]{162n}$
31. $\frac{\sqrt{72}}{3y} + \frac{5\sqrt{18}}{4y}$
32. $\frac{\sqrt[3]{8a^4}}{4a} - \frac{\sqrt[3]{27a}}{7}$
33. $\frac{\sqrt[4]{81a^7}}{3} - \frac{5a\sqrt[4]{16a^3}}{2}$
34. $\frac{5x^2\sqrt{128x^5y^3}}{\sqrt{64x^4y}} + \frac{2xy\sqrt{18x^3y}}{3\sqrt{y}}$

In Exercises 35–50, multiply.

35. $\sqrt{5}(2x + \sqrt{3})$
36. $\sqrt{2x}(\sqrt{6x} - 5)$
37. $2\sqrt{3m}(m - 3\sqrt{5m})$
38. $\sqrt[3]{x}(\sqrt[3]{54x^2} - \sqrt[3]{x})$
39. $(8\sqrt{3} - 5)(\sqrt{3} - 1)$
40. $(3 + \sqrt{5})(2 + \sqrt{7})$

41. $(\sqrt{x}+4)(\sqrt{x}+6)$
42. $(4-2\sqrt{3x})(5+\sqrt{3x})$
43. $(3-\sqrt{x})(4-\sqrt{y})$
44. $(\sqrt[3]{y}+1)(\sqrt[3]{y}-2)$
45. $(\sqrt{7z}+3)(\sqrt{7z}-3)$
46. $(\sqrt{5}+3y)^2$
47. $(\sqrt{x}+3)(3-\sqrt{x})$
48. $(\sqrt{3a}-\sqrt{2b})(\sqrt{3a}+\sqrt{2b})$
49. $(2\sqrt{x}+5)^2$
50. $(2\sqrt{m}-3\sqrt{n})^2$

In Exercises 51–64, rationalize the denominator.

51. $\dfrac{4}{\sqrt{6}}$
52. $\dfrac{\sqrt{3}}{\sqrt{5}}$
53. $\sqrt{\dfrac{9}{2x}}$
54. $\dfrac{5}{\sqrt{11x}}$
55. $\sqrt[3]{\dfrac{9}{4x^2}}$
56. $\dfrac{\sqrt[4]{3b^5}}{\sqrt[4]{25a}}$
57. $\dfrac{8}{\sqrt{12x^3y^4}}$
58. $\dfrac{\sqrt[3]{2x^2}}{\sqrt[3]{36y^5}}$
59. $\dfrac{5}{2-\sqrt{3}}$
60. $\dfrac{-4x}{5+\sqrt{6}}$
61. $\dfrac{-3}{\sqrt{x}+1}$
62. $\dfrac{\sqrt{a}}{3\sqrt{a}-\sqrt{b}}$
63. $\dfrac{\sqrt{m}-4}{\sqrt{m}-7}$
64. $\dfrac{5\sqrt{3}+2\sqrt{6}}{4\sqrt{6}-\sqrt{3}}$

R.6 Factoring Polynomials

INTRODUCTION

Read this introduction before beginning Objective 1.

OBJECTIVES

1 Factoring Out a Greatest Common Factor
2 Factoring by Grouping
3 Factoring Trinomials with a Leading Coefficient Equal to One
4 Factoring Trinomials with a Leading Coefficient Not Equal to One
5 Factoring Using Special Factoring Formulas

SECTION R.6 EXERCISES

Introduction to Section R.6

In this section, we learn techniques used to factor **polynomial expressions**. Being able to factor polynomials (and other algebraic expressions) is an essential algebraic skill and is used throughout this text to solve equations. Factoring a polynomial is basically the reverse of multiplying polynomials. The first factoring technique that we discuss is to factor out a greatest common factor.

OBJECTIVE 1 FACTORING OUT A GREATEST COMMON FACTOR

Factoring out a greatest common factor is essentially the **distributive property**. Note that the expression $AB + AC$ has two terms, AB and AC. The term AB has two factors, A and B. The term AC also has two factors, A and C. Notice that the terms

AB and AC share a common factor of A. This common factor is called the **greatest common factor (GCF)** because it is the largest factor common to both terms. We can thus rewrite the expression $AB + AC$ by "pulling out" (factoring) this greatest common factor of A to get $AB + AC = A(B + C)$. When determining the GCF of a polynomial written as the sum and/or difference of monomials, we must determine the GCF of the numeric factors and then determine the GCF of the variable factors. For example, the polynomial $4x^2 - 6x^3 + 2x$ is composed of three monomials. The number 2 is the largest integer numeric factor common to each monomial term.

The largest variable factor common to each monomial is x. Therefore, the GCF is $2x$. Using the distributive property and the GCF of $2x$, we get the following:

$$\begin{aligned} &4x^2 - 6x^3 + 2x \\ &= 2x \cdot 2x - 2x \cdot 3x^2 + 2x \cdot 1 \\ &= 2x(2x - 3x^2 + 1) \end{aligned}$$

We can always verify whether we have factored correctly by multiplying the factors together and checking to see if the product is equal to the original polynomial.

Example 1 Factor Out a Greatest Common Factor

Factor each polynomial by factoring out the GCF.

a. $5w^2y^3 - 15w^4y + 20w^3y^2$ b. $x(x^2 + 6) - 5(x^2 + 6)$

Solutions

a. The numeric GCF is 5, and the variable GCF is w^2y. Thus, the GCF is $5w^2y$.

$5w^2y^3 - 15w^4y + 20w^3y^2$	Write the original polynomial.
$= 5w^2y \cdot y^2 - 5w^2y \cdot 3w^2 + 5w^2y \cdot 4wy$	The GCF is $5w^2y$.
$= 5w^2y(y^2 - 3w^2 + 4wy)$	Factor out the GCF.

b. GCF is the binomial factor $(x^2 + 6)$.

$$x(x^2 + 6) - 5(x^2 + 6) = (x^2 + 6)(x - 5)$$

You Try It Work through the following You Try It problem.

Work Exercises 1–8 in this textbook or in the MyLab Math Study Plan.

OBJECTIVE 2 FACTORING BY GROUPING

We often encounter polynomials that, at first, appear to be difficult or impossible to factor. For example, consider the polynomial $x(x^2 + 6) - 5(x^2 + 6)$ seen in Example 1. Using the distributive property, this polynomial can be rewritten as $x^3 + 6x - 5x^2 - 30$. At first glance, it appears that we cannot factor this polynomial because the GCF is 1. However, if we group the terms in pairs and factor the GCF from each pair of terms, we see that we can indeed factor the polynomial.

$(x^3 + 6x) + (-5x^2 - 30)$	Group the terms in pairs.
$= x(x^2 + 6) - 5(x^2 + 6)$	Factor the GCF out of each pair.
$= (x^2 + 6)(x - 5)$	Factor out the GCF of $(x^2 + 6)$.

This technique is called **factoring by grouping** and often works when a polynomial has an even number of terms.

Example 2 Factor Polynomials by Grouping

Factor each polynomial by grouping.

a. $6x^2 - 2x + 9x - 3$

b. $2x^2 + 6xw - xy - 3wy$

Solutions

a. $6x^2 - 2x + 9x - 3$

$= (6x^2 - 2x) + (9x - 3)$ Group the terms in pairs.

$= 2x(3x - 1) + 3(3x - 1)$ Factor the GCF out of each pair.

$= (3x - 1)(2x + 3)$ Factor out the GCF of $(3x - 1)$.

b. $2x^2 + 6xw - xy - 3wy$

$= (2x^2 + 6xw) + (-xy - 3wy)$ Group the terms in pairs.

$= 2x(x + 3w) - y(x + 3w)$ Factor the GCF out of each pair.

$= (x + 3w)(2x - y)$ Factor out the GCF of $(x + 3w)$.

You Try It **Work through the following You Try It problem.**

Work Exercises 9–16 in this textbook or in the MyLab Math Study Plan.

OBJECTIVE 3 FACTORING TRINOMIALS WITH A LEADING COEFFICIENT EQUAL TO ONE

In **Section R.4**, the FOIL method for multiplying two binomials was introduced. For example, we can use the FOIL method to multiply the two binomials $(x + a)$ and $(x + b)$.

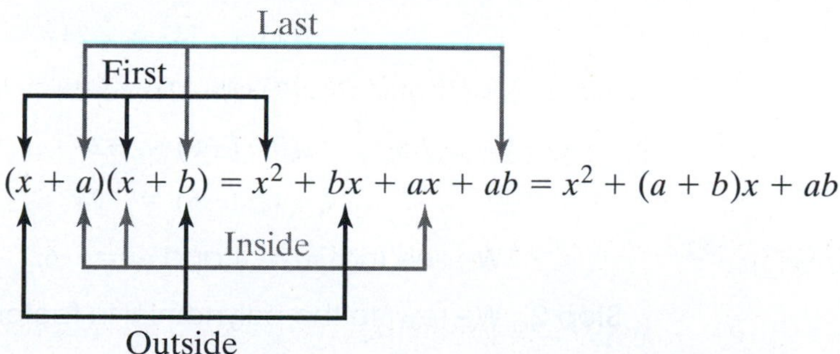

The binomial product $(x + a)(x + b)$ is said to be a factored form of the trinomial $x^2 + (a + b)x + ab$. Our goal is to start with a trinomial and try to find two binomials whose product is equal to the given trinomial. If we carefully look at the last two terms of the trinomial $x^2 + (a + b)x + ab$, we see that the middle term, $(a + b)x$, is the *sum* of the products of the inside and outside terms of the two binomials. The third term, ab, is the *product* of the last two terms of each binomial. Thus, when factoring trinomials of the form $x^2 + bx + c$, we must attempt to find two integers whose product is c and whose sum is b.

We outline this method on the following page using $x^2 - 3x - 88$ as an example.

Factoring Trinomials of the Form $x^2 + bx + c$

Example: $x^2 - 3x - 88$, $b = -3$ and $c = -88$

Step 1. Find two integers, n_1 and n_2, whose product is c and whose sum is b. So $n_1 \cdot n_2 = c$ and $n_1 + n_2 = b$. If no such pair of integers exists, the trinomial is prime.
The two integers are $n_1 = 8$ and $n_2 = -11$ because $(8)(-11) = -88$ and $8 + (-11) = -3$.

Step 2. Rewrite the trinomial in the factored form of $(x + n_1)(x + n_2)$.
$x^2 - 3x - 88 = (x + 8)(x - 11)$

Step 3. Check your answer by multiplying out the factored form.

$$(x + 8)(x - 11) = x^2 - 11x + 8x - 88$$
$$= x^2 - 3x - 88$$

Example 3 Factor Trinomials with a Leading Coefficient of One

Factor each trinomial.

a. $x^2 - 2x - 24$ **b.** $x^2 - 2x - 6$ **c.** $x^2 - 12x + 32$

Solution

a. To factor the trinomial $x^2 - 2x - 24$, we follow the steps stated above.

Step 1. We are looking for two integers n_1 and n_2 such that the product $n_1 n_2 = -24$ and the sum $n_1 + n_2 = -2$. We list all possibilities of n_1 and n_2 below:

Product	**Sum**
$n_1 \cdot n_2$	$n_1 + n_2$
$(1) \cdot (-24) = -24$	$(1) + (-24) = -23$
$(2) \cdot (-12) = -24$	$(2) + (-12) = -10$
$(3) \cdot (-8) = -24$	$(3) + (-8) = -5$
$(4) \cdot (-6) = -24$	$(4) + (-6) = -2$

We see that $n_1 = 4$ and $n_2 = -6$.

Step 2. We rewrite the polynomial in factored form as $(x + 4)(x - 6)$.

Step 3. Multiply the two binomial factors from step 2 to see if the product simplifies to the original trinomial.

$$(x + 4)(x - 6) = x^2 - 6x + 4x - 24$$
$$= x^2 - 2x - 24$$

b. To factor the trinomial $x^2 - 2x - 6$, we follow the steps stated previously.

Step 1. We are looking for two integers n_1 and n_2 such that the product $n_1 n_2 = -6$ and the sum $n_1 + n_2 = -2$. We list all possibilities of n_1 and n_2 below:

Product	**Sum**
$n_1 \cdot n_2$	$n_1 + n_2$
$(1)\cdot(-6) = -6$	$(1) + (-6) = -5$
$(2)\cdot(-3) = -6$	$(2) + (-3) = -1$
$(3)\cdot(-2) = -6$	$(3) + (-2) = 1$
$(6)\cdot(-1) = -6$	$(6) + (-1) = 5$

We see that there are no two integers whose product is -6 and whose sum is -2. Therefore, we say that the trinomial $x^2 - 2x - 6$ is not factorable with integers. A polynomial that is not factorable with integers is called a **prime polynomial**.

c. Try to factor the trinomial $x^2 - 12x + 32$ on your own. Work through the **video solution** to check your answer.

You Try It **Work through the following You Try It problem.**

Work Exercises 17–27 in this textbook or in the MyLab Math Study Plan.

In Example 3, all trinomials had a leading coefficient of one. We are now ready to explore techniques used to factor trinomials of the form $ax^2 + bx + c$, where $a \neq 1$.

OBJECTIVE 4 FACTORING TRINOMIALS WITH A LEADING COEFFICIENT NOT EQUAL TO ONE

We can try to factor trinomials of the form $ax^2 + bx + c$, where $a \neq 1$ using a **trial-and-error method**. That is, we basically "guess" until we stumble on the correct two binomial factors. For example, suppose that we want to factor the trinomial $6x^2 + 29x + 35$. The possible first terms of the two binomial factors are x and $6x$ or $2x$ and $3x$. The only integer factors of the constant, 35, are $(1)\cdot(35)$ and $(5)\cdot(7)$. This means that there are eight possible combinations of potential pairs of binomial factors for this trinomial. The eight possibilities are as follows:

1. $(x + 1)(6x + 35)$ 2. $(x + 35)(6x + 1)$ 3. $(x + 7)(6x + 5)$ 4. $(x + 5)(6x + 7)$

5. $(2x + 1)(3x + 35)$ 6. $(2x + 35)(3x + 1)$ 7. $(2x + 7)(3x + 5)$ 8. $(2x + 5)(3x + 7)$

Try to determine which of the eight possibilities above are the correct factors of $6x^2 + 29x + 35$. Carefully work through this **animation** to see how to use the trial and error method to factor trinomials. Although this "guessing" method works, it can often be a long and arduous process, especially when the leading coefficient and constant coefficient have several factors. Fortunately, there is another more methodical approach to factoring trinomials. We now outline this five-step method using the same trinomial $6x^2 + 29x + 35$ as our example.

Factoring Trinomials of the Form $ax^2 + bx + c, a \neq 0, a \neq 1$

Example: $6x^2 + 29x + 35, a = 6, b = 29$, and $c = 35$

Step 1. Multiply $a \cdot c$.

$$(6) \cdot (35) = 210$$

Step 2. Find two integers, n_1 and n_2, whose product is ac and whose sum is b. So $n_1 \cdot n_2 = ac$ and $n_1 + n_2 = b$. If no such pair of integers exists, the trinomial is prime.

The two integers are $n_1 = 15$ and $n_2 = 14$ because $(15) \cdot (14) = 210$ and $15 + 14 = 29$.

Step 3. Rewrite the middle term as the sum of two terms using the integers found in step 2. So, $ax^2 + bx + c = ax^2 + n_1x + n_2x + c$.

$$6x^2 + \underbrace{15x + 14x}_{15x + 14x = 29x} + 35$$

Step 4. Factor by grouping.

$$\begin{aligned} 6x^2 + 15x + 14x + 35 &= (6x^2 + 15x) + (14x + 35) \\ &= 3x(2x + 5) + 7(2x + 5) \\ &= (2x + 5)(3x + 7) \end{aligned}$$

Step 5. Check your answer by multiplying out the factored form.

$$(2x + 5)(3x + 7) = 6x^2 + 29x + 35$$

Example 4 Factor a Trinomial with a Leading Coefficient Not Equal to One

Factor each trinomial.

a. $4x^2 + 17x + 15$ b. $12x^2 - 25x + 12$

Solutions

a. To see how to use the ac method to factor the trinomial $4x^2 + 17x + 15$, work through this animation.

b. To factor the trinomial $12x^2 - 25x + 12$, follow the five steps outlined previously.

Step 1. $a \cdot c = 12 \cdot 12 = 144$

Step 2. $n_1 = -16$ and $n_2 = -9$

$$(-16) \cdot (-9) = 144 \quad \text{and} \quad (-16) + (-9) = -25$$

Step 3. $12x^2 - 25x + 12 = 12x^2 - 16x - 9x + 12$

Step 4. $(12x^2 - 16x) + (-9x + 12)$

$$= 4x(3x - 4) - 3(3x - 4)$$

$$= (3x - 4)(4x - 3)$$

Step 5. $(3x - 4)(4x - 3) = 12x^2 - 25x + 12$

You Try It Work through the following You Try It problem.

Work Exercises 28–34 in this textbook or in the MyLab Math Study Plan.

OBJECTIVE 5 FACTORING USING SPECIAL FACTORING FORMULAS

There are several factoring formulas that are quite useful to remember and worth mentioning at this time. In Section R.4, we saw several **special product formulas**. We now revisit three of these formulas and restate them as factoring formulas.

> **Difference of Two Squares**
>
> Let a and b represent any real number or algebraic expression, then
> $$a^2 - b^2 = (a + b)(a - b).$$

Recognizing a polynomial that can be factored using the difference of two squares pattern is fairly straightforward because such a polynomial has two terms, each of which is a perfect square. Following are three examples:

$$z^2 - 4 = (z)^2 - (2)^2 = (z + 2)(z - 2)$$
$$36y^2 - 1 = (6y)^2 - (1)^2 = (6y + 1)(6y - 1)$$
$$z^4 - 16 = (z^2)^2 - (4)^2 = (z^2 + 4)(z^2 - 4) = (z^2 + 4)(z + 2)(z - 2)$$

TIP The sum of two squares, $a^2 + b^2$, is not factorable using integers. A polynomial of this form is a prime polynomial.

In Section R.4, we introduced the following two **special product formulas** that were stated as the **square of a binomial**.

$$(a + b)^2 = a^2 + 2ab + b^2 \quad \text{and} \quad (a - b)^2 = a^2 - 2ab + b^2$$

The resulting trinomials are called *perfect square trinomials* because they can each be factored as a binomial squared.

> **Perfect Square Trinomials**
>
> Let a and b represent any real number or algebraic expression, then
> $$a^2 + 2ab + b^2 = (a + b)^2$$
> $$a^2 - 2ab + b^2 = (a - b)^2.$$

Note that the first term of a perfect square trinomial is the square of some algebraic expression a. The last term is the square of some algebraic expression b, and the middle term is the sum or difference of twice the product of a and b. Following is an example of a perfect square trinomial:

$$4x^2 - 12xy + 9y^2 = \underbrace{(2x)^2}_{a^2} - \underbrace{2(2x)(3y)}_{2ab} + \underbrace{(3y)^2}_{b^2} = \underbrace{(2x - 3y)^2}_{(a-b)^2}$$

We stated previously that the sum of two squares cannot be factored with integers. However, this is not the case for the sum of two cubes. In fact, the difference of two cubes can be factored as well. We now state these final two special factoring formulas.

Sum and Difference of Two Cubes

Let a and b represent any real number or algebraic expression, then

$$a^3 + b^3 = (a + b)(a^2 - ab + b^2)$$

$$a^3 - b^3 = (a - b)(a^2 + ab + b^2).$$

Note that the sum of two cubes and the difference of two cubes have very similar factoring patterns. The only two differences are the signs of the binomial factor and the sign of the second term of the trinomial factor. Following is an example of a sum of two cubes:

$$8y^3 + 125 = \underbrace{(2y)^3}_{a^3} + \underbrace{(5)^3}_{b^3} = \underbrace{(2y + 5)}_{(a + b)}\underbrace{(4y^2 - 10y + 25)}_{(a^2 - ab + b^2)}$$

Example 5 Use Special Factoring Formulas to Factor Polynomials

Completely factor each expression.

a. $2x^3 + 5x^2 - 12x$ **b.** $5x^4 - 45x^2$ **c.** $8z^3x - 27y + 8z^3y - 27x$

Solution Watch the **interactive video** to see each solution worked out in detail.

You Try It **Work through the following You Try It problem.**

Work Exercises 35–50 in this textbook or in the MyLab Math Study Plan.

R.6 Exercises

In Exercises 1–8, factor each polynomial by factoring out the GCF.

1. $40x + 4$
2. $wx^4 + w$
3. $3y^7 + 6xy^8$
4. $4x^4 - 20x^3 + 16x^2$
5. $39w^9y^9 - 12w^2y^2 + 12wy + 15w^2y$
6. $11(x + 7) + 3a(x + 7)$
7. $8x(z + 1) + (z + 1)$
8. $9y(x^2 + 11) - 7(x^2 + 11)$

In Exercises 9–16, factor each polynomial by grouping.

9. $xy + 5x + 3y + 15$
10. $ab + 7a - 2b - 14$
11. $9xy + 12x + 6y + 8$
12. $3xy - 2x - 12y + 8$
13. $5x^2 + 4xy + 15x + 12y$
14. $3x^2 + 3xy - 2x - 2y$
15. $x^3 + 9x^2 + 4x + 36$
16. $x^3 - x^2 - 7x + 7$

In Exercises 17–50, completely factor each polynomial or state that the polynomial is prime.

17. $x^2 + 13x + 30$
18. $y^2 + 5y - 14$
19. $w^2 - 17w + 72$
20. $x^2 - 2x - 48$
21. $x^2 - 6x + 10$
22. $x^2 - 33x - 108$

23. $x^2 + 6xy + 8y^2$
24. $24 + 2x - x^2$
25. $5x^2 + 10x - 75$
26. $4x^2w + 20xw + 24w$
27. $5x^2 - 30x - 25$
28. $9x^2 + 18x + 5$
29. $12y^2 + 29y + 15$
30. $2x^2 + 5x - 25$
31. $8x^2 - 6x - 9$
32. $10y^2 - 23y + 12$
33. $16x^2 + 36x + 18$
34. $12 - x - 20x^2$
35. $x^2 - 81$
36. $z^2 + 10z + 25$
37. $x^3 + 64$
38. $2m^2 - 48m + 288$
39. $64x^2 - 49$
40. $x^3 - 343$
41. $108t^2 - 147$
42. $(x + 8y)^2 - 25$
43. $125x^3 + 216$
44. $(2x - 5)^3 - 8$
45. $(w + 5)^2 - 100$
46. $16x^2 + 40x + 25$
47. $x^{14} - x^{11}$
48. $6 - 149x^2 - 25x^4$
49. $x^3y^4 - 27y^4$
50. $x^2 + 14x + 49 - x^4$

R.7 Rational Expressions

OBJECTIVES

1 Simplifying Rational Expressions
2 Multiplying and Dividing Rational Expressions
3 Adding and Subtracting Rational Expressions
4 Simplifying Complex Rational Expressions
5 Applications Involving Rational Expressions
SECTION R.7 EXERCISES

OBJECTIVE 1 SIMPLIFYING RATIONAL EXPRESSIONS

A **rational number** is the quotient of two integers. A rational expression is the quotient of two **polynomial expressions**.

Definition Rational Expression

A rational expression is an algebraic expression that when simplified has the form P/Q, where $P = 0$ and Q are polynomials such that the degree of Q is greater than or equal to 1.

The following are examples of rational expressions:

$$\frac{x^2 - 4x}{2x - 8}, \quad \frac{3x^2 - 2x - 8}{3x^2 - 14x - 24}, \quad \frac{x^2y^3 + 11x^5y^4 - 7xy}{xy^5 - x^2y}$$

Each of the three expressions above are quotients of polynomials such that the degree of the polynomials in each denominator is greater than or equal to one.

The expressions $\frac{x^3 - x + 1}{3}$ and $\frac{\sqrt{3xy + 2x}}{x^3y - y}$ are examples of expressions that are **not** rational expressions. Do you know **why?**

Consider the rational expression $\frac{x^2 - 4x}{2x - 8}$. If $x = 4$, then the denominator would equal zero because $2(4) - 8 = 0$. Because division by zero is never allowed, we say that this rational expression is defined for all values of x except $x = 4$. Any numeric value of the variable that causes the denominator of a rational expression to equal zero is called a **restricted value**. Therefore, $x = 4$ is a restricted value for the rational expression $\frac{x^2 - 4x}{2x - 8}$.

One way to simplify a rational expression is to factor the polynomials in the numerator and denominator and cancel any common factors. The restricted values from the original rational expression are still restricted values for the simplified expression.

For example, $\frac{x^2 - 4x}{2x - 8} = \frac{x(x - 4)}{2(x - 4)} = \frac{x\cancel{(x - 4)}}{2\cancel{(x - 4)}} = \frac{x}{2}$. Therefore, the rational expression $\frac{x^2 - 4x}{2x - 8}$ is equivalent to $\frac{x}{2}$ for all values of x not equal to 4.

Example 1 Simplify Rational Expressions

Simplify each rational expression.

a. $\frac{x^2 + x - 12}{x^2 + 9x + 20}$ b. $\frac{x^3 + 1}{x + 1}$ c. $\frac{x^2 - x - 2}{2x - x^2}$

Solutions

a. We first factor the rational expression. The numerator and denominator of this rational expression factors as follows:

$$\frac{x^2 + x - 12}{x^2 + 9x + 20} = \frac{(x + 4)(x - 3)}{(x + 5)(x + 4)}$$

This rational expression is defined for all values of x such that $x \neq -5$ and $x \neq -4$. Therefore, -5 and -4 are restricted values. This restriction must also apply to any simplified version of the original rational expression.

We now cancel all common factors within the numerator and denominator.

$$\frac{x^2 + x - 12}{x^2 + 9x + 20} = \frac{\cancel{(x + 4)}(x - 3)}{(x + 5)\cancel{(x + 4)}}$$ Cancel common factors.

Thus, $\frac{x^2 + x - 12}{x^2 + 9x + 20} = \frac{x - 3}{x + 5}$ for $x \neq -5$ and $x \neq -4$.

b. The only restricted value is -1. We can factor the numerator using the **sum of two cubes** factoring formula.

$$\frac{x^3 + 1}{x + 1} = \frac{\cancel{(x + 1)}(x^2 - x + 1)}{\cancel{(x + 1)}}$$ Factor the numerator and cancel common factors.

$$= x^2 - x + 1 \text{ for } x \neq -1$$

c. Factor the numerator and denominator.

$$\frac{x^2 - x - 2}{2x - x^2} = \frac{(x-2)(x+1)}{x(2-x)}$$

The denominator would be equal to zero for $x = 0$ and $x = 2$. Therefore, 0 and 2 are the restricted values. At first glance, it does not appear that the numerator and denominator share a common factor. However, note that if we were to multiply the factor $(2 - x)$ by -1, we get $-1 \cdot (2 - x) = -2 + x = x - 2$.

Therefore, multiply the numerator and denominator by 1 of the form $\frac{-1}{-1}$, and then cancel common factors.

$$\frac{x^2 - x - 2}{2x - x^2} = \frac{(x-2)(x+1)}{x(2-x)}$$

$$= \left(\frac{-1}{-1}\right)\frac{(x-2)(x+1)}{x(2-x)}$$ Multiply the rational expression by $\frac{-1}{-1}$.

$$= -\frac{(x-2)(x+1)}{x(x-2)}$$ $-1 \cdot (2 - x) = (x - 2)$.

$$= -\frac{\cancel{(x-2)}(x+1)}{x\cancel{(x-2)}}$$ Cancel common factors.

$$= -\frac{x+1}{x} \text{ for } x \neq 0 \text{ and } x \neq 2.$$

Watch the **interactive video** to see the solution to Example 1 worked out in detail.

You Try It Work through the following You Try It problem.

Work Exercises 1–10 in this textbook or in the MyLab Math Study Plan.

Note From this point forward in this section, we do not make any restrictions on the variables. That is, we assume that all expressions in the denominators of all rational expressions represent nonzero real numbers.

OBJECTIVE 2 MULTIPLYING AND DIVIDING RATIONAL EXPRESSIONS

We multiply rational expressions similarly to the way we multiply **rational numbers**. If $\frac{A}{B}$ and $\frac{C}{D}$ are rational expressions, then $\frac{A}{B} \cdot \frac{C}{D} = \frac{AC}{BD}$.

Example 2 Multiply Rational Expressions

Multiply $\frac{2x^2 + 3x - 2}{3x^2 - 2x - 1} \cdot \frac{3x^2 + 4x + 1}{2x^2 + x - 1}$.

Solution Factor the numerator and denominator of each rational expression. Note that each numerator and denominator are polynomials of the form $ax^2 + bx + c$, where $a \neq 1$. If you need to review how to factor polynomials of this form, watch the **video** that was introduced in Section R.6.

$$\frac{2x^2+3x-2}{3x^2-2x-1}\cdot\frac{3x^2+4x+1}{2x^2+x-1}$$

$$=\frac{(2x-1)(x+2)}{(x-1)(3x+1)}\cdot\frac{(3x+1)(x+1)}{(2x-1)(x+1)}$$ Factor the numerators and denominators.

$$=\frac{(2x-1)(x+2)(3x+1)(x+1)}{(x-1)(3x+1)(2x-1)(x+1)}$$ Multiply numerators and denominators.

$$=\frac{\cancel{(2x-1)}(x+2)\cancel{(3x+1)}\cancel{(x+1)}}{(x-1)\cancel{(3x+1)}\cancel{(2x-1)}\cancel{(x+1)}}$$ Cancel common factors.

$$=\frac{x+2}{x-1}$$

We divide two rational expressions in exactly the same manner in which we divide two **rational numbers**. Recall that if $\frac{a}{b}$ and $\frac{c}{d}$ are rational numbers such that b, c, and d are nonzero, then $\frac{a}{b}\div\frac{c}{d}=\frac{a}{b}\cdot\frac{d}{c}=\frac{ad}{bc}$. The rational numbers $\frac{c}{d}$ and $\frac{d}{c}$ are called **reciprocals**. We follow this same pattern when dividing two rational expressions. To find the quotient of two rational expressions, we find the product of the first expression times the reciprocal of the second rational expression. Thus, if $\frac{A}{B}$ and $\frac{C}{D}$ are rational expressions, then $\frac{A}{B}\div\frac{C}{D}=\frac{A}{B}\cdot\frac{D}{C}=\frac{AD}{BC}$.

Example 3 Divide Rational Expressions

Divide and simplify $\frac{x^3-8}{2x^2-x-6}\div\frac{x^2+2x+4}{6x^2+11x+3}$.

Solution

$$\frac{x^3-8}{2x^2-x-6}\div\frac{x^2+2x+4}{6x^2+11x+3}$$ Write the original expression.

$$=\frac{x^3-8}{2x^2-x-6}\cdot\frac{6x^2+11x+3}{x^2+2x+4}$$ Rewrite as a multiplication problem.

$$=\frac{(x-2)(x^2+2x+4)}{(2x+3)(x-2)}\cdot\frac{(2x+3)(3x+1)}{x^2+2x+4}$$ Factor the numerators and denominators. Use the **difference of two cubes** factoring formula to factor the first numerator.

$$=\frac{(x-2)(x^2+2x+4)(2x+3)(3x+1)}{(2x+3)(x-2)(x^2+2x+4)}$$ Multiply.

$$=\frac{\cancel{(x-2)}\cancel{(x^2+2x+4)}\cancel{(2x+3)}(3x+1)}{\cancel{(2x+3)}\cancel{(x-2)}\cancel{(x^2+2x+4)}}$$ Cancel common factors.

$$=3x+1$$

 **Example 4 Multiply and Divide Rational Expressions**

Perform the indicated operations and simplify.

$$\frac{x^2+x-6}{x^2+x-42}\cdot\frac{x^2+12x+35}{x^2-x-2}\div\frac{x+7}{x^2+8x+7}$$

Solution Watch the video to see that the expression simplifies to $\dfrac{(x+3)(x+5)}{x-6}$.

You Try It Work through the following You Try It problem.

Work Exercises 11–26 in this textbook or in the MyLab Math Study Plan.

OBJECTIVE 3 ADDING AND SUBTRACTING RATIONAL EXPRESSIONS

Recall that in order to add or subtract rational numbers, we must first rewrite both numbers in a different form using a common denominator. Once both rational numbers have a common denominator, we add or subtract the numerators, writing the result above the common denominator. For example, we subtract the rational number $\frac{1}{6}$ from $\frac{5}{8}$ as follows:

$$\frac{5}{8}-\frac{1}{6}=\left(\frac{5}{8}\right)\left(\frac{3}{3}\right)-\left(\frac{1}{6}\right)\left(\frac{4}{4}\right)=\frac{15}{24}-\frac{4}{24}=\frac{15-4}{24}=\frac{11}{24}$$

We add and subtract rational expressions in much the same way. We start by determining the **least common denominator (LCD)**. The LCD is the smallest algebraic expression divisible by all denominators.

Example 5 Add and Subtract Rational Expressions

Perform the indicated operations and simplify.

a. $\dfrac{3}{x+1}-\dfrac{2-x}{x+1}$ b. $\dfrac{3}{x^2+2x}+\dfrac{x-2}{x^2-x}$

Solutions

a. Each rational expression has the same denominator, so we can combine by subtracting the numerators.

$\dfrac{3}{x+1}-\dfrac{2-x}{x+1}$ Write the original expression.

$=\dfrac{3-(2-x)}{x+1}$ Subtract numerators.

$=\dfrac{3-2+x}{x+1}$ Use the distributive property.

$=\dfrac{x+1}{x+1}$ Combine like terms.

$=\dfrac{\cancel{x+1}}{\cancel{x+1}}=1$ Cancel common factors and simplify.

b. Factor the denominators to determine the LCD.

$$\left.\begin{aligned} x^2 + 2x &= x(x+2) \\ x^2 - \ \ x &= x(x-1) \end{aligned}\right\} \quad \text{LCD} = x(x+2)(x-1).$$

$$\frac{3}{x(x+2)} + \frac{x-2}{x(x-1)}$$ Write the original expression with the denominators factored.

$$= \frac{3}{x(x+2)} \frac{(x-1)}{(x-1)} + \frac{x-2}{x(x-1)} \frac{(x+2)}{(x+2)}$$ Multiply the first rational expression by $\frac{(x-1)}{(x-1)}$ and multiply the second rational expression by $\frac{(x+2)}{(x+2)}$.

$$= \frac{3(x-1) + (x-2)(x+2)}{x(x+2)(x-1)}$$ Add the numerators.

$$= \frac{3x - 3 + x^2 - 4}{x(x+2)(x-1)}$$ Use the **distributive property**.

$$= \frac{x^2 + 3x - 7}{x(x+2)(x-1)}$$ Combine like terms.

Try working through Example 6 on your own. Watch the **interactive video** when you are ready to see the solution.

Example 6 Add and Subtract Rational Expressions

Perform the indicated operations and simplify.

a. $\frac{3}{x-y} - \frac{x+5y}{x^2-y^2}$ **b.** $\frac{x+4}{3x^2+20x+25} + \frac{x}{3x^2+16x+5}$

Solution Watch the **interactive video** to see each solution worked out in detail.

You Try It **Work through the following You Try It problem.**

Work Exercises 27–44 in this textbook or in the MyLab Math Study Plan.

OBJECTIVE 4 SIMPLIFYING COMPLEX RATIONAL EXPRESSIONS

A complex rational expression (also called a complex fraction) is a fraction that contains a rational expression in the numerator and/or the denominator. The following are examples of complex rational expressions:

$$\frac{4 - \frac{5}{x-1}}{\frac{6}{x-1} - 7} \qquad \frac{\frac{1}{x} - \frac{3}{x+4}}{\frac{2}{x^2+4x} + \frac{2}{x}}$$

To simplify a complex rational expression, we want to rewrite the expression using only one fraction bar. There are two methods that can be used to simplify such expressions. We call these Method I and Method II. Method I involves the following three-step process.

Method I for Simplifying Complex Rational Expressions

Step 1. Simplify the rational expression in the numerator.

Step 2. Simplify the rational expression in the denominator.

Step 3. Divide the expression in the numerator by the expression in the denominator. That is, multiply the expression in the numerator by the reciprocal of the expression in the denominator. Simplify if possible.

Example 7 Simplify a Complex Rational Expression Using Method I

Simplify using Method I: $\dfrac{4 - \dfrac{5}{x-1}}{\dfrac{6}{x-1} - 7}$

Solution

Step 1. Simplify the numerator.

$$4 - \frac{5}{x-1} = 4\frac{(x-1)}{(x-1)} - \frac{5}{x-1} = \frac{4(x-1)-5}{x-1}$$

$$= \frac{4x-4-5}{x-1} = \frac{4x-9}{x-1}$$

Step 2. Simplify the denominator.

$$\frac{6}{x-1} - 7 = \frac{6}{x-1} - 7\frac{(x-1)}{(x-1)} = \frac{6-7(x-1)}{x-1}$$

$$= \frac{6-7x+7}{x-1} = \frac{13-7x}{x-1}$$

Step 3. Divide.

$$\frac{4 - \dfrac{5}{x-1}}{\dfrac{6}{x-1} - 7} = \frac{\dfrac{4x-9}{x-1}}{\dfrac{13-7x}{x-1}} = \frac{4x-9}{x-1} \cdot \frac{x-1}{13-7x}$$

$$= \frac{4x-9}{\cancel{x-1}} \cdot \frac{\cancel{x-1}}{13-7x} = \frac{4x-9}{13-7x}$$

As stated previously, there is another method that can be used to simplify a complex rational expression. Method II involves multiplying the numerator and denominator of the complex rational expression by the overall LCD. The overall LCD is the LCD of *all* fractions in both the numerator and the denominator.

Method II for Simplifying Complex Rational Expressions

Step 1. Determine the overall LCD.

Step 2. Multiply the numerator and denominator of the complex rational expression by the overall LCD.

Step 3. Simplify.

We now use Method II to simplify the same complex rational expression from Example 7.

Example 8 Simplify a Complex Rational Expression Using Method II

Simplify using Method II: $\dfrac{4 - \dfrac{5}{x-1}}{\dfrac{6}{x-1} - 7}$

Solution

Step 1. The overall LCD is $x - 1$.

Step 2. Multiply the numerator and denominator by $x - 1$.

$$\frac{4 - \dfrac{5}{x-1}}{\dfrac{6}{x-1} - 7} \cdot \frac{x-1}{x-1}$$

Step 3. Simplify.

$$\frac{4 - \dfrac{5}{x-1}}{\dfrac{6}{x-1} - 7} \cdot \frac{x-1}{x-1} = \frac{4(x-1) - \dfrac{5(x-1)}{x-1}}{\dfrac{6(x-1)}{x-1} - 7(x-1)}$$

$$= \frac{4x - 4 - \dfrac{5\cancel{(x-1)}}{\cancel{x-1}}}{\dfrac{6\cancel{(x-1)}}{\cancel{x-1}} - 7x + 7}$$

$$= \frac{4x - 4 - 5}{6 - 7x + 7} = \frac{4x - 9}{13 - 7x}$$

As you can see from Examples 7 and 8, it does not matter which method we choose. Pick the method that you are most comfortable with. See if you can simplify the complex rational expression in Example 9. You can watch the **interactive video** and choose to see the solution using either method.

Example 9 Simplify a Complex Rational Expression

Simplify the complex rational expression using Method I or Method II.

$$\frac{\dfrac{1}{x} - \dfrac{3}{x+4}}{\dfrac{2}{x^2 + 4x} + \dfrac{2}{x}}$$

Solution To see the solution, watch the **interactive video** and click on Method I or Method II.

You Try It Work through the following You Try It problem.

Work Exercises 45–54 in this textbook or in the MyLab Math Study Plan.

OBJECTIVE 5 APPLICATIONS OF RATIONAL EXPRESSIONS

Rational expressions often appear in applications as in the application shown below in Example 10 involving the area of a trapezoid.

Example 10 Using a Rational Expression to Describe the Area of a Trapezoid

The area of a trapezoid can be represented by the expression $\frac{1}{2}h(a+b)$, where a and b represent the lengths of the parallel bases of the trapezoid and h represents the height.

Suppose that $a = \frac{1}{x}$ inches, $b = \frac{1}{x+1}$ inches, and $h = x^2 - x$ inches. See Figure 6.

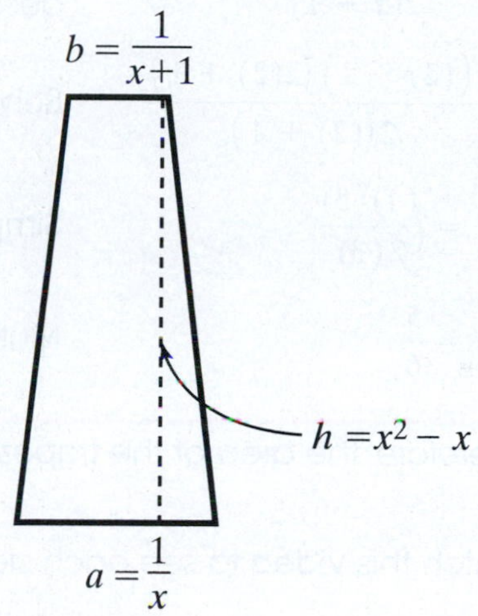

Figure 6

a. Determine the simplified rational expression that describes the area of the trapezoid.

b. What is area of the trapezoid if $x = 2$ inches?

Solution

a. $\frac{1}{2}h(a+b)$ — Write the expression that represents the area of a trapezoid.

$\frac{1}{2}(x^2 - x)\left(\frac{1}{x} + \frac{1}{x+1}\right)$ — Substitute $a = \frac{1}{x}$, $b = \frac{1}{x+1}$, and $h = x^2 - x$.

$= \frac{1}{2}(x^2 - x)\left(\frac{1}{x}\cdot\frac{(x+1)}{(x+1)} + \frac{1}{x+1}\cdot\frac{x}{x}\right)$ — Multiply the rational expression $\frac{1}{x}$ by $\frac{(x+1)}{(x+1)}$ and multiply the rational expression $\frac{1}{x+1}$ by $\frac{x}{x}$.

$= \frac{1}{2}(x^2 - x)\left(\frac{x+1+x}{x(x+1)}\right)$ — Add the numerators of the rational expressions inside the parentheses.

$= \frac{1}{2}x(x-1)\left(\frac{2x+1}{x(x+1)}\right)$ Simplify the numerator inside the parentheses and factor.

$= \frac{1}{2}\cancel{x}(x-1)\left(\frac{2x+1}{\cancel{x}(x+1)}\right)$ Cancel common factors.

$= \frac{(x-1)(2x+1)}{2(x+1)}$ Multiply numerators and multiply denominators.

The simplified rational expression that represents the area of the trapezoid is $\frac{(x-1)(2x+1)}{2(x+1)}$.

b. We can substitute $x = 2$ into the simplified rational expression from part a. to determine the area.

$\frac{(x-1)(2x+1)}{2(x+1)}$ Start with the simplified rational expression that describes the area of the trapezoid.

$\frac{((2)-1)(2(2)+1)}{2((2)+1)}$ Substitute $x = 2$.

$= \frac{(1)(5)}{2(3)}$ Simplify each parenthesis.

$= \frac{5}{6}$ Multiply.

Therefore, the area of the trapezoid is $\frac{5}{6}$ square inches.

 Watch this **video** to see each step of the solution to this example.

You Try It **Work through the following You Try It problem.**

Work Exercises 55–60 in this textbook or in the MyLab Math Study Plan.

R.7 Exercises

In Exercises 1–10, simplify each rational expression.

1. $\frac{2x - 6x^2}{2x}$

2. $\frac{x^2 - 81}{9 + x}$

3. $\frac{9a - 27}{5a - 15}$

4. $\frac{x^2 + x - 6}{x + 3}$

5. $\frac{x - 13}{13 - x}$

6. $\frac{x^2 - 4}{-2 - x}$

7. $\frac{3x^2 - 2x - 8}{3x^2 - 14x - 24}$

8. $\frac{x^3 - 125}{3x - 15}$

9. $\frac{4x^2 - 7x - 2}{8x^3 + 2x^2 + 4x + 1}$

10. $\frac{36x^2 - 42x + 49}{216x^3 + 343}$

For the remainder of this exercise set, assume that there are no restrictions on the variables. That is, assume that all expressions in the denominators of all rational expressions represent nonzero real numbers. In Exercises 11–26, perform the indicated operations and simplify.

11. $\frac{18x+18}{2x+6} \cdot \frac{x+3}{6x^2-6}$

12. $\frac{16x-12}{35} \cdot \frac{10}{3-4x}$

13. $\frac{15a-10a^2}{9a^2+6a+1} \cdot \frac{6a^2+11a+3}{4a^2-9}$

14. $\frac{4x^3-32}{2x^2+4x-16} \cdot \frac{5x+10}{4x^2+8x+16}$

15. $\frac{a^3+a^2b+a+b}{2a^3+2a} \cdot \frac{18a^2}{6a^2-6b^2}$

16. $\frac{x^2-6x-7}{2x^2-98} \cdot \frac{x^2+14x+49}{3x^2+24x+21}$

17. $\frac{x^2-4}{9} \cdot \frac{x^2-x-2}{x^2-4x+4}$

18. $\frac{4x}{9} \div \frac{8x+16}{9x+18}$

19. $\frac{a+b}{ab} \div \frac{a^2-b^2}{7a^3b}$

20. $\frac{x^2-14x+49}{x^2-x-42} \div \frac{x^2-49}{8}$

21. $\frac{x^2-6x-16}{3x^2-192} \div \frac{x^2+10x+16}{x^2+16x+64}$

22. $\frac{3x-x^2}{x^3-27} \div \frac{x}{x^2+3x+9}$

23. $\dfrac{\frac{5x^2-4x-12}{3x^2-19x-14}}{\frac{25x^2+35x+6}{3x^2+17x+10}}$

24. $\frac{8}{x} \div \frac{7xy}{x^4} \cdot \frac{5x^2}{x^7}$

25. $\frac{12x^2+8x+1}{y^2-2y-3} \cdot \frac{y^2-7y+12}{6x^2+7x+1} \div \frac{4x^2-4x-3}{3x^2-2x-5}$

26. $\frac{4a^2-64}{2a^2-4a} \div \frac{a^3+4a^2}{5a^2-10a} \cdot \frac{6a^3+4a^2}{4a^2-16a}$

27. $\frac{7}{x}+\frac{9}{x}$

28. $\frac{x+3}{x-4}+\frac{5x-2}{x-4}$

29. $\frac{x^2}{3x+1}-\frac{4}{3x+1}$

30. $\frac{6}{49x}+\frac{5}{7x^2}$

31. $\frac{9}{x-3}+\frac{x}{3-x}$

32. $\frac{8}{x-1}-\frac{6}{x+8}$

33. $\frac{x}{x+3}+\frac{2x-7}{x-3}$

34. $\frac{x-2}{x+4}-\frac{x+1}{x-4}$

35. $\frac{x}{x^2-36}+\frac{4}{x}$

36. $\frac{7}{12x^2y}-\frac{13}{4xy}$

37. $\frac{10}{a+b}+\frac{10}{a-b}$

38. $\frac{z+5}{z-9}-\frac{z-4}{z+6}$

39. $\frac{9}{x-y}-\frac{x+3y}{x^2-y^2}$

40. $\frac{x-8}{x^2+12x+27}+\frac{x-9}{x^2-9}$

41. $\frac{6x}{x^2+4x-12}-\frac{x}{x^2-36}$

42. $\frac{y-8}{2y^2+9y+7}-\frac{y+9}{2y^2-3y-35}$

43. $\frac{5}{y}-\frac{5}{y-3}+\frac{16}{(y-3)^2}$

44. $\frac{x-3}{x^2+12x+36}+\frac{1}{x+6}-\frac{2x+3}{2x^2+3x-54}$

In Exercises 45–54, simplify the complex rational expression using Method I or Method II.

45. $\dfrac{7 + \dfrac{1}{x}}{7 - \dfrac{1}{x}}$

46. $\dfrac{\dfrac{1}{x} + 3}{\dfrac{1}{x^2} - 9}$

47. $\dfrac{\dfrac{x+2}{x-6} - \dfrac{x+12}{x+5}}{x + 82}$

48. $\dfrac{\dfrac{6}{x} + \dfrac{5}{y}}{\dfrac{5}{x} - \dfrac{6}{y}}$

49. $\dfrac{\dfrac{1}{16} - \dfrac{1}{x^2}}{\dfrac{1}{4} + \dfrac{1}{x}}$

50. $\dfrac{\dfrac{8}{x+4} - \dfrac{2}{x+7}}{\dfrac{x+8}{x+4}}$

51. $\dfrac{\dfrac{2}{x+13} + \dfrac{1}{x-5}}{2 - \dfrac{x+25}{x+13}}$

52. $\dfrac{\dfrac{x-5}{x+5} + \dfrac{x-5}{x-9}}{1 + \dfrac{x+5}{x-9}}$

53. $\dfrac{1 + \dfrac{4}{x} - \dfrac{5}{x^2}}{1 - \dfrac{2}{x} - \dfrac{35}{x^2}}$

54. $\dfrac{6x^{-1} + 6y^{-1}}{xy^{-1} - x^{-1}y}$

55. A rectangle having side lengths of $\dfrac{x}{x+2}$ and $\dfrac{x}{x+5}$ is shown below.

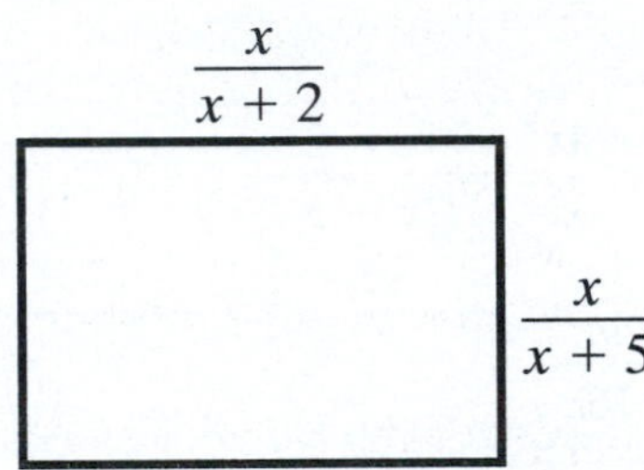

a. Determine the simplified rational expression that describes the perimeter of the rectangle.

b. What is the perimeter of the rectangle if $x = 2$ inches?

56. The length of a rectangle can be represented by the ratio $\dfrac{A}{w}$, where A is the area of the rectangle and w is the width. Suppose that the area of a rectangle is $x^2 + 5x - 6$ square feet and the width of the rectangle is $x - 1$ feet.

a. Determine the simplified rational expression that describes the length of the rectangle.

b. What is the length of the rectangle if $x = 8$ feet?

57. The height of a triangle can be represented by the ratio $\dfrac{2A}{b}$, where A is the area of the triangle and b is the base of the triangle. Suppose that the area of a triangle is $x^2 + 7x + 12$ square millimeters and the length of the base of the triangle is $x + 3$ millimeters.

a. Determine the simplified rational expression that describes the height of the triangle.

b. What is the height of the triangle if $x = 12$ millimeters?

58. The area of a trapezoid can be represented by the expression $\frac{1}{2}h(a + b)$, where a and b represent the lengths of the parallel bases of the trapezoid and h represents the height. Suppose that $b_1 = \frac{2}{x}$ inches, $b_2 = \frac{3}{x + 4}$ inches, and $h = x^2 + x - 12$ inches.

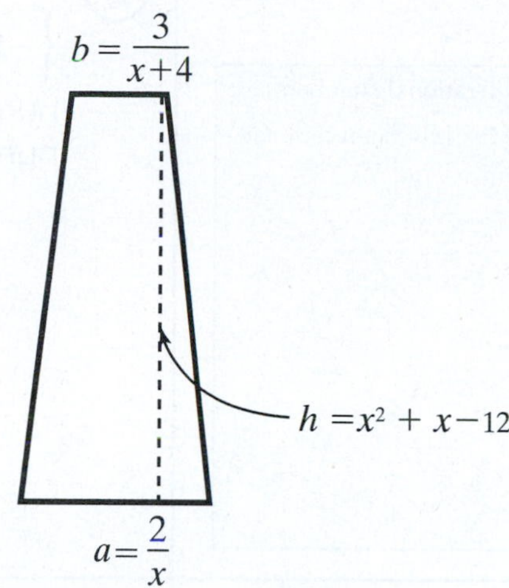

a. Determine the simplified rational expression, in factored form, that describes the area of the trapezoid.

b. What is area of the trapezoid if $x = 4$ inches?

59. On a 715-mile trip to college, Kendra's average speed was x miles per hour. On her return trip home for the holidays, she averaged 10 miles per hour less than her trip to college. The time that it took Kendra to drive to college can be represented by the rational expression $\frac{715}{x}$. The time that it took Kendra to drive home can be represented by the rational expression $\frac{715}{x - 10}$

a. Write a simplified rational expression that describes the total time that it took Kendra to drive to college and back home.

b. What was Kendra's total driving time if she averaged 65 miles per hour on her drive from home to college?

60. A small pump takes 4 more hours than a large pump to empty a city swimming pool. The rate for the large pump to drain the pool can be represented by the rational expression $\frac{1}{x}$, where x is the time that it takes for the large pump to drain the pool.

a. Write a simplified rational expression that describes the combined rate if both pumps are used to drain the pool.

b. What is the combined rate to drain the pool if it takes the larger pump 8 hours to drain the pool?

Chapter R Summary

Key Concepts	Examples/Videos
R.1 Real Numbers 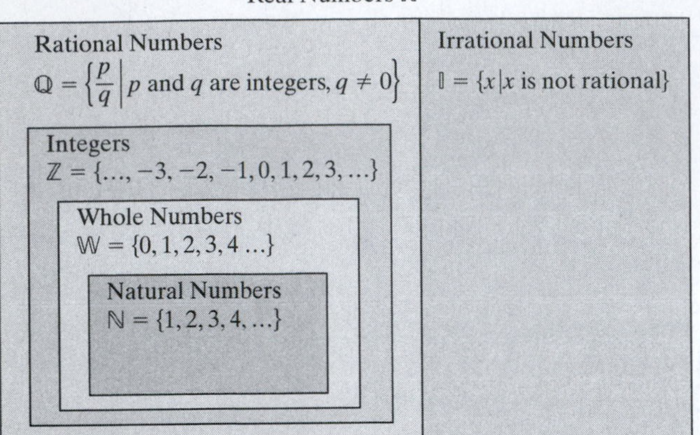	Classify each number in the set $\left\{-5, -\frac{1}{3}, 0, \sqrt{3}, 1.\overline{4}, \sqrt{36}, 2\pi, 11\right\}$ as a natural number, whole number, integer, rational number, and/or real number.
(see interval table below)	 Given the set sketched on the number line, a. Identify the type of interval. b. Write the set using set-builder notation. c. Write the set using interval notation. 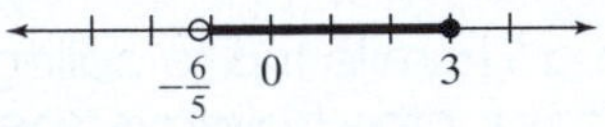 (number line: $-\frac{6}{5}$, 0, 3) Write the set $\left[\frac{-1}{3}, \infty\right)$ in set-builder notation and graph the set on a number line. Write the set $\left\{x \middle\vert -\frac{7}{2} < x \leq \pi\right\}$ in interval notation and graph the set on a number line.
For any two sets A and B: the **intersection** of A and B is given by $A \cap B$ and represents the elements that are in set A **and** set B. the **union** of A and B is given by $A \cup B$ and represents the elements that are in set A **or** in set B.	 Let $A = \{x \vert x > -2\}$ and $B = \{x \vert x \leq 5\}$. Find $A \cap B$.  Let $A = \{1, 3, 4, 5, 7, 10, 12\}$ and $B = \{2, 4, 6, 8, 10, 12\}$. Find $A \cup B$.

Type of Interval and Graph	Interval Notation	Set-Builder Notation
Open interval  (graph: a, b)	(a, b)	$\{x \vert a < x < b\}$
Closed interval  (graph: a, b)	$[a, b]$	$\{x \vert a \leq x \leq b\}$
Half-open intervals   (graphs: a, b)	$(a, b]$ $[a, b)$	$\{x \vert a < x \leq b\}$ $\{x \vert a \leq x < b\}$
Open infinite intervals   (graphs: a; b)	(a, ∞) $(-\infty, b)$	$\{x \vert x > a\}$ $\{x \vert x < b\}$
Closed infinite intervals   (graphs: a; b)	$[a, \infty)$ $(-\infty, b]$	$\{x \vert x \geq a\}$ $\{x \vert x \leq b\}$

Key Concepts		Examples/Videos
R.2 Order of Operations and Properties of Real Numbers If a, b, and c are real numbers then:		Rewrite $11 + y$ as an equivalent expression.
• $a + b = b + a$	Commutative Property of Addition	Rewrite $7(y + 4)$ as an equivalent expression. Rewrite $(2 + 3x) + 8$ as an equivalent expression.
• $ab = ba$	Commutative Property of Multiplication	Rewrite $8(pq)$ as an equivalent expression.
• $(a + b) + c = a + (b + c)$	Associative Property of Addition	Use the distributive property to multiply $-5(x + y - 4)$.
• $(a \cdot b) \cdot c = a \cdot (b \cdot c)$	Associative Property of Multiplication	Use the distributive property to write $2xw - xy + 5xz$ as a product.
• $a(b + c) = ab = ac$	Distributive Property	
$a^n = \underbrace{a \cdot a \cdot a \cdot \ldots \cdot a}_{n \text{ factors of } a}$	where a is the base and n is the exponent.	Rewrite the expression $4x \cdot 4x \cdot 4x \cdot 4x \cdot 4x \cdot 4x$ using exponential notation. Then identify the base and the exponent. Identify the base and exponent of the expression $(-3)^4$, and then evaluate. Identify the base and exponent of the expression -2^5, and then evaluate.
Order of Operations 1. Parentheses 2. Exponents 3. Multiplication and Division 4. Addition and Subtraction		Simplify each expression. a. $-2^3 + [3 - 5 \cdot (1 - 3)]$ b. $\dfrac{\lvert 2 - 3^3 \rvert + 5}{5^2 - 4^2}$

(Summary)

Key Concepts	Examples/Videos
R.3 The Laws of Exponents; Radicals **Laws of Exponents** Suppose s and t are rational numbers and a and b are real numbers, then: 1. $b^s b^t = b^{s+t}$ Product rule of exponents 2. $\frac{b^s}{b^t} = b^{s-t}$, where $b \neq 0$ Quotient rule of exponents 3. $(b^s)^t = b^{st}$ Power rule of exponents 4. $(ab)^s = a^s b^s$ Product to power rule of exponents 5. $\left(\frac{a}{b}\right)^s = \frac{a^s}{b^s}$, where $b \neq 0$ Quotient to power rule of exponents 6. $b^0 = 1$ Zero exponent rule 7. $b^{-s} = \frac{1}{b^s}$, where $b \neq 0$ Reciprocal rule of exponents	Simplify each expression using positive exponents. a. $-3^{-4} \cdot 3^4$ b. $5z^4 \cdot (-9z^{-5})$ c. $\left(\frac{a^{-3}b^4c^{-6}}{2a^5b^{-4}c}\right)^{-3}$
If $\frac{m}{n}$ is a rational number for $n \geq 2$ and if b is a real number then: $b^{1/n} = \sqrt[n]{b}$ and $b^{m/n} = \sqrt[n]{b^m} = (\sqrt[n]{b})^m$ provided that the expression is a real number.	Simplify each expression using positive exponents. a. $(9)^{3/2}$ b. $(-32)^{-3/5}$ c. $\frac{(125x^4y^{-1/4})^{2/3}}{(x^2y)^{1/3}}$
If $\sqrt[n]{a}$ and $\sqrt[n]{b}$ are real numbers, then: • $\sqrt[n]{a}\sqrt[n]{b} = \sqrt[n]{ab}$ **Product Rule for Radicals** • $\sqrt[n]{\frac{a}{b}} = \frac{\sqrt[n]{a}}{\sqrt[n]{b}}$ **Quotient Rule for Radicals**	Multiply $\sqrt[5]{9x^2} \cdot \sqrt[5]{14y^3}$. Assume all variables represent positive real numbers. Simplify $\sqrt[3]{\frac{16x^5}{27}}$. Assume all variables represent positive real numbers.
Using the Product Rule to Simplify Radical Expressions of the Form $\sqrt[n]{a}$ **Step 1** Write the radicand as a product of two factors, one being the largest possible perfect nth power. **Step 2** Use the product rule for radicals to take the nth root of each factor. **Step 3** Simplify the nth root of the perfect n th power.	Use the product rule to simplify $\sqrt{50x^4y^3}$. Assume all variables represent positive real numbers.

Key Concepts	Examples/Videos
R.4 Polynomials A **monomial** is a simplified term in which all variables are raised to non-negative integer powers and no variables appear in any denominator. The **degree of a monomial** is the sum of the exponents of the variables. A **polynomial** is the sum or difference of one or more monomials. The **degree of a polynomial** is the degree of the highest monomial term.	Determine whether each algebraic expression is a polynomial. If so, state the degree. a. $8x^3y^2 - 2$ b. $5w^5 - 3y^2 - \frac{6}{w}$ c. $7p^8m^4 - 5p^5m^3 + 2pm$
Add or subtract polynomials by combining like terms.	Add or subtract the polynomials as indicated. a. $(6x^4y^2 - 2xy^3) + (7xy^3 - 8x^4y^2)$ b. $(6x^4y^2 - 2xy^3) - (7xy^3 - 8x^4y^2)$
To multiply polynomials, use the **distributive property** multiple times. If m is a monomial and $p_1, p_2, p_3 \ldots, p_n$ are the terms of a polynomial, then $m(p_1 + p_2 + p_3 + \ldots + p_n) = mp_1 + mp_2 + mp_3 + \ldots mp_n$.	Find the product. $(2x^2 - 3)(3x^3 - x^2 + 1)$
The **FOIL** Method Last First $(x + a)(x + b) = x^2 + bx + ax + ab = x^2 + (a + b)x + ab$ Inside Outside	Find the product using the FOIL Method. $(3x + 5)(2x + 7)$
Special Products Let a and b represent any real number or algebraic expression, then: $(a + b)(a - b) = a^2 - b^2$ Difference of two squares $(a + b)^2 = a^2 + 2ab + b^2$ Squaring a binomial $(a - b)^2 = a^2 - 2ab + b^2$ $(a + b)^3 = a^3 + 3a^2b + 3ab^2 + b^3$ Cubing a binomial $(a - b)^3 = a^3 - 3a^2b + 3ab^2 - b^3$	Find each product using a special product formula. a. $(2x + 3)(2x - 3)$ b. $(3x + 2y)^2$ c. $(5 - 3z)^3$
When dividing two polynomials using **long division**, be sure to write the divisor and dividend in standard form inserting any missing terms.	Find the quotient and remainder when $3x^4 + x^3 + 7x + 4$ is divided by $x^2 - 1$.

Key Concepts	Examples/Videos
R.5 Operations with Radicals  Two radicals are **like radicals** if they have the same **index** and the same **radicand**. Add or subtract radicals by first simplifying the radicals, then add or subtract like terms and like radicals.	 Simplify. $\sqrt[3]{24} - \sqrt[3]{192} + 4\sqrt[3]{250}$  Simplify. $2a\sqrt{16ab^3} + 4\sqrt{9a^2b} - 5\sqrt{4a^3b^3}$
Multiply radical expressions using the distributive property to multiply each term in the first expression by each term in the second expression. Then combine like terms and like radicals.	Multiply. Assume variables represent non-negative values. 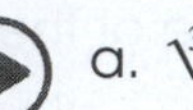  a. $\sqrt[3]{2n^2}\left(\sqrt[3]{4n} + \sqrt[3]{5n}\right)$  b. $\left(3\sqrt{x} - 2\right)^2$
Rationalizing a Denominator with One Term To rationalize a denominator with a single radical of **index** n, multiply the **numerator** and denominator by a radical of index n so that the radicand in the denominator is a **perfect** n**th power**.	 Simplify each expression and rationalize the denominator. a. $\sqrt{\frac{3x}{50}}$ b. $\frac{\sqrt{18x}}{\sqrt{27xy}}$ c. $\sqrt[3]{\frac{-4x^5}{16y^5}}$
Rationalizing a Denominator with Two Terms To rationalize a denominator with two terms involving one or more square roots, multiply the numerator and denominator by the **conjugate** of the denominator.	 Rationalize the denominator. $\frac{\sqrt{y} - 3}{\sqrt{y} + 2}$
R.6 Factoring Polynomials The **Greatest Common Factor** (GCF) is the largest factor in common to two or more terms. 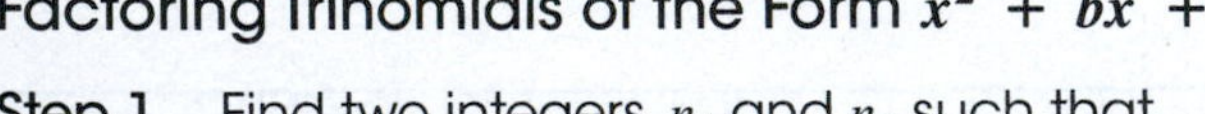 **Factoring Trinomials of the Form $x^2 + bx + c$** **Step 1** Find two integers, n_1 and n_2 such that $n_1 \cdot n_2 = c$ and $n_1 + n_2 = b$. **Step 2** Rewrite the trinomial in the factored form as $(x + n_1)(x + n_2)$. **Step 3** Check your answer by multiplying out the factored form.	 Factor out the GCF of the following expressions. $4x^2 - 6x^3 + 2x$, $5w^2y^3 - 15w^4y + 20w^3y^2$, and $x(x^2 + 6) - 5(x^2 + 6)$ Factor each trinomial or state that it is prime. a. $x^2 - 2x - 24$ b. $x^2 - 2x - 6$ c. $x^2 - 12x + 32$
Factoring Trinomials of the Form $ax^2 + bx + c, a \neq 1$ **Step 1** Multiply ac. **Step 2** Find two integers, n_1 and n_2 such that $n_1 \cdot n_2 = ac$ and $n_1 + n_2 = b$. **Step 3** Rewrite the trinomial as $ax^2 + n_1x + n_2x + c$ **Step 4** Factor by grouping. **Step 5** Check your answer by multiplying out the factored form.	 Factor the trinomial $4x^2 + 17x + 15$.

Key Concepts	Examples/Videos
Special Factoring Formulas Let a and b represent any real number or algebraic expression, then: $a^2 - b^2 = (a + b)(a - b)$ Difference of two squares $(a + b)^2 = a^2 + 2ab + b^2$ Perfect square trinomials $(a - b)^2 = a^2 - 2ab + b^2$ $a^3 + b^3 = (a + b)(a^2 - ab + b^2)$ Sum of two cubes $a^3 - b^3 = (a - b)(a^2 + ab - b^2)$ Difference of two cubes	Completely factor each expression. a. $2x^3 + 5x^2 - 12x$ b. $5x^4 - 45x^2$ c. $8z^3x - 27y + 8z^3y - 27x$
R.7 Rational Expressions A rational expression is an algebraic expression that when simplified has the form $\frac{P}{Q}$, where $P \neq 0$ and Q are polynomials such that the degree of Q is greater than or equal to 1.	
One way to simplify a rational expression of the form $\frac{P}{Q}$ is to factor P and Q and cancel any common factors.	Simplify each rational expression. a. $\frac{x^2 + x - 12}{x^2 + 9x + 20}$ b. $\frac{x^3 + 1}{x + 1}$ c. $\frac{x^2 - x - 2}{2x - x^2}$
Multiplying and Dividing Rational Expressions If $\frac{A}{B}$ and $\frac{C}{D}$ are rational expressions, then: $\frac{A}{B} \cdot \frac{C}{D} = \frac{AC}{BD}$ and $\frac{A}{B} \div \frac{C}{D} = \frac{A}{B} \cdot \frac{D}{C} = \frac{AD}{BC}$.	Perform the indicated operations and simplify. $\frac{x^2 + x - 6}{x^2 + x - 42} \cdot \frac{x^2 + 12x + 35}{x^2 - x - 2} \div \frac{x + 7}{x^2 + 8x + 7}$
Adding and Subtracting Rational Expressions To add or subtract rational expressions with different denominators, rewrite each expression using the LCD for each denominator. Then combine the numerators and simplify.	Perform the indicated operations and simplify. a. $\frac{3}{x - y} - \frac{x + 5y}{x^2 - y^2}$ b. $\frac{x + 4}{3x^2 + 20x + 25} + \frac{x}{3x^2 + 16x + 5}$

Key Concepts	Examples/Videos
A **complex rational expression** is a fraction that contains a rational expression in the numerator and/or the denominator. Simplifying Complex Rational Expressions **Method I** **Step 1** Simplify the rational expression in the numerator. **Step 2** Simplify the rational expression in the denominator. **Step 3** Multiply the expression in the numerator by the reciprocal of the expression in the denominator and simplify. **Method II** **Step 1** Determine the overall LCD. **Step 2** Multiply the numerator and denominator by the overall LCD. **Step 3** Simplify.	 Simplify the complex rational expression using Method I or Method II. $\dfrac{\dfrac{1}{x} - \dfrac{3}{x+4}}{\dfrac{2}{x^2+4x} + \dfrac{2}{x}}$

Chapter R Review Exercises

1. Given the set of real numbers, list the numbers that are a) natural numbers, b) whole numbers, c) integers, d) rational numbers, and/or e) irrational numbers.

$$\left\{-17, -\sqrt{5}, -\frac{25}{19}, 0, 0.331, 1, \frac{\pi}{2}\right\}$$

2. Given the set sketched on the number line, a) identify the type of interval, b) write the set using set-builder notation, and c) write the set using interval notation.

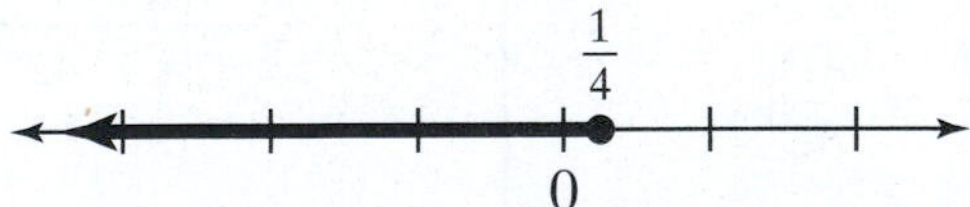

3. Write the interval $(-1, \infty)$ in set-builder notation and graph the set on a number line.

4. Write the set $\{x \mid 0 \le x < 3\}$ in interval notation and graph the set on a number line.

5. Find the union and intersection of the given sets.

$$((-\infty, 0) \cup (0, \infty)) \cap [-3, \infty)$$

6. Use the distributive property to write the given expression as a product.

$$qp - rq + qt$$

7. Write the given expression using exponential notation. Then identify the base and the exponent.

$$(-6y) \cdot (-6y) \cdot (-6y) \cdot (-6y) \cdot (-6y)$$

8. Simplify the expression using order of operations.

$$3 \cdot (18 + 3)^2 - 4 \cdot (8 - 3)^2$$

9. Evaluate the algebraic expression for the indicated variable.

$$\frac{6b - 9b^2}{b^2 - 6} \quad \text{for } b = 3$$

10. Simplify the algebraic expression.

$$56p^2 - 36 - [4(p^2 - 9) + 1]$$

For Exercises 11 and 12, use laws of exponents to simplify each expression using positive exponents only. Assume all variables represent positive real numbers.

11. $(16a^{-3}bc^{-6})(2ab)^{-5}$

12. $x^{\frac{1}{3}} \cdot x^{\frac{1}{4}} \cdot x^{\frac{-1}{2}}$

13. Use the product rule to simplify. Assume all variables represent positive real numbers.

$$3\sqrt{125y^9}$$

14. Multiply and simplify.

$$\sqrt{6} \cdot \sqrt{30}$$

15. Use the quotient rule to simplify. Assume all variables represent positive real numbers.

$$\sqrt{\frac{12x^3}{25y^2}}$$

For exercises 16–19, perform the indicated operations and simplify.

16. $(9x^2y - 6xy) + (8x^2y - xy)$

17. $2x^2(3x^5 - 2x^3 + 7x - 4)$

18. $(5x - 4)(3x + 2)$

19. $(4x - 9)^2$

20. Find the quotient and remainder when $7x^3 - 5x^2 + x + 4$ is divided by $x + 3$.

21. Subtract and simplify. $\sqrt{45} - \sqrt{80}$

22. Add and simplify. $6\sqrt{20x^3} + 4x\sqrt{45x}$

23. Multiply and simplify. $(3 + \sqrt{5})(2 + \sqrt{7})$

(Summary)

24. Rationalize the denominator. $\sqrt{\dfrac{9}{2x}}$

25. Rationalize the denominator. $\dfrac{\sqrt{a}}{3\sqrt{a} - \sqrt{b}}$

For exercises 26–30, completely factor each polynomial.

26. $4x^4 - 20x^3 + 16x^2$

27. $5x^2 + 4xy + 15x + 12y$

28. $y^2 + 5y - 14$

29. $2x^2 + 5x - 25$

30. $x^2 - 81$

31. Simplify the rational expression. $\dfrac{2x - x^2}{2x}$

For exercises 32–34, perform the indicated operations and simplify.

32. $\dfrac{12x^2 + 8x + 1}{y^2 - 2y - 3} \cdot \dfrac{y^2 - 7y + 12}{6x^2 + 7x + 1} \div \dfrac{4x^2 - 4x - 3}{3x^2 - 2x - 5}$

33. $\dfrac{8}{x - 1} - \dfrac{6}{x + 8}$

34. $\dfrac{6x}{x^2 + 4x - 12} - \dfrac{x}{x^2 - 36}$

35. $\dfrac{\dfrac{1}{16} - \dfrac{1}{x^2}}{\dfrac{1}{4} + \dfrac{1}{x}}$

CHAPTER 1

Equations, Inequalities, and Applications

CHAPTER ONE CONTENTS

1.1 Linear and Rational Equations

THINGS TO KNOW

Before working through this section, be sure that you are familiar with the following concepts:

VIDEO ANIMATION INTERACTIVE

You Try It

1. Factoring Trinomials with a Leading Coefficient Equal to 1 (Section R.6)

You Try It

2. Factoring Trinomials with a Leading Coefficient Not Equal to 1 (Section R.6)

You Try It

3. Simplifying Rational Expressions (Section R.7)

OBJECTIVES

1 Recognizing Linear Equations

2 Solving Linear Equations with Integer Coefficients

3 Solving Linear Equations Involving Fractions

4 Solving Linear Equations Involving Decimals

OBJECTIVE 1 RECOGNIZING LINEAR EQUATIONS

In the Review chapter, we manipulated **algebraic expressions**. Remember, algebraic expressions do *not* involve an equal sign. An **equation** indicates that two algebraic expressions are equal. For example, $2x + 6$ and $5x$ are two algebraic expressions. By equating these two expressions with an equal sign, we obtain the equation $2x + 6 = 5x$. This equation is an example of a **linear equation in one variable**. A linear equation in one variable involves only constants and variables that are raised to the first power. Examples of linear equations in one variable include the following:

$$7x - 4 = x - 5, \quad \frac{1}{3}y - \sqrt{2} = 11y, \quad \text{and} \quad 0.4a - 1 = 10$$

> **Definition** Linear Equation in One Variable
>
> A linear equation in one variable is an equation that can be written in the form $ax + b = 0$ where a and b are real numbers and $a \neq 0$.

Watch this **interactive video** to determine if equations are linear or nonlinear, or view this **example** of how to recognize linear and nonlinear equations.

You Try It **Work through this You Try It problem.**

Work Exercises 1–7 in this textbook or in the MyLab Math Study Plan.

OBJECTIVE 2 SOLVING LINEAR EQUATIONS WITH INTEGER COEFFICIENTS

Let's now begin to solve linear equations. A solution to an equation in one variable is a value that, when substituted for the variable, makes the equation true. For example, given the equation $2x - 1 = 5$, we see that if $x = 3$, the equation is a true statement. In other words, by substituting the number 3 into the equation for x, we get a true statement:

$2x - 1 = 5$	Write the original equation.
$2(3) - 1 \stackrel{?}{=} 5$	Substitute 3 in for x.
$6 - 1 \stackrel{?}{=} 5$	Multiply.
$5 = 5$	True statement!

We start by solving the most basic linear equations, which are linear equations with integer **coefficients**.

Example 1 Solve a Linear Equation with Integer Coefficients

Solve $5(x - 6) - 2x = 3 - (x + 1)$.

Solution The goal here is to isolate the variable x on one side of the equation. First, use the **distributive property** to remove the parentheses on the left-hand side:

$5(x - 6) - 2x = 3 - (x + 1)$	Write the original equation.
$5x - 30 - 2x = 3 - x - 1$	Use the distributive property.
$3x - 30 = 2 - x$	Combine like terms.
$3x - 30 + x = 2 - x + x$	Add x to both sides.
$4x - 30 = 2$	Simplify.
$4x - 30 + 30 = 2 + 30$	Add 30 to both sides.
$4x = 32$	Combine like terms.
$\frac{4x}{4} = \frac{32}{4}$	Divide both sides by 4.
$x = 8$	Possible solution.

You should now check by substituting $x = 8$ back into the original equation. Check your answer when finished.

Example 2 Solve a Linear Equation with Integer Coefficients

Solve $6 - 4(x + 4) = 8x - 2(3x + 5)$.

Solution To isolate the variable x, we first use the **distributive property** to remove all parentheses and then simplify:

$6 - 4(x + 4) = 8x - 2(3x + 5)$	Write the original equation.
$6 - 4x - 16 = 8x - 6x - 10$	Use the distributive property.
$-4x - 10 = 2x - 10$	Combine like terms.
$-4x - 10 + 4x = 2x - 10 + 4x$	Add $4x$ to both sides.
$-10 = 6x - 10$	Simplify.
$-10 + 10 = 6x - 10 + 10$	Add 10 to both sides.
$0 = 6x$	Combine like terms.
$\frac{0}{6} = \frac{6x}{6}$	Divide both sides by 6.
$0 = x$	Possible solution.

You should verify that $x = 0$ is the solution by substituting $x = 0$ back into the original equation.

It is important to point out that $x = 0$ can be a valid solution to an equation, as in Example 2. Once we encounter equations in which variables appear in one or more denominators, we have to be extremely careful when checking our answers because division by zero is never allowed! (See **Example 6**.)

You Try It Work through this You Try It problem.

Work Exercises 8–11 in this textbook or in the MyLab Math Study Plan.

OBJECTIVE 3 SOLVING LINEAR EQUATIONS INVOLVING FRACTIONS

To solve a linear equation involving fractions, we want to first transform the equation into a linear equation involving integer coefficients. We do this by multiplying both sides of the equation by the **least common denominator** (LCD).

Example 3 Solve a Linear Equation Involving Fractions

Solve $\frac{1}{3}(1 - x) - \frac{x+1}{2} = -2$.

Solution The first thing to do when solving equations involving fractions is to find the LCD. The LCD for this equation is $\mathbf{6}$. We can eliminate the fractions by multiplying both sides of this equation by $\mathbf{6}$.

$$6\left(\frac{1}{3}(1 - x) - \frac{x+1}{2}\right) = 6(-2) \quad \text{Multiply both sides by 6.}$$

$$6\left(\frac{1}{3}(1 - x)\right) - 6\left(\frac{x+1}{2}\right) = 6(-2) \quad \text{Use the distributive property.}$$

$$2(1 - x) - 3(x + 1) = -12 \quad \text{Multiply.}$$

Notice that the original equation containing fractions has now been transformed into a linear equation involving integer coefficients, as in Example 1. You should solve this equation and/or watch the corresponding **video** to verify that the solution set is $\left\{\frac{11}{5}\right\}$. ●

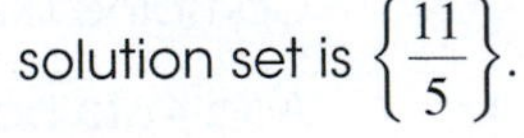

You Try It Work through this You Try It problem.

Work Exercises 12–18 in this textbook or in the MyLab Math Study Plan.

OBJECTIVE 4 SOLVING LINEAR EQUATIONS INVOLVING DECIMALS

The strategy for solving linear equations involving decimals is similar to the one used to solve linear equations involving fractions. We want to eliminate all decimals. We can eliminate decimals by multiplying both sides of the equation by the appropriate power of 10, such as $10^1 = 10$, $10^2 = 100$, $10^3 = 1{,}000$, etc. To determine the appropriate power of 10, look at the constants in the equation and choose the constant that has the greatest number of decimal places. Count those decimal places and then raise 10 to that power. Multiplying by that power of 10 will immediately eliminate all decimal places in the equation.

Example 4 Solve a Linear Equation Involving Decimals

Solve $0.1(y - 2) + 0.03(y - 4) = 0.02(10)$.

Solution The constants .03 and .02 in the equation each have *two* decimal places. We can eliminate the decimals in this equation by multiplying both sides of the equation by $10^2 = 100$, thus moving each decimal two places to the right to obtain the new equation $10(y - 2) + 3(y - 4) = 2(10)$. Watch the **video** to see this problem solved in its entirety. You should verify that the solution set is $\{4\}$. ●

You Try It Work through this You Try It problem.

Work Exercises 19–21 in this textbook or in the MyLab Math Study Plan.

OBJECTIVE 5 RECOGNIZING RATIONAL EQUATIONS

Recall that a **rational number** is the quotient of two integers. A **rational expression** is the quotient of two polynomials such that the degree of the polynomial in the denominator is greater than or equal to 1. A rational equation is an equation that contains one or more rational expressions. The precise definition of a rational equation is given below.

> **Definition** Rational Equation
>
> A rational equation is an equation consisting of one or more rational expressions with any other expressions of the equation being **polynomials**.

Examples of rational equations include the following:

$$\frac{x^2 - x - 12}{x + 5} = 1, \; \frac{1}{x} - x = \frac{2}{x + 5} - 8, \text{ and } \frac{3}{x^3 - 1} = \frac{4x - 5}{x^2 - 1}$$

Example 5 Recognizing Rational Equations

Determine which of the following equations are rational equations.

a. $\dfrac{2 - x}{x + 5} + 3 = \dfrac{4}{x + 2}$ **b.** $x^2 - 2x - 24 = \dfrac{1}{2}$

c. $\dfrac{12}{x^2 + x - 2} - \dfrac{x + 3}{x - 1} = \dfrac{1 - x}{x + 2}$

Solution Watch and work through the **video** to see the solution. ●

You Try It Work through this You Try It problem.

Work Exercises 22–26 in this textbook or in the MyLab Math Study Plan.

OBJECTIVE 6 SOLVING RATIONAL EQUATIONS THAT LEAD TO LINEAR EQUATIONS

A rational equation always contains at least one expression that involves a numerator and a non-constant denominator. The process of solving a rational equation is very similar to the process of solving linear equations containing fractions. That is, we first determine the **least common denominator** (LCD) and then multiply both sides of the equation by the LCD. We have to be extra cautious when solving rational equations because we have to be aware of the **restricted values**.

Example 6 Solve a Rational Equation

Solve $\frac{2-x}{x+2}+3=\frac{4}{x+2}$.

Solution First note that the value of $x=-2$, is a **restricted value** because this value would make the denominator equal to zero.

To solve this rational equation, first multiply both sides of the equation by the LCD, $x+2$, to eliminate the denominators.

$$(x+2)\left(\frac{2-x}{x+2}+3\right)=(x+2)\left(\frac{4}{x+2}\right) \quad \text{Multiply both sides by } x+2.$$

$$(x+2)\left(\frac{2-x}{x+2}\right)+(x+2)(3)=(x+2)\left(\frac{4}{x+2}\right) \quad \text{Use the distributive property.}$$

$$2-x+3x+6=4 \quad \text{Multiply.}$$

$$2x+8=4 \quad \text{Combine like terms.}$$

$$2x=-4 \quad \text{Subtract 8 from both sides.}$$

$$x=-2 \quad \text{Divide both sides by 2.}$$

The solution of $x=-2$ was a **restricted value**. Therefore, if we substitute $x=-2$ back into the original equation, we obtain a zero in at least one denominator. Because division by zero is never permitted, $x=-2$ is **not** a solution. Therefore, there is no solution to this rational equation. ●

When we obtain a restricted value when solving an equation, we call that restricted value an **extraneous solution.**

> **Definition** Extraneous Solution
>
> An extraneous solution is a solution to an equation obtained through algebraic manipulations that is not a solution to the original equation.

CAUTION Because rational equations often have an extraneous solution, it is imperative to first determine all restricted values before beginning the solution process.

(eText Screens 1.1-1–1.1-19)

The following steps can be used for solving rational equations.

> **Solving Rational Equations**
>
> **Step 1** Factor any denominators then list all **restricted values**.
>
> **Step 2** Determine the LCD of all denominators in the equation.
>
> **Step 3** Multiply both sides of the equation by the LCD.
>
> **Step 4** Solve the resulting equation.
>
> **Step 5** Discard any restricted values.

Example 7 Solve a Rational Equation

Solve $\dfrac{2}{x+4} + \dfrac{1}{x-5} = \dfrac{5}{x^2 - x - 20}$

Solutions

Step 1 We can factor the denominator on the right-hand side.

$$\frac{2}{x+4} + \frac{1}{x-5} = \frac{5}{(x+4)(x-5)}$$

There are two restricted values. These values are $x = -4$ and $x = 5$.

Step 2 The LCD is $(x+4)(x-5)$.

Step 3 Multiply both sides of the equation by $(x+4)(x-5)$.

$$(x+4)(x-5)\left(\frac{2}{x+4} + \frac{1}{x-5}\right) = \frac{5}{(x+4)(x-5)}(x+4)(x-5)$$ Multiply both sides by the LCD.

$$\frac{2}{x+4}(x+4)(x-5) + \frac{1}{x-5}(x+4)(x-5) = \frac{5}{(x+4)(x-5)}(x+4)(x-5)$$ Use the **distributive property**.

$$\frac{2}{\cancel{x+4}}\cancel{(x+4)}(x-5) + \frac{1}{\cancel{x-5}}(x+4)\cancel{(x-5)} = \frac{5}{\cancel{(x+4)(x-5)}}\cancel{(x+4)}\,\cancel{(x-5)}$$ Cancel common factor.

$$2(x-5) + 1(x+4) = 5$$ Rewrite.

Step 4 We now solve the linear equation.

$2(x-5) + 1(x+4) = 5$	Write the linear equation obtained in Step 3.
$2x - 10 + x + 4 = 5$	Use the **distributive property**.
$3x - 6 = 5$	Simplify the left-hand side of the equation.
$3x = 11$	Add 6 to both sides of the equation.
$x = \dfrac{11}{3}$	Divide both sides of the equation by 3.

Step 5 The two restricted values from Step 1 were $x = -4$ and $x = 5$. Thus, the solution of $x = \dfrac{11}{3}$ is a valid solution to this rational equation.

 **Example 8 Solve a Rational Equation**

Solve $\frac{12}{x^2+x-2} - \frac{x+3}{x-1} = \frac{1-x}{x+2}$.

Solution Watch and work through the video to see the solution.

You Try It Work through this You Try It problem.

Work Exercises 27–36 in this textbook or in the MyLab Math Study Plan.

1.1 Exercises

Skill Check Exercises

For exercises SCE-1 through SCE-8, determine the Least Common Denominator (LCD) of the given expression.

SCE-1. $\frac{2}{9} + \frac{1}{3} - \frac{1}{6}$

SCE-2. $\frac{1}{8}(2p-1) - \frac{7}{3}p - \frac{p-4}{6}$

SCE-3. $\frac{a-3}{6} - \frac{3(a-1)}{10} + \frac{2a+1}{5}$

SCE-4. $\frac{3x}{x+1} - \frac{5x+7}{x-1}$

SCE-5. $\frac{1}{2x} - \frac{1}{4} + \frac{6}{8x^2}$

SCE-6. $\frac{w}{w-3} - \frac{2w}{2w-1} - \frac{w-3}{2w^2-7w+3}$

SCE-7. $\frac{3}{x-1} + \frac{4}{x+1} - \frac{8x}{x^2-1}$

SCE-8. $\frac{6}{x^2-x} - \frac{2}{x} + \frac{3}{x-1}$

In Exercises 1–7, determine whether the given equation is linear or nonlinear.

1. $7x - \frac{1}{3} = 5$
2. $\frac{5}{x} + 4 = 10$
3. $\sqrt{2}x - 1 = 0$
4. $x^2 + x = 1$
5. $8x - 7 = \pi x - 3$
6. $\frac{5x-1}{x+2} = 3$
7. $5 - 0.2x = 0.1 - 4x$

In Exercises 8–21, solve each equation:

8. $5x - 8 = 3 - 7(x+1)$
9. $4(2x+5) = 10 - 2(x-5)$
10. $3(3x+4) = -(x-3)$
11. $-3(2-x) + 1 = 4 - (7x-2)$
12. $\frac{1}{5} - \frac{1}{3}x = \frac{4}{3}$
13. $\frac{1}{2}x - 3 = 7 - \frac{3}{4}x$
14. $\frac{1}{2}y - \frac{1}{3}(y-1) = 5y$
15. $\frac{1}{8}(2p-1) = \frac{7}{3}p - \frac{p-4}{6}$

16. $\dfrac{x+5}{4} - \dfrac{x-10}{5} = 2$

17. $\dfrac{x-4}{2} - \dfrac{x+1}{4} = \dfrac{2x-3}{4}$

18. $\dfrac{a-3}{6} - \dfrac{3(a-1)}{10} = \dfrac{2a+1}{5}$

19. $0.12x + 0.3(x-4) = 0.01(2x-3)$

20. $0.002(1-k) + 0.01(k-3) = 1$

21. $-0.17x + 0.01(16x+5) = 0.02(3-x)$

In Exercises 22–26, determine whether the given equation is rational or non-rational.

22. $\dfrac{1}{x} = 10$

23. $\dfrac{2x+5}{8} = x$

24. $\dfrac{3}{x^2} - 3x^2 + \dfrac{x^2}{3} = 3$

25. $\dfrac{\sqrt{x^3-2x^2}}{x^2-4} + \dfrac{1}{x^2-4} = 5$

26. $\dfrac{3x^3+5x^2-8x+4}{7x^4-11x^2-\sqrt{5}x+1} - 2 = x^2 + 11x$

In Exercises 27–36, solve each rational equation.

27. $\dfrac{5}{x+1} - 1 = \dfrac{8}{x+1}$

28. $\dfrac{3x}{x-4} + 1 = \dfrac{12}{x-4}$

29. $\dfrac{2x}{x^2-5x-6} - \dfrac{5}{x-6} = \dfrac{1}{x+1}$

30. $\dfrac{w}{w-3} - \dfrac{2w}{2w-1} = \dfrac{w-3}{2w^2-7w+3}$

31. $\dfrac{3}{x-1} + \dfrac{4}{x-1} = \dfrac{8x}{x^2-1}$

32. $\dfrac{6}{x^2-x} - \dfrac{2}{x} = \dfrac{3}{x-1}$

33. $\dfrac{1}{x-4} + \dfrac{2}{x-2} = \dfrac{2}{x^2-6x+8}$

34. $\dfrac{2x}{x^2-4} = \dfrac{4}{x^2-4} - \dfrac{1}{x+2}$

35. $\dfrac{2}{x+3} + \dfrac{1}{x+7} = \dfrac{5}{x^2+10x+21}$

36. $\dfrac{-2x^2+5x}{x^3-1} + \dfrac{2x}{x-1} = 2$

1.2 Applications of Linear Equations and Rational Equations

THINGS TO KNOW

Before working through this section, be sure that you are familiar with the following concepts:

VIDEO | ANIMATION | INTERACTIVE

You Try It

1. Solving Linear Equations with Integer Coefficients (Section 1.1)

You Try It

2. Solving Linear Equations Involving Fractions (Section 1.1)

(eText Screens 1.2-1–1.2-40)

You Try It 3. Solving Linear Equations Involving Decimals (Section 1.1) 

You Try It 4. Solving Rational Equations that Lead to Linear Equations (Section 1.1) 

INTRODUCTION

Read this introduction before beginning Objective 1.

OBJECTIVES

1 Converting Verbal Statements into Mathematical Statements

2 Solving Applications Involving Unknown Numeric Quantities

3 Solving Applications Involving Geometric Formulas

4 Solving Applications Involving Decimal Equations (Money, Mixture, Interest)

5 Solving Applications Involving Uniform Motion

6 Solving Applications Involving Rates of Work

SECTION 1.2 EXERCISES

Introduction to Section 1.2

We find algebraic applications in all sorts of subjects, including economics, physics, medicine, and countless other disciplines. In this section, we learn how to solve a variety of useful applications that involve **linear equations** and **rational equations**. Before we look at some specific examples, it is important to first practice translating or converting verbal statements into mathematical statements.

OBJECTIVE 1 CONVERTING VERBAL STATEMENTS INTO MATHEMATICAL STATEMENTS

When solving word problems, it is important to recognize key words and phrases that translate into algebraic expressions involving addition, subtraction, multiplication, and division.

Example 1 Convert Verbal Statements into Mathematical Statements

Rewrite each statement as an algebraic expression or equation:

a. 7 more than three times a number

b. 5 less than twice a number

c. Three times the quotient of a number and 11

d. The sum of a number and 9 is 1 less than half of the number.

e. The product of a number and 4 is 1 more than 8 times the difference of 10 and the number.

Solution

a. If x is the number, then three times a number is $3x$. Therefore, 7 more than three times a number is equivalent to $3x + 7$ (or $7 + 3x$).

b. Twice a number is equivalent to $2x$. So, 5 less than twice a number is written as $2x - 5$.

CAUTION **The expression $2x - 5$ is not equivalent to $5 - 2x$ because subtraction is not commutative; that is, $a - b \neq b - a$.**

c. The quotient of a number and 11 is equivalent to $\frac{x}{11}$. So, three times the quotient of a number and 11 is equivalent to $3\left(\frac{x}{11}\right)$ or $\frac{3x}{11}$.

d. Notice the key word "*is*." The word "*is*" often translates into an equal sign, thus indicating that we need to translate the phrase into an **equation**. In this case, the sum of a number and 9 translates to $x + 9$. The statement "1 less than half of the number" translates to $\frac{1}{2}x - 1$. Putting it all together, we get the following equation:

$$\underbrace{x + 9}_{\text{The sum of a number and 9}} \underset{\text{is}}{=} \underbrace{\frac{1}{2}x - 1}_{\text{one less than half the number}}$$

e. This statement translates into the following equation:

$$\underbrace{4x}_{\text{The product of a number and 4}} \underset{\text{is}}{=} \underbrace{8(10 - x) + 1}_{\text{one more than 8 times the difference of 10 and the number}}$$

You Try It **Work through this You Try It problem.**

Work Exercises 1–6 in this textbook or in the MyLab Math Study Plan.

It is important to have a strategy in place before trying to solve an applied problem. **George Polya**, a Hungarian mathematician who taught at Stanford University for many years, was famous for his book *How to Solve It* (Princeton University Press). The Mathematics Lab at the University of Idaho was named the Polya Mathematics Learning Center in honor of George Polya. His problem-solving model can be summarized into four basic steps.

Polya's Guidelines for Problem Solving

1. Understand the problem.
2. Devise a plan.
3. Carry out the plan.
4. Look back.

Following is a four-step strategy for solving applied problems with Polya's problem-solving guidelines in mind.

FOUR-STEP STRATEGY FOR PROBLEM SOLVING

Step 1.	Read the problem several times until you have an understanding of what is being asked. If possible, create diagrams, charts, or tables to assist you in your understanding.	Understand the problem.
Step 2.	Pick a variable that describes the unknown quantity that is to be found. All other quantities must be written in terms of that variable. Write an equation using the given information and the variable.	Devise a plan.
Step 3.	Carefully solve the equation.	Carry out the plan.
Step 4.	Make sure that you have answered the question and then check all answers to make sure they make sense.	Look back.

OBJECTIVE 2 SOLVING APPLICATIONS INVOLVING UNKNOWN NUMERIC QUANTITIES

Example 2 Number of Touchdowns Thrown

Roger Staubach and Terry Bradshaw were both quarterbacks in the National Football League. In 1973, Staubach threw three touchdown passes more than twice the number of touchdown passes thrown by Bradshaw. If the total number of touchdown passes between Staubach and Bradshaw was 33, how many touchdown passes did each player throw?

Solution

Step 1. After carefully reading the problem, we see that we are trying to figure out how many touchdown passes were thrown by each player.

Step 2. Let B be the number of touchdown passes thrown by Bradshaw. Because Staubach threw 3 more than twice the number of touchdowns thrown by Bradshaw, then $2B + 3$ represents the number of touchdown passes thrown by Staubach.

The sum of the number of touchdown passes is 33, so we can write the following equation:

$$\underbrace{\text{Bradshaw's touchdown passes}}_{B} + \underbrace{\text{Staubach's touchdown passes}}_{(2B+3)} = \underbrace{\text{Total}}_{33}$$

Step 3. Solve:

$B + (2B + 3) = 33$	Write the equation.
$3B + 3 = 33$	Combine like terms.
$3B = 30$	Subtract 3 from both sides.
$B = 10$	Divide both sides by 3.

To answer the question, Bradshaw threw 10 touchdown passes in 1973, and Staubach threw $2(B) + 3 = 2(10) + 3 = 23$ touchdown passes in 1973.

Step 4. Check: We see that the total number of touchdown passes is $10 + 23 = 33$. The number of touchdown passes thrown by Staubach is 3 more than two times the number of touchdown passes thrown by Bradshaw.

You Try It Work through this You Try It problem.

Work Exercises 7–9 in this textbook or in the MyLab Math Study Plan.

Example 3 Determine Unknown Numbers

One number is three times another number. Determine the two numbers if the sum of their reciprocals is 4.

Solution

Step 1. We are trying to determine the value of two numbers.

Step 2. Let n be the first number. Then $3n$ is the value of the second number. The sum of their reciprocals is 4, so we can write the following equation.

$$\underbrace{\text{The reciprocal of the first number}}_{\frac{1}{n}} + \underbrace{\text{The reciprocal of the second number}}_{\frac{1}{3n}} = \underbrace{\text{Sum}}_{4}$$

We see that this equation is a **rational equation**. To solve this rational equation, we follow the process that was discussed in **Section 1.1**.

Step 3. Solve: $\frac{1}{n} + \frac{1}{3n} = 4$ Write the equation. (The only **restricted value** is $n = 0$)

$3n\left(\frac{1}{n} + \frac{1}{3n}\right) = (4)3n$ Multiply both sides by the LCD, $3n$.

$3n\frac{1}{n} + 3n\frac{1}{3n} = 4 \cdot 3n$ Use the **distributive property**.

$3 + 1 = 12n$

$4 = 12n$ Simplify.

$\frac{4}{12} = n$ Divide both sides by 12.

$n = \frac{1}{3}$ Reduce the fraction.

Step 4. The first number is $n = \frac{1}{3}$. Therefore, the second number is $3 \cdot \frac{1}{3} = 1$. The sum of their reciprocals is

$$\frac{1}{\frac{1}{3}} + \frac{1}{1} = 3 + 1 = 4.$$

You Try It Work through this You Try It problem.

Work Exercises 10 and 11 in this textbook or in the MyLab Math Study Plan.

OBJECTIVE 3 SOLVING APPLICATIONS INVOLVING GEOMETRIC FORMULAS

In this objective, we use some common formulas from geometry. You may want to review these **formulas** and refer to them while working through the following examples.

Example 4 Determine the Dimensions of a Basketball Court

The length of a college basketball court (rectangle) is 6 feet less than twice its width. If the perimeter is 288 feet, what are the dimensions of the court?

Solution

Step 1. We need to determine the length and width of the basketball court (rectangle). We can use the formula for the **perimeter of a rectangle** to determine the equation.

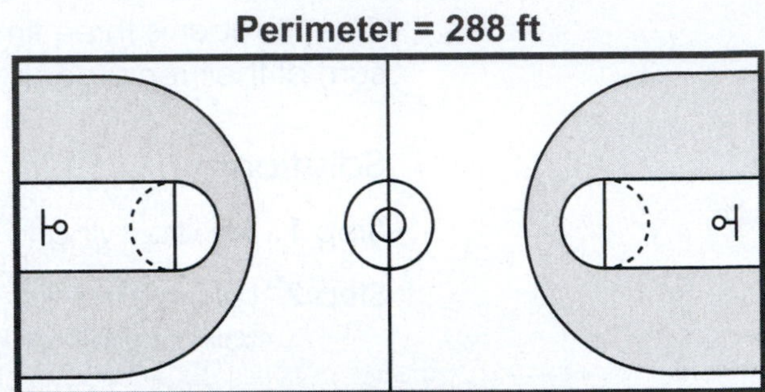

Step 2. Let w be the width of the rectangle. Then the length is $2w - 6$ (6 feet less than twice the width). The perimeter is given to be 288 feet. Thus, we use the formula for the perimeter of a rectangle to get:

Substitute $2w - 6$ for l

$$2l + 2w = P$$ Write the formula for the perimeter of a rectangle.

$$2(2w - 6) + 2w = 288$$

Substitute 288 for P

Finish solving this problem on your own. You may watch this **video** to see each step of the solution process.

You Try It **Work through this You Try It problem.**

Work Exercises 12–15 in this textbook or in the MyLab Math Study Plan.

Example 5 Determine the Side Lengths of a Trapezoid

Suppose that the area of a trapezoid can be represented by the expression $\frac{6}{x}$ square feet and suppose that the lengths of the parallel bases of the trapezoid can be represented by the expressions $\frac{2}{x}$ feet $\frac{3}{x+1}$ feet respectively. If the height is 4 feet, determine the value of x. Then determine the area and values of the lengths of the parallel bases.

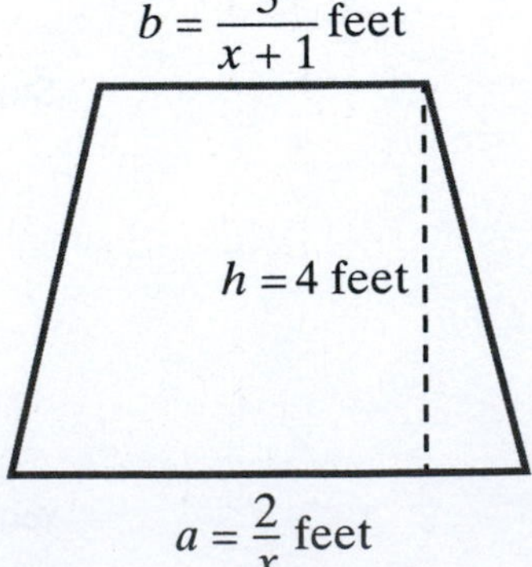

Solution

Step 1. We need to determine the value of x. Then we can find the area of the trapezoid and the lengths of the parallel bases. We can use the formula for the **area of a trapezoid** to determine the equation.

Step 2. We can substitute the given expressions into the formula for the **area of a trapezoid**.

$A = \frac{1}{2}h(a + b)$ Write the formula for the area of a trapezoid.

$\frac{6}{x} = \frac{1}{2}(4)\left(\frac{2}{x} + \frac{3}{x+1}\right)$ Substitute the given expressions in for *A*, *h*, *a*, and *b*.

$\frac{6}{x} = \frac{4}{x} + \frac{6}{x+1}$ Use the **distributive property**.

We see that this equation is a **rational equation.** To solve this rational equation, we follow the process that was discussed in **Section 1.1**.

Step 3. Solve: $\frac{6}{x} = \frac{4}{x} + \frac{6}{x+1}$ Write the equation. (The **restricted values** are $x = 0$ and $x = -1$).

$x(x+1)\frac{6}{x} = x(x+1)\left(\frac{4}{x} + \frac{6}{x+1}\right)$ Multiply both sides by the LCD, $x(x + 1)$.

$\frac{6\cancel{x}(x+1)}{\cancel{x}} = \frac{4\cancel{x}(x+1)}{\cancel{x}} + \frac{6x\cancel{(x+1)}}{\cancel{x+1}}$ Use the **distributive property** and cancel common factors.

$6x + 6 = 10x + 4$ Use the **distributive property** and collect like terms on each side.

$2 = 4x$ Subtract $6x$ from both sides and subtract 4 from both sides.

$\frac{2}{4} = x$ Divide both sides by 4.

$x = \frac{1}{2}$ Reduce the fraction.

Step 4. If $x = \frac{1}{2}$, then we get the following values:.

$Area = \frac{6}{x} = \frac{6}{\frac{1}{2}} = 12$ square feet, $a = \frac{2}{x} = \frac{2}{\frac{1}{2}} = 4$ feet, $b = \frac{3}{x+1} = \frac{3}{\frac{1}{2}+1} = \frac{3}{\frac{3}{2}} = 2$ feet

Try substituting these values into the formula for the **area of a trapezoid** to verify that we have correctly solved this problem.

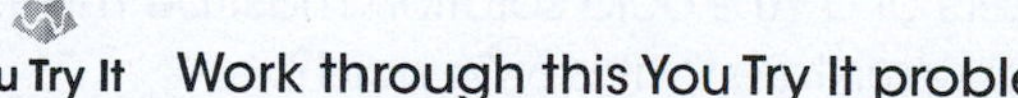

You Try It **Work through this You Try It problem.**

Work Exercise 16 in this textbook or in the MyLab Math Study Plan.

OBJECTIVE 4 SOLVING APPLICATIONS INVOLVING DECIMAL EQUATIONS (MONEY, MIXTURE, INTEREST)

In Example 2, the equation turned out to be a linear equation involving integer coefficients. In the next two examples, we see that the equations turn out to be linear equations involving decimals. You may want to review the technique used to solve **linear equations involving decimals** that was covered in Section 1.1.

Example 6 Money Application

Billy has \$16.50 in his piggy bank, consisting of nickels, dimes, and quarters. Billy notices that he has 20 fewer quarters than dimes. If the number of nickels is equal to the number of quarters and dimes combined, how many of each coin does Billy have?

Solution

Step 1. We must find out how many nickels, dimes, and quarters Billy has.

Step 2. Let d = number of dimes.

The remaining quantities must also be expressed in terms of the variable d. Because Billy has 20 fewer quarters than dimes, we can express the number of quarters in terms of the number of dimes, or

$$d - 20 = \text{number of quarters.}$$

CAUTION The expression $d - 20$ is not equivalent to $20 - d$ because subtraction is not commutative, that is, $a - b \neq b - a$.

The number of nickels is equal to the sum of the number of dimes and quarters.

$$\underbrace{d}_{\text{number of dimes}} + \underbrace{(d - 20)}_{\text{number of quarters}} = \text{number of nickels}$$

Nickels are worth \$.05, dimes are worth \$.10, quarters are worth \$.25, and the total amount is \$16.50. So, we get the following equation:

$$\underbrace{0.05(d + (d - 20))}_{\text{value of the nickels}} + \underbrace{0.10d}_{\text{value of the dimes}} + \underbrace{0.25(d - 20)}_{\text{value of the quarters}} = \underbrace{16.50}_{\text{Total}}$$

Watch the **video** to see the rest of the solution, or solve the equation yourself to verify that Billy has 80 nickels, 50 dimes, and 30 quarters.

You Try It Work through this You Try It problem.

Work Exercises 17–20 in this textbook or in the MyLab Math Study Plan.

Example 7 Mixture Application

How many milliliters of a 70% acid solution must be mixed with 30 mL of a 40% acid solution to obtain a mixture that is 50% acid?

Solution

Step 1. We are mixing two solutions together to obtain a third solution.

Step 2. The unknown quantity in this problem is the amount (in mL) of a 70% acid solution. So, let x = amount of a 70% acid solution. A diagram may help set up the required equation. See Figure 1.

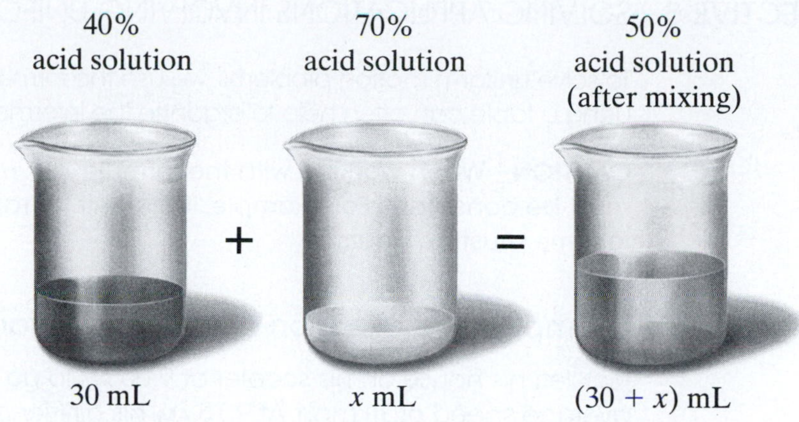

Figure 1

Because we are mixing 30 mL of one solution with x mL of another, the resulting total quantity will be $(x + 30)$ mL, as in **Figure 1**. To set up an equation, notice that the number of milliliters of pure acid in the 40% solution plus the number of milliliters of pure acid in the 70% solution must equal the number of milliliters of pure acid in the 50% solution. The equation is as follows:

$$\underbrace{0.40(30)}_{\text{amount of pure acid in the 40\% container}} + \underbrace{0.70x}_{\text{amount of pure acid in the 70\% container}} = \underbrace{0.50(30 + x)}_{\text{amount of pure acid in the 50\% container}}$$

Step 3. Solving this linear equation involving decimals, we get $x = 15$ mL. (Watch the **video** to see the problem solved in its entirety.) Therefore, we must mix 15 mL of 70% solution to reach the desired mixture. ●

You Try It Work through this You Try It problem.

Work Exercises 21–26 in this textbook or in the MyLab Math Study Plan.

Example 8 Interest Application

Kristen inherited \$20,000 from her Aunt Dawn Ann, with the stipulation that she invest part of the money in an account paying 4.5% **simple interest** and the rest in an account paying 6% simple interest locked in for 3 years. If at the end of 1 year, the total interest earned was \$982.50, how much was invested at each rate?

Solution Watch this **video** to see that Kristen invested \$14,500 at 4.5% and \$5,500 at 6%. ●

You Try It Work through this You Try It problem.

Work Exercises 27–30 in this textbook or in the MyLab Math Study Plan.

OBJECTIVE 5 SOLVING APPLICATIONS INVOLVING UNIFORM MOTION

To solve **uniform motion** problems, we use the formula $d = rt$ or distance $=$ rate $\times$ time. Using a table can often help to organize the information.

CAUTION **When working with the formula $d = rt$, the units for rate and time must be consistent. For example, if the unit for rate is miles per hour, then the unit for time must be hours.**

Example 9 Application of Uniform Motion

Rick left his house on his scooter at 9:00 AM to go fishing. He rode his scooter at an average speed of 10 mph. At 9:15 AM, his girlfriend Deb (who did not find Rick at home) pedaled after Rick on her new 10-speed bicycle at a rate of 15 mph. If Deb caught up with Rick at precisely the time they both reached the fishing hole, how far is it from Rick's house to the fishing hole? At what time did Rick and Deb arrive at the fishing hole?

Solution It is 7.5 miles to the fishing hole from Rick's house. Deb and Rick arrived at the fishing hole at 9:45 AM. Watch the **video** to see this problem worked out in detail.

Example 10 Application of Uniform Motion Involving Wind Speed

An airplane that can maintain an average velocity of 320 mph in still air is transporting smokejumpers to a forest fire. On takeoff from the airport, it encounters a headwind and takes 34 minutes to reach the jump site. The return trip from the jump site takes 30 minutes. What is the speed of the wind? How far is it from the airport to the fire?

Solution

Step 1. We are asked to find the speed of the wind, and then to find the distance from the airport to the forest fire.

Step 2. Let $w =$ speed of the wind.

The plane flies 320 mph in still air (with no wind). Therefore, when the plane encounters a headwind, the plane's rate will decrease by the amount of the wind. So, the net rate (speed) of the plane into a headwind can be described as $320 - w$. Similarly, the net speed of the plane flying with the tailwind is $320 + w$. We know that the time, in hours, is $\frac{34}{60} = \frac{17}{30}$ hours into the headwind and $\frac{30}{60} = \frac{1}{2}$ hour with the tailwind. Knowing that distance $=$ rate $\times$ time, we now create the following table.

	Rate	Time (hours)	Distance
Headwind	$320 - w$	$\frac{17}{30}$	$\frac{17}{30}(320 - w)$
Tailwind	$320 + w$	$\frac{1}{2}$	$\frac{1}{2}(320 + w)$

Because both distances are the same, we can equate the two distances:

$$\underbrace{\frac{17}{30}(320 - w)}_{\substack{\text{distance traveled}\\\text{by the plane into}\\\text{the headwind}}} = \underbrace{\frac{1}{2}(320 + w)}_{\substack{\text{distance traveled}\\\text{by the plane with}\\\text{the tailwind}}}$$

Step 3. Solve: $\frac{17}{30}(320 - w) = \frac{1}{2}(320 + w)$ Rewrite the equation.

$17(320 - w) = 15(320 + w)$ Multiply both sides by 30.

$5440 - 17w = 4800 + 15w$ Use the **distributive property**.

$640 = 32w$ Combine like terms.

$20 = w$ Divide by 32.

Step 4. To answer the question, the wind is blowing at a speed of 20 mph. We are also asked to find the distance from the airport to the fire. We can substitute $w = 20$ into either distance expression. Substituting $w = 20$ into the distance traveled with the tailwind expression, we see that the distance from the airport to the fire is as follows:

$$\frac{1}{2}(320 + 20) = \frac{1}{2}(340) = 170 \text{ mi}$$

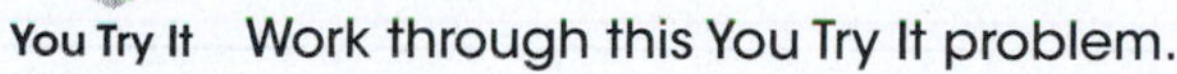

You Try It Work through this You Try It problem.

Work Exercises 31–35 in this textbook or in the MyLab Math Study Plan.

Example 11 Determine the Speed of a Boat Using a Rational Equation

Emalie can travel 16 miles upriver in the same amount of time it takes her to travel 24 miles downriver. If the speed of the current is 4 mph, how fast can her boat travel in still water?

Solution

Step 1. We want to find the speed of Emalie's boat in still water. We know she can travel 16 miles upriver in the same amount of time she can travel 24 miles downriver, and we know the speed of the current is 4 mph.

Step 2. Let $r =$ the speed of Emalie's boat in still water. Then her speed upriver is $r - 4$ (going against the current) and her speed downriver is $r + 4$ (going with the current).

Using the formula $d = rt$, we can solve for t to obtain $t = \frac{d}{r}$ or $\text{time} = \frac{\text{distance}}{\text{rate}}$.

The time to travel upriver must equal the time to travel downriver. Thus we get the following equation.

$$\text{time}_{\text{upriver}} = \text{time}_{\text{downriver}}$$

We rewrite the equation above as the following **rational equation**:

$$\underbrace{\frac{16}{r-4}}_{\text{Time to Travel Upriver}} = \underbrace{\frac{24}{r+4}}_{\text{Time to Travel Downriver}}$$

Step 3. Solve: $\frac{16}{r-4} = \frac{24}{r+4}$ Write the equation. (The restricted values are $r = 4$ and $r = -4$.)

$(r-4)(r+4)\frac{16}{r-4} = \frac{24}{r+4}(r-4)(r+4)$ Multiply both sides by the LCD, $(r-4)(r+4)$.

Finish solving the equation on your own to answer the question. View the **answer**, or watch this **video** for a complete solution.

You Try It Work through this You Try It problem.

Work Exercises 36 and 37 in this textbook or in the MyLab Math Study Plan.

OBJECTIVE 6 SOLVING APPLICATIONS INVOLVING RATES OF WORK

For work problems involving multiple workers, such as people, copiers, pumps, etc., it is often helpful to consider **rate of work**.

> **Rate of work**
>
> The **rate of work** is the number of jobs that can be completed in a given unit of time.
>
> If one job can be completed in t units of time, then the rate of work is given by $\frac{1}{t}$.

When dealing with two workers, use the following formula:

$$\underbrace{\frac{1}{t_1}}_{\text{Work rate of 1st worker}} + \underbrace{\frac{1}{t_2}}_{\text{Work rate of 2nd worker}} = \underbrace{\frac{1}{t}}_{\text{Work rate together}}$$

Here, t_1 and t_2 are the individual times to complete one job, and t is the time to complete the job when working together.

Example 12 Rate of Work Application

Brad and Michelle decide to paint the entire upstairs of their new house. Brad can do the job by himself in 8 hours. If it took them 3 hours to paint the upstairs together, how long would it have taken Michelle to paint it by herself?

Solution

Step 1. We are asked to find the time that it would take Michelle to do the job by herself.

Step 2. Let t = time it takes for Michelle to complete the job; thus, she can complete $\frac{1}{t}$ of the job in an hour. Brad can do this job in 8 hours by himself, so he can complete $\frac{1}{8}$ of the job per hour. Working together it took them 3 hours to complete the job. We can fill out the following table:

	Time needed to complete the job in hours	Portion of job completed in 1 hour (rate)
Brad	8	$\frac{1}{8}$
Michelle	t	$\frac{1}{t}$
Together	3	$\frac{1}{3}$

Step 3. There are two methods that we can use to solve this problem:

Method 1: From the table, we see that $\frac{1}{3}$ = portion of the job completed in 1 hour together. Adding Brad and Michelle's rate together, $\left(\frac{1}{8}+\frac{1}{t}\right)$, we get another expression describing the portion of the job completed together in 1 hour. Thus, equating these rates we obtain the following equation:

$$\left(\frac{1}{8}+\frac{1}{t}\right)=\frac{1}{3}$$

To solve for t, we first multiply by the **LCD** of $24t$ to clear the fractions:

$$\left(\frac{1}{8}+\frac{1}{t}\right)=\frac{1}{3} \quad \text{Rewrite the equation.}$$

$$24t\left(\frac{1}{8}+\frac{1}{t}\right)=\left(\frac{1}{3}\right)24t \quad \text{Multiply both sides by } 24t.$$

$$24t\cdot\frac{1}{8}+24t\cdot\frac{1}{t}=\frac{1}{3}\cdot 24t \quad \text{Use the \textbf{distributive property}.}$$

$$3t+24=8t \quad \text{Simplify.}$$

$$24=5t \quad \text{Subtract } 3t \text{ from both sides.}$$

$$\frac{24}{5}=t \quad \text{Divide both sides by 5.}$$

So, $t=\frac{24}{5}$ hours or $4\frac{4}{5}$ hours = 4 hours and 48 minutes $\left(\frac{4}{5}\text{ hour}\times\frac{60\text{ minutes}}{1\text{ hour}}=48\text{ minutes}\right)$. Therefore, it would take Michelle 4 hours and 48 minutes to paint the upstairs by herself.

Method 2: Another method is based on the fact that ***(time working on job)*** $\times$ ***(rate to complete job)*** $=$ ***fraction of job completed.***

In this case, it took Brad and Michelle 3 hours to complete the entire job together. Thus, the fraction of the job completed is equal to 1. The rate to complete the job together was $\left(\frac{1}{8}+\frac{1}{t}\right)$. Therefore, we can set up the following equation:

$$\underbrace{3}_{\text{time to complete the job}} \cdot \underbrace{\left(\frac{1}{8}+\frac{1}{t}\right)}_{\text{rate to complete job}} = \underbrace{1}_{\text{complete job}}$$

You should verify that $t = \frac{24}{5}$.

Step 4. The solution makes sense because Michelle's time must be less than Brad's time. Also, we can substitute $t = \frac{24}{5}$ into the Method 1 or Method 2 equation to see that a true statement results.

You Try It Work through this You Try It problem.

Work Exercises 38–43 in this textbook or in the MyLab Math Study Plan.

Example 13 Rate of Work Application

Jim and Earl were replacing the transmission on Earl's old convertible. Earl could replace the transmission by himself in 8 hours, whereas it would take Jim 6 hours to do the same job. They worked together for 2 hours, but then Jim had to go to his job at the grocery store. How long did it take Earl to finish replacing the transmission by himself?

Solution Using Method 2 as in Example 12, it would take Earl 3 hours and 20 minutes to finish replacing the transmission. Watch this **video** to see the entire solution.

You Try It Work through this You Try It problem.

Work Exercise 44 in this textbook or in the MyLab Math Study Plan.

1.2 Exercises

Skill Check Exercises

For exercises SCE-1 through SCE-4, determine the Least Common Denominator of the given rational equation.

SCE-1. $\frac{1}{8}+\frac{1}{t}=\frac{1}{3}$ **SCE-2.** $\frac{5+x}{7+x}=\frac{17}{23}$ **SCE-3.** $\frac{10}{r-5}=\frac{14}{r+5}$ **SCE-4.** $\frac{21}{x}=3\left(\frac{4}{x}+\frac{7}{x+2}\right)$

In Exercises 1–6, write the corresponding algebraic expression or equation for each verbal statement. Let n represent the unknown number.

1. 10 more than twice a number

2. 5 less than three times a number and 6

3. The quotient of three and four times a number

4. The product of a number and 2 is 3 less than the quotient of a number and 4.

5. Three less than four times the quotient of a number and 7 is equal to 8 less than twice.

6. The sum of a three times a number and the reciprocal of the number is 5.

In Exercises 7–44, solve the problems algebraically. Clearly define all variables, write an appropriate equation, and solve.

7. One number is 5 more than twice the other number. If the sum of the two numbers is 26, find the two numbers.

8. The sum of three consecutive even integers is 42 feet. Find the integers.

9. Together, Steve and Tom sold 121 raffle tickets for their school. Steve sold 1 more than twice as many raffle tickets as Tom. How many raffle tickets did each boy sell?

10. One number is four times another number. Determine the two numbers if the sum of their reciprocals is $\frac{5}{12}$.

11. When the same number is added to the numerator and denominator of the fraction $\frac{5}{7}$, the result is $\frac{17}{23}$. What is that number?

12. The perimeter of a rectangular garden is 32 feet. The length of the garden is 4 feet less than three times the width. Find the dimensions of the garden.

13. The length of a lacrosse field is 10 yards less than twice its width, and the perimeter is 340 yards. The defensive area of the field $\frac{7}{22}$ is of the total field area. Find the defensive area of the lacrosse field.

14. One length of a rectangular garden lies along a patio wall. However, the rest of the garden is enclosed by 36 feet of fencing. If the length of the garden is twice its width, what is the area of the garden?

15. A triangular-shaped deck has one side that is 5 feet longer than the shortest side and a third side that is 5 feet shorter than twice the length of the shorter side. If the perimeter of the deck is 60 feet, what are the lengths of the three sides?

16. Suppose that the area of a trapezoid can be represented by the expression $\frac{21}{x}$ square feet and lengths of the parallel bases of the trapezoid can be represented by the expressions

$\frac{4}{x}$ feet and $\frac{7}{x+2}$ feet respectively. If the height is 6 feet, determine the value of x. Then determine the area and values of the lengths of the parallel bases.

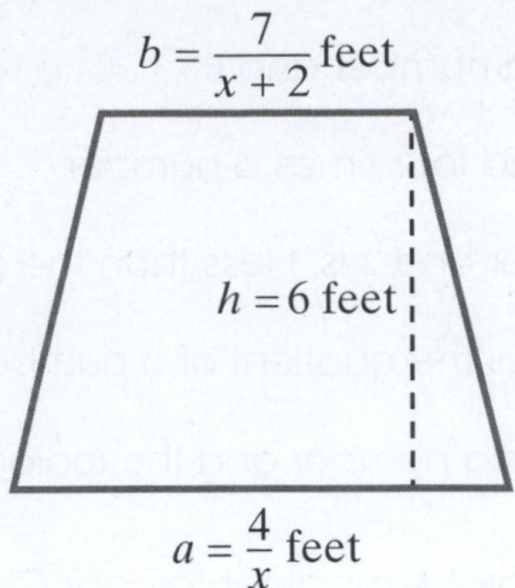

17. Jose has been collecting quarters and dimes. In his collection he has 33 coins worth $5.40. How many quarters and dimes does Jose have?

18. Emily has $18.75, consisting of only dimes, quarters, and silver dollars. She has seven times as many dimes as silver dollars and 6 more quarters than dimes. How many of each coin does she have?

19. A basketball concession stand sells a soda for $1.25, a bag of popcorn for $0.75, and a candy bar for $1.00. At one game, 14 fewer candy bars were sold than bags of popcorn. That same night, the number of sodas sold was 10 more than the number of popcorn. That same night, the number of sodas sold was 10 more than the number of candy bars or popcorn sold altogether. If the concession stand made $117 that night, how many of each item was sold?

20. Shanika has invested in three different stocks. Stock A is currently worth $4.50 per share. Stock B is worth $8.00 per share, while stock C is worth $10.00 per share. Shanika purchased half as many shares of stock B as stock A and half as many shares of stock C as stock B. If her total investments are currently worth $704, how many shares of each stock does she own?

21. A coffee shop owner decides to blend a gourmet brand of coffee with a cheaper brand. The gourmet coffee usually sells for $9.00/lb. The cheaper brand sells for $5.00/lb. How much of each type should he mix in order to have 30 lb of coffee that is worth $6.50/lb?

22. How much of an 80% orange juice drink must be mixed with 20 gallons of a 20% orange juice drink to obtain a mixture that is 50% orange juice?

23. How many liters of pure water should be mixed with a 5-L solution of 80% acid to produce a mixture that is 70% water?

24. Suppose 8 pints of a 12% alcohol solution is mixed with 2 pints of a 60% alcohol solution. What is the concentration of alcohol in the new 10-pint mixture?

25. The granular fertilizer 12-12-12 is composed of 12% nitrogen, 12% phosphate, and 12% potassium. Similarly, the fertilizer 16-20-0 is composed of 16% nitrogen, 20% phosphate, and 0% potassium. If a gardener mixes a 10 pound bag of 12-12-12 with a 15 pound bag of 16-20-0, what are the concentrations of nitrogen, phosphate, and potassium in the mixture? Express the answer as percents.

26. Ben recently moved from southern California to northern Alaska. After consulting with a local mechanic, Ben realized that his car radiator fluid should consist of 70% antifreeze. Currently, his 4.5-L radiator is full, with a concentration of 40% antifreeze. How much coolant should be drained and filled with pure antifreeze to reach the desired concentration?

27. A woman has $28,000 to invest in two investments that pay simple interest. One investment pays 4% simple annual interest, and the other pays 5% simple annual interest. How much would she have to invest in each investment if the total interest earned is to be $1,320?

28. Mark works strictly on commission of his gross sales from selling two different products for his company. Last month, his gross sales were $82,000. If he earns 6% commission on product A and 5% commission on product B, what were his gross sales for each product if he earned $4,310 in total commission?

29. Jamal borrowed $4,000 to buy a used car. He borrowed some of the money from a bank that charged 8.5% simple interest and the rest from a friend who charged 10% simple interest. If the total interest for a one-year loan was $362.50, how much did Jamal borrow at each rate?

30. How much principal should be invested in a savings account paying 6% simple interest if you want to earn $450 in interest in 2 years.

31. A freight train leaves the train station 2 hours before a passenger train. The two trains are travelling in the same direction on parallel tracks. If the rate of the passenger train is 20 mph faster than the freight train, how fast is each train travelling if the passenger train passes the freight train in 3 hours?

32. Manual traveled 31 hours nonstop to Mexico, a total of 2,300 miles. He took a train part of the way, which averaged 80 mph, and then took a bus the remaining distance, which averaged 60 mph. How long was Manuel on the train?

33. Abbas and Muhammad are at opposite ends of a 16-miles hiking trail. If Abbas' average hiking speed on the trail is 1.5 mph whereas Muhammad's average hiking speed is 2.5 mph, how long will it be before they meet?

34. A homing pigeon left his coup to deliver a message. On takeoff, the pigeon encountered a tailwind of 10 mph and made the delivery in 15 minutes. The return trip into the wind took 30 minutes. How fast could the pigeon fly in still air?

35. A grain barge travels on a river from point A to point B loading and unloading grain. The barge travels at a rate of 5 mph relative to the water. The river flows downstream at a rate of 1 mph. If the trip upstream takes 2 hours longer than the trip downstream, how far is it from point A to point B?

36. On a float trip, Kim traveled 30 miles downstream in the same amount of time that it would take her to row 12 miles upstream. If her speed in still water is 7 mph, find the speed of the current.

37. A plane can fly 500 miles against the wind in the same amount of time that it can fly 725 miles with the wind. If the wind speed is 45 mph, find the speed of the plane in still air.

38. Tommy can paint a fence in 5 hours. If his friend Huck helps, they can paint the fence in 1 hour. How fast could Huck paint the fence by himself?

39. By himself, Chris can mow his lawn in 60 minutes. If his daughter Claudia helps, they can mow the lawn together in 40 minutes. How long would it take Claudia to mow the lawn by herself?

40. It takes Joan three times longer than Jane to file the reports. Together, they can file the reports in 15 minutes. How long would it take each woman to file the reports by herself?

41. It takes Shawn 2 more hours to stain a deck than Michelle. Together it takes them 2.4 hours to complete the work. How long would it take Shawn to stain the deck by himself?

42. A garden hose can fill a hot tub in 180 minutes, whereas the drain on a hot tub can empty it in 300 minutes. It the drain is accidentally left open, how long will it take to fill the tub?

43. A leak in the hull will fill a boat's bilge in 8 hours. The boat's bilge pump can empty a full bilge in 3 hours. How long will it take the pump to empty a full bilge if the boat is leaking?

44. Two pumps were required to pump the water out of Lakeview in New Orleans after Hurricane Katrina. Pump A, the larger of the two pumps, can pump the water out in 24 hours, whereas it would take pump B 72 hours. Both pumps were working for the first 8 hours until pump A broke down. How long did it take pump B to pump out the remaining water?

1.3 Complex Numbers

THINGS TO KNOW

Before working through this section, be sure that you are familiar with the following concept:

		VIDEO	ANIMATION	INTERACTIVE
 You Try It	1. Simplify Radical Expressions Using the Product Rule (Section R.3)			

INTRODUCTION

Read this introduction before beginning Objective 1.

OBJECTIVES

1 Simplifying Powers of i

2 Adding and Subtracting Complex Numbers

3 Multiplying Complex Numbers

4 Finding the Quotient of Complex Numbers

5 Simplifying Radicals with Negative Radicands

SECTION 1.3 EXERCISES

Introduction to Section 1.3

 **THE IMAGINARY UNIT**

It is easy to verify that $x = -1$ is the solution to the equation $x + 1 = 0$. Now consider the equation $x^2 + 1 = 0$. Is $x = -1$ a solution to this equation? To check, let's substitute $x = -1$ into the equation $x^2 + 1 = 0$ to see if we get a true statement:

$x^2 + 1 = 0$	Write the original equation.
$(-1)^2 + 1 \stackrel{?}{=} 0$	Substitute $x = -1$.
$1 + 1 \neq 0$	The left-hand side does not equal the right-hand side.

In fact, the equation $x^2 + 1 = 0$ has *no real solution* because any **real number** squared added to one is greater than zero. Does this equation have a solution? If so, what is the solution? To answer this question, suppose that the number i is a solution to the equation $x^2 + 1 = 0$. In other words, $i^2 + 1 = 0$ or $i^2 = -1$. This number i, which is a solution to the equation $x^2 + 1 = 0$, is known as the **imaginary unit**.

Definition Imaginary Unit

The imaginary unit, i, is defined by $i = \sqrt{-1}$, or equivalently, $i^2 = -1$.

(eText Screens 1.3-1–1.3-15)

 OBJECTIVE 1 SIMPLIFYING POWERS OF i

Given the previous definition, we have the following:

$$i = \sqrt{-1} \qquad i^5 = i^4 \cdot i = (1)\cdot i = i$$
$$i^2 = (\sqrt{-1})^2 = -1 \qquad i^6 = i^4 \cdot i^2 = (1)\cdot(-1) = -1$$
$$i^3 = i^2 \cdot i = (-1)\cdot i = -i \qquad i^7 = i^4 \cdot i^3 = (1)\cdot(-i) = -i$$
$$i^4 = i^2 \cdot i^2 = (-1)\cdot(-1) = 1 \qquad i^8 = i^4 \cdot i^4 = (1)\cdot(1) = 1$$
$$\vdots$$

You can see the cyclic nature of i. After every fourth power, the cycle repeats itself. In fact, every integer power of i can be written as i, -1, $-i$, or 1.

 Example 1 Simplify Powers of i

Simplify each of the following:

a. i^{43} b. i^{100} c. i^{-21}

Solution Because $i^4 = 1$, we factor out the largest fourth power of i.

a. $i^{43} = (i^4)^{10} \cdot i^3 = 1^{10} \cdot i^3 = i^3 = -i$

b. $i^{100} = (i^4)^{25} = 1^{25} = 1$

c. $i^{-21} = \dfrac{1}{i^{21}} = \dfrac{1}{(i^4)^5 \cdot i} = \dfrac{1}{(1)^5 \cdot i} = \dfrac{1}{i}$

It appears that we are done and $i^{-21} = \frac{1}{i}$. However, if we multiply this expression by $\frac{i}{i}$, we get the following:

$$i^{-21} = \frac{1}{i}\cdot\frac{i}{i} = \frac{i}{i^2} = \frac{i}{-1} = -i$$

Therefore, $i^{-21} = -i$.

You Try It Work through this You Try It problem.

Work Exercises 1-5 in this textbook or in the MyLab Math Study Plan.

COMPLEX NUMBERS

Now that we have defined the imaginary unit, it is time to introduce a new type of number known as a **complex number**.

> **Definition** Complex Number
>
> A complex number is a number that can be written in the form $a + bi$, where a and b are real numbers. We often use the variable z to denote a complex number.

If $z = a + bi$ is a complex number, then a is called the **real part**, and b is called the **imaginary part**. For example, $z = \sqrt{3} - 2i$ is a complex number, where $a = \sqrt{3}$ is the real part and $b = -2$ is the imaginary part. The number 5 is also a complex number because 5 can be written as $5 + 0i$. Therefore, every real number is also a complex number.

OBJECTIVE 2 ADDING AND SUBTRACTING COMPLEX NUMBERS

To add or subtract complex numbers, simply combine the real parts and combine the imaginary parts.

Example 2 Adding and Subtracting Complex Numbers

Perform the indicated operations:

a. $(7 - 5i) + (-2 + i)$ **b.** $(7 - 5i) - (-2 + i)$

Solution

a. $(7 - 5i) + (-2 + i)$ Write the original expression.

$= 7 - 5i - 2 + i$ Remove parentheses.

$= 5 - 4i$ Combine the real parts and the imaginary parts.

b. $(7 - 5i) - (-2 + i)$ Write the original expression.

$= 7 - 5i + 2 - i$ Remove the parentheses $[-(-2 + i) = 2 - i]$.

$= 9 - 6i$ Combine the real parts and the imaginary parts.

You Try It Work through this You Try It problem.

Work Exercises 6–9 in this textbook or in the MyLab Math Study Plan.

OBJECTIVE 3 MULTIPLYING COMPLEX NUMBERS

When multiplying two complex numbers, treat the problem as if it were the multiplication of two **binomials**. Just remember that $i^2 = -1$.

Example 3 Multiplying Complex Numbers

Multiply $(4 - 3i)(7 + 5i)$.

Solution Treating this multiplication as if it were the multiplication of two binomials, we get the following:

$(4 - 3i)(7 + 5i) = 4(7) + 4(5i) - (3i)(7) - (3i)(5i)$ Use the **distributive property**.

$= 28 + 20i - 21i - 15i^2$ Multiply.

$= 28 - i - 15(-1)$ $20i - 21i = -i$ and $i^2 = -1$.

$= 28 - i + 15$ $-15(-1) = 15$.

$= 43 - i$ Combine the real parts.

Example 4 Squaring a Complex Number

Simplify $(\sqrt{3} - 5i)^2$.

Solution You should verify that the answer is $-22 - 10i\sqrt{3}$. Watch the video to see the worked out solution.

We now define the **complex conjugate**.

> **Definition** Complex Conjugate
>
> The complex conjugate of a complex number $z = a + bi$ is denoted as $\bar{z} = \overline{a + bi} = a - bi$.

For example, the complex conjugate of $z = -2 - 7i$ is $\bar{z} = -2 - (-7i) = -2 + 7i$.

Example 5 Multiplying a Complex Number by Its Complex Conjugate

Multiply the complex number $z = -2 - 7i$ by its complex conjugate $\bar{z} = -2 + 7i$.

Solution

$$z\bar{z} = (-2 - 7i)(-2 + 7i) \quad \text{Multiply.}$$
$$= 4 - 14i + 14i - 49i^2$$
$$= 4 - 49(-1) \quad \text{Simplify and } i^2 = -1.$$
$$= 4 + 49 = 53 \quad \text{Simplify.}$$

Thus, $(-2 - 7i)(-2 + 7i) = 53$.

Notice in Example 5 that the product of $z = -2 - 7i$ and its complex conjugate $\bar{z} = -2 + 7i$ is a real number. In fact, every complex number when multiplied by its complex conjugate will yield a real number. The following theorem is worth remembering:

> **Theorem**
>
> The product of a complex number $z = a + bi$ and its conjugate $\bar{z} = a - bi$ is the real number $a^2 + b^2$. In symbols, $z\bar{z} = a^2 + b^2$.

To see why this is true, read the proof of this theorem.

You Try It **Work through this You Try It problem.**

Work Exercises 10–19 in this textbook or in the MyLab Math Study Plan.

OBJECTIVE 4 FINDING THE QUOTIENT OF COMPLEX NUMBERS

The goal when dividing two complex numbers is to eliminate the imaginary part from the denominator and express the quotient in the standard form of $a + bi$. We can do this by multiplying the numerator and denominator by the complex conjugate of the denominator.

Example 6 The Quotient of Complex Numbers

Write the quotient in the form $a + bi$:

$$\frac{1 - 3i}{5 - 2i}$$

Solution

$\dfrac{1 - 3i}{5 - 2i} \cdot \dfrac{5 + 2i}{5 + 2i}$ — Multiply numerator and denominator by the complex conjugate. Note: $\dfrac{5 + 2i}{5 + 2i} = 1$.

$= \dfrac{5 - 13i - 6i^2}{(5)^2 + (2)^2}$ — Multiply the numerators and the denominators. Note: $(a + bi)(a - bi) = a^2 + b^2$.

$= \dfrac{5 - 13i - 6(-1)}{29}$ — Simplify denominator, and replace i^2 with -1.

$= \dfrac{11 - 13i}{29}$ — Simplify.

$= \dfrac{11}{29} - \dfrac{13}{29}i$ — Write in the form $a + bi$.

You Try It Work through this You Try It problem.

Work Exercises 20–23 in this textbook or in the MyLab Math Study Plan.

OBJECTIVE 5 SIMPLIFYING RADICALS WITH NEGATIVE RADICANDS

In Section 1.4, we solve equations that often lead to solutions with a negative number inside of a radical. Therefore, it is important to learn how to simplify a radical that contains a negative **radicand** such as $\sqrt{-49}$. We simplify such radicals using the following property, remembering that $\sqrt{-1} = i$.

> **Property**
>
> If M is a positive real number, then $\sqrt{-M} = \sqrt{-1} \cdot \sqrt{M} = i\sqrt{M}$.

Example 7 Simplify a Square Root with a Negative Radicand

Simplify $\sqrt{-108}$.

Solution

$\sqrt{-108} = \sqrt{-1} \cdot \sqrt{108} = i\sqrt{108} = 6i\sqrt{3}$

Thus, $\sqrt{-108} = 6i\sqrt{3}$. To review simplifying radicals, refer back to Section R.3.

CAUTION The property $\sqrt{a} \cdot \sqrt{b} = \sqrt{ab}$ is only true for real numbers $\sqrt{a}$ and $\sqrt{b}$. The expressions $\sqrt{a}$ and $\sqrt{b}$ are real numbers when both a and b are greater than or equal to zero. This property does not apply for nonreal numbers. Attempting to use this property with square roots of negative numbers will lead to false results:

$$\sqrt{-3} \cdot \sqrt{-12} = \sqrt{(-3) \cdot (-12)} = \sqrt{36} = 6 \quad \text{False!}$$

$$\sqrt{-3} \cdot \sqrt{-12} = i\sqrt{3} \cdot i\sqrt{12} = i^2\sqrt{36} = -6 \quad \text{True!}$$

Example 8 Simplifying Expressions That Contain Negative Radicands

Simplify the following expressions:

a. $\sqrt{-8} + \sqrt{-18}$

b. $\sqrt{-8} \cdot \sqrt{-18}$

c. $\dfrac{-6 + \sqrt{(-6)^2 - 4(2)(5)}}{2}$

d. $\dfrac{4 \pm \sqrt{-12}}{4}$

Solution

a. $\sqrt{-8} + \sqrt{-18} = 2i\sqrt{2} + 3i\sqrt{2} = 5i\sqrt{2}$

b. $\sqrt{-8} \cdot \sqrt{-18} = (2i\sqrt{2}) \cdot (3i\sqrt{2}) = 6i^2(2) = -12$

or

$\sqrt{-8} \cdot \sqrt{-18} = (i\sqrt{8}) \cdot (i\sqrt{18}) = i^2\sqrt{144} = -12$

c. $$\frac{-6 + \sqrt{(-6)^2 - 4(2)(5)}}{2} = \frac{-6 + \sqrt{36 - 40}}{2} = \frac{-6 + \sqrt{-4}}{2}$$

$$= \frac{-6 + 2i}{2} = \frac{\cancel{2}(-3 + i)}{\cancel{2}} = -3 + i$$

d. $$\frac{4 \pm \sqrt{-12}}{4} = \frac{4 \pm 2i\sqrt{3}}{4} = \frac{2(2 \pm i\sqrt{3})}{4} = \frac{\cancel{2}(2 \pm i\sqrt{3})}{\cancel{2} \cdot 2} = \frac{2 \pm i\sqrt{3}}{2}$$

You Try It Work through this You Try It problem.

Work Exercises 24–31 in this textbook or in the MyLab Math Study Plan.

1.3 Exercises

Skill Check Exercises

For exercises SCE-1 through SCE-6, simplify the given expression.

SCE-1. $(\sqrt{2} + 3) + (5\sqrt{2} - 5)$ **SCE-2.** $(5\sqrt{3} - 9) - (-4\sqrt{3} - 13)$ **SCE-3.** $\sqrt{20}$

SCE-4. $\sqrt{108}$ **SCE-5.** $(\sqrt{2} - 5)^2$ **SCE-6.** $(3 + \sqrt{5})^2$

In Exercises 1–5, write each power of i as $i, -1, -i,$ or 1.

1. i^{17} 2. $i^5 \cdot i^9$ 3. i^{-6} 4. i^{-59} 5. $\dfrac{i^{24}}{i^{-23}}$

In Exercises 6–9, find the desired sum or difference. Write each answer in the form $a + bi$.

6. $(3 - 2i) + (-7 + 9i)$ 7. $(3 - 2i) - (-7 + 9i)$

8. $i - (1 + i)$ 9. $(\sqrt{2} - 3i) + (2\sqrt{2} + 3i)$

In Exercises 10–14, perform the desired operations. Write each answer in the form $a + bi$.

10. $2i(4 - 3i)$ 11. $-i(1 - i)$ 12. $(-2 - i)(3 - 4i)$

13. $(6 - 2i)^2$ 14. $(\sqrt{2} - i)^2$

In Exercises 15–19, find the product of the given complex number and its conjugate. Write each answer in the form $a + bi$.

15. $5 - 2i$ 16. $1 - i$ 17. $\dfrac{1}{2} - 3i$ 18. $\sqrt{5} + i$ 19. $\dfrac{\sqrt{3}}{2} + \dfrac{1}{2}i$

In Exercises 20–23, write each quotient in the form $a + bi$.

20. $\dfrac{2 - i}{3 + 4i}$ 21. $\dfrac{1}{2 - i}$ 22. $\dfrac{3i}{2 + 2i}$ 23. $\dfrac{5 + i}{5 - i}$

In Exercises 24–31, write each expression in the form $a + bi$.

24. $\sqrt{-36} - \sqrt{49}$ 25. $\sqrt{-1} + 3 - \sqrt{-64}$ 26. $\sqrt{-2} \cdot \sqrt{-18}$ 27. $(\sqrt{-8})^2$

28. $(i\sqrt{-4})^2$ 29. $\dfrac{-4 - \sqrt{-20}}{2}$ 30. $\dfrac{-3 - \sqrt{-81}}{6}$ 31. $\dfrac{4 + \sqrt{-8}}{4}$

1.4 Quadratic Equations

THINGS TO KNOW

Before working through this section, be sure that you are familiar with the following concepts:

VIDEO ANIMATION INTERACTIVE

You Try It 1. Simplify Radical Expressions Using the Product Rule (Section R.3)

You Try It 2. Simplifying Radicals with Negative Radicands (Section 1.3)

You Try It 3. Factoring Trinomials with a Leading Coefficient Equal to 1 (Section R.6)

You Try It 4. Factoring Trinomials with a Leading Coefficient Not Equal to 1 (Section R.6)

INTRODUCTION

Read this introduction before beginning Objective 1.

OBJECTIVES

1 Solving Quadratic Equations by Factoring and the Zero Product Property

2 Solving Quadratic Equations Using the Square Root Property

3 Solving Quadratic Equations by Completing the Square

4 Solving Quadratic Equations Using the Quadratic Formula

5 Using the Discriminant to Determine the Type of Solutions of a Quadratic Equation

SECTION 1.4 EXERCISES

Introduction to Section 1.4

In Section 1.1, we studied linear equations of the form $ax + b = c, a \neq 0$. These equations are also known as first-order **polynomial** equations. In this section, we learn how to solve second-order polynomial equations. Second-order polynomial equations are called **quadratic equations**.

> **Definition** Quadratic Equation in One Variable
>
> A **quadratic equation in one variable** is an equation that can be written in the form $ax^2 + bx + c = 0, a \neq 0$.

(eText Screens 1.4-1–1.4-23)

OBJECTIVE 1 SOLVING QUADRATIC EQUATIONS BY FACTORING AND THE ZERO PRODUCT PROPERTY

Some quadratic equations can be easily solved by factoring and by using the following important property.

> **Zero Product Property**
>
> If $AB = 0$, then $A = 0$ or $B = 0$.

The zero product property says that if two factors multiplied together are equal to zero, then at least one of the factors must be zero. Example 1 shows how this property is used to solve certain quadratic equations.

Example 1 Use Factoring and the Zero Product Property to Solve a Quadratic Equation

Solve $6x^2 - 17x = -12$.

Solution First, rewrite the equation to get 0 on the right-hand side by adding the constant 12 to both sides.

$6x^2 - 17x = -12$	Write the original equation.
$6x^2 - 17x + 12 = 0$	Add 12 to both sides.
$(3x - 4)(2x - 3) = 0$	Factor the left-hand side.
$3x - 4 = 0 \quad \text{or} \quad 2x - 3 = 0$	Use the zero product property.
$3x = 4 \quad \text{or} \quad 2x = 3$	
$x = \frac{4}{3} \quad \text{or} \quad x = \frac{3}{2}$	Solve each equation for x.

The solution is $\left\{\frac{4}{3}, \frac{3}{2}\right\}$.

You Try It Work through this You Try It problem.

Work Exercises 1–14 in this textbook or in the MyLab Math Study Plan.

OBJECTIVE 2 SOLVING QUADRATIC EQUATIONS USING THE SQUARE ROOT PROPERTY

Consider the quadratic equation $x^2 - 9 = 0$. The left-hand side is a **difference of two squares**, so we can easily solve by factoring.

$x^2 - 9 = 0$	Write the original equation.
$(x - 3)(x + 3) = 0$	Factor using difference of squares formula.

$x - 3 = 0$ or $x + 3 = 0$ Use the zero product property.

$x = 3$ or $x = -3$ Solve each equation for x.

The solution is $x = \pm 3$. In fact, any quadratic equation of the form $x^2 - c = 0$ where $c > 0$ can be solved by factoring:

$$x^2 - c = 0$$

$(x - \sqrt{c})(x + \sqrt{c}) = 0$ Factor using difference of squares formula.

$x - \sqrt{c} = 0$ or $x + \sqrt{c} = 0$ Use the zero product property.

$x = \sqrt{c}$ or $x = -\sqrt{c}$ Solve each equation for x.

Although quadratic equations of this form can be solved by factoring, they can be more readily solved by using the following square root property.

> **Square Root Property**
>
> The solution to the quadratic equation $x^2 - c = 0$, or equivalently $x^2 = c$, is $x = \pm\sqrt{c}$.

Example 2 Solve Quadratic Equations Using the Square Root Property

Solve each quadratic equation.

a. $x^2 - 16 = 0$ **b.** $2x^2 + 72 = 0$ **c.** $(x - 1)^2 = 7$

Solution

a. $x^2 - 16 = 0$ Write the original equation.

$x^2 = 16$ Add 16 to both sides.

$x = \pm 4$ Use the square root property.

b. $2x^2 + 72 = 0$ Write the original equation.

$2x^2 = -72$ Subtract 72 from both sides.

$x^2 = -36$ Divide by 2.

$x = \pm\sqrt{-36}$ Use the square root property.

$x = \pm 6i$ Simplify the radical with a negative radicand.

c. $(x - 1)^2 = 7$

$x - 1 = \pm\sqrt{7}$ Use the square root property.

$x = 1 \pm \sqrt{7}$ Add 1 to both sides.

You Try It Work through this You Try It problem.

Work Exercises 15–24 in this textbook or in the MyLab Math Study Plan.

In Example 2, we see that the quadratic equation $(x - 1)^2 = 7$ can be solved using the square root property. The left side of the equation is a perfect square.

In fact, every quadratic equation can be written in the form $(x - h)^2 = k$ using a method known as **completing the square**.

OBJECTIVE 3 SOLVING QUADRATIC EQUATIONS BY COMPLETING THE SQUARE

Consider the following perfect square trinomials:

$$x^2 + 2x + 1 = (x + 1)^2 \qquad x^2 - 6x + 9 = (x - 3)^2 \qquad x^2 - 7x + \frac{49}{4} = \left(x - \frac{7}{2}\right)^2$$

$$\left(\frac{1}{2}\cdot 2\right)^2 = 1 \qquad \left(\frac{1}{2}\cdot(-6)\right)^2 = 9 \qquad \left(\frac{1}{2}\cdot(-7)\right)^2 = \frac{49}{4}$$

In each perfect square trinomial, notice the relationship between the coefficient of the linear term (x-term) and the constant term. The constant term of a perfect square trinomial is equal to the square of $\frac{1}{2}$ the linear coefficient.

Example 3 Completing the Square

What number must be added to each binomial to make it a perfect square trinomial?

a. $x^2 - 12x$ b. $x^2 + 5x$ c. $x^2 - \frac{3}{2}x$

Solution

a. The linear coefficient is -12, so we must add $\left(\frac{1}{2}(-12)\right)^2 = (-6)^2 = 36$ to complete the square. Thus, the expression $x^2 - 12x + 36$ is a perfect square trinomial and $x^2 - 12x + 36 = (x - 6)^2$.

b. $\left(\frac{1}{2}\cdot 5\right)^2 = \left(\frac{5}{2}\right)^2 = \frac{25}{4}$ must be added to complete the square:

$x^2 + 5x + \frac{25}{4} = \left(x + \frac{5}{2}\right)^2.$

c. $\left(\frac{1}{2}\left(-\frac{3}{2}\right)\right)^2 = \left(-\frac{3}{4}\right)^2 = \frac{9}{16}$ must be added to complete the square:

$x^2 - \frac{3}{2}x + \frac{9}{16} = \left(x - \frac{3}{4}\right)^2.$

You Try It Work through this You Try It problem.

Work Exercises 25–30 in this textbook or in the MyLab Math Study Plan.

To solve a quadratic equation of the form $ax^2 + bx + c = 0, a \neq 0$, by completing the square, follow these steps:

Steps for Solving $ax^2 + bx + c = 0, a \neq 0$, by Completing the Square

1. If $a \neq 1$, divide both sides of the equation by a.
2. Move all constants to the right-hand side.
3. Take half the coefficient of the x-term, square it, and add it to both sides of the equation.
4. The left-hand side is now a perfect square trinomial. Rewrite it as a binomial squared.
5. Use the square root property to solve for x.

Example 4 Solve by Completing the Square

Solve $3x^2 - 18x + 19 = 0$ by completing the square.

Solution

Step 1. Divide the equation by 3:

$$\frac{3x^2}{3} - \frac{18x}{3} + \frac{19}{3} = \frac{0}{3}$$

$$x^2 - 6x + \frac{19}{3} = 0$$

Step 2. Move all constants to the right-hand side:

$$x^2 - 6x = -\frac{19}{3}$$

Step 3. Take half the coefficient of the x-term, square it, and add it to both sides of the equation:

$$x^2 - 6x + 9 = -\frac{19}{3} + 9 \qquad \left(\frac{1}{2}\cdot(-6)\right)^2 = (-3)^2 = 9$$

$$x^2 - 6x + 9 = \frac{8}{3} \qquad \text{Simplify.}$$

Step 4. The left-hand side is now a perfect square trinomial. Rewrite it as a binomial squared.

$$(x - 3)^2 = \frac{8}{3}$$

Step 5. Use the square root property to solve for x.

$$x - 3 = \pm\sqrt{\frac{8}{3}} \qquad \text{Use the square root property.}$$

$$x - 3 = \pm\frac{\sqrt{8}}{\sqrt{3}} \qquad \sqrt{\frac{a}{b}} = \frac{\sqrt{a}}{\sqrt{b}}$$

$$x - 3 = \pm\frac{\sqrt{8}}{\sqrt{3}}\cdot\frac{\sqrt{3}}{\sqrt{3}} \qquad \text{Rationalize the denominator.}$$

$$x - 3 = \pm\frac{\sqrt{24}}{3} \qquad \text{Simplify.}$$

$$x - 3 = \pm\frac{2\sqrt{6}}{3} \qquad \sqrt{24} = 2\sqrt{6}$$

$$x = 3 \pm \frac{2\sqrt{6}}{3} \quad \text{or} \quad x = \frac{9 \pm 2\sqrt{6}}{3} \qquad \text{Add 3 to both sides and simplify.}$$

Example 5 Solve by Completing the Square

Solve $2x^2 - 10x - 6 = 0$ by completing the square.

Solution

Step 1. Divide the equation by 2:

$$\frac{2x^2}{2} - \frac{10x}{2} - \frac{6}{2} = \frac{0}{2}$$

$$x^2 - 5x - 3 = 0$$

Step 2. Move all constants to the right-hand side:

$$x^2 - 5x = 3$$

Step 3. Take half the coefficient of the x-term, square it, and add it to both sides of the equation:

$$x^2 - 5x + \frac{25}{4} = 3 + \frac{25}{4} \qquad \left(\frac{1}{2}\cdot(-5)\right)^2 = \left(-\frac{5}{2}\right)^2 = \frac{25}{4}$$

$$x^2 - 5x + \frac{25}{4} = \frac{37}{4} \qquad \text{Simplify.}$$

Step 4. The left-hand side is now a perfect square trinomial. Rewrite it as a binomial squared.

$$\left(x - \frac{5}{2}\right)^2 = \frac{37}{4}$$

Step 5. Use the square root property to solve for x.

$$x - \frac{5}{2} = \pm\sqrt{\frac{37}{4}} \qquad \text{Use the square root property.}$$

$$x - \frac{5}{2} = \pm\frac{\sqrt{37}}{\sqrt{4}} \qquad \sqrt{\frac{a}{b}} = \frac{\sqrt{a}}{\sqrt{b}}$$

$$x - \frac{5}{2} = \pm\frac{\sqrt{37}}{2} \qquad \text{Simplify.}$$

$$x = \frac{5}{2} \pm \frac{\sqrt{37}}{2} = \frac{5 \pm \sqrt{37}}{2} \qquad \text{Add } \frac{5}{2} \text{ to both sides and simplify.}$$

You Try It Work through this You Try It problem.

Work Exercises 31–39 in this textbook or in the MyLab Math Study Plan.

OBJECTIVE 4 SOLVING QUADRATIC EQUATIONS USING THE QUADRATIC FORMULA

In Example 5, we see how to solve a quadratic equation by completing the square. In fact, the method of completing the square can be used to solve *any* quadratic equation. However, this method is often long and tedious. If we solve the general quadratic equation $ax^2 + bx + c = 0, a \neq 0$, by the method of completing the square, we can obtain the following quadratic formula. The quadratic formula can be used to solve any quadratic equation and is often less tedious than the method of completing the square. Be sure to work through the **derivation of the quadratic formula** to see exactly how this formula is derived.

Quadratic Formula

The solution to the quadratic equation $ax^2 + bx + c = 0, a \neq 0$, is given by the following formula:

$$x = \frac{-b \pm \sqrt{b^2 - 4ac}}{2a}$$

Example 6 Solve a Quadratic Equation Using the Quadratic Formula

Solve $3x^2 + 2x - 2 = 0$ using the quadratic formula.

Solution This quadratic equation has values of $a = 3, b = 2$, and $c = -2$. Substitute these values into the quadratic formula.

$$x = \frac{-2 \pm \sqrt{(2)^2 - 4(3)(-2)}}{2(3)} \quad \text{Use the quadratic formula.}$$

$$= \frac{-2 \pm \sqrt{28}}{6} \quad \text{Simplify.}$$

$$= \frac{-2 \pm 2\sqrt{7}}{6} \quad \sqrt{28} = 2\sqrt{7}$$

$$= \frac{2(-1 \pm \sqrt{7})}{6} \quad \text{Factor.}$$

$$= \frac{-1 \pm \sqrt{7}}{3} \quad \text{Simplify.}$$

Example 7 Solve a Quadratic Equation Using the Quadratic Formula

Solve $4x^2 - x + 6 = 0$ using the quadratic formula.

Solution This quadratic equation has values of $a = 4, b = -1$, and $c = 6$. Substitute these values into the quadratic formula.

$$x = \frac{-(-1) \pm \sqrt{(-1)^2 - 4(4)(6)}}{2(4)} \quad \text{Use the quadratic formula.}$$

(eText Screens 1.4-1–1.4-23)

$$= \frac{1 \pm \sqrt{-95}}{8} \quad \text{Simplify.}$$

$$= \frac{1 \pm i\sqrt{95}}{8} \quad \text{Simplify the radical with a negative radicand.}$$

You Try It Work through this You Try It problem.

Work Exercises 40–47 in this textbook or in the MyLab Math Study Plan.

OBJECTIVE 5 USING THE DISCRIMINANT TO DETERMINE THE TYPE OF SOLUTIONS OF A QUADRATIC EQUATION

In Example 7, the quadratic equation $4x^2 - x + 6 = 0$ had two nonreal solutions. The solutions were nonreal because the expression $b^2 - 4ac$ under the radical was a negative number. Given a quadratic equation of the form $ax^2 + bx + c = 0$, the expression $b^2 - 4ac$ is called the **discriminant**. Knowing the value of the discriminant can help us determine the number and nature of the solutions to a quadratic equation.

> **Definition** Discriminant
>
> Given a quadratic equation $ax^2 + bx + c = 0, a \neq 0$, the expression $D = b^2 - 4ac$ is called the **discriminant**.
>
> If $D > 0$, then the quadratic equation has two real solutions.
> If $D < 0$, then the quadratic equation has two nonreal solutions.
> If $D = 0$, then the quadratic equation has exactly one real solution.

Example 8 Use the Discriminant

Use the discriminant to determine the number and nature of the solutions to each of the following quadratic equations:

a. $3x^2 + 2x + 2 = 0$ b. $4x^2 + 1 = 4x$

Solution

a. The equation $3x^2 + 2x + 2 = 0$ is in standard form with $a = 3, b = 2$, and $c = 2$. $D = 2^2 - 4(3)(2) = 4 - 24 = -20 < 0$. Because the discriminant is less than zero, there are two nonreal solutions.

b. The equation $4x^2 + 1 = 4x$ is not in the correct form. To get the equation into the correct form, subtract $4x$ from both sides: $4x^2 - 4x + 1 = 0$. $D = (-4)^2 - 4(4)(1) = 16 - 16 = 0$. Because the discriminant is zero, there is exactly one real solution.

You Try It Work through this You Try It problem.

Work Exercises 48–51 in this textbook or in the MyLab Math Study Plan.

1.4 Exercises

Skill Check Exercises

For exercises SCE-1 through SCE-8, simplify the given expression.

SCE-1. $-13 + \sqrt{529}$ **SCE-2.** $-6 + \sqrt{44}$ **SCE-3.** $4 + \sqrt{-196}$ **SCE-4.** $6 - \sqrt{-116}$

SCE-5. $\dfrac{-13 + \sqrt{529}}{6}$ **SCE-6.** $\dfrac{-6 + \sqrt{44}}{2}$ **SCE-7.** $\dfrac{4 + \sqrt{-196}}{6}$ **SCE-8.** $\dfrac{6 - \sqrt{-116}}{4}$

For exercises SCE-9 through SCE-13, factor each trinomial.

SCE-9. $x^2 + 10x + 24$ **SCE-10.** $x^2 - 7x - 18$ **SCE-11.** $x^2 - 10x + 21$

SCE-12. $2x^2 + 9x + 4$ **SCE-13.** $6x^2 - x - 15$

In Exercises 1–14, solve each equation by factoring.

1. $x^2 - 8x = 0$
2. $x^2 = -12x$
3. $6m^2 + 36m = 0$
4. $5x^2 = -15x$
5. $x^2 - x = 6$
6. $x^2 + 9x + 20 = 0$
7. $t^2 - 10t + 24 = 0$
8. $v^2 + 11v = -28$
9. $3x^2 + 8x - 3 = 0$
10. $6n^2 - 37n - 35 = 0$
11. $12x^2 + 52x + 16 = 0$
12. $8r^2 + 29r = 12$
13. $8m^2 - 15 = 14m$
14. $28z^2 - 13z - 6 = 0$

In Exercises 15–24, solve each equation using the square root property.

15. $x^2 - 64 = 0$
16. $x^2 + 64 = 0$
17. $3x^2 = 72$
18. $2x^2 - 216 = 0$
19. $3x^2 + 1350 = 0$
20. $(x + 2)^2 - 9 = 0$
21. $(x - 5)^2 - 162 = 0$
22. $(2x + 1)^2 + 4 = 0$
23. $(4x + 2)^2 - 192 = 0$
24. $(9x - 3)^2 + 180 = 0$

In Exercises 25–30, decide what number must be added to each binomial to make a perfect square trinomial.

25. $x^2 - 8x$
26. $x^2 + 10x$
27. $x^2 - 7x$
28. $x^2 + 3x$
29. $x^2 - \frac{1}{4}x$
30. $x^2 + \frac{5}{3}x$

In Exercises 31–39, solve each quadratic equation by completing the square.

SbS 31. $x^2 + 6x - 16 = 0$
SbS 32. $x^2 - 8x - 2 = 0$
SbS 33. $x^2 - 5x - 3 = 0$
SbS 34. $x^2 + 10x + 34 = 0$
SbS 35. $x^2 + 7x + 14 = 0$
SbS 36. $3x^2 + 8x - 3 = 0$
SbS 37. $2x^2 + 6x + 8 = 0$
SbS 38. $3x^2 + 24x - 7 = 0$
SbS 39. $3x^2 + 5x + 12 = 0$

(eText Screens 1.5-1–1.5-18)

In Exercises 40–47, solve each quadratic equation using the quadratic formula.

SbS 40. $x^2 + 11x - 26 = 0$

SbS 41. $3x^2 + 8x = 3$

SbS 42 $x^2 - 8x - 2 = 0$

SbS 43. $3x^2 - 1 = 4x$

SbS 44. $2x(x - 4) = 5$

SbS 45. $5x^2 - 3x = -1$

SbS 46. $5x^2 + 3x + 1 = 0$

SbS 47. $4x^2 - x + 8 = 0$

In Exercises 48–51, use the discriminant to determine the number and nature of the solutions to each quadratic equation. Do not solve the equations.

48. $x^2 + 2x + 1 = 0$

49. $4x^2 + 4x + 1 = 0$

50. $2x^2 + x - 5 = 0$

51. $3x^2 + \sqrt{12x} + 4 = 0$

Brief Exercises

In Exercises 52–60, solve each quadratic equation by completing the square.

52. $x^2 + 6x - 16 = 0$

53. $x^2 - 8x - 2 = 0$

54. $x^2 - 5x - 3 = 0$

55. $x^2 + 10x + 34 = 0$

56. $x^2 + 7x + 14 = 0$

57. $3x^2 + 8x - 3 = 0$

58. $2x^2 + 6x + 8 = 0$

59. $3x^2 + 24x - 7 = 0$

60. $3x^2 + 5x + 12 = 0$

1.5 Applications of Quadratic Equations

THINGS TO KNOW

Before working through this section, be sure that you are familiar with the following concepts:

VIDEO ANIMATION INTERACTIVE

You Try It
1. Solving Applications Involving Uniform Motion (Section 1.2)

You Try It
2. Solving Applications Involving Rates of Work (Section 1.2)

You Try It
3. Solving Quadratic Equations by Factoring and the Zero Product Property (Section 1.4)

You Try It
4. Solving Quadratic Equations by Completing the Square (Section 1.4)

You Try It
5. Solving Quadratic Equations Using the Quadratic Formula (Section 1.4)

INTRODUCTION

Read this introduction before beginning Objective 1.

OBJECTIVES

1 Solving Applications Involving Unknown Numeric Quantities

2 Using the Projectile Motion Model

3 Solving Geometric Applications

4 Solving Applications Involving Uniform Motion

5 Solving Applications Involving Rates of Work

SECTION 1.5 EXERCISES

Introduction to Section 1.5

In Section 1.2, we learn how to solve applied problems involving **linear equations**. In this section, we learn to solve applications that involve **quadratic equations**. We follow the same four-step strategy for solving applied problems that is discussed in Section 1.2. The four steps are outlined as follows.

FOUR-STEP STRATEGY FOR PROBLEM SOLVING

Step 1. Read the problem several times until you have an understanding of what is being asked. If possible, create diagrams, charts, or tables to assist you in your understanding. } Understand the problem.

Step 2. Pick a variable that describes the unknown quantity that is to be found. All other quantities must be written in terms of that variable. Write an equation using the given information and the variable. } Devise a plan.

Step 3. Carefully solve the equation. } Carry out the plan.

Step 4. Make sure that you have answered the question and then check all answers to make sure they make sense. } Look back.

OBJECTIVE 1 SOLVING APPLICATIONS INVOLVING UNKNOWN NUMERIC QUANTITIES

Example 1 Find Two Numbers

The product of a number and 1 more than twice the number is 36. Find the two numbers.

Solution

Step 1. We are looking for two numbers.

Step 2. Let x = the first number, and then $2x + 1$ = the other number. Because their product is 36, we get the equation $x(2x + 1) = 36$, which simplifies to the quadratic equation $2x^2 + x - 36 = 0$.

Step 3. Solve:

$$2x^2 + x - 36 = 0 \qquad \text{Write the equation.}$$

$$(2x + 9)(x - 4) = 0 \qquad \text{Factor.}$$

$$2x + 9 = 0 \quad \text{or} \quad x - 4 = 0 \qquad \text{Use the zero product property.}$$

$$x = -\frac{9}{2} \quad \text{or} \quad x = 4 \qquad \text{Solve.}$$

Step 4. If $x = -\frac{9}{2}$, then the other number is $2x + 1 = 2\left(-\frac{9}{2}\right) + 1 = -8$.

If $x = 4$, then the other number is $2x + 1 = 2(4) + 1 = 9$.

We can see that there are two possibilities for the two numbers, $-\frac{9}{2}$ and -8 or 4 and 9. In either case, the second number is one more than twice the first, and their products are 36. ●

You Try It Work through this You Try It problem.

Work Exercises 1–5 in this textbook or in the MyLab Math Study Plan.

OBJECTIVE 2 USING THE PROJECTILE MOTION MODEL

An object launched, thrown, or shot vertically into the air with an initial velocity of v_0 meters/second (m/s) from an initial height of h_0 meters above the ground can be modeled by the equation $h = -4.9t^2 + v_0t + h_0$. The variable h is the height (in meters) of the projectile t seconds after its departure. This equation can be used to model the height of a toy rocket at any time after liftoff, as in Example 2.

Example 2 Launch a Toy Rocket

A toy rocket is launched at an initial velocity of 14.7 m/s from a 49-m-tall platform. The height h of the object at any time t seconds after launch is given by the equation $h = -4.9t^2 + 14.7t + 49$. When will the rocket hit the ground?

Solution

Step 1. We are looking for the time when h is equal to zero.

Step 2. Set h equal to 0, and solve for t: $0 = -4.9t^2 + 14.7t + 49$.

TIP It is often helpful to rewrite a quadratic equation so that the leading coefficient is 1. For this equation, we can divide both sides by -4.9 to obtain a leading coefficient of 1.

Step 3. Solve:

$$0 = -4.9t^2 + 14.7t + 49 \qquad \text{Write the original equation.}$$

$0 = t^2 - 3t - 10$	Divide both sides by -4.9 to obtain a leading coefficient of 1.
$0 = (t - 5)(t + 2)$	Factor.
$t - 5 = 0 \quad \text{or} \quad t + 2 = 0$	Use the zero product property.
$t = 5 \quad \text{or} \quad t = -2$	Solve.

Step 4. Because t represents the time (in seconds) after launch, the value $t = -2$ seconds does not make sense. Therefore, the rocket will hit the ground 5 seconds after launch.

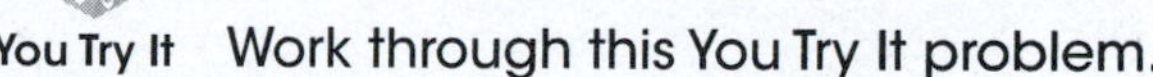

You Try It Work through this You Try It problem.

Work Exercises 6 and 7 in this textbook or in the MyLab Math Study Plan.

OBJECTIVE 3 SOLVING GEOMETRIC APPLICATIONS

Example 3 Find the Dimensions of a Rectangle

The length of a rectangle is 6 in. less than four times the width. Find the dimensions of the rectangle if the area of the rectangle is 54 in^2.

Solution Work through the **interactive video** to verify that the dimensions of the rectangle are 12 in. by 4.5 in.

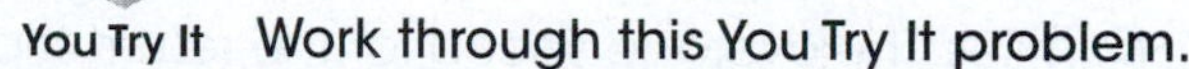

You Try It Work through this You Try It problem.

Work Exercises 8 and 9 in this textbook or in the MyLab Math Study Plan.

Before trying to work through Example 4, it is critical that you have an understanding of one of the most famous theorems in all of mathematics: The **Pythagorean** Theorem.

Example 4 Find the Width of a High-Definition Television

Jimmy bought a new 40-in. high-definition television. If the length of Jimmy's television is 8 in. longer than the width, find the width of the television.

Solution

Step 1. We are trying to find the width of the television.

Step 2. Let w = width of the television, then $w + 8$ = length of the television. The size of a television is the length of the diagonal, which in this case is 40 in. We can create an equation using the Pythagorean theorem $a^2 + b^2 = c^2$.

$w^2 + (w + 8)^2 = 40^2$	Use the **Pythagorean theorem.**

Step 3. Solve:

$w^2 + (w + 8)^2 = 40^2$	Write the equation from Step 2.
$w^2 + (w + 8)(w + 8) = 1600$	Write $(w + 8)^2$ as $(w + 8)(w + 8)$.

$w^2 + w^2 + 16w + 64 = 1{,}600$	Use the **distributive property**.
$2w^2 + 16w - 1{,}536 = 0$	Combine like terms.
$w^2 + 8w - 768 = 0$	Divide by 2.
$(w - 24)(w + 32) = 0$	Factor.
$w - 24 = 0$ or $w + 32 = 0$	Use the zero product property.
$w = 24$ or $w = -32$	Solve for w.

Step 4. Because w represents the width of the television, we can disregard the negative solution. Thus, the width of the television is 24 in.

You Try It Work through this You Try It problem.

Work Exercise 10 in this textbook or in the MyLab Math Study Plan.

OBJECTIVE 4 SOLVING APPLICATIONS INVOLVING UNIFORM MOTION

In **Section 1.2** we introduced applications involving **uniform motion**. When we solve applied problems involving distance, rate, and time, we use the formula $d = rt$ or $\text{distance} = \text{rate} \times \text{time}$ as in Example 5.

Example 5 Find the Speed of an Airplane

Kevin flew his new Cessna O-2A airplane from Jonesburg to Mountainview, a distance of 2,560 miles. The average speed for the return trip was 64 mph faster than the average outbound speed. If the total flying time for the round trip was 18 hours, what was the plane's average speed on the outbound trip from Jonesburg to Mountainview?

Solution

Step 1. We are asked to find the average speed of the plane from Jonesburg to Mountainview.

Step 2. Let r = speed of plane on the outbound trip from Jonesburg to Mountainview, and then $r + 64$ = speed of plane on the return trip from Mountainview to Jonesburg.

Step 3. Because $\text{distance} = \text{rate} \times \text{time}$, we know that $\text{time} = \dfrac{\text{distance}}{\text{rate}}$.

We can organize our data in the following table.

	Distance	Rate	Time = Distance/Rate
Outbound to Mountainview	2,560	r	$\dfrac{2{,}560}{r}$
Return to Jonesburg	2,560	$r + 64$	$\dfrac{2{,}560}{r + 64}$

The total flying time was 18 hours, so we can set up the following equation:

$$\underbrace{\frac{2{,}560}{r}}_{\text{Outbound Time}} + \underbrace{\frac{2{,}560}{r+64}}_{\text{Return Time}} = \underbrace{18}_{\text{Total Flight Time}}$$

Solving this equation for r, we get $r = 256$ or $r = -\frac{320}{9} \approx -35.56$. Watch the video to see the solution worked out in detail.

Step 4. Because the rate cannot be negative, the average outbound speed to Mountainview was 256 mph.

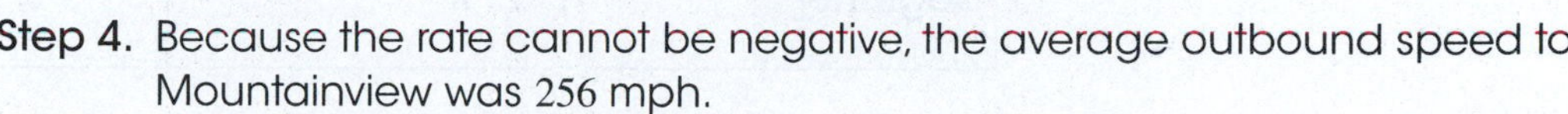

You Try It Work through this You Try It problem.

Work Exercises 11–13 in this textbook or in the MyLab Math Study Plan.

OBJECTIVE 5 SOLVING APPLICATIONS INVOLVING RATES OF WORK

In **Section 1.2** we introduced applications involving **rates of work**. Recall that when dealing with two workers that are performing a job (or task), we use the following formula:

$$\underbrace{\frac{1}{t_1}}_{\text{Work rate of 1st worker}} + \underbrace{\frac{1}{t_2}}_{\text{Work rate of 2nd worker}} = \underbrace{\frac{1}{t}}_{\text{Work rate together}}$$

Here, t_1 and t_2 are the individual times to complete one job, and t is the time to complete the job when working together.

> **Rate of Work**
>
> The rate of **work** is the number of jobs that can be completed in a given unit of time.
>
> If one job can be completed in t units of time, then the rate of work is given by $\frac{1}{t}$.

Example 6 Work Together to Create Monthly Sales Reports

Dawn can finish the monthly sales reports in 2 hours less time than it takes Adam. Working together, they were able to finish the sales reports in 8 hours. How long does it take each person to finish the monthly sales reports alone? (Round to the nearest minute.)

Solution

Step 1. We are asked to find the time that it takes each person to finish the monthly sales reports.

Step 2. Let t = Adam's time to do the job alone and $t - 2$ = Dawn's time to do the job alone. We can now set up a table with the given data.

	Time needed to complete the job, in hours	Portion of job completed in 1 hour (rate)
Adam	t	$\frac{1}{t}$
Dawn	$t-2$	$\frac{1}{t-2}$
Together	8	$\frac{1}{8}$

We now use the information from the table to set up the following equation:

$$\underbrace{\frac{1}{t}}_{\text{Adam's Rate}} + \underbrace{\frac{1}{t-2}}_{\text{Dawn's Rate}} = \underbrace{\frac{1}{8}}_{\text{Rate Together}}$$

Step 3. Solving for t, we see that Adam can do the job by himself in approximately 17 hours and 4 minutes, while it would take Dawn approximately 15 hours and 4 minutes to do the same job. To see this problem worked out in detail, watch this **video**.

Step 4. You should verify that Dawn's time is exactly 2 hours less than Adam's time and that the sum of their rates is equal to $\frac{1}{8}$.

You Try It **Work through this You Try It problem.**

Work Exercises 14–16 in this textbook or in the MyLab Math Study Plan.

1.5 Exercises

1. The product of some negative number and 5 less than three times that number is 12. Find the number.

2. The square of a number plus the number is 132. What is the number?

3. The sum of the square of a number and the square of 5 more than the number is 169. What is the number?

4. Three consecutive odd integers are such that the square of the third integer is 15 more than the sum of the squares of the first two. Find the integers.

5. Find two real numbers that have a sum of 10 and a product of 12.

6. A toy rocket is launched from a 2.8-m-high platform in such a way that its height, h (in meters), after t seconds is given by the equation $h = -4.9t^2 + 18.9t + 2.8$. How long will it take for the rocket to hit the ground?

7. Benjamin threw a rock straight up from a cliff that was 24 ft above the water. If the height of the rock h, in feet, after t seconds is given by the equation $h = -16t^2 + 20t + 24$, how long will it take for the rock to hit the water?

8. The length of a rectangle is 1 cm less than three times the width. If the area of the rectangle is 30 cm^2, find the dimensions of the rectangle.

9. The length of a rectangle is 1 in. less than twice the width. If the diagonal is 2 in. more than the length, find the dimensions of the rectangle.

10. A 35-ft by 20-ft rectangular swimming pool is surrounded by a walkway of uniform width. If the total area of the walkway is 434 ft^2, how wide is the walkway?

11. A boat traveled downstream a distance of 45 mi and then came right back. If the speed of the current is 5 mph and if the total trip took 3 hours and 45 minutes, find the average speed of the boat relative to the water.

12. Meg rowed her boat upstream a distance of 9 mi and then rowed back to the starting point. The total time of the trip was 10 hours. If the rate of the current was 4 mph, find the average speed of the boat relative to the water.

13. Imogene's car traveled 280 miles. averaging a certain speed. If the car had gone 5 mph faster, the trip would have taken 1 hour less. Find the average speed.

14. Twin brothers, Billy and Bobby, can mow their grandparent's lawn together in 56 minutes. Billy could mow the lawn by himself in 15 minutes less time than it would take Bobby. How long would it take Bobby to mow the lawn by himself?

15. Jeff and Kirk can build a 75-ft. retaining wall together in 12 hours. Because Jeff has more experience, he could build the wall by himself 4 hours quicker than Kirk. How long would it take Kirk (to the nearest minute) to build the wall by himself?

16. Homer and Mike were replacing the boards on Mike's old deck. Mike can do the job alone in 1 hour less time than Homer. They worked together for 3 hours until Homer had to go home. Mike finished the job working by himself in an additional 4 hours. How long would it have taken Homer to fix the deck by himself? (*Hint:* Use Method 2, as discussed in **Section 1.2**.)

1.6 Other Types of Equations

THINGS TO KNOW

Before working through this section, be sure that you are familiar with the following concepts:

VIDEO ANIMATION INTERACTIVE

1. Factoring Trinomials with a Leading Coefficient Equal to 1 (Section R.6)

2. Factoring Trinomials with a Leading Coefficient Not Equal to 1 (Section R.6)

3. Factoring Polynomials by Grouping (Section R.6)

4. Solving Quadratic Equations by Factoring and the Zero Product Property (Section 1.4)

5. Solving Quadratic Equations Using the Quadratic Formula (Section 1.4)

(eText Screens 1.6-1–1.6-14)

OBJECTIVES

1 Solving Higher-Order Polynomial Equations

2 Solving Equations That Are Quadratic in Form (Disguised Quadratics)

3 Solving Equations Involving Radicals

SECTION 1.6 EXERCISES

OBJECTIVE 1 SOLVING HIGHER-ORDER POLYNOMIAL EQUATIONS

So far in this text, we have learned methods for solving **linear equations and quadratic equations**. Linear equations and quadratic equations in one variable are both examples of **polynomial equations** of first and second degree, respectively. In this section, we first start by looking at certain higher-order polynomial equations that can be solved using special factoring techniques.

One useful technique is to factor out a common factor, then use the **zero product property**.

Example 1 Factoring Out a Common Factor to Solve a Polynomial Equation

Find all solutions of the equation $3x^3 - 2x = -5x^2$.

Solution First, add $5x^2$ to both sides to set the equation equal to 0.

$$3x^3 + 5x^2 - 2x = 0$$

Now, try to factor the left-hand side.

$x(3x^2 + 5x - 2) = 0$	Factor out the greatest common factor, x.
$x(3x - 1)(x + 2) = 0$	Factor the quadratic.
$x = 0$ or $3x - 1 = 0$ or $x + 2 = 0$	Use the zero product property.
$x = 0$ or $x = \frac{1}{3}$ or $x = -2$	Solve for x.

The solution set is $\left\{-2, 0, \frac{1}{3}\right\}$.

CAUTION In the equation $3x^3 + 5x^2 - 2x = 0$, do *not* divide both sides by x. This would produce the equation $3x^2 + 5x - 2 = 0$, which has only $\frac{1}{3}$ and -2 as solutions. The solution $x = 0$ would be "lost." In addition, because $x = 0$ is a solution of the original equation, dividing by x would mean dividing by 0, which of course is undefined and produces incorrect results.

Sometimes, polynomials can be solved by grouping terms and factoring (especially when the polynomial has four terms). You may want to review how to **factor by grouping**, which was discussed in **Section R.6**.

Example 2 Factor by Grouping to Solve a Polynomial Equation

Find all solutions of the equation $2x^3 - x^2 + 8x - 4 = 0$.

Solution Arrange the terms of the polynomial in descending order, and group the terms of the polynomial in pairs.

$2x^3 - x^2 + 8x - 4 = 0$	Write the original equation.
$(2x^3 - x^2) + (8x - 4) = 0$	Group the terms in pairs.
$x^2(2x - 1) + 4(2x - 1) = 0$	Factor the G.C.F from each pair.
$(2x - 1)(x^2 + 4) = 0$	Factor $2x - 1$ from the left-hand side.
$2x - 1 = 0$ or $x^2 + 4 = 0$	Use the zero product property.
$2x = 1$ or $x^2 = -4$	Solve for x.
$x = \frac{1}{2}$ or $x = \pm\sqrt{-4} = \pm 2i$	Simplify the radical with a negative radicand.

The solution set is $\left\{-2i, 2i, \frac{1}{2}\right\}$.

You Try It Work through this You Try It problem.

Work Exercises 1–6 in this textbook or in the MyLab Math Study Plan.

OBJECTIVE 2 SOLVING EQUATIONS THAT ARE QUADRATIC IN FORM (DISGUISED QUADRATICS)

Quadratic equations of the form $ax^2 + bx + c = 0, a \neq 0$, are relatively straightforward to solve because we know several methods for solving quadratics. Sometimes, equations that are not quadratic can be made into a quadratic equation by using a substitution. Equations of this type are said to be *quadratic in form* or *disguised quadratics*. These equations typically have the form $au^2 + bu + c = 0, a \neq 0$, after an appropriate substitution. Following are a few examples.

Original Equation		Make an Appropriate Substitution		New Equation is a Quadratic
$2x^4 - 11x^2 + 12 = 0$	Determine the proper substitution →	Let $u = x^2$, then $u^2 = x^4$.	New equation is a quadratic →	$2u^2 - 11u + 12 = 0$
$\left(\frac{1}{x-2}\right)^2 + \frac{2}{x-2} - 15 = 0$	Determine the proper substitution →	Let $u = \frac{1}{x-2}$, then $u^2 = \left(\frac{1}{x-2}\right)^2$.	New equation is a quadratic →	$u^2 + 2u - 15 = 0$
$x^{\frac{2}{3}} - 9x^{\frac{1}{3}} + 8 = 0$	Determine the proper substitution →	Let $u = x^{\frac{1}{3}}$, then $u^2 = (x^{\frac{1}{3}})^2 = x^{\frac{2}{3}}$.	New equation is a quadratic →	$u^2 - 9u + 8 = 0$
$3x^{-2} - 5x^{-1} - 2 = 0$	Determine the proper substitution →	Let $u = x^{-1} = \frac{1}{x}$, then $u^2 = (x^{-1})^2 = x^{-2}$.	New equation is a quadratic →	$3u^2 - 5u - 2 = 0$

We now solve each of these equations listed in Example 3:

Example 3 Solve Equations That Are Quadratic in Form

Solve each of the following:

a. $2x^4 - 11x^2 + 12 = 0$

b. $\left(\frac{1}{x-2}\right)^2 + \frac{2}{x-2} - 15 = 0$

c. $x^{2/3} - 9x^{1/3} + 8 = 0$

d. $3x^{-2} - 5x^{-1} - 2 = 0$

Solution

a. $2x^4 - 11x^2 + 12 = 0$ — Write the original equation.

Let $u = x^2$ then $u^2 = x^4$. — Determine the proper substitution.

$2u^2 - 11u + 12 = 0$ — Substitute u for x^2 and u^2 for x^4 into the original equation.

$(2u - 3)(u - 4) = 0$ — Factor.

$2u - 3 = 0$ or $u - 4 = 0$ — Use the **zero product property**.

$u = \frac{3}{2}$ or $u = 4$ — Solve for u.

Don't stop yet! We want to solve for x. Because we know that $u = x^2$, we get

$$x^2 = \frac{3}{2} \text{ or } x^2 = 4, \text{ so } x = \pm\sqrt{\frac{3}{2}} = \pm\frac{\sqrt{6}}{2} \text{ or } x = \pm 2.$$

Try to find the solutions to (b) through (d). Watch the **interactive video** to see each worked out solution.

You Try It **Work through this You Try It problem.**

Work Exercises 7–13 in this textbook or in the MyLab Math Study Plan.

OBJECTIVE 3 SOLVING EQUATIONS INVOLVING RADICALS

A radical equation is an equation that involves a variable inside a square root, cube root, or any higher root. To solve these equations, we must try to isolate the radical and then raise each side of the equation to the appropriate power to eliminate the radical. Consider the following examples:

Example 4 Solve an Equation Involving a Square Root

Solve $\sqrt{x-1} - 2 = x - 9$.

Solution First, we isolate the radical on the left side of the equation, adding 2 to both sides.

$\sqrt{x-1} = x - 7$ — Add 2 to both sides.

To eliminate the radical, square both sides of the equation.

$(\sqrt{x-1})^2 = (x-7)^2$	Square both sides of the equation.
$x - 1 = x^2 - 14x + 49$	Expand the right-hand side: $(x-7)^2 = (x-7)(x-7) = x^2 - 14x + 49$.
$0 = x^2 - 15x + 50$	Move all terms to the right-hand side.
$0 = (x-5)(x-10)$	Factor.

Setting each factor equal to zero gives us $x = 5$ or $x = 10$ as possible solutions. It is tempting to stop here. When solving equations involving radicals it is important to always check each potential solution by substituting back into the original equation.

Check $x = 10$:

$$\sqrt{x-1} - 2 = x - 9$$
$$\sqrt{10-1} - 2 \stackrel{?}{=} 10 - 9$$
$$\sqrt{9} - 2 \stackrel{?}{=} 1$$
$$3 - 2 \stackrel{?}{=} 1$$
$$1 = 1 \checkmark$$

Check $x = 5$:

$$\sqrt{x-1} - 2 = x - 9$$
$$\sqrt{5-1} - 2 \stackrel{?}{=} 5 - 9$$
$$\sqrt{4} - 2 \stackrel{?}{=} -4$$
$$2 - 2 \stackrel{?}{=} -4$$
$$0 \neq -4 \quad \times$$

We see that $x = 10$ checks, but $x = 5$ does not. Therefore, the solution set to this equation is $\{10\}$. $x = 5$ is called an **extraneous solution**.

CAUTION Because the squaring operation can make a false statement true ($-2 \neq 2$, but $(-2)^2 = (2)^2$), it is essential to always check your answers after solving an equation in which this operation was performed. The squaring operation is usually used when solving an equZation involving square roots.

CAUTION Be careful when squaring an expression of the form $(a+b)^2$ or $(a-b)^2$. Remember, $(a+b)^2 = (a+b)(a+b) = a^2 + 2ab + b^2$ and $(a-b)^2 = (a-b)(a-b) = a^2 - 2ab + b^2$.

Example 5 Solve an Equation Involving Two Square Roots

Solve $\sqrt{2x+3} + \sqrt{x-2} = 4$.

Solution When solving equations containing multiple radicals, first isolate one of the radicals and square both sides:

$$\sqrt{2x+3} = 4 - \sqrt{x-2}$$
$$(\sqrt{2x+3})^2 = (4 - \sqrt{x-2})^2$$

Watch the **video** to see how to obtain the solution of $x = 3$.

Example 6 Solve an Equation Involving a Cube Root

Solve $\sqrt[3]{1-4x} + 3 = 0$

Solution To eliminate the radical, we first isolate the radical and then cube both sides.

$\sqrt[3]{1-4x} = -3$	Subtract 3 from both sides.
$(\sqrt[3]{1-4x})^3 = (-3)^3$	Cube both sides of the equation.

$1 - 4x = -27$ Simplify.

$-4x = -28$ Combine like terms.

$x = 7$ Divide both sides by -4.

Check $\sqrt[3]{1 - 4(7)} + 3 = \sqrt[3]{-27} + 3 = -3 + 3 = 0$ ✓

You Try It **Work through this You Try It problem.**

Work Exercises 14–21 in this textbook or in the MyLab Math Study Plan.

1.6 Exercises

Skill Check Exercises

For exercises SCE-1 through SCE-14, completely factor each polynomial.

SCE-1. $x^2 + 23x + 60$ **SCE-2.** $y^2 + 5y - 14$ **SCE-3.** $w^2 - 20w + 96$

SCE-4. $x^2 - 6x - 40$ **SCE-5.** $12x^2 + 25x + 7$ **SCE-6.** $3x^2 + 4x - 15$

SCE-7. $6y^2 - 19y + 10$ **SCE-8.** $3x^2 + 24x - 99$ **SCE-9.** $5x^2w + 25xw + 30w$

SCE-10. $20x^2 + 54x + 10$ **SCE-11.** $x^2 - 121$ **SCE-12.** $108t^2 - 147$

SCE-13. $x^3 + 8x^2 + 4x + 32$ **SCE-14.** $x^3 - x^2 - 3x + 3$

For exercises SCE-15 through SCE-20, perform the indicated operation and simplify.

SCE-15. $(\sqrt{2x + 4})^2$ SCE-16. $(\sqrt{18 - 7x})^2$ SCE-17. $(2x - 1)^2$

SCE-18. $(-4\sqrt{8x + 9})^2$ SCE-19. $(3 + \sqrt{y + 4})^2$ SCE-20. $(2 - \sqrt{4x + 5})^2$

In Exercises 1–6, find all solutions.

1. $x^3 + x = 0$
2. $2x^3 + 4x^2 - 30x = 0$
3. $x^2(x^2 - 1) - 9(x^2 - 1) = 0$
4. $x^3 - x^2 + 3x = 3$
5. $2x^3 + 3x^2 = 8x + 12$
6. $x^3 - x^2 - x = -1$

In Exercises 7–13, solve the equation after making an appropriate substitution.

SbS 7. $x^4 - 6x^2 + 8 = 0$

SbS 8. $(13x - 1)^2 - 2(13x - 1) - 3 = 0$

SbS 9. $2x^{2/3} - 5x^{1/3} + 2 = 0$

SbS 10. $x^6 - 7x^3 = 8$

SbS 11. $\sqrt{x} - 3\sqrt[4]{x} - 4 = 0$

SbS 12. $3\left(\dfrac{1}{x - 1}\right)^2 - \dfrac{5}{x - 1} - 2 = 0$

SbS 13. $2x^{-2} - 3x^{-1} - 2 = 0$

In Exercises 14–21, solve each radical equation.

SbS 14. $\sqrt{2x - 1} = 3$ SbS 15. $\sqrt{32 - 4x} - x = 0$ SbS 16. $x - 1 = \sqrt{2x + 1}$

SbS 17. $\sqrt{12 - 2x} = x + 6$ SbS 18. $\sqrt{5y + 1} - \sqrt{y + 1} = 2$ SbS 19. $\sqrt{15 + 5x} + \sqrt{3x + 13} = 2$

SbS 20. $\sqrt[3]{2x - 1} = -2$ SbS 21. $\sqrt[4]{x^2 - 6x} - 2 = 0$

Brief Exercises

In Exercises 22–28, solve the equation after making an appropriate substitution.

22. $x^4 - 6x^2 + 8 = 0$

23. $(13x - 1)^2 - 2(13x - 1) - 3 = 0$

24. $2x^{2/3} - 5x^{1/3} + 2 = 0$

25. $x^6 - 7x^3 = 8$

26. $\sqrt{x} - 3\sqrt[4]{x} - 4 = 0$

27. $3\left(\frac{1}{x - 1}\right)^2 - \frac{5}{x - 1} - 2 = 0$

28. $2x^{-2} - 3x^{-1} - 2 = 0$

In Exercises 29–36, solve each radical equation.

29. $\sqrt{2x - 1} = 3$

30. $\sqrt{32 - 4x} - x = 0$

31. $x - 1 = \sqrt{2x + 1}$

32. $\sqrt{12 - 2x} = x + 6$

33. $\sqrt{5y + 1} - \sqrt{y + 1} = 2$

34. $\sqrt{15 + 5x} + \sqrt{3x + 13} = 2$

35. $\sqrt[3]{2x - 1} = -2$

36. $\sqrt[4]{x^2 - 6x} - 2 = 0$

1.7 Linear Inequalities in One Variable

THINGS TO KNOW

Before working through this section, be sure that you are familiar with the following concepts:

VIDEO ANIMATION INTERACTIVE

You Try It 1. Describing Intervals of Real Numbers (Section R.1)

You Try It 2. Determining the Intersection and Union of Sets and Intervals (Section R.1)

You Try It 3. Solving Linear Equations Involving Fractions (Section 1.1)

You Try It 4. Solving Linear Equations Involving Decimals (Section 1.1)

INTRODUCTION

Read this introduction before beginning Objective 1.

(eText Screens 1.7-1–1.7-32)

OBJECTIVES

1 Solving Linear Inequalities in One Variable

2 Solving Three-Part Inequalities in One Variable

3 Solving Compound Inequalities in One Variable

4 Using Linear Inequalities to Solve Application Problems

SECTION 1.7 EXERCISES

Introduction to Section 1.7

Unlike equations that usually have a finite number of solutions (or no solution at all), inequalities often have infinitely many solutions. For instance, the inequality $2x - 3 \le 5$ has infinitely many solutions because there are infinite values of x for which the inequality is true. Take, for example, the numbers -1, $\frac{7}{2}$, and 4. Substituting each of these values into the inequality $2x - 3 \le 5$ yields a true statement:

Inequality $2x - 3 \le 5$

$x = -1$ $\qquad 2(-1) - 3 \overset{?}{\le} 5$

$\qquad\qquad -5 \le 5$ True statement!

$x = \frac{7}{2}$ $\qquad 2\left(\frac{7}{2}\right) - 3 \overset{?}{\le} 5$

$\qquad\qquad 4 \le 5$ True statement!

$x = 4$ $\qquad 2(4) - 3 \overset{?}{\le} 5$

$\qquad\qquad 5 \le 5$ True statement!

In fact, any real number x that is less than or equal to 4 makes this inequality a true statement. We certainly cannot list all solutions to this inequality individually. Thus, there are typically three methods used to describe the solution to an inequality:

1. Graph the solution on a number line.
2. Write the solution in set-builder notation.
3. Write the solution in interval notation.

Figure 2 shows the solution to the inequality in these three ways. In this section, you learn how to solve such inequalities. You may want to review the terms set-builder notation and interval notation, which are described in **Section R.1**.

Solution graphed on a number line:

Figure 2 Solution to the inequality $2x - 3 \le 5$

Solution in set-builder notation: $\{x | x \le 4\}$

Solution in interval notation: $(-\infty, 4]$

OBJECTIVE 1 SOLVING LINEAR INEQUALITIES IN ONE VARIABLE

The inequality $2x - 3 \le 5$ is called a **linear inequality**. In this section, you learn techniques necessary to solve linear inequalities in one variable.

> **Definition** A linear inequality in one variable is an inequality that can be written in the form $ax + b < c$, where a, b, and c are real numbers and $a \neq 0$.
>
> **Note:** The inequality symbol "<" in the previous definition and the following properties can be replaced with $>$, $\le$, or $\ge$.

Before we solve a linear inequality, it is important to introduce some inequality properties:

Properties of Inequalities

Let a, b, and c be real numbers:

	Property	In Words	Example
1	If $a < b$, then $a + c < b + c$	The same number may be added to both sides of an inequality.	$-3 < 7$ $-3 + 4 < 7 + 4$ $1 < 11$
2	If $a < b$, then $a - c < b - c$	The same number may be subtracted from both sides of an inequality.	$9 \ge 2$ $9 - 6 \ge 2 - 6$ $3 \ge -4$
3	For $c > 0$, if $a < b$, then $ac < bc$	Multiplying both sides of an inequality by a *positive* number *does not* switch *the direction* of the inequality.	$3 > 2$ $(3)(5) > (2)(5)$ $15 > 10$
4	For $c < 0$, if $a < b$, then $ac > bc$	Multiplying both sides of an inequality by a *negative* number switches *the direction* of the inequality.	$3 > 2$ $(3)(-5) < (2)(-5)$ $-15 < -10$
5	For $c > 0$, if $a < b$, then $\frac{a}{c} < \frac{b}{c}$	Dividing both sides of an inequality by a *positive* number *does not* switch *the direction* of the inequality.	$6 > 4$ $\frac{6}{2} > \frac{4}{2}$ $3 > 2$
6	For $c < 0$, if $a < b$, then $\frac{a}{c} > \frac{b}{c}$	Dividing both sides of an inequality by a *negative* number switches *the direction* of the inequality.	$6 > 4$ $\frac{6}{-2} < \frac{4}{-2}$ $-3 < -2$

(eText Screens 1.7-1–1.7-32)

When multiplying or dividing both sides of an inequality by a negative number, it is important to remember to switch the direction of the inequality. Watch this **animation** to see why this is important.

Example 1 Solve a Linear Inequality

Solve the inequality $2 - 5(x - 2) < 4(3 - 2x) + 7$. Express the answer in set-builder notation.

Solution The technique to use when solving linear inequalities is to isolate the variable on one side.

$2 - 5(x - 2) < 4(3 - 2x) + 7$	Write the original inequality.
$2 - 5x + 10 < 12 - 8x + 7$	Use the **distributive property**.
$12 - 5x < 19 - 8x$	Simplify.
$12 - 5x + 8x < 19 - 8x + 8x$	Add $8x$ to both sides.
$3x + 12 < 19$	Simplify.
$3x + 12 - 12 < 19 - 12$	Subtract $12x$ from both sides.
$3x < 7$	Simplify.
$\frac{3x}{3} < \frac{7}{3}$	Divide both sides by $3x$.
$x < \frac{7}{3}$	Simplify.

The solution is the set of all values of x such that x is strictly less than $\frac{7}{3}$. We write this solution in set-builder notation as $\left\{x \middle| x < \frac{7}{3}\right\}$. Watch the **video** to see the solution worked out in its entirety.

You Try It Work through this You Try It problem.

Work Exercises 1–4 in this textbook or in the MyLab Math Study Plan.

CAUTION Remember to switch the direction of the inequality when multiplying or dividing a linear inequality by a negative number. See the solution to Example 2 below.

Example 2 Solve a Linear Inequality

Solve the inequality $-9x - 3 \geq 7 - 4x$. Graph the solution set on a number line, and express the answer in interval notation.

Solution The technique to use when solving linear inequalities is to isolate the variable on one side.

$-9x - 3 \geq 7 - 4x$	
$-9x + 4x - 3 \geq 7 - 4x + 4x$	Add $4x$ to both sides.
$-5x - 3 \geq 7$	Simplify.

$-5x - 3 + 3 \geq 7 + 3$ Add 3 to both sides.

$-5x \geq 10$ Simplify.

$\frac{-5x}{-5} \leq \frac{10}{-5}$ Divide by -5. **Switch the direction of the inequality!**

$x \leq -2$ Simplify.

The graph of the solution is depicted in Figure 3. The solution set in interval notation is $(-\infty, -2]$.

Solution graphed on number line:

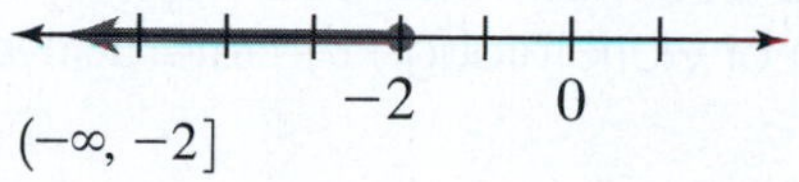

Solution in interval notation: $(-\infty, -2]$

Figure 3 Solution to the inequality $-9x - 3 \geq 7 - 4x$

You Try It Work through this You Try It problem.

Work Exercises 5–12 in this textbook or in the MyLab Math Study Plan.

When a linear inequality involves fractions, we begin the solution process in the same way as when solving linear equations involving fractions. That is, we eliminate the fractions by first multiplying both sides of the inequality by the **least common denominator** (LCD).

Similarly, if we encounter a linear inequality involving decimals, we can eliminate the decimals by multiplying both sides of the inequality by the appropriate power of 10.

Before trying Example 3, take a moment to review these two objectives by clicking on the links below:

Section 1.1: Solving Linear Equations Involving Fractions
Section 1.1: Solving Linear Equations Involving Decimals

Example 3 Solving Linear Inequalities Involving Fractions or Decimals

Solve the linear inequalities.

a. $\frac{1 - 4w}{5} - \frac{w}{2} \leq -5$ **b.** $3.7y - 6 > 6.1 + 3.45y$

Solution

a. The LCD is 10. So, start by multiplying both sides of the inequality by 10.

$10\left(\frac{1 - 4w}{5} - \frac{w}{2}\right) \leq -5 \cdot 10$ Multiply both sides of the inequality by 10.

$10 \cdot \left(\frac{1 - 4w}{5}\right) - 10 \cdot \left(\frac{w}{2}\right) \leq -5 \cdot 10$ Use the **distributive property**.

$2(1 - 4w) - 5w \leq -50$ Multiply.

Now that the fractions have been eliminated, try solving this linear inequality on your own or work through the **interactive video**.

b. There are three terms involving decimals in this linear inequality; $3.7y$, 6.1, and $3.45y$. The number 3.45 has two decimal places. Therefore, we start by multiplying both sides of the inequality by $10^2 = 100$.

$100(3.7y - 6) > (6.1 + 3.45y)100$	Multiply both sides of the inequality by 100.
$100 \cdot (3.7y) - 100 \cdot (6) > 100 \cdot (6.1) + 100 \cdot (3.45y)$	Use the **distributive property**.
$370y - 600 > 610 + 345y$	Multiply.

Now that the decimals have been eliminated, try solving this linear inequality on your own or work through the **interactive video**.

You Try It **Work through this You Try It problem.**

Work Exercises 13–20 in this textbook or in the MyLab Math Study Plan.

OBJECTIVE 2 SOLVING THREE-PART INEQUALITIES IN ONE VARIABLE

The weight of a laptop computer depends on several factors including the size of the monitor, the thickness, the type of battery, and the type of internal hard drive. The weight of a typical 14-inch laptop varies between 1.2 lbs to 8.4 pounds. The weight can be expressed using the following *three-part inequality* $1.2 \le W \le 8.4$, where W is the weight of the laptop. Notice that the variable W is between the two inequality symbols. The technique to use when solving three-part inequalities is to simplify until the variable is "sandwiched" in the middle, as in our example, $1.2 \le W \le 8.4$.

Example 4 Solve a Three-Part Inequality

Solve the inequality $-2 \le \dfrac{2 - 4x}{3} < 5$. Graph the solution set on a number line, and write the solution in set-builder notation and interval notation.

Solution We start by multiplying all three parts by 3 to eliminate the fraction.

$-2 \le \dfrac{2 - 4x}{3} < 5$	Write the original three-part inequality.
$3 \cdot (-2) \le \cancel{3} \cdot \dfrac{2 - 4x}{\cancel{3}} < 3 \cdot 5$	Multiply by 3.
$-6 \le 2 - 4x < 15$	Simplify.
$-6 - 2 \le 2 - 2 - 4x < 15 - 2$	Subtract 2.
$-8 \le -4x < 13$	Simplify.
$\dfrac{-8}{-4} \ge \dfrac{-4x}{-4} > \dfrac{13}{-4}$	Divide by -4, and reverse the direction of the inequalities.
$2 \ge x > -\dfrac{13}{4}$	Simplify.

Although this answer is adequate, it is good practice to rewrite the inequality so that the variable is in the middle and the smaller of the two outside numbers is on the left.

$$-\frac{13}{4} < x \le 2 \qquad \text{Rewrite the inequality.}$$

We can write the solution in set-builder notation as $\left\{x \,\middle|\, -\frac{13}{4} < x \le 2\right\}$. The graph of the solution is displayed in Figure 4.

Solution graphed on a number line:

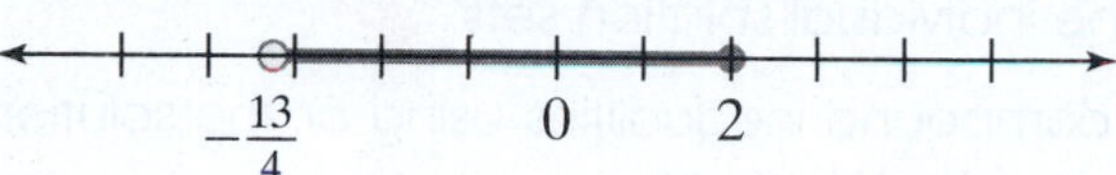

Figure 4 Solution to the three-part inequality $-2 \le \frac{2-4x}{3} < 5$

Solution in set-builder notation: $\left\{x \,\middle|\, -\frac{13}{4} < x \le 2\right\}$

Solution in interval notation: $\left(-\frac{13}{4}, 2\right]$

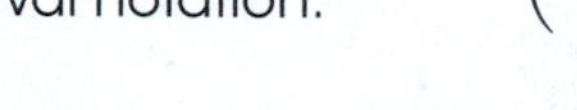

You Try It Work through this You Try It problem

Work Exercises 21–30 in this textbook or in the MyLab Math Study Plan.

OBJECTIVE 3 SOLVING COMPOUND INEQUALITIES IN ONE VARIABLE

Before we solve compound inequalities, it might be a good idea to review the concept of the **intersection** and **union** of sets. Refer to **Section R.1** to review this concept.

A **compound inequality** consists of two inequalities that are joined together using the words *and* or *or*. Two compound inequalities are shown below.

(a) $x + 2 < 5$ and $3x \ge -6$

(b) $x + 3 \le 1$ or $2x - 5 > 7$

A number is a solution to a compound inequality involving *and* if that number is a solution to **both** inequalities. In (a) above, the number 2 is a solution to both inequalities. Thus, 2 is a solution to the compound inequality. To find all solutions to a compound inequality involving the word *and*, we first find the individual solution sets to each inequality. We then determine the **intersection** of the individual solution sets.

A number is a solution to a compound inequality involving the word *or* if that number is a solution to **either** inequality. In (b) above, 8 is a solution to $2x - 5 > 7$. Therefore 8 is a solution to the compound inequality even though the number 8 is not a solution to $x + 3 \le 1$. To find all solutions to a compound inequality involving the word *or*, we first find the individual solution sets to each inequality. We then determine the **union** of the individual solution sets.

Watch this **video** to see an introduction to the concept of compound inequalities.

(eText Screens 1.7-1–1.7-32)

The following general strategy can be used when solving compound inequalities.

> **Guidelines for Solving Compound Linear Inequalities**
>
> **Step 1.** Solve each inequality separately.
>
> **Step 2.** Graph each solution set on a number line.
>
> **Step 3.** For compound inequalities using *and*, the solution set is the **intersection** of the individual solution sets.
>
> For compound inequalities using *or*, the solution set is the **union** of the individual solution sets.

Example 5 Solve a Compound Inequality with *and*

Solve the compound inequality $2x - 7 < -1$ and $3x + 5 \geq 3$. Write the solution in interval notation.

Solution

Step 1. Solve each inequality separately.

$$2x - 7 < -1 \quad \text{and} \quad 3x + 5 \geq 3$$
$$2x < 6 \quad \text{and} \quad 3x \geq -2$$
$$x < 3 \quad \text{and} \quad x \geq -\frac{2}{3}$$

(Watch this **video** to see the entire solution process.)

Step 2. Graph each solution set on a number line.

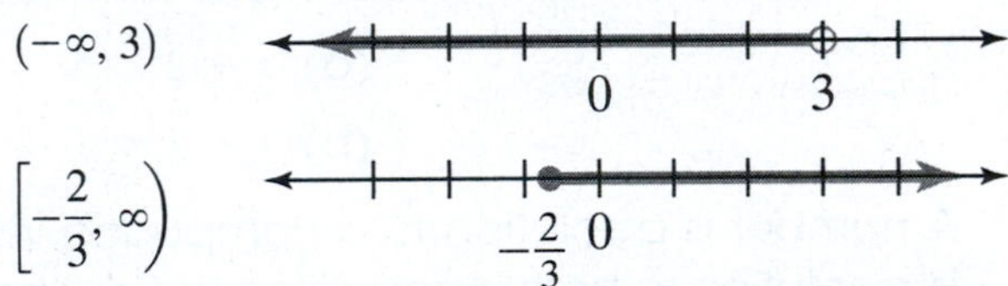

Step 3. Since the compound inequality uses the word *and*, the solution set is the **intersection** of the two graphs. We are looking for all values of x that are common to both graphs. The values in common are all x-values to the right and including $-\frac{2}{3}$ and to the left and not including 3, as depicted in Figure 5.

Solution graphed on a number line:

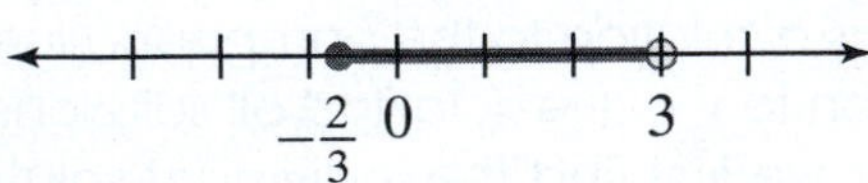

Figure 5 Solution to the compound inequality $2x - 7 < -1$ and $3x + 5 \geq 3$

Solution in interval notation: $\left[-\frac{2}{3}, 3\right)$

Example 6 Solve a Compound Inequality with *or*

Solve the compound inequality $1 - 3x \geq 7$ or $3x + 4 > 7$. Write the solution in interval notation.

Solution

Step 1. Solve each inequality separately.

Read through these **steps** or watch this **video** to verify that we get the following solutions.

$$x \leq -2 \text{ or } x > 1$$

Step 2. Graph each solution set on a number line.

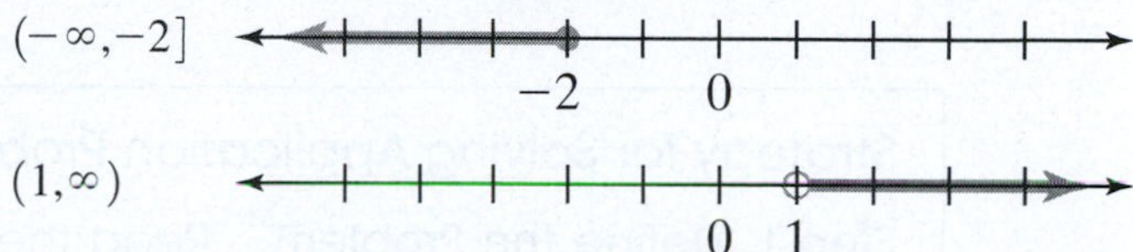

Step 3. Since the compound inequality uses the word *or*, the solution set is the union of the two graphs. This means that we take all values of x in the first graph and join them with all values of x in the second graph. To express the solution in interval notation, we join the two intervals with the *union symbol* ∪ as shown in Figure 6.

Solution graphed on a number line:

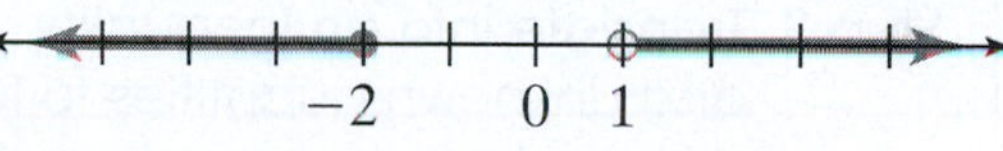

Solution in interval notation: $(-\infty, -2] \cup (1, \infty)$

Figure 6 Solution to the compound inequality $1 - 3x \geq 7$ and $3x + 4 > 7$

Example 7 Example of a Compound Inequality with No Solution

Solve $3x - 1 < -7$ and $4x + 1 > 9$.

Solution

Step 1. Solve each inequality separately.

$$3x - 1 < -7 \quad \text{and} \quad 4x + 1 > 9$$
$$3x < -6 \quad \text{and} \quad 4x > 8$$
$$x < -2 \quad \text{and} \quad x > 2$$

(Watch this **video** to see the entire solution process.)

Step 2. Graph each solution set on a number line.

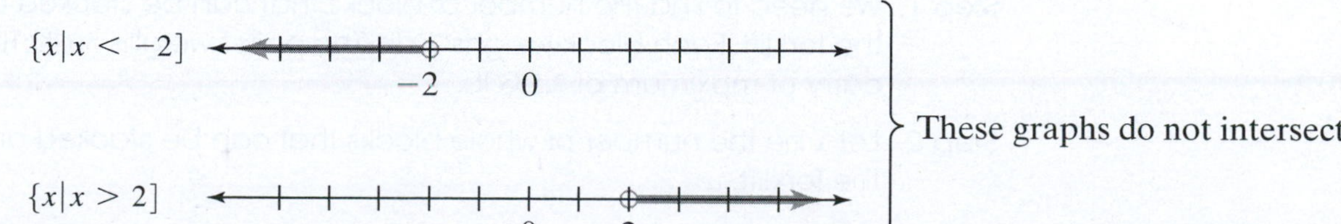

(eText Screens 1.7-1–1.7-32)

Step 3. Since the compound inequality uses the word *and*, the solution set is the **intersection** of all x-values to the left of -2 and all x-values to the right of 2. However, as you can see in Step 2 above, these two graphs do not intersect. Therefore, there is no solution to this compound inequality. We say that the solution set is empty, and use the symbol $\varnothing$ to represent the empty set.

You Try It Work through this You Try It problem.

Work Exercises 31–40 in this textbook or in the MyLab Math Study Plan.

OBJECTIVE 4 USING LINEAR INEQUALITIES TO SOLVE APPLICATION PROBLEMS

Strategy for Solving Application Problems Involving Linear Inequalities

Step 1. Define the Problem. Read the problem carefully, or multiple times if necessary. Identify what you are trying to find and determine what information is available to help you find it.

Step 2. Assign Variables. Choose a variable to assign to an unknown quantity in the problem. For example, use P for price. If other unknown quantities exist, express them in terms of the selected variable.

Step 3. Translate into an Inequality. Use the relationships among the known and unknown quantities to form an inequality. It is helpful to consider different **key words or phrases** for each inequality symbol.

Step 4. Solve the Inequality. Determine the solution set of the inequality.

Step 5. Check the Reasonableness of Your Answer. Check to see if your results make sense within the context of the problem. If not, check your work for errors and try again.

Step 6. Answer the Question. Write a clear statement that answers the question(s) posed.

Example 8 Number of Blocks that can be Lifted

Suppose you rented a forklift to move a pallet with 70-lb blocks stacked on it. The forklift can carry a maximum of 2,535 lb. If the pallet weighs 50 lb by itself with no blocks, how many whole blocks can be stacked on a pallet and lifted by the forklift?

Solution

Step 1. We need to find the number of blocks that can be stacked and lifted by the forklift. Each block weighs 70 lb. The pallet weighs 50 lb. The forklift can carry of maximum of 2,535 lb.

Step 2. Let x be the number of whole blocks that can be stacked and lifted by the forklift.

Step 3. Each block weighs 70 lb, so $70x$ is the total weight in blocks that can be lifted. The pallet weighs 50 lb, so we can set up the following inequality:

$$\underbrace{50}_{\text{50 lb for the pallet}} + \underbrace{70x}_{\text{70 lb per block}} \underbrace{\leq}_{\text{is less than or equal to}} \underbrace{2{,}535}_{\text{2,535 lb}}$$

Step 4. Solve:

$50 + 70x \leq 2{,}535$	Write the inequality.
$70x \leq 2{,}485$	Subtract 50 from both sides of the inequality.
$x \leq 35.5$	Divide both sides of the inequality by 70.

Step 5. We are trying to determine the number of whole blocks that can be lifted so we must round our answer down to 35.

Step 6. The forklift can lift no more than 35 whole blocks at a time.

Example 9 Determine the Length of a Rectangular Fence

The perimeter of a rectangular fence is to be at least 80 feet and no more than 140 feet. If the width of the fence is 12 feet, what is the range of values for the length of the fence?

Solution

Step 1. We need to determine the length range of a rectangular fence. The perimeter is at least 80 feet and no more than 140 feet. The width is fixed at 12 feet.

Step 2. Let l be the length as the rectangle as shown in Figure 7.

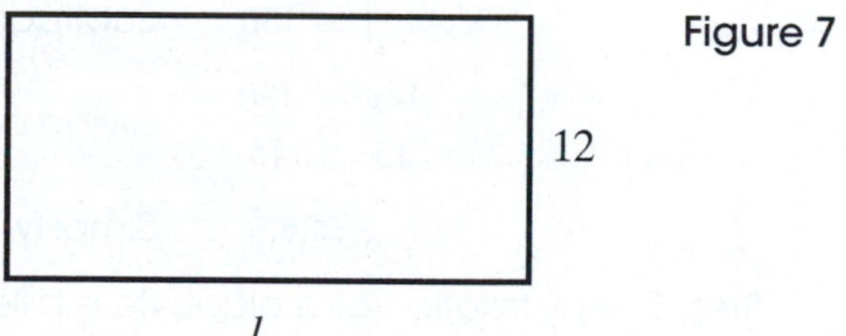

Figure 7

Step 3. The perimeter, P, of the rectangle is $P = 2(12) + 2l = 24 + 2l$. The perimeter must be greater than or equal to 80 and less than or equal to 140 or $80 \leq P \leq 140$. Because the perimeter is $24 + 2l$, we get the following three-part inequality:

$$80 \leq 24 + 2l \leq 140$$

Watch this **video** to verify that the length of the fence must be at least 28 feet and no more than 58 feet.

You Try It Work through this You Try It problem.

Work Exercises 41–46 in this textbook or in the MyLab Math Study Plan.

(eText Screens 1.7-1–1.7-32)

Example 10 Wireless Data Plan Usage

Suppose that a wireless phone company offers a monthly plan for a smartphone that includes 4 gigabytes (GB) of data for $\$110$. Each additional gigabyte of data (or fraction thereof) costs $\$15$. If Antoine subscribes to this plan, how many gigabytes of data can he use each month while keeping his total monthly cost to no more than $\$180$ (before taxes)?

Solution Use the **problem-solving strategy** for applications of linear inequalities. Follow the worked out solution below, or watch this **video**.

Step 1. We need to find the number of data gigabytes that Antoine can use each month. His monthly charge is $\$110$, which includes 4 GB of data. Also, each additional gigabyte (or fraction thereof) costs $\$15$. Antoine wants his total monthly cost to be no more than $\$180$ before taxes.

Step 2. Choose the variable g to represent the unknown number of gigabytes that Antoine can use each month. The charge for any additional gigabyte used will be $15(g - 4)$

Step 3. The sum of the monthly fee and the charge for additional gigabytes must not exceed $\$180$. So, we have the inequality:

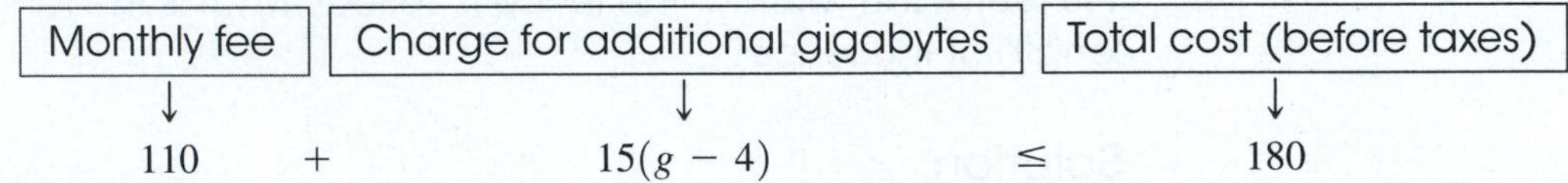

Step 4. Solve:

$110 + 15(g - 4) \le 180$	Write the inequality.
$110 + 15g - 60 \le 180$	Use the **distributive property**.
$15g + 50 \le 180$	Simplify.
$15g \le 130$	Subtract 50 from both sides of the inequality.
$\dfrac{15g}{15} \le \dfrac{130}{15}$	Divide both sides of the inequality by 15.
$g \le 8.\overline{6}$	Simplify.

Step 5. Any fraction of a gigabyte is billed as a whole gigabyte. Therefore, we must round our answer down to 8. Otherwise, the total cost would be higher than $\$180$.

Step 6. Antoine can use no more than 8 gigabytes of data per month if he wants to keep his monthly cost to no more than $\$180$. ●

You Try It Work through this You Try It problem.

Work Exercises 47–49 in this textbook or in the MyLab Math Study Plan.

Example 11 Making a profit

An online retailer sells plush toys. She purchases the toys at the wholesale price of $2.75 each and sells them online for $7.25. If her fixed costs are $900, how many plush toys must she sell in order to make a **profit**? Solve the inequality $R > C$ with R as her **revenue** and C as her **cost**.

Solution

Step 1. We need to find the number of plush toys that must be sold in order to make a **profit**. The price online is $7.25 each, and the wholesale cost to the retailer is $2.75 each. The fixed costs are $900.

Step 2. Let the variable x represent the number of plush toys that are sold. The retailer's **revenue**, R, is $7.25x$ because the number sold times the price online equals the revenue. The retailer's **cost**, C, is the sum of the fixed costs and the wholesale cost per toy. Therefore, C is $2.75x + 900$.

Step 3. To create a profit, the retailer's revenue must be greater than her cost. So, we have the following inequality:

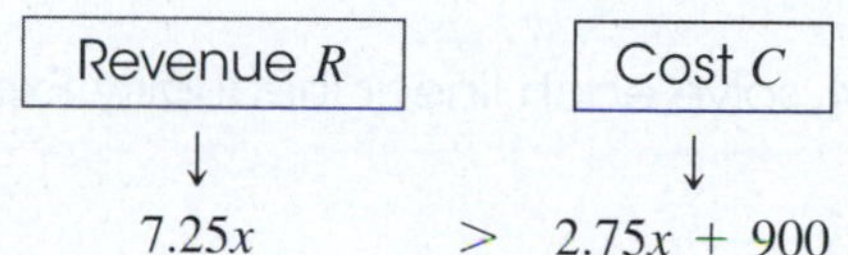

$$7.25x > 2.75x + 900$$

Step 4. The solution set is $\{x|x > 200\}$, or $(200, \infty)$ in interval notation. Watch this **video** to see the rest of the solution completed in detail. The online retailer needs to sell more than 200 plush toys to make a profit.

You Try It Work through this You Try It problem.

Work Exercises 50–52 in this textbook or in the MyLab Math Study Plan.

1.7 Exercises

Skill Check Exercises

In exercises SCE-1 through SCE-4, write each interval in set-builder notation and graph the set on a number line.

SCE-1. $\left[-\frac{1}{2}, 5\right]$ **SCE-2.** $\left(0, \frac{5}{2}\right)$ **SCE-3.** $(-\infty, 3]$ **SCE-4.** $(-1, \infty)$

In exercises SCE-5 through SCE-8, write the given set in interval notation and graph the set on a number line.

SCE-5. $\left\{x|-\frac{5}{2} \le x \le 1\right\}$ **SCE-6.** $\{x|0 \le x < 3\}$ **SCE-7.** $\left\{x|x \ge \frac{3}{4}\right\}$ **SCE-8.** $\{x|x > -4\}$

In exercises SCE-9 through SCE-12, find the intersection of sets A and B. (Find $A \cap B$.)

SCE-9. $A = \{x|x < 9\}; B = \{x|x \geq 2\}$

SCE-10. $A = \{x|x > 5\}; B = \{x|x \geq -3\}$

SCE-11. $A = \{x|x \leq -2\}; B = \{x|x \leq 0\}$

SCE-12. $A = \{x|x \leq 14\}; B = \{x|0 < x < 18\}$

In exercises SCE-13 through SCE-16, find the union of sets A and B. (Find $A \cup B$.)

SCE-13. $A = \{x|x \leq -3\}; B = \{x|x \geq 5\}$

SCE-14. $A = \{x|x \leq -8\}; B = \{x|x < 6.4\}$

SCE-15. $A = \{x|x < 10\}; B = \{x|x > -1\}$

SCE-16. $A = \{x|x \geq 0\}; B = \{x|-7 \leq x < 0\}$

In exercises SCE-17 through SCE-22, find the intersection of the given sets.

SCE-17. $(-\infty, \infty) \cap (-\infty, 3)$

SCE-18. $(-8, 5] \cap (-12, 3)$

SCE-19. $[0, \infty) \cap (-\infty, 4]$

SCE-20. $((-\infty, 1) \cup (1, \infty)) \cap (-\infty, \infty)$

SCE-21. $((-\infty, 0) \cup (0, \infty)) \cap [-3, \infty)$

SCE-22. $(-\infty, 10) \cap (-5, \infty) \cap (-\infty, 1]$

In Exercises 1–4, solve each linear inequality. Express each solution in set-builder notation.

1. $4x + 2 > -8$

2. $2(y + 1) + 3 < -3(y - 2)$

3. $11 - 2a \leq 4a + 35$

4. $\frac{1}{2}w - 1 > 4 - \frac{2}{3}w$

In Exercises 5–20, solve each linear inequality. Graph the solution set on a number line and express each solution using interval notation.

5. $x + 3 < 5$

6. $x - 1 \geq 4$

7. $5x > 10$

8. $x + 4 \leq 2x + 1$

9. $-5x - 3 \leq 2x + 4$

10. $7 - 3x \leq 1$

11. $2x + 1 > 5x + 7$

12. $3(t - 1) - 5t \geq t + 6$

13. $3x + \frac{1}{3} > \frac{3}{8}x - 1$

14. $\frac{12 - 7w}{5} - \frac{w}{5} < -8$

15. $\frac{a}{12} - \frac{7}{6} \leq 6 - 1$

16. $\frac{7}{3}(b - 12) \geq \frac{5}{2}(4b - 11)$

17. $0.75 - (0.35m + 1) < 1.5 - 0.7m$

18. $0.16x + 0.3(x - 5) \geq 0.01(2x - 2)$

19. $-0.02p - 0.04(4 - 2p) < 0.04(p - 2) - 0.12$

20. $0.0002(4 - k) + 0.09(k - 6) > 1$

In Exercises 21–30 solve each three-part inequality. Express each solution in interval notation.

21. $3 < 4x + 7 \leq 15$

22. $-2 \leq 2m - 5 \leq 2$

23. $-\frac{5}{6} \leq x + 1 < \frac{1}{2}$

24. $-10 < 1 - 2W < 4$

25. $0 \leq -6 - (a + 3) \leq 1$

26. $\frac{1}{4} < \frac{t + 1}{12} \leq 1$

27. $-1 < \frac{3 - x}{2} < 5$

28. $-7 < \frac{6 - 5x}{2} < -2$

29. $-2 \le \dfrac{5(x+1)+9}{3} < 3$

30. $-1 \le \dfrac{7-7x}{-7} \le 4$

In Exercises 31–40 solve each compound inequality. Express each solution in interval notation.

31. $x + 2 < -6$ and $2x - 3 < -9$

32. $4x + 2 \le -10$ or $3x - 4 > 8$

33. $2x - 3 \le 5$ and $5x + 1 > 6$

34. $12x - 6 \ge 2(8x - 4)$ or $5 + 13x < 15x - 9$

35. $3x - 5 \le 1$ and $1 - x < -3$

36. $3x + 1 > 7$ or $3 - 4x < 15$

37. $16x + 2 > 14x - 22$ and $4x + 2 > 7x - 4$

38. $2x - 3 \le 5$ or $3x - 3 > 3$

39. $\dfrac{x-1}{3} \ge 1$ and $\dfrac{4x-2}{2} \le 9$

40. $5x - 15 \le -20 + 4x$ or $3 + 5x > 12 - 18$

In Exercises 41–52, solve each application problem.

41. Derek can spend no more than $300 to pay for a limousine. The limousine rental company charges $200 to rent a limo, plus $20 for each hour of use. What is the maximum amount of time that he can rent the limo?

42. Jason and Sadie went to Cancun for a week. They wanted to rent a car, so they looked at two different rental car companies. Company A charges $40 per day plus 5 cents per mile. Company *B* charges $200 per week plus 30 cents per mile. How many miles must they drive in order for the price of company *A* to be less than the price of company *B*?

43. Your scores on your first three algebra exams were 91, 92, and 85. What must you score on exam four so that the average of the four test scores is at least 90?

44. The heights of the two starting guards and two starting forwards on a basketball team are 6′2″, 5′ 11″, 6′7″. and 6′9″. How tall must the starting center be in order for the starting players to have an average height of at least 6′6″ (*Hint*: First convert all heights to inches.)

45. The perimeter of a rectangular fence is to be at least 100 feet and no more than 180 feet. If the width of the fence is 22 feet, what is the range of values for the length of the fence?

46. The perimeter of a rectangular fence is to be at least 120 feet and no more than 180 feet. If the length of the fence is to be twice the width, what is the range of values for the width of the fence?

47. Suppose Verizon Wireless offers a monthly plan for a smartphone that includes 5 gigabytes of data for $120. Each additional gigabyte (or fraction thereof) costs the user $10. If Logan subscribes to this plan, how many gigabytes of data can she use each month while keeping her total cost to no more than $165 (before taxes)?

48. As of October 31, 2016, all checked baggage on American Airlines must not exceed 62 inches or an additional fee is charged. The size of baggage is determined by adding length + width + height. If a bag has a height of 14 inches and a width that is 12 inches less than the length, what lengths are acceptable for the bag to be checked free of charge? (Source: www.aa.com)

49. Beverly is planning her wedding reception at the Missouri Botanical Gardens. The rental cost is $2000 plus $76 per person for catering. How many guests can she invite if her reception budget is $10,000? (Source: www.mobot.org)

50. An online retailer sells plush toys. He purchases the toys at the wholesale price of $3.15 each and sells them online for $14.75. If his fixed costs are $1060, how many plush toys must he sell in order to make a profit? That is, solve the inequality $R > C$, with R as his revenue and C as his cost.

51. Glider Truck Rentals will rent a 24-foot truck for a daily rate of $85 plus $0.10 per mile with 200 free miles. Y'All Rentals will rent a 24-foot truck for a daily rate of $40 plus $0.20 per mile. For what number of miles driven (in one day) will Glider Truck Rentals be cheaper?

52. Nancy sells handcrafted bracelets at a flea market for $7. If her monthly fixed costs are $675 and each bracelet costs her $2.75 to make, how many bracelets must she sell in a month to make a profit?

1.8 Absolute Value Equations and Inequalities

THINGS TO KNOW

Before working through this section, be sure that you are familiar with the following concepts:

VIDEO ANIMATION INTERACTIVE

You Try It

1. Solving Three-Part Inequalities in One Variable (Section 1.7) 

You Try It

2. Solving Compound Inequalities in One Variable (Section 1.7) 

INTRODUCTION

Read this introduction before beginning Objective 1.

OBJECTIVES

1 Solving an Absolute Value Equation

2 Solving Absolute Value Inequalities

SECTION 1.8 EXERCISES

Introduction to Section 1.8

In **Section R.1**, we see that the absolute value of a number a, written as $|a|$, represents the distance from a number a to zero on the number line. Consider the equation $|x| = 5$. To solve for x, we must find all values of x that are five units away from zero on the number line. The two numbers that are five units away from zero on the number line are $x = -5$ and $x = 5$, as shown in Figure 8.

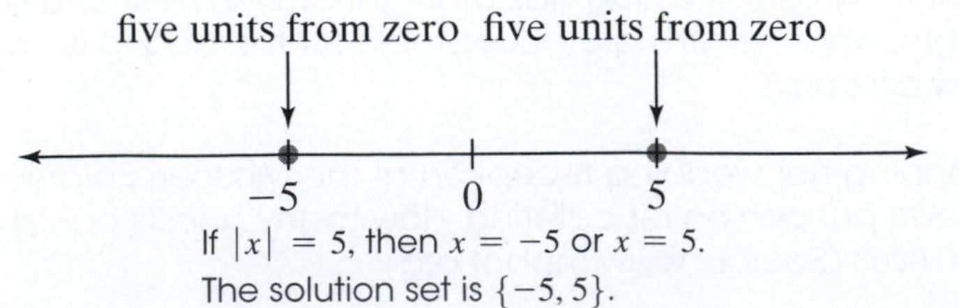

Figure 8 Solution to $|x| = 5$

The solution to the inequality $|x| < 5$ consists of all values of x whose distance from zero is less than five units on the number line. See Figure 9.

These values are all less than five units from zero.

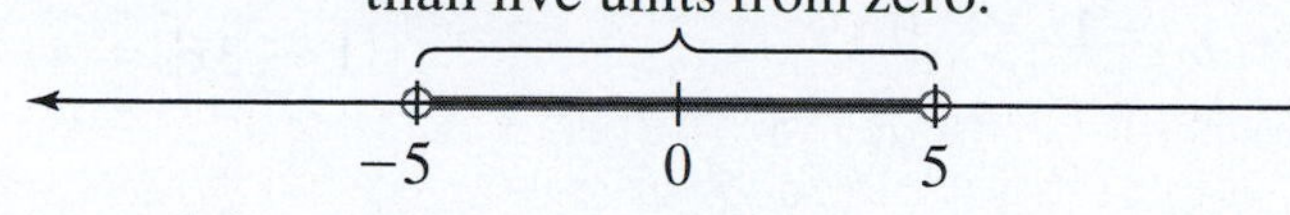

Figure 9 Solution to $|x| < 5$

If $|x| < 5$, then $-5 < x < 5$.
The solution set is $|x|-5 < x < 5\}$ in set-builder notation or $(-5, 5)$ in interval notation.

Figure 10 shows the solution to the inequality $|x| > 5$. Notice that we are now looking for all values of x that are more than five units away from zero. The solution is the set of all values of x greater than 5 combined with the set of all values of x less than -5.

These values are more than five units from zero. These values are more than five units from zero.

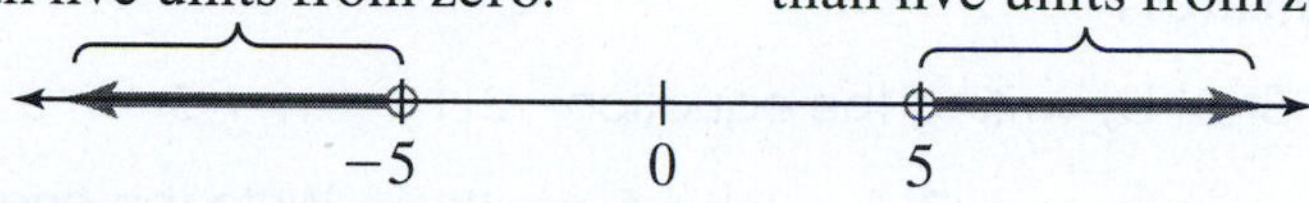

Figure 10 Solution to $|x| < 5$

If $|x| > 5$, then $x < -5$ or $x > 5$.
The solution set is $\{x|x < -5$ or $x > 5\}$ in set-builder notation or $(-\infty, -5) \cup (5, \infty)$ in interval notation.

Table 1 illustrates how to solve absolute value equations and inequalities. Each absolute value equation and inequality in Table 1 is in standard form. When solving an absolute value equation or inequality, it is necessary to first rewrite it in standard form.

Table 1 Absolute Value Equations and Inequality Properties

Let u be an algebraic expression and c be a real number such that $c > 0$, then
1. $|u| = c$ is equivalent to $u = -c$ or $u = c$.
2. $|u| < c$ is equivalent to $-c < u < c$.
3. $|u| > c$ is equivalent to $u < -c$ or $u > c$.

OBJECTIVE 1 SOLVING AN ABSOLUTE VALUE EQUATION

Example 1 Solve an Absolute Value Equation

Solve $|1 - 3x| = 4$.

Solution This equation is in standard form. Therefore, by property 1, the equation $|1 - 3x| = 4$ is equivalent to the following two equations:

$1 - 3x = 4$ or $1 - 3x = -4$ Use property 1.

$-3x = 3$ or $-3x = -5$ Subtract 1 from both sides.

$x = -1$ or $x = \frac{5}{3}$ Divide both sides by -3.

The solution set is $\left\{-1, \frac{5}{3}\right\}$. See Figure 11.

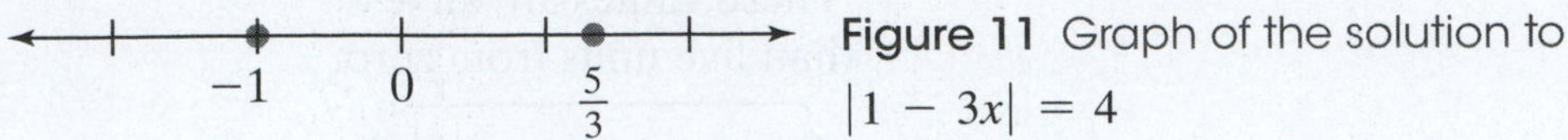

Figure 11 Graph of the solution to $|1 - 3x| = 4$

Example 2 Solving Absolute Value Equations

Solve each absolute value equation.

a. $-2|1 - 3x| + 5 = -9$ b. $\left|\frac{5x - 1}{x + 3}\right| = 11$ c. $|x^2 - x - 8| = 4$

d. $|\sqrt{2x + 9} - x| = 3$

Solution

a. Start by writing the equation $-2|1 - 3x| + 5 = -9$ in standard form.

$$-2|1 - 3x| + 5 = -9 \quad \text{Write the original equation.}$$
$$-2|1 - 3x| = -14 \quad \text{Subtract 5 from both sides.}$$
$$|1 - 3x| = 7 \quad \text{Divide both sides by } -2.$$

The absolute value equation is now in standard form. Therefore, by property 1, the equation $|1 - 3x| = 7$ is equivalent to the following two equations:

$$1 - 3x = 7 \quad \text{or} \quad 1 - 3x = -7$$

Solve these two equations on your own or work through this **interactive video** to verify that the solution set is $\left\{-2, \frac{8}{3}\right\}$.

b. The equation $\left|\frac{5x - 1}{x + 3}\right| = 11$ is in standard form. Therefore, by property 1, this absolute value equation is equivalent to the following two equations:

$$\frac{5x - 1}{x + 3} = 11 \quad \text{or} \quad \frac{5x - 1}{x + 3} = -11$$

These two equations are **rational equations**. So, we follow the **steps for solving rational equations** that were outlined in **Section 1.1**. Solve these two rational equations on your own or work through this **interactive video** to verify that the solution set is $\left\{-\frac{17}{3}, -2\right\}$.

c. The equation $|x^2 - x - 8| = 4$ is in standard form. Therefore, by property 1, this absolute value equation is equivalent to the following two equations:

$$x^2 - x - 8 = 4 \quad \text{or} \quad x^2 - x - 8 = -4$$

These two equations are **quadratic equations.** To solve these quadratic equations, start by getting 0 on the right hand side of each equation.

$x^2 - x - 8 = 4 \quad \text{or} \quad x^2 - x - 8 = -4$	Write the two quadratic equations.
$x^2 - x - 8 - 4 = 4 - 4 \quad \text{or} \quad x^2 - x - 8 + 4 = -4 + 4$	Subtract 4 from both sides of first equation and add 4 to both sides of second equation.
$x^2 - x - 12 = 0 \quad \text{or} \quad x^2 - x - 4 = 0$	Simplify.

Try solving these two quadratic equations on your own then work through this **interactive video** to verify that there are four solutions to the original absolute value equation. The solution set is $\left\{-3, 4, \frac{1+\sqrt{17}}{2}, \frac{1-\sqrt{17}}{2}\right\}$.

d. The equation $|\sqrt{2x+9} - x| = 3$ is in standard form. Therefore, by property 1, this absolute value equation is equivalent to the following equations.

$$\sqrt{2x+9} - x = 3 \quad \text{or} \quad \sqrt{2x+9} - x = -3$$

You may want to review how to solve equations involving radicals by referring back to **Section 1.6**. We start by solving the first equation, $\sqrt{2x+9} - x = 3$.

$\sqrt{2x+9} - x = 3$	Write the first equation.
$\sqrt{2x+9} = x + 3$	Add x to both sides.
$(\sqrt{2x+9})^2 = (x+3)^2$	Square both sides of the equation.
$2x + 9 = x^2 + 6x + 9$	Expand the right-hand side: $(x+3)^2 = (x+3)(x+3) = x^2 + 6x + 9$.
$0 = x^2 + 4x$	Move all terms to the right-hand side.
$0 = x(x+4)$	Factor.

Using the **zero product property**, we get $x = 0$ and -4. Recall that when solving equations involving radicals it is important to always check each potential solutions by substituting back into the original equation.

Check $x = 0$:	Check $x = -4$:
$\|\sqrt{2x+9} - x\| = 3$	$\|\sqrt{2x+9} - x\| = 3$
$\|\sqrt{2(0)+9} - (0)\| \stackrel{?}{=} 3$	$\|\sqrt{2(-4)+9} - (-4)\| \stackrel{?}{=} 3$
$\|\sqrt{0+9} - 0\| \stackrel{?}{=} 3$	$\|\sqrt{-8+9} + 4\| \stackrel{?}{=} 3$
$\|\sqrt{9} - 0\| \stackrel{?}{=} 3$	$\|\sqrt{1} + 4\| \stackrel{?}{=} 3$
$\|3 - 0\| \stackrel{?}{=} 3$	$\|1 + 4\| \stackrel{?}{=} 3$
$3 = 3$ ✓	$5 \neq 3$ ×

We see that $x = 0$ checks, but $x = -4$ does not. Thus, $x = -4$ is called an **extraneous solution**.

Try solving the equation $\sqrt{2x+9} - x = -3$ on your own. Work through this **interactive video** to verify that the solution to this radical equation is $x = 8$.

Thus, there are two unique solutions to the original absolute value equation $|\sqrt{2x+9} - x| = 3$. The solution set is $\{0, 8\}$.

You Try It Work through this You Try It problem.

Work Exercises 1–18 in this textbook or in the MyLab Math Study Plan.

OBJECTIVE 2 SOLVING ABSOLUTE VALUE INEQUALITIESS

When solving absolute value inequalities of the form $|u| < c$ where c is a positive constant, we are trying to determine all values of the algebraic expression u that are less than c units away from 0 on a number line.

In the **introduction** to this section we used the inequality $|x| < 5$ as a motivating example.

(You may want to watch this **video** that was introduced in the introduction to this section.)

The figure below illustrates that the solution to the inequality $|x| < 5$ consists of all values of x whose distance from zero is less than five units on the number line.

These values are all less than five units from zero.

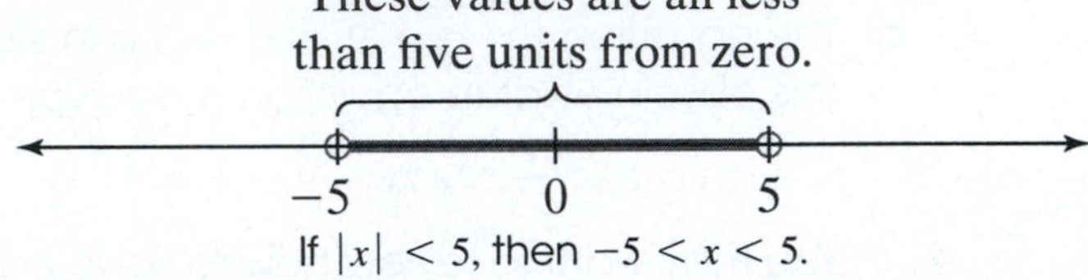

If $|x| < 5$, then $-5 < x < 5$.

When we encounter an inequality of the form $|u| < c$ where c is a positive constant, **property 2** states that the inequality is equivalent to the three-part inequality $-c < u < c$.

Example 3 Solve an Absolute Value *Less Than* Inequality

Solve $|4x - 3| + 2 \le 7$.

Solution This inequality is not in standard form. Rewrite this inequality in standard form and then use **property 2** to form a three-part inequality.

$\lvert 4x - 3\rvert + 2 \le 7$	Write the original inequality.
$\lvert 4x - 3\rvert \le 5$	Subtract 2 from both sides.
$-5 \le 4x - 3 \le 5$	Use **property 2** and rewrite as a three-part inequality.
$-2 \le 4x \le 8$	Add 3 to all three parts and simplify.
$-\frac{1}{2} \le x \le 2$	Divide all three parts by 4 and simplify.

The solution is $\left\{x \middle| -\frac{1}{2} \le x \le 2\right\}$ in set-builder notation or $\left[-\frac{1}{2}, 2\right]$ in interval notation. See Figure 12.

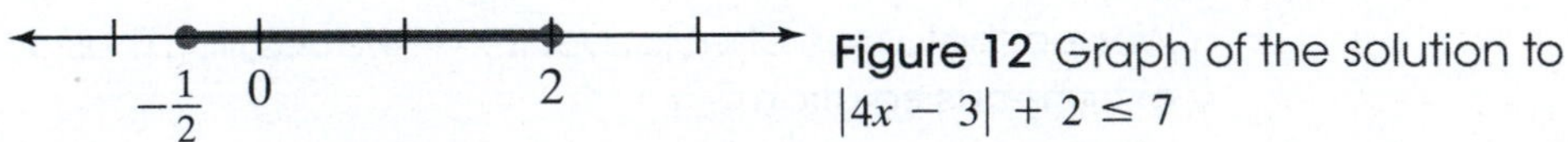

Figure 12 Graph of the solution to $|4x - 3| + 2 \le 7$

You Try It **Work through this You Try It problem.**

When solving absolute value inequalities of the form $|u| > c$ where c is a positive constant, we are trying to determine all values of the algebraic expression u that are more than c units away from 0 on a number line.

In the introduction to this section we used the inequality $|x| > 5$ as a motivating example.

(You may want to watch this video that was introduced in the introduction to this section.)

The figure below illustrates that the solution to the inequality $|x| > 5$ consists of all values of x whose distance from zero is more than five units on the number line.

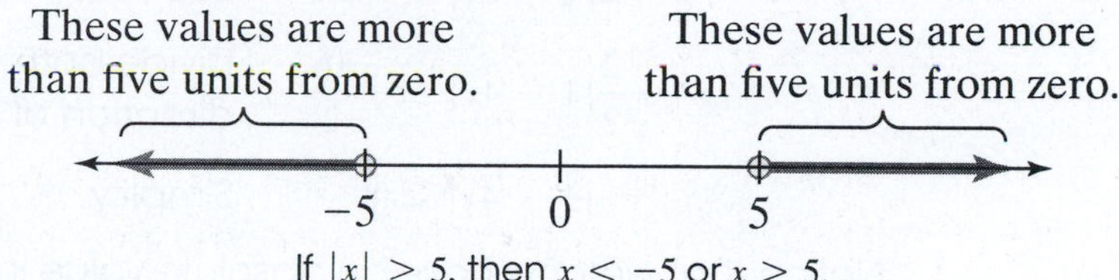
These values are more than five units from zero.
These values are more than five units from zero.
−5
0
5

If $|x| > 5$, then $x < -5$ or $x > 5$.

When we encounter an inequality of the form $|u| > c$ where c is a positive constant, **property 3** states that the inequality is equivalent to the compound inequality $u < -c$ or $u > c$.

Example 4 Solve an Absolute Value *Greater Than* Inequality

Solve $|5x + 1| > 3$.

Solution This inequality is in standard form. Therefore, we can use property 3 to rewrite the inequality as a compound inequality and solve.

$|5x + 1| > 3$	Write the original inequality.
$5x + 1 < -3 \quad \text{or} \quad 5x + 1 > 3$	Use **property 3** and rewrite as compound inequality.
$5x < -4 \quad \text{or} \quad 5x > 2$	Subtract 1 from both sides.
$x < -\dfrac{4}{5} \quad \text{or} \quad x > \dfrac{2}{5}$	Divide both sides by 5.

The solution is $\left\{x \,\middle|\, x < -\dfrac{4}{5} \text{ or } x > \dfrac{2}{5}\right\}$ in set-builder notation or $\left(-\infty, -\dfrac{4}{5}\right) \cup \left(\dfrac{2}{5}, \infty\right)$ in interval notation. See Figure 13.

$-\frac{4}{5}$ 0 $\frac{2}{5}$

Figure 13 Graph of the solution to $|5x + 1| > 3$

CAUTION In Example 4, $|5x + 1| > 3$ is *not* equivalent to $-3 > 5x + 1 > 3$. In addition, a common error in this type of problem is to write $5x + 1 > -3$ for the first inequality, instead of $5x + 1 < -3$. Think carefully about the meaning of the inequality before writing it.

You Try It Work through this You Try It problem.

Example 5 Solving an Absolute Value Inequality

Solve $7 - 2|1 - 4x| \le -9$.

Solution Start by writing the inequality $7 - 2|1 - 4x| \le -9$ in standard form.

$7 - 2|1 - 4x| \le -9$ Write the original inequality.

$-2|1 - 4x| \le -16$ Subtract 7 from both sides.

$\frac{-2}{-2}|1 - 4x| \ge \frac{-16}{-2}$ Divide both sides by -2 **and switch the direction of the inequality.**

$|1 - 4x| \ge 8$ Simplify.

Notice that we now have an absolute value inequality of the form $|u| \ge c$. Therefore, we use **property 3** to rewrite the inequality $|1 - 4x| \ge 8$ as the following compound inequality:

$$1 - 4x \le -8 \text{ or } 1 - 4x \ge 8$$

Now, solve the compound inequality.

$1 - 4x \le -8$ or $1 - 4x \ge 8$ Write the compound inequality.

$-4x \le -9$ or $-4x \ge 7$ Subtract 1 from both sides.

$\frac{-4x}{-4} \ge \frac{-9}{-4}$ or $\frac{-4x}{-4} \le \frac{7}{-4}$ Divide both sides by -4 **and switch the direction of the inequalities.**

$x \ge \frac{9}{4}$ or $x \le -\frac{7}{4}$

The solution to the inequality $7 - 2|1 - 4x| \le -9$ is $\left\{x \mid x \le -\frac{7}{4} \text{ or } x \ge \frac{9}{4}\right\}$ in set-builder notation or $\left(-\infty, -\frac{7}{4}\right] \cup \left[\frac{9}{4}, \infty\right)$ in interval notation. See Figure 14.

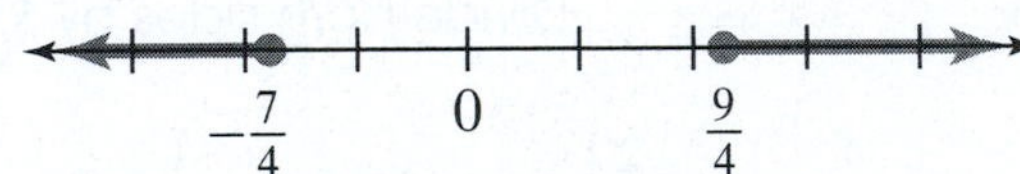

Figure 14 Graph of the solution to $7 - 2|1 - 4x| \le -9$

You Try It Work through this You Try It problem.

In every example so far in this section, we were able to write the absolute value equation or inequality into standard form. That is, we were able to write the equation or inequality in the form of $|u| = c$, $|u| < c$, or $|u| > c$ where the value of c on the right-hand side is a positive constant. Example 6 illustrates how to solve absolute value equations and inequalities when the right-hand s e is negative or zero.

Example 6 Absolute Value Equations and Inequalities Involving Zero or Negative Constants

Solve each of the following:

a. $|3x - 2| = 0$

b. $|x + 6| = -4$

c. $|7x + 5| \le 0$

d. $|3 - 4x| < -6$

e. $|8x - 3| > 0$

f. $|1 - 9x| \ge -5$

Solution

a. The expression $|3x - 2|$ will equal zero only when $3x - 2 = 0$ or when $x = \frac{2}{3}$.

The solution set is $\left\{\frac{2}{3}\right\}$.

b. The absolute value of an expression can never be less than zero. Therefore, there is no solution to this equation. The solution set is $\varnothing$.

c. The expression $|7x + 5|$ can never be less than zero. However, $|7x + 5|$ is equal to zero when $7x + 5 = 0$ or when $x = -\frac{5}{7}$. The solution set is $\left\{-\frac{5}{7}\right\}$.

d. The absolute value of an expression can never be less than zero. Therefore, there is no solution to this inequality. The solution set is $\varnothing$.

e. The expression $|8x - 3|$ is equal to zero when $x = \frac{3}{8}$ and is greater than zero for all other values of x. Therefore, the solution set in set-builder notation is $\left\{x \,\middle|\, x \neq \frac{3}{8}\right\}$ or $\left(-\infty, \frac{3}{8}\right) \cup \left(\frac{3}{8}, \infty\right)$.

f. The absolute value of an expression is always nonnegative. Therefore, $|1 - 9x| \geq -5$ is true for all values of x. The solution is all real numbers or $(-\infty, \infty)$.

For the complete solution to this entire example, work through this interactive video.

You Try It Work through this You Try It problem.

Work Exercises 19–36 in this textbook or in the MyLab Math Study Plan.

1.8 Exercises

Skill Check Exercises

In exercises SCE-1 through SCE-4, write the given set in interval notation and graph the set on a number line.

SCE-1. $\left\{x \,\middle|\, -\frac{5}{2} \leq x \leq 1\right\}$ **SCE-2.** $\{x|0 \leq x < 3\}$ **SCE-3.** $\left\{x \,\middle|\, x \geq \frac{3}{4}\right\}$ **SCE-4.** $\{x|x > -4\}$

SCE-5. If $A = \{x|x < 9\}$ and $B = \{x|x \geq 2\}$, then find the intersection of sets A and B. (Find $A \cap B$.)

SCE- 6. If $A = \{x|x \leq -3\}$ and $B = \{x|x \geq 5\}$, then find the union of sets A and B. (Find $A \cup B$.)

In exercises SCE-7 through SCE-9, solve each three-part inequality. Express each solution in interval notation.

SCE-7. $3 < 4x + 7 \le 15$ **SCE-8.** $-2 \le 2m - 5 \le 2$ **SCE-9.** $-2 \le \dfrac{5(x + 1) + 9}{3} < 3$

In exercises SCE-10 through SCE-12, solve each compound inequality. Express each solution in interval notation.

SCE-10. $4x + 2 \le -10$ or $3x - 4 > 8$

SCE-11. $2x - 3 \le 5$ or $3x - 3 > 3$

SCE-12. $5x - 15 \le -20 + 4x$ or $3 + 5x > 12 - 18$

In Exercises 1–18, solve each equation.

1. $|x| = 7$
2. $|x - 1| = 5$
3. $|x + 1| = 1$
4. $|10x + 9| = 5$
5. $|6x + 7| - 8 = 3$
6. $|8 - 3x| = 8$
7. $|7x + 3| = 0$
8. $|4x - 1| = -5$
9. $\left|\dfrac{3x + 15}{7}\right| = 0$
10. $\left|\dfrac{7x - 5}{6}\right| = \dfrac{4}{3}$
11. $|1 - 6x| = -11$
12. $|2x - 7| - 9 = -4$
13. $6|3 - 10x| = 24$
14. $3|-8x - 9| - 6 = 15$
15. $-2|3x + 2| + 7 = 5$
16. $|x^2 - 2x| = 1$
17. $\left|\dfrac{3x - 4}{x - 2}\right| = 5$
18. $|\sqrt{3x + 4} - x| = 2$

In Exercises 19–36, solve each inequality. Express each solution in interval notation.

19. $|9x| \le 3$
20. $|x - 2| < 5$
21. $|3 - 4x| < 11$
22. $|5x - 1| + 7 \le 9$
23. $|3x - 7| \le 0$
24. $|x - 1| < -2$
25. $\left|\dfrac{2x - 5}{3}\right| \le \dfrac{1}{5}$
26. $2|3 - 4x| + 1 < 4$
27. $|2x| > 6$
28. $|x + 4| > 5$
29. $|3 - 5x| \ge 2$
30. $|7x - 5| > 0$
31. $|3x + 1| \ge -5$
32. $|7x - 1| - 5 \ge 8$
33. $\left|\dfrac{2x - 6}{7}\right| \ge \dfrac{3}{14}$
34. $-|3x + 2| + 5 > -6$
35. $8 - |4 - 3x| \ge 5$
36. $-4|5x + 2| - 3 < 5$

1.9 Polynomial and Rational Inequalities

THINGS TO KNOW

Before working through this section, be sure that you are familiar with the following concepts:

VIDEO ANIMATION INTERACTIVE

 You Try It 1. Factoring Trinomials with a Leading Coefficient Equal to 1 (Section R.6)  

 You Try It 2. Factoring Trinomials with a Leading Coefficient Not Equal to 1 (Section R.6) 

 You Try It 3. Factoring Polynomials by Grouping (Section R.6) 

 You Try It 4. Solving Higher-Order Polynomial Equations (Section 1.6) 

OBJECTIVES

1 Solving Polynomial Inequalities

2 Solving Rational Inequalities

SECTION 1.9 EXERCISES

OBJECTIVE 1 SOLVING POLYNOMIAL INEQUALITIES

In Section 1.7, we learned how to solve linear inequalities. In this section we will learn how to solve **polynomial inequalities** and **rational inequalities**. We start by solving a polynomial inequality in Example 1.

Example 1 Solve a Polynomial Inequality

Solve $x^3 - 3x^2 + 2x \geq 0$.

Solution First, factor the left-hand side to get $x(x-1)(x-2) \geq 0$. Second, find all real values of x that make the expression on the left *equal to* 0. These values are called **boundary points**. To find the boundary points, set the factored polynomial on the left equal to 0 and solve for x.

$x(x-1)(x-2) = 0$ Set the factored polynomial equal to 0.

$x = 0 \quad \text{or} \quad x - 1 = 0 \quad \text{or} \quad x - 2 = 0$ Use the **zero product property**.

The boundary points are $x = 0, x = 1$, and $x = 2$.

Now plot each boundary point on a number line. Because the expression $x(x-1)(x-2)$ is equal to 0 at our three boundary points, we use a solid circle ● at each of the boundary points to indicate that the inequality $x(x-1)(x-2) \geq 0$ is satisfied at these points. (*Note:* If the inequality is a strict inequality such as $>$ or $<$, then we represent the fact that the boundary points are not part of the solution by using an open circle, ○. See Example 2.)

Notice in Figure 15 that we have naturally divided the number line into four intervals.

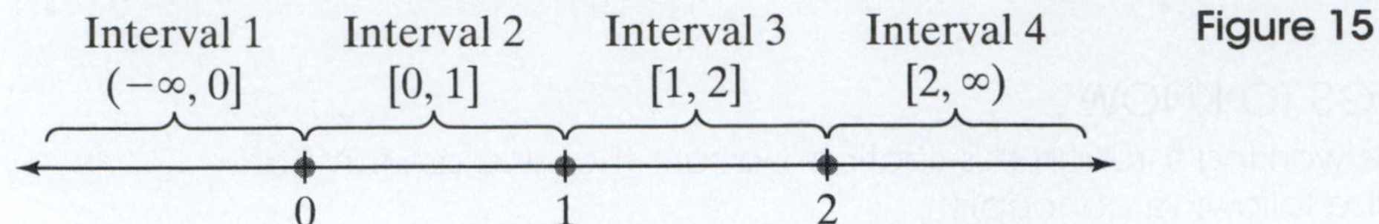

Figure 15

The expression $x(x-1)(x-2)$ is equal to zero *only* at the three boundary points 0, 1, and 2. This means that in any of the four intervals shown in Figure 15, the expression $x(x-1)(x-2)$ must be either *positive* or *negative* throughout the entire interval. To check whether this expression is positive or negative on each interval, pick a number from each interval called a **test value**. Substitute this test value into the expression $x(x-1)(x-2)$, and check to see if it yields a positive or negative value. Possible test values are plotted in red in Figure 16.

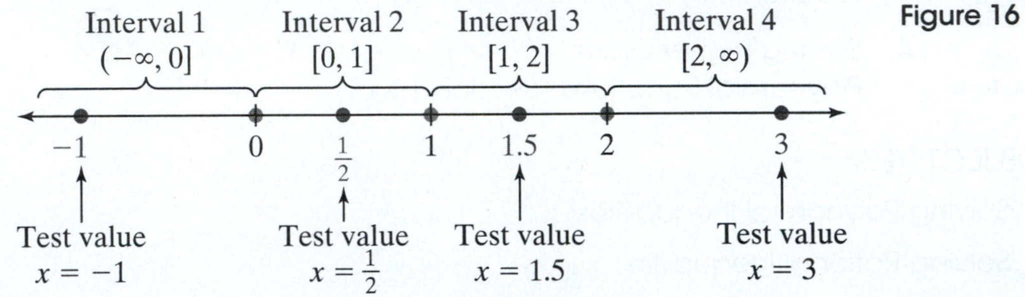

Figure 16

Interval	Test Value	Substitute Test Value into $x(x-1)(x-2)$	Comment
1. $(-\infty, 0]$	$x=-1$	$(-1)(-1-1)(-1-2) = (-1)(-2)(-3) \quad = -6 \quad \Rightarrow -$	Expression is negative on $(-\infty, 0]$
2. $[0, 1]$	$x=\frac{1}{2}$	$\left(\frac{1}{2}\right)\left(\frac{1}{2}-1\right)\left(\frac{1}{2}-2\right) = \left(\frac{1}{2}\right)\left(-\frac{1}{2}\right)\left(-\frac{3}{2}\right) = \frac{3}{8} \quad \Rightarrow +$	Expression is positive on $[0, 1]$.
3. $[1, 2]$	$x=1.5$	$(1.5)(1.5-1)(1.5-2) = (1.5)(0.5)(-0.5) \quad = -0.375 \Rightarrow -$	Expression is negative on $[1, 2]$.
4. $[2, \infty)$	$x=3$	$(3)(3-1)(3-2) \quad = (3)(2)(1) \quad = 6 \quad \Rightarrow +$	Expression is positive on $[2, \infty)$.

If the expression $x(x-1)(x-2)$ is positive on an interval, we place a "+" above the interval on the number line. If the expression $x(x-1)(x-2)$ is negative, we place a "−" above the interval. See Figure 17.

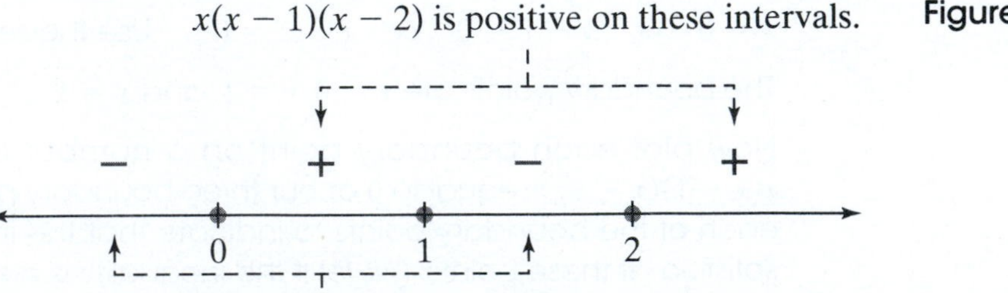

Figure 17

From the number line in **Figure 17**, we see that $x(x-1)(x-2)$ is greater than or equal to zero on the interval $[0, 1] \cup [2, \infty)$ which is precisely the solution to the original inequality $x^3 - 3x^2 + 2x \geq 0$.

 Watch this **video** to see the solution to this polynomial inequaltiy. ●

Example 1 illustrates the following steps for solving a polynomial inequality that can be factored.

Steps for Solving Polynomial Inequalities

Step 1. Move all terms to one side of the inequality leaving zero on the other side.

Step 2. Factor the nonzero side of the inequality.

Step 3. Find all boundary points by setting the factored polynomial equal to zero.

Step 4. Plot the boundary points on a number line. If the inequality is $\leq$ or $\geq$, then use a solid circle ●. If the inequality is $<$ or $>$, then use an open circle ○.

Step 5. Now that the number line is divided into intervals, pick a test value from each interval.

Step 6. Substitute the test value into the polynomial, and determine whether the expression is positive or negative on the interval.

Step 7. Determine the intervals that satisfy the inequality.

Example 2 Solve a Polynomial Inequality

Solve $x^2 + 5x < 3 - x^2$.

Solution

Step 1. Move all terms to one side of the inequality leaving zero on the other side.

$$2x^2 + 5x - 3 < 0$$

Step 2. Factor the nonzero side of the inequality.

$$(2x - 1)(x + 3) < 0$$

Step 3. Find all boundary points by setting the factored polynomial equal to zero.

The boundary points are $x = -3$ and $x = \frac{1}{2}$ because these are the values that are solutions to $(2x - 1)(x + 3) = 0$.

Step 4. Plot the boundary points on a number line.

Note that we use open circles on our boundary points because these points are *not* part of the solution. (We are looking for values that make the expression strictly less than zero.)

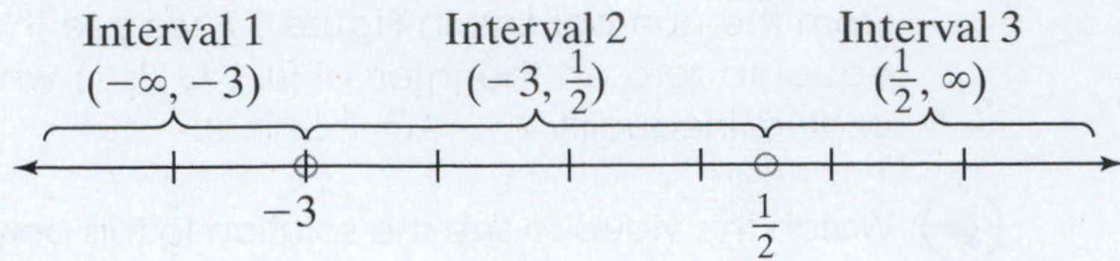

Step 5. Now that the number line is divided into three intervals, pick a test value from each interval.

Interval 1: Test value $x = -4$
Interval 2: Test value $x = 0$
Interval 3: Test value $x = 1$

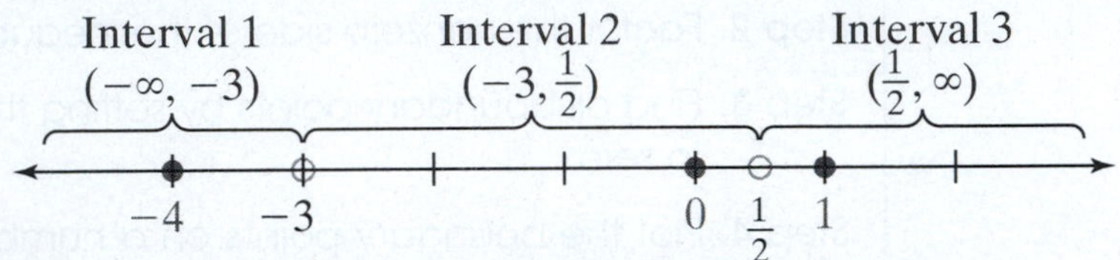

Step 6. Substitute the test value into the polynomial, and determine whether the expression is positive or negative on the interval.

$x = -4$: $(2(-4) - 1)((-4) + 3) = (-9)(-1) = 9 \Rightarrow +$ Expression is positive on $(-\infty, -3)$.

$x = 0$: $(2(0) - 1)((0) + 3) = (-1)(3) = -3 \Rightarrow -$ Expression is negative on $\left(-3, \frac{1}{2}\right)$.

$x = 1$: $(2(1) - 1)((1) + 3) = (1)(4) = 4 \Rightarrow +$ Expression is positive on $\left(\frac{1}{2}, \infty\right)$.

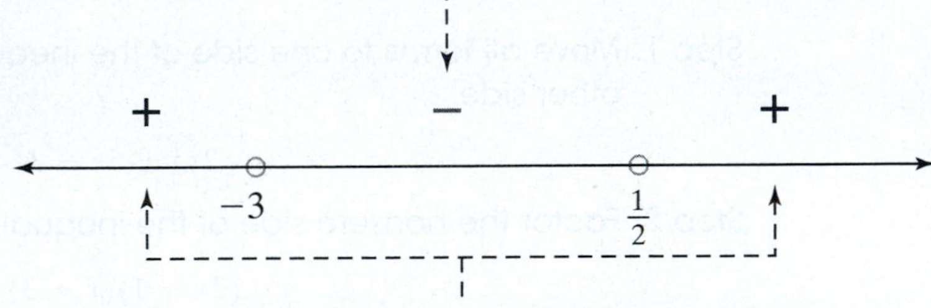

Step 7. Determine the intervals that satisfy the inequality.

Because we are looking for values of x that are less than zero (negative values), the solution must be the interval $\left(-3, \frac{1}{2}\right)$.

You Try It Work through this You Try It problem.

Work Exercise-s 1–14 in this textbook or in the MyLab Math Study Plan.

OBJECTIVE 2 SOLVING RATIONAL INEQUALITIES

A **rational inequality** can be solved using a technique similar to the one used in Examples 1 and 2, except that the boundary points are found by setting both the polynomial in the numerator and the denominator equal to zero. We will follow the seven steps outlined below.

Steps for Solving Rational Inequalities

Step 1. Move all terms to one side of the inequality leaving zero on the other side.

Step 2. Factor the numerator and denominator of the nonzero side of the inequality and cancel any common factors.

Step 3. Find all boundary points by setting the factored polynomials in the numerator and the denominator equal to zero.

Step 4. Plot the boundary points on a number line.
-For the boundary points found by setting the numerator equal to zero:
If the inequality is $\le$ or $\ge$, then use a solid circle ●.
If the inequality is $<$ or $>$, then use an open circle ○.
-Use an open circle ○ to represent all boundary points found by setting the denominator equal to zero regardless of the inequality symbol that is used.

Step 5. Now that the number line is divided into intervals, pick a test value from each interval.

Step 6. Substitute the test value into the polynomial, and determine whether the expression is positive or negative on the interval.

Step 7. Determine the intervals that satisfy the inequality.

 **Example 3 Solve a Rational Inequality**

Solve $\frac{x-4}{x+1} \ge 0$.

Solution Because the inequality is already in completely factored form, we can skip steps 1 and 2 and go right to step 3.

Step 3. Find all boundary points by setting the factored polynomial in the numerator and the denominator equal to zero.

Numerator: $x - 4 = 0$, so $x = 4$ is a boundary point.

Denominator: $x + 1 = 0$, so $x = -1$ is a boundary point.

Step 4. Plot the boundary points on a number line.

The inequality is a greater than *or equal to* inequality; thus, the boundary point $x = 4$ is represented by a closed circle. Because division by zero is *never* permitted, the boundary point, $x = -1$, that was found by setting the polynomial in the denominator equal to zero, must always be represented by an open circle on the number line.

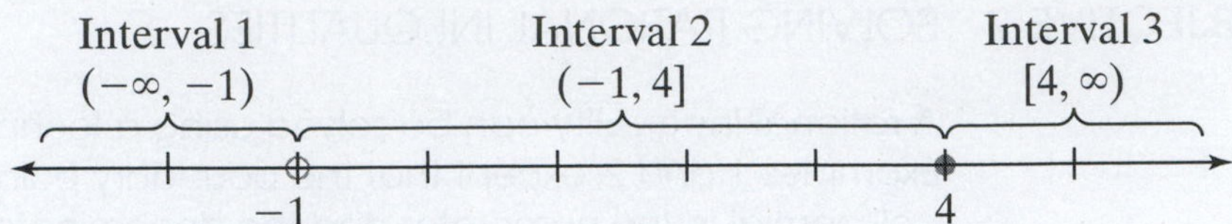

Step 5. Now that the number line is divided into three intervals, pick a test value from each interval.

Interval 1: Test value $x = -2$

Interval 2: Test value $x = 0$

Interval 3: Test value $x = 5$

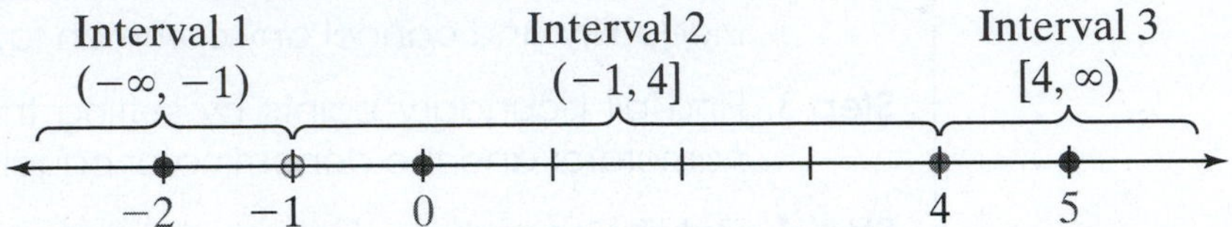

Step 6. Substitute each test value into the rational expression and determine whether the expression is positive or negative on the interval.

$x = -2$: $\dfrac{(-2-4)}{(-2+1))} = \dfrac{-6}{-1} = 6 \Rightarrow +$ Expression is positive on the interval $(-\infty, -1)$.

$x = 0$: $\dfrac{(0-4)}{(0+1)} = \dfrac{-4}{1} = -4 \Rightarrow -$ Expression is negative on the interval $(-1, 4]$.

$x = 5$: $\dfrac{(5-4)}{(5+1)} = \dfrac{1}{6} \Rightarrow +$ Expression is positive on the interval $[4, \infty)$.

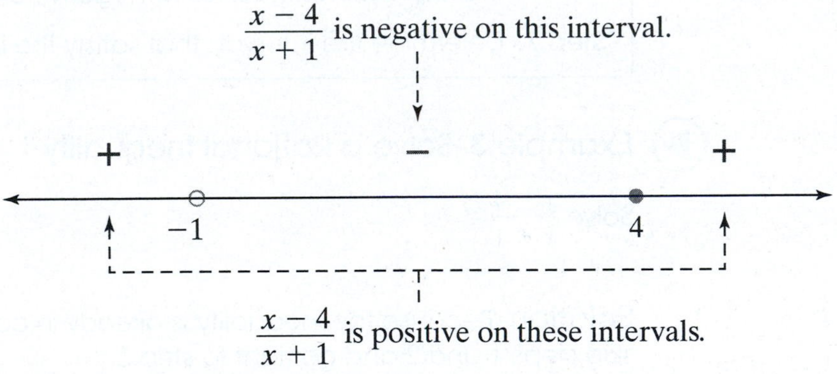

Step 7. Determine the intervals that satisfy the inequality.

Finally, because we are looking for values of x for which the rational expression is greater than or equal to zero, looking at the previous number line we see that the solution to the inequality is $(-\infty, -1) \cup [4, \infty)$.

Note that we include 4 as a solution but not -1 because $x = -1$ makes the denominator equal to 0.

Example 4 Solve a Rational Inequality with Nonzero Factors on Both Sides of the Inequality

Solve $x > \dfrac{3}{x-2}$

Solution

Step 1. Move all terms to one side of the inequality, leaving zero on the other side.

$$x - \frac{3}{x-2} > 0$$

CAUTION You cannot multiply both sides of the inequality by $x - 2$ to eliminate the fraction. This is because we do not know whether $x - 2$ is negative or positive; therefore, we do not know whether we would need to switch the direction of the inequality.

Now that all nonzero terms are on the left-hand side, we must combine the terms by getting a common denominator:

$$x - \frac{3}{x-2} > 0$$ Rewrite the inequality.

$$\frac{x(x-2)}{x-2} - \frac{3}{x-2} > 0$$ Rewrite using a common denominator of $x - 2$.

$$\frac{x^2 - 2x - 3}{x-2} > 0$$ Combine terms.

Step 2. Factor the numerator and denominator of the nonzero side of the inequality and cancel any common factors.

$$\frac{(x-3)(x+1)}{x-2} > 0$$

Finish steps 3–7 on your own, and see if you can come up with the correct solution of $(-1, 2) \cup (3, \infty)$. Watch this video to see the entire solution.

You Try It Work through this You Try It problem.

Work Exercises 15–30 in this textbook or in the MyLab Math Study Plan.

1.9 Exercises

Skill Check Exercises

For exercises SCE-1 through SCE-9, completely factor each polynomial.

SCE-1. $x^2 + 23x + 60$

SCE-2. $y^2 + 5y - 14$

SCE-3. $w^2 - 20w + 96$

SCE-4. $3x^2 + 4x - 15$

SCE-5. $12x^2 + 25x + 7$

SCE-6. $6y^2 - 19y + 10$

SCE-7. $x^2 - 121$

SCE-8. $x^3 + 8x^2 + 4x + 32$

SCE-9. $x^3 - x^2 - 3x + 3$

For exercises SCE-10 through SCE-15, perform the indicated operations and simplify.

SCE-10. $\frac{8}{x-1} - \frac{6}{x+8}$

SCE-11. $\frac{x}{x+3} + \frac{2x-7}{x-3}$

SCE-12. $\dfrac{z+5}{z-9}-\dfrac{z-4}{z+6}$

SCE-13. $\dfrac{x-8}{x^2+12x+27}+\dfrac{x-9}{x^2-9}$

SCE-14. $\dfrac{6x}{x^2+4x-12}-\dfrac{x}{x^2-36}$

SCE-15. $\dfrac{x-3}{x^2+12x+36}+\dfrac{1}{x+6}-\dfrac{2x+3}{2x^2+3x-54}$

In Exercises 1–14, determine all boundary points and solve the polynomial inequality. Express each solution in interval notation.

SbS 1. $(x-1)(x+3)\geq 0$

SbS 2. $x(3x+2)\leq 0$

SbS 3. $(x-1)(x+4)(x-3)\geq 0$

SbS 4. $-3x^2(x-5)<0$

SbS 5. $(x-1)^2(x+2)\leq 0$

SbS 6. $2x^2-4x>0$

SbS 7. $x^2+3x-21<0$

SbS 8. $x^2-8x\geq -15$

SbS 9. $x^2\leq 1$

SbS 10. $3x^2+x<3x-1$

SbS 11. $2x^3>24x-2x^2$

SbS 12. $x^3+x^2-x\leq 1$

SbS 13. $x^3\geq -2x^2-x$

SbS 14. $15x^3>3x^4+12x^2$

In Exercises 15–30, determine all boundary points and solve the rational inequality. Express each solution using interval notation.

SbS 15 $\dfrac{x+3}{x-1}\leq 0$

SbS 16. $\dfrac{x+2}{x-5}>0$

SbS 17. $\dfrac{2-x}{3x+9}\geq 0$

SbS 18 $\dfrac{x}{x-1}>0$

SbS 19. $\dfrac{x^2-9}{x+2}\leq 0$

SbS 20 $\dfrac{(x-7)^2}{x+3}>0$

SbS 21. $\dfrac{x-2}{x^2+5x-24}\leq 0$

SbS 22. $\dfrac{x+2}{x^2-x-6}\leq 0$

SbS 23. $\dfrac{x^2-7x-18}{x^2-6x+9}<0$

SbS 24 $\dfrac{x^2-10x+25}{x^2-4x-12}\geq 0$

SbS 25. $\dfrac{4}{x+1}\geq 2$

SbS 26. $\dfrac{x+5}{2x-3}>1$

SbS 27. $\dfrac{x^2-8}{x+4}\geq x$

SbS 28. $\dfrac{x}{2-x}\leq\dfrac{1}{x}$

SbS 29. $\dfrac{x-1}{x-2}\geq\dfrac{x+2}{x+3}$

SbS 30. $\dfrac{x-1}{x+1}+\dfrac{x+1}{x-1}\leq\dfrac{x+5}{x^2-1}$

Brief Exercises

In Exercises 31–44, solve each polynomial inequality. Express each solution in interval notation.

31. $(x-1)(x+3)\geq 0$

32. $x(3x+2)\leq 0$

33. $(x-1)(x+4)(x-3)\geq 0$

34. $-3x^2(x-5)<0$

35. $(x-1)^2(x+2)\leq 0$

36. $2x^2-4x>0$

37. $x^2+3x-21<0$

38. $x^2-8x\geq -15$

39. $x^2\leq 1$

40. $3x^2+x<3x-1$

41. $2x^3>24x-2x^2$

42. $x^3+x^2-x\leq 1$

43. $x^3\geq -2x^2-x$

44. $15x^3>3x^4+12x^2$

In Exercises 45–60, solve each rational inequality. Express each solution using interval notation.

45. $\frac{x+3}{x-1} \le 0$

46. $\frac{x+2}{x-5} > 0$

47. $\frac{2-x}{3x+9} \ge 0$

48. $\frac{x}{x-1} > 0$

49. $\frac{x^2-9}{x+2} \le 0$

50. $\frac{(x-7)^2}{x+3} > 0$

51. $\frac{x-2}{x^2+5x-24} \le 0$

52. $\frac{x+2}{x^2-x-6} \le 0$

53. $\frac{x^2-7x-18}{x^2-6x+9} < 0$

54. $\frac{x^2-10x+25}{x^2-4x-12} \ge 0$

55. $\frac{4}{x+1} \ge 2$

56. $\frac{x+5}{2x-3} > 1$

57. $\frac{x^2-8}{x+4} \ge x$

58. $\frac{x}{2-x} \le \frac{1}{x}$

59. $\frac{x-1}{x-2} \ge \frac{x+2}{x+3}$

60. $\frac{x-1}{x+1} + \frac{x+1}{x-1} \le \frac{x+5}{x^2-1}$

Chapter 1 Summary

Key Concepts	Examples/Videos
1.1 Linear and Rational Equations A linear equation in one variable is an equation that can be written in the form $ax + b = 0$ where a and b are real numbers and $a \neq 0$. To solve a linear equation involving fractions, first multiply both sides of the equation by the **least common denominator**. To solve a linear equation involving decimals, first multiply both sides of the equation by the appropriate power of 10.	 Which of the following is not a linear equation? a. $\sqrt{3}x - \frac{1}{4} = \pi x + 1$ b. $3\sqrt{x} - 5 = 6$ c. $\frac{2}{3}x = 0$  Solve: $\frac{1}{3}(1-x) - \frac{x+1}{2} = -2$  Solve: $1(y-2) + 0.03(y-4) = 0.02(10)$
A **rational equation** is an equation consisting of one or more **rational expressions** with any other expressions of the equation being **polynomials**.	 Which of the following are rational equations? a. $\frac{x^2-x-12}{x+5} = 1$ b. $x^2 - 2x - 24 = \frac{1}{2}$ c. $\frac{12}{x^2+x-2} - \frac{x-3}{x-1} = \frac{1-x}{x+2}$

Key Concepts	Examples/Videos
Steps for Solving Rational Equations **Step 1** Factor any denominators then list all restricted values. **Step 2** Determine the LCD of all denominators in the equation. **Step 3** Multiply both sides of the equation by the LCD. **Step 4** Solve the resulting equation. **Step 5** Discard any restricted values.	Solve $\frac{2}{x+4} + \frac{1}{x-5} = \frac{5}{x^2 - x - 20}$
1.2 Applications of Linear and Rational Equations	Determine Unknown Numbers Determine the Dimensions of a Basketball Court Money Application Mixture Application Application of Uniform Motion ($d = r \cdot t$) Rate of Work Application
1.3 Complex Numbers The **imaginary unit**, i, is defined by $i = \sqrt{-1}$, or equivalently, $i^2 = 1$. A **complex number** is a number that can be written in the form $a + bi$, where a and b are real numbers. The **complex conjugate** of a complex number $a + bi$ is $a - bi$. To divide complex numbers, multiply the numerator and denominator by the complex conjugate of the denominator.	Perform the indicated operations. **a.** $(7 - 5i) + (-2 + i)$ **b.** $(7 - 5i) - (-2 + i)$ Multiply: $(4 - 3i)(7 + 5i)$ Write the quotient in the form $a + bi$: $\frac{1 - 3i}{5 - 2i}$
1.4 Quadratic Equations A **quadratic equation in one variable** is an equation that can be written in the form $ax^2 + bx + c = 0$, $a \neq 0$. **Zero Product Property** If $AB = 0$, then $A = 0$ or $B = 0$.	Use factoring and the zero product property to solve. $6x^2 - 17x = -12$

Key Concepts	Examples/Videos
Square Root Property The solution to the quadratic equation $x^2 - c = 0$, or equivalently $x^2 = c$, is $x = \pm\sqrt{c}$.	Solve each equation using the square root property. a. $x^2 - 16 = 0$ b. $2x^2 + 72 = 0$ c. $(x - 1)^2 = 7$
Solving $ax^2 + bx + c = 0, a \neq 0$ by Completing the Square 1. If $a \neq 1$, divide both sides of the equation by a. 2. Move all constants to the right-hand side. 3. Add $\left(\frac{1}{2}b\right)^2$ to both sides. 4. Factor the left-hand side as a binomial squared. 5. Use the square root property to solve for x.	Solve $3x^2 + 2x - 2 = 0$ by completing the square.
Quadratic Formula The solution to the quadratic equation $ax^2 + bx + c = 0, a \neq 0$, is given by the formula $x = \dfrac{-b \pm \sqrt{b^2 - 4ac}}{2a}$	Solve $3x^2 + 2x - 2 = 0$ using the quadratic formula.
Given a quadratic equation $ax^2 + bx + c = 0$, $a \neq 0$, the expression $D = b^2 - 4ac$ is called the **discriminant.** If $D > 0$, then the quadratic equation has two real solutions. If $D < 0$, then the quadratic equation has two non-real solutions. If $D = 0$, then the quadratic equation has exactly one real solution.	Use the discriminant to determine the number and nature of the solutions to each of the following quadratic equations. a. $3x^2 + 2x + 2 = 0$ b. $4x^2 + 1 = 4x$
1.5 Applications of Quadratic Equations	Solve an Application Involving Unknown Quantities Use the Projectile Motion Model. Solve a Geometric Application Solve an Application Involving Uniform Motion $(d = r \cdot t)$ Solve an Application Involving Rates of Work.

Key Concepts	Examples/Videos
1.6 Quadratic Equations If a polynomial equation of degree 3 or higher has a common factor among each term, then try factoring and using the zero product property.	Find all solutions of the equation $3x^3 - 2x = -5x^2$.
If a polynomial equation has an even number of terms, try factoring by grouping.	Find all solutions of the equation $2x^3 - x^2 + 8x - 4 = 0$.
To solve equations that are quadratic in form (disguised quadratics) first make an appropriate substitution to rewrite the equation in the form $au^2 + bu + c = 0$. For example, the equation $2(x - 5)^2 - 7(x - 5) + 3 = 0$ is quadratic in form. Let $u = x - 5$. Then, rewrite the equation as $2u^2 - 7u + 3 = 0$ to solve for u.	Solve each equation that is quadratic in form. **a.** $2x^4 - 11x + 12 = 0$ **b.** $\left(\frac{1}{x - 2}\right)^2 + \frac{2}{x - 2} - 15 = 0$ **c.** $x^{2/3} - 9x^{1/3} + 8 = 0$ **d.** $3x^{-2} - 5x^{-1} - 2 = 0$
When solving radical equations, first isolate the radical. Then raise each side of the equation to the appropriate power to eliminate the radical. If the radical is an even root, make sure to check for extraneous solutions.	Solve $\sqrt{2x + 3} + \sqrt{x - 2} = 4$
1.7 Linear Inequalities **Properties** Let a, b, and c be real numbers: **1.** If $a < b$, then $a + c < b + c$ **2.** If $a < b$, then $a - c < b - c$ **3.** For $c > 0$, if $a < b$, then $ac < bc$ **4.** For $c > 0$, if $a < b$, then $ac > bc$ (Multiplying both sides of an inequality by a negative number switches the direction of the inequality.) **5.** For $c > 0$, if $a < b$, then $\frac{a}{c} < \frac{b}{c}$ **6.** For $c > 0$, if $a < b$, then $\frac{a}{c} > \frac{b}{c}$ (Dividing both sides of an inequality by a negative number switches the direction of the inequality.)	Solve the inequality $-9x - 3 \geq 7 - 4x$. Solve the inequalities: **a.** $\frac{1 - 4w}{5} - \frac{w}{2} \leq -5$ **b.** $3.7y - 6 > 6.1 + 4.35$ Solve the three-part inequality $-2 \leq \frac{2 - 4x}{3} < 3$.

Key Concepts	Examples/Videos
Guidelines for Solving Compound Inequalities **Step 1** Solve each inequality separately. **Step 2** Graph each solution set on a number line. **Step 3** For compound inequalities using *and*, the solution set is the **intersection** of the individual solution sets. For compound inequalities using or, the solution set is the union of the individual solution sets.	Solve: $2x - 7 \le -1$ *and* $3x + 5 \ge 3$ Solve: $1 - 3x \ge 7$ *or* $3x + 4 > 7$
1.8 Absolute Value Equations and Inequalities The absolute value of a number a, written as $\|a\|$, represents the distance from a number a to zero on a number line	
Let u be an algebraic expression and c be a real number such that $c > 0$, then 1. $\|u\| = c$ Is equivalent to $u = c$ or $u = -c$. 2. $\|u\| < c$ Is equivalent to $-c < u < c$. 3. $\|u\| > c$ Is equivalent to $u < -c$ or $u > c$.	Solve each absolute value equation. a. $-2\|1 - 3x\| + 5 = -9$ b. $\left\|\frac{5x - 1}{x + 3}\right\| = 11$ c. $\|x^2 - x - 8\| = 4$ d. $\|\sqrt{2x + 9} - x\| = 3$ Solve $\|4x - 3\| + 2 \le 7$. Solve $\|5x + 1\| > 3$.
1.9 Polynomial and Rational Inequalities **Steps for Solving Polynomial Inequalities** **Step 1** Move all terms to one side of the inequality leaving zero on the other side. **Step 2** Factor the nonzero side of the inequality. **Step 3** Find all boundary points by setting the factored polynomial equal to zero. **Step 4** Plot the boundary points on a number line. If the inequality is $\le$ or $\ge$, then use a solid circle ●. If the inequality is $<$ or $>$, then use an open circle ○. **Step 5** Now that the number line is divided into intervals, pick a test value from each interval.	Solve $x^3 - 3x^2 + 2x \ge 0$.

Key Concepts	Examples/Videos
Step 6 Substitute the test value into the polynomial, and determine whether the expression is positive or negative on the interval. **Step 7** Determine the intervals that satisfy the inequality.	
Steps for Solving Rational Inequalities **Step 1** Move all terms to one side of the inequality leaving zero on the other side. **Step 2** Factor the numerator and denominator of the nonzero side of the inequality and cancel any common factors. **Step 3** Find all boundary points by setting the factored polynomials in the numerator and the denominator equal to zero. **Step 4** Plot the boundary points on a number line. For the boundary points found by setting the numerator equal to zero: If the inequality is ≤ or ≥, then use a solid circle ●. If the inequality is < or >, then use an open circle ○. -Use an open circle ○ to represent all boundary points found by setting the denominator equal to zero regardless of the inequality symbol that is used. **Step 5** Now that the number line is divided into intervals, pick a test value from each interval. **Step 6** Substitute the test value into the rational expression, and determine whether the expression is positive or negative on the interval. **Step 7** Determine the intervals that satisfy the inequality.	Solve $\frac{x-4}{x+1} \geq 0$. Solve $x > \frac{3}{x-2}$.

Chapter 1 Review Exercises

1. Determine whether the equation $\sqrt{2}x - 1 = 0$ is linear or nonlinear.

2. Solve: $-3(2 - x) + 1 = 4 - (7x - 2)$

3. Solve: $\frac{x - 4}{2} - \frac{x + 1}{4} = \frac{2x - 3}{4}$

4. Solve: $0.002(1 - k) + 0.01(k - 3) = 1$

5. Determine whether the equation $\frac{2x + 5}{8} = x$ is rational or non-rational.

6. Solve: $\frac{6}{x^2 - x} - \frac{2}{x} = \frac{3}{x - 1}$

7. Together, Steve and Tom sold 121 raffle tickets for their school. Steve sold 1 more than twice as many raffle tickets as Tom. How many raffle tickets did each boy sell?

8. The perimeter of a rectangular garden is 32 feet. The length of the garden is 4 feet less than three times the width. Find the dimensions of the garden.

9. How much of an 80% orange juice drink must be mixed with 20 gallons of a 20% orange juice drink to obtain a mixture that is 50% orange juice?

10. By himself, Chris can mow his lawn in 60 minutes. If his daughter Claudia helps, they can mow the lawn together in 40 minutes. How long would it take Claudia to mow the lawn by herself?

11. Find the difference and write your answer in the form $a + bi$. $(3 - 2i) - (-7 + 9i)$

12. Multiply and write your answer in the form $a + bi$. $(-2 - i)(3 - 4i)$

13. Find the quotient and write your answer in the form $a + bi$. $\frac{2 - i}{3 + 4i}$

14. Solve the equation $3x^2 + 8x - 3 = 0$ by factoring.

15. Solve the equation $x^2 + 6x - 16 = 0$ by completing the square.

16. Solve the equation $5x^2 - 3x = -1$ using the quadratic formula.

17. Benjamin threw a rock straight up from a cliff that was 24 ft above the water. If the height of the rock *h*, in feet, after *t* seconds is given by the equation $h = -16t^2 + 20t + 24$, how long will it take for the rock to hit the water?

18. The length of a rectangle is 1 in. less than twice the width. If the diagonal is 2 in. more than the length, find the dimensions of the rectangle.

19. Imogene's car traveled 280 miles averaging a certain speed. If the car had gone 5 mph faster, the trip would have taken 1 hour less. Find the average speed.

20. Twin brothers, Billy and Bobby, can mow their grandparent's lawn together in 56 minutes. Bill could mow the lawn by himself in 15 minutes less time than it would take Bobby. How long would it take Bobby to mow the lawn by himself?

21. Solve: $x^3 - x^2 + 3x = 3$

22. Solve the equation $3\left(\frac{1}{x-1}\right)^2 - \frac{5}{x-1} - 2 = 0$ after making an appropriate substitution.

23. Solve: $x - 1 = \sqrt{2x + 1}$

24. Solve: $\sqrt{5y + 1} - \sqrt{y + 1} = 2$

For Exercises 25–30, solve each inequality. Express the solution using interval notation.

25. $3(t - 1) - 5t \geq t + 6$

26. $\frac{12 - 7w}{5} - \frac{w}{5} < -8$

27. $3 < 4x + 7 \leq 15$

28. $3x - 1 \leq 1$ *and* $1 - x < -3$

29. $3x + 1 > 7$ *or* $3 - 4x < 15$

30. $|10x + 9| = 5$

For Exercises 31–35, solve each inequality. Express the solution using interval notation.

31. $|x + 4| > 5$

32. $8 - |4 - 3x| \geq 5$

33. $x^2 + 3x - 21 < 0$

34. $\frac{x + 3}{x - 1} \leq 0$

35. $\frac{x^2 - 8}{x + 4} \geq x$

CHAPTER TWO

The Rectangular Coordinate System, Lines, and Circles

CHAPTER TWO CONTENTS

2.1 The Rectangular Coordinate System

THINGS TO KNOW

Before working through this section, be sure that you are familiar with the following concepts:

VIDEO  ANIMATION INTERACTIVE

 You Try It

1. Simplify Radical Expressions Using the Product Rule (Section R.3)   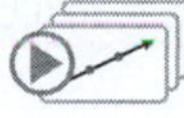

 You Try It

2. Solving Rational Equations that Lead to Linear Equations (Section 1.1) 

You Try It

3. Solving Quadratic Equations Using the Square Root Property (Section 1.4) 

 You Try It

4. Solving Equations Involving Radicals (Section 1.6) 

INTRODUCTION

Read this introduction before beginning Objective 1.

OBJECTIVES

1 Plotting Ordered Pairs

2 Graphing Equations by Plotting Points

3 Finding Intercepts of a Graph Given an Equation

4 Finding the Midpoint of a Line Segment Using the Midpoint Formula

5 Finding the Distance between Two Points Using the Distance Formula

SECTION 2.1 EXERCISES

Introduction to Section 2.1

Throughout chapter one, we solved several types of equations, including **linear equations, quadratic equations,** and **rational equations**. Each equation had something in common. They were all examples of **equations in one variable**. In this chapter, we study **equations involving two variables**. To motivate the discussion, consider the two-variable equation $x - 2y = -8$. What does it mean for this equation to have a solution? A solution to this equation consists of a *pair of numbers,* that is, an x-value and a y-value for which the equation is true.

For example, the pair of numbers $x = 2$ and $y = 5$ is a solution to this equation because substituting these two values into the equation produces a true statement:

$$x - 2y = -8 \quad \text{Write the two-variable equation.}$$
$$(2) - 2(5) \stackrel{?}{=} -8 \quad \text{Substitute } x = 2 \text{ and } y = 5.$$
$$-8 = -8 \checkmark \quad \text{The statement is true.}$$

The pair of numbers, $x = -2$ and $y = 3$ is also the solution to this equation:

$$x - 2y = -8 \quad \text{Write the two-variable equation.}$$
$$(-2) - 2(3) \stackrel{?}{=} -8 \quad \text{Substitute } x = -2 \text{ and } y = 3.$$
$$-8 = -8 \checkmark \quad \text{The statement is true.}$$

In fact, there are infinitely many pairs of numbers that satisfy this equation. Table 1 illustrates a few solutions to this equation.

Table 1

x	y	$x - 2y = -8$
−2	3	$-2 - 2(3) = -8$
0	4	$0 - 2(4) = -8$
2	5	$2 - 2(5) = -8$
4	6	$4 - 2(6) = -8$

Each pair of values is called an **ordered pair** because you can see that the order does matter. We use the notation (x, y) to represent an ordered pair. Notice that the x-coordinate or **abscissa** is first, followed by the y-coordinate or **ordinate** listed second. To represent an ordered pair graphically, we use the rectangular coordinate system, also called the Cartesian coordinate system, named after the French mathematician **René Descartes**. The plane used in this system is called the **coordinate plane** or **Cartesian plane**. The horizontal axis (x-axis) and the vertical axis (y-axis) intersect at the **origin** and naturally divide the plane into four quadrants, labeled quadrants I, II, III, and IV. See **Figure 1**.

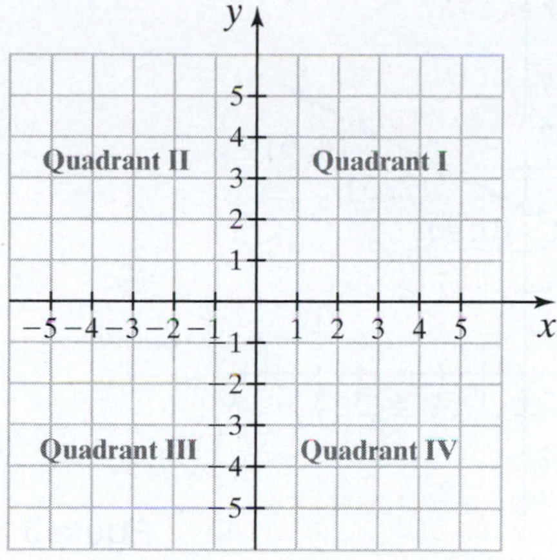

Figure 1 Rectangular coordinate system

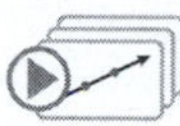

Watch this **animation** for a quick overview of the rectangular coordinate system and ordered pairs.

OBJECTIVE 1 PLOTTING ORDERED PAIRS

Example 1 Plot Ordered Pairs

Plot the ordered pairs $(-2, 3)$, $(0, 4)$, $(2, 5)$, and $(4, 6)$, and state in which quadrant or on which axis each ordered pair lies.

Solution To plot the point $(-2, 3)$, go two units to the left of the origin on the x-axis, and then move three units up, parallel to the y-axis. The point corresponding to the ordered pair $(-2, 3)$ is labeled A in Figure 2 and is located in Quadrant II. To plot the point $(0, 4)$, start at the origin and move up four units. The point corresponding to the ordered pair $(0, 4)$ is labeled B in Figure 2 and lies on the y-axis. The points $(2, 5)$ and $(4, 6)$ are labeled C and D and are both located in Quadrant I.

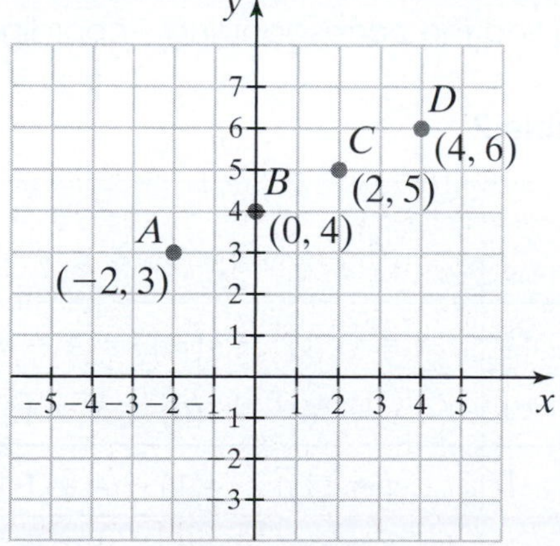

Figure 2

The four ordered pairs plotted in **Figure 2** were all solutions to the equation $x - 2y = -8$. Connecting these points forms the **graph of the equation** $x - 2y = -8$. See Figure 3. The graph of the equation $x - 2y = -8$ is a straight line. We study the equations of lines extensively in **Section 2.3**.

(eText Screens 2.1-1–2.1-31)

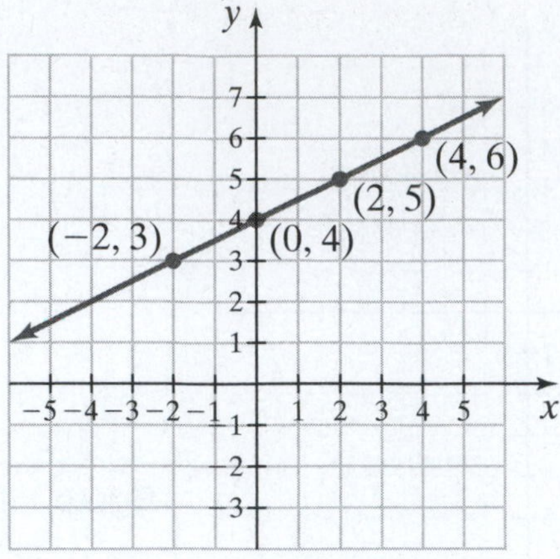

Figure 3 Graph of $x - 2y = -8$. ●

You Try It Work through this You Try It problem.

Work Exercises 1 and 2 in this textbook or in the MyLab Math Study Plan.

OBJECTIVE 2 GRAPHING EQUATIONS BY PLOTTING POINTS

Not every equation in two variables has a graph that is a straight line. One way to sketch the graph of an equation in two variables is to find several ordered pairs that satisfy the equation, plot those ordered pairs, and then connect the points with a smooth curve. We choose arbitrary values for one of the coordinates and then solve the equation for the other coordinate.

Example 2 Graph an Equation by Plotting Points

Sketch the graph of $y = x^2 - 4x + 4$.

Solution To find several ordered pairs that satisfy the equation, we arbitrarily choose several values of x and then find the corresponding values of y. Let $x = -1, 0, 1, 2, 3,$ and 4. Substitute each x-coordinate into the equation and solve for y to find the corresponding y-coordinate. The results appear in Table 2.

Table 2

x	$y = x^2 - 4x + 4$	Ordered pair that lies on the graph of $y = x^2 - 4x + 4$
-1	$y = (-1)^2 - 4(-1) + 4 = 9$	$(-1, 9)$
0	$y = (0)^2 - 4(0) + 4 = 4$	$(0, 4)$
1	$y = (1)^2 - 4(1) + 4 = 1$	$(1, 1)$
2	$y = (2)^2 - 4(2) + 4 = 0$	$(2, 0)$
3	$y = (3)^2 - 4(3) + 4 = 1$	$(3, 1)$
4	$y = (4)^2 - 4(4) + 4 = 4$	$(4, 4)$

The six ordered pairs from Table 2 are plotted and connected with a smooth curve in Figure 4. The graph in Figure 4 is called a parabola. We will explore parabolas in depth later on in this eText.

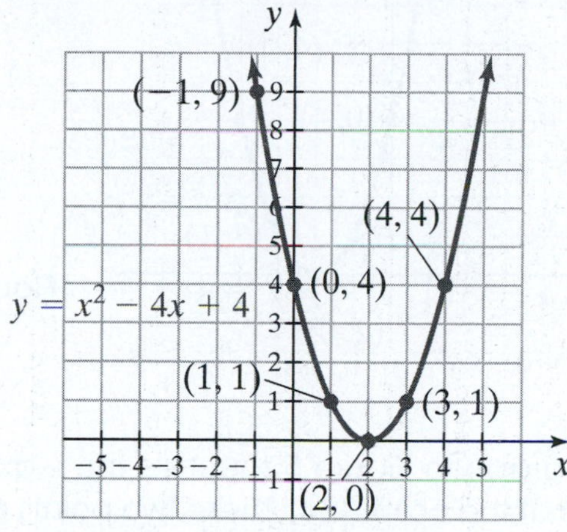

Figure 4 The graph of $y = x^2 - 4x + 4$.

You Try It Work through this You Try It problem.

Work Exercises 3–9 in this textbook or in the MyLab Math Study Plan.

Example 3 Determine Whether a Point Lies on the Graph of an Equation

Determine whether the following ordered pairs lie on the graph of the equation $x^2 + y^2 = 1$.

a. $(0, -1)$ b. $(1, 0)$ c. $\left(\frac{1}{3}, \frac{2}{3}\right)$ d. $\left(-\frac{\sqrt{2}}{2}, \frac{\sqrt{2}}{2}\right)$

Solution Watch the video to verify that the points $(0, -1)$, $(1, 0)$, and $\left(-\frac{\sqrt{2}}{2}, \frac{\sqrt{2}}{2}\right)$ lie on the graph of $x^2 + y^2 = 1$, and the point $\left(\frac{1}{3}, \frac{2}{3}\right)$ does not lie on the graph.

You Try It Work through this You Try It problem.

Work Exercises 10–12 in this textbook or in the MyLab Math Study Plan.

OBJECTIVE 3 FINDING INTERCEPTS OF A GRAPH GIVEN AN EQUATION

In **Example 2** we plotted several ordered pairs that **satisfy** the equation $y = x^2 - 4x + 4$. The graph of the equation $y = x^2 - 4x + 4$ is shown again below in Figure 5.

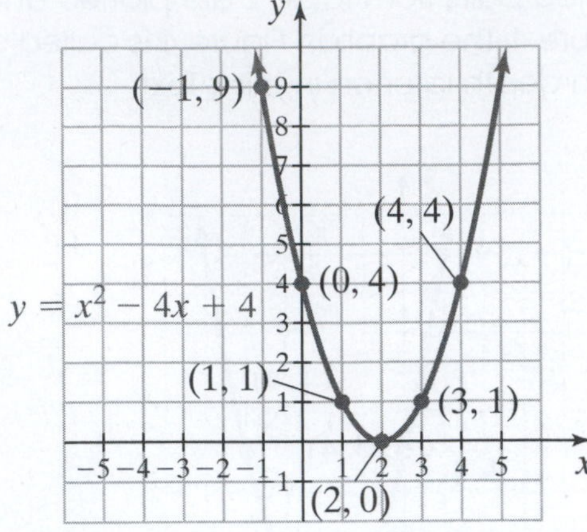

Figure 5 The graph of $y = x^2 - 4x + 4$.

Notice that the graph in Figure 5 touches the x-axis and the point $(2, 0)$ and crosses the y-axis at the point $(0, 4)$. These two points are called **intercepts**.

> **Definition** The intercepts of a graph are points where a graph crosses or touches a coordinate axis.

A **y-intercept** is the y-coordinate of a point where a graph touches or crosses the y-axis.

An **x-intercept** is the x-coordinate of a point where a graph touches or crosses the x-axis.

The intercepts of the graph in **Figure 5** are $(2, 0)$ and $(0, 4)$. The x-intercept is 2 and the y-intercept is 4.

An intercept will always have at least one coordinate that is equal to 0. Therefore, if we are given an equation in two variables x and y, it may be possible to algebraically determine an x- or y-intercept by substituting the value of 0 in for the appropriate variable. To algebraically attempt to find x- and y-intercepts, we follow the steps outlined below.

> **Algebraically Finding x- and y-Intercepts Given an Equation in Two Variables**
>
> **Finding x-intercepts:** Set all values of the variable y equal to 0 and solve for x.
>
> **Finding y-intercepts:** Set all values of the variable x equal to 0 and solve for y.

Example 4 demonstrates how to algebraically find the x- and y-intercepts of the graph of three different equations in two variables.

Example 4 Algebraically Finding x- and y-Intercepts Given an Equation

Find the x- and y-intercepts of the graphs of the given equations.

a. $y = \dfrac{2x - 1}{x + 3}$ b. $\sqrt{x + 2} + y = 3$ c. $(x - 1)^2 + (y - 3)^2 = 5$

Solution

a. To find the x-intercept(s) of the graph of $y = \frac{2x - 1}{x + 3}$, set $y = 0$ and solve for x.

$$y = \frac{2x - 1}{x + 3} \quad \text{Write the equation.}$$

$$0 = \frac{2x - 1}{x + 3} \quad \text{Set } y = 0.$$

The equation $0 = \frac{2x - 1}{x + 3}$ is a **rational equation** with a **restricted value** of $x = -3$. We can multiply both sides of this rational equation by the LCD, $x + 3$ to eliminate the fractions.

$$0 = \frac{2x - 1}{x + 3} \quad \text{Write the rational equation.}$$

$$(x + 3) \cdot 0 = \frac{2x - 1}{x + 3}(x + 3) \quad \text{Multiply both sides by the LCD.}$$

$$0 = 2x - 1 \quad \text{Simplify.}$$

$$1 = 2x \quad \text{Add 1 to both sides.}$$

$$\frac{1}{2} = x \quad \text{Divide both sides by 2.}$$

The only **restricted value** of the rational equation was $x = -3$. Thus, $x = \frac{1}{2}$ is a valid solution and is the one x-intercept of the graph of $y = \frac{2x - 1}{x + 3}$.

To find the y-intercept(s) of the graph of $y = \frac{2x - 1}{x + 3}$, set $x = 0$ and solve for y.

$$y = \frac{2x - 1}{x + 3} \quad \text{Write the equation.}$$

$$y = \frac{2(0) - 1}{(0) + 3} \quad \text{Set } x = 0.$$

$$y = -\frac{1}{3} \quad \text{Simplify.}$$

The one y-intercept of the equation $y = \frac{2x - 1}{x + 3}$ is $y = -\frac{1}{3}$.

b. To find the x-intercept(s) of the graph of $\sqrt{x + 2} + y = 3$, set $y = 0$ and solve for x.

$$\sqrt{x + 2} + y = 3 \quad \text{Write the equation.}$$

$$\sqrt{x + 2} + 0 = 3 \quad \text{Set } y = 0.$$

The equation $\sqrt{x + 2} = 3$ involves a square root. You may wish to review how to solve equations involving radicals from **Section 1.6**.

Work through this **interactive video** to verify that the x-intercept of the graph of $\sqrt{x+2}+y=3$ is $x=7$.

Any y-intercept(s) of the graph of $\sqrt{x+2}+y=3$ can be found by setting $x=0$ and solving for y.

$\sqrt{x+2}+y=3$ Write the equation.

$\sqrt{0+2}+y=3$ Set $x=0$.

$\sqrt{2}+y=3$ Simplify.

$y=3-\sqrt{2}$ Subtract $\sqrt{2}$ from both sides.

The y-intercept of the graph of $\sqrt{x+2}+y=3$ is $y=3-\sqrt{2}$.

c. To find any x-intercept(s) of the graph of $(x-1)^2+(y-3)^2=5$, set $y=0$ and solve for x.

$(x-1)^2+(y-3)^2=5$ Write the equation.

$(x-1)^2+(0-3)^2=5$ Set $y=0$.

$(x-1)^2+9=5$ Simplify.

$(x-1)^2=-4$ Subtract 9 from both sides.

$x-1=\pm\sqrt{-4}$ Use the **square root property**.

The number $\sqrt{-4}$ is **not** a real number. (There is no real number whose square is equal to -4.) Therefore, the graph of the equation $(x-1)^2+(y-3)^2=5$ has no x-intercept.

Any y-intercept(s) of the graph of $(x-1)^2+(y-3)^2=5$ can be found by setting $x=0$ and solving for y.

$(x-1)^2+(y-3)^2=5$ Write the equation.

$(0-1)^2+(y-3)^2=5$ Set $x=0$.

$1+(y-3)^2=5$ Simplify.

$(y-3)^2=4$ Subtract 1 from both sides.

$y-3=\pm 2$ Use the **square root property**.

$y=3\pm 2$ Add 3 to both sides.

The graph of the equation $(x-1)^2+(y-3)^2=5$ has two y-intercepts: $y=3+2=5$ and $y=3-2=1$.

Work through this **interactive video** to see the see the entire solution to this example worked out in detail.

You Try It **Work through this You Try It problem.**

Work Exercises 13–21 in this textbook or in the MyLab Math Study Plan.

OBJECTIVE 4 FINDING THE MIDPOINT OF A LINE SEGMENT USING THE MIDPOINT FORMULA

Suppose we want to find the midpoint $M(x, y)$ of the line segment between the points $A(x_1, y_1)$ and $B(x_2, y_2)$. See Figure 6. To find this midpoint, we simply average the

x- and y-coordinates, respectively. In other words, the x-coordinate of the midpoint is $\frac{x_1 + x_2}{2}$, whereas the y-coordinate of the midpoint is $\frac{y_1 + y_2}{2}$.

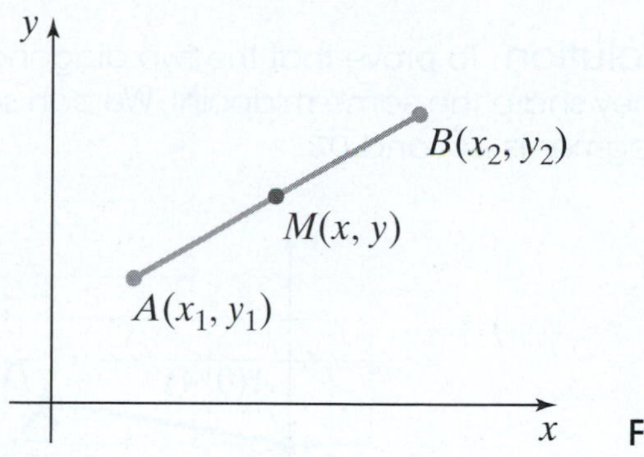

Figure 6

Midpoint Formula

The midpoint of the line segment from $A(x_1, y_1)$ to $B(x_2, y_2)$ is

$$\left(\frac{x_1 + x_2}{2}, \frac{y_1 + y_2}{2}\right).$$

Example 5 Find the Midpoint of a Line Segment

Find the midpoint of the segment whose endpoints are $(-3, 2)$ and $(4, 6)$.

Solution The midpoint of this line segment is $\left(\frac{-3 + 4}{2}, \frac{2 + 6}{2}\right) = \left(\frac{1}{2}, 4\right)$. See Figure 7.

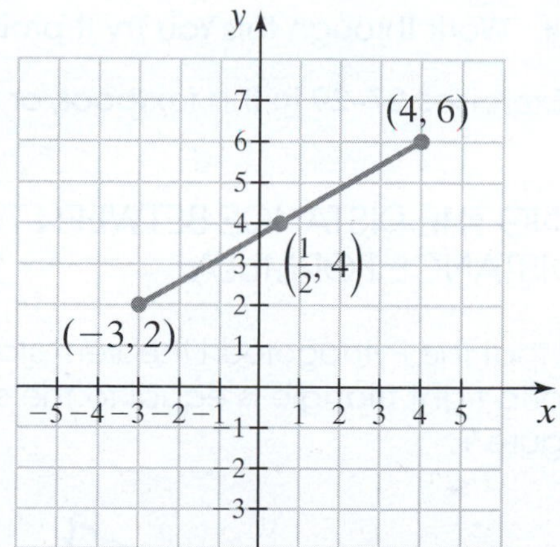

Figure 7

You Try It Work through this You Try It problem.

Work Exercises 22–26 in this textbook or in the MyLab Math Study Plan.

Example 6 Application of the Midpoint Formula

In geometry, it can be shown that four points in a plane form a **parallelogram** if the two diagonals of the quadrilateral formed by the four points **bisect** each other. Do the points $A(0, 4)$, $B(3, 0)$, $C(9, 1)$, and $D(6, 5)$ form a parallelogram?

Solution To prove that the two diagonals bisect each other, we can show that they share the same midpoint. We can see in Figure 8 that the diagonals are the segments $\overline{AC}$ and $\overline{DB}$.

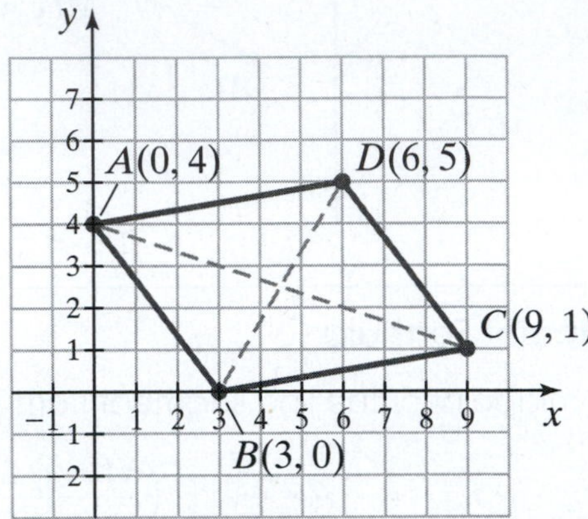

Figure 8

The midpoint of diagonal $\overline{AC}$ is $\left(\frac{0+9}{2}, \frac{4+1}{2}\right) = \left(\frac{9}{2}, \frac{5}{2}\right)$.

The midpoint of diagonal $\overline{DB}$ is $\left(\frac{6+3}{2}, \frac{5+0}{2}\right) = \left(\frac{9}{2}, \frac{5}{2}\right)$.

Because the diagonals share the same midpoint, it is implied that they bisect each other. Therefore, the quadrilateral is a parallelogram. Watch this **video** for a detailed solution.

You Try It Work through this You Try It problem.

Work Exercises 27–29 in this textbook or in the MyLab Math Study Plan.

OBJECTIVE 5 FINDING THE DISTANCE BETWEEN TWO POINTS USING THE DISTANCE FORMULA

Recall that the Pythagorean theorem states that the sum of the squares of the two sides of a right triangle is equal to the square of the hypotenuse or $a^2 + b^2 = c^2$. See Figure 9.

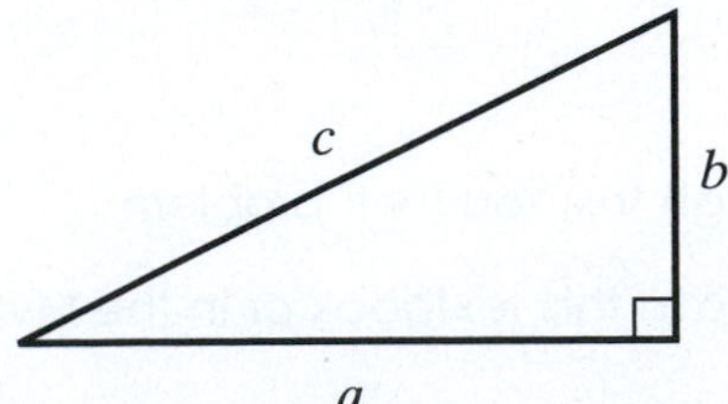

Figure 9 Pythagorean theorem: $a^2 + b^2 = c^2$

We can use the Pythagorean theorem to find the distance between any two points in a plane. For example, suppose we want to find the distance, $d(A, B)$, between the points $A(x_1, y_1)$ and $B(x_2, y_2)$. To find the length $d(A, B)$ of the line segment AB, consider the point $C(x_1, y_2)$ in **Figure 10**, which is on the same vertical line segment as point A and the same horizontal line segment as point B. The triangle formed by points A, B, and C is a right triangle whose hypotenuse has length $d(A, B)$. The horizontal leg of the triangle has length $a = |x_2 - x_1|$, whereas the vertical leg of the triangle has length $b = |y_2 - y_1|$.

Therefore, by the Pythagorean theorem, $\underbrace{|x_2 - x_1|^2}_{a^2} + \underbrace{|y_2 - y_1|^2}_{b^2} = \underbrace{[d(A, B)]^2}_{c^2}$.

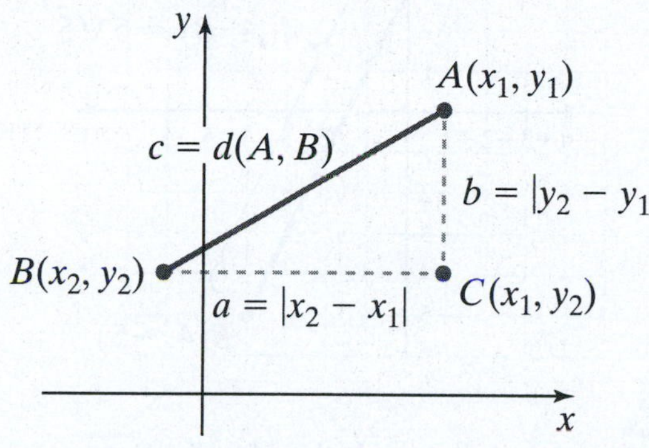

Figure 10

$[d(A, B)]^2 = |x_2 - x_1|^2 + |y_2 - y_1|^2$ Use the Pythagorean theorem.

$d(A, B) = \pm\sqrt{|x_2 - x_1|^2 + |y_2 - y_1|^2}$ Use the **square root property**.

However, because distance cannot be negative, exclude $-\sqrt{|x_2 - x_1|^2 + |y_2 - y_1|^2}$.

$d(A, B) = \sqrt{|x_2 - x_1|^2 + |y_2 - y_1|^2}$ Use the positive square root only.

$d(A, B) = \sqrt{(x_2 - x_1)^2 + (y_2 - y_1)^2}$ For any real number a, $|a|^2 = a^2$.

This formula is known as the distance formula.

Distance Formula

The distance between any two points $A(x_1, y_1)$ and $B(x_2, y_2)$ is given by the formula

$$d(A, B) = \sqrt{(x_2 - x_1)^2 + (y_2 - y_1)^2}.$$

Example 7 Use the Distance Formula to Find the Distance between Two Points

Find the distance, $d(A, B)$, between the points A and B.

$A(-1, 5); B(4, -5)$

Solution

$d(A, B) = \sqrt{(x_2 - x_1)^2 + (y_2 - y_1)^2}$ Use the distance formula.

$= \sqrt{(4 - (-1))^2 + (-5 - 5)^2}$ Substitute in the distance formula.

$= \sqrt{5^2 + (-10)^2}$ Combine terms.

$= \sqrt{25 + 100}$ Simplify.

$= \sqrt{125}$ Add.

$= 5\sqrt{5}$ Simplify the radical.

The distance between the two given points is $d(A, B) = 5\sqrt{5}$ units. See **Figure 11**.

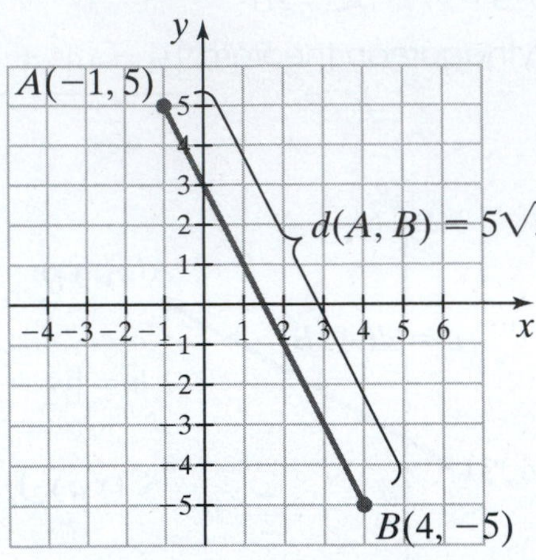

Figure 11 Distance from A to B is d, $d(A, B) = 5\sqrt{5}$.

You Try It Work through this You Try It problem.

Work Exercises 30–33 in this textbook or in the MyLab Math Study Plan.

Example 8 Application of the Distance Formula

Verify that the points $A(3, -5)$, $B(0, 6)$, and $C(5, 5)$ form a right triangle.

Solution The three points do form a right triangle. Watch the **video** to see the solution to this example.

You Try It Work through this You Try It problem.

Work Exercises 34–37 in this textbook or in the MyLab Math Study Plan.

2.1 Exercises

Skill Check Exercises

For exercises SCE-1 through SCE-3, find the average of the two indicated real numbers.

SCE-1. $-3, 8$

SCE-2. $\frac{1}{3}, -2$

SCE-3. $-\frac{2}{5}, \frac{3}{7}$

For exercises SCE-4 through SCE-7, simplify the radical.

SCE-4. $\sqrt{108}$

SCE-5. $\sqrt{(-7)^2 + (24)^2}$

SCE-6. $\sqrt{(-9)^2 + (-3)^2}$

SCE-7. $\sqrt{\left(-\frac{\sqrt{29}}{2}\right)^2 + \left(\frac{\sqrt{71}}{2}\right)^2}$

For exercises SCE-8 through SCE-13, solve each equation.

SCE-8. $\dfrac{5}{x+1} - 1 = \dfrac{8}{x+1}$

SCE-9. $\dfrac{2x}{x^2 - 5x - 6} - \dfrac{5}{x-6} = \dfrac{1}{x+1}$

SCE-10. $(x-5)^2 - 162 = 0$

SCE-11. $(2x+1)^2 + 4 = 0$

SCE-12. $x - 1 = \sqrt{2x+1}$

SCE-13. $\sqrt{12 - 2x} = x + 6$

In Exercises 1 and 2, plot each ordered pair in the Cartesian plane, and state in which quadrant or on which axis it lies.

1. a. $A(-1, 4)$ b. $B(-2, -2)$ c. $C(0, -3)$ d. $D\left(1, \frac{1}{2}\right)$ e. $E(2, 0)$

2. a. $A(2, -4)$ b. $B\left(-\frac{5}{2}, 1\right)$ c. $C(\sqrt{2}, \pi)$ d. $D\left(0, -\frac{9}{2}\right)$ e. $E(0, 0)$

In Exercises 3–9, sketch the graph for each equation by plotting points.

3. $y = 2x + 1$

4. $y = \frac{1}{4}x - 2$

5. $3x - 2y = 8$

6. $3x + y = 3$

7. $y = 3x^2$

8. $y = x^2 - 2x - 3$

9. $y = \frac{1}{2}x^3$

In Exercises 10–12, determine whether the indicated ordered pairs lie on the graph of the given equation.

10. $y = 2x^2 + 1$
    a. $(0, 1)$
    b. $(1, 1)$
    c. $(1, 3)$

11. $y = \sqrt{x} - 1$
    a. $(4, 1)$
    b. $(1, 4)$
    c. $(0, 1)$

12. $y = |x|$
    a. $(4, 1)$
    b. $(-9, 9)$
    c. $(3, 3)$

In Exercises 13–21 find the x- and y-intercepts of the graphs of the given equations.

13. $y = 3x - 5$

14. $2x - 3y = 8$

15. $y = x^2 - 2x - 8$

16. $y = x^2 - 4x + 6$

17. $y = \dfrac{x+1}{2x-3}$

18. $y = \dfrac{x^2 - x - 6}{x+4}$

19. $\sqrt{x+3} - y = 4$

20. $(x+1)^2 + (y-2)^2 = 4$

21. $(x+4)^2 + (y-1)^2 = 10$

In Exercises 22–25, find the midpoint of the line segment joining points *A* and *B*.

22. $A(1, -4)$; $B(-2, -2)$

23. $A(2, -5)$; $B(4, 1)$

24. $A(0, 1)$; $B\left(-3, \frac{1}{2}\right)$

25. $A(a, b)$; $B(c, d)$

26. A line segment has a midpoint of $\left(1, \frac{1}{2}\right)$. If one endpoint is $(7, 3)$, what is the other endpoint?

In Exercises 27–29, determine whether the points $A, B, C,$ and D form a parallelogram.

27. $A(1, 4); B(-2, -1); C(4, 2); D(7, 7)$

28. $A(-2, 4); B(-1, 2); C(1, 1); D(3, 3)$

29. $A(-2, -1); B(2, 0); C(3, 3); D(-1, 2)$

In Exercises 30–33, find the distance, $d(a, b)$, between points A and B.

30. $A(1, 5); B(-2, 1)$

31. $A(3, 5); B(-2, -2)$

32. $A(2, -4); B(-3, 6)$

33. $A\left(-\frac{\sqrt{3}}{2}, \frac{\sqrt{17}}{2}\right); B(0, 0)$

In Exercises 34–36, determine whether points $A, B,$ and C form a right triangle.

34. $A(-3, -3); B(3, -1); C(2, 2)$

35. $A(1, 4); B(-2, -1); C(4, 2)$

36. $A(1, 2); B(2, 6); C(9, 0)$

37. Find the y-coordinates of the points that are 5 units away from the point $(-1, 3)$ that have an x-coordinate of 3.

2.2 Circles

THINGS TO KNOW

Before working through this section, be sure that you are familiar with the following concepts:

VIDEO ANIMATION INTERACTIVE

You Try It

1. Solving Quadratic Equations by Completing the Square (Section 1.4)

You Try It

2. Solving Quadratic Equations Using the Square Root Property (Section 1.4)

You Try It

3. Finding the Midpoint of a Line Segment Using the Midpoint Formula (Section 2.1)

You Try It

4. Finding the Distance between Two Points Using the Distance Formula (Section 2.1)

You Try It

5. Finding Intercepts of a Graph Given an Equation (Section 2.1)

INTRODUCTION

Read this introduction before beginning Objective 1.

OBJECTIVES

1 Writing the Standard Form of an Equation of a Circle

2 Sketching the Graph of a Circle

3 Converting the General Form of a Circle into Standard Form

SECTION 2.2 EXERCISES

Introduction to Section 2.2

A **circle** is the set of all points (x, y) in the Cartesian plane that are a fixed distance, r, from a fixed point, (h, k). The fixed distance, r, is called the **radius** of the circle, and the fixed point (h, k), is called the **center** of the circle. See Figure 12. To derive the equation of a circle, we use the **distance formula** that was discussed in **Section 2.1**.

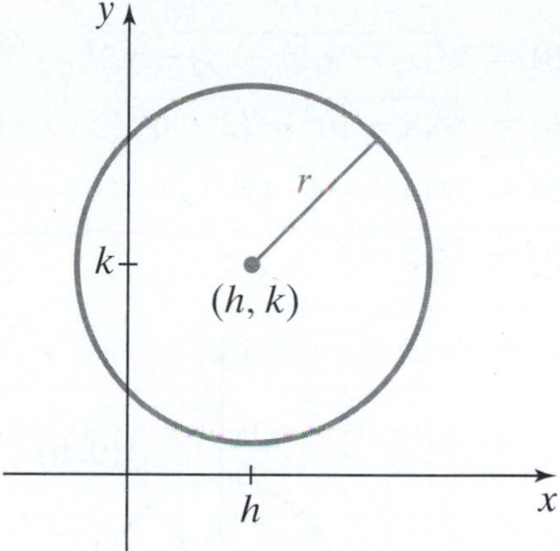

Figure 12 Circle with center, (h, k) and radius, r.

Watch this **animation** to see how the standard form of the equation of a circle is derived.

> The **standard form of an equation of a circle** with center, (h, k), and radius, r, is
>
> $$(x - h)^2 + (y - k)^2 = r^2.$$
>
> The standard form of an equation of a circle centered at the origin with radius, r, is
>
> $$x^2 + y^2 = r^2.$$

OBJECTIVE 1 WRITING THE STANDARD FORM OF AN EQUATION OF A CIRCLE

Example 1 Find the Standard Form of an Equation of a Circle

Find the standard form of the equation of the circle whose center is $(-2, 3)$ and with radius 6.

Solution We know that the standard form of the equation of a circle is $(x - h)^2 + (y - k)^2 = r^2$. We are given $h = -2$, $k = 3$, and $r = 6$. Substituting these values into the equation, we get

$$(x - (-2))^2 + (y - 3)^2 = 6^2 \quad \text{or} \quad (x + 2)^2 + (y - 3)^2 = 36.$$

You Try It Work through this You Try It problem.

Work Exercises 1–4 in this textbook or in the MyLab Math Study Plan.

Example 2 Find the Equation of a Circle Given the Center and a Point on the Circle

Find the standard form of the equation of the circle whose center is $(0, 6)$ and that passes through the point $(4, 2)$.

Solution The graph of this circle is sketched in Figure 13. The center is labeled $A(0, 6)$, and the point on the circle is labeled $B(4, 2)$. To find the equation of this circle, we must calculate the radius. The radius can be found using the distance formula.

$r = d(A, B) = \sqrt{(x_2 - x_1)^2 + (y_2 - y_1)^2}$ Write the distance formula.

$r = d(A, B) = \sqrt{(4 - 0)^2 + (2 - 6)^2}$ Substitute into the distance formula.

$r = d(A, B) = \sqrt{(4)^2 + (-4)^2}$ Combine terms.

$r = d(A, B) = \sqrt{32}$ Add.

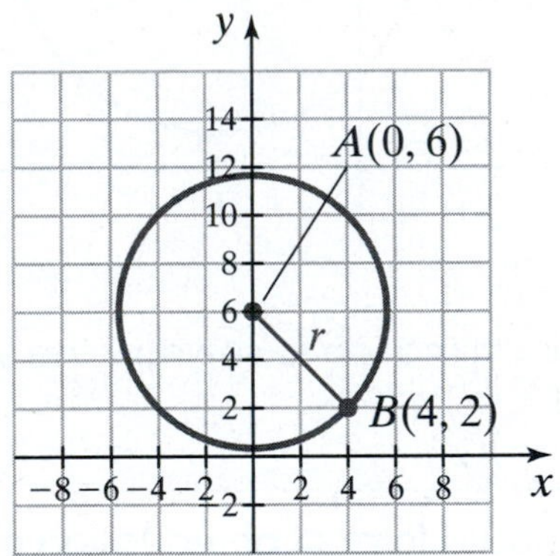

Figure 13

Now that we know the center and the radius, we can determine the equation of this circle.

$(x - h)^2 + (y - k)^2 = r^2$ Write the standard equation of a circle.

$(x - 0)^2 + (y - 6)^2 = (\sqrt{32})^2$ Substitute values for the center and the radius.

$x^2 + (y - 6)^2 = 32$ Simplify.

The equation of this circle in standard form is $x^2 + (y - 6)^2 = 32$.

Example 3 Find the Equation of a Circle Given the Endpoints of a Diameter

Find the standard form of the equation of the circle that contains endpoints of a diameter at $(-4, -3)$ and $(2, -1)$.

Solution We can use the **midpoint formula** to determine the center of this circle. The radius can be found using the **distance formula**. Watch this **video** to verify that the equation of this circle is $(x + 1)^2 + (y + 2)^2 = 10$.

You Try It Work through this You Try It problem.

Work Exercises 5–10 in this textbook or in the MyLab Math Study Plan.

OBJECTIVE 2 SKETCHING THE GRAPH OF A CIRCLE

Once you know the center, (h, k), and the radius, r, of a circle, it is fairly straightforward to sketch the graph of the circle.

To sketch the graph, first, plot the center of the circle, then locate a few points that are all r units from the center. Then, draw a circle through the points.

To illustrate how to sketch the graph of a circle, click on the guided visualization icon below and take some time to explore this guided visualization by sketching the graphs of several circles.

Sketching the Graph of a Circle

Example 4 Sketch the Graph of a Circle

Find the center and the radius, and sketch the graph of the circle $(x - 1)^2 + (y + 2)^2 = 9$. Also find any intercepts.

Solution The circle $(x - 1)^2 + (y + 2)^2 = 9$ has center $(h, k) = (1, -2)$ and $r = 3$.

CAUTION Note that the y-coordinate of the center ($k = -2$) is negative because $(y + 2)^2 = (y - (-2))^2$.

To sketch the graph of this circle, plot the center, and then locate a few points on the circle. The points that are easiest to locate are the points that are three units left and right of the center and three units up and down from the center. Complete the graph by drawing the circle through these four points, as in Figure 14.

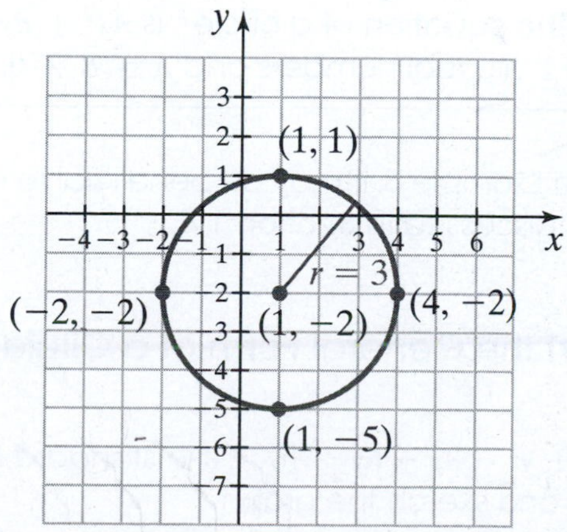

Figure 14 Graph of $(x - 1)^2 + (y + 2)^2 = 9$.

To find the **x-intercepts**, set $y = 0$ and solve for x.

$$(x - 1)^2 + (0 + 2)^2 = 9 \quad \text{Set } y = 0.$$

$$(x - 1)^2 + 4 = 9 \quad \text{Simplify.}$$

$$(x - 1)^2 = 5 \quad \text{Subtract 4 from both sides.}$$

$$x - 1 = \pm\sqrt{5} \quad \text{Use the square root property.}$$

$$x = 1 \pm \sqrt{5} \quad \text{Add 1 to both sides.}$$

The two x-intercepts are $x \approx 3.24$ and $x \approx -1.24$.

To find any **y-intercepts**, set $x = 0$ and solve for y. You should verify that the y-intercepts are $y = -2 + 2\sqrt{2} \approx .83$ and $y = -2 - 2\sqrt{2} \approx -4.83$. Watch the

video to see this example worked out in its entirety.

You Try It **Work through this You Try It problem.**

Work Exercises 11–17 in this textbook or in the MyLab Math Study Plan.

OBJECTIVE 3 CONVERTING THE GENERAL FORM OF A CIRCLE INTO STANDARD FORM

As you can see from **Example 4**, if the equation of a circle is in standard form, it is not too difficult to determine the center and the radius. For example, the center of the circle $(x + 3)^2 + (y - 1)^2 = 49$ is at $(-3, 1)$, and the radius is 7. Let's now take this same equation, eliminate both sets of parentheses, and simplify:

$$(x + 3)^2 + (y - 1)^2 = 49 \quad \text{Write the original equation in standard form.}$$

$$x^2 + 6x + 9 + y^2 - 2y + 1 = 49 \quad \text{Remove parentheses.}$$

$$x^2 + 6x + y^2 - 2y + 10 = 49 \quad \text{Combine like terms.}$$

$$x^2 + y^2 + 6x - 2y - 39 = 0 \quad \text{Subtract 49 from both sides and rearrange terms.}$$

We see that the equation $(x + 3)^2 + (y - 1)^2 = 49$ is equivalent to the equation $x^2 + y^2 + 6x - 2y - 39 = 0$; however, determining the center and the radius of this equation is less obvious. The equation $x^2 + y^2 + 6x - 2y - 39 = 0$ is said to be in **general form.**

> The **general form of the equation of a circle*** is $Ax^2 + By^2 + Cx + Dy + E = 0$, where A, B, C, D, and E are real numbers and $A = B \neq 0$.

Before working through Example 5, it may be beneficial to review how to complete the square, which was discussed in **Section 1.4**.

Example 5 Convert the General Form of a Circle into Standard Form

Write the equation $x^2 + y^2 - 8x + 6y + 16 = 0$ in standard form; find the center, radius, and intercepts, and sketch the graph.

Solution Because we are going to complete the square on both x and y, we first rearrange the terms, leaving some room to complete the square, and move any constants to the right-hand side.

$x^2 - 8x \quad + y^2 + 6y \quad = -16$ Rearrange the terms.

$x^2 - 8x + 16 + y^2 + 6y + 9 = -16 + 16 + 9$ Complete the square on x and y. Remember to add 16 and 9 to both sides.

$\left(\frac{1}{2}\cdot(-8)\right)^2 = 16 \qquad \left(\frac{1}{2}\cdot 6\right)^2 = 9$

$(x - 4)^2 + (y + 3)^2 = 9$ Factor the left side.

We have now converted the general form of the circle into standard form. The center is at $(4, -3)$, and the radius is $r = 3$. Watch the **video** to verify that there are no y-intercepts. The only x-intercept is $x = 4$. The graph of the circle is shown in Figure 15.

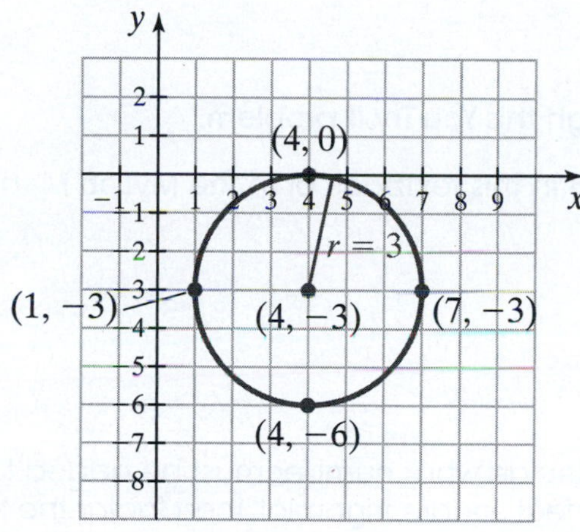

Figure 15 Graph of $x^2 + y^2 - 8x + 6y + 16 = 0$. ●

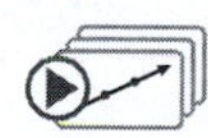

Example 6 Convert the General Form of a Circle into Standard Form, Where $A \neq 1$ and $B \neq 1$

Write the equation $4x^2 + 4y^2 + 4x - 8y + 1 = 0$ in standard form; find the center, radius, and intercepts, and sketch the graph.

Solution Work through the **animation** to see that this equation is equivalent to the following equation in standard form:

$$\left(x + \frac{1}{2}\right)^2 + (y - 1)^2 = 1;\text{ center } = \left(-\frac{1}{2}, 1\right),\text{ and } r = 1$$

Work through the **animation** to verify that the one x-intercept is found at $\left(-\frac{1}{2}, 0\right)$, whereas the y-intercepts are found at $\left(0, 1 + \frac{\sqrt{3}}{2}\right)$ and $\left(0, 1 - \frac{\sqrt{3}}{2}\right)$.

See the graph in Figure 16.

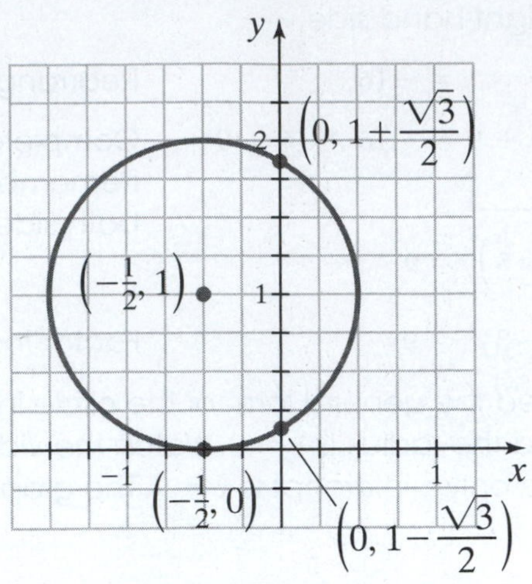

Figure 16 Graph of $4x^2 + 4y^2 + 4x - 8y + 1 = 0.$ ●

You Try It Work through this You Try It problem.

Work Exercises 18–28 in this textbook or in the MyLab Math Study Plan.

2.2 Exercises

Skill Check Exercises

For exercises SCE-1 through SCE-5, decide what number must be added to both sides of each equation to make the left side a perfect square trinomial, then factor the left side.

SCE-1. $x^2 + 10x = 3$ **SCE-2.** $x^2 - 5x = 1$ **SCE-3.** $y^2 - 8y = -7$

SCE-4. $y^2 - 9y = 2$ **SCE-5.** $x^2 + \frac{5}{3}x = -1$

For exercises SCE-6 through SCE-9, decide what numbers must be added to both sides of each equation to make the left side the sum of a perfect square trinomial in x and a perfect square trinomial in y.

SCE-6. $x^2 + 10x + y^2 - 8y = -4$ **SCE-7.** $x^2 - 5x + y^2 - 6y = 3$ **SCE-8.** $x^2 - 7x + y^2 + 11y = -2$

SCE-9. $x^2 + \frac{2}{3}x + y^2 + \frac{5}{7}y = -1$

In Exercises 1–10, write the standard form of the equation of each circle described.

1. Center $(0, 0)$, $r = 1$
2. Center $(-2, 3)$, $r = 4$
3. Center $(1, -4)$, $r = 3$
4. Center $\left(-\frac{1}{4}, -\frac{1}{3}\right)$, $r = 2$
5. Center $(0, 2)$, passes through the point $(4, -1)$
6. Center $(-4, 7)$, passes through the point $(2, 1)$

7. The endpoints of a diameter are $(0, 1)$ and $(6, -3)$

8. The endpoints of a diameter are $(2, -6)$ and $(6, 1)$

9. Center $(2, -3)$ and tangent to the x-axis

10. Center $(-4, 1)$ and tangent to the y-axis

In Exercises 11–17, find the center, radius, and intercepts of each circle and then sketch the graph.

SbS 11. $x^2 + y^2 = 1$

SbS 12. $x^2 + (y - 2)^2 = 4$

SbS 13. $(x - 1)^2 + (y + 5)^2 = 16$

SbS 14. $(x + 2)^2 + (y + 4)^2 = 36$

SbS 15. $(x - 4)^2 + (y + 7)^2 = 12$

SbS 16. $(x + 1)^2 + (y - 3)^2 = 20$

SbS 17. $\left(x - \frac{1}{4}\right)^2 + \left(y + \frac{1}{2}\right)^2 = 4$

In Exercises 18–28, find the center, radius, and intercepts of each circle and then sketch the graph.

SbS 18. $x^2 + y^2 + 2x - 4y + 1 = 0$

SbS 19. $x^2 + y^2 - 10x + 6y + 18 = 0$

SbS 20. $x^2 + y^2 - 4x - 8y + 19 = 0$

SbS 21. $x^2 + y^2 + 2y - 8 = 0$

SbS 22. $x^2 + y^2 - 6x - 12y - 5 = 0$

SbS 23. $x^2 + y^2 - 3x - y - \frac{1}{2} = 0$

SbS 24. $x^2 + y^2 + \frac{2}{3}x - \frac{1}{2}y - \frac{7}{18} = 0$

SbS 25. $2x^2 + 2y^2 - 4x + 8y + 2 = 0$

SbS 26. $16x^2 + 16y^2 - 16x + 8y - 11 = 0$

SbS 27. $144x^2 + 144y^2 - 72x - 96y - 551 = 0$

SbS 28. $36x^2 + 36y^2 + 12x + 72y - 35 = 0$

Brief Exercises

In Exercises 29–35, find the center and radius of each circle.

29. $x^2 + y^2 = 1$

30. $x^2 + (y - 2)^2 = 4$

31. $(x - 1)^2 + (y + 5)^2 = 16$

32. $(x + 2)^2 + (y + 4)^2 = 36$

33. $(x - 4)^2 + (y + 7)^2 = 12$

34. $(x + 1)^2 + (y - 3)^2 = 20$

35. $\left(x - \frac{1}{4}\right)^2 + \left(y + \frac{1}{2}\right)^2 = 4$

In Exercises 36–42, find intercepts of each circle.

36. $x^2 + y^2 = 1$

37. $x^2 + (y - 2)^2 = 4$

38. $(x - 1)^2 + (y + 5)^2 = 16$

39. $(x + 2)^2 + (y + 4)^2 = 36$

40. $(x - 4)^2 + (y + 7)^2 = 12$

41. $(x + 1)^2 + (y - 3)^2 = 20$

42. $\left(x - \frac{1}{4}\right)^2 + \left(y + \frac{1}{2}\right)^2 = 4$

In Exercises 43–49, sketch the graph of each circle.

43. $x^2 + y^2 = 1$

44. $x^2 + (y - 2)^2 = 4$

45. $(x - 1)^2 + (y + 5)^2 = 16$

46. $(x + 2)^2 + (y + 4)^2 = 36$

47. $(x - 4)^2 + (y + 7)^2 = 12$

48. $(x + 1)^2 + (y - 3)^2 = 20$

49. $\left(x - \frac{1}{4}\right)^2 + \left(y + \frac{1}{2}\right)^2 = 4$

In Exercises 50–60, find the center and radius of each circle.

50. $x^2 + y^2 + 2x - 4y + 1 = 0$

51. $x^2 + y^2 - 10x + 6y + 18 = 0$

52. $x^2 + y^2 - 4x - 8y + 19 = 0$

53. $x^2 + y^2 + 2y - 8 = 0$

54. $x^2 + y^2 - 6x - 12y - 5 = 0$

55. $x^2 + y^2 - 3x - y - \frac{1}{2} = 0$

56. $x^2 + y^2 + \frac{2}{3}x - \frac{1}{2}y - \frac{7}{18} = 0$

57. $2x^2 + 2y^2 - 4x + 8y + 2 = 0$

58. $16x^2 + 16y^2 - 16x + 8y - 11 = 0$

59. $144x^2 + 144y^2 - 72x - 96y - 551 = 0$

60. $36x^2 + 36y^2 + 12x + 72y - 35 = 0$

In Exercises 61–71, find intercepts of each circle.

61. $x^2 + y^2 + 2x - 4y + 1 = 0$

62. $x^2 + y^2 - 10x + 6y + 18 = 0$

63. $x^2 + y^2 - 4x - 8y + 19 = 0$

64. $x^2 + y^2 + 2y - 8 = 0$

65. $x^2 + y^2 - 6x - 12y - 5 = 0$

66. $x^2 + y^2 - 3x - y - \frac{1}{2} = 0$

67. $x^2 + y^2 + \frac{2}{3}x - \frac{1}{2}y - \frac{7}{18} = 0$

68. $2x^2 + 2y^2 - 4x + 8y + 2 = 0$

69. $16x^2 + 16y^2 - 16x + 8y - 11 = 0$

70. $144x^2 + 144y^2 - 72x - 96y - 551 = 0$

71. $36x^2 + 36y^2 + 12x + 72y - 35 = 0$

In Exercises 72–82, sketch the graph of each circle.

72. $x^2 + y^2 + 2x - 4y + 1 = 0$

73. $x^2 + y^2 - 10x + 6y + 18 = 0$

74. $x^2 + y^2 - 4x - 8y + 19 = 0$

75. $x^2 + y^2 + 2y - 8 = 0$

76. $x^2 + y^2 - 6x - 12y - 5 = 0$

77. $x^2 + y^2 - 3x - y - \frac{1}{2} = 0$

78. $x^2 + y^2 + \frac{2}{3}x - \frac{1}{2}y - \frac{7}{18} = 0$

79. $2x^2 + 2y^2 - 4x + 8y + 2 = 0$

80. $16x^2 + 16y^2 - 16x + 8y - 11 = 0$

81. $144x^2 + 144y^2 - 72x - 96y - 551 = 0$

82. $36x^2 + 36y^2 + 12x + 72y - 35 = 0$

2.3 Lines

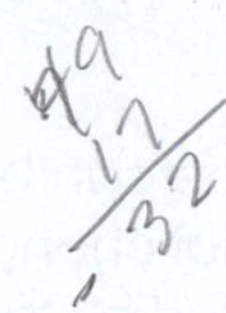

THINGS TO KNOW

Before working through this section, be sure that you are familiar with the following concepts:

VIDEO | ANIMATION | INTERACTIVE

You Try It 1. Recognizing Linear Equations (Section 1.1)

You Try It 2. Plotting Ordered Pairs (Section 2.1)

You Try It 3. Finding Intercepts of a Graph Given an Equation (Section 2.1)

OBJECTIVES

1 Determining the Slope of a Line

2 Sketching a Line Given a Point and the Slope

3 Finding the Equation of a Line Using the Point–Slope Form

4 Finding the Equation of a Line Using the Slope–Intercept Form

5 Writing the Equation of a Line in Standard Form

6 Finding the Slope and the y-Intercept of a Line in Standard Form

7 Sketching Lines by Plotting Intercepts

8 Finding the Equation of a Horizontal Line and a Vertical Line

SECTION 2.3 EXERCISES

OBJECTIVE 1 DETERMINING THE SLOPE OF A LINE

In this section, we study equations of lines that lie in the Cartesian plane. Before we learn about a line's equation, we must first establish a way to measure the steepness of a line. In mathematics, the steepness of a line can be measured by computing the line's **slope**. Every nonvertical line has slope. (Vertical lines are said to have no slope and are discussed in detail at the end of this section.) A line going up from left to right has **positive slope**, a line going down from left to right has **negative slope**, horizontal lines have **zero slope**, and vertical lines have **no slope** or **undefined slope**. See Figure 17.

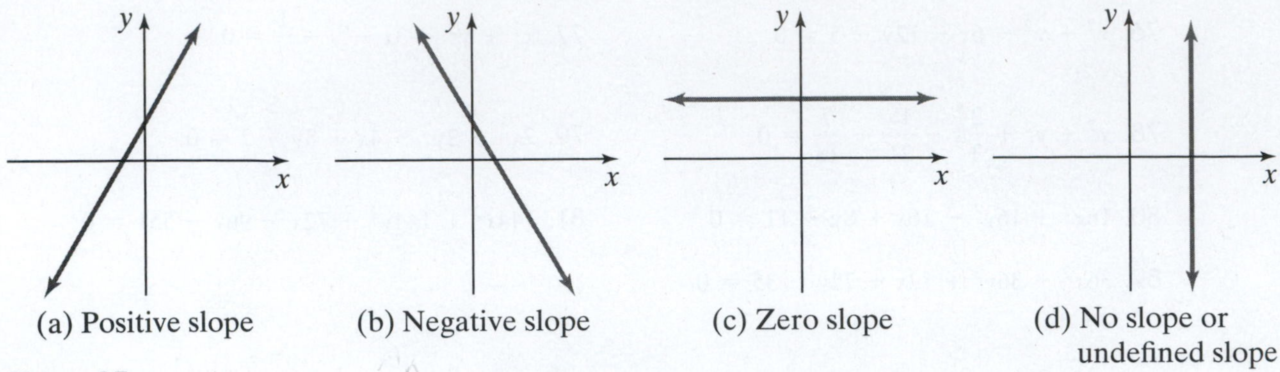

Figure 17

The slope can be computed by comparing the vertical change (the **rise**) to the horizontal change (the **run**). Given any two points on the line, the slope m, can be computed by taking the quotient of the rise over the run.

Definition Slope

If $x_1 \neq x_2$, the **slope** of a nonvertical line passing through distinct points (x_1, y_1) and (x_2, y_2) is given by

$$m = \frac{rise}{run} = \frac{\text{change in } y}{\text{change in } x} = \frac{y_2 - y_1}{x_2 - x_1}.$$

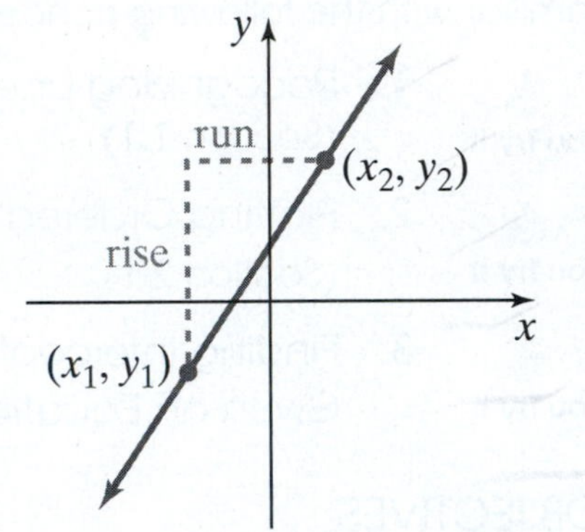

Note that any two points on the line can be used to calculate the slope. Therefore, the slope depends only on the line and not on the choice of a pair of points that lie on it.

Click on the guided visualization icon below and take some time to explore the slope of a line. Try to create lines with positive slope, negative slope, zero slope, and undefined slope.

 Determining the Slope of a Line

Example 1 Find the Slope of a Line

Find the slope of the line that passes through the indicated ordered pairs.

a. $(6, -4)$ and $(-5, 1)$ **b.** $(3, -1)$ and $(3, 6)$ **c.** $(-5, 4)$ and $(-2, 4)$

Solution

a. $m = \frac{y_2 - y_1}{x_2 - x_1} = \frac{1 - (-4)}{-5 - 6} = \frac{1 + 4}{-11} = -\frac{5}{11}$

b. $m = \frac{y_2 - y_1}{x_2 - x_1} = \frac{6 - (-1)}{3 - 3} = \frac{6 + 1}{0}.$

Since division by 0 is undefined, we say that the line that passes through the points $(3, -1)$ and $(3, 6)$ has undefined slope.

c. $m = \frac{y_2 - y_1}{x_2 - x_1} = \frac{4 - 4}{-2 - (-5)} = \frac{0}{-2 + 5} = 0.$

You Try It Work through this You Try It problem.

Work Exercises 1–6 in this textbook or in the MyLab Math Study Plan.

OBJECTIVE 2 SKETCHING A LINE GIVEN A POINT AND THE SLOPE

If we know a point on a line and the slope, we can quickly sketch the line. Watch this **video** to see how to sketch the line described in Example 2.

Example 2 Sketch a Line Given a Point and the Slope

Sketch the line with slope $m = \frac{2}{3}$ that passes through the point $(-1, -4)$.

Also, find three more points located on the line.

Solution To sketch the line, plot the ordered pair $(-1, -4)$. We can now use the slope to plot another point. Starting at $(-1, -4)$, we rise two units and then run three units. In other words, we add 2 to the y-coordinate and 3 to the x-coordinate to get $(2, -2)$. We can continue to find more points on the line by adding 2 to each successive y-coordinate and 3 to each successive x-coordinate to get the points listed in **Figure 18**.

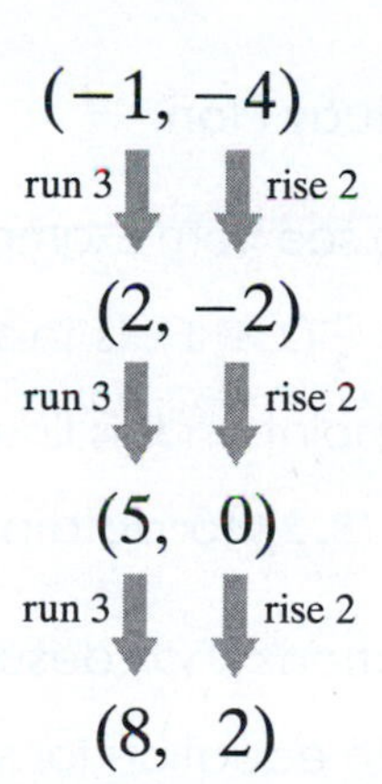

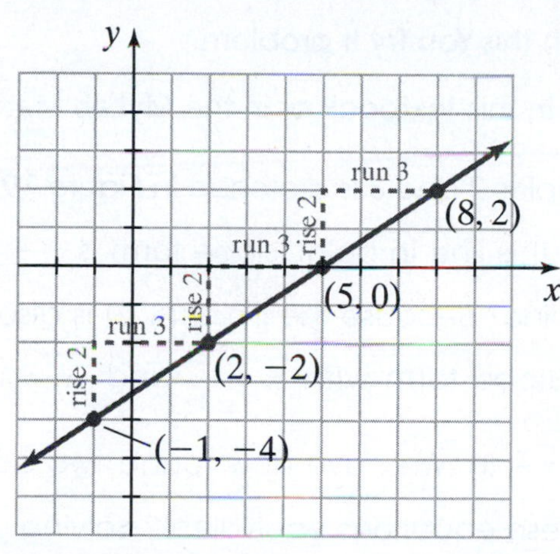

Figure 18 Line with slope passing through the point $(-1, -4)$. ●

You Try It Work through this You Try It problem.

Work Exercises 7–10 in this textbook or in the MyLab Math Study Plan.

OBJECTIVE 3 FINDING THE EQUATION OF A LINE USING THE POINT–SLOPE FORM

Given the slope m of a line and a point on the line, (x_1, y_1), we can use what is known as the **point-slope form of the equation** to determine the equation of the line. The point-slope form of a line is given in the following box. Watch the **video** to see exactly how this equation is derived.

Derivation of the Point-Slope Form of the Equation of a Line

Point-Slope Form of the Equation of a Line

Given the slope m of a line and a point on the line (x_1, y_1), the point-slope form of the equation of a line is given by $y - y_1 = m(x - x_1)$.

Example 3 Find the Equation of a Line Given a Point and the Slope

Find an equation in point-slope form of the line with slope $m = \frac{2}{3}$ that passes through the point $(-1, -4)$.

Solution This is the same line that was discussed in Example 2. Because we are given the slope of the line and a point on that line, we can use the point-slope form to produce an equation for this specific line. Substitute $m = \frac{2}{3}$ and $(x_1, y_1) = (-1, -4)$ into the point-slope form $y - y_1 = m(x - x_1)$ to get the equation

$$y - (-4) = \frac{2}{3}(x - (-1)) \quad \text{or} \quad y + 4 = \frac{2}{3}(x + 1).$$

 Watch the second half of this **video** to see the solution to this example.

Click on the guided visualization icon below and take some time to explore the point-slope form of a line. Try plotting a point with a line through it. Then change the slope and watch the equation change.

 Equations of Lines: Point-Slope

You Try It Work through this You Try It problem.

Work Exercises 11–14 in this textbook or in the MyLab Math Study Plan.

The line used in Examples 2 and 3 is sketched in **Figure 19** We see from Example 3 that the equation of this line in point-slope form is $y + 4 = \frac{2}{3}(x + 1)$. Is this the only equation of this line? Because the point $(8, 2)$ is also a point on this line, we can use the point-slope form with $m = \frac{2}{3}$ and $(x_1, y_1) = (8, 2)$ to obtain the equation $y - 2 = \frac{2}{3}(x - 8)$. We have now found two equations that describe the same line! Are these equations equivalent? Solving each equation for y, we see, in fact, that these equations are equivalent.

Equations $y + 4 = \frac{2}{3}(x + 1)$ and $y - 2 = \frac{2}{3}(x - 8)$ are equivalent

$y + 4 = \frac{2}{3}(x + 1)$	Write the original equations.	$y - 2 = \frac{2}{3}(x - 8)$
$y + 4 = \frac{2}{3}x + \frac{2}{3}$	Use the distributive property.	$y - 2 = \frac{2}{3}x - \frac{16}{3}$
$y = \frac{2}{3}x + \frac{2}{3} - 4$	Move constant term to the right.	$y = \frac{2}{3}x - \frac{16}{3} + 2$
$y = \frac{2}{3}x + \frac{2}{3} - \frac{12}{3}$	Get a common denominator.	$y = \frac{2}{3}x - \frac{16}{3} + \frac{6}{3}$
$y = \frac{2}{3}x - \frac{10}{3}$ ⟵	The two equations are equivalent. ⟶	$y = \frac{2}{3}x - \frac{10}{3}$

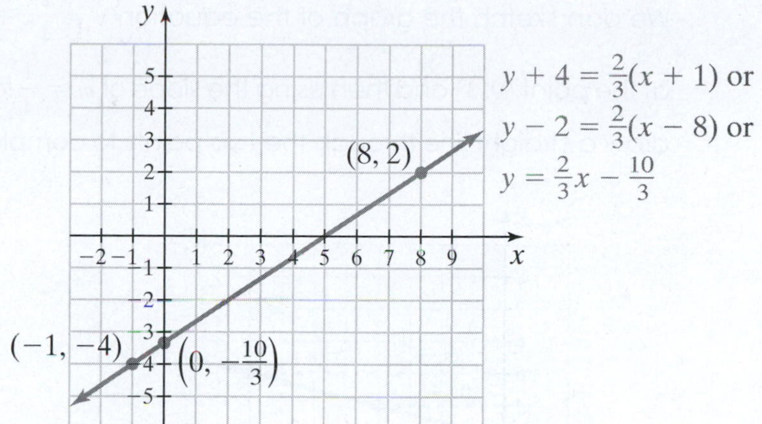

Figure 19

It does not matter which point on the line we choose to determine the equation, as long as we use the correct slope.

OBJECTIVE 4 FINDING THE EQUATION OF A LINE USING THE SLOPE–INTERCEPT FORM

The equation $y = \frac{2}{3}x - \frac{10}{3}$ sketched in **Figure 19** is said to be in **slope–intercept form** because $\frac{2}{3}$ describes the slope of the line, whereas the constant $-\frac{10}{3}$ is the y-intercept. The slope–intercept form of a line is extremely important because every nonvertical line has exactly one slope–intercept equation.

> **Slope–Intercept Form of the Equation of a Line**
>
> Given the slope m of a line and the y-intercept b, the slope–intercept form of the equation of a line is given by $y = mx + b$.

To see how we can derive the slope intercept form from the point-slope form, watch this **video**.

Derivation of the Slope–Intercept Form of the Equation of a Line

Example 4 Find the Equation of a Line Given the Slope and y-Intercept

Find the equation of the line with slope $\frac{1}{4}$ and y-intercept 3, and write the equation in slope–intercept form.

Solution To find the equation of this line, we substitute $m = \frac{1}{4}$ and $b = 3$ into the slope–intercept form.

$y = mx + b$ Write the slope–intercept form.

$y = \frac{1}{4}x + 3$ Substitute $m = \frac{1}{4}$ and $b = 3$.

(eText Screens 2.3-1–2.3-33)

We can sketch the graph of the equation $y = \frac{1}{4}x + 3$ by plotting the y-intercept at the point $(0, 3)$ and then using the slope of $m = \frac{1}{4}$ to plot one more point. We then draw a straight line through the two points to complete the sketch. See Figure 20.

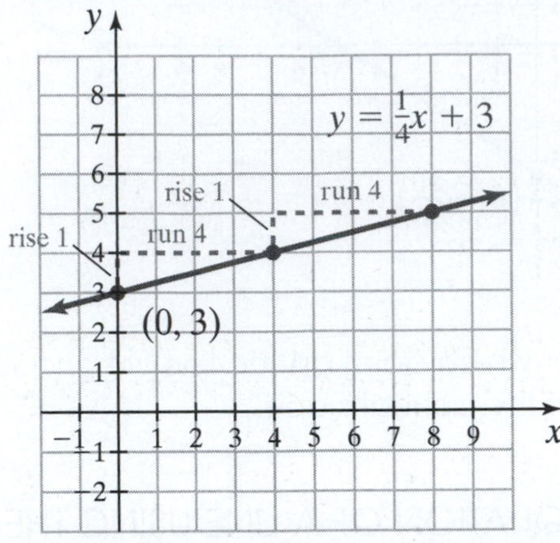

Figure 20 Graph of $y = \frac{1}{4}x + 3$.

Click on the guided visualization icon below and take some time to explore the slope-intercept form of a line. For the equation of the line with equation $y = mx + b$, try changing the values of m and b and watch the equation and the line change.

 Equations of Lines: Point-Slope

You Try It Work through this You Try It problem.

Work Exercises 15–18 in this textbook or in the MyLab Math Study Plan.

OBJECTIVE 5 WRITING THE EQUATION OF A LINE IN STANDARD FORM

The equation $y = \frac{1}{4}x + 3$ from Example 4 is in **slope-intercept form.** We can subtract $\frac{1}{4}x$ from both sides of this equation to obtain the equation $-\frac{1}{4}x + y = 3$. The equation $-\frac{1}{4}x + y = 3$ is said to be in **standard form.**

> **Standard Form Equation of a Line**
>
> The standard form of an equation of a line is given by $Ax + By = C$, where A, B, and C are real numbers such that A and B are not both zero.

Note that all coefficients of the equation $-\frac{1}{4}x + y = 3$ are **rational numbers.** When the coefficients are all rational numbers, we can eliminate any fractions by multiplying both sides of the equation by an appropriate constant. If we multiply the equation $-\frac{1}{4}x + y = 3$ by -4 we get the following.

$$-4\left(-\frac{1}{4}x + y\right) = 3(-4) \quad \text{Multiply both sides of the equation by } -4.$$

$$x - 4y = -12 \quad \text{Use the distributive property.}$$

The equations $y = \frac{1}{4}x + 3$, $-\frac{1}{4}x + y = 3$, and $x - 4y = -12$ all represent the same line. However, the equation $x - 4y = -12$ is often more desirable because of its simplicity with all integer coefficients and a positive leading coefficient.

TIP In this text (and in all exercises), whenever possible, the standard form of the line will be written using integer coefficients with $A \geq 0$.

Click on the guided visualization icon below and take some time to explore the standard form of a line. For the equation of the line with equation $Ax + By = C$, try changing the values of A, B and C and watch the equation and the line change.

Equations of Lines: Standard Form

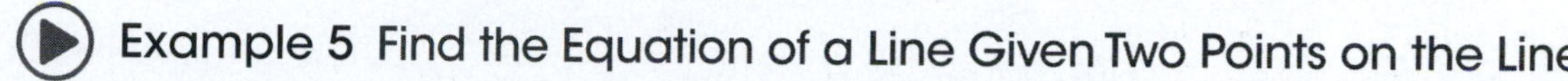

Example 5 Find the Equation of a Line Given Two Points on the Line

Find the equation of the line passing through the points $(-1, 3)$ and $(2, -4)$. Write the equation in point-slope form, slope-intercept form, and standard form.

Solution First, we always find the slope.

$$m = \frac{-4 - 3}{2 - (-1)} = \frac{-7}{2 + 1} = -\frac{7}{3}.$$

Next, select either of the given points. Let's choose the first point $(-1, 3)$. Now that we have a point and the slope, we can use the point-slope form.

$$y - y_1 = m(x - x_1) \quad \text{Write the point-slope equation.}$$

$$y - 3 = -\frac{7}{3}(x + 1) \quad \text{Substitute } m = -\frac{7}{3} \text{ and } (x_1, y_1) = (-1, 3).$$

To find the slope-intercept form of the line, we solve the equation $y - 3 = -\frac{7}{3}(x + 1)$ for y to get $y = -\frac{7}{3}x + \frac{2}{3}$. To see how this is done, read these **steps**.

Finally, write the equation in standard form by eliminating the fractions of the slope-intercept form.

$$3 \cdot y = 3 \cdot \left(-\frac{7}{3}x + \frac{2}{3}\right) \quad \text{Multiply by 3.}$$

$$3y = -7x + 2 \quad \text{Use the distributive property.}$$

$$7x + 3y = 2 \quad \text{Add } 7x \text{ to both sides.}$$

Therefore, the equation of the line that passes through the points $(-1, 3)$ and $(2, -4)$ can be represented in the following forms:

1. **Point-slope form:** $y - 3 = -\frac{7}{3}(x + 1)$ (or $y + 4 = -\frac{7}{3}(x - 2)$, or infinitely many others)

2. Slope–intercept form: $y = -\frac{7}{3}x + \frac{2}{3}$

3. Standard form: $7x + 3y = 2$

This line is sketched and labeled in Figure 21.

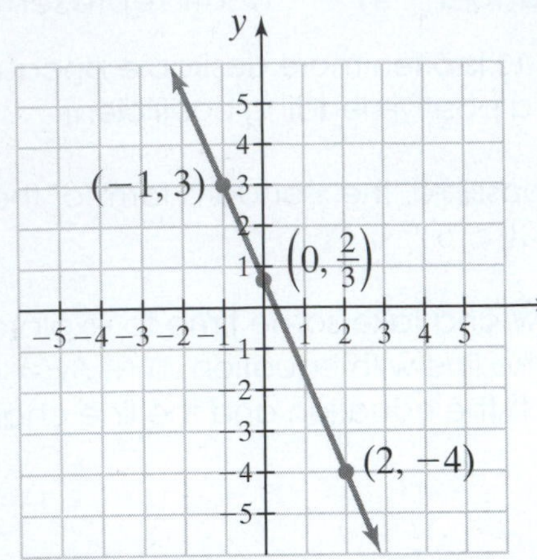

$$y - 3 = -\frac{7}{3}(x + 1) \text{ or}$$
$$y + 4 = -\frac{7}{3}(x - 2) \text{ or}$$
$$y = -\frac{7}{3}x + \frac{2}{3} \text{ or}$$
$$7x + 3y = 2$$

Figure 21 Line through the points $(-1, 3)$ and $(2, -4)$.

You Try It **Work through this You Try It problem.**

Work Exercises 19–22 in this textbook or in the MyLab Math Study Plan.

OBJECTIVE 6 FINDING THE SLOPE AND THE y-INTERCEPT OF A LINE IN STANDARD FORM

Suppose we are given the standard form of a line $Ax + By = C$ with $B \neq 0$ and want to solve for y. To solve for y, we subtract Ax from both sides and divide by B.

$Ax + By = C$	Write the standard form of a line.
$By = -Ax + C$	Subtract Ax from both sides.
$\frac{\cancel{B}y}{\cancel{B}} = -\frac{A}{B}x + \frac{C}{B}$	Divide both sides by B.

The standard form of a line $Ax + By = C$ with $B \neq 0$ is equivalent to the equation $y = \overbrace{-\frac{A}{B}}^{m}x + \overbrace{\frac{C}{B}}^{b}$, which is the equation of a line in slope–intercept form. Thus, given the standard form of a line $Ax + By = C$ with $B \neq 0$, the slope of the line is $m = -\frac{A}{B}$, and the y-intercept is $b = \frac{C}{B}$. Watch the **video** to see how to find the slope and y-intercept of a line given in standard form and to see how to work Example 6.

> **The Slope and y-Intercept of a Line Given in Standard Form**
>
> Given the standard form of a line $Ax + By = C$ with $B \neq 0$, the slope is given by $m = -\frac{A}{B}$, and the y-intercept is $\frac{C}{B}$.

Example 6 Find the Slope and y-Intercept of a Line in Standard Form

Find the slope and y-intercept and sketch the line $3x - 2y = 6$.

Solution Because $A = 3$ and $B = -2$, the slope is $m = -\frac{A}{B} = -\frac{3}{-2} = \frac{3}{2}$. The y-intercept is $\frac{C}{B} = \frac{6}{-2} = -3$. To sketch this line, plot the y-intercept, and then rise 3 and run 2 to plot the point $(2, 0)$. Now connect the points with a straight line, as in Figure 22. Watch the **video** to see this solution worked out in its entirety.

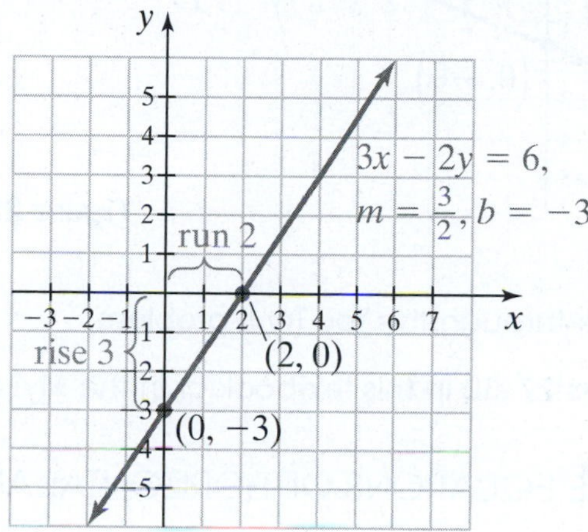

Figure 22 Line $3x - 2y = 6$.

You Try It Work through this You Try It problem.

Work Exercises 23–26 in this textbook or in the MyLab Math Study Plan.

OBJECTIVE 7 SKETCHING LINES BY PLOTTING INTERCEPTS

In Example 6, we were able to sketch the graph of the line $3x - 2y = 6$ by first plotting two points, and then drawing a straight line through them. These two points happened to be the x- and y-intercepts. The method of sketching a line by plotting intercepts is very useful and is illustrated again in Example 7.

Example 7 Sketch a Line by Plotting Intercepts

Sketch the line $2x - 5y = 8$ by plotting intercepts.

Solution Find the **x-intercept.**

The x-intercept is found by setting $y = 0$ and solving for x:

$$2x - 5(0) = 8$$
$$2x = 8$$
$$x = 4$$

The x-intercept is 4.

Find the **y-intercept.**

The y-intercept is found by setting $x = 0$ and solving for y:

$$2(0) - 5y = 8$$
$$-5y = 8$$
$$y = -\frac{8}{5}$$

The y-intercept is $-\frac{8}{5}$.

To sketch the line, plot the two intercepts by plotting the points $(4, 0)$ and $\left(0, -\frac{8}{5}\right)$ and then draw a straight line through the points, as in Figure 23. Watch this **video** to see the complete solution to this example.

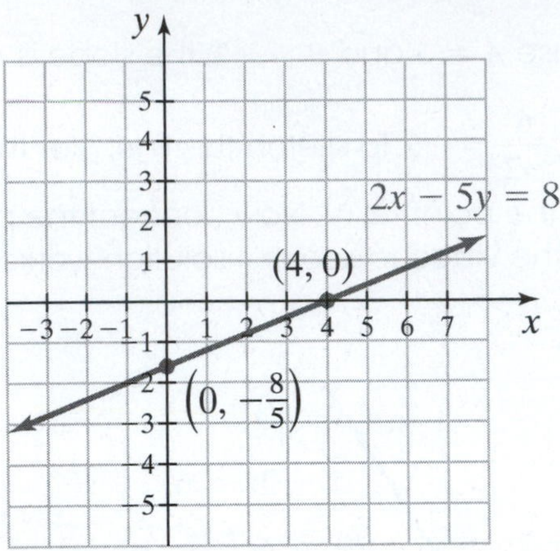

Figure 23 Line $2x - 5y = 8$. ●

You Try It Work through this You Try It problem.

Work Exercises 27–30 in this textbook or in the MyLab Math Study Plan.

OBJECTIVE 8 FINDING THE EQUATIONS OF HORIZONTAL AND VERTICAL LINES

HORIZONTAL LINES

Suppose we want to determine the equation of the horizontal line that contains the point (a, b). To find this equation, we must first determine the slope. Because the line must also pass through the point $(0, b)$, we see that the slope of this line is $m = \frac{b - b}{a - b} = \frac{0}{a} = 0$. Using the slope-intercept form of a line with $m = 0$ and y-intercept b, we see that the equation is $y = 0x + b$ or $y = b$. See Figure 24.

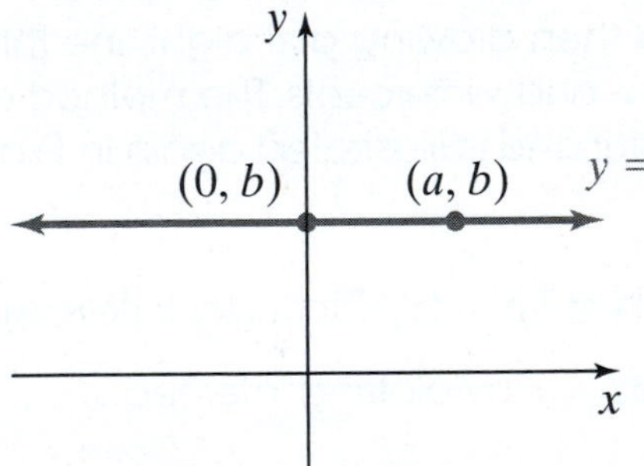

Figure 24 Equation of a horizontal line is $y = b$. ●

Therefore, we know that for any horizontal line that contains the point (a, b), the equation of that line is $y = b$, and the slope is $m = 0$.

VERTICAL LINES

Vertical lines have **no slope** (also called an **undefined slope**). We can see this by looking at the vertical line that passes through the point (a, b). Because this line also passes through the x-intercept at the point $(a, 0)$, we see that the slope of this line is $m = \frac{b - 0}{a - a} = \frac{b}{0}$, which is not a real number as division by zero is not defined.

Because the x-coordinate of this vertical line is always equal to a regardless of the y-coordinate, we say that the equation of a vertical line is $x = a$. See Figure 25.

Therefore, we know that for any vertical line that contains the point (a, b), the equation of that line is $x = a$, and the slope is undefined.

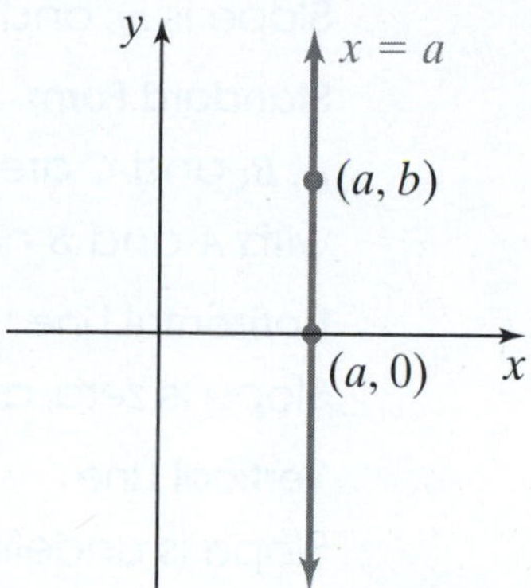

Figure 25 Equation of a vertical line is $x = a$.

Example 8 Find the Equations of Horizontal and Vertical Lines

a. Find the equation of the horizontal line passing through the point $(-1, 3)$.

b. Find the equation of the vertical line passing through the point $(-1, 3)$.

Solution

a. A horizontal line has the equation $y = b$, where b is the y-coordinate of any point on the line. Therefore, the equation is $y = 3$.

b. A vertical line has the equation $x = a$, where a is the x-coordinate of any point on the line. Therefore, the equation is $x = -1$.

Both lines are sketched in Figure 26.

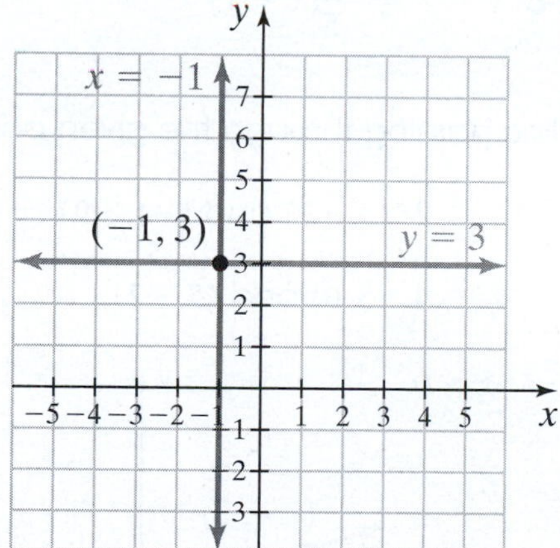

Figure 26 Graph of $y = 3$ and $x = -1$.

You Try It Work through this You Try It problem.

Work Exercises 31–34 in this textbook or in the MyLab Math Study Plan.

Summary of Forms of Equations of Lines

1. $y - y_1 = m(x - x_1)$	**Point–Slope Form** Slope is m, and (x_1, y_1) is a point on the line.
2. $y = mx + b$	**Slope–Intercept Form** Slope is m, and y-intercept is at $(0, b)$.
3. $Ax + By = C$	**Standard Form** A, B, and C are real numbers, with A and B not both zero and $A \geq 0$.
4. $y = b$	**Horizontal Line** Slope is zero, and y-intercept is at $(0, b)$.
5. $x = a$	**Vertical Line** Slope is undefined, and x-intercept is at $(a, 0)$.

2.3 Exercises

Skill Check Exercises

For exercises SCE-1 through SCE-2 solve each equation for the variable y.

SCE-1. $2x - y = 5$

SCE-2. $3x + 2y = 7$

For exercises SCE-3 through SCE-6, first use the distributive property then solve the equation for the variable y.

SCE-3. $y - 1 = 2(x + 3)$

SCE-4. $y + 3 = -5(x - 2)$

SCE-5. $y - 5 = \frac{2}{7}(x + 3)$

SCE-6. $y + 4 = -\frac{3}{5}(x - 1)$

In Exercises 1–6, find the slope of the line passing through the given points.

1. $(-5, 2)$ and $(2, -6)$
2. $(0, -7)$ and $(-4, -9)$
3. $(2, -3)$ and $(-1, -3)$
4. $(5, 6)$ and $(5, -2)$
5.
6.

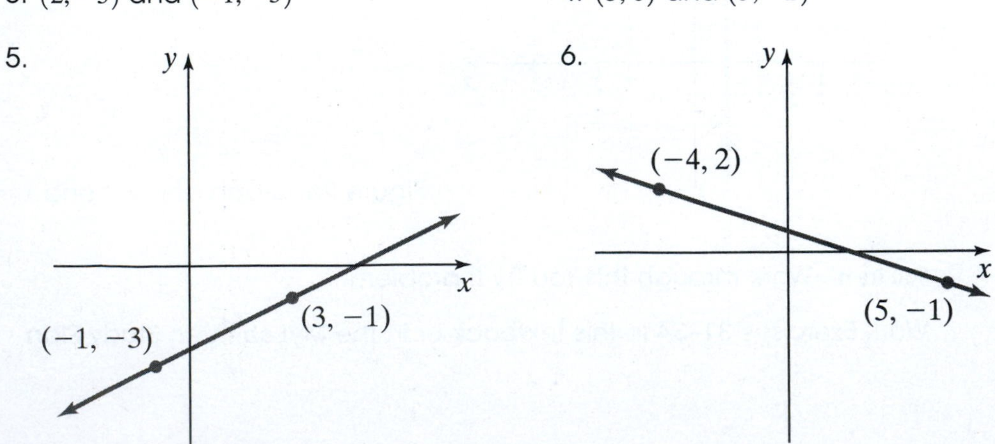

In Exercises 7–10, sketch the line with the given slope that passes through the indicated point.

7. Slope $= \frac{1}{2}$; line passes through the point $(-2, -2)$

8. Slope $= -\frac{4}{3}$; line passes through the point $(3, 8)$

9. Slope $= 0$; line passes through the point $(-5, -1)$

10. Slope is undefined; line passes through the point $(8, 0)$

In Exercises 11–14, find the point-slope form of the line with the given slope that passes through the indicated point.

11. Slope $= \frac{1}{2}$; line passes through the point $(-2, -2)$

12. Slope $= -\frac{4}{3}$; line passes through the point $(3, 8)$

13. Slope $= 3$; line passes through the point $(-5, 3)$

14. Slope $= \frac{5}{11}$; line passes through the point $(-99, 88)$

In Exercises 15–18, find the slope-intercept form of the line with the given slope and y-intercept.

15. Slope $= 1$; y-intercept $= -2$

16. Slope $= -\frac{1}{6}$; y-intercept $= \frac{1}{2}$

17. Slope $= \frac{7}{9}$; y-intercept $= -\frac{8}{7}$

18. Slope $= 0$; y-intercept $= 5$

In Exercises 19–22, find the equation of the line passing through the indicated two points. Write the equation in point-slope form, slope-intercept form, and standard form.

19. $(1, 2)$ and $(-2, 5)$

20. $(-5, 7)$ and $(3, -5)$

21. $\left(-\frac{1}{2}, 1\right)$ and $\left(-3, \frac{2}{3}\right)$

22. $(-3, 4)$ and $(2, 4)$

In Exercises 23–26, given the equation of a line in standard form, determine the slope and y-intercept, and sketch the line.

23. $4x - y = 12$

24. $x - 3y = -9$

25. $5x + 7y = 12$

26. $8x - 11y = -23$

In Exercises 27–30, sketch the given line by plotting intercepts.

27. $4x - y = 12$

28. $x - 3y = -9$

29. $4x + 3y = 6$

30. $5x + 4y = -10$

31. Find the equation of the horizontal line passing through the point $(5, -2)$.

32. Find the equation of the vertical line passing through the point $(5, -2)$.

In Exercises 33 and 34, find the equation of the given line.

33.

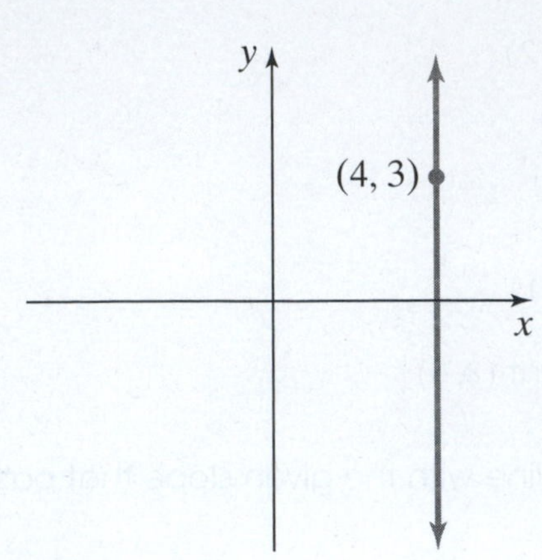

34.

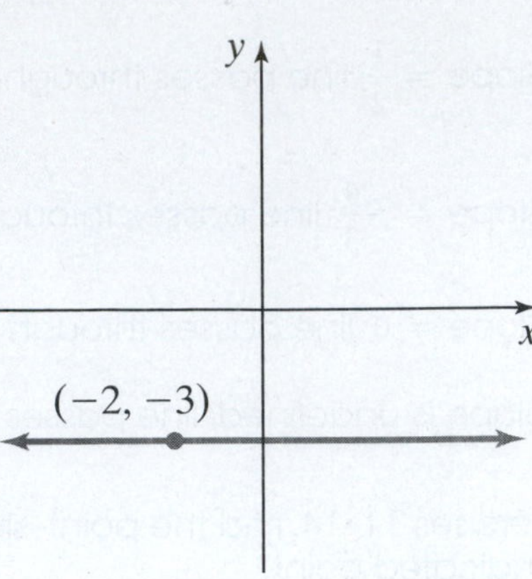

Brief Exercises

In Exercises 35–38, find the equation of the line passing through the indicated two points. Write the equation in point-slope form.

35. $(1, 2)$ and $(-2, 5)$

36. $(-5, 7)$ and $(3, -5)$

37. $\left(-\frac{1}{2}, 1\right)$ and $\left(-3, \frac{2}{3}\right)$

38. $(-3, 4)$ and $(2, 4)$

In Exercises 39–42, find the equation of the line passing through the indicated two points. Write the equation in slope-intercept form.

39. $(1, 2)$ and $(-2, 5)$

40. $(-5, 7)$ and $(3, -5)$

41. $\left(-\frac{1}{2}, 1\right)$ and $\left(-3, \frac{2}{3}\right)$

42. $(-3, 4)$ and $(2, 4)$

In Exercises 43–46, find the equation of the line passing through the indicated two points. Write the equation in standard form.

43. $(1, 2)$ and $(-2, 5)$

44. $(-5, 7)$ and $(3, -5)$

45. $\left(-\frac{1}{2}, 1\right)$ and $\left(-3, \frac{2}{3}\right)$

46. $(-3, 4)$ and $(2, 4)$

In Exercises 47–50, given the equation of a line in standard form, determine the slope and y-intercept.

47. $4x - y = 12$

48. $x - 3y = -9$

49. $5x + 7y = 12$

50. $8x - 11y = -23$

In Exercises 51–54, sketch the line whose equation is given in standard form.

51. $4x - y = 12$

52. $x - 3y = -9$

53. $5x + 7y = 12$

54. $8x - 11y = -23$

2.4 Parallel and Perpendicular Lines

THINGS TO KNOW

Before working through this section, be sure that you are familiar with the following concepts:

VIDEO | ANIMATION | INTERACTIVE

You Try It 1. Determining the Slope of a Line (Section 2.3) 

You Try It 2. Finding the Equation of a Line Using the Point–Slope Form (Section 2.3) 

You Try It 3. Writing the Equation of a Line in Standard Form (Section 2.3) 

You Try It 4. Finding the Slope and the y-Intercept of a Line in Standard Form (Section 2.3) 

INTRODUCTION

Read this introduction before beginning Objective 1.

OBJECTIVES

1 Understanding the Definition of Parallel Lines

2 Understanding the Definition of Perpendicular Lines

3 Determining Whether Two Lines Are Parallel, Perpendicular, or Neither

4 Finding the Equations of Parallel and Perpendicular Lines

5 Solving a Geometric Application of Parallel and Perpendicular Lines

SECTION 2.4 EXERCISES

Introduction to Section 2.4

Given any two *distinct* lines in the Cartesian plane, the two lines will either intersect or not. In this section, we investigate the nature of two lines that do not intersect (**parallel lines**) and then discuss the special case of two lines that intersect at a right angle (**perpendicular lines**). These two cases are interesting because we need only know the slope of the two lines to determine whether the lines are parallel, perpendicular, or neither. Let's first consider two lines in the plane that do not intersect.

OBJECTIVE 1 UNDERSTANDING THE DEFINITION OF PARALLEL LINES

As stated, two lines are parallel if they do not intersect, or in other words, the lines do not share any common points. In Section 2.3, we learned about the slope of a line. It is imperative that you can compute the slope of a line given its equation. To review this concept, refer back to **Section 2.3**. Because parallel lines do not intersect, the ratio of the vertical change (rise) to the horizontal change (run) of each line must be equivalent (Figure 27). In other words, parallel lines have the same slope!

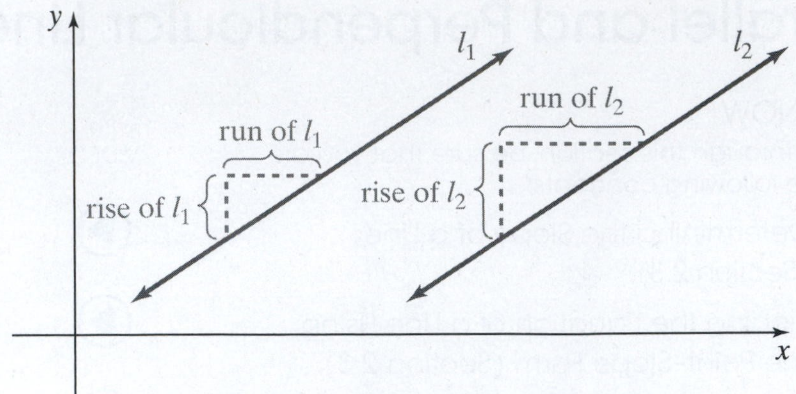

The ratios of the vertical rise to the horizontal run of any two parallel lines must be equal:

$$\frac{\text{rise of } l_1}{\text{run of } l_1} = \frac{\text{rise of } l_2}{\text{run of } l_2}.$$

Figure 27

> **Theorem**
>
> Two distinct nonvertical lines in the Cartesian plane are parallel if and only if they have the same slope.

We can prove this theorem using a little geometry. We can show that if l_1 is parallel to l_2, then triangles ΔABC and ΔDEF are similar. See Figure 28. Because the ratio of two corresponding sides of similar triangles are proportional, it follows that the slope of $l_1 = \frac{AB}{BC} = \frac{DE}{EF} =$ slope of l_2. Conversely, if the slope of l_1 is the same as the slope of l_2, then the two triangles are similar. Therefore, $\angle CAB \cong \angle FDE$ and thus the lines must be parallel.

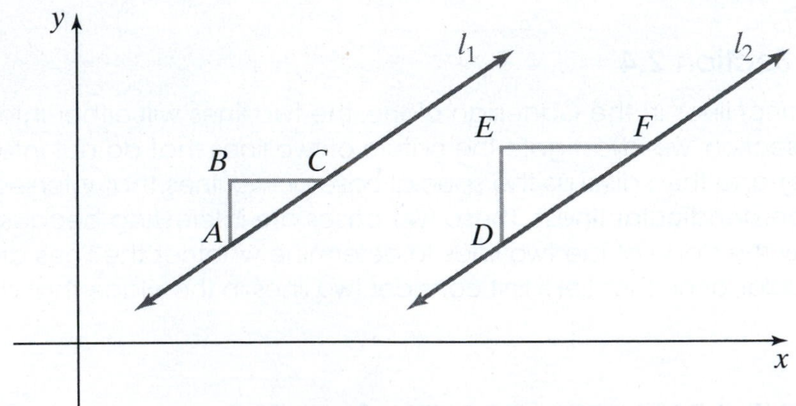

Figure 28 ΔABC and ΔDEF are similar, so

$$\frac{AB}{BC} = \frac{DE}{EF}.$$

This implies that the slopes are equal.

Example 1 Show That Two Lines Are Parallel

Show that the lines $y = -\frac{2}{3}x - 1$ and $4x + 6y = 12$ are parallel.

Solution The line $y = -\frac{2}{3}x - 1$ is in **slope–intercept form** with a slope of $m = -\frac{2}{3}$ and a y-intercept -1. The line $4x + 6y = 12$ is in **standard form** with slope $m = -\frac{A}{B} = -\frac{4}{6} = -\frac{2}{3}$ and y-intercept $\frac{C}{B} = \frac{12}{6} = 2$.

The two lines have the same slope but have different y-intercepts. Therefore, the lines must be parallel. See Figure 29.

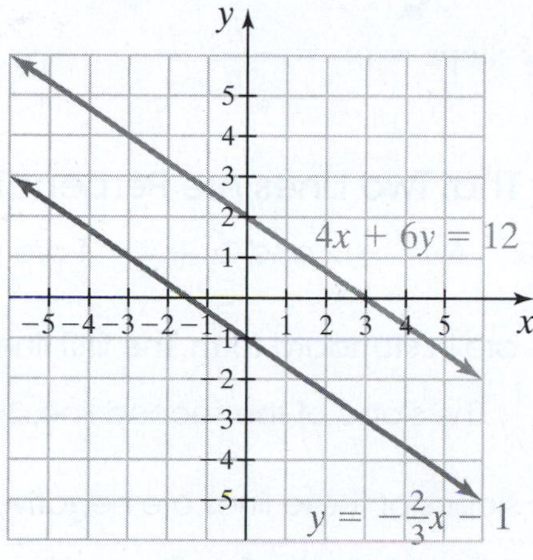

Figure 29 Parallel lines $y = -\frac{2}{3}x - 1$ and $4x + 6y = 12$.

OBJECTIVE 2 UNDERSTANDING THE DEFINITION OF PERPENDICULAR LINES

If two *distinct* lines are not parallel, then they must intersect at a single point. If the two lines intersect at a right angle (90 degrees), then the lines are said to be **perpendicular**. The slopes of perpendicular lines have a special relationship that is stated in the following theorem.

Theorem

Two nonvertical lines in the Cartesian plane are perpendicular if and only if the product of their slopes is -1.

This theorem states that if line 1 with slope m_1 is perpendicular to line 2 with slope m_2, then $m_1 m_2 = -1$. Because the product of the slopes of nonvertical perpendicular lines is -1, it follows that

$$m_1 = -\frac{1}{m_2}.$$

This means that if $m_1 = \frac{a}{b}$, then $m_2 = -\frac{b}{a}$. In other words, the slopes of nonvertical perpendicular lines are **negative reciprocals** of each other, as in Figure 30.

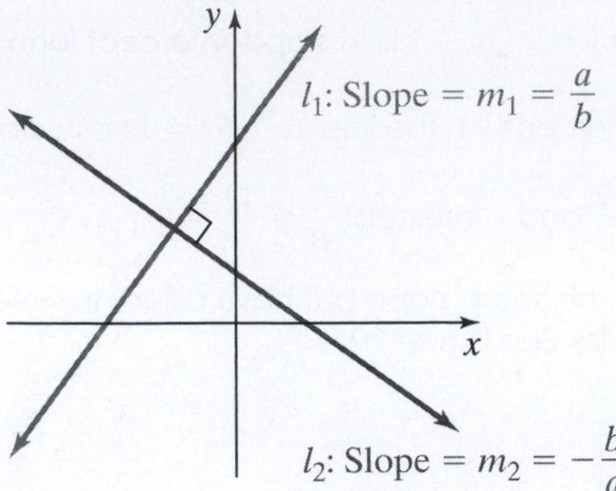

Figure 30 Slopes of perpendicular lines are negative reciprocals of each other.

Example 2 Show That Two Lines Are Perpendicular

Show that the lines $3x - 6y = -12$ and $2x + y = 4$ are perpendicular.

Solution Both lines are in **standard form**. The first line, $3x - 6y = -12$, has slope $m_1 = -\frac{A}{B} = -\frac{3}{-6} = \frac{1}{2}$. The slope of the second line, $2x + y = 4$, has slope $m_2 = -\frac{A}{B} = -\frac{2}{1}$. The slopes of these lines are negative reciprocals of each other. Therefore, the lines are perpendicular. See Figure 31.

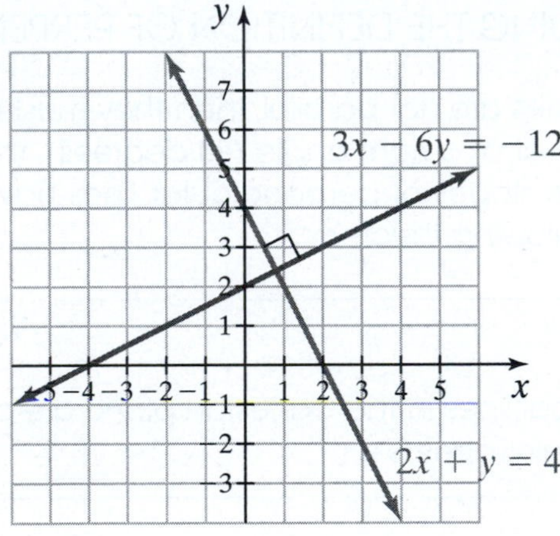

Figure 31 Perpendicular lines $3x - 6y = -12$ and $2x + y = 4$.

Summary of Parallel and Perpendicular Lines

Given two nonvertical lines l_1 and l_2 such that the slope of line l_1 is $m_1 = \frac{a}{b}$,

1. l_2 is parallel to l_1 if and only if $m_2 = \frac{a}{b}$.
2. l_2 is perpendicular to l_1 if and only if $m_2 = -\frac{b}{a}$.

TIP Any two distinct vertical lines are parallel to each other, whereas any horizontal line is perpendicular to any vertical line.

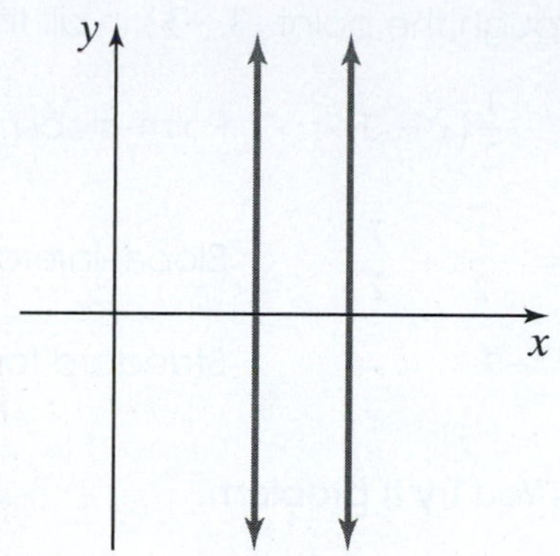

Figure 32 Any two distinct vertical lines are parallel.

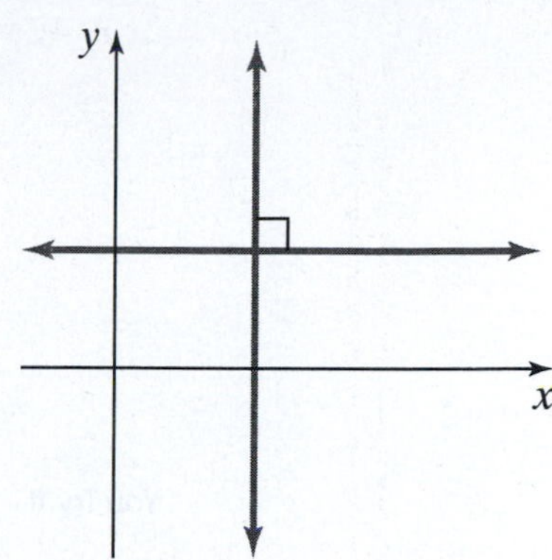

Figure 33 Horizontal lines are perpendicular to vertical lines.

OBJECTIVE 3 DETERMINING WHETHER TWO LINES ARE PARALLEL, PERPENDICULAR, OR NEITHER

Given any two distinct lines, we are now able to quickly determine whether the lines are parallel to each other, perpendicular to each other, or neither by simply evaluating their slopes.

Example 3 Determine Whether Two Lines Are Parallel, Perpendicular, or Neither

For each of the following pairs of lines, determine whether the lines are parallel, perpendicular, or neither.

a. $3x - y = 4$
$x + 3y = 7$

b. $y = \frac{1}{2}x + 3$
$x + 2y = 1$

c. $x = -1$
$x = 3$

Solution Watch the **video** to verify that

a. The lines are perpendicular.

b. The lines are neither parallel nor perpendicular.

c. The lines are parallel.

You Try It **Work through this You Try It problem.**

Work Exercises 1–6 in this textbook or in the MyLab Math Study Plan.

OBJECTIVE 4 FINDING THE EQUATIONS OF PARALLEL AND PERPENDICULAR LINES

Example 4 Find the Equation of a Parallel Line

Find the equation of the line parallel to the line $2x + 4y = 1$ that passes through the point $(3, -5)$. Write the answer in **point-slope form, slope-intercept form,** and **standard form.**

Solution Watch the video to see that the equations of the line parallel to $2x + 4y = 1$ that passes through the point $(3, -5)$ in all three forms are

$$y + 5 = -\frac{1}{2}(x - 3) \quad \text{Point–slope form}$$

$$y = -\frac{1}{2}x - \frac{7}{2} \quad \text{Slope–intercept form}$$

$$x + 2y = -7 \quad \text{Standard form}$$

You Try It Work through this You Try It problem.

Work Exercises 7–11 in this textbook or in the MyLab Math Study Plan.

Example 5 Find the Equation of a Perpendicular Line

Find the equation of the line perpendicular to the line $y = -5x + 2$ that passes through the point $(3, -1)$. Write the answer in slope–intercept form.

Solution The line $y = -5x + 2$ is in slope–intercept form with slope -5. We are looking for the line perpendicular to this line. Therefore, the slope of the new line must be the negative reciprocal of -5. Because we can rewrite -5 as $-\frac{5}{1}$, the negative reciprocal is $\frac{1}{5}$. Using a slope of $\frac{1}{5}$ and the point $(3, -1)$, we can obtain the point–slope equation $y + 1 = \frac{1}{5}(x - 3)$. Solving this equation for y, we obtain the slope–intercept form $y = \frac{1}{5}x - \frac{8}{5}$. You can see in Figure 34 how these two lines intersect at a right angle.

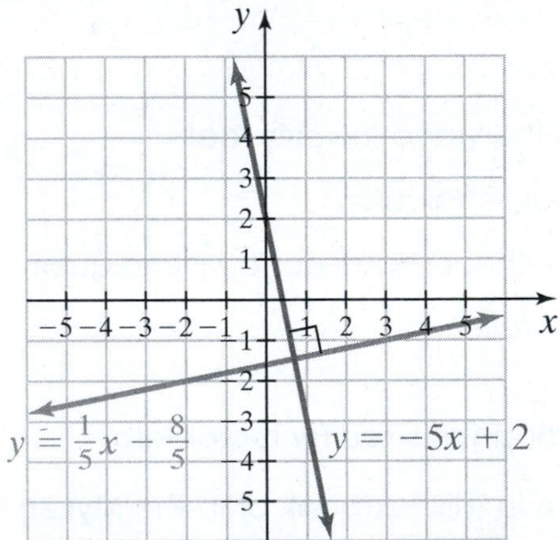

Figure 34 Graph of perpendicular lines $y = -5x + 2$ and $y = \frac{1}{5}x - \frac{8}{5}$.

You Try It Work through this You Try It problem.

Work Exercises 12–16 in this textbook or in the MyLab Math Study Plan.

OBJECTIVE 5 SOLVING A GEOMETRIC APPLICATION OF PARALLEL AND PERPENDICULAR LINES

In Section 2.1, we used the midpoint formula to determine whether four points in the plane form a parallelogram. (See **Section 2.1, Example 6.**) Another way to determine whether four points in the plane form a parallelogram is to determine the slopes of the sides opposite each other. If the slopes of the segments opposite one another are equal, then the four points must form a parallelogram. A special parallelogram in which all four sides have the same length is called a *rhombus*. In geometry, it can be shown that if the diagonals of a parallelogram are perpendicular, then the parallelogram is a rhombus. See Figures 35 and 36.

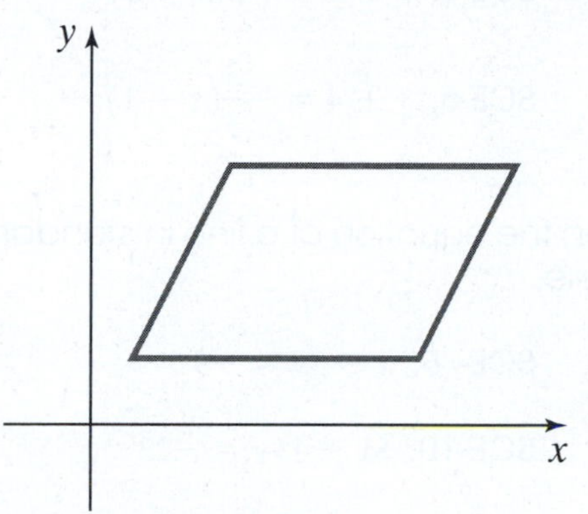

Figure 35 Parallelogram

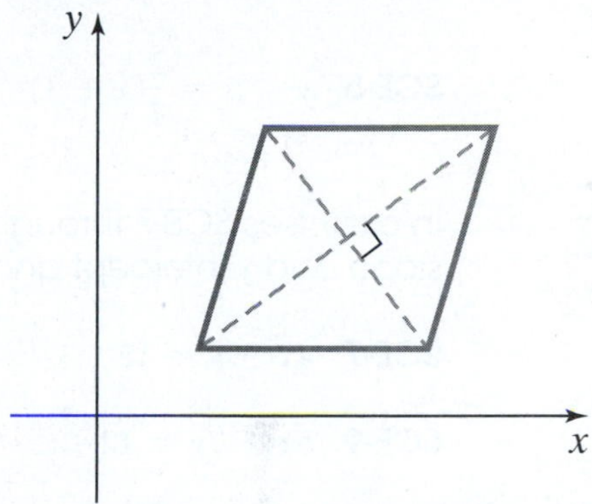

Figure 36 Rhombus

Example 6 Determine Whether Four Points in the Plane Form a Parallelogram or a Rhombus

Do the points $A(0, 4)$, $B(3, 0)$, $C(9, 1)$, and $D(6, 5)$ form a parallelogram? If the points form a parallelogram, then is the parallelogram a rhombus?

Solution These four points do form a parallelogram because the slopes of the segments opposite one another are equal. Verify this for yourself, or watch the **video** to see the worked out solution. The parallelogram formed by these four points is not a rhombus because diagonal $\overline{AC}$ has slope $-\frac{1}{3}$, whereas diagonal $\overline{BD}$ has slope $\frac{5}{3}$ and $\left(-\frac{1}{3}\right)\left(\frac{5}{3}\right) = -\frac{5}{9} \neq -1.$

Therefore, the diagonals are *not* perpendicular. Watch the **video** to see the entire solution.

You Try It Work through this You Try It problem.

Work Exercises 17–20 in this textbook or in the MyLab Math Study Plan.

2.4 Exercises

Skill Check Exercises

For exercises SCE-1 and SCE-2 solve each equation for the variable y.

SCE-1. $2x - y = 5$ **SCE-2.** $3x + 2y = 7$

For exercises SCE-3 through SCE-6, first use the distributive property then solve the equation for the variable y.

SCE-3. $y - 1 = 2(x + 3)$ **SCE-4.** $y + 3 = -5(x - 2)$

SCE-5. $y - 5 = \frac{2}{7}(x + 3)$ **SCE-6.** $y + 4 = -\frac{3}{5}(x - 1)$

In exercises SCE-7 through SCE-10, given the equation of a line in standard form, determine the slope and y-intercept and sketch the line.

SCE-7. $4x - y = 12$ **SCE- 8.** $x - 3y = -9$

SCE-9. $5x + 7y = 12$ **SCE-10.** $8x - 11y = -23$

In Exercises 1–6, decide whether the pair of lines are parallel, perpendicular, or neither.

1. $-4x - 2y = -9$
   $5x - 10y = 7$
2. $y = 3x - 2$
   $9x - 3y = 1$
3. $y = -2$
   $3y = 12$
4. $x = -1$
   $y = -1$
5. $4x - 3y = 7$
   $3x - 4y = -6$
6. $2x + 3y = 3$
   $y = -\frac{2}{3}x + 1$

In Exercises 7–9, find the equation of a line described as follows, and express your answer in point-slope form, slope-intercept form, and standard form.

SbS 7. Find the equation of the line parallel to the line $y = \frac{1}{4}x - 2$ that passes through the point $(-2, 3)$.

SbS 8. Find the equation of the line parallel to the line $3x - 5y = 1$ that passes through the point $(1, -4)$.

SbS 9. Find the equation of the line parallel to the line $y + 1 = 3(x - 5)$ that passes through the point $(-2, -5)$.

In Exercises 10 and 11, find the equation of a line described as follows, and express your answer in standard form.

10. Find the equation of the line parallel to the line $x = 7$ that passes through the point $(11, -3)$.

11. Find the equation of the line parallel to the line $y = 8$ that passes through the point $(11, -3)$.

In Exercises 12–14, find the equation of a line described as follows, and express your answer in point-slope form, slope-intercept form, and standard form.

SbS 12. Find the equation of the line perpendicular to the line $y = \frac{1}{4}x - 2$ that passes through the point $(-2, 3)$.

SbS 13. Find the equation of the line perpendicular to the line $3x - 5y = 1$ that passes through the point $(1, -4)$.

SbS 14. Find the equation of the line perpendicular to the line $y + 1 = 3(x - 5)$ that passes through the point $(-2, -5)$.

In Exercises 15 and 16, find the equation of a line described as follows, and express your answer in standard form.

15. Find the equation of the line perpendicular to the line $x = 7$ that passes through the point $(11, -3)$.

16. Find the equation of the line perpendicular to the line $y = 8$ that passes through the point $(11, -3)$.

In Exercises 17–20, use slope to determine whether the quadrilateral with the given vertices forms a parallelogram. If it is a parallelogram, determine whether it is also a rhombus.

17. $A(-1, 2), B(3, 3), C(4, 7), D(0, 6)$

18. $A(2, 2), B(3, -1), C(5, 2), D(4, 5)$

19. $A(1, 1), B(5, 2), C(8, 5), D(3, 3)$

20. $A(1, 1), B(4, -3), C(7, 1), D(4, 5)$

Brief Exercises

21. Find the equation of the line parallel to the line $y = \frac{1}{4}x - 2$ that passes through the point $(-2, 3)$. Express your answer in point-slope form.

22. Find the equation of the line parallel to the line $3x - 5y = 1$ that passes through the point $(1, -4)$. Express your answer in point-slope form.

23. Find the equation of the line parallel to the line $y + 1 = 3(x - 5)$ that passes through the point $(-2, -5)$. Express your answer in point-slope form.

24. Find the equation of the line parallel to the line $y = \frac{1}{4}x - 2$ that passes through the point $(-2, 3)$. Express your answer in slope-intercept form.

25. Find the equation of the line parallel to the line $3x - 5y = 1$ that passes through the point $(1, -4)$. Express your answer in slope-intercept form.

26. Find the equation of the line parallel to the line $y + 1 = 3(x - 5)$ that passes through the point $(-2, -5)$. Express your answer in slope-intercept form.

27. Find the equation of the line parallel to the line $y = \frac{1}{4}x - 2$ that passes through the point $(-2, 3)$. Express your answer in standard form.

28. Find the equation of the line parallel to the line $3x - 5y = 1$ that passes through the point $(1, -4)$. Express your answer in standard form.

29. Find the equation of the line parallel to the line $y + 1 = 3(x - 5)$ that passes through the point $(-2, -5)$. Express your answer in standard form.

30. Find the equation of the line perpendicular to the line $y = \frac{1}{4}x - 2$ that passes through the point $(-2, 3)$. Express your answer in point-slope form.

31. Find the equation of the line perpendicular to the line $3x - 5y = 1$ that passes through the point $(1, -4)$. Express your answer in point-slope form.

32. Find the equation of the line perpendicular to the line $y + 1 = 3(x - 5)$ that passes through the point $(-2, -5)$. Express your answer in point-slope form.

33. Find the equation of the line perpendicular to the line $y = \frac{1}{4}x - 2$ that passes through the point $(-2, 3)$. Express your answer in slope-intercept form.

34. Find the equation of the line perpendicular to the line $3x - 5y = 1$ that passes through the point $(1, -4)$. Express your answer in slope-intercept form.

35. Find the equation of the line perpendicular to the line $y + 1 = 3(x - 5)$ that passes through the point $(-2, -5)$. Express your answer in slope-intercept form.

36. Find the equation of the line perpendicular to the line $y = \frac{1}{4}x - 2$ that passes through the point $(-2, 3)$. Express your answer in standard form.

37. Find the equation of the line perpendicular to the line $3x - 5y = 1$ that passes through the point $(1, -4)$. Express your answer in standard form.

38. Find the equation of the line perpendicular to the line $y + 1 = 3(x - 5)$ that passes through the point $(-2, -5)$. Express your answer in standard form.

Chapter 2 Summary

Key Concepts	Examples/Videos
2.1 The Rectangular Coordinate System Quadrant II, Quadrant I, Quadrant III, Quadrant IV	Plot the ordered pairs $(-2, 3)$, $(0, 4)$, $(2, 5)$, and $(4, 6)$, and state in which quadrant or on which axis each ordered pair lies.

Key Concepts	Examples/Videos
One way to sketch the graph of an equation in two variables is to find several ordered pairs that **satisfy** the equation, plot those ordered pairs, and then connect the points with a smooth curve.	Sketch the graph of $y = x^2 - 4x + 4$. Determine whether the following ordered pairs lie on the graph of the equation $x^2 + y^2 = 1$. a. $(0, -1)$ b. $(1, 0)$ c. $\left(\frac{1}{3}, \frac{2}{3}\right)$ d. $\left(-\frac{\sqrt{2}}{2}, \frac{\sqrt{2}}{2}\right)$
The **intercepts** of a graph are the points where a graph crosses or touches a coordinate axis. The **x-intercept** is the x-coordinate of a point where a graph crosses or touches the x-axis. To algebraically determine the x-intercepts of the graph of an equation in two variables, set all values of the variable y equal to 0 and solve for x. The **y-intercept** is the y-coordinate of a point where a graph crosses or touches the y-axis. To algebraically determine the y-intercepts of the graph of an equation in two variables, set all values of the variable x equal to 0 and solve for y.	Find the x- and y-intercepts of the given equations. a. $y = \frac{2x - 1}{x + 3}$ b. $\sqrt{x + 2} + y = 3$ c. $(x - 1)^2 + (y - 3)^2 = 5$
Midpoint Formula The midpoint of the line segment from $A(x_1, y_1)$ to $B(x_2, y_2)$ is $\left(\frac{x_1 + x_2}{2}, \frac{y_1 + y_2}{2}\right)$.	Find the midpoint of the segment whose endpoints are $(-3, 2)$ and $(4, 6)$
Distance Formula The distance between any two point $A(x_1, y_1)$ and $B(x_2, y_2)$ is given by the formula $d(A, B) = \sqrt{(x_2 - x_1)^2 + (y_2 - y_1)^2}$.	Find the distance, $d(A, B)$, between the points A and B. $A(-1, 5)$ and $B(4, -5)$
2.2 Circles The **standard form of an equation of a circle** with center (h, k) and radius, r, is $(x - h)^2 + (y - k)^2 = r^2$	Find the standard form of the equation of the circle whose center is $(-2, 3)$ with radius 6.

Key Concepts	Examples/Videos
 Click on the **guided visualization** icon to practice sketching the graphs of circles. The **general form of an equation of a circle** is given by $Ax^2 + By^2 + Cx + Dy + E = 0$.	 Find the center, radius, intercepts and graph the circle $(x - 1)^2 + (y + 2)^2 = 9$.  Write the equation $4x^2 + 4y^2 + 4x - 8y + 1 = 0$ in standard form. Find the center, radius, intercepts, and sketch the graph.
2.3 Lines If $x_1 \neq x_2$, the **slope** of a nonvertical line passing through distinct points (x_1, y_1) and (x_2, y_2) is given by $m = \frac{rise}{run} = \frac{\text{change in } y}{\text{change in } x} = \frac{y_2 - y_1}{x_2 - x_1}$ 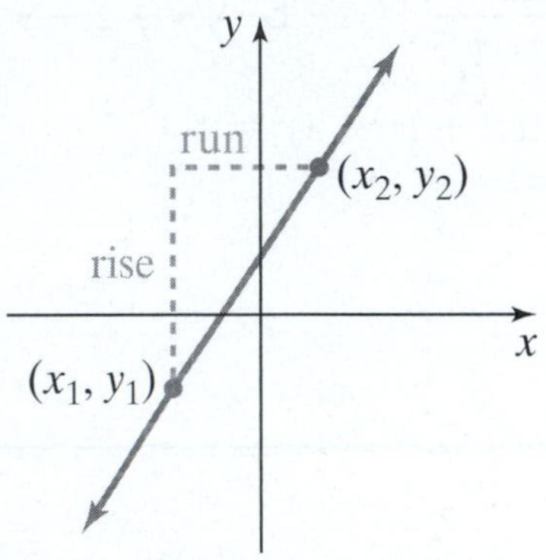	 Find the slope of the line that passes through the indicated ordered pairs. a. $(6, -4)$ and $(-5, 1)$ b. $(3, -1)$ and $(3, 6)$ c. $(-5, 4)$ and $(-2, 4)$
Summary of Equations of Lines  Point-Slope Form: $y - y_1 = m(x - x_1)$  Slope-Intercept Form: $y = mx + b$  Standard Form: $Ax + By = C$ Horizontal Line: $y = b$ Vertical Line: $x = a$	 Find the equation in point-slope form of the line with slope $m = \frac{2}{3}$ that passes through the point $(-1, -4)$.  Find the equation of the line with slope $\frac{1}{4}$ and y-intercept 3, and write the equation in slope-intercept form. Find the equation of the line passing through the points $(-1, 3)$ and $(2, -4)$. Write the equation in point-slope form, slope-intercept form, and standard form.  Horizontal/Vertical Line Example a. Find the equation of the horizontal line passing through the point $(-1, 3)$. b. Find the equation of the vertical line passing through the point $(-1, 3)$.

Key Concepts	Examples/Videos
2.4 Parallel and Perpendicular Lines Two nonvertical lines are parallel if and only if they have the same slope. $(m_1 = m_2)$ Two nonvertical lines are perpendicular if and only if the product of their slopes is -1. $\left(\text{If } m_1 = \frac{a}{b}, \text{ then } m_2 = -\frac{b}{a}\right)$ Any two distinct vertical lines are parallel. Every horizontal line is perpendicular to every vertical line.	For each of the following pairs of lines, determine the lines are parallel, perpendicular, or neither. a. $3x - y = 4$ $x + 3y = 7$ b. $y = \frac{1}{2}x + 3$ $x + 2y = 1$ c. $x = -1$ $x = 3$  Find the equation of the line parallel to the line $2x + 4y = 1$ that passes through the point $(3, -5)$.  Find the equation of the line perpendicular to the line $y = -5x + 2$ that passes through the point $(3, -1)$.

Chapter 2 Review Exercises

1. Plot each ordered pair in the Cartesian plane, and state which quadrant or on which axis it lies.

   a. $A(-1, 4)$ b. $B(-2, -2)$ c. $C(0, 3)$ d. $D\left(1, \frac{1}{2}\right)$ e. $E(2, 0)$

2. Sketch the graph of $y = 3x^2$ by plotting points.

3. Determine whether the indicated ordered pairs lie on the graph of the equation $y = \sqrt{x} - 1$.

   a. $(4, 1)$ b. $(1, 4)$ c. $(0, 1)$

4. Find the x- and y-intercepts of the graph of $y = \dfrac{x^2 - x - 6}{x + 4}$.

5. Find the midpoint of the line segments joining points *A* and *B*.

   $$A(-1, 4); B(-2, -2)$$

6. Find the distance, $d(A, B)$, between the points $A(2, -4)$ and $B(-3, 6)$.

7. Write the standard form of the equation of the circle with center $(-2, 3)$, and radius $r = 4$.

8. Write the standard form of the equation of the circle with center whose endpoints of a diameter are $(2, -6)$, and $(6, 1)$.

9. Find the center, radius, and intercepts then sketch the graph of the circle whose equation is given by $(x - 4)^2 + (y + 7)^2 = 12$.

10. Find the center, radius, and intercepts then sketch the graph of the circle whose equation is given by $x^2 + y^2 - 10x + 6y + 18 = 0$.

11. Find the center, radius, and intercepts then sketch the graph of the circle whose equation is given by $16x^2 + 16y^2 - 16x + 8y - 11 = 0$.

12. Find the slope of the line that passes through the points $(0, -7)$ and $(-4, -9)$.

13. Sketch the line that has a slope of $\frac{1}{2}$ and passes through the point $(-2, -2)$.

14. Find the point-slope equation of the line that has a slope of 3 and passes through the point $(-5, 3)$.

15. Find the slope-intercept equation of the line that has a slope of 1 and a y-intercept of -2.

16. Find the equation of the line that passes through the points $(-5, 7)$ and $(3, -5)$. Write the equation in point-slope form, slope-intercept form, and standard form.

17. Determine the slope and y-intercept of the line having the equation $x - 3y = -9$. Then sketch the line.

18. Sketch the line that has the equation $4x - y = 12$ by plotting the intercepts.

19. Find the equation of the horizontal line passing through the point $(5, -2)$.

20. Find the equation of the vertical line passing through the point $(5, -2)$.

21. Decide whether the pair of lines are parallel, perpendicular, or neither.

$$-4x - 2y = -9$$
$$5x - 10y = 7$$

22. Find the equation of the line parallel to the line $3x - 5y = 1$ that passes through the point $(1, -4)$. Express your answer in point-slope form, slope-intercept form, and standard form.

23. Find the equation of the line parallel to the line $x = 7$ that passes through the point $(11, -3)$.

24. Find the equation of the line perpendicular to the line $3x - 5y = 1$ that passes through the point $(1, -4)$. Express your answer in point-slope form, slope-intercept form, and standard form.

25. Find the equation of the line perpendicular to the line $y = 8$ that passes through the point $(11, -3)$.

CHAPTER THREE

Functions

CHAPTER THREE CONTENTS

3.1 Relations and Functions

THINGS TO KNOW

Before working through this section, be sure that you are familiar with the following concepts:

 You Try It 1. Writing Sets Using Set-Builder Notation and Interval Notation (Section R.1) 

 You Try It 2. Solving Linear Inequalities in One Variable (Section 1.7) 

 You Try It 3. Solving Polynomial Inequalities (Section 1.9) 

 You Try It 4. Solving Rational Inequalities (Section 1.9) 

 You Try It 5. Graphing Equations by Plotting Points (Section 2.1) 

INTRODUCTION

Read this introduction before beginning Objective 1.

OBJECTIVES

1 Understanding the Definitions of Relations and Functions

2 Determining Whether Equations Represent Functions

3 Using Function Notation; Evaluating Functions

4 Using the Vertical Line Test

5 Classifying Functions

6 Determining the Domain of a Function Given the Equation

SECTION 3.1 EXERCISES

Introduction to Section 3.1

Functions appear all around us and are a fundamental part of mathematics. Functions occur when one quantity depends on another. Here are just a few examples:

- The revenue generated by selling a certain product depends on the number of items sold.
- The volume of a cube depends on the length of an edge of the cube.
- For a person who works by the hour, the gross pay depends on the number of hours worked.

If you think about it, you can probably come up with countless situations where one thing, event, or occurrence depends on another. Before we officially define a function, we must first define a **relation**.

OBJECTIVE 1 UNDERSTANDING THE DEFINITIONS OF RELATIONS AND FUNCTIONS

Definition Relation

A **relation** is a correspondence between two sets A and B such that each element of set A corresponds to one or more elements of set B. Set A is called the **domain** of the relation, and set B is called the **range** of the relation.

Suppose that Adam, Patrice, and Scott are college roommates and their weights are 180 lb, 190 lb, and 180 lb, respectively. The pairing of the roommates' names with their weights is a relation. In this example, we are relating a set of names with a set of weights. The set of names {Adam, Patrice, Scott} is the domain of this relation, whereas the set of weights {180 lb, 190 lb} is the range. Notice that 180 lb is listed only once in the range. There is no need to list an element of either set more than once. See Figure 1.

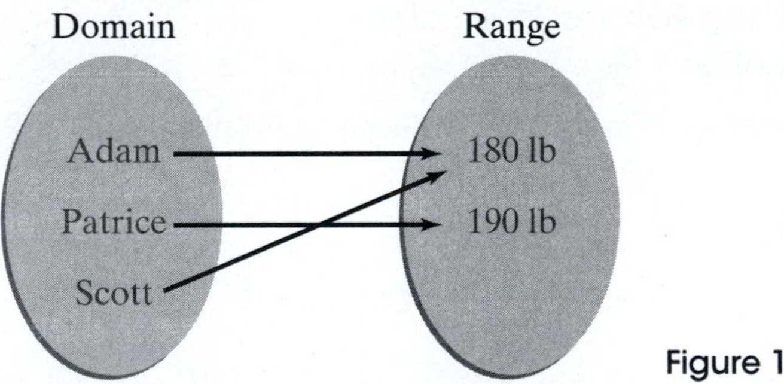

Figure 1

A relation can be represented as a set of ordered pairs with the elements of the domain listed first. The previous relation can be written as the set of ordered pairs {(Adam, 180 lb), (Patrice, 190 lb), (Scott, 180 lb)}.

Let's now switch the domain and range of this relation. The relation in Figure 2 can be written as the set of ordered pairs {(180 lb, Adam), (180 lb, Scott), (190 lb, Patrice)}.

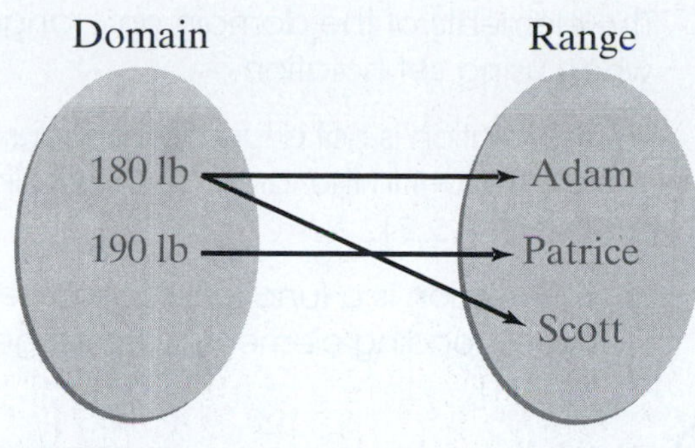

Figure 2

Suppose that a delivery man was to deliver a package to the roommate who weighs 180 lb. Because there are two people who weigh 180 lb, there is no way to know for sure who this package belongs to. This is because the element of 180 lb in the domain is associated with two distinct elements in the range. We say that this relation is not a **function**. For a relation to be a function, every element in the domain must correspond to exactly one element in the range.

> **Definition** Function
>
> A **function** is a relation such that for each element in the domain, there is *exactly one* corresponding element in the range.

The relation in **Figure 1** is a function because for every name in the domain, there is exactly one corresponding weight in the range. The relation in **Figure 2** is not a function because there is a domain value (180 lb) that corresponds to two range values (Adam and Scott).

Example 1 Determine Whether a Relation Is a Function

Determine whether each relation is a function, and then find the domain and range.

a.

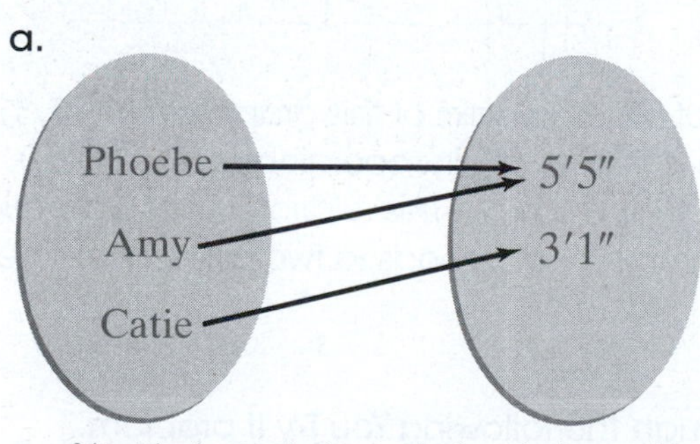

b.

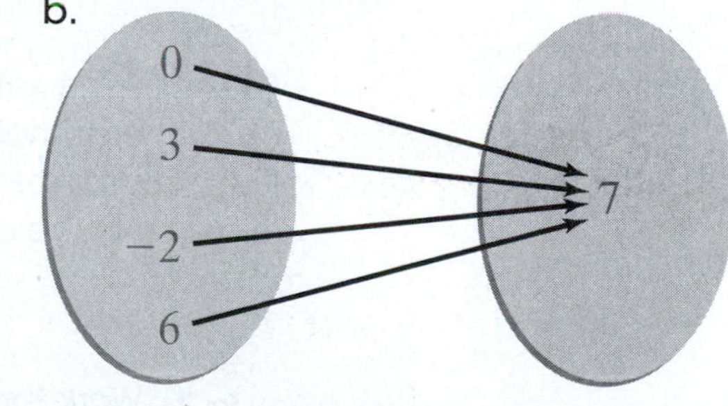

c. $\{(3, 7), (-3, 2), (-4, 5), (1, 4), (3, -4)\}$ 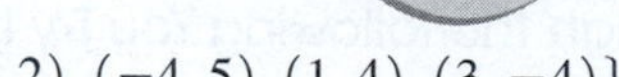 d. $\{(-1, 2), (0, 2), (1, 5), (2, 3), (3, 2)\}$

Solution

a. This relation is a function. For each element in the domain, there is *exactly one* corresponding element in the range. The domain is {Phoebe, Amy, Catie}, and the range is $\{5'5'', 3'1''\}$.

b. This relation is a function. For each element in the domain, there is *exactly one* corresponding element in the range. The domain is $\{-2, 0, 3, 6\}$, and the range is $\{7\}$.

TIP The elements of the domain and range are typically listed in ascending order when using set notation.

c. This relation is not a function because the element 3 in the domain corresponds to 7 and -4 in the range. The domain of this relation is $\{-4, -3, 1, 3\}$. The range is $\{-4, 2, 4, 5, 7\}$.

d. This relation is a function. For each element in the domain, there is *exactly one* corresponding element in the range. The domain is $\{-1, 0, 1, 2, 3\}$. The range is $\{2, 3, 5\}$.

Watch this **video** to see a complete solution to this example.

If the domain and range of a relation are sets of real numbers, then the relation can be represented by plotting the ordered pairs in the Cartesian plane. The set of all x-coordinates is the domain of the relation, and the set of all y-coordinates is the range of the relation.

Example 2 Determine Whether a Relation Is a Function

Use the domain and range of the following relation to determine whether the relation is a function.

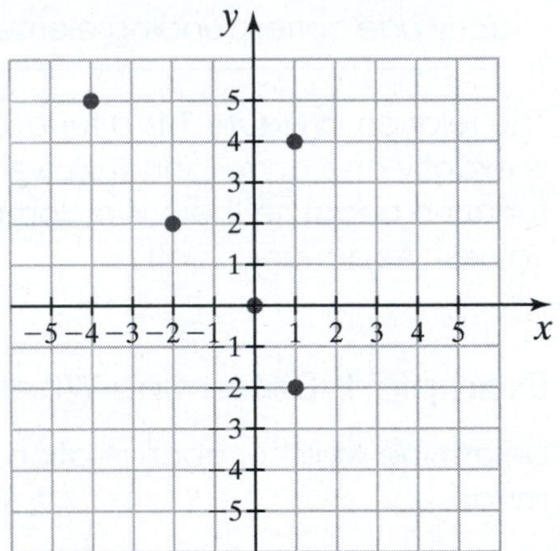

Solution The set of ordered pairs of this graph are $\{(-4, 5), (-2, 2), (0, 0), (1, -2), (1, 4)\}$. The domain is the set of all x-coordinates: $\{-4, -2, 0, 1\}$. The range is the set of all y-coordinates $\{-2, 0, 2, 4, 5\}$. This relation is *not* a function because the element 1 in the domain corresponds to two different elements (-2 and 4) in the range.

You Try It **Work through the following You Try It problem.**

Work Exercises 1–7 in this textbook or in the MyLab Math Study Plan.

OBJECTIVE 2 DETERMINING WHETHER EQUATIONS REPRESENT FUNCTIONS

In Example 2, we see that a *finite* number of ordered pairs in the Cartesian plane can represent a relation. Relations can also be described by *infinitely* many ordered pairs in the plane. Recall that the solution set of an equation in two variables is the set of all ordered pairs (x, y) for which the equation is true. For example, the ordered pairs $(1, 1)$, $(\sqrt{3}, 0)$, and $(1, -1)$ all satisfy the equation $x^2 + 2y^2 = 3$. (To verify, read this **explanation**.) Although this equation is a relation, it is not a function because $x = 1$ corresponds to both $y = 1$ and $y = -1$. To determine whether an equation represents a function, we must show that for any value in the domain, there is exactly one corresponding value in the range.

Example 3 Determine Whether an Equation Represents a Function

Determine whether the following equations represent y as a function of x:

a. $3x - 2y = -12$ **b.** $y = 3x^2 - x + 2$ **c.** $(x + 3)^2 + y^2 = 16$

Solution The equations in parts a and b both represent y as a function of x. Watch the **interactive video** to see parts a and b worked out in detail.

To determine whether the equation $(x + 3)^2 + y^2 = 16$ from part c represents y as a function of x, we can solve for y.

Solving the equation for y, we obtain $y = \pm\sqrt{16 - (x + 3)^2}$. To see how we solved for y, read these **steps**. Notice the symbol "$\pm$," which precedes the radical in the previous equation. This symbol indicates that there are two possible y-values for a given x-value. For example, if $x = -3$, then the corresponding y-values are

$$y = \pm\sqrt{16 - (-3 + 3)^2} = \pm\sqrt{16 - (0)^2} = \pm\sqrt{16} = \pm 4.$$

Therefore, the value of $x = -3$ in the domain corresponds to $y = 4$ and $y = -4$ in the range. Because there are two y-values that correspond to a single x-value, the original equation $(x + 3)^2 + y^2 = 16$ does not represent a function. Recall from **Section 2.2** that the equation $(x + 3)^2 + y^2 = 16$ is the equation of a circle in **standard form** with center $(-3, 0)$ and $r = 4$. We can sketch this circle using the techniques discussed in **Section 2.2**, or we can use a graphing utility to graph the two equations $y = \sqrt{16 - (x + 3)^2}$ and $y = -\sqrt{16 - (x + 3)^2}$. For all x-values on the interval $(-7, 1)$, there are two corresponding values of y.

Watch the **interactive video** to see all three solutions worked out in detail.

Using Technology

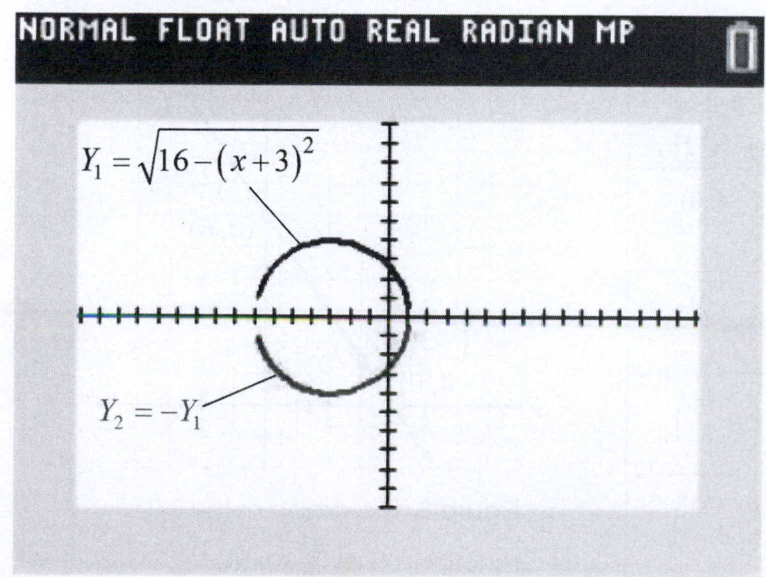

The graph of $(x + 3)^2 + y^2 = 16$ was created using a graphing utility. The top half of the circle is $y = \sqrt{16 - (x + 3)^2}$, whereas the bottom half of the circle represents the equation $y = -\sqrt{16 - (x + 3)^2}$. ●

You Try It Work through the following You Try It problem.

Work Exercises 8–12 in this textbook or in the MyLab Math Study Plan.

OBJECTIVE 3 USING FUNCTION NOTATION; EVALUATING FUNCTIONS

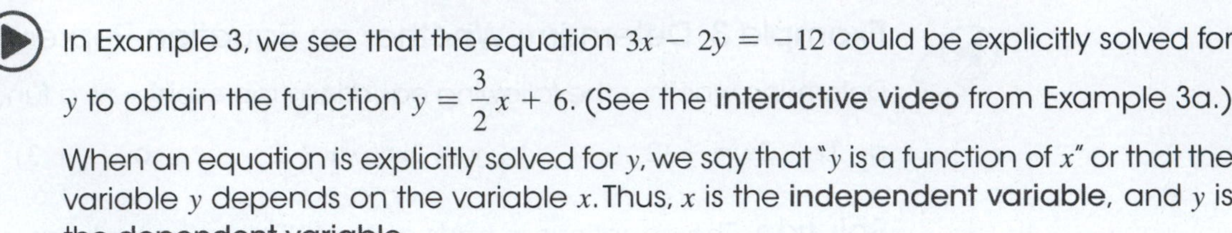

In Example 3, we see that the equation $3x - 2y = -12$ could be explicitly solved for y to obtain the function $y = \frac{3}{2}x + 6$. (See the **interactive video** from Example 3a.) When an equation is explicitly solved for y, we say that "y is a function of x" or that the variable y depends on the variable x. Thus, x is the **independent variable**, and y is the **dependent variable**.

Instead of using the variable y, letters such as f, g, or h (and others) are commonly used for functions. For example, suppose we want to name a function f. Then, for any x-value in the domain, we call the y-value (or function value) $f(x)$. The symbol $f(x)$ is read as "the value of the function f at x" or simply "f of x." In our example, the function $y = \frac{3}{2}x + 6$ can be written as $f(x) = \frac{3}{2}x + 6$. The notation $f(x)$ is called **function notation**.

In our example, if $f(x) = \frac{3}{2}x + 6$, then the expression $f(-4)$ represents the range value (y-coordinate) when the domain value (x-coordinate) is -4. To find this value, replace x with -4 in the equation $f(x) = \frac{3}{2}x + 6$ to get $f(-4) = \frac{3}{2}(-4) + 6 = -6 + 6 = 0$. Thus, the ordered pair $(-4, 0)$ must lie on the graph of f. **Table 1** shows this function evaluated at several values of x. The function $f(x) = \frac{3}{2}x + 6$ is a linear function whose graph is sketched in **Figure 3**.

CAUTION The symbol $f(x)$ does not mean f times x. The notation $f(x)$ refers to the value of the function at x.

Table 1

Domain Value x	Range Value $f(x) = \frac{3}{2}x + 6$	Ordered Pair $(x, f(x))$
-4	$f(-4) = \frac{3}{2}(-4) + 6 = 0$	$(-4, 0)$
0	$f(0) = \frac{3}{2}(0) + 6 = 6$	$(0, 6)$
1	$f(1) = \frac{3}{2}(1) + 6 = \frac{15}{2}$	$\left(1, \frac{15}{2}\right)$
-3	$f(-3) = \frac{3}{2}(-3) + 6 = \frac{3}{2}$	$\left(-3, \frac{3}{2}\right)$

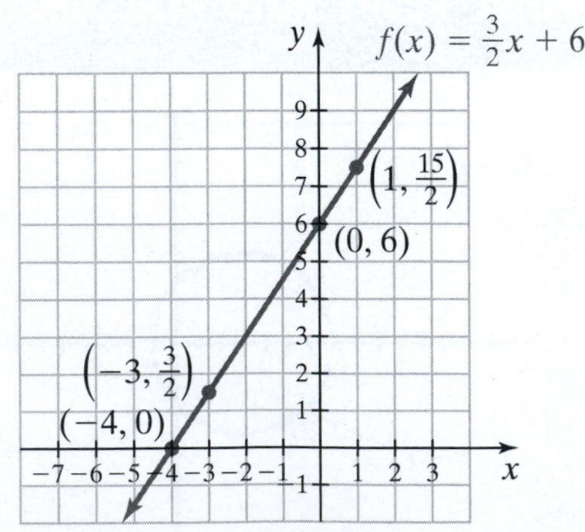

Figure 3 The graph of the function $f(x) = \frac{3}{2}x + 6$.

Example 4 Use Function Notation

Rewrite these equations using function notation where y is a function of x. Then answer the question following each equation.

a. $3x - y = 5$

What is the value of $f(4)$?

b. $x^2 - 2y + 1 = 0$

Does the point $(-2, 1)$ lie on the graph of this function?

c. $y + 7 = 0$

What is $f(x)$ when $x = 3$?

Solution You should watch the **interactive video** to verify the following:

a. This equation can be rewritten in function notation as $f(x) = 3x - 5$ and $f(4) = 7$.

b. This equation can be rewritten in function notation as $f(x) = \frac{1}{2}x^2 + \frac{1}{2}$. The point $(-2, 1)$ is *not* on the graph of this function because $f(-2) \neq 1$.

c. The equation $y + 7 = 0$ can be rewritten using function notation as $f(x) = -7$. When $x = 3$, the value of $f(x)$ is -7.

Example 5 Evaluate a Function

Given that $f(x) = x^2 + x - 1$, evaluate the following:

a. $f(0)$ b. $f(-1)$ c. $f(x + h)$ d. $\dfrac{f(x + h) - f(x)}{h}$

Solution

a. Substitute $x = 0$ into the function to get

$$f(0) = (0)^2 + (0) - 1 = -1$$

b. Substitute $x = -1$ into the function to get

$$\begin{aligned} f(-1) &= (-1)^2 + (-1) - 1 \\ &= 1 - 1 - 1 \\ &= -1 \end{aligned}$$

CAUTION The expression $(-1)^2$ does not equal -1^2.

c. Substitute $x = x + h$ into the function to get

$$\begin{aligned} f(x + h) &= (x + h)^2 + (x + h) - 1 \\ &= x^2 + 2xh + h^2 + x + h - 1 \end{aligned}$$

d. The expression $\dfrac{f(x+h)-f(x)}{h}$ is called the **difference quotient** and is very important in calculus. First, from part c, we see that $f(x+h) = x^2 + 2xh + h^2 + x + h - 1$. So,

$$\begin{aligned}\frac{f(x+h)-f(x)}{h} &= \frac{\overbrace{x^2+2xh+h^2+x+h-1}^{f(x+h)} - \overbrace{[x^2+x-1]}^{f(x)}}{h} \\ &= \frac{x^2+2xh+h^2+x+h-1-x^2-x+1}{h} \\ &= \frac{2xh+h^2+h}{h} \\ &= \frac{h(2x+h+1)}{h} \\ &= 2x+h+1\end{aligned}$$

Therefore, $\dfrac{f(x+h)-f(x)}{h} = 2x + h + 1$.

Note that function notation is not restricted to the use of x for the independent variable or to the use of f for the name of the function, which is the dependent variable. Other letters commonly used in place of x are $h, t, r, a,$ and b. Letters such as A for area, V for volume, and d for distance are often used to represent the names of functions. In fact, any letter can be used to represent values of either the domain or the range.

You Try It **Work through the following You Try It problem.**

Work Exercises 13–28 in this textbook or in the MyLab Math Study Plan.

OBJECTIVE 4 USING THE VERTICAL LINE TEST

We see in **Figure 3** that the graph of the function $f(x) = \frac{3}{2}x + 6$ is created by plotting severalordered pairs in the Cartesian plane. Although every function has a graph, it is *not* the case that every graph in the plane represents a function. A graph does not represent a function if two or more points on the graph with the same first coordinate have different second coordinates. The **vertical line test** can be used to quickly determine whether a graph represents a function. See **Figure 4**.

> **Vertical Line Test**
>
> A graph in the Cartesian plane is the graph of a function if and only if no vertical line intersects the graph more than once.

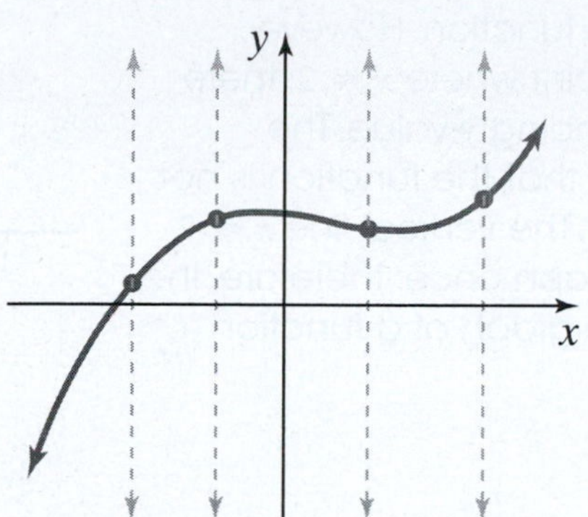

This graph is a function. (No vertical line intersects the graph more than once.)

This graph is not a function. (The graph does not pass the vertical line test.)

Figure 4

Example 6 Use the Vertical Line Test

Use the vertical line test to determine which of the following graphs represents the graph of a function.

a.

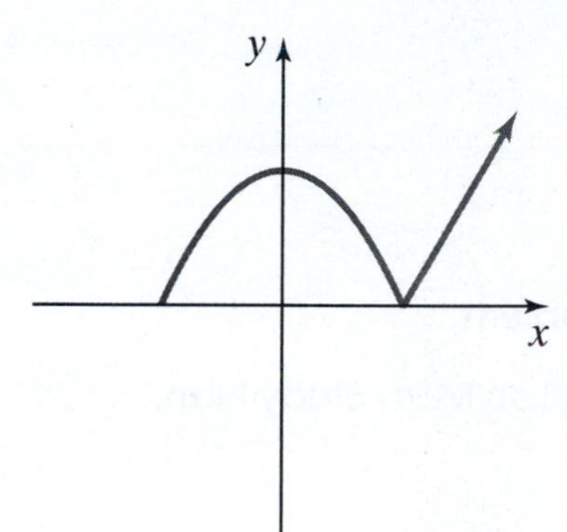

b.

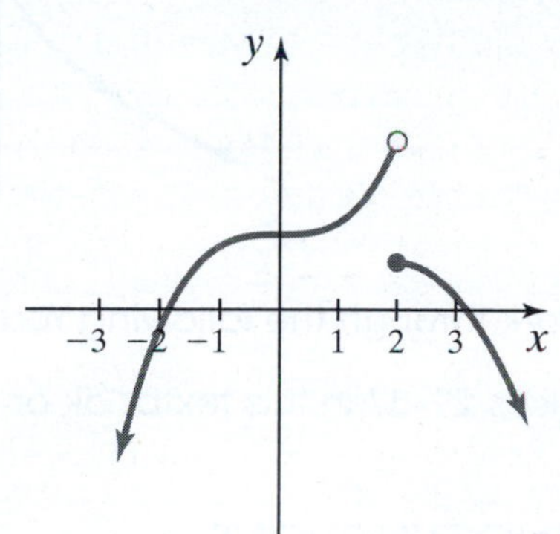

c.

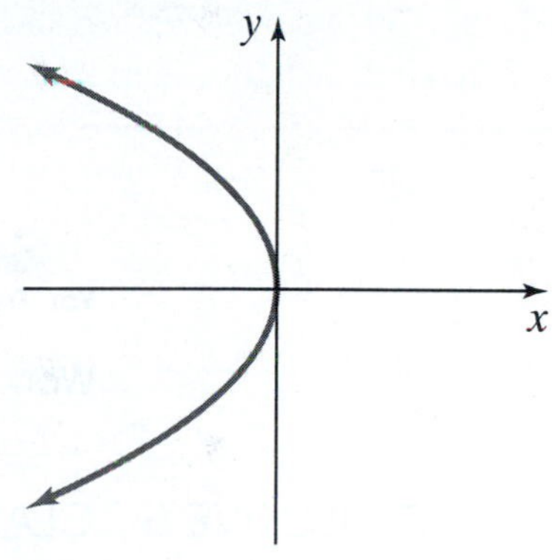

Solution

a. This graph represents a function because no vertical lines intersect the graph more than once.

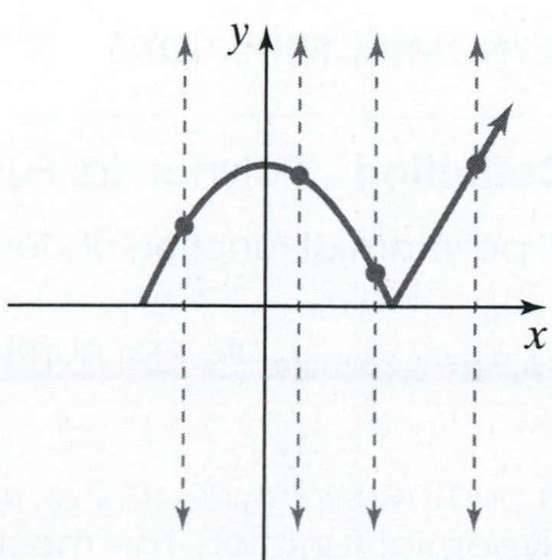

b. At first glance, it may appear that this graph does not represent a function. However, we see that at the point where $x = 2$, there is only one corresponding y-value. The "open dot" indicates that the function is not defined at that point. The vertical line $x = 2$ only intersects the graph once; therefore, the graph represents the graph of a function.

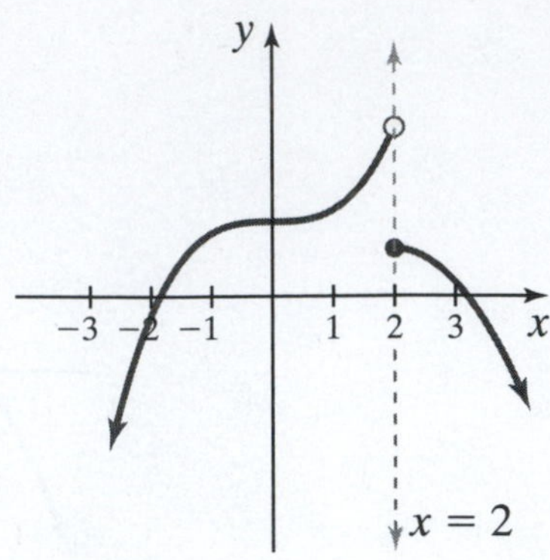

c. This graph does not represent a function because we can find several vertical lines that intersect the graph more than once.

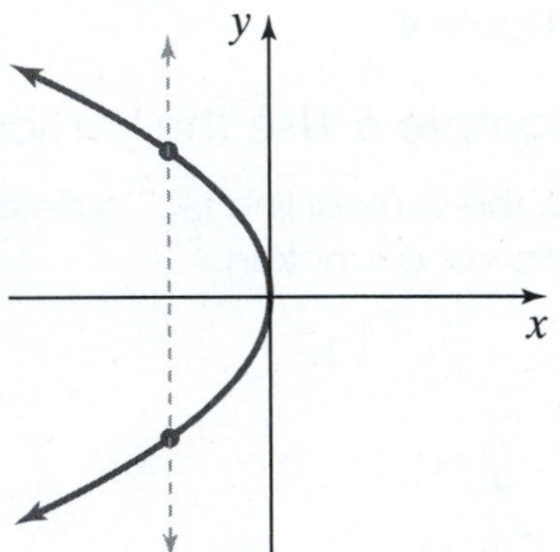

You Try It Work through the following You Try It problem.

Work Exercises 29–37 in this textbook or in the MyLab Math Study Plan.

OBJECTIVE 5 CLASSIFYING FUNCTIONS

There are many different classes of functions that you will encounter throughout this eText. Therefore, given a function, it is extremely important to learn how to classify it. We start by introducing three classifications of functions: **Polynomial functions, Rational functions,** and **Root functions.**

POLYNOMIAL FUNCTIONS

> **Definition** Polynomial Function
>
> A **polynomial function** of degree n is a function of the form $f(x) = a_n x^n + a_{n-1}x^{n-1} + a_{n-2}x^{n-2} + \ldots + a_1 x + a_0$, where n is a nonnegative integer and $a_0, a_1, a_2, \ldots, a_n$ are real numbers.

The numbers $a_0, a_1, a_2, a_3, \ldots, a_n$ in the definition above are the **coefficients** of the polynomial function. The most important thing to remember about polynomial functions is that the powers, or exponents, of each term are integer values that are greater than or equal to 0. The degree of a polynomial function is the highest power of x.

Example 7 Determine Whether a Function Is a Polynomial Function

Determine whether the following functions are polynomial functions. If the function is a polynomial function, state the degree of the polynomial function.

a. $f(x) = 7 - 3x^2 + 5x^3$ b. $h(t) = -11$ c. $f(t) = 3t^6 + \frac{1}{t}$

Solution

a. The function $f(x) = 7 - 3x^2 + 5x^3$ can be rewritten in descending order as $f(x) = 5x^3 - 3x^2 + 7$. Each term has an integer exponent that is greater than or equal to 0. The highest power of x is 3. Therefore, the function is a polynomial function of degree 3.

b. The function $h(t) = -11$ could be rewritten as $h(t) = -11t^0$. Thus, the function is considered a polynomial function of degree 0.

c. The function $f(t) = 3t^6 + \frac{1}{t}$ is **not** a polynomial function because the second term, $\frac{1}{t}$, can be written as t^{-1}. Therefore, this function has a **negative** exponent and is thus not a polynomial function.

RATIONAL FUNCTIONS

A rational function is simply the quotient of two polynomial functions.

> **Definition** Rational Function
>
> A **rational function** is a function of the form $f(x) = \frac{g(x)}{h(x)}$, where g and h are polynomial functions such that $g(x) \neq 0$ and the degree of $h(x)$ is a greater than 0.

If the degree of $h(x)$ is 0, then $h(x) = c$, where c is a real number and we consider the function $f(x) = \frac{g(x)}{h(x)} = \frac{g(x)}{c}$ to be a polynomial function.

Example 8 Determine Whether a Function Is a Rational Function

Determine whether the following functions are rational functions

a. $f(t) = \frac{t^3 - 1}{t + 1}$ b. $h(x) = (x^3 - x^2 - 9)x^{-5}$

Solution

a. The function $f(t) = \frac{t^3 - 1}{t + 1}$ is the quotient of two polynomial functions. The degree of the polynomial function in the denominator is 1. Therefore, this is a rational function.

b. The function $h(x) = (x^3 - x^2 - 9)x^{-5}$ could be rewritten as $h(x) = \dfrac{x^3 - x^2 - 9}{x^5}$.

Thus, the function is the quotient of two polynomial functions such that the degree of the polynomial function in the denominator is greater than 0. Therefore, this is a rational function. ●

ROOT FUNCTIONS

The last class of functions that we will discuss in this section are root functions.

> **Definition** Root Function
>
> The function $f(x) = \sqrt[n]{g(x)}$ is a **root function**, where n is an integer such that $n \geq 2$.

Examples of root functions are $f(x) = \sqrt{x - 1}$, $h(t) = \sqrt[3]{\dfrac{t - 4}{t^3 + 1}}$, and $V(r) = \sqrt{\pi\left(r^2 - 1\right)}$.

You Try It Work through the following You Try It problem.

Work Exercises 38–41 in this textbook or in the MyLab Math Study Plan.

OBJECTIVE 6 DETERMINING THE DOMAIN OF A FUNCTION GIVEN THE EQUATION

The domain of a function $y = f(x)$ is the set of all values of x for which the function is defined. The figure below illustrates that the input value of $x = a$ is in the domain of a function f **only if** the output value, $y = f(a)$, is a real number.

$x = a$ Input → $y = f(x)$ → $y = f(a)$ Output

- If the output value of $y = f(a)$ is a real number, then $x = a$ is in the domain of f.
- If the output value of $y = f(a)$ is a **not** a real number, then $x = a$ is **not** in the domain of f.

> **Definition** The Domain of a Function
>
> The **domain of a function** $y = f(x)$ is the set of all values of x for which the function is defined. In other words, a number $x = a$ is in the domain of f if $y = f(a)$ is a real number.

For the remainder of this section, we will focus on determining the domains of polynomial functions, rational functions, and root functions.

FINDING THE DOMAIN OF POLYNOMIAL FUNCTIONS

Consider the polynomial function $f(x) = 3x^3 - 2x^2 - x + 5$. The number $x = 0$ is in the domain because $f(0) = 3(0)^3 - 2(0)^2 - (0) + 5 = 5$ is a real number. Similarly, the number $x = -1$ is in the domain because $f(-1) = 3(-1)^3 - 2(-1)^2 - (-1) + 5 = 3(-1) - 2(1) + 1 + 5 = 1$ is also a real number. In fact, the domain of this polynomial function is all real numbers because for any real number $x = a$, the value of $f(a)$ is $f(a) = 3a^3 - 2a^2 - a + 5$, which will always be a real number. It follows that the domain of **every** polynomial function is all real numbers. We typically write the domain of a function in interval notation. Therefore, the domain of every polynomial function is $(-\infty, \infty)$.

FINDING THE DOMAIN OF RATIONAL FUNCTIONS

Consider the rational function $f(x) = \dfrac{x+5}{x-4}$. This function is defined for every real number x except $x = 4$ because the value $f(4) = \dfrac{4+5}{4-4} = \dfrac{9}{0}$ is undefined. Therefore, the domain of this function is every real number except $x = 4$. Thus, the domain of $f(x) = \dfrac{x+5}{x-4}$ in interval notation is $(-\infty, 4) \cup (4, \infty)$.

Given a rational function of the form $f(x) = \dfrac{g(x)}{h(x)}$, where the degree of $h(x)$ is greater than 0, the domain of f is the set of all real numbers such that $h(x) \neq 0$.

FINDING THE DOMAIN OF ROOT FUNCTIONS

Consider the root function $f(x) = \sqrt{x+1}$. The number $x = 3$ is in the domain because $f(3) = \sqrt{3+1} = \sqrt{4} = 2$ is a real number. However, $x = -5$ is not in the domain because $f(-5) = \sqrt{-5+1} = \sqrt{-4} = 2i$ is *not* a real number. Therefore, the domain of $f(x) = \sqrt{x+1}$ consists only of values of x for which the **radicand** is greater than or equal to zero. Thus, the domain of f is the solution to the inequality $x + 1 \geq 0$.

$x + 1 \geq 0$ — Create an inequality of the form $g(x) \geq 0$, where $g(x)$ is the radicand of the root function.

$x \geq -1$ — Subtract 1 from both sides.

Therefore, the domain of $f(x) = \sqrt{x+1}$ is $[-1, \infty)$.

The domain of every root function of the form $f(x) = \sqrt[n]{g(x)}$, where the index, n, is an **even** positive integer can be found by solving the inequality $g(x) \geq 0$.

If the index of $f(x) = \sqrt[n]{g(x)}$ is a positive **odd** integer, then the domain consists of all real numbers for which $g(x)$ is defined.

Following is a quick guide for finding the domain of the three classifications of functions that were described in this section.

Class of Function	Form	Domain
Polynomial functions	$f(x) = a_n x^n + a_{n-1}x^{n-1} + \cdots + a_1 x + a_0$	Domain is $(-\infty, \infty)$.
Rational functions	$f(x) = \dfrac{g(x)}{h(x)}$, where $g \neq 0$ and h are polynomial functions such that the degree of $h(x)$ is greater than 0.	Domain is all real numbers such that $h(x) \neq 0$.
Root functions	$f(x) = \sqrt[n]{g(x)}$, where $g(x)$ is a function and n is an integer such that $n \geq 2$.	1. If n is even, the domain is the solution to the inequality $g(x) \geq 0$. 2. If n is odd, the domain is the set of all real numbers for which g is defined.

Example 9 Find the Domain of a Function Given the Equation

Classify each function as a polynomial function, rational function, or root function, and then find the domain. Write the domain in interval notation.

a. $f(x) = 2x^2 - 5x$

b. $f(x) = \dfrac{x}{x^2 - x - 6}$

c. $h(x) = \sqrt{x^2 - 2x - 8}$

d. $f(x) = \sqrt[3]{5x - 9}$

e. $k(x) = \sqrt[4]{\dfrac{x - 2}{x^2 + x}}$

Solution Work through the **interactive video** to verify the following:

a. The function $f(x) = 2x^2 - 5x$ is a polynomial function. The domain of f is all real numbers or $(-\infty, \infty)$.

b. The function $f(x) = \dfrac{x}{x^2 - x - 6}$ is a rational function. The domain of f is $\{x \mid x \neq -2, x \neq 3\}$. The domain in interval notation is $(-\infty, -2) \cup (-2, 3) \cup (3, \infty)$.

c. The function $h(x) = \sqrt{x^2 - 2x - 8}$ is a root function. The domain of h in interval notation is $(-\infty, -2] \cup [4, \infty)$.

d. The function $f(x) = \sqrt[3]{5x - 9}$ is a root function. The domain of f is all real numbers or $(-\infty, \infty)$.

e. The function $k(x) = \sqrt[4]{\dfrac{x - 2}{x^2 + x}}$ is a root function. The domain of k in interval notation is $(-1, 0) \cup [2, \infty)$. To see the solution to this entire example, work through this **interactive video**. ●

You Try It Work through the following You Try It problem.

Work Exercises 42–60 in this textbook or in the MyLab Math Study Plan.

3.1 Exercises

Skill Check Exercises

For exercises SCE-1 through SCE-6, solve the inequality. Write your answer using interval notation.

SCE-1. $2 - 3x \geq 0$ **SCE-2.** $x^2 + 4x - 21 \geq 0$ **SCE-3.** $(x - 1)(x + 4)(x - 3) \geq 0$

SCE-4. $\dfrac{x + 3}{x - 1} \geq 0$ **SCE-5.** $\dfrac{x^2 - 9}{x + 7} \geq 0$ **SCE-6.** $\dfrac{x - 2}{x^2 + 5x - 24} \geq 0$

For exercises SCE-7 and SCE-8, simplify the expression.

SCE-7. $6 - 9(x + h)^2 - (6 - 9x^2)$ **SCE-8.** $\sqrt{x + h} - 2(x + h) - (\sqrt{x} - 2x)$

In Exercises 1–6, find the domain and range of each relation, and then determine whether each relation represents a function.

1.

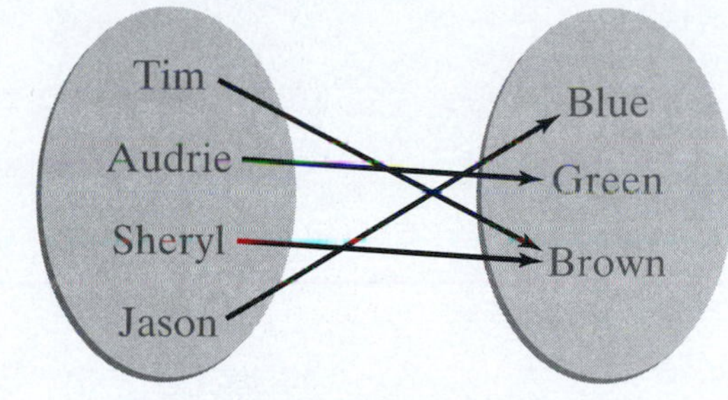

2.

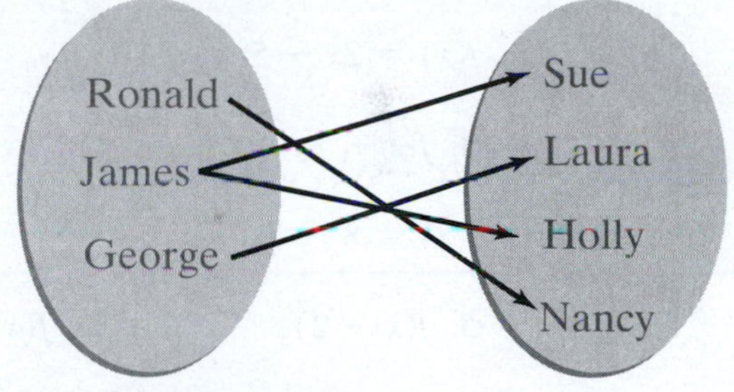

3. $\{(-1, 5), (0, 2), (1, 5), (2, 3), (1, 7)\}$

4. $\{(-1, 5), (0, 2), (1, 5), (2, 3), (3, 7)\}$

5.

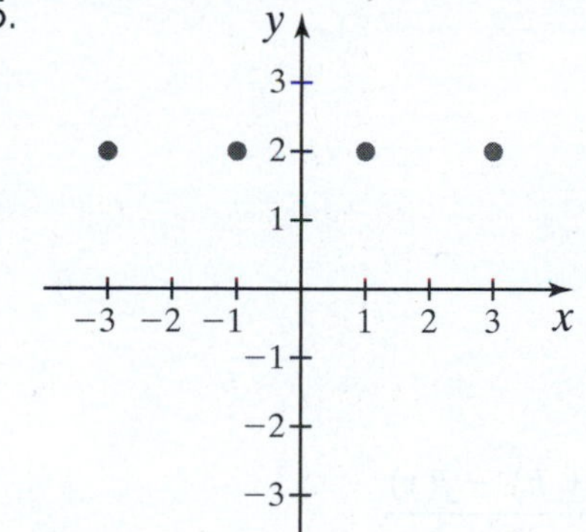

6.

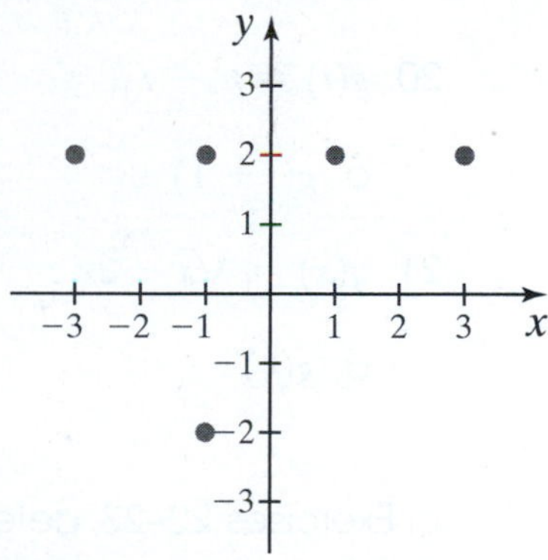

7. For the given relation, determine three additional ordered pairs which could be included such that the relation is a function. **(There are infinitely many correct answers.)**

$$\{(-2, 4), (0, 1), (1, 6), (3, 5), (7, -4), \boxed{?}, \boxed{?}, \boxed{?}\}$$

In Exercises 8–12, determine whether the equation represents y as a function of x.

8. $x^2 + y = 1$

9. $x^2 + y^2 = 1$

10. $x^3 + y^3 = 1$

11. $3xy = 6$

12. $3x - 4y = 10$

In Exercises 13–21, evaluate the function at the indicated values.

13. $f(x) = 3x - 4$

    a. $f(-3)$ b. $f\left(\frac{1}{2}\right)$ c. $f(x - 5)$

14. $f(x) = x^2 + 1$

    a. $f(2)$ b. $f(-2)$ c. $f(x - 1)$

15. $f(t) = \frac{t}{t + 1}$

    a. $f(-2)$ b. $f(7)$ c. $f(a)$

16. $V(r) = \frac{4}{3}\pi r^3$

    a. $V(3)$ b. $V(6)$ c. $V(r + 1)$

17. $f(x) = 2x - 5$

    a. $f(x + 1)$ b. $f(x + h)$

18. $f(x) = 8 - 7x$

    a. $f(x - 2)$ b. $f(x + h)$

19. $f(x) = x^2 - 2x$

    a. $f(x + 1)$ b. $f(x + h)$

20. $g(t) = t^3 - t$

    a. $g(t + 1)$ b. $g(t + h)$

21. $g(x) = \sqrt{x} - 2x$

    a. $g(4)$ b. $g(x + h)$

In Exercises 22–28, determine the difference quotient $\frac{f(x + h) - f(x)}{h}$.

SbS 22. $f(x) = 2x - 5$ SbS 23. $f(x) = 8 - 7x$ SbS 24. $f(x) = x^2 - 2x$

SbS 25. $f(x) = 6 - 9x^2$ SbS 26. $f(x) = 5x^2 - 4$ SbS 27. $f(x) = x^3 - x$

SbS 28. $f(x) = \sqrt{x} - 2x$

In Exercises 29–37, use the vertical line test to determine whether the graph represents a function.

29.

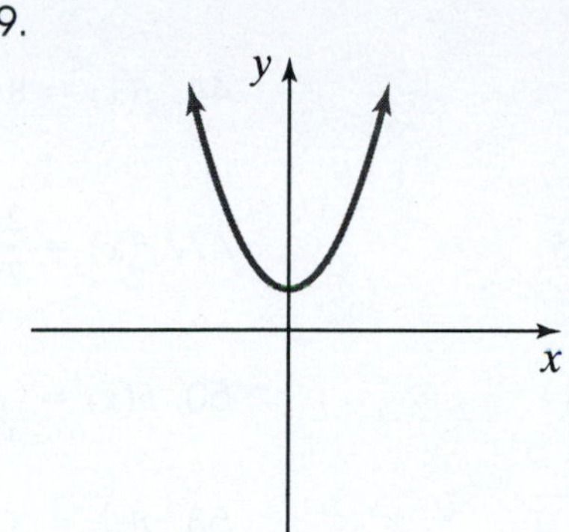

30.

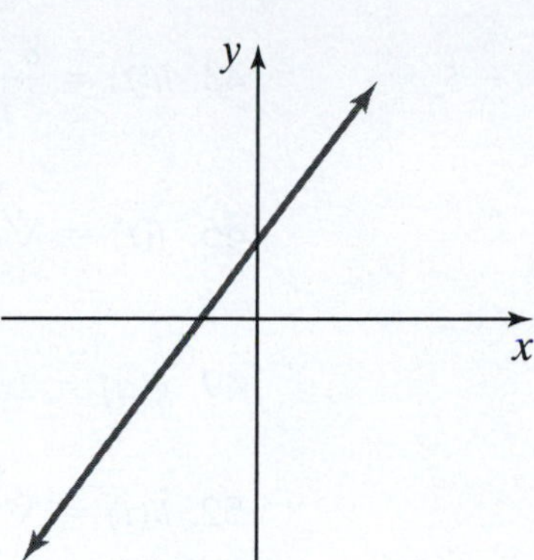

31.

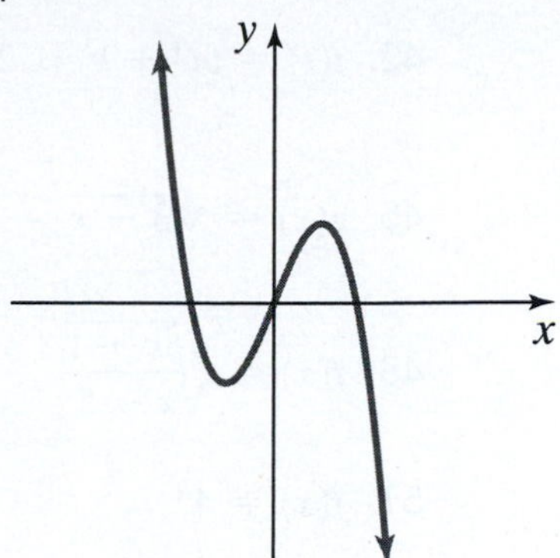

32.

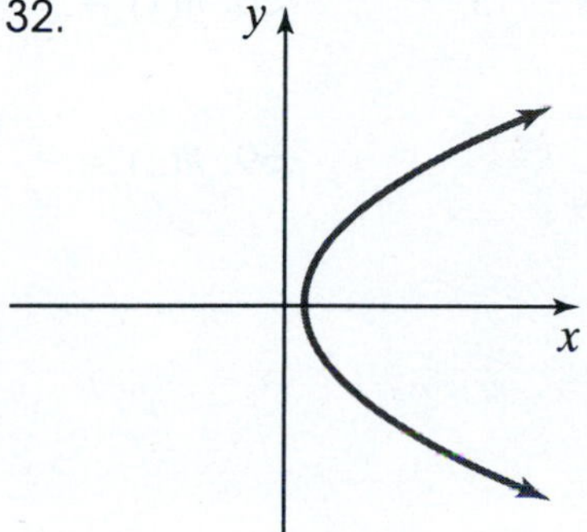

33.

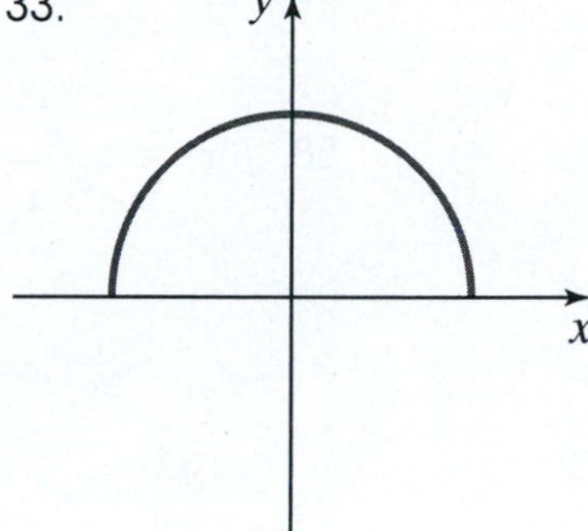

34.

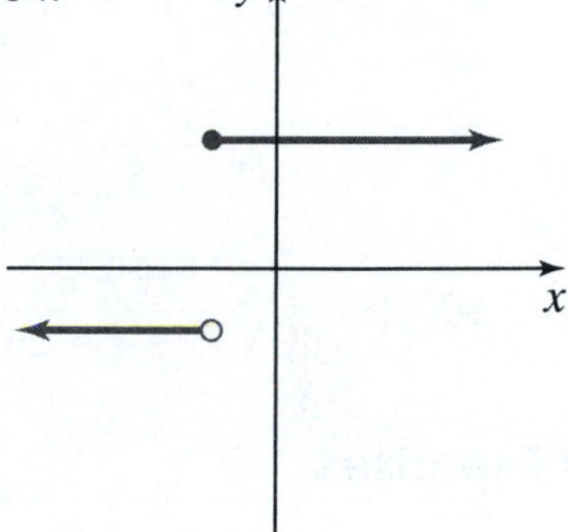

35.

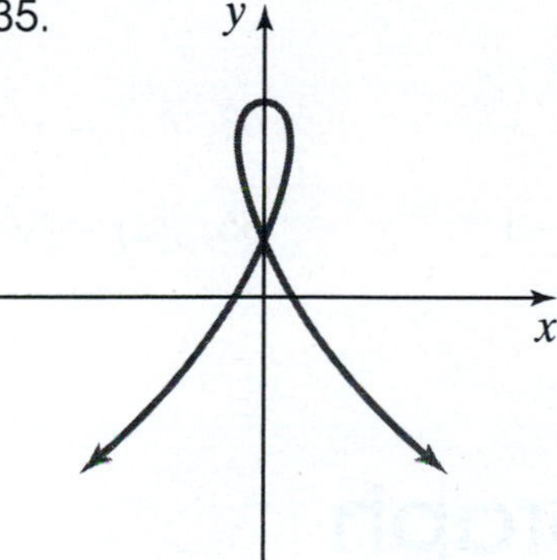

36.

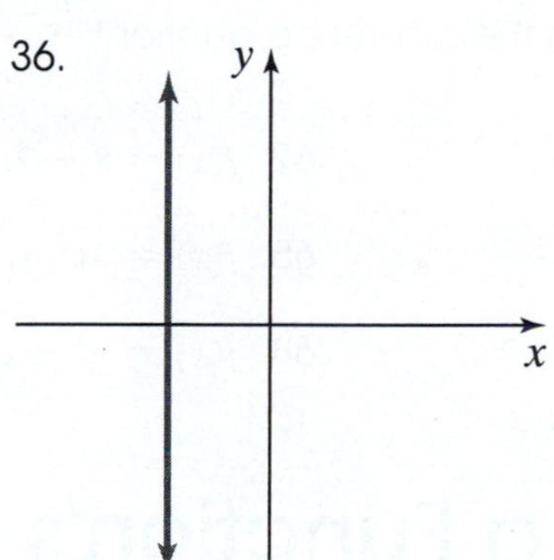

37.

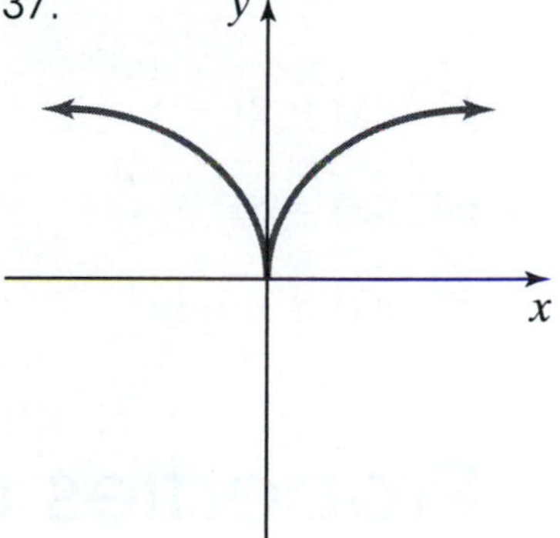

For each collection of functions in Exercises 38–41, classify each function as a polynomial function, rational function, or root function.

38. $f(x) = \dfrac{2x}{3x + 1}$, $g(t) = 3t^2 + 5$, $h(x) = -\dfrac{4}{x}$, $A(r) = \pi r^2$

39. $h(t) = 3t^{-2}$, $f(x) = \sqrt[3]{1 + x}$, $m(t) = t(t + 2)$, $H(x) = x$

40. $f(x) = -5$, $h(t) = \dfrac{3t^2 + 5}{8}$, $V(n) = \sqrt{\dfrac{n}{5}}$, $q(x) = \dfrac{8}{3x^2 + 5}$

41. $T(t) = \dfrac{\sqrt[5]{7}}{t^5}$, $f(x) = 8 - x^3$, $g(x) = \sqrt[5]{\dfrac{x^2 - 9}{2x}}$, $H(y) = 4^{-1}$

(eText Screens 3.2-1–3.2-33)

In Exercises 42–60, classify each function as a polynomial function, rational function, or root function, and then find the domain. Write the domain in interval notation.

42. $f(t) = 6t^4 + t^3 - 2t^2 + t - 5$

43. $h(t) = \dfrac{8 - t}{t^2}$

44. $f(x) = 8x^2 - x$

45. $g(x) = \sqrt{3 - x}$

46. $f(t) = \sqrt[3]{t - 5}$

47. $f(x) = \dfrac{2x - 1}{2x + 1}$

48. $f(x) = \sqrt[5]{\dfrac{x + 1}{x - 5}}$

49. $g(x) = 2x - 1$

50. $h(x) = \dfrac{x^2 + 4}{x^2 + x - 42}$

51. $f(x) = 4$

52. $h(t) = \sqrt[5]{t^4 - 1}$

53. $f(t) = \sqrt[4]{t^4 - 1}$

54. $g(t) = \dfrac{t + 5}{t^3 - 3t^2 - 4t}$

55. $f(x) = \sqrt{x^2 + 2x - 15}$

56. $h(x) = x^3 - 1$

57. $g(x) = \sqrt{\dfrac{x^2 - 9}{5}}$

58. $f(x) = \sqrt{\dfrac{5}{x^2 - 9}}$

59. $h(x) = \dfrac{1}{x^2 + 1}$

60. $f(x) = \sqrt[6]{\dfrac{2x - 6}{x^2 - x - 20}}$

Brief Exercises

In Exercises 61–68, determine the difference quotient $\dfrac{f(x + h) - f(x)}{h}$.

61. $f(x) = 2x - 5$

62. $f(x) = 8 - 7x$

63. $f(x) = x^2 - 2x$

64. $f(x) = 5x^2 - 4$

65. $f(x) = 2x^2 + 3x - 1$

66. $f(x) = \sqrt{x} - 2x$

67. $f(x) = 6 - 9x^2$

68. $f(x) = x^3 - x$

3.2 Properties of a Function's Graph

THINGS TO KNOW

Before working through this section, be sure that you are familiar with the following concepts:

VIDEO ANIMATION INTERACTIVE

You Try It

1. Solving Higher-Order Polynomial Equations (Section 1.6)

You Try It

2. Understanding the Definitions of Relations and Functions (Section 3.1)

You Try It

3. Using the Vertical Line Test (Section 3.1)

You Try It

4. Determining the Domain of a Function Given the Equation (Section 3.1)

OBJECTIVES

OBJECTIVE 1 DETERMINING THE INTERCEPTS OF A FUNCTION

In Section 3.1, we discussed that every function has a graph in the Cartesian plane. In this section, we take a closer look at a function's graph by discussing some of its characteristics. A good place to start is to talk about a function's intercepts. An **intercept** of a function is a point on the graph of a function where the graph either crosses or touches a coordinate axis. Intercepts can be represented by ordered pairs of the form $(a, 0)$ or $(0, b)$. There are two types of intercepts:

1. The y-intercept is the y-coordinate of the point where the graph crosses or touches the y-axis. If a function has an intercept at the point $(0, b)$, then the y-intercept is b.

2. The x-intercepts are the x-coordinates of the points where the graph crosses or touches the x-axis. If a function has an intercept at the point $(a, 0)$, then the x-intercept is a.

THE y-INTERCEPT

A function can have *at most* one y-intercept. The y-intercept exists if $x = 0$ is in the domain of the function. The y-intercept can be found by evaluating $f(0)$.

Example 1 Determine the y-Intercept of a Function

Find the y-intercept of the function $f(x) = -3x + 2$.

Solution Because $f(x) = -3x + 2$ is a polynomial function, we know that the domain of f is comprised of all real numbers. Thus, $x = 0$ is in the domain of f. The y-intercept is $f(0) = -3(0) + 2 = 2$. We say that the y-intercept of f is found at the point $(0, 2)$. The graph of f is the line sketched in Figure 5. You can see that the graph intersects the y-axis at $(0, 2)$.

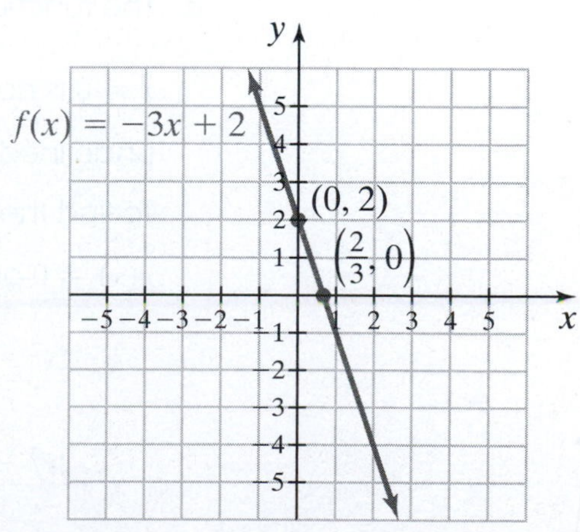

Figure 5

THE x-INTERCEPT(S)

A function may have several (even infinitely many) x-intercepts. The x-intercepts, also called **real zeros**, can be found by finding all *real solutions* to the equation $f(x) = 0$. Although a function function may have several zeros, only the real zeros are x-intercepts. The only real zero of $f(x) = -3x + 2$ is $x = \frac{2}{3}$ because the solution to the equation $-3x + 2 = 0$ is $x = \frac{2}{3}$. Notice from the graph in Figure 5 that the graph crosses the x-axis at the point $\left(\frac{2}{3}, 0\right)$.

Example 2 Find the Intercepts of a Function

Given the functions below, find the x-intercept(s) and the y-intercept (if they exist).

a. $f(x) = x^3 - 2x^2 + x - 2$

b. $g(x) = \dfrac{2x^2 - 7x - 4}{x^3}$

Solution

a. The function $f(x) = x^3 - 2x^2 + x - 2$ is a **polynomial function**. The domain of every polynomial function is all real numbers or $(-\infty, \infty)$. Therefore, $x = 0$ is in the domain of f and we can determine the y-intercept by evaluating $f(0)$.

$f(x) = x^3 - 2x^2 + x - 2$ Write the function.

$f(0) = (0)^3 - 2(0)^2 + (0) - 2$ Evaluate the function at $x = 0$.

$f(0) = -2$ Simplify.

The y-intercept is -2.

To find the x-intercepts of f, we must find the real solutions to the equation $f(x) = 0$ or $x^3 - 2x^2 + x - 2 = 0$. To solve this equation, we can factor by grouping and solve to get the solution set $\{2, i, -i\}$. To see how we obtained this solution, read these **steps** or watch this **video**. The only x-intercept of this function is 2 since 2 is the only real solution.

TIP The solutions $x = i$ and $x = -i$ are **zeros** of f because $f(i) = 0$ and $f(-i) = 0$; however, these values do not represent x-intercepts because they are not real numbers.

b. The function $g(x) = \dfrac{2x^2 - 7x - 4}{x^3}$ is a **rational function**. The number $x = 0$ is not in the domain since $g(0) = \dfrac{2(0)^2 - 7(0) - 4}{(0)^3} = \dfrac{-4}{0}$ which is undefined. Therefore, this function has no y-intercept.

To find the x-intercepts of g, we must find the real solutions to the equation $g(x) = 0$ or $\dfrac{2x^2 - 7x - 4}{x^3} = 0$.

$\dfrac{2x^2 - 7x - 4}{x^3} = 0$ Set up the equation $g(x) = 0$.

$x^3 \cdot \dfrac{2x^2 - 7x - 4}{x^3} = 0 \cdot x^3$ Multiply by the LCD to eliminate the fractions.

$2x^2 - 7x - 4 = 0$ Simplify.

(eText Screens 3.2-1–3.2-33)

Watch this video to see that the solution set to the quadratic equation $2x^2 - 7x - 4 = 0$ is $\left\{-\frac{1}{2}, 4\right\}$.

The two x-intercepts of $g(x) = \dfrac{2x^2 - 7x - 4}{x^3}$ are $\frac{1}{2}$ and 4.

Using Technology

The graphs below were created using a graphing utility. The graph on the left represents the graph of the function $f(x) = x^3 - 2x^2 + x - 2$. The graph on the right represents the function $g(x) = \dfrac{2x^2 - 7x - 4}{x^3}$.

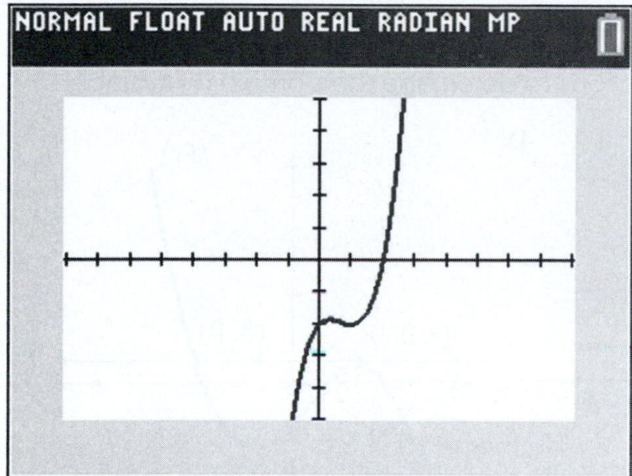

The graph of $f(x) = x^3 - 2x^2 + x - 2$ crosses the x-axis at the point $(2, 0)$ and the y-axis at the point $(0, -2)$. Thus, the x-intercept is 2 and the y-intercept is -2.

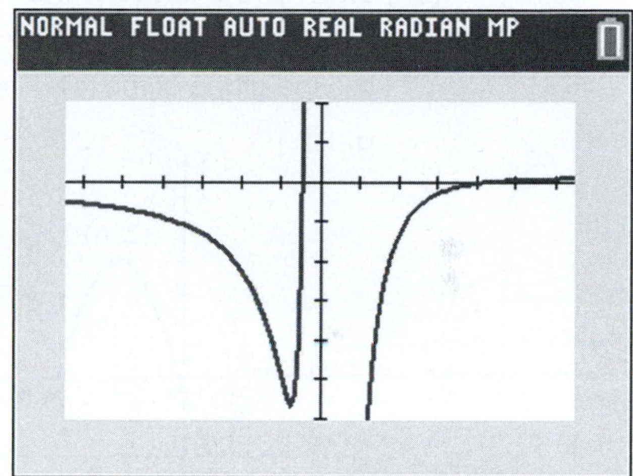

The graph of $g(x) = \dfrac{2x^2 - 7x - 4}{x^3}$ crosses and crosses the x-axis at the point $\left(-\frac{1}{2}, 0\right)$ and $(4, 0)$ The graph does not intersect the y-axis. Thus, the x-intercepts are $-\frac{1}{2}$ and 4 and there is no y-intercept.

You Try It **Work through the following You Try It problem.**

Work Exercises 1–12 in this textbook or in the MyLab Math Study Plan.

OBJECTIVE 2 DETERMINING THE DOMAIN AND RANGE OF A FUNCTION FROM ITS GRAPH

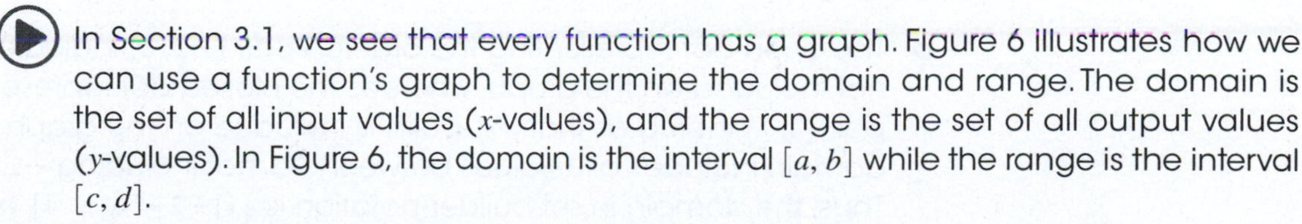

In Section 3.1, we see that every function has a graph. Figure 6 illustrates how we can use a function's graph to determine the domain and range. The domain is the set of all input values (x-values), and the range is the set of all output values (y-values). In Figure 6, the domain is the interval $[a, b]$ while the range is the interval $[c, d]$.

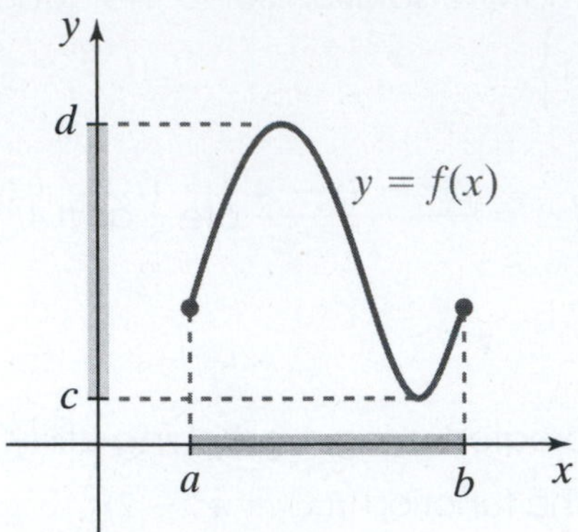

Figure 6 Domain and range of $y = f(x)$.

Example 3 Determine the Domain and Range of a Function from Its Graph

Use the graph of the following functions to determine the domain and range.

a.

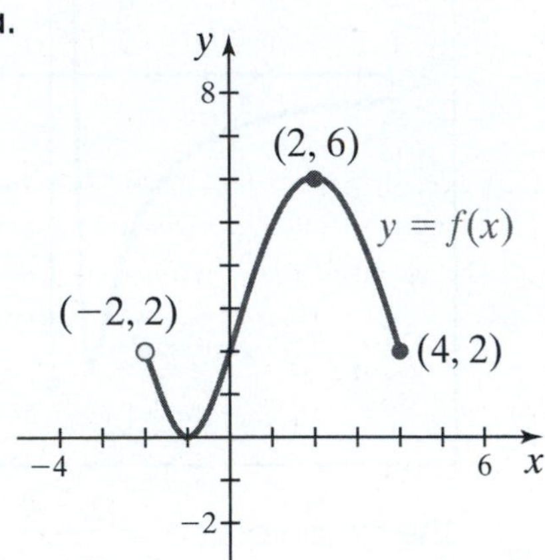

b.

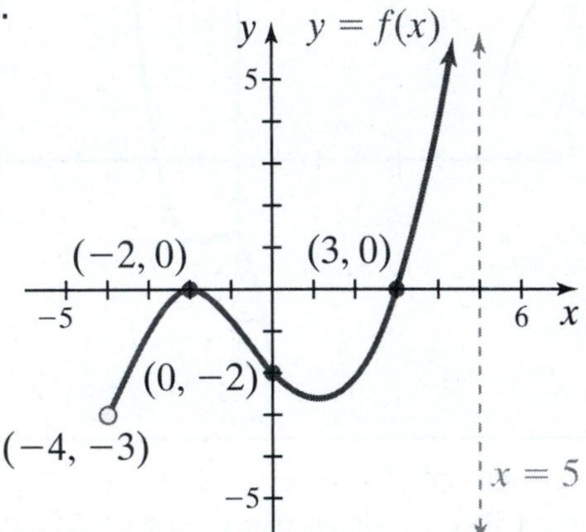

c.

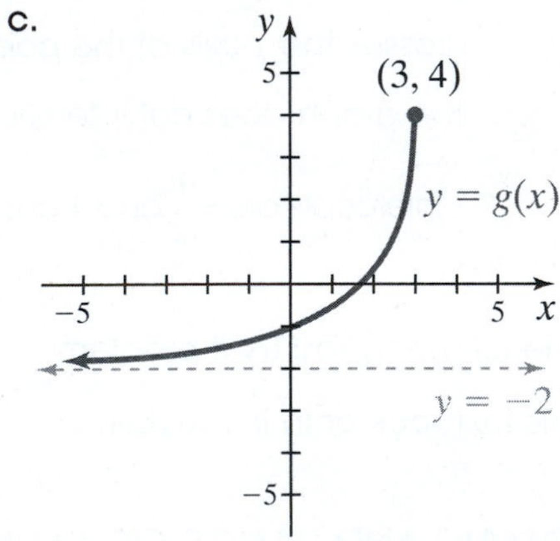

Solution

a. The open dot representing the ordered pair $(-2, 2)$ indicates that this point is not included on the graph, whereas the closed dot representing the ordered pair $(4, 2)$ indicates that this point is included on the graph. Therefore, the domain includes all x-values between -2, not including -2, and 4 inclusive. Thus, the domain in set-builder notation is $\{x \mid -2 < x \leq 4\}$ or the interval $(-2, 4]$. The range includes all y-values between 0 and 6 inclusive because the points $(-1, 0)$ and $(2, 6)$ are both included on the graph. The range is $\{y \mid 0 \leq y \leq 6\}$ or the interval $[0, 6]$.

b. The vertical dashed line at $x = 5$ is called a **vertical asymptote** and indicates that the function will get arbitrarily close to the line $x = 5$ but will never intersect the line. Therefore, the domain of f in set-builder notation is $\{x | -4 < x < 5\}$ or the interval $(-4, 5)$. The range includes all y-values strictly greater than -3. Thus, the range is $\{y \mid y > -3\}$ or the interval $(-3, \infty)$.

c. To determine the domain of g, notice that the graph extends indefinitely to the left and includes all values of x less than or equal to 3. Thus, the domain is $\{x | x \leq 3\}$ or the interval $(-\infty, 3]$. The horizontal dashed line $y = -2$ is called a **horizontal asymptote** and indicates that the function will get arbitrarily close to the line $y = -2$. Thus, the range includes all y-values greater than -2 but less than or equal to 4. So, the range of g is $\{y | -2 < y \leq 4\}$ or the interval $(-2, 4]$.

Watch this **video** to see the solution to this entre example.

You Try It **Work through the following You Try It problem.**

Work Exercises 13–18 in this textbook or in the MyLab Math Study Plan.

OBJECTIVE 3 DETERMINING WHETHER A FUNCTION IS INCREASING, DECREASING, OR CONSTANT

A function f is said to be **increasing** on an open **interval** if the value of $f(x)$ gets larger as x gets larger on the interval. The graph of f rises from left to right on the interval in which f is increasing. Likewise, a function f is said to be **decreasing** on an open interval if the value of $f(x)$ gets smaller as x gets larger on the interval. The graph of f falls from left to right on the interval in which f is decreasing. A graph is **constant** on an open interval if the values of $f(x)$ do not change as x gets larger on the interval. In this case, the graph is a horizontal line on the interval. Figure 7 illustrates all three cases.

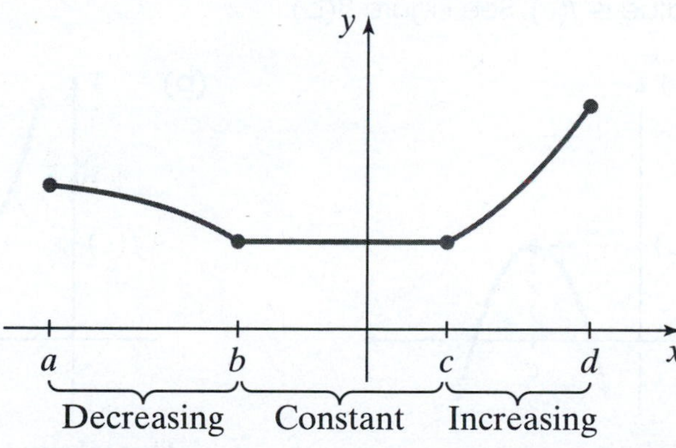

Figure 7 The function is increasing on the interval (c, d). The function is decreasing on the interval (a, b). The function is constant on the interval (b, c).

Example 4 Determine Whether a Function Is Increasing, Decreasing, or Constant from Its Graph

Given the graph of $y = f(x)$, determine where the function is increasing, decreasing, or constant.

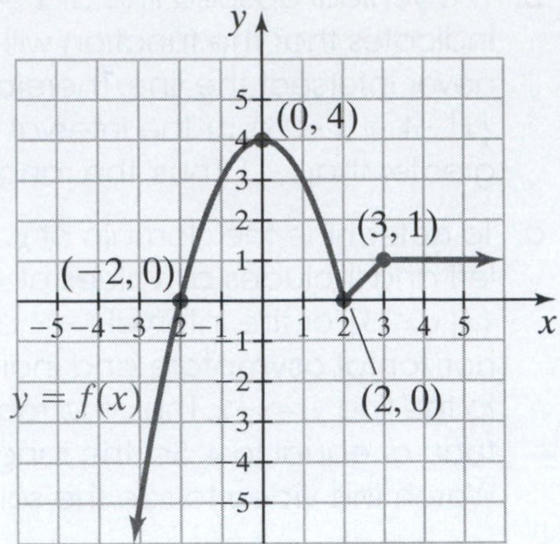

Solution The function is increasing on the interval $(-\infty, 0)$ and on the interval $(2, 3)$. The function is decreasing on the interval $(0, 2)$. The function is constant on the interval $(3, \infty)$. See the **video** to see the solution worked out in detail.

You Try It Work through the following You Try It problem.

Work Exercises 19–22 in this textbook or in the MyLab Math Study Plan.

OBJECTIVE 4 DETERMINING RELATIVE MAXIMUM AND RELATIVE MINIMUM VALUES OF A FUNCTION

When a function changes from increasing to decreasing at a point $(c, f(c))$, then f is said to have a relative maximum at $x = c$. The relative maximum value is $f(c)$. See Figure 8(a). Similarly, when a function changes from decreasing to increasing at a point $(c, f(c))$, then f is said to have a relative minimum at $x = c$. The relative minimum value is $f(c)$. See Figure 8(b).

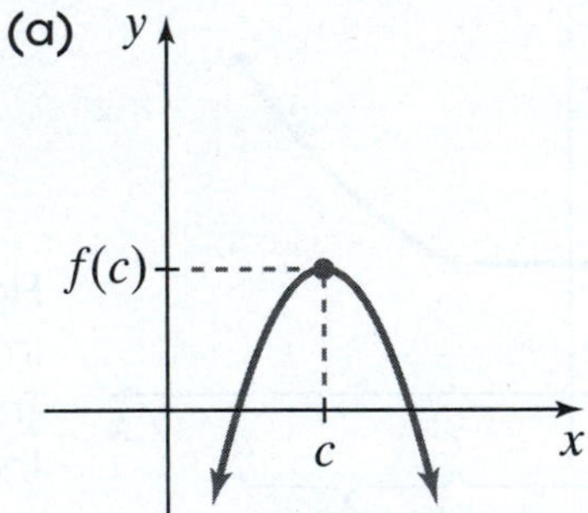

The relative maximum occurs at $x = c$, and the relative maximum value is $f(c)$.

(b) y

f(c)

c x

The relative minimum occurs at $x = c$, and the relative minimum value is $f(c)$.

Figure 8

TIP The word "relative" indicates that the function obtains a maximum or minimum value relative to some **open interval**. It is not necessarily the maximum (or minimum) value of the function on the entire domain.

Example 5 Determine Information from the Graph of a Function

Use the graph of $y = f(x)$ to answer each question.

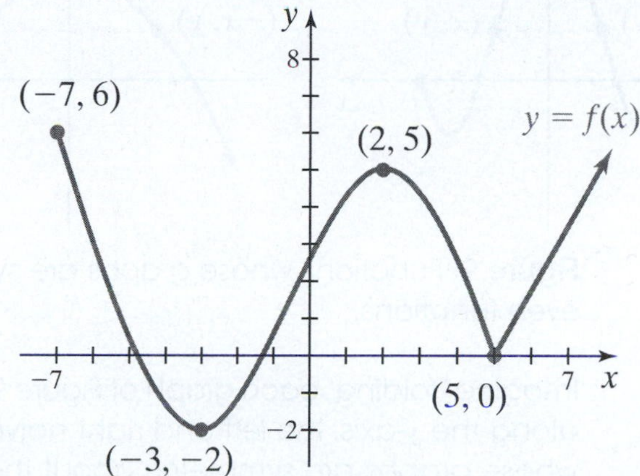

a. On what interval(s) is f increasing?

b. On what interval(s) is f decreasing?

c. For what value(s) of x does f have a relative minimum?

d. For what value(s) of x does f have a relative maximum?

e. What are the relative minima?

f. What are the relative maxima?

Solution

a. The function f is increasing on the interval $(-3, 2)$. The function is also increasing for values of x greater than 5; that is, f is also increasing on the interval $(5, \infty)$.

See the graph of $y = f(x)$.

b. The function f is decreasing on the intervals $(-7, -3)$ and $(2, 5)$.

c. The function f obtains a relative minimum at $x = -3$ and $x = 5$.

d. The function f obtains a relative maximum at $x = 2$.

e. The relative minima are $f(-3) = -2$ and $f(5) = 0$.

f. The relative maximum is $f(2) = 5$.

CAUTION In Example 5, the value $f(-7) = 6$ does not represent a relative maximum. This is because the function does not change from increasing to decreasing at the point $(-7, 6)$. Also, there is no open interval that includes $x = -7$ because the function is not defined for values of x less than -7.

You Try It Work through the following You Try It problem.

Work Exercises 23–27 in this textbook or in the MyLab Math Study Plan.

OBJECTIVE 5 DETERMINING WHETHER A FUNCTION IS EVEN, ODD, OR NEITHER

The graphs of some functions are symmetric about the y-axis. For example, the graphs of the three functions sketched in Figure 9 are all symmetric about the y-axis.

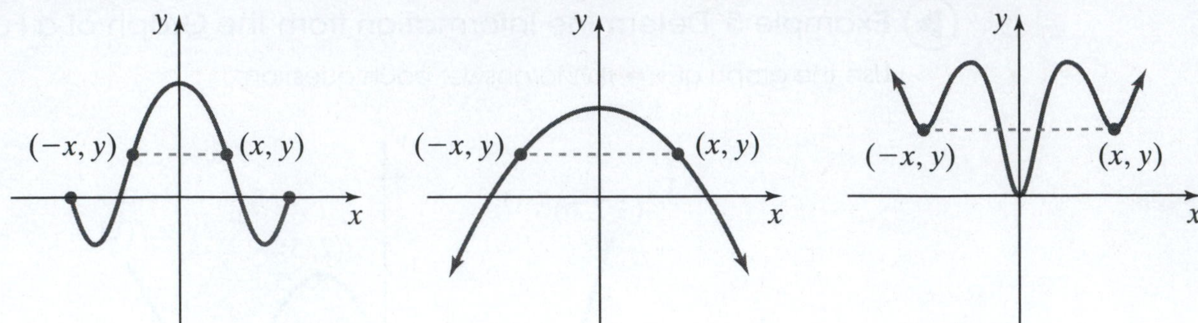

Figure 9 Functions whose graphs are symmetric about the y-axis are called even functions.

Imagine "folding" each graph of Figure 9 along the y-axis. If each graph was folded along the y-axis, the left and right halves of each graph would match. Functions whose graphs are symmetric about the y-axis are called *even functions.* For any point (x, y) on each graph, the point $(-x, y)$ also lies on the graph. Therefore, for any x-value in the domain, $f(x) = f(-x)$.

> **Definition** Even Functions
>
> A function f is **even** if for every x in the domain, $f(x) = f(-x)$. The graph of an even function is symmetric about the y-axis. For each point (x, y) on the graph, the point $(-x, y)$ is also on the graph.

The graph of a function can also be symmetric about the origin. The functions sketched in Figure 10 illustrate this type of symmetry. Functions whose graphs are symmetric about the origin are called *odd functions.* If a function is an odd function, then for any point (x_1, y_1) on the graph of f in Quadrant I, there is a corresponding point $(-x_1, -y_1)$ on the graph of f in Quadrant III. Similarly, for any point (x_2, y_2) on the graph of f in Quadrant II, there is a corresponding point $(-x_2, -y_2)$ on the graph of f in Quadrant IV.

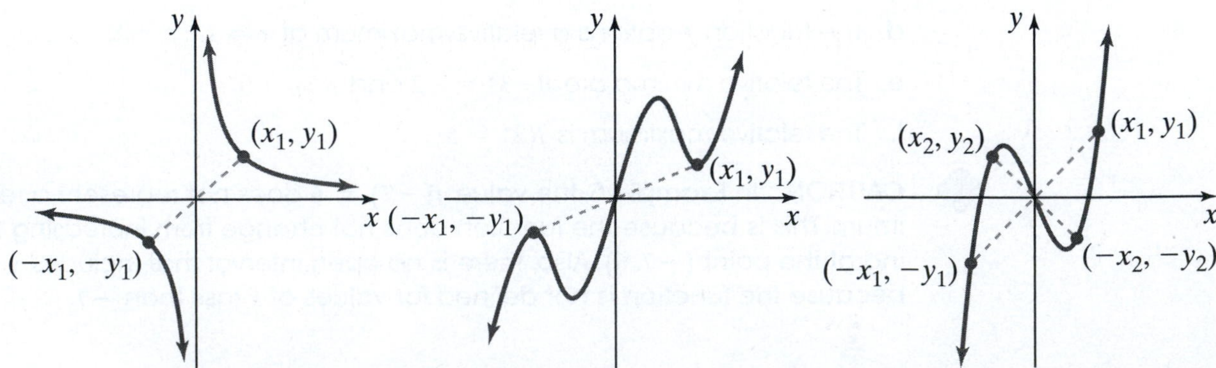

Figure 10 Functions whose graphs are symmetric about the origin are called odd functions.

For any point (x, y) on each graph in **Figure 10**, the point $(-x, -y)$ also lies on the graph. Therefore, for any x-value in the domain, $f(x) = -f(-x)$ or, equivalently, $-f(x) = f(-x)$.

Definition Odd Functions

A function f is **odd** if for every x in the domain, $-f(x) = f(-x)$. The graph of an odd function is symmetric about the origin. For each point (x, y) on the graph, the point $(-x, -y)$ is also on the graph.

Example 6 Determine Even and Odd Functions from the Graph

Determine whether each function is even, odd, or neither.

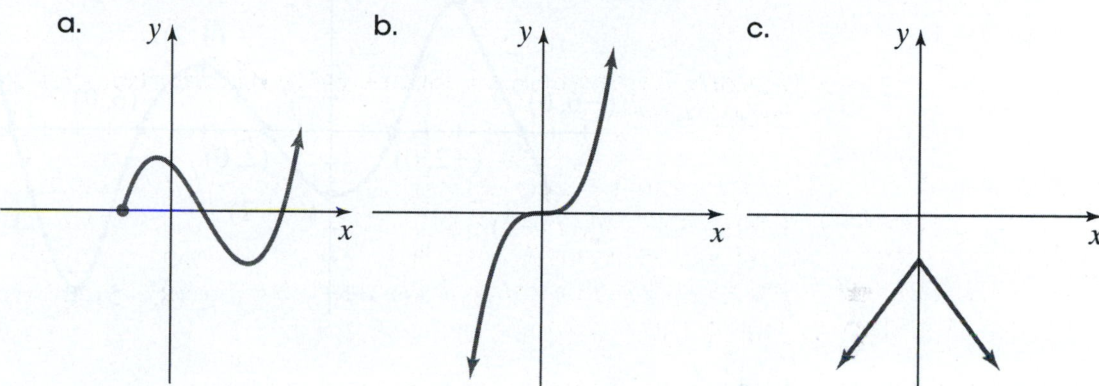

Solution

a. The graph is neither even nor odd because the graph has no symmetry.

b. The graph is symmetric about the origin; therefore, this is the graph of an odd function.

c. This graph is symmetric about the y-axis; thus, this represents the graph of an even function.

You Try It Work through the following You Try It problem.

Work Exercises 28–31 in this textbook or in the MyLab Math Study Plan.

Example 7 Determine Even and Odd Functions Algebraically

Determine whether each function is even, odd, or neither.

a. $f(x) = x^3 + x$ b. $g(x) = \frac{1}{x^2} + 7|x|$ c. $h(x) = 2x^5 - \frac{1}{x}$ d. $G(x) = x^2 + 4x$

Solution Watch the video to verify that the functions in parts a and c are odd functions. The function in part b is even, whereas the function in part d is neither even nor odd.

You Try It Work through the following You Try It problem.

Work Exercises 32–36 in this textbook or in the MyLab Math Study Plan.

OBJECTIVE 6 DETERMINING INFORMATION ABOUT A FUNCTION FROM A GRAPH

We are now ready to take the concepts learned in this section to answer questions about a function's graph. Carefully work through the animation of Example 8.

Example 8 Determine Information about a Function from a Graph

Use the graph of $y = f(x)$ to answer each question.

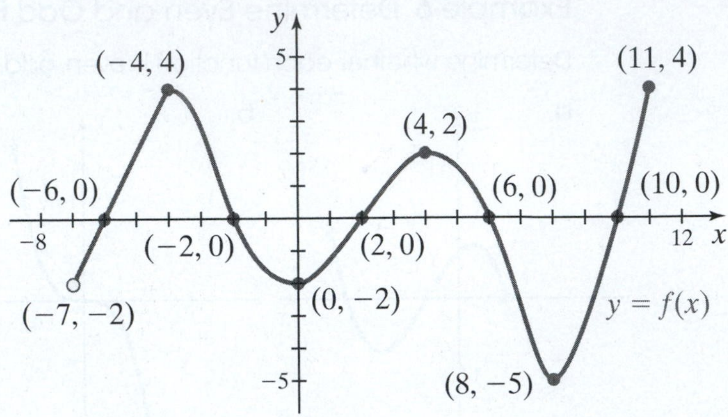

a. What is the y-intercept?

b. What are the real zeros of f?

c. Determine the domain and range of f.

d. Determine the interval(s) on which f is increasing, decreasing, and constant.

e. For what value(s) of x does f obtain a relative maximum? What are the relative maxima?

f. For what value(s) of x does f obtain a relative minimum? What are the relative minima?

g. Is f even, odd, or neither?

h. For what values of x is $f(x) = 4$?

i. For what values of x is $f(x) < 0$?

Solution Carefully work through the animation to see the solution worked out in detail.

You Try It Work through this You Try It problem.

Work Exercises 37 and 38 in this textbook or in the MyLab Math Study Plan.

3.2 Exercises

Skill Check Exercises

In exercises SCE-1 through SCE-4, for each function, determine $f(-x)$.

SCE-1. $f(x) = -3x^5 + 4x^4 + 3x^3 - 7x^2 - 21$

SCE-2. $f(x) = x^4 - 7x^2 - 11$

SCE-3. $f(x) = -3x^7 + 8x^3 - x$

SCE-4. $f(x) = \dfrac{2x^2 - 4}{x}$

In Exercise 1–12, given the function, find the x-intercept(s) and the y-intercept (if they exist).

1. $f(x) = 2x - 1$
2. $g(x) = 3$
3. $f(t) = t^2 - t - 6$
4. $F(x) = 2x^2 - 7x - 3$
5. $h(x) = x^3 + 2x^2 - x - 2$
6. $f(x) = 2x^3 + 11x^2 - 6x$
7. $f(x) = \dfrac{3x - 1}{x + 2}$
8. $g(x) = \dfrac{x}{x^2 - 4}$
9. $G(t) = \dfrac{x^2 - 25}{x}$
10. $L(x) = \dfrac{x}{x^2 + 4}$
11. $f(x) = \dfrac{x^2 + x - 12}{x^2 + x}$
12. $f(x) = \dfrac{x^3 - x^2 - 7x + 7}{x^2 + 1}$

In Exercises 13–18, use the graph to determine the domain and range of each function.

13.

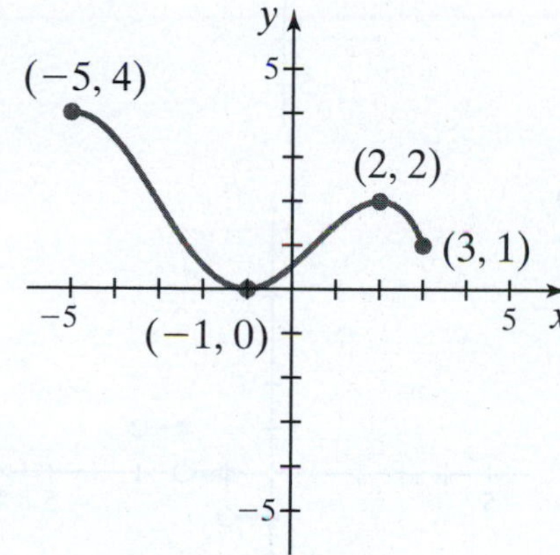

14.

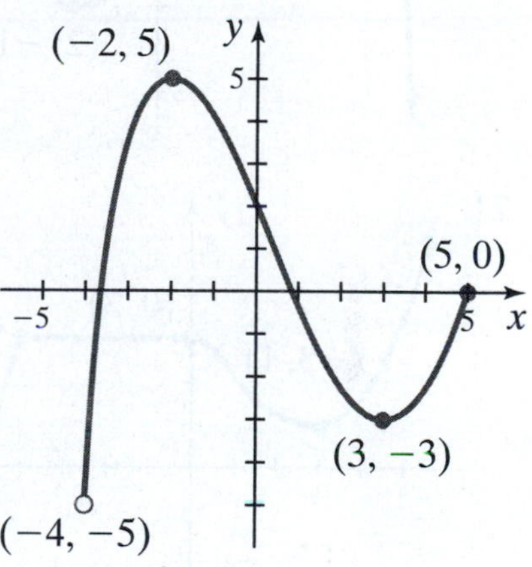

15.

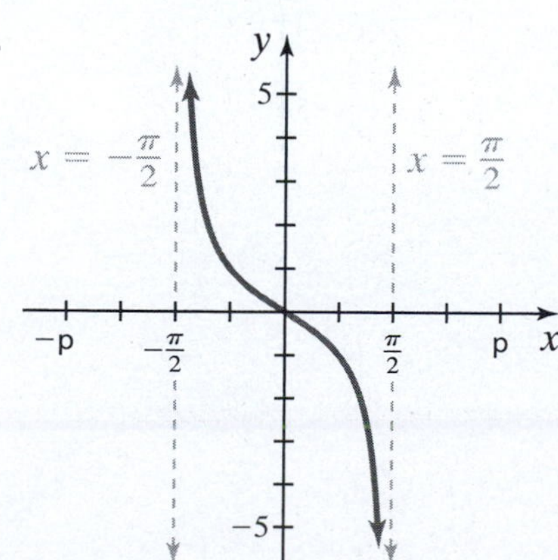

16.

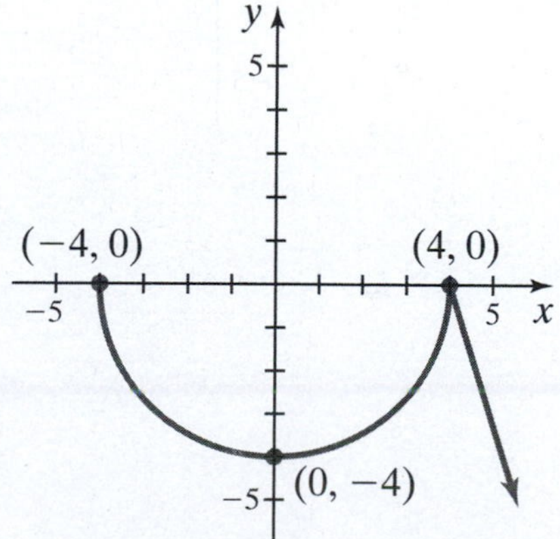

17.

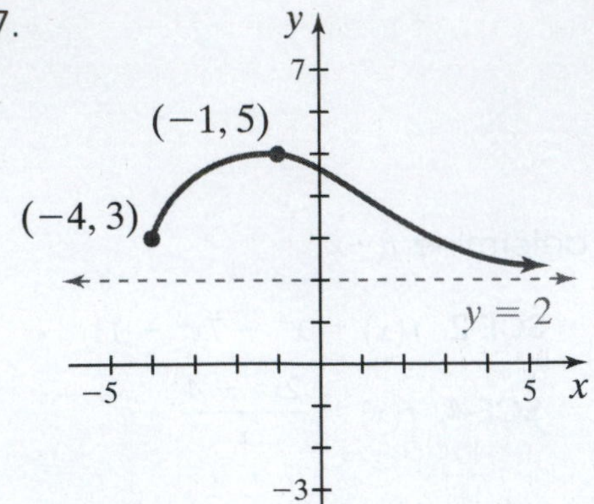

18.

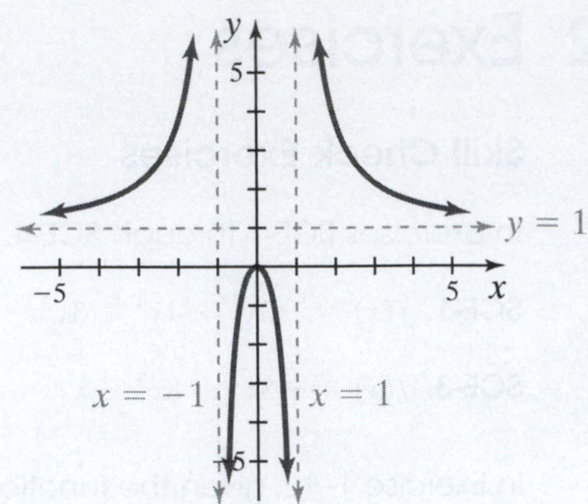

In Exercises 19–22, determine the interval(s) for which each function is (a) increasing, (b) decreasing, and (c) constant.

19.

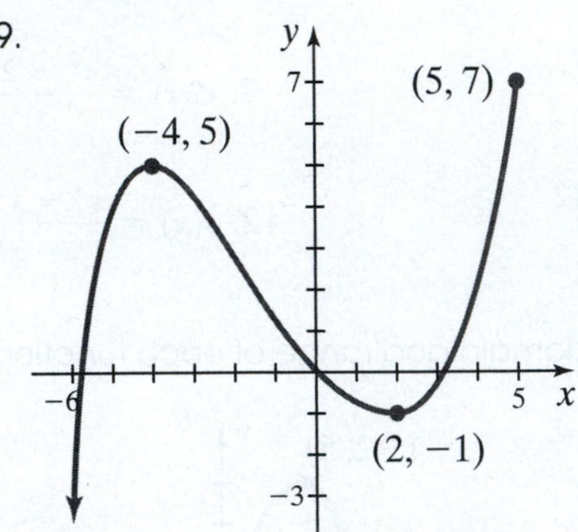

20.

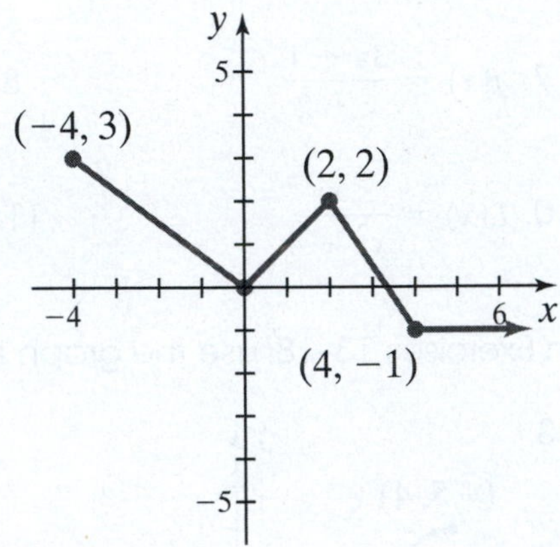

21.

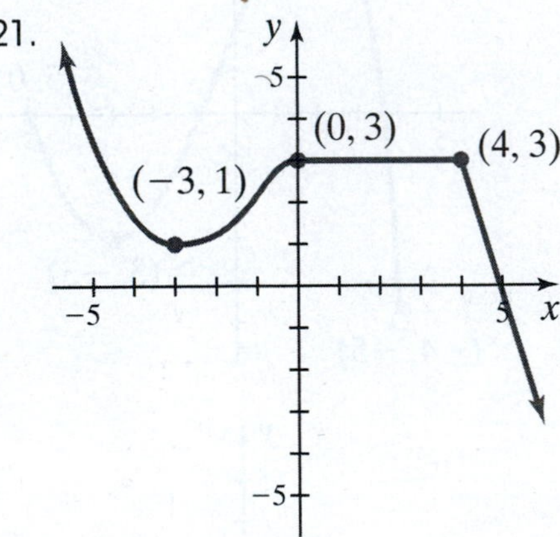

22.

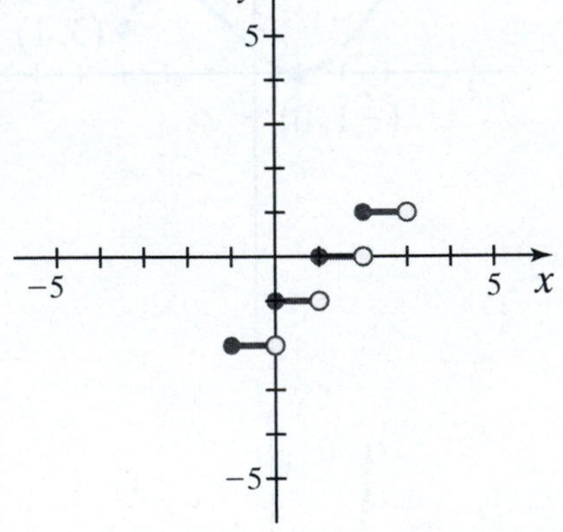

In Exercises 23–27, (a) For what value(s) of x does the function obtain a relative minimum? (b) find the relative minimum value, (c) For what value(s) of x does the function obtain a relative maximum? and (d) find the relative maximum value.

23.

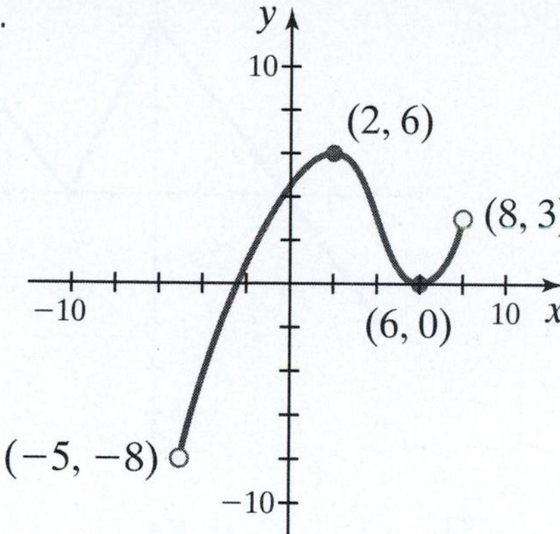

24.

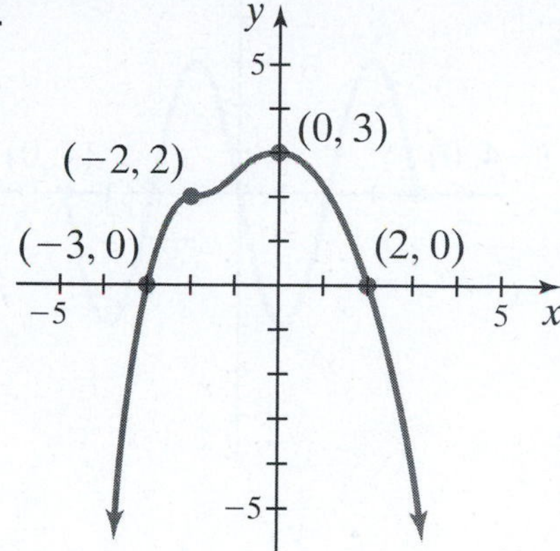

25.

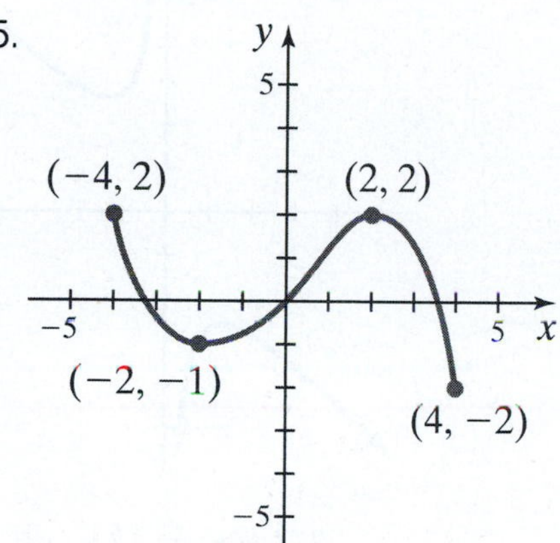

26.

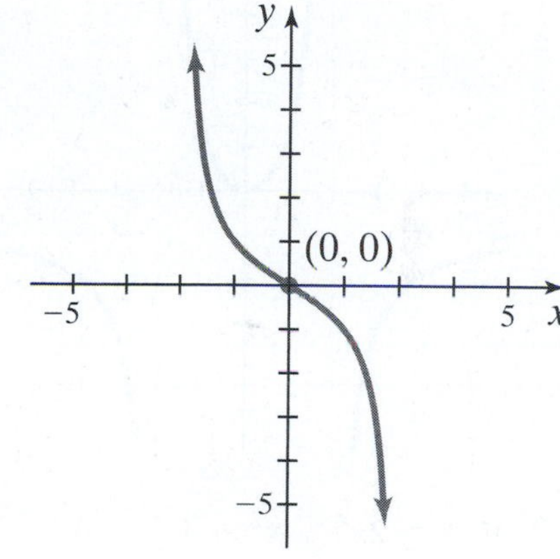

27.

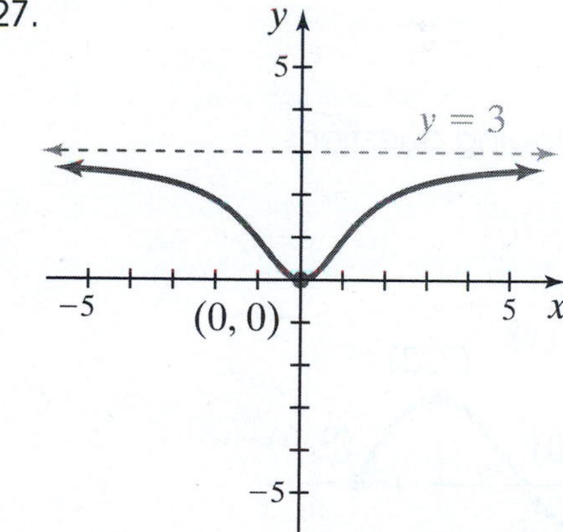

In Exercises 28–36 determine whether each function is even, odd, or neither.

28.

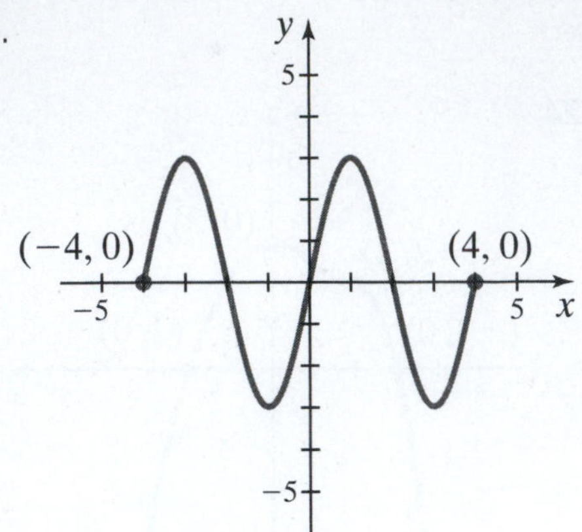

29.

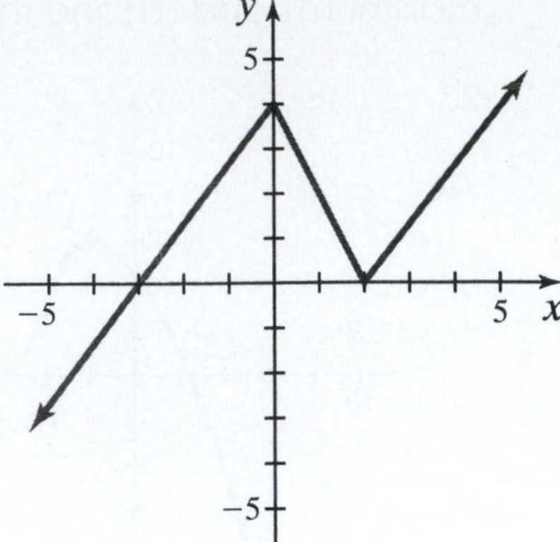

30.

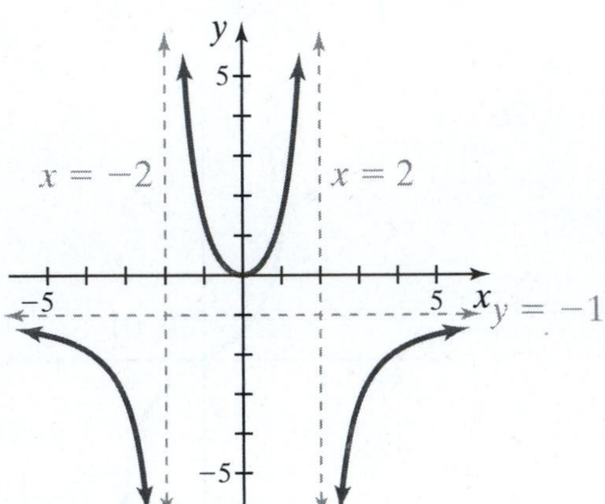

31.

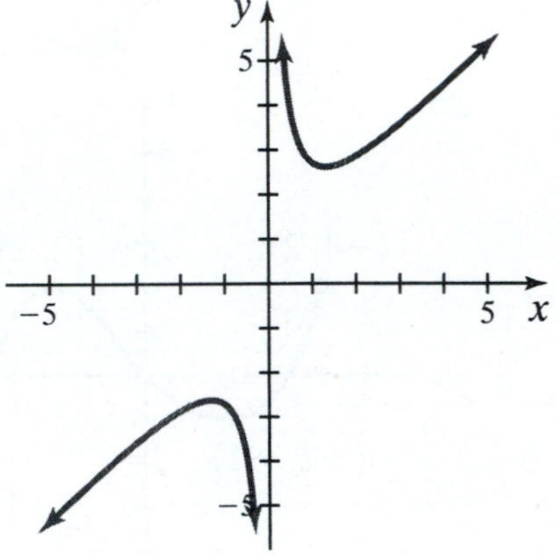

32. $f(x) = 2x^4 - x^2 + 1$

33. $g(x) = x^5 - x^2$

34. $h(x) = x^5 - x^3$

35. $F(x) = \dfrac{2x^2 - 4}{x}$

36. $G(x) = \sqrt{9 - x^2}$

37. Use the graph of $y = f(x)$ to answer each of the following questions:

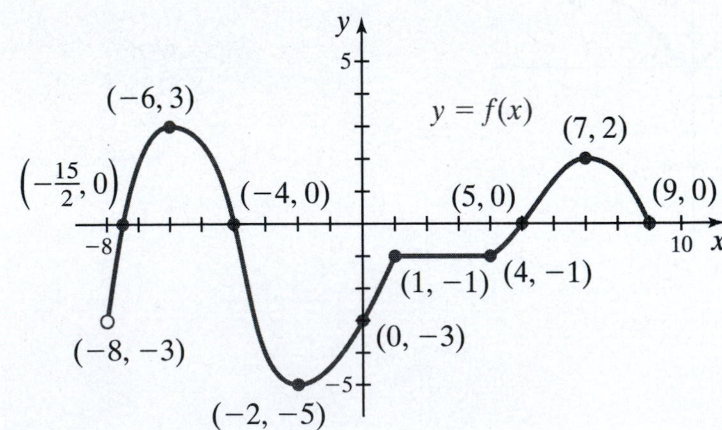

a. What is the domain of f?

b. What is the range of f?

c. On what interval(s) is f increasing, decreasing, or constant?

d. For what value(s) of x does f obtain a relative minimum? What are the relative minima?

e. For what value(s) of x does f obtain a relative maximum? What are the relative maxima?

f. What are the real zeros of f?

g. What is the y-intercept?

h. For what values of x is $f(x) \leq 0$?

i. For how many values of x is $f(x) = -3$?

j. What is the value of $f(2)$?

38. Use the graph of $y = f(x)$ to answer each of the following questions:

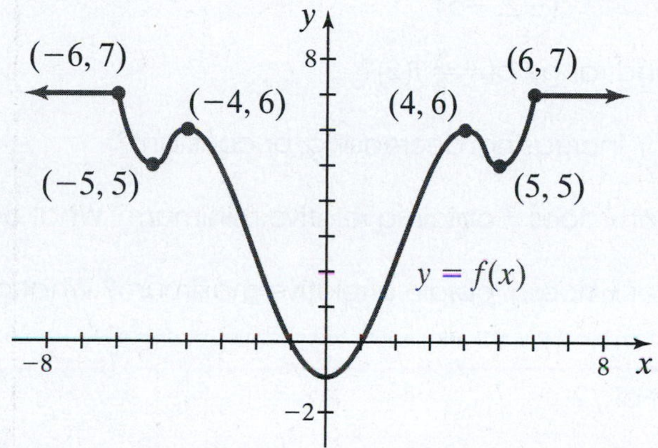

a. What is the domain of f?

b. What is the range of f?

c. On what interval(s) is f increasing, decreasing, or constant?

d. For what value(s) of x does f obtain a relative minimum? What are the relative minima?

e. For what value(s) of x does f obtain a relative maximum? What are the relative maxima?

f. What are the x-intercept(s) of f?

g. Is f even, odd, or neither?

h. For what values of x is $f(x) > 0$?

i. True or false: $f(-3) > f(2)$

j. For what value(s) of x is $f(x) = 7$?

Brief Exercises

For Exercises 39–44, use the graph of $y = f(x)$ to answer each question.

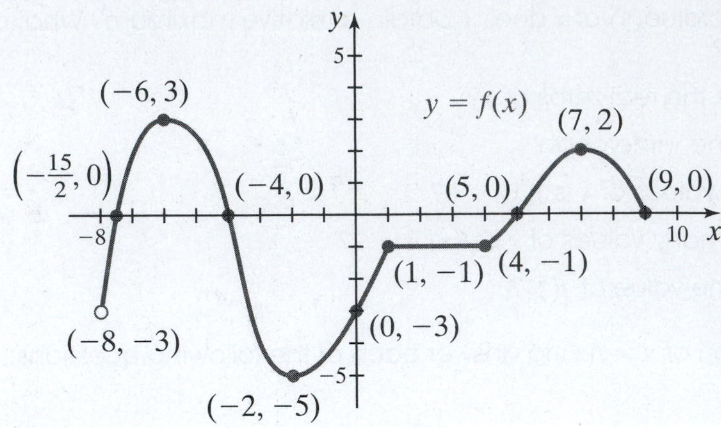

39. What is the domain and range of $y = f(x)$?

40. On what interval(s) is f increasing, decreasing, or constant?

41. a. For what value(s) of x does f obtain a relative minimum? What are the relative minima?

    b. For what value(s) of x does f obtain a relative maximum? What are the relative maxima?

42 What is the y-intercept of f?

43. For what values of x is $f(x) \leq 0$?

44. a. For how many values of x is $f(x) = 3$?

    b. What is the value of $f(2)$?

For Exercises 45–50, use the graph of $y = f(x)$ to answer each question.

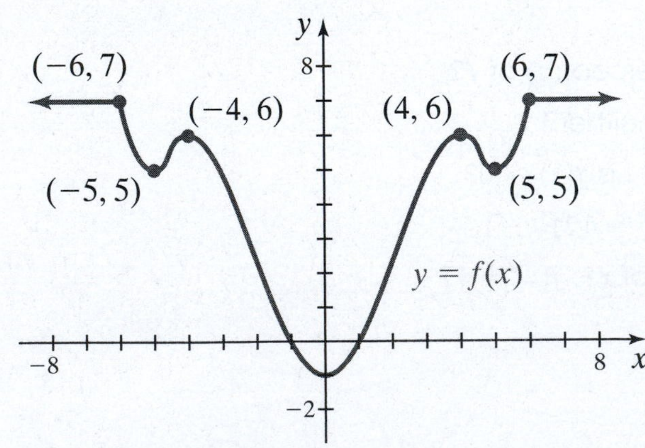

45. What is the domain and range of $y = f(x)$?

46. On what interval(s) is f increasing, decreasing, or constant?

47. a. For what value(s) of x does f obtain a relative minimum? What are the relative minima?

    b. For what value(s) of x does f obtain a relative maximum? What are the relative maxima?

48. What are the x-intercepts of f?

49. a. For what values of x is $f(x) > 0$?

    b. True or False: $f(-3) < f(-2)$?

    c. For what value(s) of x is $f(x) = 7$?

50. Is f even, odd or neither?

3.3 Graphs of Basic Functions; Piecewise Functions

THINGS TO KNOW

Before working through this section, be sure that you are familiar with the following concepts:

		VIDEO	ANIMATION	INTERACTIVE
 You Try It	1. Determining the Domain of a Function Given the Equation (Section 3.1)			
 You Try It	2. Determining the Domain and Range of a Function from Its Graph (Section 3.2)			

OBJECTIVES

OBJECTIVE 1 SKETCHING THE GRAPHS OF THE BASIC FUNCTIONS

There are several functions whose graphs are worth memorizing. In this section, you learn about the graphs of nine basic functions. We begin by discussing the graphs of two specific linear functions. Recall that a linear function has the form $f(x) = mx + b$, where m is the slope of the line and b represents the y-coordinate of the y-intercept.

1. The constant function $f(x) = b$

The **constant function** is a linear function with $m = 0$. The graph of the constant function is a horizontal line.

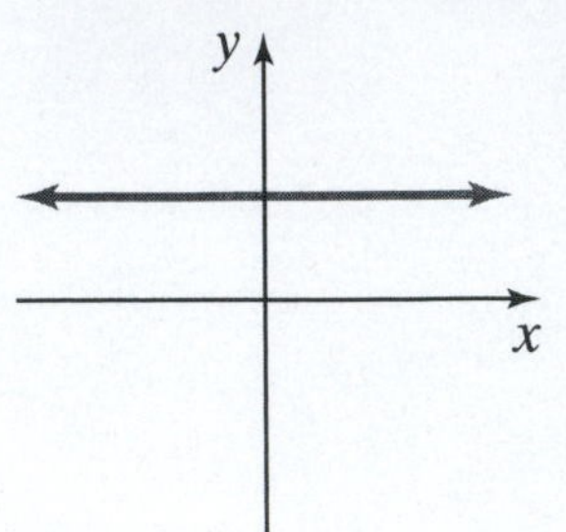

To see how to sketch the constant function, read this explanation.

2. The identity function $f(x) = x$

The **identity function** defined by $f(x) = x$ is a linear function with $m = 1$ and $b = 0$. This function assigns to each number in the domain the exact same number in the range.

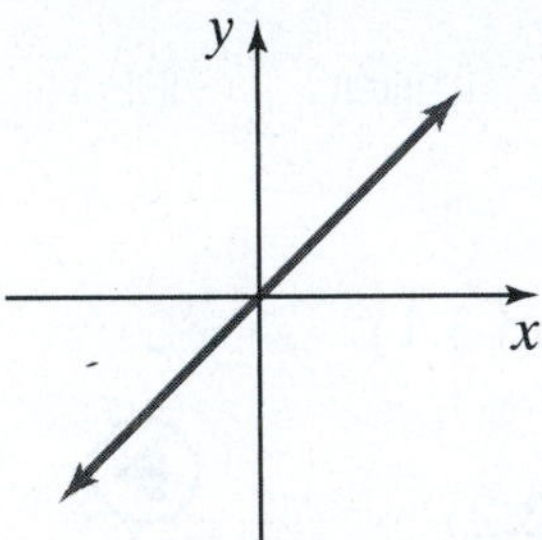

To see how to sketch the identity function, read this explanation.

Of course, you should be able to sketch the graph of any linear function of the form $f(x) = mx + b$. To review how to sketch a line that is given in **slope-intercept form**, refer to **Objective 4 in Section 2.3.**

3. The square function $f(x) = x^2$

The **square function**, $f(x) = x^2$, assigns to each real number in the domain the square of that number in the range.

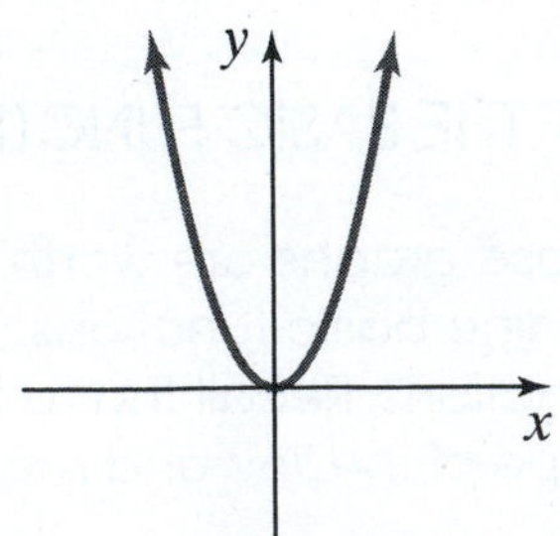

To see how to sketch the square function, read this explanation.

4. The cube function $f(x) = x^3$

The **cube function**, $f(x) = x^3$, assigns to each real number in the domain the cube of that number in the range.

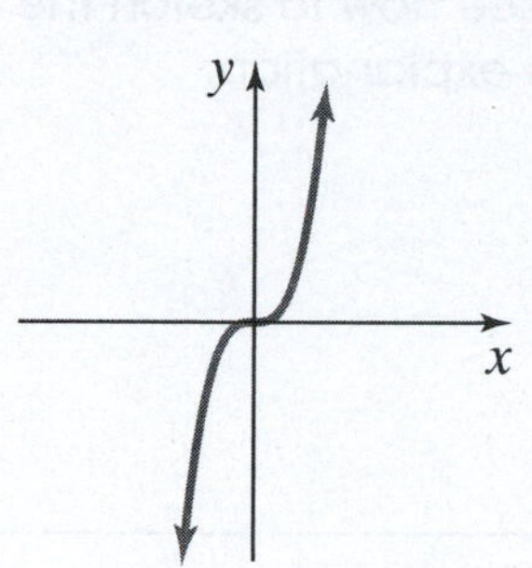

To see how to sketch the cube function, read this explanation.

5. The absolute value function $f(x) = |x|$

The **absolute value function**, $f(x) = |x|$, assigns to each real number in the domain the absolute value of that number in the range.

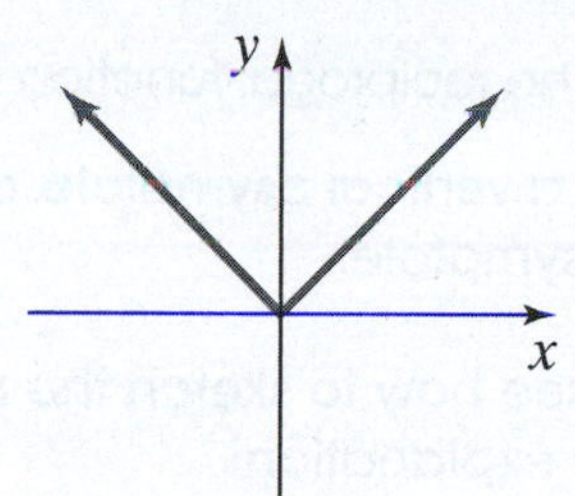

To see how to sketch the absolute value function, read this explanation.

6. The square root function $f(x) = \sqrt{x}$

The **square root function**, $f(x) = \sqrt{x}$, is only defined for values of x that are greater than or equal to zero. It assigns to each real number in the domain the square root of that number in the range.

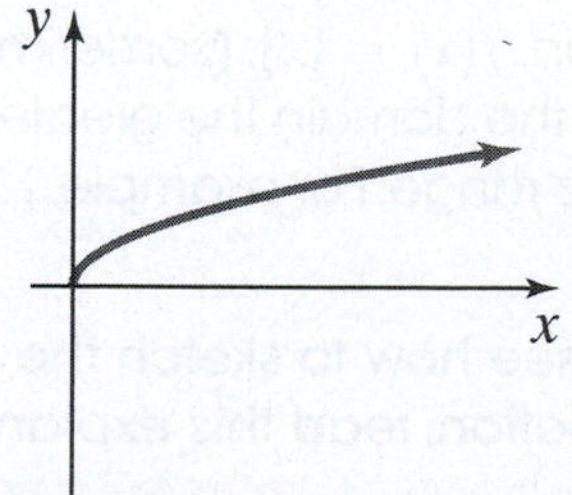

To see how to sketch the square root function, read this explanation.

7. The cube root function $f(x) = \sqrt[3]{x}$

The **cube root function**, $f(x) = \sqrt[3]{x}$, is defined for all real numbers and assigns to each number in the domain the cube root of that number in the range.

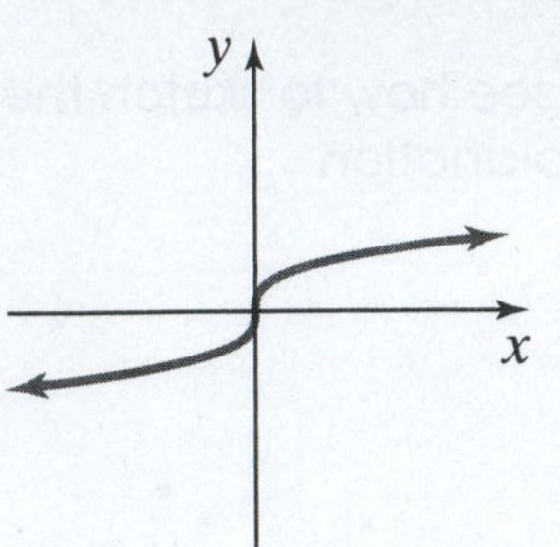

To see how to sketch the cube root function, read this explanation.

8. The reciprocal function $f(x) = \frac{1}{x}$

The **reciprocal function**, $f(x) = \frac{1}{x}$, is a **rational function** whose domain is $\{x|x \neq 0\}$. It assigns to each number a in the domain its reciprocal, $\frac{1}{a}$, in the range. The reciprocal function has two asymptotes. The y-axis (the line $x = 0$) is a **vertical asymptote**, and the x-axis (the line $y = 0$) is a **horizontal asymptote**.

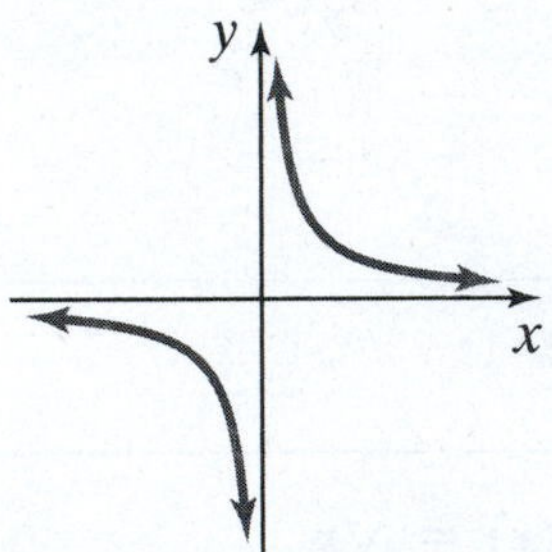

To see how to sketch the reciprocal function, read this explanation.

9. The greatest integer function $f(x) = [\![x]\!]$

The greatest integer function, $f(x) = [\![x]\!]$, (sometimes denoted $f(x) \doteq \text{int}(x)$) assigns to each number in the domain the greatest integer less than or equal to that number in the range. For example, $[\![3.001]\!] = 3$, whereas $[\![2.999]\!] = 2$.

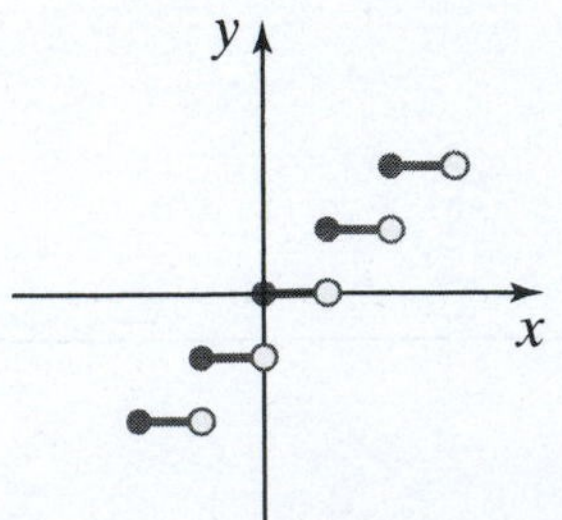

To see how to sketch the greatest integer function, read this explanation.

You Try It Work through the following You Try It problem.

Work Exercises 1–9 in this textbook or in the MyLab Math Study Plan.

OBJECTIVE 2 SKETCHING THE GRAPHS OF BASIC FUNCTIONS WITH RESTRICTED DOMAINS

The domain of the **square function**, $f(x) = x^2$, is all real numbers or $(-\infty, \infty)$. However, suppose that we were only interested in sketching the portion of the graph of $f(x) = x^2$ for positive values of x. In other words, suppose that we wanted to restrict the domain of $f(x) = x^2$ to only the values of x that are strictly greater than 0.

The function sketched below on the right is an example of a function with a restricted domain. When we define functions with restricted domains, we typically define the domain using inequalities. The restricted function on the right below is defined as $f(x) = x^2$ for $x > 0$.

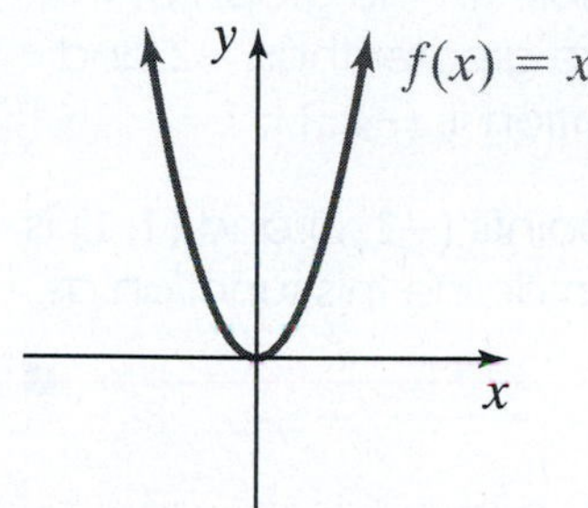

Graph of the function $f(x) = x^2$

The function is not defined at $x = 0$.
There is a "hole" in the graph at the origin.

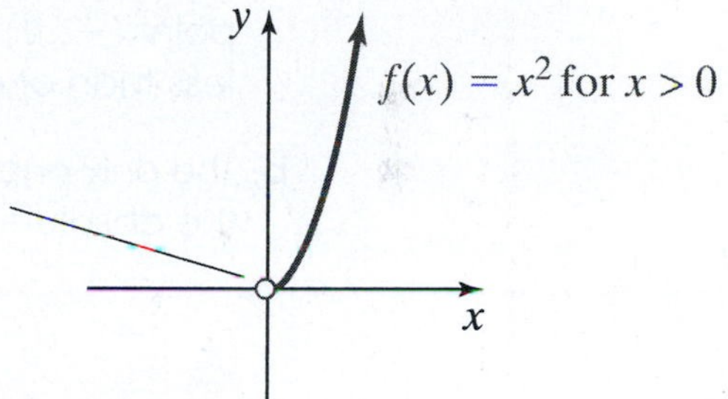

Graph of the function $f(x) = x^2$ for $x > 0$

Example 1 Sketching Basic Functions Having Restricted Domains

Sketch the function $f(x) = \sqrt{x}$ for $x \geq 1$.

Solution The graph of the function $f(x) = \sqrt{x}$ for $x \geq 1$ is shown below. Watch **this video** to see how to sketch the graph of this function.

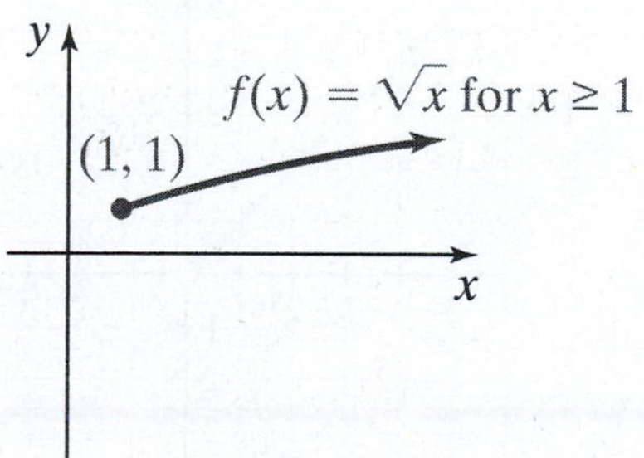

Example 2 Define a Function Given the Graph

The graph of a basic function with a restricted domain is given below.

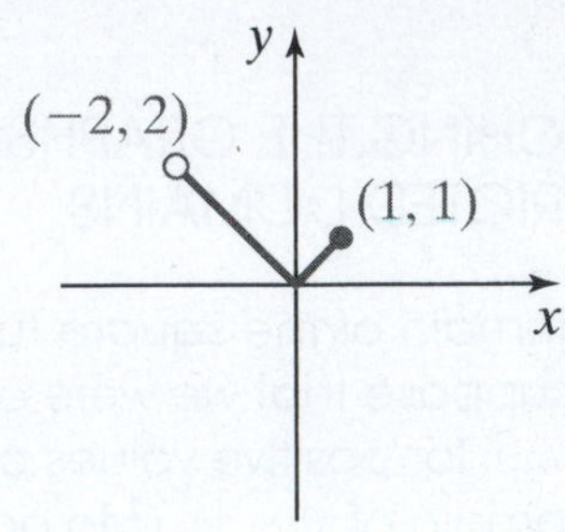

a. Write the domain of the function using interval notation.

b. Define the function using an inequalityto express the restricted domain.

Solution

a. The function is **not defined** at $x = -2$ since there is a "hole" in the graph at the point $(-2, 2)$. The graph is defined for values of x that are greater than -2 and less than or equal to 1. Thus, the domain in interval notation is $(-2, 1]$.

b. The only one of the 9 basic functions that contains the points $(-2, 2)$ and $(1, 1)$ is the **absolute value function**, $f(x) = |x|$. Therefore, we can define this function as

$$f(x) = |x| \text{ for } -2 < x \leq 1.$$

You Try It **Work through the following You Try It problem.**

Work Exercises 10–31 in this textbook or in the MyLab Math Study Plan.

OBJECTIVE 3 ANALYZING PIECEWISE-DEFINED FUNCTIONS

Consider the graphs below of the restricted functions $y = -x$ for $x < 0$ and $y = x$ for $x \geq 0$

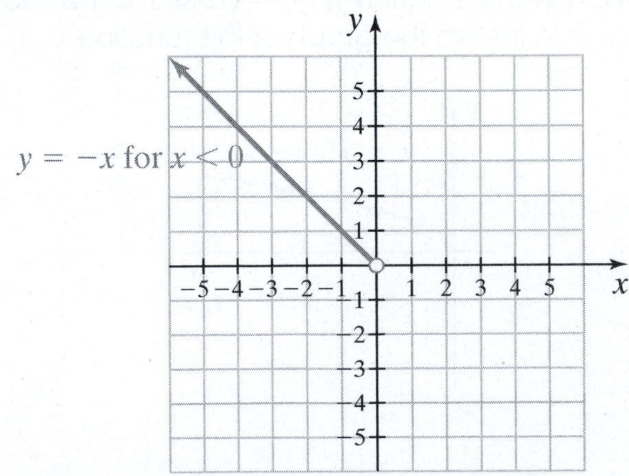

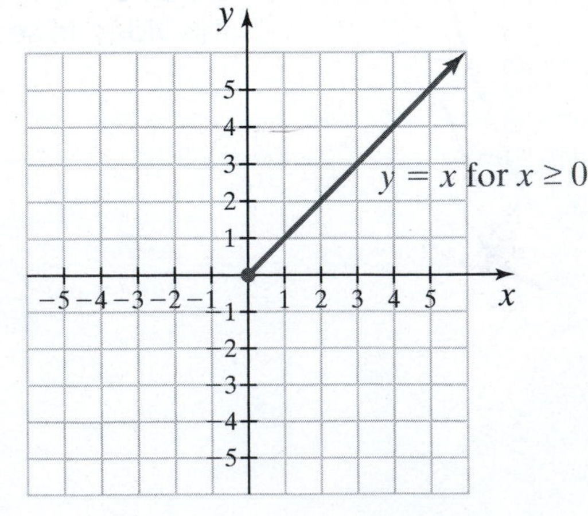

If we combine the two functions from the previous page on the same Cartesian plane, we get the following graph which is the exact same graph as the **absolute value function** $f(x) = |x|$.

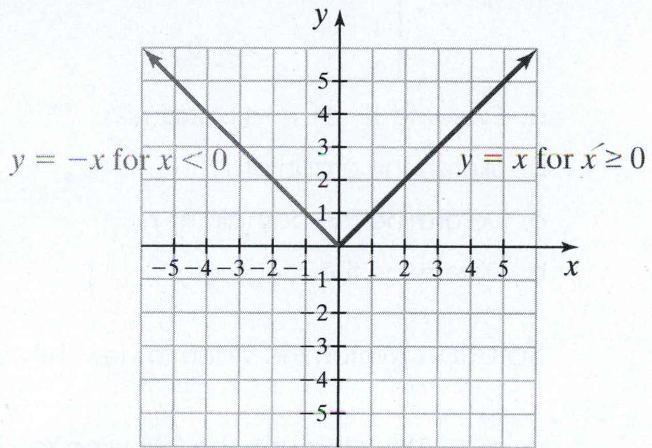

The **absolute value function** can, therefore, be defined by a rule that has two different pieces. We can define the absolute value function as $f(x) = |x| = \begin{cases} x & \text{if } x \geq 0 \\ -x & \text{if } x < 0 \end{cases}$. Functions defined by a rule that has more than one piece are called **piecewise-defined functions.**

Example 3 Sketch a Piecewise-Defined Function

Sketch the function $f(x) = \begin{cases} x^2 & \text{if } x < 1 \\ 1 - x & \text{if } x \geq 1 \end{cases}$.

Solution The graph of this piecewise function is sketched as follows. Watch the **animation** to see how to sketch this function.

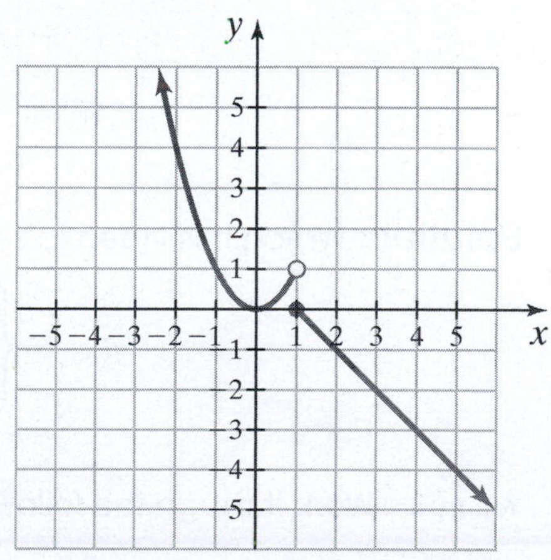

Example 4 Analyze a Piecewise-Defined Function

Let $f(x) = \begin{cases} 1 & \text{if} \quad x < -1 \\ \sqrt[3]{x} & \text{if} \quad -1 \le x < 0. \\ \dfrac{1}{x} & \text{if} \quad x > 0 \end{cases}$

a. Evaluate $f(-3), f(-1)$, and $f(2)$.

b. Sketch the graph of f.

c. Determine the domain of f.

d. Determine the range of f.

Solution Watch the **video** to view the solution.

You Try It Work through the following You Try It problem.

Work Exercises 32–38 in this textbook or in the MyLab Math Study Plan.

Example 5 Find a Rule That Defines a Piecewise Function

Find a rule that describes the following piecewise function:

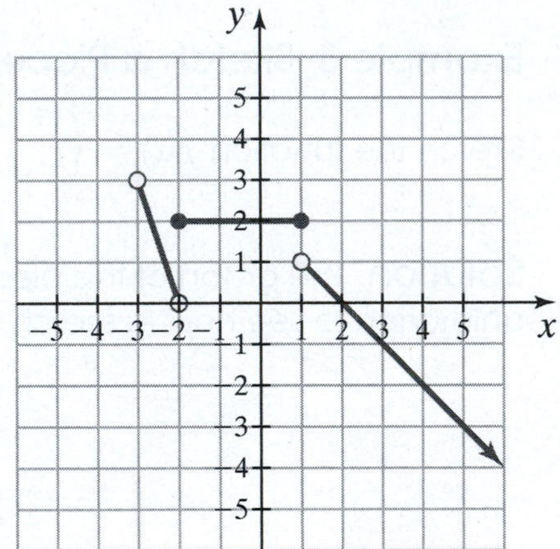

Solution Watch this **video** to see that we can define this piecewise function as

$$f(x) = \begin{cases} -3x - 6 & \text{if } -3 < x < -2 \\ 2 & \text{if } -2 \le x \le 1 \\ -x + 2 & \text{if } x > 1 \end{cases}$$

You Try It Work through the following You Try It problem.

Work Exercises 39–43 in this textbook or in the MyLab Math Study Plan.

OBJECTIVE 4 SOLVING APPLICATIONS OF PIECEWISE-DEFINED FUNCTIONS

Example 6 A Flat Tax Plan

Ted Cruz, a presidential candidate during the 2016 Republican presidential primaries, proposed a flat tax system to replace the existing U.S. income tax system. In his tax proposal, every adult would pay $0.00 in federal taxes on the first $36,000 earned. They would then pay a flat tax of 10% on everything over $36,000. Cruz's tax plan can be described using a piecewise-defined function.

a. According to Cruz's plan, how much in taxes are owed for someone earning $62,500?

b. Find the piecewise function, $T(x)$, that describes the amount of taxes paid, T, as a function of the dollars earned, x, for Cruz's plan.

c. Sketch the piecewise function, $T(x)$.

Solution

a. On $62,500, a person would pay 0% on the first $36,000 and 10% on the remaining $26,500. Thus, the federal taxes owed would be

$$0.10(62{,}500 - 36{,}000) = 0.10(26{,}500) = \$2{,}650.$$

b. For values of x greater than or equal to $0 and less than or equal to $36,000, $T(x) = 0$. When x is greater than $36,000, the tax rate is 10% or $T(x) = 0.10(x - 36{,}000)$ for $x > 36{,}000$. Thus, the rule for computing the tax rate owed is

$$T(x) = \begin{cases} 0 & \text{if } 0 \le x \le 36{,}000 \\ 0.10(x - 36{,}000) & \text{if } \quad x > 36{,}000 \end{cases}.$$

c. Watch this **video** to see how to sketch the graph of $T(x)$ which is shown below.

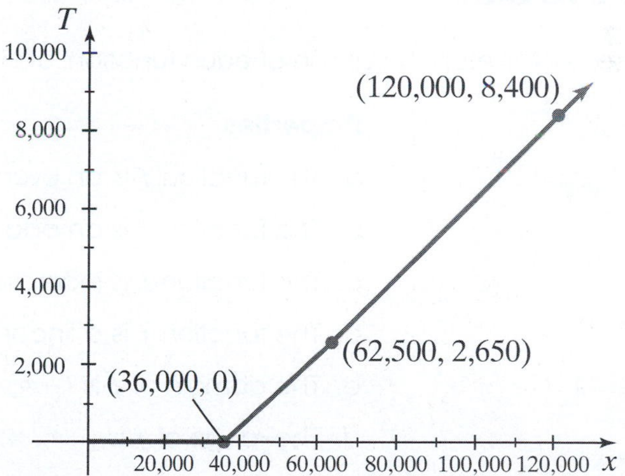

Example 7 Rent a Car

Cheapo Rental Car Co. charges a flat rate of $85 to rent a car for up to 2 days. The company charges an additional $20 for each additional day (or part of a day).

a. How much does it cost to rent a car for $2\frac{1}{2}$ days? 3 days?

b. Write and sketch the piecewise function, $C(x)$, that describes the cost of renting a car as a function of the time, x, in days rented for values of x less than or equal to 5.

Solution The rule for this piecewise function is $C(x) = \begin{cases} 85 & \text{if} \quad x \le 2 \\ 105 & \text{if} \quad 2 < x \le 3 \\ 125 & \text{if} \quad 3 < x \le 4 \\ 145 & \text{if} \quad 4 < x \le 5 \end{cases}$

and is sketched as follows. Watch the **video** for a detailed explanation.

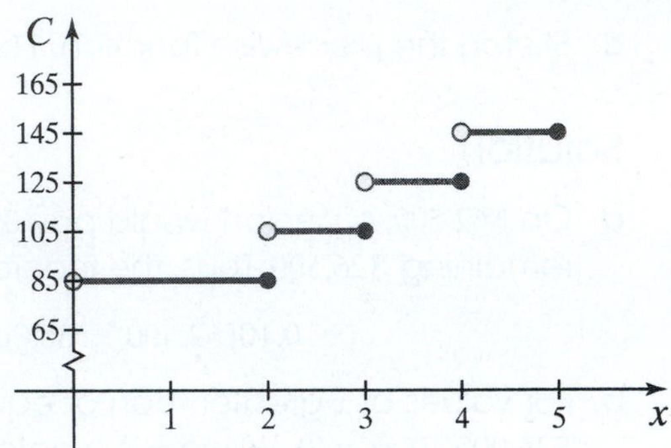

You Try It **Work through the following You Try It problem.**

Work Exercises 44–47 in this textbook or in the MyLab Math Study Plan.

3.3 Exercises

In Exercises 1–9, sketch the graph of each function, and identify all properties that apply.

1. $f(x) = x^2$
2. $f(x) = \sqrt[3]{x}$
3. $f(x) = 2x - 1$
4. $f(x) = |x|$
5. $f(x) = -4$
6. $f(x) = x$
7. $f(x) = x^3$
8. $f(x) = \sqrt{x}$
9. $f(x) = \frac{1}{x}$

Properties

a. The function f is an even function.

b. The function f is an odd function.

c. The function f is increasing on the interval $(-\infty, \infty)$.

d. The function f is a linear function.

e. The domain of f is $(-\infty, \infty)$.

f. The range of f is $(-\infty, \infty)$.

In Exercises 10–23, sketch the graph of the basic function having the indicated restricted domain.

10. $f(x) = x$ for $-3 < x \le 1$

11. $f(x) = x^2$ for $x \le 1$

12. $f(x) = \sqrt{x}$ for $0 \le x < 4$

13. $f(x) = \dfrac{1}{x}$ for $x \ge -1$

14. $f(x) = x^3$ for $-1 \le x \le 1$

15. $f(x) = \sqrt[3]{x}$ for $x < 8$

16. $f(x) = |x|$ for $-2 < x \le 1$

17. $f(x) = x$ for $x > -1$

18. $f(x) = x^2$ for $-2 < x \le 3$

19. $f(x) = \sqrt{x}$ for $x > 1$

20. $f(x) = \dfrac{1}{x}$ for $-1 \le x < 2$

21. $f(x) = x^3$ for $x \ge -2$

22. $f(x) = \sqrt[3]{x}$ for $-1 < x \le 8$

23. $f(x) = |x|$ for $x \le 3$

In Exercises 24–31, the graph of a basic function with a restricted domain is given. For each graph, (a) write the domain in interval notation; (b) Define the function using an inequality to express the restricted domain.

24.

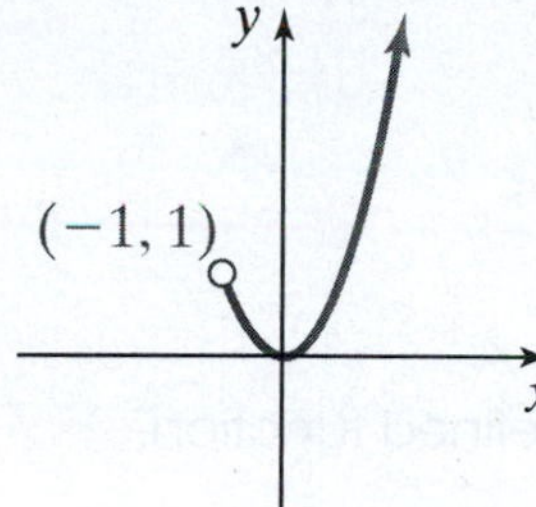

25.

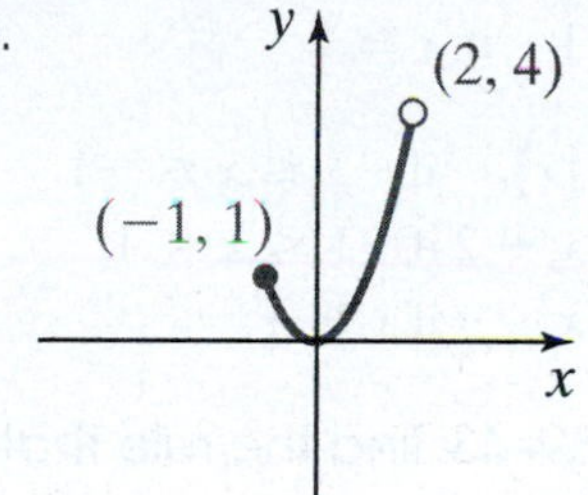

26.

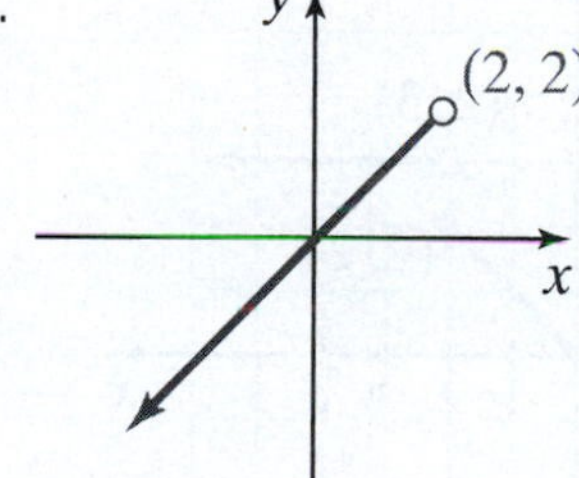

27.

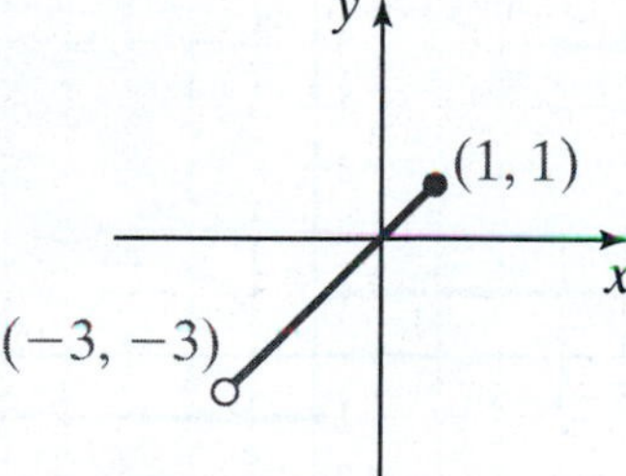

28.

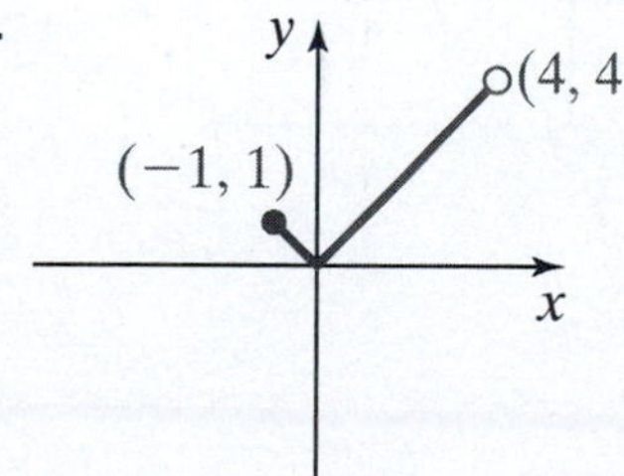

29.

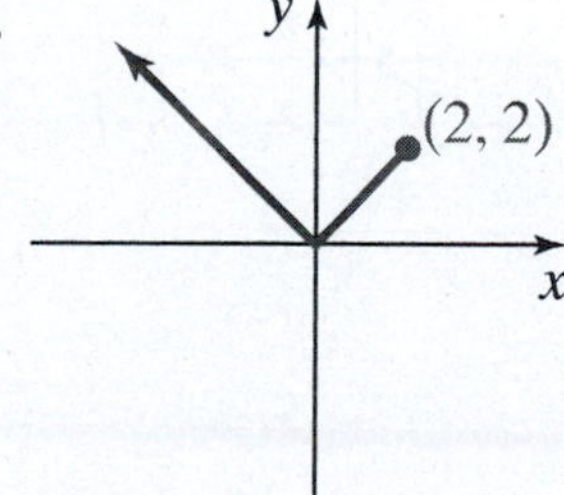

30.

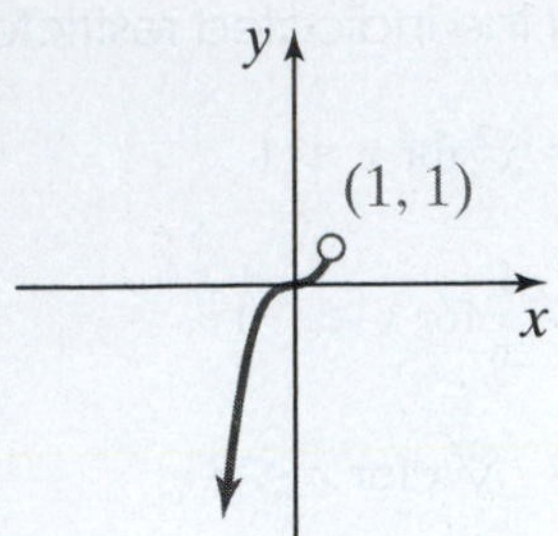

31.

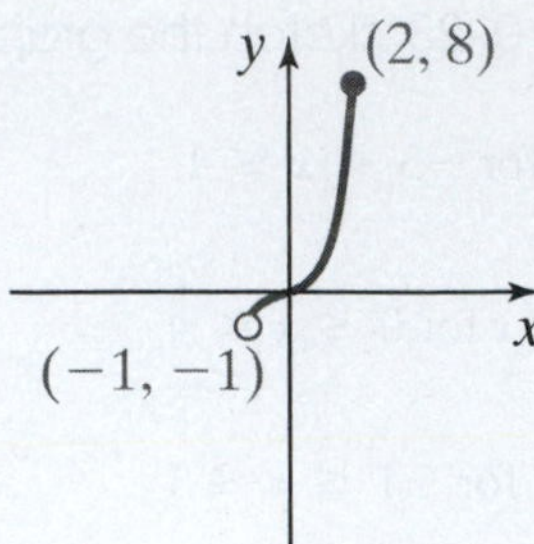

In Exercises 32–38, (a) find $f(-2), f(0)$, and $f(2)$; (b) sketch the graph of each piecewise-defined function; (c) determine the domain of f; and (d) determine the range of f.

32. $f(x) = \begin{cases} 1 & \text{if } x < 2 \\ -3 & \text{if } x \geq 2 \end{cases}$

33. $f(x) = \begin{cases} 2x - 1 & \text{if } x \leq 1 \\ x + 2 & \text{if } x > 1 \end{cases}$

34. $f(x) = \begin{cases} 2x - 1 & \text{if } -2 \leq x < 2 \\ -x + 3 & \text{if } 2 \leq x \leq 4 \end{cases}$

35. $f(x) = \begin{cases} x^2 & \text{if } x \leq 0 \\ \dfrac{1}{x} & \text{if } x > 0 \end{cases}$

36. $f(x) = \begin{cases} 3 & \text{if } x < -1 \\ |x| & \text{if } -1 \leq x < 2 \\ 1 & \text{if } x \geq 2 \end{cases}$

37. $f(x) = \begin{cases} x^2 & \text{if } x < 0 \\ 2 & \text{if } x = 0 \\ \sqrt{x} & \text{if } x > 0 \end{cases}$

38. $f(x) = \begin{cases} [\![x]\!] & \text{if } -3 \leq x \leq -1 \\ x + 2 & \text{if } -1 < x < 1 \\ x^3 & \text{if } x \geq 1 \end{cases}$

In Exercises 39–43, find the rule that describes each piecewise-defined function.

39.

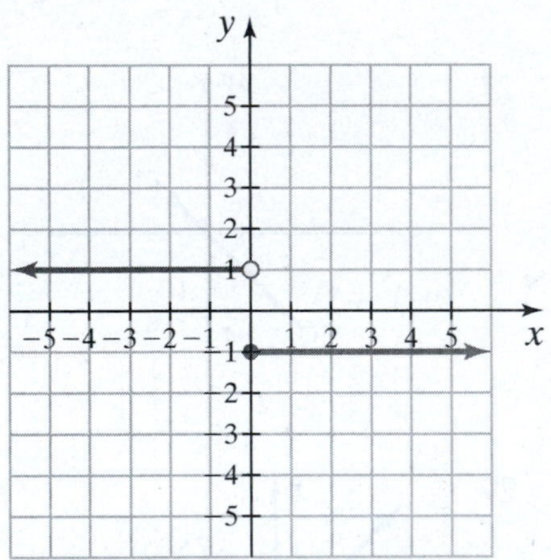

40.

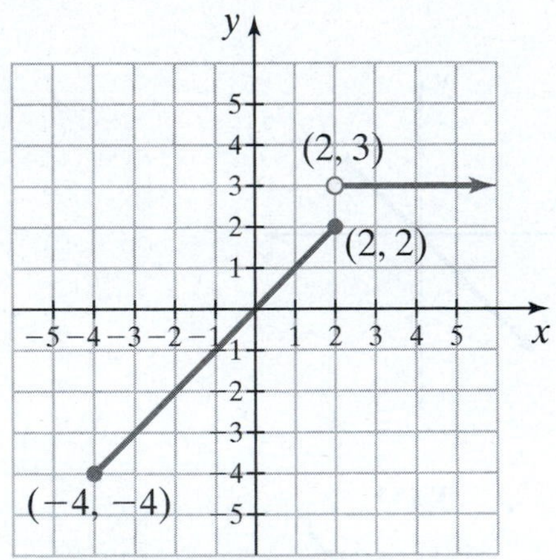

41.

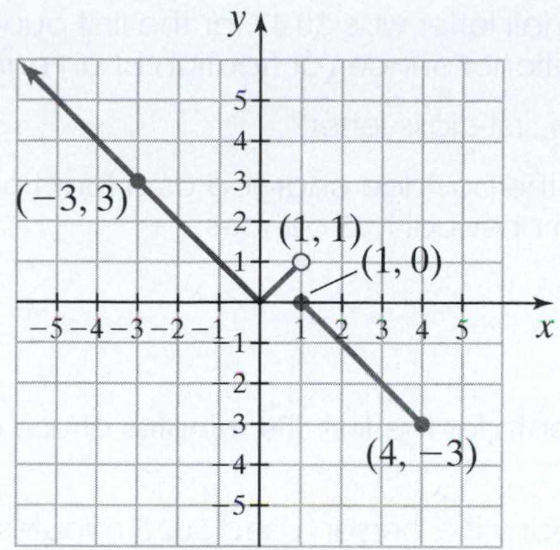

42.

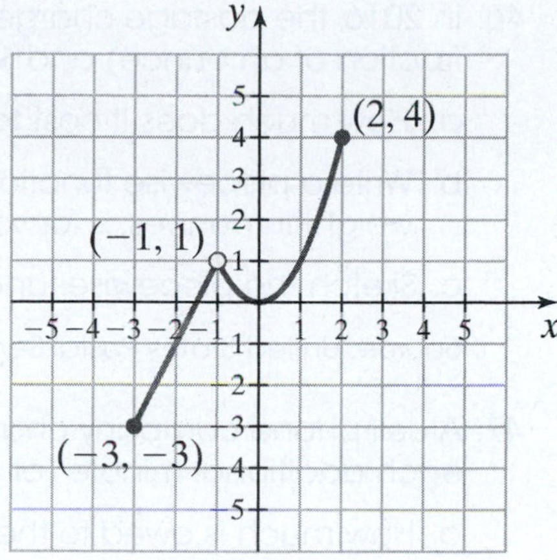

43.

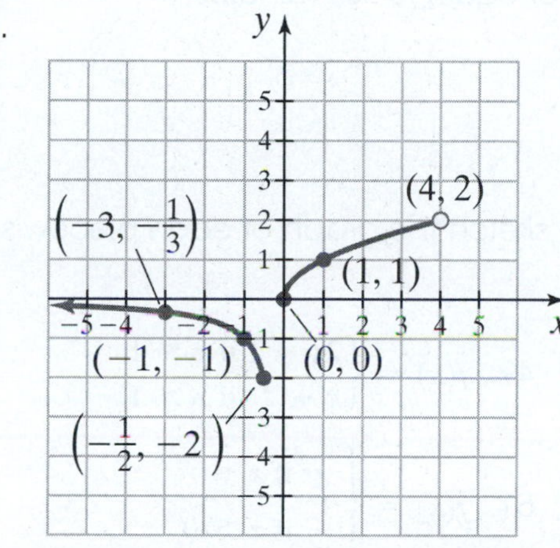

44. The currency in Elbonia is called the dilbert and is abbreviated DIL. Following is a table describing the 2018 class 1 individual tax rate on personal income in Elbonia:

Income	Tax Rate
DIL 0–26,300	0%
Over DIL 26,300	28%

a. How many dilberts in taxes are owed for an individual earning DIL 30,000?

b. Find the piecewise function, $T(x)$, that describes the amount of taxes paid, T, as a function of dilberts earned, x, for Elbonians paying class 1 personal income tax in Elbonia in 2018.

c. Sketch the piecewise function, $T(x)$.

45. An electric company charges \$9.50 per month plus \$0.07 per kilowatt-hour (kWh) for the first 350 kWh used per month and \$0.05 per kWhr for anything over 350 kWh used.

a. How much is owed for using 280 kWh per month?

b. How much is owed for using 400 kWh per month?

c. Write a piecewise function, $C(x)$, where C is the cost charged per month as a function of the monthly kilowatt-hours used, x.

d. Sketch the piecewise function.

46. In 2016, the postage charged for a first-class mail letter was \$0.47 for the first ounce (or fraction of an ounce) and \$0.21 for each additional ounce (or fraction of an ounce).

    a. How much does it cost to send a 2.4-ounce first-class letter?

    b. Write a piecewise function, $P(x)$, where P is the postage charged as a function of its weight in ounces, x, for values of x less than or equal to 5 ounces.

    c. Sketch the piecewise function.

    Source: United States Postal Service.

47. A cell phone company charges \$25.00 per month for the first 500 minutes of use and \$0.05 for each additional minute (or part of a minute).

    a. How much is owed to the cell phone company if a person used 502.5 minutes?

    b. Write a piecewise function, $C(x)$, where C is the cost charged per month as a function of the minutes, x, used for values of x less than or equal to 505 minutes.

    c. Sketch the piecewise function.

Brief Exercises

In Exercises 48–54, (a) find $f(-2), f(0)$, and $f(2)$; (b) sketch the graph of each piecewise-defined function.

48. $f(x) = \begin{cases} 1 & \text{if } x < 2 \\ -3 & \text{if } x \ge 2 \end{cases}$

49. $f(x) = \begin{cases} 2x - 1 & \text{if } x \le 1 \\ x + 2 & \text{if } x > 1 \end{cases}$

50. $f(x) = \begin{cases} 2x - 1 & \text{if } -2 \le x < 2 \\ -x + 3 & \text{if } 2 \le x \le 4 \end{cases}$

51. $f(x) = \begin{cases} x^2 & \text{if } x \le \\ \dfrac{1}{x} & \text{if } x > 0 \end{cases}$

52. $f(x) = \begin{cases} 3 & \text{if } x < -1 \\ |x| & \text{if } -1 \le x < 2 \\ 1 & \text{if } x \ge 2 \end{cases}$

53. $f(x) = \begin{cases} x^2 & \text{if } x < 0 \\ 2 & \text{if } x = 0 \\ \sqrt{x} & \text{if } x > 0 \end{cases}$

54. $f(x) = \begin{cases} [\![x]\!] & \text{if } -3 \le x \le -1 \\ x + 2 & \text{if } -1 < x < 1 \\ x^3 & \text{if } x \ge 1 \end{cases}$

3.4 Transformations of Functions

THINGS TO KNOW

Before working through this section, be sure that you are familiar with the following concepts:

VIDEO | ANIMATION | INTERACTIVE

You Try It

1. Determining the Domain of a Function Given the Equation (Section 3.1)

You Try It

2. Determining the Domain and Range of a Function from Its Graph (Section 3.2)

Introduction to Section 3.4

In this section, we learn how to sketch the graphs of new functions using the graphs of known functions. Starting with the graph of a known function, we "transform" it into a new function by applying various transformations. Before we begin our discussion about transformations, it is critical that you know the graphs of the basic functions that are discussed in Section 3.3. Take a moment to review the basic functions.

REVIEW OF THE BASIC FUNCTIONS

Click on any function to review its graph.

The **identity function**	$f(x) = x$
The **cube function**	$f(x) = x^3$
The **square root function**	$f(x) = \sqrt{x}$
The **reciprocal function**	$f(x) = \frac{1}{x}$
The **square function**	$f(x) = x^2$
The **absolute value function**	$f(x) = \lvert x \rvert$
The **cube root function**	$f(x) = \sqrt[3]{x}$

OBJECTIVE 1 USING VERTICAL SHIFTS TO GRAPH FUNCTIONS

Example 1 Vertically Shift a Function

Sketch the graphs of $f(x) = |x|$ and $g(x) = |x| + 2$.

Solution Table 2 shows that for every value of x, the y-coordinate of the function g is always 2 greater than the y-coordinate of the function f. The two functions are sketched in **Figure 11**. The graph of $g(x) = |x| + 2$ is exactly the same as the graph of $f(x) = |x|$, except the graph of g is shifted *up* two units.

Table 2

x	$f(x) = \|x\|$	$g(x) = \|x\| + 2$
−3	3	5
−2	2	4
−1	1	3
0	0	2
1	1	3
2	2	4
3	3	5

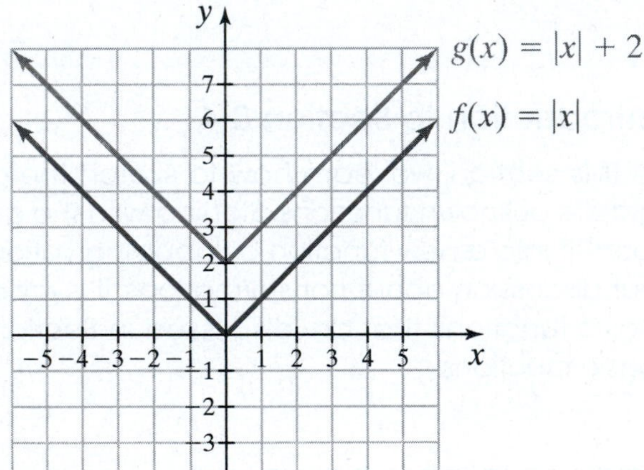

Figure 11

Watch this **video** to see the solution to this example. ●

We see from Example 1 that if c is a positive number, then $y = f(x) + c$ is the graph of f shifted *up* c units. It follows that for $c > 0$, the graph of $y = f(x) - c$ is the graph of f shifted *down* c units.

Vertical Shifts of Functions

If c is a positive real number,

The graph of $y = f(x) + c$ is obtained by shifting the graph of $y = f(x)$ vertically upward c units.

The graph of $y = f(x) - c$ is obtained by shifting the graph of $y = f(x)$ vertically downward c units.

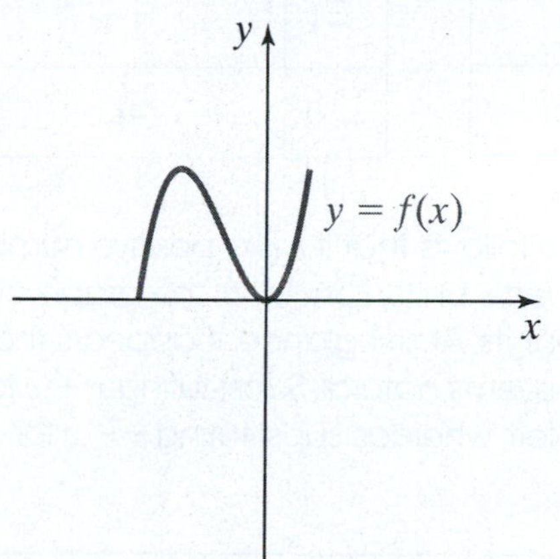

Sketch $y = f(x) + c$.

 Click icon to animate

Sketch $y = f(x) - c$.

 Click icon to animate

To illustrate how vertical transformations work, click on the guided visualization icon below and take some time to explore how to vertically shift some of the basic functions that were introduced in Section 3.3.

 Vertical Transformations of Functions

You Try It Work through the following You Try It problem.

Work Exercises 1–14 in this textbook or in the MyLab Math Study Plan.

OBJECTIVE 2 USING HORIZONTAL SHIFTS TO GRAPH FUNCTIONS

To illustrate a horizontal shift, let $f(x) = x^2$ and $g(x) = (x + 2)^2$. Tables 3 and 4 show tables of values for f and g, respectively. The graphs of f and g are sketched in Figure 12. The graph of g is the graph of f shifted to the *left* two units.

Table 3

x	$f(x) = x^2$
−2	4
−1	1
0	0
1	1
2	4

Table 4

x	$g(x) = (x + 2)^2$
−4	4
−3	1
−2	0
−1	1
0	4

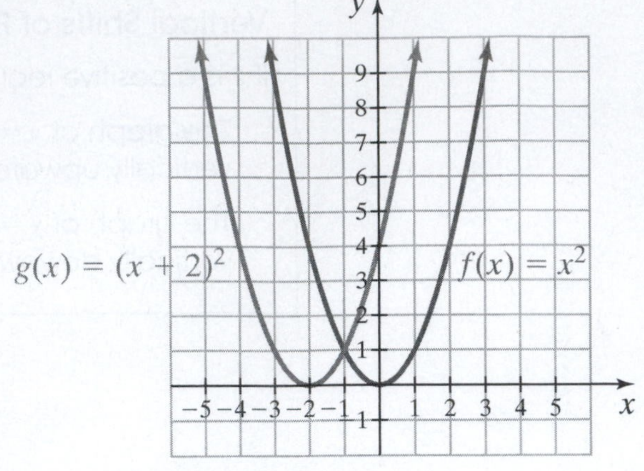

Figure 12

It follows that if c is a positive number, then $y = f(x + c)$ is the graph of f shifted to the *left* c units. For $c > 0$, the graph of $y = f(x - c)$ is the graph of f shifted to the *right* c units. At first glance, it appears that the rule for horizontal shifts is the opposite of what seems natural. Substituting $x + c$ for x causes the graph of $y = f(x)$ to be shifted to the left, whereas substituting $x - c$ for x causes the graph to shift to the right c units.

Horizontal Shifts of Functions

If c is a positive real number,

The graph of $y = f(x + c)$ is obtained by shifting the graph of $y = f(x)$ horizontally to the left c units.

The graph of $y = f(x - c)$ is obtained by shifting the graph of $y = f(x)$ horizontally to the right c units.

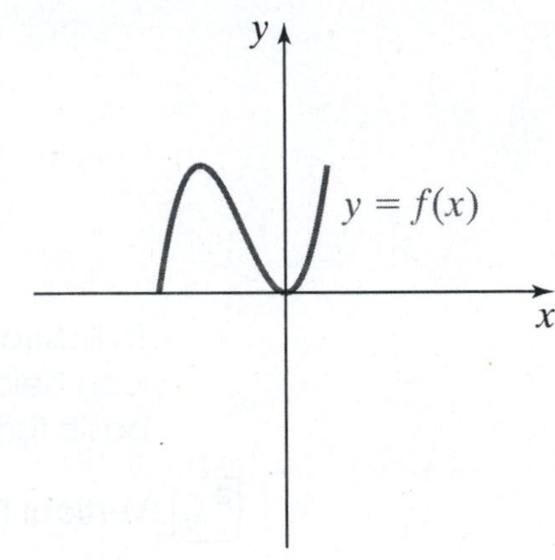

Sketch $y = f(x + c)$.

Sketch $y = f(x - c)$.

To illustrate how horizontal transformations work, click on the guided visualization icon below and take some time to explore how to horizontally shift some of the basic functions that were introduced in Section 3.3.

Horizontal Transformations of Functions

You Try It Work through the following You Try It problem.

Work Exercises 15–28 in this textbook or in the MyLab Math Study Plan.

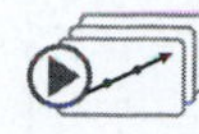

Example 2 Combine Horizontal and Vertical Shifts

Use the graph of $y = x^3$ to sketch the graph of $g(x) = (x - 1)^3 + 2$.

Solution The graph of g is obtained by shifting the graph of the basic function $y = x^3$ first horizontally to the right one unit and then vertically upward two units. When doing a problem with multiple transformations, it is good practice to always perform the vertical transformation last. Click on the animate button to see how to sketch the graph of $g(x) = (x - 1)^3 + 2$.

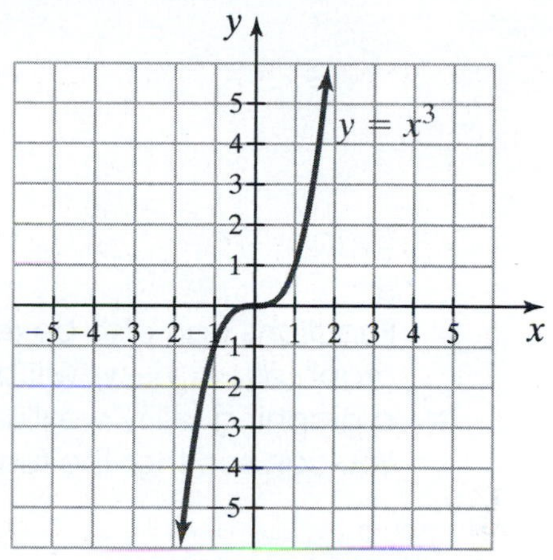

Click icon to animate

You Try It Work through the following You Try It problem.

Work Exercises 29–40 in this textbook or in the MyLab Math Study Plan.

OBJECTIVE 3 USING REFLECTIONS TO GRAPH FUNCTIONS

Given the graph of $y = f(x)$, what does the graph of $y = -f(x)$ look like?

Using a graphing utility with $y_1 = x^2$ and $y_2 = -x^2$, we can see that the graph of $y_2 = -x^2$ is the graph of $y_1 = x^2$ reflected about the ***x*-axis**.

Using Technology

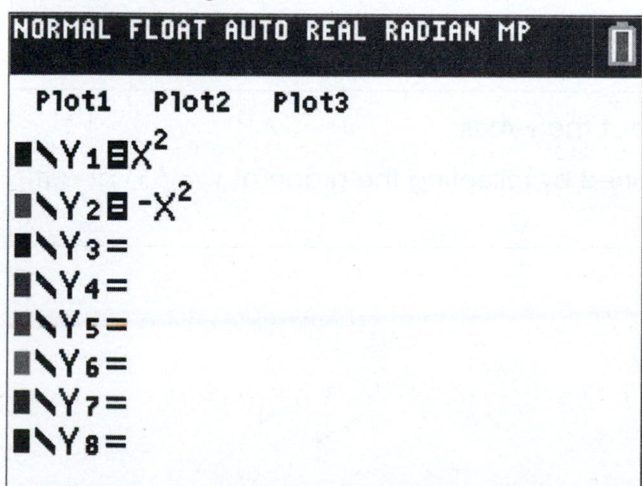

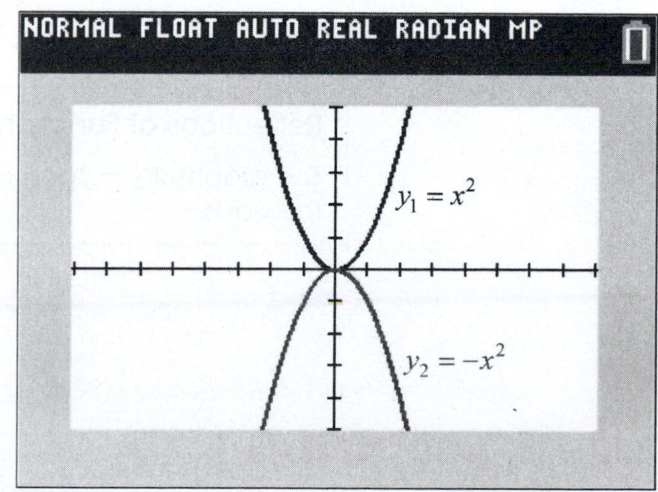

Reflections of Functions about the x-Axis

The graph of $y = -f(x)$ is obtained by reflecting the graph of $y = f(x)$ about the x-axis.

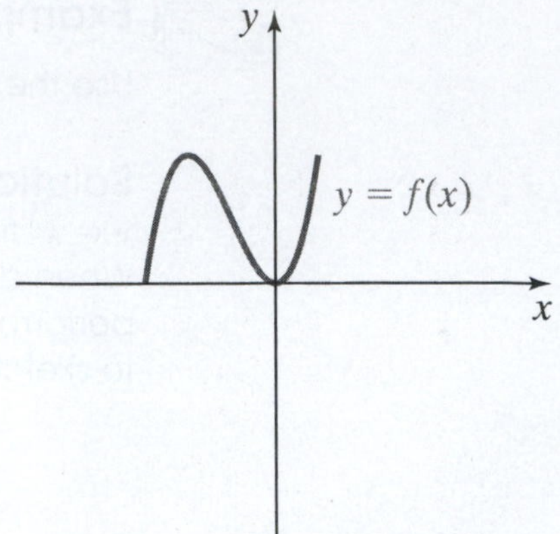

Sketch $y = -f(x)$.

Click icon to animate

Functions can also be reflected about the y-axis. Given the graph of $y = f(x)$, the graph of $y = f(-x)$ will be the graph of $y = f(x)$ reflected about the y-axis. Using a graphing utility, we illustrate a y-axis reflection by letting $y_1 = \sqrt{x}$ and $y_2 = \sqrt{-x}$. You can see that the functions are mirror images of each other about the y-axis.

Using Technology

NORMAL FLOAT AUTO REAL RADIAN MP

Plot1 Plot2 Plot3

$\backslash Y_1 = \sqrt{X}$

$\backslash Y_2 = \sqrt{-X}$

$\backslash Y_3 =$

$\backslash Y_4 =$

$\backslash Y_5 =$

$\backslash Y_6 =$

$\backslash Y_7 =$

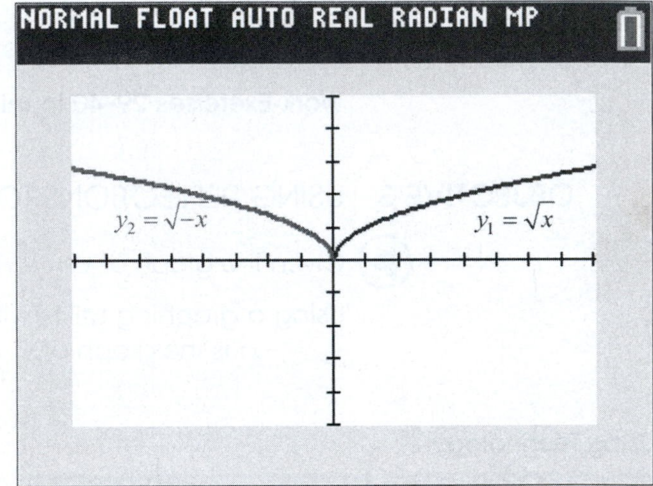

Reflections of Functions about the y-Axis

The graph of $y = f(-x)$ is obtained by reflecting the graph of $y = f(x)$ about the y-axis.

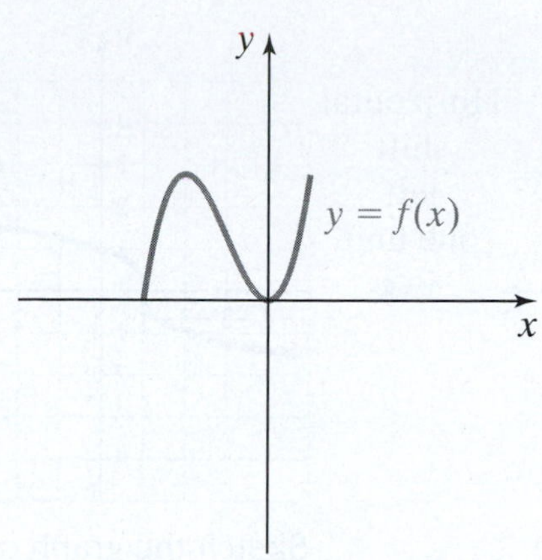

Sketch $y = f(-x)$.

Click icon to animate

Example 3 Sketch Functions Using Reflections and Shifts

Use the graph of the basic function $y = \sqrt[3]{x}$ to sketch each graph.

a. $g(x) = -\sqrt[3]{x} - 2$ b. $h(x) = \sqrt[3]{1 - x}$.

Solution

a. Starting with the graph of $y = \sqrt[3]{x}$, we can obtain the graph of $g(x) = -\sqrt[3]{x} - 2$ by performing two transformations:

1. Reflect about the x-axis.
2. Vertically shift down two units.

See Figure 13.

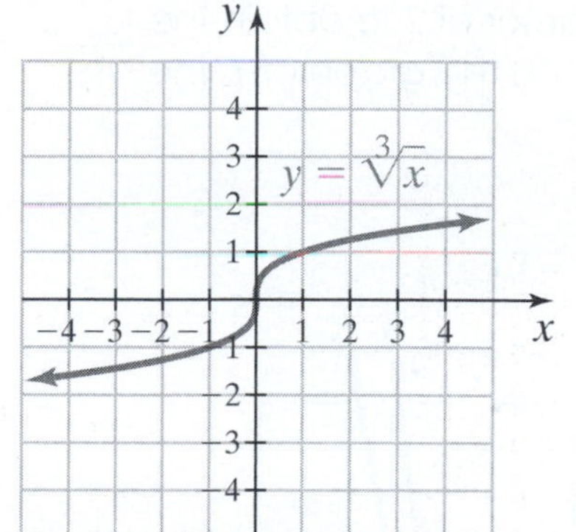

Start with the graph of the basic function $y = \sqrt[3]{x}$.

Reflect about the x-axis

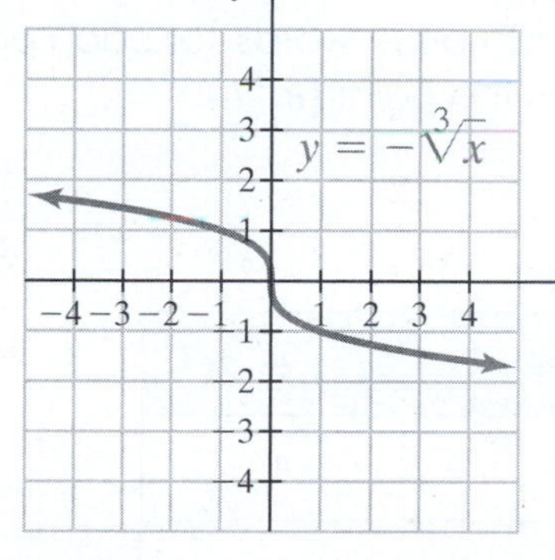

Sketch the graph of $y = -\sqrt[3]{x}$.

Vertical shift down two units

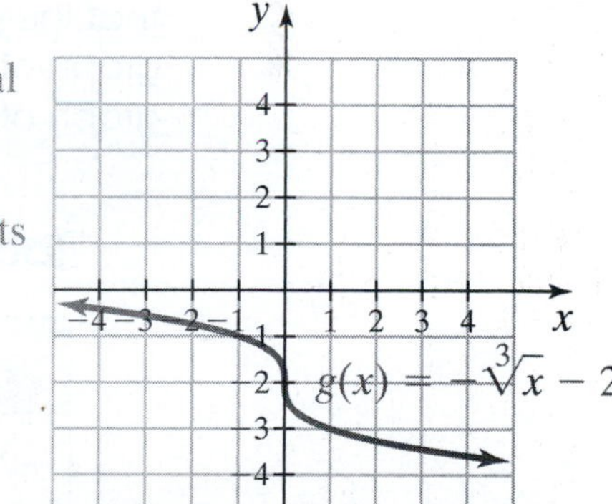

Sketch the graph of $g(x) = -\sqrt[3]{x} - 2$.

Figure 13

b. Starting with the graph of $y = \sqrt[3]{x}$, we can obtain the graph of $h(x) = \sqrt[3]{1 - x}$ by performing two transformations:

1. Horizontal shift left one unit.
2. Reflect about the y-axis.

See Figure 14.

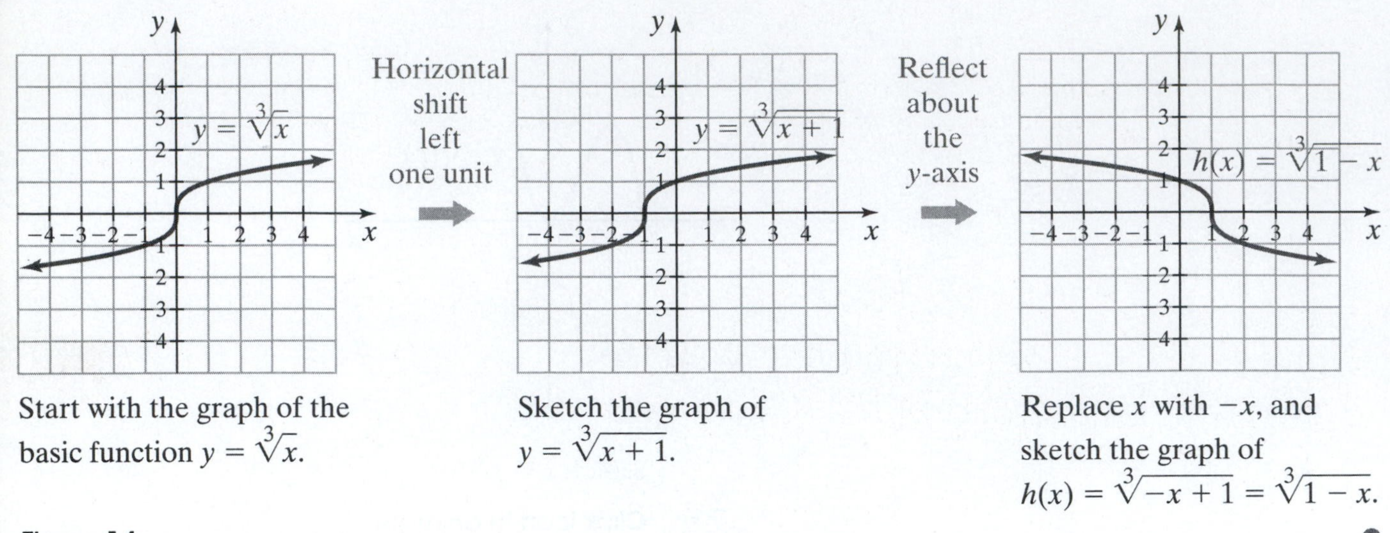

Start with the graph of the basic function $y = \sqrt[3]{x}$.

Sketch the graph of $y = \sqrt[3]{x+1}$.

Replace x with $-x$, and sketch the graph of $h(x) = \sqrt[3]{-x+1} = \sqrt[3]{1-x}$.

Figure 14

You Try It Work through the following You Try It problem.

Work Exercises 41–54 in this textbook or in the MyLab Math Study Plan.

OBJECTIVE 4 USING VERTICAL STRETCHES AND COMPRESSIONS TO GRAPH FUNCTIONS

Example 4 Vertical Stretch and Compression

Use the graph of $f(x) = x^2$ to sketch the graph of $g(x) = 2x^2$.

Solution Notice in Table 5 that for each value of x, the y-coordinate of g is two times as large as the corresponding y-coordinate of f. We can see in Figure 15 that the graph of $f(x) = x^2$ is vertically stretched by a factor of 2 to obtain the graph of $g(x) = 2x^2$. In other words, for each point (a, b) on the graph of f, the graph of g contains the point $(a, 2b)$.

Table 5

x	$f(x) = x^2$	$g(x) = 2x^2$
-2	4	8
-1	1	2
0	0	0
1	1	2
2	4	8

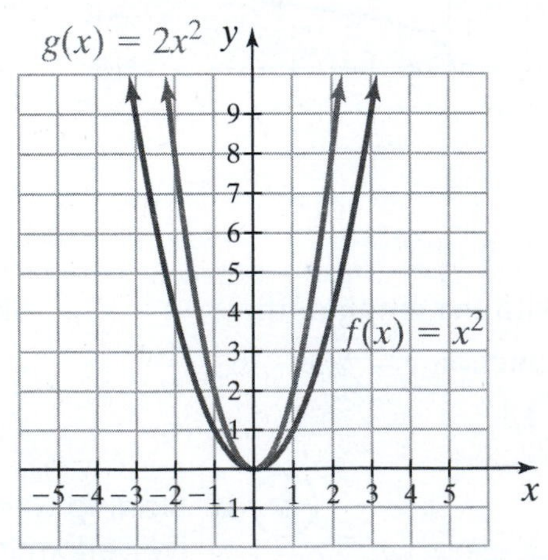

Figure 15

It follows from Example 4 that if $a > 1$, the graph of $y = af(x)$ is a **vertical stretch** of the graph of $y = f(x)$ and is obtained by multiplying each y-coordinate on the graph of f by a factor of a. If $0 < a < 1$, then the graph of $y = af(x)$ is a **vertical compression** of the graph of $y = f(x)$. Table 6 and Figure 16 show the relationship between the graphs of the functions $f(x) = x^2$ and $h(x) = \frac{1}{2}x^2$.

Table 6

x	$f(x) = x^2$	$h(x) = \frac{1}{2}x^2$
-2	4	2
-1	1	$\frac{1}{2}$
0	0	0
0	1	$\frac{1}{2}$
2	4	2

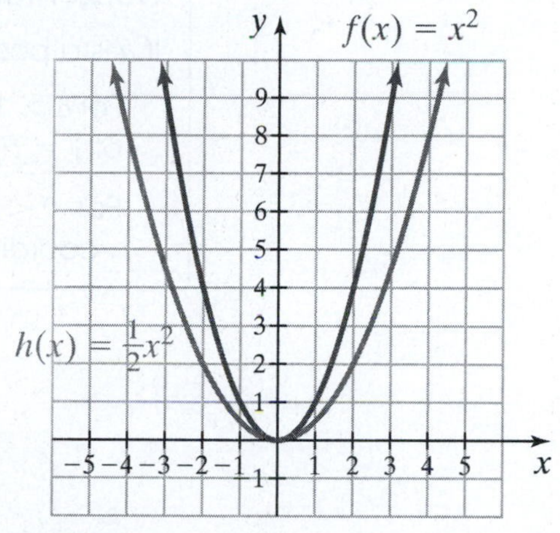

Figure 16

Vertical Stretches and Compressions of Functions

Suppose a is a positive real number:

The graph of $y = af(x)$ is obtained by multiplying each y-coordinate of $y = f(x)$ by a. If $a > 1$, the graph of $y = af(x)$ is a vertical stretch of the graph of $y = f(x)$. If $0 < a < 1$, the graph of $y = af(x)$ is a vertical compression of the graph of $y = f(x)$.

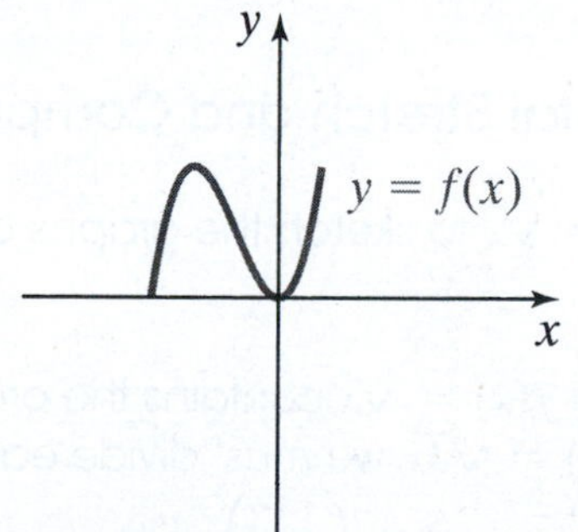

Sketch $y = af(x)$.
$(a > 1)$

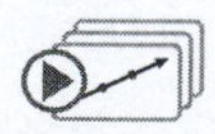 Click icon to animate

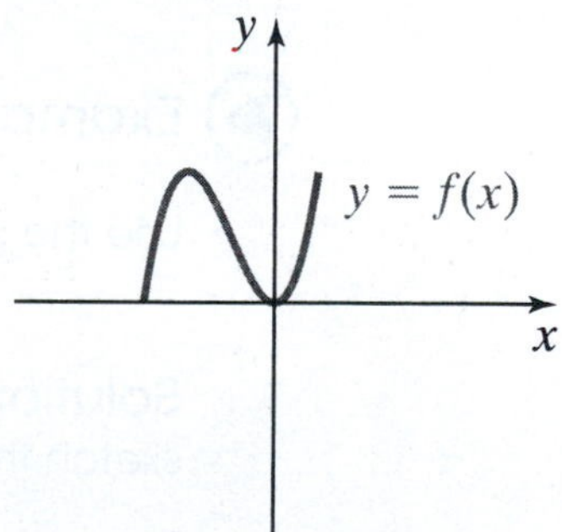

Sketch $y = af(x)$.
$(0 < a < 1)$

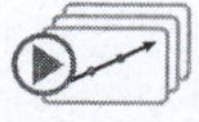 Click icon to animate

You Try It Work through the following You Try It problem.

Work Exercises 55–68 in this textbook or in the MyLab Math Study Plan.

OBJECTIVE 5 USING HORIZONTAL STRETCHES AND COMPRESSIONS TO GRAPH FUNCTIONS

The final transformation to discuss is a horizontal stretch or compression. A function, $y = f(x)$, will be horizontally stretched or compressed when x is multiplied by a positive number, $a \neq 1$, to obtain the new function, $y = f(ax)$.

Horizontal Stretches and Compressions of Functions

If a is a positive real number,

For $a > 1$, the graph of $y = f(ax)$ is obtained by dividing each x-coordinate of $y = f(x)$ by a. The resultant graph is a horizontal compression.

For $0 < a < 1$, the graph of $y = f(ax)$ is obtained by dividing each x-coordinate of $y = f(x)$ by a. The resultant graph is a horizontal stretch.

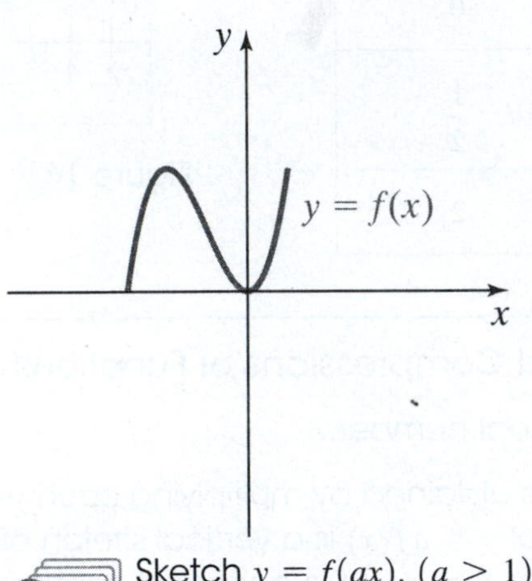

Sketch $y = f(ax)$. $(a > 1)$

Click icon to animate

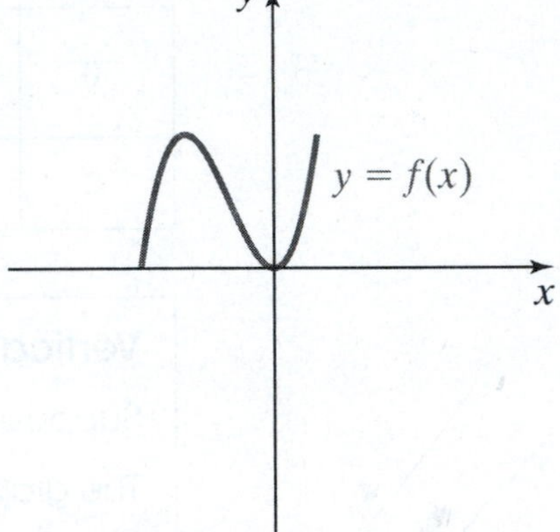

Sketch $y = f(ax)$. $(0 < a < 1)$

Click icon to animate

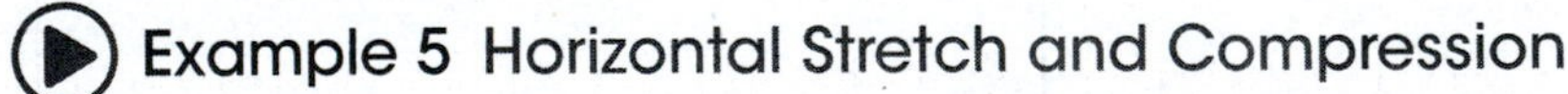

Example 5 Horizontal Stretch and Compression

Use the graph of $f(x) = \sqrt{x}$ to sketch the graphs of $g(x) = \sqrt{4x}$ and $h(x) = \sqrt{\frac{1}{4}x}$.

Solution The graph of $f(x) = \sqrt{x}$ contains the ordered pairs $(0, 0)$, $(1, 1)$, $(4, 2)$. To sketch the graph of $g(x) = \sqrt{4x}$, we must divide each previous x-coordinate by 4. Therefore, the ordered pairs $(0, 0)$, $\left(\frac{1}{4}, 1\right)$, $(1, 2)$ must lie on the graph of g. You can see that the graph of g is a horizontal compression of the graph of f. See Figure 17.

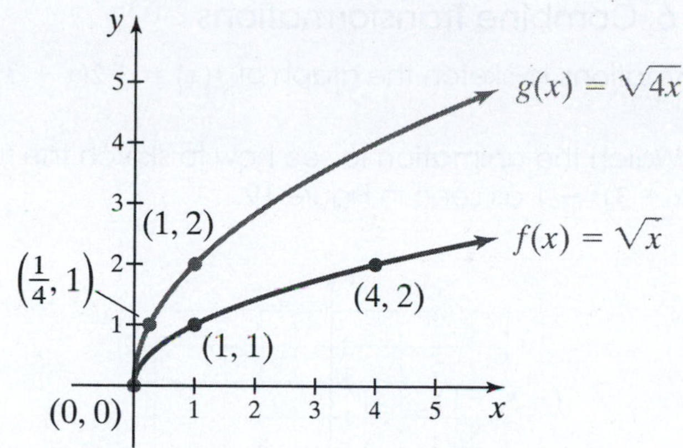

Figure 17 Graphs of $f(x) = \sqrt{x}$ and $g(x) = \sqrt{4x}$.

Similarly, to sketch the graph of $h(x) = \sqrt{\frac{1}{4}x}$, we divide the x-coordinates of the ordered pairs of f by $\frac{1}{4}$ to get the ordered pairs $(0, 0)$, $(4, 1)$, $(16, 2)$. You can see that the graph of h is a horizontal stretch of the graph of f. See Figure 18.

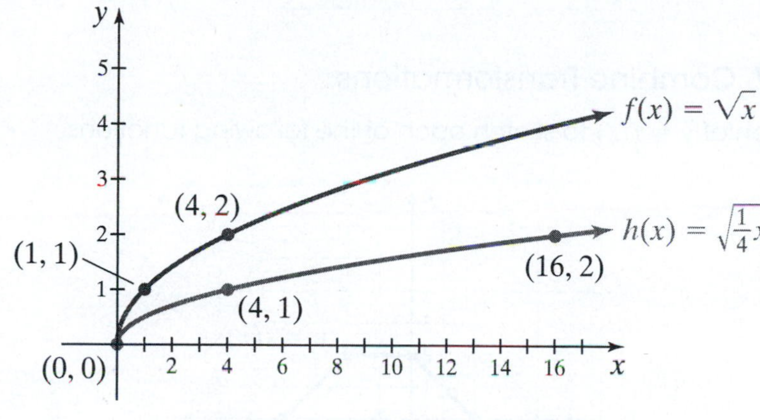

Figure 18 Graphs of $f(x) = \sqrt{x}$ and $g(x) = \sqrt{\frac{1}{4}x}$.

You Try It **Work through the following You Try It problem.**

Work Exercises 69–73 in this textbook or in the MyLab Math Study Plan.

OBJECTIVE 6 USING COMBINATIONS OF TRANSFORMATIONS TO GRAPH FUNCTIONS

You may encounter functions that combine many (if not all) of the transformations discussed in this section. When sketching a function that involves multiple transformations, it is important to follow a certain "order of operations." Following is the order in which each transformation is performed in this text:

1. Horizontal shifts
2. Horizontal stretches/compressions
3. Reflection about y-axis
4. Vertical stretches/compressions
5. Reflection about x-axis
6. Vertical shifts

Although different ordering is possible, the order above will always work.

Example 6 Combine Transformations

Use transformations to sketch the graph of $f(x) = -2(x + 3)^2 - 1$.

Solution Watch the **animation** to see how to sketch the function $f(x) = -2(x + 3)^2 - 1$ as seen in Figure 19.

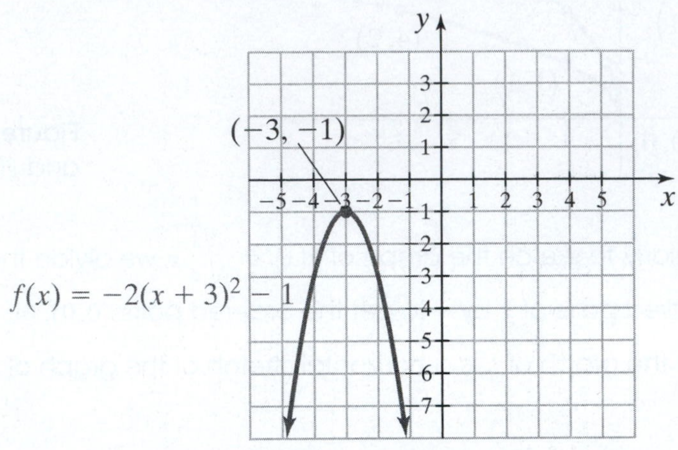

Figure 19 Graph of $f(x) = -2(x + 3)^2 - 1$. ●

Example 7 Combine Transformations

Use the graph of $y = f(x)$ to sketch each of the following functions.

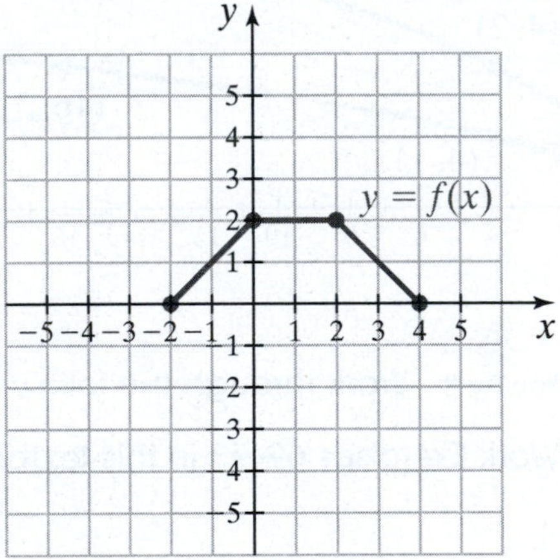

a. $y = -f(2x)$ **b.** $y = 2f(x - 3) - 1$ **c.** $y = -\frac{1}{2}f(2 - x) + 3$

Solution Watch the **interactive video** to see any one of the solutions worked out in detail.

a. The graph of $y = -f(2x)$ can be obtained from the graph of $y = f(x)$ using two transformations: (1) a horizontal compression and (2) a reflection about the x-axis. The resultant graph is sketched in Figure 20.

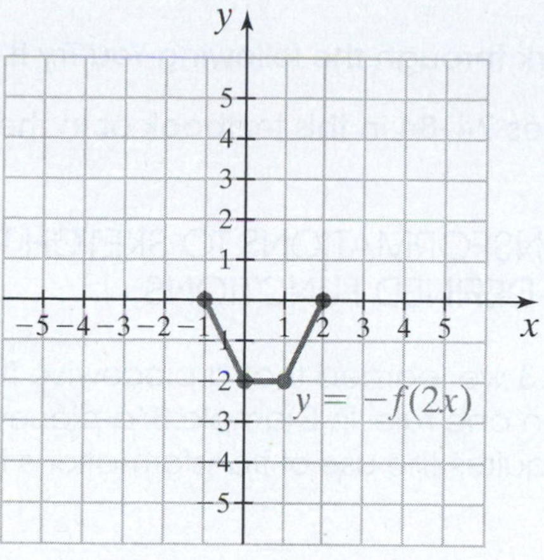

Figure 20 Graph of $y = -f(2x)$.

b. The graph of $y = 2f(x - 3) - 1$ can be obtained from the graph of $y = f(x)$ using three transformations: (1) a horizontal shift to the right three units, (2) a vertical stretch by a factor of 2, and (3) a vertical shift down one unit. The resultant graph is sketched in Figure 21.

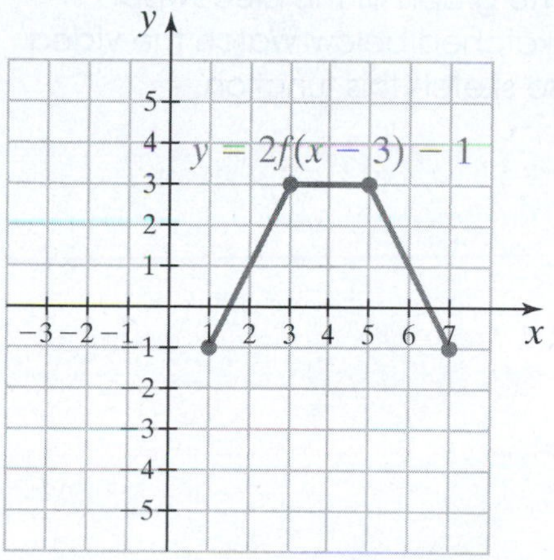

Figure 21 Graph of $y = 2f(x - 3) - 1$

c. The graph of $y = -\frac{1}{2}f(2 - x) + 3$ can be obtained from the graph of $y = f(x)$ using five transformations: (1) a horizontal shift to the left two units, (2) a reflection about the y-axis, (3) a vertical compression by a factor of $\frac{1}{2}$, (4) a reflection about the x-axis, and (5) a vertical shift up three units. The resultant graph is sketched in Figure 22.

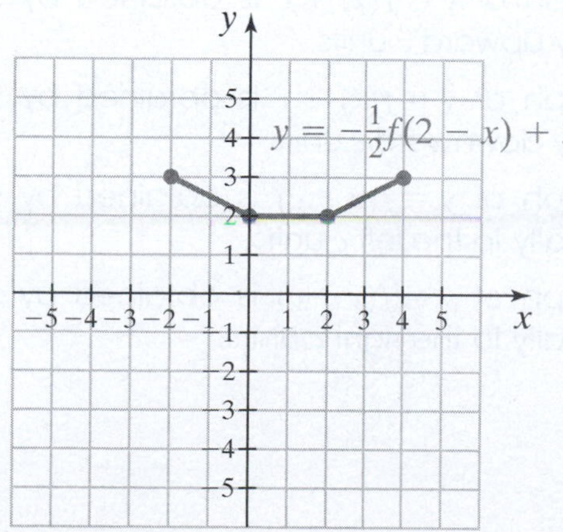

Figure 22 Graph of $y = -\frac{1}{2}f(2 - x) + 3$. ●

You Try It Work through the following You Try It problem.

Work Exercises 74–84 in this textbook or in the MyLab Math Study Plan.

OBJECTIVE 7 USING TRANSFORMATIONS TO SKETCH THE GRAPHS OF PIECEWISE-DEFINED FUNCTIONS

In Section 3.3 we learned that a piecewise function is a function that is defined by more than one rule. In Example 8, a piecewise function is defined by two rules. Each rule requires the use of transformations to sketch the graphs.

Example 8 Piecewise Functions

Sketch the graph of the function $f(x) = \begin{cases} -|x+2| & \text{if } x < -1 \\ \sqrt{x+1} & \text{if } x \geq -1 \end{cases}$

Solution The graph of this piecewise function is sketched below. Watch the **video** to see how to sketch this function.

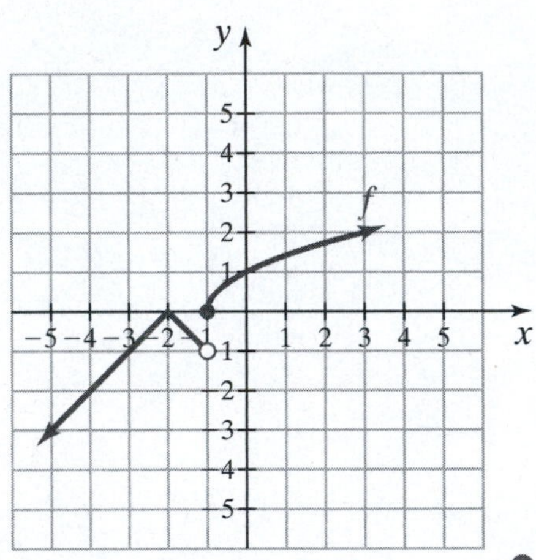

You Try It Work through the following You Try It problem.

Work Exercises 85–88 in this textbook or in the MyLab Math Study Plan.

Summary of Transformation Techniques

Given a function $y = f(x)$ and a constant $c > 0$:

1. The graph of $y = f(x) + c$ is obtained by shifting the graph of $y = f(x)$ vertically upward c units.
2. The graph of $y = f(x) - c$ is obtained by shifting the graph of $y = f(x)$ vertically downward c units.
3. The graph of $y = f(x + c)$ is obtained by shifting the graph of $y = f(x)$ horizontally to the left c units.
4. The graph of $y = f(x - c)$ is obtained by shifting the graph of $y = f(x)$ horizontally to the right c units.

5. The graph of $y = -f(x)$ is obtained by reflecting the graph of $y = f(x)$ about the x-axis.
6. The graph of $y = f(-x)$ is obtained by reflecting the graph of $y = f(x)$ about the y-axis.
7. Suppose a is a positive real number. The graph of $y = af(x)$ is obtained by multiplying each y-coordinate of $y = f(x)$ by a.

   If $a > 1$, the graph of $y = af(x)$ is a vertical stretch of the graph of $y = f(x)$.

   If $0 < a < 1$, the graph of $y = af(x)$ is a vertical compression of the graph of $y = f(x)$.
8. Suppose a is a positive real number. The graph of $y = f(ax)$ is obtained by dividing each x-coordinate of $y = f(x)$ by a.

   If $a > 1$, the graph of $y = f(ax)$ is a horizontal compression of the graph of $y = f(x)$.

   If $0 < a < 1$, the graph of $y = f(ax)$ is a horizontal stretch of the graph of $y = f(x)$.

3.4 Exercises

In Exercises 1–6, use the graph of a known basic function and vertical shifting to sketch each function.

1. $f(x) = x^2 - 1$
2. $y = \sqrt{x} + 2$
3. $g(x) = \frac{1}{x} + 3$
4. $h(x) = \sqrt[3]{x} - 2$
5. $y = |x| - 3$
6. $g(x) = x^3 + 1$
7. The graph below was created by vertically shifting the graph of the basic function $y = x^2$. Write a function that describes the graph below.

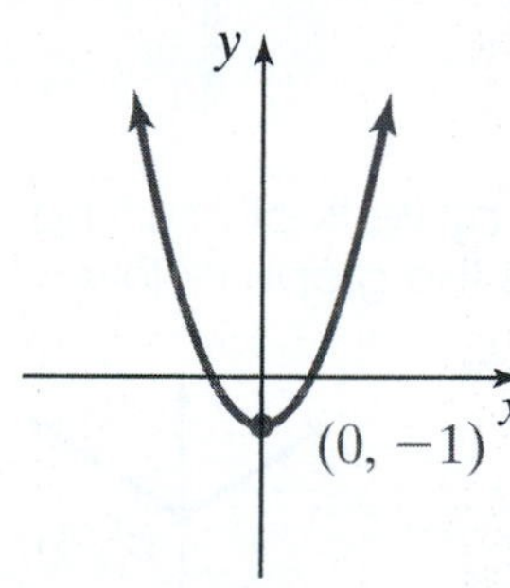

8. The graph below was created by vertically shifting the graph of the basic function $y = x^3$. Write a function that describes the graph below.

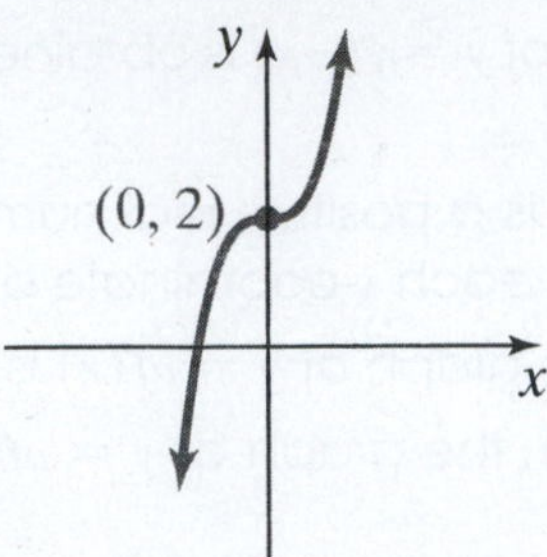

9. The graph below was created by vertically shifting the graph of the basic function $y = \sqrt{x}$. Write a function that describes the graph below.

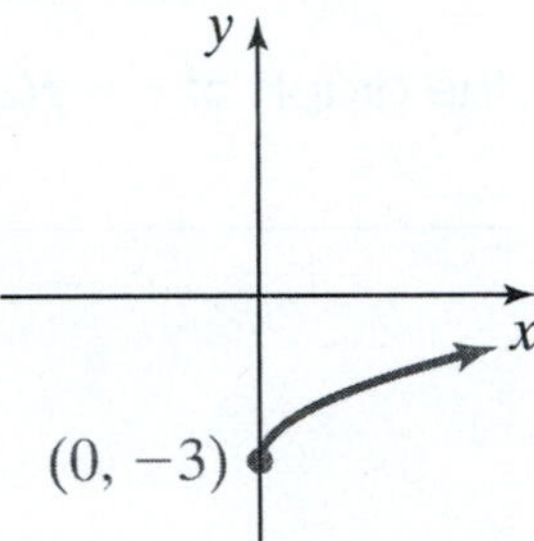

10. The graph below was created by vertically shifting the graph of the basic function $y = \frac{1}{x}$.

Write a function that describes the graph below.

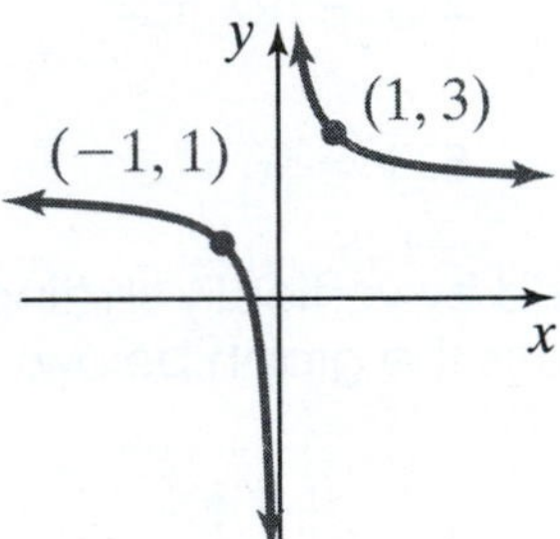

11. The graph below was created by vertically shifting the graph of the basic function $y = |x|$. Write a function that describes the graph below.

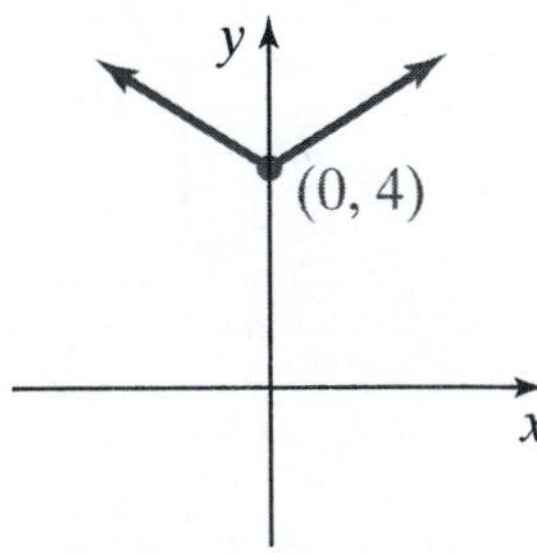

12. The graph below was created by vertically shifting the graph of the basic function $y = \sqrt[3]{x}$. Write a function that describes the graph below.

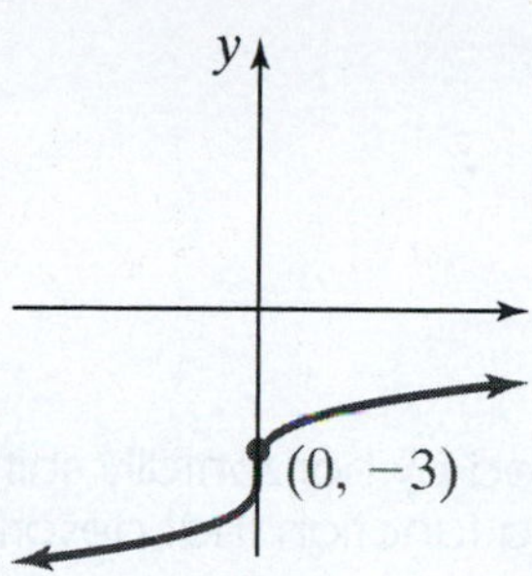

13. Use the graph of $y = f(x)$ to sketch the graph of $y = f(x) - 1$. Label at least three points on the new graph.

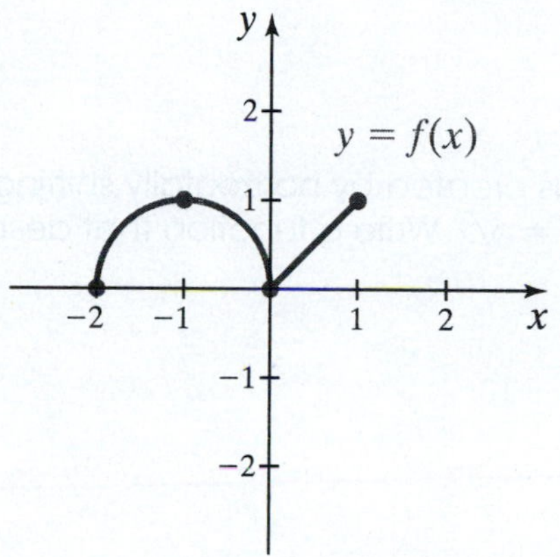

14. Use the graph of $y = f(x)$ to sketch the graph of $y = f(x) + 2$. Label at least three points on the new graph.

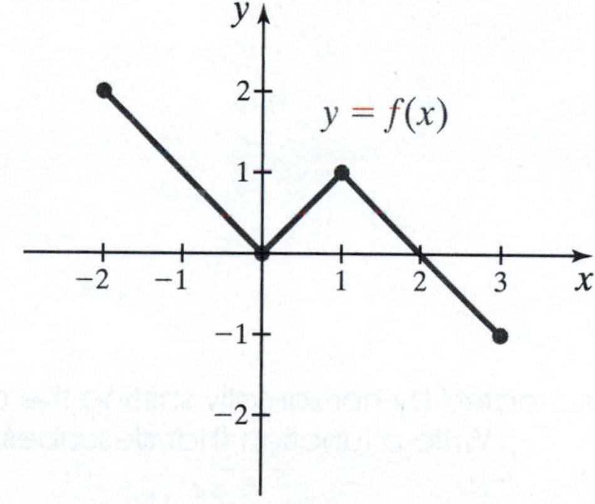

In Exercises 15–20, use the graph of a known basic function and horizontal shifts to sketch each function.

15. $f(x) = \sqrt[3]{x - 2}$

16. $g(x) = \dfrac{1}{x + 3}$

17. $y = \sqrt{x - 4}$

18. $h(x) = (x + 1)^3$

19. $k(x) = |x - 1|$

20. $y = (x - 3)^2$

21. The graph below was created by horizontally shifting the graph of the basic function $y = x^2$. Write a function that describes the graph below.

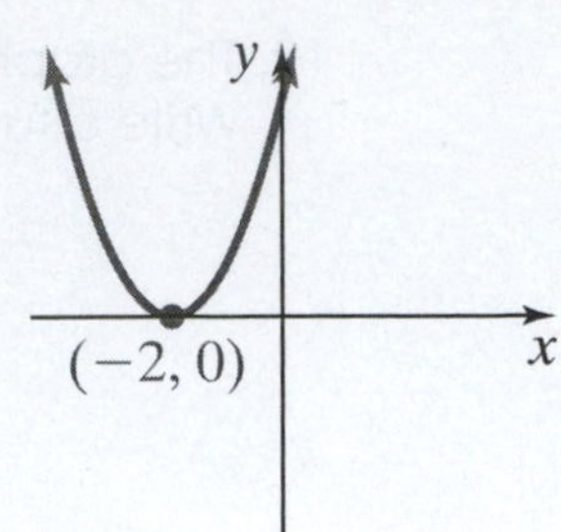

22. The graph below was created by horizontally shifting the graph of the basic function $y = x^3$. Write a function that describes the graph below.

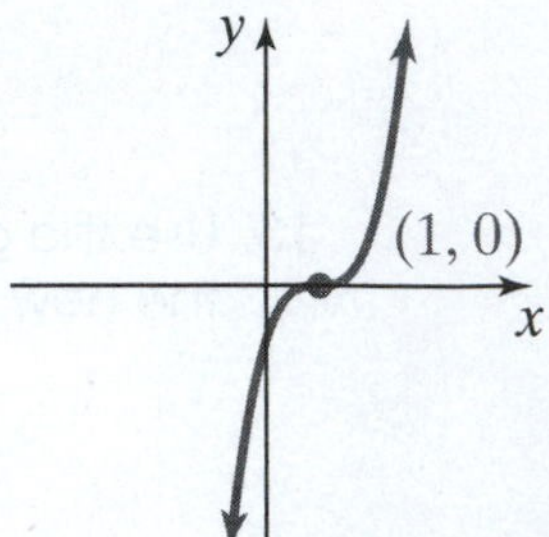

23. The graph below was created by horizontally shifting the graph of the basic function $y = \sqrt{x}$. Write a function that describes the graph below.

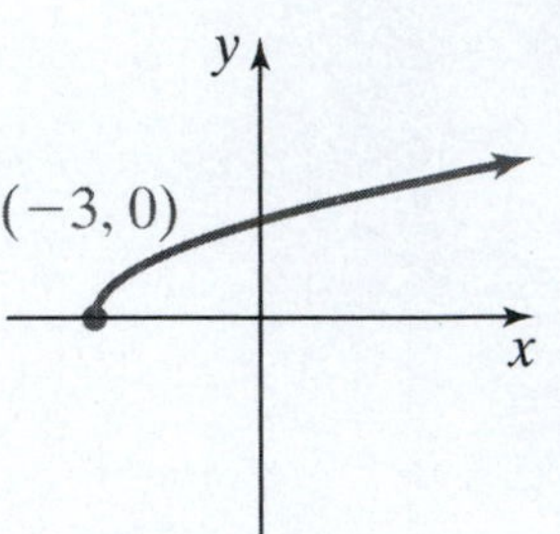

24. The graph below was created by horizontally shifting the graph of the basic function $y = \dfrac{1}{x}$. Write a function that describes the graph below.

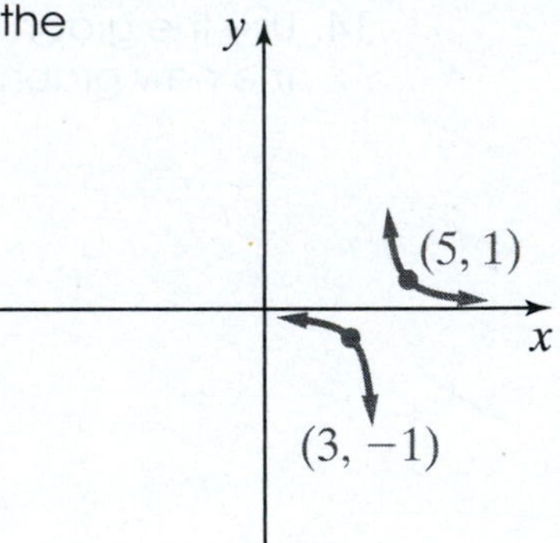

25. The graph below was created by horizontally shifting the graph of the basic function $y = |x|$. Write a function that describes the graph below.

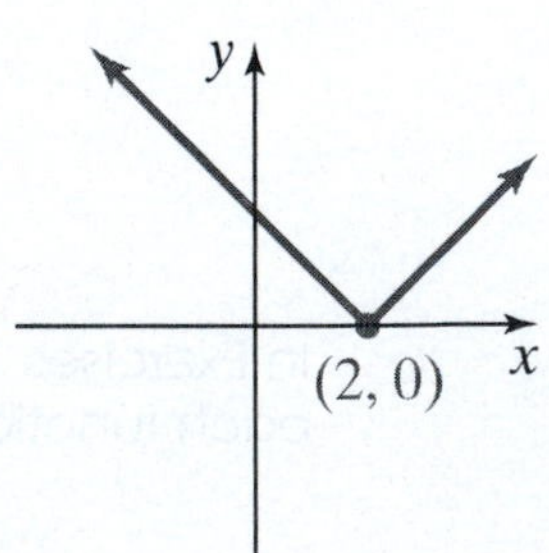

26. The graph below was created by horizontally shifting the graph of the basic function $y = \sqrt[3]{x}$. Write a function that describes the graph below.

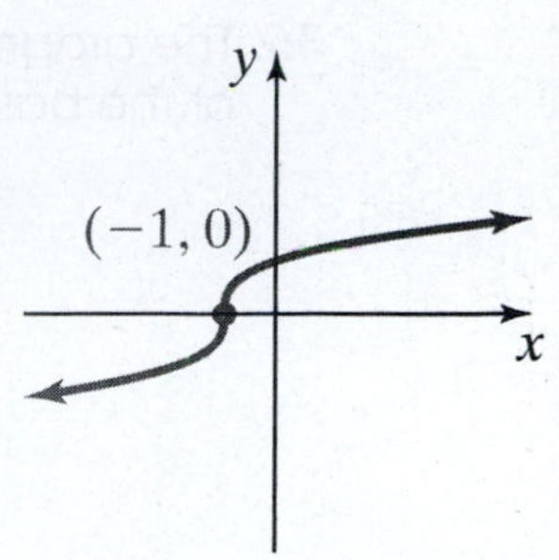

27. Use the graph of $y = f(x)$ to sketch the graph of $y = f(x - 2)$. Label at least three points on the new graph.

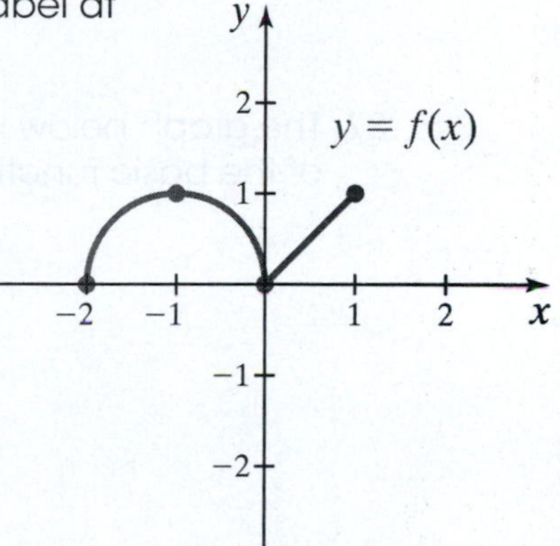

28. Use the graph of $y = f(x)$ to sketch the graph of $y = f(x + 2)$. Label at least three points on the new graph.

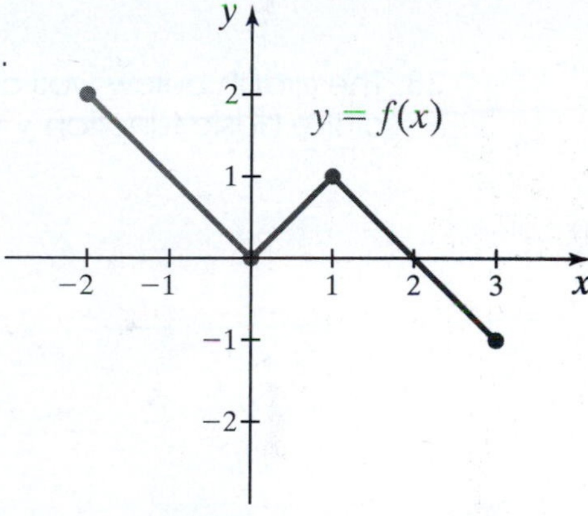

In Exercises 29–34, use the graph of a known basic function and a combination of horizontal and vertical shifts to sketch each function.

29. $y = (x + 1)^2 - 2$

30. $f(x) = (x - 3)^2 + 1$

31. $y = \sqrt{x + 3} + 2$

32. $h(x) = \dfrac{1}{x - 2} + 3$

33. $f(x) = |x + 2| + 2$

34. $y = \sqrt[3]{x + 1} - 1$

35. The graph below was created by using one horizontal shift and one vertical shift of the graph of the basic function $y = x^2$. Write a function that describes the graph below.

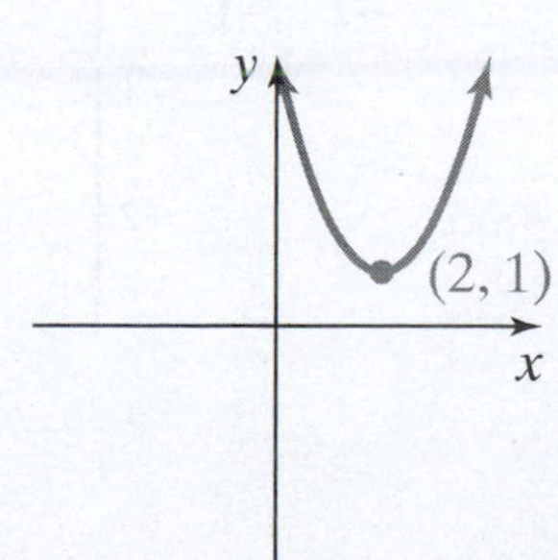

36. The graph below was created by using one horizontal shift and one vertical shift of the graph of the basic function $y = \sqrt{x}$. Write a function that describes the graph below.

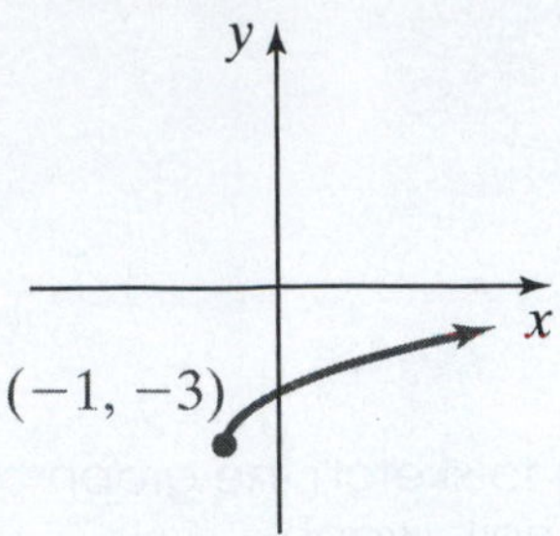

37. The graph below was created by using one horizontal shift and one vertical shift of the graph of the basic function $y = |x|$. Write a function that describes the graph below.

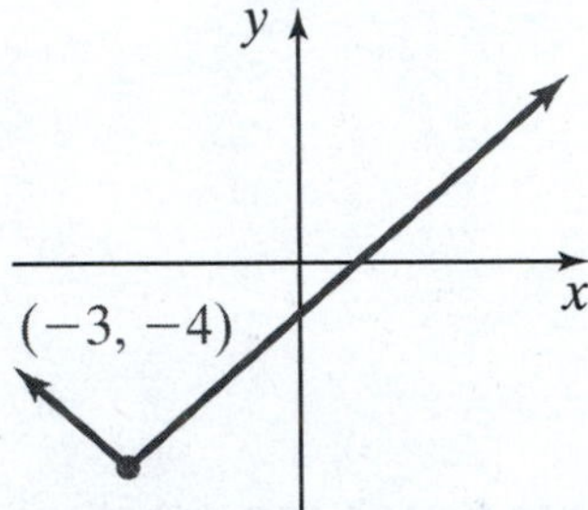

38. The graph below was created by using one horizontal shift and one vertical shift of the graph of the basic function $y = x^3$. Write a function that describes the graph below.

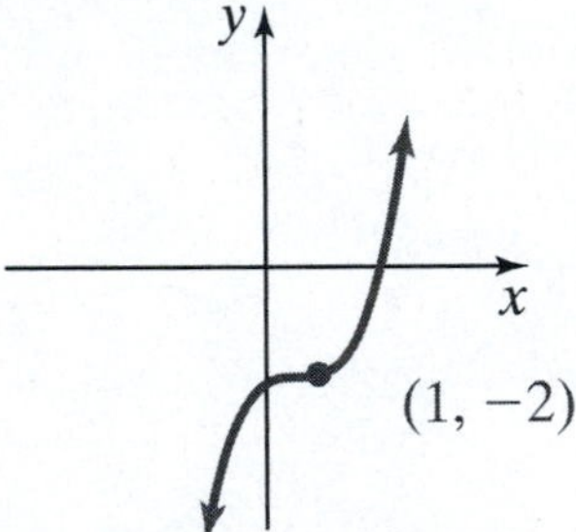

39. Use the graph of $y = f(x)$ to sketch the graph of $y = f(x - 2) - 1$. Label at least three points on the new graph.

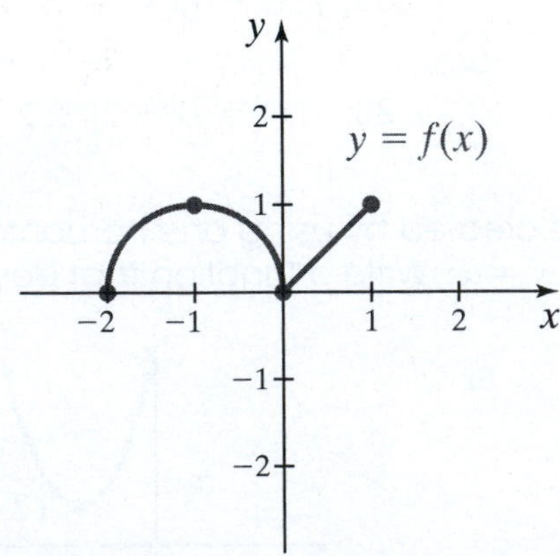

40. Use the graph of $y = f(x)$ to sketch the graph of $y = f(x + 1) + 2$. Label at least three points on the new graph.

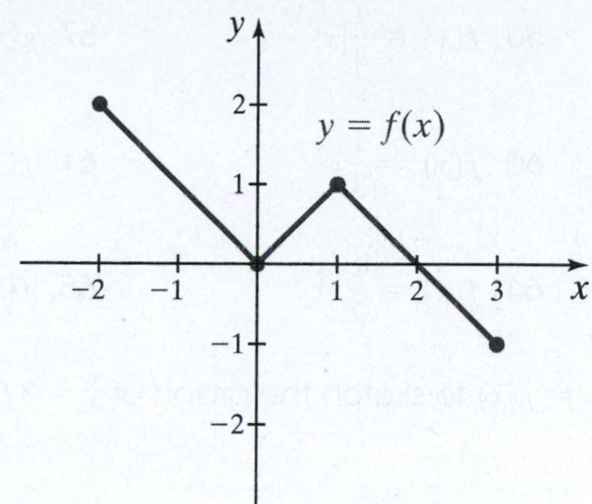

In Exercises 41–52, use the graph of a known basic function and a combination of horizontal shifts, reflections, and vertical shifts to sketch each function.

41. $g(x) = -x^2 - 2$

42. $f(x) = -\frac{1}{x} + 1$

43. $h(x) = \sqrt{2 - x}$

44. $f(x) = |-1 - x|$

45. $h(x) = \sqrt[3]{-x - 2}$

46. $g(x) = -x^3 + 1$

47. $h(x) = -\sqrt[3]{x} + 2$

48. $g(x) = (3 - x)^2$

49. $h(x) = -\sqrt{x} - 1$

50. $f(x) = -|x| + 1$

51. $g(x) = (1 - x)^3$

52. $f(x) = \frac{1}{2 - x}$

53. Use the graph of $y = f(x)$ to sketch the graph of $y = -f(x)$. Label at least three points on the new graph.

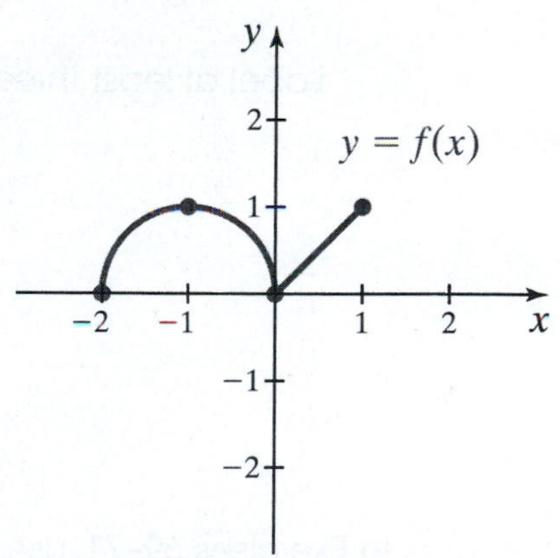

54. Use the graph of $y = f(x)$ to sketch the graph of $y = f(-x)$. Label at least three points on the new graph.

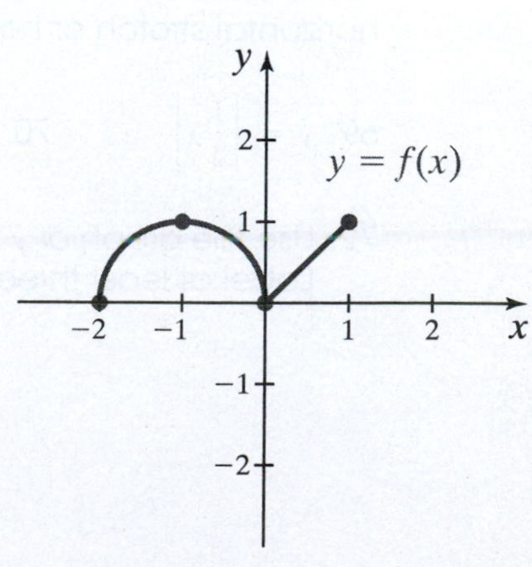

In Exercises 55–66, use the graph of a known basic function and a vertical stretch or vertical compression to sketch each function.

55. $f(x) = 3|x|$

56. $f(x) = \frac{1}{4}|x|$

57. $g(x) = 2\sqrt{x}$

58. $f(x) = \frac{1}{4}\sqrt{x}$

59. $f(x) = 3x^3$

60. $f(x) = \frac{1}{3}x^3$

61. $f(x) = 6\sqrt[3]{x}$

62. $f(x) = \frac{1}{2}\sqrt[3]{x}$

63. $g(x) = 3x^2$

64. $f(x) = \frac{1}{2}x^2$

65. $f(x) = \frac{4}{x}$

66. $y = \frac{1}{2x}$

67. Use the graph of $y = f(x)$ to sketch the graph of $y = 3f(x)$. Label at least three points on the new graph.

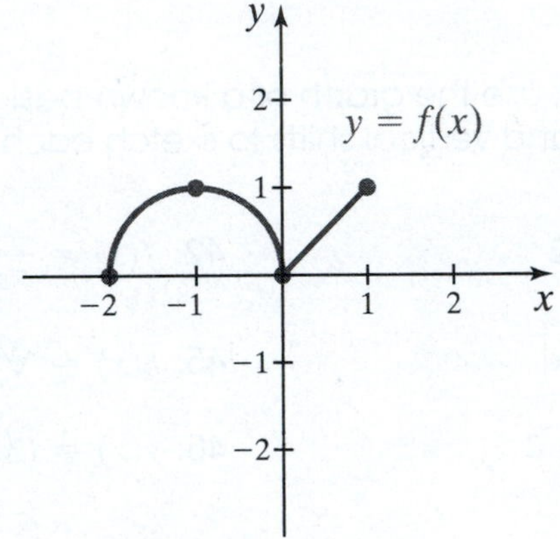

68. Use the graph of $y = f(x)$ to sketch the graph of $y = \frac{1}{2}f(x)$.

Label at least three points on the new graph.

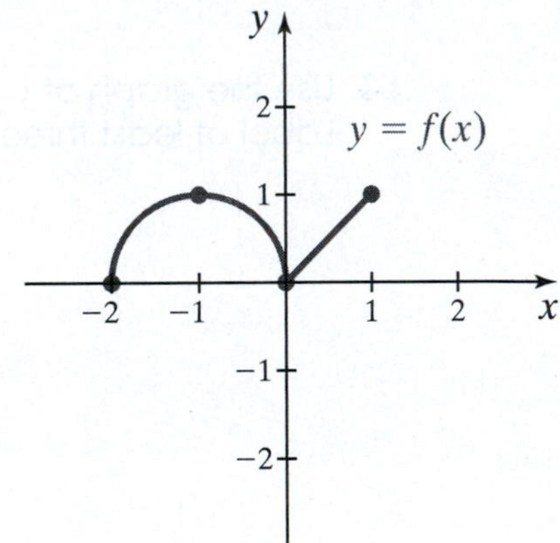

In Exercises 69–71, use the graph of a known basic function and a horizontal stretch or horizontal compression to sketch each function.

69. $y = \left|\frac{1}{4}x\right|$

70. $f(x) = \sqrt{2x}$

71. $g(x) = \sqrt[3]{3x}$

72. Use the graph of $y = f(x)$ to sketch the graph of $y = f(2x)$. Label at least three points on the new graph.

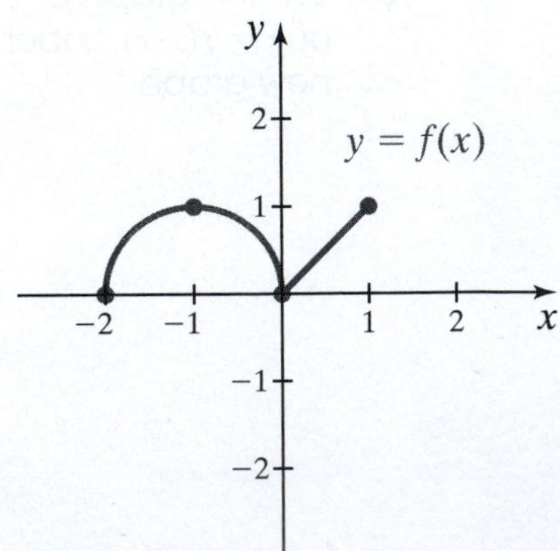

73. Use the graph of $y = f(x)$ to sketch the graph of $y = f\left(\frac{1}{2}x\right)$. Label at least three points on the new graph.

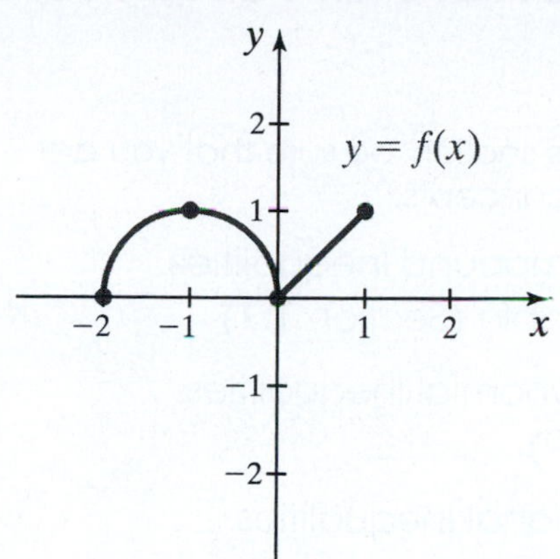

In Exercises 74–79, use the graph of a known basic function and a combination of transformations to sketch each function.

74. $f(x) = -(x - 2)^2 + 3$

75. $g(x) = \frac{1}{2}|x + 1| - 1$

76. $y = \frac{1}{x - 3} + 2$

77. $f(x) = 2\sqrt[3]{x} - 1$

78. $g(x) = -\frac{1}{2}(2 - x)^3 + 1$

79. $h(x) = 2\sqrt{4 - x} + 5$

In Exercises 80–84, use the graph of $y = f(x)$ to sketch each function. Label at least three points on each graph.

80. $y = -f(-x) - 1$

81. $y = \frac{1}{2}f(2 - x)$

82. $y = -2f(x + 1) + 2$

83. $y = 3 - 3f(x + 3)$

84. $y = -f(1 - x) - 2$

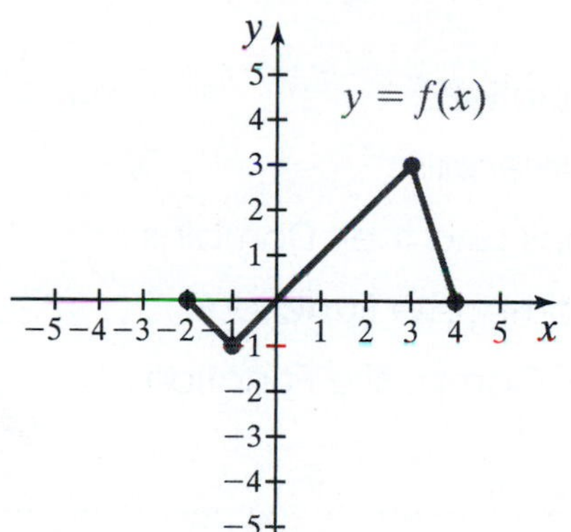

In Exercises 85–88, sketch the graph of each piecewise-defined function.

85. $f(x) = \begin{cases} |x - 2| & \text{if } x \leq 2 \\ \sqrt{x - 2} & \text{if } x > 2 \end{cases}$

86. $f(x) = \begin{cases} -x^2 + 1 & \text{if } x < 0 \\ -\sqrt{x} - 3 & \text{if } x \geq 0 \end{cases}$

87. $f(x) = \begin{cases} (x + 1)^3 & \text{if } x \leq 1 \\ |x - 2| - 2 & \text{if } x > 1 \end{cases}$

88. $f(x) = \begin{cases} \sqrt{-x} - 2 & \text{if } x \leq 0 \\ (x - 1)^2 & \text{if } x > 0 \end{cases}$

3.5 The Algebra of Functions; Composite Functions

THINGS TO KNOW

Before working through this section, be sure that you are familiar with the following concepts:

VIDEO | ANIMATION | INTERACTIVE

 You Try It

1. Solving Compound Inequalities in One Variable (Section 1.7)  

 You Try It

2. Solving Polynomial Inequalities (Section 1.9)  

 You Try It

3. Solving Rational Inequalities (Section 1.9)  

 You Try It

4. Determining the Domain of a Function Given the Equation (Section 3.1)

 You Try It

5. Determining the Domain and Range of a Function from Its Graph (Section 3.2)  

 You Try It

6. Sketching the Graphs of the Basic Functions (Section 3.3)

INTRODUCTION

Read this introduction before beginning Objective 1

OBJECTIVES

1 Evaluating a Combined Function

2 Finding the Intersection of Intervals

3 Finding Combined Functions and their Domains

4 Forming and Evaluating Composite Functions

5 Determining the Domain of Composite Functions

SECTION 3.5 EXERCISES

Introduction to Section 3.5

In this section, we learn how to create a new function by combining two or more existing functions. First, we combine functions by adding, subtracting, multiplying, or dividing two existing functions. The addition, subtraction, multiplication, and division of functions is also known as the *algebra of functions*.

Algebra of Functions

Let f and g be functions, then for all x such that both $f(x)$ and $g(x)$ are defined, the sum, difference, product, and quotient of f and g exist and are defined as follows:

1. The sum of f and g: $(f+g)(x) = f(x) + g(x)$
2. The difference of f and g: $(f-g)(x) = f(x) - g(x)$
3. The product of f and g: $(fg)(x) = f(x)\,g(x)$
4. The quotient of f and g: $\left(\frac{f}{g}\right)(x) = \frac{f(x)}{g(x)}$ for all $g(x) \neq 0$

OBJECTIVE 1 EVALUATING A COMBINED FUNCTION

Example 1 Evaluate Combined Functions

Let $f(x) = \frac{12}{2x+4}$ and $g(x) = \sqrt{x}$. Find each of the following.

a. $(f+g)(1)$ b. $(f-g)(1)$ c. $(fg)(4)$ d. $\left(\frac{f}{g}\right)(4)$

Solution

a. Because $f(1) = \frac{12}{2(1)+4} = \frac{12}{6} = 2$ and $g(1) = \sqrt{1} = 1$, it follows from the sum of f and g that

$$\begin{aligned}(f+g)(1) &= f(1) + g(1)\\ &= 2 + 1\\ &= 3\end{aligned}$$

b. $(f-g)(1) = f(1) - g(1) = 2 - 1 = 1$

c. $f(4) = \frac{12}{2(4)+4} = \frac{12}{12} = 1$ and $g(4) = \sqrt{4} = 2$, so $(fg)(4) = f(4)g(4) = (1)(2) = 2$

d. $\left(\frac{f}{g}\right)(4) = \frac{f(4)}{g(4)} = \frac{1}{2}$

You Try It Work through the following You Try It problem.

Work Exercises 1–10 in this textbook or in the MyLab Math Study Plan.

Example 2 Evaluate Combined Functions Using a Graph

Use the graph to evaluate each expression or state that it is undefined.

a. $(f + g)(1)$ b. $(f - g)(0)$

c. $(fg)(4)$ d. $\left(\frac{f}{g}\right)(2)$

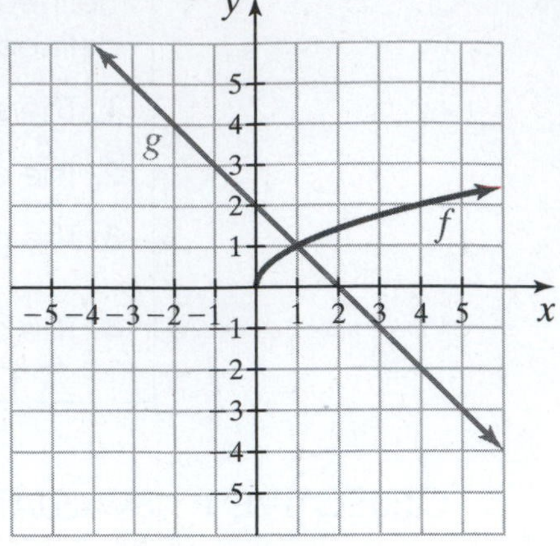

Solution Work through the **interactive video** to verify that

a. $(f + g)(1) = 2$ b. $(f - g)(0) = -2$

c. $(fg)(4) = -4$ d. $\left(\frac{f}{g}\right)(2)$ is undefined

You Try It **Work through the following You Try It problem.**

Work Exercises 11–14 in this textbook or in the MyLab Math Study Plan.

OBJECTIVE 2 FINDING THE INTERSECTION OF INTERVALS

Before learning how to find the domain of $f + g, f - g, fg$, or $\frac{f}{g}$, you must be able to find the intersection of two or more intervals. The notation $A \cap B$ is used to represent the **intersection** of sets A and B. To find the intersection of two sets, we must find the set of all numbers *common* to *both* sets.

Example 3 Find the Intersection of Sets

Find the intersection of the following sets and graph the set on a number line.

a. $[0, \infty) \cap (-\infty, 5]$ b. $((-\infty, -2) \cup (-2, \infty)) \cap [-4, \infty)$

Solution Watch the **video** to verify the solutions.

a. $[0, \infty) \cap (-\infty, 5] = [0, 5]$

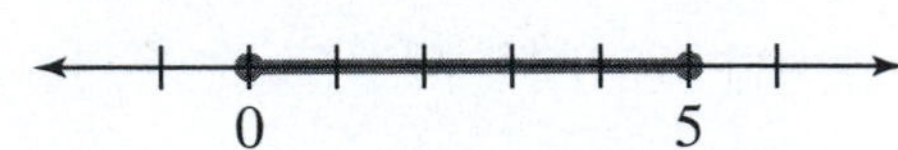

b. $((-\infty, -2) \cup (-2, \infty)) \cap [-4, \infty) = [-4, -2) \cup (-2, \infty)$

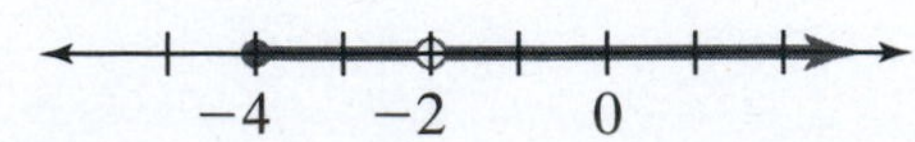

You Try It Work through the following You Try It problem.

Work Exercises 15–20 in this textbook or in the MyLab Math Study Plan.

OBJECTIVE 3 FINDING COMBINED FUNCTIONS AND THEIR DOMAINS

Consider the graph of f and g in Figure 23. Notice that the domain of f is the interval $[1, \infty)$ and the domain of g is the interval $[-1, 4)$. What is the domain of $f + g$? Because $(f + g)(x) = f(x) + g(x)$, for a number a to be in the domain of $f + g$ means that both $f(a)$ and $g(a)$ must be defined. For example, because $x = 0$ is not in the domain of f, $f(0)$ is undefined and $x = 0$ is *not* in the domain of $f + g$. Similarly, because $x = 4$ is not in the domain of g, $x = 4$ cannot be in the domain of $f + g$. Thus, for a number a to be in the domain of $f + g$, a must be in the domain of f, **and** a must be in the domain of g. In other words, the domain of $f + g$ must be the intersection of the two domains. The domain of $f + g$ in Figure 23 is $[1, 4)$.

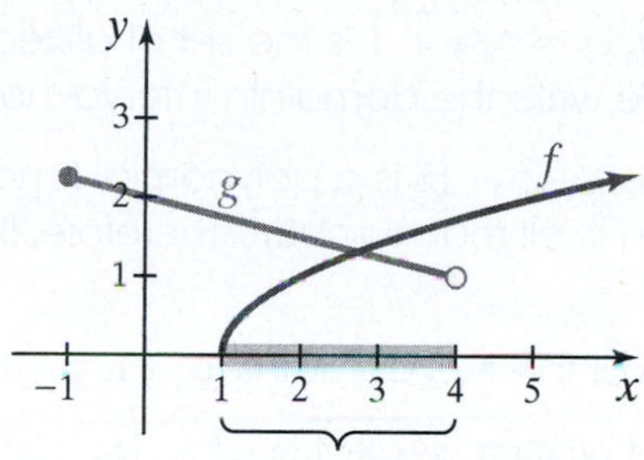

The intersection of the domain of f and the domain of g is $[1, 4)$.

Figure 23

> **Domain of $f + g$, $f - g$, fg, and $\dfrac{f}{g}$**
>
> Suppose f is a function with domain A and g is a function with domain B, then
>
> 1. The domain of the sum, $f + g$, is the set of all x in $A \cap B$.
> 2. The domain of the difference, $f - g$, is the set of all x in $A \cap B$.
> 3. The domain of the product, fg, is the set of all x in $A \cap B$.
> 4. The domain of the quotient, $\frac{f}{g}$, is the set of all x in $A \cap B$ such that $g(x) \neq 0$.

Before we look at examples of finding the domains of combined functions, it is important to remember how to find the domains of Polynomial Functions, Rational Functions, and Root Functions. The concept of finding the domains of these three classifications of functions was first introduced in **Section 3.1**. You may want to review this concept or review this **Quick Guide** for finding the domain of these three classifications of functions.

Example 4 Find $f + g$ and $\frac{f}{g}$ and Their Domains

Let $f(x) = \sqrt{x + 1}$ and $g(x) = x^2 - 16$.

a. Find $f + g$ and determine the domain of $f + g$.

b. Find $\frac{f}{g}$ and determine the domain of $\frac{f}{g}$.

Solution First, determine the domains of $f(x) = \sqrt{x + 1}$ and $g(x) = x^2 - 16$.

The function $f(x) = \sqrt{x + 1}$ is a root function of the form $f(x) = \sqrt[n]{g(x)}$, where the index, *n*, is even. Therefore, the domain is the solution to the inequality $g(x) \geq 0$ or $x + 1 \geq 0$.

$x + 1 \geq 0$ Write the inequality $g(x) \geq 0$ to determine the domain of an even root function.

$x \geq -1$ Subtract 1 from both sides of the inequality.

The domain of $f(x) = \sqrt{x + 1}$ is the set of all real numbers greater than or equal to negative 1. We write the domainin interval notation as $[-1, \infty)$.

The function $g(x) = x^2 - 16$ is a polynomial function. The domain of every polynomial function is all real numbers. Therefore, the domain of $g(x) = x^2 - 16$ is $(-\infty, \infty)$.

The intersection of the two domains is $[-1, \infty) \cap (-\infty, \infty) = [-1, \infty)$.

a. $f + g = f(x) + g(x) = \sqrt{x + 1} + x^2 - 16$.

The domain is the set of all x in the intersection of the domain of f and the domain of g. Thus, the domain of $f + g$ is $[-1, \infty)$.

b. $\frac{f}{g} = \frac{f(x)}{g(x)} = \frac{\sqrt{x + 1}}{x^2 - 16}$ or, if we factor the denominator, we get

$\frac{f(x)}{g(x)} = \frac{\sqrt{x + 1}}{(x - 4)(x + 4)}$.

The domain is the set of all x in the intersection of the domain of f and the domain of g such that $g(x) \neq 0$. Therefore, we start with the interval $[-1, \infty)$ and exclude all values of x on this interval for which $g(x) = 0$.

Note that $g(x) = 0$ when $x = 4$ and $x = -4$. The value of $x = -4$ is not in the interval $[-1, \infty)$, so in order to find the domain of $\frac{f}{g}$ we only need to exclude the value of $x = 4$ from the interval $[-1, \infty)$.

Therefore, the domain of $\frac{f}{g}$ is $[-1, 4) \cup (4, \infty)$.

Watch this video to see the entire solution to this example worked out in detail.

Example 5 Domain of $f + g$, $f - g$, and $\frac{f}{g}$ and Their Domains

Let $f(x) = \frac{x+2}{x-3}$ and $g(x) = \sqrt{4-x}$.

Find a. $f + g$, b. $f - g$, c. fg, d. $\frac{f}{g}$, and the domain of each.

Watch this **video** to verify the following:
The domain of f is $(-\infty, 3) \cup (3, \infty)$.
The domain of g is $(-\infty, 4]$.
The intersection of these two domains is $(-\infty, 3) \cup (3, 4]$.

a. $(f + g)(x) = \frac{x+2}{x-3} + \sqrt{4-x}$

The domain of $f + g$ is the set of all x in the intersection of the domain of f and the domain of g. Thus, the domain of $f + g$ is $(-\infty, 3) \cup (3, 4]$.

b. $(f - g)(x) = \frac{x+2}{x-3} - \sqrt{4-x}$

The domain of $f - g$ is the set of all x in the intersection of the domain of f and the domain of g. Thus, the domain of $f - g$ is $(-\infty, 3) \cup (3, 4]$.

c. $(fg)(x) = \frac{(x+2)\sqrt{4-x}}{x-3}$

The domain of *fg* is the set of all x in the intersection of the domain of f and the domain of g. Thus, the domain of *fg* is $(-\infty, 3) \cup (3, 4]$.

d. $\left(\frac{f}{g}\right)(x) = \frac{\frac{x+2}{x-3}}{\sqrt{4-x}} = \frac{x+2}{(x-3)\sqrt{4-x}}$

The domain of $\frac{f}{g}$ is the set of all x in the intersection of the domain of f and the domain of g such that $g(x) \neq 0$. Because $g(x) = 0$ when $x = 4$, we must exclude $x = 4$ from the domain of $\frac{f}{g}$. Thus, the domain of $\frac{f}{g}$ is $(-\infty, 3) \cup (3, 4)$.

You Try It **Work through the following You Try It problem.**

Work Exercises 21–32 in this textbook or in the MyLab Math Study Plan.

OBJECTIVE 4 FORMING AND EVALUATING COMPOSITE FUNCTIONS

Consider the function $f(x) = x^2$ and $g(x) = 2x + 1$. How could we find $f(g(x))$? To find $f(g(x))$, we substitute $g(x)$ for x in the function f to get

$$f(g(x)) = f(2x + 1) = (2x + 1)^2.$$

substitute $g(x)$ into f

The diagram in Figure 24 shows that given a number x, we first apply it to the function g to obtain $g(x)$. We then substitute $g(x)$ into f to get the result.

(eText Screens 3.5-1–3.5-26)

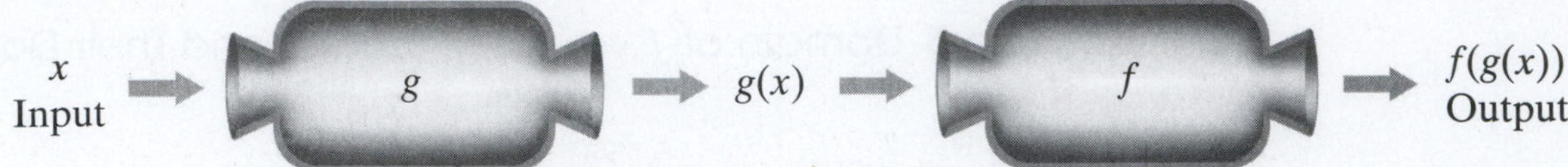

Figure 24 Composition of f and g.

The function $f(g(x))$ is called a composite function because one function is "composed" of another function.

> **Definition** Composite Function
>
> Given functions f and g, the **composite function**, $f \circ g$, (also called the **composition of f and g**) is defined by
>
> $$(f \circ g)(x) = f(g(x)),$$
>
> provided $g(x)$ is in the domain of f.

CAUTION **The composition of f and g does not equal the product of f and g: $(f \circ g)(x) \neq f(x)g(x)$. Also, the composition of f and g does not necessarily equal the composition of g and f, although this equality does exist for certain pairs of functions.**

Example 6 Form and Evaluate Composite Functions

Let $f(x) = 4x + 1$, $g(x) = \dfrac{x}{x - 2}$ and $h(x) = \sqrt{x + 3}$.

a. Find the function $f \circ g$.

b. Find the function $g \circ h$.

c. Find the function $h \circ f \circ g$.

d. Evaluate $(f \circ g)(4)$, or state that it is undefined.

e. Evaluate $(g \circ h)(1)$, or state that it is undefined.

f. Evaluate $(h \circ f \circ g)(6)$, or state that it is undefined.

Solution Work through the interactive video to verify the following:

a. $(f \circ g)(x) = 4\left(\dfrac{x}{x - 2}\right) + 1 = \dfrac{5x - 2}{x - 2}$

b. $(g \circ h)(x) = \dfrac{\sqrt{x + 3}}{\sqrt{x + 3} - 2}$

c. $(h \circ f \circ g)(x) = \sqrt{\dfrac{5x - 2}{x - 2} + 3} = \sqrt{\dfrac{8x - 8}{x - 2}}$

d. $(f \circ g)(4) = 9$

e. $(g \circ h)(1)$ is undefined

f. $(h \circ f \circ g)(6) = \sqrt{10}$

You Try It Work through the following You Try It problem.

Work Exercises 33–56 in this textbook or in the MyLab Math Study Plan.

Example 7 Evaluate Composite Functions Using a Graph

Use the graph to evaluate each expression:

a. $(f \circ g)(4)$

b. $(g \circ f)(-3)$

c. $(f \circ f)(-1)$

d. $(g \circ g)(4)$

e. $(f \circ g \circ f)(1)$

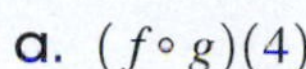

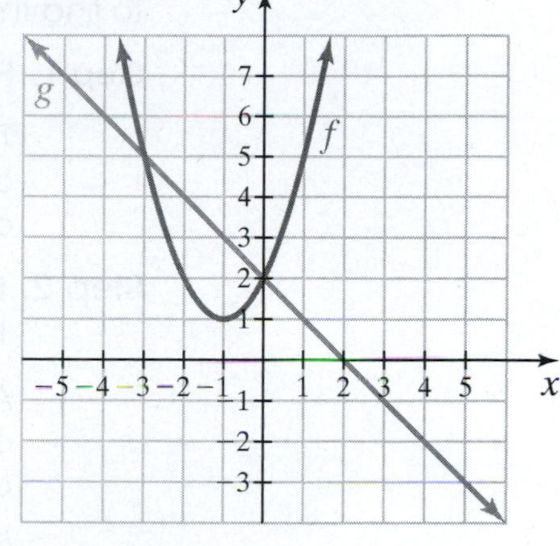

Solution Work through the interactive video to verify that

a. $(f \circ g)(4) = 2$, b. $(g \circ f)(-3) = -3$, c. $(f \circ f)(-1) = 5$,

d. $(g \circ g)(4) = 4$, and e. $(f \circ g \circ f)(1) = 5$

You Try It Work through the following You Try It problem.

Work Exercises 57 and 58 in this textbook or in the MyLab Math Study Plan.

OBJECTIVE 5 DETERMINING THE DOMAIN OF COMPOSITE FUNCTIONS

Suppose f and g are functions. For a number x to be in the domain of $f \circ g$, x must be in the domain of g *and* $g(x)$ must be in domain of f. To determine the domain of $f \circ g$, follow the two step process outlined below.

> **Determining the Domain of $f \circ g$**
>
> **Step 1.** Find the domain of g.
>
> **Step 2.** Exclude from the domain of g all values of x for which $g(x)$ is not in the domain of f.

Example 8 Find the Domain of a Composite Function

Let $f(x) = \dfrac{-10}{x-4}$, $g(x) = \sqrt{5-x}$, and $h(x) = \dfrac{x-3}{x+7}$.

a. Find the domain of $f \circ g$.

b. Find the domain of $g \circ f$.

c. Find the domain of $f \circ h$.

d. Find the domain of $h \circ f$.

Solution

a. First, form the composite function $(f \circ g)(x) = \dfrac{-10}{\sqrt{5-x}-4}$.

To find the domain of $f \circ g$, we follow these two steps:

Step 1. Find the domain of g.

The domain of $g(x) = \sqrt{5-x}$ is $(-\infty, 5]$. The domain of $f \circ g$ cannot contain any values of x that are not in this interval; in other words, the domain of $f \circ g$ is a **subset** of $(-\infty, 5]$.

Step 2. Exclude from the domain of g all values of x for which $g(x)$ is not in the domain of f.

All real numbers except 4 are in the domain of f. This implies that $g(x)$ cannot equal 4 because $g(x)$ equal to 4 would make the denominator of f, $x - 4$, equal to 0. Thus, we must exclude all values of x such that $g(x) = 4$.

$$g(x) = 4$$

$$\sqrt{5-x} = 4 \qquad \text{Substitute } \sqrt{5-x} \text{ for } g(x).$$

$$5 - x = 16 \qquad \text{Square both sides.}$$

$$x = -11 \qquad \text{Solve for } x.$$

We must *exclude* $x = -11$ from the domain of $f \circ g$. Therefore, the domain of $f \circ g$ is all values of x less than 5 such that $x \neq -11$, or the interval $(-\infty, -11) \cup (-11, 5]$.

You should carefully work through this **interactive video** to see the entire solution to part (a) and to verify the following:

b. The domain of $g \circ f$ is the set $(-\infty, 2] \cup (4, \infty)$.

c. The domain of $f \circ h$ is the set $\left(-\infty, -\dfrac{31}{3}\right) \cup \left(-\dfrac{31}{3}, -7\right) \cup (-7, \infty)$.

d. The domain of $h \circ f$ is the set $(-\infty, 4) \cup \left(4, \dfrac{38}{7}\right) \cup \left(\dfrac{38}{7}, \infty\right)$.

You Try It **Work through the following You Try It problem.**

Work Exercises 59–66 in this textbook or in the MyLab Math Study Plan.

3.5 Exercises

Skill Check Exercises

For exercises SCE-1 through SCE-7, simplify each expression and write as a single fraction. Factor the numerator and denominator whenever possible.

SCE-1. $\dfrac{\dfrac{2x+1}{x-9}}{\dfrac{5x}{5x-2}}$ SCE-2. $\dfrac{\dfrac{x}{x^2-4}}{\dfrac{1}{x-5}}$ SCE-3. $\dfrac{\dfrac{x-7}{x^2-x-6}}{\dfrac{x+8}{x^2-49}}$

SCE-4. $\dfrac{8}{x+3}+7$ SCE-5. $7\left(\dfrac{4}{x+6}\right)-6$ SCE-6. $\dfrac{3}{\dfrac{3}{x+8}+8}$

SCE-7. $2\left(\dfrac{3}{\sqrt{x+5}+2}\right)+3$

In Exercises 1–10, evaluate the following given that $f(x)=\sqrt{x+1}$, $g(x)=x^2-2$, and $h(x)=\dfrac{1}{2x+3}$.

1. $(f+g)(3)$ 2. $(f-g)(8)$ 3. $(g+h)(-2)$ 4. $\left(\dfrac{f}{g}\right)(0)$ 5. $(fh)(15)$

6. $(h-g)(4)$ 7. $\left(\dfrac{g}{f}\right)(-1)$ 8. $\left(\dfrac{f}{h}\right)(2)$ 9. $(gf)(8)$ 10. $(hh)\left(\dfrac{3}{2}\right)$

In Exercises 11–14, use the graph to evaluate the given expression, or state that it is undefined.

11.

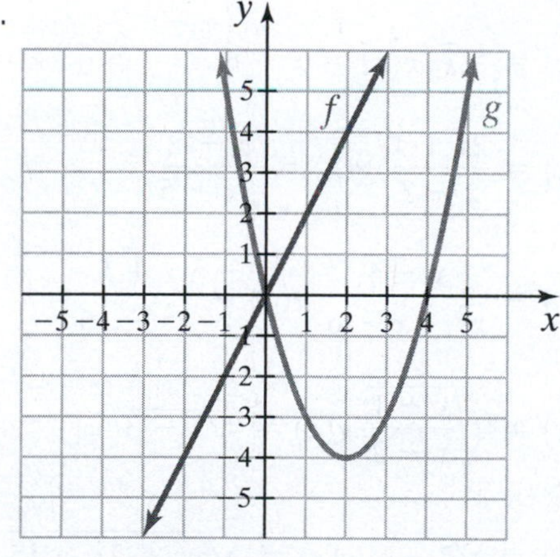

a. $(f+g)(0)$

b. $(f-g)(2)$

c. $(fg)(-1)$

d. $\left(\dfrac{f}{g}\right)(2)$

12.

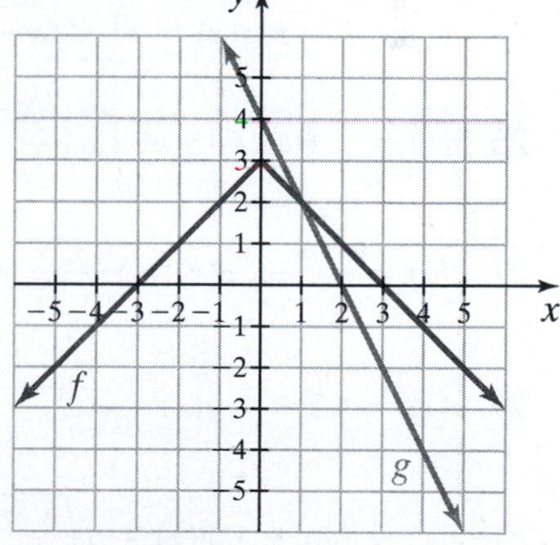

a. $(f+g)(4)$

b. $(g-f)(1)$

c. $(fg)(0)$

d. $\left(\dfrac{f}{g}\right)(2)$

13.

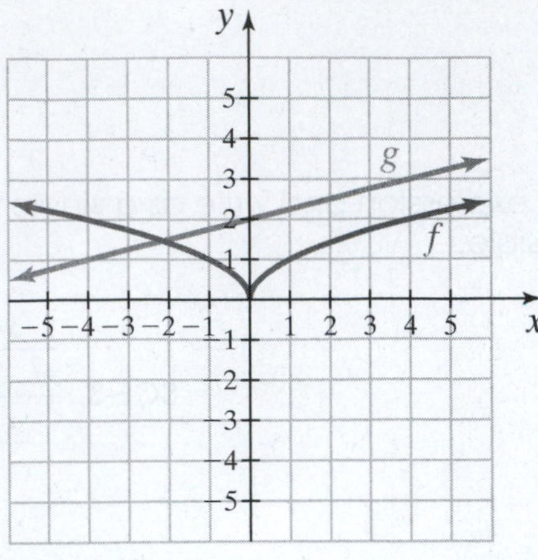

14.

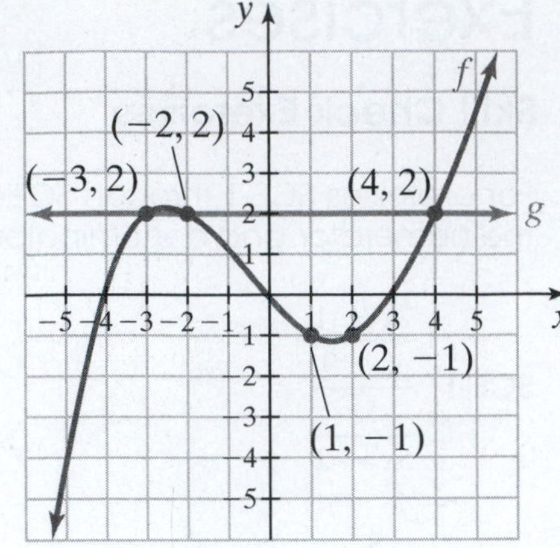

13.
a. $(f + g)(4)$
b. $(f - g)(-4)$
c. $(gg)(4)$
d. $\left(\dfrac{f}{g}\right)(4)$

14.
a. $(g + f)(4)$
b. $(g - f)(1)$
c. $(fg)(-3)$
d. $\left(\dfrac{f}{g}\right)(2)$

In Exercises 15–20, find the intersection of the given intervals and graph the set on a number line.

15. $(-\infty, \infty) \cap (-\infty, 3)$

16. $(-8, 5] \cap (-12, 3)$

17. $[0, \infty) \cap (-\infty, 4]$

18. $((-\infty, 1) \cup (1, \infty)) \cap (-\infty, \infty)$

19. $((-\infty, 0) \cup (0, \infty)) \cap [-3, \infty)$

20. $(-\infty, 10) \cap (-5, \infty) \cap (-\infty, 1]$

In Exercises 21–32, find **a.** $f + g$, **b.** $f - g$, **c.** fg, **d.** $\dfrac{f}{g}$, and the domain of each.

21. $f(x) = x^2 + 2, g(x) = x - 1$

22. $f(x) = x - 5, g(x) = x^2 - 4$

23. $f(x) = x + 5, g(x) = x^2 - 6x - 16$

24. $f(x) = \sqrt{x}, g(x) = x + 6$

25. $f(x) = \dfrac{1}{x}, g(x) = \sqrt{x - 1}$

26. $f(x) = \dfrac{2x + 1}{x + 3}, g(x) = \dfrac{x + 2}{3x + 5}$

27. $f(x) = \dfrac{x}{x^2 - 4}, g(x) = \dfrac{1}{x - 3}$

28. $f(x) = \dfrac{x - 4}{x^2 - x - 6}, g(x) = \dfrac{x + 5}{x^2 - 16}$

29. $f(x) = \sqrt{x + 2}, g(x) = \sqrt{2 - x}$

30. $f(x) = \sqrt{\dfrac{x + 1}{x + 2}}, g(x) = \sqrt{x - 3}$

31. $f(x) = \sqrt[3]{x + 1}, g(x) = \sqrt[3]{\dfrac{x - 1}{x + 2}}$

32. $f(x) = \sqrt{7 - x}, g(x) = \sqrt{x^2 - 2x - 15}$

In Exercises 33–44, let $f(x) = 3x + 1, g(x) = \dfrac{2}{x + 1}$, and $h(x) = \sqrt{x + 3}$.

33. Find the function $f \circ g$ and simplify.

34. Find the function $g \circ f$ and simplify.

35. Find the function $f \circ h$ and simplify.

36. Find the function $g \circ h$ and simplify.

37. Find the function $h \circ f$ and simplify.

38. Find the function $h \circ g$ and simplify.

39. Find the function $f \circ f$ and simplify.

40. Find the function $g \circ g$ and simplify.

41. Find the function $h \circ h$ and simplify.

42. Find the function $f \circ g \circ h$ and simplify.

43. Find the function $g \circ f \circ h$ and simplify.

44. Find the function $h \circ f \circ g$ and simplify.

In Exercises 45–56, evaluate the following composite functions given that $f(x) = 3x + 1$, $g(x) = \frac{2}{x + 1}$, and $h(x) = \sqrt{x + 3}$.

45. $(f \circ g)(0)$

46. $(f \circ h)(6)$

47. $(g \circ f)(1)$

48. $(g \circ h)(-2)$

49. $(h \circ f)(0)$

50. $(h \circ g)(3)$

51. $(f \circ f)(-1)$

52. $(g \circ g)(4)$

53. $(h \circ h)(1)$

54. $(f \circ g \circ h)(-2)$

55. $(h \circ f \circ g)(2)$

56. $(g \circ f \circ h)(6)$

In Exercises 57 and 58, use the graph to evaluate each expression.

57. a. $(f \circ g)(1)$

b. $(g \circ f)(-1)$

c. $(g \circ g)(0)$

d. $(f \circ f)(1)$

e. $(f \circ g \circ f)(0)$

f. $(g \circ f \circ g)(0)$

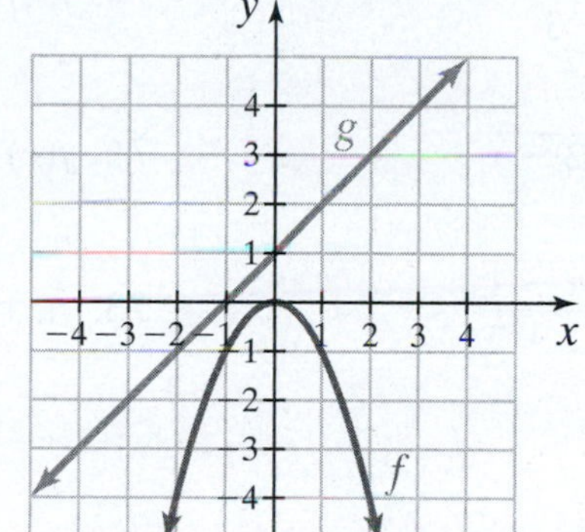

58. a. $(f \circ g)(1)$

b. $(g \circ f)(-1)$

c. $(g \circ g)(0)$

d. $(f \circ f)(1)$

e. $(f \circ g \circ f)(0)$

f. $(g \circ f \circ g)(0)$

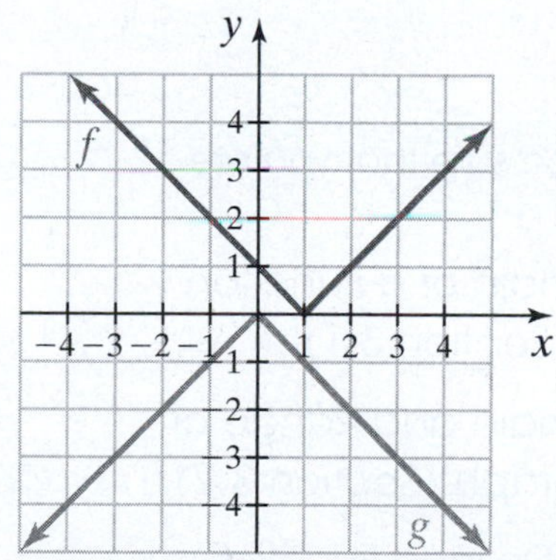

In Exercises 59–66, find the domain of $(f \circ g)(x)$ and $(g \circ f)(x)$.

59. $f(x) = x^2, g(x) = 2x - 1$

60. $f(x) = 3x - 5, g(x) = 2x^2 + 1$

61. $f(x) = x^2, g(x) = \sqrt{x}$

62. $f(x) = \frac{1}{x}, g(x) = x^2 - 4$

63. $f(x) = \frac{3}{x+1}, g(x) = \frac{x}{x-2}$

64. $f(x) = \frac{2x}{x-3}, g(x) = \frac{x+1}{x-1}$

65. $f(x) = \sqrt{x}, g(x) = \frac{x+3}{x-2}$

66. $f(x) = \frac{1}{x-2}, g(x) = \sqrt{4-x}$

Brief Exercises

In Exercises 67–78, find **a.** $f + g$, **b.** $f - g$, **c.** fg, **d.** $\frac{f}{g}$.

67. $f(x) = x^2 + 2, g(x) = x - 1$

68. $f(x) = x - 5, g(x) = x^2 - 4$

69. $f(x) = x + 5, g(x) = x^2 - 6x - 16$

70. $f(x) = \sqrt{x}, g(x) = x + 6$

71. $f(x) = \frac{1}{x}, g(x) = \sqrt{x-1}$

72. $f(x) = \frac{2x+1}{x+3}, g(x) = \frac{x+2}{3x+5}$

73. $f(x) = \frac{x}{x^2-4}, g(x) = \frac{1}{x-3}$

74. $f(x) = \frac{x-4}{x^2-x-6}, g(x) = \frac{x+5}{x^2-16}$

75. $f(x) = \sqrt{x+2}, g(x) = \sqrt{2-x}$

76. $f(x) = \sqrt{\frac{x+1}{x+2}}, g(x) = \sqrt{x-3}$

77. $f(x) = \sqrt[3]{x+1}, g(x) = \sqrt[3]{\frac{x-1}{x+2}}$

78. $f(x) = \sqrt{7-x}, g(x) = \sqrt{x^2-2x-15}$

3.6 One-to-One Functions; Inverse Functions

THINGS TO KNOW

Before working through this section, be sure that you are familiar with the following concepts:

VIDEO ANIMATION INTERACTIVE

You Try It 1. Determining the Domain of a Function Given the Equation (Section 3.1)

You Try It 2. Determining the Domain and Range of a Function from Its Graph (Section 3.2)

You Try It 3. Analyzing Piecewise-Defined Functions (Section 3.3)

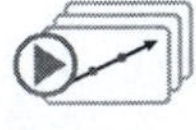

You Try It 4. Forming and Evaluating Composite Functions (Section 3.5)

INTRODUCTION

Read this introduction before beginning Objective 1.

OBJECTIVES

1 Understanding the Definition of a One-to-One Function

2 Determining If a Function Is One-to-One Using the Horizontal Line Test

3 Understanding and Verifying Inverse Functions

4 Sketching the Graphs of Inverse Functions

5 Finding the Inverse of a One-to-One Function

SECTION 3.6 EXERCISES

Introduction to Section 3.6

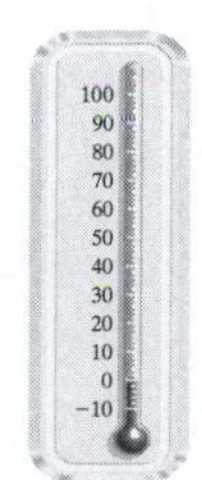

In this section, we study *inverse functions.* To illustrate an inverse function, let's consider the function $F(C) = \frac{9}{5}C + 32$. This function is used to convert a given temperature in degrees Celsius into its equivalent temperature in degrees Fahrenheit. For example,

$$F(25) = \frac{9}{5}(25) + 32 = 45 + 32 = 77.$$

Thus, a temperature of 25°C corresponds to a temperature of 77°F. To convert a temperature of 77°F back into 25°C, we need a different function. The function that is used to accomplish this task is $C(F) = \frac{5}{9}(F - 32)$.

$$\begin{aligned} C(77) &= \frac{5}{9}(77 - 32) \\ &= \frac{5}{9}(45) \\ &= 25 \end{aligned}$$

The function C is the *inverse* of the function F. In essence, these functions perform opposite actions (they "undo" each other). The first function converted 25°C into 77°F, whereas the second function converted 77°F back into 25°C.

The function F has an inverse function C. Later in this section, we examine a process for finding the inverse of a function, but keep the following in mind. We can find the inverse of many functions using this process, but that inverse will not always be a function. In this text, we are only interested in inverses that are functions, so we first develop a test to determine if a function has an inverse *function.* When the word *inverse* is used throughout the remainder of this section, we assume that we are referring to the inverse that is a function.

OBJECTIVE 1 UNDERSTANDING THE DEFINITION OF A ONE-TO-ONE FUNCTION

First, we must define the concept of **one-to-one** functions.

> **Definition** One-to-One Function
>
> A function f is **one-to-one** if for any values $a \neq b$ in the domain of f, $f(a) \neq f(b)$.

This definition suggests that a function is one-to-one if for any two *different* input values (domain values), the corresponding output values (range values) must be different. An alternate definition says that if two range values are the same, $f(u) = f(v)$, then the domain values must be the same; that is, $u = v$.

> **Alternate Definition of a One-to-One Function**
>
> A function f is **one-to-one** if for any two range values $f(u)$ and $f(v)$, $f(u) = f(v)$ implies that $u = v$.

The function sketched in Figure 25a is one-to-one because for any two distinct x-values in the domain $(a \neq b)$, the function values or range values are not equal $(f(a) \neq f(b))$. In Figure 25b, we see that the function $y = g(x)$ is *not* one-to-one because we can easily find two different domain values that correspond to the same range value.

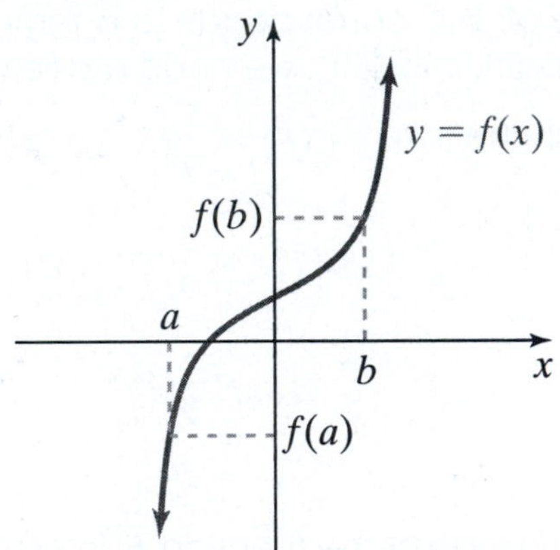

Figure 25a An example of a one-to-one function.

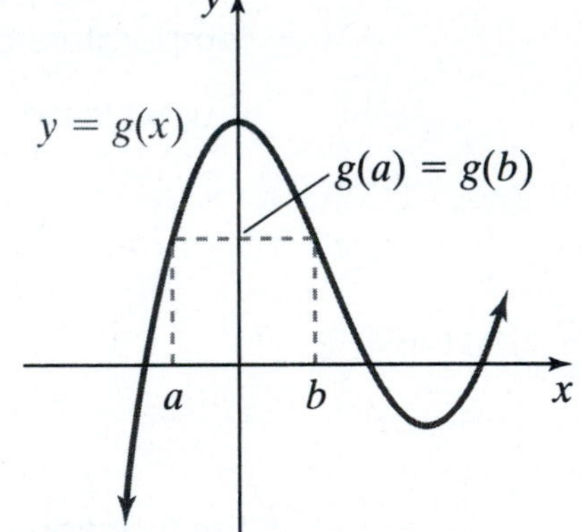

Figure 25b An example of a function that is not one-to-one.

OBJECTIVE 2 DETERMINING IF A FUNCTION IS ONE-TO-ONE USING THE HORIZONTAL LINE TEST

Notice in Figure 26 that the horizontal lines intersect the graph of $y = f(x)$ in at most one place, while we can find many horizontal lines that intersect the graph of $y = g(x)$ more than once. This gives us a visual example of how we can use horizontal lines to help us determine from the graph if a function is one-to-one. Using horizontal lines to determine if the graph of a function is one-to-one is known as the *horizontal line test*.

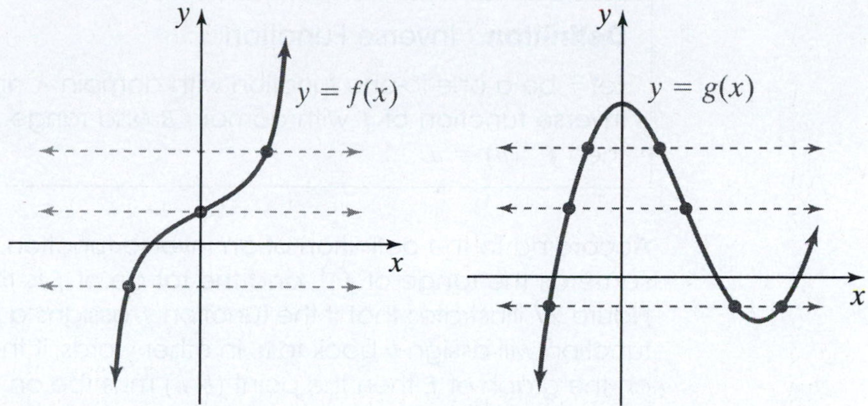

Figure 26 Drawing horizontal lines can help determine whether a graph represents a one-to-one function.

> **Horizontal Line Test**
>
> If every horizontal line intersects the graph of a function f at most once, then f is one-to-one.

Example 1 Determine If a Function Is One-to-One

Determine if each function is one-to-one.

a.

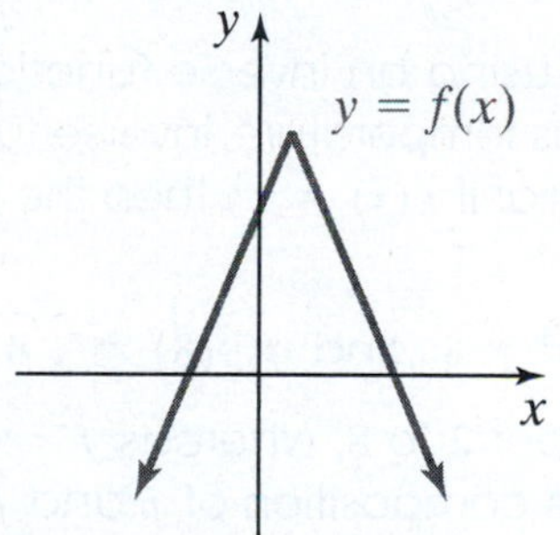

b.

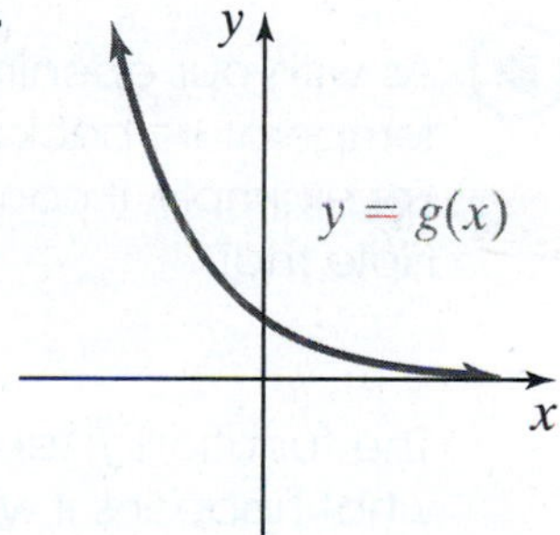

c. $f(x) = x^2 + 1, x \leq 0$

d. $f(x) = \begin{cases} 2x + 4 \text{ for } x \leq -1 \\ 2x - 6 \text{ for } x \geq 4 \end{cases}$

Solution The functions in parts b and c are one-to-one, whereas the functions in parts a and d are not one-to-one. Watch the **animation** to verify.

You Try It **Work through the following You Try It problem.**

Work Exercises 1–17 in this textbook or in the MyLab Math Study Plan.

OBJECTIVE 3 UNDERSTANDING AND VERIFYING INVERSE FUNCTIONS

We are now ready to ask ourselves the question, "Why should we be concerned with one-to-one functions?"

Answer: Every one-to-one function has an inverse function!

Definition Inverse Function

Let f be a one-to-one function with domain A and range B. Then f^{-1} is the **inverse function of** f with domain B and range A. Furthermore, if $f(a) = b$, then $f^{-1}(b) = a$.

According to the definition of an inverse function, the domain of f is exactly the same as the range of f^{-1}, and the range of f is the same as the domain of f^{-1}. Figure 27 illustrates that if the function f assigns a number a to b, then the inverse function will assign b back to a. In other words, if the point (a, b) is an ordered pair on the graph of f, then the point (b, a) must be on the graph of f^{-1}.

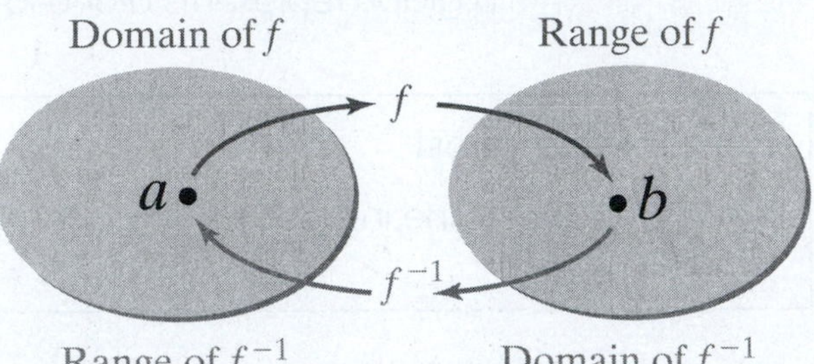

Figure 27

CAUTION Do not confuse f^{-1} with $\dfrac{1}{f(x)}$. The negative 1 in f^{-1} is ***not*** an exponent!

As with our opening example using an inverse function to convert a Fahrenheit temperature back into a Celsius temperature, inverse functions "undo" each other. For example, it can be shown that if $f(x) = x^3$, then the inverse of f is $f^{-1}(x) = \sqrt[3]{x}$. Note that

$$f(2) = (2)^3 = 8 \text{ and } f^{-1}(8) = \sqrt[3]{8} = 2.$$

The function f takes the number 2 to 8, whereas f^{-1} takes 8 back to 2. Observe what happens if we look at the composition of f and f^{-1} and the composition of f^{-1} and f at specified values:

$$\underbrace{}_{x\text{-value}\downarrow}\quad \left(f \circ f^{-1}\right)(8) = f\left(f^{-1}(8)\right) = f(2) = 8 \quad \text{(same as } x\text{-value)}$$

$$\underbrace{}_{x\text{-value}\downarrow}\quad \left(f^{-1} \circ f\right)(2) = f^{-1}\left(f(2)\right) = f^{-1}(8) = 2 \quad \text{(same as } x\text{-value)}$$

Because of the "undoing" nature of inverse functions, we get the following **composition cancellation equations:**

> **Composition Cancellation Equations**
>
> $f(f^{-1}(x)) = x$ for all x in the domain of f^{-1} and $f^{-1}(f(x)) = x$ for all x in the domain of f

These cancellation equations can be used to show if two functions are inverses of each other. We can see from our example that if $f(x) = x^3$ and $f^{-1}(x) = \sqrt[3]{x}$, then

$$f(f^{-1}(x)) = f(\sqrt[3]{x}) = (\sqrt[3]{x})^3 = x \text{ and } f^{-1}(f(x)) = f^{-1}(x^3) = \sqrt[3]{x^3} = x.$$

Example 2 Verify Inverse Functions

Show that $f(x) = \dfrac{x}{2x + 3}$ and $g(x) = \dfrac{3x}{1 - 2x}$ are inverse functions using the composition cancellation equations.

Solution To show that f and g are inverses of each other, we must show that $(f \circ g)(x) = x$ and $(g \circ f)(x) = x$. Work through the **interactive video** to verify that both composition cancellation equations are satisfied.

You Try It **Work through the following You Try It problem.**

Work Exercises 18–24 in this textbook or in the MyLab Math Study Plan.

OBJECTIVE 4 SKETCHING THE GRAPHS OF INVERSE FUNCTIONS

If f is a one-to-one function, then we know that it must have an inverse function, f^{-1}. Given the graph of a one-to-one function f, we can obtain the graph of f^{-1} by simply interchanging the coordinates of each ordered pair that lies on the graph of f. In other words, for any point (a, b) on the graph of f, the point (b, a) must lie on the graph of f^{-1}. Notice in Figure 28 that the points (a, b) and (b, a) are symmetric about the line $y = x$. Therefore, the graph of f^{-1} is a reflection of the graph of f about the line $y = x$. Figure 29 shows the graph of $f(x) = x^3$ and $f^{-1}(x) = \sqrt[3]{x}$. You can see that if the functions have any points in common, they must lie along the line $y = x$.

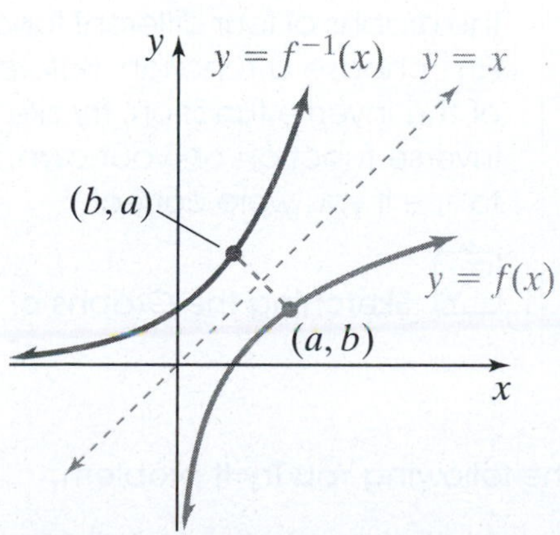

Figure 28 Graph of a one-to-one function and its inverse.

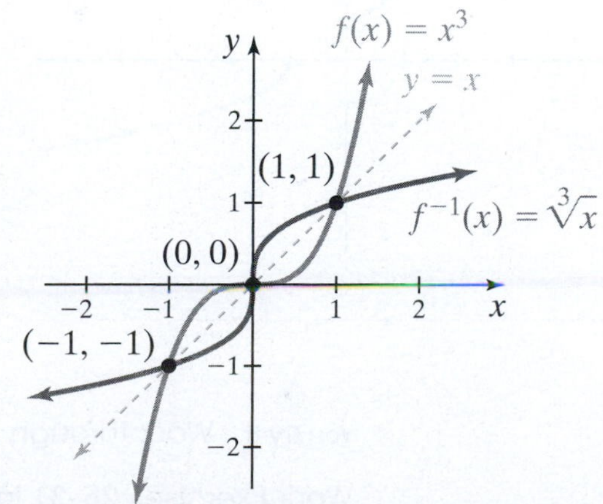

Figure 29 Graph of $f(x) = x^3$ and $f^{-1}(x) = \sqrt[3]{x}$.

Example 3 Sketch the Graph of a One-to-One Function and Its Inverse

Sketch the graph of $f(x) = x^2 + 1, x \le 0$, and its inverse. Also state the domain and range of f and f^{-1}.

Solution The graphs of f and f^{-1} are sketched in Figure 30. Notice how the graph of f^{-1} is a reflection of the graph of f about the line $y = x$. Also notice that the domain of f is the same as the range of f^{-1}. Likewise, the domain of f^{-1} is equivalent to the range of f.

View the **animation** to see exactly how to sketch f and f^{-1}. We will determine the equation of f^{-1} in Example 5.

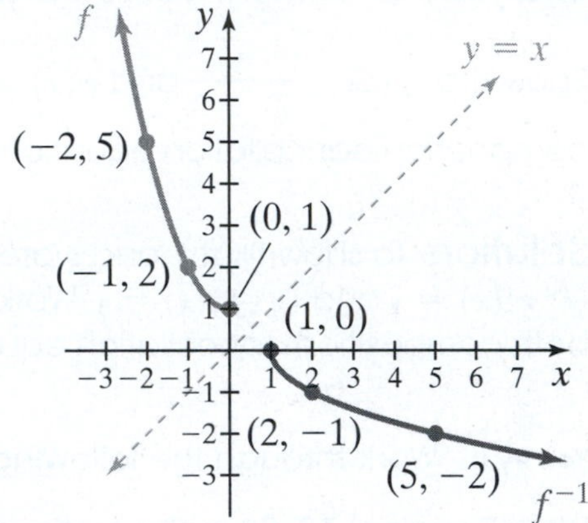

Domain of f: $(-\infty, 0]$ → Domain of f^{-1}: $[1, \infty)$

Range of f: $[1, \infty)$ → Range of f^{-1}: $(-\infty, 0]$

Figure 30 Graph of $f(x) = x^2 + 1, x \le 0$, and its inverse.

Using Technology

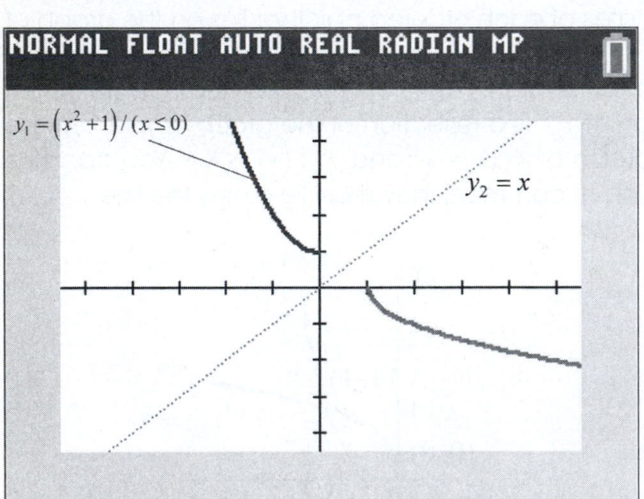

Using a TI-84+, we can sketch the functions from Example 3 by letting $y_1 = (x^2 + 1)/(x \le 0)$. We can draw the inverse function by typing the command **DrawInv Y_1** in the calculator's main viewing window. The resultant graphs are shown here.

Click on the guided visualization below and explore the graphs of four different functions and their inverses. First choose a function. Before you display the graph of the inverse function, try sketching the graph of the inverse function on your own. Then display the graph to see if you were correct.

Sketching the Graphs of Inverse Functions

You Try It Work through the following You Try It problem.

Work Exercises 25–31 in this textbook or in the MyLab Math Study Plan.

OBJECTIVE 5 FINDING THE INVERSE OF A ONE-TO-ONE FUNCTION

We are now ready to find the inverse of a one-to-one function algebraically. We know that if a point (x, y) is on the graph of a one-to-one function, then the point (y, x) is on the graph of its inverse function. We can use this information to develop a process for finding the inverse of a function algebraically simply by switching x and y in the original function to produce its inverse function.

Carefully work through Example 4 and watch the **animation** to learn the four-step process for algebraically finding inverse functions.

Example 4 Find the Inverse of a Function

Find the inverse of the function $f(x) = \dfrac{2x}{1 - 5x}$, and state the domain and range of f and f^{-1}.

Solution

Step 1. Change $f(x)$ to y: $y = \dfrac{2x}{1 - 5x}$

Step 2. Interchange x and y: $x = \dfrac{2y}{1 - 5y}$

Step 3. Solve for y:

$x = \dfrac{2y}{1 - 5y}$ Write the equation from Step 2.

$x(1 - 5y) = \dfrac{2y}{1 - 5y}(1 - 5y)$ Multiply both sides of the equation by the LCD.

$x - 5xy = 2y$ Simplify.

$x = 2y + 5xy$ Add $5xy$ to both sides.

$x = y(2 + 5x)$ Factor.

$\dfrac{x}{2 + 5x} = y$ Divide both sides by $2 + 5x$.

Step 4. Change y to $f^{-1}(x)$: $f^{-1}(x) = \dfrac{x}{2 + 5x}$ or $f^{-1}(x) = \dfrac{x}{5x + 2}$

Carefully work through this **animation** to verify that the domain of f is $\left(-\infty, \frac{1}{5}\right) \cup \left(\frac{1}{5}, \infty\right)$, whereas the domain of f^{-1} is $\left(-\infty, -\frac{2}{5}\right) \cup \left(-\frac{2}{5}, \infty\right)$. Because the range of f must be the domain of f^{-1} and the range of f^{-1} must be the domain of f, we get the following result:

$f(x) = \dfrac{2x}{1 - 5x}$ $\qquad$ $f^{-1}(x) = \dfrac{x}{5x + 2}$

Domain of f: $\left(-\infty, \frac{1}{5}\right) \cup \left(\frac{1}{5}, \infty\right)$ $\qquad$ Domain of f^{-1}: $\left(-\infty, -\frac{2}{5}\right) \cup \left(-\frac{2}{5}, \infty\right)$

Range of f: $\left(-\infty, -\frac{2}{5}\right) \cup \left(-\frac{2}{5}, \infty\right)$ $\qquad$ Range of f^{-1}: $\left(-\infty, \frac{1}{5}\right) \cup \left(\frac{1}{5}, \infty\right)$

In Example 5, use the **Four-Step Process for Algebraically Finding Inverse Functions** that was laid out in the previous example to determine the inverse of the function $f(x) = x^2 + 1,\ x \le 0$ which was discussed in **Example 3**.

Example 5 Find the Inverse of a Function

Find the inverse of the function $f(x) = x^2 + 1,\ x \le 0$ that was discussed in Example 3.

Solution

Step 1. Change $f(x)$ to y: $\quad y = x^2 + 1$

Step 2. Interchange x and y: $\quad x = y^2 + 1$

Step 3. Solve for y:

$x = y^2 + 1$ — Write the equation from Step 2.

$x - 1 = y^2$ — Subtract 1 from both sides.

$\pm\sqrt{x - 1} = y$ — Use the square root property.

(Because the domain of f is $(-\infty, 0]$, the range of f^{-1} must be $(-\infty, 0]$. Therefore, we must use the negative square root or $y = -\sqrt{x - 1}$.)

Step 4. Change y to $f^{-1}(x)$: $\quad f^{-1}(x) = -\sqrt{x - 1}$

Figure 31 shows the graphs of $f(x) = x^2 + 1,\ x \le 0$ and $f^{-1}(x) = -\sqrt{x - 1}$. Notice that the two graphs are symmetric about the line $y = x$ and that the domain of f is exactly the range of f^{-1}. Similarly, the range of f is exactly the domain of f^{-1}.

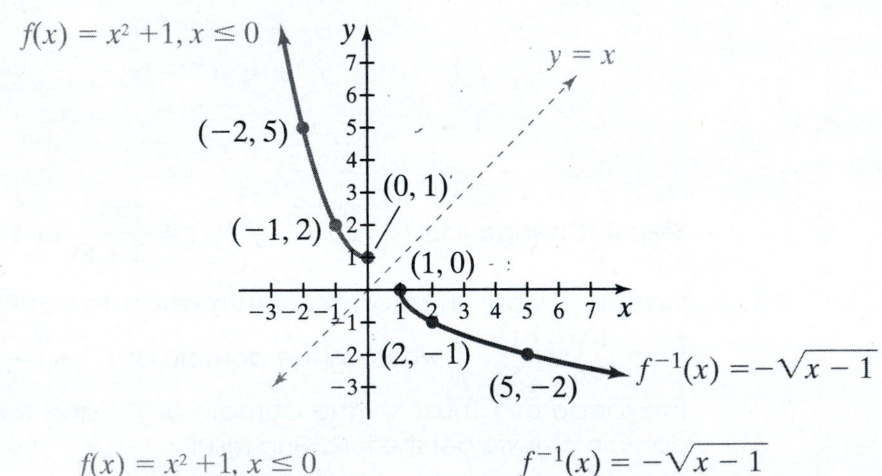

$f(x) = x^2 + 1,\ x \le 0$	$f^{-1}(x) = -\sqrt{x - 1}$
Domain of f: $(-\infty, 0]$	Domain of f^{-1}: $[1, \infty)$
Range of f: $[1, \infty)$	Range of f^{-1}: $(-\infty, 0]$

Figure 31 The graphs of $f(x) = x^2 + 1,\ x \le 0$ and $f^{-1}(x) = -\sqrt{x - 1}$.

You Try It Work through the following You Try It problem.

Work Exercises 32–46 in this textbook or in the MyLab Math Study Plan.

3.6 Exercises

Skill Check Exercises

For exercises SCE-1 and SCE-2, simplify each expression and write as a single fraction.

SCE-1. $\dfrac{2\left(\dfrac{x+5}{x-1}\right)-3}{\dfrac{x+5}{x-1}}$

SCE-2. $\dfrac{11\left(\dfrac{2x-4}{3+5x}\right)+5}{4-7\left(\dfrac{2x-4}{3+5x}\right)}$

For exercises SCE-3 through SCE-8, solve each equation for the variable y.

SCE-3. $x = \frac{1}{4}y - 5$

SCE-4. $x = \dfrac{3y+9}{7}$

SCE-5. $x(7-5y) = 8y - 1$

SCE-6. $x = \dfrac{4y-5}{8y+4}$

SCE-7. $x = 6 - \sqrt[3]{y-3}$

SCE-8. $x = (y-4)^2 + 9$ for $y \le 4$

In Exercises 1–17, determine if each function is one-to-one.

1. $f(x) = 3x - 1$
2. $f(x) = 2x^2$
3. $f(x) = (x-1)^2, x \ge 1$
4. $f(x) = (x-1)^2, x \ge -1$
5. $f(x) = \frac{1}{x} - 2$
6. $f(x) = 4\sqrt{x}$
7. $f(x) = -2|x|$
8. $f(x) = 2$
9. $f(x) = (x+1)^3 - 2$
10. $f(x) = \begin{cases} x+4 & \text{if } x < -1 \\ 2x+2 & \text{if } x \ge \frac{1}{2} \end{cases}$
11. $f(x) = \begin{cases} -3x & \text{if } x \le 0 \\ 2-x & \text{if } x \ge 2 \end{cases}$

12.

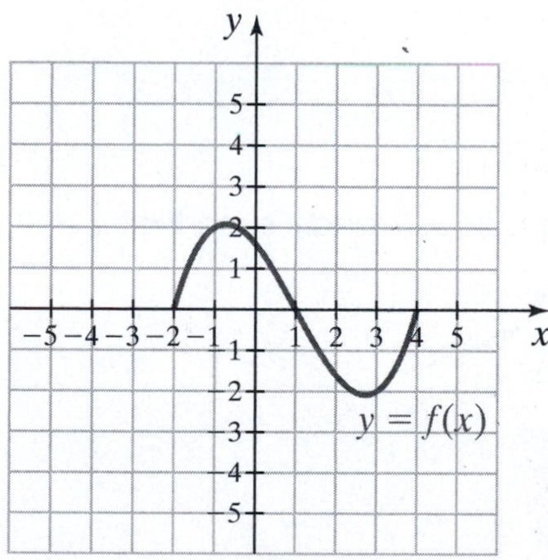

13.

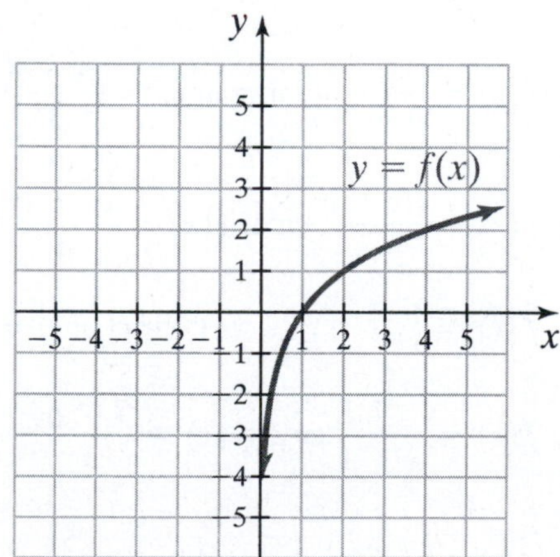

14.

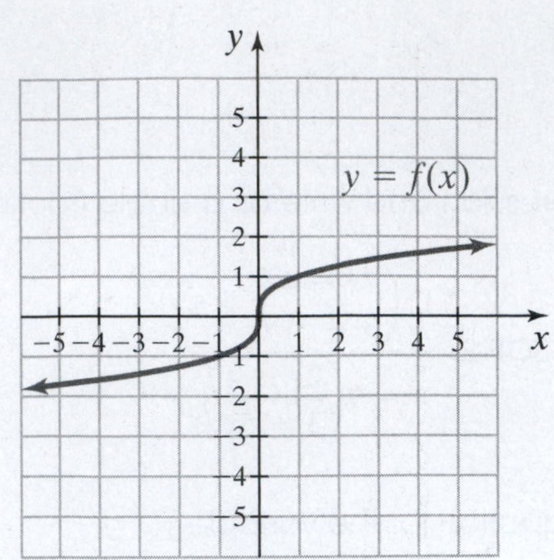

15.

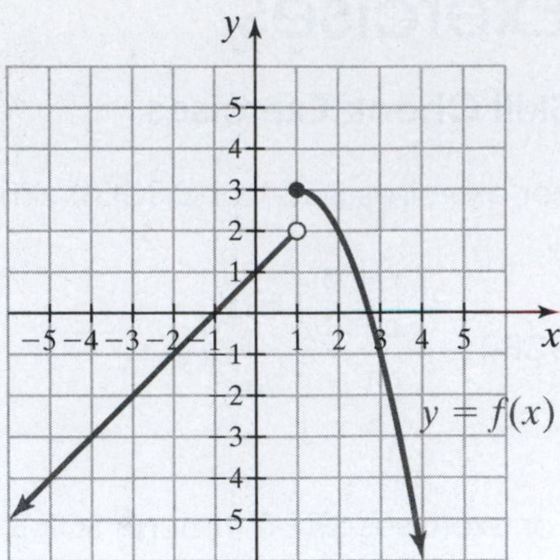

16.

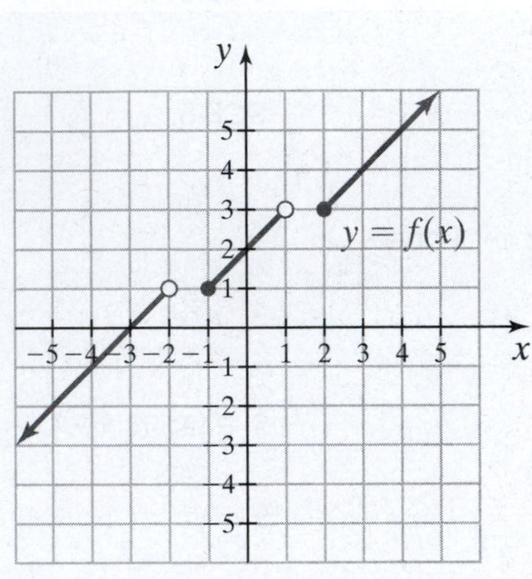

17.

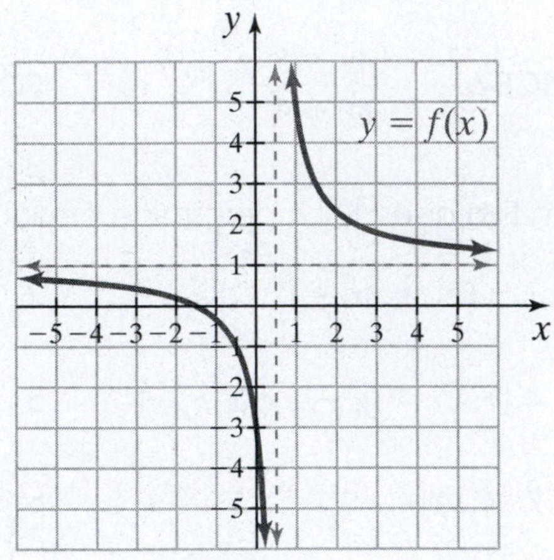

In Exercises 18–24, use the composition cancellation equations to verify that f and g are inverse functions.

18. $f(x) = \frac{3}{2}x - 4$ and $g(x) = \frac{2x + 8}{3}$

19. $f(x) = \frac{ax + b}{c}$ and $g(x) = \frac{cx - b}{a}$ for all real numbers a, b, and c such that $a \neq 0, c \neq 0$

20. $f(x) = (x - 1)^2, x \geq 1$, and $g(x) = \sqrt{x} + 1$

21. $f(x) = \frac{7}{x + 1}$ and $g(x) = \frac{7 - x}{x}$

22. $f(x) = \frac{x}{5 + 3x}$ and $g(x) = \frac{5x}{1 - 3x}$

23. $f(x) = 2\sqrt[3]{x - 1} + 3$ and $g(x) = \frac{(x - 3)^3}{8} + 1$

24. $f(x) = \frac{11x + 5}{4 - 7x}$ and $g(x) = \frac{4x - 5}{7x + 11}$

In Exercises 25–30, use the graph of f to sketch the graph of f^{-1}. Use the graphs to determine the domain and range of each function.

25.

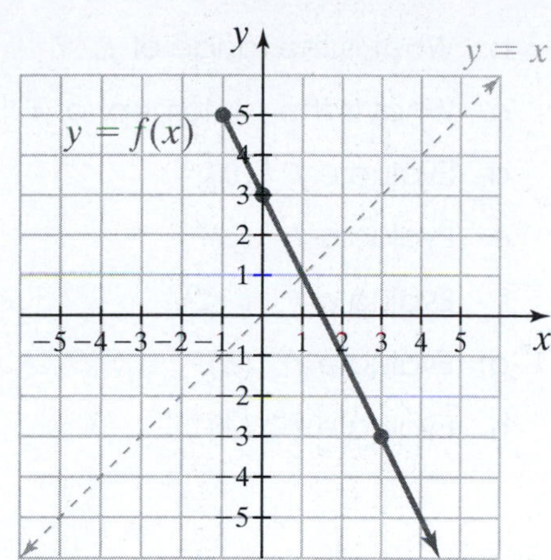

26.

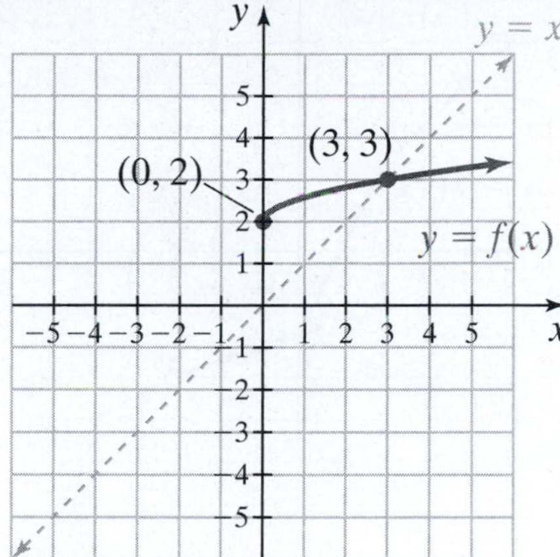

27.

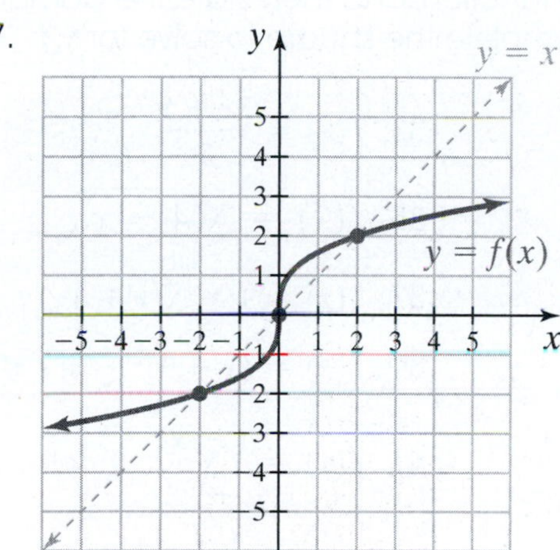

28.

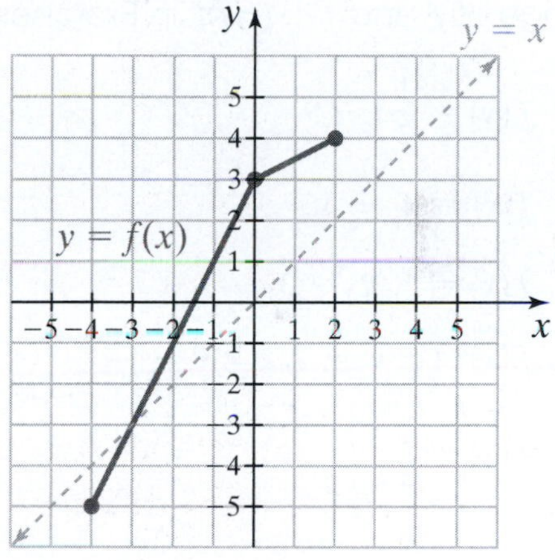

29.

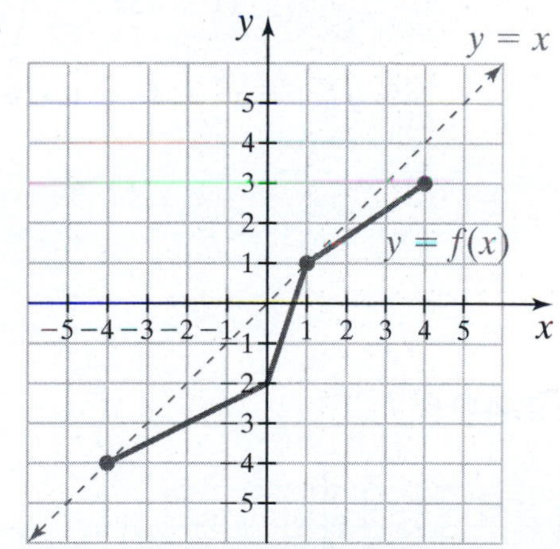

30.

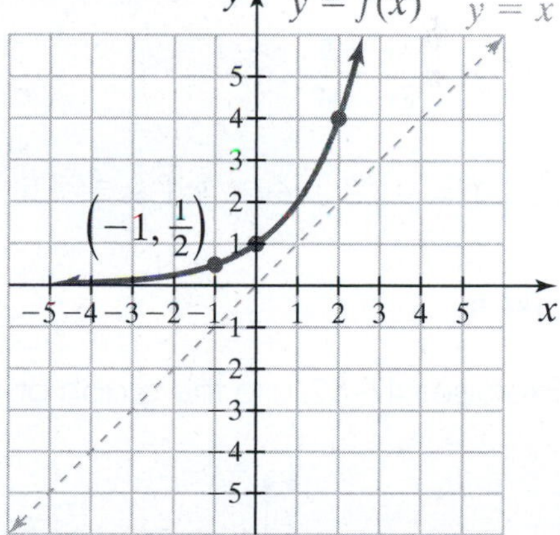

In Exercise 31, refer to the graph of the function $y = f(x)$.

31.

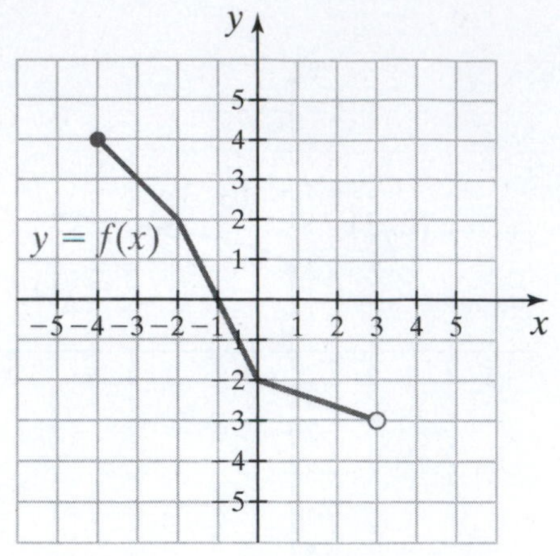

a. What is the domain of f^{-1}?

b. What is the range of f^{-1}?

c. What is the y-intercept of f^{-1}?

d. Evaluate $f^{-1}(0)$.

e. Evaluate $f^{-1}(2)$.

f. Evaluate $f^{-1}(-2)$.

g. Evaluate $f^{-1}(3)$.

h. Evaluate $f^{-1}(4)$.

In Exercises 32–46, write an equation for the inverse function, and then state the domain and range of f and f^{-1}. (*Hint:* In Exercises 43 and 44, complete the square to solve for y.)

32. $f(x) = \frac{1}{3}x - 5$

33. $f(x) = \frac{3x + 9}{7}$

34. $f(x) = x^3 + 5$

35. $f(x) = \sqrt[3]{4 - x}$

36. $f(x) = \sqrt[3]{2x - 3}$

37. $f(x) = 1 - \sqrt[5]{x + 4}$

38. $f(x) = -x^2 - 2, x \geq 0$

39. $f(x) = (x + 3)^2 - 5, x \leq -3$

40. $f(x) = \frac{3}{x}$

41. $f(x) = \frac{1 - x}{2x}$

42. $f(x) = \frac{8x - 1}{7 - 5x}$

43. $f(x) = \frac{1 - 4x}{11 + 13x}$

44. $f(x) = \frac{4x - 5}{8x + 4}$

45. $f(x) = x^2 + 4x - 1, x \geq -2$

46. $f(x) = -2x^2 + 6x + 8, x \leq \frac{3}{2}$

Brief Exercises

In Exercises 47–52, use the graph of f to sketch the graph of f^{-1}.

47.

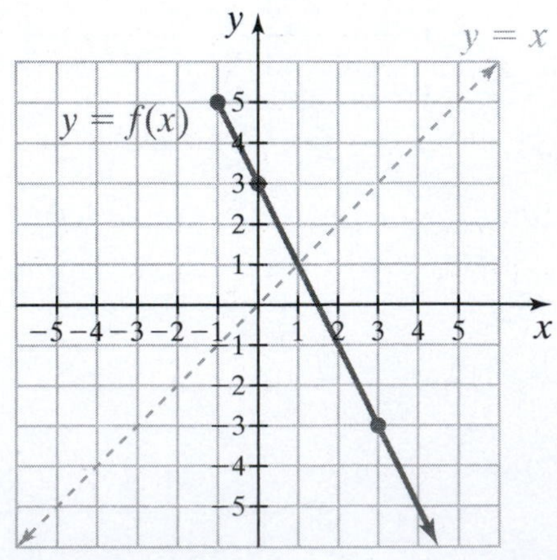

48.

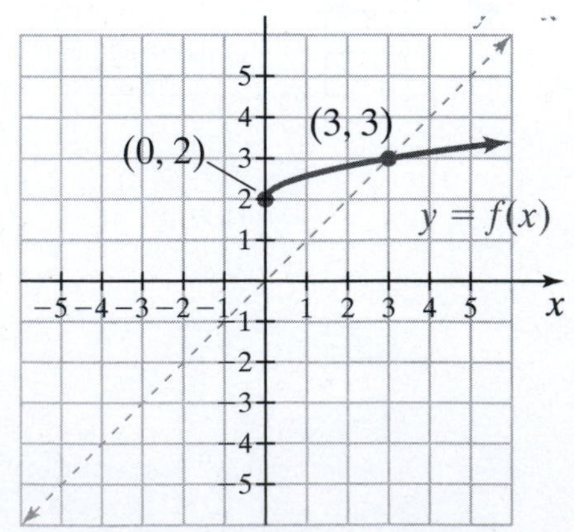

49.

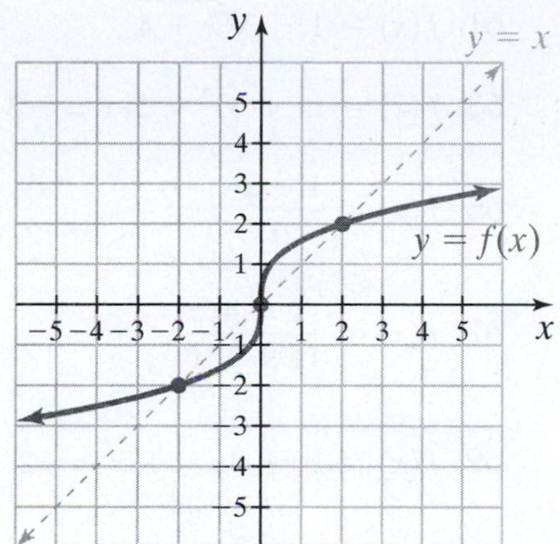

50.

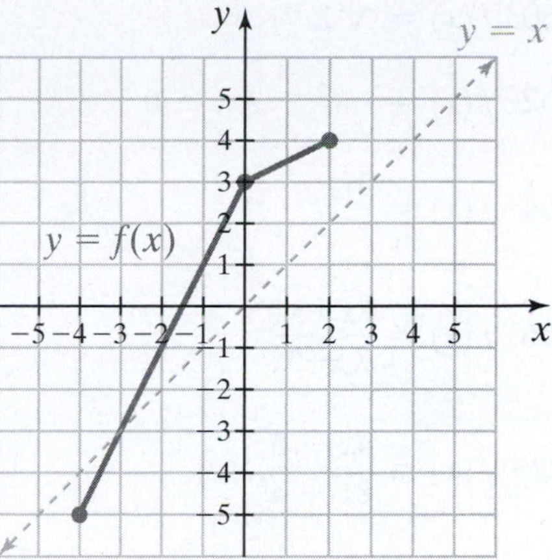

51.

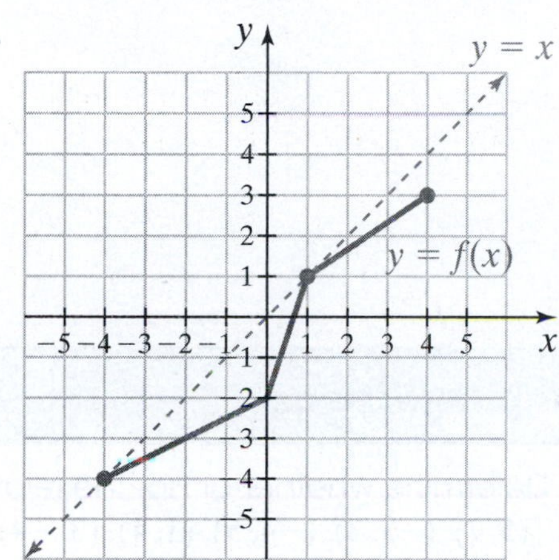

52.

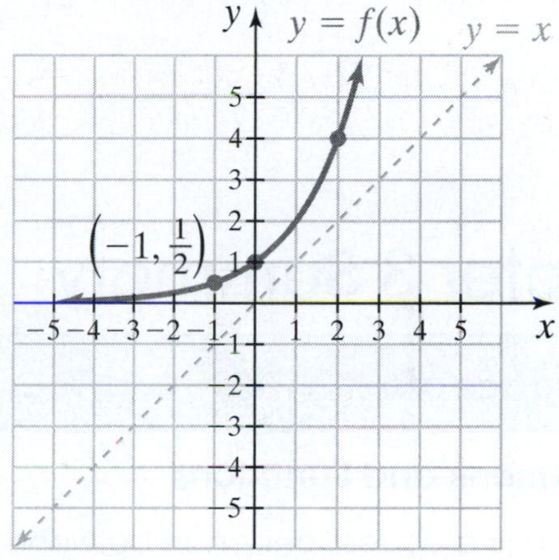

For Exercises 53–55, use the graph of the function $y = f(x)$ to answer each question.

53. a. What is the domain of f^{-1}?

b. What is the range of f^{-1}?

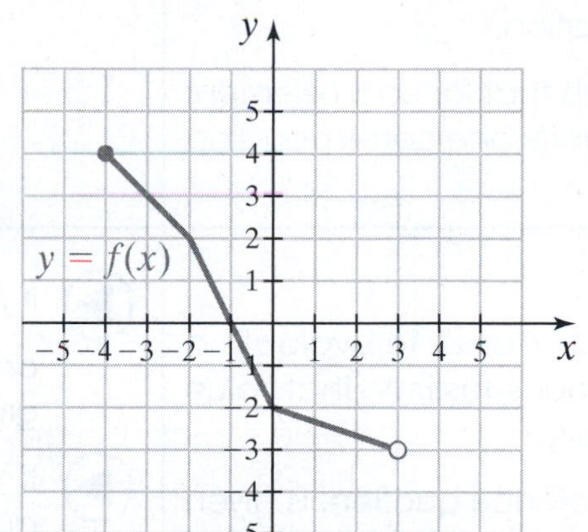

54. What is the y-intercept of f^{-1}?

55. Evaluate each of the following:

a. $f^{-1}(0)$ b. $f^{-1}(2)$ c. $f^{-1}(-2)$ d. $f^{-1}(3)$ e. $f^{-1}(4)$

In Exercises 56–70, write an equation for the inverse of the given one-to-one function. (*Hint:* In Exercises 69 and 70, complete the square to solve for y.)

56. $f(x) = \frac{1}{3}x - 5$

57. $f(x) = \frac{3x + 9}{7}$

58. $f(x) = x^3 + 5$

59. $f(x) = \sqrt[3]{4 - x}$

60. $f(x) = \sqrt[3]{2x - 3}$

61. $f(x) = 1 - \sqrt[5]{x + 4}$

62. $f(x) = -x^2 - 2, x \geq 0$

63. $f(x) = (x + 3)^2 - 5, x \leq -3$

64. $f(x) = \frac{3}{x}$

65. $f(x) = \frac{1 - x}{2x}$

66. $f(x) = \frac{8x - 1}{7 - 5x}$

67. $f(x) = \frac{1 - 4x}{11 + 13x}$

68. $f(x) = \frac{4x - 5}{8x + 4}$

69. $f(x) = x^2 + 4x - 1, x \geq -2$

70. $f(x) = -2x^2 + 6x + 8, x \leq \frac{3}{2}$

Chapter 3 Summary

Key Concepts	Examples/Videos
3.1 Relations and Functions A **relation** is a correspondence between two sets *A* and *B* such that each element of set *A* corresponds to one or more elements is set *B*. Set *A* is called the **domain** of the relation and set *B* is called the **range** of the relation. A **function** is a relation such that for each element in the domain, there is*exactly one* corresponding element in the range.	Determine whether or not the relation $\{(3, 7), (-3, 2), (-4, 5), (1, 4), (3, -4)\}$ is a function, and then find the domain and range.
Function Notation The symbol $f(x)$ is read as "f of x". To evaluate a function at a specific number, substitute that value into the function and simplify. Given a function f, the **difference quotient** is given by the expression $\frac{f(x + h) - f(x)}{h}$.	If $f(x) = \frac{3}{2}x + 6$, evaluate $f(-4), f(0)$, and $f(1)$ and state three ordered pairs that lie on the graph of f. If $f(x) = x^2 + x - 1$, determine the difference quotient.

Key Concepts	Examples/Videos
The Vertical Line Test A graph in the Cartesian plane is the graph of a function if and only if no vertical line intersects the graph more than once.	Use the vertical line test to determine which of the following graphs represents the graph of a function. a. 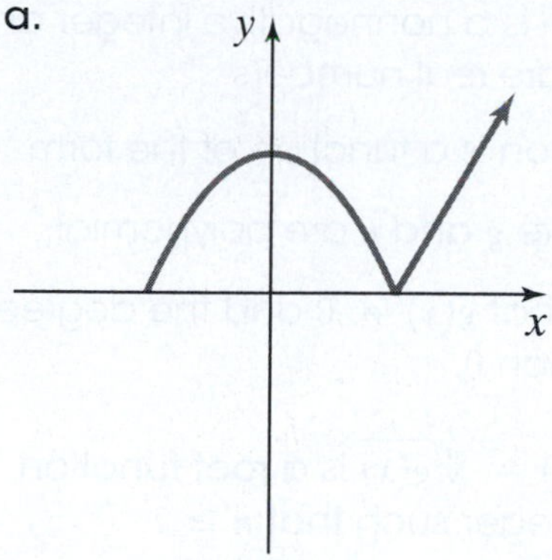 b. 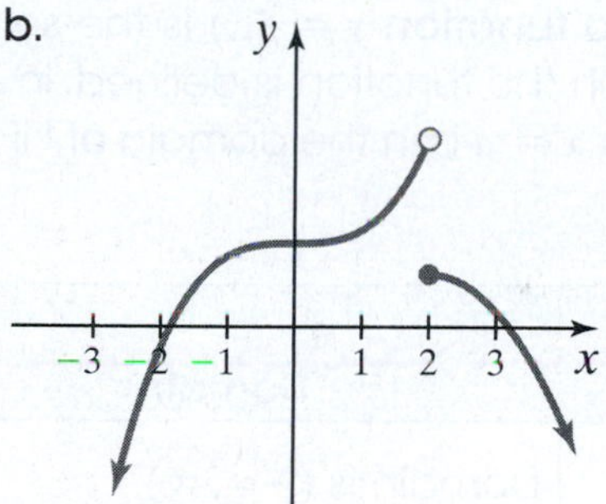 c. 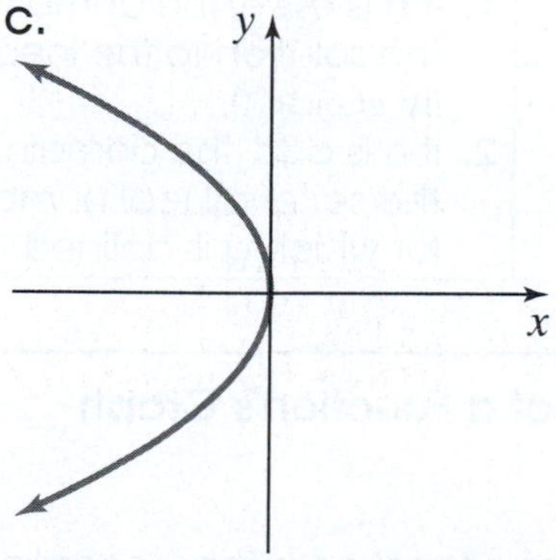

Key Concepts	Examples/Videos
Classifying Functions A **polynomial function** of degree n is a function of the form $f(x) = a_n x^n + a_{n-1}x^{n-1} + a_{n-2}x^{n-2} + \ldots + a_1 x + a_0$, where n is a nonnegative integer and $a_0, a_1, a_2, \ldots, a_n$ are real numbers. A **rational function** is a function of the form $f(x) = \frac{g(x)}{h(x)}$, where g and h are polynomial functions such that $g(x) \neq 0$ and the degree of $h(x)$ is greater than 0. The function $f(x) = \sqrt[n]{g(x)}$ is a **root function**, where n is an integer such that $n \geq 2$.	 Determine whether the following functions are polynomial functions, If so, state the degree. a. $f(x) = 7 - 3x^2 + 5x^3$ b. $h(t) = -11$ c. $f(t) = 3t^6 + \frac{1}{t}$  Determine whether the following functions are rational functions. a. $f(t) = \frac{t^3 - 1}{t + 1}$ b. $h(x) = (x^3 - x^2 - 9)x^{-5}$
The **domain of a function** $y = f(x)$ is the set of all values x for which the function is defined. In ot her words, a number $x = a$ is in the domain of f if $f(a)$ is a real number. (see table below)	 Classify each function as a polynomial function, rational function, or root function, and then find the domain. Write the domain in interval notation. a. $f(x) = 2x^2 - 5x$ b. $f(x) = \frac{x}{x^2 - x - 6}$ c. $h(x) = \sqrt{x^2 - 2x - 8}$ d. $f(x) = \sqrt[3]{5x - 9}$ e. $k(x) = \sqrt[4]{\frac{x - 2}{x^2 + x}}$
3.2 Properties of a Function's Graph **Intercepts** The ***y*-intercept** of a function is the y-coordinate of the point where the graph crosses or touches the y-axis. – The y-intercept can be found by evaluating $f(0)$. An ***x*-intercept** of a function is the x-coordinate of the point where the graph crosses or touches the x-axis. The x-intercepts of a function are also called real zeros. – The x-intercepts can be found by finding all real solutions to the equation $f(x) = 0$.	 Given the functions below, find the x-intercept(s) and the y-intercept (if they exist). a. $f(x) = x^3 - 2x^2 + x - 2$ b. $g(x) = \frac{(2x^2 - 7x - 4)}{x^3}$

Class of Function	Domain
Polynomial functions	Domain is $(-\infty, \infty)$.
Rational functions	Domain is all real numbers such that $h(x) \neq 0$.
Root functions	1. If n is even, the domain is the solution to the inequality $g(x) \geq 0$. 2. If n is odd, the domain is the set of all real numbers for which g is defined.

Key Concepts	Examples/Videos
Determining the Domain and Range from a Graph Given the graph of a function, the domain is the set of all x-values and the range is the set of all y-values.	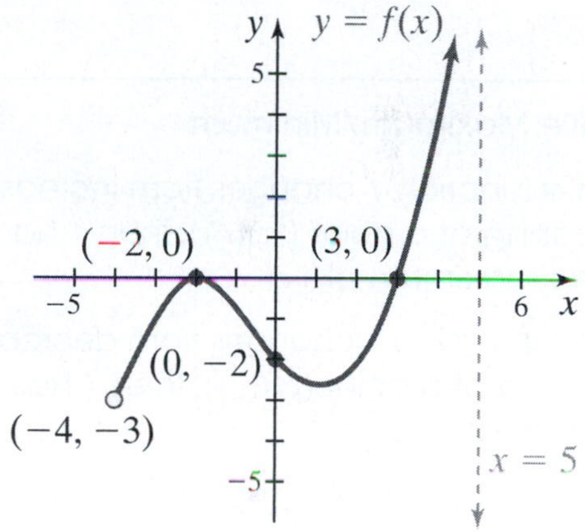 Use the graph of the following functions to determine the domain and range. a. 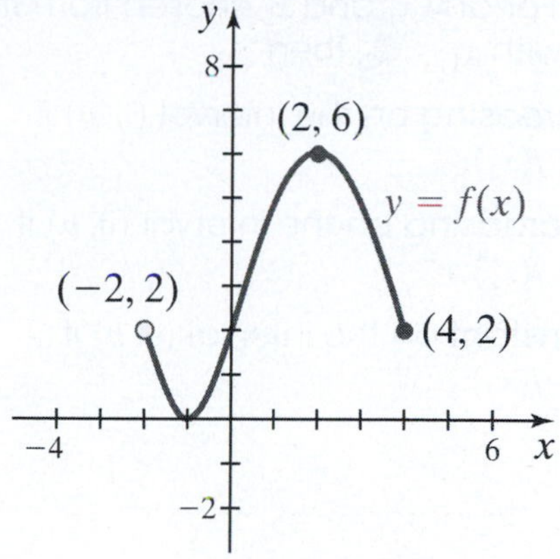 y; 8; (2, 6); $y = f(x)$; (−2, 2); (4, 2); −4; 6 x; −2 b. y; $y = f(x)$; 5; (−2, 0); (3, 0); −5; 6 x; (0, −2); (−4, −3); −5; $x = 5$ c. y; 5; (3, 4); $y = g(x)$; −5; 5 x; $y = -2$; −5 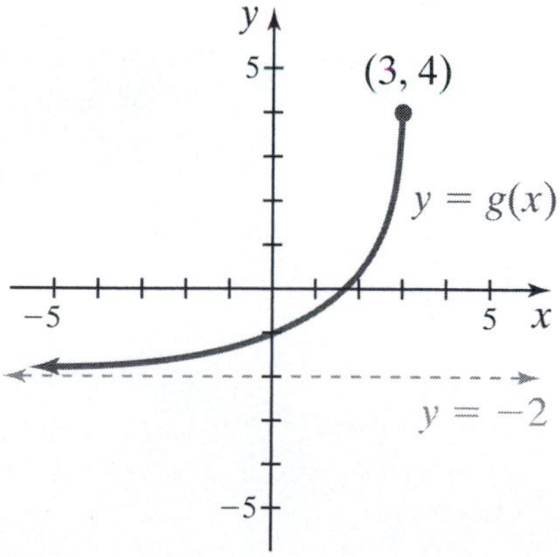

Key Concepts	Examples/Videos
Increasing/Decreasing/Constant Suppose that a function f is defined on the interval (a, b). For any x_1 and x_2 chosen from the interval (a, b) with $x_1 < x_2$, then f is **increasing** on the interval (a, b) if $f(x_1) < f(x_2)$; f is **decreasing** on the interval (a, b) if $f(x_1) > f(x_2)$; f is **constant** on the interval (a, b) if $f(x_1) = f(x_2)$.	 Given the function $y = f(x)$, determine where the function is increasing, decreasing, or constant. 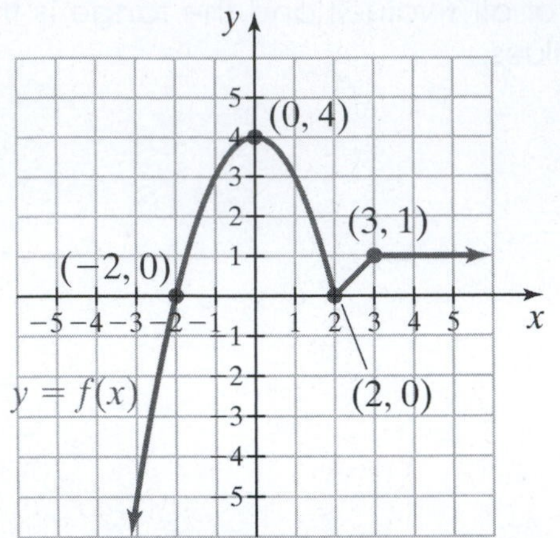
Relative Maximum/Minimum When a function f changes from increasing to decreasing at a point $(c, f(c))$, then f has a **relative maximum** at $x = c$. When a function f changes from decreasing to increasing at a point $(c, f(c))$, then f has a **relative minimum** at $x = c$.	 Use the graph to answer each question. 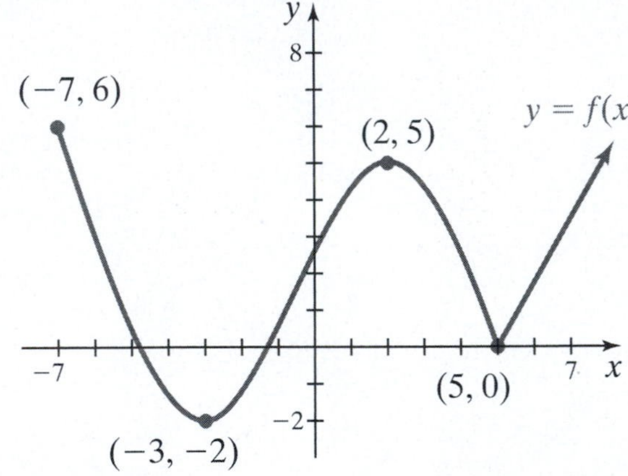 a. On what interval(s) is f increasing? b. On what interval(s) is f decreasing? c. For what value(s) of x does f have a relative minimum? d. For what value(s) of x does f have a relative maximum? e. What are the relative minima? f. What are the relative maxima?

Key Concepts	Examples/Videos

Even/Odd Functions

A function f is **even** if for every x in the domain, $f(x) = f(-x)$. Even functions are symmetric about the y-axis.

A function f is **odd** if for every x in the domain, $-f(x) = f(-x)$. Odd functions are symmetric about the *origin*.

Determine whether each function is even, odd or neither.

a. $f(x) = x^3 + x$

b. $f(x) = \frac{1}{x^2} + 7|x|$

c. $f(x) = 2x^5 - \frac{1}{x}$

d. $f(x) = x^2 + 4x$

Use the graph of $y = f(x)$ to answer each question on the right.

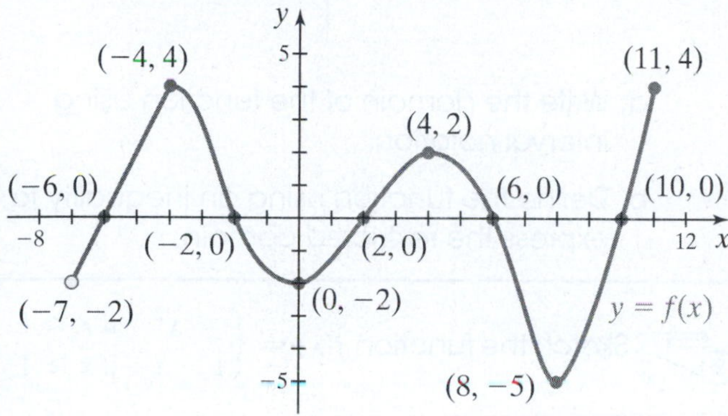

a. What is the y-intercept?

b. What are the real zeros of f?

c. Determine the domain and range of f.

d. Determine the interval(s) on which f is increasing, decreasing, and constant.

e. For what value(s) of x does f obtain a relative maximum? What are the relative maxima?

f. For what value(s) of x does f obtain a relative minimum? What are the relative minima?

g. Is f even, odd, or neither?

h. For what values of x is $f(x) = 4$?

i. For what values of x is $f(x) < 0$?

3.3 Graphs of Basic Functions; Piecewise Functions

Review of the Basic Functions

Click on any function to review its graph.

The identity function $f(x) = x$

The cube function $f(x) = x^3$

The square root function $f(x) = \sqrt{x}$

The reciprocal function $f(x) = \frac{1}{x}$

The square function $f(x) = x^2$

The absolute value function $f(x) = |x|$

The cube root function $f(x) = \sqrt[3]{x}$

Key Concepts	Examples/Videos
Sketching Basic Functions Having Restricted Domains To sketch a basic function with a restricted domain, first lightly sketch the entire function. Then, erase the portion of the graph that is not defined.	Sketch the function $f(x) = \sqrt{x}$ for $x \geq 1$. The graph of a basic function with a restricted domain is given below. 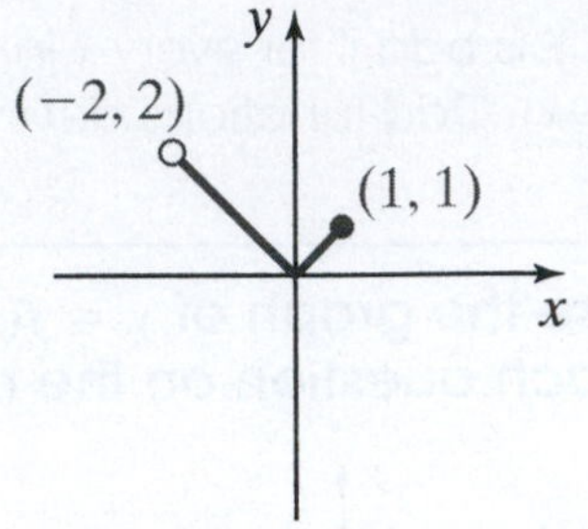 y, $(-2, 2)$, $(1, 1)$, x a. Write the domain of the function using interval notation. b. Define the function using an inequality to express the restricted domain.
Piecewise-Defined Functions Functions defined by a rule that has more than one piece are called **piecewise-defined functions.** To sketch the graph of a piecewise-defined function, sketch each function independently paying close attention to any restricted domains.	Sketch the function $f(x) = \begin{cases} x^2 & \text{if } x < 1 \\ 1 - x & \text{if } x \geq 1 \end{cases}$. Find a rule that describes the following piecewise function: 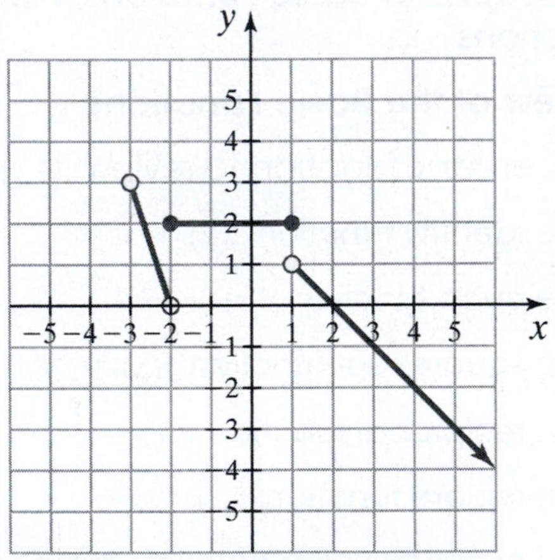 y, x, −5 −4 −3 −2 −1 1 2 3 4 5, 5 4 3 2 1 −1 −2 −3 −4 −5
	Cheapo Rental Car Co. charges a flat rate of \$85 to rent a car for up to 2 days. The company charges an additional \$20 for each additional day (or part of a day).

Key Concepts	Examples/Videos
	a. How much does it cost to rent a car for $2\frac{1}{2}$ days? 3 days? b. Write and sketch a piecewise function, $C(x)$, that describes the cost of renting a car as a function of time, x, in days rented for values of x less than or equal to 5.
3.4 Transformations of Functions Vertical Shift Up Vertical Shift Down Horizontal Shift Right Horizontal Shift Left	Use the graph of $y = x^3$ to sketch the graph of $g(x) = (x - 1)^3 + 2$.
Reflection About the x-axis Reflection About the y-axis	Use the graph of the basic function $y = \sqrt[3]{x}$ to sketch each graph. a. $g(x) = -\sqrt[3]{x} - 2$ b. $h(x) = \sqrt[3]{1 - x}$
Vertical Stretch Vertical Compression	Use the graph of $y = x^2$ to sketch the graph of $g(x) = 2x^2$.
Horizontal Stretch Horizontal Compression	Use the graph of $f(x) = \sqrt{x}$ to sketch the graphs of $g(x) = \sqrt{4x}$ and $h(x) = \sqrt{\frac{1}{4}x}$.
Order of Operations for Transformations 1. Horizontal shifts 2. Horizontal stretches/compressions 3. Reflection about y-axis 4. Vertical stretches/compressions 5. Reflection about x-axis 6. Vertical shifts Although different ordering is possible, the order above will always work.	Use transformations to sketch the graph of $f(x) = -2(x + 3)^2 - 1$.

<table>
<tr><th>Key Concepts</th><th>Examples/Videos</th></tr>
<tr><td>

3.5 The Algebra of Functions; Composite Functions

Algebra of Functions

Let f and g be functions, then for all x such that both $f(x)$ and $g(x)$ are defined, the sum, difference, product, and quotient of f and g exist and are defined as follows:

1. The sum of f and g: $(f + g)(x) = f(x) + g(x)$
2. The difference of f and g: $(f - g)(x) = f(x) - g(x)$
3. The product of f and g: $(fg)(x) = f(x)g(x)$
4. The quotient of f and g: $\left(\frac{f}{g}\right)(x) = \frac{f(x)}{g(x)}$ for all $g(x) \neq 0$

</td><td>

Let $f(x) = \frac{12}{2x + 4}$ and $g(x) = \sqrt{x}$. Find each of the following.

a. $(f + g)(1)$ b. $(f - g)(1)$

c. $(fg)(4)$ d. $\left(\frac{f}{g}\right)(4)$

</td></tr>
<tr><td>

The notation $A \cap B$ is used to represent the intersection of sets A and B. When finding the intersection of two sets, we must find the set of all numbers common to *both* sets.

</td><td>

Find the intersection of the following sets and graph the set on a number line.

a. $[0, \infty) \cap (-\infty, 5]$

b. $((-\infty, -2) \cup (-2, \infty)) \cap [-4, \infty)$

</td></tr>
<tr><td>

Finding the Domain of Combined Functions

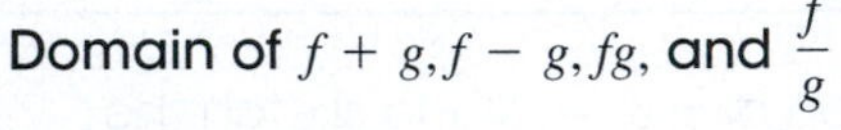

Domain of $f + g, f - g, fg$, and $\frac{f}{g}$

Suppose f is a function with domain A and g is a function with domain A, then

1. The domain of the sum, $f + g$, is the set of all x in $A \cap B$.
2. The domain of the difference, $f - g$, is the set of all x in $A \cap B$.
3. The domain of the product, fg, is the set of all x in $A \cap B$.
4. The domain of the quotient, $\frac{f}{g}$, is the set of all x in $A \cap B$ such that $g(x) \neq 0$.

</td><td>

Let $f(x) = \frac{x + 2}{x - 3}$ and $g(x) = \sqrt{4 - x}$.

Find a. $f + g$ b. $f - g$ c. fg d. $\frac{f}{g}$, and the domains of each.

</td></tr>
</table>

Key Concepts	Examples/Videos
Given functions f and g, the **composite function**, $f \circ g$, (also called the **composition of f and g**) is defined by $$(f \circ g)(x) = f(g(x)),$$ provided $g(x)$ is in the domain of f.	 Let $f(x) = 4x + 1$, $g(x) = \frac{x}{x-2}$ and $h(x) = \sqrt{x+3}$. a. Find the function $f \circ g$. b. Find the function $g \circ h$. c. Find the function $h \circ f \circ g$. d. Evaluate $(f \circ g)(4)$, or state that it is undefined. e. Evaluate $(g \circ h)(1)$, or state that it is undefined. f. Evaluate $(h \circ f \circ g)(6)$, or state that it is undefined.
Determining the Domain of $f \circ g$ **Step 1.** Find the domain of g. **Step 2.** Exclude from the domain of g all values of x for which $g(x)$ is not in the domain of f.	 Let $f(x) = \frac{-10}{x-4}$, $g(x) = \sqrt{5-x}$, and $h(x) = \frac{x-3}{x+7}$. a. Find the domain of $f \circ g$. b. Find the domain of $g \circ f$. c. Find the domain of $f \circ h$. d. Find the domain of $h \circ f$.
3.6 One-to-One Functions; Inverse Functions A function f is **one-to-one** if for any values $a \neq b$ in the domain of f, $f(a) \neq f(b)$. **Horizontal Line Test** If every horizontal line intersects the graph of a function f at most once, then f is one-to-one.	 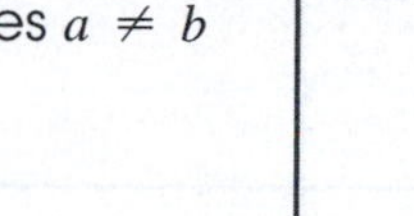 Determine if each function is one-to-one. a. 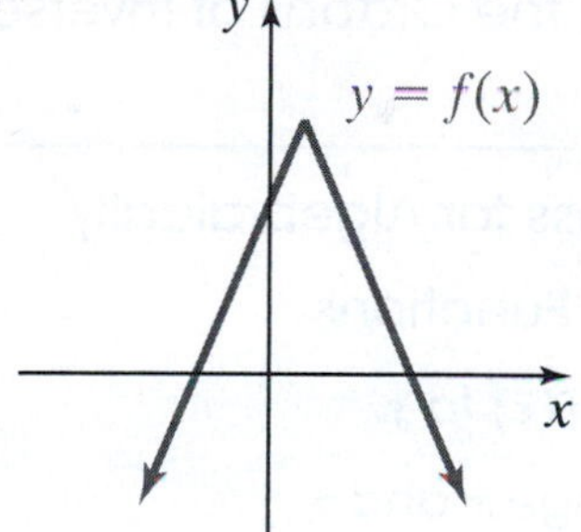 b. 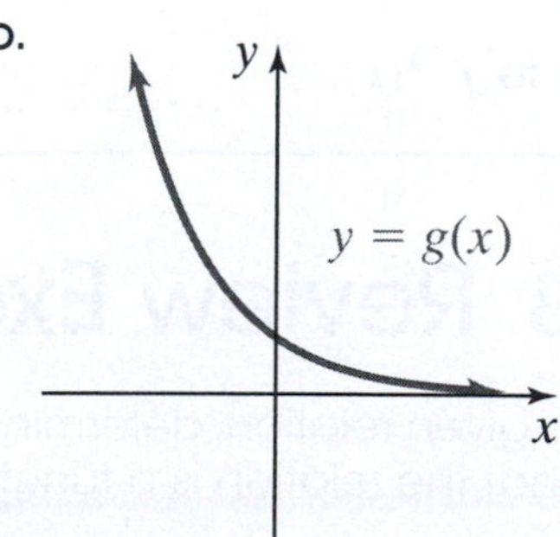 c. $f(x) = x^2 + 1, x \leq 0$ d. $f(x) = \begin{cases} 2x + 4 \text{ for } x \leq -1 \\ 2x - 6 \text{ for } x \geq 4 \end{cases}$

Key Concepts	Examples/Videos
Let f be a one-to-one function with domain A and range B. Then f^{-1} is the **inverse function of** f with domain B and range A. Furthermore, if $f(a) = b$, then $f^{-1}(b) = a$. Composition Cancellation Equations $f(f^{-1}(x)) = x$ for all x in the domain of f^{-1} and $f^{-1}(f(x)) = x$ for all x in the domain of f.	Show that $f(x) = \frac{x}{2x+3}$ and $g(x) = \frac{3x}{1-2x}$ are inverse functions using the composition cancellation equations.
Inverse Functions • Every one-to-one function f has an inverse function f^{-1}. • The domain of f is the same as the range of f^{-1}. • The range of f is the same as the domain of f^{-1}. • If $f(a) = b$ then $f^{-1}(b) = a$. • Given the graph of f, the graph of f^{-1} can be obtained by interchanging the coordinates of the ordered pairs that lie on the graph of f. • The graphs of f and f^{-1} are symmetric about the line $y = x$. Sketching the Graphs of Inverse Functions	Sketch the graph of $f(x) = x^2 + 1, x \leq 0$, and its inverse. Also state the domain and range of f and f^{-1}.
Four-Step Process for Algebraically Finding Inverse Functions Step 1. Change $f(x)$ to y. Step 2. Interchange x and y. Step 3. Solve for y. Step 4. Change y to $f^{-1}(x)$.	Find the inverse of the function $f(x) = \frac{2x}{1-5x}$, and state the domain and range of f and f^{-1}.

Chapter 3 Review Exercises

1. For the given relation, determine three additional ordered pairs which could be included such that the relation is a function. (There are infinitely many correct answers.)

$$\{(-2, 4), (0, 1), (1, 6), (3, 5), (7, -4), \boxed{?}, \boxed{?}, \boxed{?}\}$$

2. Evaluate the function at the indicated values.

$$f(t) = \frac{t}{t+1}$$

a. $f(-2)$ b. $f(7)$ c. $f(a)$

3. Determine the difference quotient $\frac{f(x+h)-f(x)}{h}$ for $f(x) = x^2 - 2x$.

4. Classify the function $f(x) = 8x^2 - x$ as a polynomial function, rational function, or root function, and then find the domain. Write the domain in interval notation.

5. Classify the function $h(x) = \frac{x^2+4}{x^2+x-42}$ as a polynomial function, rational function, or root function, and then find the domain. Write the domain in interval notation.

6. Find the x-intercept(s) and the y-intercept (if they exist) of the function $f(x) = \frac{x^2+x-12}{x^2+x}$.

7. Use the graph to determine the domain and range.

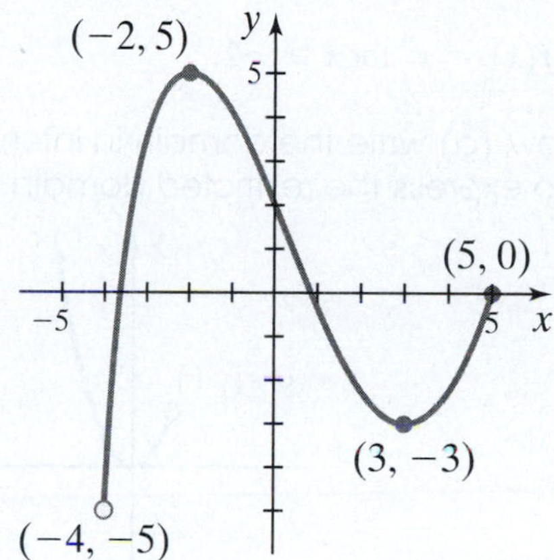

8. Use the graph to determine the interval(s) for which the function is (a) increasing, (b) decreasing, (c) constant.

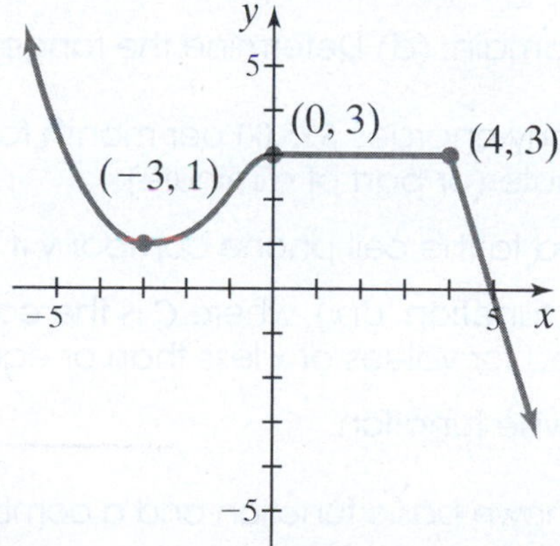

9. Determine whether the function $f(x) = x^5 - x^3$ is even, odd, or neither.

10. Use the graph of $y = f(x)$ to answer each of the following questions.
    a. What is the domain of f?
    b. What is the range of f?

c. On what interval(s) is f increasing, decreasing, or constant?

d. For what value(s) of x does f obtain a relative minimum? What are the relative minima?

e. For what value(s) of x does f obtain a relative maximum? What are the relative maxima?

f. What are the real zeros of f?

g. What is the y-intercept?

h. For what values of x is $f(x) \leq 0$?

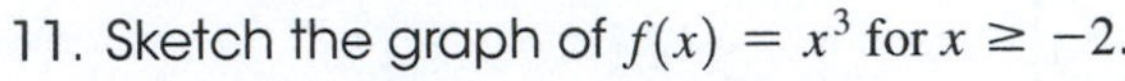

11. Sketch the graph of $f(x) = x^3$ for $x \geq -2$.

12. Given the graph below, (a) write the domain in interval notation; (b) Define the function using an inequality to express the restricted domain.

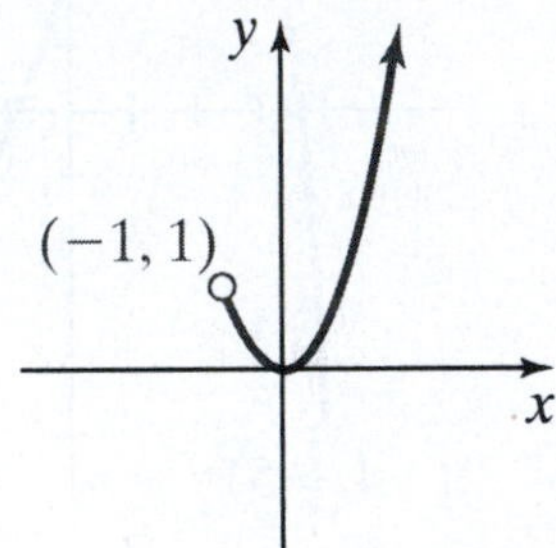

13. Let $f(x) = \begin{cases} 3 & \text{if } x < -1 \\ |x| & \text{if } -1 \leq x < 2 \\ 1 & \text{if } x \geq 2 \end{cases}$. (a) Find $f(-2), f(0)$, and $f(2)$; (b) Sketch the graph; (c) Determine the domain; (d) Determine the range

14. A cell phone company charges \$25.00 per month for the first 500 minutes of use and \$0.05 for each additional minute (or part of a minute).

    a. How much is owed to the cell phone company if a person used 502.5 minutes?

    b. Write a piecewise function. $C(x)$. where C is the cost charged per month as a function of the minutes, x, used for values of x less than or equal to 505 minutes.

    c. Sketch the piecewise function.

15. Use the graph of a known basic function and a combination of horizontal and vertical shifts to sketch the graph of $f(x) = (x + 1)^2 - 2$

16. The graph below was created by using one horizontal shift and one vertical shift of the graph of the basic function $y = x^3$. Write a function that describes the graph below.

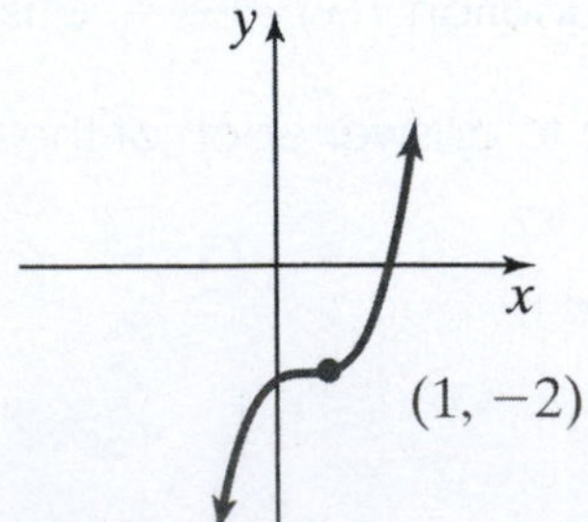

17. Use the graph of $y = f(x)$ to sketch the graph of $y = -2f(x + 1) + 2$

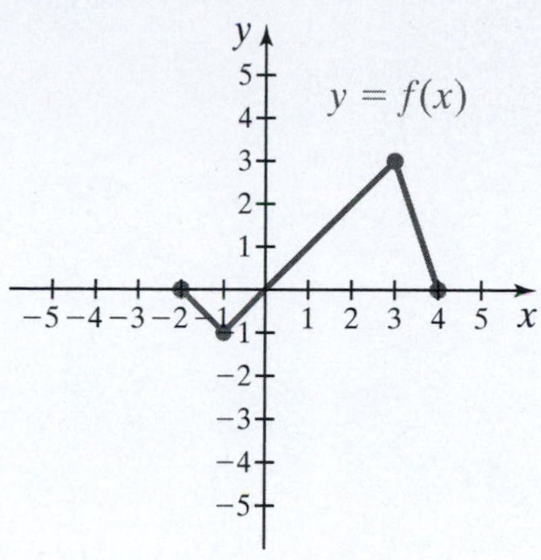

18. Sketch the graph of the following piecewise-defined function.

$$f(x) = \begin{cases} |x - 2| & \text{if } x \leq 2 \\ \sqrt{x - 2} & \text{if } x > 2 \end{cases}$$

19. Given that $f(x) = \sqrt{x + 1}$ and $g(x) = x^2 - 2$, evaluate $(f + g)(3)$.

20. Given that $f(x) = \dfrac{x}{x^2 - 4}$ and $g(x) = \dfrac{1}{x - 3}$, find (a) $f + g$, (b) $f - g$, (c) fg, (d) $\dfrac{f}{g}$, and find the domains of each.

21. Evaluate $(f \circ g)(6)$ given that $f(x) = 3x + 1$ and $g(x) = \sqrt{x + 3}$.

22. Given that $f(x) = \dfrac{1}{x - 2}$ and $g(x) = \sqrt{4 - x}$, find the domain of $(f \circ g)(x)$ and $(g \circ f)(x)$.

23. Determine if the function $f(x) = (x - 1)^2,\ x \geq 1$ is one-to-one.

24. Use the graph of f to sketch the graph of f^{-1}. Use the graphs to determine the domain and range of each function.

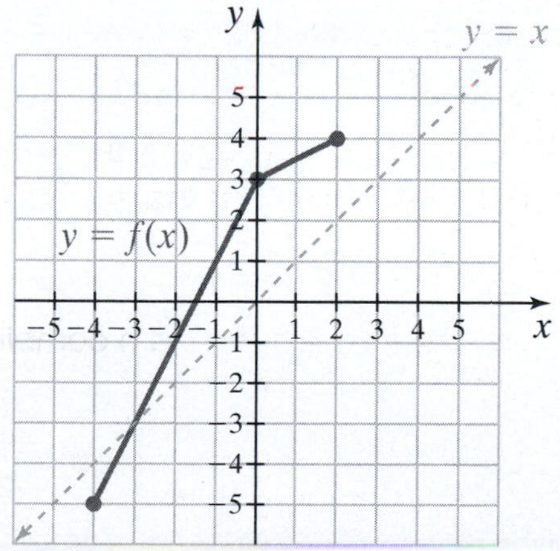

25. Write an equation for the inverse of $f(x) = \dfrac{8x - 1}{7 - 5x}$. Then state the domain and range of f and f^{-1}.

CHAPTER FOUR

Polynomial and Rational Functions

CHAPTER FOUR CONTENTS

4.1 Quadratic Functions

THINGS TO KNOW

Before working through this section, be sure that you are familiar with the following concepts:

VIDEO ANIMATION INTERACTIVE

You Try It

1. Solving Quadratic Equations by Factoring and the Zero Product Property (Section 1.4) 

You Try It

2. Solving Quadratic Equations by Completing the Square (Section 1.4) 

You Try It

3. Solving Quadratic Equations Using the Quadratic Formula (Section 1.4) 

You Try It

4. Determining the Domain and Range of a Function From Its Graph (Section 3.2) 

You Try It

5. Using Combinations of Transformations to Graph Functions (Section 3.4) 

OBJECTIVES

1 Understanding the Definition of a Quadratic Function and Its Graph

2 Graphing Quadratic Functions Written in Vertex Form

3 Graphing Quadratic Functions by Completing the Square

(eText Screens 4.1-1–4.1-35)

4 Graphing Quadratic Functions Using the Vertex Formula

5 Determining the Equation of a Quadratic Function Given Its Graph

SECTION 4.1 EXERCISES

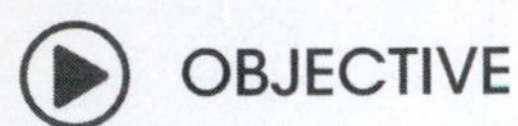

OBJECTIVE 1 UNDERSTANDING THE DEFINITION OF A QUADRATIC FUNCTION AND ITS GRAPH

In Section 1.4, we learned how to solve quadratic equations of the form $ax^2 + bx + c = 0, a \neq 0$. In this section, we learn about *quadratic functions*.

> **Definition** Quadratic Function
>
> A **quadratic function** is a function that can be written in the form $f(x) = ax^2 + bx + c$, where a, b, and c are real numbers with $a \neq 0$. Every quadratic function has a graph called a *parabola*.

The function $f(x) = x^2$ is a quadratic function with $a = 1, b = 0,$ and $c = 0$ whose graph is shown in **Figure 1a**. The function $g(x) = -x^2$ is a quadratic function with $a = -1, b = 0,$ and $c = 0$ whose graph is shown in **Figure 1b**. Notice that both graphs are parabolas.

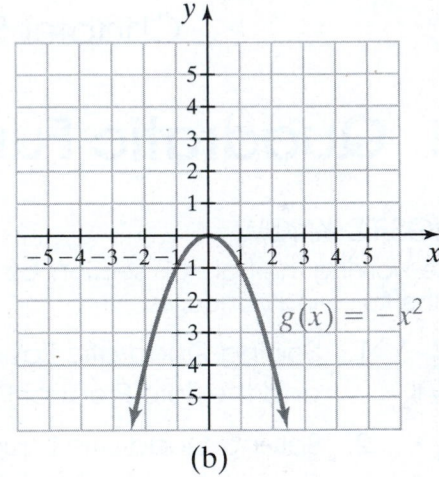

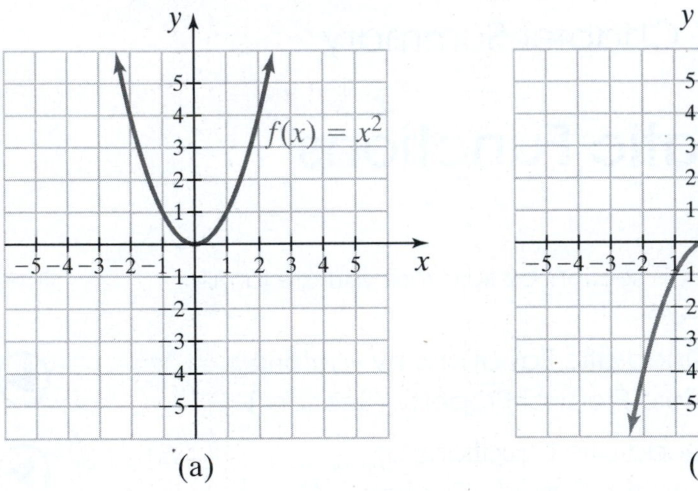

Figure 1 Graphs of $f(x) = x^2$ and $g(x) = -x^2$.

A parabola either opens up or opens down, depending on the **leading coefficient**, a. If $a > 0$, as in Figure 1a, the parabola will "open up." If $a < 0$, as in Figure 1b, the parabola will "open down."

Example 1 Determine Whether the Graph of a Quadratic Function Opens Up or Opens Down

Without graphing, determine whether the graph of the quadratic function $f(x) = -3x^2 + 6x + 1$ opens up or opens down.

Solution Because the leading coefficient is $a = -3 < 0$, the graph of the quadratic function $f(x) = -3x^2 + 6x + 1$ must open down. ●

Using Technology

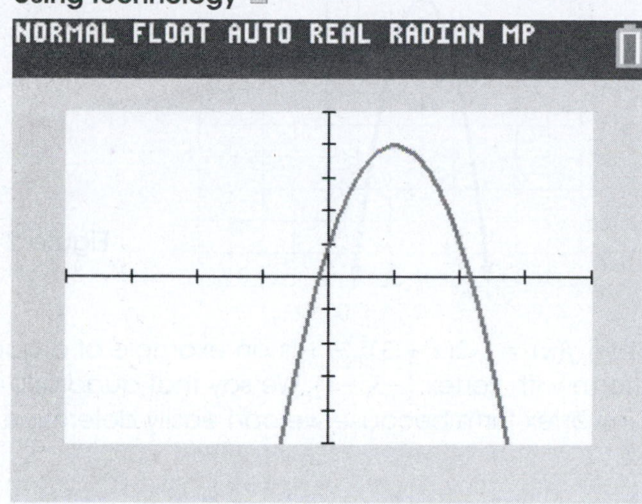

The graph of the quadratic function $f(x) = -3x^2 + 6x + 1$ is seen here using a graphing utility. You can see that the graph "opens down."

You Try It Work through the following You Try It problem.

Work Exercises 1–4 in this textbook or in the MyLab Math Study Plan.

CHARACTERISTICS OF A PARABOLA

Before we learn how to sketch the graphs of other quadratic functions, it is important to identify the five basic characteristics of a parabola. Work through the following **animation**, and click on each characteristic to get a detailed description.

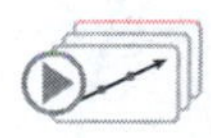

1. Vertex
2. Axis of symmetry
3. y-Intercept
4. x-Intercept(s) or real zeros
5. Domain and range

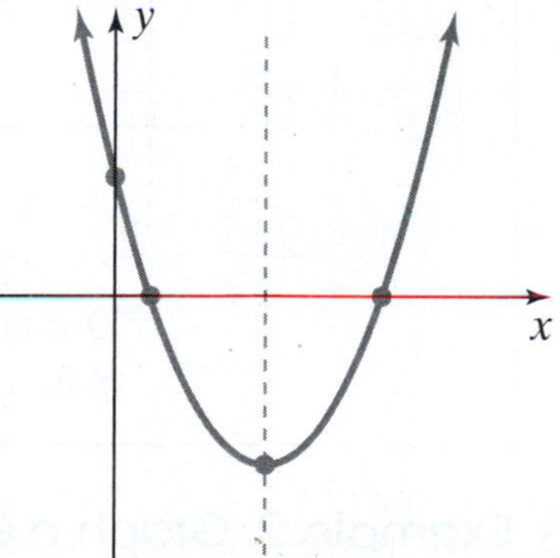

OBJECTIVE 2 GRAPHING QUADRATIC FUNCTIONS WRITTEN IN VERTEX FORM

In **Section 3.4, Example 6**, we sketched the graph of $f(x) = -2(x+3)^2 - 1$ using **transformation techniques**. Watch the **animation** to see how to sketch the function $f(x) = -2(x + 3)^2 - 1$ using transformations. The graph of $f(x) = -2(x + 3)^2 - 1$ is sketched in Figure 2. Note that the domain is $(-\infty, \infty)$ and the range is $(-\infty, -1]$.

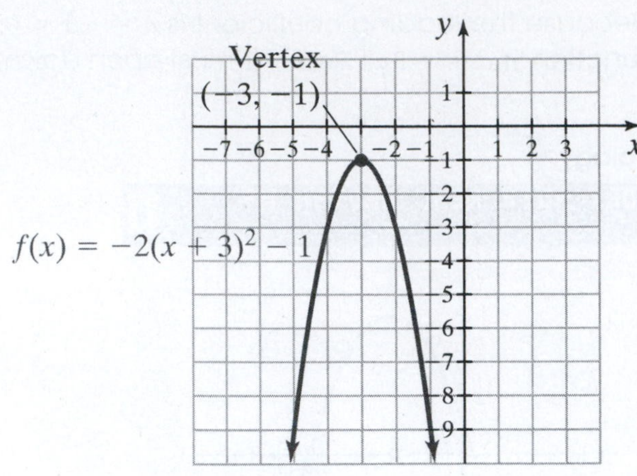

Figure 2 Graph of $f(x) = -2(x + 3)^2 - 1$.

The function $f(x) = -2(x + 3)^2 - 1$ is an example of a quadratic function that is in **vertex form** with vertex $(-3, -1)$. We say that quadratic functions written in this form are in vertex form because we can easily determine the coordinates of the vertex.

Definition Vertex Form of a Quadratic Function

A quadratic function is in vertex form if it is written as $f(x) = a(x - h)^2 + k$. The graph is a parabola with vertex (h, k). The parabola "opens up" if $a > 0$. The parabola "opens down" if $a < 0$.

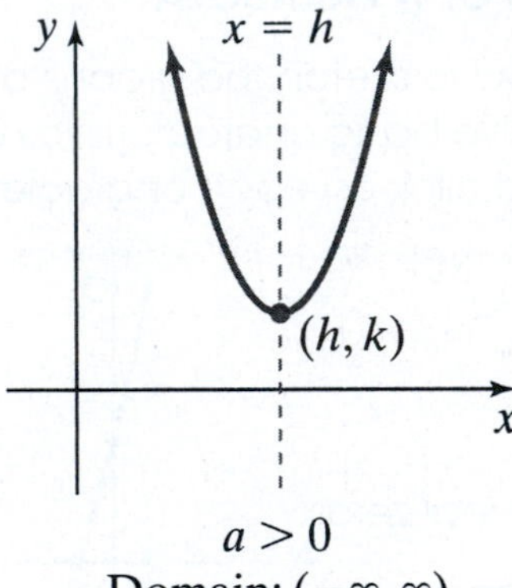

$a > 0$
Domain: $(-\infty, \infty)$
Range: $[k, \infty)$

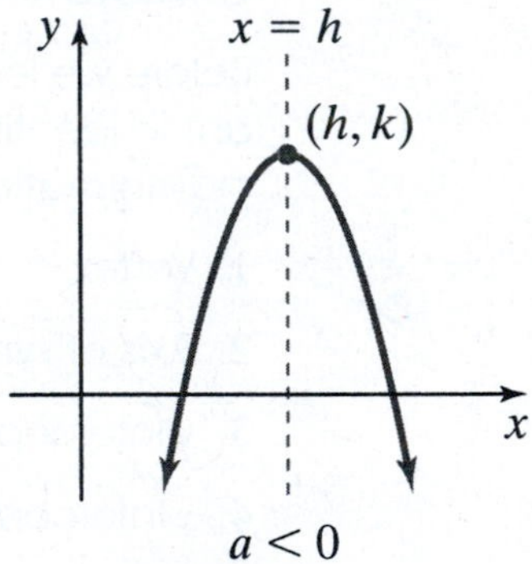

$a < 0$
Domain: $(-\infty, \infty)$
Range: $(-\infty, k]$

Example 2 Graph a Quadratic Function that Is in Vertex Form

Given that the quadratic function $f(x) = -(x - 2)^2 - 4$ is in vertex form, address the following:

a. What are the coordinates of the vertex?

b. Does the graph "open up" or "open down"?

c. What is the equation of the axis of symmetry?

d. Find any x-intercepts.

e. Find the y-intercept.

f. Sketch the graph.

g. State the domain and range in interval notation.

Solution

a. The quadratic function $f(x) = -(x-2)^2 - 4$ is in vertex form. Therefore, we can find the vertex by determining the values of h and k.

$f(x) = a(x-h)^2 + k$ Quadratic function in vertex form

$a = -1$ $h = 2$ $k = -4$

$f(x) = -(x-2)^2 - 4$ Given function

Because $h = 2$ and $k = -4$, the vertex of this parabola is $(2, -4)$.

b. The leading coefficient is $a = -1 < 0$. Therefore, the parabola opens down.

c. The equation of the axis of symmetry is $x = 2$.

d. To find any x-intercepts, we must determine the real solution(s) to the equation $f(x) = 0$.

$f(x) = -(x-2)^2 - 4$ Write the original function.

$0 = -(x-2)^2 - 4$ Replace $f(x)$ with 0 in $f(x) = -(x-2)^2 - 4$.

$(x-2)^2 = -4$ Add $(x-2)^2$ to both sides.

$x - 2 = \pm\sqrt{-4}$ Use the square root property.

$x - 2 = \pm 2i$ Recall that $\sqrt{-1} = i$ so $\sqrt{-4} = \sqrt{4}\sqrt{-1} = 2i$.

$x = 2 \pm 2i$ Add 2 to both sides.

Because the equation $f(x) = 0$ has no real solutions, the parabola has no x-intercepts.

e. The y-intercept is $f(0) = -(0-2)^2 - 4 = -4 - 4 = -8$.

f. The vertex is $(2, -4)$, the parabola opens down, the equation of the axis of symmetry is $x = 2$, and the y-intercept is -8. Using this information, we can sketch the parabola as shown in **Figure 3**.

g. The domain of f is the interval $(-\infty, \infty)$. The range of f is the interval $(-\infty, -4]$.

Watch the **video** to see each step worked out in detail.

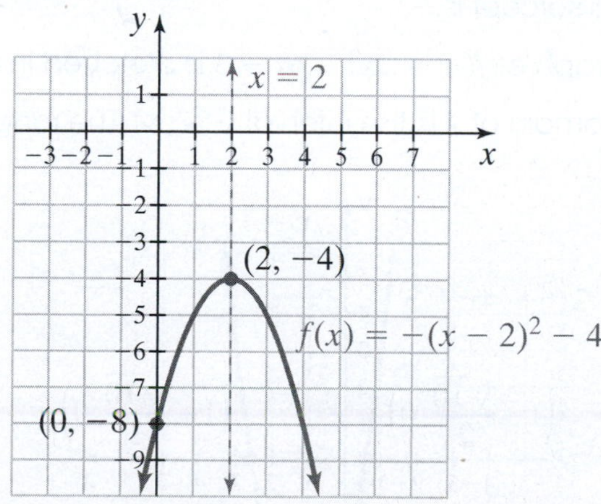

Figure 3 Graph of $f(x) = -(x-2)^2 - 4$.

You Try It Work through the following You Try It problem.

Work Exercises 5–10 in this textbook or in the MyLab Math Study Plan.

OBJECTIVE 3 GRAPHING QUADRATIC FUNCTIONS BY COMPLETING THE SQUARE

We can see from Example 2 that if a quadratic function is in vertex form $f(x) = a(x - h)^2 + k$, it is fairly straightforward to determine its graph. Therefore, to graph a quadratic function of the form $f(x) = ax^2 + bx + c$, we convert it into vertex form by completing the square as in Example 3.

Example 3 Write a Quadratic Function in Vertex Form by Completing the Square and Sketch the Graph

Rewrite the quadratic function $f(x) = 2x^2 - 4x - 3$ in vertex form, and then complete a) through g), as in Example 2.

Solution Our goal is to rewrite $f(x) = 2x^2 - 4x - 3$ in the form $f(x) = a(x - h)^2 + k$. To do this, we must complete the square.

$f(x) = 2x^2 - 4x - 3$	Write the original function.
$f(x) = (2x^2 - 4x) - 3$	Isolate the constant.
$f(x) = 2(x^2 - 2x) - 3$	Factor out the leading coefficient 2 from $2x^2 - 4x$.
$f(x) = 2(x^2 - 2x + 1) - 3 - 2$	Complete the square of $x^2 - 2x$. Note that we are adding $2 \cdot 1 = 2$, so we must subtract 2 as well!
$f(x) = 2(x - 1)^2 - 5$	Rewrite $(x^2 - 2x + 1)$ as a binomial squared.

Thus, the quadratic function $f(x) = 2x^2 - 4x - 3$ written in vertex form is $f(x) = 2(x - 1)^2 - 5$. Watch the **video** to verify the following:

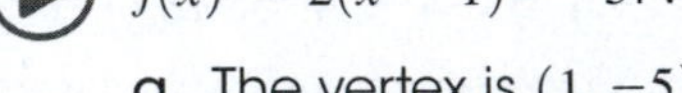

a. The vertex is $(1, -5)$.

b. The parabola opens up.

c. The equation of the axis of symmetry is $x = 1$.

d. The two x-intercepts are $x = 1 + \frac{\sqrt{10}}{2} \approx 2.5811$ and $x = 1 - \frac{\sqrt{10}}{2} \approx -0.5811$.

e. The y-intercept is -3.

f. The graph of $f(x) = 2x^2 - 4x - 3$ is sketched in **Figure 4**.

g. The domain of f is the interval $(-\infty, \infty)$. The range of f is the interval $[-5, \infty)$.

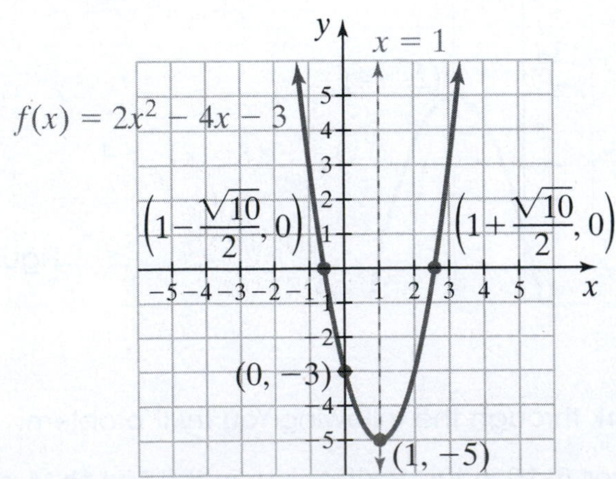

Figure 4 Graph of $f(x) = 2x^2 - 4x - 3$. ●

Try sketching the graphs of some quadratic functions on your own by exploring the guided visualization below. Click on the icon below to sketch quadratic functions of the form $f(x) = a(x - h)^2 + k$. Choose your own values of a, h, and k.

Sketching Quadratic Functions

You Try It **Work through the following You Try It problem.**

Work Exercises 11–20 in this textbook or in the MyLab Math Study Plan.

OBJECTIVE 4 GRAPHING QUADRATIC FUNCTIONS USING THE VERTEX FORMULA

We can establish a "formula" for the vertex by completing the square on the quadratic function $f(x) = ax^2 + bx + c$. Work through the **video** to verify that the function $f(x) = ax^2 + bx + c$ is equivalent to $f(x) = a\left(x + \frac{b}{2a}\right)^2 + c - \frac{b^2}{4a}$.

We can now compare the function $f(x) = a\left(x + \frac{b}{2a}\right)^2 + c - \frac{b^2}{4a}$ to $f(x) = a(x - h)^2 + k$.

$$f(x) = a\left(x - \left(-\frac{b}{2a}\right)\right)^2 + \underbrace{c - \frac{b^2}{4a}}$$

$$\downarrow \qquad \downarrow$$

$$f(x) = a(x - \quad h)^2 \quad + \quad k$$

We see that the coordinates of the vertex must be $\left(-\frac{b}{2a}, c - \frac{b^2}{4a}\right)$. Because the y-coordinate of the vertex can be found more easily by evaluating the function at the x-coordinate of the vertex than by memorizing the expression $c - \frac{b^2}{4a}$, we write the coordinates of the vertex as $\left(-\frac{b}{2a}, f\left(-\frac{b}{2a}\right)\right)$.

> **Formula for the Vertex of a Parabola**
>
> Given a quadratic function of the form $f(x) = ax^2 + bx + c$, $a \neq 0$, the vertex of the parabola is $\left(-\frac{b}{2a}, f\left(-\frac{b}{2a}\right)\right)$.

Example 4 Graph a Quadratic Function Using the Vertex Formula and Intercepts

Given the quadratic function $f(x) = -2x^2 - 4x + 5$, address the following:

a. Use the vertex formula to determine the vertex.

b. Does the graph "open up" or "open down"?

c. What is the equation of the axis of symmetry?

d. Find any x-intercepts.

e. Find the y-intercept.

f. Sketch the graph.

g. State the domain and range in interval notation.

Solution

a. The x-coordinate of the vertex is $-\frac{b}{2a} = -\frac{(-4)}{2(-2)} = -1$. The y-coordinate of the vertex is $f\left(-\frac{b}{2a}\right) = f(-1) = -2(-1)^2 - 4(-1) + 5 = -2 + 4 + 5 = 7$. Therefore, the vertex is $(-1, 7)$.

b. Because $a = -2 < 0$, the parabola must open down.

c. The axis of symmetry is $x = -1$.

d. To find the x-intercepts, we solve the quadratic equation $-2x^2 - 4x + 5 = 0$. Read these steps to verify that the x-intercepts are $x = -1 + \frac{\sqrt{14}}{2} \approx 0.8708$ and $x = -1 - \frac{\sqrt{14}}{2} \approx -2.8708$.

e. The y-intercept is $f(0) = 5$.

f. Plotting the vertex and all intercepts, we get the graph sketched in **Figure 5**.

g. The domain of f is the interval $(-\infty, \infty)$. The range of f is the interval $(-\infty, 7]$.

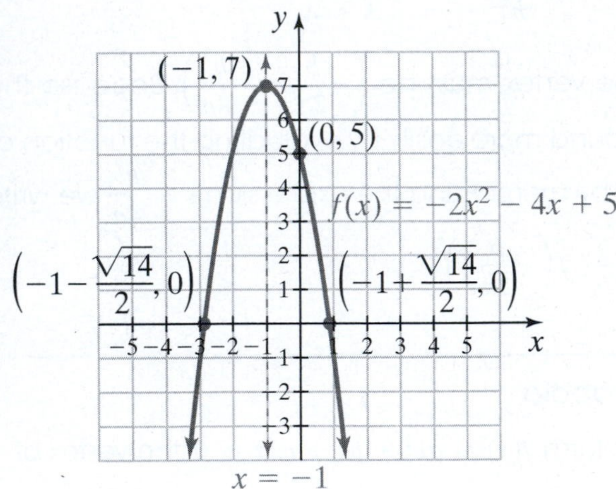

Figure 5 Graph of $f(x) = -2x^2 - 4x + 5$.

Using Technology 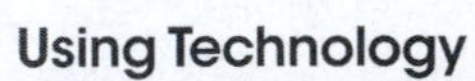

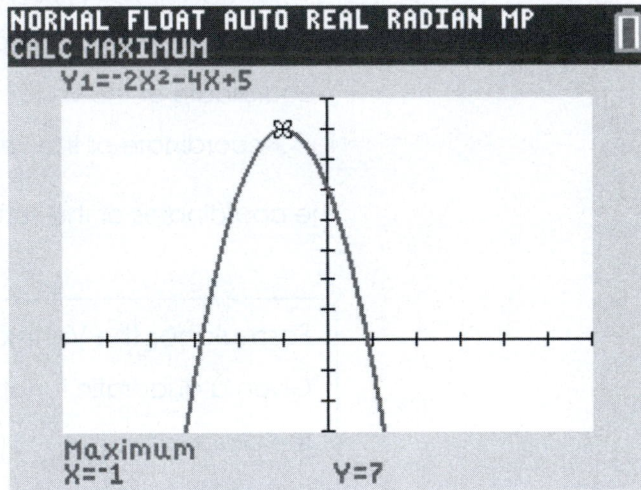

Using a graphing utility and the maximum function, we can verify that the vertex is $(-1, 7)$.

You Try It Work through the following You Try It problem.

Work Exercises 21–30 in this textbook or in the MyLab Math Study Plan.

OBJECTIVE 5 DETERMINING THE EQUATION OF A QUADRATIC FUNCTION GIVEN ITS GRAPH

Given the graph of a parabola, it may be necessary to find the equation of the quadratic function that it represents.

Example 5 Determine the Equation of a Quadratic Function Given Its Graph

Analyze the graph to address the following about the quadratic function it represents.

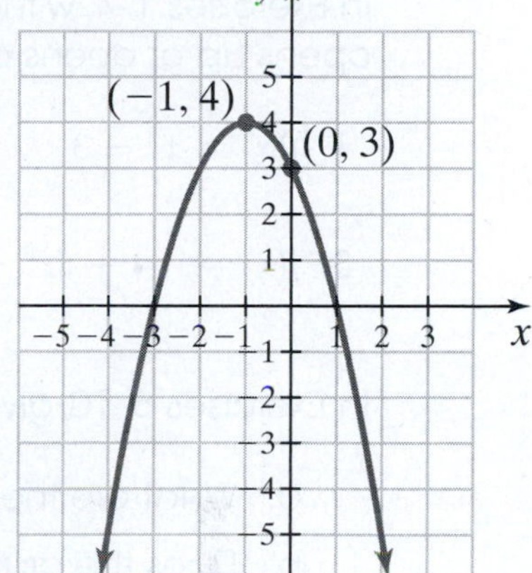

a. Is the leading coefficient positive or negative?

b. What is the value of h? What is the value of k?

c. What is the value of the leading coefficient a?

d. Write the equation of the function in vertex form $f(x) = a(x - h)^2 + k$.

e. Write the equation of the function in the form $f(x) = ax^2 + bx + c$.

Solution

a. The leading coefficient is negative because the parabola opens down.

b. The value of h is the x-coordinate of the vertex, so $h = -1$. The value of k is the y-coordinate of the vertex, so $k = 4$.

c. Substitute the values of $h = -1$ and $k = 4$ into the vertex form $f(x) = a(x - h)^2 + k$ to get $f(x) = a(x - (-1))^2 + 4$ or $f(x) = a(x + 1)^2 + 4$. To determine the value of a, we can use the fact that the graph contains the point $(0, 3)$, and therefore, $f(0) = 3$.

$f(x) = a(x + 1)^2 + 4$	Use the vertex form with $h = -1$ and $k = 4$.
$f(0) = a(0 + 1)^2 + 4$	Evaluate $f(0)$.
$f(0) = 3$	The graph contains the point $(0, 3)$.
$a(0 + 1)^2 + 4 = 3$	Equate the two expressions for $f(0)$.
$a = -1$	Solve for a.

d. Substituting the values of $a = -1$, $h = -1$, and $k = 4$ into the equation $f(x) = a(x - h)^2 + k$, we get $f(x) = -(x + 1)^2 + 4$.

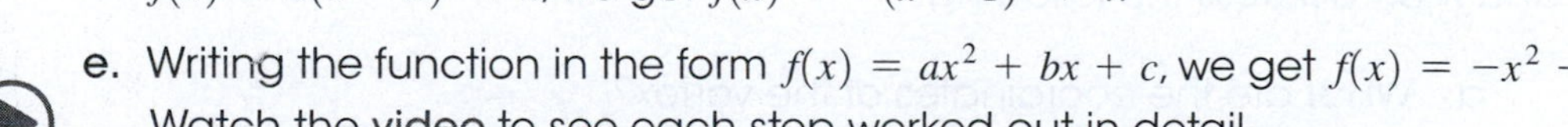

e. Writing the function in the form $f(x) = ax^2 + bx + c$, we get $f(x) = -x^2 - 2x + 3$.

Watch the video to see each step worked out in detail.

You Try It Work through the following You Try It problem.

Work Exercises 31–36 in this textbook or in the MyLab Math Study Plan.

4.1 Exercises

Skill Check Exercises

For exercises SCE-1 through SCE-3, decide what number must be added to each binomial to make a perfect square trinomial.

SCE-1. $x^2 + 10x$ **SCE-2.** $x^2 - 5x$ **SCE-3.** $x^2 + \frac{5}{3}x$

In Exercises 1–4, without graphing, determine whether the graph of each quadratic function opens up or opens down.

1. $f(x) = x^2 - 3$
2. $f(x) = -2x^2 + 4x + 1$
3. $f(x) = -4 + 5x^2 - 2x$
4. $f(x) = -\frac{2}{3}x^2 + x + 7$

In Exercises 5–10, given the quadratic function in vertex form, address the following.

a. What are the coordinates of the vertex?

b. Does the graph "open up" or "open down"?

c. What is the equation of the axis of symmetry?

d. Find any x-intercepts.

e. Find the y-intercept.

f. Sketch the graph.

g. State the domain and range.

SbS 5. $f(x) = (x - 2)^2 - 4$

SbS 6. $f(x) = -(x + 1)^2 - 9$

SbS 7. $f(x) = -2(x - 3)^2 + 2$

SbS 8. $f(x) = \frac{1}{2}(x + 2)^2 + 2$

SbS 9. $f(x) = -\frac{1}{4}(x - 4)^2 + 2$

SbS 10. $f(x) = 3\left(x + \frac{1}{3}\right)^2 - 4$

In Exercises 11–20, first rewrite the quadratic function in vertex form by completing the square, and then address the following.

a. What are the coordinates of the vertex?

b. Does the graph "open up" or "open down"?

c. What is the equation of the axis of symmetry?

d. Find any x-intercepts.

e. Find the y-intercept.

f. Sketch the graph.

g. State the domain and range in interval notation.

SbS 11. $f(x) = x^2 - 8x$

SbS 12. $f(x) = -x^2 - 4x + 12$

SbS 13. $f(x) = 3x^2 + 6x - 4$

SbS 14. $f(x) = 2x^2 - 5x - 3$

SbS 15. $f(x) = -x^2 + 2x - 6$

SbS 16. $f(x) = \frac{1}{2}x^2 + 6x + 1$

SbS 17. $f(x) = -\frac{1}{3}x^2 - 2x + 5$

SbS 18. $f(x) = -3x^2 + 7x + 5$

SbS 19. $f(x) = x^2 + \frac{8}{3}x - 1$

SbS 20. $f(x) = -\frac{1}{4}x^2 + 6x - 1$

In Exercises 21–30, use the quadratic function to address the following.

a. Use the vertex formula to determine the vertex.
b. Does the graph "open up" or "open down"?
c. What is the equation of the axis of symmetry?
d. Find any x-intercepts.
e. Find the y-intercept.
f. Sketch the graph.
g. State the domain and range in interval notation.

SbS 21. $f(x) = x^2 - 8x$

SbS 22. $f(x) = -x^2 - 4x + 8$

SbS 23. $f(x) = 3x^2 + 6x - 4$

SbS 24. $f(x) = 2x^2 - 5x - 3$

SbS 25. $f(x) = -x^2 + 2x - 6$

SbS 26. $f(x) = \frac{1}{2}x^2 + 6x + 1$

SbS 27. $f(x) = -\frac{1}{3}x^2 - 9x + 5$

SbS 28. $f(x) = -3x^2 + 7x + 5$

SbS 29. $f(x) = x^2 + \frac{8}{3}x - 1$

SbS 30. $f(x) = -\frac{1}{4}x^2 + 6x - 1$

In Exercises 31–36, analyze the graph to address the following about the quadratic function it represents.

a. Is the leading coefficient positive or negative?
b. What is the value of h? What is the value of k?
c. What is the value of the leading coefficient a?
d. Write the equation of the function in vertex form $f(x) = a(x - h)^2 + k$.
e. Write the equation of the function in the form $f(x) = ax^2 + bx + c$.

SbS 31.

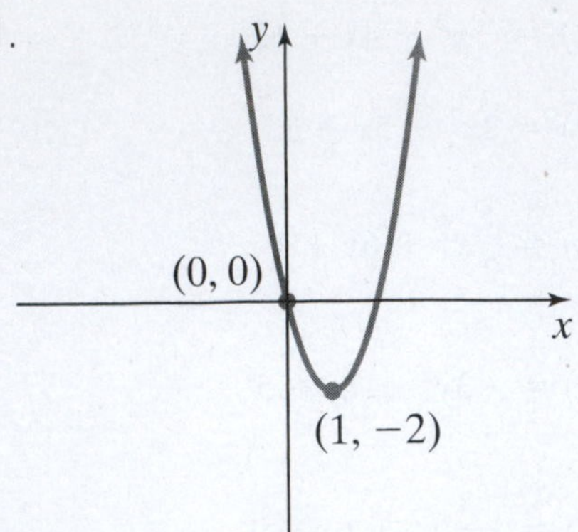

SbS 32.

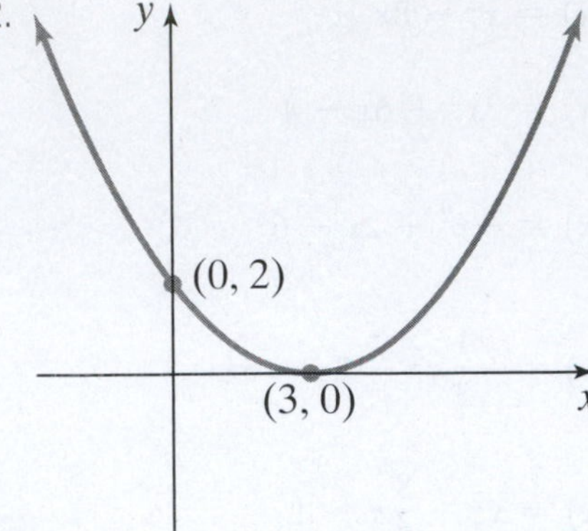

SbS 33.

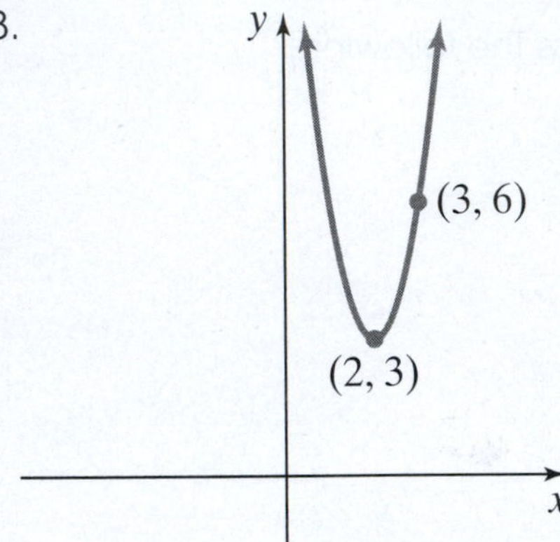

SbS 34.

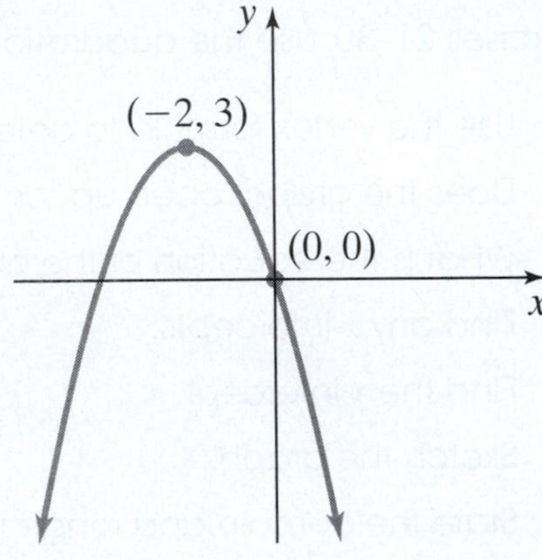

SbS 35.

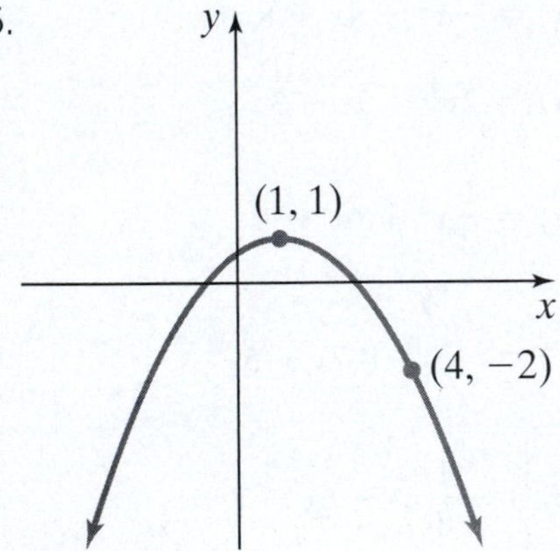

SbS 36.

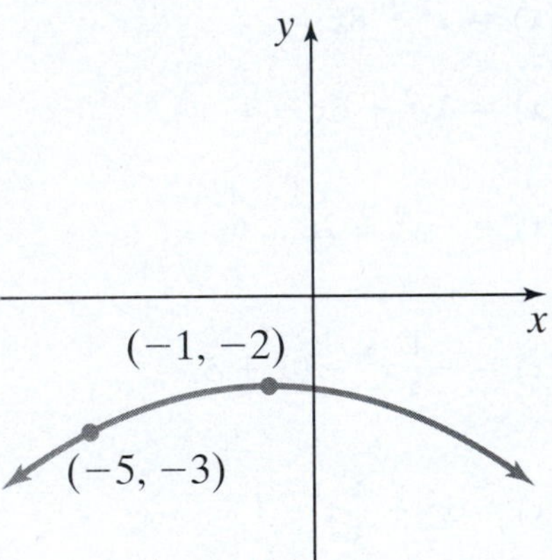

Brief Exercises

In Exercises 37–42, given the quadratic function in vertex form, address the following.

a. What are the coordinates of the vertex.

b. Does the graph "open up" or "open down"?

c. What is the equation of the axis of symmetry?

37. $f(x) = (x - 2)^2 - 4$

38. $f(x) = -(x + 1)^2 - 9$

39. $f(x) = -2(x - 3)^2 + 2$

40. $f(x) = \frac{1}{2}(x + 2)^2 + 2$

41. $f(x) = -\frac{1}{4}(x - 4)^2 + 2$

42. $f(x) = 3\left(x + \frac{1}{3}\right)^2 - 4$

In Exercises 43–48, given the quadratic function in vertex form, address the following.

a. Find any x-intercepts.

b. Find the y-intercept.

43. $f(x) = (x - 2)^2 - 4$

44. $f(x) = -(x + 1)^2 - 9$

45. $f(x) = -2(x - 3)^2 + 2$

46. $f(x) = \frac{1}{2}(x + 2)^2 + 2$

47. $f(x) = -\frac{1}{4}(x - 4)^2 + 2$

48. $f(x) = 3\left(x + \frac{1}{3}\right)^2 - 4$

In Exercises 49–54, given the quadratic function in vertex form, address the following.

a. Sketch the graph.

b. State the domain and range in interval notation.

49. $f(x) = (x - 2)^2 - 4$

50. $f(x) = -(x + 1)^2 - 9$

51. $f(x) = -2(x - 3)^2 + 2$

52. $f(x) = \frac{1}{2}(x + 2)^2 + 2$

53. $f(x) = -\frac{1}{4}(x - 4)^2 + 2$

54. $f(x) = 3\left(x + \frac{1}{3}\right)^2 - 4$

In Exercises 55–59, first rewrite the quadratic function in vertex form by completing the square, and then address the following.

a. What are the coordinates of the vertex.

b. Does the graph "open up" or "open down"?

c. What is the equation of the axis of symmetry?

55. $f(x) = 3x^2 + 6x - 4$

56. $f(x) = -x^2 + 2x - 6$

57. $f(x) = \frac{1}{2}x^2 + 6x + 1$

58. $f(x) = -3x^2 + 7x + 5$

59. $f(x) = -\frac{1}{4}x^2 + 6x - 1$

In Exercises 60–64, first rewrite the quadratic function in vertex form by completing the square, and then address the following.

a. Find any x-intercepts.

b. Find the y-intercept.

60. $f(x) = 3x^2 + 6x - 4$

61. $f(x) = -x^2 + 2x - 6$

62. $f(x) = \frac{1}{2}x^2 + 6x + 1$

63. $f(x) = -3x^2 + 7x + 5$

64. $f(x) = -\frac{1}{4}x^2 + 6x - 1$

In Exercises 65–69, first rewrite the quadratic function in vertex form by completing the square, and then address the following.

a. Sketch the graph.

b. State the domain and range in interval notation.

65. $f(x) = 3x^2 + 6x - 4$

66. $f(x) = -x^2 + 2x - 6$

67. $f(x) = \frac{1}{2}x^2 + 6x + 1$

68. $f(x) = -3x^2 + 7x + 5$

69. $f(x) = -\frac{1}{4}x^2 + 6x - 1$

In Exercises 70–74, use the quadratic function to address the following.

a. Use the vertex formula to determine the coordinates of the vertex.

b. Does the graph "open up" or "open down"?

c. What is the equation of the axis of symmetry?

70. $f(x) = 3x^2 + 6x - 4$

71. $f(x) = -x^2 + 2x - 6$

72. $f(x) = \frac{1}{2}x^2 + 6x + 1$

73. $f(x) = -3x^2 + 7x + 5$

74. $f(x) = -\frac{1}{4}x^2 + 6x - 1$

In Exercise 75–79, use the quadratic function to address the following.

a. Find any x-intercepts.

b. Find any y-intercepts.

75. $f(x) = 3x^2 + 6x - 4$

76. $f(x) = -x^2 + 2x - 6$

77. $f(x) = \frac{1}{2}x^2 + 6x + 1$

78. $f(x) = -3x^2 + 7x + 5$

79. $f(x) = -\frac{1}{4}x^2 + 6x - 1$

In Exercises 80–84, use the quadratic function to address the following.

a. Sketch the graph

b. State the domain and range the interval notation.

80. $f(x) = 3x^2 + 6x - 4$

81. $f(x) = -x^2 + 2x - 6$

82. $f(x) = \frac{1}{2}x^2 + 6x + 1$

83. $f(x) = -3x^2 + 7x + 5$

84. $f(x) = -\frac{1}{4}x^2 + 6x - 1$

In Exercises 85–94, use the quadratic function to address the following.

a. Does the function have a maximum or minimum value?

b. What is this maximum or minimum value?

85. $f(x) = x^2 - 8x$

86. $f(x) = -x^2 - 4x + 8$

87. $f(x) = 3x^2 + 6x - 4$

88. $f(x) = 2x^2 - 5x - 3$

89. $f(x) = -x^2 + 2x - 6$

90. $f(x) = \frac{1}{2}x^2 + 6x + 1$

91. $f(x) = -\frac{1}{2}x^2 - 9x + 5$

92. $f(x) = -3x^2 + 7x + 5$

93. $f(x) = x^2 + \frac{8}{3}x - 1$

94. $f(x) = -\frac{1}{4}x^2 + 6x - 1$

In Exercises 95–100, given the graph of a quadratic function, write the equation of the function is vertex form $f(x) = a(x - h)^2 + k$.

95.

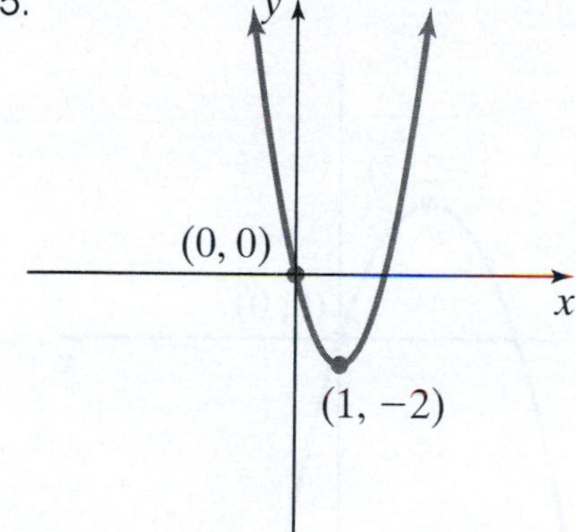

96.

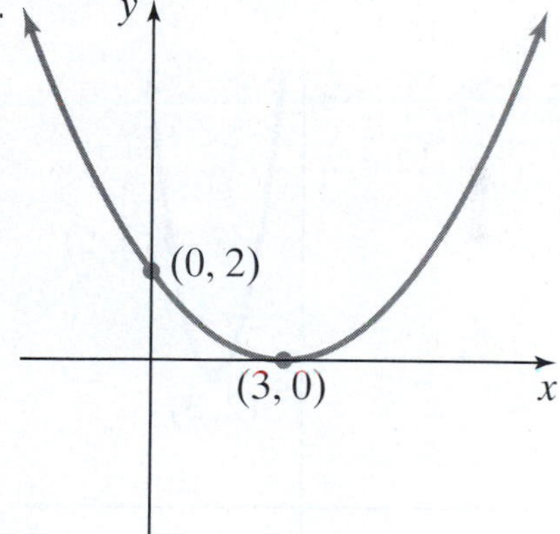

97.

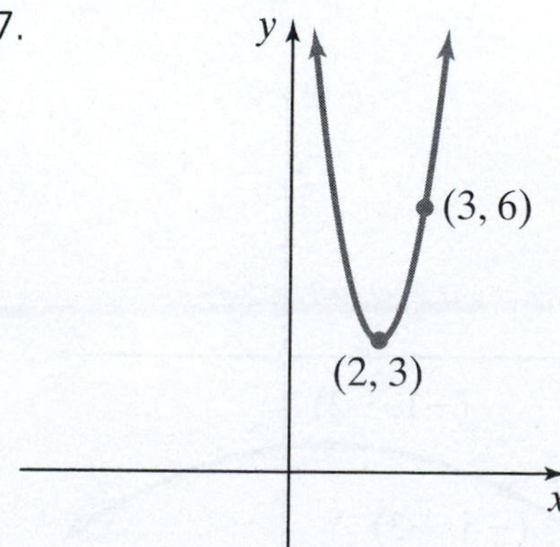

98.

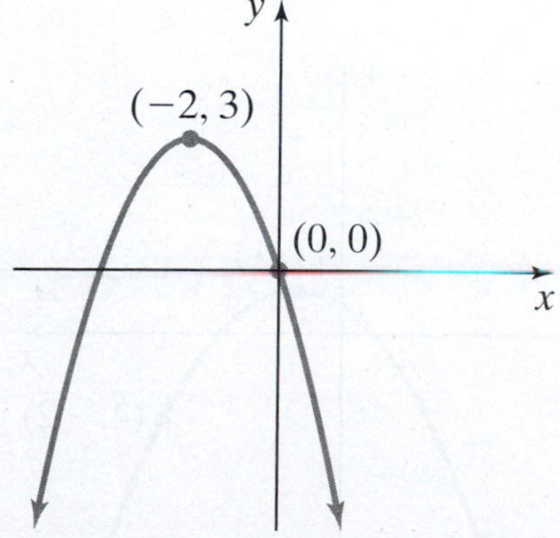

99.

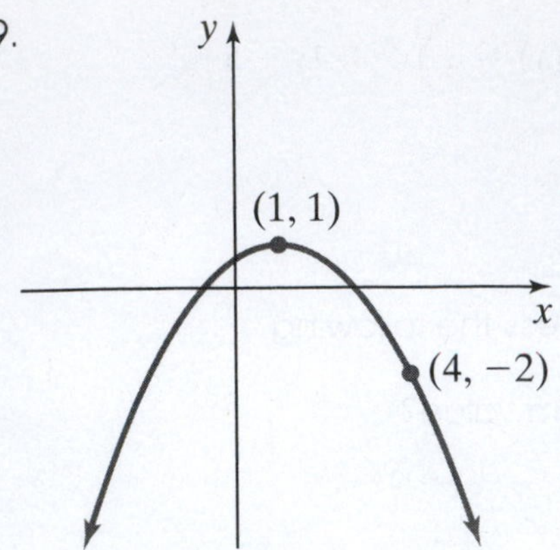

100.

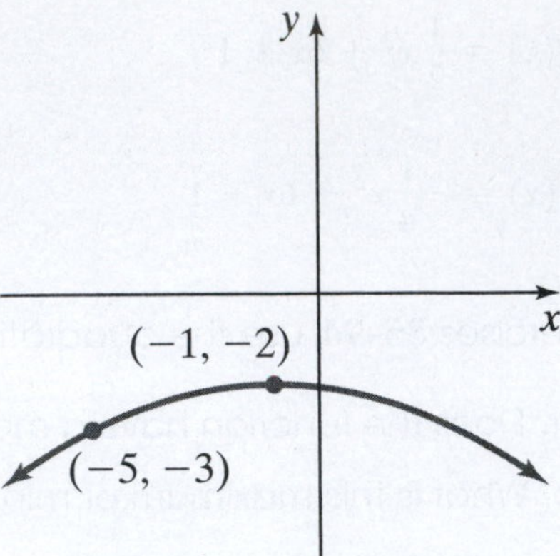

In Exercises 101–106, given the graph of a quadratic function, write the equation of the function in the form $f(x) = ax^2 + bx + c$.

101.

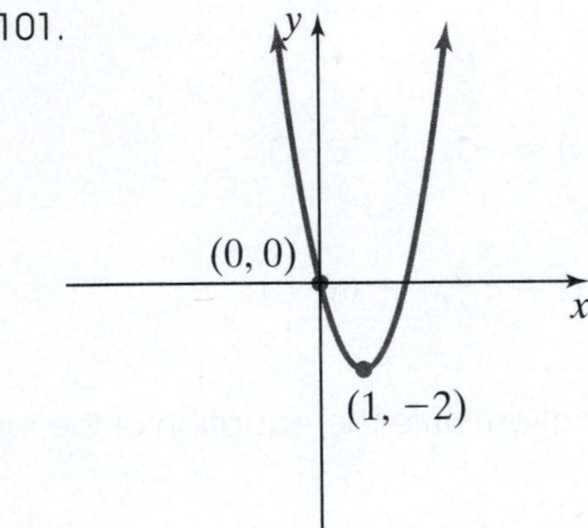

102.

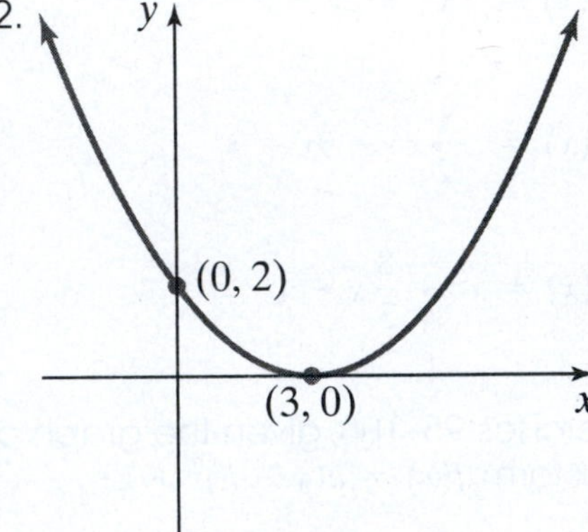

103.

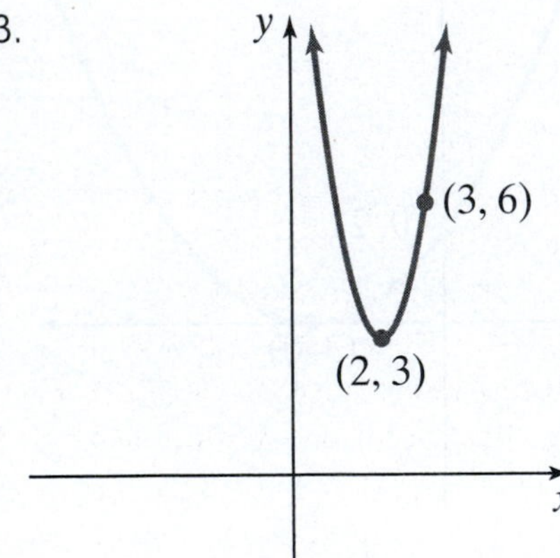

104.

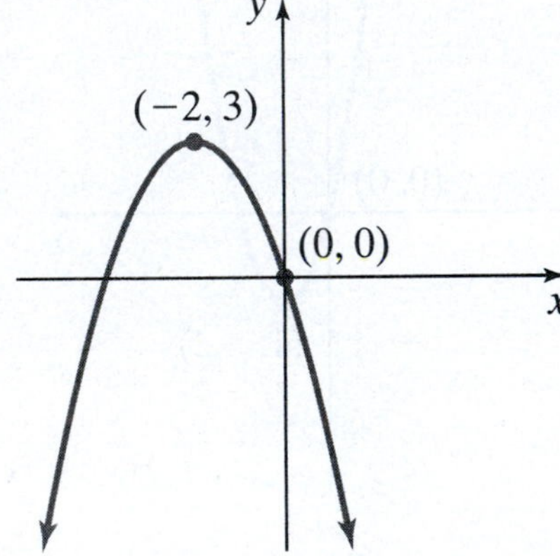

105.

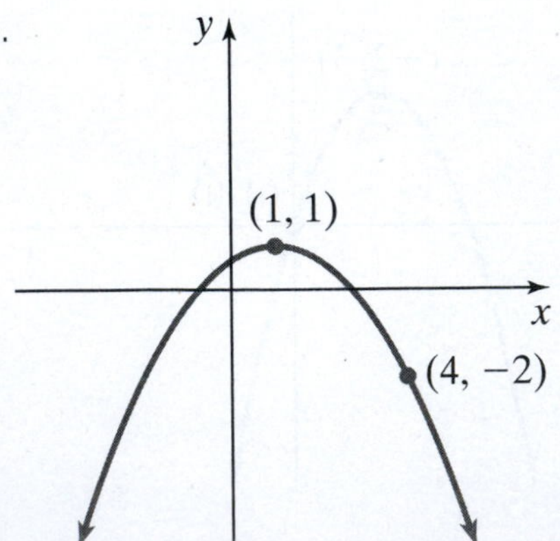

106.

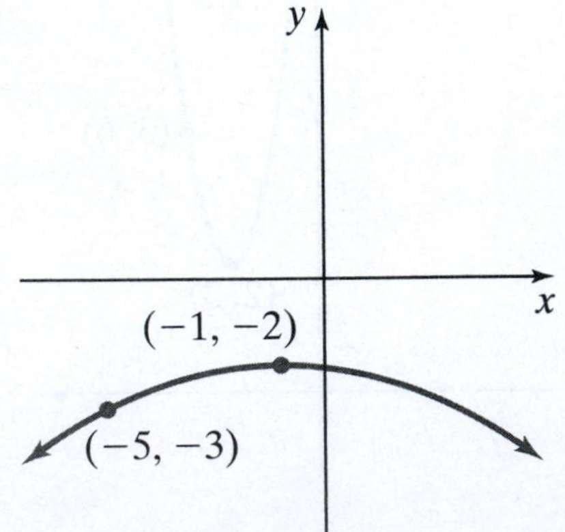

4.2 Applications and Modeling of Quadratic Functions

THINGS TO KNOW

Before working through this section, be sure that you are familiar with the following concepts:

VIDEO ANIMATION INTERACTIVE

You Try It

1. Using the Projectile Motion Model (Section 1.5)

You Try It

2. Graphing Quadratic Functions Using the Vertex Formula (Section 4.1)

INTRODUCTION

Read this introduction before beginning Objective 1.

OBJECTIVES

1 Maximizing Projectile Motion Functions

2 Maximizing Functions in Economics

3 Maximizing Area Functions

SECTION 4.2 EXERCISES

Introduction to Section 4.2

With application problems involving functions, we are often interested in finding the maximum or minimum value of the function. For example, a professional golfer may want to control the trajectory of a struck golf ball. Thus, he may be interested in determining the maximum height of the ball given certain parameters such as swing velocity and club angle. An economist may want to minimize a cost function or maximize a revenue or profit function. A builder with a fixed amount of fencing may want to maximize an area function. In a calculus course, you learn how to maximize or minimize a wide variety of functions. In this section, we concentrate only on quadratic functions. Quadratic functions are relatively easy to maximize or minimize because we know a formula for finding the coordinates of the vertex. Recall that if $f(x) = ax^2 + bx + c, a \neq 0$, we know that the coordinates of the vertex are $\left(-\frac{b}{2a}, f\left(-\frac{b}{2a}\right)\right)$. If $a > 0$, the parabola *opens up* and has a minimum value at the y-coordinate of the vertex, $f\left(\frac{-b}{2a}\right)$. If $a < 0$, the parabola *opens down* and has a maximum value at the y-coordinate of the vertex, $f\left(\frac{-b}{2a}\right)$. See **Figure 6**.

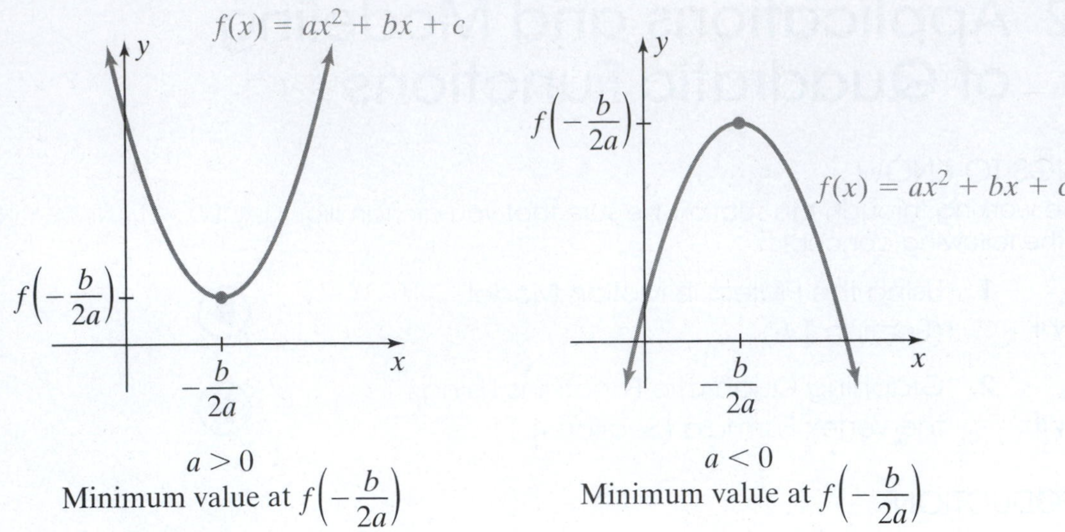

Figure 6

OBJECTIVE 1 MAXIMIZING PROJECTILE MOTION FUNCTIONS

We start our discussion by looking at the projectile motion model that was introduced in **Section 1.5**. An object launched, thrown, or shot vertically into the air with an initial velocity of v_0 meters/second from an initial height of h_0 meters above the ground can be modeled by the **function*** $h(t) = -4.9t^2 + v_0 t + h_0$, where $h(t)$ is the height of the projectile t seconds after its departure.

Example 1 Launch a Toy Rocket

A toy rocket is launched with an initial velocity of 44.1 meters per second from a 1-meter-tall platform. The height h of the rocket at any time t seconds after launch is given by the function $h(t) = -4.9t^2 + 44.1t + 1$. How long after launch did it take the rocket to reach its maximum height? What is the maximum height obtained by the toy rocket?

Solution Because the function is quadratic with $a = -4.9 < 0$, we know that the graph is a parabola that opens down. Thus, the function obtains a maximum value at the vertex.

The t-coordinate of the vertex is $t = -\frac{b}{2a} = -\frac{44.1}{2(-4.9)} = \frac{-44.1}{-9.8} = 4.5$ seconds.

Therefore, the rocket reaches its maximum height 4.5 seconds after launch. The maximum height obtained by the rocket is $h(4.5) = -4.9(4.5)^2 + 44.1(4.5) + 1 = 100.225$ meters.

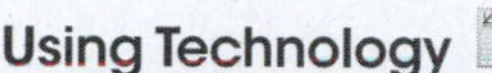

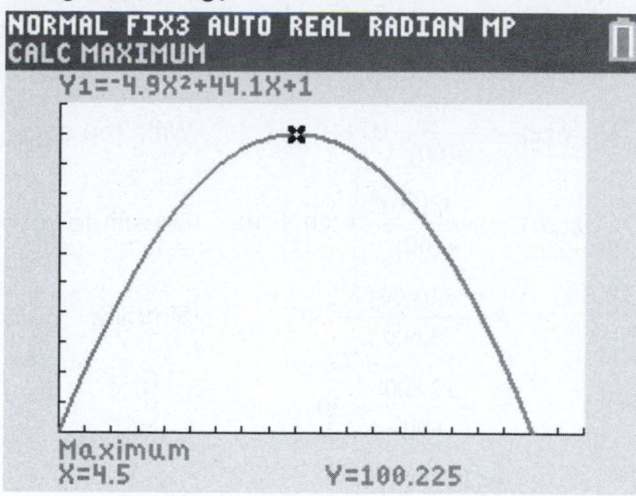

The graph of the quadratic function $h(t) = -4.9t^2 + 44.1t + 1$ is seen here using a graphing utility. Using the Maximum function, we can see that the maximum height is 100.225 meters when time is 4.5 seconds.

You Try It Work through the following You Try It problem.

Work Exercises 1–3 in this textbook or in the MyLab Math Study Plan.

Example 2 Horizontal and Vertical Projectile Motion

If an object is launched at an angle of 45 degrees from a 10-foot platform at 60 feet per second, it can be shown that the height of the object in feet is given by the quadratic function $h(x) = -\frac{32x^2}{(60)^2} + x + 10$, where x is the horizontal distance of the object from the platform. See Figure 7.

a. What is the height of the object when its horizontal distance from the platform is 20 feet? Round to two decimal places.

b. What is the horizontal distance from the platform when the object is at its maximum height?

c. What is the maximum height of the object?

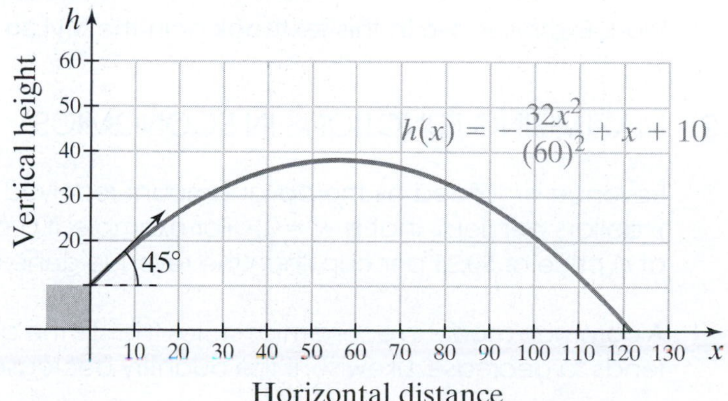

Figure 7

a. To find the height of the object when the horizontal distance from the platform is 20 feet, we must find the value of $h(20)$.

$$h(x) = -\frac{32x^2}{(60)^2} + x + 10 \quad \text{Write the original function.}$$

$$h(20) = -\frac{32(20)^2}{(60)^2} + 20 + 10 \quad \text{Substitute 20 for } x.$$

$$= -\frac{32(400)}{3{,}600} + 30 \quad \text{Simplify.}$$

$$= -\frac{12{,}800}{3{,}600} + 30$$

$$\approx 26.44$$

When the horizontal distance from the platform is 20 feet, the height of the object is approximately 26.44 feet.

b. To find the horizontal distance of the object when it is at its maximum height, we must find the x-coordinate of the vertex. The x-coordinate of the vertex is found using the vertex formula $x = -\frac{b}{2a}$. Since $a = -\frac{32}{60^2}$ and $b = 1$, we get

$$x = -\frac{1}{2\left(\frac{-32}{60^2}\right)} = \frac{-1}{\frac{-64}{3{,}600}} = -1 \cdot \frac{3{,}600}{-64} = 56.25 \text{ feet.}$$

Thus, the object obtains its maximum height at a horizontal distance of 56.25 feet from the base of the platform.

c. From part b, the object obtains its maximum height at a horizontal distance of 56.25 feet from the base of the platform. Therefore, the maximum height is

$$h(56.25) = -\frac{32(56.25)^2}{(60)^2} + 56.25 + 10 = 38.125 \text{ feet.}$$

You Try It Work through the following You Try It problem.

Work Exercises 4–6 in this textbook or in the MyLab Math Study Plan.

OBJECTIVE 2 MAXIMIZING FUNCTIONS IN ECONOMICS

Revenue is defined as the dollar amount received by selling x items at a price of p dollars per item; that is, $R = xp$. For example, if a child sells 50 cups of lemonade at a price of $0.25 per cup, then the revenue generated is $R = \underbrace{(50)}_{x}\underbrace{(0.25)}_{p} = \$12.50$.

A common model in economics states that as the quantity, x, increases, the price, p, tends to decrease. Likewise, if the quantity decreases, the price tends to increase.

Example 3 Maximize Revenue

Records can be kept on the price of shoes and the number of pairs sold in order to gather enough data to reasonably model shopping trends for a particular type of shoe. Demand functions of this type are often linear and can be developed using knowledge of the slope and equations of lines. Suppose that the marketing and research department of a shoe company determined that the price of a certain type of basketball shoe obeys the demand equation $p = -\frac{1}{50}x + 110$.

a. According to the demand equation, how much should the shoes sell for if 500 pairs of shoes are sold? 1,200 pairs of shoes?

b. What is the revenue if 500 pairs of shoes are sold? 1,200 pairs of shoes?

c. How many pairs of shoes should be sold in order to maximize revenue? What is the maximum revenue?

d. What price should be charged in order to maximize revenue?

Solution

a. The demand equation $p = -\frac{1}{50}x + 110$ is sketched in Figure 8. If 500 pairs of shoes are sold, the price should be $p(500) = -\frac{1}{50}(500) + 110 = -10 + 110 = \$100$. If 1,200 pairs of shoes are sold, the price should be $p(1{,}200) = -\frac{1}{50}(1{,}200) + 110 = -24 + 110 = \$86$.

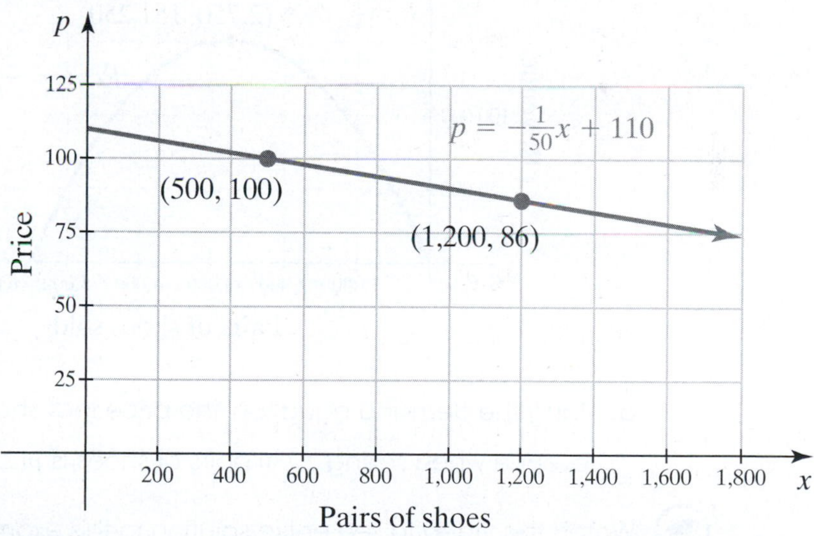

Figure 8

b. Because $R = xp$, we can substitute the demand equation, $p = -\frac{1}{50}x + 110$, for p to obtain the function $R(x) = x\underbrace{\left(-\frac{1}{50}x + 110\right)}_{p}$ or $R(x) = -\frac{1}{50}x^2 + 110x$. The graph of R is shown in Figure 9. The revenue generated by selling 500 pairs of shoes is $R(500) = \$50{,}000$, whereas the revenue generated by selling 1,200 pairs of shoes is $R(1{,}200) = \$103{,}200$.

(eText Screens 4.2-1–4.2-34)

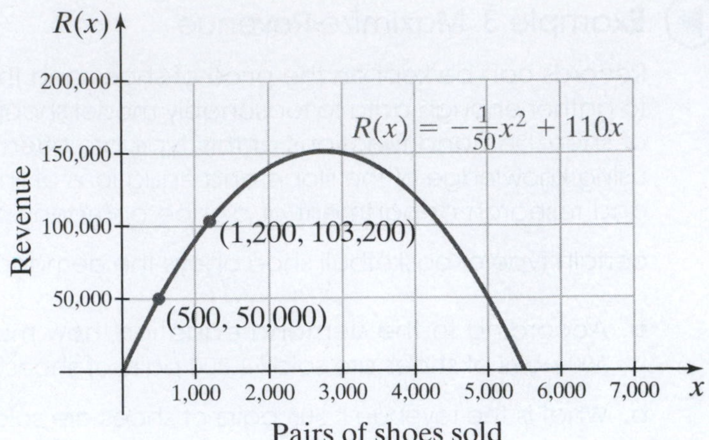

Figure 9

c. $R(x) = -\frac{1}{50}x^2 + 110x$ is a quadratic function with $a = -\frac{1}{50}$ and $b = 110$. Because a is less than 0, the function has a *maximum* value at the vertex. Therefore, the value of x that produces the maximum revenue is $x = -\frac{b}{2a} = -\frac{110}{2\left(-\frac{1}{50}\right)} = \frac{-110}{-\frac{1}{25}} = 2{,}750$ pairs of shoes. The maximum revenue is $R(2{,}750) = \$151{,}250$. See Figure 10.

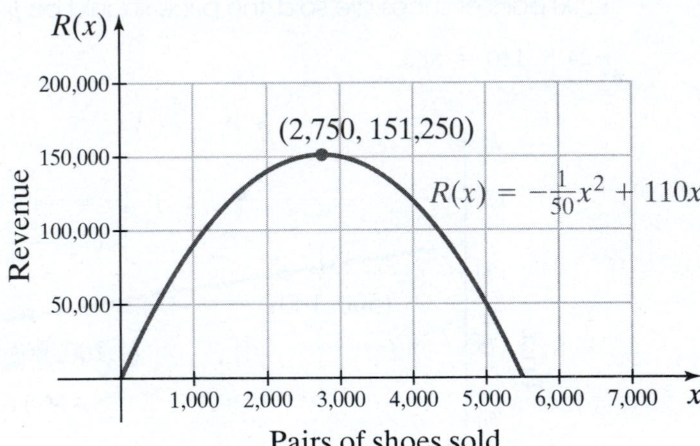

Figure 10

d. Using the demand equation, the price that should be charged to maximize revenue when selling 2,750 pairs of shoes is $p(2{,}750) = -\frac{1}{50}(2{,}750) + 110 = \$55.$

Watch this **video** to see entire solution to this example.

You Try It Work through the following You Try It problem.

Work Exercises 7 and 8 in this textbook or in the MyLab Math Study Plan.

Example 4 Maximize Profit

To sell x waterproof CD alarm clocks, WaterTime, LLC, has determined that the price in dollars must be $p = 250 - 2x$, which is the demand equation. Each clock costs $2 to produce, with fixed costs of $4,000, producing the cost function of $C(x) = 2x + 4{,}000$.

a. Express the revenue R as a function of x.

b. Express the profit P as a function of x.

c. Find the value of x that maximizes profit. What is the maximum profit?

d. What is the price of the alarm clock that will maximize profit?

Solution

a. The equation for revenue is $R = xp$. Therefore, we can substitute the demand equation $p = 250 - 2x$ for p to obtain the function $R(x) = x\underbrace{(250 - 2x)}_{p}$ or $R(x) = -2x^2 + 250x$.

b. Profit is equal to revenue minus cost.

$$\begin{aligned} P(x) &= R(x) - C(x) \\ &= -2x^2 + 250x - (2x + 4{,}000) \\ &= -2x^2 + 248x - 4{,}000 \end{aligned}$$

c. The value of x that will produce the maximum profit is

$$x = -\frac{b}{2a} = -\frac{248}{2(-2)} = \frac{-248}{-4} = 62 \text{ alarm clocks}$$

The maximum profit is $P(62) = -2(62)^2 + 248(62) - 4{,}000 = \$3{,}688$.

d. Using the demand equation $p = 250 - 2x$ and the fact that profit will be maximized when 62 alarm clocks are produced, we obtain a price of $p = 250 - 2(62) = \$126$.

You Try It Work through the following You Try It problem.

Work Exercises 9–12 in this textbook or in the MyLab Math Study Plan.

Example 5 Maximize Revenue at a Country Club

A country club currently has 400 members who pay $500 per month for membership dues. The country club's board members want to increase monthly revenue by *lowering* the monthly dues in hopes of attracting new members. A market research study has shown that for each $1 decrease in the monthly membership price, two additional people will join the club. What price should the club charge to maximize revenue? What is the maximum revenue?

Solution We know that $R = xp$, where x is the number of members and p is the price of monthly membership. Currently, $R = (400)(\$500) = \$200{,}000$. If the price is reduced by $1, two additional people will join the club, so the revenue generated would be $R = (402)(\$499) = \$200{,}598$. We can see that by reducing monthly dues to $499, it can be anticipated that revenue will increase by $598. We want to find the optimal price to charge in order to maximize revenue.

(eText Screens 4.2-1–4.2-34)

Because the number of members, x, depends on price, we can find an expression for x in terms of p:

$$x = \underbrace{400}_{\substack{\text{Number}\\ \text{of current}\\ \text{members}}} + \underbrace{2(500 - p)}_{\substack{\text{Additional two members}\\ \text{for each dollar decrease in}\\ \text{membership monthly dues}}}$$

We can now write the revenue as a function of p:

$$R = xp$$

$$R(p) = \underbrace{(400 + 2(500 - p))}_{x}p \quad \text{or}$$

$$R(p) = -2p^2 + 1{,}400p$$

The revenue is maximized when the price is $p = -\dfrac{b}{2a} = -\dfrac{1{,}400}{2(-2)} = \$350$.

The maximum revenue is $R(350) = -2(350)^2 + 1{,}400(350) = \$245{,}000$. Watch the interactive video to see this problem worked out using an alternate method.

You Try It Work through the following You Try It problem.

Work Exercises 13–20 in this textbook or in the MyLab Math Study Plan.

OBJECTIVE 3 MAXIMIZING AREA FUNCTIONS

Suppose you are asked to build a rectangular fence that borders a river but you have only 3,000 feet of fencing available. No fencing will be required along the side of the river. How many ways can you build this fence? As it turns out, there are infinitely many ways to construct such a fence.

Although there are infinitely many ways in which you could build this fence, there is only one way to build this fence in order to *maximize* the area! What should the length and width of the fence be in order to maximize the area? Watch this video to see how to determine the length and width of the fence that will maximize the area.

Example 6 Maximize Area

Mark has 100 feet of fencing available to build a rectangular pen for his hens and roosters. One side of the pen will border a river and requires no fencing. He wants to separate the hens and roosters by dividing the pen into two equal areas. Let x represent the length of the center partition that is perpendicular to the river.

a. Create a function $A(x)$ that describes the total area of the rectangular enclosure as a function of x where x is the length of the center partition.

b. Find the length of the center partition that will yield the maximum area.

c. Find the length of the side of the fence parallel to the river that will yield the maximum area.

d. What is the maximum area?

Solution

a. The image below illustrates the rectangular enclosure where x represents the length of the center partition and y represents the length of the side parallel to the river.

The area of the rectangular figure in two variables is $A = xy$. Our task is to write a function that that describes the area using only the variable x.

There is 100 feet of fencing available. So, the equation that describes the total fencing is given by $3x + y = 100$.

Solve this equation for y.

$3x + y = 100$ Write the equation that describes the total amount of fencing.

$y = 100 - 3x$ Subtract $3x$ from both sides.

We can substitute the expression $100 - 3x$ into the area equation to write a function in one variable.

$A = xy$ Write the equation in two variables that describes the area.

$A = x(100 - 3x)$ Substitute $100 - 3x$ for y.

Therefore, the function that describes the total area of the enclosure as a function of the length of the partition is $A(x) = x(100 - 3x)$ or $A(x) = -3x^2 + 100x$.

b. The function $A(x) = -3x^2 + 100x$ is a quadratic function that has a maximum at the vertex. The x-coordinate of the vertex represents the length of the partition that will yield the maximum area. Therefore, the value of x that will produce the maximum area is

$$x = -\frac{b}{2a} = -\frac{100}{2(-3)} = \frac{100}{6} = 16\frac{2}{3} \text{ feet.}$$

Watch the **video** to verify that the length of the side of the fence parallel to the river that will yield the maximum area is $y = 50$ feet. The maximum area is $416\frac{2}{3}$ square feet.

You Try It **Work through the following You Try It problem.**

Work Exercises 21–26 in this textbook or in the MyLab Math Study Plan.

4.2 Exercises

1. A baseball player swings and hits a pop fly straight up in the air to the catcher. The height of the baseball in meters t seconds after it is hit is given by the quadratic function $h(t) = -4.9t^2 + 34.3t + 1$. How long does it take for the baseball to reach its maximum height? What is the maximum height obtained by the baseball?

2. An object is launched vertically in the air at 36.75 meters per second from a 10-meter-tall platform. Using the projectile motion model, determine how long it will take for the object to reach its maximum height. What is the maximum height?

3. A toy rocket is shot vertically into the air from a 5-foot-tall launching pad with an initial velocity of 112 feet per second. Suppose the height of the rocket in feet t seconds after being launched can be modeled by the function $h(t) = -16t^2 + v_0t + h_0$, where v_0 is the initial velocity of the rocket and h_0 is the initial height of the rocket. How long will it take for the rocket to reach its maximum height? What is the maximum height?

4. A football player punts a football at an inclination of 45 degrees to the horizontal at an initial velocity of 75 feet per second. The height of the football in feet is given by the quadratic function $h(x) = -\frac{32x^2}{(75)^2} + x + 4$, where x is the horizontal distance of the football from the point at

which it was kicked. What was the horizontal distance of the football from the point at which it was kicked when the height of the ball is at a maximum? What is the maximum height of the football?

5. A golf ball is struck by a 60-degree golf club at an initial velocity of 88 feet per second. The height of the golf ball in feet is given by the quadratic function $h(x) = -\frac{16x^2}{(44)^2} + \frac{76.2}{44}x$, where x is the horizontal distance of the golf ball from the point of impact. What is the horizontal distance of the golf ball from the point of impact when the ball is at its maximum height? What is the maximum height obtained by the golf ball?

6. If a projectile is launched upward at an acute angle, then the height of the projectile at a horizontal distance of x units from the launch site is given by the quadratic function $h(x) = -\frac{1}{2}g\left(\frac{x}{w_0}\right)^2 + \left(\frac{v_0}{w_0}\right)x + h_0$, where g is acceleration due to gravity, w_0 is horizontal initial velocity, v_0 is vertical initial velocity, and h_0 is initial height of the object.

   a. Find a formula describing the horizontal distance from the launch point where the projectile is at a maximum height in terms of g, w_0, and v_0.

   b. Find a formula for the maximum height of the projectile in terms of g, v_0, and h_0.

   c. Suppose $g = 32$, $w_0 = 64$, $v_0 = 64$, and $h_0 = 1$, use the formulas from parts a and b to find the horizontal distance from the launch point where the projectile is at a maximum height and find the maximum height.

7. The price p and the quantity x sold of a small flat screen television set obeys the demand equation $p = -0.15x + 300$.

   a. How much should be charged for the television set if there are 50 television sets in stock?

   b. What quantity x will maximize revenue? What is the maximum revenue?

   c. What price should be charged in order to maximize revenue?

8. The dollar price for a barrel of oil sold at a certain oil refinery tends to follow the demand equation $p = -\frac{1}{10}x + 72$, where x is the number of barrels of oil on hand (in millions).

   a. How much should be charged for a barrel of oil if there are 4 million barrels on hand?

   b. What quantity x will maximize revenue?

   c. What price should be charged in order to maximize revenue?

In Exercises 9–12, use the fact that profit is defined as revenue minus cost or $P(x) = R(x) - C(x)$.

9. Rite-Cut riding lawnmowers obey the demand equation $p = -\frac{1}{20}x + 1{,}000$. The cost of producing x lawnmowers is given by the function $C(x) = 100x + 5{,}000$.

   a. Express the revenue R as a function of x.

   b. Express the profit P as a function of x.

   c. Find the value of x that maximizes profit. What is the maximum profit?

   d. What price should be charged in order to maximize profit?

10. The CarryItAll minivan, a popular vehicle among soccer moms, obeys the demand equation $p = -\frac{1}{40}x + 8{,}000$. The cost of producing x vans is given by the function $C(x) = 4{,}000x + 20{,}000$.
    a. Express the revenue R as a function of x.
    b. Express the profit P as a function of x.
    c. Find the value of x that maximizes profit. What is the maximum profit?
    d. What price should be charged in order to maximize profit?

11. Silver Scooter, Inc., finds that it costs $\$200$ to produce each motorized scooter and that the fixed costs are $\$1{,}500$. The price is given by $p = 600 - 5x$, where p is the price in dollars at which exactly x scooters will be sold. Find the quantity of scooters that Silver Scooter, Inc., should produce and the price it should charge to maximize profit. Find the maximum profit.

12. Amy, the owner of Amy's Pottery, can produce china pitchers at a cost of $\$5$ each. She estimates her price function to be $p = 17 - 5x$, where p is the price at which exactly x pitchers will be sold per week. Find the number of pitchers that she should produce and the price that she should charge in order to maximize profit. Also find the maximum profit.

13. Lewiston Golf and Country Club currently has 300 members who pay $450 per month for membership dues. The country club's board members want to increase monthly revenue by *lowering* the monthly dues in hopes of attracting new members. A market research study has shown that for each $1 decrease in the monthly membership price, three additional people will join the club. What price should the club charge to maximize revenue? What is the maximum revenue?

14. An orange grove produces a profit of $90 per tree when there are 1,200 trees planted. Because of overcrowding, the profit per tree (for every tree in the grove) is reduced by 5 cents per tree for each additional tree planted. How many trees should be planted in order to maximize the total profit of the orange grove? What is the maximum profit?

15. All-Nite Fitness Health Club currently charges its 1,500 clients monthly membership dues of $45. The board of directors decides to increase the monthly membership dues. Market research states that each $1 increase in dues will result in the loss of five clients. How much should the club charge each month to optimize the revenue from monthly dues?

16. The marketing department of a local shoe company found that approximately 600 pairs of running shoes are sold monthly when the price of each pair is $90. It was also observed that for each $1 reduction in price, an additional 20 pairs of running shoes are sold monthly. What price should the shoe store charge for a pair of running shoes in order to maximize revenue?

17. Pearson Travel, Inc., charges $10 for a trip to a concert if 30 people travel in a group. But for each 1 person above the 30-person level, the charge will be reduced by $.20. How many people will maximize the total revenue for the company?

18. Sunshine Bicycle Rental Company at a California resort rents 100 bicycles per day at a flat rate of $10 per day for each bicycle. For each $1 increase in the rate, 5 fewer bikes are rented. At what rate should the bicycles be rented to produce the maximum daily revenue? What is the maximum daily revenue?

19. Cherrypickers, Inc., estimates from past records that if 30 trees are planted per acre, each tree will yield an average of 50 pounds of cherries per season. If for each additional tree planted per acre the average yield per tree is reduced by 1 pound, how many trees should be planted per acre to obtain the maximum yield per acre?

20. An orchard contains 30 apple trees, each of which yields about 400 apples over the growing season. The owner plans to add more trees, but each new tree will reduce the average yield per tree by about 10 apples over the season. How many trees should be added to maximize the total yield of apples? What would be the maximum yield for the apple orchard for the season?

21. A farmer has 1,800 feet of fencing available to enclose a rectangular area bordering a river. No fencing is required along the river. Let x represent the length of the side of the rectangular enclosure that is **perpendicular** to the river.

a. Create a function $A(x)$ that describes the total area of the rectangular enclosure as a function of x where x is the length of the rectangular enclosure that is perpendicular to the river.

b. Find the dimensions of the fence that will maximize the area.

c. What is the maximum area?

22. A farmer has 2,000 feet of fencing available to enclose a rectangular area bordering a river. No fencing is required along the river. Let x represent the length of the fence that is **parallel** to the river

a. Create a function $A(x)$ that describes the total area of the rectangular enclosure as a function of x where x is the length of the rectangular enclosure that is parallel to the river.

b. Find the dimensions of the fence that will maximize the area.

c. What is the maximum area?

23. Jim wants to build a rectangular parking lot along a busy street but has only 2.500 feet of fencing available. If no fencing is required along the street, find the maximum area of the parking lot.

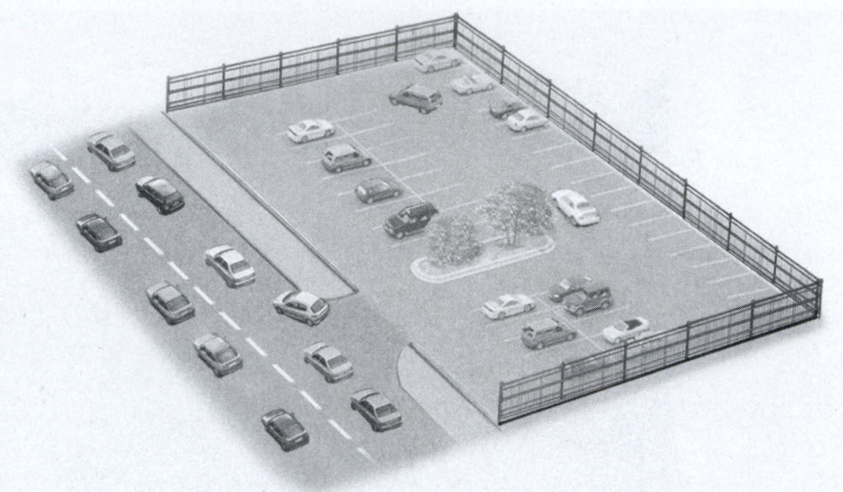

24. A rancher has 4,950 feet of fencing available to enclose a rectangular area bordering a river. She wants to use part of the fencing to create a partition to separate her cows from her horses by dividing the enclosure into two equal areas. No fencing is required along the river. Let x represent the length of the center partition.

a. Create a function $A(x)$ that describes the total area of the rectangular enclosure as a function of x where x is the length of the center partition.

b. Find the length of the center partition that will yield the maximum area.

c. Find the length of the side of the fence parallel to the river that will yield the maximum area.

d. What is the maximum area?

25. A rancher has 5,200 feet of fencing available to enclose a rectangular area bordering a river. He wants to use part of the fencing to create two partitions to separate his cows, horses, and pigs by dividing the enclosure into three equal areas. No fencing is required along the river. Let x represent the length of the partitions.

a. Create a function $A(x)$ that describes the total area of the rectangular enclosure as a function of x where x is the length of a partition.

b. Find the length of a partition that will yield the maximum area.

c. Find the length of the side of the fence parallel to the river that will yield the maximum area.

d. What is the maximum area?

26. A 32-inch wire is to be cut. One piece is to be bent into the shape of a square, whereas the other piece is to be bent into the shape of a rectangle whose length is twice the width. Find the width of the rectangle that will minimize the total area.

4.3 The Graphs of Polynomial Functions

THINGS TO KNOW

Before working through this section, be sure that you are familiar with the following concepts:

VIDEO ANIMATION INTERACTIVE

You Try It

1. Determining the Intercepts of a Function (Section 3.2) 

You Try It

2. Sketching the Graphs of the Basic Functions (Section 3.3)

You Try It

3. Using Combinations of Transformations to Graph Functions (Section 3.4)  

OBJECTIVES

1 Understanding the Definition of a Polynomial Function

2 Sketching the Graphs of Power Functions

3 Determining the End Behavior of Polynomial Functions

4 Determining the Intercepts of a Polynomial Function

5 Determining the Real Zeros of Polynomial Functions and Their Multiplicities

6 Sketching the Graph of a Polynomial Function

7 Determining a Possible Equation of a Polynomial Function Given Its Graph

SECTION 4.3 EXERCISES

OBJECTIVE 1 UNDERSTANDING THE DEFINITION OF A POLYNOMIAL FUNCTION

We have already studied linear functions of the form $f(x) = mx + b$, where $m \neq 0$, and quadratic functions of the form $f(x) = ax^2 + bx + c$, where $a \neq 0$. Both functions are examples of **polynomial functions**.

Definition Polynomial Function

The function $f(x) = a_n x^n + a_{n-1}x^{n-1} + a_{n-2}x^{n-2} + \cdots + a_1 x + a_0$ is a polynomial function of degree n, where n is a nonnegative integer. The numbers $a_0, a_1, a_2, \ldots, a_n$ are called the **coefficients** of the polynomial function. The number a_n is called the **leading coefficient**, and a_0 is called the **constant coefficient**.

The important thing to remember from the definition is that the power (exponent) of each term must be an integer greater than or equal to zero. Although the coefficients can be any number (real or imaginary), we focus primarily on polynomial functions with real coefficients.

Example 1 Determine Whether a Function Is a Polynomial Function

Determine which functions are polynomial functions. If the function is a polynomial function, identify the degree, the leading coefficient, and the constant coefficient.

a. $f(x) = \sqrt{3}x^3 - 2x^2 - \frac{1}{6}$

b. $g(x) = 4x^5 - 3x^3 + x^2 + \frac{7}{x} - 3$

c. $h(x) = \frac{3x - x^2 + 7x^4}{9}$

Solution

a. The function $f(x) = \sqrt{3}x^3 - 2x^2 - \frac{1}{6}$ is a polynomial function because all powers of x are integers greater than or equal to 0. The highest power is 3, so this polynomial is of degree 3. The leading coefficient is $\sqrt{3}$, and the constant coefficient is $-\frac{1}{6}$.

b. The function $g(x) = 4x^5 - 3x^3 + x^2 + \frac{7}{x} - 3$ is *not* a polynomial function because the term $\frac{7}{x}$ can be rewritten as $7x^{-1}$ and -1 is *not* greater than or equal to 0.

c. The function $h(x) = \frac{3x - x^2 + 7x^4}{9}$ is a polynomial function of degree 4. Dividing each term by 9 and reordering the terms, we can write this polynomial function as $h(x) = \frac{7}{9}x^4 - \frac{1}{9}x^2 + \frac{1}{3}x$. The leading coefficient is $\frac{7}{9}$, and the constant coefficient is 0.

You Try It **Work through the following You Try It problem.**

Work Exercises 1–6 in this textbook or in the MyLab Math Study Plan.

OBJECTIVE 2 SKETCHING THE GRAPHS OF POWER FUNCTIONS

The most basic polynomial functions are **monomial functions** of the form $f(x) = ax^n$ where n is a positive integer. Polynomial functions of this form are called **power functions. Figure 11** shows the graphs of power functions for $a = 1$ and for $n = 1, 2, 3, 4,$ and 5.

You can see from **Figure 11** that the graph of $f(x) = x^4$ closely resembles the graph of $f(x) = x^2$. The difference in the two graphs is that the graph of $f(x) = x^4$ flattens out near the origin and gets steep quicker than the graph of $f(x) = x^2$. Likewise, the graph of $f(x) = x^5$ resembles the graph of $f(x) = x^3$. In general if n is even, the graph of $f(x) = x^n$ will resemble the graph of $f(x) = x^2$. If n is odd, the graph of $f(x) = x^n$ will resemble the graph of $f(x) = x^3$.

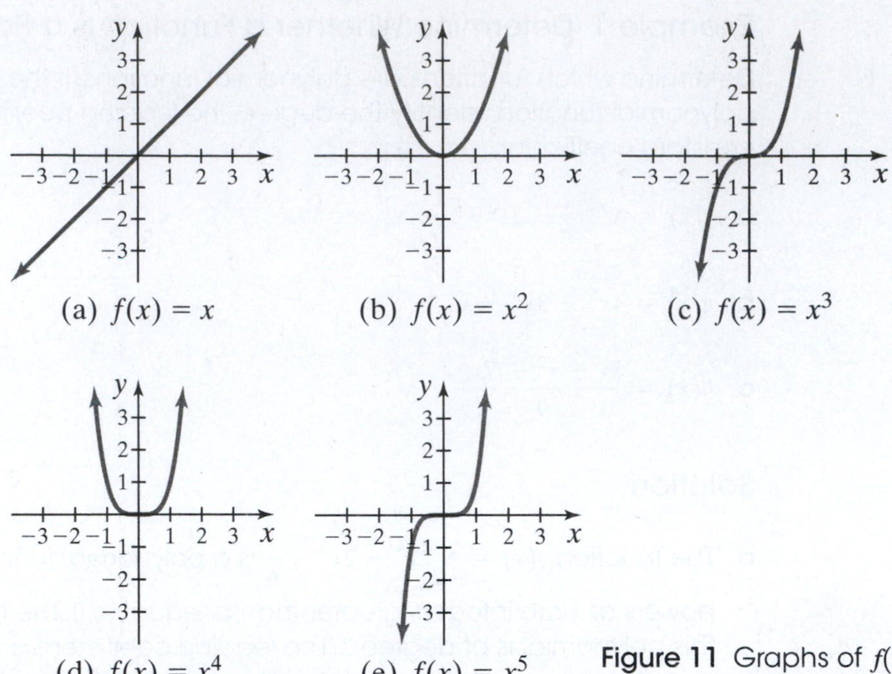

(a) $f(x) = x$ (b) $f(x) = x^2$ (c) $f(x) = x^3$

(d) $f(x) = x^4$ (e) $f(x) = x^5$

Figure 11 Graphs of $f(x) = x^n$, where $n = 1, 2, 3, 4,$ and 5.

Because we know the basic shape of $f(x) = x^n$, we can use the transformation techniques discussed in Section 3.4 to sketch more complicated polynomial functions. Read this **summary** for a quick review of the transformation techniques or refer to **Section 3.4**.

Example 2 Use Transformations to Sketch Polynomial Functions

Use transformations to sketch the following functions:

a. $f(x) = -x^6$ **b.** $f(x) = (x + 1)^5 + 2$ **c.** $f(x) = 2(x - 3)^4$

Solution

a. The graph of $f(x) = -x^6$ is obtained by reflecting the graph of $y = x^6$ about the x-axis. See **Figure 12(a)**.

b. Beginning with the graph of $y = x^5$, we can obtain the graph of $f(x) = (x + 1)^5 + 2$ by horizontally shifting $y = x^5$ to the *left* one unit and then *up* two units. See **Figure 12(b)**.

c. To sketch the graph of $f(x) = 2(x - 3)^4$, we start with the graph of $y = x^4$. We then horizontally shift the graph to the *right* three units and vertically stretch the graph by a factor of 2. See **Figure 12(c)**.

Watch the **interactive video** for a detailed solution.

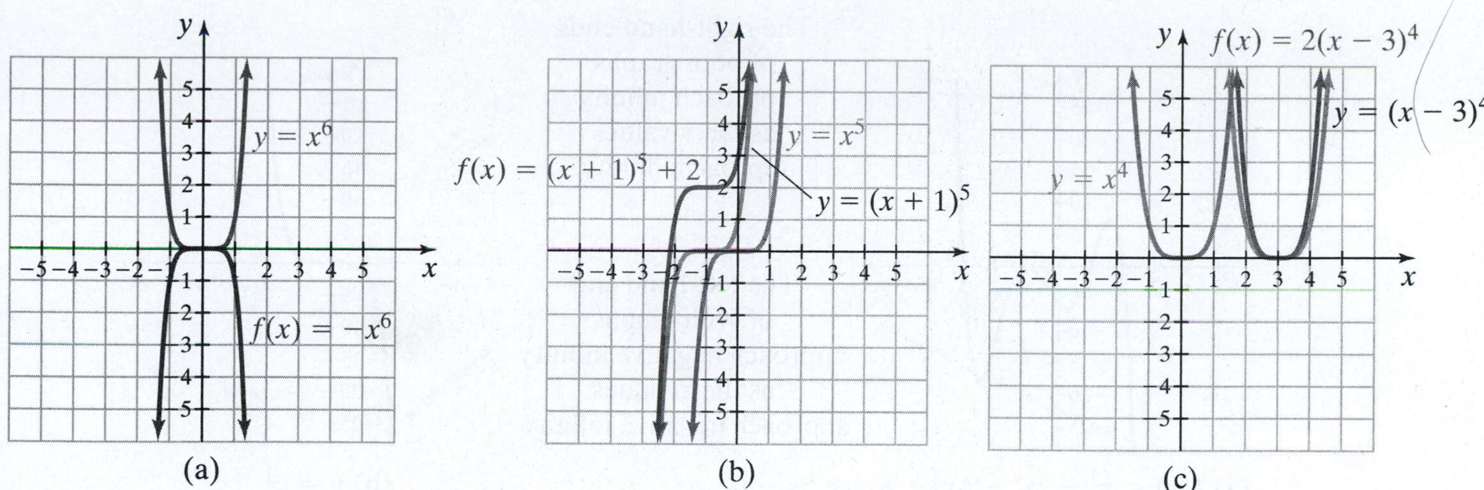

Figure 12

You Try It Work through the following You Try It problem.

Work Exercises 7–15 in this textbook or in the MyLab Math Study Plan.

OBJECTIVE 3 DETERMINING THE END BEHAVIOR OF POLYNOMIAL FUNCTIONS

To begin to sketch the graph of a more complicated polynomial function, it is necessary to understand the behavior of the graph as the x-coordinate gets large in the positive direction (approaches infinity) and as the x-coordinate gets small in the negative direction (approaches negative infinity). The nature of the graph of a polynomial function for large values of x in the positive and negative direction is known as the **end behavior**. The end behavior of the graph of a polynomial function depends on the leading term, that is, the term of highest degree. The graph of a polynomial function $f(x) = a_nx^n + a_{n-1}x^{n-1} + a_{n-2}x^{n-2} + \cdots + a_1x + a_0$ has the same end behavior as the graph of the power function $f(x) = a_nx^n$.

To illustrate end behavior, consider the polynomial function $f(x) = x^5 - 2x^4 - 6x^3 + 8x^2 + 5x - 6$ sketched in **Figure 13a** and compare it to the graph of $f(x) = x^5$ sketched in **Figure 13b**. Notice that the "ends" of both graphs show similar behavior. The right-hand ends of both graphs increase without bound; that is, the y-values approach infinity as the x-values approach infinity. The left-hand ends of both graphs approach negative infinity as the x-values approach negative infinity.

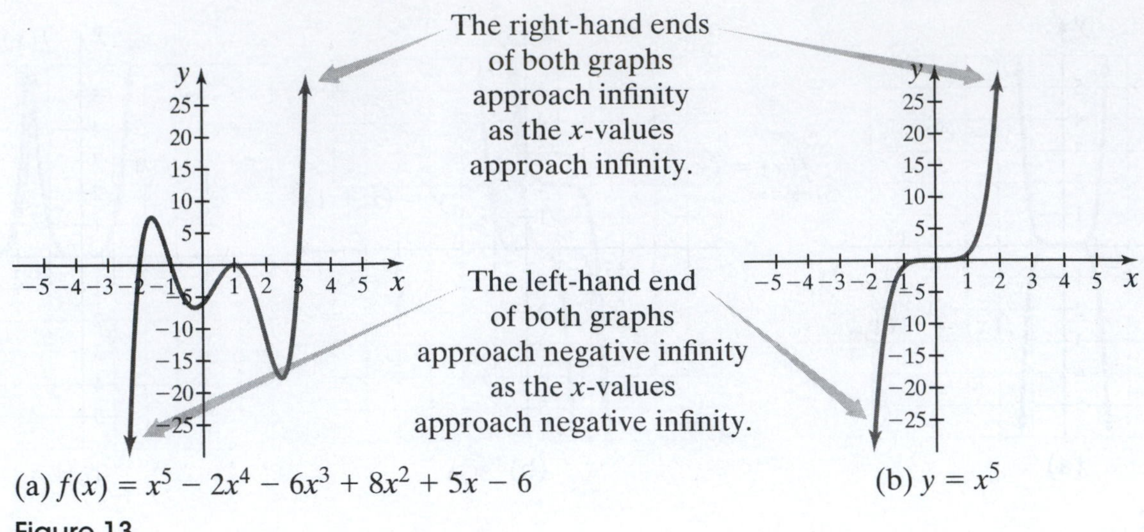
The right-hand ends of both graphs approach infinity as the x-values approach infinity.
The left-hand end of both graphs approach negative infinity as the x-values approach negative infinity.
(a) $f(x) = x^5 - 2x^4 - 6x^3 + 8x^2 + 5x - 6$
(b) $y = x^5$

Figure 13

To determine the end behavior of a polynomial function $f(x) = a_n x^n + a_{n-1}x^{n-1} + a_{n-2}x^{n-2} + \ldots + a_1 x + a_0$, we look at the leading term $a_n x^n$ and follow a two-step process.

Two-Step Process for Determining the End Behavior of a Polynomial Function $f(x) = a_n x^n + a_{n-1}x^{n-1} + a_{n-2}x^{n-2} + \cdots + a_1 x + a_0$

Step 1. Determine the sign of the leading coefficient a_n.

If $a_n > 0$, the right-hand behavior "finishes up."

If $a_n < 0$, the right-hand behavior "finishes down."

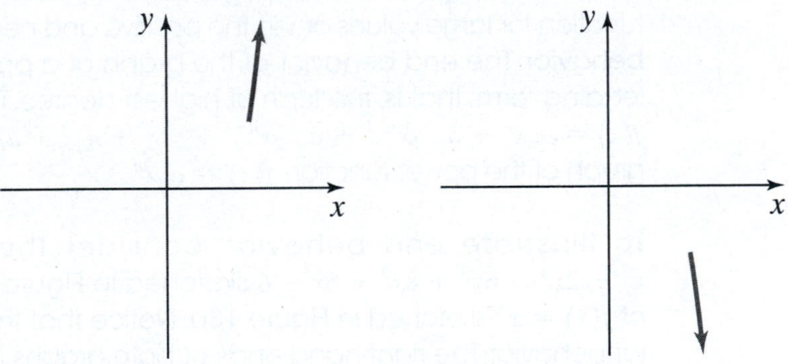

Step 2. Determine the degree.

If the degree n is odd, the graph has opposite left-hand and right-hand end behavior; that is, the graph "starts" and "finishes" in *opposite* directions.

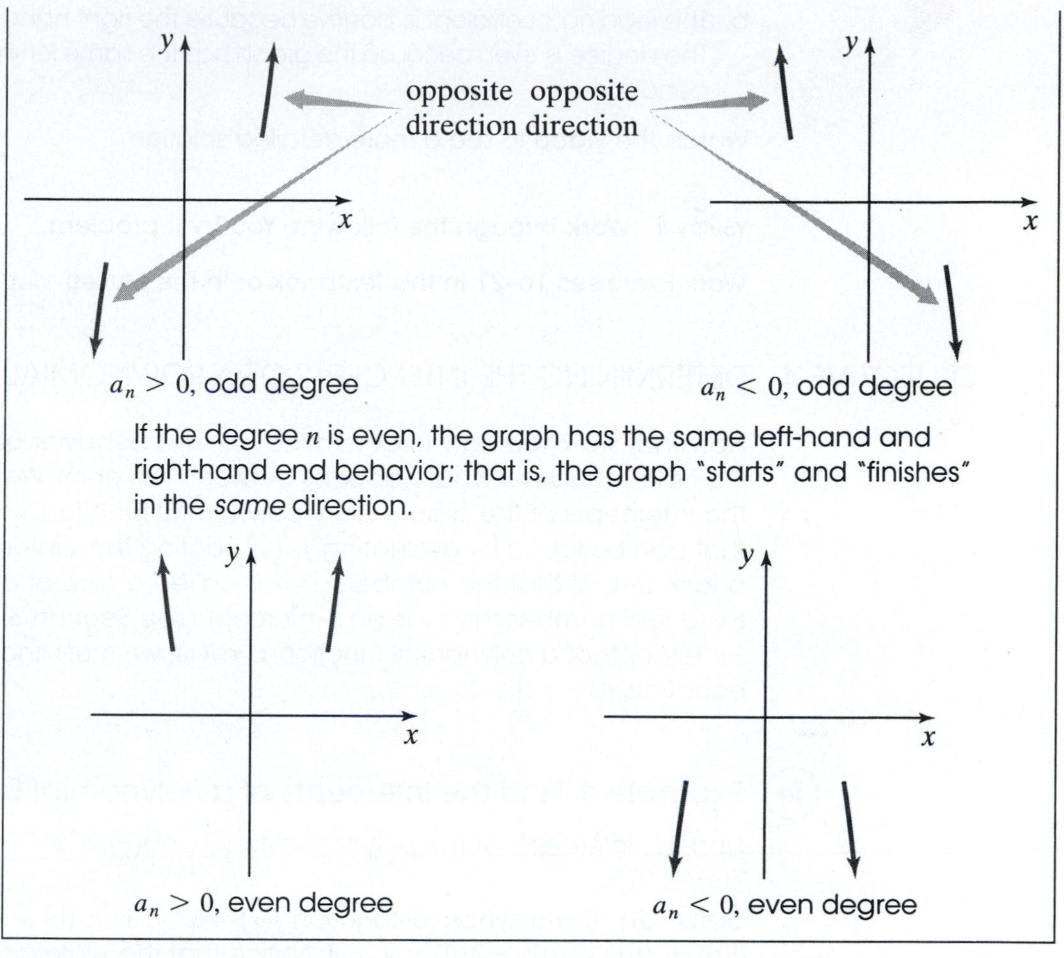

Example 3 Use End Behavior to Determine the Degree and Leading Coefficient

Use the end behavior of each graph to determine whether the degree is even or odd and whether the leading coefficient is positive or negative.

a.

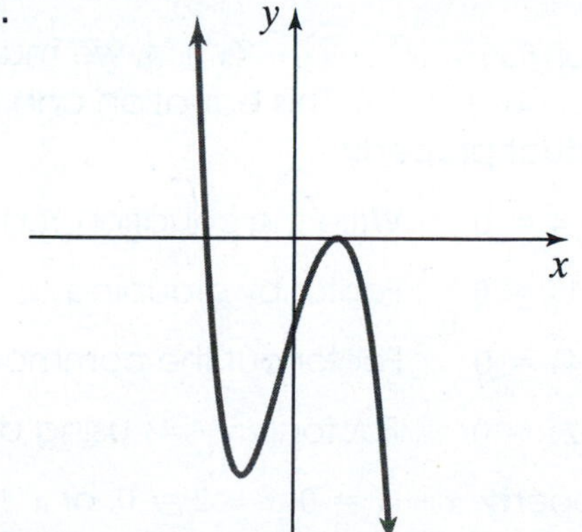

b.

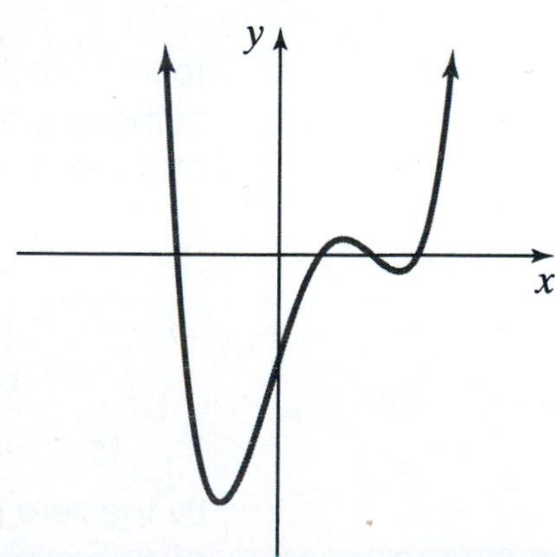

Solution

a. The leading coefficient is negative because the right-hand end behavior "ends down." The degree is *odd* because the graph has opposite left-hand and right-hand end behavior.

b. The leading coefficient is *positive* because the right-hand end behavior "ends up." The degree is *even* because the graph has the same left-hand and right-hand end behavior.

Watch the **video** to see a more detailed solution. ●

You Try It **Work through the following You Try It problem.**

Work Exercises 16–21 in this textbook or in the MyLab Math Study Plan.

OBJECTIVE 4 DETERMINING THE INTERCEPTS OF A POLYNOMIAL FUNCTION

Now that we know how to determine the end behavior of a polynomial function, it is time to find out what happens between the ends. We start by trying to locate the **intercepts** of the graph. Every polynomial function, $y = f(x)$, has a y-intercept that can be found by evaluating $f(0)$. Locating the x-intercepts is not that easy of a task. Recall that the number $x = c$ is called a **zero** of a function f if $f(c) = 0$. If c is a real number, then c is an x-intercept (see **Section 3.2**). Therefore, to find the x-intercepts of a polynomial function $y = f(x)$, we must find the real solutions of the equation $f(x) = 0$.

Example 4 Find the Intercepts of a Polynomial Function

Find the intercepts of the polynomial function $f(x) = x^3 - x^2 - 4x + 4$.

Solution The polynomial function $f(x) = x^3 - x^2 - 4x + 4$ has a y-intercept at $f(0) = (0)^3 - (0)^2 - 4(0) + 4 = 4$. Notice that the y-intercept is the same as the constant coefficient.

In general, every polynomial function of the form

$$f(x) = a_n x^n + a_{n-1}x^{n-1} + a_{n-2}x^{n-2} + \cdots + a_1 x + a_0$$

has a y-intercept at the constant coefficient a_0.

To find the x-intercepts of $f(x) = x^3 - x^2 - 4x + 4$, we must find the real solutions of the equation $x^3 - x^2 - 4x + 4 = 0$. This equation can be solved by factoring and using the **zero product property**.

$x^3 - x^2 - 4x + 4 = 0$	Write the equation $f(x) = 0$.
$x^2(x - 1) - 4(x - 1) = 0$	Factor by grouping.
$(x - 1)(x^2 - 4) = 0$	Factor out the common factor of $(x - 1)$.
$(x - 1)(x - 2)(x + 2) = 0$	Factor $(x^2 - 4)$ using difference of squares.

By the **zero product property**, $x - 1 = 0$, $x - 2 = 0$, or $x + 2 = 0$. Thus, the zeros of $f(x) = x^3 - x^2 - 4x + 4$ are $x = 1$, $x = 2$, and $x = -2$. Because the zeros are all real numbers, $x = 1$, $x = 2$, and $x = -2$ are the x-intercepts. By determining the end behavior and plotting the intercepts, we begin to get an idea of what the  graph of $f(x) = x^3 - x^2 - 4x + 4$ looks like. See Figure 14. Watch this **video** to see this entire example worked out in detail.

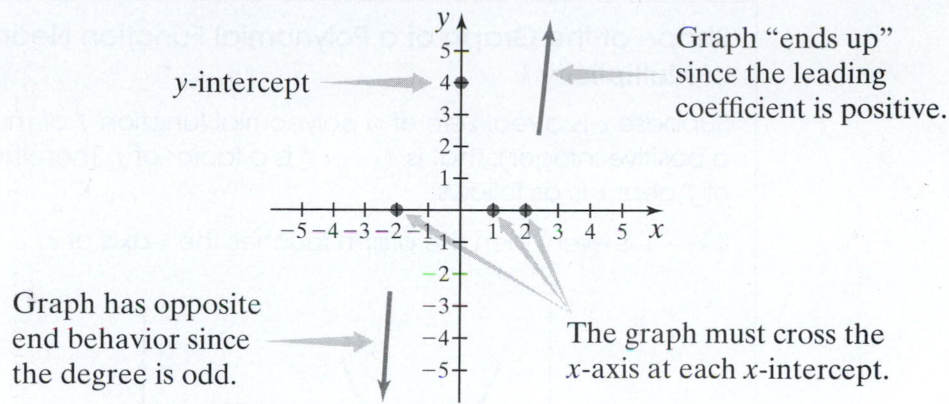

Figure 14 A portion of the graph of $f(x) = x^3 - x^2 - 4x + 4$.

You Try It Work through the following You Try It problem.

Work Exercises 22–25 in this textbook or in the MyLab Math Study Plan.

OBJECTIVE 5 DETERMINING THE REAL ZEROS OF POLYNOMIAL FUNCTIONS AND THEIR MULTIPLICITIES

In Example 4, we were able to factor the polynomial function $f(x) = x^3 - x^2 - 4x + 4$ into $f(x) = (x - 1)(x - 2)(x + 2)$. Once the polynomial function was in factored form, we were able to determine the three zeros $x = 1, x = 2$, and $x = -2$. In general, if $(x - c)$ is a factor of a polynomial function, then $x = c$ is a zero. The converse of the previous statement is also true. (If f is a polynomial function and $x = c$ is a zero, then $(x - c)$ is a factor.)

For example, consider the polynomial functions $f(x) = (x - 1)^2$ and $g(x) = (x - 1)^3$. Both functions have a real zero at $x = 1$ because $(x - 1)$ is a factor of both polynomials. For the function $f(x) = (x - 1)^2, x = 1$ is called a zero of multiplicity 2 because the factor $(x - 1)$ occurs two times in the factorization of f. Likewise, for $g(x) = (x - 1)^3, x = 1$ is a zero of multiplicity 3. Notice in **Figure 15** that the graph of $f(x) = (x - 1)^2$ is *tangent* to (touches) the x-axis at $x = 1$, whereas the graph of $g(x) = (x - 1)^3$ *crosses* the x-axis at $x = 1$. The graph of a polynomial function will touch the x-axis at a real zero of even multiplicity and will cross the x-axis at a real zero of odd multiplicity.

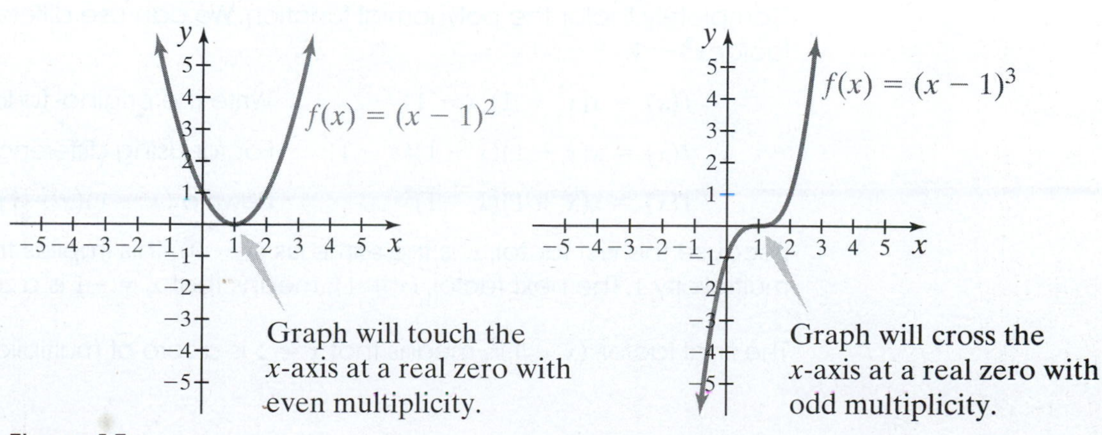

Figure 15

Shape of the Graph of a Polynomial Function Near a Zero of Multiplicity k

Suppose c is a real zero of a polynomial function f of multiplicity k (where k is a positive integer); that is, $(x - c)^k$ is a factor of f. Then the shape of the graph of f near c is as follows:

If $k > 1$ is even, then the graph touches the x-axis at c.

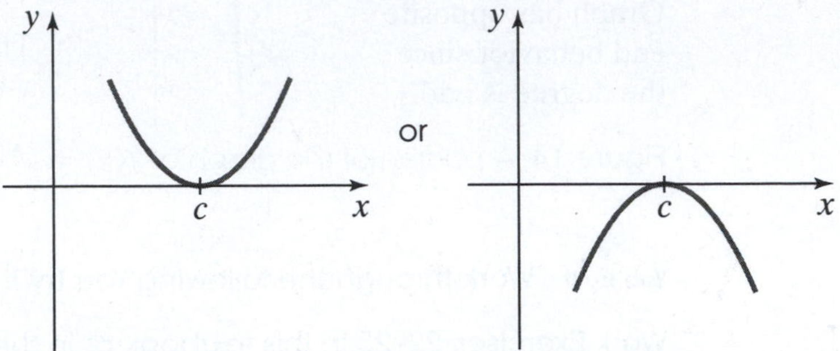

If $k \geq 1$ is odd, then the graph crosses the x-axis at c.

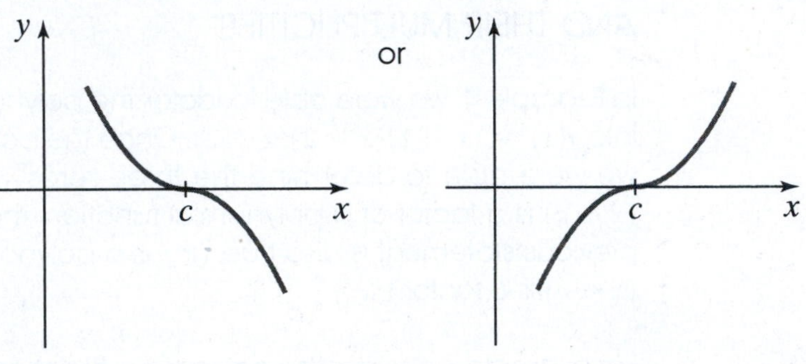

Example 5 Determine the Real Zeros (and Their Multiplicities) of a Polynomial Function

Find all real zeros of $f(x) = x(x^2 - 1)(x - 1)$. Determine the multiplicities of each zero, and decide whether the graph touches or crosses at each zero.

Solution The solution to this example can be seen at the end of this **video**.

Completely factor the polynomial function. We can use difference of squares to factor $x^2 - 1$.

$f(x) = x(x^2 - 1)(x - 1)$	Write the original function.
$f(x) = x(x + 1)(x - 1)(x - 1)$	Factor using difference of squares.
$f(x) = x(x + 1)(x - 1)^2$	Rewrite $(x - 1)(x - 1)$ as $(x - 1)^2$.

Because the first factor, x, is the same as $(x - 0)$, this implies that $x = 0$ is a zero of multiplicity 1. The next factor, $(x + 1)$, means that $x = -1$ is a zero of multiplicity 1.

The final factor, $(x - 1)^2$, means that $x = 1$ is a zero of multiplicity 2.

The graph of $f(x) = x(x + 1)(x - 1)^2$ must cross the x-axis at $x = 0$ and $x = -1$ and must touch the x-axis at $x = 1$. See the following calculator graph.

Using Technology

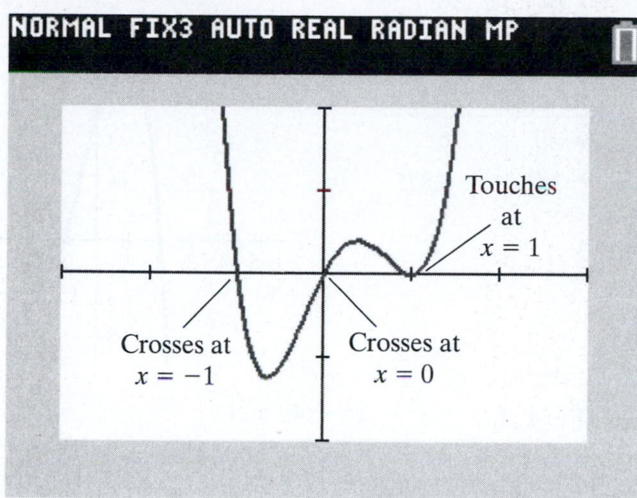

The graph of the function $f(x) = x(x + 1)(x - 1)^2$ is seen here using a graphing utility. Notice that the graph crosses the x-axis at $x = -1$ and $x = 0$. The graph touches the x-axis at $x = 1$. ●

You Try It Work through the following You Try It problem.

Work Exercises 26–33 in this textbook or in the MyLab Math Study Plan.

OBJECTIVE 6 SKETCHING THE GRAPH OF A POLYNOMIAL FUNCTION

If we are able to find the zeros of a polynomial function, we will be able to sketch an approximate graph. Earlier in this section, we were able to factor the polynomial function $f(x) = x^3 - x^2 - 4x + 4$ into $f(x) = (x - 1)(x - 2)(x + 2)$. Note that the three zeros $x = 1, x = 2,$ and $x = -2$ have odd multiplicities. Therefore, the graph of f must cross the x-axis at each x-intercept. By determining the end behavior of this function and by plotting the x-intercepts and y-intercept, we can begin to sketch the graph as seen in Figure 16.

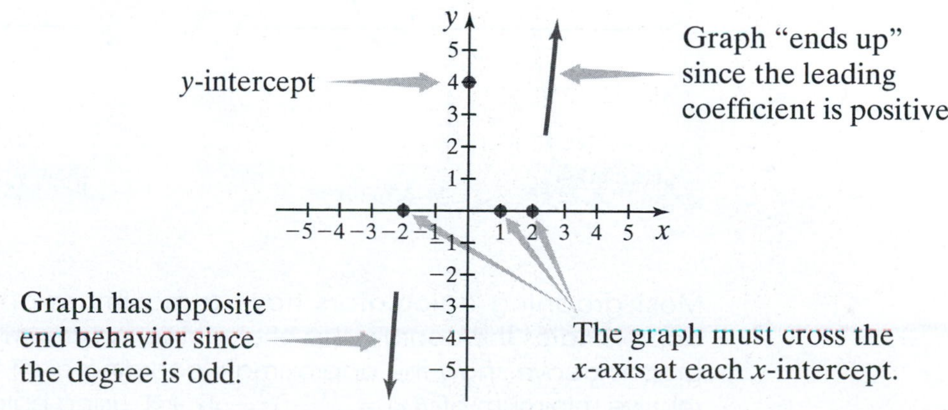

Figure 16 A portion of the graph of $f(x) = x^3 - x^2 - 4x + 4$.

To complete the graph, we can create a table of values to plot additional points by choosing a **test value** between each of the zeros. Here, we choose $x = -1$ and $x = 1.5$. The approximate graph is sketched in Figure 17.

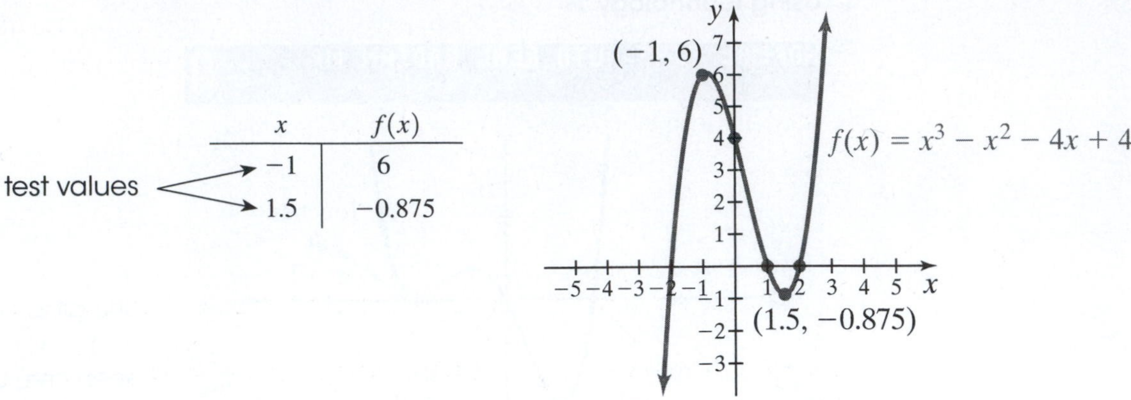

x	$f(x)$
-1	6
1.5	-0.875

Figure 17 The graph of $f(x) = x^3 - x^2 - 4x + 4$.

We see from the graph in Figure 17 that there are two **turning points**. A polynomial function of degree n has at most $n - 1$ turning points. A turning point in which the graph changes from increasing to decreasing is also called a **relative maximum**. A turning point in which the graph changes from decreasing to increasing is called a **relative minimum**.

Without the use of calculus (or a graphing utility), there is no way to determine the precise coordinates of the relative minima and relative maxima. Using a graphing utility, we can use the Maximum and Minimum features to determine the coordinates of the relative maxima and minima as follows.

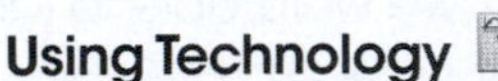

Using Technology

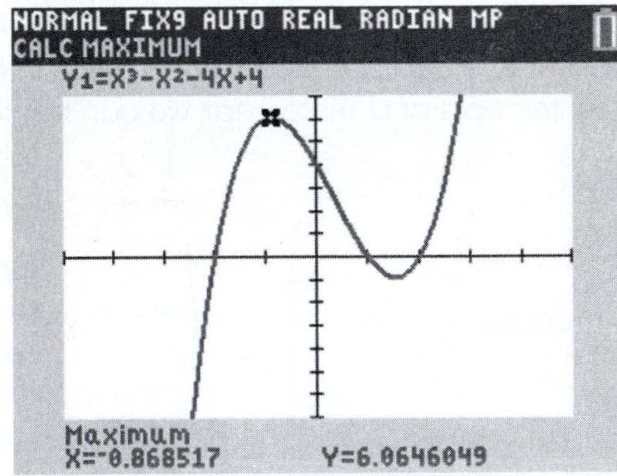

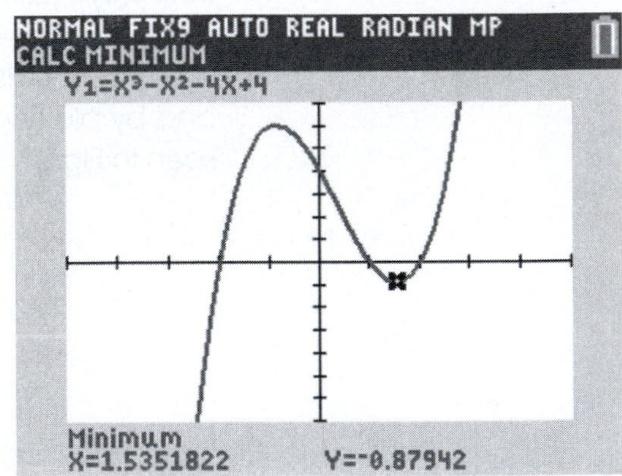

Most graphing calculators have a maximum and minimum feature that approximates the coordinates of a relative maximum or a relative minimum. The graphs above show the approximate coordinates of the relative maximum and relative minimum of $f(x) = x^3 - x^2 - 4x + 4$. Using calculus, it can be shown that the exact coordinates are $\left(\frac{1 - \sqrt{13}}{3}, \frac{70 + 26\sqrt{13}}{27}\right)$ and $\left(\frac{1 + \sqrt{13}}{3}, \frac{70 - 26\sqrt{13}}{27}\right)$, respectively.

Some texts use the terms *local maximum* and *local minimum* to describe a turning point.

We can summarize the sketching procedure with the following four-step process.

Four-Step Process for Sketching the Graph of a Polynomial Function

Step 1. Determine the end behavior.

Step 2. Plot the y-intercept $f(0) = a_0$.

Step 3. Completely factor f to find all real zeros and their **multiplicities**.

Step 4. Choose a test value between each real zero and sketch the graph.

(Remember that without calculus, there is no way to determine the precise coordinates of the turning points.)

Example 6 Use the Four-Step Process to Sketch the Graph of a Polynomial Function

Use the four-step process to sketch the graphs of the following polynomial functions:

a. $f(x) = -2(x + 2)^2(x - 1)$

b. $f(x) = x^4 - 2x^3 - 3x^2$

Solution The graphs of each polynomial are sketched in Figure 18. Work through the **interactive video** to see how to sketch these polynomial functions.

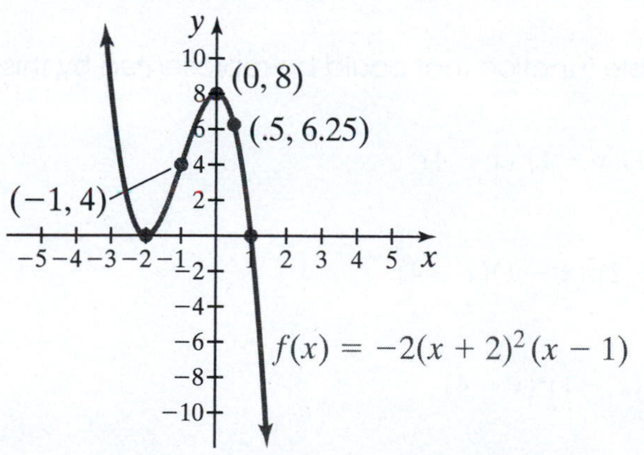

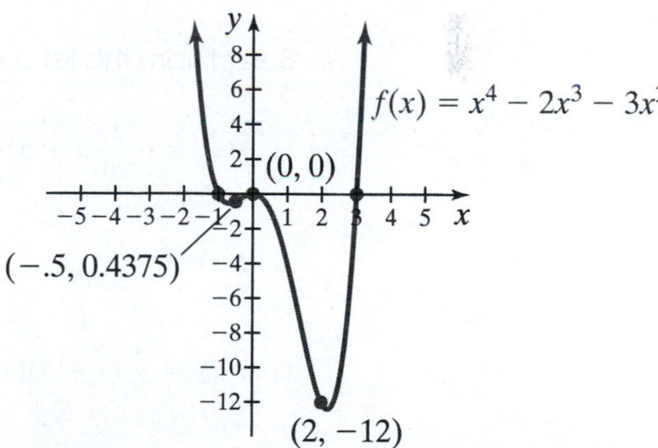

Figure 18 (a) Graph of $f(x) = -2(x + 2)^2(x - 1)$. (b) Graph of $f(x) = x^4 - 2x^3 - 3x^2$.

You Try It Work through the following You Try It problem.

Work Exercises 34–44 in this textbook or in the MyLab Math Study Plan.

OBJECTIVE 7 DETERMINING A POSSIBLE EQUATION OF A POLYNOMIAL FUNCTION GIVEN ITS GRAPH

Now that we have identified the characteristics of a polynomial function and can create an approximate sketch of its graph, we should be able to analyze the graph of a polynomial function and describe features of its equation. The end behavior of the graph gives us information about the degree and leading coefficient, the y-intercept gives us the value of the constant coefficient, and the behavior of the graph at the x-intercepts gives us information about the multiplicity of the zeros. We can use this information to identify possible equations that the graph of a polynomial function might represent.

Example 7 Determine a Possible Equation of a Polynomial Function Given Its Graph

Analyze the graph to address the following about the polynomial function it represents.

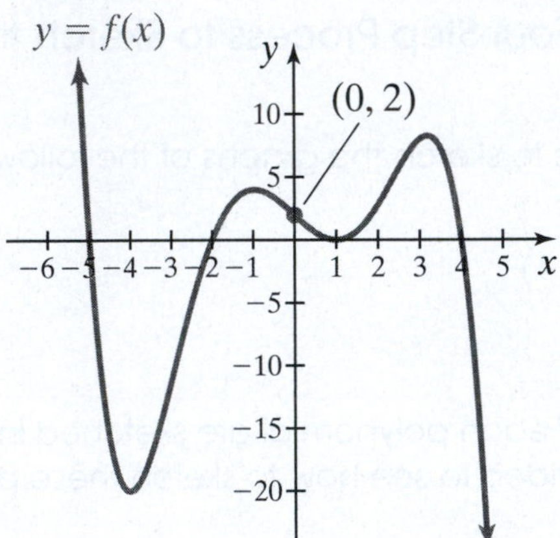

a. Is the degree of the polynomial function even or odd?

b. Is the leading coefficient positive or negative?

c. What is the value of the constant coefficient?

d. Identify the real zeros, and state the multiplicity of each.

e. Select from this list a possible function that could be represented by this graph.

i. $f(x) = -\frac{1}{20}(x + 5)(x + 2)(x - 1)^2(x - 4)$

ii. $f(x) = -\frac{1}{800}(x + 5)^2(x + 2)^2(x - 1)(x - 4)^2$

iii. $f(x) = \frac{1}{20}(x + 5)(x + 2)(x - 1)^2(x - 4)$

iv. $f(x) = -\frac{1}{10}(x + 5)(x + 2)(x - 1)^2(x - 4)$

Solution

a. The degree of the polynomial function is odd because the graph "starts" and "finishes" in opposite directions.

b. The leading coefficient is negative because the right-hand behavior "finishes down."

c. The y-intercept is 2, so $f(0) = 2$.

d. The real zeros are $-5, -2, 1$, and 4. The zeros that have odd multiplicity are $-5, -2$, and 4 because the graph crosses the x-axis at these zeros. The only zero of even multiplicity is 1 because the graph touches the x-axis at $x = 1$.

e. The best choice is item i. Watch the **interactive video** for a more detailed explanation.

You Try It **Work through the following You Try It problem.**

Work Exercises 45–47 in this textbook or in the MyLab Math Study Plan.

4.3 Exercises

In Exercises 1–6, determine whether the function is a polynomial function. For each function that is a polynomial, identify the degree, the leading coefficient, and the constant coefficient.

1. $f(x) = \frac{1}{3}x^7 - 8x^3 + \sqrt{5}x - 8$
2. $g(x) = \dfrac{8x^3 - 5x + 3}{7x^2}$
3. $h(x) = 2x^{\frac{2}{3}} - 8x^2 + 5$
4. $f(x) = \dfrac{3x - 5x^3}{11}$
5. $G(x) = 7x^4 + 1 + 3x - 9x^5$
6. $f(x) = 12$

In Exercises 7–15, use the graph of a power function and transformations to sketch the graph of each polynomial function.

7. $f(x) = -x^5$
8. $f(x) = x^6 + 1$
9. $f(x) = -x^5 - 2$
10. $f(x) = \frac{1}{2}(x - 1)^4$
11. $f(x) = 2(x + 1)^5$
12. $f(x) = -(x - 2)^6$
13. $f(x) = (x - 2)^5 - 3$
14. $f(x) = -(x + 2)^4 + 1$
15. $f(x) = -(x - 1)^5 + 2$

In Exercises 16–21, use the end behavior of the graph of the polynomial function to

a. determine whether the degree is even or odd and

b. determine whether the leading coefficient is positive or negative.

16.

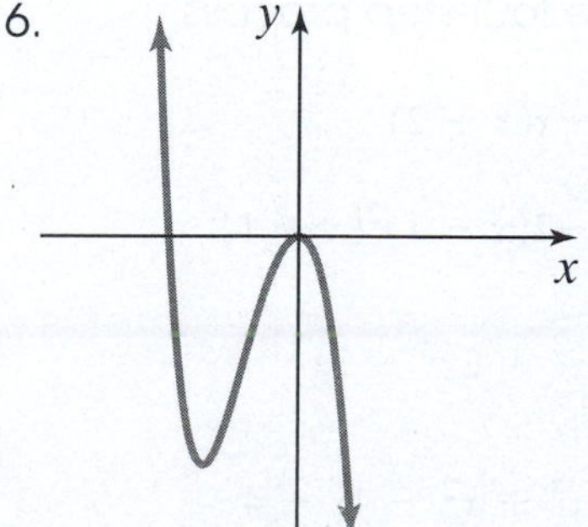

17.

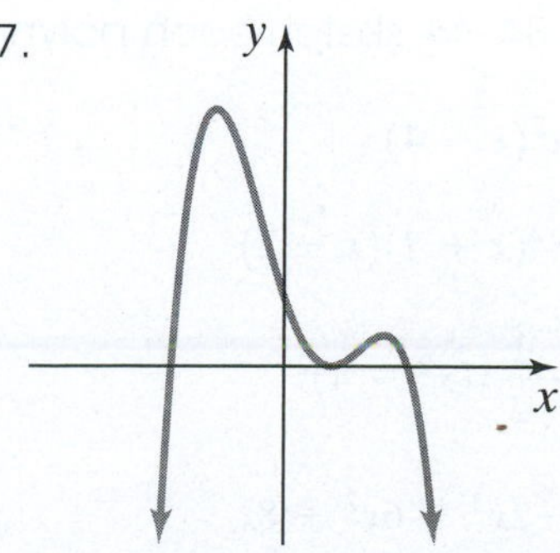

18. 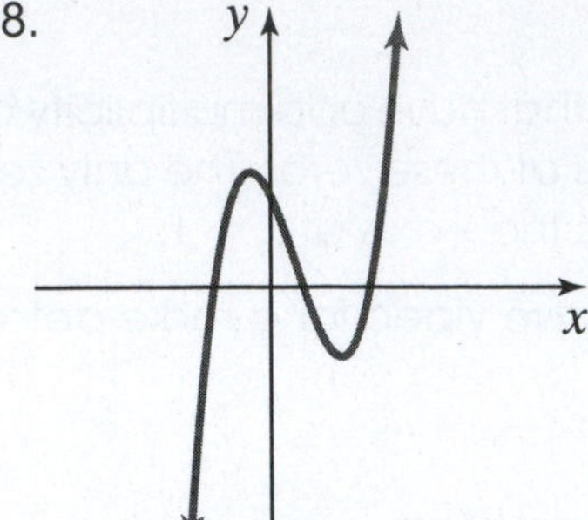

19. 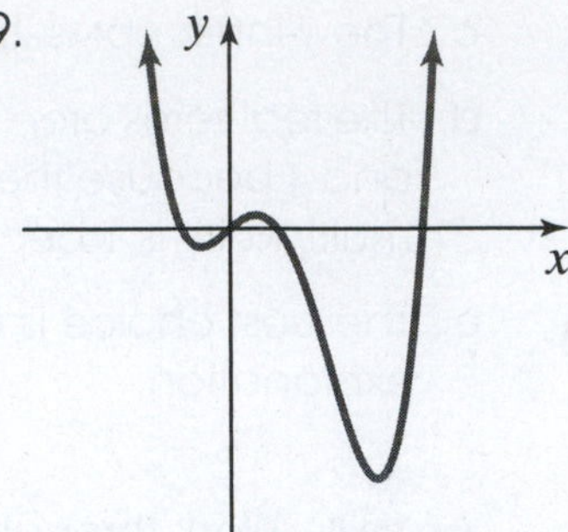

20. 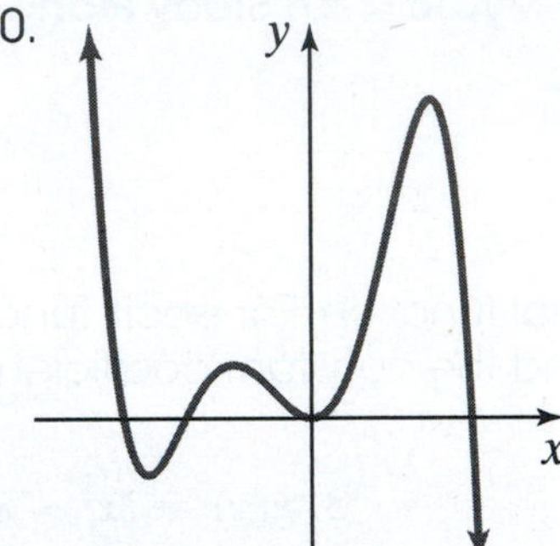

21. 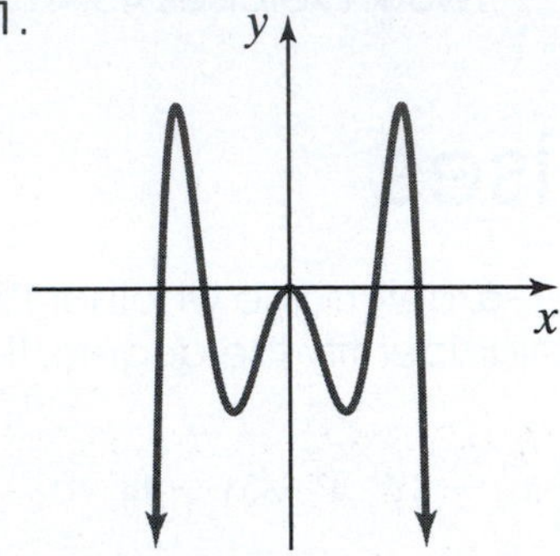

In Exercises 22–25, find the intercepts of each polynomial function.

22. $f(x) = x^2 - 3x - 10$

23. $f(x) = x^3 + 3x^2 - 16x - 48$

24. $f(x) = (x^2 + 1)\left(x - \frac{1}{2}\right)(x + 6)$

25. $f(x) = (x^2 - 9)(2 - x)$

In Exercises 26–33, determine the real zeros of each polynomial and their multiplicities. Then decide whether the graph touches or crosses the x-axis at each zero.

SbS 26. $f(x) = (x - 2)^2(x + 1)^3$

SbS 27. $f(x) = -(x + 1)(x + 3)^2$

SbS 28. $f(x) = x(x - 3)(x - 4)^4$

SbS 29. $f(x) = 3(x^2 - 2)^2$

SbS 30. $f(x) = -5(x - 2)(x^2 - 4)^3$

SbS 31. $f(x) = x^2(x^2 - x - 3)$

SbS 32. $f(x) = (x^2 - 3)(x + 1)^3$

SbS 33. $f(x) = -(x^2 - 7)^2(x + \sqrt{7})^3$

In Exercises 34–44, sketch each polynomial function using the four-step process.

SbS 34. $f(x) = x^2(x - 4)$

SbS 35. $f(x) = -x(x + 2)^2$

SbS 36. $f(x) = x^2(x + 2)(x - 2)$

SbS 37. $f(x) = -2(x - 1)^2(x + 1)$

SbS 38. $f(x) = -\frac{1}{2}x(x^2 - 4)^2$

SbS 39. $f(x) = x^4 + 2x^3 - 3x^2$

SbS 40. $f(x) = -2x^3 + 6x^2 + 8x$

SbS 41. $f(x) = x^3 + x^2 - 4x - 4$

SbS 42. $f(x) = -2x^3 - 4x^2 + 2x + 4$

SbS 43. $f(x) = x^4 + 2x^3 - 9x^2 - 18x$

SbS 44. $f(x) = -x^5 + 3x^4 + 9x^3 - 27x^2$

SbS 45. Analyze the graph to address the following about the polynomial function it represents.

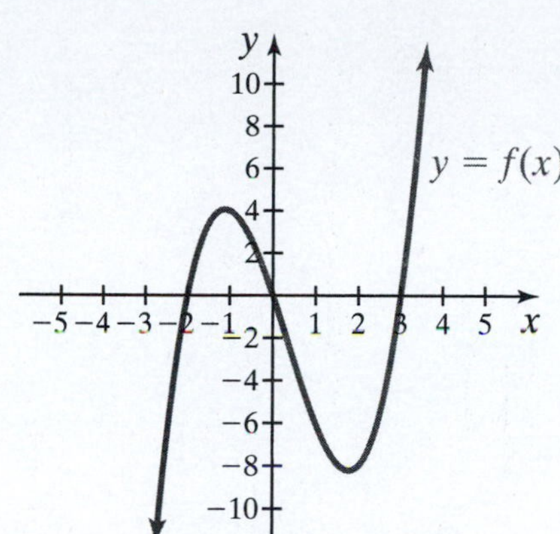

a. Is the degree of the polynomial function even or odd?

b. Is the leading coefficient positive or negative?

c. What is the value of the constant coefficient?

d. Identify the real zeros, and state the multiplicity of each.

e. Select from this list a possible function that could be represented by this graph.

i. $f(x) = -x(x + 2)(x - 3)$

ii. $f(x) = x(x + 2)(x - 3)$

iii. $f(x) = -x^3(x + 2)(x - 3)$

iv. $f(x) = (x + 2)(x - 3)$

SbS 46. Analyze the graph to address the following about the polynomial function it represents.

a. Is the degree of the polynomial function even or odd?

b. Is the leading coefficient positive or negative?

c. What is the value of the constant coefficient?

d. Identify the real zeros, and state the multiplicity of each.

e. Select from this list a possible function that could be represented by this graph.

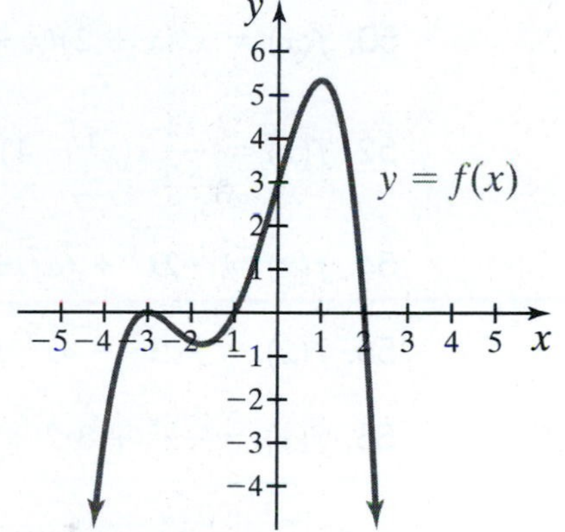

i. $f(x) = \frac{1}{4}(x + 3)(x + 1)^2(x - 2)^2$

ii. $f(x) = -\frac{1}{18}(x + 3)^3(x + 1)(x - 2)$

iii. $f(x) = -\frac{1}{6}(x + 3)^2(x - 1)(x + 2)$

iv. $f(x) = -\frac{1}{6}(x + 3)^2(x + 1)(x - 2)$

SbS 47. Analyze the graph to address the following about the polynomial function it represents.

a. Is the degree of the polynomial function even or odd?

b. Is the leading coefficient positive or negative?

c. What is the value of the constant coefficient?

d. Identify the real zeros, and state the multiplicity of each.

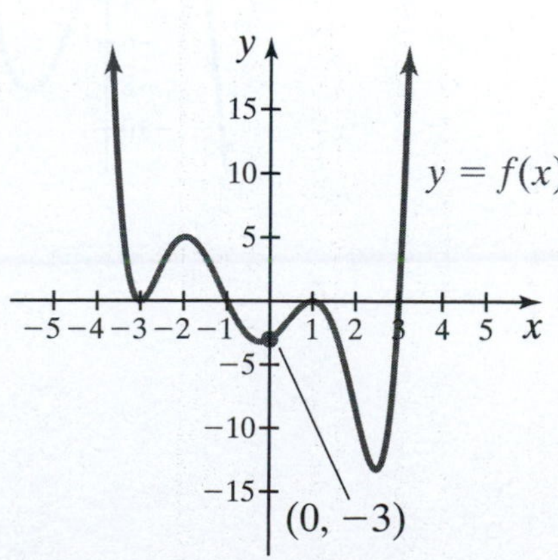

e. Select from this list a possible function that could be represented by this graph.

i. $f(x) = \frac{1}{6}(x + 3)^2(x + 1)(x - 1)^2(x - 3)$

ii. $f(x) = \frac{1}{9}(x + 3)^2(x + 1)(x - 1)^2(x - 3)$

iii. $f(x) = \frac{1}{3}(x + 3)^2(x + 1)(x - 1)^2(x - 3)$

iv. $f(x) = \frac{1}{12}(x + 3)^2(x + 1)(x - 1)^2(x - 3)$

Brief Exercises

In Exercises 48–58, sketch each polynomial function using the four-step process.

48. $f(x) = x^2(x - 4)$

49. $f(x) = -x(x + 2)^2$

50. $f(x) = x^2(x + 2)(x - 2)$

51. $f(x) = -2(x - 1)^2(x + 1)$

52. $f(x) = -\frac{1}{2}x(x^2 - 4)^2$

53. $f(x) = x^4 + 2x^3 - 3x^2$

54. $f(x) = -2x^3 + 6x^2 + 8x$

55. $f(x) = x^3 + x^2 - 4x - 4$

56. $f(x) = -2x^3 - 4x^2 + 2x + 4$

57. $f(x) = x^4 + 2x^3 - 9x^2 - 18x$

58. $f(x) = -x^5 + 3x^4 + 9x^3 - 27x^2$

In Exercises 59–61, select from a list a possible function that could be represented by the given graph.

59.

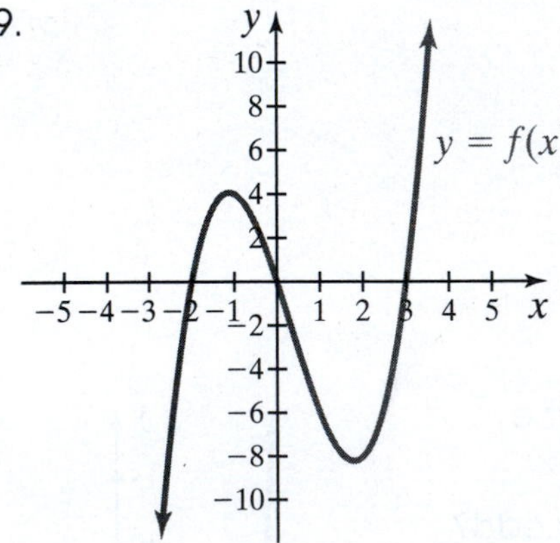

i. $f(x) = -x(x + 2)(x - 3)$

ii. $f(x) = x(x + 2)(x - 3)$

iii. $f(x) = -x^3(x + 2)(x - 3)$

iv. $f(x) = (x + 2)(x - 3)$

60.

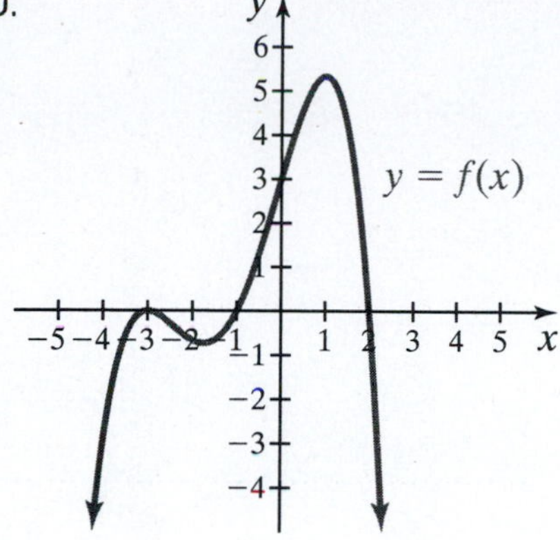

i. $f(x) = \frac{1}{4}(x + 3)(x + 1)^2(x - 2)^2$

ii. $f(x) = -\frac{1}{18}(x + 3)^3(x + 1)(x - 2)$

iii. $f(x) = -\frac{1}{6}(x - 3)^2(x - 1)(x + 2)$

iv. $f(x) = -\frac{1}{6}(x + 3)^2(x + 1)(x - 2)$

61.

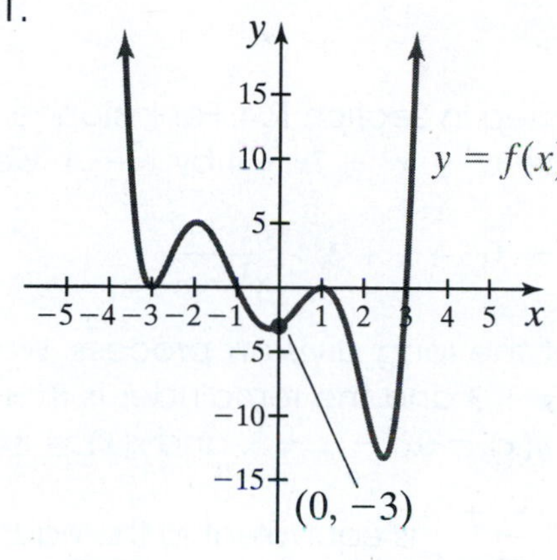

i. $f(x) = \frac{1}{6}(x + 3)^2(x + 1)(x - 1)^2(x - 3)$

ii. $f(x) = \frac{1}{9}(x + 3)^2(x + 1)(x - 1)^2(x - 3)$

iii. $f(x) = \frac{1}{3}(x + 3)^2(x + 1)(x - 1)^2(x - 3)$

iv. $f(x) = \frac{1}{12}(x + 3)^2(x + 1)(x - 1)^2(x - 3)$

4.4 Synthetic Division; The Remainder and Factor Theorems

THINGS TO KNOW

Before working through this section, be sure that you are familiar with the following concepts:

VIDEO ANIMATION INTERACTIVE

 You Try It

1. Dividing Polynomials Using Long Division (Section R.4) 

 You Try It

2. Simplifying Powers of i (Section 1.3) 

 You Try It

3. Determining the Intercepts of a Function (Section 3.2) 

 You Try It

4. Using Combinations of Transformations to Graph Functions (Section 3.4) 

 You Try It

5. Sketching the Graph of a Polynomial Function (Section 4.3) 

OBJECTIVES

OBJECTIVE 1 USING THE DIVISION ALGORITHM

Long division of polynomials was discussed in **Section R.4**. For instance, we saw in Example 8 of Section R.4 that if we divide $3x^4 + x^3 + 7x + 4$ by $x^2 - 1$, we get

$$\frac{3x^4 + x^3 + 7x + 4}{x^2 - 1} = 3x^2 + x + 3 + \frac{8x + 7}{x^2 - 1}.$$

Watch the **video** to see each step of the long division process. We see from this division that the quotient is $3x^2 + x + 3$ and the remainder is $8x + 7$. Letting $f(x) = 3x^4 + x^3 + 7x + 4$, $d(x) = x^2 - 1$, $q(x) = 3x^2 + x + 3$, and $r(x) = 8x + 7$, then

$$\frac{3x^4 + x^3 + 7x + 4}{x^2 - 1} = 3x^2 + x + 3 + \frac{8x + 7}{x^2 - 1}$$ is equivalent to the equation

$$\frac{f(x)}{d(x)} = q(x) + \frac{r(x)}{d(x)}.$$

If we multiply both sides of these equations by the divisor $d(x)$, we get

$$(x^2 - 1)\frac{3x^4 + x^3 + 7x + 4}{x^2 - 1} = (x^2 - 1)(3x^2 + x + 3) + \frac{8x + 7}{x^2 - 1}(x^2 - 1),$$

which is equivalent to

$$d(x)\frac{f(x)}{d(x)} = d(x)\,q(x) + \frac{r(x)}{d(x)}d(x) \quad \text{or}$$

$3x^4 + x^3 + 7x + 4 = (x^2 - 1)(3x^2 + x + 3) + (8x + 7)$, which is equivalent to

$$f(x) = d(x)q(x) + r(x).$$

The equation $f(x) = d(x)q(x) + r(x)$ is used to check that the long division was done properly. The original polynomial should equal the product of the divisor and the quotient plus the remainder. This process is known as the **division algorithm**.

Division Algorithm

If $f(x)$ and $d(x)$ are polynomial functions with $d(x) \neq 0$ and the degree of $d(x)$ is less than or equal to the degree of $f(x)$, then unique polynomial functions $q(x)$ and $r(x)$ exist such that

$$f(x) = d(x)q(x) + r(x)$$

or equivalently

$$\frac{f(x)}{d(x)} = q(x) + \frac{r(x)}{d(x)},$$

where the remainder $r(x) = 0$ or is of degree less than the degree of $d(x)$.

Example 1 Use the Division Algorithm

Given the polynomials $f(x) = 4x^4 - 3x^3 + 5x - 6$ and $d(x) = x - 2$, find polynomial functions $q(x)$ and $r(x)$, and write $f(x)$ in the form of $f(x) = d(x)q(x) + r(x)$.

Solution We use long division inserting the term $0x^2$ into the dividend as a placeholder to make sure that all columns of the long division process line up properly.

$$\begin{array}{r} 4x^3 + 5x^2 + 10x + 25 \\ x - 2\overline{)4x^4 - 3x^3 + 0x^2 + 5x - 6} \\ \underline{4x^4 - 8x^3} \qquad\qquad\qquad \\ 5x^3 + 0x^2 \qquad\qquad \\ \underline{5x^3 - 10x^2} \qquad\qquad \\ 10x^2 + 5x \qquad \\ \underline{10x^2 - 20x} \qquad \\ 25x - 6 \\ \underline{25x - 50} \\ 44 \end{array}$$

The process is complete because 44 is of lesser degree than the divisor, $x - 2$. Therefore, $q(x) = 4x^3 + 5x^2 + 10x + 25$ and $r(x) = 44$, so

$$4x^4 - 3x^3 + 5x - 6 = (x - 2)(4x^3 + 5x^2 + 10x + 25) + 44.$$

Notice in Example 1 that the remainder, $r(x)$, was a constant. We know that the remainder must always be of degree less than the degree of the divisor. Therefore, if a polynomial $f(x)$ is divided by a first-degree polynomial, then the remainder must be a constant.

Corollary to the Division Algorithm

If a polynomial $f(x)$ of degree greater than or equal to 1 is divided by a polynomial $d(x)$ where $d(x)$ is of degree 1, then there exists a unique polynomial function $q(x)$ and a constant r such that

$$f(x) = d(x)q(x) + r.$$

OBJECTIVE 2 USING SYNTHETIC DIVISION

If a polynomial $f(x)$ is divided by $d(x) = x - c$, then we can use a "shortcut method" called **synthetic division** to find the quotient, $q(x)$, and the remainder, r. To illustrate synthetic division, we do a side-by-side comparison of long division and synthetic division using the polynomial function of Example 1.

Long Division

$$\begin{array}{r}
4x^3 + 5x^2 + 10x + 25 \\
x - 2\overline{)4x^4 - 3x^3 + 0x^2 + 5x - 6} \\
\underline{4x^4 - 8x^3} \qquad\qquad\qquad \\
5x^3 + 0x^2 \qquad\qquad \\
\underline{5x^3 - 10x^2} \qquad\qquad \\
10x^2 + 5x \qquad \\
\underline{10x^2 - 20x} \qquad \\
25x - 6 \\
\underline{25x - 50} \\
44
\end{array}$$

Synthetic Division

$$\begin{array}{r|rrrrr}
2 & 4 & -3 & 0 & 5 & -6 \\
 & & 8 & 10 & 20 & 50 \\
\hline
 & 4 & 5 & 10 & 25 & 44
\end{array}$$

You can see how much more compact the synthetic division process is compared to long division. This is how it works.

Step 1. The constant coefficient, c, of the divisor $d(x) = x - c$ is written to the far left, whereas all coefficients of the polynomial $f(x)$ are written inside the symbol $\underline{|\quad\quad}$. Once again, be sure to include the 0 for $0x^2$.

This is c in $x - c$. ⟶ $2\,\underline{|\,4 \quad -3 \quad 0 \quad 5 \quad -6}$

These are the coefficients of $f(x) = 4x^4 - 3x^3 + 5x - 6$.

Step 2. Bring down the leading coefficient 4.

$$\begin{array}{r|rrrrr}
2 & 4 & -3 & 0 & 5 & -6 \\
 & \downarrow & & & & \\
\hline
 & 4 & & & &
\end{array}$$

Step 3. Multiply c times the leading coefficient that was just brought down. In this case, we multiply $2 \cdot 4$. The product (8 in this case) is written in the next column in the second row.

Multiply $2 \cdot 4 = 8$

$$\begin{array}{r|rrrrr}
2 & 4 & -3 & 0 & 5 & -6 \\
 & \downarrow & 8 & & & \\
\hline
 & 4 & & & &
\end{array}$$

Step 4. Add down the column and write the sum (5 in this case) in the bottom row.

$$\begin{array}{r|rrrrr}
2 & 4 & -3 & 0 & 5 & -6 \\
 & & 8 & & & \\
\hline
 & 4 & 5 & & &
\end{array}$$

We now repeat this process by multiplying $c = 2$ times the value in the last row, always adding down the columns.

Multiply $2 \cdot 5 = 10$

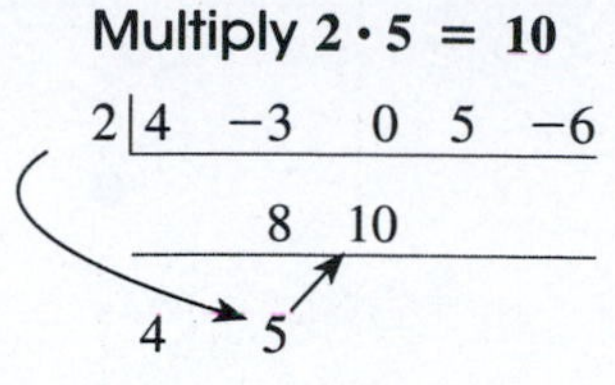

Add $0 + 10 = 10$

$$\begin{array}{r|rrrrr} 2 & 4 & -3 & 0 & 5 & -6 \\ & & 8 & 10 & & \\ \hline & 4 & 5 & 10 & & \end{array}$$

Multiply $2 \cdot 10 = 20$

$$\begin{array}{r|rrrrr} 2 & 4 & -3 & 0 & 5 & -6 \\ & & 8 & 10 & 20 & \\ \hline & 4 & 5 & 10 & & \end{array}$$

Add $5 + 20 = 25$

$$\begin{array}{r|rrrrr} 2 & 4 & -3 & 0 & 5 & -6 \\ & & 8 & 10 & 20 & \\ \hline & 4 & 5 & 10 & 25 & \end{array}$$

Multiply $2 \cdot 25 = 50$

$$\begin{array}{r|rrrrr} 2 & 4 & -3 & 0 & 5 & -6 \\ & & 8 & 10 & 20 & 50 \\ \hline & 4 & 5 & 10 & 25 & \end{array}$$

Add $-6 + 50 = 44$

$$\begin{array}{r|rrrrr} 2 & 4 & -3 & 0 & 5 & -6 \\ & & 8 & 10 & 20 & 50 \\ \hline & 4 & 5 & 10 & 25 & 44 \end{array}$$

The last row now represents the quotient and the remainder. The last entry of the bottom row (44 in this case) is the remainder. The other numbers of the last row represent the coefficients of the quotient.

$$\begin{array}{r|rrrrr} 2 & 4 & -3 & 0 & 5 & -6 \\ & & 8 & 10 & 20 & 50 \\ \hline & 4 & 5 & 10 & 25 & 44 \end{array}$$

remainder $r = 44$

coefficients of $q(x) = 4x^3 + 5x^2 + 10x + 25$

You can watch the **video** to see this example worked out. Whenever possible, synthetic division is used throughout the rest of this chapter.

CAUTION **Synthetic division can only be used when the divisor $d(x)$ has the form $(x - c)$.**

Example 2 Use Synthetic Division to Find the Quotient and Remainder

Use synthetic division to divide $f(x)$ by $(x - c)$, and then write $f(x)$ in the form $f(x) = (x - c)q(x) + r$ for $f(x) = -2x^4 + 3x^3 + 7x^2 - x + 5$ divided by $x + 1$.

Solution The divisor $x + 1$ or $x - (-1)$ is of the form $x - c$ with $c = -1$ so we can use synthetic division. We start by writing $c = -1$ on the far left, putting the coefficients of f on the inside and bringing down the -2:

$$\begin{array}{r|rrrrr} -1 & -2 & 3 & 7 & -1 & 5 \\ & \downarrow & & & & \\ \hline & -2 & & & & \end{array}$$

Using a series of multiplications and additions, we can complete the synthetic division process:

$$\begin{array}{r|rrrrr} -1 & -2 & 3 & 7 & -1 & 5 \\ & \downarrow & 2 & -5 & -2 & 3 \\ \hline & -2 & 5 & 2 & -3 & 8 \end{array}$$

Watch the **video** to see each step. We can see from the last entry of the bottom row that the remainder is 8. Thus,

$$q(x) = -2x^3 + 5x^2 + 2x - 3 \quad \text{and} \quad r = 8, \text{ so}$$
$$f(x) = -2x^4 + 3x^3 + 7x^2 - x + 5$$
$$= (x + 1)(-2x^3 + 5x^2 + 2x - 3) + 8$$

You Try It Work through the following You Try It problem.

Work Exercises 1–10 in this textbook or in the MyLab Math Study Plan.

OBJECTIVE 3 USING THE REMAINDER THEOREM

In Example 2, we see that the remainder is 8 when $f(x) = -2x^4 + 3x^3 + 7x^2 - x + 5$ is divided by $x + 1$. If we evaluate f at $x = -1$, we get

$$f(-1) = -2(-1)^4 + 3(-1)^3 + 7(-1)^2 - (-1) + 5$$
$$= -2(1) + 3(-1) + 7(1) + 1 + 5$$
$$= -2 - 3 + 7 + 1 + 5 = 8$$

It is no coincidence that $f(-1)$ is equal to the remainder. If we rewrite $f(x) = -2x^4 + 3x^3 + 7x^2 - x + 5$ in the form of $f(x) = (x + 1)(-2x^3 + 5x^2 + 2x - 3) + 8$ as in Example 2, we see that

$$f(-1) = \underbrace{(-1 + 1)(-2(-1)^3 + 5(-1)^2 + 2(-1) - 3)}_{=\;0} + 8 = 8$$

Therefore, $f(-1) = 8$, and we can conclude that if a polynomial is divided by $x - c$, then the remainder is the value of $f(c)$. This is known as the **remainder theorem.**

> **Remainder Theorem**
>
> If a polynomial $f(x)$ is divided by $x - c$, then the remainder is $f(c)$.

To see why the remainder theorem is true, read this **proof**.

Example 3 Use the Remainder Theorem

Use the remainder theorem to find the remainder when $f(x)$ is divided by $x - c$.

a. $f(x) = 5x^4 - 8x^2 + 3x - 1;\ x - 2$

b. $f(x) = 3x^3 + 5x^2 - 5x - 6;\ x + 2$

Solution

a. We could use long division or synthetic division to find the remainder, but the remainder theorem says that the remainder must be $f(2)$.

$$f(2) = 5(2)^4 - 8(2)^2 + 3(2) - 1 = 80 - 32 + 6 - 1 = 53$$

Thus, the remainder is 53.

b. To find the remainder when $f(x) = 3x^3 + 5x^2 - 5x - 6$ is divided by $x + 2 = x - (-2)$, we must evaluate $f(-2)$.

$$\begin{aligned} f(-2) &= 3(-2)^3 + 5(-2)^2 - 5(-2) - 6 \\ &= -24 + 20 + 10 - 6 = 0 \end{aligned}$$

Therefore, the remainder is 0. ●

You Try It Work through the following You Try It problem.

Work Exercises 11–16 in this textbook or in the MyLab Math Study Plan.

OBJECTIVE 4 USING THE FACTOR THEOREM

In Example 3, we see that when $f(x) = 3x^3 + 5x^2 - 5x - 6$ is divided by $x + 2$, then the remainder is 0. The remainder of 0 implies that $x + 2$ divides $f(x)$ *even*, and therefore, $x + 2$ is a **factor** of f. In fact, by the remainder theorem, if $f(c) = 0$, then the remainder when a polynomial $f(x)$ is divided by $x - c$ must be zero. Conversely, if $x - c$ is a factor of $f(x)$, then $f(c)$ must equal zero. The last two statements can be rewritten as the following important theorem known as the **factor theorem.**

> **Factor Theorem**
>
> The polynomial $x - c$ is a factor of the polynomial $f(x)$ **if and only if*** $f(c) = 0$.

Example 4 Use the Factor Theorem to Determine whether $x - c$ Is a Factor of $f(x)$

Determine whether $x + 3$ is a factor of $f(x) = 2x^3 + 7x^2 + 2x - 3$.

Solution By the factor theorem, $x + 3$ is a factor of $f(x)$ if and only if $f(-3) = 0$. We can use synthetic division to determine the value of $f(-3)$.

$$\begin{array}{r|rrrr} -3 & 2 & 7 & 2 & -3 \\ & \downarrow & -6 & -3 & 3 \\ \hline & 2 & 1 & -1 & 0 \end{array} \longleftarrow f(-3) = 0$$

The remainder $f(-3)$ is equal to zero, and therefore, $x + 3$ is a factor of $f(x)$. Furthermore, using the bottom row of the synthetic division, we can rewrite $f(x)$ as $f(x) = (x + 3)(2x^2 + x - 1)$. ●

You Try It Work through the following You Try It problem.

Work Exercises 17–26 in this textbook or in the MyLab Math Study Plan.

OBJECTIVE 5 SKETCHING THE GRAPH OF A POLYNOMIAL FUNCTION

In Example 4, we show that $x + 3$ is a factor of $f(x) = 2x^3 + 7x^2 + 2x - 3$. Using synthetic division and the division algorithm, we were able to rewrite $f(x)$ as

$f(x) = (x + 3)(2x^2 + x - 1)$. Because the quadratic function $2x^2 + x - 1$ can also be factored, we can now completely factor $f(x)$ as

$f(x) = 2x^3 + 7x^2 + 2x - 3$	Write the original polynomial.
$f(x) = (x + 3)(2x^2 + x - 1)$	Because $f(-3) = 0$, $x + 3$ is a factor (See Example 4).
$f(x) = (x + 3)(2x - 1)(x + 1)$	Factor $2x^2 + x - 1$ as $(2x - 1)(x + 1)$.
$f(x) = 2(x + 3)\left(x - \frac{1}{2}\right)(x + 1)$	Factor $(2x - 1)$ as $2\left(x - \frac{1}{2}\right)$.

The polynomial $f(x) = 2(x + 3)\left(x - \frac{1}{2}\right)(x + 1)$ is now said to be written in **completely factored form**. Because the polynomial is in completely factored form, we can follow the four-step process discussed in **Section 4.3** to sketch the graph of $f(x)$. The graph of $f(x) = 2(x + 3)\left(x - \frac{1}{2}\right)(x + 1)$ is sketched in **Figure 19**. Watch this **video** to see how to use the four-step process to sketch this polynomial.

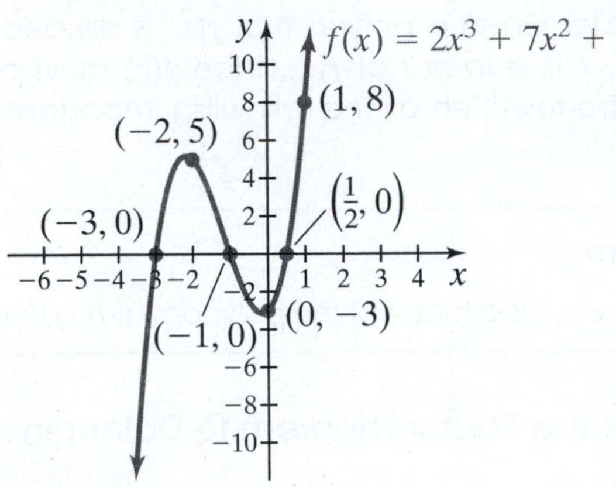

Figure 19 Graph of $f(x) = 2x^3 + 7x^2 + 2x - 3$.

Example 5 Factor Completely and Sketch the Graph of a Polynomial Function

Given that $x = 2$ is a zero of $f(x) = x^3 - 6x + 4$, completely factor f and sketch its graph.

Solution We begin by using the synthetic division process to divide $f(x) = x^3 - 6x + 4$ by $x - 2$, remembering to use a 0 for the coefficient of the x^2-term.

$$\begin{array}{r|rrrr} 2 & 1 & 0 & -6 & 4 \\ & \downarrow & 2 & 4 & -4 \\ \hline & 1 & 2 & -2 & 0 \end{array}$$

We see from the synthetic division process that we can rewrite $f(x)$ as $f(x) = (x - 2)(x^2 + 2x - 2)$. Because the quadratic $x^2 + 2x - 2$ does not factor with integer coefficients, we can use the quadratic formula or complete the square to find the zeros of $g(x) = x^2 + 2x - 2$. The zeros of $g(x) = x^2 + 2x - 2$ are $x = -1 + \sqrt{3} \approx .732$ and $x = -1 - \sqrt{3} \approx -2.732$. Because $x = -1 + \sqrt{3}$ is a zero of $g(x) = x^2 + 2x - 2$, this implies that $x - (-1 + \sqrt{3})$ is a factor of $x^2 + 2x - 2$. Similarly, $x - (-1 - \sqrt{3})$ is also a factor of $x^2 + 2x - 2$.

Thus, the complete factorization of f is $f(x) = (x - 2)\big(x - (-1 + \sqrt{3})\big)\big(x - (-1 - \sqrt{3})\big)$. Once again, we can now use the four-step process to sketch the graph seen in Figure 20. Watch the **video** to see this example worked out in detail.

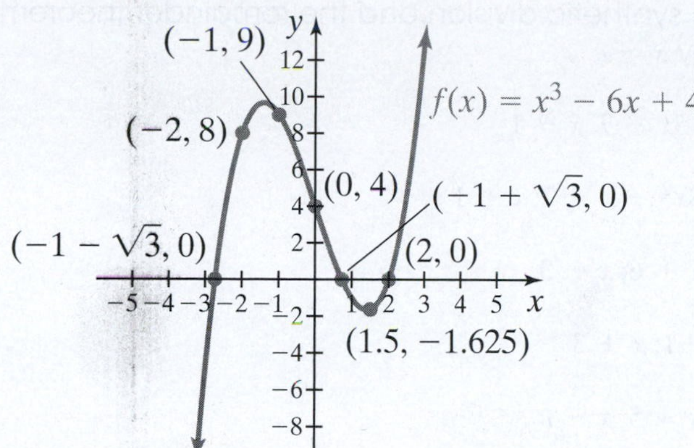

Figure 20 Graph of $f(x) = x^3 - 6x + 4$.

You Try It Work through the following You Try It problem.

Work Exercises 27–34 in this textbook or in the MyLab Math Study Plan.

4.4 Exercises

Skill Check Exercises

SCE-1. Find the quotient and remainder when $f(x) = 9x^4 - 2x^2 + 2x + 9$ is divided by $x^2 + 5$.

SCE-2. Find the quotient and remainder when $6x^5 - 5x^2 + 7x + 6$ is divided by $2x^3 - 1$.

SCE-3. Find the quotient and remainder when $7 - x^2 + x^4$ is divided by $x^2 + x + 9$.

In Exercises 1–10, use synthetic division to divide $f(x)$ by $x - c$, and then write $f(x)$ in the form $f(x) = (x - c)q(x) + r$.

1. $f(x) = x^3 - 3x^2 + x - 5; x - 1$

2. $f(x) = x^3 + 5x^2 - 3x - 1; x + 1$

3. $f(x) = 6x^3 - 2x + 3; x - 2$

4. $f(x) = 2x^3 - x^2 + 4; x + 3$

5. $f(x) = x^4 + x^3 - x^2 - x - 4; x - 2$

6. $f(x) = -2x^4 + 3x^3 - x^2 + 7x + 2; x + 1$

7. $f(x) = -x^4 - x^3 - 5x; x + 4$

8. $f(x) = 3x^5 - x^2 - 2; x + 1$

9. $f(x) = x^3 - 1; x - 1$

10. $f(x) = x^5 + 1; x + 1$

In Exercises 11–16, use synthetic division and the remainder theorem to find the remainder when $f(x)$ is divided by $x - c$.

11. $f(x) = 2x^3 - 3x^2 - 5x - 7; x - 1$

12. $f(x) = x^4 + 3x^3 - 8x^2 - x + 2; x + 1$

13. $f(x) = x^5 - 2x^2 + x + 6; x - 3$

14. $f(x) = -4x^3 + 2x - 1; x + 3$

15. $f(x) = x^3 - 3x^2 + x - 5; x - i$

16. $f(x) = x^4 - x^3 - 2x^2 - x - 3; x + i$

In Exercises 17–26, use synthetic division and the factor theorem to determine whether $x - c$ is a factor of $f(x)$.

17. $f(x) = 2x^3 - 5x^2 + 7x - 4; x - 1$

18. $f(x) = 2x^3 - 5x^2 + 7x - 4; x + 1$

19. $f(x) = x^5 + x^4 - 3x^3 - 3x^2 - 4x - 4; x - 2$

20. $f(x) = x^5 + x^4 - 3x^3 - 3x^2 - 4x - 4; x + 2$

21. $f(x) = 2x^3 + 5x^2 - 8x - 5; x - \frac{1}{2}$

22. $f(x) = 2x^3 + 5x^2 - 8x - 5; x + \frac{1}{2}$

23. $f(x) = x^4 + x^3 + x^2 - 2x - 6; x - \sqrt{2}$

24. $f(x) = x^4 + x^3 + x^2 - 2x - 6; x + \sqrt{2}$

25. $f(x) = 3x^4 - x^3 - x^2 - x - 4; x - i$

26. $f(x) = x^3 + x^2 + 3x - 5; x - (1 + 2i)$

In Exercises 27–34, find the remaining zeros of $f(x)$ given that c is a zero. Then rewrite $f(x)$ in completely factored form and sketch its graph.

SbS 27. $f(x) = x^3 + 2x^2 - x - 2; c = 1$ is a zero

SbS 28. $f(x) = -x^3 + 5x^2 - 3x - 9; c = -1$ is a zero

SbS 29. $f(x) = 2x^3 + 7x^2 + 2x - 3; c = \frac{1}{2}$ is a zero

SbS 30. $f(x) = \frac{1}{2}x^3 - \frac{3}{2}x - 1; c = 2$ is a zero

SbS 31. $f(x) = -\frac{1}{3}x^3 - \frac{4}{3}x^2 - \frac{1}{3}x + 2$; $c = -3$ is a zero

SbS 32. $f(x) = 3x^3 + 11x^2 + 8x - 4$; $c = -2$ is a zero of multiplicity 2

SbS 33. $f(x) = 4x^4 + 5x^3 - 3x^2 - 5x - 1$; $c = -1$ is a zero of multiplicity 2

SbS 34. $f(x) = 3x^5 - 5x^4 - 2x^3 + 6x^2 - x - 1$; $c = 1$ is a zero of multiplicity 3

Brief Exercises

In Exercises 35–42, find the remaining zeros of $f(x)$ given that c is a zero. Then rewrite $f(x)$ in completely factored form.

35. $f(x) = x^3 + 2x^2 - x - 2$; $c = 1$ is a zero

36. $f(x) = -x^3 + 5x^2 - 3x - 9$; $c = -1$ is a zero

37. $f(x) = 2x^3 + 7x^2 + 2x - 3$; $c = \frac{1}{2}$ is a zero

38. $f(x) = \frac{1}{2}x^3 - \frac{3}{2}x - 1$; $c = 2$ is a zero

39. $f(x) = -\frac{1}{3}x^3 - \frac{4}{3}x^2 - \frac{1}{3}x + 2$; $c = -3$ is a zero

40. $f(x) = 3x^3 + 11x^2 + 8x - 4$; $c = -2$ is a zero of multiplicity 2

41. $f(x) = 4x^4 + 5x^3 - 3x^2 - 5x - 1$; $c = -1$ is a zero of multiplicity 2

42. $f(x) = 3x^5 - 5x^4 - 2x^3 + 6x^2 - x - 1$; $c = 1$ is a zero of multiplicity 3

4.5 The Zeros of Polynomial Functions; The Fundamental Theorem of Algebra

THINGS TO KNOW

Before working through this section, be sure that you are familiar with the following concepts:

You Try It

1. Sketching the Graph of a Polynomial Function (Section 4.3)

You Try It

2. Using the Remainder Theorem (Section 4.4)

You Try It

3. Using the Factor Theorem (Section 4.4)

You Try It

4. Sketching the Graph of a Polynomial Function (Section 4.4)

(eText Screens 4.5-1–4.5-31)

INTRODUCTION

Read this introduction before beginning Objective 1.

OBJECTIVES

1 Using the Rational Zeros Theorem

2 Using Descartes' Rule of Signs

3 Finding the Zeros of a Polynomial Function

4 Solving Polynomial Equations

5 Using the Complex Conjugate Pairs Theorem

6 Using the Intermediate Value Theorem

7 Sketching the Graphs of Polynomial Functions

SECTION 4.5 EXERCISES

Introduction to Section 4.5

The goal of this section is to be able to find the zeros of polynomial functions, or equivalently, we are interested in solving polynomial equations of the form $f(x) = 0$. To begin our discussion, we must take a look at a very important theorem that will help us create a list of potential zeros of a polynomial function.

OBJECTIVE 1 USING THE RATIONAL ZEROS THEOREM

In Example 5 of Section 4.4, we were able to find all real zeros of $f(x) = x^3 - 6x + 4$ given that $x = 2$ was a zero. You may want to work through this **video** to review this example. Because we knew that $x = 2$ was a zero, we were able to use the **factor theorem** and synthetic division to rewrite f as $f(x) = (x - 2)(x^2 + 2x - 2)$. We then solved the quadratic equation $x^2 + 2x - 2 = 0$ to find the other two zeros.

But what if we were not given the fact that $x = 2$ was a zero? How do we start looking for the zeros? Is there a systematic way to determine possible zeros of a polynomial function? The answer to this question is yes *if* we are given a polynomial with **integer** coefficients. If a polynomial has integer coefficients, then we are able to create a list of the **potential rational zeros** using the **rational zeros theorem**.

Rational Zeros Theorem

Let f be a polynomial function of the form $f(x) = a_n x^n + a_{n-1}x^{n-1} + a_{n-2}x^{n-2} + \cdots + a_1 x + a_0$ of degree $n \geq 1$, where each **coefficient** is an integer. If $\frac{p}{q}$ is a **rational zero** of f (where $\frac{p}{q}$ is written in lowest terms), then p must be a factor of the constant coefficient, a_0, and q must be a factor of the leading coefficient, a_n.

CAUTION The rational zeros theorem can only provide us with a list of possible rational zeros. It ***does not*** guarantee that the polynomial will have a zero from the list. Simply stated, if a polynomial with integer coefficients has a rational zero, then it must be on the list created using this theorem. Example 1 illustrates how this theorem works.

Example 1 Use the Rational Zeros Theorem

Use the rational zeros theorem to determine the potential rational zeros of the polynomial function $f(x) = 4x^4 - 7x^3 + 9x^2 - x - 10$.

Solution Because f has integer coefficients, we can use the rational zeros theorem to create a list of **potential** rational zeros. The factors of the constant coefficient, -10, are $p = \pm 1, \pm 2, \pm 5, \pm 10$, whereas the factors of the leading coefficient, 4, are $q = \pm 1, \pm 2, \pm 4$. Listing all possibilities of $\frac{p}{q}$, we get

$$\frac{p}{q}: \pm 1, \ \pm 2, \ \pm 5, \ \pm 10, \ \pm\frac{1}{2}, \ \pm\frac{1}{4}, \ \pm\frac{5}{2}, \ \pm\frac{5}{4}.$$

Thus, if f has a rational zero, then it must appear on this list.

You Try It Work through the following You Try It problem.

Work Exercises 1–6 in this textbook or in the MyLab Math Study Plan.

OBJECTIVE 2 USING DESCARTES' RULE OF SIGNS

The French Mathematician **René Descartes** first described a rule for determining the number of positive or negative real zeros of a polynomial function. The rule, which is known as **Descartes' Rule of Signs**, is based on the number of variations in signs in a given polynomial function. A variation in sign occurs when successive nonzero terms of a polynomial function change from positive to negative or from negative to positive. Note that the term *nonzero* indicates that we do not consider terms with a zero coefficient.

Descartes' Rule of Signs

Let f be a polynomial function with real coefficients written in descending order.

1. The number of *positive* real zeros of f is equal to the number of variations in sign of $f(x)$ or is less than the number of variations in sign of $f(x)$ by a positive even integer.

   If $f(x)$ has one variation in sign, then f has exactly one positive real zero.

2. The number of *negative* real zeros of f is equal to the number of variations in sign of $f(-x)$ or is less than the number of variations in sign of $f(-x)$ by a positive even integer.

   If $f(-x)$ has one variation in sign, then f has exactly one negative real zero.

Example 2 Use Descartes' Rule of Signs

Determine the possible number of positive real zeros and negative real zeros of each polynomial function.

a. $f(x) = 2x^5 + 3x^4 + 8x^2 + 2x$ b. $f(x) = 6x^4 + 13x^3 + 61x^2 + 8x - 10$

Solution

a. The polynomial function $f(x) = 2x^5 + 3x^4 + 8x^2 + 2x$ is written in descending order and has real coefficients, thus we can apply Descartes' Rule of Signs.

1. All coefficients of $f(x)$ are positive. Therefore, there are no variations in sign. Therefore, there are no positive real zeros.

2. To determine the number of possible negative real zeros, we must determine the number of variations in sign of $f(-x)$. We obtain $f(-x)$ by replacing x with $-x$ in the given polynomial function.

$$f(-x) = 2(-x)^5 + 3(-x)^4 + 8(-x)^2 + 2(-x) = -2x^5 + 3x^4 + 8x^2 - 2x$$

We now count the number of variations in sign of $f(-x)$.

$$f(-x) = \underbrace{-2x^5 + 3x^4}_{\text{variation in sign}} + \underbrace{8x^2 - 2x}_{\text{variation in sign}}$$

There are two variations in sign. Therefore there are 2 or $2 - 2 = 0$ possible negative real zeros.

b. The polynomial function $f(x) = 6x^4 + 13x^3 + 61x^2 + 8x - 10$ is written in descending order and has real coefficients, thus we can apply Descartes' Rule of Signs.

1. Count the number of variations in sign of $f(x)$.

$$f(x) = 6x^4 + 13x^3 + 61x^2 \underbrace{+ 8x - 10}_{\text{variation in sign}}$$

There is one variation in sign. Therefore, f has exactly one positive real zero.

2. We now count the number of variations in sign of $f(-x)$.

$$f(-x) = 6(-x)^4 + 13(-x)^3 + 61(-x)^2 + 8(-x) - 5$$
$$= \underbrace{6x^4 -}_{\text{variation in sign}} \underbrace{13x^3 +}_{\text{variation in sign}} \underbrace{61x^2 -}_{\text{variation in sign}} 8x - 5$$

There are three variations in sign. Therefore, there are 3 or $3 - 2 = 1$ possible negative real zeros.

You Try It Work through this You Try It problem.

Work Exercises 7–10 in this textbook or in the MyLab Math Study Plan.

OBJECTIVE 3 FINDING THE ZEROS OF A POLYNOMIAL FUNCTION

Before we begin to find the zeros of polynomial functions, we should know how many zeros to expect. The fundamental theorem of algebra says that every polynomial of degree $n \geq 1$ has at least one **complex zero**.

> **Fundamental Theorem of Algebra**
>
> Every polynomial function of degree $n \geq 1$ has at least one **complex zero**.

Suppose $f(x)$ is a polynomial function of degree $n \geq 1$. By the fundamental theorem of algebra, f has at least one complex zero, call it c_1. By the **factor theorem**, $x - c_1$ is a factor of $f(x)$, and it follows that we can rewrite $f(x)$ as $f(x) = (x - c_1)q_1(x)$, where $q_1(x)$ is another polynomial. If $q_1(x)$ is of degree 1 or more, then we repeat the process again and rewrite $f(x)$ as $f(x) = (x - c_1)(x - c_2)q_2(x)$, where c_2 is a zero of $q_1(x)$. If $f(x)$ is a degree n polynomial, then we can repeat this process a total of n times to rewrite $f(x)$ as $f(x) = a(x - c_1)(x - c_2)(x - c_3) \cdots (x - c_n)$, where $c_1, c_2, c_3 \ldots c_n$ are zeros of $f(x)$ and a is the leading coefficient.

Therefore, every polynomial of degree n has n **complex zeros** and can be written in **completely factored form**. Note that some of the zeros could be the same and need to be counted each time, as indicated by the following theorem.

> **Number of Zeros Theorem**
>
> Every polynomial of degree n has n **complex zeros** provided each zero of multiplicity greater than 1 is counted accordingly.

Example 3 Find the Zeros of a Polynomial Function

Find all zeros of $f(x) = 6x^4 + 13x^3 + 61x^2 + 8x - 10$ and rewrite $f(x)$ in completely factored form.

Solution This is a degree 4 polynomial so there must be four complex zeros. The factors of the constant coefficient, -10, are $p = \pm 1, \pm 2, \pm 5, \pm 10$. The factors of the leading coefficient, 6, are $q = \pm 1, \pm 2, \pm 3, \pm 6$. By the rational zeros theorem, the list of possible rational zeros are

$$\frac{p}{q}: \pm 1, \pm 2, \pm 5, \pm 10, \pm\frac{1}{2}, \pm\frac{1}{3}, \pm\frac{1}{6}, \pm\frac{2}{3}, \pm\frac{5}{2}, \pm\frac{5}{3}, \pm\frac{5}{6}, \pm\frac{10}{3}.$$

Using Descartes' Rule of Signs we know that there must be exactly one positive real zero and either three or one negative real zeros. See **Example 2b**.

We go through the list of possible rational zeros looking for the one positive real zero. Remember, if $f\left(\frac{p}{q}\right) = 0$, then $\frac{p}{q}$ is a zero.

Eventually getting to $\frac{1}{3}$ and using synthetic division, we see that we obtain a zero.

$$\begin{array}{r|rrrrr} \frac{1}{3} & 6 & 13 & 61 & 8 & -10 \\ & \downarrow & 2 & 5 & 22 & 10 \\ \hline & 6 & 15 & 66 & 30 & 0 \end{array} \longleftarrow f\left(\tfrac{1}{3}\right) = 0, \text{ so } \tfrac{1}{3} \text{ is a zero.}$$

Thus, we can rewrite f as $f(x) = (x - \frac{1}{3})(6x^3 + 15x^2 + 66x + 30)$. We now try to find a zero of the polynomial $q_1(x) = 6x^3 + 15x^2 + 66x + 30$. There is no need to search for any more positive zeros since we have already found the one and only positive real zero of $f(x)$. Therefore, we have eliminated half of our list of possible rational zeros. The new list becomes

$$\frac{p}{q}: -1, -2, -5, -10, -\frac{1}{2}, -\frac{1}{3}, -\frac{1}{6}, -\frac{2}{3}, -\frac{5}{2}, -\frac{5}{3}, -\frac{5}{6}, -\frac{10}{3}.$$

Eventually getting to $-\frac{1}{2}$ and using synthetic division, we see that we obtain another zero.

$$\begin{array}{r|rrrr} -\frac{1}{2} & 6 & 15 & 66 & 30 \\ & \downarrow & -3 & -6 & -30 \\ \hline & 6 & 12 & 60 & 0 \end{array} \longleftarrow q_1\left(-\tfrac{1}{2}\right) = 0, \text{ so } -\tfrac{1}{2} \text{ is a zero.}$$

The remainder when $q_1(x) = 6x^3 + 15x^2 + 66x + 30$ is divided by $x + \frac{1}{2}$ is zero.

This implies that $-\frac{1}{2}$ is a zero. Using the bottom row of the synthetic division we see that $f(x) = \left(x - \frac{1}{3}\right)(6x^3 + 15x^2 + 66x + 30) = \left(x - \frac{1}{3}\right)\left(x + \frac{1}{2}\right)(6x^2 + 12x + 60)$.

Factoring out a 6 from $6x^2 + 12x + 60$ we can rewrite $f(x)$ as $f(x) = 6\left(x - \frac{1}{3}\right)\left(x + \frac{1}{2}\right)(x^2 + 2x + 10)$. We must now solve the quadratic equation $x^2 + 2x + 10 = 0$ to find the remaining two zeros. Solving either by **completing the square** or by using the **quadratic formula** we get $x = -1 \pm 3i$. Thus the four zeros are $\frac{1}{3}, -\frac{1}{2}, -1 + 3i$ and $-1 - 3i$. The completely factored form of this polynomial is

$$f(x) = 6\left(x - \frac{1}{3}\right)\left(x + \frac{1}{2}\right)(x - (-1 + 3i))(x - (-1 - 3i)).$$

To see the complete solution, watch this **video**. ●

TIP Once we factor a polynomial into the product of linear factors and a quadratic function of the form $f(x) = (x - c_1)(x - c_2)(x - c_3) \cdots (x - c_k)(ax^2 + bx + c)$, we need simply to solve the quadratic equation $ax^2 + bx + c = 0$ to find the remaining two zeros. For example, if we want to find all zeros of $g(x) = 2x^3 - 3x^2 + 4x - 3$, we need only try to find *one* zero from the list created by the rational zeros theorem and then find the two zeros of the remaining quadratic function. See if you can find the zeros of $g(x) = 2x^3 - 3x^2 + 4x - 3$. Watch the **video** to see if you are correct!

(eText Screens 4.5-1–4.5-31)

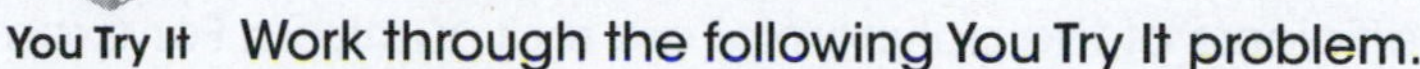

You Try It Work through the following You Try It problem.

Work Exercises 11–19 in this textbook or in the MyLab Math Study Plan.

OBJECTIVE 4 SOLVING POLYNOMIAL EQUATIONS

Solving polynomial equations of the form $f(x) = 0$ is equivalent to finding the zeros of $f(x)$. For example, the solutions to the equation $6x^4 + 13x^3 + 61x^2 + 8x - 10 = 0$ are exactly the zeros of the polynomial function $f(x) = 6x^4 + 13x^3 + 61x^2 + 8x - 10$. Using the result from Example 2, we conclude that the real solutions of this equation are $-\frac{1}{2}$ and $\frac{1}{3}$, whereas the two nonreal solutions are $-1 + 3i$ and $-1 - 3i$.

Example 4 Solve a Polynomial Equation

Solve $5x^5 - 9x^4 + 23x^3 - 35x^2 + 12x + 4 = 0$.

Solution Let $f(x) = 5x^5 - 9x^4 + 23x^3 - 35x^2 + 12x + 4$. The solutions to this equation must be the zeros of f. Using the rational zeros theorem, we see that the potential rational zeros are $\pm 1, \pm 2, \pm 4, \pm\frac{1}{5}, \pm\frac{2}{5}, \pm\frac{4}{5}$. Our goal is to attempt to find three zeros from this list and then solve the remaining quadratic to find the last two zeros. Watch the **video** to verify that the rational zeros of f are $x = 1$ (of multiplicity 2) and $-\frac{1}{5}$. Using synthetic division and factoring out the leading coefficient, we can rewrite f as $f(x) = 5(x - 1)^2\left(x + \frac{1}{5}\right)(x^2 + 4)$. The solutions to the quadratic equation $x^2 + 4 = 0$ are $\pm 2i$. Thus, the equation $5x^5 - 9x^4 + 23x^3 - 35x^2 + 12x + 4 = 0$ is equivalent to $5(x - 1)^2\left(x + \frac{1}{5}\right)(x - 2i)(x + 2i) = 0$. Therefore, the solutions are $\left\{1, -\frac{1}{5}, 2i, -2i\right\}$.

You Try It Work through the following You Try It problem.

Work Exercises 20–25 in this textbook or in the MyLab Math Study Plan.

OBJECTIVE 5 USING THE COMPLEX CONJUGATE PAIRS THEOREM

In Example 2, the two nonreal zeros of $f(x) = 6x^4 + 13x^3 + 61x^2 + 8x - 10$ are $-1 + 3i$ and $-1 - 3i$. In Example 3, $2i$ and $-2i$ are the two nonreal solutions of $5x^5 - 9x^4 + 23x^3 - 35x^2 + 12x + 4 = 0$. Notice that the nonreal zeros in both examples are **complex conjugates** of each other. This is no coincidence! In fact, the following theorem explains that complex zeros must occur in pairs.

Complex Conjugate Pairs Theorem

If $a + bi$ is a zero of a polynomial function with real coefficients, then the complex conjugate $a - bi$ is also a zero.

CAUTION To use the complex conjugate pairs theorem, it is imperative that the coefficients of the polynomial are real numbers. For example, the polynomial function $f(x) = x - i$ has a constant coefficient i. Note that $f(i) = i - i = 0$, and therefore, i is a zero of the polynomial function $f(x) = x - i$. However, the complex conjugate, $-i$, is not a zero because $f(-i) = -i - i = -2i$.

Example 5 Use the Complex Conjugate Pairs Theorem

Find the equation of a polynomial function f with real coefficients that satisfies the given conditions:

a. Fourth-degree polynomial function such that 1 is a zero of multiplicity 2 and $2 - i$ is also a zero.

b. Fifth-degree polynomial function sketched in Figure 21 given that $1 + 3i$ is a zero.

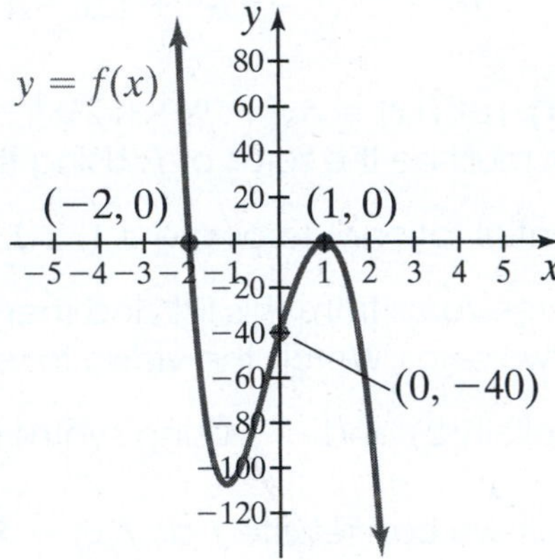

Figure 21

Solution

a. By the complex conjugate pairs theorem, because $2 - i$ is a zero, this implies that $2 + i$ is also a zero. The four zeros of f are $1, 1, 2 - i$, and $2 + i$. Using the factor theorem, we can write f as $f(x) = a(x - 1)(x - 1)(x - (2 - i))(x - (2 + i))$, where a is a constant. We can actually let a equal any number here because there are infinitely many polynomial functions that meet the stated conditions, and they all differ only by the leading coefficient. So, when no specific information is given to allow the exact leading coefficient to be determined, we choose a equal to 1 just for the sake of simplicity. Letting $a = 1$ and multiplying the factors together, we get $f(x) = x^4 - 6x^3 + 14x^2 - 14x + 5$. To see this multiplication worked out, read these **steps** or watch the **interactive video**.

b. Because we are looking for a fifth-degree polynomial, there must be five zeros (including multiplicities). From the graph, we see that $x = -2$ is a zero with odd multiplicity because the graph crosses the x-axis at $x = -2$. Similarly, $x = 1$ is a zero with even multiplicity because the graph is tangent to the x-axis at $x = 1$. By the complex conjugate pairs theorem, $1 + 3i$ and $1 - 3i$ are zeros. Thus, the completely factored form of f is $f(x) = a(x + 2)^m(x - 1)^n(x - (1 + 3i))(x - (1 - 3i))$, where m is odd and n is even.

Because f is a fifth-degree polynomial, m must equal 1 and n must equal 2. Therefore, $f(x) = a(x + 2)(x - 1)^2(x - (1 + 3i))(x - (1 - 3i))$. Multiplying the factors, we get $f(x) = a(x^5 - 2x^4 + 7x^3 + 8x^2 - 34x + 20)$. To see this multiplication worked out, read these **steps** or watch the **interactive video**. Unlike in part a, we cannot randomly assign a value for the leading coefficient because specific information has been given. To determine the value of a, we can use the fact that the graph contains the point $(0, -40)$ and, therefore, $f(0) = -40$.

$f(x) = a(x^5 - 2x^4 + 7x^3 + 8x^2 - 34x + 20)$	Write the polynomial function.
$f(0) = a(0^5 - 2(0)^4 + 7(0)^3 + 8(0)^2 - 34(0) + 20)$	Evaluate $f(0)$.
$f(0) = a(20)$	Simplify.
$f(0) = -40$	The graph contains the point $(0, -40)$.
$a(20) = -40$	Equate the two expressions for $f(0)$.
$a = -2$	Solve for a.

Thus,

$f(x) = -2(x^5 - 2x^4 + 7x^3 + 8x^2 - 34x + 20)$ or $f(x) = -2x^5 + 4x^4 - 14x^3 - 16x^2 + 68x - 40$.

You Try It Work through the following You Try It problem.

Work Exercises 26–31 in this textbook or in the MyLab Math Study Plan.

OBJECTIVE 6 USING THE INTERMEDIATE VALUE THEOREM

A direct consequence of the complex conjugate pairs theorem is the fact that every *odd degree* polynomial with real coefficients has at least one real zero. This is true because complex zeros always occur in pairs; thus, the number of nonreal zeros of a polynomial with real coefficients is always *even*. Therefore, if f is a polynomial function of odd degree, then at least one of its zeros must be real. For example, the polynomial function $f(x) = x^3 + 2x - 1$ is of odd degree, so we know that there exists at least one real zero. By the rational zeros theorem, the only possible rational zeros are ± 1. But

$$f(1) = 1 + 2 - 1 = 2 \quad \text{and}$$
$$f(-1) = -1 - 2 - 1 = -4$$

Neither 1 nor -1 are zeros of f, so there are no real **rational zeros**. Therefore, the real zero(s) of f must be **irrational**. So how do we find the irrational zero(s)? The following theorem will help us approximate the irrational zeros to as many decimal places as we want.

Intermediate Value Theorem

Let f be a polynomial function with $a < b$. If $f(a)$ and $f(b)$ have opposite signs, then there exists at least one real zero strictly between a and b.

Although the proof of the intermediate value theorem involves calculus, you can see from Figure 22 that if $f(a)$ is negative and $f(b)$ is positive, then the graph of f *must* cross the x-axis somewhere between a and b.

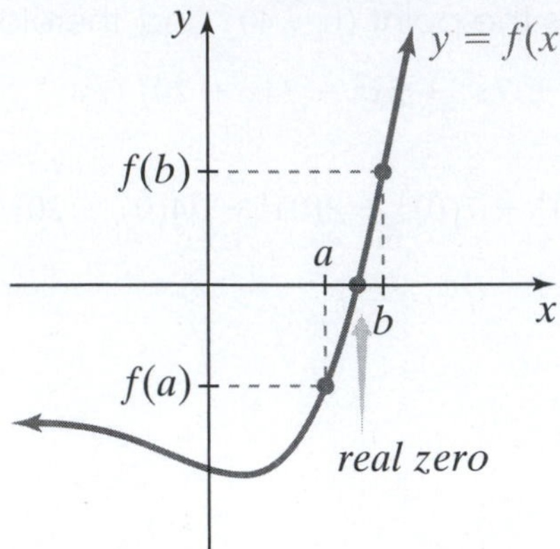

Figure 22 For a polynomial $y = f(x)$, if $f(a)$ is negative and $f(b)$ is positive, then there is a real zero somewhere between a and b.

Example 6 Use the Intermediate Value Theorem to Approximate a Real Zero

Find the real zero of $f(x) = x^3 + 2x - 1$ **correct to two decimal places.**

Solution Because f is of odd degree, there is at least one real zero. To approximate this zero, we must first find values of a and b such that $f(a)$ and $f(b)$ have opposite signs.

Note that for $a = 0$, $f(a) = f(0) = -1$. We must now find a number b such that $f(b)$ is positive. Using trial and error, we see that for $b = 1$, $f(b) = f(1) = 2$. Therefore, by the intermediate value theorem, there exists at least one real zero strictly between 0 and 1. We now evaluate the function at the decimal values $0.1, 0.2, 0.3, \ldots$ until we obtain another sign change:

$$f(0.1) = -0.7990$$
$$f(0.2) = -0.5920$$
$$f(0.3) = -0.3730$$
$$f(0.4) = -0.1360$$
$$f(0.5) = 0.1250$$

Sign change!

Notice that $f(0.4)$ and $f(0.5)$ have opposite signs. Therefore, by the intermediate value theorem, there exists a zero strictly between 0.4 and 0.5. We repeat the process again, this time evaluating the function at $0.41, 0.42, 0.43, \cdots$ until we obtain another sign change.

$$f(0.45) = -0.0089$$
$$f(0.46) = 0.01734$$

Sign change!

We conclude that the real zero lies between $x = 0.45$ and $x = 0.46$. Therefore, the real zero correct to two decimal places is $x = 0.45$. Using a graphing utility, we can see that the zero carried out to seven decimal places is $x = 0.4533977$.

Watch this **video** to see the entire solution to this example.

Using Technology

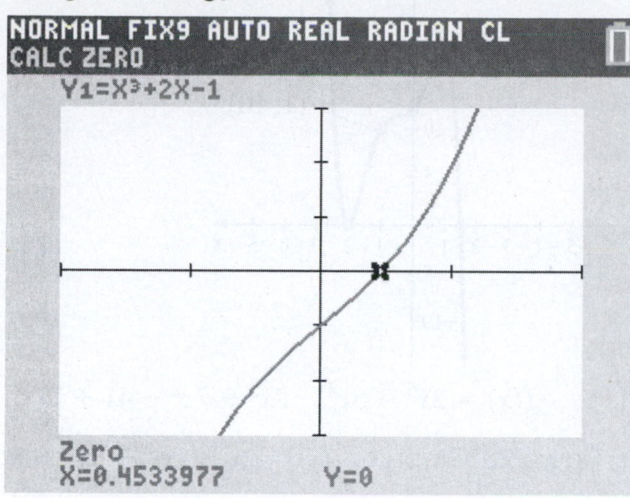

Use a graphing utility and the zero feature to approximate the real zero of $f(x) = x^3 + 2x - 1$.

You Try It Work through the following You Try It problem.

Work Exercises 32–37 in this textbook or in the MyLab Math Study Plan.

OBJECTIVE 7 SKETCHING THE GRAPHS OF POLYNOMIAL FUNCTIONS

Now that we have a procedure for finding the zeros of a polynomial function, we can modify the four-step process that was discussed in Section 4.3 and put it all together to sketch the graphs of polynomial functions of the form $f(x) = a_n x^n + a_{n-1}x^{n-1} + a_{n-2}x^{n-2} + \cdots + a_1 x + a_0$.

Steps for Sketching the Graphs of Polynomial Functions

Step 1. Determine the end behavior.

Step 2. Plot the y-intercept $f(0) = a_0$.

Step 3. Use the rational zeros theorem, the factor theorem and synthetic division, or the intermediate value theorem to find all zeros and completely factor f.

Step 4. Choose a test value between each real zero and complete the graph.

Example 7 Sketch a Polynomial Function

Sketch the graph of $f(x) = 2x^5 - 5x^4 - 2x^3 + 7x^2 - 4x + 12$.

Solution Work through the video to see how to sketch this polynomial function. The polynomial is sketched in Figure 23.

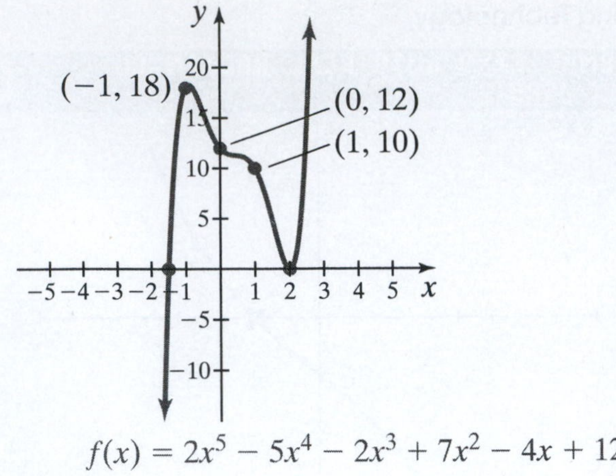

Figure 23 Graph of $f(x) = 2x^5 - 5x^4 - 2x^3 + 7x^2 - 4x + 12$.

You Try It Work through the following You Try It problem.

Work Exercises 38–43 in this textbook or in the MyLab Math Study Plan.

4.5 Exercises

Skill Check Exercises

In exercises SCE-1 through SCE-3, for each function, determine $f(-x)$.

SCE-1. $f(x) = -3x^5 + 4x^4 + 3x^3 - 7x^2 - 21$

SCE-2. $f(x) = x^4 - 7x^2 - 11$

SCE-3. $f(x) = -3x^7 + 8x^3 - x$

In Exercises 1–6, use the rational zeros theorem to determine the potential rational zeros of each polynomial function. Do *not* find the zeros.

1. $f(x) = x^4 - 2$
2. $f(x) = x^3 - 3x + 16$
3. $f(x) = x^5 - 2x^4 + 8x^3 - x^2 + 2x - 12$
4. $f(x) = 3x^4 - 7x^2 - x + 8$
5. $f(x) = x^4 + 4 + 3x - 9x^5$
6. $f(x) = 5x^3 - 27x^6 + 10x^2 + 16 - 4x$

In Exercises 7–10, determine the possible number of positive real zeros and negative real zeros of each polynomial function using Descarates' rule of signs.

7. $f(x) = 2x^5 - 13x^4 + 16x^3 - 5x^2 + 8x - 10$
8. $f(x) = -8x^4 + 13x^3 - 11x^2 + 3x - 1$
9. $f(x) = 7x^4 - 11x^2 - 8x - 10$
10. $f(x) = -2x^6 - 11x^4 - 8x^2 - 10$

In Exercises 11–19, find all complex zeros of the given polynomial function, and write the polynomial in completely factored form.

11. $f(x) = 4x^3 - 5x^2 - 23x + 6$

12. $f(x) = 2x^3 + 7x^2 - 6x - 21$

13. $f(x) = -2x^3 - 7x^2 - 36x + 20$

14. $f(x) = -6x^4 + 19x^3 - 15x^2 - 3x + 5$

15. $f(x) = 3x^4 + 4x^3 + 73x^2 - 54x - 26$

16. $f(x) = -2x^5 - 11x^4 - 19x^3 - 8x^2 + 7x + 5$

17. $f(x) = 8x^5 + 36x^4 + 70x^3 - 5x^2 - 68x + 24$

18. $f(x) = 63x^5 + 33x^4 - 125x^3 - 67x^2 - 2x + 2$

19. $f(x) = 5x^6 - 18x^5 + 29x^4 - 32x^3 + 27x^2 - 14x + 3$

In Exercises 20–25, solve each polynomial equation in the complex numbers.

20. $x^3 - 1 = 0$

21. $x^4 - 1 = 0$

22. $x^3 - 4x^2 + x + 6 = 0$

23. $2x^4 + x^3 - 4x^2 + 11x - 10 = 0$

24. $3x^4 + 4x^3 + 73x^2 - 54x - 26 = 0$

25. $12x^4 + 56x^3 + 3x^2 - 13x - 2 = 0$

In Exercises 26–31, form a polynomial function $f(x)$ with real coefficients using the given information.

26. Form a third-degree polynomial function with real coefficients such that $5 + i$ and 2 are zeros.

27. Form a fourth-degree polynomial function with real coefficients such that $3 - 2i$ is a zero and 1 is a zero of multiplicity 2.

28. Form a fifth-degree polynomial function with real coefficients such that $3i$, $1 - 5i$, and -1 are zeros and $f(0) = -3$.

29. Form the third-degree polynomial function with real coefficients sketched here given that $2i$ is a zero.

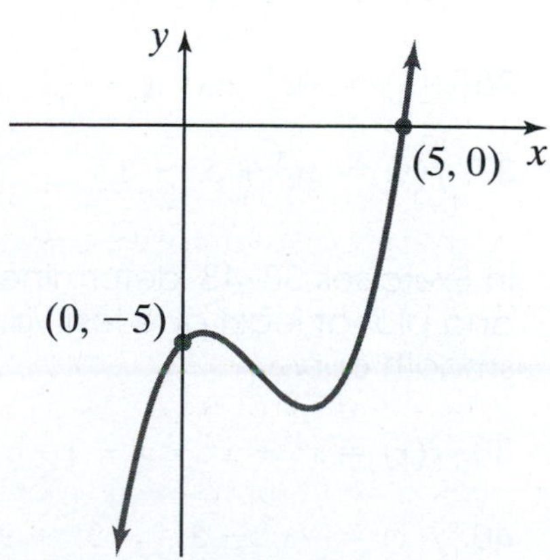

30. Form the fifth-degree polynomial function with real coefficients sketched here given that $1 + i$ is a zero.

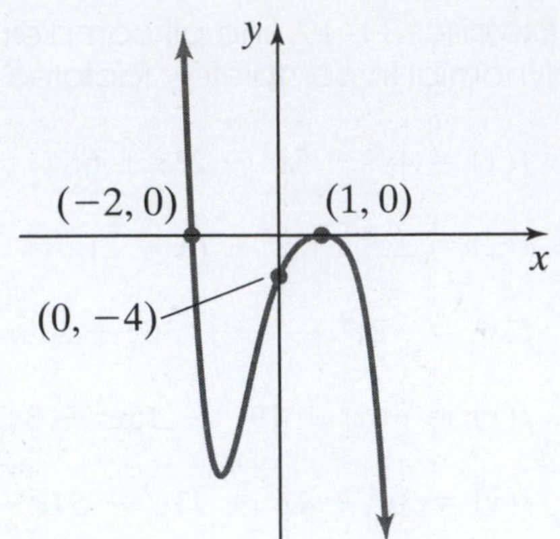

31. Form the fourth-degree polynomial function with real coefficients sketched here given that $-1 - 2i$ is a zero.

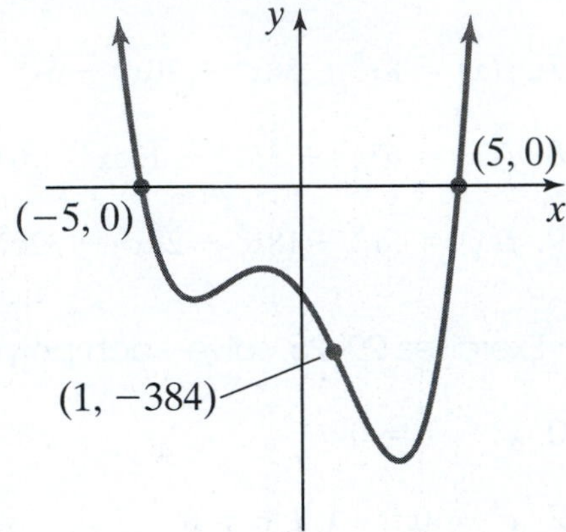

In Exercises 32–34, use the intermediate value theorem to show that the polynomial has a real zero on the given interval.

32. $f(x) = x^3 + 2x - 5; [1, 2]$

33. $f(x) = 3x^4 - x^3 + 8x^2 + x - 1; [0, 1]$

34. $f(x) = 2x^5 + 5x + 1; [-1, 0]$

In Exercises 35–37, use the intermediate value theorem to find the real zero correct to two decimal places.

35. $f(x) = x^3 + 2x - 5$

36. $f(x) = -x^3 + x^2 + x - 1$

37. $f(x) = 2x^5 + 5x + 1$

In Exercises 38–43, determine the end behavior, plot the y-intercept, find and plot all real zeros, and plot at least one test value between each intercept. Then connect the points with a smooth curve.

SbS 38. $f(x) = x^3 + x^2 - x - 1$

SbS 39. $f(x) = 4x^3 - 5x^2 - 23x + 6$

SbS 40. $f(x) = -x^3 - 5x^2 - 3x + 9$

SbS 41. $f(x) = -6x^4 + 19x^3 - 15x^2 - 3x + 5$

SbS 42. $f(x) = 2x^5 + 11x^4 + 18x^3 + 8x^2 - 4x - 3$

SbS 43. $f(x) = -6x^4 + x^3 + 53x^2 - 3x - 105$

Brief Exercises

In Exercises 44–49, sketch the graph of the given function.

44. $f(x) = x^3 + x^2 - x - 1$

45. $f(x) = 4x^3 - 5x^2 - 23x + 6$

46. $f(x) = -x^3 - 5x^2 - 3x + 9$

47. $f(x) = -6x^4 + 19x^3 - 15x^2 - 3x + 5$

48. $f(x) = 2x^5 + 11x^4 + 18x^3 + 8x^2 - 4x - 3$

49. $f(x) = -6x^4 + x^3 + 53x^2 - 3x - 105$

4.6 Rational Functions and Their Graphs

THINGS TO KNOW

Before working through this section, be sure that you are familiar with the following concepts:

VIDEO ANIMATION INTERACTIVE

You Try It

1. Determining the Domain of a Function Given the Equation (Section 3.1)

You Try It

2. Determining Whether a Function Is Even, Odd, or Neither (Section 3.2)

You Try It

3. Using Combinations of Transformations to Graph Functions (Section 3.4)

INTRODUCTION

Read this introduction before beginning Objective 1.

OBJECTIVES

1 Finding the Domain and Intercepts of Rational Functions

2 Identifying Vertical Asymptotes

3 Identifying Horizontal Asymptotes

4 Using Transformations to Sketch the Graphs of Rational Functions

5 Sketching Rational Functions Having Removable Discontinuities

6 Identifying Slant Asymptotes

7 Sketching Rational Functions

SECTION 4.6 EXERCISES

Introduction to Section 4.6

In this section, we investigate the properties and graphs of rational functions. Rational functions were first introduced in **Section 3.1**. We now redefine a rational function.

> **Definition** Rational Function
>
> A **rational function** is a function of the form $f(x) = \frac{g(x)}{h(x)}$, where g and h are **polynomial functions** such that $g(x) \neq 0$ and the degree of $h(x)$ is greater than 0.

If the degree of $h(x)$ is 0, then $h(x) = c$, where c is a real number and we consider the function $f(x) = g(x)/h(x) = g(x)/c$ to be a polynomial function.

OBJECTIVE 1 FINDING THE DOMAIN AND INTERCEPTS OF RATIONAL FUNCTIONS

We start our investigation of rational functions by discussing the domain and intercepts. Let $f(x)$ be a rational function of the form $f(x) = \frac{g(x)}{h(x)}$ where $g \neq 0$ and h are polynomial functions and the degree of $h(x)$ is greater than 0.

DOMAIN

Recall that the domain of every polynomial function is all real numbers. Therefore, it follows that the domain of $f(x) = \frac{g(x)}{h(x)}$ is also all real numbers **except** for any values of x for which the denominator $h(x)$ is equal to zero.

THE y-INTERCEPT

If $f(x)$ has a y-intercept, then it can be found by evaluating $f(0)$, provided that $f(0)$ is defined. The rational function $f(x) = \frac{g(x)}{h(x)}$ can have at most one y-intercept.

THE x-INTERCEPT(S)

A rational function can have several x-intercepts (or no x-intercepts at all). If $f(x) = \frac{g(x)}{h(x)}$ has any x-intercepts, then they can be found by solving the equation $g(x) = 0$ (provided that the polynomial functions g and h do not share a non-constant common factor).

Example 1 Find the Domain and Intercepts of a Rational Function

Let $f(x) = \frac{x - 4}{x^2 + x - 6}$.

a. Determine the domain of f.

b. Determine the y-intercept (if any).

c. Determine any x-intercepts.

Solution Factor the denominator, and rewrite f as $f(x) = \dfrac{x-4}{(x+3)(x-2)}$.

a. The domain of f is the set of all real numbers except those for which the denominator is equal to zero.

The denominator, $(x+3)(x-2)$, is equal to zero when $x = -3$ or $x = 2$. Therefore, the domain is the set of all real numbers except -3 and 2 or $\{x \mid x \neq -3, x \neq 2\}$. In interval notation, the domain is $(-\infty, -3) \cup (-3, 2) \cup (2, \infty)$.

b. The y-intercept is $f(0) = \dfrac{-4}{-6} = \dfrac{2}{3}$.

c. To find any x-intercepts, we must solve the equation $x - 4 = 0$. The solution to this equation is $x = 4$. Thus, the only x-intercept is 4.

Watch the **video** to see the solution to Example 1 worked out in detail.

You Try It Work through the following You Try It problem.

Work Exercises 1–6 in this textbook or in the MyLab Math Study Plan.

OBJECTIVE 2 IDENTIFYING VERTICAL ASYMPTOTES

To begin our discussion of the graphs of rational functions, we first look at a basic function that was introduced in Section 3.3, the **reciprocal function,** $f(x) = \dfrac{1}{x}$. The function $f(x) = \dfrac{1}{x}$ is defined everywhere except at $x = 0$; therefore, the domain of $f(x) = \dfrac{1}{x}$ is $(-\infty, 0) \cup (0, \infty)$. You can see from the graph of $f(x) = \dfrac{1}{x}$ sketched in **Figure 24** that as the values of x get closer to 0 from the right side of 0, the values of $f(x)$ increase without bound. Mathematically, we say, "as x approaches zero from the right, $f(x)$ approaches infinity." In symbols, this is written as

$$f(x) \to \infty \quad \text{as} \quad x \to 0^+.$$

The "+" symbol indicates that we are only looking at values of x that are located to the *right* of zero. Similarly, as x gets closer to 0 from the left side of 0, the values of $f(x)$ approach negative infinity. Symbolically, we can write the following:

$$f(x) \to -\infty \quad \text{as} \quad x \to 0^-.$$

The "−" symbol indicates that we are only looking at values of x that are located to the *left* of zero.

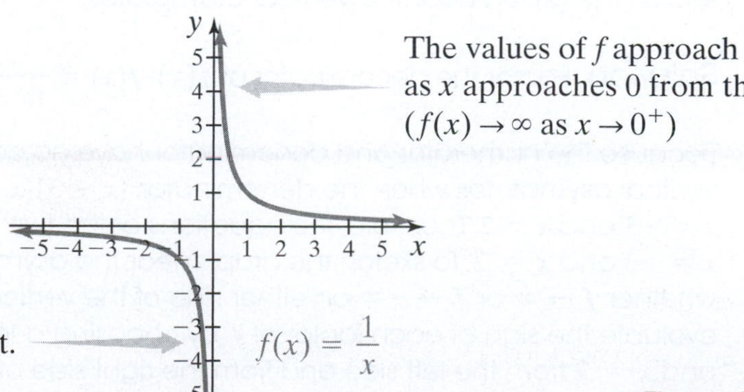
The values of f approach ∞ as x approaches 0 from the right. ($f(x) \to \infty$ as $x \to 0^+$)
The values of f approach $-\infty$ as x approaches 0 from the left. ($f(x) \to -\infty$ as $x \to 0^-$)
$f(x) = \dfrac{1}{x}$
y
x
-5 -4 -3 -2 -1 1 2 3 4 5
5 4 3 2 1 -1 -2 -3 -4 -5

Figure 24 Graph of $f(x) = \dfrac{1}{x}$.

The graph of the function $f(x) = \frac{1}{x}$ gets closer and closer to the line $x = 0$ (the y-axis) as x gets closer and closer to 0 from either side of 0, but the graph never touches the line $x = 0$. The line $x = 0$ is said to be a **vertical asymptote** of the graph of $f(x) = \frac{1}{x}$. Many rational functions have vertical asymptotes. The vertical asymptotes occur when the graph of a function approaches positive or negative infinity as x approaches some finite number a as stated in the following definition.

Definition Vertical Asymptote

The vertical line $x = a$ is a **vertical asymptote** of a function $y = f(x)$ if *at least* one of the following occurs:

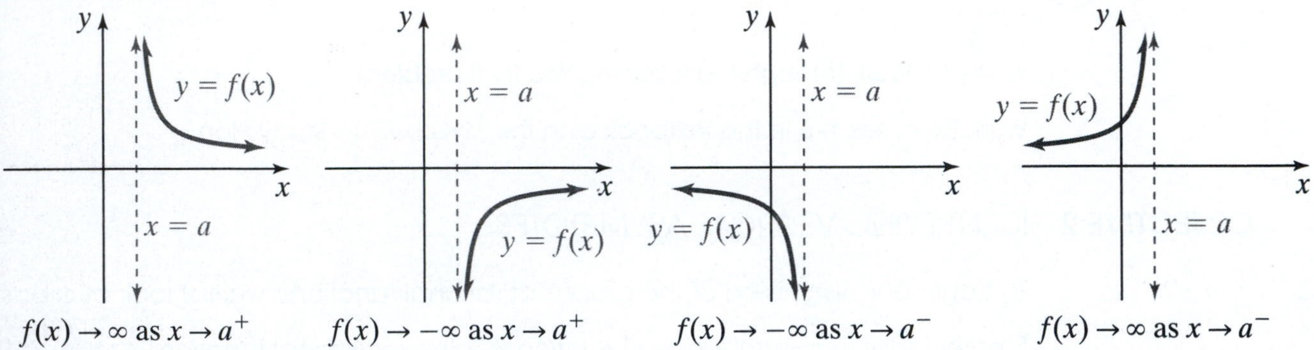

A rational function of the form $f(x) = \frac{g(x)}{h(x)}$, where $g(x)$ and $h(x)$ have no common factors, will have a vertical asymptote at $x = a$ if $h(a) = 0$.

CAUTION It is essential to cancel any common factors before locating the vertical asymptotes.

The situation that arises when $g(x)$ and $h(x)$ share a common factor is discussed in objective 5 of this section.

Example 2 Find the Vertical Asymptotes of a Rational Function

Find the vertical asymptotes (if any) of the function $f(x) = \frac{x-3}{x^2 + x - 6}$, and then sketch the graph near the vertical asymptotes.

Solution Factor the denominator of $f(x)$: $f(x) = \frac{x-3}{(x+3)(x-2)}$.

Because the numerator and denominator have no common factors, f will have vertical asymptotes when the denominator, $(x + 3)(x - 2)$, is equal to zero or when $x = -3$ and $x = 2$. Therefore, the equations of the two vertical asymptotes are $x = -3$ and $x = 2$. To sketch the graph near the asymptotes, we must determine whether $f \to \infty$ or $f \to -\infty$ on either side of the vertical asymptote. To do this, we evaluate the sign of each factor of f by choosing a test value "close to" $x = -3$ and $x = 2$ from the left side and from the right side of $x = -3$ and $x = 2$.

Vertical Asymptote	$x = -3$		$x = 2$	
Approach from	Left	Right	Left	Right
Test Value	$x = -3.1$	$x = -2.9$	$x = 1.9$	$x = 2.1$
Substitute Test Value into $f(x) = \frac{x-3}{(x+3)(x-2)}$	$f(-3.1) = \frac{(-3.1-3)}{(-3.1+3)(-3.1-2)} = \frac{(-)}{(-)(-)} = -$	$f(-2.9) = \frac{(-2.9-3)}{(-2.9+3)(-2.9-2)} = \frac{(-)}{(+)(-)} = +$	$f(1.9) = \frac{(1.9-3)}{(1.9+3)(1.9-2)} = \frac{(-)}{(+)(-)} = +$	$f(2.1) = \frac{(2.1-3)}{(2.1+3)(2.1-2)} = \frac{(-)}{(+)(+)} = -$
Result	Sign is negative. $f \to -\infty$	Sign is positive. $f \to \infty$	Sign is positive. $f \to \infty$	Sign is negative. $f \to -\infty$

CAUTION **If there is an x-intercept near the vertical asymptote, it may be beneficial to choose a test value that is between the x-intercept and the vertical asymptote.**

We see from the table that as the values of x approach -3 from the left side of -3, the values of f approach $-\infty$. Similarly, as the values of x approach -3 from the right side of -3, the values of f approach ∞. As x approaches 2 from the left side of 2, the values of f approach ∞. Finally, as the values of x approach 2 from the right side of 2, the values of f approach $-\infty$. Notice that we chose 2.1 as our test value for the case when x approaches 2 from the right. If we would have picked a test value greater than 3, our conclusions would have been incorrect. This is because $x = 3$ is an x-intercept and the graph to the right of 3 is above the x-axis. It is absolutely crucial to choose test values that are strictly between the x-intercept and the vertical asymptote.

Watch the **video** to see this entire example worked out in detail. The graph of f near both vertical asymptotes is sketched in **Figure 25**.

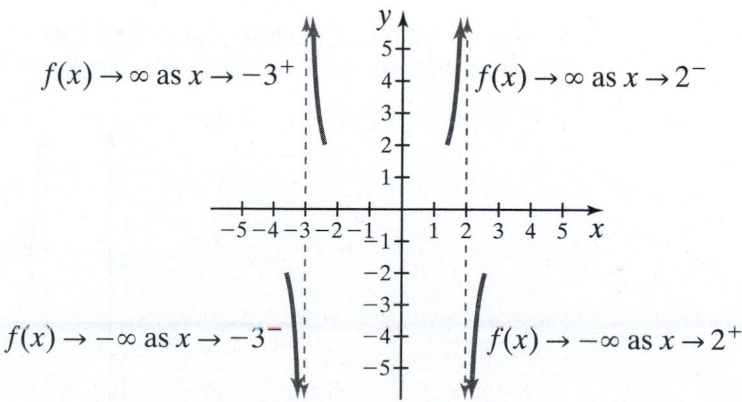

Figure 25 Graph of $f(x) = \frac{x-3}{x^2+x-6}$ near the vertical asymptotes. ●

Example 3 Find the Vertical Asymptotes of a Rational Function

Find the vertical asymptotes (if any) of the function $f(x) = \dfrac{x+3}{x^2+x-6}$, and then sketch the graph near the vertical asymptotes.

Solution We start by factoring the denominator: $f(x) = \dfrac{x+3}{(x+3)(x-2)}$. Notice that the numerator and denominator share a common factor of $x+3$. Canceling the common factor, we get $f(x) = \dfrac{x+3}{(x+3)(x-2)} = \dfrac{1}{x-2}$ for $x \neq -3$. After cancellation, the only zero in the denominator is $x = 2$. Thus, the line $x = 2$ is the only vertical asymptote of the graph of f. To sketch the graph near the asymptote, we must determine the sign of the denominator as x approaches 2 from the left side and the right side of $x = 2$.

Vertical Asymptote	$x = 2$	
Approach from	**Left**	**Right**
Test Value	$x = 1.9$	$x = 2.1$
Substitute Test Value into $f(x) = \dfrac{1}{x-2}$ for $x \neq -3$.	$f(1.9) = \dfrac{1}{(1.9-2)} = \dfrac{1}{(-)} = -$	$f(2.1) = \dfrac{1}{(2.1-2)} = \dfrac{1}{(+)} = +$
Result	Sign is negative. $f \to -\infty$	Sign is positive. $f \to \infty$

We see that as the values of x approach 2 from the left side of 2, the values of f approach $-\infty$. Similarly, as the values of x approach 2 from the right side of 2, the values of f approach ∞. Watch the **video** to see this entire example worked out in detail.

A rough sketch of the graph of f near the vertical asymptote can be seen in Figure 26. We sketch the complete graph of f in Example 6.

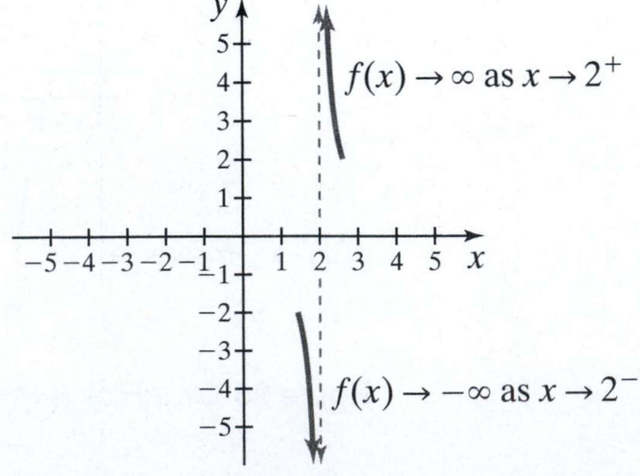

Figure 26 Graph of $f(x) = \dfrac{x+3}{x^2+x-6}$ near the vertical asymptote.

You Try It Work through the following You Try It problem.

Work Exercises 7–12 in this textbook or in the MyLab Math Study Plan.

OBJECTIVE 3 IDENTIFYING HORIZONTAL ASYMPTOTES

The graph of $f(x) = \frac{1}{x}$ is sketched again in Figure 27. The graph shows that as the values of x increase without bound (as $x \to \infty$), the values of f approach 0 $(f(x) \to 0)$. Similarly, as the values of x decrease without bound (as $x \to -\infty$), the values of f approach 0 $(f(x) \to 0)$. In this case, the line $y = 0$ is called a **horizontal asymptote** of the graph of $f(x) = \frac{1}{x}$.

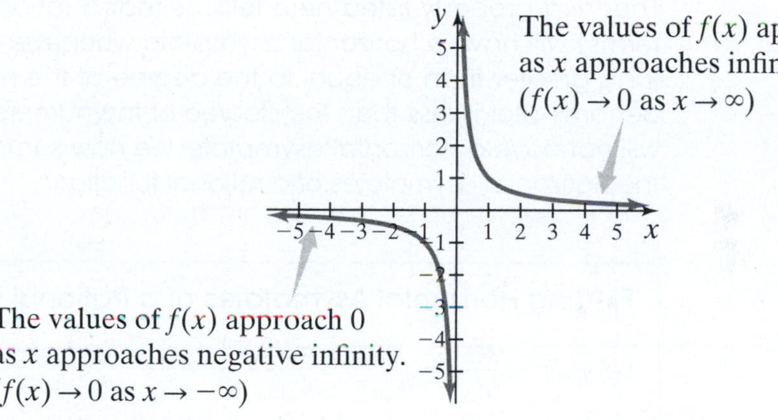

The values of $f(x)$ approach 0 as x approaches infinity. $(f(x) \to 0$ as $x \to \infty)$
The values of $f(x)$ approach 0 as x approaches negative infinity. $(f(x) \to 0$ as $x \to -\infty)$

Figure 27 The line $y = 0$ is a horizontal asymptote of the graph of $f(x) = \frac{1}{x}$.

Definition Horizontal Asymptote

A horizontal line $y = H$ is a **horizontal asymptote** of a function f if the values of $f(x)$ approach some fixed number H as the values of x approach ∞ or $-\infty$.

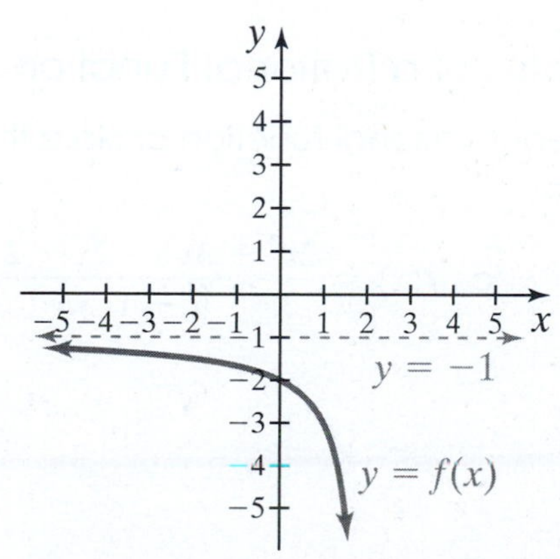

The line $y = -1$ is a horizontal asymptote because the values of $f(x)$ approach -1 as x approaches $-\infty$.

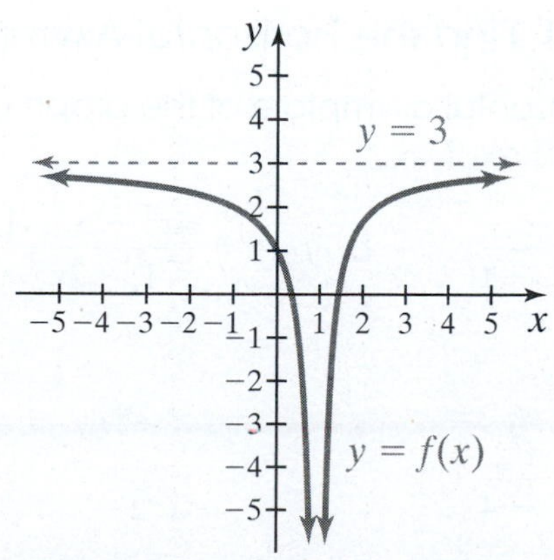

The line $y = 3$ is a horizontal asymptote because the values of $f(x)$ approach 3 as x approaches $\pm\infty$.

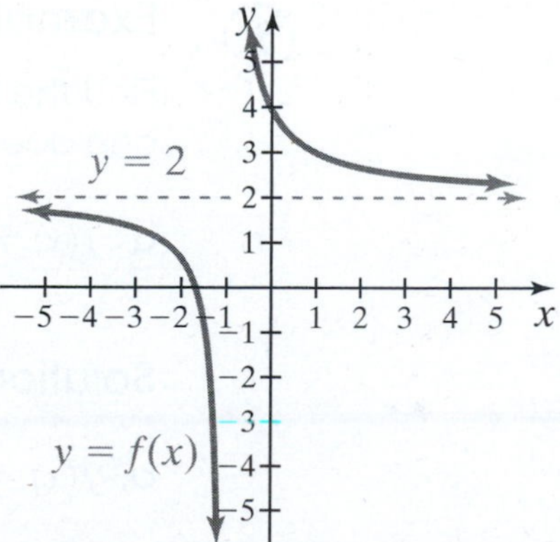

The line $y = 2$ is a horizontal asymptote because the values of $f(x)$ approach 2 as x approaches $\pm\infty$.

Properties of Horizontal Asymptotes of Rational Functions

- Although a rational function can have many vertical asymptotes, it can have, at most, one horizontal asymptote.
- The graph of a rational function will never intersect a vertical asymptote but may intersect a horizontal asymptote.
- A rational function $f(x) = \dfrac{g(x)}{h(x)}$ that is written in lowest terms (all common factors of the numerator and denominator have been canceled) will have a horizontal asymptote whenever the degree of $h(x)$ is greater than or equal to the degree of $g(x)$.

The third property listed here tells us that a rational function (written in lowest terms) will have a horizontal asymptote whenever the degree of the denominator is greater than or equal to the degree of the numerator. If the degree of the denominator is less than the degree of the numerator, then the rational function will not have a horizontal asymptote. We now summarize a technique for finding the horizontal asymptotes of a rational function.

Finding Horizontal Asymptotes of a Rational Function

Let $f(x) = \dfrac{g(x)}{h(x)} = \dfrac{a_nx^n + a_{n-1}x^{n-1} + a_{n-2}x^{n-2} + \cdots + a_1x + a_0}{b_mx^m + b_{m-1}x^{m-1} + b_{m-2}x^{m-2} + \cdots + b_1x + b_0}, a_n \neq 0, b_m \neq 0,$

where f is written in lowest terms, n is the degree of g, and m is the degree of h.

- If $m > n$, then $y = 0$ is the horizontal asymptote.
- If $m = n$, then the horizontal asymptote is $y = \dfrac{a_n}{b_m}$, the ratio of the leading coefficients.
- If $m < n$, then there are no horizontal asymptotes.

Example 4 Find the Horizontal Asymptotes of a Rational Function

Find the horizontal asymptote of the graph of each rational function or state that one does not exist.

a. $f(x) = \dfrac{x}{x^2 - 4}$ b. $f(x) = \dfrac{4x^2 - x + 1}{1 - 2x^2}$ c. $f(x) = \dfrac{2x^3 + 3x^2 - 2x - 2}{x - 1}$

Solution

a. $f(x) = \dfrac{x}{x^2 - 4}$

The degree of the denominator is *greater than* the degree of the numerator. Therefore, the graph has a horizontal asymptote whose equation is $y = 0$ (the x-axis). View the **graph** of f or work through the **interactive video**.

b. $f(x) = \dfrac{4x^2 - x + 1}{1 - 2x^2}$

The degree of the denominator is *equal* to the degree of the numerator. Thus, the graph has a horizontal asymptote. Note that the leading coefficient of the polynomial in the numerator is 4. The leading coefficient of the polynomial in the denominator is -2 because -2 is the coefficient of the term with the highest degree. Thus, the equation of the horizontal asymptote is $y = \dfrac{4}{-2}$ or $y = -2$.

View the **graph** of f or work through the **interactive video.**

c. $f(x) = \dfrac{2x^3 + 3x^2 - 2x - 2}{x - 1}$

The degree of the denominator is *less than* the degree of the numerator. Therefore, the graph has no horizontal asymptotes. View the **graph** of f or work through the **interactive video.**

You Try It Work through the following You Try It problem.

Work Exercises 13–20 in this textbook or in the MyLab Math Study Plan.

OBJECTIVE 4 USING TRANSFORMATIONS TO SKETCH THE GRAPHS OF RATIONAL FUNCTIONS

We now introduce another basic rational function, $f(x) = \dfrac{1}{x^2}$. Like the **reciprocal function,** the domain of $f(x) = \dfrac{1}{x^2}$ is $(-\infty, 0) \cup (0, \infty)$. The denominator, x^2, is always greater than zero, and this implies that the range includes all values of y greater than zero. The graph has a y-axis vertical asymptote $(x = 0)$ and an x-axis horizontal asymptote $(y = 0)$. Following are the graphs of $f(x) = \dfrac{1}{x}$ and $f(x) = \dfrac{1}{x^2}$, along with some important properties of each. Knowing these two basic graphs will help us sketch more complicated rational functions.

Graphs of $f(x) = \dfrac{1}{x}$ and $f(x) = \dfrac{1}{x^2}$

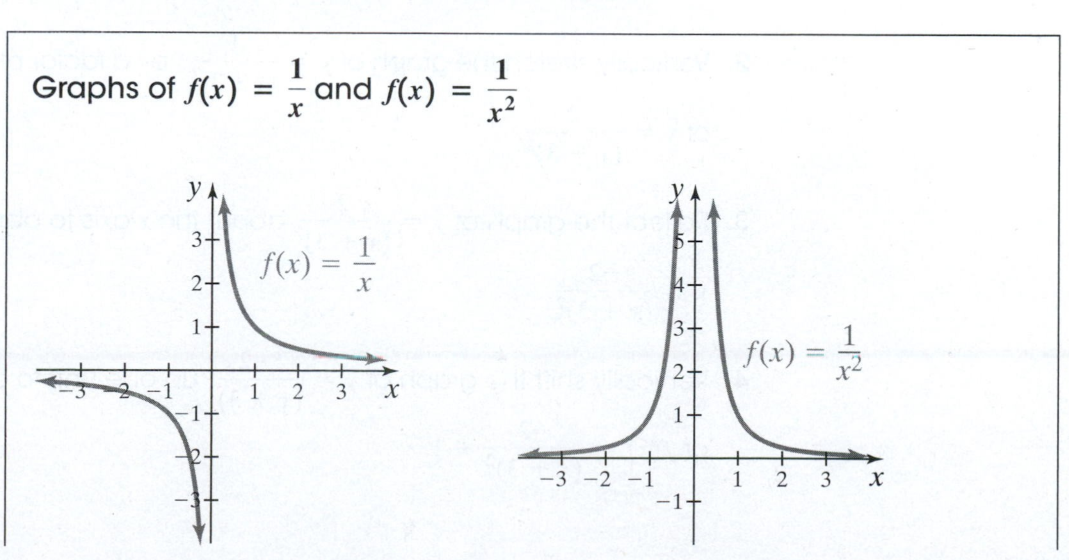

Properties of the graph of $f(x) = \frac{1}{x}$	Properties of the graph of $f(x) = \frac{1}{x^2}$
Domain: $(-\infty, 0) \cup (0, \infty)$	Domain: $(-\infty, 0) \cup (0, \infty)$
Range: $(-\infty, 0) \cup (0, \infty)$	Range: $(0, \infty)$
No intercepts	No intercepts
Vertical asymptote: $x = 0$	Vertical asymptote: $x = 0$
Horizontal asymptote: $y = 0$	Horizontal asymptote: $y = 0$
Odd function $f(-x) = -f(x)$	Even function $f(x) = f(-x)$
The graph is symmetric about the origin.	The graph is symmetric about the y-axis.

Sometimes we are able to use transformations to sketch the graphs of rational functions. To review this idea, read this **summary** of the transformation techniques that are discussed in **Section 3.4**.

Example 5 Use Transformations to Sketch the Graph of a Rational Function

Use transformations to sketch the graph of $f(x) = \frac{-2}{(x+3)^2} + 1$.

Solution Because $x + 3$ is raised to the second power in the denominator, start with the graph of $y = \frac{1}{x^2}$. We can obtain the graph of $f(x) = \frac{-2}{(x+3)^2} + 1$ using the following sequence of transformations:

1. Horizontally shift the graph of $y = \frac{1}{x^2}$ to the left three units to obtain the graph of $y = \frac{1}{(x+3)^2}$.
2. Vertically stretch the graph of $y = \frac{1}{(x+3)^2}$ by a factor of 2 to obtain the graph of $y = \frac{2}{(x+3)^2}$.
3. Reflect the graph of $y = \frac{2}{(x+3)^2}$ about the x-axis to obtain the graph of $y = \frac{-2}{(x+3)^2}$.
4. Vertically shift the graph of $y = \frac{-2}{(x+3)^2}$ up one unit to obtain the final graph of $f(x) = \frac{-2}{(x+3)^2} + 1$.

You can see from the final graph in Figure 28 that the horizontal asymptote is now shifted up one unit to $y = 1$. The x-intercepts of $-3 - \sqrt{2}$ and $-3 + \sqrt{2}$ can be found by setting $f(x) = 0$ and solving for x. The y-intercept is $f(0) = \frac{7}{9}$. Watch the video to see each step worked out in detail.

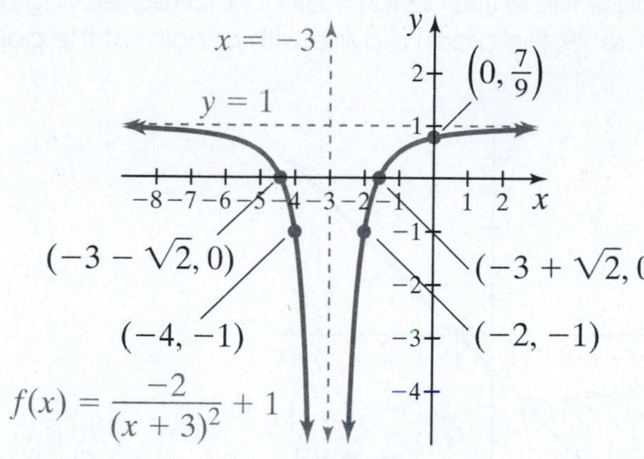

Figure 28 Graph of $f(x) = \dfrac{-2}{(x + 3)^2} + 1$. ●

You Try It **Work through the following You Try It problem.**

Work Exercises 21–32 in this textbook or in the MyLab Math Study Plan.

OBJECTIVE 5 SKETCHING RATIONAL FUNCTIONS HAVING REMOVABLE DISCONTINUITIES

A rational function $f(x) = \dfrac{g(x)}{h(x)}$ may sometimes have a "hole" in its graph. In calculus, these holes are called **removable discontinuities**. For a rational function to have a removable discontinuity, $g(x)$ and $h(x)$ must share a common factor as in Example 6.

Example 6 Sketch a Rational Function Having a Removable Discontinuity

Sketch the graph of $f(x) = \dfrac{x^2 - 1}{x + 1}$, and find the coordinates of all removable discontinuities.

Solution The domain of f is $\{x | x \neq -1\}$. Factoring the numerator, we can rewrite f as $f(x) = \dfrac{(x + 1)(x - 1)}{x + 1}$. Canceling the common factor of $x + 1$, we get $f(x) = x - 1$ for $x \neq -1$. Notice that the domain of this simplified function is still all real numbers *except* $x = -1$. Because there are no longer any factors containing the variable x in the denominator of the simplified function, there are no vertical asymptotes.

The graph of $f(x) = \dfrac{x^2 - 1}{x + 1}$ behaves exactly like the graph of the simplified function $f(x) = x - 1$ for $x \neq -1$. This graph will look like the graph of $y = x - 1$ except we must remove the point on the graph when $x = -1$. We can find this point by evaluating $y = x - 1$ at $x = -1$. We get $y = (-1) - 1 = -2$. Thus, the removable discontinuity is the point $(-1, -2)$. The graph no longer looks like a typical rational function because the entire denominator canceled leaving only $f(x) = x - 1$ for $x \neq -1$. In Figure 29, the graph is a line with a "hole" at the point $(-1, -2)$.

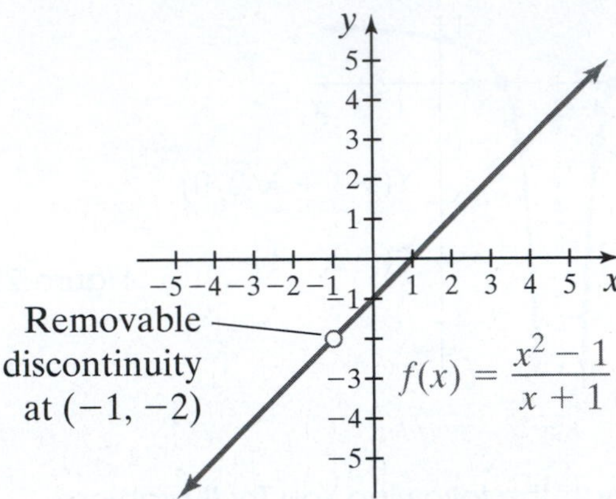

Figure 29 Graph of $f(x) = \dfrac{x^2 - 1}{x + 1}$.

It is possible to have a rational function with a common factor in the numerator and denominator whose graph still looks like the graph of a rational function but has a "hole" in it, as in Example 7.

Example 7 Sketch a Rational Function Having a Removable Discontinuity

Sketch the graph of $f(x) = \dfrac{x + 3}{x^2 + x - 6}$, and find the coordinates of all removable discontinuities.

Solution Notice that this is the function that was discussed in **Example 3**. Factoring the denominator, we get $f(x) = \dfrac{x + 3}{(x + 3)(x - 2)}$. The domain of $f(x)$ is all real numbers, except $x = 2$ and $x = -3$. Canceling the common factor of $x + 3$, we get $f(x) = \dfrac{1}{x - 2}$ for $x \neq -3$. Therefore, the graph of $f(x) = \dfrac{x + 3}{(x + 3)(x - 2)}$ will behave exactly like the graph of $f(x) = \dfrac{1}{x - 2}$ for $x \neq -3$. Notice that this graph will look like the graph of $y = \dfrac{1}{x - 2}$, except we must remember that $x = -3$ is *not* in the domain! There will be a removable discontinuity at the point where $x = -3$. We can find this point by evaluating $y = \dfrac{1}{x - 2}$ at $x = -3$. We get $y = \dfrac{1}{(-3) - 2} = -\dfrac{1}{5}$. Using transformations, we can shift the graph of $y = \dfrac{1}{x}$ to the right two units to obtain the graph of $y = \dfrac{1}{x - 2}$, remembering to remove the point $\left(-3, -\dfrac{1}{5}\right)$. See **Figure 30**.

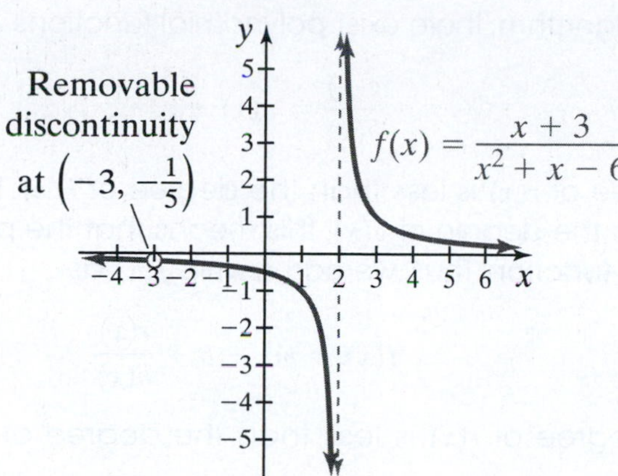

Figure 30 Graph of $f(x) = \dfrac{x+3}{x^2+x-6}$.

CAUTION Be careful when using graphing devices. Most graphing calculators will not show removable discontinuities!

Using Technology

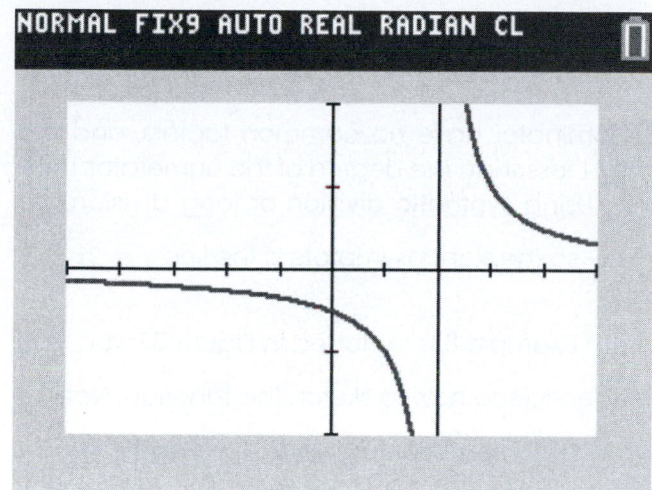

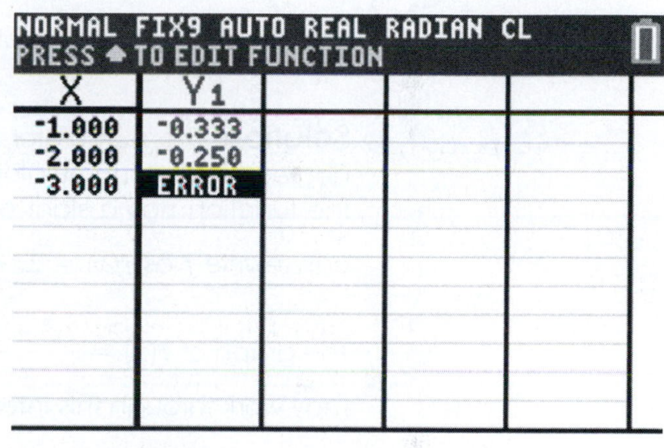

The graph of $y_1 = \dfrac{x+3}{x^2+x-6}$ is shown using a graphing utility. The graph does not suggest that the function is undefined at $x = -3$. However, the table of values does illustrate this fact.

You Try It Work through the following You Try It problem.

Work Exercises 33–38 in this textbook or in the MyLab Math Study Plan.

OBJECTIVE 6 IDENTIFYING SLANT ASYMPTOTES

Recall that the graph of a rational function $f(x) = \dfrac{g(x)}{h(x)}$, where f is written in lowest terms has a horizontal asymptote whenever the degree of $h(x)$ is greater than or equal to the degree of $g(x)$. But what happens if the degree of $h(x)$ is less than the degree of $g(x)$? If the degree of $h(x)$ is exactly 1 less than the degree of $g(x)$, then there will be a **slant asymptote**.

By the **division algorithm**, there exist polynomial functions $q(x)$ and $r(x)$ such that

$$\frac{g(x)}{h(x)} = q(x) + \frac{r(x)}{h(x)},$$

where the degree of $r(x)$ is less than the degree of $h(x)$. Because the degree of $h(x)$ is 1 less than the degree of $g(x)$, this means that the polynomial function $q(x)$ must be a **linear function**. Thus, we can rewrite $f(x)$ as

$$f(x) = mx + b + \frac{r(x)}{h(x)}.$$

Because the degree of $r(x)$ is less than the degree of $h(x)$, we know that as $x \to \pm\infty$, $\frac{r(x)}{h(x)} \to 0$. Therefore, as $x \to \pm\infty$, the graph of $f(x) = mx + b + \frac{r(x)}{h(x)}$ approaches the line $y = mx + b$. This line is the **slant asymptote**.

Example 8 Find the Slant Asymptote of a Rational Function

Find the slant asymptote of $f(x) = \frac{2x^2 + 3x - 2}{x - 1}$.

Solution The numerator and denominator have no common factors, and the degree of the denominator is exactly 1 less than the degree of the numerator. Thus, the function has a slant asymptote. Using **synthetic division** or **long division**, we can rewrite f as $f(x) = 2x + 5 + \frac{3}{x - 1}$ so the slant asymptote is the line $y = 2x + 5$.

The graph of $f(x) = \frac{2x^2 + 3x - 2}{x - 1}$ from Example 8 is sketched in **Figure 31**. You may work through this **interactive video** to see how to sketch this function. Notice in Figure 31 that the y-intercept is $f(0) = 2$. The x-intercepts are $x = \frac{1}{2}$ and $x = -2$, which are the solutions to the equation $2x^2 + 3x - 2 = 0$. The vertical asymptote is the line $x = 1$, and the slant asymptote is the line $y = 2x + 5$.

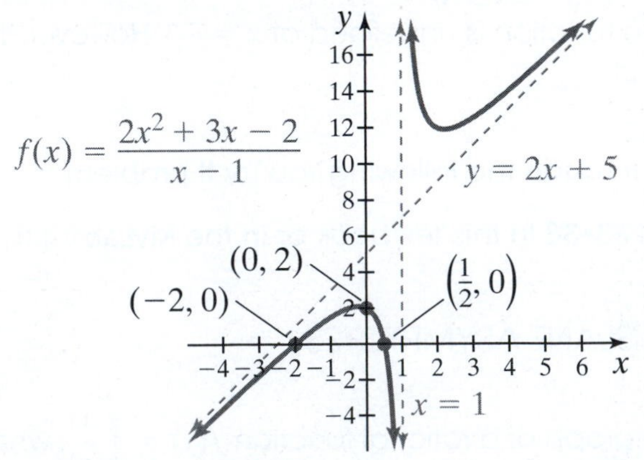

Figure 31 Graph of $f(x) = \frac{2x^2 + 3x - 2}{x - 1}$.

You Try It Work through the following You Try It problem.

Work Exercises 39–42 in this textbook or in the MyLab Math Study Plan.

OBJECTIVE 7 SKETCHING RATIONAL FUNCTIONS

Given a rational function of the form $f(x) = \dfrac{g(x)}{h(x)}$, we can sketch the graph of f by carefully following the nine-step procedure outlined here.

Steps for Graphing Rational Functions of the Form $f(x) = \dfrac{g(x)}{h(x)}$

Step 1. Find the domain.

Step 2. If $g(x)$ and $h(x)$ have common factors, cancel all common factors determining the x-coordinates of any removable discontinuities and rewrite f in lowest terms.

Step 3. Check for symmetry.

If $f(-x) = -f(x)$, then the graph of $f(x)$ is *odd* and thus symmetric about the origin.

If $f(x) = f(-x)$, then the graph of $f(x)$ is *even* and thus symmetric about the y-axis.

Step 4. Find the y-intercept by evaluating $f(0)$.

Step 5. Find the x-intercepts by finding the zeros of the numerator of f, being careful to use the new numerator if a common factor has been removed.

Step 6. Find the vertical asymptotes by finding the zeros of the denominator of f, being careful to use the new denominator if a common factor has been removed. Use test values to determine the behavior of the graph on each side of the vertical asymptotes.

Step 7. Determine whether the graph has any horizontal or slant asymptotes.

Step 8. Plot points, choosing values of x between each intercept and values of x on either side of all vertical asymptotes.

Step 9. Complete the sketch.

Example 9 Sketch a Rational Function

Sketch the graph of $f(x) = \dfrac{x^3 + 2x^2 - 9x - 18}{x^3 + 6x^2 + 5x - 12}$.

Solution Watch the video to see each of the following steps worked out in detail:

1. We can use the **rational zeros theorem** and synthetic division to factor and rewrite f as $f(x) = \dfrac{(x+3)(x-3)(x+2)}{(x+3)(x-1)(x+4)}$. The domain is $\{x \mid x \neq -4, x \neq -3, x \neq 1\}$.
2. Because the numerator and denominator share a common factor of $x + 3$, there will be a removable discontinuity at $x = -3$. The function simplifies

to $f(x) = \dfrac{(x-3)(x+2)}{(x-1)(x+4)}$ for $x \neq -3$ or $f(x) = \dfrac{x^2 - x - 6}{x^2 + 3x - 4}$ for $x \neq -3$. The removable discontinuity is the point $\left(-3, -\dfrac{3}{2}\right)$.

3. This function is neither odd nor even because $f(-x) \neq -f(x)$ and $f(x) \neq f(-x)$. Therefore, the graph is not symmetric about the origin or the y-axis.

4. The y-intercept is $f(0) = \dfrac{3}{2}$.

5. The x-intercepts occur when $(x-3)(x+2) = 0$. The x-intercepts are $x = -2$ and $x = 3$.

6. The equations of the two vertical asymptotes are $x = -4$ and $x = 1$. Using test values, we can determine the behavior of the graph near the vertical asymptotes.

   As $x \to -4^-$, $f(x) \to \infty$. The graph approaches positive infinity as x approaches -4 from the left.

   As $x \to -4^+$, $f(x) \to -\infty$. The graph approaches negative infinity as x approaches -4 from the right.

   As $x \to 1^-$, $f(x) \to \infty$. The graph approaches positive infinity as x approaches 1 from the left.

   As $x \to 1^+$, $f(x) \to -\infty$. The graph approaches negative infinity as x approaches 1 from the right.

7. Because the degree of the denominator is equal to the degree of the numerator, the graph of f has a horizontal asymptote at $y = \dfrac{1}{1} = 1$.

8. We can evaluate the function at several values of x. Note that $f(-5) = 4$, $f\left(\dfrac{1}{2}\right) = \dfrac{25}{9}$, and $f(2) = -\dfrac{2}{3}$. Hence, the graph must pass through the points $(-5, 4)$, $\left(\dfrac{1}{2}, \dfrac{25}{9}\right)$, and $\left(2, -\dfrac{2}{3}\right)$.

9. The completed graph is sketched in **Figure 32**. Watch the **video** to see each step worked out in detail.

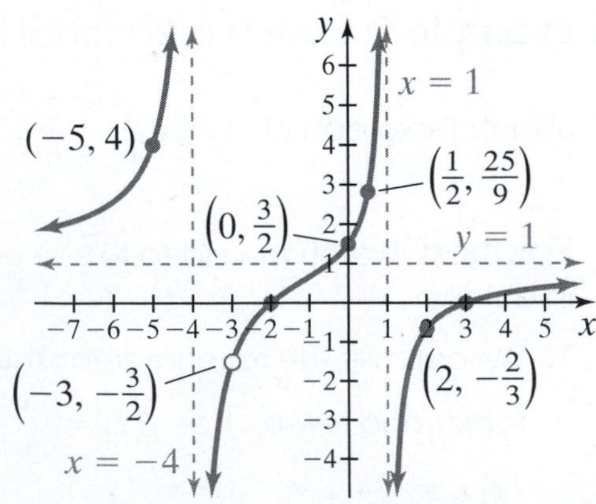

Figure 32 Graph of $f(x) = \dfrac{x^3 + 2x^2 - 9x - 18}{x^3 + 6x^2 + 5x - 12}$.

The graph seen in Figure 32 does not have any symmetry. Some rational functions, however, do have symmetry so it is important not to skip step 3. For example, the rational function $f(x) = \dfrac{x}{x^2 - 4}$ is symmetric about the origin because $f(-x) = -f(x)$. See if you can follow the nine-step graphing process to sketch this graph. To see how

to sketch the function $f(x) = \dfrac{x}{x^2 - 4}$, watch this video.

You Try It **Work through the following You Try It problem.**

Work Exercises 43–54 in this textbook or in the MyLab Math Study Plan.

4.6 Exercises

In Exercises 1–6, for each rational function, answer parts a to c.

a. Determine the domain.

b. Determine the y-intercept (if any).

c. Determine any x-intercepts.

1. $f(x) = \dfrac{x}{x^2 - 1}$

2. $f(x) = \dfrac{x}{x^2 + 1}$

3. $f(x) = \dfrac{x^2 - 1}{x}$

4. $f(x) = \dfrac{3x - 1}{x + 2}$

5. $f(x) = \dfrac{x^2 + x - 12}{x^2 + x}$

6. $f(x) = \dfrac{x^3 - 2x^2 - 5x + 6}{x^2 - 4x - 5}$

In Exercises 7–12, find all vertical asymptotes, and create a rough sketch of the graph near each asymptote.

7. $f(x) = \dfrac{3}{x + 1}$

8. $f(x) = \dfrac{x}{x - 2}$

9. $f(x) = \dfrac{x - 1}{x^2 - 1}$

10. $f(x) = \dfrac{x + 3}{x^2 - 6x + 8}$

11. $f(x) = \dfrac{x^2 - 2x - 15}{x^2 + x - 2}$

12. $f(x) = \dfrac{x^2 - 9}{x^3 - 2x^2 - 5x + 6}$

In Exercises 13–20, find the equation of all horizontal asymptotes (if any) of each rational function.

13. $f(x) = \dfrac{3}{x + 1}$

14. $f(x) = \dfrac{x^2 - 1}{x + 1}$

15. $f(x) = \dfrac{3x - 1}{5x + 2}$

16. $f(x) = \dfrac{2x^3 - x^2 + 1}{x^2 + 1}$

17. $f(x) = \dfrac{3x^2 - 4x^3}{7x^3 + x^2 - x + 1}$

18. $f(x) = \dfrac{9x^2}{4 - x^2}$

19. $f(x) = \dfrac{1}{x^3 + x^2 - 1}$

20. $f(x) = \dfrac{3x^2 - 2x - 9}{11x^2 + x - 2}$

In Exercises 21–32, use transformations of $y = \dfrac{1}{x}$ or $y = \dfrac{1}{x^2}$ to sketch each rational function. Label all intercepts, and find the equation of all asymptotes.

SbS 21. $f(x) = \dfrac{1}{x} - 2$

SbS 22. $f(x) = \dfrac{1}{x^2} + 2$

SbS 23. $f(x) = \dfrac{1}{x - 1}$

SbS 24. $f(x) = \dfrac{1}{(x - 1)^2}$

SbS 25. $f(x) = \dfrac{1}{x + 1} - 3$

SbS 26. $f(x) = \dfrac{1}{(x + 3)^2} - 2$

SbS 27. $f(x) = \dfrac{-1}{x - 2} - 3$

SbS 28. $f(x) = \dfrac{-1}{(x + 1)^2} + 2$

SbS 29. $f(x) = \dfrac{2}{x + 1}$

SbS 30. $f(x) = \dfrac{2}{(x - 3)^2}$

SbS 31. $f(x) = \dfrac{-2}{x + 1} - 1$

SbS 32. $f(x) = \dfrac{-2}{(x - 1)^2} + 3$

In Exercises 33–38, identify the coordinates of all removable discontinuities, and sketch the graph of each rational function. Label all intercepts, and find the equations of all asymptotes.

SbS 33. $f(x) = \dfrac{x^2 - 9}{x - 3}$

SbS 34. $f(x) = \dfrac{x^2 + 3x - 4}{x + 4}$

SbS 35. $f(x) = \dfrac{x - 1}{x^2 - x}$

SbS 36. $f(x) = \dfrac{x + 2}{x^2 - x - 6}$

SbS 37. $f(x) = \dfrac{x + 2}{(x + 2)(x - 1)^2}$

SbS 38. $f(x) = \dfrac{x - 1}{x^3 + 3x^2 - 4}$

In Exercises 39–42, find the slant asymptote of the graph of each rational function.

39. $f(x) = \dfrac{2x^2 - 1}{x}$

40. $f(x) = \dfrac{x^2 - 2x + 5}{x + 1}$

41. $f(x) = \dfrac{x^3 - 1}{x^2 + x}$

42. $f(x) = \dfrac{2x^4 - 1}{x^3 - x}$

In Exercises 43–54, follow the nine-step graphing strategy to sketch the graph of each rational function. Be sure to label all intercepts and find the equations of all asymptotes. Also label any removable discontinuities.

SbS 43. $f(x) = \dfrac{2x + 4}{x - 1}$

SbS 44. $f(x) = \dfrac{2x + 4}{x^2 - x - 2}$

SbS 45. $f(x) = \dfrac{x^2 - 1}{x^2 + 1}$

SbS 46. $f(x) = \dfrac{2x^2 - 5x - 3}{x + 2}$

SbS 47. $f(x) = \dfrac{1}{x^2 - 9}$

SbS 48. $f(x) = \dfrac{x}{x^2 - 9}$

SbS 49. $f(x) = \dfrac{x^2 + 2x - 3}{x + 1}$

SbS 50. $f(x) = \dfrac{2x^2 + 4x - 6}{x^2 - 4x + 3}$

SbS 51. $f(x) = \dfrac{x^3 - 3x^2 - x + 3}{x^2 - x - 6}$

SbS 52. $f(x) = \dfrac{x^3 - 7x + 6}{x^3 - 5x^2 + 2x + 8}$

SbS 53. $f(x) = \dfrac{2x^3 - 3x^2 - 3x + 2}{x^3 - 3x^2 - 6x + 8}$

SbS 54. $f(x) = \dfrac{2x^3 - 3x^2 - 3x + 2}{x^2 - 9}$

Brief Exercises

In Exercises 55–60, determine the domain of each rational function.

55. $f(x) = \dfrac{x}{x^2 - 1}$

56. $f(x) = \dfrac{x}{x^2 + 1}$

57. $f(x) = \dfrac{x^2 - 1}{x}$

58. $f(x) = \dfrac{3x - 1}{x + 2}$

59. $f(x) = \dfrac{x^2 + x - 12}{x^2 + x}$

60. $f(x) = \dfrac{x^3 - 2x^2 - 5x + 6}{x^2 - 4x - 5}$

In Exercises 61–66, determine the vertical asymptotes of each rational function.

61. $f(x) = \dfrac{3}{x + 1}$

62. $f(x) = \dfrac{x}{x - 2}$

63. $f(x) = \dfrac{x - 1}{x^2 - 1}$

64. $f(x) = \dfrac{x + 3}{x^2 - 6x + 8}$

65. $f(x) = \dfrac{x^2 - 2x - 15}{x^2 + x - 2}$

66. $f(x) = \dfrac{x^2 - 9}{x^3 - 2x^2 - 5x + 6}$

In Exercise 67–96, sketch the graph of each rational function.

67. $f(x) = \dfrac{1}{x} - 2$

68. $f(x) = \dfrac{1}{x^2} + 2$

69. $f(x) = \dfrac{1}{x - 1}$

70. $f(x) = \dfrac{1}{(x - 1)^2}$

71. $f(x) = \dfrac{1}{x + 1} - 3$

72. $f(x) = \dfrac{1}{(x + 3)^2} - 2$

73. $f(x) = \dfrac{-1}{x - 2} - 3$

74. $f(x) = \dfrac{-1}{(x + 1)^2} + 2$

75. $f(x) = \dfrac{2}{x+1}$

76. $f(x) = \dfrac{2}{(x-3)^2}$

77. $f(x) = \dfrac{-2}{x+1} - 1$

78. $f(x) = \dfrac{-2}{(x-1)^2} + 3$

79. $f(x) = \dfrac{x^2-9}{x-3}$

80. $f(x) = \dfrac{x^2+3x-4}{x+4}$

81. $f(x) = \dfrac{x-1}{x^2-x}$

82. $f(x) = \dfrac{x+2}{x^2-x-6}$

83. $f(x) = \dfrac{x+2}{(x+2)(x-1)^2}$

84. $f(x) = \dfrac{x-1}{x^3+3x^2-4}$

85. $f(x) = \dfrac{2x+4}{x-1}$

86. $f(x) = \dfrac{2x+4}{x^2-x-2}$

87. $f(x) = \dfrac{x^2-1}{x^2+1}$

88. $f(x) = \dfrac{2x^2-5x-3}{x+2}$

89. $f(x) = \dfrac{1}{x^2-9}$

90. $f(x) = \dfrac{x}{x^2-9}$

91. $f(x) = \dfrac{x^2+2x-3}{x+1}$

92. $f(x) = \dfrac{2x^2+4x-6}{x^2-4x+3}$

93. $f(x) = \dfrac{x^3-3x^2-x+3}{x^2-x-6}$

94. $f(x) = \dfrac{x^3-7x+6}{x^3-5x^2+2x+8}$

95. $f(x) = \dfrac{2x^3-3x^2-3x+2}{x^3-3x^2-6x+8}$

96. $f(x) = \dfrac{2x^3-3x^2-3x+2}{x^2-9}$

4.7 Variation

THINGS TO KNOW

Before working through this section, be sure that you are familiar with the following concepts:

VIDEO ANIMATION INTERACTIVE

You Try It

1. Converting Verbal Statements into Mathematical Statements (Section 1.2)

OBJECTIVES

1 Solve Application Problems Involving Direct Variation

2 Solve Application Problems Involving Inverse Variation

3 Solve Application Problems Involving Combined Variation

SECTION 4.7 EXERCISES

OBJECTIVE 1 SOLVE APPLICATION PROBLEMS INVOLVING DIRECT VARIATION

In application problems, we are often concerned with exploring how one quantity varies with respect to other quantities. Variation equations allow us to show how one quantity changes with respect to one or more additional quantities. In this section we will discuss direct variation, inverse variation, and combined variation. We start by introducing **direct variation**.

Direct Variation

For an equation of the form

$$y = kx,$$

we say that y **varies directly** with x, or y is **proportional to** x. The constant k is called the **constant of variation** or the **proportionality constant**.

Example 1 Direct Variation

Suppose y varies directly with x, and $y = 20$ when $x = 8$.

a. Find the equation that relates x and y.

b. Find y when $x = 12$.

Solutions Read the following, or watch this **video** for a complete solution.

a. Because y **varies directly** with x, the equation has the form $y = kx$. Use the fact that $y = 20$ when $x = 8$ to find k.

$y = kx$	Write the direct variation equation.
$20 = k(8)$	Substitute $y = 20$ and $x = 8$.
$\frac{20}{8} = \frac{k(8)}{8}$	Divide both sides by 8.
$2.5 = k$	Simplify.

The **constant of variation** is 2.5. so the equation is $y = 2.5x$.

b. The variables y and x are related by the equation $y = 2.5x$. To find y when $x = 12$, substitute 12 for x in the equation and simplify:

$$y = 2.5(12) = 30$$

Therefore, $y = 30$ when $x = 12$.

You Try It Work through this You Try It problem.

Work Exercises 1 and 2 in this textbook or in the MyLab Math Study Plan.

In Example 1, the model $y = kx$ is a **linear equation**. However, direct variation equations will not always be linear. For example, if y varies directly with the **cube** of x, then the equation will be $y = kx^3$, where it is the constant of variation.

(eText Screens 4.7-1–4.7-26)

Example 2 Direct Variation

Suppose y **varies directly** with the cube of x, and $y = 375$ when $x = 5$.

a. Find the equation that relates x and y.

b. Find y when $x = 2$.

Solutions Because y varies directly with the cube of x, or x^3, the variation equation has the form $y = kx^3$. Use the fact that $y = 375$ when $x = 5$ to find k, and then the equation. Try to complete this problem on your own. View the **answer** or watch this **video** for a complete solution.

You Try It Work through this You Try It problem.

Work Exercises 3–6 in this textbook or in the MyLab Math Study Plan.

For direct variation, the **ratio** of the two quantities is constant (the **constant of variation**). For example, consider $y = kx$ and $y = kx^3$.

$$\underbrace{y = kx}_{\substack{y \text{ varies}\\ \text{directly}\\ \text{with } x}} \;\rightarrow\; \underbrace{\frac{y}{x}}_{\substack{\text{Ratio of the}\\ \text{quantities}\\ y \text{ and } x}} = \underbrace{k}_{\substack{\text{Constant of}\\ \text{variation}}} \qquad \underbrace{y = kx^3}_{\substack{y \text{ varies}\\ \text{directly with}\\ \text{the cube}\\ \text{of } x}} \;\rightarrow\; \underbrace{\frac{y}{x^3}}_{\substack{\text{Ratio of the}\\ \text{quantities}\\ y \text{ and } x^3}} = \underbrace{k}_{\substack{\text{Constant of}\\ \text{variation}}}$$

Problems involving variation can generally be solved using the following guidelines.

> **Solving Variation Problems**
>
> **Step 1.** Translate the problem into an equation that models the situation.
>
> **Step 2.** Substitute given values for the variables into the equation and solve for the constant of variation, k.
>
> **Step 3.** Substitute the value for k into the equation to form the general model.
>
> **Step 4.** Use the general model to answer the question posed in the problem.

Example 3 Kinetic Energy

The kinetic energy of an object in motion varies directly with the square of its speed. If a van traveling at a speed of 30 meters per second has 945,000 joules of kinetic energy, how much kinetic energy does it have if it is traveling at a speed of 20 meters per second?

Solution Follow the **guidelines for solving variation problems.**

Step 1. We are told that the kinetic energy of an object in motion varies **directly** with the square of its speed. If we let $K = \text{kinetic energy}$ and $s = \text{speed}$, we can translate the problem statement into the model

$$K = ks^2,$$

where k is the **constant of variation.**

Step 2. To determine the value of k, we use the fact that the kinetic energy is 945,000 joules when the velocity is 30 meters per second.

$945{,}000 = k(30)^2$	Substitute 945,000 for K and 30 for s.
$945{,}000 = 900k$	Simplify 30^2.
$1050 = k$	Divide both sides by 900.

Step 3. The constant of variation is 1050, so the general model is $K = 1050s^2$.

Step 4. We want to determine the kinetic energy of the van if its speed is 20 meters per second. Substituting 20 for s, we find

$$K = 1050(20)^2 = 1050(400) = 420{,}000.$$

The van will have 420,000 joules of kinetic energy if it is traveling at a speed of 20 meters per second. ●

You Try It Work through this You Try It problem.

Work Exercises 7 and 8 in this textbook or in the MyLab Math Study Plan.

Example 4 Measuring Leanness

The Ponderal Index measure of leanness states that body mass varies directly with the cube of height. If a "normal" person who is 1.2 m tall has a body mass of 21.6 kg, then what is the body mass of a "normal" person that is 1.8 m tall?

Solution Follow the guidelines for solving variation problems.

Step 1. We are told that weight varies directly with the cube of height. If we let $w =$ weight and $h =$ height, we can translate the problem statement into the model

$$w = kh^3,$$

where k is the constant of variation.

Step 2. To determine the value of k, we use the fact that a normal person who is 1.2 meters tall has a mass of 21.6 kg.

$21.6 = k(1.2)^3$	Substitute 1.2 for h and 21.6 for w.
$21.6 = 1.728k$	Simplify $(1.2)^3$
$12.5 = k$	Divide both sides by 1.728.

Step 3. The constant of variation is 12.5, so the general model is $w = 12.5h^3$.

Use the general model to determine the body mass of a normal person who is 1.8 m tall. View the solution, or watch this video to see the entire detailed solution ●

You Try It Work through this You Try It problem.

Work Exercises 9 and 10 in this textbook or in the MyLab Math Study Plan.

OBJECTIVE 2 SOLVE APPLICATION PROBLEMS INVOLVING INVERSE VARIATION

We now discuss **inverse variation**. Inverse variation means that one variable is a constant multiple of the **reciprocal** of another variable.

> **Inverse Variation**
>
> For equations of the form
>
> $$y = \frac{k}{x} \quad \text{or} \quad y = k \cdot \frac{1}{x},$$
>
> we say that y **varies inversely** with x, or y is **inversely proportional to** x. The constant k is called the **constant of variation**.

For inverse variation, the product of the two quantities is constant (the constant of variation). For example, consider $y = \frac{k}{x}$ and $y = \frac{k}{x^2}$.

$$\underbrace{y = \frac{k}{x}}_{y \text{ varies inversely with } x} \rightarrow \underbrace{xy}_{\text{Product of the quantities } y \text{ and } x} = \underbrace{k}_{\text{Constant of variation}} \qquad \underbrace{y = \frac{k}{x^2}}_{y \text{ varies inversely with the square of } x} \rightarrow \underbrace{x^2y}_{\text{Product of the quantities } y \text{ and } x^2} = \underbrace{k}_{\text{Constant of variation}}$$

Example 5 Inverse Variation

Suppose y **varies inversely** with x, and $y = 72$ when $x = 50$.

a. Find the equation that relates x and y.

b. Find y when $x = 45$.

Solutions

a. Because y varies inversely with x, the equation has the form $y = \frac{k}{x}$. We use the fact that $y = 72$ when $x = 50$ to find k.

$y = \frac{k}{x}$ Write the inverse variation equation.

$72 = \frac{k}{50}$ Substitute $y = 72$ and $x = 50$.

$3600 = k$ Multiply both sides by 50.

The **constant of variation** is 3600, so the equation is $y = \frac{3600}{x}$.

b. Try to work this part on your own. View the **answer** or watch this **video** for a complete solution to both parts.

You Try It Work through this You Try It problem.

Work Exercises 11–14 in this textbook or in the MyLab Math Study Plan.

Example 6 Density of an Object

For a given mass, the density of an object is **inversely proportional** to its volume. If 50 cubic centimeters of an object with a density of $28\frac{\text{g}}{\text{cm}^3}$ is compressed to 40 cubic centimeters, what would be its new density?

Solution Follow the **guidelines for solving variation problems.**

Step 1. We are told that the density of an object with a given mass varies inversely with its volume. If we let $D =$ density and $V =$ volume, we can translate the problem statement into the model

$$D = \frac{k}{V},$$

where k is the **constant of variation.**

Step 2. To determine the value of k, we use the fact that the density is $28\text{ g}/\text{cm}^3$ when the volume is 50 cm^3.

$$28 = \frac{k}{50} \quad \text{Substitute 28 for } D \text{ and 50 for } V.$$

$$(28)(50) = k \quad \text{Multiply both sides by 50.}$$

$$1400 = k \quad \text{Simplify.}$$

Step 3. The constant of variation is 1400, so the general model is $D = \frac{1400}{V}$.

Step 4. We want to determine the density of the object if the volume is compressed to 40 cm^3. Substituting 40 for V, we find

$$D = \frac{1400}{40} = 35\text{ g}/\text{cm}^3.$$

The density of the compressed object would be $35\text{ g}/\text{cm}^3$. ●

You Try It **Work through this You Try It problem.**

Work Exercises 15 and 16 in this textbook or in the MyLab Math Study Plan.

Example 7 Shutter Speed

The shutter speed, S, of a camera varies inversely as the square of the aperture setting, f. If the shutter speed is 125 for an aperture of 5.6, what is the shutter speed if the aperture is 1.4?

Solution Follow the **guidelines for solving variation problems.**

Step 1. We are told that shutter speed varies **inversely** with the square of the aperture setting. Letting $S =$ shutter speed and $f =$ aperture setting, we can translate the problem statement into the model

$$S = \frac{k}{f^2},$$

where k is the **constant of variation.**

Step 2. To determine the value of k, we use the fact that an aperture setting of 5.6 corresponds to a shutter speed of 125.

$$125 = \frac{k}{(5.6)^2} \quad \text{Substitute 5.6 for } f \text{ and 125 for } S.$$

$$125 = \frac{k}{31.36} \quad \text{Simplify } (5.6)^2.$$

$$3920 = k \quad \text{Multiply both sides by 31.36.}$$

Step 3. The constant of variation is 3920, so the general model is $S = \dfrac{3920}{f^2}$.

Use the general model to determine the shutter speed for an aperture setting of 1.4. View the **solution**, or watch this **video** to see the entire detailed solution. ●

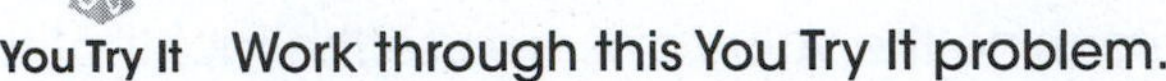

You Try It **Work through this You Try It problem.**

Work Exercises 17 and 18 in this textbook or in the MyLab Math Study Plan.

OBJECTIVE 3 SOLVE APPLICATION PROBLEMS INVOLVING COMBINED VARIATION

When a variable is related to more than one other variable, we call this **combined variation**. Some examples of **combined variation** are

$$y = k\frac{x}{z} \quad y = kxz \quad y = k\frac{w^2x}{z}.$$

In the first example we say that y varies directly as x and inversely as z. In the second example we say that y varies directly as x and z. In the third example we say that y varies directly as x and the square of w, and inversely as z. In general, variables **directly related** to the dependent variable occur in the numerator while variables **inversely related** occur in the denominator.

$$y = k\frac{x \leftarrow \text{directly related}}{z \leftarrow \text{inversely related}} \qquad y = kxz \leftarrow \text{directly related} \qquad y = k\frac{w^2x \leftarrow \text{directly related}}{z \leftarrow \text{inversely related}}$$

We solve these problems the same way we solved other variation problems. First we find the **constant of variation**, then use it to help answer the question.

When a variable is directly proportional to the product of two or more other variables, such as $y = kxz$, this is often called **joint variation**. In this case we would say that y varies **jointly** as x and z.

Example 8 Combined Variation

Suppose y varies directly with w and inversely with x, and $y = 240$ when $w = 20$ and $x = 4$.

a. Find the equation that relates $w, x,$ and y.

b. Find y when $w = 10$ and $x = 3$.

Solutions Read the following, or watch this **video** for a complete solution.

a. Because y varies directly with w and inversely with x, the equation has the form $y = k\frac{w}{x}$. We use the fact that $y = 240$ when $w = 20$ and $x = 4$ to find k.

Write the combined variation equation: $y = k\frac{w}{x}$

Substitute $y = 240$, $w = 20$, and $x = 4$: $240 = k\frac{20}{4}$

Simplify: $240 = 5k$

Divide both sides by 5: $48 = k$

The **constant of variation** is 48, so the equation is $y = 48\frac{w}{x}$, or $y = \frac{48w}{x}$

b. To find y when $w = 10$ and $x = 3$, substitute the values into the equation and simplify:

$$y = \frac{48(10)}{3} = 160$$

Therefore, $y = 160$ when $w = 10$ and $x = 3$.

You Try It Work through this You Try It problem.

Work Exercises 19–22 in this textbook or in the MyLab Math Study Plan.

Example 9 Volume of a Conical Tank

The number of gallons of a liquid that can be stored in a conical tank is directly proportional to the area of the base of the tank and its height (**joint variation**). A tank with a base area of 1200 square feet and a height of 15 feet holds 45,000 gallons of liquid. How tall must the tank be to hold 75,000 gallons of liquid if its base area is 1500 square feet?

Solution Follow the **guidelines for solving variation problems**.

Step 1. We are told that the volume is **directly** related to the area of the base of the tank and its height. If we let $V = \text{volume}$, $B = \text{base area}$, and $h = \text{height}$, we can translate the problem statement into the model

$$V = kBh \leftarrow \text{joint variation } (B \text{ and } h \text{ both direct}),$$

where k is the constant of variation.

Step 2. To determine the value of k, we use the fact that a tank with a base area of 1200 square feet and a height of 15 feet holds 45,000 gallons.

$45{,}000 = k(1200)(15)$ Substitute 1200 for B, 15 for h, and 45,000 for V.

$45{,}000 = 18{,}000k$ Simplify.

$2.5 = k$ Divide both sides by 18,000.

(eText Screens 4.7-1–4.7-26)

Step 3. The constant of variation is 2.5, so the general model is $V = 2.5Bh$.

Step 4. We want to determine the height of a tank with a base area of 1500 square feet that holds 75,000 gallons. Substituting 75,000 for V and 1500 for B, we then solve for h.

$75{,}000 = 2.5(1500)h$	Substitute 1500 for B and 75,000 for V.
$75{,}000 = 3750h$	Simplify.
$20 = h$	Divide both sides by 3750.

The tank would need to be 20 feet tall.

You Try It **Work through this You Try It problem.**

Work Exercises 23 and 24 in this textbook or in the MyLab Math Study Plan.

Example 10 Electrical Resistance

The resistance of a wire varies directly with the length of the wire and inversely with the square of its radius. A wire with a length of 500 cm and a radius of 0.5 cm has a resistance of 15 ohms. Determine the resistance of an 800 cm piece of similar wire with a radius of 0.8 cm.

Solution

Follow the **guidelines for solving variation problems.**

Step 1. We are told that resistance varies **directly** with the length of the wire and **inversely** with the square of its radius. Letting $R = \text{resistance}$, $L = \text{length}$, and $r = \text{radius}$, we can translate the problem statement into the model

$$R = k\frac{L}{r^2} \quad \begin{matrix} \leftarrow \text{length (direct)} \\ \leftarrow \text{square of the radius (inverse)} \end{matrix}$$

where k is the **constant of variation.**

Step 2. To determine the value of k, we use the fact that a wire of length 500 cm and radius 0.5 cm has a resistance of 15 ohms.

$15 = k\dfrac{(500)}{(0.5)^2}$	Substitute 500 for L, 0.5 for r, and 15 for R.
$15 = 2000k$	Simplify.
$0.0075 = k$	Divide both sides by 2000.

Use the constant of variation to write a general model for this situation. Then use the model to determine the resistance of an 800 cm piece of similar wire with a radius of 0.8 cm. View the **solution**, or watch this **video** to see the entire detailed solution.

You Try It **Work through this You Try It problem.**

Work Exercises 25 and 26 in this textbook or in the MyLab Math Study Plan.

 **Example 11 Burning Calories**

For a fixed speed, the number of calories burned while jogging varies **jointly** with the mass of the jogger (in kg) and the time spent jogging (in minutes). If a 100-kg man jogs for 40 minutes and burns 490 calories, how many calories will a 130-kg man burn if he jogs for 60 minutes at the same speed?

Solution Letting $C =$ calories burned, $W =$ weight in kg, and $T =$ time jogging in minutes, we have the model $C = kWT$, where k is the **constant of variation**. Solve the problem on your own. View the **solution**, or watch this **video** to see the entire detailed solution.

You Try It Work through this You Try It problem.

Work Exercises 27 and 28 in this textbook or in the MyLab Math Study Plan.

4.7 Exercises

In Exercises 1–28, solve each variation problem.

1. Suppose y varies directly with x, and $y = 14$ when $x = 4$.

   a. Find the equation that relates x and y.

   b. Find y when $x = 7$.

2. Suppose S in directly proportional to t, and $S = 900$ when $t = 25$.

   a. Find the equation that relates t and S.

   b. Find S when $t = 4$.

3. Suppose y varies directly with the square of x, and $y = 216$ when $x = 3$.

   a. Find the equation that relates x and y.

   b. Find y when $x = 5$.

4. Suppose M is directly proportional to the square of r, and $M = 14$ when $r = \frac{1}{6}$.

   a. Find the equation that relates r and M.

   b. Find M when $r = \frac{2}{3}$.

5. Suppose p is directly proportional to the cube of n, and $p = 135$ when $n = 6$.

   a. Find the equation that relates n and p.

   b. Find p when $n = 10$.

6. Suppose A varies directly with the cube of m, and $A = 48$ when $m = 4$.

   a. Find the equation that relates m and A.

   b. Find A when $m = 8$.

7. The water pressure on a scuba diver is directly proportional to the depth of the diver. If the pressure on a diver is 13.5 psi when she is 30 feet below the surface, how far below the surface will she be when the pressure is 18 psi?

8. The length of a simple pendulum varies directly with the square of its period. If a pendulum of length 2.25 meters has a period of 3 seconds, how long is a pendulum with a period of 8 seconds?

9. For a fixed water flow rate, the amount of water that can be pumped through a pipe varies directly as the square of the diameter of the pipe. In one hour a pipe with an 8-inch diameter can pump 400 gallons of water. Assuming the same water flow rate, how much water could be pumped through a pipe that is 12 inches in diameter?

10. The distance an object falls varies directly with the square of the time it spends falling. If a ball falls 19.6 meters after falling for 2 seconds, how far will it fall after 9 seconds?

11. Suppose y varies inversely with x, and $y = 24$ when $x = 25$.

    a. Find the equation that relates x and y.

    b. Find y when $x = 30$.

12. Suppose p is inversely proportional to q, and $p = 3.5$ when $q = 4.6$.

    a. Find the equation that relates q and p.

    b. Find p when $q = 1.4$.

13. Suppose n varies inversely with the square of m, and $n = 6.4$ when $m = 9$.

    a. Find the equation that relates m and n.

    b. Find n when $m = 4$.

14. Suppose d is inversely proportional to the cube of t, and $d = 76.8$ when $t = 2.5$.

    a. Find the equation that relates t and d.

    b. Find d when $t = 5$.

15. The value of a car is inversely proportional to its age. If a car is worth \$8100 when it is 4 years old, how old will it be when it is worth \$3600?

16. For a given voltage, the resistance of a circuit is inversely related to its current. If a circuit has a resistance of 5 ohms and a current of 12 amps, what is the resistance if the current is 20 amps?

17. The weight of an object with in Earth's atmosphere varies inversely with the square of the distance of the object from Earth's center. If a low Earth orbit satellite weighs 100 kg on Earth's surface (6400 km), how much will it weigh in its orbit 800 km above Earth?

18. The intensity of a light varies inversely as the square of the distance from the light source. If the intensity from a light source 3 feet away is 8 lumens, what is the intensity at a distance of 2 feet?

19. Suppose y varies directly with x and inversely with z, and $y = 48$ when $x = 8$ and $z = 2.5$.

    a. Find the equation that relates y, x, and z.

    b. Find y when $x = 24$ and $z = 18$.

20. Suppose A varies jointly with b and c, and $A = 15$ when $b = 10$ and $c = 3$.

    a. Find the equation that relates A, b, and c.

    b. Find b when $A = 21$ and $c = 9$.

21. Suppose P varies directly with n and the square of m, and $P = 43.2$ when $n = 4$ and $m = 1.5$.

    a. Find the equation that relates P, n and m.

    b. Find P when $n = 3.5$ and $m = 2$.

22. Suppose A varies directly with the square of t and inversely with the cube of r, and $A = 54$ when $t = 6$ and $r = 2$.

    a. Find the equation that relates A, t, and r.

    b. Find A when $t = 16$ and $r = 4$.

23. For a car loan using simple interest at a given rate, the amount of interest charged varies jointly with the loan amount and the time of the loan (in years). If a \$15,000 car loan earns \$2175 in simple interest over 5 years, how much interest will a \$32,000 car loan earn over 6 years?

24. The horsepower of a water pump varies jointly with the weight of the water moved (in lb) and the length of the discharge head (in feet). A 3.1-horsepawer pump can move 341 lb of water against a discharge head of 300 feet. What is the horsepower of a pump that moves 429 lb of water against a discharge head of 400 feet?

25. The pressure of a gas in a container varies directly with its temperature and inversely with its volume. At a temperature of 90 Kelvin and a volume of 10 cubic meters, the pressure is 32.4 kilograms per square meter. What is the pressure if the temperature is increased to 100 Kelvin and the volume is decreased to 8 cubic meters?

26. In milling operations, the spindle speed S (in revolutions per minute) is directly related to the cutting speed C (in feet per minute) and inversely related to the tool diameter D (in inches). A milling cut taken with a 2-inch high-speed drill and a cutting speed of 70 feet per minute has a spindle speed of 133.7 revolutions per minute. What is the spindle speed for a cut taken with a 4-inch high-speed drill and a cutting speed of 50 feet per minute?

27. The load that can be supported by a rectangular beam varies jointly as the width of the beam and the square of its height and inversely as the length of the beam. A beam 12 feet long, with a width of 6 inches and a height of 4 inches can support a maximum load of 900 pounds. If a similar board has a width of 8 inches and a height of 5 inches. how long must it be to support 1200 pounds?

28. The power produced by a windmill varies directly with the square of its diameter and the cube of the wind speed. A fan with a diameter of 2.5 meters produces 270 watts of power if the wind speed is 3 m/s. How much power would a similar fan produce if it had a diameter of 4 meters and the wind speed was 4 m/s?

Chapter 4 Summary

Key Concepts	Examples/Videos
4.1 Quadratic Functions A **quadratic function** is a function that can be written in the form $f(x) = ax^2 + bx + c$, where $a, b,$ and c are real numbers with $a \neq 0$. The graph of every quadratic function is a parabola. Characteristics of a Parabola A quadratic function is in **vertex form** if it is written as $f(x) = a(x - h)^2 + k$. The graph is a parabola with vertex (h, k). The parabola "opens up" if $a > 0$. The parabola "opens down" if $a < 0$.	Given the quadratic function $f(x) = -(x - 2)^2 - 4$ is in vertex form, address the following: **a.** What are the coordinates of the vertex. **b.** Does the graph open "up" or "open down". **c.** What is the equation of the axis of symmetry? **d.** Find any x-intercepts. **e.** Find the y-intercept. **f.** Sketch the graph. **g.** State the domain and range in interval notation.
To rewrite a quadratic function from the form $f(x) = ax^2 + bx + c$ into vertex form, factor out the leading coefficient and complete the square. Practice sketching quadratic functions.	Rewrite the quadratic function $f(x) = 2x^2 - 4x - 3$ in vertex form and answer questions a-g as in the example above.
Formula for the Vertex of a Parabola Given a quadratic function of the form $f(x) = ax^2 + bx + c, a \neq 0$, the vertex of the parabola is $\left(-\frac{b}{2a}, f\left(-\frac{b}{2a}\right)\right)$.	Given the quadratic function $f(x) = -2x^2 - 4x + 5$, use the vertex formula to determine the vertex then answer questions b-g as in the examples on the previous page.
4.2 Applications and Modeling Quadratic Functions If a quadratic function of the form $f(x) = ax^2 + bx + c, a \neq 0$ is used to maximize or minimize applied problems, then we are interested in determining the values of the coordinates of the vertex. The x-coordinate of the vertex is $-\frac{b}{2a}$ and the y-coordinate of the vertex is $f\left(-\frac{b}{2a}\right)$.	A toy rocket is launched with an initial velocity of 44.1 meters per second from a 1-meter-tall platform. The height h of the object at anytime t seconds after launch is given by the function $h(t) = -4.9t^2 + 44.1t + 1$. How long after launch did it take the rocket to reach its maximum height? What is the maximum height obtained by the toy rocket?

Key Concepts	Examples/Videos
	To sell x waterproof CD alarm clocks, WaterTime, LLC, has determined that the price in dollars must be $p = 250 - 2x$, which is the demand equation. Each clock costs \$2 to produce, with fixed costs of \$4,000, producing the cost function of $C(x) = 2x + 4{,}000$. **a.** Express the revenue R as a function of x. **b.** Express the profit P as a function of x. **c.** Find the value of x that maximizes profit, What is the maximum profit? **d.** What is the price of the alarm clock that will maximize profit?
	Mark has 100 feet of fencing available to build a rectangular pen for his hens and roosters. He wants to separate the hens and roosters by dividing the pen into two equal areas. Let x represent the length of the center partition. **a.** Create a function $A(x)$ that describes the total area of the rectangular enclosure as a function of x where x is the length of the center partition. **b.** Find the length of the center partition that will yield the maximum area. **c.** Find the length of the side of the fence parallel to the river that will yield the maximum area. **d.** What is the maximum area?
4.3 The Graphs of Polynomial Functions **Four-Step Process for Sketching the Graph of a Polynomial Function** **Step 1.** Determine the end behavior. **Step 2.** Plot the y-intercept $f(0) = a_0$. **Step 3.** Completely factor f to find all real zeros and their **multiplicities.** **Step 4.** Choose a test value between each real zero and sketch the graph.	Use the four-step process to sketch the graphs of the following polynomial functions: **a.** $f(x) = -2(x + 2)^2(x - 1)$ **b.** $f(x) = x^4 - 2x^3 - 3x^2$

Key Concepts	Examples/Videos
4.4 Synthetic Division; The Remainder and Factor Theorems **Synthetic Division** is a "shortcut method" to long division of polynomials but only works when the divisor is of the form $x - c$.	Use synthetic division to divide $f(x)$ by $(x - c)$, and then write $f(x)$ in the form $f(x) = (x - c)q(x) + r$ for $f(x) = -2x^4 + 3x^3 + 7x^2 - x + 5$ divided by $x + 1$.
Remainder Theorem If a polynomial $f(x)$ is divided by $x - c$, then the remainder is $f(c)$.	Use the remainder theorem to find the remainder when $f(x)$ is divided by $x - c$. a. $f(x) = 5x^4 - 8x^2 + 3x - 1;\quad x - 2$ b. $f(x) = 3x^3 + 5x^2 - 5x - 6 - 1;\quad x + 2$
Factor Theorem The polynomial $x - c$ is a factor of the polynomial $f(x)$ if and only if $f(c) = 0$.	Determine whether $x + 3$ is a factor of $f(x) = 2x^3 + 7x^2 + 2x - 3$. Given that $x = 2$ is a zero of $f(x) = x^3 - 6x + 4$, completely factor f and sketch its graph.
4.5 The Zeros of Polynomial Functions; The Fundamental Theorem of Algebra **Rational Zeros Theorem** If a polynomial function with integer coefficients has a rational zero of the form $\frac{p}{q}$, then p must be a factor of the constant coefficient and q must be a factor of the leading coefficient.	Use the rational zeros theorem to determine the potential rational zeros of the polynomial function $f(x) = 4x^4 - 7x^3 + 9x^2 - x - 10$.
Descartes' Rule of Signs Let f be a polynomial function with real coefficients written in descending order. 1. The number of *positive* real zeros of f is equal to the number of variations in sign of $f(x)$ or is less than the number of variations in sign by a positive even integer. If $f(x)$ has one variation in sign, then f has exactly one positive real zero. 2. The number of *negative* real zeros of f is equal to the number of variations in sign of $f(-x)$ or is less than the number of variations in sign of $f(-x)$ by a positive even integer. If $f(-x)$ has one variation in sign, then f has exactly one negative real zero.	Determine the possible number of positive and negative real zeros of each polynomial function. a. $f(x) = 2x^5 + 3x^4 + 8x^2 + 2x$ b. $f(x) = 6x^4 + 13x^3 + 61x^2 + 8x - 10$

Key Concepts	Examples/Videos
The Fundamental Theorem of Algebra Every polynomial function of degree $n \geq 1$ has at least one complex zero. **Number of Zeros Theorem** Every polynomial of degree n has n complex zeros provided each zero of multiplicity greater than 1 is counted accordingly.	Find all zeros of $f(x) = 6x^4 + 13x^3 + 61x^2 + 8x - 10$ and rewrite $f(x)$ in completely factored form.
The solution to a polynomial equation of the form $f(x) = 0$ is equivalent to finding the zeros of $f(x)$.	Solve $5x^5 - 9x^4 + 23x^3 - 35x^2 + 12x + 4 = 0$.
Complex Conjugate Pairs Theorem If $a + bi$ is a zero of a polynomial function with real coefficients, then the complex conjugate $a - bi$ is also a zero.	Find a fourth-degree polynomial function with real coefficients such That 1 is a zero of multiplicity 2 and $2 - i$ is also a zero.
Intermediate Value Theorem Let f be a polynomial function with $a < b$. If $f(a)$ and $f(b)$ have opposite signs, then there exists at least one real zero strictly between a and b.	Find the real zeros of $f(x) = x^3 + 2x - 1$ correct to two decimal places.
Steps for Sketching the Graphs of Polynomial Functions **Step 1.** Determine the end behavior. **Step 2.** Plot the y-intercept $f(0) = a_0$. **Step 3.** Use the rational zeros theorem, the factor theorem and synthetic division, or the intermediate value theorem to find all zeros and completely factor f. **Step 4.** Choose a test value between each real zero and complete the graph.	Sketch the graph of $f(x) = 2x^5 - 5x^4 - 2x^3 + 7x^2 - 4x + 12$.

Key Concepts	Examples/Videos
4.6 Rational Functions and their Graphs A **rational function** is a function of the form $f(x) = \frac{g(x)}{h(x)}$, where $g \neq 0$ and h are polynomial functions such that the degree of $h(x)$ is greater than 0. A rational function of the form $f(x) = \frac{g(x)}{h(x)}$, where g and h have no common factors, will have a **vertical asymptote** at $x = a$ if $h(a) = 0$.	 Find the vertical asymptotes (if any) of the function $f(x) = \frac{x - 3}{x^2 + x - 6}$, and then sketch the graph near the vertical asymptotes.
A rational function of the form $f(x) = \frac{g(x)}{h(x)}$ that is written in lowest terms (all common factors of the numerator and denominator have been canceled) will have a **horizontal asymptote** whenever the degree of $h(x)$ is greater than or equal the degree of $g(x)$. **Finding Horizontal Asymptotes of a Rational Function** Let $f(x) = \frac{g(x)}{h(x)} = \frac{a_nx^n + a_{n-1}x^{n-1} + a_{n-2}x^{n-2} + \ldots + a_1x + a_0}{b_mx^m + b_{m-1}x^{m-1} + b_{m-2}x^{m-2} + \ldots + b_1x + b_0}$, $a_n \neq 0, b_m \neq 0$, where f is written in lowest terms, n is the degree of g, and m is the degree of h. • If $m > n$, then $y = 0$ is the horizontal asymptote. • If $m = n$, then the horizontal asymptote is $y = \frac{a_n}{b_m}$. • If $m < n$, then there are no horizontal asymptotes.	Find the horizontal asymptotes of the graph of each rational function or state that one does not exist. a. $f(x) = \frac{x}{x^2 - 4}$ b. $f(x) = \frac{4x^2 - x + 1}{1 - 2x^2}$ c. $f(x) = \frac{2x^3 + 3x^2 - 2x - 2}{x - 1}$
A rational function of the form $f(x) = \frac{g(x)}{h(x)}$ will have a **removable discontinuity** (a "hole" in the graph) if $g(x)$ and $h(x)$ share a common factor.	 Sketch the graph of $f(x) = \frac{x + 3}{x^2 + x - 6}$, and find the coordinates of all removable discontinuities.
If a rational function of the form $f(x) = \frac{g(x)}{h(x)}$ where the degree of $h(x)$ is exactly 1 less than the degree of $g(x)$, then f can be rewritten as $f(x) = mx + b + \frac{r(x)}{h(x)}$. The line $y = mx + b$ is called the **slant asymptote**.	 Find the slant asymptotes of $f(x) = \frac{2x^2 + 3x - 2}{x - 1}$.

Key Concepts	Examples/Videos
Steps for Graphing Rational Functions of the Form $f(x) = \frac{g(x)}{h(x)}$	Sketch the graph of $f(x) = \frac{x^3 + 2x^2 - 9x - 18}{x^3 + 6x^2 + 5x - 12}$.
4.7 Variation Let k be a constant called the **constant of variation** or the **proportionality constant.** For an equation of the form $y = kx$, we say that y varies **directly** with x. For an equation of the form $y = \frac{k}{x}$, we say that y varies **inversely** with x. For an equation of the form $y = kxz$, we say that y varies **jointly** as x and z.	
Solving Variation Problems **Step 1** Translate the problem into an equation that models the situation. **Step 2** Substitute the given values of the variables into the equation and solve for the constant of variation, k. **Step 3** Substitute the value for k into the equation to form the general model. **Step 4** Use the general model to answer the question posed in the problem.	**Direct Variation Example** The Ponderal Index measure of leanness states that the body mass varies directly with the cube of height. If a "normal" person who is 1.2 m tall has a body mass of 21.6 kg, then what is the body mass of a "normal" person that is 1.8 m tall? **Inverse Variation Example** The shutter speed, S, of a camera varies inversely as the square of the aperture setting, f. If the shutter speed is 125 for an aperture of 5.6, what is the shutter speed if the aperture is 1.4? **Combined Variation Example** The resistance of a wire varies directly with the length of the wire and inversely with the square of its radius. A wire with a length of 500 cm and a radius of 0.5 cm has a resistance of 15 ohms. Determine the resistance of an 800 cm piece of similar wire with a radius of 0.8 cm.

Chapter 4 Review Exercises

1. Given the quadratic function $f(x) = -2(x - 3)^2 + 2$, address the following:

   a. What are the coordinates of the vertex?

   b. Does the graph "open up" or "open down"?

   c. What is the equation of the axis of symmetry?

   d. Find any x-intercepts.

   e. Find the y-intercept.

   f. Sketch the graph.

   g. State the domain and range.

2. Given the quadratic function $f(x) = -x^2 - 4x + 8$, address the following:

   a. Use the vertex formula to determine the vertex?

   b. Does the graph "open up" or "open down"?

   c. What is the equation of the axis of symmetry?

   d. Find any x-intercepts.

   e. Find the y-intercept.

   f. Sketch the graph.

   g. State the domain and range.

3. Analyze the graph to address the following about the quadratic function it represents.

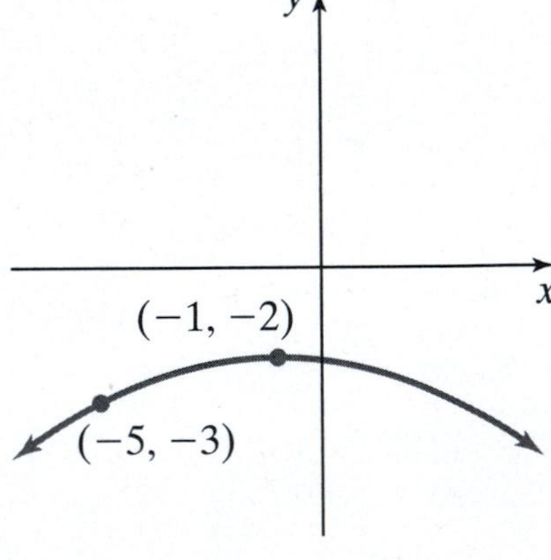

   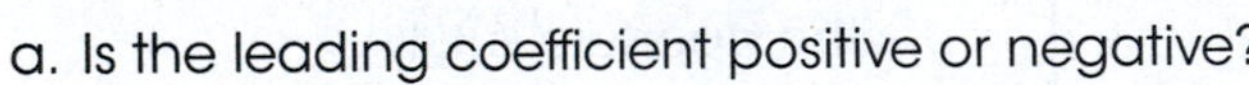

   a. Is the leading coefficient positive or negative?

   b. What is the value of h? What is the value of k?

   c. What is the value of the leading coefficient?

   d. Write the equation of the function in vertex form.

   e. Write the equation of the function in the form $f(x) = ax^2 + bx + c$.

4. A baseball player swings and hits a pop fly straight up in the air. The height of the baseball in meters, t seconds after it is hit, is given by the quadratic function $h(t) = -4.9t^2 + 34.3t + 1$. How long does it take for the baseball to reach its maximum height? What is the maximum height obtained by the baseball?

5. The price p and the quantity x sold of a small flat screen television set obeys the demand equation $p = -0.15x + 300$.

   a. How much should be charged for the television set if there are 50 television sets in stock?

   b. What quantity x will maximize revenue? What is the maximum revenue?

   c. What price should be charged in order to maximize revenue?

6. A farmer has 2,000 feet of fencing available to enclose a rectangular area bordering a river. No fencing is required along the river. Let x represent the length of the fence that is **parallel** to the river.

   a. Create a function $A(x)$ that describes the total area of the rectangular enclosure as a function of x where x is the length of the rectangular enclosure that is parallel to the river.

   b. Find the dimensions of the fence that will maximize the area.

   c. What is the maximum area?

7. Determine whether the function $G(x) = 7x^4 + 1 + 3x - 9x^5$ is a polynomial function. If it is a polynomial function, identify the degree, the leading coefficient, and the constant coefficient.

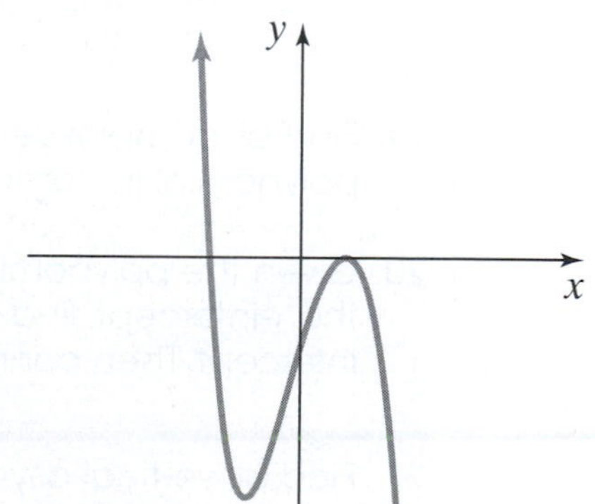

8. Use the end behavior of the graph of the polynomial function to

   a. determine whether the degree is even or odd and

   b. determine whether the leading coefficient is positive or negative.

9. Find the intercepts of the polynomial function $f(x) = x^3 + 3x^2 - 16x - 48$.

10. Determine the real zeros of $f(x) = x(x - 3)(x - 4)^4$ and state their multiplicities. Then decide whether the graph touches or crosses the x-axis at each zero.

11. Analyze the graph to address the following about the polynomial function it represents

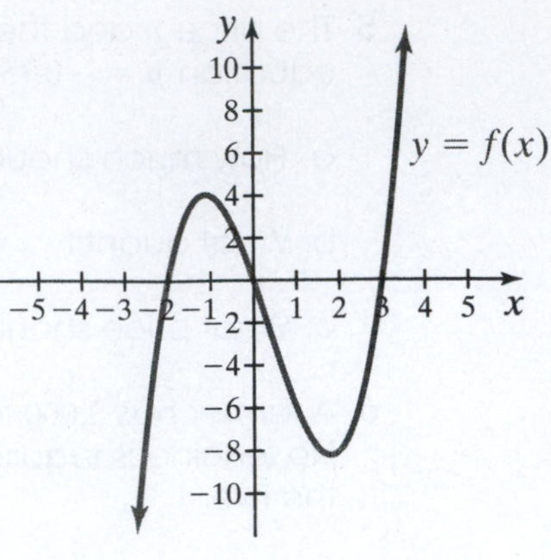

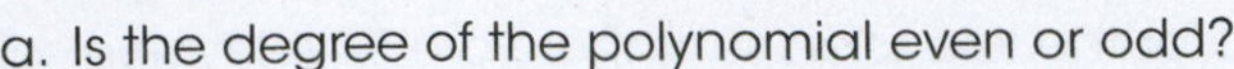

a. Is the degree of the polynomial even or odd?

b. Is the leading coefficient positive or negative?

c. What is the value of the constant coefficient?

d. Identify the real zeros, and state the multiplicity of each.

e. Select from this list a possible function that could be represented by this graph.

i. $f(x) = -x(x + 2)(x - 3)$

ii. $f(x) = x(x + 2)(x - 3)$

iii. $f(x) = -x^3(x + 2)(x - 3)$

iv. $f(x) = (x + 2)(x - 3)$

12. Sketch the graph of $f(x) = -2(x - 1)^2(x + 1)$ using the four-step process.

13. Use synthetic division to divide $f(x) = 6x^3 - 2x + 3$ by $x - 2$, and then rewrite $f(x)$ in the form $f(x) = (x - 2)q(x) + r$.

14. Use synthetic division and the remainder theorem to find the remainder when $f(x) = -4x^3 + 2x - 1$ is divided by $x + 3$.

15. Use synthetic division and the factor theorem to determine whether $x - 2$ is a factor of $f(x) = x^5 + x^4 - 3x^3 - 3x^2 - 4x - 4$.

16. Find the remaining zeros of $f(x) = x^3 + 2x^2 - x - 2$ given that $c = 1$ is a zero. Then rewrite $f(x)$ in completely factored form and sketch its graph.

17. Use the rational zeros theorem to determine the potential rational zeros of the polynomial $f(x) = 5x^3 - 27x^6 + 10x^2 + 16 - 4x$. Do not find the zeros.

18. Determine the possible number of positive real zeros and negative real zeros of the function $f(x) = 2x^5 - 13x^4 + 16x^3 - 5x^2 + 8x - 10$ using Descartes' rule of signs. Do not find the zeros.

19. Find all complex zeros of the polynomial function $f(x) = 4x^3 - 5x^2 - 23x + 6$, and write the polynomial in completely factored form.

20. Given the polynomial function $f(x) = x^3 + x^2 - x - 1$, determine the end behavior, plot the y-intercept, find and plot all real zeros, and plot at least one test value between each intercept. Then connect the points with a smooth curve.

21. Find all vertical asymptotes of the graph of $f(x) = \dfrac{x + 3}{x^2 - 6x + 8}$ and create a rough sketch of the graph near each asymptote.

22. Find the equation of all horizontal asymptotes (if any) of the rational function $f(x) = \dfrac{3x - 1}{5x + 2}$.

23. Follow the **nine-step strategy** to sketch the graph of the rational function $f(x) = \frac{2x + 4}{x^2 - x - 2}$. Be sure to label all intercepts and find the equation of all asymptotes. Also label any removable discontinuities.

24. The water pressure on a scuba diver is directly proportional to the depth of the diver. If the pressure on the diver is 13.5 psi when she is 30 psi?

25. The weight of an object within the Earth's atmosphere varies inversely with the square of the distance of the object from the Earth's center. If a low Earth orbit satellite weighs 100 kg on the Earth's surface (6400 km), how much will it weigh in its orbit 800 km above Earth?

CHAPTER FIVE

Exponential and Logarithmic Functions and Equations

CHAPTER FIVE CONTENTS

5.1 Exponential Functions

THINGS TO KNOW

Before working through this section, be sure that you are familiar with the following concepts:

VIDEO | ANIMATION | INTERACTIVE

You Try It

1. Using Combinations of Transformations to Graph Functions (Section 3.4)

You Try It

2. Determining if a Function Is One-to-One Using the Horizontal Line Test (Section 3.6)

OBJECTIVES

1 Understanding the Characteristics of Exponential Functions

2 Sketching the Graphs of Exponential Functions Using Transformations

3 Solving Exponential Equations by Relating the Bases

4 Solving Applications of Exponential Functions

SECTION 5.1 EXERCISES

OBJECTIVE 1 UNDERSTANDING THE CHARACTERISTICS OF EXPONENTIAL FUNCTIONS

Many natural phenomena and real-life applications can be modeled using exponential functions. Before we define the exponential function, it is important to remember how to manipulate exponential expressions because this skill is necessary when solving certain equations involving exponents. In **Section R.3**, expressions of the form b^r were evaluated for **rational numbers** r. For example,

$$3^2 = 9, \quad 4^{-2} = \frac{1}{4^2} = \frac{1}{16}, \quad \text{and} \quad 27^{-2/3} = \frac{1}{27^{2/3}} = \frac{1}{(\sqrt[3]{27})^2} = \frac{1}{(3)^2} = \frac{1}{9}.$$

In this section, we extend the meaning of b^r to include all **real** values of r by defining the exponential function $f(x) = b^x$. Watch this **video** for an explanation of the exponential function.

> **Definition** Exponential Function
>
> An **exponential function** is a function of the form $f(x) = b^x$, where x is any real number and $b > 0$ such that $b \neq 1$.
>
> The constant, b, is called the base of the exponential function.

Notice in the definition that the base, b, must be positive and must not equal 1. If $b = 1$, then the function $f(x) = 1^x$ is equal to 1 for all x and is, hence, equivalent to the constant function $f(x) = 1$. If b were negative, then $f(x) = b^x$ would not be defined for all real values of x.

For example, if $b = -4$, then $f\left(\frac{1}{2}\right) = (-4)^{1/2} = \sqrt{-4} = 2i$, which is *not* a positive real number.

To illustrate the basic shapes of exponential functions we create a table of values and sketch the graph of $y = b^x$ for $b = 2, 3, \frac{1}{2}$ and $\frac{1}{3}$. See Table 1 and Figure 1.

Table 1

x	$y = 2^x$	$y = 3^x$	$y = \left(\frac{1}{2}\right)^x$	$y = \left(\frac{1}{3}\right)^x$
-2	$2^{-2} = \frac{1}{2^2} = \frac{1}{4}$	$3^{-2} = \frac{1}{3^2} = \frac{1}{9}$	$\left(\frac{1}{2}\right)^{-2} = 2^2 = 4$	$\left(\frac{1}{3}\right)^{-2} = 3^2 = 9$
-1	$2^{-1} = \frac{1}{2^1} = \frac{1}{2}$	$3^{-1} = \frac{1}{3^1} = \frac{1}{3}$	$\left(\frac{1}{2}\right)^{-1} = 2^1 = 2$	$\left(\frac{1}{3}\right)^{-1} = 3^1 = 3$
0	$2^0 = 1$	$3^0 = 1$	$\left(\frac{1}{2}\right)^0 = 1$	$\left(\frac{1}{3}\right)^0 = 1$
1	$2^1 = 2$	$3^1 = 3$	$\left(\frac{1}{2}\right)^1 = \frac{1}{2}$	$\left(\frac{1}{3}\right)^1 = \frac{1}{3}$
2	$2^2 = 4$	$3^2 = 9$	$\left(\frac{1}{2}\right)^2 = \frac{1}{4}$	$\left(\frac{1}{3}\right)^2 = \frac{1}{9}$

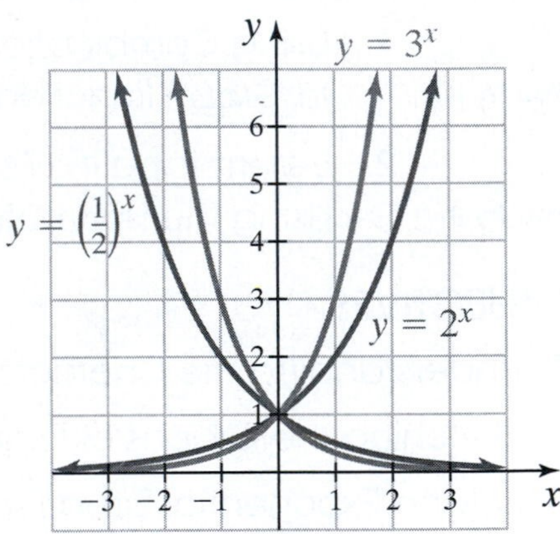

Figure 1 Graphs of $y = 2^x$, $y = 3^x$, $y = \left(\frac{1}{2}\right)^x$, and $y = \left(\frac{1}{3}\right)^x$.

Any positive number b, where $b \neq 1$, can be used as the base of an **exponential function**. However, there is one number that appears as the base in exponential applications more than any other number. This number is called the **natural base** and is symbolized using the letter e. The number e is an **irrational number** that is defined as the value of the expression $\left(1 + \frac{1}{n}\right)^n$ as n approaches infinity. Table 2 shows the values of the expression $\left(1 + \frac{1}{n}\right)^n$ for increasingly large values of n.

You can see from Table 2 that as the values of n get large, the value e (rounded to six decimal places) is 2.718282. The function $f(x) = e^x$ is called the **natural exponential function.**

Table 2

n	$\left(1 + \frac{1}{n}\right)^n$
1	2
2	2.25
10	2.5937424601
100	2.7048138294
1,000	2.7169239322
10,000	2.7181459268
100,000	2.7182682372
1,000,000	2.7182804693
10,000,000	2.7182816925
100,000,000	2.7182818149

Because $2 < e < 3$, it follows that the graph of $f(x) = e^x$ lies between the graphs of $y = 2^x$ and $y = 3^x$, as seen in Figure 2.

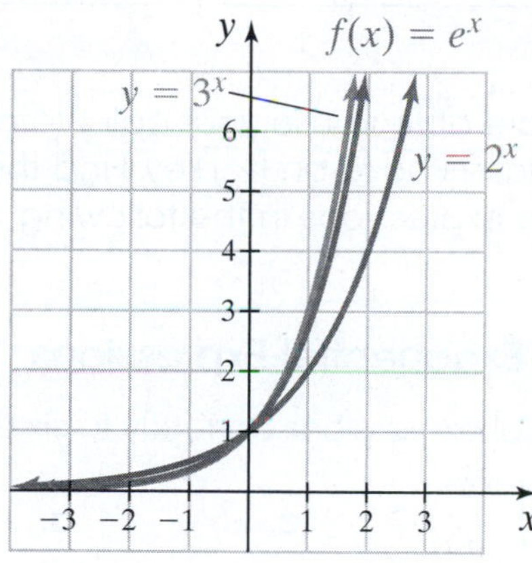

Figure 2 The graphs of $y = 2^x$, $y = 3^x$, and the graph of the natural exponential function $f(x) = e^x$.

You can see from the graphs sketched in **Figure 1** and **Figure 2** that all of the exponential functions intersect the y-axis at the point $(0, 1)$. This is true because $b^0 = 1$ for all nonzero values of b. For values of $b > 1$, the graph of $y = b^x$ increases rapidly as the values of x approach positive infinity ($b^x \to \infty$ as $x \to \infty$). In fact, the larger the base, the faster the graph will grow. Also, for $b > 1$, the graph of $y = b^x$ decreases quickly, approaching 0 as the values of x approach negative infinity ($b^x \to 0$ as $x \to -\infty$). Thus, the x-axis (the line $y = 0$) is a **horizontal asymptote**.

However, for $0 < b < 1$, the graph decreases quickly, approaching the horizontal asymptote $y = 0$ as the values of x approach positive infinity ($b^x \to 0$ as $x \to \infty$), whereas the graph increases rapidly as the values of x approach negative infinity ($b^x \to \infty$ as $x \to -\infty$). The preceding statements, along with some other characteristics of the graphs of exponential functions, are outlined on the following page.

Characteristics of Exponential Functions

For $b > 0, b \neq 1$, the exponential function with base b is defined by $f(x) = b^x$.

The domain of $f(x) = b^x$ is $(-\infty, \infty)$, and the range is $(0, \infty)$. The graph of $f(x) = b^x$ has one of the following two shapes:

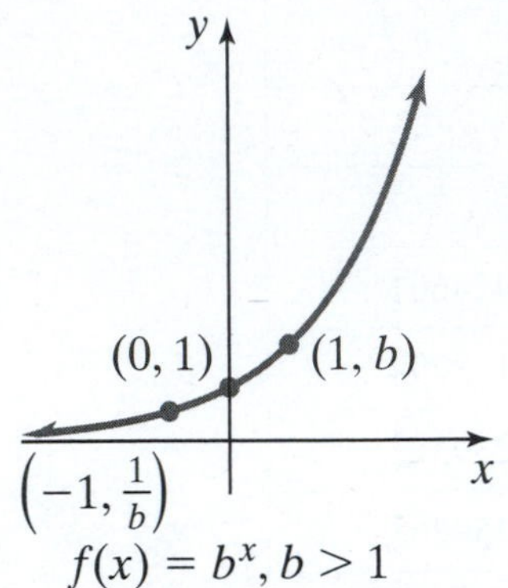

$f(x) = b^x, b > 1$

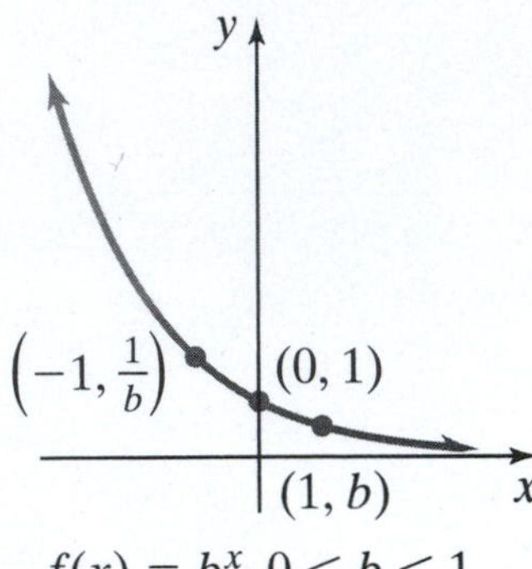

$f(x) = b^x, 0 < b < 1$

The graph intersects the y-axis at $(0, 1)$.
The graph contains the points $\left(-1, \frac{1}{b}\right)$ and $(1, b)$.
$b^x \to \infty$ as $x \to \infty$
$b^x \to 0$ as $x \to -\infty$
The line $y = 0$ is a **horizontal asymptote.**
The function is **one to one.**

The graph intersects the y-axis at $(0, 1)$.
The graph contains the points $\left(-1, \frac{1}{b}\right)$ and $(1, b)$.
$b^x \to 0$ as $x \to \infty$
$b^x \to \infty$ as $x \to -\infty$
The line $y = 0$ is a **horizontal asymptote.**
The function is **one to one.**

It is important that you are able to use your calculator to evaluate exponential expressions. Most calculators have an $\boxed{e^x}$ key. Find this special key on your calculator and evaluate the expressions in the following example.

Example 1 Evaluate Exponential Expressions

Evaluate each exponential expressions correctly to six decimal places.

a. $\left(\frac{3}{7}\right)^{0.6}$ b. $9\left(\frac{5}{8}\right)^{-0.375}$ c. e^2 d. $e^{-0.534}$ e. $1{,}000e^{0.013}$

Solution

a. $\left(\frac{3}{7}\right)^{0.6} \approx 0.601470$ b. $9\left(\frac{5}{8}\right)^{-0.375} \approx 10.734640$

For parts c and e, we use the $\boxed{e^x}$ key on a calculator to get

c. $e^2 \approx 7.389056$ d. $e^{-0.534} \approx 0.586255$ e. $1{,}000e^{0.013} \approx 1{,}013.084867$ ●

You Try It Work through this You Try It problem.

Work Exercises 1–7 in this textbook or in the MyLab Math Study Plan.

Example 2 Sketch the Graph of an Exponential Function

Sketch the graph of each exponential function.

a. $f(x) = \left(\frac{2}{3}\right)^x$ b. $f(x) = -2e^x$

Solution

a. Because the base of the exponential function is $\frac{2}{3}$, which is between 0 and 1, the graph must approach the x-axis as the value of x approaches positive infinity. The graph intersects the y-axis at $(0, 1)$. We can find a few more points by choosing some negative and positive values of x:

$$f(-2) = \left(\frac{2}{3}\right)^{-2} = \left(\frac{3}{2}\right)^2 = \frac{9}{4}$$

$$f(-1) = \left(\frac{2}{3}\right)^{-1} = \left(\frac{3}{2}\right)^1 = \frac{3}{2}$$

$$f(1) = \left(\frac{2}{3}\right)^1 = \frac{2}{3}$$

$$f(2) = \left(\frac{2}{3}\right)^2 = \frac{4}{9}$$

We can complete the graph by connecting the points with a smooth curve. Watch this **video** to see how to sketch the complete function as seen in Figure 3.

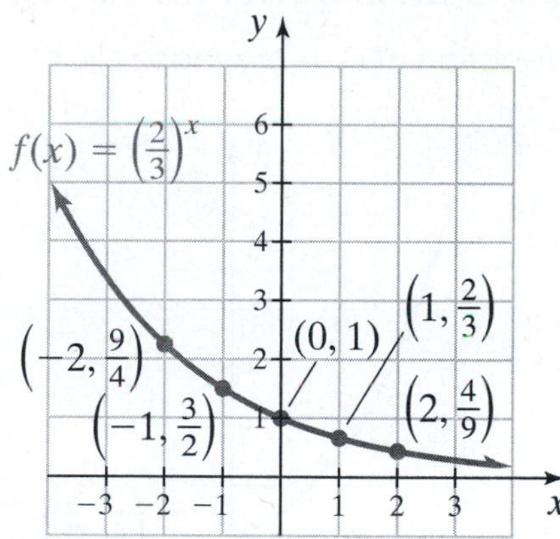

Figure 3 The graph of $f(x) = \left(\frac{2}{3}\right)^x$. ●

b. We know that the value of e^x is greater than zero for all values of x. Thus, the values of $-2e^x$ must be less than zero for all values of x. This means that the graph of the function $f(x) = -2e^x$ must be below the x-axis for all values of x.

The y-intercept is $f(0) = -2e^0 = -2(1) = -2$. Therefore, the graph intersects the y-axis at $(0, -2)$. We can find a few more points by choosing some arbitrary values of x.

$$f(-2) = -2e^{-2} = -2\left(\frac{1}{e^2}\right) \approx -0.27067$$

$$f(-1) = -2e^{-1} = -2\left(\frac{1}{e}\right) \approx -0.73576$$

$$f(1) = -2e^1 = -2(e) \approx -5.43656$$

We can complete the graph by connecting the points with a smooth curve. Watch this **interactive video** to see how to sketch the function seen in Figure 4. Note that you could also use transformation techniques to sketch the graph of $f(x) = -2e^x$. We will use transformation techniques to sketch the graphs of exponential functions in **Objective 2**.

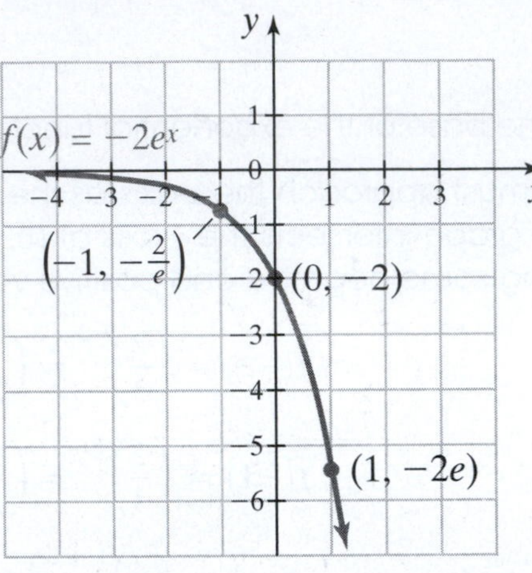

Figure 4 The graph of $f(x) = -2e^x$.

You Try It Work through this You Try It problem.

Work Exercises 8–14 in this textbook or in the MyLab Math Study Plan.

Example 3 Determine an Exponential Function Given the Graph

Find the exponential function $f(x) = b^x$ whose graph is given as follows.

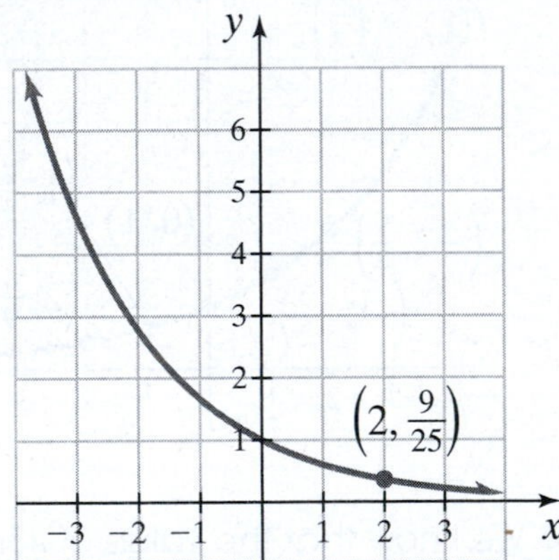

Solution From the point $\left(2, \frac{9}{25}\right)$, we see that $f(2) = \frac{9}{25}$. Thus,

$f(x) = b^x$ — Write the exponential function $f(x) = b^x$.

$f(2) = b^2$ — Evaluate $f(2)$.

$f(2) = \frac{9}{25}$ — The graph contains the point $\left(2, \frac{9}{25}\right)$.

$b^2 = \frac{9}{25}$ — Equate the two expressions for $f(2)$.

Therefore, we are looking for a constant b such that $b^2 = \frac{9}{25}$. Using the **square root property**, we get

$$\sqrt{b^2} = \pm\sqrt{\frac{9}{25}}$$

$$b = \pm\frac{3}{5}.$$

By definition of an exponential function, $b > 0$; thus, $b = \frac{3}{5}$.

Therefore, this is the graph of $f(x) = \left(\frac{3}{5}\right)^x$.

You Try It **Work through this You Try It problem.**

Work Exercises 15–21 in this textbook or in the MyLab Math Study Plan.

OBJECTIVE 2 SKETCHING THE GRAPHS OF EXPONENTIAL FUNCTIONS USING TRANSFORMATIONS

Often we can use the **transformation techniques** that are discussed in **Section 3.4** to sketch the graph of an exponential function. For example, the graph of $f(x) = 3^x - 1$ can be obtained by vertically shifting the graph of $y = 3^x$ down one unit. You can see in Figure 5 that the y-intercept of $f(x) = 3^x - 1$ is $(0, 0)$ and the horizontal asymptote is the line $y = -1$.

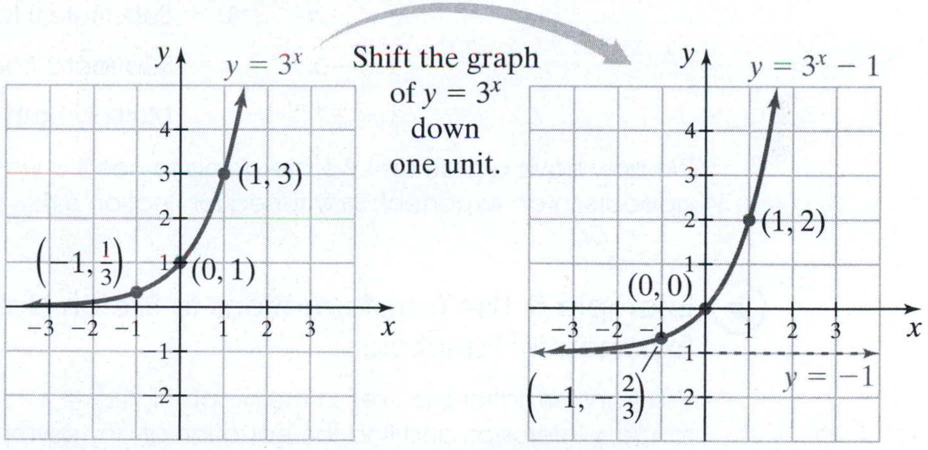

Figure 5 The graph of $f(x) = 3^x - 1$ can be obtained by vertically shifting the graph of $y = 3^x$ down one unit.

Example 4 Use Transformations to Sketch an Exponential Function

Use transformations to sketch the graph of $f(x) = -2^{x+1} + 3$.

Solution Starting with the graph of $y = 2^x$, we can obtain the graph of $f(x) = -2^{x+1} + 3$ through a series of three transformations:

1. Horizontally shift the graph of $y = 2^x$ to the left one unit, producing the graph of $y = 2^{x+1}$.

2. Reflect the graph of $y = 2^{x+1}$ about the x-axis, producing the graph of $y = -2^{x+1}$.

3. Vertically shift the graph of $y = -2^{x+1}$ up three units, producing the graph of $f(x) = -2^{x+1} + 3$.

The graph of $f(x) = -2^{x+1} + 3$ is shown in Figure 6. Watch the **video** to see every step.

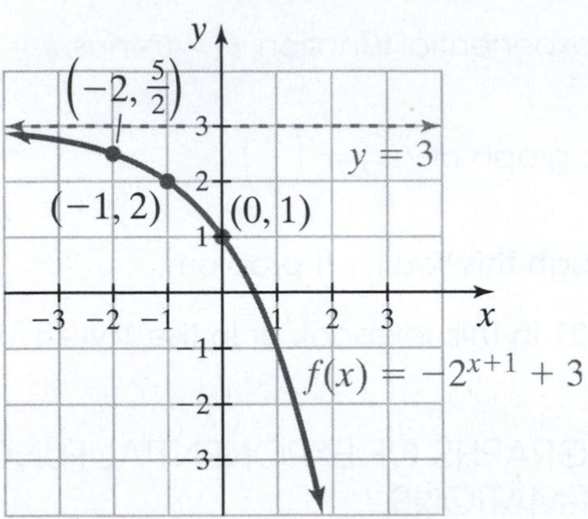

Figure 6 Graph of $f(x) = -2^{x+1} + 3$.

Notice that the graph of $f(x) = -2^{x+1} + 3$ in **Figure 6** has a y-intercept. We can find the y-intercept by evaluating $f(0) = -2^{0+1} + 3 = -2 + 3 = 1$. Also notice that the graph has an x-intercept. Can you find it? Recall that to find an x-intercept, we need to set $f(x) = 0$ and solve for x.

$f(x) = -2^{x+1} + 3$	Write the original function.
$0 = -2^{x+1} + 3$	Substitute 0 for $f(x)$.
$-3 = -2^{x+1}$	Subtract 3 from both sides.
$3 = 2^{x+1}$	Multiply both sides by -1.

We now have a problem! We are going to need a way to solve for a variable that appears in an exponent. Stay tuned for Section 5.4.

Example 5 Use Transformations to Sketch Natural Exponential Functions

Use transformations to sketch the graph of $f(x) = e^{-x} - 2$. Determine the domain, range, y-intercept, and find the equation of any asymptotes.

Solution First sketch the graph of $y = e^x$ and label the points $\left(-1, \frac{1}{e}\right)$, $(0, 1)$ and $(1, e)$.

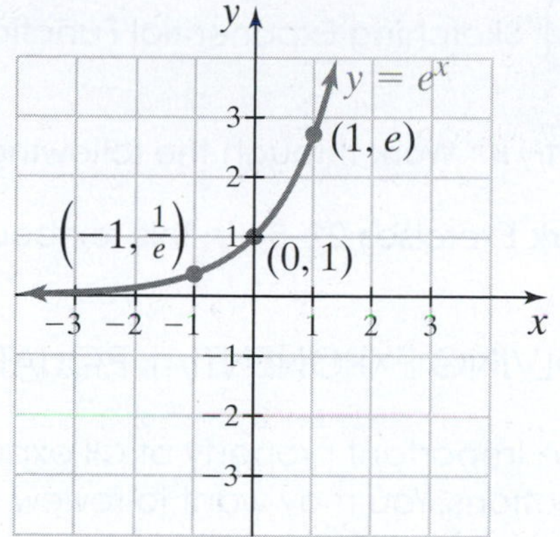

We can sketch the graph of $f(x) = e^{-x} - 2$ using the following two transformations.

1. Reflect the graph of $y = e^x$ about the y-axis, producing the graph of $y = e^{-x}$.
2. Vertically shift the graph of $y = e^{-x}$ down two units, producing the graph of $f(x) = e^{-x} - 2$.

Watch this **video** to see each step of this graphing process. The original three points that lie on the graph of $y = e^x$ and the corresponding points that lie on the graph of $f(x) = e^{-x} - 2$ are shown below.

Points that lie on the graph of $y = e^x$:	$\left(-1, \frac{1}{e}\right)$	$(0, 1)$	$(1, e)$
	↓	↓	↓
Corresponding points that lie on the graph of $f(x) = e^{-x} - 2$:	$\left(1, \frac{1}{e} - 2\right)$	$(0, -1)$	$(-1, e - 2)$

The graph of $f(x) = e^{-x} - 2$ is shown below in Figure 7. You can see from the graph that the domain of $f(x) = e^{-x} - 2$ is the interval $(-\infty, \infty)$ and the range of $f(x) = e^{-x} - 2$ is the interval $(-2, \infty)$. The y-intercept is -1 and the equation of the horizontal asymptote is $y = -2$.

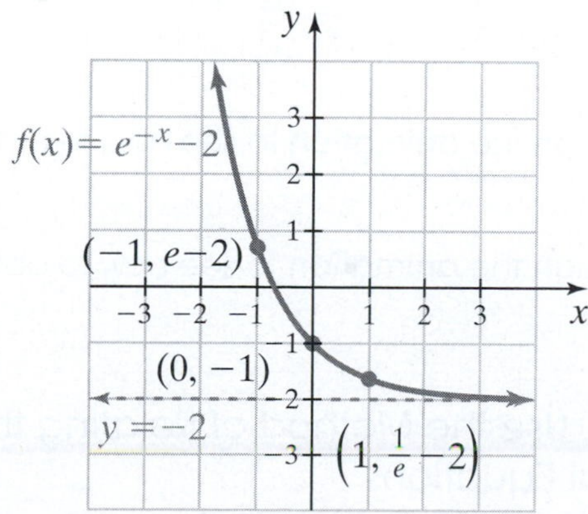

Figure 7 Graph of $f(x) = e^{-x} - 2$.

Try sketching the graphs of some exponential functions on your own by exploring the guided visualization below. Click on the icon below to sketch exponential functions of the form $f(x) = ab^{x+c} + d$. Choose your own values of a, b, c, and d.

(eText Screens 5.1-1–5.1-51)

 Sketching Exponential Functions

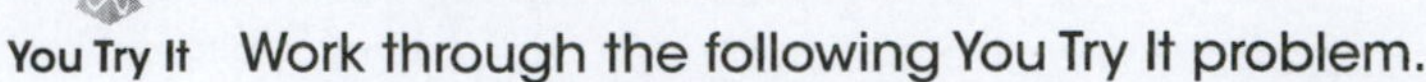

You Try It Work through the following You Try It problem.

Work Exercises 22–34 in this textbook or in the MyLab Math Study Plan.

OBJECTIVE 3 SOLVING EXPONENTIAL EQUATIONS BY RELATING THE BASES

One important property of all exponential functions is that they are **one to one functions**. You may want to review **Section 3.6**, which discusses one to one functions in detail. The function $f(x) = b^x$ is one to one because the graph of f passes the **horizontal line test**. In Section 3.6, the alternate definition of one to one is stated as follows:

A function f is one to one if for any two range values $f(u)$ and $f(v)$, $f(u) = f(v)$ implies that $u = v$.

Using this definition and letting $f(x) = b^x$, we can say that if $b^u = b^v$, then $u = v$. In other words, if the bases of an exponential equation of the form $b^u = b^v$ are the same, then the exponents must be the same. Solving exponential equations with this property is known as the **method of relating the bases** for solving exponential equations.

> **Method of Relating the Bases**
>
> If an exponential equation can be written in the form $b^u = b^v$, then $u = v$.

Example 6 Use the Method of Relating the Bases to Solve Exponential Equations

Solve the following equations:

a. $8 = \dfrac{1}{16^x}$ b. $\dfrac{1}{27^x} = (\sqrt[4]{3})^{x-2}$

Solution

a. Work through the **animation** to see how to obtain a solution of $x = -\dfrac{3}{4}$.

b. Work through the **animation** to see how to obtain a solution of $x = \dfrac{2}{13}$. ●

Example 7 Use the Method of Relating the Bases to Solve Natural Exponential Equations

Use the method of relating the bases to solve each exponential equation:

a. $e^{3x-1} = \dfrac{1}{\sqrt{e}}$ b. $\dfrac{e^{x^2}}{e^{10}} = (e^x)^3$

Solution Work through the interactive video to verify that the solutions are as follows:

a. $x = \frac{1}{6}$ b. $x = -2$ or $x = 5$

You Try It Work through this You Try It problem.

Work Exercises 35–52 in this textbook or in the MyLab Math Study Plan.

OBJECTIVE 4 SOLVING APPLICATIONS OF EXPONENTIAL FUNCTIONS

Exponential functions are used to describe many real-life situations and natural phenomena. We now look at some examples.

Example 8 Learn to Hit a 3-Wood on a Golf Driving Range

Most golfers find that their golf skills improve dramatically at first and then level off rather quickly. For example, suppose that the distance (in yards) that a typical beginning golfer can hit a 3-wood after t weeks of practice on the driving range is given by the exponential function $d(t) = 225 - 100e^{-0.7t}$. This function has been developed after many years of gathering data on beginning golfers.

How far can a typical beginning golfer initially hit a 3-wood? How far can a typical beginning golfer hit a 3-wood after 1 week of practice on the driving range? After 5 weeks? After 9 weeks? Round to the nearest hundredth yard.

Solution Initially, when $t = 0$, $d(0) = 225 - 100e^{0} = 225 - 100 = 125$ yards. Therefore, a typical beginning golfer can hit a 3-wood 125 yards.

After 1 week of practice on the driving range, a typical beginning golfer can hit a 3-wood $d(1) = 225 - 100e^{-0.7(1)} \approx 175.34$ yards.

After 5 weeks of practice, $d(5) = 225 - 100e^{-0.7(5)} \approx 221.98$ yards.
After 9 weeks of practice, $d(9) = 225 - 100e^{-0.7(9)} \approx 224.82$ yards.

Using a graphing utility, we can sketch the graph of $d(t) = 225 - 100e^{-0.7t}$. You can see from the graph in Figure 8 that the distance increases rather quickly and then tapers off toward a **horizontal asymptote** of 225 yards.

Using Technology

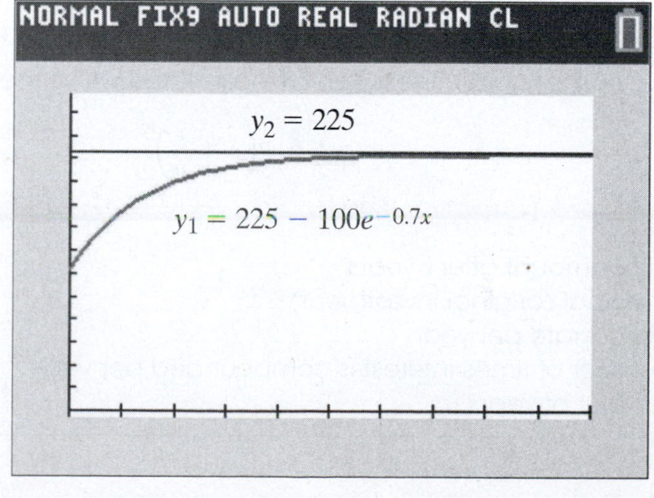

Figure 8 A TI-83 Plus was used to sketch the function $d(t) = 225 - 100e^{-0.7t}$ and the **horizontal asymptote** $y = 225$.

You Try It Work through this You Try It problem.

Work Exercises 53–56 in this textbook or in the MyLab Math Study Plan.

COMPOUND INTEREST

A real-life application of exponential functions is the concept of **compound interest**, that is, interest that is paid on both principal and interest. First, we take a look at how **simple interest** is accrued. If an investment of P dollars is invested at r percent annually (written as a decimal) using simple interest, then the interest earned after 1 year is $\boldsymbol{Pr}$ dollars. Adding this interest to the original investment yields a total amount, A, of

$$A = \underbrace{P}_{\text{Original investment}} + \underbrace{Pr}_{\text{Interest earned}} = P(1 + r).$$

If this amount is reinvested at the same interest rate, then the total amount after 2 years becomes

$$A = \underbrace{P(1 + r)}_{\text{Total investment after 1 year}} + \underbrace{P(1 + r)r}_{\text{Interest earned}} = P(1 + r)(1 + r) = P(1 + r)^2.$$

Reinvesting this amount for a third year gives an amount of $P(1 + r)^3$. Continuing this process for k years, we can see that the amount becomes $A = P(1 + r)^k$. This is an exponential function with base $1 + r$.

We can now modify this formula to obtain another formula that will model interest that is compounded periodically throughout the year(s). When interest is compounded periodically, then k no longer represents the number of years but rather the number of pay periods. If interest is paid n times per year for t years, then $k = nt$ pay periods. Thus, in the formula $A = P(1 + r)^k$, we substitute nt for k and get $A = P(1 + r)^{nt}$.

In the earlier simple interest model, the variable r was used to represent annual interest. In the periodically compound interest model being developed here with n pay periods per year, the interest rate per pay period is no longer r but rather $\frac{r}{n}$.

Thus, in the formula $A = P(1 + r)^{nt}$, we replace r with $\frac{r}{n}$ and get the periodic compound interest formula $A = P\left(1 + \frac{r}{n}\right)^{nt}$.

Periodic Compound Interest Formula

Periodic compound interest can be calculated using the formula

$$A = P\left(1 + \frac{r}{n}\right)^{nt},$$

where

A = Total amount after t years
P = Principal (original investment)
r = Interest rate per year
n = Number of times interest is compounded per year
t = Number of years

Example 9 Calculate Compound Interest

Which investment results in the greatest total amount after 25 years?

Investment A: $12,000 invested at 3% compounded monthly
Investment B: $10,000 invested at 3.9% compounded quarterly

Solution Investment A: $P = 12{,}000, r = 0.03, n = 12, t = 25$:

$$A = 12{,}000\left(1 + \frac{0.03}{12}\right)^{12(25)} \approx \$25{,}380.23$$

Investment B: $P = 10{,}000, r = 0.039, n = 4, t = 25$:

$$A = 10{,}000\left(1 + \frac{0.039}{4}\right)^{4(25)} \approx \$26{,}386.77$$

Investment B will yield the most money after 25 years.

You Try It Work through this You Try It problem.

Work Exercises 57–59 in this textbook or in the MyLab Math Study Plan.

CONTINUOUS COMPOUND INTEREST

Some banks use **continuous compounding**; that is, they compound the interest every fraction of a second every day! If we start with the formula for periodic compound interest, $A = P\left(1 + \frac{r}{n}\right)^{nt}$, and let n (the number of times the interest is compounded each year) approach infinity, we can derive the formula $A = Pe^{rt}$, which is the formula for continuous compound interest. Work through this **animation** to see exactly how this formula is derived.

> **Continuous Compound Interest Formula**
>
> Continuous compound interest can be calculated using the formula
>
> $$A = Pe^{rt},$$
>
> where
>
> A = Total amount after t years
> P = Principal
> r = Interest rate per year
> t = Number of years

Example 10 Calculate Continuous Compound Interest

How much money would be in an account after 5 years if an original investment of $6,000 was compounded continuously at 4.5%? Compare this amount to the same investment that was compounded daily. Round to the nearest cent.

Solution First, the amount after 5 years compounded continuously is $A = Pe^{rt} = 6{,}000e^{0.045(5)} \approx \$7{,}513.94$. The same investment compounded daily yields an amount of $A = P\left(1 + \frac{r}{n}\right)^{nt} = 6{,}000\left(1 + \frac{0.045}{365}\right)^{365(5)} \approx \$7{,}513.83$. Continuous compound interest yields $.11 more interest after 5 years!

You Try It Work through this You Try It problem.

Work Exercises 60–62 in this textbook or in the MyLab Math Study Plan.

PRESENT VALUE FOR PERIODIC COMPOUND INTEREST

Sometimes investors want to know how much money to invest now to reach a certain investment goal in the future. This amount of money, P, is known as the **present value** of A dollars. To find a formula for present value, start with the formula for periodic compound interest and solve the formula for P:

$$A = P\left(1+\frac{r}{n}\right)^{nt}$$ Use the periodic compound interest formula.

$$\frac{A}{\left(1+\frac{r}{n}\right)^{nt}} = \frac{P\left(1+\frac{r}{n}\right)^{nt}}{\left(1+\frac{r}{n}\right)^{nt}}$$ Divide both sides by $\left(1+\frac{r}{n}\right)^{nt}$.

$$A\left(1+\frac{r}{n}\right)^{-nt} = P$$ Rewrite $\frac{1}{\left(1+\frac{r}{n}\right)^{nt}}$ as $\left(1+\frac{r}{n}\right)^{-nt}$.

The formula $P = A\left(1+\frac{r}{n}\right)^{-nt}$ is known as the **present value formula for periodic compound interest**.

> **Present Value Formula for Periodic Compound Interest**
>
> Present value can be calculated using the formula
>
> $$P = A\left(1+\frac{r}{n}\right)^{-nt},$$
>
> where
>
> P = Principal (original investment)
> A = Total amount after t years
> r = Interest rate per year
> n = Number of times interest is compounded per year
> t = Number of years

PRESENT VALUE FOR CONTINUOUS COMPOUND INTEREST

To find a formula for present value on money that is compounded continuously, we start with the formula for continuous compound interest and solve for P.

$$A = Pe^{rt}$$ Write the continuous compound interest formula.

$$\frac{A}{e^{rt}} = P\frac{e^{rt}}{e^{rt}}$$ Divide both sides by e^{rt}.

$$Ae^{-rt} = P$$ Rewrite $\frac{1}{e^{rt}}$ as e^{-rt}.

> **Present Value Formula for Continuous Compound Interest**
>
> The present value of A dollars after t years of continuous compound interest, with interest rate r, is given by the formula
>
> $$P = Ae^{-rt}.$$

Example 11 Determine Present Value

a. Find the present value of $\$8{,}000$ if interest is paid at a rate of 5.6% compounded quarterly for 7 years. Round to the nearest cent.

b. Find the present value of $\$18{,}000$ if interest is paid at a rate of 8% compounded continuously for 20 years. Round to the nearest cent.

Solution

a. Using the present value formula for periodic compound interest $P = A\left(1 + \frac{r}{n}\right)^{-nt}$ with $A = \$8{,}000$, $r = 0.056$, $n = 4$, and $t = 7$, we get

$$P = A\left(1 + \frac{r}{n}\right)^{-nt}$$ Write the present value formula for periodic compound interest.

$$P = 8{,}000\left(1 + \frac{0.056}{4}\right)^{-(4)(7)} \approx 5{,}420.35$$ Substitute and use a calculator to simplify.

Therefore, the present value of $\$8{,}000$ in 7 years at 5.6% compounded quarterly is $\$5{,}420.35$.

b. Using the present value formula for continuous compound interest $P = Ae^{-rt}$ with $A = \$18{,}000$, $r = 0.08$, and $t = 20$, we get

$$P = Ae^{-rt}$$ Write the present value formula for continuous compound interest.

$$P = (18{,}000)e^{-(0.08)(20)} \approx \$3{,}634.14.$$ Substitute and use a calculator to simplify.

You Try It Work through this You Try It problem.

Work Exercises 63–67 in this textbook or in the MyLab Math Study Plan.

EXPONENTIAL GROWTH MODEL

You have probably heard that some populations grow exponentially. Most populations grow at a rate proportional to the size of the population. In other words, the larger the population, the faster the population grows. With this in mind, it can be shown in a more advanced math course that the mathematical model that can describe population growth is given by the function $P(t) = P_0e^{kt}$.

> **Exponential Growth**
>
> A model that describes the population, P, after a certain time, t, is
>
> $$P(t) = P_0e^{kt},$$
>
> where $P_0 = P(0)$ is the initial population and $k > 0$ is a constant called the **relative growth rate**. (*Note*: k may be given as a percent.)

The graph of the exponential growth model is shown in Figure 9. Notice that the graph has a y-intercept of P_0.

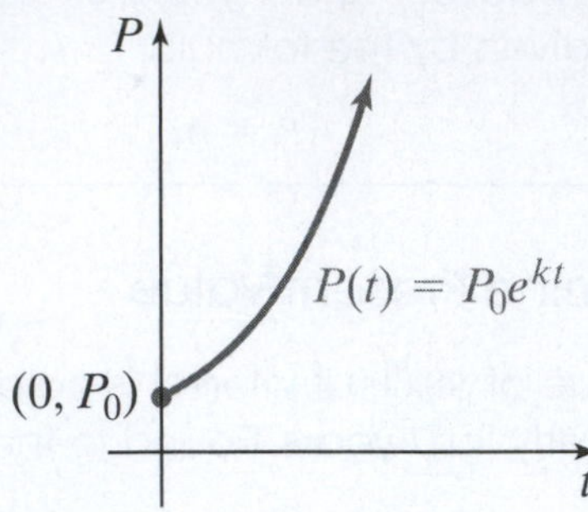

Figure 9 Graph of the exponential growth model $P(t) = P_0e^{kt}$.

Example 12 Population Growth

The population of a small town follows the exponential growth model $P(t) = 900e^{0.015t}$, where t is the number of years after 1900.

Answer the following questions, rounding each answer to the nearest whole number:

a. What was the population of this town in 1900?

b. What was the population of this town in 1950?

c. Use this model to predict the population of this town in 2012.

Solution

a. The initial population was $P(0) = 900e^{0.015(0)} = 900$.

b. Because 1950 is 50 years after 1900, we must evaluate $P(50)$. $P(50) = 900e^{0.015(50)} \approx 1{,}905$.

c. In the year 2012, we can predict that the population will be $P(112) = 900e^{0.015(112)} \approx 4{,}829$.

Example 13 Determine the Initial Population

Twenty years ago, the Idaho Fish and Game Department introduced a new breed of wolf into a certain Idaho forest. The current wolf population in this forest is now estimated at 825, with a relative growth rate of 12%.

Use the exponential growth model $P(t) = P_0e^{kt}$ to answer the following questions. Round each answer to the nearest whole number.

a. How many wolves did the Idaho Fish and Game Department initially introduce into this forest?

b. How many wolves can be expected after another 20 years?

Solution

a. The relative growth rate is 0.12, so we use the exponential growth model $P(t) = P_0e^{0.12t}$. Because $P(20) = 825$, we get

$$P(20) = P_0e^{0.12(20)} \quad \text{Substitute 20 for } t.$$
$$825 = P_0e^{0.12(20)} \quad \text{Substitute 825 for } P(20).$$
$$P_0 = \frac{825}{e^{0.12(20)}} \approx 75 \quad \text{Solve for } P_0.$$

Therefore, the Idaho Fish and Game Department initially introduced 75 wolves into the forest.

b. Because $P_0 = 75$, we can use the exponential growth model $P(t) = 75e^{0.12t}$. In another 20 years, the value of t will be 40. Thus, we must evaluate $P(40)$.

$$P(40) = 75e^{0.12(40)} \approx 9{,}113$$

Therefore, we can expect the wolf population to be approximately 9,113 in another 20 years. Watch this **video** to see the entire solution to this example. ●

You Try It Work through this You Try It problem.

Work Exercises 68–71 in this textbook or in the MyLab Math Study Plan.

5.1 Exercises

Skill Check Exercises

For exercises SCE-1 through SCE-5, rewrite each expression in the form 2^u, 3^u, or 5^u where *u* is a constant or an algebraic expression.

SCE-1. 27 **SCE-2.** $\frac{1}{2^3}$ **SCE-3.** $\sqrt[4]{3}$ **SCE-4.** $\frac{1}{5^{3x}}$ **SCE-5.** $(\sqrt[3]{3})^x$

SCE-6. $\left(\frac{1}{2}\right)^{x-1}$ **SCE-7.** $\frac{25}{\sqrt[4]{5}}$ **SCE-8.** $\frac{81}{\sqrt[5]{3^x}}$ **SCE-9.** $\frac{3^{x^2}}{9^x}$ **SCE-10.** $\frac{4^x}{2^{-x^2}}$

In Exercises 1–7, use a calculator to approximate each exponential expression to six decimal places.

1. $\left(\frac{2}{11}\right)^{0.7}$ 2. $4\left(\frac{3}{13}\right)^{-0.254}$ 3. e^3 4. $e^{-0.2}$ 5. $e^{1/3}$ 6. $100e^{-.123}$ 7. $\sqrt{2}e^{\pi}$

In Exercises 8–14, sketch the graph of each exponential function. Label the y-intercept and at least two other points on the graph using both positive and negative values of x.

8. $f(x) = 4^x$ 9. $f(x) = \left(\frac{1}{4}\right)^x$ 10. $f(x) = \left(\frac{3}{2}\right)^x$ 11. $f(x) = (.4)^x$

12. $f(x) = (2.7)^x$ 13. $f(x) = -3e^x$ 14. $f(x) = 2\left(\frac{1}{e}\right)^x$

In Exercises 15–21, determine the correct exponential function of the form $f(x) = b^x$ whose graph is given.

15.

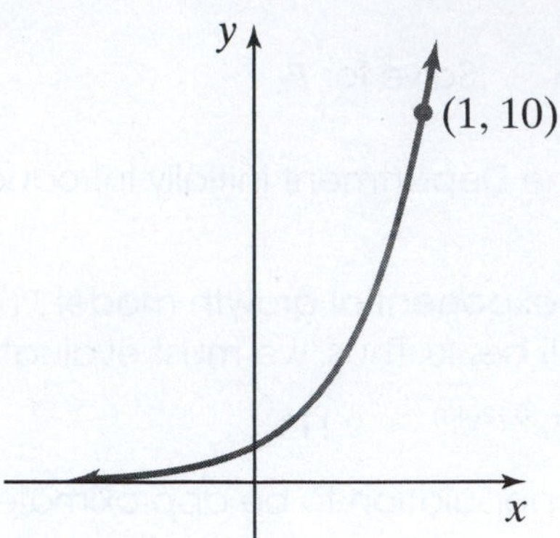

16.

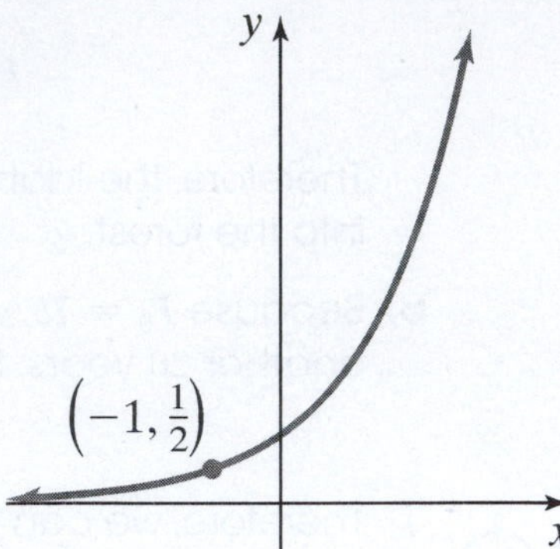

17.

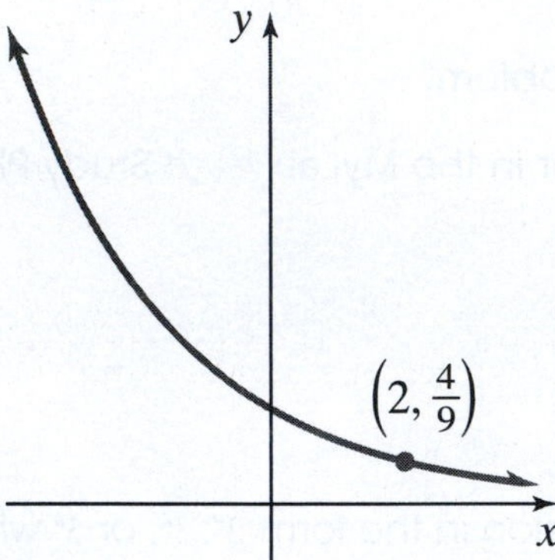

18.

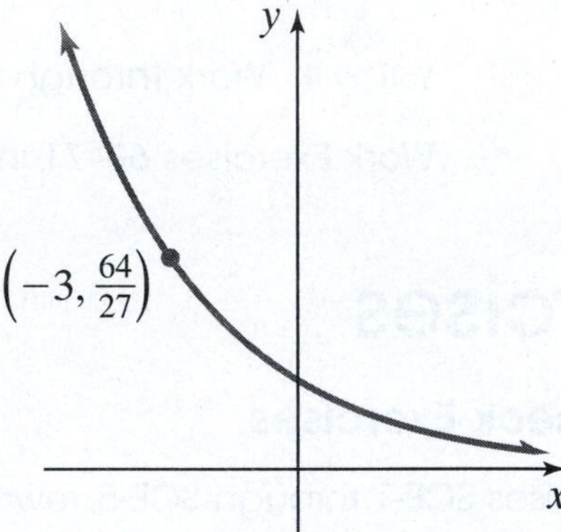

19.

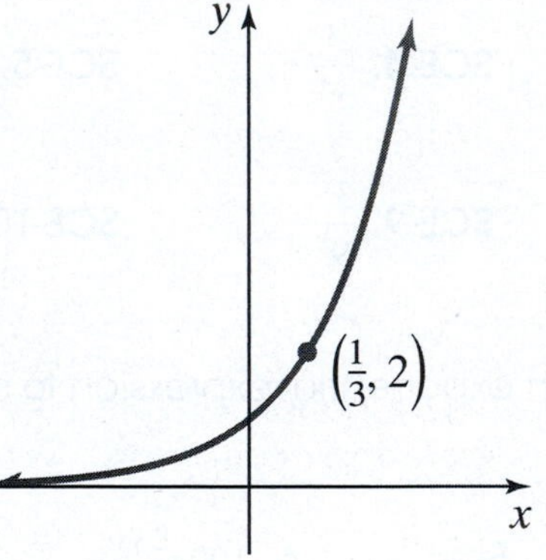

20.

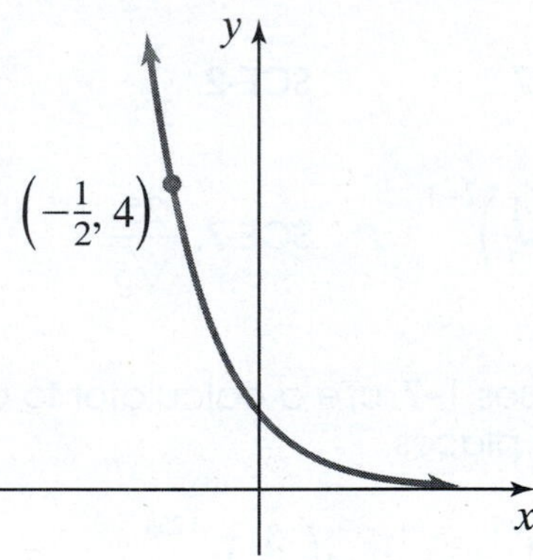

21.

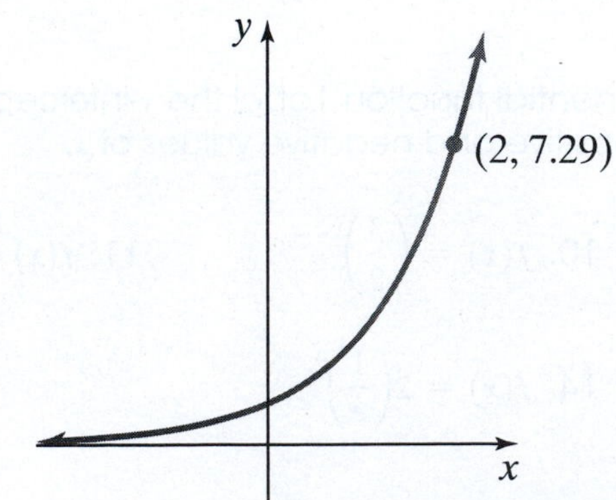

In Exercises 22–34, use the graph of $y = 2^x$, $y = e^x$, or $y = 3^x$ and transformations to sketch each exponential function. Determine the domain and range. Also, determine the y-intercept, and find the equation of the horizontal asymptote.

SbS 22. $f(x) = 2^{x-1}$ SbS 23. $f(x) = 3^x - 1$ SbS 24. $f(x) = -3^{x+2}$

SbS 25. $f(x) = -2^{x+1} - 1$ SbS 26. $f(x) = \left(\frac{1}{2}\right)^{x+1}$ SbS 27. $f(x) = \left(\frac{1}{3}\right)^x - 3$

SbS 28. $f(x) = 2^{-x} + 1$ SbS 29. $f(x) = 3^{1-x} - 2$

SbS 30. $f(x) = e^{x-1}$ SbS 31. $f(x) = e^x - 1$ SbS 32. $f(x) = -e^{x+2}$

SbS 33. $f(x) = -e^{x+1} - 1$ SbS 34. $f(x) = e^{-x} - 2$

In Exercises 35–52, solve each exponential equation using the method of "relating the bases" by first rewriting the equation in the form $b^u = b^v$.

35. $2^x = 16$ 36. $3^{x-1} = \frac{1}{9}$ 37. $\sqrt{5} = 25^x$ 38. $(\sqrt[3]{3})^x = 9$

39. $\frac{1}{\sqrt[5]{8}} = 2^x$ 40. $\frac{9}{\sqrt[4]{3}} = \left(\frac{1}{27}\right)^x$ 41. $(49)^x = \left(\frac{1}{7}\right)^{x-1}$ 42. $\frac{125}{\sqrt[3]{5^x}} = \left(\frac{1}{25^x}\right)$

43. $\frac{3^{x^2}}{9^x} = 27$ 44. $2^{x^3} = \frac{4^x}{2^{-x^2}}$

45. $e^x = \frac{1}{e^2}$ 46. $e^{5x+2} = \sqrt[3]{e}$ 47. $\frac{1}{e^x} = \frac{\sqrt{e}}{e^{1-x}}$ 48. $(e^{x^2})^2 = e^8$

49. $e^{x^2} = (e^x) \cdot e^{12}$ 50. $e^{x^2} = \frac{e^3}{(e^x)^5}$ 51. $\frac{e^{x^3}}{e^x} = \frac{e^{2x^2}}{e^2}$ 52. $e^{x^3} \cdot e^6 = \frac{(e^{2x^2})^2}{e^x}$

53. Typically, weekly sales will drop off rather quickly after the end of an advertising campaign. This drop in sales is known as *sales decay*. Suppose that the gross sales S, in hundreds of dollars, of a certain product is given by the exponential function $S(t) = 3{,}000(1.5^{-0.3t})$, where t is the number of weeks after the end of the advertising campaign.

    Answer the following questions, rounding each answer to the nearest whole number:

    a. What was the level of sales immediately after the end of the advertising campaign when $t = 0$?

    b. What was the level of sales 1 week after the end of the advertising campaign?

    c. What was the level of sales 5 weeks after the end of the advertising campaign?

54. Most people who start a serious weight-lifting regimen initially notice a rapid increase in the maximum amount of weight that they can bench press. After a few weeks, this increase starts to level off. The following function models the maximum weight, w, that a particular person can bench press in pounds at the end of t weeks of working out.

$$w(t) = 250 - 120e^{-0.3t}$$

    Answer the following questions, rounding each answer to the nearest whole number:

    a. What is the maximum weight that this person can bench press initially?

    b. What is the maximum weight that this person can bench press after 3 weeks of weight lifting?

c. What is the maximum weight that this person can bench press after 7 weeks of weight lifting?

55. *Escherichia coli* bacteria reproduce by simple cell division, which is known as binary fission. Under ideal conditions, a population of *E. coli* bacteria can double every 20 minutes. This behavior can be modeled by the exponential function $N(t) = N_0(2^{0.05t})$, where t is in minutes and N_0 is the initial number of *E. coli* bacteria.

    Answer the following questions, rounding each answer to the nearest bacteria:

    a. If the initial number of *E. coli* bacteria is five, how many bacteria will be present in 3 hours?

    b. If the initial number of *E. coli* bacteria is eight, how many bacteria will be present in 3 hours?

    c. If the initial number of *E. coli* bacteria is eight, how many bacteria will be present in 10 hours?

56. A wildlife-management research team noticed that a certain forest had no rabbits, so they decided to introduce a rabbit population into the forest for the first time. The rabbit population will be controlled by wolves and other predators. This rabbit population can be modeled by the function $R(t) = \dfrac{960}{0.6 + 23.4e^{-0.045t}}$, where t is the number of weeks after the research team first introduced the rabbits into the forest.

    Answer the following questions, rounding each answer to the nearest whole number:

    a. How many rabbits did the wildlife management research team bring into the forest?

    b. How many rabbits can be expected after 10 weeks?

    c. How many rabbits can be expected after the first year?

    d. What is the expected rabbit population after 4 years? 5 years?
    What can the expected rabbit population approach as time goes on?

Use the **periodic compound interest** formula to solve Exercises 57–59.

57. Suppose that \$9,000 is invested at 3.5% compounded quarterly. Find the total amount of this investment after 10 years. Round to the nearest cent.

58. Suppose that you have \$5,000 to invest. Which investment yields the greater return over a 10-year period: 7.35% compounded daily or 7.4% compounded quarterly?

59. Which investment yields the greatest total amount?

    Investment A: \$4,000 invested for 5 years compounded semiannually at 8%
    Investment B: \$5,000 invested for 4 years compounded quarterly at 4.5%

Use the **continuous compound interest** formula to solve Exercises 60–62.

60. An original investment of \$6,000 earns 6.25% interest compounded continuously. What will the investment be worth in 2 years? 20 years? Round to the nearest cent.

61. How much more will an investment of \$10,000 earning 5.5% compounded continuously for 9 years earn compared to the same investment at 5.5% compounded quarterly for 9 years? Round to the nearest cent.

62. Suppose your great-great grandfather invested $\$500$ earning 6.5% interest compounded continuously 100 years ago. How much would his investment be worth today? Round to the nearest cent.

Use the **present value formula for periodic compound interest** to solve Exercises 63–65.

63. Find the present value of $\$10{,}000$ if interest is paid at a rate of 4.5% compounded semiannually for 12 years. Round to the nearest cent.

64. Find the present value of $\$1{,}000{,}000$ if interest is paid at a rate of 9.5% compounded monthly for 8 years. Round to the nearest cent.

65. How much money would you have to invest at 10% compounded semiannually so that the total investment has a value of $\$2{,}205$ after 1 year? Round to the nearest cent.

Use the **present value formula for continuous compound interest** to solve Exercises 66 and 67.

66. Find the present value of $\$16{,}000$ if interest is paid at a rate of 4.5% compounded continuously for 10 years. Round to the nearest cent.

67. Which has the lower present value:

    a. $\$20{,}000$ if interest is paid at a rate of 5.18% compounded continuously for 2 years or

    b. $\$25{,}000$ if interest is paid at a rate of 3.8% compounded continuously for 30 months?

68. The population of a rural city follows the exponential growth model $P(t) = 2{,}000e^{0.035t}$, where t is the number of years after 1995.

    a. What was the population of this city in 1995?

    b. What is the relative growth rate as a percent?

    c. Use this model to approximate the population in 2030, rounding to the nearest whole number.

69. The relative growth rate of a certain bacteria colony is 25%.

    Suppose there are 10 bacteria initially.

    Use the exponential growth model $P(t) = P_0e^{kt}$ to answer the following.

    a. Find a function that describes the population of the bacteria after t hours.

    b. How many bacteria should we expect after 1 day? Round to the nearest whole number.

70. In 2006, the population of a certain American city was $18{,}221$. If the relative growth rate has been 6% since 1986, what was the population of this city in 1986? Round to the nearest whole number. Use the exponential growth model $P(t) = P_0e^{kt}$.

71. In 1970, a wildlife resource management team introduced a certain rabbit species into a forest for the first time. In 2004, the rabbit population had grown to $7{,}183$. The relative growth rate for this rabbit species is 20%.

    Use the exponential growth model $P(t) = P_0e^{kt}$ to answer the following questions. Round each answer to the nearest whole number.

    a. How many rabbits did the wildlife resource management team introduce into the forest in 1970?

    b. How many rabbits can be expected in the year 2025?

Brief Exercises

In Exercises 72–84, use the graph of $y = 2^x$, $y = e^x$, or $y = 3^x$ and transformations to sketch each exponential function.

72. $f(x) = 2^{x-1}$
73. $f(x) = 3^x - 1$
74. $f(x) = -3^{x+2}$
75. $f(x) = -2^{x+1} - 1$
76. $f(x) = \left(\frac{1}{2}\right)^{x+1}$
77. $f(x) = \left(\frac{1}{3}\right)^x - 3$
78. $f(x) = 2^{-x} + 1$
79. $f(x) = 3^{1-x} - 2$
80. $f(x) = e^{x-1}$
81. $f(x) = e^x - 1$
82. $f(x) = -e^{x+2}$
83. $f(x) = -e^{x+1} - 1$
84. $f(x) = e^{-x} - 2$
85. The population of a rural city follows the exponential growth model $P(t) = 2{,}000e^{0.035t}$, where t is the number of years after 1995.

    Use this model to approximate the population in 2030, rounding to the nearest whole number.
86. The relative growth rate of a certain bacteria colony is 25%. Suppose there are 10 bacteria initially. Use the exponential growth model $P(t) = P_0e^{kt}$ to determine the number of bacteria that should be expected after 1 day.
87. In 1970, a wildlife resource management team introduced a certain rabbit species into a forest for the first time. In 2004, the rabbit population had grown to 7,183. The relative growth rate for this rabbit species is 20%. Use the exponential growth model $P(t) = P_0e^{kt}$ to determine the number of rabbits that can be expected in the year 2025. Round the answer to the nearest whole number.

5.2 Logarithmic Functions

THINGS TO KNOW

Before working through this section, be sure that you are familiar with the following concepts: VIDEO ANIMATION INTERACTIVE

You Try It
1. Solving Polynomial Inequalities (Section 1.9)

You Try It
2. Solving Rational Inequalities (Section 1.9)

You Try It
3. Determining If a Function Is One to One Using the Horizontal Line Test (Section 3.6)

You Try It
4. Sketching the Graph of an Inverse Function (Section 3.6)

You Try It
5. Using the Composition Cancellation Properties of Inverse Functions (Section 3.6)

You Try It
6. Finding the Equation of an Inverse Function (Section 3.6)

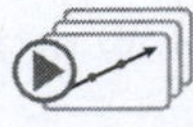

7. Sketching the Graphs of Exponential Functions (Section 5.1)

8. Sketching the Graphs of Exponential Functions Using Transformations (Section 5.1)

9. Solving Exponential Equations by Relating the Bases (Section 5.1)

OBJECTIVES

1 Understanding the Definition of a Logarithmic Function

2 Evaluating Logarithmic Expressions

3 Understanding the Properties of Logarithms

4 Using the Common and Natural Logarithms

5 Understanding the Characteristics of Logarithmic Functions

6 Sketching the Graphs of Logarithmic Functions Using Transformations

7 Finding the Domain of Logarithmic Functions

SECTION 5.2 EXERCISES

OBJECTIVE 1 UNDERSTANDING THE DEFINITION OF A LOGARITHMIC FUNCTION

Every exponential function of the form $f(x) = b^x$, where $b > 0$ and $b \neq 1$, is **one to one** and thus has an **inverse function**. (You may want to refer to **Section 3.6** to review one to one functions and inverse functions.) Remember, given the graph of a one to one function f, the graph of the inverse function is a reflection about the line $y = x$. That is, for any point (a, b) that lies on the graph of f, the point (b, a) must lie on the graph of f^{-1}. In other words, the graph of f^{-1} can be obtained by simply interchanging the x and y coordinates of the ordered pairs of $f(x) = b^x$. Watch this **video** to see how to sketch the graph of $f(x) = b^x$ and its inverse.

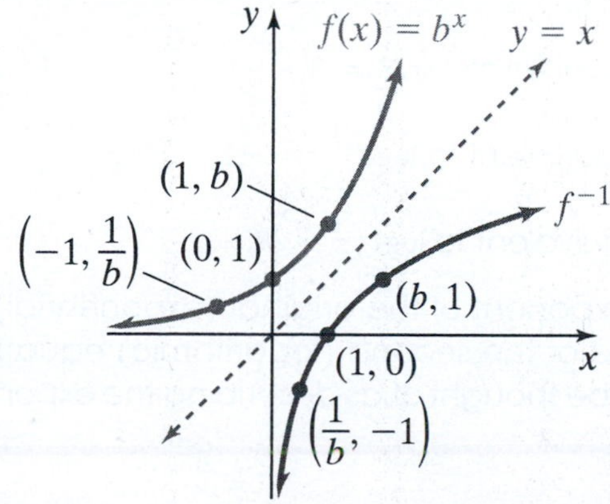

Figure 10 Graph of $f(x) = b^x, b > 1$, and its inverse function.

The graphs of f and f^{-1} are sketched in **Figure 10**, but what is the equation of the inverse of $f(x) = b^x$? To find the equation of f^{-1}, we follow the **four-step process** for finding inverse functions that is discussed in **Section 3.6**.

(eText Screens 5.2-1–5.2-28)

Step 1. Change $f(x)$ to y: $y = b^x$

Step 2. Interchange x and y: $x = b^y$

Step 3. Solve for y: ??

Before we can solve for y we must introduce the following definition:

> **Definition** Logarithmic Function
>
> For $x > 0, b > 0$, and $b \neq 1$, the **logarithmic function** with base b is defined by
>
> $y = \log_b x$ if and only if $x = b^y$.

The equation $y = \log_b x$ is said to be in **logarithmic form,** whereas the equation $x = b^y$ is in **exponential form.** We can now continue to find the inverse of $f(x) = b^x$ by completing steps 3 and 4.

Step 3. Solve for y: $x = b^y$ can be written as $y = \log_b x$

Step 4. Change y to $f^{-1}(x)$: $f^{-1}(x) = \log_b x$

In general, if $f(x) = b^x$ for $b > 0$ and $b \neq 1$, then the inverse function is $f^{-1}(x) = \log_b x$. For example, the inverse of $f(x) = 2^x$ is $f^{-1}(x) = \log_2 x$, which is read as "the log base 2 of x." We revisit the graphs of logarithmic functions later on in this section, but first it is very important to understand the definition of the logarithmic function and practice how to go back and forth writing exponential equations as logarithmic equations and vice versa.

Example 1 Change from Exponential Form to Logarithmic Form

Write each exponential equation as an equation involving a logarithm.

a. $2^3 = 8$ **b.** $5^{-2} = \frac{1}{25}$ **c.** $1.1^M = z$

Solution We use the fact that the equation $x = b^y$ is equivalent to the equation $y = \log_b x$.

a. $2^3 = 8$ is equivalent to $\log_2 8 = 3$.

b. $5^{-2} = \frac{1}{25}$ is equivalent to $\log_5 \frac{1}{25} = -2$.

c. $1.1^M = z$ is equivalent to $\log_{1.1} z = M$.

Note that the exponent of the original (exponential) equation ends up by itself on the right side of the second (logarithmic) equation. Therefore, a logarithmic expression can be thought of as describing the exponent of a certain exponential equation.

Watch the video to see this example worked out in more detail.

You Try It Work through this You Try It problem.

Work Exercises 1–6 in this textbook or in the MyLab Math Study Plan.

Example 2 Change from Logarithmic Form to Exponential Form

Write each logarithmic equation as an equation involving an exponent.

a. $\log_3 81 = 4$ b. $\log_4 16 = y$ c. $\log_{3/5} x = 2$

Solution We use the fact that the equation $y = \log_b x$ is equivalent to the equation $x = b^y$.

a. $\log_3 81 = 4$ is equivalent to $3^4 = 81$.

b. $\log_4 16 = y$ is equivalent to $4^y = 16$.

c. $\log_{3/5} x = 2$ is equivalent to $\left(\frac{3}{5}\right)^2 = x$.

Watch the **video** to see this example worked out in more detail.

You Try It **Work through this You Try It problem.**

Work Exercises 7–11 in this textbook or in the MyLab Math Study Plan.

OBJECTIVE 2 EVALUATING LOGARITHMIC EXPRESSIONS

Because a logarithmic expression represents the exponent of an exponential equation, it is possible to evaluate many logarithms by inspection or by creating the corresponding exponential equation. Remember that the expression $\log_b x$ is the exponent to which b must be raised to in order to get x. For example, suppose we are to evaluate the expression $\log_4 64$. To evaluate this expression, we must ask ourselves, "4 raised to what power is 64?" Because $4^3 = 64$, we conclude that $\log_4 64 = 3$. For some logarithmic expressions, it is often convenient to create an exponential equation and use the **method of relating the bases** for solving exponential equations. For more complex logarithmic expressions, additional techniques are required. These techniques are discussed in Sections 5.4 and 5.5.

Example 3 Evaluate Logarithmic Expressions

Evaluate each logarithm:

a. $\log_5 25$ b. $\log_3 \frac{1}{27}$ c. $\log_{\sqrt{2}} \frac{1}{4}$

Solution

a. To evaluate $\log_5 25$, we must ask, "5 raised to what exponent is 25?" Because $5^2 = 25$, $\log_5 25 = 2$.

b. The expression $\log_3 \frac{1}{27}$ requires more analysis. In this case, we ask, "3 raised to what exponent is $\frac{1}{27}$?" Suppose we let y equal this exponent. Then $3^y = \frac{1}{27}$. To solve for y, we can use the **method of relating the bases** for solving exponential equations.

$3^y = \frac{1}{27}$ Write the exponential equation.

$3^y = \frac{1}{3^3}$ Rewrite 27 as 3^3.

$3^y = 3^{-3}$ Use $\frac{1}{b^n} = b^{-n}$.

$y = -3$ Use the method of relating the bases. (If $b^u = b^v$, then $u = v$.)

Thus, $\log_3 \frac{1}{27} = -3$.

c. Watch the **interactive video** to verify that $\log_{\sqrt{2}} \frac{1}{4} = -4$ and to see each solution worked out in detail.

You Try It Work through this You Try It problem.

Work Exercises 12–18 in this textbook or in the MyLab Math Study Plan.

OBJECTIVE 3 UNDERSTANDING THE PROPERTIES OF LOGARITHMS

Because $b^1 = b$ for any real number b, we can use the definition of the logarithmic function ($y = \log_b x$ if and only if $x = b^y$) to rewrite this expression as $\log_b b = 1$. Similarly, because $b^0 = 1$ for any real number b, we can rewrite this expression as $\log_b 1 = 0$. These two general properties are summarized as follows.

General Properties of Logarithms

For $b > 0$ and $b \neq 1$,

1. $\log_b b = 1$
2. $\log_b 1 = 0$.

In **Section 3.6**, we see that a function f and its inverse function f^{-1} satisfy the following two composition cancellation equations:

$$f(f^{-1}(x)) = x \text{ for all } x \text{ in the domain of } f^{-1} \text{ and}$$

$$f^{-1}(f(x)) = x \text{ for all } x \text{ in the domain of } f$$

If $f(x) = b^x$, then $f^{-1}(x) = \log_b x$. Applying the two composition cancellation equations, we get

$$f(f^{-1}(x)) = b^{\log_b x} = x \text{ and}$$

$$f^{-1}(f(x)) = \log_b b^x = x.$$

Cancellation Properties of Exponentials and Logarithms

For $b > 0$ and $b \neq 1$,

1. $b^{\log_b x} = x$
2. $\log_b b^x = x$.

Example 4 Use the Properties of Logarithms

Use the properties of logarithms to evaluate each expression.

a. $\log_3 3^4$ b. $\log_{12} 12$ c. $7^{\log_7 13}$ d. $\log_8 1$

Solution

a. By the second cancellation property, $\log_3 3^4 = 4$.

b. Because $\log_b b = 1$ for all $b > 0$ and $b \neq 1$, $\log_{12} 12 = 1$.

c. By the first cancellation property, $7^{\log_7 13} = 13$.

d. Because $\log_b 1 = 0$ for all $b > 0$ and $b \neq 1$, $\log_8 1 = 0$. ●

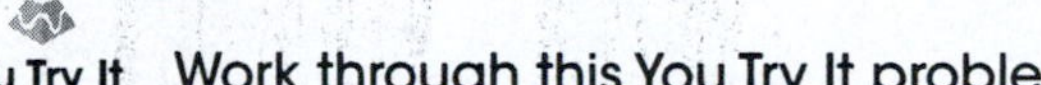

You Try It Work through this You Try It problem.

Work Exercises 19–26 in this textbook or in the MyLab Math Study Plan.

OBJECTIVE 4 USING THE COMMON AND NATURAL LOGARITHMS

There are two bases that are used more frequently than any other base. They are base 10 and base e. (Refer to **Section 5.1** to review the natural base e). Because our counting system is based on the number 10, the base 10 logarithm is known as the common logarithm. Instead of using the notation $\log_{10} x$ to denote the common logarithm, it is usually abbreviated without the subscript 10 as simply $\log x$. The base e logarithm is called the natural logarithm and is abbreviated as $\ln x$ instead of $\log_e x$. Most scientific calculators are equipped with a $\boxed{\log}$ key and a $\boxed{\ln}$ key. We can apply the definition of the logarithmic function for the base 10 and for the base e logarithm as follows.

> **Definition** Common Logarithmic Function
>
> For $x > 0$, the **common logarithmic function** is defined by
>
> $y = \log x$ if and only if $x = 10^y$.

> **Definition** Natural Logarithmic Function
>
> For $x > 0$, the **natural logarithmic function** is defined by
>
> $y = \ln x$ if and only if $x = e^y$.

Example 5 Change from Exponential Form to Logarithmic Form

Write each exponential equation as an equation involving a common logarithm or natural logarithm.

a. $e^0 = 1$ b. $10^{-2} = \dfrac{1}{100}$ c. $e^K = w$

Solution

a. $e^0 = 1$ is equivalent to $\ln 1 = 0$.

b. $10^{-2} = \frac{1}{100}$ is equivalent to $\log\left(\frac{1}{100}\right) = -2$.

c. $e^K = w$ is equivalent to $\ln w = K$.

Watch the **video** to see this example worked out in more detail.

You Try It Work through this You Try It problem.

Work Exercises 27–31 in this textbook or in the MyLab Math Study Plan.

Example 6 Change from Logarithmic Form to Exponential Form

Write each logarithmic equation as an equation involving an exponent.

a. $\log 10 = 1$ b. $\ln 20 = Z$ c. $\log(x-1) = T$

Solution

a. $\log 10 = 1$ is equivalent to $10^1 = 10$.

b. $\ln 20 = Z$ is equivalent to $e^Z = 20$.

c. $\log(x-1) = T$ is equivalent to $10^T = x - 1$.

Watch the **video** to see this example worked out in more detail.

You Try It Work through this You Try It problem.

Work Exercises 32–35 in this textbook or in the MyLab Math Study Plan.

Example 7 Evaluate Common and Natural Logarithmic Expressions

Evaluate each expression without the use of a calculator.

a. $\log 100$ b. $\ln\sqrt{e}$ c. $e^{\ln 51}$ d. $\log 1$

Solution

a. $\log 100 = 2$ because $10^2 = 100$ by the **definition of the logarithmic function** or $\log 100 = \log 10^2 = 2$ by **cancellation property (2)**.

b. $\ln\sqrt{e} = \ln e^{1/2} = \frac{1}{2}$ by **cancellation property (2)**.

c. $e^{\ln 51} = 51$ by **cancellation property (1)**.

d. $\log 1 = 0$ by **general property (1)**.

Watch the **video** to see this example worked out in more detail.

You Try It Work through this You Try It problem.

Work Exercises 36–43 in this textbook or in the MyLab Math Study Plan.

OBJECTIVE 5 UNDERSTANDING THE CHARACTERISTICS OF LOGARITHMIC FUNCTIONS

To sketch the graph of a logarithmic function of the form $f(x) = \log_b x$, where $b > 0$ and $b \neq 1$, follow these three steps:

Steps for Sketching Logarithmic Functions of the Form $f(x) = \log_b x$

Step 1. Start with the graph of the exponential function $y = b^x$, labeling several ordered pairs.

Step 2. Because $f(x) = \log_b x$ is the inverse of $y = b^x$, we can find several points on the graph of $f(x) = \log_b x$ by reversing the coordinates of the ordered pairs of $y = b^x$.

Step 3. Plot the ordered pairs from step 2, and complete the graph of $f(x) = \log_b x$ by connecting the ordered pairs with a smooth curve. The graph of $f(x) = \log_b x$ is a reflection of the graph of $y = b^x$ about the line $y = x$.

Example 8 Sketch the Graph of a Logarithmic Function

Sketch the graph of $f(x) = \log_3 x$.

Solution

Step 1. The graph of $y = 3^x$ passes through the points $\left(-1, \frac{1}{3}\right)$, $(0, 1)$, and $(1, 3)$. See Figure 11.

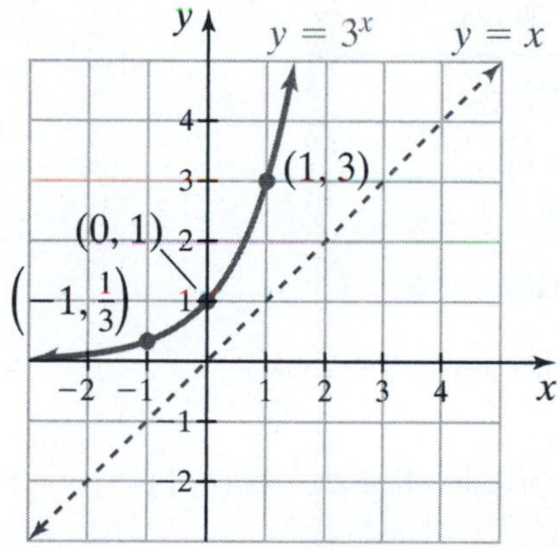

Figure 11 Graph of $y = 3^x$.

Step 2. We reverse the three ordered pairs from step 1 to obtain the following three points: $\left(\frac{1}{3}, -1\right)$, $(1, 0)$, and $(3, 1)$.

Step 3. Plot the points $\left(\frac{1}{3}, -1\right)$, $(1, 0)$, and $(3, 1)$, and connect them with a smooth curve to obtain the graph of $f(x) = \log_3 x$. See Figure 12.

(eText Screens 5.2-1–5.2-28)

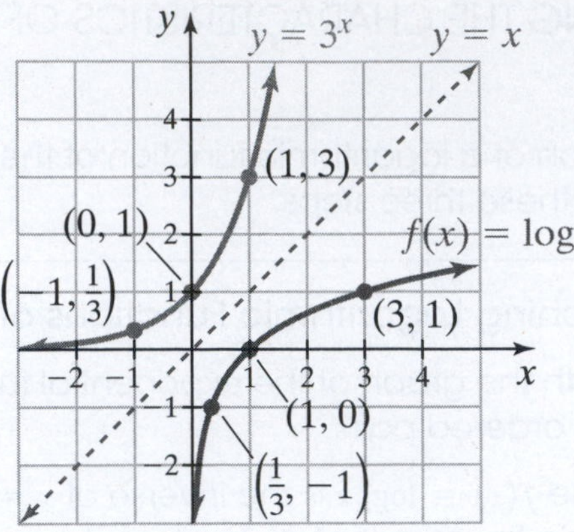

Figure 12 Graph of $y = 3^x$ and $f(x) = \log_3 x$.

Notice in Figure 12 that the y-axis is a **vertical asymptote** of the graph of $f(x) = \log_3 x$. Every logarithmic function of the form $y = \log_b x$, where $b > 0$ and $b \neq 1$ has a vertical asymptote at the y-axis. The graphs and the characteristics of logarithmic functions are outlined as follows. Watch this **video** to see each step of this graphing process.

Characteristics of Logarithmic Functions

For $b > 0$, $b \neq 1$, the logarithmic function with base b is defined by $y = \log_b x$. The domain of $f(x) = \log_b x$ is $(0, \infty)$, and the range is $(-\infty, \infty)$. The graph of $f(x) = \log_b x$ has one of the following two shapes.

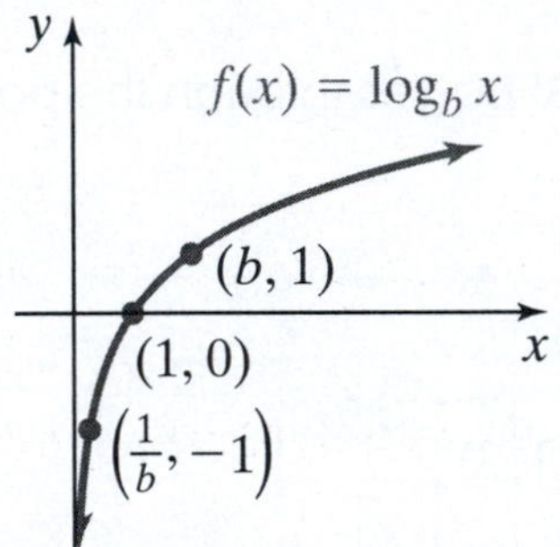

$f(x) = \log_b x, b > 1$

- The graph intersects the x-axis at $(1, 0)$.
- The graph contains the points $(b, 1)$ and $\left(\frac{1}{b}, -1\right)$.
- The graph is **increasing** on the interval $(0, \infty)$.
- The y-axis $(x = 0)$ is a **vertical asymptote**.
- The function is **one to one**.

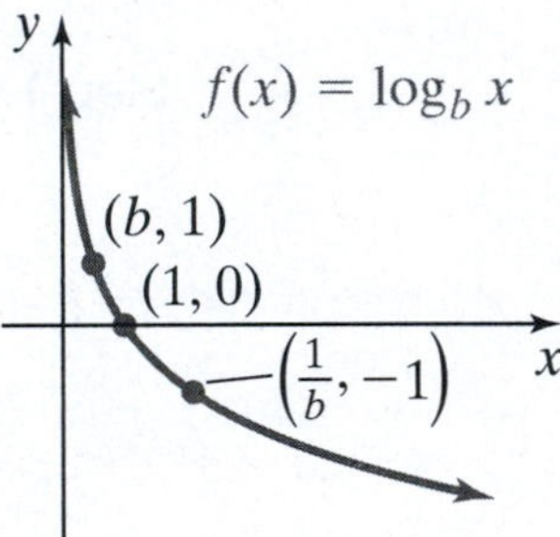

$f(x) = \log_b x, 0 < b < 1$

- The graph intersects the x-axis at $(1, 0)$.
- The graph contains the points $(b, 1)$ and $\left(\frac{1}{b}, -1\right)$.
- The graph is **decreasing** on the interval $(0, \infty)$.
- The y-axis $(x = 0)$ is a **vertical asymptote**.
- The function is **one to one**.

You Try It Work through this You Try It problem.

Work Exercises 44 and 45 in this textbook or in the MyLab Math Study Plan.

OBJECTIVE 6 SKETCHING THE GRAPHS OF LOGARITHMIC FUNCTIONS USING TRANSFORMATIONS

Often we can use the **transformation techniques** that are discussed in **Section 3.4** to sketch the graph of logarithmic functions.

Example 9 Sketch the graph of $f(x) = -\ln(x + 2) - 1$.

Solution Recall that the function $y = \ln x$ has a base of e, where $2 < e < 3$. This means that the graph of $y = \ln x$ is increasing on the interval $(0, \infty)$ and contains the three points $\left(\frac{1}{e}, -1\right)$, $(1, 0)$, and $(e, 1)$.

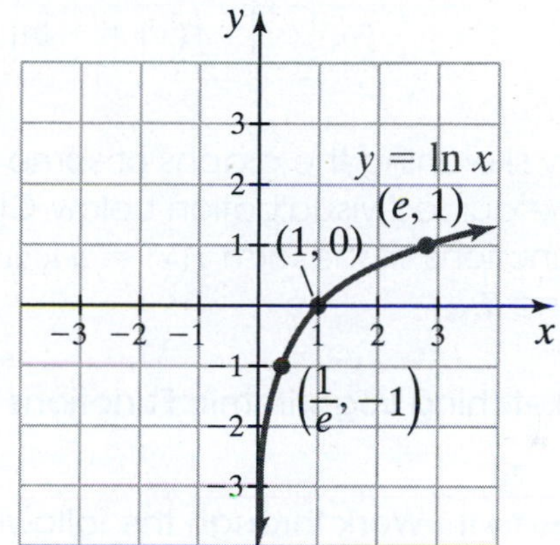

Starting with the graph of $y = \ln x$, we can obtain the graph of $f(x) = -\ln(x + 2) - 1$ through the following series of transformations:

1. Shift the graph of $y = \ln x$ horizontally to the left two units to obtain the graph of $y = \ln(x + 2)$.
2. Reflect the graph of $y = \ln(x + 2)$ about the x-axis to obtain the graph of $y = -\ln(x + 2)$.
3. Shift the graph of $y = -\ln(x + 2)$ vertically down one unit to obtain the final graph of $f(x) = -\ln(x + 2) - 1$.

Watch this **video** to see each step of this graphing process. The original three points that lie on the graph of $y = \ln x$ and the corresponding points that lie on the graph of $f(x) = -\ln(x + 2) - 1$ are shown below.

Points that lie on the graph of $y = \ln x$:

Corresponding points that lie on the graph of $f(x) = -\ln(x + 2) - 1$:

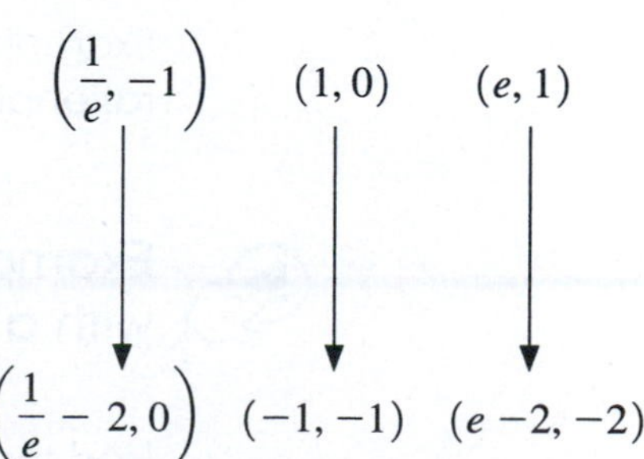

The graph of $f(x) = -\ln(x + 2) - 1$ is sketched in Figure 13. You can see from the graph that the domain of $f(x) = -\ln(x + 2) - 1$ is $(-2, \infty)$. The vertical asymptote is the line $x = -2$.

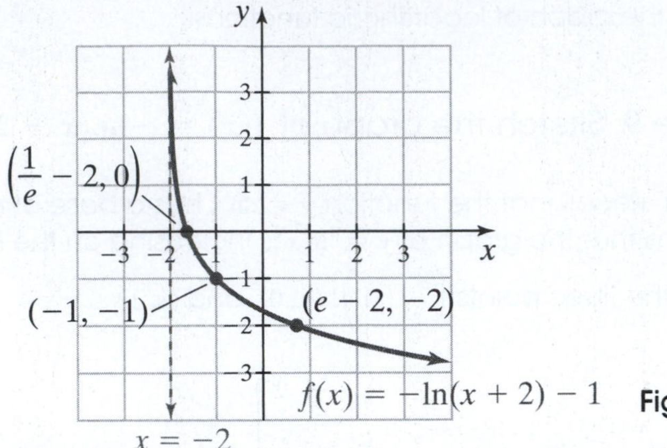

Figure 13 Graph of $f(x) = -\ln(x + 2) - 1$. ●

Try sketching the graphs of some logarithmic functions on your own by exploring the guided visualization below. Click on the icon below to sketch logarithmic functions of the form $f(x) = a \log_b(x + c) + d$. Choose your own values of a, b, c, and d.

 Sketching Logarithmic Functions

You Try It Work through the following You Try It problem.

Work Exercises 46–55 in this textbook or in the MyLab Math Study Plan.

OBJECTIVE 7 FINDING THE DOMAIN OF LOGARITHMIC FUNCTIONS

In Example 9, we sketch the function $f(x) = -\ln(x + 2) - 1$ and observe that the domain was $(-2, \infty)$. We do not have to sketch the graph of a logarithmic function to determine the domain. The domain of a logarithmic function consists of all values of x for which the **argument** of the logarithm is greater than zero. In other words, if $f(x) = \log_b[g(x)]$, then the domain of f can be found by solving the inequality $g(x) > 0$. For example, given the function $f(x) = -\ln(x + 2) - 1$ from Example 9, we can determine the domain by solving the linear inequality $x + 2 > 0$. Solving this inequality for x, we obtain $x > -2$. Thus, the domain of $f(x) = -\ln(x + 2) - 1$ is $(-2, \infty)$.

Example 10 is a bit more challenging because the argument of the logarithm is a rational expression.

Example 10 Find the Domain of a Logarithmic Function with a Rational Argument

Find the domain of $f(x) = \log_5\left(\frac{2x - 1}{x + 3}\right)$.

Solution To find the domain of f, we must find all values of x for which the argument $\frac{2x-1}{x+3}$ is greater than zero. That is, you must solve the rational inequality $\left(\frac{2x-1}{x+3}\right) > 0$. See **Section 1.9** if you need help remembering how to solve this inequality. By the techniques discussed in Section 1.9, we find that the solution to $\left(\frac{2x-1}{x+3}\right) > 0$ is $x < -3$ or $x > \frac{1}{2}$. Therefore, the domain of $f(x) = \log_5\left(\frac{2x-1}{x+3}\right)$ in set notation is $\left\{x \middle| x < -3 \text{ or } x > \frac{1}{2}\right\}$. In interval notation, the domain is $(-\infty, -3) \cup \left(\frac{1}{2}, \infty\right)$. Watch the **interactive video** to see this problem worked out in detail.

You Try It **Work through the following You Try It problem.**

Work Exercises 56–62 in this textbook or in the MyLab Math Study Plan.

5.2 Exercises

In Exercises 1–6, write each exponential equation as an equation involving a logarithm.

1. $3^2 = 9$
2. $16^{1/2} = 4$
3. $2^{-3} = \frac{1}{8}$
4. $\sqrt{2}^{\pi} = W$
5. $\left(\frac{1}{3}\right)^t = 27$
6. $7^{5k} = L$

In Exercises 7–11, write each logarithmic equation as an exponential equation.

7. $\log_5 1 = 0$
8. $\log_7 343 = 3$
9. $\log_{\sqrt{2}} 8 = 6$
10. $\log_4 K = L$
11. $\log_a (x - 1) = 3$

In Exercises 12–18, evaluate each logarithm without the use of a calculator.

12. $\log_2 8$
13. $\log_6 \sqrt{6}$
14. $\log_3 \frac{1}{9}$
15. $\log_{\sqrt{5}} 25$
16. $\log_4\left(\frac{1}{\sqrt[5]{64}}\right)$
17. $\log_{1/7} \sqrt[3]{7}$
18. $\log_{0.1} 100$

In Exercises 19–26, use the properties of logarithms to evaluate each expression without the use of a calculator.

19. $2^{\log_2 11}$
20. $\log_4 4$
21. $\log_9 1$
22. $\log_7 7^{-3}$
23. $\log_a a,\ a > 1$
24. $5^{\log_5 M},\ M > 0$
25. $\log_y 1,\ y > 0$
26. $\log_x x^{20},\ x > 1$

(eText Screens 5.2-1–5.2-28)

In Exercises 27–31, write each exponential equation as an equation involving a common logarithm or a natural logarithm.

27. $10^3 = 1{,}000$

28. $e^{-1} = \dfrac{1}{e}$

29. $e^k = 2$

30. $10^e = M$

31. $e^{10} = Z$

In Exercises 32–35, write each logarithmic equation as an exponential equation.

32. $\ln 1 = 0$

33. $\log(1{,}000{,}000) = 6$

34. $\log K = L$

35. $\ln Z = 4$

In Exercises 36–43, evaluate each expression without the use of a calculator, and then verify your answer using a calculator.

36. $\log 10{,}000$

37. $\log\left(\dfrac{1}{1{,}000}\right)$

38. $\ln 1$

39. $\ln \sqrt[3]{e^2}$

40. $10^{\log e}$

41. $e^{\ln 49}$

42. $\log 10^6$

43. $\ln e + \ln e^3$

In Exercises 44–55, sketch each logarithmic function. Label three points that lie on the graph, and determine the domain and the equation of any vertical asymptotes.

SbS 44. $h(x) = \log_4 x$

SbS 45. $g(x) = \log_{\frac{1}{3}} x$

SbS 46. $f(x) = \log_2(x) - 1$

SbS 47. $f(x) = \log_5(x - 1)$

SbS 48. $f(x) = \ln(x + 1)$

SbS 49. $f(x) = -\ln(x)$

SbS 50. $f(x) = -\ln(x - 2)$

SbS 51. $f(x) = -2 + \ln x$

SbS 52. $f(x) = 3 - \ln x$

SbS 53. $y = \log_{1/2}(x + 1) + 2$

SbS 54. $y = \log_3(1 - x)$

SbS 55. $h(x) = -\dfrac{1}{2}\log_3(x + 3) + 1$

In Exercises 56–62, find the domain of each logarithmic function.

56. $f(x) = \log(-x)$

57. $f(x) = \log_{1/4}(2x + 6)$

58. $f(x) = \ln(1 - 3x)$

59. $f(x) = \log_2(x^2 - 9)$

60. $f(x) = \log_7(x^2 - x - 20)$

61. $f(x) = \ln\left(\dfrac{x + 5}{x - 8}\right)$

62. $f(x) = \log\left(\dfrac{x^2 - x - 6}{x + 10}\right)$

Brief Exercises

In Exercises 63–74, sketch each logarithmic function. Label three points that lie on the graph.

63. $h(x) = \log_4 x$

64. $g(x) = \log_{\frac{1}{3}} x$

65. $f(x) = \log_2 (x) - 1$

66. $f(x) = \log_5 (x - 1)$

67. $f(x) = \ln (x + 1)$

68. $f(x) = -\ln (x)$

69. $f(x) = -\ln (x - 2)$

70. $f(x) = -2 + \ln x$

71. $f(x) = 3 - \ln x$

72. $y = \log_{1/2} (x + 1) + 2$

73. $y = \log_3 (1 - x)$

74. $h(x) = -\dfrac{1}{2}\log_3 (x + 3) + 1$

5.3 Properties of Logarithms

THINGS TO KNOW

Before working through this section, be sure that you are familiar with the following concepts: VIDEO ANIMATION INTERACTIVE

You Try It 1. Solving Exponential Equations by Relating the Bases (Section 5.1)

You Try It 2. Change from Exponential Form to Logarithmic Form (Section 5.2)

You Try It 3. Change from Logarithmic Form to Exponential Form (Section 5.2)

You Try It 4. Evaluating Logarithmic Expressions (Section 5.2)

You Try It 5. Using the Common and Natural Logarithms (Section 5.2)

You Try It 6. Finding the Domain of Logarithmic Functions (Section 5.2)

OBJECTIVES

1 Using the Product Rule, Quotient Rule, and Power Rule for Logarithms

2 Expanding and Condensing Logarithmic Expressions

3 Solving Logarithmic Equations Using the Logarithm Property of Equality

4 Using the Change of Base Formula

SECTION 5.3 EXERCISES

OBJECTIVE 1 USING THE PRODUCT RULE, QUOTIENT RULE, AND POWER RULE FOR LOGARITHMS

In this section, we learn how to manipulate logarithmic expressions using properties of logarithms. Understanding how to use these properties will help us solve exponential and logarithmic equations that are encountered in the next section. Recall from Section 5.3 the **general properties** and **cancellation properties** of logarithms. We now look at three additional properties of logarithms.

Properties of Logarithms

If $b > 0, b \neq 1, u$ and v represent positive numbers and r is any real number, then

$\log_b uv = \log_b u + \log_b v$ product rule for logarithms 

$\log_b \frac{u}{v} = \log_b u - \log_b v$ quotient rule for logarithms 

$\log_b u^r = r \log_b u$ power rule for logarithms 

To prove the product rule and quotient rule for logarithms, we use properties of exponents and the **method of relating the bases** to solve exponential equations. The power rule for logarithms is a direct result of the product rule. Click on a video proof link above to see a proof of one or more of these properties.

Example 1 Use the Product Rule

Use the product rule for logarithms to expand each expression. Assume $x > 0$.

a. $\ln(5x)$ **b.** $\log_2(8x)$

Solution

a. $\ln(5x) = \ln 5 + \ln x$ Use the product rule for logarithms.

b. $\log_2(8x) = \log_2 8 + \log_2 x$ Use the product rule for logarithms

$= 3 + \log_2 x$ Use the **definition of the logarithmic function** to rewrite $\log_2 8$ as 3 because $2^3 = 8$.

CAUTION $\log_b(u + v)$ **is *not* equivalent to** $\log_b u + \log_b v$**.**

You Try It Work through this You Try It problem.

Example 2 Use the Quotient Rule

Use the quotient rule for logarithms to expand each expression. Assume $x > 0$.

a. $\log_5\left(\frac{12}{x}\right)$ **b.** $\ln\left(\frac{x}{e^5}\right)$

Solution

a. $\log_5\left(\dfrac{12}{x}\right) = \log_5 12 - \log_5 x$ Use the quotient rule for logarithms.

b. $\ln\left(\dfrac{x}{e^5}\right) = \ln x - \ln e^5$ Use the quotient rule for logarithms

$= \ln x - 5$ Use cancellation property (2) to rewrite $\ln e^5$ as 5.

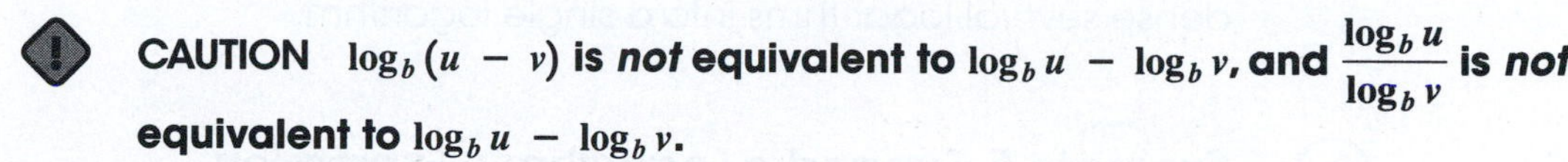

CAUTION **$\log_b(u - v)$ is *not* equivalent to $\log_b u - \log_b v$, and $\dfrac{\log_b u}{\log_b v}$ is *not* equivalent to $\log_b u - \log_b v$.**

You Try It Work through this You Try It problem.

Example 3 Use the Power Rule

Use the power rule for logarithms to rewrite each expression. Assume $x > 0$.

a. $\log 6^3$ b. $\log_{1/2} \sqrt[4]{x}$

Solution

a. $\log 6^3 = 3 \log 6$ Use the power rule for logarithms.

b. $\log_{1/2} \sqrt[4]{x} = \log_{1/2} x^{1/4}$ Rewrite the fourth root of x using a rational exponent.

$= \dfrac{1}{4}\log_{1/2} x$ Use the power rule for logarithms.

The process of using the power rule to simplify a logarithmic expression is often casually referred to as "bringing down the exponent."

CAUTION **$(\log_b u)^r$ is *not* equivalent to $r \log_b u$.**

You Try It Work through this You Try It problem.

Work Exercises 12–15 in this eText or in the MyLab Math Study Plan.

Example 4 Use Properties of Logarithms to Evaluate Expressions

Use properties of logarithms to evaluate each expression without the use of a calculator.

a. $7^{\log_7 6 + \log_7 3}$ b. $e^{2\ln 5 - \frac{1}{3}\ln 64 + \ln 1}$

Solution

a. $7^{\log_7 6 + \log_7 3}$ Write the original expression.

$= 7^{\log_7 (6 \cdot 3)}$ Use the product rule for logarithms.

$= 7^{\log_7 18}$ Simplify.

$= 18$ Use cancellation property (1).

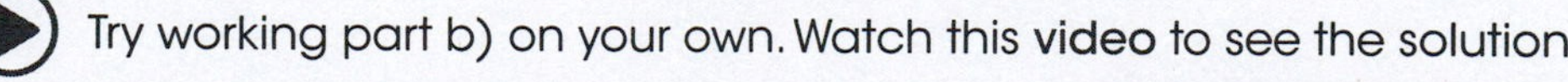

Try working part b) on your own. Watch this video to see the solution.

You Try It Work through this You Try It problem.

Work Exercises 12–15 in this eText or in the MyLab Math Study Plan.

OBJECTIVE 2 EXPANDING AND CONDENSING LOGARITHMIC EXPRESSIONS

Sometimes it is necessary to combine several properties of logarithms to expand a logarithmic expression into the sum and/or difference of logarithms or to condense several logarithms into a single logarithm.

Example 5 Expand a Logarithmic Expression

Use properties of logarithms to expand each logarithmic expression as much as possible.

a. $\log_7 (49x^3\sqrt[5]{y^2})$ b. $\ln\left(\frac{(x^2-4)}{9e^{x^3}}\right)$

Solution

a. $\log_7 (49x^3\sqrt[5]{y^2})$	Write the original expression.
$= \log_7 49 + \log_7 x^3\sqrt[5]{y^2}$	Use the product rule.
$= \log_7 49 + \log_7 x^3 + \log_7 \sqrt[5]{y^2}$	Use the product rule again.
$= \log_7 49 + \log_7 x^3 + \log_7 y^{2/5}$	Rewrite $\sqrt[5]{y^2}$ using a rational exponent.
$= 2 + 3\log_7 x + \frac{2}{5}\log_7 y$	Rewrite $\log_7 49$ as 2 and use the power rule.
b. $\ln\left(\frac{(x^2-4)}{9e^{x^3}}\right) = \ln\left(\frac{(x-2)(x+2)}{9e^{x^3}}\right)$	Factor the expression in the numerator.
$= \ln(x-2)(x+2) - \ln 9e^{x^3}$	Use the quotient rule.
$= \ln(x-2) + \ln(x+2) - \left[\ln 9 + \ln e^{x^3}\right]$	Use the product rule twice.
$= \ln(x-2) + \ln(x+2) - \left[\ln 9 + x^3\right]$	Use cancellation property (2) to rewrite $\ln e^{x^3}$ as x^3.
$= \ln(x-2) + \ln(x+2) - \ln 9 - x^3$	Simplify.

Watch the **interactive video** to see this example worked out in detail.

You Try It Work through this You Try It problem.

Work Exercises 12–15 in this eText or in the MyLab Math Study Plan.

Example 6 Condense a Logarithmic Expression

Use properties of logarithms to rewrite each expression as a single logarithm.

a. $\frac{1}{2}\log(x-1) - 3\log z + \log 5$ b. $\frac{1}{3}(\log_3 x - 2\log_3 y) + \log_3 10$

Solution

a. $\frac{1}{2}\log(x-1) - 3\log z + \log 5$ Write the original expression.

$= \log(x-1)^{1/2} - \log z^3 + \log 5$ Use the power rule twice.

$= \log\frac{(x-1)^{1/2}}{z^3} + \log 5$ Use the quotient rule.

$= \log\frac{5(x-1)^{1/2}}{z^3}$ or $\log\frac{5\sqrt{x-1}}{z^3}$ Use the product rule.

b. $\frac{1}{3}(\log_3 x - 2\log_3 y) + \log_3 10$ Write the original expression.

$= \frac{1}{3}(\log_3 x - \log_3 y^2) + \log_3 10$ Use the power rule.

$= \frac{1}{3}\log_3\frac{x}{y^2} + \log_3 10$ Use the quotient rule.

$= \log_3\left(\frac{x}{y^2}\right)^{1/3} + \log_3 10$ Use the power rule.

$= \log_3\left[10\left(\frac{x}{y^2}\right)^{1/3}\right]$ or $\log_3\left[10\sqrt[3]{\frac{x}{y^2}}\right]$ Use the product rule. ●

Watch the **interactive video** to see this example worked out in detail.

You Try It Work through this You Try It problem.

Work Exercises 12–15 in this eText or in the MyLab Math Study Plan.

OBJECTIVE 3 SOLVING LOGARITHMIC EQUATIONS USING THE LOGARITHM PROPERTY OF EQUALITY

Remember that all logarithmic functions of the form $f(x) = \log_b x$ for $b > 0$ and $b \neq 1$ are **one to one**. In Section 3.6, the alternate definition of **one to one** stated that

A function f is one to one if, for any two range values $f(u)$ and $f(v)$, $f(u) = f(v)$ implies that $u = v$.

Using this definition and letting $f(x) = \log_b x$, we can say that if $\log_b u = \log_b v$, then $u = v$. In other words, if the bases of a logarithmic equation of the form $\log_b u = \log_b v$ are equal, then the **arguments** must be equal. This is known as the **logarithm property of equality**.

> **Logarithm Property of Equality**
> If a logarithmic equation can be written in the form $\log_b u = \log_b v$, then $u = v$. Furthermore, if $u = v$, then $\log_b u = \log_b v$.

The second statement of the logarithm property of equality says that if we start with the equation $u = v$, then we can rewrite the equation as $\log_b u = \log_b v$. This process is often casually referred to as "taking the log of both sides."

Example 7 Use the Logarithm Property of Equality to Solve Logarithmic Equations

Solve the following equations:

a. $\log_7 (x - 1) = \log_7 12$ b. $2 \ln x = \ln 16$

Solution

a. Because the base of each logarithm is 7, we can use the logarithm property of equality to eliminate the logarithms.

$\log_7 (x - 1) = \log_7 12$	Write the original equation.
$(x - 1) = 12$	If $\log_b u = \log_b v$, then $u = v$.
$x = 13$	Solve for x.

b.

$2 \ln x = \ln 16$	Write the original expression.
$\ln x^2 = \ln 16$	Use the power rule.
$x^2 = 16$	If $\log_b u = \log_b v$, then $u = v$.
$x = \pm 4$	Use the square root property.

The domain of $\ln x$ is $x > 0$; this implies that $x = -4$ is an **extraneous solution**, and hence, we must discard it. Therefore, this equation has only one solution, $x = 4$.

Watch this **interactive video** to see the entire solution to this example. ●

You Try It **Work through the following You Try It problem.**

Work Exercises 38–43 in this textbook or in the MyLab Math Study Plan.

OBJECTIVE 4 USING THE CHANGE OF BASE FORMULA

Most scientific calculators are equipped with a [log] key and a [ln] key to evaluate common logarithms and natural logarithms. But how do we use a calculator to evaluate logarithmic expressions having a base other than 10 or e? The answer is to use the following **change of base formula**.

> **Change of Base Formula**
>
> For any positive base $b \neq 1$ and for any positive real number u, then
>
> $$\log_b u = \frac{\log_a u}{\log_a b},$$
>
> where a is any positive number such that $a \neq 1$.

 To see the proof of the change of base formula, watch this **video**.

The change of base formula allows us to change the base of a logarithmic expression into a ratio of two logarithms using any base we choose. For example, suppose we are given the logarithmic expression $\log_3 10$. We can use the change of base formula to write this logarithm as a quotient of logarithms involving any base we choose:

$$\log_3 10 = \frac{\log_7 10}{\log_7 3} \quad \text{or} \quad \log_3 10 = \frac{\log_2 10}{\log_2 3} \quad \text{or} \quad \log_3 10 = \frac{\log 10}{\log 3} \quad \text{or} \quad \log_3 10 = \frac{\ln 10}{\ln 3}$$

 In each of the previous four cases, we introduced a new base (7, 2, 10, and e, respectively). However, if we want to use a calculator to get a numerical approximation of $\log_3 10$, then it really only makes sense to change $\log_3 10$ into an expression involving base 10 or base e because these are the only two bases most calculators can handle.

$$\log_3 10 = \frac{\log 10}{\log 3} \approx 2.0959 \text{ or } \log_3 10 = \frac{\ln 10}{\ln 3} \approx 2.0959$$

Example 8 Use the Change of Base Formula

Approximate the following expressions. Round each to four decimal places.

a. $\log_9 200$ b. $\log_{\sqrt{3}} \pi$

Solution

a. $\log_9 200 = \dfrac{\log 200}{\log 9} \approx 2.4114$ or $\log_9 200 = \dfrac{\ln 200}{\ln 9} \approx 2.4114$

b. $\log_{\sqrt{3}} \pi = \dfrac{\log \pi}{\log \sqrt{3}} \approx 2.0840$ or $\log_{\sqrt{3}} \pi = \dfrac{\ln \pi}{\ln \sqrt{3}} \approx 2.0840$

You Try It Work through the following You Try It problem.

Work Exercises 44–47 in this textbook or in the MyLab Math Study Plan.

Example 9 Use the Change of Base Formula and Properties of Logarithms

Use the change of base formula and the properties of logarithms to rewrite as a single logarithm involving base 2.

$$\log_4 x + 3 \log_2 y$$

(eText Screens 5.3-1–5.3-21)

Solution To use properties of logarithms, the base of each logarithmic expression must be the same. We use the change of base formula to rewrite $\log_4 x$ as a logarithmic expression involving base 2:

$\log_4 x + 3\log_2 y = \dfrac{\log_2 x}{\log_2 4} + 3\log_2 y$	Use the change of base formula $\log_4 x = \dfrac{\log_2 x}{\log_2 4}$.
$= \dfrac{\log_2 x}{2} + 3\log_2 y$	Rewrite $\log_2 4$ as 2 because $2^2 = 4$.
$= \dfrac{1}{2}\log_2 x + 3\log_2 y$	Rewrite $\dfrac{\log_2 x}{2}$ as $\dfrac{1}{2}\log_2 x$.
$= \log_2 x^{1/2} + \log_2 y^3$	Use the power rule.
$= \log_2 x^{1/2}y^3 \quad \text{or} \quad \log_2 \sqrt{x}y^3$	Use the product rule.

Therefore, the expression $\log_4 x + 3\log_2 y$ is equivalent to $\log_2 \sqrt{x}y^3$. Note that we could have chosen to rewrite the original expression as a single logarithm involving base 4. Watch the video to see that the expression $\log_4 x + 3\log_2 y$ is also equivalent to $\log_4 xy^6$.

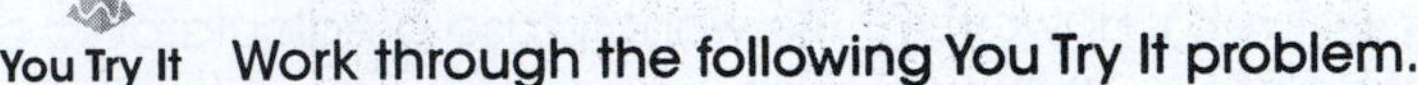

You Try It Work through the following You Try It problem.

Work Exercises 48–51 in this textbook or in the MyLab Math Study Plan.

Example 10 Use the Change of Base Formula to Solve Logarithmic Equations

Use the change of base formula and the properties of logarithms to solve the following equation:

$$2\log_3 x = \log_9 16$$

Solution Watch the video to see how the change of base formula and the power rule for logarithms can be used to solve this equation.

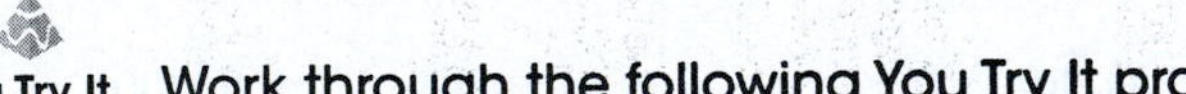

You Try It Work through the following You Try It problem.

Work Exercises 52–55 in this textbook or in the MyLab Math Study Plan.

Below is a summary of the logarithm properties that have been presented thus far in the eText. At this time, it would be beneficial to write these properties down in a notebook or on note cards so that you can easily refer to these properties throughout the remaining portion of this chapter.

Summary of Logarithm Property

If $b > 0, b \neq 1, u$ and v, represent positive numbers and r is any real number, then

$\log_b b = 1$ and $\log_b 1 = 0$	general properties of logarithms
$b^{\log_b u} = u$ and $\log_b b^u = u$	cancellation properties
$\log_b uv = \log_b u + \log_b v$	product rule for logarithms

$\log_b \frac{u}{v} = \log_b u - \log_b v$	quotient rule for logarithms
$\log_b u^r = r \log_b u$	product rule for logarithms
If $\log_b u = \log_b v$, then $u = v$.	logarithm property of equality
If $u = v$, then $\log_b u = \log_b v$.	logarithms property of equality
$\log_b u = \frac{\log_a u}{\log_a b}$, where a is any positive number such that $a \neq 1$.	change of base formula

5.3 Exercises

In Exercises 1–10, use the product rule, quotient rule, or power rule to expand each logarithmic expression. Wherever possible, evaluate logarithmic expressions.

1. $\log_4 (xy)$
2. $\log \left(\frac{9}{t}\right)$
3. $\log_5 y^3$
4. $\log_3 (27w)$
5. $\ln 5e^2$
6. $\log_9 \sqrt[4]{k}$
7. $\log 100P$
8. $\ln \left(\frac{e^5}{r}\right)$
9. $\log_{\sqrt{2}} 8x$
10. $\log_2 \left(\frac{M}{32}\right)$

In Exercises 11–14, use properties of logarithms to evaluate each expression without the use of a calculator.

11. $9^{\log_9 5 + \log_9 8 - \log_9 1}$
12. $5^{\log_5 3 - \log_5 11}$
13. $e^{\ln 5 + 2 \ln 2}$
14. $10^{\log 1 + \log 3 - 2 \log 11}$

In Exercises 15–24, use the properties of logarithms to expand each logarithmic expression. Wherever possible, evaluate logarithmic expressions.

15. $\log_7 x^2 y^3$
16. $\ln \frac{a^2 b^3}{c^4}$
17. $\log \frac{\sqrt{x}}{10y^3}$
18. $\log_3 9(x^2 - 25)$
19. $\log_2 \sqrt{4xy}$
20. $\log_5 \frac{\sqrt{5x^5}}{\sqrt[3]{25y^4}}$
21. $\ln \frac{\sqrt[5]{ez}}{\sqrt{x - 1}}$
22. $\log_3 \sqrt[4]{\frac{x^2 y^5}{9}}$
23. $\ln \left[\frac{x + 1}{(x^2 - 1)^3}\right]^{2/3}$
24. $\log \frac{(10x)^3 \sqrt{x - 4}}{(x^2 - 16)^5}$

In Exercises 25–37, use properties of logarithms to rewrite each expression as a single logarithm. Wherever possible, evaluate logarithmic expressions.

25. $\log_b A + \log_b C$
26. $\log_4 M - \log_4 N$
27. $2 \log_8 x + \frac{1}{3} \log_8 y$
28. $\log 20 + \log 5$
29. $\log_2 80 - \log_2 5$
30. $\ln \sqrt{x} - \frac{1}{3} \ln x + \ln \sqrt[4]{x}$
31. $\log_5 (x - 2) + \log_5 (x + 2)$
32. $\log_3 (x + 1) - \log_3 (x + 4) - \log_3 \sqrt{6x}$
33. $\ln(x - 1) - \ln\sqrt{x + 5} - \ln 4x$

34. $\ln(x-3)^2 + \ln 7x^2$

35. $\log_9(x^2-5x+6) - \log_9(x^2-4) + \log_9(x+2)$

36. $\log(x-3) + 2\log(x+3) - \log(x^3+x^2-9x-9)$

37. $\frac{1}{2}[\ln(x-1)^2 - \ln(2x^2-x-1)^4] + 2\ln(2x+1)$

In Exercises 38–43, use the properties of logarithms and the logarithm property of equality to solve each logarithmic equation.

38. $\log_3(2x+1) = \log_3 11$

39. $\log_{11}\sqrt{x} = \log_{11} 6$

40. $2\log(x+5) = \log 12 + \log 3$

41. $\ln 5 + \ln x = \ln 7 + \ln(3x-2)$

42. $\log_7(x+6) - \log_7(x+2) = \log_7 x$

43. $\log_2(x+3) + \log_2(x-4) = \log_2(x+12)$

In Exercises 44–47, use the change of base formula and a calculator to approximate each logarithmic expression. Round your answers to four decimal places.

44. $\log_4 51$

45. $\log_7 0.8$

46. $\log_{1/5} 72$

47. $\log_{\sqrt{7}} 100$

In Exercises 48–51, use the change of base formula and the properties of logarithms to rewrite each expression as a single logarithm in the indicated base.

48. $\log_3 x + 4\log_9 w$, base 3

49. $\log_5 x + \log_{1/5} x^3$, base 5

50. $\log_{16} x^4 + \log_8 y^3 + \log_4 w^2$, base 2

51. $\log_{e^2} x^5 + \log_{e^3} x^6 + \log_{e^4} x^{12}$, base e

In Exercises 52–55, solve each logarithmic equation.

52. $\log_2 x = \log_4 25$

53. $\log_{1/3} x = \log_3 20$

54. $\log_5 x = \log_{\sqrt{5}} 6$

55. $2\ln x = \log_{e^3} 125$

5.4 Exponential and Logarithmic Equations

THINGS TO KNOW

Before working through this section, be sure that you are familiar with the following concepts:

VIDEO | ANIMATION | INTERACTIVE

You Try It

1. Solving Exponential Equations by Relating the Bases (Section 5.1)

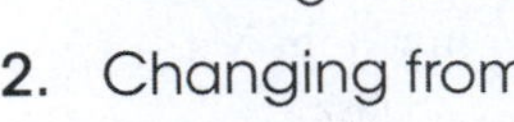
You Try It

2. Changing from Exponential to Logarithmic Form (Section 5.2)

You Try It
3. Changing from Logarithmic to Exponential Form (Section 5.2) 

You Try It
4. Using the Cancellation Properties of Exponentials and Logarithms (Section 5.2)

You Try It
5. Expanding and Condensing Logarithmic Expressions (Section 5.3) 

You Try It
6. Solving Logarithmic Equations Using the Logarithm Property of Equality (Section 5.3) 

OBJECTIVES

1 Solving Exponential Equations

2 Solving Logarithmic Equations

SECTION 5.4 EXERCISES

In this section, we learn how to solve exponential and logarithmic equations. The techniques and strategies learned in this section help us solve applied problems that are discussed in Section 5.5. We start by developing a strategy to solve exponential equations.

OBJECTIVE 1 SOLVING EXPONENTIAL EQUATIONS

We have already solved exponential equations using the **method of relating the bases**. For example, we can solve the equation $4^{x+3} = \frac{1}{2}$ by converting the base on both sides of the equation to base 2. To see how to solve this equation, read these **steps**.

But suppose we are given an exponential equation in which the bases cannot be related, such as $2^{x+1} = 3$. Remember in **Section 5.1, Example 4**, we wanted to find the one x-intercept of the graph of $f(x) = -2^{x+1} + 3$ (Figure 14).

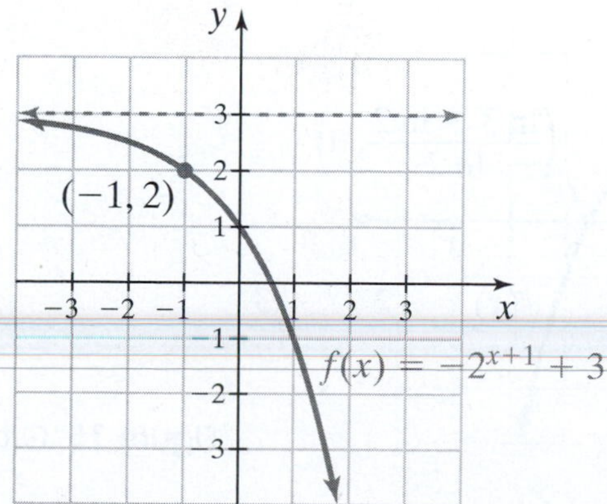

Figure 14 Graph of $f(x) = -2^{x+1} + 3$.

(eText Screens 5.4-1–5.4-19)

To find the x-intercept of $f(x) = -2^{x+1} + 3$, we need to set $f(x) = 0$ and solve for x.

$$f(x) = -2^{x+1} + 3$$
$$0 = -2^{x+1} + 3$$
$$2^{x+1} = 3$$

In Section 5.1, we could not solve this equation for x because we had not yet defined the logarithm. We can now use some properties of logarithms to solve this equation. Recall the following logarithmic properties.

If $u = v$, then $\log_b u = \log_b v$. **logarithm property of equality**

$\log_b u^r = r \log_b u$ **power rule for logarithms**

We can use these two properties to solve the equation $2^{x+1} = 3$ and thus determine the x-intercept of $f(x) = -2^{x+1} + 3$. We solve the equation $2^{x+1} = 3$ in Example 1.

Example 1 Solve an Exponential Equation

Solve $2^{x+1} = 3$.

Solution

$2^{x+1} = 3$	Write the original equation.
$\ln 2^{x+1} = \ln 3$	Use the **logarithm property of equality.**
$(x + 1)\ln 2 = \ln 3$	Use the power rule for logarithms.
$x\ln 2 + \ln 2 = \ln 3$	Use the distributive property.
$x\ln 2 = \ln 3 - \ln 2$	Subtract ln 2 from both sides.
$x = \dfrac{\ln 3 - \ln 2}{\ln 2}$	Divide both sides by ln 2. ●

The solution to Example 1 verifies that the x-intercept of $f(x) = -2^{x+1} + 3$ is $x = \dfrac{\ln 3 - \ln 2}{\ln 2} \approx 0.5850$. See Figure 15.

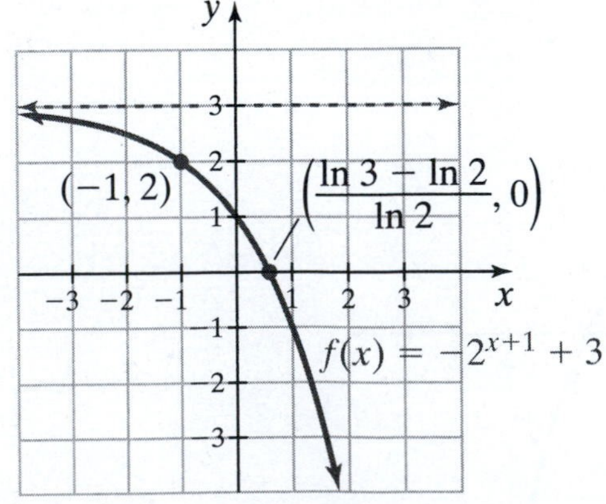

Figure 15 Graph of $f(x) = -2^{x+1} + 3$.

When we cannot easily relate the bases of an exponential equation, as in Example 1, we use logarithms and their properties to solve them. The methods used to solve exponential equations are outlined as follows.

> **Solving Exponential Equations**
>
> - If the equation can be written in the form $b^u = b^v$, then solve the equation $u = v$.
> - If the equation cannot easily be written in the form $b^u = b^v$,
>   1. Use the logarithm property of equality to "take the log of both sides" (typically using base 10 or base e).
>   2. Use the power rule of logarithms to "bring down" any exponents.
>   3. Solve for the given variable.

Example 2 Solve Exponential Equations

Solve each equation. For part b, round to four decimal places.

a. $3^{x-1} = \left(\frac{1}{27}\right)^{2x+1}$ **b.** $7^{x+3} = 4^{2-x}$

Solution

a. Watch the **interactive video**, or read these **steps** to see that the solution is $x = -\frac{2}{7}$.

b. We cannot easily use the method of relating the bases because we cannot easily write both 7 and 4 using a common base. Therefore, we use logarithms to solve.

$7^{x+3} = 4^{2-x}$	Write the original equation.
$\ln 7^{x+3} = \ln 4^{2-x}$	Use the **logarithm property of equality**.
$(x+3)\ln 7 = (2-x)\ln 4$	Use the power rule for logarithms.
$x\ln 7 + 3\ln 7 = 2\ln 4 - x\ln 4$	Use the distributive property.
$x\ln 7 + x\ln 4 = 2\ln 4 - 3\ln 7$	Add $x\ln 4$ to both sides, and subtract $3\ln 7$ from both sides.
$x(\ln 7 + \ln 4) = 2\ln 4 - 3\ln 7$	Factor out an x from the left-hand side.
$x = \frac{2\ln 4 - 3\ln 7}{\ln 7 + \ln 4}$	Divide both sides by $\ln 7 + \ln 4$.
$= \frac{\ln 16 - \ln 343}{\ln 28}$	Use the power rule for logarithms in the numerator, and use the product rule for logarithms in the denominator.
$= \frac{\ln\left(\frac{16}{343}\right)}{\ln 28}$	Use the quotient rule for logarithms to rewrite $\ln 16 - \ln 343$ as $\ln\left(\frac{16}{343}\right)$.
≈ -0.9199	Use a calculator to round to four decimal places.

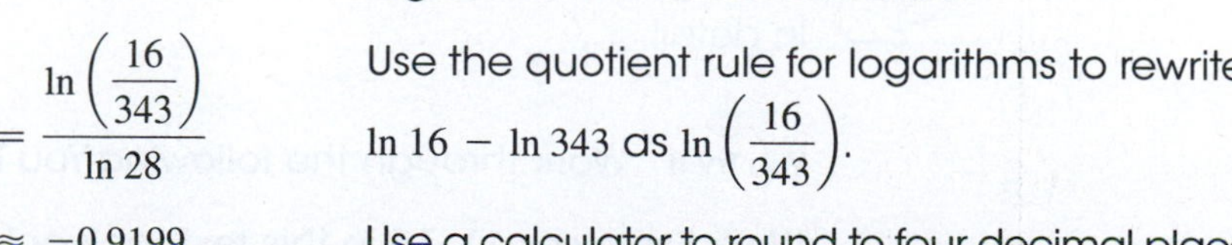

Watch this **interactive video** to see the entire solution to this example. ●

(eText Screens 5.4-1–5.4-19)

You Try It **Work through the following You Try It problem.**

Work Exercises 1–10 in this textbook or in the MyLab Math Study Plan.

Example 3 Solve Exponential Equations Involving the Natural Exponential Function

Solve each equation. Round to four decimal places.

a. $25e^{x-5} = 17$ b. $e^{2x-1} \cdot e^{x+4} = 11$

Solution

a. First Isolate the exponential term on the left by dividing both sides of the equation by 25.

$25e^{x-5} = 17$ Write the original equation.

$e^{x-5} = \frac{17}{25}$ Divide both sides by 25.

Now use the natural logarithm and the logarithm property of equality to solve for x.

$\ln e^{x-5} = \ln \frac{17}{25}$ Use the **logarithm property of equality**.

$x - 5 = \ln \frac{17}{25}$ Use **cancellation property (2)** to rewrite $\ln e^{x-5}$ as $x - 5$.

$x = \ln \frac{17}{25} + 5$ Add 5 to both sides.

≈ 4.6143 Use a calculator to round to four decimal places.

b. $e^{2x-1} \cdot e^{x+4} = 11$ Write the original equation.

$e^{(2x-1)+(x+4)} = 11$ Use $b^m \cdot b^n = b^{m+n}$.

$e^{3x+3} = 11$ Combine like terms in the exponent.

$\ln e^{3x+3} = \ln 11$ Use the **logarithm property of equality**.

$3x + 3 = \ln 11$ Use **cancellation property (2)** to rewrite $\ln e^{3x+3}$ as $3x + 3$.

$3x = \ln 11 - 3$ Subtract 3 from both sides.

$x = \frac{\ln 11 - 3}{3}$ Divide both sides by 3.

≈ -0.2007 Use a calculator to round to four decimal places.

Watch the **interactive video** to see the solutions to this example worked out in detail.

You Try It **Work through the following You Try It problem.**

Work Exercises 11–15 in this textbook or in the MyLab Math Study Plan.

OBJECTIVE 2 SOLVING LOGARITHMIC EQUATIONS

We now turn our attention to solving logarithmic equations. In **Section 5.3**, we learned how to solve certain logarithmic equations by using the **logarithm property of equality**. That is, if we can write a logarithmic equation in the form $\log_b u = \log_b v$, then $u = v$. Before we look at an example, let's review three of the properties of logarithms.

> **Properties of Logarithms**
>
> If $b > 0, b \neq 1, u$ and v represent positive numbers and r is any real number, then
>
> $\log_b uv = \log_b u + \log_b v$ — product rule for logarithms
>
> $\log_b \dfrac{u}{v} = \log_b u - \log_b v$ — quotient rule for logarithms
>
> $\log_b u^r = r \log_b u$ — power rule for logarithms

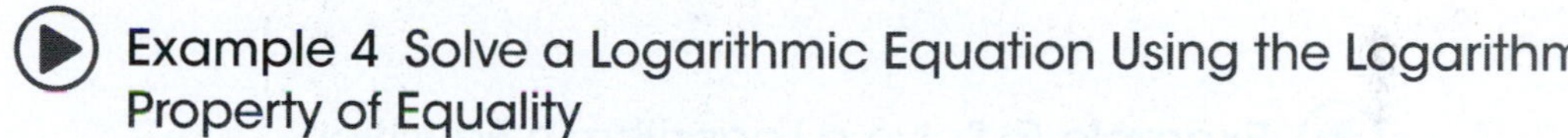

Example 4 Solve a Logarithmic Equation Using the Logarithm Property of Equality

Solve $2 \log_5 (x - 1) = \log_5 64$.

Solution We can use the power rule for logarithms and the logarithmic property of equality to solve.

$2 \log_5 (x - 1) = \log_5 64$	Write the original equation.
$\log_5 (x - 1)^2 = \log_5 64$	Use the power rule.
$(x - 1)^2 = 64$	Use the logarithm property of equality.
$x - 1 = \pm 8$	Use the square root property.
$x = 1 \pm 8$	Solve for x.
$x = 9$ or $x = -7$	Simplify.

Recall that the domain of a logarithmic function consists of all values of x for which the **argument** of the logarithm is greater than zero; thus, $x - 1$ must be positive. Therefore, the solution of $x = -7$ must be discarded. The only solution is $x = 9$. You may want to review how to determine the domain of a logarithmic function, which is discussed in **Section 5.2**.

CAUTION When solving logarithmic equations, it is important to always verify the solutions. Logarithmic equations often lead to extraneous solutions, as in Example 4.

When a logarithmic equation cannot be written in the form $\log_b u = \log_b v$, as in Example 4, we adhere to the steps outlined as follows:

Solving Logarithmic Equations

1. Determine the domain of the variable.
2. Use properties of logarithms to combine all logarithms, and write as a single logarithm, if needed.
3. Eliminate the logarithm by rewriting the equation in exponential form. To review how to change from logarithmic form to exponential form, view Example 2 from Section 5.2.
4. Solve for the given variable.
5. Check for any extraneous solutions. Verify that each solution is in the domain of the variable.

You Try It **Work through the following You Try It problem.**

Work Exercises 16–19 in this textbook or in the MyLab Math Study Plan.

Example 5 Solve a Logarithmic Equation

Solve $\log_4(2x - 1) = 2$.

Solution The domain of the variable in this equation is the solution to the inequality $2x - 1 > 0$ or $x > \frac{1}{2}$. Thus, our solution must be greater than $\frac{1}{2}$. Because the equation involves a single logarithm, we can proceed to the third step.

$\log_4(2x - 1) = 2$	Write the original equation.
$4^2 = 2x - 1$	Rewrite in exponential form.
$16 = 2x - 1$	Simplify.
$17 = 2x$	Add 1 to both sides.
$x = \frac{17}{2}$	Divide by 2.

Because the solution satisfies the inequality, there are no extraneous solutions. We can verify the solution by substituting $x = \frac{17}{2}$ into the original equation.

Check:

$\log_4(2x - 1) = 2$	Write the original equation.
$\log_4\left(2\left(\frac{17}{2}\right) - 1\right) \stackrel{?}{=} 2$	Substitute $x = \frac{17}{2}$.
$\log_4(17 - 1) \stackrel{?}{=} 2$	Simplify.
$\log_4(16) = 2$	This is a true statement because $4^2 = 16$. ●

You Try It Work through the following You Try It problem.

Work Exercises 20–24 in this textbook or in the MyLab Math Study Plan.

Example 6 Solve a Logarithmic Equation

Solve $\log_2(x + 10) + \log_2(x + 6) = 5$.

Solution The domain of the variable in this equation is the solution to the compound inequality $x + 10 > 0$ and $x + 6 > 0$. The solution to this compound inequality is $x > -6$. (You may want to review compound inequalities from **Section 1.7**.)

$\log_2(x + 10) + \log_2(x + 6) = 5$	Write the original equation.
$\log_2(x + 10)(x + 6) = 5$	Use the product rule.
$(x + 10)(x + 6) = 2^5$	Rewrite in exponential form.
$x^2 + 16x + 60 = 32$	Simplify.
$x^2 + 16x + 28 = 0$	Subtract 32 from both sides.
$(x + 14)(x + 2) = 0$	Factor.
$x = -14$ or $x = -2$	Use the **zero product property** to solve.

Because the domain of the variable is $x > -6$, we must *exclude* the solution $x = -14$. Therefore, the only solution to this logarithmic equation is $x = -2$. Work through the **interactive video** to see this solution worked out in detail. ●

You Try It Work through the following You Try It problem.

Work Exercises 25–33 in this textbook or in the MyLab Math Study Plan.

Example 7 Solve a Logarithmic Equation

Solve $\ln(x - 4) - \ln(x - 5) = 2$. Round to four decimal places.

Solution The domain of the variable is the solution to the compound inequality $x - 4 > 0$ and $x - 5 > 0$. The solution to this compound inequality is $x > 5$. (You may want to review compound inequalities from **Section 1.7**.)

$\ln(x - 4) - \ln(x - 5) = 2$	Write the original equation.
$\ln\left(\frac{x - 4}{x - 5}\right) = 2$	Use the quotient rule.
$e^2 = \frac{x - 4}{x - 5}$	Rewrite in exponential form.
$e^2(x - 5) = x - 4$	Multiply both sides by $x - 5$.
$e^2x - 5e^2 = x - 4$	Use the distributive property.
$e^2x - x = 5e^2 - 4$	Add $5e^2$ to both sides and subract x from both sides.

$x(e^2 - 1) = 5e^2 - 4$ Factor out an x from the left-hand side.

$x = \dfrac{5e^2 - 4}{e^2 - 1} \approx 5.1565$ Solve for x. Use a calculator to round to four decimal places.

We approximate the exact answer $x = \dfrac{5e^2 - 4}{e^2 - 1}$ in order to verify that the solution is in the domain of the variable. In this example, we see that 5.1565 is clearly greater than 5. In some cases, we may need to use the exact answer to verify a solution to a logarithmic equation. To see this verification for Example 7, read these steps.

You Try It Work through the following You Try It problem.

Work Exercises 34 and 35 in this textbook or in the MyLab Math Study Plan.

5.4 Exercises

Skill Check Exercises

For exercises SCE-1 through SCE-8, evaluate the expression using a calculator. Round your answer to 4 decimal places.

SCE-1. $\ln\left(\dfrac{7}{2}\right)$ **SCE-2.** $\dfrac{5e^2 - 4}{e^2 - 1}$ **SCE-3.** $\dfrac{\ln 3 + \ln 5}{\ln 5}$ **SCE-4.** $\dfrac{\ln \pi - 3\ln 4}{\ln \pi - 2\ln 4}$

SCE-5. $\dfrac{\ln\left(\dfrac{17}{3}\right)}{0.00235}$ **SCE-6.** $\dfrac{\ln\left(\dfrac{77}{131}\right)}{\ln\left(\dfrac{120}{131}\right)}$ **SCE-7.** $\dfrac{\ln 2}{12\ln\left(1 + \dfrac{0.06}{12}\right)}$ **SCE-8.** $\dfrac{-\ln 65}{\left(\dfrac{\ln\left(\dfrac{37}{117}\right)}{20}\right)}$

In Exercises 1–15, solve each exponential equation. For **irrational solutions**, round to four decimal places.

1. $3^x = 5$
2. $2^{x/3} = 19$
3. $4^{x^2-2x} = 64$
4. $3^{x+7} = -20$
5. $(1.52)^{-3x/7} = 11$
6. $\left(\dfrac{1}{5}\right)^{x-1} = 25^x$
7. $8^{4x-7} = 11^{5+x}$
8. $3(9)^{x-1} = (81)^{2x+1}$
9. $(3.14)^x = \pi^{1-2x}$
10. $7(2 - 10^{4x-2}) = 8$
11. $e^x = 2$
12. $150e^{x-4} = 5$
13. $e^{x-3} \cdot e^{3x+7} = 24$
14. $2(e^{x-1})^2 \cdot e^{3-x} = 80$
15. $8e^{-x/3} \cdot e^x = 1$

In Exercises 16–33, solve each logarithmic equation.

16. $\log_4(x+1) = \log_4(6x-5)$

17. $\log_3(x^2-21) = \log_3 4x$

18. $2\log_5(3-x) - \log_5 2 = \log_5 18$

19. $2\ln x - \ln(2x-3) = \ln 2x - \ln(x-1)$

20. $\log_2(4x-7) = 3$

21. $\log(1-5x) = 2$

22. $\log_3(2x-5) = -2$

23. $\log_x 3 = -1$

24. $\log_{x/2} 16 = 2$

25. $\log_2(x-2) + \log_2(x+2) = 5$

26. $\log_7(x+9) + \log_7(x+15) = 1$

27. $\log_3(3x+1) - \log_3(x-2) = 2$

28. $\log_6(x-8) = 2 - \log_6(x+8)$

29. $\log_4(x+21) - 2 = -\log_4(x+6)$

30. $2 - \log_5(5x+3) + \log_5(x-1) = 0$

31. $\log_4(x-7) + \log_4 x = \frac{3}{2}$

32. $\ln 3 + \ln\left(x^2 + \frac{2x}{3}\right) = 0$

33. $\log_2(x-4) + \log_2(x+6) = 2 + \log_2 x$

In Exercises 34 and 35, solve each logarithmic equation. Round to four decimal places.

34. $\ln x - \ln(x+6) = 1$

35. $\ln(x+3) - \ln(x-2) = 4$

5.5 Applications of Exponential and Logarithmic Functions

THINGS TO KNOW

Before working through this section, be sure that you are familiar with the following concepts:

		VIDEO	ANIMATION	INTERACTIVE
 You Try It	1. Solving Applications of Exponential Functions (Compound Interest) (Section 5.1) 			
 You Try It	2. Solving Applications of Exponential Functions (Exponential Growth) (Section 5.1)			
 You Try It	3. Solving Exponential Equations (Section 5.4)			

INTRODUCTION

Read this introduction before beginning Objective 1.

OBJECTIVES

Introduction to Section 5.5

We have seen that exponential functions appear in a wide variety of settings, including biology, chemistry, physics, and business. In this section, we revisit some applications that are discussed previously in this chapter and then introduce several new applications. The difference between the applications presented earlier and the applications presented in this section is that we are now equipped with the tools necessary to solve for variables that appear as exponents. We start with applications involving **compound interest**. You may want to review **periodic compound interest** and **continuous compound interest** from Section 5.1 before proceeding.

OBJECTIVE 1 SOLVING COMPOUND INTEREST APPLICATIONS

In Section 5.1, the formulas for compound interest and continuous compound interest are defined as follows.

Compound Interest Formulas

Periodic Compound Interest Formula

$$A = P\left(1 + \frac{r}{n}\right)^{nt}$$

Continuous Compound Interest Formula

$$A = Pe^{rt},$$

where

A = Total amount after t years
P = Principal (original investment)
r = Interest rate per year
n = Number of times interest is compounded per year
t = Number of years

Example 1 Find the Doubling Time

How long will it take (in years and months) for an investment to double if it earns 7.5% compounded monthly?

Solution We use the periodic compound interest formula $A = P\left(1 + \frac{r}{n}\right)^{nt}$ with $r = 0.075$ and $n = 12$ and solve for t. Notice that the principal is not given. As it turns out, any value of P will suffice. If the principal is P, then the amount needed to double the investment is $A = 2P$. We now have all of the information necessary to solve for t:

$$A = P\left(1 + \frac{r}{n}\right)^{nt}$$ Use the periodic compound interest formula.

$$2P = P\left(1 + \frac{0.075}{12}\right)^{12t}$$ Substitute the appropriate values.

$$2 = \left(1 + \frac{0.075}{12}\right)^{12t}$$ Divide both sides by P.

$$2 = (1.00625)^{12t}$$ Simplify within the parentheses but **do not round**.

$$\ln 2 = \ln (1.00625)^{12t}$$ Use the logarithm property of equality.

$$\ln 2 = 12t \ln (1.00625)$$ Use the power rule: $\log_b u^r = r \log_b u$.

$$t = \frac{\ln 2}{12 \ln (1.00625)}$$ Divide both sides by $12 \ln (1.00625)$.

$$t \approx 9.27 \text{ years}$$ Round to two decimal places.

Note that $0.27 \text{ years} = 0.27 \text{ years} \times \frac{12 \text{ months}}{1 \text{ year}} = 3.24 \text{ months}$. Because the interest is compounded at the end of each month, the investment will not double until 9 years and 4 months.

Example 2 Continuous Compound Interest

Suppose an investment of $\$5{,}000$ compounded continuously grew to an amount of $\$5{,}130.50$ in 6 months. Find the interest rate, and then determine how long it will take for the investment to grow to $\$6{,}000$. Round the interest rate to the nearest hundredth of a percent and the time to the nearest hundredth of a year.

Solution Because the investment is compounded continuously, we use the formula $A = Pe^{rt}$.

We are given that $P = 5{,}000$, so $A = 5{,}000e^{rt}$. In 6 months, or when $t = 0.5$ years, we know that $A = 5{,}130.50$. Substituting these values into the compound interest formula will enable us to solve for r:

$$5{,}130.50 = 5{,}000e^{r(0.5)}$$ Substitute the appropriate values.

$$\frac{5{,}130.50}{5{,}000} = e^{0.5r}$$ Divide by 5,000.

$$\ln\left(\frac{5{,}130.50}{5{,}000}\right) = \ln e^{0.5r}$$ Use the **logarithm property of equality**.

$$\ln\left(\frac{5{,}130.50}{5{,}000}\right) = 0.5r$$ Use **cancellation property (2)** to rewrite $\ln e^{0.5r}$ as $0.5r$.

$$r = \frac{\ln\left(\frac{5{,}130.50}{5{,}000}\right)}{0.5} \approx .051530$$ Divide by 0.5.

Therefore, the interest rate is 5.15%. To find the time that it takes for the investment to grow to $\$6{,}000$, we use the formula $A = Pe^{rt}$, with $A = 6{,}000$, $P = 5{,}000$, and $r = 0.0515$, and solve for t. Watch the **video** to verify that it will take approximately 3.54 years.

You Try It **Work through the following You Try It problem.**

Work Exercises 1–5 in this textbook or in the MyLab Math Study Plan.

OBJECTIVE 2 EXPONENTIAL GROWTH AND DECAY

In **Section 5.1**, the exponential growth model is introduced. This model is used when a population grows at a rate proportional to the size of its current population. This model is often called the **uninhibited growth** model. We review this exponential growth model and sketch the graph in Figure 16.

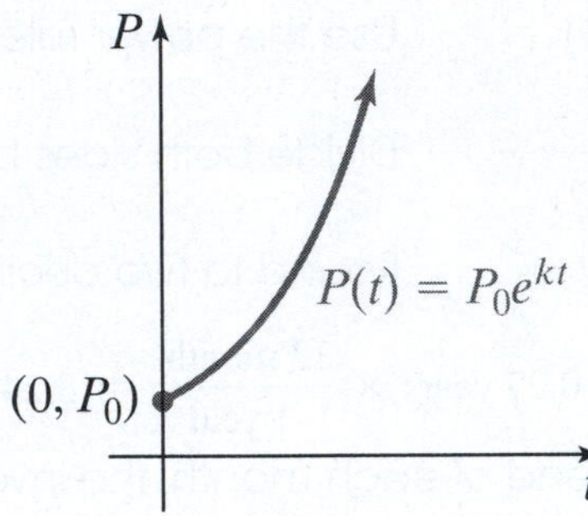

Figure 16 Graph of $P(t) = P_0e^{kt}$ for $k > 0$.

> **Exponential Growth**
>
> A model that describes the exponential uninhibited growth of a population, P, after a certain time, t, is
>
> $$P(t) = P_0e^{kt},$$
>
> where $P_0 = P(0)$ is the initial population and $k > 0$ is a constant called the **relative growth rate**. (*Note*: k is sometimes given as a percent.)

Example 3 Population Growth

The population of a small town grows at a rate proportional to its current size. In 1900, the population was 900. In 1920, the population had grown to $1{,}600$. What was the population of this town in 1950? Round to the nearest whole number.

Solution Using the model $P(t) = P_0e^{kt}$, we must first determine the constants P_0 and k. The initial population was 900 in 1900 so $P_0 = 900$. Therefore, $P(t) = 900e^{kt}$. To find k, we use the fact that in 1920, or when $t = 20$, the population was $1{,}600$; thus,

$$P(20) = 900e^{k(20)} = 1{,}600 \qquad \text{Substitute } P(20) = 1{,}600.$$

$$900e^{20k} = 1{,}600$$

$$e^{20k} = \frac{16}{9} \qquad \text{Divide by 900 and simplify.}$$

$$\ln e^{20k} = \ln \frac{16}{9}$$ Use the **logarithm property of equality**.

$$20k = \ln \frac{16}{9}$$ Use **cancellation property (2)** to rewrite $\ln e^{20k}$ as $20k$.

$$k = \frac{\ln\left(\frac{16}{9}\right)}{20}$$ Divide by 20.

The function that models the population of this town at any time t is given by $P(t) = 900e^{\frac{\ln(16/9)}{20}t}$. To determine the population in 1950, or when $t = 50$, we evaluate $P(50)$:

$$P(50) = 900e^{\frac{\ln(16/9)}{20}(50)} \approx 3{,}793$$

Some populations exhibit *negative exponential growth*. In other words, the population, quantity, or amount *decreases* over time. Such models are called **exponential decay** models. The only difference between an exponential growth model and an exponential decay model is that the constant, k, is less than zero. See Figure 17.

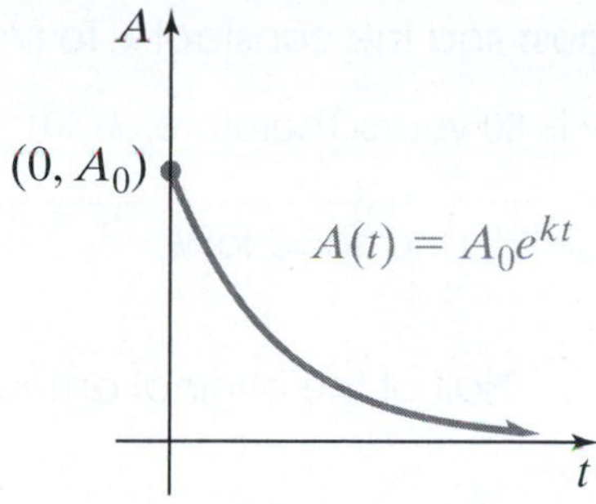

Figure 17 Graph of $A(t) = A_0e^{kt}$ for $k < 0$.

> **Exponential Decay**
>
> A model that describes the exponential decay of a population, quantity, or amount A, after a certain time, t, is
>
> $$A(t) = A_0e^{kt},$$
>
> where $A_0 = A(0)$ is the initial quantity and $k < 0$ is a constant called the **relative decay constant**. (*Note*: k is sometimes given as a percent.)

HALF-LIFE

Every radioactive element has a half-life, which is the required time for a given quantity of that element to decay to half of its original mass. For example, the half-life of cesium-137 is 30 years. Thus, it takes 30 years for any amount of cesium-137 to decay to $\frac{1}{2}$ of its original mass. It takes an additional 30 years to decay to $\frac{1}{4}$ of its original mass and so on. See Figure 18, and view the **animation** that illustrates the half-life of cesium-137.

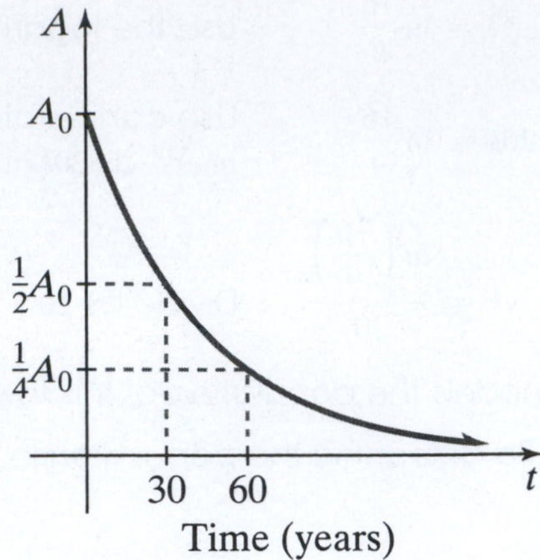

Figure 18 Half-life of cesium-137

Example 4 Radioactive Decay

Suppose that a meteorite is found containing 4% of its original krypton-99. If the half-life of krypton-99 is 80 years, how old is the meteorite? Round to the nearest year.

Solution We use the formula $A(t) = A_0e^{kt}$, where A_0 is the original amount of krypton-99. We first must find the constant k. To find k, we use the fact that the half-life of krypton-99 is 80 years. Therefore, $A(80) = \frac{1}{2}A_0$. Because $A(80) = A_0e^{k(80)}$, we can set $\frac{1}{2}A_0 = A_0e^{k(80)}$ and solve for k.

$\frac{1}{2}A_0 = A_0e^{k(80)}$ Half of the original amount will be present in 80 years.

$\frac{1}{2} = e^{80k}$ Divide both sides by A_0.

$\ln\frac{1}{2} = \ln e^{80k}$ Use the **logarithm property of equality**.

$\ln\frac{1}{2} = 80k$ Use **cancellation property (2)** to rewrite $\ln e^{80k}$ as $80k$.

$\frac{\ln\frac{1}{2}}{80} = k$ Divide both sides by 80.

$\frac{-\ln 2}{80} = k$ $\ln\frac{1}{2} = \ln 1 - \ln 2 = 0 - \ln 2 = -\ln 2$

Now that we know $k = \frac{-\ln 2}{80}$, our function becomes $A(t) = A_0e^{\frac{-\ln 2}{80}t}$. To find out the age of the meteorite, we set $A(t) = .04A_0$ because the meteorite now contains 4% of the original amount of krypton-99.

$.04A_0 = A_0e^{\frac{-\ln 2}{80}t}$ Substitute $.04A_0$ for $A(t)$.

$.04 = e^{\frac{-\ln 2}{80}t}$ Divide both sides by A_0.

$\ln .04 = \ln e^{\frac{-\ln 2}{80}t}$ Use the **logarithm property of equality**.

$\ln .04 = \frac{-\ln 2}{80}t$ Use **cancellation property (2)** to rewrite $\ln e^{\frac{-\ln 2}{80}t}$ as $\frac{-\ln 2}{80}t$.

$\frac{\ln .04}{\left(\frac{-\ln 2}{80}\right)} = t \approx 372 \text{ years}$ Divide both sides by $\frac{-\ln 2}{80}$.

The meteorite is about 372 years old.

Watch this **video** to see the solution to this example worked out in detail.

You Try It **Work through the following You Try It problem.**

Work Exercises 6–12 in this textbook or in the MyLab Math Study Plan.

OBJECTIVE 3 SOLVING LOGISTIC GROWTH APPLICATIONS

The uninhibited exponential growth model $P(t) = P_0e^{kt}$ for $k > 0$ is used when there are no outside limiting factors such as predators or disease that affect the population growth. When such outside factors exist, scientists often use a **logistic model** to describe the population growth. One such logistic model is described and sketched in Figure 19.

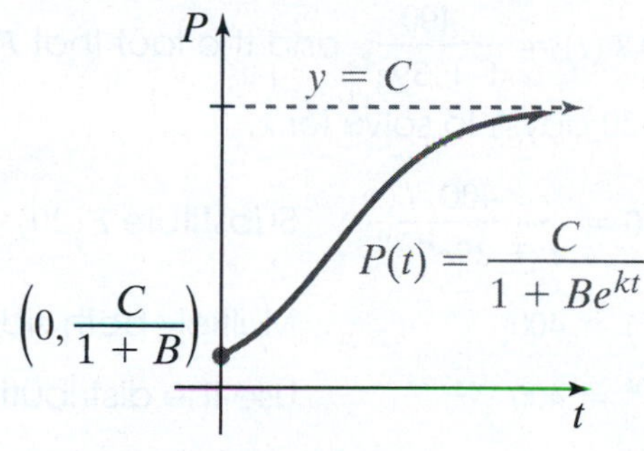

Figure 19 Graph of $P(t) = \frac{C}{1 + Be^{kt}}$.

> **Logistic Growth**
>
> A model that describes the logistic growth of a population P at any time t is given by the function
>
> $$P(t) = \frac{C}{1 + Be^{kt}},$$
>
> where B, C, and k are constants with $C > 0$ and $k < 0$.

The number C is called the **carrying capacity**. In the logistic model, the population will approach the value of the carrying capacity over time but never exceed it. You can see in the graph sketched in Figure 19 that the graph of the logistic growth model approaches the **horizontal asymptote** $y = C$.

Example 5 Logistic Growth

Ten goldfish were introduced into a small pond. Because of limited food, space, and oxygen, the carrying capacity of the pond is 400 goldfish. The goldfish population at any time t, in days, is modeled by the logistic growth function $F(t) = \dfrac{C}{1 + Be^{kt}}$.

If 30 goldfish are in the pond after 20 days,

a. Find B.

b. Find k.

c. When will the pond contain 250 goldfish? Round to the nearest whole number.

Solution

a. The carrying capacity is 400; thus, $C = 400$. Also, initially (at $t = 0$) there were 10 goldfish, so $F(0) = 10$. Therefore,

$$10 = \frac{400}{1 + Be^{k(0)}} \quad \text{Substitute } C = 400, t = 0, \text{ and } f(0) = 10.$$

$$10 = \frac{400}{1 + B} \quad \text{Evaluate } e^0 = 1.$$

$$10 + 10B = 400 \quad \text{Multiply both sides by } 1 + B.$$

$$B = 39 \quad \text{Solve for } B.$$

b. Use the function $F(t) = \dfrac{400}{1 + 39e^{kt}}$ and the fact that $F(20) = 30$ (there are 30 goldfish after 20 days) to solve for k.

$$30 = \frac{400}{1 + 39e^{k(20)}} \quad \text{Substitute } F(20) = 30.$$

$$30(1 + 39e^{20k}) = 400 \quad \text{Multiply both sides by } 1 + 39e^{20k}.$$

$$30 + 1{,}170e^{20k} = 400 \quad \text{Use the distributive property.}$$

$$1{,}170e^{20k} = 370 \quad \text{Subtract 30 from both sides.}$$

$$e^{20k} = \frac{370}{1{,}170} \quad \text{Divide both sides by 1,170.}$$

$$e^{20k} = \frac{37}{117} \quad \text{Simplify.}$$

$$\ln e^{20k} = \ln \frac{37}{117} \quad \text{Use the logarithm property of equality.}$$

$$20k = \ln \frac{37}{117} \quad \text{Use cancellation property (2) to rewrite } \ln e^{20k} \text{ as } 20k.$$

$$k = \frac{\ln \frac{37}{117}}{20} \quad \text{Divide both sides by 20.}$$

c. Use the function $F(t) = \dfrac{400}{1 + 39e^{\frac{\ln \frac{37}{117}}{20}t}}$, and then find t when $F(t) = 250$.

By repeating the exact same process as in part b, we find that it will take approximately 73 days until there are 250 goldfish in the pond.

Watch the **interactive video** to verify the solution.

You Try It Work through the following You Try It problem.

Work Exercises 13–15 in this textbook or in the MyLab Math Study Plan.

OBJECTIVE 4 USING NEWTON'S LAW OF COOLING

Newton's law of cooling states that the temperature of an object changes at a rate proportional to the difference between its temperature and that of its surroundings. It can be shown in a more advanced course that the function describing Newton's law of cooling is given by the following.

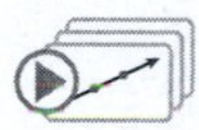

Newton's Law of Cooling

The temperature T of an object at any time t is given by

$$T(t) = S + (T_0 - S)e^{kt},$$

where T_0 is the original temperature of the object, S is the constant temperature of the surroundings, and k is the cooling constant.

View the **animation** to see how this function behaves.

Example 6 Newton's Law of Cooling

Suppose that the temperature of a cup of hot tea obeys Newton's law of cooling. If the tea has a temperature of 200°F when it is initially poured and 1 minute later has cooled to 189°F in a room that maintains a constant temperature of 69°F, determine when the tea reaches a temperature of 146°F. Round to the nearest minute.

Solution We start using the formula for Newton's law of cooling with $T_0 = 200$ and $S = 69$.

$T(t) = S + (T_0 - S)e^{kt}$	Use Newton's law of cooling formula.
$T(t) = 69 + (200 - 69)e^{kt}$	Substitute $T_0 = 200$ and $S = 69$.
$T(t) = 69 + 131e^{kt}$	Simplify.

We now proceed to find k. The object cools to 189°F in 1 minute; thus, $T(1) = 189$. Therefore,

$189 = 69 + 131e^{k(1)}$	Substitute $t = 1$ and $T(1) = 189$.
$120 = 131e^{k}$	Subtract 69 from both sides.
$\dfrac{120}{131} = e^{k}$	Divide both sides by 131.

$\ln \frac{120}{131} = \ln e^k$ Use the **logarithm property of equality**.

$\ln \frac{120}{131} = k$ Use **cancellation property (2)** to rewrite $\ln e^k$ as k.

Now that we know the cooling constant $k = \ln \frac{120}{131}$, we can use the function $T(t) = 69 + 131e^{\ln \frac{120}{131} t}$ and determine the value of t when $T(t) = 146$.

$146 = 69 + 131e^{\ln \frac{120}{131} t}$ Set $T(t) = 146$.

$77 = 131e^{\ln \frac{120}{131} t}$ Subtract 69 from both sides.

$\frac{77}{131} = e^{\ln \frac{120}{131} t}$ Divide both sides by 131.

$\ln \frac{77}{131} = \ln e^{\ln \frac{120}{131} t}$ Use the **logarithm property of equality**.

$\ln \frac{77}{131} = \ln \frac{120}{131} t$ Use **cancellation property (2)** to rewrite $\ln e^{\ln \frac{120}{131} t}$ as $\ln \frac{120}{131} t$.

$\frac{\ln \frac{77}{131}}{\ln \frac{120}{131}} = t$ Divide both sides by $\ln \frac{120}{131}$.

$t \approx 6$ minutes Use a calculator to approximate the time rounded to the nearest minute.

So, it takes approximately 6 minutes for the tea to cool to 146°F.

Using Technology

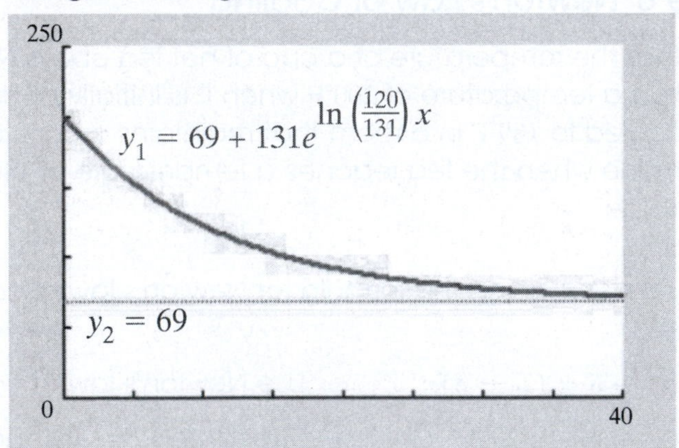

The graph of $T(t) = 69 + 131e^{\ln \frac{120}{131} t}$, which describes the temperature of the tea t minutes after being poured, was created using a graphing utility. Note that the line $y = 69$, which represents the temperature of the surroundings, is a **horizontal asymptote**.

You Try It Work through the following You Try It problem.

Work Exercises 16–18 in this textbook or in the MyLab Math Study Plan.

5.5 Exercises

1. Jimmy invests $15,000 in an account that pays 6.25% compounded quarterly. How long (in years and months) will it take for his investment to reach $20,000?

2. How long (in years and months) will it take for an investment to double at 9% compounded monthly?

3. How long will it take for an investment to triple if it is compounded continuously at 8%? Round to 2 decimal places.

4. What is the interest rate necessary for an investment to quadruple after 8 years of continuous compound interest? Round to the nearest hundredth of a percent.

5. Marsha and Jan both invested money on March 1, 2005. Marsha invested $5,000 at Bank A, where the interest was compounded quarterly. Jan invested $3,000 at Bank B, where the interest was compounded continuously. On March 1, 2007, Marsha had a balance of $5,468.12, whereas Jan had a balance of $3,289.09. What was the interest rate at each bank? Round to the nearest tenth of a percent.

6. The population of Adamsville grew from 9,000 to 15,000 in 6 years. Assuming uninhibited exponential growth, what is the expected population in an additional 4 years? Round to the nearest whole number.

7. During a research experiment, it was found that the number of bacteria in a culture grew at a rate proportional to its size. At 8:00 AM, there were 2,000 bacteria present in the culture. At noon, the number of bacteria grew to 2,400. How many bacteria will there be at midnight? Round to the nearest whole number.

8. A skull cleaning factory cleans animal skulls such as deer, buffalo, and other types of animal skulls using flesh-eating beetles to clean the skulls. The factory owner started with only 10 adult beetles. After 40 days, the beetle population grew to 30 adult beetles. How long did it take before the beetle population reached 10,000 beetles? Round to the nearest whole number.

9. The population of a Midwest industrial town decreased from 210,000 to 205,000 in just 3 years. Assuming that this trend continues, what will the population be after an additional 3 years? Round to the nearest whole number.

10. A certain radioactive isotope is leaked into a small stream. Three hundred days after the leak, 2% of the original amount of the substance remained. Determine the half-life of this radioactive isotope. Round to the nearest whole number.

11. Radioactive iodine-131 is a by-product of certain nuclear reactors. On April 26, 1986, one of the nuclear reactors in Chernobyl, Ukraine, a republic of the former Soviet Union, experienced a massive release of radioactive iodine. Fortunately, iodine-131 has a very short half-life of 8 days. Estimate the percentage of the original amount of iodine-131 released by the Chernobyl explosion, 5 days after the explosion. Round to 2 decimal places.

12. Superman is rendered powerless when exposed to 50 or more grams of kryptonite. A 500-year-old rock that originally contained 300 grams of kryptonite was recently stolen from a rock museum by Superman's enemies. The half-life of kryptonite is known to be 200 years.

    a. How many grams of kryptonite are still contained in the stolen rock? Round to two decimal places.

    b. For how many years can this rock be used by Superman's enemies to render him powerless? Round to the nearest whole number.

SbS 13. The logistic growth model $H(t) = \frac{6{,}000}{1 + 2e^{-.65t}}$ represents the number of families that own a home in a certain small (but growing) Idaho city t years after 1980.

a. What is the maximum number of families that will own a home in this city?

b. How many families owned a home in 1980?

c. In what year did 5,920 families own a home?

SbS 14. The number of students that hear a rumor on a small college campus t days after the rumor starts is modeled by the logistic function $R(t) = \frac{3{,}000}{1 + Be^{kt}}$. Determine the following if 8 students initially heard the rumor and 100 students heard the rumor after 1 day.

a. What is the carrying capacity for the number of students who will hear the rumor?

b. Find B.

c. Find k.

d. How long will it take 2,900 students to hear the rumor?

SbS 15. In 1999, 1,500 runners entered the inaugural Run-for-Your-Life marathon in Joppetown, USA. In 2005, 21,500 runners entered the race. Because of the limited number of hotels, restaurants, and portable toilets in the area, the carrying capacity for the number of racers is 61,500. The number of racers at any time, t, in years, can be modeled by the logistic function $P(t) = \frac{C}{1 + Be^{kt}}$.

a. What is the value of C?

b. Find B.

c. Find k.

d. In what year should at least 49,500 runners be expected to run in the race? Round to the nearest year.

16. Estabon poured himself a hot beverage that had a temperature of 198°F and then set it on the kitchen table to cool. The temperature of the kitchen was a constant 75°F. If the drink cooled to 180°F in 2 minutes, how long will it take for the drink to cool to 100°F?

17. Police arrive at a murder scene at 1:00 AM and immediately record the body's temperature, which was 92°F. At 2:30 AM, after thoroughly inspecting and fingerprinting the area, they again took the temperature of the body, which had dropped to 85°F. The temperature of the crime scene has remained at a constant 60°F. Determine when the person was murdered. (Assume that the victim was healthy at the time of death. That is, assume that the temperature of the body at the time of death was 98.6°F.)

18. Jodi poured herself a cold soda that had an initial temperature of 40°F and immediately went outside to sunbathe where the temperature was a steady 99°F. After 5 minutes, the temperature of the soda was 47°F. Jodi had to run back into the house to answer the phone. What is the expected temperature of the soda after an additional 10 minutes?

Brief Exercises

19. The logistic growth model $H(t) = \frac{6{,}000}{1 + 2e^{-.65t}}$ represents the number of families that own a home in a certain small (but growing) Idaho city t years after 1980. In what year did 5,920 families own a home?

20. The number of students that hear a rumor on a small college campus t days after the rumor starts is modeled by the logistic function $R(t) = \dfrac{3{,}000}{1 + Be^{kt}}$. If 8 students initially heard the rumor and 100 students heard the after 1 day, then how long will it take 2,900 students to hear the rumor?

21. In 1999, 1,500 runners entered the inaugural Run-for-Your-Life marathon in Joppetown, USA. In 2005, 21,500 runners entered the race. Because of the limited number of hotels, restaurants, and portable toilets in the area, the carrying capacity for the number of racers is 61,500. The number of racers at any time, t, in years, can be modeled by the logistic function $P(t) = \dfrac{C}{1 + Be^{kt}}$. In what year should at least 49,500 runners be expected to run in the race? Round to the nearest year.

Chapter 5 Summary

Key Concepts

5.1 Exponential Functions

An exponential **function** is a function of the form $f(x) = b^x$, where x is any real number and $b > 0$ such that $b \neq 1$. The constant, b, is called the base of the exponential function.

Below are three exponential functions having base 2, 3, and e.

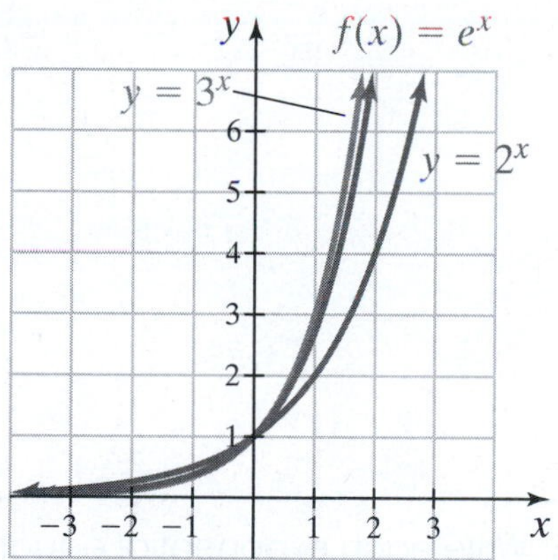

Sketching Exponential Functions

Examples/Videos

Sketch the graph of each exponential function.

a. $f(x) = \left(\dfrac{2}{3}\right)^x$ b. $f(x) = -2e^x$

Use transformations to sketch the graph of $f(x) = -2^{x+1} + 3$

Method of Relating the Bases

If an exponential equation can be written in the form $b^u = b^v$, then $u = v$.

Solve the following equations using the Method of Relating the Bases.

a. $8 = \dfrac{1}{16^x}$

b. $\dfrac{1}{27^x} = \left(\sqrt[4]{3}\right)^{x-2}$

Key Concepts	Examples/Videos
Periodic Compound Interest Formula Periodic compound interest can be calculated using the formula $$A = P\left(1 + \frac{r}{n}\right)^{nt},$$ where A = Total amount after t years P = Principal (original investment) r = Interest rate per year n = Number of times interest is compounded per year t = Number of years **Continuous Compound Interest Formula** Continuous compound interest can be calculated using the formula $$A = Pe^{rt}$$ **Present Value Formula for Periodic Compound Interest** Present value can be calculated using the formula $$P = A\left(1 + \frac{r}{n}\right)^{-nt}$$ **Present Value Formula for Continuous Compound Interest** $$P = Ae^{-rt}$$	 How much money would be in an account after 5 years if an original investment of \$6,000 was compounded continuously at 4.5%? Compare this amount to the same investment that was compounded daily. Round to the nearest cent.  Find the present value of \$8,000 if the interest is paid at a rate of 5% compounded quarterly for 7 years. Round to the nearest cent. Find the present value of \$18,000 if interest is paid at a rate of 8% compounded continuously for 20 years. Round to the nearest cent.
5.2 Logarithmic Functions For $x > 0, b > 0,$ and $b \neq 1$, the **logarithmic function** with base b is defined by $y = \log_b x$ if and only if $x = b^y$.	 Write each exponential equation as an equation involving a logarithm. **a.** $2^3 = 8$ **b.** $5^{-2} = \frac{1}{25}$ **c.** $1.1^M = z$  Write each logarithmic equation as an equation involving an exponent. **a.** $\log_3 81 = 4$ **b.** $\log_4 16 = y$ **c.** $\log_{3/5} x = 2$  Evaluate each logarithm: **a.** $\log_5 25$ **b.** $\log_3 \frac{1}{27}$ **c.** $\log_{\sqrt{2}} \frac{1}{4}$

Key Concepts	Examples/Videos
For $x > 0$, the **common logarithmic function** is defined by $y = \log x$ if and only if $x = 10^y$. For $x > 0$, the **natural logarithmic function** is defined by $y = \ln x$ if and only if $x = e^y$.	Write each exponential equation as an equation involving a common logarithm or natural logarithm. a. $e^0 = 1$ b. $10^{-2} = \frac{1}{100}$ c. $e^K = w$ Write each logarithmic equation as an equation involving an exponent. a. $\log 10 = 1$ b. $\ln 20 = Z$ c. $\log (x - 1) = T$
General Properties of Logarithms For $b > 0$ and $b \neq 1$, 1. $\log_b b = 1$ 2. $\log_b 1 = 0$	Evaluate each expression without the use of a calculator. a. $\log 100$ b. $\ln \sqrt{e}$ c. $e^{\ln 51}$ d. $\log 1$
Cancellation Properties of Exponentials and Logarithms For $b > 0$ and $b \neq 1$, 1. $b^{\log_b x} = x$ 2. $\log_b b^x = x$	
Steps for Sketching Logarithmic Functions of the Form $f(x) = \log_b x$ **Step 1.** Start with the graph of the exponential function $y = b^x$, labeling several ordered pairs. **Step 2.** Because $f(x) = \log_b x$ is the inverse of $y = b^x$, we can find several points on the graph of $f(x) = \log_b x$ by reversing the coordinates of the ordered pairs of $y = b^x$. **Step 3.** Plot the ordered pairs from step 2, and complete the graph of $f(x) = \log_b x$ by connecting the ordered pairs with a smooth curve. The graph of $f(x) = \log_b x$ is a reflection of the graph of $y = b^x$ about the line $y = x$. G/V Sketching Logarthmic Functions	Sketch the graph of $f(x) = \log_3 x$. Sketch the graph of $f(x) = -\ln(x + 2) - 1$.

Key Concepts	Examples/Videos
5.3 Properties of Logarithms If $b > 0, b \neq 1, u$ and v represent positive numbers and r is any real number, then $\log_b uv = \log_b u + \log_b v$ $\log_b \frac{u}{v} = \log_b u - \log_b v$ $\log_b u^r = r \log_b u$	 Use properties of logarithms to expand each logarithmic expression as much as possible. **a.** $\log_7 (49x^3\sqrt[5]{y^2})$ **b.** $\ln\left(\frac{(x^2-4)}{9e^{x^3}}\right)$  Use properties of logarithms to rewrite each expression as a single logarithm. **a.** $\frac{1}{2}\log(x-1) - 3\log z + \log 5$ **b.** $\frac{1}{3}(\log_3 x - 2\log_3 y) + \log_3 10$
Logarithm Property of Equality If a logarithmic equation can be written in the form $\log_b u = \log_b v$, then $u = v$. Furthermore, if $u = v$, then $\log_b u = \log_b v$	 Solve the following equations. **a.** $\log_7(x-1) = \log_7 12$ **b.** $2\ln x = \ln 16$
Change of Base Formula For any positive base $b \neq 1$ and for any positive real number u, then $\log_b u = \frac{\log_a u}{\log_a b}$ where a is any positive number such that $a \neq 1$.	 Use the change of base formula and the properties of logarithms to rewrite as a single logarithm involving base 2. $\log_4 x + 3\log_2 y$
5.4 Exponential and Logarithmic Equations **Solving Exponential Equations** • If the equation can be written in the form $b^u = b^v$, then solve the equation $u = v$. • If the equation cannot easily be written in the form $b^u = b^v$. **1.** Use the logarithm property of equality to "take the log of both sides" (typically using base 10 or base e). **2.** Use the power rule of logarithms to "bring down" any exponents. **3.** Solve for the given variable.	 Solve each equation. Round to four decimal places. **a.** $3^{x-1} = \left(\frac{1}{27}\right)^{2x+1}$ **b.** $7^{x+3} = 4^{2-x}$  Solve each equation. Round to four decimal places. **a.** $25e^{x-5} = 17$ **b.** $e^{2x-1} \cdot e^{x+4} = 11$

Key Concepts	Examples/Videos
Solving Logarithmic Equations 1. Determine the domain of the variable. 2. Use properties of logarithms to combine all logarithms, and write as a single logarithm, if needed. 3. Eliminate the logarithm by rewriting the equation in exponential form. 4. Solve for the given variable. 5. Check for any extraneous solutions. Verity that each solution is in the domain of the variable.	 Solve $\log_4(2x - 1) = 2$.  Solve $\log_2(x + 10) + \log_2(x + 6) = 5$.
5.5 Applications of Exponential and Logarithmic Functions **Exponential Growth and Decay** 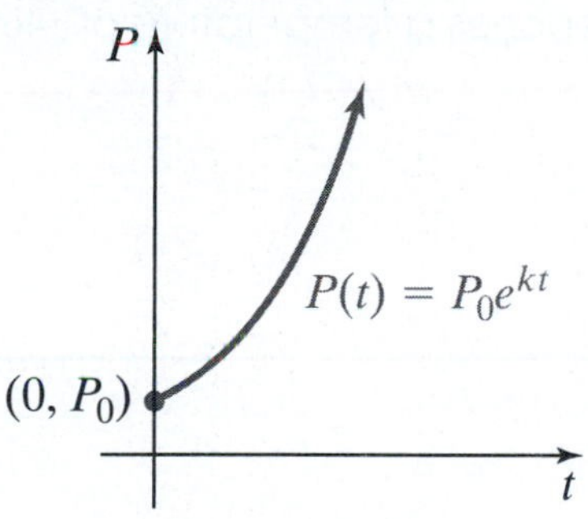 Graph of $P(t) = P_0e^{kt}$ for $k > 0$ 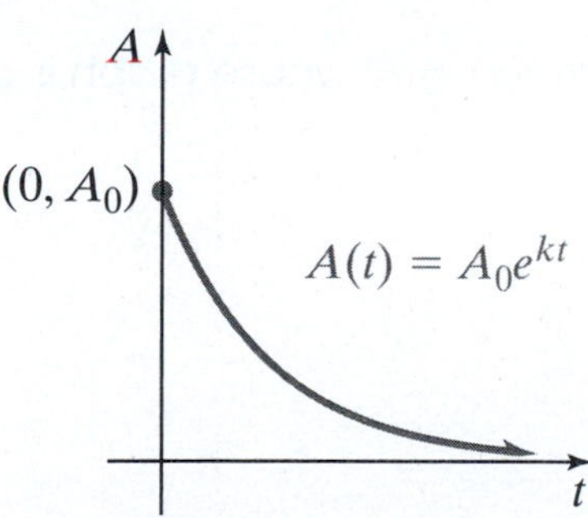 Graph of $A(t) = A_0e^{kt}$ for $k < 0$  **Half-Life**	 How long will it take (in years and months) for an investment to double if it earns 7.5% compounded monthly?  The population of a small town grown at a rate proportional to its current size. In 1900, the population was 900. In 1920, the population had grown to 1600. What was the population of this town in 1950?  Suppose that a meteorite is found containing 4% of its original krypton-99. If the half-life of krypton-99 is 80 years, how old is the meteorite?

Key Concepts	Examples/Videos
Logistic Growth P, $y = C$, $P(t) = \dfrac{C}{1 + Be^{kt}}$, $\left(0, \dfrac{C}{1 + B}\right)$, t Graph of $P(t) = \dfrac{C}{1 - Be^{kt}}$	Ten goldfish were introduced into a small pond. Because of limited food, space, and oxygen, the carrying capacity of the pond is 400 goldfish. The goldfish population at any time t, in days, is modeled by the logistic growth function $F(t) = \dfrac{C}{1 + Be^{kt}}$. If 30 goldfish are in the pond after 20 days, a. Find B. b. Find k. c. When will the pond contain 250 goldfish?
Newton's Law of Cooling T, T_0, $T(t) = S + (T_0 - S)e^{kt}$, $y = S$, t Graph of $T(t) = S + (T_0 - S)e^{kt}$	Suppose that the temperature of a cup of hot tea obeys Newton's law of cooling. If the tea has a temperature of 200°F when it is initially poured and 1 minute later has cooled to 189°F in a room that maintains a constant temperature of 69°F, determine when the tea reaches a temperature of 146°F.

Chapter 5 Review Exercises

1. Determine the correct exponential function of the form $f(x) = b^x$ whose graph is given.

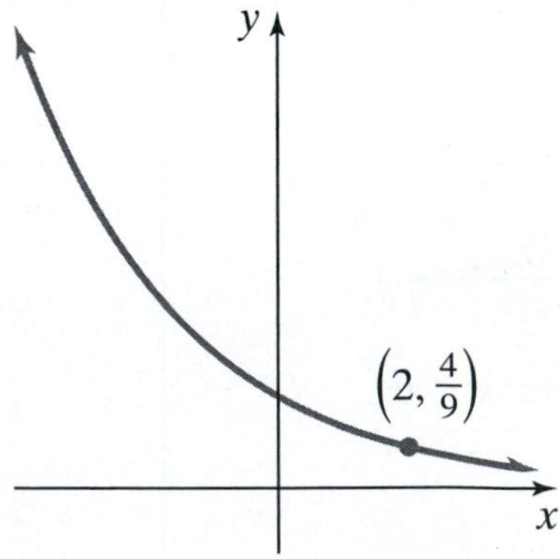

2. Use transformations to sketch the graph of $f(x) = -2^{x+1} - 1$.

In Review Exercises 3 and 4, Solve using the method of relating the bases.

3. $3^{x-1} = \frac{1}{9}$

4. $\frac{125}{\sqrt[3]{5^x}} = \left(\frac{1}{25^x}\right)$

5. Write the equation $\left(\frac{1}{9}\right)^t = 27$ as an equation involving a logarithm.

6. Write the equation $\log_4 K = L$ as an exponential equation.

7. Evaluate the expression $\log_4\left(\frac{1}{\sqrt[5]{64}}\right)$ without the use of a calculator.

8. Use the properties of logarithms to evaluate the expression $\log_9 1$ without the use of a calculator.

9. Sketch the logarithmic function $f(x) = \log_5(x - 1)$. Label three points that lie on the graph and determine the domain and the equation of any asymptotes.

In 10–12, use the product rule, quotient rule, or power rule to expand each logarithmic expression. Whenever possible, evaluate logarithmic expressions.

10. $\log_5 y^3$

11. $\log_3(27w)$

12. $\log \frac{\sqrt{x}}{10y^3}$

13. Use properties of logarithms to evaluate the expression $5^{\log_5 3 - \log_5 11}$ without the use of a calculator.

14. Use properties of logarithms to rewrite the expression $2\log_8 x + \frac{1}{3}\log_8 y$ as a single logarithm.

15. Use properties of logarithms and the logarithmic property of equality to solve the equation $\log_7(x + 6) - \log_7(x + 2) = \log_7 x$.

16. Use the change of base formula and a calculator to approximate the value of the expression $\log_4 51$.

In 17–19, solve each exponential equation. Round your answer to four decimal places.

17. $3^x = 5$

18. $8^{4x-7} = 11^{5+x}$

19. $150e^{x-4} = 5$

In 20–21, solve each logarithmic equation.

20. $2\ln x - \ln(2x - 3) = \ln 2x - \ln(x - 1)$

21. $\log_6(x + 14) + \log_6(x + 9) = 1$

22. How long (in years and months) will it take for an investment to double at 9% compounded monthly?

23. During a research experiment, it was found that the number of bacteria in a culture grew at a rate proportional to its size. At 8:00 AM, there were 2,000 bacteria present in the culture. At noon, the number of bacteria grew to 2,400. How many bacteria will there be at midnight? Round to the nearest whole number.

24. A certain radioactive isotope is leaked into a small stream. Three hundred days after the leak, 2% of the original amount of the substance remained. Determine the half-life of this radioactive isotope. Round to the nearest whole number.

25. Estabon poured himself a hot beverage that had a temperature of $198°F$ and then set it on the kitchen table to cool. The temperature of the kitchen was a constant $75°F$. If the drink cooled to $180°F$ in 2 minutes, how long will it take for the drink to cool to $100°F$?

CHAPTER SIX

An Introduction to Trigonometric Functions

CHAPTER SIX CONTENTS

6.1 An Introduction to Angles: Degree and Radian Measure

THINGS TO KNOW

Before working through this section, be sure that you are familiar with the following concepts:

VIDEO ANIMATION INTERACTIVE

You Try It

1. Sketching the Graph of a Circle (Section 2.2) 

INTRODUCTION

Read this introduction before beginning Objective 1.

OBJECTIVES

1 Understanding Degree Measure

2 Finding Coterminal Angles Using Degree Measure

3 Understanding Radian Measure

4 Converting between Degree Measure and Radian Measure

5 Finding Coterminal Angles Using Radian Measure

SECTION 6.1 EXERCISES

Introduction to Section 6.1

An angle is made up of two **rays** that share a common endpoint called the **vertex**. An angle is created by rotating one ray away from a fixed ray. The fixed ray is called the **initial side** of the angle and the rotated ray is called the terminal side of the angle. When we draw an angle, we show the direction in which the **terminal side** is rotated and the amount of rotation. Greek letters such as θ, α, and β (theta, alpha, and beta) or capital letters such as A, B, and C are used to label angles. When we rotate the ray in a *counterclockwise* fashion, we say that the angle has a *positive amount of rotation* (the angle has positive measure). An angle formed from a *clockwise rotation* is said to *have negative measure*. In Figure 1a, angle θ has positive measure. In Figure 1b, angle α has negative measure.

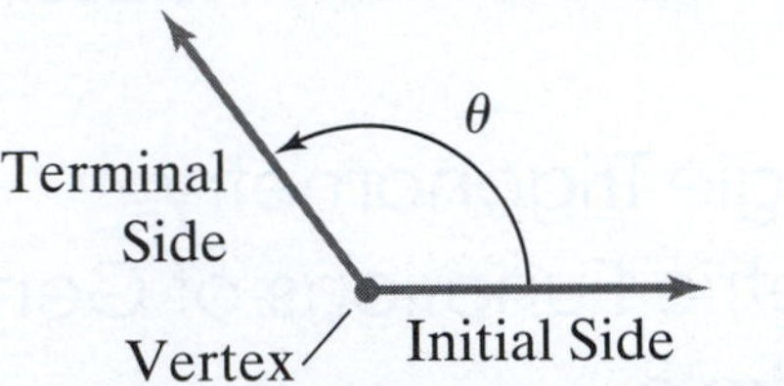

(a) An angle with positive measure

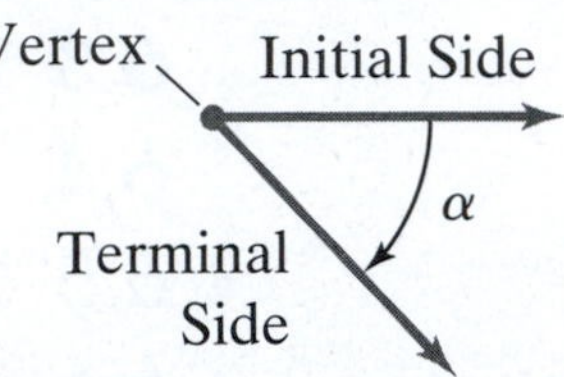

(b) An angle with negative measure

Figure 1

An angle is in **standard position** if the vertex is at the **origin** of a rectangular coordinate system and the initial side of the angle is along the positive x-axis. The terminal side of the angle will always lie in one of the **four quadrants** or on either axis. Angle θ and angle α are sketched again in Figure 2. These angles are now said to be in standard position.

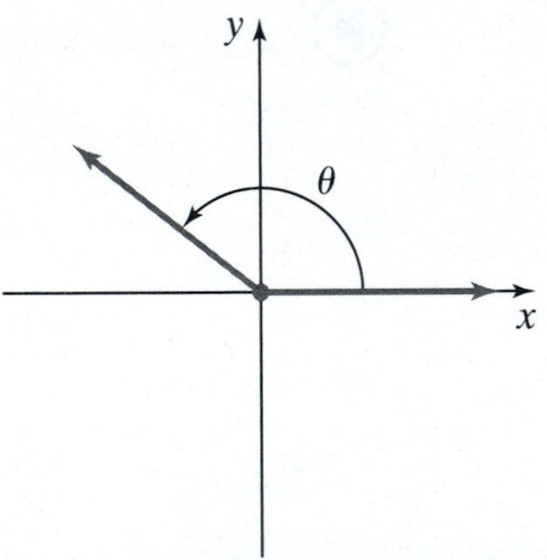

(a) Angle θ has **positive measure** because the direction of rotation is **counterclockwise**. The terminal side of θ lies in Quadrant II.

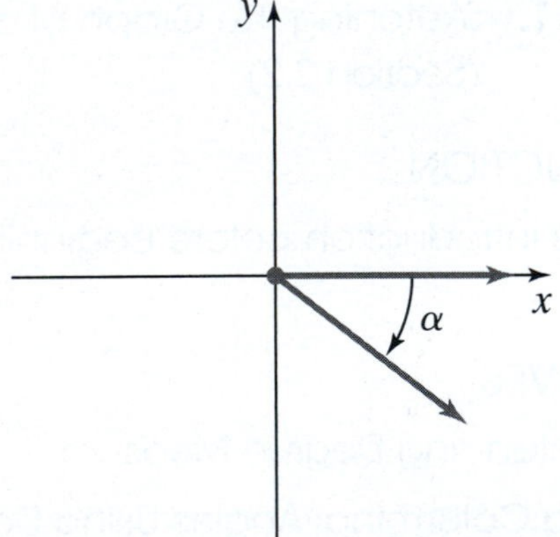

(b) Angle α has **negative measure** because the direction of rotation is **clockwise**. The terminal side of α lies in Quadrant IV.

Figure 2 Angles in standard position

OBJECTIVE 1 UNDERSTANDING DEGREE MEASURE

DEGREE MEASURE

There are two ways that we will measure angles. We start by introducing **degree measure**. The notation for degrees is the ° symbol. The angle formed by rotating the terminal side one complete counterclockwise rotation has a measure of $360°$. See **Figure 3**. We can draw an angle by knowing the desired amount of rotation necessary. For example, an angle of $90°$ is equal to $\frac{1}{4}$ of one complete rotation because $\frac{90°}{360°} = \frac{1}{4}$. So, we can draw an angle of $90°$ by rotating the terminal side of an angle $\frac{1}{4}$ of one complete counterclockwise rotation. See **Figure 4**. Similarly, an angle of $-45°$ is $\frac{1}{8}$ of one complete clockwise rotation because $\frac{45°}{360°} = \frac{1}{8}$. See **Figure 5**.

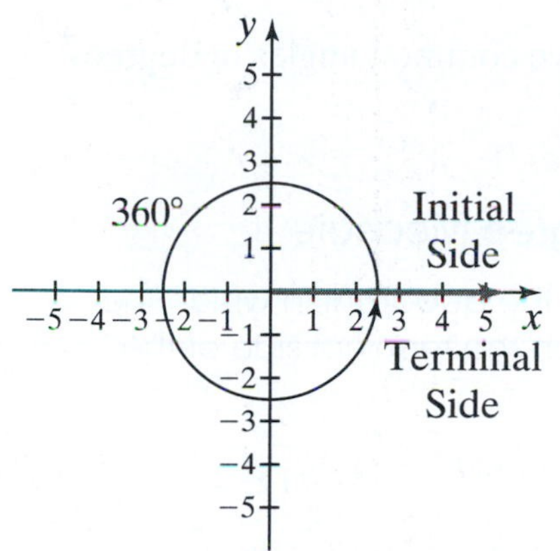

Figure 3 A $360°$ angle is one complete counterclockwise rotation.

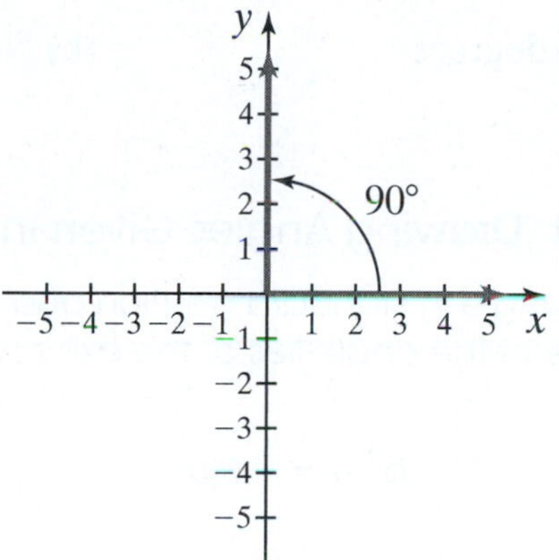

Figure 4 An angle of $90°$ is $\frac{1}{4}$ of one complete counterclockwise rotation.

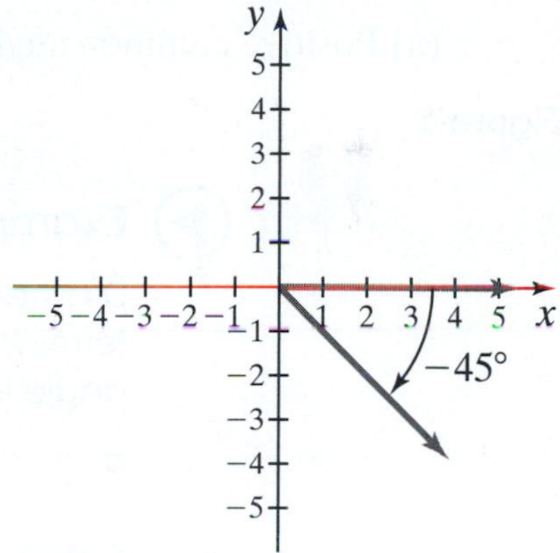

Figure 5 An angle of $-45°$ is $\frac{1}{8}$ of one complete clockwise rotation.

Eventually you should be comfortable sketching the positive and negative angles measured in degrees, as seen in **Figure 6**. Notice in Figure 6a that all angles that appear in Quadrant I are between $0°$ and $90°$. These are called **acute angles**. All angles that appear in Quadrant II are between $90°$ and $180°$ and are called **obtuse angles**. Angles whose terminal sides lie along an axis are called **quadrantal angles**. The quadrantal angle of exactly $90°$ is called a **right angle**. The quadrantal angle of exactly $180°$ is called a **straight angle**.

The angles sketched in Figure 6a are formed by rotating the terminal side of the angle in a counterclockwise direction and are thus positive angles. The angles sketched in Figure 6b are negative angles because they are formed by rotating the terminal side of the angle in a clockwise fashion. It is important to point out that even though two angles in standard position may have identical terminal sides, the measure of the two angles may be different. For example, notice in Figure 6a that the angle with a measure of $45°$ has the exact same terminal side as an angle of $-315°$ sketched in Figure 6b. Angles in standard position having the same terminal side are called **coterminal angles**. Coterminal angles will be discussed in detail in **Objective 2** of this section.

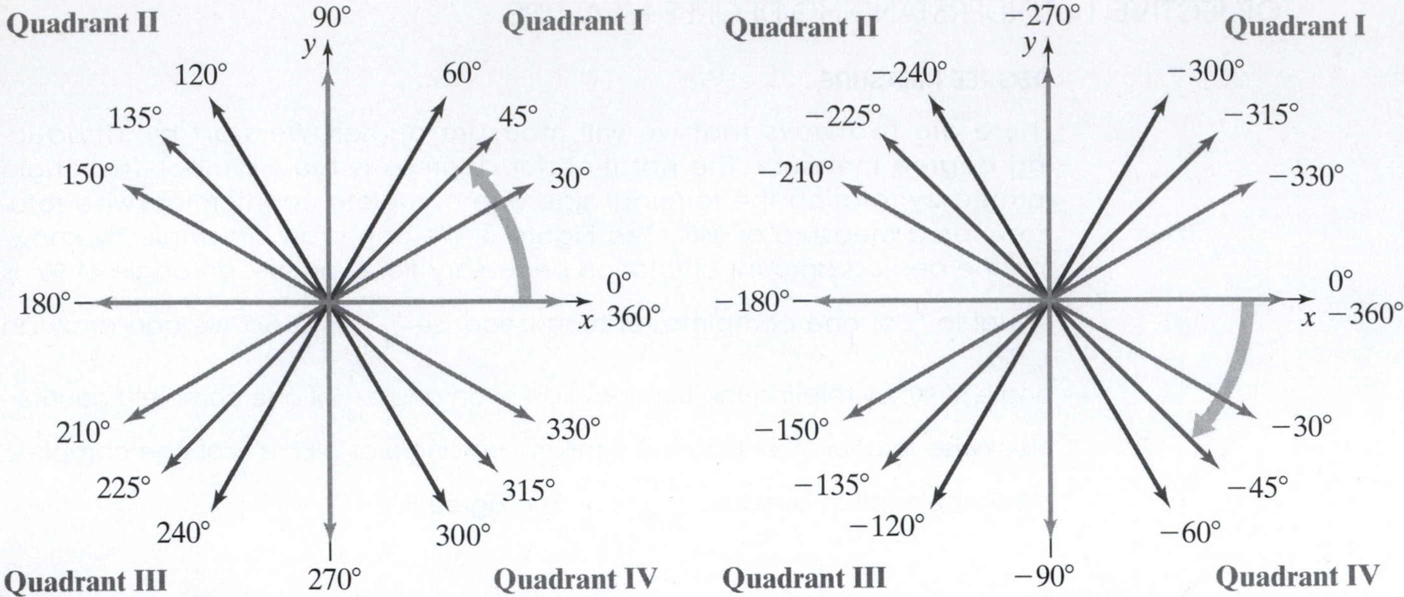

(a) Positive common angles in degrees (b) Negative common angles in degrees

Figure 6

Example 1 Drawing Angles Given in Degree Measure

Draw each angle in standard position and state the quadrant in which the terminal side of the angle lies or the axis on which the terminal side of the angle lies.

a. $\theta = 60°$ **b.** $\alpha = -270°$ **c.** $\beta = 420°$

Solution Watch this **video** to see how to draw each angle seen below.

a.

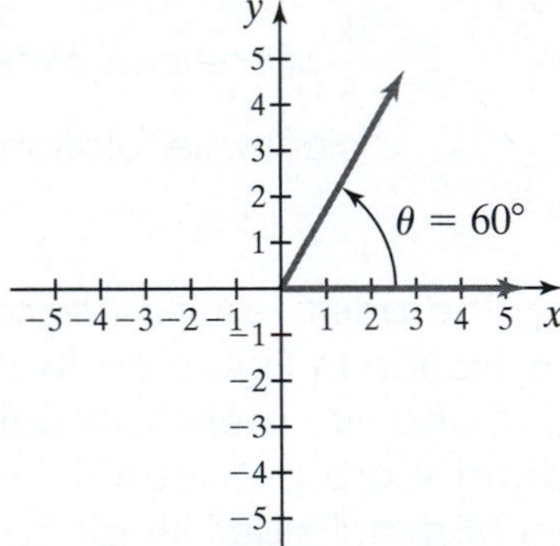

Terminal side lies in Quadrant I.

b.

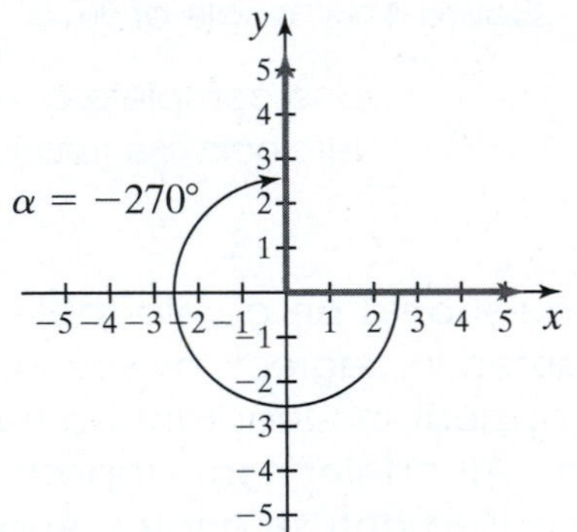

Terminal side lies on the positive y-axis.

c.

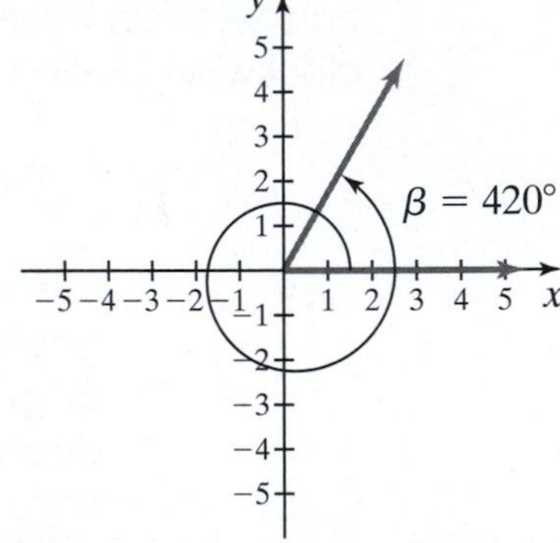

Terminal side lies in Quadrant I.

TIP Not every angle written in degrees has nice integer measurements as in Example 1. An angle can have any decimal measurement, such as $\theta = 41.23°$ or $\alpha = -0.375°$. These angles are said to be written in **degree decimal form.** Angles can also be written in **degrees, minutes, seconds form.** These two forms are discussed in **Appendix A.**

You Try It Work through this You Try It problem.

Work Exercises 1–5 in this textbook or in the MyLab Math Study Plan.

OBJECTIVE 2 FINDING COTERMINAL ANGLES USING DEGREE MEASURE

Consider the $45°$, $405°$, and $-315°$ angles in standard position seen in Figure 7. Notice that each angle has the exact same terminal side. Angles in standard position having the same terminal side are called **coterminal angles**.

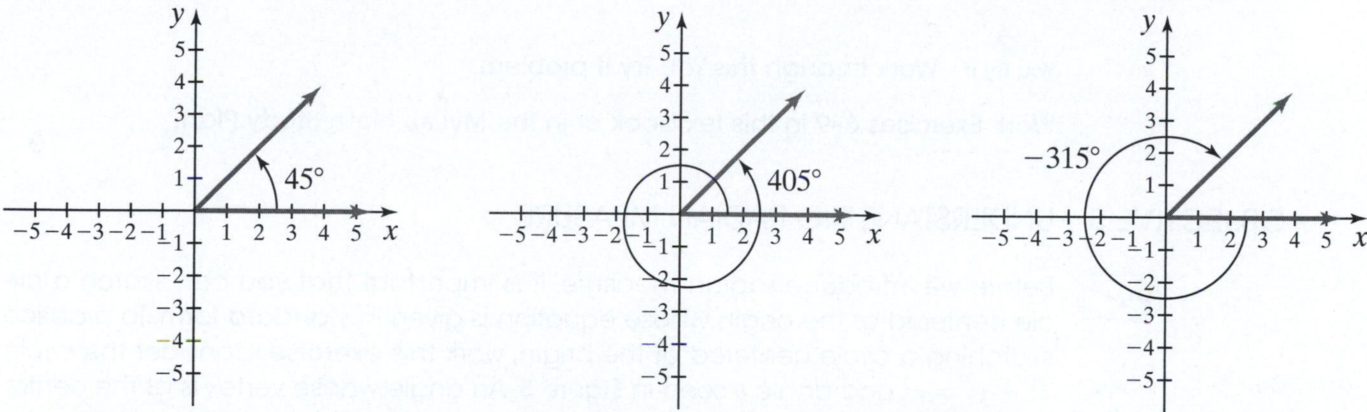

Figure 7 Coterminal angles in standard position have the same terminal side.

> **Definition** Coterminal Angles
>
> **Coterminal angles** are angles in standard position having the same terminal side but different measures.
>
> Coterminal angles can be obtained by adding any nonzero integer multiple of $360°$ to a given angle. Given any angle θ and any nonzero integer k, the angles θ and $\theta + k \cdot 360°$ are coterminal angles.

Below are several examples of angles that are coterminal with the angle $45°$.

$$k = 1: \quad 45° + 1 \cdot 360° = 405°$$
$$k = -1: \quad 45° + (-1) \cdot 360° = -315°$$
$$k = 2: \quad 45° + 2 \cdot 360° = 765°$$
$$k = -2: \quad 45° + (-2) \cdot 360° = -675°$$

The number of angles coterminal to a given angle is infinite. However, we will often be interested in determining the angle of least nonnegative measure that is coterminal with a given angle. Every angle has a coterminal angle of least nonnegative measure. If θ is a given angle, then we will use the notation θ_C to denote the angle of least nonnegative measure coterminal with θ. Note that $0° \leq \theta_C < 360°$ (or $0 \leq \theta_C < 2\pi$). See Example 2.

Example 2 Finding Coterminal Angles Using Degree Measure

Find the angle of least nonnegative measure, θ_C, that is coterminal with $\theta = -697°$.

Solution The angles coterminal with $\theta = -697°$ have the form $-697° + k(360°)$, where k is an integer. Because θ is a negative angle, we choose positive values of k until we find the angle of least nonnegative measure.

$$k = 1: \; -697° + (1)(360°) = -337°$$

The angle $-337°$ has negative measure, so we now choose $k = 2$.

$$k = 2: \ -697° + (2)(360°) = 23°$$

The angle $23°$ is the smallest *nonnegative* angle coterminal with $\theta = -697°$. Therefore, the angle of least nonnegative measure that is coterminal with $\theta = -697°$ is $\theta_C = 23°$. For the solution to this example, watch this **video**. ●

You Try It **Work through this You Try It problem.**

Work Exercises 6-9 in this textbook or in the MyLab Math Study Plan.

OBJECTIVE 3 UNDERSTANDING RADIAN MEASURE

You Try It

Before we introduce radian measure, it is important that you can sketch a circle centered at the **origin** whose equation is given in **standard form**. To practice sketching a circle centered at the origin, work this **exercise**. Consider the circle $x^2 + y^2 = r^2$ and angle θ seen in **Figure 8**. An angle whose vertex is at the center of a circle is called a **central angle**. Every central angle intercepts a portion of a circle called an **intercepted arc**. We typically use the variable s to represent the length of an intercepted arc.

When a central angle has an intercepted arc whose length is equal to the radius of the circle, then the central angle is said to have a measurement equal to **1 radian.** It is worth pointing out that 1 radian has a degree measure of about $57.3°$. See **Figure 9**. Watch this **animation** for a further explanation of radian measure and to see how to obtain a relationship between radians and degrees.

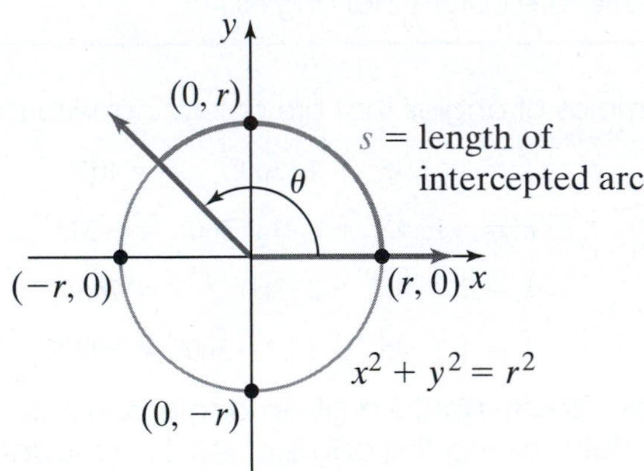

Figure 8 A central angle θ and the corresponding intercepted arc

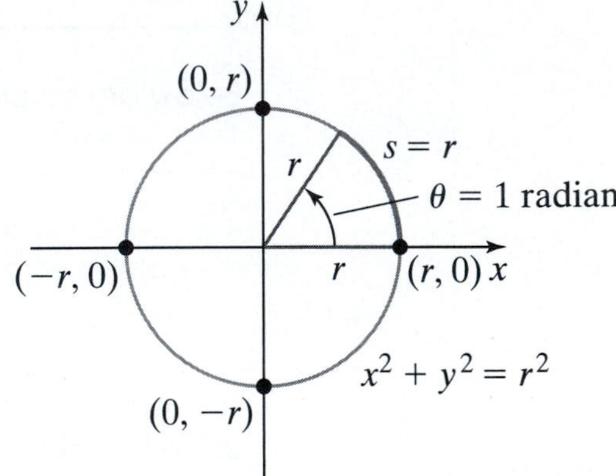

Figure 9 A central angle of $\theta = 1$ radian has a corresponding intercepted arc equal to the length of the radius. An angle of 1 radian is approximately $57.3°$.

> **Definition** Radian
>
> One **radian** is the measure of a central angle that has an intercepted arc equal in length to the radius of the circle.

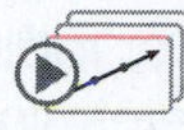

Make sure that you have carefully watched this **animation** to verify that there are exactly 2π radians in a circle. This allows us to establish the relationship that $360° = 2\pi$ radians or $180° = \pi$ radians.

Relationship Between Degrees and Radians

$$360° = 2\pi \text{ radians}$$

$$180° = \pi \text{ radians}$$

Although we can draw angles that are given in radians by converting the radian measure to degrees, it is important to be able to draw angles given in radian measure by knowing the amount of rotation necessary to sketch them. For example, the angle formed by rotating the terminal side of an angle one complete counterclockwise rotation has a measure of 2π radians. See **Figure 10**. An angle formed by rotating its terminal side $\frac{1}{4}$ of one complete counterclockwise rotation has a radian measure of $\theta = \frac{\pi}{2}$ because $\frac{1}{4}(2\pi) = \frac{\pi}{2}$. So, we draw an angle of $\theta = \frac{\pi}{2}$ by rotating the terminal side $\frac{1}{4}$ of one complete counterclockwise rotation. See **Figure 11**. Similarly, an angle formed by rotating its terminal side $\frac{1}{8}$ of one complete clockwise rotation has a radian measure of $\theta = -\frac{\pi}{4}$ because $\frac{1}{8}(-2\pi) = -\frac{\pi}{4}$. See **Figure 12**.

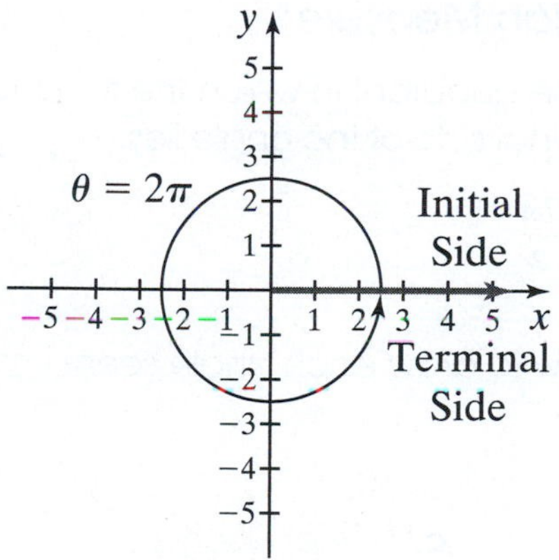

Figure 10 An angle of 2π radians is one complete counterclockwise rotation.

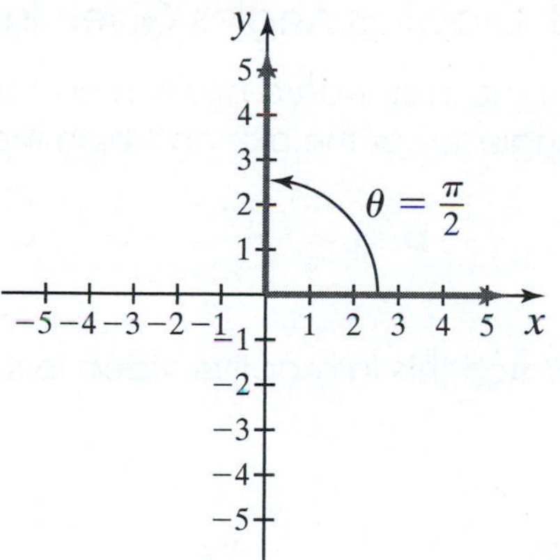

Figure 11 An angle of $\frac{\pi}{2}$ radians is $\frac{1}{4}$ of one complete counterclockwise rotation.

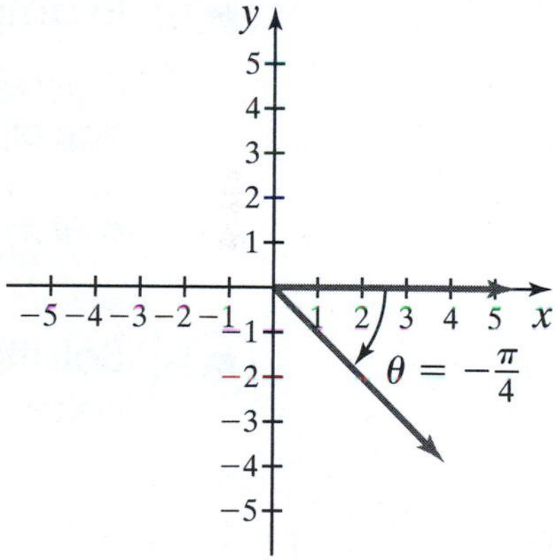

Figure 12 An angle of $-\frac{\pi}{4}$ is $\frac{1}{8}$ of one complete clockwise rotation.

It is important to point out that there is no symbol to indicate radian measure. Although we may use the abbreviation "rad" to represent radians, we will typically omit this abbreviation when angles measured in radians are written in terms of π. For example, we will simply use $\frac{\pi}{4}$ to represent $\frac{\pi}{4}$ radians. At this point, you should be comfortable sketching angles that are given in degrees or radians. You should be able to sketch each angle seen in **Figure 13** as well as determine the quadrant or axis on which the terminal side of the angle lies.

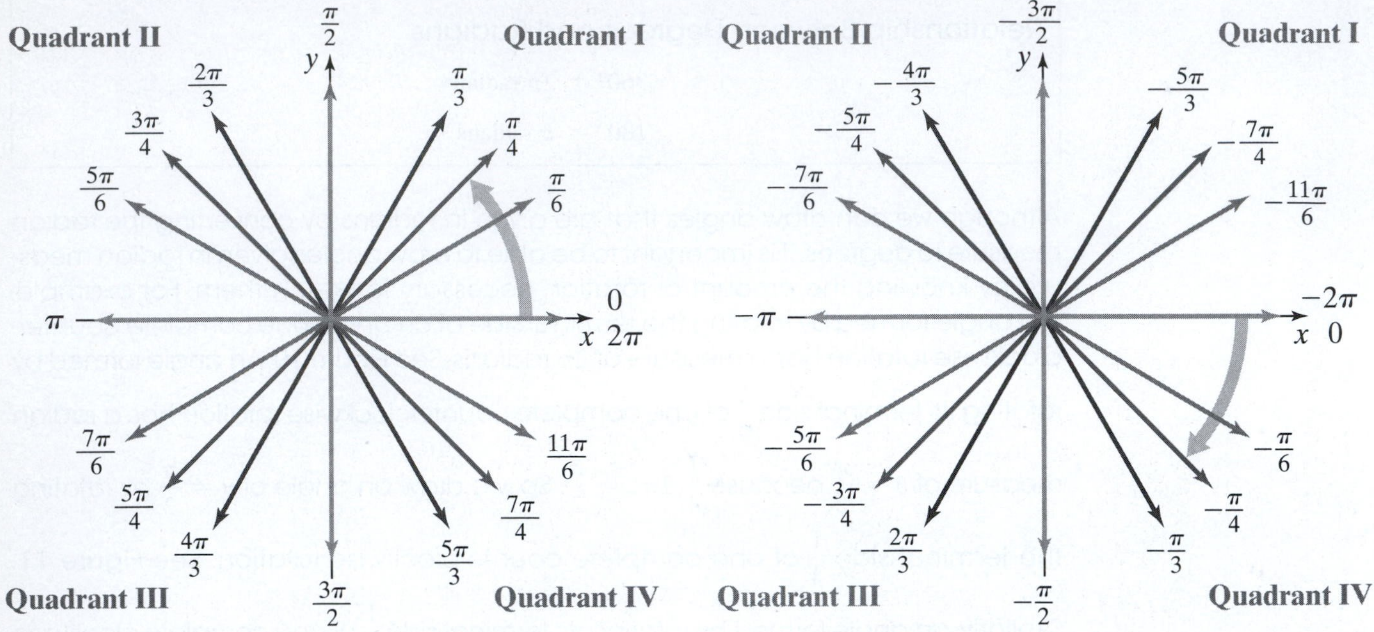

(a) Positive common angles in radians (b) Negative common angles in radians

Figure 13

Example 3 Drawing Angles Given in Radian Measure

Draw each angle in standard position and state the quadrant in which the terminal side of the angle lies or the axis on which the terminal side of the angle lies.

a. $\theta = \frac{\pi}{3}$ **b.** $\alpha = -\frac{3\pi}{2}$ **c.** $\beta = \frac{7\pi}{3}$

Solution Watch this **interactive video** to see how to draw each angle seen below.

a.

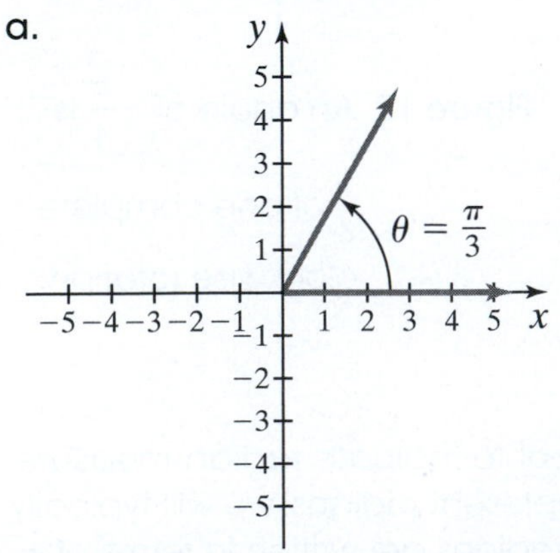

Terminal side lies in Quadrant I.

b.

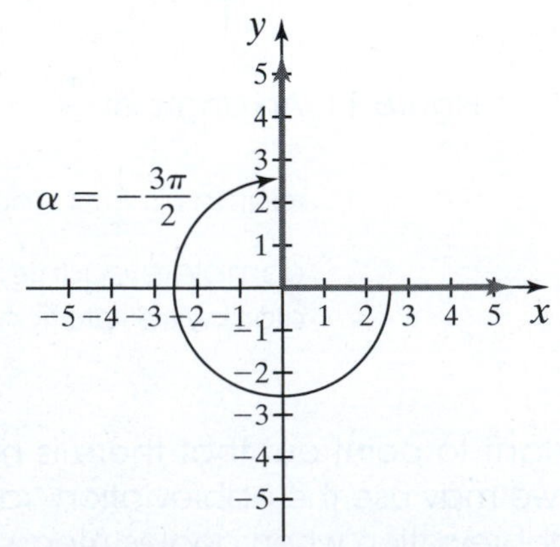

Terminal side lies on the positive y-axis.

c.

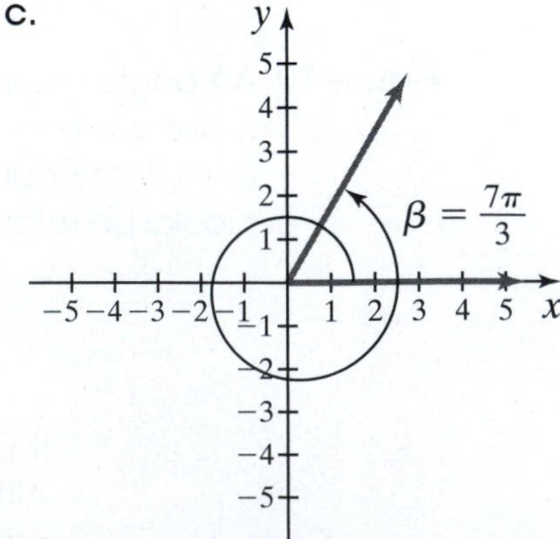

Terminal side lies in Quadrant I.

At first glance it may appear that the angles $\theta = \frac{\pi}{3}$ and $\beta = \frac{7\pi}{3}$ from Example 3 are the same. Both angles are in standard position and have the same terminal side. However, the angles are quite different because different amounts of rotation are required to sketch the two angles. Recall that angles in standard position that have the same terminal side are called **coterminal angles**. We discuss how to find coterminal angles using radian measure at length in **Objective 5.**

You Try It Work through the following You Try It problem.

Work Exercises 10–15 in this textbook or in the MyLab Math Study Plan.

OBJECTIVE 4 CONVERTING BETWEEN DEGREE MEASURE AND RADIAN MEASURE

We established in this **animation** that $180° = \pi$ radians. Dividing both sides of this equation 180° or by π radians gives the following rules for converting between degrees and radians.

Relationship Between Degrees and Radians

To convert degrees to radians, multiply by $\frac{\pi \text{ radians}}{180°}$.

To convert radians to degrees, multiply by $\frac{180°}{\pi \text{ radians}}$.

Example 4 Converting from Degree Measure to Radian Measure

Convert each angle given in degree measure into radians.

a. 45° **b.** −150° **c.** 56°

Solution To convert from degrees to radians, we multiply by $\frac{\pi \text{ radians}}{180°}$.

a. $45° = 45° \cdot \frac{\pi \text{ radians}}{180°} = \frac{45\pi}{180} \text{ radians} = \frac{\pi}{4} \text{ radians}$

b. $-150° = -150° \cdot \frac{\pi \text{ radians}}{180°} = \frac{150\pi}{180} \text{ radians} = \frac{5\pi}{6} \text{ radians}$

c. $56° = 56° \cdot \frac{\pi \text{ radians}}{180°} = \frac{56\pi}{180} \text{ radians} = \frac{14\pi}{45} \text{ radians}$

You Try It Work through this You Try It problem.

Work Exercises 16–20 in this textbook or in the MyLab Math Study Plan.

Example 5 Converting from Radian Measure to Degree Measure

Convert each angle given in radian measure into degrees. Round to two decimal places if needed.

a. $\frac{2\pi}{3}$ radians b. $-\frac{11\pi}{6}$ radians c. 3 radians

Solution To convert from degrees to radians, we multiply by $\frac{180°}{\pi \text{ radians}}$.

a. $\frac{2\pi}{3}\text{ radians} = \frac{2\cancel{\pi}}{3}\cancel{\text{radians}}\cdot\frac{180°}{\cancel{\pi}\cancel{\text{radians}}} = \frac{360°}{3} = 120°$

b. $-\frac{11\pi}{6}\text{ radians} = -\frac{11\cancel{\pi}}{6}\cancel{\text{radians}}\cdot\frac{180°}{\cancel{\pi}\cancel{\text{radians}}} = -\frac{1980°}{6} = -330°$

c. $3\text{ radians} = 3\,\cancel{\text{radians}}\cdot\frac{180°}{\pi\ \cancel{\text{radians}}} = \frac{540°}{\pi} \approx 171.89°$

You Try It Work through this You Try It problem.

Work Exercises 21–26 in this textbook or in the MyLab Math Study Plan.

OBJECTIVE 5 FINDING COTERMINAL ANGLES USING RADIAN MEASURE

Recall that angles in standard position having the same terminal side are called **coterminal angles**. Coterminal angles in degrees can be obtained by adding any nonzero integer multiple of 360° to a given angle. Similarly, coterminal angles given in radians can be obtained by adding a nonzero integer multiple of 2π to a given angle. Therefore, for any angle θ and for any nonzero integer k, we can find a coterminal angle using the expression below:

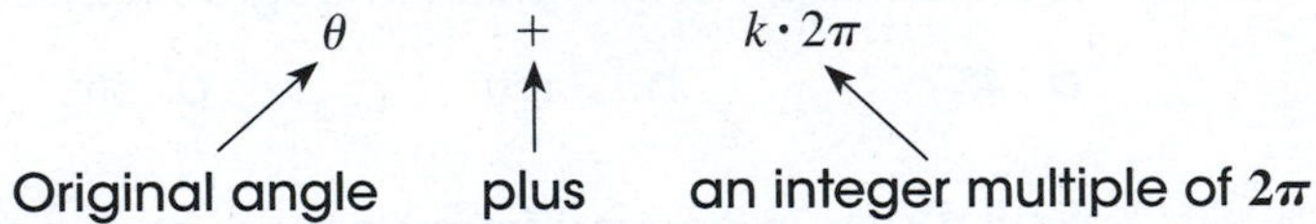

Example 6 Finding Coterminal Angles Using Radian Measure

Find three angles that are coterminal with $\theta = \frac{\pi}{3}$ using $k = 1, k = -1$, and $k = -2$.

Solution Using the expression $\frac{\pi}{3} + k\cdot 2\pi$ with $k = 1, k = -1$, and $k = -2$, we get the following.

$$k = 1: \frac{\pi}{3} + (1)2\pi = \frac{\pi}{3} + 2\pi = \frac{\pi}{3} + \frac{6\pi}{3} = \frac{7\pi}{3}$$

$$k = -1: \frac{\pi}{3} + (-1)2\pi = \frac{\pi}{3} - 2\pi = \frac{\pi}{3} - \frac{6\pi}{3} = -\frac{5\pi}{3}$$

$$k = -2: \frac{\pi}{3} + (-2)2\pi = \frac{\pi}{3} - 4\pi = \frac{\pi}{3} - \frac{12\pi}{3} = -\frac{11\pi}{3}$$

These angles are sketched in Figure 14.

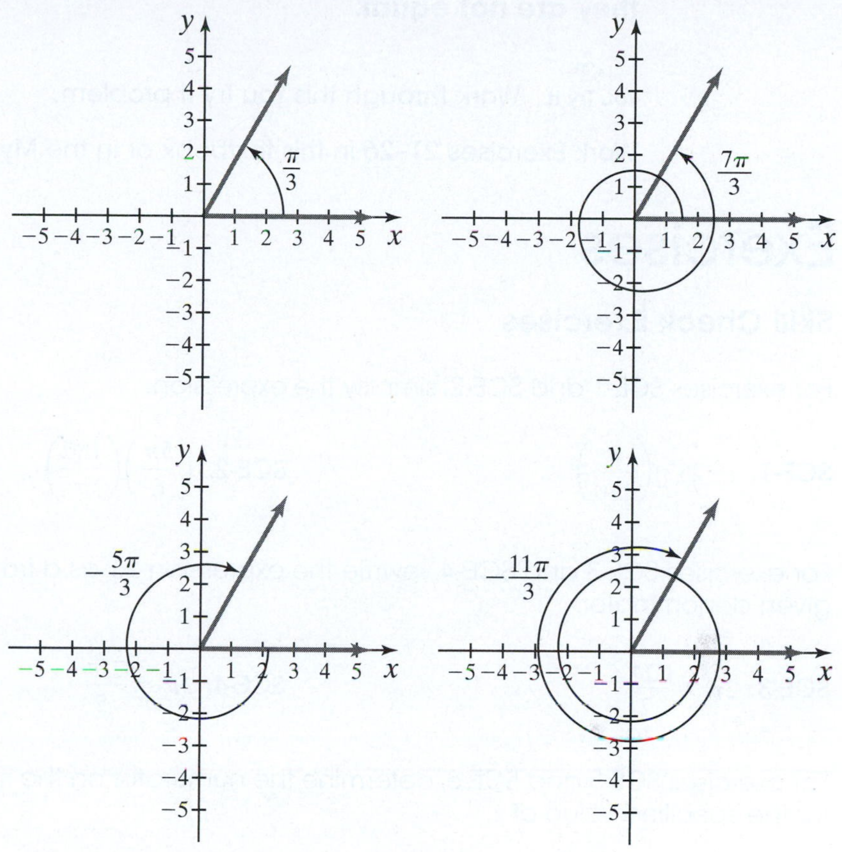

Figure 14 The angles $\frac{\pi}{3}, \frac{7\pi}{3}, -\frac{5\pi}{3}$, and $-\frac{11\pi}{3}$ are coterminal angles.

Example 7 Finding Coterminal Angles Using Radian Measure

Find the angle of least nonnegative measure, θ_C, that is coterminal with $\theta = -\frac{21\pi}{4}$.

Solution The angles coterminal with $\theta = -\frac{21\pi}{4}$ have the form $-\frac{21\pi}{4} + k \cdot 2\pi$, where k is an integer. Because θ is a negative angle, we choose positive integer values of k until we find the angle of least nonnegative measure.

$$k = 1: -\frac{21\pi}{4} + (1)2\pi = -\frac{21\pi}{4} + 2\pi = -\frac{21\pi}{4} + \frac{8\pi}{4} = -\frac{13\pi}{4}$$

$$k = 2: -\frac{21\pi}{4} + (2)2\pi = -\frac{21\pi}{4} + 4\pi = -\frac{21\pi}{4} + \frac{16\pi}{4} = -\frac{5\pi}{4}$$

$$k = 3: -\frac{21\pi}{4} + (3)2\pi = -\frac{21\pi}{4} + 6\pi = -\frac{21\pi}{4} + \frac{24\pi}{4} = \frac{3\pi}{4}$$

The angle of least nonnegative measure that is coterminal with $\theta = -\frac{21\pi}{4}$ is $\theta_C = \frac{3\pi}{4}$.

(eText Screens 6.1-1–6.1-27)

CAUTION Do not write $-\frac{21\pi}{4} = \frac{3\pi}{4}$. These angles are coterminal but they are not equal.

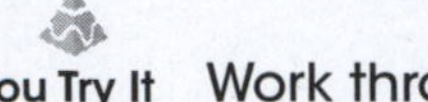

You Try It Work through this You Try It problem.

Work Exercises 21–26 in this textbook or in the MyLab Math Study Plan.

6.1 Exercises

Skill Check Exercises

For exercises SCE-1 and SCE-2, simplify the expression.

SCE-1. $(-252)\left(\frac{\pi}{180}\right)$

SCE-2. $\left(\frac{5\pi}{6}\right)\left(\frac{180}{\pi}\right)$

For exercises SCE-3 and SCE-4, rewrite the expression 2π as a fraction using the given denominator.

SCE-3. $2\pi = \frac{\square\pi}{2}$

SCE-4. $2\pi = \frac{\square\pi}{5}$

For exercises SCE-5 and SCE-6, determine the numerator on the right-hand side of the equation for the specified value of k.

SCE-5. $\frac{8\pi}{4} \cdot k = \frac{\square}{4}$ when $k = 2$

SCE-6. $\frac{10\pi}{5} \cdot k = \frac{\square}{5}$ when $k = -3$

In Exercises 1–5, draw each angle in standard position and state the quadrant in which the terminal side of the angle lies or the axis on which the terminal side of the angle lies.

1. $15°$ 2. $270°$ 3. $-150°$ 4. $480°$ 5. $-570°$

6. Find the angle of least nonnegative measure, θ_C, that is coterminal with $\theta = 848°$.

7. Find the angle of least nonnegative measure, θ_C, that is coterminal with $\theta = -622°$.

8. Find the angle of least nonnegative measure that is coterminal with $-51°$. Then find the measure of the negative angle that is coterminal with $-51°$ such that the angle lies between $-720°$ and $-360°$.

9. Find two positive angles and two negative angles that are coterminal with the quadrantal angle $\theta = -630°$ such that each angle lies between $-1440°$ and $720°$.

In Exercises 10–19, draw each angle in standard position and state the quadrant in which the terminal side of the angle lies or the axis on which the terminal side of the angle lies.

10. 5π 11. -4π 12. $\frac{11\pi}{6}$ 13. $-\frac{13\pi}{6}$ 14. $\frac{9\pi}{4}$

15. $-\frac{5\pi}{4}$ 16. $\frac{5\pi}{2}$ 17. $-\frac{7\pi}{2}$ 18. $\frac{4\pi}{3}$ 19. $-\frac{10\pi}{3}$

In Exercises 20–24, convert each angle given in degree measure into radians.

20. $15°$ 21. $150°$ 22. $-540°$ 23. $96°$ 24. $-357°$

In Exercises 25–30, convert each angle given in radian measure into degrees. Round to two decimal places if needed.

25. $\frac{\pi}{12}$ radians 26. $\frac{3\pi}{5}$ radians 27. $-\frac{5\pi}{2}$ radians

28. $\frac{8\pi}{7}$ radians 29. $-\frac{29\pi}{15}$ radians 30. 3 radians

31. Find the angle of least nonnegative measure, θ_C, that is coterminal with $\theta = \frac{27\pi}{8}$.

32. Find the angle of least nonnegative measure, θ_C, that is coterminal with $\theta = \frac{41\pi}{6}$.

33. Find the angle of least nonnegative measure, θ_C, that is coterminal with $\theta = -\frac{29\pi}{3}$.

34. Find the angle of least nonnegative measure, θ_C, that is coterminal with $\theta = -\frac{\pi}{70}$.

35. Find the angle of least nonnegative measure, θ_C, that is coterminal with $-\frac{11\pi}{12}$. Then find the measure of the negative angle that is coterminal with $-\frac{11\pi}{12}$ such that the angle lies between -4π and -2π.

36. Find two positive angles and two negative angles that are coterminal with the quadrantal angle $\theta = -\frac{7\pi}{2}$ such that each angle lies between -8π and 4π.

6.2 Applications of Radian Measure

THINGS TO KNOW

Before working through this section, be sure that you are familiar with the following concepts:

VIDEO ANIMATION INTERACTIVE

1. Converting between Degree Measure and Radian Measure (Section 6.1)

INTRODUCTION

Read this introduction before beginning Objective 1.

OBJECTIVES

1 Determining the Area of a Sector of a Circle

2 Computing the Arc Length of a Sector of a Circle

3 Understanding Linear Velocity and Angular Velocity

SECTION 6.2 EXERCISES

Introduction to Section 6.2

In this section, we will look at various applications of radian measure. You may want to watch this **animation**, which was introduced in the previous section, to review the definition of a radian and to verify that you understand the relationship between radians and degrees.

OBJECTIVE 1 DETERMINING THE AREA OF A SECTOR OF A CIRCLE

A **sector** of a circle is a portion of a circle bounded by two radii and an arc of the circle. See Figure 15.

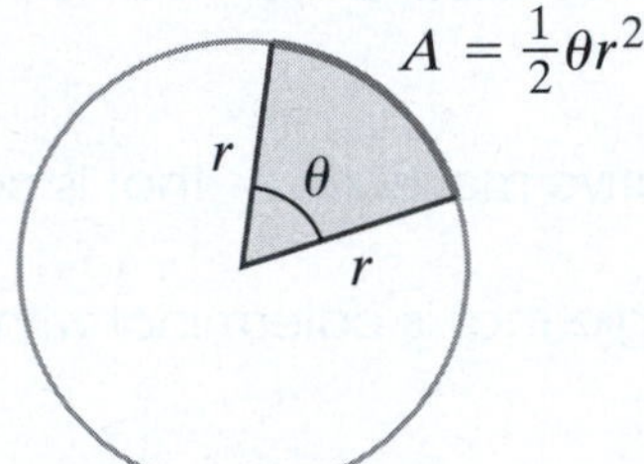

Figure 15 A sector of a circle

Recall that the area of a circle is $A = \pi r^2$. The area of a sector of a circle is simply a portion of the total area. To establish a formula for the area of a sector of a circle, let's look at a concrete example. Suppose that the central angle of a sector of a circle is $\theta = \frac{\pi}{4}$. We know that there are 2π radians in an entire circle. Thus, the sector with a central angle of $\theta = \frac{\pi}{4}$ represents a portion of the circle equal to $\frac{\frac{\pi}{4}}{2\pi} = \frac{\cancel{\pi}}{4} \cdot \frac{1}{2\cancel{\pi}} = \frac{1}{8}$. Therefore, the area of the sector of the circle must be $\frac{1}{8}$ of the area of the entire circle. So, the area of a sector of a circle with a central angle of $\theta = \frac{\pi}{4}$ is $A = \frac{1}{8}\pi r^2$.

In general, if θ is the central angle of a sector of a circle in radians, then the sector represents a portion of the circle equal to $\frac{\theta}{2\pi}$ of the entire circle. Thus, the area, A, of a sector of a circle is given by $A = \left(\frac{\theta}{2\cancel{\pi}}\right)(\cancel{\pi} r^2) = \frac{1}{2}\theta r^2$.

Area of a Sector of a Circle

The area, A, of a sector of a circle with radius, r, and central angle θ radians is given by $A = \dfrac{\theta}{2\pi}(\pi r^2) = \dfrac{1}{2}\theta r^2$.

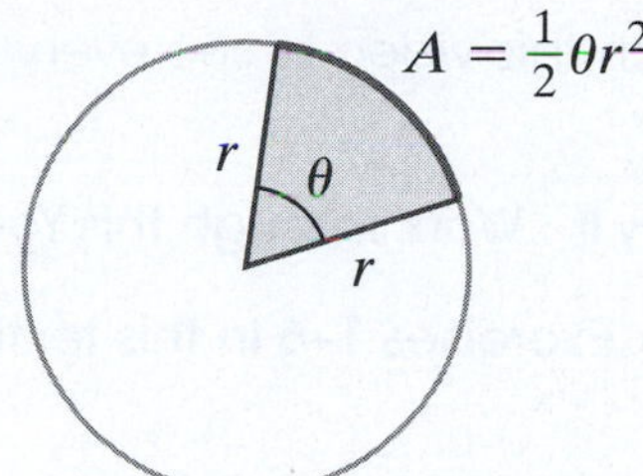

Example 1 Computing the Area of a Sector of a Circle

Find the area of the sector of a circle of radius 15 inches formed by a central angle of $\theta = \dfrac{\pi}{10}$ radians. Round the answer to two decimal places.

Solution We can use the formula $A = \dfrac{1}{2}\theta r^2$ with $\theta = \dfrac{\pi}{10}$ and $r = 15$ inches.

$A = \dfrac{1}{2}\theta r^2$	Write the formula for the area of a sector of a circle.
$= \dfrac{1}{2}\left(\dfrac{\pi}{10}\right)(15 \text{ in.})^2$	Substitute $\theta = \dfrac{\pi}{10}$ and $r = 15$ in.
$= \left(\dfrac{\pi}{20}\right)(225 \text{ in.}^2)$	Multiply.
$= \dfrac{45\pi}{4} \text{ in.}^2$	Simplify.
$\approx 35.34 \text{ in.}^2$	Round to two decimal places.

The area of the sector of the circle is approximately 35.34 square inches.

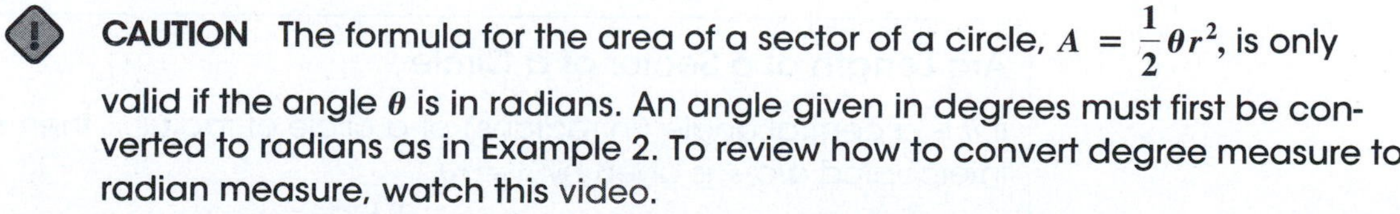

CAUTION **The formula for the area of a sector of a circle, $A = \dfrac{1}{2}\theta r^2$, is only valid if the angle θ is in radians. An angle given in degrees must first be converted to radians as in Example 2. To review how to convert degree measure to radian measure, watch this video.**

Example 2 Computing the Area of a Sector of a Circle

Find the area of the sector of a circle of diameter 21 meters formed by a central angle of $135°$. Round the answer to two decimal places.

Solution Because the radius of the circle is half the diameter, we know that $r = \dfrac{21}{2} = 10.5$ meters. We must now convert the central angle of $135°$ to radians.

$$135° = 135\not{°}\,\frac{\pi \text{ radians}}{180\not{°}} = \frac{135\pi}{180} \text{ radians} = \frac{3\pi}{4} \text{ radians}$$

We can now find the area of the sector of the circle using $r = 10.5$ and $\theta = \dfrac{3\pi}{4}$.

Therefore, the area of the sector of the circle is

$$A = \frac{1}{2}\theta r^2 = \frac{1}{2}\left(\frac{3\pi}{4}\right)(10.5)^2 \approx 129.89 \text{ square meters.}$$

Watch this **video** to see every step of this solution.

You Try It **Work through this You Try It problem.**

Work Exercises 1–6 in this textbook or in the MyLab Math Study Plan.

OBJECTIVE 2 COMPUTING THE ARC LENGTH OF A SECTOR OF A CIRCLE

The arc length of a sector of a circle depends on the corresponding **central angle** that intercepts the arc and the length of the radius of the circle. See Figure 16.

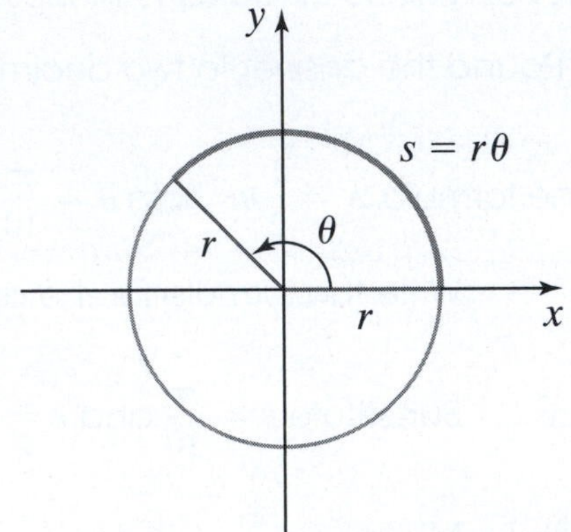

Figure 16 The arc length of a sector of a circle s, depends on the corresponding central angle θ and the length of the radius r.

Recall that the circumference of an entire circle is equal to $2\pi r$. The arc length of a sector of a circle is simply a portion of the circumference. We established that a sector represents a portion of the circle equal to $\dfrac{\theta}{2\pi}$ of the entire circle. Thus, the arc length, s, of a sector of a circle is given by $s = \left(\dfrac{\theta}{\cancel{2\pi}}\right)(\cancel{2\pi}r) = r\theta$.

Arc Length of a Sector of a Circle

If θ is a central angle (in radians) of a circle of radius r, then the length of the intercepted arc, s, is given by $s = r\theta$.

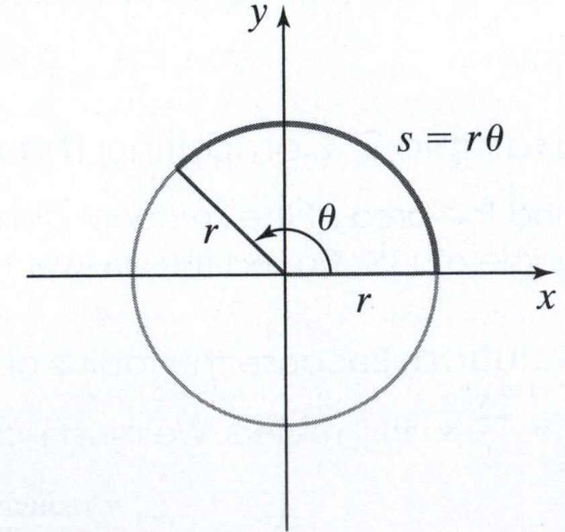

Example 3 Computing the Arc Length of a Sector of a Circle

Find the length of the arc intercepted by a central angle of $\theta = \frac{\pi}{6}$ in a circle of radius $r = 22$ centimeters. Round the answer to two decimal places.

Solution We can use the arc length formula $s = r\theta$ with $\theta = \frac{\pi}{6}$ and $r = 22$ cm. Therefore, the length of the arc is $s = r\theta = (22)\left(\frac{\pi}{6}\right) = \frac{22\pi}{6} = \frac{11\pi}{3} \approx 11.52$ cm. ●

CAUTION The arc length formula $s = r\theta$ is only valid if the angle θ is in radians. An angle given in degrees must first be converted to radians. To review how to convert degree measure to radian measure, watch this video.

Example 4 Calculate the Radius of a Circle

A $120°$ central angle intercepts an arc of 23.4 inches. Calculate the radius of the circle. Round the answer to two decimal places.

Solution First, we must convert the angle to radian measure.

$$120° = 120\cancel{°} \cdot \frac{\pi \text{ radians}}{180\cancel{°}} = \frac{120\pi}{180}\text{ radians} = \frac{2\pi}{3}\text{ radians}$$

Now, substitute $s = 23.4$ in. and $\theta = \frac{2\pi}{3}$ into the arc length formula $s = r\theta$ and solve for r.

$s = r\theta$	Write the formula for the arc length of a sector of a circle.
$23.4 \text{ in.} = r\left(\frac{2\pi}{3}\right)$	Substitute $s = 23.4$ in. and $\theta = \frac{2\pi}{3}$.
$\frac{3}{2\pi} \cdot 23.4 \text{ in.} = r\left(\frac{\cancel{2\pi}}{\cancel{3}}\right) \cdot \frac{\cancel{3}}{\cancel{2\pi}}$	Multiply both sides by the reciprocal of $\frac{2\pi}{3}$.
$11.17 \text{ in.} \approx r$	Simplify and round to two decimal places.

The radius is $r \approx 11.17$ inches. For the solution to this example, watch this **video**. ●

Example 5 Calculate the Angle of Rotation of a Mechanical Gear

Two gears are connected so that the smaller gear turns the larger gear. When the smaller gear with a radius of 3.6 centimeters rotates $300°$, how many degrees will the larger gear with a radius of 7.5 cm rotate?

Solution We will first determine how far the smaller gear rotates. The smaller gear has a radius of $r_1 = 3.6$ cm and an angle of rotation of $\theta_1 = 300°$, which is equivalent to $\theta_1 = \frac{5\pi}{3}$. (Watch this **video** to verify.) We can use the arc length formula $s = r_1\theta_1$ to find the distance that the small gear rotates.

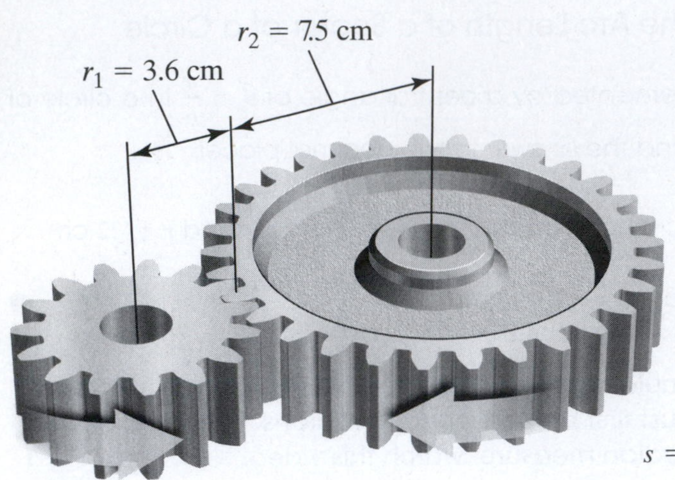

$s = r_1\theta_1$ Write the arc length formula.

$= (3.6\text{ cm})\left(\frac{5\pi}{3}\right)$ Substitute $r_1 = 3.6\text{ cm}$ and $\theta_1 = \frac{5\pi}{3}$.

$= 6\pi\text{ cm}$ Multiply.

Because the gears are connected, the larger gear will also rotate a distance of 6π cm. Therefore, we can use the radius of the larger gear, $r_2 = 7.5\text{ cm}$, and the arc length, $s = 6\pi\text{ cm}$, to find the amount of rotation of the larger gear.

$s = r_2\theta_2$ Write the arc length formula.

$6\pi\text{ cm} = (7.5\text{ cm})(\theta_2)$ Substitute $s = 6\pi\text{ cm}$ and $r_2 = 7.5\text{ cm}$.

$\frac{6\pi}{7.5} = \theta_2$ Divide both sides by 7.5 cm.

$\theta_2 = \frac{4\pi}{5}$ Simplify and rewrite.

The larger gear rotates $\theta_2 = \frac{4\pi}{5}$ radians which can be **converted** to $\theta_2 = 144°$. You may want to watch this **video** to see the solution to this example worked out in detail.

You Try It Work through this You Try It problem.

Work Exercises 7–14 in this textbook or in the MyLab Math Study Plan.

OBJECTIVE 3 UNDERSTANDING ANGULAR VELOCITY AND LINEAR VELOCITY

Suppose that an object is moving along a circular path such as the rotor blade of a helicopter, children riding on a merry-go-round, or a tire on a moving bicycle. In the following discussion, we assume that all objects are moving along a circular path at a constant speed.

There are two different measures of the speed or velocity of an object traveling on a circular path. **Angular velocity** measures the rotation of the object over time and is usually measured in radians per unit of time. **Linear velocity** measures the distance that a particular point on the object travels over time and is usually measured in distance (such as miles or meters) per unit of time. Typically, the Greek letter ω (omega) is used to describe angular velocity and the letter v is used to describe linear velocity.

> **Angular Velocity**
>
> The angular velocity, ω, of an object moving along a circular path is the amount of rotation, θ, in radians of the object per unit of time, t, and is given by $\omega = \frac{\theta}{t}$.

Linear Velocity

The linear velocity, v, of an object moving along a circular path is the distance traveled by the object, s, per unit of time, t, and is given by $v = \frac{s}{t}$.

Note that $s = r\theta$, where θ is given in radians.

Example 6 Computing the Angular Velocity and the Linear Velocity of the Earth

Determine the angular velocity and the linear velocity of a point on the equator of the Earth. Assume that the radius of the Earth at the equator is 3963 miles.

Solution To determine the angular velocity of a point on the equator of the Earth, we first use the fact that the Earth makes one complete revolution every 24 hours. Therefore, any point on the surface of the Earth, not at a pole, rotates $\theta = 2\pi$ radians every 24 hours. Thus, a point on the equator of the Earth has an angular velocity of

$$\omega = \frac{\theta}{t} = \frac{2\pi \text{ rad}}{24 \text{ hr}} = \frac{\pi}{12} \frac{\text{rad}}{\text{hr}}.$$

To determine the linear velocity of a point on the equator of the Earth, we use the fact that an object at the equator travels a distance, s, equal to the circumference of the Earth, in 24 hours. Therefore, an object at the equator travels $s = 2\pi r = (2\pi)(3963 \text{ miles}) = 7926\pi$ miles every 24 hours. Thus, a point on the equator of the Earth has a linear velocity of $v = \frac{s}{t} = \frac{7926\pi \text{ miles}}{24 \text{ hr}} \approx 1038$ mph. (See **Figure 17**.)

Figure 17 The angular velocity of an object located on the equator of the Earth is $\omega = \frac{\pi}{12} \frac{\text{rad}}{\text{hr}}$. The linear velocity of the same object is $v \approx 1038$ mph.

Example 7 Computing Angular Velocity

The propeller of a small airplane is rotating 1500 revolutions per minute. Find the angular velocity of the propeller in radians per minute.

Solution The propeller rotates 1500 revolutions per minute. In one minute, the propeller rotates through an angle of $\theta = \frac{2\pi \text{ rad}}{\cancel{\text{rev}}} \cdot 1500 \cancel{\text{rev}} = 3000\pi \text{ rad}$. We now use the formula $\omega = \frac{\theta}{t}$ with $\theta = 3000\pi \text{ rad}$ and $t = 1 \text{ min}$ to get $\omega = \frac{3000\pi \text{ rad}}{1 \text{ min}} = 3000\pi \frac{\text{rad}}{\text{min}}$.

Example 8 Computing Linear Velocity

In 2008, the old Ferris wheel at the world-famous Santa Monica pier in Santa Monica, California, was replaced with a new solar-powered Ferris wheel that contains thousands of energy-efficient LED (light-emitting diode) lights that can illuminate into many different colors and designs. The wheel is approximately 85 feet in diameter and rotates 2.5 revolutions per minute. Find the linear velocity (in mph) of this new Ferris wheel at the outer edge. Round to two decimal places.

Solution The linear velocity is given by the formula $v = \frac{s}{t}$, where s is the distance traveled and t is time.

First, we will find the distance travelled, s. Recall that $s = r\theta$, where θ is the central angle and r is the radius. There are 2π radians in one complete rotation. The wheel rotates 2.5 revolutions per minute, so in 1 minute the wheel rotates through an angle of $\theta = (2.5)(2\pi) = 5\pi$ radians. We can now use one-half of the diameter, or $r = 42.5$ feet, to get the distance traveled in 1 minute.

$s = r\theta$	Write the formula for arc length.
$s = (42.5 \text{ ft})(5\pi)$	Substitute $r = 42.5 \text{ ft}$ and $\theta = 5\pi$.
$s = (212.5)\pi \text{ ft}$	Multiply.

Now we find the linear velocity, v. A point on the outer edge of the Ferris wheel travels $(212.5)\pi$ feet in 2.5 revolutions (or in 1 minute). Therefore,

$v = \frac{s}{t}$	Write the formula for linear velocity.
$v = \frac{(212.5)\pi \text{ ft}}{1 \text{ min}}$	Substitute $s = (212.5)\pi \text{ ft}$ and $t = 1 \text{ minute}$.

To convert the linear velocity to miles per hour, we use the fact that there are 60 minutes in 1 hour and that there are 5280 feet in 1 mile.

$$v = \frac{(212.5)\pi \cancel{\text{ft}}}{1 \cancel{\text{min}}} \cdot \frac{60 \cancel{\text{min}}}{1 \text{ hr}} \cdot \frac{1 \text{ mile}}{5280 \cancel{\text{ft}}} \approx 7.59 \text{ mph}$$

Therefore, the outer edge of the Ferris wheel has a linear velocity of approximately 7.59 miles per hour.

Because angular velocity and linear velocity are related, we can develop a formula to convert easily from one to the other.

$v = \frac{s}{t}$ Write the formula for linear velocity.

$v = \frac{r\theta}{t}$ Use the arc length formula $s = r\theta$ and substitute $r\theta$ for s.

$v = r\frac{\theta}{t}$ $\frac{r\theta}{t} = r\frac{\theta}{t}$

$v = r\omega$ Use the formula for angular velocity $\omega = \frac{\theta}{t}$ and substitute ω for $\frac{\theta}{t}$.

Relationship Between Linear Velocity and Angular Velocity

If ω is the angular velocity of an object moving along a circular path of radius r, then the linear velocity, v, is given by $v = r\omega$.

To illustrate the relationship between linear velocity and angular velocity, let's revisit Example 6, where we computed the angular velocity and linear velocity of an object located on the equator of the Earth.

Object Located on the Equator of the Earth:

$$\omega = \frac{\pi}{12}\frac{\text{rad}}{\text{hr}}$$

$$v = r\omega = (3963 \text{ miles})\left(\frac{\pi}{12}\frac{\text{rad}}{\text{hr}}\right) \approx 1038 \text{ mph}^*$$

You can see that if we know the angular velocity and the radius of the circular object, then we can easily determine the linear velocity.

*An angle measured in radians can be thought of as the ratio of an arc length of a sector of a circle to the radius, or $\theta = \frac{s}{r}$. Because s and r have the same unit of measure, the ratios of those units cancel each other out. Thus, a radian has no unit of measure and is said to be "dimensionless." When multiplying an expression involving radians with another expression involving any unit of measure, the radians can always be dropped from the final expression.

Example 9 Computing Linear Velocity

Suppose that the propeller blade from **Example 7** is 2.5 meters in diameter. Find the linear velocity in meters per minute for a point located on the tip of the propeller.

Solution Using the formula $v = r\omega$ with $r = \frac{2.5}{2} = 1.25$ meters and $\omega = 3000\pi\frac{\text{rad}}{\text{min}}$, we get

$$v = r\omega = (1.25 \text{ m})\left(3000\pi\frac{\text{rad}}{\text{min}}\right) = 3750\pi\frac{\text{m}}{\text{min}} \approx 11{,}781\frac{\text{m}}{\text{min}}.$$

You Try It Work through this You Try It problem.

Work Exercises 15–21 in this textbook or in the MyLab Math Study Plan.

6.2 Exercises

1. Find the area of the sector of a circle of radius 12 cm formed by a central angle of $\frac{1}{12}$ radian.

2. Find the area of the sector of a circle of radius 30 inches formed by a central angle of $\frac{\pi}{5}$ radians.

3. Find the area of the sector of a circle of radius 8 meters formed by a central angle of 60°.

4. A 16-inch-diameter pizza is cut into various sizes. What is the area of a piece that was cut with a central angle of 32°?

5. A sector of a circle has an area of 30 square feet. Find the central angle that forms the sector if the radius is 7 feet.

6. A farmer uses an irrigation sprinkler system that pivots about the center of a circular field. Determine the length of the sprinkler if an area of 2140 square feet is watered after a pivot of 50°.

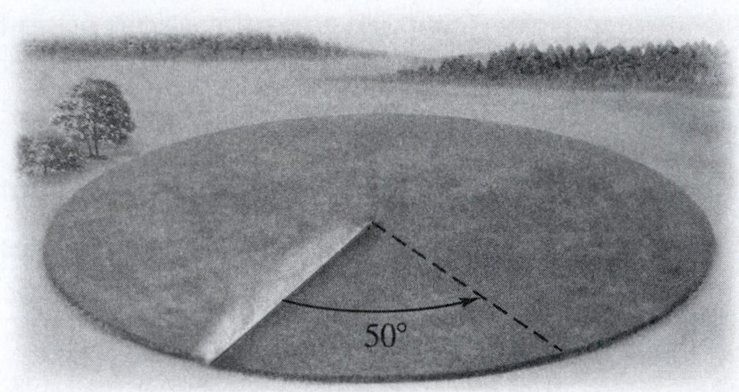

7. Find the length of the arc intercepted by a central angle of $\theta = \frac{\pi}{10}$ in a circle of radius $r = 30$ inches.

8. Find the length of the arc intercepted by a central angle of $\theta = \frac{4\pi}{3}$ in a circle of radius $r = 54$ centimeters.

9. Find the length of the arc intercepted by a central angle of $\theta = \frac{5\pi}{6}$ in a circle of radius $r = 31$ meters.

10. Find the length of the arc intercepted by a central angle of 305° in a circle of radius $r = 40$ kilometers.

11. Find the measure of the central angle (in radians) of a circle of radius $r = 25$ inches that intercepts an arc of length $s = 40$ inches.

12. On the planet Mercury, a 45° central angle intercepts an arc of approximately 379π miles. What is the radius of the planet mercury at the equator?

13. Find the central angle in degrees of an arc of 5000 km on the surface of Neptune if the radius of Neptune is 24,553 km.

14. Two gears of a small mechanical device are connected so that the smaller gear turns the larger gear. If the smaller gear with a radius of 12.2 millimeters rotates 225°, how many degrees will the larger gear with a radius of 22.5 millimeters rotate?

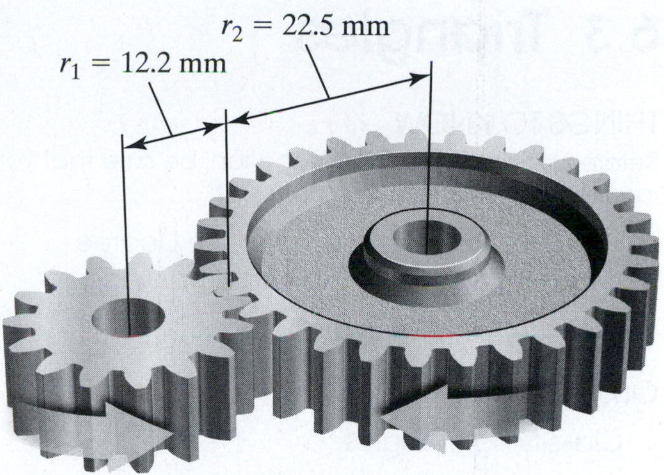

15. Find the angular velocity in radians per minute of a microwave turntable that turns through an angle of 36° each second.

16. The use of wind-powered energy is becoming more and more prevalent around the world to produce clean energy. A windmill with a blade 40 feet long rotates at a rate of 50 revolutions per minute. Find the angular velocity of the blade.

17. What is the linear velocity of the tip of the windmill blade described in Exercise 16?

18. In 2009, the Singapore Flyer was the largest Ferris wheel in the world at approximately 75 meters in diameter. The wheel rotates once every 30 minutes. Find the linear speed of the Singapore Flyer at the outer edge of the wheel.

19. The blade of a circular saw rotates at a rate of 3000 revolutions per minute. What is the linear velocity in miles per hour of a point on the tip of the outer edge of a $7\frac{1}{4}$-inch-diameter blade?

20. Fishing line is being pulled from a circular fishing reel at a rate of 54 cm per second. Find the radius of the reel if it makes 160 revolutions per minute.

21. Suppose that two 3-blade windmills rotate 80 times per minute. What is the difference in linear velocity, in miles per hour, of the tips of the windmill blades if the larger windmill has 75-foot blades and the smaller windmill has 60-foot blades?

6.3 Triangles

THINGS TO KNOW

Before working through this section, be sure that you are familiar with the following concept:

VIDEO ANIMATION INTERACTIVE

You Try It

1. Converting between Degree Measure and Radian Measure (Section 6.1)

OBJECTIVES

OBJECTIVE 1 CLASSIFYING TRIANGLES

The word *trigonometry* comes from the Greek words *trigonon*, meaning "triangle," and *metron*, meaning "measure." In the first section of Chapter 6, we introduced angle measure. In this section, we introduce some fundamentals of triangles. You have undoubtedly worked with triangles in previous courses. Thus, much of this section may be review for you.

Every triangle has three angles and three sides. The sum of the measures of the three angles is equal to $180°$ or π radians. Two angles of a triangle are **congruent** if the measures of the two angles are the same. Similarly, two sides of a triangle are congruent if the lengths of the two sides are equal.

We can classify triangles based on the measures of their angles or the lengths of their sides. We start by classifying triangles based on their angle measures. We will use the notation $\measuredangle$ to represent an angle. The notation $\measuredangle A$ refers to angle A. A triangle with three **acute angles** is called an **acute triangle**. A triangle with one **obtuse angle** is called an **obtuse triangle**. A triangle with one **right angle** is called a **right triangle**. The side opposite the right angle of a right triangle is called the **hypotenuse** and the two other sides of a right triangle are called **legs**. Also, note that the hypotenuse and both legs can be referred to as "sides" but that the hypotenuse is never referred to as a "leg." **Figure 18** illustrates the three classifications of triangles based on the measures of their angles.

Acute Triangle

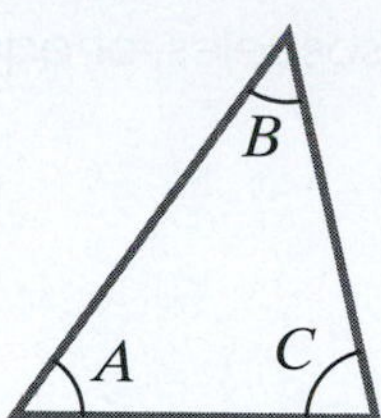

Acute triangles have three acute angles.
$\measuredangle A < 90°$, $\measuredangle B < 90°$, and $\measuredangle C < 90°$

Obtuse Triangle

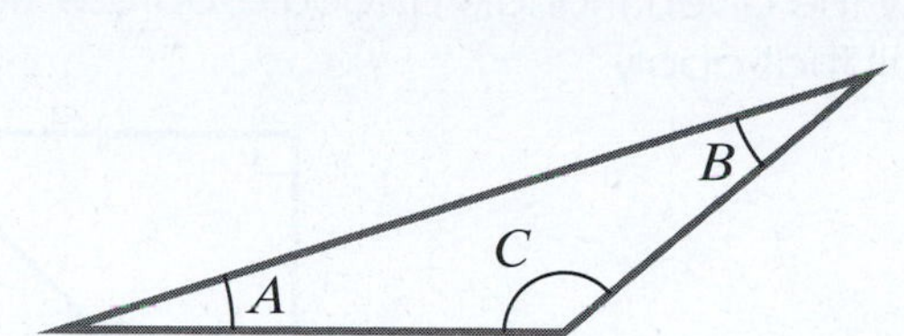

Obtuse triangles have one obtuse angle.
$\measuredangle A < 90°$, $\measuredangle B < 90°$, and $\measuredangle C > 90°$

Right Triangle

hypotenuse
B
leg
A
C
leg

Right triangles have one right angle.
$\measuredangle A < 90°$, $\measuredangle B < 90°$, and $\measuredangle C = 90°$

Figure 18 Classification of triangles based on their angle measures

We can also classify triangles based on their side lengths. We can represent that two sides of a triangle are congruent by drawing the same number of "tick marks" through the congruent sides. A triangle with no congruent sides is called a **scalene triangle**. See **Figure 19a**. A triangle with two congruent sides is called an **isosceles triangle**. See **Figure 19b**. The angles opposite the congruent sides of an isosceles triangle are also congruent. A special triangle with three congruent sides is called an **equilateral triangle**. Note that all three angles of an equilateral triangle are congruent with a measure of 60° or $\frac{\pi}{3}$ radians. See **Figure 19c**.

Scalene Triangle

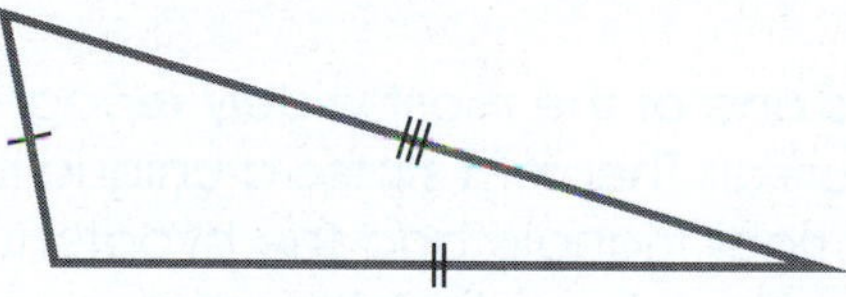

(a) Scalene triangles have no congruent sides. Each side is represented by a different number of "tick marks" that represents that the length of each side is different.

Isosceles Triangle

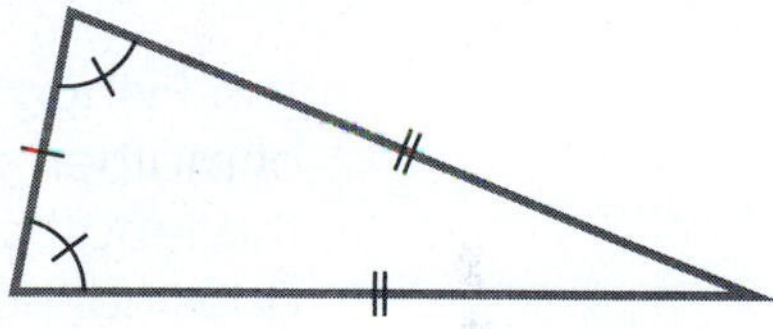

(b) Isosceles triangles have two congruent sides. The angles opposite the two congruent sides are congruent.

Equilateral Triangle

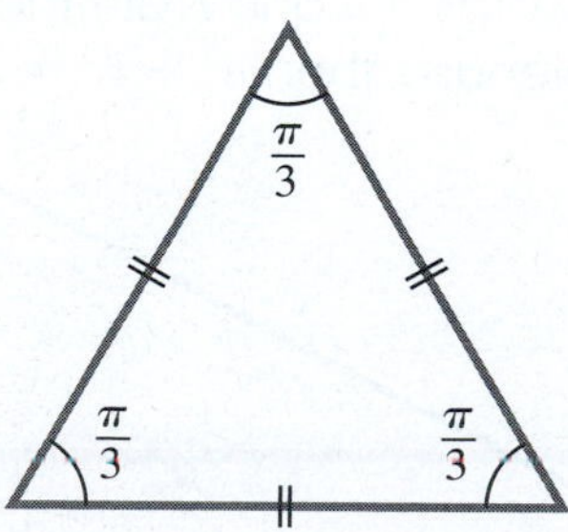

(c) Equilateral triangles have three congruent sides and three congruent angles. Each angle has a measure of 60° or $\frac{\pi}{3}$ radians.

Figure 19 The classification of triangles based on their side lengths

Example 1 Classifying a Triangle

Classify the given triangle as acute, obtuse, right, scalene, isosceles, or equilateral. State all that apply.

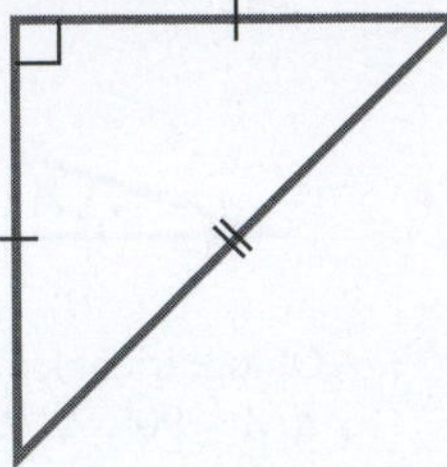

Solution The triangle is a right triangle because it has a right angle. The two legs of the triangle are congruent, as indicated by the single tick marks. Therefore, this is an isosceles right triangle. ●

The triangle in Example 1 is an isosceles right triangle. It is important to point out that every isosceles right triangle has two acute angles that have a measure of $\frac{\pi}{4}$ radians or 45°. Do you know why? Watch this **animation** for an explanation.

You Try It **Work through this You Try It problem.**

Work Exercises 1–3 in this textbook or in the MyLab Math Study Plan.

OBJECTIVE 2 USING THE PYTHAGOREAN THEOREM

The Pythagorean Theorem is one of the most widely recognized theorems in all of mathematics. The Pythagorean Theorem states a unique relationship between the lengths of the legs of a right triangle and the hypotenuse. This relationship does not hold for acute triangles or for obtuse triangles.

The Pythagorean Theorem

Given any right triangle, the sum of the squares of the lengths of the legs is equal to the square of the length of the hypotenuse.

In other words, if a and b are the lengths of the two legs and if c is the length of the hypotenuse, then $a^2 + b^2 = c^2$. See Figure 20.

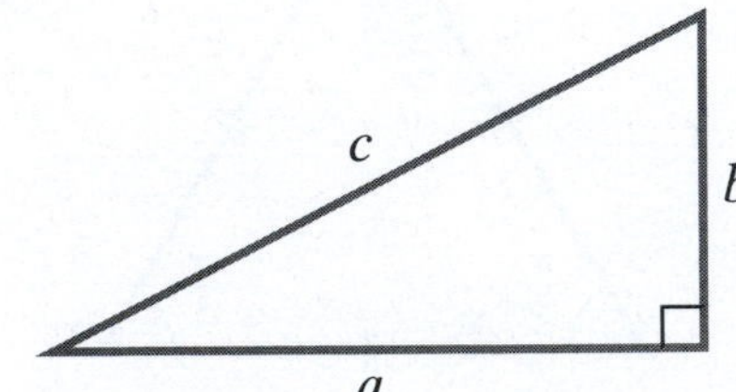

Figure 20 The Pythagorean Theorem: $a^2 + b^2 = c^2$

Example 2 Using the Pythagorean Theorem to Find the Lengths of the Missing Sides

Use the Pythagorean Theorem to find the length of the missing side of the given right triangle.

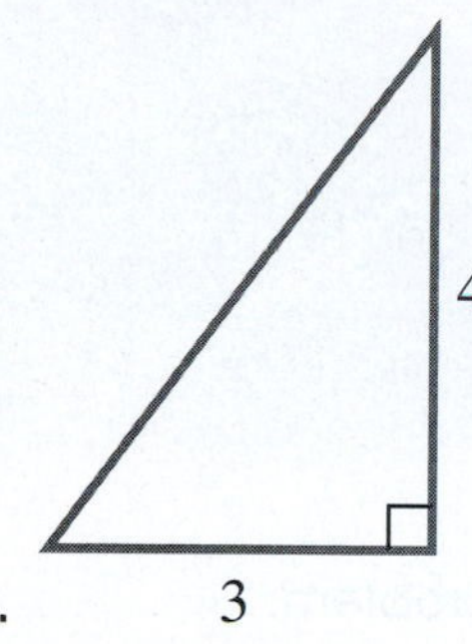

a.

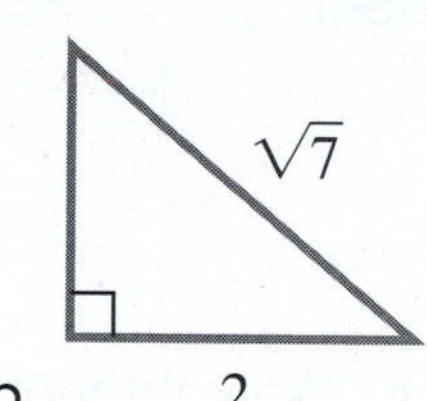

b.

Solution

a. We are given the length of both legs. Using the equation $a^2 + b^2 = c^2$ with $a = 3$ and $b = 4$, we can solve for c.

$a^2 + b^2 = c^2$	Write the equation representing the Pythagorean Theorem.
$(3)^2 + (4)^2 = c^2$	Substitute $a = 3$ and $b = 4$.
$9 + 16 = c^2$	Simplify.
$25 = c^2$	Add.
$\pm 5 = c$	Use the square root property.

Because the length of the hypotenuse must be positive, the length is 5.

b. See if you can determine the length of the missing leg. When you think you are correct, view the **solution** or watch this **video**.

Notice in Example 2a that all three sides of the right triangle have a length represented by an integer. This is very special, and these numbers are called Pythagorean triples. In addition to 3-4-5, other examples of Pythagorean triples are 5-12-13 and 8-15-17. Multiples of these numbers are also Pythagorean triples, such as 6-8-10, 9-12-15, 12-16-20, and so on. Being familiar with these Pythagorean triples can often save you time and effort in working trigonometric exercises, so you might want to become very familiar with them. You may want to view this interesting technique used for **generating Pythagorean triples**.

You Try It **Work through this You Try It problem.**

Work Exercises 4–8 in this textbook or in the MyLab Math Study Plan.

Example 3 Using the Pythagorean Theorem to Find the Lengths of the Missing Sides

A Major League baseball "diamond" is really a square. The distance between each consecutive base is 90 feet. What is the distance between home plate and second base? Round to two decimal places.

Solution Watch this **video** to see how to use the Pythagorean Theorem to determine that the distance between home plate and second base is approximately 127.28 feet.

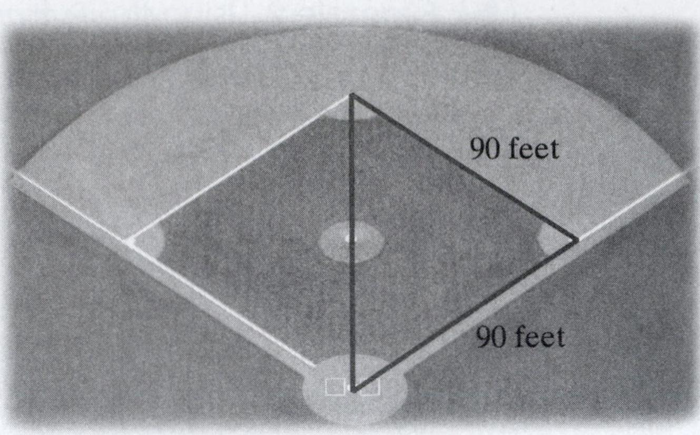

You Try It Work through this You Try It problem.

Work Exercises 9–12 in this textbook or in the MyLab Math Study Plan.

OBJECTIVE 3 UNDERSTANDING SIMILAR TRIANGLES

Imagine drawing a triangle on a piece of paper and placing the paper in a copy machine. If you increase (or decrease) the size of the image to be copied, the resulting image will be a triangle of exactly the same shape as the original triangles but it will be a different size. Triangles that have the same shape but not necessarily the same size are called **similar triangles**. Triangles ABC and XYZ in Figure 21 are similar.

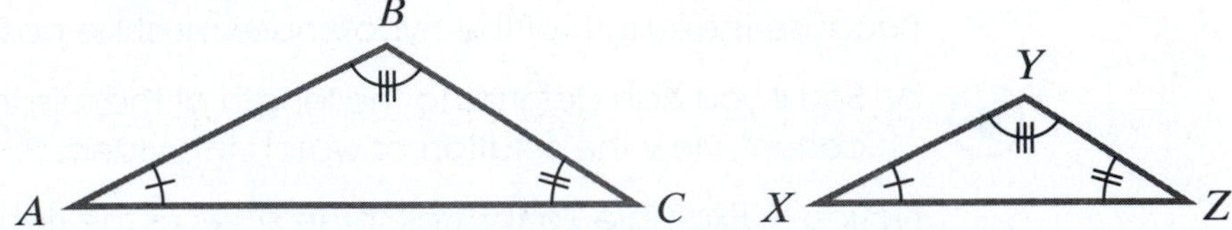

Figure 21 Triangle ABC and triangle XYZ are similar. Similar triangles have the same shape but not necessarily the same size.

The measures of the corresponding angles of the triangles in Figure 21 are equal (as indicated by the small arcs with tick marks) but the lengths of the corresponding sides are obviously different. However, the ratio of the lengths of any two sides of triangle ABC is equal to the ratio of the lengths of the corresponding sides of triangle XYZ. For example, the ratio of side AB to side AC must be equal to the ratio of side XY to side XZ. This can be written as $\frac{AB}{AC} = \frac{XY}{XZ}$.

Properties of Similar Triangles

1. The corresponding angles have the same measure.
2. The ratio of the lengths of any two sides of one triangle is equal to the ratio of the lengths of the corresponding sides of the other triangle.

Example 4 Determining the Side Lengths of Similar Triangles

Triangles ABC and XYZ are similar. Find the lengths of the missing sides of triangle ABC.

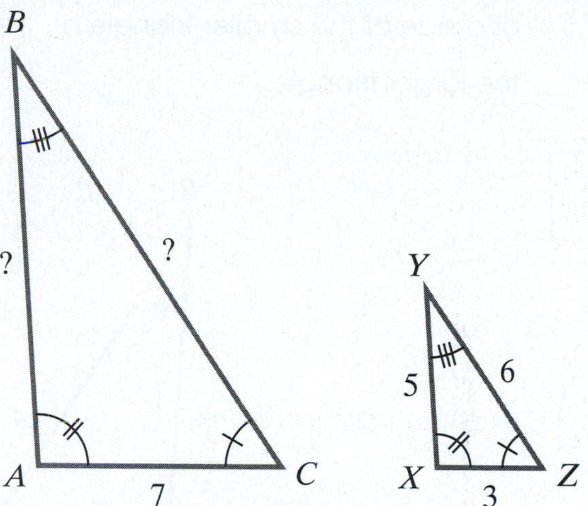

Solution We start by finding the length of side AB. Side AB corresponds to side XY. Also, side AC corresponds to side XZ. Therefore, we can set up the proportion $\frac{AB}{AC} = \frac{XY}{XZ}$.

$\frac{AB}{AC} = \frac{XY}{XZ}$	The lengths of corresponding sides of similar triangles are proportional.
$\frac{AB}{7} = \frac{5}{3}$	Substitute $AC = 7, XY = 5,$ and $XZ = 3$.
$3AB = 35$	Multiply both sides by the LCD, which is $(3)(7) = 21$.
$AB = \frac{35}{3}$	Divide both sides by 3.

The length of side AB is $\frac{35}{3}$.

To find the length of side BC, we can use the proportion $\frac{BC}{AC} = \frac{YZ}{XZ}$ to determine that the length of side BC is 14. Watch this **video** to see the entire solution to this example.

In Figure 22, we reconsider similar triangles ABC and XYZ from Example 4. Side AB corresponds to side XY, side BC corresponds to side YZ, and side AC corresponds to side XZ. Looking at the ratios of the lengths of these corresponding sides, we get $\frac{AB}{XY} = \frac{35/3}{5} = \frac{7}{3}, \frac{BC}{YZ} = \frac{14}{6} = \frac{7}{3}$, and $\frac{AC}{XZ} = \frac{7}{3}$. The number $\frac{7}{3}$ is called the **proportionality constant**. Note that the length of each side of triangle ABC is $\frac{7}{3}$ times the length of the corresponding side of triangle XYZ. Note also that the length of each side of triangle XYZ is $\frac{3}{7}$ times the length of the corresponding side of triangle ABC. For consistency, we will define the proportionality constant to always be greater than or equal to 1. Therefore, if two triangles are similar, then the length of a side of the larger triangle is k times the length of the corresponding side of the smaller triangle, where k is the proportionality constant. Also, the length of a side of the smaller triangle is $\frac{1}{k}$ times the length of the corresponding side of the larger triangle.

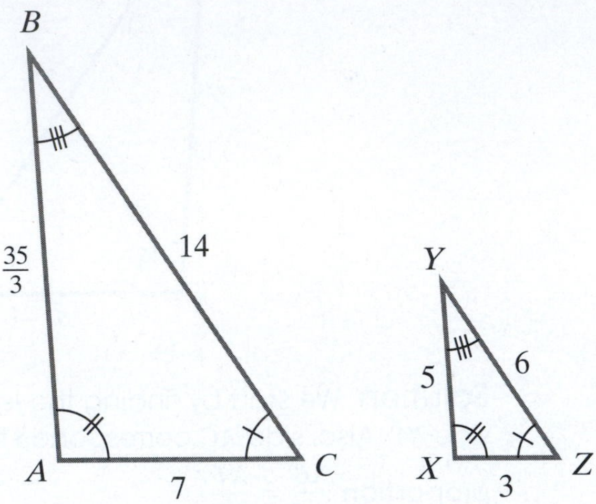

Figure 22 The lengths of the sides of triangle ABC are $k = \frac{7}{3}$ times the lengths of the corresponding sides of triangle XYZ.

Definition Proportionality Constant of Similar Triangles

If two triangles are similar, there exists a constant $k \geq 1$ called the **proportionality constant of similar triangles** equal to the ratio of the lengths of corresponding sides.

Given the similar triangles shown below, $k = \frac{a}{x} = \frac{b}{y} = \frac{c}{z}$, where $a \geq x, b \geq y$, and $c \geq z$.

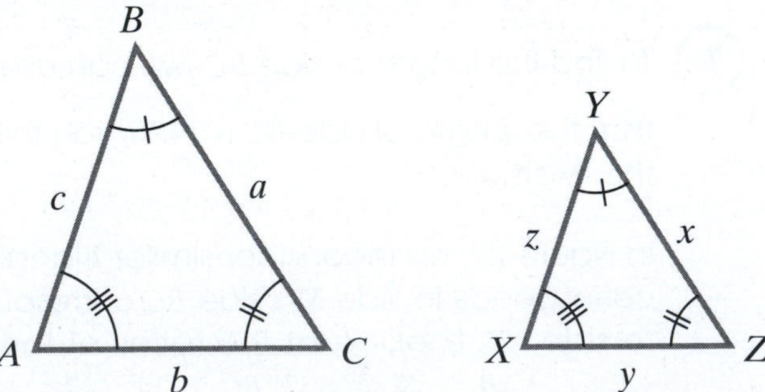

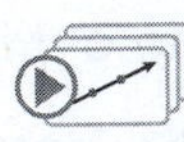

Example 5 Determining the Proportionality Constant of Similar Triangles

The triangles below are similar. Find the proportionality constant. Then find the lengths of the missing sides.

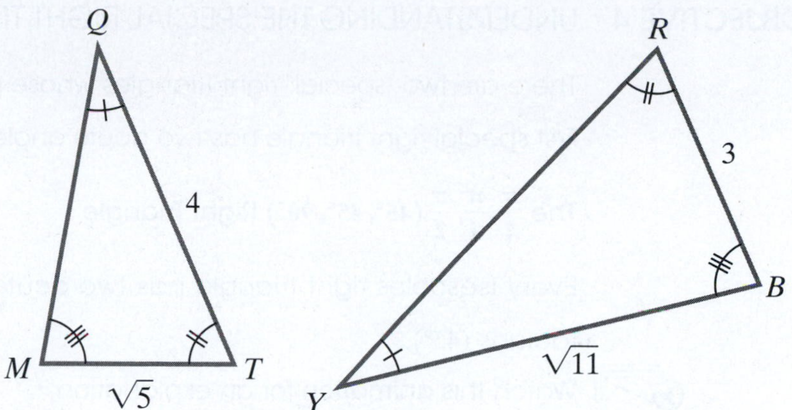

Solution Work through this animation to verify that $k = \frac{3}{\sqrt{5}}$ and that the lengths of missing sides YR and QM are $\frac{12}{\sqrt{5}}$ and $\frac{\sqrt{55}}{3}$, respectively.

You Try It Work through this You Try It problem.

Work Exercises 13–16 in this textbook or in the MyLab Math Study Plan.

Example 6 Determining the Side Lengths of Similar Right Triangles

The right triangles below are similar. Determine the lengths of the missing sides.

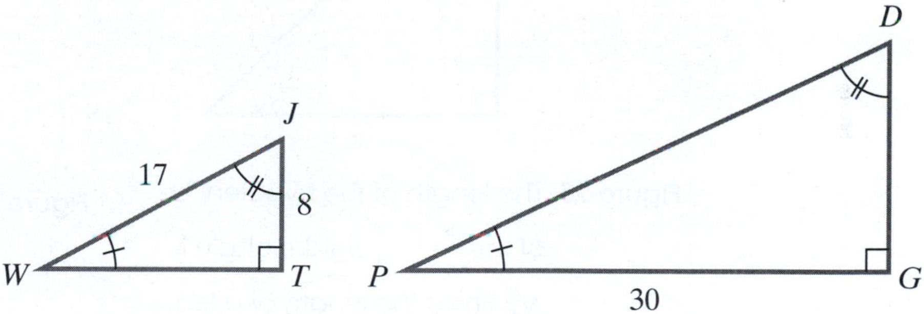

Solution Using the Pythagorean Theorem, we can determine the length of side WT. Once we have the length of side WT, we can find the proportionality constant, which will allow us to find the lengths of the missing sides of triangle PDG. Try to find the lengths of the missing sides yourself, and then watch this video to see if you are correct.

You Try It Work through this You Try It problem.

Work Exercises 17–20 in this textbook or in the MyLab Math Study Plan.

OBJECTIVE 4 UNDERSTANDING THE SPECIAL RIGHT TRIANGLES

There are two "special" right triangles whose properties are worth memorizing. The first special right triangle has two acute angles of $\frac{\pi}{4}$ radians $(45°)$.

The $\frac{\pi}{4}, \frac{\pi}{4}, \frac{\pi}{2}$ (45°, 45°, 90°) Right Triangle

Every isosceles right triangle has two acute angles that have a measure of $\frac{\pi}{4}$ radians $(45°)$.

Watch this **animation** for an explanation.

Because the triangle is isosceles, the length of the two legs must be congruent. If we let the length of a leg be a units, then we can use the **Pythagorean Theorem** to determine that the length of the hypotenuse is $a\sqrt{2}$ units. See **Figure 23**. The lengths of the sides of every $\frac{\pi}{4}, \frac{\pi}{4}, \frac{\pi}{2}$ right triangle has this relationship. For example, if we let $a = 1$, then the length of the hypotenuse is $\sqrt{2}$. See **Figure 24**.

Watch this **animation** for a detailed explanation of the relationship between the lengths of the sides of the $\frac{\pi}{4}, \frac{\pi}{4}, \frac{\pi}{2}$ special right triangle.

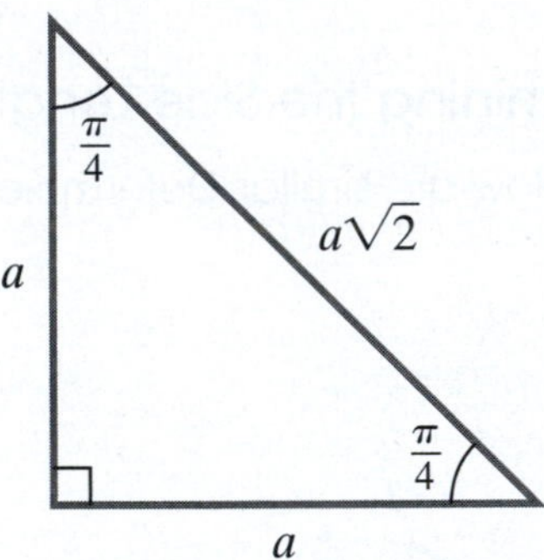

Figure 23 The length of the hypotenuse of the $\frac{\pi}{4}, \frac{\pi}{4}, \frac{\pi}{2}$ right triangle is $\sqrt{2}$ times the length of a leg.

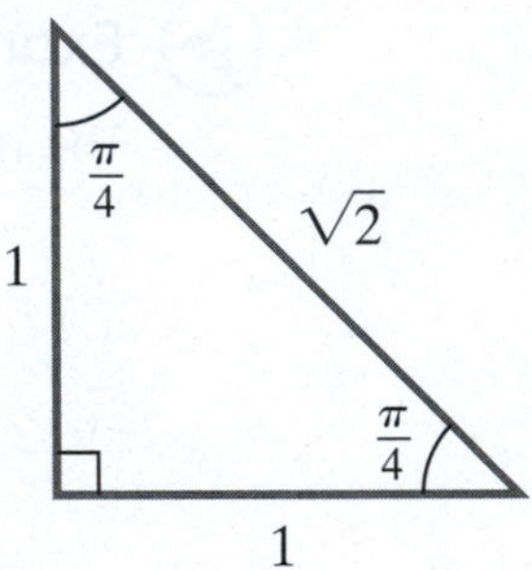

Figure 24 The $\frac{\pi}{4}, \frac{\pi}{4}, \frac{\pi}{2}$ right triangle with sides of lengths $1, 1,$ and $\sqrt{2}$

The $\frac{\pi}{6}, \frac{\pi}{3}, \frac{\pi}{2}$ (30°, 60°, 90°) Right Triangle

The second special right triangle has acute angles of $\frac{\pi}{6}$ radians $(30°)$ and $\frac{\pi}{3}$ radians $(60°)$. We can construct this triangle by starting with an equilateral triangle whose angles all measure $\frac{\pi}{3}$ radians $(60°)$. See **Figure 25**. We can then draw a perpendicular line segment that bisects one of the angles and one of the sides to create two $\frac{\pi}{6}, \frac{\pi}{3}, \frac{\pi}{2}$ right triangles. See **Figure 26**.

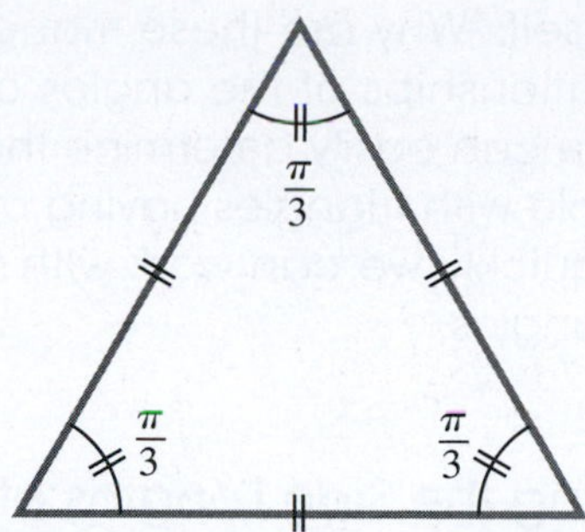

Figure 25 An equilateral triangle whose angles all measure $\frac{\pi}{3}$ radians and whose sides are all equal in length

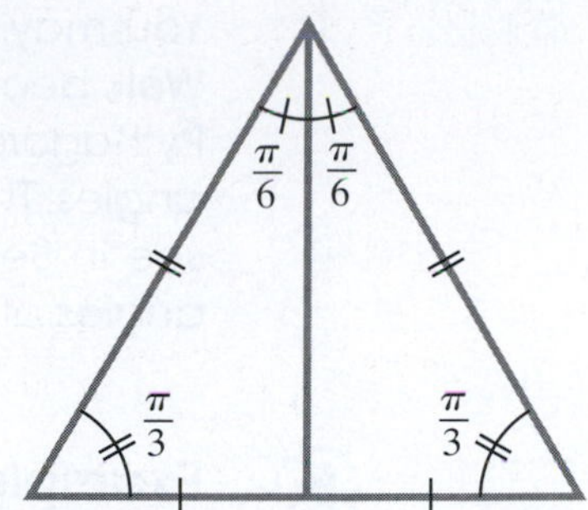

Figure 26 Bisecting an angle and a side of an equilateral triangle produces two $\frac{\pi}{6}, \frac{\pi}{3}, \frac{\pi}{2}$ right triangles.

Because we bisected one of the sides of the equilateral triangle to create two $\frac{\pi}{6}, \frac{\pi}{3}, \frac{\pi}{2}$ triangles, the length of the shortest side of a $\frac{\pi}{6}, \frac{\pi}{3}, \frac{\pi}{2}$ triangle must be exactly half of the length of the hypotenuse. (Or, the length of the hypotenuse must be twice as long as the length of the shortest side.) Therefore, if the length of the shortest leg (the leg opposite the $\frac{\pi}{6}$ angle) is a units, then the length of the hypotenuse is $2a$ units. We can use the **Pythagorean Theorem** to determine that the length of the other leg is $a\sqrt{3}$ units.

See Figure 27. If we let $a = 1$, then the length of the hypotenuse is 2 and the length of the side opposite the $\frac{\pi}{3}$ angle is $\sqrt{3}$. See Figure 28. Watch this **animation** for a complete explanation of the $\frac{\pi}{6}, \frac{\pi}{3}, \frac{\pi}{2}$ special right triangle.

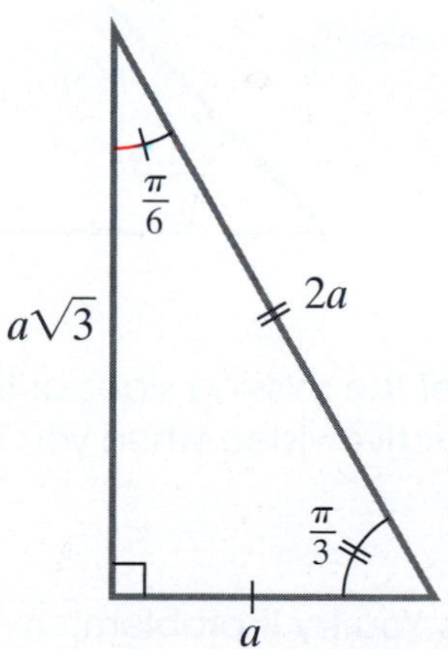

Figure 27 The length of the hypotenuse of the $\frac{\pi}{6}, \frac{\pi}{3}, \frac{\pi}{2}$ right triangle is 2 times the length of the shortest leg. The length of the longer leg is $\sqrt{3}$ times the length of the shorter leg.

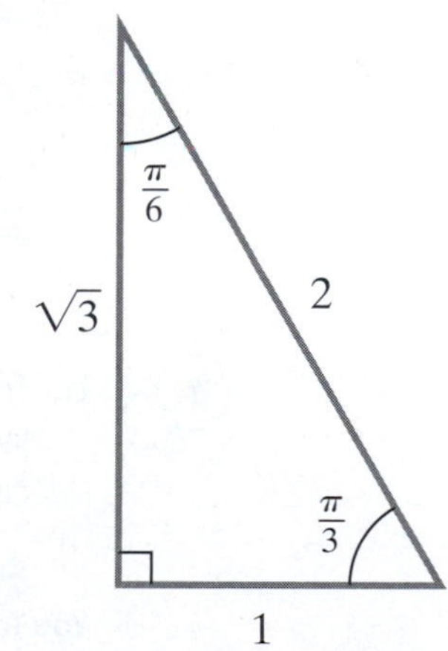

Figure 28 The $\frac{\pi}{6}, \frac{\pi}{3}, \frac{\pi}{2}$ right triangle with sides of lengths 1, $\sqrt{3}$, and 2

You may be asking yourself, "Why are these triangles and angles so special?" Well, because of the relationships of the angles of the right triangles and the Pythagorean Theorem, we can easily determine the ratios of the sides of the triangles. This is not so simple with triangles having other angle measures. You will see in Section 6.4 how quickly we can work with these angles as opposed to angles of non-special triangles.

Example 7 Determining the Side Lengths of Special Triangles

Determine the lengths of the missing sides of each right triangle.

a. 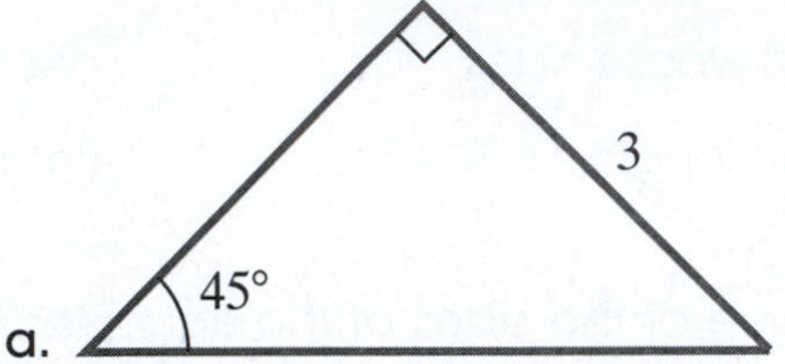
3
45°

b. 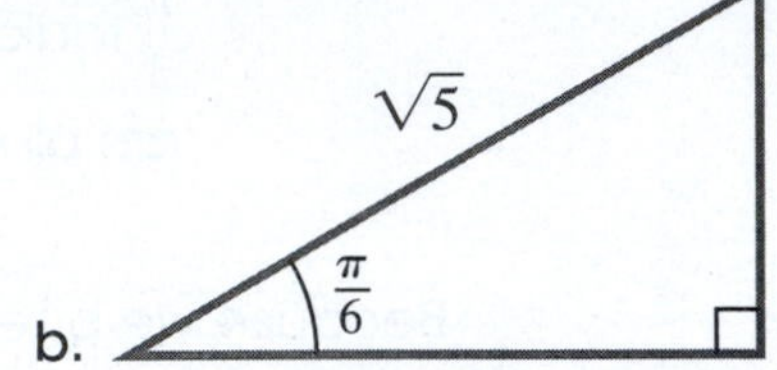
√5
π/6

Solution

a. This is a $\frac{\pi}{4}, \frac{\pi}{4}, \frac{\pi}{2}$ right triangle similar to the $\frac{\pi}{4}, \frac{\pi}{4}, \frac{\pi}{2}$ right triangle having side lengths of $1, 1, \sqrt{2}$. The proportionality constant is $\frac{3}{1} = 3$. Therefore, the lengths of the sides must be 3 times the lengths of the corresponding sides of the triangle with lengths $1, 1, \sqrt{2}$. Thus, the missing lengths are 3 and $3\sqrt{2}$.

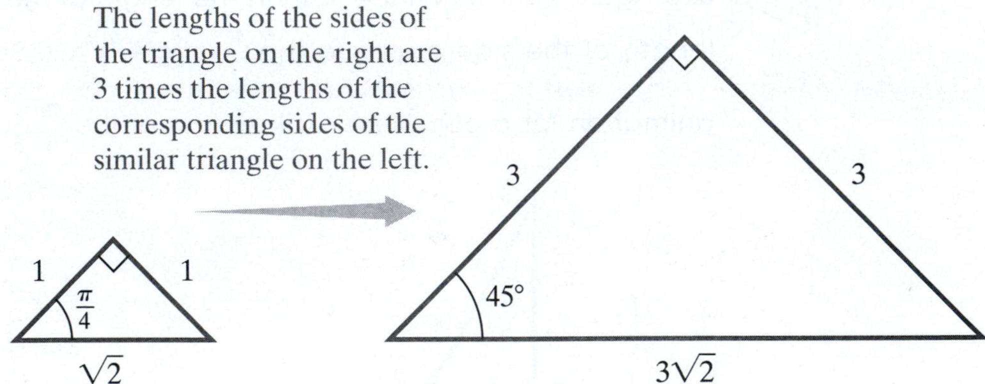
The lengths of the sides of the triangle on the right are 3 times the lengths of the corresponding sides of the similar triangle on the left.
1
1
π/4
√2
3
3
45°
3√2

b. Try finding the lengths of the missing sides of this triangle on your own. Work through this **interactive video** when you have finished to see if you are correct. ●

You Try It **Work through this You Try It problem.**

Work Exercises 21–30 in this textbook or in the MyLab Math Study Plan.

OBJECTIVE 5 USING SIMILAR TRIANGLES TO SOLVE APPLIED PROBLEMS

Understanding similar triangles and the special right triangles can help us solve application problems involving triangles.

Example 8 Determining the Height of a Cell Tower

The shadow of a cell tower is 80 feet long. A boy 3 feet 9 inches tall is standing next to the tower. If the boy's shadow is 6 feet long, find the height of the cell tower.

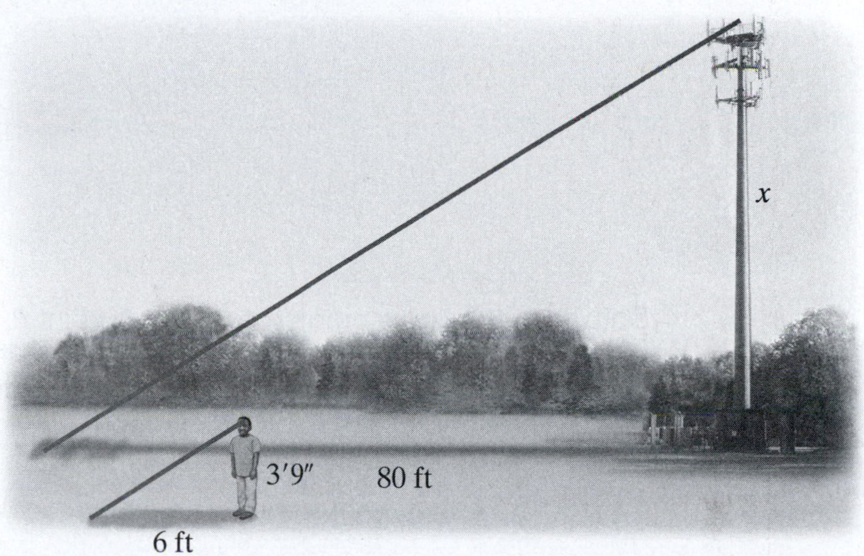

Solution First, convert 9 inches to feet.

$$9\,\cancel{\text{in}}\cdot\frac{1\text{ ft}}{12\,\cancel{\text{in}}} = \frac{9}{12}\text{ ft} = \frac{3}{4}\text{ ft} = 0.75\text{ ft}$$

Therefore, the height of the boy is 3.75 feet (*not* 3.9 feet). We can now draw two similar right triangles representing this problem. Let x represent the height of the cell tower. We can then write a **proportion** to solve for x.

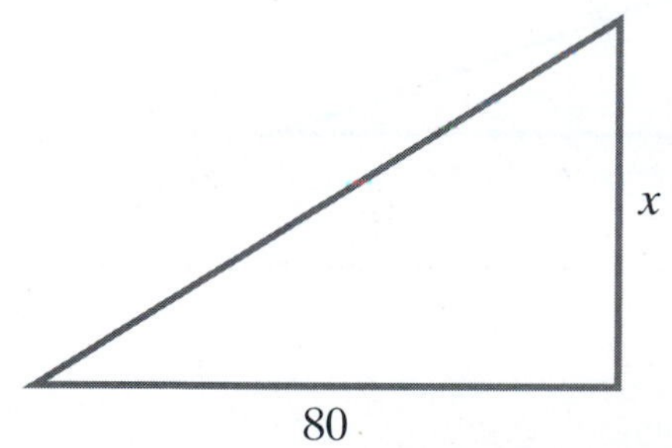

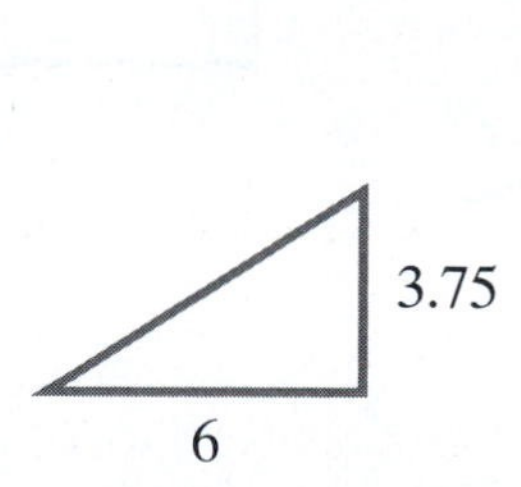

$$\frac{x}{3.75} = \frac{80}{6}$$ Write a proportion using the lengths of corresponding sides of similar triangles

$$6x = 300$$ Multiply both sides by the LCD, which is $(6)(3.75)$.

$$x = 50$$ Divide both sides by 6.

The height of the cell tower is 50 feet.

Example 9 Determining the Length of a Bridge

Two people are standing on opposite sides of a small river. One person is located at point Q, a distance of 20 feet from a bridge. The other person is standing on the southeast corner of the bridge at point P. The angle between the bridge and the line of sight from P to Q is 30°. Use this information to determine the length of the bridge and the distance between the two people. Round your answers to two decimal places as needed.

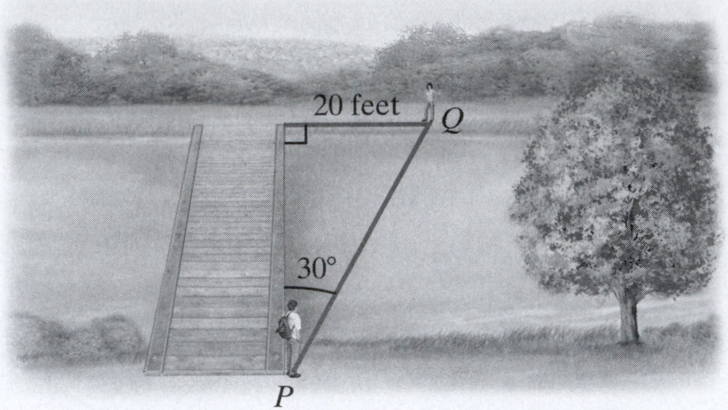

Solution The figure created by point Q, point P, and the bridge forms a $30°, 60°, 90°$ special right triangle that is similar to the $30°, 60°, 90°$ triangle with sides of lengths $1, \sqrt{3}$, and 2. Try finding the length of the bridge on your own. Watch this **video** to see if you are correct.

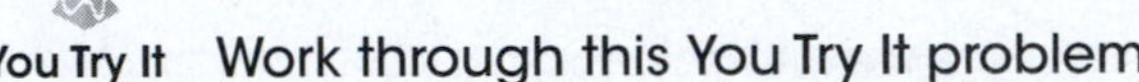

You Try It Work through this You Try It problem.

Work Exercises 31–36 in this textbook or in the MyLab Math Study Plan.

6.3 Exercises

In Exercises 1–3, classify each triangle as acute, obtuse, right, scalene, isosceles, or equilateral. State all that apply.

1.

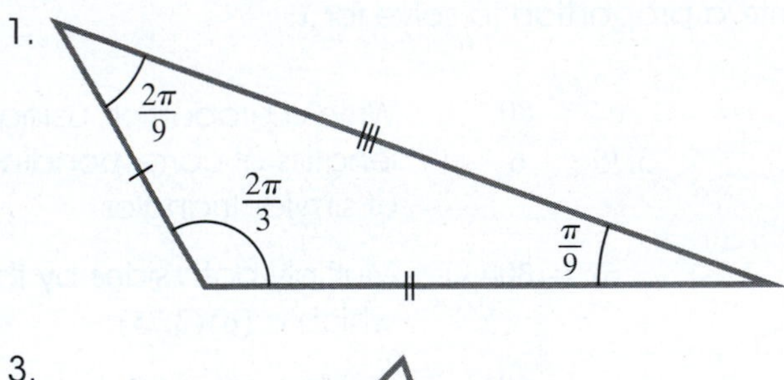

2.

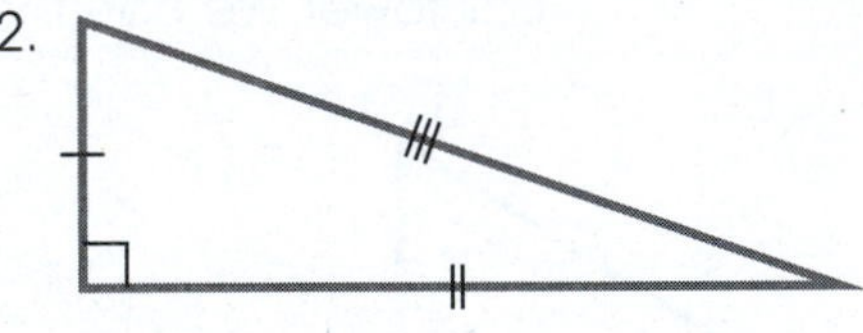

3.

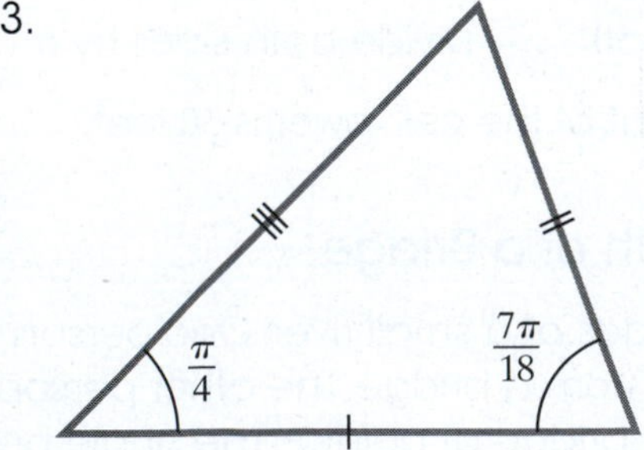

In Exercises 4–8, use the Pythagorean Theorem to find the missing side of each right triangle.

4.

7

12

5.

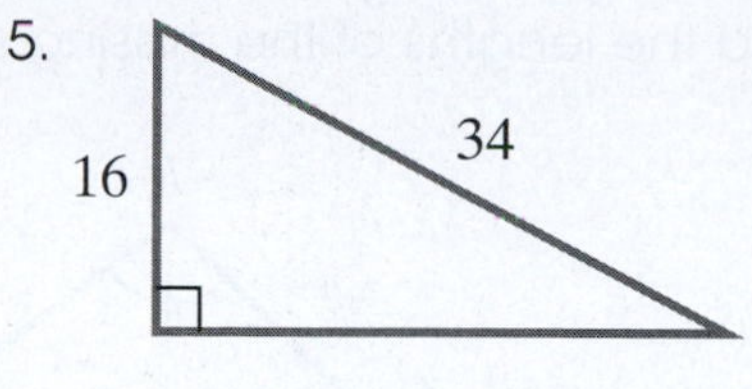

6.

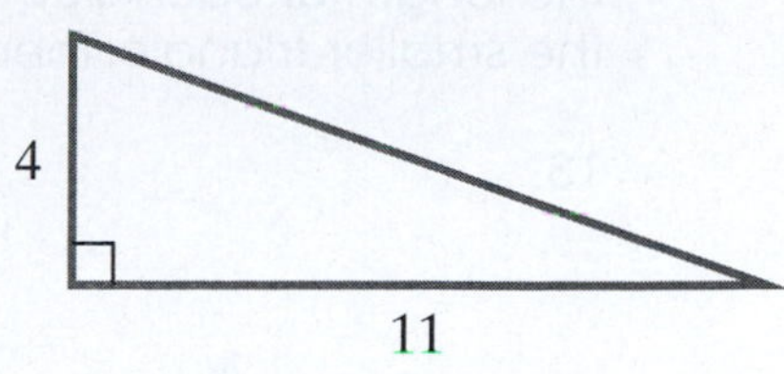

7.

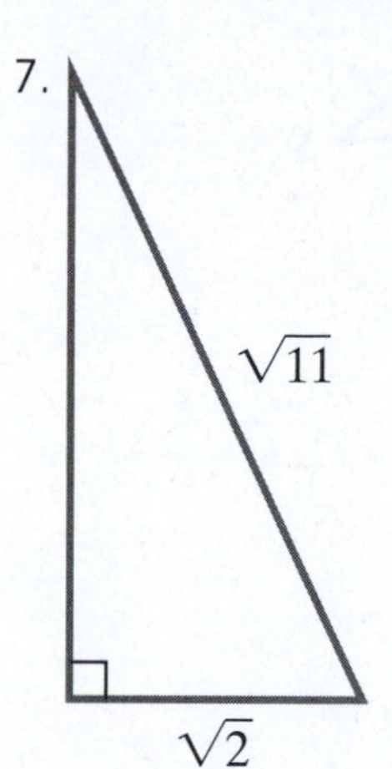

8.

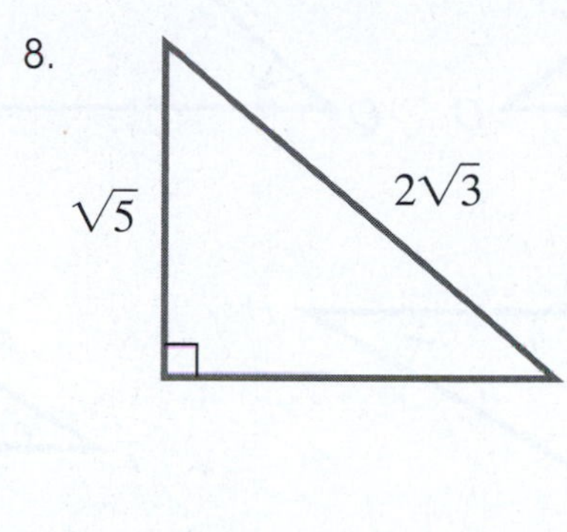

9. An official National Collegiate Athletic Association (NCAA) women's softball "diamond" is really a square. The distance between each consecutive base is 60 feet. What is the distance between home plate and second base?

10. How far up the side of a building will a 9-meter ladder reach if the foot of the ladder is 2 meters from the base of the building?

11. The length of a rectangle is 1 inch less than twice the width. If the diagonal is 2 inches more than the length, find the dimensions of the rectangle.

12. The cross section of a camping tent forms a triangle. The slanted sides measure 7 feet and the base of the tent is 8 feet. What is the height of the tent?

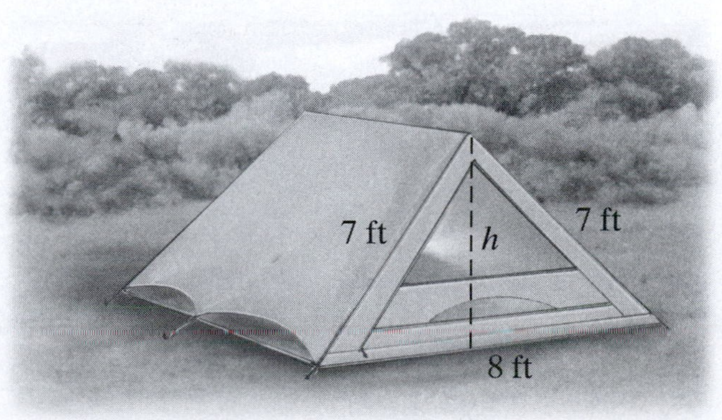

In Exercises 13–16, two similar triangles are given. Find the proportionality constant k such that the length of each side of the larger triangle is k times the length of the corresponding sides of the smaller triangle. Then find the lengths of the missing sides of each triangle.

13.

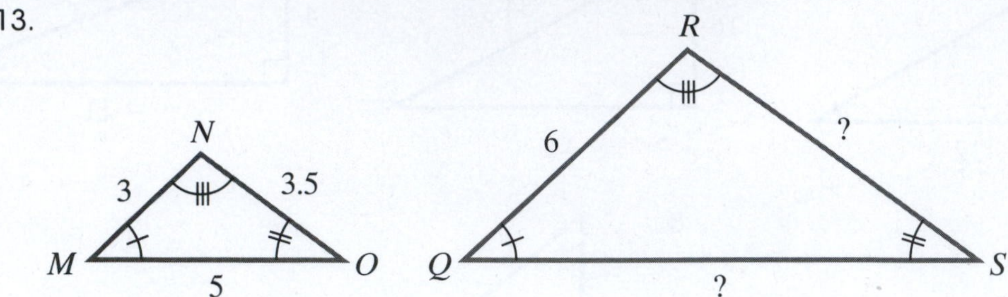

14.

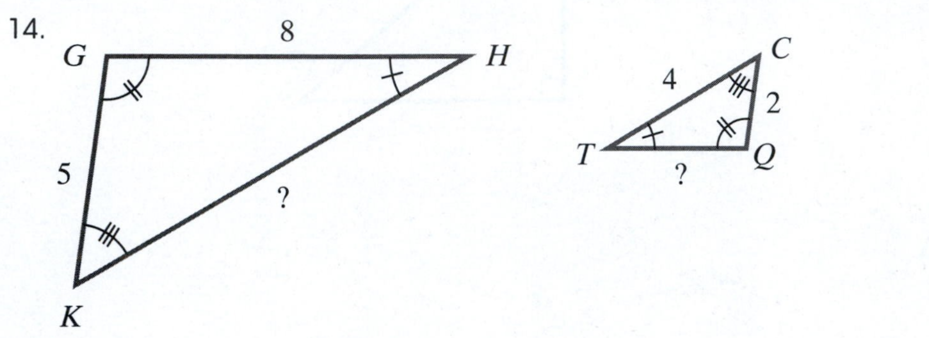

15.

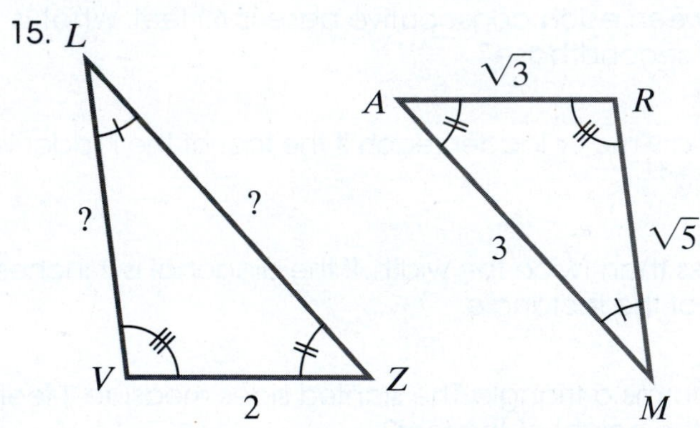

16.

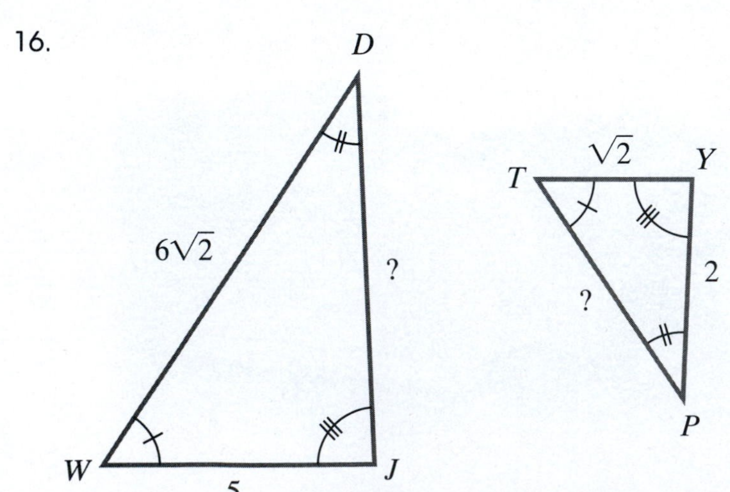

In Exercises 17–20, find the missing sides of the given similar right triangles.

17.

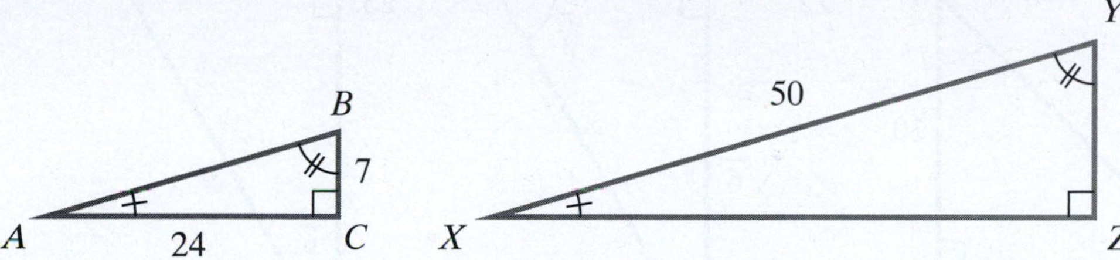

18.

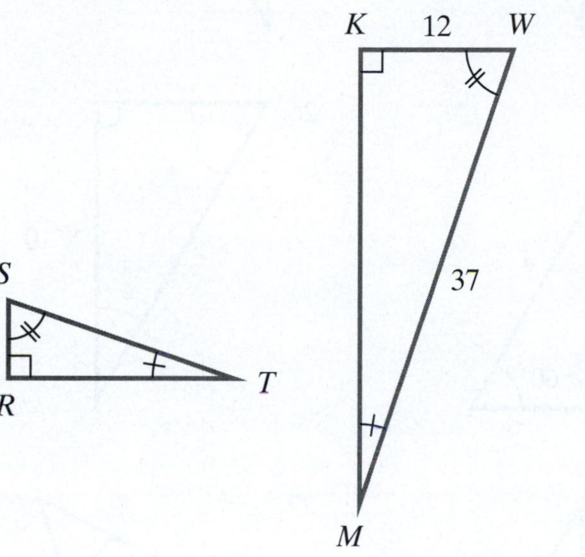

19.

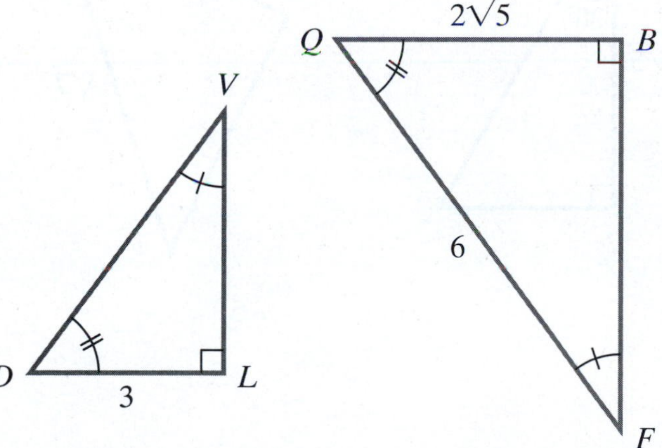

20.

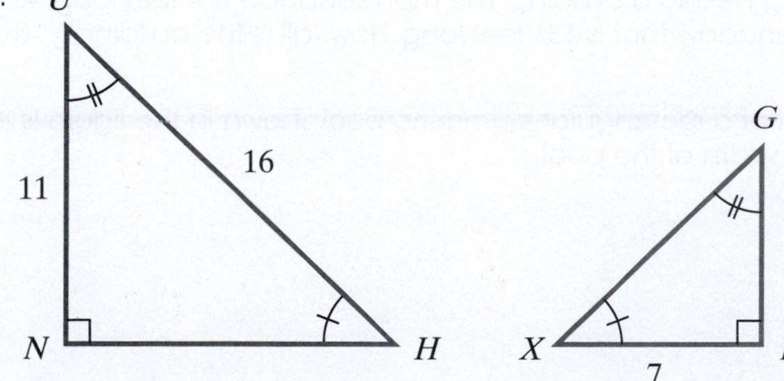

In Exercises 21–30, find the missing sides of each special right triangle.

21.

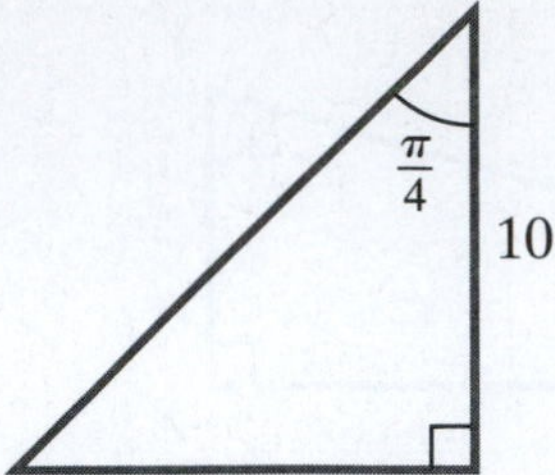

22.

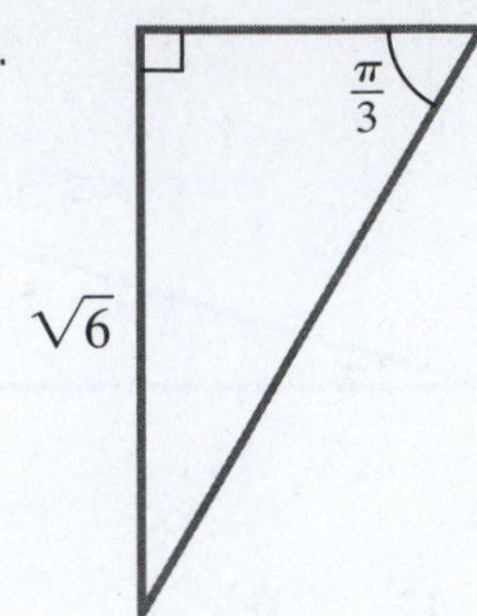

23.

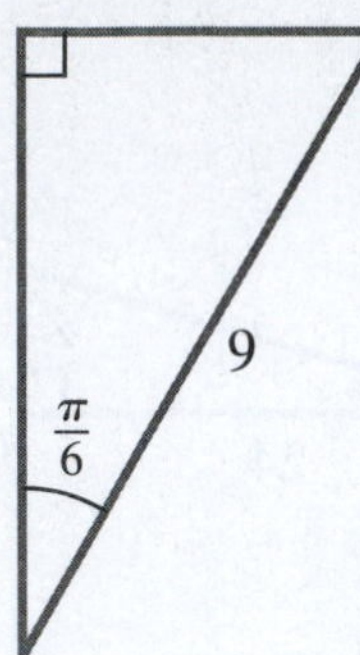

24.

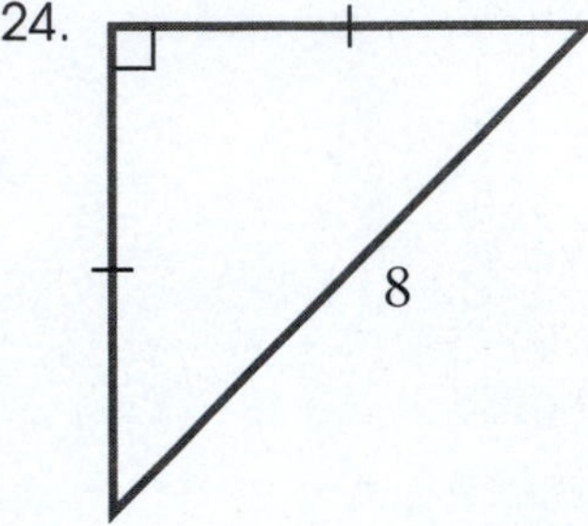

25.

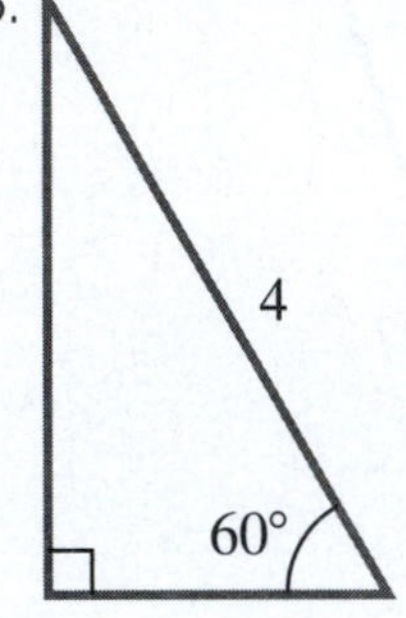

26.

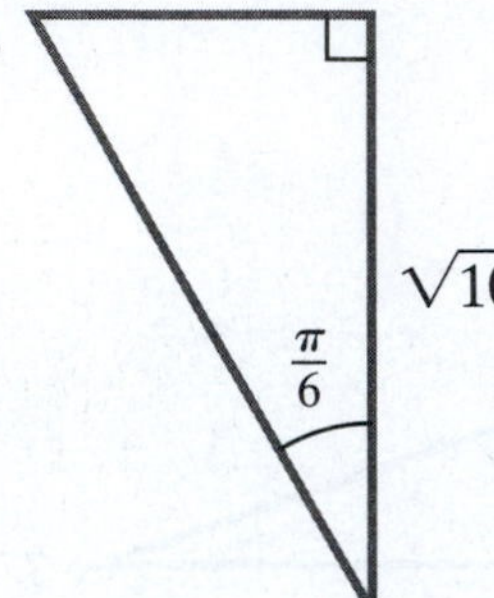

27.

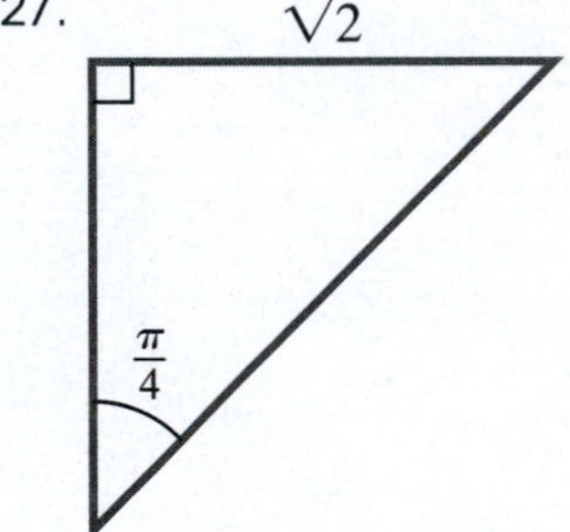

28.

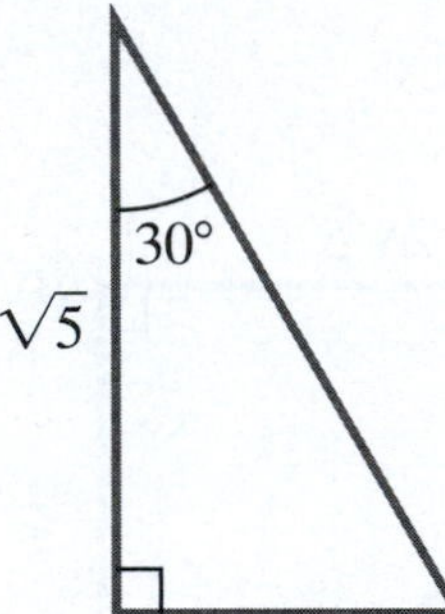

29.

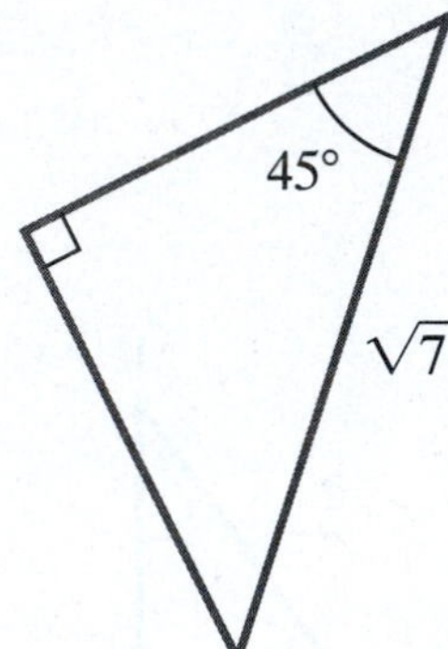

30.

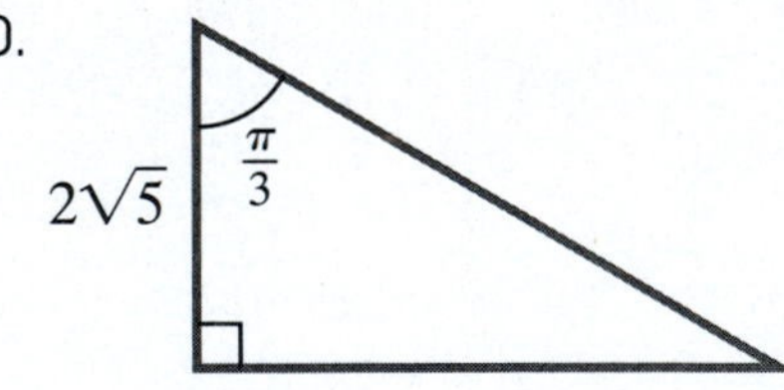

31. A 6-foot-tall man is standing next to a building. The man's shadow is 4 feet long. At the same time, the building casts a shadow that is 135 feet long. How tall is the building?

32. The length of the diagonal of a rectangular swimming pool shown in the figure is 40 feet. Determine the length and width of the pool.

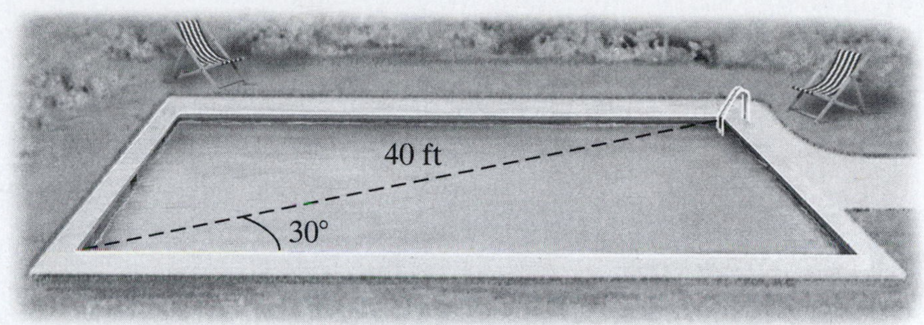

33. Two ladders lean against a wall in such a way that the angle that they make with the ground is the same. The 15-foot ladder rests 8 feet up the wall. How much further up the wall will the 20-foot ladder reach?

34. The **angle of elevation** is the angle from the horizontal looking up at an object. The angle of elevation of an airplane is $\frac{\pi}{4}$ radians, and the altitude of the plane is 12,000 feet. How far away is the plane "as the crow flies," which means through the air rather than on the ground?

35. Two people are standing on opposite sides of a small river. One person is located at point Q, a distance of 35 meters from a bridge. The other person is standing on the southeast corner of the bridge at point P. The angle between the bridge and the line of sight from P to Q is $\frac{\pi}{3}$ radians. Use this information to determine the length of the bridge and the distance between the two people. Round your answers to two decimal places as needed.

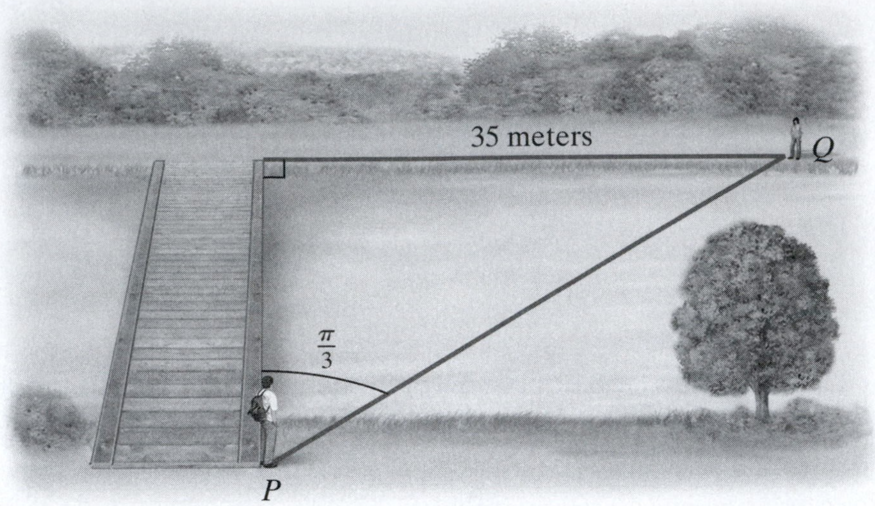

36. A team of forestry graduate students wants to determine the height of a giant sequoia tree using lasers. The laser is placed on top of a 2-foot tripod and shines a beam through the top of a 20-foot pole positioned 10 feet from the tripod and 140 feet from the approximate center of the base of the tree. Determine the height of the tree.

6.4 Right Triangle Trigonometry

THINGS TO KNOW

Before working through this section, be sure that you are familiar with the following concepts:

VIDEO ANIMATION INTERACTIVE

You Try It 1. Converting between Degree Measure and Radian Measure (Section 6.1)  

You Try It 2. Understanding Similar Triangles (Section 6.3)  

You Try It 3. Understanding the Special Right Triangles (Section 6.3) 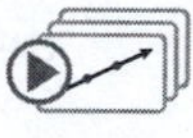

OBJECTIVES

1 Understanding the Right Triangle Definitions of the Trigonometric Functions

2 Using the Special Right Triangles

3 Understanding the Fundamental Trigonometric Identities

4 Understanding Cofunctions

5 Evaluating Trigonometric Functions Using a Calculator

SECTION 6.4 EXERCISE

OBJECTIVE 1 UNDERSTANDING THE RIGHT TRIANGLE DEFINITIONS OF THE TRIGONOMETRIC FUNCTIONS

Consider the right triangles in Figure 29 with the given **acute angle** θ. The side opposite the right angle is called the **hypotenuse**. The hypotenuse is always the longest side of a right triangle. We label the two legs of the right triangle relative to the given angle θ. The leg opposite angle θ is appropriately named the **opposite side**. The other leg is called the **adjacent side**.

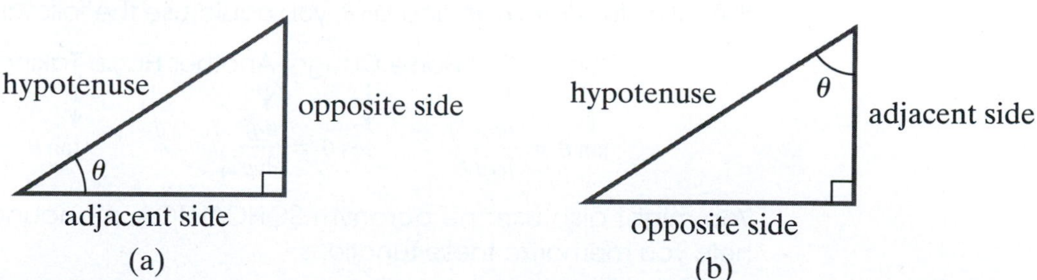

Figure 29 The sides of two right triangles relative to the given acute angle θ

We will be concerned about the *lengths* of each of these three sides. For simplicity, we will abbreviate the length of the hypotenuse as "*hyp*" The length of the opposite side will be referred to as "*opp*" and the length of the adjacent side will be denoted as "*adj*". See **Figure 30**.

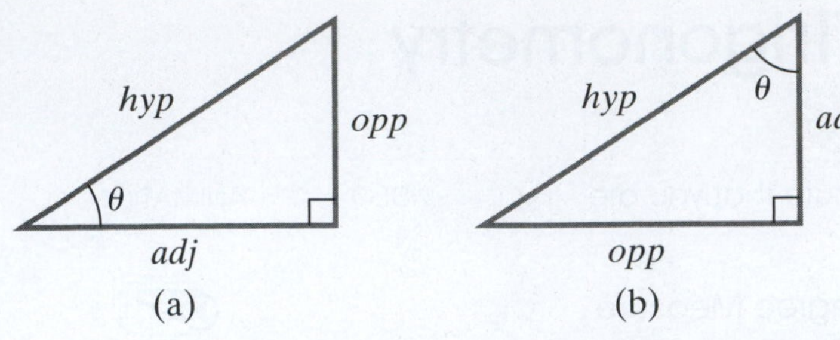

Figure 30 Two right triangles and a given acute angle θ with side lengths hyp, opp, and adj

Choosing side lengths two at a time, we can create a total of six ratios. These six ratios are $\frac{opp}{hyp}, \frac{adj}{hyp}, \frac{opp}{adj}, \frac{hyp}{opp}, \frac{hyp}{adj}$, and $\frac{adj}{opp}$. The value of each of these six ratios depends on the measure of the acute angle θ. Thus, the six ratios are functions of the variable θ and are called the **trigonometric functions of the acute angle** θ. For convenience, the six ratios have been given names. Historically, these six trigonometric functions have been named **sine** of theta, **cosine** of theta, **tangent** of theta, **cosecant** of theta, **secant** of theta, and **cotangent** of theta. The six functions are abbreviated as $\sin\theta$, $\cos\theta$, $\tan\theta$, $\csc\theta$, $\sec\theta$, and $\cot\theta$.

The Right Triangle Definitions of the Trigonometric Functions

Given a right triangle with acute angle θ and side lengths of hyp, opp, and adj, the six trigonometric functions of angle θ are defined as follows:

$$\sin\theta = \frac{opp}{hyp} \qquad \csc\theta = \frac{hyp}{opp}$$

$$\cos\theta = \frac{adj}{hyp} \qquad \sec\theta = \frac{hyp}{adj}$$

$$\tan\theta = \frac{opp}{adj} \qquad \cot\theta = \frac{adj}{opp}$$

hyp
opp
θ
adj

It is extremely important to memorize the six trigonometric functions. To memorize the ratios for $\sin\theta$, $\cos\theta$, and $\tan\theta$, you could use the following silly phrase:

"**S**ome **O**ld **H**orse **C**aught **A**nother **H**orse **T**aking **O**ats **A**way"

$$\sin\theta = \frac{opp}{hyp} \qquad \cos\theta = \frac{adj}{hyp} \qquad \tan\theta = \frac{opp}{adj}$$

You might also use the **acronym** SOHCAHTOA, pronounced "So-Kah-Toe-Ah," to help you memorize these functions.

Once you have memorized the ratios for $\sin\theta$, $\cos\theta$, and $\tan\theta$, you can easily obtain the ratios of $\csc\theta$, $\sec\theta$, and $\cot\theta$. Note that the ratios for $\csc\theta$, $\sec\theta$, and $\cot\theta$ are simply the **reciprocals** of the ratios for $\sin\theta$, $\cos\theta$, and $\tan\theta$, respectively.

Recall from **Section 6.3** that if two triangles are similar, then the ratio of the lengths of any two sides of one triangle is equal to the ratio of the lengths of the corresponding sides of the other triangle. Thus, the value of the six trigonometric functions for a specific acute angle θ will be exactly the same regardless of the size of the triangle. To illustrate this, consider the three similar triangles in Figure 31 below. Note that the values of θ and $\sin\theta$ are the same for each triangle even though the sides are of different length. This is because the ratio of any two sides of a right triangle is a function of the measure of the acute angle θ.

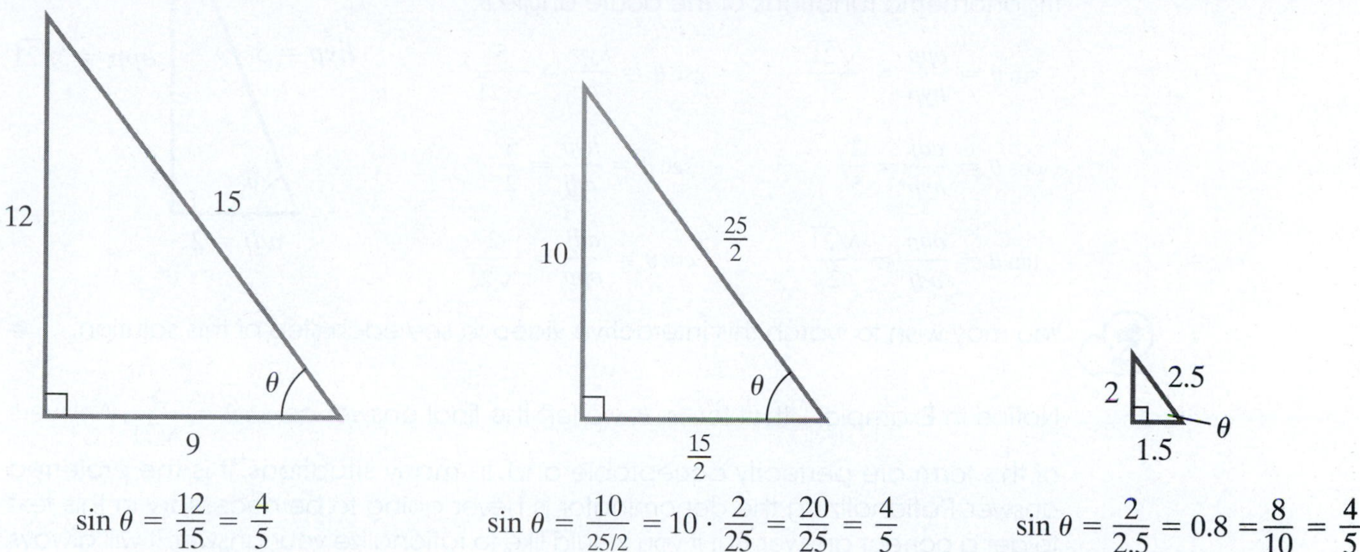

Figure 31 In each triangle, though the corresponding sides all have different lengths, the ratio of the length of the opposite side to the length of the hypotenuse is the same.

You should verify that the remaining five trigonometric functions of angle θ for each of the three similar right triangles are also equivalent. (To see the values of the six trigonometric functions for the three right triangles shown previously, consult this **table**.) It follows that the values of the six trigonometric functions of an acute angle θ depend only on the measure of θ, not on the size of the right triangle.

Example 1 Evaluating the Trigonometric Functions Given a Right Triangle

Given the right triangle, evaluate the six trigonometric functions of the acute angle θ.

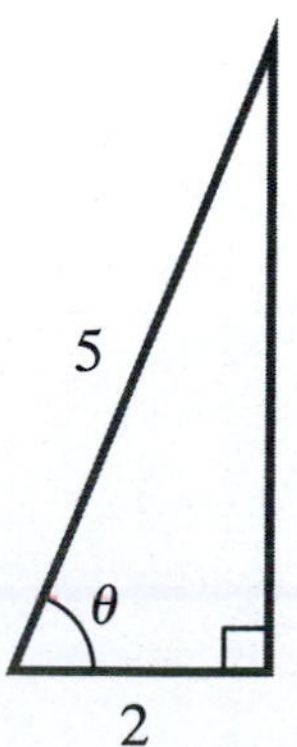

Solution We are given that $hyp = 5$ and $adj = 2$. We must first determine the length of the opposite side (opp) using the **Pythagorean Theorem** before we can evaluate the six trigonometric functions.

$(adj)^2 + (opp)^2 = (hyp)^2$	Write an equation representing the Pythagorean Theorem.
$2^2 + (opp)^2 = 5^2$	Substitute $adj = 2$ and $hyp = 5$.
$4 + (opp)^2 = 25$	Simplify.
$(opp)^2 = 21$	Subtract 4 from both sides.
$opp = \pm\sqrt{21}$	Take the square root of both sides.

Because opp represents the length of a side of a triangle, we use only the positive square root value. Therefore, $opp = \sqrt{21}$. We can now evaluate the six trigonometric functions of the acute angle θ.

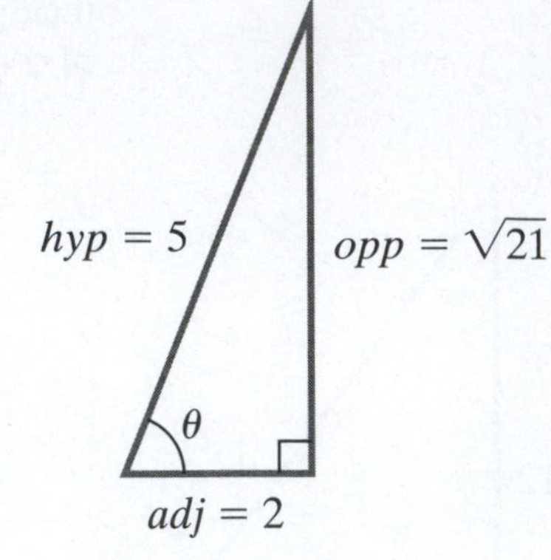

$$\sin\theta = \frac{opp}{hyp} = \frac{\sqrt{21}}{5} \qquad \csc\theta = \frac{hyp}{opp} = \frac{5}{\sqrt{21}}$$

$$\cos\theta = \frac{adj}{hyp} = \frac{2}{5} \qquad \sec\theta = \frac{hyp}{adj} = \frac{5}{2}$$

$$\tan\theta = \frac{opp}{adj} = \frac{\sqrt{21}}{2} \qquad \cot\theta = \frac{adj}{opp} = \frac{2}{\sqrt{21}}$$

You may wish to watch this **interactive video** to see each step of this solution. ●

Notice in Example 1 that for $\csc\theta$, we left the final answer as $\csc\theta = \frac{5}{\sqrt{21}}$. Answers of this form are perfectly acceptable and, in many situations, it is the preferred answer. Rationalizing the denominator is never going to be necessary in this text to get a correct answer. But if you would like to rationalize your answer, it will always be counted as correct as long as your work contains no errors. If you would like a quick review on how to rationalize the denominator, watch this **video**.

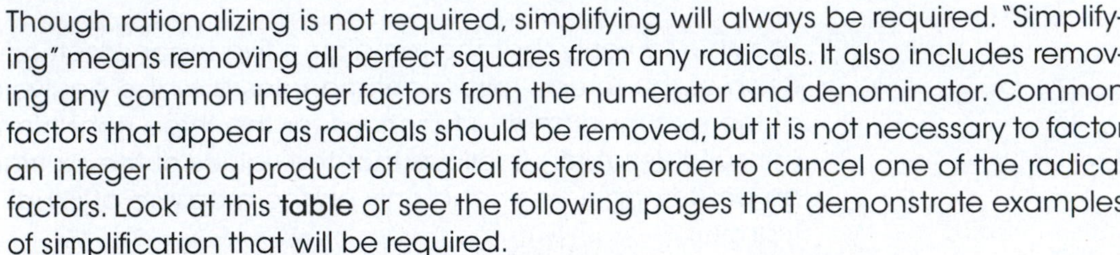

Though rationalizing is not required, simplifying will always be required. "Simplifying" means removing all perfect squares from any radicals. It also includes removing any common integer factors from the numerator and denominator. Common factors that appear as radicals should be removed, but it is not necessary to factor an integer into a product of radical factors in order to cancel one of the radical factors. Look at this **table** or see the following pages that demonstrate examples of simplification that will be required.

Examples of Simplifying Radical Expressions

Calculated result	Simplification required	Correct simplified, unrationalized answer
$\frac{5}{\sqrt{21}}$	None	$\frac{5}{\sqrt{21}}$
$\frac{1}{\sqrt{2}}$	None	$\frac{1}{\sqrt{2}}$
$\frac{1}{\sqrt{24}}$	$\frac{1}{\sqrt{24}} = \frac{1}{\sqrt{4}\sqrt{6}} = \frac{1}{2\sqrt{6}}$	$\frac{1}{2\sqrt{6}}$
$\frac{3}{\sqrt{3}}$	None	$\frac{3}{\sqrt{3}}$
$\frac{3}{\sqrt{63}}$	$\frac{3}{\sqrt{63}} = \frac{3}{\sqrt{9}\sqrt{7}} = \frac{\cancel{3}}{\cancel{3}\sqrt{7}} = \frac{1}{\sqrt{7}}$	$\frac{1}{\sqrt{7}}$
$\frac{\sqrt{3}}{\sqrt{10}}$	None	$\frac{\sqrt{3}}{\sqrt{10}}$

(eText Screens 6.4-1–6.4-35)

Calculated result	Simplification required	Correct simplified, unrationalized answer
$\dfrac{\sqrt{2}}{\sqrt{10}}$	$\dfrac{\sqrt{2}}{\sqrt{10}} = \dfrac{\cancel{\sqrt{2}}}{\cancel{\sqrt{2}}\sqrt{5}} = \dfrac{1}{\sqrt{5}}$	$\dfrac{1}{\sqrt{5}}$
$\dfrac{\sqrt{3}}{\sqrt{24}}$	$\dfrac{\sqrt{3}}{\sqrt{24}} = \dfrac{\cancel{\sqrt{3}}}{\cancel{\sqrt{3}}\sqrt{8}} = \dfrac{1}{\sqrt{8}} = \dfrac{1}{\sqrt{4}\sqrt{2}} = \dfrac{1}{2\sqrt{2}}$	$\dfrac{1}{2\sqrt{2}}$
$\dfrac{\sqrt{98}}{\sqrt{2}}$	$\dfrac{\sqrt{98}}{\sqrt{2}} = \dfrac{\cancel{\sqrt{2}}\sqrt{49}}{\cancel{\sqrt{2}}} = \sqrt{49} = 7$	7
$\dfrac{\sqrt{72}}{5}$	$\dfrac{\sqrt{72}}{5} = \dfrac{\sqrt{36}\sqrt{2}}{5} = \dfrac{6\sqrt{2}}{5}$	$\dfrac{6\sqrt{2}}{5}$

You Try It Work through this You Try It problem.

Work Exercises 1–5 in this textbook or in the MyLab Math Study Plan.

Example 2 Finding the Values of the Remaining Trigonometric Functions

If θ is an acute angle of a right triangle and if $\sin\theta = \dfrac{3}{4}$, then find the values of the remaining five trigonometric functions for angle θ.

Solution We know that $\sin\theta = \dfrac{opp}{hyp} = \dfrac{3}{4}$, where $opp = 3$ is the length of the side of a right triangle opposite an acute angle θ and $hyp = 4$ is the length of the hypotenuse. Using the **Pythagorean Theorem**, we find that the length of the adjacent side is $\sqrt{7}$.

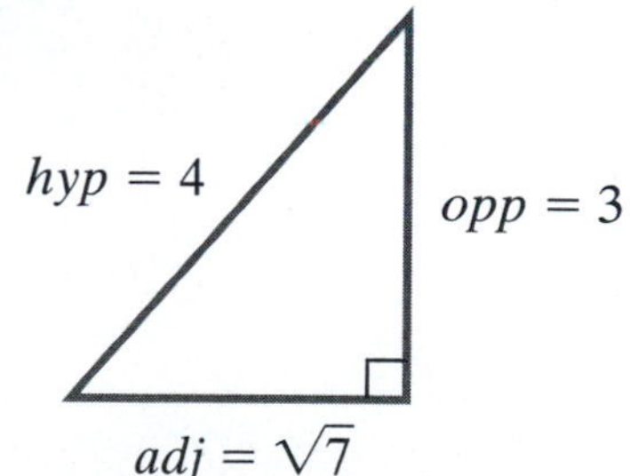

Once we know the lengths of all three sides of the right triangle, we can determine the values of the remaining trigonometric functions. Try finding these values on your own. Once you have completed this work, watch this **video** to see if you are correct.

You Try It Work through this You Try It problem.

Work Exercises 6–11 in this textbook or in the MyLab Math Study Plan.

OBJECTIVE 2 USING THE SPECIAL RIGHT TRIANGLES

In **Section 6.3**, we introduced two special right triangles, the $\frac{\pi}{4}, \frac{\pi}{4}, \frac{\pi}{2}$ $(45°, 45°, 90°)$ triangle and the $\frac{\pi}{6}, \frac{\pi}{3}, \frac{\pi}{2}$ $(30°, 60°, 90°)$. You may want to watch this **animation** for a quick review of these two special right triangles. These two special triangles are seen in Figure 32.

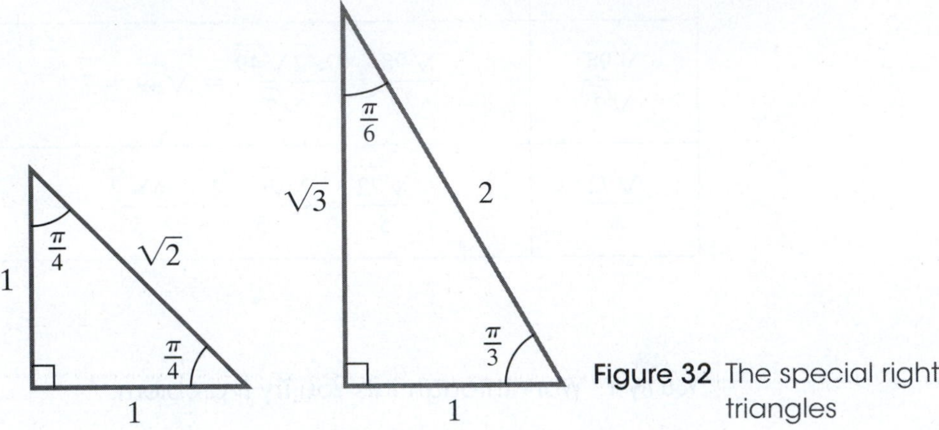

Figure 32 The special right triangles

Because we know the relationship between the sides of these two special right triangles, we can determine the exact values of the six trigonometric functions of θ by letting θ take on the values of $\frac{\pi}{6}, \frac{\pi}{4}$, and $\frac{\pi}{3}$. **Table 1** shows the values of the sine, cosine, and tangent functions for these three acute angles. Notice that these values do not have to be rationalized.

That is, we don't have to write $\frac{1}{\sqrt{2}}$ as $\frac{\sqrt{2}}{2}$. It would be in your best interest to become very familiar with Table 1.

Table 1 The Trigonometric Functions for Acute Angles $\frac{\pi}{6}, \frac{\pi}{4}$, and $\frac{\pi}{3}$

θ	$\frac{\pi}{6}(30°)$	$\frac{\pi}{4}(45°)$	$\frac{\pi}{3}(60°)$
$\sin\theta$	$\frac{1}{2}$	$\frac{1}{\sqrt{2}}$	$\frac{\sqrt{3}}{2}$
$\cos\theta$	$\frac{\sqrt{3}}{2}$	$\frac{1}{\sqrt{2}}$	$\frac{1}{2}$
$\tan\theta$	$\frac{1}{\sqrt{3}}$	1	$\sqrt{3}$

To find the values of $\csc\theta, \sec\theta$, and $\cot\theta$ for the angles in Table 1, we can use the fact that the ratios for $\csc\theta, \sec\theta$, and $\cot\theta$ are simply the **reciprocals** of the ratios for $\sin\theta, \cos\theta$, and $\tan\theta$, respectively. Try to determine the values of $\csc\theta, \sec\theta$, and $\cot\theta$ for each of the angles in Table 1. When finished, check your **answers**.

Example 3 Evaluating Trigonometric Functions Using the Special Right Triangles

Determine the value of $\csc \frac{\pi}{6} + \cot \frac{\pi}{4}$.

Solution Draw the two special right triangles and then find the value of $\csc \frac{\pi}{6}$ and $\cot \frac{\pi}{4}$.

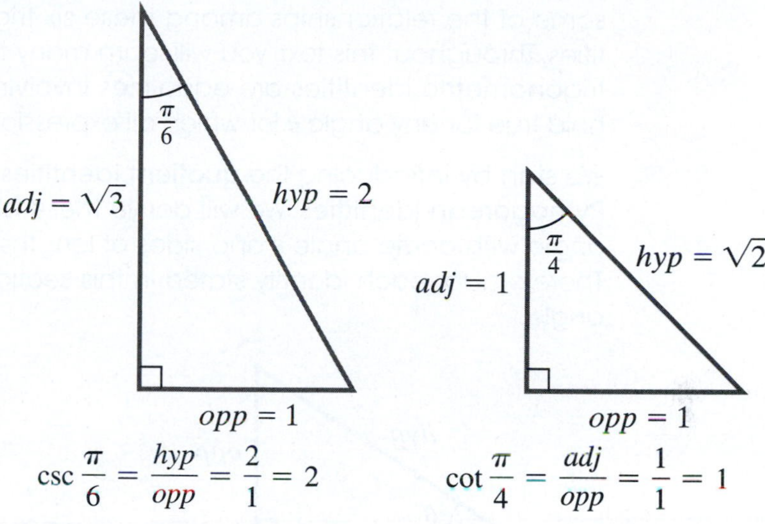

$$\csc \frac{\pi}{6} = \frac{hyp}{opp} = \frac{2}{1} = 2 \qquad \cot \frac{\pi}{4} = \frac{adj}{opp} = \frac{1}{1} = 1$$

Now, add the two expressions together.

$$\csc \frac{\pi}{6} + \cot \frac{\pi}{4} = 2 + 1 = 3$$

You Try It Work through this You Try It problem.

Work Exercises 12–18 in this textbook or in the MyLab Math Study Plan.

Example 4 Determining the Measure of an Acute Angle

Determine the measure of the acute angle θ for which $\sec \theta = 2$.

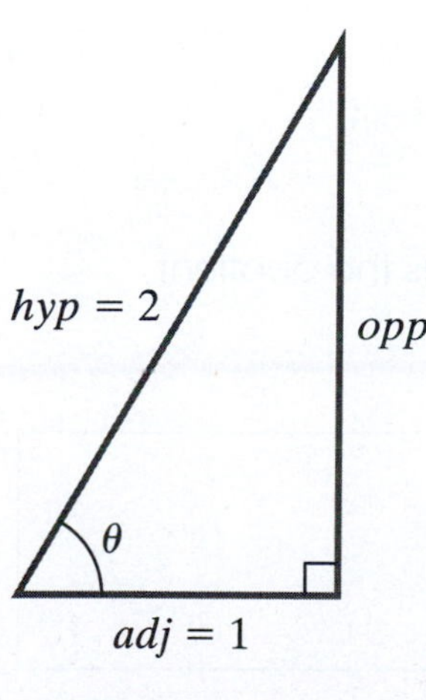

Solution By definition, the secant of an acute angle of a right triangle is equal to the ratio of the length of the hypotenuse to the length of the side adjacent to the angle. This ratio is given to be 2. Therefore, $\sec \theta = \frac{hyp}{adj} = \frac{2}{1}$. Note that we need to insert the 1 in the denominator of the whole number 2 to "see" the length of the adjacent side, adj. We can use this information to draw the following right triangle. We can use the Pythagorean Theorem to determine that the length of the opposite side is $opp = \sqrt{3}$. We can see that this right triangle is a special $\frac{\pi}{6}, \frac{\pi}{3}, \frac{\pi}{2}$ triangle. The angle opposite the side of length $\sqrt{3}$ is always $\frac{\pi}{3}$. Therefore, $\theta = \frac{\pi}{3}$. You may wish to watch this **video** to see a complete solution to this example.

(eText Screens 6.4-1–6.4-35)

You Try It Work through this You Try It problem.

Work Exercises 19–24 in this textbook or in the MyLab Math Study Plan.

OBJECTIVE 3 UNDERSTANDING THE FUNDAMENTAL TRIGONOMETRIC IDENTITIES

Now that you know the ratios of the lengths of the sides of a right triangle and their correspondingly named trigonometric functions of θ, it is time to recognize some of the relationships among these six trigonometric functions, called identities. Throughout this text, you will learn many trigonometric identities. Simply put, trigonometric identities are equalities involving trigonometric expressions that hold true for any angle θ for which all expressions are defined.

We start by introducing the **quotient identities**, the **reciprocal identities**, and the **Pythagorean identities**. We will derive these identities by referring to the right triangle with acute angle θ and sides of lengths opp, adj, and hyp seen in Figure 33. Therefore, for each identity stated in this section, it is assumed that θ is an **acute angle**.

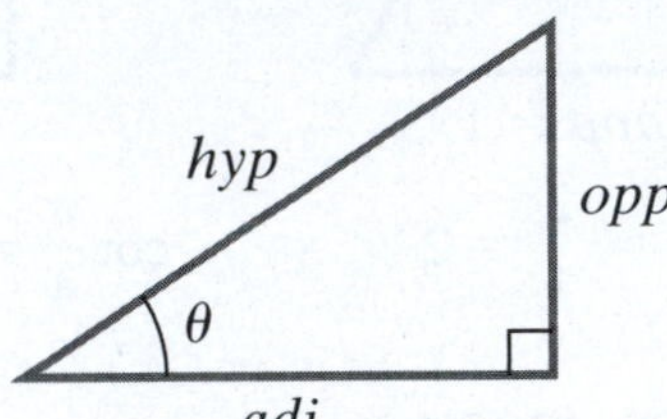

Figure 33 A right triangle and acute angle θ with side lengths hyp, opp, and adj

THE QUOTIENT IDENTITIES

Recall that $\sin\theta = \dfrac{opp}{hyp}$ and $\cos\theta = \dfrac{adj}{hyp}$. Therefore, the quotient of $\sin\theta$ and $\cos\theta$ is

$$\frac{\sin\theta}{\cos\theta} = \frac{\frac{opp}{hyp}}{\frac{adj}{hyp}} = \frac{opp}{\cancel{hyp}} \cdot \frac{\cancel{hyp}}{adj} = \frac{opp}{adj} = \tan\theta.$$

Similarly, the quotient of $\cos\theta$ and $\sin\theta$ is

$$\frac{\cos\theta}{\sin\theta} = \frac{\frac{adj}{hyp}}{\frac{opp}{hyp}} = \frac{adj}{\cancel{hyp}} \cdot \frac{\cancel{hyp}}{opp} = \frac{adj}{opp} = \cot\theta.$$

The two equations $\tan\theta = \dfrac{\sin\theta}{\cos\theta}$ and $\cot\theta = \dfrac{\cos\theta}{\sin\theta}$ are known as the quotient identities.

The Quotient Identities

$$\tan\theta = \frac{\sin\theta}{\cos\theta} \qquad \cot\theta = \frac{\cos\theta}{\sin\theta}$$

THE RECIPROCAL IDENTITIES

In algebra, the reciprocal of a real number a is defined as $\frac{1}{a}$. Similarly, the reciprocal of $\sin\theta$ is $\frac{1}{\sin\theta}$. If θ is the acute angle of the right triangle from **Figure 33**, then $\frac{1}{\sin\theta} = \frac{1}{\frac{opp}{hyp}} = \frac{hyp}{opp} = \csc\theta$.

The identity $\csc\theta = \frac{1}{\sin\theta}$ is one of six reciprocal identities. Watch this **video** to see how to derive each reciprocal identity.

> **The Reciprocal Identities**
>
> $$\sin\theta = \frac{1}{\csc\theta} \qquad \csc\theta = \frac{1}{\sin\theta}$$
>
> $$\cos\theta = \frac{1}{\sec\theta} \qquad \sec\theta = \frac{1}{\cos\theta}$$
>
> $$\tan\theta = \frac{1}{\cot\theta} \qquad \cot\theta = \frac{1}{\tan\theta}$$

Example 5 Using the Quotient and Reciprocal Identities

Given that $\sin\theta = \frac{5}{7}$ and $\cos\theta = \frac{2\sqrt{6}}{7}$, find the values of the remaining four trigonometric functions using identities.

Solution We can use the **quotient identity** $\tan\theta = \frac{\sin\theta}{\cos\theta}$ to find the value of $\tan\theta$.

$$\tan\theta = \frac{\sin\theta}{\cos\theta} = \frac{\frac{5}{7}}{\frac{2\sqrt{6}}{7}} = \frac{5}{\cancel{7}}\cdot\frac{\cancel{7}}{2\sqrt{6}} = \frac{5}{2\sqrt{6}}$$

Now that we know the values of $\sin\theta$, $\cos\theta$, and $\tan\theta$, we can use the **reciprocal identities** to find the values of the remaining three trigonometric functions.

$$\csc\theta = \frac{1}{\sin\theta} = \frac{1}{\frac{5}{7}} = \frac{7}{5}$$

$$\sec\theta = \frac{1}{\cos\theta} = \frac{1}{\frac{2\sqrt{6}}{7}} = \frac{7}{2\sqrt{6}}$$

$$\cot\theta = \frac{1}{\tan\theta} = \frac{1}{\frac{5}{2\sqrt{6}}} = \frac{2\sqrt{6}}{5}$$

(eText Screens 6.4-1–6.4-35)

You Try It Work through this You Try It problem.

Work Exercises 25–28 in this textbook or in the MyLab Math Study Plan.

THE PYTHAGOREAN IDENTITIES

Given the right triangle from **Figure 33**, we can use the **Pythagorean Theorem** to establish the equation $(opp)^2 + (adj)^2 = (hyp)^2$.

If we start with this equation and then divide both sides by $(hyp)^2$, we can establish the first of three Pythagorean identities.

$$(opp)^2 + (adj)^2 = (hyp)^2$$ Start with the equation representing the Pythagorean Theorem.

$$\frac{(opp)^2}{(hyp)^2} + \frac{(adj)^2}{(hyp)^2} = \frac{(hyp)^2}{(hyp)^2}$$ Divide both sides of the equation by $(hyp)^2$.

$$\left(\frac{opp}{hyp}\right)^2 + \left(\frac{adj}{hyp}\right)^2 = 1$$ Simplify.

$$(\sin\theta)^2 + (\cos\theta)^2 = 1$$ Substitute $\sin\theta = \dfrac{opp}{hyp}$ and $\cos\theta = \dfrac{adj}{hyp}$.

We will typically write $(\sin\theta)^2$ as $\sin^2\theta$ and write $(\cos\theta)^2$ as $\cos^2\theta$. Therefore, the first Pythagorean identity is $\sin^2\theta + \cos^2\theta = 1$. Watch this **video** to see how to derive the remaining two Pythagorean identities.

> **The Pythagorean Identities**
>
> $\sin^2\theta + \cos^2\theta = 1$ $\quad$ $1 + \tan^2\theta = \sec^2\theta$ $\quad$ $1 + \cot^2\theta = \csc^2\theta$

Example 6 Using Identities to Find the Exact Value of a Trigonometric Expression

Use identities to find the exact value of each trigonometric expression.

a. $\tan 37° - \dfrac{\sin 37°}{\cos 37°}$ $\quad$ **b.** $\dfrac{1}{\cos^2\frac{\pi}{9}} - \dfrac{1}{\cot^2\frac{\pi}{9}}$

Solution

a. We can rewrite $\dfrac{\sin 37°}{\cos 37°}$ as $\tan 37°$ using the **quotient identity** $\tan\theta = \dfrac{\sin\theta}{\cos\theta}$.

Therefore, $\tan 37° - \dfrac{\sin 37°}{\cos 37°} = \tan 37° - \tan 37° = 0$.

b. Try using two **reciprocal identities** and a **Pythagorean identity** to find the exact value of this trigonometric expression. Watch this **interactive video** when you have finished to see if you are correct.

You Try It Work through this You Try It problem.

Work Exercises 29–33 in this textbook or in the MyLab Math Study Plan.

OBJECTIVE 4 UNDERSTANDING COFUNCTIONS

Two positive angles are said to be **complementary** if the sum of their measures is $\frac{\pi}{2}$ radians (or $90°$). The two acute angles of every right triangle are complementary. To see why, read this **explanation**. Thus, if θ is an acute angle of a right triangle, then the measure of the other acute angle must be $\frac{\pi}{2} - \theta$ (or $90° - \theta$ if θ is given in degrees). For example, if one acute angle of a right triangle is $\frac{\pi}{5}$, then the other acute angle is $\frac{\pi}{2} - \frac{\pi}{5} = \frac{5\pi}{10} - \frac{2\pi}{10} = \frac{3\pi}{10}$.

Consider the right triangle in Figure 34 with complementary acute angles of θ and $\frac{\pi}{2} - \theta$, and side lengths a, b, and c.

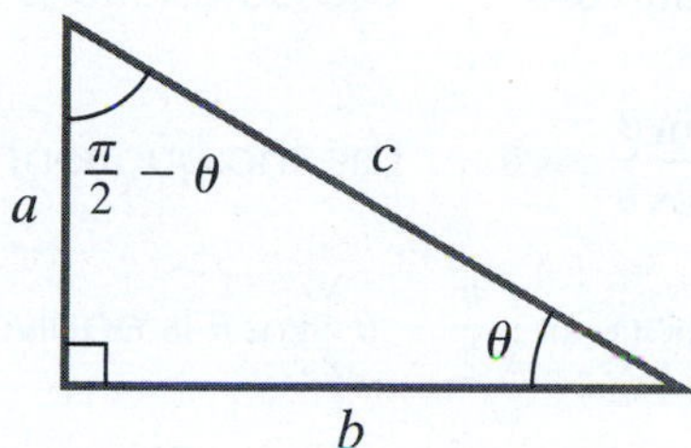

Figure 34 A right triangle with acute angles θ and $\frac{\pi}{2} - \theta$

Looking at **Figure 34**, we see that

$$\sin\theta = \frac{\text{length of side opposite } \theta}{\text{length of the hypotenuse}} = \frac{a}{c} \text{ and } \cos\left(\frac{\pi}{2} - \theta\right) = \frac{\text{length of side adjacent to } \left(\frac{\pi}{2} - \theta\right)}{\text{length of the hypotenuse}} = \frac{a}{c}.$$

Therefore, $\sin\theta = \cos\left(\frac{\pi}{2} - \theta\right)$. In words we say, "the sine of angle θ is equal to the **co**sine of the complement of angle θ." Trigonometric functions that have this relationship are called **cofunctions.** There are three pair of cofunctions: sine and **co**sine, tangent and **co**tangent, and secant and **co**secant. We now state the cofunction identities.

Cofunction Identities

$$\sin\theta = \cos\left(\frac{\pi}{2} - \theta\right) \qquad \cos\theta = \sin\left(\frac{\pi}{2} - \theta\right)$$

$$\tan\theta = \cot\left(\frac{\pi}{2} - \theta\right) \qquad \cot\theta = \tan\left(\frac{\pi}{2} - \theta\right)$$

$$\sec\theta = \csc\left(\frac{\pi}{2} - \theta\right) \qquad \csc\theta = \sec\left(\frac{\pi}{2} - \theta\right)$$

If angle θ is given in degrees, replace $\frac{\pi}{2}$ with $90°$.

(eText Screens 6.4-1–6.4-35)

Example 7 Using the Cofunction Identities to Rewrite or Evaluate Expressions

a. Rewrite the expression $\left(\cot\left(\frac{\pi}{2}-\theta\right)\right)\cos\theta$ as one of the six trigonometric functions of acute angle θ.

b. Determine the exact value of $\sec 55° \csc 35° - \tan 55° \cot 35°$.

Solution

a. Using cofunctions, we can rewrite $\cot\left(\frac{\pi}{2}-\theta\right)$ as $\tan\theta$ and then simplify.

$\left(\cot\left(\frac{\pi}{2}-\theta\right)\right)\cos\theta$	Start with the original expression.
$=\tan\theta\cos\theta$	Use cofunctions to rewrite $\cot\left(\frac{\pi}{2}-\theta\right)$ as $\tan\theta$.
$=\frac{\sin\theta}{\cancel{\cos\theta}}\cancel{\cos\theta}$	Use the **quotient identity** $\tan\theta=\frac{\sin\theta}{\cos\theta}$.

Thus, the expression $\cot\left(\frac{\pi}{2}-\theta\right)\cos\theta$ is equivalent to $\sin\theta$.

b. Using cofunctions, we can rewrite $\sec 55°$ as $\csc(90° - 55°)$, which is equivalent to $\csc 35°$. Similarly, we can rewrite $\tan 55°$ as $\cot(90° - 55°)$, which is equivalent to $\cot 35°$. We now use these cofunction identities to simplify the original expression.

$\sec 55° \csc 35° - \tan 55° \cot 35°$	Start with the original expression.
$=\csc 35° \csc 35° - \cot 35° \cot 35°$	Substitute $\csc 35°$ for $\sec 55°$ and $\cot 35°$ for $\tan 55°$.
$=\csc^2 35° - \cot^2 35°$	Simplify.

Note that the **Pythagorean identity** $1+\cot^2\theta=\csc^2\theta$ is equivalent to $1=\csc^2\theta-\cot^2\theta$. Therefore, $\csc^2 35° - \cot^2 35° = 1$. Hence, the exact value of the original expression, $\sec 55° \csc 35° - \tan 55° \cot 35°$, is 1.

Watch this **interactive video** to see this solution worked out in detail.

You Try It **Work through this You Try It problem.**

Work Exercises 34–38 in this textbook or in the MyLab Math Study Plan.

OBJECTIVE 5 EVALUATING TRIGONOMETRIC FUNCTIONS USING A CALCULATOR

We now know how to find the exact values of trigonometric functions of the acute special angles using our knowledge of the special right triangles. But what about evaluating trigonometric functions of non-special angles? For this, we need to find decimal approximations of these values using a calculator. Take a few moments to locate the [sin] key, [cos] key, and [tan] key on your calculator. More importantly, make sure that you know how to set your calculator to degree mode or radian mode. Then try working through Example 8.

Example 8 Approximate Trigonometric Expressions Using a Calculator

Evaluate each trigonometric expression using a calculator. Round each answer to four decimal places.

a. $\sin \frac{8\pi}{7}$ b. $\cot 70°$ c. $\sec \frac{\pi}{5}$

Solution

a. First, make sure that your calculator is in radian mode.

$$\sin \frac{8\pi}{7} \approx -0.4339$$

b. First, make sure that your calculator is in degree mode. Since most calculators do not have a cotangent key, we will use the **reciprocal identity** $\cot \theta = \frac{1}{\tan \theta}$ and the [tan] key to evaluate the expression.

$$\cot 70° = \frac{1}{\tan 70°} \approx 0.3640$$

c. First, make sure that your calculator is in radian mode. Because most calculators do not have a secant key, we will use the **reciprocal identity** $\sec \theta = \frac{1}{\cos \theta}$ and the [cos] key to evaluate the expression.

$$\sec \frac{\pi}{5} = \frac{1}{\cos \frac{\pi}{5}} \approx 1.2361$$

Watch this **video** to see how to enter these expressions into your calculator.

CAUTION It is extremely important to make sure that your calculator is set to the correct mode. In Example 8a, your calculator must be set to radian mode. Failing to set your calculator to the correct mode will yield erroneous answers, as illustrated below.

Using Technology

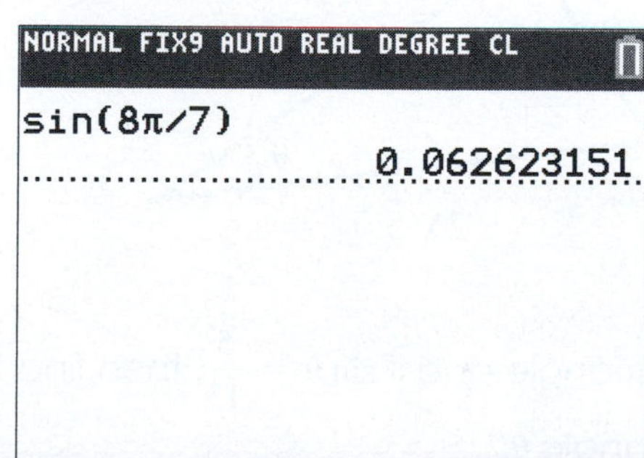

NORMAL FIX9 AUTO REAL RADIAN CL
sin(8π/7)
-0.433883739

The screenshot on the left shows the value of $\sin \frac{8\pi}{7}$ using a calculator set to degree mode. The screenshot on the right shows the correct value of $\sin \frac{8\pi}{7}$ using the same calculator set to radian mode. You can see the importance of setting your calculator to the correct mode!

You should quickly test yourself by clicking on the You Try It icon below to verify that you can evaluate trigonometric expressions using your calculator.

You Try It Work through this You Try It problem.

Work Exercises 39–42 in this textbook or in the MyLab Math Study Plan.

6.4 Exercises

Skill Check Exercises

For exercises SCE-1 and SCE-2, simplify each expression by writing as a single term using a common denominator.

SCE-1. $\frac{\pi}{2} - \frac{\pi}{6}$ **SCE-2.** $\frac{\pi}{2} - \frac{5\pi}{12}$

In Exercises 1–5, a right triangle with acute angle θ is given. Evaluate the six trigonometric functions of the acute angle θ.

1.
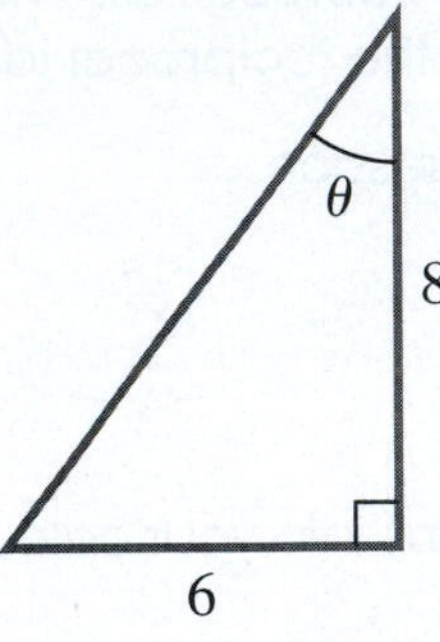

2.
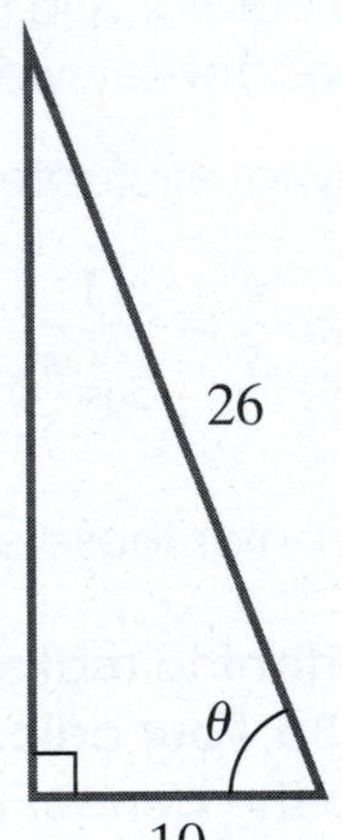

3.
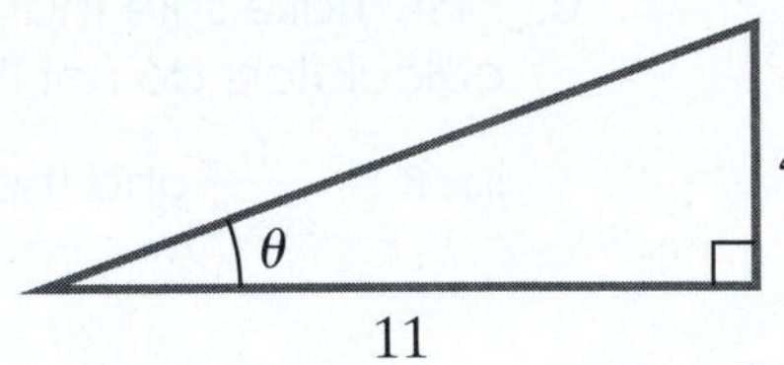

4.
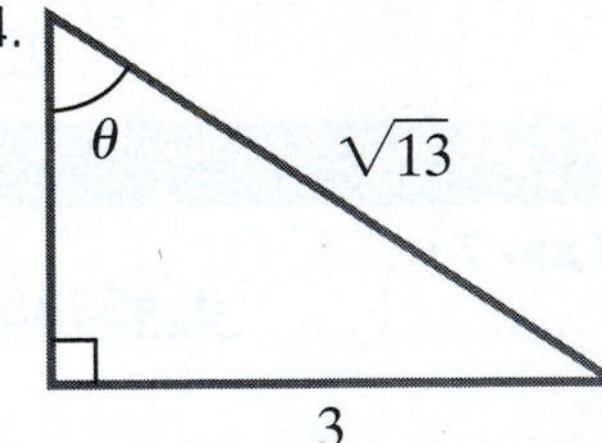

5.
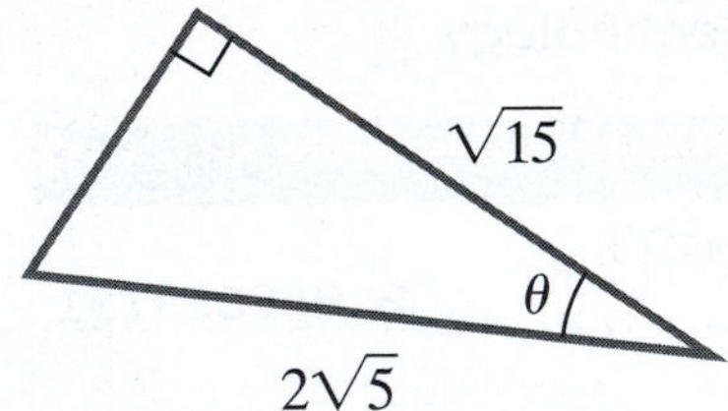

6. If θ is an acute angle of a right triangle and if $\sin\theta = \frac{5}{13}$, then find the values of the remaining five trigonometric functions for angle θ.

7. If θ is an acute angle of a right triangle and if $\cos\theta = \frac{60}{61}$, then find the values of the remaining five trigonometric functions for angle θ.

8. If θ is an acute angle of a right triangle and if $\tan\theta = \frac{1}{2}$, then find the values of the remaining five trigonometric functions for angle θ.

9. If θ is an acute angle of a right triangle and if $\csc\theta = \sqrt{6}$, then find the values of the remaining five trigonometric functions for angle θ.

10. If θ is an acute angle of a right triangle and if $\sec\theta = \frac{7}{\sqrt{5}}$, then find the values of the remaining five trigonometric functions for angle θ.

11. If θ is an acute angle of a right triangle and if $\cot\theta = 3$, then find the values of the remaining five trigonometric functions for angle θ.

In Exercises 12–18, use special right triangles to evaluate each expression.

12. $\sin\frac{\pi}{6}$

13. $\sec 45^\circ$

14. $\csc\frac{\pi}{3}$

15. $\csc 45^\circ - \cot 30^\circ$

16. $\sqrt{3}\sec\frac{\pi}{6} + \sqrt{2}\csc\frac{\pi}{4}$

17. $\sec^2 45^\circ + \csc^2 60^\circ$

18. $\dfrac{\sin^2\frac{\pi}{3} + \cos^2\frac{\pi}{6}}{\sec^2\frac{\pi}{4}}$

19. Determine the measure of the acute angle θ (in radians) for which $\sin\theta = \frac{1}{\sqrt{2}}$.

20. Determine the measure of the acute angle θ (in radians) for which $\cos\theta = \frac{\sqrt{3}}{2}$.

21. Determine the measure of the acute angle θ (in radians) for which $\tan\theta = \frac{1}{\sqrt{3}}$.

22. Determine the measure of the acute angle θ (in radians) for which $\csc\theta = 2$.

23. Determine the measure of the acute angle θ (in radians) for which $\cot\theta = \sqrt{3}$.

24. Determine the measure of the acute angle θ (in radians) for which $\sec\theta = \sqrt{2}$.

25. Given that $\sin\theta = \frac{7}{25}$ and $\cos\theta = \frac{24}{25}$, find the values of the remaining four trigonometric functions using identities.

26. Given that $\sin\theta = \frac{3}{\sqrt{11}}$ and $\cos\theta = \frac{\sqrt{2}}{\sqrt{11}}$, find the values of the remaining four trigonometric functions using identities.

27. Given that $\csc\theta = \frac{29}{21}$ and $\sec\theta = \frac{29}{20}$, find the values of the remaining four trigonometric functions using identities.

28. Given that $\csc\theta = \frac{\sqrt{23}}{4}$ and $\sec\theta = \frac{\sqrt{23}}{\sqrt{7}}$, find the values of the remaining four trigonometric functions using identities.

In Exercises 29–31, use identities to find the exact value of each trigonometric expression. Assume that θ is an acute angle.

29. $\cot 19^\circ - \frac{\cos 19^\circ}{\sin 19^\circ}$

30. $\frac{1}{\csc^2 \frac{\pi}{11}} + \frac{1}{\sec^2 \frac{\pi}{11}}$

31. $\tan\frac{5\pi}{12}\left(\frac{1}{\cos^2\frac{5\pi}{12}} - \frac{1}{\cot^2\frac{5\pi}{12}}\right) - \tan\frac{5\pi}{12}$

32. Use identities to rewrite the following expression as $\sin\theta$, $\cos\theta$, $\tan\theta$, $\csc\theta$, $\sec\theta$, or $\cot\theta$.

$$\sin\theta\sec^2\theta\cot^2\theta$$

33. Use identities to rewrite the following expression as $\sin\theta$, $\cos\theta$, $\tan\theta$, $\csc\theta$, $\sec\theta$, or $\cot\theta$.

$$\frac{\sin^2\theta + \tan^2\theta + \cos^2\theta}{\sec\theta}$$

34. Find a cofunction equivalent to $\sin\frac{5\pi}{7}$.

35. Find a cofunction equivalent to $\cot 39^\circ$.

36. Determine the exact value of $\tan\frac{\pi}{7}\cot\frac{5\pi}{14} - \sec\frac{\pi}{7}\csc\frac{5\pi}{14}$.

37. Rewrite the expression $\cos(90^\circ - \theta)\sec\theta$ as one of the six trigonometric functions of acute angle θ.

38. Given that θ is an acute angle, rewrite the $\tan(\theta - 15^\circ)$ as an equivalent value using its cofunction.

In Exercises 39–42, use a calculator to evaluate each trigonometric expression. Round your answer to four decimal places.

39. $\cos\frac{2\pi}{9}$

40. $\csc 81^\circ$

41. $\sec\frac{4\pi}{13}$

42. $\sin 61^\circ$

6.5 Trigonometric Functions of General Angles

THINGS TO KNOW

Before working through this section, be sure that you are familiar with the following concepts:

 You Try It
1. Understanding Radian Measure (Section 6.1) 

 You Try It
2. Converting between Degree Measure and Radian Measure (Section 6.1)  

You Try It
3. Finding Coterminal Angles Using Radian Measure (Section 6.1)

 You Try It
4. Understanding the Special Right Triangles (Section 6.3) 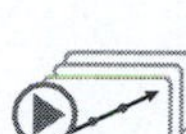

 You Try It
5. Understanding the Right Triangle Definitions of the Trigonometric Functions (Section 6.4) 

OBJECTIVES

1 Understanding the Four Families of Special Angles

2 Understanding the Definitions of the Trigonometric Functions of General Angles

3 Finding the Values of the Trigonometric Functions of Quadrantal Angles

4 Understanding the Signs of the Trigonometric Functions

5 Determining Reference Angles

6 Evaluating Trigonometric Functions of Angles Belonging to the $\frac{\pi}{3}$, $\frac{\pi}{6}$, or $\frac{\pi}{4}$ Families

SECTION 6.5 EXERCISES

OBJECTIVE 1 UNDERSTANDING THE FOUR FAMILIES OF SPECIAL ANGLES

In Section 6.4, we defined the six trigonometric functions of an angle θ where θ was an **acute angle** of a right triangle. In this section, we will extend the definitions of the six trigonometric functions to include *all* angles for which each function is defined. We start by introducing four groups, or families, of special angles. These four families are known as **the quadrantal family, the $\frac{\pi}{3}$ family, the $\frac{\pi}{6}$ family, and the $\frac{\pi}{4}$ family.**

THE QUADRANTAL FAMILY OF ANGLES

Recall that a quadrantal angle is an angle in standard position whose terminal side lies along either axis. Although there are infinitely many quadrantal angles, there are only four possible positions for the terminal side of the quadrantal angles.

Any angle belonging to the quadrantal family must be **coterminal** with $0, \frac{\pi}{2}, \pi,$ or $\frac{3\pi}{2}$. See Figure 35. Watch this **video** for a detailed explanation of the quadrantal family of angles.

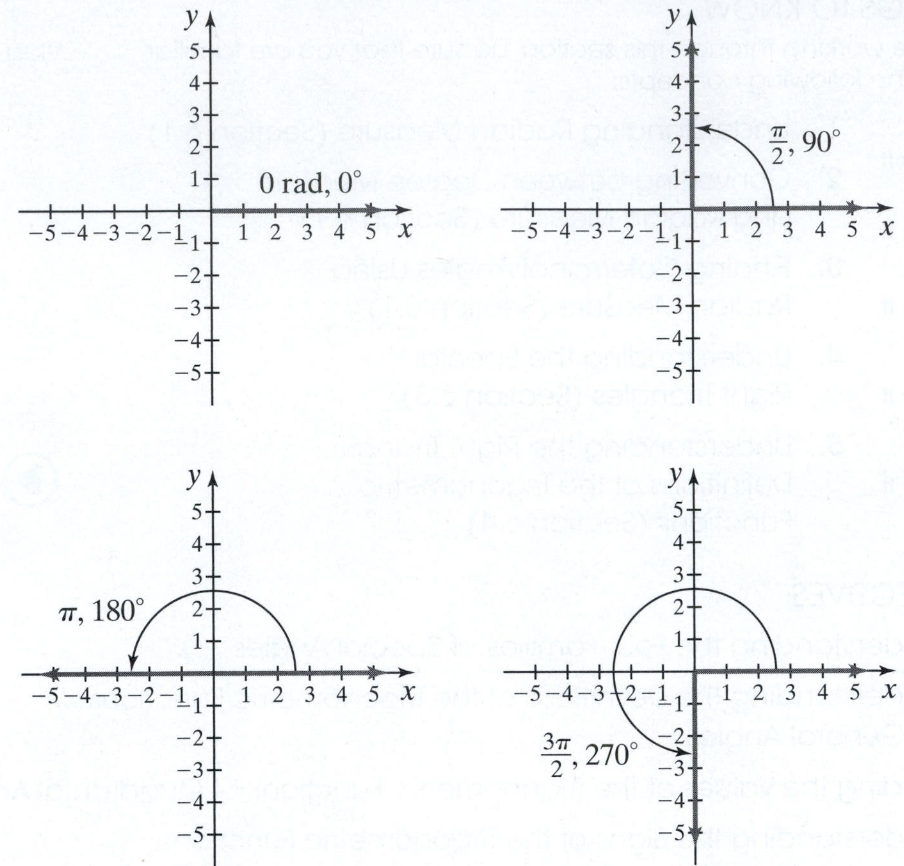

Figure 35 Angles belonging to the quadrantal family must be coterminal with $0, \frac{\pi}{2}, \pi,$ or $\frac{3\pi}{2}$.

THE $\frac{\pi}{3}$ FAMILY OF ANGLES

An angle in standard position belongs to the $\frac{\pi}{3}$ family if the angle is **coterminal** with $\frac{\pi}{3}, \frac{2\pi}{3}, \frac{4\pi}{3},$ or $\frac{5\pi}{3}$. See Figure 36. Angles belonging to the $\frac{\pi}{3}$ family must be an integer multiple of $\frac{\pi}{3}$ but must **not** be a **quadrantal angle**. For example, the angle $\theta = \frac{6\pi}{3}$ does not belong to the $\frac{\pi}{3}$ family because $\theta = \frac{6\pi}{3} = 2\pi$ and $\theta = 2\pi$ is a quadrantal angle. Watch this **video** for a detailed explanation of the $\frac{\pi}{3}$ family of angles.

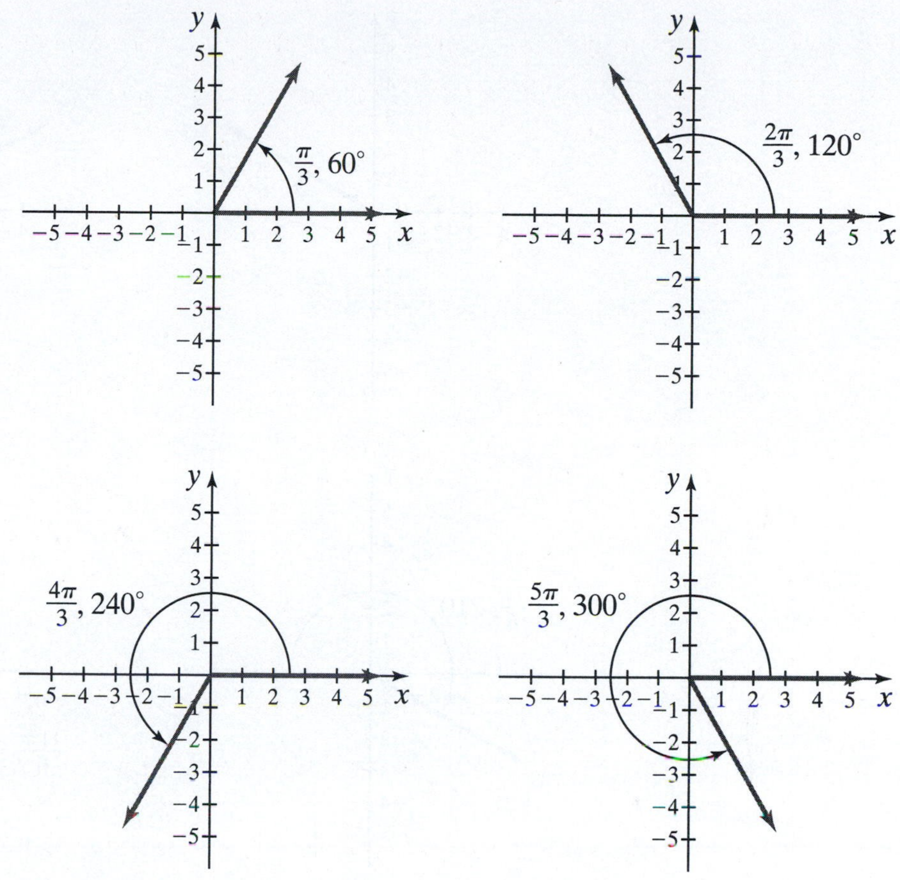

Figure 36 Angles belonging to the $\frac{\pi}{3}$ family must be coterminal with $\frac{\pi}{3}, \frac{2\pi}{3}, \frac{4\pi}{3}$, or $\frac{5\pi}{3}$.

THE $\frac{\pi}{6}$ FAMILY OF ANGLES

An angle in standard position belongs to the $\frac{\pi}{6}$ family if the angle is **coterminal** with $\frac{\pi}{6}, \frac{5\pi}{6}, \frac{7\pi}{6}$, or $\frac{11\pi}{6}$. See Figure 37. Angles belonging to the $\frac{\pi}{6}$ family must be an integer multiple of $\frac{\pi}{6}$ but must **not** belong to the **quadrantal** family or the $\frac{\pi}{3}$ family. For example, the angle $\theta = \frac{4\pi}{6}$ does not belong to the $\frac{\pi}{6}$ family because $\theta = \frac{4\pi}{6} = \frac{2\pi}{3}$ and $\theta = \frac{2\pi}{3}$ belongs to the $\frac{\pi}{3}$ family of angles. Watch this **video** for a detailed explanation of the $\frac{\pi}{6}$ family of angles.

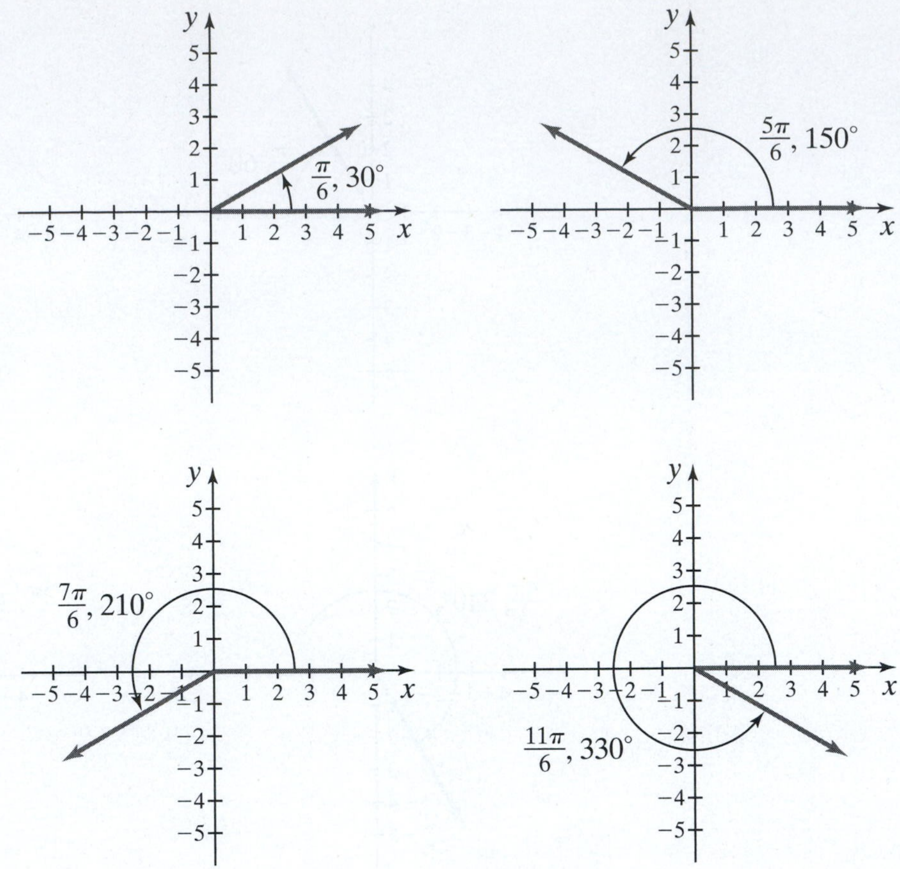

Figure 37 Angles belonging to the $\frac{\pi}{6}$ family must be coterminal with $\frac{\pi}{6}, \frac{5\pi}{6}, \frac{7\pi}{6}$, or $\frac{11\pi}{6}$.

THE $\frac{\pi}{4}$ FAMILY OF ANGLES

An angle in standard position belongs to the $\frac{\pi}{4}$ family if the angle is **coterminal** with $\frac{\pi}{4}, \frac{3\pi}{4}, \frac{5\pi}{4}$, or $\frac{7\pi}{4}$. See Figure 38. Angles belonging to the $\frac{\pi}{4}$ family must be an integer multiple of $\frac{\pi}{4}$ but must **not** belong to the **quadrantal** family. For example, the angle $\theta = \frac{10\pi}{4}$ does not belong to the $\frac{\pi}{4}$ family because $\theta = \frac{10\pi}{4} = \frac{5\pi}{2}$ and $\theta = \frac{5\pi}{2}$ belongs to the quadrantal family. Watch this **video** for a detailed explanation of the $\frac{\pi}{4}$ family of angles.

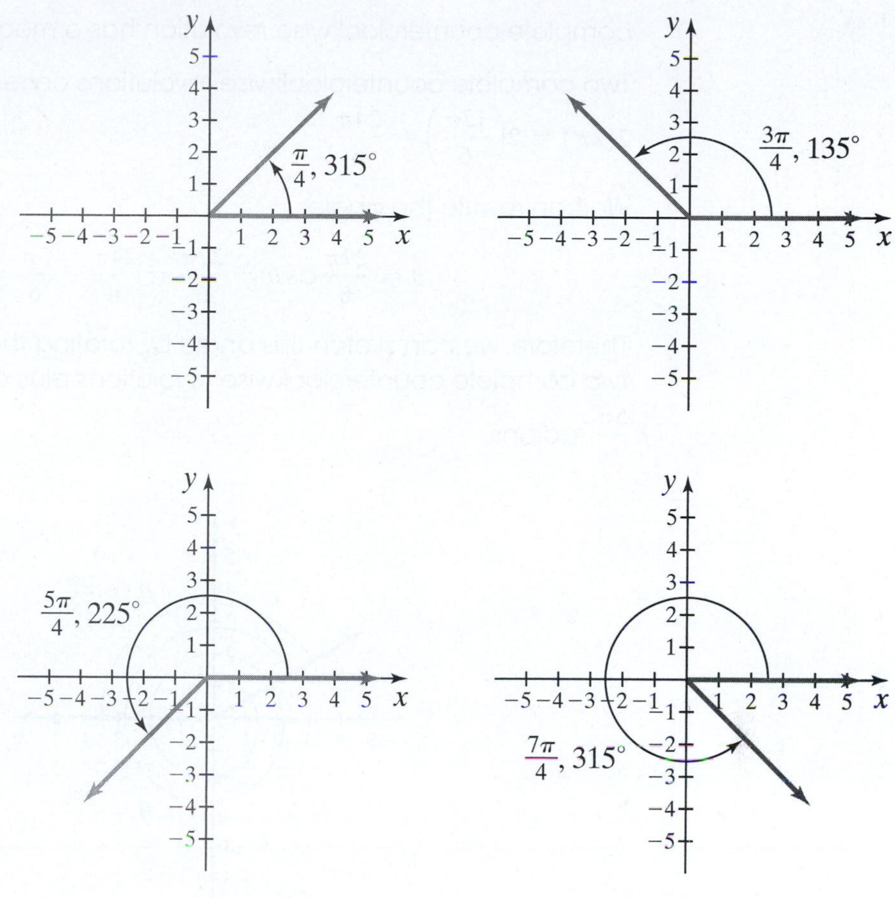

Figure 38 Angles belonging to the $\frac{\pi}{4}$ family must be coterminal with $\frac{\pi}{4}, \frac{3\pi}{4}, \frac{5\pi}{4}$, or $\frac{7\pi}{4}$.

Example 1 Working with Angles Belonging to the Families of Special Angles

Each of the given angles belongs to one of the four families of special angles. Determine the family of angles to which it belongs, sketch the angle, and then determine the angle of least nonnegative measure, θ_C, coterminal with the given angle.

a. $\theta = \frac{29\pi}{6}$ **b.** $\theta = \frac{14\pi}{2}$ **c.** $\theta = -\frac{18\pi}{4}$ **d.** $\theta = \frac{11\pi}{4}$

e. $\theta = \frac{14\pi}{6}$ **f.** $\theta = 420°$ **g.** $\theta = -495°$

Solution

a. The angle $\theta = \frac{29\pi}{6}$ is written in lowest terms and is 29 times $\frac{\pi}{6}$. Therefore, $\theta = \frac{29\pi}{6}$ belongs to the $\frac{\pi}{6}$ family. Recall that the angle created from one complete counterclockwise revolution has a measure of $2\pi = \frac{12\pi}{6}$ radians. Two complete counterclockwise revolutions correspond to the angle $2(2\pi) = 2\left(\frac{12\pi}{6}\right) = \frac{24\pi}{6}$.

We can rewrite the angle

$$\theta = \frac{29\pi}{6} \text{ as } \theta = \frac{29\pi}{6} = \frac{24\pi}{6} + \frac{5\pi}{6} = 2(2\pi) + \frac{5\pi}{6}.$$

Therefore, we can sketch this angle by rotating the terminal side of the angle two complete counterclockwise revolutions plus an additional rotation of $\frac{5\pi}{6}$ radians.

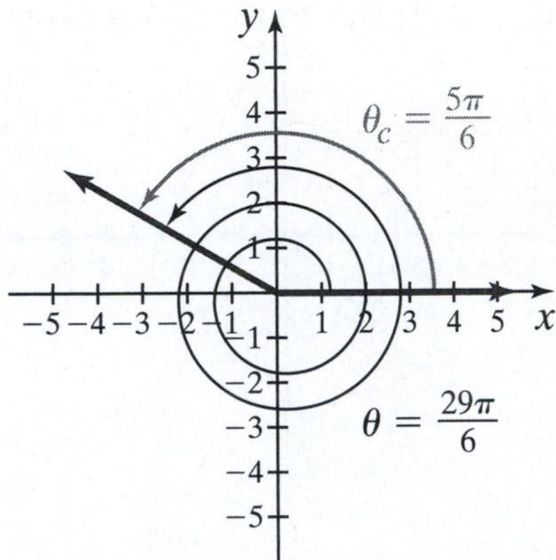

Figure 39

See Figure 39. The angle of least nonnegative measure coterminal with $\theta = \frac{29\pi}{6}$ is $\theta_C = \frac{5\pi}{6}$.

b. The angle $\theta = \frac{14\pi}{2}$ is not written in lowest terms. We can rewrite this angle in lowest terms as $\theta = \frac{14\pi}{2} = \frac{\cancel{2} \cdot 7\pi}{\cancel{2}} = 7\pi$. Any angle that is an integer multiple of π belongs to the quadrantal family. Therefore, this angle belongs to the quadrantal family. The angle $\theta = 7\pi$ can be rewritten as $\theta = 3(2\pi) + \pi$. Therefore, this angle can be drawn by rotating the terminal side counterclockwise three complete revolutions plus an additional rotation of π radians. See **Figure 40**.

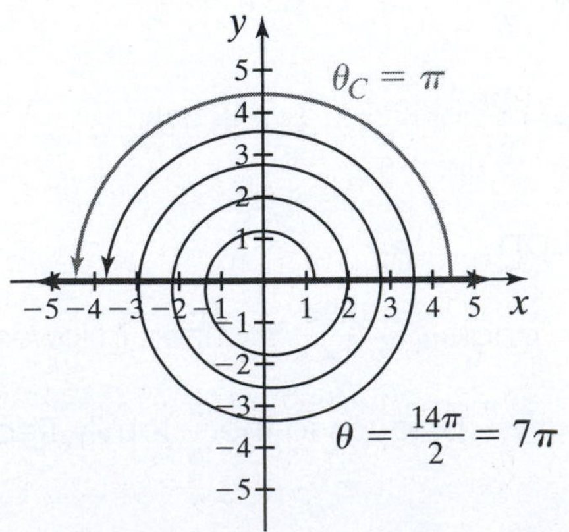

Figure 40

The terminal side of this angle lies along the negative x-axis. Therefore, the angle of least nonnegative measure coterminal with $\theta = 7\pi$ is $\theta_C = \pi$.

c. The angle $\theta = -\frac{18\pi}{4}$ is not written in lowest terms. We can rewrite this angle in lowest terms as $\theta = -\frac{18\pi}{4} = -\frac{\cancel{2} \cdot 9\pi}{\cancel{2} \cdot 2} = -\frac{9\pi}{2}$. Any angle that is an integer multiple of $\frac{\pi}{2}$ belongs to the quadrantal family. Therefore, this angle belongs to the quadrantal family. The angle $\theta = -\frac{9\pi}{2}$ can be rewritten as $\theta = -\frac{8\pi}{2} - \frac{\pi}{2} = 2(-2\pi) - \frac{\pi}{2}$. Therefore, this angle can be drawn by rotating the terminal side clockwise two complete revolutions plus an additional rotation of $\frac{\pi}{2}$ radians in a clockwise direction. See **Figure 41**.

The terminal side of this angle lies along the negative y-axis. Therefore, the angle of least nonnegative measure coterminal with $\theta = -\frac{9\pi}{2}$ is $\theta_C = \frac{3\pi}{2}$.

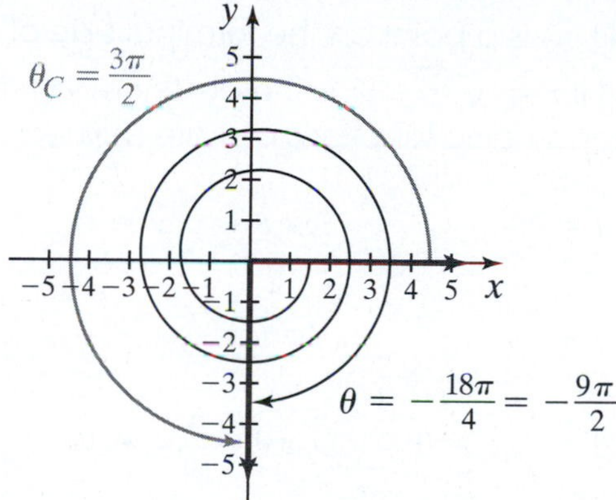

Figure 41

Try completing parts d–g on your own. Then watch this **interactive video** to see the remainder of this solution.

You Try It **Work through this You Try It problem.**

Work Exercises 1–12 in this textbook or in the MyLab Math Study Plan.

OBJECTIVE 2 UNDERSTANDING THE DEFINITIONS OF THE TRIGONOMETRIC FUNCTIONS OF GENERAL ANGLES

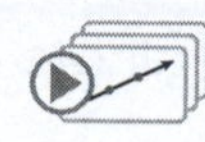

In **Section 6.4**, we presented the definitions of the trigonometric functions using the lengths of the sides and the hypotenuse of a right triangle. You may wish to review these **definitions**. The right triangle definitions of the trigonometric functions only applies to **acute angles**. We now want to define the trigonometric functions of *general angles*. The term *general angle* is used here to indicate that these angles are not restricted in size and can be either positive angles, negative angles, or zero.

(eText Screens 6.5-1–6.5-55)

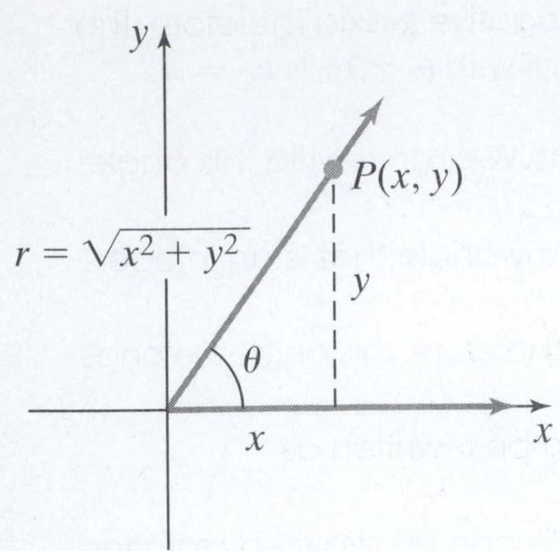

Figure 42 This illustrates an acute angle θ in Quadrant I.

In order to develop the definitions of trigonometric functions of general angles, let $P(x, y)$ be a point lying on the terminal side of an acute angle θ in standard position and let $r > 0$ represent the distance from the **origin** to point P. By the **distance formula**, $r = \sqrt{x^2 + y^2}$. We use the right triangle definitions of the trigonometric functions to write the trigonometric ratios in terms of x, y, and r. See Figure 42.

$$\sin\theta = \frac{opp}{hyp} = \frac{y}{r} \qquad \csc\theta = \frac{hyp}{opp} = \frac{r}{y}$$

$$\cos\theta = \frac{adj}{hyp} = \frac{x}{r} \qquad \sec\theta = \frac{hyp}{adj} = \frac{r}{x}$$

$$\tan\theta = \frac{opp}{adj} = \frac{y}{x} \qquad \cot\theta = \frac{adj}{opp} = \frac{x}{y}$$

We now use this representation to create a second set of definitions of trigonometric functions. These definitions will apply to general angles. Notice that the new definitions are consistent with our previous right triangle definitions.

The General Angle Definitions of the Trigonometric Functions

If $P(x, y)$ is a point on the terminal side of *any* angle θ in standard position and if $r = \sqrt{x^2 + y^2}$ is the distance from the origin to point P, then the six trigonometric functions of θ are defined as follows:

$$\sin\theta = \frac{y}{r} \qquad \csc\theta = \frac{r}{y}, y \neq 0$$

$$\cos\theta = \frac{x}{r} \qquad \sec\theta = \frac{r}{x}, x \neq 0$$

$$\tan\theta = \frac{y}{x}, x \neq 0 \qquad \cot\theta = \frac{x}{y}, y \neq 0$$

Because division by zero is undefined, it is important to note that four of the trigonometric functions will not be defined for certain angles. For example, $\tan\theta = \frac{y}{x}$ and $\sec\theta = \frac{r}{x}$ will be undefined when $x = 0$. A point having an x-coordinate of zero must lie along the y-axis. Therefore, the tangent and secant functions are undefined for any angle whose **terminal side lies along the y-axis**. Similarly, $\csc\theta = \frac{r}{y}$ and $\cot\theta = \frac{x}{y}$ are undefined when $y = 0$. A point having a y-coordinate of zero must lie along the x-axis. Thus, the cosecant and cotangent functions are undefined for any angle whose **terminal side lies along the x-axis**. We will explore this further when we discuss the trigonometric functions of **quadrantal angles**. The sine $\left(\sin\theta = \frac{y}{r}\right)$ and cosine $\left(\cos\theta = \frac{x}{r}\right)$ functions are defined for any angle because the value of r, which is in the denominator of each ratio, will never be zero.

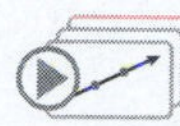

You may want to watch this **animation** to see the development of the definitions of the trigonometric functions of general angles.

Example 2 Finding the Values of the Six Trigonometric Functions for a General Angle

Suppose that the point $(-4, -6)$ is on the terminal side of an angle θ. Find the six trigonometric functions of θ.

Solution Always start by plotting the point, constructing the angle θ, and identifying the values of x, y, and r. We are given that $x = -4$ and $y = -6$. Using the equation $r = \sqrt{x^2 + y^2}$, we can find the value of r.

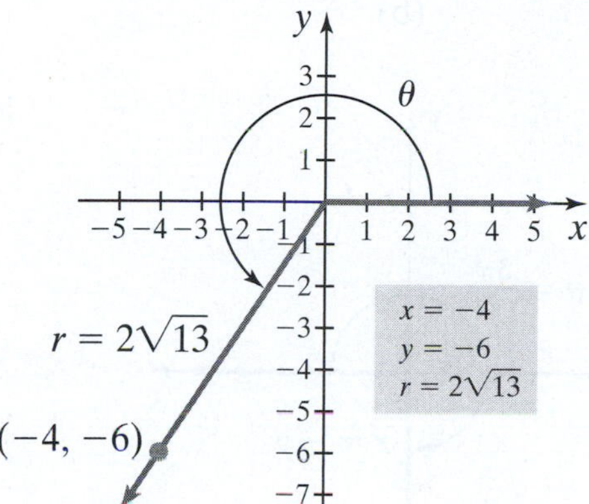

$r = \sqrt{x^2 + y^2}$ Write the equation representing the value of r.

$= \sqrt{(-4)^2 + (-6)^2}$ Substitute $x = -4$ and $y = -6$.

$= \sqrt{52} = 2\sqrt{13}$ Simplify the radical.

Now that we know the values of x, y, and r, we can use the general angle definitions of the trigonometric functions to find the values of $\sin\theta$, $\cos\theta$, and $\tan\theta$.

$$\sin\theta = \frac{y}{r} = \frac{-6}{2\sqrt{13}} = -\frac{3}{\sqrt{13}} \qquad \csc\theta = \frac{r}{y} = \frac{2\sqrt{13}}{-6} = -\frac{\sqrt{13}}{3}$$

$$\cos\theta = \frac{x}{r} = \frac{-4}{2\sqrt{13}} = -\frac{2}{\sqrt{13}} \qquad \sec\theta = \frac{r}{x} = \frac{2\sqrt{13}}{-4} = -\frac{\sqrt{13}}{2}$$

$$\tan\theta = \frac{y}{x} = \frac{-6}{-4} = \frac{3}{2} \qquad \cot\theta = \frac{x}{y} = \frac{-4}{-6} = \frac{2}{3}$$

Notice that you may leave a radical in the denominator as long as the radical is simplified and that all factors common to the numerator and denominator are removed. This includes common integer factors as well as radical factors. To see some examples of simplification that will be required, view this **table**. Watch this

video to see a complete solution to this example.

You Try It Work through this You Try It problem.

Work Exercises 13–18 in this textbook or in the MyLab Math Study Plan.

OBJECTIVE 3 FINDING THE VALUES OF THE TRIGONOMETRIC FUNCTIONS OF QUADRANTAL ANGLES

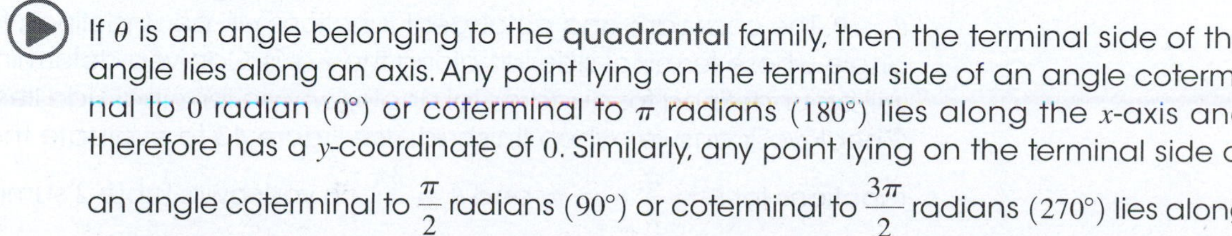

If θ is an angle belonging to the **quadrantal** family, then the terminal side of the angle lies along an axis. Any point lying on the terminal side of an angle coterminal to 0 radian $(0°)$ or coterminal to π radians $(180°)$ lies along the x-axis and therefore has a y-coordinate of 0. Similarly, any point lying on the terminal side of an angle coterminal to $\frac{\pi}{2}$ radians $(90°)$ or coterminal to $\frac{3\pi}{2}$ radians $(270°)$ lies along the y-axis and has an x-coordinate of 0. To determine the values of the trigonometric functions of angles belonging to the quadrantal family, we can use *any* point lying on the terminal side and apply the definitions of the six trigonometric functions. See **Figure 43**.

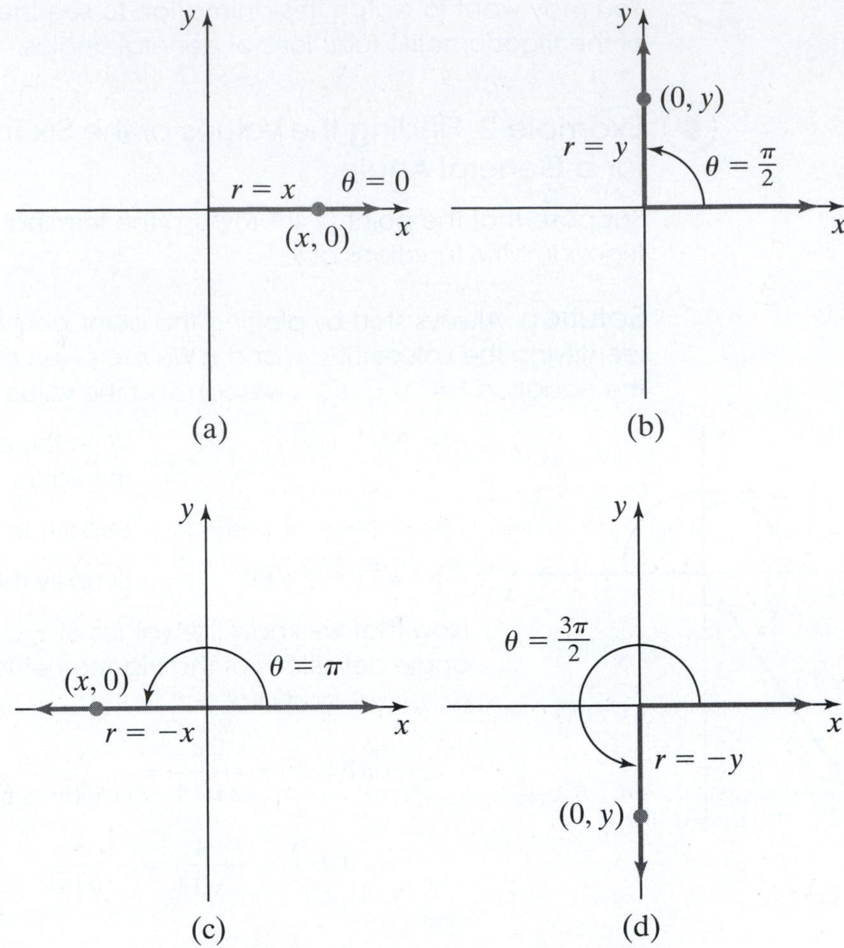

Figure 43 Quadrantal angles with an arbitrary point located on the terminal side of each angle.

Looking at Figure 43a, we can use $x, y = 0$, and $r = x$ to find the values of the six trigonometric functions for the angle $\theta = 0$ radian (or $\theta = 0°$).

$$\sin 0 = \frac{y}{r} = \frac{0}{x} = 0 \qquad \csc 0 = \frac{r}{y} = \frac{x}{0}\,(\text{undefined})$$

$$\cos 0 = \frac{x}{r} = \frac{x}{x} = 1 \qquad \sec 0 = \frac{r}{x} = \frac{x}{x} = 1$$

$$\tan 0 = \frac{y}{x} = \frac{0}{x} = 0 \qquad \cot 0 = \frac{x}{y} = \frac{x}{0}\,(\text{undefined})$$

We can see that the cosecant function and cotangent functions are undefined when $\theta = 0$. The cosecant and cotangent functions will be undefined for any quadrantal angle whose terminal side lies along the x-axis. Can you determine which functions will be undefined for quadrantal angles whose terminal side lies along the y-axis? Check your **answer** when finished. Use **Figure 43** to evaluate the six trigonometric functions for $\theta = \frac{\pi}{2}$, $\theta = \pi$, and $\theta = \frac{3\pi}{2}$ on your own. Table 2 summarizes the values of the six trigonometric functions of quadrantal angles.

Table 2 Trigonometric Functions of Quadrantal Angles

θ	$\sin\theta$	$\cos\theta$	$\tan\theta$	$\csc\theta$	$\sec\theta$	$\cot\theta$
0	0	1	0	Undefined	1	Undefined
$\frac{\pi}{2}$	1	0	Undefined	1	Undefined	0
π	0	-1	0	Undefined	-1	Undefined
$\frac{3\pi}{2}$	-1	0	Undefined	-1	Undefined	0

You will find that, rather than trying to memorize this table, determining these values by drawing the graphs seen in **Figure 43** is much easier. Also, the values seen in Table 2 will become clear once you become familiar with the graphs of the trigonometric functions. The graphs of the trigonometric functions will be studied in depth in Chapter 7.

Example 3 Finding the Values of the Six Trigonometric Functions of Quadrantal Angles

Without using a calculator, determine the value of the trigonometric function or state that the value is undefined.

a. $\cos(-11\pi)$ **b.** $\csc(-270°)$ **c.** $\tan\left(\frac{13\pi}{2}\right)$ **d.** $\sin(540°)$ **e.** $\cot\left(-\frac{7\pi}{2}\right)$

Solution Draw each angle and choose an arbitrary point (other than the origin) on the terminal side of the angle. Then use the general angle definitions of the trigonometric functions to find each value or state that the value is undefined. Once you have completed this, watch this **interactive video** to see if you are correct.

You Try It Work through this You Try It problem.

Work Exercises 19–30 in this textbook or in the MyLab Math Study Plan.

OBJECTIVE 4 UNDERSTANDING THE SIGNS OF THE TRIGONOMETRIC FUNCTIONS

Suppose $0 < \theta < 2\pi$ and $\sin\theta = \frac{3}{5}$. What is the value of $\cos\theta$? By definition, $\sin\theta = \frac{y}{r}$, so possible choices for y and r are $y = 3$ and $r = 5$. (We could have picked $y = 6$ and $r = 10$ or $y = 9$ and $r = 15$ because $\frac{6}{10} = \frac{3}{5}$ and $\frac{9}{15} = \frac{3}{5}$.) Using the equation $r = \sqrt{x^2 + y^2}$, we can solve for x to get $x = \pm 4$. (View these **steps** to see how.) Looking at Figure 44, we see that there are two possible angles (between 0 and 2π) for which $\sin\theta = \frac{3}{5}$. One possible angle appears in Quadrant I with $x = 4$, $y = 3$, and $r = 5$. See Figure 44a. The other possible angle appears in Quadrant II with $x = -4$, $y = 3$, and $r = 5$. See Figure 44b.

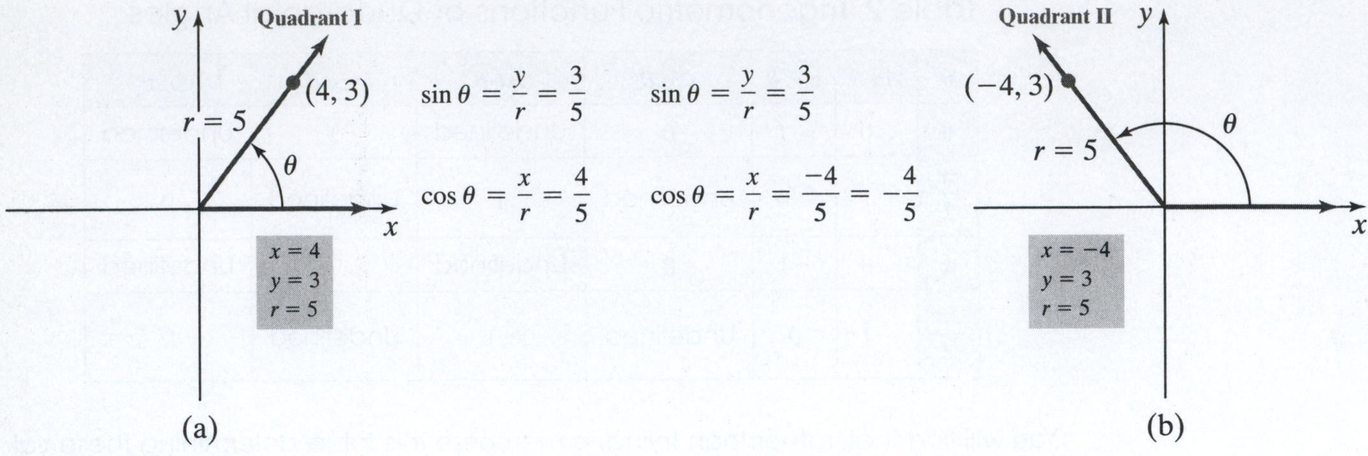

Figure 44 There are two possible values of θ between 0 and 2π for which $\sin\theta = \frac{3}{5}$.

In **Figure 44a**, $\cos\theta = \frac{4}{5}$. In **Figure 44b**, $\cos\theta = -\frac{4}{5}$. Thus, it is impossible to determine the value of $\cos\theta$ unless we know the quadrant in which the terminal side of the angle lies. It is crucial to understand that the sign of each trigonometric function is determined by the quadrant in which the terminal side of the angle lies. The value of r is always positive. Thus, it follows that the sign of a trigonometric function depends on the sign of the x- and y-coordinates of the ordered pair $P(x, y)$ lying on the terminal side of the angle. The signs of x and y depends on the quadrant in which the point is located. We now explore how to determine the sign of each trigonometric function given a point on the terminal side of an angle in each of the four quadrants.

QUADRANT I

In Quadrant I, $r > 0, x > 0$, and $y > 0$. Therefore, the ratios $\frac{y}{r}, \frac{x}{r}$, and $\frac{y}{x}$ (and their reciprocals) are all positive. Thus, for any angle θ whose terminal side lies in Quadrant I, $\sin\theta = \frac{y}{r} > 0, \cos\theta = \frac{x}{r} > 0$, and $\tan\theta = \frac{y}{x} > 0$. Therefore, *all* six trigonometric functions are positive for *all* angles whose terminal side lies in Quadrant I. See Figure 45.

$$\sin\theta = \frac{y}{r} = \frac{+}{+} = +$$

$$\cos\theta = \frac{x}{r} = \frac{+}{+} = +$$

$$\tan\theta = \frac{y}{x} = \frac{+}{+} = +$$

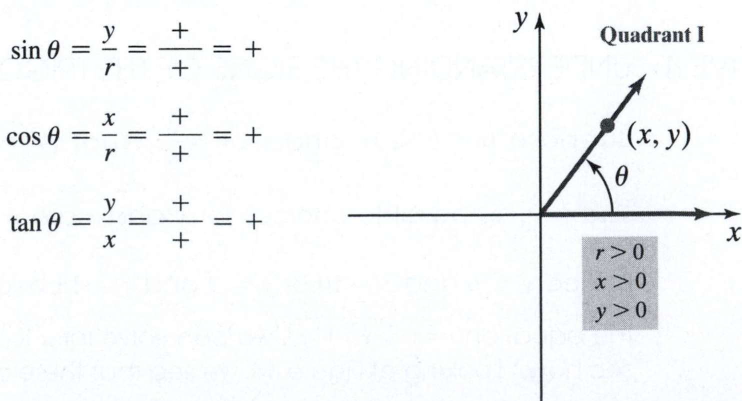

Figure 45 All trigonometric functions are positive for all angles in Quadrant I.

QUADRANT II

In Quadrant II, $r > 0, x < 0,$ and $y > 0$. Therefore, $\frac{y}{r} > 0$ but $\frac{x}{r} < 0$ and $\frac{y}{x} < 0$. Thus, for any angle θ whose terminal side lies in Quadrant II, $\sin\theta = \frac{y}{r} > 0, \cos\theta = \frac{x}{r} < 0,$ and $\tan\theta = \frac{y}{x} < 0$. We only need to remember that $\sin\theta$ (and its reciprocal $\csc\theta$) are positive for all angles with a terminal side lying in Quadrant II. All other trigonometric functions are negative. See Figure 46.

$$\sin\theta = \frac{y}{r} = \frac{+}{+} = +$$

$$\cos\theta = \frac{x}{r} = \frac{-}{+} = -$$

$$\tan\theta = \frac{y}{x} = \frac{+}{-} = -$$

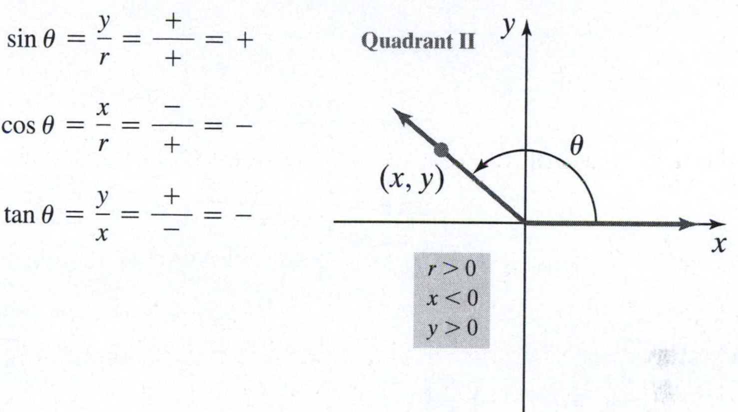

Figure 46 The sine function is positive for all angles in Quadrant II.

QUADRANT III

In Quadrant III, $r > 0, x < 0,$ and $y < 0$. Therefore, $\frac{y}{r} < 0$ and $\frac{x}{r} < 0$. The ratio $\frac{y}{x}$ is positive because both x and y are negative values. Thus, for any angle θ whose terminal side lies in Quadrant III, $\tan\theta = \frac{y}{x} > 0,\ \sin\theta = \frac{y}{r} < 0,$ and $\cos\theta = \frac{x}{r} < 0$. We need only remember that $\tan\theta$ (and its reciprocal $\cot\theta$) are positive for all angles whose terminal side lies in Quadrant III. All other trigonometric functions are negative. See Figure 47.

$$\sin\theta = \frac{y}{r} = \frac{-}{+} = -$$

$$\cos\theta = \frac{x}{r} = \frac{-}{+} = -$$

$$\tan\theta = \frac{y}{x} = \frac{-}{-} = +$$

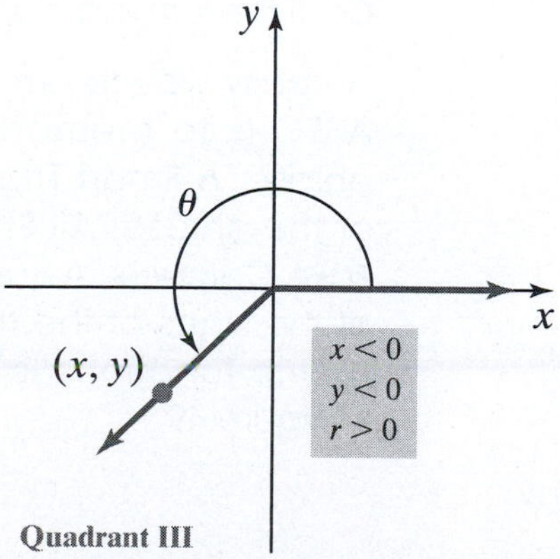

y
θ
x
(x, y)
x < 0
y < 0
r > 0
Quadrant III

Figure 47 The tangent function is positive for all angles in Quadrant III.

QUADRANT IV

In Quadrant IV, $r > 0, x > 0,$ and $y < 0.$ Therefore, $\frac{x}{r} > 0$ but $\frac{y}{r} < 0$ and $\frac{y}{x} < 0.$

Thus, for any angle θ whose terminal side lies in Quadrant IV, $\cos\theta = \frac{x}{r} > 0,$ $\sin\theta = \frac{y}{r} < 0,$ and $\tan\theta = \frac{y}{x} < 0.$ We need to remember that only $\cos\theta$ (and its reciprocal $\sec\theta$) are positive for all angles with a terminal side lying in Quadrant IV. All other trigonometric functions are negative. See Figure 48.

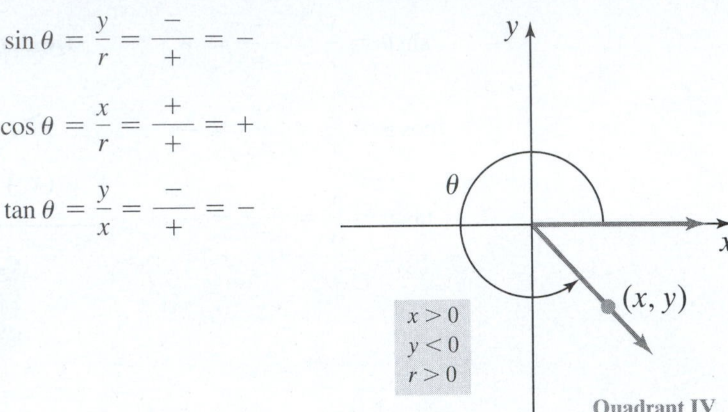

Figure 48 The cosine function is positive for all angles in Quadrant IV.

We can memorize the signs of the trigonometric functions for angles whose terminal side lies in one of the four quadrants by remembering the acronym **ASTC**. These four letters stand for the following:

A: If the terminal side of an angle lies in **Quadrant I**, then **A**ll trigonometric functions are positive.

S: If the terminal side of an angle lies in **Quadrant II**, then the **S**ine function is positive.

T: If the terminal side of an angle lies in **Quadrant III**, then the **T**angent function is positive.

C: If the terminal side of an angle lies in **Quadrant IV**, the **C**osine function is positive.

An easy way to remember ASTC is to memorize the phrase "**A** **S**mart **T**rig **C**lass" or the phrase "**A**ll **S**tudents **T**ake **C**alculus." A summary of the signs of the trigonometric functions is shown in Figure 49.

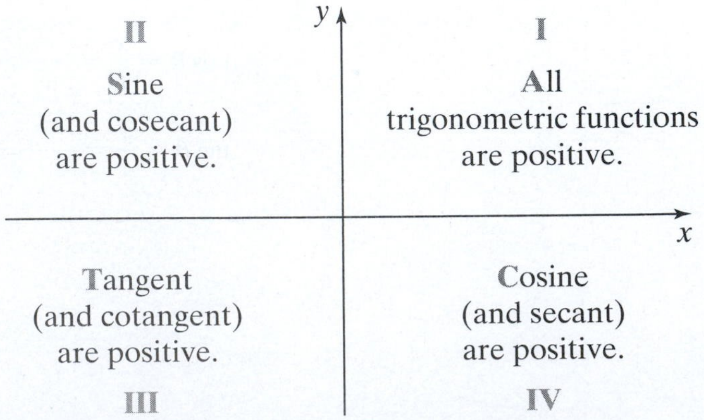

II
Sine (and cosecant) are positive.

I
All trigonometric functions are positive.

III
Tangent (and cotangent) are positive.

IV
Cosine (and secant) are positive.

Figure 49

The signs of the trigonometric functions $y = \sin x, y = \cos x$, and $y = \tan x$ for angles having terminal sides in each of the four quadrants are displayed below in Figure 50.

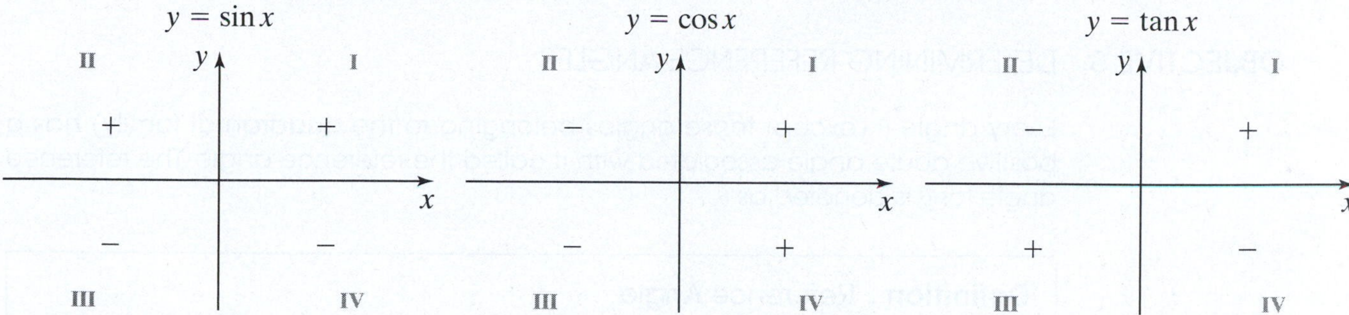

The sine function is positive for angles with terminal sides lying in Quadrants I or II.

The cosine function is positive for angles with terminal sides lying in Quadrants I or IV.

The tangent function is positive for angles with terminal sides lying in Quadrants I or III.

Figure 50

Example 4 Determining the Quadrant of an Angle and Evaluating a Trigonometric Function

Suppose θ is a positive angle in standard position such that $\sin \theta < 0$ and $\sec \theta > 0$.

a. Determine the quadrant in which the terminal side of angle θ lies.

b. Find the value of $\tan \theta$ if $\sec \theta = \sqrt{5}$.

Solution

a. The sine function is negative in Quadrant III and Quadrant IV. The secant function is the reciprocal of the cosine function which is positive in Quadrant I and Quadrant IV. The terminal side of angle θ must lie in the quadrant common to these two pairs. Therefore, the terminal side of angle θ must lie in Quadrant IV.

b. The terminal side of angle θ lies in Quadrant IV, so the value of y must be negative. Because $\sec \theta = \frac{\sqrt{5}}{1} = \frac{r}{x}$, we know that $r = \sqrt{5}$ and $x = 1$. We can use the equation $r = \sqrt{x^2 + y^2}$ to solve for y. Read these **steps** to verify that $y = -2$. We can now draw angle θ, which passes through the point $(1, -2)$. See Figure 51. By the definition of the tangent function, we get

$$\tan \theta = \frac{y}{x} = \frac{-2}{1} = -2.$$

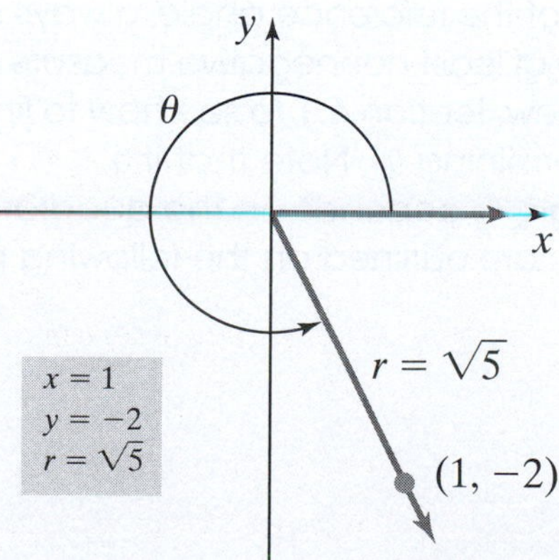

Figure 51 Angle θ lies in Quadrant IV and passes through the point $(1, -2)$.

You Try It Work through this You Try It problem.

Work Exercises 31–40 in this textbook or in the MyLab Math Study Plan.

OBJECTIVE 5 DETERMINING REFERENCE ANGLES

Every angle θ (except those angles belonging to the **quadrantal** family) has a positive acute angle associated with it called the reference angle. The reference angle for θ is denoted as θ_R.

Definition Reference Angle

The reference angle, θ_R, is the positive acute angle associated with a given angle θ. The reference angle is formed by the "nearest" x-axis and the terminal side of θ. Below are angles sketched in each of the four quadrants along with their corresponding reference angles.

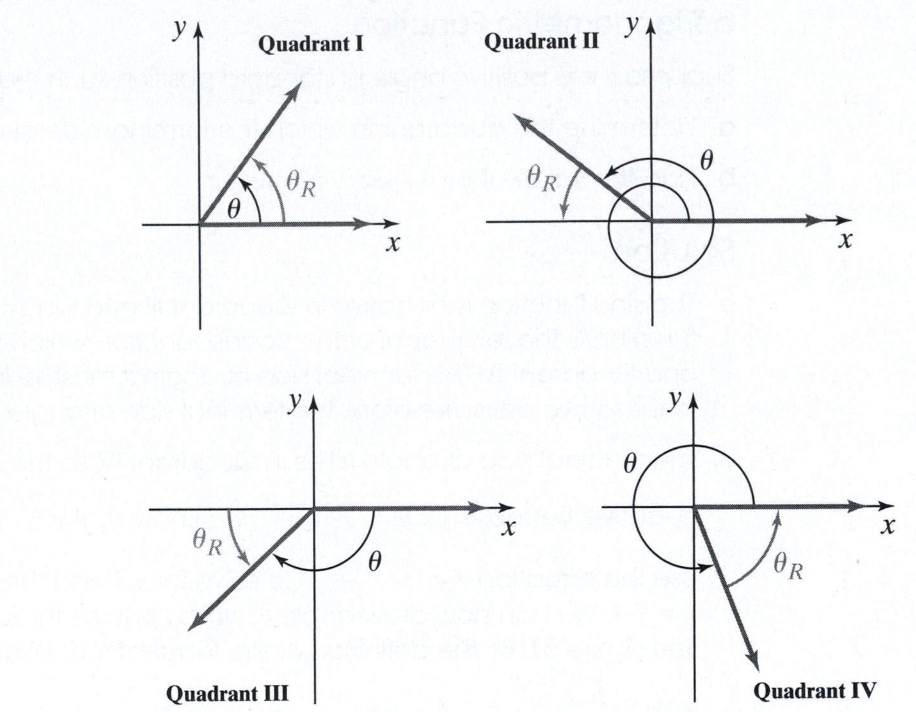

To find the measure of the reference angle, always start by drawing a picture. First, determine the angle of least nonnegative measure coterminal to θ, denoted as θ_C. You may want to review **Section 6.1** to see how to find θ_C or try working through this **exercise** about determining θ_C. Note that if $0 < \theta < 2\pi$, then $\theta = \theta_C$. The measure of the reference angle θ_R depends on the quadrant in which the terminal side of θ_C lies. The four cases are outlined on the following page.

Determining Reference Angles

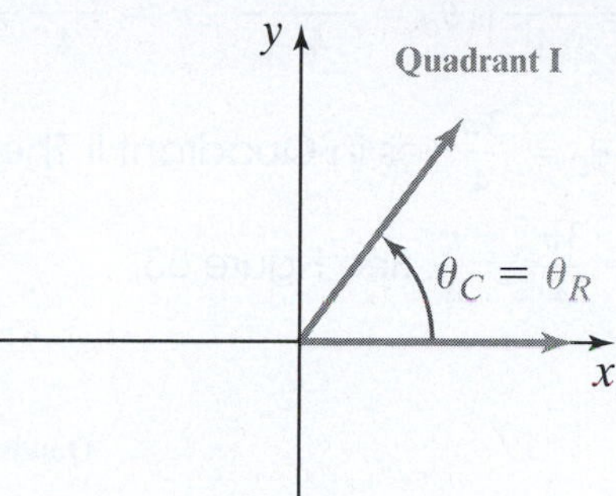

Case 1: If the terminal side of θ_C lies in Quadrant I, then $\theta_R = \theta_C$.

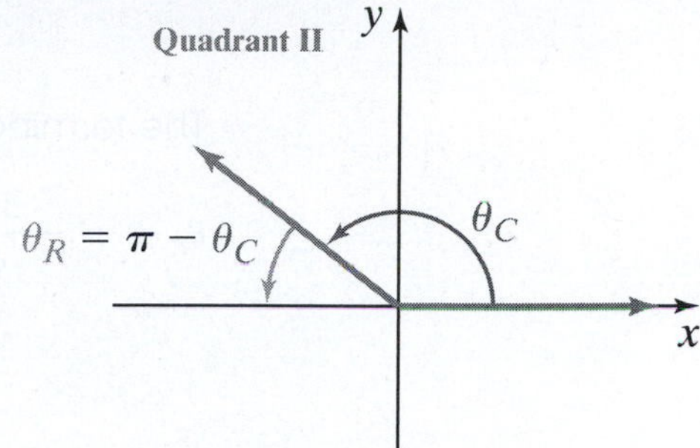

Case 2: If the terminal side of θ_C lies in Quadrant II, then $\theta_R = \pi - \theta_C$ (or $\theta_R = 180° - \theta_C$).

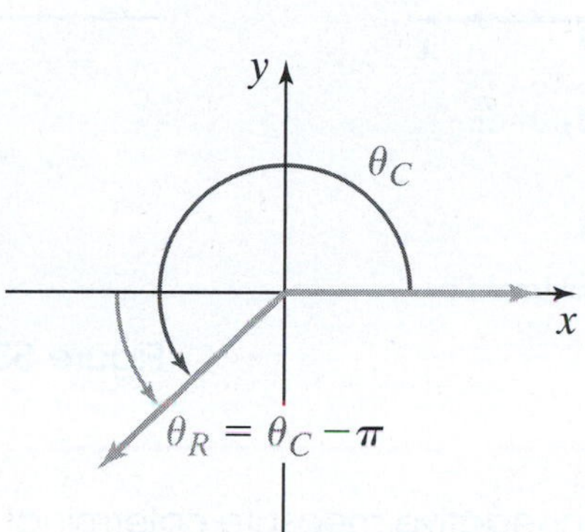

Case 3: If the terminal side of θ_C lies in Quadrant III, then $\theta_R = \theta_C - \pi$ (or $\theta_R = \theta_C - 180°$).

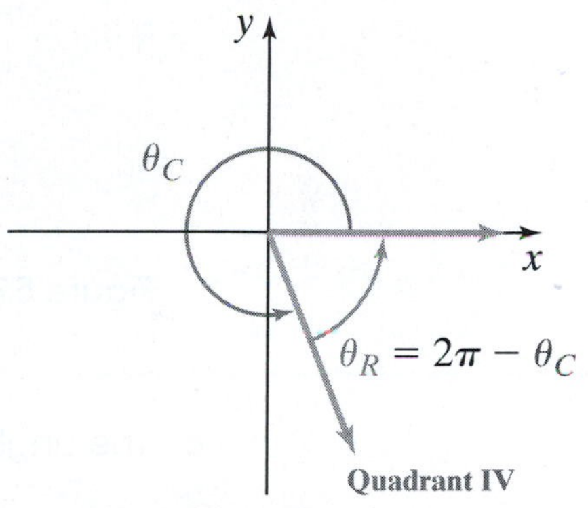

Case 4: If the terminal side of θ_C lies in Quadrant IV, then $\theta_R = 2\pi - \theta_C$ (or $\theta_R = 360° - \theta_C$).

Example 5 Determining Reference Angles for Angles in Radians Belonging to the $\frac{\pi}{3}, \frac{\pi}{6}$, or $\frac{\pi}{4}$ Families

For each of the given angles, determine the reference angle.

a. $\theta = \frac{5\pi}{3}$ b. $\theta = \frac{11\pi}{4}$ c. $\theta = -\frac{25\pi}{6}$ d. $\theta = \frac{16\pi}{6}$

Solution

a. The angle $\theta = \frac{5\pi}{3}$ lies between 0 and 2π. Therefore, $\theta = \theta_C = \frac{5\pi}{3}$. The terminal side of $\theta_C = \frac{5\pi}{3}$ lies in Quadrant IV. Therefore, $\theta_R = 2\pi - \theta_C$. Thus, the reference angle is $\theta_R = 2\pi - \frac{5\pi}{3} = \frac{6\pi}{3} - \frac{5\pi}{3} = \frac{\pi}{3}$. See Figure 52.

b. The angle of least nonnegative measure coterminal with

$$\theta = \frac{11\pi}{4} \text{ is } \theta_C = \frac{11\pi}{4} - 2\pi = \frac{11\pi}{4} - \frac{8\pi}{4} = \frac{3\pi}{4}.$$

The terminal side of $\theta_C = \frac{3\pi}{4}$ lies in Quadrant II. Therefore, $\theta_R = \pi - \theta_C$. Thus, $\theta_R = \pi - \frac{3\pi}{4} = \frac{4\pi}{4} - \frac{3\pi}{4} = \frac{\pi}{4}$. See Figure 53.

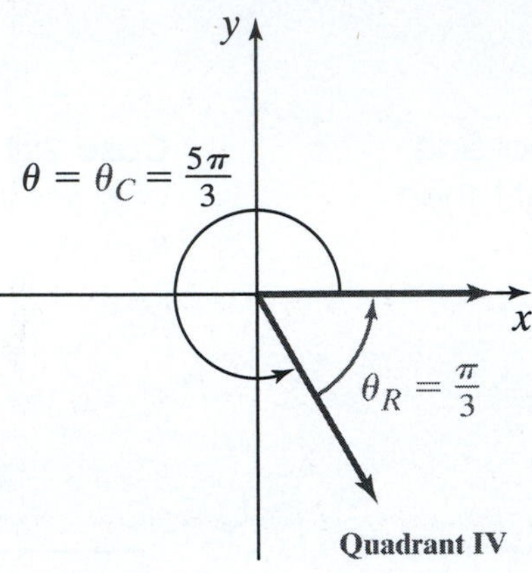

Figure 52

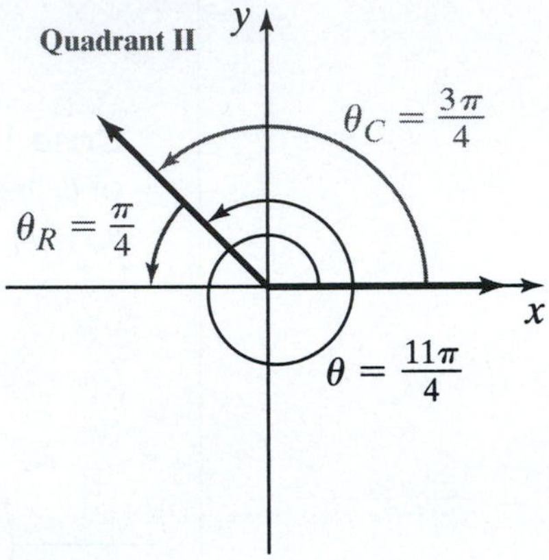

Figure 53

c. The angle of least nonnegative measure coterminal with $\theta = -\frac{25\pi}{6}$ is $\theta_C = \frac{11\pi}{6}$.

To see how to find θ_C, read this **explanation**. The terminal side of $\theta_C = \frac{11\pi}{6}$ lies in Quadrant IV. Therefore, $\theta_R = 2\pi - \theta_C$. Thus, $\theta_R = 2\pi - \theta_C = 2\pi - \frac{11\pi}{6} = \frac{12\pi}{6} - \frac{11\pi}{6} = \frac{\pi}{6}$. See Figure 54.

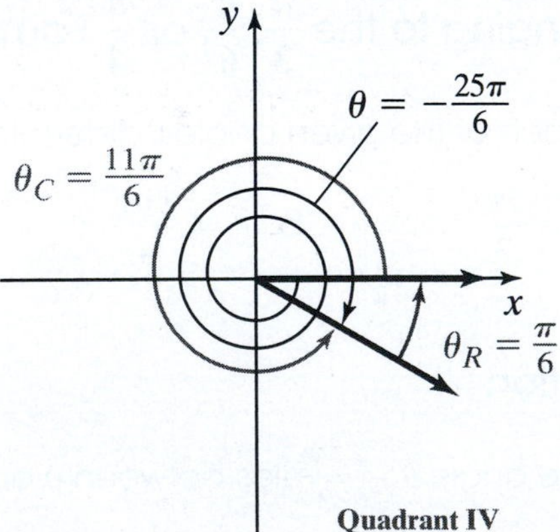

Figure 54

At this point, we can observe from Example 5a, 5b, and 5c that angles belonging to the $\frac{\pi}{6}, \frac{\pi}{3}$, or $\frac{\pi}{4}$ families will always have a reference angle of $\frac{\pi}{6}, \frac{\pi}{3}$, or $\frac{\pi}{4}$, respectively. However, when determining reference angles, make sure that the angle is written in lowest terms.

d. For the angle $\theta = \frac{16\pi}{6}$, do not assume that the reference angle is $\frac{\pi}{6}$. Note that $\theta = \frac{16\pi}{6}$ can be reduced to $\theta = \frac{16\pi}{6} = \frac{\cancel{2} \cdot 8\pi}{\cancel{2} \cdot 3} = \frac{8\pi}{3}$. This is a special angle belonging to the $\frac{\pi}{3}$ family. Therefore, the reference angle is $\theta_R = \frac{\pi}{3}$.

You may want to work through this **interactive video** to see the solution to each part of Example 5.

Try to determine angles of least positive measure and reference angles (θ_C and θ_R) of angles from the various families of angles. Click on the Guided Visualization icon below to experiment with different angles from all of the families of angles.

Sketching General Angles

We now know that if we are given an angle belonging to the $\frac{\pi}{6}, \frac{\pi}{3}$, or $\frac{\pi}{4}$ families, then the reference angle will be $\frac{\pi}{6}, \frac{\pi}{3}$, or $\frac{\pi}{4}$, respectively. But how do we find the reference angle of a given angle that does not belong to one of these families? In this situation, we must use the **four cases** outlined earlier to determine the reference angle.

Example 6 Determining Reference Angles for Angles in Radians Not Belonging to the $\frac{\pi}{3}, \frac{\pi}{6}$, or $\frac{\pi}{4}$ Families

For each of the given angles, determine the reference angle.

a. $\theta = \frac{5\pi}{8}$ **b.** $\theta = \frac{22\pi}{9}$ **c.** $\theta = -\frac{5\pi}{7}$

Solution

a. The angle $\theta = \frac{5\pi}{8}$ lies between 0 and 2π. Therefore, $\theta = \theta_C = \frac{5\pi}{8}$. The terminal side of $\theta_C = \frac{5\pi}{8}$ lies in Quadrant II. Therefore, $\theta_R = \pi - \theta_C$. Thus, the reference angle is $\theta_R = \pi - \frac{5\pi}{8} = \frac{8\pi}{8} - \frac{5\pi}{8} = \frac{3\pi}{8}$. See Figure 55.

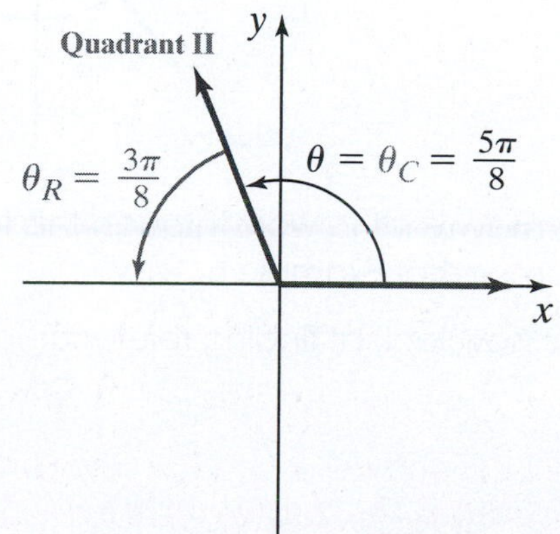

Figure 55

b. The angle of least nonnegative measure coterminal with $\theta = \frac{22\pi}{9}$ is

$$\theta_C = \frac{22\pi}{9} - 2\pi = \frac{22\pi}{9} - \frac{18\pi}{9} = \frac{4\pi}{9}.$$

Because $\frac{4\pi}{9} < \frac{\pi}{2}$, the terminal side of $\theta_C = \frac{4\pi}{9}$ lies in Quadrant I. Therefore, $\theta_R = \theta_C$. Thus, $\theta_R = \theta_C = \frac{4\pi}{9}$. See Figure 56.

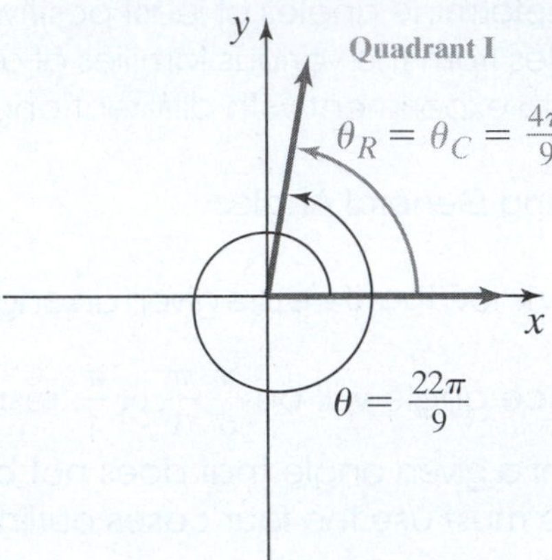

Figure 56

c. The angle of least nonnegative measure coterminal with

$$\theta = -\frac{5\pi}{7} \text{ is } \theta_C = -\frac{5\pi}{7} + 2\pi = -\frac{5\pi}{7} + \frac{14\pi}{7} = \frac{9\pi}{7}.$$

The terminal side of $\theta_C = \frac{9\pi}{7}$ lies in Quadrant III. Therefore, $\theta_R = \theta_C - \pi$.

Thus, $\theta_R = \theta_C - \pi = \frac{9\pi}{7} - \pi = \frac{9\pi}{7} - \frac{7\pi}{7} = \frac{2\pi}{7}$. See Figure 57.

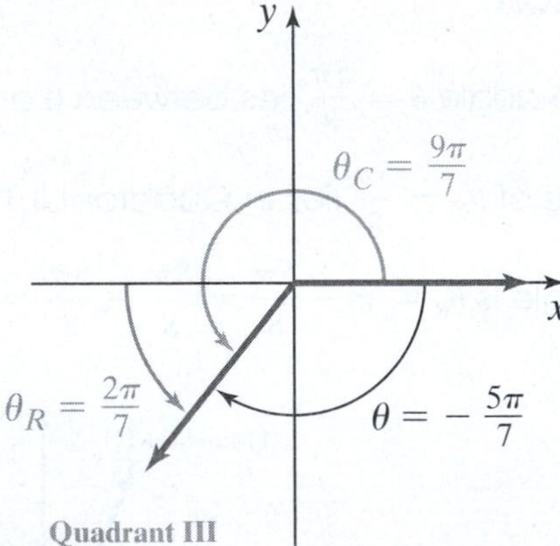

Figure 57

You may want to work through this **interactive video** to see the solution to each part of Example 6.

Let's now look at finding reference angles for angles given in degree measure.

Example 7 Determining Reference Angles for Angles Given in Degrees

For each of the given angles, determine the reference angle.

a. $\theta = 225°$ b. $\theta = -233°$ c. $\theta = 510°$

Solution

a. The angle $\theta = 225°$ belongs to the $\frac{\pi}{4}$ (or $45°$) family of angles because $\theta = 225° = 5(45°)$. Therefore, the reference angle is $\theta_R = 45°$. See Figure 58.

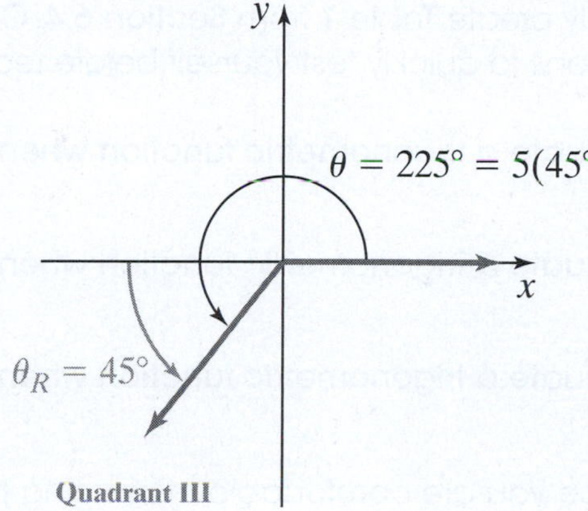

Figure 58

b. The angle $\theta = -233°$ does not belong to one of the special families of angles because $-233°$ is not an integer multiple of $30°$, $45°$, or $60°$. The angle of least nonnegative measure coterminal with $\theta = -233°$ is $\theta_C = -233° + 360° = 127°$. The terminal side of angle $\theta_C = 127°$ lies in Quadrant II. Therefore, $\theta_R = 180° - \theta_C$. Thus, $\theta_R = 180° - \theta_C = 180° - 127° = 53°$. See Figure 59.

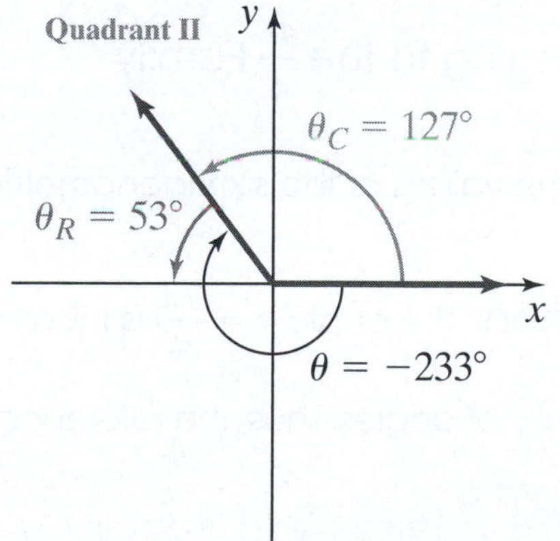

Figure 59

Try working part (c) on your own. Then work through this **interactive video** to see all solutions to Example 7.

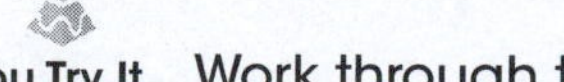

You Try It Work through this You Try It problem.

Work Exercises 41–50 in this textbook or in the MyLab Math Study Plan.

OBJECTIVE 6 EVALUATING TRIGONOMETRIC FUNCTIONS OF ANGLES BELONGING TO THE $\frac{\pi}{3}, \frac{\pi}{6}$, OR $\frac{\pi}{4}$ FAMILIES

We are now ready to evaluate the trigonometric functions for any angle belonging to the $\frac{\pi}{3}, \frac{\pi}{6}$, or $\frac{\pi}{4}$ families. Before we continue, make sure that you have a good understanding of the **two special right triangles** discussed in **Section 6.4**. It is essential that you can use these special right triangles to evaluate all trigonometric functions for the acute angles of $\frac{\pi}{3}, \frac{\pi}{6}$, or $\frac{\pi}{4}$. At this point, you should be able to easily create **Table 1** from **Section 6.4**. Click on one of the following three You Try It buttons to quickly test yourself before reading on.

You Try It Evaluate a trigonometric function when $\boldsymbol{\theta = \frac{\pi}{6}}$ **(or 30°)**.

You Try It Evaluate a trigonometric function when $\boldsymbol{\theta = \frac{\pi}{3}}$ **(or 60°)**.

You Try It Evaluate a trigonometric function when $\boldsymbol{\theta = \frac{\pi}{4}}$ **(or 45°)**.

Once you are comfortable evaluating trigonometric functions at angles of $\frac{\pi}{3}, \frac{\pi}{6}$, or $\frac{\pi}{4}$ using right triangles, you are ready to evaluate trigonometric functions for any angle belonging to the $\frac{\pi}{3}, \frac{\pi}{6}$, or $\frac{\pi}{4}$ families.

Example 8 Finding Trigonometric Function Values for an Angle Belonging to the $\frac{\pi}{4}$ Family

Find the values of the six trigonometric functions for $\theta = \frac{7\pi}{4}$.

Solution The angle $\theta = \frac{7\pi}{4}$ is located in Quadrant IV. The angle belongs to the $\frac{\pi}{4}$ family of angles. Thus, the reference angle is $\theta_R = \frac{\pi}{4}$. See Figure 60.

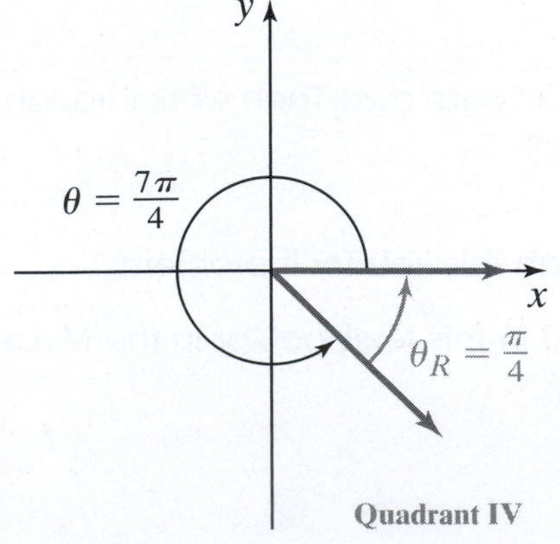

Figure 60 This illustrates the angle $\theta = \frac{7\pi}{4}$ and its reference angle $\theta_R = \frac{\pi}{4}$.

To find the values of the six trigonometric functions, we choose any point P on the terminal side of angle θ. The simplest point to choose is $P(1, -1)$, where $r = \sqrt{2}$. See Figure 61. To see why this point lies on the terminal side of angle theta, read this explanation.

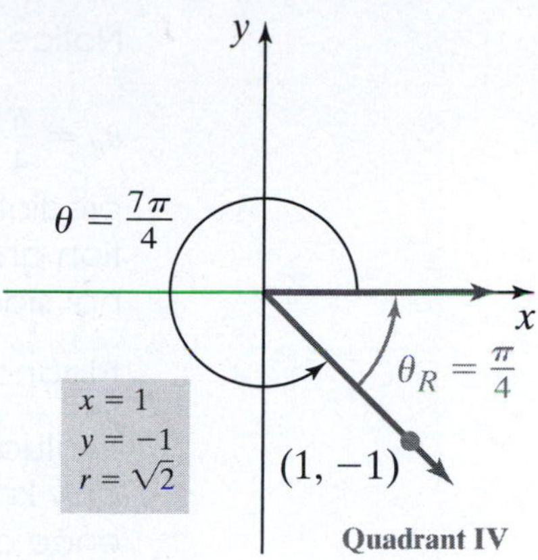

Figure 61

We can now use the general angle definitions to evaluate each trigonometric function.

$$\sin\left(\frac{7\pi}{4}\right) = \frac{y}{r} = \frac{-1}{\sqrt{2}} = -\frac{1}{\sqrt{2}} \qquad \csc\left(\frac{7\pi}{4}\right) = \frac{r}{y} = \frac{\sqrt{2}}{-1} = -\sqrt{2}$$

$$\cos\left(\frac{7\pi}{4}\right) = \frac{x}{r} = \frac{1}{\sqrt{2}} \qquad \sec\left(\frac{7\pi}{4}\right) = \frac{r}{x} = \frac{\sqrt{2}}{1} = \sqrt{2}$$

$$\tan\left(\frac{7\pi}{4}\right) = \frac{y}{x} = \frac{-1}{1} = -1 \qquad \cot\left(\frac{7\pi}{4}\right) = \frac{x}{y} = \frac{1}{-1} = -1$$

In Example 8, note that the absolute values of the trigonometric functions of $\theta = \frac{7\pi}{4}$ are the same as the absolute values of the trigonometric functions of the reference angle, $\theta_R = \frac{\pi}{4}$. In other words, the trigonometric functions of $\theta = \frac{7\pi}{4}$ are exactly the same as the trigonometric functions of $\theta_R = \frac{\pi}{4}$, except that they may have opposite signs.

$$\sin\left(\frac{7\pi}{4}\right) = -\frac{1}{\sqrt{2}} \xleftrightarrow{\text{opposite signs}} \frac{1}{\sqrt{2}} = \sin\left(\frac{\pi}{4}\right)$$

$$\cos\left(\frac{7\pi}{4}\right) = \frac{1}{\sqrt{2}} \xleftrightarrow{\text{both are positive}} \frac{1}{\sqrt{2}} = \cos\left(\frac{\pi}{4}\right)$$

$$\tan\left(\frac{7\pi}{4}\right) = -1 \xleftrightarrow{\text{opposite signs}} 1 = \tan\left(\frac{\pi}{4}\right)$$

$$\csc\left(\frac{7\pi}{4}\right) = -\sqrt{2} \xleftrightarrow{\text{opposite signs}} \sqrt{2} = \csc\left(\frac{\pi}{4}\right)$$

$$\sec\left(\frac{7\pi}{4}\right) = \sqrt{2} \xleftrightarrow{\text{both are positive}} \sqrt{2} = \sec\left(\frac{\pi}{4}\right)$$

$$\cot\left(\frac{7\pi}{4}\right) = -1 \xleftrightarrow{\text{opposite signs}} 1 = \cot\left(\frac{\pi}{4}\right)$$

Notice that only the cosine and secant functions are positive for both $\theta = \frac{7\pi}{4}$ and $\theta_R = \frac{\pi}{4}$. All other trigonometric functions of $\theta = \frac{7\pi}{4}$ are negative. We could have predicted this because we know by the acronym ASTC that only the cosine function and its reciprocal, the secant function, are positive for an angle whose terminal side lies in Quadrant IV. This example suggests a shortcut for evaluating the trigonometric functions for any angle belonging to the $\frac{\pi}{3}, \frac{\pi}{6}$, or $\frac{\pi}{4}$ families. To evaluate a trigonometric function belonging to one of these families, we need only know the quadrant in which the terminal side of the angle lies and the reference angle. Thus, we can then use one of the special right triangles to evaluate the function. We now summarize this process.

Steps for Evaluating Trigonometric Functions of Angles Belonging to the $\frac{\pi}{3}, \frac{\pi}{6}$, or $\frac{\pi}{4}$ Families

Step 1. Draw the angle and determine the quadrant in which the terminal side of the angle lies.

Step 2. Determine if the sign of the function is positive or negative in that quadrant.

Step 3. Determine if the reference angle, θ_R, is $\frac{\pi}{3}, \frac{\pi}{6}$, or $\frac{\pi}{4}$.

Step 4. Use the appropriate special right triangle or your knowledge of **Table 1** to determine the value of the trigonometric function.

Example 9 Evaluating Trigonometric Functions of Angles Belonging to the $\frac{\pi}{3}, \frac{\pi}{6}$, or $\frac{\pi}{4}$ Families

Find the exact value of each trigonometric expression without using a calculator.

a. $\sin\left(\frac{7\pi}{6}\right)$ **b.** $\cot\left(-\frac{22\pi}{3}\right)$ **c.** $\tan\left(\frac{11\pi}{4}\right)$

d. $\cos\left(\frac{11\pi}{3}\right)$ **e.** $\sec\left(\frac{5\pi}{6}\right)$ **f.** $\csc\left(-\frac{7\pi}{6}\right)$

Solution

a. Step 1. The angle $\theta = \frac{7\pi}{6}$ is sketched below with the terminal side of the angle located in Quadrant III.

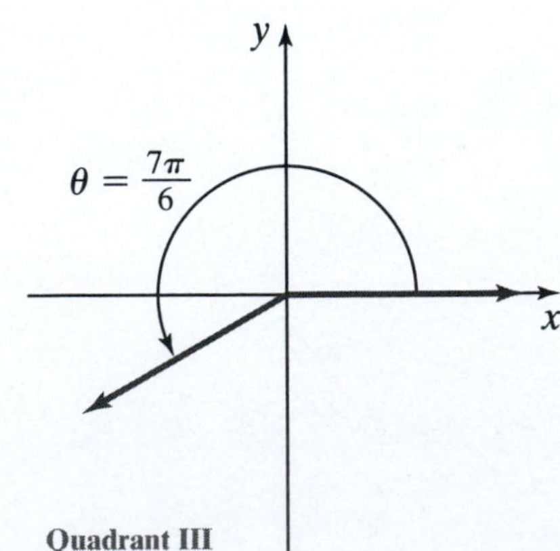

Step 2. Because the terminal side of the angle lies in Quadrant III, we know that this quadrant corresponds with "T" from the acronym ASTC, meaning that *only* the tangent and its reciprocal, cotangent, are positive for angles lying in this quadrant. Thus, the value of $\sin\left(\frac{7\pi}{6}\right)$ must be negative.

Step 3. The angle belongs to the $\frac{\pi}{6}$ family so the reference angle is $\theta_R = \frac{\pi}{6}$.

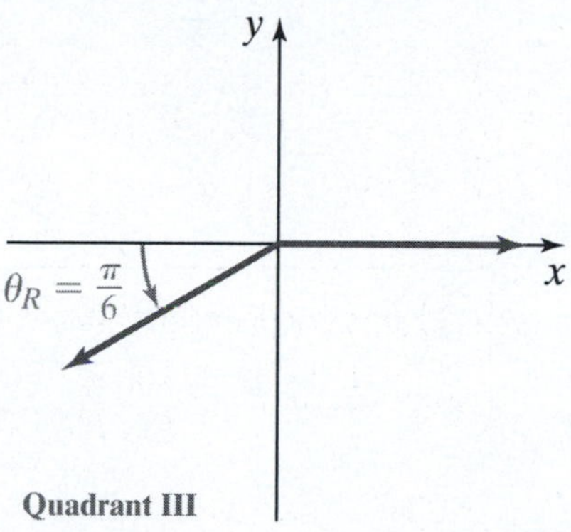

Step 4. We can use the special $\frac{\pi}{6}, \frac{\pi}{3}, \frac{\pi}{2}$ **right triangle** or **Table 1** to get $\sin\left(\frac{\pi}{6}\right) = \frac{1}{2}$.

Therefore, $\sin\left(\frac{7\pi}{6}\right) = \underbrace{\boxed{-}}_{\text{The sine function is negative in Quadrant III.}} \; \underbrace{\boxed{\tfrac{1}{2}}}_{\sin\left(\frac{\pi}{6}\right) = \frac{1}{2}} = -\frac{1}{2}$.

b. To evaluate $\cot\left(-\frac{22\pi}{3}\right)$, we follow the four-step process:

Step 1. The angle $\theta = -\frac{22\pi}{3}$ is sketched below with the terminal side of the angle located in Quadrant II.

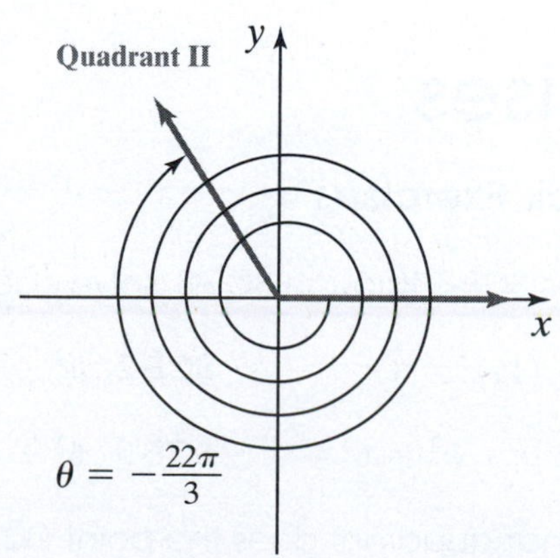

Step 2. Because the terminal side of the angle lies in Quadrant II, we know that this quadrant corresponds with "S" from the acronym ASTC, meaning that *only* the sine and its reciprocal, cosecant, are positive for angles lying in this quadrant. Thus, the value of $\cot\left(-\frac{22\pi}{3}\right)$ must be negative.

Step 3. The angle $\theta = -\frac{22\pi}{3}$ belongs to the $\frac{\pi}{3}$ family so the reference angle is $\theta_R = \frac{\pi}{3}$.

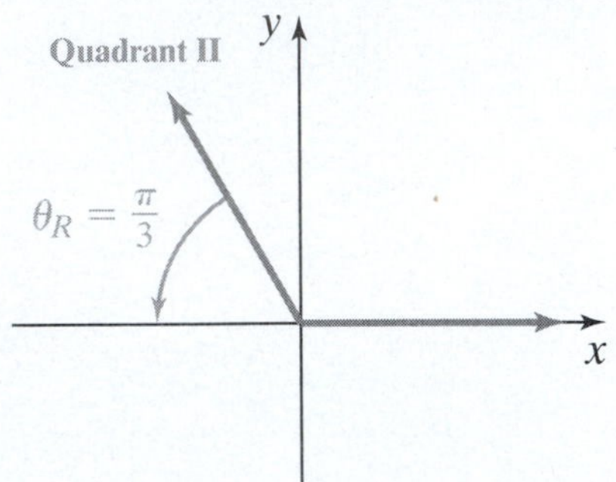

Step 4. We can use the special $\frac{\pi}{6}, \frac{\pi}{3}, \frac{\pi}{2}$ right triangle or Table 1 to get

$\cot\left(\frac{\pi}{3}\right) = \frac{1}{\sqrt{3}}$.

Therefore, $\cot\left(-\frac{22\pi}{3}\right) = \underbrace{\boxed{-}}_{\text{The cotangent function is negative in Quadrant II.}} \underbrace{\boxed{\frac{1}{\sqrt{3}}}}_{\cot\left(\frac{\pi}{3}\right)} = -\frac{1}{\sqrt{3}}$.

Watch this **interactive video** to see all solutions to Example 9.

You Try It **Work through this You Try It problem.**

Work Exercises 51–66 in this textbook or in the MyLab Math Study Plan.

6.5 Exercises

Skill Check Exercises

For exercises SCE-1 through SCE-8, solve each equation.

SCE-1. $a^2 + (15)^2 = 17^2$

SCE-2. $a^2 + (\sqrt{3})^2 = 5^2$

SCE-3. $(24)^2 + b^2 = 25^2$

SCE-4. $(\sqrt{5})^2 + b^2 = 6^2$

SCE-5. $6^2 + 8^2 = c^2$

SCE-6. $2^2 + (\sqrt{7})^2 = c^2$

SCE-7. In what quadrant does the point $(12, -10)$ lie?

SCE-8. On which axis does the point $(-4, 0)$ lie?

1. List all angles written in degrees in the interval $[0°, 360°)$ that are members of the quadrantal family. List answers in order from smallest degree measure to largest degree measure.

2. List all angles written in radians in the interval $[0, 2\pi)$ that are members of the $\frac{\pi}{6}$ family. List answers in order from smallest radian measure to largest radian measure.

In Exercises 3–12, each angle belongs to one of the four families of special angles. Determine the family of angles to which it belongs, sketch the angle, and then determine the angle of least nonnegative measure, θ_C, coterminal with the given angle.

3. $\theta = \frac{5\pi}{2}$
4. $\theta = 480°$
5. $\theta = \frac{8\pi}{3}$
6. $\theta = \frac{14\pi}{4}$
7. $\theta = -315°$
8. $\theta = 6\pi$
9. $\theta = -\frac{17\pi}{6}$
10. $\theta = -420°$
11. $\theta = -\frac{21\pi}{4}$
12. $\theta = -\frac{22\pi}{6}$

In Exercises 13–18, a point lying on the terminal side of an angle θ is given. Find the exact value of the six trigonometric functions of θ.

13. $(4, 3)$
14. $(-5, 12)$
15. $(-2, -5)$
16. $(2, -8)$
17. $\left(-\frac{1}{2}, \frac{\sqrt{3}}{2}\right)$
18. $\left(-\frac{1}{\sqrt{2}}, -\frac{1}{\sqrt{2}}\right)$

In Exercises 19–30, without using a calculator, determine the value of each trigonometric function or state that the value is undefined.

SbS 19. $\sin(8\pi)$
SbS 20. $\cos(-270°)$
SbS 21. $\tan\left(\frac{7\pi}{2}\right)$
SbS 22. $\csc(-540°)$
SbS 23. $\sec(-5\pi)$
SbS 24. $\cot(180°)$
SbS 25. $\sin(450°)$
SbS 26. $\cos\left(-\frac{3\pi}{2}\right)$
SbS 27. $\tan(-90°)$
SbS 28. $\csc(9\pi)$
SbS 29. $\sec(360°)$
SbS 30. $\cot\left(\frac{13\pi}{2}\right)$

31. List the quadrants where the sine function has positive values.

32. List the quadrants where the secant function has positive values.

33. List the quadrants where the tangent function has negative values.

34. List the quadrants where the cosecant function has negative values.

35. If $\sin\theta < 0$ and $\cos\theta > 0$, determine the quadrant in which the terminal side of angle θ lies.

36. If $\tan\theta > 0$ and $\csc\theta > 0$, determine the quadrant in which the terminal side of angle θ lies.

37. If $\cos\theta > 0$ and $\tan\theta > 0$, determine the quadrant in which the terminal side of angle θ lies.

38. Find the value of $\csc\theta$ if $\cos\theta = -\frac{7}{25}$ and given that the terminal side of θ lies in Quadrant II.

39. Find the value of $\sin\theta$ if $\tan\theta = \frac{1}{4}$ and $\sec\theta < 0$.

40. Find the value of $\csc\theta$ if $\sec\theta = -\frac{5}{2}$ and $\tan\theta > 0$.

In Exercises 41–50, determine the reference angle for each of the given angles.

SbS 41. $\theta = \frac{7\pi}{3}$ SbS 42. $\theta = \frac{27\pi}{4}$ SbS 43. $\theta = -\frac{37\pi}{3}$ SbS 44. $\theta = -\frac{19\pi}{4}$ SbS 45. $\theta = \frac{20\pi}{6}$

SbS 46. $\theta = \frac{16\pi}{9}$ SbS 47. $\theta = -\frac{32\pi}{7}$

SbS 48. $\theta = 570°$ SbS 49. $\theta = -675°$ SbS 50. $\theta = -341°$

In Exercises 51–66, find the exact value of each trigonometric expression without using a calculator.

SbS 51. $\sin\left(\frac{5\pi}{6}\right)$ SbS 52. $\cos\left(\frac{3\pi}{4}\right)$ SbS 53. $\cot\left(\frac{4\pi}{3}\right)$ SbS 54. $\csc\left(\frac{7\pi}{6}\right)$

SbS 55. $\tan\left(-\frac{2\pi}{3}\right)$ SbS 56. $\cos\left(-\frac{7\pi}{6}\right)$ SbS 57. $\sec\left(-\frac{\pi}{4}\right)$ SbS 58. $\cot\left(-\frac{5\pi}{4}\right)$

SbS 59. $\sin\left(\frac{17\pi}{6}\right)$ SbS 60. $\tan\left(\frac{11\pi}{3}\right)$ SbS 61. $\sec\left(\frac{23\pi}{6}\right)$ SbS 62. $\csc\left(\frac{10\pi}{3}\right)$

SbS 63. $\cos\left(-\frac{15\pi}{4}\right)$ SbS 64. $\sin\left(-\frac{17\pi}{6}\right)$ SbS 65. $\cot\left(-\frac{11\pi}{3}\right)$ SbS 66. $\sec\left(-\frac{10\pi}{3}\right)$

Brief Exercises

In Exercises 67–94, without using a calculator, determine the value of each trigonometric function or state that the value is undefined.

67. $\sin(8\pi)$ 68. $\cos(-270°)$ 69. $\tan\left(\frac{7\pi}{2}\right)$ 70. $\csc(-540°)$ 71. $\sec(-5\pi)$

72. $\cot(180°)$ 73. $\sin(450°)$ 74. $\cos\left(-\frac{3\pi}{2}\right)$ 75. $\tan(-90°)$ 76. $\csc(9\pi)$

77. $\sec(360°)$ 78. $\cot\left(\frac{13\pi}{2}\right)$ 79. $\sin\left(\frac{5\pi}{6}\right)$ 80. $\cos\left(\frac{3\pi}{4}\right)$ 81. $\cot\left(\frac{4\pi}{3}\right)$

82. $\csc\left(\frac{7\pi}{6}\right)$ 83. $\tan\left(-\frac{2\pi}{3}\right)$ 84. $\cos\left(-\frac{7\pi}{6}\right)$ 85. $\sec\left(-\frac{\pi}{4}\right)$ 86. $\cot\left(-\frac{5\pi}{4}\right)$

87. $\sin\left(\frac{17\pi}{6}\right)$ 88. $\tan\left(\frac{11\pi}{3}\right)$ 89. $\sec\left(\frac{23\pi}{6}\right)$ 90. $\csc\left(\frac{10\pi}{3}\right)$ 91. $\cos\left(-\frac{15\pi}{4}\right)$

92. $\sin\left(-\frac{17\pi}{6}\right)$ 93. $\cot\left(-\frac{11\pi}{3}\right)$ 94. $\sec\left(-\frac{10\pi}{3}\right)$

6.6 The Unit Circle

THINGS TO KNOW

Before working through this section, be sure that you are familiar with the following concepts:

VIDEO | ANIMATION | INTERACTIVE

You Try It

1. Converting between Degree Measure and Radian Measure (Section 6.1)

2. Understanding the Special Right Triangles (Section 6.3)

You Try It

3. Understanding the Right Triangle Definitions of the Trigonometric Functions (Section 6.4)

You Try It

4. Understanding the Four Families of Special Angles (Section 6.5)

You Try It

5. Understanding the Definitions of the Trigonometric Functions of General Angles (Section 6.5)

You Try It

6. Evaluating Trigonometric Functions of Angles Belonging to the $\frac{\pi}{3}$, $\frac{\pi}{6}$, or $\frac{\pi}{4}$ Families (Section 6.5)

OBJECTIVES

1 Understanding the Definition of the Unit Circle

2 Using Symmetry to Determine Points on the Unit Circle

3 Understanding the Unit Circle Definitions of the Trigonometric Functions

4 Using the Unit Circle to Evaluate Trigonometric Functions at Increments of $\frac{\pi}{2}$

5 Using the Unit Circle to Evaluate Trigonometric Functions for Increments of $\frac{\pi}{6}$, $\frac{\pi}{4}$, and $\frac{\pi}{3}$

SECTION 6.6 EXERCISES

OBJECTIVE 1 UNDERSTANDING THE DEFINITION OF THE UNIT CIRCLE

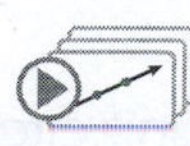

Recall that the standard form of the equation of a circle with center (h, k) and radius r is given by $(x - h)^2 + (y - k)^2 = r^2$. A circle centered at the origin with a radius length of 1 unit is called the **unit circle**. The standard form of the equation of the unit circle is $x^2 + y^2 = 1$. Watch this **animation** for a further explanation of the unit circle.

The Unit Circle

A circle centered at the origin with a radius of 1 unit is called the **unit circle** whose equation is given by $x^2 + y^2 = 1$.

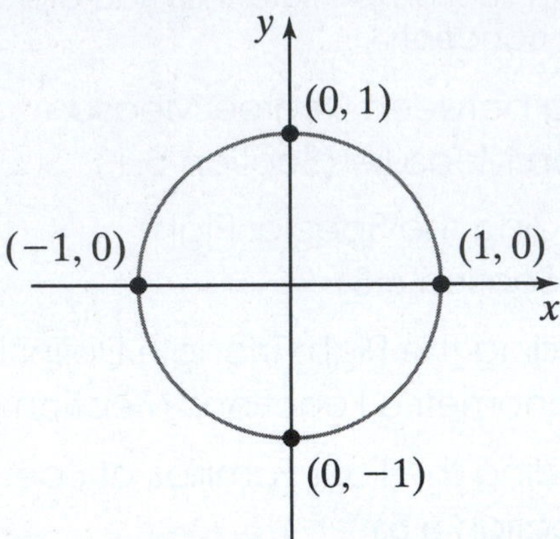

Note that the points $(1, 0)$, $(0, 1)$, $(-1, 0)$, and $(0, -1)$ all lie on the graph of the unit circle. These are the only four points that lie on the unit circle whose coordinates are integers. From the graph, you can see that the x- and y-coordinates of all other points that lie on the unit circle must be between -1 and 1 and cannot equal 0. A point (a, b) lies on the graph of the unit circle if and only if $a^2 + b^2 = 1$. In Example 1, one of the coordinates of a point that lies on the graph of the unit circle is given. Work through Example 1 to see if you can determine the value of the missing coordinate.

Example 1 Determine the Missing Coordinate of a Point That Lies on the Unit Circle

Determine the missing coordinate of a point that lies on the graph of the unit circle given the quadrant in which the point is located.

a. $\left(-\frac{1}{8}, y\right)$; Quadrant III

b. $\left(x, -\frac{\sqrt{3}}{2}\right)$; Quadrant IV

c. $\left(-\frac{1}{\sqrt{2}}, y\right)$; Quadrant II

Solution

a. We are given that $x = -\frac{1}{8}$. Substitute $x = -\frac{1}{8}$ into the equation of the unit circle and solve for y.

$x^2 + y^2 = 1$ Write the equation of the unit circle.

$\left(-\frac{1}{8}\right)^2 + y^2 = 1$ Substitute $x = -\frac{1}{8}$.

$\frac{1}{64} + y^2 = 1$ Simplify.

$y^2 = 1 - \frac{1}{64}$ Subtract $\frac{1}{64}$ from both sides.

$y^2 = \frac{63}{64}$ Simplify. $\left(1 - \frac{1}{64} = \frac{64}{64} - \frac{1}{64} = \frac{63}{64}\right)$

$y = \pm\sqrt{\frac{63}{64}} = \pm\frac{3\sqrt{7}}{8}$ Use the **square root property** and simplify.

Because the y-coordinates of all points that lie in Quadrant III are *negative*, use $y = -\frac{3\sqrt{7}}{8}$.

Try finding the missing coordinates for parts (b) and (c) on your own. Then watch this **interactive video** to see if you are correct.

You Try It **Work through this You Try It problem.**

Work Exercises 1–5 in this textbook or in the MyLab Math Study Plan.

OBJECTIVE 2 USING SYMMETRY TO DETERMINE POINTS THAT LIE ON THE UNIT CIRCLE

It is important to point out that the graph of the unit circle is symmetric about both axes and the origin. In Example 1b, we found that the point $\left(\frac{1}{2}, -\frac{\sqrt{3}}{2}\right)$ lies on the graph of the unit circle. Because the graph of the unit circle is **symmetric about the y-axis**, it follows that the point $\left(-\frac{1}{2}, -\frac{\sqrt{3}}{2}\right)$ also lies on the graph of the unit circle. Now that we know two points lying on the unit circle located *below* the x-axis, we can use the fact that the unit circle is **symmetric about the x-axis** to find two more points lying on the unit circle *above* the x-axis. These two points are $\left(\frac{1}{2}, \frac{\sqrt{3}}{2}\right)$ and $\left(-\frac{1}{2}, \frac{\sqrt{3}}{2}\right)$. Note that we also could have used **origin symmetry** to locate these two points. In general, for any point (a, b) lying on the unit circle (not on an axis), the points $(-a, b)$, $(a, -b)$, and $(-a, -b)$ must also lie on the unit circle. See Figure 62.

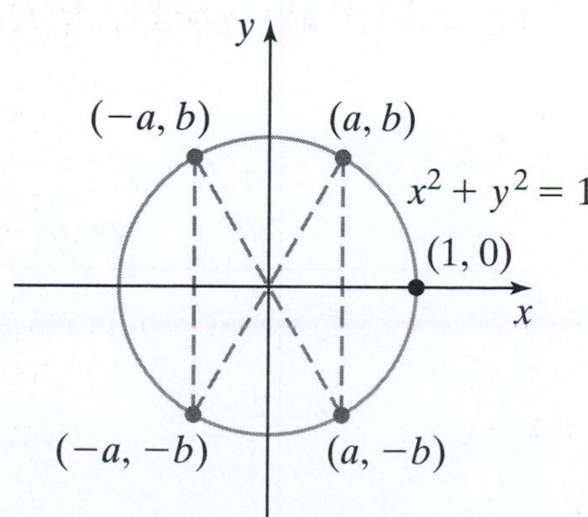

Figure 62 For any point (a, b) on the unit circle not lying on an axis, symmetry can be used to locate three more points.

Example 2 Finding Points Lying on the Graph of the Unit Circle

Verify that the point $\left(-\frac{1}{8}, -\frac{3\sqrt{7}}{8}\right)$ lies on the graph of the unit circle. Then use symmetry to find three other points that also lie on the graph of the unit circle.

Solution We can verify that the point $\left(-\frac{1}{8}, -\frac{3\sqrt{7}}{8}\right)$ lies on the graph of the unit circle by substituting $x = -\frac{1}{8}$ and $y = -\frac{3\sqrt{7}}{8}$ into the equation $x^2 + y^2 = 1$ to see if a true statement is produced. To see this verification, read these **steps**.

Because the unit circle has **y-axis symmetry**, we are guaranteed that the point $\left(\frac{1}{8}, -\frac{3\sqrt{7}}{8}\right)$ also lies on the graph of the unit circle. We now know two points on the graph of the unit circle that lie *below* the x-axis. We can use **x-axis symmetry** to find the corresponding points on the graph of the unit circle that lie *above* the x-axis. These two points are $\left(\frac{1}{8}, \frac{3\sqrt{7}}{8}\right)$ and $\left(-\frac{1}{8}, \frac{3\sqrt{7}}{8}\right)$. Watch this **video** to see this solution worked out in detail. ●

You Try It **Work through this You Try It problem.**

Work Exercises 6–8 in this textbook or in the MyLab Math Study Plan.

OBJECTIVE 3 UNDERSTANDING THE UNIT CIRCLE DEFINITIONS OF THE TRIGONOMETRIC FUNCTIONS

So far in this text, we have seen two groups of definitions for the trigonometric functions. In **Section 6.4,** we saw the **right triangle definitions** of the trigonometric functions of acute angles. Then, in **Section 6.5** we saw the **definitions of the trigonometric functions of general angles.**

We now turn our attention to the third set of definitions of the trigonometric functions that involve the unit circle. Suppose that t is the measure (in radians) of a central angle of a unit circle with a corresponding arc length, s. See Figure 63.

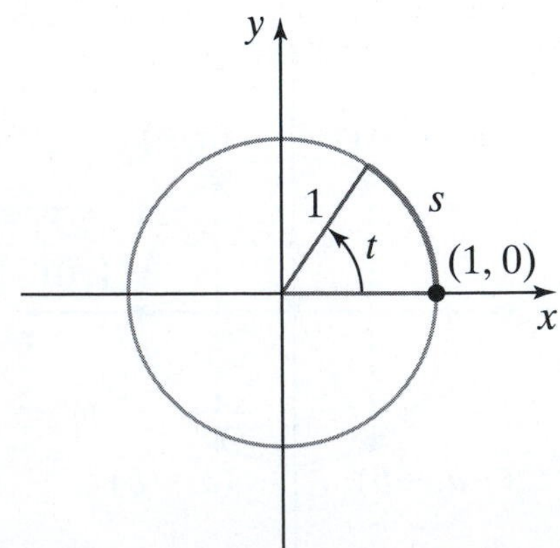

Figure 63 This illustrates the unit circle with a central angle of t radians and a corresponding arc of length s.

We can use the **formula for the arc length of a sector of a circle**, which we derived in **Section 6.2**, to find the arc length, s, of a unit circle that corresponds to a central angle of t radians.

$s = r\theta$	Write the formula for the arc length of a sector of a circle.
$s = 1 \cdot t$	Substitute $r = 1$ and $\theta = t$.
$s = t$	Simplify.

We see that the length of the intercepted arc seen in Figure 63 is $s = t$. **This means that the arc length of a sector of the** ***unit circle*** **is exactly equal to the measure of the central angle!** The arc length and the angle are represented by the same real number t.

We can now create the unit circle definitions of the trigonometric functions. To do this, let t be any real number and let $P(x, y)$ be the point on the unit circle that has an arc length of t units from the point $(1, 0)$. The measure of the central angle (in radians) is exactly the same as the arc length, t. If $t > 0$, then point P is obtained by rotating in a *counterclockwise* direction. See **Figure 64a**. If $t < 0$, then point P is obtained by rotating in a *clockwise* direction. See **Figure 64b**.

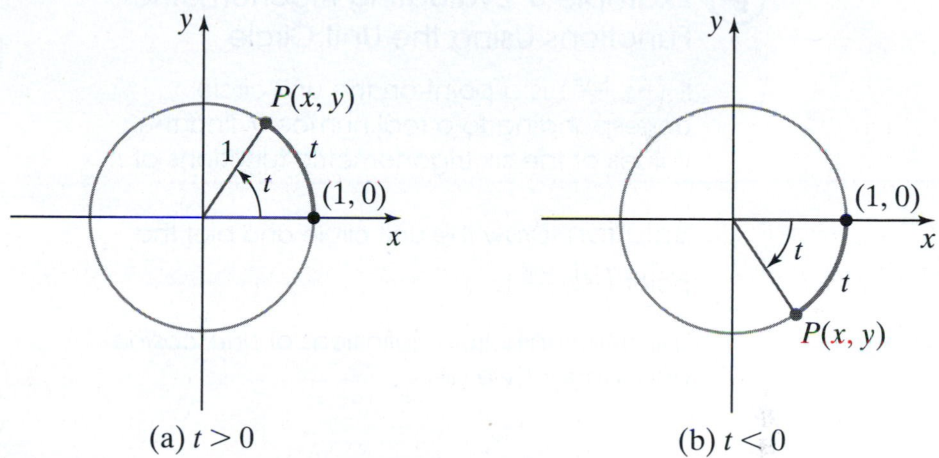

Figure 64

For the real number t and the corresponding point $P(x, y)$ lying on the graph of the unit circle, we define the cosine of t as the x-coordinate of P and the sine of t as the y-coordinate of P. Therefore, $\cos t = x$ and $\sin t = y$. This choice of x for the cosine of t and y for the sine of t is not made arbitrarily. If you remember the **definitions of the trigonometric functions of general angles**, the sine of an angle theta was defined as $\frac{y}{r}$ and the cosine of theta was defined as $\frac{x}{r}$. Therefore, it seems logical and consistent that we choose $\sin t$ to be y (which is $\frac{y}{r}$ when $r = 1$) and $\cos t$ to be x (which is $\frac{x}{r}$ when $r = 1$).

We now define all six of the trigonometric functions using the unit circle.

The Unit Circle Definitions of the Trigonometric Functions

For any real number t, if $P(x, y)$ is a point on the unit circle corresponding to t, then

$$\sin t = y \qquad \csc t = \frac{1}{y}, y \neq 0$$

$$\cos t = x \qquad \sec t = \frac{1}{x}, x \neq 0$$

$$\tan t = \frac{y}{x}, x \neq 0 \qquad \cot t = \frac{x}{y}, y \neq 0$$

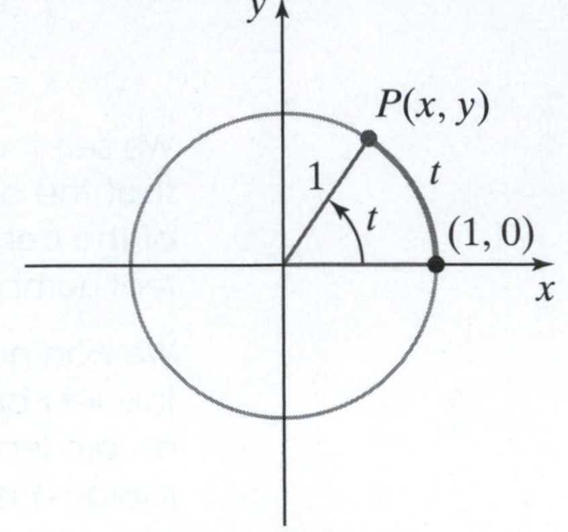

Example 3 Evaluating Trigonometric Functions Using the Unit Circle

If $\left(-\frac{1}{4}, \frac{\sqrt{15}}{4}\right)$ is a point on the unit circle corresponding to a real number t, find the values of the six trigonometric functions of t.

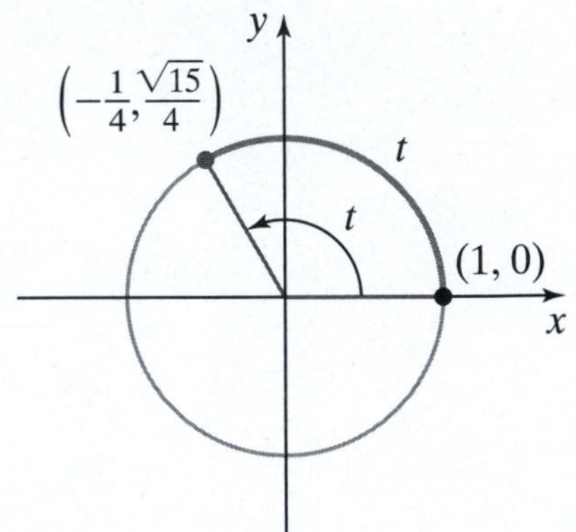

Solution Draw the unit circle and plot the point $\left(-\frac{1}{4}, \frac{\sqrt{15}}{4}\right)$.

Using the unit circle definitions of *sine, cosine,* and *tangent,* we get

$$\sin t = y = \frac{\sqrt{15}}{4}, \quad \cos t = x = -\frac{1}{4}, \quad \text{and} \quad \tan t = \frac{y}{x} = \frac{\frac{\sqrt{15}}{4}}{-\frac{1}{4}} = \frac{\sqrt{15}}{4}\cdot\left(-\frac{4}{1}\right) = -\sqrt{15}$$

We can find the cosecant, secant, and cotangent of t using the reciprocals of the sine, cosine, and tangent functions. Thus,

$$\csc t = \frac{4}{\sqrt{15}}, \quad \sec t = -4, \quad \text{and} \quad \cot t = -\frac{1}{\sqrt{15}}.$$

You Try It Work through this You Try It problem.

Work Exercises 9–11 in this textbook or in the MyLab Math Study Plan.

OBJECTIVE 4 USING THE UNIT CIRCLE TO EVALUATE TRIGONOMETRIC FUNCTIONS AT INCREMENTS OF $\frac{\pi}{2}$

Understanding the relationship between certain values of t and specific points on the graph of the unit circle will help you quickly evaluate the trigonometric functions of these values. Undoubtedly, the easiest points on the unit circle to remember are the points $(1, 0)$, $(0, 1)$, $(-1, 0)$, and $(0, -1)$. The arc length from the point $(1, 0)$ to itself is 0 unit. Thus, the real number $t = 0$ corresponds to the point $(1, 0)$. See **Figure 65a**. To determine the values of t that correspond to the points $(0, 1)$, $(-1, 0)$, and $(0, -1)$, we use the fact that the circumference of a circle of radius, r, is $2\pi r$. Therefore, the circumference of the unit circle is $2\pi(1) = 2\pi$. The arc length from the point $(1, 0)$ to the point $(0, 1)$ is one-quarter of the circumference of the unit circle. Therefore, $t = \frac{1}{4} \cdot 2\pi = \frac{\pi}{2}$ corresponds to the point $(0, 1)$. See **Figure 65b**. The arc length from the point $(1, 0)$ to the point $(-1, 0)$ is one-half of the circumference of the unit circle. Thus, $t = \frac{1}{2} \cdot 2\pi = \pi$ corresponds to the point $(-1, 0)$. See **Figure 65c**. Finally, the arc length from the point $(1, 0)$ to the point $(0, -1)$ is three-quarters of the circumference of the unit circle. So, $t = \frac{3}{4} \cdot 2\pi = \frac{3\pi}{2}$ corresponds to the point $(0, -1)$. See **Figure 65d**.

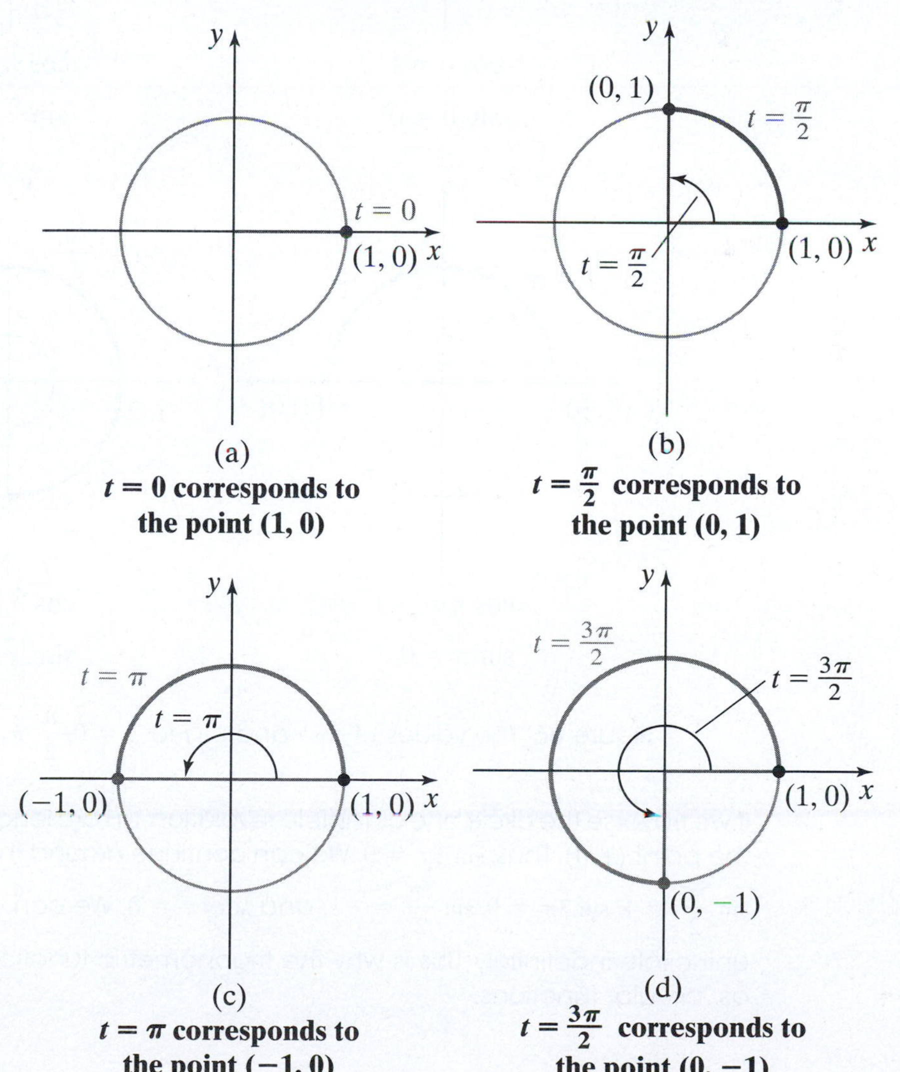

(a) **$t = 0$ corresponds to the point (1, 0)**

(b) **$t = \frac{\pi}{2}$ corresponds to the point (0, 1)**

(c) **$t = \pi$ corresponds to the point (−1, 0)**

(d) **$t = \frac{3\pi}{2}$ corresponds to the point (0, −1)**

Figure 65

We can now use the unit circle definitions to evaluate the trigonometric functions for values of t equal to $0, \frac{\pi}{2}, \pi$, and $\frac{3\pi}{2}$. For example, using the fact that $\cos t = x$ and $\sin t = y$, we can use **Figure 66** to find the values of $\cos t$ and $\sin t$ for $0, \frac{\pi}{2}, \pi$, and $\frac{3\pi}{2}$. Note in **Figure 66** that the numbers in blue represent the real number t and the x-coordinates of the corresponding points on the unit circle in green represent the value of $\cos t$. Similarly, the numbers in red represent the values of $\sin t$.

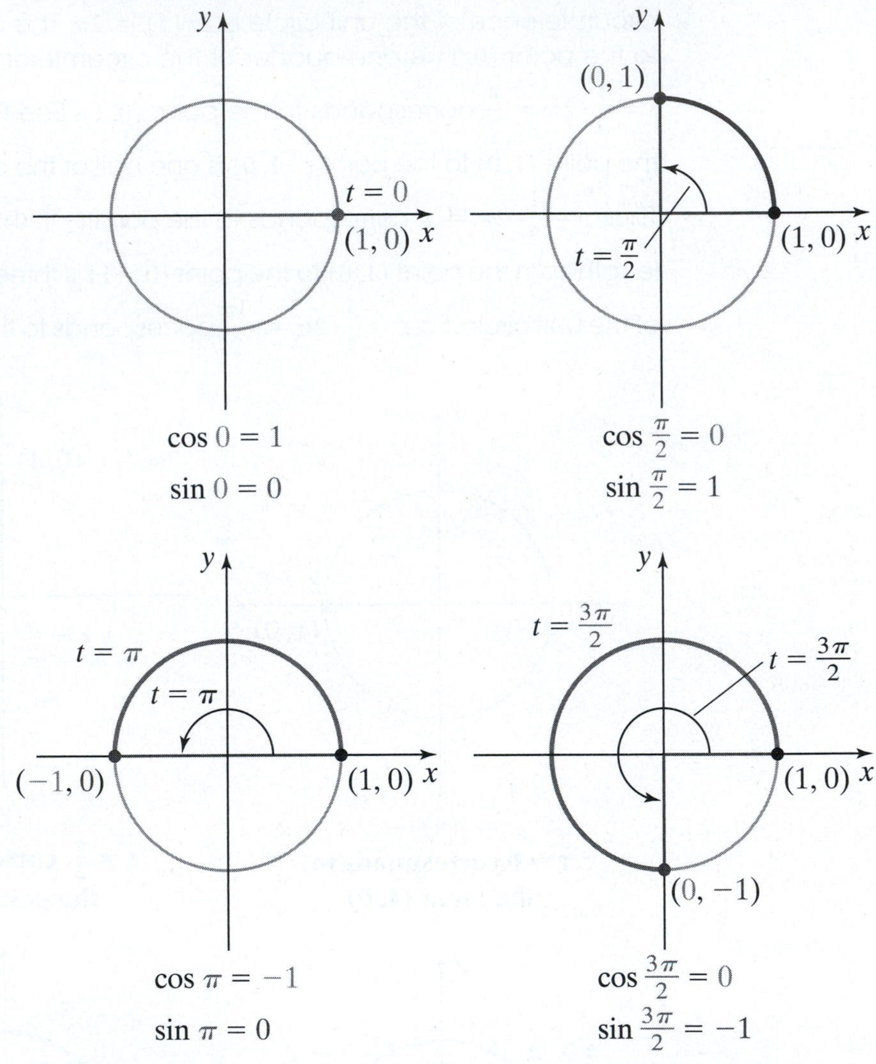

Figure 66 The values of $\cos t$ and $\sin t$ for $t = 0, \frac{\pi}{2}, \pi$, and $\frac{3\pi}{2}$.

If we traverse the circle one complete revolution, an arc length of 2π, we are back at the point $(1, 0)$. Thus, $\sin 2\pi = 0$. We can continue around the circle again to obtain $\sin \frac{5\pi}{2} = 1, \sin 3\pi = 0, \sin \frac{7\pi}{2} = -1$, and $\sin 4\pi = 0$. We can continue traversing the unit circle indefinitely. This is why the trigonometric functions are often referred to as "circular functions."

Because division by zero is undefined, the values of trigonometric functions may be undefined. For example, when $t = \frac{\pi}{2}$, the corresponding point on the unit circle is $(0, 1)$. Thus, $\tan\frac{\pi}{2} = \frac{y}{x} = \frac{1}{0}$, which is undefined. Table 3 shows the values of the six trigonometric functions for values of $0, \frac{\pi}{2}, \pi, \frac{3\pi}{2}$, and 2π. Use Figure 67 and watch this **video** for a complete explanation on how to fill out Table 3.

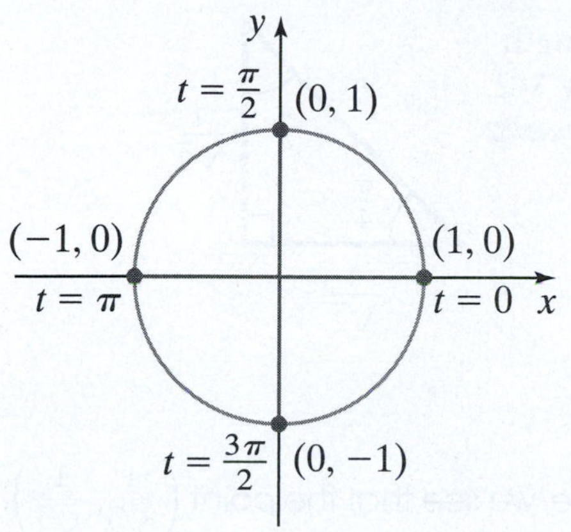

Figure 67

Table 3

t	$\sin t$	$\cos t$	$\tan t$	$\csc t$	$\sec t$	$\cot t$
$0, 2\pi$	0	1	0	Undefined	1	Undefined
$\frac{\pi}{2}$	1	0	Undefined	1	Undefined	0
π	0	−1	0	Undefined	−1	Undefined
$\frac{3\pi}{2}$	−1	0	Undefined	−1	Undefined	0

Example 4 Evaluating Trigonometric Functions Using the Unit Circle

Use the unit circle to determine the value of each expression or state that it is undefined.

a. $\cos 11\pi$ b. $\tan\frac{5\pi}{2}$ c. $\csc\left(-\frac{3\pi}{2}\right)$

Solution

a. The number 11π is equal to $10\pi + \pi$, which is equivalent to the distance around the unit circle five times plus an additional half of a revolution. Thus, 11π corresponds to the point $(-1, 0)$ on the unit circle. Therefore, $\cos 11\pi = -1$. Work through parts (b) and (c) on your own. Then watch this **interactive video** to see if you are correct.

You Try It Work through this You Try It problem.

Work Exercises 12–19 in this textbook or in the MyLab Math Study Plan.

OBJECTIVE 5 USING THE UNIT CIRCLE TO EVALUATE TRIGONOMETRIC FUNCTIONS FOR INCREMENTS OF $\frac{\pi}{6}, \frac{\pi}{4}$, AND $\frac{\pi}{3}$

In **Section 6.3**, we introduced two special right triangles, the $\frac{\pi}{4}, \frac{\pi}{4}, \frac{\pi}{2}$ right triangle and the $\frac{\pi}{3}, \frac{\pi}{6}, \frac{\pi}{2}$ right triangle. We can use our knowledge of these two special right triangles to determine several more points that lie on the unit circle. If we look at the $\frac{\pi}{4}, \frac{\pi}{4}, \frac{\pi}{2}$ right triangle with side lengths of $1, 1, \sqrt{2}$ and divide the length of each side by $\sqrt{2}$, we can form a similar right triangle with side lengths of $\frac{1}{\sqrt{2}}, \frac{1}{\sqrt{2}}, 1$. See Figure 68.

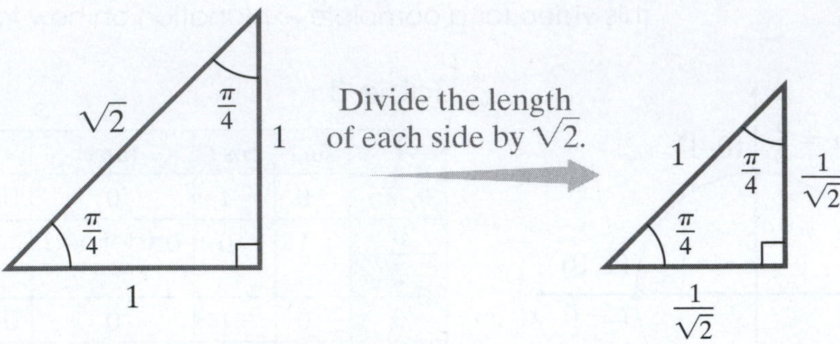

Figure 68

If we superimpose this triangle onto the unit circle, we see that the point $\left(\frac{1}{\sqrt{2}}, \frac{1}{\sqrt{2}}\right)$ lies on the unit circle. See Figure 69a. Similarly, we can use our knowledge of the $\frac{\pi}{3}, \frac{\pi}{6}, \frac{\pi}{2}$ right triangle to show that the points $\left(\frac{1}{2}, \frac{\sqrt{3}}{2}\right)$ and $\left(\frac{\sqrt{3}}{2}, \frac{1}{2}\right)$ also lie on the unit circle. See Figure 69b and Figure 69c. For a complete explanation, watch this interactive video.

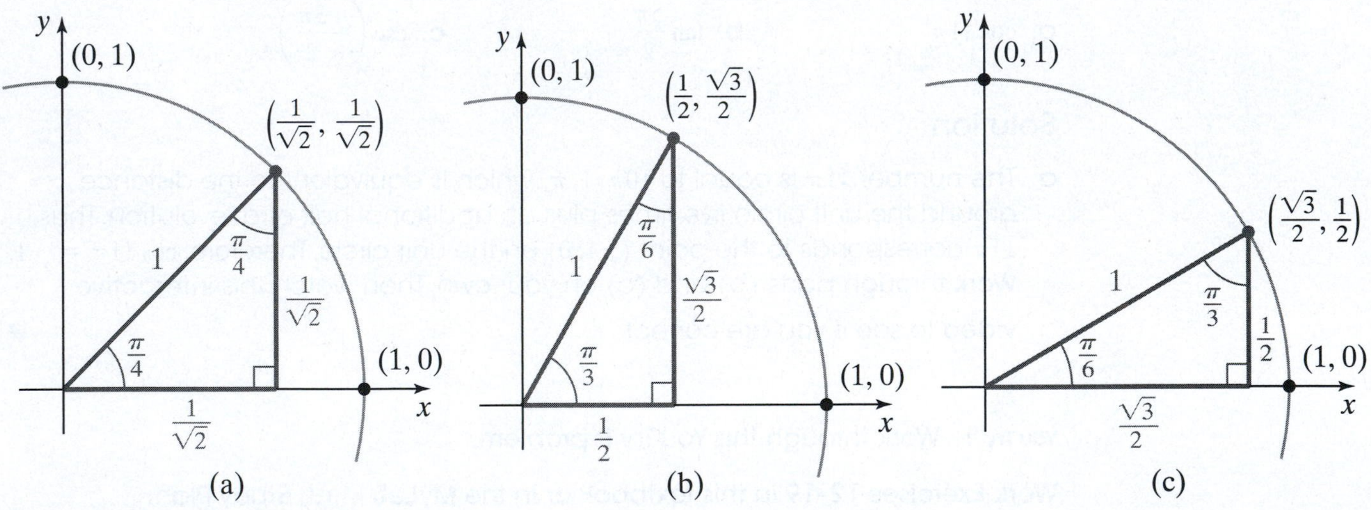

Figure 69

Now that we know that the points $\left(\frac{1}{\sqrt{2}}, \frac{1}{\sqrt{2}}\right)$, $\left(\frac{1}{2}, \frac{\sqrt{3}}{2}\right)$, and $\left(\frac{\sqrt{3}}{2}, \frac{1}{2}\right)$ lie on the unit circle, we can use the symmetry properties of the unit circle to complete the unit circle, as shown in **Figure 70**. This unit circle contains the "special points" that correspond with "special values of t." For convenience, both the radian measure and degree measure is given for each corresponding point. You should become familiar with the unit circle shown in **Figure 70**.

(eText Screens 6.6-1–6.6-26)

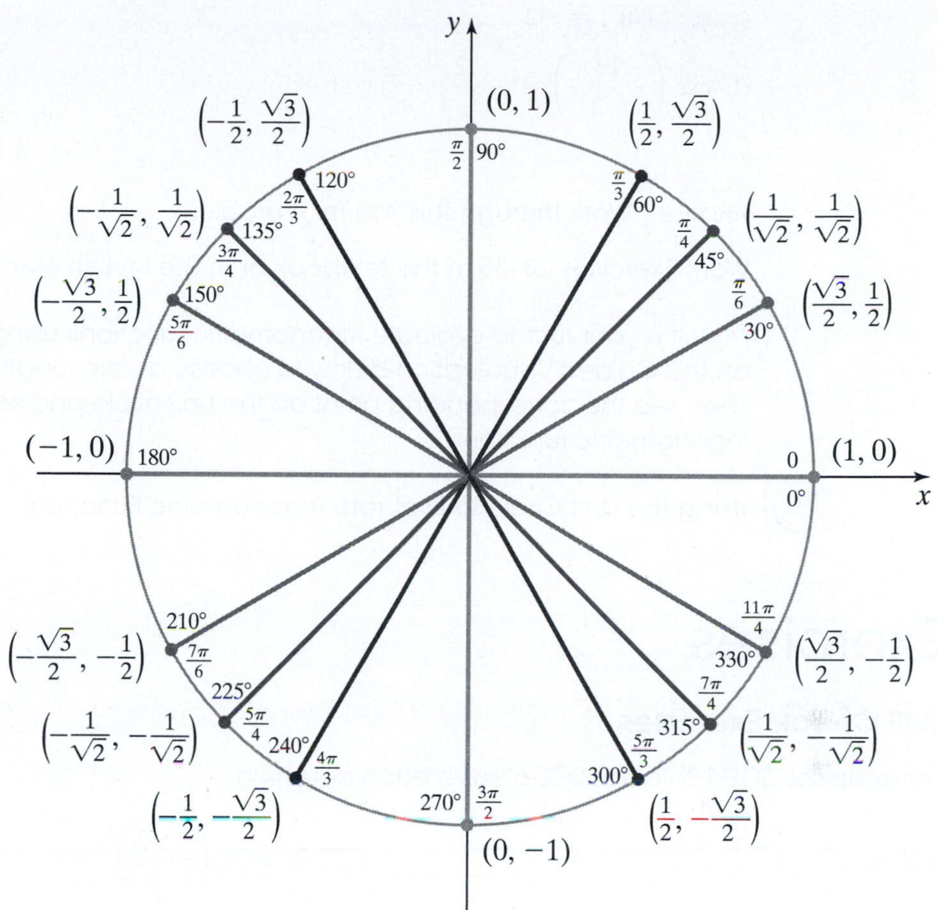

Figure 70 The Unit Circle

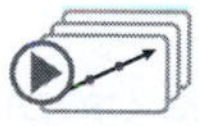

Example 5 Evaluating Trigonometric Functions Using the Unit Circle

Use the unit circle to determine the following values.

a. $\tan\left(\frac{7\pi}{3}\right)$ **b.** $\sin\left(-\frac{3\pi}{4}\right)$ **c.** $\sec(480°)$ **d.** $\csc\left(-\frac{13\pi}{3}\right)$

Solution

a. We first need to identify the ordered pair on the unit circle that corresponds to $\frac{7\pi}{3}$. Watch this **animation** to see how to determine that this point is $\left(\frac{1}{2}, \frac{\sqrt{3}}{2}\right)$. Using the unit circle definition of the tangent function, we get

$\tan\frac{7\pi}{3} = \frac{y}{x} = \frac{\frac{\sqrt{3}}{2}}{\frac{1}{2}} = \left(\frac{\sqrt{3}}{\cancel{2}}\right)(\cancel{2}) = \sqrt{3}$. Work through the entire **animation** to verify the following:

b. $\sin\left(-\frac{3\pi}{4}\right) = -\frac{1}{\sqrt{2}}$

c. $\sec(480^\circ) = -2$

d. $\csc\left(-\frac{13\pi}{3}\right) = -\frac{2}{\sqrt{3}}$

You Try It Work through this You Try It problem.

Work Exercises 20–35 in this textbook or in the MyLab Math Study Plan.

Now it is your turn to evaluate trigonometric functions using the unit circle. Click on the Guided Visualization below to choose an arc length (or central angle). Then see the corresponding point on the unit circle and see the values of the six trigonometric functions.

 Using the Unit Circle to Evaluate Trigonometric Functions

6.6 Exercises

Skill Check Exercises

For exercises SCE-1 through SCE-6, solve each equation.

SCE-1. $x^2 + \left(\frac{4}{5}\right)^2 = 1$

SCE-2. $x^2 + \left(-\frac{\sqrt{3}}{2}\right)^2 = 1$

SCE-3. $\left(-\frac{3}{7}\right)^2 + y^2 = 1$

SCE-4. $\left(\frac{1}{\sqrt{2}}\right)^2 + y^2 = 1$

In Exercises 1–5, determine the missing coordinate of a point that lies on the graph of the unit circle, given the quadrant in which it is located.

1. $\left(\frac{2}{3}, y\right)$; Quadrant IV

2. $\left(-\frac{1}{5}, y\right)$; Quadrant II

3. $\left(x, -\frac{1}{7}\right)$; Quadrant III

4. $\left(x, \frac{2}{9}\right)$; Quadrant II

5. $\left(\frac{1}{3\sqrt{5}}, y\right)$; Quadrant IV

In Exercises 6–8, verify that the given point lies on the graph of the unit circle. Then use symmetry to find three other points that also lie on the graph of the unit circle.

6. $\left(\frac{1}{7}, \frac{4\sqrt{3}}{7}\right)$

7. $\left(-\frac{1}{2}, -\frac{\sqrt{3}}{2}\right)$

8. $\left(-\frac{4}{\sqrt{17}}, -\frac{1}{\sqrt{17}}\right)$

In Exercises 9–11, the given point lies on the graph of the unit circle and corresponds to a real number t. Find the values of the six trigonometric functions of t.

9. $\left(\frac{1}{7}, -\frac{4\sqrt{3}}{7}\right)$

10. $\left(-\frac{4}{\sqrt{17}}, -\frac{1}{\sqrt{17}}\right)$

11. $\left(-\frac{1}{2}, -\frac{\sqrt{3}}{2}\right)$

In Exercises 12–19, use the unit circle to determine the value of the given expression or state that it is undefined.

SbS 12. $\cos 4\pi$ SbS 13. $\cot \frac{7\pi}{2}$ SbS 14. $\tan\left(-\frac{11\pi}{2}\right)$ SbS 15. $\sec(-3\pi)$

SbS 16. $\csc 5\pi$ SbS 17. $\sin(-6\pi)$ SbS 18. $\cos\left(\frac{5\pi}{2}\right)$ SbS 19. $\csc\left(-\frac{17\pi}{2}\right)$

In Exercises 20–35, use the unit circle to determine the value of the given expression or state that it is undefined.

SbS 20. $\sin\left(\frac{5\pi}{6}\right)$ SbS 21. $\cos\left(\frac{3\pi}{4}\right)$ SbS 22. $\cot\left(\frac{4\pi}{3}\right)$ SbS 23. $\csc(315^\circ)$

SbS 24. $\tan\left(-\frac{2\pi}{3}\right)$ SbS 25. $\cos\left(-\frac{7\pi}{6}\right)$ SbS 26. $\sec\left(-\frac{\pi}{4}\right)$ SbS 27. $\cot\left(-\frac{5\pi}{4}\right)$

SbS 28. $\sin\left(\frac{17\pi}{6}\right)$ SbS 29. $\tan\left(\frac{11\pi}{3}\right)$ SbS 30. $\sec\left(\frac{23\pi}{6}\right)$ SbS 31. $\csc\left(\frac{10\pi}{3}\right)$

SbS 32. $\cos(-390^\circ)$ SbS 33. $\sin\left(-\frac{17\pi}{6}\right)$ SbS 34. $\cot\left(-\frac{11\pi}{3}\right)$ SbS 35. $\sec(-510^\circ)$

Brief Exercises

In Exercises 36–59, use the unit circle to determine the value of the given expression or state that it is undefined.

36. $\cos 4\pi$ 37. $\cot \frac{7\pi}{2}$ 38. $\tan\left(-\frac{11\pi}{2}\right)$ 39. $\sec(-3\pi)$

40. $\csc 5\pi$ 41. $\sin(-6\pi)$ 42. $\cos\left(\frac{5\pi}{2}\right)$ 43. $\csc\left(-\frac{17\pi}{2}\right)$

44. $\sin\left(\frac{5\pi}{6}\right)$ 45. $\cos\left(\frac{3\pi}{4}\right)$ 46. $\cot\left(\frac{4\pi}{3}\right)$ 47. $\csc(315^\circ)$

48. $\tan\left(-\frac{2\pi}{3}\right)$ 49. $\cos\left(-\frac{7\pi}{6}\right)$ 50. $\sec\left(-\frac{\pi}{4}\right)$ 51. $\cot\left(-\frac{5\pi}{4}\right)$

52. $\sin\left(\frac{17\pi}{6}\right)$ 53. $\tan\left(\frac{11\pi}{3}\right)$ 54. $\sec\left(\frac{23\pi}{6}\right)$ 55. $\csc\left(\frac{10\pi}{3}\right)$

56. $\cos(-390^\circ)$ 57. $\sin\left(-\frac{17\pi}{6}\right)$ 58. $\cot\left(-\frac{11\pi}{3}\right)$ 59. $\sec(-510^\circ)$

Chapter 6 Summary

<table>
<tr><th>Key Concepts</th><th>Examples/Videos</th></tr>
<tr><td>

6.1 An Introduction to Angles: Degree and Radian Measure

An angle is in **standard position** if the vertex is at the origin of a rectangular coordinate system and the initial side is along the positive x-axis.

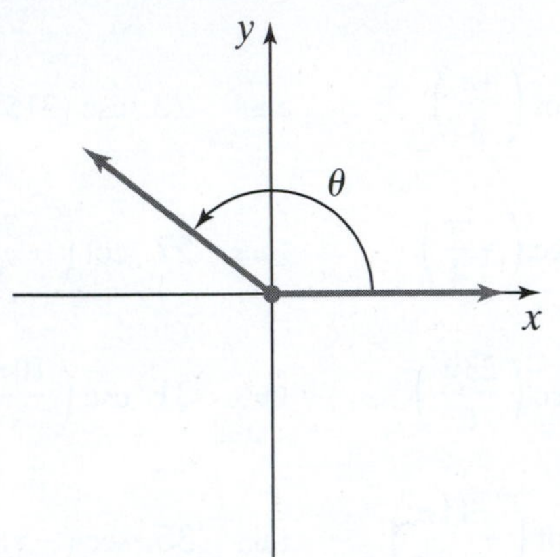

Angle θ has **positive measure** because the direction of rotation is **counterclockwise**. The terminal side of θ lies in Quadrant II.

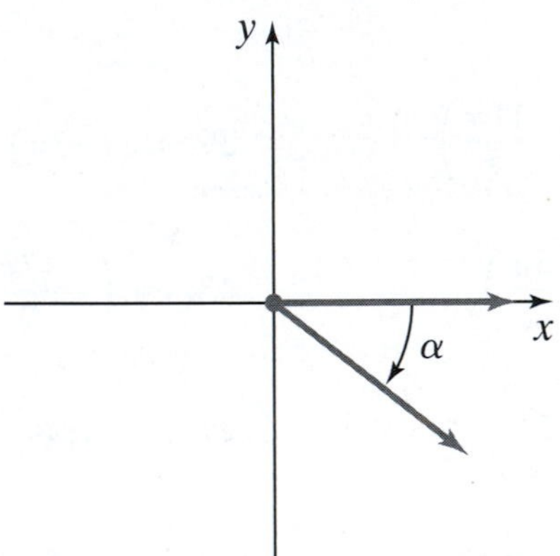

Angle α has **negative measure** because the direction of rotation is **clockwise**. The terminal side of α lies in Quadrant IV.

</td><td>

 Draw each angle in standard position and state the quadrant in which the terminal side of the angle lies or the axis on which the terminal side of the angle lies.

a. $\theta = 60°$

b. $\alpha = -270°$

c. $\beta = 420°$

</td></tr>
</table>

Key Concepts	Examples/Videos
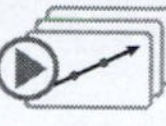 One **radian** is the measure of a central angle that has an intercepted arc equal in length to the radius of the circle. 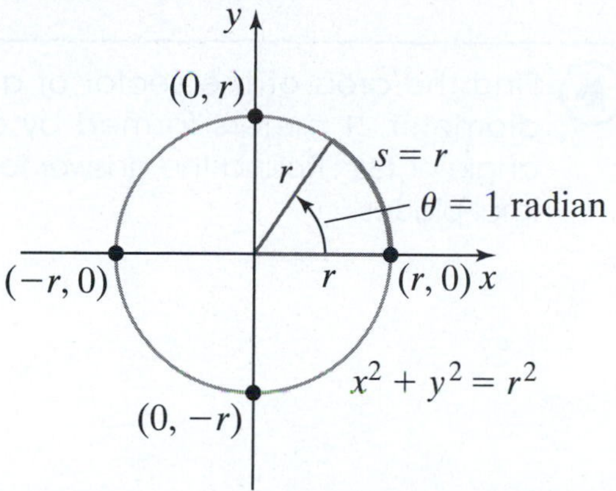 A central angle of $\theta = 1$ radian has a corresponding intercepted arc equal to the length of the radius. An angle of 1 radian is approximately $57.3°$.	
Relationship between Degrees and Radians $360° = 2\pi$ radians $180° = \pi$ radians	 Draw each angle in standard position and state the quadrant in which the terminal side of the angle lies or the axis on which the terminal side of the angle lies. **a.** $\theta = \frac{\pi}{3}$ **b.** $\alpha = -\frac{3\pi}{2}$ **c.** $\beta = \frac{7\pi}{3}$
Converting between Degrees and Radians To convert degrees to radians, multiply by $\frac{\pi \text{ radians}}{180°}$. To convert radians to degrees, multiply by $\frac{180°}{\pi \text{ radians}}$.	 Convert each angle from degrees into radians. **a.** $45°$ **b.** $-150°$ **c.** $56°$  Convert each angle from radians to degrees. Round to two decimal places if needed. **a.** $\frac{2\pi}{3}$ radians **b.** $\frac{11\pi}{6}$ radians **c.** 3 radians
Two angles in standard position having the same terminal side are **coterminal angles.** Coterminal angles can be obtained by adding nonzero integer multiples of $360°$ (or 2π) to a given angle θ. Coterminal Angle $= \theta + k \cdot 360°$ or Coterminal Angle $= \theta + k \cdot 2\pi$ where θ is any given angle and k is any integer.	 Find the angle of least nonnegative measure, θ_C, that is coterminal with $\theta = -697°$. Find the angle of least nonnegative measure, θ_C, that is coterminal with $\theta = -\frac{21\pi}{4}$.

Key Concepts	Examples/Videos
The **angle of least nonnegative measure,** θ_C, is the angle of smallest angle that is coterminal to a given angle θ such that $0 \leq \theta < 360°$ if θ is given in degrees or $0 \leq \theta < 2\pi$ if θ is given in radians.	
6.2 Applications of Radian Measure **Area of a Sector of a Circle** The area A, of a sector of a circle with radius r, and a central angle of θ radians is given by $A = \frac{1}{2}\theta r^2$. 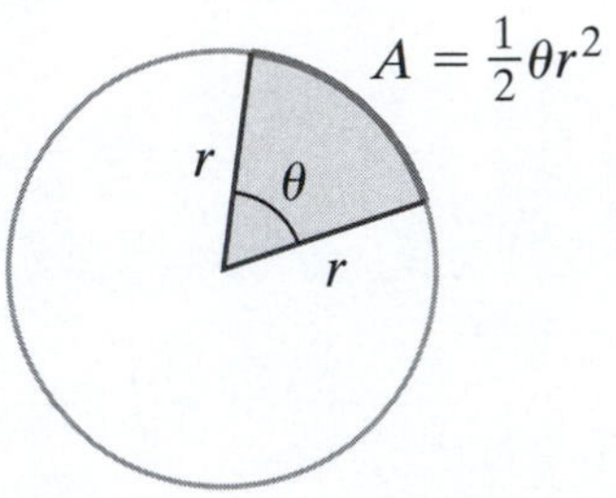 $A = \frac{1}{2}\theta r^2$ r θ r	Find the area of the sector of a circle of diameter 21 meters formed by a central angle of 135°. Round the answer to two decimal places.
Arc Length of a Sector of a Circle If θ is a central angle (in radians) of a circle with radius r, then the length of the intercepted arc, s, is given by $s = r\theta$. 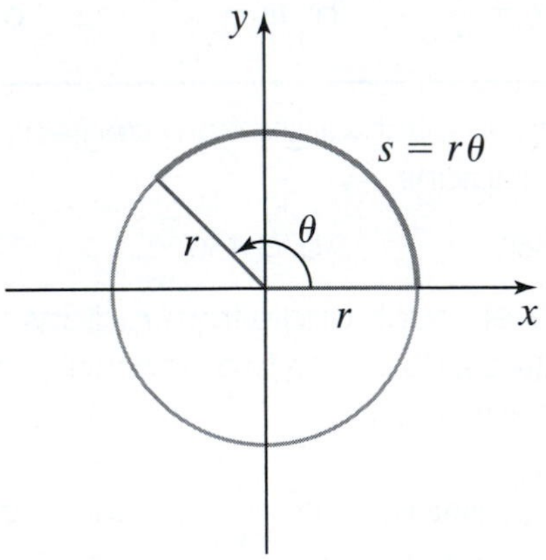 y $s = r\theta$ r θ r x	A 120° central angle intercepts an arc of 23.4 inches. Calculate the radius of the circle. Round to two decimal places.
Angular Velocity The angular velocity, ω, of an object moving along a circular path is the amount of rotation, θ, in radians of the object per unit of time, t, and is given by $\omega = \frac{\theta}{t}$.	

Key Concepts	Examples/Videos
Linear Velocity The linear velocity, v, of an object along a circular path is the distance traveled by the object, s, per unit of time, t, and is given by $v = \frac{s}{t}$. Note that $s = r\theta$, where θ is given in radians.	Determine the angular velocity and the linear velocity of a point on the equator of the Earth. Assume that the radius of the Earth at the equator is 3963 miles.
6.3 Triangles Acute Triangle 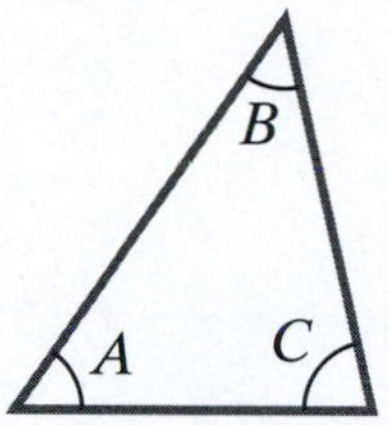 Acute triangles have three acute angles. $\measuredangle A < 90°$, $\measuredangle B < 90°$, and $\measuredangle C < 90°$ Obtuse Triangle 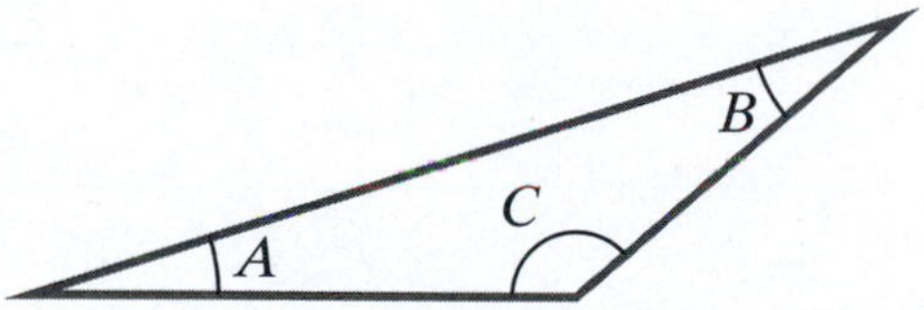 Obtuse triangles have one obtuse angle. $\measuredangle A < 90°$, $\measuredangle B < 90°$, and $\measuredangle C > 90°$ Right Triangle 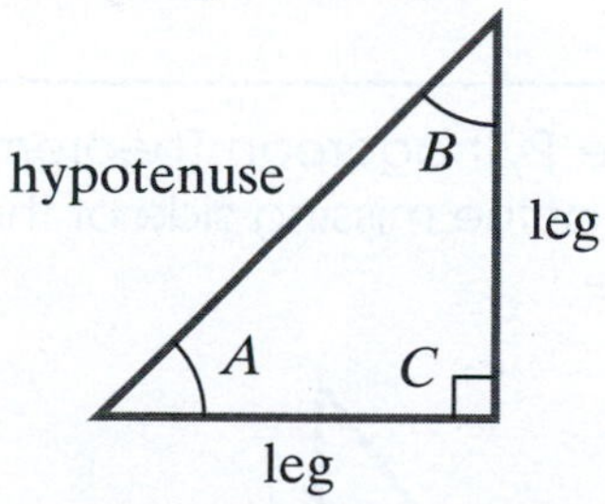 hypotenuse, leg, leg Right triangles have one right angle. $\measuredangle A < 90°$, $\measuredangle B < 90°$, and $\measuredangle C = 90°$	

Key Concepts	Examples/Videos
Scalene Triangle 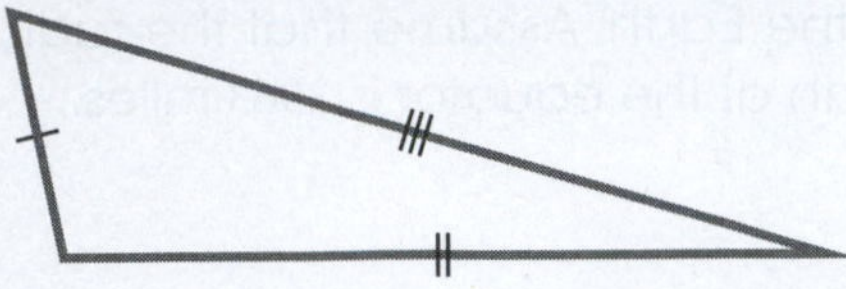 Scalene triangles have no congruent sides. Isosceles Triangle 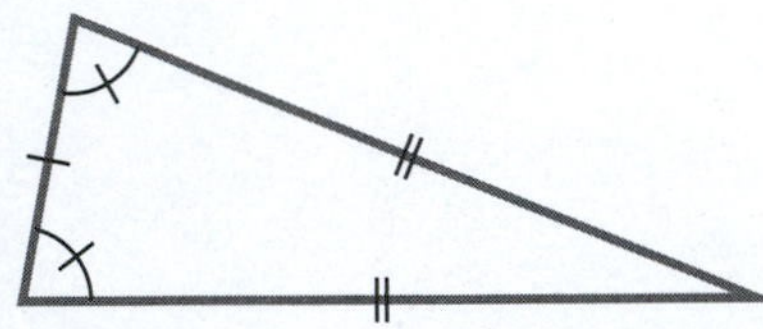 Isosceles triangles have two congruent sides. Equilateral Triangle 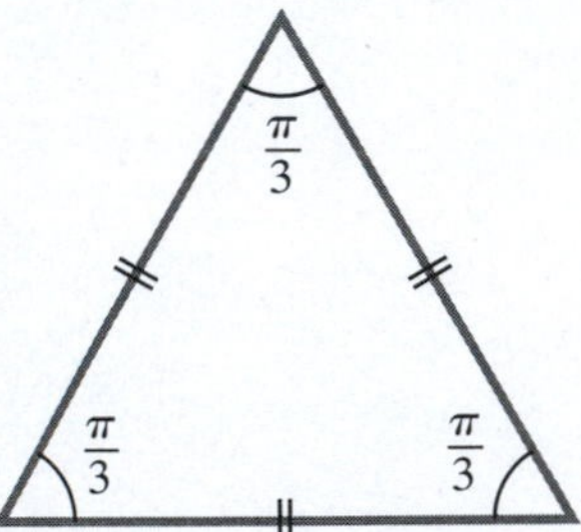 Equilateral triangles have three congruent sides and three congruent angles. Each angle has a measure of $60°$ or $\frac{\pi}{3}$ radians.	
The Pythagorean Theorem Given any right triangle, the sum of the squares of the lengths of the legs is equal to the square of the length of the hypotenuse. 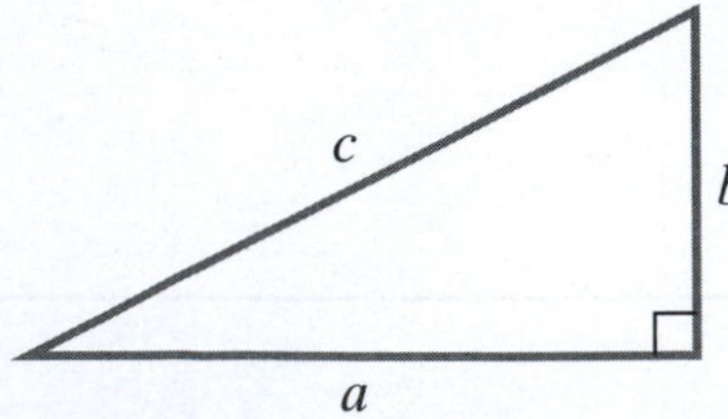 The Pythagorean Theorem: $a^2 + b^2 = c^2$	 Use the Pythagorean Theorem to find the length of the missing side of the given right triangle. 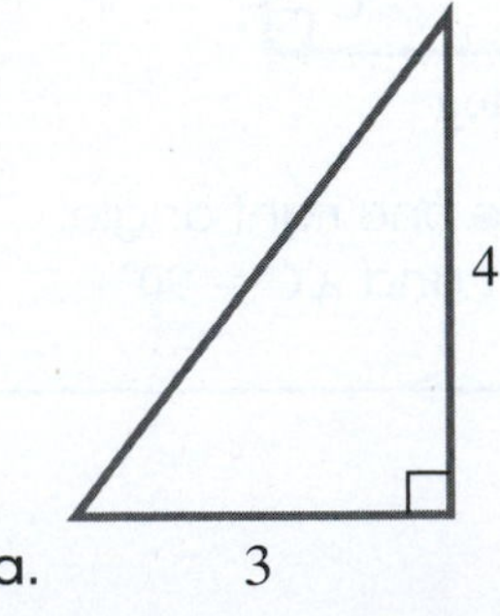 a. 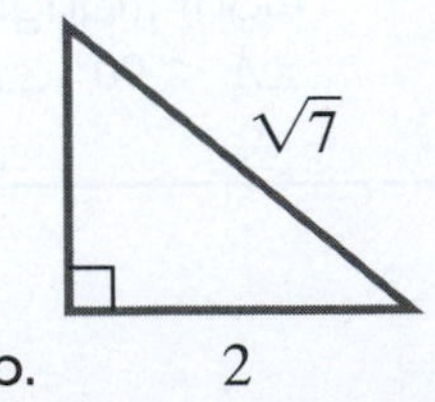 b.

Key Concepts	Examples/Videos

Key Concepts

Properties of Similar Triangles

1. The corresponding angles have the same measure.
2. The ratio of the lengths of any two sides of one triangle is equal to the ratio of the lengths of the corresponding sides of the other triangle.

Proportionality Constant of Similar Triangles

If two triangles are similar, there exists a constant $k \geq 1$ called the **proportionality constant of similar triangles** equal to the ratio of the lengths of corresponding sides. Given the similar triangles shown below, $k = \frac{a}{x} = \frac{b}{y} = \frac{c}{z}$, where $a \geq x, b \geq y$, and $c \geq z$.

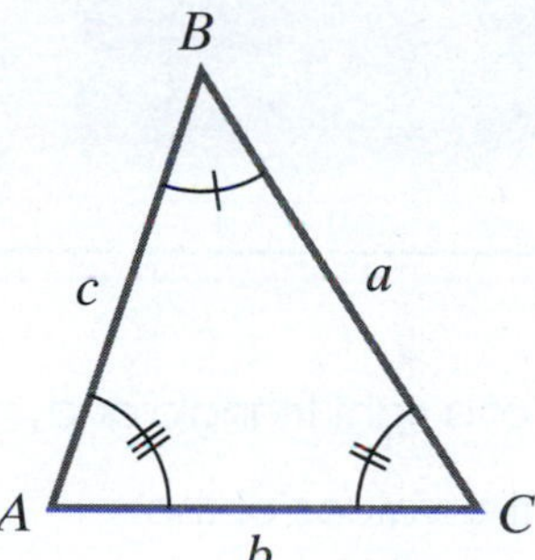

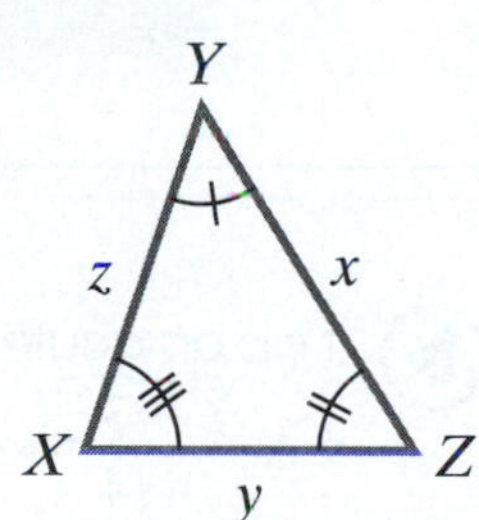

Examples/Videos

Triangles ABC and XYZ are similar. Find the lengths of the missing sides of triangle ABC.

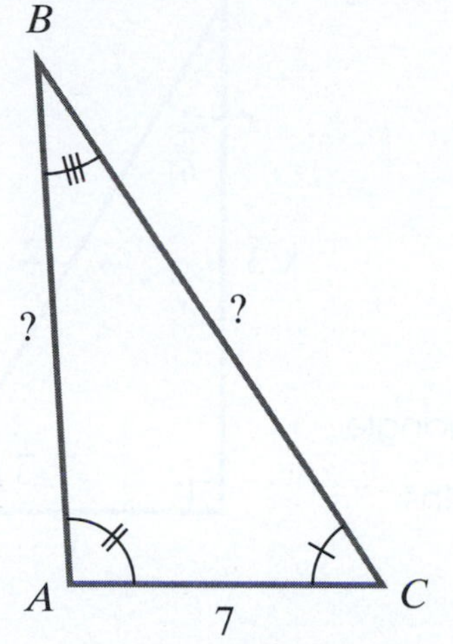

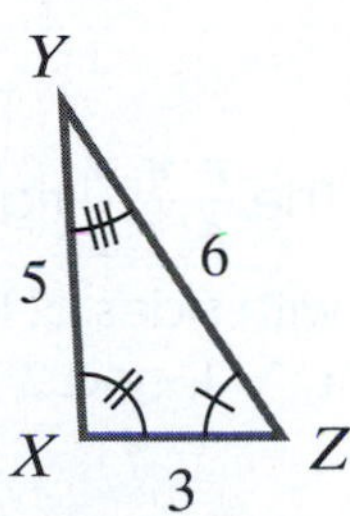

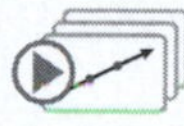

The triangles below are similar. Find the proportionality constant. Then find the lengths of the missing sides.

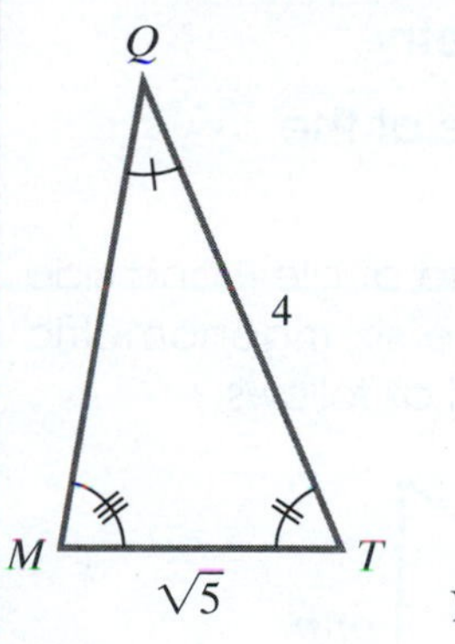

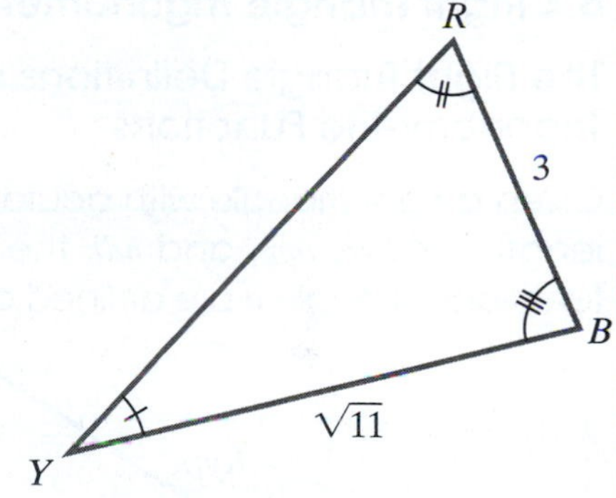

Key Concepts

Special Right Triangles

The $\frac{\pi}{4}, \frac{\pi}{4}, \frac{\pi}{2}$ $(45°, 45°, 90°)$ Right Triangle

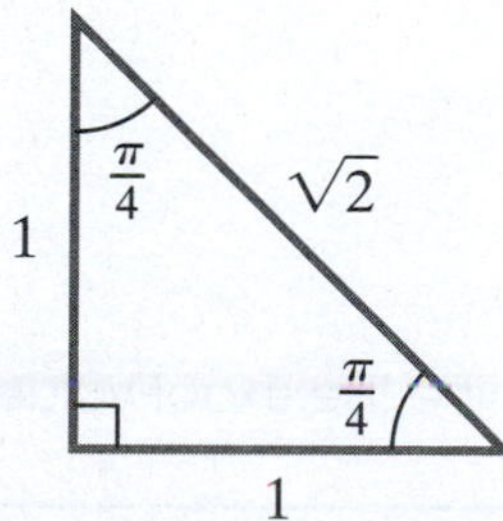

The $\frac{\pi}{4}, \frac{\pi}{4}, \frac{\pi}{2}$ $(45°, 45°, 90°)$ right triangle with sides of lengths $1, 1$, and $\sqrt{2}$

Examples/Videos

Determine the lengths of the missing sides of each triangle.

a.

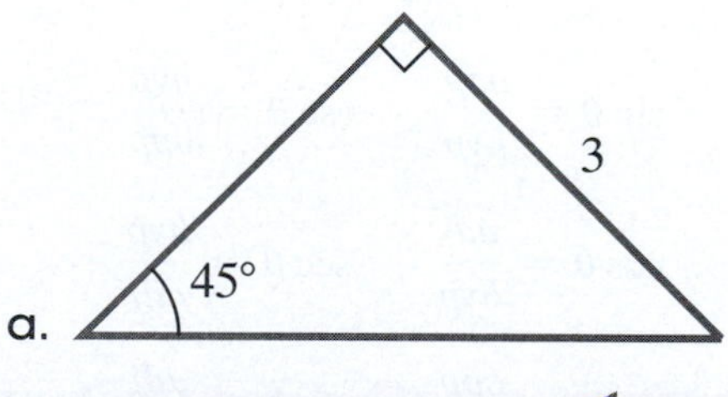

b.

$\sqrt{5}$

$\frac{\pi}{6}$

Key Concepts	Examples/Videos
 The $\frac{\pi}{6}, \frac{\pi}{3}, \frac{\pi}{2}$ $(30°, 60°, 90°)$ Right Triangle 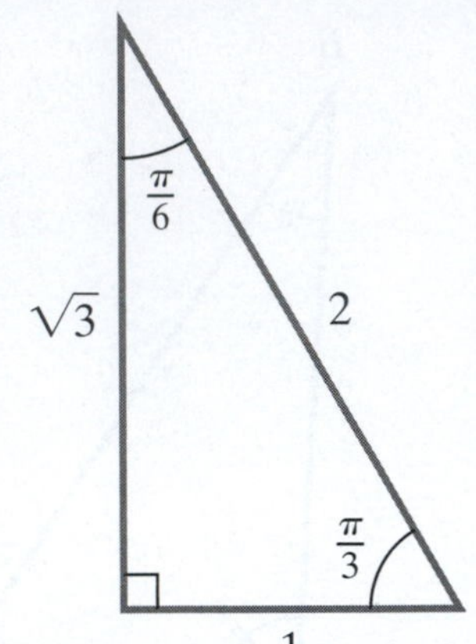 The $\frac{\pi}{6}, \frac{\pi}{3}, \frac{\pi}{2}$ right triangle with sides of lengths $1, \sqrt{3}$, and 2	Two people are standing on opposite sides of a small river. One person is located at point Q, a distance of 20 feet from a bridge. The other person is standing on the southeast corner of the bridge at point P. The angle between the bridge and the line of sight from P to Q is $30°$. Use this information to determine the length of the bridge and the distance between the two people. Round your answers to two decimal places as needed. 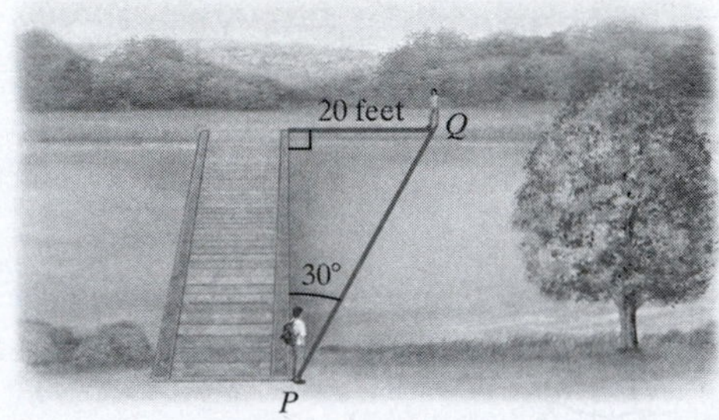
6.4 Right Triangle Trigonometry **The Right Triangle Definitions of the Trigonometric Functions** Given a right triangle with acute angle θ and side lengths of *hyp, opp,* and *adj,* the six trigonometric functions of angle θ are defined as follows: 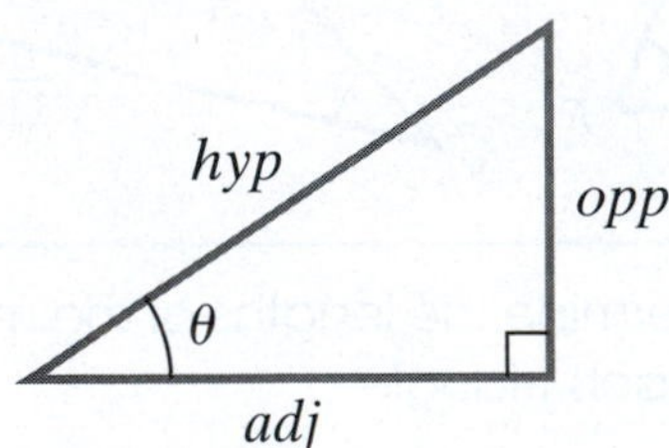 $\sin\theta = \frac{opp}{hyp}$ $\csc\theta = \frac{hyp}{opp}$ $\cos\theta = \frac{adj}{hyp}$ $\sec\theta = \frac{hyp}{adj}$ $\tan\theta = \frac{opp}{adj}$ $\cot\theta = \frac{adj}{opp}$	 If θ is an acute angle of a right triangle and if $\sin\theta = \frac{3}{4}$, then find the values of the remaining five trigonometric functions for angle θ.  Determine the exact value of $\csc\frac{\pi}{6} + \cot\frac{\pi}{4}$.

Key Concepts	Examples/Videos
The Fundamental Trigonometric Identities **Quotient Identities** $\tan\theta = \dfrac{\sin\theta}{\cos\theta}$ $\cot\theta = \dfrac{\cos\theta}{\sin\theta}$ **Reciprocal Identities** $\sin\theta = \dfrac{1}{\csc\theta}$ $\csc\theta = \dfrac{1}{\sin\theta}$ $\cos\theta = \dfrac{1}{\sec\theta}$ $\sec\theta = \dfrac{1}{\cos\theta}$ $\tan\theta = \dfrac{1}{\cot\theta}$ $\cot\theta = \dfrac{1}{\tan\theta}$ **Pythagorean Identities** $\sin^2\theta + \cos^2\theta = 1$ $1 + \tan^2\theta = \sec^2\theta$ $1 + \cot^2\theta = \csc^2\theta$	 Given that $\sin\theta = \dfrac{5}{7}$ and $\cos\theta = \dfrac{2\sqrt{6}}{7}$, find the values of the remaining four trigonometric functions using identities.  Use identities to find the exact value of each trigonometric expression. a. $\tan 37° - \dfrac{\sin 37°}{\cos 37°}$ b. $\dfrac{1}{\cos^2\frac{\pi}{9}} - \dfrac{1}{\cot^2\frac{\pi}{9}}$
Cofunction Identities $\sin\theta = \cos\left(\dfrac{\pi}{2} - \theta\right)$ $\cos\theta = \sin\left(\dfrac{\pi}{2} - \theta\right)$ $\tan\theta = \cot\left(\dfrac{\pi}{2} - \theta\right)$ $\cot\theta = \tan\left(\dfrac{\pi}{2} - \theta\right)$ $\sec\theta = \csc\left(\dfrac{\pi}{2} - \theta\right)$ $\csc\theta = \sec\left(\dfrac{\pi}{2} - \theta\right)$ If angle θ is given in degrees, replace $\dfrac{\pi}{2}$ with $90°$.	 a. Rewrite the expression $\left(\cot\left(\dfrac{\pi}{2} - \theta\right)\right)\cos\theta$ as one of the six trigonometric functions of angle θ. b. Determine the exact value of $\sec 55° \csc 55° - \tan 55° \cot 55°$
6.5 Trigonometric Functions of General Angles **Families of Special Angles**  Quadrantal Family  $\dfrac{\pi}{3}$ Family  $\dfrac{\pi}{6}$ Family  $\dfrac{\pi}{4}$ Family	 Each of the given angles belongs to one of the four families of special angles. Determine the family of angles to which it belongs, sketch the angle, then determine the angle of least nonnegative measure, θ_c, coterminal with the given angle. a. $\theta = \dfrac{29\pi}{6}$ b. $\theta = \dfrac{14\pi}{2}$ c. $\theta = \dfrac{18\pi}{4}$ d. $\theta = \dfrac{11\pi}{4}$ e. $\theta = \dfrac{14\pi}{6}$ f. $\theta = 420°$ g. $\theta = -495°$

Key Concepts	Examples/Videos

The General Angle Definitions of the Trigonometric Functions

If $P(x, y)$ is a point on the terminal side of *any* angle θ in standard position and if $r = \sqrt{x^2 + y^2}$ is the distance from the origin to point P, then the six trigonometric functions of θ are defined as follows:

$$\sin\theta = \frac{y}{r} \qquad \csc\theta = \frac{r}{y}, y \neq 0$$

$$\cos\theta = \frac{x}{r} \qquad \sec\theta = \frac{r}{x}, x \neq 0$$

$$\tan\theta = \frac{y}{x}, x \neq 0 \qquad \cot\theta = \frac{x}{y}, y \neq 0$$

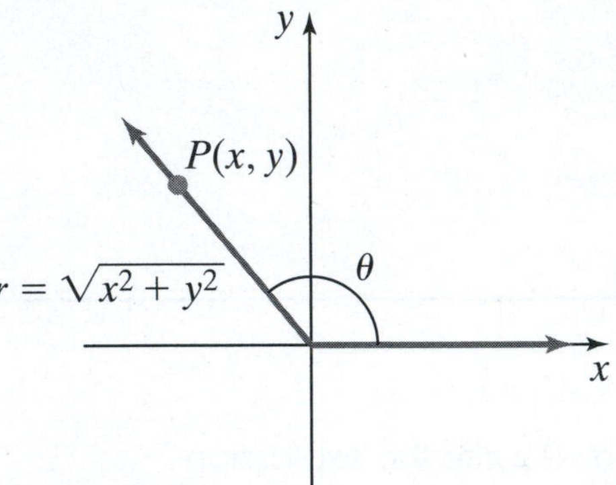

Suppose that the point $(-4, -6)$ is on the terminal side of an angle θ. Find the six trigonometric functions of θ.

Trigonometric Functions of Quadrantal Families

θ	$\sin\theta$	$\cos\theta$	$\tan\theta$	$\csc\theta$	$\sec\theta$	$\cot\theta$
0	0	1	a	Undefined	1	Undefined
$\frac{\pi}{2}$	1	0	Undefined	1	Undefined	0
π	0	−1	0	Undefined	−1	Undefined
$\frac{3\pi}{2}$	−1	0	Undefined	−1	Undefined	0

Without using a calculator, determine the value of the trigonometric function or state that the value is undefined.

a. $\cos(-11\pi)$

b. $\csc(-270°)$

c. $\tan\left(\frac{13\pi}{2}\right)$

d. $\sin(540°)$

e. $\cot\left(-\frac{7\pi}{2}\right)$

Key Concepts	Examples/Videos

The Signs of the Trigonometric Functions

II: Sine (and cosecant) are positive.

I: All trigonometric functions are positive.

III: Tangent (and cotangent) are positive.

IV: Cosine (and secant) are positive.

Suppose θ is a positive angle in standard position such that $\sin\theta < 0$ and $\sec\theta > 0$.

a. Determine the quadrant in which the terminal side of angle θ lies.

b. Find the value of $\tan\theta$ if $\sec\theta = \sqrt{5}$.

The **reference angle**, θ_R, is the positive acute angle associated with a given angle θ. The reference angle is formed by the "nearest" x-axis and the terminal side of angle θ. Below are angles sketched in each of the four quadrants along with their corresponding reference angles.

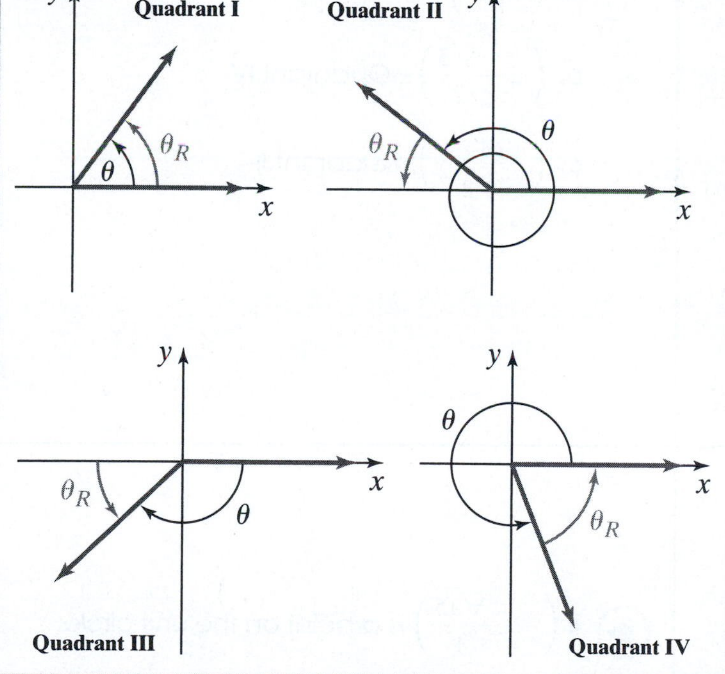

For each of the given angles, determine the reference angle.

a. $\theta = \frac{5\pi}{3}$ b. $\theta = \frac{11\pi}{4}$

c. $\theta = \frac{25\pi}{6}$ d. $\theta = \frac{16\pi}{6}$

For each of the given angles, determine the reference angle.

a. $\theta = \frac{5\pi}{8}$ b. $\theta = \frac{22\pi}{9}$ c. $\theta = \frac{5\pi}{7}$

G/V Sketching General Angles

Key Concepts	Examples/Videos

Steps for Evaluating Trigonometric Functions of Angles Belonging to the $\frac{\pi}{3}, \frac{\pi}{6}$, or $\frac{\pi}{4}$ Families

Step 1. Draw the angle and determine the quadrant in which the terminal side of the angle lies.

Step 2. Determine if the sign of the function is positive or negative in that quadrant.

Step 3. Determine if the reference angle, θ_R, is $\frac{\pi}{3}, \frac{\pi}{6}$, or $\frac{\pi}{4}$.

Step 4. Use the appropriate special right triangle or your knowledge of Table 1 to determine the value of the trigonometric function.

Find the exact value of each trigonometric expression without the use of a calculator.

a. $\sin\left(\frac{7\pi}{6}\right)$ b. $\cot\left(-\frac{22\pi}{3}\right)$

c. $\tan\left(\frac{11\pi}{4}\right)$ d. $\cos\left(\frac{11\pi}{3}\right)$

e. $\sec\left(\frac{5\pi}{6}\right)$ f. $\csc\left(-\frac{7\pi}{6}\right)$

6.6 The Unit Circle

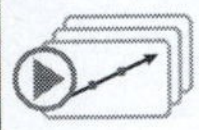
A circle centered at the origin with a radius of 1 unit is called the **unit circle** whose equation is given by $x^2 + y^2 = 1$.

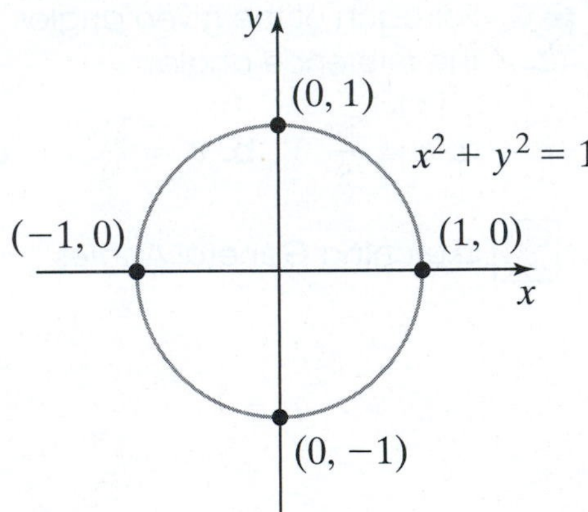

Determine the missing coordinate of a point that lies on the graph of the unit circle given the quadrant in which the point is located.

a. $\left(-\frac{1}{8}, y\right)$; Quadrant III

b. $\left(x, -\frac{\sqrt{3}}{2}\right)$; Quadrant IV

c. $\left(-\frac{1}{\sqrt{2}}, y\right)$; Quadrant II

The Unit Circle Definitions of the Trigonometric Functions

For any real number t, if $P(x, y)$ is a point on the unit circle corresponding to t, then

$\sin t = y$ $\csc t = \frac{1}{y}, y \neq 0$

$\cos t = x$ $\sec t = \frac{1}{x}, x \neq 0$

$\tan t = \frac{y}{x}, x \neq 0$ $\cot t = \frac{x}{y}, y \neq 0$

If $\left(-\frac{1}{4}, \frac{\sqrt{15}}{4}\right)$ is a point on the unit circle corresponding to a real number t, find the values of the six trigonometric functions of t.

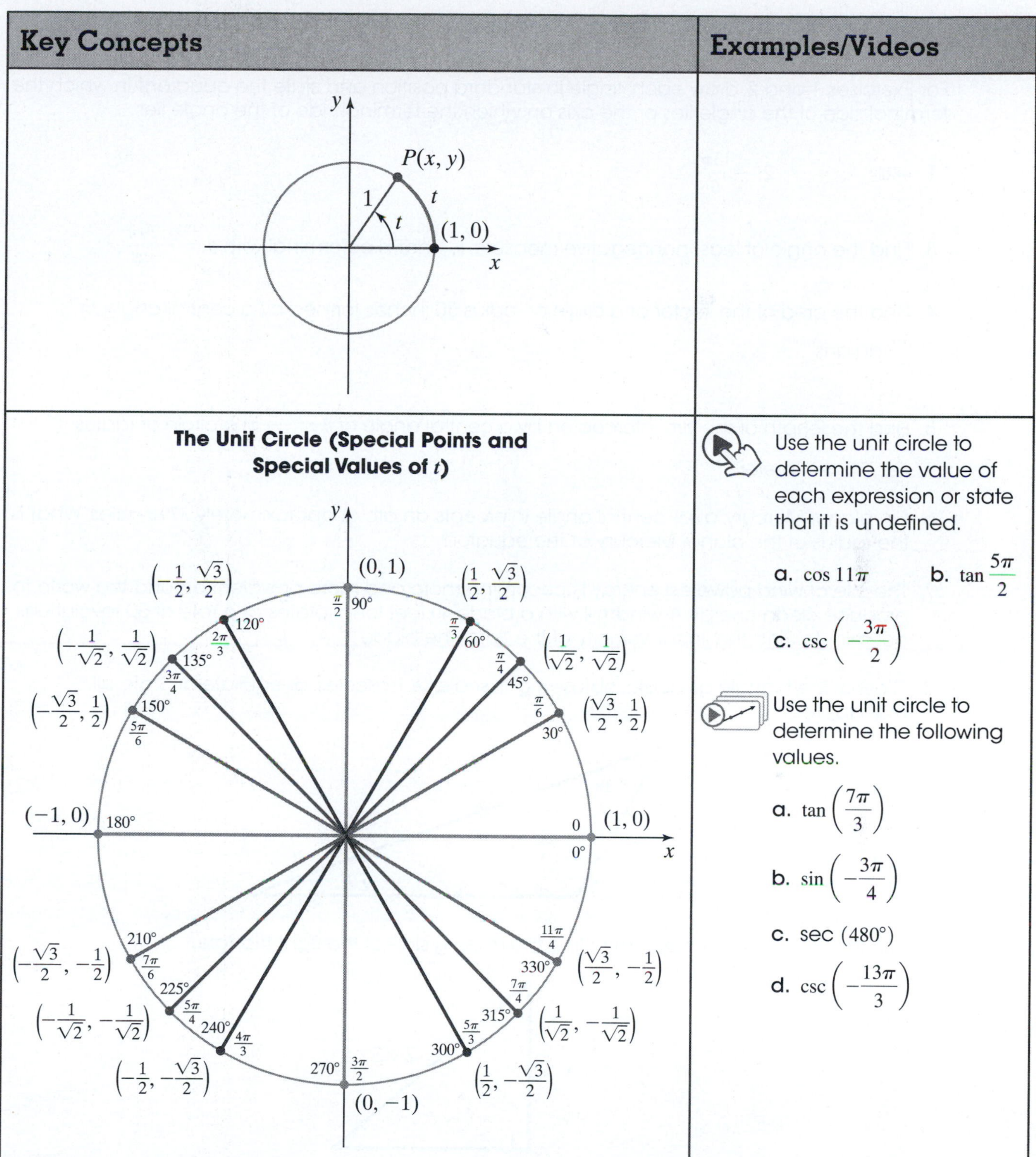
Key Concepts
Examples/Videos
y
$P(x, y)$
1
t
t
(1, 0)
x
The Unit Circle (Special Points and Special Values of t)
y
(0, 1)
$\frac{\pi}{2}$ 90°
$\left(-\frac{1}{2}, \frac{\sqrt{3}}{2}\right)$
$\left(\frac{1}{2}, \frac{\sqrt{3}}{2}\right)$
120° $\frac{2\pi}{3}$
$\frac{\pi}{3}$ 60°
$\left(-\frac{1}{\sqrt{2}}, \frac{1}{\sqrt{2}}\right)$
$\left(\frac{1}{\sqrt{2}}, \frac{1}{\sqrt{2}}\right)$
135° $\frac{3\pi}{4}$
$\frac{\pi}{4}$ 45°
$\left(-\frac{\sqrt{3}}{2}, \frac{1}{2}\right)$
$\left(\frac{\sqrt{3}}{2}, \frac{1}{2}\right)$
150° $\frac{5\pi}{6}$
$\frac{\pi}{6}$ 30°
(−1, 0) 180°
0 (1, 0)
0°
x
210° $\frac{7\pi}{6}$
$\left(-\frac{\sqrt{3}}{2}, -\frac{1}{2}\right)$
$\frac{11\pi}{6}$ 330°
$\left(\frac{\sqrt{3}}{2}, -\frac{1}{2}\right)$
225° $\frac{5\pi}{4}$
$\left(-\frac{1}{\sqrt{2}}, -\frac{1}{\sqrt{2}}\right)$
$\frac{7\pi}{4}$ 315°
$\left(\frac{1}{\sqrt{2}}, -\frac{1}{\sqrt{2}}\right)$
240° $\frac{4\pi}{3}$
$\left(-\frac{1}{2}, -\frac{\sqrt{3}}{2}\right)$
$\frac{5\pi}{3}$ 300°
$\left(\frac{1}{2}, -\frac{\sqrt{3}}{2}\right)$
270° $\frac{3\pi}{2}$
(0, −1)
Use the unit circle to determine the value of each expression or state that it is undefined.
a. $\cos 11\pi$
b. $\tan \frac{5\pi}{2}$
c. $\csc\left(-\frac{3\pi}{2}\right)$
Use the unit circle to determine the following values.
a. $\tan\left(\frac{7\pi}{3}\right)$
b. $\sin\left(-\frac{3\pi}{4}\right)$
c. $\sec(480°)$
d. $\csc\left(-\frac{13\pi}{3}\right)$

Chapter 6 Review Exercises

For Exercises 1 and 2, draw each angle in standard position and state the quadrant in which the terminal side of the angle lies or the axis on which the terminal side of the angle lies.

1. $480°$ 2. $-\frac{13\pi}{6}$

3. Find the angle of least nonnegative measure, θ_C, that is coterminal with $\theta = -\frac{29\pi}{3}$.

4. Find the area of the sector of a circle or radius 30 inches formed by a central angle of $\frac{\pi}{5}$ radians.

5. Find the length of the arc intercepted by a central angle of $\theta = \frac{4\pi}{3}$ in a circle of radius $r = 54$ centimeters.

6. On planet Mercury, a $45°$ central angle intercepts an arc of approximately 379π miles. What is the radius of the planet Mercury at the equator?

7. The use of wind-powered energy is becoming more and more prevalent around the world to produce clean energy. A windmill with a blade 40 feet long rotates at a rate of 50 revolutions per minute. Find the linear velocity of the tip of the blade.

8. Classify the triangle as acute, obtuse, right, scalene, isosceles, or equilateral. State all that apply.

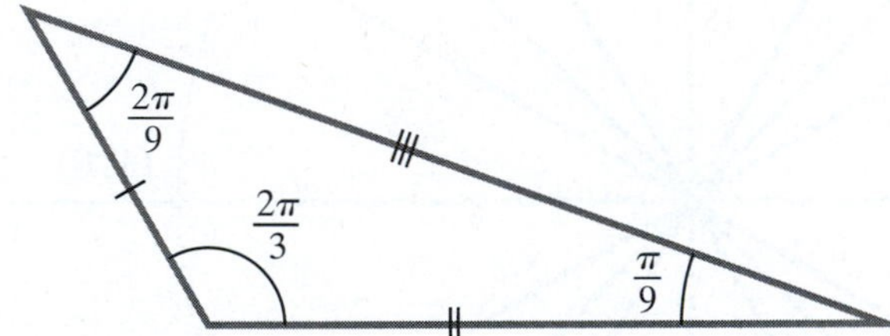

9. Use the Pythagorean Theorem to find the missing side of the right triangle.

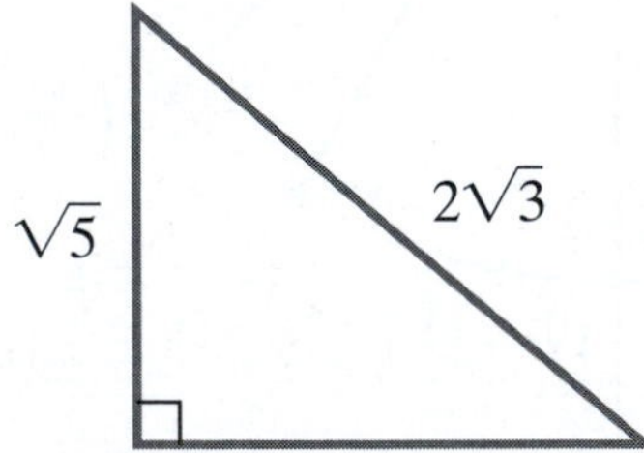

10. How far up the side of a building will a 9-meter ladder reach if the foot of the ladder is 2 meters from the base of the building?

11. The triangles below are similar. Find the proportionality constant k such that the length of each side of the larger triangle is k times the length of the corresponding sides of the smaller triangle. Then find the lengths of the missing sides of each triangle.

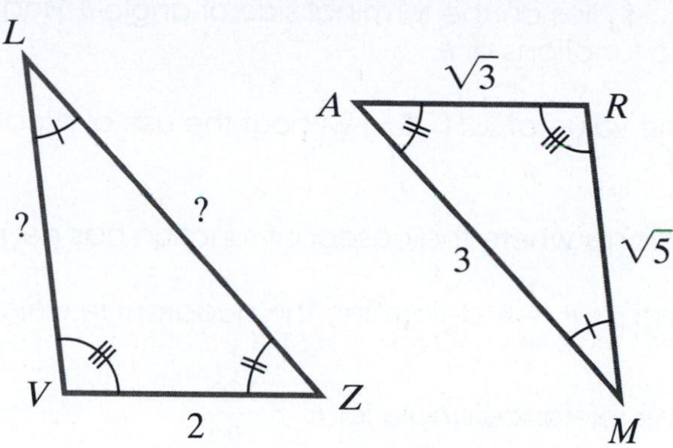

For Exercises 12 and 13, find the missing sides of each special right triangle.

12.

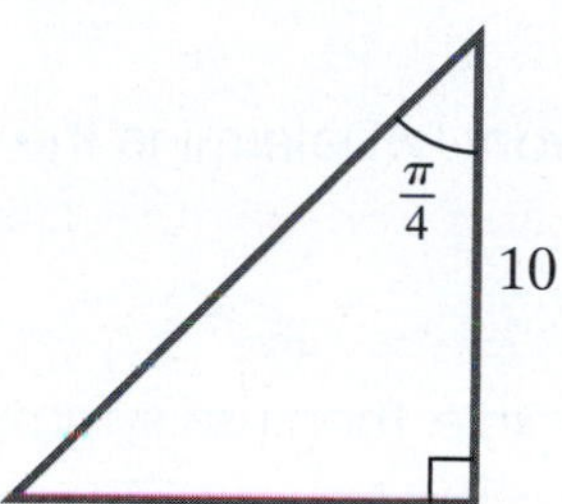

13.

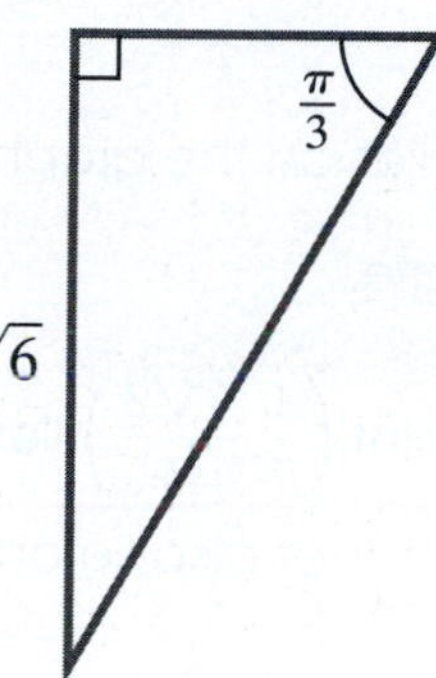

14. For the given right triangle, evaluate the six trigonometric functions of the acute angle θ.

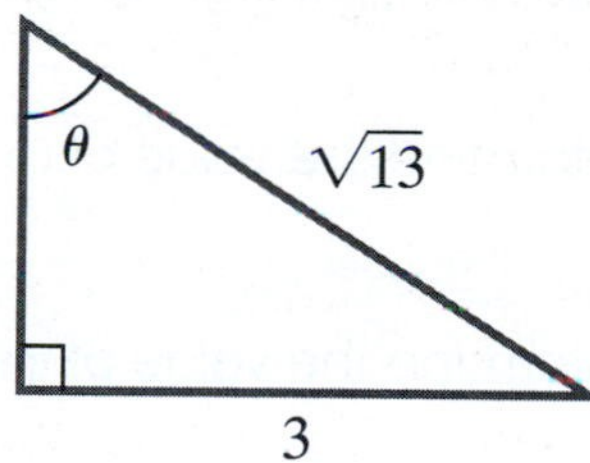

15. If θ is an acute angle of a right triangle and if $\sec\theta = \dfrac{7}{\sqrt{5}}$, then find the values of the remaining five trigonometric functions for angle θ.

16. Determine the measure of the acute angle θ for which $\cos\theta = \dfrac{\sqrt{3}}{2}$.

17. Given that $\sin\theta = \dfrac{7}{25}$ and $\cos\theta = \dfrac{24}{25}$, find the values of the remaining four trigonometric functions using identities.

18. Use identities to rewrite the following expression as $\sin\theta$, $\cos\theta$, $\tan\theta$, $\csc\theta$, $\sec\theta$, or $\cot\theta$.

$$\sin\theta\sec^2\theta\cot^2\theta$$

19. Find a cofunction equivalent to $\sin \frac{5\pi}{7}$.

20. The point $(2, -8)$ lies on the terminal side of angle θ. Find the exact value of the six trigonometric functions of θ.

21. Determine the value of $\sec(-5\pi)$ without the use of a calculator or state that the value is undefined.

22. List the quadrants where the cosecant function has negative values.

23. If $\sin\theta < 0$ and $\cos\theta > 0$, determine the quadrant in which the terminal side of angle θ lies.

24. Determine the reference angle for $\theta = \frac{16\pi}{9}$.

25. Find the exact value of $\cos\left(-\frac{7\pi}{6}\right)$ without the use of a calculator.

26. The point $\left(\frac{2}{3}, y\right)$ lies on the graph of the unit circle in Quadrant IV. Determine the missing coordinate.

27. Verify that the point $\left(\frac{1}{7}, \frac{4\sqrt{3}}{7}\right)$ lies on the graph of the unit circle. Then use symmetry to find three other points that also lie on the graph of the unit circle.

28. The point $\left(-\frac{4}{\sqrt{17}}, -\frac{1}{\sqrt{17}}\right)$ lies on the graph of the unit circle and corresponds to a real number t. Find the values of the six trigonometric functions of t.

29. Use the unit circle to determine the value of $\cos\frac{7\pi}{2}$ or state that the value is undefined.

30. Use the unit circle to determine the value of $\tan\frac{11\pi}{3}$ or state that the value is undefined.

CHAPTER SEVEN

The Graphs of Trigonometric Functions

CHAPTER SEVEN CONTENTS

7.1 The Graphs of Sine and Cosine

THINGS TO KNOW

Before working through this section, be sure that you are familiar with the following concepts:

VIDEO ANIMATION INTERACTIVE

1. Determining Whether a Function Is Even, Odd, or Neither (Section 3.2) 

2. Using Vertical Stretches and Compressions to Graph Functions (Section 3.4) 

3. Using Horizontal Stretches and Compressions to Graph Functions (Section 3.4) 

4. Using the Special Right Triangles (Section 6.4)  

5. Finding the Values of the Trigonometric Functions of Quadrantal Angles (Section 6.5) 

6. Evaluating Trigonometric Functions of Angles Belonging to the $\frac{\pi}{3}, \frac{\pi}{6},$ or $\frac{\pi}{4}$ Families (Section 6.5)  

INTRODUCTION

Read this introduction before beginning Objective 1.

OBJECTIVES

1 Understanding the Graph of the Sine Function and Its Properties

2 Understanding the Graph of the Cosine Function and Its Properties

3 Sketching Graphs of the Form $y = A \sin x$ and $y = A \cos x$

4 Sketching Graphs of the Form $y = \sin(Bx)$ and $y = \cos(Bx)$

5 Sketching Graphs of the Form $y = A \sin(Bx)$ and $y = A \cos(Bx)$

6 Determine the Equation of a Function of the Form $y = A \sin(Bx)$ or $y = A \cos(Bx)$ Given Its Graph

SECTION 7.1 EXERCISES

Introduction to Section 7.1

Recall that in Chapter 6, we sketched an angle θ in standard position in the rectangular coordinate system. We used the ordered pair (x, y) to represent a point P lying on the terminal side of θ. From there, we were able to find the trigonometric values of that angle using the **general angle definitions of the trigonometric functions.** We now review these definitions.

The General Angle Definitions of the Trigonometric Functions

If $P(x, y)$ is a point on the terminal side of *any* angle θ in standard position and if $r = \sqrt{x^2 + y^2}$ is the distance from the origin to point P, then the six trigonometric functions of θ are defined as follows:

$\sin\theta = \frac{y}{r}$ $\quad$ $\csc\theta = \frac{r}{y},\ y \neq 0$

$\cos\theta = \frac{x}{r}$ $\quad$ $\sec\theta = \frac{r}{x},\ x \neq 0$

$\tan\theta = \frac{y}{x},\ x \neq 0$ $\quad$ $\cot\theta = \frac{x}{y},\ y \neq 0$

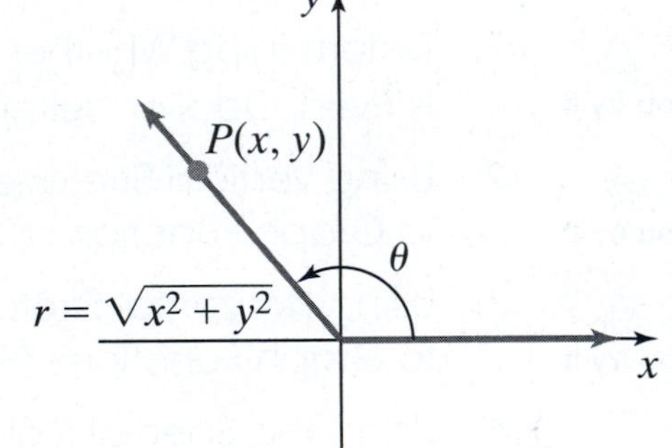

Now, in Chapter 7, we will use a rectangular coordinate system for a different purpose. We will plot points and label them (x, y), but the x-value (or **independent variable**) will represent an angle measured in radians and the y-value (or **dependent variable**) will represent a trigonometric function of the angle, either $y = \sin x$, $y = \cos x$, $y = \tan x$, $y = \csc x$, $y = \sec x$, or $y = \cot x$. For example, consider the function $y = \sin x$. When $x = \frac{\pi}{4}$, $y = \sin\frac{\pi}{4} = \frac{1}{\sqrt{2}}$. Thus, the ordered pair $\left(\frac{\pi}{4}, \frac{1}{\sqrt{2}}\right)$ represents a point that lies on the graph of $y = \sin x$.

Before we establish the graphs of $y = \sin x$ and $y = \cos x$, it is imperative that you can evaluate trigonometric functions at certain values. Take a moment to test yourself by clicking on any of the three You Try It exercises below.

You Try It Evaluating Trigonometric Functions for Angles Belonging to the **Quadrantal Family**

You Try It Using Special Right Triangles to Evaluate Trigonometric Functions

You Try It Evaluating Trigonometric Functions of Angles Belonging to the $\frac{\pi}{6}, \frac{\pi}{4}$, or $\frac{\pi}{3}$ Families

OBJECTIVE 1 UNDERSTANDING THE GRAPH OF THE SINE FUNCTION AND ITS PROPERTIES

We can sketch the graph of $y = \sin x$ by setting up a table (or tables) of values and plotting points. Take some time to fill in the four tables shown below. Table 1 includes values of x belonging to the **quadrantal family**. Table 2 includes values of x belonging to the $\frac{\pi}{6}$ **family**. Table 3 includes values of x belonging to the $\frac{\pi}{4}$ **family**. Table 4 includes values of x belonging to the $\frac{\pi}{3}$ **family**. Once these four tables are complete, we can plot the corresponding ordered pairs to sketch the graph of $y = \sin x$ on the interval $[0, 2\pi]$. See **Figure 1**. Watch this **video** to see how to complete each of the four tables and to see how to sketch the graph of $y = \sin x$.

Table 1 (View the completed Table 1.)

x	0	$\frac{\pi}{2}$	π	$\frac{3\pi}{2}$	2π
$y = \sin x$					

Table 2 (View the completed Table 2.)

x	$\frac{\pi}{6}$	$\frac{5\pi}{6}$	$\frac{7\pi}{6}$	$\frac{11\pi}{6}$
$y = \sin x$				

Table 3 (View the completed Table 3.)

x	$\frac{\pi}{4}$	$\frac{3\pi}{4}$	$\frac{5\pi}{4}$	$\frac{7\pi}{4}$
$y = \sin x$				

Table 4 (View the completed Table 4.)

x	$\frac{\pi}{3}$	$\frac{2\pi}{3}$	$\frac{4\pi}{3}$	$\frac{5\pi}{3}$
$y = \sin x$				

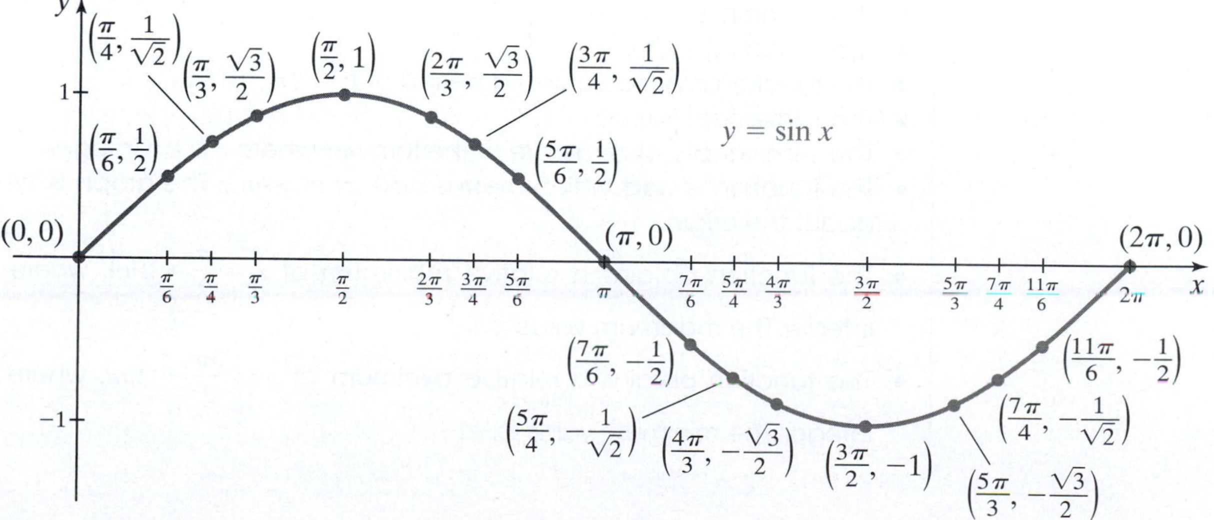

Figure 1 The graph of $y = \sin x$ on the interval $[0, 2\pi]$

(eText Screens 7.1-1–7.1-67)

The graph of $y = \sin x$ in Figure 1 is sketched on the interval $[0, 2\pi]$ and shows only one cycle of the graph. If we were to create another table of values choosing values of x between 2π and 4π (or between -2π and 0) to plot several more ordered pairs, we would see that the graph repeats itself. In fact, on every interval of length 2π, the graph of $y = \sin x$ repeats itself. We say that the graph of $y = \sin x$ is a **periodic function** with a period of 2π.

> **Definition** Periodic Function
>
> A function f is said to be **periodic** if there is a positive number P such that $f(x + P) = f(x)$ for all x in the domain of f. The smallest number P for which f is periodic is called the **period** of f.

In Figure 2, we see three complete cycles of the graph of $y = \sin x$. As you can see in Figure 2, the graph of $y = \sin x$ repeats itself after every interval of length 2π. That is, for any value of x, $\sin x = \sin(x + 2\pi n)$, where n is any integer.

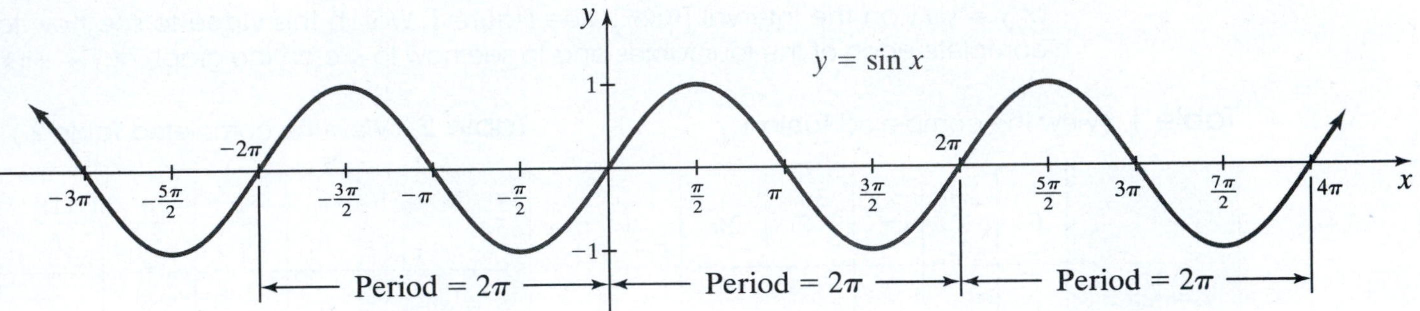

Figure 2 Three complete cycles of $y = \sin x$

Note that the domain of $y = \sin x$ includes all real numbers, whereas the range includes all values between and including -1 and 1. We now list these and several other characteristics of the sine function. Watch this **video** for a detailed description of the characteristics of the sine function.

> **Characteristics of the Sine Function**
>
> - The domain is $(-\infty, \infty)$.
> - The range is $[-1, 1]$.
> - The function is periodic with a period of $P = 2\pi$.
> - The y-intercept is 0.
> - The x-intercepts, or **zeros**, are of the form $n\pi$ where n is an integer.
> - The function is **odd**, which means $\sin(-x) = -\sin x$. The graph is symmetric about the origin.
> - The function obtains a **relative maximum** at $x = \frac{\pi}{2} + 2\pi n$, where n is an integer. The maximum value is 1.
> - The function obtains a **relative minimum** at $x = \frac{3\pi}{2} + 2\pi n$, where n is an integer. The minimum value is -1.

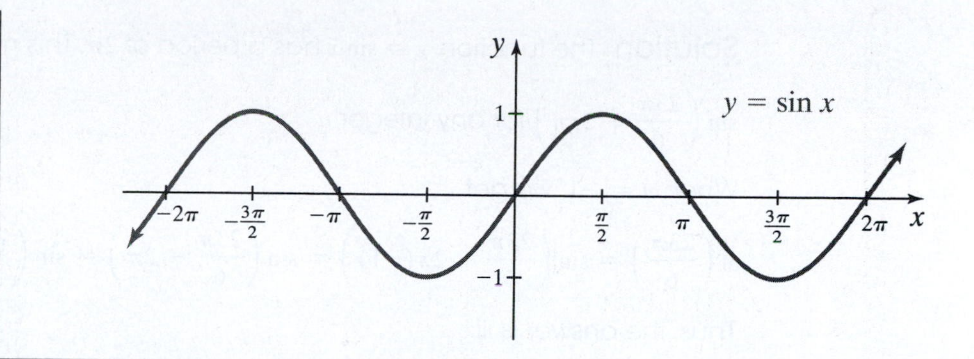

Example 1 Understanding the Graph of $y = \sin x$

Using the graph of $y = \sin x$, list all values of x on the interval $\left[-3\pi, \frac{7\pi}{4}\right]$ that satisfy the ordered pair $(x, 0)$.

Solution First, sketch the graph of $y = \sin x$ on the interval $\left[-3\pi, \frac{7\pi}{4}\right]$.

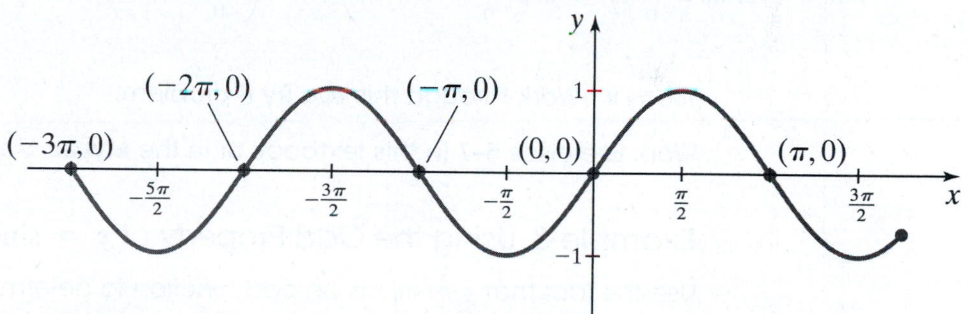

The values of x that satisfy the ordered pair $(x, 0)$ are precisely the x-intercepts of $y = \sin x$. The x-intercepts have the form $n\pi$, where n is an integer. We see from the graph that the x-intercepts on the interval $\left[-3\pi, \frac{7\pi}{4}\right]$ are $-3\pi, -2\pi, -\pi, 0$, and π. ●

You Try It Work through this You Try It problem.

Work Exercises 1–4 in this textbook or in the MyLab Math Study Plan.

Example 2 Using the Periodic Property of $y = \sin x$

Use the periodic property of $y = \sin x$ to determine which of the following expressions is equivalent to $\sin\left(\frac{23\pi}{6}\right)$.

i. $\sin\left(\frac{\pi}{6}\right)$ ii. $\sin\left(\frac{5\pi}{6}\right)$ iii. $\sin\left(\frac{11\pi}{6}\right)$ iv. $\sin\left(\frac{13\pi}{6}\right)$

Solution The function $y = \sin x$ has a period of 2π. This means that $\sin\left(\frac{23\pi}{6}\right) = \sin\left(\frac{23\pi}{6} + 2\pi n\right)$ for any integer n.

When $n = -1$, we get

$$\sin\left(\frac{23\pi}{6}\right) = \sin\left(\frac{23\pi}{6} + 2\pi(-1)\right) = \sin\left(\frac{23\pi}{6} - 2\pi\right) = \sin\left(\frac{23\pi}{6} - \frac{12\pi}{6}\right) = \sin\left(\frac{11\pi}{6}\right).$$

Thus, the answer is iii. ●

In the previous example, we could find infinitely many more expressions that are equivalent to $\sin\left(\frac{23\pi}{6}\right)$ using the periodic property of $y = \sin x$. Below are just a few.

For $n = 1$, $\sin\left(\frac{23\pi}{6}\right) = \sin\left(\frac{23\pi}{6} + 2\pi(1)\right) = \sin\left(\frac{23\pi}{6} + \frac{12\pi}{6}\right) = \sin\left(\frac{35\pi}{6}\right)$.

For $n = -2$, $\sin\left(\frac{23\pi}{6}\right) = \sin\left(\frac{23\pi}{6} + 2\pi(-2)\right) = \sin\left(\frac{23\pi}{6} - 4\pi\right) = \sin\left(\frac{23\pi}{6} - \frac{24\pi}{6}\right) = \sin\left(-\frac{\pi}{6}\right)$.

For $n = 2$, $\sin\left(\frac{23\pi}{6}\right) = \sin\left(\frac{23\pi}{6} + 2\pi(2)\right) = \sin\left(\frac{23\pi}{6} + 4\pi\right) = \sin\left(\frac{23\pi}{6} + \frac{24\pi}{6}\right) = \sin\left(\frac{47\pi}{6}\right)$.

You Try It **Work through this You Try It problem.**

Work Exercises 5–7 in this textbook or in the MyLab Math Study Plan.

Example 3 Using the Odd Property of $y = \sin x$

Use the fact that $y = \sin x$ is an odd function to determine which of the following expressions is equivalent to $-\sin\left(\frac{9\pi}{16}\right)$.

i. $\sin\left(-\frac{9\pi}{16}\right)$ ii. $\sin\left(\frac{9\pi}{16}\right)$ iii. $-\sin\left(-\frac{9\pi}{16}\right)$

Solution Because $y = \sin x$ is an **odd function**, we know that $-\sin x = \sin(-x)$. Therefore, if $x = \frac{9\pi}{16}$, then $-\sin\left(\frac{9\pi}{16}\right) = \sin\left(-\frac{9\pi}{16}\right)$. Thus, the answer is i. ●

You Try It **Work through this You Try It problem.**

Work Exercise 8 in this textbook or in the MyLab Math Study Plan.

OBJECTIVE 2 UNDERSTANDING THE GRAPH OF THE COSINE FUNCTION AND ITS PROPERTIES

To sketch the graph of $y = \cos x$, we once again create several tables of values and plot the corresponding ordered pairs. You should take some time and complete Table 5, Table 6, Table 7, and Table 8 shown below. Watch this **video** to see how to complete each of these four tables and to see how to sketch the graph of $y = \cos x$. One complete cycle of the graph of $y = \cos x$ is sketched in **Figure 3**.

Table 5 (View the completed **Table 5**.)

x	0	$\frac{\pi}{2}$	π	$\frac{3\pi}{2}$	2π
$y = \cos x$					

Table 6 (View the completed **Table 6**.)

x	$\frac{\pi}{6}$	$\frac{5\pi}{6}$	$\frac{7\pi}{6}$	$\frac{11\pi}{6}$
$y = \cos x$				

Table 7 (View the completed **Table 7**.)

x	$\frac{\pi}{4}$	$\frac{3\pi}{4}$	$\frac{5\pi}{4}$	$\frac{7\pi}{4}$
$y = \cos x$				

Table 8 (View the completed **Table 8**.)

x	$\frac{\pi}{3}$	$\frac{2\pi}{3}$	$\frac{4\pi}{3}$	$\frac{5\pi}{3}$
$y = \cos x$				

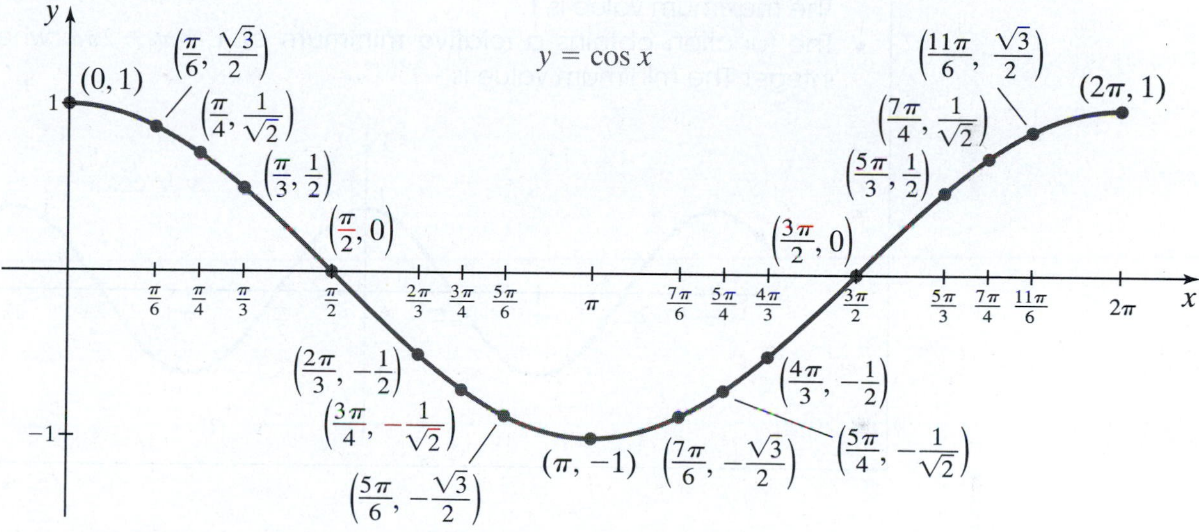

Figure 3 The graph of $y = \cos x$ on the interval $[0, 2\pi]$

Just like $y = \sin x$, the function $y = \cos x$ is periodic with a period of 2π. In Figure 4, we see three complete cycles of the graph of $y = \cos x$.

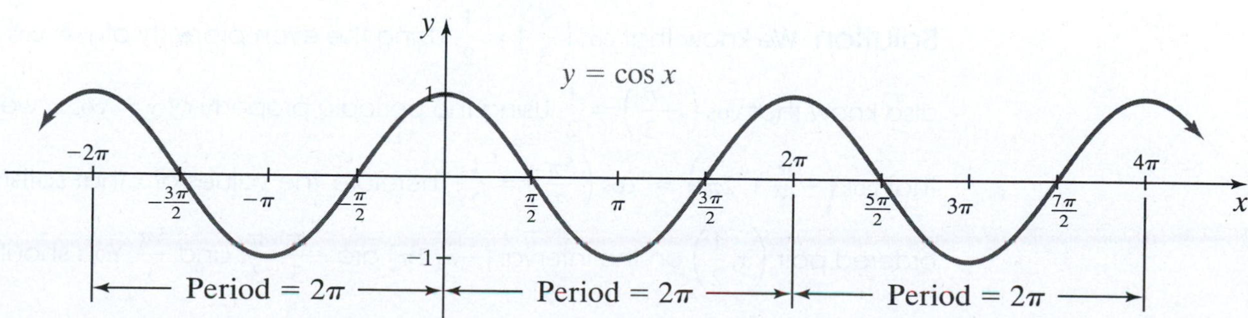

Figure 4 Three complete cycles of $y = \cos x$

As you can see in Figure 4, the domain of $y = \cos x$ includes all real numbers, whereas the range includes all values between and including -1 and 1. We now list these and several other characteristics of the cosine function. Watch this **video** for a detailed description of the characteristics of the cosine function.

Characteristics of the Cosine Function

- The domain is $(-\infty, \infty)$.
- The range is $[-1, 1]$.
- The function is periodic with a period of $P = 2\pi$.
- The y-intercept is 1.
- The x-intercepts, or **zeros**, are of the form $(2n + 1) \cdot \frac{\pi}{2}$, where n is an integer.
- The function is **even**, which means $\cos(-x) = \cos x$. The graph is symmetric about the y-axis.
- The function obtains a **relative maximum** at $x = 2\pi n$, where n is an integer. The maximum value is 1.
- The function obtains a **relative minimum** at $x = \pi + 2\pi n$, where n is an integer. The minimum value is -1.

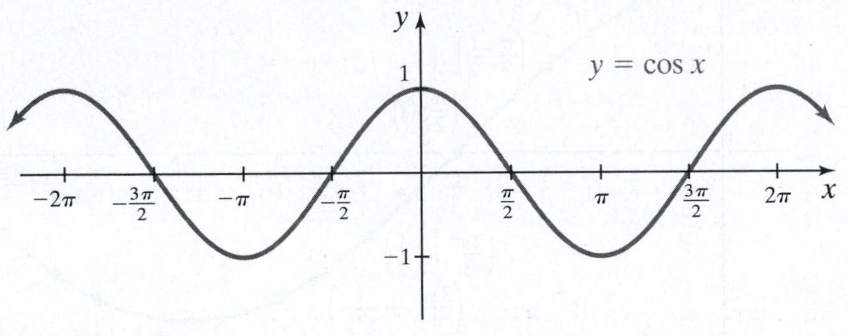

Example 4 Understanding the Graph of $y = \cos x$

Using the graph of $y = \cos x$, list all values of x on the interval $[-\pi, 2\pi]$ that satisfy the ordered pair $\left(x, \frac{1}{2}\right)$.

Solution We know that $\cos\left(\frac{\pi}{3}\right) = \frac{1}{2}$. Using the **even** property of $y = \cos x$, we also know that $\cos\left(-\frac{\pi}{3}\right) = \frac{1}{2}$. Using the periodic property of $y = \cos x$, we know that $\cos\left(-\frac{\pi}{3} + 2\pi\right) = \cos\left(\frac{5\pi}{3}\right) = \frac{1}{2}$. Therefore, the values of x that satisfy the ordered pair $\left(x, \frac{1}{2}\right)$ on the interval $[-\pi, 2\pi]$ are $-\frac{\pi}{3}, \frac{\pi}{3}$, and $\frac{5\pi}{3}$. You should verify that we have accurately found all x- values by sketching the graph of $y = \cos x$ on the interval $[-\pi, 2\pi]$.

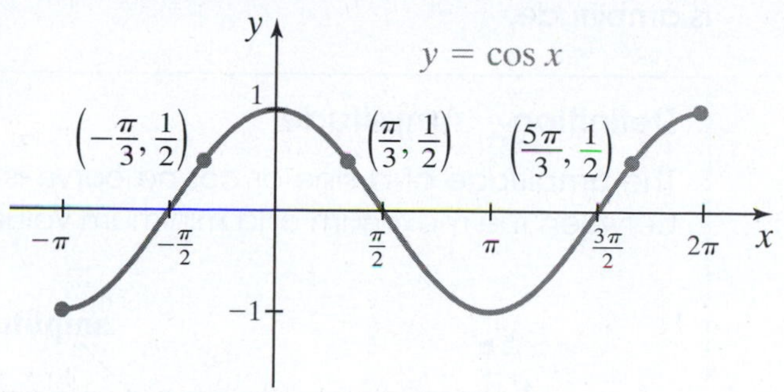

You Try It Work through this You Try It problem.

Work Exercises 9–16 in this textbook or in the MyLab Math Study Plan.

Now that you have a basic understanding of the graphs of $y = \sin x$ and $y = \cos x$, we will introduce variations of these graphs. As we go forward, we will focus on sketching one cycle of each graph with the understanding that this cycle repeats itself indefinitely. We will focus on sketching five points on every graph. These five points evenly divide one cycle of the sine or cosine curve into fourths, or quarters. Therefore, we will call these five points **quarter points**. You should memorize the coordinates of these quarter points because they will help you sketch complicated sine and cosine curves. We now sketch one cycle of the sine and cosine curves using these five quarter points over the interval $[0, 2\pi]$.

The Five Quarter Points of $y = \sin x$

1. $(0, 0)$ **2.** $\left(\frac{\pi}{2}, 1\right)$ **3.** $(\pi, 0)$ **4.** $\left(\frac{3\pi}{2}, -1\right)$
5. $(2\pi, 0)$

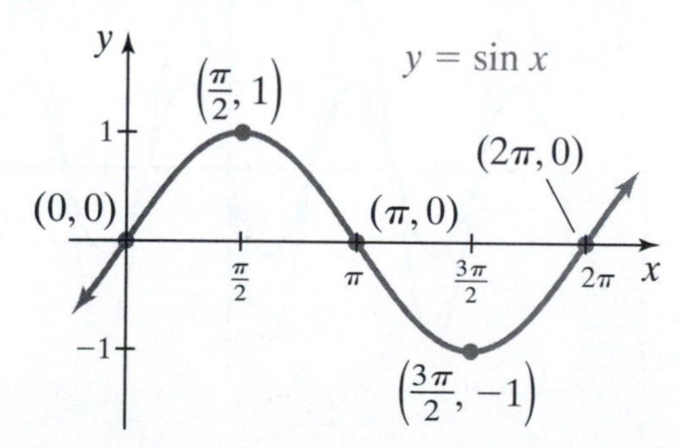

The Five Quarter Points of $y = \cos x$

1. $(0, 1)$ **2.** $\left(\frac{\pi}{2}, 0\right)$ **3.** $(\pi, -1)$ **4.** $\left(\frac{3\pi}{2}, 0\right)$
5. $(2\pi, 1)$

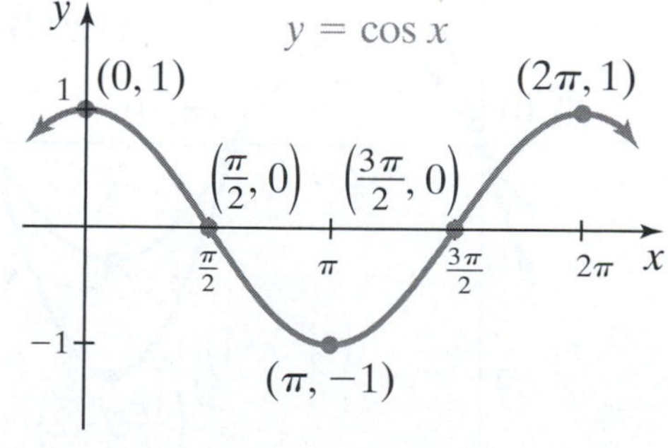

OBJECTIVE 3 SKETCHING GRAPHS OF THE FORM $y = A\sin x$ AND $y = A\cos x$

There are four factors that affect the graph of a sine or cosine curve. The first is **amplitude**.

Definition Amplitude

The **amplitude** of a sine or cosine curve is the measure of half the distance between the maximum and minimum values.

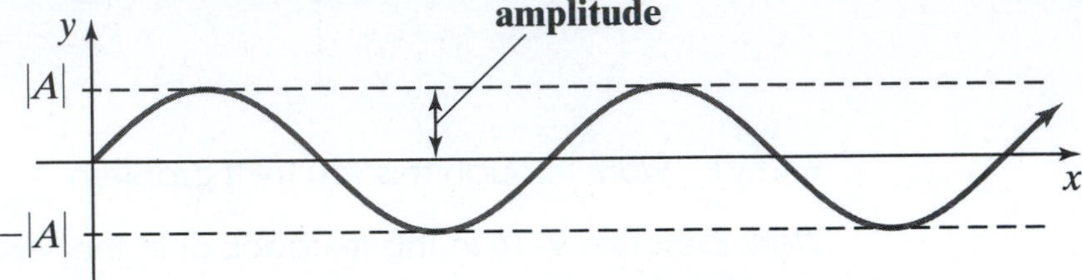

Trigonometric functions of the form $y = A\sin x$ and $y = A\cos x$ have an amplitude of $|A|$ and a range of $[-|A|, |A|]$.

Knowing the amplitude will help us to determine the maximum and minimum values of functions of the form $y = A\sin x$ or $y = A\cos x$. For example, the function $y = 2\sin x$ has an amplitude of $|A| = |2| = 2$ and a range of $[-2, 2]$. Thus, the maximum value of $y = 2\sin x$ is 2 and the minimum value is -2. The graphs of $y = \sin x$ and $y = 2\sin x$ are shown on the same rectangular coordinate system in **Figure 5**.

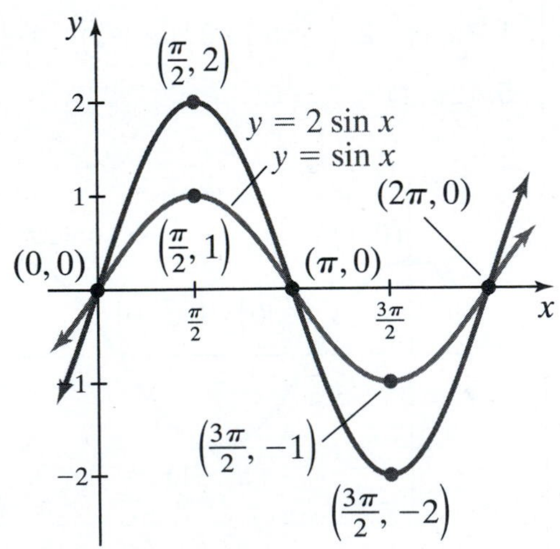

Figure 5 This illustrates one cycle of the graphs of $y = \sin x$ and $y = 2\sin x$.

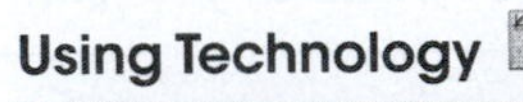

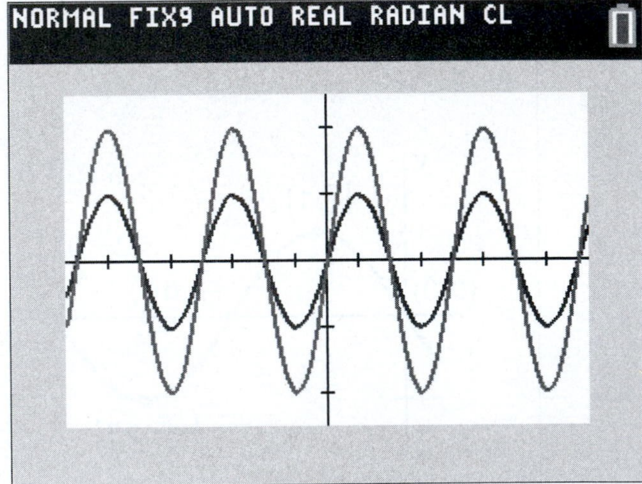

Figure 6 This illustrates several cycles of the graphs of $y = \sin x$ and $y = 2\sin x$ using a graphing utility.

Notice in Figure 5 that the x-coordinates of all quarter points on each graph are the same. This is because the period of each function is 2π. We simply multiplied each y-coordinate of the quarter points of $y = \sin x$ by 2 to obtain the five quarter points used to sketch the graph of $y = 2\sin x$. Because the graph of $y = \sin x$ has three quarter points with y-coordinates of 0, multiplying those values by 2 causes no change in the values of the corresponding y-coordinates on the graph of $y = 2\sin x$.

Example 5 Sketching a graph of the form $y = A\cos x$

Determine the amplitude and range of $y = -\frac{2}{3}\cos x$ and then sketch the graph.

Solution The amplitude of $y = -\frac{2}{3}\cos x$ is $A = \left|-\frac{2}{3}\right| = \frac{2}{3}$. Thus the range is $\left[-\frac{2}{3}, \frac{2}{3}\right]$.

To sketch the graph, we multiply the y-coordinates of the quarter points of $y = \cos x$ by $-\frac{2}{3}$ to obtain the quarter points for $y = -\frac{2}{3}\cos x$.

Quarter points for $y = \cos x$: $(0, 1)$, $\left(\frac{\pi}{2}, 0\right)$, $(\pi, -1)$, $\left(\frac{3\pi}{2}, 0\right)$, $(2\pi, 1)$

Quarter points for $y = -\frac{2}{3}\cos x$: $\left(0, -\frac{2}{3}\right)$, $\left(\frac{\pi}{2}, 0\right)$, $\left(\pi, \frac{2}{3}\right)$, $\left(\frac{3\pi}{2}, 0\right)$, $\left(2\pi, -\frac{2}{3}\right)$

One cycle of the graphs of $y = \cos x$ and $y = -\frac{2}{3}\cos x$ is sketched in Figure 7.

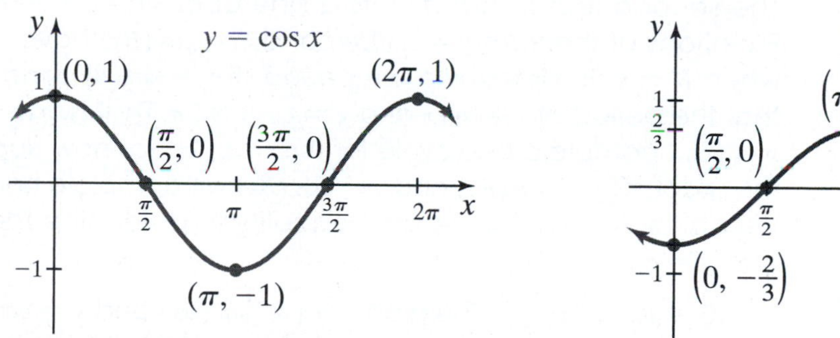

Figure 7 One cycle of the graphs of $y = \cos x$ and $y = -\frac{2}{3}\cos x$

In the previous example, we sketched the graph of $y = -\frac{2}{3}\cos x$ using amplitude and the five quarter points. Note that we could have used two transformations to sketch the graph of $y = -\frac{2}{3}\cos x$. First, recall that if $A > 1$, then the graph of a function $y = Af(x)$ can be obtained by **vertically stretching** the graph of the function $y = f(x)$ by a factor of A. If $0 < A < 1$, the graph of $y = Af(x)$ can be obtained by **vertically compressing** the graph of $y = Af(x)$ by a factor of A.

You may want to review the vertical stretch transformation or vertical compression transformation that was discussed in **Section 3.4** or click on the animations below to review this concept. (Note that in Section 3.4 and in the animations we used a lowercase "a" and in this section we are using a capital "A.")

To review the concept of a vertical stretch, watch this **animation**.

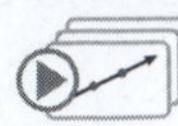

To review the concept of a vertical compression, watch this **animation**.

Also, recall that the graph of $y = -f(x)$ can be obtained by reflecting the graph of $y = f(x)$ about the x-axis. To review the concept of a reflection about the x-axis, watch this **animation** or read about this transformation that was discussed in **Section 3.4**.

Therefore, using transformations, we can sketch the graph of $y = -\frac{2}{3}\cos x$ by first vertically compressing the graph of $y = \cos x$ by a factor of $\frac{2}{3}$ to obtain the graph of $y = \frac{2}{3}\cos x$. Then we can reflect the graph of $y = \frac{2}{3}\cos x$ about the x-axis, producing the graph of $y = -\frac{2}{3}\cos x$. To see how to sketch the graph of $y = -\frac{2}{3}\cos x$ using transformations, read these **steps**.

You Try It **Work through this You Try It problem.**

Work Exercises 17–26 in this textbook or in the MyLab Math Study Plan.

OBJECTIVE 4 SKETCHING GRAPHS OF THE FORM $y = \sin(Bx)$ AND $y = \cos(Bx)$

The second factor that affects a sine or cosine curve is a change in period. Functions of the form $y = \sin(Bx)$ and $y = \cos(Bx)$ have a period other than 2π when $B \neq 1$. To determine the period of $y = \sin(Bx)$ and $y = \cos(Bx)$, first recall that the period of $y = \sin x$ and $y = \cos x$ is 2π. That is, the graphs of $y = \sin x$ and $y = \cos x$ complete one cycle for $0 \le x \le 2\pi$. For now, suppose that $B > 0$. Then the graph of $y = \sin(Bx)$ and $y = \cos(Bx)$ will complete one cycle for $0 \le Bx \le 2\pi$. We can solve this three-part inequality to determine the period of $y = \sin(Bx)$ and $y = \cos(Bx)$.

$0 \le Bx \le 2\pi$	The graph of $y = \sin(Bx)$ and $y = \cos(Bx)$ will complete one cycle for values of x satisfying this inequality.
$\frac{0}{B} \le \frac{Bx}{B} \le \frac{2\pi}{B}$	Divide each part by B. The direction of the inequality remains the same because $B > 0$.
$0 \le x \le \frac{2\pi}{B}$	Simplify.

The graph of the functions $y = \sin(Bx)$ and $y = \cos(Bx)$ will complete one cycle for $0 \le x \le \frac{2\pi}{B}$, which means that the period, P, of $y = \sin(Bx)$ and $y = \cos(Bx)$ is $P = \frac{2\pi}{B}$.

Determining the Period of $y = \sin(Bx)$ and $y = \cos(Bx)$

The **period** of a sine or cosine curve is equal to $P = \frac{2\pi}{B}$, where $B > 0$.

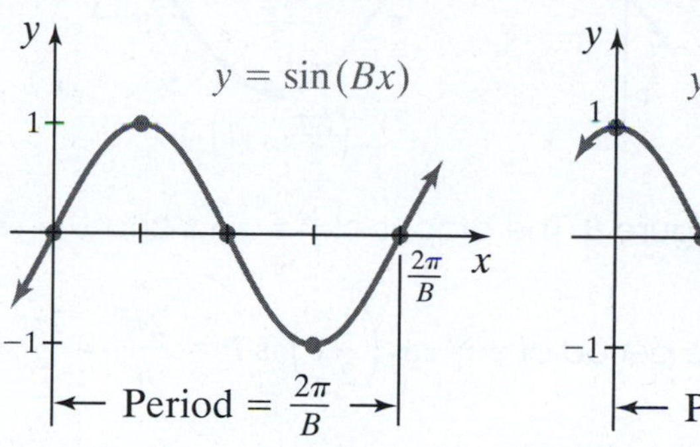

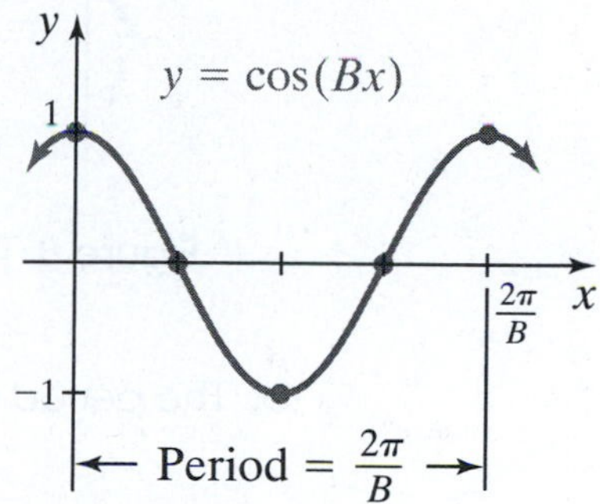

For $B > 0$, the graphs of $y = \sin(Bx)$ and $y = \cos(Bx)$ can be obtained by subdividing the interval $\left[0, \frac{2\pi}{B}\right]$ into four equal subintervals of length $\frac{2\pi}{B} \div 4$ and by plotting **five quarter points.** The y-coordinate of each quarter point of $y = \sin(Bx)$ and $y = \cos(Bx)$ has the same value as the y-coordinate of each quarter point of $y = \sin x$ and by $y = \cos x$, respectively.

Example 6 Sketching graphs of the form $y = \sin(Bx)$ and $y = \cos(Bx)$

Determine the period and sketch the graph of each function.

a. $y = \sin(2x)$ **b.** $y = \cos\left(\frac{1}{2}x\right)$ **c.** $y = \sin(\pi x)$

Solution

a. The period of $y = \sin(2x)$ is $P = \frac{2\pi}{B} = \frac{2\pi}{2} = \pi$. Next, we divide the interval $[0, \pi]$ into four equal subintervals of length $\pi \div 4 = \frac{\pi}{4}$ by starting with 0 and adding $\frac{\pi}{4}$ to the x-coordinate of each successive quarter point. Thus, the x-coordinates of the five quarter points are $0, \frac{\pi}{4}, \frac{\pi}{2}, \frac{3\pi}{4}$, and π. The y-coordinates of the five quarter points of $y = \sin(2x)$ will be the same as the y-coordinates of the five quarter points of $y = \sin x$.

Quarter points for $y = \sin x$: $(0, 0)$, $\left(\frac{\pi}{2}, 1\right)$, $(\pi, 0)$, $\left(\frac{3\pi}{2}, -1\right)$, $(2\pi, 0)$

Quarter points for $y = \sin(2x)$: $(0, 0)$, $\left(\frac{\pi}{4}, 1\right)$, $\left(\frac{\pi}{2}, 0\right)$, $\left(\frac{3\pi}{4}, -1\right)$, $(\pi, 0)$

One cycle of the graphs of $y = \sin x$ and $y = \sin(2x)$ is sketched in **Figure 8**. Watch this **interactive video** to see how to sketch the graph of $y = \sin(2x)$.

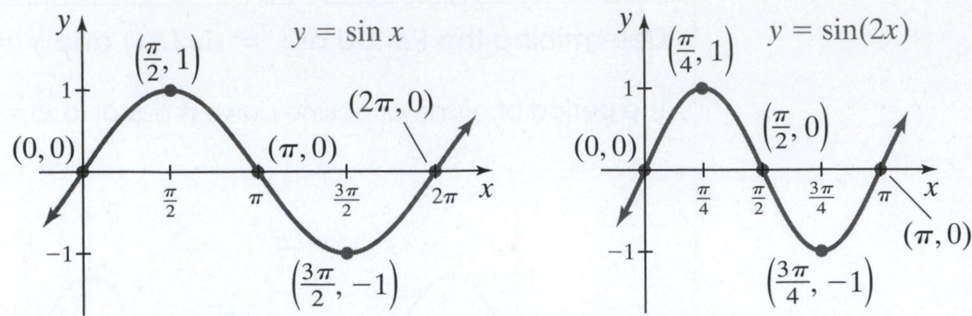

Figure 8 The graphs of $y = \sin x$ and $y = \sin(2x)$

b. The period of $y = \cos\left(\frac{1}{2}x\right)$ is $P = \frac{2\pi}{B} = \frac{2\pi}{\frac{1}{2}} = 2\pi \cdot \frac{2}{1} = 4\pi$. Next, we divide the interval $[0, 4\pi]$ into four equal subintervals of length $4\pi \div 4 = \pi$ by starting with 0 and adding π to the x-coordinate of each successive quarter point. The x-coordinates of the five quarter points are $0,\ \pi,\ 2\pi,\ 3\pi$, and 4π. The y-coordinates of the five quarter points of $y = \cos\left(\frac{1}{2}x\right)$ will be the same as the y-coordinates of the five quarter points of $y = \cos x$.

Quarter points for $y = \cos x$: $(0, 1),\ \left(\frac{\pi}{2}, 0\right),\ (\pi, -1),\ \left(\frac{3\pi}{2}, 0\right),\ (2\pi, 1)$

Quarter points for $y = \cos\left(\frac{1}{2}x\right)$: $(0, 1),\ (\pi, 0),\ (2\pi, -1),\ (3\pi, 0),\ (4\pi, 1)$

Watch this **interactive video** to see how to sketch the graph of $y = \cos\left(\frac{1}{2}x\right)$ as seen below in Figures 9 and 10.

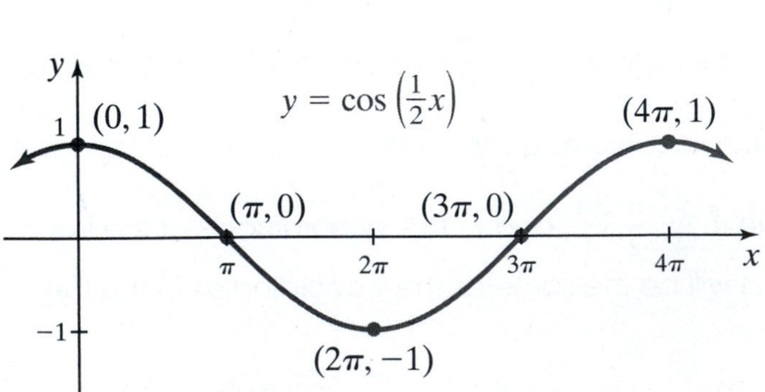

Figure 9 This illustrates one cycle of the graphs of $y = \cos\left(\frac{1}{2}x\right)$.

Using Technology 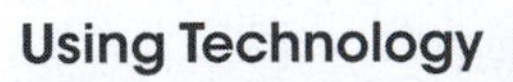

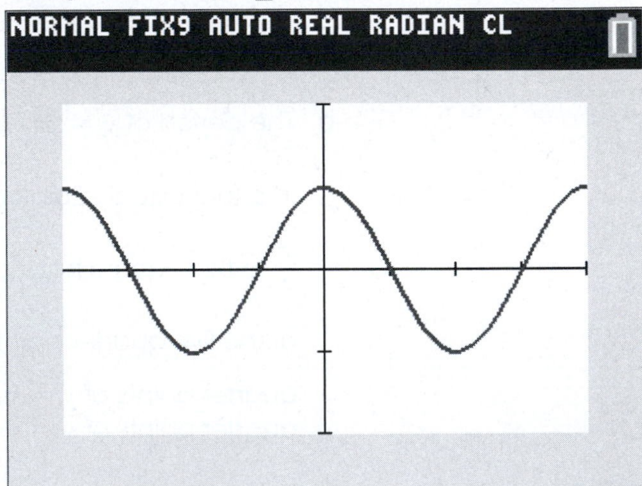

Figure 10 This illustrates two cycles of the graphs of $y = \cos\left(\frac{1}{2}x\right)$ using a graphing utility.

c. The period of $y = \sin(\pi x)$ is $P = \frac{2\pi}{B} = \frac{2\pi}{\pi} = 2$. Next, we divide the interval $[0, 2]$ into four equal subintervals of length $\frac{2}{4} = \frac{1}{2}$ by starting with 0 and adding $\frac{1}{2}$ to the x-coordinate of each successive quarter point. Thus, the x-coordinates of the five quarter points are $0, \frac{1}{2}, 1, \frac{3}{2}$, and 2. The y-coordinates of the five quarter points of $y = \sin(\pi x)$ will be the same as the y-coordinates of the five quarter points of $y = \sin x$. The graph of $y = \sin(\pi x)$ can be seen in **Figures 11 and 12**.

Quarter points for $y = \sin x$: $(0, 0)$, $\left(\frac{\pi}{2}, 1\right)$, $(\pi, 0)$, $\left(\frac{3\pi}{2}, -1\right)$, $(2\pi, 0)$

Quarter points for $y = \sin(\pi x)$: $(0, 0)$, $\left(\frac{1}{2}, 1\right)$, $(1, 0)$, $\left(\frac{3}{2}, -1\right)$, $(2, 0)$

Watch this **interactive video** see the complete solution to this example.

Using Technology

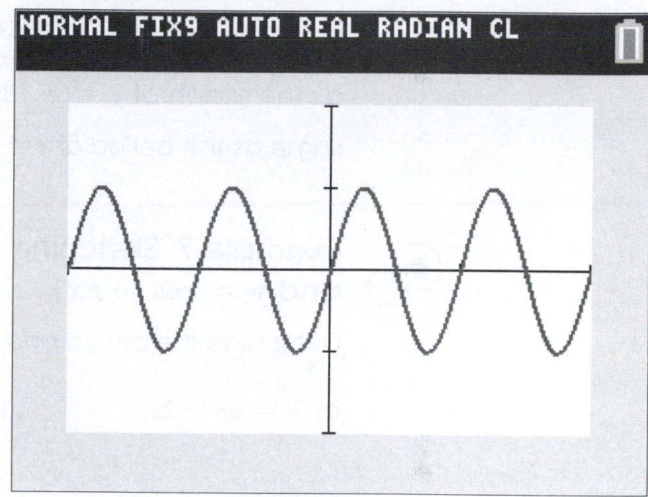

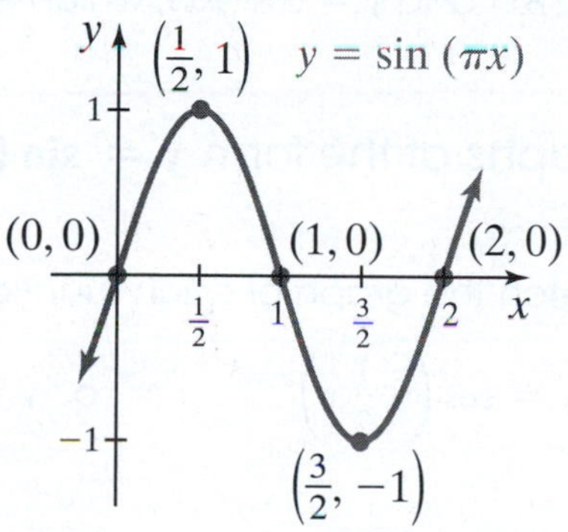

Figure 11 This illustrates one cycle of the graph of $y = \sin(\pi x)$.

Figure 12 This illustrates several cycles of the graph of $y = \sin(\pi x)$ using a graphing utility.

It is noteworthy that we could have used transformation techniques to sketch the three graphs from Example 6. In Example 6a we sketched a function of the form $y = \sin(Bx)$, where $B = 2$. Because $B > 1$, the graph of a function $y = f(Bx)$ can be obtained by **horizontally compressing** the graph of the function $y = f(x)$ by dividing each x-coordinate of $y = f(x)$ by B.

Similarly, if $0 < B < 1$, the graph of $y = f(Bx)$ can be obtained by **horizontally stretching** the graph of $y = f(x)$ by a factor of B.

You may want to review the horizontal stretch transformation or horizontal compression transformation that was discussed in **Section 3.4** or click on the animations below to review this concept. (Note that in Section 3.4 and in the animations we used a lowercase "a" and in this section we are using a capital "B.")

To review the concept of a horizontal compression, watch this **animation**.

To review the concept of a horizontal stretch, watch this **animation**.

To see how to sketch each of the functions from Example 6 using transformations, click on the Show Graph link located under the following three functions.

$y = \sin(2x)$	$y = \cos\left(\frac{1}{2}x\right)$	$y = \sin(\pi x)$
Show Graph	Show Graph	Show Graph

Now that we know how to sketch the graphs of $y = \sin(Bx)$ and $y = \cos(Bx)$, where $B > 0$, we can use the even and odd properties of cosine and sine to sketch the graphs of $y = \sin(-Bx)$ and $y = \cos(-Bx)$.

Recall that $y = \sin x$ is an **odd** function. This means that $\sin(-x) = -\sin x$ for all x in the domain of $y = \sin x$. It follows that

$$\sin(-Bx) = -\sin(Bx).$$

Similarly, $y = \cos x$ is an **even** function. This means that $\cos(-x) = \cos x$ for all x in the domain of $y = \cos x$. It follows that

$$\cos(-Bx) = \cos(Bx).$$

Therefore, when $B > 0$, we can sketch the graph of $y = \sin(-Bx)$ by simply sketching the graph of $y = -\sin(Bx)$. This graph can be obtained by reflecting the graph of $y = \sin(Bx)$ about the x-axis. The graph of $y = \cos(-Bx)$ is the exact same graph as the graph of $y = \cos(Bx)$. The period of $y = \sin(-Bx)$ and $y = \cos(-Bx)$ is the same as the period of $y = \sin(Bx)$ and $y = \cos(Bx)$, which is $P = \frac{2\pi}{B}$.

Example 7 Sketching graphs of the form $y = \sin(-Bx)$ and $y = \cos(-Bx)$

Determine the period and sketch the graph of each function.

a. $y = \sin(-2x)$ **b.** $y = \cos\left(-\frac{1}{2}x\right)$ **c.** $y = \sin(-\pi x)$

Solution

a. We use the odd property of the sine function to rewrite $y = \sin(-2x)$ as $y = -\sin(2x)$. We have already sketched the graph of $y = \sin(2x)$ in **Example 6a**. See **Figure 8**. We did this by finding the period of $y = \sin(2x)$ to be $P = \frac{2\pi}{B} = \frac{2\pi}{2} = \pi$ and then dividing the interval $[0, \pi]$ into four equal subintervals of length $\pi \div 4 = \frac{\pi}{4}$. This produced x-coordinates of the five quarter points of $y = \sin(2x)$, namely $0, \frac{\pi}{4}, \frac{\pi}{2}, \frac{3\pi}{4}$, and π.

To sketch the graph of $y = -\sin(2x)$ using the graph of $y = \sin(2x)$, we can multiply the y-coordinates of the quarter points of $y = \sin(2x)$ by -1 to obtain the quarter points for $y = -\sin(2x)$ or we can simply reflect the graph of $y = \sin(2x)$ about the x-axis. See Figure 13.

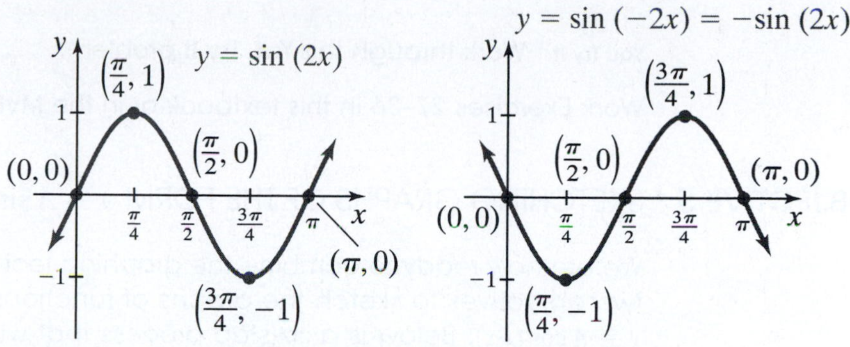

Figure 13 The graphs of $y = \sin(2x)$ and $y = \sin(-2x)$

b. Because the cosine function is even, the graph of $y = \cos\left(-\frac{1}{2}x\right)$ is the same as the graph of $y = \cos\left(\frac{1}{2}x\right)$, which we have already sketched in Example 6. See **Figure 9**.

We sketch the graph of $y = \cos\left(-\frac{1}{2}x\right) = \cos\left(\frac{1}{2}x\right)$ again in Figure 14.

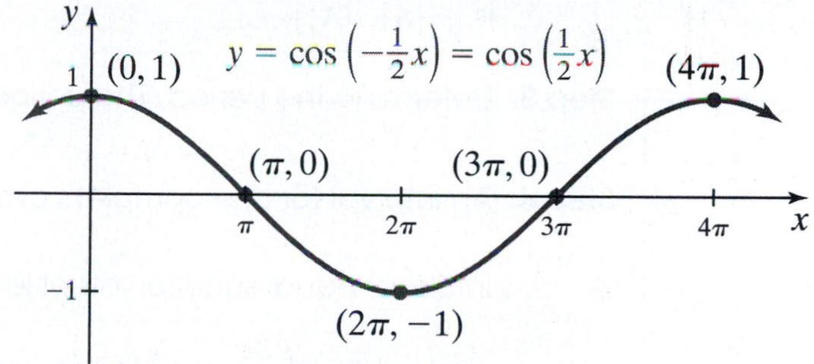

Figure 14 One cycle of the graph of $y = \cos\left(-\frac{1}{2}x\right) = \cos\left(\frac{1}{2}x\right)$

c. Try sketching the graph of $y = \sin(-\pi x)$ on your own. Watch this **interactive video** when you have finished to see if you are correct. ●

We could have used transformation techniques to sketch the graphs of the two functions from Example 7.

Recall that the graph of $y = f(-x)$ can be obtained by reflecting the graph of $y = f(x)$ about the y-axis. To review this transformation, watch this short **animation**.

In Example 7 we could have sketched the graphs of $y = \sin(-2x)$, $y = \cos\left(-\frac{1}{2}x\right)$, and $y = \sin(-\pi x)$ by first using a horizontal compression (or stretch) followed by a reflection about the y-axis to sketch both of these graphs. To see how to sketch these two functions using transformations, click on the Show Graph links below.

$y = \sin(-2x)$	$y = \cos\left(-\frac{1}{2}x\right)$	$y = \sin(-\pi x)$
Show Graph	Show Graph	Show Graph

You Try It Work through this You Try It problem.

Work Exercises 27–36 in this textbook or in the MyLab Math Study Plan.

OBJECTIVE 5 SKETCHING GRAPHS OF THE FORM $y = A\sin(Bx)$ AND $y = A\cos(Bx)$

We are now ready to combine the graphing techniques learned in the previous two objectives to sketch the graphs of functions of the form $y = A\sin(Bx)$ and $y = A\cos(Bx)$. Below is a six-step process that will help you sketch functions of this form.

Steps for Sketching Functions of the Form $y = A\sin(Bx)$ and $y = A\cos(Bx)$

Step 1. If $B < 0$, use the even and odd properties of the sine and cosine function to rewrite the function in an equivalent form such that $B > 0$.

We now use this new form to determine A and B.

Step 2. Determine the amplitude and range. The amplitude is $|A|$. The range is $[-|A|, |A|]$.

Step 3. Determine the period. The period is $P = \frac{2\pi}{B}$.

Step 4. An interval for one complete cycle is $\left[0, \frac{2\pi}{B}\right]$. Subdivide this interval into four equal subintervals of length $\frac{2\pi}{B} \div 4$ by starting with 0 and adding $\left(\frac{2\pi}{B} \div 4\right)$ to the x-coordinate of each successive quarter point.

Step 5. Multiply the y-coordinates of the quarter points of $y = \sin x$ or $y = \cos x$ by A to determine the y-coordinates of the corresponding quarter points for the new graph.

Step 6. Connect the quarter points to obtain one complete cycle.

Example 8 Sketching graphs of the form $y = A\sin(Bx)$ and $y = A\cos(Bx)$

Use the six-step process outlined in this section to sketch each graph.

a. $y = 3\sin(4x)$ **b.** $y = -2\cos\left(\frac{1}{3}x\right)$ **c.** $y = -6\sin\left(-\frac{\pi x}{2}\right)$

Solution

a. Step 1. For the function $y = 3\sin(4x)$, we see that $B = 4 > 0$. Therefore, we do not need to use the odd property of sine to rewrite the function, and we continue to step 2.

Step 2. The amplitude of $y = 3\sin(4x)$ is $|A| = |3| = 3$. Therefore, the range is $[-3, 3]$. The maximum value is 3 and the minimum value is -3.

Step 3. The period is $P = \frac{2\pi}{B} = \frac{2\pi}{4} = \frac{\pi}{2}$.

Step 4. An interval for one complete cycle of the graph is $\left[0, \frac{\pi}{2}\right]$. We divide this interval into four equal subintervals of length $\frac{\pi}{2} \div 4 = \frac{\pi}{2} \cdot \frac{1}{4} = \frac{\pi}{8}$ by starting with 0 and adding $\frac{\pi}{8}$ to the x-coordinate of each successive quarter point. Thus, the x-coordinates of the five quarter points are $0, \frac{\pi}{8}, \frac{\pi}{4}, \frac{3\pi}{8}$, and $\frac{\pi}{2}$.

Step 5. Multiply the y-coordinate of each of the five quarter points of $y = \sin x$ by 3 to obtain the y-coordinates of the corresponding quarter points of $y = 3\sin(4x)$. The y-coordinates of the five quarter points are $0, 3, 0, -3$, and 0.

Step 6. From the information gathered from the previous steps, we get the following.

Quarter points for $y = \sin x$:	$(0, 0)$,	$\left(\frac{\pi}{2}, 1\right)$,	$(\pi, 0)$,	$\left(\frac{3\pi}{2}, -1\right)$,	$(2\pi, 0)$
Quarter points for $y = 3\sin(4x)$:	$(0, 0)$,	$\left(\frac{\pi}{8}, 3\right)$,	$\left(\frac{\pi}{4}, 0\right)$,	$\left(\frac{3\pi}{8}, -3\right)$,	$\left(\frac{\pi}{2}, 0\right)$

We connect these quarter points with a smooth curve to complete the sketch of $y = 3\sin(4x)$. See Figures 15 and 16.

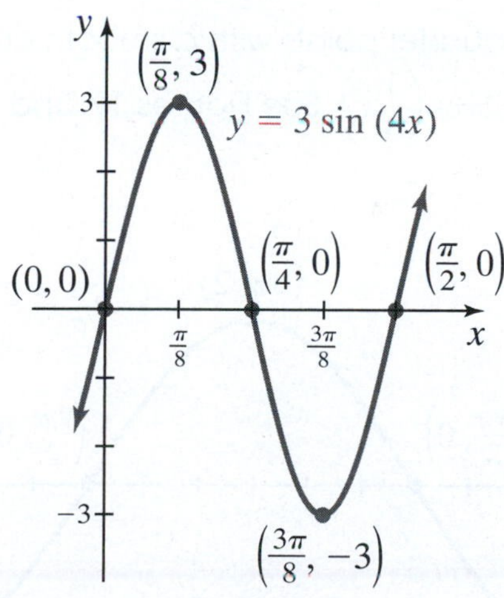

Figure 15 This illustrates one cycle of the graph of $y = 3\sin(4x)$.

Using Technology

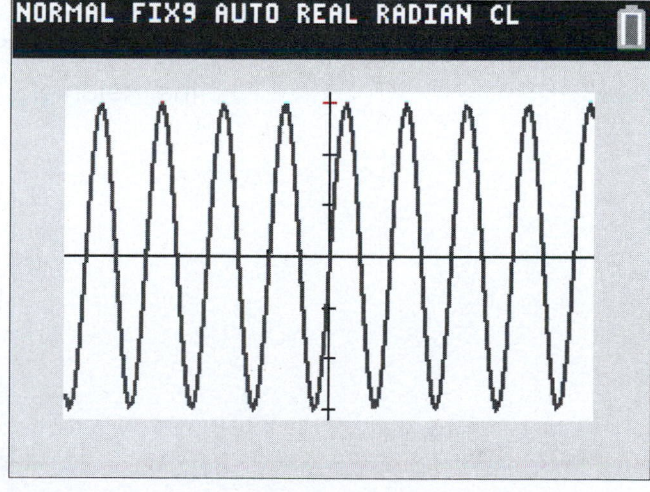

Figure 16 This illustrates several cycles of the graph of $y = 3\sin(4x)$ using a graphing utility.

You may want to work through this **interactive video** to see how to sketch the graph of $y = 3\sin(4x)$.

b. Step 1. For the function $y = -2\cos\left(\frac{1}{3}x\right)$, we see that $B = \frac{1}{3} > 0$. Therefore, we do not need to use the even property of cosine to rewrite the function, and we continue to step 2.

Step 2. The amplitude of $y = -2\cos\left(\frac{1}{3}x\right)$ is $|A| = |-2| = 2$. Therefore, the range is $[-2, 2]$. The maximum value is 2 and the minimum value is -2.

Step 3. The period is $P = \frac{2\pi}{B} = \frac{2\pi}{\frac{1}{3}} = 2\pi \cdot 3 = 6\pi$.

Step 4. An interval for one complete cycle of the graph is $[0, 6\pi]$. We divide this interval into four equal subintervals of length $6\pi \div 4 = \frac{6\pi}{4} = \frac{3\pi}{2}$ by starting with 0 and adding $\frac{3\pi}{2}$ to the x-coordinate of each successive quarter point. Thus, the x-coordinates of the five quarter points are $0, \frac{3\pi}{2}, 3\pi, \frac{9\pi}{2}$, and 6π.

Step 5. Multiply the y-coordinate of each of the five quarter points of $y = \cos x$ by -2 to obtain the y-coordinates of the corresponding quarter points of $y = -2\cos\left(\frac{1}{3}x\right)$. The y-coordinates of the five quarter points are $-2, 0, 2, 0$, and -2.

Step 6. From the information gathered from the previous steps, we get the following.

Quarter points for $y = \cos x$: $(0, 1)$, $\left(\frac{\pi}{2}, 0\right)$, $(\pi, -1)$, $\left(\frac{3\pi}{2}, 0\right)$, $(2\pi, 1)$

Quarter points for $y = -2\cos\left(\frac{1}{3}x\right)$: $(0, -2)$, $\left(\frac{3\pi}{2}, 0\right)$, $(3\pi, 2)$, $\left(\frac{9\pi}{2}, 0\right)$, $(6\pi, -2)$

We connect these quarter points with a smooth curve to complete the sketch of $y = -2\cos\left(\frac{1}{3}x\right)$. See **Figures 17 and 18**.

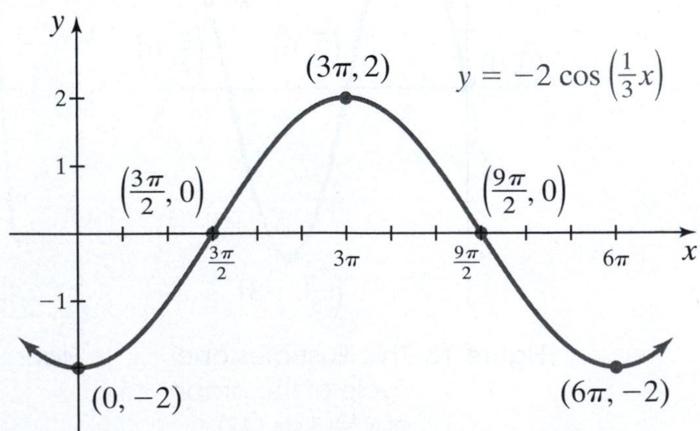

Figure 17 This illustrates one cycle of the graph of $y = -2\cos\left(\frac{1}{3}x\right)$.

(eText Screens 7.1-1–7.1-67)

Using Technology

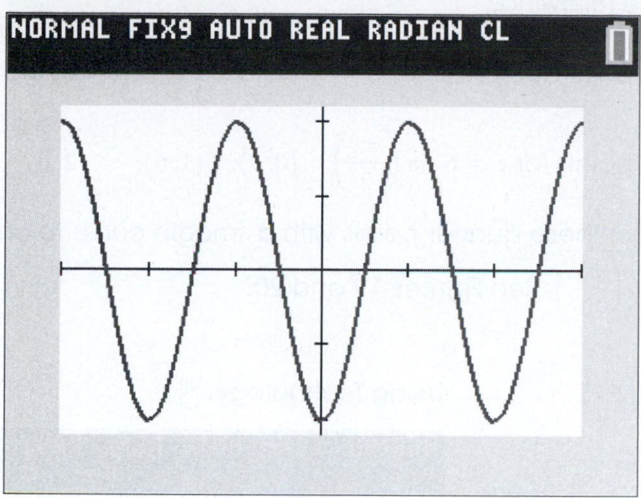

Figure 18 This illustrates several cycles of the graph of $y = -2\cos\left(\frac{1}{3}x\right)$ using a graphing utility.

You may want to work through this **interactive video** to see how to sketch the graph of $y = -2\cos\left(\frac{1}{3}x\right)$.

c. Step 1. For the function $y = -6\sin\left(-\frac{\pi x}{2}\right)$, we see that $B = -\frac{\pi}{2} < 0$. Therefore, we use the odd property of sine to rewrite the function using a positive value of B.

$y = -6\sin\left(-\frac{\pi x}{2}\right)$ Write the original function.

$y = 6\sin\left(\frac{\pi x}{2}\right)$ Use the odd property of $y = \sin x$: $\sin(-x) = -\sin x$.

We now continue through the steps using the function $y = 6\sin\left(\frac{\pi x}{2}\right)$ with $A = 6$ and $B = \frac{\pi}{2}$.

Step 2. The amplitude of $y = 6\sin\left(\frac{\pi x}{2}\right)$ is $|A| = |6| = 6$. Therefore, the range is $[-6, 6]$. The maximum value is 6 and the minimum value is -6.

Step 3. The period of $y = 6\sin\left(\frac{\pi x}{2}\right)$ is $P = \frac{2\pi}{B} = 2\pi \div \frac{\pi}{2} = 2\pi \cdot \frac{2}{\pi} = 4$.

Step 4. An interval for one complete cycle of the graph is $[0, 4]$. We divide this interval into four equal subintervals of length $\frac{4}{4} = 1$ by starting with 0 and adding 1 to the x-coordinate of each successive quarter point. Thus, the x-coordinates of the five quarter points are $0, 1, 2, 3,$ and 4.

Step 5. Multiply the y-coordinates of the five quarter points of $y = \sin x$ by 6 to obtain the y-coordinates of the corresponding quarter points of $y = 6\sin\left(\frac{\pi x}{2}\right)$. The y-coordinates of the five quarter points are $0, 6, 0, -6,$ and 0.

(eText Screens 7.1-1–7.1-67)

Step 6. From the information gathered from the previous steps, we get the following.

Quarter points for $y = \sin x$: $(0, 0)$, $\left(\frac{\pi}{2}, 1\right)$, $(\pi, 0)$, $\left(\frac{3\pi}{2}, -1\right)$, $(2\pi, 0)$

Quarter points for $y = 6\sin\left(\frac{\pi x}{2}\right)$: $(0, 0)$, $(1, 6)$, $(2, 0)$, $(3, -6)$, $(4, 0)$

We connect these quarter points with a smooth curve to complete the sketch of $y = 6\sin\left(\frac{\pi x}{2}\right)$. See **Figures 19** and **20**.

$y = -6\sin\left(-\frac{\pi x}{2}\right)$ or $y = 6\sin\left(\frac{\pi x}{2}\right)$

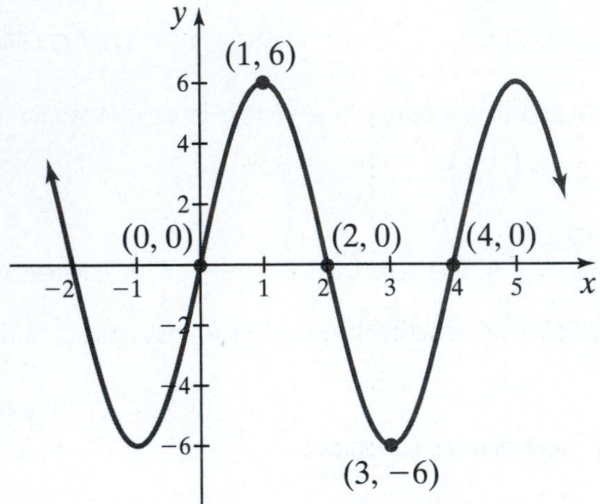

Figure 19 This illustrates the graph of $y = -6\sin\left(-\frac{\pi x}{2}\right)$ or $y = 6\sin\left(\frac{\pi x}{2}\right)$.

Using Technology

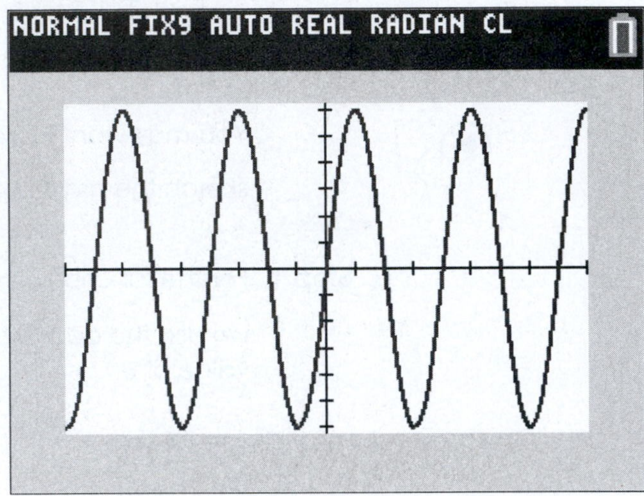

Figure 20 This illustrates several cycles of the graph of $y = -6\sin\left(-\frac{\pi x}{2}\right)$ or $y = 6\sin\left(\frac{\pi x}{2}\right)$ using a graphing utility. ●

You may want to work through this **interactive video** to see how to sketch the graph of $y = -6\sin\left(-\frac{\pi x}{2}\right)$.

You Try It Work through this You Try It problem.

Work Exercises 37–52 in this textbook or in the MyLab Math Study Plan.

Take some time sketching functions of the form $y = A\sin(Bx)$ and $y = A\cos(Bx)$ by experimenting with the Guided Visualization below. Click on the Guided Visualization and choose your own values of A and B.

Sketching Functions of the form $y = A\sin(Bx)$ and $y = A\cos(Bx)$

OBJECTIVE 6 DETERMINE THE EQUATION OF A FUNCTION OF THE FORM $y = A\sin(Bx)$ OR $y = A\cos(Bx)$ GIVEN ITS GRAPH

We now know how to sketch the graphs of $y = A\sin(Bx)$ and $y = A\cos(Bx)$. Suppose that we are given a graph whose function is given by $y = A\sin(Bx)$ or $y = A\cos(Bx)$ for $B > 0$. (We will assume that the value of B is positive. Otherwise, there could be more than one correct answer because of the even nature of the cosine function.) To determine the proper function we must determine three things:

1. Is the given graph a representation of a function of the form $y = A\sin(Bx)$ or $y = A\cos(Bx)$?
2. What is the value of B?
3. What is the value of A?

Therefore, we can establish the following three steps for determining the proper function.

Steps for Determining the Equation of a Function of the Form $y = A\sin(Bx)$ or $y = A\cos(Bx)$ Given the Graph

Step 1. Determine whether the given graph is a representation of a function of the form $y = A\sin(Bx)$, or $y = A\cos(Bx)$. Choose $y = A\sin(Bx)$ if the graph passes through the origin or choose $y = A\cos(Bx)$ if the graph does not pass through the origin.*

Step 2. Determine the period $P = \frac{2\pi}{B}$ then use the period to determine the value of $B > 0$.

Step 3. Use the information from the previous two steps then use one of the given points on the graph to solve for A.

Example 9 Determining the Equation of a Function Given the Graph that it Represents

One cycle of the graphs of three trigonometric functions of the from $y = A\sin(Bx)$ or $y = A\cos(Bx)$ for $B > 0$ are given below. Determine the equation of the function represented by each graph.

a.

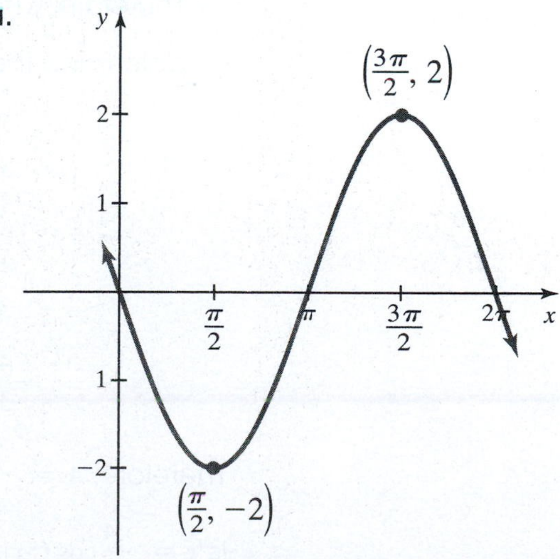

*As long as we restrict the choices of the equations to $y = A\sin(Bx)$ and $y = A\cos(Bx)$ *where* $B > 0$, this will be true. Without these restrictions, there would be more than one correct equation represented by the given graph.

b.

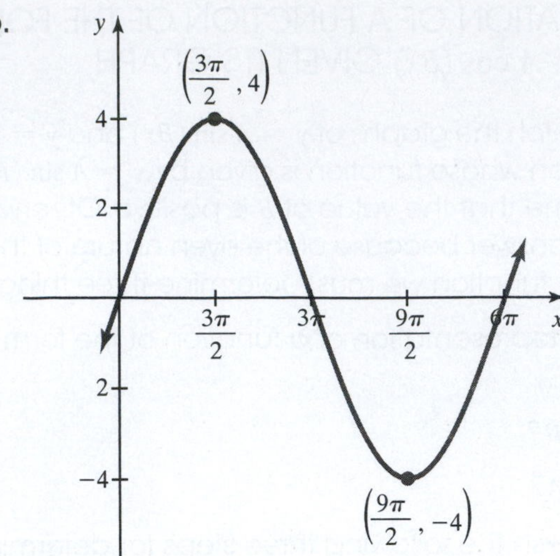

c.

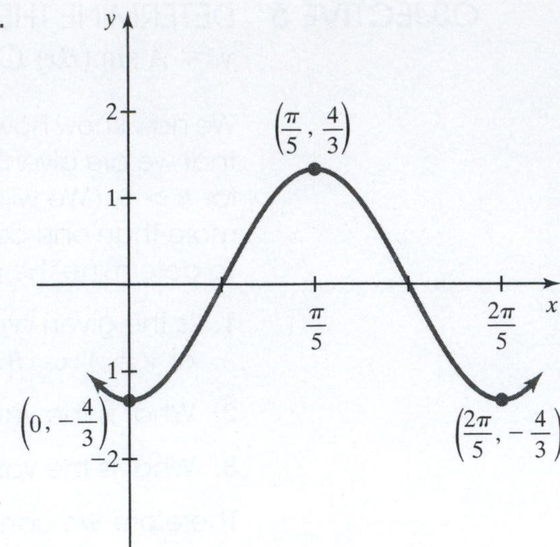

Solution Work through the **interactive video** to verify that the function whose graph is given in part a) is $y = -2\sin x$.

Work through the **interactive video** to verify that the function whose graph is given in part b) is $y = 4\sin\left(\frac{1}{3}x\right)$.

c. **Step 1.** The function whose graph is given in part (c) does **not** pass through the origin. Therefore, a possible representation of this graph is a function of the form $y = A\cos(Bx)$.

Step 2. The graph completes one cycle on the interval $\left[0, \frac{2\pi}{5}\right]$. Thus, the period is $P = \frac{2\pi}{5} = \frac{2\pi}{B}$. Therefore, $B = 5$.

Step 3. The function is of the form $y = A\cos(5x)$ and the graph passes through the point $\left(0, -\frac{4}{3}\right)$.

Therefore, when $x = 0$, the value of y is $-\frac{4}{3}$. We can use this information to determine the value of A.

$y = A\cos(5x)$	Start with the equation $y = A\cos(5x)$.
$-\frac{4}{3} = A\cos\big(5(0)\big)$	Substitute $x = 0$ and $y = -\frac{4}{3}$.
$-\frac{4}{3} = A\cos 0$	Simplify.
$-\frac{4}{3} = A$	Evaluate $\cos 0 = 1$.

Therefore, $A = -\frac{4}{3}$ and the function whose graph is represented in part (c) is $y = -\frac{4}{3}\cos(5x)$.

You Try It Work through this You Try It problem.

Work Exercises 53–64 in this textbook or in the MyLab Math Study Plan.

7.1 Exercises

Skill Check Exercises

In exercises SCE-1 and SCE-2, perform the indicated operations and simplify.

SCE-1. $\frac{2\pi}{5} \div 4$ SCE-2. $\frac{2\pi}{\frac{\pi}{3}}$

1. Sketch the graph of $y = \sin x$, and identify the properties that apply.
   The function is an even function.
   The function is an odd function.
   The function is increasing on the interval $\left(-\pi, -\frac{\pi}{2}\right)$.
   The function is decreasing on the interval $\left(\pi, \frac{3\pi}{2}\right)$.
   The domain is $(-\infty, \infty)$.
   The range is $(-1, 1)$.
   The range is $[-1, 1]$.
   The y-intercept is 0.
   The y-intercept is 1.
   The zeros are of the form $(2n + 1) \cdot \frac{\pi}{2}$, where n is any integer.
   The zeros are of the form $n\pi$, where n is any integer.
   The function obtains a relative maximum at $x = \frac{\pi}{2} + 2\pi n$, where n is an integer.
   The function obtains a relative maximum at $x = 2\pi n$, where n is an integer.
   The function obtains a relative minimum at $x = \frac{3\pi}{2} + 2\pi n$, where n is an integer.
   The function obtains a relative minimum at $x = \pi n$, where n is an integer.

2. Using the graph of $y = \sin x$, list all values of x on the interval $\left[-2\pi, \frac{5\pi}{2}\right]$ that satisfy the ordered pair $\left(x, \frac{1}{2}\right)$.

3. Using the graph of $y = \sin x$, list all values of x on the interval $\left[-\frac{11\pi}{4}, \frac{\pi}{2}\right]$ that satisfy the ordered pair $\left(x, \frac{1}{\sqrt{2}}\right)$.

4. Using the graph of $y = \sin x$, list all values of x on the interval $\left[-\frac{7\pi}{2}, \frac{4\pi}{3}\right]$ that satisfy the ordered pair $\left(x, -\frac{\sqrt{3}}{2}\right)$.

5. Use the periodic property of $y = \sin x$ to determine which of the following expressions is equivalent to $\sin\left(\frac{7\pi}{3}\right)$.

i. $\sin\left(-\frac{\pi}{3}\right)$ ii. $\sin\left(\frac{4\pi}{3}\right)$ iii. $\sin\left(\frac{\pi}{3}\right)$ iv. $\sin\left(\frac{5\pi}{3}\right)$

6. Use the periodic property of $y = \sin x$ to determine which of the following expressions is equivalent to $\sin\left(-\frac{5\pi}{4}\right)$.

i. $\sin\left(-\frac{13\pi}{4}\right)$ ii. $\sin\left(-\frac{\pi}{4}\right)$ iii. $\sin\left(\frac{5\pi}{4}\right)$ iv. $\sin\left(-\frac{9\pi}{4}\right)$

7. Use the periodic property of $y = \sin x$ to determine which of the following expressions is equivalent to $\sin\left(-\frac{25\pi}{6}\right)$.

i. $\sin\left(\frac{\pi}{6}\right)$ ii. $\sin\left(\frac{23\pi}{6}\right)$ iii. $\sin\left(\frac{29\pi}{6}\right)$ iv. $\sin\left(-\frac{31\pi}{6}\right)$

8. Use the fact that $y = \sin x$ is an odd function to determine which of the following expressions is equivalent to $-\sin\left(\frac{9\pi}{16}\right)$.

i. $\sin\left(-\frac{9\pi}{16}\right)$ ii. $\sin\left(\frac{9\pi}{16}\right)$ iii. $-\sin\left(-\frac{9\pi}{16}\right)$ iv. $-\sin\left(-\frac{\pi}{16}\right)$

9. Sketch the graph of $y = \cos x$, and identify the properties that apply.

The function is an even function.

The function is an odd function.

The function is increasing on the interval $\left(-\pi, -\frac{\pi}{2}\right)$.

The function is decreasing on the interval $\left(\pi, \frac{3\pi}{2}\right)$.

The domain is $(-\infty, \infty)$.

The range is $(-1, 1)$.

The range is $[-1, 1]$.

The y-intercept is 0.

The y-intercept is 1.

The zeros are of the form $(2n + 1) \cdot \frac{\pi}{2}$, where n is any integer.

The zeros are of the form $n\pi$, where n is any integer.

The function obtains a relative maximum at $x = \frac{\pi}{2} + 2\pi n$, where n is an integer.

The function obtains a relative maximum at $x = 2\pi n$, where n is an integer.

The function obtains a relative minimum at $x = \frac{3\pi}{2} + 2\pi n$, where n is an integer.

The function obtains a relative minimum at $x = \pi n$, where n is an integer.

10. Using the graph of $y = \cos x$, list all values of x on the interval $\left[-3\pi, \frac{5\pi}{4}\right]$ that satisfy the ordered pair $(x, 0)$.

11. Using the graph of $y = \cos x$, list all values of x on the interval $\left[-\frac{11\pi}{4}, \frac{\pi}{2}\right]$ that satisfy the ordered pair $\left(x, \frac{1}{\sqrt{2}}\right)$.

12. Using the graph of $y = \cos x$, list all values of x on the interval $\left[-\frac{7\pi}{2}, \frac{4\pi}{3}\right]$ that satisfy the ordered pair $\left(x, -\frac{\sqrt{3}}{2}\right)$.

13. Use the periodic property of $y = \cos x$ to determine which of the following expressions is equivalent to $\cos\left(\frac{25\pi}{6}\right)$.

i. $\cos\left(\frac{5\pi}{6}\right)$ ii. $\cos\left(\frac{7\pi}{6}\right)$ iii. $\cos\left(\frac{13\pi}{6}\right)$ iv. $\cos\left(-\frac{7\pi}{6}\right)$

14. Use the periodic property of $y = \cos x$ to determine which of the following expressions is equivalent to $\cos\left(-\frac{4\pi}{3}\right)$.

i. $\cos\left(\frac{2\pi}{3}\right)$ ii. $\cos\left(-\frac{\pi}{3}\right)$ iii. $\cos\left(\frac{7\pi}{3}\right)$ iv. $\cos\left(\frac{5\pi}{3}\right)$

15. Use the periodic property of $y = \cos x$ to determine which of the following expressions is equivalent to $\cos\left(-\frac{17\pi}{4}\right)$.

i. $\cos\left(\frac{5\pi}{4}\right)$ ii. $\cos\left(-\frac{9\pi}{4}\right)$ iii. $\cos\left(\frac{3\pi}{4}\right)$ iv. $\cos\left(-\frac{11\pi}{4}\right)$

16. Use the fact that $y = \cos x$ is an even function to determine which of the following expressions is equivalent to $-\cos\left(\frac{7\pi}{13}\right)$.

i. $-\cos\left(-\frac{7\pi}{13}\right)$ ii. $\cos\left(-\frac{7\pi}{13}\right)$ iii. $\cos\left(\frac{7\pi}{13}\right)$ iv. $\cos\left(-\frac{\pi}{13}\right)$

In Exercises 17–26, determine the amplitude and range of each function and sketch its graph.

SbS 17. $y = 3\sin x$ SbS 18. $y = 4\cos x$ SbS 19. $y = \frac{1}{3}\sin x$ SbS 20. $y = \frac{1}{4}\cos x$

SbS 21. $y = \frac{3}{2}\sin x$ SbS 22. $y = -2\sin x$ SbS 23. $y = -5\cos x$ SbS 24. $y = -\frac{3}{4}\sin x$

SbS 25. $y = -\frac{1}{2}\cos x$ SbS 26. $y = -\frac{5}{4}\cos x$

In Exercises 27–36, determine the period and sketch the graph of each function.

SbS 27. $y = \sin(3x)$ SbS 28. $y = \cos\left(\frac{1}{3}x\right)$ SbS 29. $y = \sin\left(\frac{3}{2}x\right)$

SbS 30. $y = \cos(3\pi x)$ SbS 31. $y = \sin\left(\frac{\pi}{3}x\right)$ SbS 32. $y = \cos(-4x)$

SbS 33. $y = \sin\left(-\frac{1}{4}x\right)$ SbS 34. $y = \cos\left(-\frac{5}{2}x\right)$

SbS 35. $y = \sin(-3\pi x)$ SbS 36. $y = \cos\left(-\frac{\pi}{2}x\right)$

In Exercises 37–52, determine the amplitude, range, and period of each function and sketch the graph.

SbS 37. $y = 2\sin(3x)$ SbS 38. $y = -\frac{5}{3}\cos(4x)$ SbS 39. $y = \frac{3}{4}\sin(-2x)$

SbS 40. $y = -4\cos(-3x)$ SbS 41. $y = 5\cos\left(\frac{1}{2}x\right)$ SbS 42. $y = \frac{1}{2}\sin\left(-\frac{1}{4}x\right)$

SbS 43. $y = -\cos\left(-\frac{1}{5}x\right)$ SbS 44. $y = 6\sin\left(\frac{3}{2}x\right)$ SbS 45. $y = 2\sin\left(-\frac{4}{3}x\right)$

SbS 46. $y = -2\cos\left(\frac{5}{4}x\right)$ SbS 47. $y = -\sin\left(-\frac{5}{2}x\right)$ SbS 48. $y = 3\cos\left(\frac{\pi}{2}x\right)$

SbS 49. $y = -4\sin\left(-\frac{\pi}{3}x\right)$ SbS 50. $y = \frac{5}{4}\cos(\pi x)$ SbS 51. $y = -\frac{1}{2}\sin(-\pi x)$

SbS 52. $y = -6\cos\left(-\frac{2\pi}{3}x\right)$

In Exercises 53–62, one cycle of the graph a trigonometric function of the from $y = A\sin(Bx)$ or $y = A\cos(Bx)$ for $B > 0$ is given. Determine the equation of the function represented by each graph.

SbS 53.

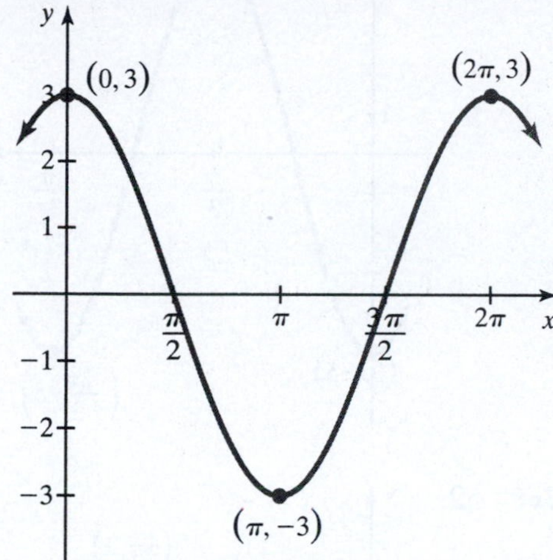

SbS 54.

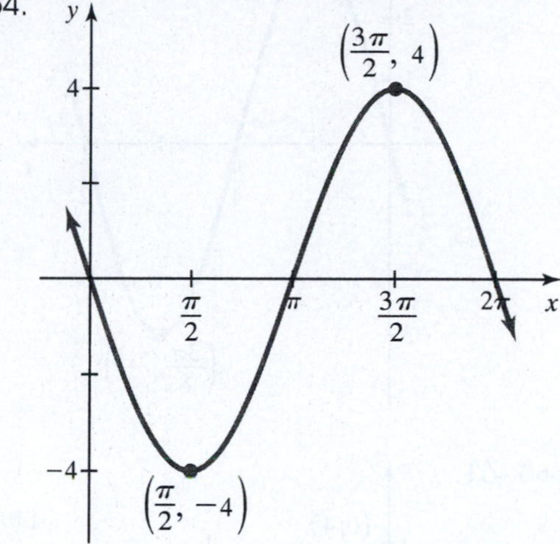

SbS 55.

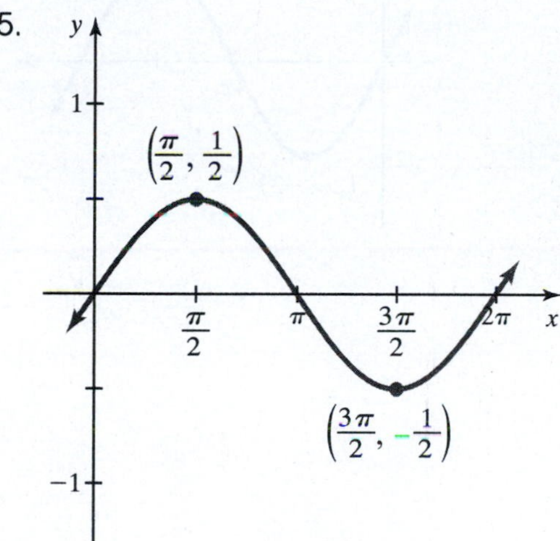

SbS 56.

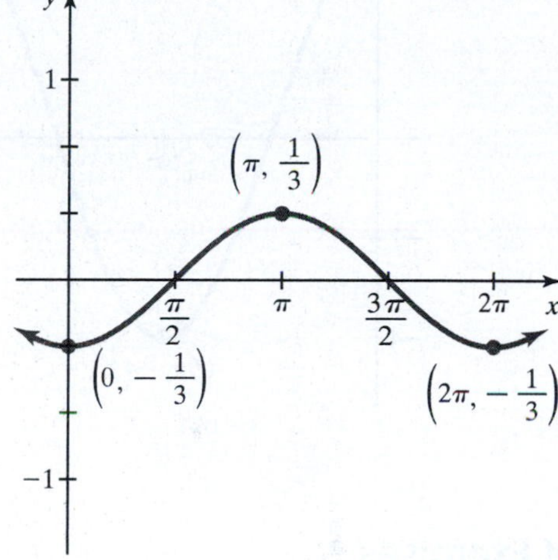

SbS 57.

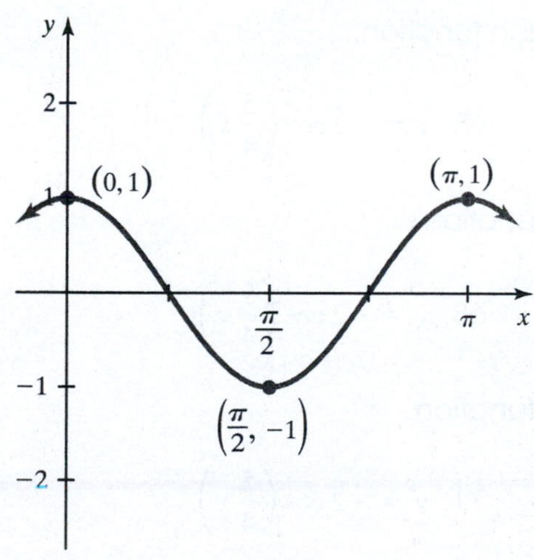

SbS 58.

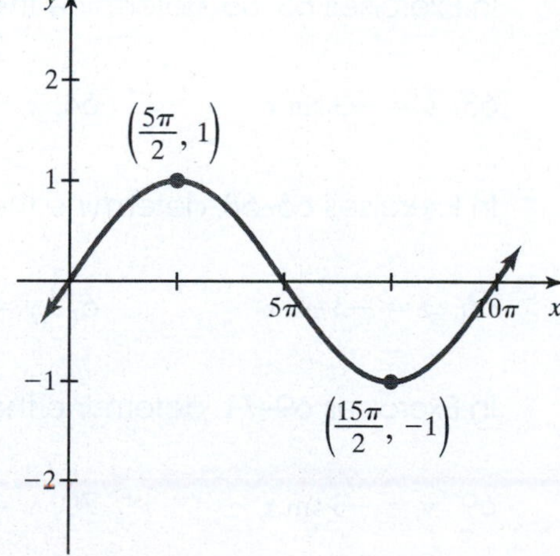

SbS 59.

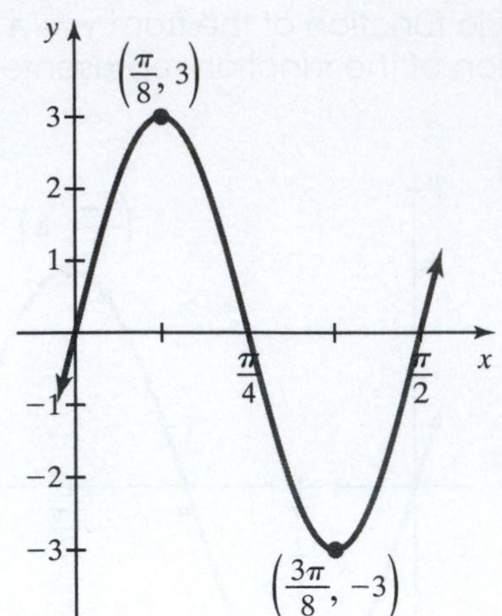

SbS 60.

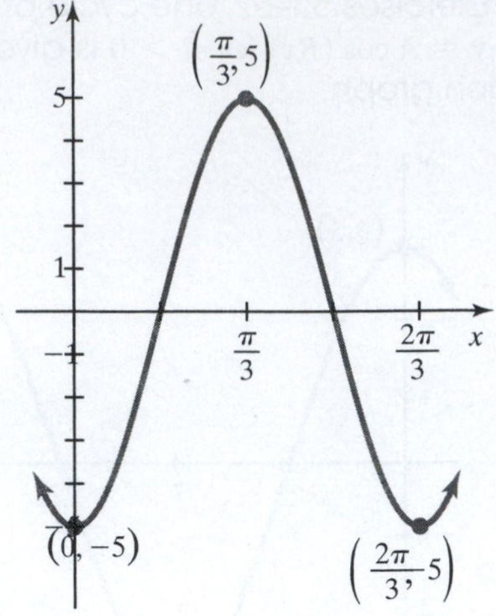

SbS 61.

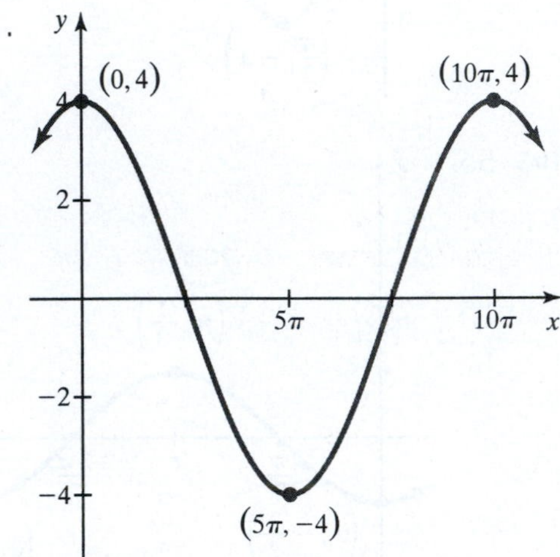

SbS 62.

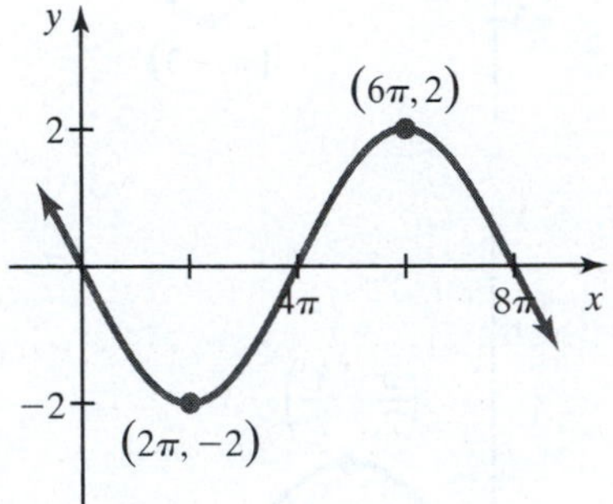

Brief Exercises

In Exercises 63–65, determine the amplitude of each function.

63. $y = -3 \sin x$ 64. $y = \cos(-3x)$ 65. $y = -2\cos\left(\frac{5}{4}x\right)$

In Exercises 66–68, determine the range of each function.

66. $y = -3 \sin x$ 67. $y = \cos(-3x)$ 68. $y = -2\cos\left(\frac{5}{4}x\right)$

In Exercises 69–71, determine the period of each function.

69. $y = -3 \sin x$ 70. $y = \cos(-3x)$ 71. $y = -2\cos\left(\frac{5}{4}x\right)$

In Exercises 72–74, determine the coordinates for the five quarter points of each function that correspond to the five quarter points of $y = \sin x$ or $y = \cos x$.

72. $y = -3 \sin x$ 73. $y = \cos(-3x)$ 74. $y = -2\cos\left(\frac{5}{4}x\right)$

In Exercises 75–90, sketch the graph.

75. $y = 2\sin(3x)$

76. $y = -\frac{5}{3}\cos(4x)$

77. $y = \frac{3}{4}\sin(-2x)$

78. $y = -4\cos(-3x)$

79. $y = 5\cos\left(\frac{1}{2}x\right)$

80. $y = \frac{1}{2}\sin\left(-\frac{1}{4}x\right)$

81. $y = -\cos\left(-\frac{1}{5}x\right)$

82. $y = 6\sin\left(\frac{3}{2}x\right)$

83. $y = 2\sin\left(-\frac{4}{3}x\right)$

84. $y = -2\cos\left(\frac{5}{4}x\right)$

85. $y = -\sin\left(-\frac{5}{2}x\right)$

86. $y = 3\cos\left(\frac{\pi}{2}x\right)$

87. $y = -4\sin\left(-\frac{\pi}{3}x\right)$

88. $y = \frac{5}{4}\cos(\pi x)$

89. $y = -\frac{1}{2}\sin(-\pi x)$

90. $y = -6\cos\left(-\frac{2\pi}{3}x\right)$

In Exercises 91–100, one cycle of the graph a trigonometric function of the from $y = A\sin(Bx)$ or $y = A\cos(Bx)$ for $B > 0$ is given. Determine the equation of the function represented by each graph.

91.

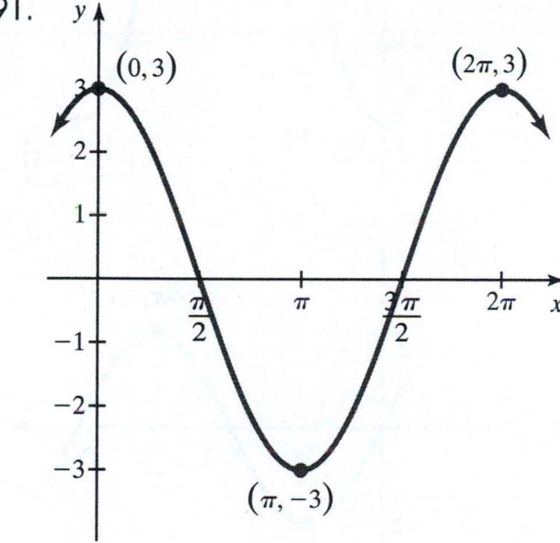

92.

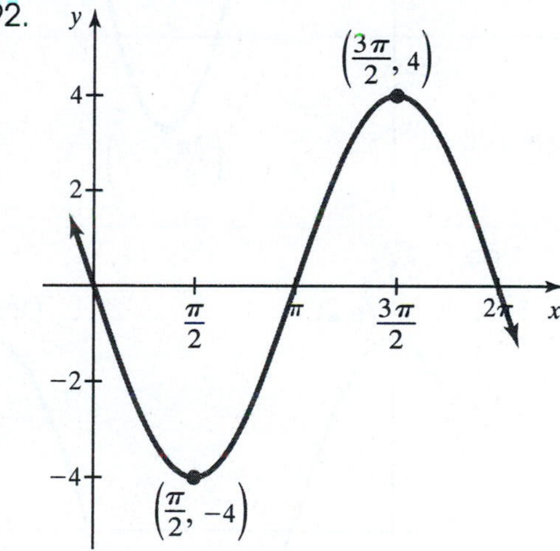

93.

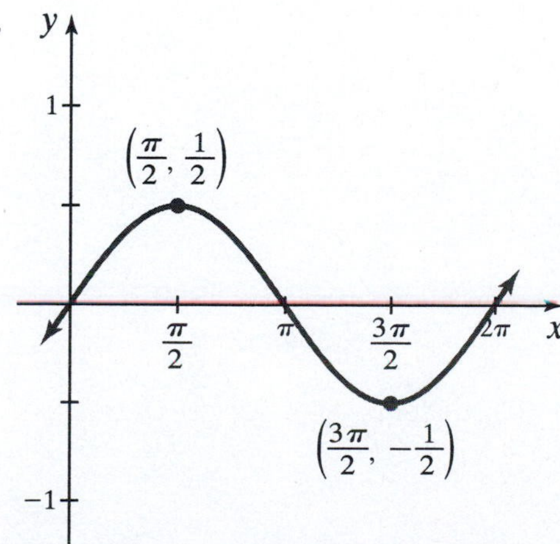

94.

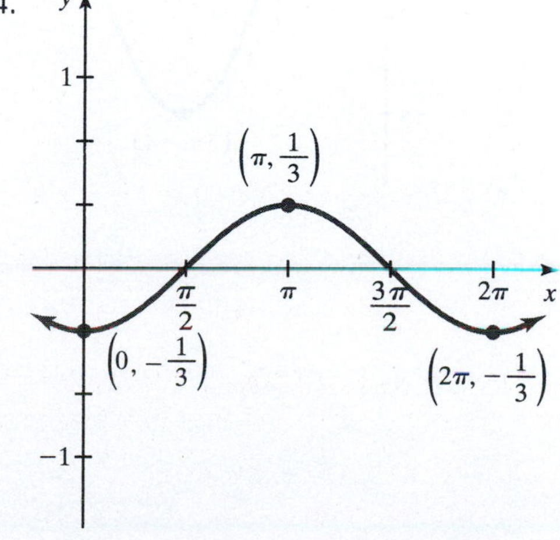

95.

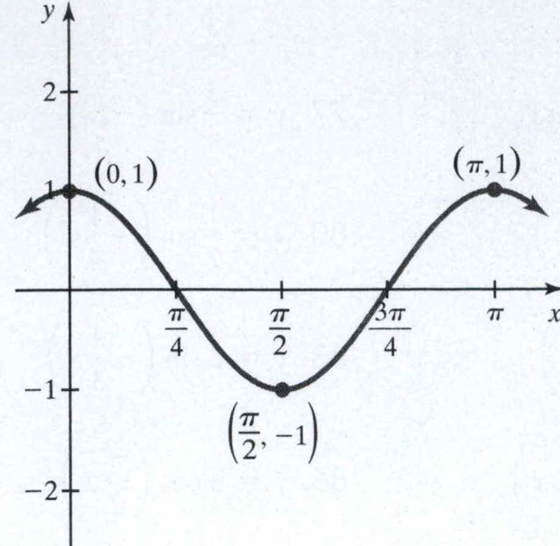

96.

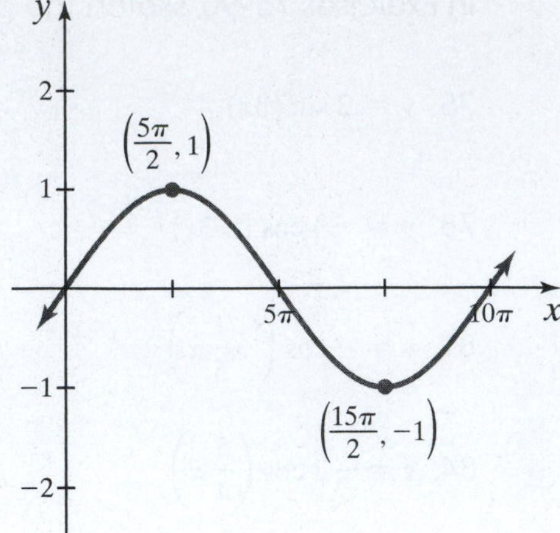

97.

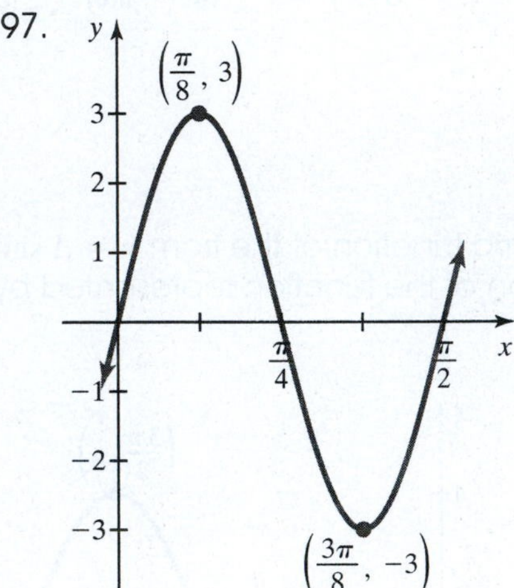

98.

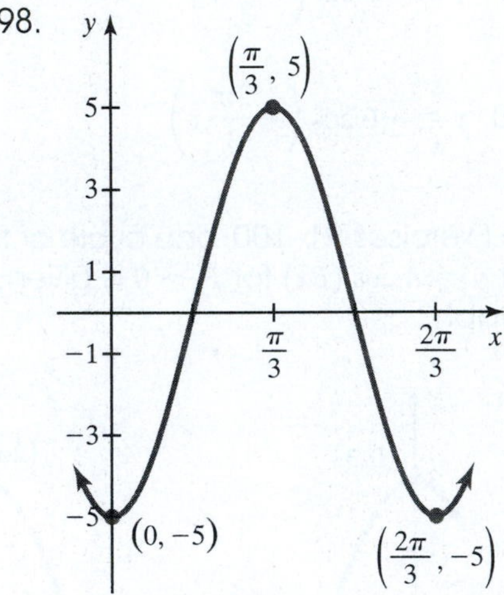

99.

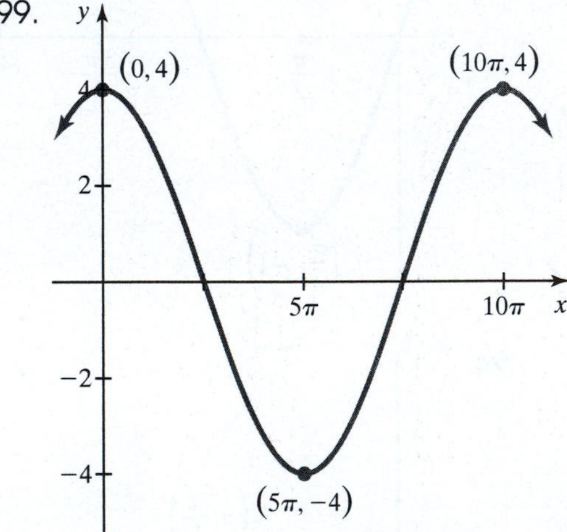

100.

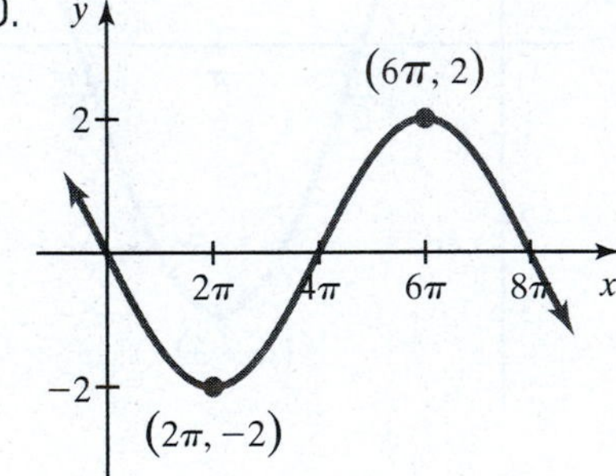

7.2 More on the Graphs of Sine and Cosine

THINGS TO KNOW

Before working through this section, be sure that you are familiar with the following concepts:

VIDEO | ANIMATION | INTERACTIVE

 You Try It 1. Using Vertical Shifts to Graph Functions (Section 3.4) 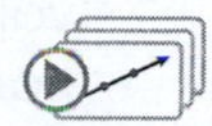 

You Try It 2. Using Horizontal Shifts to Graph Functions (Section 3.4)  

You Try It 3. Sketching Graphs of the Form $y = A \sin x$ and $y = A \cos x$ (Section 7.1) 

 You Try It 4. Sketching Graphs of the Form $y = \sin (Bx)$ and $y = \cos (Bx)$ (Section 7.1) 

 You Try It 5. Sketching Graphs of the Form $y = A \sin (Bx)$ and $y = A \cos (Bx)$ (Section 7.1) 

INTRODUCTION

Read this introduction before beginning Objective 1.

OBJECTIVES

1 Sketching Graphs of the Form $y = \sin (x - C)$ and $y = \cos (x - C)$

2 Sketching Graphs of the Form $y = A \sin (Bx - C)$ and $y = A \cos (Bx - C)$

3 Sketching Graphs of the Form $y = A \sin (Bx - C) + D$ and $y = A \cos (Bx - C) + D$

4 Determine the Equation of a Function of the Form $y = A \sin (Bx - C) + D$ or $y = A \cos (Bx - C) + D$ Given Its Graph

SECTION 7.2 EXERCISES

Introduction to Section 7.2

In Section 7.1, we mentioned that there are four factors that affect the graph of a sine or cosine curve. The first factor was **amplitude** and the second factor was **period**. The next two are **phase shift** and **vertical shift**. Before we learn about these last two factors that can affect a sine or cosine curve, take a moment and work through the following You Try It exercises to make sure that you have a complete understanding of amplitude and period. Do this before starting Objective 1.

 You Try It Sketching a Graph of the Form $y = A \sin x$

You Try It Sketching a Graph of the Form $y = A \cos x$

You Try It Sketching a Graph of the Form $y = \sin (Bx)$

You Try It Sketching a Graph of the Form $y = \cos (Bx)$

You Try It Sketching a Graph of the Form $y = A \sin (Bx)$

You Try It Sketching a Graph of the Form $y = A \cos (Bx)$

(eText Screens 7.2-1–7.2-62)

OBJECTIVE 1 SKETCHING GRAPHS OF THE FORM $y = \sin(x - C)$ AND $y = \cos(x - C)$

The third factor that can affect the graph of a sine or cosine curve is known as **phase shift**. In this objective, for functions of the form $y = A\sin(Bx - C)$ and $y = A\cos(Bx - C)$, we consider only the case where $A = 1$ and $B = 1$. We will look at functions of the form $y = A\sin(Bx - C)$ and $y = A\cos(Bx - C)$ for any nonzero values of A and B in Objective 2. Before we define *phase shift*, recall from **Section 3.4** that if C is a positive constant, then the graph of a function $y = f(x - C)$ can be obtained by horizontally shifting the graph of $y = f(x)$ to the *right* C units.

Watch this **animation** that illustrates a horizontal shift to the right.

Similarly, if C is a positive constant, the graph of $y = f(x + C)$ can be obtained by horizontally shifting the graph of $y = f(x)$ to the *left* C units. Watch this **animation** that illustrates a horizontal shift to the left.

It follows that if $C > 0$, the graph of $y = \sin(x - C)$ can be obtained by horizontally shifting each **quarter point of $y = \sin x$** to the *right* C units and the graph of $y = \sin(x + C)$ can be obtained by horizontally shifting each **quarter point of $y = \sin x$** to the *left* C units. If $C > 0$ the graph of $y = \cos(x - C)$ can be obtained by horizontally shifting each **quarter point of $y = \cos x$** to the *right* C units and the graph of $y = \cos(x + C)$ can be obtained by horizontally shifting each **quarter point of $y = \cos x$** to the *left* C units. See **Figure 21**.

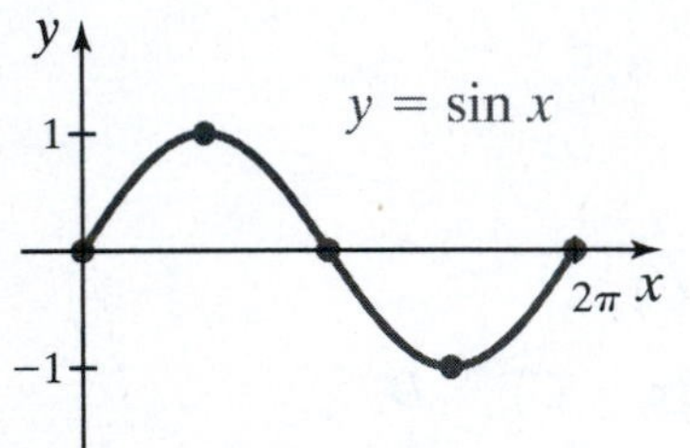

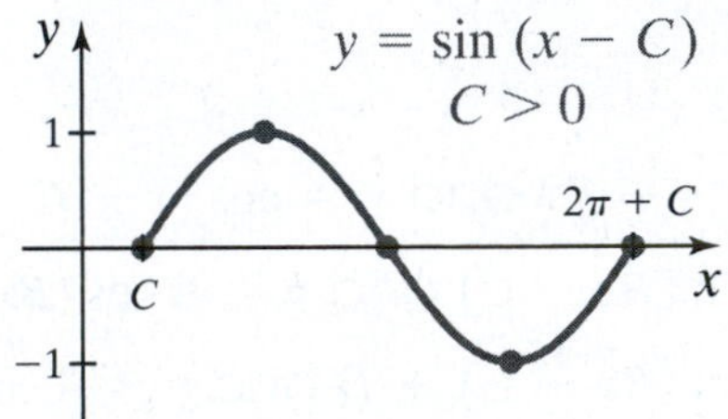

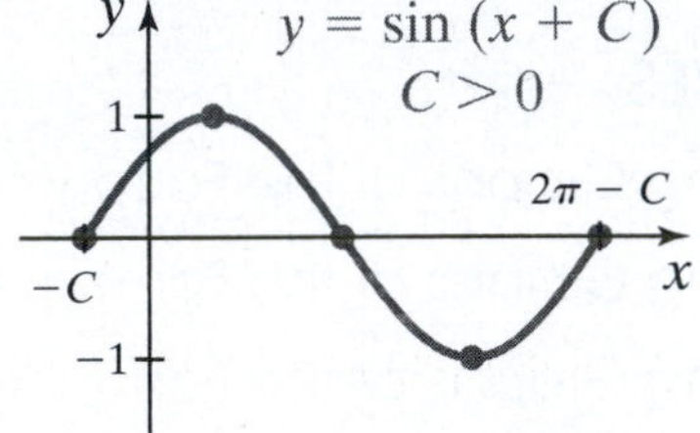

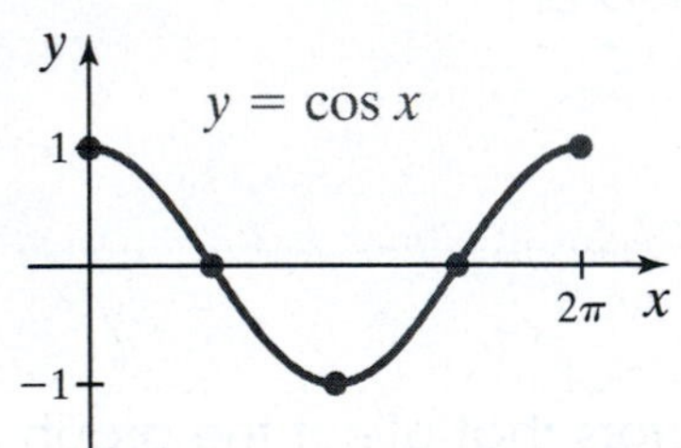

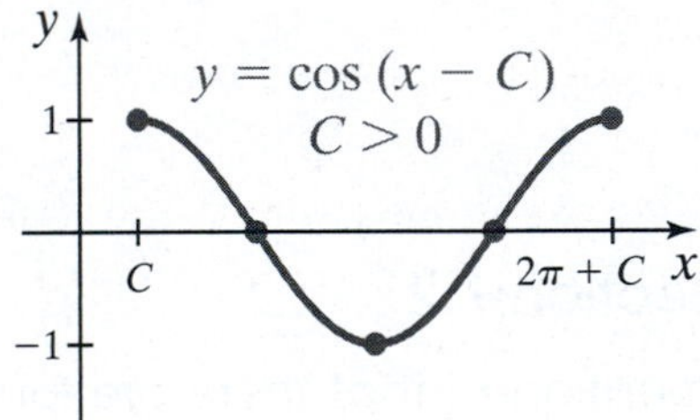

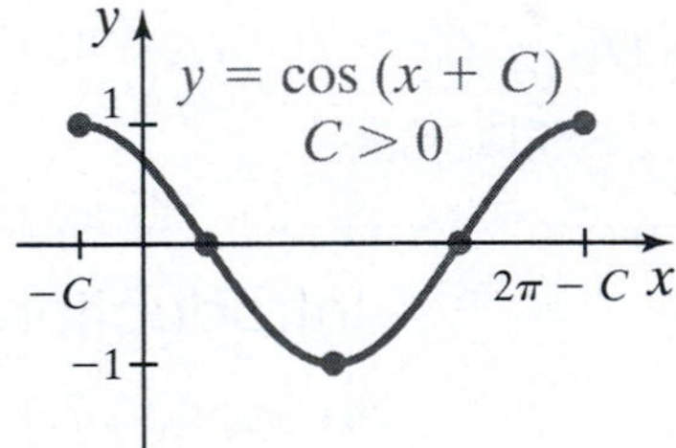

Figure 21

As you can see in Figure 21, the amplitude for each graph is $|A| = |1| = 1$ and the range is $[-1, 1]$. The period of each graph is $P = \frac{2\pi}{B} = \frac{2\pi}{1} = 2\pi$. Notice that the increments between the x-coordinates of each quarter point of all graphs remain unchanged at $\frac{2\pi}{4} = \frac{\pi}{2}$. Only the x-coordinates of each quarter point of the graph of $y = \sin(x - C)$ and $y = \cos(x - C)$ are different from the x-coordinates of the quarter points of one cycle of the graph of $y = \sin x$ and $y = \cos x$, respectively. The x-coordinates of the quarter points of $y = \sin(x - C)$ and $y = \cos(x - C)$ depend on the constant C. In general, the number $\frac{C}{B}$ is known as the **phase shift**. Because $B = 1$, we call the number $\frac{C}{B} = \frac{C}{1} = C$ the **phase shift** for functions of the form $y = \sin(x - C)$ or $y = \cos(x - C)$.

The Graphs of $y = \sin(x - C)$ or $y = \cos(x - C)$

The amplitude is 1 (because $A = 1$) and the range is $[-1, 1]$.

The period is $P = 2\pi\left(\text{Because } B = 1 \text{ and } P = \dfrac{2\pi}{B}\right)$.

The phase shift is $C\left(\text{Because } B = 1 \text{ and the phase shift is } \dfrac{C}{B}\right)$.

The x-coordinates of the quarter points are $C, C + \frac{\pi}{2}, C + \pi, C + \frac{3\pi}{2}$, and $C + 2\pi$.

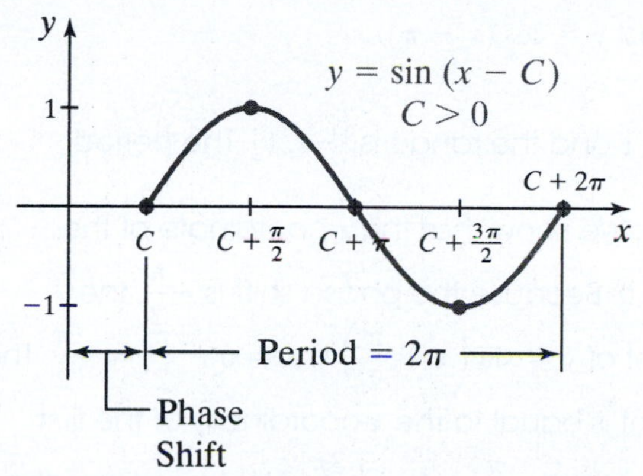

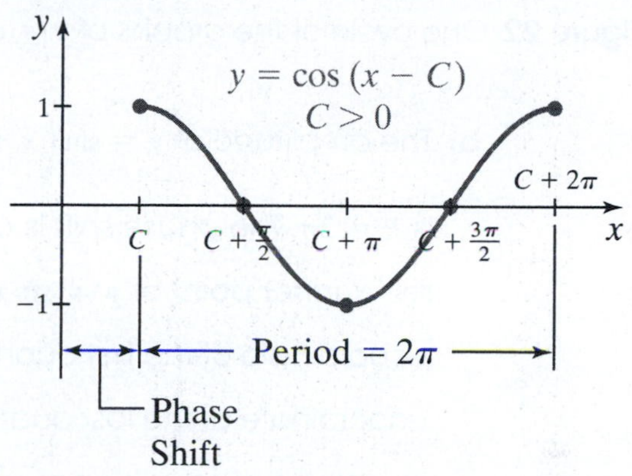

Example 1 Sketching graphs of the form $y = \sin(x - C)$ and $y = \cos(x - C)$

Determine the phase shift and sketch the graph of each function.

a. $y = \cos(x - \pi)$ b. $y = \sin\left(x + \frac{\pi}{2}\right)$

Solution

a. The amplitude of $y = \cos(x - \pi)$ is 1 and the range is $[-1, 1]$. The period is $P = 2\pi$. The phase shift is $C = \pi$. We know that the x-coordinate of the first quarter point of $y = \cos x$ is $x = 0$. Because the phase shift is π, the x-coordinate of the first quarter point of $y = \cos(x - \pi)$ is $x = 0 + \pi = \pi$. The x-coordinate of the last quarter point is equal to the x-coordinate of the first quarter point plus the period, which is $x = \pi + 2\pi = 3\pi$. Next, we divide the interval $[\pi, 3\pi]$ into four equal subintervals of length $P \div 4 = 2\pi \div 4 = \frac{\pi}{2}$ by starting with π and adding $\frac{\pi}{2}$ to the x-coordinate of each successive quarter point. Thus, the x-coordinates of the five quarter points are $\pi, \frac{3\pi}{2}, 2\pi, \frac{5\pi}{2}$, and 3π.

The y-coordinates of the five quarter points of $y = \cos(x - \pi)$ will be exactly the same as the y-coordinates of the five quarter points of $y = \cos x$. We now state the quarter points of $y = \cos x$ and $y = \cos(x - \pi)$. The graphs of these functions can be seen in **Figure 22**.

Quarter points for $y = \cos x$: $(0, 1)$, $\left(\frac{\pi}{2}, 0\right)$, $(\pi, -1)$, $\left(\frac{3\pi}{2}, 0\right)$, $(2\pi, 1)$

Quarter points for $y = \cos(x - \pi)$: $(\pi, 1)$, $\left(\frac{3\pi}{2}, 0\right)$, $(2\pi, -1)$, $\left(\frac{5\pi}{2}, 0\right)$, $(3\pi, 1)$

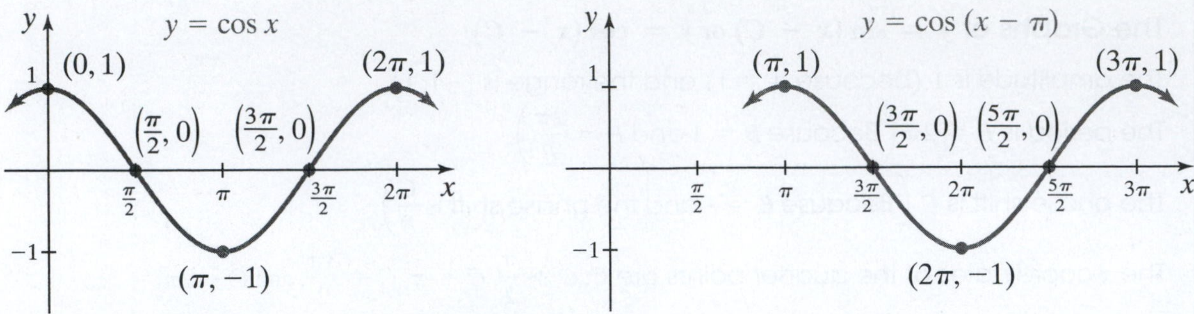

Figure 22 One cycle of the graphs of $y = \cos x$ and $y = \cos(x - \pi)$.

b. The amplitude of $y = \sin\left(x + \frac{\pi}{2}\right)$ is 1 and the range is $[-1, 1]$. The period is $P = 2\pi$. The phase shift is $C = -\frac{\pi}{2}$. We know that the x-coordinate of the first **quarter point of $y = \sin x$** is $x = 0$. Because the phase shift is $-\frac{\pi}{2}$, the x-coordinate of the first quarter point of $y = \sin\left(x + \frac{\pi}{2}\right)$ is $x = 0 - \frac{\pi}{2} = -\frac{\pi}{2}$. The x-coordinate of the last quarter point is equal to the x-coordinate of the first quarter point plus the period, which is $x = -\frac{\pi}{2} + 2\pi = \frac{3\pi}{2}$. Next, we divide the interval $\left[-\frac{\pi}{2}, \frac{3\pi}{2}\right]$ into four equal subintervals of length $P \div 4 = 2\pi \div 4 = \frac{\pi}{2}$ by starting with $-\frac{\pi}{2}$ and adding $\frac{\pi}{2}$ to the x-coordinate of each successive quarter point. Thus, the x-coordinates of the five quarter points are $-\frac{\pi}{2}, 0, \frac{\pi}{2}, \pi$, and $\frac{3\pi}{2}$. The y-coordinates of the five quarter points of $y = \sin\left(x + \frac{\pi}{2}\right)$ will be exactly the same as the y-coordinates of the five quarter points of $y = \sin x$. We now state the quarter points of $y = \sin x$ and $y = \sin\left(x + \frac{\pi}{2}\right)$. The graphs of these functions can be seen in **Figure 23**.

Quarter points for $y = \sin x$: $(0, 0)$, $\left(\frac{\pi}{2}, 1\right)$, $(\pi, 0)$, $\left(\frac{3\pi}{2}, -1\right)$, $(2\pi, 0)$

Quarter points for $y = \sin\left(x + \frac{\pi}{2}\right)$: $\left(-\frac{\pi}{2}, 0\right)$, $(0, 1)$, $\left(\frac{\pi}{2}, 0\right)$, $(\pi, -1)$, $\left(\frac{3\pi}{2}, 0\right)$

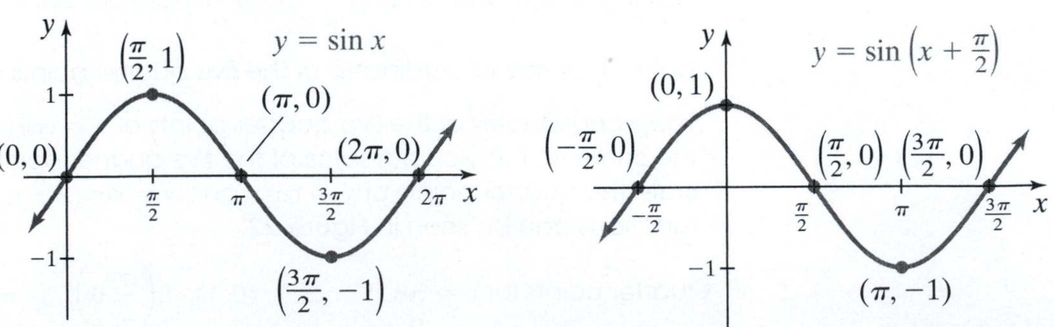

Figure 23 One cycle of the graphs of $y = \sin x$ and $y = \sin\left(x + \frac{\pi}{2}\right)$

Functions of the form $y = \sin(x - C)$ and $y = \cos(x - C)$ (such as $y = \cos(x - \pi)$ and $y = \sin\left(x + \frac{\pi}{2}\right)$ from Example 1) are fairly easy to sketch using transformations. Using transformations, we can sketch the graph of $y = \cos(x - \pi)$ by horizontally shifting the graph of $y = \cos x$ to the right π units. Similarly, we can sketch the graph of $y = \sin\left(x + \frac{\pi}{2}\right)$ by horizontally shifting the graph of $y = \sin\left(x + \frac{\pi}{2}\right)$ to the left $\frac{\pi}{2}$ units. To see how to sketch each of the functions from Example 1 using transformations, click on the Show Graph link located under the following two functions.

$y = \cos(x - \pi)$ Show Graph

$y = \sin\left(x + \frac{\pi}{2}\right)$ Show Graph

If we take a closer look at the graph of $y = \sin\left(x + \frac{\pi}{2}\right)$, we see that it is identical to the graph of $y = \cos x$. See Figure 24. Therefore, we can establish the trigonometric identity, $\sin\left(x + \frac{\pi}{2}\right) = \cos x$.

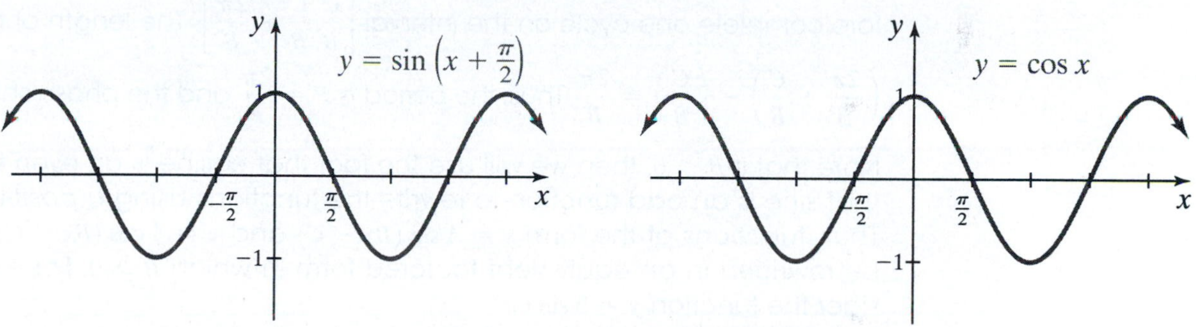

Figure 24 The graph of $y = \sin\left(x + \frac{\pi}{2}\right)$ is the same as the graph of $y = \cos x$.

A trigonometric identity is an equality involving trigonometric functions that are true for every single value of the given variable for which both sides of the equality are defined. We have been exposed to a few trigonometric identities earlier in the text such as the **reciprocal identities** and the **Pythagorean identities**. Trigonometric identities will be discussed extensively in **Chapter 8**.

If we **sketch the graph of** $\boldsymbol{y = \cos\left(x - \frac{\pi}{2}\right)}$, we see that this graph is exactly the same as the graph of $y = \sin x$. This leads to the following two trigonometric identities.

Relationship between Graphs of Sine and Cosine Functions

$$\sin\left(x + \frac{\pi}{2}\right) = \cos x \qquad \cos\left(x - \frac{\pi}{2}\right) = \sin x$$

You Try It Work through this You Try It problem.

Work Exercises 1–5 in this textbook or in the MyLab Math Study Plan.

OBJECTIVE 2 SKETCHING GRAPHS OF THE FORM $y = A\sin(Bx - C)$ AND $y = A\cos(Bx - C)$

We now want to discuss trigonometric functions of the form $y = A\sin(Bx - C)$ and $y = A\cos(Bx - C)$. The amplitude is $|A|$ and the range is $[-|A|, |A|]$. At first glance, it may appear that these graphs have a phase shift of C units. This is *not* true unless $B = 1$! To determine the period and phase shift of functions of this form, we use the fact that the period of $y = A\sin x$ and $y = A\cos x$ is $P = 2\pi$. That is, the graphs of $y = A\sin x$ and $y = A\cos x$ complete one cycle for $0 \le x \le 2\pi$. For now, suppose that $B > 0$. Then the graph of $y = A\sin(Bx - C)$ and $y = A\cos(Bx - C)$ will complete one cycle for $0 \le Bx - C \le 2\pi$. We can solve this three-part inequality to determine the period and phase shift of $y = A\sin(Bx - C)$ and $y = A\cos(Bx - C)$.

$0 \le Bx - C \le 2\pi$	The graph of $y = A\sin(Bx - C)$ and $y = A\cos(Bx - C)$ will complete one cycle for values of x satisfying this inequality.
$C \le Bx \le 2\pi + C$	Add C to each part of the inequality.
$\dfrac{C}{B} \le x \le \dfrac{2\pi}{B} + \dfrac{C}{B}$	Divide each part by B and simplify. The direction of the inequality remains the same because $B > 0$.

The graph of the functions $y = A\sin(Bx - C)$ and $y = A\cos(Bx - C)$ will therefore complete one cycle on the interval $\left[\dfrac{C}{B}, \dfrac{C}{B} + \dfrac{2\pi}{B}\right]$. The length of this interval is $\left(\dfrac{2\pi}{B} + \dfrac{C}{B}\right) - \left(\dfrac{C}{B}\right) = \dfrac{2\pi}{B}$. Thus, the period is $P = \dfrac{2\pi}{B}$ and the phase shift is $\dfrac{C}{B}$.

Note that if $B < 0$, then we will use the fact that cosine is an **even function** and that sine is an **odd function** to rewrite the functions using a *positive* value of B. Thus, functions of the form $y = A\sin(Bx - C)$ and $y = A\cos(Bx - C)$ can *always* be rewritten in an equivalent factored form in which $B > 0$. For example, consider the function $y = 3\sin(\pi - 2x)$.

$y = 3\sin(\pi - 2x)$	Write the original function.
$y = 3\sin(-2x + \pi)$	Reorder inside the parentheses.
$y = 3\sin(-(2x - \pi))$	Factor out a negative within the parentheses.
$y = -3\sin(2x - \pi)$	Use the odd property of $y = \sin x$: $\sin(-x) = -\sin x$.

As you can see, we have rewritten the original function into an equivalent form such that $B > 0$. When sketching trigonometric functions, we will always rewrite the functions using a positive value of B.

When $B \neq 1$, we may choose to factor out B from the **argument** of the function so that the function's characteristics are more apparent.

$y = -3\sin(2x - \pi)$	Start with the form of the function where $B > 0$.
$y = -3\sin\left(2x - \dfrac{2\pi}{2}\right)$	Rewrite the second term within the parentheses so that $B = 2$ is in the numerator.
$y = -3\sin\left(2\left(x - \dfrac{\pi}{2}\right)\right)$	Factor out 2.

Amplitude: $|A| = |-3| = 3$ Period: $P = \dfrac{2\pi}{B} = \dfrac{2\pi}{2} = \pi$ Phase Shift: $\dfrac{C}{B} = \dfrac{\pi}{2}$

We will sketch the graph of $y = 3\sin(\pi - 2x)$ in Example 2c. We now list a seven-step process for sketching the graphs of functions of the form $y = A\sin(Bx - C)$ and $y = A\cos(Bx - C)$.

Steps for Sketching Functions of the Form $y = A\sin(Bx - C)$ and $y = A\cos(Bx - C)$

Step 1. If $B < 0$, rewrite the function in an equivalent form such that $B > 0$. Use the odd property of the sine function or the even property of the cosine function.

It is often helpful to factor out B when $B \neq 1$ such that

$$y = A\sin(Bx - C) = A\sin\left(B\left(x - \frac{C}{B}\right)\right)$$

or

$$y = A\cos(Bx - C) = A\cos\left(B\left(x - \frac{C}{B}\right)\right).$$

In the factored form, the amplitude, period, and phase shift are more apparent.

Step 2. The amplitude is $|A|$. The range is $[-|A|, |A|]$.

Step 3. The period is $P = \frac{2\pi}{B}$.

Step 4. The phase shift is $\frac{C}{B}$.

Step 5. The x-coordinate of the first quarter point is $\frac{C}{B}$. The x-coordinate of the last quarter point is $\frac{C}{B} + P$. An interval for one complete cycle is $\left[\frac{C}{B}, \frac{C}{B} + P\right]$. Subdivide this interval into four equal subintervals of length $P \div 4$ by starting with $\frac{C}{B}$ and adding $(P \div 4)$ to the x-coordinate of each successive quarter point.

Step 6. Multiply the y-coordinates of the quarter points of $y = \sin x$ or $y = \cos x$ by A to determine the y-coordinates of the corresponding quarter points for $y = A\sin(Bx - C)$ and $y = A\cos(Bx - C)$.

Step 7. Connect the quarter points to obtain one complete cycle.

Example 2 Sketching graphs of the form $y = A\sin(Bx - C)$ and $y = A\cos(Bx - C)$

Sketch the graph of each function.

a. $y = 3\sin(2x - \pi)$

b. $y = -2\cos\left(3x + \frac{\pi}{2}\right)$

c. $y = 3\sin(\pi - 2x)$

d. $y = -2\cos\left(-3x + \frac{\pi}{2}\right)$

Solution

a. **Step 1.** We see that $B = 2 > 0$. Therefore, there is no need to use the odd property of sine to rewrite the function.

We can factor this function as $y = 3\sin(2x - \pi) = 3\sin\left(2\left(x - \frac{\pi}{2}\right)\right)$ to determine more easily the function's characteristics. Note that $A = 3, B = 2$, and $\frac{C}{B} = \frac{\pi}{2}$.

Step 2. The amplitude is $|A| = |3| = 3$. Thus, the range is $[-3, 3]$.

Step 3. The period is $P = \frac{2\pi}{B} = \frac{2\pi}{2} = \pi$.

Step 4. The phase shift is $\frac{C}{B} = \frac{\pi}{2}$.

Step 5. The x-coordinate of the first quarter point is $x = \frac{C}{B} = \frac{\pi}{2}$. The x-coordinate of the last quarter point is equal to the x-coordinate of the first quarter point plus the period, which is $x = \frac{\pi}{2} + \pi = \frac{3\pi}{2}$. Next, we divide the interval $\left[\frac{\pi}{2}, \frac{3\pi}{2}\right]$ into four equal subintervals of length $P \div 4 = \pi \div 4 = \frac{\pi}{4}$ by starting with $\frac{\pi}{2}$ and adding $\frac{\pi}{4}$ to the x-coordinate of each successive quarter point. Thus, the x-coordinates of the five quarter points are $\frac{\pi}{2}, \frac{3\pi}{4}, \pi, \frac{5\pi}{4}$, and $\frac{3\pi}{2}$.

Step 6. The y-coordinates of the five quarter points will be $A = 3$ times the corresponding y-coordinates of the five **quarter points of $y = \sin x$**. We now state the quarter points of $y = \sin x$ and $y = 3\sin(2x - \pi) = 3\sin\left(2\left(x - \frac{\pi}{2}\right)\right)$.

Quarter points for $y = \sin x$:	$(0, 0)$,	$\left(\frac{\pi}{2}, 1\right)$,	$(\pi, 0)$,	$\left(\frac{3\pi}{2}, -1\right)$,	$(2\pi, 0)$
	$(3)\cdot 0 = 0$	$(3)\cdot 1 = 3$	$(3)\cdot 0 = 0$	$(3)\cdot(-1) = -3$	$(3)\cdot 0 = 0$
Quarter points for $y = 3\sin(2x - \pi)$ $= 3\sin\left(2\left(x - \frac{\pi}{2}\right)\right)$	$\left(\frac{\pi}{2}, 0\right)$,	$\left(\frac{3\pi}{4}, 3\right)$,	$(\pi, 0)$,	$\left(\frac{5\pi}{4}, -3\right)$,	$\left(\frac{3\pi}{2}, 0\right)$

Step 7. We connect these quarter points with a smooth curve to complete one cycle of the graph of $y = 3\sin(2x - \pi)$. See **Figure 25**. Several cycles of the graph can be seen in **Figure 26**.

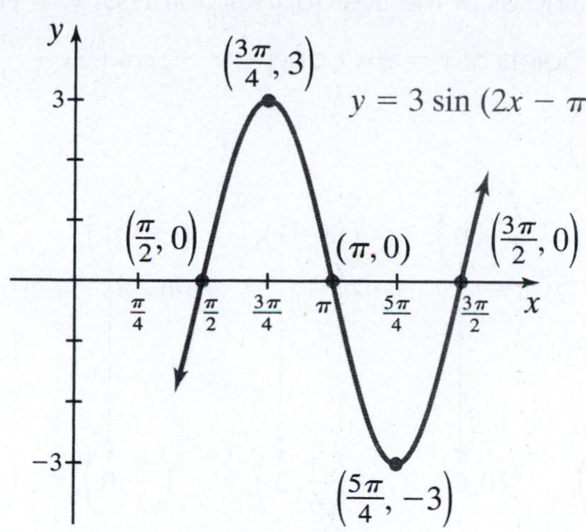

Figure 25 One complete cycle of the graph of $y = 3\sin(2x - \pi)$

Using Technology

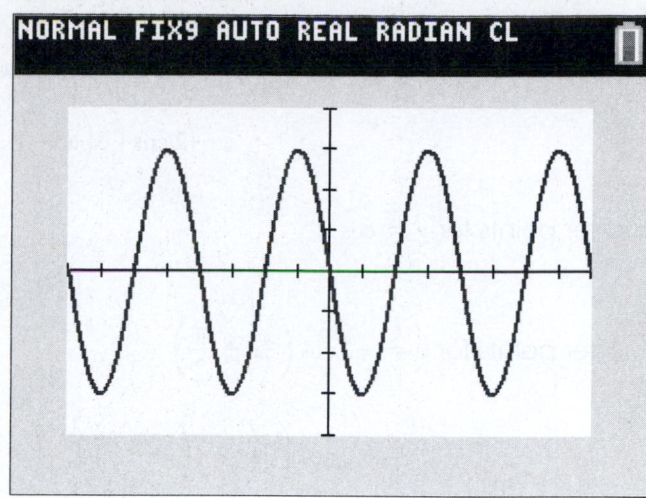

Figure 26 Several cycles of the graph of $y = 3\sin(2x - \pi)$ using a graphing utility

To see how to sketch the function $y = 3\sin(2x - \pi)$, watch this **interactive video**.

b. Step 1. For the function $y = -2\cos\left(3x + \frac{\pi}{2}\right)$, we see that $B = 3 > 0$. Therefore, there is no need to use the even property of cosine to rewrite the function. We can factor this function as $y = -2\cos\left(3x + \frac{\pi}{2}\right) = -2\cos\left(3\left(x - \left(-\frac{\pi}{6}\right)\right)\right)$ to more easily determine the function's characteristics. Note that $A = -2, B = 3,$ and $\frac{C}{B} = -\frac{\pi}{6}$.

Step 2. The amplitude is $|A| = |-2| = 2$. Thus, the range is $[-2, 2]$.

Step 3. The period is $P = \frac{2\pi}{B} = \frac{2\pi}{3}$.

Step 4. The phase shift is $\frac{C}{B} = -\frac{\pi}{6}$.

Step 5. The x-coordinate of the first quarter point is $x = \frac{C}{B} = -\frac{\pi}{6}$. The x-coordinate of the last quarter point is equal to the x-coordinate of the first quarter point plus the period, which is $x = -\frac{\pi}{6} + \frac{2\pi}{3} = -\frac{\pi}{6} + \frac{4\pi}{6} = \frac{3\pi}{6} = \frac{\pi}{2}$.

Next, we divide the interval $\left[-\frac{\pi}{6}, \frac{\pi}{2}\right]$ into four equal subintervals of length $P \div 4 = \frac{2\pi}{3} \div 4 = \frac{2\pi}{3} \cdot \frac{1}{4} = \frac{2\pi}{12} = \frac{\pi}{6}$ by starting with $-\frac{\pi}{6}$ and adding $\frac{\pi}{6}$ to the x-coordinate of each successive quarter point. Thus, the x-coordinates of the five quarter points are $-\frac{\pi}{6}, 0, \frac{\pi}{6}, \frac{\pi}{3},$ and $\frac{\pi}{2}$.

Step 6. The y-coordinates of the five quarter points will be $A = -2$ times the corresponding y-coordinates of the five **quarter points of $y = \cos x$**. We now state the quarter points of $y = \cos x$ and $y = -2\cos\left(3x + \frac{\pi}{2}\right) = -2\cos\left(3\left(x - \left(-\frac{\pi}{6}\right)\right)\right)$.

Quarter points for $y = \cos x$: $(0, 1)$, $\left(\frac{\pi}{2}, 0\right)$, $(\pi, -1)$, $\left(\frac{3\pi}{2}, 0\right)$, $(2\pi, 1)$

$(-2)\cdot 1 = -2$ | $(-2)\cdot 0 = 0$ | $(-2)\cdot(-1) = 2$ | $(-2)\cdot 0 = 0$ | $(-2)\cdot 1 = -2$

Quarter points for $y = -2\cos\left(3x + \frac{\pi}{2}\right)$

$= -2\cos\left(3\left(x - \left(-\frac{\pi}{6}\right)\right)\right)$: $\left(-\frac{\pi}{6}, -2\right)$, $(0, 0)$, $\left(\frac{\pi}{6}, 2\right)$, $\left(\frac{\pi}{3}, 0\right)$, $\left(\frac{\pi}{2}, -2\right)$

Step 7. The graph of $y = -2\cos\left(3x + \frac{\pi}{2}\right)$ can be seen in Figure 27 and Figure 28.

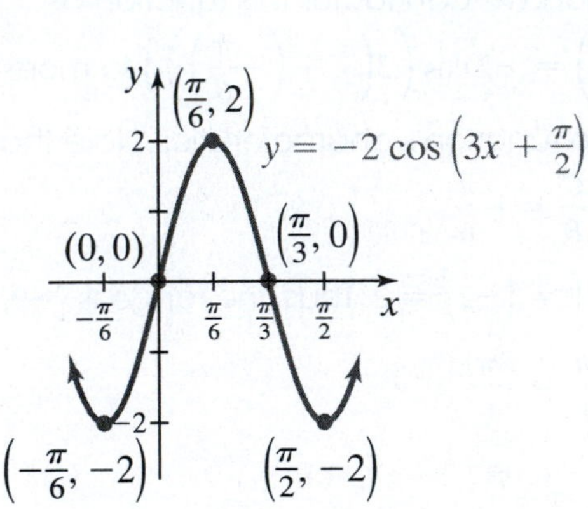

Figure 27 One complete cycle of the graph of $y = -2\cos\left(3x + \frac{\pi}{2}\right)$

Using Technology

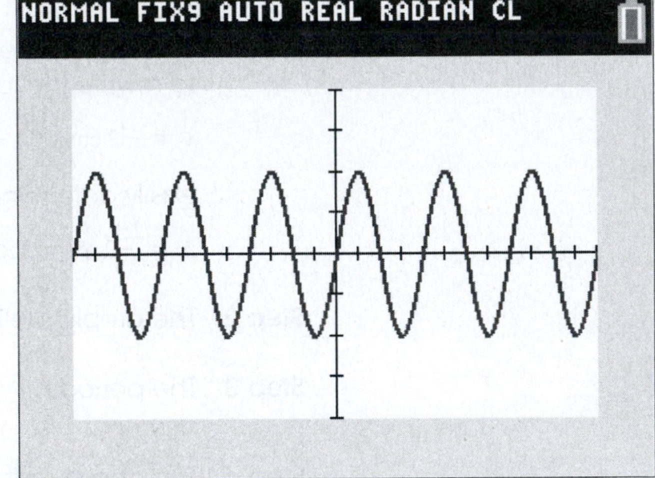

Figure 28 Several cycles of the graph of $y = -2\cos\left(3x + \frac{\pi}{2}\right)$ using a graphing utility

To see how to sketch the function $y = -2\cos\left(3x + \frac{\pi}{2}\right)$, watch this **interactive** video.

c. Step 1. For the function $y = 3\sin(\pi - 2x)$, we see that $B = -2 < 0$. Therefore, we use the odd property of the sine function to rewrite the function as an equivalent function using a positive value of B.

$y = 3\sin(\pi - 2x)$	Write the original function.
$y = 3\sin(-2x + \pi)$	Reorder inside the parentheses.
$y = 3\sin(-(2x - \pi))$	Factor out a negative within the parentheses.
$y = -3\sin(2x - \pi)$	Use the odd property of $y = \sin x$: $\sin(-x) = -\sin x$.

We can now factor this function as $y = -3\sin(2x - \pi) = -3\sin\left(2\left(x - \frac{\pi}{2}\right)\right)$ to determine more easily the function's characteristics.

Note that $A = -3$, $B = 2$, and $\frac{C}{B} = \frac{\pi}{2}$.

Step 2. The amplitude is $|A| = |-3| = 3$. Thus, the range is $[-3, 3]$.

Step 3. The period is $P = \frac{2\pi}{B} = \frac{2\pi}{2} = \pi$.

Step 4. The phase shift is $\frac{C}{B} = \frac{\pi}{2}$.

Step 5. The x-coordinate of the first quarter point is $x = \frac{C}{B} = \frac{\pi}{2}$. The x-coordinate of the last quarter point is equal to the x-coordinate of the first quarter point plus the period, which is $x = \frac{\pi}{2} + \pi = \frac{\pi}{2} + \frac{2\pi}{2} = \frac{3\pi}{2}$. Next, we divide the interval $\left[\frac{\pi}{2}, \frac{3\pi}{2}\right]$ into four equal subintervals of length $P \div 4 = \pi \div 4 = \frac{\pi}{4}$ by starting with $\frac{\pi}{2}$ and adding $\frac{\pi}{4}$ to the x-coordinate of each successive quarter point. Thus, the x-coordinates of the five quarter points are $\frac{\pi}{2}, \frac{3\pi}{4}, \pi, \frac{5\pi}{4}$, and $\frac{3\pi}{2}$.

Step 6. The y-coordinates of the five quarter points of $y = -3\sin\left(2\left(x - \frac{\pi}{2}\right)\right)$ will be $A = -3$ times the corresponding y-coordinates of the five quarter points of $y = \sin x$. We now state the quarter points of $y = \sin x$ and $y = 3\sin(\pi - 2x) = -3\sin(2x - \pi) = -3\sin\left(2\left(x - \frac{\pi}{2}\right)\right)$.

Quarter points for $y = \sin x$:	$(0, 0)$,	$\left(\frac{\pi}{2}, 1\right)$,	$(\pi, 0)$,	$\left(\frac{3\pi}{2}, -1\right)$,	$(2\pi, 0)$
	$(-3)\cdot 0 = 0$	$(-3)\cdot 1 = -3$	$(-3)\cdot 0 = 0$	$(-3)\cdot(-1) = 3$	$(-3)\cdot 0 = 0$
Quarter points for $y = 3\sin(\pi - 2x)$ $= -3\sin(2x - \pi)$ $= -3\sin\left(2\left(x - \frac{\pi}{2}\right)\right)$:	$\left(\frac{\pi}{2}, 0\right)$,	$\left(\frac{3\pi}{4}, -3\right)$,	$(\pi, 0)$,	$\left(\frac{5\pi}{4}, 3\right)$,	$\left(\frac{3\pi}{2}, 0\right)$

Step 7. We can now connect the quarter points with a smooth curve to create one complete cycle of the graph of $y = 3\sin(\pi - 2x)$. See **Figure 29**. Several cycles of the graph can be seen in **Figure 30**.

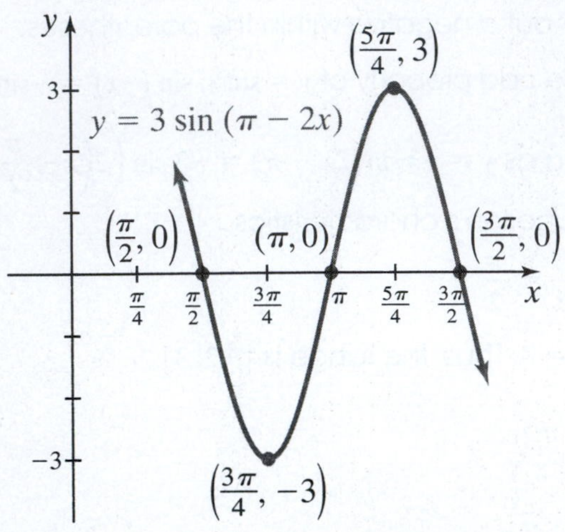

Figure 29 One complete cycle of the graph of $y = 3\sin(\pi - 2x)$

Using Technology

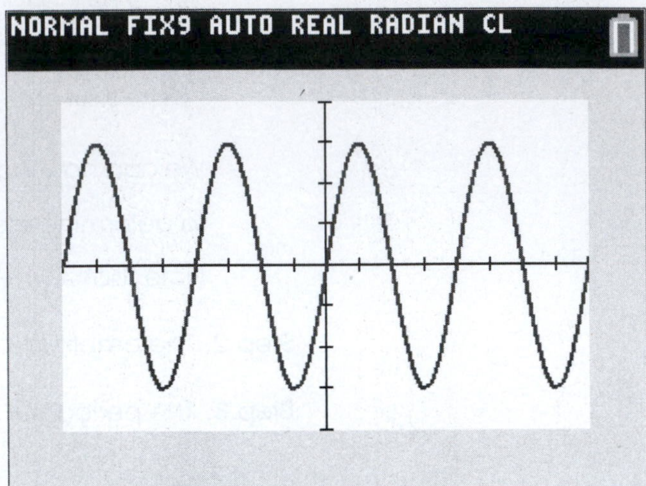

Figure 30 Several cycles of the graph of $y = 3\sin(\pi - 2x)$ using a graphing utility

To see how to sketch the function $y = 3\sin(\pi - 2x)$, watch this **interactive video**.

d. Step 1. For the function $y = -2\cos\left(-3x + \frac{\pi}{2}\right)$, we see that $B = -3 < 0$. Therefore, we use the even property of the cosine function to rewrite the function as an equivalent function using a positive value of B.

$y = -2\cos\left(-3x + \frac{\pi}{2}\right)$ Write the original function.

$y = -2\cos\left(-\left(3x - \frac{\pi}{2}\right)\right)$ Factor out the negative within the parentheses.

$y = -2\cos\left(3x - \frac{\pi}{2}\right)$ Use the even property of $y = \cos x$: $\cos(-x) = \cos x$.

We can now factor this function as $y = -2\cos\left(3x - \frac{\pi}{2}\right) = -2\cos\left(3\left(x - \frac{\pi}{6}\right)\right)$ to more easily determine the function's characteristics. Note that $A = -2$, $B = 3$, and $\frac{C}{B} = \frac{\pi}{6}$.

Step 2. The amplitude is $|A| = |-2| = 2$. Thus, the range is $[-2, 2]$.

Step 3. The period is $P = \frac{2\pi}{B} = \frac{2\pi}{3}$.

Step 4. The phase shift is $\frac{C}{B} = \frac{\pi}{6}$.

Step 5. The x-coordinate of the first quarter point is $x = \frac{C}{B} = \frac{\pi}{6}$. The x-coordinate of the last quarter point is equal to the x-coordinate of the first quarter point plus the period, which is $x = \frac{\pi}{6} + \frac{2\pi}{3} = \frac{\pi}{6} + \frac{4\pi}{6} = \frac{5\pi}{6}$. Next, we divide the interval $\left[\frac{\pi}{6}, \frac{5\pi}{6}\right]$ into four equal subintervals of length $P \div 4 = \frac{2\pi}{3} \div 4 = \frac{2\pi}{3} \cdot \frac{1}{4} = \frac{2\pi}{12} = \frac{\pi}{6}$. We do this by starting with $\frac{\pi}{6}$ and adding $\frac{\pi}{6}$ to the x-coordinate of each successive quarter point. Thus, the x-coordinates of the five quarter points are $\frac{\pi}{6}, \frac{\pi}{3}, \frac{\pi}{2}, \frac{2\pi}{3}$, and $\frac{5\pi}{6}$.

Step 6. The y-coordinates of the five quarter points will be $A = -2$ times the corresponding y-coordinates of the five quarter points of $y = \cos x$. We now state the quarter points of $y = \cos x$ and $y = -2\cos\left(-3x + \frac{\pi}{2}\right) = -2\cos\left(3\left(x - \frac{\pi}{6}\right)\right)$.

Quarter points for $y = \cos x$: $(0, 1)$, $\left(\frac{\pi}{2}, 0\right)$, $(\pi, -1)$, $\left(\frac{3\pi}{2}, 0\right)$, $(2\pi, 1)$

$(-2)\cdot 1 = -2$ $\quad (-2)\cdot 0 = 0$ $\quad (-2)\cdot -1 = 2$ $\quad (-2)\cdot 0 = 0$ $\quad (-2)\cdot 1 = -2$

Quarter points for $y = -2\cos\left(-3x + \frac{\pi}{2}\right)$
$= -2\cos\left(3\left(x - \frac{\pi}{6}\right)\right)$: $\left(\frac{\pi}{6}, -2\right)$, $\left(\frac{\pi}{3}, 0\right)$, $\left(\frac{\pi}{2}, 2\right)$, $\left(\frac{2\pi}{3}, 0\right)$, $\left(\frac{5\pi}{6}, -2\right)$

Step 7. The graph of $y = -2\cos\left(-3x + \frac{\pi}{2}\right)$ can be seen in Figures 31 and 32.

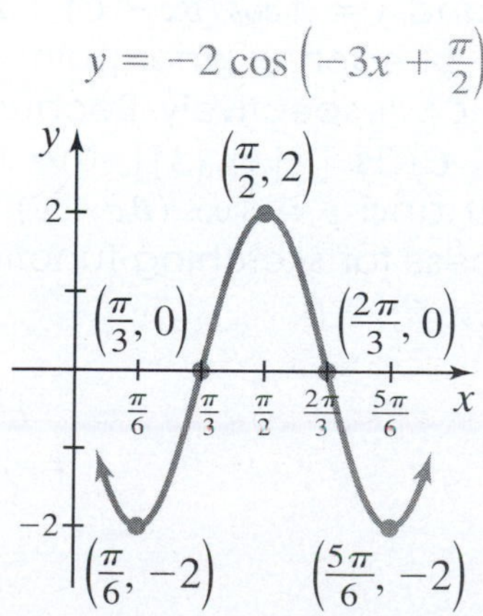

Figure 31 One complete cycle of the graph of $y = -2\cos\left(-3x + \frac{\pi}{2}\right)$

Using Technology

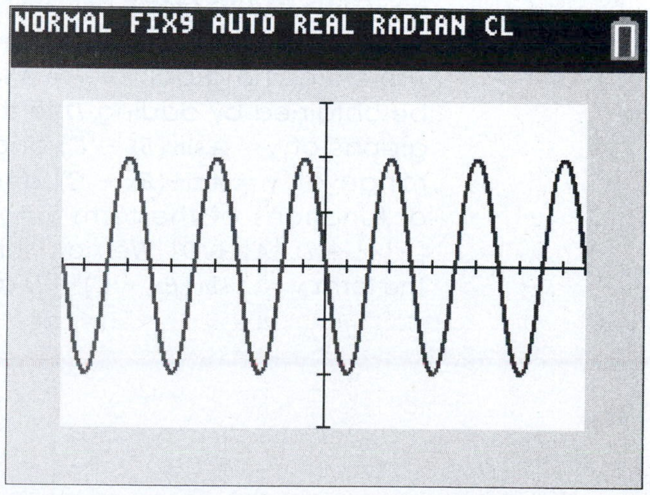

Figure 32 Several cycles of the graph of $y = -2\cos\left(-3x + \frac{\pi}{2}\right)$ using a graphing utility

To see how to sketch the function $y = -2\cos\left(-3x + \frac{\pi}{2}\right)$, watch this **interactive video**.

It is possible to use transformations to sketch functions of the form $y = A\sin(Bx - C)$ and $y = A\cos(Bx - C)$ as seen in Example 2. Such functions require several transformations. When sketching functions that require multiple transformations, we will always use the following order of transformations.

1. Horizontal shifts
2. Horizontal stretches/compressions
3. Reflection about the y-axis
4. Vertical stretches/compressions
5. Reflection about the x-axis
6. Vertical shifts

Functions of the form $y = A\sin(Bx - C)$ and $y = A\cos(Bx - C)$ never require a vertical shift. However, we will encounter functions requiring vertical shifts in Example 3. To see how to sketch each of the functions from Example 2 using transformations, click on the Show Graph link below the following four functions.

$y = 3\sin(2x - \pi)$	$y = -2\cos\left(3x + \frac{\pi}{2}\right)$	$y = 3\sin(-2x - \pi)$	$y = -2\cos\left(-3x + \frac{\pi}{2}\right)$
Show Graph	Show Graph	Show Graph	Show Graph

You Try It Work through this You Try It problem.

Work Exercises 6–15 in this textbook or in the MyLab Math Study Plan.

OBJECTIVE 2 SKETCHING GRAPHS OF THE FORM $y = A\sin(Bx - C) + D$ AND $y = A\cos(Bx - C) + D$

Recall that when we add a nonzero constant to a function $y = f(x)$, we get $y = f(x) + D$. The graph of $y = f(x) + D$ can be obtained by adding D to the corresponding y-coordinates of each ordered pair lying on the graph of $y = f(x)$. Therefore, the graph of $y = A\sin(Bx - C) + D$ and $y = A\cos(Bx - C) + D$ can be obtained by adding D to the y-coordinate of each quarter point of the graphs of $y = A\sin(Bx - C)$ and $y = A\cos(Bx - C)$, respectively. Because the range of $y = A\sin(Bx - C)$ and $y = A\cos(Bx - C)$ is $[-|A|, |A|]$, the range of functions of the form $y = A\sin(Bx - C) + D$ and $y = A\cos(Bx - C) + D$ is $[-|A| + D, |A| + D]$. We now summarize a process for sketching functions of the form $y = A\sin(Bx - C) + D$ and $y = A\cos(Bx - C) + D$.

Steps for Sketching Functions of the Form $y = A\sin(Bx - C) + D$ and $y = A\cos(Bx - C) + D$

Step 1. If $B < 0$, rewrite the function in an equivalent form such that $B > 0$. Use the odd property of the sine function or the even property of the cosine function.

It is often helpful to factor out B when $B \neq 1$ such that $y = A\sin(Bx - C) + D = A\sin\left(B\left(x - \frac{C}{B}\right)\right) + D$ or $y = A\cos(Bx - C) + D = A\cos\left(B\left(x - \frac{C}{B}\right)\right) + D$. In the factored form, the amplitude, period, and phase shift are more apparent.

Step 2. The amplitude is $|A|$. The range is $[-|A| + D, |A| + D]$.

Step 3. The period is $P = \frac{2\pi}{B}$.

Step 4. The phase shift is $\frac{C}{B}$.

Step 5. The x-coordinate of the first quarter point is $\frac{C}{B}$. The x-coordinate of the last quarter point is $\frac{C}{B} + P$. An interval for one complete cycle is $\left[\frac{C}{B}, \frac{C}{B} + P\right]$. Subdivide this interval into four equal subintervals of length $P \div 4$ by starting with $\frac{C}{B}$ and adding $(P \div 4)$ to the x-coordinate of each successive quarter point.

Step 6. Multiply the y-coordinates of the quarter points of $y = \sin x$ or $y = \cos x$ by A and then add D to determine the y-coordinates of the corresponding quarter points for $y = A\sin(Bx - C) + D$ and $y = A\cos(Bx - C) + D$.

Step 7. Connect the quarter points to obtain one complete cycle.

Example 3 Sketching graphs of the form $y = A\sin(Bx - C) + D$ and $y = A\cos(Bx - C) + D$

Sketch the graph of each function.

a. $y = 3\sin\left(2x - \frac{\pi}{2}\right) - 1$ **b.** $y = 4 - \cos(-\pi x + 2)$

Solution

a. Step 1. For the function $y = 3\sin\left(2x - \frac{\pi}{2}\right) - 1$, we see that $B = 2 > 0$. Therefore, there is no need to use the odd properly of sine to rewrite the function. We can now factor this function as $y = 3\sin\left(2x - \frac{\pi}{2}\right) - 1 = 3\sin\left(2\left(x - \frac{\pi}{4}\right)\right) - 1$ to determine more easily the function's characteristics. Note that $A = 3, B = 2, \frac{C}{B} = \frac{\pi}{4}$, and $D = -1$.

Step 2. The amplitude is $|A| = |3| = 3$.
The range is $[-|A| + D, |A| + D] = [-3 + (-1), 3 + (-1)] = [-4, 2]$.

Step 3. The period is $P = \frac{2\pi}{B} = \frac{2\pi}{2} = \pi$.

Step 4. The phase shift is $\frac{C}{B} = \frac{\pi}{4}$.

Step 5. The x-coordinate of the first quarter point is $x = \frac{C}{B} = \frac{\pi}{4}$. The x-coordinate of the last quarter point is equal to the x-coordinate of the first quarter point plus the period, which is $x = \frac{\pi}{4} + \pi = \frac{5\pi}{4}$. An interval for one complete cycle of the graph is $\left[\frac{\pi}{4}, \frac{5\pi}{4}\right]$. We divide this interval into four equal subintervals of length $\frac{P}{4} = \frac{\pi}{4}$ by starting with $\frac{\pi}{4}$ and adding $\frac{\pi}{4}$ to the x-coordinate of each successive quarter point. Thus, the x-coordinates of the five quarter points are $\frac{\pi}{4}, \frac{\pi}{2}, \frac{3\pi}{4}, \pi$, and $\frac{5\pi}{4}$.

Step 6. Multiply each y-coordinate of the **quarter points of $y = \sin x$** by $A = 3$ and then add $D = -1$ to obtain the quarter points of

$$y = 3\sin\left(2x - \frac{\pi}{2}\right) - 1 = 3\sin\left(2\left(x - \frac{\pi}{4}\right)\right) - 1.$$

Quarter points for $y = \sin x$:	$(0, 0)$,	$\left(\frac{\pi}{2}, 1\right)$,	$(\pi, 0)$,	$\left(\frac{3\pi}{2}, -1\right)$,	$(2\pi, 0)$
	$(3)\cdot 0 + (-1) = -1$	$(3)\cdot 1 + (-1) = 2$	$(3)\cdot 0 + (-1) = -1$	$(3)\cdot(-1) + (-1) = -4$	$(3)\cdot 0 + (-1) = -1$
Quarter points for $y = 3\sin\left(2x - \frac{\pi}{2}\right) - 1 = 3\sin\left(2\left(x - \frac{\pi}{4}\right)\right) - 1$:	$\left(\frac{\pi}{4}, -1\right)$,	$\left(\frac{\pi}{2}, 2\right)$,	$\left(\frac{3\pi}{4}, -1\right)$,	$(\pi, -4)$,	$\left(\frac{5\pi}{4}, -1\right)$

Step 7. We connect these quarter points with a smooth curve to complete one cycle of the graph of $y = 3\sin\left(2x - \frac{\pi}{2}\right) - 1$. See Figures 33 and 34.

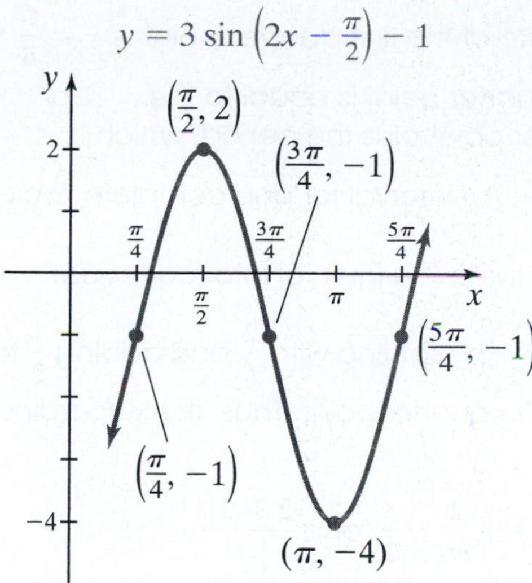

Figure 33 One complete cycle of the graph of $y = 3\sin\left(2x - \frac{\pi}{2}\right) - 1$

Using Technology

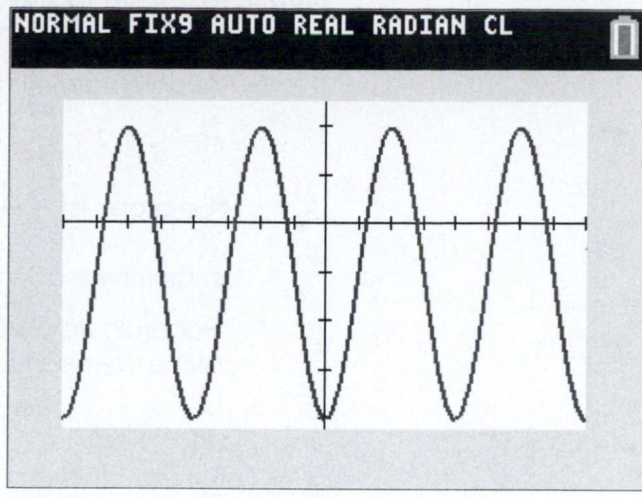

Figure 34 Several cycles of the graph of $y = 3\sin\left(2x - \frac{\pi}{2}\right) - 1$ using a graphing utility

To see how to sketch the graph of $y = 3\sin\left(2x - \frac{\pi}{2}\right) - 1$, watch this **interactive video**.

b. Step 1. For the function $y = 4 - \cos(-\pi x + 2)$, we see that $B = -\pi < 0$. Therefore, we use the even property of the cosine function to rewrite the function as an equivalent function using a positive value of B.

$y = 4 - \cos(-\pi x + 2)$	Write the original function.
$y = -\cos(-\pi x + 2) + 4$	Reorder the terms.
$y = -\cos(-(\pi x - 2)) + 4$	Factor out the negative inside the parentheses.
$y = -\cos(\pi x - 2) + 4$	Use the even property of $y = \cos x$: $\cos(-x) = \cos x$.

We now factor this function as $y = -\cos(\pi x - 2) + 4 = -\cos\left(\pi\left(x - \frac{2}{\pi}\right)\right) + 4$ to determine more easily the function's characteristics. Note that $A = -1, B = \pi, \frac{C}{B} = \frac{2}{\pi}$, and $D = 4$.

Step 2. The amplitude is $|A| = |-1| = 1$.
The range is $[-|A| + D, |A| + D] = [-1 + 4, 1 + 4] = [3, 5]$.

Step 3. The period is $P = \frac{2\pi}{B} = \frac{2\pi}{\pi} = 2$.

Step 4. The phase shift is $\frac{C}{B} = \frac{2}{\pi}$.

Step 5. We know that the x-coordinate of the first quarter point is $x = \frac{C}{B} = \frac{2}{\pi}$. The x-coordinate of the last quarter point is equal to the x-coordinate of the first quarter point plus the period, which is $x = \frac{2}{\pi} + 2 = \frac{2}{\pi} + \frac{2\pi}{\pi} = \frac{2 + 2\pi}{\pi}$. An interval for one complete cycle of the graph is $\left[\frac{2}{\pi}, \frac{2 + 2\pi}{\pi}\right]$. We divide this interval into four equal subintervals of length $\frac{P}{4} = \frac{2}{4} = \frac{1}{2}$ by starting with $\frac{2}{\pi}$ and adding $\frac{1}{2}$ to the x-coordinate of each successive quarter point. Thus, the x-coordinates of the five quarter points are

$$\frac{2}{\pi}, \frac{2}{\pi} + \frac{1}{2}, \frac{2}{\pi} + 1, \frac{2}{\pi} + \frac{3}{2}, \text{ and } \frac{2 + 2\pi}{\pi}.$$

These x-coordinates can be simplified and rewritten as

$$\frac{2}{\pi}, \frac{4 + \pi}{2\pi}, \frac{2 + \pi}{\pi}, \frac{4 + 3\pi}{2\pi}, \text{ and } \frac{2 + 2\pi}{\pi}.$$

Step 6. Multiply each y-coordinate of the **quarter points of $y = \cos x$** by $A = -1$ and then add $D = 4$ to obtained the corresponding y-coordinates of the quarter points of

$$y = 4 - \cos(-\pi x + 2) = -\cos\left(\pi\left(x - \frac{2}{\pi}\right)\right) + 4.$$

Quarter points for $y = \cos x$:	$(0, 1)$,	$\left(\frac{\pi}{2}, 0\right)$,	$(\pi, -1)$,	$\left(\frac{3\pi}{2}, 0\right)$,	$(2\pi, 1)$
	$(-1)\cdot 1 + 4 = 3$	$(-1)\cdot 0 + 4 = 4$	$(-1)\cdot(-1) + 4 = 5$	$(-1)\cdot 0 + 4 = 4$	$(-1)\cdot 1 + 4 = 3$
Quarter points for $y = 4 - \cos(-\pi x + 2)$ $= -\cos\left(\pi\left(x - \frac{2}{\pi}\right)\right) + 4$:	$\left(\frac{2}{\pi}, 3\right)$,	$\left(\frac{4 + \pi}{2\pi}, 4\right)$,	$\left(\frac{2 + \pi}{\pi}, 5\right)$,	$\left(\frac{4 + 3\pi}{2\pi}, 4\right)$,	$\left(\frac{2 + 2\pi}{\pi}, 3\right)$

Step 7. We connect these quarter points with a smooth curve to complete one cycle of the graph of $y = 4 - \cos(-\pi x + 2)$. See Figures 35 and 36.

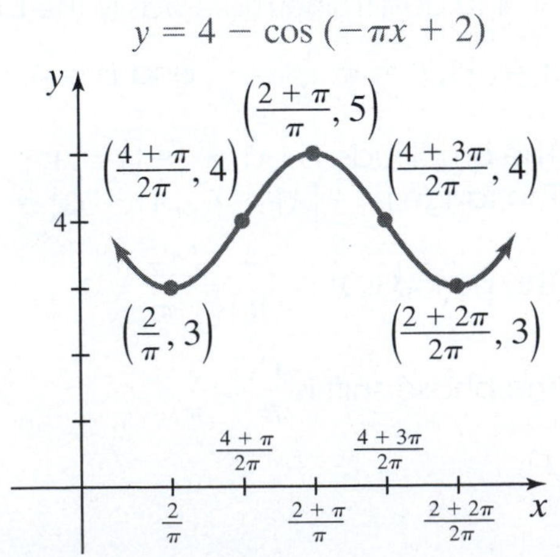

Figure 35 One complete cycle of the graph of $y = 4 - \cos(-\pi x + 2)$

Using Technology

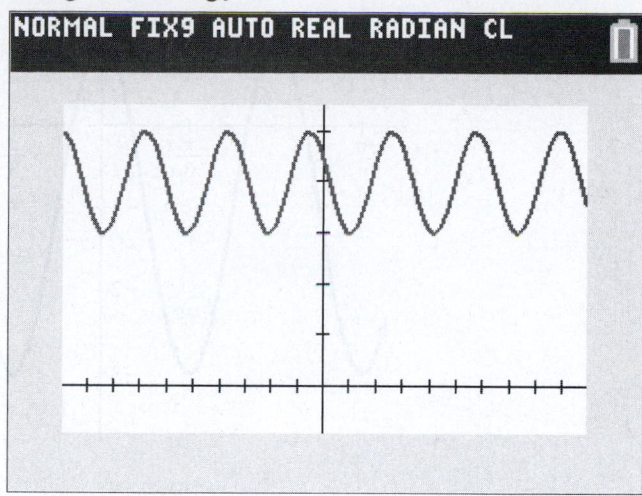

Figure 36 Several cycles of the graph of $y = 4 - \cos(-\pi x + 2)$ using a graphing utility

To see how to sketch the graph of $y = 4 - \cos(-\pi x + 2)$, watch this interactive video.

We can sketch the graphs seen in Example 3 using transformations if we carefully follow the **order of transformations**. Try sketching the functions $y = 3\sin\left(2x - \frac{\pi}{2}\right) - 1$ and $y = 4 - \cos(-\pi x + 2)$ using transformations and then click on the Show Graph links below to see if you are correct.

$y = 3\sin\left(2x - \frac{\pi}{2}\right) - 1$ Show Graph

$y = 4 - \cos(-\pi x + 2)$ Show Graph

You Try It Work through this You Try It problem.

Work Exercises 16–25 in this textbook or in the MyLab Math Study Plan.

Take some time sketching functions of the form $y = A\sin(Bx - C) + D$ and $y = A\cos(Bx - C) + D$ by experimenting with the Guided Visualization below. Click on the Guided Visualization and choose your own values of $A, B, C,$ and D.

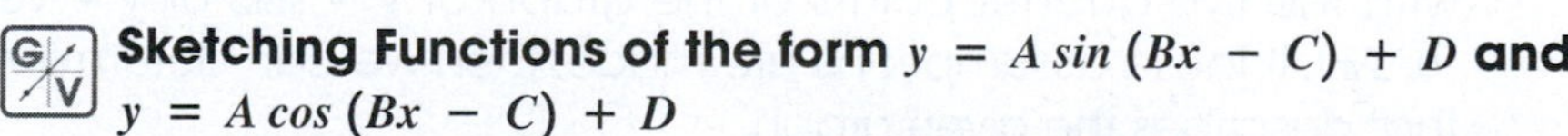

Sketching Functions of the form $y = A\sin(Bx - C) + D$ and $y = A\cos(Bx - C) + D$

OBJECTIVE 4 DETERMINE THE EQUATION OF A FUNCTION OF THE FORM $y = A\sin(Bx - C) + D$ OR $y = A\cos(Bx - C) + D$ GIVEN ITS GRAPH

Suppose that you were asked to determine the equation of a function that describes the graph seen in Figure 37 where the function is of the form $y = A\sin(Bx - C) + D$ or $y = A\cos(Bx - C) + D$.

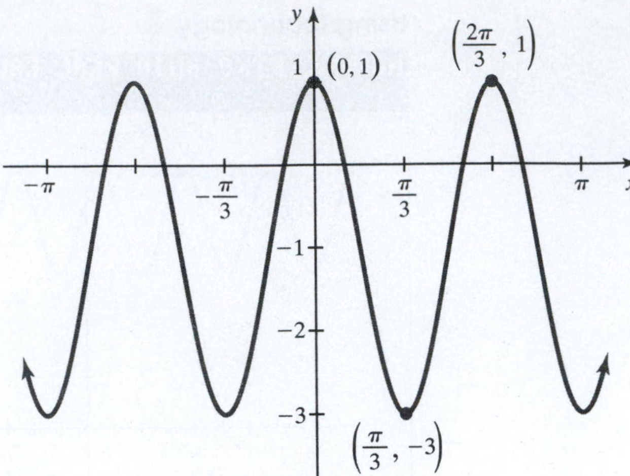

Figure 37

Furthermore, suppose that you were asked to choose a function that describes the graph seen in **Figure 37** where the function is chosen from the five functions listed below.

$$y = 2\cos(3x) - 1$$

$$y = 2\sin\left(3x + \frac{\pi}{2}\right) - 1$$

$$y = -2\cos(3x + \pi) - 1$$

$$y = -2\sin\left(3x - \frac{\pi}{2}\right) - 1$$

$$y = 2\cos(-3x) - 1$$

Which of the five functions above would you choose to accurately describe the graph displayed in **Figure 37**? The answer is **all five functions describe the graph seen in Figure 37**! You might want to take a few minutes and indeed verify that all of these functions describe the same graph.

Therefore, given the graph of a sine curve or cosine curve, it is impossible to determine a unique function that describes the graph unless certain assumptions are made. In this objective, when given the graph of a function of the form $y = A\sin(Bx - C) + D$ or $y = A\cos(Bx - C) + D$, we will first state whether the given graph is a sine curve or a cosine curve and we will assume that $B > 0$. We will also always label the five quarter points of one cycle of the given graph to correspond with the five quarter points of the graph of $y = \sin x$ or $y = \cos x$ over the interval $[0, 2\pi]$. If these assumptions are made, then we can determine a unique function that describes the given graph.

We can follow the six steps outlined below to determine the function that describes a given sine or cosine curve.

Steps for Determining an Equation of a Function of the Form $y = A\sin(Bx - C) + D$ or $y = A\cos(Bx - C) + D$ Given the Graph

Step 1. Subtract the x-coordinate of the first quarter point from the x-coordinate of the fifth quarter point to determine the period.

Step 2. Use the equation $P = \frac{2\pi}{B}$ to determine the value of $B > 0$.

Step 3. The x-coordinate of the first quarter point represents the phase shift, $\frac{C}{B}$. Use this information to determine the value of C.

Step 4. The amplitude is $|A| = \frac{|b - a|}{2}$ where $[a, b]$ is the range of the given graph.

Step 5. If the given graph is a sine curve, then $A = |A|$ if the graph is *increasing* from the first quarter point to the second quarter point. Otherwise, $A = -|A|$.

If the given graph is a cosine curve, then $A = |A|$ if the graph is *decreasing* from the first quarter point to the second quarter point. Otherwise, $A = -|A|$.

Step 6. Choose the first quarter point (x_1, y_1) that lies on the given graph and the corresponding first quarter point of the graph of $y = \sin x$ or $y = \cos x$. Note that the first quarter point of $y = \sin x$ is $(0, 0)$ and the first quarter point of $y = \cos x$ is $(0, 1)$.

The relationship between y_1, the y-coordinate of the first quarter point of the given graph, and the y-coordinate of the first quarter point of $y = \sin x$ or $y = \cos x$ is given by the following:

If the graph is a sine curve, then $y_1 = (0)A + D$.

If the graph is a cosine curve, then $y_1 = (1)A + D$.

Use this information to determine the value of D.

Example 4 Determining an Equation of a Function Given the Graph that it Represents

a. The graph of a function of the form $y = A\cos(Bx - C) + D$, where $B > 0$ is given below. The five quarter points of one cycle of the graph are labeled. These five quarter points on the graph correspond to the five quarter points of the graph of $y = \cos x$ over the interval $[0, 2\pi]$. Determine the specific function that is represented by the given graph based on the association of the labeled quarter points and the quarter points of the graph of $y = \cos x$ over the interval $[0, 2\pi]$.

$y = A\cos(Bx - C) + D$

b. The graph of a function of the form $y = A\sin(Bx - C) + D$ where $B > 0$ is given below. The five quarter points of one cycle of the graph are labeled. These five quarter points on the graph correspond to the five quarter points of the graph of $y = \sin x$ over the interval $[0, 2\pi]$. Determine the specific function that is represented by the given graph based on the association of the labeled quarter points and the quarter points of the graph of $y = \sin x$ over the interval $[0, 2\pi]$.

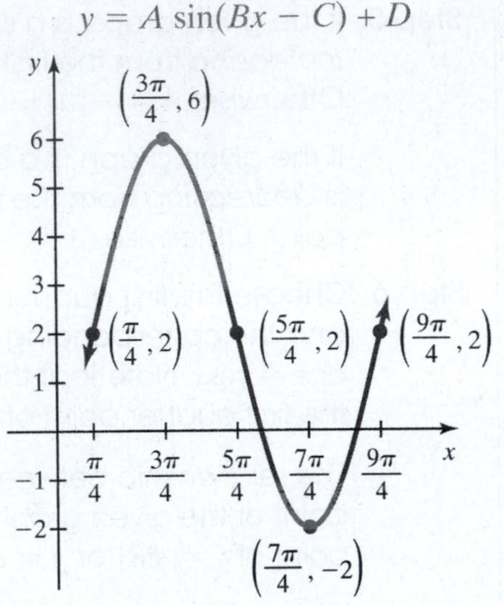

Solution

a. The given graph is shown on the right. Before completing Step 1, we compare the coordinates of the five quarter points of the graph of $y = \cos x$ over the interval $[0, 2\pi]$ with the five quarter points of the given graph.

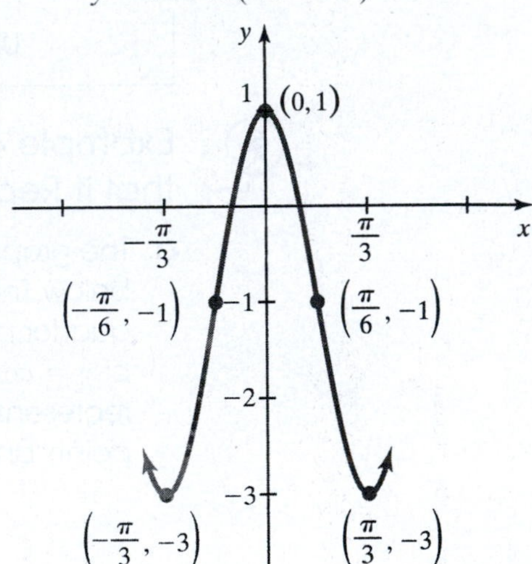

	First	Second	Third	Fourth	Fifth
Quarter Points of $y = \cos x$:	$(0, 1)$	$\left(\frac{\pi}{2}, 0\right)$	$(\pi, -1)$	$\left(\frac{3\pi}{2}, 0\right)$	$(2\pi, 1)$.
Quarter Points of the Given Graph:	$\left(-\frac{\pi}{3}, -3\right)$	$\left(-\frac{\pi}{6}, -1\right)$	$(0, 1)$	$\left(\frac{\pi}{6}, -1\right)$	$\left(\frac{\pi}{3}, -3\right)$

Step 1. Looking at the graph we see that the x-coordinate of the first quarter point is $-\frac{\pi}{3}$ and the x-coordinate of the fifth quarter point is $\frac{\pi}{3}$.

Therefore, the period is $\frac{\pi}{3} - \left(-\frac{\pi}{3}\right) = \frac{\pi}{3} + \frac{\pi}{3} = \frac{2\pi}{3}$.

Step 2. The period is $P = \frac{2\pi}{B} = \frac{2\pi}{3}$. Therefore, $B = 3$.

Step 3. The x-coordinate of the first quarter point is $-\frac{\pi}{3}$. Therefore, the phase shift is $\frac{C}{B} = -\frac{\pi}{3}$. From Step 2, we know that $B = 3$. Thus, $\frac{C}{B} = -\frac{\pi}{3}$ so $C = -\pi$.

Step 4. The range of this graph is $[-3, 1]$. Therefore, the amplitude is $|A| = \frac{|1-(-3)|}{2} = \frac{|4|}{2} = 2$.

Step 5. The given graph is a cosine curve that *increases* between the first quarter point and the second quarter point. Therefore, the value of A must be negative. Hence, $A = -|A| = -2$.

Step 6. The first quarter point of the given graph is $(x_1, y_1) = \left(-\frac{\pi}{3}, -3\right)$. The first quarter point of the graph of $y = \cos x$ is $(0,1)$. Therefore, we can use the equation $y_1 = (1)A + D$ to solve for D.

$y_1 = (1)A + D$ Start with the equation describing the y-coordinate of the first quarter point of the given graph.

$-3 = (1)(-2) + D$ Substitute $y_1 = -3$, and $A = -2$.

Solve this equation for D to get $D = -1$.

We see that $A = -2$, $B = 3$, $C = -\pi$, and $D = -1$. Therefore, substituting these values into the function of the form $y = A\cos(Bx - C) + D$, we see that the function which describes the given graph using the association with the quarter points of the function $y = \cos x$ over the interval $[0, 2\pi]$ is $y = -2\cos(3x - (-\pi)) - 1$ or $y = -2\cos(3x + \pi) - 1$. You should sketch the graph of $y = -2\cos(3x + \pi) - 1$ to make sure that it matches the given graph.

b. Work through this interactive video to verify that the function of the form $y = A\sin(Bx - C) + D$ whose graph is shown on the right is $y = 4\sin\left(x - \frac{\pi}{4}\right) + 2$.

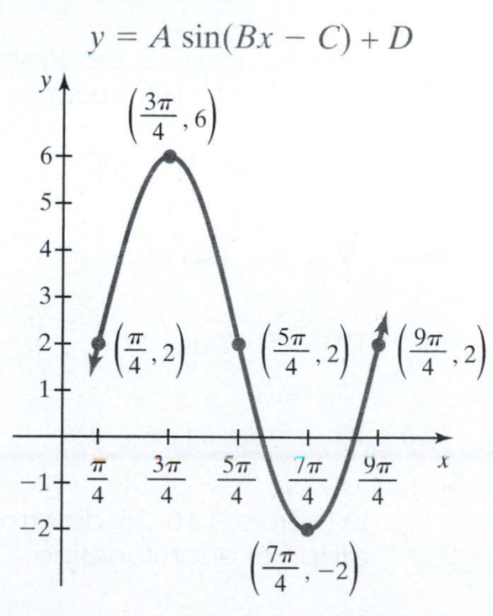

You Try It Work through the following You Try It problem.

Work Exercises 26–33 in this textbook or in the MyLab Math Study Plan.

7.2 Exercises

Skill Check Exercises

In exercises SCE-1 and SCE-3, solve the inequality. Write your answer in interval notation form.

SCE-1. $0 \le 2x - \pi \le 2\pi$ **SCE-2.** $0 \le 3x - \frac{\pi}{6} \le 2\pi$ **SCE-3.** $0 \le \pi x - 3 \le 2\pi$

In exercises SCE-4 and SCE-5, perform the indicated operations and simplify.

SCE-4. $\left(\frac{3\pi}{2} - \frac{\pi}{2}\right) \div 4$ **SCE-5.** $\left(\frac{3\pi}{4} - \frac{\pi}{4}\right) \div 4$

In exercises SCE-6 through SCE-11, an algebraic expression is given. On the second line, complete the factorization by correctly filling in the parentheses. The factored expression on the second line should be equivalent to the expression on the first line.

SCE-6. $(4x - \pi)$
$= 4(\qquad)$

SCE-7. $\left(5x - \frac{\pi}{3}\right)$
$= 5(\qquad)$

SCE-8. $(-3x + \pi)$
$= -3(\qquad)$

SCE-9. $\left(-4x + \frac{\pi}{3}\right)$
$= -4(\qquad)$

SCE-10. $(\pi - 2x)$
$= -2(\qquad)$

SCE-11. $(\pi x - 3)$
$= \pi(\qquad)$

In Exercises 1–5, determine the amplitude, range, period, and phase shift and then sketch the graph.

SbS 1. $y = \sin(x - \pi)$ SbS 2. $y = \cos\left(x + \frac{\pi}{2}\right)$ SbS 3. $y = \sin\left(x - \frac{\pi}{6}\right)$

SbS 4. $y = \sin\left(x + \frac{2\pi}{3}\right)$ SbS 5. $y = \cos(x - 2)$

In Exercises 6–15, determine the amplitude, range, period, and phase shift and then sketch the graph of each function.

SbS 6. $y = \cos(3x + \pi)$ SbS 7. $y = -\sin(2x - \pi)$ SbS 8. $y = 3\cos(4x - \pi)$

SbS 9. $y = -2\sin(6x + \pi)$ SbS 10. $y = 4\cos(-2x - \pi)$ SbS 11. $y = 3\sin\left(-2x - \frac{\pi}{2}\right)$

SbS 12. $y = -2\sin\left(3x - \frac{\pi}{2}\right)$ SbS 13. $y = -4\cos\left(-x - \frac{\pi}{3}\right)$ SbS 14. $y = -\cos(\pi - 2x)$

SbS 15. $y = 4\sin(\pi x - 3)$

In Exercises 16–25, determine the amplitude, range, period, and phase shift and then sketch the graph of each function.

SbS 16. $y = 2\cos(3x + \pi) + 1$ SbS 17. $y = -\sin(2x - \pi) - 2$ SbS 18. $y = 3\cos(4x - \pi) - 1$

SbS 19. $y = 1 - 2\sin(6x + \pi)$ SbS 20. $y = 4\cos(-2x - \pi) + 3$ SbS 21. $y = -3\sin\left(-2x + \frac{\pi}{4}\right) - 4$

SbS 22. $y = 5 - 2\sin\left(3x - \frac{\pi}{2}\right)$ SbS 23. $y = -4\cos\left(-x - \frac{\pi}{3}\right) + 3$ SbS 24. $y = -\cos(\pi - 2x) + 3$

SbS 25. $y = 4\sin(\pi x - 3) - 2$

In Exercises 26–33, the graph is of a function of the form $y = A\cos(Bx - C) + D$ or $y = A\sin(Bx - C) + D$ where $B > 0$ is given. The five quarter points of one cycle of the graph are labeled. These five quarter points on the graph correspond to the five quarter points of the graph of either $y = \cos x$ or $y = \sin x$ over the interval $[0, 2\pi]$. Determine the equation of the specific function that is represented by the given graph based on the association of the labeled quarter points and the quarter points of the graph of either $y = \sin x$ or $y = \cos x$ over the interval $[0, 2\pi]$.

SbS 26.

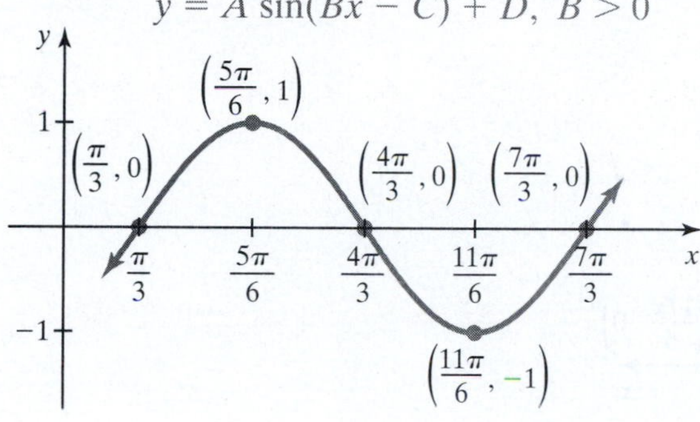

SbS 27.

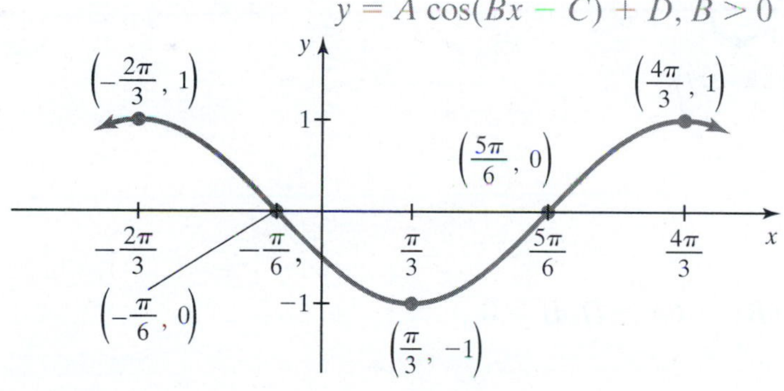

SbS 28.

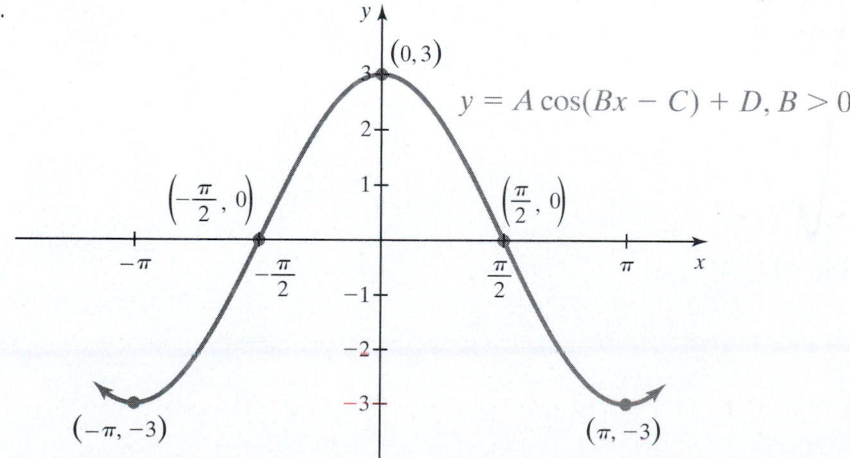

29.

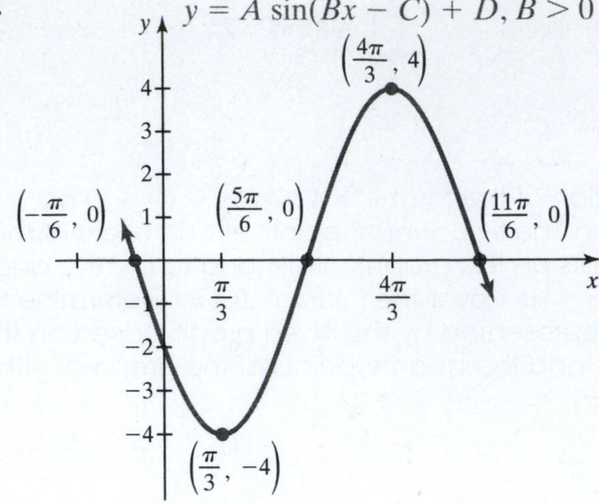

SbS 30.

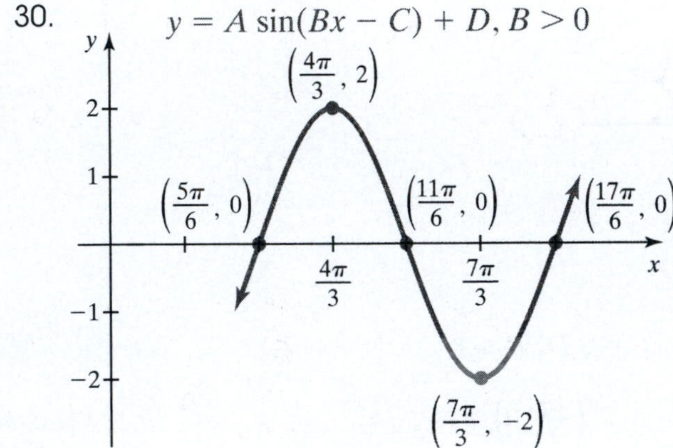

SbS 31.

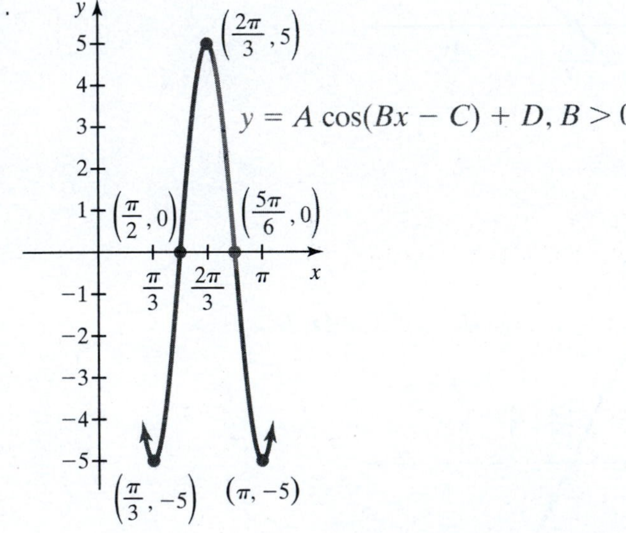

SbS 32.

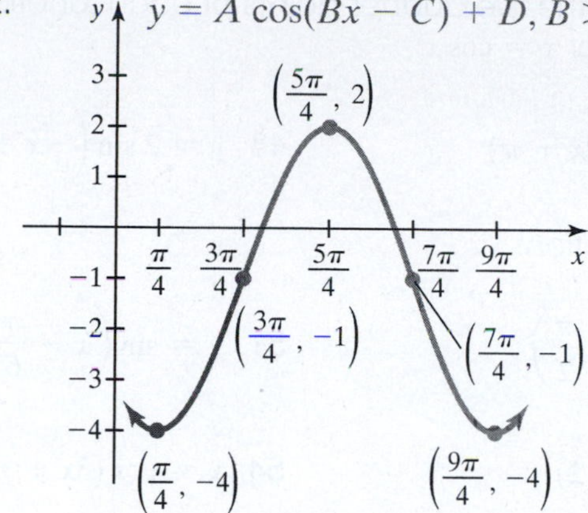

SbS 33.

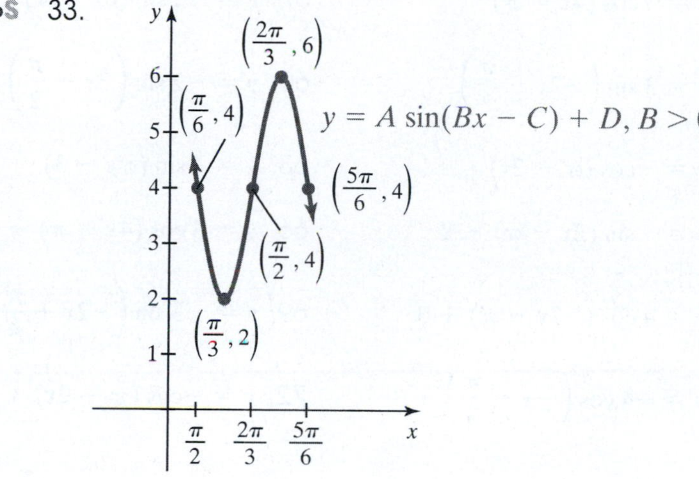

Brief Exercises

In Exercises 34–36, determine the amplitude of each function.

34. $y = \sin\left(x - \frac{\pi}{6}\right)$

35. $y = -3\cos(2x + \pi)$

36. $y = 2\sin\left(-x + \frac{\pi}{3}\right) - 2$

In Exercises 37–39, determine the range of each function.

37. $y = \sin\left(x - \frac{\pi}{6}\right)$

38. $y = -3\cos(2x + \pi)$

39. $y = 2\sin\left(-x + \frac{\pi}{3}\right) - 2$

In Exercises 40–42, determine the period of each function.

40. $y = \sin\left(x - \frac{\pi}{6}\right)$

41. $y = -3\cos(2x + \pi)$

42. $y = 2\sin\left(-x + \frac{\pi}{3}\right) - 2$

In Exercises 43–45, determine the phase shift of each function.

43. $y = \sin\left(x - \frac{\pi}{6}\right)$

44. $y = -3\cos(2x + \pi)$

45. $y = 2\sin\left(-x + \frac{\pi}{3}\right) - 2$

In Exercises 46–48, determine the coordinates for the five quarter points of each function that correspond to the five quarter points of $y = \sin x$ or $y = \cos x$.

46. $y = \sin\left(x - \frac{\pi}{6}\right)$

47. $y = -3\cos(2x + \pi)$

48. $y = 2\sin\left(-x + \frac{\pi}{3}\right) - 2$

In Exercises 49–72, sketch the graph of each function.

49. $y = \sin(x - \pi)$

50. $y = \cos\left(x + \frac{\pi}{2}\right)$

51. $y = \sin\left(x - \frac{\pi}{6}\right)$

52. $y = \sin\left(x + \frac{2\pi}{3}\right)$

53. $y = \cos(x - 2)$

54. $y = \cos(3x + \pi)$

55. $y = -\sin(2x - \pi)$

56. $y = 3\cos(4x - \pi)$

57. $y = -2\sin(6x + \pi)$

58. $y = 4\cos(-2x - \pi)$

59. $y = 3\sin\left(-2x - \frac{\pi}{2}\right)$

60. $y = -2\sin\left(3x - \frac{\pi}{2}\right)$

61. $y = -4\cos\left(-x - \frac{\pi}{3}\right)$

62. $y = -\cos(\pi - 2x)$

63. $y = 4\sin(\pi x - 3)$

64. $y = 2\cos(3x + \pi) + 1$

65. $y = -\sin(2x - \pi) - 2$

66. $y = 3\cos(4x - \pi) - 1$

67. $y = 1 - 2\sin(6x + \pi)$

68. $y = 4\cos(-2x - \pi) + 3$

69. $y = -3\sin\left(-2x + \frac{\pi}{4}\right) - 4$

70. $y = 5 - 2\sin\left(3x - \frac{\pi}{2}\right)$

71. $y = -4\cos\left(-x - \frac{\pi}{3}\right) + 3$

72. $y = -\cos(\pi - 2x) + 3$

In Exercises 73–80, the graph of a function of the form $y = A\cos(Bx - C) + D$ or $y = A\sin(Bx - C) + D$, where $B > 0$ is given. The five quarter points of one cycle of the graph are labeled. These five quarter points on the graph correspond to the five quarter points of the graph of either $y = \cos x$ or $y = \sin x$ over the interval $[0, 2\pi]$. Determine the equation of the specific function that is represented by the given graph based on the association of the labeled quarter points and the quarter points of the graph of either $y = \sin x$ or $y = \cos x$ over the interval $[0, 2\pi]$.

73.

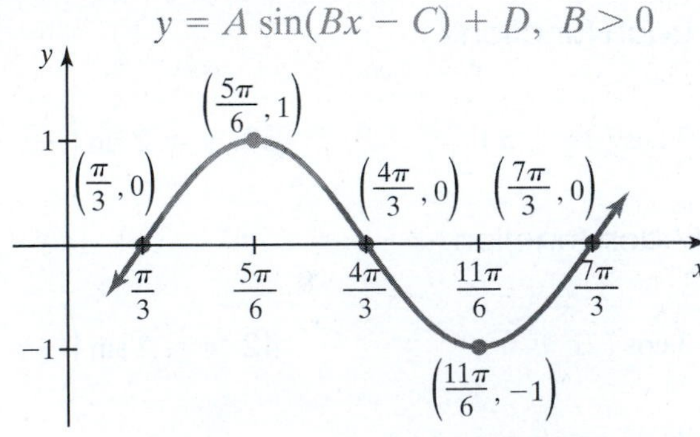

74.

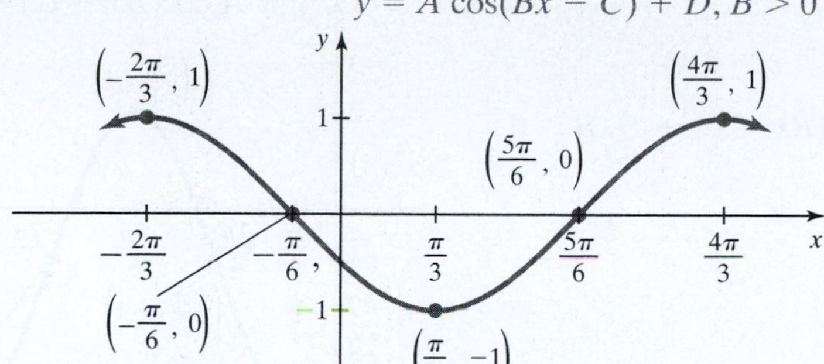
$y = A\cos(Bx - C) + D, B > 0$
$\left(-\frac{2\pi}{3}, 1\right)$
$\left(\frac{4\pi}{3}, 1\right)$
$\left(\frac{5\pi}{6}, 0\right)$
$\left(-\frac{\pi}{6}, 0\right)$
$\left(\frac{\pi}{3}, -1\right)$
$-\frac{2\pi}{3}$
$-\frac{\pi}{6}$
$\frac{\pi}{3}$
$\frac{5\pi}{6}$
$\frac{4\pi}{3}$
1
−1
x
y

75.

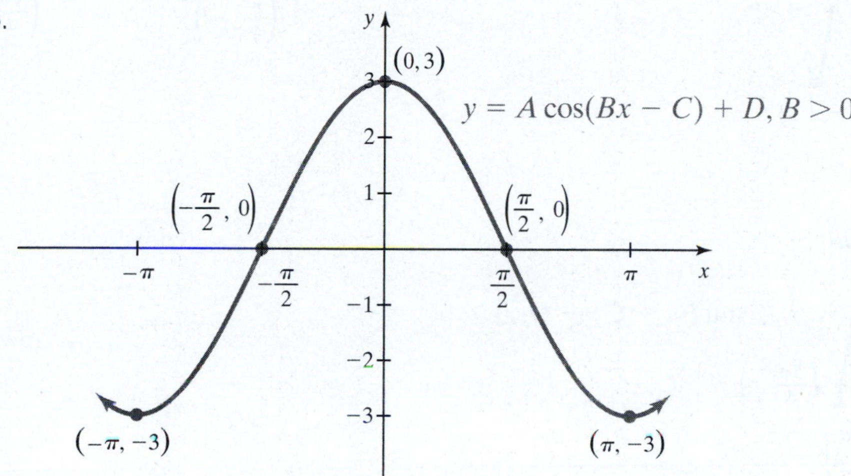
$y = A\cos(Bx - C) + D, B > 0$
(0, 3)
$\left(-\frac{\pi}{2}, 0\right)$
$\left(\frac{\pi}{2}, 0\right)$
$(-\pi, -3)$
$(\pi, -3)$
$-\pi$
$-\frac{\pi}{2}$
$\frac{\pi}{2}$
π
3
2
1
−1
−2
−3
x
y

76.

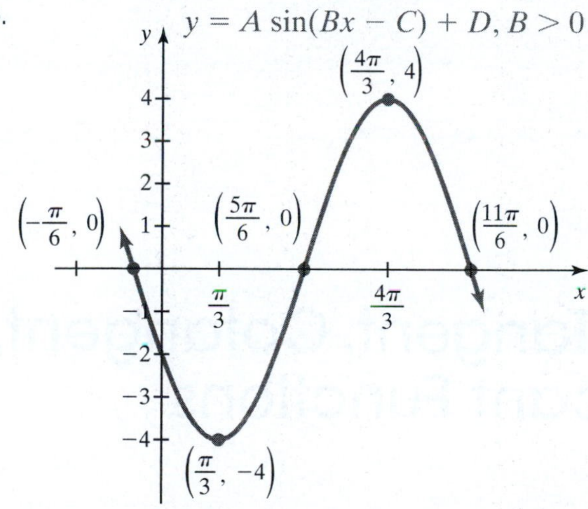
$y = A\sin(Bx - C) + D, B > 0$
$\left(\frac{4\pi}{3}, 4\right)$
$\left(-\frac{\pi}{6}, 0\right)$
$\left(\frac{5\pi}{6}, 0\right)$
$\left(\frac{11\pi}{6}, 0\right)$
$\left(\frac{\pi}{3}, -4\right)$
$\frac{\pi}{3}$
$\frac{4\pi}{3}$
4
3
2
1
−1
−2
−3
−4
x
y

77.

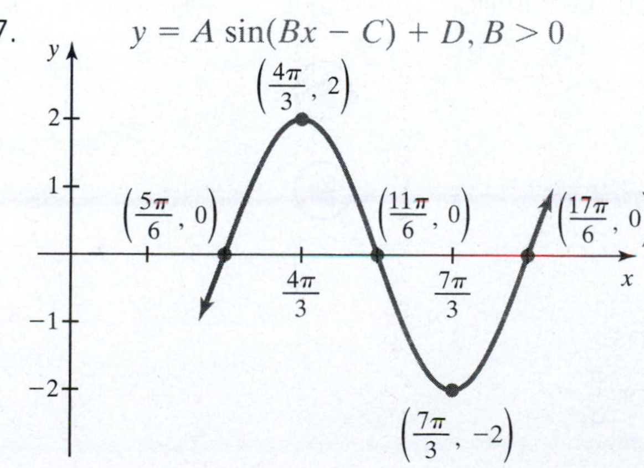
$y = A\sin(Bx - C) + D, B > 0$
$\left(\frac{4\pi}{3}, 2\right)$
$\left(\frac{5\pi}{6}, 0\right)$
$\left(\frac{11\pi}{6}, 0\right)$
$\left(\frac{17\pi}{6}, 0\right)$
$\left(\frac{7\pi}{3}, -2\right)$
$\frac{4\pi}{3}$
$\frac{7\pi}{3}$
2
1
−1
−2
x
y

78.

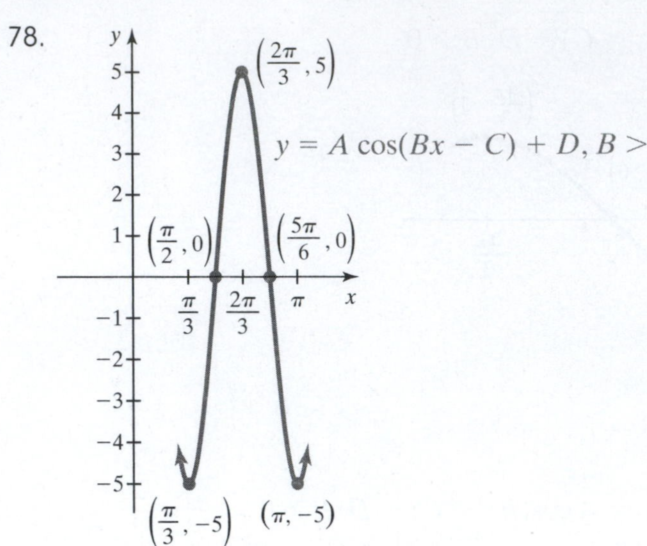

79.

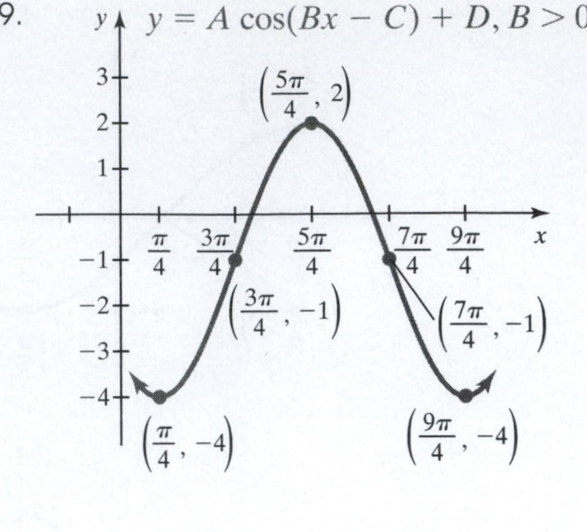

80.

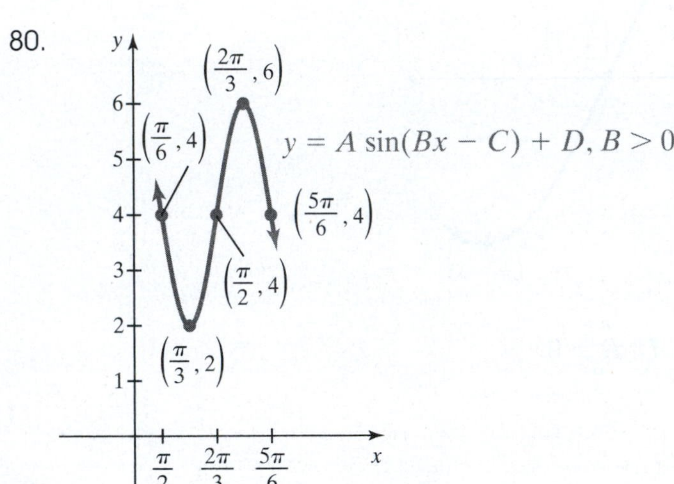

7.3 The Graphs of the Tangent, Cotangent, Cosecant, and Secant Functions

THINGS TO KNOW

Before working through this section, be sure that you are familiar with the following concepts:

VIDEO  

You Try It

1. Using the Special Right Triangles (Section 6.4)

You Try It

2. Finding the Values of the Trigonometric Functions of Quadrantal Angles (Section 6.5)

You Try It

3. Evaluating Trigonometric Functions of Angles Belonging to the $\frac{\pi}{3}$, $\frac{\pi}{6}$, or $\frac{\pi}{4}$ Families (Section 6.5)

You Try It

4. Sketching Graphs of the Form $y = A\sin(Bx)$ and $y = A\cos(Bx)$ (Section 7.1)

You Try It

5. Sketching Graphs of the Form $y = A\sin(Bx - C) + D$ and $y = A\cos(Bx - C) + D$ (Section 7.2)

OBJECTIVES

1 Understanding the Graph of the Tangent Function and Its Properties

2 Sketching Graphs of the Form $y = A\tan(Bx - C) + D$

3 Understanding the Graph of the Cotangent Function and Its Properties

4 Sketching Graphs of the Form $y = A\cot(Bx - C) + D$

5 Determine the Equation of a Function of the Form $y = A\tan(Bx - C) + D$ or $y = A\cot(Bx - C) + D$ Given Its Graph

6 Understanding the Graph of the Cosecant and Secant Functions and Their Properties

7 Sketching Graphs of the Form $y = A\csc(Bx - C) + D$ and $y = A\sec(Bx - C) + D$

SECTION 7.3 EXERCISES

OBJECTIVE 1 UNDERSTANDING THE GRAPH OF THE TANGENT FUNCTION AND ITS PROPERTIES

In Sections 7.1 and 7.2, we sketched the sine and cosine functions, as well as variations of them. In this section, we will sketch the tangent, cotangent, secant, and cosecant functions, as well as variations of them. We start by introducing the graph of $y = \tan x$. Watch this **video** to see how the graph of the tangent function is established.

Unlike the functions $y = \sin x$ and $y = \cos x$, the domain of $y = \tan x$ is *not* all real numbers. To determine the domain of $y = \tan x$, we start by using a **quotient identity** to rewrite $y = \tan x$ as $y = \dfrac{\sin x}{\cos x}$. Because division by zero is never allowed, we see that the tangent function is not defined when the denominator, $\cos x$, is equal to zero. The **zeros of $y = \cos x$** are of the form $(2n + 1)\cdot\dfrac{\pi}{2}$, where n is an integer. Therefore, the domain of $y = \tan x$ is all real numbers *except* values of x of the form $(2n + 1)\cdot\dfrac{\pi}{2}$, where n is an integer. Thus, the tangent function is undefined for values of x belonging to the set

$$\left\{\ldots, -\frac{5\pi}{2}, -\frac{3\pi}{2}, -\frac{\pi}{2}, \frac{\pi}{2}, \frac{3\pi}{2}, \frac{5\pi}{2}, \ldots\right\}$$

We will see that the graph approaches a **vertical asymptote** at these values.

We can determine the x-intercepts of the tangent function by observing that $y = \tan x = \dfrac{\sin x}{\cos x}$ will equal zero only when the numerator, $\sin x$, is equal to zero. Therefore, the x-intercepts, or zeros, of $y = \tan x = \dfrac{\sin x}{\cos x}$ will be precisely the zeros of $y = \sin x$. Thus, the x-intercepts of $y = \tan x$ are of the form $n\pi$, where n is an integer. Therefore, the x-intercepts belong to the set $\{\ldots, -3\pi, -2\pi, \pi, 0, \pi, 2\pi, 3\pi, \ldots\}$.

We begin by creating a table of values and plotting points to sketch one cycle of the graph of $y = \tan x$. We start by choosing $x = 0, x = \frac{\pi}{6}, x = \frac{\pi}{4}$, and $x = \frac{\pi}{3}$. See Table 9 and Figure 38.

Table 9

x	$y = \tan x$
0	0
$\frac{\pi}{6}$	$\frac{1}{\sqrt{3}} \approx 0.577$
$\frac{\pi}{4}$	1
$\frac{\pi}{3}$	$\sqrt{3} \approx 1.732$

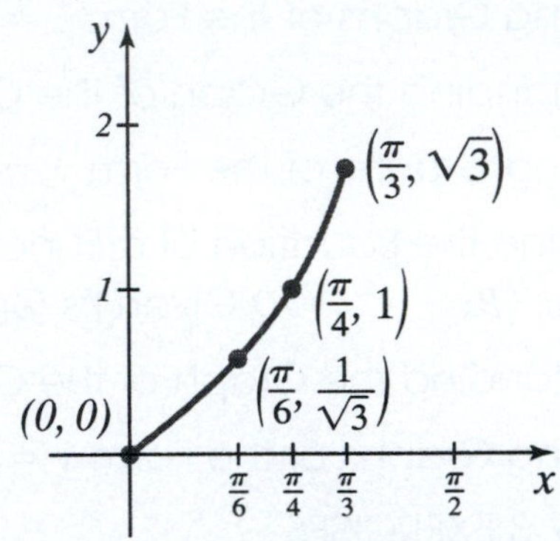

Figure 38 A portion of the graph $y = \tan x$

If we continue to choose values of x closer and closer to $x = \frac{\pi}{2}$, we see that the values of $y = \tan x$ increase without bound and the graph approaches the vertical asymptote $x = \frac{\pi}{2}$. See Table 10 and Figure 39.

Table 10

x	$y = \tan x$
$\frac{4\pi}{10}$	≈ 3.078
$\frac{49\pi}{100}$	≈ 31.821
$\frac{499\pi}{1000}$	≈ 318.309
$\frac{4999\pi}{10{,}000}$	≈ 3183.099
$\frac{5000\pi}{10{,}000} = \frac{\pi}{2}$	Undefined

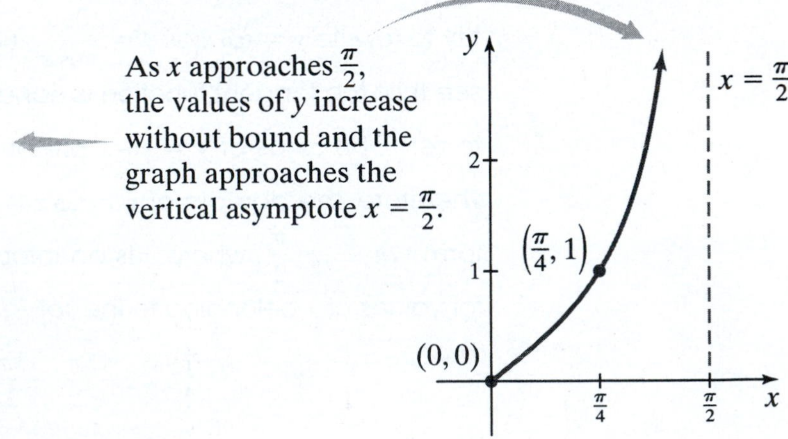

Figure 39 The graph of $y = \tan x$ on the interval $\left[0, \frac{\pi}{2}\right)$

The tangent function is an **odd function**, which means that the graph is symmetric about the origin. Therefore, we can reflect the portion of the graph seen in **Figure 39** about the origin to get one complete cycle of the graph of $y = \tan x$ on the interval $\left(-\frac{\pi}{2}, \frac{\pi}{2}\right)$. See Figure 40.

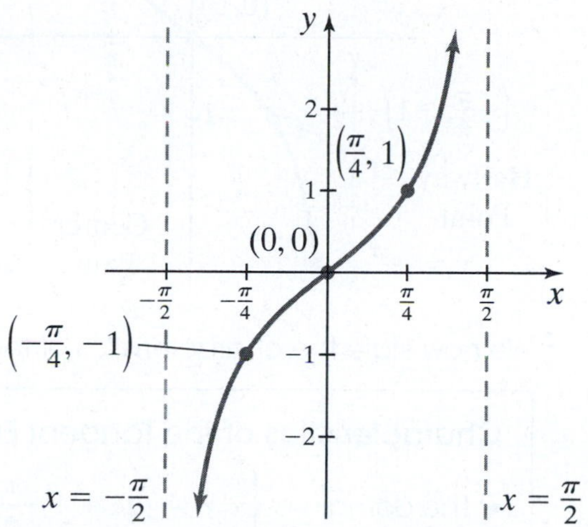

Figure 40 One cycle of the graph of $y = \tan x$ on the interval $\left(-\frac{\pi}{2}, \frac{\pi}{2}\right)$

We will call the graph seen in **Figure 40** the **principal cycle** of the graph of $y = \tan x$. Notice that the graph of $y = \tan x$ completes this principal cycle on the interval $\left(-\frac{\pi}{2}, \frac{\pi}{2}\right)$. The length of this interval is π units. Therefore, the tangent function is **periodic** with a period of $P = \pi$. Each cycle, or period, of the graph lies between consecutive values of x for which the function is undefined. Also note that each cycle of the graph of $y = \tan x$ is **one-to-one**. We can obtain a complete graph of $y = \tan x$ by repeating the graph seen in **Figure 40** infinitely many times in either direction.

For each cycle of the graph of $y = \tan x$, there are three special points that will help us sketch the graph. The first point is called the **center point**. The center point is located at the center of the graph of each cycle. The center point of the principal cycle of the graph of $y = \tan x$ is located at the origin. Note that when we sketch the graphs of variations of the tangent function in Objective 2, this center point will not necessarily be located at the origin.

The two other special points are called **halfway points**. One halfway point is located halfway between the x-coordinate of the center point and the vertical asymptote to the left of the center point and has a y-coordinate of -1. The other halfway point is located halfway between the x-coordinate of the center point and the vertical asymptote to the right of the center point and has a y-coordinate of 1. See Figure 41.

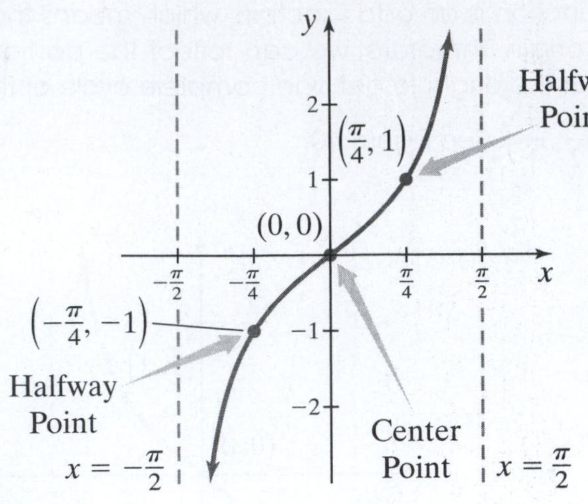

Figure 41 The principal cycle of the graph of $y = \tan x$ with the center point and the two halfway points labeled.

We now state the characteristics of the tangent function.

Characteristics of the Tangent Function

- The domain is $\left\{x \mid x \neq (2n + 1) \cdot \frac{\pi}{2}, \text{ where } n \text{ is an integer}\right\}$.
- The range is $(-\infty, \infty)$.
- The function is periodic with a period of $P = \pi$. The principal cycle of the graph occurs on the interval $\left(-\frac{\pi}{2}, \frac{\pi}{2}\right)$.
- The function has infinitely many vertical asymptotes with equations $x = (2n + 1) \cdot \frac{\pi}{2}$, where n is an integer.
- The y-intercept is 0.
- For each cycle there is one center point. The x-coordinates of the center points are also the x-intercepts, or **zeros**, and are of the form $n\pi$, where n is an integer.
- For each cycle there are two halfway points. The halfway point to the left of the x-intercept has a y-coordinate of -1. The halfway point to the right of the x-intercept has a y-coordinate of 1.
- The function is **odd**, which means $\tan(-x) = -\tan x$. The graph is symmetric about the origin.
- The graph of each cycle of $y = \tan x$ is **one-to-one**.

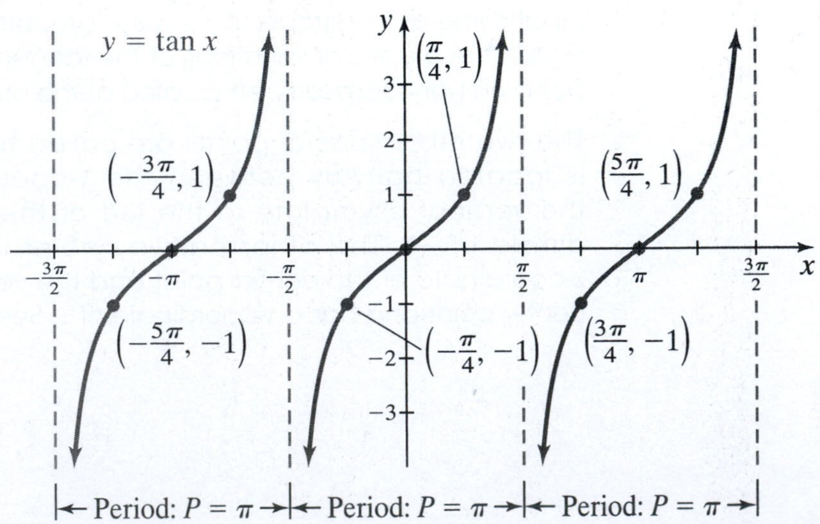

Example 1 Understanding the Graph of $y = \tan x$

List all halfway points of $y = \tan x$ on the interval $\left[-\pi, \frac{5\pi}{2}\right]$ that have a y-coordinate of -1.

Solution From the graph of $y = \tan x$ on the previous page, or using the fact that the graph of each cycle of the graph of $y = \tan x$ is one-to-one, we know that there is exactly one point from each cycle that has a y-coordinate of -1. The only halfway point of the principal cycle of $y = \tan x$ with a y-coordinate of -1 is $\left(-\frac{\pi}{4}, -1\right)$. To find all other halfway points with a y-coordinate of -1, we use the fact that the tangent function is periodic with a period of $P = \pi$. Therefore, every halfway point with a y-coordinate of -1 must be a distance of plus or minus π units horizontally from the previous halfway point with a y-coordinate of -1.

We can express all halfway points with a y-coordinate of -1 as $\left(-\frac{\pi}{4} + k\pi, -1\right)$, where k is an integer. We now choose integer values of k, making sure that the resulting x-coordinate is in the interval $\left[-\pi, \frac{5\pi}{2}\right]$.

$$k = -1: \left(-\frac{\pi}{4} + (-1)\pi, -1\right) \rightarrow \cancel{\left(-\frac{5\pi}{4}, -1\right)} \quad \left(x = -\frac{5\pi}{4} \text{ is not in the interval } \left[-\pi, \frac{5\pi}{2}\right]\right)$$

$$k = 1: \left(-\frac{\pi}{4} + (1)\pi, -1\right) \rightarrow \left(\frac{3\pi}{4}, -1\right)$$

$$k = 2: \left(-\frac{\pi}{4} + (2)\pi, -1\right) \rightarrow \left(\frac{7\pi}{4}, -1\right)$$

$$k = 3: \left(-\frac{\pi}{4} + (3)\pi, -1\right) \rightarrow \cancel{\left(\frac{11\pi}{4}, -1\right)} \quad \left(x = \frac{11\pi}{4} \text{ is not in the interval } \left[-\pi, \frac{5\pi}{2}\right]\right)$$

See Figure 42.

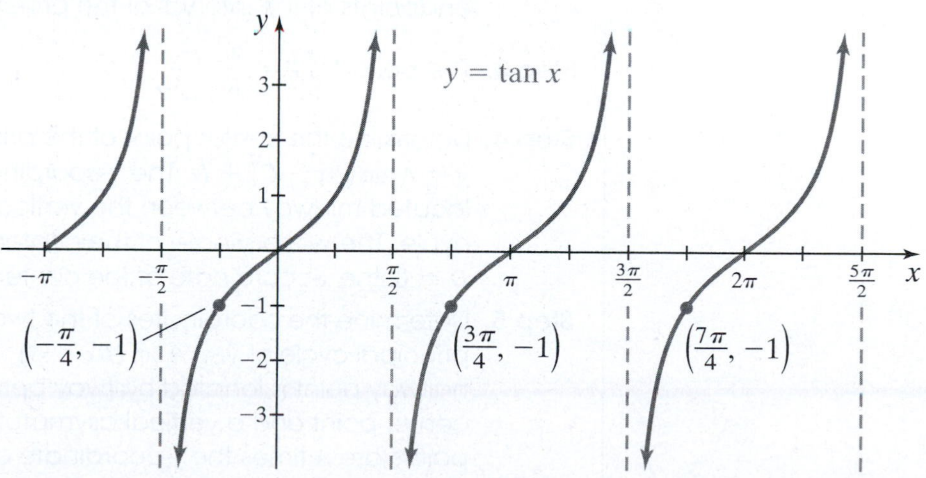

Figure 42 All halfway points of $y = \tan x$ on the interval $\left[-\pi, \frac{5\pi}{2}\right]$ that have a y-coordinate of -1

(eText Screens 7.3-1–7.3-72)

 The three halfway points of $y = \tan x$ that have a y-coordinate of -1 on the interval $\left[-\pi, \frac{5\pi}{2}\right]$ are

$$\left(-\frac{\pi}{4}, -1\right), \left(\frac{3\pi}{4}, -1\right), \text{ and } \left(\frac{7\pi}{4}, -1\right).$$

You may want to watch this **video** to see this solution worked out in detail. ●

 **You Try It** **Work through this You Try It problem.**

Work Exercises 1–5 in this textbook or in the MyLab Math Study Plan.

OBJECTIVE 2 SKETCHING FUNCTIONS OF THE FORM $y = A\tan(Bx - C) + D$

We will now sketch variations of the tangent function. These functions have the form $y = A\tan(Bx - C) + D$. Just like the graph of $y = \tan x$, the graph of $y = A\tan(Bx - C) + D$ has infinitely many cycles. We will focus on sketching the principal cycle. The principal cycle of $y = A\tan(Bx - C) + D$ corresponds to the principal cycle of $y = \tan x$ and has two vertical asymptotes, one center point, and two halfway points. To sketch the principal cycle of functions of the form $y = A\tan(Bx - C) + D$, we follow the process outlined below.

Steps for Sketching Functions of the Form $y = A\tan(Bx - C) + D$

Step 1. If $B < 0$, rewrite the function in an equivalent form such that $B > 0$. Use the odd property of the tangent function.

We now use this new form to determine A, B, C, and D.

Step 2. Determine the interval and the equations of the vertical asymptotes of the principal cycle. The interval for the principal cycle can be found by solving the inequality $-\frac{\pi}{2} < Bx - C < \frac{\pi}{2}$. The vertical asymptotes of the principal cycle occur at the **endpoints** of the interval of the principal cycle.

Step 3. The period is $P = \frac{\pi}{B}$.

Step 4. Determine the center point of the principal cycle of $y = A\tan(Bx - C) + D$. The x-coordinate of the center point is located midway between the vertical asymptotes of the principal cycle. The y-coordinate of the center point is D. Note that when $D = 0$, the x-coordinate of the center point is the x-intercept.

Step 5. Determine the coordinates of the two halfway points of the principal cycle of $y = A\tan(Bx - C) + D$. Each x-coordinate of a halfway point is located halfway between the x-coordinate of the center point and a vertical asymptote. The y-coordinates of these points are A times the y-coordinate of the corresponding **halfway point of $y = \tan x$** plus D.

Step 6. Sketch the vertical asymptotes, plot the center point, and plot the two halfway points. Connect these points with a smooth curve. Complete the sketch, showing appropriate behavior of the graph as it approaches each asymptote.

Example 2 Sketching graphs of the form $y = A\tan(Bx - C) + D$

For each function, determine the interval for the principal cycle. Then for the principal cycle, determine the equations of the vertical asymptotes, the coordinates of the center point, and the coordinates of the halfway points. Sketch the graph.

a. $y = \tan\left(x - \frac{\pi}{6}\right)$ **b.** $y = 4\tan(\pi - 2x) + 3$ **c.** $y = \frac{1}{2}\tan(3x) - 1$

Solution

a. Step 1. For the function $y = \tan\left(x - \frac{\pi}{6}\right)$, we see that $B > 0$.

Therefore, there is no need to rewrite the function and we can skip **step 1.**

Note that $A = 1, B = 1, C = \frac{\pi}{6}$, and $D = 0$.

Step 2. The interval for the principal cycle can be found by solving the inequality $-\frac{\pi}{2} < x - \frac{\pi}{6} < \frac{\pi}{2}$.

$-\frac{\pi}{2} < x - \frac{\pi}{6} < \frac{\pi}{2}$	Write the inequality.
$-\frac{\pi}{2} + \frac{\pi}{6} < x < \frac{\pi}{2} + \frac{\pi}{6}$	Add $\frac{\pi}{6}$ to all parts of the inequality.
$-\frac{3\pi}{6} + \frac{\pi}{6} < x < \frac{3\pi}{6} + \frac{\pi}{6}$	Get a common denominator.
$-\frac{2\pi}{6} < x < \frac{4\pi}{6}$	Combine terms.
$-\frac{\pi}{3} < x < \frac{2\pi}{3}$	Simplify.

The principal cycle occurs on the interval $\left(-\frac{\pi}{3}, \frac{2\pi}{3}\right)$. Therefore, the equations of the vertical asymptotes of the principal cycle are $x = -\frac{\pi}{3}$ and $x = \frac{2\pi}{3}$.

Step 3. The period is $P = \frac{\pi}{B} = \frac{\pi}{1} = \pi$.

Step 4. The x-coordinate of the center point of $y = \tan\left(x - \frac{\pi}{6}\right)$ is located midway between the two vertical asymptotes. Therefore, the x-coordinate of the center point is $x = \frac{-\frac{\pi}{3} + \frac{2\pi}{3}}{2} = \frac{\frac{\pi}{3}}{2} = \frac{\pi}{3} \cdot \frac{1}{2} = \frac{\pi}{6}$. The y-coordinate of the center point is $D = 0$. Thus, the coordinates of the center point are $\left(\frac{\pi}{6}, 0\right)$.

Step 5. The x-coordinate of the halfway point located to the *left* of the center point is $-\frac{\pi}{12}$ and the x-coordinate of the halfway point located to the *right* of the center point is $\frac{5\pi}{12}$. To see how to obtain these values, read these **steps**. The y-coordinates of these two halfway points will be $A = 1$ times the corresponding y-coordinate of the **halfway point of the principal cycle of $y = \tan x$** plus $D = 0$.

Halfway points of the principal cycle of $y = \tan x$: $\left(-\frac{\pi}{4}, -1\right)$, $\left(\frac{\pi}{4}, 1\right)$

$1 \cdot (-1) + 0 = -1$ $\quad$ $1 \cdot 1 + 0 = 1$

Halfway points of the principal cycle of $y = \tan\left(x - \frac{\pi}{6}\right)$: $\left(-\frac{\pi}{12}, -1\right)$, $\left(\frac{5\pi}{12}, 1\right)$

Step 6. We now sketch the vertical asymptotes,

$$x = -\frac{\pi}{3} \text{ and } x = \frac{2\pi}{3};$$

plot the center point $\left(\frac{\pi}{6}, 0\right)$; and plot the two halfway points,

$$\left(-\frac{\pi}{12}, -1\right) \text{ and } \left(\frac{5\pi}{12}, 1\right).$$

We connect these three special points with a smooth curve showing the appropriate behavior of the graph as it approaches each asymptote. Watch this **interactive video** to see how to sketch the graph of $y = \tan\left(x - \frac{\pi}{6}\right)$.

The graph of $y = \tan\left(x - \frac{\pi}{6}\right)$ can be seen in **Figures 43 and 44.**

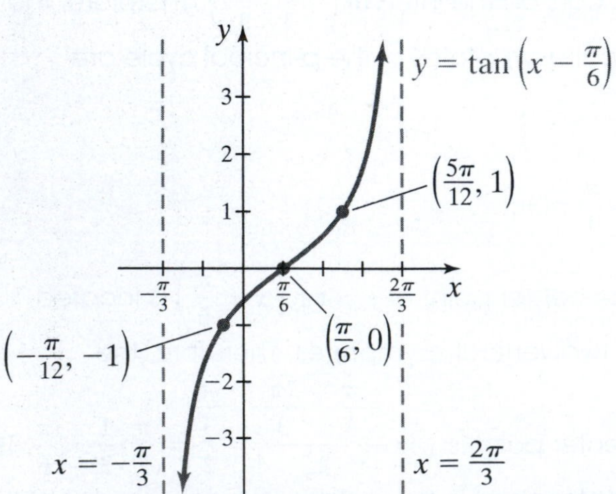

Figure 43 The principal cycle of the graph of $y = \tan\left(x - \frac{\pi}{6}\right)$

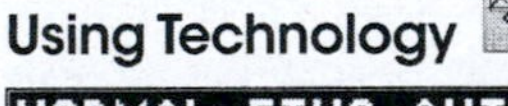

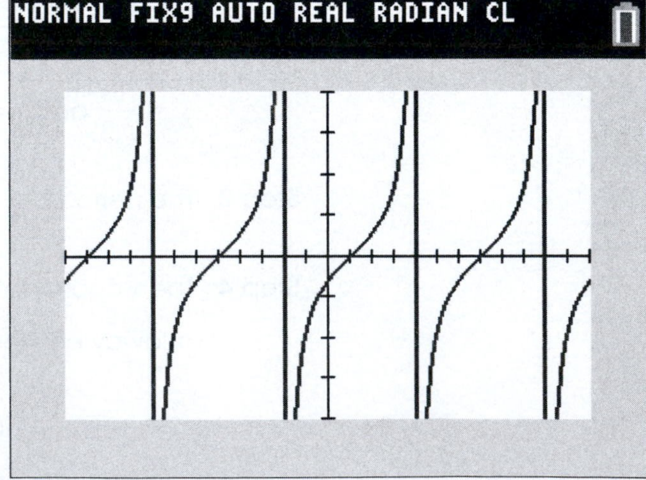

Figure 44 Several cycles of the graph of $y = \tan\left(x - \frac{\pi}{6}\right)$ using a graphing utility

b. Step 1. For the function $y = 4\tan(\pi - 2x) + 3$, we see that $B = -2$, which is less than zero. Therefore, we will rewrite the function as an equivalent function using a positive value of B.

$y = 4\tan(\pi - 2x) + 3$ Write the original function.

$y = 4\tan(-2x + \pi) + 3$ Reorder the terms.

$y = 4\tan(-(2x - \pi)) + 3$ Factor out a negative inside the parentheses.

$y = -4\tan(2x - \pi) + 3$ Use the odd property of $y = \tan x$: $\tan(-x) = -\tan x$.

We see that the function $y = 4\tan(\pi - 2x) + 3$ is equivalent to $y = -4\tan(2x - \pi) + 3$. We now continue using the function $y = -4\tan(2x - \pi) + 3$ with $A = -4, B = 2, C = \pi$, and $D = 3$.

Step 2. The interval for the principal cycle can be found by solving the inequality $-\frac{\pi}{2} < 2x - \pi < \frac{\pi}{2}$.

$-\frac{\pi}{2} < 2x - \pi < \frac{\pi}{2}$ Write the inequality.

$-\frac{\pi}{2} + \pi < 2x < \frac{\pi}{2} + \pi$ Add π to all parts of the inequality.

$-\frac{\pi}{2} + \frac{2\pi}{2} < 2x < \frac{\pi}{2} + \frac{2\pi}{2}$ Get a common denominator.

$\frac{\pi}{2} < 2x < \frac{3\pi}{2}$ Simplify.

$\frac{\pi}{4} < x < \frac{3\pi}{4}$ Divide all parts of the inequality by 2.

The principal cycle occurs on the interval $\left(\frac{\pi}{4}, \frac{3\pi}{4}\right)$. Thus, the equations of the vertical asymptotes of the principal cycle are $x = \frac{\pi}{4}$ and $x = \frac{3\pi}{4}$.

Step 3. The period is $P = \frac{\pi}{B} = \frac{\pi}{2}$.

Step 4. The x-coordinate of the center point of $y = -4\tan(2x - \pi) + 3$ is located midway between the two vertical asymptotes. Thus, the x-coordinate of the center point is

$$x = \frac{\frac{\pi}{4} + \frac{3\pi}{4}}{2} = \frac{\frac{4\pi}{4}}{2} = \frac{\pi}{2}.$$

The y-coordinate of the center point is $D = 3$. Therefore, the coordinates of the center point of

$y = -4\tan(2x - \pi) + 3$ are $\left(\frac{\pi}{2}, 3\right)$.

Step 5. The x-coordinates of the two halfway points are $x = \frac{3\pi}{8}$ and $x = \frac{5\pi}{8}$.

To see how to obtain these values, read these **steps**. The y-coordinates of these two halfway points will be $A = -4$ times the corresponding y-coordinate of the **halfway point of the principal cycle of $y = \tan x$** plus $D = 3$.

Halfway points of the principal cycle of $y = \tan x$: $\left(-\frac{\pi}{4}, -1\right)$ $\left(\frac{\pi}{4}, 1\right)$

$(-4)\cdot(-1) + 3 = 7$ $(-4)\cdot(1) + 3 = -1$

Halfway points of the principal cycle of $y = -4\tan(2x - \pi) + 3$: $\left(\frac{3\pi}{8}, 7\right)$ $\left(\frac{5\pi}{8}, -1\right)$

Step 6. We now sketch the vertical asymptotes,

$$x = \frac{\pi}{4} \text{ and } x = \frac{3\pi}{4};$$

plot the center point $\left(\frac{\pi}{2}, 3\right)$; and plot the halfway points $\left(\frac{3\pi}{8}, 7\right)$ and $\left(\frac{5\pi}{8}, -1\right)$. We connect these three special points with a smooth curve showing the appropriate behavior of the graph as it approaches each asymptote. Watch this **interactive video** to see how to sketch the graph of $y = 4\tan(\pi - 2x) + 3$. The graph of $y = 4\tan(\pi - 2x) + 3$ can be seen in **Figures 45** and **46**.

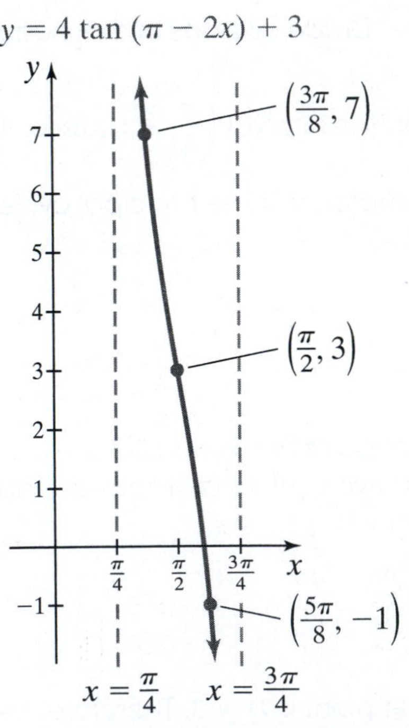

Figure 45 The principal cycle of the graph of $y = 4\tan(\pi - 2x) + 3$

Using Technology

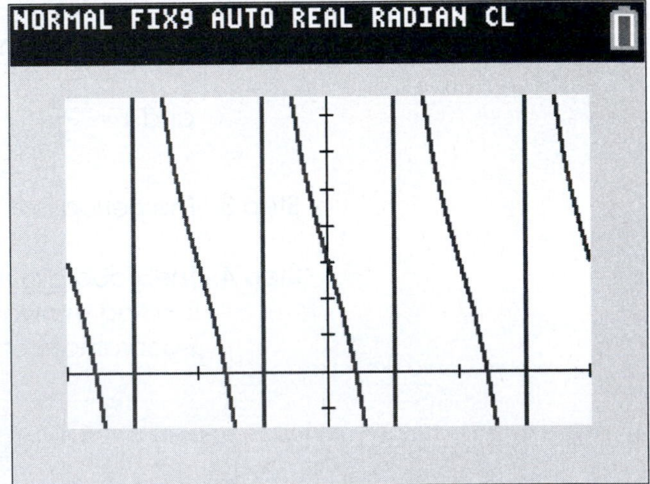

Figure 46 Several cycles of the graph of $y = 4\tan(\pi - 2x) + 3$ using a graphing utility

c. See if you can sketch the graph of $y = \frac{1}{2}\tan(3x) - 1$ on your own. Watch this **interactive video** to see if you are correct.

We could have used transformations to sketch the functions from Example 2. Click on the Show Graph link under the following functions to see how to sketch the function using transformations.

$y = \tan\left(x - \frac{\pi}{6}\right)$	$y = 4\tan(\pi - 2x) + 3$	$y = \frac{1}{2}\tan(3x) - 1$
Show Graph	Show Graph	Show Graph

You Try It Work through this You Try It problem.

Work Exercises 6–19 in this textbook or in the MyLab Math Study Plan.

OBJECTIVE 3 UNDERSTANDING THE GRAPH OF THE COTANGENT FUNCTION AND ITS PROPERTIES

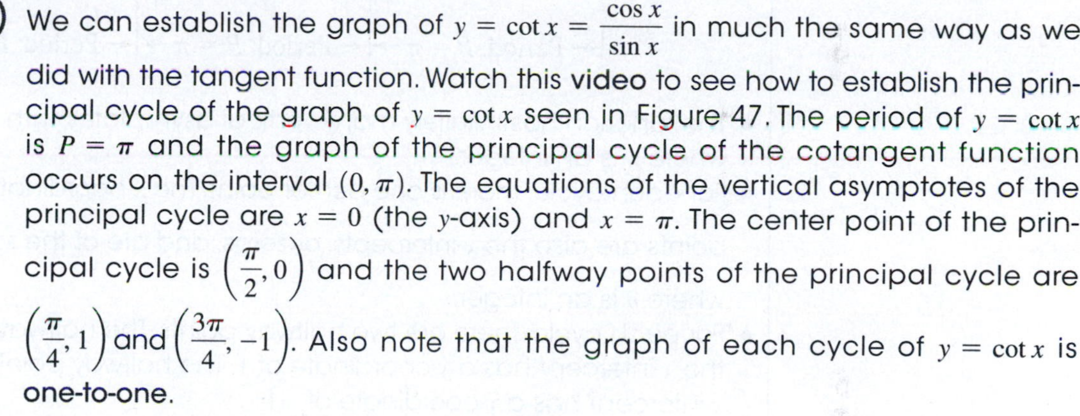

We can establish the graph of $y = \cot x = \dfrac{\cos x}{\sin x}$ in much the same way as we did with the tangent function. Watch this **video** to see how to establish the principal cycle of the graph of $y = \cot x$ seen in Figure 47. The period of $y = \cot x$ is $P = \pi$ and the graph of the principal cycle of the cotangent function occurs on the interval $(0, \pi)$. The equations of the vertical asymptotes of the principal cycle are $x = 0$ (the y-axis) and $x = \pi$. The center point of the principal cycle is $\left(\frac{\pi}{2}, 0\right)$ and the two halfway points of the principal cycle are $\left(\frac{\pi}{4}, 1\right)$ and $\left(\frac{3\pi}{4}, -1\right)$. Also note that the graph of each cycle of $y = \cot x$ is one-to-one.

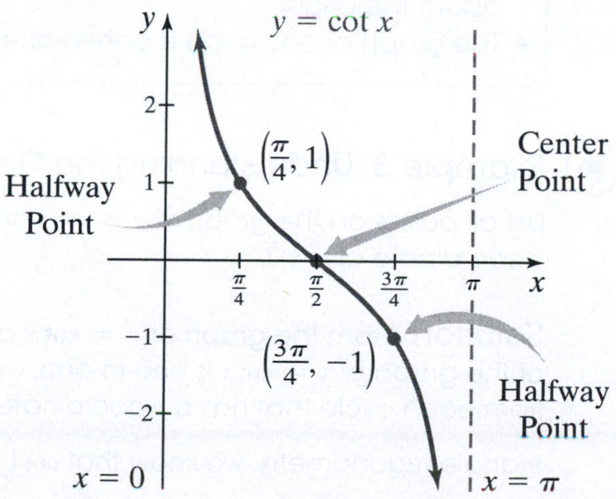

Figure 47 The principal cycle of the graph of $y = \cot x$ on the interval $(0, \pi)$ with the center point and the two halfway points labeled.

We now state the characteristics of the cotangent function.

Characteristics of the Cotangent Function

- The domain is $\{x \mid x \neq n\pi, \text{ where } n \text{ is an integer}\}$.
- The range is $(-\infty, \infty)$.
- The function is periodic with a period of $P = \pi$.
- The principal cycle of the graph occurs on the interval $(0, \pi)$.

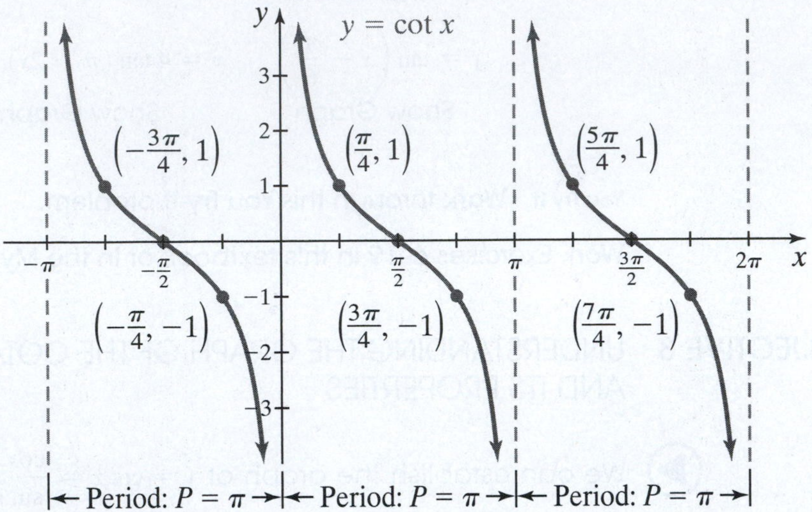

- The function has infinitely many vertical asymptotes with equations $x = n\pi$, where n is an integer.
- For each cycle, there is one center point. The x-coordinates of the center points are also the x-intercepts, or **zeros**, and are of the form $(2n + 1)\cdot\frac{\pi}{2}$, where n is an integer.
- For each cycle, there are two halfway points. The halfway point to the left of the x-intercept has a y-coordinate of 1. The halfway point to the right of the x-intercept has a y-coordinate of -1.
- The function is **odd**, which means $\cot(-x) = -\cot x$. The graph is symmetric about the origin.
- The graph of one cycle is **one-to-one**.

Example 3 Understanding the Graph of $y = \cot x$

List all points on the graph of $y = \cot x$ on the interval $[-2\pi, 2\pi]$ that have a y-coordinate of $-\sqrt{3}$.

Solution From the graph of $y = \cot x$ and using the property that each cycle of the graph of $y = \cot x$ is **one-to-one**, we know that there is exactly one point from each cycle that has a y-coordinate of $-\sqrt{3}$. Using our knowledge of right triangle trigonometry, we know that $\cot\left(\frac{\pi}{6}\right) = \sqrt{3}$. To see why this is true, read this explanation.

Therefore, the point $\left(\frac{\pi}{6}, \sqrt{3}\right)$ lies on the graph of $y = \cot x$. Using the fact that the cotangent function is **odd**, we know that the point $\left(-\frac{\pi}{6}, -\sqrt{3}\right)$ also lies on the graph. This point lies on the given interval $[-2\pi, 2\pi]$. To find all other points with a y-coordinate of $-\sqrt{3}$, we use the fact that the cotangent function is periodic with a period of $P = \pi$. Therefore, every point with a y-coordinate of $-\sqrt{3}$ must be of the form $\left(-\frac{\pi}{6} + k\pi, -\sqrt{3}\right)$, where k is an integer. We now choose negative integer values of k and positive integer values of k, making sure that the resulting x-coordinate is in the interval $[-2\pi, 2\pi]$.

$$k = -2: \left(-\frac{\pi}{6} + (-2)\pi, -\sqrt{3}\right) \rightarrow \cancel{\left(-\frac{13\pi}{6}, -\sqrt{3}\right)} \quad \left(x = -\frac{13\pi}{6} \text{ is not in the interval } [-2\pi, 2\pi]\right)$$

$$k = -1: \left(-\frac{\pi}{6} + (-1)\pi, -\sqrt{3}\right) \rightarrow \left(-\frac{7\pi}{6}, -\sqrt{3}\right)$$

$$k = 1: \left(-\frac{\pi}{6} + (1)\pi, -\sqrt{3}\right) \rightarrow \left(\frac{5\pi}{6}, -\sqrt{3}\right)$$

$$k = 2: \left(-\frac{\pi}{6} + (2)\pi, -\sqrt{3}\right) \rightarrow \left(\frac{11\pi}{6}, -\sqrt{3}\right)$$

$$k = 3: \left(-\frac{\pi}{6} + (3)\pi, -\sqrt{3}\right) \rightarrow \cancel{\left(\frac{17\pi}{6}, -\sqrt{3}\right)} \quad \left(x = \frac{17\pi}{6} \text{ is not in the interval } [-2\pi, 2\pi]\right)$$

See **Figure 48**.

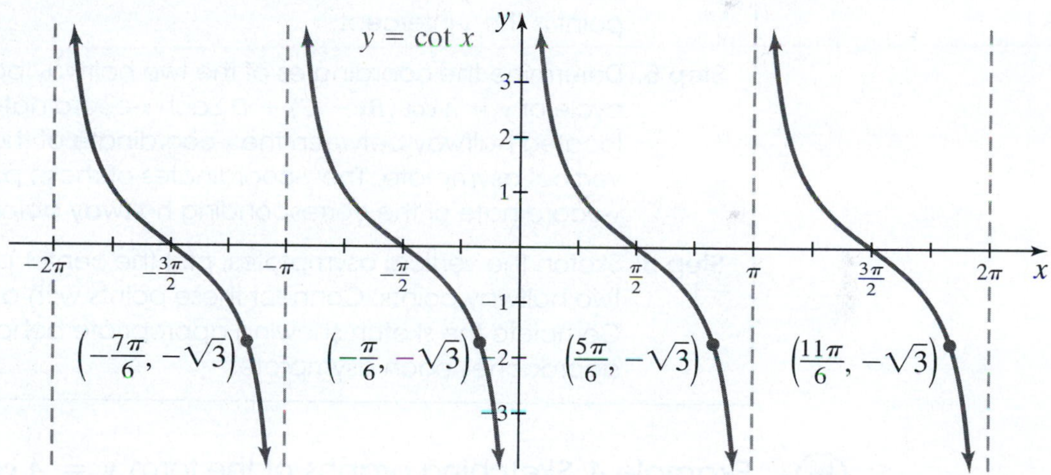

Figure 48 All points of $y = \cot x$ on the interval $[-2\pi, 2\pi]$ that have a y-coordinate of $-\sqrt{3}$

The four points of $y = \cot x$ on the interval $[-2\pi, 2\pi]$ that have a y-coordinate of $-\sqrt{3}$ are

$$\left(-\frac{7\pi}{6}, -\sqrt{3}\right), \left(-\frac{\pi}{6}, -\sqrt{3}\right), \left(\frac{5\pi}{6}, -\sqrt{3}\right) \text{ and } \left(\frac{11\pi}{6}, -\sqrt{3}\right).$$

You may want to watch this **video** to see this solution worked out in detail.

You Try It Work through this You Try It problem.

Work Exercises 20–24 in this textbook or in the MyLab Math Study Plan.

OBJECTIVE 4 SKETCHING FUNCTIONS OF THE FORM $y = A\cot(Bx - C) + D$

To sketch the principal cycle of functions of the form $y = A\cot(Bx - C) + D$, we follow the process outlined below.

Steps for Sketching Functions of the Form $y = A\cot(Bx - C) + D$

Step 1. If $B < 0$, rewrite the function in an equivalent form such that $B > 0$. Use the odd property of the cotangent function.

We now use this new form to determine A, B, C, and D.

Step 2. Determine the interval and the equations of the vertical asymptotes of the principal cycle. The interval for the principal cycle can be found by solving the inequality $0 < Bx - C < \pi$. The vertical asymptotes of the principal cycle occur at the **endpoints** of the interval of the principal cycle.

Step 3. The period is $P = \dfrac{\pi}{B}$.

Step 4. Determine the center point of the principal cycle of $y = A\cot(Bx - C) + D$. The x-coordinate of the center point is located midway between the vertical asymptotes of the principal cycle. The y-coordinate of the center point is D. Note that when $D = 0$ the x-coordinate of the center point is the x-intercept.

Step 5. Determine the coordinates of the two halfway points of the principal cycle of $y = A\cot(Bx - C) + D$ Each x-coordinate of a halfway point is located halfway between the x-coordinate of the center point and a vertical asymptote. The y-coordinates of these points are A times the y-coordinate of the corresponding **halfway point of $y = \cot x$** plus D.

Step 6. Sketch the vertical asymptotes, plot the center point, and plot the two halfway points. Connect these points with a smooth curve. Complete the sketch showing appropriate behavior of the graph as it approaches each asymptote.

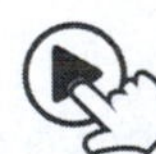

Example 4 Sketching graphs of the form $y = A\cot(Bx - C) + D$

For each function, determine the interval for the principal cycle. Then for the principal cycle, determine the equations of the vertical asymptotes, the coordinates of the center point, and the coordinates of the halfway points. Sketch the graph.

a. $y = \cot(2x + \pi) + 1$ **b.** $y = -3\cot\left(x - \dfrac{\pi}{4}\right)$

Solution

a. Step 1. For the function $y = \cot(2x + \pi) + 1$, we see that $B > 0$. Therefore, there is no need to rewrite the function, and we can skip **step 1.** Note that $A = 1$, $B = 2$, $C = -\pi$, and $D = 1$.

Step 2. The interval for the principal cycle can be found by solving the inequality $0 < 2x + \pi < \pi$.

$0 < 2x + \pi < \pi$	Write the inequality.
$0 - \pi < 2x < \pi - \pi$	Subtract π from all parts of the inequality.
$-\pi < 2x < 0$	Simplify.
$-\frac{\pi}{2} < x < 0$	Divide all parts of the inequality by 2.

The principal cycle occurs on the interval $\left(-\frac{\pi}{2}, 0\right)$. The equations of the vertical asymptotes of the principal cycle are $x = -\frac{\pi}{2}$ and $x = 0$.

Step 3. The period is $P = \frac{\pi}{B} = \frac{\pi}{2}$.

Step 4. The x-coordinate of the center point of $y = \cot(2x + \pi) + 1$ is located midway between the two vertical asymptotes. Therefore, the x-coordinate of the center point is $x = \dfrac{-\frac{\pi}{2} + 0}{2} = -\frac{\pi}{2} \div 2 = -\frac{\pi}{2} \cdot \frac{1}{2} = -\frac{\pi}{4}$. The y-coordinate of the center point is $D = 1$. Thus, the coordinates of the center point are $\left(-\frac{\pi}{4}, 1\right)$.

Step 5. There are two halfway points in the interval $\left(-\frac{\pi}{2}, 0\right)$. The x-coordinate of the halfway point located to the *left* of the center point is $-\frac{3\pi}{8}$ and the x-coordinate of the halfway point located to the *right* of the center point is $-\frac{\pi}{8}$. To see how to obtain these values, read these steps. Now, list the y-coordinates of the two **halfway points of the principal cycle of $y = \cot x$**. Next, multiply these values by $A = 1$ and add $D = 1$ to get the y-coordinates of the halfway points.

Halfway points of the principal cycle of $y = \cot x$: $\left(\frac{\pi}{4}, 1\right)$ $\left(\frac{3\pi}{4}, -1\right)$

$(1) \cdot (1) + 1 = 2$ $(1) \cdot (-1) + 1 = 0$

Halfway points of the principal cycle of $y = \cot(2x + \pi)$: $\left(-\frac{3\pi}{8}, 2\right)$ $\left(-\frac{\pi}{8}, 0\right)$

Step 6. We now sketch the vertical asymptotes, $x = -\frac{\pi}{2}$ and $x = 0$; plot the center point $\left(-\frac{\pi}{4}, 1\right)$; and plot the halfway points $\left(-\frac{3\pi}{8}, 2\right)$ and $\left(-\frac{\pi}{8}, 0\right)$. We connect these points with a smooth curve showing the appropriate behavior of the graph as it approaches each asymptote.

Watch this **interactive video** to see how to sketch the graph of $y = \cot(2x + \pi) + 1$. The graph of $y = \cot(2x + \pi) + 1$ can be seen in Figures 49 and 50.

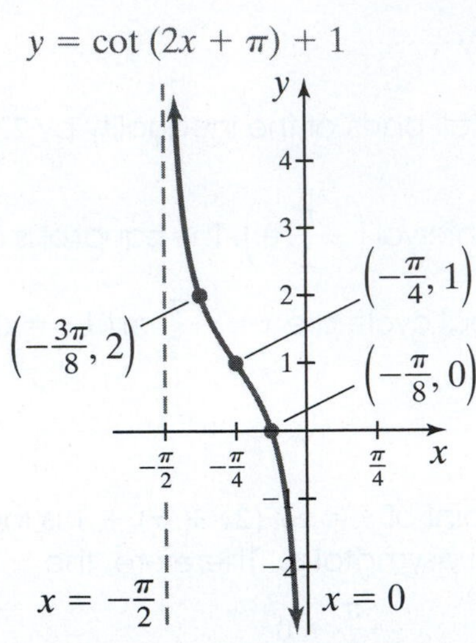

Figure 49 The principal cycle of the graph of $y = \cot(2x + \pi) + 1$

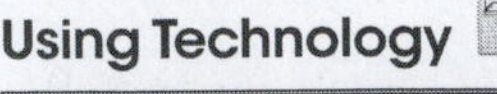

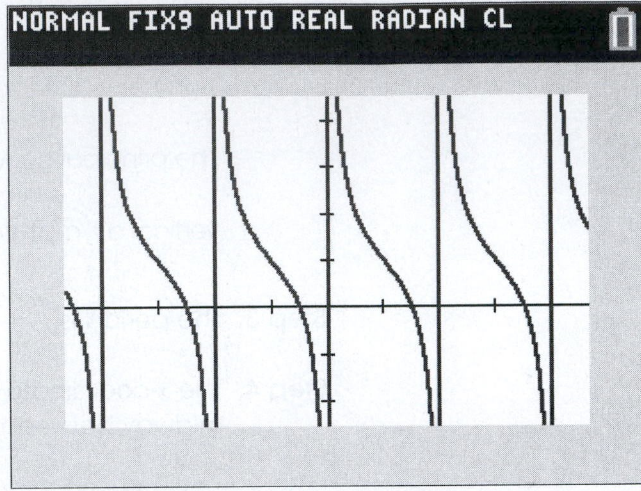

Figure 50 Several cycles of the graph of $y = \cot(2x + \pi) + 1$ using a graphing utility

b. See if you can sketch the graph of $y = -3\cot\left(x - \frac{\pi}{4}\right)$ on your own. To see how to sketch this graph, watch this **interactive video**.

We could have used transformations to sketch the functions from Example 4. Click on the Show Graph link under the following functions to see how to sketch the function using transformations.

$y = \cot(2x + \pi) + 1$ $y = -3\cot\left(x - \frac{\pi}{4}\right)$

Show Graph   Show Graph

You Try It Work through this You Try It problem.

Work Exercises 25–38 in this textbook or in the MyLab Math Study Plan.

OBJECTIVE 5 DETERMINE THE EQUATION OF A FUNCTION OF THE FORM $y = A\tan(Bx - C) + D$ OR $y = A\cot(Bx - C) + D$ GIVEN ITS GRAPH

We now know how to sketch the graphs of $y = A\tan(Bx - C) + D$ and $y = A\cot(Bx - C) + D$. Suppose that we are given the principal cycle of the graph whose function is given by $y = A\tan(Bx - C) + D$ or $y = A\cot(Bx - C) + D$ for $B > 0$. (We will assume that the value of B is positive. Otherwise, there could be more than one correct answer because of the odd nature of the tangent and cotangent functions.) To determine the proper function we must first identify the following five characteristics of the graph of the given principal cycle.

1. $x = a$, the equation of the left-most vertical asymptote
2. $x = b$, the equation of the right-most vertical asymptote
3. (x_1, y_1), the coordinates of the left-most halfway point
4. (x_2, y_2), the coordinates of the center point
5. (x_3, y_3), the coordinates of the right-most halfway point

We will use these five characteristics to help us to determine the values of A, B, C and D. Figure 51 on the following page illustrates the principal cycle of a possible tangent or cotangent curve having the five characteristics above.

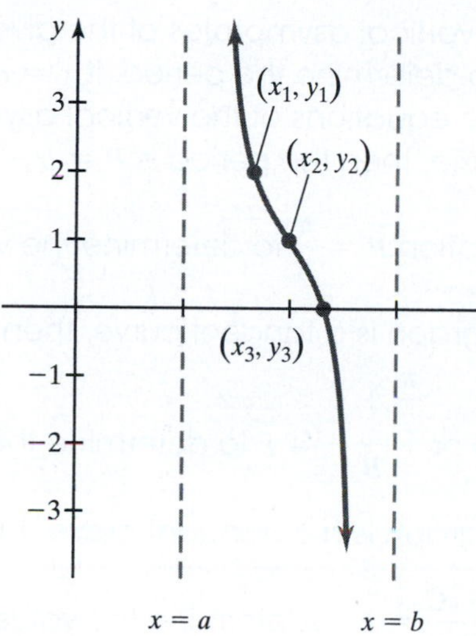

Figure 51 The principal cycle of the graph of a tangent or cotangent function of the form $y = A\tan(Bx - C) + D$ or $y = A\cot(Bx - C) + D$

Note that the period of the graph seen in Figure 51 must be $b - a$, the distance between the right-most vertical asymptote and the left-most vertical asymptote. Using the fact that the period of a tangent curve or cotangent curve is $P = \frac{\pi}{B}$, we can easily determine the value of B.

Recall that the interval for the principal cycle of a tangent curve is $-\frac{\pi}{2} < Bx - C < \frac{\pi}{2}$ whereas the interval for the principal cycle of a cotangent curve is $0 < Bx - C < \pi$.

If we solve these two inequalities, we get

$$\underbrace{\frac{-\frac{\pi}{2} + C}{B}}_{a} < x < \underbrace{\frac{\frac{\pi}{2} + C}{B}}_{b} \quad \text{and} \quad \underbrace{\frac{C}{B}}_{a} < x < \underbrace{\frac{\pi + C}{B}}_{b} \quad \text{respectively.}$$

Therefore, if we are given the principal cycle of the graph of a tangent function or a cotangent function whose vertical asymptotes are $x = a$, $x = b$ where $a < b$, then we can establish the following equations:

If the graph is a tangent curve, then $\dfrac{-\frac{\pi}{2} + C}{B} = a$ and $\dfrac{\frac{\pi}{2} + C}{B} = b$.

If the graph is a cotangent curve, then $\dfrac{C}{B} = a$ and $\dfrac{\pi + C}{B} = b$.

These equations can be used to determine the value of C. We can determine the values of A and D by comparing the given half-way points and the center point with the known half-way points and center point of the graph of $y = \tan x$ or $y = \cot x$. We can follow the five steps outlined on the following page to determine the equation of the function that describes a given tangent or cotangent curve.

Steps for Determining the Equation of a Function of the Form $y = A\tan(Bx - C) + D$ or $y = A\cot(Bx - C) + D$ Given the Graph

Step 1. Use the two vertical asymptotes of the given principal cycle of the graph to determine the period. If $x = a$ and $x = b$ where $a < b$ are the equations of the vertical asymptotes of the principal cycle, then the period is $P = b - a$.

Step 2. Use the equation $P = \frac{\pi}{B}$ to determine the value of $B > 0$.

Step 3. If the given graph is a tangent curve, then use the equation

$$\frac{-\frac{\pi}{2} + C}{B} = a \text{ or } \frac{\frac{\pi}{2} + C}{B} = b \text{ to determine the value of } C.$$

If the given graph is a cotangent curve, then use the equation

$$\frac{C}{B} = a \text{ or } \frac{\pi + C}{B} = b \text{ to determine the value of } C.$$

Step 4. The y-coordinate of the center point represents the value of D.

Step 5. Identify the halfway points (x_1, y_1) and (x_3, y_3) that lie on the given graph. If the given graph is a **tangent curve**, then use the relationship between (x_3, y_3), the coordinates of the **right-most** halfway point of the given graph, and $\left(\frac{\pi}{4}, 1\right)$, the **right-most** halfway point of the principal cycle of the graph of $y = \tan x$ to get the equation $y_3 = A + D$. Use this information to determine the value of A.

If the given graph is a **cotangent curve**, then use the relationship between (x_1, y_1), the coordinates of the **left-most** halfway point of the given graph, and $\left(\frac{\pi}{4}, 1\right)$, the **left-most** halfway point of the principal cycle of the graph of $y = \cot x$ to get the equation $y_1 = A + D$. Use this information to determine the value of A.

Example 5 Determining the Equation of a Function Given the Graph that it Represents

The principal cycle of the graphs of two trigonometric functions of the from $y = A\tan(Bx - C) + D$ or $y = A\cot(Bx - C) + D$ for $B > 0$ are given below. Determine the equation of the function represented by each graph.

a. $y = A\tan(Bx - C) + D$

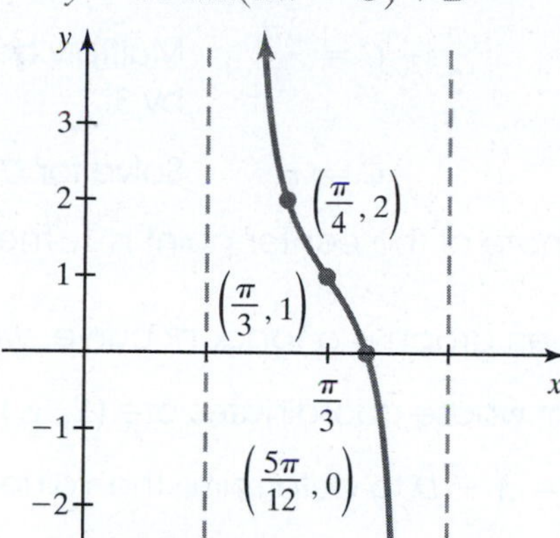

b. $y = A\cot(Bx - C) + D$

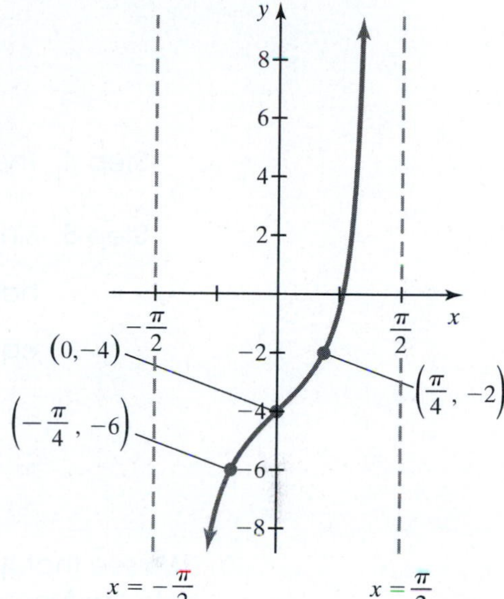

Solution

a. First, use the **given graph** to identify the following.

1. The equation of the left-most vertical asymptote is $x = \frac{\pi}{6}$.

2. The equation of the right-most vertical asymptote is $x = \frac{\pi}{2}$.

3. The coordinates of the left-most halfway point are $(x_1, y_1) = \left(\frac{\pi}{4}, 2\right)$.

4. The coordinates of the center point are $(x_2, y_2) = \left(\frac{\pi}{3}, 1\right)$.

5. The coordinates of the right-most halfway point are $(x_3, y_3) = \left(\frac{5\pi}{12}, 0\right)$.

Step 1. The period is the distance between the two vertical asymptotes of the principal cycle of the given graph. Thus, the period is

$$P = b - a = \frac{\pi}{2} - \frac{\pi}{6} = \frac{\pi}{3}.$$

Step 2. The period is $P = \frac{\pi}{B} = \frac{\pi}{3}$. Therefore, $B = 3$.

Step 3. The given graph is a tangent curve. Therefore, we can use the equation

$$\frac{-\frac{\pi}{2} + C}{B} = a \text{ or } \frac{\frac{\pi}{2} + C}{B} = b$$ to determine the value of C.

$$\frac{\frac{\pi}{2} + C}{B} = b \qquad \text{Start with one of the equations.}$$

$$\frac{\frac{\pi}{2} + C}{3} = \frac{\pi}{2} \qquad \text{Substitute } B = 3 \text{ and } b = \frac{\pi}{2}.$$

$$\frac{\pi}{2} + C = \frac{3\pi}{2} \qquad \text{Multiply both sides of the equation by 3.}$$

$$C = \pi \qquad \text{Solve for } C.$$

Step 4. The y-coordinate of the center point is 1. Therefore, $D = 1$.

Step 5. Since the given graph is a tangent curve, we identify the right-most halfway point whose coordinates are $(x_3, y_3) = \left(\frac{5\pi}{12}, 0\right)$ and use the equation $y_3 = A + D$ to determine the value of A.

$y_3 = A + D$ Write the equation relating y_3, A, and D.

$0 = A + 1$ Substitute $y_3 = 0$ and $D = 1$.

$A = -1$ Solve for A.

We see that $A = -1, B = 3, C = \pi$, and $D = 1$. Therefore, substituting these values into the function of the form $y = A \tan(Bx - C) + D$, we see that the function which describes the principal cycle of the given graph is $y = -\tan(3x - \pi) + 1$. You should sketch the graph of $y = -\tan(3x - \pi) + 1$ on your own to make sure that it matches the given graph. The graph of several cycles of the function $y = -\tan(3x - \pi) + 1$ is shown in Figure 52.

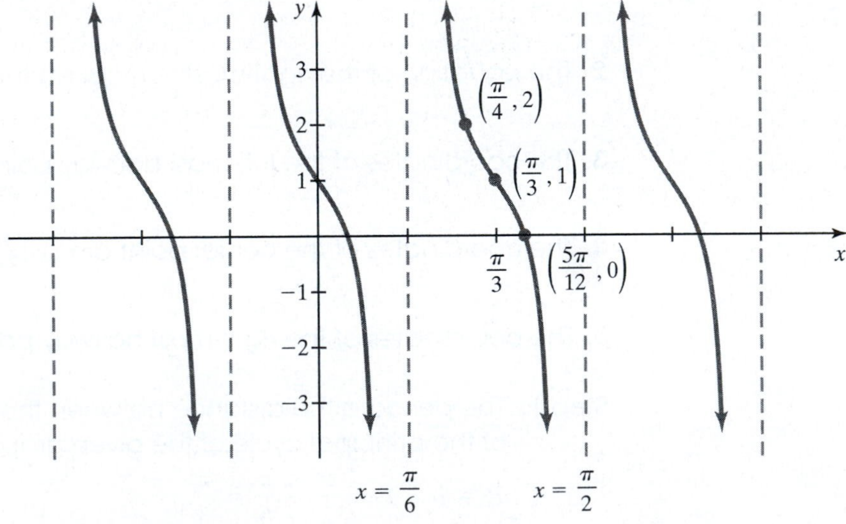

Figure 52 Several cycles of the graph of $y = -\tan(3x - \pi) + 1$

b. Work through this **interactive video** to verify that the function of the form $y = A \cot(Bx - C) + D$ whose graph is shown below is $y = -2 \cot\left(x + \frac{\pi}{2}\right) - 4$.

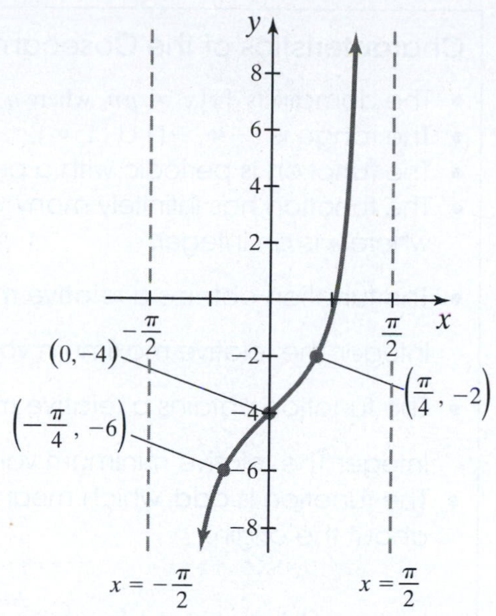

You Try It Work through this You Try It problem.

Work Exercises 39–44 in this textbook or in the MyLab Math Study Plan.

OBJECTIVE 6 UNDERSTANDING THE GRAPHS OF THE COSECANT AND SECANT FUNCTIONS AND THEIR PROPERTIES

To sketch the graph of the cosecant function, we use a **reciprocal identity** to rewrite $y = \csc x$ as $y = \frac{1}{\sin x}$ and use our knowledge of the graph of the sine function to help sketch the graph of the cosecant function. Note that if x is any value such that $\sin x \neq 0$, then the value of $y = \csc x$ will be the reciprocal of the value of $\sin x$. If $\sin x = 0$, then the cosecant function is undefined. Therefore, the domain of the cosecant function is all real numbers except values of x for which x is a **zero of $y = \sin x$**. Thus, the domain is all real numbers except integer multiples of π. The graph of $y = \csc x$ has infinitely many vertical asymptotes of the form $x = n\pi$, where n is an integer. Recall that the graph of $y = \sin x$ obtains a **relative maximum** value of 1 at $x = \frac{\pi}{2} + 2\pi n$, where n is an integer. Because of the reciprocal relationship between $y = \sin x$ and $y = \csc x$, these values will be relative *minimums* of the graph of $y = \csc x$. Likewise, the **relative minimum** values of $y = \sin x$ are -1, which occur at $x = \frac{3\pi}{2} + 2\pi n$, where n is an integer. These values will be relative *maximums* of the graph of $y = \csc x$. Watch this **interactive video** to see how to establish the graph of $y = \csc x$ and its characteristics.

Characteristics of the Cosecant Function

- The domain is $\{x \mid x \neq n\pi, \text{ where } n \text{ is an integer}\}$.
- The range is $(-\infty, -1] \cup [1, \infty)$.
- The function is periodic with a period of $P = 2\pi$.
- The function has infinitely many vertical asymptotes with equations $x = n\pi$, where n is an integer.
- The function obtains a **relative maximum** at $x = \frac{3\pi}{2} + 2\pi n$, where n is an integer. The relative maximum value is -1.
- The function obtains a **relative minimum** at $x = \frac{\pi}{2} + 2\pi n$, where n is an integer. The relative minimum value is 1.
- The function is **odd**, which means $\csc(-x) = -\csc x$. The graph is symmetric about the origin.

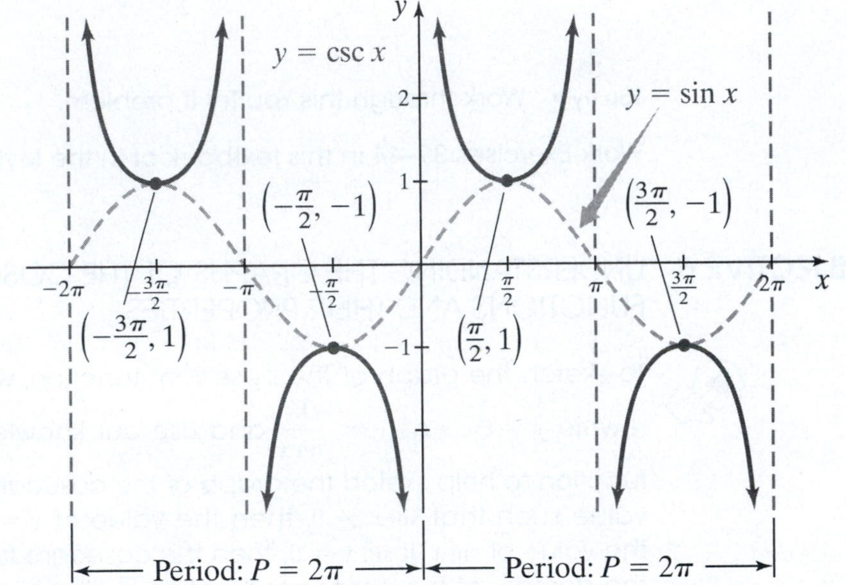

We can sketch the graph of $y = \sec x$ in much the same way as we did with the graph of $y = \csc x$. Because we can use a **reciprocal identity** to rewrite $y = \sec x$ as $y = \frac{1}{\cos x}$, we can use our knowledge of the graph of the cosine function to help sketch the graph of the secant function. Watch this **interactive video** to see how to establish the graph of $y = \sec x$ and its characteristics that are shown on the following page.

Characteristics of the Secant Function

- The domain is $\left\{x \mid x \neq (2n + 1)\cdot\frac{\pi}{2}, \text{ where } n \text{ is an integer}\right\}$.
- The range is $(-\infty, -1] \cup [1, \infty)$.
- The function is periodic with a period of $P = 2\pi$.
- The function has infinitely many vertical asymptotes with equations $x = (2n + 1)\cdot\frac{\pi}{2}$, where n is an integer.
- The function obtains a **relative maximum** at $x = (2n + 1)\pi$, where n is an integer. The relative maximum value is -1.
- The function obtains a **relative minimum** at $2\pi n$, where n is an integer. The relative minimum value is 1.
- The function is **even**, which means $\sec(-x) = \sec x$. The graph is symmetric about the y-axis.

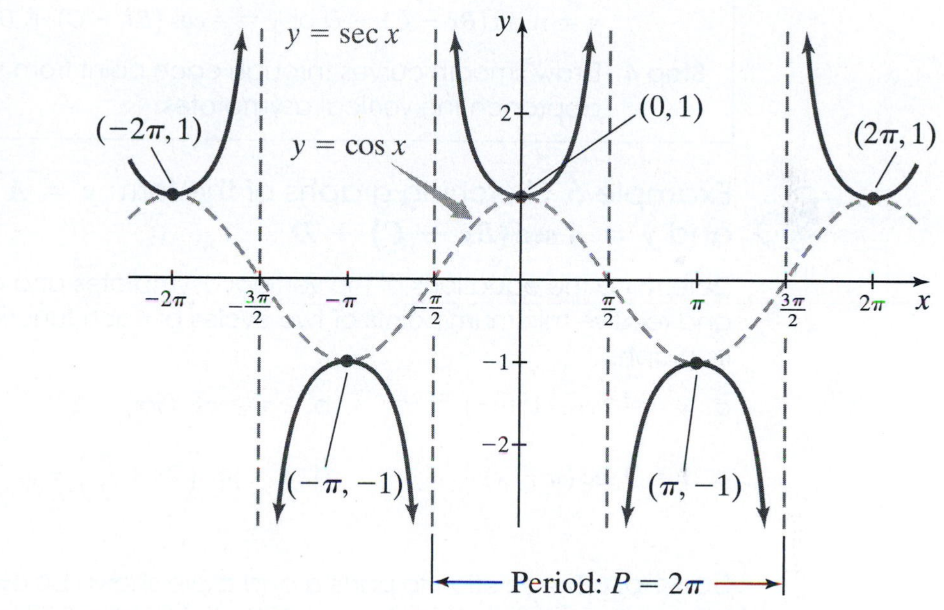

You Try It Work through this You Try It problem.

Work Exercises 45 and 46 in this textbook or in the MyLab Math Study Plan.

OBJECTIVE 7 SKETCHING FUNCTIONS OF THE FORM $y = A\csc(Bx - C) + D$ AND $y = A\sec(Bx - C) + D$

To sketch functions of the form $y = A\csc(Bx - C) + D$ and $y = A\sec(Bx - C) + D$, we must first sketch the graph of the corresponding reciprocal function. The corresponding reciprocal function of $y = A\csc(Bx - C) + D$ is $y = A\sin(Bx - C) + D$ and the corresponding reciprocal function of $y = A\sec(Bx - C) + D$ is $y = A\cos(Bx - C) + D$. We sketch the corresponding reciprocal functions by following the steps that were outlined in **Section 7.2**. Before going on, make sure that you can sketch functions of the form $y = A\sin(Bx - C) + D$ and $y = A\cos(Bx - C) + D$. Click on the You Try It icon below to see if you understand how to sketch such functions.

You Try It An example of sketching functions of the form $y = A\sin(Bx - C) + D$ and $y = A\cos(Bx - C) + D$

To sketch functions of the form $y = A\csc(Bx - C) + D$ and $y = A\sec(Bx - C) + D$, we follow the steps outlined below.

Steps for Sketching Functions of the Form $y = A\csc(Bx - C) + D$ and $y = A\sec(Bx - C) + D$

Step 1. Lightly sketch at least two cycles of the corresponding reciprocal function using the process outlined in **Section 7.2**. If $D \neq 0$, lightly sketch two reciprocal functions, one with $D = 0$ and one with $D \neq 0$.

Step 2. Sketch the vertical asymptotes. The vertical asymptotes will correspond to the x-intercepts of the reciprocal function $y = A\sin(Bx - C)$ or $y = A\cos(Bx - C)$.

Step 3. Plot all maximum and minimum points on the graph of $y = A\sin(Bx - C) + D$ or $y = A\cos(Bx - C) + D$.

Step 4. Draw smooth curves through each point from step 3, making sure to approach the vertical asymptotes.

Example 6 Sketching graphs of the form $y = A\csc(Bx - C) + D$ and $y = A\sec(Bx - C) + D$

Determine the equations of the vertical asymptotes and all relative maximum and relative minimum points of two cycles of each function and then sketch its graph.

a. $y = -2\csc(x + \pi)$ **b.** $y = -\csc(\pi x) - 2$

c. $y = 3\sec(\pi - x)$ **d.** $y = \sec\left(2x + \frac{\pi}{2}\right) + 1$

Solution The solution to parts a and d are shown below. If you would like to see the solution to parts b and c, watch this **interactive video**.

a. Step 1. Sketch the corresponding reciprocal function, $y = -2\sin(x + \pi)$. Two complete cycles of $y = -2\sin(x + \pi)$ are sketched in Figure 53.

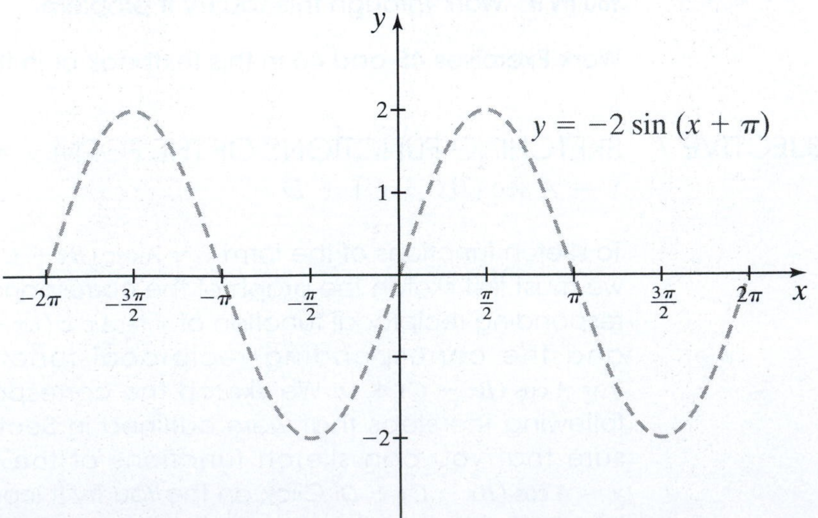

Figure 53 Two complete cycles of the graph of $y = -2\sin(x + \pi)$

Step 2. As you can see in **Figure 53**, the zeros of $y = -2\sin(x+\pi)$ on the interval $[-2\pi, 2\pi]$ are $-2\pi, -\pi, 0, \pi,$ and 2π. Therefore, the equations of the vertical asymptotes of the graph of $y = -2\csc(x+\pi)$ are $x = -2\pi, x = -\pi, x = 0, x = \pi,$ and $x = 2\pi$.

Step 3. The relative maximum points of the graph of $y = -2\csc(x+\pi)$ are $\left(-\frac{\pi}{2}, -2\right)$ and $\left(\frac{3\pi}{2}, -2\right)$. The relative minimum points are $\left(-\frac{3\pi}{2}, 2\right)$ and $\left(\frac{\pi}{2}, 2\right)$.

Step 4. We draw smooth curves through the points from step 3, making sure to approach the vertical asymptotes. The graph of $y = -2\csc(x+\pi)$ can be seen in **Figure 54** and **Figure 55**.

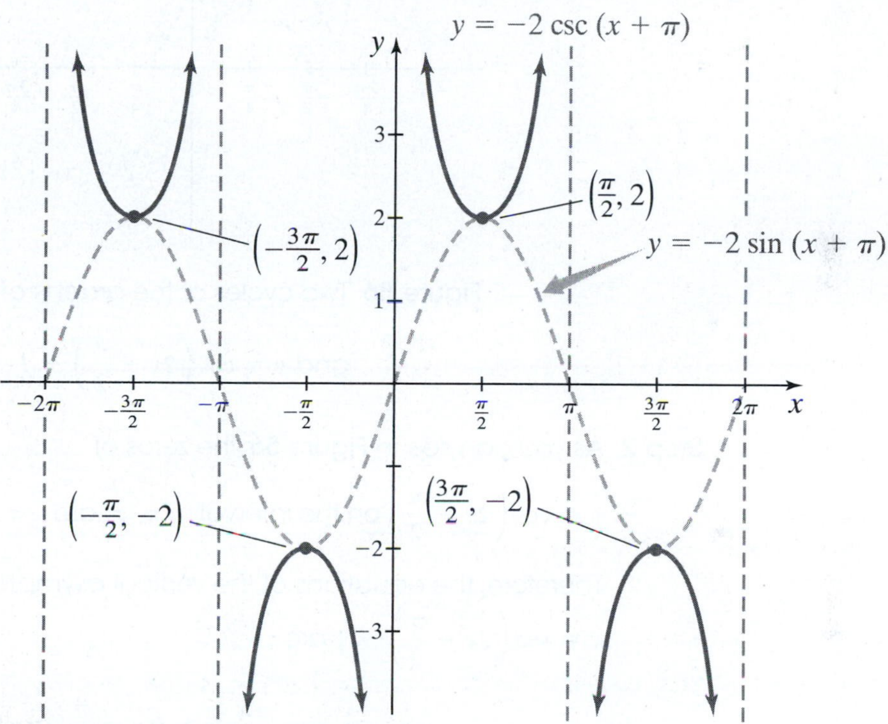

Figure 54 Two cycles of the graph of $y = -2\csc(x+\pi)$

Using Technology

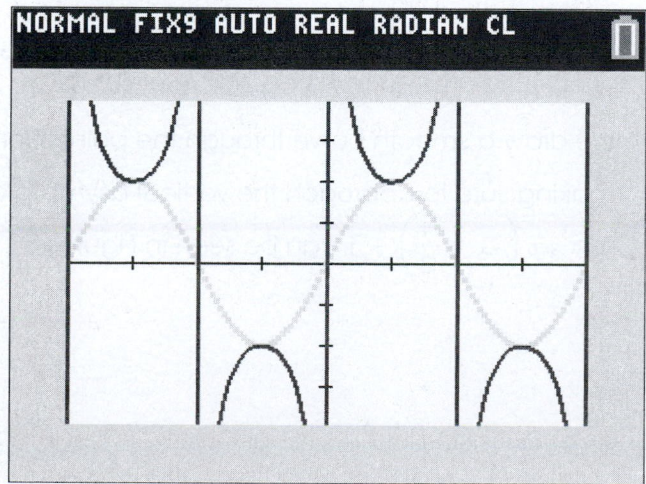

Figure 55 Two cycles of the graph of $y = -2\csc(x+\pi)$ using a graphing utility.

b. Step 1. Sketch the corresponding reciprocal functions,

$$y = \cos\left(2x + \frac{\pi}{2}\right) \text{ and } y = \cos\left(2x + \frac{\pi}{2}\right) + 1.$$

Two complete cycles of $y = \cos\left(2x + \frac{\pi}{2}\right)$ and $y = \cos\left(2x + \frac{\pi}{2}\right) + 1$ are sketched in Figure 56.

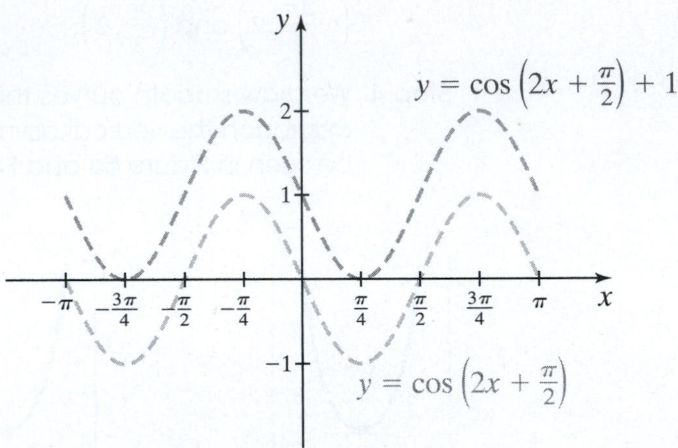

Figure 56 Two cycles of the graphs of $y = \cos\left(2x + \frac{\pi}{2}\right)$ and $y = \cos\left(2x + \frac{\pi}{2}\right) + 1$

Step 2. As you can see in **Figure 56**, the zeros of

$y = \cos\left(2x + \frac{\pi}{2}\right)$ on the interval $[-\pi, \pi]$ are $-\pi, -\frac{\pi}{2}, 0, \frac{\pi}{2},$ and π.

Therefore, the equations of the vertical asymptotes of the graph of $y = \sec\left(2x + \frac{\pi}{2}\right) + 1$ are

$$x = -\pi, x = -\frac{\pi}{2}, x = 0\, x = \frac{\pi}{2}, \text{and } x = \pi.$$

Step 3. The relative maximum and relative minimum points of the graph of $y = \sec\left(2x + \frac{\pi}{2}\right) + 1$ on the interval $[-\pi, \pi]$ are

$$\left(-\frac{3\pi}{4}, 0\right), \left(-\frac{\pi}{4}, 2\right), \left(\frac{\pi}{4}, 0\right), \text{and } \left(\frac{3\pi}{4}, 2\right).$$

Step 4. We draw a smooth curve through the points from step 3, making sure to approach the vertical asymptotes. The graph of $y = \sec\left(2x + \frac{\pi}{2}\right) + 1$ can be seen in Figures 57 and 58.

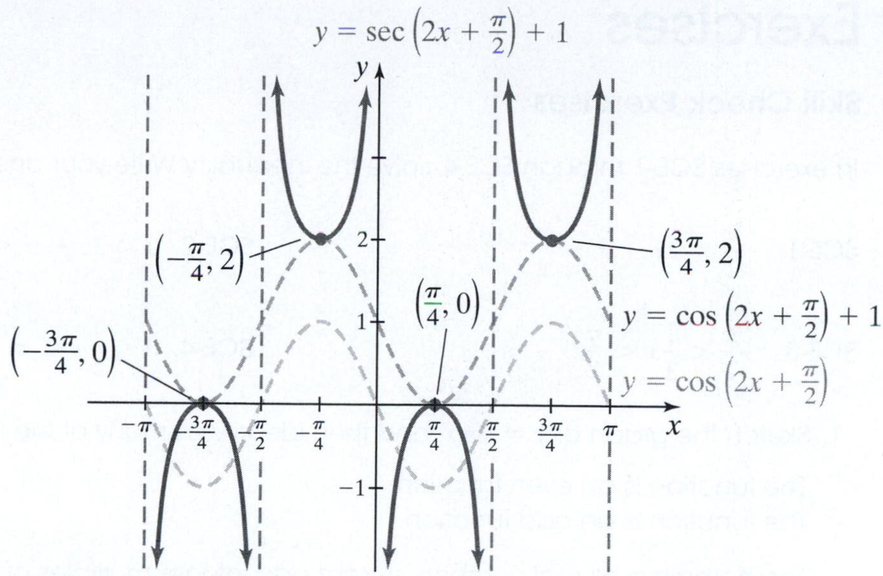

Figure 57 Two cycles of the graphs of $y = \sec\left(2x + \frac{\pi}{2}\right) + 1$

Using Technology

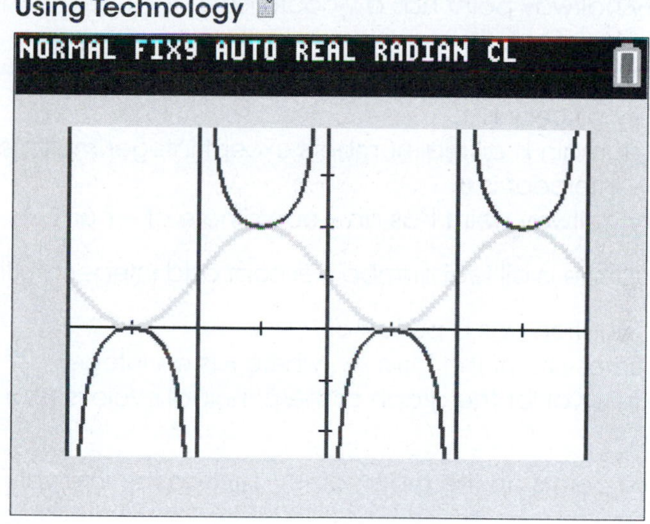

Figure 58 Two cycles of the graph of $y = \sec\left(2x + \frac{\pi}{2}\right) + 1$ using a graphing utility.

To see how to sketch the functions from Example 5 using transformations, click on the Show Graph link under each function below.

$y = -2\csc(x + \pi)$	$y = \csc(\pi x) - 2$	$y = 3\sec(\pi - x)$	$y = \sec\left(2x + \frac{\pi}{2}\right) + 1$
Show Graph	Show Graph	Show Graph	Show Graph ●

You Try It Work through this You Try It problem.

Work Exercises 47–56 in this textbook or in the MyLab Math Study Plan.

7.3 Exercises

Skill Check Exercises

In exercises SCE-1 through SCE-4, solve the inequality. Write your answer in interval notation form.

SCE-1. $-\frac{\pi}{2} < 2x - \pi < \frac{\pi}{2}$

SCE-2. $0 < 3x + \pi < \pi$

SCE-3. $-\frac{\pi}{2} < \frac{1}{3}x < \frac{\pi}{2}$

SCE-4. $0 < \frac{1}{5}x < \pi$

1. Sketch the graph of $y = \tan x$, and then identify as many of the following properties that apply.

   The function is an even function.
   The function is an odd function.
   The domain is all real numbers except odd integer multiples of $\frac{\pi}{2}$.
   The range is $(-\infty, \infty)$.
   The function has a period of $P = 2\pi$.
   The zeros are of the form $\frac{n\pi}{2}$, where n is an odd integer.
   Every halfway point has a y-coordinate of -1 or 1.
   The interval for the graph of the principal cycle is $\left(-\frac{\pi}{2}, \frac{\pi}{2}\right)$.
   The y-intercept is 1.
   The domain is all real numbers except integer multiples of π.
   The y-intercept is 0.
   Every halfway point has an x-coordinate of -1 or 1.
   The range is all real numbers except odd integer multiples of $\frac{\pi}{2}$.
   The function has a period of $P = \pi$.
   The zeros are of the form $n\pi$, where n is an integer.
   The interval for the graph of the principal cycle is $(0, \pi)$.

2. List all points on the graph of $y = \tan x$ on the interval $\left[-2\pi, \frac{5\pi}{2}\right]$ that have a y-coordinate of $\sqrt{3}$.

3. List all points on the graph of $y = \tan x$ on the interval $\left[-3\pi, \frac{\pi}{2}\right]$ that have a y-coordinate of $-\frac{1}{\sqrt{3}}$.

4. List all halfway points of $y = \tan x$ on the interval $[-2\pi, \pi]$.

5. Use the periodic property of $y = \tan x$ to determine which of the following expressions is equivalent to $\tan\left(\frac{4\pi}{3}\right)$.

   i. $\tan\left(\frac{2\pi}{3}\right)$ ii. $\tan\left(-\frac{\pi}{3}\right)$ iii. $\tan\left(-\frac{2\pi}{3}\right)$ iv. $\tan\left(\frac{5\pi}{3}\right)$

For Exercises 6–19, given each function, determine the interval for the principal cycle. Determine the period. Then for the principal cycle, determine the equations of the vertical asymptotes, the coordinates of the center point, and the coordinates of the halfway points. Sketch the graph.

SbS 6. $y = \tan\left(x + \frac{\pi}{4}\right)$ SbS 7. $y = 3\tan\left(x - \frac{\pi}{6}\right)$ SbS 8. $y = -2\tan\left(x - \frac{\pi}{4}\right)$

SbS 9. $y = \tan(3x + \pi)$ SbS 10. $y = \tan(2x)$ SbS 11. $y = \tan\left(\frac{1}{2}x\right)$

SbS 12. $y = 2\tan(-3x)$ SbS 13. $y = \tan\left(\frac{1}{2}x + \pi\right)$ SbS 14. $y = \tan(\pi - 2x)$

SbS 15. $y = 3\tan(2x + \pi) - 1$ SbS 16. $y = 4\tan(-2x + \pi) + 1$ SbS 17. $y = \frac{1}{2}\tan(-3x) + 4$

SbS 18. $y = 1 - \tan\left(x - \frac{\pi}{2}\right)$ SbS 19. $y = -3\tan(\pi - 2x) - 1$

20. Sketch the graph of $y = \cot x$, and then identify as many of the following properties that apply.

The function is an even function.
The function is an odd function.
The domain is all real numbers except odd integer multiples of $\frac{\pi}{2}$.
The range is $(-\infty, \infty)$.
The function has a period of $P = 2\pi$.
The zeros are of the form $\frac{n\pi}{2}$, where n is an odd integer.
Every halfway point has a y-coordinate of -1 or 1.
The interval for the graph of the principal cycle is $\left(-\frac{\pi}{2}, \frac{\pi}{2}\right)$.
The y-intercept is 1.
The domain is all real numbers except integer multiples of π.
The y-intercept is 0.
Every halfway point has an x-coordinate of -1 or 1.
The range is all real numbers except odd integer multiples of $\frac{\pi}{2}$.
The function has a period of $P = \pi$.
The zeros are of the form $n\pi$, where n is an integer.
The interval for the graph of the principal cycle is $(0, \pi)$.

21. List all points on the graph of $y = \cot x$ on the interval $\left[-2\pi, \frac{5\pi}{2}\right]$ that have a y-coordinate of $\sqrt{3}$.

22. List all points on the graph of $y = \cot x$ on the interval $\left[-3\pi, \frac{\pi}{2}\right]$ that have a y-coordinate of $-\frac{1}{\sqrt{3}}$.

23. List all halfway points of $y = \cot x$ on the interval $[-2\pi, \pi]$.

24. Use the periodic property of $y = \cot x$ to determine which of the following expressions is equivalent to $\cot\left(\frac{13\pi}{6}\right)$.

i. $\cot\left(-\frac{5\pi}{6}\right)$ ii. $\cot\left(-\frac{\pi}{6}\right)$ iii. $\cot\left(\frac{11\pi}{6}\right)$ iv. $\cot\left(\frac{5\pi}{6}\right)$

For Exercises 25–38, given each function, determine the interval for the principal cycle. Determine the period. Then for the principal cycle, determine the equations of the vertical asymptotes, the coordinates of the center point, and the coordinates of the halfway points. Sketch the graph.

SbS 25. $y = \cot\left(x + \frac{\pi}{4}\right)$ SbS 26. $y = 3\cot\left(x - \frac{\pi}{2}\right)$ SbS 27. $y = -2\cot\left(x - \frac{\pi}{3}\right)$

SbS 28. $y = \cot(3x + \pi)$ SbS 29. $y = \cot(2x)$ SbS 30. $y = \cot\left(\frac{1}{3}x\right)$

SbS 31. $y = 2\cot(-3x)$ SbS 32. $y = -3\cot(-2x + \pi)$ SbS 33. $y = \cot(\pi - 2x)$

SbS 34. $y = \cot\left(\frac{1}{2}x + \pi\right)$ SbS 35. $y = 3\cot(2x + \pi) - 1$ SbS 36. $y = \frac{1}{2}\cot(-3x) + 4$

SbS 37. $y = 1 - \cot\left(x - \frac{\pi}{2}\right)$ SbS 38. $y = -3\cot(\pi - 2x) - 1$

In Exercises 39–44, the principal cycle of the graph of a trigonometric function of the from $y = A\tan(Bx - C) + D$ or $y = A\cot(Bx - C) + D$ for $B > 0$ is given. Determine the equation of the function represented by each graph.

SbS 39. 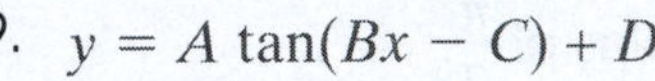
$y = A\tan(Bx - C) + D$

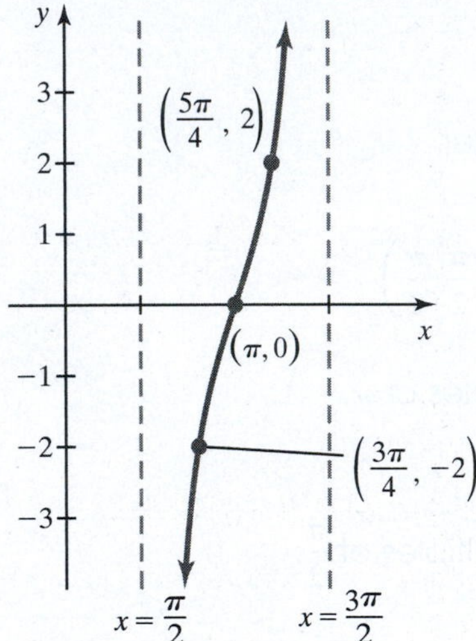

SbS 40. 
$y = A\cot(Bx - C) + D$

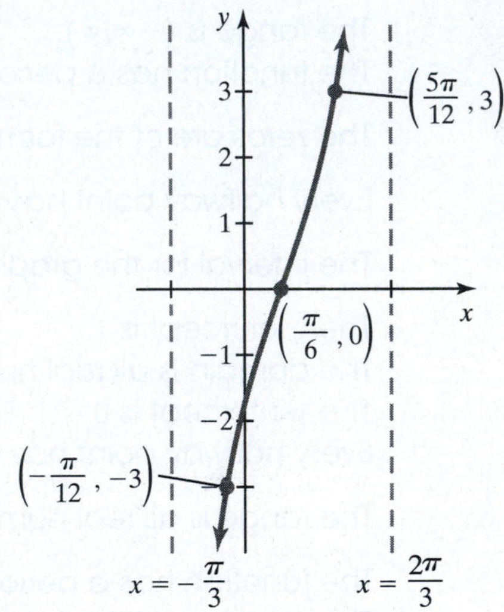

SbS 41. $y = A\tan(Bx - C) + D$

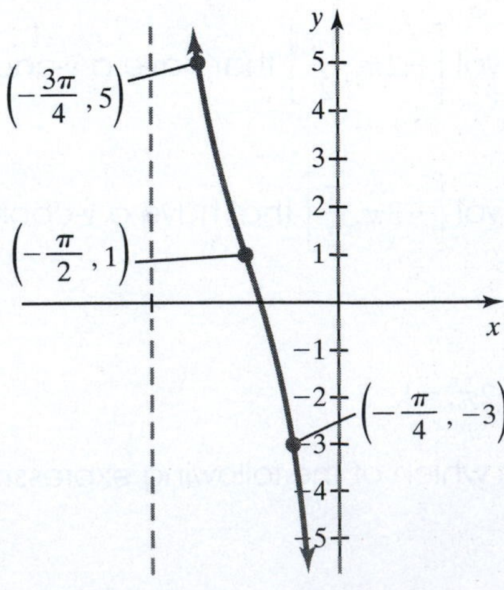

SbS 42. $y = A\cot(Bx - C) + D$

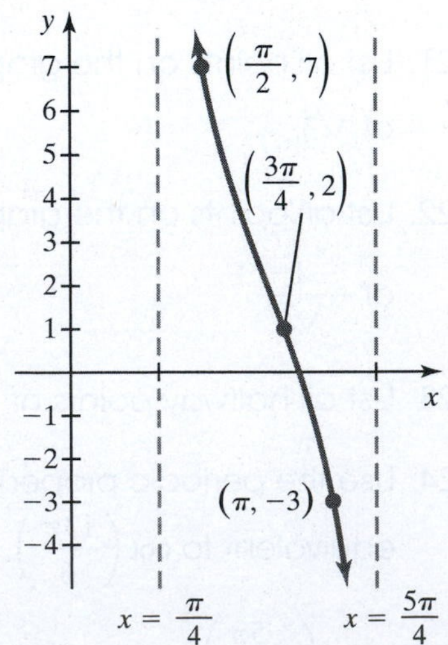

SbS 43. $y = A\cot(Bx - C) + D$

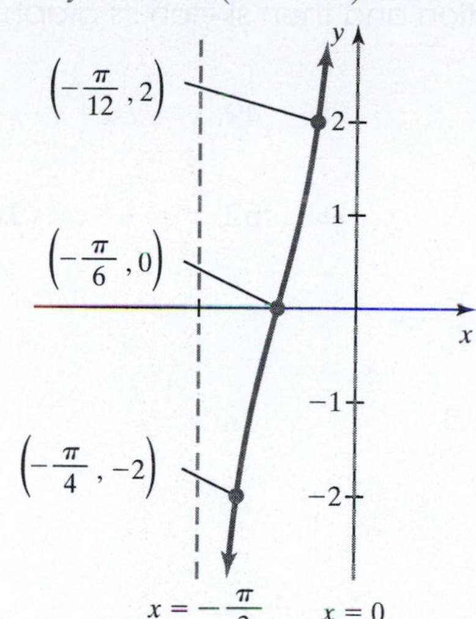

SbS 44. $y = A\tan(Bx - C) + D$

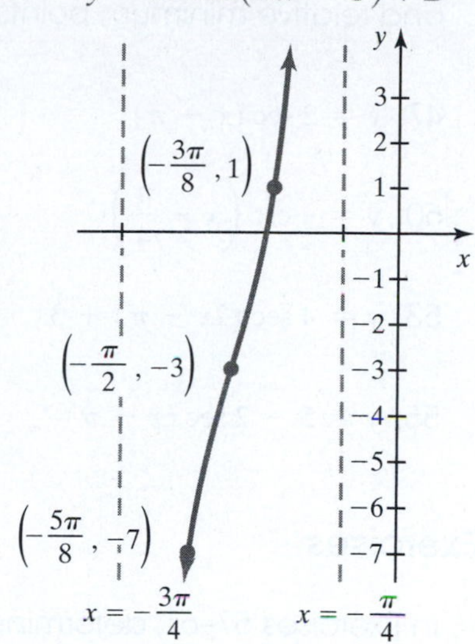

45. Sketch the graph of $y = \csc x$, and then identify as many of the following properties that apply

The function is an even function.
The function is an odd function.
The domain is all real numbers except integer multiples of π.
The range is $(-\infty, \infty)$.
The function has a period of $P = 2\pi$.
The equations of the vertical asymptotes are of the form $x = n\pi$, where n is an integer.
The function obtains a relative maximum at $x = -\frac{\pi}{2} + 2\pi n$, where n is an integer.
The domain is all real numbers except odd integer multiples of $\frac{\pi}{2}$.
The range is $(-\infty, -1] \cup [1, \infty)$.
The function obtains a relative maximum at $x = \pi n$, where n is an odd integer.
The equations of the vertical asymptotes are of the form $x = \frac{n\pi}{2}$, where n is an odd integer.
The function has a period of $P = \pi$.

46. Sketch the graph of $y = \sec x$, and then identify as many of the following properties that apply.

The function is an even function.
The function is an odd function.
The domain is all real numbers except integer multiples of π.
The range is $(-\infty, \infty)$.
The function has a period of $P = 2\pi$.
The equations of the vertical asymptotes are of the form $x = n\pi$, where n is an integer.
The function obtains a relative maximum at $x = -\frac{\pi}{2} + 2\pi n$, where n is an integer.
The domain is all real numbers except odd integer multiples of $\frac{\pi}{2}$.
The range is $(-\infty, -1] \cup [1, \infty)$.
The function obtains a relative maximum at $x = \pi n$, where n is an odd integer.
The equations of the vertical asymptotes are of the form $x = \frac{n\pi}{2}$, where n is an odd integer.
The function has a period of $P = \pi$.

In Exercises 47–56, determine the equations of the vertical asymptotes and all relative maximum and relative minimum points of two cycles of each function and then sketch its graph.

SbS 47. $y = 2\sec(x - \pi)$

SbS 48. $y = -2\sec(x - \pi)$

SbS 49. $y = \csc\left(x - \frac{\pi}{4}\right)$

SbS 50. $y = -\csc\left(x - \frac{\pi}{4}\right)$

SbS 51. $y = 2\sec(3x + \pi)$

SbS 52. $y = -3\csc(2x + \pi)$

SbS 53. $y = 4\sec(2x - \pi) + 3$

SbS 54. $y = -3\csc\left(x - \frac{\pi}{4}\right) - 4$

SbS 55. $y = 5 - 2\sec(x - \pi)$

SbS 56. $y = \csc(\pi - 2x) + 3$

Brief Exercises

In Exercises 57–61, determine the interval of the principal cycle of each function.

57. $y = 3\tan\left(x - \frac{\pi}{6}\right)$

58. $y = \cot(2x)$

59. $y = \tan(3x + \pi)$

60. $y = \cot(\pi - 2x)$

61. $y = -3\tan(-2x + \pi) + 1$

In Exercises 62–66, determine the period of each function.

62. $y = 3\tan\left(x - \frac{\pi}{6}\right)$

63. $y = \cot(2x)$

64. $y = \tan(3x + \pi)$

65. $y = \cot(\pi - 2x)$

66. $y = -3\tan(-2x + \pi) + 1$

In Exercises 67–71, determine the equations of the vertical asymptotes of the principal cycle of each function.

67. $y = 3\tan\left(x - \frac{\pi}{6}\right)$

68. $y = \cot(2x)$

69. $y = \tan(3x + \pi)$

70. $y = \cot(\pi - 2x)$

71. $y = -3\tan(-2x + \pi) + 1$

In Exercises 72–76, determine the coordinates of the center point of the principal cycle of each function.

72. $y = 3\tan\left(x - \frac{\pi}{6}\right)$

73. $y = \cot(2x)$

74. $y = \tan(3x + \pi)$

75. $y = \cot(\pi - 2x)$

76. $y = -3\tan(-2x + \pi) + 1$

In Exercises 77–81, determine the coordinates of the two halfway points of the principal cycle of each function.

77. $y = 3\tan\left(x - \frac{\pi}{6}\right)$

78. $y = \cot(2x)$

79. $y = \tan(3x + \pi)$

80. $y = \cot(\pi - 2x)$

81. $y = -3\tan(-2x + \pi) + 1$

In Exercises 82–87, the principal cycle of the graph of a trigonometric function of the from $y = A\tan(Bx - C) + D$ or $y = A\cot(Bx - C) + D$ for $B > 0$ is given. Determine the equation of the function represented by each graph.

82. $y = A\tan(Bx - C) + D$

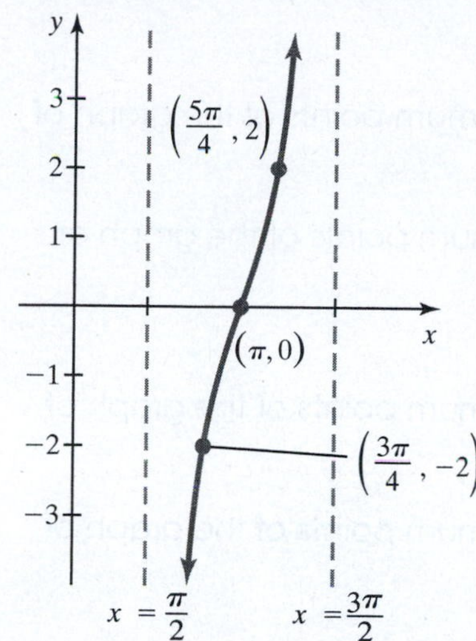

83. $y = A\cot(Bx - C) + D$

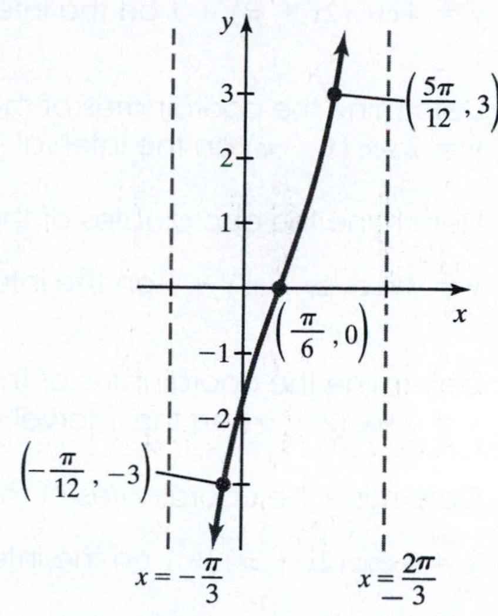

84. $y = A\tan(Bx - C) + D$

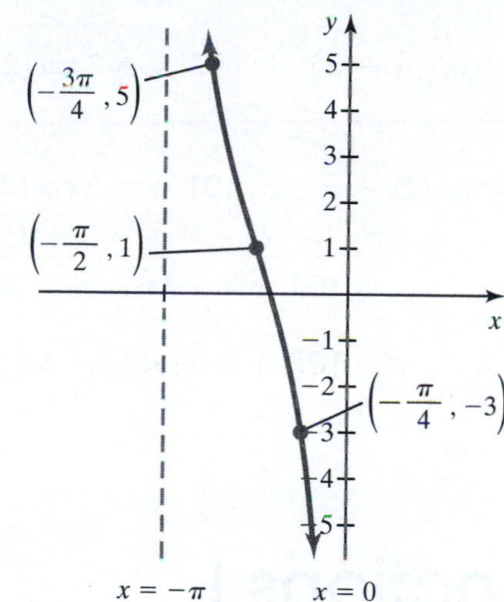

85. $y = A\cot(Bx - C) + D$

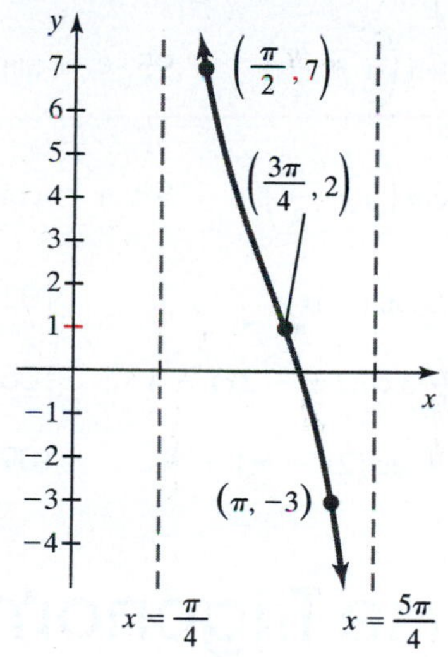

86. $y = A\cot(Bx - C) + D$

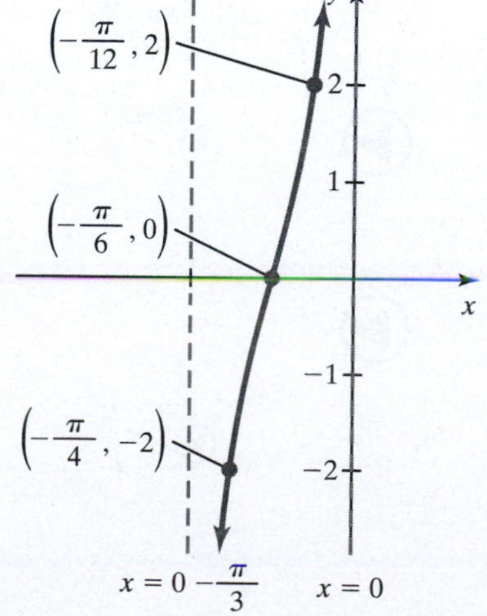

87. $y = A\tan(Bx - C) + D$

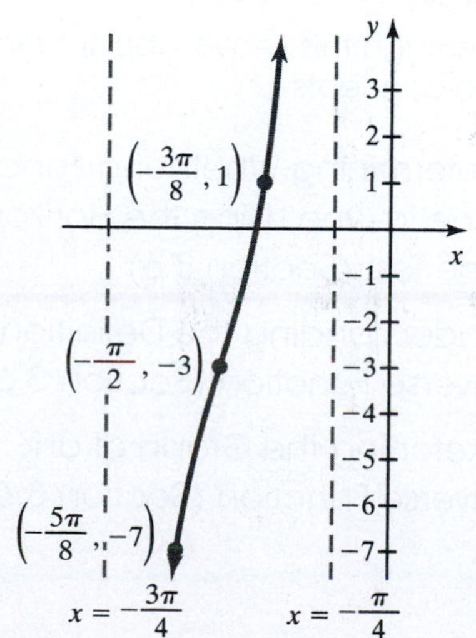

88. Determine the equations of the four vertical asymptotes of the graph of $y = 2\sec(x - \pi)$ on the interval $[-\pi, 3\pi]$.

89. Determine the equations of the four vertical asymptotes of the graph of $y = 4\csc(2x + \pi) + 1$ on the interval $\left[-\frac{\pi}{2}, \frac{3\pi}{2}\right]$.

90. Determine the coordinates of the three relative minimum points of the graph of $y = 2\sec(x - \pi)$ on the interval $[-\pi, 3\pi]$.

91. Determine the coordinates of the two relative minimum points of the graph of $y = 4\csc(2x + \pi) + 1$ on the interval $\left[-\frac{\pi}{2}, \frac{3\pi}{2}\right]$.

92. Determine the coordinates of the two relative maximum points of the graph of $y = 2\sec(x - \pi)$ on the interval $[-\pi, 3\pi]$.

93. Determine the coordinates of the two relative maximum points of the graph of $y = 4\csc(2x + \pi) + 1$ on the interval $\left[-\frac{\pi}{2}, \frac{3\pi}{2}\right]$.

In Exercises 94–109, sketch each function.

94. $y = \tan\left(x + \frac{\pi}{4}\right)$ 95. $y = \tan(2x)$ 96. $y = \tan(3x + \pi)$ 97. $y = 3\tan(2x + \pi) - 1$

98. $y = \cot\left(x - \frac{\pi}{3}\right)$ 99. $y = \cot(3x)$ 100. $y = \cot(2x - \pi)$ 101. $y = 3\cot(2x + \pi) - 1$

102. $y = 2\tan(-3x)$ 103. $y = 2\cot(-3x)$ 104. $y = -3\tan(\pi - 2x) - 1$

105. $y = -3\cot(\pi - 2x) - 1$ 106. $y = \csc\left(x - \frac{\pi}{4}\right)$ 107. $y = 2\sec(x - \pi)$

108. $y = 4\sec(2x - \pi) + 3$ 109. $y = \csc(\pi - 2x) + 3$

7.4 Inverse Trigonometric Functions I

THINGS TO KNOW

Before working through this section, be sure that you are familiar with the following concepts:

VIDEO ANIMATION INTERACTIVE

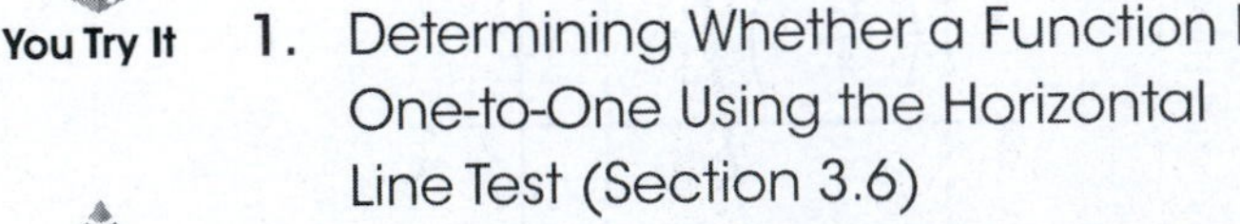

You Try It 1. Determining Whether a Function Is One-to-One Using the Horizontal Line Test (Section 3.6)

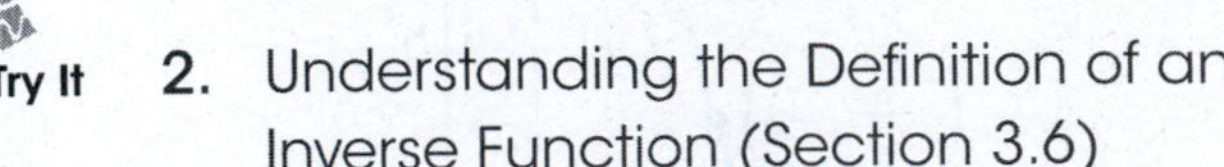

You Try It 2. Understanding the Definition of an Inverse Function (Section 3.6)

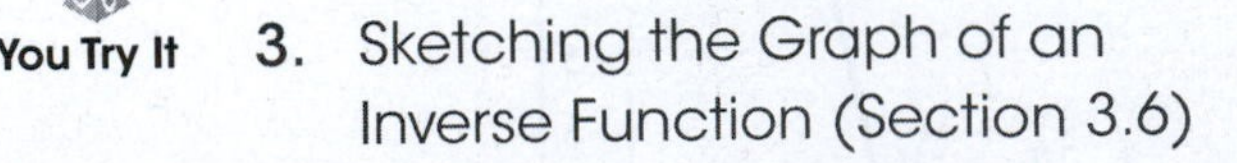

You Try It 3. Sketching the Graph of an Inverse Function (Section 3.6)

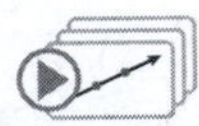

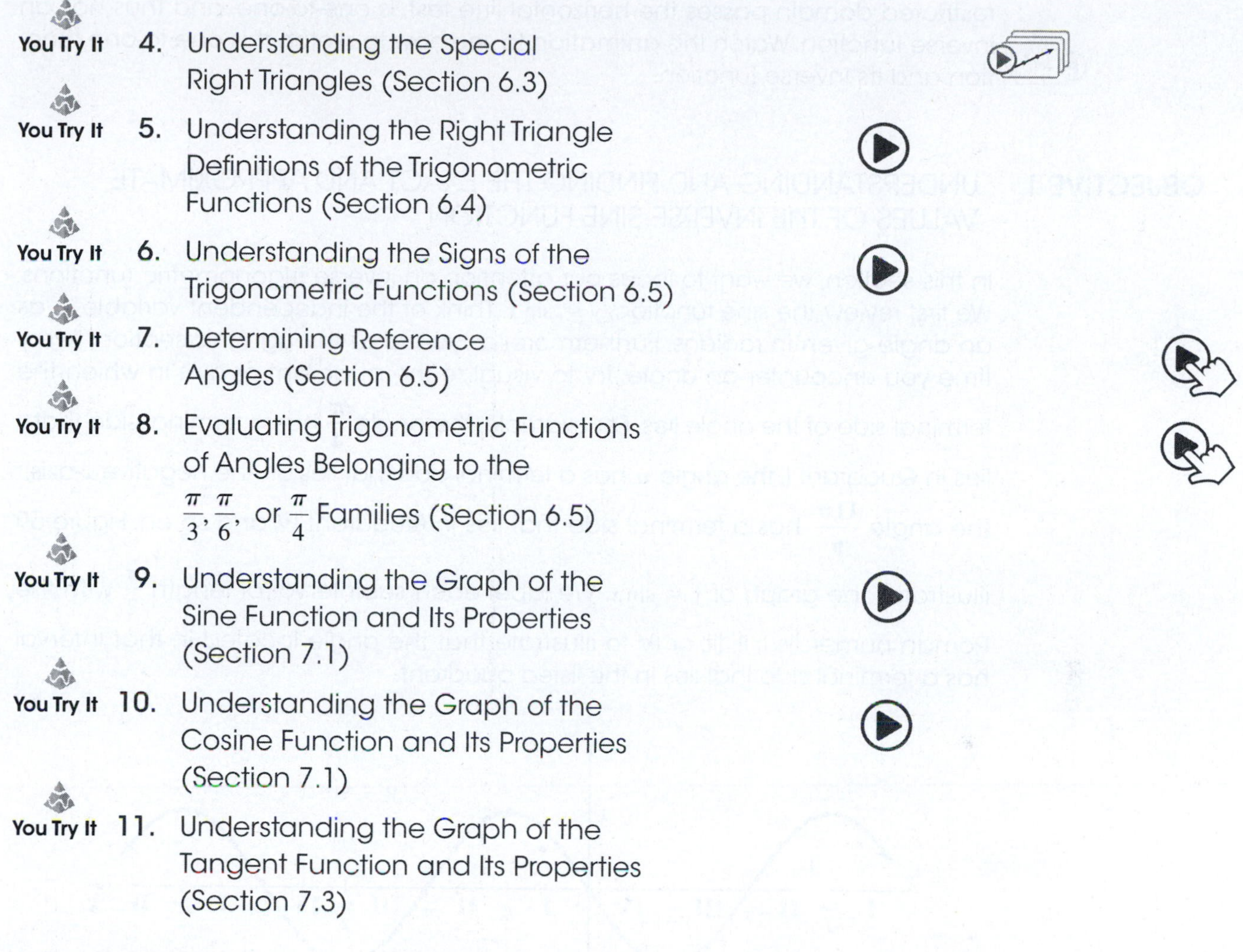

You Try It 4. Understanding the Special Right Triangles (Section 6.3)

You Try It 5. Understanding the Right Triangle Definitions of the Trigonometric Functions (Section 6.4)

You Try It 6. Understanding the Signs of the Trigonometric Functions (Section 6.5)

You Try It 7. Determining Reference Angles (Section 6.5)

You Try It 8. Evaluating Trigonometric Functions of Angles Belonging to the $\frac{\pi}{3}, \frac{\pi}{6},$ or $\frac{\pi}{4}$ Families (Section 6.5)

You Try It 9. Understanding the Graph of the Sine Function and Its Properties (Section 7.1)

You Try It 10. Understanding the Graph of the Cosine Function and Its Properties (Section 7.1)

You Try It 11. Understanding the Graph of the Tangent Function and Its Properties (Section 7.3)

INTRODUCTION

Read this introduction before beginning Objective 1.

OBJECTIVES

1 Understanding and Finding the Exact and Approximate Values of the Inverse Sine Function

2 Understanding and Finding the Exact and Approximate Values of the Inverse Cosine Function

3 Understanding and Finding the Exact and Approximate Values of the Inverse Tangent Function

SECTION 7.4 EXERCISES

Introduction to Section 7.4

In **Section 3.6**, we introduced **one-to-one** functions. Recall that if a graph of a function passes the **horizontal line test**, then the graph represents a one-to-one function. *Only* one-to-one functions have **inverse functions**. However, it is possible to restrict the domain of a function that is not one-to-one to produce a one-to-one function. For example, the function $f(x) = x^2 + 1$ **is not one-to-one** because its graph does not pass the horizontal line test. However, if we **restrict the domain of** $f(x) = x^2 + 1$ to the interval $(-\infty, 0]$, then the graph of f on this

restricted domain passes the horizontal line test, is one-to-one, and thus has an inverse function. Watch this **animation** to see how to sketch this one-to-one function and its inverse function.

OBJECTIVE 1 UNDERSTANDING AND FINDING THE EXACT AND APPROXIMATE VALUES OF THE INVERSE SINE FUNCTION

In this section, we want to focus our attention on inverse trigonometric functions. We first review the sine function, $y = \sin x$. Think of the independent variable, x, as an angle given in radians. Furthermore, as you read through this section, every time you encounter an angle, try to visualize the quadrant or axis in which the terminal side of the angle lies. For example, **the angle** $\frac{\pi}{4}$ has a terminal side that lies in Quadrant I, **the angle** π has a terminal side that lies on the negative x-axis, **the angle** $\frac{11\pi}{6}$ has a terminal side that lies in Quadrant IV, and so on. Figure 59 illustrates the graph of $y = \sin x$. We label each subinterval of length $\frac{\pi}{2}$ with the Roman numerals I, II, III, or IV to illustrate that the angle located in that interval has a terminal side that lies in the listed quadrant.

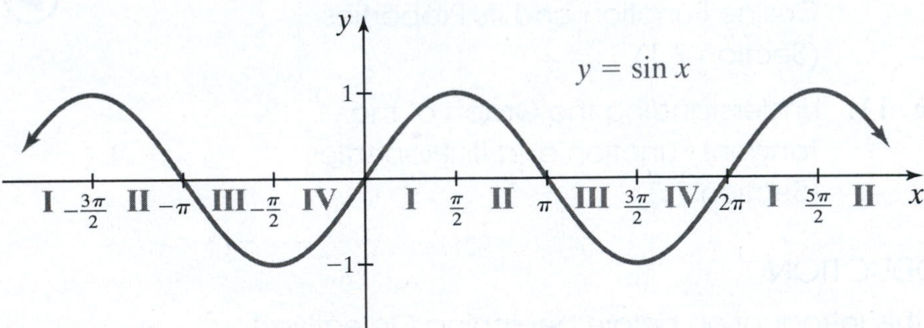

Figure 59 The Roman numerals under each subinterval indicate that any angle located within the interval has a terminal side that lies in that quadrant.

The domain of $y = \sin x$ is $(-\infty, \infty)$. The sine function is clearly not one-to-one because the graph fails the horizontal line test. See Figure 60. However, perhaps we can restrict the domain of $y = \sin x$ by limiting it to a **closed interval** to produce a graph that is one-to-one. But what interval do we choose? There are actually infinitely many ways to restrict the domain of $y = \sin x$ to produce a graph that is one-to-one. Traditionally, it has been agreed upon to use the interval $-\frac{\pi}{2} \leq x \leq \frac{\pi}{2}$ as the restricted domain. Note that this interval shows the entire range of $[-1, 1]$. See Figure 61. This graph is a one-to-one function and therefore has an inverse function. Watch the first part of this **animation** for further explanation.

(eText Screens 7.4-1–7.4-46)

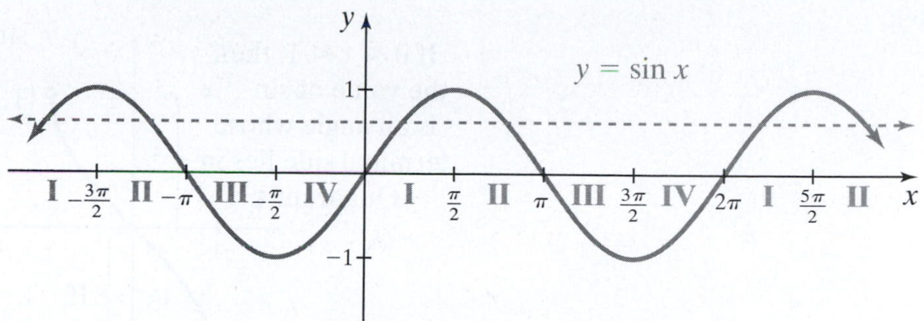

Figure 60 The graph of the function $y = \sin x$ fails the horizontal line test and is therefore not a one-to-one function. This function does not have an inverse function.

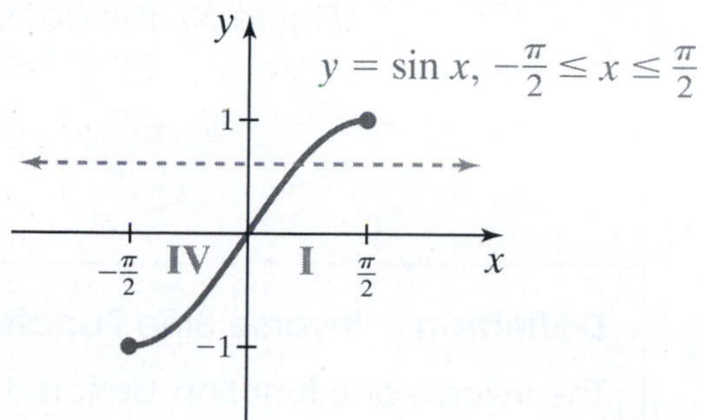

Figure 61 The graph of the function $y = \sin x$ on the restricted domain $-\frac{\pi}{2} \leq x \leq \frac{\pi}{2}$ passes the horizontal line test and is therefore a one-to-one function. This function has an inverse function.

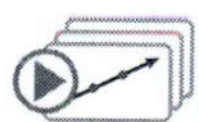

We now want to establish the graph of the inverse sine function. In **Section 3.6**, we learned that if f is a one-to-one function, then the graph of the inverse function, f^{-1}, can be obtained by reversing the coordinates of each of the ordered pairs lying on the graph of f. In other words, for any point (a, b) lying on the graph of f, the point (b, a) must lie on the graph of f^{-1}.

Therefore, starting with the graph of the restricted sine function, we can obtain the graph of $y = \sin^{-1} x$ by reversing the coordinates of the ordered pairs that lie on the graph of the restricted sine function. We then plot these ordered pairs and connect them with a smooth curve to obtain the graph of $y = \sin^{-1} x$. This function is called the **inverse sine function**. Be sure to work through this **animation** to see exactly how to obtain the graphs shown in Figures 62 and 63.

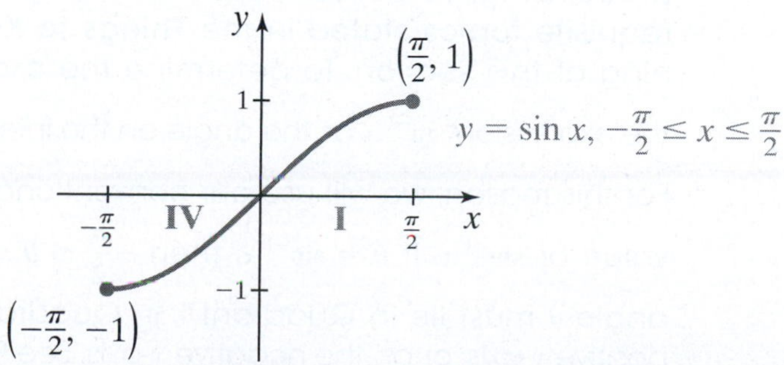

Figure 62 The graph of $y = \sin x$ on the restricted domain $-\frac{\pi}{2} \leq x \leq \frac{\pi}{2}$ with range $-1 \leq y \leq 1$

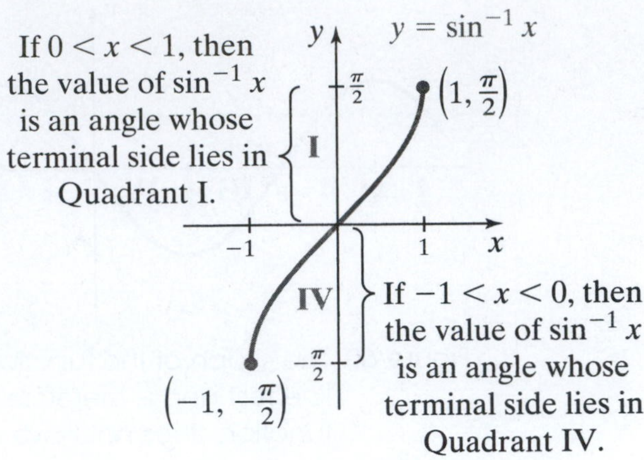
If $0 < x < 1$, then the value of $\sin^{-1} x$ is an angle whose terminal side lies in Quadrant I.
$y = \sin^{-1} x$
$\left(1, \frac{\pi}{2}\right)$
I
IV
$\left(-1, -\frac{\pi}{2}\right)$
If $-1 < x < 0$, then the value of $\sin^{-1} x$ is an angle whose terminal side lies in Quadrant IV.

Figure 63 The domain of $y = \sin^{-1}(x)$ is $-1 \le x \le 1$ and the range is $-\frac{\pi}{2} \le y \le \frac{\pi}{2}$.

Definition Inverse Sine Function

The **inverse sine function**, denoted as $y = \sin^{-1} x$, is the inverse of $y = \sin x, -\frac{\pi}{2} \le x \le \frac{\pi}{2}$.

The domain of $y = \sin^{-1} x$ is $-1 \le x \le 1$ and the range is $-\frac{\pi}{2} \le y \le \frac{\pi}{2}$.

(Note that an alternative notation for $\sin^{-1} x$ is $\arcsin x$.)

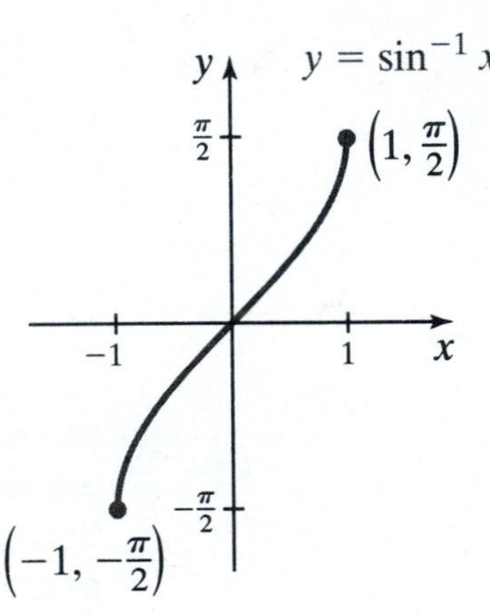
$y = \sin^{-1} x$
$\left(1, \frac{\pi}{2}\right)$
$\left(-1, -\frac{\pi}{2}\right)$

CAUTION Do not confuse the notation $\sin^{-1} x$ with $(\sin x)^{-1} = \frac{1}{\sin x} = \csc x$.

The negative 1 is not an exponent! Thus, $\sin^{-1} x \ne \frac{1}{\sin x}$.

We are often interested in determining the exact value of $\sin^{-1} x$. Before proceeding, you must have a complete understanding of all of the prerequisite topics stated in the **Things to Know** pages located at the beginning of this section. To determine the exact value of $\sin^{-1} x$, we will think of the expression $\sin^{-1} x$ as the angle on the interval $\left[-\frac{\pi}{2}, \frac{\pi}{2}\right]$ whose sine is equal to x. For this reason, we will use our conventional angle notation, θ, to represent the value of $\sin^{-1} x$. If $\theta = \sin^{-1} x$, then $-\frac{\pi}{2} \le \theta \le \frac{\pi}{2}$. Therefore, the terminal side of angle θ must lie in Quadrant I, in Quadrant IV, on the positive x-axis, on the positive y-axis, or on the negative y-axis. See Figure 64.

(eText Screens 7.4-1–7.4-46)

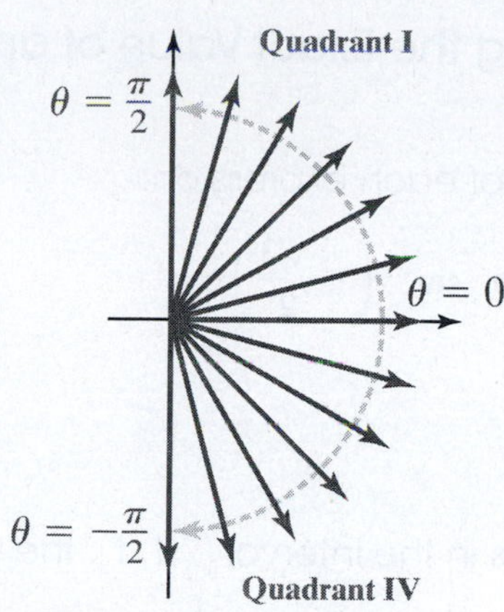

Figure 64 If $\theta = \sin^{-1} x$, then $-\frac{\pi}{2} \le \theta \le \frac{\pi}{2}$ and the terminal side of angle θ lies in Quadrant I, in Quadrant IV, on the positive x-axis, on the positive y-axis, or on the negative y-axis.

To determine the exact value of $\sin^{-1} x$, we use the four-step process outlined on the following page.

Steps for Determining the Exact Value of $\sin^{-1} x$

Step 1. If x is in the interval $[-1, 1]$, then the value of $\sin^{-1}x$ must be an angle in the interval $\left[-\frac{\pi}{2}, \frac{\pi}{2}\right]$.

Step 2. Let $\sin^{-1}x = \theta$ such that $\sin \theta = x$.

Step 3. If $\sin \theta = 0$, then $\theta = 0$ and the terminal side of angle θ lies on the positive x-axis.

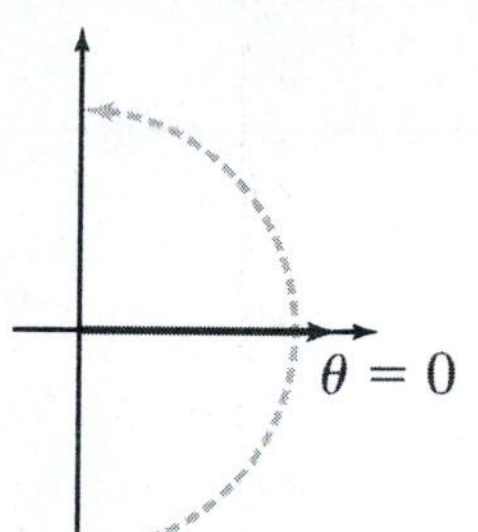

If $\sin \theta > 0$, then $0 < \theta \le \frac{\pi}{2}$ and the terminal side of angle θ lies in Quadrant I or on the positive y-axis.

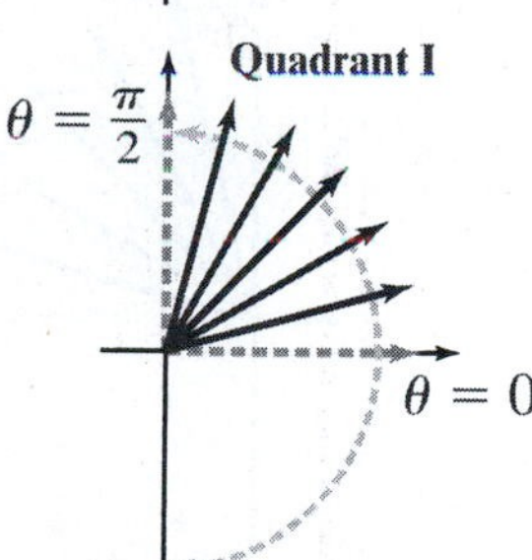

If $\sin \theta < 0$, then $-\frac{\pi}{2} \le \theta < 0$ and the terminal side of angle θ lies in Quadrant IV or on the negative y-axis.

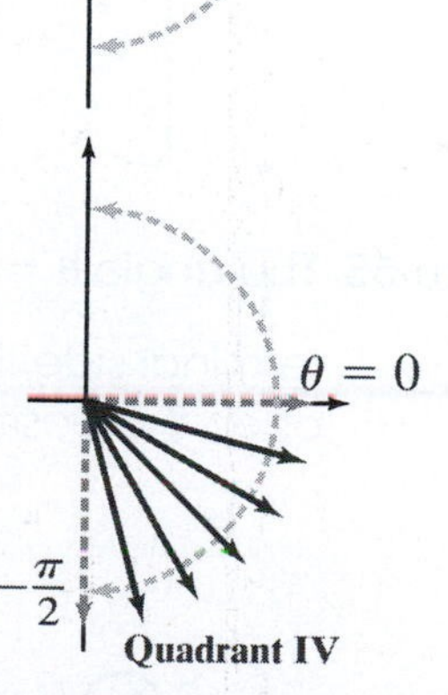

Step 4. Use your knowledge of the two **special right triangles**, the graphs of the trigonometric functions, **Table 1** from Section 6.4, or **Table 2** from Section 6.5 to determine the angle in the correct quadrant whose sine is x.

Example 1 Determining the Exact Value of an Inverse Sine Expression

Determine the exact value of each expression.

a. $\sin^{-1}\left(\frac{1}{2}\right)$ **b.** $\sin^{-1}\left(-\frac{\sqrt{3}}{2}\right)$

Solution

a. Step 1. Because $x = \frac{1}{2}$ is in the interval $[-1, 1]$, the value of $\sin^{-1}\left(\frac{1}{2}\right)$ must be an angle in the interval $\left[-\frac{\pi}{2}, \frac{\pi}{2}\right]$.

Step 2. Let $\sin^{-1}\left(\frac{1}{2}\right) = \theta$ such that $\sin\theta = \frac{1}{2}$.

Step 3. Because $\sin\theta = \frac{1}{2} > 0$, it follows that $0 < \theta \le \frac{\pi}{2}$. Therefore, we are looking for an angle whose terminal side lies in Quadrant I or on the positive y-axis. See Figure 65.

Step 4. Using our knowledge of the special $\frac{\pi}{6}, \frac{\pi}{3}, \frac{\pi}{2}$ right triangle or Table 1, we know that $\sin\frac{\pi}{6} = \frac{1}{2}$. The angle $\theta = \frac{\pi}{6}$ is the **only** angle for which $0 < \theta \le \frac{\pi}{2}$ whose terminal side lies in Quadrant I or on the positive y-axis such that $\sin\frac{\pi}{6} = \frac{1}{2}$. See Figure 66. Therefore, $\sin^{-1}\left(\frac{1}{2}\right) = \frac{\pi}{6}$.

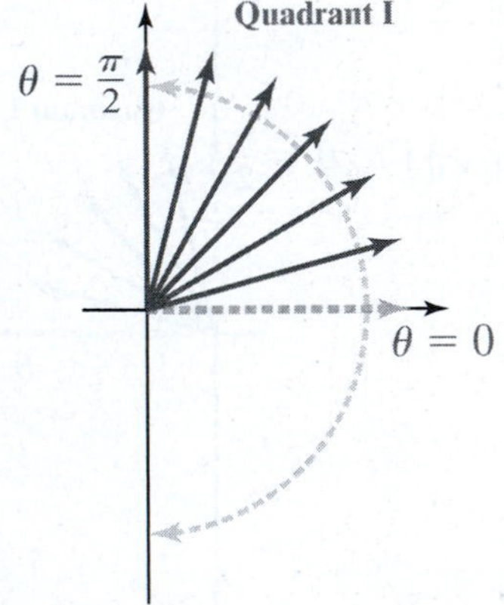

Figure 65 The angle $\theta = \sin^{-1}\left(\frac{1}{2}\right)$ must have a terminal side that lies in Quadrant I or on the positive y-axis.

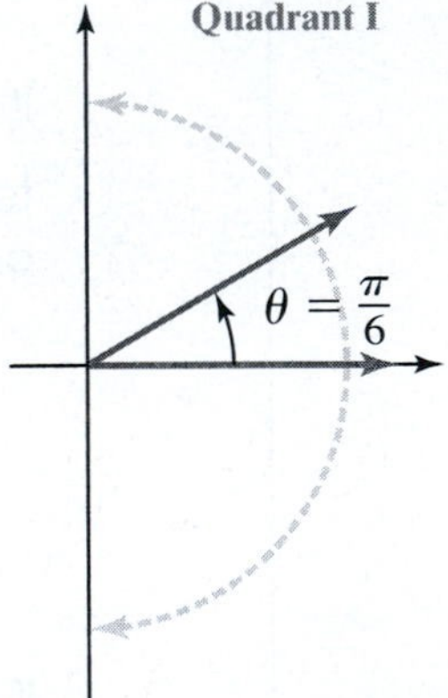

Figure 66 The angle $\theta = \frac{\pi}{6}$ is the only angle in the interval $0 < \theta \le \frac{\pi}{2}$ whose terminal side lies in Quadrant I or on the positive y-axis such that $\sin\theta = \frac{1}{2}$.

Work through this **interactive video** to see each s tep of this solution.

b. To find the value of $\sin^{-1}\left(-\frac{\sqrt{3}}{2}\right)$, we follow the four-step process.

Step 1. Because $x = -\frac{\sqrt{3}}{2}$ is in the interval $[-1, 1]$, the value of $\sin^{-1}\left(-\frac{\sqrt{3}}{2}\right)$ must be an angle in the interval $\left[-\frac{\pi}{2}, \frac{\pi}{2}\right]$.

Step 2. Let $\sin^{-1}\left(-\frac{\sqrt{3}}{2}\right) = \theta$ such that $\sin\theta = -\frac{\sqrt{3}}{2}$.

Step 3. Because $\sin\theta = -\frac{\sqrt{3}}{2} < 0$, it follows that $-\frac{\pi}{2} \le \theta < 0$. Therefore, we are looking for an angle whose terminal side lies in Quadrant IV or on the negative y-axis. See Figure 67.

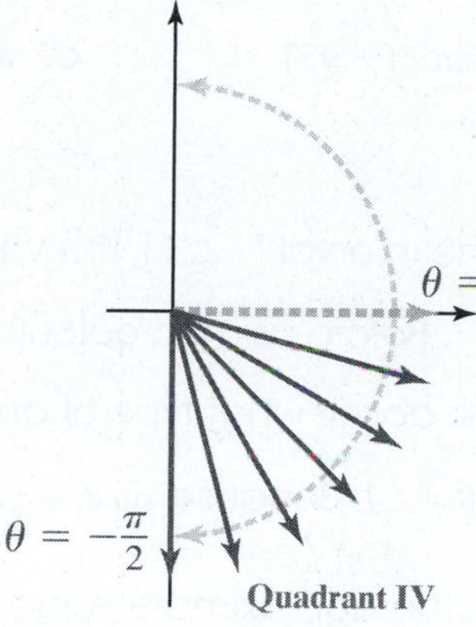

Figure 67 The angle $\theta = \sin^{-1}\left(-\frac{\sqrt{3}}{2}\right)$ must have a terminal side that lies in Quadrant IV or on the negative y-axis.

Step 4. Using our knowledge of the special $\frac{\pi}{6}, \frac{\pi}{3}, \frac{\pi}{2}$ **right triangle** or **Table 1**, we know that $\sin\frac{\pi}{3} = \frac{\sqrt{3}}{2}$. However, $\theta \neq \frac{\pi}{3}$ because the terminal side of this angle does **not** lie in Quadrant IV or on the negative y-axis. We must identify the angle θ that satisfies the following three conditions:

- The reference angle is $\theta_R = \frac{\pi}{3}$.
- The terminal side lies in Quadrant IV or on the negative y-axis.
- $-\frac{\pi}{2} \le \theta < 0$.

The only such angle is $\theta = -\frac{\pi}{3}$. See Figure 68. Therefore, $\sin^{-1}\left(-\frac{\sqrt{3}}{2}\right) = -\frac{\pi}{3}$.

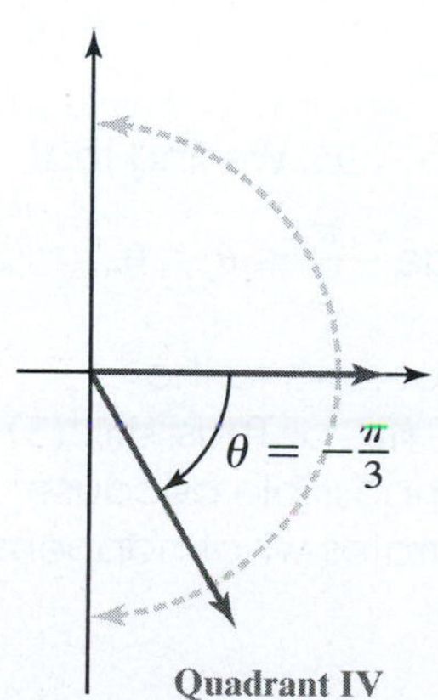

Figure 68

The angle $\theta = -\frac{\pi}{3}$ is the only angle in the interval $-\frac{\pi}{2} \le \theta < 0$ whose terminal side lies in Quadrant IV or on the negative y-axis such that $\sin\theta = -\frac{\sqrt{3}}{2}$.

(eText Screens 7.4-1–7.4-46)

Work through this **interactive video** to see each step of this solution.

You Try It Work through this You Try It problem.

Work Exercises 1–3 in this textbook or in the MyLab Math Study Plan.

The **steps for determining the exact value of $\sin^{-1}x$** will not apply when the **argument**, x, is not a recognized value associated with special right triangles or quadrantal angles. In this situation, the use of a calculator will be necessary to find an approximate value of $\sin^{-1} x$.

Example 2 Finding the Approximate Value of an Inverse Sine Expression

Use a calculator to approximate each value, or state that the value does not exist.

a. $\sin^{-1}(.7)$ **b.** $\sin^{-1}(-.95)$ **c.** $\sin^{-1}(3)$

Solution

a. Because $x = .7$ is in the interval $[-1, 1]$, the value of $\sin^{-1}(.7)$ must be an angle in the interval $\left[-\frac{\pi}{2}, \frac{\pi}{2}\right]$. Before using a calculator to approximate this value, it is good practice to think about what type of answer we can expect. If we let $\sin^{-1}(.7) = \theta$, then $\sin\theta = .7$. Because $\sin\theta = .7 > 0$, then $0 < \theta \leq \frac{\pi}{2}$, where $\frac{\pi}{2} \approx 1.5708$. Therefore, the approximate value of $\sin^{-1}(.7)$ must be between 0 and 1.5708.

Using a calculator in radian mode, type 2nd sin and then .7. We find that $\sin^{-1}(.7) \approx 0.7754$ radian. The value of $\theta \approx 0.7754$ seems reasonable because $0 < 0.7754 \leq \frac{\pi}{2}$.

b. Because $x = -.95$ is in the interval $[-1, 1]$, the value of $\sin^{-1}(-.95)$ must be an angle in the interval $\left[-\frac{\pi}{2}, \frac{\pi}{2}\right]$. If $\sin^{-1}(-.95) = \theta$, then $\sin\theta = -.95 < 0$. Thus, $-\frac{\pi}{2} \leq \theta < 0$. Therefore, the approximate value of $\sin^{-1}(-.95)$ must be negative but greater than or equal to $-\frac{\pi}{2} \approx -1.5708$.

Using a calculator in radian mode, type 2nd sin and then $-.95$. We find that $\sin^{-1}(-.95) \approx -1.2532$ radians, which is certainly in the range $-\frac{\pi}{2} \leq \theta < 0$.

c. Looking at the expression, $\sin^{-1}(3)$, we see that 3 does not belong to the interval $-1 \leq x \leq 1$. Therefore, 3 is not in the domain of $y = \sin^{-1} x$. Thus, $\sin^{-1}(3)$ does not exist. Also, if $\theta = \sin^{-1}(3)$, then $\sin\theta = 3$, which is impossible because the maximum value of the sine function is 1. Figure 69 illustrates what happens when we input $\sin^{-1}(3)$ into a calculator.

Using Technology

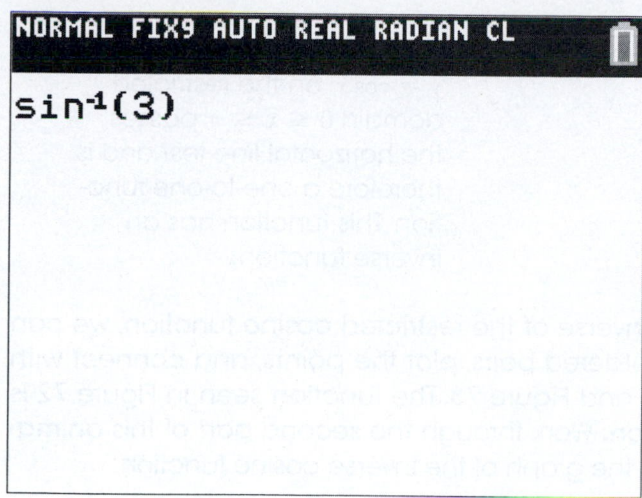

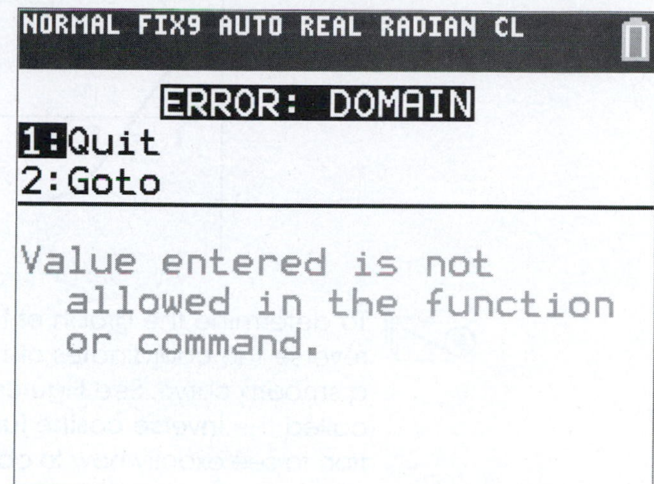

Figure 69 An input of $\sin^{-1}(3)$ using a graphing utility results in a domain error.

You Try It Work through this You Try It problem.

Work Exercises 4-6 in this textbook or in the MyLab Math Study Plan.

OBJECTIVE 2 UNDERSTANDING AND FINDING THE EXACT AND APPROXIMATE VALUES OF THE INVERSE COSINE FUNCTION

We establish the inverse cosine function in much the same way as we did with the inverse sine function. First, the cosine function is defined for all real numbers and is not one-to-one because the graph does not pass the horizontal line test. See **Figure 70**. Note that we label each subinterval of length $\frac{\pi}{2}$ with the Roman numerals I, II, III, or IV to illustrate that the angle located in that interval has a terminal side that lies in the listed quadrant.

As with the sine function, we can restrict the domain of the cosine function to produce a graph that passes the horizontal line test and hence is one-to-one. Although there are infinitely many ways to restrict the domain of the cosine function to produce a graph that is one-to-one, most mathematicians have agreed on the interval $0 \le x \le \pi$ as the common restricted interval. The graph of the restricted cosine function on the interval $0 \le x \le \pi$ passes the horizontal line test, is one-to-one, and hence has an inverse function. See **Figure 71**. Work through the first part of this **animation** for further explanation.

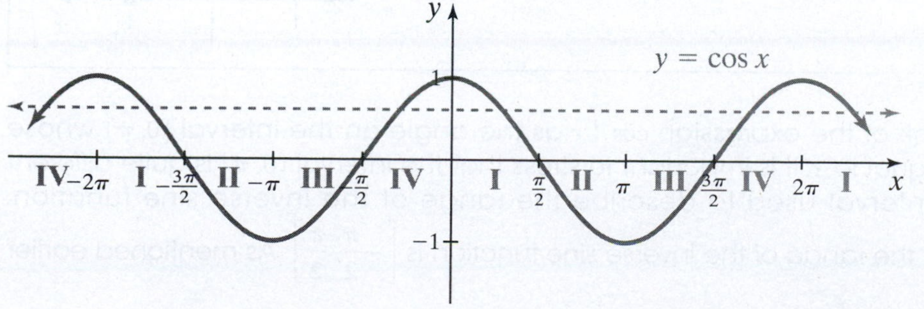

Figure 70 The graph of the function $y = \cos x$ fails the horizontal line test and is therefore not a one-to-one function. This function does not have an inverse function.

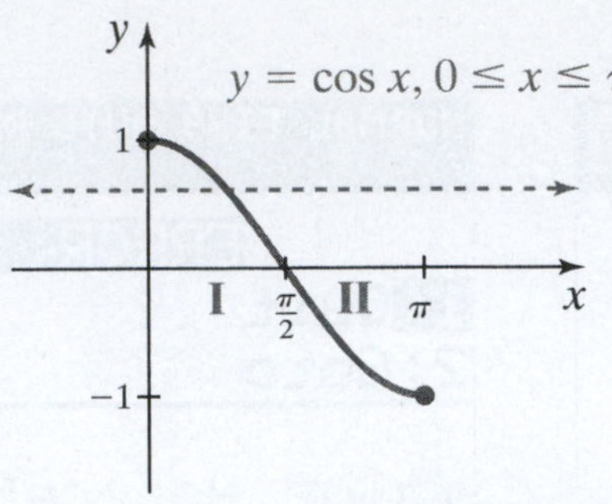

Figure 71 The graph of the function $y = \cos x$ on the restricted domain $0 \le x \le \pi$ passes the horizontal line test and is therefore a one-to-one function. This function has an inverse function.

To determine the graph of the inverse of the restricted cosine function, we can reverse the coordinates of the ordered pairs, plot the points, and connect with a smooth curve. See **Figures 72** and **Figure 73**. The function seen in Figure 72 is called the **inverse cosine function.** Work through the second part of this **animation** to see exactly how to obtain the graph of the inverse cosine function.

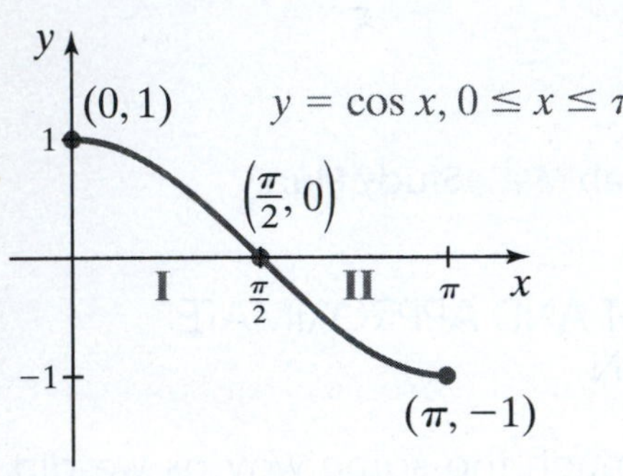

Figure 72 The graph of $y = \cos x$ on the restricted domain $0 \le x \le \pi$ with range $-1 \le y \le 1$.

Start with the restricted cosine function and reverse the coordinates of each ordered pair to produce the graph of the inverse cosine function.

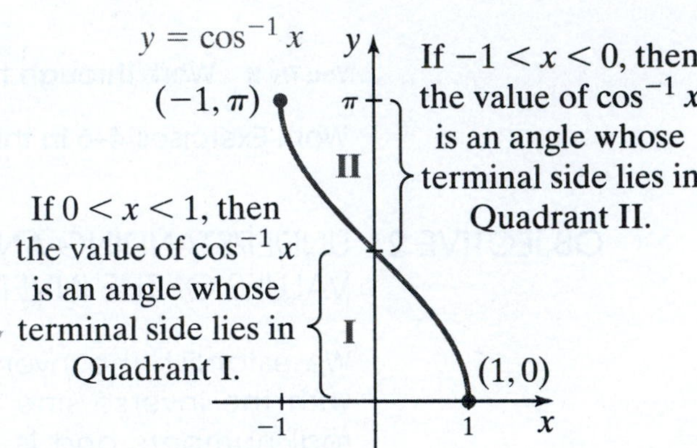

Figure 73 The domain of $y = \cos^{-1} x$ is $-1 \le x \le 1$ and the range is $0 \le y \le \pi$.

Definition Inverse Cosine Function

The **inverse cosine function**, denoted as $y = \cos^{-1} x$, is the inverse of $y = \cos x, 0 \le x \le \pi$.

The domain of $y = \cos^{-1} x$ is $-1 \le x \le 1$ and the range is $0 \le y \le \pi$.

(Note that an alternative notation for $\cos^{-1} x$ is $\arccos x$.)

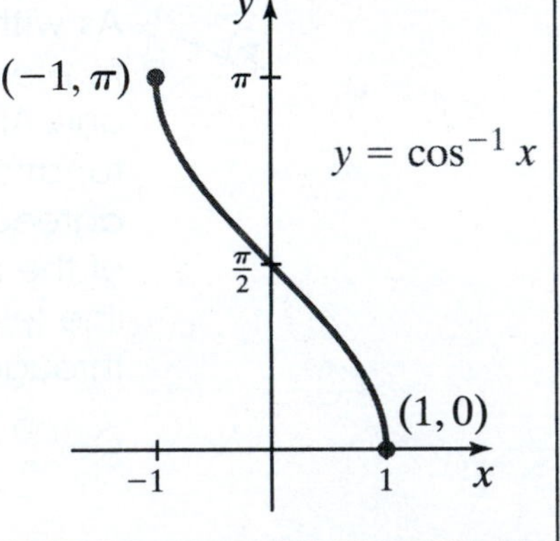

We will think of the expression $\cos^{-1} x$ as the angle on the interval $[0, \pi]$ whose cosine is equal to x. It is important to stress that the interval $[0, \pi]$ is quite different from the interval used to describe the range of the inverse sine function. Recall that the range of the inverse sine function is $\left[-\frac{\pi}{2}, \frac{\pi}{2}\right]$. As mentioned earlier when we described the inverse sine function, it is customary to use the angle notation, θ, to represent the value of $\cos^{-1} x$. Therefore, if $\theta = \cos^{-1} x$, then $0 \le \theta \le \pi$. Thus, the terminal side of angle θ must lie in Quadrant I, in Quadrant II, on the positive y-axis, on the positive x-axis, or on the negative x-axis. See **Figure 74**.

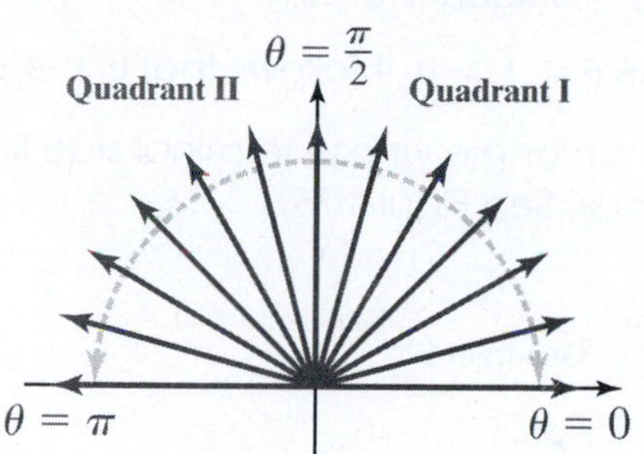

Figure 74 If $\theta = \cos^{-1} x$, then $0 \le \theta \le \pi$ and the terminal side of angle θ lies in Quadrant I, in Quadrant II, on the positive y-axis, on the positive x-axis, or on the negative x-axis.

To determine the exact value of an inverse cosine expression, we follow the four-step process outlined on the following page.

Steps for Determining the Exact Value of $\cos^{-1} x$

Step 1. If x is in the interval $[-1, 1]$, then the value of $\cos^{-1} x$ must be an angle in the interval $[0, \pi]$.

Step 2. Let $\cos^{-1} x = \theta$ such that $\cos \theta = x$.

Step 3. If $\cos \theta = 0$, then $\theta = \frac{\pi}{2}$ and the terminal side of angle θ lies on the positive y-axis.

If $\cos \theta > 0$, then $0 \le \theta < \frac{\pi}{2}$ and the terminal side of angle θ lies in Quadrant I or on the positive x-axis.

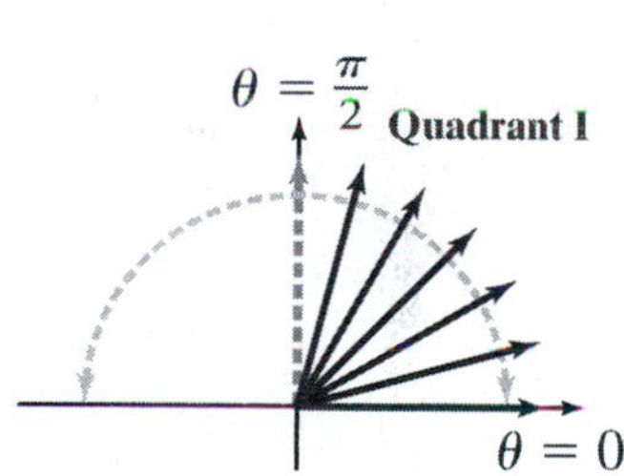

If $\cos \theta < 0$, then $\frac{\pi}{2} < \theta \le \pi$ and the terminal side of angle θ lies in Quadrant II or on the negative x-axis.

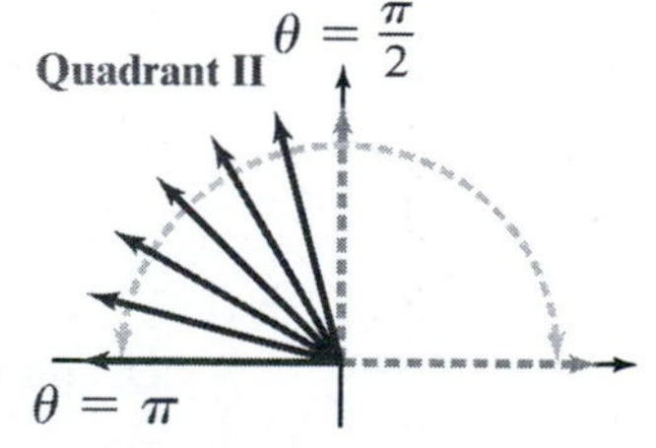

Step 4. Use your knowledge of the two special **right triangles**, the graphs of the trigonometric functions, **Table 1** from Section 6.4, or **Table 2** from Section 6.5 to determine the angle in the correct quadrant whose cosine is x.

Example 3 Finding the Exact Value of an Inverse Cosine Expression

Determine the exact value of each expression.

a. $\cos^{-1}(1)$ **b.** $\cos^{-1}\left(-\frac{1}{\sqrt{2}}\right)$

Solution

a. Step 1. Because $x = 1$ is in the interval $[-1, 1]$, the value of $\cos^{-1}(1)$ must be an angle in the interval $[0, \pi]$.

Step 2. Let $\cos^{-1}(1) = \theta$ such that $\cos\theta = 1$.

Step 3. Because $\cos\theta = 1 > 0$, it follows that $0 \le \theta < \frac{\pi}{2}$. Therefore, we are looking for an angle whose terminal side lies in Quadrant I or on the positive x-axis. See Figure 75.

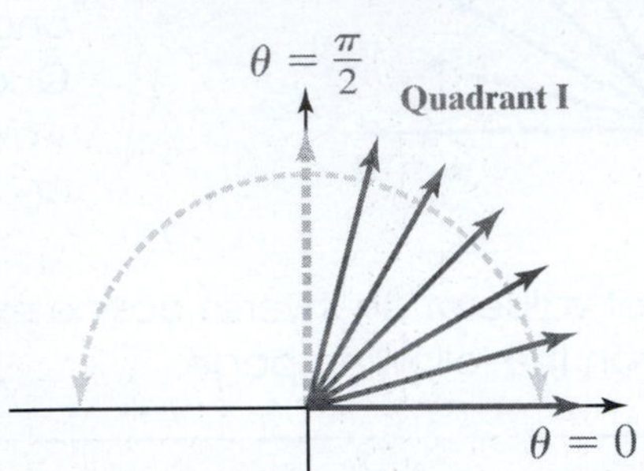

Figure 75 The angle $\theta = \cos^{-1}(1)$ must have a terminal side that lies in Quadrant I or on the positive x-axis.

Step 4. From the **graph of $y = \cos\theta$** or from Section 6.5 **Table 2**, we know that if $\theta = 0$, then $\cos 0 = 1$. The angle $\theta = 0$ is the *only* angle for which $0 \le \theta < \frac{\pi}{2}$ whose terminal side lies in Quadrant I or on the positive x-axis such that $\cos 0 = 1$. See Figure 76. Therefore, $\cos^{-1}(1) = 0$.

Work through this **interactive video** to see each step of this solution.

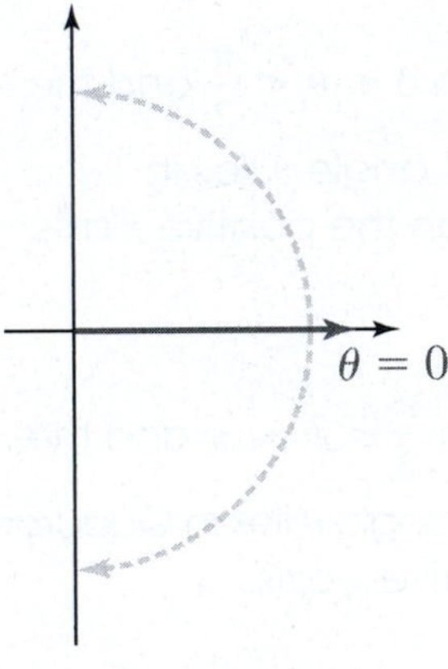

Figure 76 The angle $\theta = 0$ is the only angle in the interval $0 \le \theta < \frac{\pi}{2}$ whose terminal side lies in Quadrant I or on the positive x-axis such that $\cos\theta = 1$.

b. To find the value of $\cos^{-1}\left(-\frac{1}{\sqrt{2}}\right)$, we follow the four-step process.

Step 1. Because $x = -\frac{1}{\sqrt{2}}$ is in the interval $[-1, 1]$, the value of $\cos^{-1}\left(-\frac{1}{\sqrt{2}}\right)$ must be an angle in the interval $[0, \pi]$.

Step 2. Let $\cos^{-1}\left(-\frac{1}{\sqrt{2}}\right) = \theta$ such that $\cos\theta = -\frac{1}{\sqrt{2}}$.

Step 3. Because $\cos\theta = -\frac{1}{\sqrt{2}} < 0$, it follows that $\frac{\pi}{2} < \theta \le \pi$. Thus, we are looking for an angle whose terminal side lies in Quadrant II or on the negative x-axis. See Figure 77.

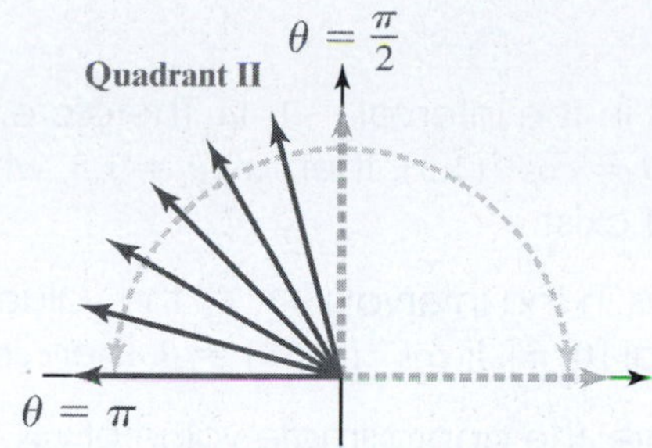

Figure 77 The angle $\theta = \cos^{-1}\left(-\frac{1}{\sqrt{2}}\right)$ must have a terminal side that lies in Quadrant II or on the negative x-axis.

Step 4. Using our knowledge of the special $\frac{\pi}{4}, \frac{\pi}{4}, \frac{\pi}{2}$ right triangle or Table 1 from Section 6.4, we know that $\cos\frac{\pi}{4} = \frac{1}{\sqrt{2}}$. However, $\theta \neq \frac{\pi}{4}$ because the terminal side of this angle does **not** lie in Quadrant II or on the negative x-axis. We must identify the angle θ that satisfies the following three conditions:

- The reference angle is $\theta_R = \frac{\pi}{4}$.
- The terminal side lies in Quadrant II or on the negative x-axis.
- $\frac{\pi}{2} < \theta \leq \pi$.

The only such angle is $\theta = \frac{3\pi}{4}$. See Figure 78. Therefore, $\cos^{-1}\left(-\frac{1}{\sqrt{2}}\right) = \frac{3\pi}{4}$.

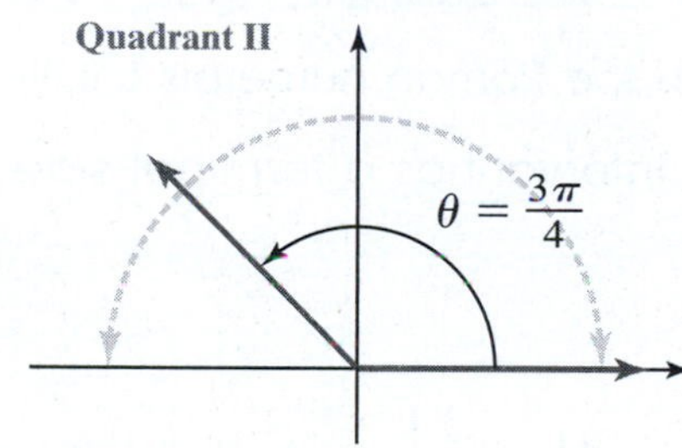

Figure 78 The angle $\theta = \frac{3\pi}{4}$ is the only angle in the interval $\frac{\pi}{2} < \theta \leq \pi$ whose terminal side lies in Quadrant II or on the negative x-axis such that $\cos\theta = -\frac{1}{\sqrt{2}}$.

Work through this interactive video to see each step of this solution.

You Try It Work through this You Try It problem.

Work Exercises 7–9 in this textbook or in the MyLab Math Study Plan.

Example 4 Finding the Approximate Value of an Inverse Cosine Expression

Use a calculator to approximate each value, or state that the value does not exist.

a. $\cos^{-1}(1.5)$ **b.** $\cos^{-1}(-.25)$

Solution

a. Note that 1.5 is not in the interval $[-1, 1]$. Therefore, 1.5 is not in the domain of $y = \cos^{-1} x$. Also, if $\theta = \cos^{-1}(1.5)$, then $\cos\theta = 1.5$, which is not possible. Thus, $\cos^{-1}(1.5)$ does not exist.

b. Because $x = -.25$ is in the interval $[-1, 1]$, the value of $\cos^{-1}(-.25)$ must be an angle in the interval $[0, \pi]$. If $\cos^{-1}(-.25) = \theta$, then $\cos\theta = -.25 < 0$. Thus, $\frac{\pi}{2} < \theta \le \pi$. Therefore, the approximate value of $\cos^{-1}(-.25)$ must be a decimal value greater than $\frac{\pi}{2} \approx 1.5708$ but less than or equal to $\pi \approx 3.1416$. Using a calculator in radian mode, type [2nd] [COS] and then $-.25$. We find that $\cos^{-1}(-.25) \approx 1.8235$ radians. This value represents an angle in Quadrant II between $\frac{\pi}{2}$ and π, as predicted.

You Try It Work through this You Try It problem.

Work Exercises 10–12 in this textbook or in the MyLab Math Study Plan.

OBJECTIVE 3 UNDERSTANDING AND FINDING THE EXACT AND APPROXIMATE VALUES OF THE INVERSE TANGENT FUNCTION

Recall that the function $y = \tan x$ is defined for all values except odd integer multiples of $\frac{\pi}{2}$. The tangent function on its entire domain is not one-to-one because the graph does not pass the horizontal line test. See **Figure 79**. However, we established in Section 7.3 that the **principal cycle of the tangent function is one-to-one**. See **Figure 80**. Once again, in Figure 79 and 80, we label each subinterval of length $\frac{\pi}{2}$ with the Roman numerals I, II, III, or IV to illustrate that the angle located in that interval has a terminal side that lies in the listed quadrant.

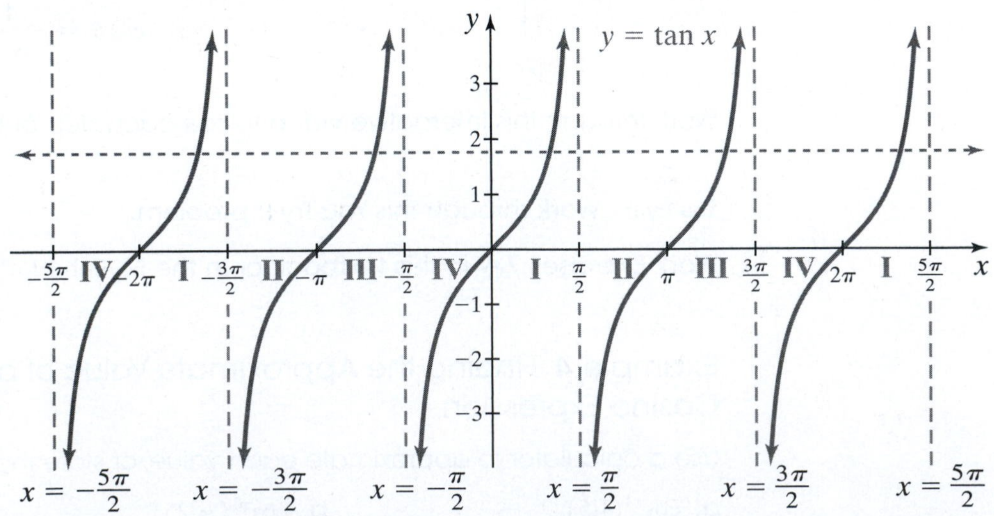

Figure 79 The tangent function fails the horizontal line test and is not one-to-one. This function does not have an inverse function.

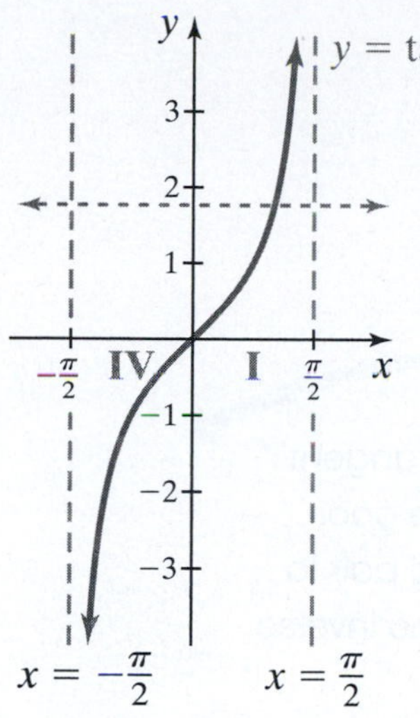

Figure 80 The principal cycle of the tangent function on the interval $-\frac{\pi}{2} < x < \frac{\pi}{2}$ passes the horizontal line test and is one-to-one. This function has an inverse function.

Although there are infinitely many ways to restrict the domain of the tangent function to produce a graph that is one-to-one, it is customary to use the principal cycle of the tangent function, the portion of the graph restricted to the domain $-\frac{\pi}{2} < x < \frac{\pi}{2}$. Because this restricted tangent function is one-to-one, it must have an inverse function. To determine the graph of the inverse of the restricted tangent function, we can reverse the coordinates of the ordered pairs, plot the points, and connect with a smooth curve. See **Figures 81 and 82**. The inverse of the restricted tangent function is called the **inverse tangent function** and is denoted as $y = \tan^{-1} x$.

Notice that the domain of the restricted tangent function is $\left(-\frac{\pi}{2}, \frac{\pi}{2}\right)$. This is precisely the range of $y = \tan^{-1} x$. Likewise, the range of the restricted tangent function is $(-\infty, \infty)$. This is exactly the domain of $y = \tan^{-1} x$. The restricted tangent function has two **vertical asymptotes** at $x = -\frac{\pi}{2}$ and $x = \frac{\pi}{2}$. The inverse tangent function has two **horizontal asymptotes** at $y = -\frac{\pi}{2}$ and $y = \frac{\pi}{2}$. Work through this **animation** to see how to obtain the inverse tangent function.

Definition Inverse Tangent Function

The **inverse tangent function**, denoted as $y = \tan^{-1} x$, is the inverse of $y = \tan x$, $-\frac{\pi}{2} < x < \frac{\pi}{2}$.

The domain of $y = \tan^{-1} x$ is all real numbers and the range is $-\frac{\pi}{2} < y < \frac{\pi}{2}$.

(Note that an alternative notation for $\tan^{-1} x$ is $\arctan x$.)

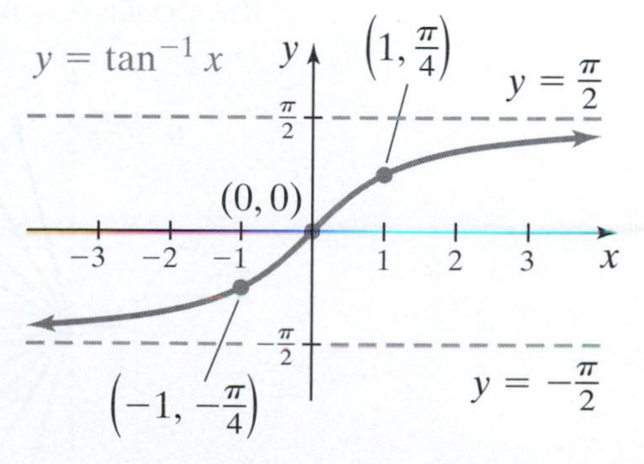

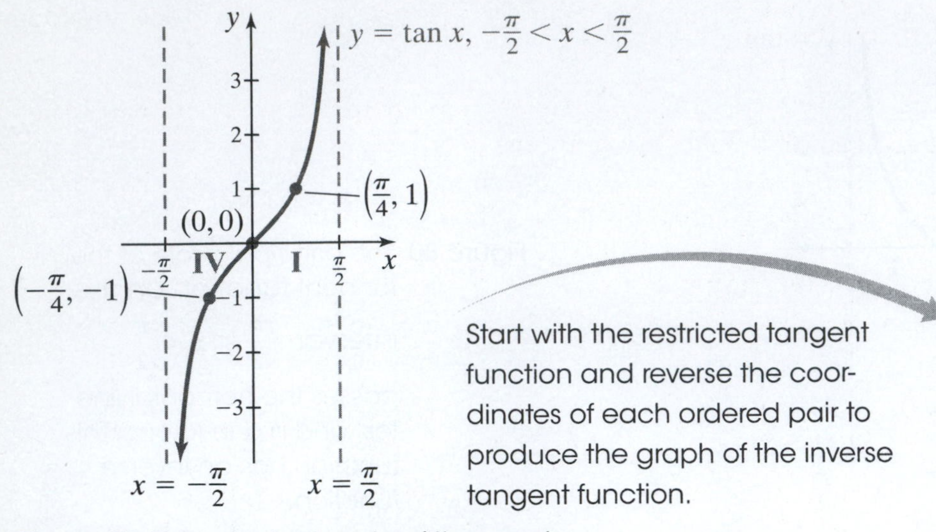

Figure 81 The principal cycle of the graph of $y = \tan x$ on the restricted domain $-\frac{\pi}{2} < x < \frac{\pi}{2}$ with range $-\infty < y < \infty$

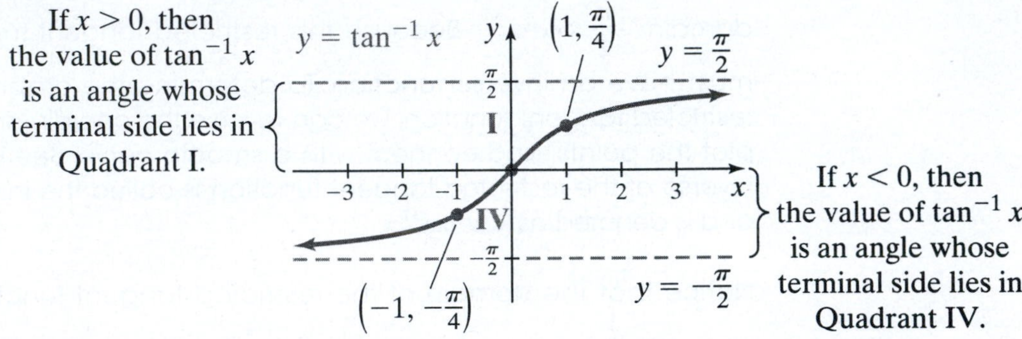

If $x > 0$, then the value of $\tan^{-1} x$ is an angle whose terminal side lies in Quadrant I.
$y = \tan^{-1} x$
$\left(1, \frac{\pi}{4}\right)$
$y = \frac{\pi}{2}$
I
IV
$\left(-1, -\frac{\pi}{4}\right)$
$y = -\frac{\pi}{2}$
If $x < 0$, then the value of $\tan^{-1} x$ is an angle whose terminal side lies in Quadrant IV.

Figure 82 The domain of $y = \tan^{-1} x$ is $-\infty < x < \infty$ and the range is $-\frac{\pi}{2} < y < \frac{\pi}{2}$.

We will think of the expression $\tan^{-1}x$ as the angle on the interval $\left(-\frac{\pi}{2}, \frac{\pi}{2}\right)$ whose tangent is equal to x. For this reason, we continue to use our conventional angle notation, θ, to represent the value of $\tan^{-1}x$. If $\theta = \tan^{-1} x$, then $-\frac{\pi}{2} < \theta < \frac{\pi}{2}$. Therefore, the terminal side of angle θ must lie in Quadrant I, in Quadrant IV, or on the positive x-axis. See **Figure 83**.

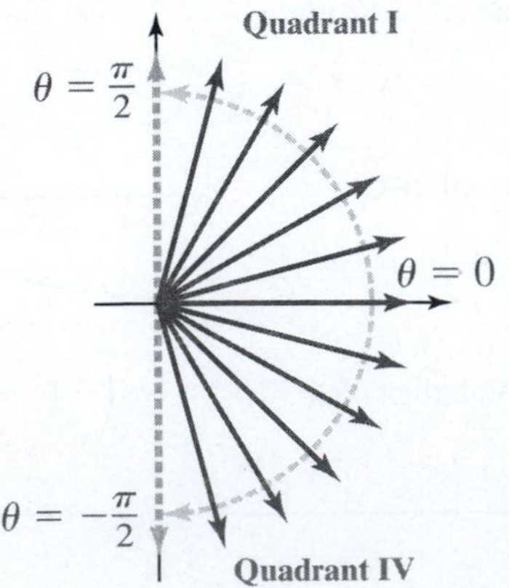

Figure 83 If $\theta = \tan^{-1} x$, then $-\frac{\pi}{2} < \theta < \frac{\pi}{2}$ and the terminal side of angle θ lies in Quadrant I, in Quadrant IV, or on the positive x-axis.

To determine the exact value of an inverse tangent expression, we follow the four-step process below.

Steps for Determining the Exact Value of $\tan^{-1} x$

Step 1. The value of $\tan^{-1}x$ must be an angle in the interval $\left(-\frac{\pi}{2}, \frac{\pi}{2}\right)$.

Step 2. Let $\tan^{-1}x = \theta$ such that $\tan\theta = x$.

Step 3. If $\tan\theta = 0$, then $\theta = 0$ and the terminal side of angle θ lies on the positive x-axis.

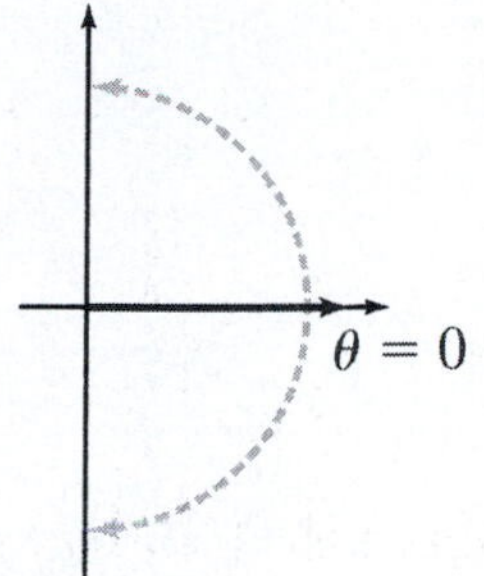

If $\tan\theta > 0$, then $0 < \theta < \frac{\pi}{2}$ and the terminal side of angle θ lies in Quadrant I.

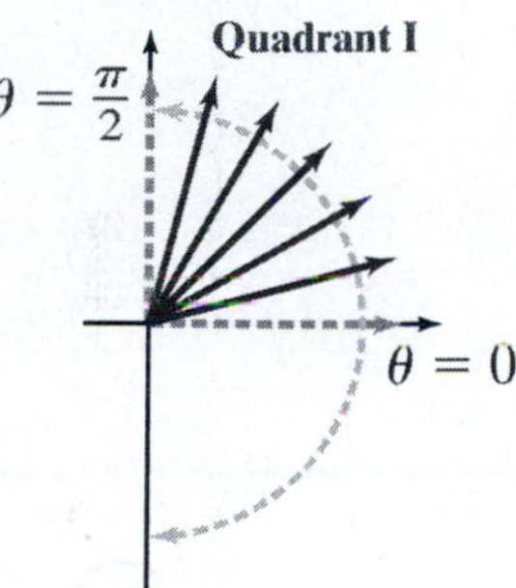

If $\tan\theta < 0$, then $-\frac{\pi}{2} < \theta < 0$ and the terminal side of angle θ lies in Quadrant IV.

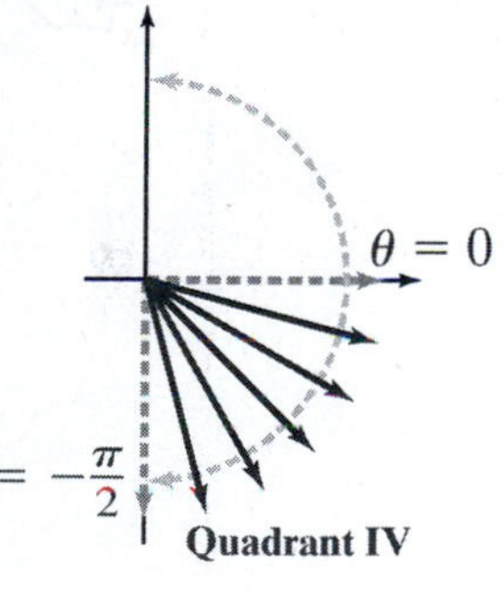

Step 4. Use your knowledge of the two **special right triangles**, the graphs of the trigonometric functions, **Table 1** from Section 6.4, or **Table 2** from Section 6.5 to determine the angle in the correct quadrant whose tangent is x.

Example 5 Finding the Exact Value of an Inverse Tangent Expression

Determine the exact value of each expression.

a. $\tan^{-1}\left(\frac{1}{\sqrt{3}}\right)$ **b.** $\tan^{-1}\left(-\frac{1}{\sqrt{3}}\right)$

Solution

a. Step 1. The value of $\tan^{-1}\left(\frac{1}{\sqrt{3}}\right)$ must be an angle in the interval $\left(-\frac{\pi}{2}, \frac{\pi}{2}\right)$.

Step 2. Let $\tan^{-1}\left(\frac{1}{\sqrt{3}}\right) = \theta$ such that $\tan\theta = \frac{1}{\sqrt{3}}$.

Step 3. Because $\tan \theta = \frac{1}{\sqrt{3}} > 0$, then $0 < \theta < \frac{\pi}{2}$. Thus, we are looking for an angle whose terminal side lies in Quadrant I. See Figure 84.

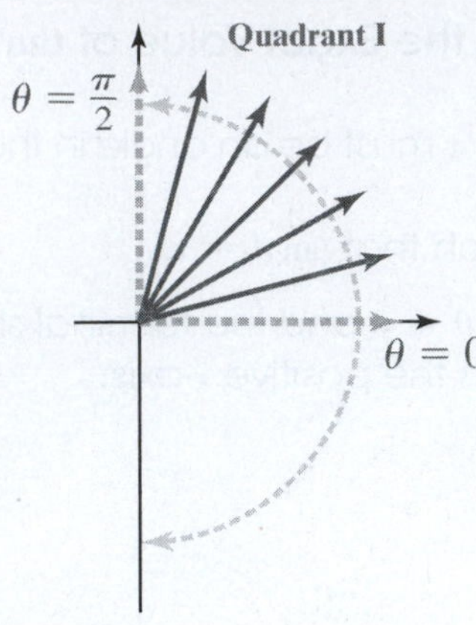

Figure 84 The angle $\theta = \tan^{-1}\left(\frac{1}{\sqrt{3}}\right)$ must have a terminal side that lies in Quadrant I.

Step 4. Using our knowledge of the **special** $\boldsymbol{\frac{\pi}{6}, \frac{\pi}{3}, \frac{\pi}{2}}$ **right triangle** or **Table 1**, we know that $\tan \frac{\pi}{6} = \frac{1}{\sqrt{3}}$. The angle $\theta = \frac{\pi}{6}$ is the **only** angle for which $0 < \theta < \frac{\pi}{2}$ whose terminal side lies in Quadrant I such that $\tan \theta = \frac{1}{\sqrt{3}}$. See Figure 85. Therefore, $\tan^{-1}\left(\frac{1}{\sqrt{3}}\right) = \frac{\pi}{6}$.

Work through this **interactive video** to see each step of this solution.

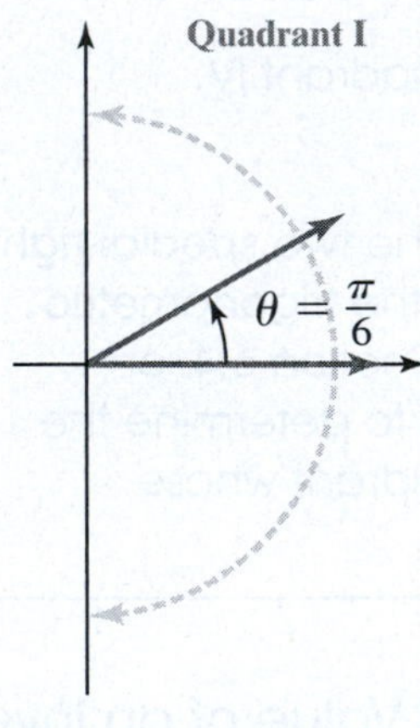

Figure 85 The angle $\theta = \frac{\pi}{6}$ is the only angle in the interval $0 < \theta < \frac{\pi}{2}$ whose terminal side lies in Quadrant I such that $\tan \theta = \frac{1}{\sqrt{3}}$.

b. To find the value of $\tan^{-1}\left(-\frac{1}{\sqrt{3}}\right)$, we follow the four-step process.

Step 1. The value of $\tan^{-1}\left(-\frac{1}{\sqrt{3}}\right)$ must be an angle in the interval $\left(-\frac{\pi}{2}, \frac{\pi}{2}\right)$.

Step 2. Let $\tan^{-1}\left(-\frac{1}{\sqrt{3}}\right) = \theta$ such that $\tan \theta = -\frac{1}{\sqrt{3}}$.

Step 3. Because $\tan \theta = -\frac{1}{\sqrt{3}} < 0$, then $-\frac{\pi}{2} < \theta < 0$. Thus, we are looking for an angle whose terminal side lies in Quadrant IV. See Figure 86.

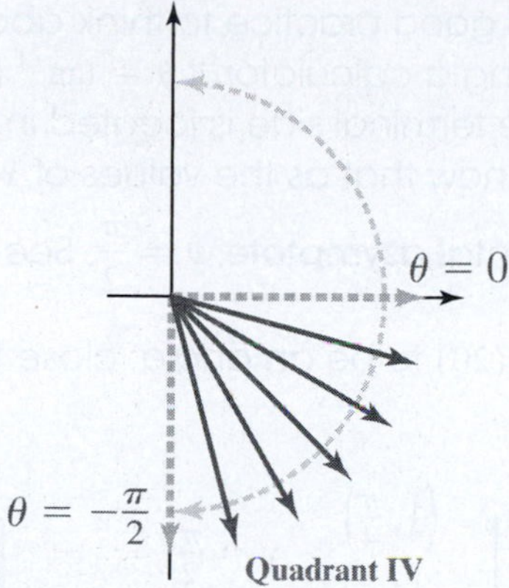

Figure 86 The angle $\theta = \tan^{-1}\left(-\frac{1}{\sqrt{3}}\right)$ must have a terminal side that lies in Quadrant IV.

Step 4. Using our knowledge of the special $\frac{\pi}{6}, \frac{\pi}{3}, \frac{\pi}{2}$ right triangle or **Table 1**, we know that $\tan\frac{\pi}{6} = \frac{1}{\sqrt{3}}$. Therefore, we must identify the angle θ that satisfies the following three conditions:

- The reference angle is $\theta_R = \frac{\pi}{6}$.
- The terminal side lies in Quadrant IV.
- $-\frac{\pi}{2} < \theta < 0$.

The only such angle is $\theta = -\frac{\pi}{6}$. See Figure 87. Therefore, $\tan^{-1}\left(-\frac{1}{\sqrt{3}}\right) = -\frac{\pi}{6}$.

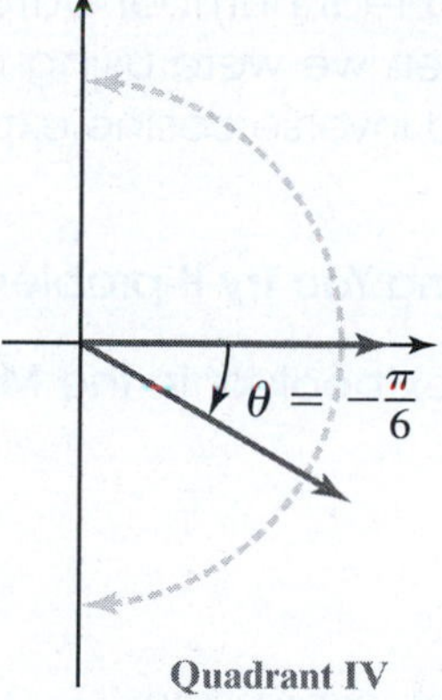

Figure 87 The angle $\theta = -\frac{\pi}{6}$ is the only angle in the interval $-\frac{\pi}{2} < \theta < 0$ whose terminal side lies in Quadrant IV such that $\tan\theta = -\frac{1}{\sqrt{3}}$.

Work through this **interactive video** to see each step of this solution.

You Try It Work through this You Try It problem.

Work Exercises 13–15 in this textbook or in the MyLab Math Study Plan.

Example 6 Finding the Approximate Value of an Inverse Tangent Expression

Use a calculator to approximate the value of $\tan^{-1}(20)$, or state that the value does not exist.

Solution As always, it is good practice to think about the type of result that we can expect when using a calculator. If $\theta = \tan^{-1}(20)$, then $\tan\theta = 20 > 0$. Thus, θ is an angle whose terminal side is located in Quadrant I. If we think of the graph of $y = \tan^{-1} x$, we know that as the values of x get large, then the value of y approaches the **horizontal asymptote**, $y = \frac{\pi}{2}$. See Figure 88. Therefore, we can expect the value of $\tan^{-1}(20)$ to be an angle "close to" $\frac{\pi}{2} \approx 1.5708$.

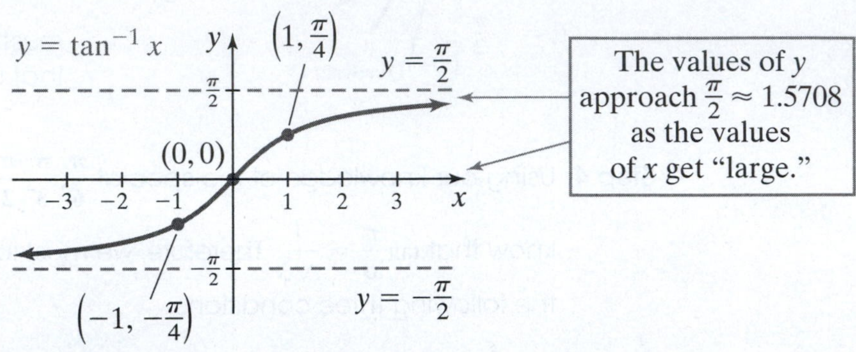

Figure 88 The values of y approach $\frac{\pi}{2} \approx 1.5708$ as the values of x get "large."

Using a calculator in radian mode, type 2nd tan and then 20. We find that $\tan^{-1}(20) \approx 1.521$ radians. Notice that this value is "close to" $\frac{\pi}{2} \approx 1.5708$, as expected.

Note Because the domain of $y = \tan^{-1} x$ is all real numbers, every real number input value will produce a real number output value, unlike with **Example 2c** or **Example 4a**, when we were using a calculator to evaluate an inverse sine expression and inverse cosine expression.

You Try It Work through the following You Try It problem.

Work Exercises 16 and 17 in this textbook or in the MyLab Math Study Plan.

7.4 Exercises

In Exercises 1–3, determine the exact value of each expression.

SbS 1. $\sin^{-1}\left(\frac{\sqrt{3}}{2}\right)$ SbS 2. $\sin^{-1}\left(-\frac{1}{\sqrt{2}}\right)$ SbS 3. $\sin^{-1}(1)$

In Exercises 4–6, use a calculator to approximate each value, or state that the value does not exist.

SbS 4. $\sin^{-1}(.33)$ SbS 5. $\sin^{-1}(-.87)$ SbS 6. $\sin^{-1}(-2.5)$

In Exercises 7–9, determine the exact value of each expression.

SbS 7. $\cos^{-1}\left(\frac{\sqrt{3}}{2}\right)$ SbS 8. $\cos^{-1}\left(-\frac{1}{\sqrt{2}}\right)$ SbS 9. $\cos^{-1}(1)$

In Exercises 10–12, use a calculator to approximate each value, or state that the value does not exist.

SbS 10. $\cos^{-1}(.35)$

SbS 11. $\cos^{-1}(-.92)$

SbS 12. $\cos^{-1}(3.1)$

In Exercises 13–15, determine the exact value of each expression.

SbS 13. $\tan^{-1}\left(\frac{1}{\sqrt{3}}\right)$

SbS 14. $\tan^{-1}(-\sqrt{3})$

SbS 15. $\tan^{-1}(1)$

In Exercises 16–17, use a calculator to approximate each value, or state that the value does not exist.

SbS 16. $\tan^{-1}(.76)$

SbS 17. $\tan^{-1}(5.4)$

Brief Exercises

18. The value of $\sin^{-1} x$ represents an angle in what interval?

19. The value of $\cos^{-1} x$ represents an angle in what interval?

20. The value of $\tan^{-1} x$ represents an angle in what interval?

21. If $\theta = \sin^{-1}\left(\frac{\sqrt{3}}{2}\right)$, then determine the quadrant in which or the axis on which the terminal side of θ lies.

22. If $\theta = \sin^{-1}\left(-\frac{1}{\sqrt{2}}\right)$, then determine the quadrant in which or the axis on which the terminal side of θ lies.

23. If $\theta = \sin^{-1}(1)$, then determine the quadrant in which or the axis on which the terminal side of θ lies.

24. If $\theta = \cos^{-1}\left(\frac{\sqrt{3}}{2}\right)$, then determine the quadrant in which or the axis on which the terminal side of θ lies.

25. If $\theta = \cos^{-1}\left(-\frac{1}{\sqrt{2}}\right)$, then determine the quadrant in which or the axis on which the terminal side of θ lies.

26. If $\theta = \cos^{-1}(1)$, then determine the quadrant in which or the axis on which the terminal side of θ lies.

27. If $\theta = \tan^{-1}\left(\frac{1}{\sqrt{3}}\right)$, then determine the quadrant in which or the axis on which the terminal side of θ lies.

28. If $\theta = \tan^{-1}(-\sqrt{3})$, then determine the quadrant in which or the axis on which the terminal side of θ lies.

29. If $\theta = \tan^{-1}(1)$, then determine the quadrant in which or the axis on which the terminal side of θ lies.

In Exercises 30–38, determine the exact value of the given expression.

30. $\sin^{-1}\left(\frac{\sqrt{3}}{2}\right)$ 31. $\sin^{-1}\left(-\frac{1}{\sqrt{2}}\right)$ 32. $\sin^{-1}(1)$

33. $\cos^{-1}\left(\frac{\sqrt{3}}{2}\right)$ 34. $\cos^{-1}\left(-\frac{1}{\sqrt{2}}\right)$ 35. $\cos^{-1}(1)$

36. $\tan^{-1}\left(\frac{1}{\sqrt{3}}\right)$ 37. $\tan^{-1}(-\sqrt{3})$ 38. $\tan^{-1}(1)$

7.5 Inverse Trigonometric Functions II

THINGS TO KNOW

Before working through this section, be sure that you are familiar with the following concepts:

VIDEO ANIMATION INTERACTIVE

You Try It 1. Understanding the Composition Cancellation Equations (Section 3.6)   

You Try It 2. Understanding the Special Right Triangles (Section 6.3)

You Try It 3. Understanding the Right Triangle Definitions of the Trigonometric Functions (Section 6.4)  

You Try It 4. Understanding the Signs of the Trigonometric Functions (Section 6.5)  

You Try It 5. Determining Reference Angles (Section 6.5) 

You Try It 6. Evaluating Trigonometric Functions of Angles Belonging to the $\frac{\pi}{3}$, $\frac{\pi}{6}$, or $\frac{\pi}{4}$ Families (Section 6.5) 

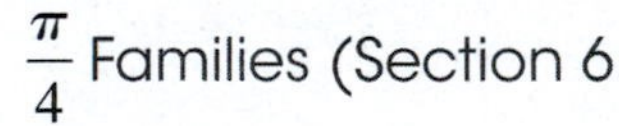

You Try It 7. Understanding the Inverse Sine Function (Section 7.4)  

You Try It 8. Understanding the Inverse Cosine Function (Section 7.4)  

You Try It 9. Understanding the Inverse Tangent Function (Section 7.4)  

INTRODUCTION

Read this introduction before beginning Objective 1.

OBJECTIVES

1 Evaluating Composite Functions Involving Inverse Trigonometric Functions of the Form $f \circ f^{-1}$ and $f^{-1} \circ f$

(eText Screens 7.5-1–7.5-33)

2 Evaluating Composite Functions Involving Inverse Trigonometric Functions of the Form $f \circ g^{-1}$ and $f^{-1} \circ g$

3 Understanding the Inverse Cosecant, Inverse Secant, and Inverse Cotangent Functions

4 Writing Trigonometric Expressions as Algebraic Expressions

SECTION 7.5 EXERCISES

Introduction to Section 7.5

Before reading this section, it is crucial that you understand the three inverse trigonometric functions that were discussed in **Section 7.4**. These functions are $y = \sin^{-1}x$, $y = \cos^{-1}x$, and $y = \tan^{-1}x$. Below is a quick overview of each of these functions.

The inverse sine function, $y = \sin^{-1}x$

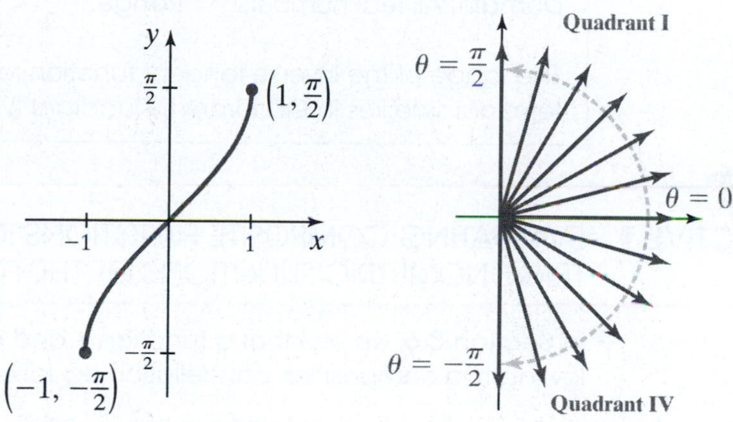

Domain: $-1 \le x \le 1$ **Range:** $-\frac{\pi}{2} \le y \le \frac{\pi}{2}$

The range of the inverse sine function represents an angle whose terminal side lies in Quadrant I, Quadrant IV, on the positive x-axis, on the positive y-axis, or on the negative y-axis.

The inverse cosine function, $y = \cos^{-1}x$

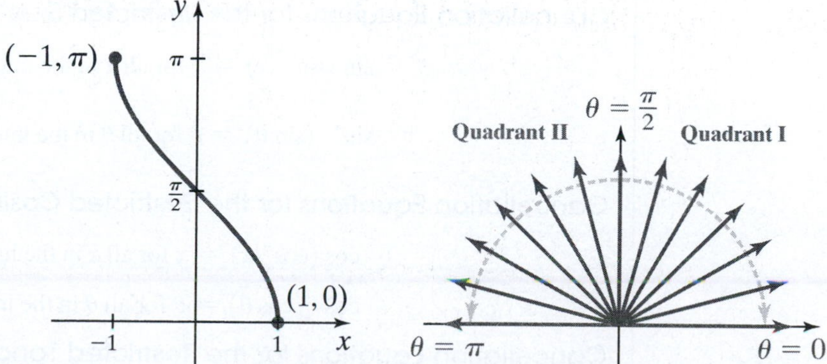

Domain: $-1 \le x \le 1$ **Range:** $0 \le y \le \pi$

The range of the inverse cosine function represents an angle whose terminal side lies in Quadrant I, Quadrant II, on the positive y-axis, on the positive x-axis, or on the negative x-axis.

The inverse tangent function, $y = \tan^{-1}x$

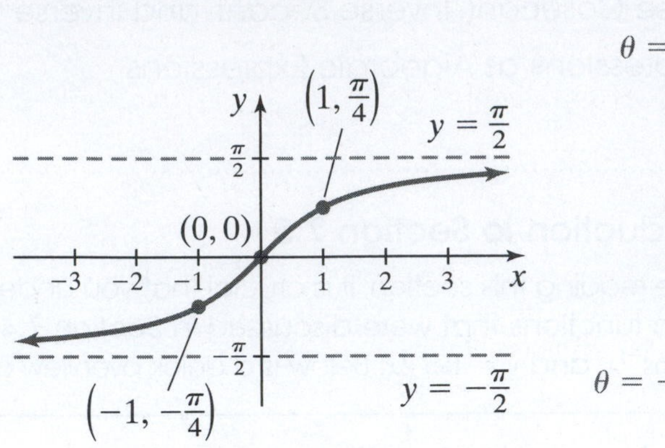

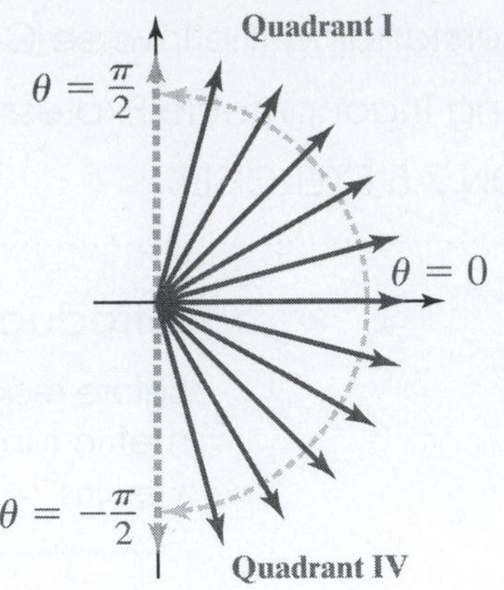

Domain: All real numbers **Range:** $-\frac{\pi}{2} < y < \frac{\pi}{2}$

The range of the inverse tangent function represents an angle whose terminal side lies in Quadrant I, Quadrant IV, or on the positive x-axis.

OBJECTIVE 1 EVALUATING COMPOSITE FUNCTIONS INVOLVING INVERSE TRIGONOMETRIC FUNCTIONS OF THE FORM $f \circ f^{-1}$ AND $f^{-1} \circ f$

In **Section 3.6**, we see that a function f and its inverse function f^{-1} satisfy the following two composition cancellation equations:

$$f(f^{-1}(x)) = x \text{ for all } x \text{ in the domain of } f^{-1} \text{ and } f^{-1}(f(x)) = x \text{ for all } x \text{ in the domain of } f$$

Therefore, if we encounter a trigonometric expression of the form $(f \circ f^{-1})(x)$ or $(f^{-1} \circ f)\,(x)$, we can use the cancellation equations *only* if we verify that x is in the domain of the "inner" function. This gives the following cancellation equations for compositions of inverse trigonometric functions.

Cancellation Equations for Compositions of Inverse Trigonometric Functions

Cancellation Equations for the Restricted Sine Function and Its Inverse

$$\sin(\sin^{-1} x) = x \text{ for all } x \text{ in the interval } [-1, 1]$$

$$\sin^{-1}(\sin\theta) = \theta \text{ for all } \theta \text{ in the interval } \left[-\frac{\pi}{2}, \frac{\pi}{2}\right]$$

Cancellation Equations for the Restricted Cosine Function and Its Inverse

$$\cos(\cos^{-1}x) = x \text{ for all } x \text{ in the interval } [-1, 1]$$

$$\cos^{-1}(\cos\theta) = \theta \text{ for all } \theta \text{ in the interval } [0, \pi]$$

Cancellation Equations for the Restricted Tangent Function and Its Inverse

$$\tan(\tan^{-1}x) = x \text{ for all } x \text{ in the interval } (-\infty, \infty)$$

$$\tan^{-1}(\tan\theta) = \theta \text{ for all } \theta \text{ in the interval } \left(-\frac{\pi}{2}, \frac{\pi}{2}\right)$$

Example 1 Finding the Exact Value of a Composite Expression of the Form $f \circ f^{-1}$

Find the exact value of each expression or state that it does not exist.

a. $\sin\left(\sin^{-1}\frac{1}{2}\right)$ **b.** $\cos\left(\cos^{-1}\frac{3}{2}\right)$ **c.** $\tan(\tan^{-1}(8.2))$ **d.** $\sin(\sin^{-1}(1.3))$

Solution

a. Here, we have an expression of the form $\sin(\sin^{-1}x)$. The value of $x = \frac{1}{2}$ is in the interval $[-1, 1]$. Thus we can use the cancellation equation, $\sin(\sin^{-1}x) = x$. Therefore, $\sin\left(\sin^{-1}\frac{1}{2}\right) = \frac{1}{2}$.

We do not have to use a cancellation equation to evaluate the expression $\sin\left(\sin^{-1}\frac{1}{2}\right)$, as shown in the following alternate method for evaluating $\sin\left(\sin^{-1}\frac{1}{2}\right)$:

$\sin\left(\sin^{-1}\frac{1}{2}\right)$ Write the original expression.

$= \sin\left(\frac{\pi}{6}\right)$ Evaluate $\sin^{-1}\frac{1}{2}$ using the **four-step process for determining the exact value of $\sin^{-1}x$.**

$= \frac{1}{2}$ Evaluate $\sin\left(\frac{\pi}{6}\right)$ using a **special right triangle** or **Table 1** from Section 6.4.

b. To evaluate the expression $\cos\left(\cos^{-1}\frac{3}{2}\right)$, we cannot use a cancellation equation because $x = \frac{3}{2}$ is *not* in the domain of the inverse cosine function. In fact, because $x = \frac{3}{2}$ is not in the domain of the inverse cosine function, the expression $\cos^{-1}\frac{3}{2}$ does not exist. $\left(\text{There is no angle whose cosine is equal to } \frac{3}{2}.\right)$ Therefore, $\cos\left(\cos^{-1}\frac{3}{2}\right)$ is undefined.

CAUTION **Do not get into the habit of using a calculator to evaluate the composition of trigonometric expressions because it is possible to get false results. Displays of two different graphing devices are shown on the following page. The screenshot on the far right of Figure 89 displays an "incorrect" result because this particular calculator is able to handle imaginary and complex numbers.**

Using Technology

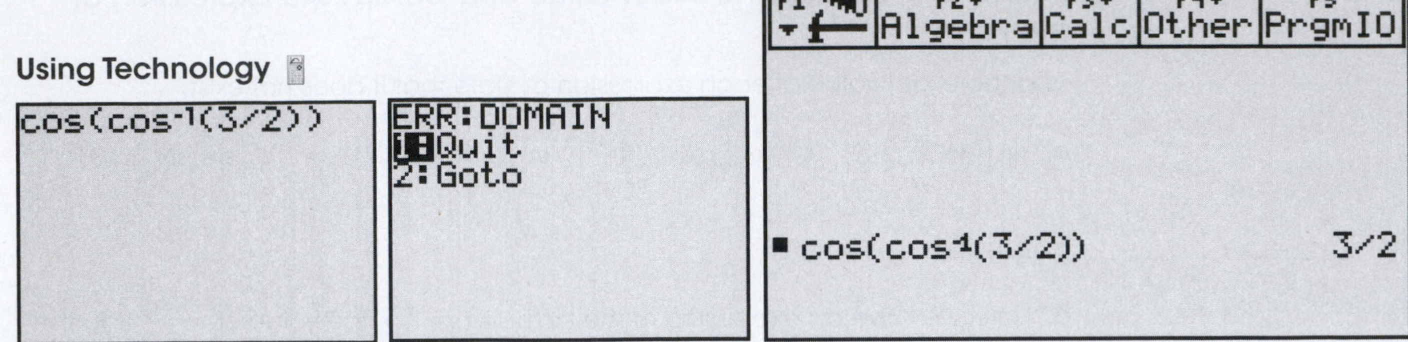

Figure 89 An input of $\cos\left(\cos^{-1}\left(\frac{3}{2}\right)\right)$ using a graphing utility results in a domain error on some calculators but may give a false result on others.

Try working parts c and d on your own. When you have completed your work, view the **solution**, or watch this **interactive video** to see the worked-out solution.

You Try It Work through this You Try It problem.

Work Exercises 1–9 in this textbook or in the MyLab Math Study Plan.

Example 2 Finding the Exact Value of a Composite Expression of the Form $f^{-1} \circ f$

Find the exact value of each expression or state that it does not exist.

a. $\sin^{-1}\left(\sin\frac{\pi}{6}\right)$ b. $\cos^{-1}\left(\cos\frac{2\pi}{3}\right)$ c. $\sin^{-1}\left(\sin\frac{4\pi}{3}\right)$ d. $\tan^{-1}\left(\tan\frac{7\pi}{10}\right)$

Solution

a. The terminal side of $\theta = \frac{\pi}{6}$ lies in Quadrant I and is thus is in the domain of the restricted sine function, $\left[-\frac{\pi}{2}, \frac{\pi}{2}\right]$. Therefore, we can use the cancellation equation $\sin^{-1}(\sin\theta) = \theta$. It follows that, $\sin^{-1}\left(\sin\frac{\pi}{6}\right) = \frac{\pi}{6}$.

b. The terminal side of $\theta = \frac{2\pi}{3}$ lies in Quadrant II and is thus in the domain of the restricted cosine function, $[0, \pi]$. Therefore, we can use the cancellation equation $\cos^{-1}(\cos\theta) = \theta$. It follows that $\cos^{-1}\left(\cos\frac{2\pi}{3}\right) = \frac{2\pi}{3}$.

c. To evaluate the expression $\sin^{-1}\left(\sin\frac{4\pi}{3}\right)$, we first note that the terminal side of $\theta = \frac{4\pi}{3}$ lies in Quadrant III and is **not** in the interval $\left[-\frac{\pi}{2}, \frac{\pi}{2}\right]$. Therefore, we cannot directly use a cancellation equation. However, we can first evaluate the "inner expression" $\sin\frac{4\pi}{3}$.

$\sin^{-1}\left(\sin\frac{4\pi}{3}\right)$ Write the original expression.

$= \sin^{-1}\left(-\frac{\sqrt{3}}{2}\right)$ Evaluate the inner expression, $\sin\frac{4\pi}{3}$, using the steps for **evaluating trigonometric functions of general angles.**

$= -\frac{\pi}{3}$ Evaluate the inverse sine expression using the **four-step process for determining the exact value of $\sin^{-1}x$.**

Therefore, $\sin^{-1}\left(\sin\frac{4\pi}{3}\right) = -\frac{\pi}{3}$.

Work through this **interactive video** to see every step of this solution.

d. To evaluate the expression $\tan^{-1}\left(\tan\frac{7\pi}{10}\right)$, we first note that $\frac{7\pi}{10}$ is **not** in the interval $\left(-\frac{\pi}{2}, \frac{\pi}{2}\right)$. We can first try to evaluate the "inner expression" $\tan\frac{7\pi}{10}$. However, $\frac{7\pi}{10}$ does not belong to one of the **four families of special angles.** Therefore, we cannot determine the *exact* value of $\tan\frac{7\pi}{10}$. So we must develop an alternate strategy to evaluate $\tan^{-1}\left(\tan\frac{7\pi}{10}\right)$. Let $\tan^{-1}\left(\tan\frac{7\pi}{10}\right) = \theta$. Because the value of the inverse tangent function is an angle that must be in the interval $\left(-\frac{\pi}{2}, \frac{\pi}{2}\right)$, we are looking for an angle whose terminal side lies in Quadrant I or Quadrant IV.

The terminal side of $\frac{7\pi}{10}$ lies in Quadrant II and has a reference angle of $\frac{3\pi}{10}$. The tangent function is negative in Quadrant II. The tangent function is also negative in Quadrant IV. The tangent of any angle in Quadrant IV having a reference angle of $\theta_R = \frac{3\pi}{10}$ will be equivalent to the value of $\tan\frac{7\pi}{10}$. The only angle θ in the interval $\left(-\frac{\pi}{2}, \frac{\pi}{2}\right)$ whose terminal side lies in Quadrant IV and has a reference angle of $\theta_R = \frac{3\pi}{10}$ is $\theta = -\frac{3\pi}{10}$. Therefore, $\tan\frac{7\pi}{10} = \tan\left(-\frac{3\pi}{10}\right)$. See Figure 90.

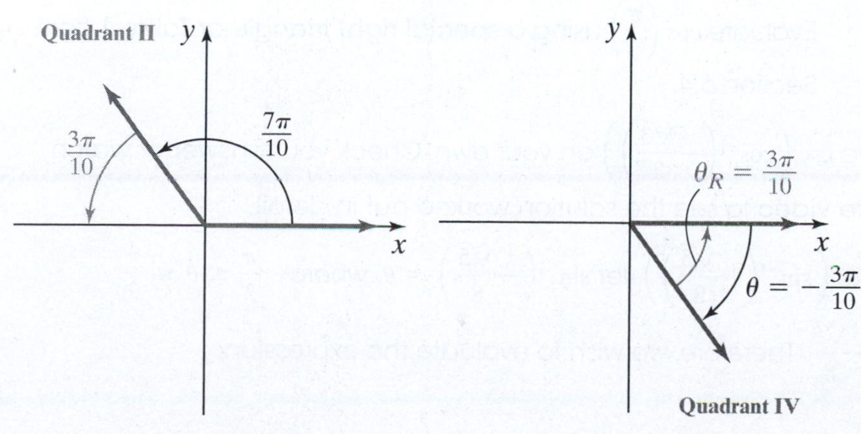

Figure 90 The angles $\frac{7\pi}{10}$ and $-\frac{3\pi}{10}$ both have a reference angle of $\frac{3\pi}{10}$. Since the tangent function is negative in Quadrants II and IV, it follows that $\tan\frac{7\pi}{10} = \tan\left(-\frac{3\pi}{10}\right)$.

Therefore, we can rewrite the expression $\tan^{-1}\left(\tan\dfrac{7\pi}{10}\right)$ as $\tan^{-1}\left(\tan\left(-\dfrac{3\pi}{10}\right)\right)$.

Because $\theta = -\dfrac{3\pi}{10}$ is in the interval $\left(-\dfrac{\pi}{2}, \dfrac{\pi}{2}\right)$, we can use the cancellation equation $\tan^{-1}(\tan\theta) = \theta$. It follows that $\tan^{-1}\left(\tan\dfrac{7\pi}{10}\right) = \tan^{-1}\left(\tan\left(-\dfrac{3\pi}{10}\right)\right) = -\dfrac{3\pi}{10}$.

Work through this **interactive video** to see this solution worked out in its entirety. ●

You Try It **Work through this You Try It problem.**

Work Exercises 10–28 in this textbook or in the MyLab Math Study Plan.

OBJECTIVE 2 EVALUATING COMPOSITE FUNCTIONS INVOLVING INVERSE TRIGONOMETRIC FUNCTIONS OF THE FORM $f \circ g^{-1}$ AND $f^{-1} \circ g$

When we encounter composite functions involving inverse trigonometric functions that are not of the form $f \circ f^{-1}$ or $f^{-1} \circ f$, as in Examples 1 and 2, then the **cancellation equations for inverse trigonometric functions** *cannot* be directly used. To evaluate composite functions of the form $f \circ g^{-1}$ or $f^{-1} \circ g$, where f and g are trigonometric functions and f^{-1} and g^{-1} are inverse trigonometric functions, first try to evaluate the "inner function" and then evaluate the "outer function."

Example 3 Finding the Exact Value of a Composite Expression of the Form $f \circ g^{-1}$

Find the exact value of each expression or state that it does not exist.

a. $\cos(\tan^{-1}\sqrt{3})$ **b.** $\csc\left(\cos^{-1}\left(-\dfrac{\sqrt{3}}{2}\right)\right)$ **c.** $\sec\left(\sin^{-1}\left(-\dfrac{\sqrt{5}}{8}\right)\right)$

Solution

a. To evaluate the expression $\cos(\tan^{-1}\sqrt{3})$, first evaluate the "inner expression" $\tan^{-1}\sqrt{3}$, and then evaluate the "outer expression."

$\cos(\tan^{-1}\sqrt{3})$ Write the original expression.

$= \cos\left(\dfrac{\pi}{3}\right)$ Evaluate the inverse tangent expression using the **four-step process for determining the exact value of $\tan^{-1}x$.**

$= \dfrac{1}{2}$ Evaluate $\cos\left(\dfrac{\pi}{3}\right)$ using a **special right triangle** or **Table 1** from Section 6.4.

b. Try to evaluate $\csc\left(\cos^{-1}\left(-\dfrac{\sqrt{3}}{2}\right)\right)$ on your own. Check your **answer** or watch this **interactive video** to see the solution worked out in detail.

c. To evaluate $\sec\left(\sin^{-1}\left(-\dfrac{\sqrt{5}}{8}\right)\right)$, let $\sin^{-1}\left(-\dfrac{\sqrt{5}}{8}\right) = \theta$, where $-\dfrac{\pi}{2} \le \theta \le \dfrac{\pi}{2}$ and $\sin\theta = -\dfrac{\sqrt{5}}{8}$. Therefore, we wish to evaluate the expression

$\sec\left(\sin^{-1}\left(-\frac{\sqrt{5}}{8}\right)\right) = \sec\theta$. The terminal side of θ lies in Quadrant IV because $\sin\theta = -\frac{\sqrt{5}}{8} < 0$. By the definition of the sine function for general angles, we know that $\sin\theta = \frac{y}{r} = -\frac{\sqrt{5}}{8}$. Thus, the terminal side of angle θ lies in Quadrant IV and passes through the point having a y-coordinate of $-\sqrt{5}$, a distance of 8 units from the origin. See Figure 91.

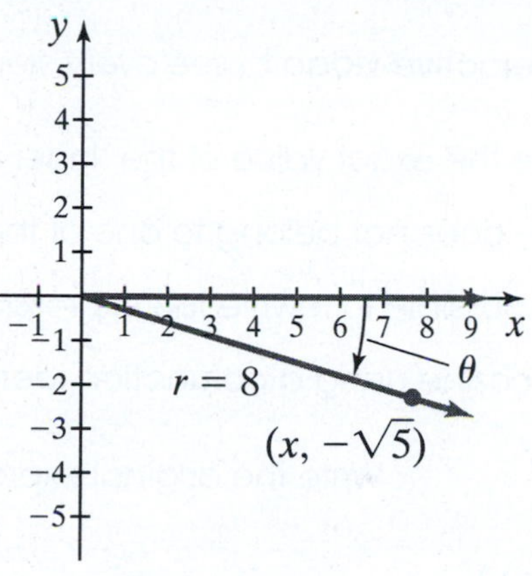

Figure 91 The terminal side of angle θ lies in Quadrant IV and passes through the point $(x, -\sqrt{5})$, a distance of 8 units from the origin.

To find the value of x, we use the formula $r = \sqrt{x^2 + y^2}$ to get $x = \pm\sqrt{59}$. Because the point $(x, -\sqrt{5})$ lies in Quadrant IV, we know that x must be **positive.** Therefore, $x = \sqrt{59}$.

Using the general angle definition of the secant function, we get

$$\sec\left(\sin^{-1}\left(-\frac{\sqrt{5}}{8}\right)\right) = \sec\theta = \frac{r}{x} = \frac{8}{\sqrt{59}}.$$

Watch this **interactive video** to see this solution worked out in detail.

You Try It Work through this You Try It problem.

Work Exercises 29–40 in this textbook or in the MyLab Math Study Plan.

Example 4 Finding the Exact Value of a Composite Expression of the Form $f^{-1} \circ g$

Find the exact value of each expression or state that it does not exist.

a. $\sin^{-1}\left(\cos\left(-\frac{2\pi}{3}\right)\right)$ **b.** $\cos^{-1}\left(\sin\frac{\pi}{7}\right)$

Solution

a. To evaluate the expression $\sin^{-1}\left(\cos\left(-\frac{2\pi}{3}\right)\right)$, first evaluate the "inner expression" $\cos\left(-\frac{2\pi}{3}\right)$, and then evaluate the "outer expression."

$\sin^{-1}\left(\cos\left(-\frac{2\pi}{3}\right)\right)$ Write the original expression.

$= \sin^{-1}\left(-\frac{1}{2}\right)$ Evaluate the inner expression, $\cos\left(-\frac{2\pi}{3}\right)$ using the steps for **evaluating trigonometric functions of general angles.**

$= -\frac{\pi}{6}$ Evaluate the inverse sine expression using the **four-step process for determining the exact value of $\sin^{-1}x$.**

Work through this **interactive video** to see every step of this solution.

b. We cannot determine the exact value of the "inner expression" of $\cos^{-1}\left(\sin\frac{\pi}{7}\right)$ because the angle $\frac{\pi}{7}$ does not belong to one of the **four families of special angles.** However, it is possible to rewrite the expression $\sin\frac{\pi}{7}$ as an equivalent expression involving cosine using a **cofunction identity.**

$\cos^{-1}\left(\sin\frac{\pi}{7}\right)$ Write the original expression.

$= \cos^{-1}\left(\cos\left(\frac{\pi}{2}-\frac{\pi}{7}\right)\right)$ Use the cofunction identity $\sin\theta = \cos\left(\frac{\pi}{2}-\theta\right)$.

$= \cos^{-1}\left(\cos\frac{5\pi}{14}\right)$ Simplify.

Therefore, we see that we can rewrite the expression $\cos^{-1}\left(\sin\frac{\pi}{7}\right)$ as $\cos^{-1}\left(\cos\frac{5\pi}{14}\right)$. Because $\frac{5\pi}{14}$ is on the interval $[0, \pi]$, we can use the cancellation equation $\cos^{-1}(\cos\theta) = \theta$. It follows that $\cos^{-1}\left(\sin\frac{\pi}{7}\right) = \cos^{-1}\left(\cos\frac{5\pi}{14}\right) = \frac{5\pi}{14}$.

Work through this **interactive video** to this example worked out in detail.

You Try It **Work through this You Try It problem.**

Work Exercises 41–48 in this textbook or in the MyLab Math Study Plan.

OBJECTIVE 3 UNDERSTANDING THE INVERSE COSECANT, INVERSE SECANT, AND INVERSE COTANGENT FUNCTIONS

The remaining three inverse trigonometric functions are $y = \csc^{-1}x$, $y = \sec^{-1}x$, and $y = \cot^{-1}x$. Their graphs and definitions are shown below. Click on the video icon next to each definition to see how each graph is obtained.

Definition Inverse Cosecant Function

The **inverse cosecant function**, denoted as $y = \csc^{-1} x$, is the inverse of $\mathbf{y = \csc x, \left[-\frac{\pi}{2}, 0\right) \cup \left(0, \frac{\pi}{2}\right]}$.

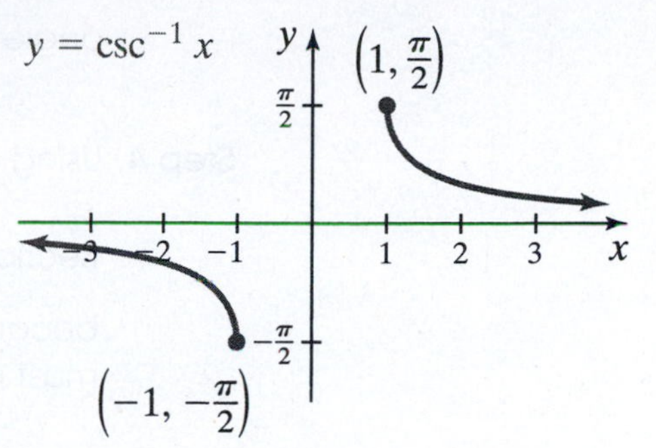

The domain of $y = \csc^{-1} x$ is $(-\infty, -1] \cup [1, \infty)$ and the range is $\left[-\frac{\pi}{2}, 0\right) \cup \left(0, \frac{\pi}{2}\right]$.

Definition Inverse Secant Function

The **inverse secant function**, denoted as $y = \sec^{-1} x$, is the inverse of $\mathbf{y = \sec x, \left[0, \frac{\pi}{2}\right) \cup \left(\frac{\pi}{2}, \pi\right]}$.

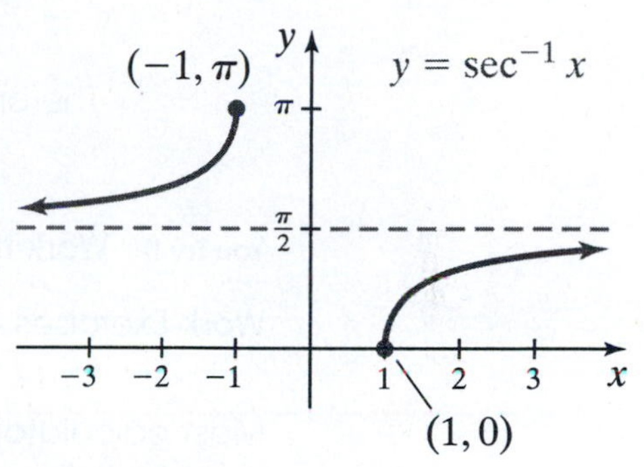

The domain of $y = \sec^{-1} x$ is $(-\infty, -1] \cup [1, \infty)$ and the range is $\left[0, \frac{\pi}{2}\right) \cup \left(\frac{\pi}{2}, \pi\right]$.

Definition Inverse Cotangent Function*

The **inverse cotangent function**, denoted as $y = \cot^{-1} x$, is the inverse of $\mathbf{y = \cot x}, \left(-\frac{\pi}{2}, 0\right) \cup \left(0, \frac{\pi}{2}\right]$.

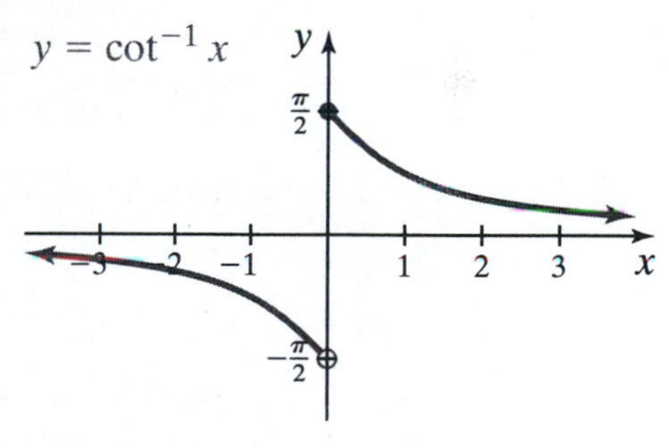

The domain of $y = \cot^{-1} x$ is $(-\infty, \infty)$ and the range is $\left(-\frac{\pi}{2}, 0\right) \cup \left(0, \frac{\pi}{2}\right]$.

*Some texts define the inverse cotangent function as the inverse of $y = \cot x, 0 < x < \pi$.

Example 5 Finding the Exact Value of an Inverse Secant Expression

Find the exact value of $\sec^{-1}(-\sqrt{2})$ or state that it does not exist.

Solution We follow steps similar to the steps used for **determining the exact value of $\cos^{-1} x$.**

Step 1. Let $\theta = \sec^{-1}(-\sqrt{2})$.

Step 2. Rewrite $\theta = \sec^{-1}(-\sqrt{2})$ as $\sec\theta = -\sqrt{2}$ (or $\cos\theta = -\frac{1}{\sqrt{2}}$), where $0 \le \theta < \frac{\pi}{2}$ or $\frac{\pi}{2} < \theta \le \pi$.

Step 3. Because $\sec\theta = -\sqrt{2} < 0$ (or $\cos\theta = -\frac{1}{\sqrt{2}} < 0$), we are looking for an angle whose terminal side lies in Quadrant II such that $\frac{\pi}{2} < \theta \le \pi$.

Step 4. Using our knowledge of the **special** $\boldsymbol{\frac{\pi}{4}, \frac{\pi}{4}, \frac{\pi}{2}}$ **right triangle** or **Table 1** from Section 6.4, we know that $\sec\frac{\pi}{4} = \sqrt{2}$ (or $\cos\frac{\pi}{4} = \frac{1}{\sqrt{2}}$.) However, $\theta \ne \frac{\pi}{4}$ because the terminal side of this angle does **not** lie in Quadrant II. We must identify the angle that satisfies the following three conditions:

- The reference angle is $\theta_R = \frac{\pi}{4}$.
- The terminal side lies in Quadrant II.
- $\frac{\pi}{2} < \theta \le \pi$.

The only such angle is $\theta = \frac{3\pi}{4}$. Therefore, $\sec^{-1}\left(-\sqrt{2}\right) = \frac{3\pi}{4}$.

You Try It **Work through this You Try It problem.**

Work Exercises 49–51 in this textbook or in the MyLab Math Study Plan.

Most calculators do not have inverse cosecant, inverse secant, or inverse cotangent keys. To use a calculator, we need to rewrite the given inverse cosecant, inverse secant, or inverse cotangent expression as an expression involving the inverse sine, inverse cosine, or inverse tangent, respectively. We illustrate this by rewriting $y = \sec^{-1} x$ as a function involving inverse cosine.

$y = \sec^{-1} x$	Start with the inverse secant function.
$\sec y = x$	$y = \sec^{-1} x$ means that $x = \sec y$.
$\frac{1}{\cos y} = x$	Use the reciprocal identity $\sec\theta = \frac{1}{\cos\theta}$.
$1 = x\cos y$	Multiply both sides by $\cos y$.
$\frac{1}{x} = \cos y$	Divide both sides by x.
$\cos^{-1}\left(\frac{1}{x}\right) = y$	Rewrite as an expression involving inverse cosine.

Therefore, $y = \sec^{-1} x = \cos^{-1}\left(\frac{1}{x}\right)$, except where either expression is undefined. Example 6 illustrates how a calculator can be used to approximate an inverse cosecant, inverse secant, or inverse cotangent expression.

Example 6 Finding the Approximate Value of an Inverse Trigonometric Expression

Use a calculator to approximate each value or state that the value does not exist.

a. $\sec^{-1}(5)$ **b.** $\cot^{-1}(-10)$ **c.** $\csc^{-1}(0.4)$

Solution

a. We first note that 5 is in the interval $(-\infty, -1] \cup [1, \infty)$. Now, rewrite the expression $\sec^{-1}(5)$ as $\cos^{-1}\left(\frac{1}{5}\right)$. Using a calculator in radian mode, type 2nd cos and then $\frac{1}{5}$. We find that $\sec^{-1}(5) = \cos^{-1}\left(\frac{1}{5}\right) \approx 1.3694$ radians.

b. We know that $\cot^{-1}(-10)$ exists because the domain of the inverse cotangent function is all real numbers. Now, rewrite the expression $\cot^{-1}(-10)$ as $\tan^{-1}\left(-\frac{1}{10}\right)$. Using a calculator in radian mode, type 2nd tan and then $-\frac{1}{10}$. We find that $\cot^{-1}(-10) = \tan^{-1}\left(-\frac{1}{10}\right) \approx -0.9967$ radian.

c. The domain of the cosecant function is $(-\infty, -1] \cup [1, \infty)$. Because 0.4 is not on this interval, this means that the value of $\csc^{-1}(0.4)$ does not exist.

You Try It Work through this You Try It problem.

Work Exercises 52–54 in this textbook or in the MyLab Math Study Plan.

OBJECTIVE 4 WRITING TRIGONOMETRIC EXPRESSIONS AS ALGEBRAIC EXPRESSIONS

Given an expression involving inverse trigonometric functions, it is often useful (especially in calculus) to rewrite the expression as an **algebraic expression** involving the given variable, as in Example 7.

Example 7 Rewrite a Trigonometric Expression as an Algebraic Expression

Rewrite the trigonometric expression $\sin(\tan^{-1}u)$ as an algebraic expression involving the variable u. Assume that $\tan^{-1}u$ represents an angle whose terminal side is located in Quadrant I.

Solution Let $\theta = \tan^{-1}u$. Then θ is the angle in Quadrant I whose tangent is u or $\frac{u}{1}$. Hence, $\tan\theta = \frac{u}{1}$. Because $u > 1$, we know that the terminal side of θ lies in Quadrant I. We can construct a right triangle with acute angle θ with sides of lengths u and 1. See Figure 92.

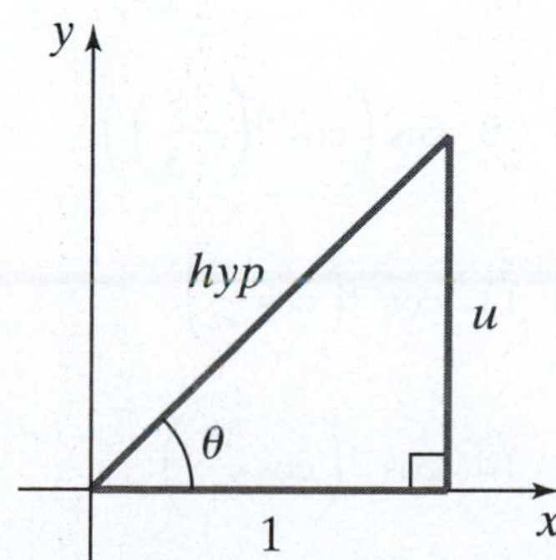

Figure 92

By the Pythagorean Theorem, we can find the length of the hypotenuse, *hyp*, in terms of u.

$a^2 + b^2 = c^2$	Write the Pythagorean Theorem.
$1^2 + u^2 = (hyp)^2$	Substitute $a = 1, b = u$, and $c = hyp$.
$1 + u^2 = (hyp)^2$	Simplify.
$\sqrt{1 + u^2} = hyp$	Take square roots of both sides. (Ignore the negative square root.)

Thus, the length of the hypotenuse is $\sqrt{1 + u^2}$. See Figure 93.

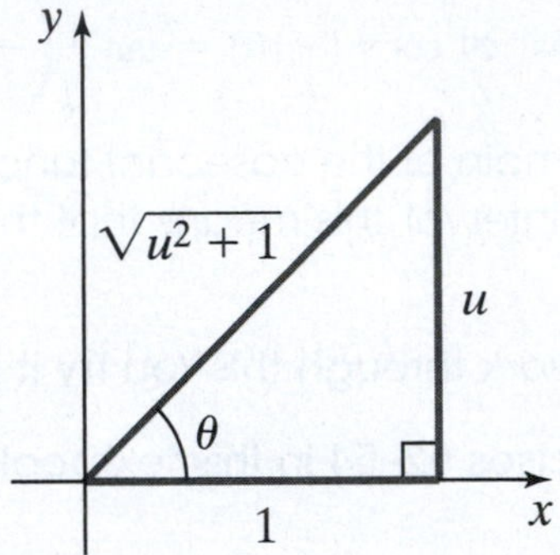

Figure 93

Therefore, $\sin(\tan^{-1}u) = \sin\theta = \dfrac{opp}{hyp} = \dfrac{u}{\sqrt{u^2 + 1}}$.

You Try It **Work through this You Try It problem.**

Work Exercises 55–59 in this textbook or in the MyLab Math Study Plan.

7.5 Exercises

In Exercises 1–51, find the exact value of each expression or state that it does not exist.

SbS 1. $\sin\left(\sin^{-1}\dfrac{1}{\sqrt{2}}\right)$ SbS 2. $\cos\left(\cos^{-1}\dfrac{\sqrt{3}}{2}\right)$ SbS 3. $\tan\left(\tan^{-1}\left(\dfrac{1}{\sqrt{3}}\right)\right)$

SbS 4. $\sin\left(\sin^{-1}\left(-\dfrac{\sqrt{3}}{2}\right)\right)$ SbS 5. $\cos\left(\cos^{-1}\left(-\dfrac{1}{2}\right)\right)$ SbS 6. $\tan(\tan^{-1}(-\sqrt{3}))$

SbS 7. $\sin\left(\sin^{-1}\dfrac{7}{9}\right)$ SbS 8. $\cos\left(\cos^{-1}\left(-\dfrac{8}{5}\right)\right)$ SbS 9. $\tan(\tan^{-1}35.4)$

SbS 10. $\sin^{-1}\left(\sin\dfrac{\pi}{7}\right)$ SbS 11. $\cos^{-1}\left(\cos\dfrac{\pi}{5}\right)$ SbS 12. $\tan^{-1}\left(\tan\dfrac{\pi}{3}\right)$

SbS 13. $\sin^{-1}\left(\sin\left(-\dfrac{\pi}{4}\right)\right)$ SbS 14. $\cos^{-1}\left(\cos\dfrac{3\pi}{4}\right)$ SbS 15. $\tan^{-1}\left(\tan\dfrac{7\pi}{6}\right)$

SbS 16. $\sin^{-1}\left(\sin\dfrac{14\pi}{3}\right)$ SbS 17. $\cos^{-1}\left(\cos\dfrac{9\pi}{4}\right)$ SbS 18. $\sin^{-1}\left(\sin\left(-\dfrac{7\pi}{2}\right)\right)$

SbS 19. $\cos^{-1}(\cos 22\pi)$

SbS 20. $\tan^{-1}\left(\tan \frac{8\pi}{3}\right)$

SbS 21. $\sin^{-1}\left(\sin \frac{13\pi}{4}\right)$

SbS 22. $\cos^{-1}\left(\cos \frac{19\pi}{6}\right)$

SbS 23. $\tan^{-1}\left(\tan \frac{12\pi}{5}\right)$

SbS 24. $\sin^{-1}\left(\sin \frac{15\pi}{7}\right)$

SbS 25. $\cos^{-1}\left(\cos \frac{17\pi}{9}\right)$

SbS 26. $\sin^{-1}\left(\sin \frac{13\pi}{12}\right)$

SbS 27. $\cos^{-1}\left(\cos \frac{13\pi}{10}\right)$

SbS 28. $\tan^{-1}\left(\tan \frac{11\pi}{7}\right)$

SbS 29. $\sin(\tan^{-1}\sqrt{3})$

SbS 30. $\tan(\cos^{-1}(0))$

SbS 31 $\cos\left(\sin^{-1}\left(\frac{\sqrt{3}}{2}\right)\right)$

SbS 32. $\sec\left(\sin^{-1}\left(\frac{1}{2}\right)\right)$

SbS 33. $\csc\left(\tan^{-1}\left(\frac{1}{\sqrt{3}}\right)\right)$

SbS 34. $\cos(\tan^{-1}(-1))$

SbS 35. $\sin\left(\cos^{-1}\left(-\frac{1}{2}\right)\right)$

SbS 36. $\tan\left(\sin^{-1}\left(-\frac{1}{\sqrt{2}}\right)\right)$

SbS 37. $\sec\left(\cos^{-1}\left(\frac{4}{7}\right)\right)$

SbS 38. $\csc\left(\tan^{-1}\left(\frac{7}{5}\right)\right)$

SbS 39. $\tan\left(\sin^{-1}\left(\frac{\sqrt{3}}{4}\right)\right)$

SbS 40. $\csc\left(\cos^{-1}\left(-\frac{2}{\sqrt{11}}\right)\right)$

SbS 41. $\sin^{-1}\left(\cos\left(\frac{\pi}{3}\right)\right)$

SbS 42. $\cos^{-1}\left(\tan\left(-\frac{5\pi}{3}\right)\right)$

SbS 43. $\tan^{-1}\left(\sin\left(-\frac{21\pi}{2}\right)\right)$

SbS 44 $\sin^{-1}\left(\tan\left(\frac{11\pi}{3}\right)\right)$

SbS 45. $\cos^{-1}\left(\sin\left(\frac{5\pi}{4}\right)\right)$

SbS 46. $\tan^{-1}\left(\cos\left(-\frac{19\pi}{2}\right)\right)$

SbS 47. $\sin^{-1}\left(\cos\left(\frac{\pi}{11}\right)\right)$

SbS 48. $\cos^{-1}\left(\sin\left(\frac{19\pi}{9}\right)\right)$

SbS 49. $\sec^{-1}\left(-\frac{2}{\sqrt{3}}\right)$

SbS 50. $\csc^{-1}(2)$

SbS 51. $\cot^{-1}\left(-\frac{1}{\sqrt{3}}\right)$

In Exercises 52–54, use a calculator to approximate each value or state that the value does not exist.

52. $\cot^{-1}(15)$

53. $\sec^{-1}(-.92)$

54. $\csc^{-1}(3.1)$

In Exercises 55–59, rewrite each trigonometric expression as an algebraic expression involving the variable u. Assume that $u > 0$ and that the value of the "inner" inverse trigonometric expression represents an angle θ such that $0 < \theta < \frac{\pi}{2}$.

55. $\cos(\sin^{-1}u)$

56. $\tan(\cos^{-1}2u)$

57. $\cos\left(\sin^{-1}\frac{3}{u}\right)$

58. $\cot\left(\tan^{-1}\frac{u}{\sqrt{11}}\right)$

59. $\sec\left(\sin^{-1}\frac{u}{\sqrt{u^2 + 121}}\right)$

Brief Exercises

In Exercises 60–107, find the exact value of each expression or state that it does not exist.

60. $\sin\left(\sin^{-1}\frac{1}{\sqrt{2}}\right)$

61. $\cos\left(\cos^{-1}\frac{\sqrt{3}}{2}\right)$

62. $\tan\left(\tan^{-1}\left(\frac{1}{\sqrt{3}}\right)\right)$

63. $\sin\left(\sin^{-1}\left(-\frac{\sqrt{3}}{2}\right)\right)$

64. $\cos\left(\cos^{-1}\left(-\frac{1}{2}\right)\right)$

65. $\tan\left(\tan^{-1}(-\sqrt{3})\right)$

66. $\sin\left(\sin^{-1}\frac{7}{9}\right)$

67. $\cos\left(\cos^{-1}\left(-\frac{8}{5}\right)\right)$

68. $\tan\left(\tan^{-1}35.4\right)$

69. $\sin^{-1}\left(\sin\frac{\pi}{7}\right)$

70. $\cos^{-1}\left(\cos\frac{\pi}{5}\right)$

71. $\tan^{-1}\left(\tan\frac{\pi}{3}\right)$

72. $\sin^{-1}\left(\sin\left(-\frac{\pi}{4}\right)\right)$

73. $\cos^{-1}\left(\cos\frac{3\pi}{4}\right)$

74. $\tan^{-1}\left(\tan\frac{7\pi}{6}\right)$

75. $\sin^{-1}\left(\sin\frac{14\pi}{3}\right)$

76. $\cos^{-1}\left(\cos\frac{9\pi}{4}\right)$

77. $\sin^{-1}\left(\sin\left(-\frac{7\pi}{2}\right)\right)$

78. $\cos^{-1}(\cos 22\pi)$

79. $\tan^{-1}\left(\tan\frac{8\pi}{3}\right)$

80. $\sin^{-1}\left(\sin\frac{13\pi}{4}\right)$

81. $\cos^{-1}\left(\cos\frac{19\pi}{6}\right)$

82. $\tan^{-1}\left(\tan\frac{12\pi}{5}\right)$

83. $\sin^{-1}\left(\sin\frac{15\pi}{7}\right)$

84. $\cos^{-1}\left(\cos\frac{17\pi}{9}\right)$

85. $\sin^{-1}\left(\sin\frac{13\pi}{12}\right)$

86. $\cos^{-1}\left(\cos\frac{13\pi}{10}\right)$

87. $\tan^{-1}\left(\tan\frac{11\pi}{7}\right)$

88. $\sin\left(\tan^{-1}\sqrt{3}\right)$

89. $\tan\left(\cos^{-1}(0)\right)$

90. $\cos\left(\sin^{-1}\left(\frac{\sqrt{3}}{2}\right)\right)$

91. $\sec\left(\sin^{-1}\left(\frac{1}{2}\right)\right)$

92. $\csc\left(\tan^{-1}\left(\frac{1}{\sqrt{3}}\right)\right)$

93. $\cos\left(\tan^{-1}(-1)\right)$

94. $\sin\left(\cos^{-1}\left(-\frac{1}{2}\right)\right)$

95. $\tan\left(\sin^{-1}\left(-\frac{1}{\sqrt{2}}\right)\right)$

96. $\sec\left(\cos^{-1}\left(\frac{4}{7}\right)\right)$

97. $\csc\left(\tan^{-1}\left(\frac{7}{5}\right)\right)$

98. $\tan\left(\sin^{-1}\left(\frac{\sqrt{3}}{4}\right)\right)$

99. $\csc\left(\cos^{-1}\left(-\frac{2}{\sqrt{11}}\right)\right)$

100. $\sin^{-1}\left(\cos\left(\frac{\pi}{3}\right)\right)$

101. $\cos^{-1}\left(\tan\left(-\frac{5\pi}{3}\right)\right)$

102. $\tan^{-1}\left(\sin\left(-\frac{21\pi}{2}\right)\right)$

103. $\sin^{-1}\left(\tan\left(\frac{11\pi}{3}\right)\right)$

104. $\cos^{-1}\left(\sin\left(\frac{5\pi}{4}\right)\right)$

105. $\tan^{-1}\left(\cos\left(-\frac{19\pi}{2}\right)\right)$

106. $\sin^{-1}\left(\cos\left(\frac{\pi}{11}\right)\right)$

107. $\cos^{-1}\left(\sin\left(\frac{19\pi}{9}\right)\right)$

Chapter 7 Summary

Key Concepts	Examples/Videos
7.1 The Graphs of Trigonometric Functions Characteristics of the Sine Function 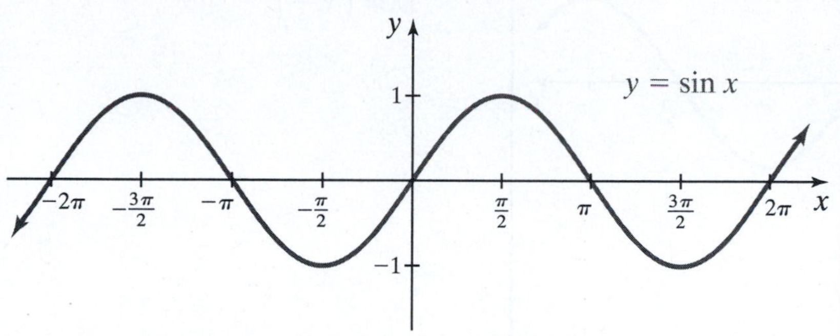 • The domain is $(-\infty, \infty)$. • The range is $[-1, 1]$. • The function is periodic with a period of $P = 2\pi$. • The y-intercept is 0. • The x-intercepts, or **zeros**, are of the form $n\pi$ where n is an integer. • The function is **odd**, which means $\sin(-x) = -\sin x$. The graph is symmetric about the origin. • The function obtains a **relative maximum** at $x = \frac{\pi}{2} + 2\pi n$, where n is an integer. The maximum value is 1. • The function obtains a **relative minimum** at $x = \frac{3\pi}{2} + 2\pi n$, where n is an integer. The minimum value is -1. The Five Quarter Points of $y = \sin x$ 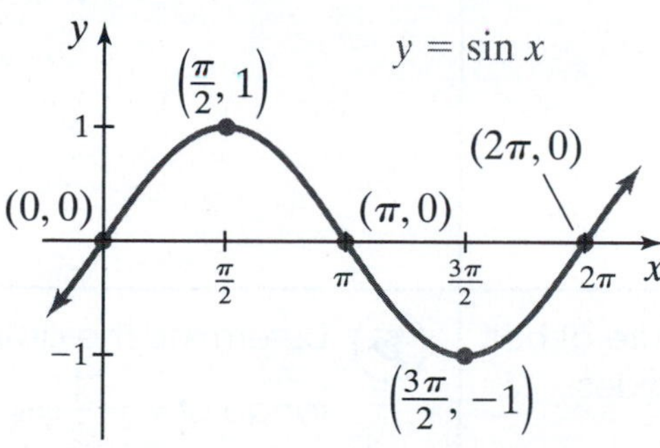	 Using the graph of $y = \sin x$, list all values of x on the interval $\left[-3\pi, \frac{7\pi}{4}\right]$ that satisfy the ordered pair $(x, 0)$.  Use the periodic property of $y = \sin x$ to determine which of the following expressions is equivalent to $\sin\left(\frac{23\pi}{6}\right)$. i. $\sin\left(\frac{\pi}{6}\right)$ ii. $\sin\left(\frac{5\pi}{6}\right)$ iii. $\sin\left(\frac{11\pi}{6}\right)$ iv. $\sin\left(\frac{13\pi}{6}\right)$

Key Concepts	Examples/Videos
Characteristics of the Cosine Function 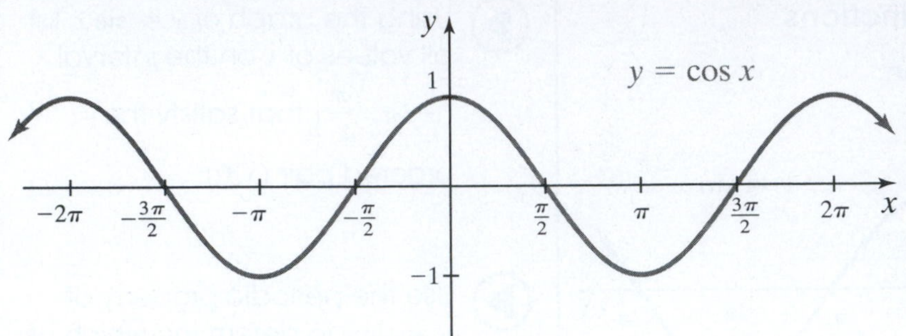 • The domain is $(-\infty, \infty)$. • The range is $[-1, 1]$. • The function is periodic with a period of $P = 2\pi$. • The y-intercept is 1. • The x-intercepts, or **zeros**, are of the form $(2n + 1) \cdot \frac{\pi}{2}$, where n is an integer. • The function is **even**, which means $\cos(-x) = \cos x$. The graph is symmetric about the y-axis. • The function obtains a **relative maximum** at $x = 2\pi n$, where n is an integer. The maximum value is 1. • The function obtains a **relative minimum** at $x = \pi + 2\pi n$, where n is an integer. The minimum value is -1. **The Five Quarter Points of $y = \cos x$** 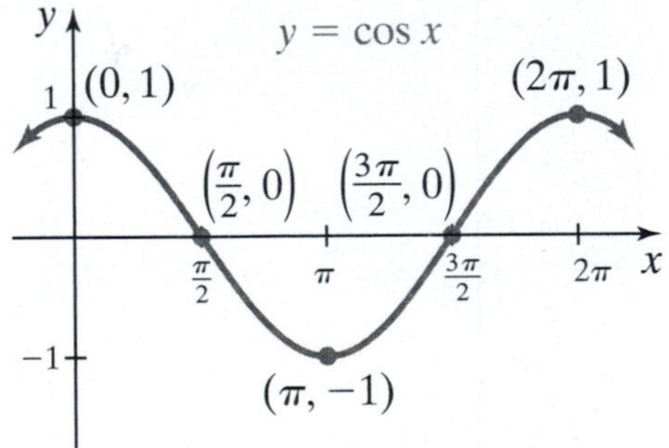 	 Using the graph of $y = \cos x$, list all values of x on the interval $[\pi, 2\pi]$ that satisfy the ordered pair $\left(x, \frac{1}{2}\right)$.
The **amplitude** of a sine or cosine curve is the measure of half the distance between the maximum and minimum values. 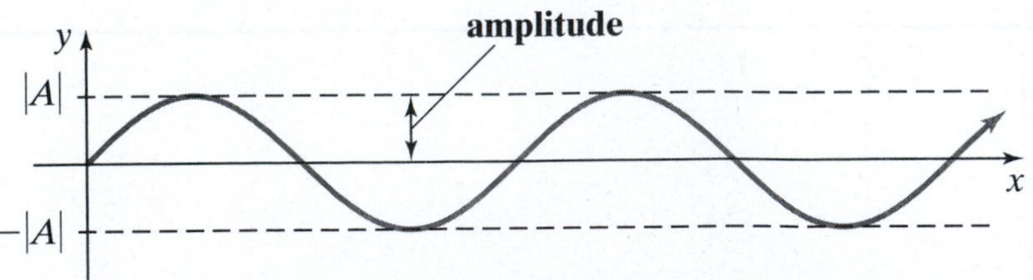 Trigonometric functions of the form $y = A \sin x$ and $y = A \cos x$ have an amplitude of $\|A\|$ and a range of $[-\|A\|, \|A\|]$.	 Determine the amplitude and range of $y = \frac{2}{3} \cos x$ and then sketch the graph.

(Summary)

Key Concepts	Examples/Videos

Determining the Period of $y = \sin(Bx)$ and $y = \cos(Bx)$

The **period** of a sine or cosine curve is equal to $P = \frac{2\pi}{B}$, where $B > 0$.

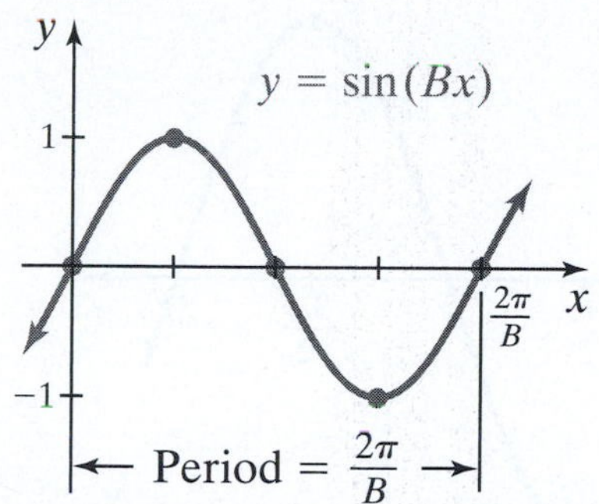

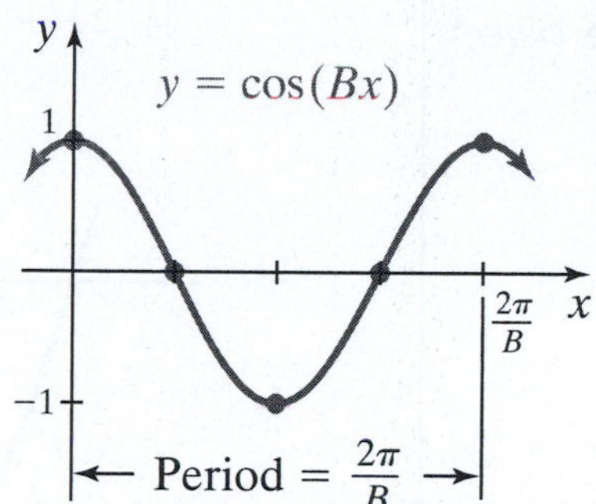

Determine the period and sketch the graph of each function.

a. $y = \sin(2x)$

b. $y = \cos\left(\frac{1}{2}x\right)$

c. $y = \sin(\pi x)$

Steps for Sketching Functions of the Form $y = A\sin(Bx)$ and $y = A\cos(Bx)$

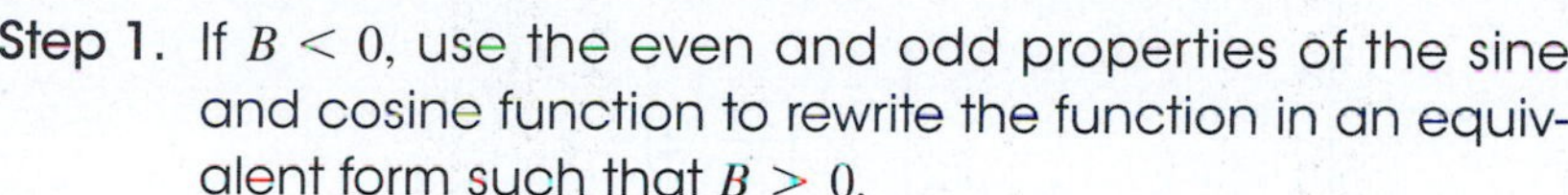

Step 1. If $B < 0$, use the even and odd properties of the sine and cosine function to rewrite the function in an equivalent form such that $B > 0$.

We now use this new form to determine A and B.

Step 2. Determine the amplitude and range. The amplitude is $|A|$. The range is $[-|A|, |A|]$.

Step 3. Determine the period. The period is $P = \frac{2\pi}{B}$.

Step 4. An interval for one complete cycle is $\left[0, \frac{2\pi}{B}\right]$. Subdivide this interval into four equal subintervals of length $\frac{2\pi}{B} \div 4$ by starting with 0 and adding $\left(\frac{2\pi}{B} \div 4\right)$ to the x-coordinate of each successive quarter point.

Step 5. Multiply the y-coordinates of the quarter points of $y = \sin x$ or $y = \cos x$ by A to determine the y-coordinates of the corresponding quarter points for the new graph.

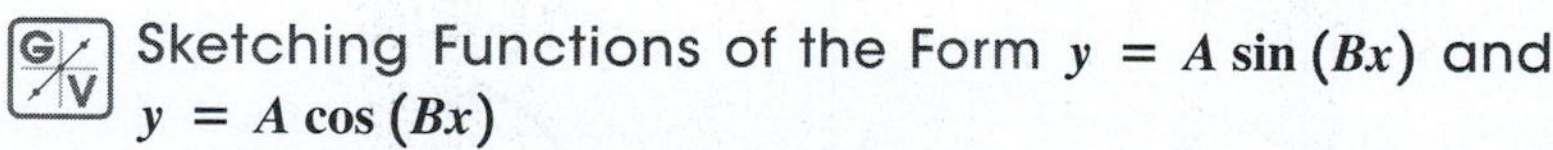

Sketching Functions of the Form $y = A\sin(Bx)$ and $y = A\cos(Bx)$

Use the six-step process to sketch each graph.

a. $y = 3\sin(4x)$

b. $y = -2\cos\left(\frac{1}{3}x\right)$

c. $y = -6\sin\left(-\frac{\pi x}{2}\right)$

Steps for Determining the Equation of a Function of the Form $y = A\sin(Bx)$ or $y = A\cos(Bx)$

Given the Graph

Step 1. Determine whether the given graph is a representation of a function of the form $y = A\sin(Bx)$ or $y = A\cos(Bx)$. Choose $y = A\sin(Bx)$ if the graph passes through the origin or choose $y = A\cos(Bx)$ if the graph does not pass through the origin.

One cycle of the graphs of three trigonometric functions of the form $y = A\sin(Bx)$ or $y = A\cos(Bx)$ for $B > 0$ are given below.

Key Concepts	Examples/Videos
Step 2. Determine the period $P = \frac{2\pi}{B}$ then use the period to determine the value of $B > 0$. **Step 3.** Use the information from the previous two steps then use one of the given points on the graph to solve for A.	Determine the equation of the function represented by each graph. **a.** 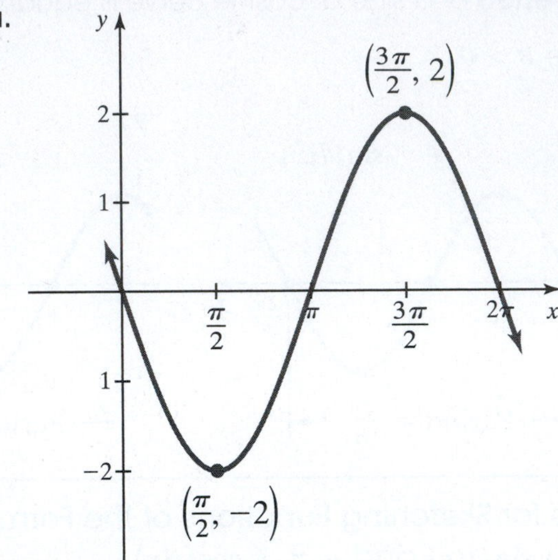 **b.** 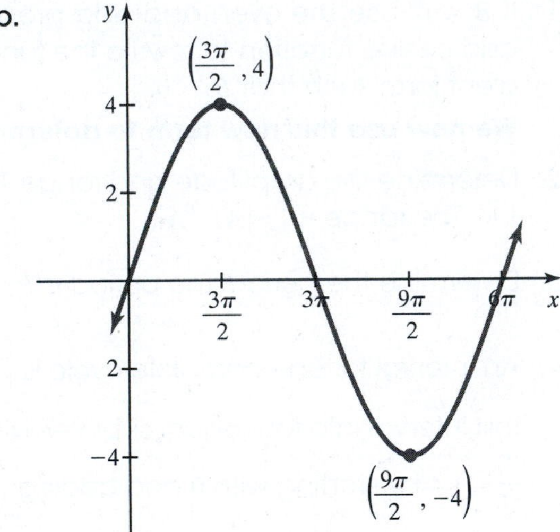 **c.** 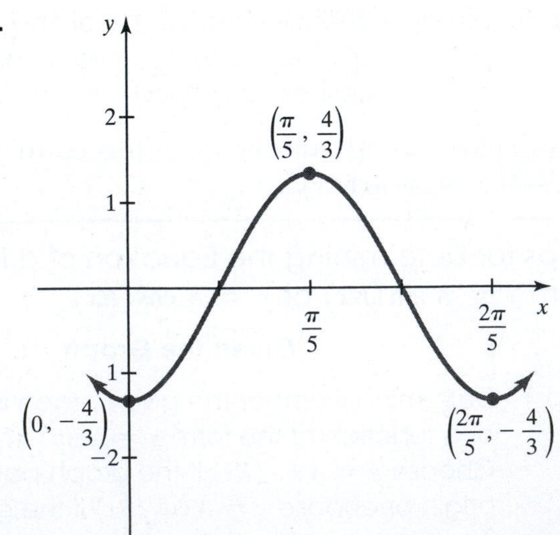

Key Concepts	Examples/Videos
7.2 More on the Graphs of Sine and Cosine For functions of the form $y = A\sin(Bx - C)$ or $y = A\cos(Bx - C)$, the value of $\frac{C}{B}$ is called the **phase shift.** **Steps for Sketching Functions of the Form $y = A\sin(Bx - C) + D$ and $y = A\cos(Bx - C) + D$** Step 1. If $B < 0$, rewrite the function in an equivalent form such that $B > 0$. Use the odd property of the sine function or the even property of the cosine function. **It is often helpful to factor out B when $B \neq 1$ such that $y = A\sin(Bx - C) + D = A\sin\left(B\left(x - \frac{C}{B}\right)\right) + D$ or $y = A\cos(Bx - C) + D = A\cos\left(B\left(x - \frac{C}{B}\right)\right) + D$. In the factored form, the amplitude, period, and phase shift are more apparent.** Step 2. The amplitude is $\lvert A\rvert$. The range is $[-\lvert A\rvert + D, \lvert A\rvert + D]$. Step 3. The period is $P = \frac{2\pi}{B}$. Step 4. The phase shift is $\frac{C}{B}$. Step 5. The x-coordinate of the first quarter point is $\frac{C}{B}$. The x-coordinate of the last quarter point is $\frac{C}{B} + P$. Subdivide this interval into four equal subintervals of length $P \div 4$ by starting with $\frac{C}{B}$ and adding $(P \div 4)$ to the x-coordinate of each successive quarter point. Step 6. Multiply the y-coordinates of the quarter points of $y = \sin x$ or $y = \cos x$ by A and then add D to determine the y-coordinates of the corresponding quarter points for $y = A\sin(Bx - C) + D$ and $y = A\cos(Bx - C) + D$. Step 7. Connect the quarter points to obtain one complete cycle. 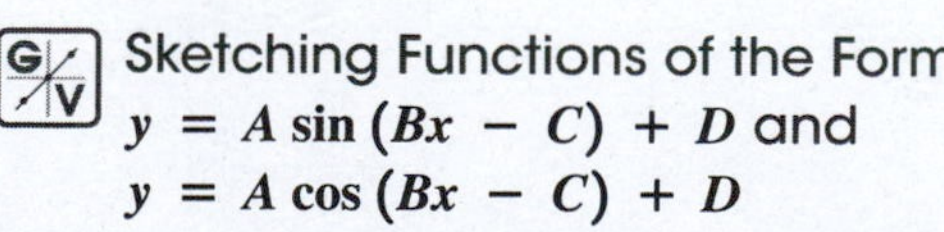	 Determine the phase shift and sketch the graph of each function. a. $y = \cos(x - \pi)$ b. $y = \sin\left(x + \frac{\pi}{2}\right)$  Sketch the graph of each function. a. $y = 3\sin\left(2x - \frac{\pi}{2}\right) - 1$ b. $y = 4 - \cos(-\pi + 2)$

Key Concepts	Examples/Videos

Key Concepts

Steps for Determining an Equation of a Function of the Form $y = A\sin(Bx - C) + D$ or $y = A\cos(Bx - C) + D$

Given the Graph

Step 1. Subtract the x-coordinate of the first quarter point from the x-coordinate of the fifth quarter point to determine the period.

Step 2. Use the equation $P = \dfrac{2\pi}{B}$ to determine the value of $B > 0$.

Step 3. The x-coordinate of the first quarter point represents the phase shift, $\dfrac{C}{B}$. Use this information to determine the value of C.

Step 4. The amplitude is $|A| = \dfrac{|b - a|}{2}$, where $[a, b]$ is the range of the given graph.

Step 5. If the given graph is a sine curve, then $A = |A|$ if the graph is *increasing* from the first quarter point to the second quarter point. Otherwise, $A = -|A|$.

If the given graph is a cosine curve, then $A = |A|$ if the graph is *decreasing* from the first quarter point to the second quarter point. Otherwise, $A = -|A|$.

Step 6. Choose the first quarter point (x_1, y_1) that lies on the given graph and the corresponding first quarter point of the graph of $y = \sin x$ or $y = \cos x$. Note that the first quarter point of $y = \sin x$ is $(0, 0)$ and the first quarter point of $y = \cos x$ is $(0, 1)$.

The relationship between y_1, the y-coordinate of the first quarter point of the given graph, and the y-coordinate of the first quarter point of $y = \sin x$ or $y = \cos x$ is given by the following:

If the graph is a sine curve, then $y_1 = (0)A + D$.

If the graph is a cosine curve, then $y_1 = (1)A + D$.

Use this information to determine the value of D.

Examples/Videos

Determine the values of A, B, C, and D, where $B > 0$.

a. $y = A\cos(Bx - C) + D$

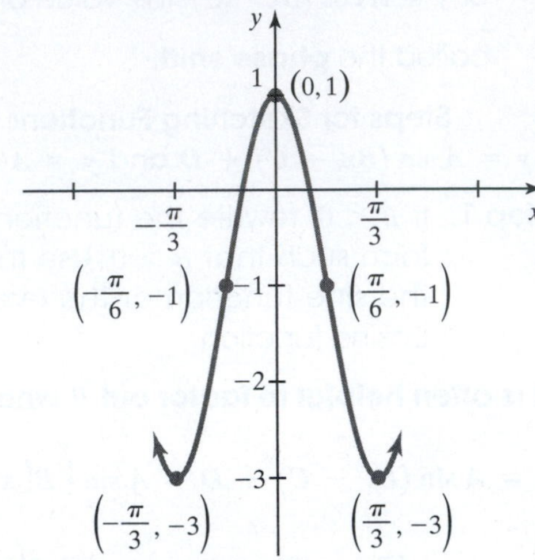

b. $y = A\sin(Bx - C) + D$

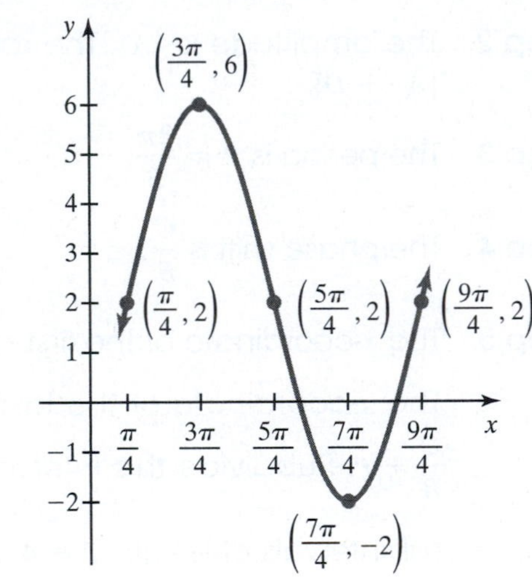

Key Concepts	Examples/Videos
7.3 The Graphs of the Tangent, Cotangent, Cosecant, and Secant Functions **Characteristics of the Tangent Function** 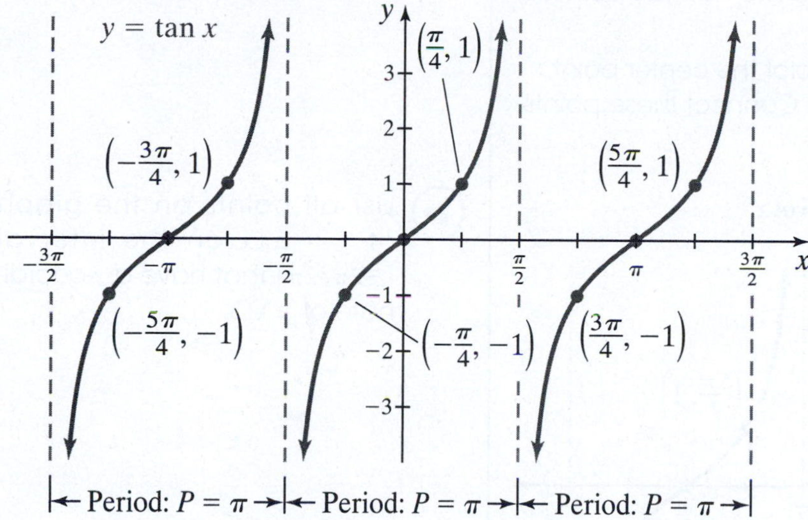 • The domain is $\left\{x \mid x \neq (2n+1) \cdot \frac{\pi}{2}, \text{ where } n \text{ is an integer}\right\}$. • The range is $(-\infty, \infty)$. • The function is periodic with a period of $P = \pi$. The principal cycle of the graph occurs on the interval $\left(-\frac{\pi}{2}, \frac{\pi}{2}\right)$. • The function has infinitely many vertical asymptotes with equations $x = (2n+1), \frac{\pi}{2}$, where n is an integer. • The y-intercept is 0. • For each cycle there is one center point. The x-coordinates of the center points are also the x-intercepts, or **zeros**, and are of the form $n\pi$, where n is an integer. • For each cycle there are two halfway points. The halfway point to the left of the x-intercept has a y-coordinate of -1. The halfway point to the right of the x-intercept has a y-coordinate of 1. • The function is **odd**, which means $\tan(-x) = -\tan x$. The graph is symmetric about the origin. • The graph of each cycle of $y = \tan x$ is **one-to-one**.	 List all halfway points of $y = \tan x$ on the interval $\left[-\pi, \frac{5\pi}{2}\right]$ that have a y-coordinate of -1.
Steps for Sketching Functions of the Form $y = A\tan(Bx - C) + D$ **Step 1.** If $B < 0$, rewrite the function in an equivalent form such that $B > 0$. Use the odd property of the tangent function. **We now use this new form to determine A, B, C, and D.** **Step 2.** Determine the interval and the equations of the vertical asymptotes of the principal cycle.	 Sketch the graph of each function. **a.** $y = \tan\left(x - \frac{\pi}{6}\right)$ **b.** $y = 4\tan(\pi - 2x) + 3$ **c.** $y = \frac{1}{2}\tan(3x) - 1$

Key Concepts	Examples/Videos
Step 3. The period is $P = \frac{\pi}{B}$. **Step 4.** Determine the center point of the principal cycle. **Step 5.** Determine the coordinates of the two halfway points of the principal cycle. **Step 6.** Sketch the vertical asymptotes, plot the center point, and plot the two halfway points. Connect these points with a smooth curve. **The Graph of $y = A \cot x$** $y = \cot x$; points $\left(-\frac{3\pi}{4}, 1\right)$, $\left(-\frac{\pi}{4}, -1\right)$, $\left(\frac{\pi}{4}, 1\right)$, $\left(\frac{3\pi}{4}, -1\right)$, $\left(\frac{5\pi}{4}, 1\right)$, $\left(\frac{7\pi}{4}, -1\right)$ Period: $P = \pi$ Period: $P = \pi$ Period: $P = \pi$	List all points on the graph of $y = \cot x$ on the interval $[-2\pi, 2\pi]$ that have a y-coordinate of $-\sqrt{3}$.
Steps for Sketching Functions of the Form $y = A \cot(Bx - C) + D$ **Step 1.** If $B < 0$, rewrite the function in an equivalent form such that $B > 0$. Use the odd property of the cotangent function. **We now use this new form to determine A, B, C, and D.** **Step 2.** Determine the interval and the equations of the vertical asymptotes of the principal cycle. **Step 3.** The period is $P = \frac{\pi}{B}$. **Step 4.** Determine the center point of the principal cycle. **Step 5.** Determine the coordinates of the two halfway points of the principal cycle. **Step 6.** Sketch the vertical asymptotes, plot the center point, and plot the two halfway points. Connect these points with a smooth curve.	Sketch the graph of each function. a. $y = \cot(2x + \pi) + 1$ b. $y = -3\cot\left(x - \frac{\pi}{4}\right)$

Key Concepts	Examples/Videos
The graph of $y = \csc x$	

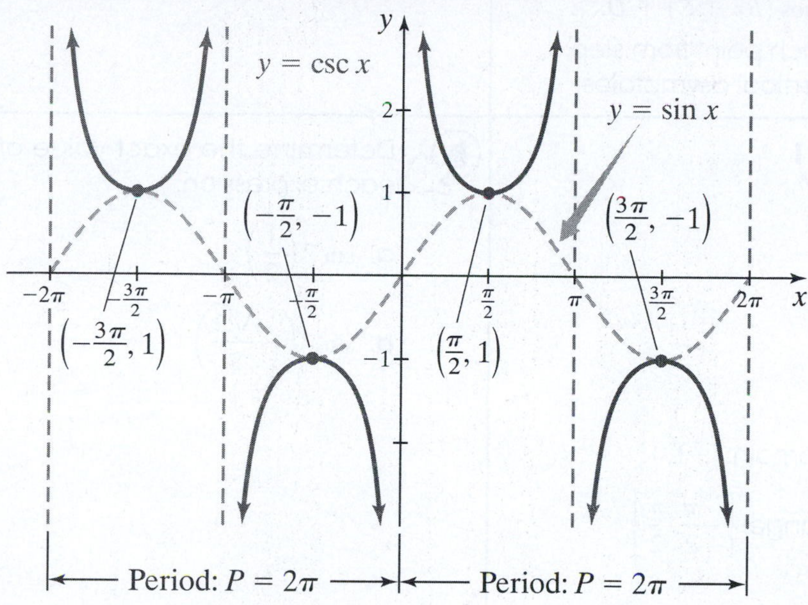

The graph of $y = \sec x$

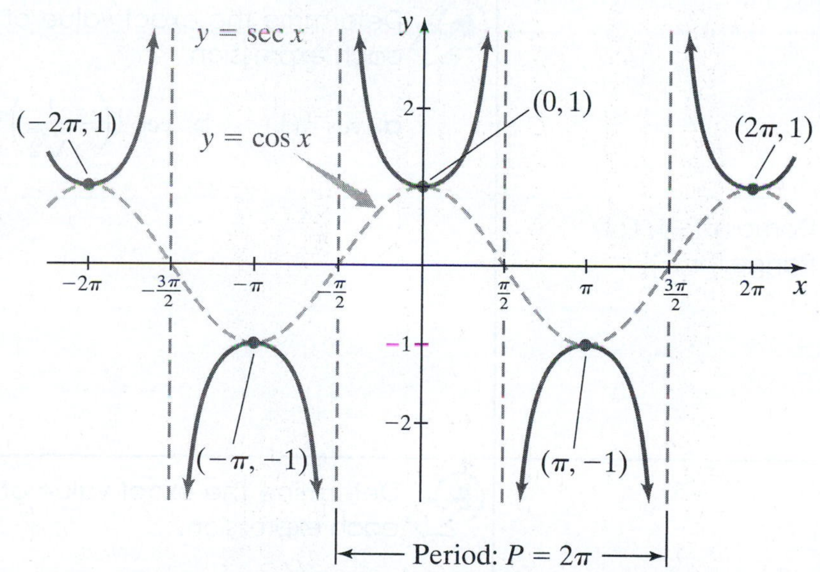

Steps for Sketching Functions of the Form
$y = A\csc(Bx - C) + D$ and $y = A\csc(Bx - C) + D$

Step 1. Lightly sketch at least two cycles of the corresponding reciprocal function. If $D \neq 0$, lightly sketch two reciprocal functions, one with $D = 0$ and one with $D \neq 0$.

Step 2. Sketch the vertical asymptotes. The vertical asymptotes will correspond to the x-intercepts of the reciprocal function $y = A\sin(Bx - C)$ or $y = A\cos(Bx - C)$.

Sketch the graph of each function.

a. $y = -2\csc(x + \pi)$

b. $y = -3\csc(\pi x) - 2$

c. $y = 3\sec(\pi - x)$

d. $y = \sec\left(2x + \frac{\pi}{2}\right) + 1$

Key Concepts	Examples/Videos
Step 3. Plot all maximum and minimum points on the graph of $y = A\sin(Bx - C) + D$ or $y = A\cos(Bx - C) + D$. **Step 4.** Draw smooth curves through each point from step 3, making sure to approach the vertical asymptotes.	
7.4 Inverse Trigonometric Functions I **The Graph of $y = \sin^{-1}x$** 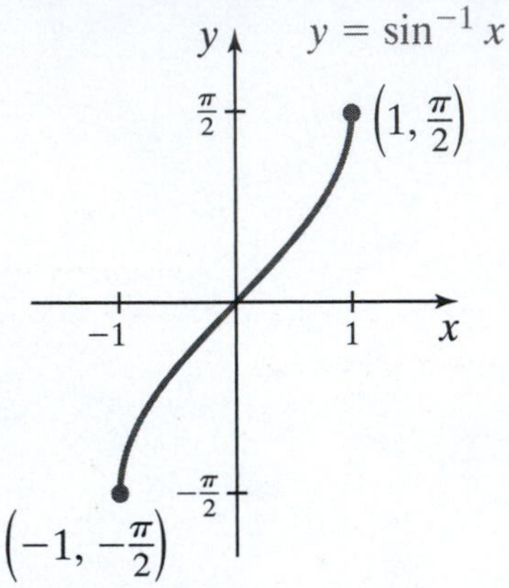 $y = \sin^{-1}x$; points $(1, \frac{\pi}{2})$, $(-1, -\frac{\pi}{2})$ Domain: $[-1, 1]$ Range: $\left[-\frac{\pi}{2}, \frac{\pi}{2}\right]$	Determine the exact value of each expression. **a.** $\sin^{-1}\left(\frac{1}{2}\right)$ **b.** $\sin^{-1}\left(-\frac{\sqrt{3}}{2}\right)$
The Graph of $y = \cos^{-1}x$ 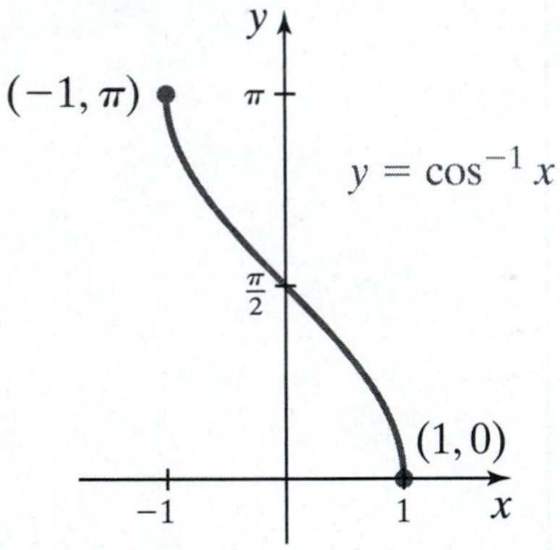 $y = \cos^{-1}x$; points $(-1, \pi)$, $(1, 0)$ Domain: $[-1, 1]$ Range: $[0, \pi]$	Determine the exact value of each expression. **a.** $\cos^{-1}(1)$ **b.** $\cos^{-1}\left(-\frac{1}{\sqrt{2}}\right)$
The Graph of $y = \tan^{-1}x$ 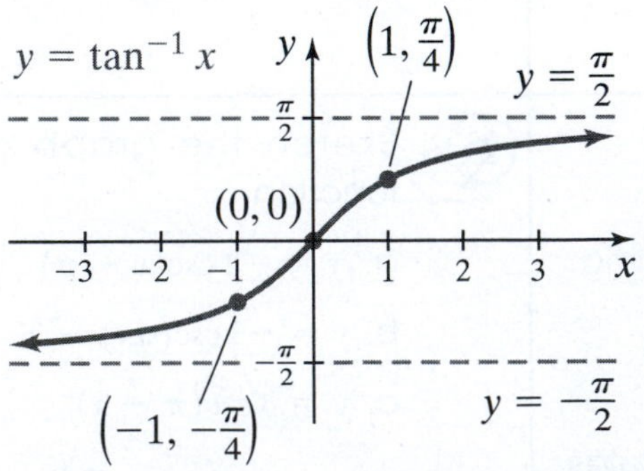 $y = \tan^{-1}x$; points $(1, \frac{\pi}{4})$, $(0, 0)$, $(-1, -\frac{\pi}{4})$; asymptotes $y = \frac{\pi}{2}$, $y = -\frac{\pi}{2}$ Domain: $(-\infty, \infty)$ Range: $\left(-\frac{\pi}{2}, \frac{\pi}{2}\right)$	Determine the exact value of each expression. **a.** $\tan^{-1}\left(\frac{1}{\sqrt{3}}\right)$ **b.** $\tan^{-1}\left(-\frac{1}{\sqrt{3}}\right)$

Key Concepts	Examples/Videos
7.5 Inverse Trigonometric Functions II **Cancellation Equations for the Restricted Sine Function and Its Inverse** $\sin(\sin^{-1} x) = x$ for all x in the interval $[-1, 1]$ $\sin^{-1}(\sin\theta) = \theta$ for all θ in the interval $\left[-\frac{\pi}{2}, \frac{\pi}{2}\right]$. **Cancellation Equations for the Restricted Cosine Function and Its Inverse** $\cos(\cos^{-1} x) = x$ for all x in the interval $[-1, 1]$ $\cos^{-1}(\cos\theta) = \theta$ for all θ in the interval $[0, \pi]$	 Find the exact value of each expression or state that it does not exist. a. $\sin\left(\sin^{-1}\frac{1}{2}\right)$ b. $\cos\left(\cos^{-1}\frac{3}{2}\right)$ c. $\tan(\tan^{-1}(8.2))$ d. $\sin(\sin^{-1}(1.3))$
Cancellation Equations for the Restricted Tangent Function and Its Inverse $\tan(\tan^{-1} x) = x$ for all x in the interval $(-\infty, \infty)$ $\tan^{-1}(\tan\theta) = \theta$ for all θ in the interval $\left(-\frac{\pi}{2}, \frac{\pi}{2}\right)$	 Find the exact value of each expression or state that it does not exist. a. $\sin^{-1}\left(\sin\frac{\pi}{6}\right)$ b. $\cos^{-1}\left(\cos\frac{2\pi}{3}\right)$ c. $\sin^{-1}\left(\sin\frac{4\pi}{3}\right)$ d. $\tan^{-1}\left(\tan\frac{7\pi}{10}\right)$
Evaluating Functions of the Form $f \circ g^{-1}$ or $f^{-1} \circ g$ 1. Evaluate the "inner expression" 2. Evaluate the "outer expression"	 Find the exact value of each expression or state that it does not exist. a. $\cos(\tan^{-1}\sqrt{3})$ b. $\csc\left(\cos^{-1}\left(-\frac{\sqrt{3}}{2}\right)\right)$ c. $\sec\left(\sin^{-1}\left(-\frac{\sqrt{5}}{8}\right)\right)$  Find the exact value of each expression or state that it does not exist. a. $\sin^{-1}\left(\cos\left(-\frac{2\pi}{3}\right)\right)$ b. $\cos^{-1}\left(\sin\frac{\pi}{7}\right)$
 **The Graph of $y = \csc^{-1}x$** 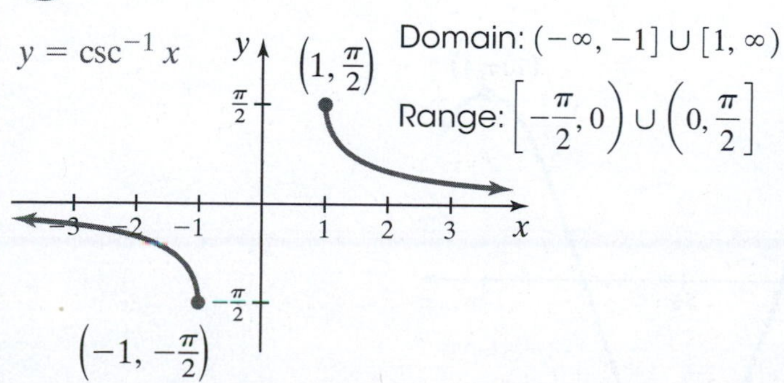 	

Key Concepts	Examples/Videos
The Graph of $y = \sec^{-1} x$ $y = \sec^{-1} x$ $(-1, \pi)$ $(1, 0)$ Domain: $(-\infty, -1] \cup [1, \infty)$ Range: $\left[0, \frac{\pi}{2}\right) \cup \left(\frac{\pi}{2}, \pi\right]$	
The Graph of $y = \cot^{-1} x$ $y = \cot^{-1} x$ Domain: $(-\infty, \infty)$ Range: $\left(-\frac{\pi}{2}, 0\right) \cup \left(0, \frac{\pi}{2}\right]$	

Chapter 7 Review Exercises

1. Determine the amplitude of $y = -3 \sin x$.

2. Determine the period of $y = \cos(-4x)$.

3. Determine the amplitude, range, and period of $y = 2 \sin(3x)$ and sketch the graph.

4. One cycle of the graph of a trigonometric function of the form $y = A \sin(Bx)$ or $y = A \cos(Bx)$ for $B > 0$ is given. Determine the equation of the function represented by the graph.

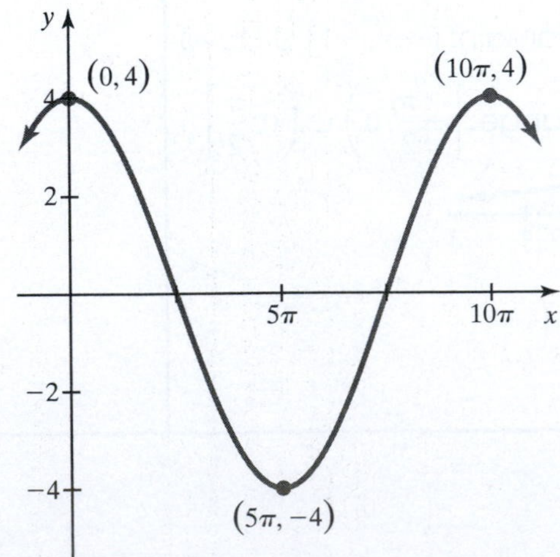

5. Determine the phase shift of $y = -3\cos(2x + \pi)$.

6. Determine the amplitude, range, period, and phase shift of $y = \cos\left(x + \frac{\pi}{2}\right)$ and sketch the graph.

7. Determine the amplitude, range, period, and phase shift of $y = -\sin(2x - \pi) - 2$ and sketch the graph.

8. Determine the interval of the principal cycle of $y = \cot(2x)$.

9. Determine the equations of the vertical asymptotes of the principal cycle of $y = 3\tan\left(x - \frac{\pi}{6}\right)$.

10. For the function $y = \tan(3x + \pi)$, determine the interval for the principal cycle. Determine the period. Then for the principal cycle, determine the equations of the vertical asymptotes, the coordinates of the center point, and the coordinates of the halfway points. Then sketch the graph.

11. The principal cycle of the graph of a function of the form $y = A\cot(Bx - C) + D$ for $B > 0$ is given. Determine the equation of the function represented by the graph.

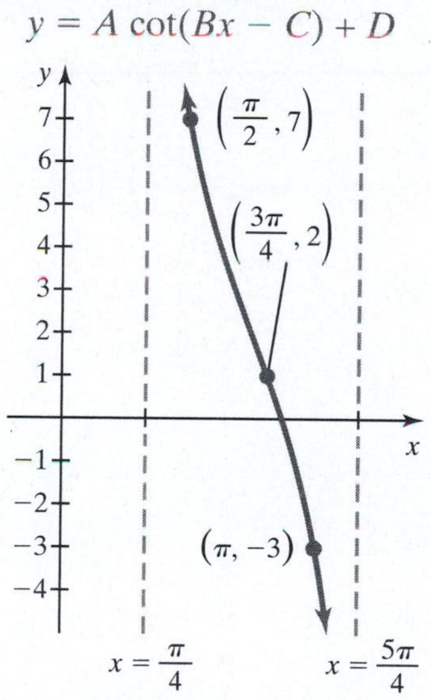

12. For the function $y = 2\sec(x - \pi)$, determine the equations of the vertical asymptotes and all relative maximum and relative minimum point of two cycles. Then sketch the graph.

13. Determine the exact value of $y = \sin^{-1}\left(\frac{\sqrt{3}}{2}\right)$.

14. If $\theta = \cos^{-1}\left(-\frac{1}{\sqrt{2}}\right)$, then determine the quadrant in which or the axis on which the terminal side of θ lies.

15. Determine the exact value of $y = \tan^{-1}(-\sqrt{3})$.

For 16–19, find the exact value of each expression or state that it does not exist.

16. $\sin\left(\sin^{-1}\left(\frac{1}{\sqrt{2}}\right)\right)$

17. $\cos^{-1}\left(\cos\frac{9\pi}{4}\right)$

18. $\sec\left(\sin^{-1}\left(\frac{1}{2}\right)\right)$

19. $\cos^{-1}\left(\sin\frac{19\pi}{9}\right)$

20. Write the expression $\cos\left(\sin^{-1}\frac{3}{u}\right)$ as an algebraic expression involving the variable u.

Assume that $u > 0$ and that the value of the "inner" inverse trigonometric expression represents an angle θ such that $0 < \theta < \frac{\pi}{2}$.

CHAPTER EIGHT

Trigonometric Identities

CHAPTER EIGHT CONTENTS

8.1 Trigonometric Identities

THINGS TO KNOW

Before working through this section, be sure that you are familiar with the following concepts:

VIDEO | ANIMATION | INTERACTIVE

1. Understanding the Quotient Identities for Acute Angles (Section 6.4)

2. Understanding the Reciprocal Identities for Acute Angles (Section 6.4)

3. Understanding the Pythagorean Identities for Acute Angles (Section 6.4)

OBJECTIVES

1 Reviewing the Fundamental Identities

2 Substituting Known Identities to Verify an Identity

3 Changing to Sines and Cosines to Verify an Identity

4 Factoring to Verify an Identity

5 Separating a Single Quotient into Multiple Quotients to Verify an Identity

6 Combining Fractional Expressions to Verify an Identity

7 Multiplying by Conjugates to Verify an Identity

8 Summarizing the Techniques for Verifying Identities

SECTION 8.1 EXERCISES

(eText Screens 8.1-1–8.1-28)

OBJECTIVE 1 REVIEWING THE FUNDAMENTAL IDENTITIES

In **Section 6.4**, we established some fundamental trigonometric identities that were valid for all **acute** angles. We used right triangles to prove many of these identities. We now review these fundamental identities (the quotient identities, the reciprocal identities, and the Pythagorean identities), which are true for *all* values of θ for which each trigonometric expression is defined.

The Quotient Identities

$$\tan\theta = \frac{\sin\theta}{\cos\theta} \qquad \cot\theta = \frac{\cos\theta}{\sin\theta}$$

The Reciprocal Identities

$$\sin\theta = \frac{1}{\csc\theta} \qquad \csc\theta = \frac{1}{\sin\theta}$$

$$\cos\theta = \frac{1}{\sec\theta} \qquad \sec\theta = \frac{1}{\cos\theta}$$

$$\tan\theta = \frac{1}{\cot\theta} \qquad \cot\theta = \frac{1}{\tan\theta}$$

The Pythagorean Identities

$$\sin^2\theta + \cos^2\theta = 1$$

$$1 + \tan^2\theta = \sec^2\theta$$

$$1 + \cot^2\theta = \csc^2\theta$$

Note that there are several variations of the Pythagorean identities such as $\sin^2\theta = 1 - \cos^2\theta$ that are obtained by a simple algebraic manipulation of one of the three listed identities. You should become very familiar with the various forms of the Pythagorean identities.

We have also explored the graphs and properties of each of the trigonometric functions in depth in Chapter 7. Recall that one of the properties of the sine, tangent, cosecant, and cotangent functions is that each of these functions is an **odd function**. In contrast, the cosine and secant functions are **even functions**. These facts give rise to the following **odd and even properties**.

The Odd Properties

$$\sin(-\theta) = -\sin\theta \qquad \tan(-\theta) = -\tan\theta$$

$$\csc(-\theta) = -\csc\theta \qquad \cot(-\theta) = -\cot\theta$$

The Even Properties

$$\cos(-\theta) = \cos\theta$$

$$\sec(-\theta) = \sec\theta$$

You Try It Work through this You Try It problem.

Work Exercises 1–7 in this textbook or in the MyLab Math Study Plan.

OBJECTIVE 2 SUBSTITUTING KNOWN IDENTITIES TO VERIFY AN IDENTITY

We can use the fundamental identities to determine other trigonometric identities. When given a trigonometric identity, it is important to remember that the identity is valid for *all* values of the **independent variable** for which both sides of the identity are defined. In this section, we will explore several different techniques that can be used to verify identities. Verifying trigonometric identities is the process of showing that the expression on one side of the identity can be simplified to be the exact expression that appears on the other side. We typically start with what appears to be the more complicated side of the identity and then attempt to use known identities to transform that side into the trigonometric expression that appears on the other side of the identity. There is no one correct way to verify identities. In fact, there can be several ways to verify an identity. We start with a technique that first involves substituting one or more known trigonometric identities and then simplifying. Be especially aware of squared trigonometric expressions because they may indicate the use of a Pythagorean identity.

Example 1 Verifying an Identity

Verify each identity.

a. $\tan x \cot x = 1$

b. $\sec^2 3x + \cot^2 3x - \tan^2 3x = \csc^2 3x$

c. $(5 \sin y + 2 \cos y)^2 + (5 \cos y - 2 \sin y)^2 = 29$

Solution

a. The left-hand side of the identity is more complicated. Note that we can first use a **reciprocal identity** to rewrite $\cot x$ and then simplify.

$\tan x \cot x \stackrel{?}{=} 1$	Write the original identity.
$\tan x \cdot \dfrac{1}{\tan x} \stackrel{?}{=} 1$	Use the quotient identity $\cot x = \dfrac{1}{\tan x}$.
$\cancel{\tan x} \cdot \dfrac{1}{\cancel{\tan x}} \stackrel{?}{=} 1$	Cancel the common factors.
$1 = 1$	The left-hand side is identical to the right-hand side.

The left-hand side is now identical to the right-hand side. Therefore, the identity is verified. You may wish to work through this **interactive video** to see each step of this solution. Note that if we multiply any trigonometric function by its reciprocal function, then the result will always be 1. To see these products worked out, read these **steps**.

b. The left-hand side appears to be more difficult than the right-hand side. We see that the left-hand side contains three squared trigonometric expressions. This is an indication that one or more Pythagorean identities might be used.

$\sec^2 3x + \cot^2 3x - \tan^2 3x \stackrel{?}{=} \csc^2 3x$	Write the original identity.
$\cot^2 3x + (\sec^2 3x - \tan^2 3x) \stackrel{?}{=} \csc^2 3x$	Rearrange the terms.
$\cot^2 3x + 1 \stackrel{?}{=} \csc^2 3x$	Use the Pythagorean identity $\sec^2\theta - \tan^2\theta = 1$.
$\csc^2 3x = \csc^2 3x$	Use the Pythagorean identity $\cot^2\theta + 1 = \csc^2\theta$.

The left-hand side is now identical to the right-hand side. Therefore, the identity is verified. You may wish to work through this **interactive video** to see each step of this solution.

CAUTION In the previous example, note that we treated the argument, $3x$, just as if it were a single symbol like θ. Never separate an argument:

$$\sec^2 3x = \sec^2(3x) \text{ but } \sec^2 3x \neq \sec^2(3) \cdot x$$

c. It does not appear that we can directly substitute a known identity. However, squaring the two expressions will produce squared trigonometric expressions that may allow us to use a Pythagorean identity.

$(5\sin y + 2\cos y)^2 + (5\cos y - 2\sin y)^2 \stackrel{?}{=} 29$	Write the original identity.
$29\sin^2 y + 29\cos^2 y \stackrel{?}{=} 29$	Square each expression on the left-hand side. To see this squaring process, read these **steps**.
$29(\sin^2 y + \cos^2 y) \stackrel{?}{=} 29$	Factor out the common factor of 29.
$29(1) \stackrel{?}{=} 29$	Use the Pythagorean identity $\sin^2\theta + \cos^2\theta = 1$.
$29 = 29$	Multiply.

The left-hand side is now identical to the right-hand side. Therefore, the identity is verified. You may wish to work through this **interactive video** to see each step of this solution. ●

You Try It Work through this You Try It problem.

Work Exercises 8–12 in this textbook or in the MyLab Math Study Plan.

OBJECTIVE 3 CHANGING TO SINES AND COSINES TO VERIFY AN IDENTITY

When we have simple forms of multiple trigonometric functions on both sides of the identity, it is often efficient first to rewrite all of the functions on one side of the identity in terms of sines and cosines and then simplify.

Example 2 Verifying an Identity

Verify each identity.

a. $\sin^2 t = \tan t \cot t - \cos^2 t$ b. $\dfrac{\sec\theta\csc\theta}{\cot\theta} = \sec^2\theta$ c. $\dfrac{\cos(-\theta)}{\sec\theta} + \sin(-\theta)\csc\theta = -\sin^2\theta$

Solution

a. The right-hand side of the identity $\sin^2 t = \tan t \cot t - \cos^2 t$ looks like the more complicated side. Therefore, we will rewrite the right-hand side in terms of sines and cosines.

$\sin^2 t \stackrel{?}{=} \tan t \cot t - \cos^2 t$	Write the original identity.
$\sin^2 t \stackrel{?}{=} \dfrac{\sin t}{\cos t} \cdot \dfrac{\cos t}{\sin t} - \cos^2 t$	Use the quotient identities $\tan t = \dfrac{\sin t}{\cos t}$ and $\cot t = \dfrac{\cos t}{\sin t}$.
$\sin^2 t \stackrel{?}{=} \dfrac{\cancel{\sin t}}{\cancel{\cos t}} \cdot \dfrac{\cancel{\cos t}}{\cancel{\sin t}} - \cos^2 t$	Cancel common factors.
$\sin^2 t \stackrel{?}{=} 1 - \cos^2 t$	Simplify.
$\sin^2 t = \sin^2 t$	Use the Pythagorean identity $1 - \cos^2\theta = \sin^2\theta$.

The right-hand side is now identical to the left-hand side. Therefore, the identity is verified. Note that we could have used an alternate method to verify this identity without changing the right-hand side to sines and cosines. To see an alternate solution, read these **steps** or work through this **interactive video**.

b. It appears that the trigonometric expression on the left-hand side of the identity $\frac{\sec\theta\csc\theta}{\cot\theta} = \sec^2\theta$ is more complicated than the expression on the right-hand side. Therefore, we will start by rewriting the left-hand side in terms of sines and cosines.

$\frac{\sec\theta\csc\theta}{\cot\theta} \stackrel{?}{=} \sec^2\theta$ Write the original identity.

$\frac{\frac{1}{\cos\theta}\cdot\frac{1}{\sin\theta}}{\frac{\cos\theta}{\sin\theta}} \stackrel{?}{=} \sec^2\theta$ Use the reciprocal identities $\sec\theta = \frac{1}{\cos\theta}$ and $\csc\theta = \frac{1}{\sin\theta}$ and the quotient identity $\cot\theta = \frac{\cos\theta}{\sin\theta}$.

$\frac{\frac{1}{\sin\theta\cos\theta}}{\frac{\cos\theta}{\sin\theta}} \stackrel{?}{=} \sec^2\theta$ Multiply the expressions in the numerator.

$\frac{1}{\sin\theta\cos\theta} \div \frac{\cos\theta}{\sin\theta} \stackrel{?}{=} \sec^2\theta$ Rewrite as the division of two expressions.

$\frac{1}{\sin\theta\cos\theta}\cdot\frac{\sin\theta}{\cos\theta} \stackrel{?}{=} \sec^2\theta$ Multiply the first expression by the reciprocal of the second expression.

$\frac{1}{\cancel{\sin\theta}\cos\theta}\cdot\frac{\cancel{\sin\theta}}{\cos\theta} \stackrel{?}{=} \sec^2\theta$ Cancel the common factors.

$\frac{1}{\cos^2\theta} \stackrel{?}{=} \sec^2\theta$ Multiply.

$\left(\frac{1}{\cos\theta}\right)^2 \stackrel{?}{=} \sec^2\theta$ Rewrite $\frac{1}{\cos^2\theta}$ as $\left(\frac{1}{\cos\theta}\right)^2$.

$(\sec\theta)^2 \stackrel{?}{=} \sec^2\theta$ Use the reciprocal identity $\frac{1}{\cos\theta} = \sec\theta$.

$\sec^2\theta = \sec^2\theta$ Rewrite $(\sec\theta)^2$ as $\sec^2\theta$.

The left-hand side is now identical to the right-hand side. Therefore, the identity is verified. Work through this **interactive video** to see each step of this solution.

c. Try to verify the identity $\frac{\cos(-\theta)}{\sec\theta} + \sin(-\theta)\csc\theta = -\sin^2\theta$ on your own. To see the solution, read these **steps** or work through part c of this **interactive video**. ●

You Try It Work through this You Try It problem.

Work Exercises 13–18 in this textbook or in the MyLab Math Study Plan.

OBJECTIVE 4 FACTORING TO VERIFY AN IDENTITY

In the solution to **Example 1c**, we factored out a common factor of 29 to help us verify the identity. We may encounter identities where it may be necessary to use a more complicated factoring technique. We now quickly review some factoring techniques.

You should also be very comfortable factoring trinomials of the form $ax^2 + bx + c$. Watch this **video** to review how to factor trinomials that have a leading coefficient equal to 1. You might also want to watch this **video** to review how to factor trinomials that have a leading coefficient not equal to 1.

It is also worthwhile to remember the following special factoring formulas. These formulas hold true for all algebraic and trigonometric expressions a and b.

Difference of Two Squares	$a^2 - b^2 = (a + b)(a - b)$
Perfect Square Formulas	$a^2 + 2ab + b^2 = (a + b)^2$ and $a^2 - 2ab + b^2 = (a - b)^2$
Sum of Two Cubes	$a^3 + b^3 = (a + b)(a^2 - ab + b^2)$
Difference of Two Cubes	$a^3 - b^3 = (a - b)(a^2 + ab + b^2)$

Example 3 Verifying a Trigonometric Identity by Factoring

Verify each trigonometric identity.

a. $\sin x - \cos^2 x \sin x = \sin^3 x$

b. $\dfrac{\tan^3\alpha - 1}{\tan\alpha - 1} = \sec^2\alpha + \tan\alpha$

c. $\dfrac{8\sin^2\theta - 2\sin\theta - 3}{1 + 2\sin\theta} = 4\sin\theta - 3$

Solution

a. Note that both terms of the left-hand side of the identity share a common factor of $\sin x$. Therefore, we start by first factoring out $\sin x$ from both terms on the left-hand side and then substituting with a known identity as we first did in Objective 2.

$\sin x - \cos^2 x \sin x \stackrel{?}{=} \sin^3 x$	Write the original trigonometric identity.
$\sin x\,(1 - \cos^2 x) \stackrel{?}{=} \sin^3 x$	Factor out the common factor of $\sin x$.
$\sin x\,(\sin^2 x) \stackrel{?}{=} \sin^3 x$	Use the Pythagorean identity $1 - \cos^2\theta = \sin^2\theta$.
$\sin^3 x = \sin^3 x$	Multiply.

The left-hand side is now identical to the right-hand side. Therefore, the identity is verified. Watch this **interactive video** to see each step of this verification process.

b. The numerator of the left-hand expression is a difference of two cubes of the form $a^3 - b^3$, where $a = \tan\alpha$ and $b = 1$. Start by factoring the difference of two cubes and then look for a known identity to substitute and simplify.

$$\frac{\tan^3\alpha - 1}{\tan\alpha - 1} \stackrel{?}{=} \sec^2\alpha + \tan\alpha$$ Write the original trigonometric identity.

$$\frac{(\tan\alpha - 1)(\tan^2\alpha + \tan\alpha + 1)}{\tan\alpha - 1} \stackrel{?}{=} \sec^2\alpha + \tan\alpha$$ Use the difference of two cubes formula $a^3 - b^3 = (a - b)(a^2 + ab + b^2)$ with $a = \tan\alpha$ and $b = 1$.

$$\frac{\cancel{(\tan\alpha - 1)}(\tan^2\alpha + \tan\alpha + 1)}{\cancel{\tan\alpha - 1}} \stackrel{?}{=} \sec^2\alpha + \tan\alpha$$ Cancel common factors.

$$(1 + \tan^2\alpha) + \tan\alpha \stackrel{?}{=} \sec^2\alpha + \tan\alpha$$ Rearrange the terms.

$$\sec^2\alpha + \tan\alpha = \sec^2\alpha + \tan\alpha$$ Use the Pythagorean identity $1 + \tan^2\theta = \sec^2\theta$.

The left-hand side is now identical to the right-hand side. Therefore, the identity is verified. Watch this **interactive video** to see each step of this verification process.

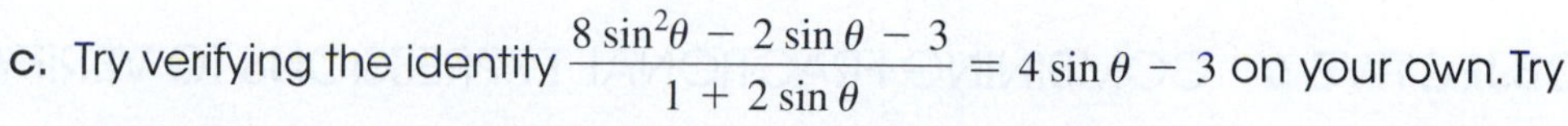

c. Try verifying the identity $\frac{8\sin^2\theta - 2\sin\theta - 3}{1 + 2\sin\theta} = 4\sin\theta - 3$ on your own. Try

factoring the numerator on the left-hand side by treating it as a trinomial. Check your work by watching this **interactive video**.

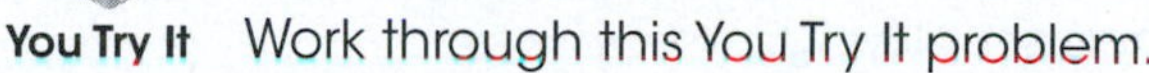

You Try It Work through this You Try It problem.

Work Exercises 19–23 in this textbook or in the MyLab Math Study Plan.

OBJECTIVE 5 SEPARATING A SINGLE QUOTIENT INTO MULTIPLE QUOTIENTS TO VERIFY AN IDENTITY

When one side of a trigonometric identity is a quotient of the form $\frac{A + B}{C}$, where C is a single trigonometric expression, then it is often advantageous to begin by separating the quotient into multiple quotients. We do this using the algebraic property $\frac{A + B}{C} = \frac{A}{C} + \frac{B}{C}$.

Example 4 Verifying a Trigonometric Identity by Separating a Single Quotient into Multiple Quotients

Verify the trigonometric identity $\frac{\sin\alpha + \cos\alpha}{\cos\alpha} - \frac{\sin\alpha + \cos\alpha}{\sin\alpha} = \tan\alpha - \cot\alpha$.

Solution The left-hand side appears to be more complex than the right-hand side. Note that the denominator of each quotient on the left-hand side of the identity contains one term. We can therefore separate each quotient into two separate quotients, substitute, and then simplify.

$$\frac{\sin\alpha + \cos\alpha}{\cos\alpha} - \frac{\sin\alpha + \cos\alpha}{\sin\alpha} \stackrel{?}{=} \tan\alpha - \cot\alpha$$ Write the original trigonometric identity.

$$\frac{\sin\alpha}{\cos\alpha} + \frac{\cos\alpha}{\cos\alpha} - \frac{\sin\alpha}{\sin\alpha} - \frac{\cos\alpha}{\sin\alpha} \stackrel{?}{=} \tan\alpha - \cot\alpha$$ Separate each single quotient into two quotients.

$$\tan\alpha + 1 - 1 - \cot\alpha \stackrel{?}{=} \tan\alpha - \cot\alpha$$ Use two quotient identities and rewrite the quotients $\frac{\cos\alpha}{\cos\alpha}$ and $\frac{\sin\alpha}{\sin\alpha}$ as 1.

$$\tan\alpha - \cot\alpha = \tan\alpha - \cot\alpha$$ Combine like terms.

The left-hand side is now identical to the right-hand side. Therefore, the identity is verified. Watch this **video** to see each step of this verification process. ●

You Try It Work through this You Try It problem.

Work Exercises 24–26 in this textbook or in the MyLab Math Study Plan.

OBJECTIVE 6 COMBINING FRACTIONAL EXPRESSIONS TO VERIFY AN IDENTITY

When two or more fractional expressions appear on one side of an identity and the other side contains only one term, it may be useful to begin by combining all fractions into a single quotient using a common denominator, substituting if necessary, and then simplifying.

Example 5 Verifying a Trigonometric Identity by Combining Fractional Expressions

Verify the trigonometric identity $\frac{1 - \csc\theta}{\cot\theta} - \frac{\cot\theta}{1 - \csc\theta} = 2\tan\theta$.

Solution The left-hand side looks more complicated than the right-hand side. So we will focus on transforming the left-hand side.

$$\frac{1 - \csc\theta}{\cot\theta} - \frac{\cot\theta}{1 - \csc\theta} \stackrel{?}{=} 2\tan\theta$$ Write the original trigonometric identity.

$$\frac{(1 - \csc\theta)(1 - \csc\theta) - (\cot\theta)(\cot\theta)}{(\cot\theta)(1 - \csc\theta)} \stackrel{?}{=} 2\tan\theta$$ Combine the two expressions to write as a single quotient using the LCD $(\cot\theta)(1 - \csc\theta)$.

$$\frac{1 - 2\csc\theta + \csc^2\theta - \cot^2\theta}{(\cot\theta)(1 - \csc\theta)} \stackrel{?}{=} 2\tan\theta$$ Multiply.

$$\frac{1 - 2\csc\theta + 1}{(\cot\theta)(1 - \csc\theta)} \stackrel{?}{=} 2\tan\theta$$ Use the Pythagorean identity $\csc^2\theta - \cot^2\theta = 1$.

$$\frac{2 - 2\csc\theta}{(\cot\theta)(1 - \csc\theta)} \stackrel{?}{=} 2\tan\theta$$ Combine like terms.

$$\frac{2(1 - \csc\theta)}{(\cot\theta)(1 - \csc\theta)} \stackrel{?}{=} 2\tan\theta$$ Factor out the common factor of 2.

$$\frac{2\cancel{(1 - \csc\theta)}}{(\cot\theta)\cancel{(1 - \csc\theta)}} \stackrel{?}{=} 2\tan\theta$$ Cancel common factors.

$\dfrac{2}{\cot\theta} \stackrel{?}{=} 2\tan\theta$ Simplify.

$2\tan\theta = 2\tan\theta$ Use the reciprocal identity $\dfrac{1}{\cot\theta} = \tan\theta$.

The left-hand side is now identical to the right-hand side. Therefore, the identity is verified. Watch this **video** to see each step of this verification process.

You Try It Work through this You Try It problem.

Work Exercises 27–30 in this textbook or in the MyLab Math Study Plan.

OBJECTIVE 7 MULTIPLYING BY CONJUGATES TO VERIFY IDENTITIES

Given an expression of the form $A + B$, we define its conjugate as the expression $A - B$. When verifying an identity, if the numerator or denominator of one of the expressions is of the form $A + B$, try first multiplying the numerator and denominator of the expression by $\dfrac{A - B}{A - B} = 1$. We illustrate this technique in Example 6.

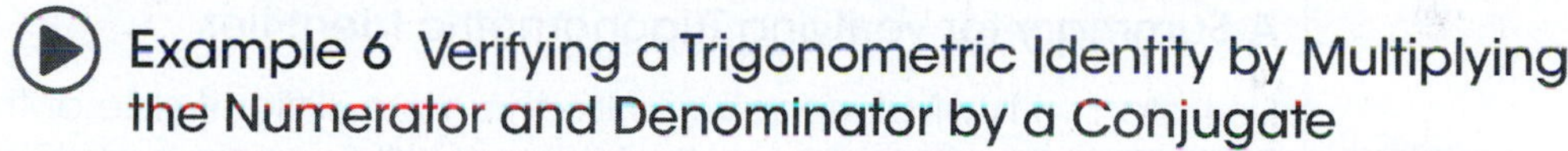

Example 6 Verifying a Trigonometric Identity by Multiplying the Numerator and Denominator by a Conjugate

Verify the trigonometric identity $\dfrac{\sin\theta}{\csc\theta + 1} = \dfrac{1 - \sin\theta}{\cot^2\theta}$.

Solution Neither side of the identity looks more complicated than the other. We could try to separate the right-hand side into two terms. However, this is not an efficient first step because the left-hand side does not contain two separate terms. So, multiply the left-hand side by the conjugate of its denominator.

$\dfrac{\sin\theta}{\csc\theta + 1} \stackrel{?}{=} \dfrac{1 - \sin\theta}{\cot^2\theta}$ Write the original trigonometric identity.

$\dfrac{\sin\theta}{\csc\theta + 1} \cdot \dfrac{\csc\theta - 1}{\csc\theta - 1} \stackrel{?}{=} \dfrac{1 - \sin\theta}{\cot^2\theta}$ Multiply the numerator and denominator by $1 = \dfrac{\csc\theta - 1}{\csc\theta - 1}$.

$\dfrac{\sin\theta\csc\theta - \sin\theta}{\csc^2\theta - 1} \stackrel{?}{=} \dfrac{1 - \sin\theta}{\cot^2\theta}$ Multiply the terms in the numerator and denominator.

$\dfrac{\sin\theta \cdot \dfrac{1}{\sin\theta} - \sin\theta}{\cot^2\theta} \stackrel{?}{=} \dfrac{1 - \sin\theta}{\cot^2\theta}$ Use the reciprocal identity $\csc\theta = \dfrac{1}{\sin\theta}$ and the Pythagorean identity $\csc^2\theta - 1 = \cot^2\theta$.

$\dfrac{\cancel{\sin\theta} \cdot \dfrac{1}{\cancel{\sin\theta}} - \sin\theta}{\cot^2\theta} \stackrel{?}{=} \dfrac{1 - \sin\theta}{\cot^2\theta}$ Cancel common factors.

$$\dfrac{1 - \sin\theta}{\cot^2\theta} = \dfrac{1 - \sin\theta}{\cot^2\theta}$$

The left-hand side is now identical to the right-hand side. Therefore, the identity is verified. Watch this **video** to see each step of this verification process.

(eText Screens 8.1-1–8.1-28)

You Try It Work through this You Try It problem.

Work Exercises 31 and 32 in this textbook or in the MyLab Math Study Plan.

OBJECTIVE 8 SUMMARIZING THE TECHNIQUES FOR VERIFYING IDENTITIES

As stated earlier, there is no one correct way to verify an identity. This fact makes it sometimes difficult to know how to begin. But it seems that often there are one or two initial techniques that will allow for the most efficient verification process, and the key is to recognize those before you begin.

We have seen six different techniques that can be used as the initial step for verifying trigonometric identities, and we have discussed how to recognize when to use each technique. We have also seen that often other techniques along with simplification may be needed in subsequent steps to complete the verification process.

We now state some generic guidelines for verifying trigonometric identities, and we summarize the six techniques learned in this section along with tips on how to recognize when each should be used.

A Summary for Verifying Trigonometric Identities

First, start with what appears to be the more difficult side of the given identity. Try to transform this side so that it eventually matches identically to the other side. Don't hesitate to start over and work with the other side of the identity if you are having trouble. Try using one of the following techniques to begin and then use others if necessary to continue the verification process.

1. Look for ways to use known identities such as the **reciprocal identities, quotient identities,** and **even/odd properties**. If the identity includes a squared trigonometric expression, try using a variation of a **Pythagorean identity**.
2. Try rewriting each trigonometric expression in terms of sines and cosines.
3. Factor out a greatest common factor and use algebraic factoring techniques such as factoring the **difference of two squares** or the **sum/difference of two cubes**.
4. If a single term appears in the denominator of a quotient, try separating the quotient into two or more quotients:
$$\frac{A+B}{C}=\frac{A}{C}+\frac{B}{C}$$
5. If there are two or more fractional expressions, try combining the expressions using a common denominator:
$$\frac{A}{B}+\frac{C}{D}=\frac{AD+BC}{BD}$$
6. If the numerator or denominator of one or more quotients contains an expression of the form $A+B$, try multiplying the numerator and denominator by its conjugate $A-B$.
$$\frac{C}{A+B}=\frac{C}{A+B}\cdot\frac{A-B}{A-B} \quad \text{or} \quad \frac{A+B}{C}=\frac{A+B}{C}\cdot\frac{A-B}{A-B}$$

Example 7 Verifying Trigonometric Identities

Verify each trigonometric identity.

a. $\dfrac{\sin^2 t + 6\sin t + 9}{\sin t + 3} = \dfrac{3\csc t + 1}{\csc t}$ b. $\dfrac{2\csc\theta}{\sec\theta} + \dfrac{\cos\theta}{\sin\theta} = 3\cot\theta$

c. $\dfrac{1 - \sin\theta}{\cos\theta} + \dfrac{\cos\theta}{1 - \sin\theta} = 2\sec\theta$

Solution

a. It appears that the left-hand side is more complicated than the right-hand side. It also appears that we can factor the numerator on the left-hand side.

$$\frac{\sin^2 t + 6\sin t + 9}{\sin t + 3} \stackrel{?}{=} \frac{3\csc t + 1}{\csc t}$$ Write the original trigonometric identity.

$$\frac{\cancel{(\sin t + 3)}(\sin t + 3)}{\cancel{\sin t + 3}} \stackrel{?}{=} \frac{3\csc t + 1}{\csc t}$$ Factor the numerator and cancel common factors.

$$\sin t + 3 \stackrel{?}{=} \frac{3\csc t + 1}{\csc t}$$ Simplify.

The left-hand side is now more simplified but it is not yet identical to the right-hand side. The denominator on the right-hand side is $\csc t$. Therefore, we will use a reciprocal identity to rewrite $\sin t$ as $\dfrac{1}{\csc t}$. We can then apply the technique of combining fractional expressions.

$$\frac{1}{\csc t} + 3 \stackrel{?}{=} \frac{3\csc t + 1}{\csc t}$$ Use the reciprocal identity $\sin t = \dfrac{1}{\csc t}$.

$$\frac{1 + 3\csc t}{\csc t} \stackrel{?}{=} \frac{3\csc t + 1}{\csc t}$$ Combine using the LCD $\csc t$.

$$\frac{3\csc t + 1}{\csc t} = \frac{3\csc t + 1}{\csc t}$$ Rearrange the terms.

The left-hand side is now identical to the right-hand side. As you can see, we used two different techniques (factoring and combining fractional expressions) to verify this identity.

Try to verify the identities for parts b and c on your own. Use the **summary for verifying trigonometric identities** to assist you. Work through this **interactive video** when you are through to see the verification process. Note that this **interactive video** illustrates two different ways to verify the identities in parts *b* and *c*.

You Try It Work through this You Try It problem.

Work Exercises 33–43 in this textbook or in the MyLab Math Study Plan.

8.1 Exercises

1. Complete the quotient identity: $\tan\theta = \dfrac{\sin\theta}{\square}$.

2. Complete the quotient identity: $\dfrac{\cos\theta}{\sin\theta} = \square$.

3. Complete the reciprocal identity: $\csc\theta = \dfrac{1}{\square}$.

4. Complete the reciprocal identity: $\square = \dfrac{1}{\sec\theta}$.

5. Complete the Pythagorean identity: $\square + \cos^2\theta = 1$.

6. Complete the Pythagorean identity: $\sec^2\theta - \square = \square$.

7. Use fundamental identities to complete the identity: $1 - \dfrac{1}{\cos^2\theta} = \square$.

In Exercises 8–43, verify each identity.

8. $1 + \cot^2(-\theta) = \csc^2\theta$

9. $\dfrac{\cot^2\beta + 1}{\csc\beta} = \csc\beta$

10. $\tan^2 4x + \csc^2 4x - \cot^2 4x = \sec^2 4x$

11. $(3\cos\theta - 4\sin\theta)^2 + (4\cos\theta + 3\sin\theta)^2 = 25$

12. $7\sin^2\theta + 4\cos^2\theta = 4 + 3\sin^2\theta$

13. $\cos\theta\csc\theta = \cot\theta$

14. $\tan(-x)\cos x = -\sin x$

15. $\cos\theta\tan\theta\csc\theta = 1$

16. $\sec\theta - \cos\theta = \sin\theta\tan\theta$

17. $\csc t\sin t - \sin^2 t = \cos^2 t$

18. $\dfrac{\csc(\theta)}{\sin\theta} - \cos(-\theta)\sec(-\theta) = \cot^2\theta$

19. $\sec x + \tan^2 x\sec x = \sec^3 x$

20. $\dfrac{\sin^2\beta - \cos^2\beta}{\sin\beta - \cos\beta} = \sin\beta + \cos\beta$

21. $\dfrac{\cot^3\theta + 1}{\cot\theta + 1} = \csc^2\theta - \cot\theta$

22. $\dfrac{6\csc^2\theta - 7\csc\theta - 3}{1 + 3\csc\theta} = 2\csc\theta - 3$

23. $\sin^4\theta - \cos^4\theta = 2\sin^2\theta - 1$

24. $\dfrac{1 + 2\sec\theta}{\sec\theta} = 2 + \cos\theta$

25. $\dfrac{\cot\theta + 1}{\csc\theta} = \sin\theta + \cos\theta$

26. $\dfrac{\tan\alpha + \cot\alpha}{\tan\alpha} - \dfrac{\cot\alpha + \tan\alpha}{\cot\alpha} = \cot^2\alpha - \tan^2\alpha$

27. $\dfrac{\sin x + \tan x + 1}{\cos x} = \sec x + \tan x + \sin x\sec^2 x$

28. $\dfrac{1-\cos\theta}{\sin\theta}+\dfrac{\sin\theta}{1-\cos\theta}=2\csc\theta$

29. $\dfrac{\sin t}{1+\cos t}+\cot t=\csc t$

30. $\dfrac{\sec\beta}{\sin\beta}-\dfrac{\sin\beta}{\sec\beta}=\dfrac{\tan^2\beta+\cos^2\beta}{\tan\beta}$

31. $\dfrac{\sin\theta}{1-\cos\theta}=\dfrac{1+\cos\theta}{\sin\theta}$

32. $\dfrac{\sec t-1}{\tan t}=\dfrac{\tan t}{\sec t+1}$

33. $\cot^2 3x+\sec^2 3x-\tan^2 3x=\csc^2 3x$

34. $1+\cot^2(-\theta)=\csc^2\theta$

35. $\tan(-x)\cos x=-\sin x$

36. $\dfrac{\csc\theta+1}{\cot\theta}=\sec\theta+\tan\theta$

37. $\dfrac{1-\sec\theta}{\tan\theta}-\dfrac{\tan\theta}{1-\sec\theta}=2\cot\theta$

38. $\dfrac{\csc\theta}{\sec\theta}+\dfrac{4\cos\theta}{\sin\theta}=5\cot\theta$

39. $\dfrac{\cos^2 t+3\cos t-10}{\cos t+5}=\dfrac{1-2\sec t}{\sec t}$

40. $1+\dfrac{1-\cot^2 x}{1+\cot^2 x}=2\sin^2 x$

41. $\sec^4\theta-\tan^4\theta=2\sec^2\theta-1$

42. $\dfrac{\tan(-\theta)}{\cot(-\theta)}-\sin\theta\csc(-\theta)=\sec^2\theta$

43. $\dfrac{\cos^3\theta+\sin^3\theta}{\cos\theta+\sin\theta}=1-\cos\theta\sin\theta$

8.2 The Sum and Difference Formulas

THINGS TO KNOW

Before working through this section, be sure that you are familiar with the following concepts:

VIDEO | ANIMATION | INTERACTIVE

You Try It

1. Understanding Cofunctions (Section 6.4)

You Try It

2. Evaluating Trigonometric Functions of Angles Belonging to the $\frac{\pi}{3}$, $\frac{\pi}{6}$, or $\frac{\pi}{4}$ Families (Section 6.5)

You Try It

3. Finding the Exact and Approximate Values of an Inverse Sine Expression (Section 7.4)

You Try It

4. Finding the Exact and Approximate Values of an Inverse Cosine Expression (Section 7.4)

OBJECTIVES

1 Understanding the Sum and Difference Formulas for the Cosine Function

2 Understanding the Sum and Difference Formulas for the Sine Function

3 Understanding the Sum and Difference Formulas for the Tangent Function

(eText Screens 8.2-1–8.2-42)

4 Using the Sum and Difference Formulas to Verify Identities

5 Using the Sum and Difference Formulas to Evaluate Expressions Involving Inverse Trigonometric Functions

SECTION 8.2 EXERCISES

OBJECTIVE 1 UNDERSTANDING THE SUM AND DIFFERENCE FORMULAS FOR THE COSINE FUNCTION

In the previous section, we verified trigonometric identities using a wide variety of techniques. In this section we introduce several new trigonometric identities. Unlike the identities in the previous section, where our identities contained only one variable, many of the identities in this section contain two variables. In this section, we will focus on formulas involving the sum and difference of two angles. We begin by introducing the sum and difference formulas for the cosine function.

The Sum and Difference Formulas for the Cosine Function

$\cos(\alpha + \beta) = \cos\alpha\cos\beta - \sin\alpha\sin\beta$ **Cosine of the Sum of Two Angles Formula**

$\cos(\alpha - \beta) = \cos\alpha\cos\beta + \sin\alpha\sin\beta$ **Cosine of the Difference of Two Angles Formula**

These sum and difference formulas are true for *all* real numbers α and β, although we typically think of α and β as angles. We start by proving the cosine of the difference of two angles formula. To prove this formula, we will assume that α and β are angles such that $0 < \beta < \alpha < 2\pi$. It is a very good exercise to work carefully through the **animation** proving this formula. Before you work through the animation proof, be prepared to draw two angles in standard position on a piece of paper. You will also need to remember the **distance formula**. As you work through the animation, you will need to use the pause button to draw and label your angles and points. Also, don't be afraid to rewind when necessary!

The cosine of the sum of two angles formula can be derived using the cosine of the difference of two angles along with the **even and odd properties**. To see how this is done, read this **proof**.

Example 1 Finding the Exact Value of a Trigonometric Expression Involving Cosine

Find the exact value of each trigonometric expression without the use of a calculator.

a. $\cos\left(\frac{2\pi}{3} + \frac{3\pi}{4}\right)$

b. $\cos(225° - 150°)$

Solution

a. To evaluate the expression $\cos\left(\frac{2\pi}{3}+\frac{3\pi}{4}\right)$, we can use the **cosine of the sum of two angles formula.**

$\cos\left(\frac{2\pi}{3}+\frac{3\pi}{4}\right)$ — Write the original expression.

$= \cos\frac{2\pi}{3}\cos\frac{3\pi}{4} - \sin\frac{2\pi}{3}\sin\frac{3\pi}{4}$ — Use the cosine of the sum of two angles formula.

$= \left(-\frac{1}{2}\right)\left(-\frac{1}{\sqrt{2}}\right) - \left(\frac{\sqrt{3}}{2}\right)\left(\frac{1}{\sqrt{2}}\right)$ — Evaluate each trigonometric function. (See how to evaluate each function by watching the **interactive video.)**

$= \frac{1}{2\sqrt{2}} - \frac{\sqrt{3}}{2\sqrt{2}} = \frac{1-\sqrt{3}}{2\sqrt{2}}$ — Simplify.

b. To evaluate the expression $\cos(225° - 150°)$, we can use the **cosine of the difference of two angles formula.** Try to find the exact value of this expression on your own. When you're finished, check your **solution** or work through this **interactive video** to see each step worked out in detail.

You Try It Work through this You Try It problem.

Work Exercises 1–16 in this textbook or in the MyLab Math Study Plan.

Example 2 Finding the Exact Value of a Trigonometric Expression Involving Cosine

Find the exact value of each trigonometric expression without the use of a calculator.

a. $\cos\left(\frac{7\pi}{12}\right)\cos\left(\frac{5\pi}{12}\right)+\sin\left(\frac{7\pi}{12}\right)\sin\left(\frac{5\pi}{12}\right)$ **b.** $\cos\left(\frac{7\pi}{12}\right)$ **c.** $\cos(-75°)$

Solution

a. To evaluate the expression $\cos\left(\frac{7\pi}{12}\right)\cos\left(\frac{5\pi}{12}\right)+\sin\left(\frac{7\pi}{12}\right)\sin\left(\frac{5\pi}{12}\right)$, we recognize that this expression represents one side of the **cosine of the difference of two angles formula** with $\alpha = \frac{7\pi}{12}$ and $\beta = \frac{5\pi}{12}$. Therefore,

$$\cos\left(\frac{7\pi}{12}\right)\cos\left(\frac{5\pi}{12}\right)+\sin\left(\frac{7\pi}{12}\right)\sin\left(\frac{5\pi}{12}\right) = \cos\left(\frac{7\pi}{12}-\frac{5\pi}{12}\right) = \cos\left(\frac{2\pi}{12}\right) = \cos\frac{\pi}{6} = \frac{\sqrt{3}}{2}.$$

b. To evaluate the expression $\cos\left(\frac{7\pi}{12}\right)$, we first need to write the angle $\frac{7\pi}{12}$ as the sum or difference of angles whose cosine is known. We know the exact value of $\cos\left(\frac{k\pi}{6}\right)$, $\cos\left(\frac{k\pi}{4}\right)$, $\cos\left(\frac{k\pi}{3}\right)$, and $\cos\left(\frac{k\pi}{2}\right)$, where k is an integer.

So, perhaps we can rewrite $\frac{7\pi}{12}$ as the sum or difference of angles of the form $\frac{k\pi}{6}, \frac{k\pi}{4}, \frac{k\pi}{3}$, and $\frac{k\pi}{2}$.

Note that $\frac{7\pi}{12} = \frac{3\pi}{12} + \frac{4\pi}{12} = \frac{\pi}{4} + \frac{\pi}{3}$. Therefore, using the **cosine of the sum of two angles formula**, we get

$$\cos\left(\frac{7\pi}{12}\right) = \cos\left(\frac{\pi}{4} + \frac{\pi}{3}\right) = \cos\frac{\pi}{4}\cos\frac{\pi}{3} - \sin\frac{\pi}{4}\sin\frac{\pi}{3}.$$

Try evaluating this expression on your own. When you're finished, check your **answer** or watch this **interactive video** to see the video solution.

c. To evaluate the expression $\cos(-75°)$, we first need to rewrite $-75°$ as the sum or difference of angles whose cosine is known. We know that we can determine the cosine of any angle that is an integer multiple of $30°, 45°, 60°$, or $90°$. So, we try the sum or difference of integer multiples of these angles until we find a combination that works. Note that $3 \cdot (45°) = 135°$ and $-75° = 60° - 135°$. Therefore, using the **cosine of the difference of two angles formula**, we get

$$\cos(-75°) = \cos(60° - 135°) = \cos 60° \cos(135°) + \sin 60° \sin(135°).$$

Try evaluating this expression on your own. When you're finished, check your **answer** or watch this **interactive video** to see the video solution.

You Try It Work through this You Try It problem.

Work Exercises 17–32 in this textbook or in the MyLab Math Study Plan.

Example 3 Finding the Exact Value of a Trigonometric Expression

Suppose that the terminal side of angle α lies in Quadrant IV and the terminal side of angle β lies in Quadrant III. If $\cos\alpha = \frac{4}{7}$ and $\sin\beta = -\frac{8}{13}$, find the exact value of $\cos(\alpha + \beta)$.

Solution To evaluate the expression $\cos(\alpha + \beta)$ using the **cosine of the sum of two angles formula**, we must first determine the value of $\sin\alpha$ and $\cos\beta$.

To find the value of $\sin\alpha$, sketch an angle α with the terminal side of α lying in Quadrant IV. Because $\cos\alpha = \frac{4}{7} = \frac{x}{r}$, we can choose the point $(4, y)$ lying on the terminal side of α, a distance of $r = 7$ units from the origin. See **Figure 1**.

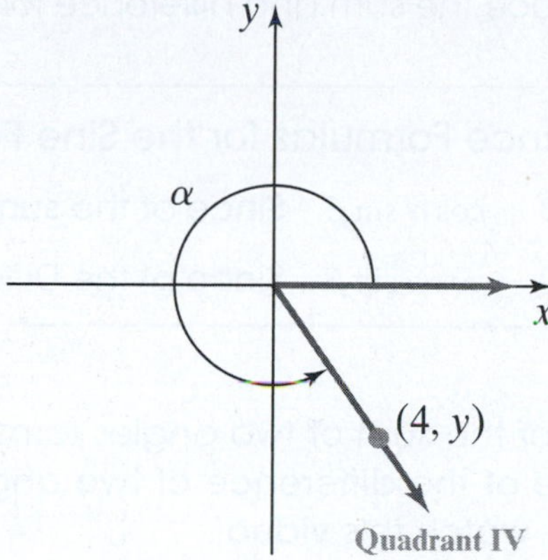

Figure 1 Angle α and the point $(4, y)$ located a distance of $r = 7$ units from the origin.

Because the point $(4, y)$ lies in Quadrant IV, we know that the value of y must be negative. Using the relationship $x^2 + y^2 = r^2$, we can determine that $y = -\sqrt{33}$. To see how we obtain this y-coordinate, read through these **steps**. Therefore, $\sin \alpha = \frac{y}{r} = -\frac{\sqrt{33}}{7}$.

We can determine the value of $\cos \beta$ in a similar fashion. Read through these **steps** to see how we determine that $\cos \beta = -\frac{\sqrt{105}}{13}$. Now that we know the values of $\sin \alpha$, $\cos \alpha$, $\sin \beta$, and $\cos \beta$, we can evaluate $\cos(\alpha + \beta)$.

$$\cos(\alpha + \beta) = \cos \alpha \cos \beta - \sin \alpha \sin \beta$$ Write the cosine of the sum of two angles formula.

$$= \left(\frac{4}{7}\right)\left(-\frac{\sqrt{105}}{13}\right) - \left(-\frac{\sqrt{33}}{7}\right)\left(-\frac{8}{13}\right)$$ Substitute the appropriate values.

$$= \frac{-4\sqrt{105} - 8\sqrt{33}}{91}$$ Simplify.

You Try It Work through this You Try It problem.

Work Exercises 33–36 in this textbook or in the MyLab Math Study Plan.

OBJECTIVE 2 UNDERSTANDING THE SUM AND DIFFERENCE FORMULAS FOR THE SINE FUNCTION

Before we introduce the sum and difference formulas for the sine function, it is important to revisit the **cofunction identities** because these identities are necessary to prove the sine of the sum of two angles formula. In **Section 6.4** we proved that the cofunction identities were true for **acute angles**. Using formulas from this section, we can **verify** that these cofunction identities are true for *all* angles. We now restate the cofunction identities for sine and cosine.

Cofunction Identities for Sine and Cosine

$$\cos\left(\frac{\pi}{2} - \theta\right) = \sin \theta \qquad \sin\left(\frac{\pi}{2} - \theta\right) = \cos \theta$$

We can now introduce the sum and difference formulas for the sine function.

The Sum and Difference Formulas for the Sine Function

$\sin(\alpha + \beta) = \sin\alpha\cos\beta + \cos\alpha\sin\beta$ **Since of the sum of Two Angles Formula**

$\sin(\alpha - \beta) = \sin\alpha\cos\beta - \cos\alpha\sin\beta$ **Since of the Difference of Two Angles Formula**

To derive the sine of the sum of two angles formula, we use a **cofunction identity** and the **cosine of the difference of two angles formula**. To see the derivation of this formula, watch this **video**.

We derive the sine of the difference of two angles formula by rewriting the left-hand side as $\sin(\alpha + (-\beta))$ and then use the sine of the sum of two angles formula and **even and odd properties of cosine and sine**. To see the derivation of this formula, watch this **video**.

Example 4 Finding the Exact Value of a Trigonometric Expression Involving Sine

Find the exact value of each trigonometric expression without the use of a calculator.

a. $\sin\left(-\frac{\pi}{3} + \frac{5\pi}{4}\right)$ **b.** $\sin(12°)\cos(78°) + \cos(12°)\sin(78°)$ **c.** $\sin(15°)$

Solution

a. To evaluate the expression $\sin\left(-\frac{\pi}{3} + \frac{5\pi}{4}\right)$, we can use the **sine of the sum of two angles formula**.

$\sin\left(-\frac{\pi}{3} + \frac{5\pi}{4}\right)$ Write the original expression.

$= \sin\left(-\frac{\pi}{3}\right)\cos\left(\frac{5\pi}{4}\right) + \cos\left(-\frac{\pi}{3}\right)\sin\left(\frac{5\pi}{4}\right)$ Use the sine of the sum of two angles formula.

$= \left(-\frac{\sqrt{3}}{2}\right)\left(-\frac{1}{\sqrt{2}}\right) + \left(\frac{1}{2}\right)\left(-\frac{1}{\sqrt{2}}\right)$ Evaluate each trigonometric function. (See how to evaluate each function by watching the **interactive video.**)

$= \frac{\sqrt{3}}{2\sqrt{2}} - \frac{1}{2\sqrt{2}} = \frac{\sqrt{3} - 1}{2\sqrt{2}}$ Simplify.

Note that we could have rearranged the order of the angles and rewritten the expression $\sin\left(-\frac{\pi}{3} + \frac{5\pi}{4}\right)$ as $\sin\left(\frac{5\pi}{4} - \frac{\pi}{3}\right)$ and then used the **sine of the difference of two angles formula** to simplify the expression. To see this alternate method worked out in detail, read these **steps**.

b. To evaluate the expression $\sin(12°)\cos(78°) + \cos(12°)\sin(78°)$, we recognize that this expression represents one side of the **sine of the sum of two angles formula** with $\alpha = 12°$ and $\beta = 78°$. Therefore,

$$\sin(12°)\cos(78°) + \cos(12°)\sin(78°) = \sin(12° + 78°) = \sin 90° = 1.$$

c. To evaluate the expression $\sin(15°)$, we first need to write the angle $15°$ as the sum or difference of angles whose sine is known. Note that $15° = 60° - 45°$. Therefore, using the **sine of the difference of two angles formula**, we get

$$\sin(15°) = \sin(60° - 45°) = \sin 60° \cos 45° - \cos 60° \sin 45°.$$

Try evaluating this expression on your own. When you're finished, check your answer or watch this **interactive video** to see the video solution.

You Try It Work through this You Try It problem.

Work Exercises 37–68 in this textbook or in the MyLab Math Study Plan.

Example 5 Finding the Exact Value of a Trigonometric Expression

Suppose that α is an angle such that $\tan\alpha = \frac{5}{7}$ and $\cos\alpha < 0$. Also, suppose that β is an angle such that $\sec\beta = -\frac{4}{3}$ and $\csc\beta > 0$. Find the exact value of $\sin(\alpha + \beta)$.

Solution To evaluate the expression $\sin(\alpha + \beta)$ using the **sine of the sum of two angles formula**, we must first determine the value of $\sin\alpha$, $\cos\alpha$, $\sin\beta$, and $\cos\beta$.

To find the value of $\sin\alpha$ and $\cos\alpha$, we must first determine the quadrant in which the terminal side of angle α lies. Because $\tan\alpha > 0$, we know that the terminal side of α could lie in either Quadrant I or Quadrant III. Also, because $\cos\alpha < 0$, we know that the terminal side of α could lie in Quadrant II or Quadrant III. Therefore, angle α lies in Quadrant III.

Because $\tan\alpha = \frac{5}{7} = \frac{y}{x}$, we can choose the point $(-7, -5)$ that lies on the terminal side of α. Using the relationship $x^2 + y^2 = r^2$, we can determine that $r = \sqrt{74}$. To see exactly how we obtained this value of r, read these **steps**. See **Figure 2**.

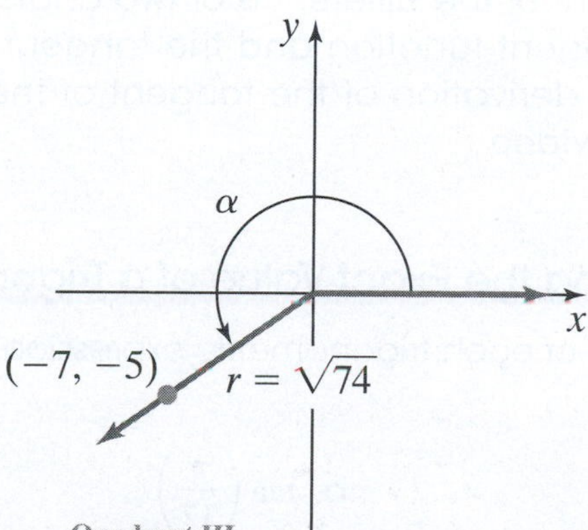

Figure 2 Angle α and the point $(-7, -5)$, which is located a distance of $r = \sqrt{74}$ units from the origin.

Thus, $\sin\alpha = \frac{y}{r} = \frac{-5}{\sqrt{74}}$ and $\cos\alpha = \frac{x}{r} = \frac{-7}{\sqrt{74}}$. Watch this **video** to see that $\sin\beta = \frac{\sqrt{7}}{4}$ and $\cos\beta = -\frac{3}{4}$. Now that we know the values of $\sin\alpha$, $\cos\alpha$, $\sin\beta$, and $\cos\beta$, we can evaluate $\sin(\alpha+\beta)$.

$\sin(\alpha+\beta) = \sin\alpha\cos\beta + \cos\alpha\sin\beta$ — Write the sine of the sum of two angles formula.

$= \left(\frac{-5}{\sqrt{74}}\right)\left(-\frac{3}{4}\right) + \left(\frac{-7}{\sqrt{74}}\right)\left(\frac{\sqrt{7}}{4}\right)$ — Substitute the appropriate values.

$= \frac{15 - 7\sqrt{7}}{4\sqrt{74}}$ — Simplify.

You Try It Work through this You Try It problem.

Work Exercises 69–72 in this textbook or in the MyLab Math Study Plan.

OBJECTIVE 3 UNDERSTANDING THE SUM AND DIFFERENCE FORMULAS FOR THE TANGENT FUNCTION

We now introduce the final sum and difference formulas, which are the sum and difference formulas for the tangent function.

The Sum and Difference Formulas for the Tangent Function

$\tan(\alpha+\beta) = \frac{\tan\alpha + \tan\beta}{1 - \tan\alpha\tan\beta}$ **Tangent of the sum of Two Angles Formula**

$\tan(\alpha-\beta) = \frac{\tan\alpha - \tan\beta}{1 + \tan\alpha\tan\beta}$ **Tangent of the Difference of Two Angles Formula**

To derive the tangent of the sum of two angles formula, we use the identity $\tan(\alpha+\beta) = \frac{\sin(\alpha+\beta)}{\cos(\alpha+\beta)}$ and then use the **sum of two angles formulas for sine and cosine.** To see the derivation of the tangent of the sum of two angles formula, watch this **video**.

To derive the tangent of the difference of two angles formula, we use the **odd property** of the tangent function and the tangent of the sum of two angles formula. To see the derivation of the tangent of the difference of two angles formula, watch this **video**.

Example 6 Finding the Exact Value of a Trigonometric Expression

Find the exact value of each trigonometric expression without the use of a calculator.

a. $\tan\left(\frac{5\pi}{6} + \frac{3\pi}{4}\right)$ **b.** $\tan\left(\frac{\pi}{12}\right)$

Solution

a. To evaluate the expression $\tan\left(\frac{5\pi}{6}+\frac{3\pi}{4}\right)$, we can use the **tangent of the sum of two angles formula.**

$\tan\left(\frac{5\pi}{6}+\frac{3\pi}{4}\right)$ Write the original expression.

$= \dfrac{\tan\left(\frac{5\pi}{6}\right)+\tan\left(\frac{3\pi}{4}\right)}{1-\tan\left(\frac{5\pi}{6}\right)\tan\left(\frac{3\pi}{4}\right)}$ Use the tangent of the sum of two angles formula.

$= \dfrac{-\frac{1}{\sqrt{3}}+(-1)}{1-\left(-\frac{1}{\sqrt{3}}\right)(-1)}$ Evaluate each trigonometric function. (See how to evaluate each function by watching the **interactive video.**)

$= \dfrac{-\frac{1}{\sqrt{3}}-1}{1-\frac{1}{\sqrt{3}}}$ Simplify.

$= \dfrac{\frac{-1-\sqrt{3}}{\sqrt{3}}}{\frac{\sqrt{3}-1}{\sqrt{3}}}$ Combine terms in the numerator and denominator.

$= \dfrac{-1-\sqrt{3}}{\sqrt{3}} \div \dfrac{\sqrt{3}-1}{\sqrt{3}}$ Rewrite as the division of two expressions.

$= \dfrac{-1-\sqrt{3}}{\sqrt{3}} \cdot \dfrac{\sqrt{3}}{\sqrt{3}-1}$ Multiply the first expression by the reciprocal of the second expression.

$= \dfrac{-1-\sqrt{3}}{\cancel{\sqrt{3}}} \cdot \dfrac{\cancel{\sqrt{3}}}{\sqrt{3}-1}$ Cancel common factors.

$= \dfrac{-1-\sqrt{3}}{\sqrt{3}-1}$ Simplify.

Recall that **rationalizing the denominator** is never required in this text. Therefore, the expression $\frac{-1-\sqrt{3}}{\sqrt{3}-1}$ is a perfectly acceptable answer. However, if we do rationalize the denominator, we see that an equivalent answer is $-2-\sqrt{3}$. To see how we obtain this, read these **steps.** Thus, $\tan\left(\frac{5\pi}{6}+\frac{3\pi}{4}\right)=\frac{-1-\sqrt{3}}{\sqrt{3}-1}$ or $\tan\left(\frac{5\pi}{6}+\frac{3\pi}{4}\right)=-2-\sqrt{3}$.

b. To evaluate the expression $\tan\left(\frac{\pi}{12}\right)$, we first need to write the angle $\frac{\pi}{12}$ as the sum or difference of angles whose tangent is known. Note that

$\frac{\pi}{12} = \frac{4\pi}{12} - \frac{3\pi}{12} = \frac{\pi}{3} - \frac{\pi}{4}$. Therefore, using the **tangent of the difference of two angles formula**, we get

$$\tan\left(\frac{\pi}{12}\right) = \tan\left(\frac{\pi}{3} - \frac{\pi}{4}\right) = \frac{\tan\left(\frac{\pi}{3}\right) - \tan\left(\frac{\pi}{4}\right)}{1 + \tan\left(\frac{\pi}{3}\right)\tan\left(\frac{\pi}{4}\right)}.$$

Try evaluating this expression on your own. When you're finished, check your answer or watch this **interactive video** to see the video solution. ●

You Try It Work through this You Try It problem.

Work Exercises 73–106 in this textbook or in the MyLab Math Study Plan.

OBJECTIVE 4 USING THE SUM AND DIFFERENCE FORMULAS TO VERIFY IDENTITIES

Recall that verifying a trigonometric identity is the process of showing that the expression on one side of the identity can be simplified to be the exact expression that appears on the other side. We typically start with what appears to be the more complicated side of the identity and then attempt to use known identities to transform that side into the trigonometric expression that appears on the other side of the identity. However, if one side of the identity includes an expression of the form $\cos(\alpha \pm \beta)$, $\sin(\alpha \pm \beta)$, or $\tan(\alpha \pm \beta)$, then first substitute one of the sum or difference formulas discussed in this section. Next, use the strategies we developed in Section 8.1 for verifying identities. To review these strategies, read this **summary** of techniques that were introduced in Section 8.1.

Example 7 Verifying a Trigonometric Identity

Verify the trigonometric identity $\sin(2\theta) = 2\sin\theta\cos\theta$.

Solution We start by writing the left-hand side as $\sin(\theta + \theta)$ and then using the **sine of the sum of two angles formula**.

$\sin(2\theta) \stackrel{?}{=} 2\sin\theta\cos\theta$	Write the original trigonometric identity.
$\sin(\theta + \theta) \stackrel{?}{=} 2\sin\theta\cos\theta$	Write 2θ as $\theta + \theta$.
$\sin\theta\cos\theta + \cos\theta\sin\theta \stackrel{?}{=} 2\sin\theta\cos\theta$	Use the sine of the sum of two angles formula.
$2\sin\theta\cos\theta = 2\sin\theta\cos\theta$	Combine like terms.

The left-hand side is now identical to the right-hand side. Therefore, the identity is verified. The identity $\sin(2\theta) = 2\sin\theta\cos\theta$ is called a **double angle identity** and will be discussed in **Section 8.3**. ●

Example 8 Verifying a Trigonometric Identity

Verify the trigonometric identity $\csc(\alpha - \beta) = \frac{\sin\alpha\cos\beta + \cos\alpha\sin\beta}{\sin^2\alpha - \sin^2\beta}$.

Solution Although the right-hand side of the identity may look more complicated, we will start with the left-hand side because we can first use a **reciprocal identity** to rewrite $\csc(\alpha - \beta)$ as $\frac{1}{\sin(\alpha - \beta)}$. Then we can use the **sine of the difference of two angles formula** to rewrite the denominator.

$$\csc(\alpha - \beta) \stackrel{?}{=} \frac{\sin\alpha\cos\beta + \cos\alpha\sin\beta}{\sin^2\alpha - \sin^2\beta}$$ Write the original trigonometric identity.

$$\frac{1}{\sin(\alpha - \beta)} \stackrel{?}{=} \frac{\sin\alpha\cos\beta + \cos\alpha\sin\beta}{\sin^2\alpha - \sin^2\beta}$$ Use the reciprocal identity $\csc(\alpha - \beta) = \frac{1}{\sin(\alpha - \beta)}$.

$$\frac{1}{\sin\alpha\cos\beta - \cos\alpha\sin\beta} \stackrel{?}{=} \frac{\sin\alpha\cos\beta + \cos\alpha\sin\beta}{\sin^2\alpha - \sin^2\beta}$$ Use the sine of the difference of two angles formula.

$$\frac{1}{\sin\alpha\cos\beta - \cos\alpha\sin\beta} \cdot \frac{\sin\alpha\cos\beta + \cos\alpha\sin\beta}{\sin\alpha\cos\beta + \cos\alpha\sin\beta} \stackrel{?}{=} \frac{\sin\alpha\cos\beta + \cos\alpha\sin\beta}{\sin^2\alpha - \sin^2\beta}$$ Multiply the numerator and denominator by the conjugate of the denominator.

$$\frac{\sin\alpha\cos\beta + \cos\alpha\sin\beta}{\sin^2\alpha\cos^2\beta - \cos^2\alpha\sin^2\beta} \stackrel{?}{=} \frac{\sin\alpha\cos\beta + \cos\alpha\sin\beta}{\sin^2\alpha - \sin^2\beta}$$ Multiply the terms in the numerator and denominator.

$$\frac{\sin\alpha\cos\beta + \cos\alpha\sin\beta}{\sin^2\alpha(1 - \sin^2\beta) - (1 - \sin^2\alpha)\sin^2\beta} \stackrel{?}{=} \frac{\sin\alpha\cos\beta + \cos\alpha\sin\beta}{\sin^2\alpha - \sin^2\beta}$$ Substitute $1 - \sin\beta$ for $\cos^2\beta$

$$\frac{\sin\alpha\cos\beta + \cos\alpha\sin\beta}{\sin^2\alpha - \sin^2\alpha\sin^2\beta - \sin^2\beta + \sin^2\alpha\sin^2\beta} \stackrel{?}{=} \frac{\sin\alpha\cos\beta + \cos\alpha\sin\beta}{\sin^2\alpha - \sin^2\beta}$$ Multiply.

$$\frac{\sin\alpha\cos\beta + \cos\alpha\sin\beta}{\sin^2\alpha - \sin^2\beta} = \frac{\sin\alpha\cos\beta + \cos\alpha\sin\beta}{\sin^2\alpha - \sin^2\beta}$$ Combine like terms.

The left-hand side is now identical to the right-hand side. Therefore, the identity is verified. Watch this **video** to see each step of this verification process.

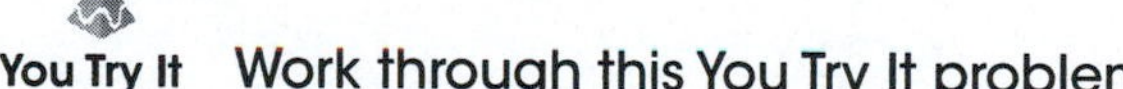
You Try It Work through this You Try It problem.

Work Exercises 107–112 in this textbook or in the MyLab Math Study Plan.

OBJECTIVE 5 USING THE SUM AND DIFFERENCE FORMULAS TO EVALUATE EXPRESSIONS INVOLVING INVERSE TRIGONOMETRIC FUNCTIONS

The sum and difference formulas discussed in this section may be helpful when we encounter trigonometric expressions involving one or more inverse trigonometric functions. Before working through the next example, you may want to review the inverse trigonometric functions that were first introduced in **Section 7.4**. Click on one of the following links for a quick review of these functions.

A Quick Review of Inverse Trigonometric Functions

$y = \sin^{-1}x$ $y = \cos^{-1}x$ $y = \tan^{-1}x$

(eText Screens 8.2-1–8.2-42)

Example 9 Finding the Exact Value of a Trigonometric Expression Involving Inverse Trigonometric Expressions

Find the exact value of the expression $\cos\left(\sin^{-1}\left(\frac{1}{5}\right) + \cos^{-1}\left(-\frac{3}{4}\right)\right)$ without using a calculator.

Solution If we let $\sin^{-1}\left(\frac{1}{5}\right) = \alpha$ and $\cos^{-1}\left(-\frac{3}{4}\right) = \beta$, then we are trying to find the value of the cosine of the sum of α and β.

Remember that if $\mathbf{y = \sin^{-1}x}$, then $\sin y = x$, where $-\frac{\pi}{2} \le y \le \frac{\pi}{2}$. This means that the terminal side of angle y must lie in Quadrant I or Quadrant IV. Because $\sin^{-1}\left(\frac{1}{5}\right) = \alpha$, then $\sin\alpha = \frac{1}{5}$. Because $\sin\alpha = \frac{1}{5} > 0$, we know that α is an angle whose terminal side lies in Quadrant I. Therefore, $\cos\alpha > 0$ because *all* trigonometric functions are positive for angles whose terminal side lies in Quadrant I. To find $\cos\alpha$, we use the Pythagorean identity $\sin^2\alpha + \cos^2\alpha = 1$ or $\cos\alpha = \sqrt{1 - \sin^2\alpha}$. Read these **steps** to see how to obtain $\cos\alpha = \frac{2\sqrt{6}}{5}$.

Similarly, if $\mathbf{y = \cos^{-1}x}$, then $\cos y = x$, where $0 \le y \le \pi$. This means that the terminal side of angle y must lie in Quadrant I or Quadrant II. Because $\cos^{-1}\left(-\frac{3}{4}\right) = \beta$, then $\cos\beta = -\frac{3}{4}$. Because $\cos\beta = -\frac{3}{4} < 0$, we know that β is an angle whose terminal side lies in Quadrant II. Therefore, $\sin\alpha > 0$ because the sine function is positive for all angles whose terminal side lies in Quadrant II. We can use the identity $\sin^2\beta + \cos^2\beta = 1$ or $\sin\beta = \sqrt{1 - \cos^2\beta}$ to get $\sin\beta = \frac{\sqrt{7}}{4}$. To see how we obtained this, read these **steps**.

By the **cosine of the sum of two angles formula** we can find the exact value of $\cos\left(\sin^{-1}\left(\frac{1}{5}\right) + \cos^{-1}\left(-\frac{3}{4}\right)\right)$.

$\cos\left(\sin^{-1}\left(\frac{1}{5}\right) + \cos^{-1}\left(-\frac{3}{4}\right)\right)$	Write the original expression.
$= \cos(\alpha + \beta)$	Write as the cosine of the sum of two angles.
$= \cos\alpha\cos\beta - \sin\alpha\sin\beta$	Use the cosine of the sum of two angles formula.
$= \left(\frac{2\sqrt{6}}{5}\right)\left(-\frac{3}{4}\right) - \left(\frac{1}{5}\right)\left(\frac{\sqrt{7}}{4}\right)$	Substitute the appropriate values.
$= \frac{-6\sqrt{6} - \sqrt{7}}{20}$	Simplify.

You Try It Work through this You Try It problem.

Work Exercises 113–116 in this textbook or in the MyLab Math Study Plan.

8.2 Exercises

Skill Check Exercises

For exercises SCE-1 through SCE-4 simplify each expression.

SCE-1. $\left(\frac{1}{\sqrt{2}}\cdot\frac{1}{2}+\frac{1}{\sqrt{2}}\cdot\frac{\sqrt{3}}{2}\right)$ **SCE-2.** $\left(\frac{-5}{\sqrt{61}}\right)\left(\frac{-4}{\sqrt{41}}\right)+\left(\frac{-6}{\sqrt{61}}\right)\left(\frac{5}{\sqrt{41}}\right)$

SCE-3. $\dfrac{(-1)+(-\sqrt{3})}{1-(-1)(-\sqrt{3})}$ **SCE-4.** $\dfrac{-\frac{1}{\sqrt{3}}+(-1)}{1-\left(-\frac{1}{\sqrt{3}}\right)(-1)}$

SCE-5. Write the expression $\frac{5\pi}{12}$ as the **sum** of two angles that each belong to one of the four special angle families. (The quadrantal family, the $\frac{\pi}{3}$ family, the $\frac{\pi}{4}$ family, or the $\frac{\pi}{6}$ family.)

SCE-6. Write the expression $\frac{7\pi}{12}$ as the **difference** of two angles that each belong to one of the four special angle families. (The quadrantal family, the $\frac{\pi}{3}$ family, the $\frac{\pi}{4}$ family, or the $\frac{\pi}{6}$ family.)

In Exercises 1–8, find the exact value of each trigonometric expression without the use of a calculator. **Note to instructors:** The final answer to these exercise is **not** fully simplified.

SbS 1. $\cos\left(\frac{\pi}{3}+\frac{\pi}{4}\right)$ SbS 2. $\cos(30^\circ+45^\circ)$ SbS 3. $\cos\left(\frac{\pi}{4}-\frac{\pi}{6}\right)$ SbS 4. $\cos(60^\circ-45^\circ)$

SbS 5. $\cos\left(\frac{7\pi}{6}+\frac{5\pi}{4}\right)$ SbS 6. $\cos(210^\circ+135^\circ)$ SbS 7. $\cos\left(\frac{5\pi}{4}-\frac{7\pi}{6}\right)$ SbS 8. $\cos(120^\circ-225^\circ)$

In Exercises 9–20, find the exact value of each trigonometric expression without the use of a calculator. **Note to instructors:** The final answer to these exercise is fully simplified.

SbS 9. $\cos\left(\frac{\pi}{3}+\frac{\pi}{4}\right)$ SbS 10. $\cos(30^\circ+45^\circ)$ SbS 11. $\cos\left(\frac{\pi}{4}-\frac{\pi}{6}\right)$ SbS 12. $\cos(60^\circ-45^\circ)$

SbS 13. $\cos\left(\frac{7\pi}{6}+\frac{5\pi}{4}\right)$ SbS 14. $\cos(210^\circ+135^\circ)$ SbS 15. $\cos\left(\frac{5\pi}{4}-\frac{7\pi}{6}\right)$ SbS 16. $\cos(120^\circ-225^\circ)$

SbS 17. $\cos\left(\frac{11\pi}{12}\right)\cos\left(\frac{7\pi}{12}\right)+\sin\left(\frac{11\pi}{12}\right)\sin\left(\frac{7\pi}{12}\right)$ SbS 18. $\cos(88^\circ)\cos(58^\circ)+\sin(88^\circ)\sin(58^\circ)$

SbS 19. $\cos\left(\frac{4\pi}{9}\right)\cos\left(\frac{2\pi}{9}\right)-\sin\left(\frac{4\pi}{9}\right)\sin\left(\frac{2\pi}{9}\right)$ SbS 20. $\cos(66^\circ)\cos(69^\circ)-\sin(66^\circ)\sin(69^\circ)$

In Exercises 21–23, use the cosine of the **sum** of two angles formula to find the exact value of each trigonometric expression without the use of a calculator. **Note to instructors:** The final answer to these exercises is **not** fully simplified.

SbS 21. $\cos\left(\frac{5\pi}{12}\right)$ SbS 22. $\cos(75°)$ SbS 23. $\sec\left(\frac{11\pi}{12}\right)$

In Exercises 24–26, use the cosine of the **difference** of two angles formula to find the exact value of each trigonometric expression without the use of a calculator. **Note to instructors:** The final answer to these exercises is **not** fully simplified.

SbS 24. $\cos\left(\frac{7\pi}{12}\right)$ SbS 25. $\cos(165°)$ SbS 26. $\sec\left(\frac{13\pi}{12}\right)$

In Exercises 27–29, use the cosine of the **sum** of two angles formula to find the exact value of each trigonometric expression without the use of a calculator. **Note to instructors:** The final answer to these exercises is fully simplified.

SbS 27. $\cos\left(\frac{5\pi}{12}\right)$ SbS 28. $\cos(75°)$ SbS 29. $\sec\left(\frac{11\pi}{12}\right)$

In Exercises 30–32, use the cosine of the **difference** of two angles formula to find the exact value of each trigonometric expression without the use of a calculator. **Note to instructors:** The final answer to these exercises is fully simplified.

SbS 30. $\cos\left(\frac{7\pi}{12}\right)$ SbS 31. $\cos(165°)$ SbS 32. $\sec\left(\frac{13\pi}{12}\right)$

Note to instructors: The final answer to Exercises 33 and 34 is **not** fully simplified.

SbS 33. Suppose that the terminal side of angle α lies in Quadrant I and the terminal side of angle β lies in Quadrant IV. If $\sin \alpha = \frac{8}{17}$ and $\cos \beta = \frac{3}{\sqrt{11}}$, find the exact value of $\cos(\alpha + \beta)$.

SbS 34. Suppose that the terminal side of angle α lies in Quadrant II and the terminal side of angle β lies in Quadrant I. If $\tan \alpha = -\frac{8}{15}$ and $\cos \beta = \frac{3}{4}$, find the exact value of $\cos(\alpha - \beta)$.

Note to instructors: The final answer to Exercises 35 and 36 is fully simplified.

SbS 35. Suppose that the terminal side of angle α lies in Quadrant I and the terminal side of angle β lies in Quadrant IV. If $\sin \alpha = \frac{8}{17}$ and $\cos \beta = \frac{3}{\sqrt{11}}$, find the exact value of $\cos(\alpha + \beta)$.

SbS 36. Suppose that the terminal side of angle α lies in Quadrant II and the terminal side of angle β lies in Quadrant I. If $\tan \alpha = -\frac{8}{15}$ and $\cos \beta = \frac{3}{4}$, find the exact value of $\cos(\alpha - \beta)$.

In Exercises 37–44, find the exact value of each trigonometric expression without the use of a calculator. **Note to instructors:** The final answer to these exercises is **not** fully simplified.

SbS 37. $\sin\left(\frac{\pi}{6} - \frac{\pi}{4}\right)$ SbS 38. $\sin(30° + 45°)$ SbS 39. $\sin\left(\frac{\pi}{4} - \frac{\pi}{6}\right)$ SbS 40. $\sin(60° - 45°)$

SbS 41. $\sin\left(\frac{7\pi}{6} + \frac{5\pi}{4}\right)$ SbS 42. $\sin(210^\circ + 135^\circ)$ SbS 43. $\sin\left(\frac{5\pi}{4} - \frac{7\pi}{6}\right)$ SbS 44. $\sin(120^\circ - 225^\circ)$

In Exercises 45–56, find the exact value of each trigonometric expression without the use of a calculator. **Note to instructors:** The final answer to these exercises is fully simplified.

SbS 45. $\sin\left(\frac{\pi}{6} + \frac{\pi}{4}\right)$ SbS 46. $\sin(30^\circ + 45^\circ)$ SbS 47. $\sin\left(\frac{\pi}{4} - \frac{\pi}{6}\right)$ SbS 48. $\sin(60^\circ - 45^\circ)$

SbS 49. $\sin\left(\frac{7\pi}{6} + \frac{5\pi}{4}\right)$ SbS 50. $\sin(210^\circ + 135^\circ)$ SbS 51. $\sin\left(\frac{5\pi}{4} - \frac{7\pi}{6}\right)$ SbS 52. $\sin(120^\circ - 225^\circ)$

SbS 53. $\sin\left(\frac{\pi}{12}\right)\cos\left(\frac{2\pi}{3}\right) + \cos\left(\frac{\pi}{12}\right)\sin\left(\frac{2\pi}{3}\right)$ SbS 54. $\sin(25^\circ)\cos(5^\circ) + \cos(25^\circ)\sin(5^\circ)$

SbS 55. $\sin\left(\frac{7\pi}{9}\right)\cos\left(\frac{\pi}{9}\right) - \cos\left(\frac{7\pi}{9}\right)\sin\left(\frac{\pi}{9}\right)$ SbS 56. $\sin(275^\circ)\cos(65^\circ) - \cos(275^\circ)\sin(65^\circ)$

In Exercises 57–59, use the sine of the **sum** of two angles formula to find the exact value of each trigonometric expression without the use of a calculator. **Note to instructors:** The final answer to these exercises is **not** fully simplified.

SbS 57. $\sin\left(\frac{5\pi}{12}\right)$ SbS 58. $\sin(75^\circ)$ SbS 59. $\csc\left(\frac{11\pi}{12}\right)$

In Exercises 60–62, use the sine of the **difference** of two angles formula to find the exact value of each trigonometric expression without the use of a calculator. **Note to instructors:** The final answer to these exercises is **not** fully simplified.

SbS 60. $\sin\left(\frac{7\pi}{12}\right)$ SbS 61. $\sin(165^\circ)$ SbS 62. $\csc\left(\frac{13\pi}{12}\right)$

In Exercises 63–65, use the sine of the **sum** of two angles formula to find the exact value of each trigonometric expression without the use of a calculator. **Note to instructors:** The final answer to these exercises is fully simplified.

SbS 63. $\sin\left(\frac{5\pi}{12}\right)$ SbS 64. $\sin(75^\circ)$ SbS 65. $\csc\left(\frac{11\pi}{12}\right)$

In Exercises 66–68, use the sine of the **difference** of two angles formula to find the exact value of each trigonometric expression without the use of a calculator. **Note to instructors:** The final answer to these exercises is fully simplified.

SbS 66. $\sin\left(\frac{7\pi}{12}\right)$ SbS 67. $\sin(165^\circ)$ SbS 68. $\csc\left(\frac{13\pi}{12}\right)$

Note to instructors: The final answer to Exercises 69 and 70 is **not** fully simplified.

SbS 69. Suppose that the terminal side of angle α lies in Quadrant I and the terminal side of angle β lies in Quadrant IV. If $\sin\alpha = \frac{8}{17}$ and $\cos\beta = \frac{3}{\sqrt{11}}$, find the exact value of $\sin(\alpha + \beta)$.

SbS 70. Suppose that α is an angle such that $\tan\alpha = \frac{4}{9}$ and $\sin\alpha < 0$. Also, suppose that β is an angle such that $\cot\beta = -\frac{2}{7}$ and $\sec\beta > 0$. Find the exact value of $\sin(\alpha - \beta)$.

Note to instructors: The final answer to Exercises 71 and 72 is fully simplified.

SbS 71. Suppose that the terminal side of angle α lies in Quadrant I and the terminal side of angle β lies in Quadrant IV. If $\sin\alpha = \frac{8}{17}$ and $\cos\beta = \frac{3}{\sqrt{11}}$, find the exact value of $\sin(\alpha + \beta)$.

SbS 72. Suppose that α is an angle such that $\tan\alpha = \frac{4}{9}$ and $\sin\alpha < 0$. Also, suppose that β is an angle such that $\cot\beta = -\frac{2}{7}$ and $\sec\beta > 0$. Find the exact value of $\sin(\alpha - \beta)$.

In Exercises 73–80, find the exact value of each trigonometric expression without the use of a calculator. **Note to instructors:** The final answer to these exercises is **not** fully simplified.

SbS 73. $\tan\left(\frac{\pi}{6} + \frac{\pi}{4}\right)$

SbS 74. $\tan(60° + 45°)$

SbS 75. $\tan\left(\frac{\pi}{3} - \frac{\pi}{4}\right)$

SbS 76. $\tan(45° - 60°)$

SbS 77. $\tan\left(\frac{5\pi}{3} + \frac{7\pi}{4}\right)$

SbS 78. $\tan(240° + 315°)$

SbS 79. $\tan\left(\frac{7\pi}{6} - \frac{9\pi}{4}\right)$

SbS 80. $\tan(225° - 300°)$

In Exercises 81–90, find the exact value of each trigonometric expression without the use of a calculator. **Note to instructors:** The final answer to these exercises is fully simplified.

SbS 81. $\tan\left(\frac{\pi}{6} + \frac{\pi}{4}\right)$

SbS 82. $\tan(60° + 45°)$

SbS 83. $\tan\left(\frac{\pi}{3} - \frac{\pi}{4}\right)$

SbS 84. $\tan(45° - 60°)$

SbS 85. $\tan\left(\frac{5\pi}{3} + \frac{7\pi}{4}\right)$

SbS 86. $\tan(240° + 315°)$

87. $\tan\left(\frac{7\pi}{6} - \frac{9\pi}{4}\right)$

88. $\tan(225° - 300°)$

89. $\dfrac{\tan 40° + \tan 20°}{1 - \tan 40° \tan 20°}$

90. $\dfrac{\tan\left(\frac{4\pi}{5}\right) - \tan\left(\frac{2\pi}{15}\right)}{1 + \tan\left(\frac{4\pi}{5}\right)\tan\left(\frac{2\pi}{15}\right)}$

In Exercises 91–93, use the tangent of the **sum** of two angles formula to find the exact value of each trigonometric expression without the use of a calculator. **Note to instructors:** The final answer to these exercises is **not** fully simplified.

SbS 91. $\tan\left(\frac{17\pi}{12}\right)$

SbS 92. $\tan(75°)$

SbS 93. $\cot\left(\frac{11\pi}{12}\right)$

In Exercises 94–96, use the tangent of the **difference** of two angles formula to find the exact value of each trigonometric expression without the use of a calculator. **Note to instructors:** The final answer to these exercises is **not** fully simplified.

SbS 94. $\tan\left(\frac{7\pi}{12}\right)$ SbS 95. $\tan(165°)$ SbS 96. $\cot\left(\frac{5\pi}{12}\right)$

In Exercises 97–99, use the tangent of the **sum** of two angles formula to find the exact value of each trigonometric expression without the use of a calculator. **Note to instructors:** The final answer to these exercises is fully simplified.

SbS 97. $\tan\left(\frac{17\pi}{12}\right)$ SbS 98. $\tan(75°)$ SbS 99. $\cot\left(\frac{11\pi}{12}\right)$

In Exercises 100–102, use the tangent of the **difference** of two angles formula to find the exact value of each trigonometric expression without the use of a calculator. **Note to instructors:** The final answer to these exercises is fully simplified.

SbS 100. $\tan\left(\frac{7\pi}{12}\right)$ SbS 101. $\tan(165°)$ SbS 102. $\cot\left(\frac{5\pi}{12}\right)$

Note to instructors: The final answer to Exercises 103 and 104 is **not** fully simplified.

SbS 103. Suppose that the terminal side of angle α lies in Quadrant II and the terminal side of angle β lies in Quadrant III. If $\sin\alpha = \frac{1}{4}$ and $\cos\beta = -\frac{3}{5}$, find the exact value of $\tan(\alpha + \beta)$.

SbS 104. Suppose that the terminal side of angle α lies in Quadrant IV and the terminal side of angle β lies in Quadrant III. If $\cos\alpha = \frac{5}{\sqrt{61}}$ and $\sin\beta = -\frac{7}{\sqrt{91}}$, find the exact value of $\tan(\alpha - \beta)$.

Note to instructors: The final answer to Exercises 105 and 106 is fully simplified.

SbS 105. Suppose that the terminal side of angle α lies in Quadrant II and the terminal side of angle β lies in Quadrant III. If $\sin\alpha = \frac{1}{4}$ and $\cos\beta = -\frac{3}{5}$, find the exact value of $\tan(\alpha + \beta)$.

SbS 106. Suppose that the terminal side of angle α lies in Quadrant IV and the terminal side of angle β lies in Quadrant III. If $\cos\alpha = \frac{5}{\sqrt{61}}$ and $\sin\beta = -\frac{7}{\sqrt{91}}$, find the exact value of $\tan(\alpha - \beta)$.

For Exercises 107–112, verify each identity.

107. $\sin\left(\frac{\pi}{2} + \theta\right) = \cos\theta$

108. $\cos(\alpha + \beta) + \cos(\alpha - \beta) = 2\cos\alpha\cos\beta$

109. $\frac{\cos(\alpha + \beta)}{\sin\alpha\cos\beta} = \cot\alpha - \tan\beta$

110. $\frac{\cos(\alpha - \beta)}{\cos(\alpha + \beta)} = \frac{1 + \tan\alpha\tan\beta}{1 - \tan\alpha\tan\beta}$

111. $\tan(\alpha - \beta) = \frac{\cot\beta - \cot\alpha}{\cot\alpha\cot\beta + 1}$

112. $\csc(\alpha - \beta) = \frac{\csc\alpha\csc\beta}{\cot\beta - \cot\alpha}$

For Exercises 113–116, find the exact value of each expression without the use of a calculator.

113. $\sin\left(\tan^{-1}1 + \cos^{-1}\frac{\sqrt{3}}{2}\right)$

114. $\cos\left(\sin^{-1}\left(\frac{5}{13}\right) + \cos^{-1}\left(-\frac{12}{13}\right)\right)$

115. $\sin\left(\sin^{-1}\left(\frac{1}{6}\right) - \cos^{-1}\left(-\frac{3}{7}\right)\right)$

116. $\tan\left(\sin^{-1}\left(\frac{1}{\sqrt{2}}\right) + \cos^{-1}\left(-\frac{1}{2}\right)\right)$

Brief Exercises

In Exercises 117–128, find the exact value of each trigonometric expression without the use of a calculator. **Note to instructors:** The final answer to these exercise is fully simplified.

117. $\cos\left(\frac{\pi}{3} + \frac{\pi}{4}\right)$

118. $\cos(30° + 45°)$

119. $\cos\left(\frac{\pi}{4} - \frac{\pi}{6}\right)$

120. $\cos(60° - 45°)$

121. $\cos\left(\frac{7\pi}{6} + \frac{5\pi}{4}\right)$

122. $\cos(210° + 135°)$

123. $\cos\left(\frac{5\pi}{4} - \frac{7\pi}{6}\right)$

124. $\cos(120° - 225°)$

125. $\cos\left(\frac{11\pi}{12}\right)\cos\left(\frac{7\pi}{12}\right) + \sin\left(\frac{11\pi}{12}\right)\sin\left(\frac{7\pi}{12}\right)$

126. $\cos(88°)\cos(58°) + \sin(88°)\sin(58°)$

127. $\cos\left(\frac{4\pi}{9}\right)\cos\left(\frac{2\pi}{9}\right) - \sin\left(\frac{4\pi}{9}\right)\sin\left(\frac{2\pi}{9}\right)$

128. $\cos(66°)\cos(69°) - \sin(66°)\sin(69°)$

In Exercises 129–131, use the cosine of the **sum** of two angles formula to find the exact value of each trigonometric expression without the use of a calculator. **Note to instructors:** The final answer to these exercises is fully simplified.

129. $\cos\left(\frac{5\pi}{12}\right)$

130. $\cos(75°)$

131. $\sec\left(\frac{11\pi}{12}\right)$

In Exercises 132–134, use the cosine of the **difference** of two angles formula to find the exact value of each trigonometric expression without the use of a calculator. **Note to instructors:** The final answer to these exercises is fully simplified.

132. $\cos\left(\frac{7\pi}{12}\right)$

133. $\cos(165°)$

134. $\sec\left(\frac{13\pi}{12}\right)$

In Exercises 135–146, find the exact value of each trigonometric expression without the use of a calculator. **Note to instructors:** The final answer to these exercises is fully simplified.

135. $\sin\left(\frac{\pi}{6} + \frac{\pi}{4}\right)$

136. $\sin(30° + 45°)$

137. $\sin\left(\frac{\pi}{4} - \frac{\pi}{6}\right)$

138. $\sin(60° - 45°)$

139. $\sin\left(\frac{7\pi}{6} + \frac{5\pi}{4}\right)$

140. $\sin(210° + 135°)$

141. $\sin\left(\frac{5\pi}{4} - \frac{7\pi}{6}\right)$

142. $\sin(120° - 225°)$

143. $\sin\left(\frac{\pi}{12}\right)\cos\left(\frac{2\pi}{3}\right) + \cos\left(\frac{\pi}{12}\right)\sin\left(\frac{2\pi}{3}\right)$

144. $\sin(25°)\cos(5°) + \cos(25°)\sin(5°)$

145. $\sin\left(\frac{7\pi}{9}\right)\cos\left(\frac{\pi}{9}\right) - \cos\left(\frac{7\pi}{9}\right)\sin\left(\frac{\pi}{9}\right)$

146. $\sin(275°)\cos(65°) - \cos(275°)\sin(65°)$

In Exercises 147–149, use the sine of the **sum** of two angles formula to find the exact value of each trigonometric expression without the use of a calculator. **Note to instructors:** The final answer to these exercises is fully simplified.

147. $\sin\left(\frac{5\pi}{12}\right)$ 148. $\sin(75°)$ 149. $\csc\left(\frac{11\pi}{12}\right)$

In Exercises 150–152, use the sine of the **difference** of two angles formula to find the exact value of each trigonometric expression without the use of a calculator. **Note to instructors:** The final answer to these exercises is fully simplified.

150. $\sin\left(\frac{7\pi}{12}\right)$ 151. $\sin(165°)$ 152. $\csc\left(\frac{13\pi}{12}\right)$

In Exercises 153–162, find the exact value of each trigonometric expression without the use of a calculator. **Note to instructors:** The final answer to these exercises is fully simplified.

153. $\tan\left(\frac{\pi}{6}+\frac{\pi}{4}\right)$ 154. $\tan(60° + 45°)$ 155. $\tan\left(\frac{\pi}{3}-\frac{\pi}{4}\right)$

156. $\tan(45° - 60°)$ 157. $\tan\left(\frac{5\pi}{3}+\frac{7\pi}{4}\right)$ 158. $\tan(240° + 315°)$

159. $\tan\left(\frac{7\pi}{6}-\frac{9\pi}{4}\right)$ 160. $\tan(225° - 300°)$ 161. $\frac{\tan 40° + \tan 20°}{1 - \tan 40° \tan 20°}$

162. $\frac{\tan\left(\frac{4\pi}{5}\right) - \tan\left(\frac{2\pi}{15}\right)}{1 + \tan\left(\frac{4\pi}{5}\right)\tan\left(\frac{2\pi}{15}\right)}$

In Exercises 163–165, use the tangent of the **sum** of two angles formula to find the exact value of each trigonometric expression without the use of a calculator. **Note to instructors:** The final answer to these exercises is fully simplified.

163. $\tan\left(\frac{17\pi}{12}\right)$ 164. $\tan(75°)$ 165. $\cot\left(\frac{11\pi}{12}\right)$

In Exercises 166–168, use the tangent of the **difference** of two angles formula to find the exact value of each trigonometric expression without the use of a calculator. **Note to instructors:** The final answer to these exercises is fully simplified.

166. $\tan\left(\frac{7\pi}{12}\right)$ 167. $\tan(165°)$ 168. $\cot\left(\frac{5\pi}{12}\right)$

Note to instructors: The final answer to Exercises 169–174 is **not** fully simplified.

169. Suppose that the terminal side of angle α lies in Quadrant I and the terminal side of angle β lies in Quadrant IV. If $\sin\alpha = \frac{8}{17}$ and $\cos\beta = \frac{3}{\sqrt{11}}$, find the exact value of $\cos(\alpha + \beta)$.

170. Suppose that the terminal side of angle α lies in Quadrant II and the terminal side of angle β lies in Quadrant I. If $\tan \alpha = -\frac{8}{15}$ and $\cos \beta = \frac{3}{4}$, find the exact value of $\cos(\alpha - \beta)$.

171. Suppose that the terminal side of angle α lies in Quadrant I and the terminal side of angle β lies in Quadrant IV. If $\sin \alpha = \frac{8}{17}$ and $\cos \beta = \frac{3}{\sqrt{11}}$, find the exact value of $\sin(\alpha + \beta)$.

172. Suppose that α is an angle such that $\tan \alpha = \frac{4}{9}$ and $\sin \alpha < 0$. Also, suppose that β is an angle such that $\cot \beta = -\frac{2}{7}$ and $\sec \beta > 0$. Find the exact value of $\sin(\alpha - \beta)$.

173. Suppose that the terminal side of angle α lies in Quadrant II and the terminal side of angle β lies in Quadrant III. If $\sin \alpha = \frac{1}{4}$ and $\cos \beta = -\frac{3}{5}$, find the exact value of $\tan(\alpha + \beta)$.

174. Suppose that the terminal side of angle α lies in Quadrant IV and the terminal side of angle β lies in Quadrant III. If $\cos \alpha = \frac{5}{\sqrt{61}}$ and $\sin \beta = -\frac{7}{\sqrt{91}}$, find the exact value of $\tan(\alpha - \beta)$.

Note to instructors: The final answer to Exercises 175–180 is fully simplified.

175. Suppose that the terminal side of angle α lies in Quadrant I and the terminal side of angle β lies in Quadrant IV. If $\sin \alpha = \frac{8}{17}$ and $\cos \beta = \frac{3}{\sqrt{11}}$, find the exact value of $\cos(\alpha + \beta)$.

176. Suppose that the terminal side of angle α lies in Quadrant II and the terminal side of angle β lies in Quadrant I. If $\tan \alpha = -\frac{8}{15}$ and $\cos \beta = \frac{3}{4}$, find the exact value of $\cos(\alpha - \beta)$.

177. Suppose that the terminal side of angle α lies in Quadrant I and the terminal side of angle β lies in Quadrant IV. If $\sin \alpha = \frac{8}{17}$ and $\cos \beta = \frac{3}{\sqrt{11}}$, find the exact value of $\sin(\alpha + \beta)$.

178. Suppose that α is an angle such that $\tan \alpha = \frac{4}{9}$ and $\sin \alpha < 0$. Also, suppose that β is an angle such that $\cot \beta = -\frac{2}{7}$ and $\sec \beta > 0$. Find the exact value of $\sin(\alpha - \beta)$.

179. Suppose that the terminal side of angle α lies in Quadrant II and the terminal side of angle β lies in Quadrant III. If $\sin \alpha = \frac{1}{4}$ and $\cos \beta = -\frac{3}{5}$, find the exact value of $\tan(\alpha + \beta)$.

180. Suppose that the terminal side of angle α lies in Quadrant IV and the terminal side of angle β lies in Quadrant III. If $\cos \alpha = \frac{5}{\sqrt{61}}$ and $\sin \beta = -\frac{7}{\sqrt{91}}$, find the exact value of $\tan(\alpha - \beta)$.

8.3 The Double-Angle and Half-Angle Formulas

THINGS TO KNOW

Before working through this section, be sure that you are familiar with the following concepts:

VIDEO ANIMATION INTERACTIVE

You Try It

1. Evaluating Trigonometric Functions of Angles Belonging to the $\frac{\pi}{3}$, $\frac{\pi}{6}$, or $\frac{\pi}{4}$ Families (Section 6.5)

You Try It

2. Finding the Exact and Approximate Values of an Inverse Sine Expression (Section 7.4)

You Try It

3. Finding the Exact and Approximate Values of an Inverse Cosine Expression (Section 7.4)

You Try It

4. Finding the Exact and Approximate Values of an Inverse Tangent Expression (Section 7.4)

You Try It

5. Understanding the Sum and Difference Formulas for the Sine Function (Section 8.2)

You Try It

6. Understanding the Sum and Difference Formulas for the Cosine Function (Section 8.2)

OBJECTIVES

1 Understanding the Double-Angle Formulas

2 Understanding the Power Reduction Formulas

3 Understanding the Half-Angle Formulas

4 Using the Double-Angle, Power Reduction, and Half-Angle Formulas to Verify Identities

5 Using the Double-Angle and Half-Angle Formulas to Evaluate Expressions Involving Inverse Trigonometric Functions

SECTION 8.3 EXERCISES

OBJECTIVE 1 UNDERSTANDING THE DOUBLE-ANGLE FORMULAS

In Section 8.2 we introduced the sum and difference formulas for sine, cosine, and tangent. We can use these formulas to derive the following **double-angle formulas**.

Double-Angle Formulas

$$\sin 2\theta = 2\sin\theta\cos\theta \qquad \cos 2\theta = \cos^2\theta - \sin^2\theta \qquad \tan 2\theta = \frac{2\tan\theta}{1-\tan^2\theta}$$

Work through this **interactive video** to see how to use the **sum formulas** to derive each of the three double-angle formulas. Using a form of the Pythagorean identity $\sin^2\theta + \cos^2\theta = 1$, we can rewrite the formula for $\cos 2\theta$ in two additional ways.

$\cos 2\theta = \cos^2\theta - \sin^2\theta$	Write the double-angle formula for cosine.
$= (1 - \sin^2\theta) - \sin^2\theta$	Use the identity $\cos^2\theta = 1 - \sin^2\theta$.
$= 1 - 2\sin^2\theta$	Simplify.
$\cos 2\theta = \cos^2\theta - \sin^2\theta$	Write the double-angle formula for cosine.
$= \cos^2\theta - (1 - \cos^2\theta)$	Use the identity $\sin^2\theta = 1 - \cos^2\theta$.
$= \cos^2\theta - 1 + \cos^2\theta$	Distribute.
$= 2\cos^2\theta - 1$	Simplify.

We now state all three forms of the double-angle formula for cosine.

The Double-Angle Formulas for Cosine

$$\cos 2\theta = \cos^2\theta - \sin^2\theta$$
$$\cos 2\theta = 1 - 2\sin^2\theta$$
$$\cos 2\theta = 2\cos^2\theta - 1$$

Example 1 Evaluating Trigonometric Expressions Using Double-Angle Formulas

Rewrite each expression as the sine, cosine, or tangent of a double angle. Then evaluate the expression without using a calculator.

a. $\cos^2\left(\frac{11\pi}{12}\right) - \sin^2\left(\frac{11\pi}{12}\right)$

b. $2\sin 67.5^\circ \cos 67.5^\circ$

c. $\dfrac{2\tan\left(-\frac{\pi}{8}\right)}{1 - \tan^2\left(-\frac{\pi}{8}\right)}$

d. $2\cos^2 105^\circ - 1$

Solution

a. To evaluate the expression $\cos^2\left(\frac{11\pi}{12}\right) - \sin^2\left(\frac{11\pi}{12}\right)$, we should recognize that this expression represents the right-hand side of the double-angle formula $\cos 2\theta = \cos^2\theta - \sin^2\theta$, where $\theta = \frac{11\pi}{12}$.

$\cos^2\left(\frac{11\pi}{12}\right) - \sin^2\left(\frac{11\pi}{12}\right)$	Write the original expression.
$= \cos\left(2\cdot\frac{11\pi}{12}\right)$	Use the double-angle formula $\cos 2\theta = \cos^2\theta - \sin^2\theta$.
$= \cos\left(\frac{11\pi}{6}\right)$	Simplify.
$= \frac{\sqrt{3}}{2}$	Evaluate the trigonometric function.

Therefore, $\cos^2\left(\frac{11\pi}{12}\right) - \sin^2\left(\frac{11\pi}{12}\right) = \frac{\sqrt{3}}{2}$.

Try working parts b–d on your own using one of the double-angle formulas. When finished, check your **solutions** or work through this **interactive video** to see each step worked out in detail.

You Try It Work through this You Try It problem.

Work Exercises 1–10 in this textbook or in the MyLab Math Study Plan.

Example 2 Evaluating Trigonometric Expressions Using Double-Angle Formulas

Suppose that the terminal side of angle θ lies in Quadrant II such that $\sin\theta = \frac{5}{7}$. Find the values of $\sin 2\theta$, $\cos 2\theta$, and $\tan 2\theta$.

Solution We start by sketching an angle whose terminal side lies in Quadrant II such that $\sin\theta = \frac{5}{7} = \frac{y}{r}$. See Figure 3.

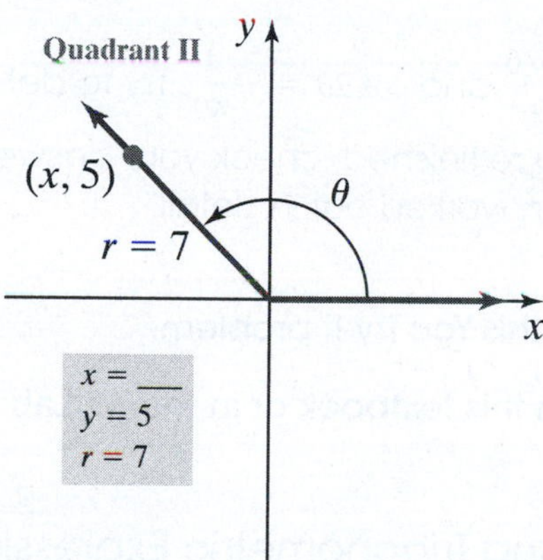

Figure 3 Angle θ and the point $(x, 5)$ located in Quadrant II, a distance of $r = 7$ units from the origin.

Because the point $(x, 5)$ lies in Quadrant II, we know that the value of x must be negative. We can use the relationship $x^2 + y^2 = r^2$ to determine the value of x. Read through these **steps** to see how we determine that $x = -2\sqrt{6}$. See Figure 4.

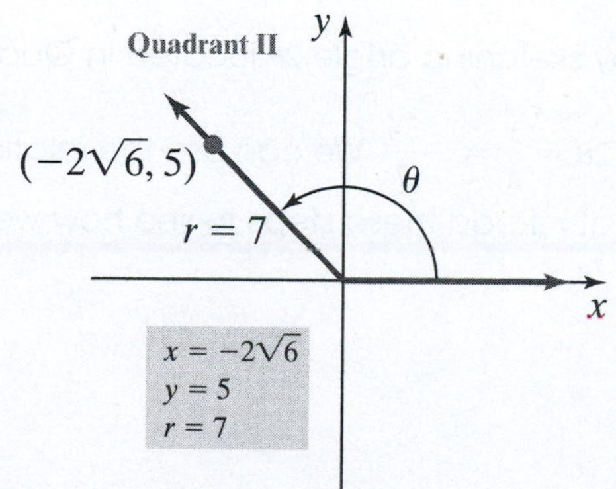

Figure 4 Angle θ lies in Quadrant II and passes through the point $(-2\sqrt{6}, 5)$.

We now summarize the information gathered thus far about angle θ.

$$\sin\theta = \frac{5}{7} \qquad \cos\theta = \frac{x}{r} = -\frac{2\sqrt{6}}{7} \qquad \tan\theta = \frac{y}{x} = \frac{5}{-2\sqrt{6}} = -\frac{5}{2\sqrt{6}}$$

Use the double-angle formulas to determine the values of $\sin 2\theta$, $\cos 2\theta$, and $\tan 2\theta$.

$\sin 2\theta = 2\sin\theta\cos\theta$	Write the double-angle formula for sine.
$= 2\left(\frac{5}{7}\right)\left(-\frac{2\sqrt{6}}{7}\right)$	Substitute $\sin\theta = \frac{5}{7}$ and $\cos\theta = -\frac{2\sqrt{6}}{7}$.
$= -\frac{20\sqrt{6}}{49}$	Multiply.
$\cos 2\theta = \cos^2\theta - \sin^2\theta$	Write a double-angle formula for cosine.
$= \left(-\frac{2\sqrt{6}}{7}\right)^2 - \left(\frac{5}{7}\right)^2$	Substitute $\sin\theta = \frac{5}{7}$ and $\cos\theta = -\frac{2\sqrt{6}}{7}$.
$= \frac{24}{49} - \frac{25}{49}$	Square both terms.
$= -\frac{1}{49}$	Subtract.

Therefore, $\sin 2\theta = -\frac{20\sqrt{6}}{49}$ and $\cos 2\theta = -\frac{1}{49}$. Try to determine the value of $\tan 2\theta$ on your own. When you're finished, check your **answer** or watch this **interactive video** to see the solution worked out in detail.

You Try It Work through this You Try It problem.

Work Exercises 11–34 in this textbook or in the MyLab Math Study Plan.

Example 3 Evaluating Trigonometric Expressions Using Double-Angle Formulas

If $\tan 2\theta = -\frac{24}{7}$ for $\frac{3\pi}{2} < 2\theta < 2\pi$, then find the values of $\sin\theta$, $\cos\theta$, and $\tan\theta$.

Solution We start by sketching angle 2θ located in Quadrant IV between $\frac{3\pi}{2}$ and 2π such that $\tan 2\theta = \frac{y}{x} = -\frac{24}{7}$. We can use the relationship $x^2 + y^2 = r^2$ to determine the value of r. Read these **steps** to see how we determine that $r = 25$. See Figure 5.

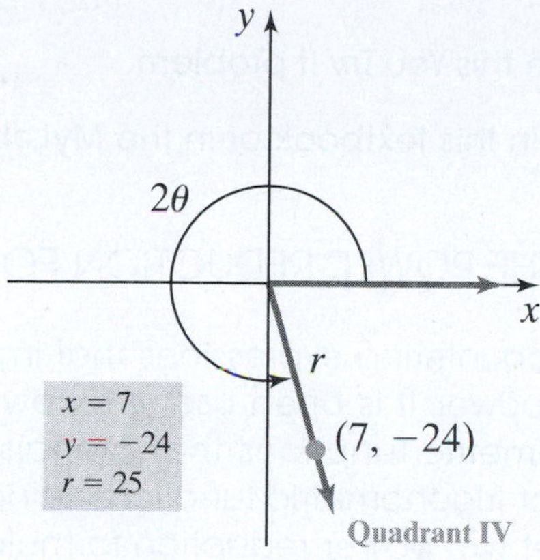

Figure 5 This illustrates angle 2θ and the point $(7, -24)$ located in Quadrant IV, a distance of $r = 25$ from the origin.

We now summarize the information gathered thus far about angle 2θ.

$$\sin 2\theta = \frac{y}{r} = -\frac{24}{25} \qquad \cos 2\theta = \frac{x}{r} = \frac{7}{25} \qquad \tan 2\theta = -\frac{24}{7}$$

Before we determine the value of $\sin\theta$, $\cos\theta$, and $\tan\theta$, we need to determine the quadrant in which the terminal side of θ lies. Dividing each part of this inequality $\frac{3\pi}{2} < 2\theta < 2\pi$ by 2, we get $\frac{3\pi}{4} < \theta < \pi$. Thus, the terminal side of angle θ lies in Quadrant II.

We can determine the value of $\sin\theta$ by using a double-angle formula. We choose $\cos 2\theta = 1 - 2\sin^2\theta$ because this formula contains only $\cos 2\theta$ and $\sin\theta$ and allows us to solve easily for $\sin\theta$.

$\cos 2\theta = 1 - 2\sin^2\theta$ Start with a double-angle formula for cosine that involves $\sin\theta$.

$\frac{7}{25} = 1 - 2\sin^2\theta$ Substitute $\cos 2\theta = \frac{7}{25}$.

$\frac{7}{25} - 1 = -2\sin^2\theta$ Subtract 1 from both sides.

$\frac{7}{25} - \frac{25}{25} = -2\sin^2\theta$ Use 25 as the LCD.

$-\frac{18}{25} = -2\sin^2\theta$ Subtract.

$\frac{9}{25} = \sin^2\theta$ Divide both sides by -2.

$\pm\frac{3}{5} = \sin\theta$ Take square roots of both sides.

Because the terminal side of θ lies in Quadrant II, we know that $\sin\theta$ must be positive. Therefore, $\sin\theta = \frac{3}{5}$. We can now determine the values of $\cos\theta$ and $\tan\theta$.

 Watch this **video** to verify that $\cos\theta = -\frac{4}{5}$ and $\tan\theta = -\frac{3}{4}$.

(eText Screens 8.3-1–8.3-51)

You Try It Work through this You Try It problem.

Work Exercises 35–38 in this textbook or in the MyLab Math Study Plan.

OBJECTIVE 2 UNDERSTANDING THE POWER REDUCTION FORMULAS

In calculus, when encountering expressions that involve trigonometric functions that are raised to a power, it is often useful to rewrite the expression so that it involves only trigonometric functions that are raised to a power of 1. We can rewrite *even* powers of trigonometric functions using the following **power reduction formulas.** The first two power reduction formulas are alternate forms of the double-angle formulas for cosine, and the third formula is the quotient of the first two.

The Power Reduction Formulas

$$\sin^2\theta = \frac{1-\cos 2\theta}{2} \qquad \cos^2\theta = \frac{1+\cos 2\theta}{2} \qquad \tan^2\theta = \frac{1-\cos 2\theta}{1+\cos 2\theta}$$

Work through this **interactive video** to see how we can use the double-angle formulas to establish the three power reduction formulas.

Example 4 Reducing the Power of a Trigonometric Function Using a Power Reduction Formula

Rewrite the function $f(x) = 6\sin^4 x$ as an equivalent function containing only cosine terms raised to a power of 1.

Solution Start by rewriting $6\sin^4 x$ as $6(\sin^2 x)^2$, and then make the substitution $\sin^2 x = \frac{1-\cos 2x}{2}$.

$6\sin^4 x = 6(\sin^2 x)^2$ — Write $\sin^4 x$ as $(\sin^2 x)^2$.

$= 6\left(\frac{1-\cos 2x}{2}\right)^2$ — Use the power reduction formula $\sin^2 x = \frac{1-\cos 2x}{2}$.

$= 6\left(\frac{1-2\cos 2x+\cos^2 2x}{4}\right)$ — Square the numerator and denominator.

$= \frac{3}{2}(1-2\cos 2x+\cos^2 2x)$ — Write $\frac{6}{4}$ as $\frac{3}{2}$.

$= \frac{3}{2} - 3\cos 2x + \frac{3}{2}\cos^2 2x$ — Distribute.

$= \frac{3}{2} - 3\cos 2x + \frac{3}{2}\left(\frac{1+\cos 2(2x)}{2}\right)$ — Use the power reduction formula $\cos^2\theta = \frac{1+\cos 2\theta}{2}$, where $\theta = 2x$.

$= \frac{3}{2} - 3\cos 2x + \frac{3}{4}(1 + \cos 4x)$ Simplify.

$= \frac{3}{2} - 3\cos 2x + \frac{3}{4} + \frac{3}{4}\cos 4x$ Distribute.

$= \frac{9}{4} - 3\cos 2x + \frac{3}{4}\cos 4x$ Combine like terms.

Therefore, we can rewrite $f(x) = 6\sin^4 x$ as $f(x) = \frac{9}{4} - 3\cos 2x + \frac{3}{4}\cos 4x$.

You Try It Work through this You Try It problem.

Work Exercises 39–43 in this textbook or in the MyLab Math Study Plan.

OBJECTIVE 3 UNDERSTANDING THE HALF-ANGLE FORMULAS

Starting with the power reduction formula $\sin^2\theta = \frac{1 - \cos 2\theta}{2}$, we can take the square root of both sides of the equation to get $\sin\theta = \sqrt{\frac{1 - \cos 2\theta}{2}}$ or $\sin\theta = -\sqrt{\frac{1 - \cos 2\theta}{2}}$. The choice of which root to use depends on the quadrant in which the terminal side of θ lies Similarly, we can determine that $\cos\theta = \sqrt{\frac{1 + \cos 2\theta}{2}}$ or $\cos\theta = -\sqrt{\frac{1 + \cos 2\theta}{2}}$. Using the substitution $\theta = \frac{\alpha}{2}$, we obtain the following **half-angle formulas** for sine and cosine.

The Half-Angle Formulas for Sine and Cosine

$\sin\left(\frac{\alpha}{2}\right) = \sqrt{\frac{1 - \cos\alpha}{2}}$ for $\frac{\alpha}{2}$ in Quadrant I or Quadrant II.

$\sin\left(\frac{\alpha}{2}\right) = -\sqrt{\frac{1 - \cos\alpha}{2}}$ for $\frac{\alpha}{2}$ in Quadrant III or Quadrant IV.

$\cos\left(\frac{\alpha}{2}\right) = \sqrt{\frac{1 + \cos\alpha}{2}}$ for $\frac{\alpha}{2}$ in Quadrant I or Quadrant IV.

$\cos\left(\frac{\alpha}{2}\right) = -\sqrt{\frac{1 + \cos\alpha}{2}}$ for $\frac{\alpha}{2}$ in Quadrant II or Quadrant III.

Note that the choice of which form of $\sin\left(\frac{\alpha}{2}\right)$ and $\cos\left(\frac{\alpha}{2}\right)$ to use depends on the quadrant in which the terminal side of angle $\frac{\alpha}{2}$ lies. We use the acronym ASTC first discussed in **Section 6.5** to determine the signs of trigonometric functions for angles lying in each of the four quadrants. Refer to this **figure** for a quick review of the signs of the trigonometric functions.

To derive the half-angle formula for tangent, we start with the power reduction formula $\tan^2\theta = \frac{1 - \cos 2\theta}{1 + \cos 2\theta}$ and take the square root of both sides to get $\tan\theta = \sqrt{\frac{1 - \cos 2\theta}{1 + \cos 2\theta}}$ or $\tan\theta = -\sqrt{\frac{1 - \cos 2\theta}{1 + \cos 2\theta}}$. Again, the choice of which root to use depends on the quadrant in which the terminal side of θ lies. Using the substitution $\theta = \frac{\alpha}{2}$, we obtain the half-angle formulas $\tan\left(\frac{\alpha}{2}\right) = \sqrt{\frac{1 - \cos\alpha}{1 + \cos\alpha}}$ or $\tan\left(\frac{\alpha}{2}\right) = -\sqrt{\frac{1 - \cos\alpha}{1 + \cos\alpha}}$. We can derive two other formulas for $\tan\left(\frac{\alpha}{2}\right)$ that do not include any radicals and can be used regardless of the quadrant in which the terminal side of $\frac{\alpha}{2}$ lies. Watch this **interactive video** to see how to derive the half-angle formulas for tangent listed below.

The Half-Angle Formulas for Tangent

$$\tan\left(\frac{\alpha}{2}\right) = \sqrt{\frac{1 - \cos\alpha}{1 + \cos\alpha}} \quad \text{for } \frac{\alpha}{2} \text{ in Quadrant I or Quadrant III.}$$

$$\tan\left(\frac{\alpha}{2}\right) = -\sqrt{\frac{1 - \cos\alpha}{1 + \cos\alpha}} \quad \text{for } \frac{\alpha}{2} \text{ in Quadrant II or Quadrant IV.}$$

$$\tan\left(\frac{\alpha}{2}\right) = \frac{1 - \cos\alpha}{\sin\alpha} \quad \text{for } \frac{\alpha}{2} \text{ in any quadrant.}$$

$$\tan\left(\frac{\alpha}{2}\right) = \frac{\sin\alpha}{1 + \cos\alpha} \quad \text{for } \frac{\alpha}{2} \text{ in any quadrant.}$$

Example 5 Evaluating Trigonometric Expressions Using Half-Angle Formulas

Use a half-angle formula to evaluate each expression without using a calculator.

a. $\sin\left(-\frac{7\pi}{8}\right)$ **b.** $\cos(-15°)$ **c.** $\tan\left(\frac{11\pi}{12}\right)$

Solution

a. To evaluate the expression $\sin\left(-\frac{7\pi}{8}\right)$ using a half-angle formula, we first identify $\frac{\alpha}{2}$ and α.

$$\frac{\alpha}{2} = -\frac{7\pi}{8} \quad \text{Let } \frac{\alpha}{2} \text{ represent the given angle.}$$

$$2\cdot\frac{\alpha}{2} = 2\cdot\left(-\frac{7\pi}{8}\right) \quad \text{Multiply by 2.}$$

$$\alpha = -\frac{7\pi}{4} \quad \text{Simplify.}$$

The terminal side of $\frac{\alpha}{2} = -\frac{7\pi}{8}$ lies in Quadrant III. See Figure 6.

The sine function is negative for all angles lying in Quadrant III, which indicates that $\sin\left(-\frac{7\pi}{8}\right)$ is a negative value. Thus, we must use the half-angle formula $\sin\left(\frac{\alpha}{2}\right) = -\sqrt{\frac{1-\cos\alpha}{2}}$ with $\frac{\alpha}{2} = -\frac{7\pi}{8}$ and $\alpha = -\frac{7\pi}{4}$.

$$\sin\left(\frac{\alpha}{2}\right) = -\sqrt{\frac{1-\cos\alpha}{2}}$$ Write the half-angle formula for sine using the negative square root.

$$\sin\left(-\frac{7\pi}{8}\right) = -\sqrt{\frac{1-\cos\left(-\frac{7\pi}{4}\right)}{2}}$$ Substitute $\frac{\alpha}{2} = -\frac{7\pi}{8}$ and $\alpha = -\frac{7\pi}{4}$.

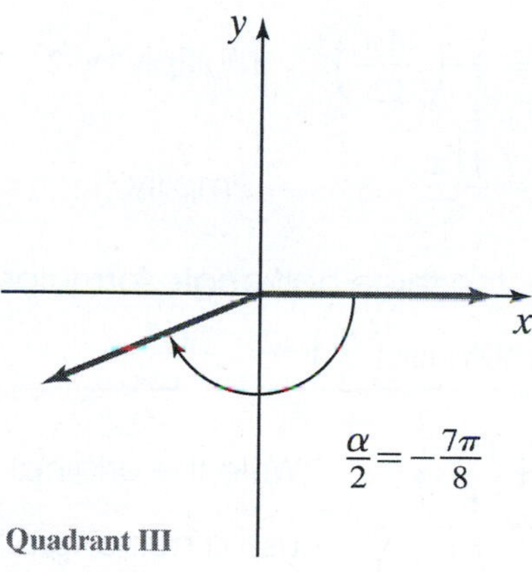

Figure 6

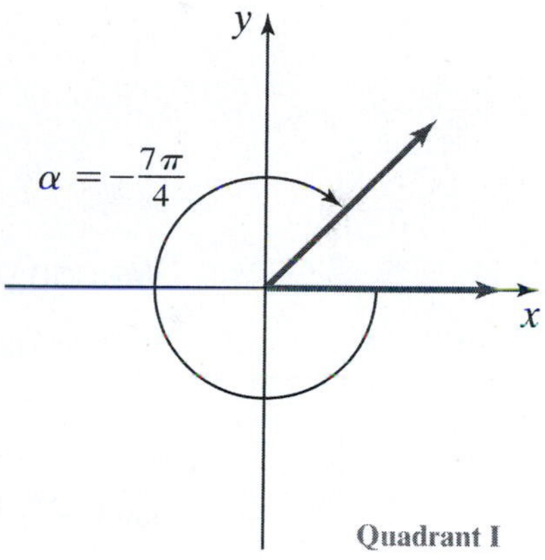

Figure 7

The terminal side of $\alpha = -\frac{7\pi}{4}$ lies in Quadrant I. See Figure 7. The cosine of any angle whose terminal side lies in Quadrant I is *positive*. When we evaluate the expression $\cos\left(-\frac{7\pi}{4}\right)$, we get $\cos\left(-\frac{7\pi}{4}\right) = \frac{1}{\sqrt{2}}$.

$$\sin\left(-\frac{7\pi}{8}\right) = -\sqrt{\frac{1-\cos\left(-\frac{7\pi}{4}\right)}{2}}$$

$$= -\sqrt{\frac{1-\frac{1}{\sqrt{2}}}{2}}$$ Evaluate: $\cos\left(-\frac{7\pi}{4}\right) = \frac{1}{\sqrt{2}}$.

$$= -\sqrt{\frac{\frac{\sqrt{2}-1}{\sqrt{2}}}{2}}$$ Combine the expressions in the numerator.

$$= -\sqrt{\frac{\sqrt{2}-1}{2\sqrt{2}}}$$ Simplify: $-\sqrt{\frac{\frac{\sqrt{2}-1}{\sqrt{2}}}{2}} = -\sqrt{\frac{\sqrt{2}-1}{\sqrt{2}} \div 2} = -\sqrt{\frac{\sqrt{2}-1}{\sqrt{2}} \cdot \frac{1}{2}} = -\sqrt{\frac{\sqrt{2}-1}{2\sqrt{2}}}$.

We see that $\sin\left(-\frac{7\pi}{8}\right) = -\sqrt{\frac{\sqrt{2}-1}{2\sqrt{2}}}$. Watch this **interactive video** to see every step of this solution. Note that if we choose to rationalize the expression under the radical (which is not required), we can obtain an equivalent solution of $\sin\left(-\frac{7\pi}{8}\right) = -\frac{\sqrt{2-\sqrt{2}}}{2}$.

b. Try to evaluate the expression $\cos(-15°)$ on your own. Watch this **interactive video** to check your work.

c. To evaluate the expression $\tan\left(\frac{11\pi}{12}\right)$ using a half-angle formula, we first identify $\frac{\alpha}{2}$ and α.

$$\frac{\alpha}{2} = \frac{11\pi}{12} \qquad \text{Let } \frac{\alpha}{2} \text{ represent the given angle.}$$

$$2\cdot\frac{\alpha}{2} = 2\cdot\left(\frac{11\pi}{12}\right) \qquad \text{Multiply by 2.}$$

$$\alpha = \frac{11\pi}{6} \qquad \text{Simplify.}$$

We can use any of the three **half-angle formulas for tangent**. For this example, we will use the formula $\tan\left(\frac{\alpha}{2}\right) = \frac{\sin\alpha}{1+\cos\alpha}$.

$$\tan\left(\frac{11\pi}{12}\right) = \frac{\sin\left(\frac{11\pi}{6}\right)}{1+\cos\left(\frac{11\pi}{6}\right)} \qquad \text{Write the original expression then use a half-angle formula for } \tan\left(\frac{\alpha}{2}\right).$$

$$= \frac{-\frac{1}{2}}{1+\frac{\sqrt{3}}{2}} \qquad \text{Evaluate: } \sin\left(\frac{11\pi}{6}\right) = -\frac{1}{2} \text{ and } \cos\left(\frac{11\pi}{6}\right) = \frac{\sqrt{3}}{2}.$$

$$= \frac{-\frac{1}{2}}{\frac{2+\sqrt{3}}{2}} \qquad \text{Combine the expressions in the denominator.}$$

$$= -\frac{1}{2} \div \frac{2+\sqrt{3}}{2} \qquad \text{Rewrite as the division of two expressions.}$$

$$= -\frac{1}{2}\cdot\frac{2}{2+\sqrt{3}} \qquad \text{Multiply the first expression by the reciprocal of the second expression.}$$

$$= -\frac{1}{\cancel{2}}\cdot\frac{\cancel{2}}{2+\sqrt{3}} \qquad \text{Cancel common factors.}$$

$$= -\frac{1}{2+\sqrt{3}} \qquad \text{Simplify.}$$

Therefore, $\tan\left(\frac{11\pi}{12}\right) = -\frac{1}{2+\sqrt{3}}$. You may wish to work through this **interactive video** to see each step of this solution. Note that if we choose to rationalize this answer (which is not required), we can obtain an equivalent solution $\tan\left(\frac{11\pi}{12}\right) = \sqrt{3} - 2.$

You Try It Work through this You Try It problem.

Work Exercises 44–59 in this textbook or in the MyLab Math Study Plan.

Example 6 Evaluating Trigonometric Expressions Using Half-Angle Formulas

Suppose that $\csc\alpha = \frac{8}{3}$ such that $\frac{\pi}{2} < \alpha < \pi$.

Find the values of $\sin\left(\frac{\alpha}{2}\right)$, $\cos\left(\frac{\alpha}{2}\right)$, and $\tan\left(\frac{\alpha}{2}\right)$.

Solution To determine the values of $\sin\left(\frac{\alpha}{2}\right)$, $\cos\left(\frac{\alpha}{2}\right)$, and $\tan\left(\frac{\alpha}{2}\right)$, we must first find the values of $\sin\alpha$ and $\cos\alpha$. The value of $\sin\alpha$ is the reciprocal of $\csc\alpha = \frac{8}{3}$. Thus, $\sin\alpha = \frac{3}{8}$. To determine the value of $\cos\alpha$, we start by sketching an angle whose terminal side lies in Quadrant II and that illustrates $\csc\alpha = \frac{8}{3} = \frac{r}{y}$. See Figure 8.

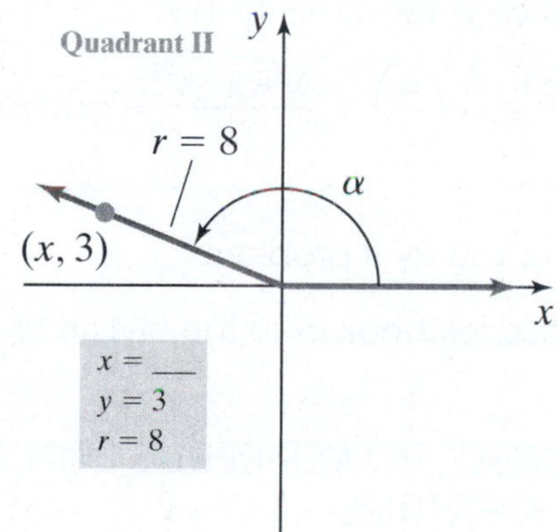

Figure 8 This illustrates angle α and the point $(x, 3)$, located in Quadrant II a distance of $r = 8$ units from the origin.

The point $(x, 3)$ lies in Quadrant II; thus we know that the value of x must be negative. We can use the relationship $x^2 + y^2 = r^2$ to determine the value of x. Read these **steps** to see how we determine that $x = -\sqrt{55}$. See Figure 9.

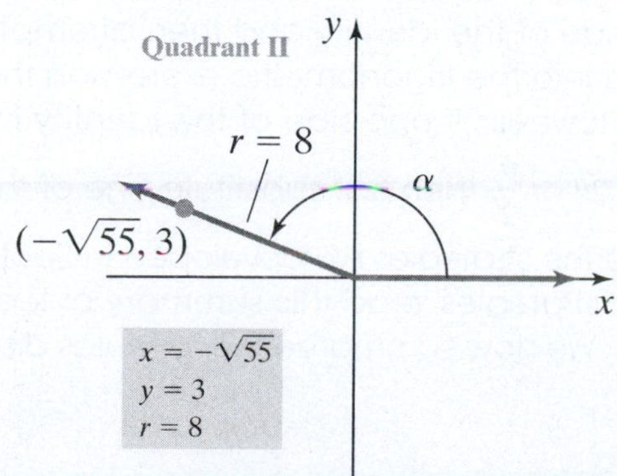

Figure 9 This illustrates angle α and the point $(-\sqrt{55}, 3)$ located in Quadrant II, a distance of $r = 8$ from the origin.

(eText Screens 8.3-1–8.3-51)

We now summarize the information gathered thus far about angle α.

$$\sin\alpha = \frac{y}{r} = \frac{3}{8} \qquad \cos\alpha = \frac{x}{r} = -\frac{\sqrt{55}}{8}$$

Now we consider the angle $\frac{\alpha}{2}$. We were given that $\frac{\pi}{2} < \alpha < \pi$.

Dividing each part of this inequality by 2, we get

$$\frac{\pi}{4} < \frac{\alpha}{2} < \frac{\pi}{2}.$$

This indicates that the terminal side of $\frac{\alpha}{2}$ lies in Quadrant I. Therefore, the values of $\sin\left(\frac{\alpha}{2}\right)$, $\cos\left(\frac{\alpha}{2}\right)$, and $\tan\left(\frac{\alpha}{2}\right)$ must all be positive because all trigonometric functions are positive for any angle whose terminal side lies in Quadrant I. We can now use the formulas

$$\sin\left(\frac{\alpha}{2}\right) = \sqrt{\frac{1-\cos\alpha}{2}},\ \cos\left(\frac{\alpha}{2}\right) = \sqrt{\frac{1+\cos\alpha}{2}},\ \text{and } \tan\left(\frac{\alpha}{2}\right) = \frac{\sin\alpha}{1+\cos\alpha}$$

with $\sin\alpha = \frac{3}{8}$ and $\cos\alpha = \frac{-\sqrt{55}}{8}$ to determine the values of each half-angle.

Work through this **interactive video** to verify that

$$\sin\left(\frac{\alpha}{2}\right) = \frac{\sqrt{8+\sqrt{55}}}{4},\ \cos\left(\frac{\alpha}{2}\right) = \frac{\sqrt{8-\sqrt{55}}}{4},\ \text{and } \tan\left(\frac{\alpha}{2}\right) = \frac{3}{8-\sqrt{55}}.$$

You Try It **Work through this You Try It problem.**

Work Exercises 60–95 in this textbook or in the MyLab Math Study Plan.

OBJECTIVE 4 USING THE DOUBLE-ANGLE, POWER REDUCTION, AND HALF-ANGLE FORMULAS TO VERIFY IDENTITIES

We can use formulas discussed in this section, along with any identity discussed so far in this text, to verify trigonometric identities. As always, we will verify an identity by trying to transform one side of the identity into the exact trigonometric expression that appears on the opposite side. We typically start with what appears to be the more complicated side of the identity and then attempt to use known identities to transform that side into the trigonometric expression that appears on the other side of the identity. However, if one side of the identity includes a trigonometric expression involving 2θ or $\frac{\theta}{2}$, then first substitute one of the formulas discussed in this section. Next, use the strategies we developed in Section 8.1 for verifying identities. To review these strategies, read this **summary** of techniques that were introduced in Section 8.1. We now summarize the formulas discussed in this section.

Double-Angle Formulas	Half-Angle Formulas	
$\sin 2\theta = 2\sin\theta\cos\theta$	$\sin\left(\frac{\alpha}{2}\right) = \sqrt{\frac{1-\cos\alpha}{2}}$	for $\frac{\alpha}{2}$ in Quadrant I or Quadrant II.
$\cos 2\theta = \cos^2\theta - \sin^2\theta$	$\sin\left(\frac{\alpha}{2}\right) = -\sqrt{\frac{1-\cos\alpha}{2}}$	for $\frac{\alpha}{2}$ in Quadrant III or Quadrant IV.
$\cos 2\theta = 1 - 2\sin^2\theta$	$\cos\left(\frac{\alpha}{2}\right) = \sqrt{\frac{1+\cos\alpha}{2}}$	for $\frac{\alpha}{2}$ in Quadrant I or Quadrant IV.
$\cos 2\theta = 2\cos^2\theta - 1$	$\cos\left(\frac{\alpha}{2}\right) = -\sqrt{\frac{1+\cos\alpha}{2}}$	for $\frac{\alpha}{2}$ in Quadrant II or Quadrant III.
$\tan 2\theta = \frac{2\tan\theta}{1-\tan^2\theta}$	$\tan\left(\frac{\alpha}{2}\right) = \sqrt{\frac{1-\cos\alpha}{1+\cos\alpha}}$	for $\frac{\alpha}{2}$ in Quadrant I or Quadrant III.
Power Reduction Formulas	$\tan\left(\frac{\alpha}{2}\right) = -\sqrt{\frac{1-\cos\alpha}{1+\cos\alpha}}$	for $\frac{\alpha}{2}$ in Quadrant II or Quadrant IV.
$\sin^2\theta = \frac{1-\cos 2\theta}{2}$	$\tan\left(\frac{\alpha}{2}\right) = \frac{1-\cos\alpha}{\sin\alpha}$	for $\frac{\alpha}{2}$ in any quadrant.
$\cos^2\theta = \frac{1+\cos 2\theta}{2}$	$\tan\left(\frac{\alpha}{2}\right) = \frac{\sin\alpha}{1+\cos\alpha}$	for $\frac{\alpha}{2}$ in any quadrant.
$\tan^2\theta = \frac{1-\cos 2\theta}{1+\cos 2\theta}$		

Example 7 Verifying a Trigonometric Identity

Verify the trigonometric identity $\frac{\cos(2\theta)}{1+\sin(2\theta)} = \frac{\cos\theta - \sin\theta}{\cos\theta + \sin\theta}$.

Solution We start by rewriting the left-hand side using the **double-angle** formulas for sine and cosine.

$\frac{\cos(2\theta)}{1+\sin(2\theta)} \stackrel{?}{=} \frac{\cos\theta - \sin\theta}{\cos\theta + \sin\theta}$	Write the original trigonometric identity.
$\frac{\cos^2\theta - \sin^2\theta}{1 + 2\sin\theta\cos\theta} \stackrel{?}{=} \frac{\cos\theta - \sin\theta}{\cos\theta + \sin\theta}$	Choose $\cos 2\theta = \sin^2\theta - \cos^2\theta$ because it is a good match for the right-hand side. Use the double-angle formula for sine.
$\frac{\cos^2\theta - \sin^2\theta}{\sin^2\theta + \cos^2\theta + 2\sin\theta\cos\theta} \stackrel{?}{=} \frac{\cos\theta - \sin\theta}{\cos\theta + \sin\theta}$	Use the Pythagorean identity $1 = \sin^2\theta + \cos^2\theta$ to eliminate the 1.
$\frac{\cos^2\theta - \sin^2\theta}{\cos^2\theta + 2\sin\theta\cos\theta + \sin^2\theta} \stackrel{?}{=} \frac{\cos\theta - \sin\theta}{\cos\theta + \sin\theta}$	Rearrange the terms in the denominator.
$\frac{(\cos\theta + \sin\theta)(\cos\theta - \sin\theta)}{(\cos\theta + \sin\theta)(\cos\theta + \sin\theta)} \stackrel{?}{=} \frac{\cos\theta - \sin\theta}{\cos\theta + \sin\theta}$	Factor the numerator and denominator.
$\frac{\cancel{(\cos\theta + \sin\theta)}(\cos\theta - \sin\theta)}{\cancel{(\cos\theta + \sin\theta)}(\cos\theta + \sin\theta)} \stackrel{?}{=} \frac{\cos\theta - \sin\theta}{\cos\theta + \sin\theta}$	Cancel common factors.
$\frac{\cos\theta - \sin\theta}{\cos\theta + \sin\theta} \stackrel{?}{=} \frac{\cos\theta - \sin\theta}{\cos\theta + \sin\theta}$	Simplify.

The left-hand side is now identical to the right-hand side. Therefore, the identity is verified.

You Try It Work through this You Try It problem.

Work Exercises 96–102 in this textbook or in the MyLab Math Study Plan.

OBJECTIVE 5 USING THE DOUBLE-ANGLE AND HALF-ANGLE FORMULAS TO EVALUATE EXPRESSIONS INVOLVING INVERSE TRIGONOMETRIC FUNCTIONS

The double-angle and half-angle formulas discussed in this section may be helpful when we encounter trigonometric expressions involving one or more inverse trigonometric functions. Before working through the next example, you should review the inverse trigonometric functions that were first introduced in **Section 7.4**. Click on one of the following links for a quick review of these functions.

A Quick Review of Inverse Trigonometric Functions

$y = \sin^{-1}x \qquad y = \cos^{-1}x \qquad y = \tan^{-1}x$

Example 8 Finding the Exact Value of a Trigonometric Expression Involving Inverse Trigonometric Expressions

Find the exact value of the expression $\cos\left(\frac{1}{2}\sin^{-1}\left(-\frac{3}{11}\right)\right)$ without the use of a calculator.

Solution If we let $\sin^{-1}\left(-\frac{3}{11}\right) = \alpha$, then we are trying to find the value of $\cos\left(\frac{\alpha}{2}\right)$.

Because $\sin^{-1}\left(-\frac{3}{11}\right) = \alpha$, then $\sin\alpha = -\frac{3}{11}$ and $-\frac{\pi}{2} \le \alpha \le \frac{\pi}{2}$. Thus the terminal side of α lies in either Quadrant I or Quadrant IV. Because $\sin\alpha < 0$, we know that α must lie in Quadrant IV. To find the value of $\cos\alpha$, we use the Pythagorean identity $\sin^2\alpha + \cos^2\alpha = 1$ or $\cos\alpha = \sqrt{1 - \sin^2\alpha}$. Read these **steps** to see how to obtain $\cos\alpha = \frac{4\sqrt{7}}{11}$.

Because the terminal side of the angle lies in Quadrant IV, we know that $-\frac{\pi}{2} < \alpha < 0$. Dividing each part of this inequality by 2, we get $-\frac{\pi}{4} < \frac{\alpha}{2} < 0$. Therefore, the terminal side of $\frac{\alpha}{2}$ also lies in Quadrant IV. Thus, $\cos\left(\frac{\alpha}{2}\right) > 0$ because the cosine function is positive for all angles whose terminal side lies in Quadrant IV.

We therefore use the **positive form** of the **half-angle formula for cosine** to find the exact value of $\cos\left(\frac{1}{2}\sin^{-1}\left(-\frac{3}{11}\right)\right)$.

$\cos\left(\frac{1}{2}\sin^{-1}\left(-\frac{3}{11}\right)\right)$ Write the original expression.

$= \cos\left(\frac{\alpha}{2}\right)$ Write as the cosine of a half-angle, where $\alpha = \sin^{-1}\left(-\frac{3}{11}\right)$.

$= \sqrt{\dfrac{1 + \cos\alpha}{2}}$ Use the positive form of the half-angle formula for cosine.

$= \sqrt{\dfrac{1 + \frac{4\sqrt{7}}{11}}{2}}$ Substitute $\cos\alpha = \dfrac{4\sqrt{7}}{11}$.

$= \sqrt{\dfrac{\frac{11 + 4\sqrt{7}}{11}}{2}}$ Combine the expressions in the denominator.

$= \sqrt{\dfrac{11 + 4\sqrt{7}}{22}}$ Simplify.

Therefore, $\cos\left(\dfrac{1}{2}\sin^{-1}\left(-\dfrac{3}{11}\right)\right) = \sqrt{\dfrac{11 + 4\sqrt{7}}{22}}$.

You Try It Work through this You Try It problem.

Work Exercises 103–122 in this textbook or in the MyLab Math Study Plan.

8.3 Exercises

Skill Check Exercises

In exercises SCE-1 through SCE-5 simplify each expression.

SCE-1. $\sqrt{\dfrac{1 - \frac{1}{\sqrt{2}}}{2}}$

SCE-2. $\sqrt{\dfrac{1 - \left(-\frac{1}{\sqrt{2}}\right)}{2}}$

SCE-3. $-\sqrt{\dfrac{1 - \frac{5}{11}}{2}}$

SCE-4. $\sqrt{\dfrac{1 - \frac{5}{23}}{1 + \frac{5}{23}}}$

SCE-5. $\sqrt{\dfrac{1 - \left(-\frac{1}{\sqrt{2}}\right)}{1 + \left(-\frac{1}{\sqrt{2}}\right)}}$

In Exercises 1–10, rewrite each expression as the sine, cosine, or tangent of a double-angle. Then find the exact value of the trigonometric expression without the use of a calculator.

1. $2\sin\left(\dfrac{\pi}{8}\right)\cos\left(\dfrac{\pi}{8}\right)$

2. $2\cos^2 75° - 1$

3. $\dfrac{2\tan\left(\frac{7\pi}{8}\right)}{1 - \tan^2\left(\frac{7\pi}{8}\right)}$

4. $\cos^2\left(-\dfrac{5\pi}{12}\right) - \sin^2\left(-\dfrac{5\pi}{12}\right)$

5. $2\sin 202.5°\cos 202.5°$

6. $1 - 2\sin^2\left(\dfrac{-5\pi}{8}\right)$

7. $\dfrac{2\tan(-67.5°)}{1 - \tan^2(-67.5°)}$

8. $\cos^2 105° - \sin^2 105°$

9. $1 - 2\sin^2(-157.5°)$

10. $2\cos^2\left(\dfrac{-9\pi}{8}\right) - 1$

Note to instructors: The final answer to Exercises 11–22 is **not** fully simplified.

SbS 11. Suppose that the terminal side of angle θ lies in Quadrant II such that $\sin\theta = \frac{2}{13}$. Determine the value of $\sin 2\theta$.

SbS 12. Suppose that the terminal side of angle θ lies in Quadrant II such that $\sin\theta = \frac{19}{20}$. Determine the value of $\cos 2\theta$.

SbS 13. Suppose that the terminal side of angle θ lies in Quadrant II such that $\sin\theta = \frac{7}{9}$. Determine the value of $\tan 2\theta$.

SbS 14. Suppose that the terminal side of angle θ lies in Quadrant IV such that $\cos\theta = \frac{21}{29}$. Determine the value of $\sin 2\theta$.

SbS 15. Suppose that the terminal side of angle θ lies in Quadrant IV such that $\cos\theta = \frac{8}{17}$. Determine the value of $\cos 2\theta$.

SbS 16. Suppose that the terminal side of angle θ lies in Quadrant IV such that $\cos\theta = \frac{12}{13}$. Determine the value of $\tan 2\theta$.

SbS 17. Suppose that the terminal side of angle θ lies in Quadrant III such that $\sin\theta = -\frac{8}{17}$. Determine the value of $\sin 2\theta$.

SbS 18. Suppose that the terminal side of angle θ lies in Quadrant III such that $\sin\theta = -\frac{12}{13}$. Determine the value of $\cos 2\theta$.

SbS 19. Suppose that the terminal side of angle θ lies in Quadrant III such that $\sin\theta = -\frac{5}{13}$. Determine the value of $\tan 2\theta$.

SbS 20. Suppose that the terminal side of angle θ lies in Quadrant III such that $\cot\theta = 10$. Determine the value of $\sin 2\theta$.

SbS 21. Suppose that the terminal side of angle θ lies in Quadrant III such that $\cot\theta = 8$. Determine the value of $\cos 2\theta$.

SbS 22. Suppose that the terminal side of angle θ lies in Quadrant III such that $\cot\theta = 15$. Determine the value of $\tan 2\theta$.

Note to instructors: The final answer to Exercises 23 and 34 is fully simplified.

SbS 23. Suppose that the terminal side of angle θ lies in Quadrant II such that $\sin\theta = \frac{2}{13}$. Determine the value of $\sin 2\theta$.

SbS 24. Suppose that the terminal side of angle θ lies in Quadrant II such that $\sin\theta = \frac{19}{20}$. Determine the value of $\cos 2\theta$.

SbS 25. Suppose that the terminal side of angle θ lies in Quadrant II such that $\sin\theta = \frac{7}{9}$. Determine the value of $\tan 2\theta$.

SbS 26. Suppose that the terminal side of angle θ lies in Quadrant IV such that $\cos\theta = \frac{21}{29}$. Determine the value of $\sin 2\theta$.

SbS 27. Suppose that the terminal side of angle θ lies in Quadrant IV such that $\cos\theta = \frac{8}{17}$. Determine the value of $\cos 2\theta$.

SbS 28. Suppose that the terminal side of angle θ lies in Quadrant IV such that $\cos\theta = \frac{12}{13}$. Determine the value of $\tan 2\theta$.

SbS 29. Suppose that the terminal side of angle θ lies in Quadrant III such that $\sin\theta = -\frac{8}{17}$. Determine the value of $\sin 2\theta$.

SbS 30. Suppose that the terminal side of angle θ lies in Quadrant III such that $\sin\theta = -\frac{12}{13}$. Determine the value of $\cos 2\theta$.

SbS 31. Suppose that the terminal side of angle θ lies in Quadrant III such that $\sin\theta = -\frac{5}{13}$. Determine the value of $\tan 2\theta$.

SbS 32. Suppose that the terminal side of angle θ lies in Quadrant III such that $\cot\theta = 10$. Determine the value of $\sin 2\theta$.

SbS 33. Suppose that the terminal side of angle θ lies in Quadrant III such that $\cot\theta = 8$. Determine the value of $\cos 2\theta$.

SbS 34. Suppose that the terminal side of angle θ lies in Quadrant III such that $\cot\theta = 15$. Determine the value of $\tan 2\theta$.

In Exercises 35–38, use the given information to determine the values of $\sin\theta$, $\cos\theta$, and $\tan\theta$.

35. $\sin 2\theta = \frac{5}{13}; \frac{\pi}{2} < 2\theta < \pi$

36. $\cos 2\theta = \frac{3}{5}; \frac{3\pi}{2} < 2\theta < 2\pi$

37. $\tan 2\theta = \frac{2}{7}; \pi < 2\theta < \frac{3\pi}{2}$

38. $\csc 2\theta = -\frac{7}{5}; \frac{3\pi}{2} < 2\theta < 2\pi$

In Exercises 39–43, rewrite the given function as an equivalent function containing only cosine terms raised to a power of 1.

39. $f(x) = -3\sin^2 x$

40. $f(x) = 7\cos^2 x$

41. $f(x) = 2\sin^4 x$

42. $f(x) = -3\cos^4 x$

43. $f(x) = -2\sin^2 x\cos^2 x$

In Exercises 44–51, use a half-angle formula to evaluate each expression without using a calculator. **Note to instructors:** The final answer to these exercises is **not** fully simplified.

SbS 44. $\sin\left(-\frac{3\pi}{8}\right)$ SbS 45. $\cos(-22.5°)$ SbS 46. $\tan 157.5°$ SbS 47. $\sin 67.5°$

SbS 48. $\cos\left(\frac{\pi}{12}\right)$ SbS 49. $\tan\left(\frac{5\pi}{12}\right)$ SbS 50. $\csc\left(\frac{9\pi}{8}\right)$ SbS 51. $\sec\left(\frac{13\pi}{12}\right)$

In Exercises 52–59, use a half-angle formula to evaluate each expression without using a calculator. **Note to instructors:** The final answer to these exercises is completely simplified.

SbS 52. $\sin\left(-\frac{3\pi}{8}\right)$ SbS 53. $\cos(-22.5°)$ SbS 54. $\tan 157.5°$ SbS 55. $\sin 67.5°$

SbS 56. $\cos\left(\frac{\pi}{12}\right)$ SbS 57. $\tan\left(\frac{5\pi}{12}\right)$ SbS 58. $\csc\left(\frac{9\pi}{8}\right)$ SbS 59. $\sec\left(\frac{13\pi}{12}\right)$

Note to instructors: The final answer to Exercises 60–77 is **not** fully simplified.

SbS 60. Suppose that $\sin\alpha = \frac{12}{13}$ such that $\frac{\pi}{2} < \alpha < \pi$. Determine the value of $\sin\left(\frac{\alpha}{2}\right)$.

SbS 61. Suppose that $\sin\alpha = \frac{4}{5}$ such that $\frac{\pi}{2} < \alpha < \pi$. Determine the value of $\cos\left(\frac{\alpha}{2}\right)$.

SbS 62. Suppose that $\sin\alpha = \frac{15}{17}$ such that $\frac{\pi}{2} < \alpha < \pi$. Determine the value of $\tan\left(\frac{\alpha}{2}\right)$.

SbS 63. Suppose that $\cos\alpha = -\frac{21}{29}$ such that $\frac{\pi}{2} < \alpha < \pi$. Determine the value of $\sin\left(\frac{\alpha}{2}\right)$.

SbS 64. Suppose that $\cos\alpha = -\frac{5}{13}$ such that $\frac{\pi}{2} < \alpha < \pi$. Determine the value of $\cos\left(\frac{\alpha}{2}\right)$.

SbS 65. Suppose that $\cos\alpha = -\frac{15}{17}$ such that $\frac{\pi}{2} < \alpha < \pi$. Determine the value of $\tan\left(\frac{\alpha}{2}\right)$.

SbS 66. Suppose that $\tan\alpha = \frac{12}{5}$ such that $\pi < \alpha < \frac{3\pi}{2}$. Determine the value of $\sin\left(\frac{\alpha}{2}\right)$.

SbS 67. Suppose that $\tan\alpha = \frac{15}{8}$ such that $\pi < \alpha < \frac{3\pi}{2}$. Determine the value of $\cos\left(\frac{\alpha}{2}\right)$.

SbS 68. Suppose that $\tan\alpha = \frac{4}{3}$ such that $\pi < \alpha < \frac{3\pi}{2}$. Determine the value of $\tan\left(\frac{\alpha}{2}\right)$.

SbS 69. Suppose that $\sec\alpha = -\frac{5}{4}$ such that $\pi < \alpha < \frac{3\pi}{2}$. Determine the value of $\sin\left(\frac{\alpha}{2}\right)$.

SbS 70. Suppose that $\sec\alpha = -\frac{13}{5}$ such that $\pi < \alpha < \frac{3\pi}{2}$. Determine the value of $\cos\left(\frac{\alpha}{2}\right)$.

SbS 71. Suppose that $\sec\alpha = -\frac{17}{5}$ such that $\pi < \alpha < \frac{3\pi}{2}$. Determine the value of $\tan\left(\frac{\alpha}{2}\right)$.

SbS 72. Suppose that $\cot\alpha = \frac{1}{3}$ such that $\pi < \alpha < \frac{3\pi}{2}$. Determine the value of $\sin\left(\frac{\alpha}{2}\right)$.

SbS 73. Suppose that $\cot\alpha = \frac{1}{17}$ such that $\pi < \alpha < \frac{3\pi}{2}$. Determine the value of $\cos\left(\frac{\alpha}{2}\right)$.

SbS 74. Suppose that $\cot\alpha = \frac{1}{13}$ such that $\pi < \alpha < \frac{3\pi}{2}$. Determine the value of $\tan\left(\frac{\alpha}{2}\right)$.

SbS 75. Suppose that $\cos\alpha = \frac{5}{11}$ such that $\frac{3\pi}{2} < \alpha < 2\pi$. Determine the value of $\sin\left(\frac{\alpha}{2}\right)$.

SbS 76. Suppose that $\cos\alpha = \frac{7}{13}$ such that $\frac{3\pi}{2} < \alpha < 2\pi$. Determine the value of $\cos\left(\frac{\alpha}{2}\right)$.

SbS 77. Suppose that $\cos\alpha = \frac{5}{19}$ such that $\frac{3\pi}{2} < \alpha < 2\pi$. Determine the value of $\tan\left(\frac{\alpha}{2}\right)$.

Note to instructors: The final answer to Exercises 78–95 is fully simplified.

SbS 78. Suppose that $\sin\alpha = \frac{12}{13}$ such that $\frac{\pi}{2} < \alpha < \pi$. Determine the value of $\sin\left(\frac{\alpha}{2}\right)$.

SbS 79. Suppose that $\sin\alpha = \frac{4}{5}$ such that $\frac{\pi}{2} < \alpha < \pi$. Determine the value of $\cos\left(\frac{\alpha}{2}\right)$.

SbS 80. Suppose that $\sin\alpha = \frac{15}{17}$ such that $\frac{\pi}{2} < \alpha < \pi$. Determine the value of $\tan\left(\frac{\alpha}{2}\right)$.

SbS 81. Suppose that $\cos\alpha = -\frac{21}{29}$ such that $\frac{\pi}{2} < \alpha < \pi$. Determine the value of $\sin\left(\frac{\alpha}{2}\right)$.

SbS 82. Suppose that $\cos\alpha = -\frac{5}{13}$ such that $\frac{\pi}{2} < \alpha < \pi$. Determine the value of $\cos\left(\frac{\alpha}{2}\right)$.

SbS 83. Suppose that $\cos\alpha = -\frac{15}{17}$ such that $\frac{\pi}{2} < \alpha < \pi$. Determine the value of $\tan\left(\frac{\alpha}{2}\right)$.

SbS 84. Suppose that $\tan\alpha = \frac{12}{5}$ such that $\pi < \alpha < \frac{3\pi}{2}$. Determine the value of $\sin\left(\frac{\alpha}{2}\right)$.

SbS 85. Suppose that $\tan\alpha = \frac{15}{8}$ such that $\pi < \alpha < \frac{3\pi}{2}$. Determine the value of $\cos\left(\frac{\alpha}{2}\right)$.

SbS 86. Suppose that $\tan\alpha = \frac{4}{3}$ such that $\pi < \alpha < \frac{3\pi}{2}$. Determine the value of $\tan\left(\frac{\alpha}{2}\right)$.

SbS 87. Suppose that $\sec\alpha = -\frac{5}{4}$ such that $\pi < \alpha < \frac{3\pi}{2}$. Determine the value of $\sin\left(\frac{\alpha}{2}\right)$.

SbS 88. Suppose that $\sec\alpha = -\frac{13}{5}$ such that $\pi < \alpha < \frac{3\pi}{2}$. Determine the value of $\cos\left(\frac{\alpha}{2}\right)$.

SbS 89. Suppose that $\sec\alpha = -\frac{17}{5}$ such that $\pi < \alpha < \frac{3\pi}{2}$. Determine the value of $\tan\left(\frac{\alpha}{2}\right)$.

SbS 90. Suppose that $\cot\alpha = \frac{1}{3}$ such that $\pi < \alpha < \frac{3\pi}{2}$. Determine the value of $\sin\left(\frac{\alpha}{2}\right)$.

SbS 91. Suppose that $\cot\alpha = \frac{1}{17}$ such that $\pi < \alpha < \frac{3\pi}{2}$. Determine the value of $\cos\left(\frac{\alpha}{2}\right)$.

SbS 92. Suppose that $\cot\alpha = \frac{1}{13}$ such that $\pi < \alpha < \frac{3\pi}{2}$. Determine the value of $\tan\left(\frac{\alpha}{2}\right)$.

SbS 93. Suppose that $\cos\alpha = \frac{5}{11}$ such that $\frac{3\pi}{2} < \alpha < 2\pi$. Determine the value of $\sin\left(\frac{\alpha}{2}\right)$.

SbS 94. Suppose that $\cos\alpha = \frac{7}{13}$ such that $\frac{3\pi}{2} < \alpha < 2\pi$. Determine the value of $\cos\left(\frac{\alpha}{2}\right)$.

SbS 95. Suppose that $\cos\alpha = \frac{5}{19}$ such that $\frac{3\pi}{2} < \alpha < 2\pi$. Determine the value of $\tan\left(\frac{\alpha}{2}\right)$.

In Exercises 96–102, verify each identity.

96. $\frac{\tan^2\theta - 1}{1 + \tan^2\theta} = -\cos 2\theta$

97. $\cot 2\theta = \frac{1}{2}\sec\theta\csc\theta - \tan\theta$

98. $\sec 2\theta = \frac{\csc^2\theta}{\csc^2\theta - 2}$

99. $\tan\theta = \frac{1 - \cos 2\theta}{\sin 2\theta}$

100. $\sin^2\frac{\theta}{2} = \frac{\tan\theta - \sin\theta}{2\tan\theta}$

101. $\tan\frac{\theta}{2} = \frac{1}{\csc\theta + \cot\theta}$

102. $\cot\frac{\theta}{2} - \cot\theta = \csc\theta$

In Exercises 103–106, use a double angle formula to find the exact value of each expression without using a calculator.

103. $\sin\left(2\cos^{-1}\frac{\sqrt{3}}{2}\right)$

104. $\cos\left(2\sin^{-1}\left(-\frac{1}{\sqrt{2}}\right)\right)$

105. $\sin\left(2\tan^{-1}\frac{5}{2}\right)$

106. $\cos\left(2\sin^{-1}\left(-\frac{5}{9}\right)\right)$

In Exercises 107–114, use a half-angle formula to evaluate each expression without using a calculator. **Note to instructors:** The final answer to these exercises is **not** simplified.

SbS 107. $\sin\left(\frac{1}{2}\cos^{-1}\left(-\frac{1}{2}\right)\right)$

SbS 108. $\cos\left(\frac{1}{2}\cos^{-1}\left(-\frac{1}{\sqrt{2}}\right)\right)$

SbS 109. $\sin\left(\frac{1}{2}\cos^{-1}\left(-\frac{3}{7}\right)\right)$

SbS 110. $\cos\left(\frac{1}{2}\cos^{-1}\left(\frac{5}{11}\right)\right)$ SbS 111. $\sin\left(\frac{1}{2}\tan^{-1}(-1)\right)$ SbS 112. $\cos\left(\frac{1}{2}\tan^{-1}\left(\frac{1}{\sqrt{3}}\right)\right)$

SbS 113. $\sin\left(\frac{1}{2}\tan^{-1}\left(\frac{5}{7}\right)\right)$ SbS 114. $\cos\left(\frac{1}{2}\tan^{-1}(-5)\right)$

In Exercises 115–122, use a half-angle formula to evaluate each expression without using a calculator. **Note to instructors:** The final answer to these exercises is completely simplified.

SbS 115. $\sin\left(\frac{1}{2}\cos^{-1}\left(-\frac{1}{2}\right)\right)$ SbS 116. $\cos\left(\frac{1}{2}\cos^{-1}\left(-\frac{1}{\sqrt{2}}\right)\right)$ SbS 117. $\sin\left(\frac{1}{2}\cos^{-1}\left(-\frac{3}{7}\right)\right)$

SbS 118. $\cos\left(\frac{1}{2}\cos^{-1}\left(\frac{5}{11}\right)\right)$ SbS 119. $\sin\left(\frac{1}{2}\tan^{-1}(-1)\right)$ SbS 120. $\cos\left(\frac{1}{2}\tan^{-1}\left(\frac{1}{\sqrt{3}}\right)\right)$

SbS 121. $\sin\left(\frac{1}{2}\tan^{-1}\left(\frac{5}{7}\right)\right)$ SbS 122. $\cos\left(\frac{1}{2}\tan^{-1}(-5)\right)$

Brief Exercises

Note to instructors: The final answer to Exercises 123–134 is fully simplified.

123. Suppose that the terminal side of angle θ lies in Quadrant II such that $\sin\theta = \frac{2}{13}$. Determine the value of $\sin 2\theta$.

124. Suppose that the terminal side of angle θ lies in Quadrant II such that $\sin\theta = \frac{19}{20}$. Determine the value of $\cos 2\theta$.

125. Suppose that the terminal side of angle θ lies in Quadrant II such that $\sin\theta = \frac{7}{9}$. Determine the value of $\tan 2\theta$.

126. Suppose that the terminal side of angle θ lies in Quadrant IV such that $\cos\theta = \frac{21}{29}$. Determine the value of $\sin 2\theta$.

127. Suppose that the terminal side of angle θ lies in Quadrant IV such that $\cos\theta = \frac{8}{17}$. Determine the value of $\cos 2\theta$.

128. Suppose that the terminal side of angle θ lies in Quadrant IV such that $\cos\theta = \frac{12}{13}$. Determine the value of $\tan 2\theta$.

129. Suppose that the terminal side of angle θ lies in Quadrant III such that $\sin\theta = -\frac{8}{17}$. Determine the value of $\sin 2\theta$.

130. Suppose that the terminal side of angle θ lies in Quadrant III such that $\sin\theta = -\frac{12}{13}$. Determine the value of $\cos 2\theta$.

131. Suppose that the terminal side of angle θ lies in Quadrant III such that $\sin\theta = -\frac{5}{13}$. Determine the value of tan 2θ.

132. Suppose that the terminal side of angle θ lies in Quadrant III such that $\cot\theta = 10$. Determine the value of sin 2θ.

133. Suppose that the terminal side of angle θ lies in Quadrant III such that $\cot\theta = 8$. Determine the value of cos 2θ.

134. Suppose that the terminal side of angle θ lies in Quadrant III such that $\cot\theta = 15$. Determine the value of tan 2θ.

In Exercises 135–142, use a half-angle formula to evaluate each expression without using a calculator. **Note to instructors:** The final answer to these exercises is **not** completely simplified.

135. $\sin\left(-\frac{3\pi}{8}\right)$ 136. $\cos(-22.5°)$ 137. $\tan 157.5°$ 138. $\sin 67.5°$

139. $\cos\left(\frac{\pi}{12}\right)$ 140. $\tan\left(\frac{5\pi}{12}\right)$ 141. $\csc\left(\frac{9\pi}{8}\right)$ 142. $\sec\left(\frac{13\pi}{12}\right)$

In Exercises 143–150, use a half-angle formula to evaluate each expression without using a calculator. **Note to instructors:** The final answer to these exercises is completely simplified.

143. $\sin\left(-\frac{3\pi}{8}\right)$ 144. $\cos(-22.5°)$ 145. $\tan 157.5°$ 146. $\sin 67.5°$

147. $\cos\left(\frac{\pi}{12}\right)$ 148. $\tan\left(\frac{5\pi}{12}\right)$ 149. $\csc\left(\frac{9\pi}{8}\right)$ 150. $\sec\left(\frac{13\pi}{12}\right)$

Note to instructors: The final answer to Exercises 151–168 is **not** fully simplified.

151. Suppose that $\sin\alpha = \frac{12}{13}$ such that $\frac{\pi}{2} < \alpha < \pi$. Determine the value of $\sin\left(\frac{\alpha}{2}\right)$.

152. Suppose that $\sin\alpha = \frac{4}{5}$ such that $\frac{\pi}{2} < \alpha < \pi$. Determine the value of $\cos\left(\frac{\alpha}{2}\right)$.

153. Suppose that $\sin\alpha = \frac{15}{17}$ such that $\frac{\pi}{2} < \alpha < \pi$. Determine the value of $\tan\left(\frac{\alpha}{2}\right)$.

154. Suppose that $\cos\alpha = -\frac{21}{29}$ such that $\frac{\pi}{2} < \alpha < \pi$. Determine the value of $\sin\left(\frac{\alpha}{2}\right)$.

155. Suppose that $\cos\alpha = -\frac{5}{13}$ such that $\frac{\pi}{2} < \alpha < \pi$. Determine the value of $\cos\left(\frac{\alpha}{2}\right)$.

156. Suppose that $\cos\alpha = -\frac{15}{17}$ such that $\frac{\pi}{2} < \alpha < \pi$. Determine the value of $\tan\left(\frac{\alpha}{2}\right)$.

157. Suppose that $\tan\alpha = \frac{12}{5}$ such that $\pi < \alpha < \frac{3\pi}{2}$. Determine the value of $\sin\left(\frac{\alpha}{2}\right)$.

158. Suppose that $\tan \alpha = \frac{15}{8}$ such that $\pi < \alpha < \frac{3\pi}{2}$. Determine the value of $\cos\left(\frac{\alpha}{2}\right)$.

159. Suppose that $\tan \alpha = \frac{4}{3}$ such that $\pi < \alpha < \frac{3\pi}{2}$. Determine the value of $\tan\left(\frac{\alpha}{2}\right)$.

160. Suppose that $\sec \alpha = -\frac{5}{4}$ such that $\pi < \alpha < \frac{3\pi}{2}$. Determine the value of $\sin\left(\frac{\alpha}{2}\right)$.

161. Suppose that $\sec \alpha = -\frac{13}{5}$ such that $\pi < \alpha < \frac{3\pi}{2}$. Determine the value of $\cos\left(\frac{\alpha}{2}\right)$.

162. Suppose that $\sec \alpha = -\frac{17}{5}$ such that $\pi < \alpha < \frac{3\pi}{2}$. Determine the value of $\tan\left(\frac{\alpha}{2}\right)$.

163. Suppose that $\cot \alpha = \frac{1}{3}$ such that $\pi < \alpha < \frac{3\pi}{2}$. Determine the value of $\sin\left(\frac{\alpha}{2}\right)$.

164. Suppose that $\cot \alpha = \frac{1}{17}$ such that $\pi < \alpha < \frac{3\pi}{2}$. Determine the value of $\cos\left(\frac{\alpha}{2}\right)$.

165. Suppose that $\cot \alpha = \frac{1}{13}$ such that $\pi < \alpha < \frac{3\pi}{2}$. Determine the value of $\tan\left(\frac{\alpha}{2}\right)$.

166. Suppose that $\cos \alpha = \frac{5}{11}$ such that $\frac{3\pi}{2} < \alpha < 2\pi$. Determine the value of $\sin\left(\frac{\alpha}{2}\right)$.

167. Suppose that $\cos \alpha = \frac{7}{13}$ such that $\frac{3\pi}{2} < \alpha < 2\pi$. Determine the value of $\cos\left(\frac{\alpha}{2}\right)$.

168. Suppose that $\cos \alpha = \frac{5}{19}$ such that $\frac{3\pi}{2} < \alpha < 2\pi$. Determine the value of $\tan\left(\frac{\alpha}{2}\right)$.

Note to instructors: The final answer to Exercises 169–186 is fully simplified.

169. Suppose that $\sin \alpha = \frac{12}{13}$ such that $\frac{\pi}{2} < \alpha < \pi$. Determine the value of $\sin\left(\frac{\alpha}{2}\right)$.

170. Suppose that $\sin \alpha = \frac{4}{5}$ such that $\frac{\pi}{2} < \alpha < \pi$. Determine the value of $\cos\left(\frac{\alpha}{2}\right)$.

171. Suppose that $\sin \alpha = \frac{15}{17}$ such that $\frac{\pi}{2} < \alpha < \pi$. Determine the value of $\tan\left(\frac{\alpha}{2}\right)$.

172. Suppose that $\cos \alpha = -\frac{21}{29}$ such that $\frac{\pi}{2} < \alpha < \pi$. Determine the value of $\sin\left(\frac{\alpha}{2}\right)$.

173. Suppose that $\cos \alpha = -\frac{5}{13}$ such that $\frac{\pi}{2} < \alpha < \pi$. Determine the value of $\cos\left(\frac{\alpha}{2}\right)$.

174. Suppose that $\cos \alpha = -\frac{15}{17}$ such that $\frac{\pi}{2} < \alpha < \pi$. Determine the value of $\tan\left(\frac{\alpha}{2}\right)$.

175. Suppose that $\tan \alpha = \frac{12}{5}$ such that $\pi < \alpha < \frac{3\pi}{2}$. Determine the value of $\sin\left(\frac{\alpha}{2}\right)$.

176. Suppose that $\tan \alpha = \frac{15}{8}$ such that $\pi < \alpha < \frac{3\pi}{2}$. Determine the value of $\cos\left(\frac{\alpha}{2}\right)$.

177. Suppose that $\tan \alpha = \frac{4}{3}$ such that $\pi < \alpha < \frac{3\pi}{2}$. Determine the value of $\tan\left(\frac{\alpha}{2}\right)$.

178. Suppose that $\sec \alpha = -\frac{5}{4}$ such that $\pi < \alpha < \frac{3\pi}{2}$. Determine the value of $\sin\left(\frac{\alpha}{2}\right)$.

179. Suppose that $\sec \alpha = -\frac{13}{5}$ such that $\pi < \alpha < \frac{3\pi}{2}$. Determine the value of $\cos\left(\frac{\alpha}{2}\right)$.

180. Suppose that $\sec \alpha = -\frac{17}{5}$ such that $\pi < \alpha < \frac{3\pi}{2}$. Determine the value of $\tan\left(\frac{\alpha}{2}\right)$.

181. Suppose that $\cot \alpha = \frac{1}{3}$ such that $\pi < \alpha < \frac{3\pi}{2}$. Determine the value of $\sin\left(\frac{\alpha}{2}\right)$.

182. Suppose that $\cot \alpha = \frac{1}{17}$ such that $\pi < \alpha < \frac{3\pi}{2}$. Determine the value of $\cos\left(\frac{\alpha}{2}\right)$.

183. Suppose that $\cot \alpha = \frac{1}{13}$ such that $\pi < \alpha < \frac{3\pi}{2}$. Determine the value of $\tan\left(\frac{\alpha}{2}\right)$.

184. Suppose that $\cos \alpha = \frac{5}{11}$ such that $\frac{3\pi}{2} < \alpha < 2\pi$. Determine the value of $\sin\left(\frac{\alpha}{2}\right)$.

185. Suppose that $\cos \alpha = \frac{7}{13}$ such that $\frac{3\pi}{2} < \alpha < 2\pi$. Determine the value of $\cos\left(\frac{\alpha}{2}\right)$.

186. Suppose that $\cos \alpha = \frac{5}{19}$ such that $\frac{3\pi}{2} < \alpha < 2\pi$. Determine the value of $\tan\left(\frac{\alpha}{2}\right)$.

In Exercises 187–194, use a half-angle formula to evaluate each expression without using a calculator. **Note to instructors:** The final answer to these exercises is not completely simplified.

187. $\sin\left(\frac{1}{2}\cos^{-1}\left(-\frac{1}{2}\right)\right)$

188. $\cos\left(\frac{1}{2}\cos^{-1}\left(-\frac{1}{\sqrt{2}}\right)\right)$

189. $\sin\left(\frac{1}{2}\cos^{-1}\left(-\frac{3}{7}\right)\right)$

190. $\cos\left(\frac{1}{2}\cos^{-1}\left(\frac{5}{11}\right)\right)$

191. $\sin\left(\frac{1}{2}\tan^{-1}(-1)\right)$

192. $\cos\left(\frac{1}{2}\tan^{-1}\left(-\frac{1}{\sqrt{3}}\right)\right)$

193. $\sin\left(\frac{1}{2}\tan^{-1}\left(\frac{5}{7}\right)\right)$

194. $\cos\left(\frac{1}{2}\tan^{-1}(-5)\right)$

In Exercises 195–202, use a half-angle formula to evaluate each expression without using a calculator. **Note to instructors:** The final answer to these exercises is completely simplified.

195. $\sin\left(\frac{1}{2}\cos^{-1}\left(-\frac{1}{2}\right)\right)$

196. $\cos\left(\frac{1}{2}\cos^{-1}\left(-\frac{1}{\sqrt{2}}\right)\right)$

197. $\sin\left(\frac{1}{2}\cos^{-1}\left(-\frac{3}{7}\right)\right)$

198. $\cos\left(\frac{1}{2}\cos^{-1}\left(\frac{5}{11}\right)\right)$

199. $\sin\left(\frac{1}{2}\tan^{-1}(-1)\right)$

200. $\cos\left(\frac{1}{2}\tan^{-1}\left(-\frac{1}{\sqrt{3}}\right)\right)$

201. $\sin\left(\frac{1}{2}\tan^{-1}\left(\frac{5}{7}\right)\right)$

202. $\cos\left(\frac{1}{2}\tan^{-1}(-5)\right)$

8.4 The Product-to-Sum and Sum-to-Product Formulas

THINGS TO KNOW

Before working through this section, be sure that you are familiar with the following concepts:

VIDEO | ANIMATION | INTERACTIVE

You Try It

1. Evaluating Trigonometric Functions 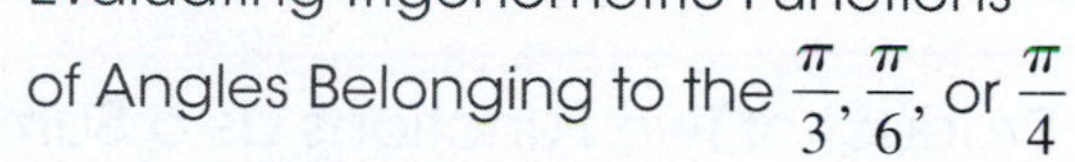 of Angles Belonging to the $\frac{\pi}{3}$, $\frac{\pi}{6}$, or $\frac{\pi}{4}$ Families (Section 6.5)

You Try It

2. Understanding the Sum and Difference Formulas for the Sine Function (Section 8.2)

You Try It

3. Understanding the Sum and Difference 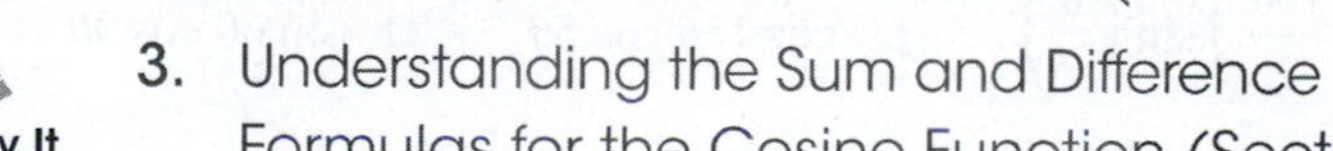 Formulas for the Cosine Function (Section 8.2)

INTRODUCTION

Read this introduction before beginning Objective 1.

OBJECTIVES

1 Understanding the Product-to-Sum Formulas

2 Understanding the Sum-to-Product Formulas

3 Using the Product-to-Sum and Sum-to-Product Formulas to Verify Identities

SECTION 8.4 EXERCISES

Introduction to Section 8.4

In this section, we will learn formulas that will allow us to convert the product of trigonometric functions into the sum or difference of trigonometric functions and vice versa. We start by introducing the **product-to-sum formulas.**

OBJECTIVE 1 UNDERSTANDING THE PRODUCT-TO-SUM FORMULAS

There are four product-to-sum formulas. That is, there are four different formulas that can be used to change the product of two trigonometric expressions into the sum or difference of two trigonometric expressions. These formulas are especially useful in calculus. Although the formulas appear to be difficult to memorize, they are actually fairly easy to derive provided that you can recall the **sum and difference formulas for sine and cosine**.

Product-to-Sum Formulas

$$\sin\alpha\sin\beta = \frac{1}{2}\left[\cos(\alpha-\beta) - \cos(\alpha+\beta)\right]$$

$$\cos\alpha\cos\beta = \frac{1}{2}\left[\cos(\alpha-\beta) + \cos(\alpha+\beta)\right]$$

$$\sin\alpha\cos\beta = \frac{1}{2}\left[\sin(\alpha+\beta) + \sin(\alpha-\beta)\right]$$

$$\cos\alpha\sin\beta = \frac{1}{2}\left[\sin(\alpha+\beta) - \sin(\alpha-\beta)\right]$$

Work through this **interactive video** to see how to use the **sum and difference formulas for sine and cosine** to derive each of the four product-to-sum formulas.

Example 1 Writing the Product of Two Functions as a Sum or Difference of Two Functions

Write each product as a sum or difference containing only sines or cosines.

a. $\sin 4\theta \sin 2\theta$ **b.** $\cos\left(\frac{19\theta}{2}\right)\sin\left(\frac{\theta}{2}\right)$ **c.** $\cos 11\theta \cos 5\theta$ **d.** $\sin 6\theta \cos 3\theta$

Solution

a. To write the product $\sin(4\theta)\sin(2\theta)$ as a sum or difference of sines and cosines, we must first identify the correct **product-to-sum formula**. Because we are given the product of two sine expressions, we will use the formula

$$\sin\alpha\sin\beta = \frac{1}{2}\left[\cos(\alpha-\beta) - \cos(\alpha+\beta)\right] \text{ with } \alpha = 4\theta \text{ and } \beta = 2\theta.$$

$$\sin\alpha\sin\beta = \frac{1}{2}\left[\cos(\alpha-\beta) - \cos(\alpha+\beta)\right]$$ Write the appropriate product-to-sum formula.

$$\sin(4\theta)\sin(2\theta) = \frac{1}{2}\left[\cos(4\theta-2\theta) - \cos(4\theta+2\theta)\right]$$ Substitute $\alpha = 4\theta$ and $\beta = 2\theta$.

$$= \frac{1}{2}\left[\cos(2\theta) - \cos(6\theta)\right]$$ Simplify the **arguments**.

$$= \frac{1}{2}\cos(2\theta) - \frac{1}{2}\cos(6\theta)$$ Distribute.

b. To write the product $\cos\left(\frac{19\theta}{2}\right)\sin\left(\frac{\theta}{2}\right)$ as a sum or difference of sines or cosines, we must first identify the correct **product-to-sum formula**. Because we are given the product of a cosine expression and a sine expression, we will use the formula $\cos\alpha\sin\beta = \frac{1}{2}\left[\sin(\alpha+\beta) - \sin(\alpha-\beta)\right]$ with $\alpha = \frac{19\theta}{2}$ and $\beta = \frac{\theta}{2}$.

$$\cos\alpha\sin\beta = \frac{1}{2}\left[\sin(\alpha+\beta) - \sin(\alpha-\beta)\right]$$ Write the appropriate product-to-sum formula.

$$\cos\left(\frac{19\theta}{2}\right)\sin\left(\frac{\theta}{2}\right) = \frac{1}{2}\left[\sin\left(\frac{19\theta}{2}+\frac{\theta}{2}\right) - \sin\left(\frac{19\theta}{2}-\frac{\theta}{2}\right)\right]$$ Substitute $\alpha = \frac{19\theta}{2}$ and $\beta = \frac{\theta}{2}$.

$$= \frac{1}{2}\left[\sin 10\theta - \sin 9\theta\right]$$ Simplify the **arguments**.

$$= \frac{1}{2}\sin 10\theta - \frac{1}{2}\sin 9\theta$$ Distribute.

Try to determine the solution to part c and part d on your own by first choosing the appropriate **product-to-sum formula**. When you're finished, check your **solutions**, or watch this **interactive video** to see the solutions to all four examples.

You Try It Work through this You Try It problem.

Work Exercises 1–4 in this textbook or in the MyLab Math Study Plan.

Example 2 Evaluating the Product of a Trigonometric Expression

Determine the exact value of the expression $\sin\left(\frac{3\pi}{8}\right)\cos\left(\frac{\pi}{8}\right)$ without the use of a calculator.

Solution To evaluate the expression $\sin\left(\frac{3\pi}{8}\right)\cos\left(\frac{\pi}{8}\right)$, first use the **product-to-sum formula** $\sin\alpha\cos\beta = \frac{1}{2}\left[\sin(\alpha+\beta) + \sin(\alpha-\beta)\right]$ with $\alpha = \frac{3\pi}{8}$ and $\beta = \frac{\pi}{8}$.

$$\sin\left(\frac{3\pi}{8}\right)\cos\left(\frac{\pi}{8}\right)$$ Write the original expression.

$$= \frac{1}{2}\left[\sin\left(\frac{3\pi}{8}+\frac{\pi}{8}\right) + \sin\left(\frac{3\pi}{8}-\frac{\pi}{8}\right)\right]$$ Use the formula $\sin\alpha\cos\beta = \frac{1}{2}[\sin(\alpha+\beta) + \sin(\alpha-\beta)]$ with $\alpha = \frac{3\pi}{8}$ and $\beta = \frac{\pi}{8}$.

$$= \frac{1}{2}\left[\sin\frac{\pi}{2} + \sin\frac{\pi}{4}\right]$$ Simplify the **arguments**.

$$= \frac{1}{2}\left[1 + \frac{1}{\sqrt{2}}\right]$$ Substitute $\sin\frac{\pi}{2} = 1$ and $\sin\frac{\pi}{4} = \frac{1}{\sqrt{2}}$.

$$= \frac{1}{2}\left[\frac{\sqrt{2}+1}{\sqrt{2}}\right] = \frac{\sqrt{2}+1}{2\sqrt{2}}$$ Rewrite using a common denominator and multiply.

Therefore, $\sin\left(\frac{3\pi}{8}\right)\cos\left(\frac{\pi}{8}\right) = \frac{\sqrt{2}+1}{2\sqrt{2}}$. Watch this **video** to see every step of this solution. Note that the expression $\frac{\sqrt{2}+1}{2\sqrt{2}}$ can be written as $\frac{2+\sqrt{2}}{4}$ if we choose to rationalize the denominator. ●

You Try It Work through this You Try It problem.

Work Exercises 5–8 in this textbook or in the MyLab Math Study Plan.

OBJECTIVE 2 UNDERSTANDING THE SUM-TO-PRODUCT FORMULAS

We can use the four product-to-sum formulas to verify each of the four sum-to-product formulas. These formulas are particularly useful when solving trigonometric equations. (See **Section 8.5, Example 4d.**)

Sum-to-Product Formulas

$$\sin\alpha + \sin\beta = 2\sin\left(\frac{\alpha+\beta}{2}\right)\cos\left(\frac{\alpha-\beta}{2}\right)$$

$$\sin\alpha - \sin\beta = 2\sin\left(\frac{\alpha-\beta}{2}\right)\cos\left(\frac{\alpha+\beta}{2}\right)$$

$$\cos\alpha + \cos\beta = 2\cos\left(\frac{\alpha+\beta}{2}\right)\cos\left(\frac{\alpha-\beta}{2}\right)$$

$$\cos\alpha - \cos\beta = -2\sin\left(\frac{\alpha+\beta}{2}\right)\sin\left(\frac{\alpha-\beta}{2}\right)$$

Work through this **interactive video** to see how to use the **product-to-sum formulas** to verify each of the sum-to-product formulas.

Example 3 Writing the Sum or Difference of Two Functions as a Product of Two Functions

Write each sum or difference as a product of sines and/or cosines.

a. $\sin 5\theta + \sin 3\theta$ **b.** $\cos\left(\frac{3\theta}{2}\right) - \cos\left(\frac{17\theta}{2}\right)$

Solution

a. To write the sum $\sin 5\theta + \sin 3\theta$ as a product of sines and/or cosines, we must first identify the correct **sum-to-product formula**. Because we are given the sum of two sine expressions, we will use the formula

$$\sin\alpha + \sin\beta = 2\sin\left(\frac{\alpha+\beta}{2}\right)\cos\left(\frac{\alpha-\beta}{2}\right) \text{ with } \alpha = 5\theta \text{ and } \beta = 3\theta.$$

$$\sin\alpha + \sin\beta = 2\sin\left(\frac{\alpha+\beta}{2}\right)\cos\left(\frac{\alpha-\beta}{2}\right)$$ Write the appropriate product-to-sum formula.

$$\sin 5\theta + \sin 3\theta = 2\sin\left(\frac{5\theta+3\theta}{2}\right)\cos\left(\frac{5\theta-3\theta}{2}\right)$$ Substitute $\alpha = 5\theta$ and $\beta = 3\theta$.

$$= 2\sin 4\theta\cos\theta$$ Simplify the **arguments**.

Watch this **interactive video** to see each step of this solution.

b. To write the difference $\cos\left(\frac{3\theta}{2}\right) - \cos\left(\frac{17\theta}{2}\right)$ as a product of sines and/or cosines, we must first identify the correct **sum-to-product formula**. Because we are given the difference of two cosine expressions, we will use the formula

$$\cos\alpha - \cos\beta = -2\sin\left(\frac{\alpha+\beta}{2}\right)\sin\left(\frac{\alpha-\beta}{2}\right)$$ with $\alpha = \frac{3\theta}{2}$ and $\beta = \frac{17\theta}{2}$.

$$\cos\alpha - \cos\beta = -2\sin\left(\frac{\alpha+\beta}{2}\right)\sin\left(\frac{\alpha-\beta}{2}\right)$$ Write the appropriate product-to-sum formula.

$$\cos\left(\frac{3\theta}{2}\right) - \cos\left(\frac{17\theta}{2}\right) = -2\sin\left(\frac{\frac{3\theta}{2}+\frac{17\theta}{2}}{2}\right)\sin\left(\frac{\frac{3\theta}{2}-\frac{17\theta}{2}}{2}\right)$$ Substitute $\alpha = \frac{3\theta}{2}$ and $\beta = \frac{17\theta}{2}$.

$$= -2\sin\left(\frac{\frac{20\theta}{2}}{2}\right)\sin\left(\frac{-\frac{14\theta}{2}}{2}\right)$$ Combine the expressions in the numerator of each **argument**.

$$= -2\sin 5\theta\sin\left(\frac{-7\theta}{2}\right)$$ Simplify each **argument**.

 Watch this **interactive video** to see each step of this solution.

You Try It 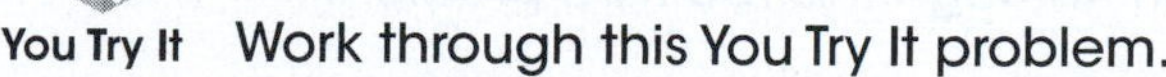 Work through this You Try It problem.

Work Exercises 9–12 in this textbook or in the MyLab Math Study Plan.

Example 4 Evaluating the Difference of Two Trigonometric Expressions

Determine the exact value of the expression $\sin\left(\frac{\pi}{12}\right) - \sin\left(\frac{17\pi}{12}\right)$ without the use of a calculator.

Solution To evaluate the expression $\sin\left(\frac{\pi}{12}\right) - \sin\left(\frac{17\pi}{12}\right)$, first use the **sum-to-product formula** $\sin\alpha - \sin\beta = 2\sin\left(\frac{\alpha-\beta}{2}\right)\cos\left(\frac{\alpha+\beta}{2}\right)$ with $\alpha = \frac{\pi}{12}$ and $\beta = \frac{17\pi}{12}$.

$\sin\left(\frac{\pi}{12}\right) - \sin\left(\frac{17\pi}{12}\right)$ Write the original expression.

$= 2\sin\left(\dfrac{\frac{\pi}{12} - \frac{17\pi}{12}}{2}\right)\cos\left(\dfrac{\frac{\pi}{12} + \frac{17\pi}{12}}{2}\right)$ Use the formula $\sin\alpha - \sin\beta = 2\sin\left(\frac{\alpha-\beta}{2}\right)\cos\left(\frac{\alpha+\beta}{2}\right)$ with $\alpha = \frac{\pi}{12}$ and $\beta = \frac{17\pi}{12}$.

Watch this **interactive video** to see how this expression simplifies to $2\sin\left(-\frac{2\pi}{3}\right)\cos\left(\frac{3\pi}{4}\right)$. Using our knowledge of the trigonometric functions of general angles, we can determine the values of $\sin\left(-\frac{2\pi}{3}\right)$ and $\cos\left(\frac{3\pi}{4}\right)$.

$2\sin\left(-\frac{2\pi}{3}\right)\cos\left(\frac{3\pi}{4}\right)$ Write the new expression that represents $\sin\left(\frac{\pi}{12}\right) - \sin\left(\frac{17\pi}{12}\right)$.

$= 2\left(-\frac{\sqrt{3}}{2}\right)\left(-\frac{1}{\sqrt{2}}\right)$ Substitute $\sin\left(-\frac{2\pi}{3}\right) = -\frac{\sqrt{3}}{2}$ and $\cos\left(\frac{3\pi}{4}\right) = -\frac{1}{\sqrt{2}}$.

$= \frac{\sqrt{3}}{\sqrt{2}}$ Multiply.

Therefore, the exact value of $\sin\left(\frac{\pi}{12}\right) - \sin\left(\frac{17\pi}{12}\right)$ is $\frac{\sqrt{3}}{\sqrt{2}}$. This expression can also be written as $\sqrt{\frac{3}{2}}$ or $\frac{\sqrt{6}}{2}$. Watch this **interactive video** to see each step of this solution.

You Try It **Work through this You Try It problem.**

Work Exercises 13–16 in this textbook or in the MyLab Math Study Plan.

OBJECTIVE 3 USING THE PRODUCT-TO-SUM AND SUM-TO-PRODUCT FORMULAS TO VERIFY IDENTITIES

We can use formulas discussed in this section along with any identity discussed so far in this text to verify trigonometric identities. As always, we will verify an identity by trying to transform one side of the identity into the exact trigonometric expression that appears on the opposite side. We typically start with what appears to be the more complicated side of the identity and then attempt to use known identities to transform that side into the trigonometric expression that appears on the other side of the identity. However, if one side of the identity includes a trigonometric expression involving the sum or difference of sine and cosine, then first substitute the appropriate **sum-to-product formula**. Likewise, if one side of the identity includes the product of sines and/or cosines, then first substitute the appropriate **product-to-sum formula**. Next, use the strategies we developed in Section 8.1 for verifying identities. To review these strategies, read this **summary** of techniques that were introduced in Section 8.1.

Example 5 Verifying a Trigonometric Identity

Verify the trigonometric identity $\dfrac{\cos\theta + \cos 3\theta}{2\cos 2\theta} = \cos\theta$.

Solution We start by rewriting the numerator of the left-hand side using a sum-to-product formula.

$$\frac{\cos\theta + \cos 3\theta}{2\cos 2\theta} \stackrel{?}{=} \cos\theta$$

Write the original trigonometric identity.

$$\frac{2\cos\left(\frac{\theta + 3\theta}{2}\right)\cos\left(\frac{\theta - 3\theta}{2}\right)}{2\cos 2\theta} \stackrel{?}{=} \cos\theta$$

Use the sum-to-product formula $\cos\alpha + \cos\beta = 2\cos\left(\frac{\alpha+\beta}{2}\right)\cos\left(\frac{\alpha-\beta}{2}\right)$ with $\alpha = \theta$ and $\beta = 3\theta$.

$$\frac{2\cos 2\theta\cos(-\theta)}{2\cos 2\theta} \stackrel{?}{=} \cos\theta$$

Simplify each **argument**.

$$\frac{\cancel{2\cos 2\theta}\cos(-\theta)}{\cancel{2\cos 2\theta}} \stackrel{?}{=} \cos\theta$$

Cancel common factors.

$$\cos(-\theta) \stackrel{?}{=} \cos\theta$$

Simplify.

$$\cos\theta = \cos\theta$$

Use the even property of cosine: $\cos(-\theta) = \cos\theta$.

The left-hand side is now identical to the right-hand side. Therefore, the identity is verified. ●

You Try It Work through this You Try It problem.

Work Exercises 17–21 in this textbook or in the MyLab Math Study Plan.

8.4 Exercises

In Exercises 1–4, write each product as the sum or difference containing only sines or cosines.

1. $\sin(8\theta)\sin(4\theta)$
2. $\sin(7\theta)\cos(3\theta)$
3. $\cos(\theta)\cos(5\theta)$
4. $\sin\left(\frac{5\theta}{2}\right)\cos\left(\frac{\theta}{2}\right)$

In Exercises 5–8, determine the exact value of each expression without the use of a calculator.

5. $\sin\left(\frac{5\pi}{24}\right)\cos\left(\frac{\pi}{24}\right)$
6. $(\sin 75°)(\sin 15°)$
7. $\cos\left(\frac{5\pi}{8}\right)\cos\left(\frac{3\pi}{8}\right)$
8. $(\cos 67.5°)(\sin 22.5°)$

In Exercises 9–12, write each sum or difference as a product containing only sines and/or cosines.

9. $\sin(9\theta) - \sin(7\theta)$

10. $\cos(5\theta) + \cos(3\theta)$

11. $\sin(11\theta) + \sin(9\theta)$

12. $\cos\left(\dfrac{5\theta}{2}\right) - \cos\left(\dfrac{11\theta}{2}\right)$

In Exercises 13–16, determine the exact value of each expression without the use of a calculator.

13. $\sin\left(\dfrac{5\pi}{12}\right) - \sin\left(\dfrac{11\pi}{12}\right)$

14. $\sin 225^\circ + \sin 135^\circ$

15. $\cos\left(\dfrac{19\pi}{12}\right) - \cos\left(\dfrac{\pi}{12}\right)$

16. $\cos 225^\circ + \cos 135^\circ$

In Exercises 17–21, verify each identity.

17. $\dfrac{\sin\theta - \sin 3\theta}{\cos\theta + \cos 3\theta} = -\tan\theta$

18. $\dfrac{\cos\theta + \cos 3\theta}{\sin\theta + \sin 3\theta} = \cot 2\theta$

19. $\dfrac{\sin x + \sin y}{\cos x + \cos y} = \tan\dfrac{x+y}{2}$

20. $\dfrac{\cos x + \cos y}{\cos x - \cos y} = -\cot\dfrac{x+y}{2}\cot\dfrac{x-y}{2}$

21. $\dfrac{\sin(6\theta) + \sin(8\theta)}{\sin(6\theta) - \sin(8\theta)} = -\dfrac{\tan(7\theta)}{\tan\theta}$

8.5 Trigonometric Equations

THINGS TO KNOW

Before working through this section, be sure that you are familiar with the following concepts:

VIDEO | ANIMATION | INTERACTIVE

You Try It

1. Solving Equations That Are Quadratic in Form (Section 1.6)

You Try It

2. Evaluating Trigonometric Functions of Angles Belonging to the $\dfrac{\pi}{3}, \dfrac{\pi}{6}$, or $\dfrac{\pi}{4}$ Families (Section 6.5)

You Try It

3. Finding the Exact and Approximate Values of an Inverse Sine Expression (Section 8.2)

You Try It

4. Finding the Exact and Approximate Values of an Inverse Cosine Expression (Section 8.2)

INTRODUCTION

Read this introduction before beginning Objective 1.

OBJECTIVES

1 Solving Trigonometric Equations That Are Linear in Form

2 Solving Trigonometric Equations That Are Quadratic in Form

3 Solving Trigonometric Equations Using Identities

4 Solving Other Types of Trigonometric Equations

5 Solving Trigonometric Equations Using a Calculator

SECTION 8.5 EXERCISES

Introduction to Section 8.5

Some trigonometric equations are true for *all* values of the variable for which each trigonometric function is defined. We call these **identities.** We have verified many identities throughout this chapter. Other trigonometric equations are only true for *specific* values of the variable (or no values at all). These are called **conditional trigonometric equations** or simply trigonometric equations.

Because of the periodic nature of trigonometric functions, we will often find that trigonometric equations have infinitely many solutions. We obviously cannot make a list of the infinitely many solutions. Therefore, we will describe these solutions using a formula (or formulas.) The most concise formula(s) describing the infinite set of solutions to a trigonometric equation is called the **general solution(s)**. We will also be interested in finding the **specific solution(s)** on a given restricted interval. We will typically try to find the solutions on the interval $[0, 2\pi)$.

To motivate our discussion of trigonometric equations, consider the equation $\sin\theta = b$ for $-1 \leq b \leq 1$. In this equation, we are trying to determine all values θ for which the sine of θ is equal to the constant b. We can get a visual representation of the solutions to this equation by sketching the graphs $y = \sin\theta$ and $y = b$ on the same coordinate plane. The solutions to the equation $\sin\theta = b$ will be the first coordinate of all points of intersection of the two graphs. Notice in **Figure 10** that the graph of $y = b$ for $-1 \leq b \leq 1$ will intersect the graph of $y = \sin\theta$ an infinite number of times. Thus, if $-1 \leq b \leq 1$, then the equation $\sin\theta = b$ has infinitely many solutions. Note that for $b > 1$ or for $b < -1$ the equation $\sin\theta = b$ has no solution because the line $y = b$ would not intersect the graph of $y = \sin\theta$. Work through this **animation** to see this visual representation of the solutions to this equation.

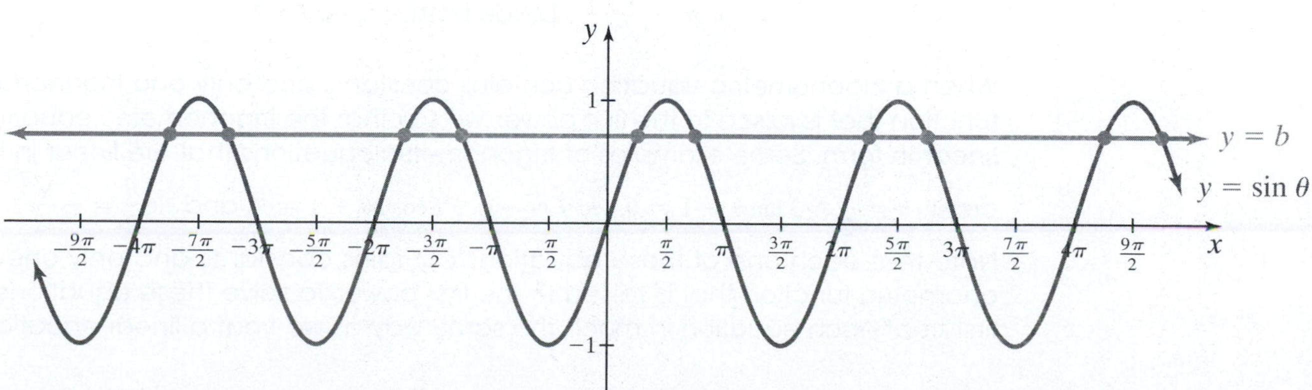

Figure 10 The graphs of $y = \sin\theta$ and $y = b$ for $-1 \leq b \leq 1$ intersect infinitely many times. Thus, the equation $\sin\theta = b$ has infinitely many solutions for $-1 \leq b \leq 1$.

If we choose to restrict θ to angles on the interval $[0, 2\pi)$, then we see that there are exactly two points of intersection of $\sin\theta = b$ and $y = b$. See **Figure 11**. Thus, the equation $\sin\theta = b$ has exactly two solutions on this interval. This is **not** always the case. It is possible to have many more solutions on a restricted interval or have no solutions at all.

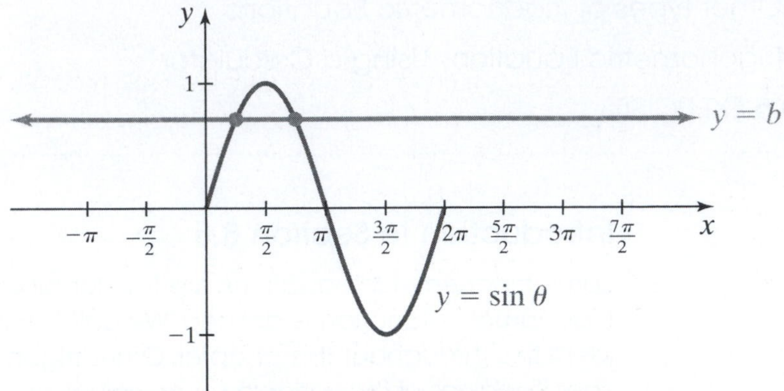

Figure 11 The graphs of $y = \sin\theta$ and $y = b$ for $-1 \le b \le 1$ intersect exactly twice on the interval $[0, 2\pi)$. Thus, the equation $\sin\theta = b$ has two solutions on the restricted interval $[0, 2\pi)$ for $-1 \le b \le 1$.

Throughout this section, we will learn a variety of techniques used to solve trigonometric equations. First, we will look at trigonometric equations that are linear in form.

OBJECTIVE 1 SOLVING TRIGONOMETRIC EQUATIONS THAT ARE LINEAR IN FORM

Recall that a linear equation in one variable is an equation that involves constants and **only one** variable that is raised to the first power. For example, the equation $\sqrt{2}x + 1 = 0$ is a linear equation in one variable. We solve this equation by isolating the variable x. We do this by subtracting 1 from both sides and then dividing both sides by $\sqrt{2}$.

$\sqrt{2}x + 1 = 0$ Write the original linear equation.

$\sqrt{2}x = -1$ Subtract 1 from both sides.

$x = -\dfrac{1}{\sqrt{2}}$ Divide both sides by $\sqrt{2}$.

When a trigonometric equation contains constants and **only one** trigonometric function that is raised to the first power, we say that the trigonometric equation is **linear in form.** Some examples of trigonometric equations that are linear in form are $\sin\theta = \frac{1}{2}$, $\sqrt{3}\tan\theta + 1 = 0$, $\sec\theta = -1$, $\sqrt{2}\cos 2\theta + 1 = 0$, and $\sin\frac{\theta}{2} = -\frac{\sqrt{3}}{2}$. Note that each one of these equations contains constants and only one trigonometric function that is raised to the first power. To solve these equations, we first treat each equation in much the same way as we treat a linear equation in one variable. That is, we use algebraic manipulations to isolate the trigonometric function. In the following two examples, we will follow a four-step procedure to solve each of these equations.

Steps for Solving Trigonometric Equations That Are Linear in Form

Step 1. Isolate the trigonometric function on one side of the equation.

Step 2. Determine the quadrants in which the terminal side of the argument of the function lies or determine the axis on which the terminal side of the argument of the function lies.

Step 3. If the terminal side of the argument of the function lies within a quadrant, then determine the reference angle and the value(s) of the argument on the interval $[0, 2\pi)$.

If the terminal side of the argument of the function lies along an axis, then determine the angle associated with it on the interval $[0, 2\pi)$, choosing from $0, \frac{\pi}{2}, \pi$, or $\frac{3\pi}{2}$.

Step 4. Use the period of the given function to determine the solutions.

Example 1 Solving Trigonometric Equations That Are Linear in Form

Determine a general formula (or formulas) for all solutions to each equation. Then, determine the specific solutions (if any) on the interval $[0, 2\pi)$.

a. $\sin\theta = \frac{1}{2}$ b. $\sqrt{3}\tan\theta + 1 = 0$ c. $\sec\theta = -1$

Solution

a. **Step 1.** For the equation $\sin\theta = \frac{1}{2}$, the trigonometric function is already isolated. Therefore, we may skip to step 2.

Step 2. Because $\sin\theta = \frac{1}{2} > 0$, we know that the terminal side of θ must lie in Quadrant I or Quadrant II.

Step 3. All values of θ must have the same reference angle, which is $\frac{\pi}{6}$ because $\sin\frac{\pi}{6} = \frac{1}{2}$. It follows that

$\theta = \text{Reference Angle} = \frac{\pi}{6}$ in Quadrant I and

$\theta = \pi - \text{Reference Angle} = \pi - \frac{\pi}{6} = \frac{5\pi}{6}$ in Quadrant II. See Figure 12.

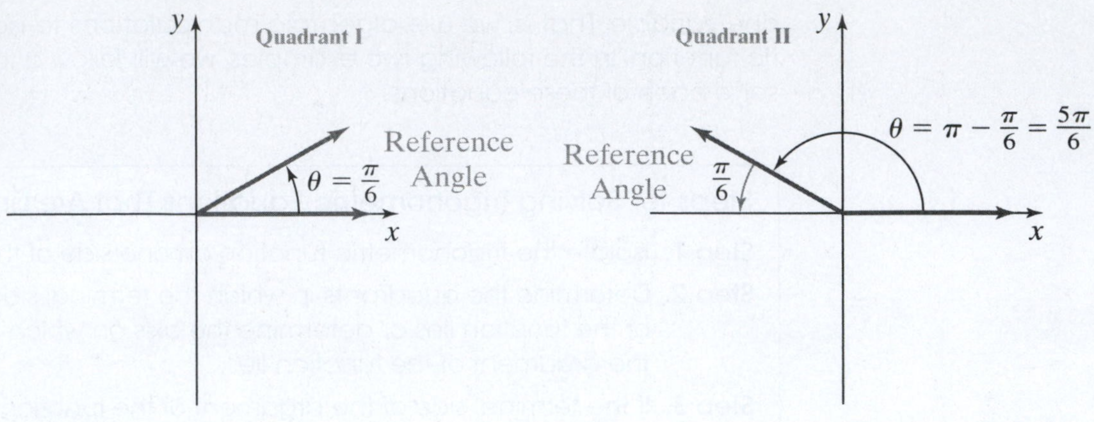

Figure 12 The reference angle is $\frac{\pi}{6}$. In Quadrant I, $\theta = \frac{\pi}{6}$. In Quadrant II, $\theta = \frac{5\pi}{6}$.

Step 4. The sine function has a period of 2π. Therefore, the possible general solutions are of the form

$$\theta = \frac{\pi}{6} + 2\pi k \quad \text{or} \quad \theta = \frac{5\pi}{6} + 2\pi k, \text{ where } k \text{ is any integer.}$$

These two equations produce uniquely different values and thus represent the general solutions to the equation $\sin\theta = \frac{1}{2}$.

In order to find the specific solutions between 0 and 2π, we must carefully observe the value of the angles to be certain that we have all angles in the interval and that we do not list angles outside of that interval. Using the formulas for the general solutions with $k = 1$, we get

$$\theta = \frac{\pi}{6} + 2\pi = \frac{13\pi}{6} \quad \text{or} \quad \theta = \frac{5\pi}{6} + 2\pi = \frac{17\pi}{6}.$$

Both of these angles lie outside of the interval $[0, 2\pi)$. Therefore, the only solutions on the interval $[0, 2\pi)$ are $\theta = \frac{\pi}{6}$ or $\theta = \frac{5\pi}{6}$. Work through this **interactive video** to see the video solution to this equation.

b. Step 1. Given the equation $\sqrt{3}\tan\theta + 1 = 0$, we start by isolating the function $\tan\theta$.

$\sqrt{3}\tan\theta + 1 = 0$	Write the original equation.
$\sqrt{3}\tan\theta = -1$	Subtract 1 from both sides.
$\tan\theta = -\frac{1}{\sqrt{3}}$	Divide both sides by $\sqrt{3}$.

Step 2. Because $\tan\theta = -\frac{1}{\sqrt{3}} < 0$, we know that the terminal side of θ must lie in Quadrant II or Quadrant IV.

Step 3. All values of θ must have the same reference angle of $\frac{\pi}{6}$ because $\tan\frac{\pi}{6} = \frac{1}{\sqrt{3}}$. It follows that

$$\theta = \pi - \text{Reference Angle} = \pi - \frac{\pi}{6} = \frac{5\pi}{6} \text{ or}$$

$$\theta = 2\pi - \text{Reference Angle} = 2\pi - \frac{\pi}{6} = \frac{11\pi}{6}. \text{ See Figure 13.}$$

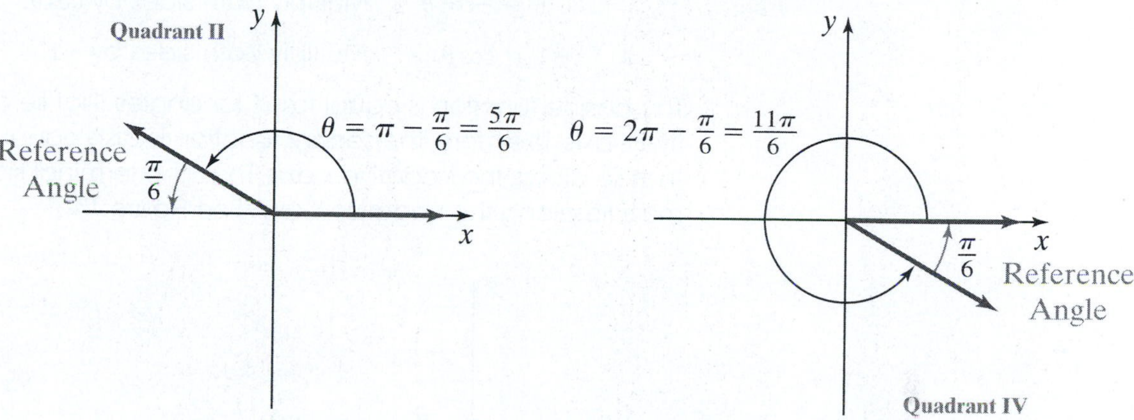

Figure 13 The reference angle is $\frac{\pi}{6}$. In Quadrant II, $\theta = \frac{5\pi}{6}$. In Quadrant IV, $\theta = \frac{11\pi}{6}$.

Step 4. The tangent function has a period of π. Therefore, the possible general solutions are of the form

$$\theta = \frac{5\pi}{6} + \pi k \quad \text{or} \quad \theta = \frac{11\pi}{6} + \pi k, \text{ where } k \text{ is any integer.}$$

However, these two equations do **not** produce unique values and thus using both formulas is not the most concise way to describe the general solution.

Note that the formula $\theta = \frac{5\pi}{6} + \pi k$ produces the exact same solutions as the formula $\theta = \frac{11\pi}{6} + \pi k$. Therefore, only one of these equations is needed to describe the general solution. We will choose the first equation. Thus, the general solution to the equation $\sqrt{3}\tan\theta + 1 = 0$ is $\theta = \frac{5\pi}{6} + \pi k$, where k is any integer.

As we observed in step 3, the only angles that satisfy the equation $\sqrt{3}\tan\theta + 1 = 0$ on the interval $[0, 2\pi)$ are $\theta = \frac{5\pi}{6}$ and $\theta = \frac{11\pi}{6}$.

Work through this **interactive video** to see the video solution to this equation.

c. **Step 1.** For the equation $\sec\theta = -1$, the trigonometric function is already isolated. Therefore, we may skip to step 2.

Step 2. To determine the quadrant in which the terminal side of θ lies or the axis on which the terminal side of θ lies, we can use the reciprocal identity $\sec\theta = \dfrac{1}{\cos\theta}$ and then use our knowledge of the more familiar cosine function.

$\sec\theta = -1$	Write the original equation.
$\dfrac{1}{\cos\theta} = -1$	Use the reciprocal identity $\sec\theta = \dfrac{1}{\cos\theta}$.
$1 = -\cos\theta$	Multiply both sides by $\cos\theta$.
$-1 = \cos\theta$	Multiply both sides by -1.

The cosine function is equal to -1 for angles that lie along the negative x-axis. Therefore, the secant function is also equal to -1 for angles that lie along the negative x-axis. Thus, the terminal side of angle θ must lie along the negative x-axis. See Figure 14.

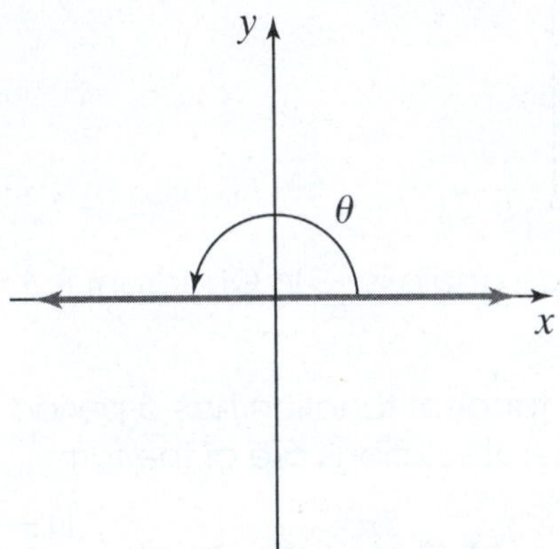

Figure 14 For the equation $\sec\theta = -1$, the terminal side of θ lies along the negative x-axis.

Step 3. The angle on the interval $[0, 2\pi)$ that lies along the negative x-axis is $\theta = \pi$.

Step 4. The secant (and cosine) function has a period of 2π. Therefore, the general solutions are of the form $\theta = \pi + 2\pi k$. The only solution to the equation $\sec\theta = -1$ on the interval $[0, 2\pi)$ is $\theta = \pi$.

Work through this **interactive video** to see the video solution to this equation.

You Try It **Work through this You Try It problem.**

Work Exercises 1–4 in this textbook or in the MyLab Math Study Plan.

In Example 1, the **argument** of the trigonometric function in each equation was θ. In Example 2, we look at equations in which the argument is something other than just θ.

Example 2 Solving Trigonometric Equations That Are Linear in Form

Determine a general formula (or formulas) for all solutions to each equation. Then, determine the specific solutions (if any) on the interval $[0, 2\pi)$.

a. $\sqrt{2}\cos 2\theta + 1 = 0$ b. $\sin\dfrac{\theta}{2} = -\dfrac{\sqrt{3}}{2}$ c. $\tan\left(\theta + \dfrac{\pi}{6}\right) + 1 = 0$

Solution

a. **Step 1.** To isolate the trigonometric function, we first subtract 1 from both sides then divide both sides by $\sqrt{2}$.

$\sqrt{2}\cos 2\theta + 1 = 0$ Write the original equation.

$\sqrt{2}\cos 2\theta = -1$ Subtract 1 from both sides.

$\cos 2\theta = -\dfrac{1}{\sqrt{2}}$ Divide both sides by $\sqrt{2}$.

Step 2. Because $\cos 2\theta = -\dfrac{1}{\sqrt{2}} < 0$, we know that the terminal side of the argument, 2θ, must lie in Quadrant II or Quadrant III.

Step 3. All values of 2θ must have the reference angle of $\dfrac{\pi}{4}$ because $\cos\dfrac{\pi}{4} = \dfrac{1}{\sqrt{2}}$. It follows that

$$2\theta = \pi - \text{Reference Angle} = \pi - \frac{\pi}{4} = \frac{3\pi}{4} \text{ or}$$

$$2\theta = \text{Reference Angle} + \pi = \frac{\pi}{4} + \pi = \frac{5\pi}{4}.$$ See **Figure 15**.

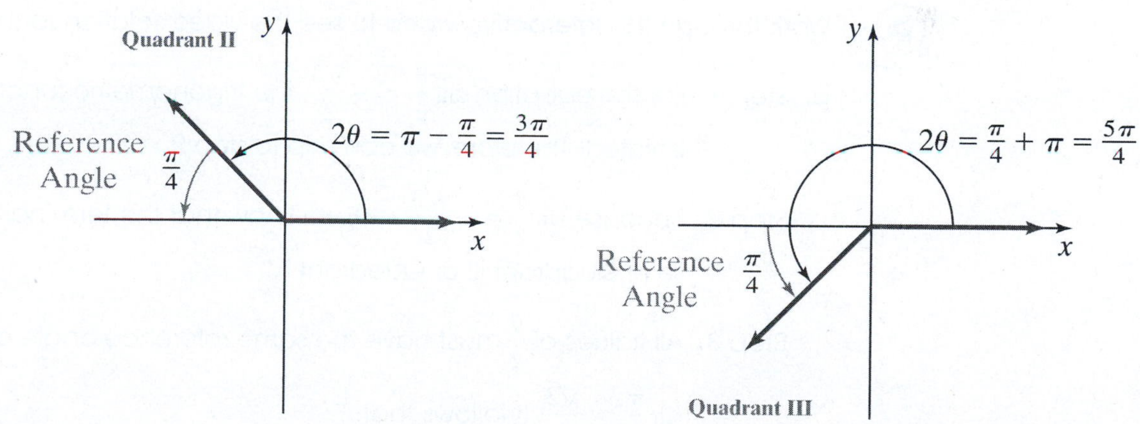

Figure 15 The reference angle is $\dfrac{\pi}{4}$. In Quadrant II, $2\theta = \dfrac{3\pi}{4}$. In Quadrant III, $2\theta = \dfrac{5\pi}{4}$.

Step 4. The cosine function has a period of 2π. Therefore, the solutions for 2θ are of the form

$$2\theta = \frac{3\pi}{4} + 2\pi k \quad \text{or} \quad 2\theta = \frac{5\pi}{4} + 2\pi k, \text{ where } k \text{ is any integer.}$$

To determine the general solutions, solve for θ by dividing both sides of each of these equations by 2.

$$2\theta = \frac{3\pi}{4} + 2\pi k \quad \text{or} \quad 2\theta = \frac{5\pi}{4} + 2\pi k$$ Write the two equations representing 2θ.

$$\theta = \frac{3\pi}{8} + \pi k \quad \text{or} \quad \theta = \frac{5\pi}{8} + \pi k$$ Divide both sides of each equation by 2.

These two equations produce uniquely different values. Thus, the formulas describing the general solutions are $\theta = \frac{3\pi}{8} + \pi k$ or $\theta = \frac{5\pi}{8} + \pi k$, where k is any integer.

In order to find the specific solutions on the interval $[0, 2\pi)$, we must carefully observe the value of the angles to be certain that we have all angles in the interval and that we do not list angles outside of that interval.

Using $k = 1$ for each formula above, we get $\theta = \frac{3\pi}{8} + \pi = \frac{11\pi}{8}$ or $\theta = \frac{5\pi}{8} + \pi = \frac{13\pi}{8}$. Notice that both of these new solutions are within the interval $[0, 2\pi)$.

Using $k = 2$, we get $\theta = \frac{3\pi}{8} + 2\pi = \frac{19\pi}{8}$ or $\theta = \frac{5\pi}{8} + 2\pi = \frac{21\pi}{8}$ and we observe that these angles are outside of the interval $[0, 2\pi)$.

Thus, there are four specific solutions to the equation $\sqrt{2}\cos 2\theta + 1 = 0$ on the interval $[0, 2\pi)$. These solutions are $\theta = \frac{3\pi}{8}, \theta = \frac{5\pi}{8}, \theta = \frac{11\pi}{8}$ and $\theta = \frac{13\pi}{8}$.

 Work through this **interactive video** to see the video solution to this equation.

b. Step 1. For the equation $\sin\frac{\theta}{2} = -\frac{\sqrt{3}}{2}$, the trigonometric function is already isolated. Therefore, we can skip to step 2.

Step 2. Because $\sin\frac{\theta}{2} = -\frac{\sqrt{3}}{2} < 0$, we know that the terminal side of $\frac{\theta}{2}$ must lie in Quadrant III or Quadrant IV.

Step 3. All values of $\frac{\theta}{2}$ must have the same reference angle of $\frac{\pi}{3}$ because $\sin\frac{\pi}{3} = \frac{\sqrt{3}}{2}$. It follows that

$$\frac{\theta}{2} = \text{Reference Angle} + \pi = \frac{\pi}{3} + \pi = \frac{4\pi}{3} \text{ or}$$

$$\frac{\theta}{2} = 2\pi - \text{Reference Angle} = 2\pi - \frac{\pi}{3} = \frac{5\pi}{3}.$$ See **Figure 16**.

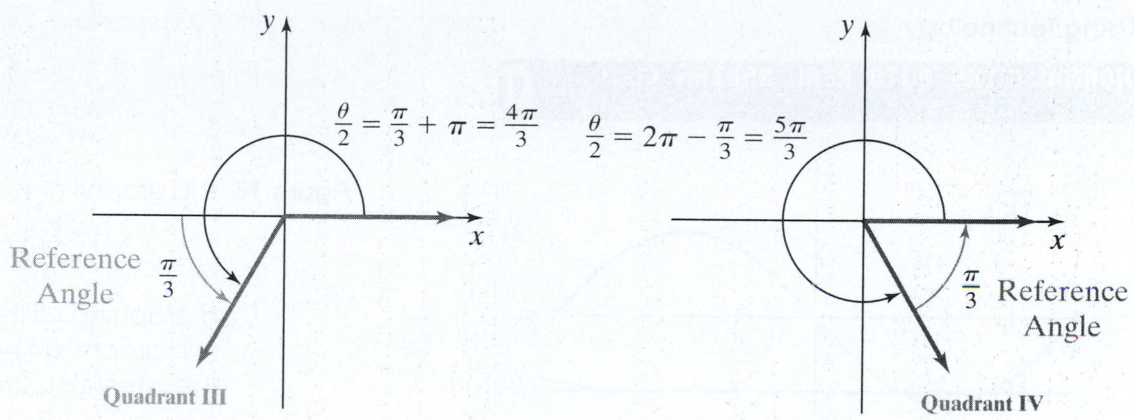

Figure 16 The reference angle is $\frac{\pi}{3}$. In Quadrant III, $\frac{\theta}{2} = \frac{4\pi}{3}$. In Quadrant IV, $\frac{\theta}{2} = \frac{5\pi}{3}$.

Step 4. The sine function has a period of 2π. Therefore, the solutions for $\frac{\theta}{2}$ are of the form $\frac{\theta}{2} = \frac{4\pi}{3} + 2\pi k$ or $\frac{\theta}{2} = \frac{5\pi}{3} + 2\pi k$, where k is any integer.

To determine the general solutions, solve each of these equations for θ by multiplying both sides of each equation by 2.

$$\frac{\theta}{2} = \frac{4\pi}{3} + 2\pi k \quad \text{or} \quad \frac{\theta}{2} = \frac{5\pi}{3} + 2\pi k$$ Write the two equations representing $\frac{\theta}{2}$.

$$\theta = \frac{8\pi}{3} + 4\pi k \quad \text{or} \quad \theta = \frac{10\pi}{3} + 4\pi k$$ Multiply both sides of each equation by 2.

These two equations produce uniquely different values. Thus, the general solutions are $\theta = \frac{8\pi}{3} + 4\pi k$ or $\theta = \frac{10\pi}{3} + 4\pi k$, where k is any integer.

Note that for any integer value of k, the angles $\theta = \frac{8\pi}{3} + 4\pi k$ and $\theta = \frac{10\pi}{3} + 4\pi k$ lie outside of the interval $[0, 2\pi)$. Therefore, the equation $\sin\frac{\theta}{2} = -\frac{\sqrt{3}}{2}$ has no solution on the interval $[0,2\pi)$. Work through this interactive video to see the video solution to this equation.

CAUTION It is possible for a trigonometric equation to have general solutions but not have a solution on the interval $[0, 2\pi)$ as in the previous example. Therefore, it is extremely important to check carefully to see if a solution exists on an indicated interval. Using a graphing utility, we can see why there is no solution to the equation $\sin\frac{\theta}{2} = -\frac{\sqrt{3}}{2}$ on the interval $[0, 2\pi)$. See Figure 17.

Using Technology

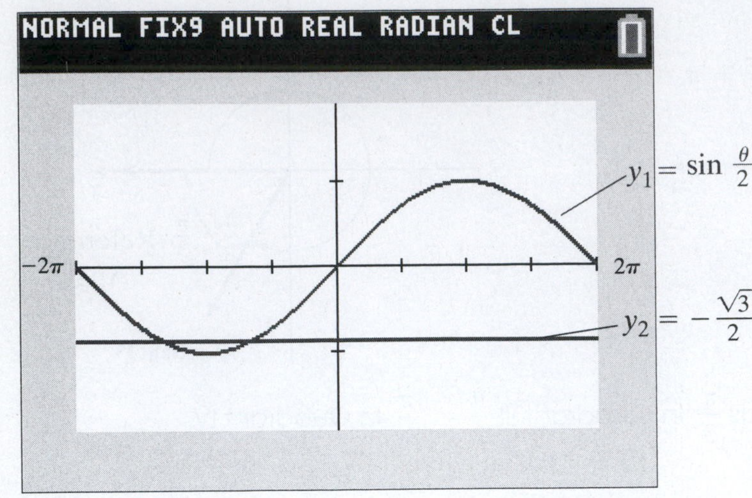

Figure 17 The graphs of $y_1 = \sin\frac{\theta}{2}$ and $y_2 = -\frac{\sqrt{3}}{2}$ are shown here using a graphing utility. Notice that the two graphs do not intersect on the interval $[0, 2\pi)$. Thus, the equation $\sin\frac{\theta}{2} = -\frac{\sqrt{3}}{2}$ has no solution on the restricted interval $[0, 2\pi)$.

c. **Step 1.** For the equation $\tan\left(\theta + \frac{\pi}{6}\right) + 1 = 0$, we isolate the trigonometric function by subtracting 1 from both sides to obtain the equation $\tan\left(\theta + \frac{\pi}{6}\right) = -1$.

Step 2. Because $\tan\left(\theta + \frac{\pi}{6}\right) = -1 < 0$, we know that the terminal side of the argument, $\left(\theta + \frac{\pi}{6}\right)$, must lie in Quadrant II or Quadrant IV.

Step 3. All values of the angle $\left(\theta + \frac{\pi}{6}\right)$ must have a reference angle of $\frac{\pi}{4}$ because $\tan\frac{\pi}{4} = 1$. It follows that

$$\left(\theta + \frac{\pi}{6}\right) = \pi - \text{Reference Angle} = \pi - \frac{\pi}{4} = \frac{3\pi}{4} \text{ or}$$

$$\left(\theta + \frac{\pi}{6}\right) = 2\pi - \text{Reference Angle} = 2\pi - \frac{\pi}{4} = \frac{7\pi}{4}.$$ See Figure 18.

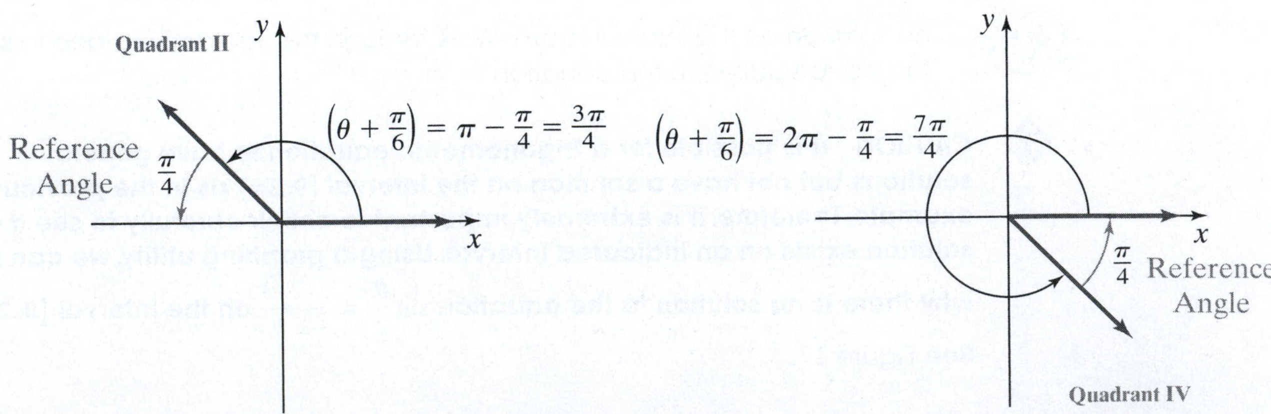

Figure 18 The reference angle is $\frac{\pi}{4}$. In Quadrant II, $\left(\theta + \frac{\pi}{6}\right) = \frac{3\pi}{4}$. In Quadrant IV, $\left(\theta + \frac{\pi}{6}\right) = \frac{7\pi}{4}$.

Step 4. The tangent function has a period of π. Therefore, the solutions for $\left(\theta + \frac{\pi}{6}\right)$ are of the form

$$\left(\theta + \frac{\pi}{6}\right) = \frac{3\pi}{4} + \pi k \text{ or } \left(\theta + \frac{\pi}{6}\right) = \frac{7\pi}{4} + \pi k, \text{ where } k \text{ is any integer.}$$

Note that both equations produce exactly the same values and thus using both equations is not the most concise way to describe the general solution. Therefore, only one equation is needed. Thus, we will only use $\left(\theta + \frac{\pi}{6}\right) = \frac{3\pi}{4} + \pi k$. This situation also exists in **Example 1b** and often occurs when solving a trigonometric equation involving a tangent function because the period of the tangent function is π.

To determine the general solution, remove the parentheses and solve for θ.

$\theta + \frac{\pi}{6} = \frac{3\pi}{4} + \pi k$	Write the equation representing $\theta + \frac{\pi}{6}$.
$\theta = \frac{3\pi}{4} - \frac{\pi}{6} + \pi k$	Subtract $\frac{\pi}{6}$ from both sides.
$\theta = \frac{9\pi}{12} - \frac{2\pi}{12} + \pi k$	Write each fraction using a common denominator.
$\theta = \frac{7\pi}{12} + \pi k$	Simplify.

Thus, the general solution to the equation $\tan\left(\theta + \frac{\pi}{6}\right) + 1 = 0$ is $\theta = \frac{7\pi}{12} + \pi k$. Using the formula for the general solutions with $k = 0$ and $k = 1$, we get

$$\theta = \frac{7\pi}{12} + \pi(0) = \frac{7\pi}{12} \text{ and}$$

$$\theta = \frac{7\pi}{12} + \pi(1) = \frac{19\pi}{12}$$

Both of these angles lie in the interval $[0, 2\pi)$. Any other values of k will produce angles outside of the interval $[0, 2\pi)$. Therefore, the only solutions on the interval $[0, 2\pi)$ are $\theta = \frac{7\pi}{12}$ and $\theta = \frac{19\pi}{12}$.

Work through this **interactive video** to see the video solution to this equation. ●

You Try It Work through this You Try It problem.

Work Exercises 5–13 in this textbook or in the MyLab Math Study Plan.

OBJECTIVE 2 SOLVING TRIGONOMETRIC EQUATIONS THAT ARE QUADRATIC IN FORM

Recall that a quadratic equation has the form $ax^2 + bx + c = 0,\ a \neq 0$. These equations are relatively straightforward to solve because we know several methods for solving these types of equations. If $f(\theta)$ is a trigonometric function, then the equation $a(f(\theta))^2 + b(f(\theta)) + c = 0, a \neq 0$ is said to be quadratic in form because we can transform it into a quadratic equation using the substitution $u = f(\theta)$. Example 3 illustrates two trigonometric equations that are quadratic in form.

Example 3 Solving a Trigonometric Equation That Is Quadratic in Form

Determine a general formula (or formulas) for all solutions to each equation. Then, determine the specific solutions (if any) on the interval $[0, 2\pi)$.

a. $\sin^2\theta - 4\sin\theta + 3 = 0$ b. $4\cos^2\theta - 3 = 0$

Solution

a. $\sin^2\theta - 4\sin\theta + 3 = 0$ Write the original equation.

Let $u = \sin\theta$, then $u^2 = \sin^2\theta$. Determine the proper substitution.

$u^2 - 4u + 3 = 0$ Substitute u for $\sin\theta$ and u^2 for $\sin^2\theta$ into the original equation.

$(u - 3)(u - 1) = 0$ Factor.

$u - 3 = 0 \text{ or } u - 1 = 0$ Use the **zero product property**.

$u = 3 \text{ or } u = 1$ Solve for u.

$\sin\theta = 3 \text{ or } \sin\theta = 1$ Substitute $\sin\theta$ for u.

We now have two trigonometric functions that are linear in form similar to the equations from **Example 1**. Note that the range of $y = \sin\theta$ is $[-1, 1]$; therefore, it is impossible for the sine of an angle to equal 3. Thus, the equation $\sin\theta = 3$ has no solution. We can solve the equation $\sin\theta = 1$ using the **steps for solving trigonometric equations that are linear in form** discussed earlier. Try solving the equation $\sin\theta = 1$ on your own. When you're finished, check your **answer** or watch this **interactive video** to see each step of the solution process.

b. To solve the equation $4\cos^2\theta - 3 = 0$, we first make the substitution $u = \cos\theta$.

$4\cos^2\theta - 3 = 0$ Write the original equation.

Let $u = \cos\theta$, then $u^2 = \cos^2\theta$. Determine the proper substitution.

$4u^2 - 3 = 0$ Substitute u^2 for $\cos^2\theta$ in the original equation.

$4u^2 = 3$ Add 3 to both sides.

$u^2 = \frac{3}{4}$ Divide both sides by 4.

$u = \pm\sqrt{\frac{3}{4}} = \pm\frac{\sqrt{3}}{2}$ Use the **square root property**.

$\cos\theta = \pm\frac{\sqrt{3}}{2}$ Substitute $\cos\theta$ for u.

We now see that $\cos\theta = \frac{\sqrt{3}}{2}$ or $\cos\theta = -\frac{\sqrt{3}}{2}$. To solve each equation, we follow the **steps for solving trigonometric equations that are linear in form**.

Follow these steps or work through this **interactive video** to verify the following general solutions:

General solutions to the equation $\cos\theta = \frac{\sqrt{3}}{2}$: $\quad \theta = \frac{\pi}{6} + 2\pi k$ or $\theta = \frac{11\pi}{6} + 2\pi k$

General solutions to the equation $\cos\theta = -\frac{\sqrt{3}}{2}$: $\quad \theta = \frac{5\pi}{6} + 2\pi k$ or $\theta = \frac{7\pi}{6} + 2\pi k$

All four of these equations could be used to describe the general solution to the original equation $4\cos^2\theta - 3 = 0$. However, this is not the most concise way to describe the general solution. Note that the two equations $\theta = \frac{\pi}{6} + \pi k$ and $\theta = \frac{5\pi}{6} + \pi k$ describe the exact same values as the four equations describe. Therefore, the most concise way to describe the general solutions to the equation $4\cos^2\theta - 3 = 0$ is $\theta = \frac{\pi}{6} + \pi k$ or $\theta = \frac{5\pi}{6} + \pi k$.

Use the general formulas $\theta = \frac{\pi}{6} + \pi k$ and $\theta = \frac{5\pi}{6} + \pi k$ using different integer values of k until we find all specific solutions to the equation $4\cos^2\theta - 3 = 0$ on the interval $[0, 2\pi)$. We start with $k = -1$ as sometimes a specific solution can be found using negative values of k. If a specific solution is found using $k = -1$, then try $k = -2$. Continue this process until the angles found are outside the interval $[0, 2\pi)$.

$k = -1$: $\theta = \frac{\pi}{6} + \pi(-1) = -\frac{5\pi}{6}$ and $\theta = \frac{5\pi}{6} + \pi(-1) = -\frac{\pi}{6}$ (Neither angle lies within the interval 30, 2π).)

There were no specific solutions found using the negative interger $k = -1$, so we now try $k = 0, k = 1$, etc.

$k = 0$: $\theta = \frac{\pi}{6} + \pi(0) = \frac{\pi}{6}$ and $\theta = \frac{5\pi}{6} + \pi(0) = \frac{5\pi}{6}$ (Both angles lie within the interval 30, 2π).)

$k = 1$: $\theta = \frac{\pi}{6} + \pi(1) = \frac{7\pi}{6}$ and $\theta = \frac{5\pi}{6} + \pi(1) = \frac{11\pi}{6}$ (Both angles lie within the interval 30, 2π).)

$k = 2$: $\theta = \frac{\pi}{6} + \pi(2) = \frac{13\pi}{6}$ and $\theta = \frac{5\pi}{6} + \pi(2) = \frac{17\pi}{6}$ (Neither angle lies within the interval 30, 2π).)

We see that when $k = 2$, the angles produced do not lie within the interval $[0, 2\pi)$. Thus, there are no more specific solutions on the interval $[0, 2\pi)$.

Therefore, $\theta = \frac{\pi}{6}, \theta = \frac{5\pi}{6}, \theta = \frac{7\pi}{6}$, and $\theta = \frac{11\pi}{6}$ are the four specific solutions to the equation $4\cos^2\theta - 3 = 0$ on the interval $[0, 2\pi)$.

Work through this **interactive video** to see the video solution to this equation. ●

You Try It Work through this You Try It problem.

Work Exercises 14–19 in this textbook or in the MyLab Math Study Plan.

OBJECTIVE 3 SOLVING TRIGONOMETRIC EQUATIONS USING IDENTITIES

We will often encounter trigonometric equations that cannot be solved directly. However, it might be possible to use a trigonometric identity to transform the equation into a more manageable equation. Then, we can use solving techniques already discussed. Before we look at such equations, take a moment to refresh your memory by clicking on the identity and formula links below.

Quotient Identities	Half-Angle Formulas
Reciprocal Identities	Double Angle Formulas
Pythagorean Identities	Product-to-Sum Formulas
Cofunction Identities	Sum-to-Product Formulas

Example 4 Solving Trigonometric Equation Using Identities

Determine a general formula (or formulas) for all solutions to each equation. Then, determine the specific solutions (if any) on the interval $[0, 2\pi)$.

a. $2\sin^2\theta = 3\cos\theta + 3$ b. $\sin\theta\cos\theta = -\frac{1}{2}$

c. $\cos 2\theta + 4\sin^2\theta = 2$ d. $\sin 5\theta + \sin 3\theta = 0$

Solution

a. It appears that the equation $2\sin^2\theta = 3\cos\theta + 3$ might be quadratic in form. But, because there are two different trigonometric functions, there is no immediate substitution that will allow us to transform the equation into a quadratic equation. However, we can first use the Pythagorean identity of the form $\sin^2\theta = 1 - \cos^2\theta$ to produce an equation that contains all functions in terms of cosine.

$2\sin^2\theta = 3\cos\theta + 3$	Write the original equation.
$2(1 - \cos^2\theta) = 3\cos\theta + 3$	Use the substitution $\sin^2\theta = 1 - \cos^2\theta$.
$2 - 2\cos^2\theta = 3\cos\theta + 3$	Distribute.
$0 = 2\cos^2\theta + 3\cos\theta + 1$	Add $2\cos^2\theta$ to both sides and subtract 2 from both sides.
$2\cos^2\theta + 3\cos\theta + 1 = 0$	Rewrite the equation so that 0 is on the right-hand side.
Let $u = \cos\theta$, then $u^2 = \cos^2\theta$	Determine the proper substitution.
$2u^2 + 3u + 1 = 0$	Substitute u for $\cos\theta$ and u^2 for $\cos^2\theta$.
$(2u + 1)(u + 1) = 0$	Factor.
$2u + 1 = 0$ or $u + 1 = 0$	Use the zero product property.
$u = -\frac{1}{2}$ or $u = -1$	Solve for u.
$\cos\theta = -\frac{1}{2}$ or $\cos\theta = -1$	Substitute $\cos\theta$ for u.

We now see that $\cos\theta = -\frac{1}{2}$ or $\cos\theta = -1$. To solve each equation, follow the **steps for solving trigonometric equations that are linear in form.**

The general solutions to the equation $\cos\theta = -\frac{1}{2}$ are $\theta = \frac{2\pi}{3} + 2\pi k$ or $\theta = \frac{4\pi}{3} + 2\pi k$. The general solution to the equation $\cos\theta = -1$ is $\theta = \pi + 2\pi k$.

Therefore, the general solutions to the equation $2\sin^2\theta = 3\cos\theta + 3$ are $\theta = \frac{2\pi}{3} + 2\pi k$ or $\theta = \frac{4\pi}{3} + 2\pi k$ or $\theta = \pi + 2\pi k$.

There are three specific solutions to the equation $2\sin^2\theta = 3\cos\theta + 3$ on the interval $[0, 2\pi)$. These solutions are $\theta = \frac{2\pi}{3}$, $\theta = \pi$, and $\theta = \frac{4\pi}{3}$.

Work through this **interactive video** to see every step of the solution to this equation.

b. Recall that the double-angle formula for sine is given by $\sin 2\theta = 2\sin\theta\cos\theta$. If we multiply both sides of the equation $\sin\theta\cos\theta = -\frac{1}{2}$ by 2, we will be able to use this double-angle formula to transform the equation into a trigonometric equation that is linear in form.

$\sin\theta\cos\theta = -\frac{1}{2}$	Write the original equation.
$2\sin\theta\cos\theta = -1$	Multiply both sides by 2.
$\sin 2\theta = -1$	Substitute $\sin 2\theta$ for $2\sin\theta\cos\theta$.

We now have transformed the original equation into a trigonometric equation that is linear in form. Follow the **steps for solving trigonometric equations that are linear in form** to solve this equation.

Step 1. For the equation $\sin 2\theta = -1$, the trigonometric function is already isolated. Therefore, we can skip to step 2.

Step 2. Because $\sin 2\theta = -1$, we know that the terminal side of the **argument**, 2θ, must lie along the negative y-axis.

Step 3. The angle on the interval $[0, 2\pi)$ that lies along the negative y-axis is $2\theta = \frac{3\pi}{2}$. See **Figure 19**.

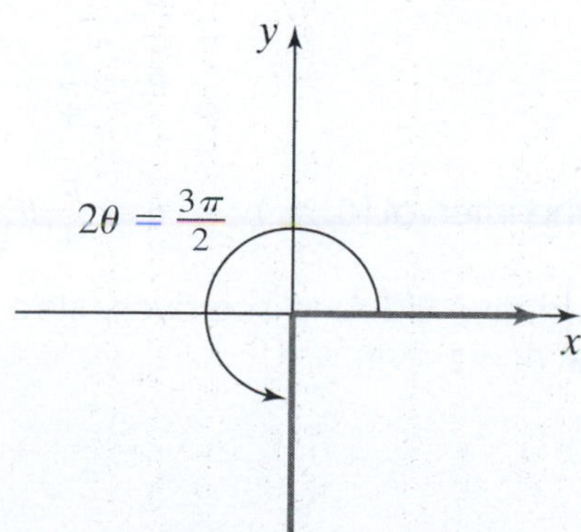

Figure 19 The sine function is equal to -1 for angles that lie on the negative y-axis. The angle on the interval $[0, 2\pi)$ that lies along the negative y-axis is $2\theta = \frac{3\pi}{2}$.

Step 4. The sine function has a period of 2π. Therefore, the solutions for 2θ are of the form $2\theta = \frac{3\pi}{2} + 2\pi k$, where k is any integer. We can divide both sides this equations by 2 to solve for θ.

$2\theta = \frac{3\pi}{2} + 2\pi k$ Write the equation representing 2θ.

$\theta = \frac{3\pi}{4} + \pi k$ Divide both sides by 2.

Therefore, the general solution to the equation $\sin\theta\cos\theta = -\frac{1}{2}$ is $\theta = \frac{3\pi}{4} + \pi k$, where k is any integer.

We can use the general solution $\theta = \frac{3\pi}{4} + \pi k$ using different values of k until we find all specific solutions on the interval $[0, 2\pi)$. We start with $k = -1$.

$k = -1$: $\theta = \frac{3\pi}{4} + \pi(-1) = \frac{3\pi}{4} - \pi = \frac{3\pi}{4} - \frac{4\pi}{4} = -\frac{\pi}{4}$ (This angle does **not** lie within the interval $[0, 2\pi)$.)

$k = 0$: $\theta = \frac{3\pi}{4} + \pi(0) = \frac{3\pi}{4}$ (This angle lies within the interval $[0, 2\pi)$.)

$k = 1$: $\theta = \frac{3\pi}{4} + \pi(1) = \frac{3\pi}{4} + \pi = \frac{3\pi}{4} + \frac{4\pi}{4} = \frac{7\pi}{4}$ (This angle lies within the interval $[0, 2\pi)$.)

$k = 2$: $\theta = \frac{3\pi}{4} + \pi(2) = \frac{3\pi}{4} + 2\pi = \frac{3\pi}{4} + \frac{8\pi}{4} = \frac{11\pi}{4}$ (This angle does **not** lie within the interval $[0, 2\pi)$.)

The specific solutions to the equation $\sin\theta\cos\theta = -\frac{1}{2}$ on the interval $[0, 2\pi)$ are $\theta = \frac{3\pi}{4}$ and $\theta = \frac{7\pi}{4}$.

Work through this **interactive video** to every step of the solution to the equation $\sin\theta\cos\theta = -\frac{1}{2}$.

c. For the equation $\cos 2\theta + 4\sin^2\theta = 2$, use a **double-angle formula for cosine** to rewrite the function $\cos 2\theta$ in terms of only sine. This will transform the equation into a trigonometric equation that is quadratic in form. Try solving this equation on your own to see if you obtain the following solutions.

General solution: $\theta = \frac{\pi}{4} + \frac{\pi}{2}k$

Specific solutions on the interval $[0, 2\pi)$: $\theta = \frac{\pi}{4}, \theta = \frac{3\pi}{4}, \theta = \frac{5\pi}{4}$ and $\theta = \frac{7\pi}{4}$

Work through this **interactive video** to see every step of the solution to the equation $\cos 2\theta + 4\sin^2\theta = 2$.

d. For the equation $\sin 5\theta + \sin 3\theta = 0$, first use a **sum-to-product formula.** (See Section 8.4, **Example 3a.**) This will transform the equation into the product of two trigonometric functions that will allow us to use the **zero product property**.

$\sin 5\theta + \sin 3\theta = 0$	Write the original equation.
$2\sin\left(\frac{5\theta + 3\theta}{2}\right)\cos\left(\frac{5\theta - 3\theta}{2}\right) = 0$	Use the sum-to-product formula $\sin\alpha + \sin\beta = 2\sin\left(\frac{\alpha+\beta}{2}\right)\cos\left(\frac{\alpha-\beta}{2}\right)$.
$2\sin 4\theta\cos\theta = 0$	Simplify the arguments.
$\sin 4\theta\cos\theta = 0$	Divide both sides by 2.
$\sin 4\theta = 0$ or $\cos\theta = 0$	Use the **zero product property**.

Solve these two equations on your own. Check your work by watching this interactive video.

You Try It Work through this You Try It problem.

Work Exercises 20–29 in this textbook or in the MyLab Math Study Plan.

OBJECTIVE 4 SOLVING OTHER TYPES OF TRIGONOMETRIC EQUATIONS

In addition to the techniques for solving trigonometric equations discussed so far in this section, there are still other techniques that can be used. These include but are not limited to the use of factoring techniques and creative algebraic manipulations.

Example 5 Solving Other Types of Trigonometric Equations

Determine a general formula (or formulas) for all solutions to each equation. Then, determine the specific solutions (if any) on the interval $[0, 2\pi)$.

a. $\sin 2\theta + 2\cos\theta\sin 2\theta = 0$ **b.** $\cos^2\theta = \sin\theta\cos\theta$ **c.** $\sin\theta + \cos\theta = 1$

Solution

a. Observe that the two terms of the equation $\sin 2\theta + 2\cos\theta\sin 2\theta = 0$ on the left-hand side contain the common factor $\sin 2\theta$. We start by factoring out this common factor.

$\sin 2\theta + 2\cos\theta\sin 2\theta = 0$	Write the original equation.
$\sin 2\theta(1 + 2\cos\theta) = 0$	Factor out the common factor of $\sin 2\theta$.
$\sin 2\theta = 0$ or $1 + 2\cos\theta = 0$	Use the **zero product property**.

We now solve the equations $\sin 2\theta = 0$ and $1 + 2\cos\theta = 0$ independently. Try solving these two equations on your own. When you have solved them, work through this **interactive video** to see if you are correct.

b. To solve the equation $\cos^2\theta = \sin\theta\cos\theta$, it is tempting to divide both sides of the equation by $\cos\theta$ and then cancel a factor of $\cos\theta$ on each side of the equation. However, if $\cos\theta = 0$, that would mean we divided by zero (which is undefined). This would result in the "loss" of any solutions that come from the equation $\cos\theta = 0$. Instead, follow the procedure below.

$\cos^2\theta = \sin\theta\cos\theta$	Write the original equation.
$\cos^2\theta - \sin\theta\cos\theta = 0$	Subtact $\sin\theta\cos\theta$ from both sides.
$\cos\theta\,(\cos\theta - \sin\theta) = 0$	Factor out the common factor of $\cos\theta$.
$\cos\theta = 0$ or $\cos\theta - \sin\theta = 0$	Use the **zero product property**.

We now solve the equations $\cos\theta = 0$ and $\cos\theta - \sin\theta = 0$ independently.

To solve the equation $\cos\theta = 0$, follow the **steps for solving trigonometric equations that are linear in form.** Read these steps to verify that the general solutions to the equation $\cos\theta = 0$ are of the form $\theta = \frac{\pi}{2} + \pi k$, where k is any integer.

To solve the equation $\cos\theta - \sin\theta = 0$, start by adding $\sin\theta$ to both sides of the equation.

$\cos\theta - \sin\theta = 0$	Write the equation.
$\cos\theta = \sin\theta$	Add $\sin\theta$ to both sides.
$\sin\theta = \cos\theta$	Rearrange the order of the equation.

Note that we are now able to divide both sides of the equation $\sin\theta = \cos\theta$ by $\cos\theta$ without dividing by zero and without losing any solutions. To see why this is true, read this **explanation**.

$\sin\theta = \cos\theta$	Write the equation.
$\frac{\sin\theta}{\cos\theta} = 1$	Divide both sides by $\cos\theta$.
$\tan\theta = 1$	Substitute $\frac{\sin\theta}{\cos\theta} = \tan\theta$.

Work through this **interactive video** to verify that the general solution to the equation $\tan\theta = 1$ is $\theta = \frac{\pi}{4} + \pi k$, where k is any integer.

Thus, the general solutions to the equation $\cos^2\theta = \sin\theta\cos\theta$ are the two formulas $\theta = \frac{\pi}{2} + \pi k$ or $\theta = \frac{\pi}{4} + \pi k$, where k is any integer.

The solutions to the equation $\cos^2\theta = \sin\theta\cos\theta$ on the interval $[0, 2\pi)$ are $\theta = \frac{\pi}{4}, \theta = \frac{\pi}{2}, \theta = \frac{5\pi}{4}$, and $\theta = \frac{3\pi}{2}$.

CAUTION In the previous example, it was not possible to divide both sides of the original equation $\cos^2\theta = \sin\theta\cos\theta$ by $\cos\theta$ because, for this equation, the expression $\cos\theta$ could equal 0 (and division by zero is never allowed.) However, we saw that it was possible to divide both sides of the equation $\sin\theta = \cos\theta$ by $\cos\theta$ because we verified in this equation that $\cos\theta$ could not equal 0. Therefore, be extremely careful when attempting to multiply or divide both sides of a trigonometric equation by a function. In order to perform this operation, you must first verify that the function cannot equal 0 for the given equation.

c. At first glance it appears that the equation $\sin\theta + \cos\theta = 1$ is a **Pythagorean identity**. However, this is not the case because neither trigonometric function is squared. If we square both sides of this equation, it may be possible to use a Pythagorean identity.

CAUTION **Great care must be taken when introducing the squaring operation because doing so may produce** extraneous solutions**.**

$\sin\theta + \cos\theta = 1$	Write the original equation.
$(\sin\theta + \cos\theta)^2 = (1)^2$	Square both sides.
$\sin^2\theta + 2\sin\theta\cos\theta + \cos^2\theta = 1$	Perform the squaring operations.
$\sin^2\theta + \cos^2\theta + 2\sin\theta\cos\theta = 1$	Rearrange the terms.
$1 + 2\sin\theta\cos\theta = 1$	Substitute $\sin^2\theta + \cos^2\theta = 1$.
$2\sin\theta\cos\theta = 0$	Subtract 1 from both sides.
$\sin 2\theta = 0$	Substitute $\sin 2\theta$ for $2\sin\theta\cos\theta$. (This is a **double-angle formula**.)

To solve the equation $\sin 2\theta = 0$, use the **steps for solving trigonometric equations that are linear in form**. Using these steps, we find that the general solution to the equation $\sin 2\theta = 0$ is $\theta = \frac{k\pi}{2}$, where k is *any* integer. It is possible that $\theta = \frac{k\pi}{2}$ is not a solution to the original equation for all integer values of k because we introduced the squaring operation during the solution process. Thus, we must check all possibilities of k for $\theta = \frac{k\pi}{2}$ to determine the general solution to the original equation $\sin\theta + \cos\theta = 1$.

k	$\sin\left(\frac{k\pi}{2}\right) + \cos\left(\frac{k\pi}{2}\right) = 1$	
$k = 0$:	$\sin(0) + \cos(0) = 0 + 1 = 1$	✓
$k = 1$:	$\sin\left(\frac{\pi}{2}\right) + \cos\left(\frac{\pi}{2}\right) = 1 + 0 = 1$	✓
$k = 2$:	$\sin(\pi) + \cos(\pi) = 0 - 1 = -1$	×
$k = 3$:	$\sin\left(\frac{3\pi}{2}\right) + \cos\left(\frac{3\pi}{2}\right) = -1 + 0 = -1$	×

We do not have to check any more values of k because the period of sine and cosine is 2π. In other words, $k = 4$ will produce the same result as $k = 0$. Similarly, $k = 5$ will produce the same result as $k = 1$, and so on.

We can see that $\theta = 0$ and $\theta = \frac{\pi}{2}$ are solutions to the equation $\sin\theta + \cos\theta = 1$. The period of sine and cosine is 2π. Therefore, integer multiples of 2π added to each of these values must also be solutions. Thus, the general solutions to the equation $\sin\theta + \cos\theta = 1$ are of the form

$$\theta = 2\pi k \quad \text{or} \quad \theta = \frac{\pi}{2} + 2\pi k.$$

The only two solutions to the equation $\sin\theta + \cos\theta = 1$ on the interval $[0, 2\pi)$ are $\theta = 0$ and $\theta = \frac{\pi}{2}$, as illustrated by the points of intersection of the graphs in Figure 20.

Watch this **interactive video** to see each step of this solution and to check your work.

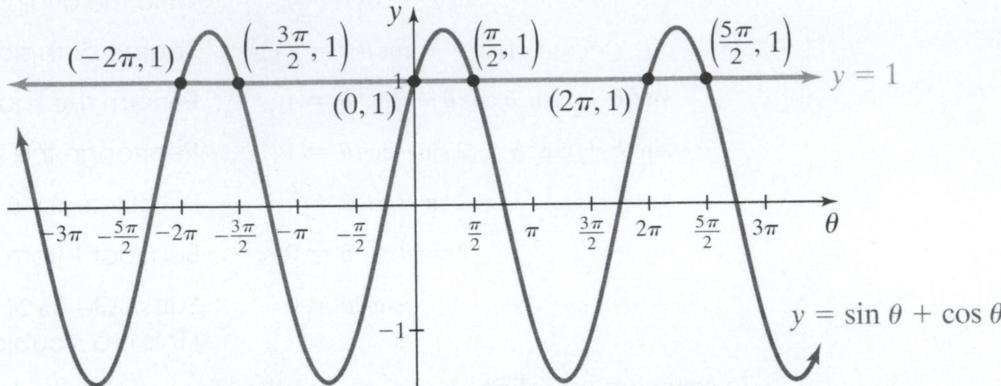

Figure 20 The solutions to the equation $\sin\theta + \cos\theta = 1$ are the first coordinates of the points of intersection of the graphs $y = \sin\theta + \cos\theta$ and $y = 1$. The two graphs intersect at all values of $\theta = 2\pi k$ and $\theta = \frac{\pi}{2} + 2\pi k$ where k is an integer.

You Try It Work through this You Try It problem.

Work Exercises 30–33 in this textbook or in the MyLab Math Study Plan.

OBJECTIVE 5 SOLVING TRIGONOMETRIC EQUATIONS USING A CALCULATOR

In the previous examples, we were able to find an exact solution to each trigonometric equation. This is not always the case. When an exact solution cannot be obtained, it may be necessary to approximate the solution using a calculator. Before trying the following example, you should review inverse trigonometric functions that were discussed in **Section 7.4**.

Example 6 Solving Trigonometric Equations Using a Calculator

Approximate all solutions to the trigonometric equation $\cos\theta = -0.4$ on the interval $[0, 2\pi)$. Round your answer to four decimal places.

Solution The equation $\cos\theta = -0.4$ is a trigonometric equation that is linear in form. Therefore, we can follow the **steps for solving trigonometric equations that are linear in form** to solve this equation.

Step 1. For the equation $\cos\theta = -0.4$, the trigonometric function is already isolated. Therefore, we may skip to step 2.

Step 2. Because $\cos\theta = -0.4 < 0$, we know that the terminal side of θ must lie in Quadrant II or Quadrant III.

Step 3. All values of θ must have the same reference angle. The reference angle is the positive acute angle whose cosine is equal to 0.4. Thus, the reference angle is $\cos^{-1}(0.4)$. It follows that

$$\theta = \pi - \text{Reference Angle} = \pi - \cos^{-1}(0.4) \approx 1.9823 \text{ in Quadrant II and}$$

$$\theta = \pi + \text{Reference Angle} = \pi + \cos^{-1}(0.4) \approx 4.3009 \text{ in Quadrant III. See Figure 21.}$$

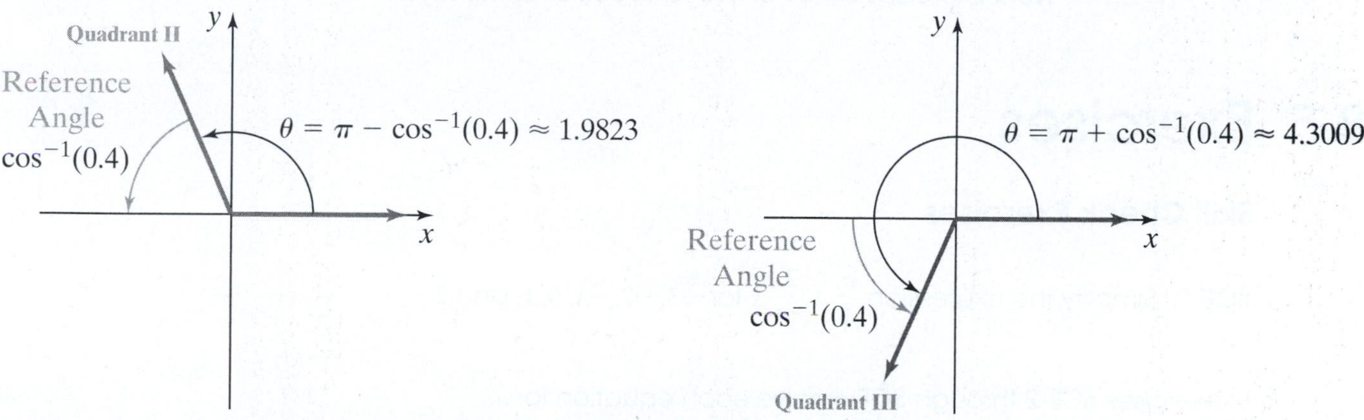

Figure 21 The reference angle is $\cos^{-1}(0.4)$. In Quadrant II, $\theta = \pi - \cos^{-1}(0.4) \approx 1.9823$. In Quadrant III, $\theta = \pi + \cos^{-1}(0.4) \approx 4.3009$.

Step 4. The period of the cosine function is 2π. Adding or subtracting any integer multiple of 2π to $\theta \approx 1.9823$ or $\theta \approx 4.3009$ will yield an angle that is outside of the interval $[0, 2\pi)$. Therefore, the only two solutions to the equation $\cos\theta = -0.4$ on the interval $[0, 2\pi)$ are $\theta \approx 1.9823$ and $\theta \approx 4.3009$.

Using the Intersect feature of a graphing utility, we can verify that our answers are correct. See Figure 22.

Using Technology 

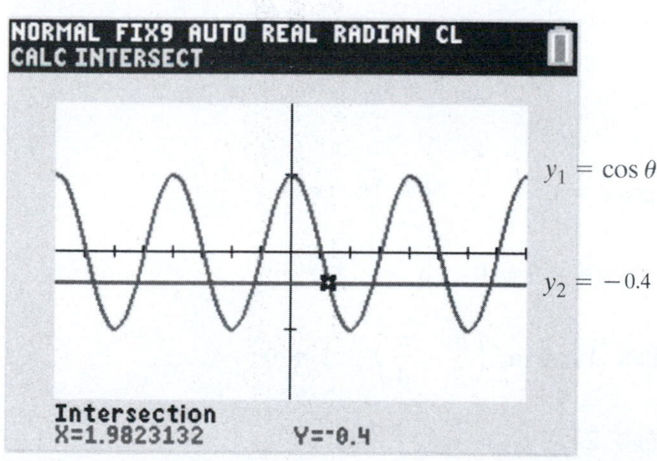

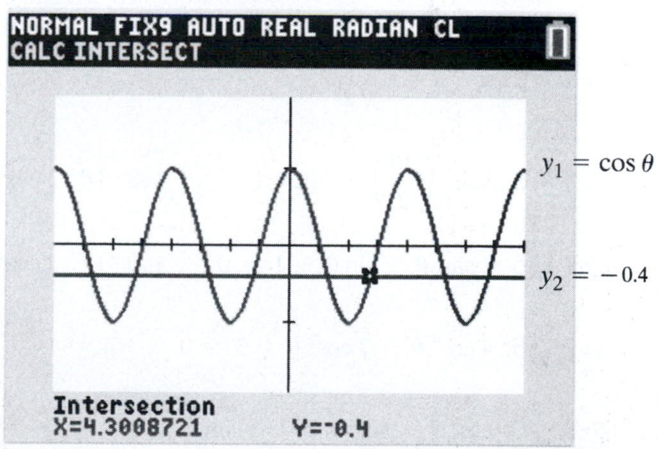

Figure 22 Using the intersect feature of a graphing utility with $y_1 = \cos\theta$ and $y_2 = -0.4$ we see that the graphs intersect at $\theta \approx 1.9823$ and $\theta \approx 4.3009$.

CAUTION **You cannot simply use an inverse trigonometric function key on your calculator to solve this trigonometric equation. Most calculators are programmed to restrict the interval of the solution of an inverse trigonometric function to the same intervals we used in Section 7.4 when we first defined $y = \sin^{-1}x, y = \cos^{-1}x$, and $y = \tan^{-1}x$. Therefore, using the inverse function**

key will never give you more than one solution, and sometimes even the one solution it does give you will not be a correct solution to the given equation.

You Try It Work through this You Try It problem.

Work Exercises 34–39 in this textbook or in the MyLab Math Study Plan.

8.5 Exercises

Skill Check Exercises

SCE-1. Simplify the expression $\frac{\pi}{4} + \frac{2\pi}{3}k$ for $-3, -2, -1, 0, 1,$ and 2

In exercises SCE-2 through SCE-4, solve each equation for u.

SCE-2. $2u^2 + 3u + 1 = 0$ SCE-3. $u + u^2 = 0$ SCE-4. $2(1 - u^2) - 3u - 3 = 0$

In Exercises 1–33, determine a general formula (or formulas) for the solution to each equation. Then, determine the specific solutions (if any) on the interval $[0, 2\pi)$.

SbS 1. $\cos\theta = \frac{1}{2}$ SbS 2. $\tan\theta + \sqrt{3} = 0$ SbS 3. $\sin\theta = 0$

SbS 4. $5\csc\theta - 3 = 2$ SbS 5. $\sin 2\theta = -1$ SbS 6. $\cos 2\theta = 0$

SbS 7. $\sqrt{2}\sin 2\theta + 1 = 0$ SbS 8. $2\cos 3\theta + \sqrt{2} = 0$ SbS 9. $\tan\left(\frac{\theta}{2}\right) = -1$

SbS 10. $\sin\frac{\theta}{2} = -\frac{1}{\sqrt{2}}$ SbS 11. $\cos\frac{\theta}{2} = \frac{1}{2}$ SbS 12. $\cos\left(\theta - \frac{\pi}{3}\right) - 1 = 0$

SbS 13. $5\sec\left(\frac{3\theta}{2}\right) + 10 = 0$ SbS 14. $2\cos^2\theta + \cos\theta = 0$ SbS 15. $\tan^2\theta = \frac{1}{3}$

SbS 16. $2\sin^2\theta - \sin\theta - 1 = 0$ SbS 17. $2\cos^2\theta - 9\cos\theta - 5 = 0$

SbS 18. $4\cos^4\theta - 7\cos^2\theta + 3 = 0$ SbS 19. $2\sin^2\left(\frac{\theta}{3} - \frac{\pi}{4}\right) - 1 = 0$

SbS 20. $\cos^2\theta - \sin^2\theta = 1 - \sin\theta$ SbS 21. $3\cos\theta + 3 = 2\sin^2\theta$

SbS 22. $\sec^2\theta - \tan\theta = 1$ SbS 23. $\cos 2\theta + 3 = 5\cos\theta$

SbS 24. $\cos 2\theta + 10\sin^2\theta = 7$ SbS 25. $\sin 2\theta\sin\theta = \cos\theta$

SbS 26. $\tan\theta = 2\sin\theta$ SbS 27. $4\cos\theta = 3\sec\theta$

SbS 28. $\sin(4\theta) - \sin(2\theta) = 0$

SbS 29. $\cos(2\theta) - \cos(6\theta) = 0$

SbS 30. $3\sin\theta = -3\cos\theta$

SbS 31. $\sin\theta + \cos\theta = -\sqrt{2}$

SbS 32. $\sin\theta\cos\theta = -\dfrac{1}{2}$

SbS 33. $\sin^2\theta + \sqrt{3}\sin\theta\cos\theta = 0$

In Exercises 34–39, approximate the solutions to each equation on the interval $[0, 2\pi)$. Round your answer to four decimal places.

34. $\sin\theta = 0.72$

35. $\cos\theta = 0.63$

36. $\tan\theta = 0.5$

37. $\cos\theta = -0.8$

38. $\sin\theta = -0.2$

39. $\tan\theta = -4$

Brief Exercises

In Exercises 40–72, determine the specific solutions (if any) to each equation on the interval $[0, 2\pi)$.

40. $\cos\theta = \dfrac{1}{2}$

41. $\tan\theta + \sqrt{3} = 0$

42. $\sin\theta = 0$

43. $5\csc\theta - 3 = 2$

44. $\sin 2\theta = -1$

45. $\cos 2\theta = 0$

46. $\sqrt{2}\sin 2\theta + 1 = 0$

47. $2\cos 3\theta + \sqrt{2} = 0$

48. $\tan\left(\dfrac{\theta}{2}\right) = -1$

49. $\sin\dfrac{\theta}{2} = -\dfrac{1}{\sqrt{2}}$

50. $\cos\dfrac{\theta}{2} = \dfrac{1}{2}$

51. $\cos\left(\theta - \dfrac{\pi}{3}\right) - 1 = 0$

52. $5\sec\left(\dfrac{3\theta}{2}\right) + 10 = 0$

53. $2\cos^2\theta + \cos\theta = 0$

54. $\tan^2\theta = \dfrac{1}{3}$

55. $2\sin^2\theta - \sin\theta - 1 = 0$

56. $2\cos^2\theta - 9\cos\theta - 5 = 0$

57. $4\cos^4\theta - 7\cos^2\theta + 3 = 0$

58. $2\sin^2\left(\dfrac{\theta}{3} - \dfrac{\pi}{4}\right) - 1 = 0$

59. $\cos^2\theta - \sin^2\theta = 1 - \sin\theta$

60. $3\cos\theta + 3 = 2\sin^2\theta$

61. $\sec^2\theta - \tan\theta = 1$

62. $\cos 2\theta + 3 = 5\cos\theta$

63. $\cos 2\theta + 10\sin^2\theta = 7$

64. $\sin 2\theta\sin\theta = \cos\theta$

65. $\tan\theta = 2\sin\theta$

66. $4\cos\theta = 3\sec\theta$

67. $\sin(4\theta) - \sin(2\theta) = 0$

68. $\cos(2\theta) - \cos(6\theta) = 0$

69. $3\sin\theta = -3\cos\theta$

70. $\sin\theta + \cos\theta = -\sqrt{2}$

71. $\sin\theta\cos\theta = -\dfrac{1}{2}$

72. $\sin^2\theta + \sqrt{3}\sin\theta\cos\theta = 0$

Chapter 8 Summary

Key Concepts	Examples/Videos

8.1 Trigonometric Identities

Summary for Verifying Trigonometric Identities

First, start with what appears to be the more difficult side of the given identity. Try to transform this side so that it eventually matches identically to the other side. Don't hesitate to start over and work with the other side of the identity if you are having trouble. Try using one of the following techniques to begin and then use others if necessary to continue the verification process.

1. Look for ways to use known identities such as the **reciprocal identities**, **quotient identities**, and **even/odd properties**. If the identity includes a squared trigonometric expression, try using a variation of a **Pythagorean identity**.
2. Try rewriting each trigonometric expression in terms of sines and cosines.
3. Factor out a greatest common factor and use algebraic factoring techniques such as factoring the **difference of two squares** or the **sum/difference of two cubes**.
4. If a single term appears in the denominator of a quotient, try separating the quotient into two or more quotients:

$$\frac{A+B}{C} = \frac{A}{C} + \frac{B}{C}$$

5. If there are two or more fractional expressions, try combining the expressions using a common denominator:

$$\frac{A}{B} + \frac{C}{D} = \frac{AD+BC}{BD}$$

6. If the numerator or denominator of one or more quotients contains an expression of the form $A + B$, try multiplying the numerator and denominator by its conjugate $A - B$.

$$\frac{C}{A+B} = \frac{C}{A+B}\cdot\frac{A-B}{A-B} \quad \text{or} \quad \frac{A+B}{C} = \frac{A+B}{C}\cdot\frac{A-B}{A-B}$$

Examples/Videos

Verify each trigonometric identity.

a. $\dfrac{\sin^2 t + 6\sin t + 9}{\sin t + 3} = \dfrac{3\csc t + 1}{\csc t}$

b. $\dfrac{2\csc\theta}{\sec\theta} + \dfrac{\cos\theta}{\sin\theta} = 3\cot\theta$

c. $\dfrac{1-\sin\theta}{\cos\theta} + \dfrac{\cos\theta}{1-\sin\theta} = 2\sec\theta$

Key Concepts	Examples/Videos
8.2 The Sum and Difference Formulas **The Sum and Difference Formulas for the Cosine Function** $\cos(\alpha + \beta) = \cos\alpha\cos\beta - \sin\alpha\sin\beta$ $\cos(\alpha - \beta) = \cos\alpha\cos\beta + \sin\alpha\sin\beta$	Find the exact value of each trigonometric expression without the use of a calculator. **a.** $\cos\left(\frac{7\pi}{12}\right)\cos\left(\frac{5\pi}{12}\right) + \sin\left(\frac{7\pi}{12}\right)\sin\left(\frac{5\pi}{12}\right)$ **b.** $\cos\left(\frac{7\pi}{12}\right)$ **c.** $\cos(-75°)$
The Sum and Difference Formulas for the Sine Function $\sin(\alpha + \beta) = \sin\alpha\cos\beta + \cos\alpha\sin\beta$ $\sin(\alpha - \beta) = \sin\alpha\cos\beta - \cos\alpha\sin\beta$	Suppose that α is an angle such that $\tan\alpha = \frac{5}{7}$ and $\cos\alpha < 0$. Also, suppose that β is an angle such that $\sec\beta = -\frac{4}{3}$ and $\csc\beta > 0$. Find the exact value of $\sin(\alpha + \beta)$.
The Sum and Difference Formulas for the Tangent Function $\tan(\alpha + \beta) = \frac{\tan\alpha + \tan\beta}{1 - \tan\alpha\tan\beta}$ $\tan(\alpha - \beta) = \frac{\tan\alpha - \tan\beta}{1 + \tan\alpha\tan\beta}$	Find the exact value of each trigonometric expression without the use of a calculator. **a.** $\tan\left(\frac{5\pi}{6} + \frac{3\pi}{4}\right)$ **b.** $\tan\left(\frac{\pi}{12}\right)$
8.3 The Double-Angle and Half-Angle Formulas **Double-Angle Formulas** $\sin 2\theta = 2\sin\theta\cos\theta$ $\cos 2\theta = \cos^2\theta - \sin^2\theta$ $\cos 2\theta = 1 - 2\sin^2\theta$ $\cos 2\theta = 2\cos^2\theta - 1$ $\tan 2\theta = \frac{2\tan\theta}{1 - \tan^2\theta}$	Rewrite each expression as the sine, cosine, or tangent of a double angle. Then evaluate the expression without using a calculator. **a.** $\cos^2\left(\frac{11\pi}{12}\right) - \sin^2\left(\frac{11\pi}{12}\right)$ **b.** $2\sin 67.5°\cos 67.5°$ **c.** $\frac{2\tan\left(-\frac{\pi}{8}\right)}{1 - \tan^2\left(-\frac{\pi}{8}\right)}$ **d.** $2\cos^2 105° - 1$ Suppose that the terminal side of angle θ lies in Quadrant II such that $\sin\theta = \frac{5}{7}$. Find the values of $\sin 2\theta$, $\cos 2\theta$, and $\tan 2\theta$.

Key Concepts	Examples/Videos
Power Reduction Formulas $\sin^2\theta = \dfrac{1 - \cos 2\theta}{2}$ $\cos^2\theta = \dfrac{1 + \cos 2\theta}{2}$ $\tan^2\theta = \dfrac{1 - \cos 2\theta}{1 + \cos 2\theta}$	Rewrite the function $f(x) = 6\sin^4 x$ as an equivalent function containing only cosine terms raised to a power of 1.
Half-Angle Formulas $\sin\left(\dfrac{\alpha}{2}\right) = \sqrt{\dfrac{1 - \cos\alpha}{2}}$ for $\dfrac{\alpha}{2}$ in Quadrant I or Quadrant II. $\sin\left(\dfrac{\alpha}{2}\right) = -\sqrt{\dfrac{1 - \cos\alpha}{2}}$ for $\dfrac{\alpha}{2}$ in Quadrant III or Quadrant IV. $\cos\left(\dfrac{\alpha}{2}\right) = \sqrt{\dfrac{1 + \cos\alpha}{2}}$ for $\dfrac{\alpha}{2}$ in Quadrant I or Quadrant IV. $\cos\left(\dfrac{\alpha}{2}\right) = -\sqrt{\dfrac{1 + \cos\alpha}{2}}$ for $\dfrac{\alpha}{2}$ in Quadrant II or Quadrant III. $\tan\left(\dfrac{\alpha}{2}\right) = \sqrt{\dfrac{1 - \cos\alpha}{1 + \cos\alpha}}$ for $\dfrac{\alpha}{2}$ in Quadrant I or Quadrant III. $\tan\left(\dfrac{\alpha}{2}\right) = -\sqrt{\dfrac{1 - \cos\alpha}{1 + \cos\alpha}}$ for $\dfrac{\alpha}{2}$ in Quadrant II or Quadrant IV. $\tan\left(\dfrac{\alpha}{2}\right) = \dfrac{1 - \cos\alpha}{\sin\alpha}$ for $\dfrac{\alpha}{2}$ in any quadrant. $\tan\left(\dfrac{\alpha}{2}\right) = \dfrac{\sin\alpha}{1 + \cos\alpha}$ for $\dfrac{\alpha}{2}$ in any quadrant.	Use a half-angle formula to evaluate each expression without using a calculator. a. $\sin\left(-\dfrac{7\pi}{8}\right)$ b. $\cos(-15°)$ c. $\tan\left(\dfrac{11\pi}{12}\right)$ Suppose that $\csc\alpha = \dfrac{8}{3}$ such that $\dfrac{\pi}{2} < \alpha < \pi$. Find the values of $\sin\left(\dfrac{\alpha}{2}\right)$, $\cos\left(\dfrac{\alpha}{2}\right)$, and $\tan\left(\dfrac{\alpha}{2}\right)$.
8.4 The Product-to-Sum and Sum-to-Product Formulas **The Product-to-Sum Formulas** $\sin\alpha\sin\beta = \dfrac{1}{2}[\cos(\alpha - \beta) - \cos(\alpha + \beta)]$ $\cos\alpha\cos\beta = \dfrac{1}{2}[\cos(\alpha - \beta) + \cos(\alpha + \beta)]$ $\sin\alpha\cos\beta = \dfrac{1}{2}[\sin(\alpha + \beta) + \sin(\alpha - \beta)]$ $\cos\alpha\sin\beta = \dfrac{1}{2}[\sin(\alpha + \beta) - \sin(\alpha - \beta)]$	 Write each product as a sum or difference containing only sines or cosines. a. $\sin 4\theta \sin 2\theta$ b. $\cos\left(\dfrac{19\theta}{2}\right)\sin\left(\dfrac{\theta}{2}\right)$ c. $\cos 11\theta \cos 5\theta$ d. $\sin 6\theta \cos 3\theta$ Determine the exact value of the 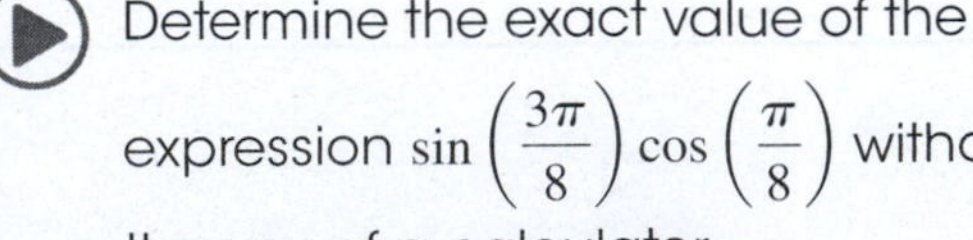 expression $\sin\left(\dfrac{3\pi}{8}\right)\cos\left(\dfrac{\pi}{8}\right)$ without the use of a calculator.

Key Concepts	Examples/Videos
The Sum-to-Product Formulas $\sin\alpha + \sin\beta = 2\sin\left(\frac{\alpha+\beta}{2}\right)\cos\left(\frac{\alpha-\beta}{2}\right)$ $\sin\alpha - \sin\beta = 2\sin\left(\frac{\alpha-\beta}{2}\right)\cos\left(\frac{\alpha+\beta}{2}\right)$ $\cos\alpha + \cos\beta = 2\cos\left(\frac{\alpha+\beta}{2}\right)\cos\left(\frac{\alpha-\beta}{2}\right)$ $\cos\alpha - \cos\beta = -2\sin\left(\frac{\alpha+\beta}{2}\right)\sin\left(\frac{\alpha-\beta}{2}\right)$	 Write each sum or difference as a product of sines and/or cosines. a. $\sin 5\theta + \sin 3\theta$ b. $\cos\left(\frac{3\theta}{2}\right) - \cos\left(\frac{17\theta}{2}\right)$ Determine the exact value of the expression $\sin\left(\frac{\pi}{12}\right) - \sin\left(\frac{17\pi}{12}\right)$ without the use of a calculator.
8.5 Trigonometric Equations A trigonometric equation that contains constants and only one trigonometric function that is raised to the first power is a **trigonometric equation that is linear in form**. **Steps for Solving Trigonometric Equations that Are Linear in Form** **Step 1.** Isolate the trigonometric function on one side of the equation. **Step 2.** Determine the quadrants in which the terminal side of the argument of the function lies or determine the axis on which the terminal side of the argument of the function lies. **Step 3.** If the terminal side of the argument of the function lies within a quadrant, then determine the reference angle and the value(s) of the argument on the Interval $[0, 2\pi)$. If the terminal side of the argument of the function lies along an axis, then determine the angle associated with it on the interval $(0, 2\pi)$, choosing from $0, \frac{\pi}{2}, \pi, or \frac{3\pi}{2}$. **Step 4.** Use the period of the given function to determine the solutions.	 Determine a general formula (or formulas) for all solutions to each equation. Then, determine the specific solutions (if any) on the interval $[0, 2\pi]$. a. $\sqrt{2}\cos 2\theta + 1 = 0$ b. $\sin\frac{\theta}{2} = -\frac{\sqrt{3}}{2}$ c. $\tan\left(\theta + \frac{\pi}{6}\right) + 1 = 0$
If $f(\theta)$ is a trigonometric function, then an equation of the form $a(f(\theta))^2 + b(f(\theta)) + c = 0$, $a \neq 0$ is said to be a **trigonometric equation that is quadratic in form**. To solve these equations, transform the equation into an algebraic equation using the substitution $u = f(\theta)$.	 Determine a general formula (or formulas) for all solutions to each equation. Then, determine the specific solutions (if any) on the interval $[0, 2\pi)$. a. $\sin^2\theta - 4\sin\theta + 3 = 0$ b. $4\cos^2\theta - 3 = 0$

Key Concepts	Examples/Videos
When a trigonometric equation cannot be solved directly, it may be possible to first use a trigonometric identity to transform the equation into one that can be solved directly. The identities below are useful to remember when solving trigonometric equations. Quotient Identities — Half-Angle Formulas Reciprocal Identities — Double-Angle Formulas Pythagorean Identities — Product-to-Sum Formulas Cofunction Identities — Sum-to-Product Formulas	Determine a general formula (or formulas) for all solutions to each equation. Then, determine the specific solutions (if any) on the interval $[0, 2\pi)$. a. $2\sin^2\theta - 3\cos\theta + 3$ b. $\sin\theta\cos\theta = -\frac{1}{2}$ c. $\cos 2\theta + 4\sin^2\theta = 2$ d. $\sin 5\theta + \sin 3\theta - 0$

Chapter 8 Review Exercises

In 1–3, verify the identity.

1. $\dfrac{\cot\theta + 1}{\csc\theta} = \sin\theta + \cos\theta$

2. $\dfrac{\sec\beta}{\sin\beta} - \dfrac{\sin\beta}{\sec\beta} = \dfrac{\tan^2\beta + \cos^2\beta}{\tan\beta}$

3. $1 + \dfrac{1 - \cot^2 x}{1 + \cot^2 x} = 2\sin^2 x$

4. Find the exact value of the expression $\cos\left(\frac{11\pi}{12}\right)\cos\left(\frac{7\pi}{12}\right) + \sin\left(\frac{11\pi}{12}\right)\sin\left(\frac{7\pi}{12}\right)$ without the use of a calculator. The final answer must be fully simplified.

5. Use the cosine of the sum of two angles formula to find the exact value of the expression $\cos\left(\frac{5\pi}{12}\right)$ without the use of a calculator. The final answer must be fully simplified.

6. Suppose that the terminal side of angle α lies in Quadrant I and the terminal side of angle β lies in Quadrant IV. If $\sin\alpha = \frac{8}{17}$ and $\cos\beta = -\frac{3}{\sqrt{11}}$, find the exact value of $\sin(\alpha + \beta)$. The final answer must be fully simplified.

7. Find the exact value of the expression $\tan\left(\frac{7\pi}{6} - \frac{9\pi}{4}\right)$ without the use of a calculator. The final answer must be fully simplified.

8. Find the exact value of the expression $\cos\left(\sin^{-1}\left(\frac{5}{13}\right) + \cos^{-1}\left(-\frac{12}{13}\right)\right)$ without the use of a calculator.

9. Rewrite the expression $2\sin\left(\frac{\pi}{8}\right)\cos\left(\frac{\pi}{8}\right)$ as the sine, cosine, or tangent of a double-angle. Then find the exact value of the trigonometric expression without the use of a calculator.

10. Rewrite the expression $\cos^2\left(-\frac{5\pi}{12}\right) - \sin^2\left(-\frac{5\pi}{12}\right)$ as the sine, cosine, or tangent of a double-angle. Then find the exact value of the trigonometric expression without the use of a calculator.

11. Suppose that the terminal side of angle θ lies in Quadrant II such that $\sin\theta = \frac{2}{13}$. Determine the value of $\sin 2\theta$. The final answer must be fully simplified.

12. Suppose that the terminal side of angle θ lies in Quadrant IV such that $\cos\theta = \frac{12}{13}$. Determine the value of $\tan 2\theta$. The final answer must be fully simplified.

13. Use a half-angle formula to evaluate the expression $\sin\left(-\frac{3\pi}{8}\right)$ without using a calculator. The final answer must be fully simplified.

14. Suppose that $\tan\alpha = \frac{15}{8}$ such that $\pi < \alpha < \frac{3\pi}{2}$. Determine the value of $\cos\left(\frac{\alpha}{2}\right)$. The final answer must be fully simplified.

15. Find the exact value of the expression $\cos\left(2\sin^{-1}\left(-\frac{1}{\sqrt{2}}\right)\right)$ without the use of a calculator.

16. Write the product as sum or difference containing only sines or cosines.

$$\sin(8\theta)\sin(4\theta)$$

17. Determine the exact value of the expression $\cos\left(\frac{19\pi}{12}\right) - \cos\left(\frac{\pi}{12}\right)$ without the use of a calculator.

In 18–20, determine a general formula (or formulas) for the solutions to each equation. Then determine the specific solutions (if any) on the interval $[0, 2\pi)$.

18. $\tan\theta + \sqrt{3} = 0$

19. $2\cos^2\theta + \cos\theta = 0$

20. $3\cos\theta + 3 = 2\sin^2\theta$

CHAPTER NINE

Applications of Trigonometry

CHAPTER NINE CONTENTS

9.1 Right Triangle Applications

THINGS TO KNOW

Before working through this section, be sure that you are familiar with the following concepts:

VIDEO ANIMATION INTERACTIVE

You Try It

1. Using the Pythagorean Theorem (Section 6.3) 

You Try It

2. Evaluating Trigonometric Functions Using a Calculator (Section 6.4) 

OBJECTIVES

1 Solving Right Triangles

2 Applications Using Right Triangles

SECTION 9.1 EXERCISES

OBJECTIVE 1 SOLVING RIGHT TRIANGLES

Right triangles abound in our everyday world, and so it is appropriate at this point to look at just a few situations where the right triangle definitions of trigonometric functions can simplify our lives. We start with a concept called "solving right triangles." The goal of solving a right triangle is to determine the measure of all angles and the length of all sides, given certain information. In Section 9.2 and Section 9.3, we will solve triangles that do not have a right angle.

Example 1 Solving a Right Triangle

Solve the given right triangle. Round the length of side a and the measure of the two acute angles to two decimal places.

Solution Observe that we are given the following information.

Angles	Sides
$A =$ ______	$a =$ ______
$B =$ ______	$b = 11.4$ cm
$C = 90°$	$c = 12.1$ cm

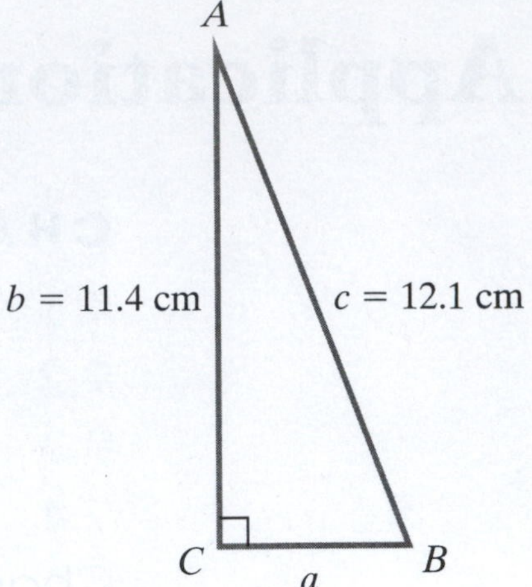

We can determine the length of side a using the **Pythagorean Theorem**.

$a^2 + b^2 = c^2$	Write the Pythagorean Theorem.
$a^2 + (11.4)^2 = (12.1)^2$	Substitute $b = 11.4$ and $c = 12.1$.
$a^2 + 129.96 = 146.41$	Square.
$a^2 = 16.45$	Subtract.
$a = \pm\sqrt{16.45} \approx \pm 4.06$	Solve for a.
$a \approx 4.06$	Use the positive square root since a represents the length of a side of a triangle.

We insert the new information.

Angles	Sides
$A =$ ______	$a \approx \underline{4.06}$ cm
$B =$ ______	$b = 11.4$ cm
$C = 90°$	$c = 12.1$ cm

Now that we know the lengths of all sides, we can determine the measure of the two **acute angles**.

It does not matter which angle to determine first. We will choose to first determine the measure of angle B. In **Figure 1** we see that the length of the side opposite of angle B is 11.4 cm. The length of the hypotenuse is 12.1 cm. Therefore, we can use the trigonometric ratio, $\sin B = \frac{11.4}{12.1}$.

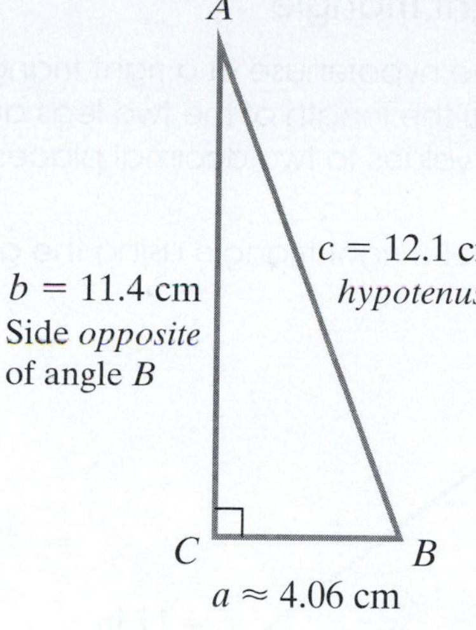

Figure 1 The length of the side opposite of B is 11.4 cm and the length of the hypotenuse is 12.1 cm. Thus, $\sin B = \frac{11.4}{12.1}$.

$\sin B = \frac{11.4}{12.1}$ The sine of angle B is equal to the ratio of the length of the side opposite angle B to the length of the hypotenuse.

$B = \sin^{-1}\left(\frac{11.4}{12.1}\right)$ Use the inverse sine function to solve for B.

$\approx 70.42°$ Use a calculator (in degree mode) to evaluate $\sin^{-1}\left(\frac{11.4}{12.1}\right)$.

Again, we insert the new information.

Angles	Sides
$A =$ ______	$a \approx 4.06$ cm
$B \approx 70.42°$	$b = 11.4$ cm
$C = 90°$	$c = 12.1$ cm

Since angle A and angle B are **complementary angles**, we can easily determine the measure of angle A.

$A \approx 90° - 70.42° \approx 19.58°$ Use the fact that angle A and B are complementary angles.

We have now solved the right triangle.

Angles	Sides
$A \approx 19.58°$	$a \approx 4.06$ cm
$B \approx 70.42°$	$b = 11.4$ cm
$C = 90°$	$c = 12.1$ cm

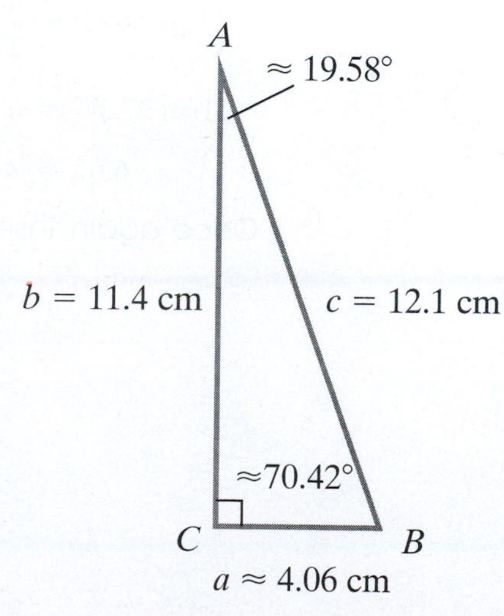

Example 2 Solving a Right Triangle

Suppose that the length of the hypotenuse of a right triangle is 11 inches. If one of the acute angles is 37.34°, find the length of the two legs and the measure of the other acute angle. Round all values to two decimal places.

Solution First, draw and label a right triangle using the given information. Let $A = 37.34°$ and let $c = 11$ in.

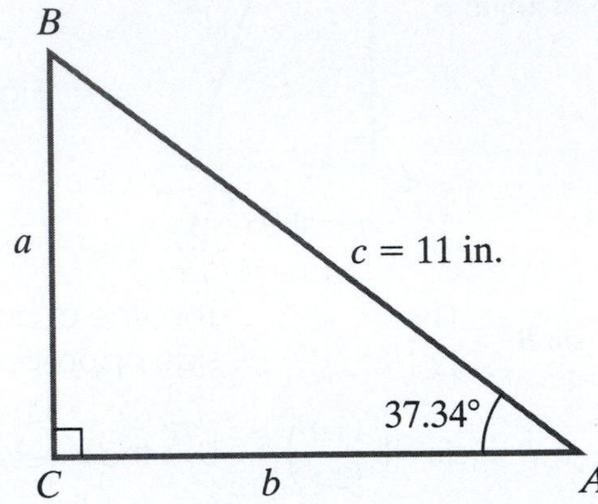

Since angle A and angle B are **complementary angles,** we can easily determine the measure of angle B.

$B = 90° - 37.34° = 52.66°$ Use the fact that angle A and B are complementary angles.

Insert the information.

Angles	Sides
$A = 37.34°$	$a \approx$ ____
$B = \underline{52.66°}$	$b =$ ____
$C = 90°$	$c = 11$ in

Side a is opposite of angle A so we can use the sine function because we are given the length of the hypotenuse.

$\sin 37.34° = \frac{a}{11}$ By definition, $\sin\theta = \frac{opp}{hyp}$.

$11\sin 37.34° = a$ Multiply both sides by 11.

$6.67 \approx a$ Use a calculator (in degree mode) to evaluate $11\sin 37.34°$.

Once again, insert the information.

Angles	Sides
$A = 37.34°$	$a \approx \underline{6.67 \text{ in}}$
$B = \underline{52.66°}$	$b \approx$ ____
$C = 90°$	$c = 11$ in

We can determine the length of the side b in a variety of ways. Below are three different ways that can be used to solve for b.

$(6.67)^2 + b^2 = (11)^2$ Use the Pythagorean Theorem with $a \approx 6.67$ and $c = 11$.

$\cos 37.34° = \dfrac{b}{11}$ Use the cosine function, $\cos\theta = \dfrac{adj}{hyp}$.

$\sin 52.66° = \dfrac{b}{11}$ Use the sine function, $\sin\theta = \dfrac{opp}{hyp}$.

You should try solving for b in each of the three equations above to verify that $b \approx 8.75$ inches. We have now solved the right triangle.

Angles	Sides
$A = 37.34°$	$a \approx \underline{6.67}$ in
$B = \underline{52.66°}$	$b \approx \underline{8.75}$ in
$C = 90°$	$c = 11$ in

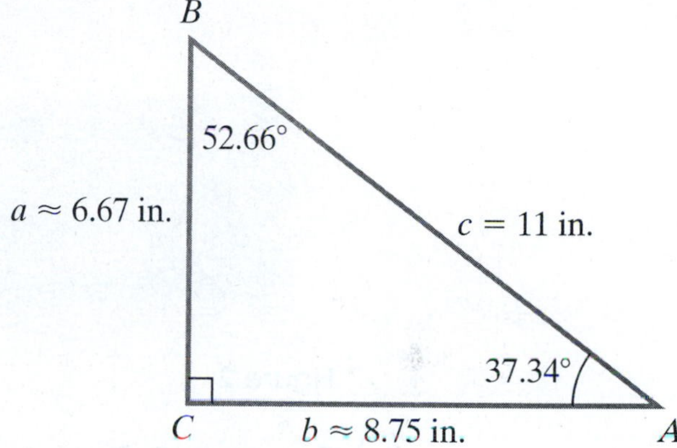

In order to solve a right triangle, we must be given at least two pieces of information. In Example 1 we were given the length of one side and the measure of an acute angle. In Example 2, we were given the length of two sides but no angles. The only two pieces of information that will not help to solve a right triangle are the measures of two angles. In this case we can easily determine the measure of the third angle but it is impossible to determine the length of the three sides. This is because there are infinitely many similar right triangles with the same three angle measures.

Work through the Guided Visualization below to practice solving right triangles.

 Solving Right Triangles

You Try It **Work through this You Try It problem.**

Work Exercises 1–7 in this textbook or in the MyLab Math Study Plan.

OBJECTIVE 2 APPLICATIONS USING RIGHT TRIANGLES

Applications involving right triangles often involve angles created by an observer looking up or looking down at an object. The angle between the horizontal and the line of sight of an object *above* the horizontal is called the **angle of elevation**. The angle between the horizontal and the line of sight of an object below the horizontal is called the **angle of depression**. See Figure 2.

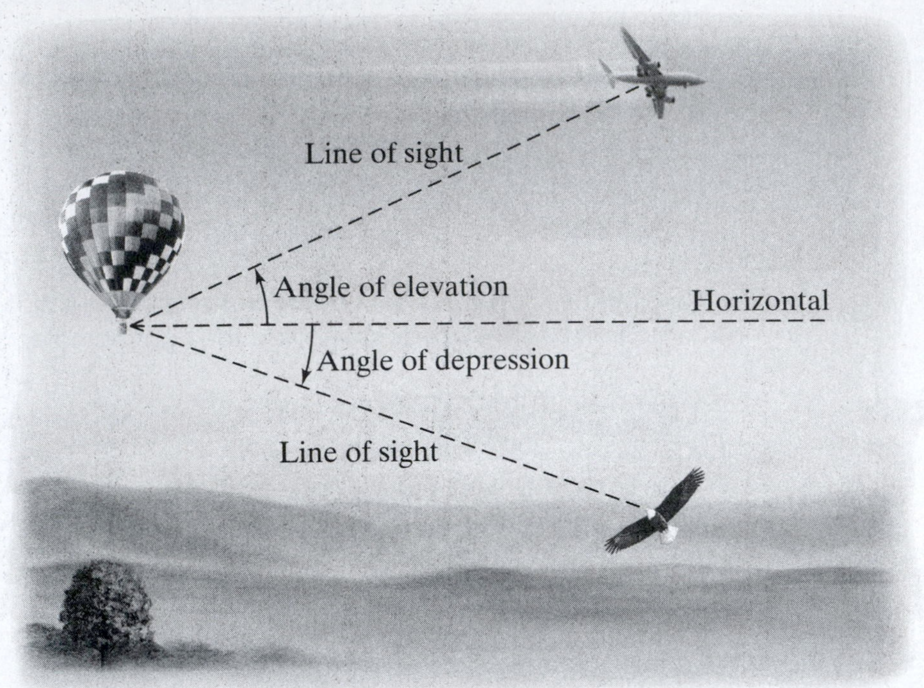

Figure 2

Example 3 Determining the Height of a Flagpole

The angle of elevation to the top of a flagpole measured by a digital protractor is $20°$ from a point on the ground 90 feet away from its base. Find the height of the flagpole. Round to two decimal places.

Solution Let h represent the height of the flagpole, as in Figure 3. Then choose the correct trigonometric function that relates the given information. Determine the value of h on your own, and then watch this **video** to see if you are correct.

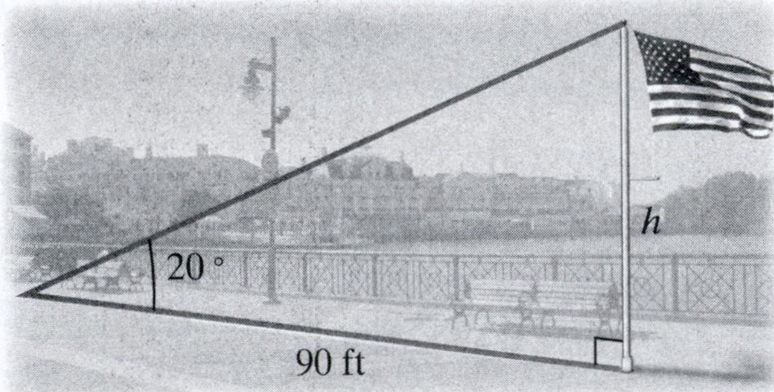

Figure 3

Example 4 Determining the Altitude of an Airplane

At the same instant, two observers 5 miles apart are looking at the same airplane. (See Figure 4.) The angle of elevation (from the ground) of the observer closest to the plane is $71°$. The angle of elevation (from the ground) of the person furthest from the plane is $39°$. Find the altitude of the plane (to the nearest foot) at the instant the two people observe the plane.

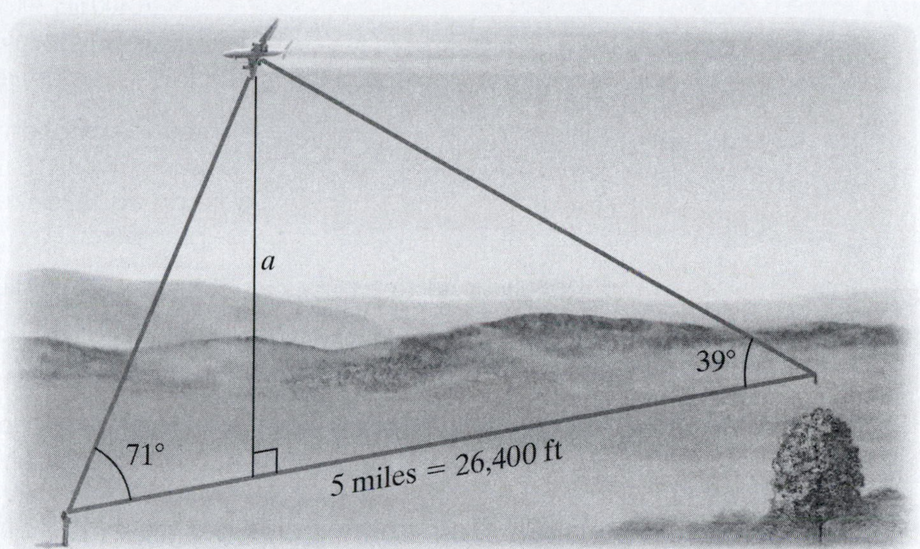

Figure 4

Solution Let a represent the altitude of the plane in feet. The 5 miles must be converted to feet to be consistent with units. There are 5280 feet in one mile. Therefore, the observers are $5\,\cancel{\text{mi}} \cdot 5280\,\frac{\text{ft}}{\cancel{\text{mi}}} = 26{,}400\text{ ft}$ apart. Let x represent the distance from the observer nearest to the plane to the base of the altitude. Then $26{,}400 - x$ is the distance from the observer furthest from the plane to the base of the altitude. We can now draw and label the two right triangles shown below.

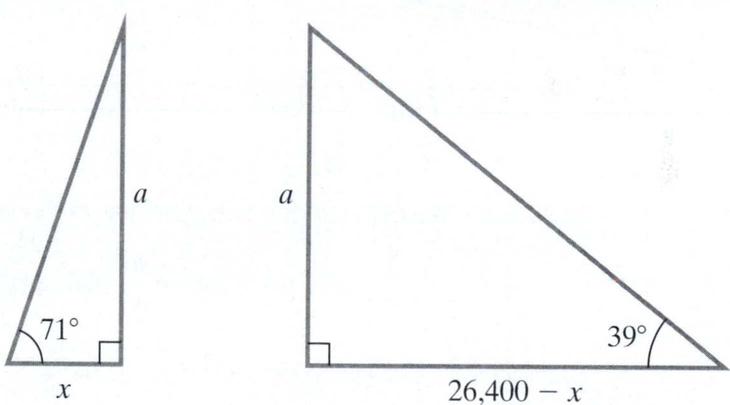

From the two right triangles, we can use the tangent function to write the following two equations:

$$\tan 71^\circ = \frac{a}{x} \quad \tan 39^\circ = \frac{a}{26{,}400 - x}$$

Multiply both sides of each equation by their respective denominators to solve each equation for a.

$$a = x \tan 71^\circ \quad a = (26{,}400 - x) \tan 39^\circ$$

The expressions $x \tan 71^\circ$ and $(26{,}400 - x) \tan 39^\circ$ both represent the altitude of the plane. Thus they are equal to each other.

$$x \tan 71^\circ = (26{,}400 - x) \tan 39^\circ$$

Solving this equation for x, we get $x = \dfrac{26{,}400 \tan 39^\circ}{\tan 71^\circ + \tan 39^\circ}$. Read this **overview** to see how to solve for x or watch this **video** to see the entire solution. We can solve for a by substituting $x = \dfrac{26{,}400 \tan 39^\circ}{\tan 71^\circ + \tan 39^\circ}$ into the equation $a = x \tan 71^\circ$.

$a = x \tan 71^\circ$ — Write an equation representing the altitude of the plane.

$a = \left(\dfrac{26{,}400 \tan 39^\circ}{\tan 71^\circ + \tan 39^\circ}\right) \tan 71^\circ$ — Substitute $\dfrac{26{,}400 \tan 39^\circ}{\tan 71^\circ + \tan 39^\circ}$ for x.

$a \approx 16{,}717 \text{ ft}$ — Evaluate by using a calculator (in degree mode).

The altitude of the airplane is approximately 16,717 feet.

Example 5 Determining the Height of the Eiffel Tower

A tourist visiting Paris determines that the angle of elevation from point A to the top of the Eiffel tower is 12.19°. She then walks 1 km on a straight line toward the tower to point B and determines that the angle of elevation to the top of the tower is now 32.94°. Determine the height of the Eiffel Tower. Round to the nearest meter.

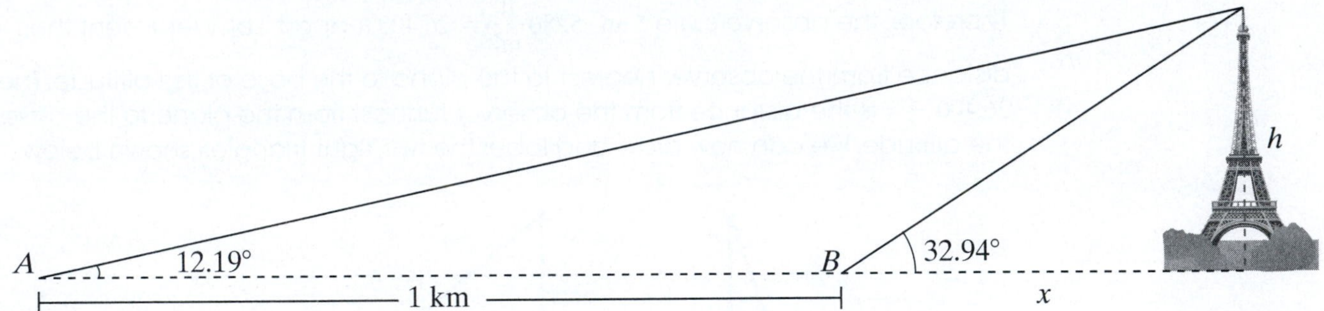

Solution We can use the two triangles in the figure to write the following equations.

$$\tan 32.94^\circ = \frac{h}{x} \quad \text{and} \quad \tan 12.19^\circ = \frac{h}{x + 1}$$

Solving for x using the equation, $\tan 32.94^\circ = \dfrac{h}{x}$, we get $x = \dfrac{h}{\tan 32.94^\circ}$. Substituting $x = \dfrac{h}{\tan 32.94^\circ}$ into the second equation we get $\tan 12.19^\circ = \dfrac{h}{\dfrac{h}{\tan 32.94^\circ} + 1}$.

Now, solve for h.

$\tan 12.19^\circ = \dfrac{h}{\dfrac{h}{\tan 32.94^\circ} + 1}$ — Write the equation found in the previous step.

$\tan 12.19^\circ = \dfrac{h}{\dfrac{h + \tan 32.94^\circ}{\tan 32.94^\circ}}$ — Simplify the denominator of the complex fraction using a common denominator.

$$\tan 12.19° = \frac{h \tan 32.94°}{h + \tan 32.94°}$$ Simplify.

$$\tan 12.19°(h + \tan 32.94°) = h \tan 32.94°$$ Multiply both sides by $h + \tan 32.94°$.

$$h \tan 12.19° + (\tan 12.19°)(\tan 32.94°) = h \tan 32.94°$$ Use the distributive property.

$$h \tan 12.19° - h \tan 32.94° = -(\tan 12.19°)(\tan 32.94°)$$ Collect the terms with the variable h on the left-hand side.

$$h(\tan 12.19° - \tan 32.94°) = -(\tan 12.19°)(\tan 32.94°)$$ Factor out an h.

$$h = \frac{-(\tan 12.19°)(\tan 32.94°)}{\tan 12.19° - \tan 32.94°} \text{ km}$$ Solve for h to get an expression that describes the height in kilometers.

To determine the height in meters, we multiply this expression by 1000. Therefore,

$$h = \frac{-(\tan 12.19°)(\tan 32.94°)}{\tan 12.19° - \tan 32.94°}(1000) \approx 324 \text{ m}$$

You Try It Work through this You Try It problem.

Work Exercises 8–15 in this textbook or in the MyLab Math Study Plan.

9.1 Exercises

Skill Check Exercises

For exercises SCE-1 through SCE-3, use a calculator in degree mode to approximate the expression. Round to two decimal places.

SCE-1. $\sin^{-1}\left(\frac{35.1}{75.8}\right)$ **SCE-2.** $\cos^{-1}\left(\frac{8.1}{31.9}\right)$ **SCE-3.** $\tan^{-1}\left(\frac{14.8}{16.2}\right)$

In Exercises 1–5, solve each right triangle. Round the length of the sides to two decimal places and round the measure of the angles to two decimal places.

1. A, $b = 24$ in., c, C, $a = 14$ in., B

2. 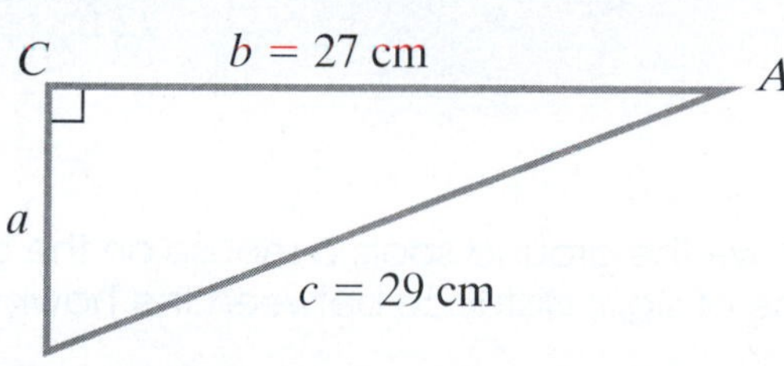

3. 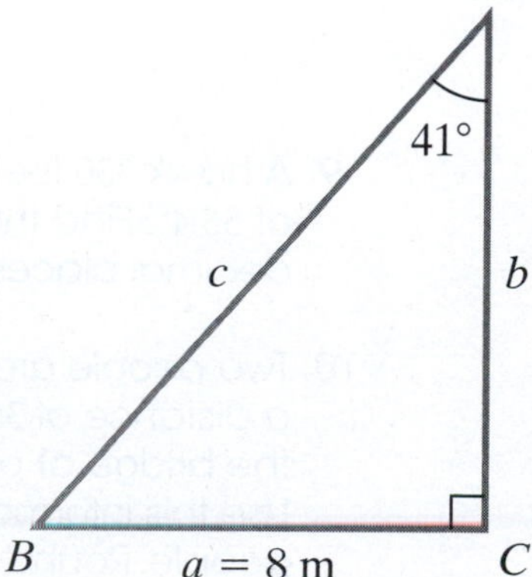

4.

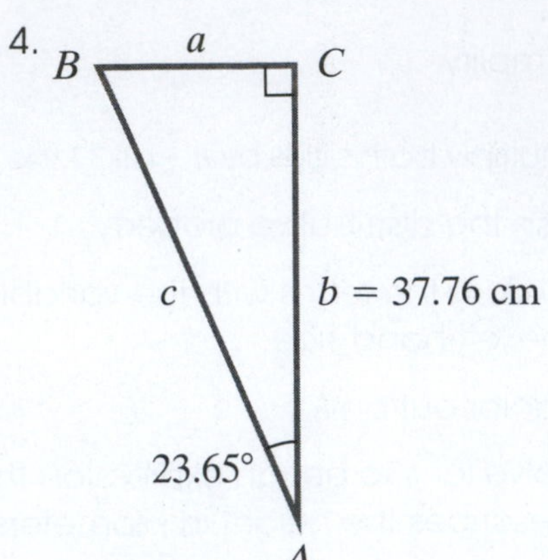

5.

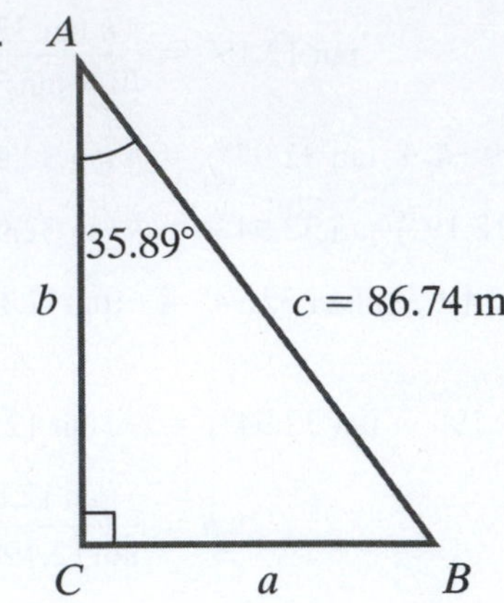

6. Find the length of each leg of a right triangle given that one angle is 27° and the length of the hypotenuse is 14 inches. Round the length of the sides to two decimal places.

7. Find the length of the two missing sides of a right triangle given that one angle is 83.1° and the length of the side adjacent to the angle is 12.5 centimeters. Round the length of the sides to two decimal places.

8. The angle of elevation to the top of a flagpole measured by a digital protractor is 54.7° from a point on the ground 72 feet away from its base. Find the height of the flagpole. Round to two decimal places.

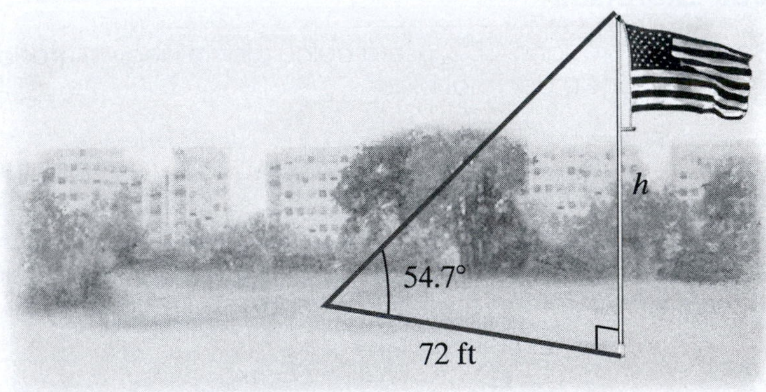

9. A hawk 330 feet above the ground spots a mouse on the ground at an angle of depression of 55.4°. Find the line of sight distance between the hawk and the mouse. Round to two decimal places.

10. Two people are standing on opposite sides of a small river. One person is located at point Q, a distance of 35 meters from a bridge. The other person is standing on the southeast corner of the bridge at point P. The angle between the bridge and the line of sight from P to 10 is 72.4°. Use this information to determine the length of the bridge and the distance between the two people. Round your answers to two decimal places as needed.

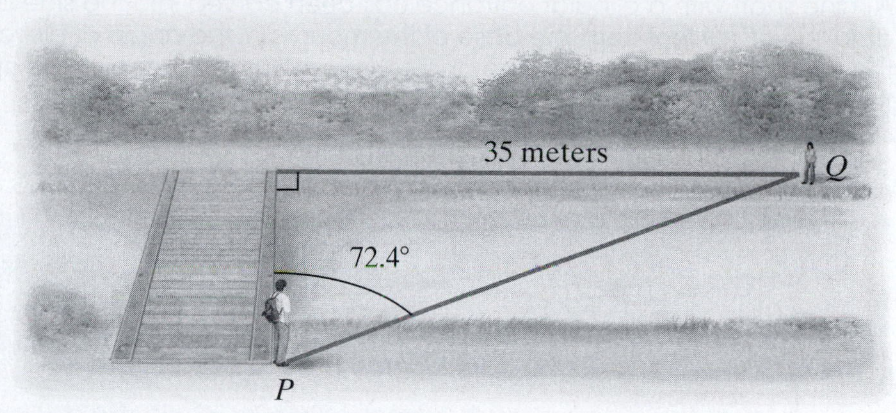

11. A basketball player spots the front of the rim of a basketball hoop at an angle of elevation of 25.7°. If the player's eyes are located 6 feet above the ground, find the horizontal distance between the basketball player and the front of the rim. (Assume that the rim of the basket is 10 feet from the floor.) Round to two decimal places.

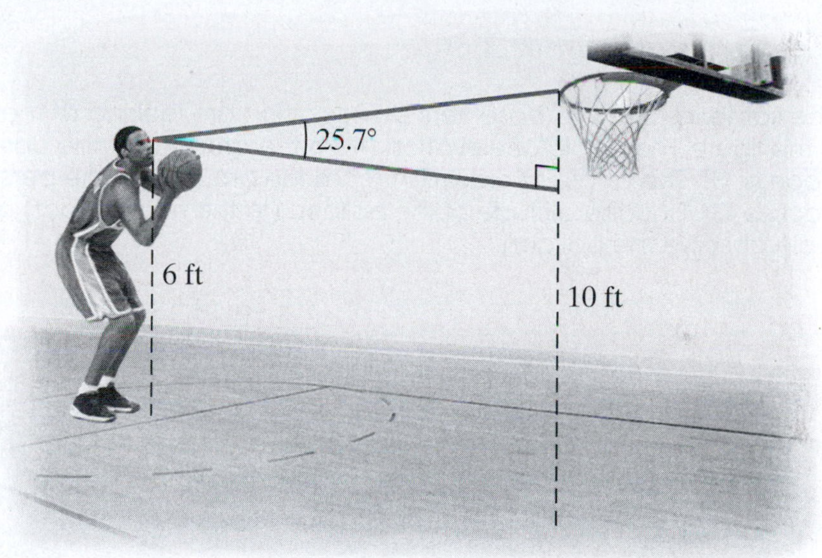

12. During the first quarter of its moon, the planet Quagnon, its moon, and its sun create a right triangle as shown. The distance between Quagnon and its moon is approximately 146,000 km. Determine the distance between the moon of Quagnon and its sun. Round to the nearest kilometer.

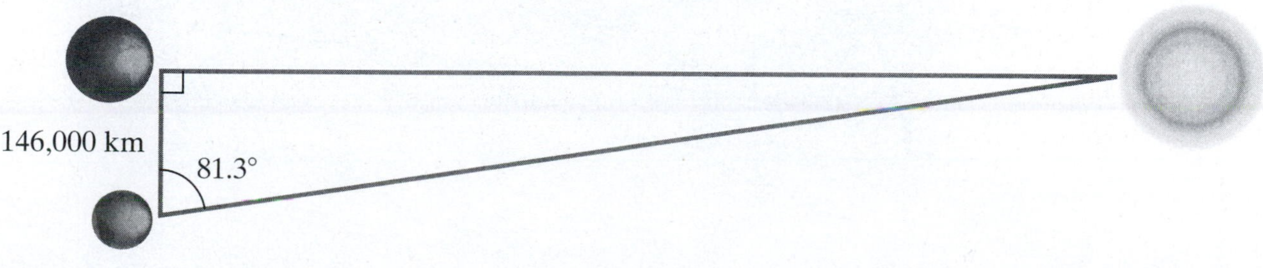

13. A mine shaft with a circular entrance has been carved into the side of a mountain. From a distance of 100 feet from the base of the mountain, the angle of elevation to the bottom of the circular opening is $27.7°$. The angle of elevation to the top of the opening is $33°$. Determine the diameter of the circular entrance. Round to two decimal places.

14. At the same instant, two observers 2 miles apart are looking at the same hot air balloon. See the figure. The angle of elevation (from the ground) of the observer closest to the balloon is $77°$. The angle of elevation (from the ground) of the person furthest from the balloon is $43°$. Find the altitude of the balloon (to the nearest foot) at the instant the two people observe the balloon.

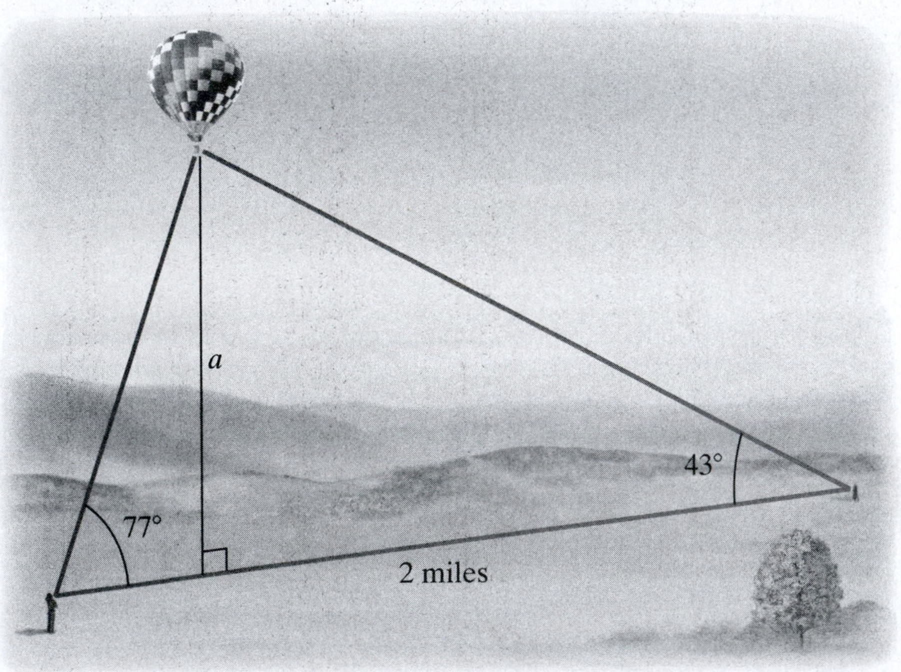

15. A surveyor wanted to determine the height of a mountain. To do this, she determined that the angle of elevation from point A to the top of the mountain is $31°$. She then drove 1500 meters toward the mountain to point B where she determined that the angle of elevation to the top of the mountain was $44°$. How tall is the mountain? Round to the nearest meter.

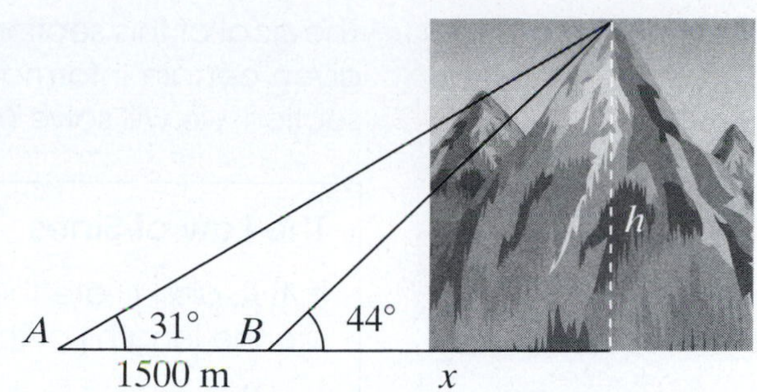

9.2 The Law of Sines

THINGS TO KNOW

Before working through this section, be sure that you are familiar with the following concepts:

VIDEO | ANIMATION | INTERACTIVE

You Try It

1. Understanding the Inverse Sine Function (Section 7.4)

OBJECTIVES

1 Determining If the Law of Sines Can Be Used to Solve an Oblique Triangle

2 Using the Law of Sines to Solve the SAA Case or ASA Case

3 Using the Law of Sines to Solve the SSA Case

4 Using the Law of Sines to Solve Applied Problems Involving Oblique Triangles

SECTION 9.2 EXERCISES

OBJECTIVE 1 DETERMINING IF THE LAW OF SINES CAN BE USED TO SOLVE AN OBLIQUE TRIANGLE

Most triangles that we have worked with thus far in this text have been right triangles. We now turn our attention to triangles that do not include a right angle. These triangles are called **oblique triangles**. There are two types of oblique triangles. The first type is an oblique triangle with three acute angles. The second type is an oblique triangle with one obtuse angle and two acute angles. Figure 5 illustrates two such oblique triangles. Note that the angles in each triangle in Figure 5 are labeled with capital letters and the sides opposite the angles are labeled with the respective lowercase letter.

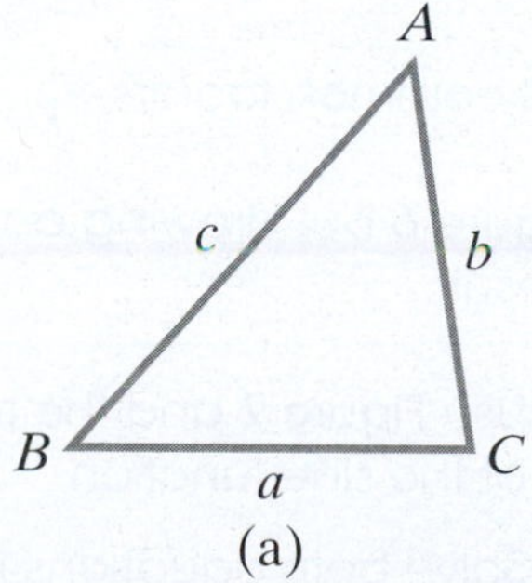

(a)

An oblique triangle with three acute angles

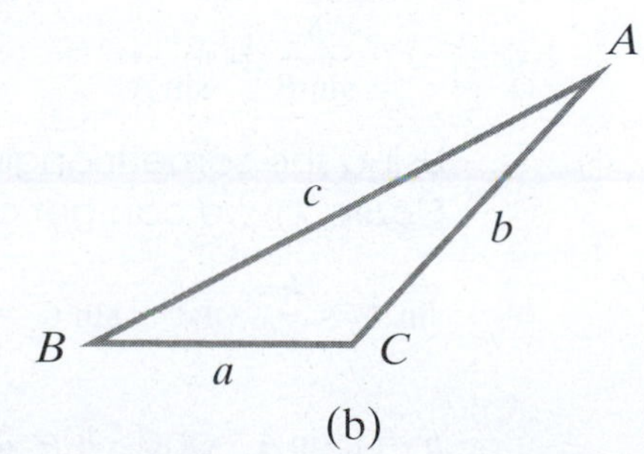

(b)

An oblique triangle with one obtuse angle and two acute angles

Figure 5

The goal of this section is to determine all angles and all sides of oblique triangles, given certain information. This process is called **solving oblique triangles.** In this section, we will solve (or attempt to solve) oblique triangles using the **Law of Sines**.

The Law of Sines

If $A, B,$ and C are the measures of the angles of any triangle and if $a, b,$ and c are the lengths of the sides opposite the corresponding angles, then

$$\frac{a}{\sin A} = \frac{b}{\sin B} = \frac{c}{\sin C} \quad \text{or} \quad \frac{\sin A}{a} = \frac{\sin B}{b} = \frac{\sin C}{c}.$$

The Law of Sines states that the ratio of the length of a side of a triangle to the sine of the angle opposite the side is equal to the ratio of the length of any other side to the sine of the angle opposite that side. We can develop a proof of the Law of Sines for the case of an oblique triangle having three acute angles by first drawing an **altitude** of length h from angle C to point D, as shown in Figure 6.

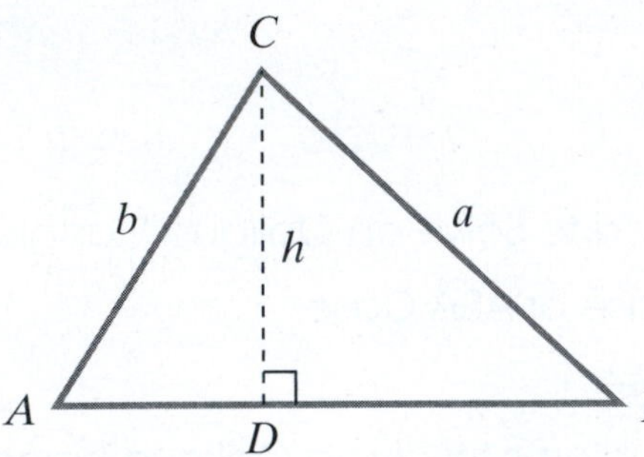

Figure 6 Oblique triangle ABC with three acute angles and an altitude of length h.

Note in **Figure 6** that triangle ACD and BCD are right triangles. Using the **right triangle definition of the sine function**, we get

$$\sin A = \frac{h}{b} \quad \text{and} \quad \sin B = \frac{h}{a}.$$

Solving both of these equations for h, we get

$$h = b \sin A \quad \text{and} \quad h = a \sin B.$$

We can now set the expressions $a \sin A$ and $b \sin B$ equal to each other because they both represent the length of the **altitude** of the triangle in **Figure 6**.

$b \sin A = a \sin B$ — Set the expressions representing h equal to each other.

$\dfrac{b \sin A}{\sin A \sin B} = \dfrac{a \sin B}{\sin A \sin B}$ — Divide both sides by $\sin A \sin B$.

$\dfrac{b}{\sin B} = \dfrac{a}{\sin A}$ — Cancel common factors.

Using the same triangle from **Figure 6** but drawing an **altitude** from angle B (see **Figure 7**), we can get a similar result.

$\sin A = \dfrac{h}{c}$ and $\sin C = \dfrac{h}{a}$ — Use **Figure 7** and the right triangle definition of the sine function.

$h = c \sin A$ and $h = a \sin C$ — Solve both equations for h.

$c \sin A = a \sin C$ — Set the expressions representing h equal to each other.

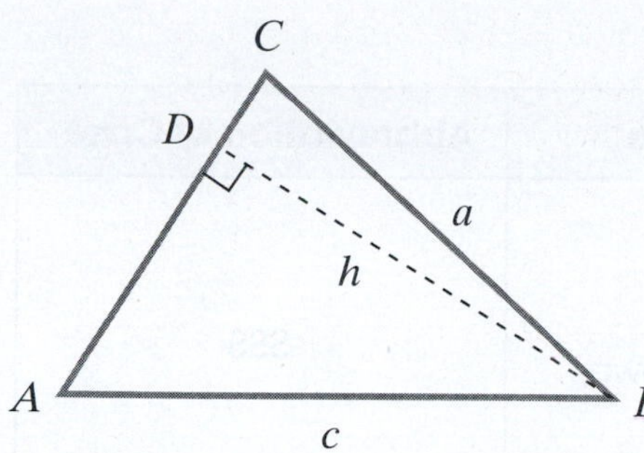

Figure 7

$\dfrac{c \sin A}{\sin A \sin C} = \dfrac{a \sin C}{\sin A \sin C}$ Divide both sides by $\sin A \sin C$.

$\dfrac{c}{\sin C} = \dfrac{a}{\sin A}$ Cancel common factors.

Thus we now have the two results,

$$\frac{b}{\sin B} = \frac{a}{\sin A} \text{ and } \frac{c}{\sin C} = \frac{a}{\sin A}.$$

Therefore, $\dfrac{a}{\sin A} = \dfrac{b}{\sin B} = \dfrac{c}{\sin C}$ and we have proved the Law of Sines for the case of an oblique triangle with three acute angles. The right triangle case and the oblique triangle with an obtuse angle case can be proved in a similar manner.

Before we use the Law of Sines to solve oblique triangles, it is necessary to consider when the Law of Sines can be used. To solve any oblique triangle, at least three pieces of information must be known. If S represents a known side of a triangle and if A represents a known angle of a triangle, then there are six possible situations where only three pieces of information can be known. The six possible cases are shown in **Table 1**.

Table 1

Triangle	Description of Case	Abbreviation of Case
	Side–Angle–Angle Two angles and a side opposite one of the angles are known.	SAA
	Angle–Side–Angle Two angles and the side between the angles are known.	ASA
	Side–Side–Angle Two sides and an angle opposite one of the sides are known.	SSA
	Side–Angle–Side Two sides and an angle between the sides are known.	SAS

(*Continued*)

Table 1 (*Continued*)

Triangle	Description of Case	Abbreviation of Case
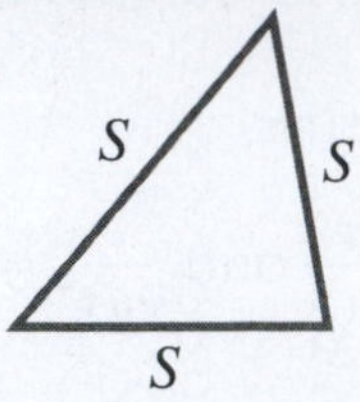	Side–Side–Side All three sides are known.	SSS
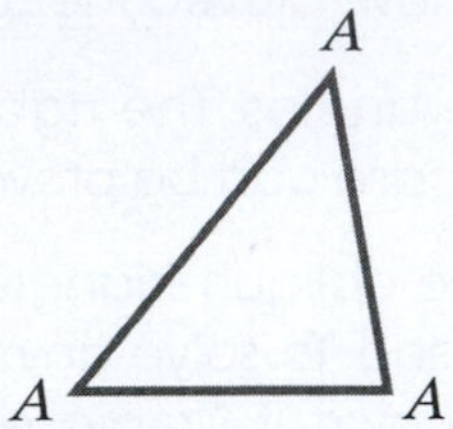	Angle–Angle–Angle All three angles are known.	AAA

Because the Law of Sines uses proportions that involve both angles and sides, the following pieces of information are needed to solve an oblique triangle using the Law of Sines:

1. The measure of an angle must be known.
2. The length of the side opposite the known angle must be known.
3. At least one more side or one more angle must be known.

The first three cases listed in **Table 1** involve situations where this information is known. Therefore, the Law of Sines can be used to solve the SAA, ASA, and SSA cases.

Example 1 Determining If the Law of Sines Can Be Used to Solve an Oblique Triangle

Decide whether or not the Law of Sines can be used to solve each triangle. Do not attempt to solve the triangle.

a.

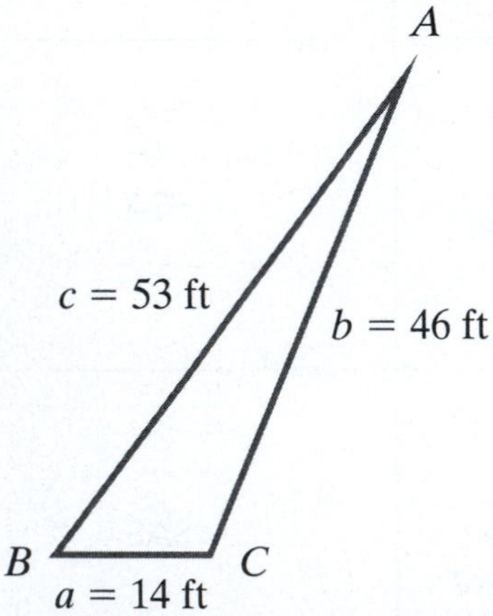

b.

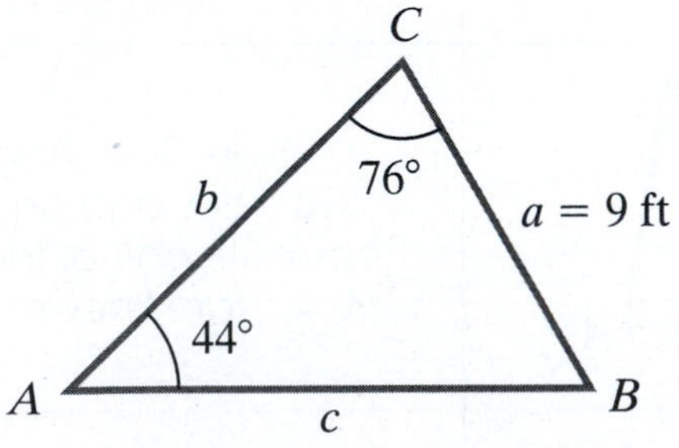

c.

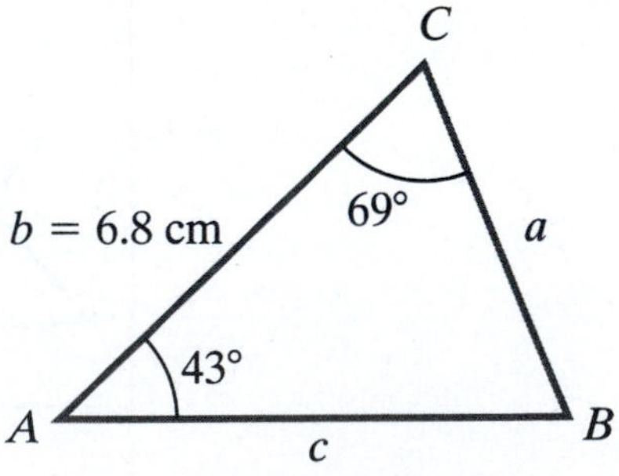

Solution

a. First, organize the given information.

Angles	Sides
$A =$ ____	$a = 14$ ft
$B =$ ____	$b = 46$ ft
$C =$ ____	$c = 53$ ft

The Law of Sines *cannot* be used because the first criterion is not satisfied. We are not given the measure of any angle. This is an example of the SSS case, which will be discussed in Section 9.2.

b. First, organize the given information.

Angles	Sides
$A = 44°$	$a = 9$ ft
$B =$ ____	$b =$ ____
$C = 76°$	$c =$ ____

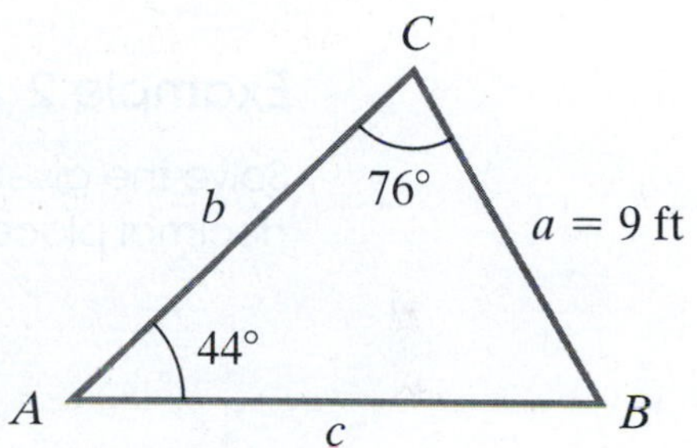

The Law of Sines *can* be used to solve the triangle because we are given an angle $A = 44°$ and the length of the side opposite of angle A ($a = 9$ ft). We are also given the measure of another angle $C = 76°$. This is an example of the SAA case.

c. First, organize the given information.

Angles	Sides
$A = 43°$	$a =$ ____
$B =$ ____	$b = 6.8$ cm
$C = 69°$	$c =$ ____

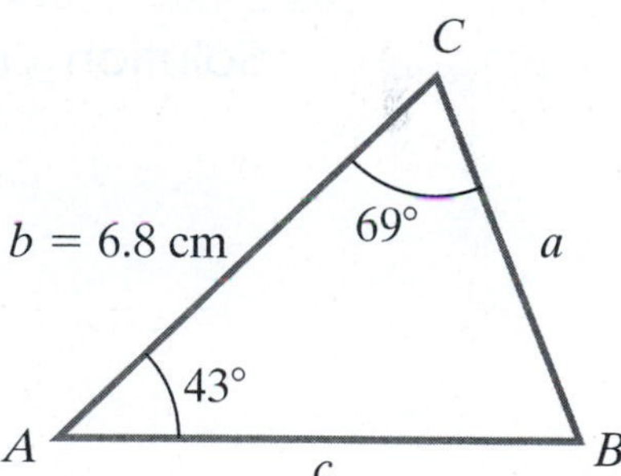

At first glance it appears that we cannot use the Law of Sines to solve this triangle because we are given the length of side b but we are not given the measure of angle B. However, because the sum of the measures of the angles of any triangle is $180°$, we can find the measure of angle B.

$B + 69° + 43° = 180°$ The sum of the measures of the angles of any triangle is $180°$.

$B + 112° = 180°$ Add.

$B = 68°$ Solve for B.

We now know the measure of an angle $B = 68°$, the length of the side opposite the angle, $b = 6.8$ cm and another angle $A = 43°$ or $C = 69°$. Therefore, the Law of Sines can be used to solve the triangle. This is an example of the ASA case.

You Try It Work through this You Try It problem.

Work Exercises 1–6 in this textbook or in the MyLab Math Study Plan.

OBJECTIVE 2 USING THE LAW OF SINES TO SOLVE THE SAA CASE OR ASA CASE

When the measure of any two angles of an oblique triangle are known and the length of any side is known, always start by determining the measure of the unknown angle. Then use appropriate Law of Sines proportions to solve for the lengths of the remaining unknown sides. Whenever possible, we will avoid using rounded information to solve for the remaining parts of the triangle. When this cannot be avoided, we will agree to use information rounded to one decimal place unless some other guideline is stated.

Example 2 Solving a SAA Triangle Using the Law of Sines

Solve the given oblique triangle. Round the lengths of the sides to one decimal place.

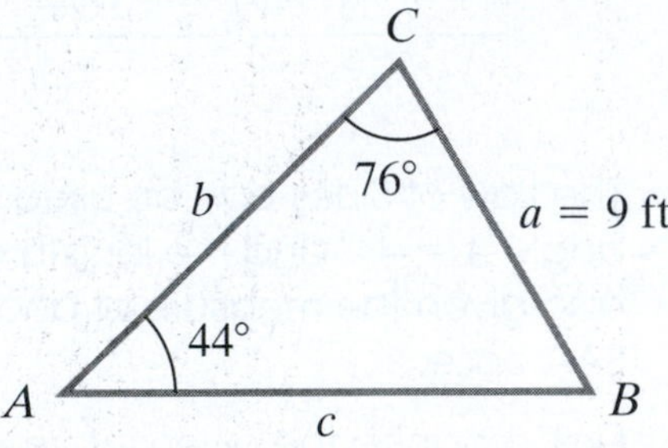

Solution Observe that we are given the following information.

Angles	Sides
$A = 44°$	$a = \underline{9\text{ ft}}$
$B = \_\_\_\_$	$b = \_\_\_\_$
$C = 76°$	$c = \_\_\_\_$

We start by determining the measure of angle B using the fact that the sum of the measures of the three angles of a triangle is $180°$.

$B + 44° + 76° = 180°$ The sum of the measures of the angles of any triangle is $180°$.

$B + 120° = 180°$ Add.

$B = 60°$ Solve for B.

We insert the new information.

Angles	Sides
$A = 44°$	$a = 9$ ft
$B = \underline{60°}$	$b = \underline{\quad}$
$C = 76°$	$c = \underline{\quad}$

We will use the known ratio $\dfrac{a}{\sin A}$ to determine the lengths of the other two sides. First find b.

$\dfrac{b}{\sin B} = \dfrac{a}{\sin A}$ Use the Law of Sines to write a proportion.

$\dfrac{b}{\sin 60°} = \dfrac{9}{\sin 44°}$ Substitute the known information.

$b = \dfrac{9 \sin 60°}{\sin 44°}$ Multiply both sides by $\sin 60°$ to solve for b.

$b \approx 11.2$ ft Use a calculator in degree mode and round to one decimal place.

Again, we insert the new information.

Angles	Sides
$A = 44°$	$a = 9$ ft
$B = \underline{60°}$	$b \approx \underline{11.2\text{ ft}}$
$C = 76°$	$c = \underline{\quad}$

Lastly, we find the length of side C.

$\dfrac{c}{\sin C} = \dfrac{a}{\sin A}$ Use the Law of Sines to write a proportion.

$\dfrac{c}{\sin 76°} = \dfrac{9}{\sin 44°}$ Substitute the known information.

$c = \dfrac{9 \sin 76°}{\sin 44°}$ Multiply both sides by $\sin 76°$ to solve for c.

$c \approx 12.6$ ft Use a calculator in degree mode and round to one decimal place.

We have now solved the triangle.

Angles	Sides
$A = 44°$	$a = 9$ ft
$B = \underline{60°}$	$b \approx \underline{11.2\text{ ft}}$
$C = 76°$	$c \approx \underline{12.6\text{ ft}}$

Example 3 Solving an ASA Triangle Using the Law of Sines

Solve oblique triangle ABC if $B = 38°$, $C = 72°$, and $a = 7.5$ cm. Round the lengths of the sides to one decimal place.

Solution Start by sketching a triangle and label the angles and sides based on the given information. See Figure 8.

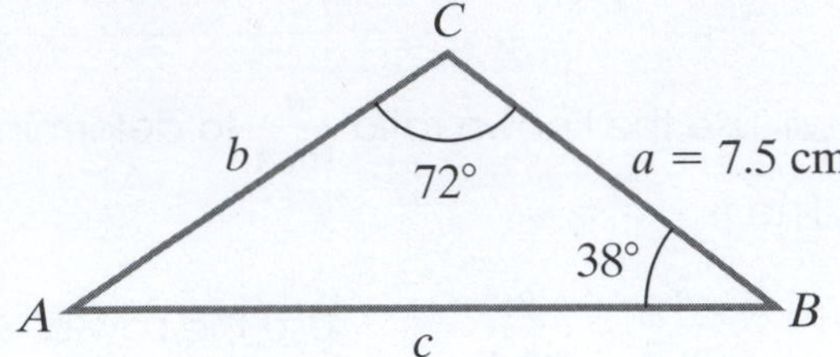

Figure 8

We can determine the measure of angle A by subtracting the measures of angle B and angle C from $180°$. Therefore, $A = 180° - 38° - 72° = 70°$.

We now organize all of the information.

Angles	Sides
$A = 70°$	$a = 7.5$ cm
$B = 38°$	$b =$ ____
$C = 72°$	$c =$ ____

Try using the Law of Sines to determine the lengths of sides b and c on your own. Work through this **video** to see the solution. ●

You Try It **Work through this You Try It problem.**

Work Exercises 7–12 in this textbook or in the MyLab Math Study Plan.

OBJECTIVE 3 USING THE LAW OF SINES TO SOLVE THE SSA CASE

When given the SAA case or the ASA case, as in Examples 2 and 3, we used the Law of Sines of the form $\dfrac{a}{\sin A} = \dfrac{b}{\sin B} = \dfrac{c}{\sin C}$ to determine the missing sides of a triangle. In the SAA case or the ASA case, a unique triangle is always formed.

We now want to look at the SSA case. This is the case when we are given two sides and the angle opposite one of the sides. In the SSA case, we will use the Law of Sines of the form $\dfrac{\sin A}{a} = \dfrac{\sin B}{b} = \dfrac{\sin C}{c}$ to first determine the measure of an angle. For example, suppose that we are given the measure of angle A, the length of side a, and the length of side b. In order to solve for angle B, we first use the Law of Sines and the proportion $\dfrac{\sin B}{b} = \dfrac{\sin A}{a}$ to solve for $\sin B$ as follows:

$$\frac{\sin B}{b} = \frac{\sin A}{a} \qquad \text{Use the Law of Sines to write a proportion.}$$

$$\sin B = \frac{b \sin A}{a} \qquad \text{Multiply both sides by } b \text{ to solve for } \sin B.$$

Because the values of b, a, and $\sin A$ are always positive, it follows that the value of $\sin B$ is always positive. Therefore, there are three possibilities for the value of $\sin B$.

1. $\sin B > 1$
2. $\sin B = 1$
3. $0 < \sin B < 1$

For the first possibility, recall that the sine of an angle cannot exceed a value of 1. Thus, if $\sin B > 1$, then there exists no angle B for which $\sin B > 1$ and the result is no triangle.

For the second possibility, where $\sin B = 1$, then $B = 90°$ and the result is a right triangle.

For the third possibility, where $0 < \sin B < 1$, then there are two potential solutions for B because the sine function is positive for angles having a terminal side in Quadrant I or in Quadrant II. Thus, the result is either one triangle or two triangles.

As you can see, unlike the SAA case or ASA case, where one unique triangle is always formed, the SSA case can result in zero, one, or two triangles. For this reason, the SSA case is often referred to as the **ambiguous case** of the Law of Sines.

Watch this **animation** to see all of the possible triangles that can result when given the SSA case.

The number of triangles depends on the value of $\sin B$ and whether or not there is exactly one solution for B, two solutions for B, or no solution for B. All possible scenarios of the SSA case are outlined in **Table 2** assuming that we are given the measure of angle A and the lengths of sides a and b.

Table 2

Possible Triangles	Number of Triangles	Description	The Value of $\sin B$
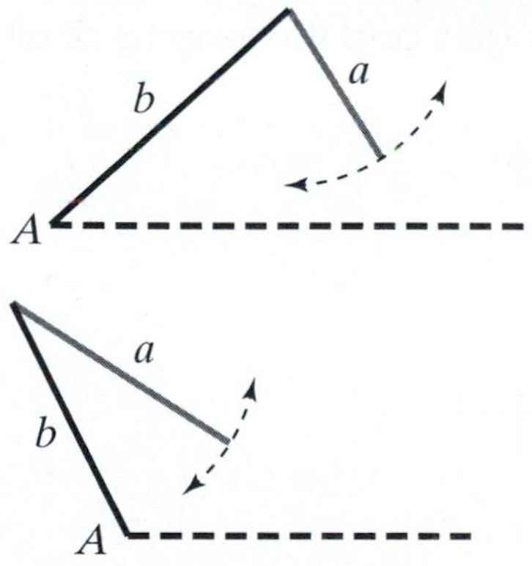	No Triangle	No angle B exists and side a is too short to reach the opposite side.	$\sin B > 1$
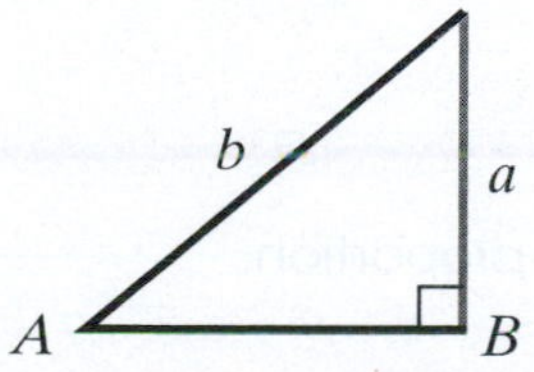	One Triangle	The measure of angle B is 90° and the result is one right triangle.	$\sin B = 1$

(*Continued*)

Table 2 *(Continued)*

Possible Triangles	Number of Triangles	Description	The Value of $\sin B$
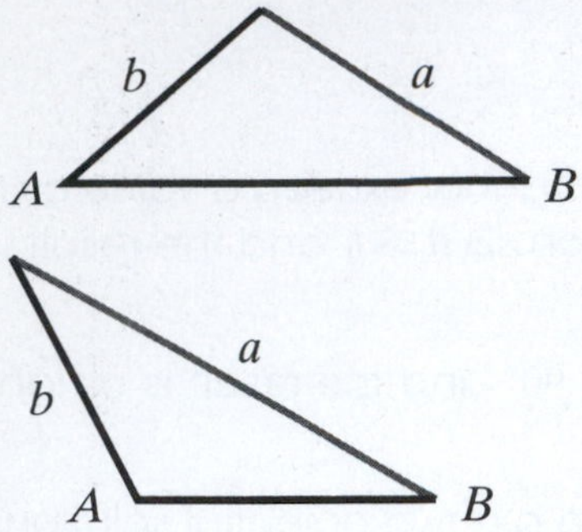	One Triangle	If there is one solution for B, then the traingle is oblique with three acute angles or the triangle is oblique with one obtuse angle.	$0 < \sin B < 1$
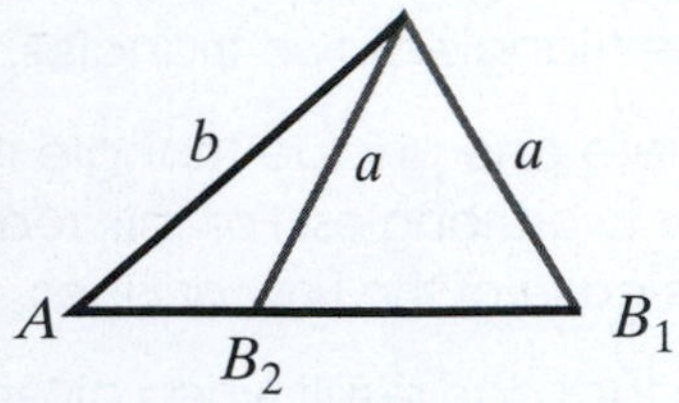	Two Triangles	If there are two solutions for B (B_1 and B_2), then the result is two triangles.	$0 < \sin B < 1$

Watch this **animation** to see all of the possible triangles that can result when given the SSA case.

Example 4 Solving a SSA Triangle Using the Law of Sines (No Triangle)

Two sides and an angle are given below. Determine whether the information results in no triangle, one right triangle, or one or two oblique triangles. Solve each resulting triangle. Round the measures of all angles and the lengths of all sides to one decimal place.

$$a = 10 \text{ ft}, \quad b = 28 \text{ ft}, \quad A = 29°$$

Solution First, organize the given information.

Angles	Sides
$A = 29°$	$a = 10$ ft
$B =$ ____	$b = 28$ ft
$C =$ ____	$c =$ ____

We use the Law of Sines to first determine the value of $\sin B$.

$$\frac{\sin B}{b} = \frac{\sin A}{a}$$ Use the Law of Sines to write a proportion.

$$\frac{\sin B}{28} = \frac{\sin 29°}{10}$$ Substitute the known information.

$$\sin B = \frac{28 \sin 29°}{10}$$ Multiply both sides by 28 to isolate $\sin B$.

The value of $\sin B = \dfrac{28 \sin 29°}{10}$ is approximately 1.3575. Because the value of $\sin B$ cannot exceed 1, we know that there is no angle B. Thus, there is no triangle that satisfies the given conditions. Figure 9 illustrates that the length of side a is too short to reach the opposite side.

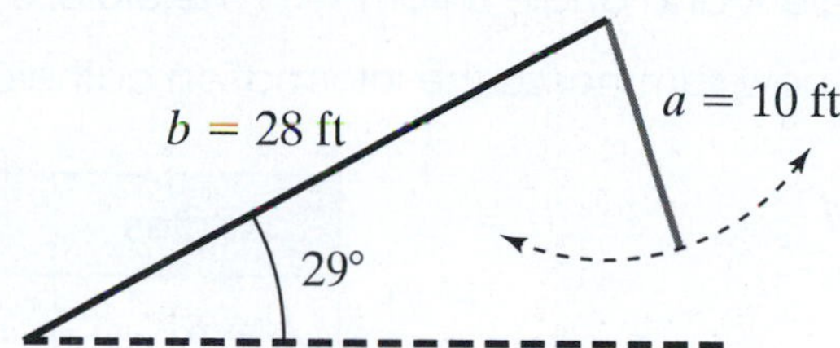

Figure 9

Example 5 Solving a SSA Triangle Using the Law of Sines (Exactly One Triangle)

Two sides and an angle are given below. Determine whether the information results in no triangle, one right triangle, or one or two oblique triangles. Solve each resulting triangle. Round the measures of all angles and the lengths of all sides to one decimal place.

$$a = 13 \text{ cm}, \quad b = 7.8 \text{ cm}, \quad A = 67°$$

Solution First, organize the given information.

Angles	Sides
$A = 67°$	$a = 13$ cm
$B =$ ____	$b = 7.8$ cm
$C =$ ____	$c =$ ____

We use the Law of Sines to first determine the value of $\sin B$.

$$\frac{\sin B}{b} = \frac{\sin A}{a}$$ Use the Law of Sines to write a proportion.

$$\frac{\sin B}{7.8} = \frac{\sin 67°}{13}$$ Substitute the known information.

$$\sin B = \frac{7.8 \sin 67°}{13}$$ Multiply both sides by 7.8 to isolate $\sin B$.

The value of $\sin B = \dfrac{7.8 \sin 67°}{13}$ is approximately 0.5523. Therefore, $0 < \sin B < 1$, and the given information results in one or two oblique triangles.

There are two possible choices for B between 0° and 180°. One choice is an angle in standard position having a terminal side that lies in Quadrant I. This angle is $B_1 = \sin^{-1}\left(\dfrac{7.8 \sin 67°}{13}\right) \approx 33.5°$. The other choice is an angle in standard position having a terminal side lying in Quadrant II. This angle is $B_2 = 180° - \sin^{-1}\left(\dfrac{7.8 \sin 67°}{13}\right) \approx 146.5°$. If $B_2 \approx 146.5°$, then the sum of the measures of the angles in the triangle would be $A + B_2 + C \approx 67° + 146.5° + C \approx 213.5° + C$. This cannot happen because the sum of the measures of the three angles cannot exceed $180°$. Therefore, we exclude B_2 as a possible choice. Thus, $B_1 = B \approx 33.5°$.

We can determine the measure of angle C by subtracting the measures of angle A and angle B from $180°$. Therefore, $C \approx 180° - 67° - 33.5° \approx 79.5°$.

We now summarize the information gathered thus far.

Angles	Sides
$A = 67°$	$a = 13$ cm
$B \approx \underline{33.5°}$	$b = 7.8$ cm
$C \approx \underline{79.5°}$	$c =$ ____

To find c we use the Law of Sines. Note that we will use the approximate measure of angle C rounded to one decimal place.

$$\frac{c}{\sin C} = \frac{a}{\sin A}$$ Use the Law of Sines to write a proportion.

$$\frac{c}{\sin 79.5°} \approx \frac{13}{\sin 67°}$$ Substitute the known information.

$$c \approx \frac{13 \sin 79.5°}{\sin 67°}$$ Multiply both sides by $\sin 79.5°$ to solve for c.

$$c \approx 13.9 \text{ cm}$$ Use a calculator in degree mode and round to one decimal place.

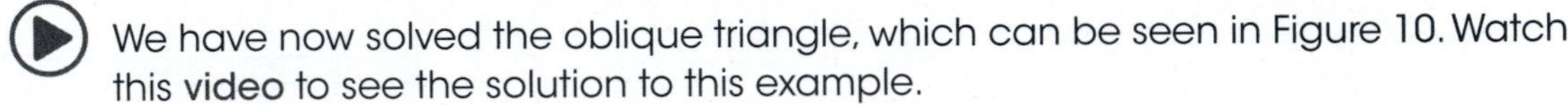
We have now solved the oblique triangle, which can be seen in Figure 10. Watch this **video** to see the solution to this example.

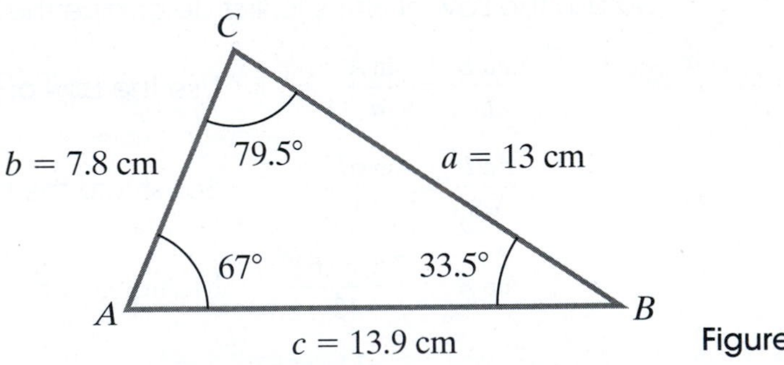

Figure 10

Example 6 Solving a SSA Triangle Using the Law of Sines (Two Triangles)

Two sides and an angle are given below. Determine whether the information results in no triangle, one right triangle, or one or two oblique triangles. Solve each resulting triangle. Round the measures of all angles and the lengths of all sides to one decimal place.

$$b = 11.3 \text{ in.}, \quad c = 15.5 \text{ in.}, \quad B = 34.7°$$

Solution First organize the given information.

Angles	Sides
$A =$ ____	$a =$ ____
$B = 34.7°$	$b = 11.3$ in.
$C =$ ____	$c = 15.5$ in.

We use the Law of Sines to first determine the value of $\sin C$.

$$\frac{\sin C}{c} = \frac{\sin B}{b}$$ Use the Law of Sines to write a proportion.

$$\frac{\sin C}{15.5} = \frac{\sin 34.7°}{11.3}$$ Substitute the known information.

$$\sin C = \frac{15.5 \sin 34.7°}{11.3}$$ Multiply both sides by 15.5 to isolate $\sin C$.

The value of $\sin C = \dfrac{15.5 \sin 34°}{11.3}$ is approximately 0.7809. Thus, there are two possible choices for C between $0°$ and $180°$. One choice is $C_1 = \sin^{-1}\left(\dfrac{15.5 \sin 34.7°}{11.3}\right) \approx 51.3°$. The other choice is $C_2 = 180° - \sin^{-1}\left(\dfrac{15.5 \sin 34.7°}{11.3}\right) \approx 128.7°$.

If $C_1 \approx 51.3°$, then the sum of the measures of the angles in the triangle would be

$$A_1 + B + C_1 \approx A_1 + 34.7° + 51.3° \approx A_1 + 86.0°.$$

If $C_2 \approx 128.7°$, then the sum of the measures of the angles in the triangle would be

$$A_2 + B + C_2 \approx A_2 + 34.7° + 128.7° \approx A_2 + 163.4°.$$

Both C_1 and C_2 are valid choices because the sum of the measures of the three angles does not yet exceed $180°$.

We now summarize the information thus far for both triangles.

Triangle 1

Angles	Sides
$A_1 =$ ____	$a_1 =$ ____
$B = 34.7°$	$b = 11.3$ in.
$C_1 \approx \underline{51.3°}$	$c = 15.5$ in.

Triangle 2

Angles	Sides
$A_2 =$ ____	$a_2 =$ ____
$B = 34.7°$	$b = 11.3$ in.
$C_2 \approx \underline{128.7°}$	$c = 15.5$ in.

 See if you can determine the values of A_1, A_2, a_1, and a_2 on your own. Watch this **video** to verify that $A_1 \approx 94°$, $A_2 \approx 16.6°$, $a_1 \approx 19.8$ in., and $a_2 \approx 5.7$ in. The two triangles are displayed in Figure 11.

(eText Screens 9.2-1–9.2-42)

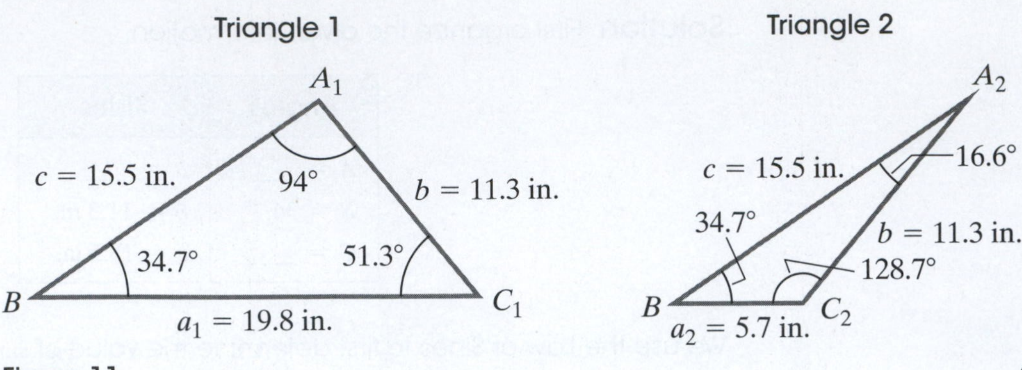

Figure 11

You Try It Work through this You Try It problem.

Work Exercises 13–19 in this textbook or in the MyLab Math Study Plan.

We now summarize the three cases for which the Law of Sines can be used.

Using the Law of Sines

If S represents a given side of a triangle and if A represents a given angle of a triangle, then the Law of Sines can be used to solve the three oblique triangle cases shown below.

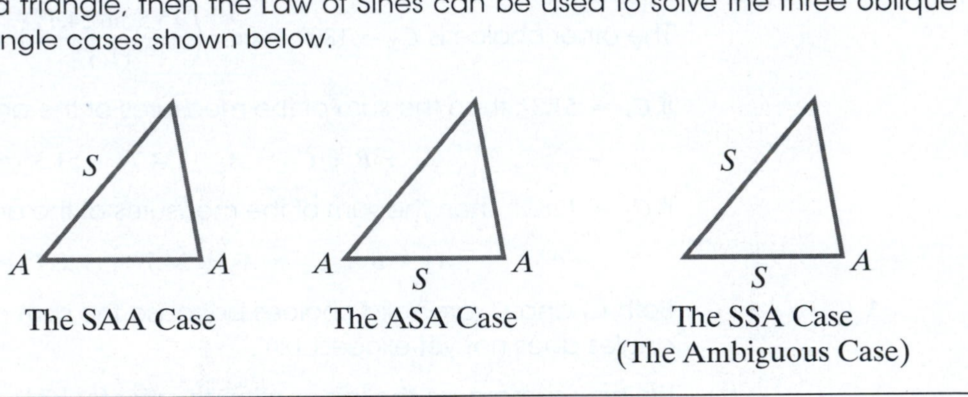

TIP In Section 9.3, we will see that the Law of Cosines can be used to solve the SAS case and the SSS case. Neither the Law of Sines nor the Law of Cosines can be used to solve the AAA case because the three given angles do not define a unique triangle. Do you know why? View this **explanation** to see why.

OBJECTIVE 4 USING THE LAW OF SINES TO SOLVE APPLIED PROBLEMS INVOLVING OBLIQUE TRIANGLES

The Law of Sines can be a useful tool to help solve many applications that arise involving triangles that are not right triangles. Many fields, such as surveying, engineering, and navigation, require the use of the Law of Sines.

Example 7 Determining the Width of a River

To determine the width of a river, forestry workers place markers on opposite sides of the river at points A and B. A third marker is placed at point C, 200 feet away from point A, forming triangle ABC. If the angle in triangle ABC at point C

is $51°$ and if the angle in triangle ABC at point A is $110°$, then determine the width of the river rounded to the nearest tenth of a foot.

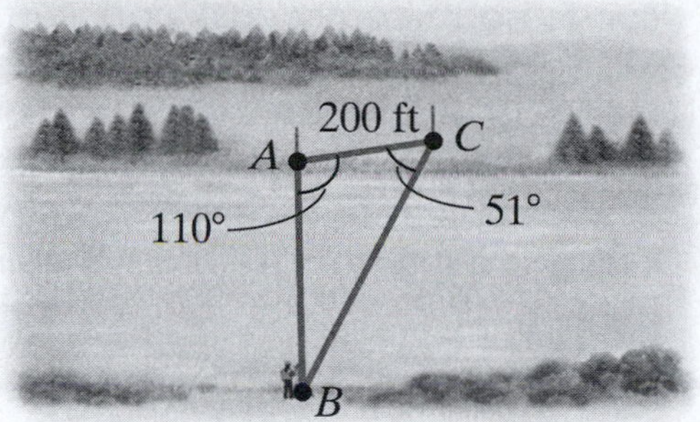

Solution First organize the given information.

Angles	Sides
$A = 110°$	$a = \_\_\_\_$
$B = \_\_\_\_$	$b = 200$ ft
$C = 51°$	$c = \_\_\_\_$

Observe that we want to find the length of side c, which represents the width of the river.

Start by determining the measure of angle B using the fact that the sum of the measures of the three angles of a triangle is $180°$.

$B + 51° + 110° = 180°$ The sum of the measures of the angles of any triangle is $180°$.

$B + 161° = 180°$ Add.

$B = 19°$ Solve for B.

Now insert the new information.

Angles	Sides
$A = 110°$	$a = \_\_\_\_$
$B = \underline{19°}$	$b = 200$ ft
$C = 51°$	$c = \_\_\_\_$

We can now use the Law of Sines of the form $\frac{c}{\sin C} = \frac{b}{\sin B}$ to solve for c.

$\frac{c}{\sin C} = \frac{b}{\sin B}$ Use the Law of Sines to write a proportion.

$\frac{c}{\sin 51°} = \frac{200}{\sin 19°}$ Substitute the known information.

$c = \frac{200 \sin 51°}{\sin 19°}$ Multiply both sides by $\sin 51°$ to solve for c.

$c \approx 477.4$ ft Use a calculator in degree mode.

Therefore, the width of the river is approximately 477.4 ft.

Many applications involving navigation use the concept of **bearing.** There are several different ways to denote a bearing. In the examples that follow, a bearing will be described as the direction that one object is from another object in relation to north, south, east, and west. Two directions and a degree measurement will be given to describe a bearing. For example, a bearing of N 45° E (read as "45 degrees east of north") can be sketched by drawing the initial side of an angle along the positive y-axis, which represents due north. The terminal side of the angle is then rotated away from the initial side in an "easterly direction" 45° toward the positive x-axis. See **Figure 12a**. Three other bearings all having an angle of 45° are sketched in **Figures 12b–d**.

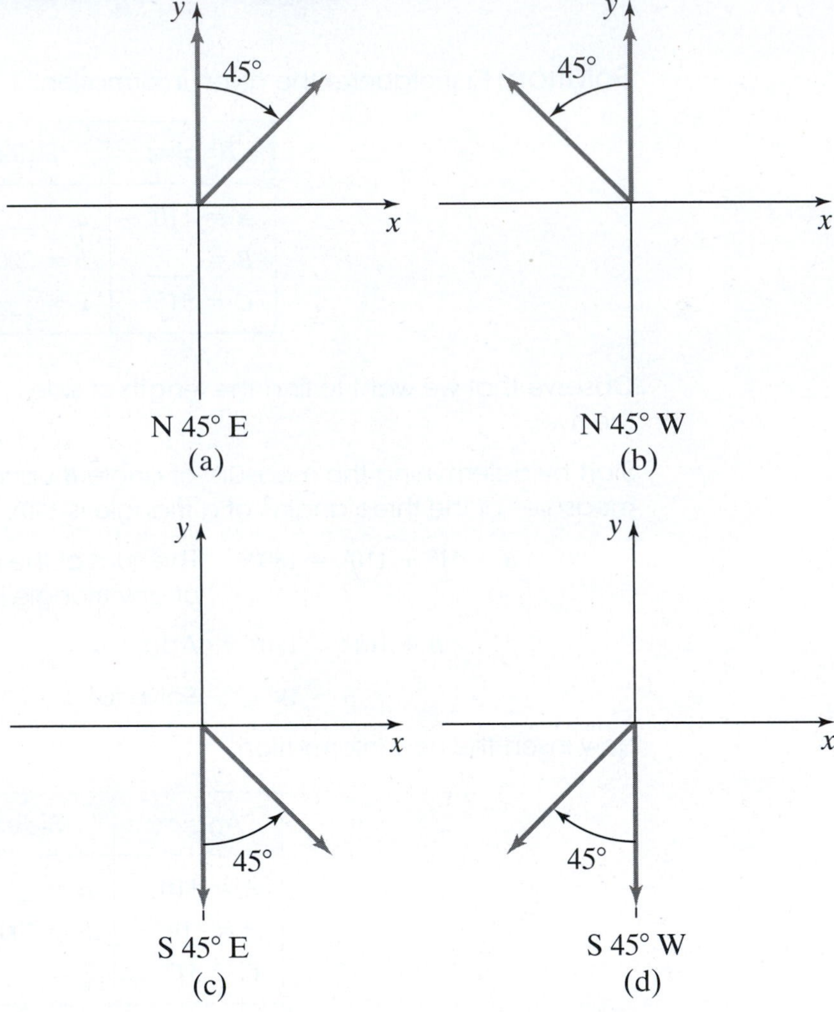

Figure 12

Example 8 Determining the Distance a Ship Is from Port

A ship set sail from port at a bearing of N 53° E and sailed 63 km to point B. The ship then turned and sailed an additional 69 km to point C. Determine the distance from port to point C if the ship's final bearing is N 74° E. Round to the nearest tenth of a kilometer.

Solution First, draw a diagram with the port located at the origin, as shown below in **Figure 13**.

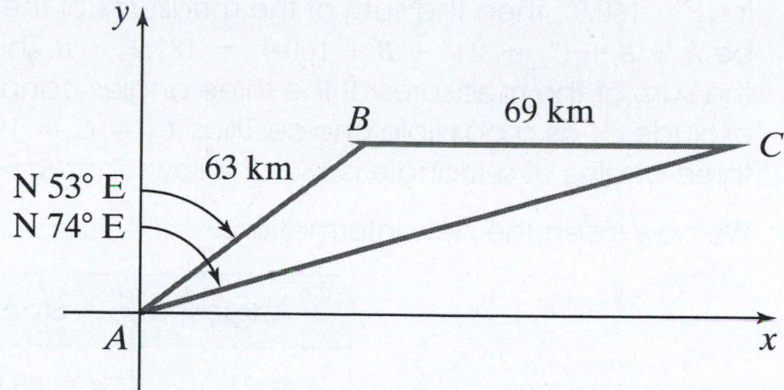

Figure 13

To determine the measure of angle A, we subtract the original bearing from the final bearing to get $A = 74° - 53° = 21°$. This gives the triangle seen in Figure 14.

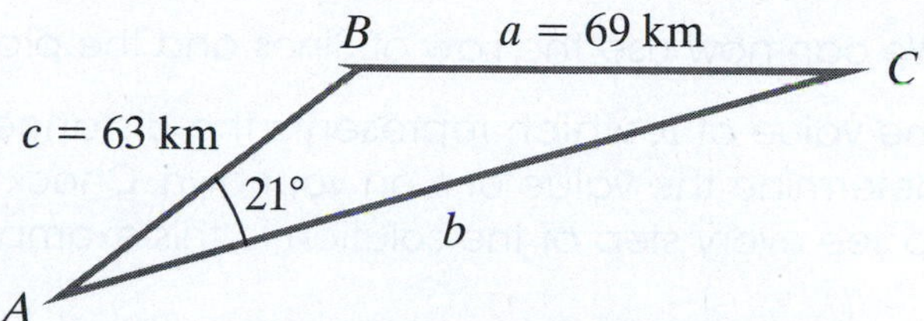

Figure 14

We now organize the information that we have thus far.

Angles	Sides
$A = 21°$	$a = 69$ km
$B =$ ____	$b =$ ____
$C =$ ____	$c = 63$ km

The goal is to find the length of side b. To do this, we must first find the measures of angles C and B. We can use the Law of Sines and the proportion $\frac{\sin C}{c} = \frac{\sin A}{a}$ to first determine the value of $\sin C$.

$$\frac{\sin C}{c} = \frac{\sin A}{a} \quad \text{Use the Law of Sines to write a proportion.}$$

$$\frac{\sin C}{63} = \frac{\sin 21°}{69} \quad \text{Substitute the known information.}$$

$$\sin C = \frac{63 \sin 21°}{69} \quad \text{Multiply both sides by 63 to isolate } \sin C.$$

The value of $\sin C = \frac{63 \sin 21°}{69}$ is approximately 0.3272. Thus, there are two possible choices for C between $0°$ and $180°$. One choice is an angle in standard position having a terminal side that lies in Quadrant I. This angle is $C_1 = \sin^{-1}\left(\frac{63 \sin 21°}{69}\right) \approx 19.1°$. The other choice is an angle in standard position having a terminal side lying in Quadrant II. This angle is $C_2 = 180° - \sin^{-1}\left(\frac{63 \sin 21°}{69}\right) \approx 160.9°$.

If $C_2 \approx 160.9^\circ$, then the sum of the measures of the angles in the triangle would be $A + B + C_2 \approx 21^\circ + B + 160.9^\circ \approx 181.9^\circ + B$. This cannot happen because the sum of the measures of the three angles cannot exceed 180°. Therefore, we exclude C_2 as a possible choice. Thus, $C_1 = C \approx 19.1^\circ$. Because the sum of the three angles of a triangle is 180°, it follows that $B \approx 139.9^\circ$.

We now insert the new information.

Angles	Sides
$A = 21^\circ$	$a = 69$ km
$B \approx \underline{139.9^\circ}$	$b =$ ____
$C \approx \underline{19.1^\circ}$	$c = 63$ km

We can now use the Law of Sines and the proportion $\dfrac{b}{\sin B} = \dfrac{a}{\sin A}$ to determine the value of b, which represents the distance from the port to the ship. Try to determine the value of b on your own. Check your **answer**, or watch this **video** to see every step of the solution to this example. ●

You Try It Work through this You Try It problem.

Work Exercises 20–25 in this textbook or in the MyLab Math Study Plan.

9.2 Exercises

Skill Check Exercises

For exercises SCE-1 and SCE-2, use a calculator to evaluate each expression. Round to one decimal place.

SCE-1. $\dfrac{7 \sin 63^\circ}{\sin 43^\circ}$ **SCE-2.** $\dfrac{23.3 \sin 84.2^\circ}{\sin 45.9^\circ}$

In Exercises 1–6, decide whether or not the Law of Sines can be used to solve each triangle. Do not attempt to solve the triangle.

1. B, a, $c = 33$ cm, 76°, 22°, C, A, b

2.

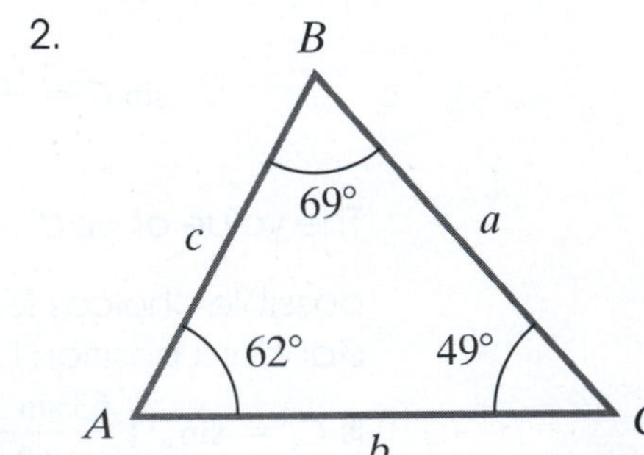

3.

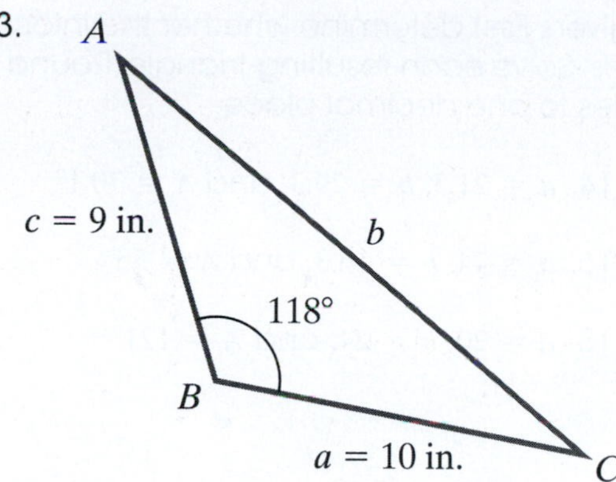

4.

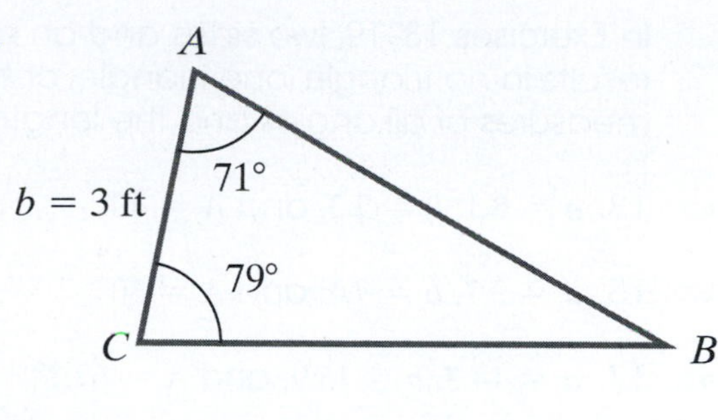

5.

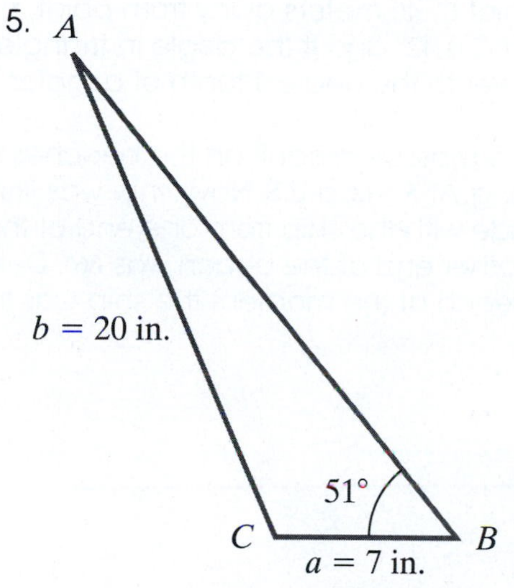

6.

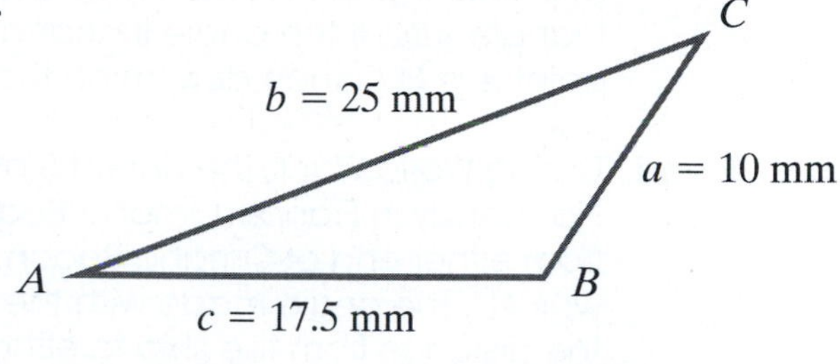

In Exercises 7–12, solve each oblique triangle. Round the measures of all angles and the lengths of all sides to one decimal place.

7.

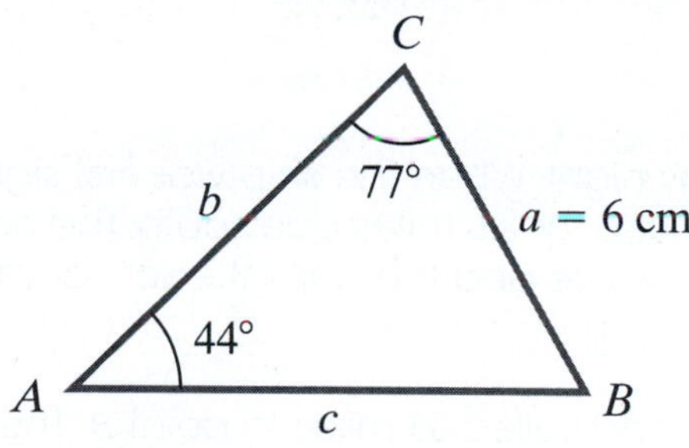

8.

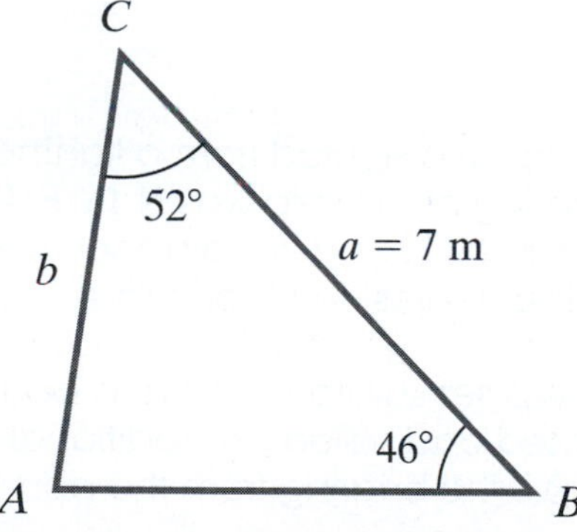

9.

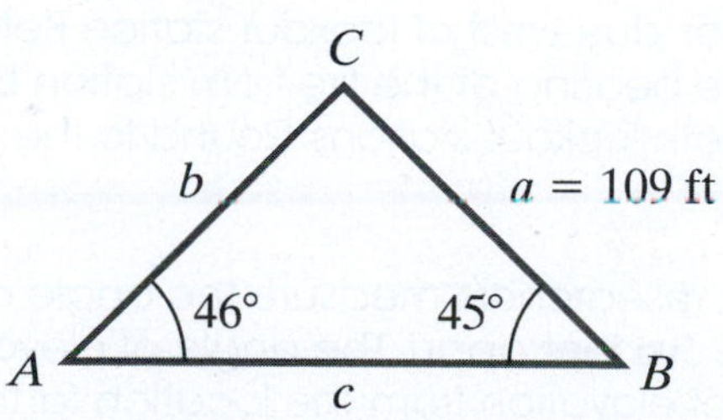

10. $A = 48°$, $B = 53°$, and $b = 6$

11. $A = 106°$, $C = 21°$, and $a = 4$

12. $B = 58°$, $C = 47°$, and $c = 9$

In Exercises 13–19, two sides and an angle are given. First determine whether the information results in no triangle, one triangle, or two triangles. Solve each resulting triangle. Round the measures of all angles and the lengths of all sides to one decimal place.

SbS 13. $a = 8.1$, $b = 7.3$, and $A = 40°$

SbS 14. $a = 21.5$, $b = 29.3$, and $A = 70.1°$

SbS 15. $a = 3.7$, $b = 7.4$, and $A = 30°$

SbS 16. $a = 14$, $b = 16.3$, and $A = 55°$

SbS 17. $a = 14.8$, $b = 15.9$, and $A = 67.7°$

SbS 18. $a = 20$, $b = 24$, and $A = 121°$

SbS 19. $a = 13.3$, $b = 13.6$, and $A = 57.1°$

20. To determine the width of a river, forestry workers place markers on opposite sides of the river at points A and B. A third marker is placed at point C, 40 meters away from point A, forming triangle ABC. If the angle in triangle ABC at point C is 42° and if the angle in triangle ABC at point A is 112°, then determine the width of the river to the nearest tenth of a meter.

21. During World War II, the United States military led a massive assault on the beaches of Normandy in France. Omaha Beach is 5 miles long. At 5 A.M, a U.S. Navy ship was first spotted from either end of Omaha Beach. The angle made with the ship from one end of the beach was 41°. The angle made with the ship from the other end of the beach was 66°. Determine the distance from the ship to either end of the beach at the moment the ship was first spotted. Round to the nearest tenth of a mile.

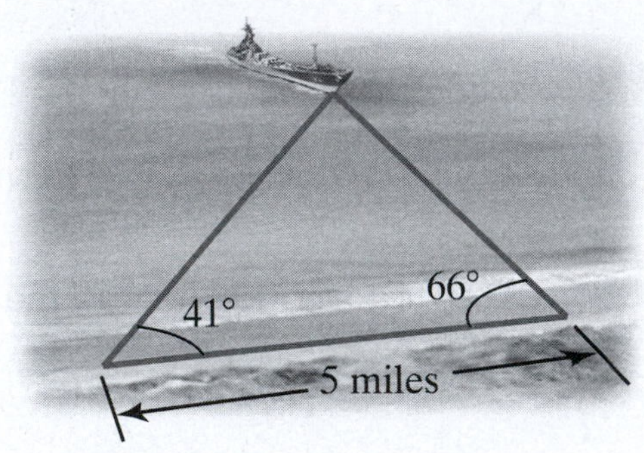

22. A ship was sighted from a lighthouse on a foggy night. When the ship was first sighted, the bearing of the ship was N 16° E. The ship then sailed for 8.6 miles due south. The new bearing was N 27° E. Find the distance between the lighthouse and the ship at each location. Round to the nearest tenth of a mile.

23. A ship set sail from port at a bearing of N 20° W and sailed 35 miles to point B. The ship then turned and sailed an additional 40 miles to point C. Determine the distance from port to the ship if the bearing from the port to point C is N 51° W. Round to the nearest tenth of a mile.

24. Forest fire lookout station Alpha is located 20 miles due west of lookout station Beta. The bearing of a fire from station Alpha is S 11° W. The bearing of the fire from station Beta is S 29° W. Determine the distance from the fire to both lookout stations. Round to the nearest tenth of a mile.

25. To determine the height of a giant Sequoia tree, researchers measure the angle of elevation to the top of the tree from two locations that are 500 feet apart. The angle of elevation from the location closest to the tree is 22°. The angle of elevation from the location farthest from the tree is 16°. What is the height of the tree to the nearest tenth of a degree?

Brief Exercises

In Exercises 26–32, two sides and an angle are given. Determine whether the information results in no triangle, one triangle, or two triangles.

26. $a = 8.1$, $b = 7.3$, and $A = 40°$

27. $a = 21.5$, $b = 29.3$, and $A = 70.1°$

28. $a = 3.7$, $b = 7.4$, and $A = 30°$

29. $a = 14$, $b = 16.3$, and $A = 55°$

30. $a = 14.8$, $b = 15.9$, and $A = 67.7°$

31. $a = 20$, $b = 24$, and $A = 121°$

32. $a = 13.3$, $b = 13.6$, and $A = 57.1°$

9.3 The Law of Cosines

THINGS TO KNOW

Before working through this section, be sure that you are familiar with the following concepts:

VIDEO ANIMATION INTERACTIVE

You Try It

1. Understanding the Inverse Sine Function (Section 7.4)

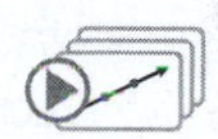

You Try It

2. Understanding the Inverse Cosine Function (Section 7.4)

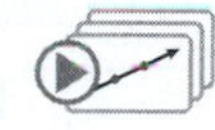

OBJECTIVES

1 Determining If the Law of Sines or the Law of Cosines Should Be Used to Begin to Solve an Oblique Triangle

2 Using the Law of Cosines to Solve the SAS Case

3 Using the Law of Cosines to Solve the SSS Case

4 Using the Law of Cosines to Solve Applied Problems Involving Oblique Triangles

SECTION 9.3 EXERCISES

OBJECTIVE 1 DETERMINING IF THE LAW OF SINES OR THE LAW OF COSINES SHOULD BE USED TO BEGIN TO SOLVE AN OBLIQUE TRIANGLE

Recall that it takes at least three pieces of known information to solve an oblique triangle. In Section 9.2, we used the **Law of Sines** to solve oblique triangles. We saw that the Law of Sines can be used when we are given the measure of an angle, the length of the side opposite that angle, and at least one other side or one other angle. Therefore, the Law of Sines can be used for the following cases.

Using the Law of Sines

If s represents a given side of a triangle and if A represents a given angle of a triangle, then the Law of Sines can be used to solve the three triangle cases shown below.

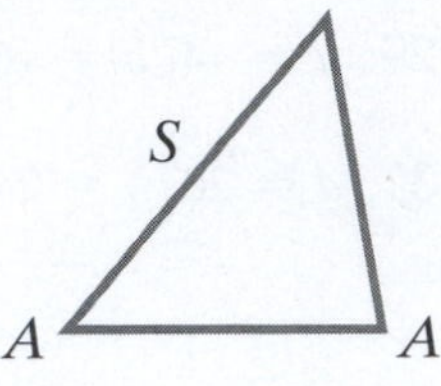

The SAA Case

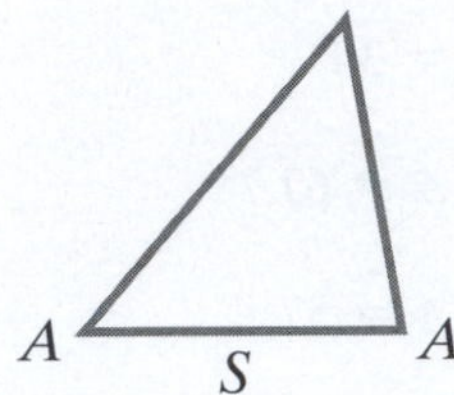

The ASA Case

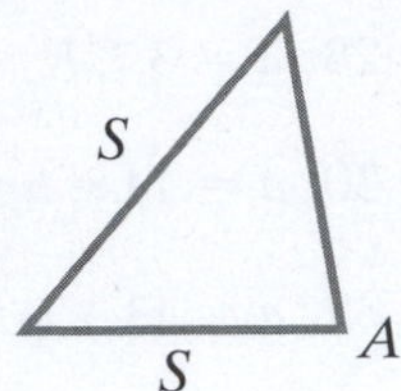

The SSA Case (The Ambiguous Case)

The other cases in which three pieces of information can be known are the ASA, SSS, and AAA cases. We will ignore the AAA case because the AAA case never defines a unique triangle. Do you know why? View this **explanation** to see why. The Law of Sines cannot be used to begin to solve the SAS or SSS cases because in either case, no angle opposite a known side is given. However, we can use the **Law of Cosines** to obtain this needed information.

The Law of Cosines

If A, B, and C are the measures of the angles of any triangle and if a, b, and c are the lengths of the sides opposite the corresponding angles, then

$$a^2 = b^2 + c^2 - 2bc\cos A,$$

$$b^2 = a^2 + c^2 - 2ac\cos B, \text{ and}$$

$$c^2 = a^2 + b^2 - 2ab\cos C$$

Note that each of the three Law of Cosines equations relates all three sides of a triangle to an angle. In words, the Law of Cosines says that the square of any side of a triangle is equal to the sum of the squares of the remaining two sides, minus twice the product of the two remaining sides and the cosine of the angle between them. Note that the angle used in each of the three equations is the angle opposite the side of the triangle that is isolated on the left-hand side of the equation.

To derive the first equation, $a^2 = b^2 + c^2 - 2bc\cos A$, start by drawing a triangle so that the vertex of angle A is at the origin of a Cartesian plane. See Figure 15a.

Next, draw an **altitude** of length y from the vertex of angle C to the opposite side at point $(x, 0)$, thus creating the two right triangles shown in Figure 15b. Note that the coordinates of the vertex of angle C are (x, y). Using the **right triangle definitions** of cosine and sine, we see that $\cos A = \frac{x}{b}$ and $\sin A = \frac{y}{b}$. Therefore, we can determine that the x-coordinate of the vertex of angle C is $x = b\cos A$. Likewise, the y-coordinate of the vertex of angle C is $y = b\sin A$. See Figure 15c.

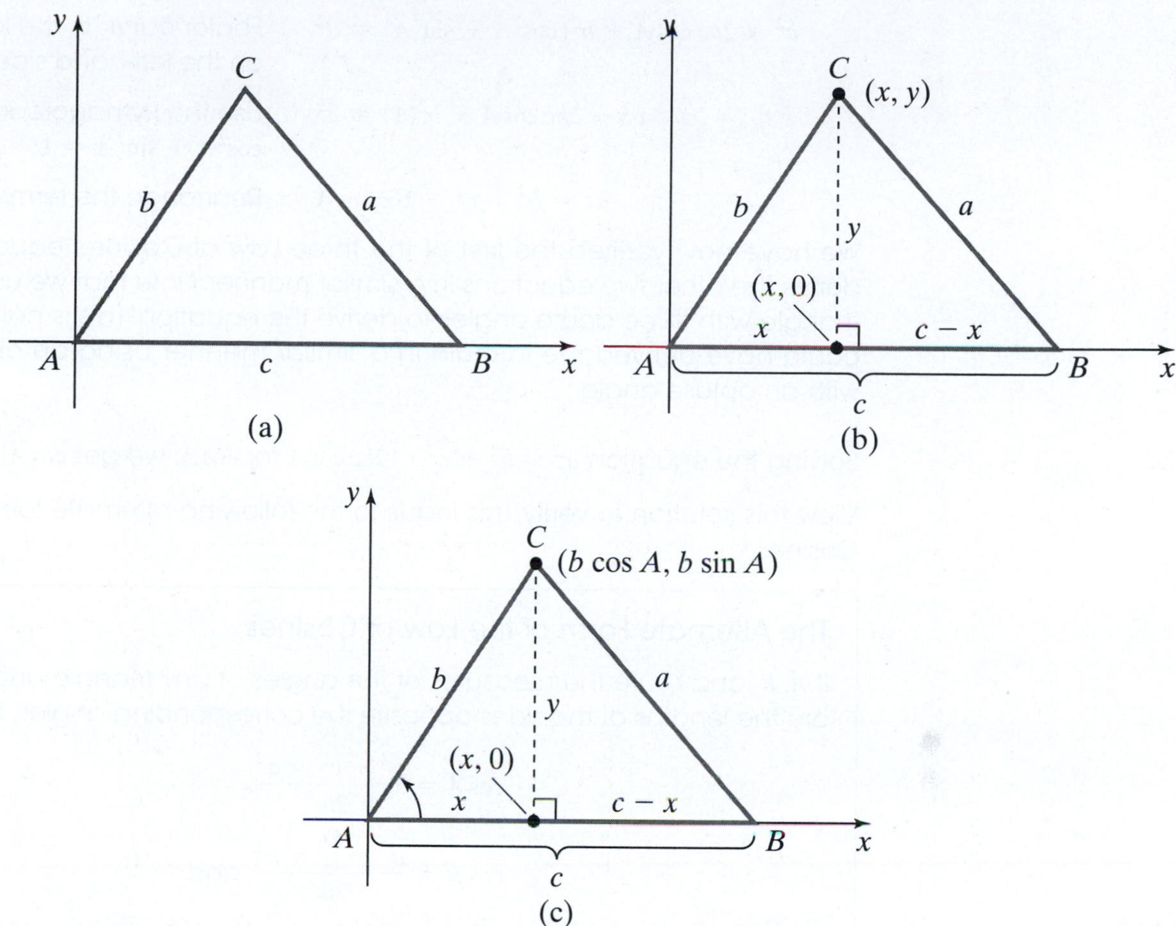

Figure 15

Now, isolate the triangle with side lengths of $c - x$, y, and a (see Figure 16) and use the Pythagorean Theorem to solve for a^2.

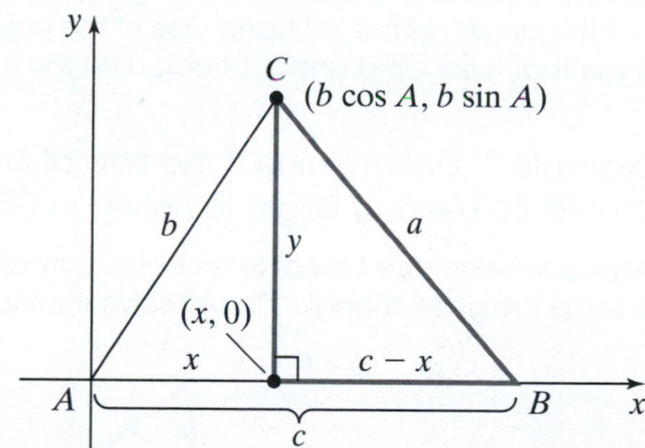

Figure 16

$(c - x)^2 + y^2 = a^2$	Write an equation representing the Pythagorean Theorem.
$c^2 - 2xc + x^2 + y^2 = a^2$	Square: $(c - x)^2 = c^2 - 2xc + x^2$.
$c^2 - 2(b\cos A)c + (b\cos A)^2 + (b\sin A)^2 = a^2$	Substitute $x = b\cos A$ and $y = b\sin A$.
$c^2 - 2bc\cos A + b^2\cos^2 A + b^2\sin^2 A = a^2$	Perform the operations and simplify.

$c^2 - 2bc\cos A + b^2(\cos^2 A + \sin^2 A) = a^2$ Factor out b^2 in the last two terms on the left-hand side.

$c^2 - 2bc\cos A + b^2(1) = a^2$ Use the Pythagorean identity: $\cos^2 A + \sin^2 A = 1$.

$a^2 = b^2 + c^2 - 2bc\cos A$ Rearrange the terms.

We have now verified the first of the three Law of Cosines equations. We can derive the other two equations in a similar manner. Note that we used an oblique triangle with three acute angles to derive the equation. This is not necessary. We could have derived the formula in a similar manner using an oblique triangle with an obtuse angle.

Solving the equation $a^2 = b^2 + c^2 - 2bc\cos A$ for $\cos A$, we get $\cos A = \dfrac{b^2 + c^2 - a^2}{2bc}$.

View this **solution** to verify. This leads to the following alternate form of the Law of Cosines.

The Alternate Form of the Law of Cosines

If A, B, and C are the measures of the angles of any triangle and if a, b, and c are the lengths of the sides opposite the corresponding angles, then

$$\cos A = \frac{b^2 + c^2 - a^2}{2bc},$$

$$\cos B = \frac{a^2 + c^2 - b^2}{2ac}, \text{ and}$$

$$\cos C = \frac{a^2 + b^2 - c^2}{2ab}$$

The three equations seen above are useful when determining the measure of a missing angle provided that the length of each of the three sides is known. Note that the angle on the left-hand side of the equation is opposite the side of the triangle that is squared and subtracted on the right-hand side of the equation.

Example 1 Determining If the Law of Sines or the Law of Cosines Should Be Used to Begin to Solve an Oblique Triangle

Decide whether the Law of Sines or the Law of Cosines should be used to begin to solve the given triangle. Do not solve the triangle.

a.

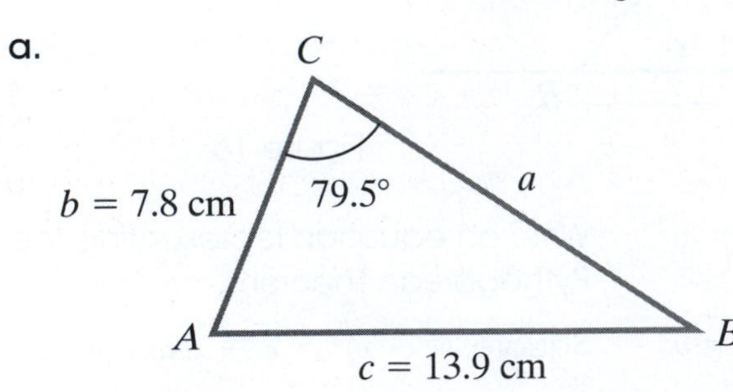

b.

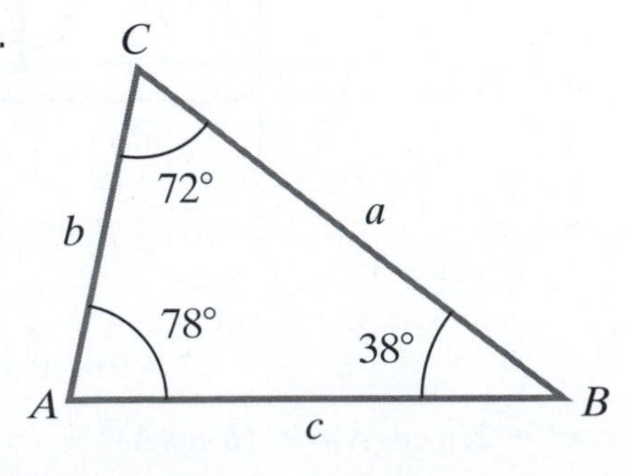

c.

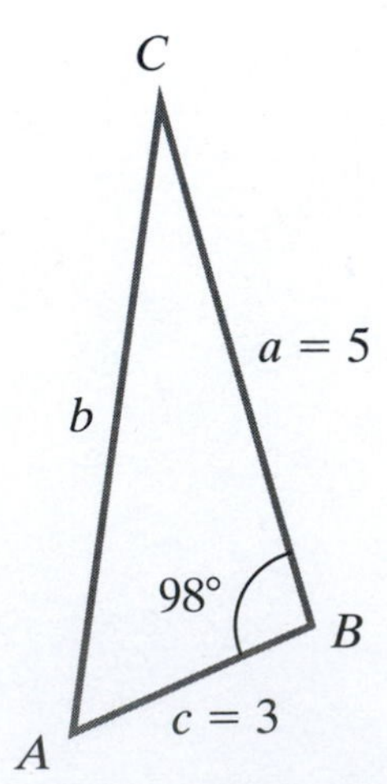

Solution

a. First, organize the given information.

Angles	Sides
$A =$ ___	$a =$ ___
$B =$ ___	$b = 7.8$ cm
$C = 79.5^\circ$	$c = 13.9$ cm

The lengths of two sides and the measure of an angle opposite one of the sides is given. This is the SSA case. Therefore, the Law of Sines should be used to begin to solve this triangle.

b. For this triangle, we see that we are given the measure of all three angles but are not given the length of any of the sides. This is the AAA case. Neither the Law of Sines nor the Law of Cosines can be used to solve this triangle because the AAA case does not result in a unique triangle.

c. First, organize the given information.

Angles	Sides
$A =$ ___	$a = 5$
$B = 98^\circ$	$b =$ ___
$C =$ ___	$c = 3$

The lengths of two sides and the measure of the angle between the two sides are given. This is the SAS case. Therefore, the Law of Cosines should be used to begin to solve this triangle.

You Try It Work through this You Try It problem.

Work Exercises 1–6 in this textbook or in the MyLab Math Study Plan.

OBJECTIVE 2 USING THE LAW OF COSINES TO SOLVE THE SAS CASE

The SAS case is the situation where we are given the lengths of two sides of a triangle and the measure of the included angle. To solve such a triangle, first use the Law of Cosines to determine the length of the missing side (the side opposite the given angle.) Once we know the lengths of the three sides, we can use the Law of Sines or the Law of Cosines to determine the measure of another angle. Note that the longer the side length, the larger the angle and vice versa. This information can often be useful when solving triangles. We now summarize a technique for solving an SAS oblique triangle.

Solving an SAS Oblique Triangle

Step 1. Use the Law of Cosines to determine the length of the missing side.

Step 2. Determine the measure of the smaller of the remaining two angles using the **Law of Sines** or using the **alternate form of the Law of Cosines**. This angle is the angle opposite of the shortest side and will always be acute.

Step 3. Use the fact that the sum of the measures of the three angles of a triangle is 180° to determine the measure of the remaining angle.

Whenever possible, we will avoid using rounded information to solve for the remaining parts of the triangle. When this cannot be avoided, we will agree to use information rounded to one decimal place unless some other guideline is stated.

Example 2 Solving an SAS Triangle

Solve the given oblique triangle. Round the measures of all angles and the lengths of all sides to one decimal place.

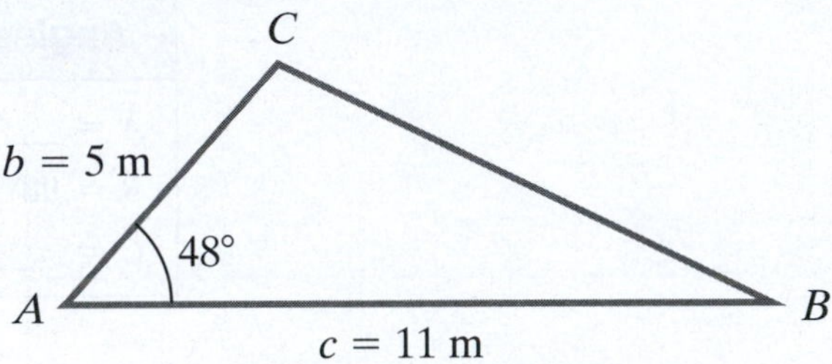

Solution Observe that we are given the following information.

Angles	Sides
$A = 48°$	$a =$ ____
$B =$ ____	$b = 5$ m
$C =$ ____	$c = 11$ m

Step 1. Use the Law of Cosines to determine the length of side a because we are given the measure of angle A.

$a^2 = b^2 + c^2 - 2bc\cos A$ — Use the Law of Cosines equation that relates $a, b, c,$ and A.

$a^2 = (5)^2 + (11)^2 - 2(5)(11)\cos 48°$ — Substitute $b = 5, c = 11,$ and $A = 48°$.

$a = \sqrt{(5)^2 + (11)^2 - 2(5)(11)\cos 48°}$ — Take the square root of both sides.

$a \approx 8.5$ — Use a calculator and round to one decimal place.

Insert the new information.

Angles	Sides
$A = 48°$	$a \approx 8.5$ m
$B =$ ____	$b = 5$ m
$C =$ ____	$c = 11$ m

Step 2. The smaller of the two missing angles is angle B because it is opposite the shortest side. We can now use either the Law of Sines or the alternate form of the Law of Cosines to determine the measure of angle B. Here, we use the alternate form of the Law of Cosines.

$$\cos B = \frac{a^2 + c^2 - b^2}{2ac}$$ Use the alternate form of the Law of Cosines equation that relates a, b, c, and $\cos B$.

$$\cos B \approx \frac{(8.5)^2 + (11)^2 - (5)^2}{2(8.5)(11)}$$ Substitute $a \approx 8.5$, $b = 5$, and $c = 11$.

$$\cos B \approx \frac{168.25}{187}$$ Perform the operations on the right-hand side and simplify.

Therefore, $B \approx \cos^{-1}\left(\frac{168.25}{187}\right) \approx 25.9°$.

Step 3. We can determine the measure of angle C by subtracting the measures of angle A and angle B from $180°$. Therefore, $C \approx 180° - 48° - 25.9° \approx 106.1°$.

We have now solved the oblique triangle, which can be seen in Figure 17.

Watch this **interactive video** to see the solution to this example.

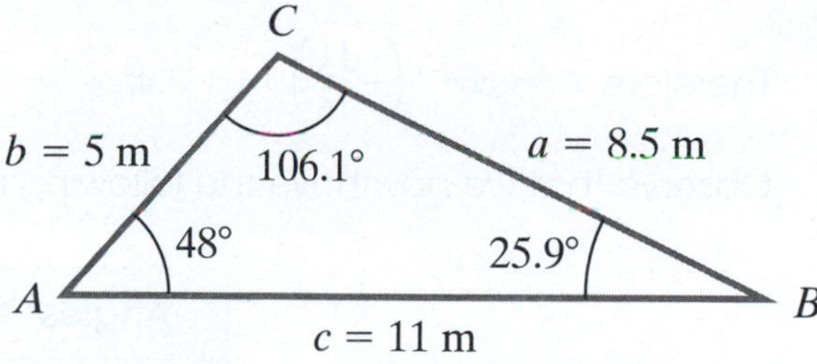

Figure 17

You Try It Work through this You Try It problem.

Work Exercises 7–11 in this textbook or in the MyLab Math Study Plan.

OBJECTIVE 3 USING THE LAW OF COSINES TO SOLVE THE SSS CASE

The SSS case is the situation where we are given the lengths of all three sides of a triangle. Thus, we need to determine the measures of all three angles. To do this, we use the appropriate alternate form of the Law of Cosines to find the measure of the largest angle. Finding the largest angle first guarantees that the remaining two angles will always be acute. Next, use the Law of Sines or the alternate form of the Law of Cosines to find the measure of either of the two remaining acute angles. Finally, use the fact that the sum of the measures of the three angles of a triangle is $180°$ to find the measure of the third angle. We now summarize a technique for solving a SSS oblique triangle.

Solving a SSS Oblique Triangle

Step 1. Use the **alternate form of the Law of Cosines** to determine the measure of the largest angle. This is the angle opposite the longest side.

Step 2. Determine the measure of one of the remaining two angles using the **Law of Sines** or the **alternate form of the Law of Cosines.** (This angle will always be acute.)

Step 3. Use the fact that the sum of the measures of the three angles of a triangle is $180°$ to determine the measure of the remaining angle.

Example 3 Solving a SSS Triangle

Solve oblique triangle ABC if $a = 5$ ft, $b = 8$ ft, and $c = 12$ ft.

Solution

Step 1. Use the alternative form of the Law of Cosines to determine the measure of the angle C. This angle has the largest measure because it is opposite the longest side.

$$\cos C = \frac{a^2 + b^2 - c^2}{2ab}$$ Use the alternate form of the Law of Cosines equation that relates a, b, c, and $\cos C$.

$$\cos C = \frac{(5)^2 + (8)^2 - (12)^2}{2(5)(8)}$$ Substitute $a = 5, b = 8$, and $c = 12$.

$$\cos C = -\frac{11}{16}$$ Perform the operations on the right-hand side and simplify.

Therefore, $C = \cos^{-1}\left(-\frac{11}{16}\right) \approx 133.4°$.

Observe that we now have the following information.

Angles	Sides
$A =$ ____	$a = 5$ ft
$B =$ ____	$b = 8$ ft
$C \approx \underline{133.4°}$	$c = 12$ ft

Step 2. We can use either the Law of Sines or the alternate form of the Law of Cosines to determine the measure of angle B. Here, we use the Law of Sines.

$$\frac{\sin B}{b} = \frac{\sin C}{c}$$ Use the Law of Sines to write a proportion.

$$\frac{\sin B}{8} \approx \frac{\sin 133.4°}{12}$$ Substitute the known information.

$$\sin B \approx \frac{8 \sin 133.4°}{12}$$ Multiply both sides by 8 to isolate $\sin B$.

Therefore, $B \approx \sin^{-1}\left(\dfrac{8 \sin 133.4°}{12}\right) \approx 29.0°$.

Note that there is another solution to the equation $\sin B \approx \dfrac{8 \sin 133.4°}{12}$, where $0° < B < 180°$. This solution is $B \approx 180° - \sin^{-1}\left(\dfrac{8 \sin 133.4°}{12}\right) \approx 151.0°$. However, because we determined that the largest angle was $C \approx 133.4°$, we can ignore this solution.

Step 3. To find the measure of angle A, subtract the measures of angles C and B from 180°. Thus, $A \approx 180° - 133.4° - 29.0° \approx 17.6°$ and the triangle is solved. See Figure 18.

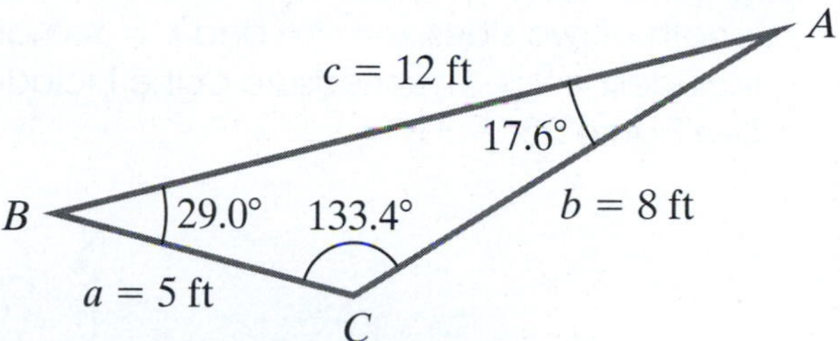

Figure 18

You Try It Work through this You Try It problem.

Work Exercises 12–16 in this textbook or in the MyLab Math Study Plan.

OBJECTIVE 4 USING THE LAW OF COSINES TO SOLVE APPLIED PROBLEMS INVOLVING OBLIQUE TRIANGLES

As with the Law of Sines, the Law of Cosines can be a useful tool to help solve many applications that arise involving triangles that are not right triangles. You may want to review the concept of bearing that was introduced in **Section 9.2**.

Example 4 Determining the Distance between Two Airplanes

Two planes take off from different runways at the same time. One plane flies at an average speed of 350 mph with a bearing of N 21° E. The other plane flies at an average speed of 420 mph with a bearing of S 84° W. How far are the planes from each other 2 hours after takeoff? Round to the nearest tenth of a mile.

Solution After 2 hours, the plane flying with a bearing of N 21° E is 700 miles from the airport and the plane flying with a bearing of S 84° W is 840 miles from the airport. Draw a diagram with the airport located at the origin, as shown in Figure 19.

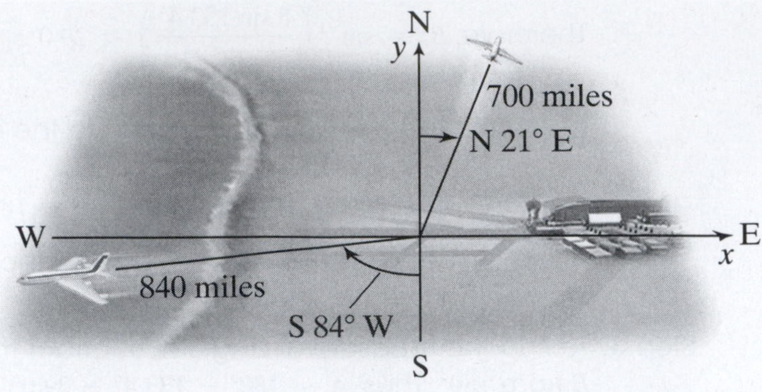

Figure 19

Let d represent the distance between the two planes. We have been given the length of two sides, $a = 700$ and $b = 840$, and the information needed to find the included angle. The measure of the included angle D is $D = 21° + 90° + 6° = 117°$. See Figure 20.

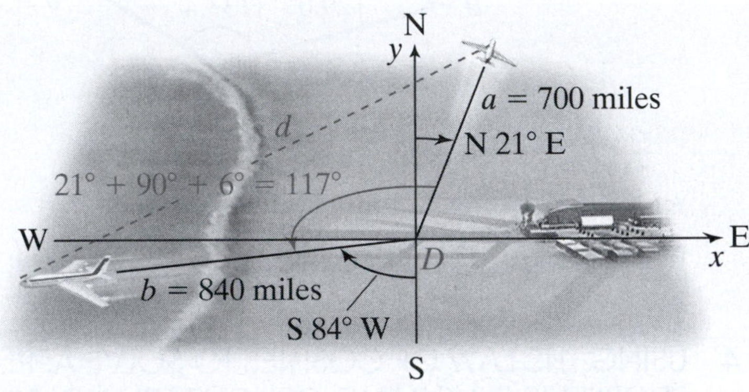

Figure 20

Now that we know the two sides and the angle between them, we use the Law of Cosines with $a = 700$, $b = 840$, and $D = 117°$ to solve for d.

$d^2 = a^2 + b^2 - 2ab\cos D$ — Use the alternate form of the Law of Cosines equation that relates $a, b, d,$ and $\cos D$.

$d^2 = (700)^2 + (840)^2 - 2(700)(840)\cos 117°$ — Substitute $a = 700,\ b = 840,$ and $D = 117°$.

$d = \sqrt{(700)^2 + (840)^2 - 2(700)(840)\cos 117°}$ — Take the square root of both sides.

$d \approx 1315.1$ — Round to the nearest tenth.

Therefore, the planes are approximately 1315.1 miles apart after 2 hours.

Example 5 Determining the Central Angle Given the Radius and the Length of a Chord

A **chord** of a circle is a line segment with endpoints that both lie on the circumference of a circle. Determine the measure of the central angle (in degrees) if the length of the chord intercepted by the central angle of a circle of radius 10 inches is 16.5 inches. Round the measure of the central angle to one decimal place.

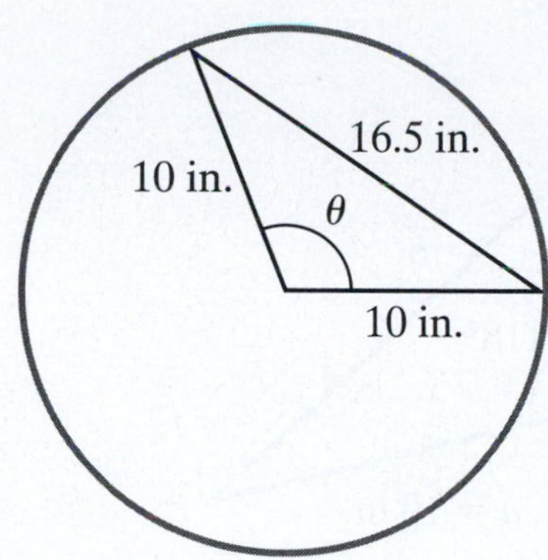

Solution To determine the measure of θ, we use the alternate form of the Law of Cosines with $a = 10$, $b = 10$, and $c = 16.5$, where c is the side opposite angle θ.

$$\cos\theta = \frac{a^2 + b^2 - c^2}{2ab}$$ Use the alternate form of the Law of Cosines equation that relates a, b, c, and $\cos\theta$.

$$\cos\theta = \frac{(10)^2 + (10)^2 - (16.5)^2}{2(10)(10)}$$ Substitute $a = 10, b = 10$, and $c = 16.5$.

$$\cos\theta = -\frac{72.25}{200}$$ Perform the operations on the right-hand side and simplify.

Therefore, $\theta = \cos^{-1}\left(-\frac{72.25}{200}\right) \approx 111.2°$.

You Try It Work through this You Try It problem.

Work Exercises 17–25 in this textbook or in the MyLab Math Study Plan.

9.3 Exercises

Skill Check Exercises

For exercises SCE-1 and SCE-2, use a calculator in degree mode to evaluate each expression. Round to one decimal place.

SCE-1. $\sqrt{(3)^2 + (5)^2 - 2(3)(5)\cos 36°}$

SCE-2. $\cos^{-1}\left(\frac{(3.1)^2 + (5)^2 - (3)^2}{2(3.1)(5)}\right)$.

In Exercises 1–6, decide whether the Law of Sines or the Law of Cosines should be used to begin to solve the given triangle. Do not solve the triangle.

1.
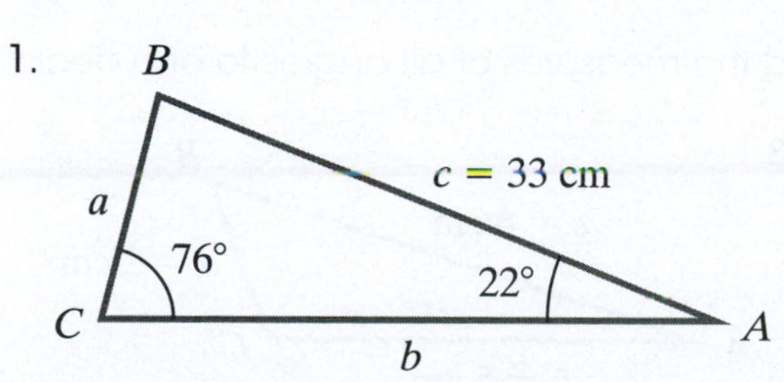

2.
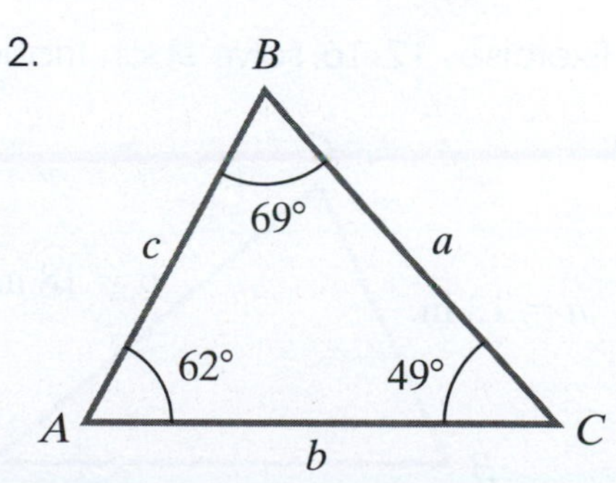

3.

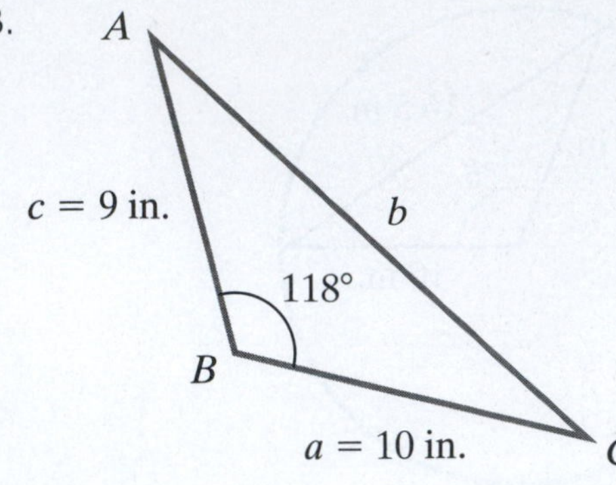

4.

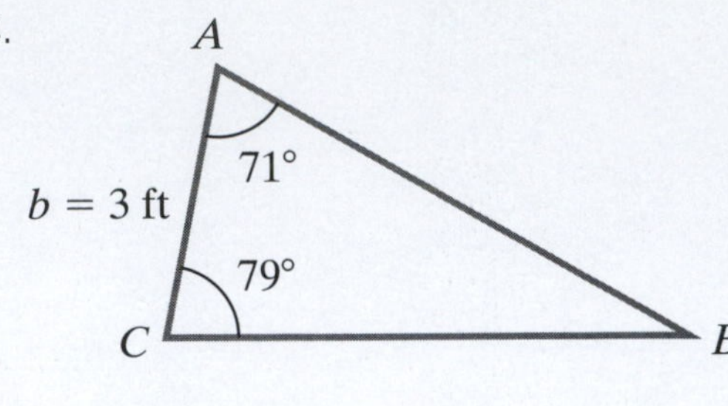

5.

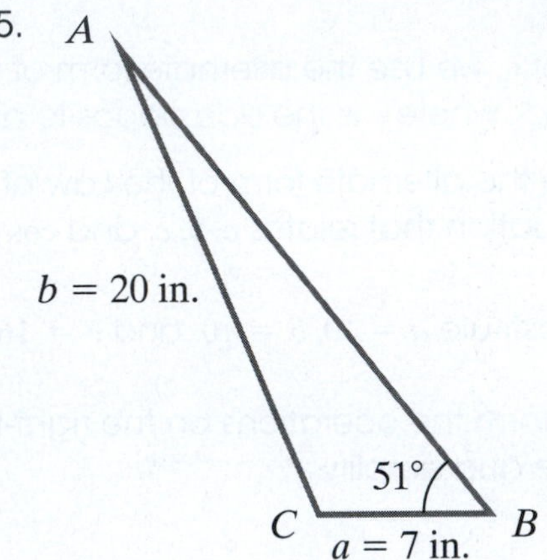

6.

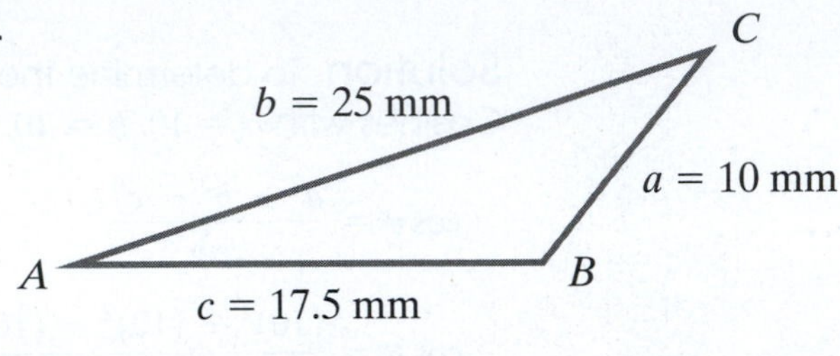

In Exercises 7–11, solve each oblique triangle. Round the measures of all angles and the lengths of all sides to one decimal place.

7.

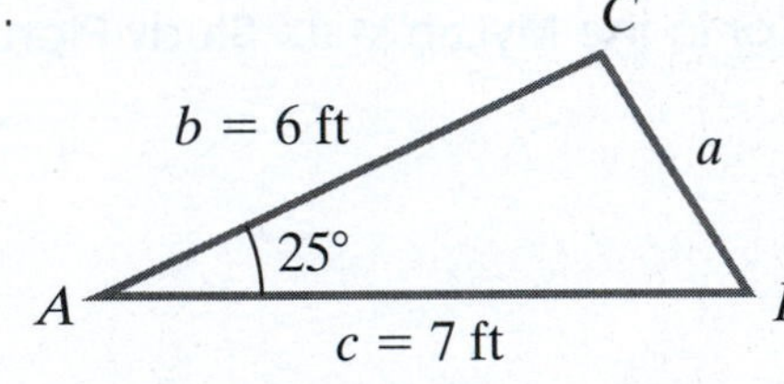

8.

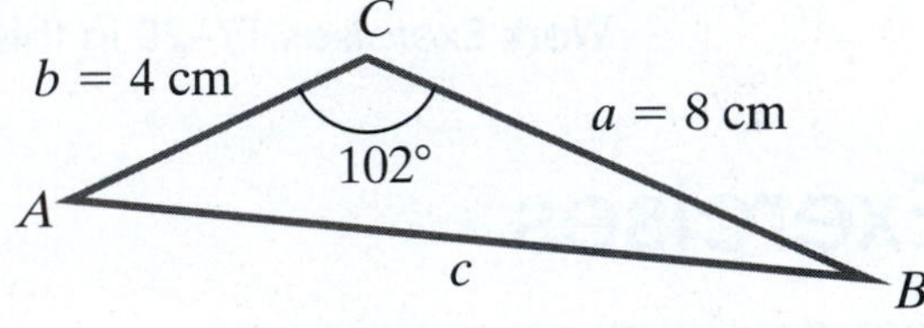

9.

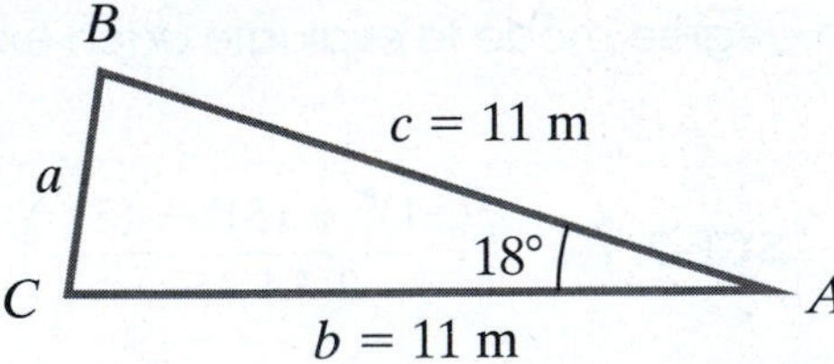

10. $a = 6, b = 8$, and $C = 20°$

11. $b = 2, c = 9$, and $A = 130°$

In Exercises 12–16, solve each triangle. Round the measures of all angles to one decimal place.

SbS 12.

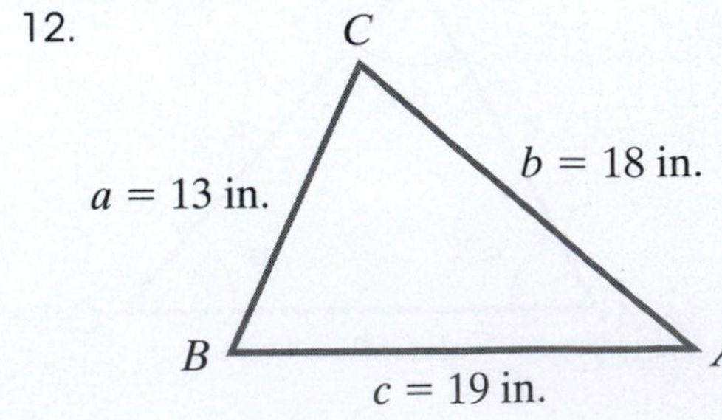

SbS 13.

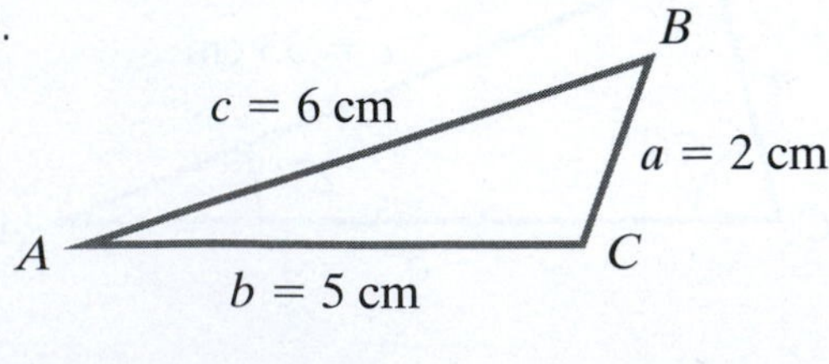

SbS 14.

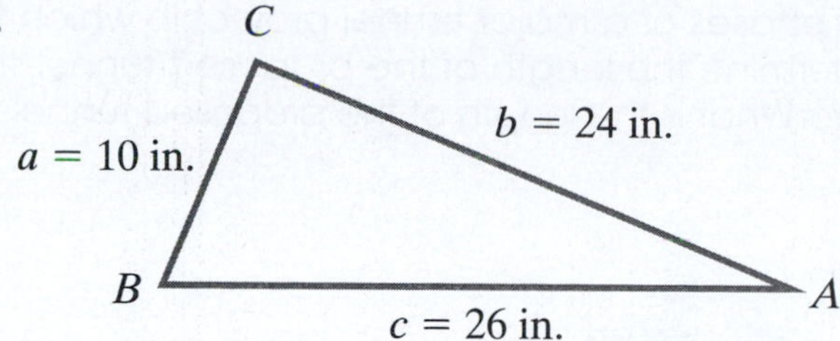

SbS 15. $a = 11, b = 18$, and $c = 20$

SbS 16. $a = 18, b = 14$, and $c = 13$

17. Two planes take off from the same airport at the same time using different runways. One plane flies at an average speed of 300 mph with a bearing of N 29° E. The other plane flies at an average speed of 350 mph with a bearing of S 47° W. How far are the planes from each other 3 hours after takeoff? Round to the nearest tenth of a mile.

18. Two planes take off from the same airport at the same time using different runways. One plane flies at an average speed of 500 mph with a bearing of N 32° W. The other plane flies at an average speed of 420 mph with a bearing of S 20° W. How far are the planes from each other 2 hours after takeoff? Round to the nearest tenth of a mile.

19. A cruise ship leaves port *A* on a four-day cruise and visits port *B* and port *C* as shown in the figure. On what bearing should the cruise ship navigate to return from port *C* back to port *A*? Round each angle to one decimal place.

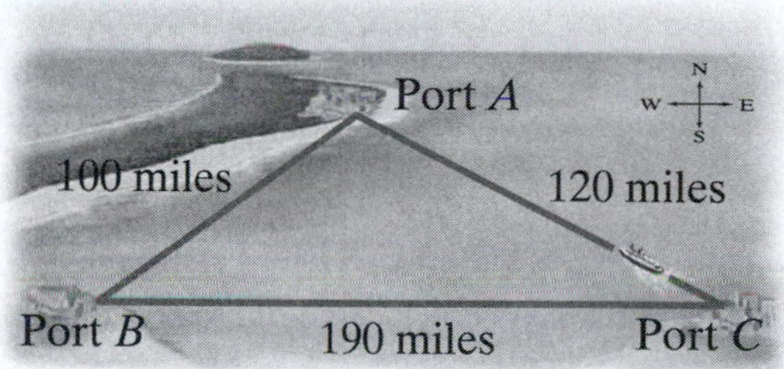

20. Determine the measure of the central angle (in degrees) if the length of the chord intercepted by the central angle of a circle of radius 35 inches is 14 inches. Round the measure of the central angle to one decimal place.

21. Determine the length of the chord intercepted by a central angle of 54° in a circle with a radius of 18 centimeters. Round to the nearest tenth of a centimeter.

22. Determine the lengths of the diagonals of a parallelogram if the adjacent sides are 13 cm and 11 cm and if the angle between them is 42°. Round the lengths of the diagonals to the nearest tenth of a centimeter.

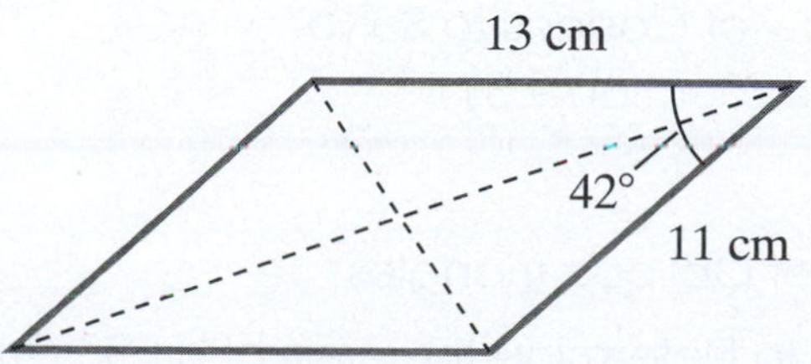

23. Three points in a plane determine a unique triangle. Solve this triangle formed by the points $A(-4, -1)$, $B(-1, 3)$, and $C(5, -1)$. Round the measures of all angles and the lengths of all sides to one decimal place.

24. Civil engineers are in the planning phases of a major tunnel project in which they plan to bore through a mountain. To determine the length of the proposed tunnel, they take measurements as shown in the figure. What is the length of the proposed tunnel to the nearest tenth of a foot?

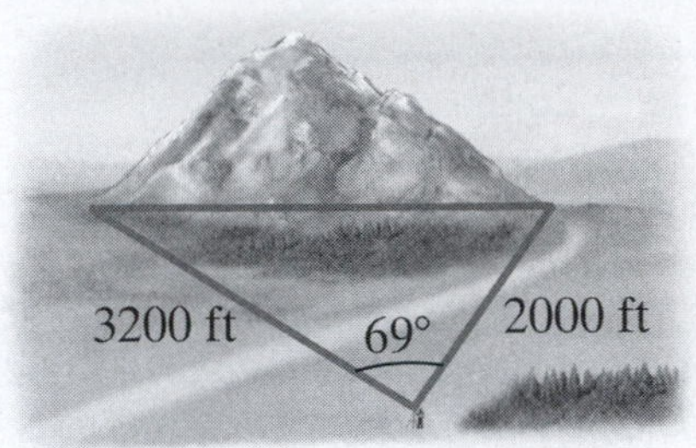

25. A common design for a mountain cabin is an A-frame cabin. A-frame cabins are fairly easy to construct, and the steeply pitched roof line is perfect for helping snow fall to the ground during heavy winterstorms. Determine the angle between the two sides of the roof of an A-frame cabin if the sides are both 22 feet long and the base of the cabin is 20 feet wide. Round to the nearest tenth of a degree.

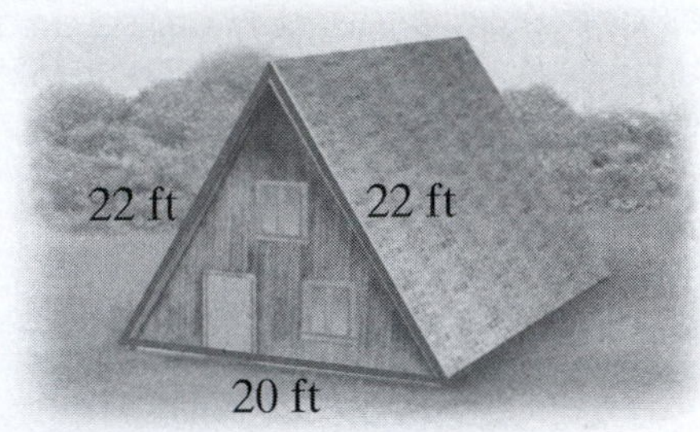

9.4 Area of Triangles

THINGS TO KNOW

Before working through this section, be sure that you are familiar with the following concepts:

	VIDEO	ANIMATION	INTERACTIVE
1. Understanding the Inverse Sine Function (Section 7.4)			
2. Understanding the Inverse Cosine Function (Section 7.4)			
3. Using the Law of Sines to Solve the SAA Case or the ASA Case (Section 9.2)	▶		
4. Using the Law of Cosines to Solve the SSS Case (Section 9.3)			

OBJECTIVES

1 Determining the Area of Oblique Triangles

2 Using Heron's Formula to Determine the Area of an SSS Triangle

3 Solving Applied Problems Involving the Area of Triangles

SECTION 9.4 EXERCISES

OBJECTIVE 1 DETERMINING THE AREA OF OBLIQUE TRIANGLES

A familiar formula for the area of a triangle is $\text{Area} = \frac{1}{2}bh$, where b is the length of the base of the triangle and h is the length of the height, or **altitude**, of the triangle.

Area of a Triangle

In any triangle, the area is given by $\text{Area} = \frac{1}{2}bh$, where b is the length of the base of the triangle and h is the length of the altitude drawn to that base (or drawn to an extension of that base).

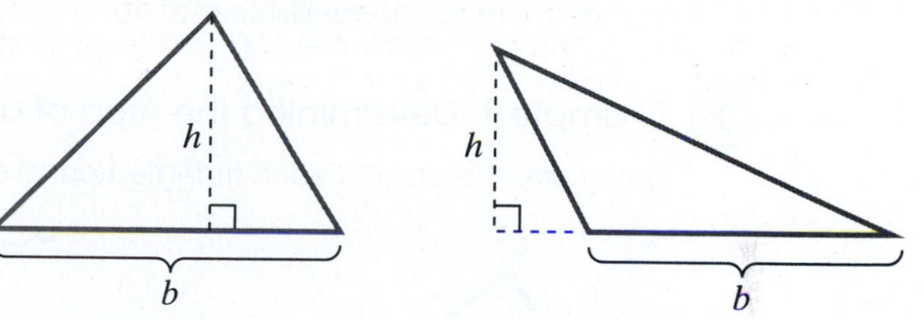

If the height of a triangle is not known, then the formula $\text{Area} = \frac{1}{2}bh$ cannot always be used readily. In this section we will develop alternate formulas to find the area of a triangle.

To develop the first set of formulas, start by drawing an oblique triangle ABC with an altitude of length h drawn from the vertex of B. See Figure 21.

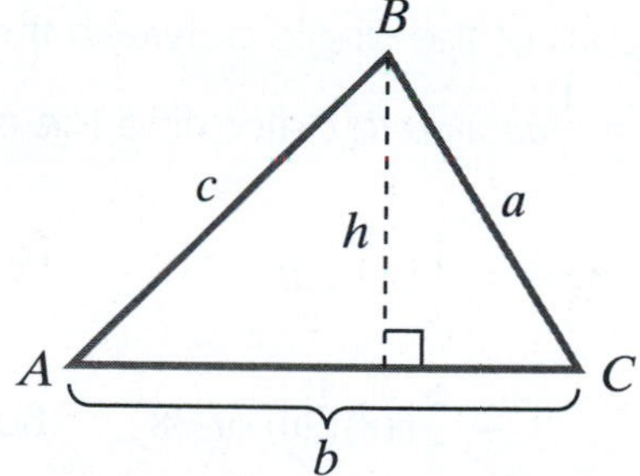

Figure 21

By the **right triangle definition** of the sine function, we have $\sin A = \frac{h}{c}$ or, equivalently, $h = c \sin A$. Substituting $h = c \sin A$ into the area formula, we get $\text{Area} = \frac{1}{2}bh = \frac{1}{2}b(c \sin A)$. Note that we could have **derived this same formula** had we started with an oblique triangle with an obtuse angle. The formula $\text{Area} = \frac{1}{2}bc \sin A$ is just one of three similar area formulas. The other two formulas can be derived the exact same way by first drawing the altitude from another vertex of the triangle.

> **Area of a Triangle**
>
> If $A, B,$ and C are the measures of the angles of any triangle and if $a, b,$ and c are the lengths of the sides opposite the corresponding angles, then the area of triangle ABC is given by
>
> $$\text{Area} = \frac{1}{2}bc\sin A, \text{ or Area} = \frac{1}{2}ac\sin B, \text{ or Area} = \frac{1}{2}ab\sin C.$$

Note that the three area formulas above require that the lengths of two sides of a triangle are known along with the measure of the angle between them. In other words, we must have a SAS (side-angle-side) triangle before using these formulas. If not enough information is known, use the **Law of Sines** or the **Law of Cosines** to determine the needed information.

Example 1 Determining the Area of an Oblique Triangle

Determine the area of each triangle. Round each answer to two decimal places.

a.

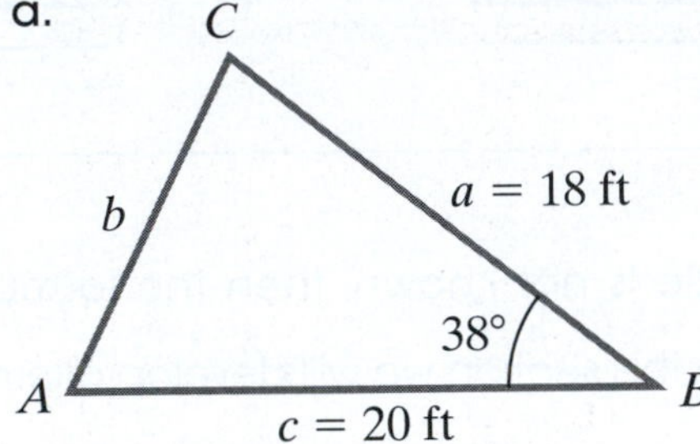

b.

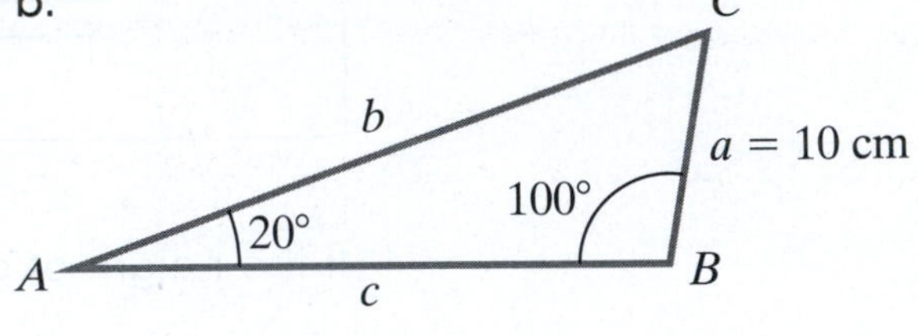

Solution

a. We are given the lengths of two sides, $a = 18$ feet and $c = 20$ feet, and the measure of the angle between them, $B = 38°$. Therefore, we use the formula $\text{Area} = \frac{1}{2}ac\sin B$ to determine the area.

$\text{Area} = \frac{1}{2}ac\sin B$	Write the appropriate area formula.
$= \frac{1}{2}(18)(20)\sin 38°$	Substitute the known information.
$\approx 110.82\text{ ft}^2$	Use a calculator in degree mode and round to two decimal places.

b. Before we can use one of the area formulas, we must know the lengths of two sides and the measure of the angle between them. Because we know the length of side a and we know the measure of angle B, we need to find the length of side c, thus giving us a SAS triangle.

To determine the length of side c, first determine the measure of angle C using the fact that the sum of the measures of the three angles must be $180°$. Then, we will use the Law of Sines to find the length of side c.

$20° + 100° + C = 180°$ The sum of the measures of the angles of any triangle is $180°$.

$120° + C = 180°$ Add.

$C = 60°$ Solve for C.

Organize the known information.

Angles	Sides
$A = 20°$	$a = 10$ cm
$B = 100°$	$b =$ ____
$C = 60°$	$c =$ ____

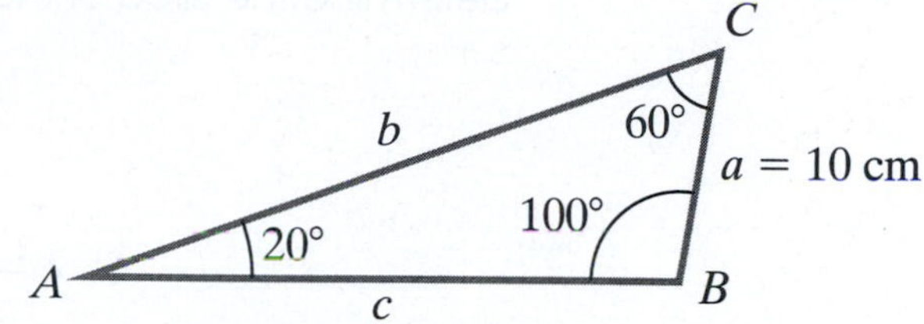

Use the **Law of Sines** to determine the length of side c.

$\dfrac{c}{\sin C} = \dfrac{a}{\sin A}$ Use the Law of Sines to write a proportion.

$\dfrac{c}{\sin 60°} = \dfrac{10}{\sin 20°}$ Substitute the known information.

$c = \dfrac{10 \sin 60°}{\sin 20°}$ Multiply both sides by $\sin 60°$ to solve for c.

 **CAUTION To avoid rounding errors, we will use *exact* values found in intermediate steps. Therefore, we will use the value $c = \dfrac{10 \sin 60°}{\sin 20°}$ when substituting into an area formula.**

Now use the formula $\text{Area} = \dfrac{1}{2}ac \sin B$ to determine the area.

$\text{Area} = \dfrac{1}{2}ac \sin B$ Write the appropriate area formula.

$= \dfrac{1}{2}(10)\left(\dfrac{10 \sin 60°}{\sin 20°}\right) \sin 100°$ Substitute the known information using exact values.

$\approx 124.68 \text{ cm}^2$ Use a calculator in degree mode and round to two decimal places.

 Work through this **video** to see the entire worked-out solution.

 **You Try It** Work through this **You Try It** problem.

Work Exercises 1–12 in this textbook or in the MyLab Math Study Plan.

OBJECTIVE 2 USING HERON'S FORMULA TO DETERMINE THE AREA OF AN SSS TRIANGLE

Suppose that we wish to find the area of a triangle in which the lengths of the three sides are known. One way to determine this area is to use the Law of Cosines to determine the measure of any angle, then use one of the three previously stated area formulas. We illustrate this procedure in Example 2.

Example 2 Determining the Area of an SSS Oblique Triangle

Determine the area of the given triangle. Round to two decimal places.

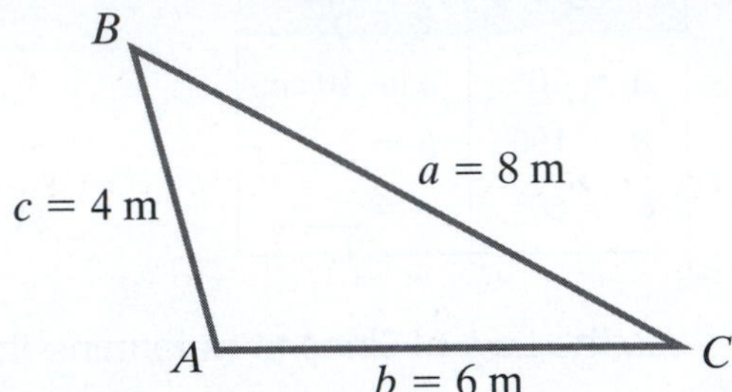

Solution To use an area formula, first determine the measure of any of the three angles. Here, we determine the measure of angle A using an **alternate form of the Law of Cosines.**

$\cos A = \dfrac{b^2 + c^2 - a^2}{2bc}$ Use the alternate form of the Law of Cosines equation that relates $a, b, c,$ and $\cos A$.

$\cos A = \dfrac{(6)^2 + (4)^2 - (8)^2}{2(6)(4)}$ Substitute $a = 8, b = 6,$ and $c = 4$.

$\cos A = -\dfrac{1}{4}$ Perform the operations on the right-hand side and simplify.

Therefore, the exact value of A is $A = \cos^{-1}\left(-\dfrac{1}{4}\right)$.

 **CAUTION Once again, to avoid rounding errors, we will substitute the exact value, $A = \cos^{-1}\left(-\dfrac{1}{4}\right)$, when using the appropriate area formula.**

Now use the formula $\text{Area} = \dfrac{1}{2}bc\sin A$ to determine the area.

$\text{Area} = \dfrac{1}{2}bc\sin A$ Write the appropriate area formula.

$= \dfrac{1}{2}(6)(4)\sin\left(\cos^{-1}\left(-\dfrac{1}{4}\right)\right)$ Substitute the known information using exact values.

$\approx 11.62\text{ m}^2$ Use a calculator in degree mode and round to two decimal places.

The procedure for determining the area of an SSS triangle can be computed much more efficiently using a formula that does not require the knowledge of the measure of any angle. This formula is known as **Heron's formula** and is named after the Greek mathematician Heron of Alexandria. Heron's formula requires that we first compute the **semiperimeter**, which is exactly equal to half of the sum of the lengths of the three sides of a triangle. Heron's formula is stated below.

> **Heron's Formula**
>
> Suppose that a triangle has side lengths of $a, b,$ and c. If the semiperimeter is $s = \frac{1}{2}(a + b + c)$, then the area of the triangle is
>
> $$\text{Area} = \sqrt{s(s - a)(s - b)(s - c)}.$$

There are many ways to prove Heron's formula. One way requires the area formula $\text{Area} = \frac{1}{2}ab\sin C$, an alternate form of the Law of Cosines, the semiperimeter formula, and a Pythagorean identity of the form $\sin C = \sqrt{1 - \cos^2 C}$. To see the complete proof of Heron's formula, watch this **video**.

In Example 3, we will use the same triangle that was used in Example 2 to illustrate the use of Heron's formula.

Example 3 Determining the Area of an SSS Oblique Triangle Using Heron's Formula

Use Heron's formula to determine the area of the given triangle. Round to two decimal places.

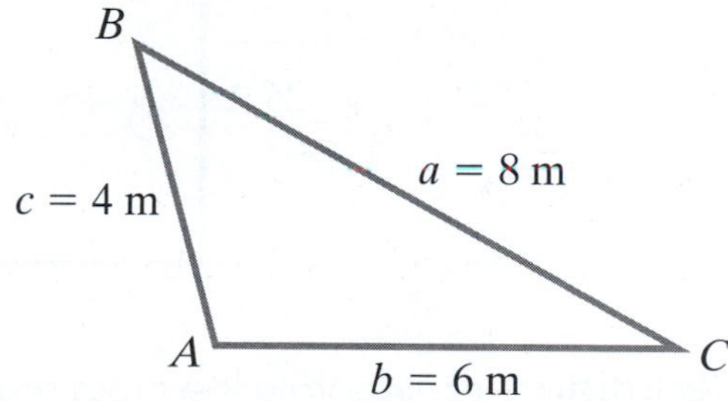

Solution First, determine the semiperimeter.

$s = \frac{1}{2}(a + b + c)$ Write the formula for the semiperimeter.

$= \frac{1}{2}(8 + 6 + 4) = 9$ m Substitute the known information and simplify.

Now use Heron's formula with $a = 8, b = 6, c = 4$, and $s = 9$.

$$\begin{aligned} \text{Area} &= \sqrt{s(s-a)(s-b)(s-c)} && \text{Write Heron's formula.} \\ &= \sqrt{9(9-8)(9-6)(9-4)} && \text{Substitute the known information.} \\ &= \sqrt{9(1)(3)(5)} && \text{Simplify inside the radical.} \\ &\approx 11.62 \text{ m}^2 && \text{Use a calculator and round to two decimal places.} \end{aligned}$$

Notice that the area computed using Heron's formula in Example 3 is the same as the area found in Example 2! ●

You Try It **Work through this You Try It problem.**

Work Exercises 13–15 in this textbook or in the MyLab Math Study Plan.

OBJECTIVE 3 SOLVING APPLIED PROBLEMS INVOLVING THE AREA OF TRIANGLES

The formulas discussed in this section can be used to solve applications that involve finding the area of triangles.

Example 4 Determining the Area of the Cross Section of a House

A painter who is painting a house has only one side of the house left to paint. He has enough paint to cover 1200 square feet. A cross section of the unpainted side of the house is shown in the figure. What is the area of the unpainted side of the house? Does he have enough paint to finish the job?

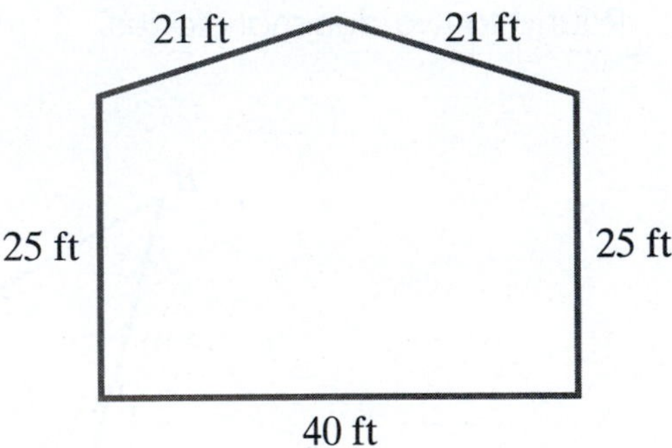

Solution First, separate the cross section of the side of the house into a rectangle and a triangle, as shown in **Figure 22**.

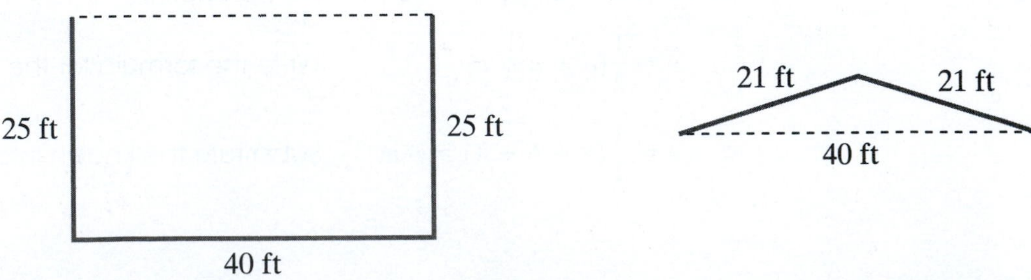

Figure 22

The total area of the cross section is the sum of the areas of the rectangular portion and the triangular portion seen in Figure 22. The area of the rectangular portion is $(25)(40) = 1000 \text{ ft}^2$.

Using Heron's formula with a semiperimeter of $s = \frac{1}{2}(21 + 21 + 40) = 41$ feet, we can determine the area of the triangular portion of the cross section. The area of the triangular portion of the side of the house is

$$\text{Area} = \sqrt{s(s-a)(s-b)(s-c)} = \sqrt{41(41-21)(41-21)(41-40)}$$
$$= \sqrt{41(20)(20)(1)} = \sqrt{16{,}400} \text{ ft}^2.$$

Thus, the total area of the unpainted side of the house is $1000 \text{ ft}^2 + \sqrt{16{,}400} \text{ ft}^2 \approx 1128 \text{ ft}^2$. The painter has enough paint to finish the job. ●

You Try It **Work through this You Try It problem.**

Work Exercises 16–21 in this textbook or in the MyLab Math Study Plan.

9.4 Exercises

Skill Check Exercises

For SCE-1, use a calculator to evaluate the expression $\sqrt{12.5(12.5-11)(12.5-9)(12.5-5)}$. Round to two decimal places.

In Exercises 1–12, determine the area of each triangle. Round your answer to two decimal places.

1.

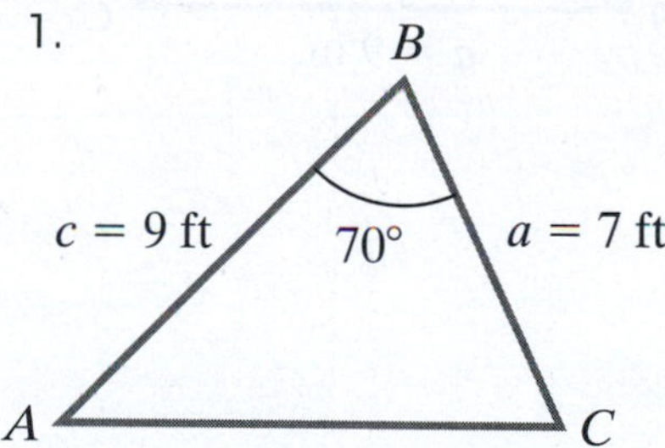

2.

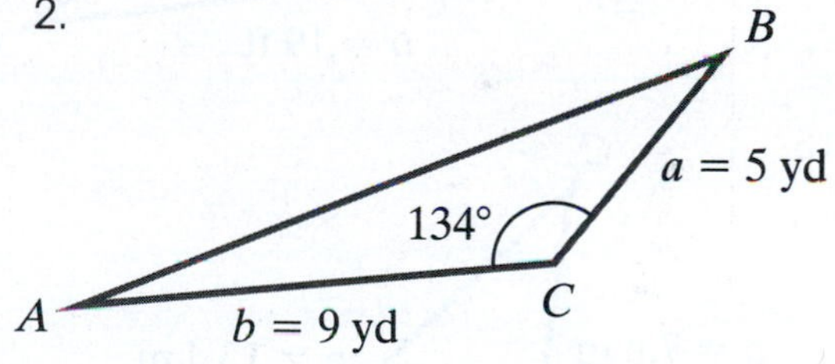

3.

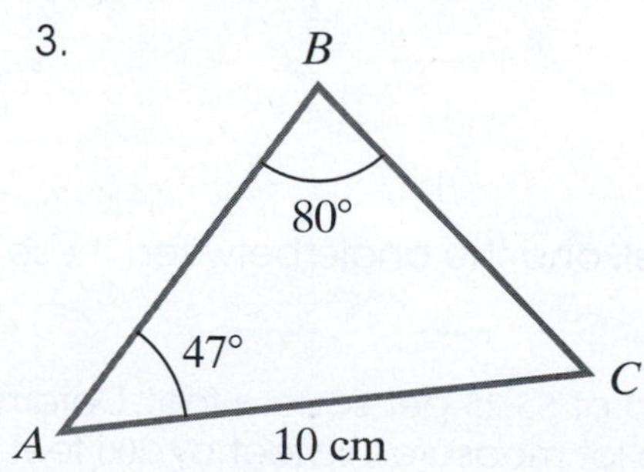

4.

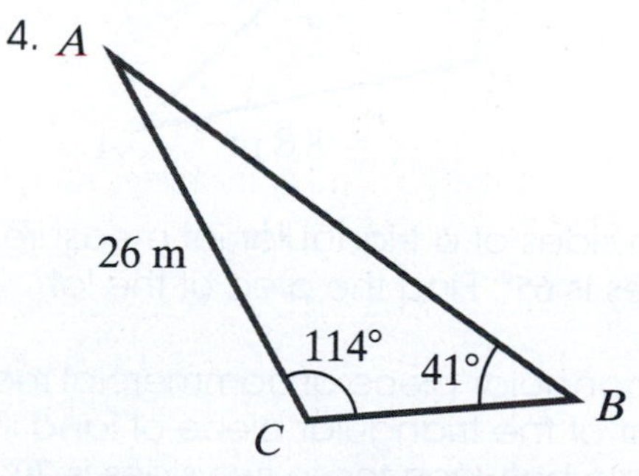

5.

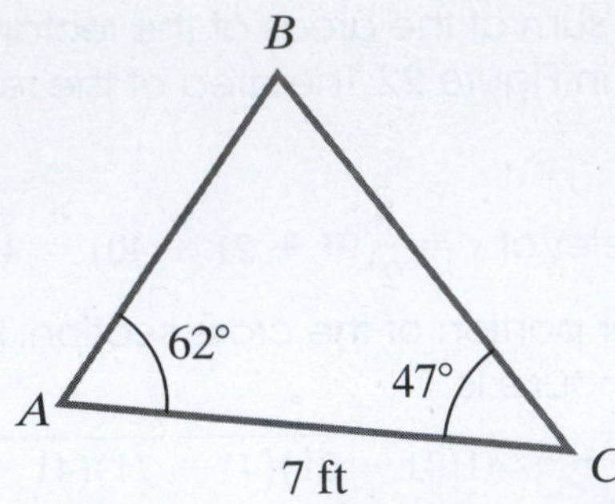

6.

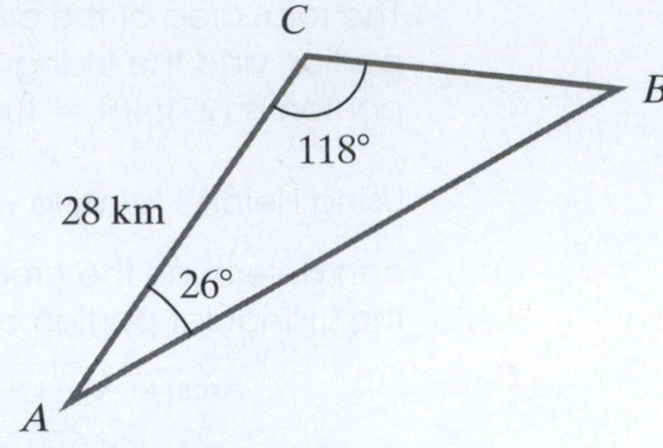

7. $a = 12$ cm, $c = 15$ cm, $B = 56°$

8. $b = 11$ yd, $c = 7$ yd, $B = 102°$

9. $a = 7$ ft, $A = 72°$, $B = 51°$

10. $b = 4$ m, $B = 21°$, $C = 131°$

11. $a = 19$ km, $B = 74°$, $C = 81°$

12. $c = 20$ cm, $B = 59°$, $A = 93°$

In Exercises 13–15, use Heron's formula to determine the area of each triangle.

Round your answer to two decimal places.

13.

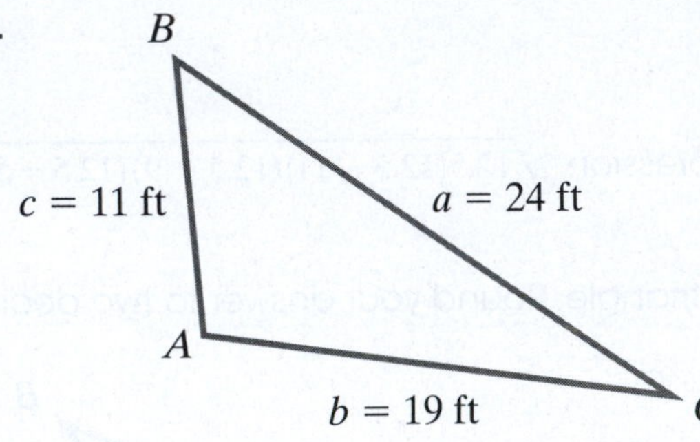

14.

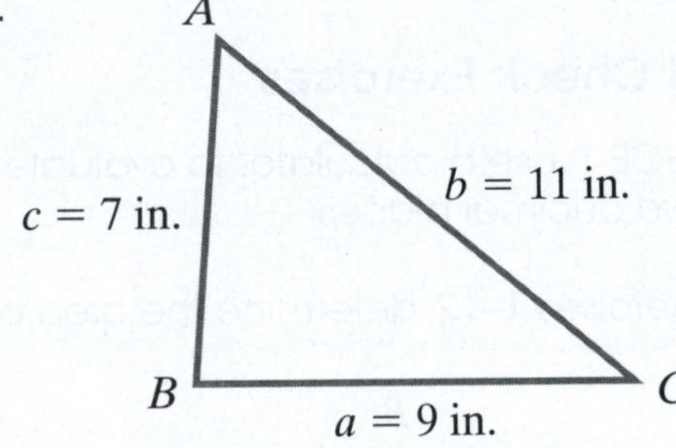

15.

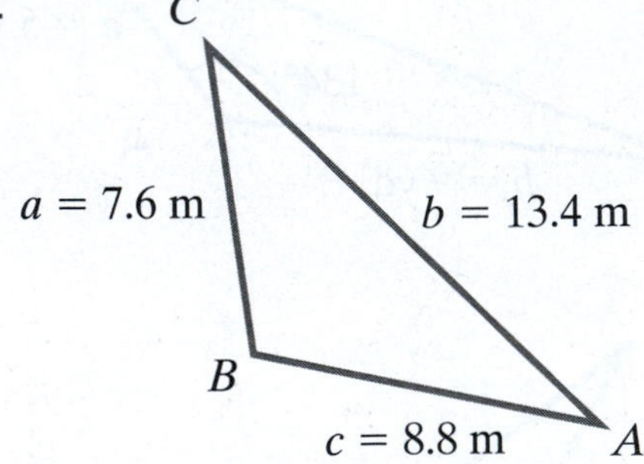

16. Two sides of a triangular lot measure 58 feet by 72 feet, and the angle between these two sides is 65°. Find the area of the lot.

17. A triangular piece of commercial real estate is priced at \$5.25 per square foot. Determine the cost of the triangular piece of land if two sides of the lot measure 180 feet by 200 feet and the angle between these two sides is 70°. Round to the nearest dollar.

18. A local merchant currently has a triangular gravel parking lot for her customers. The measurements of the lot are 125 feet by 70 feet by 65 feet. Determine the total cost to pave the lot if the price is \$1.15 per square foot.

19. A painter who is painting a house has only one side of the house left to paint. He has enough paint to cover 1000 square feet. A cross section of the unpainted side of the house is shown in the figure. What is the area of the unpainted side of the house? Does the painter have enough paint to finish the job?

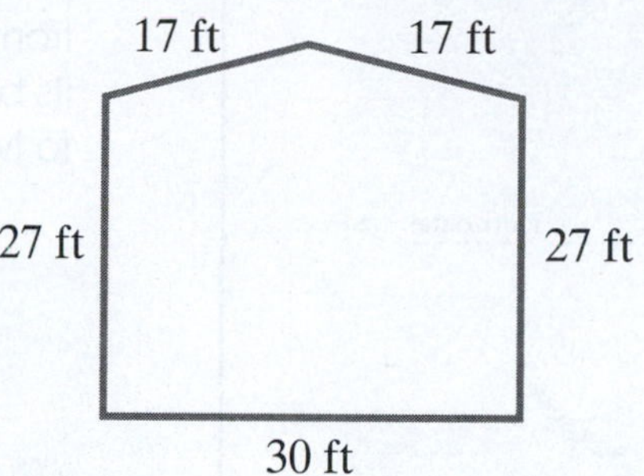

20. Determine the area of the cross section of a house shown in the figure below.

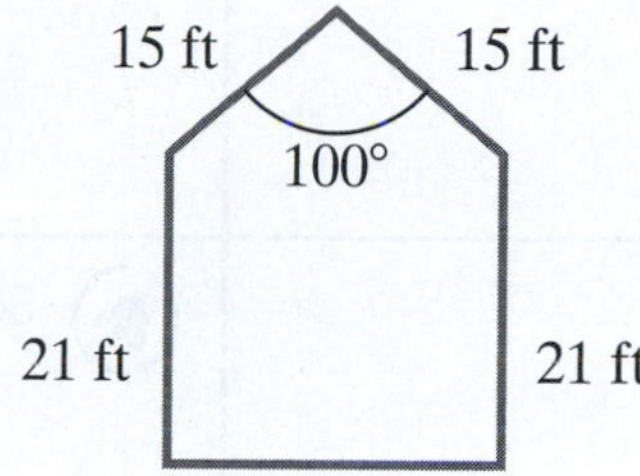

21. A couple is planning on carpeting their upstairs bonus room, which has a floor in the shape of a triangle. Two of the adjacent sides of the room are 12 feet and 11 feet. The angle opposite the 11-foot side is 52°. If the cost of the carpet is $1.25 per square foot, how much, to the nearest dollar, will it cost to carpet the room?

Chapter 9 Summary

Key Concepts	Examples/Videos
9.1 Right Triangle Applications Solving Right Triangles	Solve the right triangle. B a $c = 11$ in. $37.34°$ C b A

Key Concepts	Examples/Videos

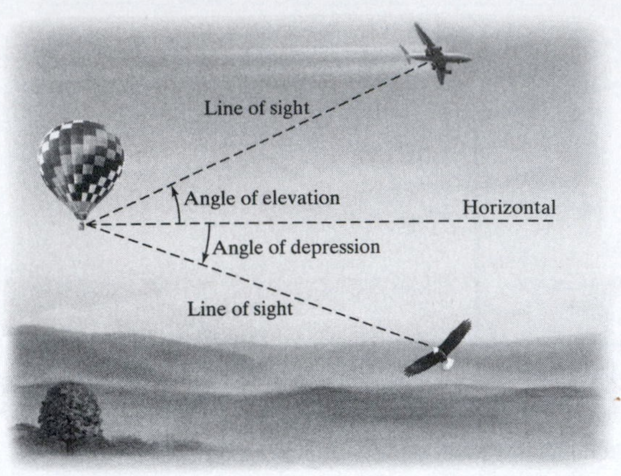

 The angle of elevation to the top of a flagpole measured by a digital protractor is 20° from a point on the ground 90 feet away from its base. Find the height of the flagpole. Round to two decimal places.

9.2 The Law of Sines

The Law of Sines

If A, B, and C are the measures of the angles of any triangle and if a, b, and c are the lengths of the sides opposite the corresponding angles, then

$$\frac{a}{\sin A} = \frac{b}{\sin B} = \frac{c}{\sin C} \quad \text{or} \quad \frac{\sin A}{a} = \frac{\sin B}{b} = \frac{\sin C}{c}.$$

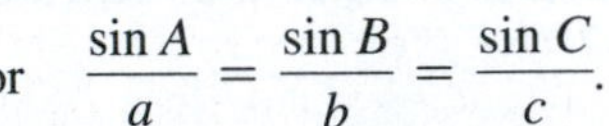

Using the Law of Sines

If S represents a given side of a triangle and if A represents a given angle of a triangle, then the Law of Sines can be used to solve the three oblique triangle cases shown below.

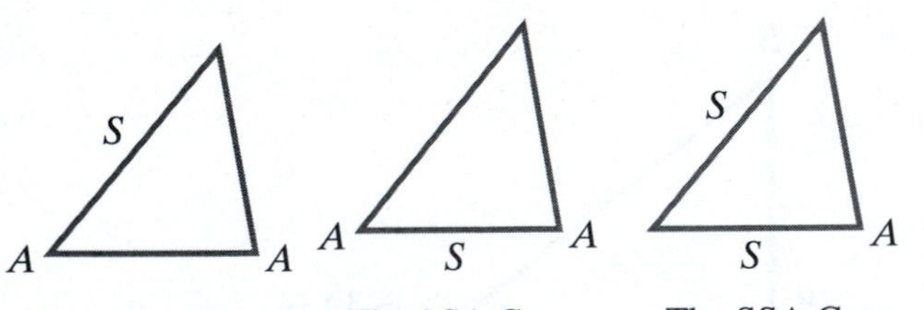

The SAA Case The ASA Case The SSA Case (The Ambiguous Case)

 Solve the oblique triangle.

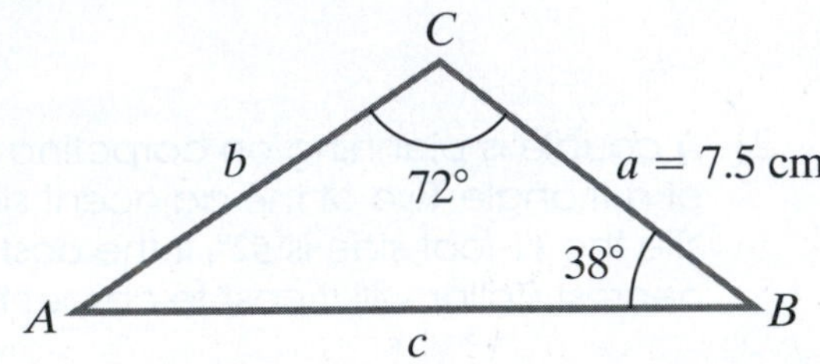

Two sides and an angle are given below. Determine whether the information results in no triangle, one right triangle, or one or two oblique triangles. Solve each resulting triangle. Round the measures of all angles and the lengths of all sides to one decimal place.

$a = 13 \text{ cm}, \quad b = 7.8 \text{ cm}, \quad A = 67°$

Two sides and an angle are given below. Determine whether the information results in no triangle, one right triangle, or one or two oblique triangles. Solve each resulting triangle. Round the measures of all angles and the lengths of all sides to one decimal place.

$b = 11.3 \text{ in.}, \quad c = 15.5 \text{ in.}, \quad B = 34.7°$

<table>
<tr><th>Key Concepts</th><th>Examples/Videos</th></tr>
<tr><td>

9.3 The Law of Cosines

The Law of Cosines

If A, B, and C are the measures of the angles of any triangle and if a, b, and c are the lengths of the sides opposite the corresponding angles, then

$$a^2 = b^2 + c^2 - 2bc\cos A,$$
$$b^2 = a^2 + c^2 - 2ac\cos B, \text{ and}$$
$$c^2 = a^2 + b^2 - 2ab\cos C$$

The Alternate Form of the Law of Cosines

If A, B, and C are the measures of the angles of any triangle and if a, b, and c are the lengths of the sides opposite the corresponding angles, then

$$\cos A = \frac{b^2 + c^2 - a^2}{2bc},$$
$$\cos B = \frac{a^2 + c^2 - b^2}{2ac}, \text{ and}$$
$$\cos C = \frac{a^2 + b^2 - c^2}{2ab}$$

</td><td></td></tr>
<tr><td>

Solving an SAS Oblique Triangle

Step 1 Use the Law of Cosines to determine the length of the missing side.

Step 2 Determine the measure of the smaller of the remaining two angles using the Law of Sines or using the alternate form of the Law of Cosines. This angle is the angle opposite of the shortest side and will always be acute.

Step 3 Use the fact that the sum of the measures of the three angles of a triangle is $180°$ to determine the measure of the remaining angle.

</td><td>

Solve the given oblique triangle. Round the measures of all angles and the lengths of all sides to one decimal place.

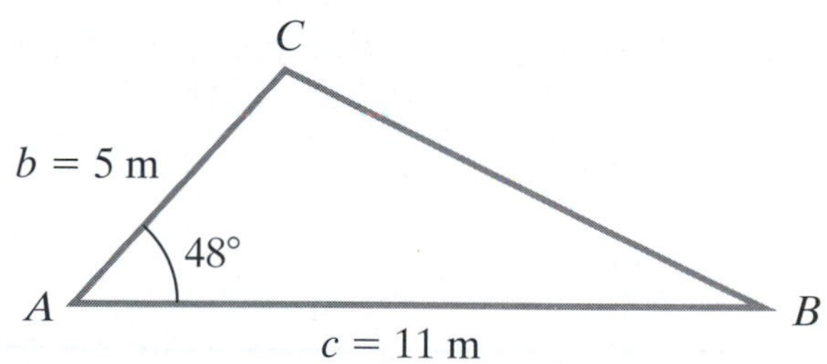

</td></tr>
</table>

Key Concepts	Examples/Videos

Solving the SSS Oblique Triangle

Step 1 Use the **alternate form of the Law of Cosines** to determine the measure of the largest angle. This is the angle opposite the longest side.

Step 2 Determine the measure of one of the remaining two angles using the **Law of Sines** or the **alternate form of the Law of Cosines**. (This angle will always be acute.)

Step 3 Use the fact that the sum of the measures of the three angles of a triangle is $180°$ to determine the measure of the remaining angle.

Solve oblique triangle ABC if $a = 5$ ft, $b = 8$ ft, and $c = 12$ ft.

9.4 Area of Triangles

Area of a Triangle

If A, B, and C are the measures of the angles of any triangle and if a, b, and c are the lengths of the sides opposite the corresponding angles, then the area of triangle ABC is given by

$$\text{Area} = \frac{1}{2}bc\sin A, \text{ or Area} = \frac{1}{2}ac\sin B, \text{ or Area} = \frac{1}{2}ab\sin C.$$

Determine the area of each triangle. Round each answer to two decimal places.

a.

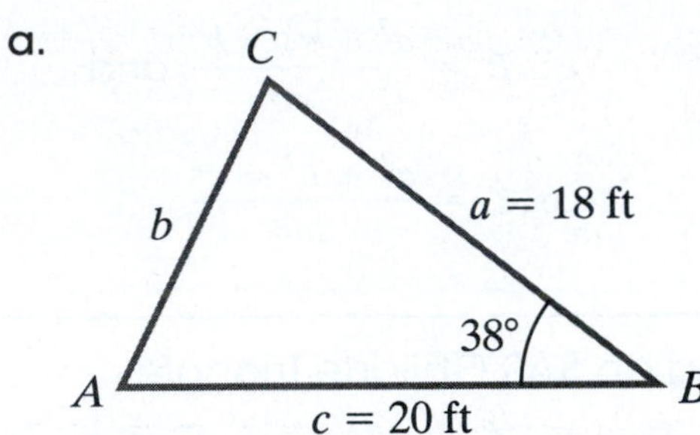

b.

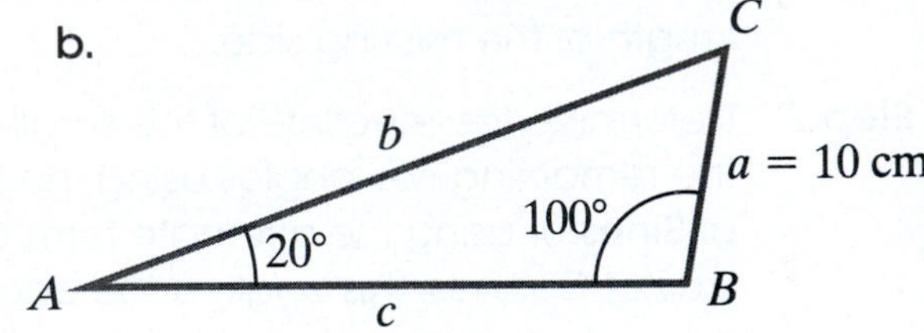

Heron's Formula

Suppose that a triangle has side lengths of a, b, and c. If the semi-perimeter is $s = \frac{1}{2}(a + b + c)$, then the area of the triangle is $Area = \sqrt{s(s - a)(s - b)(s - c)}$.

Use Heron's formula to determine the area of the given triangle. Round to two decimal places.

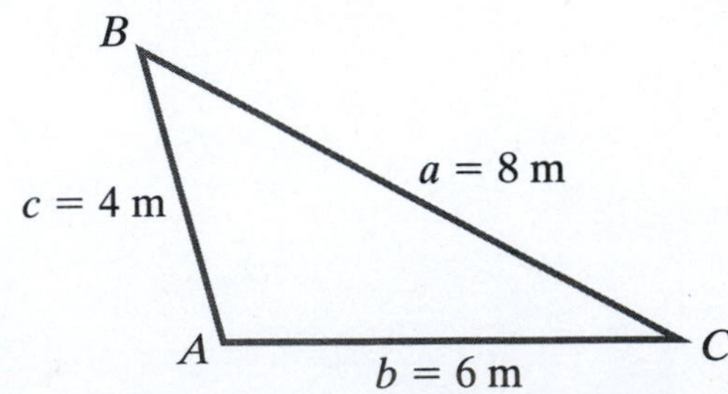

Chapter 9 Review Exercises

In 1–2, solve each right triangle.

1.

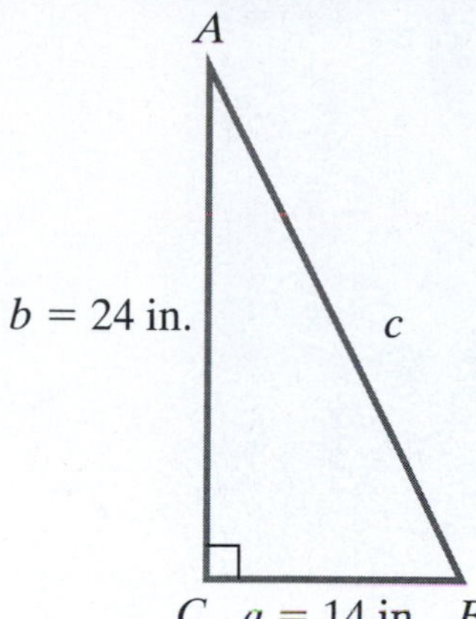

2.

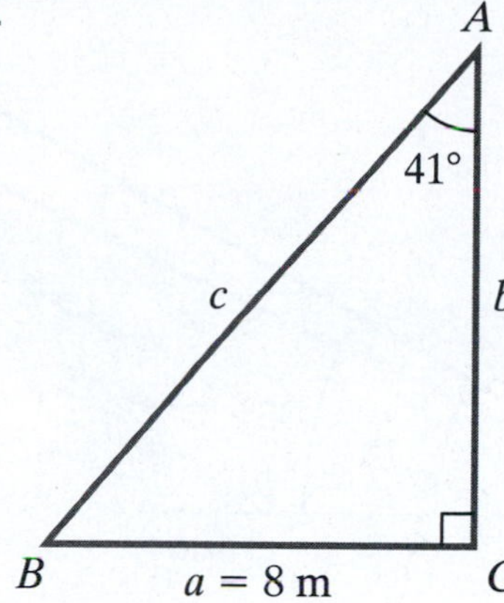

3. Find the length of the two missing sides of a right triangle given that one angle is $83.1°$ and the length of the side adjacent to the angle is 12.5 cm. Round the length of the sides to two decimal places.

4. The angle of elevation to the top of a flagpole measured by a digital protractor is $54.7°$ from a point on the ground 72 feet away from its base. Find the height of the flagpole. Round to two decimal places.

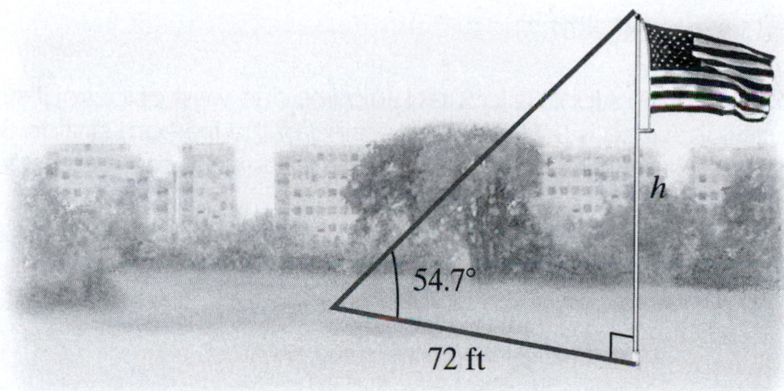

5. A mine shaft with a circular entrance has been carved into the side of a mountain. From a distance of 100 feet from the base of the mountain, the angle of elevation to the bottom of the circular opening is $27.7°$. The angle of elevation to the top of the opening is $33°$. Determine the diameter of the circular entrance. Round to two decimal places.

(Summary)

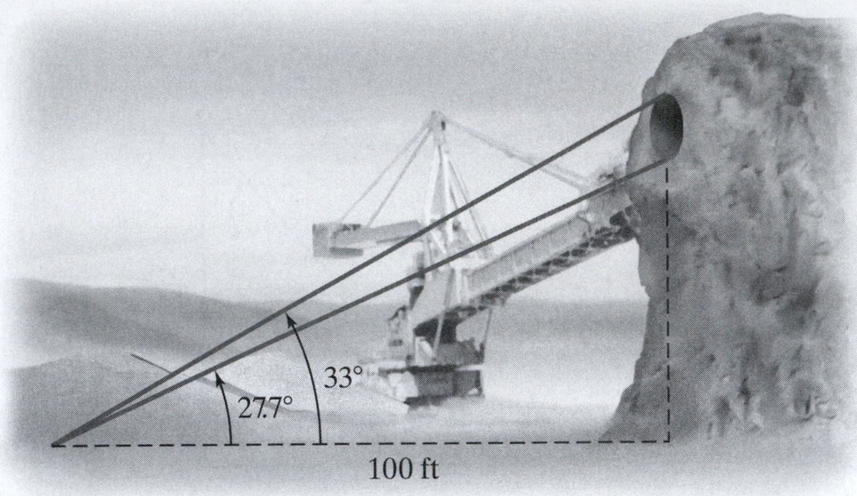

6. Solve the oblique triangle such that $A = 48°$, $B = 53°$, and $b = 6$. Round the measures of all angles and the lengths of all sides to one decimal place.

In 7–9, two sides and an angle are given. First determine whether the given information results in no triangle, one triangle, or two triangles. Solve each resulting triangle. Round the measure of all angles and the lengths of all sides to one decimal place.

7. $a = 8.1$, $b = 7.3$, and $A = 40°$

8. $a = 14$, $b = 16.3$, and $A = 55°$

9. $a = 14.8$, $b = 15.9$, and $A = 67.7°$

10. Forest fire lookout station Alpha is located 20 miles due west of lookout station Beta. The bearing of a fire from station Alpha is S 11° W. The bearing of the fire from station Beta is S 29° W. Determine the distance from the fire to both lookout stations. Round to the nearest tenth of a mile.

11. Decide whether the Law of Sines or the Law of Cosines should be used to begin to solve the given triangle. **Do not solve** the triangle.

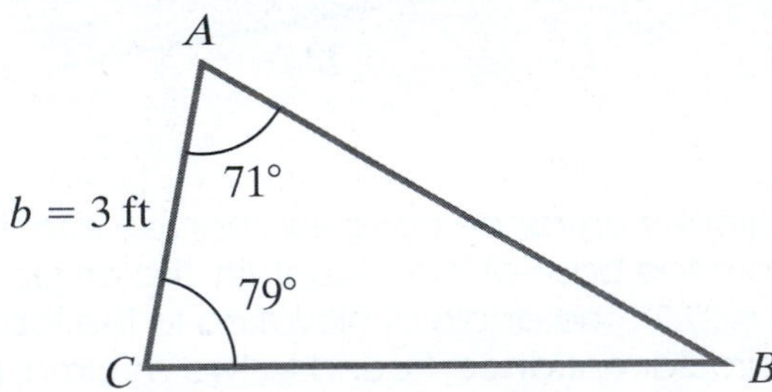

12. Solve the triangle. Round the measures of all angles to one decimal place.

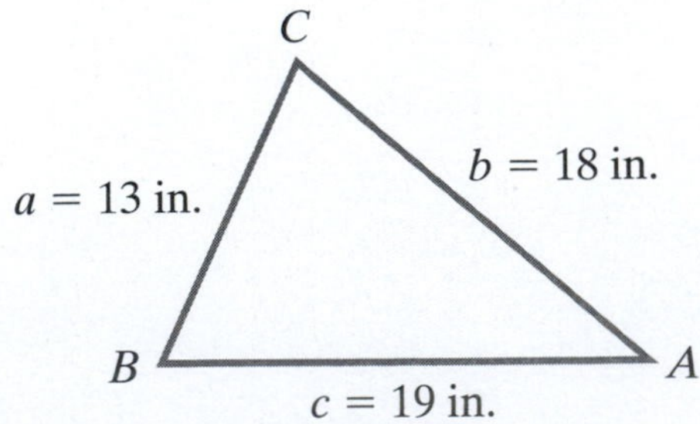

13. Solve the triangle having side lengths $a = 11$, $b = 18$, and $c = 20$.

14. Civil engineers are in the planning phases of a major tunnel project in which they plan to bore through a mountain. To determine the length of the proposed tunnel, they take measurements as shown in the figure. What is the length of the proposed tunnel to the nearest tenth of a foot?

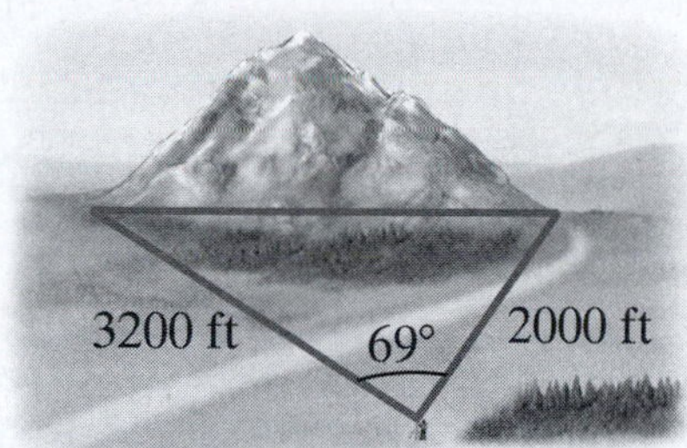

15. A common design for a mountain cabin is an A-frame cabin. A-frame cabins are fairly easy to construct, and the steeply pitched roof line is perfect for helping snow fall to the ground during heavy winterstorms. Determine the angle between the two sides of the roof of an A-frame cabin if the sides are both 22 feet long and the base of the cabin is 20 feet wide. Round to the nearest tenth of a degree.

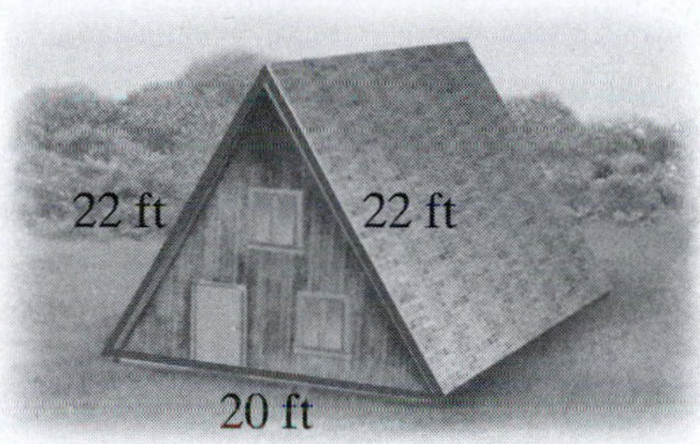

In 16–18, determine the area of the given triangle.

16.

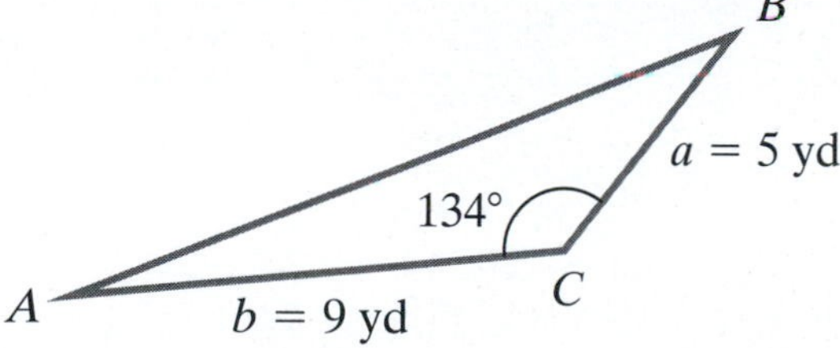

17.

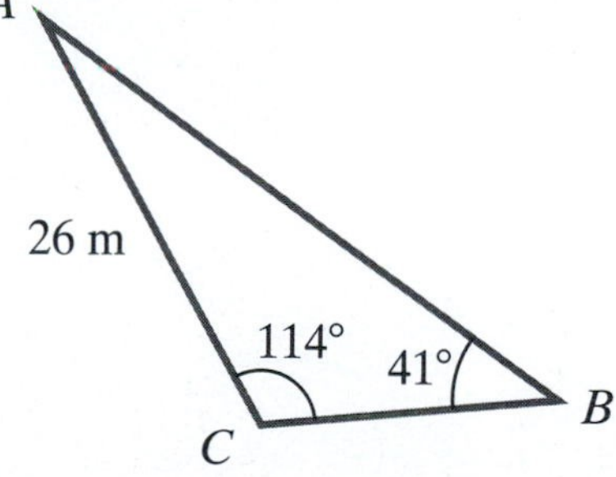

18. $a = 7$ ft, $A = 72°$, and $B = 51°$

19. Use Heron's formula to determine the area of the triangle. Round your answer to two decimal places.

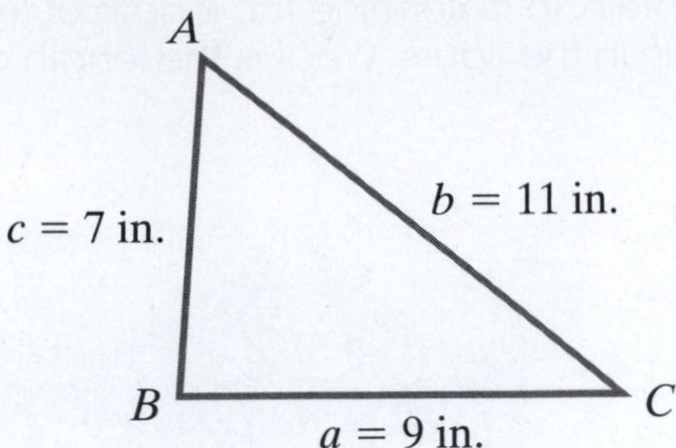

20. A triangular piece of commercial real estate is priced at \$5.25 per square foot. Determine the cost of the triangular piece of land if two sides of the lot measure 180 feet by 200 feet and the angle between these two sides is 70°. Round to the nearest dollar.

CHAPTER TEN

Polar Equations, Complex Numbers, and Vectors

CHAPTER TEN CONTENTS

10.1 Polar Coordinates and Polar Equations

THINGS TO KNOW

Before working through this section, be sure that you are familiar with the following concepts:

VIDEO ANIMATION INTERACTIVE

You Try It

1. Understanding the Four Families of Special Angles (Section 6.5) 

You Try It

2. Understanding the Definitions of the Trigonometric Functions of General Angles (Section 6.5)  

You Try It

3. Understanding the Signs of the Trigonometric Functions (Section 6.5) 

You Try It

4. Evaluating Trigonometric Functions of Angles Belonging to the $\frac{\pi}{3}, \frac{\pi}{6}$, or $\frac{\pi}{4}$ Families (Section 6.5) 

You Try It

5. Solving Trigonometric Equations That Are Linear in Form (Section 8.5) 

OBJECTIVES

1 Plotting Points Using Polar Coordinates

2 Determining Different Representations of a Point (r, θ)

3 Converting a Point from Polar Coordinates to Rectangular Coordinates

(eText Screens 10.1-1–10.1-47)

OBJECTIVE 1 PLOTTING POINTS USING POLAR COORDINATES

So far in this eText we have used the rectangular coordinate system (also called the Cartesian coordinate system) when plotting points and sketching the graphs of functions. This is not the only system that can be used to plot points and sketch functions. In this section we begin our study of the **polar coordinate system.** This system is based on a fixed point called the **pole** and a **ray** with the vertex at the pole. This ray is called the **polar axis** and is shown in Figure 1.

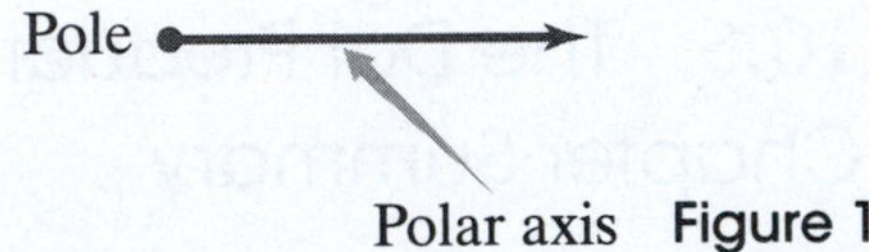

Figure 1

A point P in the polar coordinate system is represented by the ordered pair $P(r, \theta)$. The angle θ is the angle (measured in radians or degrees) between the polar axis and the line segment from the pole to point P. This angle can be positive, negative, or zero. The value r is the **directed distance** from the pole to point P. This directed distance can be positive, negative, or zero. The following box shows the three cases of r with a positive angle θ.

Definition Directed Distance

Given an ordered pair $P(r, \theta)$ in the polar coordinate system, the **directed distance** r can be positive, negative, or zero.

- If $r > 0$, then P lies on the terminal side of angle θ.

$r > 0$

$P(r, \theta)$ θ r

- If $r < 0$, then P lies on the ray opposite of the terminal side of angle θ.

$r < 0$

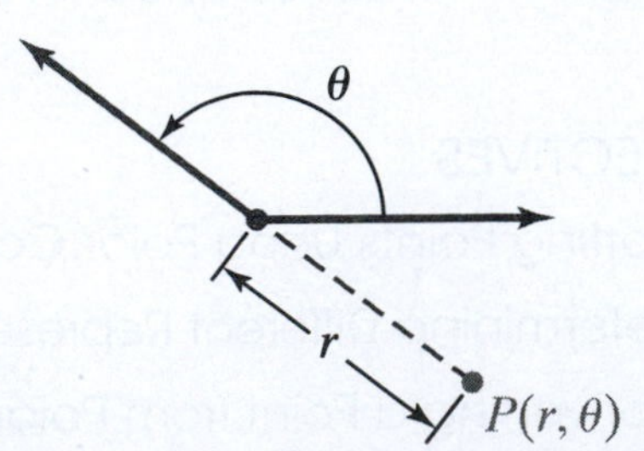

- If $r = 0$, then P lies on the pole regardless of the measure of angle θ.

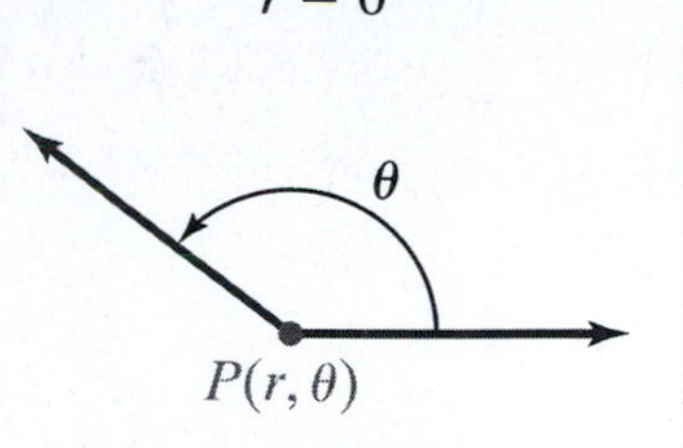

When plotting polar coordinates and sketching polar equations, we will often use a polar grid. A polar grid consists of a series of **concentric circles** of different radii and pre-sketched angles in **standard position**. See Figure 2.

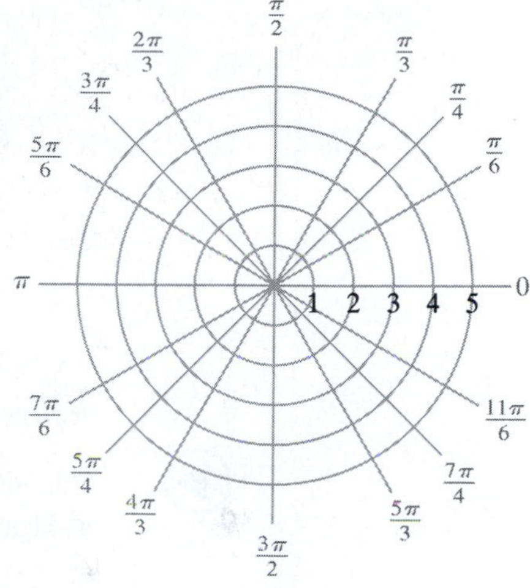

Figure 2 A polar grid

When plotting points in the polar coordinate system, we first sketch the angle in standard position. Then, we locate the point P using the appropriate directed distance for r.

Example 1 Plotting Points Using Polar Coordinates

Plot the following points in a polar coordinate system.

a. $A\left(3, \frac{\pi}{4}\right)$ **b.** $B(-2, 120°)$ **c.** $C\left(1.5, -\frac{7\pi}{6}\right)$ **d.** $D\left(-3, -\frac{3\pi}{4}\right)$

Solution

a. To plot the point $A\left(3, \frac{\pi}{4}\right)$, we identify that $r = 3$ and $\theta = \frac{\pi}{4}$. Thus, start by drawing the angle $\theta = \frac{\pi}{4}$ on a polar grid. Because the value of r is positive, the point A lies along the terminal side of angle θ, 3 units from the pole. See Figure 3.

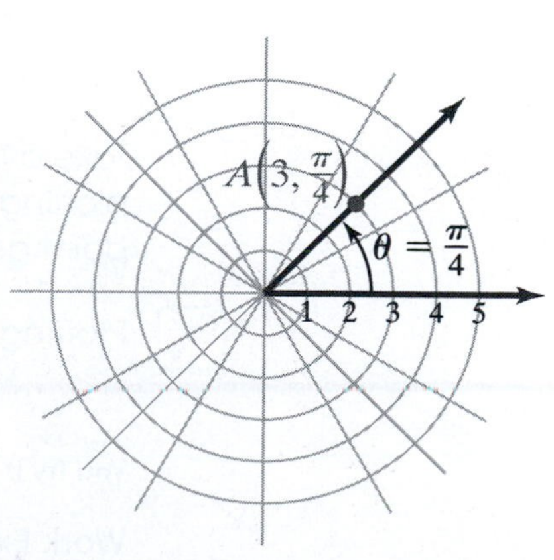

Figure 3 The point $A\left(3, \frac{\pi}{4}\right)$ lies along the terminal side of $\theta = \frac{\pi}{4}$ a distance of 3 units from the pole.

b. To plot the point $B(-2, 120°)$, we identify that $r = -2$ and $\theta = 120°$. Thus, start by drawing the angle $\theta = 120°$ on a polar grid. Because the value of r is negative, the point B lies along the ray directly opposite from the terminal side of angle θ, 2 units from the pole. See Figure 4.

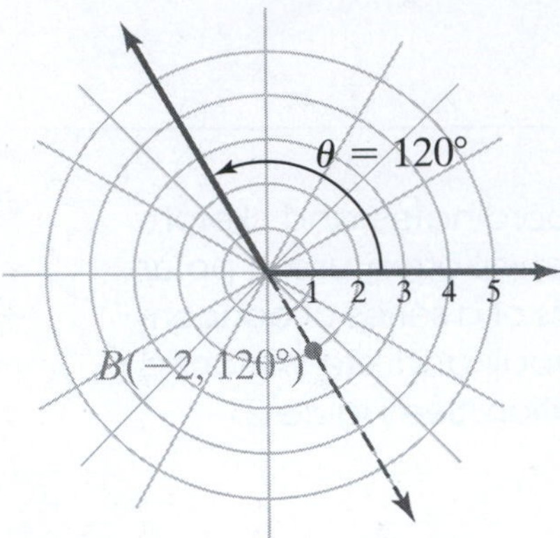

Figure 4 The point $B(-2, 120°)$ lies along the ray directly opposite the terminal side of $\theta = 120°$ a distance of 2 units from the pole.

Try plotting the points $C\left(1.5, -\frac{7\pi}{6}\right)$ and $D\left(-3, -\frac{3\pi}{4}\right)$ on your own. Then watch this **video** to see how to plot these points correctly. The two points are plotted in Figure 5.

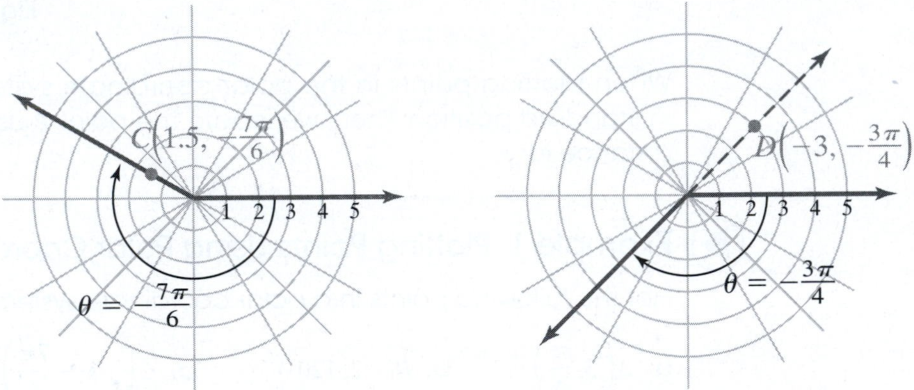

Figure 5 The points $C\left(1.5, -\frac{7\pi}{6}\right)$ and $D\left(-3, -\frac{3\pi}{4}\right)$ are shown plotted in the polar coordinate system.

Take a moment to work through the Guided Visualization below to practice plotting polar coordinates. Choose your own values of r and θ, then watch the point get plotted onto a polar coordinate system.

Plotting Polar Coordinates

You Try It Work through this You Try It problem.

Work Exercises 1–10 in this textbook or in the MyLab Math Study Plan.

OBJECTIVE 2 DETERMINING DIFFERENT REPRESENTATIONS OF THE POINT (r, θ)

In Example 1 we plotted the points $A\left(3, \frac{\pi}{4}\right)$ and $D\left(-3, -\frac{3\pi}{4}\right)$. Watch this **video** to see how to plot these two points. These two points are shown again in Figure 6. Notice that points A and D are positioned at the exact same location.

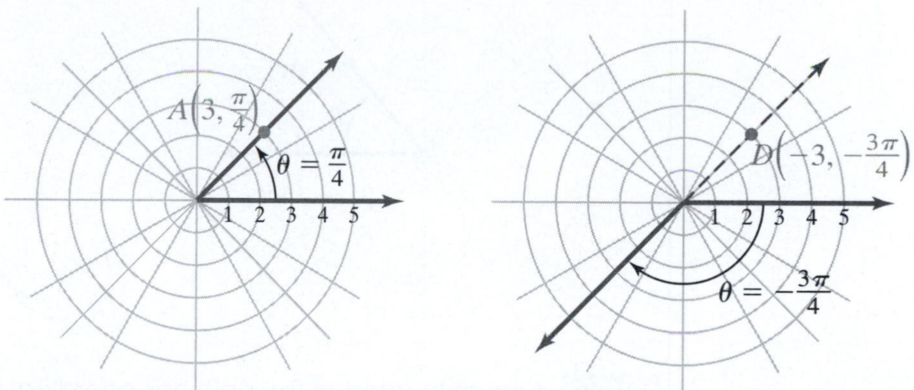

Figure 6 The points $A\left(3, \frac{\pi}{4}\right)$ and $D\left(-3, -\frac{3\pi}{4}\right)$ have the exact same location in the polar coordinate system.

Unlike points in the rectangular coordinate system, which have exactly one representation, points in the polar coordinate system can have infinitely many representations. Given a point in the polar coordinate system having coordinates (r, θ), we can determine a different representation of this point by keeping r the same and by choosing a different angle that is coterminal to the given angle. Or, we can change the sign of r and choose an angle whose terminal side is located one-half of a rotation from angle θ, as in the case of the two points shown in **Figure 6**. We now summarize how to obtain different representations of the same point.

Determining Different Representations of a Point (r, θ)

- Use the same value of r but choose an angle coterminal to θ. The coordinates will be of the form $(r, \theta + 2\pi k)$, where k is any integer.

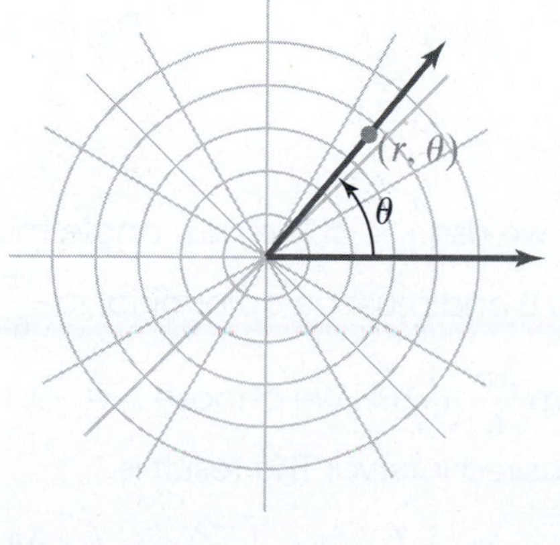

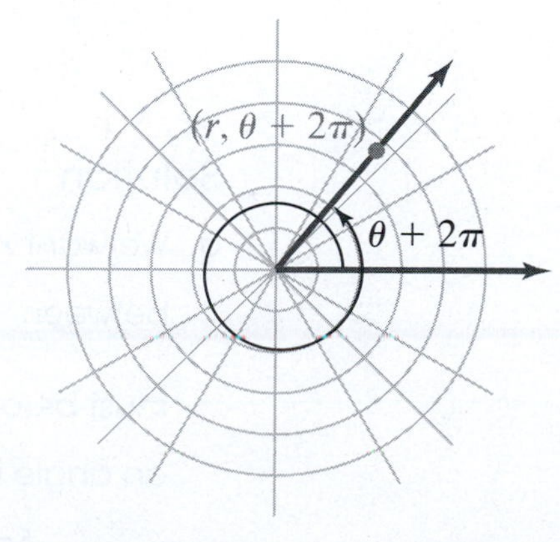

(continued)

- Use the opposite value of r but choose an angle coterminal to the angle located one-half of a rotation from angle θ. The coordinates will be of the form $(-r, \theta + \pi + 2\pi k)$, where k is any integer.

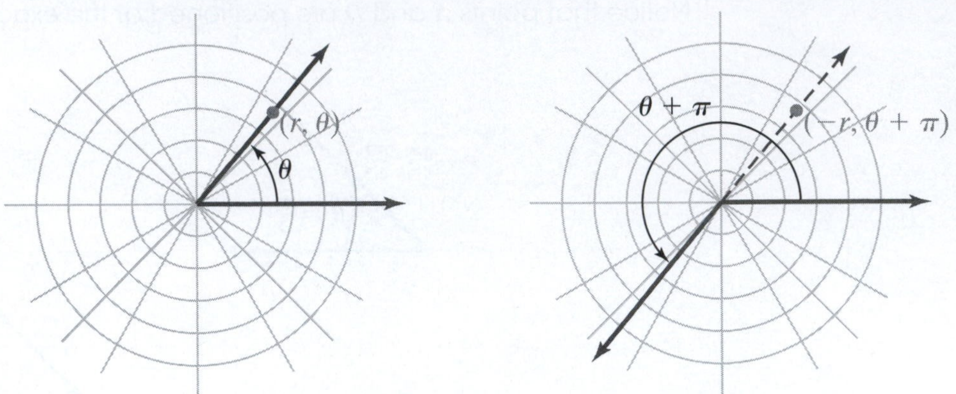

Note A point located at the pole has coordinates $(0, \theta)$, where θ is any angle.

Example 2 Determining Different Representations of the Same Point

The point $P\left(4, \frac{5\pi}{6}\right)$ is shown in **Figure 7**. Determine three different representations of point P that have the specified conditions.

a. $r > 0, -2\pi \leq \theta < 0$ **b.** $r < 0, 0 \leq \theta < 2\pi$ **c.** $r > 0, 2\pi \leq \theta < 4\pi$

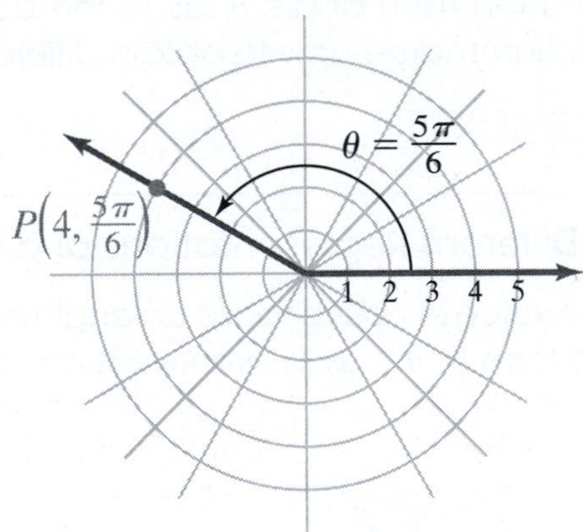

Figure 7 The point $P\left(4, \frac{5\pi}{6}\right)$

Solution

a. We want $r > 0$, so we use $r = 4$. The new angle must be a negative angle between -2π and 0 and must be coterminal to $\frac{5\pi}{6}$. Therefore, the new angle must be of the form $\frac{5\pi}{6} + 2\pi k$. We choose $k = -1$ to guarantee that we obtain an angle in the desired interval. The result is

$$P\left(4, \frac{5\pi}{6} + 2\pi(-1)\right) = P\left(4, \frac{5\pi}{6} - 2\pi\right) = P\left(4, \frac{5\pi}{6} - \frac{12\pi}{6}\right) = P\left(4, -\frac{7\pi}{6}\right).$$

b. We want $r < 0$, so we use $r = -4$. The new angle must be a positive angle between 0 and 2π and must be coterminal to $\frac{5\pi}{6} + \pi$. Therefore, the new angle must be of the form $\frac{5\pi}{6} + \pi + 2\pi k$. We choose $k = 0$ to guarantee that we obtain an angle in the desired interval. The result is

$$P\left(-4, \frac{5\pi}{6} + \pi + 2\pi(0)\right) = P\left(-4, \frac{5\pi}{6} + \pi\right) = P\left(-4, \frac{5\pi}{6} + \frac{6\pi}{6}\right) = P\left(-4, \frac{11\pi}{6}\right).$$

Try to determine the representation of point P as described in part c on your own. Then watch this **video** to see if you are correct.

You Try It Work through this You Try It problem.

Work Exercises 11–15 in this textbook or in the MyLab Math Study Plan.

OBJECTIVE 3 CONVERTING A POINT FROM POLAR COORDINATES TO RECTANGULAR COORDINATES

We can place the pole at the origin of the rectangular coordinate system so that the polar axis coincides with the positive x-axis. See Figure 8. Therefore, we can describe point P using either rectangular coordinates or polar coordinates.

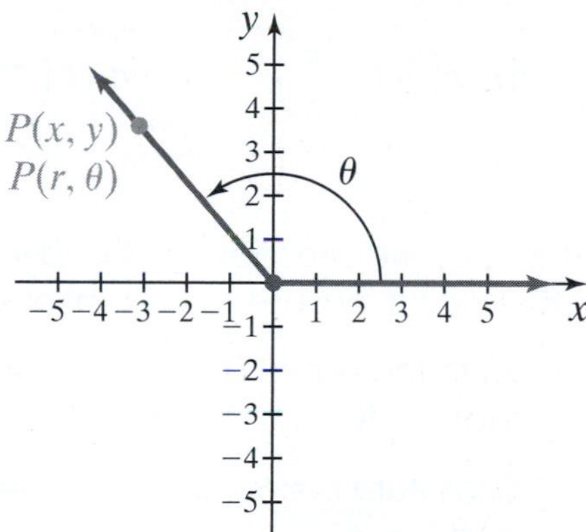

Figure 8 Rectangular and polar coordinate systems

The relationships between rectangular coordinates and polar coordinates are based on the **general angle definitions of the trigonometric functions**. If we consider point P seen in Figure 8, then by the general angle definitions of the trigonometric functions we know the following:

$$\cos\theta = \frac{x}{r} \quad \text{and} \quad \sin\theta = \frac{y}{r}$$

Multiplying both sides of the equations $\cos\theta = \frac{x}{r}$ and $\sin\theta = \frac{y}{r}$ by r gives $x = r\cos\theta$ and $y = r\sin\theta$. In these new forms, the value of r can be zero.

Relationships Used when Converting a Point from Polar Coordinates to Rectangular Coordinates

$x = r\cos\theta$ and $y = r\sin\theta$

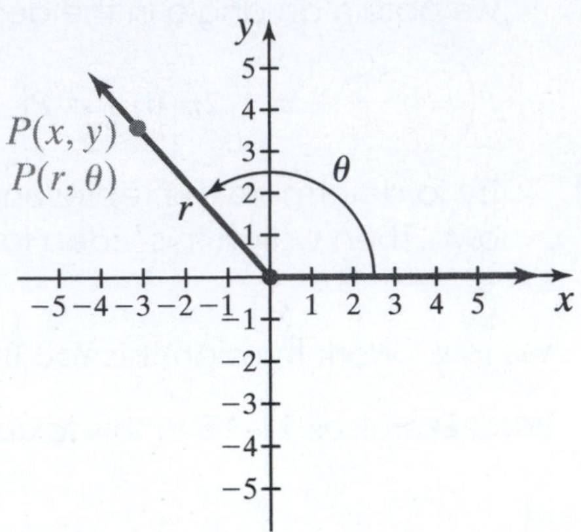

Example 3 Converting a Point from Polar Coordinates to Rectangular Coordinates

Determine the rectangular coordinates for the points with the given polar coordinates.

a. $A(5, \pi)$ **b.** $B\left(-7, -\frac{\pi}{3}\right)$ **c.** $C\left(3\sqrt{2}, \frac{5\pi}{4}\right)$

Solution

a. For the point $A(5, \pi)$, we see that $r = 5$ and $\theta = \pi$. To determine the rectangular coordinates for point A, we use the equations $x = r\cos\theta$ and $y = r\sin\theta$.

$x = r\cos\theta$	Write the equation that relates x, r, and θ.	$y = r\sin\theta$	Write the equation that relates y, r, and θ.
$x = 5\cos\pi$	Substitute $r = 5$ and $\theta = \pi$.	$y = 5\sin\pi$	Substitute $r = 5$ and $\theta = \pi$.
$x = 5(-1)$	Evaluate $\cos\pi = -1$.	$y = 5(0)$	Evaluate $\sin\pi = 0$.
$x = -5$	Multiply.	$y = 0$	Multiply.

We see that $x = -5$ and $y = 0$. Therefore, the rectangular coordinates of $A(5, \pi)$ are $A(-5, 0)$. See Figure 9.

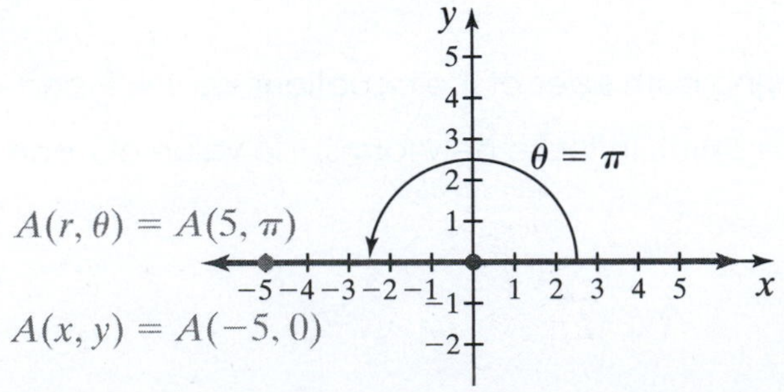

Figure 9 The rectangular coordinates of the point $A(5, \pi)$ are $A(-5, 0)$.

b. For the point $B\left(-7, -\frac{\pi}{3}\right)$, we see that $r = -7$ and $\theta = -\frac{\pi}{3}$. To determine the rectangular coordinates for point B, we use the equations $x = r\cos\theta$ and $y = r\sin\theta$.

$x = r\cos\theta$	Write the equation that relates x, r, and θ.	$y = r\sin\theta$	Write the equation that relates y, r, and θ.
$x = -7\cos\left(-\frac{\pi}{3}\right)$	Substitute $r = -7$ and $\theta = -\frac{\pi}{3}$.	$y = -7\sin\left(-\frac{\pi}{3}\right)$	Substitute $r = -7$ and $\theta = -\frac{\pi}{3}$.
$x = -7\left(\frac{1}{2}\right)$	Evaluate $\cos\left(-\frac{\pi}{3}\right) = \frac{1}{2}$.	$y = -7\left(-\frac{\sqrt{3}}{2}\right)$	Evaluate $\sin\left(-\frac{\pi}{3}\right) = -\frac{\sqrt{3}}{2}$.
$x = -\frac{7}{2}$	Multiply.	$y = \frac{7\sqrt{3}}{2}$	Multiply.

We see that $x = -\frac{7}{2}$ and $y = \frac{7\sqrt{3}}{2}$. Therefore, the rectangular coordinates of

$$B\left(-7, -\frac{\pi}{3}\right) \text{ are } B\left(-\frac{7}{2}, \frac{7\sqrt{3}}{2}\right).$$

c. Try to determine the rectangular coordinates that correspond to the polar coordinates $C\left(3\sqrt{2}, \frac{5\pi}{4}\right)$ on your own. View the **solution** to see if you are correct or watch this **video** to see the worked-out solution.

You Try It **Work through this You Try It problem.**

Work Exercises 16–25 in this textbook or in the MyLab Math Study Plan.

OBJECTIVE 4 CONVERTING A POINT FROM RECTANGULAR COORDINATES TO POLAR COORDINATES

Converting from rectangular coordinates to polar coordinates is a bit more involved than converting from polar coordinates to rectangular coordinates, because there are infinitely many representations for any given point in polar coordinates. For simplicity and consistency, we will always determine the polar coordinates with the conditions that $r \geq 0$ and $0 \leq \theta < 2\pi$.

First, consider points in the rectangular coordinate system lying along an axis and having coordinates $(a, 0)$, $(0, a)$, $(-a, 0)$, and $(0, -a)$, where $a > 0$. Each of the four cases is described in the following box.

Converting Rectangular Coordinates to Polar Coordinates for Points Lying Along an Axis

In each case, assume that $a > 0$.

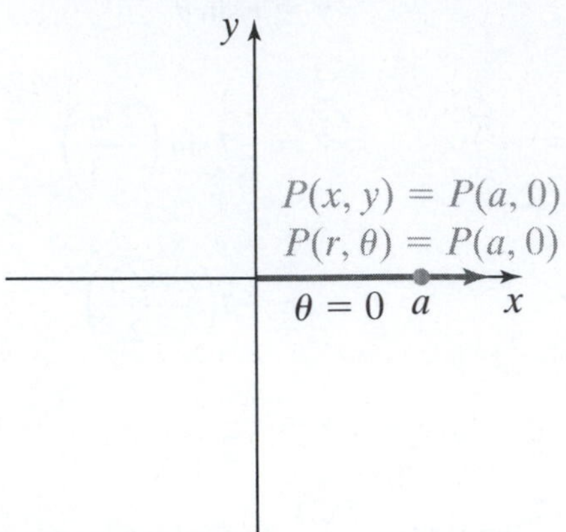

The point $P(x, y) = P(a, 0)$ lies along the positive x-axis and has polar coordinates of $P(r, \theta) = P(a, 0)$.

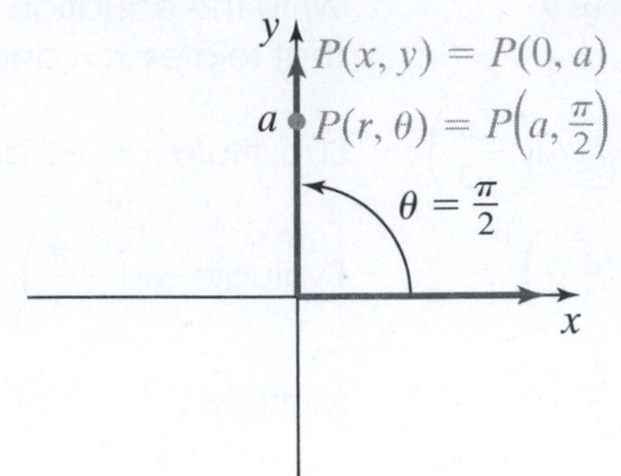

The point $P(x, y) = P(0, a)$ lies along the positive y-axis and has polar coordinates of $P(r, \theta) = P\left(a, \frac{\pi}{2}\right)$.

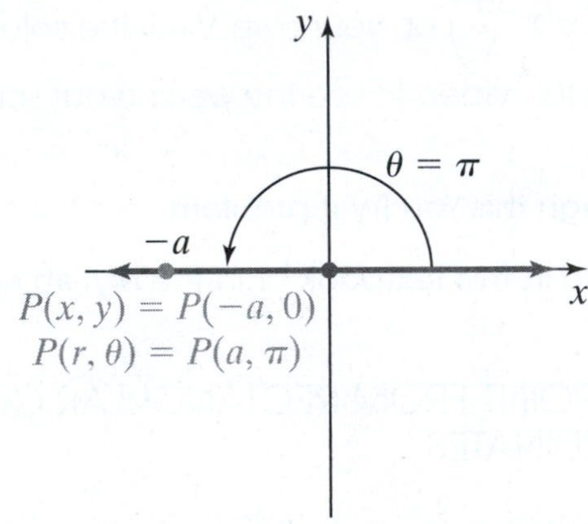

The point $P(x, y) = P(-a, 0)$ lies along the negative x-axis and has polar coordinates of $P(r, \theta) = P(a, \pi)$.

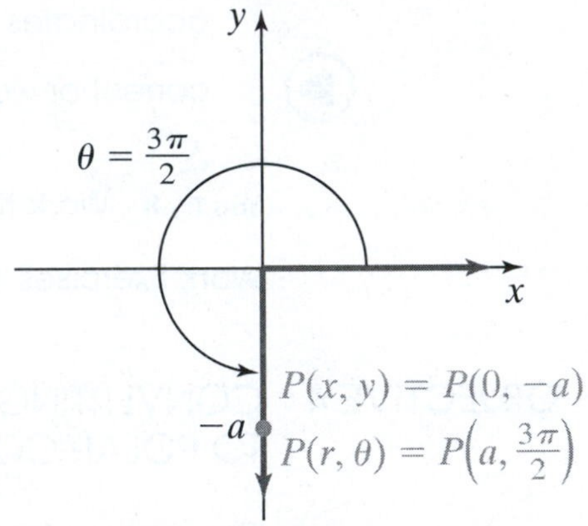

The point $P(x, y) = P(0, -a)$ lies along the negative y-axis and has polar coordinates of $P(r, \theta) = P\left(a, \frac{3\pi}{2}\right)$.

Example 4 Converting Rectangular Coordinates to Polar Coordinates for Points Lying Along an Axis

Determine the polar coordinates for the points with the given rectangular coordinates.

a. $A(-3.5, 0)$

b. $B(0, -\sqrt{7})$

Solution

a. The point $A(-3.5, 0)$ lies along the negative x-axis. Therefore, $\theta = \pi$. The distance from the origin to point A is 3.5 units, so $r = 3.5$. Thus, the polar coordinates of point A are $A(3.5, \pi)$. See Figure 10.

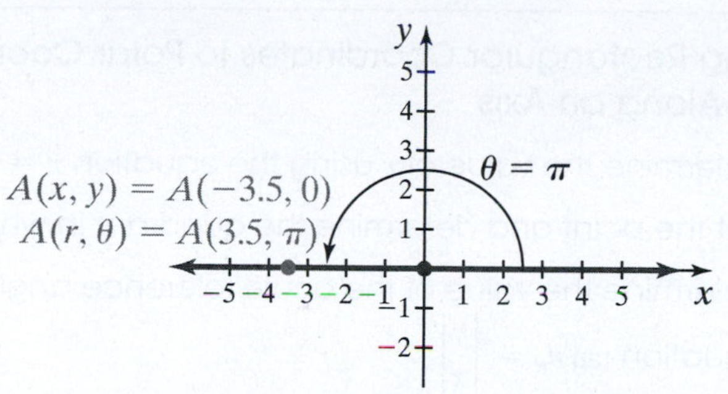

Figure 10 The polar coordinates of the point $A(-3.5, 0)$ are $A(3.5, \pi)$.

b. The point $B(0, -\sqrt{7})$ lies along the negative y-axis. Therefore, $\theta = \dfrac{3\pi}{2}$. The distance from the origin to point B is $\sqrt{7}$ units, so $r = \sqrt{7}$. Thus, the polar coordinates of point B are $B\left(\sqrt{7}, \dfrac{3\pi}{2}\right)$. See Figure 11.

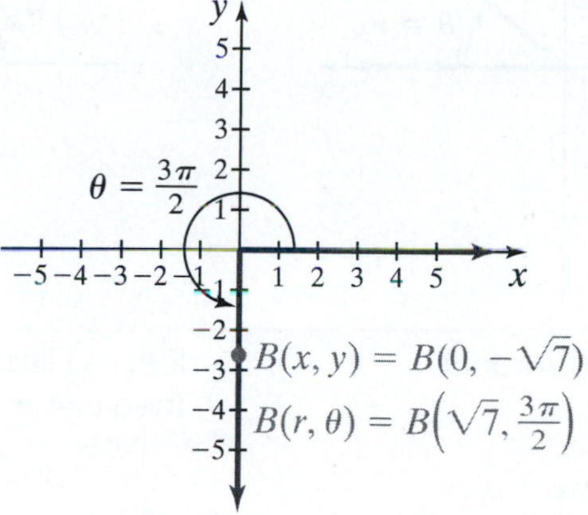

Figure 11 The polar coordinates of the point $B(0, -\sqrt{7})$ are $B\left(\sqrt{7}, \dfrac{3\pi}{2}\right)$.

Work through this **video** to see the complete solution to Example 4. ●

You Try It Work through this You Try It problem.

Work Exercises 26–29 in this textbook or in the MyLab Math Study Plan.

When a point $P(x, y)$ given in rectangular coordinates does not lie along an axis, the conversion to polar coordinates is a bit more complicated. As stated earlier, we will always determine the polar coordinates $P(r, \theta)$ with the conditions that $r \geq 0$ and $0 \leq \theta < 2\pi$.

By the **general angle definitions of the trigonometric functions** we know the following:

$$r = \sqrt{x^2 + y^2} \quad \text{and} \quad \tan\theta = \frac{y}{x}$$

First, we will determine the value of r using the fact that $r = \sqrt{x^2 + y^2}$. Second, we will determine the value of θ, which is much more involved. In order to find θ, we first have to find the reference angle θ_R. Recall that the reference angle is an acute angle. We know that the tangent of any acute angle is a positive value. Therefore, using the fact that $\tan\theta = \dfrac{y}{x}$, we can find θ_R by solving the equation $\tan\theta_R = \left|\dfrac{y}{x}\right|$. The appropriate value of θ depends on the quadrant in which the point $P(x, y)$ lies.

Converting Rectangular Coordinates to Polar Coordinates for Points Not Lying Along an Axis

Step 1. Determine the value of r using the equation $r = \sqrt{x^2 + y^2}$.

Step 2. Plot the point and determine the quadrant in which it lies.

Step 3. Determine the value of the acute reference angle θ_R by solving the equation $\tan \theta_R = \left|\frac{y}{x}\right|$.

Step 4. Determine the value of θ using θ_R and the quadrant in which the point lies. Each case is outlined below.

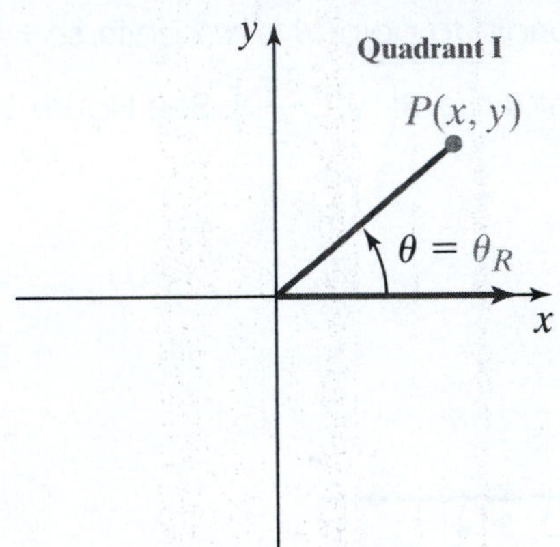

If $P(x, y)$ lies in Quadrant I, then $\theta = \theta_R$.

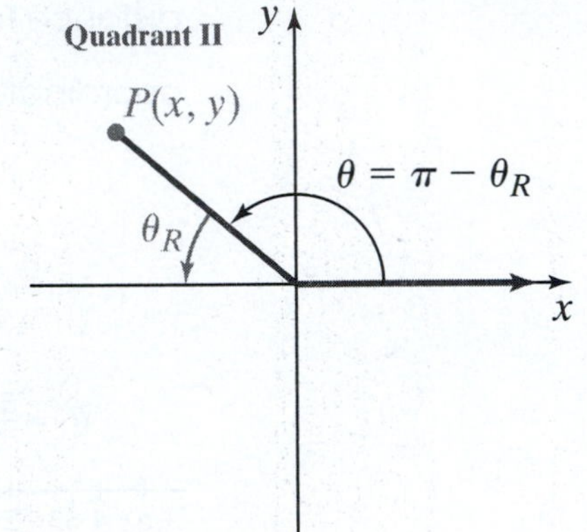

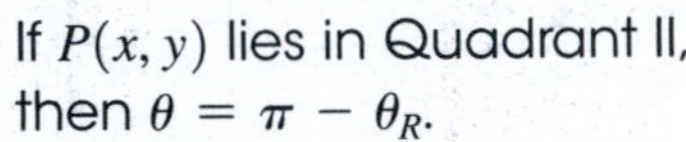

If $P(x, y)$ lies in Quadrant II, then $\theta = \pi - \theta_R$.

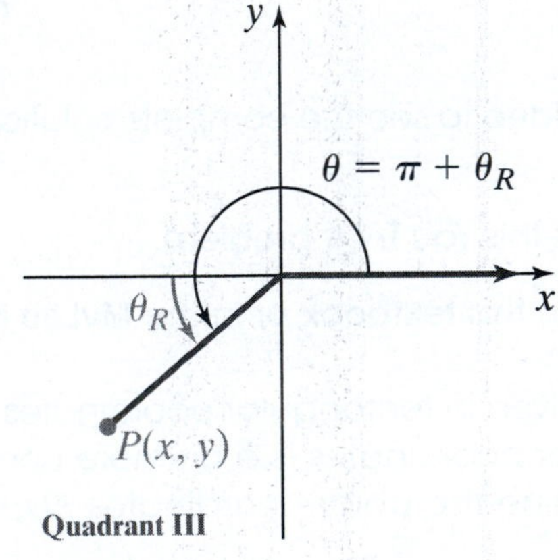

If $P(x, y)$ lies in Quadrant III, then $\theta = \pi + \theta_R$.

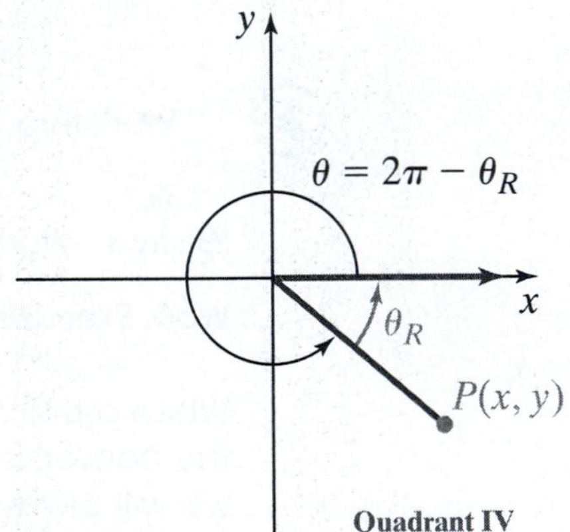

If $P(x, y)$ lies in Quadrant IV, then $\theta = 2\pi - \theta_R$.

Example 5 Converting a Point from Rectangular Coordinates to Polar Coordinates

Determine the polar coordinates for the points with the given rectangular coordinates such that $r \geq 0$ and $0 \leq \theta < 2\pi$. Round the values of r and θ to two decimal places if necessary.

a. $A(-4, -4)$ **b.** $B(-2\sqrt{3}, 2)$ **c.** $C(4, -3)$

Solution

a. **Step 1.** Given the rectangular coordinates $A(-4, -4)$, find the value of r using the equation $r = \sqrt{x^2 + y^2}$.

$r = \sqrt{x^2 + y^2}$	Write the equation relating r, x, and y.
$r = \sqrt{(-4)^2 + (-4)^2}$	Substitute $x = -4$ and $y = -4$.
$r = \sqrt{16 + 16}$	Square each term.
$r = \sqrt{32} = 4\sqrt{2}$	Simplify.

Step 2. Plot the point $A(-4, -4)$ and recognize that the point lies in Quadrant III. See Figure 12.

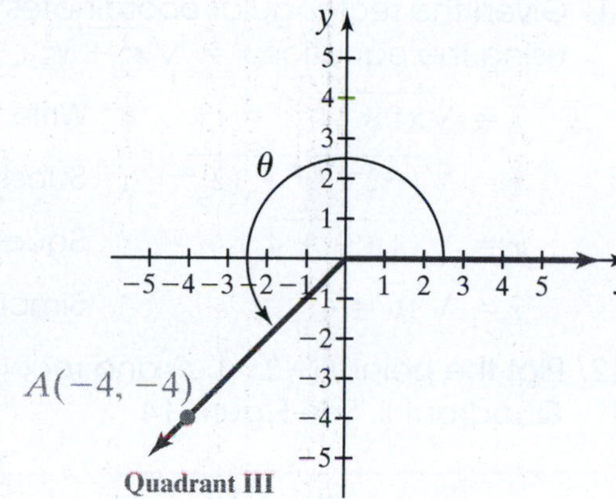

Figure 12 The point $A(-4, -4)$ lies in Quadrant III.

Step 3. To determine the value of θ, we first find the reference angle, θ_R, by solving the equation $\tan \theta_R = \left|\frac{y}{x}\right|$.

$\tan \theta_R = \left|\frac{y}{x}\right|$	Write the equation relating θ_R, x, and y.
$\tan \theta_R = \left|\frac{-4}{-4}\right|$	Substitute $x = -4$ and $y = -4$.
$\tan \theta_R = 1$	Simplify.

The acute angle whose tangent is 1 is $\theta_R = \frac{\pi}{4}$.

Step 4. Because the point lies in Quadrant III, we know that $\theta = \theta_R + \pi$. Thus, $\theta = \frac{\pi}{4} + \pi = \frac{5\pi}{4}$.

Therefore, the polar coordinates of point A for $r \geq 0$ and $0 \leq \theta < 2\pi$ are $A\left(4\sqrt{2}, \frac{5\pi}{4}\right)$. See **Figure 13**. Watch this **interactive video** to see the worked-out solution.

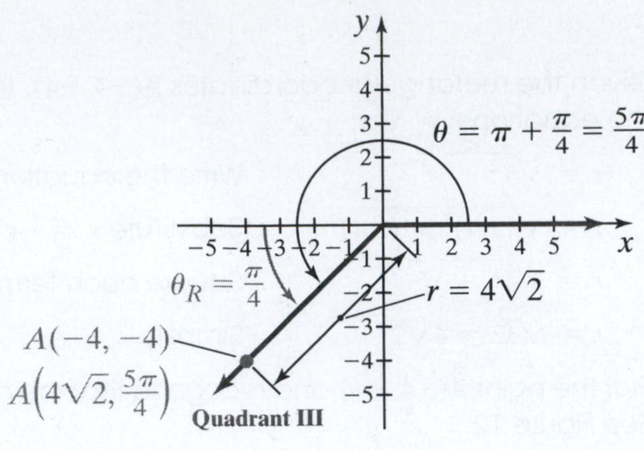

Figure 13 The polar cooridinates of the point $A(-4, -4)$ for $r \geq 0$ and $0 \leq \theta < 2\pi$ are $A\left(4\sqrt{2}, \frac{5\pi}{4}\right)$.

b. Step 1. Given the rectangular coordinates $B(-2\sqrt{3}, 2)$, find the value of r using the equation $r = \sqrt{x^2 + y^2}$.

$r = \sqrt{x^2 + y^2}$ Write the equation relating r, x, and y.

$r = \sqrt{(-2\sqrt{3})^2 + (2)^2}$ Substitute $x = -2\sqrt{3}$ and $y = 2$.

$r = \sqrt{(4)(3) + 4}$ Square each term.

$r = \sqrt{16} = 4$ Simplify.

Step 2. Plot the point $B(-2\sqrt{3}, 2)$ and recognize that the point lies in Quadrant II. See **Figure 14**.

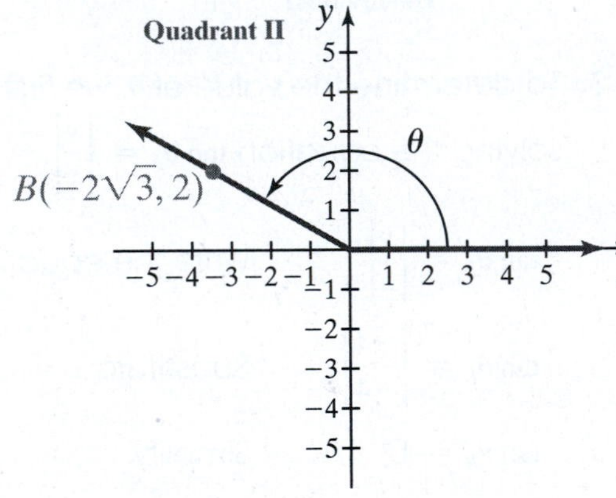

Figure 14 The point $B(-2\sqrt{3}, 2)$ lies in Quadrant II.

Step 3. Given the rectangular coordinates $B(-2\sqrt{3}, 2)$, to determine θ we first find the value of θ_R by solving the equation $\tan \theta_R = \left|\frac{y}{x}\right|$.

$\tan \theta_R = \left|\frac{y}{x}\right|$ Write the equation relating θ_R, x, and y.

$\tan \theta_R = \left|\frac{2}{-2\sqrt{3}}\right|$ Substitute $y = 2$ and $x = -2\sqrt{3}$.

$\tan \theta_R = \frac{1}{\sqrt{3}}$ Simplify.

The acute angle whose tangent is $\frac{1}{\sqrt{3}}$ is $\theta_R = \frac{\pi}{6}$.

Step 4. Because the point lies in Quadrant II, we get $\theta = \pi - \theta_R = \pi - \frac{\pi}{6} = \frac{5\pi}{6}$. Therefore, the polar coordinates of point B for $r \geq 0$ and $0 \leq \theta < 2\pi$ are $B\left(4, \frac{5\pi}{6}\right)$. See Figure 15. Watch this **interactive video** to see the worked-out solution.

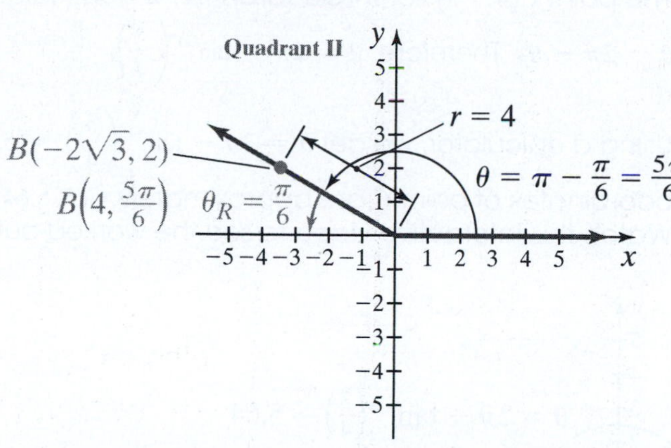

Figure 15 The polar coordinates of the point $B(-2\sqrt{3}, 2)$ for $r \geq 0$ and $0 \leq \theta < 2\pi$ are $B\left(4, \frac{5\pi}{6}\right)$.

c. Step 1. Given the rectangular coordinates $C(4, -3)$, find the value of r using the equation $r = \sqrt{x^2 + y^2}$.

$r = \sqrt{x^2 + y^2}$	Write the equation relating r, x, and y.
$r = \sqrt{(4)^2 + (-3)^2}$	Substitute $x = 4$ and $y = -3$.
$r = \sqrt{16 + 9}$	Square each term.
$r = \sqrt{25} = 5$	Simplify.

Step 2. Plot the point $C(4, -3)$ and recognize that the point lies in Quadrant IV. See Figure 16.

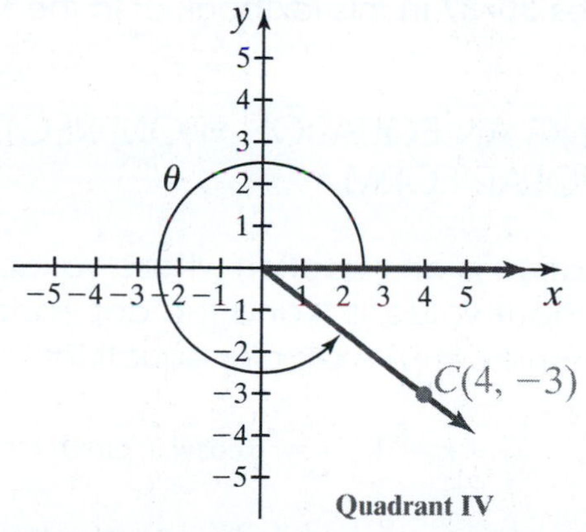

Figure 16 The point $C(4, -3)$ lies in Quadrant IV.

Step 3. Given the rectangular coordinates $C(4, -3)$, to determine θ we first find the value of θ_R by solving the equation $\tan\theta_R = \left|\frac{y}{x}\right|$.

$\tan\theta_R = \left|\frac{y}{x}\right|$	Write the equation relating θ_R, x, and y.
$\tan\theta_R = \left|\frac{-3}{4}\right|$	Substitute $x = 4$ and $y = -3$.
$\tan\theta_R = \frac{3}{4}$	Simplify.

The angle whose tangent is $\frac{3}{4}$ does not belong to one of the special angle families. However, we can use the inverse tangent function to solve for θ_R. Thus, $\theta_R = \tan^{-1}\left(\frac{3}{4}\right)$.

Step 4. The point $C(4, -3)$ lies in Quadrant IV, which indicates that $\theta = 2\pi - \theta_R$. Therefore, $\theta = 2\pi - \tan^{-1}\left(\frac{3}{4}\right)$.

Using a calculator, we get $\theta = 2\pi - \tan^{-1}\left(\frac{3}{4}\right) \approx 5.64$. Thus, the polar coordinates of point C are approximately $C(5, 5.64)$. See Figure 17.

Watch this **interactive video** to see the worked-out solution.

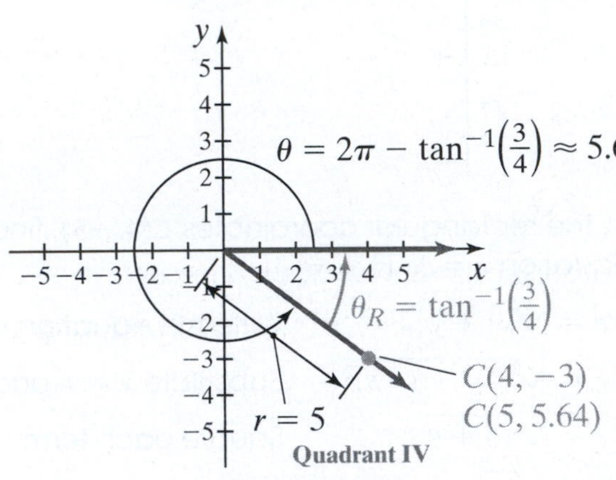

Figure 17 The polar coordinates of the point $C(4, -3)$ for $r \geq 0$ and $0 \leq \theta < 2\pi$ are approximately $C(5, 5.64)$.

You Try It Work through this You Try It problem.

Work Exercises 30–37 in this textbook or in the MyLab Math Study Plan.

OBJECTIVE 5 CONVERTING AN EQUATION FROM RECTANGULAR FORM TO POLAR FORM

A **polar equation** is an equation whose variables are r and θ. The independent variable, or input value, is θ and the dependent variable, or output value, is r. Below are some examples of polar equations:

$$r = 3, \quad r = 6\cos\theta, \quad \text{and } r = \frac{8}{2\cos\theta - \sin\theta}$$

The equations above are said to be in polar form. To convert an equation in x and y (rectangular form) to an equation in r and θ (polar form), we must revisit some relationships previously stated. Recall that $x = r\cos\theta$, $y = r\sin\theta$, and $r = \sqrt{x^2 + y^2}$. When converting rectangular to polar coordinates, we chose, for convenience, to state our answer with $r \geq 0$. But those same points could be written using a different representation where r is negative. Therefore, we can square both sides of the equation $r = \sqrt{x^2 + y^2}$ and obtain an equation where r can take on any value. This equation is $r^2 = x^2 + y^2$. Thus, to convert an equation from rectangular form to polar form, replace x with $r\cos\theta$ and y with $r\sin\theta$. If the equation in rectangular form contains the expression $x^2 + y^2$, then replace $x^2 + y^2$ with r^2. We will always attempt to write the final polar equation in the form where either r or r^2 is isolated.

Example 6 Converting an Equation from Rectangular Form to Polar Form

Convert each equation given in rectangular form into polar form.

a. $x = 7$ **b.** $2x - y = 8$ **c.** $4x^2 + 4y^2 = 3$ **d.** $x^2 + y^2 = 9y$

Solution

a. To convert the equation $x = 7$ into polar form, we replace x with $r\cos\theta$.

$x = 7$ Write the original equation in rectangular form.

$r\cos\theta = 7$ Replace x with $r\cos\theta$.

$r = \dfrac{7}{\cos\theta}$ Divide both sides by $\cos\theta$.

Therefore, given the equation $x = 7$ in rectangular form, we see that the equivalent equation in polar form is $r = \dfrac{7}{\cos\theta}$. Watch this **video** to see the worked-out solution.

b. To convert the equation $2x - y = 8$ into polar form, we replace x with $r\cos\theta$ and replace y with $r\sin\theta$.

$2x - y = 8$ Write the original equation in rectangular form.

$2r\cos\theta - r\sin\theta = 8$ Replace x with $r\cos\theta$ and replace y with $r\sin\theta$.

$r(2\cos\theta - \sin\theta) = 8$ Factor.

$r = \dfrac{8}{2\cos\theta - \sin\theta}$ Divide both sides by $2\cos\theta - \sin\theta$.

Therefore, given the equation $2x - y = 8$ in rectangular form, we see that the equivalent equation in polar form is

$$r = \frac{8}{2\cos\theta - \sin\theta}.$$

Watch this **video** to see the worked-out solution.

c. To convert the equation $4x^2 + 4y^2 = 3$ into polar form, we can factor out a 4 and then replace $x^2 + y^2$ with r^2.

$4x^2 + 4y^2 = 3$ Write the original equation in rectangular form.

$4(x^2 + y^2) = 3$ Factor.

$4r^2 = 3$ Replace $x^2 + y^2$ with r^2.

$r^2 = \dfrac{3}{4}$ Divide both sides by 4.

$r = \pm\sqrt{\dfrac{3}{4}}$ Use the square root property to take the square root of both sides.

$r = \pm\dfrac{\sqrt{3}}{2}$ Simplify.

As you will see when we graph polar equations, the graphs of $r = \frac{\sqrt{3}}{2}$ and $r = -\frac{\sqrt{3}}{2}$ are represented by the same set of points. Thus, it is unnecessary to include both equations as part of the final polar form answer. For convenience, we will only use the positive value of r. Therefore, given the equation $4x^2 + 4y^2 = 3$ in rectangular form, we see that the equivalent equation in polar form is of $r = \frac{\sqrt{3}}{2}$.

 Watch this **video** to see the worked-out solution.

d. To convert the equation $x^2 + y^2 = 9y$ into polar form, we start by replacing $x^2 + y^2$ with r^2 and y with $r \sin \theta$.

$x^2 + y^2 = 9y$	Write the original equation in rectangular form.
$r^2 = 9r \sin \theta$	Replace $x^2 + y^2$ with r^2 and replace y with $r \sin \theta$.
$r^2 - 9r \sin \theta = 0$	Subtract $9r \sin \theta$ from both sides.
$r(r - 9 \sin \theta) = 0$	Factor.
$r = 0$ or $r - 9 \sin \theta = 0$	Use the **zero product property**.
$r = 0$ or $r = 9 \sin \theta$	Solve the second equation for r.

Note that for the equation $r = 9 \sin \theta$, if k is an integer, then $r = 9 \sin (k\pi) = 0$. Therefore, it is unnecessary to include the equation $r = 0$ as part of the final polar form answer.

Thus, the equivalent equation in polar form is $r = 9 \sin \theta$.

 Watch this **video** to see the worked-out solution.

You Try It **Work through this You Try It problem.**

Work Exercises 38–45 in this textbook or in the MyLab Math Study Plan.

OBJECTIVE 6 CONVERTING AN EQUATION FROM POLAR FORM TO RECTANGULAR FORM

To convert an equation in r and θ (polar form) to an equation in x and y (rectangular form), we will use the equations $x = r \cos \theta$, $y = r \sin \theta$, $r^2 = x^2 + y^2$, and $\tan \theta = \frac{y}{x}$ that have been previously stated. To use the equations $x = r \cos \theta$ and $y = r \sin \theta$, we may be required to first multiply both sides of the original polar equation by r. To use the equation $r^2 = x^2 + y^2$, we may need to square both sides of the original polar equation. To use the equation $\tan \theta = \frac{y}{x}$, we may need to rewrite the original equation in a form that involves the expression $\tan \theta$. This will allow us to substitute $\frac{y}{x}$ for $\tan \theta$.

Example 7 Converting an Equation from Polar Form to Rectangular Form

Convert each equation given in polar form into rectangular form.

a. $3r\cos\theta - 4r\sin\theta = -1$ b. $r = 6\cos\theta$ c. $r = 3$ d. $\theta = \frac{\pi}{6}$

Solution

a. To convert the equation $3r\cos\theta - 4r\sin\theta = -1$ into rectangular form, we replace $r\cos\theta$ with x and replace $r\sin\theta$ with y.

$3r\cos\theta - 4r\sin\theta = -1$ Write the original equation in polar form.

$3x - 4y = -1$ Replace $r\cos\theta$ with x and $r\sin\theta$ with y.

Therefore, given the equation $3r\cos\theta - 4r\sin\theta = -1$ in polar form, we see that the equivalent equation in rectangular form is $3x - 4y = -1$. Note that we could solve this equation for y to obtain an alternate answer of $y = \frac{3}{4}x + \frac{1}{4}$.

Watch this **video** to see the worked-out solution.

b. To convert the equation $r = 6\cos\theta$ into rectangular form, first multiply both sides of the equation by r.

$r = 6\cos\theta$ Write the original equation in polar form.

$r^2 = 6r\cos\theta$ Multiply both sides by r.

$x^2 + y^2 = 6x$ Replace r^2 with $x^2 + y^2$ and replace $r\cos\theta$ with x.

Therefore, given the equation $r = 6\cos\theta$ in polar form, we see that the equivalent equation in rectangular form is $x^2 + y^2 = 6x$. Watch this **video** to see the worked-out solution.

c. To convert the equation $r = 3$ into rectangular form, first square both sides of the equation so that we can use the equation $r^2 = x^2 + y^2$.

$r = 3$ Write the original equation in polar form.

$r^2 = 9$ Square both sides.

$x^2 + y^2 = 9$ Replace r^2 with $x^2 + y^2$.

Therefore, given the equation $r = 3$ in polar form, we see that the equivalent equation in rectangular form is $x^2 + y^2 = 9$. Watch this **video** to see the worked-out solution.

d. To convert the equation $\theta = \frac{\pi}{6}$ into rectangular form, first rewrite the equation to involve the expression $\tan\theta$. This will allow us to use the equation $\tan\theta = \frac{y}{x}$.

$\theta = \frac{\pi}{6}$ Write the original equation in polar form.

$\tan\theta = \tan\frac{\pi}{6}$ Rewrite the original equation to involve the expression $\tan\theta$.

$\tan\theta = \frac{1}{\sqrt{3}}$ Evaluate $\tan\frac{\pi}{6} = \frac{1}{\sqrt{3}}$.

$\frac{y}{x} = \frac{1}{\sqrt{3}}$ Replace $\tan\theta$ with $\frac{y}{x}$.

$y = \frac{1}{\sqrt{3}}x$ Multiply both sides by x.

Therefore, given the equation $\theta = \frac{\pi}{6}$ in polar form, we see that the equivalent equation in rectangular form is $y = \frac{1}{\sqrt{3}}x$. Watch this **video** to see the worked-out solution.

It is important to point out that in the previous example we were given polar equations and were able to convert them into recognizable rectangular equations. Note that the graphs of the equations $3x - 4y = -1$ and $y = \frac{1}{\sqrt{3}}x$ are lines and the graphs of the equations $x^2 + y^2 = 6x$ and $x^2 + y^2 = 9$ are circles. In Section 10.2 we will sketch the graphs of polar equations.

You Try It Work through this You Try It problem.

Work Exercises 46–53 in this textbook or in the MyLab Math Study Plan.

10.1 Exercises

In Exercises 1–10, plot each point in a polar coordinate system.

1. $A\left(2, \frac{\pi}{3}\right)$ 2. $B(3, 150°)$ 3. $C\left(-1, \frac{5\pi}{4}\right)$ 4. $D(1.5, \pi)$ 5. $E\left(4, -\frac{5\pi}{6}\right)$

6. $F(-2.5, -305°)$ 7. $G\left(-2, -\frac{3\pi}{2}\right)$ 8. $H\left(-3, -\frac{7\pi}{4}\right)$ 9. $J\left(5, \frac{13\pi}{6}\right)$ 10. $K\left(-4, \frac{11\pi}{4}\right)$

In Exercises 11–15, a point $P(r, \theta)$ is given. Plot the point and then find three different representations of point P that have the specified conditions.

a. $r > 0, -2\pi \le \theta < 0$

b. $r < 0, 0 \le \theta < 2\pi$

c. $r > 0, 2\pi \le \theta < 4\pi$

11. $P\left(5, \frac{\pi}{6}\right)$ 12. $P\left(3, \frac{4\pi}{3}\right)$ 13. $P\left(4, \frac{3\pi}{2}\right)$ 14. $P\left(5, \frac{3\pi}{4}\right)$ 15. $P\left(3, \frac{11\pi}{6}\right)$

In Exercises 16–25, a point $P(r, \theta)$ in polar coordinates is given. Determine the rectangular coordinates $P(x, y)$.

16. $P\left(3, \frac{\pi}{4}\right)$ 17. $P\left(2, \frac{2\pi}{3}\right)$ 18. $P(5, 270°)$ 19. $P\left(1, \frac{7\pi}{6}\right)$ 20. $P\left(4, \frac{5\pi}{3}\right)$

21. $P(-2, 30°)$ 22. $P\left(-5, \frac{3\pi}{4}\right)$ 23. $P(-7, \pi)$ 24. $P\left(-3, \frac{4\pi}{3}\right)$ 25. $P\left(-7, \frac{11\pi}{6}\right)$

In Exercises 26–37, a point $P(x, y)$ in rectangular coordinates is given. Determine the polar coordinates $P(r, \theta)$ such that $r \geq 0$ and $0 \leq \theta < 2\pi$.

26. $P(3, 0)$ 27. $P(0, -\sqrt{2})$ 28. $P(-0.5, 0)$ 29. $P(0, 1.5)$

SbS 30. $P(3, -3)$ SbS 31. $P(-5, 5\sqrt{3})$ SbS 32. $P(-4\sqrt{3}, -4)$ SbS 33. $P(\sqrt{5}, \sqrt{5})$

SbS 34. $P(2, 2\sqrt{3})$ SbS 35. $P(-6\sqrt{3}, 6)$ SbS 36. $P(2, -7)$ SbS 37. $P(\sqrt{2}, -1)$

In Exercises 38–45, convert each equation given in rectangular form into polar form.

38. $7x - 4y = 1$ 39. $x^2 + y^2 = 25$ 40. $x = -3$ 41. $y = 5$

42. $x^2 + y^2 = -4y$ 43. $x^2 + y^2 = 6x$ 44. $(x + 1)^2 + y^2 = 1$ 45. $x^2 + (y - 4)^2 = 16$

In Exercises 46–53, convert each equation given in polar form into rectangular form.

46. $2r\cos\theta + 5r\sin\theta = 3$ 47. $r = 3\sin\theta$ 48. $r = 4$

49. $r = -2\cos\theta$ 50. $\theta = \frac{\pi}{3}$ 51. $r = 6\sin\theta - 2\cos\theta$

52. $r = 2\sec\theta$ 53. $r = -3\csc\theta$

Brief Exercises

In Exercises 54–59, a point $P(x, y)$ in rectangular coordinates is given. Determine the polar coordinates $P(r, \theta)$ such that $r \geq 0$ and $0 \leq \theta < 2\pi$.

54. $P(2, 2)$ 55. $P(-5, 5\sqrt{3})$ 56. $P(-2\sqrt{3}, 2)$

57. $P(-\sqrt{5}, \sqrt{5})$ 58. $P(-1, -\sqrt{3})$ 59. $P(-6\sqrt{3}, -6)$

10.2 Graphing Polar Equations

THINGS TO KNOW

Before working through this section, be sure that you are familiar with the following concepts:

VIDEO ANIMATION INTERACTIVE

You Try It

1. Evaluating Trigonometric Functions of Angles Belonging to the $\frac{\pi}{3}$, $\frac{\pi}{6}$, or $\frac{\pi}{4}$ Families (Section 6.5)

You Try It

2. Solving Trigonometric Equations That Are Linear in Form (Section 8.5)

You Try It

3. Plotting Points Using Polar Coordinates (Section 10.1)

You Try It

4. Converting an Equation from Polar Form to Rectangular Form (Section 10.1) 

INTRODUCTION

Read this introduction before beginning Objective 1.

OBJECTIVES

1 Sketching Equations of the Form $\theta = \alpha$, $r\cos\theta = a$, $r\sin\theta = a$, and $ar\cos\theta + br\sin\theta = c$

2 Sketching Equations of the Form $r = a$, $r = a\sin\theta$, and $r = a\cos\theta$

3 Sketching Equations of the Form $r = a + b\sin\theta$ and $r = a + b\cos\theta$

4 Sketching Equations of the Form $r = a\sin n\theta$ and $r = a\cos n\theta$

5 Sketching Equations of the Form $r^2 = a^2\sin 2\theta$ and $r^2 = a^2\cos 2\theta$

SECTION 10.2 EXERCISES

Introduction to Section 10.2

In our study of algebra, we learned to recognize the shape of the graph of some equations just by looking at the equation rather than just plotting a myriad of points. We know that the equation $x^2 + y^2 = 4$ is represented by the graph of a circle with center at the origin and radius of 2, that the equation $x = 2$ is represented by the graph of a vertical line 2 units to the right of the y-axis, and that the equation $y = -3$ is represented by the graph of a horizontal line 3 units below the x-axis. Recognizing these shapes and considering other factors specific to the shape reduces the work needed to get an accurate sketch of the graph.

As with equations in rectangular form, plotting points to sketch a polar graph does not always satisfactorily yield an accurate picture of the graph. Therefore, we will learn to recognize the shapes that represent various polar equations to reduce the work needed to accurately sketch their graphs. Other factors that may also be considered are symmetry, finding the greatest value r attains and the associated values of θ, and finding the value(s) of θ when $r = 0$. Though some polar equations may look quite complex at first, knowing this information can make their graphs easier to sketch using the polar coordinate system than graphing in the rectangular coordinate system.

OBJECTIVE 1 SKETCHING EQUATIONS OF THE FORM $\theta = \alpha$, $r\cos\theta = a$, $r\sin\theta = a$, AND $ar\cos\theta + br\sin\theta = c$

We begin our study of polar equations by looking at equations of the form $\theta = \alpha$, $r\cos\theta = a$, $r\sin\theta = a$, and $ar\cos\theta + br\sin\theta = c$. We will see that these equations all have something in common. We start by analyzing the graph of an equation of the form $\theta = \alpha$.

The graph of $\theta = \alpha$ is the set of all ordered pairs (r, α), where r is any value. When $r > 0$, the point lies along the terminal side of angle α. When $r < 0$, the point lies along the ray opposite the terminal side of angle α. The graph of $\theta = \alpha$ is a line through the pole that makes an angle of α with the polar axis. Figure 18 illustrates the graph of $\theta = \alpha$ for $0 < \alpha < \frac{\pi}{2}$.

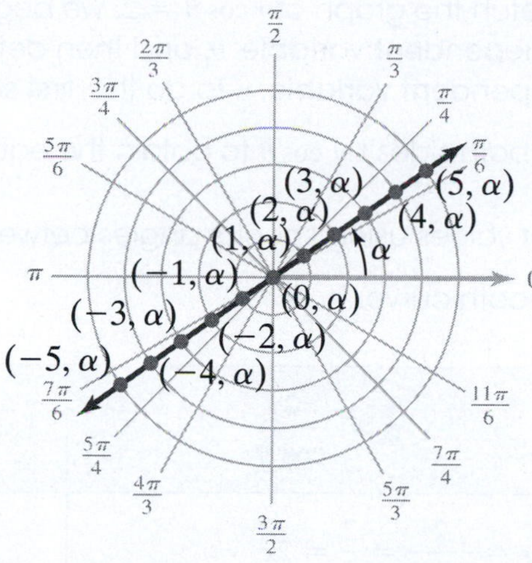

Figure 18 The graph of $\theta = a$ for $0 < \alpha < \frac{\pi}{2}$.

Note that it is possible to convert the equation $\theta = \alpha$ in polar form into rectangular form, but there is no advantage in doing so.

Example 1 Sketching the Graph of a Polar Equation of the Form $\theta = \alpha$

Sketch the graph of the polar equation $\theta = \frac{2\pi}{3}$.

Solution We recognize that the equation is of the form $\theta = \alpha$, where $\alpha = \frac{2\pi}{3}$. Thus, the graph of $\theta = \frac{2\pi}{3}$ is a line passing through the pole that makes an angle of $\frac{2\pi}{3}$ with the polar axis.

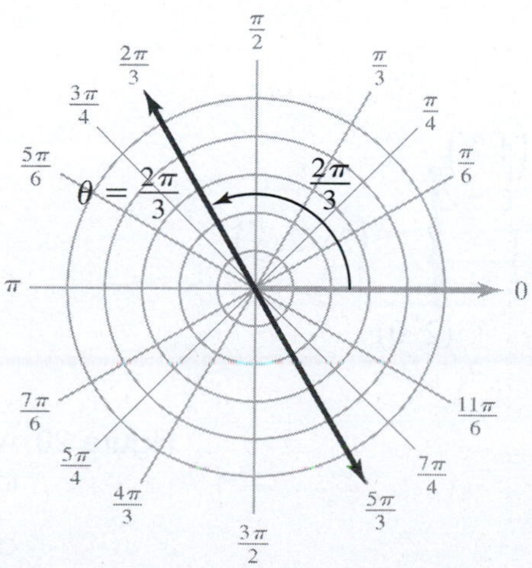

Figure 19 The graph of $\theta = \frac{2\pi}{3}$.

We now consider equations of the form $r\cos\theta = a$ as in Example 2.

Example 2 Sketching the Graph of an Equation of the Form $r\cos\theta = a$

Sketch the graph of the polar equation $r\cos\theta = 2$.

Solution To sketch the graph of $r\cos\theta = 2$, we begin by arbitrarily choosing values of the **independent variable**, θ, and then determining the corresponding values of the **dependent variable**, r. To do this, first solve the equation $r\cos\theta = 2$ for r by dividing both sides by $\cos\theta$ to obtain the equation $r = \dfrac{2}{\cos\theta}$. We now create a table of values using special angles between 0 and $\dfrac{\pi}{2}$ and connect the points with a smooth curve.

$\boldsymbol{\theta}$	$\boldsymbol{r = \dfrac{2}{\cos\theta}}$	$\boldsymbol{(r,\theta)}$
0	$r = \dfrac{2}{\cos 0} = \dfrac{2}{1} = 2x$	$(2, 0)$
$\dfrac{\pi}{6}$	$r = \dfrac{2}{\cos\dfrac{\pi}{6}} = \dfrac{2}{\dfrac{\sqrt{3}}{2}} = \dfrac{4}{\sqrt{3}}$	$\left(\dfrac{4}{\sqrt{3}}, \dfrac{\pi}{6}\right)$
$\dfrac{\pi}{4}$	$r = \dfrac{2}{\cos\dfrac{\pi}{4}} = \dfrac{2}{\dfrac{1}{\sqrt{2}}} = 2\sqrt{2}$	$\left(2\sqrt{2}, \dfrac{\pi}{4}\right)$
$\dfrac{\pi}{3}$	$r = \dfrac{2}{\cos\dfrac{\pi}{3}} = \dfrac{2}{\dfrac{1}{2}} = 4$	$\left(4, \dfrac{\pi}{3}\right)$
$\dfrac{\pi}{2}$	$r = \dfrac{2}{\cos\dfrac{\pi}{2}} = \dfrac{2}{0} =$ undefined	There is no point that corresponds to $\theta = \dfrac{\pi}{2}$.

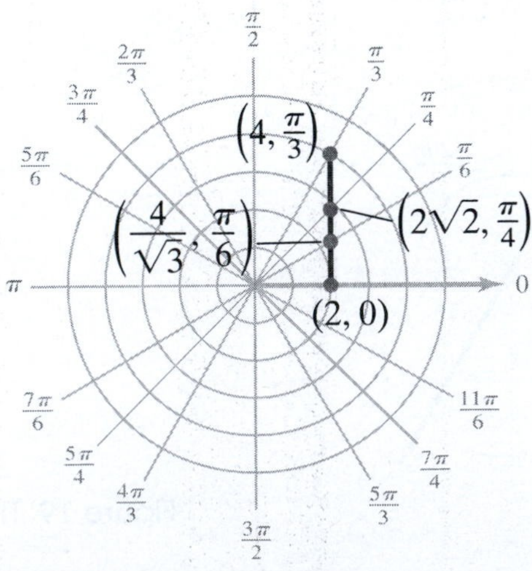

Figure 20 A partial graph of $r\cos\theta = 2$ for special angles between 0 and $\dfrac{\pi}{2}$.

The partial graph of $r\cos\theta = 2$ shown in Figure 20 appears to be taking on the shape of a straight line. This should not be surprising at all. Recall from Section 10.1 that one of the relationships used when converting an equation from polar form to rectangular form is $x = r\cos\theta$. Therefore, rather than plotting points, we could have used this relationship to convert the original equation into rectangular coordinates.

$r\cos\theta = 2$ Write the original equation in polar form.

$x = 2$ Replace $r\cos\theta$ with x.

The equation $r\cos\theta = 2$ in polar form is equivalent to the equation $x = 2$ in rectangular form. We should recognize that the graph of the equation $x = 2$ is a vertical line 2 units to the right of the y-axis. Thus, the graph of the equation $r\cos\theta = 2$ must also be a vertical line 2 units to the right of the vertical line $\theta = \frac{\pi}{2}$ on a polar grid. See Figure 21.

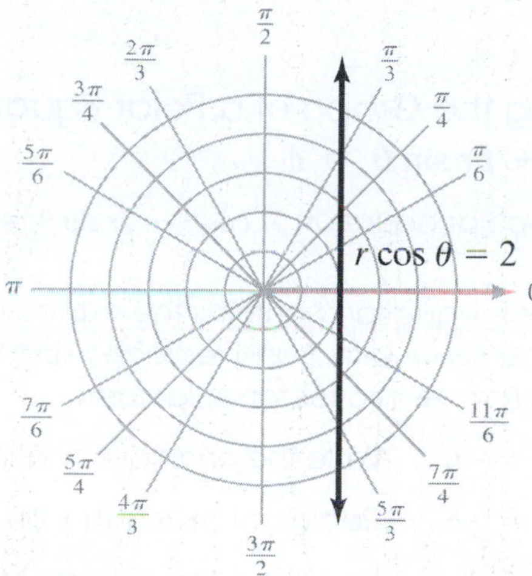

Figure 21 The graph of $r\cos\theta = 2$ is a vertical line located two units to the right of the vertical line $\theta = \frac{\pi}{2}$.

When a is a constant, the graph of the equation $r\cos\theta = a$ is a vertical line. ●

We now analyze the graph of an equation of the form $r\sin\theta = a$. We can convert the equation $r\sin\theta = a$ from polar form to rectangular form.

$r\sin\theta = a$ Write the original equation in polar form.

$y = a$ Replace $r\sin\theta$ with y.

The graph of the equation $y = a$ in rectangular coordinates is a horizontal line. Thus, when a is a constant, the graph of the equation $r\sin\theta = a$ in polar form is a horizontal line.

Example 3 Sketching the Graph of an Equation of the Form $r\sin\theta = a$

Sketch the graph of the polar equation $r\sin\theta = -3$.

Solution There is no need to plot points to determine the graph of the equation $r\sin\theta = -3$ because we should immediately recognize that this equation is equivalent to the equation $y = -3$ in rectangular form. Thus, the graph of $r\sin\theta = -3$ is a horizontal line located 3 units below the line $\theta = 0$ on a polar grid. See Figure 22.

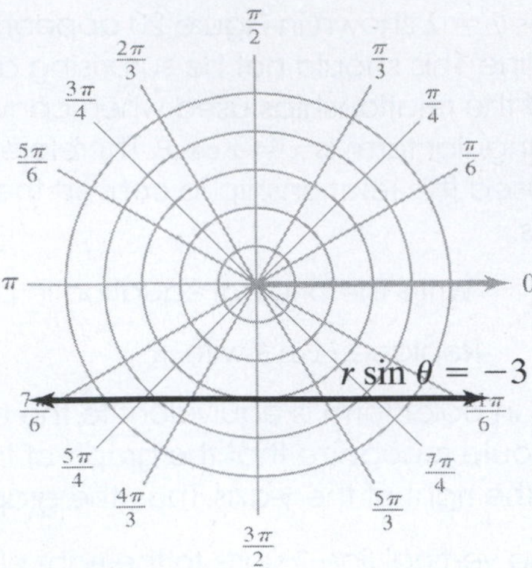

Figure 22 The graph of $r\sin\theta = -3$ is a horizontal line located three units below the line $\theta = 0$.

We now analyze polar equations containing both $r\cos\theta$ and $r\sin\theta$. These equations are represented by graphs of lines that are neither vertical nor horizontal.

Example 4 Sketching the Graph of a Polar Equation of the Form $ar\cos\theta + br\sin\theta = c$

Sketch the graph of the polar equation $3r\cos\theta - 2r\sin\theta = -6$.

Solution We see that this equation contains the expressions $r\cos\theta$ and $r\sin\theta$. Therefore, we can replace $r\cos\theta$ with x and replace $r\sin\theta$ with y to convert the equation $3r\cos\theta - 2r\sin\theta = -6$ into rectangular form.

$$3r\cos\theta - 2r\sin\theta = -6 \quad \text{Write the original equation in polar form.}$$

$$3x - 2y = -6 \quad \text{Replace } r\cos\theta \text{ with } x \text{ and } r\sin\theta \text{ with } y.$$

The graph of the equation $3x - 2y = -6$ is a line. We can solve this equation for y to write the equation in slope-intercept form as $y = \frac{3}{2}x + 3$, where $m = \frac{3}{2}$ and $b = 3$. We can therefore sketch the graph of the equation $3r\cos\theta - 2r\sin\theta = -6$ in polar form by sketching the graph of the linear equation $y = \frac{3}{2}x + 3$ in rectangular form. Watch this **video** to see how to sketch the graph shown in Figure 23.

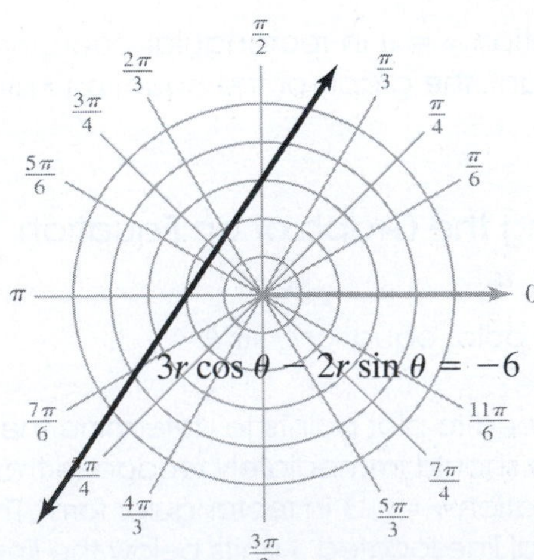

Figure 23 The graph of $3r\cos\theta - 2r\sin\theta = -6$.

When a, b, and c are constants, the graph of the equation $ar\cos\theta + br\sin\theta = c$ is a line with slope $m = -\frac{a}{b}$ and y-intercept $\frac{c}{b}$.

We now summarize some polar equations whose graphs are lines.

Graphs of Polar Equations of the Form $\theta = \alpha$, $r\cos\theta = a$, $r\sin\theta = a$, and $ar\cos\theta + br\sin\theta = c$, where a, b, and c are Constants

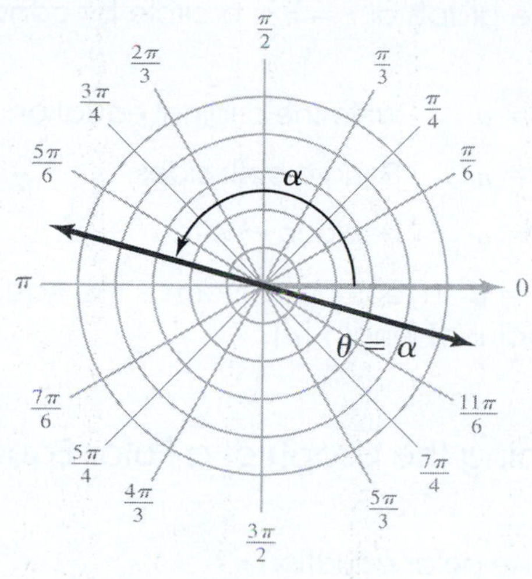

The graph of $\theta = \alpha$ is a line through the pole that makes an angle of α with the polar axis.

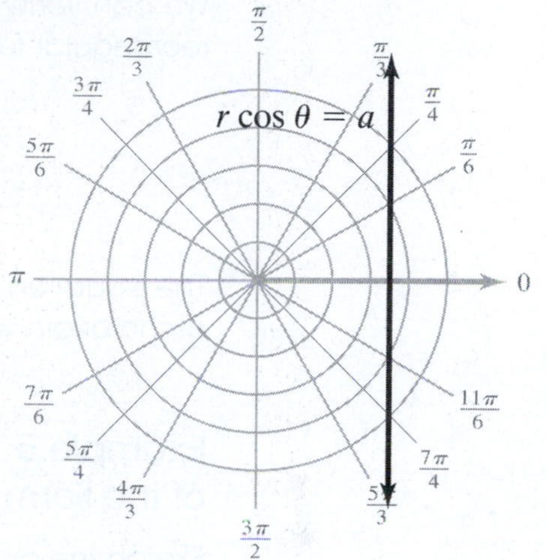

The graph of $r\cos\theta = a$ is a vertical line.

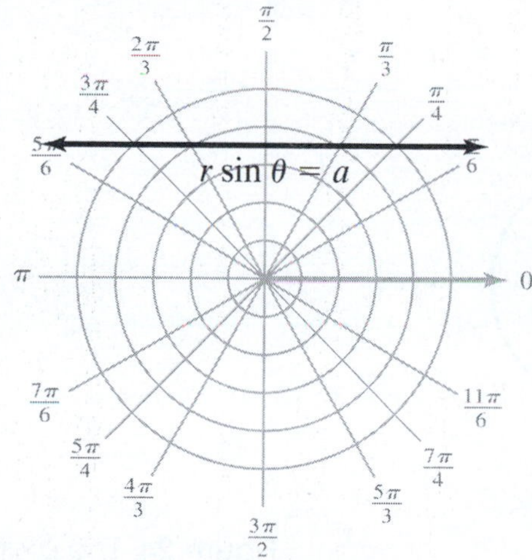

The graph of $r\sin\theta = a$ is a horizontal line.

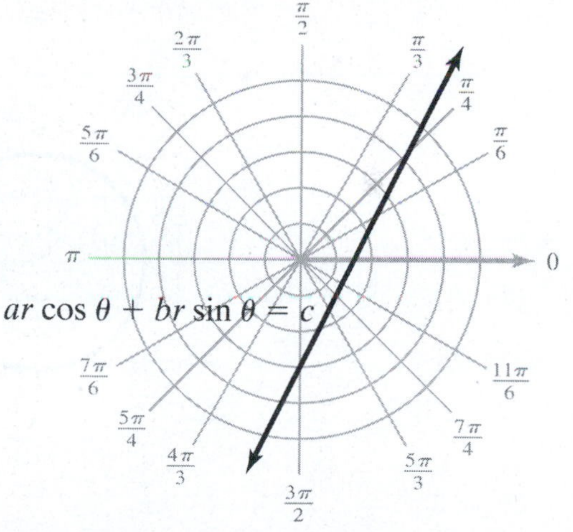

The graph of $ar\cos\theta + br\sin\theta = c$ is a line with slope $m = -\frac{a}{b}$ and y-intercept $\frac{c}{b}$.

You Try It Work through this You Try It problem.

Work Exercises 1–8 in this textbook or in the MyLab Math Study Plan.

OBJECTIVE 2 SKETCHING EQUATIONS OF THE FORM $r = a$, $r = a \sin \theta$, AND $r = a \cos \theta$

Consider the polar equation $r = a$, where a is a constant. Note that this equation does not contain the independent variable θ. This means that regardless of angle θ, the value of r is always equal to a. Thus, the graph of $r = a$ is a circle centered at the pole with a radius of length $|a|$. If $a = 0$, then the graph is a single point located at the pole.

We can verify that the graph of $r = a$ is a circle by converting the equation into rectangular form.

$r = a$	Write the original equation in polar form.
$r^2 = a^2$	Square both sides.
$x^2 + y^2 = a^2$	Replace r^2 with $x^2 + y^2$.

The equation $x^2 + y^2 = a^2$ in rectangular form is the equation of a circle centered at the origin with a radius of length $|a|$.

Example 5 Sketching the Graph of a Polar Equation of the Form $r = a$

Sketch the graph of the polar equation $r = -3$.

Solution The graph is a circle centered at the pole with a radius of length $|-3| = 3$. See Figure 24.

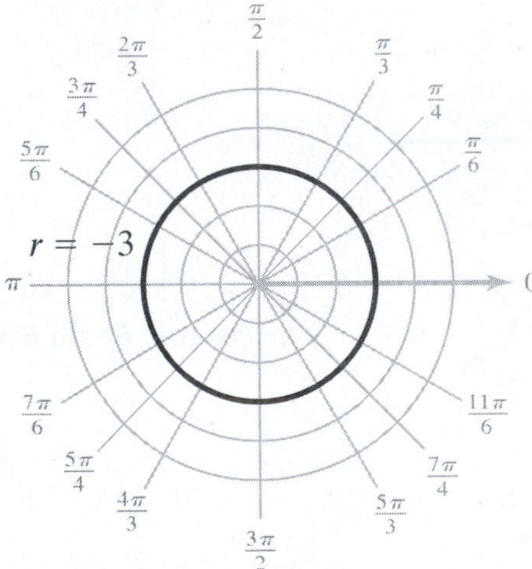

Figure 24 The graph of $r = -3$.

We now consider two more equations whose graphs are circles. These equations have the form $r = a \sin \theta$ and $r = a \cos \theta$. (Remember that if $a = 0$, then the graph is a single point located at the pole.) To sketch the graphs of equations of this form, we can convert the equations into rectangular form as in Example 6.

Example 6 Sketching the Graph of a Polar Equation of the Form $r = a \sin \theta$ and $r = a \cos \theta$

Sketch the graph of each polar equation.

a. $r = 4 \sin \theta$

b. $r = -2 \cos \theta$

Solution

a. First, convert the equation $r = 4\sin\theta$ from polar form to rectangular form.

$r = 4\sin\theta$	Write the original equation in polar form.
$r^2 = 4r\sin\theta$	Multiply both sides by r.
$x^2 + y^2 = 4y$	Replace r^2 with $x^2 + y^2$ and replace $r\sin\theta$ with y.

Now, write the equation $x^2 + y^2 = 4y$ in standard form by subtracting $4y$ from both sides and completing the square.

$x^2 + y^2 = 4y$	Write the equation in rectangular form.
$x^2 + y^2 - 4y = 0$	Subtract $4y$ from both sides.
$x^2 + y^2 - 4y + 4 = 4$	Complete the square on the y-terms by adding 4 to both sides.
$x^2 + (y - 2)^2 = 4$	Factor.

We see that the equation $r = 4\sin\theta$ in polar form is equivalent to the equation $x^2 + (y - 2)^2 = 4$ in rectangular form. The graph of the equation $x^2 + (y - 2)^2 = 4$ is a circle centered at the point $(0, 2)$ with a radius of 2 units. Therefore, the equation $r = 4\sin\theta$ is a circle with a center located 2 units directly above the pole with a radius of 2 units. See **Figure 25**.

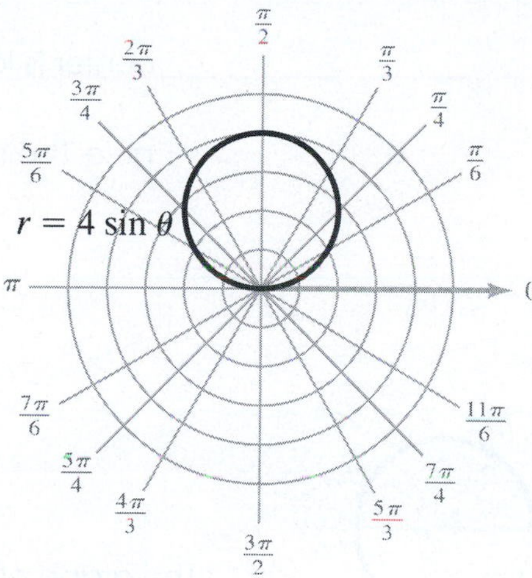

Figure 25 The graph of $r = 4\sin\theta$ is a circle centered two units above the pole with a radius of 2 units.

b. Try sketching the graph of $r = -2\cos\theta$ on your own. Watch this **interactive video** to see the complete solution or view the **graph**.

A summary of the graphs of $r = a$, $r = a\sin(\theta)$, and $r = a\cos(\theta)$ is displayed on the following page. You should now take some time to experiment with the Guided Visualization below to sketch graphs of the form $r = a$, $r = a\sin(\theta)$, or $r = a\cos(\theta)$ by choosing different values of a.

Graphs of Polar Equations of the Form $r = a$, $r = a\sin(\theta)$ or $r = a\cos(\theta)$

Graphs of Polar Equations of the Form $r = a, r = a\sin(\theta)$ or $r = a\cos(\theta)$

Graphs of Polar Equations of the Form $r = a, r = a\sin\theta$, and $r = a\cos\theta$, where $a \neq 0$ is a Constant

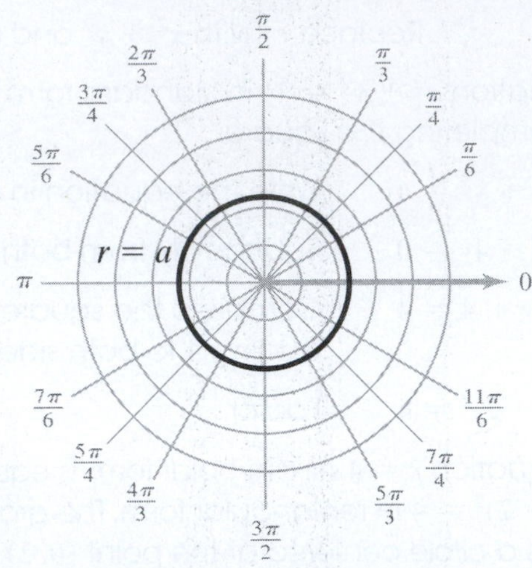

The graph of $r = a$ is a circle centered at the pole with radius of length $|a|$.

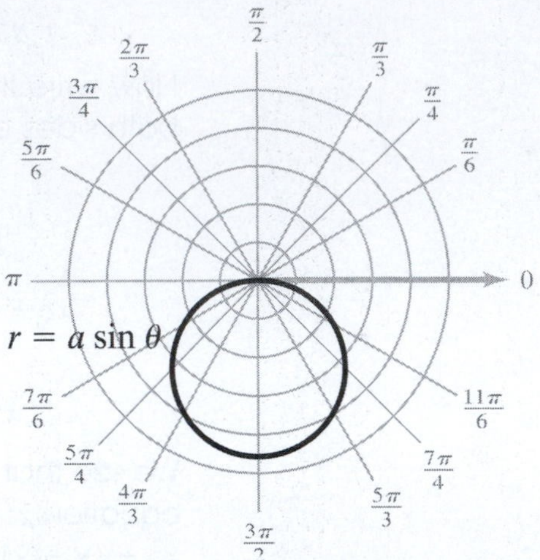

The graph of $r = a\sin\theta$ is a circle centered along the line $\theta = \frac{\pi}{2}$. The center is located $\frac{|a|}{2}$ units from the pole. The radius is $\frac{|a|}{2}$ units.

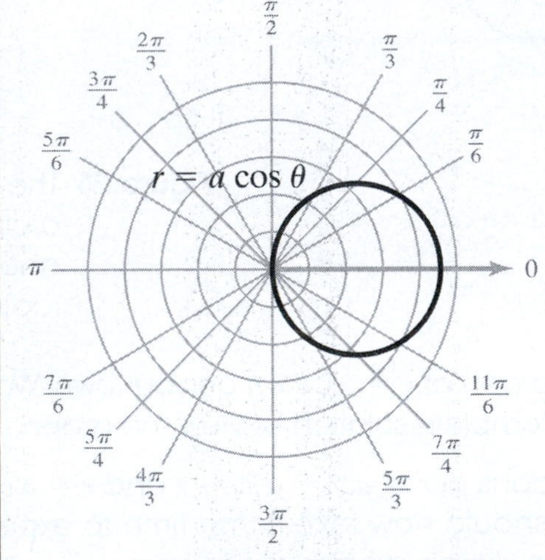

The graph of $r = a\cos\theta$ is a circle centered along the line $\theta = 0$. The center is located $\frac{|a|}{2}$ units from the pole. The radius is $\frac{|a|}{2}$ units.

You Try It **Work through this You Try It problem.**

Work Exercises 9–14 in this textbook or in the MyLab Math Study Plan.

OBJECTIVE 3 SKETCHING EQUATIONS OF THE FORM $r = a + b\sin\theta$ AND $r = a + b\cos\theta$

So far we have sketched polar equations whose graphs had familiar shapes, namely lines and circles. We now begin to explore polar equations whose graphs may be a bit unfamiliar. The next category of graphs that we will study are called **limacons**. The word *limacon* is derived from the Latin word *limax*, which means "snail." All limacons have the form $r = a + b\sin\theta$ or $r = a + b\cos\theta$, where a and b are constants. Although we can convert these equations into rectangular coordinates, the rectangular form will not help to sketch the graph. We will see that the ratio $\left|\frac{a}{b}\right|$ will determine the shape of the limacon. We start with an example where the ratio $\left|\frac{a}{b}\right|$ is equal to 1.

Example 7 Sketching the Graph of a Polar Equation of the Form $r = a + b\sin\theta$, where $\left|\frac{a}{b}\right| = 1$

Sketch the graph of the polar equation $r = 3 - 3\sin\theta$.

Solution To sketch the graph of $r = 3 - 3\sin\theta$, begin by choosing values of θ and then determine the corresponding values of r. We start by choosing the quadrantal angles $\theta = 0, \theta = \frac{\pi}{2}, \theta = \pi$, and $\theta = \frac{3\pi}{2}$. See Figure 26.

θ	$r = 3 - 3\sin\theta$	(r, θ)
0	$r = 3 - 3\sin 0 = 3 - 0 = 3$	$(3, 0)$
$\frac{\pi}{2}$	$r = 3 - 3\sin\frac{\pi}{2} = 3 - 3 = 0$	$\left(0, \frac{\pi}{2}\right)$
π	$r = 3 - 3\sin\pi = 3 - 0 = 3$	$(3, \pi)$
$\frac{3\pi}{2}$	$r = 3 - 3\sin\frac{3\pi}{2} = 3 + 3 = 6$	$\left(6, \frac{3\pi}{2}\right)$

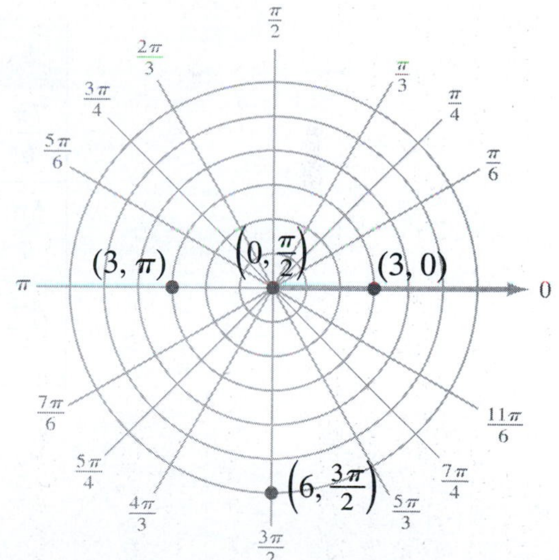

Figure 26 Points lying on the graph of $r = 3 - 3\sin\theta$ whose angles are quadrantal angles.

If we were to connect the points plotted in **Figure 26**, it would appear that the graph has the shape of a "T." This is not the actual shape of the graph, so we continue to plot several more points. We now choose the remaining special angles between 0 and π and connect the points with a smooth curve. See Figure 27.

θ	$r = 3 - 3\sin\theta$	(r, θ)
$\frac{\pi}{6}$	$r = 3 - 3\sin\frac{\pi}{6} = 3 - \frac{3}{2} = 1.5$	$\left(1.5, \frac{\pi}{6}\right)$
$\frac{\pi}{4}$	$r = 3 - 3\sin\frac{\pi}{4} = 3 - \frac{3}{\sqrt{2}} \approx 0.88$	$\left(0.88, \frac{\pi}{4}\right)$
$\frac{\pi}{3}$	$r = 3 - 3\sin\frac{\pi}{3} = 3 - \frac{3\sqrt{3}}{2} \approx 0.40$	$\left(0.40, \frac{\pi}{3}\right)$
$\frac{2\pi}{3}$	$r = 3 - 3\sin\frac{2\pi}{3} = 3 - \frac{3\sqrt{3}}{2} \approx 0.40$	$\left(0.40, \frac{2\pi}{3}\right)$
$\frac{3\pi}{4}$	$r = 3 - 3\sin\frac{3\pi}{4} = 3 - \frac{3}{\sqrt{2}} \approx 0.88$	$\left(0.88, \frac{3\pi}{4}\right)$
$\frac{5\pi}{6}$	$r = 3 - 3\sin\frac{5\pi}{6} = 3 - \frac{3}{2} = 1.5$	$\left(1.5, \frac{5\pi}{6}\right)$

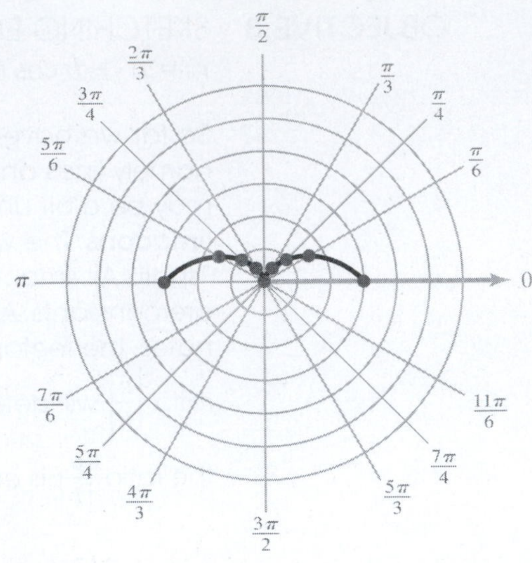

Figure 27 The graph of $r = 3 - 3\sin\theta$ for $0 \le \theta \le \pi$.

As you can see in **Figure 27**, the graph of $r = 3 - 3\sin\theta$ appears to be symmetric about the vertical line $\theta = \frac{\pi}{2}$. Every equation of the form $r = a + b\sin\theta$ will be symmetric about this vertical line. Thus, in order to complete the graph, we need only to plot points having values of θ between π and $\frac{3\pi}{2}$ and then use symmetry to finish the graph. See Figure 28.

θ	$r = 3 - 3\sin\theta$	(r, θ)
$\frac{7\pi}{6}$	$r = 3 - 3\sin\frac{7\pi}{6} = 3 + \frac{3}{2} = 4.5$	$\left(4.5, \frac{7\pi}{6}\right)$
$\frac{5\pi}{4}$	$r = 3 - 3\sin\frac{5\pi}{4} = 3 + \frac{3}{\sqrt{2}} \approx 5.12$	$\left(5.12, \frac{5\pi}{4}\right)$
$\frac{4\pi}{3}$	$r = 3 - 3\sin\frac{4\pi}{3} = 3 + \frac{3\sqrt{3}}{2} \approx 5.60$	$\left(5.60, \frac{4\pi}{3}\right)$

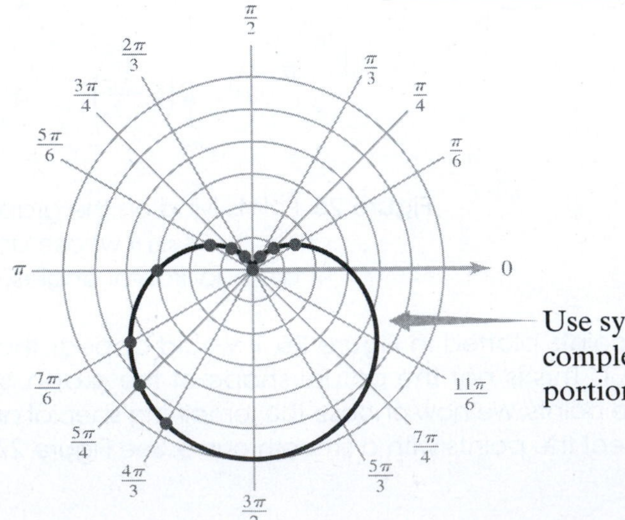
Use symmetry to complete the right-hand portion of the graph.

Figure 28 The complete graph of $r = 3 - 3\sin\theta$.

Observe that the graph of $r = 3 - 3\sin\theta$ has a "heart shape." For this reason, graphs with this shape are called **cardioids.** All polar equations of the form $r = a + b\sin\theta$ or $r = a + b\cos\theta$, where $\left|\frac{a}{b}\right| = 1$, have graphs that are cardioids. Equations of this form involving sine will have graphs that are symmetric about the vertical line $\theta = \frac{\pi}{2}$. Equations of this form involving cosine will have graphs that are symmetric about the horizontal line $\theta = 0$. In fact, the graphs of all equations of the form $r = a + b\sin\theta$ or $r = a + b\cos\theta$ follow these symmetric properties.

We now investigate equations of the form $r = a + b\sin\theta$ or $r = a + b\cos\theta$, where $\left|\frac{a}{b}\right| < 1$.

Example 8 Sketching the Graph of a Polar Equation of the Form $r = a + b\cos\theta$, where $\left|\frac{a}{b}\right| < 1$

Sketch the graph of the polar equation $r = -1 + 2\cos\theta$.

Solution The equation is of the form $r = a + b\cos\theta$, so the graph will be symmetric about the horizontal line $\theta = 0$. Plotting points for values of $0 \le \theta \le \pi$, we get exactly "one-half" of the graph. See **Figure 29**.

θ	0	$\frac{\pi}{6}$	$\frac{\pi}{4}$	$\frac{\pi}{3}$	$\frac{\pi}{2}$	$\frac{2\pi}{3}$	$\frac{3\pi}{4}$	$\frac{5\pi}{6}$	π
r	1	0.73	0.41	0	-1	-2	-2.41	-2.73	-3
(r, θ)	$(1, 0)$	$\left(0.73, \frac{\pi}{6}\right)$	$\left(0.41, \frac{\pi}{4}\right)$	$\left(0, \frac{\pi}{3}\right)$	$\left(-1, \frac{\pi}{2}\right)$	$\left(-2, \frac{2\pi}{3}\right)$	$\left(-2.41, \frac{3\pi}{4}\right)$	$\left(-2.73, \frac{5\pi}{6}\right)$	$(-3, \pi)$

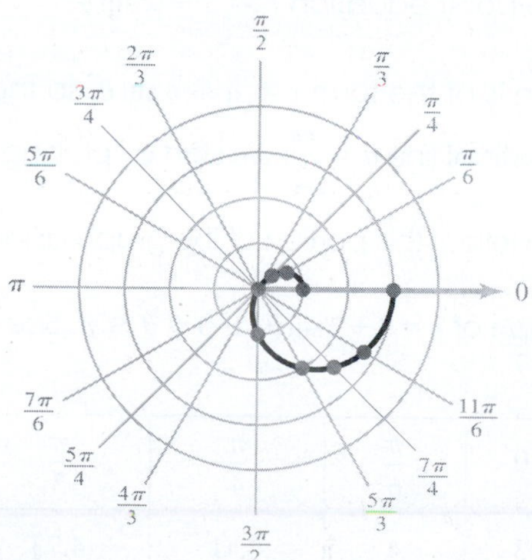

Figure 29 The graph of $r = -1 + 2\cos\theta$ for $0 \le \theta \le \pi$.

Because the graph is symmetric about the horizontal line $\theta = 0$, the graph can be completed by drawing the reflection of **Figure 29** about the line $\theta = 0$. See Figure 30.

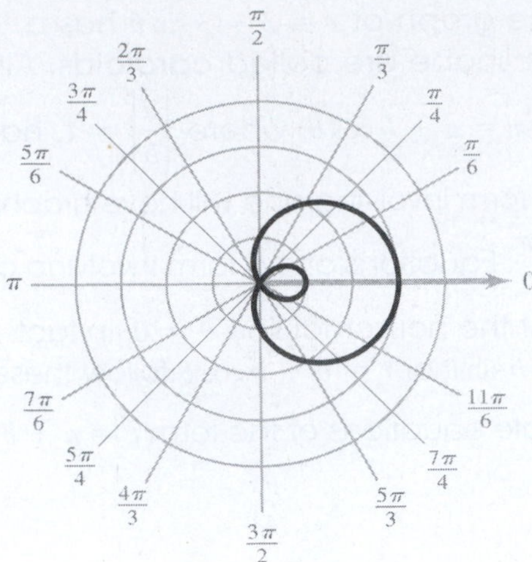

Figure 30 The complete graph of $r = -1 + 2\cos\theta$.

Notice that the graph has an "inner loop" that intersects the pole. The graph actually intersects the pole twice; once at the point $\left(0, \frac{\pi}{3}\right)$ and again at the point $\left(0, \frac{5\pi}{3}\right)$.

Graphs of the form $r = a + b\sin\theta$ or $r = a + b\cos\theta$ for $\left|\frac{a}{b}\right| < 1$ are limacons that have an innerloop that intersects the pole twice.

The next example illustrates the graph of $r = a + b\sin\theta$ for the case where $1 < \left|\frac{a}{b}\right| < 2$.

Example 9 Sketching the Graph of a Polar Equation of the Form $r = a + b\sin\theta$, where $1 < \left|\frac{a}{b}\right| < 2$

Sketch the graph of the polar equation $r = 3 + 2\sin\theta$.

Solution The equation is of the form $r = a + b\sin\theta$, so the graph will be symmetric about the vertical line $\theta = \frac{\pi}{2}$. We start by plotting points for values of $0 \le \theta \le \frac{\pi}{2}$. We can then reflect this portion of the graph about the vertical line $\theta = \frac{\pi}{2}$ to obtain the graph of $r = 3 + 2\sin\theta$ for $0 \le \theta \le \pi$. See Figure 31.

$\boldsymbol{\theta}$	0	$\frac{\pi}{6}$	$\frac{\pi}{4}$	$\frac{\pi}{3}$	$\frac{\pi}{2}$
$\boldsymbol{r}$	3	4	4.41	4.73	5
$\boldsymbol{(r, \theta)}$	$(3, 0)$	$\left(4, \frac{\pi}{6}\right)$	$\left(4.41, \frac{\pi}{4}\right)$	$\left(4.73, \frac{\pi}{3}\right)$	$\left(5, \frac{\pi}{2}\right)$

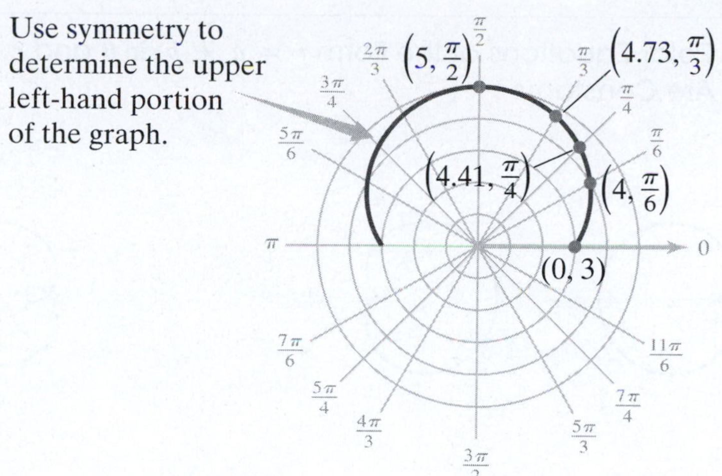
Use symmetry to determine the upper left-hand portion of the graph.
(5, π/2)
(4.73, π/3)
(4.41, π/4)
(4, π/6)
(0, 3)

Figure 31 The graph of $r = 3 + 2\sin\theta$ for $0 \le \theta \le \pi$.

We can now plot points for values of $\pi < \theta \le \frac{3\pi}{2}$ and then once again use symmetry to complete the graph. See Figure 32.

θ	$\frac{7\pi}{6}$	$\frac{5\pi}{4}$	$\frac{4\pi}{3}$	$\frac{3\pi}{2}$
r	2	1.59	1.27	1
(r, θ)	$\left(2, \frac{7\pi}{6}\right)$	$\left(1.59, \frac{5\pi}{4}\right)$	$\left(1.27, \frac{4\pi}{3}\right)$	$\left(1, \frac{3\pi}{2}\right)$

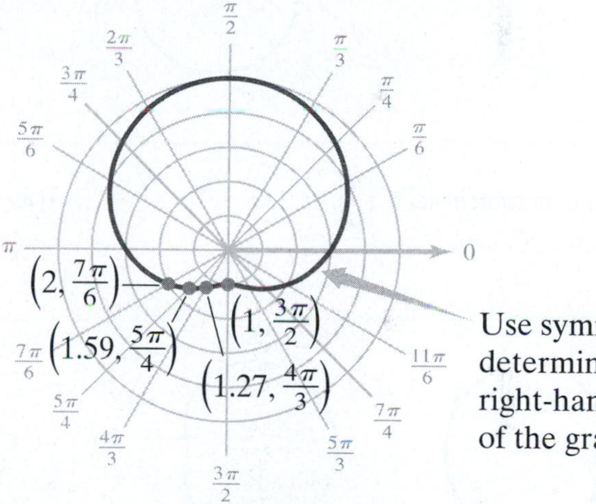
(2, 7π/6)
(1.59, 5π/4)
(1, 3π/2)
(1.27, 4π/3)
Use symmetry to determine the lower right-hand portion of the graph.

Figure 32 The complete graph of $r = 3 + 2\sin\theta$.

The graph of $r = 3 + 2\sin\theta$ is known as a limacon with a dimple. All equations of the form $r = a + b\sin\theta$ or $r = a + b\cos\theta$ for $1 < \left|\frac{a}{b}\right| < 2$ have graphs that are limacons with dimples.

The final type of limacon has no inner loop and no dimple. These equations are of the form $r = a + b\sin\theta$ or $r = a + b\cos\theta$ for $\left|\frac{a}{b}\right| \ge 2$. Example 10 shows such an equation

We now summarize each of the four types of limacons and create a step-by-step procedure for sketching their graphs.

Graphs of Polar Equations of the Form $r = a + b\sin\theta$ and $r = a + b\cos\theta$, where $a \neq 0$ and $b \neq 0$ Are Constants

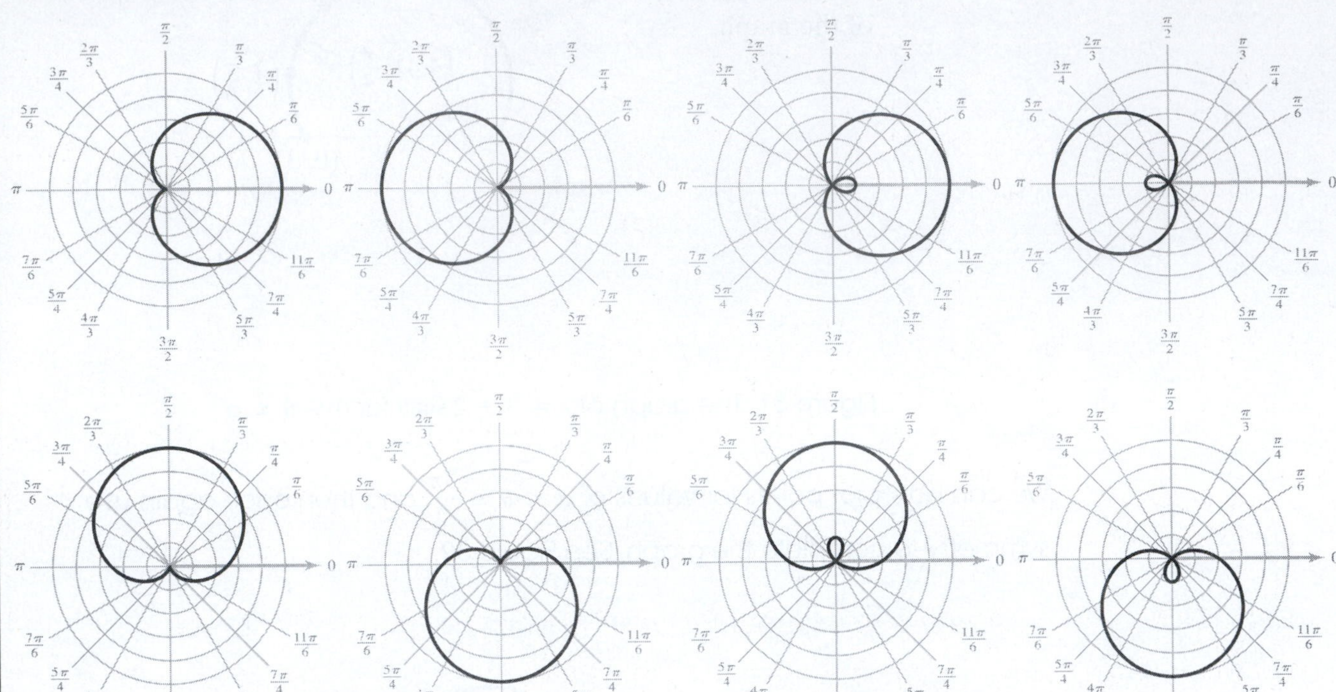

The graph is a cardioid if $\left|\frac{a}{b}\right| = 1.$

The graph is a limacon with an inner loop if $\left|\frac{a}{b}\right| < 1.$

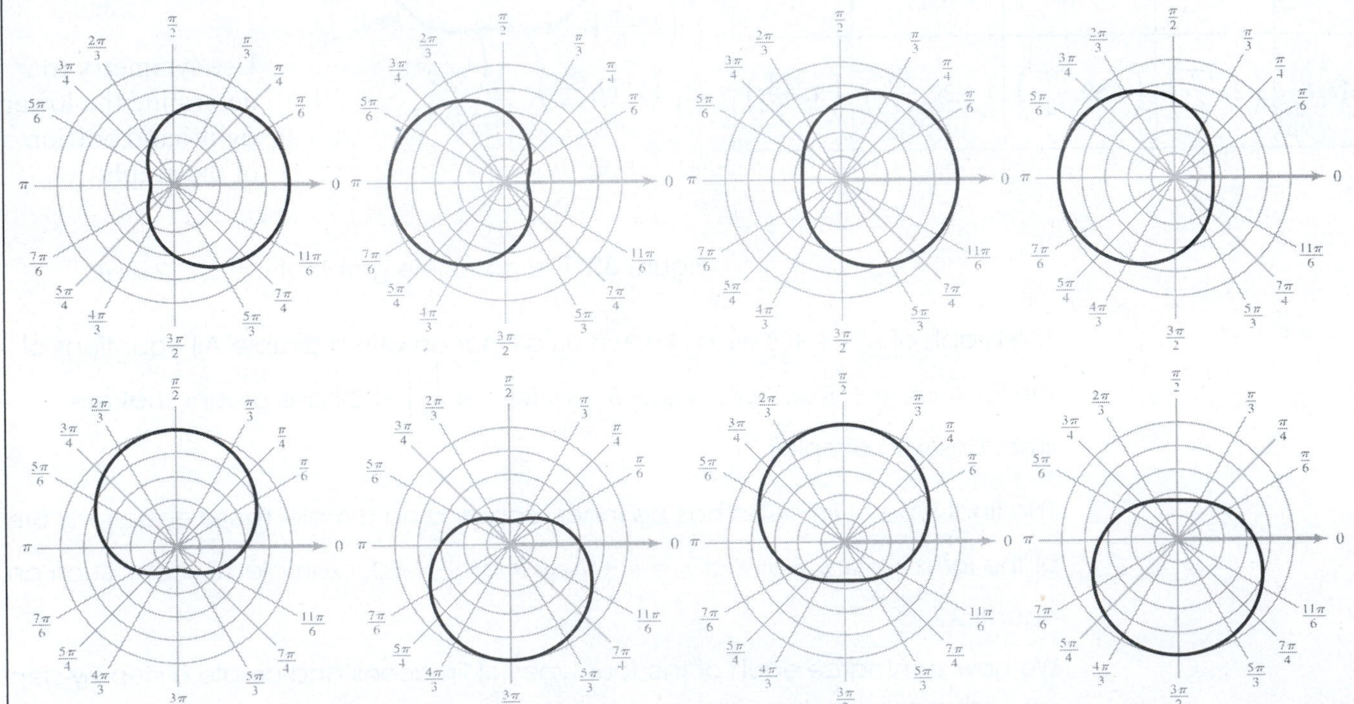

The graph is a limacon with a dimple if $1 < \left|\frac{a}{b}\right| < 2.$

The graph is a limacon with no inner loop and no dimple if $\left|\frac{a}{b}\right| \geq 2.$

Steps for Sketching Polar Equations (Limacons) of the Form $r = a + b\sin\theta$ and $r = a + b\cos\theta$

Step 1. Identify the general shape using the ratio $\left|\frac{a}{b}\right|$.

- If $\left|\frac{a}{b}\right| = 1$, then the graph is a cardioid.
- If $\left|\frac{a}{b}\right| < 1$, then the graph is a limacon with an inner loop that intersects the pole.
- If $1 < \left|\frac{a}{b}\right| < 2$, then the graph is a limacon with a dimple.
- If $\left|\frac{a}{b}\right| \geq 2$, then the graph is a limacon with no inner loop and no dimple.

Step 2. Determine the symmetry.

- If the equation is of the form $r = a + b\sin\theta$, then the graph must be symmetric about the line $\theta = \frac{\pi}{2}$.
- If the equation is of the form $r = a + b\cos\theta$, then the graph must be symmetric about the line $\theta = 0$.

Step 3. Plot the points corresponding to the quadrantal angles $\theta = 0, \theta = \frac{\pi}{2}, \theta = \pi$, and $\theta = \frac{3\pi}{2}$.

Step 4. If necessary, plot a few more points until symmetry can be used to complete the graph.

Before working through Example 10, take some time to interact with the Guided Visualization below. In this Guided Visualization, choose a function of the form $r = a + b\sin\theta$ or $r = a + b\cos\theta$. Then choose different values of a and b. Try to first predict what the graph will look like before animating the graph. Once you have had a chance to interact with the Guided Visualization, then work through Example 10.

 Graphs of polar equations of the form $r = a + b\sin\theta$ or $r = a + b\cos\theta$

Example 10 Sketching the Graph of a Polar Equation of the Form $r = a + b\sin\theta$ and $r = a + b\cos\theta$

Sketch the graph of each polar equation.

a. $r = 4 - 3\cos\theta$

b. $r = 2 + \sin\theta$

c. $r = -2 + 2\cos\theta$

d. $r = 3 - 4\sin\theta$

Solution Follow the four-step process to sketch each graph. Carefully work through this **interactive video** to see how to use this four-step process, or, once you have sketched a graph, click on the corresponding link below to see if your graph is correct.

View the **graph** of $r = 4 - 3\cos\theta$.

View the **graph** of $r = 2 + \sin\theta$.

View the **graph** of $r = -2 + 2\cos\theta$.

View the **graph** of $r = 3 - 4\sin\theta$.

You Try It Work through this You Try It problem.

Work Exercises 15–26 in this textbook or in the MyLab Math Study Plan.

OBJECTIVE 4 SKETCHING EQUATIONS OF THE FORM $r = a\sin n\theta$ AND $r = a\cos n\theta$

We now consider polar equations of the form $r = a\sin n\theta$ and $r = a\cos n\theta$, where $a \neq 0$ is a constant and $n \neq 1$ is a positive integer. The graphs of these equations are called roses because the shapes of the graphs resemble the petals of a flower. The number of petals depends on the value of n. Again, converting these types of equations into rectangular form would be of no advantage. We start by sketching a graph of the form $r = a\sin n\theta$, where n is even.

Example 11 Sketching the Graph of a Polar Equation of the Form $r = a\sin n\theta$

Sketch the graph of $r = 3\sin 2\theta$.

Solution The equation $r = 3\sin 2\theta$ is of the form $r = a\sin n\theta$, where $a = 3$ and $n = 2$. We start by plotting several points to complete one petal of a rose. We will use values of θ such that the angle 2θ will yield a special angle, as seen in the table below. Note that decimal values of r are approximations.

θ	0	$\frac{\pi}{12}$	$\frac{\pi}{6}$	$\frac{\pi}{4}$	$\frac{\pi}{3}$	$\frac{5\pi}{12}$	$\frac{\pi}{2}$
2θ	0	$\frac{\pi}{6}$	$\frac{\pi}{3}$	$\frac{\pi}{2}$	$\frac{2\pi}{3}$	$\frac{5\pi}{6}$	π
r	0	$\frac{3}{2}$	2.6	3	2.6	$\frac{3}{2}$	0
(r, θ)	$(0, 0)$	$\left(\frac{3}{2}, \frac{\pi}{12}\right)$	$\left(2.6, \frac{\pi}{6}\right)$	$\left(3, \frac{\pi}{4}\right)$	$\left(2.6, \frac{\pi}{3}\right)$	$\left(\frac{3}{2}, \frac{5\pi}{12}\right)$	$\left(0, \frac{\pi}{2}\right)$

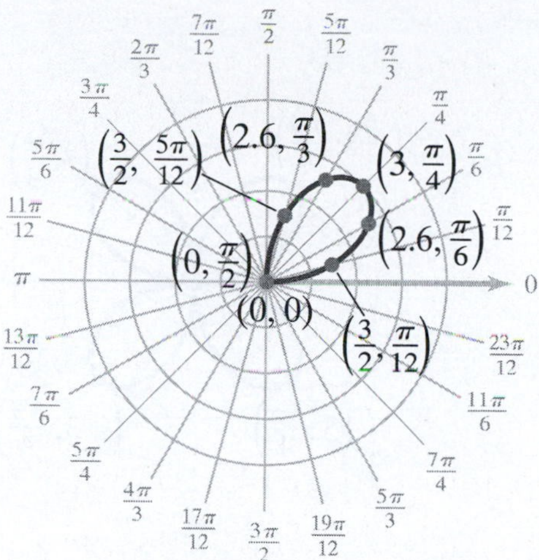

Figure 33 One petal of the graph of $r = 3 \sin 2\theta$.

It is not necessary to continue to plot more points to complete the graph. Each petal of the rose will have the exact same shape as the petal sketched in Figure 33. Note that the length of the petal is 3 units. If we can determine the endpoints of the remaining petals, then we can easily complete the graph. These endpoints occur for values of θ for which $r = 3$ or $r = -3$. Thus, we must solve the two trigonometric equations $3 \sin 2\theta = 3$ and $3 \sin 2\theta = -3$. These equations are equivalent to the equations $\sin 2\theta = 1$ and $\sin 2\theta = -1$. Both of these equations can be solved using the **steps for solving trigonometric equations that are linear in form** that were discussed in **Section 8.5**.

The **solutions to the equation $\sin 2\theta = 1$** on the interval $[0, 2\pi)$ are $\theta = \frac{\pi}{4}$ and $\theta = \frac{5\pi}{4}$.

The **solutions to the equation $\sin 2\theta = -1$** on the interval $[0, 2\pi)$ are $\theta = \frac{3\pi}{4}$ and $\theta = \frac{7\pi}{4}$.

Therefore, the endpoints of the four petals occur at the points and $\left(3, \frac{\pi}{4}\right), \left(3, \frac{5\pi}{4}\right), \left(-3, \frac{3\pi}{4}\right)$, and $\left(-3, \frac{7\pi}{4}\right)$. See Figure 34. Now that we know the endpoints of the four petals, we can complete the graph of $r = 3 \sin 2\theta$. See Figure 35.

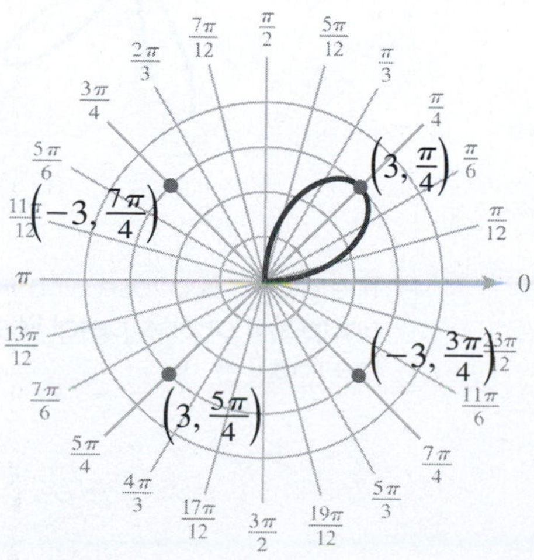

Figure 34 The endpoints of the four petals of the graph of $r = 3 \sin 2\theta$ occur at the points $\left(3, \frac{\pi}{4}\right), \left(3, \frac{5\pi}{4}\right), \left(-3, \frac{3\pi}{4}\right)$, and $\left(-3, \frac{7\pi}{4}\right)$.

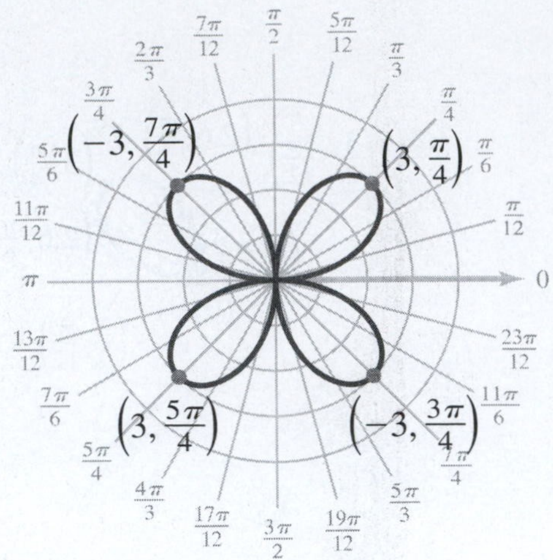

Figure 35 The complete graph of $r = 3 \sin 2\theta$

Note that the graph of the equation $r = 3 \sin 2\theta$ has 4 petals. The length of each petal is 3 units. In general, equations of the form $r = a \sin n\theta$ and $r = a \cos n\theta$ have petals of length $|a|$ units. After further examination we will be able to generalize the following:

- If n is even, then there will be $2n$ petals.
- If n is odd, there will be n petals.

We now summarize each type of rose and create a step-by-step procedure for sketching their graphs.

Graphs of Polar Equations of the Form $r = a \sin n\theta$ and $r = a \cos n\theta$, where $a \neq 0$ Is a Constant and $n \neq 1$ Is a Positive Integer

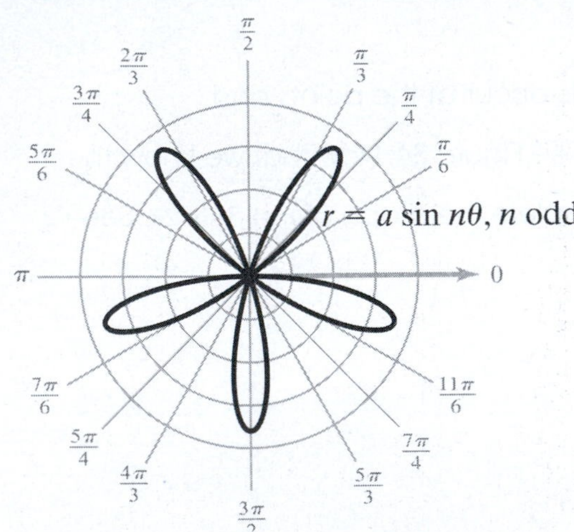

The graph is a rose with n petals. The endpoint of one petal lies along the vertical line $\theta = \frac{\pi}{2}$.

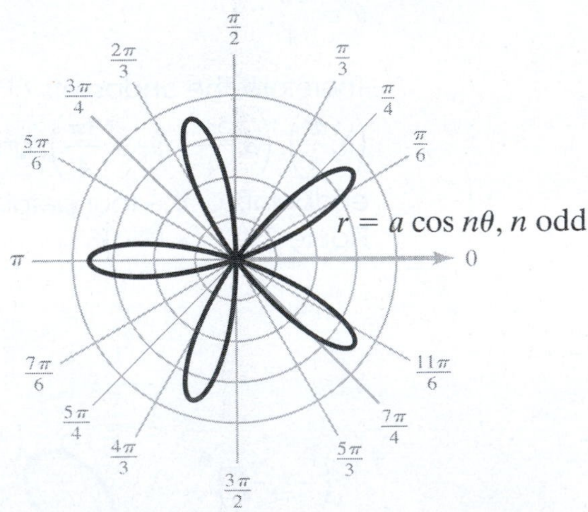

The graph is a rose with n petals. The endpoint of one petal lies along the line $\theta = 0$.

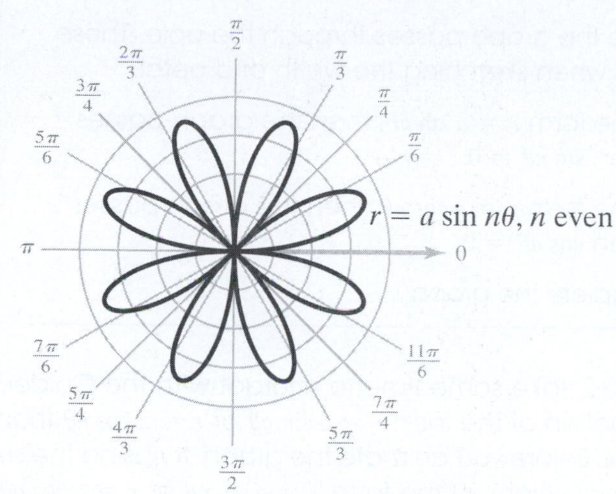

The graph is a rose with $2n$ petals. None of the petals has an endpoint lying on either the line $\theta = 0$ or the line $\theta = \frac{\pi}{2}$.

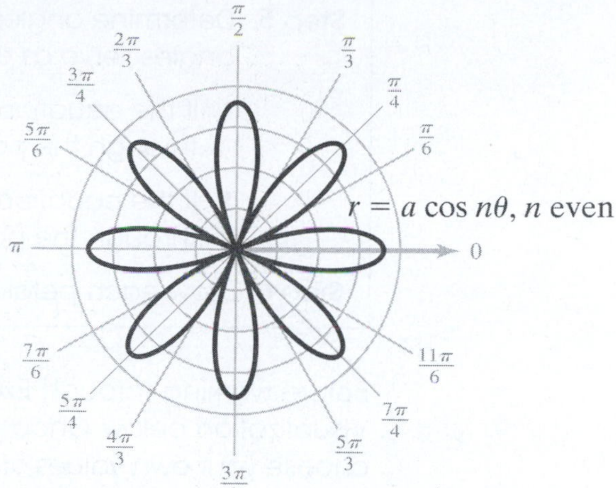

The graph is a rose with $2n$ petals. Two petals have endpoints lying along the vertical line $\theta = \frac{\pi}{2}$. Two petals have endpoints lying along the horizontal line $\theta = 0$.

Steps for Sketching Polar Equations (Roses) of the Form $r = a \sin n\theta$ and $r = a \cos n\theta$, where $a \neq 0$ and $n \neq 1$ Is a Positive Integer

Step 1. Identify the number of petals.

- If n is even, then there are $2n$ petals.
- If n is odd, then there are n petals.

Step 2. Determine the length of each petal.

- The length of each petal is $|a|$ units.

Step 3. Determine all angles where an endpoint of a petal lies.

- If the equation is of the form $r = a \sin n\theta$, then the endpoints occur for angles on the interval $[0, 2\pi)$* that satisfy the equations $\sin n\theta = 1$ and $\sin n\theta = -1$.
- If the equation is of the form $r = a \cos n\theta$, then the endpoints occur for angles on the interval $[0, 2\pi)$* that satisfy the equations $\cos n\theta = 1$ and $\cos n\theta = -1$.

***Note that when n is odd, it is only necessary to consider angles on the interval $[0, \pi)$. A complete graph is obtained on this interval because the graph will completely traverse itself on the interval $[\pi, 2\pi)$.**

Step 4. Substitute each angle determined in Step 3 back into the original equation to obtain the appropriate values of r for each angle. The ordered pairs obtained represent the endpoints of the rose petals. Plot these points on the graph.

(Continued)

Step 5. Determine angles where the graph passes through the pole. These angles serve as a guide when sketching the width of a petal.

- If the equation is of the form $r = a \sin n\theta$, then the graph passes through the pole when $\sin n\theta = 0$.
- If the equation is of the form $r = a \cos n\theta$, then the graph passes through the pole when $\cos n\theta = 0$.

Step 6. Draw each petal to complete the graph.

Before working through Example 12, take some time to interact with the Guided Visualization below. Choose a function of the form $r = a \sin n\theta$ or $r = a \cos n\theta$ then choose your own values of a and n. Before you animate the graph, try using the six step process for sketching polar equations of the form $r = a \sin n\theta$ or $r = a \cos n\theta$. After you have had a chance to interact with the Guided Visualization, work through Example 12.

 Graphs of polar equations of the form $r = a \sin n\theta$ or $r = a \cos n\theta$

Example 12 Sketching the Graph of a Polar Equation of the Form $r = a \sin n\theta$ and $r = a \cos n\theta$

Sketch the graph of each polar equation.

a. $r = -4 \cos 3\theta$ **b.** $r = -2 \sin 5\theta$ **c.** $r = 5 \cos 4\theta$

Solution

a. To sketch the graph of $r = -4 \cos 3\theta$, we follow the five-step process for sketching roses.

Step 1. For the equation $r = -4 \cos 3\theta$, $n = 3$, which is odd. So there are 3 petals.

Step 2. The length of each petal is $|a| = |-4| = 4$ units.

Step 3. Solve the equations $\cos 3\theta = 1$ and $\cos 3\theta = -1$ to determine all angles where an endpoint of a petal lies. View these **steps** to verify that the solutions to the equation $\cos 3\theta = 1$ on the interval $[0, 2\pi)$ are $\theta = 0, \theta = \frac{2\pi}{3}$ and $\theta = \frac{4\pi}{3}$. View these **steps** to verify that the solutions to the equation $\cos 3\theta = -1$ on the interval $[0, 2\pi)$ are $\theta = \frac{\pi}{3}, \theta = \pi$, and $\theta = \frac{5\pi}{3}$.

Step 4. Although there are only three petals, we found six values of θ in step 3. This is because when n is odd, we obtain a complete graph on the interval $[0, \pi)$. The graph will completely traverse itself on the interval $[\pi, 2\pi)$.

For this reason, we choose the three values of θ from step 3 that lie on the interval $[0, \pi)$. These values are $\theta = 0, \theta = \frac{\pi}{3}$, and $\theta = \frac{2\pi}{3}$.

Thus, the coordinates of the three endpoints of the petals are $(-4, 0)$, $\left(4, \frac{\pi}{3}\right)$, and $\left(-4, \frac{2\pi}{3}\right)$. See Figure 36.

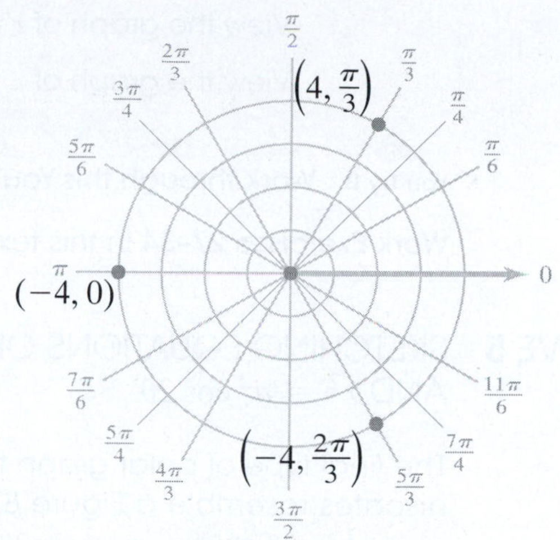

Figure 36 The endpoints of the 3 petals of the graph of $r = -4\cos 3\theta$.

Step 5. The graph passes through the pole when $\cos 3\theta = 0$. View these **steps** to verify that these angles on the interval $[0, \pi)$ are $\theta = \frac{\pi}{6}, \theta = \frac{\pi}{2}$, and $\theta = \frac{5\pi}{6}$. These three angles can be used as a guide when determining the width of each petal.

Step 6. We now draw the three petals of the rose, making certain that the graph never intersects the lines $\theta = \frac{\pi}{6}, \theta = \frac{\pi}{2}$, and $\theta = \frac{5\pi}{6}$ except at the pole. See the complete graph in Figure 37.

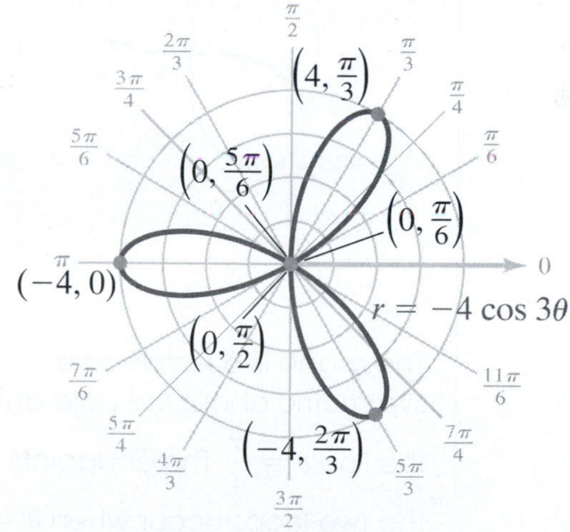

Figure 37 The complete graph of $r = -4\cos 3\theta$.

For parts b and c, carefully work through this **interactive video** to see how to sketch each graph or click on the corresponding link below to see if your graph is correct.

(eText Screens 10.2-1–10.2-50)

View the **graph** of $r = -2 \sin 5\theta$.

View the **graph** of $r = 5 \cos 4\theta$.

You Try It **Work through this You Try It problem.**

Work Exercises 27–34 in this textbook or in the MyLab Math Study Plan.

OBJECTIVE 5 SKETCHING EQUATIONS OF THE FORM $r^2 = a^2 \sin 2\theta$ AND $r^2 = a^2 \cos 2\theta$

The final type of polar graph that we will consider is called a **lemniscate.** Lemniscates resemble a Figure 8, or a propeller, and have equations of the form $r^2 = a^2 \sin 2\theta$ and $r^2 = a^2 \cos 2\theta$. The endpoints of the two loops of lemniscates of the form $r^2 = a^2 \sin 2\theta$ occur when $\theta = \frac{\pi}{4}$ and $\theta = \frac{5\pi}{4}$, whereas the endpoints of the two loops of lemniscates of the form $r^2 = a^2 \cos 2\theta$ occur when $\theta = 0$ and $\theta = \pi$. The length of each loop of a lemniscate is $|a|$ units. The two types of lemniscates are shown on the following page.

Graphs of Polar Equations of the Form $r^2 = a^2 \sin 2\theta$, and $r^2 = a^2 \cos 2\theta$, where $a \neq 0$ Is a Constant

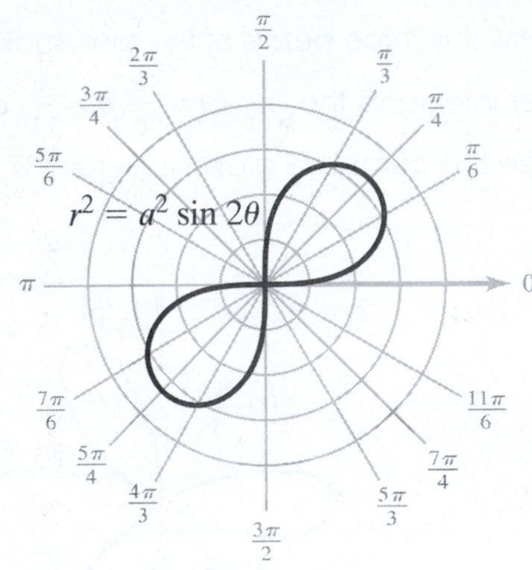

The graph is a lemniscate symmetric about the pole and the line $\theta = \frac{\pi}{4}$. The endpoints of the two loops occur when $\theta = \frac{\pi}{4}$ and $\theta = \frac{5\pi}{4}$.

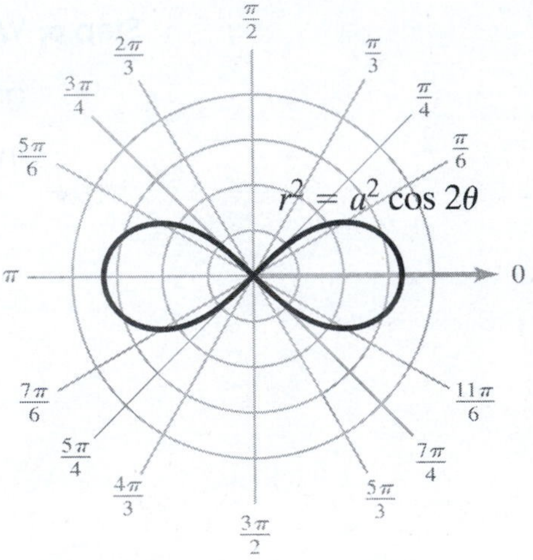

The graph is a lemniscate symmetric about the pole, the horizontal line $\theta = 0$, and the vertical line $\theta = \frac{\pi}{2}$. The endpoints of the two loops occur when $\theta = 0$ and $\theta = \pi$.

Example 13 Sketching the Graph of a Polar Equation of the Form $r^2 = a^2 \sin 2\theta$ and $r^2 = a^2 \cos 2\theta$

Sketch the graph of each polar equation.

a. $r^2 = 9 \cos 2\theta$ **b.** $r^2 = 16 \sin 2\theta$

Solution

a. The equation is of the form $r^2 = a^2 \cos 2\theta$, where $a^2 = 9$. So the graph is a lemniscate. The length of each loop is 3 units. The graph is symmetric about the pole, the horizontal line $\theta = 0$, and the vertical line $\theta = \frac{\pi}{2}$. See Figure 38.

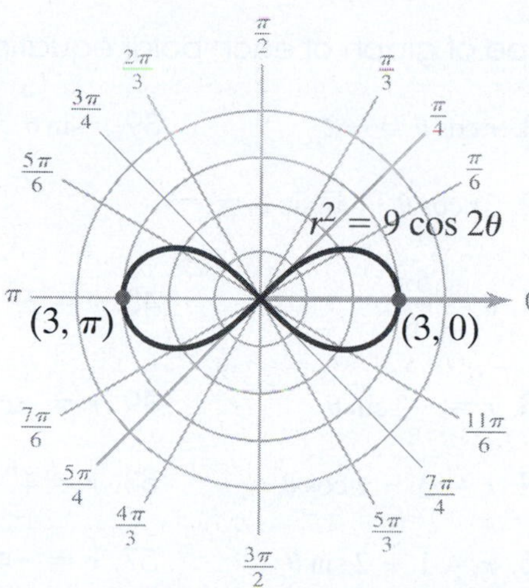

Figure 38 The complete graph of $r^2 = 9\cos 2\theta$.

b. Try sketching the graph of the equation $r^2 = 16 \sin 2\theta$ on your own. When you have finished sketching the graph, view the completed graph to see if you are correct or watch this video.

You Try It Work through this You Try It problem.

Work Exercises 35 and 36 in this textbook or in the MyLab Math Study Plan.

10.2 Exercises

Skill Check Exercises

For exercise SCE-1, simplify the expression $\frac{\pi}{4} + \frac{2\pi}{3}k$ for $k = -3, -2, -1, 0, 1,$ and 2.

In Exercises 1–36, sketch the graph of the polar equation.

SbS 1. $r\cos\theta = 5$ SbS 2. $r\cos\theta = -2$ SbS 3. $r\sin\theta = 3$ SbS 4. $r\sin\theta = -4$

SbS 5. $2r\cos\theta + 5r\sin\theta = 10$ SbS 6. $3r\cos\theta - 4r\sin\theta = -5$

SbS 7. $\theta = -\frac{\pi}{4}$ SbS 8. $\theta = \frac{5\pi}{6}$ 9. $r = 3$ 10. $r = -2$

SbS 11. $r = 2\cos\theta$ SbS 12. $r = -3\sin\theta$ 13. $r = -\cos\theta$ 14. $r = 5\sin\theta$

SbS 15. $r = 1 + \sin\theta$ SbS 16. $r = 3 - 4\cos\theta$ SbS 17. $r = 4 + 3\sin\theta$ SbS 18. $r = 3 - 3\cos\theta$

SbS 19. $r = 4 + 2\sin\theta$ SbS 20. $r = 1 + 2\sin\theta$ SbS 21. $r = -4 - 4\sin\theta$ SbS 22. $r = 5 - 3\cos\theta$

SbS 23. $r = -2 + 5\cos\theta$ SbS 24. $r = -2 + 2\cos\theta$ SbS 25. $r = -2 - 3\sin\theta$ SbS 26. $r = 3 - \cos\theta$

SbS 27. $r = -4\cos 3\theta$ 28. $r = 2\sin 2\theta$ SbS 29. $r = 5\cos 4\theta$ SbS 30. $r = -3\sin 5\theta$

SbS 31. $r = 5\sin 3\theta$ SbS 32. $r = 3\cos 5\theta$ SbS 33. $r = -4\sin 4\theta$ SbS 34. $r = -2\cos 2\theta$

SbS 35. $r^2 = 16\sin 2\theta$ SbS 36. $r^2 = 9\cos 2\theta$

Brief Exercises

In Exercises 37–72, identify the type of graph of each polar equation.

37. $r\cos\theta = 5$ 38. $r\cos\theta = -2$ 39. $r\sin\theta = 3$ 40. $r\sin\theta = -4$

41. $2r\cos\theta + 5r\sin\theta = 10$ 42. $3r\cos\theta - 4r\sin\theta = -5$

43. $\theta = -\frac{\pi}{4}$ 44. $\theta = \frac{5\pi}{6}$ 45. $r = 3$ 46. $r = -2$

47. $r = 2\cos\theta$ 48. $r = -3\sin\theta$ 49. $r = -\cos\theta$ 50. $r = 5\sin\theta$

51. $r = 1 + \sin\theta$ 52. $r = 3 - 4\cos\theta$ 53. $r = 4 + 3\sin\theta$ 54. $r = 3 - 3\cos\theta$

55. $r = 4 + 2\sin\theta$ 56. $r = 1 + 2\sin\theta$ 57. $r = -4 - 4\sin\theta$ 58. $r = 5 - 3\cos\theta$

59. $r = -2 + 5\cos\theta$ 60. $r = -2 + 2\cos\theta$ 61. $r = -2 - 3\sin\theta$ 62. $r = 3 - \cos\theta$

63. $r = -4\cos 3\theta$ 64. $r = 2\sin 2\theta$ 65. $r = 5\cos 4\theta$ 66. $r = -3\sin 5\theta$

67. $r = 5\sin 3\theta$ 68 $r = 3\cos 5\theta$ 69. $r = -4\sin 4\theta$ 70. $r = -2\cos 2\theta$

71. $r^2 = 16\sin 2\theta$ 72. $r^2 = 9\cos 2\theta$

In Exercises 73–108, sketch the graph of the polar equation.

73. $r\cos\theta = 5$ 74. $r\cos\theta = -2$ 75. $r\sin\theta = 3$ 76. $r\sin\theta = -4$

77. $2r\cos\theta + 5r\sin\theta = 10$ 78. $3r\cos\theta - 4r\sin\theta = -5$

79. $\theta = -\frac{\pi}{4}$ 80. $\theta = \frac{5\pi}{6}$ 81. $r = 3$ 82. $r = -2$

83. $r = 2\cos\theta$ 84. $r = -3\sin\theta$ 85. $r = -\cos\theta$ 86. $r = 5\sin\theta$

87. $r = 1 + \sin\theta$ 88. $r = 3 - 4\cos\theta$ 89. $r = 4 + 3\sin\theta$ 90. $r = 3 - 3\cos\theta$

91. $r = 4 + 2\sin\theta$ 92. $r = 1 + 2\sin\theta$ 93. $r = -4 - 4\sin\theta$ 94. $r = 5 - 3\cos\theta$

95. $r = -2 + 5\cos\theta$ 96. $r = -2 + 2\cos\theta$ 97. $r = -2 - 3\sin\theta$ 98. $r = 3 - \cos\theta$

99. $r = -4\cos 3\theta$ 100. $r = 2\sin 2\theta$ 101. $r = 5\cos 4\theta$ 102. $r = -3\sin 5\theta$

103. $r = 5\sin 3\theta$ 104. $r = 3\cos 5\theta$ 105. $r = -4\sin 4\theta$ 106. $r = -2\cos 2\theta$

107. $r^2 = 16\sin 2\theta$ 108. $r^2 = 9\cos 2\theta$

10.3 Complex Numbers in Polar Form; De Moivre's Theorem

THINGS TO KNOW

Before working through this section, be sure that you are familiar with the following concepts:

		VIDEO	ANIMATION	INTERACTIVE
 You Try It	1. Understanding the Four Families of Special Angles (Section 6.5)			
 You Try It	2. Understanding the Definitions of the Trigonometric Functions of General Angles (Section 6.5)			
 You Try It	3. Evaluating Trigonometric Functions of Angles Belonging to the $\frac{\pi}{3}$, $\frac{\pi}{6}$, or $\frac{\pi}{4}$ Families (Section 6.5) 			
 You Try It	4. Solving Trigonometric Equations That Are Linear in Form (Section 8.5) 			
 You Try It	5. Plotting Points Using Polar Coordinates (Section 10.1)			
 You Try It	6. Converting a Point from Rectangular Coordinates to Polar Coordinates (Section 10.1)			

OBJECTIVES

1 Understanding the Rectangular Form of a Complex Number

2 Understanding the Polar Form of a Complex Number

3 Converting a Complex Number from Polar Form to Rectangular Form

4 Converting a Complex Number from Rectangular Form to Standard Polar Form

5 Determining the Product or Quotient of Complex Numbers in Polar Form

6 Using De Moivre's Theorem to Raise a Complex Number to a Power

7 Using De Moivre's Theorem to Find the Roots of a Complex Number

SECTION 10.3 EXERCISES

OBJECTIVE 1 UNDERSTANDING THE RECTANGULAR FORM OF A COMPLEX NUMBER

Complex numbers are numbers of the form $a + bi$, where a and b are real numbers and i is the **imaginary unit**. We often use the variable z to denote a complex number. Complex numbers of the form $z = a + bi$ are said to be in **rectangular form.**

Definition The Rectangular Form of a Complex Number

A complex number is said to be in **rectangular form** if it is written as $z = a + bi$, where a and b are real numbers and i is the imaginary unit.

Complex numbers have no order. That is, there is no way to compare two complex numbers and determine if one is greater than the other. However, we can characterize complex numbers by representing them geometrically on a graph.

Every complex number can be represented by a point in the **complex plane.** The complex plane looks similar to a rectangular coordinate system. Just like in the rectangular coordinate system, the point of intersection of the two axes is called the **origin** of the complex plane. However, in the complex plane the horizontal axis is called the **real** axis and the vertical axis is called the **imaginary axis.** We plot the complex number $z = a + bi$ in the same way as we would plot the point (a, b) in the rectangular coordinate system. See **Figure 39**.

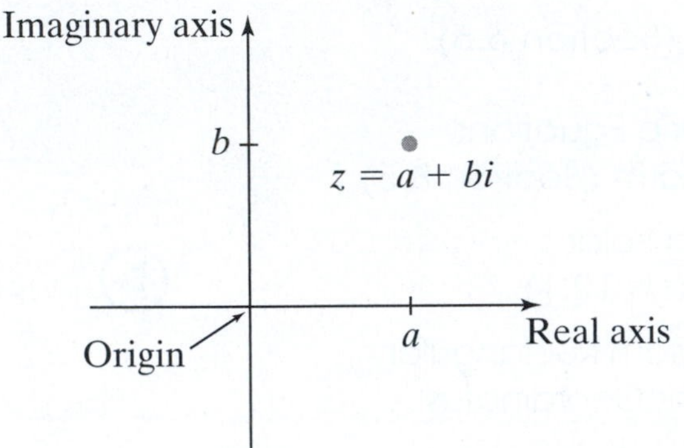

Figure 39 A point in the complex plane

Recall that the absolute value of a real number a is defined as its distance from 0 on a number line and is denoted as $|a|$. Similarly, **the absolute value of a complex number** $z = a + bi$ is the distance from the origin to the point z in the complex plane and is denoted by $|z|$. Therefore, we can use the **distance formula** to determine the absolute value of a complex number by finding the distance between the origin and the point (a, b) in a rectangular coordinate system.

Definition The Absolute Value of a Complex Number

The **absolute value of a complex number** $z = a + bi$ is denoted by $|z|$ and is the distance from the origin to z in the complex plane and is given by $|z| = \sqrt{a^2 + b^2}$.

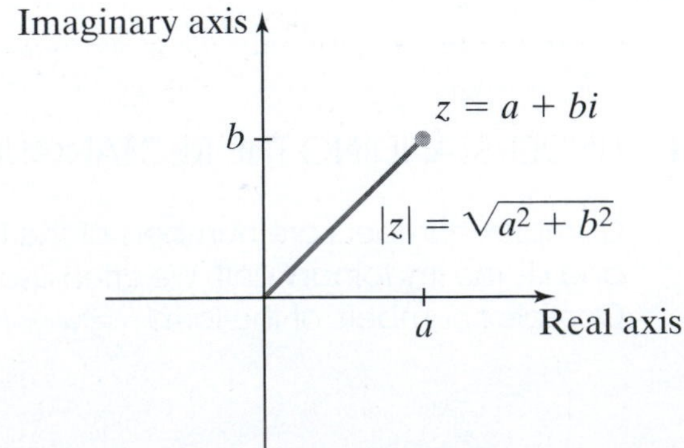

Example 1 Plotting Complex Numbers in the Complex Plane

Plot each complex number in the complex plane and determine its absolute value.

a. $z_1 = 3 - 4i$ b. $z_2 = -2 + 5i$ c. $z_3 = 3$ d. $z_4 = -2i$

Solution Each complex number is plotted in Figure 40.

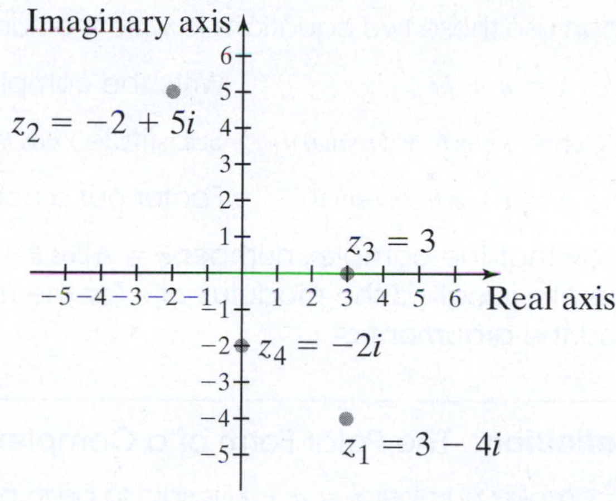

Figure 40

The absolute values of each number are as follows:

a. For $z_1 = 3 - 4i$, $a = 3$ and $b = -4$. Thus, $|z_1| = \sqrt{3^2 + (-4)^2} = \sqrt{9 + 16} = \sqrt{25} = 5$.

b. For $z_2 = -2 + 5i$, $a = -2$ and $b = 5$. Thus, $|z_2| = \sqrt{(-2)^2 + (5)^2} = \sqrt{4 + 25} = \sqrt{29}$.

c. For $z_3 = 3$, $a = 3$ and $b = 0$. Thus, $|z_3| = \sqrt{3^2 + 0^2} = \sqrt{9} = 3$.

d. For $z_4 = -2i$, $a = 0$ and $b = -2$. Thus, $|z_4| = \sqrt{0^2 + (-2)^2} = \sqrt{4} = 2$.

You Try It Work through this You Try It problem.

Work Exercises 1–4 in this textbook or in the MyLab Math Study Plan.

OBJECTIVE 2 UNDERSTANDING THE POLAR FORM OF A COMPLEX NUMBER

Suppose that we represent a complex number $z = a + bi$ using the polar coordinates (r, θ). Notice in Figure 41 that the absolute value of z, $|z| = \sqrt{a^2 + b^2}$, is the same as the value of r. Therefore, $|z| = r = \sqrt{a^2 + b^2}$.

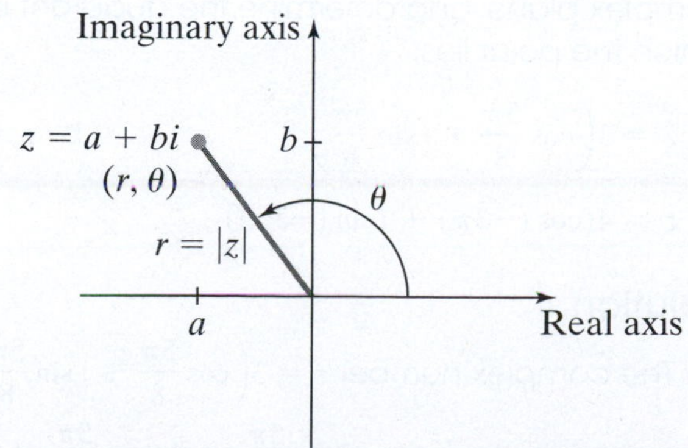

Figure 41

Considering the point sketched in Figure 41 and using the **general angle definitions of the trigonometric functions,** we know the following:

$$\cos\theta = \frac{a}{r},\ \sin\theta = \frac{b}{r},\ \text{and } \tan\theta = \frac{b}{a}$$

Multiplying both sides of the equations $\cos\theta = \frac{a}{r}$ and $\sin\theta = \frac{b}{r}$ by r gives $a = r\cos\theta$ and $b = r\sin\theta$.

We can use these two equations to write the number $z = a + bi$ in an alternate form.

$z = a + bi$	Write the complex number in rectangular form.
$z = r\cos\theta + (r\sin\theta)i$	Substitute $r\cos\theta$ for a and $r\cos\theta$ for b.
$z = r(\cos\theta + i\sin\theta)$	Factor out a common factor of r.

We say that the complex number $z = r(\cos\theta + i\sin\theta)$ is written in **polar form.** The value of r is called the **modulus** of z (or the magnitude of z) and the angle θ is called the **argument** of z.

Definition **The Polar Form of a Complex Number**

A complex number $z = a + bi$ is said to be in **polar form** if it is written as

$z = r(\cos\theta + i\sin\theta)$, where $a = r\cos\theta$, $b = r\sin\theta$, $\tan\theta = \frac{b}{a}$, and $r = \sqrt{a^2 + b^2}$.

Because of the periodic nature of sine and cosine, there are infinitely many representations of every complex number in polar form. A complex number $z = r(\cos\theta + i\sin\theta)$ is equivalent to a complex number of the form $z = r(\cos(\theta + 2\pi k) + i\sin(\theta + 2\pi k))$, where k is any integer. However, we will always write the polar form of a complex number with $0 \le \theta < 2\pi$. Complex numbers in polar form, where $0 \le \theta < 2\pi$ (or $0 \le \theta < 360°$), are said to be in **standard polar form.**

Definition **The Standard Polar Form of a Complex Number**

A complex number $z = r(\cos\theta + i\sin\theta)$ is said to be in **standard polar form** when $0 \le \theta < 2\pi$ (or $0 \le \theta < 360°$).

Example 2 Using the Standard Polar Form of a Complex Number

Rewrite the complex number in standard polar form, plot the number in the complex plane, and determine the quadrant in which the point lies or the axis on which the point lies.

a. $z = 3\left(\cos\frac{5\pi}{8} + i\sin\frac{5\pi}{8}\right)$

b. $z = 2\left(\cos\frac{23\pi}{4} + i\sin\frac{23\pi}{4}\right)$

c. $z = 4(\cos(-3\pi) + i\sin(-3\pi))$

Solution

a. The complex number $z = 3\left(\cos\frac{5\pi}{8} + i\sin\frac{5\pi}{8}\right)$ is already written in standard polar form because $\theta = \frac{5\pi}{8}$ and $0 \le \frac{5\pi}{8} < 2\pi$. To plot z, we plot a point that lies on the terminal side of $\theta = \frac{5\pi}{8}$ and is a distance of $r = 3$ units from the origin. We observe that this complex number lies in Quadrant II. See Figure 42.

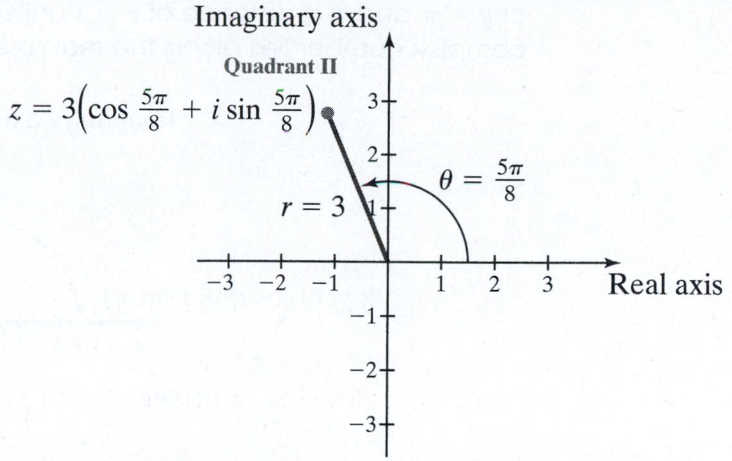

Figure 42 The complex number $z = 3\left(\cos \frac{5\pi}{8} + i \sin \frac{5\pi}{8}\right)$ is located in Quadrant II in the complex plane.

b. To write the complex number $z = 2\left(\cos \frac{23\pi}{4} + i \sin \frac{23\pi}{4}\right)$ in standard polar form, we must determine the angle on the interval $[0, 2\pi)$ that is coterminal with $\frac{23\pi}{4}$.

This angle is $\theta = \frac{23\pi}{4} - 4\pi = \frac{23\pi}{4} - \frac{16\pi}{4} = \frac{7\pi}{4}$. Therefore, the standard polar form is $z = 2\left(\cos \frac{7\pi}{4} + i \sin \frac{7\pi}{4}\right)$. To plot z, we plot a point that lies on the terminal side of $\theta = \frac{7\pi}{4}$ and is a distance of $r = 2$ units from the origin. We observe that this complex number lies in Quadrant IV. Figure 43.

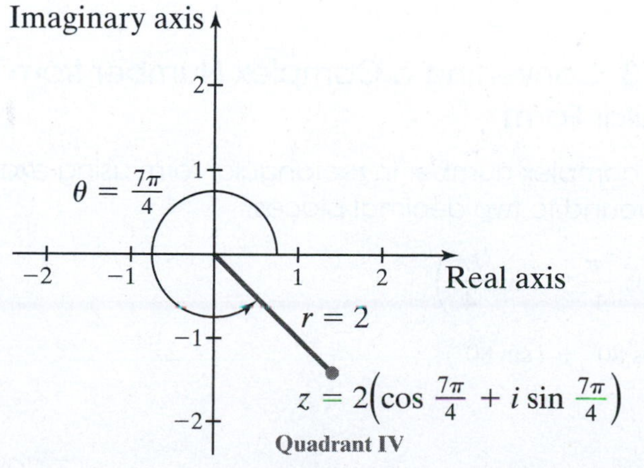

Figure 43 The complex number $z = 2\left(\cos \frac{23\pi}{4} + i \sin \frac{23\pi}{4}\right)$ written in standard polar form is $z = 2\left(\cos \frac{7\pi}{4} + i \sin \frac{7\pi}{4}\right)$ and is located in Quadrant IV in the complex plane.

c. To write the complex number $z = 4(\cos(-3\pi) + i\sin(-3\pi))$ in standard polar form, we must determine the angle on the interval $[0, 2\pi)$ that is coterminal with -3π. This angle is $\theta = -3\pi + 4\pi = \pi$. Therefore, the standard polar form is $z = 4(\cos\pi + i\sin\pi)$. To plot z, we plot a point that lies on the terminal side of $\theta = \pi$ and is a distance of $r = 4$ units from the origin. We observe that this complex number lies along the real axis. See Figure 44.

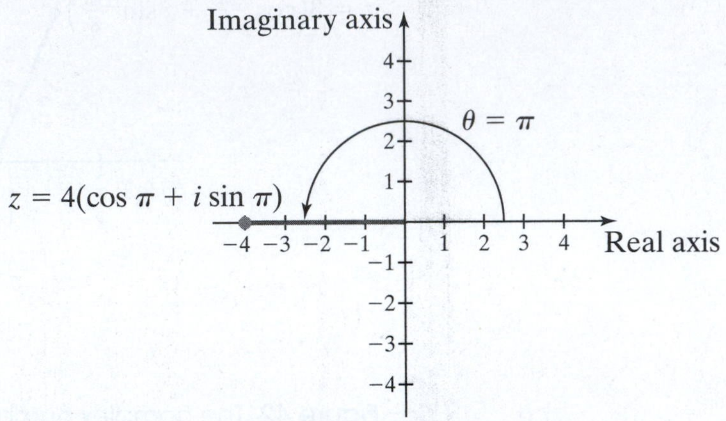

Figure 44 The complex number $z = 4(\cos(-3\pi) + i\sin(-3\pi))$ written in standard polar form is $z = 4(\cos\pi + i\sin\pi)$ and is located along the real axis in the complex plane.

You Try It Work through this You Try It problem.

Work Exercises 5–10 in this textbook or in the MyLab Math Study Plan.

OBJECTIVE 3 CONVERTING A COMPLEX NUMBER FROM POLAR FORM TO RECTANGULAR FORM

If we are given a complex number in polar form, it is very easy to convert the complex number into rectangular form. We simply evaluate $\cos\theta$ and $\sin\theta$ within the expression $z = r(\cos\theta + i\sin\theta)$.

Example 3 Converting a Complex Number from Polar Form to Rectangular Form

Write each complex number in rectangular form using exact values if possible. Otherwise, round to two decimal places.

a. $z = 3\left(\cos\frac{7\pi}{4} + i\sin\frac{7\pi}{4}\right)$

b. $z = 4(\cos 80° + i\sin 80°)$

Solution

a. To convert $z = 3\left(\cos\frac{7\pi}{4} + i\sin\frac{7\pi}{4}\right)$ to rectangular form, we must evaluate $\cos\frac{7\pi}{4}$ and $\sin\frac{7\pi}{4}$.

$z = 3\left(\cos\frac{7\pi}{4} + i\sin\frac{7\pi}{4}\right)$	Write the original complex number in polar form.
$z = 3\left(\frac{1}{\sqrt{2}} + i\left(-\frac{1}{\sqrt{2}}\right)\right)$	Evaluate $\cos\frac{7\pi}{4} = \frac{1}{\sqrt{2}}$ and $\sin\frac{7\pi}{4} = -\frac{1}{\sqrt{2}}$.
$z = \frac{3}{\sqrt{2}} - \frac{3}{\sqrt{2}}i$	Multiply.

Therefore, the rectangular form of the complex number $z = 3\left(\cos\frac{7\pi}{4} + i\sin\frac{7\pi}{4}\right)$ is $z = \frac{3}{\sqrt{2}} - \frac{3}{\sqrt{2}}i$. See Figure 45.

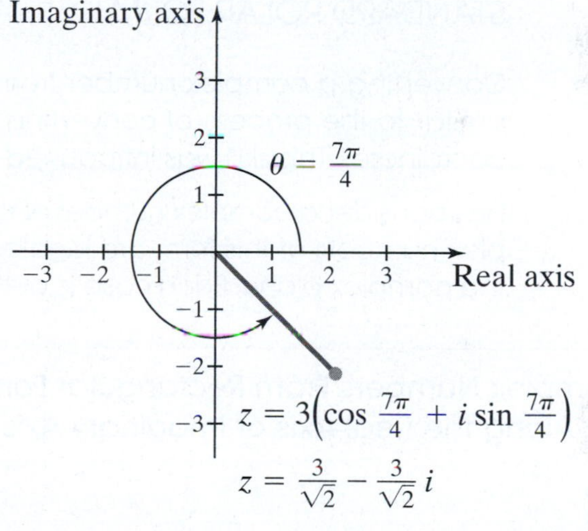

Figure 45 The rectangular form of $z = 3\left(\cos\frac{7\pi}{4} + i\sin\frac{7\pi}{4}\right)$ is $z = \frac{3}{\sqrt{2}} - \frac{3}{\sqrt{2}}i$.

b. To convert $z = 4(\cos 80° + i\sin 80°)$ to rectangular form, we observe that $\theta = 80°$ does not belong to one of the special angle families. Thus, we will use a calculator set in degree mode to approximate values.

$z = 4(\cos 80° + i\sin 80°)$	Write the original complex number in polar form.
$z = 4\cos 80° + 4i\sin 80°$	Multiply.
$z \approx 0.69 + 3.94i$	Approximate $4\cos 80° \approx 0.69$ and $4\sin 80° \approx 3.94$.

Therefore, the rectangular form of the complex number $z = 4(\cos 80° + i\sin 80°)$ is approximately $z \approx 0.69 + 3.94i$. See Figure 46.

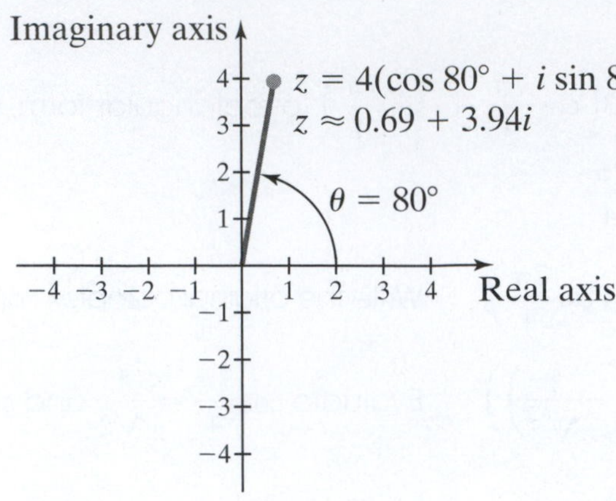

Figure 46 The rectangular form of $z = 4(\cos 80° + i \sin 80°)$ is approximately $z \approx 0.69 + 3.94i$. ●

You Try It Work through this You Try It problem.

Work Exercises 11–16 in this textbook or in the MyLab Math Study Plan.

OBJECTIVE 4 CONVERTING A COMPLEX NUMBER FROM RECTANGULAR FORM TO STANDARD POLAR FORM

Converting a complex number from rectangular form to standard polar form is similar to the process of converting a point in rectangular coordinates to polar coordinates. This skill was introduced in **Section 10.1**.

First consider a complex number of the form $z = a + bi$ where $a = 0$ or $b = 0$. Complex numbers of this form are located along the real axis or the imaginary axis in the complex plane. Each case is outlined below.

Converting Complex Numbers From Rectangular Form to Standard Polar Form for Complex Numbers Lying Along the Real Axis or Imaginary Axis

Imaginary axis
$z = a, a > 0$
$z = a(\cos 0 + i \sin 0)$
$\theta = 0$ a Real axis

The complex number $z = a, a > 0$ lies along the positive real axis so $\theta = 0$. The standard polar form is $z = a(\cos 0 + i \sin 0)$.

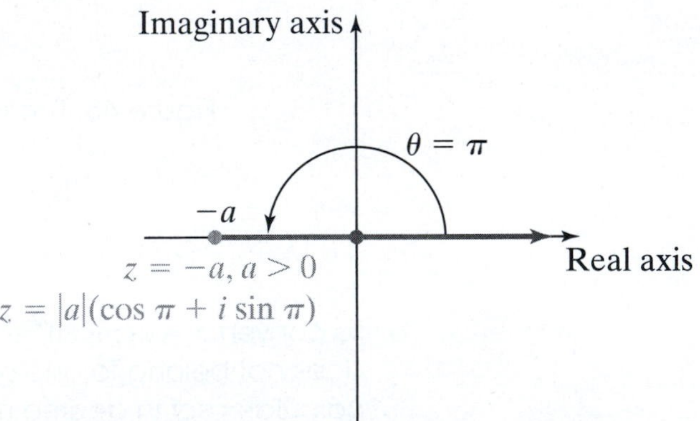

The complex number $z = -a, a > 0$ lies along the negative real axis so $\theta = \pi$. The standard polar form is $z = |a| (\cos \pi + i \sin \pi)$.

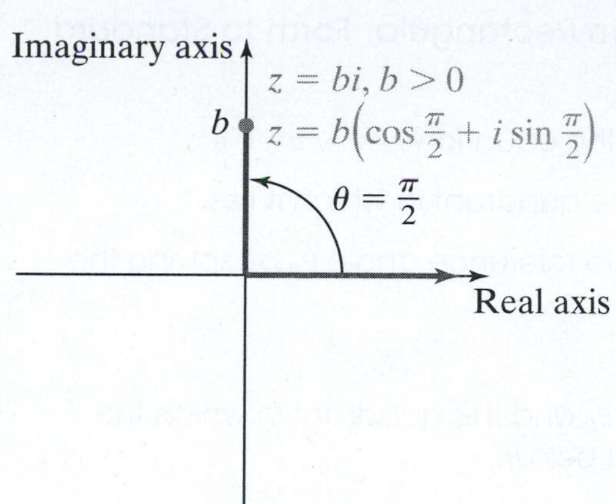

The complex number $z = bi, b > 0$ lies along the positive imaginary axis so $\theta = \frac{\pi}{2}$. The standard polar form is $z = b\left(\cos\frac{\pi}{2} + i\sin\frac{\pi}{2}\right)$.

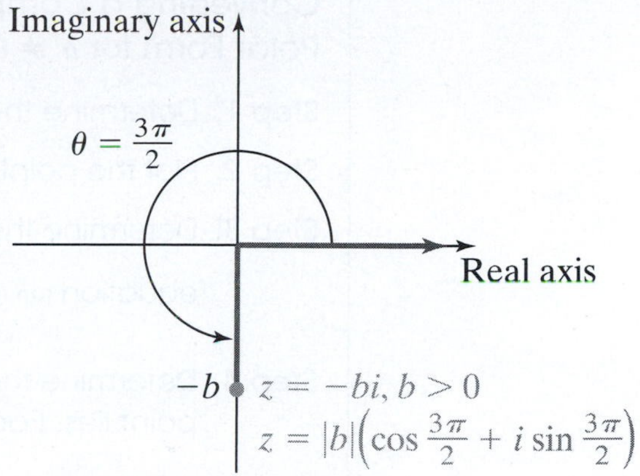

The complex number $z = -bi, b > 0$ lies along the negative imaginary axis so $\theta = \frac{3\pi}{2}$. The standard polar form is $z = |b|\left(\cos\frac{3\pi}{2} + i\sin\frac{3\pi}{2}\right)$.

Example 4 Converting a Complex Number from Rectangular Form to Standard Polar Form

Determine the standard polar form of each complex number. Write the argument using radians.

a. $z = 5$ **b.** $z = -3i$ **c.** $z = -\sqrt{7}$ **d.** $z = \frac{7}{2}i$

Solution Work through the video to verify the following:

a. The standard polar form of $z = 5$ is $z = 5(\cos 0 + i\sin 0)$.

b. The standard polar form of $z = -3i$ is $z = 3\left(\cos\frac{3\pi}{2} + i\sin\frac{3\pi}{2}\right)$.

c. The standard polar form of $z = -\sqrt{7}$ is $z = \sqrt{7}\,(\cos\pi + i\sin\pi)$.

d. The standard polar form of $z = \frac{7}{2}i$ is $z = \frac{7}{2}\left(\cos\frac{\pi}{2} + i\sin\frac{\pi}{2}\right)$.

In order to convert a complex number $z = a + bi$ into the form $z = r(\cos\theta + i\sin\theta)$, for $a \neq 0$ and $b \neq 0$, we must determine the values of r and θ. Determining the value of the modulus r is straightforward because we know that $r = \sqrt{a^2 + b^2}$. Determining the value of θ is much more involved. In order to find θ, we first have to find the reference angle θ_R. Recall that the reference angle is an acute angle. We know that the tangent of any acute angle is a positive value. Therefore, using the fact that $\tan\theta = \frac{b}{a}$, we can find θ_R by solving the equation $\tan\theta_R = \left|\frac{b}{a}\right|$. The appropriate value of θ depends on the quadrant in which the complex number $z = a + bi$ lies in the complex plane.

Converting a Complex Number from Rectangular Form to Standard Polar Form for $a \neq 0$ and $b \neq 0$

Step 1. Determine the value of r using the equation $r = \sqrt{a^2 + b^2}$.

Step 2. Plot the point and determine the quadrant in which it lies.

Step 3. Determine the value of the acute reference angle θ_R by solving the equation $\tan\theta_R = \left|\frac{b}{a}\right|$.

Step 4. Determine the value of θ using θ_R and the quadrant in which the point lies. Each case is outlined below.

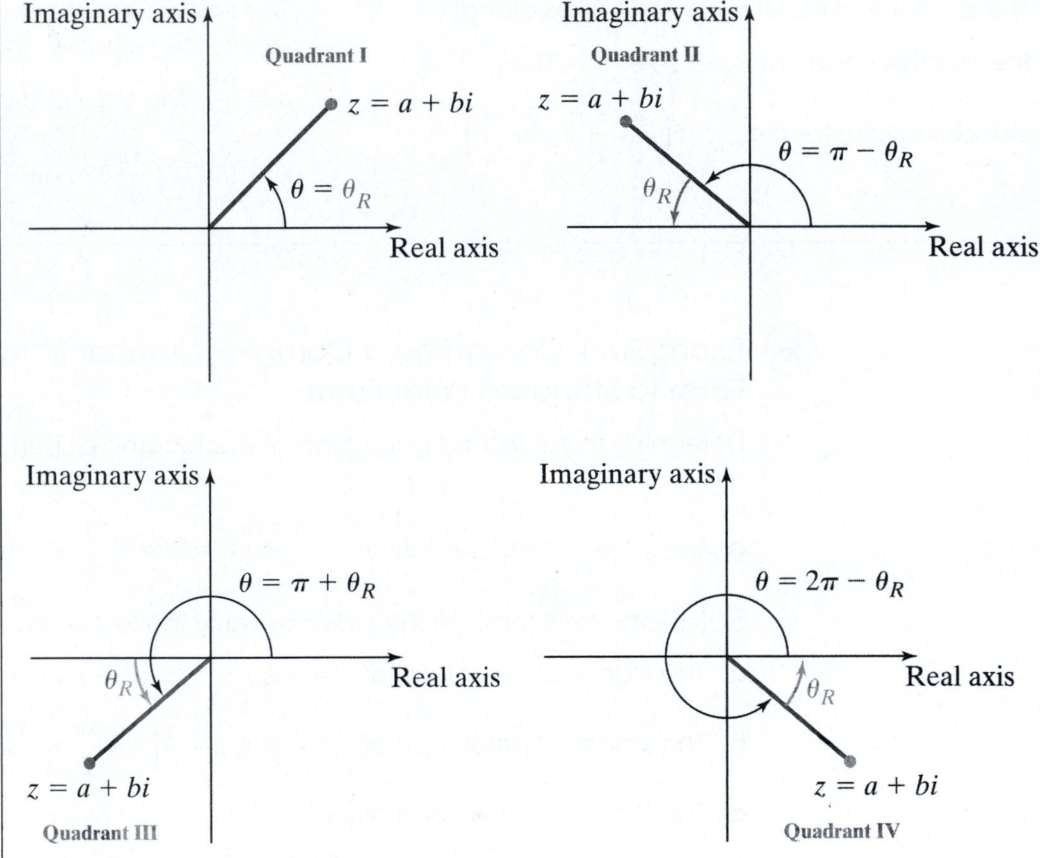

Example 5 Converting a Complex Number from Rectangular Form to Standard Polar Form

Determine the standard polar form of each complex number. Write the argument in radians using exact values if possible. Otherwise, round the argument to two decimal places.

a. $z = -2\sqrt{3} + 2i$

b. $z = 4 - 3i$

Solution

a. Step 1. To find the value of r, use the equation $r = \sqrt{a^2 + b^2}$.

$r = \sqrt{a^2 + b^2}$	Write the equation relating r, a, and b.
$r = \sqrt{(-2\sqrt{3})^2 + (2)^2}$	Substitute $a = -2\sqrt{3}$ and $b = 2$.
$r = \sqrt{12 + 4}$	Square each term.
$r = \sqrt{16} = 4$	Simplify.

Step 2. Plot the complex number $z = -2\sqrt{3} + 2i$ and recognize that it lies in Quadrant II, a distance of 4 units from the origin. See Figure 47.

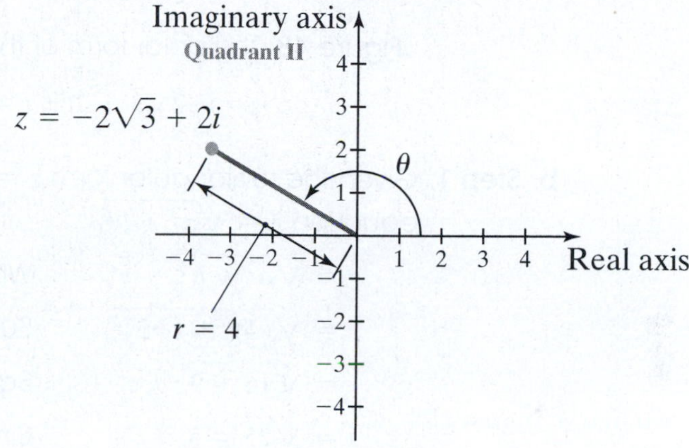

Figure 47 The complex number $z = -2\sqrt{3} + 2i$ lies in Quadrant II, a distance of four units from the origin.

Step 3. Given the rectangular form $z = -2\sqrt{3} + 2i$, to determine θ we first find the value of θ_R by solving the equation $\tan\theta_R = \left|\frac{b}{a}\right|$.

$\tan\theta_R = \left\lvert\frac{b}{a}\right\rvert$	Write the equation relating θ_R, a, and b.
$\tan\theta_R = \left\lvert\frac{2}{-2\sqrt{3}}\right\rvert$	Substitute $a = -2\sqrt{3}$ and $b = 2$.
$\tan\theta_R = \frac{1}{\sqrt{3}}$	Simplify.

The acute angle whose tangent is $\frac{1}{\sqrt{3}}$ is $\theta_R = \frac{\pi}{6}$.

Step 4. Because the point lies in Quadrant II, we know that $\theta = \pi - \theta_R$. Thus, $\theta = \pi - \frac{\pi}{6} = \frac{5\pi}{6}$.

Therefore, the standard polar form of the complex number is $z = -2\sqrt{3} + 2i$ is $z = 4\left(\cos\frac{5\pi}{6} + i\sin\frac{5\pi}{6}\right)$. See Figure 48. Watch this **interactive video** to see the worked-out solution.

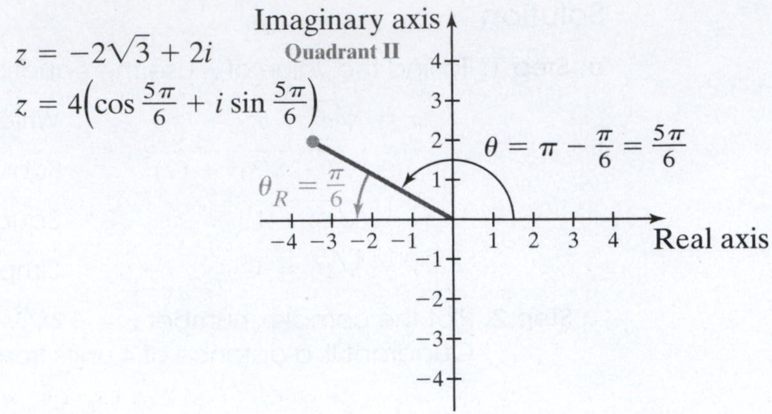

Figure 48 The polar form of the complex number $z = -2\sqrt{3} + 2i$ is $z = 4\left(\cos\frac{5\pi}{6} + i\sin\frac{5\pi}{6}\right)$.

b. Step 1. Given the rectangular form $z = 4 - 3i$, find the value of r using the equation $r = \sqrt{a^2 + b^2}$.

$r = \sqrt{a^2 + b^2}$ Write the equation relating r, a, and b.

$r = \sqrt{(4)^2 + (-3)^2}$ Substitute $a = 4$ and $b = -3$.

$r = \sqrt{16 + 9}$ Square each term.

$r = \sqrt{25} = 5$ Simplify.

Step 2. Plot the complex number $z = 4 - 3i$ and recognize that it lies in Quadrant IV, a distance of 5 units from the origin. See Figure 49.

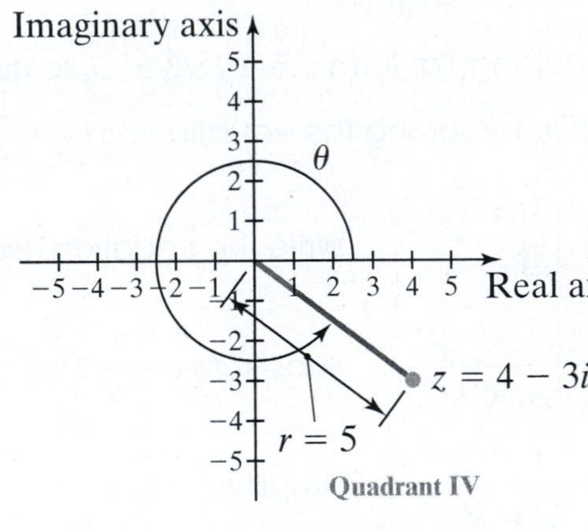

Figure 49 The complex number $z = 4 - 3i$ lies in Quadrant IV, a distance of five units from the origin.

Step 3. Given the rectangular form $z = 4 - 3i$, to determine θ we first find the value of θ_R by solving the equation $\tan\theta_R = \left|\frac{b}{a}\right|$.

$\tan\theta_R = \left|\frac{b}{a}\right|$ Write the equation relating θ_R, a, and b.

$\tan\theta_R = \left|\frac{-3}{4}\right|$ Substitute $a = 4$ and $b = -3$.

$\tan\theta_R = \frac{3}{4}$ Simplify.

The angle whose tangent is $\frac{3}{4}$ is not an angle that is a member of one of the special families of angles. However, using the inverse tangent function, we get $\theta_R = \tan^{-1}\left(\frac{3}{4}\right)$.

Step 4. The complex number $z = 4 - 3i$ lies in Quadrant IV, which indicates that $\theta = 2\pi - \theta_R$. Therefore, $\theta = 2\pi - \tan^{-1}\left(\frac{3}{4}\right)$. Using a calculator set in radian mode, we get $\theta = 2\pi - \tan^{-1}\left(\frac{3}{4}\right) \approx 5.64$.

Therefore, using $r = 5$ and an approximate value of $\theta \approx 5.64$, we see that the standard polar form of the complex number $z = 4 - 3i$ is approximately $z = 4(\cos 5.64 + i \sin 5.64)$. See Figure 50. Watch this interactive video to see the worked-out solution.

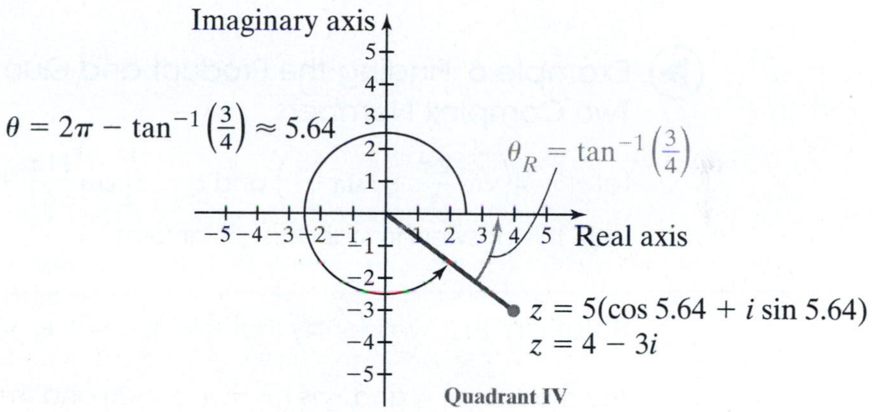

Figure 50 The standard polar form of the complex number $z = 4 - 3i$ is approximately $z = 5(\cos 5.64 + i \sin 5.64)$.

You Try It Work through this You Try It problem.

Work Exercises 17–23 in this textbook or in the MyLab Math Study Plan.

OBJECTIVE 5 DETERMINING THE PRODUCT OR QUOTIENT OF COMPLEX NUMBERS IN POLAR FORM

Given two complex numbers of the form $z_1 = r_1(\cos \theta_1 + i \sin \theta_1)$ and $z_2 = r_2(\cos \theta_2 + i \sin \theta_2)$, we wish to determine a formula for the product $z_1 z_2$ and the quotient $\frac{z_1}{z_2}$. To derive these formulas, we must recall the sum and difference formulas for sine and cosine that were first introduced in Section 8.2. Watch this video to see how to derive the product and quotient of complex numbers whose formulas are given below.

The Product and Quotient of Two Complex Numbers Written in Polar Form

If $z_1 = r_1(\cos\theta_1 + i\sin\theta_1)$ and $z_2 = r_2(\cos\theta_2 + i\sin\theta_2)$ are two complex numbers, then

$$z_1z_2 = r_1r_2(\cos(\theta_1 + \theta_2) + i\sin(\theta_1 + \theta_2))$$

and

$$\frac{z_1}{z_2} = \frac{r_1}{r_2}[\cos(\theta_1 - \theta_2) + i\sin(\theta_1 - \theta_2)]$$

Simply stated, we can say that to find the product of complex numbers we "multiply the r's and add the angles." Similarly, to divide complex numbers we "divide the r's and subtract the angles."

Example 6 Finding the Product and Quotient of Two Complex Numbers

Let $z_1 = 4\left(\cos\frac{2\pi}{3} + i\sin\frac{2\pi}{3}\right)$ and $z_2 = 5\left(\cos\frac{11\pi}{6} + i\sin\frac{11\pi}{6}\right)$. Find z_1z_2 and $\frac{z_1}{z_2}$ and write the answers in standard polar form.

Solution First, we identify that $r_1 = 4$, $r_2 = 5$, $\theta_1 = \frac{2\pi}{3}$, and $\theta_2 = \frac{11\pi}{6}$.

The product of r_1 and r_2 is $r_1r_2 = 4\cdot 5 = 20$ and the sum of θ_1 and θ_2 is

$$\theta_1 + \theta_2 = \frac{2\pi}{3} + \frac{11\pi}{6} = \frac{15\pi}{6} = \frac{5\pi}{2}.$$

To write the answer in standard polar form, we find the angle on the interval $[0, 2\pi)$ that is coterminal with $\frac{5\pi}{2}$ is $\frac{\pi}{2}$. Thus, the product written in standard polar form is

$$z_1z_2 = 20\left(\cos\frac{\pi}{2} + i\sin\frac{\pi}{2}\right).$$

The quotient of r_1 and r_2 is $\frac{r_1}{r_2} = \frac{4}{5}$ and the difference of θ_1 and θ_2 is

$$\theta_1 - \theta_2 = \frac{2\pi}{3} - \frac{11\pi}{6} = -\frac{7\pi}{6}.$$

The angle on the interval $[0, 2\pi)$ that is coterminal with $-\frac{7\pi}{6}$ is $\frac{5\pi}{6}$.

Therefore, the quotient written in standard polar form is $\frac{z_1}{z_2} = \frac{4}{5}\left(\cos\frac{5\pi}{6} + i\sin\frac{5\pi}{6}\right)$.

Note that because the arguments of the product and quotient are both angles that belong to a special family of angles, it is easy to determine the corresponding rectangular form. We could have rewritten the product and quotient as

$$z_1z_2 = 20\left(\cos\frac{\pi}{2} + i\sin\frac{\pi}{2}\right) = 20(0 + i(1)) = 20i$$

and

$$\frac{z_1}{z_2} = \frac{4}{5}\left(\cos\frac{5\pi}{6} + i\sin\frac{5\pi}{6}\right) = \frac{4}{5}\left(-\frac{\sqrt{3}}{2} + i\left(\frac{1}{2}\right)\right) = -\frac{2\sqrt{3}}{5} + \frac{2}{5}i.$$

You Try It Work through this You Try It problem.

Work Exercises 24–27 in this textbook or in the MyLab Math Study Plan.

OBJECTIVE 6 USING DE MOIVRE'S THEOREM TO RAISE A COMPLEX NUMBER TO A POWER

If $z = r(\cos\theta + i\sin\theta)$ is a complex number, then we can use the product of two complex numbers formula to find z^2.

$z^2 = z \cdot z = r(\cos\theta + i\sin\theta) \cdot r(\cos\theta + i\sin\theta)$ — Write z^2 as the product of $z = r(\cos\theta + i\sin\theta)$ and $z = r(\cos\theta + i\sin\theta)$.

$= r \cdot r(\cos(\theta + \theta) + i\sin(\theta + \theta))$ — Use the formula for the product of complex numbers written in polar form.

$z^2 = r^2(\cos 2\theta + i\sin 2\theta)$ — Write $r \cdot r$ as r^2 and $\theta + \theta$ as 2θ.

We can use this result to find z^3.

$z^3 = z^2 \cdot z = r^2(\cos 2\theta + i\sin 2\theta) \cdot r(\cos\theta + i\sin\theta)$ — Write z^3 as the product of $z^2 = r^2(\cos 2\theta + i\sin 2\theta)$ and $z = r(\cos\theta + i\sin\theta)$.

$= r^2 \cdot r(\cos(2\theta + \theta) + i\sin(2\theta + \theta))$ — Use the formula for the product of complex numbers written in polar form.

$z^3 = r^3(\cos 3\theta + i\sin 3\theta)$ — Write $r^2 \cdot r$ as r^3 and $2\theta + \theta$ as 3θ.

We can use this same procedure to get

$$z^4 = r^4(\cos 4\theta + i\sin 4\theta) \text{ and } z^5 = r^5(\cos 5\theta + i\sin 5\theta).$$

In general, if $z = r(\cos\theta + i\sin\theta)$ is a complex number and if n is a positive integer, then $z^n = r^n(\cos n\theta + i\sin n\theta)$. This formula for the nth power of a complex number is known as **De Moivre's Theorem,** named after the French mathematician Abraham de Moivre.

De Moivre's Theorem for Finding Powers of a Complex Number

If $z = r(\cos\theta + i\sin\theta)$ is a complex number written in polar form and if n is a positive integer, then z^n is given by the formula

$$z^n = [r(\cos\theta + i\sin\theta)]^n = r^n(\cos n\theta + i\sin n\theta).$$

Example 7 Using De Moivre's Theorem to Raise a Complex Number to a Power

a. Find $\left[5\left(\cos\frac{3\pi}{4} + i\sin\frac{3\pi}{4}\right)\right]^3$ and write your answer in standard polar form.

b. Find $(\sqrt{3} - i)^4$ and write your answer in rectangular form.

Solution

a. Apply De Moivre's Theorem with $r = 5, \theta = \frac{3\pi}{4}$, and $n = 3$.

$5\left[\left(\cos\frac{3\pi}{4} + i\sin\frac{3\pi}{4}\right)\right]^3$ Write the original expression.

$= 5^3\left(\cos\left(3\cdot\frac{3\pi}{4}\right) + i\sin\left(3\cdot\frac{3\pi}{4}\right)\right)$ Use De Moivre's Theorem with $r = 5, \theta = \frac{3\pi}{4}$, and $n = 3$.

$= 125\left(\cos\frac{9\pi}{4} + i\sin\frac{9\pi}{4}\right)$ Multiply.

$= 125\left(\cos\frac{\pi}{4} + i\sin\frac{\pi}{4}\right)$ Substitute the angle $\frac{\pi}{4}$ for its coterminal angle $\frac{9\pi}{4}$ to write in standard polar form.

b. To evaluate the expression $(\sqrt{3} - i)^4$, first write the complex number $z = \sqrt{3} - i$ in standard polar form. View these **steps** to verify that the standard polar form of

$$z = \sqrt{3} - i \text{ is } z = 2\left(\cos\frac{11\pi}{6} + i\sin\frac{11\pi}{6}\right).$$

Next, apply De Moivre's Theorem with $r = 2, \theta = \frac{11\pi}{6}$, and $n = 4$.

$\left[2\left(\cos\frac{11\pi}{6} + i\sin\frac{11\pi}{6}\right)\right]^4$ Write the original expression.

$= 2^4\left(\cos\left(4\cdot\frac{11\pi}{6}\right) + i\sin\left(4\cdot\frac{11\pi}{6}\right)\right)$ Use De Moivre's Theorem with $r = 2, \theta = \frac{11\pi}{6}$, and $n = 4$.

$= 16\left(\cos\frac{44\pi}{6} + i\sin\frac{44\pi}{6}\right)$ Multiply.

$= 16\left(\cos\frac{22\pi}{3} + i\sin\frac{22\pi}{3}\right)$ Simplify: $\frac{44\pi}{6} = \frac{22\pi}{3}$.

$= 16\left(\cos\frac{4\pi}{3} + i\sin\frac{4\pi}{3}\right)$ Substitute the angle $\frac{4\pi}{3}$ for its coterminal angle $\frac{22\pi}{3}$ to write in standard polar form.

$= 16\left(-\frac{1}{2} + i\left(-\frac{\sqrt{3}}{2}\right)\right)$ Evaluate $\cos\frac{4\pi}{3}$ and $\sin\frac{4\pi}{3}$.

$= -8 - 8\sqrt{3}i$ Multiply.

You Try It Work through this You Try It problem.

Work Exercises 28–37 in this textbook or in the MyLab Math Study Plan.

OBJECTIVE 7 USING DE MOIVRE'S THEOREM TO FIND THE ROOTS OF A COMPLEX NUMBER

We now want to investigate the nth roots of a complex number. There are exactly n distinct nth roots of any nonzero complex number. For example, there are 2 distinct 2nd roots (square roots) of a complex number, there are 3 distinct 3rd roots (cube roots) of a complex number, there are 4 distinct 4th roots of a complex number, and so on.

To derive a formula for the nth roots of a complex number, start with De Moivre's Theorem for positive integer powers of n and substitute $\frac{1}{n}$ for n.

$$z^n = [r(\cos\theta + i\sin\theta)]^n = r^n(\cos n\theta + i\sin n\theta)$$

Write De Moivre's Theorem for integer powers of a complex number.

$$z^{\frac{1}{n}} = [r(\cos\theta + i\sin\theta)]^{\frac{1}{n}} = r^{\frac{1}{n}}\left(\cos\frac{1}{n}\theta + i\sin\frac{1}{n}\theta\right)$$

Substitute $\frac{1}{n}$ for n.

Writing $z^{\frac{1}{n}}$ as $\sqrt[n]{z}$ and $r^{\frac{1}{n}}$ as $\sqrt[n]{r}$, we get the formula $\sqrt[n]{z} = \sqrt[n]{r}\left(\cos\frac{1}{n}\theta + i\sin\frac{1}{n}\theta\right)$. However, this formula only determines *one* of the n distinct nth roots. To find all n roots, we start overusing the fact that $z = r(\cos\theta + i\sin\theta)$ is equivalent to

$$z = r(\cos(\theta + 2\pi k) + i\sin(\theta + 2pk)),\text{ where } k \text{ is any integer.}$$

$$z^n = [r(\cos(\theta + 2\pi k) + i\sin(\theta + 2\pi k))]^n = r^n(\cos(n(\theta + 2\pi k)) + i\sin(n(\theta + 2\pi k)))$$

Write De Moivre's Theorem for integer powers of a complex number using $\theta + 2\pi k$ as the argument.

$$z^{\frac{1}{n}} = [r(\cos(\theta + 2\pi k) + i\sin(\theta + 2\pi k))]^{\frac{1}{n}} = r^{\frac{1}{n}}\left[\cos\left(\frac{1}{n}(\theta + 2\pi k)\right) + i\sin\left(\frac{1}{n}(\theta + 2\pi k)\right)\right]$$

Substitute $\frac{1}{n}$ for n.

We can write this formula as $\sqrt[n]{z} = \sqrt[n]{r}\left[\cos\left(\frac{\theta + 2\pi k}{n}\right) + i\sin\left(\frac{\theta + 2\pi k}{n}\right)\right]$.

This formula determines the n distinct nth roots of a complex number for $k = 0, 1, 2, \cdots, (n - 1)$.

De Moivre's Theorem for Finding the *n*th Roots of a Complex Number

If $z = r(\cos\theta + i\sin\theta)$ is a complex number written in polar form and if n is a positive integer, then z has exactly n complex roots determined by the formula

$$\sqrt[n]{z} = \sqrt[n]{r}\left[\cos\left(\frac{\theta + 2\pi k}{n}\right) + i\sin\left(\frac{\theta + 2\pi k}{n}\right)\right]$$

or

$$\sqrt[n]{z} = \sqrt[n]{r}\left[\cos\left(\frac{\theta + (360°)k}{n}\right) + i\sin\left(\frac{\theta + (360°)k}{n}\right)\right]$$

for $k = 0, 1, 2, \cdots, (n - 1)$. We will label the n roots as $z_0, z_1, z_2, \cdots, z_{n-1}$.

Example 8 Finding Complex Roots of a Real Number

Find the complex cube roots of 8. Write your answers in rectangular form.

Solution First write 8 in polar form as $8 = 8(\cos 0 + i \sin 0)$.

Next, apply De Moivre's Theorem for finding nth roots with $r = 8, \theta = 0$, and $n = 3$.

$$\sqrt[3]{8} = \sqrt[3]{8(\cos 0 + i \sin 0)} = \sqrt[3]{8}\left[\cos\left(\frac{0 + 2\pi k}{3}\right) + i \sin\left(\frac{0 + 2\pi k}{3}\right)\right] = \sqrt[3]{8}\left[\cos\left(\frac{2\pi k}{3}\right) + i \sin\left(\frac{2\pi k}{3}\right)\right] \text{ for } k = 0, 1, 2.$$

$$\text{If } k = 0\text{: } z_0 = \sqrt[3]{8}\left[\cos\left(\frac{2\pi \cdot 0}{3}\right) + i \sin\left(\frac{2\pi \cdot 0}{3}\right)\right] = 2[\cos 0 + i \sin 0] = 2(1 + i \cdot 0) = 2$$

$$\text{If } k = 1\text{: } z_1 = \sqrt[3]{8}\left[\cos\left(\frac{2\pi \cdot 1}{3}\right) + i \sin\left(\frac{2\pi \cdot 1}{3}\right)\right] = 2\left[\cos\frac{2\pi}{3} + i \sin\frac{2\pi}{3}\right] = 2\left(-\frac{1}{2} + i \cdot \frac{\sqrt{3}}{2}\right) = -1 + \sqrt{3}i$$

$$\text{If } k = 2\text{: } z_2 = \sqrt[3]{8}\left[\cos\left(\frac{2\pi \cdot 2}{3}\right) + i \sin\left(\frac{2\pi \cdot 2}{3}\right)\right] = 2\left[\cos\frac{4\pi}{3} + i \sin\frac{4\pi}{3}\right] = 2\left(-\frac{1}{2} + i \cdot \left(-\frac{\sqrt{3}}{2}\right)\right) = -1 - \sqrt{3}i$$

If we plot the three distinct cube roots of 8, we notice that the three roots are equally spaced around a circle centered at the origin of radius 2. See Figure 51. Each successive root is located exactly $\frac{2\pi}{3}$ radians away from the previous root.

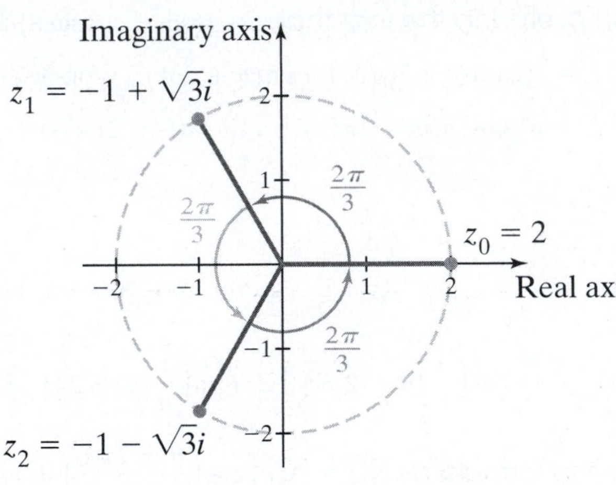

Figure 51 The 3 cube roots of 8 lie on the graph of a circle of radius 2 in the complex plane. Each successive root is located $\frac{2\pi}{3}$ radians from the previous root.

In Example 8 we saw that the three complex cube roots of 8 were all located on a circle of radius 2, where each root was exactly $\frac{2\pi}{3}$ radians from the previous root. See Figure 51. All complex nth roots will have this same graphical representation. **If we can find z_0, then we can easily determine the remaining $n - 1$ roots by adding $\frac{2\pi}{n}$ $\left(\text{or } \frac{360°}{n}\right)$ to each successive argument.** We demonstrate this in Example 9.

Example 9 Finding Complex Roots

a. Find the complex fourth roots of $z = 81\left(\cos\frac{3\pi}{5} + i \sin\frac{3\pi}{5}\right)$. Write your answers in standard polar form.

b. Find the complex square roots of $z = -2\sqrt{3} + 2i$. Write your answer in rectangular form with each part rounded to two decimal places.

Solution

a. We start by finding z_0 using De Moivre's Theorem for finding nth roots with $r = 81, \theta = \frac{3\pi}{5}$, and $n = 4$.

$$z_0 = \sqrt[4]{81}\left[\cos\left(\frac{\frac{3\pi}{5} + 2\pi \cdot 0}{4}\right) + i\sin\left(\frac{\frac{3\pi}{5} + 2\pi \cdot 0}{4}\right)\right] = 3\left[\cos\frac{3\pi}{20} + i\sin\frac{3\pi}{20}\right]$$

To find z_1, z_2, and z_3, we must add $\frac{2\pi}{4} = \frac{\pi}{2}$ to the argument of the previous root. Therefore, we get

$$z_1 = 3\left[\cos\left(\frac{3\pi}{20} + \frac{\pi}{2}\right) + i\sin\left(\frac{3\pi}{20} + \frac{\pi}{2}\right)\right] = 3\left[\cos\frac{13\pi}{20} + i\sin\frac{13\pi}{20}\right]$$

$$z_2 = 3\left[\cos\left(\frac{13\pi}{20} + \frac{\pi}{2}\right) + i\sin\left(\frac{13\pi}{20} + \frac{\pi}{2}\right)\right] = 3\left[\cos\frac{23\pi}{20} + i\sin\frac{23\pi}{20}\right]$$

$$z_3 = 3\left[\cos\left(\frac{23\pi}{20} + \frac{\pi}{2}\right) + i\sin\left(\frac{23\pi}{20} + \frac{\pi}{2}\right)\right] = 3\left[\cos\frac{33\pi}{20} + i\sin\frac{33\pi}{20}\right]$$

The four roots are shown in Figure 52. Notice that each successive root is located exactly $\frac{\pi}{2}$ radians away from the previous root in the complex plane.

Watch this **interactive video** to see the solution to this example.

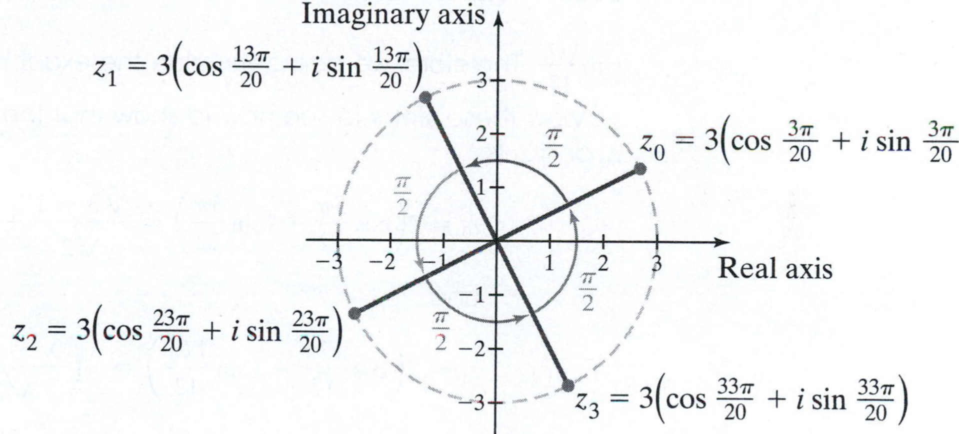

Figure 52 The 4 fourth roots of $z = 81\left(\cos\frac{3\pi}{5} + i\sin\frac{3\pi}{5}\right)$ lie on the graph of a circle of radius 3. Each successive root is located $\frac{2\pi}{4} = \frac{\pi}{2}$ radians from the previous root in the complex plane.

b. To find the two square roots of $z = -2\sqrt{3} + 2i$, first write the complex number in standard polar form. See **Example 5** to verify that the standard polar form of $z = -2\sqrt{3} + 2i$ is $z = 4\left(\cos\frac{5\pi}{6} + i\sin\frac{5\pi}{6}\right)$. We find z_0 using De Moivre's Theorem for finding nth roots with $r = 4, \theta = \frac{5\pi}{6}$, and $n = 2$.

$$z_0 = \sqrt{4}\left[\cos\left(\frac{\frac{5\pi}{6} + 2\pi \cdot 0}{2}\right) + i\sin\left(\frac{\frac{5\pi}{6} + 2\pi \cdot 0}{2}\right)\right] = 2\left(\cos\frac{5\pi}{12} + i\sin\frac{5\pi}{12}\right)$$

To find z_1, we add $\frac{2\pi}{2} = \pi$ to the argument of z_0.

$$z_1 = 2\left[\cos\left(\frac{5\pi}{12} + \pi\right) + i\sin\left(\frac{5\pi}{12} + \pi\right)\right] = 2\left(\cos\frac{17\pi}{12} + i\sin\frac{17\pi}{12}\right)$$

Using a calculator set in radian mode, we can evaluate $\cos\frac{5\pi}{12}$, $\sin\frac{5\pi}{12}$, $\cos\frac{17\pi}{12}$, and $\sin\frac{17\pi}{12}$ to approximate the rectangular form of z_0 and z_1. Thus, the approximate rectangular forms for the complex square roots of $z = -2\sqrt{3} + 2i$ are

$$z_0 = 2\left(\cos\frac{5\pi}{12} + i\sin\frac{5\pi}{12}\right) \approx 0.52 + 1.93i$$

and

$$z_1 = 2\left(\cos\frac{17\pi}{12} + i\sin\frac{17\pi}{12}\right) \approx -0.52 - 1.93i$$

Note Because the argument of each root has a denominator of 12, it is actually possible to use the sum formulas for sine and cosine discussed in Section 8.2 to determine the exact values of $\cos\frac{5\pi}{12}$, $\sin\frac{5\pi}{12}$, $\cos\frac{17\pi}{12}$, and $\sin\frac{17\pi}{12}$. Therefore, we can determine the exact rectangular form of z_0 and z_1. View these steps to see how to show that the exact rectangular form of z_0 and z_1 is

$$z_0 = 2\left(\cos\frac{5\pi}{12} + i\sin\frac{5\pi}{12}\right) = \frac{\sqrt{3} - 1}{\sqrt{2}} + \frac{\sqrt{3} + 1}{\sqrt{2}}i$$

and

$$z_1 = 2\left(\cos\frac{17\pi}{12} + i\sin\frac{17\pi}{12}\right) = -\left(\frac{\sqrt{3} - 1}{\sqrt{2}}\right) - \left(\frac{\sqrt{3} + 1}{\sqrt{2}}\right)i$$

You Try It Work through this You Try It problem.

Work Exercises 38–46 in this textbook or in the MyLab Math Study Plan.

10.3 Exercises

In Exercises 1–4, plot each complex number in the complex plane and determine its absolute value.

1. $z = -1 + 2i$
2. $z = -5 - 4i$
3. $z = 5$
4. $z = -3i$

In Exercises 5–10, rewrite the complex number in standard polar form, plot the number in the complex plane, and determine the quadrant in which the point lies or the axis on which the point lies.

SbS 5. $z = 3\left(\cos\frac{11\pi}{7} + i\sin\frac{11\pi}{7}\right)$

SbS 6. $z = 2(\cos 280° + i\sin 280°)$

SbS 7. $z = 5(\cos \pi + i \sin \pi)$

SbS 8. $z = 3\left(\cos \frac{25\pi}{3} + i \sin \frac{25\pi}{3}\right)$

SbS 9. $z = 2(\cos 390° + i \sin 390°)$

SbS 10. $z = 5\left(\cos \frac{5\pi}{2} + i \sin \frac{5\pi}{2}\right)$

In Exercises 11–16, write each complex number in rectangular form using exact values if possible. Otherwise, round to two decimal places.

11. $z = 5\left(\cos \frac{7\pi}{4} + i \sin \frac{7\pi}{4}\right)$

12. $z = 4(\cos 240° + i \sin 240°)$

13. $z = 8\left(\cos \frac{3\pi}{2} + i \sin \frac{3\pi}{2}\right)$

14. $z = 10\left(\cos \frac{5\pi}{3} + i \sin \frac{5\pi}{3}\right)$

15. $z = 10\left(\cos \frac{\pi}{10} + i \sin \frac{\pi}{10}\right)$

16. $z = 18(\cos 208° + i \sin 208°)$

In Exercises 17–23, write each complex number in standard polar form. Write the argument in radians using exact values if possible. Otherwise, round the argument to two decimal places.

SbS 17. $z = -4 + 4i$

SbS 18. $z = 5\sqrt{3} - 5i$

SbS 19. $z = -2 - 2\sqrt{3}i$

SbS 20. $z = -5$

SbS 21. $z = 7i$

SbS 22. $z = -3 + 2i$

SbS 23. $z = 1 - 4i$

24. Find $z_1 z_2$ and $\frac{z_1}{z_2}$ if $z_1 = 5\left(\cos \frac{\pi}{6} + i \sin \frac{\pi}{6}\right)$ and $z_2 = 4\left(\cos \frac{\pi}{7} + i \sin \frac{\pi}{7}\right)$. Write your answers in standard polar form with the argument in radians.

25. Find $z_1 z_2$ and $\frac{z_1}{z_2}$ if $z_1 = 7(\cos 80° + i \sin 80°)$ and $z_2 = 6(\cos 30° + i \sin 30°)$. Write your answers in standard polar form with the argument in degrees.

26. Find $z_1 z_2$ and $\frac{z_1}{z_2}$ if $z_1 = 2\left(\cos \frac{3\pi}{4} + i \sin \frac{3\pi}{4}\right)$ and $z_2 = 3\left(\cos \frac{11\pi}{6} + i \sin \frac{11\pi}{6}\right)$. Write your answers in standard polar form with the argument in radians.

27. Find $z_1 z_2$ and $\frac{z_1}{z_2}$ if $z_1 = 11(\cos 150° + i \sin 150°)$ and $z_2 = 9(\cos 310° + i \sin 310°)$. Write your answers in standard polar form with the argument in degrees.

In Exercises 28–30, determine the given complex number raised to a power and write the answer in standard polar form.

28. $\left[2\left(\cos \frac{\pi}{7} + i \sin \frac{\pi}{7}\right)\right]^5$

29. $[3(\cos 71° + i \sin 71°)]^4$

30. $\left[7\left(\cos \frac{16\pi}{9} + i \sin \frac{16\pi}{9}\right)\right]^2$

In Exercises 31–37, determine the given complex number raised to a power and write the answer in rectangular form.

31. $\left[3\left(\cos \frac{\pi}{9} + i \sin \frac{\pi}{9}\right)\right]^3$

32. $[2(\cos 80° + i \sin 80°)]^3$

33. $\left[4\left(\cos \frac{\pi}{5} + i \sin \frac{\pi}{5}\right)\right]^5$

34. $\left[\sqrt{3}(\cos 20^\circ + i\sin 20^\circ)\right]^6$ 35. $\left[\sqrt{5}\left(\cos\frac{7\pi}{6} + i\sin\frac{7\pi}{6}\right)\right]^4$ 36. $(-1 - i)^7$ 37. $(3 - 3\sqrt{3}i)^6$

38. Find all complex square roots of $z = 36\left(\cos\frac{\pi}{6} + i\sin\frac{\pi}{6}\right)$. Write each root in standard polar form with the argument in radians.

39. Find all complex cube roots of $z = 27(\cos 250^\circ + i\sin 250^\circ)$. Write each root in standard polar form with the argument in degrees.

40. Find all complex fourth roots of $z = -16$. Write each root in standard polar form with the argument in radians.

41. Find all complex fifth roots of $z = -4i$. Write each root in standard polar form with the argument in radians.

42. Find all complex cube roots of $z = 6 + 6\sqrt{3}i$. Write each root in standard polar form with the argument in radians.

43. Find all complex fourth roots of $z = 256$. Write each root in rectangular form.

44. Find all complex cube roots of $z = -125$. Write each root in rectangular form.

45. Find all complex fourth roots of $z = 64\left(\cos\frac{5\pi}{3} + i\sin\frac{5\pi}{3}\right)$. Write each root in rectangular form, rounding to two decimal places.

46. Find all complex cube roots of $z = 1 - 4i$. Write each root in rectangular form, rounding to two decimal places.

10.4 Vectors

THINGS TO KNOW

Before working through this section, be sure that you are familiar with the following concepts:

VIDEO ANIMATION INTERACTIVE

You Try It 1. Converting a Point from Polar Coordinates to Rectangular Coordinates (Section 10.1)  

You Try It 2. Converting a Point from Rectangular Coordinates to Polar Coordinates (Section 10.1) 

You Try It 3. Converting a Complex Number from Rectangular Form to Standard Polar Form (Section 10.3) 

INTRODUCTION

Read this introduction before beginning Objective 1.

OBJECTIVES

1 Understanding the Geometric Representation of a Vector
2 Understanding Operations on Vectors Represented Geometrically
3 Understanding Vectors in Terms of Components
4 Understanding Vectors in Terms of **i** and **j**
5 Finding a Unit Vector
6 Determining the Direction Angle of a Vector
7 Representing a Vector in Terms of **i** and **j** Given Its Magnitude and Direction Angle
8 Using Vectors to Solve Applications Involving Velocity
9 Using Vectors to Solve Applications Involving Force

SECTION 10.4 EXERCISES

Introduction to Section 10.4

Many physical quantities can be described by a single real number. For example, the length of a fence, the circumference of a circle, the temperature of a room, or the area of a farmer's field can all be described by a single real number. The size of these numbers is called the **magnitude** of the quantity. Quantities that can be described using only magnitude are called **scalar quantities**. Some quantities, such as velocity and force, require that we know both the magnitude of the quantity and the direction of the quantity. For example, the two cars in Figure 53 left from the same initial point and are both traveling at a speed of 60 mph. However, one car is moving in a Northwest direction and the other is moving in a Northeast direction. To describe the position of a car, it is not enough to know the speed of the car (the magnitude); it is also necessary to know the direction in which the car is moving.

Figure 53

The speed of both cars in Figure 53 is the same (magnitude of 60 mph), but they are traveling in different directions. Quantities that are described using both magnitude and direction are called **vector quantities** or **vectors**. In this section we will study vectors and we will learn how to represent vectors in three different ways. We start by introducing a geometric representation of vectors.

(eText Screens 10.4-1–10.4-50)

OBJECTIVE 1 UNDERSTANDING THE GEOMETRIC REPRESENTATION OF A VECTOR

We can represent a vector geometrically using an arrow of finite length drawn in a particular direction. Vectors can be thought of as a directed line segment that begins at an **initial point** (sometimes called the tail of the vector) and ends at a **terminal point** (sometimes called the head or tip of the vector).

Definition Geometric Representation of a Vector

A **vector v can be represented geometrically** as a directed line segment in a plane having an initial point P and a terminal point Q.

The vector can be denoted as the boldface letter $\mathbf{v}$, $\overrightarrow{PQ}$, or $\vec{v}$.

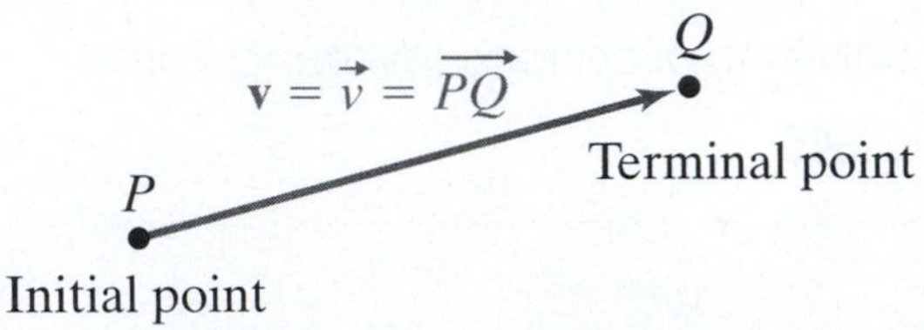

Throughout Sections 10.4 and 10.5, we will denote vectors using a boldface letter. Because we cannot write boldface letters using pencil and paper, vectors in all videos will be denoted by an arrow above a single letter.

The magnitude of a vector represented geometrically is the length of the arrow. A vector whose magnitude is 0 is called the **zero vector** and is denoted by $\vec{0}$ or $\mathbf{0}$. If we know the coordinates of the initial point and the terminal point of a vector in a rectangular coordinate system, then we can use the distance formula to determine the magnitude, as stated in the following definition.

Definition The Magnitude of a Vector Represented Geometrically

The **magnitude** of a vector $\mathbf{v}$ is the distance between the initial point and the terminal point and is denoted by $\|\mathbf{v}\|$. If the initial point has coordinates $P(x_1, y_1)$ and the terminal point has coordinates $Q(x_2, y_2)$, then

$$\|\mathbf{v}\| = \sqrt{(x_2 - x_1)^2 + (y_2 - y_1)^2}.$$

Example 1 Determining the Magnitude of a Vector Represented Geometrically

Determine the magnitude of $\mathbf{v}$.

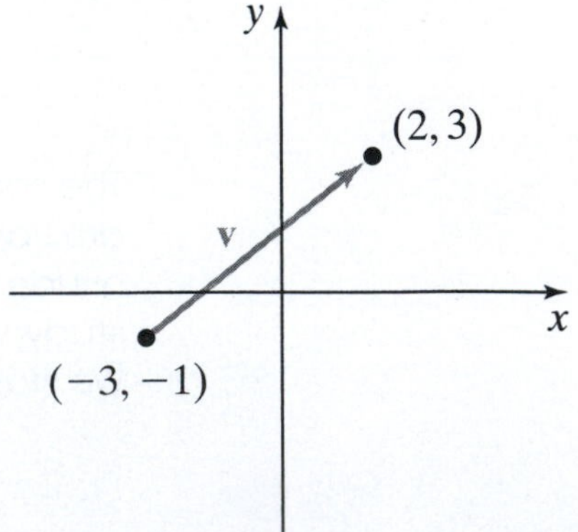

Solution

Vector $\mathbf{v}$ has an initial point $(x_1, y_1) = (-3, -1)$ and a terminal point $(x_2, y_2) = (2, 3)$. Therefore, the magnitude of $\mathbf{v}$ is

$$\|\mathbf{v}\| = \sqrt{(x_2 - x_1)^2 + (y_2 - y_1)^2} = \sqrt{(2 - (-3))^2 + (3 - (-1))^2} = \sqrt{5^2 + 4^2} = \sqrt{41}.$$

You Try It Work through this You Try It problem.

Work Exercises 1–3 in this textbook or in the MyLab Math Study Plan.

OBJECTIVE 2 UNDERSTANDING OPERATIONS ON VECTORS REPRESENTED GEOMETRICALLY

SCALAR MULTIPLICATION

Recall that a quantity described using only magnitude is called a **scalar quantity**. When working in the context of vectors, real numbers are called **scalars.** Consider the two vectors in **Figure 54a**. If we multiply vector $\mathbf{v}$ by a scalar k, we obtain the product $k\mathbf{v}$. Multiplying a vector by a scalar is called **scalar multiplication.** Suppose that k is a positive scalar such that $k \neq 1$. Then the vector $k\mathbf{v}$ has the same direction as $\mathbf{v}$ but has a magnitude that is k times the magnitude of $\mathbf{v}$. See **Figure 54b**. Similarly, if vector $\mathbf{u}$ is multiplied by $k < 0$, then $k\mathbf{u}$ is a vector in the exact opposite direction as $\mathbf{u}$ with a magnitude of $|k|\|\mathbf{u}\|$. See **Figure 54c**. Watch this **video** to see a description of scalar multiplication.

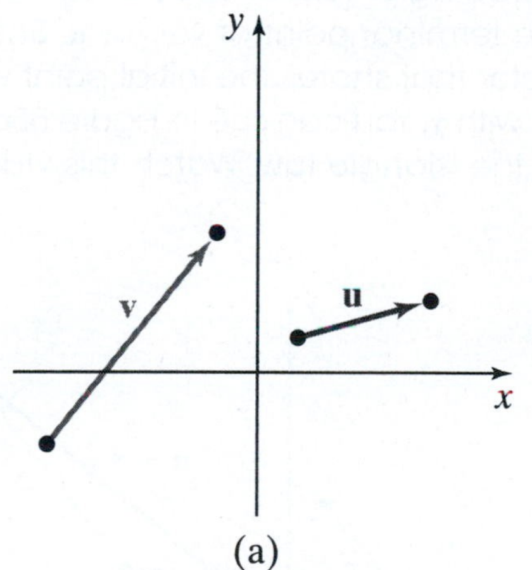

Two vectors $\mathbf{u}$ and $\mathbf{v}$

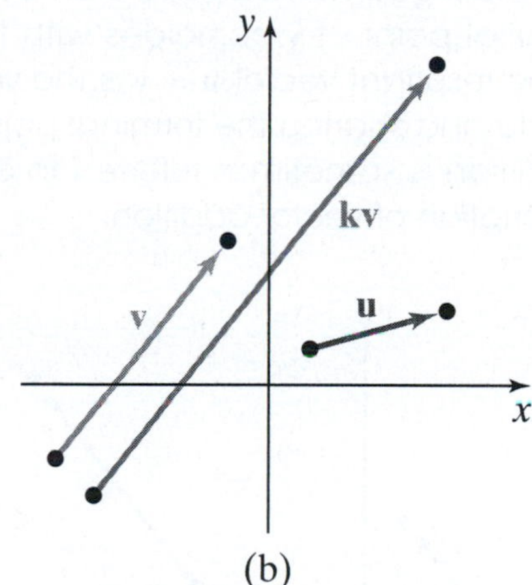

If $k > 0$ such that $k \neq 1$, then $k\mathbf{v}$ is a vector in the same direction as $\mathbf{v}$ and has a magnitude of $k\|\mathbf{v}\|$.

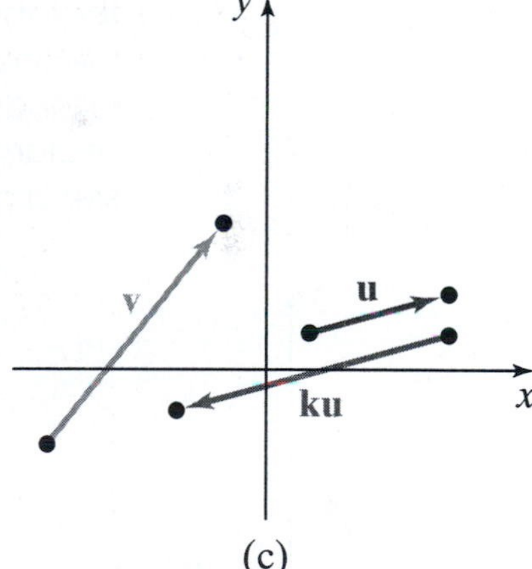

If $k < 0$, then $k\mathbf{u}$ is a vector in the exact opposite direction as $\mathbf{u}$ and has a magnitude of $|k|\ \|\mathbf{u}\|$.

Figure 54

Note that if $k = 0$, then $k\mathbf{v}$ is the zero vector, $\mathbf{0}$.

Example 2 Multiplying a Vector by a Scalar

Given the vector $\mathbf{v}$, draw the vectors $2\mathbf{v}$ and $-\frac{1}{2}\mathbf{v}$.

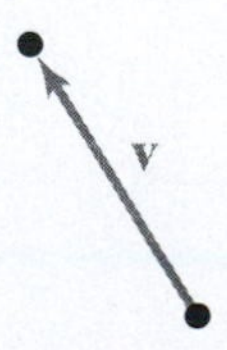

Solution The vector $2\mathbf{v}$ is twice as long as vector $\mathbf{v}$ and is in the same direction as $\mathbf{v}$. The vector $-\frac{1}{2}\mathbf{v}$ is half as long as vector $\mathbf{v}$ and is in the exact opposite direction as $\mathbf{v}$.

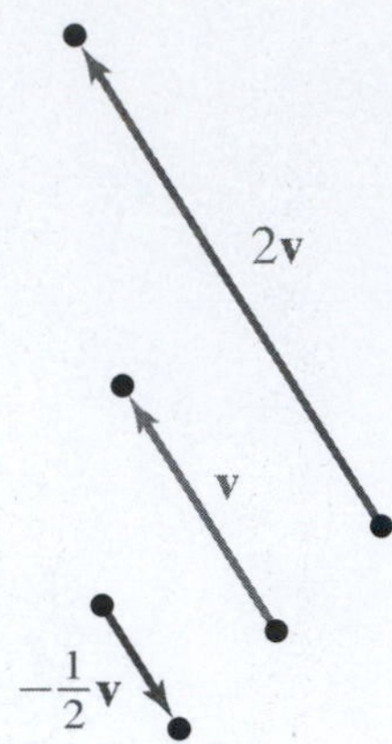

Note that the three vectors in Example 2 are **parallel** to each other. Any nonzero vector $\mathbf{u}$ that is a scalar multiple of a vector $\mathbf{v}$ is said to be parallel to $\mathbf{v}$.

> **Definition** Parallel Vectors
>
> Two vectors $\mathbf{u}$ and $\mathbf{v}$ are **parallel vectors** if there is a nonzero scalar k such that $\mathbf{u} = k\mathbf{v}$.

VECTOR ADDITION

Suppose that we wish to add the two nonzero vectors $\mathbf{u}$ and $\mathbf{v}$ shown in Figure 55a. The sum of $\mathbf{u}$ and $\mathbf{v}$ is denoted by $\mathbf{u} + \mathbf{v}$ and is called the **resultant vector.** To add the two vectors geometrically, we start by drawing an exact copy of vector $\mathbf{v}$ so that the initial point of $\mathbf{v}$ coincides with the terminal point of vector $\mathbf{u}$. See Figure 55b. The resultant vector $\mathbf{u} + \mathbf{v}$ is the vector that shares the initial point with $\mathbf{u}$ extending to and sharing the terminal point with $\mathbf{v}$. You can see in Figure 55c why vector addition is sometimes referred to as the **triangle law.** Watch this **video** to see a description of vector addition.

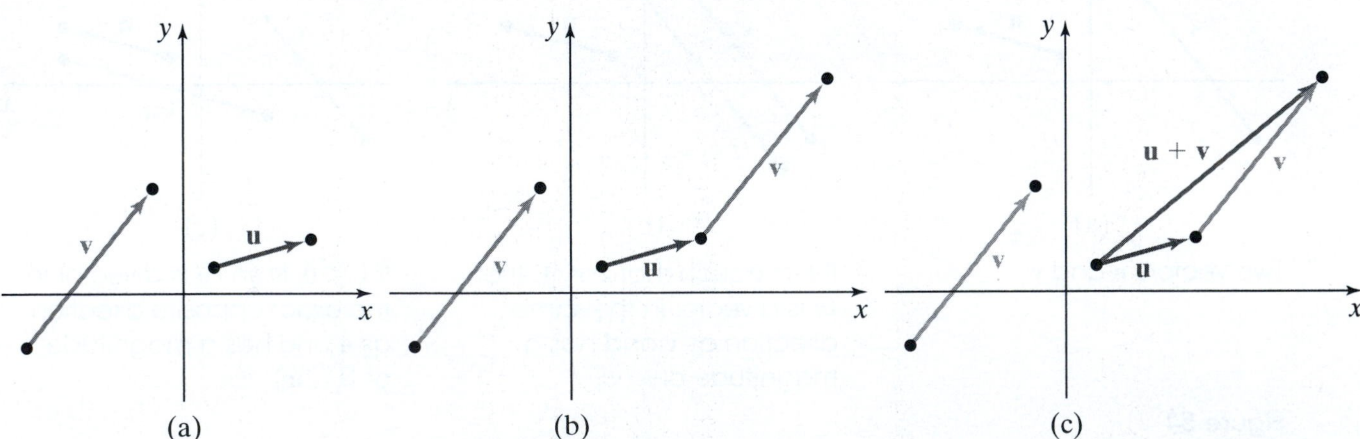

(a) Two vectors $\mathbf{u}$ and $\mathbf{v}$

(b) To find $\mathbf{u} + \mathbf{v}$, start by placing the initial point of $\mathbf{v}$ on the terminal point of $\mathbf{u}$.

(c) The resultant vector $\mathbf{u} + \mathbf{v}$ extends from the initial point of $\mathbf{u}$ to the terminal point of $\mathbf{v}$. The three vectors form a triangle.

Figure 55

THE PARALLELOGRAM LAW FOR VECTOR ADDITION

Another way to illustrate vector addition is to think of the resultant vector $\mathbf{u} + \mathbf{v}$ as the diagonal of a parallelogram. We do this by starting with the vectors $\mathbf{u}$ and $\mathbf{v}$, as shown in Figure 56a. We can complete a parallelogram by drawing copies of vectors $\mathbf{u}$ and $\mathbf{v}$ on opposite sides of each other. As you can see in Figure 56b, we can represent the resultant vector $\mathbf{u} + \mathbf{v}$ as the diagonal of a parallelogram. This is known as the **parallelogram law for vector addition.** The parallelogram law is also a nice way to illustrate that **vector addition is commutative**, which means that $\mathbf{u} + \mathbf{v} = \mathbf{v} + \mathbf{u}$. The commutative property of vector addition is just one of many properties of vectors. We will state several more properties later in this section.

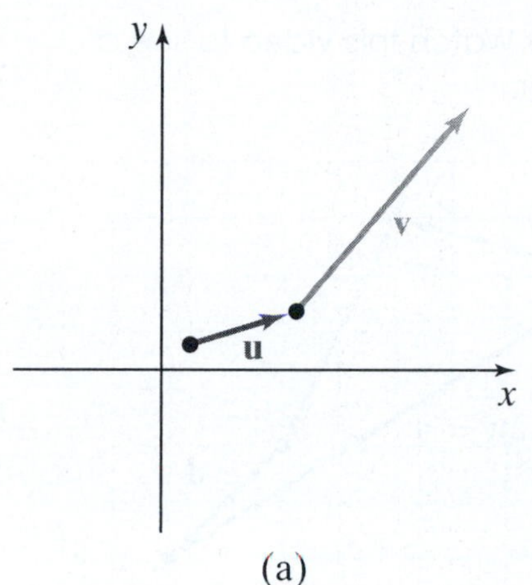

(a) Start by placing the initial point of $\mathbf{v}$ on the terminal point of $\mathbf{u}$.

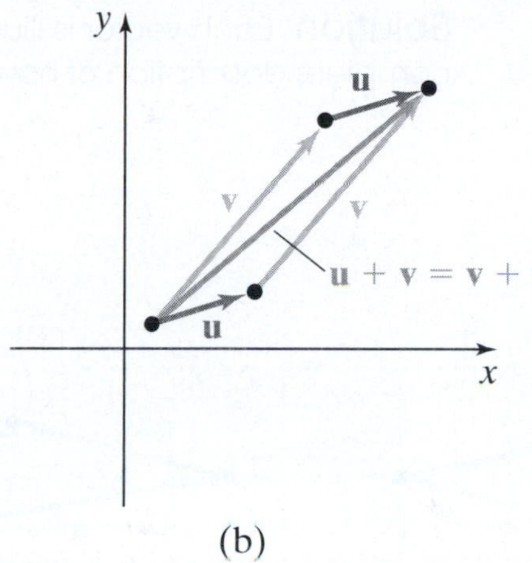

(b) Construct a parallelogram. The diagonal is the vector $\mathbf{u} + \mathbf{v} = \mathbf{v} + \mathbf{u}$.

Figure 56

VECTOR SUBTRACTION

The difference of two vectors $\mathbf{u} - \mathbf{v}$ is defined as $\mathbf{u} - \mathbf{v} = \mathbf{u} + (-\mathbf{v})$, where $-\mathbf{v}$ is obtained by multiplying $\mathbf{v}$ by the scalar -1. Therefore, to find the difference of two vectors $\mathbf{u} - \mathbf{v}$, first draw the vector $-\mathbf{v}$ in such a way that the initial point of $-\mathbf{v}$ coincides with the terminal point of $\mathbf{u}$. Then use vector addition to add $-\mathbf{v}$ to $\mathbf{u}$. See Figure 57b. The resultant vector $\mathbf{u} - \mathbf{v} = \mathbf{u} + (-\mathbf{v})$ is the vector that shares the initial point, with $\mathbf{u}$ extending to and sharing the terminal point with $\mathbf{v}$. See Figure 57c. Watch this video to see a description of vector subtraction.

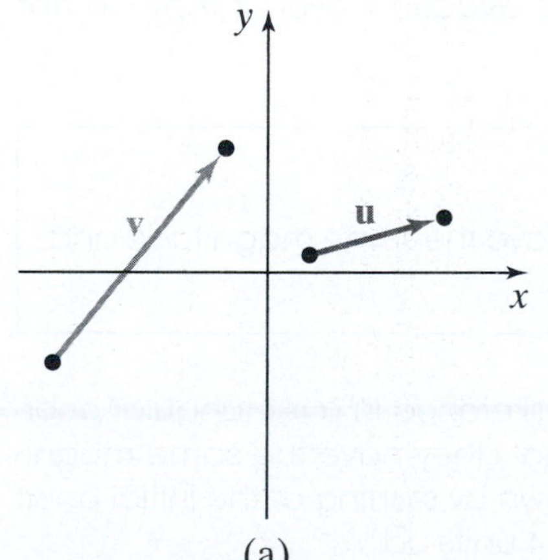

(a) Two vectors $\mathbf{u}$ and $\mathbf{v}$

(b) To find $\mathbf{u} - \mathbf{v}$, start by creating $-\mathbf{v}$ and place $-\mathbf{v}$ on the terminal point of $\mathbf{u}$.

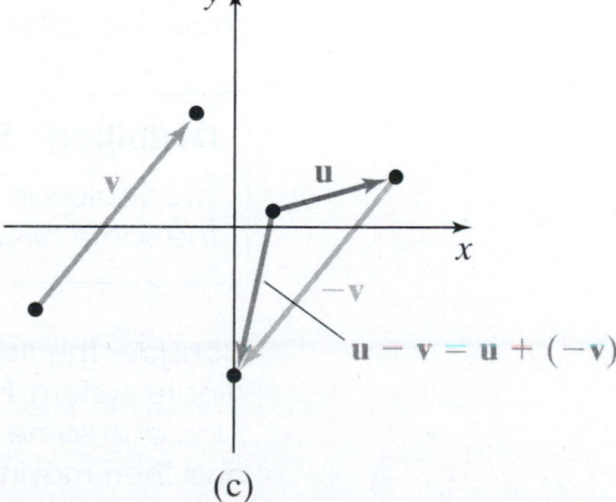

(c) The resultant vector $\mathbf{u} - \mathbf{v}$ extends from the initial point of $\mathbf{u}$ to the terminal point of $-\mathbf{v}$.

Figure 57

Example 3 Performing Operations on Vectors Represented Geometrically

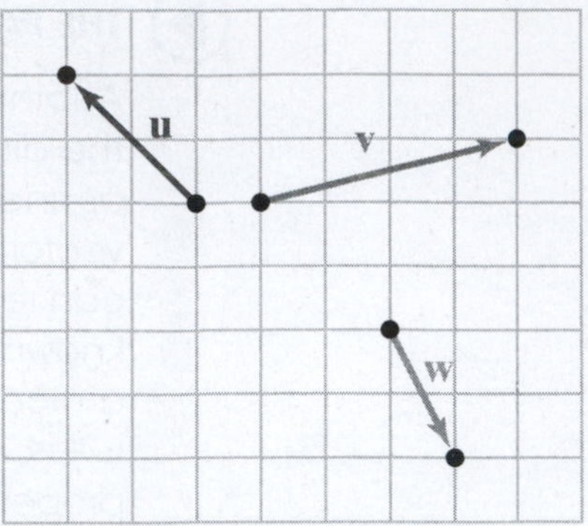

Given the vectors **u**, **v**, and **w**, draw each of the following vectors.

a. **w** + **v**

b. **v** − **u**

c. **v** + 2**w** − **u**

Solution Each vector is illustrated in green below. Watch this **video** to see a complete description of how to sketch each vector.

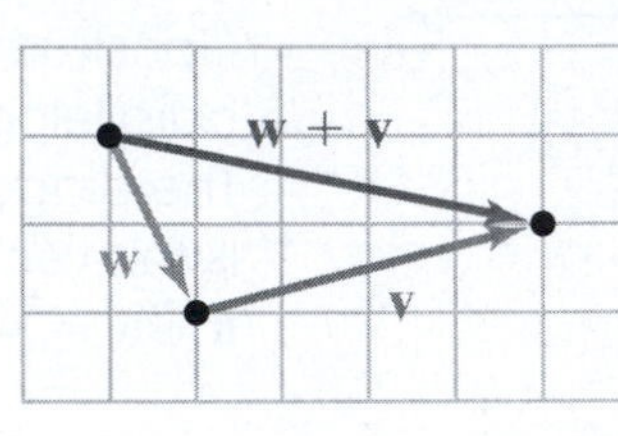

a. **w** + **v**

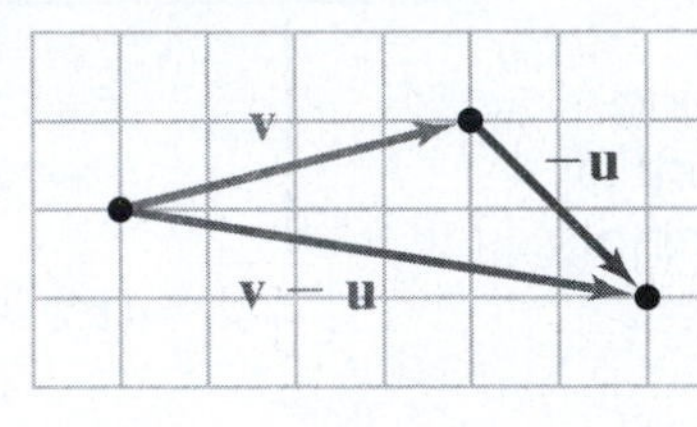

b. **v** − **u**

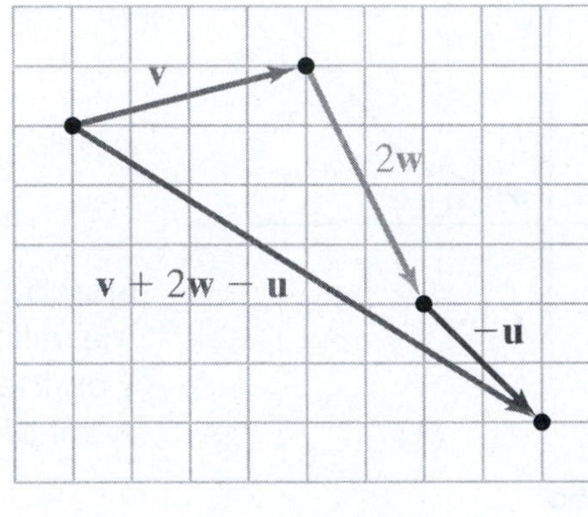

c. **v** + 2**w** − **u**

You Try It Work through this You Try It problem.

Work Exercises 4–10 in this textbook or in the MyLab Math Study Plan.

OBJECTIVE 3 UNDERSTANDING VECTORS IN TERMS OF COMPONENTS

Every vector has magnitude and direction. Two vectors in a plane are **equal vectors** if they have the same magnitude and the same direction, even if they do not share the same initial point and terminal point.

> **Definition** Equal Vectors
>
> Two vectors in a plane are **equal vectors** if they have the same magnitude and the same direction.

Consider the three vectors in Figure 58, which are sketched in a rectangular coordinate system. Note that all three vectors are equal (they have the same magnitude and same direction) and they can all be drawn by starting at the initial point and then moving 3 units right followed by moving 4 units up.

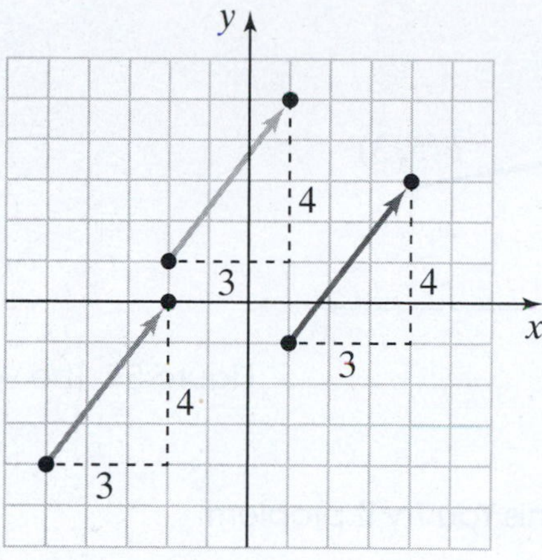

Figure 58 Each vector can be obtained by moving horizontally to the right of the initial point 3 units followed by moving vertically up 4 units.

The three equal vectors in Figure 58 can each be thought of as the "3 units right and 4 units up vector." This suggests another way to represent vectors. For each of the vectors in Figure 58, we can use the number 3 to describe the horizontal displacement from the initial point and the number 4 to describe the vertical displacement. These displacement values are called the **components** of the vector. Vectors represented using components a and b can be written in the form $\mathbf{v} = \langle a, b \rangle$, where a is the horizontal component and b is the vertical component. Each vector in Figure 58 can be represented as $\mathbf{v} = \langle 3, 4 \rangle$.

Definition Representing a Vector in Terms of Components a and b

If the initial point of a vector $\mathbf{v}$ is $P(x_1, y_1)$ and if the terminal point is $Q(x_2, y_2)$, then $\mathbf{v}$ is a **vector represented by components a and b**, where $\mathbf{v} = \langle x_2 - x_1, y_2 - y_1 \rangle = \langle a, b \rangle$.

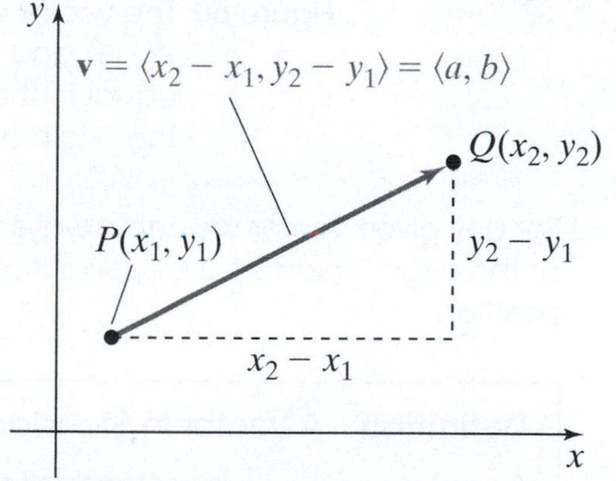

Example 4 Determining the Components and Magnitude of a Vector

Determine the component representation and the magnitude of a vector $\mathbf{v}$ having an initial point $P(5, 3)$ and a terminal point $Q(-6, 5)$.

Solution The component representation is $\mathbf{v} = \langle x_2 - x_1, y_2 - y_1 \rangle = \langle -6 - 5, 5 - 3 \rangle = \langle -11, 2 \rangle$ and the magnitude of $\mathbf{v}$ is $\|\mathbf{v}\| = \|\langle -11, 2 \rangle\| = \sqrt{(-11)^2 + 2^2} = \sqrt{121 + 4} = \sqrt{125} = 5\sqrt{5}$. See Figure 59.

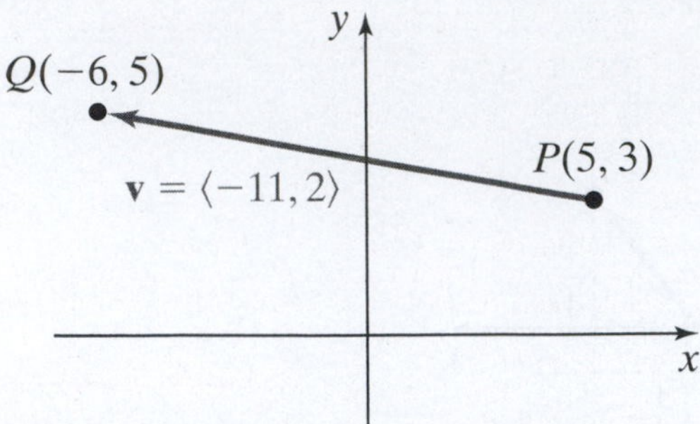

Figure 59 The vector $\mathbf{v} = \langle -11, 2 \rangle$. ●

You Try It Work through this You Try It problem.

Work Exercises 11–13 in this textbook or in the MyLab Math Study Plan.

We have previously stated that two vectors are equal if they have the same magnitude and direction. Notice that the vector $\mathbf{v} = \langle -11, 2 \rangle$ in Example 4 is equal to the vector $\mathbf{u}$ that has an initial point at the origin and a terminal point of $(-11, 2)$. See Figure 60.

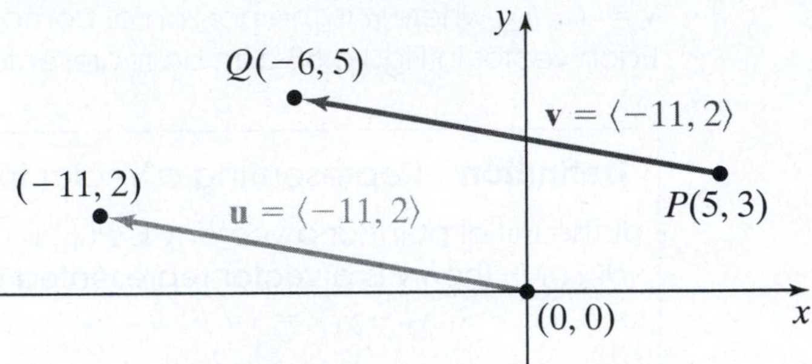

Figure 60 The vector $\mathbf{v} = \langle -11, 2 \rangle$ with initial point $P(5, 3)$ and terminal point $Q(-6, 5)$ is equal to the vector $\mathbf{u}$ with initial point at the origin and terminal point at $(-11, 2)$.

For any given vector, we can always find an equivalent vector whose initial point is at the origin. Vectors having an initial point at the origin are said to be in **standard position**.

Definition A Vector in Standard Position

A vector $\mathbf{v} = \langle a, b \rangle$ is in **standard position** if the initial point coincides with the origin. The magnitude is $\|\mathbf{v}\| = \|\langle a, b \rangle\| = \sqrt{a^2 + b^2}$.

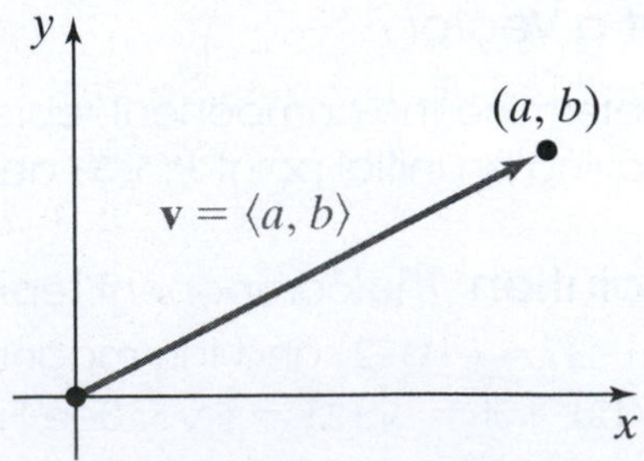

For simplicity, unless otherwise stated, we will consider all vectors discussed in the remainder of this section to be in standard position. This is possible because any vector can be represented in the form $\mathbf{v} = \langle a, b \rangle$, where the initial point is the origin and the terminal point is (a, b).

OBJECTIVE 4 UNDERSTANDING VECTORS REPRESENTED IN TERMS OF i AND j

So far we have seen a geometric representation of vectors and vectors represented using components. There is one other very useful way to represent vectors. Before we discuss this third representation, we must introduce the concept of unit vectors.

A **unit vector** is a vector that has a magnitude of 1 unit.

> **Definition** Unit Vector
>
> A **unit vector** is a vector that has a magnitude of 1 unit.

There are two unit vectors that are particularly important. The first is the unit vector whose direction is along the positive x-axis. This vector is named $\mathbf{i}$, where $\mathbf{i} = \langle 1, 0 \rangle$. The other is the unit vector whose direction is along the positive y-axis. This vector is named $\mathbf{j}$, where $\mathbf{j} = \langle 0, 1 \rangle$.

> **Definition** Unit Vectors $\mathbf{i}$ and $\mathbf{j}$
>
> The unit vector whose direction is along the positive x-axis is $\mathbf{i} = \langle 1, 0 \rangle$.
>
> The unit vector whose direction is along the positive y-axis is $\mathbf{j} = \langle 0, 1 \rangle$.

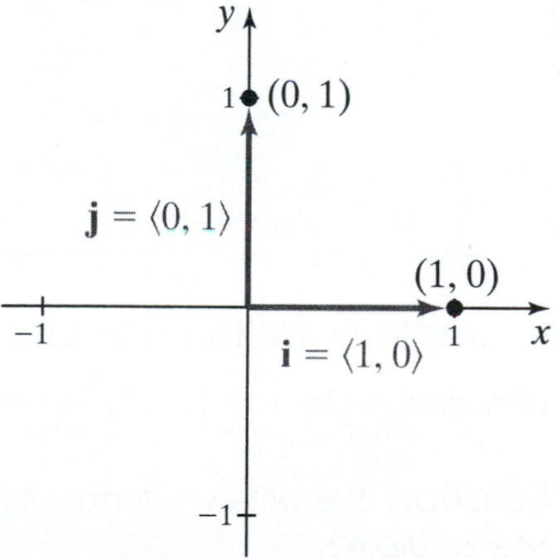

To sketch the vector $\mathbf{v} = a\mathbf{i} + b\mathbf{j}$, start by placing the initial point of $b\mathbf{j}$ on the terminal point of $a\mathbf{i}$. See Figure 61a. The resultant vector $\mathbf{v} = a\mathbf{i} + b\mathbf{j}$ extends from the origin to the terminal point of $b\mathbf{j}$. See Figure 61b.

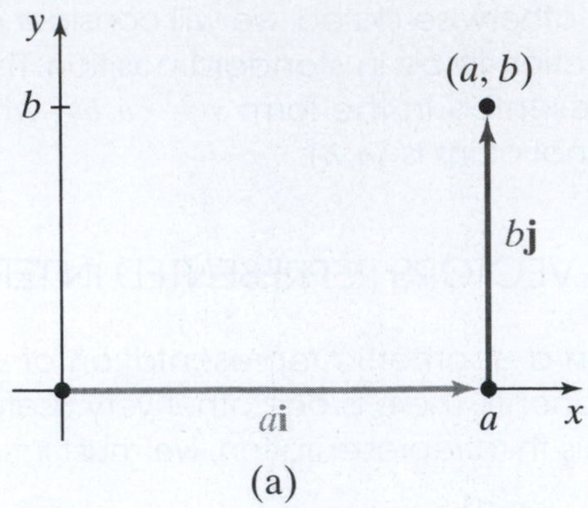

(a)

Place the initial point of $b\mathbf{j}$ on the terminal point of $a\mathbf{i}$.

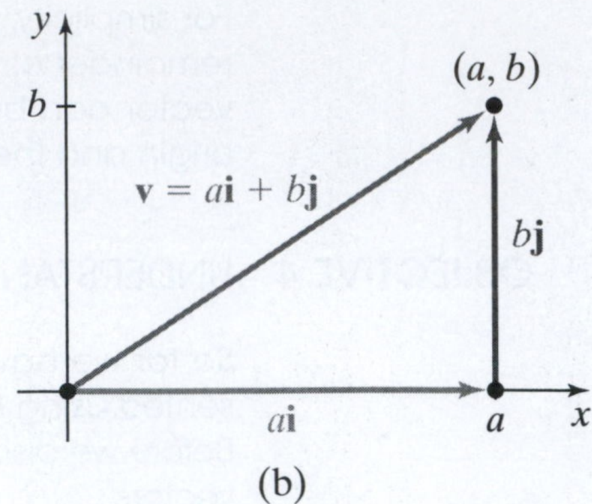

(b)

The resultant vector $\mathbf{v} = a\mathbf{i} + b\mathbf{j}$ extends from the origin to the terminal point of $b\mathbf{j}$.

Figure 61

As you can see in Figure 61b, the vector $\mathbf{v} = a\mathbf{i} + b\mathbf{j}$ is equivalent to the vector $\mathbf{v} = \langle a, b \rangle$. Therefore, any vector can be written in terms of unit vectors $\mathbf{i}$ and $\mathbf{j}$, which gives us a third way of representing vectors.

Definition A Vector Represented in Terms of $\mathbf{i}$ and $\mathbf{j}$

Any vector $\mathbf{v} = \langle a, b \rangle$ can be **represented in terms of the unit vectors $\mathbf{i}$ and $\mathbf{j}$**, where $\mathbf{v} = \langle a, b \rangle = a\mathbf{i} + b\mathbf{j}$.

The magnitude is $\|\mathbf{v}\| = \sqrt{a^2 + b^2}$.

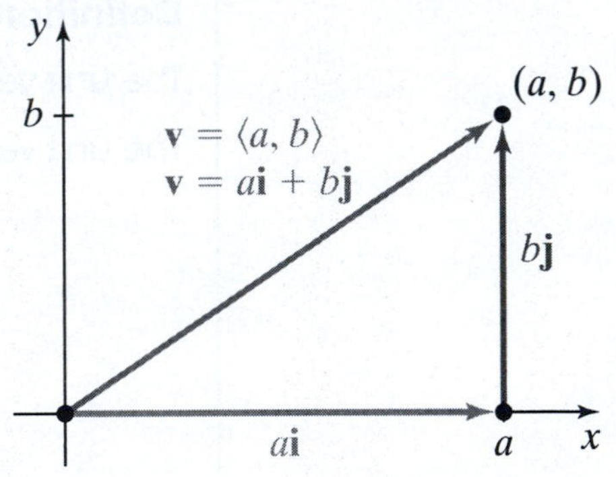

Example 5 Writing a Vector in Terms of $\mathbf{i}$ and $\mathbf{j}$

Write the vector $\mathbf{v} = \langle -5, 2 \rangle$ in terms of the unit vectors $\mathbf{i}$ and $\mathbf{j}$.

Solution We write $\mathbf{v}$ in terms of the unit vectors $\mathbf{i}$ and $\mathbf{j}$ as $\mathbf{v} = \langle -5, 2 \rangle = -5\mathbf{i} + 2\mathbf{j}$. See Figure 62.

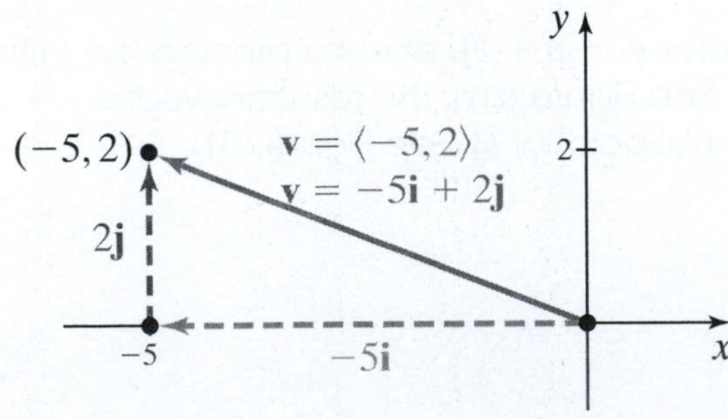

Figure 62 The vector $\mathbf{v} = \langle -5, 2 \rangle = -5\mathbf{i} + 2\mathbf{j}$. ●

You Try It Work through this You Try It problem.

Work Exercises 14–16 in this textbook or in the MyLab Math Study Plan.

When vectors are represented in terms of $\mathbf{i}$ and $\mathbf{j}$, then the operations of addition, subtraction, and scalar multiplication are very straightforward.

Operations with Vectors in Terms of i and j

If $\mathbf{u} = a\mathbf{i} + b\mathbf{j}$, $\mathbf{v} = c\mathbf{i} + d\mathbf{j}$, and k is a scalar, then

$$\mathbf{u} + \mathbf{v} = (a + c)\mathbf{i} + (b + d)\mathbf{j}$$
$$\mathbf{u} - \mathbf{v} = (a - c)\mathbf{i} + (b - d)\mathbf{j}$$
$$k\mathbf{u} = (ka)\mathbf{i} + (kb)\mathbf{j}$$

Example 6 Performing Operations on Vectors in Terms of i and j

Let $\mathbf{u} = -3\mathbf{i} + 7\mathbf{j}$ and $\mathbf{v} = 5\mathbf{i} - \mathbf{j}$. Find each vector in terms of $\mathbf{i}$ and $\mathbf{j}$ and determine the magnitude of each vector.

a. $-\frac{1}{2}\mathbf{u}$

b. $\mathbf{u} + \mathbf{v}$

c. $\mathbf{u} - \mathbf{v}$

d. $3\mathbf{u} - 5\mathbf{v}$

Solution Work through this **video** to verify the following:

a. $-\frac{1}{2}\mathbf{u} = \left(\left(-\frac{1}{2}\right)(-3)\right)\mathbf{i} + \left(\left(-\frac{1}{2}\right)(7)\right)\mathbf{j} = \frac{3}{2}\mathbf{i} - \frac{7}{2}\mathbf{j};$

$$\left\|-\frac{1}{2}\mathbf{u}\right\| = \sqrt{\left(\frac{3}{2}\right)^2 + \left(-\frac{7}{2}\right)^2} = \sqrt{\frac{9}{4} + \frac{49}{4}} = \sqrt{\frac{58}{4}} = \frac{\sqrt{58}}{2}$$

b. $\mathbf{u} - \mathbf{v} = (-3 - 5)\mathbf{i} + (7 - (-1))\mathbf{j} = -8\mathbf{i} + 8\mathbf{j};$

$$\|\mathbf{u} - \mathbf{v}\| = \sqrt{(-8)^2 + 8^2} = \sqrt{64 + 64} = \sqrt{128} = 8\sqrt{2}$$

Work through this **video** to see the solutions to parts c and d.

You Try It Work through this You Try It problem.

Work Exercises 17–22 in this textbook or in the MyLab Math Study Plan.

We have previously used the parallelogram law to introduce the fact that vector addition is commutative. We now restate the commutative property and list several other properties of vectors.

Properties of Vectors

If $\mathbf{u}$, $\mathbf{v}$, and $\mathbf{w}$ are vectors and if A and B are scalars, then the following properties are true.

1. $\mathbf{u} + \mathbf{v} = \mathbf{v} + \mathbf{u}$	Commutative Property for Vector Addition
2. $(\mathbf{u} + \mathbf{v}) + \mathbf{w} = \mathbf{u} + (\mathbf{v} + \mathbf{w})$	Associative Property for Vector Addition
3. $\mathbf{u} + \mathbf{0} = \mathbf{u}$	Additive Identity Property
4. $\mathbf{u} + (-\mathbf{u}) = \mathbf{0}$	Additive Inverse Property
5. $(AB)\mathbf{u} = A(B\mathbf{u})$	Associative Property for Scalar Multiplication
6. $A(\mathbf{u} + \mathbf{v}) = A\mathbf{u} + A\mathbf{v}$	Distributive Property
7. $(A + B)\mathbf{u} = A\mathbf{u} + B\mathbf{u}$	Distributive Property
8. $1\mathbf{u} = \mathbf{u}$	Multiplicative Identity Property
9. $0\mathbf{u} = \mathbf{0}$	Zero Multiplication Property
10. $\|A\mathbf{u}\| = \|A\|\,\|\mathbf{u}\|$	Magnitude of Scalar Multiplication Property

OBJECTIVE 5 FINDING A UNIT VECTOR

Recall that a unit vector has a magnitude of 1 unit. Given a vector $\mathbf{v} = a\mathbf{i} + b\mathbf{j}$, it is often useful to find the unit vector $\mathbf{u}$ that has the same direction as $\mathbf{v}$. We can find this unit vector $\mathbf{u}$ by dividing $\mathbf{v}$ by its magnitude $\|\mathbf{v}\|$.

Definition The Unit Vector in the Same Direction of a Given Vector

Given a nonzero vector $\mathbf{v} = a\mathbf{i} + b\mathbf{j}$, the **unit vector $\mathbf{u}$ in the same direction as $\mathbf{v}$** is $\mathbf{u} = \dfrac{\mathbf{v}}{\|\mathbf{v}\|}$.

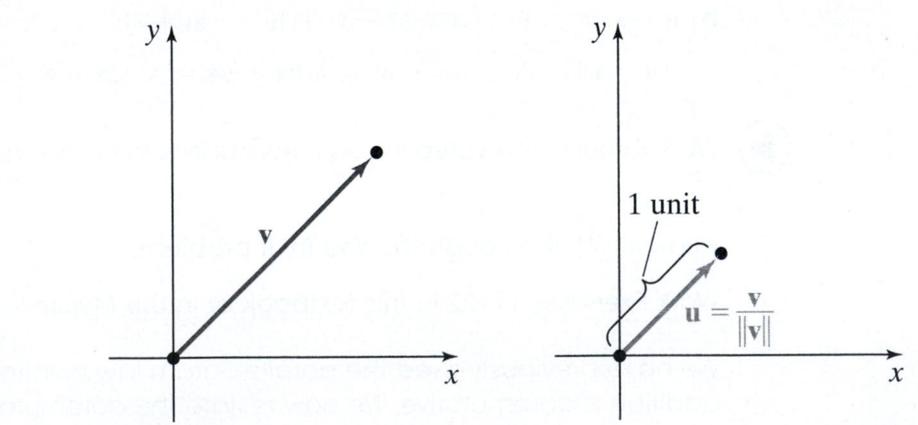

Example 7 Finding a Unit Vector

Find the unit vector that has the same direction as $\mathbf{v} = 6\mathbf{i} - 8\mathbf{j}$.

Solution First, determine the magnitude of $\mathbf{v}$.

$$\|\mathbf{v}\| = \|6\mathbf{i} - 8\mathbf{j}\| = \sqrt{6^2 + (-8)^2} = \sqrt{36 + 64} = \sqrt{100} = 10$$

The unit vector in the same direction as $\mathbf{v}$ is $\mathbf{u} = \dfrac{\mathbf{v}}{\|\mathbf{v}\|} = \dfrac{6\mathbf{i} - 8\mathbf{j}}{10} = \dfrac{6}{10}\mathbf{i} - \dfrac{8}{10}\mathbf{j} = \dfrac{3}{5}\mathbf{i} - \dfrac{4}{5}\mathbf{j}$.

We can verify that the length of $\mathbf{u}$ is 1 unit by determining the magnitude of $\mathbf{u}$.

$$\|\mathbf{u}\| = \left\|\frac{3}{5}\mathbf{i} - \frac{4}{5}\mathbf{j}\right\| = \sqrt{\left(\frac{3}{5}\right)^2 + \left(\frac{-4}{5}\right)^2} = \sqrt{\frac{9}{25} + \frac{16}{25}} = \sqrt{\frac{25}{25}} = \sqrt{1} = 1$$

You Try It Work through this You Try It problem.

Work Exercises 23–26 in this textbook or in the MyLab Math Study Plan.

OBJECTIVE 6 DETERMINING THE DIRECTION ANGLE OF A VECTOR

Recall that every vector has magnitude and direction. We can describe the direction using the positive angle formed between the positive x-axis and a vector in standard position. This positive angle is called the **direction angle**.

Definition The Direction Angle of a Vector

Given a vector $\mathbf{v} = \langle a, b \rangle = a\mathbf{i} + b\mathbf{j}$ in standard position, the **direction angle** of $\mathbf{v}$ is the positive angle θ between the positive x-axis and vector that satisfies the equation $\tan\theta = \dfrac{b}{a}$, where $a \neq 0$.

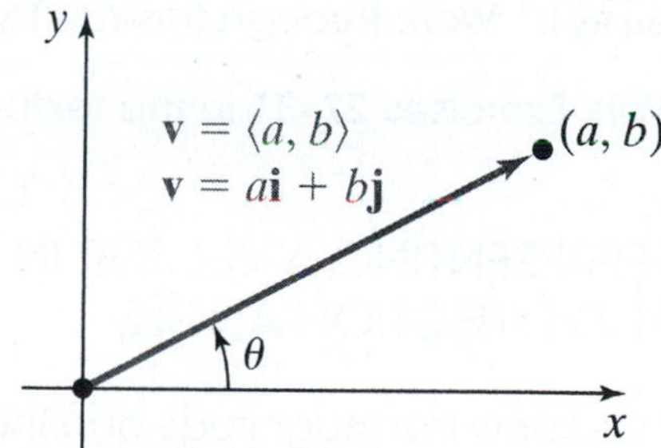

Example 8 Determining the Direction Angle of a Vector

Determine the direction angle of the vector $\mathbf{v} = -3\mathbf{i} + 2\mathbf{j}$.

Solution To find the direction angle θ, start by determining the reference angle θ_R. Using the fact that $\tan\theta = \dfrac{b}{a}$, we can find θ_R by solving the equation $\tan\theta_R = \left|\dfrac{b}{a}\right|$. This technique was first introduced in **Section 10.1**.

$\tan\theta_R = \left|\frac{b}{a}\right|$ Write the equation relating θ_R, a, and b.

$\tan\theta_R = \left|\frac{2}{-3}\right|$ Substitute $a = -3$ and $b = 2$.

$\tan\theta_R = \frac{2}{3}$ Simplify.

The angle whose tangent is $\frac{2}{3}$ does not belong to one of the special angle families. However, we can use the inverse tangent function to solve for θ_R. Thus, $\theta_R = \tan^{-1}\left(\frac{2}{3}\right)$.

The terminal side of θ lies in Quadrant II, which indicates that $\theta = 180° - \theta_R$. Using a calculator set in degree mode, we get $\theta = 180° - \tan^{-1}\left(\frac{2}{3}\right) \approx 146.3°$. See Figure 63.

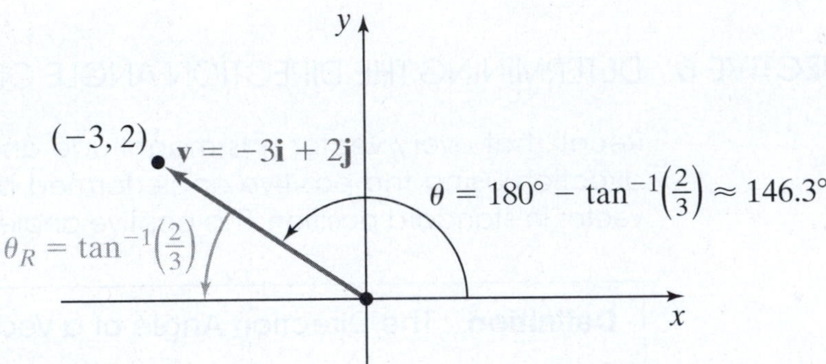

Figure 63 The vector $\mathbf{v} = -3\mathbf{i} + 2\mathbf{j}$ has a direction angle of $\theta \approx 146.3°$.

You Try It **Work through this You Try It problem.**

Work Exercises 27–31 in this textbook or in the MyLab Math Study Plan.

OBJECTIVE 7 REPRESENTING A VECTOR IN TERMS OF I AND J GIVEN ITS MAGNITUDE AND DIRECTION ANGLE

If we know the magnitude and the direction angle of a vector, we can find the horizontal component a and the vertical component b. We do this in much the same way as we did when we converted polar coordinates to rectangular coordinates.

Consider the nonzero vector $\mathbf{v}$ in Figure 64 with magnitude $\|\mathbf{v}\|$ and direction angle θ. The goal is to determine the values of a and b.

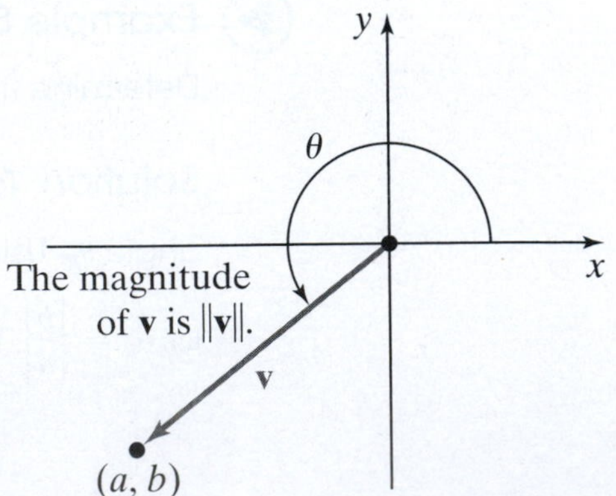

Figure 64

By the **general angle definition** of sine and cosine, we know that $\cos\theta = \frac{a}{\|\mathbf{v}\|}$ and $\sin\theta = \frac{b}{\|\mathbf{v}\|}$. Multiplying both sides of these two equations by $\|\mathbf{v}\|$, we get $a = \|\mathbf{v}\|\cos\theta$ and $b = \|\mathbf{v}\|\sin\theta$. Therefore, we can write $\mathbf{v}$ as follows:

$\mathbf{v} = a\mathbf{i} + b\mathbf{j}$ Write the vector in terms of unit vectors $\mathbf{i}$ and $\mathbf{j}$.

$\mathbf{v} = (\|\mathbf{v}\|\cos\theta)\mathbf{i} + (\|\mathbf{v}\|\sin\theta)\mathbf{j}$ Substitute $b = \|\mathbf{v}\|\cos\theta$ and $b = \|\mathbf{v}\|\sin\theta$.

Representing a Vector in Terms of i and j Given its Magnitude and Direction Angle

If θ is the direction angle measured from the positive x-axis to a vector $\mathbf{v}$, then the vector can be expressed as $\mathbf{v} = (\|\mathbf{v}\|\cos\theta)\mathbf{i} + (\|\mathbf{v}\|\sin\theta)\mathbf{j}$.

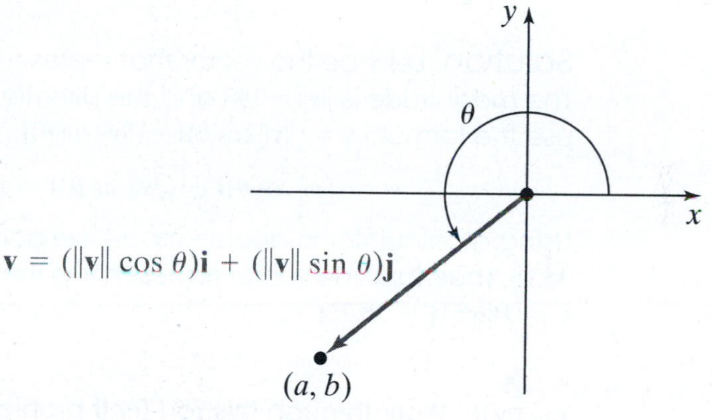

Example 9 Representing a Vector in Terms of i and j Given Its Magnitude and Direction Angle

The vector $\mathbf{v}$ has a magnitude of 20 units and direction angle of $\theta = 50°$. Represent this vector in the form $\mathbf{v} = a\mathbf{i} + b\mathbf{j}$. Round a and b to two decimal places.

Solution First sketch the vector.

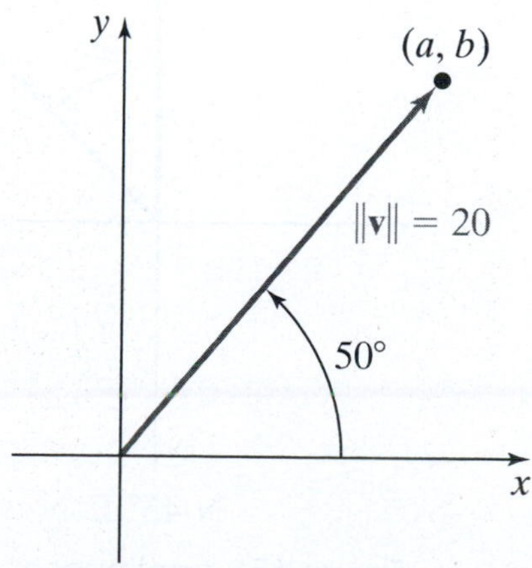

The horizontal component of the vector is $a = \|\mathbf{v}\|\cos\theta = 20\cos 50° \approx 12.86$.
The vertical component of the vector is $b = \|\mathbf{v}\|\sin\theta = 20\sin 50° \approx 15.32$.
Thus, the vector is approximately $\mathbf{v} = 12.86\mathbf{i} + 15.32\mathbf{j}$.

You Try It Work through this You Try It problem.

Work Exercises 32–36 in this textbook or in the MyLab Math Study Plan.

OBJECTIVE 8 USING VECTORS TO SOLVE APPLICATIONS INVOLVING VELOCITY

Velocity has both magnitude (speed) and direction. Thus, given the speed and direction of an object in motion, we can use a vector to represent the velocity of the object.

Example 10 Using a Vector to Represent the Velocity of an Airplane

An airplane takes off from a runway at a speed of 190 mph at an angle of $11°$. Express the velocity of the plane at takeoff as a vector in terms of $\mathbf{i}$ and $\mathbf{j}$. Round a and b to two decimal places.

Solution Let $\mathbf{v}$ be the vector that represents the velocity of the plane at takeoff. The magnitude is $\|\mathbf{v}\| = 190$ and the direction angle is $\theta = 11°$. To determine $\mathbf{v}$, we use the formula $\mathbf{v} = (\|\mathbf{v}\| \cos \theta)\mathbf{i} + (\|\mathbf{v}\| \sin \theta)\mathbf{j}$.

$$\mathbf{v} = (\|\mathbf{v}\| \cos \theta)\mathbf{i} + (\|\mathbf{v}\| \sin \theta)\mathbf{j} = (190 \cos 11°)\mathbf{i} + (190 \sin 11°)\mathbf{j}$$

Using a calculator in degree mode, we get $190 \cos 11° \approx 186.51$ and $190 \sin 11° \approx 36.25$. Therefore, the vector representing the velocity of the plane is approximately $\mathbf{v} = 186.51\mathbf{i} + 36.25\mathbf{j}$.

You Try It Work through this You Try It problem.

Work Exercises 37 and 38 in this textbook or in the MyLab Math Study Plan.

Many applications give the direction of a vector in terms of its bearing. The concept of a bearing was first introduced in **Section 9.2**. Figure 65 illustrates a review of how to sketch the bearing of an object. Each object has a bearing of $45°$ in one of four directions.

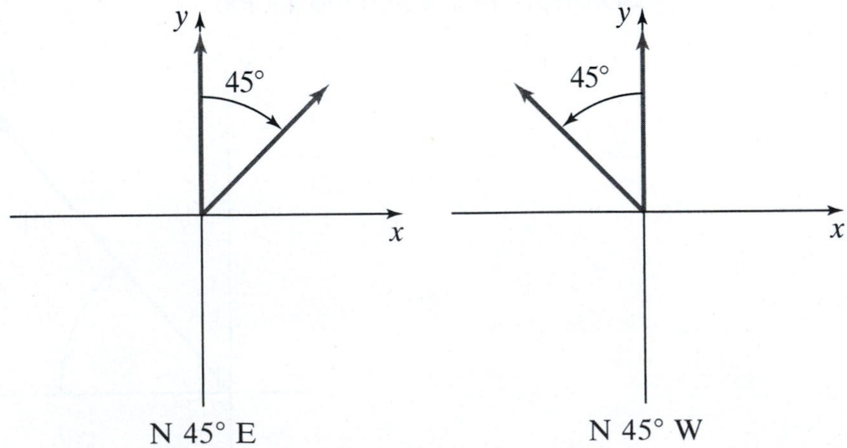

Figure 65—continues

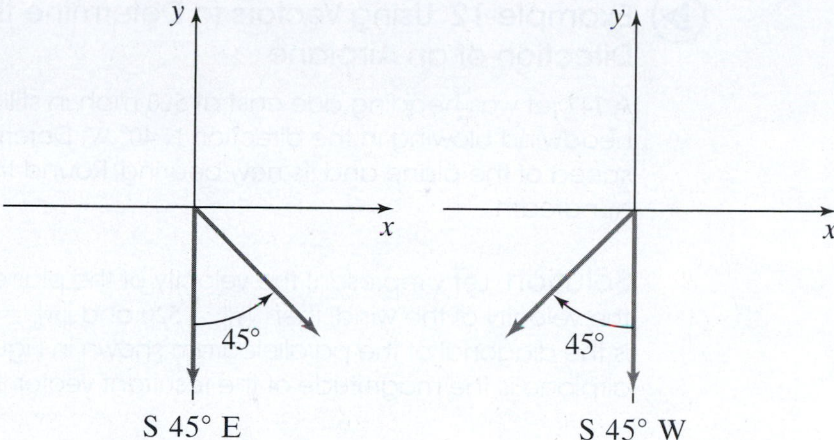

Figure 65

Recall that a bearing of N 45° E is read as "45° east of north."

For the rest of the application examples in this section, the vectors drawn in the corresponding figures will not include the initial point or terminal point. Remember that all vectors have finite length and that the arrows do not mean that the vector extends indefinitely. The arrowheads in the figures simply represent the direction of each vector.

Example 11 Using a Vector to Represent Wind Velocity

The wind is blowing at a speed of 35 mph in a direction of N 30° E. Express the velocity of the wind as a vector in terms of $\mathbf{i}$ and $\mathbf{j}$.

Solution Let $\mathbf{w}$ be the vector having a magnitude of $\|\mathbf{w}\| = 35$ and a direction angle of $\theta = 60°$. See Figure 66.

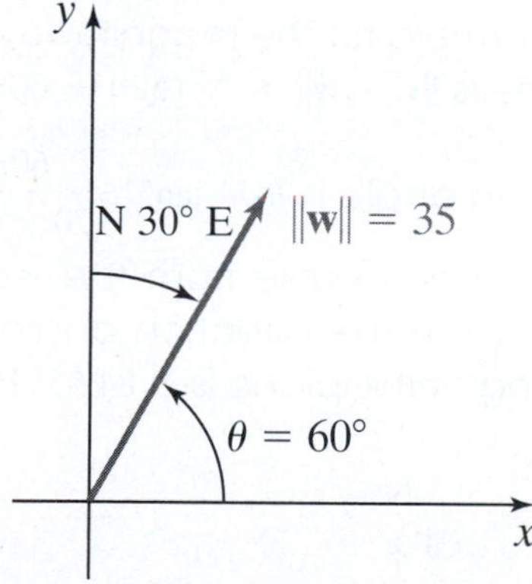

Figure 66

To represent the velocity of the wind in terms of $\mathbf{i}$ and $\mathbf{j}$, we use the formula $\mathbf{w} = (\|\mathbf{w}\|\cos\theta)\mathbf{i} + (\|\mathbf{w}\|\sin\theta)\mathbf{j}$.

$$\mathbf{w} = (\|\mathbf{w}\|\cos\theta)\mathbf{i} + (\|\mathbf{w}\|\sin\theta)\mathbf{j} = (35\cos 60°)\mathbf{i} + (35\sin 60°)\mathbf{j} = \left(35\cdot\frac{1}{2}\right)\mathbf{i} + \left(35\cdot\frac{\sqrt{3}}{2}\right)\mathbf{j} = \frac{35}{2}\mathbf{i} + \frac{35\sqrt{3}}{2}\mathbf{j}$$

You Try It Work through this You Try It problem.

Work Exercises 39 and 40 in this textbook or in the MyLab Math Study Plan.

Example 12 Using Vectors to Determine the Ground Speed and Direction of an Airplane

A 747 jet was heading due east at 520 mph in still air and encountered a 60 mph headwind blowing in the direction N 40° W. Determine the resulting ground speed of the plane and its new bearing. Round the ground speed to the nearest hundredth.

Solution Let $\mathbf{v}$ represent the velocity of the plane in still air and let $\mathbf{w}$ represent the velocity of the wind; then $\|\mathbf{v}\| = 520$ and $\|\mathbf{w}\| = 60$. The resultant vector $\mathbf{v} + \mathbf{w}$ is the diagonal of the parallelogram shown in Figure 67. The ground speed of the airplane is the magnitude of the resultant vector $\|\mathbf{v} + \mathbf{w}\|$.

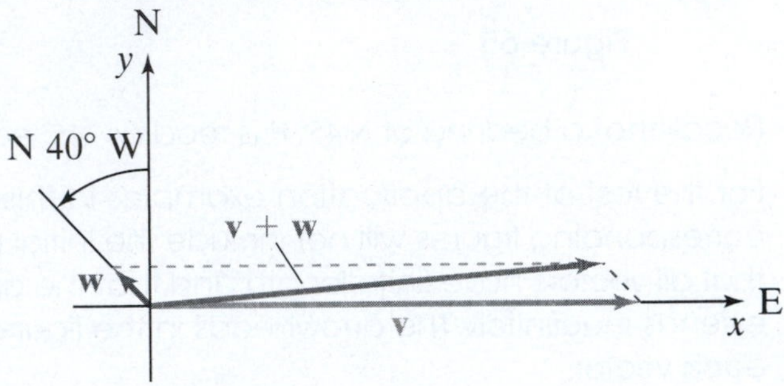

Figure 67

We first find $\mathbf{v}$ using the formula $\mathbf{v} = (\|\mathbf{v}\| \cos\theta)\mathbf{i} + (\|\mathbf{v}\| \sin\theta)\mathbf{j}$, where $\theta = 0°$ (the plane was heading due east) and $\|\mathbf{v}\| = 520$. Thus, $\mathbf{v} = (520 \cos 0°)\mathbf{i} + (520 \sin 0°)\mathbf{j} = 520(1)\mathbf{i} + 520(0)\mathbf{j} = 520\mathbf{i}$.

To find $\mathbf{w}$, we use the formula $\mathbf{w} = (\|\mathbf{w}\| \cos\theta)\mathbf{i} + (\|\mathbf{w}\| \sin\theta)\mathbf{j}$, where $\theta = 40° + 90° = 130°$ and $\|\mathbf{w}\| = 60$. Therefore, $\mathbf{w} = (60 \cos 130°)\mathbf{i} + (60 \sin 130°)\mathbf{j}$.

The resultant vector is $\mathbf{v} + \mathbf{w} = 520\mathbf{i} + (60 \cos 130°)\mathbf{i} + (60 \sin 130°)\mathbf{j} = (520 + 60 \cos 130°)\mathbf{i} + (60 \sin 130°)\mathbf{j}$.

Using the formula for the magnitude of a vector, we see that the ground speed of the plane is $\|\mathbf{v} + \mathbf{w}\| = \sqrt{(520 + 60 \cos 130°)^2 + (60 \sin 130°)^2} \approx 483.62$ mph.

The direction angle is $\theta = \tan^{-1}\dfrac{60 \sin 130°}{520 + 60 \cos 130°} \approx 5.45°$. The bearing of the plane is measured clockwise from the y-axis to $\mathbf{v} + \mathbf{w}$. Thus we must determine the complement of the direction angle, which is $90° - 5.45° = 84.55°$. Therefore, the final bearing of the plane is N 84.55° E. See Figure 68.

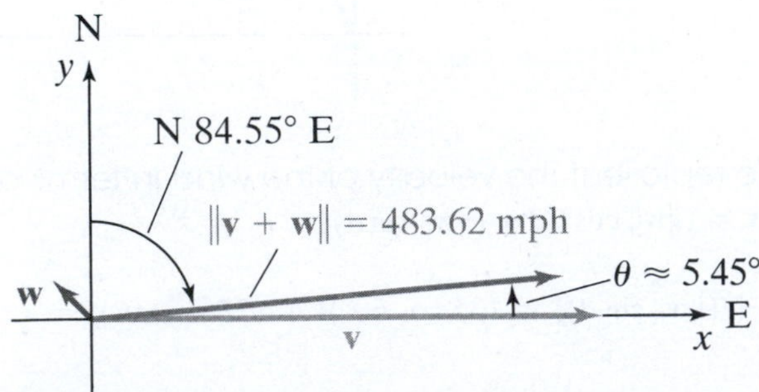

Figure 68

You Try It Work through this You Try It problem.

Work Exercises 39 and 40 in this textbook or in the MyLab Math Study Plan.

OBJECTIVE 9 USING VECTORS TO SOLVE APPLICATIONS INVOLVING FORCE

Vectors can be used to solve many applied problems involving the concept of **force.** This is because a force has *both* magnitude and direction. You can think of force as the push or pull on an object that causes the object to move. Examples include the pushing of a chair across a room or the downward gravitational pull on a ball. If more than one force is acting on an object, then the resultant force experienced by the object is the vector sum of these forces.

Isaac Newton's second law of motion states that for every action, there is an equal and opposite reaction. In the context of force, we can think of this law as saying that "forces always come in pairs." For example, when you place a 5-pound bag of groceries on your kitchen table, there is a 5-pound force of gravity acting on the bag of groceries and pulling it downward. We can represent this force using the vector $\mathbf{F}_1 = -5\mathbf{j}$. However, the bag does not move because the kitchen table stops it. This is because there is a force of equal magnitude (5 pounds) exerted by the table in the exact opposite direction.

We can represent the force of the table acting on the bag by the vector $\mathbf{F}_2 = 5\mathbf{j}$. The resultant force is $\mathbf{F}_1 + \mathbf{F}_2 = -5\mathbf{j} + 5\mathbf{j} = \mathbf{0}$, where $\mathbf{0}$ is the zero vector. Thus, the magnitude of the total force acting on the grocery bag is 0, indicating that the bag will not move. See Figure 69.

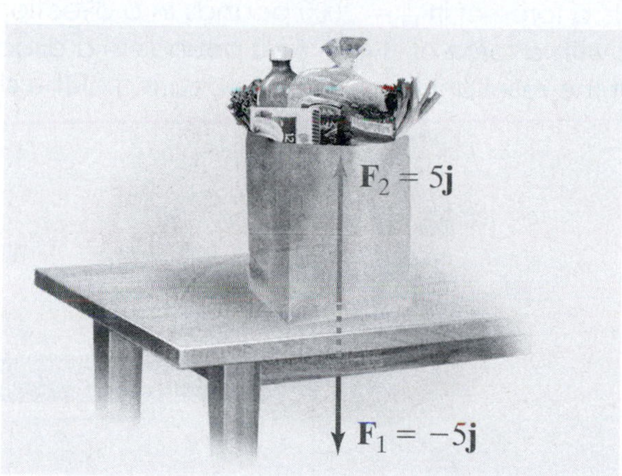

Figure 69 A 5-pound bag of groceries on a kitchen table does not move because the downward force of gravity is equal to the upward force of the table acting on the bag.

When the net result of the total force acting on an object is the zero vector, we say that the object is in **static equilibrium.** In applied problems involving force, it is often necessary to determine the forces acting on an object that will result in static equilibrium.

> **Definition** Static Equilibrium
>
> If $\mathbf{F}_1, \mathbf{F}_2, \cdots, \mathbf{F}_n$ are forces acting on an object, then the object is in **static equilibrium** if the resultant force $\mathbf{F} = \mathbf{F}_1 + \mathbf{F}_2 + \cdots + \mathbf{F}_n$ is equal to $\mathbf{0}$ (the zero vector).

Example 13 Determining the Force Necessary for an Object to Be in Static Equilibrium

The forces $\mathbf{F}_1 = 6\mathbf{i} - 8\mathbf{j}$ and $\mathbf{F}_2 = 3\mathbf{i} + 2\mathbf{j}$ are acting on an object. What additional force is required for the object to be in static equilibrium?

Solution The resultant force acting on the object is $\mathbf{F} = \mathbf{F}_1 + \mathbf{F}_2 = (6 + 3)\mathbf{i} + (-8 + 2)\mathbf{j} = 9\mathbf{i} - 6\mathbf{j}$. The additional force $\mathbf{F}_3$ required for the object to be in equilibrium is $\mathbf{F}_3 = -\mathbf{F} = -(9\mathbf{i} - 6\mathbf{j}) = -9\mathbf{i} + 6\mathbf{j}$.

You Try It Work through this You Try It problem.

Work Exercises 43–45 in this textbook or in the MyLab Math Study Plan.

If the sum of the forces acting on an object is not $\mathbf{0}$, then the resultant forces cause the object to move.

Example 14 Finding the Magnitude and Bearing of a Ship Being Towed

Two tugboats are towing a large ship out of port and into the open sea. One tugboat exerts a force of $\|\mathbf{F}_1\| = 2000$ pounds in a direction N 35° W. The other tugboat pulls with a force of $\|\mathbf{F}_2\| = 1400$ pounds in a direction S 55° W. Find the magnitude of the resultant force and the bearing of the ship.

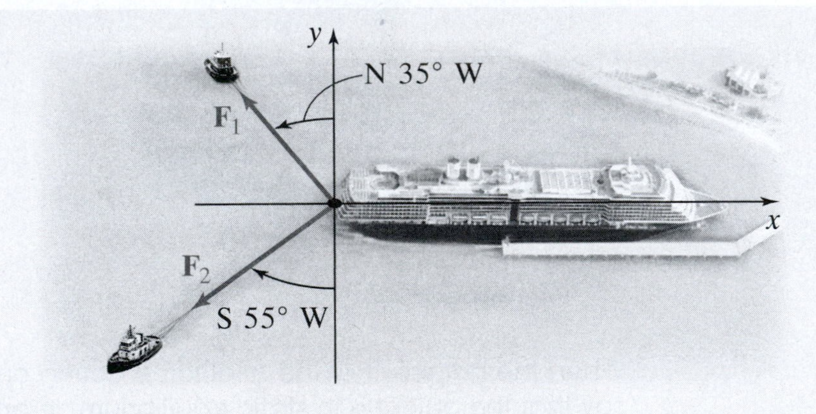

Solution We first find $\mathbf{F}_1$ using the formula $\mathbf{F}_1 = (\|\mathbf{F}_1\|\cos\theta)\mathbf{i} + (\|\mathbf{F}_1\|\sin\theta)\mathbf{j}$, where $\|\mathbf{F}_1\| = 2000$ and $\theta = 90° + 35° = 125°$. Thus, $\mathbf{F}_1 = (2000\cos 125°)\mathbf{i} + (2000\sin 125°)\mathbf{j}$.

Similarly, we find $\mathbf{F}_2$ using the formula $\mathbf{F}_2 = (\|\mathbf{F}_2\|\cos\theta)\mathbf{i} + (\|\mathbf{F}_2\|\sin\theta)\mathbf{j}$, where $\|\mathbf{F}_2\| = 1400$ and $\theta = 270° - 55° = 215°$. Thus, $\mathbf{F}_2 = (1400\cos 215°)\mathbf{i} + (1400\sin 215°)\mathbf{j}$.

The resultant force is $\mathbf{F} = \mathbf{F}_1 + \mathbf{F}_2 = (2000\cos 125° + 1400\cos 215°)\mathbf{i} + (2000\sin 125° + 1400\sin 215°)\mathbf{j} \approx -2293.97\mathbf{i} + 835.30\mathbf{j}$.

The magnitude of the resultant force is approximately

$$\|\mathbf{F}\| = \sqrt{(-2293.97)^2 + (835.30)^2} \approx 2441.32.$$

To determine the ship's bearing, we must first determine the direction angle θ for $\mathbf{F} = -2293.97\mathbf{i} + 835.30\mathbf{j}$.

To find the direction angle θ, start by determining the reference angle θ_R. Using the fact that $\tan\theta = \frac{b}{a}$, we can find θ_R by solving the equation $\tan\theta_R = \left|\frac{b}{a}\right|$.

$\tan\theta_R = \left|\frac{b}{a}\right|$ Write the equation relating θ_R, a, and b.

$\tan\theta_R = \left|\frac{835.30}{-2293.97}\right|$ Substitute $a = -2293.97$ and $b = 835.30$.

$\tan\theta_R = \frac{835.30}{2293.97}$ Simplify.

We can use the inverse tangent function to solve for θ_R. Thus, $\theta_R = \tan^{-1}\left(\frac{835.30}{2293.97}\right) \approx 20.00°$.

The bearing of the ship is measured from the positive y-axis to $\mathbf{F}$. Thus we must determine the complement of θ_R, which is $90° - 20.00° \approx 70.00°$. Therefore, the ship's bearing is N 70.00° W. See Figure 70.

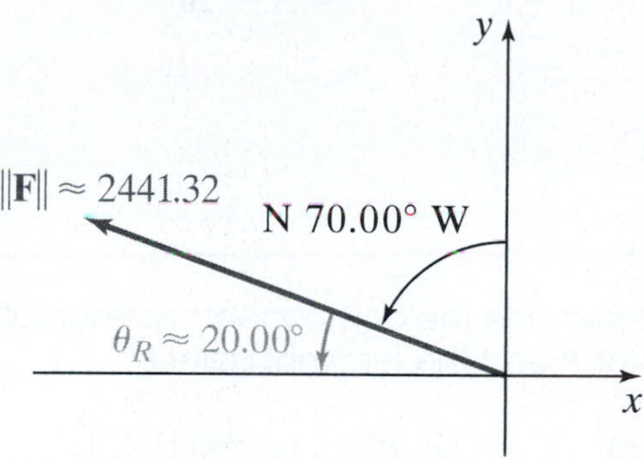

Figure 70

You Try It Work through this You Try It problem.

Work Exercises 46 and 47 in this textbook or in the MyLab Math Study Plan.

10.4 Exercises

In Exercises 1–3, determine the magnitude of $\mathbf{v}$.

1.

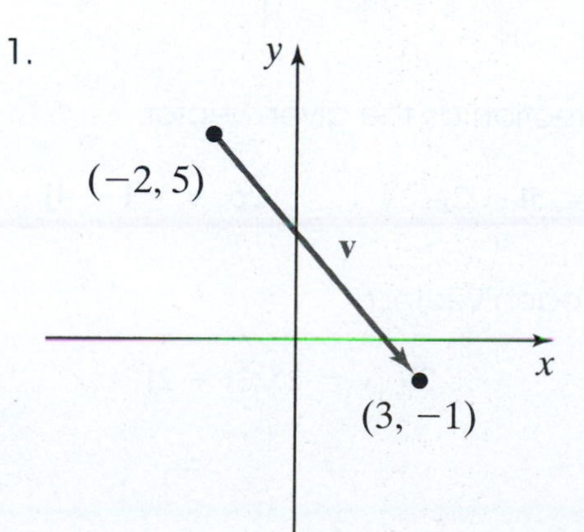

2.

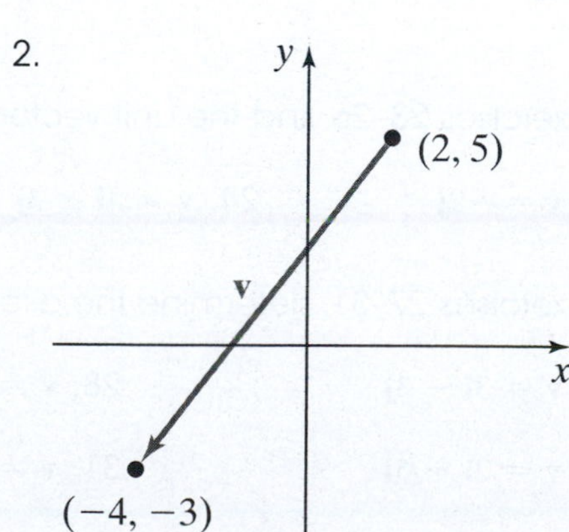

3.

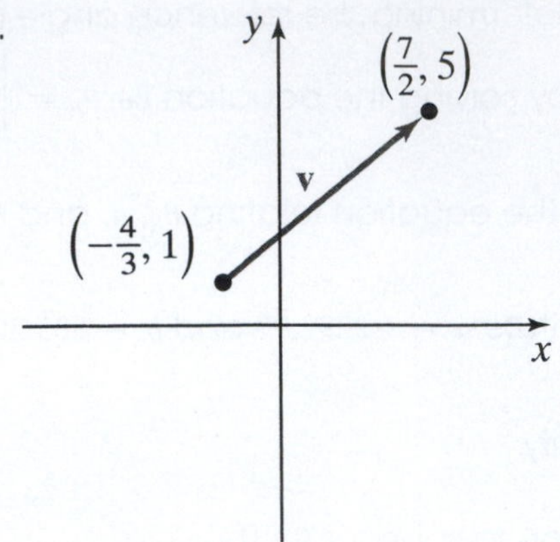

In Exercises 4–10, use the given vectors $\mathbf{u}$, $\mathbf{v}$, and $\mathbf{w}$ to draw each vector.

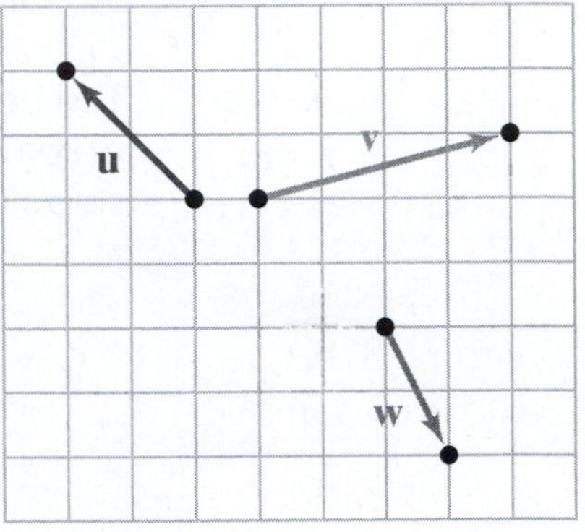

4. $2\mathbf{w}$ 5. $\mathbf{u} + \mathbf{v}$ 6. $-2\mathbf{v}$

7. $\mathbf{w} - \mathbf{v}$ 8. $\frac{1}{2}\mathbf{u}$ 9. $\mathbf{v} - 2\mathbf{u}$

10. $2\mathbf{w} + 2\mathbf{u} - \mathbf{v}$

In Exercises 11–13, determine the component representation and magnitude of the vector having the initial point P and the terminal point Q.

11. $P(2, 3)$ and $Q(6, 6)$ 12. $P(-3, 1)$ and $Q(1, 4)$ 13. $P(1, -4)$ and $Q(-5, 7)$

In Exercises 14–16, write each vector in terms of the unit vectors $\mathbf{i}$ and $\mathbf{j}$.

14. $\mathbf{v} = \langle 3, 5 \rangle$ 15. $\mathbf{v} = \langle -4, 1 \rangle$ 16. $\mathbf{v} = \langle -6, -7 \rangle$

In Exercises 17–22, if $\mathbf{u} = -3\mathbf{i} + 7\mathbf{j}$ and $\mathbf{v} = 5\mathbf{i} - \mathbf{j}$, then find the indicated vectors in terms of $\mathbf{i}$ and $\mathbf{j}$ and find the magnitude of each vector.

17. $3\mathbf{u}$ 18. $\mathbf{u} + \mathbf{v}$ 19. $\mathbf{u} - \mathbf{v}$

20. $4\mathbf{u} + 3\mathbf{v}$ 21. $-\frac{1}{3}\mathbf{u}$ 22. $\frac{1}{2}\mathbf{u} - \frac{1}{4}\mathbf{v}$

In Exercises 23–26, find the unit vector in the same direction as the given vector.

23. $\mathbf{v} = -2\mathbf{j}$ 24. $\mathbf{v} = 4\mathbf{i} - 3\mathbf{j}$ 25. $\mathbf{v} = 5\mathbf{i} - 2\mathbf{j}$ 26. $\mathbf{v} = \mathbf{i} - 4\mathbf{j}$

In Exercises 27–31, determine the direction angle for each vector.

27. $\mathbf{v} = 3\mathbf{i} - 3\mathbf{j}$ 28. $\mathbf{v} = -5\mathbf{i} + 5\sqrt{3}\mathbf{j}$ 29. $\mathbf{v} = 2\sqrt{3}\mathbf{i} + 2\mathbf{j}$

30. $\mathbf{v} = 2\mathbf{i} + 5\mathbf{j}$ 31. $\mathbf{v} = 2\mathbf{i} - 7\mathbf{j}$

In Exercises 32–36, represent the vector $\mathbf{v}$ in the form $\mathbf{v} = a\mathbf{i} + b\mathbf{j}$ whose magnitude and direction angle are given.

32. $\|\mathbf{v}\| = 15; \theta = 150°$

33. $\|\mathbf{v}\| = 22; \theta = 315°$

34. $\|\mathbf{v}\| = 7; \theta = 300°$

35. $\|\mathbf{v}\| = 10; \theta = 100°$ (Round a and b to two decimal places.)

36. $\|\mathbf{v}\| = \frac{2}{3}; \theta = 212°$ (Round a and b to two decimal places.)

37. An outfielder releases a baseball with a speed of 100 feet per second at an angle of 30° with the horizontal. Express the velocity of the ball as a vector in terms of $\mathbf{i}$ and $\mathbf{j}$.

38. An airplane approaches a runway at 180 miles per hour at an angle of 6° with the runway. Express the velocity of the plane as a vector in terms of $\mathbf{i}$ and $\mathbf{j}$. Round a and b to two decimal places.

39. The wind is blowing at a speed of 27 mph in a direction of N 50° E. Express the velocity of the wind as a vector in terms of $\mathbf{i}$ and $\mathbf{j}$.

40. The wind is blowing at a speed of 40 mph in a direction of S 20° W. Express the velocity of the wind as a vector in terms of $\mathbf{i}$ and $\mathbf{j}$.

41. An airplane was heading due east at 280 mph in still air and encountered a 30 mph headwind blowing in the direction N 40° W. Determine the resulting ground speed of the plane and its new bearing. Round the ground speed to the nearest hundredth.

42. An airplane was heading due west at 280 mph in still air and encountered a 50 mph tailwind blowing in the direction N 40° W. Determine the ground speed of the plane and its new bearing. Round the ground speed to the nearest hundredth.

43. The forces $\mathbf{F}_1 = -5\mathbf{i} + 4\mathbf{j}$ and $\mathbf{F}_2 = 6\mathbf{i} - 11\mathbf{j}$ are acting on an object. What additional force is required for the object to be in static equilibrium?

44. The forces $\mathbf{F}_1 = 8\mathbf{i} - 11\mathbf{j}$, $\mathbf{F}_2 = -12\mathbf{i} - 13\mathbf{j}$, and $\mathbf{F}_3 = -7\mathbf{i} + 8\mathbf{j}$ are acting on an object. What additional force is required for the object to be in static equilibrium?

45. At the county Fourth of July fair, a local strong man challenges contestants two at a time to a tug-of-war contest. Contestant A can tug with a force of 200 pounds. Contestant B can tug with a force of 350 pounds. The angle between the ropes of the two contestants is 40°. With how much force must the local strong man tug so that the rope does not move?

46. Two tugboats are towing a Naval ship out of port and into the open sea. One tugboat exerts a force of $\|\mathbf{F}_1\| = 1200$ pounds in a direction N 45° W. The other tugboat pulls with a force of $\|\mathbf{F}_2\| = 800$ pounds in a direction S 60° W. Find the magnitude of the resultant force and bearing of the Naval ship.

47. Two tugboats are towing a grain barge out of dangerous shallow water. One tugboat exerts a force of $\|\mathbf{F}_1\| = 500$ pounds in a direction N 25° E. The other tugboat pulls with a force of $\|\mathbf{F}_2\| = 900$ pounds in a direction S 20° E. Find the magnitude of the resultant force and bearing of the grain barge.

(eText Screens 10.5-1–10.5-31)

10.5 The Dot Product

THINGS TO KNOW

Before working through this section, be sure that you are familiar with the following concepts:

VIDEO ANIMATION INTERACTIVE

 1. Understanding Vectors in Terms of $\mathbf{i}$ and $\mathbf{j}$ (Section 10.4) 

 2. Finding a Unit Vector (Section 10.4) 

You Try It 3. Determining the Direction Angle of a Vector (Section 10.4) 

 4. Representing a Vector in Terms of $\mathbf{i}$ and $\mathbf{j}$ Given Its Magnitude and Direction (Section 10.4) 

OBJECTIVES

1 Understanding the Dot Product and Its Properties

2 Using the Dot Product to Determine the Angle between Two Vectors

3 Using the Dot Product to Determine If Two Vectors Are Orthogonal or Parallel

4 Decomposing a Vector into Two Orthogonal Vectors

5 Solving Applications Involving Forces on an Inclined Plane

6 Solving Applications Involving Work

SECTION 10.5 EXERCISES

In Section 10.4 we introduced three different ways to represent vectors. A vector $\mathbf{v}$ can be represented geometrically by an arrow of finite length in a plane, it can be represented using the component notation $\mathbf{v} = \langle a, b \rangle$, or it can be represented using the unit vectors $\mathbf{i}$ and $\mathbf{j}$ as $\mathbf{v} = a\mathbf{i} + b\mathbf{j}$. We will use this third representation of vectors throughout this section.

OBJECTIVE 1 UNDERSTANDING THE DOT PRODUCT AND ITS PROPERTIES

In Section 10.4 we learned how to add and subtract two vectors. We also learned how to multiply a vector by a scalar. We now introduce an operation called the **dot product** of two vectors. The dot product of two vectors results in a scalar. As you will see at the end of this section, the dot product is useful in many applications.

Definition The Dot Product of Two Vectors

If $\mathbf{u} = a\mathbf{i} + b\mathbf{j}$ and $\mathbf{v} = c\mathbf{i} + d\mathbf{j}$, then the **dot product** is $\mathbf{u} \cdot \mathbf{v} = ac + bd$.

Note that the dot product is often referred to as the **inner product** or the **scalar product** of two vectors.

Example 1 Determining the Dot Product of Two Vectors

If $\mathbf{u} = -3\mathbf{i} + 5\mathbf{j}$ and $\mathbf{v} = 7\mathbf{i} - 4\mathbf{j}$, then find $\mathbf{u} \cdot \mathbf{v}$ and $\mathbf{v} \cdot \mathbf{u}$.

Solution $\mathbf{u} \cdot \mathbf{v} = (-3)(7) + (5)(-4) = -21 - 20 = -41$

$\mathbf{v} \cdot \mathbf{u} = (7)(-3) + (-4)(5) = -21 - 20 = -41$

It is no coincidence that the values of $\mathbf{u} \cdot \mathbf{v}$ and $\mathbf{v} \cdot \mathbf{u}$ in Example 1 are equal. In fact, for any two vectors $\mathbf{u}$ and $\mathbf{v}$, $\mathbf{u} \cdot \mathbf{v} = \mathbf{v} \cdot \mathbf{u}$. This is just one of five properties of the dot product.

Dot Product Properties

If $\mathbf{u}$, $\mathbf{v}$, and $\mathbf{w}$ are vectors and if k is a scalar, then the following properties are true.

1. $\mathbf{u} \cdot \mathbf{v} = \mathbf{v} \cdot \mathbf{u}$
2. $\mathbf{u} \cdot (\mathbf{v} + \mathbf{w}) = \mathbf{u} \cdot \mathbf{v} + \mathbf{u} \cdot \mathbf{w}$
3. $k(\mathbf{u} \cdot \mathbf{v}) = (k\mathbf{u}) \cdot \mathbf{v} = \mathbf{u} \cdot (k\mathbf{v})$
4. $\mathbf{0} \cdot \mathbf{v} = 0$
5. $\mathbf{v} \cdot \mathbf{v} = \|\mathbf{v}\|^2$

Example 2 Finding Dot Products

If $\mathbf{u} = 4\mathbf{i} + 6\mathbf{j}$, $\mathbf{v} = -2\mathbf{i} + 8\mathbf{j}$, and $\mathbf{w} = -3\mathbf{i} - \mathbf{j}$, then find each of the following.

a. $\mathbf{u} \cdot \mathbf{v}$ b. $\mathbf{u} \cdot (\mathbf{v} + \mathbf{w})$ c. $\mathbf{u} \cdot (-5\mathbf{v})$ d. $\|\mathbf{w}\|^2$

Solution

a. $\mathbf{u} \cdot \mathbf{v} = (4)(-2) + (6)(8) = -8 + 48 = 40$

b. $\mathbf{u} \cdot (\mathbf{v} + \mathbf{w})$ Write the original expression.

$= \mathbf{u} \cdot \mathbf{v} + \mathbf{u} \cdot \mathbf{w}$ Use Property 2.

$= 40 + (4)(-3) + (6)(-1)$ Use the result from part a and determine $\mathbf{u} \cdot \mathbf{w}$.

$= 22$ Simplify.

c. $\mathbf{u} \cdot (-5\mathbf{v})$ Write the original expression.

$= -5(\mathbf{u} \cdot \mathbf{v})$ Use Property 3.

$= -5(40)$ Use the result from part a.

$= -200$ Simplify.

d. $\|\mathbf{w}\|^2$ Write the original expression.

$= \mathbf{w} \cdot \mathbf{w}$ Use Property 5.

$= (-3)(-3) + (-1)(-1)$ Substitute $a = -3, c = -3, b = -1$, and $d = -1$ into the dot product formula.

$= 10$ Simplify

You Try It Work through this You Try It problem.

Work Exercises 1–8 in this textbook or in the MyLab Math Study Plan.

OBJECTIVE 2 USING THE DOT PRODUCT TO DETERMINE THE ANGLE BETWEEN TWO VECTORS

One of the important applications of the dot product is that it is useful in determining the angle between two non-zero vectors. Suppose that $\mathbf{u}$ and $\mathbf{v}$ are non-zero vectors that share the same initial point. Let θ be the angle between the two vectors such that $0 \le \theta \le 180°$. We can determine the angle between the two vectors using the following formula.

The Angle between Two Vectors

If $\mathbf{u}$ and $\mathbf{v}$ are non-zero vectors and if θ is the angle between $\mathbf{u}$ and $\mathbf{v}$, then $\cos\theta = \dfrac{\mathbf{u}\cdot\mathbf{v}}{\|\mathbf{u}\|\|\mathbf{v}\|}$.

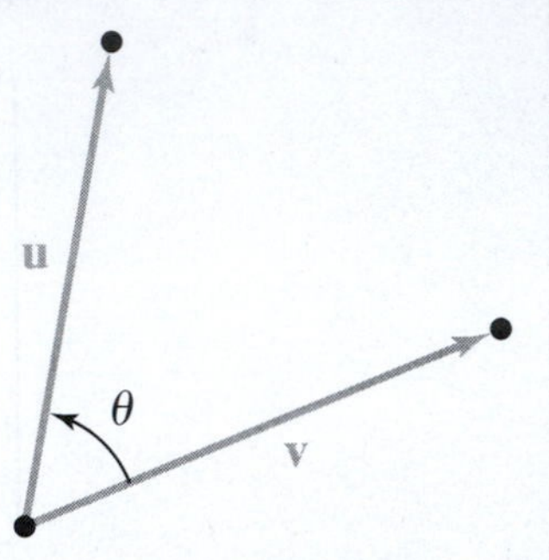

To prove the formula $\cos\theta = \dfrac{\mathbf{u}\cdot\mathbf{v}}{\|\mathbf{u}\|\,\|\mathbf{v}\|}$, start by coinciding the initial point of vector $-\mathbf{v}$ with the terminal point of $\mathbf{u}$. The angle between $\mathbf{u}$ and $-\mathbf{v}$ is θ because $\mathbf{v}$ and $-\mathbf{v}$ are parallel and alternate interior angles are congruent. See Figure 71a. The vector that extends from the initial point of $\mathbf{u}$ to the terminal point of $-\mathbf{v}$ is $\mathbf{u} - \mathbf{v}$. See Figure 71b. The triangle formed in Figure 71b has sides of lengths $\|\mathbf{u}\|$, $\|-\mathbf{v}\| = \|\mathbf{v}\|$, and $\|\mathbf{u} - \mathbf{v}\|$. See Figure 71c.

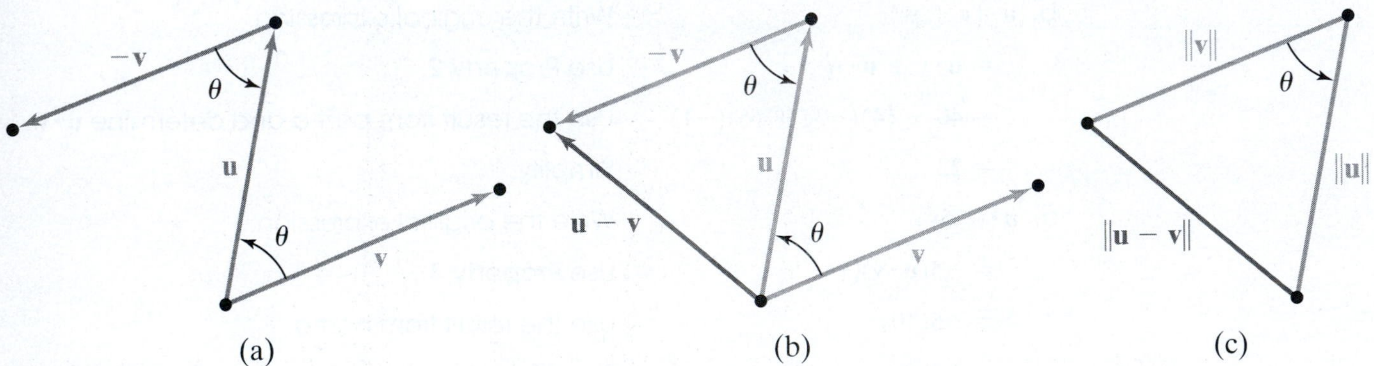

Figure 71

We use the triangle in Figure 71c and the **Law of Cosines** to get $\|\mathbf{u} - \mathbf{v}\|^2 = \|\mathbf{u}\|^2 + \|\mathbf{v}\|^2 - 2\|\mathbf{u}\|\,\|\mathbf{v}\|\cos\theta$. Carefully watch this **video** to see how to use **dot product properties** to rewrite this equation as $\cos\theta = \dfrac{\mathbf{u}\cdot\mathbf{v}}{\|\mathbf{u}\|\,\|\mathbf{v}\|}$.

Example 3 Determining the Angle between Two Vectors

Determine the angle between each pair of vectors. Give the angle in degrees rounded to the nearest hundredth of a degree.

a. $\mathbf{u} = \mathbf{i} + 4\mathbf{j}, \mathbf{v} = -2\mathbf{i} + 5\mathbf{j}$

b. $\mathbf{u} = 3\mathbf{i} - 2\mathbf{j}, \mathbf{v} = 4\mathbf{i} + 6\mathbf{j}$

Solution

a. To determine the angle between $\mathbf{u} = \mathbf{i} + 4\mathbf{j}$ and $\mathbf{v} = -2\mathbf{i} + 5\mathbf{j}$, we use the formula $\cos\theta = \dfrac{\mathbf{u}\cdot\mathbf{v}}{\|\mathbf{u}\|\,\|\mathbf{v}\|}$.

$$\cos\theta = \frac{(1)(-2) + (4)(5)}{\sqrt{17}\sqrt{29}} = \frac{18}{\sqrt{493}}$$

To find θ, we use a calculator set in degree mode and use the inverse cosine function. Recall that the range of the inverse cosine function is $0 \le \theta \le 180°$.

$$\theta = \cos^{-1}\left(\frac{18}{\sqrt{493}}\right) \approx 35.84°$$

b. To determine the angle between $\mathbf{u} = 3\mathbf{i} - 2\mathbf{j}$ and $\mathbf{v} = 4\mathbf{i} + 6\mathbf{j}$, we use the formula $\cos\theta = \dfrac{\mathbf{u}\cdot\mathbf{v}}{\|\mathbf{u}\|\,\|\mathbf{v}\|}$.

$$\cos\theta = \frac{(3)(4) + (-2)(6)}{\sqrt{13}\sqrt{52}} = \frac{0}{\sqrt{676}} = 0$$

The only angle on the interval $0 \le \theta \le 180°$ such that $\cos\theta = 0$ is $\theta = 90°$.

Note that if $\mathbf{u}$ and $\mathbf{v}$ are non-zero vectors, then we can multiply both sides of the equation $\cos\theta = \dfrac{\mathbf{u}\cdot\mathbf{v}}{\|\mathbf{u}\|\,\|\mathbf{v}\|}$ by $\|\mathbf{u}\|\,\|\mathbf{v}\|$ to obtain the equivalent equation $\|\mathbf{u}\|\,\|\mathbf{v}\|\cos\theta = \mathbf{u}\cdot\mathbf{v}$. This is an alternate form of the dot product.

Definition The Alternate Form of the Dot Product of Two Vectors

If $\mathbf{u}$ and $\mathbf{v}$ are non-zero vectors, then the **alternate form of the dot product** is

$$\mathbf{u}\cdot\mathbf{v} = \|\mathbf{u}\|\,\|\mathbf{v}\|\cos\theta.$$

You Try It Work through this You Try It problem.

Work Exercises 9–13 in this textbook or in the MyLab Math Study Plan.

OBJECTIVE 3 USING THE DOT PRODUCT TO DETERMINE IF TWO VECTORS ARE ORTHOGONAL OR PARALLEL

ORTHOGONAL VECTORS

In Example 3b, we see that the angle between the vectors $\mathbf{u} = 3\mathbf{i} - 2\mathbf{j}$ and $\mathbf{v} = 4\mathbf{i} + 6\mathbf{j}$ is $\theta = 90°$. When the angle between two non-zero vectors is $\theta = 90°$, then the vectors are said to be **orthogonal** vectors.

Definition Orthogonal Vectors

Two non-zero vectors $\mathbf{u}$ and $\mathbf{v}$ are **orthogonal** if the angle between them is $\theta = 90°$.

Because we know from the definition of orthogonal vectors that the angle between **u** and **v** is $\theta = 90°$, then by the alternate definition of the dot product we get $\mathbf{u} \cdot \mathbf{v} = \|\mathbf{u}\|\|\mathbf{v}\| \cos 90° = \|\mathbf{u}\|\|\mathbf{v}\|(0) = 0$. This gives us the following test to determine if two non-zero vectors are orthogonal.

Test for Orthogonal Vectors

Two non-zero vectors **u** and **v** are orthogonal if and only if $\mathbf{u} \cdot \mathbf{v} = 0$.

Note Every vector is orthogonal to the zero vector.

Example 4 Finding a Component of an Orthogonal Vector

Determine the value of b so that the vectors $\mathbf{u} = \mathbf{i} + b\mathbf{j}$. and $\mathbf{v} = 3\mathbf{i} + 10\mathbf{j}$. are orthogonal.

Solution We know that when two vectors are orthogonal, then the dot product must be zero. Thus, we are looking for the value of b such that $\mathbf{u} \cdot \mathbf{v} = 0$. The dot product of **u** and **v** is $\mathbf{u} \cdot \mathbf{v} = (1)(3) + (b)(10) = 3 + 10b$.

We now set the expression $3 + 10b$ equal to zero to obtain $b = -\frac{3}{10}$.

We can verify that the vectors $\mathbf{u} = \mathbf{i} - \frac{3}{10}\mathbf{j}$ and $\mathbf{v} = 3\mathbf{i} + 10\mathbf{j}$ are orthogonal by showing that $\mathbf{u} \cdot \mathbf{v} = 0$.

$$\mathbf{u} \cdot \mathbf{v} = (1)(3) + \left(-\frac{3}{10}\right)(10) = 3 - 3 = 0$$

You Try It Work through this You Try It problem.

Work Exercises 14 and 15 in this textbook or in the MyLab Math Study Plan.

PARALLEL VECTORS

Recall that two non-zero vectors **u** and **v** are parallel vectors if there is a non-zero scalar k such that $\mathbf{v} = k\mathbf{u}$. To establish a test to determine if two vectors are parallel, we can examine the angle between them. We start by finding the dot product.

$\mathbf{u} \cdot \mathbf{v}$	Start with the dot product of **u** and **v**.
$\mathbf{u} \cdot (k\mathbf{u})$	Vector **v** must be a scalar multiple of **u**.
$= k(\mathbf{u} \cdot \mathbf{u})$	Use Dot Product Property 3.
$= k\|\mathbf{u}\|^2$	Use Dot Product Property 5.

Therefore, the cosine of the angle between parallel vectors **u** and $\mathbf{v} = k\mathbf{u}$ is

$$\cos\theta = \frac{\mathbf{u} \cdot k\mathbf{u}}{\|\mathbf{u}\|\|k\mathbf{u}\|} = \frac{k\|\mathbf{u}\|^2}{|k|\|\mathbf{u}\|\|\mathbf{u}\|} = \frac{k\|\mathbf{u}\|^2}{|k|\|\mathbf{u}\|^2} = \frac{k}{|k|}.$$

If $k > 0$, then $\cos\theta = \frac{k}{|k|} = \frac{k}{k} = 1$ and hence $\theta = 0°$.

If $k < 0$, then $\cos\theta = \frac{k}{|k|} = \frac{k}{-k} = -1$ and hence $\theta = 180°$.

This gives us the following test to determine if two non-zero vectors are parallel.

> **Test for Parallel Vectors**
>
> Two non-zero vectors $\mathbf{u}$ and $\mathbf{v}$ are parallel if $\frac{\mathbf{u}\cdot\mathbf{v}}{\|\mathbf{u}\|\,\|\mathbf{v}\|} = \mathbf{1}$ or $\frac{\mathbf{u}\cdot\mathbf{v}}{\|\mathbf{u}\|\,\|\mathbf{v}\|} = -1$.
>
> Note that the angle between parallel vectors is $\theta = 0°$ if $\frac{\mathbf{u}\cdot\mathbf{v}}{\|\mathbf{u}\|\,\|\mathbf{v}\|} = 1$.
>
> The angle between parallel vectors is $\theta = 180°$ if $\frac{\mathbf{u}\cdot\mathbf{v}}{\|\mathbf{u}\|\,\|\mathbf{v}\|} = -1$.

Example 5 Determining If Two Vectors Are Orthogonal, Parallel, or Neither

Determine if $\mathbf{u} = -\frac{1}{2}\mathbf{i} - \mathbf{j}$ and $\mathbf{v} = 2\mathbf{i} + 4\mathbf{j}$ are orthogonal, parallel, or neither.

Solution We start by determining the dot product $\mathbf{u}\cdot\mathbf{v}$. The dot product of $\mathbf{u}$ and $\mathbf{v}$ is

$$\mathbf{u}\cdot\mathbf{v} = \left(-\frac{1}{2}\right)(2) + (-1)(4) = -1 - 4 = -5$$

The vectors are not orthogonal because the value of $\mathbf{u}\cdot\mathbf{v}$. is non-zero.

Next, determine the value of $\frac{\mathbf{u}\cdot\mathbf{v}}{\|\mathbf{u}\|\,\|\mathbf{v}\|}$.

$$\frac{\mathbf{u}\cdot\mathbf{v}}{\|\mathbf{u}\|\,\|\mathbf{v}\|} = \frac{-5}{\sqrt{\left(\frac{-1}{2}\right)^2 + (-1)^2}\sqrt{2^2 + 4^2}} = \frac{-5}{\sqrt{\frac{5}{4}}\sqrt{20}} = \frac{-5}{\sqrt{\frac{100}{4}}} = \frac{-5}{\sqrt{25}} = \frac{-5}{5} = -1$$

Therefore, the vectors are parallel. Furthermore, the angle between them is $\theta = 180°$ because $\cos\theta = \frac{\mathbf{u}\cdot\mathbf{v}}{\|\mathbf{u}\|\,\|\mathbf{v}\|} = -1$.

You Try It Work through this You Try It problem.

Work Exercises 16–19 in this textbook or in the MyLab Math Study Plan.

OBJECTIVE 4 DECOMPOSING A VECTOR INTO TWO ORTHOGONAL VECTORS

An important use of dot products is the idea of **projections**. Consider the non-zero vectors $\mathbf{v}$ and $\mathbf{w}$ that share the same initial point P. See Figure 72a. Draw a line segment from the terminal point of $\mathbf{v}$ that is perpendicular to $\mathbf{w}$. Label this point Q, thus creating a right triangle. See Figure 72b. The green vector $\overrightarrow{PQ}$, which is the side of the right triangle adjacent to θ in Figure 72c, is called the **vector projection of $\mathbf{v}$ onto $\mathbf{w}$**. We will denote the vector projection of $\mathbf{v}$ onto $\mathbf{w}$ as $\text{proj}_{\mathbf{w}}\mathbf{v}$.

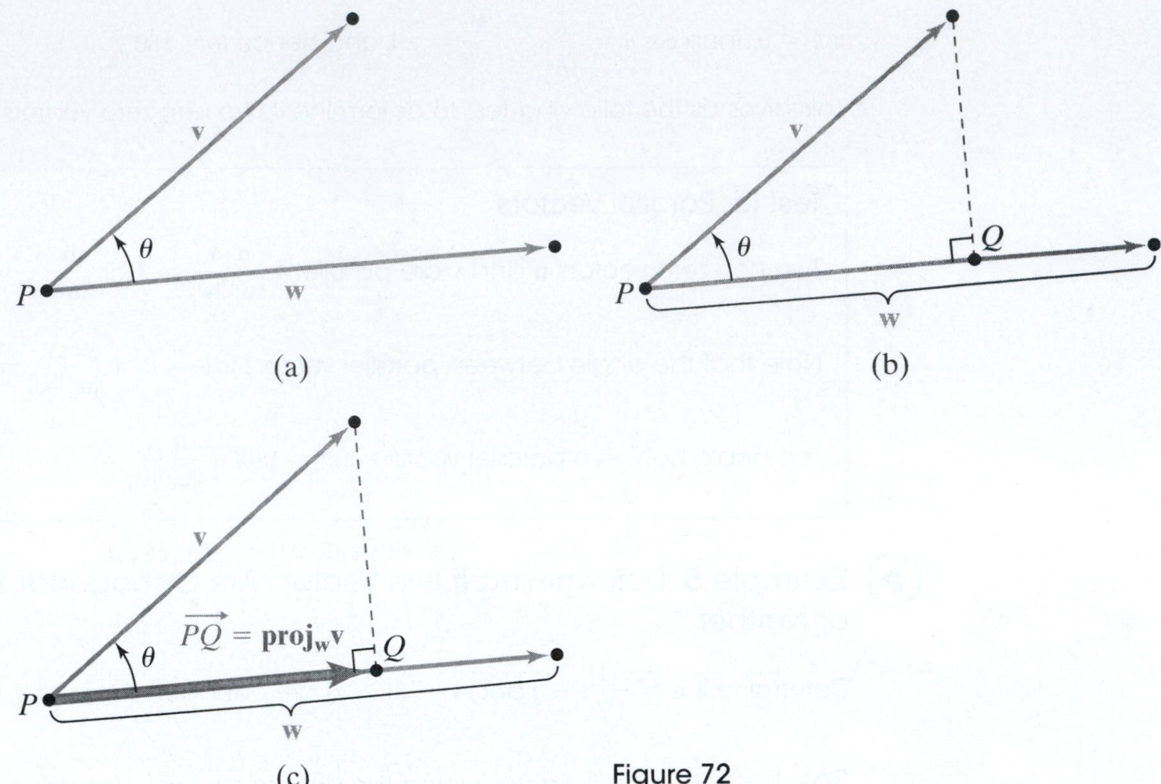

Figure 72

Our goal is to represent the vector $\text{proj}_{\mathbf{w}}\mathbf{v}$ in terms of $\mathbf{v}$ and $\mathbf{w}$. We start by determining the magnitude of $\text{proj}_{\mathbf{w}}\mathbf{v}$. Looking at the right triangle in **Figure 72c**, we see that the length of the hypotenuse of the right triangle is the magnitude of $\mathbf{v}$ and the length of the side adjacent to θ is the magnitude of $\text{proj}_{\mathbf{w}}\mathbf{v}$. Using the right triangle definition of the cosine function, we get

$$\cos\theta = \frac{adj}{hyp} = \frac{\|\text{proj}_{\mathbf{w}}\mathbf{v}\|}{\|\mathbf{v}\|}.$$

We can multiply both sides of this equation by $\|\mathbf{v}\|$ because $\mathbf{v}$ is a non-zero vector. The result is $\|\text{proj}_{\mathbf{w}}\mathbf{v}\| = \|\mathbf{v}\|\cos\theta$, which is called the **scalar component of $\mathbf{v}$ in the direction of $\mathbf{w}$**.

Definition The Scalar Component of $\mathbf{v}$ in the Direction of $\mathbf{w}$

If θ is the angle between $\mathbf{v}$ and $\mathbf{w}$, then the **scalar component of $\mathbf{v}$ in the direction of $\mathbf{w}$** is $\|\text{proj}_{\mathbf{w}}\mathbf{v}\| = \|\mathbf{v}\|\cos\theta$.

Now that we know the magnitude of $\text{proj}_{\mathbf{w}}\mathbf{v}$, we can develop a formula that represents $\text{proj}_{\mathbf{w}}\mathbf{v}$ in terms of $\mathbf{v}$ and $\mathbf{w}$. Recall that the **unit vector in the direction of $\mathbf{w}$** is defined as $\frac{\mathbf{w}}{\|\mathbf{w}\|}$. The vector $\text{proj}_{\mathbf{w}}\mathbf{v}$ is equal to its magnitude (given by $\|\mathbf{v}\|\cos\theta$) times the unit vector in the direction of $\mathbf{w}$ (given by $\frac{\mathbf{w}}{\|\mathbf{w}\|}$). We can thus represent the vector $\text{proj}_{\mathbf{w}}\mathbf{v}$ as

$$\text{proj}_{\mathbf{w}}\mathbf{v} = \|\mathbf{v}\|\cos\theta\frac{\mathbf{w}}{\|\mathbf{w}\|}.$$

We will multiply the right-hand side of $\text{proj}_{\mathbf{w}}\mathbf{v} = \|\mathbf{v}\| \cos\theta \dfrac{\mathbf{w}}{\|\mathbf{w}\|}$ by $1 = \dfrac{\|\mathbf{w}\|}{\|\mathbf{w}\|}$ and then use the **alternate definition of the dot product** to represent the vector in terms of $\mathbf{v}$ and $\mathbf{w}$.

$\text{proj}_{\mathbf{w}}\mathbf{v} = \|\mathbf{v}\| \cos\theta \dfrac{\mathbf{w}}{\|\mathbf{w}\|}$ Start with the vector representation of $\text{proj}_{\mathbf{w}}\mathbf{v} = \|\mathbf{v}\| \cos\theta \dfrac{w}{\|\mathbf{w}\|}$.

$= \|\mathbf{v}\| \cos\theta \dfrac{\mathbf{w}}{\|\mathbf{w}\|} \dfrac{\|\mathbf{w}\|}{\|\mathbf{w}\|}$ Multiply the right-hand side by $1 = \dfrac{\|\mathbf{w}\|}{\|\mathbf{w}\|}$.

$= \|\mathbf{v}\|\|\mathbf{w}\| \cos\theta \dfrac{\mathbf{w}}{\|\mathbf{w}\|^2}$ Rearrange the order of the factors and write $\|\mathbf{w}\|\,\|\mathbf{w}\|$ as $\|\mathbf{w}\|^2$.

$= \dfrac{\mathbf{v} \cdot \mathbf{w}}{\|\mathbf{w}\|^2}\mathbf{w}$ Substitute $\mathbf{v} \cdot \mathbf{w}$ for $\|\mathbf{v}\|\,\|\mathbf{w}\| \cos\theta$ using the **alternate definition of the dot product**.

We now define $\text{proj}_{\mathbf{w}}\mathbf{v}$, the **vector projection of v onto w**, in terms of $\mathbf{v}$ and $\mathbf{w}$.

> **Definition** The Vector Projection of $\mathbf{v}$ onto $\mathbf{w}$
>
> Given two non-zero vectors $\mathbf{v}$ and $\mathbf{w}$, the **vector projection of v onto w** is given by $\text{proj}_{\mathbf{w}}\mathbf{v} = \dfrac{\mathbf{v} \cdot \mathbf{w}}{\|\mathbf{w}\|^2}\mathbf{w}$.

Example 6 Determining the Vector Projection of v onto w

If $\mathbf{v} = \mathbf{i} + 2\mathbf{j}$ and $\mathbf{w} = 4\mathbf{i} + \mathbf{j}$, determine the vector $\text{proj}_{\mathbf{w}}\mathbf{v}$.

Solution To determine the vector $\text{proj}_{\mathbf{w}}\mathbf{v}$, we use the formula $\text{proj}_{\mathbf{w}}\mathbf{v} = \dfrac{\mathbf{v} \cdot \mathbf{w}}{\|\mathbf{w}\|^2}\mathbf{w}$.

$$\text{proj}_{\mathbf{w}}\mathbf{v} = \frac{\mathbf{v} \cdot \mathbf{w}}{\|\mathbf{w}\|^2}\mathbf{w} = \frac{(1)(4) + (2)(1)}{(4)^2 + (1)^2}(4\mathbf{i} + \mathbf{j}) = \frac{6}{17}(4\mathbf{i} + \mathbf{j}) = \frac{24}{17}\mathbf{i} + \frac{6}{17}\mathbf{j}$$ ●

The vectors $\mathbf{v}$, $\mathbf{w}$, and $\text{proj}_{\mathbf{w}}\mathbf{v}$ are shown in Figure 73.

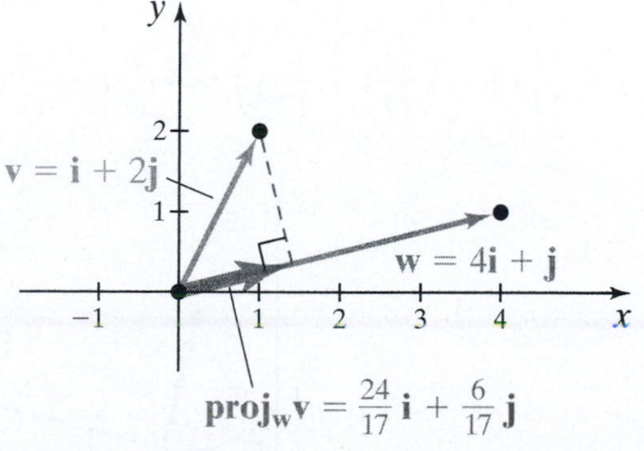

Figure 73

Every non-zero vector can be written as the sum of two orthogonal vectors. To illustrate this, consider the vectors $\mathbf{v}$, $\mathbf{w}$, $\mathbf{v}_1 = \text{proj}_{\mathbf{w}}\mathbf{v}$, and $\mathbf{v}_2$, where $\mathbf{v}_1$ is orthogonal to $\mathbf{v}_2$. See Figure 74.

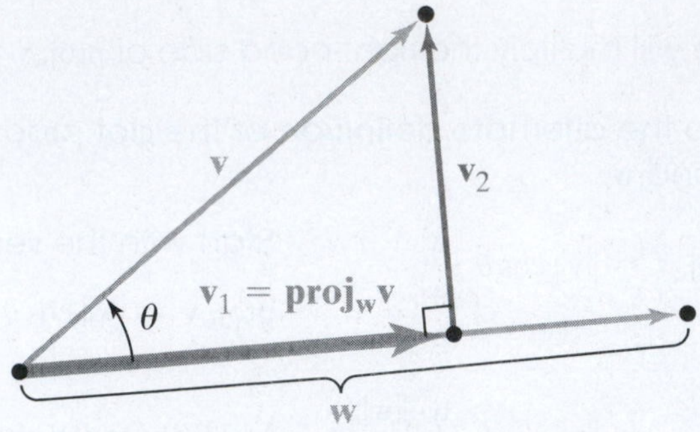

Figure 74

Using the addition of vectors represented geometrically, we see that $\mathbf{v} = \mathbf{v}_1 + \mathbf{v}_2$, where $\mathbf{v}_1 = \text{proj}_{\mathbf{w}}\mathbf{v}$ is parallel to $\mathbf{w}$ and $\mathbf{v}_2$ is orthogonal to $\mathbf{w}$. Note that $\mathbf{v}_2 = \mathbf{v} - \mathbf{v}_1$. The vector sum $\mathbf{v} = \mathbf{v}_1 + \mathbf{v}_2$ is called the **vector decomposition of v into orthogonal components.**

Definition The Vector Decomposition of $\mathbf{v}$ into Orthogonal Components

Let $\mathbf{v}$ and $\mathbf{w}$ be non-zero vectors. Then $\mathbf{v}$ can be written as the sum of orthogonal vectors $\mathbf{v}_1$ and $\mathbf{v}_2$, where $\mathbf{v}_1$ is parallel to $\mathbf{w}$ and $\mathbf{v}_2$ is orthogonal to $\mathbf{w}$ such that

$$\mathbf{v}_1 = \text{proj}_{\mathbf{w}}\mathbf{v} = \frac{\mathbf{v} \cdot \mathbf{w}}{\|\mathbf{w}\|^2}\mathbf{w} \text{ and } \mathbf{v}_2 = \mathbf{v} - \mathbf{v}_1.$$

The process of expressing $\mathbf{v}$ as the sum of $\mathbf{v}_1$ and $\mathbf{v}_2$ is called the **vector decomposition of v into orthogonal components $\mathbf{v}_1$ and $\mathbf{v}_2$.**

Example 7 Decomposing a Vector into Orthogonal Components

Let $\mathbf{v} = \mathbf{i} + 2\mathbf{j}$ and $\mathbf{w} = 4\mathbf{i} + \mathbf{j}$. Determine the vector decomposition of $\mathbf{v}$ into orthogonal components $\mathbf{v}_1$ and $\mathbf{v}_2$, where $\mathbf{v}_1$ is parallel to $\mathbf{w}$ and $\mathbf{v}_2$ is orthogonal to $\mathbf{w}$.

Solution In Example 6 we determined that $\text{proj}_{\mathbf{w}}\mathbf{v} = \frac{\mathbf{v} \cdot \mathbf{w}}{\|\mathbf{w}\|^2}\mathbf{w} = \frac{24}{17}\mathbf{i} + \frac{6}{17}\mathbf{j}$.

Therefore, $\mathbf{v}_1 = \frac{24}{17}\mathbf{i} + \frac{6}{17}\mathbf{j}$ and $\mathbf{v}_2 = \mathbf{v} - \mathbf{v}_1 = (\mathbf{i} + 2\mathbf{j}) - \left(\frac{24}{17}\mathbf{i} + \frac{6}{17}\mathbf{j}\right) =$ $\left(\frac{17}{17}\mathbf{i} + \frac{34}{17}\mathbf{j}\right) - \left(\frac{24}{17}\mathbf{i} + \frac{6}{17}\mathbf{j}\right) = -\frac{7}{17}\mathbf{i} + \frac{28}{17}\mathbf{j}$. See Figure 75.

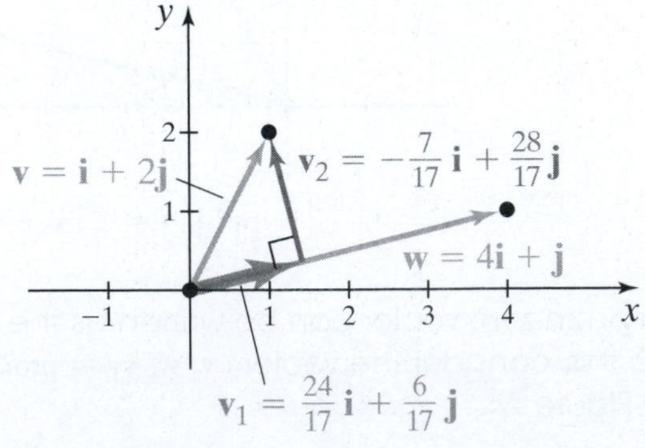

Figure 75

Note that $\mathbf{v}_1 \cdot \mathbf{v}_2 = \left(\frac{24}{17}\right)\left(-\frac{7}{17}\right) + \left(\frac{6}{17}\right)\left(\frac{28}{17}\right) = -\frac{168}{289} + \frac{168}{289} = 0$, which verifies that $\mathbf{v}_1$ and $\mathbf{v}_2$ are orthogonal. ●

You Try It **Work through this You Try It problem.**

Work Exercises 20–22 in this textbook or in the MyLab Math Study Plan.

OBJECTIVE 5 SOLVING APPLICATIONS INVOLVING FORCES ON AN INCLINED PLANE

Projections and the decomposition of vectors into orthogonal components can be used to solve applications when the force of gravity is acting on an object that is on an inclined plane, such as a hill or a ramp. The steeper the incline, the more force is required to pull or push the object up the incline.

Suppose that an object is placed on a ramp that has an incline angle of α. Let $\mathbf{w}$ represent a vector parallel to the incline. Let $\mathbf{G}$ represent the force due to gravity pulling the object directly downward. There are two other forces involved, $\mathbf{F}_1$ and $\mathbf{F}_2$, where $\mathbf{F}_1$ represents the force exerted by the object pushing down the incline and $\mathbf{F}_2$ represents the force of the object pushing against the ramp at a right angle. See Figure 76.

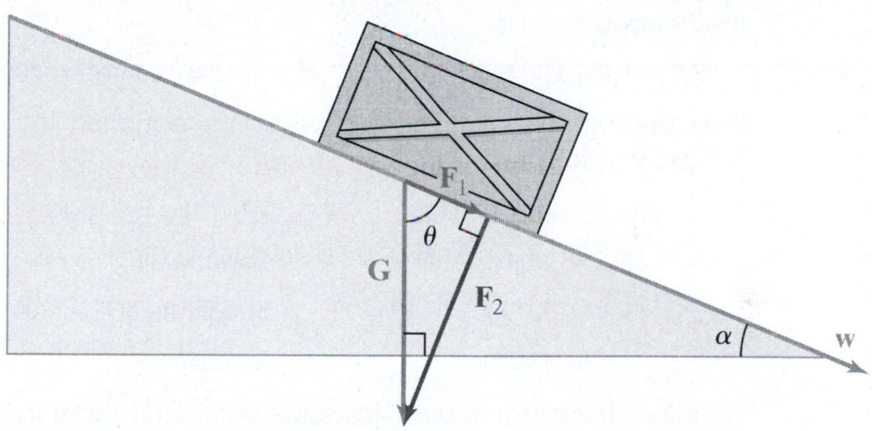

Figure 76

Note that a portion of the vector $\mathbf{G}$ forms a right angle with the base of the incline. The right triangle formed contains the **complementary angles** α and θ. The force of gravity $\mathbf{G}$ can be written as the vector decomposition of the orthogonal components $\mathbf{F}_1$ and $\mathbf{F}_2$. Thus, $\mathbf{G} = \mathbf{F}_1 + \mathbf{F}_2$, where $\mathbf{F}_1 = \text{proj}_{\mathbf{w}}\mathbf{G}$. Therefore, the force required to prevent the object from sliding down the ramp must be equal in magnitude to $\mathbf{F}_1$. The magnitude of $\mathbf{F}_1$ is precisely the scalar component of $\mathbf{G}$ in the direction of $\mathbf{w}$, which is

$$\|\mathbf{F}_1\| = \|\text{proj}_{\mathbf{w}}\mathbf{G}\| = \|\mathbf{G}\| \cos \theta.$$

Example 8 Determining the Magnitude of Forces of an Object on an Incline

A 200-pound object is placed on a ramp that is inclined at 22°. What is the magnitude of the force needed to hold the box in a stationary position to prevent the box from sliding down the ramp? What is the magnitude of the force pushing against the ramp?

Solution We can draw a diagram with vectors $\mathbf{w}$, $\mathbf{G}$, $\mathbf{F_1}$, and $\mathbf{F_2}$, where $\mathbf{w}$ is a vector positioned parallel to the incline, $\mathbf{G} = -200\mathbf{j}$ is the force due to gravity, $\mathbf{F_1} = \text{proj}_\mathbf{w}\mathbf{G}$ is the force pushing down the incline, and $\mathbf{F_2}$ is the force of the object pushing against the ramp at a right angle. The angle between $\mathbf{G}$ and $\mathbf{w}$ is 68°. See Figure 77.

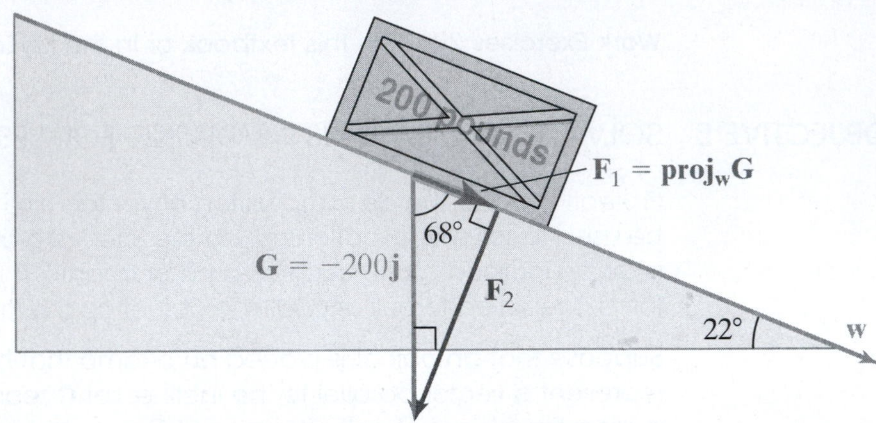

Figure 77

The force needed to hold the object in a stationary position is the magnitude of $\mathbf{F_1}$, which is the scalar component of $\mathbf{G}$ in the direction of $\mathbf{w}$. So, the magnitude of the needed force is

$$\|\mathbf{F_1}\| = \|\text{proj}_\mathbf{w}\mathbf{G}\| = \|\mathbf{G}\|\cos\theta = \sqrt{0^2 + (-200)^2}\cos 68° = 200\cos 68° \approx 74.92 \text{ pounds.}$$

From the Pythagorean Theorem, we get the equation $\|\mathbf{G}\|^2 = \|\mathbf{F_1}\|^2 + \|\mathbf{F_2}\|^2$, which will allow us to solve for $\|\mathbf{F_2}\|$.

$\|\mathbf{G}\|^2 = \|\mathbf{F_1}\|^2 + \|\mathbf{F_2}\|^2$ — Write the Pythagorean Theorem.

$\|\mathbf{F_2}\| = \sqrt{\|\mathbf{G}\|^2 - \|\mathbf{F_1}\|^2}$ — Solve for $\|\mathbf{F_2}\|$.

$\|\mathbf{F_2}\| = \sqrt{(200)^2 - (74.92)^2}$ — Substitute $\|\mathbf{G}\| = 200$ and $\|\mathbf{F_1}\| = 74.92$.

≈ 185.44 — Simplify.

Therefore, the magnitude of the force pushing against the ramp is approximately 185.44 pounds. ●

You Try It Work through this You Try It problem.

Work Exercises 23–26 in this textbook or in the MyLab Math Study Plan.

OBJECTIVE 6 SOLVING APPLICATIONS INVOLVING WORK

In physics, the concept of **work** refers to the transfer of energy from one object to another that causes the object to move a certain distance. So, work is done by a force whenever the force moves an object. The simplest formula for work is $W = Fd$, where F is the magnitude of the force and d is the distance traveled. However, this formula is true only when the force is applied parallel to the line of motion. See Figure 78.

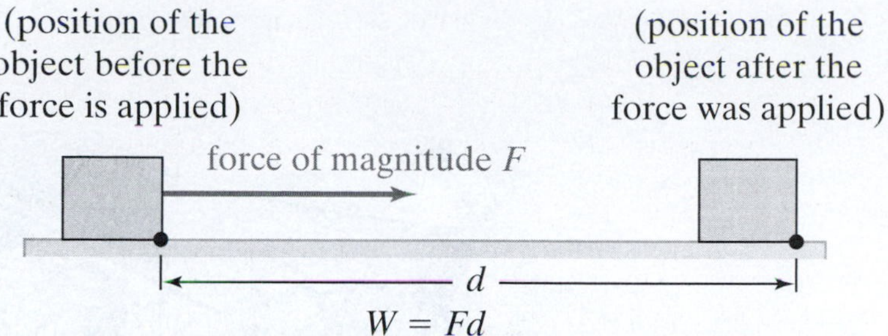

Figure 78

For example, the amount of work required to lift a 10-pound weight vertically 3 feet is

$$W = Fd = (10 \text{ lbs})(3 \text{ feet}) = 30 \text{ foot-pounds.}$$

Note that the units used for work in this scenario is foot-pounds.

Suppose that the constant force applied to an object is not parallel to the line of motion but instead is applied to the object at an angle θ between the object and the horizontal. In this situation, let $\mathbf{F}$ represent the force of an object with a direction angle θ relative to the line of motion. If the object moves from a point P to a point Q, then let $\mathbf{D} = \overrightarrow{PQ}$. The force of $\mathbf{F}$ in the direction of $\mathbf{D}$ is the vector $\text{proj}_{\mathbf{D}}\mathbf{F}$. See Figure 79.

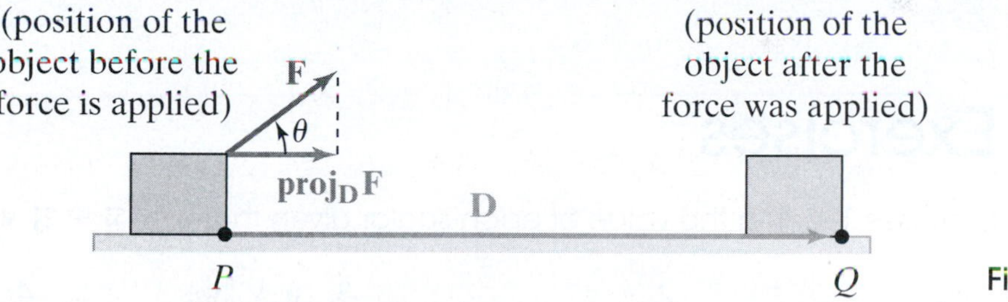

Figure 79

The work done by $\mathbf{F}$ is equal to the magnitude of the force in the direction of $\mathbf{D}$ times the distance from P to Q. This gives the following.

$W = $ (Magnitude of the force in the direction of the line of motion)(Distance from P to Q)

$= $ (Scalar Projection of $\mathbf{F}$ onto $\mathbf{D}$)(Magnitude of $\mathbf{D}$)

$= (\|\text{proj}_{\mathbf{D}}\mathbf{F}\|)(\|\mathbf{D}\|)$ Substitute the appropriate expressions.

$= (\|\mathbf{F}\| \cos \theta)(\|\mathbf{D}\|)$ Use the definition of $\text{proj}_{\mathbf{D}}\mathbf{F}$.

$= \|\mathbf{F}\| \|\mathbf{D}\| \cos \theta$ Rearrange the order of the factors.

$= \mathbf{F} \cdot \mathbf{D}$ Substitute using the alternate definition of the dot product.

Thus, the work is equal to the dot product of $\mathbf{F}$ and $\mathbf{D}$. If the angle of direction is given, then the most direct way to compute work is to use the alternate form of the dot product $\mathbf{F} \cdot \mathbf{D} = \|\mathbf{F}\| \|\mathbf{D}\| \cos \theta$.

Example 9 Calculating Work

A horse is pulling a plow with a force of 400 pounds. The angle between the harness and the ground is 20°. How much work is done to pull the plow 50 feet?

Solution We can calculate the work done as follows.

$$W = \mathbf{F} \cdot \mathbf{D}$$
$$= \|\mathbf{F}\|\,\|\mathbf{D}\| \cos \theta$$
$$= (400)(50) \cos 20° \approx 18{,}793.85 \text{ foot-pounds.}$$

You Try It Work through this You Try It problem.

Work Exercises 27–30 in this textbook or in the MyLab Math Study Plan.

10.5 Exercises

In Exercises 1–8, find the value of each scalar given that $\mathbf{u} = 5\mathbf{i} - 2\mathbf{j}$, $\mathbf{v} = 6\mathbf{i} + \mathbf{j}$, and $\mathbf{w} = -\mathbf{i} - 7\mathbf{j}$.

1. $\mathbf{u} \cdot \mathbf{v}$
2. $\mathbf{v} \cdot \mathbf{w}$
3. $(\mathbf{u} \cdot \mathbf{v})\|\mathbf{w}\|$
4. $\mathbf{u} \cdot (\mathbf{v} + \mathbf{w})$
5. $(\mathbf{u} + \mathbf{v}) \cdot \mathbf{w}$
6. $(-7\mathbf{u}) \cdot \mathbf{w}$
7. $\|\mathbf{w}\|^2$
8. $\|\mathbf{u}\|\,\|\mathbf{v}\|$

In Exercises 9–13, determine the angle between $\mathbf{u}$ and $\mathbf{v}$. Give the angle in degrees rounded to the nearest hundredth of a degree.

9. $\mathbf{u} = 5\mathbf{i} + 2\mathbf{j}, \mathbf{v} = \mathbf{i} + 4\mathbf{j}$
10. $\mathbf{u} = 2\mathbf{i} + 3\mathbf{j}, \mathbf{v} = -\mathbf{i} + 2\mathbf{j}$
11. $\mathbf{u} = 7\mathbf{i} + 3\mathbf{j}, \mathbf{v} = -5\mathbf{i} + 2\mathbf{j}$
12. $\mathbf{u} = -4\mathbf{i} + 4\mathbf{j}, \mathbf{v} = 3\mathbf{i} - 6\mathbf{j}$
13. $\mathbf{u} = \frac{1}{2}\mathbf{i} - \frac{1}{3}\mathbf{j}, \mathbf{v} = -3\mathbf{i} + 2\mathbf{j}$

14. Determine the value of b so that the vectors $\mathbf{u} = 4\mathbf{i} + b\mathbf{j}$ and $\mathbf{v} = -\mathbf{i} + 2\mathbf{j}$ are orthogonal.

15. Determine the value of a so that the vectors $\mathbf{u} = -5\mathbf{i} - 7\mathbf{j}$ and $\mathbf{v} = a\mathbf{i} + 3\mathbf{j}$ are orthogonal.

In Exercises 16–19, determine if the given vectors are orthogonal, parallel, or neither.

SbS 16. $\mathbf{u} = 2\mathbf{i} - 3\mathbf{j}, \mathbf{v} = 6\mathbf{i} + 4\mathbf{j}$

SbS 17. $\mathbf{u} = -\mathbf{i} + 2\mathbf{j}, \mathbf{v} = 2\mathbf{i} + 4\mathbf{j}$

SbS 18. $\mathbf{u} = -\mathbf{i} + \sqrt{3}\mathbf{j}, \mathbf{v} = \sqrt{3}\mathbf{i} - 3\mathbf{j}$

SbS 19. $\mathbf{u} = -\sqrt{2}\mathbf{i} + \mathbf{j}, \mathbf{v} = \sqrt{2}\mathbf{i} + 3\mathbf{j}$

In Exercises 20–22, non-zero vectors $\mathbf{v}$ and $\mathbf{w}$ are given. Determine the vector decomposition of $\mathbf{v}$ into orthogonal components $\mathbf{v}_1$ and $\mathbf{v}_2$, where $\mathbf{v}_1$ is parallel to $\mathbf{w}$ and $\mathbf{v}_2$ is orthogonal to $\mathbf{w}$.

20. $\mathbf{v} = -5\mathbf{i} - 9\mathbf{j}$; $\mathbf{w} = -\mathbf{i} + \mathbf{j}$

21. $\mathbf{v} = 3\mathbf{i} + 4\mathbf{j}$; $\mathbf{w} = -2\mathbf{i} + 7\mathbf{j}$

22. $\mathbf{v} = \frac{1}{2}\mathbf{i} - 8\mathbf{j}$; $\mathbf{w} = 5\mathbf{i} + \frac{1}{3}\mathbf{j}$

23. How much force is required to keep a 1000-pound block from sliding down an incline of 36°?

24. A mover in a moving truck is using a rope to pull a 400-pound box up a ramp that has an incline of 25°. What is the force needed to hold the box in a stationary position to prevent the box from sliding down the ramp? What is the magnitude of the force pushing against the ramp?

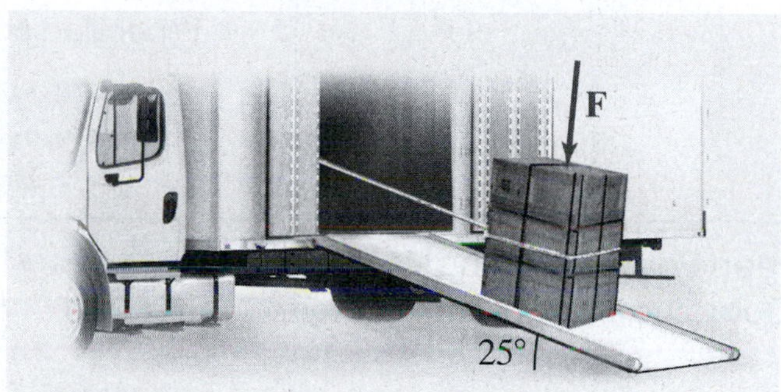

25. A car is being winched out of a ditch at an angle of 40°. Determine the weight of the car if the winch is pulling with a maximum capacity of 1200 pounds.

26. An object weighing 800 pounds is being hauled up a ramp. The object is held still by a force of 350 pounds. Determine the ramp's incline angle.

27. How much work is required to lift a 50-pound weight vertically 4 feet?

28. A horse is pulling a plow with a force of 500 pounds. The angle between the harness and the ground is 25°. How much work is done to pull the plow 60 feet?

29. A car is being winched out of a ditch at an angle of 40°. The winch exerts a constant force of 1000 pounds. What is the total work done by the winch in order to pull the car a horizontal distance of 35 feet?

30. Ten thousand foot-pounds of work was done by a force of 200 pounds to move the object 75 feet. If the force was exerted on the object at an angle θ between the object and the horizontal, then determine the angle θ. Write the angle in degrees rounded to the nearest tenth of a degree.

Chapter 10 Summary

Key Concepts	Examples/Videos

10.1 Polar Coordinates and Polar Equations

To sketch a point (r, θ) in the polar coordinate system, first sketch the angle θ in standard position. Then locate the point by using the appropriate directed distance for r.

 Plotting Polar Coordinates

 Plot the following points in a polar coordinate system.

a. $A\left(3, \frac{\pi}{4}\right)$ b. $B(-2, 120°)$

c. $C\left(1.5, -\frac{7\pi}{6}\right)$ d. $D\left(-3, -\frac{3\pi}{4}\right)$

 Determine three different representations of the point $P\left(4, \frac{5\pi}{6}\right)$ that have the specified conditions.

a. $r > 0, -2\pi \le \theta < 0$
b. $r < 0, 0 \le \theta < 2\pi$
c. $r > 0, 2\pi \le \theta < 4\pi$

Relationships Between Polar Coordinates and Rectangular Coordinates

$x = r\cos\theta$

$y = r\sin\theta$

$r = \sqrt{x^2 + y^2}$

$\tan\theta = \frac{y}{x}$

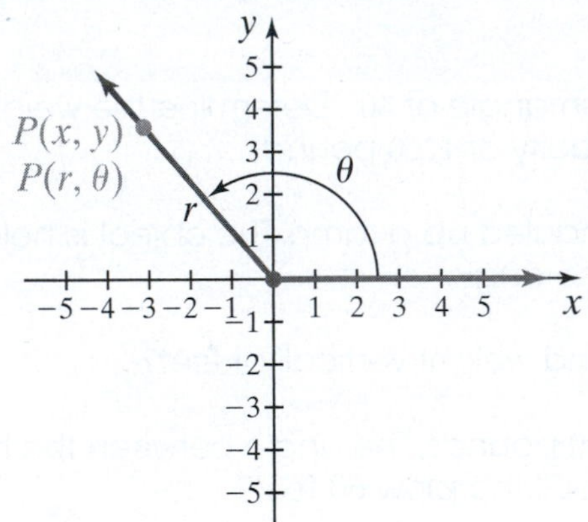

 Determine the rectangular coordinates for the points with the given polar coordinates.

a. $A(5, \pi)$ b. $B\left(-7, -\frac{\pi}{3}\right)$

c. $C\left(3\sqrt{2}, \frac{5\pi}{4}\right)$

 Determine the polar coordinates for the points with the given rectangular coordinates such that $r \ge 0$ and $0 \le \theta < 2\pi$. Round the values of r and θ to two decimal places if necessary.

a. $A(-4, -4)$ b. $B(-2\sqrt{3}, 2)$

c. $C(4, -3)$

 Convert each equation given in rectangular form into polar form.

a. $x = 7$ b. $2x - y = 8$

c. $4x^2 + 4y^2 = 3$ d. $x^2 + y^2 = 9y$

 Convert each equation given in polar form into rectangular form.

a. $3r\cos\theta - 4r\sin\theta = -1$

b. $r = 6\cos\theta$ c. $r = 3$

d. $\theta = \frac{\pi}{6}$

Key Concepts	Examples/Videos

10.2 Graphing Polar Equations

Graphs of Polar Equations of the Form $r = a, r = a\sin\theta$, and $r = a\cos\theta$

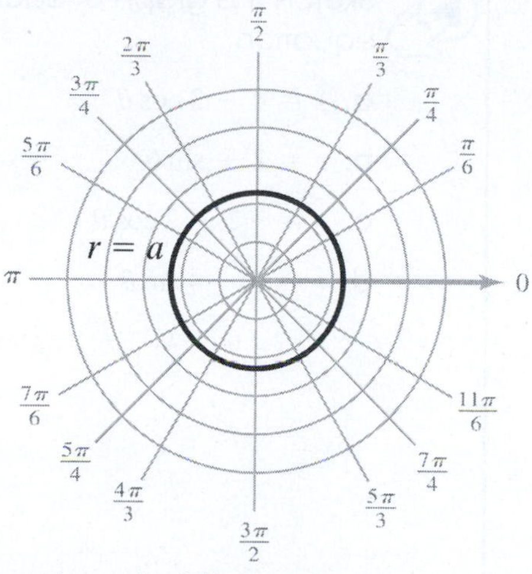

Circle centered at the pole

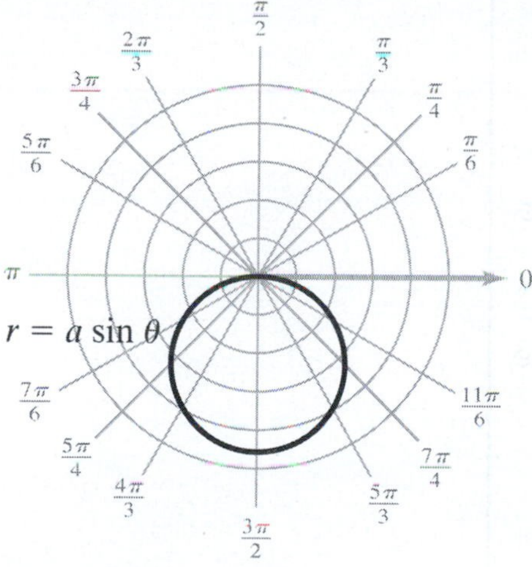

Circle centered along the line $\theta = \frac{\pi}{2}$

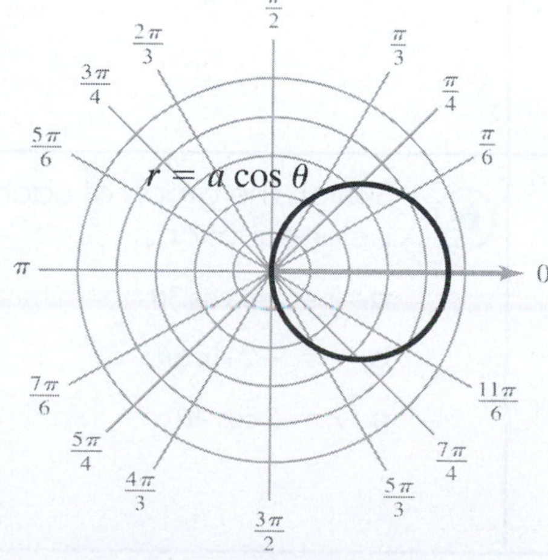

Circle centered along the line $\theta = 0$

 Sketch the graph of each polar equation.

a. $r = 4\sin\theta$

b. $r = -2\cos\theta$

Key Concepts	Examples/Videos
 Graphs of Polar Equations of the Form $r = a, r = a \sin\theta$, and $r = a \cos\theta$	

Key Concepts

Steps for Sketching Polar Equations of the Form $r = a + b\sin\theta$ and $r = a + b\cos\theta$

Step 1. Identify the general shape using the ratio $\left\|\frac{a}{b}\right\|$.

- If $\left|\frac{a}{b}\right| = 1$, then the graph is a cardioid.
- If $\left|\frac{a}{b}\right| < 1$, then the graph is a limacon with an inner loop that intersects the pole.
- If $1 < \left|\frac{a}{b}\right| < 2$, then the graph is a limacon with a dimple.
- if $\left|\frac{a}{b}\right| \geq 2$, then the graph is a limacon with no inner loop and no dimple.

Step 2. Determine the symmetry.

- If the equation is of the form $r = a + b\sin\theta$, then the graph must be symmetric about the line $\theta = \frac{\pi}{2}$.
- If the equation is of the form $r = a + b\cos\theta$, then the graph must be symmetric about the line $\theta = 0$.

Step 3. Plot the points corresponding to the quadrantal angles $\theta = 0, \theta = \frac{\pi}{2}, \theta = \pi$, and $\theta = \frac{3\pi}{2}$.

Step 4. If necessary, plot a few more points until symmetry can be used to complete the graph.

Graphs of Polar Equations of the Form $r = a + b\sin\theta$ and $r = a + b\cos\theta$

Examples/Videos

Sketch the graph of each polar equation.

a. $r = 4 - 3\cos\theta$

b. $r = 2 + \sin\theta$

c. $r = -2 + 2\cos\theta$

d. $r = 3 - 4\sin\theta$

Key Concepts

Steps for Sketching Polar Equations of the Form

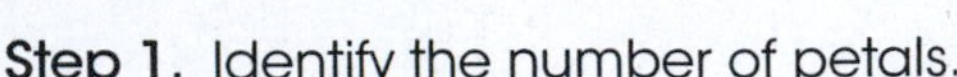

$r = a\sin n\theta$ and $r = a\cos n\theta$

Step 1. Identify the number of petals.

- If n is even, then there are $2n$ petals.
- If n is odd, then there are n petals.

Step 2. Determine the length of each petal.

- The length of each petal is $|a|$ units.

Examples/Videos

Sketch the graph of each polar equation.

a. $r = -4\cos 3\theta$

b. $r = -2\sin 5\theta$

c. $r = 5\cos 4\theta$

Key Concepts	Examples/Videos
Step 3. Determine all angles where an endpoint of a petal lies. • If the equation is of the form $r = a \sin n\theta$, then the endpoints occur for angles on the interval $[0, 2\pi)$* that satisfy the equations $\sin n\theta = 1$ and $\sin n\theta = -1$. • If the equation is of the form $r = a \cos n\theta$, then the endpoints occur for angles on the interval $[0, 2\pi)$* that satisfy the equations $\cos n\theta = 1$ and $\cos n\theta = -1$. ***Note that when n is odd, it is only necessary to consider angles on the interval $[0, \pi)$. A complete graph is obtained on this interval because the graph will completely traverse itself on the interval $[\pi, 2\pi]$.** **Step 4.** Substitute each angle determined in Step 3 back into the original equation to obtain the appropriate values of r for each angle. The ordered pairs obtained represent the endpoints of the rose petals. Plot these points on the graph. **Step 5.** Determine angles where the graph passes through the pole. These angles serve as a guide when sketching the width of a petal. • If the equation is of the form $r = a \sin n\theta$, then the graph passes through the pole when $\sin n\theta = 0$. • If the equation is of the form $r = a \cos n\theta$, then the graph passes through the pole when $\cos n\theta = 0$. **Step 6.** Draw each petal to complete the graph. G/V Graphs of Polar Equations of the Form $r = a \sin n\theta$ and $r = a \cos n\theta$	

Key Concepts	Examples/Videos
Graphs of Polar Equations of the Form $r^2 = a^2 \sin 2\theta$ and $r^2 = a^2 \cos 2\theta$ 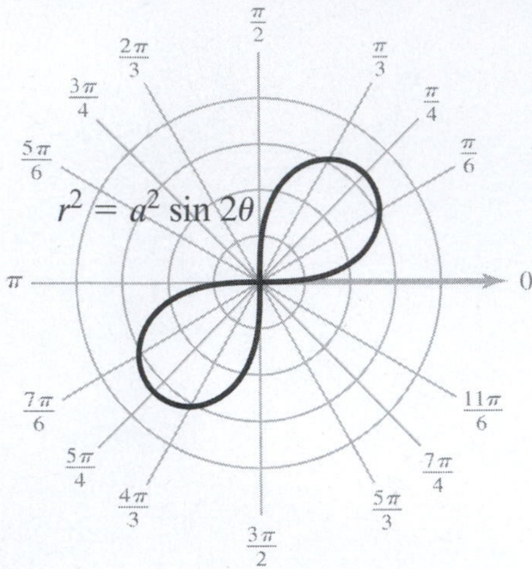 The graph is a lemniscate symmetric about the pole and the line $\theta = \frac{\pi}{4}$. The endpoints of the two loops occur when $\theta = \frac{\pi}{4}$ and $\theta = \frac{5\pi}{4}$. 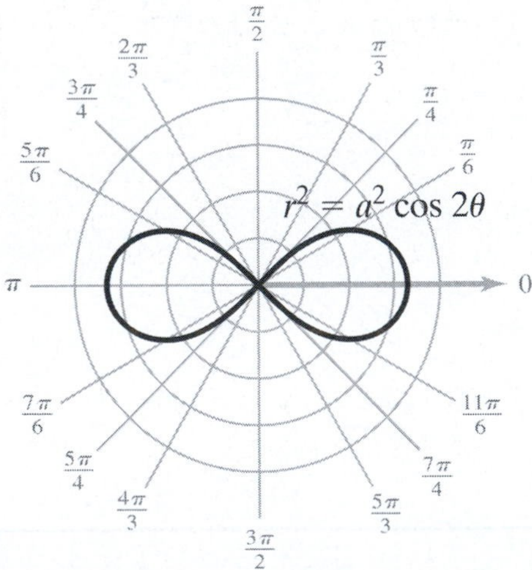 The graph is a lemniscate symmetric about the pole, the horizontal line $\theta = 0$, and the vertical line $\theta = \frac{\pi}{2}$. The endpoints of the two loops occur when $\theta = 0$ and $\theta = \pi$.	 Sketch the graph of each polar equation. a. $r^2 = 9 \cos 2\theta$ b. $r^2 = 16 \sin 2\theta$

Key Concepts	Examples/Videos
10.3 Complex Numbers in Polar Form; De Moivre's Theorem A complex number is said to be in **rectangular form** if it is written as $z = a + bi$, where a and b are real numbers and i is the imaginary unit. The **absolute value of a complex number** $z = a + bi$ is denoted by $\|z\|$ and is the distance from the origin to z in the complex plane and is given by $\|z\| = \sqrt{a^2 + b^2}$. 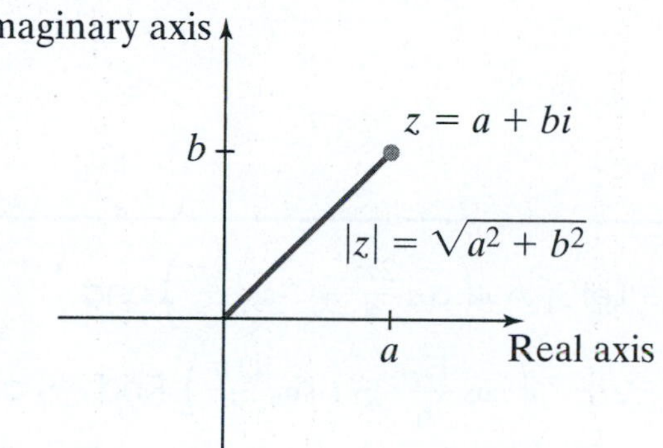	 Plot each complex number in the complex plane and determine its absolute value. **a.** $z_1 = 3 - 4i$ **b.** $z_2 = -2 + 5i$ **c.** $z_3 = 3$ **d.** $z_4 = -2i$
A complex number $z = a + bi$ is said to be in **polar form** if it is written as $z = r(\cos\theta + i\sin\theta)$, where $a = r\cos\theta, b = r\sin\theta, \tan\theta = \frac{b}{a}$, and $r = \sqrt{a^{2+} b^2}$. A complex number $z = r(\cos\theta + i\sin\theta)$ is said to be in **standard polar form** when $0 \le \theta < 2\pi$ (or $0 \le \theta < 360°$).	 Rewrite the complex number in standard polar form, plot the number in the complex plane, and determine the quadrant in which the point lies or the axis on which the point lies. **a.** $z = 3\left(\cos\frac{5\pi}{8} + i\sin\frac{5\pi}{8}\right)$ **b.** $z = 2\left(\cos\frac{23\pi}{4} + i\sin\frac{23\pi}{4}\right)$ **c.** $z = 4(\cos(-3\pi) + i\sin(-3\pi))$  Write each complex number in rectangular form using exact values if possible. Otherwise, round to two decimal places. **a.** $z = 3\left(\cos\frac{7\pi}{4} + i\sin\frac{7\pi}{4}\right)$ **b.** $z = 4(\cos 80° + i\sin 80°)$

Key Concepts	Examples/Videos
Converting a Complex Number from Rectangular Form to Standard Polar Form for $a \neq 0$ and $b \neq 0$ **Step 1.** Determine the value of r using the equation $r = \sqrt{a^2 + b^2}$. **Step 2.** Plot the point and determine the quadrant in which it lies. **Step 3.** Determine the value of the acute reference angle θ_R by solving the equation $\tan \theta_R = \left\lvert \frac{b}{a} \right\rvert$. **Step 4.** Determine the value of θ using θ_R and the quadrant in which the point lies.	 Determine the standard polar form of each complex number. Write the argument in radians using exact values if possible. Otherwise, round the argument to two decimal places. **a.** $z = -2\sqrt{3} + 2i$ **b.** $z = 4 - 3i$
The Product and Quotient of Two Complex Numbers Written in Polar Form If $z_1 = r_1 (\cos \theta_1 + i \sin \theta_1)$ and $z_2 = r_2 (\cos \theta_2 + i \sin \theta_2)$ are two complex numbers, then $z_1 z_2 = r_1 r_2 (\cos(\theta_1 + \theta_2) + i \sin(\theta_1 + \theta_2))$ and $\frac{z_1}{z_2} = \frac{r_1}{r_2}[\cos(\theta_1 - \theta_2) + i \sin(\theta_1 - \theta_2)]$	 Let $z_1 = 4\left(\cos \frac{2\pi}{3} + i \sin \frac{2\pi}{3}\right)$ and $z_2 = 5\left(\cos \frac{11\pi}{6} + i \sin \frac{11\pi}{6}\right)$. Find $z_1 z_2$ and $\frac{z_1}{z_2}$ and write the answers in standard polar form.
De Moivre's Theorem for Finding Powers of a Complex Number If $z = r(\cos \theta + i \sin \theta)$ is a complex number written in polar form and if n is a positive integer, then z^n is given by the formula $z^n = [r(\cos \theta + i \sin \theta)]^n = r^n (\cos n\theta + i \sin n\theta)$.	 **a.** Find $\left[5\left(\cos \frac{3\pi}{4} + i \sin \frac{3\pi}{4}\right)\right]^3$ and write your answer in standard polar form. **b.** Find $(\sqrt{3} - i)^4$ and write your answer in rectangular form.
De Moivre's Theorem for Finding the nth Roots of a Complex Number If $z = r(\cos \theta + i \sin \theta)$ is a complex number written in polar form and if n is a positive integer, then z has exactly n complex roots determined by the formula $\sqrt[n]{z} = \sqrt[n]{r}\left[\cos\left(\frac{\theta + 2\pi k}{n}\right) + i \sin\left(\frac{\theta + 2\pi k}{n}\right)\right]$ or $\sqrt[n]{z} = \sqrt[n]{r}\left[\cos\left(\frac{\theta + (360^\circ)k}{n}\right) + i \sin\left(\frac{\theta + (360^\circ)k}{n}\right)\right]$ for $k = 0, 1, 2, \cdots, (n - 1)$. We will label the n roots as $z_0, z_1, z_2, \cdots, z_{n-1}$.	 **a.** Find the complex fourth roots of $z = 81\left(\cos \frac{3\pi}{5} + i \sin \frac{3\pi}{5}\right)$. Write your answers in standard polar form. **b.** Find the complex square roots of $z = -2\sqrt{3} + 2i$. Write your answer in rectangular form with each part rounded to two decimal places.

Key Concepts	Examples/Videos
10.4 Vectors A **vector v can be represented geometrically** as a directed line segment in a plane having an initial point P and a terminal point Q. The vector can be denoted as the boldface letter $\mathbf{v}$, $\overrightarrow{PQ}$, or $\vec{v}$. 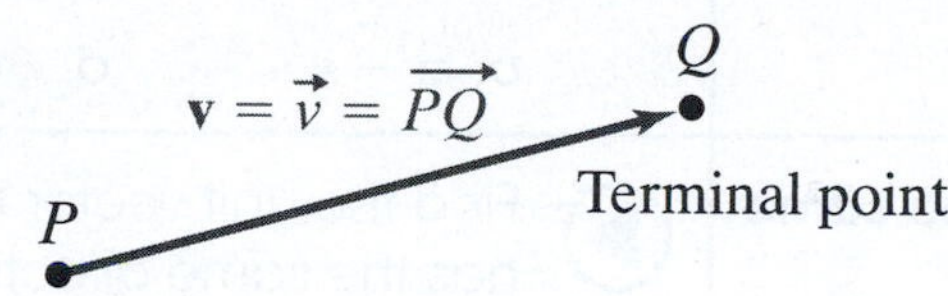 The **magnitude** of a vector $\mathbf{v}$ is the distance between the initial point and the terminal point and is denoted by $\lVert\mathbf{v}\rVert$. If the initial point has coordinates $P(x_1, y_1)$, and the terminal point has coordinates $Q(x_2, y_2)$, then $\lVert\mathbf{v}\rVert = \sqrt{(x_2 - x_1)^2 + (y_2 - y_1)^2}.$	 Determine the magnitude of $\mathbf{v}$. 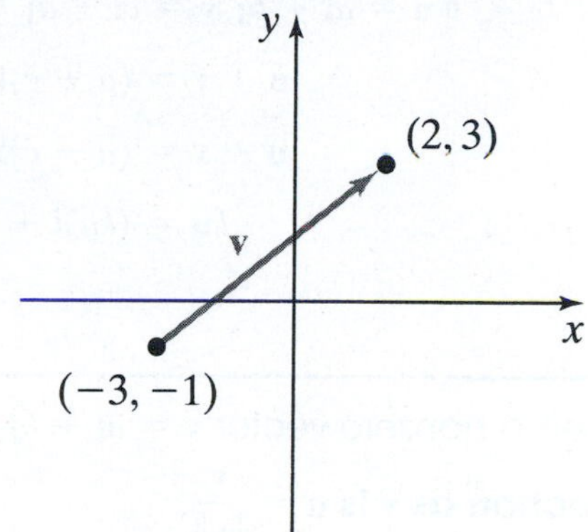
 **Scalar Multiplication**  **Vector Addition**  **The Parellelogram Law for Vector Addition** **Vector Subtraction**	 Given the vectors $\mathbf{u}$, $\mathbf{v}$, and $\mathbf{w}$, draw each of the following vectors. 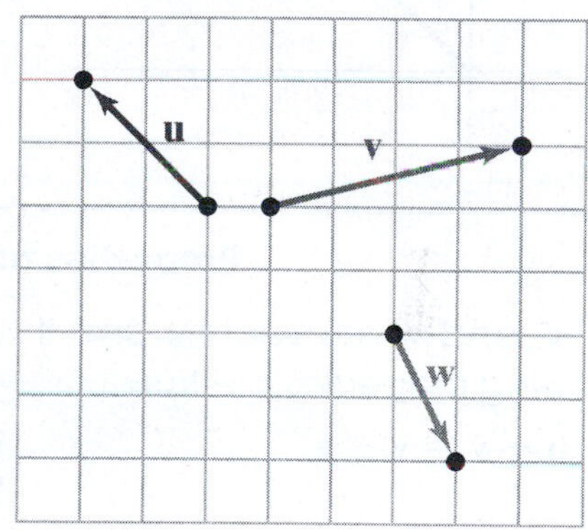 a. $\mathbf{w} + \mathbf{v}$ b. $\mathbf{v} - \mathbf{u}$ c. $\mathbf{v} + 2\mathbf{w} - \mathbf{u}$
A **unit vector** is a vector that has a magnitude of 1 unit. The unit vector whose direction is along the positive x-axis is $\mathbf{i} = \langle 1, 0 \rangle$. The unit vector whose direction is along the positive y-axis is $\mathbf{j} = \langle 0, 1 \rangle$. 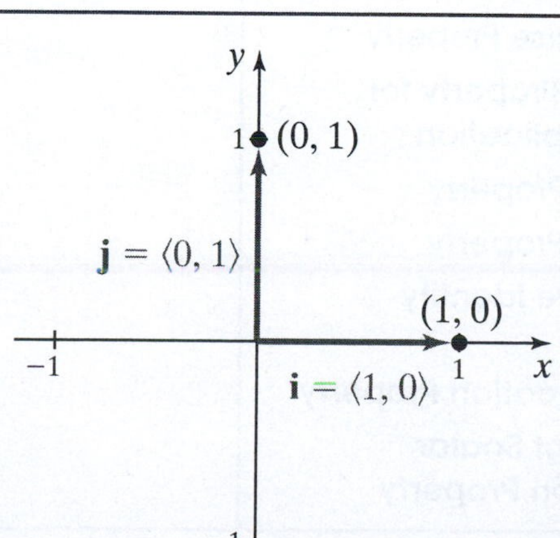	

Key Concepts	Examples/Videos

Operations with Vectors in Terms of i and j

If $\mathbf{u} = a\mathbf{i} + b\mathbf{j}$, $\mathbf{v} = c\mathbf{i} + d\mathbf{j}$, and k is a scalar, then

$$\mathbf{u} + \mathbf{v} = (a + c)\mathbf{i} + (b + d)\mathbf{j}$$

$$\mathbf{u} - \mathbf{v} = (a - c)\mathbf{i} + (b - d)\mathbf{j}$$

$$k\mathbf{u} = (ka)\mathbf{i} + (kb)\mathbf{j}$$

Let $\mathbf{u} = -3\mathbf{i} + 7\mathbf{j}$ and $\mathbf{v} = 5\mathbf{i} - \mathbf{j}$. Find each vector in terms of **i** and **j** and determine the magnitude of each vector.

a. $-\frac{1}{2}\mathbf{u}$ b. $\mathbf{u} + \mathbf{v}$

c. $\mathbf{u} - \mathbf{v}$ d. $3\mathbf{u} - 5\mathbf{v}$

Given a nonzero vector $\mathbf{v} = a\mathbf{i} + b\mathbf{j}$, the **unit vector u in the same direction as v is** $\mathbf{u} = \frac{\mathbf{v}}{\|\mathbf{v}\|}$.

Find the unit vector that has the same direction as $\mathbf{v} = 6\mathbf{i} - 8\mathbf{j}$.

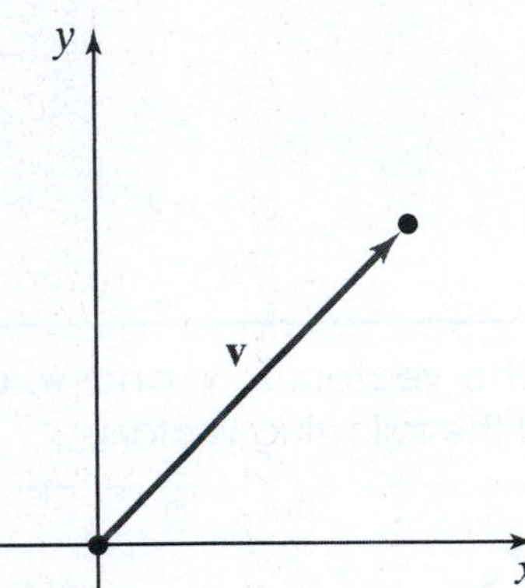

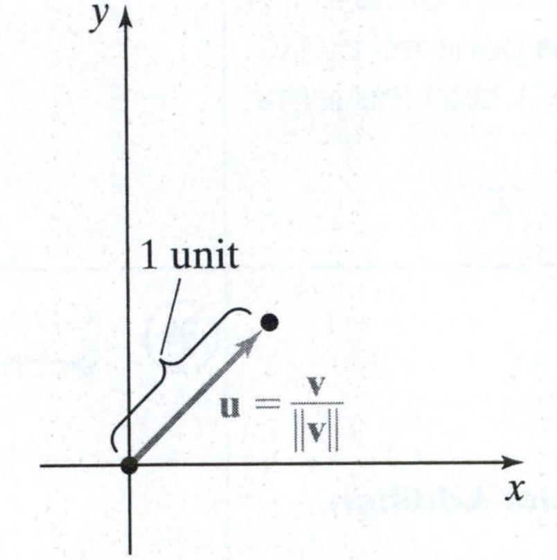

Properties of Vectors

If **u**, **v**, and **w** are vectors and if A and B are scalars, then the following properties are true.

1. $\mathbf{u} + \mathbf{v} = \mathbf{v} + \mathbf{u}$ — **Commutative Property for Vector Addition**
2. $(\mathbf{u} + \mathbf{v}) + \mathbf{w} = \mathbf{u} + (\mathbf{v} + \mathbf{w})$ — **Associative Property for vector Addition**
3. $\mathbf{u} + \mathbf{0} = \mathbf{u}$ — **Additive Identity Property**
4. $\mathbf{u} + (-\mathbf{u}) = \mathbf{0}$ — **Additive Inverse Property**
5. $(AB)\mathbf{u} = A(B\mathbf{u})$ — **Associative Property for Scalar Multiplication**
6. $A(\mathbf{u} + \mathbf{v}) = A\mathbf{u} + A\mathbf{v}$ — **Distributive Property**
7. $(A + B)\mathbf{u} = A\mathbf{u} + B\mathbf{u}$ — **Distributive Property**
8. $1\mathbf{u} = \mathbf{u}$ — **Multiplicative Identity Property**
9. $0\mathbf{u} = \mathbf{0}$ — **Zero Multiplication Property**
10. $\|A\mathbf{u}\| = \|A\| \, \|\mathbf{u}\|$ — **Magnitude of Scalar Multiplication Property**

Key Concepts	Examples/Videos
Given a vector $\mathbf{v} = \langle a, b \rangle = a\mathbf{i} + b\mathbf{j}$ in standard position, the **direction angle of v** is the positive angle θ between the positive x-axis and vector that satisfies the equation $\tan\theta = \frac{b}{a}$, where $a \neq 0$. 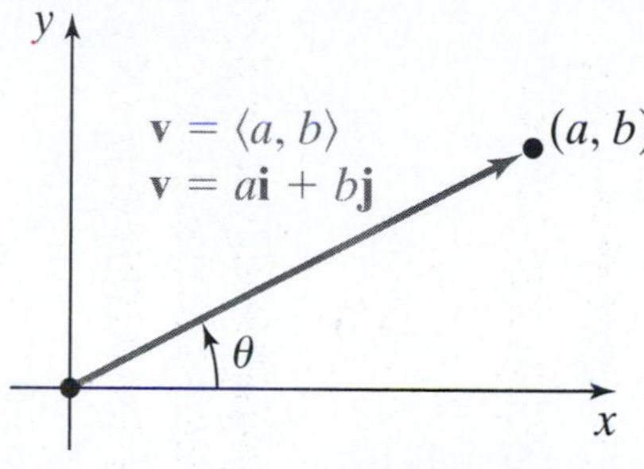	 Determine the direction angle of the vector $\mathbf{v} = -3\mathbf{i} + 2\mathbf{j}$.
If θ is the direction angle measured from the positive x-axis to a vector $\mathbf{v}$, then the vector can be expressed as $\mathbf{v} = (\|\mathbf{v}\| \cos\theta)\mathbf{i} + (\|\mathbf{v}\| \sin\theta)\mathbf{j}$. 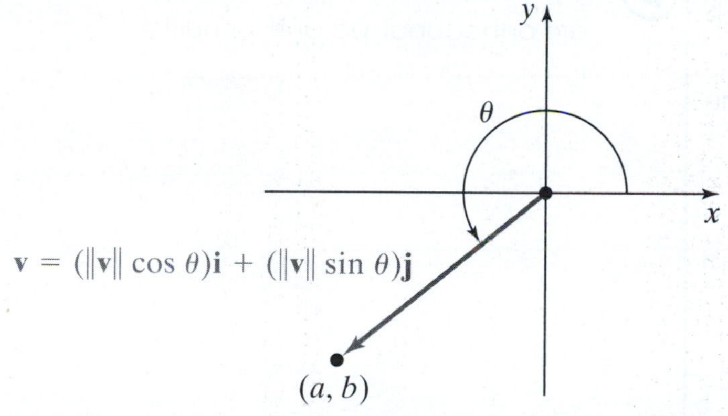	 The vector $\mathbf{v}$ has a magnitude of 20 units and direction angle of $\theta = 50°$. Represent this vector in the form $\mathbf{v} = a\mathbf{i} + b\mathbf{j}$. Round a and b to two decimal places. An airplane takes off from a runway at a speed of 190 mph at an angle of $11°$. Express the velocity of the plane at takeoff as a vector in terms of $\mathbf{i}$ and $\mathbf{j}$. Round a and b to two decimal places.
10.5 The Dot Product If $\mathbf{u} = a\mathbf{i} + b\mathbf{j}$ and $\mathbf{v} = c\mathbf{i} + d\mathbf{j}$, then the **dot product** is $\mathbf{u}\cdot\mathbf{v} = ac + bd$. **Dot Product Properties** If $\mathbf{u}$, $\mathbf{v}$, and $\mathbf{w}$ are vectors and if k is a scalar, then the following properties are true. 1. $\mathbf{u}\cdot\mathbf{v} = \mathbf{v}\cdot\mathbf{u}$ 2. $\mathbf{u}\cdot(\mathbf{v} + \mathbf{w}) = \mathbf{u}\cdot\mathbf{v} + \mathbf{u}\cdot\mathbf{w}$ 3. $k(\mathbf{u}\cdot\mathbf{v}) = (k\mathbf{u})\cdot\mathbf{v} = \mathbf{u}\cdot(k\mathbf{v})$ 4. $\mathbf{0}\cdot\mathbf{v} = 0$ 5. $\mathbf{v}\cdot\mathbf{v} = \|\mathbf{v}\|^2$	 If $\mathbf{u} = 4\mathbf{i} + 6\mathbf{j}$, $\mathbf{v} = -2\mathbf{i} + 8\mathbf{j}$, and $\mathbf{w} = -3\mathbf{i} - \mathbf{j}$, then find each of the following. a. $\mathbf{u}\cdot\mathbf{v}$ b. $\mathbf{u}\cdot(\mathbf{v} + \mathbf{w})$ c. $\mathbf{u}\cdot(-5\mathbf{v})$ d. $\|\mathbf{w}\|^2$

Key Concepts	Examples/Videos
The Angle Between Two Vectors If $\mathbf{u}$ and $\mathbf{v}$ are non-zero vectors and if θ is the angle between $\mathbf{u}$ and $\mathbf{v}$, then $\cos\theta = \frac{\mathbf{u}\cdot\mathbf{v}}{\lVert\mathbf{u}\rVert\,\lVert\mathbf{v}\rVert}$. 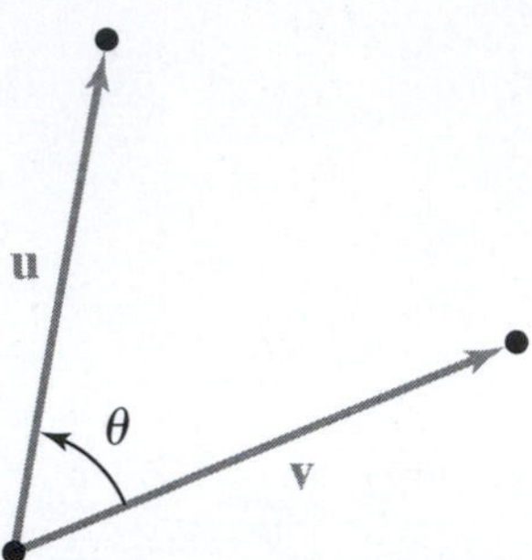 If $\mathbf{u}$ and $\mathbf{v}$ are non-zero vectors, then the **alternate form of the dot product** is $\mathbf{u}\cdot\mathbf{v} = \lVert\mathbf{u}\rVert\,\lVert\mathbf{v}\rVert\cos\theta$.	 Determine the angle between each pair of vectors. Give the angle in degrees rounded to the nearest hundredth of a degree. a. $\mathbf{u} = \mathbf{i} + 4\mathbf{j}, \mathbf{v} = -2\mathbf{i} + 5\mathbf{j}$ b. $\mathbf{u} = 3\mathbf{i} - 2\mathbf{j}, \mathbf{v} = 4\mathbf{i} + 6\mathbf{j}$
Parallel and Orthogonal Vectors Two vectors $\mathbf{u}$ and $\mathbf{v}$ are **parallel vectors** if there is a nonzero scalar k such that $\mathbf{u} = k\mathbf{v}$. Two non-zero vectors $\mathbf{u}$ and $\mathbf{v}$ are **orthogonal** if the angle between them is $\theta = 90°$. **Test for Orthogonal Vectors** Two non-zero vectors $\mathbf{u}$ and $\mathbf{v}$ are orthogonal if and only if $\mathbf{u}\cdot\mathbf{v} = 0$. **Test for Parallel Vectors** Two non-zero vectors $\mathbf{u}$ and $\mathbf{v}$ are parallel if $\frac{\mathbf{u}\cdot\mathbf{v}}{\lVert\mathbf{u}\rVert\,\lVert\mathbf{v}\rVert} = 1$ or $\frac{\mathbf{u}\cdot\mathbf{v}}{\lVert\mathbf{u}\rVert\,\lVert\mathbf{v}\rVert} = -1$.	 Determine if $\mathbf{u} = -\frac{1}{2}\mathbf{i} - \mathbf{j}$ and $\mathbf{v} = 2\mathbf{i} + 4\mathbf{j}$ are orthogonal, parallel, or neither.
If θ is the angle between $\mathbf{v}$ and $\mathbf{w}$, then the **scalar component of v in the direction of** $\mathbf{w}$ is $\lVert\text{proj}_{\mathbf{w}}\mathbf{v}\rVert = \lVert\mathbf{v}\rVert\cos\theta$. 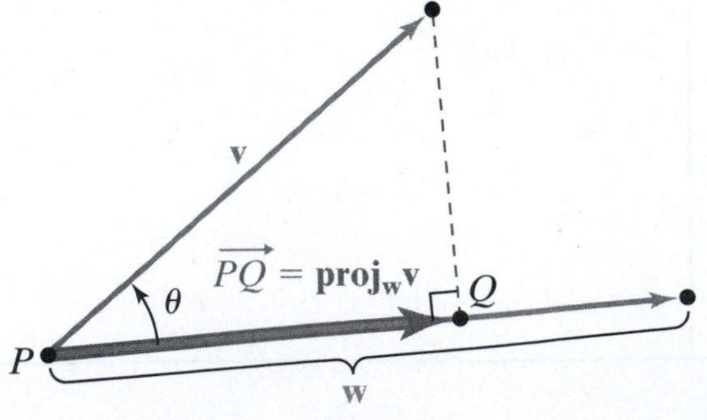 	

Key Concepts	Examples/Videos
Let $\mathbf{v}$ and $\mathbf{w}$ be non-zero vectors. Then $\mathbf{v}$ can be written as the sum of orthogonal vectors $\mathbf{v}_1$ and $\mathbf{v}_2$, where $\mathbf{v}_1$ is parallel to $\mathbf{w}$ and $\mathbf{v}_2$ is orthogonal to $\mathbf{w}$ such that $\mathbf{v}_1 = \text{proj}_{\mathbf{w}}\mathbf{v} = \frac{\mathbf{v} \cdot \mathbf{w}}{\lVert\mathbf{w}\rVert^2}\mathbf{w}$ and $\mathbf{v}_2 = \mathbf{v} - \mathbf{v}_1$. The process of expressing $\mathbf{v}$ as the sum of $\mathbf{v}_1$ and $\mathbf{v}_2$ is called the **vector decomposition of v into orthogonal components** $\mathbf{v}_1$ and $\mathbf{v}_2$.	Let $\mathbf{v} = \mathbf{i} + 2\mathbf{j}$ and $\mathbf{w} = 4\mathbf{i} + \mathbf{j}$. Determine the vector decomposition of $\mathbf{v}$ into orthogonal components $\mathbf{v}_1$ and $\mathbf{v}_2$, where $\mathbf{v}_1$ is parallel to $\mathbf{w}$ and $\mathbf{v}_2$ is orthogonal to $\mathbf{w}$.

Chapter 10 Review Exercises

1. Plot the point $P\left(4, -\frac{5\pi}{6}\right)$ in the polar coordinate system.

2. Plot the point $P\left(3, \frac{4\pi}{3}\right)$ in the polar coordinate system and then find three different representations of point P that have the specified conditions.
   a. $r > 0, -2\pi \le \theta < 0$
   b. $r < 0, 0 \le \theta < 2\pi$
   c. $r > 0, 2\pi \le \theta < 4\pi$

3. Determine the rectangular coordinates of the point $P\left(-5, \frac{3\pi}{4}\right)$.

4. Determine the polar coordinates for the point $P(-4\sqrt{3}, -4)$ such that $r > 0$ and $0 \le \theta < 2\pi$.

5. Convert the equation $(x + 1)^2 + y^2 = 1$ into polar form.

For 6–9, sketch the graph of the polar equation.

6. $r \cos\theta = -2$.

7. $r = 5 \sin\theta$

8. $r = 3 - 4\cos\theta$

9. $r = -4\cos 3\theta$

10. Plot the point $z = -1 + 2i$ in the complex plane and determine its absolute value.

11. Rewrite the complex number $z = 3\left(\cos\frac{25\pi}{3} + i\sin\frac{25\pi}{3}\right)$ in standard polar form, plot the number in the complex plane, and determine the quadrant in which the point lies or the axis on which the point lies.

12. Write the complex number $z = 10\left(\cos\frac{5\pi}{3} + i\sin\frac{5\pi}{3}\right)$ in rectangular form using exact values if possible. Otherwise, round to two decimal places.

13. Write the complex number $z = 5\sqrt{3} - 5i$ in standard polar form. Write the argument in radians using exact values if possible. Otherwise, round to two decimal places.

14. Find z_1z_2 and $\frac{z_1}{z_2}$ if $z_1 = 2\left(\cos\frac{3\pi}{4} + i\sin\frac{3\pi}{4}\right)$ and $z_2 = 3\left(\cos\frac{11\pi}{6} + i\sin\frac{11\pi}{6}\right)$. Write your answers in standard polar form with the argument in radians.

15. Determine the given complex number raised to a power and write the answer in rectangular form.

$$\left[3\left(\cos\frac{\pi}{9} + i\sin\frac{\pi}{9}\right)\right]^3$$

16. Find all complex square roots of $z = 36\left(\cos\frac{\pi}{6} + i\sin\frac{\pi}{6}\right)$. Write each root in standard polar form with the argument in radians.

17. Determine the magnitude of $\mathbf{v}$.

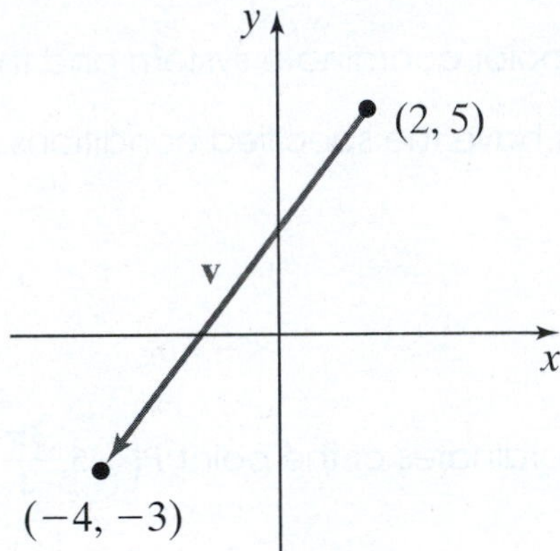

18. Use the given vectors $\mathbf{u}$, $\mathbf{v}$, and $\mathbf{w}$ to draw the vector $2\mathbf{w} + 2\mathbf{u} - \mathbf{v}$.

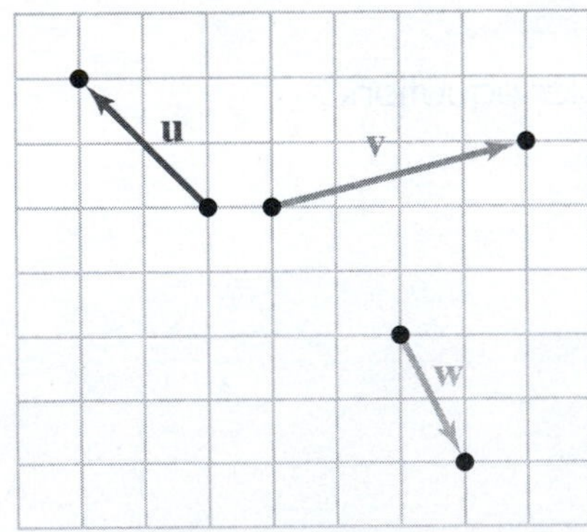

19. If $\mathbf{u} = -3\mathbf{i} + 7\mathbf{j}$ and $\mathbf{v} = 5\mathbf{i} - \mathbf{j}$, find $4\mathbf{u} + 3\mathbf{v}$ in terms of $\mathbf{i}$ and $\mathbf{j}$, then find the magnitude of $4\mathbf{u} + 3\mathbf{v}$.

20. Determine the direction angle for the vector $\mathbf{v} = 2\sqrt{3}\mathbf{i} + 2\mathbf{j}$.

21. Represent the vector $\mathbf{v}$ in the form $\mathbf{v} = a\mathbf{i} + b\mathbf{j}$ whose magnitude is $\|\mathbf{v}\| = 15$ and direction angle is $\theta = 150°$.

22. Find the value of the scalar $(\mathbf{u} + \mathbf{v}) \cdot \mathbf{w}$ given that $\mathbf{u} = 5\mathbf{i} - 2\mathbf{j}$, $\mathbf{v} = 6\mathbf{i} + \mathbf{j}$ and $\mathbf{w} = -\mathbf{i} - 7\mathbf{j}$.

23. Determine the angle between $\mathbf{u} = 5\mathbf{i} + 2\mathbf{j}$ and $\mathbf{v} = \mathbf{i} + 4\mathbf{j}$. Give the angle in degrees rounded to the nearest hundredth of a degree.

24. Determine if the vectors $\mathbf{u} = 2\mathbf{i} - 3\mathbf{j}$ and $\mathbf{v} = 6\mathbf{i} + 4\mathbf{j}$ are orthogonal, parallel, or neither.

25. Given the vectors $\mathbf{v} = -5\mathbf{i} - 9\mathbf{j}$ and $\mathbf{w} = -\mathbf{i} + \mathbf{j}$, determine the vector decomposition of $\mathbf{v}$ into orthogonal components $\mathbf{v}_1$ and $\mathbf{v}_2$, where $\mathbf{v}_1$ is parallel to $\mathbf{w}$ and $\mathbf{v}_2$ is orthogonal to $\mathbf{w}$.

CHAPTER ELEVEN

Conic Sections and Parametric Equations

CHAPTER ELEVEN CONTENTS

INTRODUCTION TO CONIC SECTIONS

In this chapter, we focus on the geometric study of conic sections. Conic sections (or conics) are formed when a plane intersects a pair of **right circular cones.** The surface of the cones comprises the set of all line segments that intersect the outer edges of the circular bases of the cones and pass through a fixed point. The fixed point is called the **vertex** of the cone, and the line segments are called the **elements**. See Figure 1.

When a plane intersects a right circular cone, a conic section is formed. The four conic sections that can be formed are circles, ellipses, parabolas, and hyperbolas. Click on one of the following four animations to see how each conic section is formed.

axis

Vertex

element

Figure 1 Pair of right circular cones

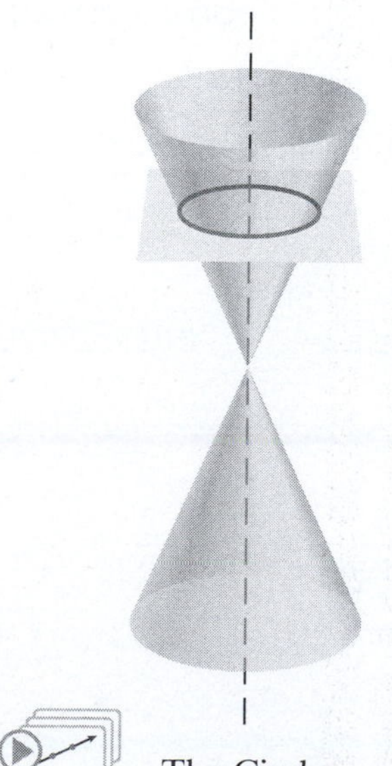

The Circle

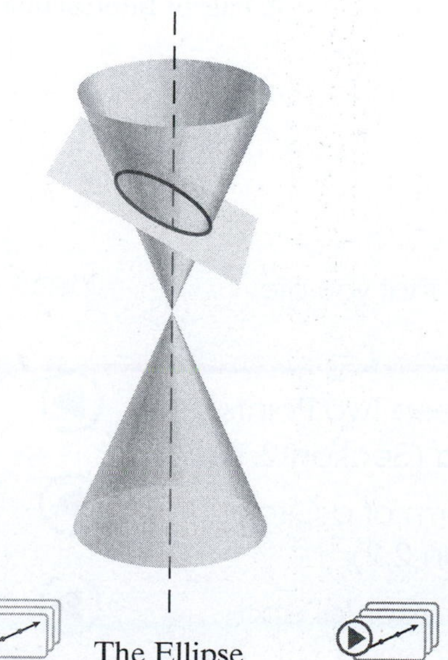

The Ellipse

The Parabola

The Hyperbola

Because we studied circles in **Section 2.2**, they are not covered again in this chapter. It may, however, be useful to review circles before going on. Watch the **video** to review how to write the equation of a circle in standard form by completing the square. We also need to be able to complete the square to write each of the other conic sections in standard form. These conic sections are introduced in the following order:

Section 11.1 The Parabola

Section 11.2 The Ellipse

Section 11.3 The Hyperbola

DEGENERATE CONIC SECTIONS

It is worth noting that when the circle, ellipse, parabola, or hyperbola is formed, the intersecting plane does not pass through the vertex of the cones. When a plane does intersect the vertex of the cones, a degenerate conic section is formed. The three degenerate conic sections are a point, a line, and a pair of intersecting lines. See **Figure 2**. We do not concern ourselves with degenerate conic sections in this chapter.

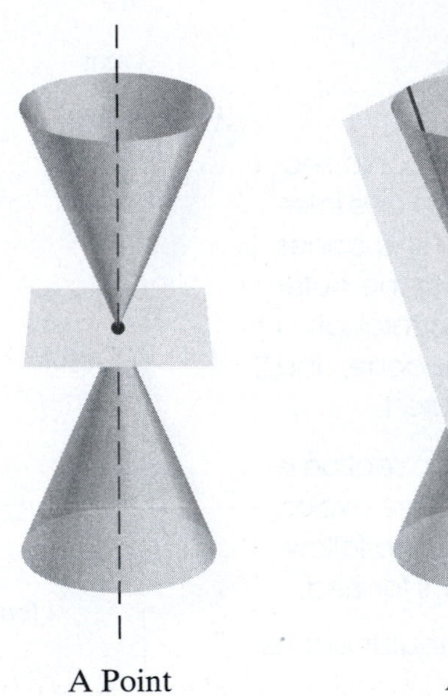

A Point

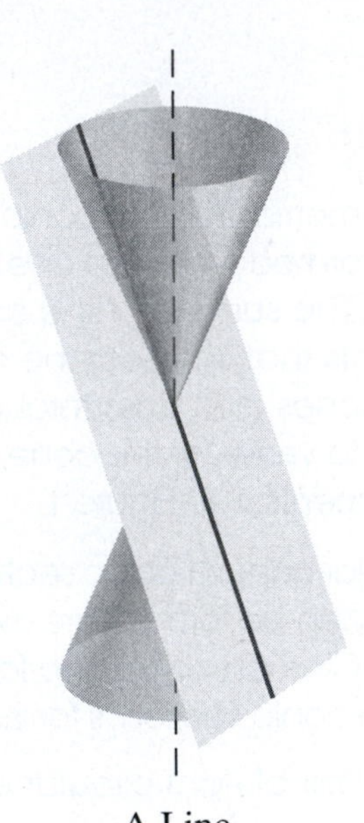

A Line

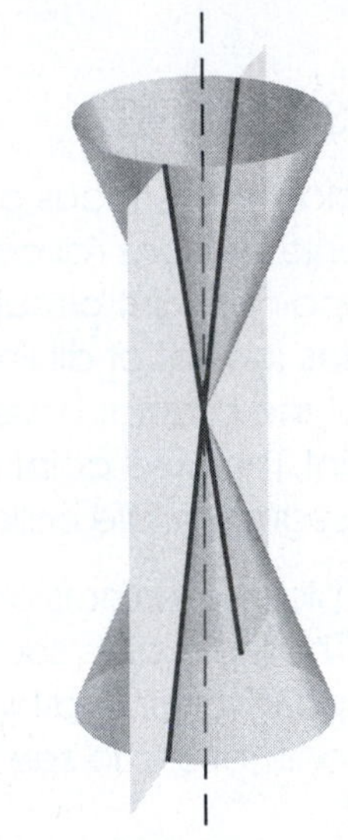

A Pair of Intersecting Lines

Figure 2 Degenerate conic sections

11.1 The Parabola

THINGS TO KNOW

Before working through this section, be sure that you are familiar with the following concepts:

		VIDEO	ANIMATION	INTERACTIVE
 You Try It	1. Finding the Distance between Two Points Using the Distance Formula (Section 2.1)	 		
 You Try It	2. Converting the General Form of a Circle into Standard Form (Section 2.2)			
 You Try It	3. Finding the Equations of Horizontal and Vertical Lines (Section 2.3)			

OBJECTIVES

OBJECTIVE 1 DETERMINING THE EQUATION OF A PARABOLA WITH A VERTICAL AXIS OF SYMMETRY

In Section 4.1, we studied quadratic functions of the form $f(x) = ax^2 + bx + c$, $a \neq 0$. We learned that every quadratic function has a U-shaped graph called a *parabola*. You may want to review the different characteristics of a parabola. Work through the following animation, and click on each characteristic to get a detailed description.

CHARACTERISTICS OF A PARABOLA

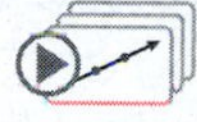

1. Vertex
2. Axis of symmetry
3. y-Intercept
4. x-Intercept(s) or real zeros
5. Domain and range

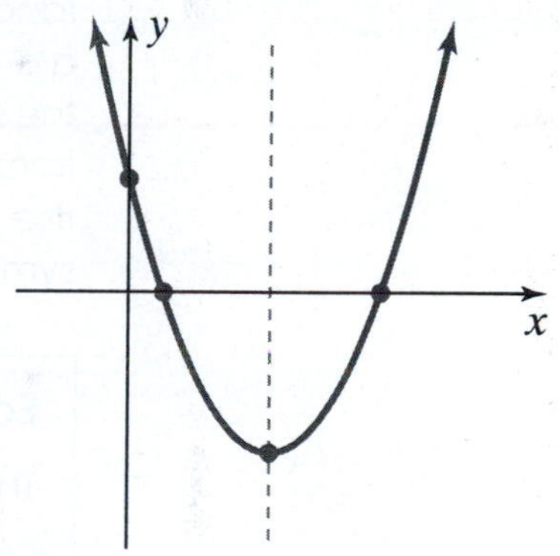

In Section 4.1, we studied quadratic functions and parabolas from an algebraic point of view. We now look at parabolas from a geometric perspective. We see in the introduction to this chapter that when a plane is parallel to an **element** of the cone, the plane will intersect the cone in a parabola. (Click on the animation.)

The set of points that define the parabola formed by the intersection described previously is stated in the following geometric definition of the parabola.

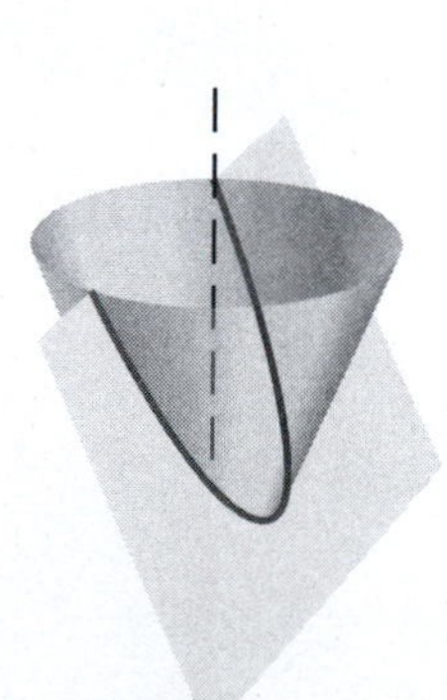

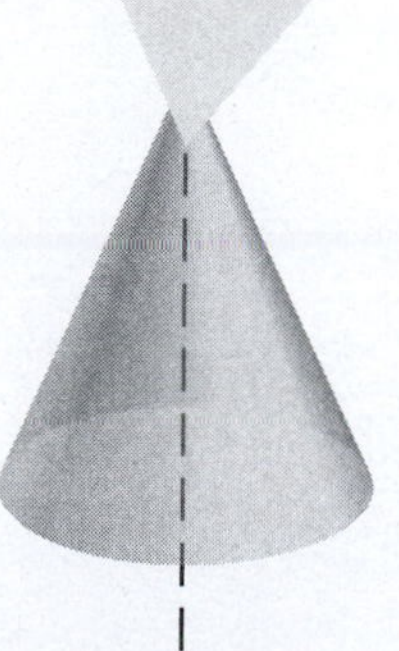

The Parabola

> **Geometric Definition of the Parabola**
>
> A **parabola** is the set of all points in a plane equidistant from a fixed point F and a fixed line D. The fixed point is called the **focus**, and the fixed line is called the **directrix**.

Watch the **video** to see how to sketch the parabola seen in **Figure 3** using the geometric definition of the parabola.

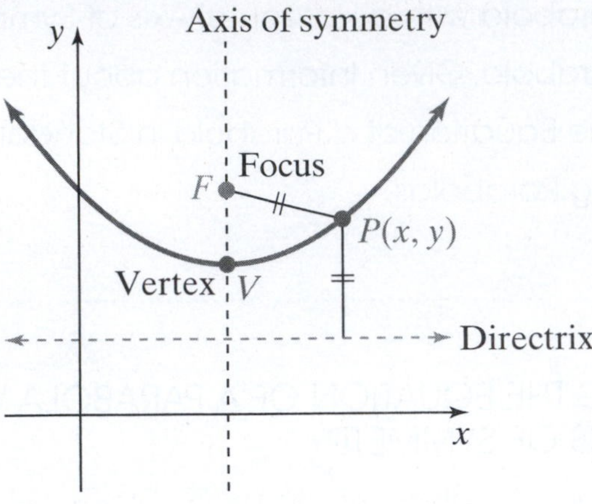

Figure 3 The distance from any point P on the parabola to the focus is the same as the distance from point P to the directrix.

In Figure 3, we can see that for any point $P(x, y)$ that lies on the graph of the parabola, the distance from point P to the focus is exactly the same as the distance from point P to the directrix. Similarly, because the vertex, V, lies on the graph of the parabola, the distance from V to the focus must also be the same as the distance from V to the directrix. Therefore, if the distance from V to F is p units, then the distance from V to the directrix is also p units. If the coordinates of the vertex in Figure 3 are (h, k), then the coordinates of the focus must be $(h, k + p)$ and the equation of the directrix is $y = k - p$. We can use this information and the fact that the distance from $P(x, y)$ to the focus is equal to the distance from $P(x, y)$ to the directrix to derive the equation of a parabola. The equation of the parabola with a vertical axis of symmetry is **derived in Appendix B**.

Equation of a Parabola in Standard Form with a Vertical Axis of Symmetry

The equation of a parabola with a vertical axis of symmetry is $(x - h)^2 = 4p(y - k)$, where

the vertex is $V(h, k)$,

$|p|$ = distance from the vertex to the focus = distance from the vertex to the directrix,

the focus is $F(h, k + p)$, and

the equation of the directrix is $y = k - p$.

The parabola opens *upward* if $p > 0$ or *downward* if $p < 0$.

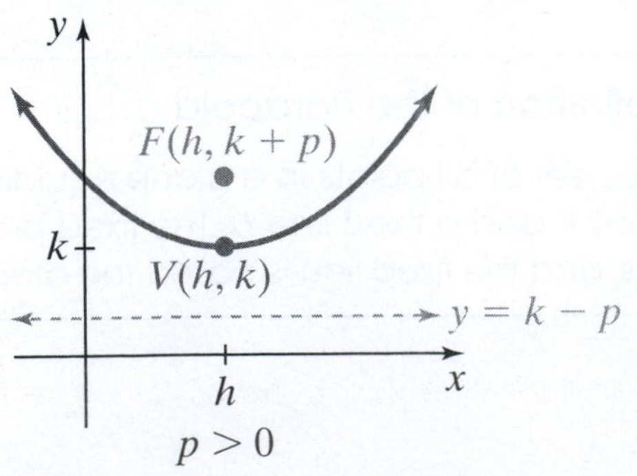

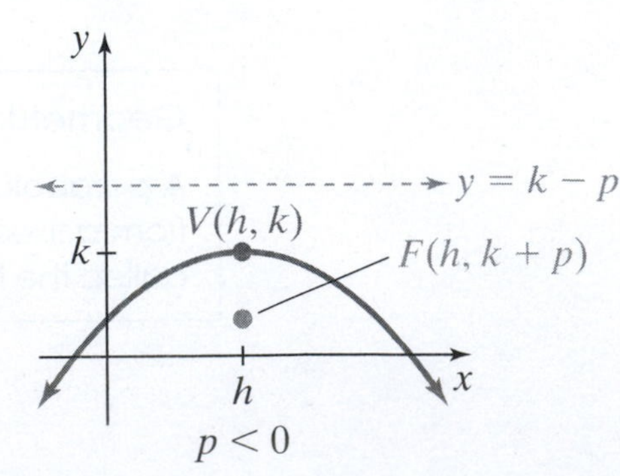

Example 1 Find the Vertex, Focus, and Directrix of a Parabola and Sketch Its Graph

Find the vertex, focus, and directrix of the parabola $x^2 = 8y$ and sketch its graph.

Solution Notice that we can rewrite the equation $x^2 = 8y$ as $(x - 0)^2 = 8(y - 0)$. We can now compare the equation $(x - 0)^2 = 8(y - 0)$ to the standard form equation $(x - h)^2 = 4p(y - k)$ to see that $h = 0$ and $k = 0$; hence, the vertex is at the origin, $(0, 0)$. To find the focus and directrix, we need to find p.

$$4p = 8$$

$$p = 2 \qquad \text{Divide both sides by 4.}$$

Because the value of p is positive, the parabola opens upward and the focus is located two units vertically *above* the vertex, whereas the directrix is the horizontal line located two units vertically *below* the vertex. The focus has coordinates $(0, 2)$, and the equation of the directrix is $y = -2$. The graph is shown in Figure 4.

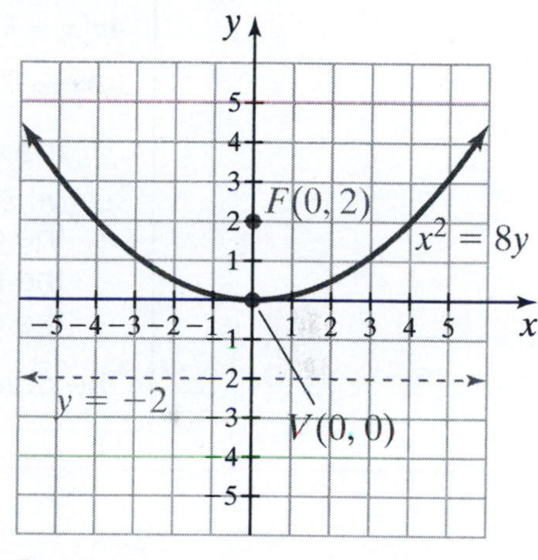

Figure 4

Example 2 Find the Vertex, Focus, and Directrix of a Parabola and Sketch Its Graph

Find the vertex, focus, and directrix of the parabola $-(x + 1)^2 = 4(y - 3)$ and sketch its graph.

Solution Watch the video to verify that the vertex has coordinates $(-1, 3)$, the focus has coordinates $(-1, 2)$, and the equation of the directrix is $y = 4$. The graph of this parabola is sketched in Figure 5.

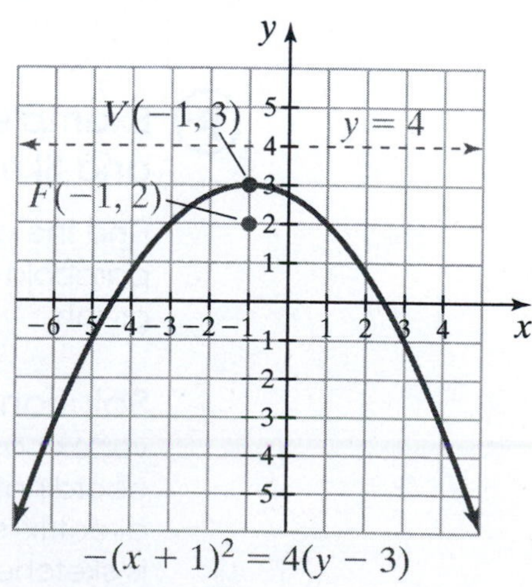

Figure 5

You Try It Work Exercises 1–5 in this textbook or in the MyLab Math Study Plan.

OBJECTIVE 2 DETERMINING THE EQUATION OF A PARABOLA WITH A HORIZONTAL AXIS OF SYMMETRY

In Examples 1 and 2, the graphs of both parabolas had vertical axes of symmetry. The graph of a parabola could also have a horizontal axis of symmetry and open "sideways." We derive the standard form of the parabola with a horizontal axis of symmetry in much the same way as we did with the parabola with a vertical axis of symmetry. This equation is **derived in Appendix B.**

Equation of a Parabola in Standard Form with a Horizontal Axis of Symmetry

The equation of a parabola with a horizontal axis of symmetry is $(y - k)^2 = 4p(x - h)$,

where

the vertex is $V(h, k)$,
$|p|$ = distance from the vertex to the focus = distance from the vertex to the directrix,
the focus is $F(h + p, k)$, and
the equation of the directrix is $x = h - p$.

The parabola opens *right* if $p > 0$ or *left* if $p < 0$.

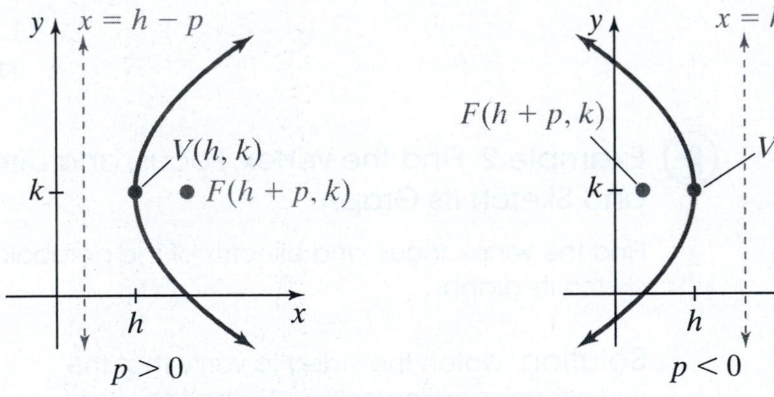

Example 3 Find the Vertex, Focus, and Directrix of a Parabola and Sketch Its Graph

Find the vertex, focus, and directrix of the parabola $(y - 3)^2 = 8(x + 2)$ and sketch its graph.

Solution Watch the video to verify that the vertex has coordinates $(-2, 3)$, the focus has coordinates $(0, 3)$, and the equation of the directrix is $x = -4$. The graph of this parabola is sketched in Figure 6.

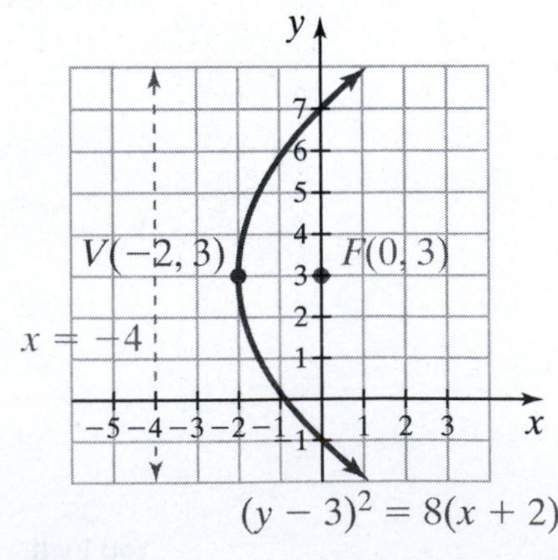

Figure 6

You Try It Work Exercises 6–10 in this textbook or in the MyLab Math Study Plan.

OBJECTIVE 3 DETERMINING THE EQUATION OF A PARABOLA, GIVEN INFORMATION ABOUT THE GRAPH

It is often necessary to determine the equation of a parabola, given certain information. It is always useful to first determine whether the parabola has a vertical axis of symmetry or a horizontal axis of symmetry. Try to work through Examples 4 and 5. Then watch the corresponding video solutions to determine whether you are correct.

Example 4 Find the Equation of a Parabola

Find the standard form of the equation of the parabola with focus $\left(-3, \frac{5}{2}\right)$ and directrix $y = \frac{11}{2}$.

Solution Watch the video to see that the equation of this parabola is $(x + 3)^2 = -6(y - 4)$. The graph is shown in Figure 7.

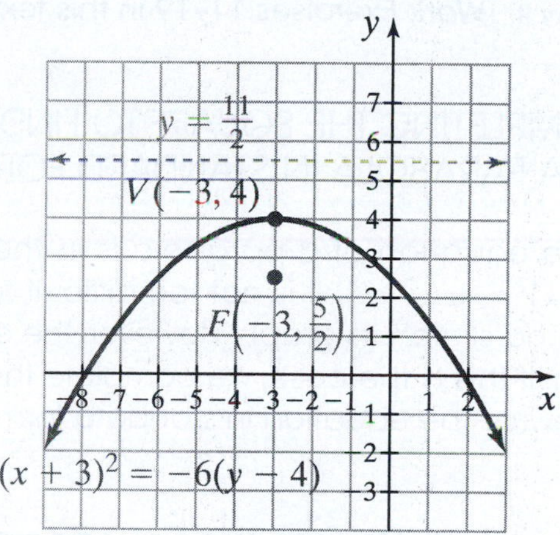

Figure 7

Example 5 Find the Vertex, Focus, and Directrix of a Parabola and Sketch Its Graph

Find the standard form of the equation of the parabola with focus $(4, -2)$ and vertex $\left(\frac{13}{2}, -2\right)$.

Solution Watch the video to see that the equation of this parabola is $(y + 2)^2 = -10\left(x - \frac{13}{2}\right)$. The graph is shown in Figure 8.

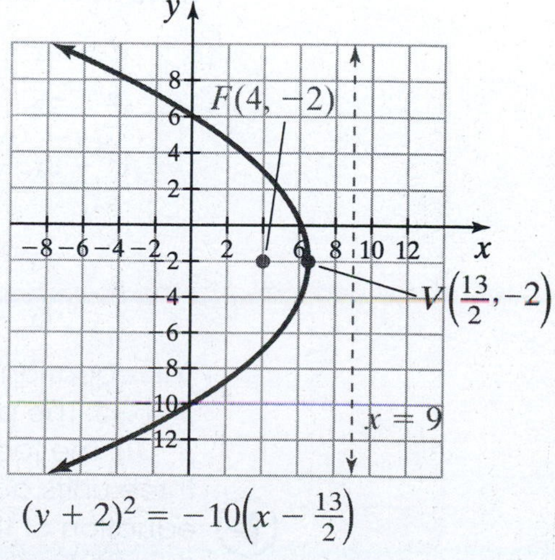

Figure 8

Using Technology

We can use a graphing utility to graph the parabola from Example 5 by solving the equation for y.

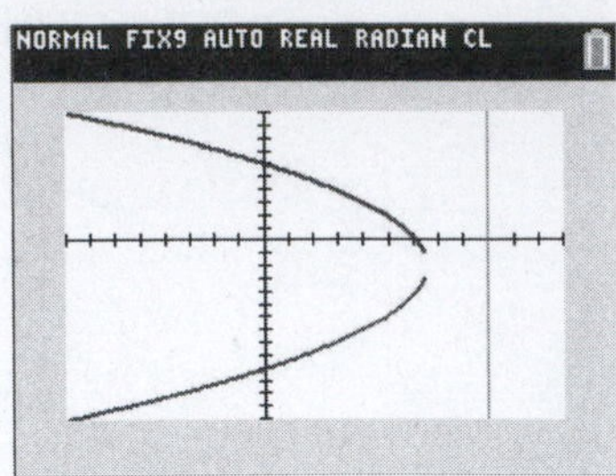

Figure 9

$$(y + 2)^2 = -10\left(x - \frac{13}{2}\right)$$

$$y + 2 = \pm\sqrt{-10\left(x - \frac{13}{2}\right)}$$

$$y = -2 \pm \sqrt{-10\left(x - \frac{13}{2}\right)}$$

Using $y_1 = -2 + \sqrt{-10\left(x - \frac{13}{2}\right)}$, and $y_2 = -2 - \sqrt{-10\left(x - \frac{13}{2}\right)}$, we obtain the graph seen in Figure 9.

Note The directrix was created using the calculator's DRAW feature.

You Try It **Work Exercises 11–19 in this textbook or in the MyLab Math Study Plan.**

OBJECTIVE 4 COMPLETING THE SQUARE TO FIND THE EQUATION OF A PARABOLA IN STANDARD FORM

If the equation of a parabola is in the standard form of $(x - h)^2 = 4p(y - k)$ or $(y - k)^2 = 4p(x - h)$, it is not too difficult to determine the vertex, focus, and directrix and sketch its graph. However, the equation might not be given in standard form. If this is the case, we complete the square on the variable that is squared to rewrite the equation in standard form as in Example 6.

Example 6 Rewrite the Equation of a Parabola in Standard Form by Completing the Square

Find the vertex, focus, and directrix and sketch the graph of the parabola $x^2 - 8x + 12y = -52$.

Solution Because x is squared, we will complete the square on the variable x.

$x^2 - 8x + 12y = -52$	Write the original equation.
$x^2 - 8x = -12y - 52$	Subtract $12y$ from both sides.
$x^2 - 8x + 16 = -12y - 52 + 16$	Complete the square by adding 16 to both sides.
$(x - 4)^2 = -12y - 36$	Factor and simplify.
$(x - 4)^2 = -12(y + 3)$	Factor.

The equation is now in standard form with vertex $(4, -3)$ and $4p = -12$ so $p = -3$. The parabola must open down because the variable x is squared and $p < 0$. The focus is located three units below the vertex, whereas the directrix is three units above the vertex. Thus, the focus has coordinates $(4, -6)$, and the

equation of the directrix is $y = 0$ or the x-axis. You can watch the **video** to see this solution worked out in detail. The graph is shown in Figure 10.

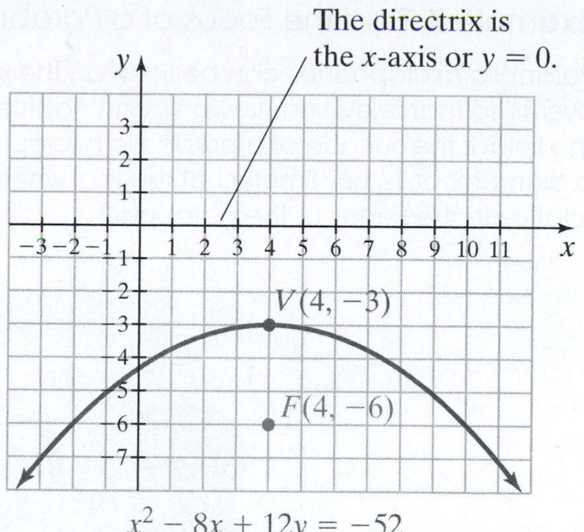

Figure 10

You Try It Work Exercises 20–23 in this textbook or in the MyLab Math Study Plan.

OBJECTIVE 5 SOLVING APPLIED PROBLEMS INVOLVING PARABOLAS

The Romans were one of the first civilizations to use the engineering properties of parabolic structures in their creation of **arch bridges**. The cables of many suspension bridges, such as the **Golden Gate Bridge** in San Francisco, span from tower to tower in the shape of a parabola.

Parabolic surfaces are used in the manufacture of many satellite dishes, search lights, car headlights, telescopes, lamps, heaters, and other objects. This is because parabolic surfaces have the property that incoming rays of light or radio waves traveling parallel to the axis of symmetry of a parabolic reflector or receiver will reflect off the parabolic surface and travel directly toward the antenna that is placed at the focus. See **Figure 11**. When a light source such as the headlight of a car is placed at the focus of a parabolic reflector, the light reflects off the surface outward, producing a narrow beam of light and thus maximizing the output of illumination.

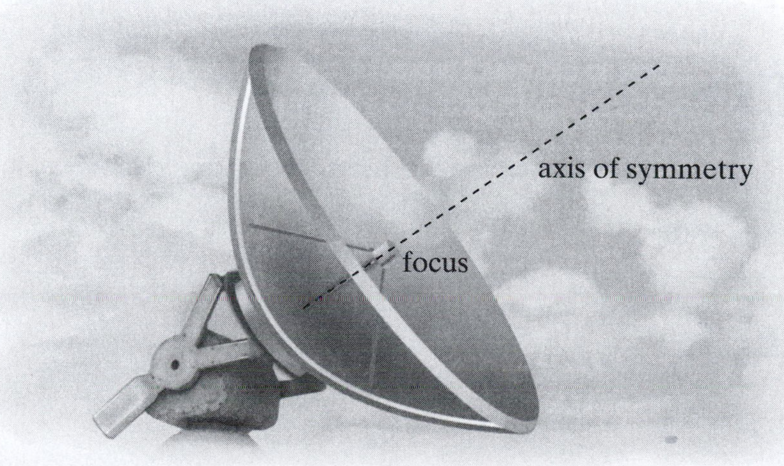

Figure 11 Incoming rays reflect off the parabolic surface toward the antenna placed at the focus.

Example 7 Find the Focus of a Parabolic Microphone

Parabolic microphones can be seen on the sidelines of professional sporting events so that television networks can capture audio sounds from the players on the field. If the surface of a parabolic microphone is 27 centimeters deep and has a diameter of 72 centimeters at the top, where should the microphone be placed relative to the vertex of the parabola?

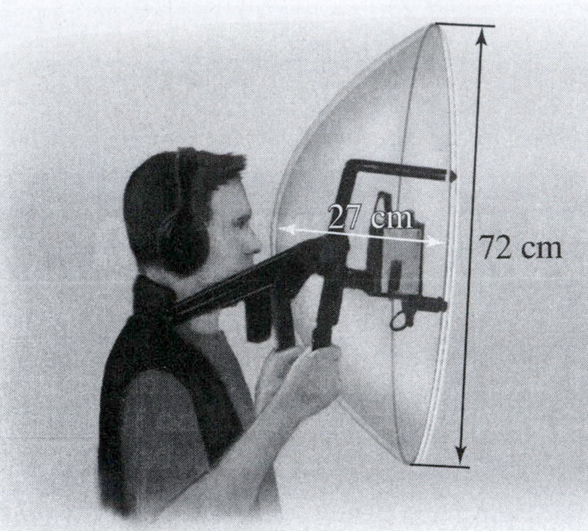

Solution We can draw a parabola with the vertex at the origin representing the center cross section of the parabolic microphone. The equation of this parabola in standard form is $x^2 = 4py$. Substitute the point $(36, 27)$ into the equation to get

$$x^2 = 4py$$
$$(36)^2 = 4p(27)$$
$$1{,}296 = 108p$$
$$p = 12.$$

The microphone must be placed 12 centimeters from the vertex. ●

You Try It Work Exercises 24–28 in this textbook or in the MyLab Math Study Plan.

11.1 Exercises

In Exercises 1–10, determine the vertex, focus, and directrix of the parabola and sketch its graph.

SbS 1. $x^2 = 16y$

SbS 2. $x^2 = -8y$

SbS 3. $(x - 1)^2 = -12(y - 4)$

SbS 4. $(x + 3)^2 = 6(y - 1)$

SbS 5. $(x + 2)^2 = 5(y + 6)$

SbS 6. $y^2 = 4x$

SbS 7. $y^2 = -8x$

SbS 8. $(y - 5)^2 = -4(x - 2)$

SbS 9. $(y + 3)^2 = 20(x - 4)$

SbS 10. $(y + 4)^2 = 9(x + 3)$

In Exercises 11–19, find the equation in standard form of the parabola described.

11. The focus has coordinates $(2, 0)$, and the equation of the directrix is $x = -2$.

12. The focus has coordinates $\left(0, -\frac{1}{2}\right)$, and the equation of the directrix is $y = \frac{1}{2}$.

13. The focus has coordinates $(3, -5)$, and the equation of the directrix is $y = -1$.

14. The focus has coordinates $(2, 4)$, and the equation of the directrix is $x = -4$.

15. The vertex has coordinates $\left(-\frac{11}{4}, -2\right)$, and the focus has coordinates $(-3, -2)$.

16. The vertex has coordinates $\left(4, -\frac{1}{4}\right)$, and the focus has coordinates $\left(4, \frac{1}{4}\right)$.

17. The vertex has coordinates $(-3, 4)$, and the equation of the directrix is $x = -7$.

18. Find the equations of the two parabolas in standard form that have a horizontal axis of symmetry and a focus at the point $(0, 4)$, and that pass through the origin.

19. Find the equations of the two parabolas in standard form that have a vertical axis of symmetry and a focus at the point $\left(2, \frac{3}{2}\right)$, and that pass through the origin.

In Exercises 20–23, determine the vertex, focus, and directrix of the parabola and sketch its graph.

SbS 20. $x^2 + 10x = 5y - 10$

SbS 21. $y^2 - 12y + 6x + 30 = 0$

SbS 22. $x^2 - 2x = -y + 1$

SbS 23. $y^2 + x + 6y = -10$

24. A parabolic eavesdropping device is used by a CIA agent to record terrorist conversations. The parabolic surface measures 120 centimeters in diameter at the top and is 90 centimeters deep at its center. How far from the vertex should the microphone be located?

25. A parabolic space heater is 18 inches in diameter and 8 inches deep. How far from the vertex should the heat source be located to maximize the heating output?

26. A large NASA parabolic satellite dish is 22 feet across and has a receiver located at the focus 4 feet from its base. The satellite dish should be how deep?

27. A parabolic arch bridge spans 160 feet at the base and is 40 feet above the water at the center. Find the equation of the parabola if the vertex is placed at the point $(0, 40)$. Can a sailboat that is 35 feet tall fit under the bridge 30 feet from the center?

28. The cable between two 40-meter towers of a suspension bridge is in the shape of a parabola that just touches the bridge halfway between the towers. The two towers are 100 meters apart. Vertical cables are spaced every 10 meters along the bridge. What are the lengths of the vertical cables located 30 meters from the center of the bridge?

Brief Exercises

In Exercises 29–42, sketch the graph of each parabola.

29. $x^2 = 16y$

30. $x^2 = -8y$

31. $(x - 1)^2 = -12(y - 4)$

32. $(x + 3)^2 = 6(y - 1)$

33. $(x + 2)^2 = 5(y + 6)$

34. $y^2 = 4x$

35. $y^2 = -8x$

36. $(y - 5)^2 = -4(x - 2)$

37. $(y + 3)^2 = 20(x - 4)$

38. $(y + 4)^2 = 9(x + 3)$

39. $x^2 + 10x = 5y - 10$

40. $y^2 - 12y + 6x + 30 = 0$

41. $x^2 - 2x = -y + 1$

42. $y^2 + x + 6y = -10$

11.2 The Ellipse

THINGS TO KNOW

Before working through this section, be sure that you are familiar with the following concepts:

VIDEO ANIMATION INTERACTIVE

You Try It 1. Finding the Distance between Two Points Using the Distance Formula (Section 2.1)

You Try It 2. Completing the Square to Find the Equation of a Parabola in Standard Form (Section 11.1)

INTRODUCTION

Read this introduction before beginning Objective 1.

OBJECTIVES

1 Sketching the Graph of an Ellipse

2 Determining the Equation of an Ellipse, Given Information about the Graph

3 Completing the Square to Find the Equation of an Ellipse in Standard Form

4 Solving Applied Problems Involving Ellipses

SECTION 11.2 EXERCISES

Introduction to Section 11.2

When a plane intersects a **right circular cone** at an angle between 0 and 90 degrees to the axis of the cone, the conic section formed is an ellipse. (Click on the animation.)

The set of points in the plane that defines an ellipse formed by the intersection of a plane and a cone described previously is stated in the following geometric definition.

> **Definition** Geometric Definition of the Ellipse
>
> An **ellipse** is the set of all points in a plane, the sum of whose distances from two fixed points is a positive constant. The two fixed points, F_1 and F_2, are called the *foci.*

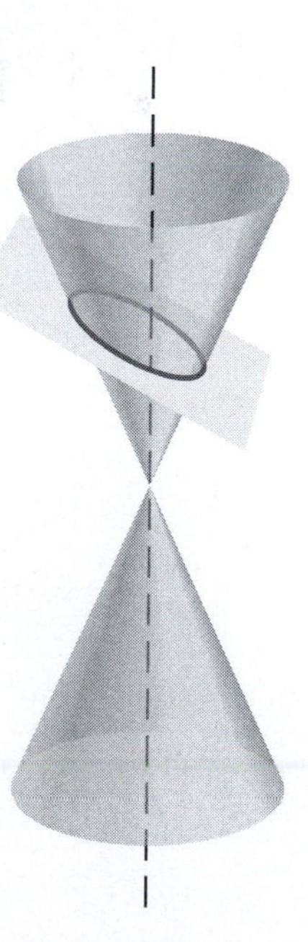

The Ellipse

The previous geometric definition implies that for any two points P and Q that lie on the graph of the ellipse, the sum of the distance between P and F_1 plus the distance between P and F_2 is equal to the sum of the distance between Q and F_1 plus the distance between Q and F_2. In symbols, we write $d(P, F_1) + d(P, F_2) = d(Q, F_1) + d(Q, F_2)$. See Figure 12.

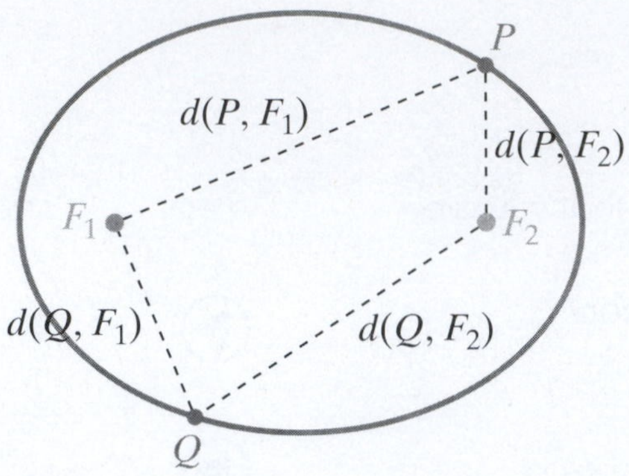

Figure 12 For any points P and Q on an ellipse, $d(P, F_1) + d(P, F_2) = d(Q, F_1) + d(Q, F_2)$.

OBJECTIVE 1 SKETCHING THE GRAPH OF AN ELLIPSE

An ellipse has two **axes of symmetry**. The longer axis, which is the line segment that connects the two vertices, is called the *major axis*. The foci are always located along the major axis. The shorter axis is called the *minor axis*. It is the line segment perpendicular to the major axis that passes through the center having endpoints that lie on the ellipse. Watch the **video** to see how to sketch the two ellipses seen in Figure 13. In Figure 13a, the ellipse has a **horizontal major axis**. The ellipse in Figure 13b has a **vertical major axis**.

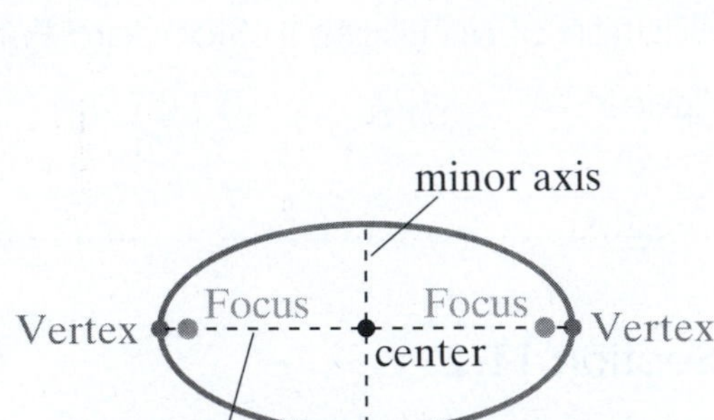

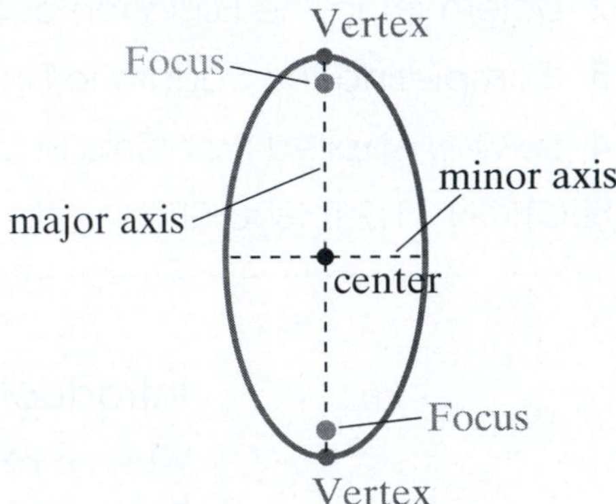

Figure 13 (a) Ellipse with Horizontal Major Axis (b) Ellipse with Vertical Major Axis

Consider the ellipse with a **horizontal major axis** centered at (h, k) as shown in Figure 14, where $c > 0$ is the distance between the center and a focus, $a > 0$ is the distance between the center and one of the vertices, and $b > 0$ is the distance from the center to an endpoint of the **minor axis**.

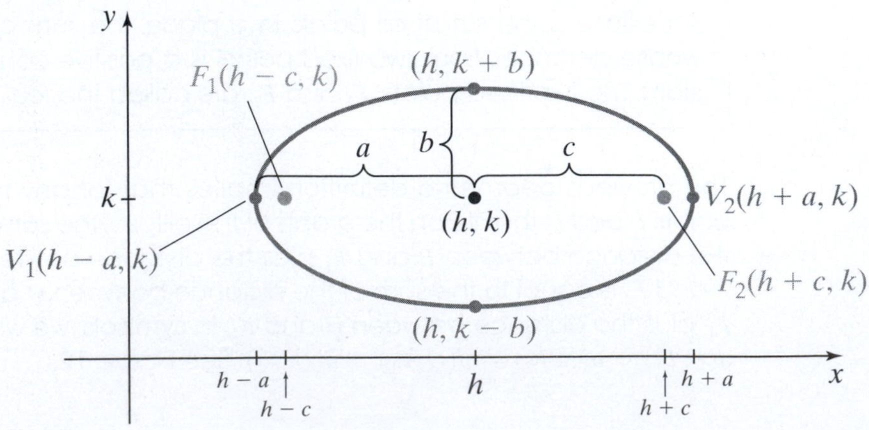

Figure 14

Before we can derive the equation of this ellipse, we must establish the following two facts:

Fact 1: The sum of the distances from any point on the ellipse to the two foci is $2a$.
Fact 2: $b^2 = a^2 - c^2$, or equivalently, $c^2 = a^2 - b^2$.

Click on **Fact 1** or **Fact 2** to see how these two facts are established in Appendix B.

Once we have established these two facts, we can derive the equation of the ellipse. The equation of an ellipse is **derived in Appendix B**. We now state the two standard equations of an ellipse.

Equation of an Ellipse in Standard Form with Center (h, k)

Horizontal Major Axis

$$\frac{(x-h)^2}{a^2} + \frac{(y-k)^2}{b^2} = 1$$

- $a > b > 0$
- Foci: $F_1(h - c, k)$ and $F_2(h + c, k)$
- Vertices: $V_1(h - a, k)$ and $V_2(h + a, k)$
- Endpoints of minor axis: $(h, k - b)$ and $(h, k + b)$
- $c^2 = a^2 - b^2$

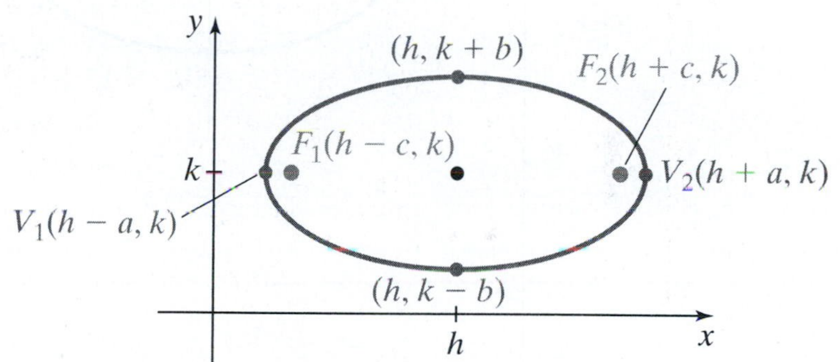

Equation of an Ellipse in Standard Form with Center (h, k)

Vertical Major Axis

$$\frac{(x-h)^2}{b^2} + \frac{(y-k)^2}{a^2} = 1$$

- $a > b > 0$
- Foci: $F_1(h, k - c)$ and $F_2(h, k + c)$
- Vertices: $V_1(h, k - a)$ and $V_2(h, k + a)$
- Endpoints of minor axis: $(h - b, k)$ and $(h + b, k)$
- $c^2 = a^2 - b^2$

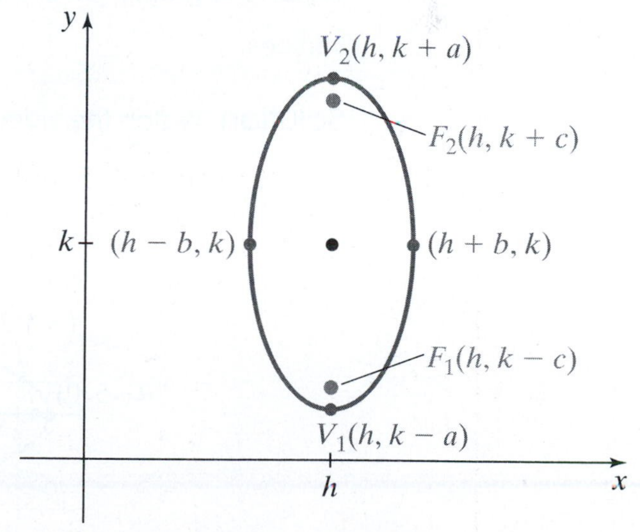

Notice that the two equations of the ellipse are identical except for the placement of a^2. If the equation of an ellipse is in standard form, we can quickly determine whether the ellipse has a horizontal major axis or a vertical major axis by looking at the denominator. If the larger denominator, a^2, appears under the expression

$(x - h)^2$, then the ellipse has a horizontal major axis. If the larger denominator appears under the expression $(y - k)^2$, then the ellipse has a vertical major axis. If the denominators are equal $(a^2 = b^2)$, then the ellipse is a circle!

If $h = 0$ and $k = 0$, then the ellipse is centered at the **origin**. Ellipses centered at the origin have the following equations.

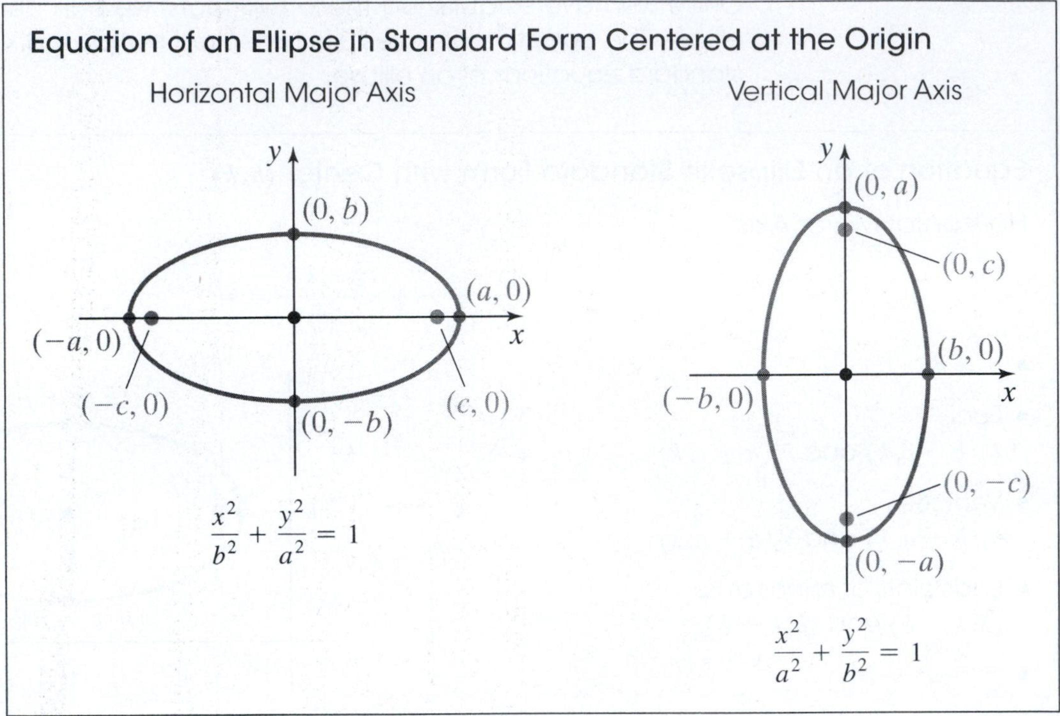

Example 1 Sketch the Graph of an Ellipse Centered at the Origin

Sketch the graph of the ellipse $\frac{x^2}{25} + \frac{y^2}{4} = 1$, and label the center, foci, and vertices.

Solution Watch the **video** to see how to sketch the ellipse shown in Figure 15.

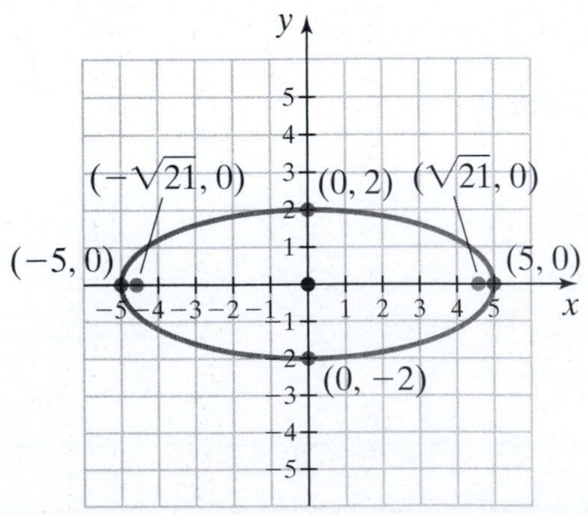

Figure 15

You Try It Work Exercises 1–4 in this textbook or in the MyLab Math Study Plan.

Example 2 Sketch the Graph of an Ellipse

Sketch the graph of the ellipse $\frac{(x+2)^2}{20}+\frac{(y-3)^2}{36}=1$, and label the center, foci, and vertices.

Solution Note that the larger denominator appears under the y-term. This indicates that the ellipse has a vertical major axis. Watch the **video** to verify that the center of the ellipse is $(-2, 3)$. The foci have coordinates $(-2, -1)$ and $(-2, 7)$, the vertices have coordinates $(-2, -3)$ and $(-2, 9)$, and the coordinates of the endpoints of the minor axis are $(-2-2\sqrt{5}, 3)$ and $(-2+2\sqrt{5}, 3)$. The graph is shown in Figure 16.

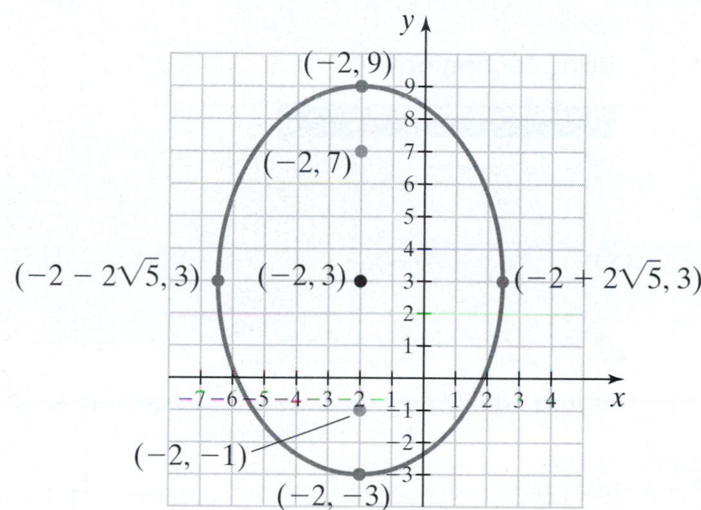

Figure 16

You Try It Work Exercises 5–8 in this textbook or in the MyLab Math Study Plan.

OBJECTIVE 2 DETERMINING THE EQUATION OF AN ELLIPSE, GIVEN INFORMATION ABOUT THE GRAPH

It is often necessary to determine the equation of an ellipse given certain information. It is always useful to first determine whether the ellipse has a horizontal major axis or a vertical major axis. Try to work through Examples 3 and 4. Then watch the solutions to the corresponding videos to determine whether you are correct.

Example 3 Determine the Equation of an Ellipse

Find the standard form of the equation of the ellipse with foci at $(-6, 1)$ and $(-2, 1)$ such that the length of the major axis is eight units.

Solution Watch the **video** to verify that this is an ellipse centered at $(-4, 1)$ with a horizontal major axis such that $a = 4$ and $b = \sqrt{12}$. The equation in standard form is

$$\frac{(x+4)^2}{4^2}+\frac{(y-1)^2}{(\sqrt{12})^2}=1 \quad \text{or} \quad \frac{(x+4)^2}{16}+\frac{(y-1)^2}{12}=1.$$

The graph of this ellipse is shown in **Figure 17**.

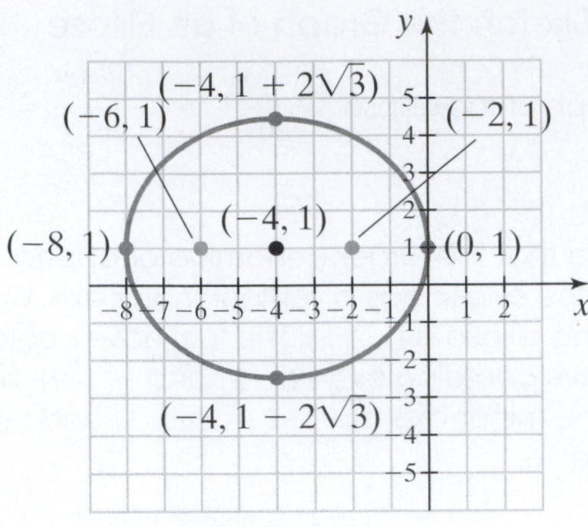

Figure 17 $\frac{(x+4)^2}{16} + \frac{(y-1)^2}{12} = 1.$ ●

Using Technology

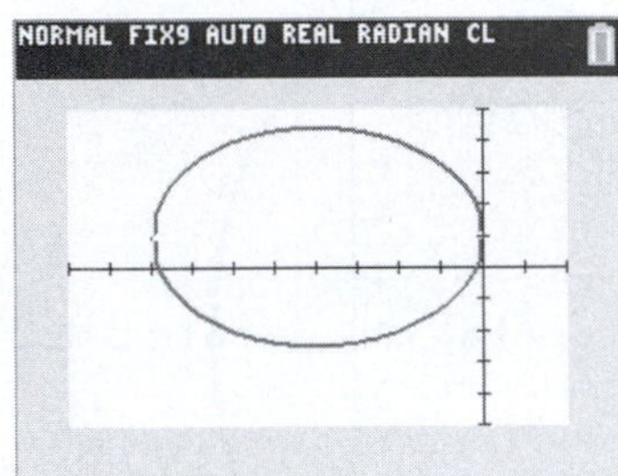

Figure 18 We can use a graphing utility to graph the ellipse from Example 3 by solving the equation for y:

$$y = 1 \pm \sqrt{12\left(1 - \frac{(x+4)^2}{16}\right)}$$

(View these **steps** to see how to solve for y.) Using

$$y_1 = 1 + \sqrt{12\left(1 - \frac{(x+4)^2}{16}\right)} \quad \text{and} \quad y_2 = 1 - \sqrt{12\left(1 - \frac{(x+4)^2}{16}\right)},$$

we obtain the graph seen in Figure 18. ●

You Try It Work Exercises 9–18 in this textbook or in the MyLab Math Study Plan.

Example 4 Determine the Equation of an Ellipse

Determine the equation of the ellipse with foci located at $(0, 6)$ and $(0, -6)$ that passes through the point $(-5, 6)$.

Solution To find the equation, we can use **Fact 1** and **Fact 2** of ellipses. Watch the **video** to see how we can use these two facts to determine that the equation of this ellipse is $\frac{x^2}{45} + \frac{y^2}{81} = 1.$ ●

You Try It Work Exercises 19 and 20 in this textbook or in the MyLab Math Study Plan.

OBJECTIVE 3 COMPLETING THE SQUARE TO FIND THE EQUATION OF AN ELLIPSE IN STANDARD FORM

If the equation of an ellipse is in the standard form of

$$\frac{(x-h)^2}{a^2}+\frac{(y-k)^2}{b^2}=1 \text{ or } \frac{(x-h)^2}{b^2}+\frac{(y-k)^2}{a^2}=1,$$

it is not too difficult to determine the center and foci and sketch its graph. However, the equation might not be given in standard form. If this is the case, we complete the square on both variables to rewrite the equation in standard form as in Example 5.

Example 5 Rewrite the Equation of an Ellipse in Standard Form by Completing the Square

Find the center and foci and sketch the ellipse $36x^2 + 20y^2 + 144x - 120y - 396 = 0$.

Solution Rearrange the terms, leaving some room to complete the square and move any constants to the right-hand side:

$36x^2 + 144x \quad + 20y^2 - 120y \quad = 396$ Rearrange the terms.

Then factor and complete the square.

$36(x^2 + 4x \quad) + 20(y^2 - 6y \quad) = 396$ Factor out 36 and 20.

$36(x^2 + 4x + 4) + 20(y^2 - 6y + 9) = 396 + 144 + 180$ Complete the square on x and y. Remember to add $36 \cdot 4 = 144$ and $20 \cdot 9 = 180$ to the right side.

$36(x+2)^2 + 20(y-3)^2 = 720$ Factor the left side, and simplify the right side.

$$\frac{36(x+2)^2}{720}+\frac{20(y-3)^2}{720}=\frac{720}{720}$$ Divide both sides by 720.

$$\frac{(x+2)^2}{20}+\frac{(y-3)^2}{36}=1$$ Simplify.

The equation is now in standard form. Watch the **video** to see the solution to this example worked out in detail. Notice that this is the exact same ellipse that we sketched in Example 2. To see how to sketch this ellipse, refer to Example 2 or watch the last part of this **video**.

You Try It **Work Exercises 21–24 in this textbook or in the MyLab Math Study Plan.**

OBJECTIVE 4 SOLVING APPLIED PROBLEMS INVOLVING ELLIPSES

Ellipses have many applications. The planets of our solar system travel around the Sun in an elliptical orbit with the Sun at one focus. Some comets, such as Halley's comet, also travel in elliptical orbits. See Figure 19. (Some comets are only seen in our solar system once because they travel in hyperbolic orbits.) We saw in Section 11.1 that the parabola had a special reflecting property where incoming rays of light or radio waves traveling parallel to the axis of symmetry of a parabolic reflector or receiver will reflect off the parabolic surface and travel directly toward the antenna that is placed at the focus.

Figure 19 Planets travel around the Sun in an elliptical orbit.

Ellipses have a similar reflecting property. When light or sound waves originate from one focus of an ellipse, the waves will reflect off the surface of the ellipse and travel directly toward the other focus. See Figure 20. This reflecting property is used in a medical procedure called *sound wave lithotripsy* in which the patient is placed in an elliptical tank with the kidney stone placed at one focus. An ultrasound wave emitter is positioned at the other focus. The sound waves reflect off the walls of the tank directly to the kidney stone, thus obliterating the stone into fragments that are easily passed naturally through the patient's body. See Example 6.

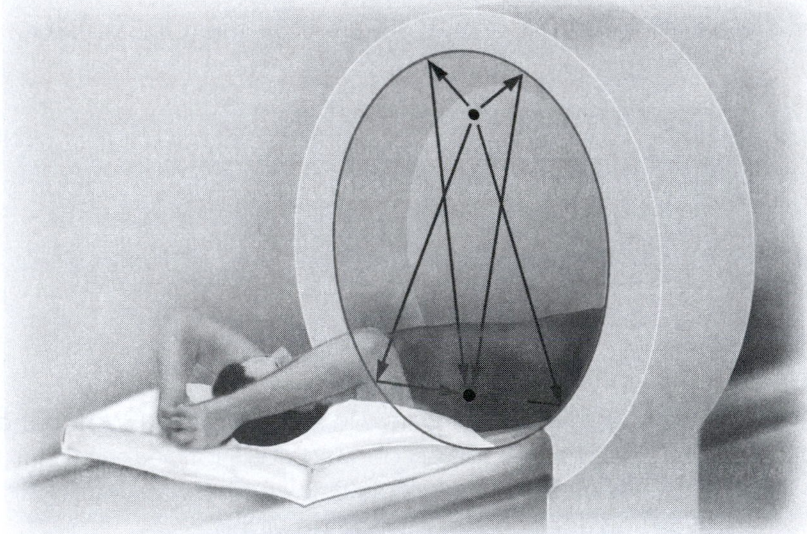

Figure 20 Sound waves or light rays emitted from one focus of an ellipse reflect off the surface directly to the other focus.

Example 6 Position a Patient During Kidney Stone Treatment

A patient is placed in an elliptical tank that is 200 centimeters long and 80 centimeters wide to undergo sound wave lithotripsy treatment for kidney stones. Determine where the sound emitter and the stone should be positioned relative to the center of the ellipse.

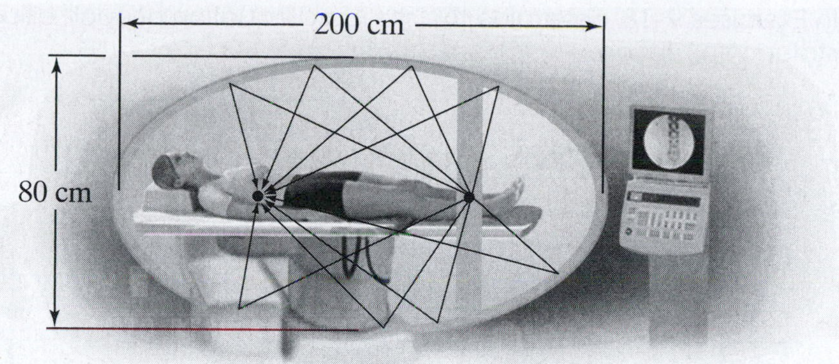

Solution The kidney stone and the sound emitter must be placed at the foci of the ellipse. We know that the major axis is 200 centimeters. Therefore, the vertices must be 100 centimeters from the center. Similarly, the endpoints of the minor axis are located 40 centimeters from the center.

To find c, we use the fact that $c^2 = a^2 - b^2$.

$$c^2 = 100^2 - 40^2$$

$$c^2 = 8{,}400$$

$$c = \sqrt{8{,}400} \approx 91.65$$

The stone and the sound emitter should be positioned approximately 91.65 centimeters from the center of the tank on the major axis.

You Try It Work Exercises 25–31 in this textbook or in the MyLab Math Study Plan.

11.2 Exercises

In Exercises 1–8, determine the center, foci, and vertices of the ellipse and sketch its graph.

SbS 1. $\frac{x^2}{16} + \frac{y^2}{4} = 1$

SbS 2. $\frac{x^2}{12} + \frac{y^2}{25} = 1$

SbS 3. $4x^2 + 9y^2 = 36$

SbS 4. $20x^2 + 5y^2 = 100$

SbS 5. $\frac{(x-2)^2}{36} + \frac{(y-4)^2}{25} = 1$

SbS 6. $\frac{(x-1)^2}{9} + \frac{(y+4)^2}{49} = 1$

SbS 7. $9(x-5)^2 + 25(y+1)^2 = 225$

SbS 8. $(x+7)^2 + 9(y-2)^2 = 9$

In Exercises 9–18, determine the standard equation of each ellipse using the given graph or the stated information.

SbS 9.

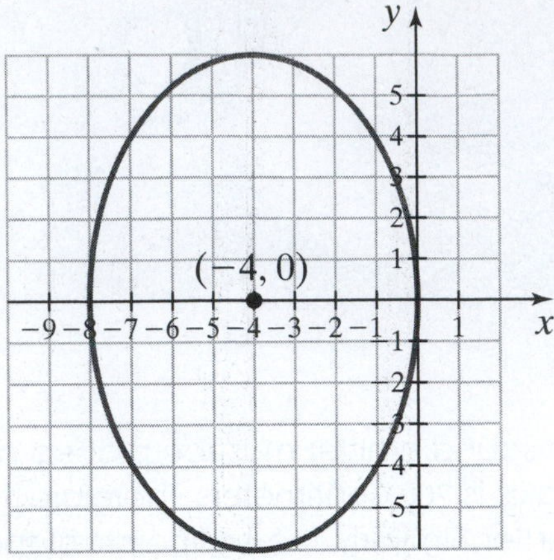

SbS 10.

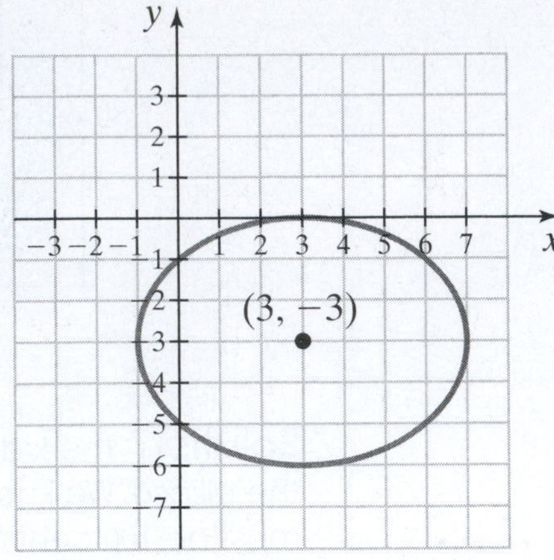

11. Foci at $(-5, 0)$ and $(5, 0)$; length of the major axis is fourteen units.

12. Foci at $(8, -1)$ and $(-2, -1)$; length of the major axis is twelve units.

13. Vertices at $(-6, 10)$ and $(-6, 0)$; length of the minor axis is eight units.

14. Center at $(4, 5)$; vertical minor axis with length sixteen units; $c = 6$.

15. Center at $(-1, 4)$; vertex at $(-1, 8)$; focus at $(-1, 7)$.

16. Vertices at $(2, -7)$ and $(2, 5)$; focus at $(2, 3)$.

17. Center at $(4, 1)$; focus at $(4, 8)$; ellipse passes through the point $(6, 1)$.

18. Center at $(1, 3)$; focus at $(1, 6)$; ellipse passes through the point $(2, 3)$.

In Exercises 19 and 20, determine the standard equation of each ellipse by first using **Fact 1** to find a and then by using **Fact 2** to find b.

19. The foci have coordinates $(-2, 0)$ and $(2, 0)$. The ellipse contains the point $(2, 3)$.

20. The foci have coordinates $(0, -8)$ and $(0, 8)$. The ellipse contains the point $(-2, 6)$.

In Exercises 21–24, complete the square to write each equation in the form

$$\frac{(x-h)^2}{a^2} + \frac{(y-k)^2}{b^2} = 1 \quad \text{or} \quad \frac{(x-h)^2}{b^2} + \frac{(y-k)^2}{a^2} = 1.$$

In Exercises 21–24, determine the center, foci, and vertices of the ellipse and sketch its graph.

SbS 21. $16x^2 + 20y^2 + 64x - 40y - 236 = 0$

SbS 22. $9y^2 + 16x^2 + 224x + 54y + 721 = 0$

SbS 23. $x^2 + 4x + 16y^2 - 32y + 4 = 0$

SbS 24. $50x^2 + y^2 + 20y = 0$

25. An elliptical arch railroad tunnel 16 feet high at the center and 30 feet wide is cut through the side of a mountain. Find the equation of the ellipse if a vertex is represented by the point $(0, 16)$.

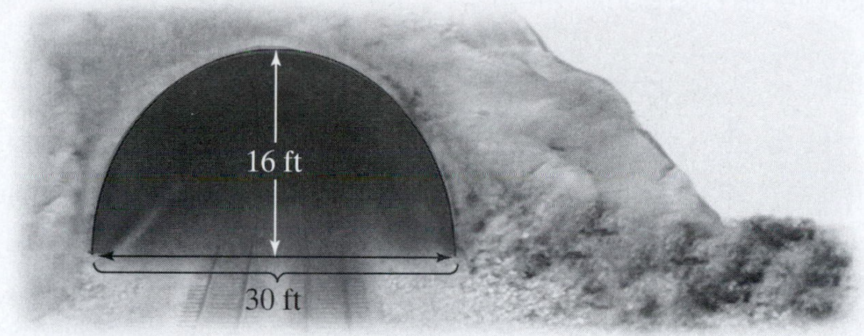

26. A 78-inch by 36-inch door contains a decorative glass elliptical pattern. One vertex of the ellipse is located 9 inches from the top of the door. The other vertex is located 27 inches from the bottom of the door. The endpoints of the minor axis are located 9 inches from each side. Find the equation of the elliptical pattern. (Assume that the center of the elliptical pattern is at the origin.)

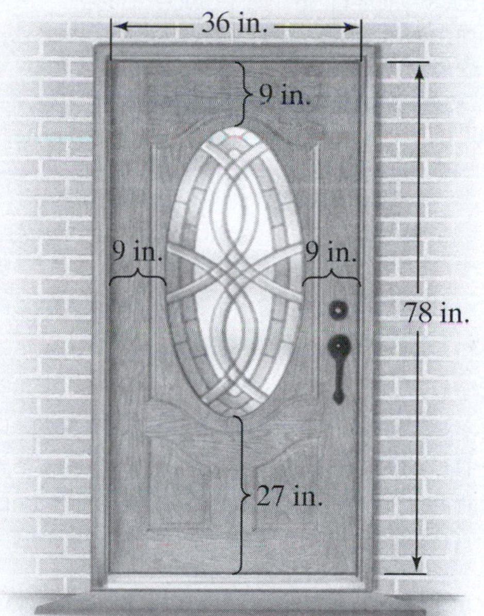

27. A rectangular playing field lies in the interior of an elliptical track that is 50 yards wide and 140 yards long. What is the width of the rectangular playing field if the width is located 10 yards from either vertex?

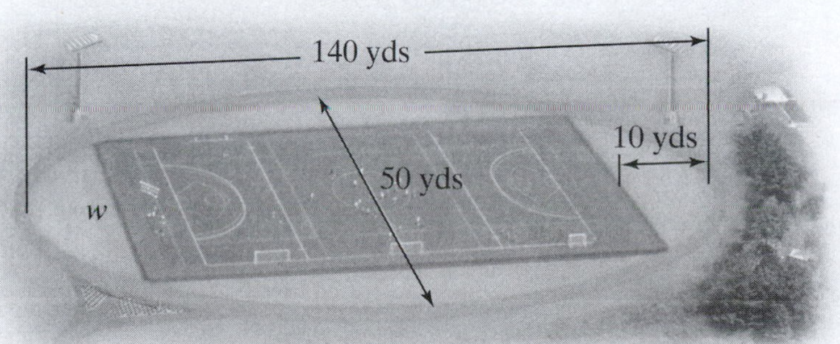

28. A patient is placed in an elliptical tank that is 180 centimeters long and 100 centimeters wide to undergo sound wave lithotripsy treatment for kidney stones. Determine where the sound emitter and the stone should be positioned relative to the center of the tank. (Round to the nearest hundredth of a centimeter.)

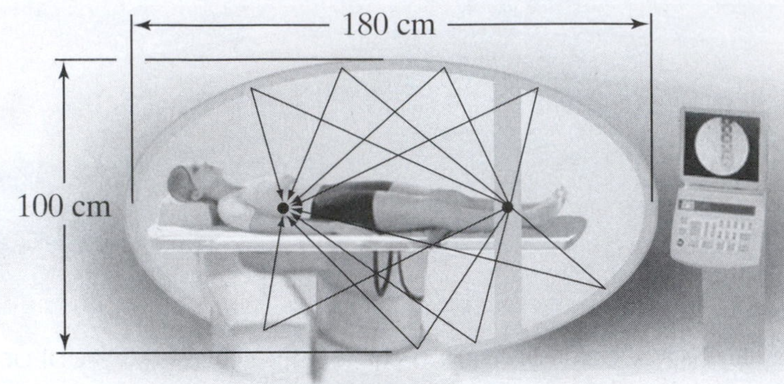

29. A window is constructed with the top half of an ellipse on top of a square. The square portion of the window has a 36-inch base. If the window is 50 inches tall at its highest point, find the height, h, of the window 14 inches from the center of the base. (Round the height to the nearest hundredth of an inch.)

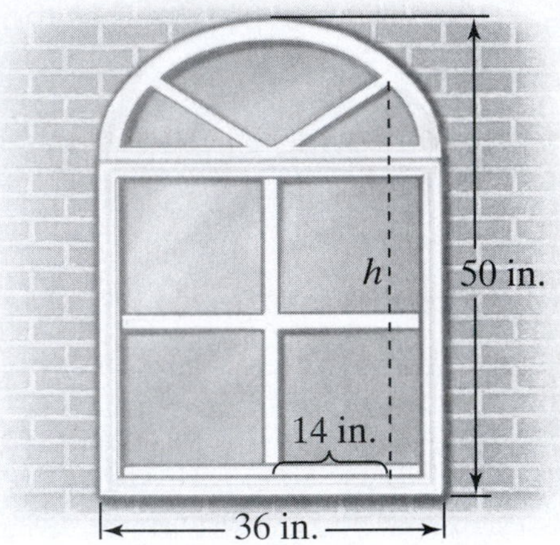

30. A government spy satellite is in an elliptical orbit around the Earth with the center of the Earth at a focus. If the satellite is 150 miles from the surface of the Earth at one vertex of the orbit and 500 miles from the surface of the Earth at the other vertex, find the equation of the elliptical orbit. (Assume that the Earth is a sphere with a diameter of 8,000 miles.)

31. An elliptical arch bridge spans 160 feet. The elliptical arch has a maximum height of 40 feet. What is the height of the arch at a distance of 10 feet from the center?

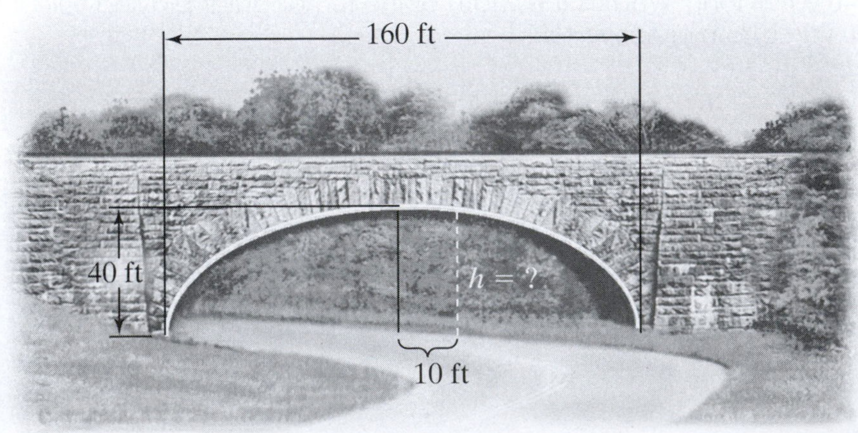

Brief Exercises

In Exercises 32–43, sketch the graph of each ellipse.

32. $\dfrac{x^2}{16} + \dfrac{y^2}{4} = 1$

33. $\dfrac{x^2}{12} + \dfrac{y^2}{25} = 1$

34. $4x^2 + 9y^2 = 36$

35. $20x^2 + 5y^2 = 100$

36. $\dfrac{(x-2)^2}{36} + \dfrac{(y-4)^2}{25} = 1$

37. $\dfrac{(x-1)^2}{9} + \dfrac{(y+4)^2}{49} = 1$

38. $9(x-5)^2 + 25(y+1)^2 = 225$

39. $(x+7)^2 + 9(y-2)^2 = 9$

40. $16x^2 + 20y^2 + 64x - 40y - 236 = 0$

41. $9y^2 + 16x^2 + 224x + 54y + 721 = 0$

42. $x^2 + 4x + 16y^2 - 32y + 4 = 0$

43. $50x^2 + y^2 + 20y = 0$

11.3 The Hyperbola

THINGS TO KNOW

Before working through this section, be sure that you are familiar with the following concepts:

VIDEO ANIMATION INTERACTIVE

You Try It

1. Finding the Distance between Two Points Using the Distance Formula (Section 2.1)

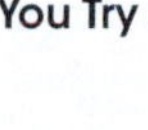
You Try It

2. Finding the Equation of a Line Using the Point–Slope Form (Section 2.3)

You Try It

3. Completing the Square to Find the Equation of an Ellipse in Standard Form (Section 11.2)

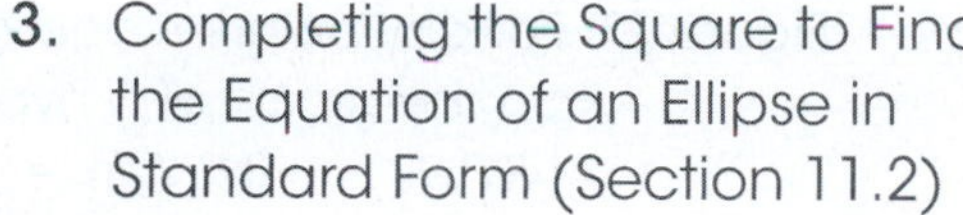

INTRODUCTION

Read this introduction before beginning Objective 1.

OBJECTIVES

1 Sketching the Graph of a Hyperbola

2 Determining the Equation of a Hyperbola in Standard Form

3 Completing the Square to Find the Equation of a Hyperbola in Standard Form

4 Solving Applied Problems Involving Hyperbolas

SECTION 11.3 EXERCISES

Introduction to Section 11.3

When a plane intersects two **right circular cones** at the same time, the conic section formed is a hyperbola. (Click on the animation.)

The set of points in the plane that defines a hyperbola formed by the intersection of a plane and the cones described previously is stated in the following geometric definition.

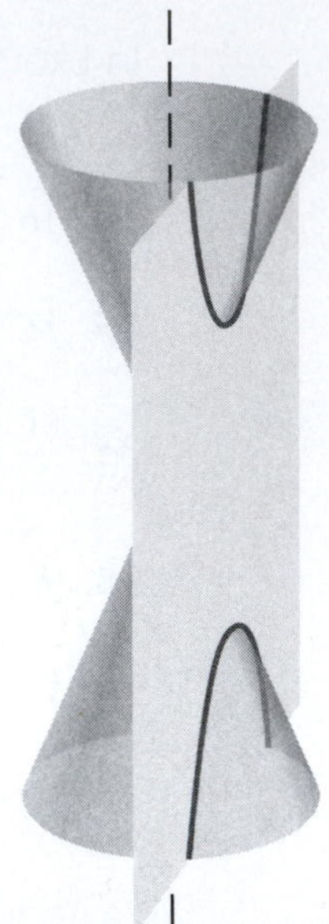

The Hyperbola

> **Definition** Geometric Definition of the Hyperbola
>
> A **hyperbola** is the set of all points in a plane, the difference of whose distances from two fixed points is a positive constant. The two fixed points, F_1 and F_2, are called the foci.

Notice that the previous geometric definition is very similar to the **geometric definition of the ellipse**. Recall that for a point to lie on the graph of an ellipse, the **sum** of the distances from the point to the two foci is constant. For a point to lie on the graph of a hyperbola, the **difference** between the distances from the point to the two foci is constant. Because subtraction is not **commutative**, we consider the absolute value of the difference in the distances between a point on the hyperbola and the foci to ensure that the constant is positive. Thus, for any two points P and Q that lie on the graph of a hyperbola, $|d(P, F_1) - d(P, F_2)| = |d(Q, F_1) - d(Q, F_2)|$. See Figure 21.

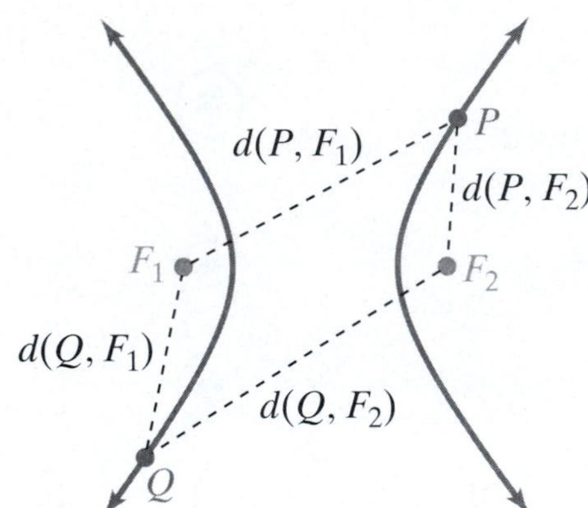

Figure 21 For any points P and Q that lie on the graph of a hyperbola, $|d(P, F_1) - d(P, F_2)| = |d(Q, F_1) - d(Q, F_2)|$.

OBJECTIVE 1 SKETCHING THE GRAPH OF A HYPERBOLA

The graph of a hyperbola has two branches. These branches look somewhat like parabolas, but they are certainly not because the branches do not satisfy the **geometric definition of the parabola**. Every hyperbola has a center, two vertices, and two foci. The vertices are located at the endpoints of an invisible line segment called the **transverse axis.** The transverse axis is either parallel to the x-axis (horizontal transverse axis) or parallel to the y-axis (vertical transverse axis). The center of a hyperbola is located midway between the two vertices (or two foci). The hyperbola has another invisible line segment called the **conjugate axis** that passes through the center and lies perpendicular to the transverse axis. Each branch of the hyperbola approaches (but never intersects) a pair of lines called *asymptotes*. A **reference rectangle** is typically used as a guide to help sketch the asymptotes. The reference rectangle is a rectangle whose midpoints of each side are the vertices of the hyperbola or the endpoints of the conjugate axis.

The asymptotes pass diagonally through opposite corners of the reference rectangle. Watch the **video** to see how to sketch the graphs of the two hyperbolas shown in Figure 22.

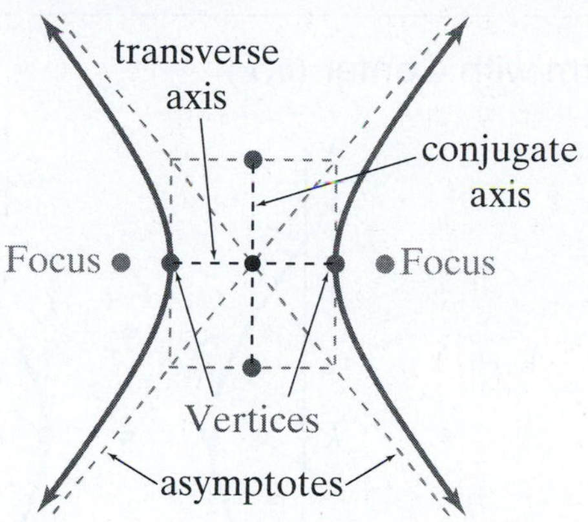

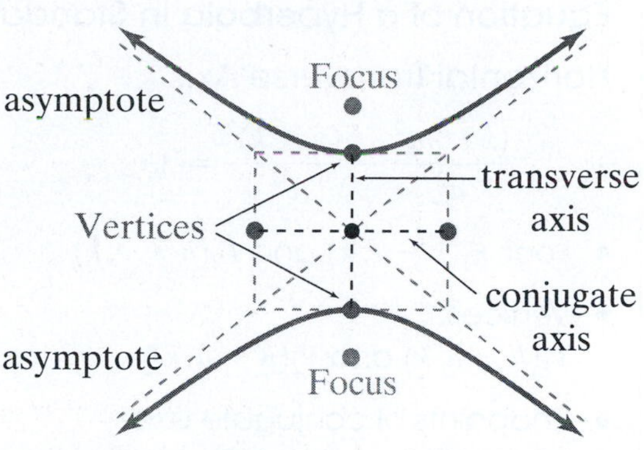

Figure 22 (a) Hyperbola with a Horizontal Transverse Axis (b) Hyperbola with a Vertical Transverse Axis

To derive the equation of a hyperbola, consider a hyperbola with a horizontal transverse axis centered at (h, k). If the distance between the center and either vertex is $a > 0$, then the coordinates of the vertices are $V_1(h - a, k)$ and $V_2(h + a, k)$, and the length of the transverse axis is equal to $2a$. If $c > 0$ is the distance between the center and either foci, then the foci have coordinates $F_1(h - c, k)$ and $F_2(h + c, k)$. See Figure 23.

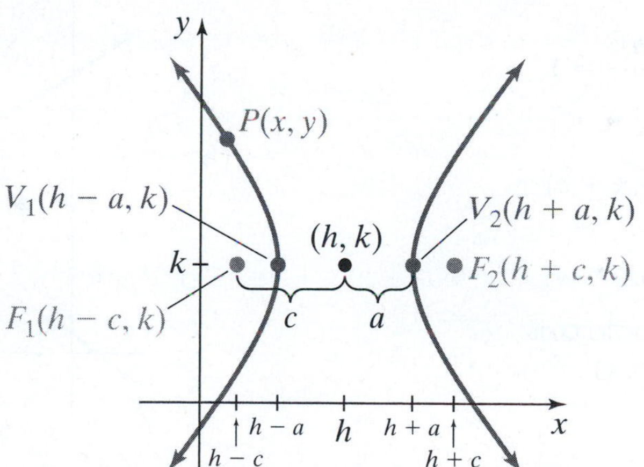

Figure 23

By the geometric definition of a hyperbola, we know that for any point $P(x, y)$ that lies on the hyperbola, the difference of the distances from P to the two foci is a constant. It can be shown that this constant is equal to $2a$, the length of the transverse axis. We now state this fact and denote it as Fact 1 for Hyperbolas.

FACT 1 FOR HYPERBOLAS

Given the foci of a hyperbola F_1 and F_2 and any point P that lies on the graph of the hyperbola, the difference of the distances between P and the foci is equal to $2a$. In other words, $|d(P, F_1) - d(P, F_2)| = 2a$. The constant $2a$ represents the length of the transverse axis.

We **prove Fact 1** for Hyperbolas in Appendix B.

Once we know that the constant stated in the geometric definition is equal to $2a$, we can derive the equation of a hyperbola. The equation of a hyperbola is **derived in Appendix B**. We now state the two standard equations of hyperbolas.

Equation of a Hyperbola in Standard Form with Center (h, k)

Horizontal Transverse Axis

$$\frac{(x-h)^2}{a^2} - \frac{(y-k)^2}{b^2} = 1$$

- Foci: $F_1(h-c, k)$ and $F_2(h+c, k)$
- Vertices: $V_1(h-a, k)$ and $V_2(h+a, k)$
- Endpoints of conjugate axis: $(h, k-b)$ and $(h, k+b)$
- $b^2 = c^2 - a^2$
- Asymptotes: $y - k = \pm\frac{b}{a}(x-h)$

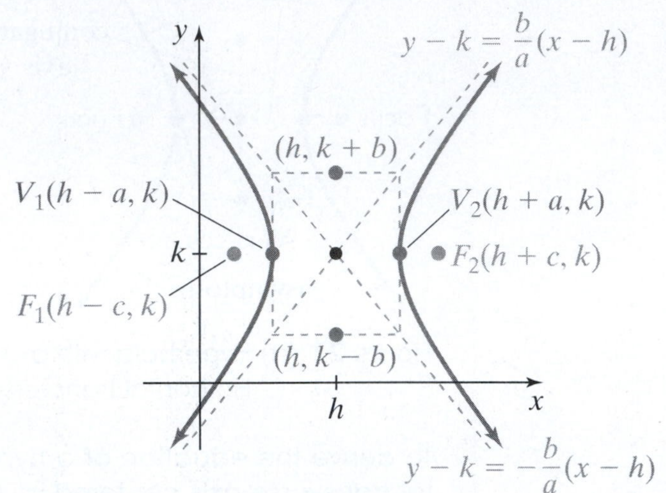

Equation of a Hyperbola in Standard Form with Center (h, k)

Vertical Transverse Axis

$$\frac{(y-k)^2}{a^2} - \frac{(x-h)^2}{b^2} = 1$$

- Foci: $F_1(h, k-c)$ and $F_2(h, k+c)$
- Vertices: $V_1(h, k-a)$ and $V_2(h, k+a)$
- Endpoints of conjugate axis: $(h-b, k)$ and $(h+b, k)$
- $b^2 = c^2 - a^2$
- Asymptotes: $y - k = \pm\frac{a}{b}(x-h)$

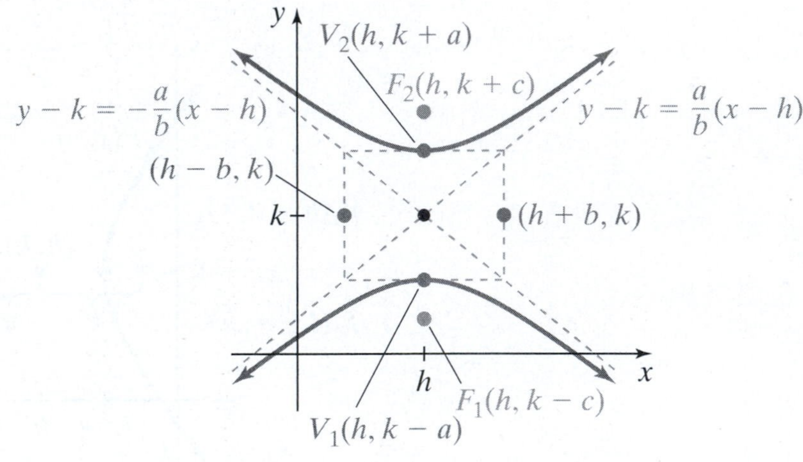

The two hyperbola equations are nearly identical except for two differences. The term involving $(x-h)^2$ is positive and the term involving $(y-k)^2$ is negative in the equation of a hyperbola with the horizontal transverse axis. The signs are reversed in the equation with a vertical transverse axis. Also note that a^2 appears in the denominator of the positive squared term in each equation. We can derive the equations of the asymptotes by using the **point-slope form of the equation of a line**.

Note If $h = 0$ and $k = 0$, then the hyperbola is centered at the **origin**. Hyperbolas centered at the origin have the following equations.

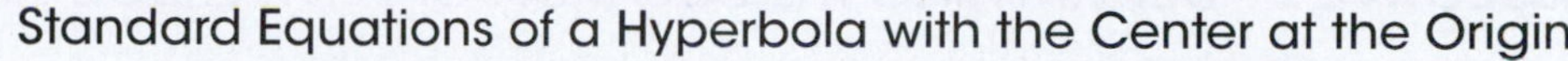

Standard Equations of a Hyperbola with the Center at the Origin

$$\frac{x^2}{a^2} - \frac{y^2}{b^2} = 1 \qquad \frac{y^2}{a^2} - \frac{x^2}{b^2} = 1$$

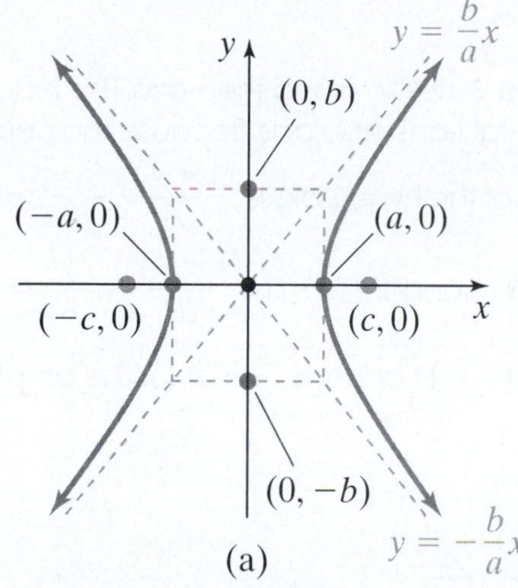

(a)

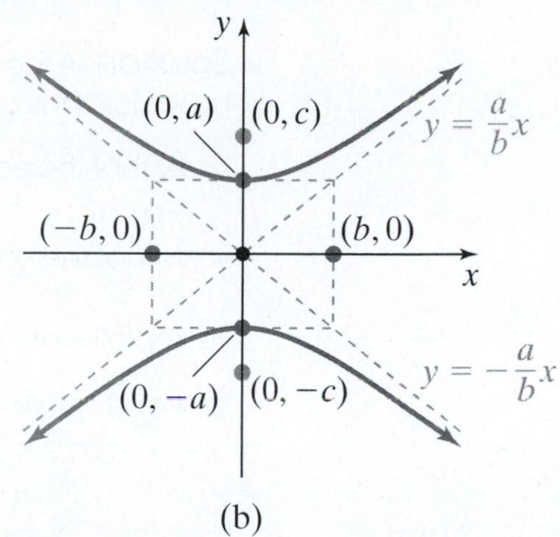

(b)

Example 1 Sketch the Graph of a Hyperbola in Standard Form

Sketch the following hyperbolas. Determine the center, transverse axis, vertices, and foci and find the equations of the asymptotes.

a. $\frac{(y-4)^2}{36} - \frac{(x+5)^2}{9} = 1$

b. $25x^2 - 16y^2 = 400$

Solution Watch the interactive video to see how to sketch each hyperbola shown in Figure 24(a) and Figure 24(b).

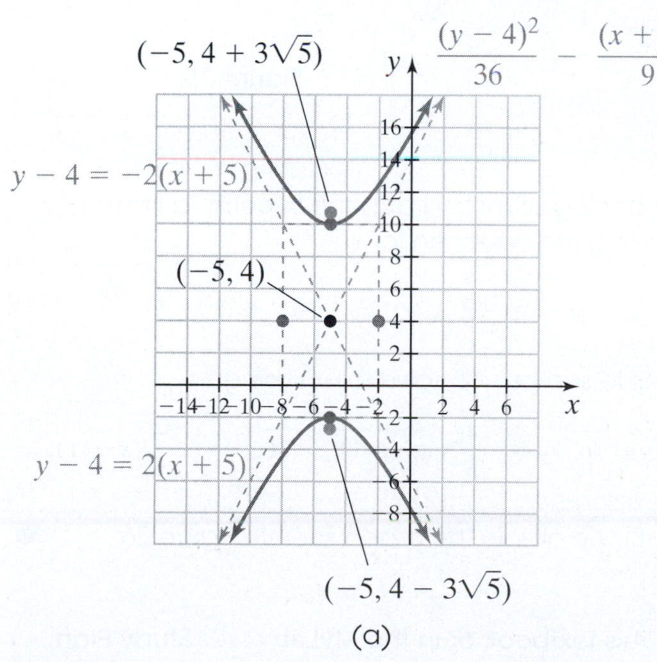

(a)

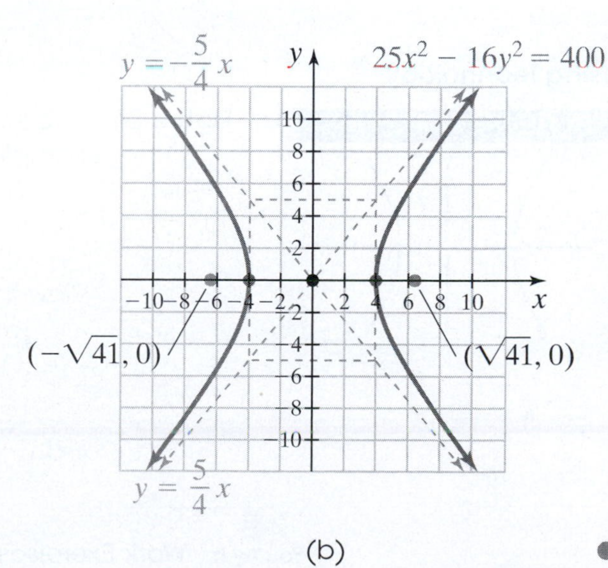

(b)

Figure 24

You Try It Work Exercises 1–6 in this textbook or in the MyLab Math Study Plan.

OBJECTIVE 2 DETERMINING THE EQUATION OF A HYPERBOLA IN STANDARD FORM

Example 2 Find the Equation of a Hyperbola

Find the equation of the hyperbola with the center at $(-1, 0)$, a focus at $(-11, 0)$, and a vertex at $(5, 0)$.

Solution A focus and a vertex lie along the x-axis. This indicates that the hyperbola has a horizontal transverse axis. Because the center is at $(-1, 0)$, we know that the equation of the hyperbola is $\frac{(x+1)^2}{a^2} - \frac{y^2}{b^2} = 1$. Watch the **video** to verify that the equation in standard form is $\frac{(x+1)^2}{36} - \frac{y^2}{64} = 1$. The equations of the asymptotes are $y = -\frac{4}{3}(x+1)$ and $y = \frac{4}{3}(x+1)$. The graph of this hyperbola is shown in Figure 25.

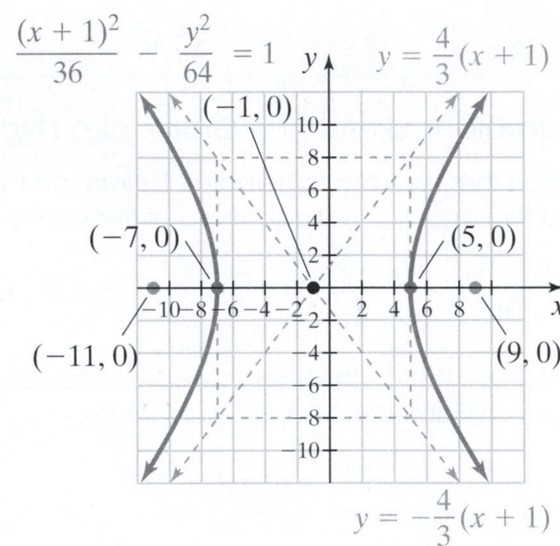

Figure 25

Using Technology

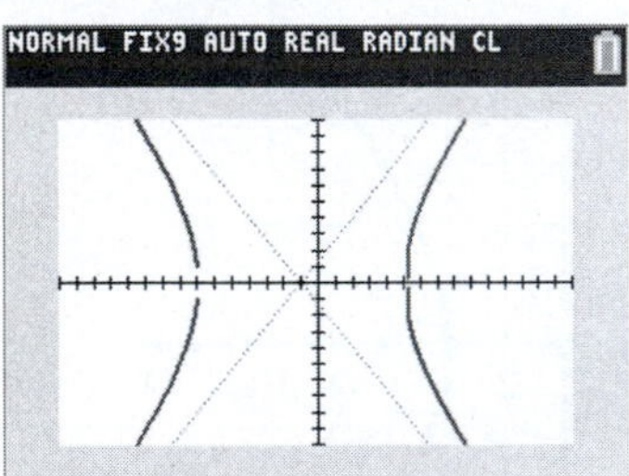

Figure 26 We can use a graphing utility to graph the hyperbola from Example 2 by solving the equation for y:

$$y = \pm\frac{4}{3}\sqrt{(x+1)^2 - 36}$$

(View these **steps** to see how to solve for y.) Using

$$y_1 = \frac{4}{3}\sqrt{(x+1)^2 - 36},\ y_2 = -\frac{4}{3}\sqrt{(x+1)^2 - 36},\ y_3 = -\frac{4}{3}(x+1),$$

and $y_4 = \frac{4}{3}(x+1)$, we obtain the graph seen in Figure 26. ●

You Try It Work Exercises 7–14 in this textbook or in the MyLab Math Study Plan.

OBJECTIVE 3 COMPLETING THE SQUARE TO FIND THE EQUATION OF A HYPERBOLA IN STANDARD FORM

If the equation of a hyperbola is in the standard form of

$$\frac{(x-h)^2}{a^2} - \frac{(y-k)^2}{b^2} = 1 \quad \text{or} \quad \frac{(y-k)^2}{a^2} - \frac{(x-h)^2}{b^2} = 1,$$

then it is not too difficult to determine the center, vertices, foci, and asymptotes and sketch its graph. However, the equation might not be given in standard form. If this is the case, we may need to complete the square on both variables as in Example 3.

Example 3 Rewrite the Equation of a Hyperbola in Standard Form by Completing the Square

Find the center, vertices, foci, and equations of asymptotes and sketch the hyperbola $12x^2 - 4y^2 - 72x - 16y + 140 = 0$.

Solution Rearrange the terms leaving some room to complete the square, and move any constants to the right-hand side:

$12x^2 - 72x \qquad - 4y^2 - 16y \qquad = -140$ Rearrange the terms.

Then factor and complete the square.

$12(x^2 - 6x \qquad) - 4(y^2 + 4y \qquad) = -140$ Factor out 12 and -4.

$12(x^2 - 6x + 9) - 4(y^2 + 4y + 4) = -140 + 108 - 16$ Complete the square on x and y. Remember to add $12 \cdot 9 = 108$ and $-4 \cdot 4 = -16$ to the right side.

$12(x-3)^2 - 4(y+2)^2 = -48$ Factor the left side, and simplify the right side.

$\frac{12(x-3)^2}{-48} - \frac{4(y+2)^2}{-48} = \frac{-48}{-48}$ Divide both sides by -48.

$-\frac{(x-3)^2}{4} + \frac{(y+2)^2}{12} = 1$ Simplify.

$\frac{(y+2)^2}{12} - \frac{(x-3)^2}{4} = 1$ Rewrite the equation.

The equation is now in standard form. Watch the video to see this example worked out in detail. You should verify that this is the equation of a hyperbola with a vertical transverse axis with center $(3, -2)$. The vertices have coordinates $(3, -2 - 2\sqrt{3})$ and $(3, -2 + 2\sqrt{3})$. The foci have coordinates $(3, -6)$ and $(3, 2)$. The equations of the asymptotes are $y + 2 = -\sqrt{3}(x - 3)$ and $y + 2 = \sqrt{3}(x - 3)$. ●

You Try It **Work Exercises 15–20 in this textbook or in the MyLab Math Study Plan.**

OBJECTIVE 4 SOLVING APPLIED PROBLEMS INVOLVING HYPERBOLAS

Hyperbolas have many applications. We saw in **Section 11.2** that the planets in our solar system and some comets, such as Halley's comet, travel through our solar system in elliptical orbits. However, some comets are only seen once in our solar system because they travel through the solar system on the **path of a hyperbola** with the Sun at a focus. On October 14, 1947, Chuck Yeager became the first person to break the sound barrier. As an airplane moves faster than the speed of sound, a **cone-shaped shock wave** is produced. The cone intersects the ground in the shape of a hyperbola. When two rocks are simultaneously tossed into a calm pool of water, ripples move outward in the form of **concentric circles**. These circles intersect in points that form a hyperbola. Hyperbolas can be used to locate ships by sending radio signals simultaneously from radio transmitters placed at some fixed distance apart. A device measures the difference in the time it takes the radio signals to reach the ship. The equation of a hyperbola can then be determined to describe the current path of the ship. If three transmitters are used, two hyperbolic equations can be determined. The precise location of the ship can be determined by finding the intersection of the two hyperbolas. This system of locating ships is known as long-range navigation or LORAN. See Example 4.

Example 4 Use a Hyperbola to Locate a Ship

One transmitting station is located 100 miles due east from another transmitting station. Each station simultaneously sends out a radio signal. The signal from the west tower is received by a ship $\frac{1{,}600}{3}$ microseconds after the signal from the east tower. If the radio signal travels at 0.18 miles per microsecond, find the equation of the hyperbola on which the ship is presently located.

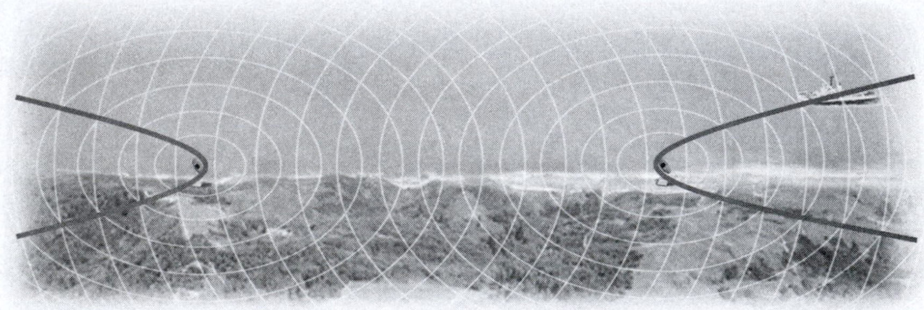

Solution Start by plotting the two foci of the hyperbola at points $F_1(-50, 0)$ and $F_2(50, 0)$. These two points represent the position of the two transmitting towers. Note that $c = 50$. Because the hyperbola is centered at the origin with a horizontal transverse axis, the equation must be of the form $\frac{x^2}{a^2} - \frac{y^2}{b^2} = 1$. The difference in the distances from the two transmitters to the ship is $\left(\frac{1{,}600}{3} \text{ microseconds}\right) \cdot \left(0.18 \frac{\text{miles}}{\text{microsecond}}\right) = 96$ miles. This distance represents the constant stated in the **Fact 1 for Hyperbolas**. Therefore, $2a = 96$, so $a = 48$ or $a^2 = 2{,}304$. To find b^2, we use the fact that $b^2 = c^2 - a^2$.

$$b^2 = c^2 - a^2$$
$$b^2 = 50^2 - 48^2$$
$$b^2 = 196$$

We now substitute the values of $a^2 = 2{,}304$ and $b^2 = 196$ into the previous equation to obtain the equation $\frac{x^2}{2{,}304} - \frac{y^2}{196} = 1.$

You Try It **Work Exercises 21–25 in this textbook or in the MyLab Math Study Plan.**

11.3 Exercises

In Exercises 1–6, determine the center, transverse axis, vertices, foci, and the equations of the asymptotes and sketch the hyperbola.

SbS 1. $\frac{x^2}{16} - \frac{y^2}{9} = 1$

SbS 2. $\frac{y^2}{9} - \frac{x^2}{16} = 1$

SbS 3. $\frac{(y-4)^2}{25} - \frac{(x-2)^2}{36} = 1$

SbS 4. $\frac{(x+1)^2}{9} - \frac{(y+3)^2}{49} = 1$

SbS 5. $20x^2 - 5y^2 = 100$

SbS 6. $20(x-1)^2 - 16(y-3)^2 = -320$

In Exercises 7–12, determine the standard equation of the hyperbola with the given characteristics and sketch the graph.

7. The center is at $(0, 0)$, a focus is at $(5, 0)$, and a vertex is at $(3, 0)$.

8. The center is at $(0, 0)$, a focus is at $(0, 10)$, and a vertex is at $(0, -6)$.

9. The center is at $(4, -4)$, a focus is at $(6, -4)$, and a vertex is at $(5, -4)$.

10. The center is at $(-6, -1)$, a focus is at $(-6, -9)$, and a vertex is at $(-6, -5)$.

11. The foci are at $(9, 3)$ and $(9, 9)$; the vertex is at $(9, 8)$.

12. The vertices are at $(-2, -3)$ and $(10, -3)$; an asymptote has equation $y + 3 = \frac{7}{6}(x - 4)$.

In Exercises 13 and 14, determine the equation of the hyperbola with the given characteristics and sketch the graph. *Hint:* $|d(P, F_1) - d(P, F_2)| = 2a$.

13. The hyperbola has foci with coordinates $F_1(0, -6)$ and $F_2(0, 6)$ and passes through the point $P(8, 10)$.

14. The hyperbola has foci with coordinates $F_1(-6, 1)$ and $F_2(10, 1)$ and passes through the point $P(10, 13)$.

In Exercises 15–20, complete the square to write each equation in the form

$$\frac{(x-h)^2}{a^2} - \frac{(y-k)^2}{b^2} = 1 \text{ or } \frac{(y-k)^2}{a^2} - \frac{(x-h)^2}{b^2} = 1.$$

In Exercises 15–20, determine the center, vertices, foci, endpoints of the conjugate axis, and the equations of the asymptotes of the hyperbola, and sketch its graph.

SbS 15. $x^2 - y^2 - 4x + 2y - 1 = 0$

SbS 16. $x^2 - y^2 + 8x - 6y + 9 = 0$

SbS 17. $y^2 - 9x^2 - 12y - 36x - 9 = 0$

SbS 18. $x^2 - 16y^2 + 10x + 64y - 55 = 0$

SbS 19. $25y^2 - 144x^2 + 1{,}728x + 400y + 16 = 0$

SbS 20. $49y^2 - 576x^2 - 98y - 1{,}152x - 28{,}751 = 0$

21. A light on a wall produces a shadow in the shape of a hyperbola. If the distance between the two vertices of the hyperbola is 14 inches and the distance between the two foci is 16 inches, find the equation of the hyperbola.

22. This figure shows the hyperbolic orbit of a comet with the center of the Sun positioned at a focus, 100 million miles from the origin. (The units are in millions of miles.) The comet will be 50 million miles from the center of the Sun at its nearest point during the orbit. Find the equation of the hyperbola describing the comet's orbit.

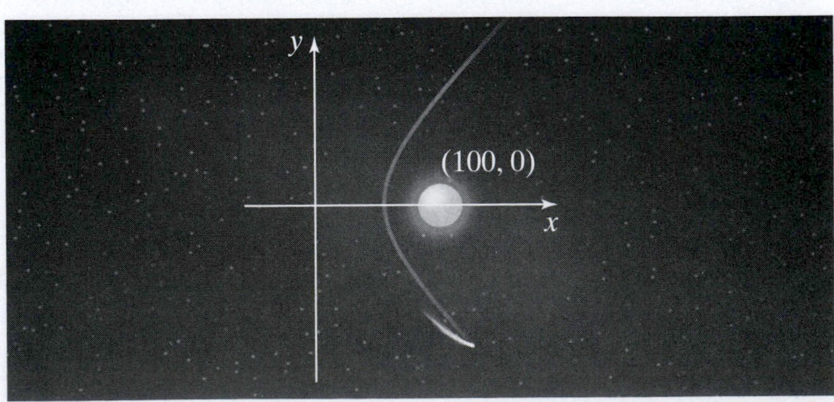

23. A nuclear power plant has a large cooling tower with sides curved in the shape of a hyperbola. The radius of the base of the tower is 60 meters. The radius of the top of the tower is 50 meters. The sides of the tower are 60 meters apart at the closest point located 90 meters above the ground.

 a. Find the equation of the hyperbola that describes the sides of the cooling tower. (Assume that the center is at the origin.)

 b. Determine the height of the tower. (Round your answer to the nearest meter.)

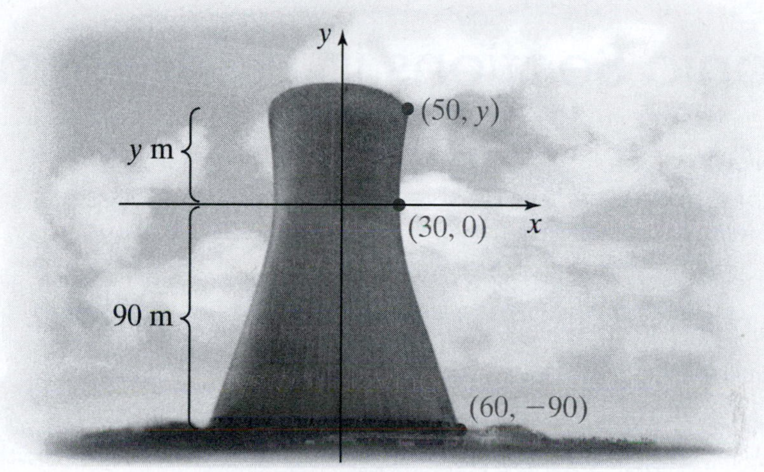

24. One transmitting station is located 80 miles north of another transmitting station. Each station simultaneously sends out a radio signal. The signal from the north station is received 200 microseconds after the signal from the south station. If the radio signal travels at 0.18 miles per microsecond, find the equation of the hyperbola on which the ship is presently located.

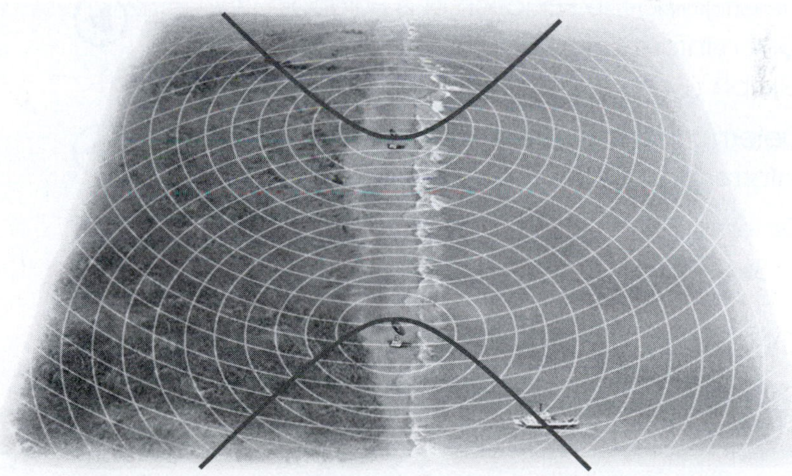

Brief Exercises

In Exercises 25–36, sketch the graph of each hyperbola.

25. $\frac{x^2}{16} - \frac{y^2}{9} = 1$

26. $\frac{y^2}{9} - \frac{x^2}{16} = 1$

27. $\frac{(y-4)^2}{25} - \frac{(x-2)^2}{36} = 1$

28. $\frac{(x+1)^2}{9} - \frac{(y+3)^2}{49} = 1$

29. $20x^2 - 5y^2 = 100$

30. $20(x-1)^2 - 16(y-3)^2 = -320$

31. $x^2 - y^2 - 4x + 2y - 1 = 0$

32. $x^2 - y^2 + 8x - 6y + 9 = 0$

33. $y^2 - 9x^2 - 12y - 36x - 9 = 0$

34. $x^2 - 16y^2 + 10x + 64y - 55 = 0$

35. $25y^2 - 144x^2 + 1{,}728x + 400y + 16 = 0$

36. $49y^2 - 576x^2 - 98y - 1{,}152x - 28{,}751 = 0$

11.4 Conic Sections in Polar Form

THINGS TO KNOW

Before working through this section, be sure that you are familiar with the following concepts:

VIDEO ANIMATION INTERACTIVE

You Try It

1. Finding the Values of the Trigonometric Functions of Quadrantal Angles (Section 6.5)

You Try It

2. Evaluating Trigonometric Functions of Angles Belonging to the $\frac{\pi}{3}, \frac{\pi}{6}$, or $\frac{\pi}{4}$ Families (Section 6.5)

You Try It

3. Plotting Points Using Polar Coordinates (Section 10.1)

You Try It

4. Converting an Equation from Polar Form to Rectangular Form (Section 10.1)

You Try It

5. Determining the Equation of a Parabola Given Information about the Graph (Section 11.1)

You Try It

6. Determining the Equation of an Ellipse Given Information about the Graph (Section 11.2)

You Try It

7. Determining the Equation of a Hyperbola in Standard Form (Section 11.3)

INTRODUCTION

Read this introduction before beginning Objective 1.

OBJECTIVES

1 Identifying Conic Sections Given in Polar Form

2 Sketching the Graphs of Conic Sections Given in Polar Form

3 Converting an Equation of a Conic Section from Polar Form to Rectangular Form

4 Determining the Equation of a Conic Section in Polar Form Given the Eccentricity and the Equation of the Directrix

SECTION 11.4 EXERCISES

Introduction to Section 11.4

Earlier in this chapter we learned how to sketch the graphs of parabolas, ellipses, and hyperbolas, which are called **conic sections**. A conic section is formed when a plane intersects a pair of right circular cones. You may want to watch the animations below to review how these conic sections are formed.

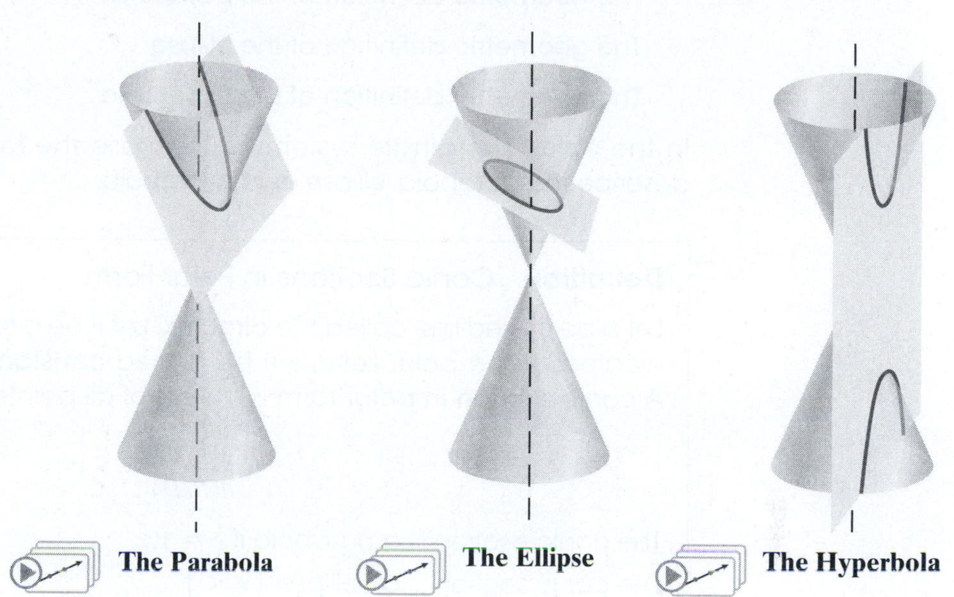

In this section, we will learn how to identify and sketch these conic sections using polar coordinates. Before we learn how to sketch conics that are in polar form, it is worthwhile to remember how to work with conic sections that are in rectangular form. Click on the links below to review how to find the equations of conic sections in rectangular form when certain information is given.

You Try It — Find the equation of a parabola such that the focus is at the origin and the equation of the directrix is $y = -2$.

You Try It — Find the equation of a parabola such that the focus is at the origin and the equation of the directrix is $x = 1$.

You Try It — Find the equation of an ellipse with foci at $(0, 0)$ and $(10, 0)$ such that the length of the major axis is fourteen units.

You Try It — Find the equation of an ellipse with foci at $(0, 0)$ and $(0, -4)$ such that the length of the major axis is eight units.

You Try It — Find the equation of a hyperbola such that a focus is at the origin, the center is at $(-3, 0)$, and a vertex is at $(-5, 0)$.

You Try It — Find the equation of a hyperbola such that a focus is at the origin, the center is at $(0, -2)$, and a vertex is at $(0, 1)$.

(eText Screens 11.4-1–11.4-35)

OBJECTIVE 1 IDENTIFYING CONIC SECTIONS GIVEN IN POLAR FORM

In Sections 11.1–11.3, we saw that the parabola, ellipse, and hyperbola each had a geometric definition that was used to derive the corresponding rectangular equation. To review these geometric definitions, click on the definition links below.

The geometric definition of the parabola

The geometric definition of the ellipse

The geometric definition of the hyperbola

In the polar coordinate system, we can use the following single definition to describe the parabola, ellipse, and hyperbola.

Definition Conic Sections in Polar Form

Let D be a fixed line called the **directrix**. Let F be a fixed point called the **focus** located at the pole. Let $e > 0$ be a fixed constant called the **eccentricity**. A **conic section in polar form** is the set of all points P in the plane such that

$$\frac{d(P, F)}{d(P, D)} = e.$$

The conic section is a parabola if $e = 1$.

The conic section is an ellipse if $e < 1$.

The conic section is a hyperbola if $e > 1$.

Note in the definition above that if $e = 1$, then this definition is consistent with the **geometric definition of the parabola** that was discussed in Section 11.1.

CAUTION Do not confuse the variable e that is used to describe eccentricity with the irrational number $e \approx 2.71828$.

For each of the three conic sections (parabola, ellipse, hyperbola) in polar form, the directrix lies in one of four positions:

Case 1. **Parallel** to the polar axis, located p units **below** the polar axis

Case 2. **Parallel** to the polar axis, located p units **above** the polar axis

Case 3. **Perpendicular** to the polar axis, located p units to the **left** of the polar axis

Case 4. **Perpendicular** to the polar axis, located p units to the **right** of the polar axis

Figure 27 illustrates an example of Case 1. Each of the three conic sections shown in Figure 27 have a directrix **parallel** to the polar axis located p units **below** the polar axis. Each conic section in Figure 27 passes through an arbitrary point $P(r, \theta)$.

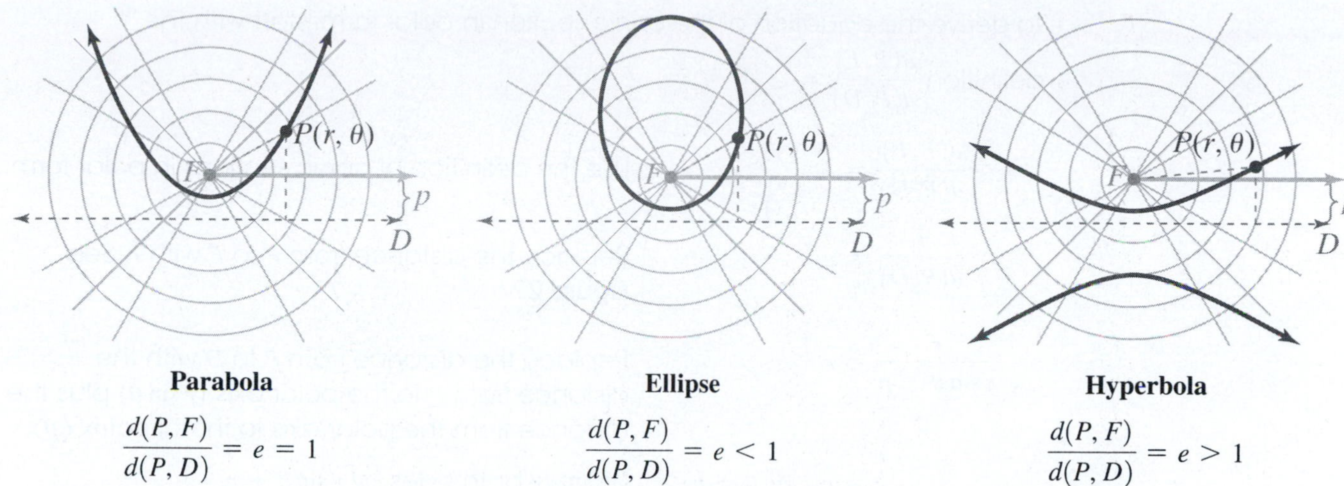

Figure 27 Conic sections in the polar coordinate system having a focus at the pole, having a directrix parallel to the polar axis located p units below the polar axis, and containing the point $P(r, \theta)$.

We will now derive the equation of a conic section in polar form for this first case where the directrix is parallel to the polar axis located p units below the polar axis. Note that $p > 0$.

We start by positioning the focus at the pole. Suppose that the conic section passes through the point $P(r, \theta)$. See Figure 28.

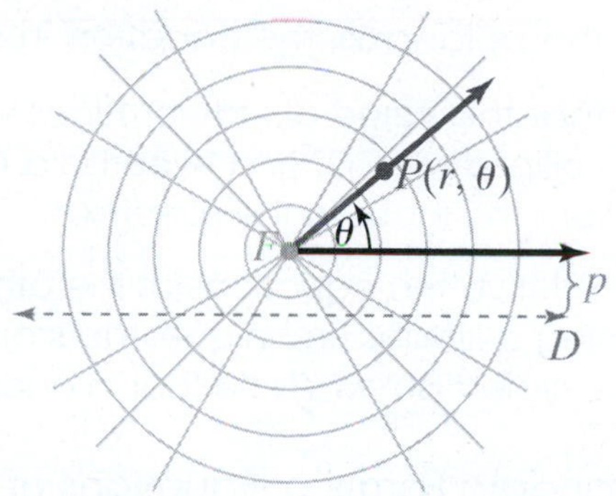

Figure 28

We can construct the right triangle with the hypotenuse extending from the focus to point P having the acute angle θ. The length of the hypotenuse is r units. The length of the side adjacent to θ is $r\cos\theta$ and length opposite of θ is $r\sin\theta$. See Figure 29.

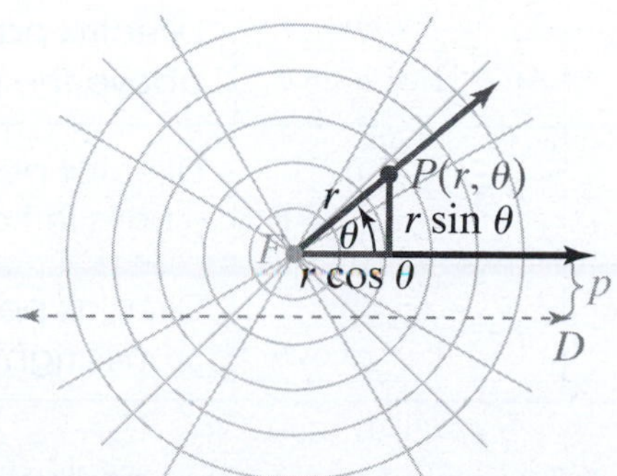

Figure 29

To derive the equation of the conic section in polar form, start with the definition $\dfrac{d(P, F)}{d(P, D)} = e.$

$\dfrac{d(P, F)}{d(P, D)} = e$	Use the definition of conic sections in polar form.
$\dfrac{r}{d(P, D)} = e$	Replace the distance from P to F with r. See Figure 29.
$\dfrac{r}{r \sin \theta + p} = e$	Replace the distance from P to D with the distance from P to the polar axis $(r \sin \theta)$ plus the distance from the polar axis to the directrix (p).
$r = e(r \sin \theta + p)$	Multiply both sides by $r \sin \theta + p$.
$r = er \sin \theta + ep$	Use the distributive property.
$r - er \sin \theta = ep$	Subtract $er \sin \theta$ from both sides.
$r(1 - e \sin \theta) = ep$	Factor out r from the two terms on the left-hand side.
$r = \dfrac{ep}{1 - e \sin \theta}$	Divide both sides by $1 - e \sin \theta$.

Therefore, for Case 1 in which the directrix is **parallel** to the polar axis located p units **below** the polar axis, the equation in standard form is given by $r = \dfrac{ep}{1 - e \sin \theta}$. Recall that the value of e determines whether the conic section is a parabola $(e = 1)$, ellipse $(e < 1)$, or a hyperbola $(e > 1)$. The value of p determines the distance from the focus to the directrix.

We can derive the equations of the other three cases of conic sections in polar form using a similar technique. The standard forms of all conic sections in polar form are stated below. (Note that in each equation $e > 0$ and $p > 0$.)

The Standard Forms of Equations of Conic Sections in Polar Form

Case	Equation	Description
1	$r = \dfrac{ep}{1 - e \sin \theta}$	Directrix **parallel** to the polar axis located p units **below** the polar axis.
2	$r = \dfrac{ep}{1 + e \sin \theta}$	Directrix **parallel** to the polar axis located p units **above** the polar axis.
3	$r = \dfrac{ep}{1 - e \cos \theta}$	Directrix **perpendicular** to the polar axis located p units **left** of the pole
4	$r = \dfrac{ep}{1 + e \cos \theta}$	Directrix **perpendicular** to the polar axis located p units **right** the pole.

On the following pages we give an illustration of each of the four cases.

Case 1

Equation of Conic Section in Standard Form

$$r = \frac{ep}{1 - e\sin\theta}$$

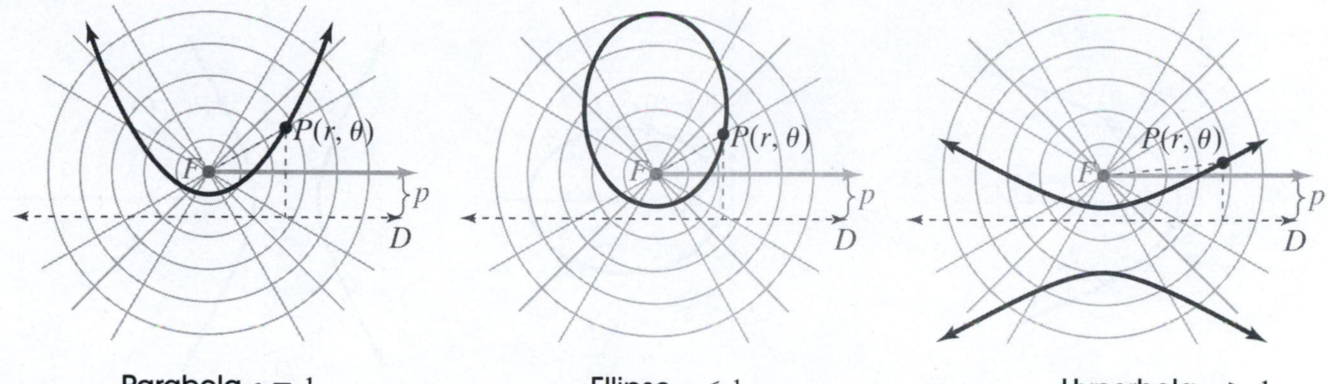

Parabola $e = 1$ Ellipse $e < 1$ Hyperbola $e > 1$

Description

Directrix **parallel** to the polar axis located p units **below** the polar axis.

Case 2

Equation of Conic Section in Standard Form

$$r = \frac{ep}{1 + e\sin\theta}$$

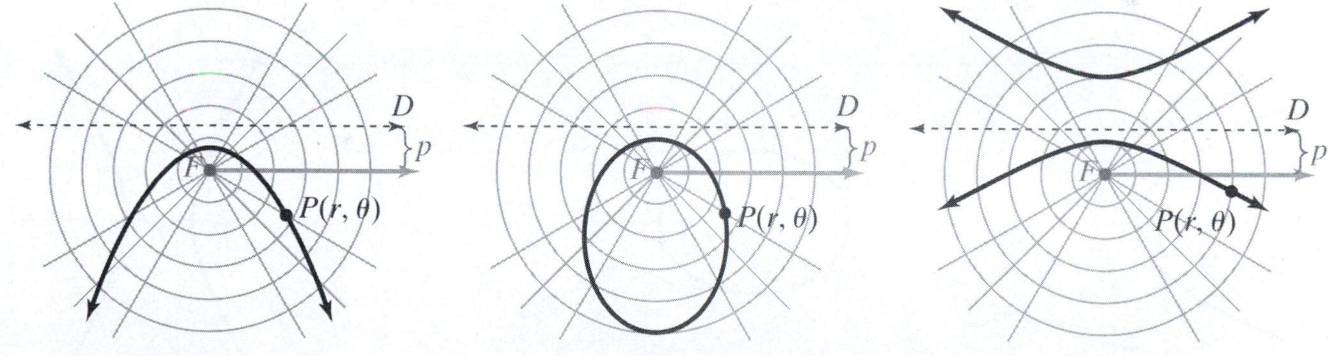

Parabola $e = 1$ Ellipse $e < 1$ Hyperbola $e > 1$

Description

Directrix **parallel** to the polar axis located p units **above** the polar axis.

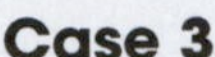

Case 3

Equation of Conic Section in Standard Form

$$r = \frac{ep}{1 - e\cos\theta}$$

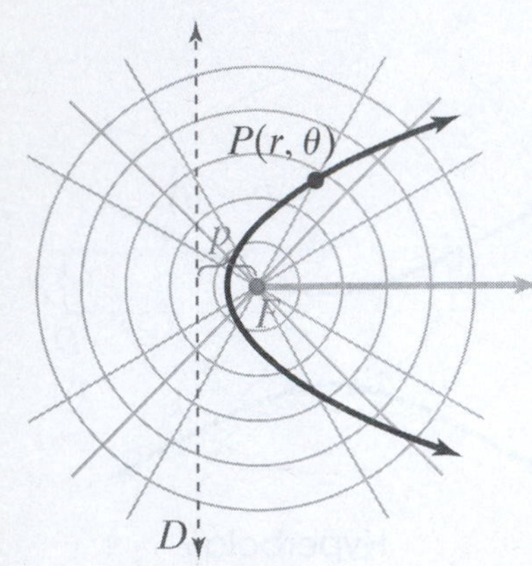

Parabola $e = 1$

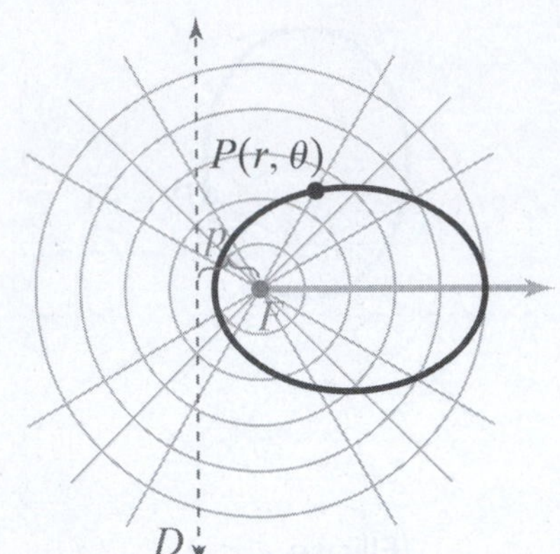

Ellipse $e < 1$

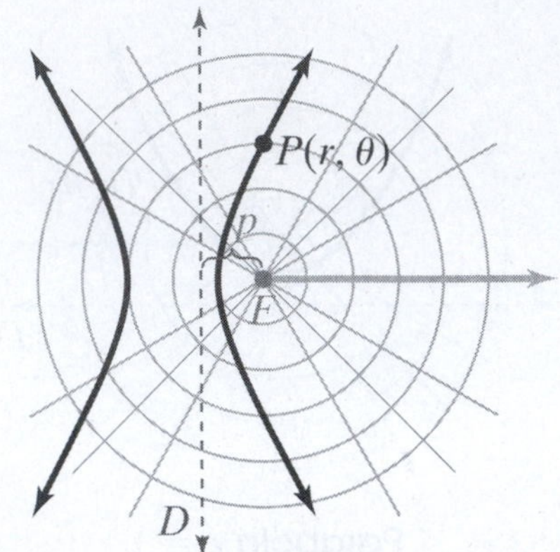

Hyperbola $e > 1$

Description

Directrix **perpendicular** to the polar axis located p units to the **left** of the pole.

Case 4

Equation of Conic Section in Standard Form

$$r = \frac{ep}{1 + e\cos\theta}$$

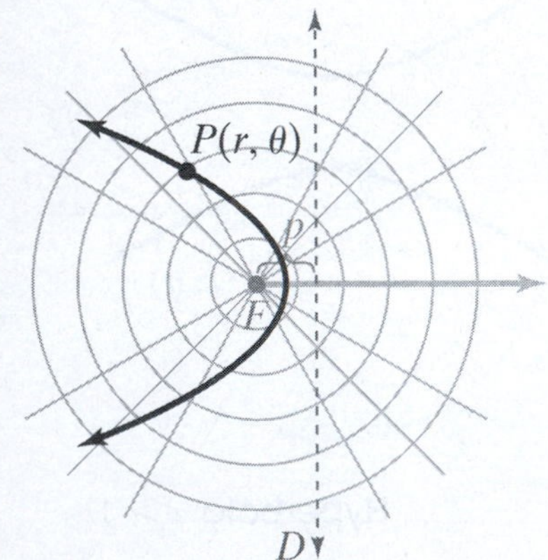

Parabola $e = 1$

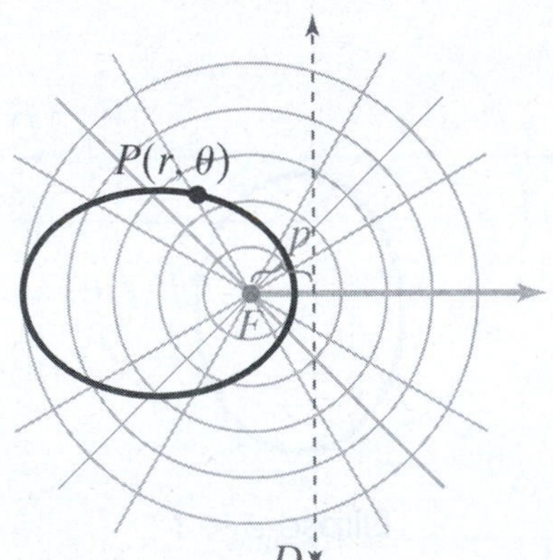

Ellipse $e < 1$

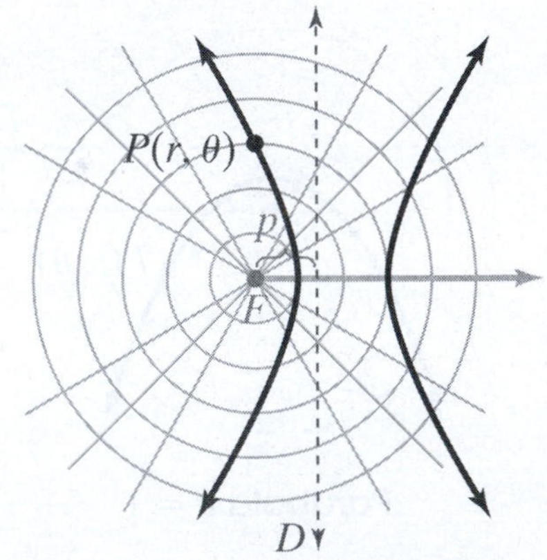

Hyperbola $e > 1$

Description

Directrix **perpendicular** to the polar axis located p units to the **right** of the pole.

Example 1 Indentifying Conic Sections Given in Polar Form

Each equation represents a conic section in polar form. Identify the conic section, the position of the directrix, and the distance between the focus and directrix.

a. $r = \dfrac{2}{1 - \cos\theta}$ b. $r = \dfrac{4}{1 + 3\sin\theta}$

Solution

a. The equation $r = \dfrac{2}{1 - \cos\theta}$ is in standard form where $r = \dfrac{ep}{1 - e\cos\theta}$. Therefore, $e = 1$ thus the graph is a parabola.

Since the equation $r = \dfrac{2}{1 - \cos\theta}$ is of the form $r = \dfrac{ep}{1 - e\cos\theta}$, this represents

Case 3 in which the directrix is **perpendicular** to the polar axis located to the **left** of the pole.

In the equation $r = \dfrac{2}{1 - \cos\theta}$ which is of the form $r = \dfrac{ep}{1 - e\cos\theta}$, we see that $e = 1$ and $ep = 2$. Therefore, $p = 2$. Thus, the distance between the focus and directrix is 2 units.

b. The equation $r = \dfrac{2}{1 + 3\sin\theta}$ is in standard form where $r = \dfrac{ep}{1 + e\sin\theta}$. Therefore, $e = 3 > 1$ thus the graph is a hyperbola.

Since the equation $r = \dfrac{2}{1 + 3\sin\theta}$ is of the form $r = \dfrac{ep}{1 + e\sin\theta}$, this represents Case 2 in which the directrix is **parallel** to the polar axis located **above** of the polar axis.

In the equation $r = \dfrac{2}{1 + 3\sin\theta}$, which is of the form $r = \dfrac{ep}{1 + e\sin\theta}$, we see that

$e = 3$ and $ep = 4$. Therefore, $p = \dfrac{4}{3}$. Thus, the distance between the focus and

directrix is $\dfrac{4}{3}$ units.

You Try It **Work through this You Try It problem.**

Work Exercises 1–6 in this textbook or in the MyLab Math Study Plan.

In Example 1 the equations of the conic sections were given in **standard form**. If the equation of the conic is not given in standard form, then we must use algebraic properties to rewrite the equation in standard form.

Example 2 Indentifying Conic Sections Given in Polar Form

The equation $3r\cos\theta + 4r = 5$ represents a conic section in polar form. Rewrite the equation in standard form. Identify the conic section, the position of the directrix, and the distance between the focus and directrix.

Solution First, solve the equation for r.

$3r\cos\theta + 4r = 5$ Start with the given equation.

$r(3\cos\theta + 4) = 5$ Factor out r from the left-hand side.

$r = \dfrac{5}{3\cos\theta + 4}$ Divide both sides by $3\cos\theta + 4$

$r = \dfrac{5}{4 + 3\cos\theta}$ Rearrange the terms in the denominator.

In order to write the equation $r = \dfrac{5}{4 + 3\cos\theta}$ in standard form, we must have a 1 as the first term in the denominator. To accomplish this, we can multiply the numerator and denominator by $\dfrac{1}{4}$.

$r = \dfrac{5}{4 + 3\cos\theta}$ Start with the equation found above.

$r = \dfrac{5}{4 + 3\cos\theta} \cdot \dfrac{\frac{1}{4}}{\frac{1}{4}}$ Multiply the numerator and denominator by $\dfrac{1}{4}$.

$r = \dfrac{\frac{5}{4}}{1 + \frac{3}{4}\cos\theta}$ Simplify.

We see that the equation of the conic section in standard form is $r = \dfrac{\frac{5}{4}}{1 + \frac{3}{4}\cos\theta}$.

This equation is of the form $r = \dfrac{ep}{1 + e\cos\theta}$ with $e = \dfrac{3}{4} < 1$. Thus, the graph is a hyperbola.

Since the equation $r = \dfrac{\frac{5}{4}}{1 + \frac{3}{4}\cos\theta}$ is of the form $r = \dfrac{ep}{1 + e\cos\theta}$, this represents Case 4 in which the directrix is **perpendicular** to the polar axis located to the **right** of the pole.

In the equation $r = \dfrac{\frac{5}{4}}{1 + \frac{3}{4}\cos\theta}$, which is of the form $r = \dfrac{ep}{1 + e\cos\theta}$, we see that $e = \dfrac{3}{4}$ and $ep = \dfrac{5}{4}$.

We can use this information to solve for p.

$ep = \frac{5}{4}$ Start with the equation $ep = \frac{5}{4}$.

$\left(\frac{3}{4}\right)p = \frac{5}{4}$ Substitute for $\frac{3}{4}$ for e.

$\frac{\not{4}}{\not{3}} \cdot \frac{\not{3}}{\not{4}} p = \frac{5}{\not{4}} \cdot \frac{\not{4}}{3}$ Multiply both sides by $\frac{4}{3}$.

$p = \frac{5}{3}$ Simplify.

Therefore, $p = \frac{5}{3}$. Thus, the distance between the focus and directrix is $\frac{5}{3}$ units.

You Try It Work through this You Try It problem.

Work Exercises 7–12 in this textbook or in the MyLab Math Study Plan.

OBJECTIVE 2 SKETCHING THE GRAPHS OF CONIC SECTIONS GIVEN IN POLAR FORM

When sketching the graphs of conic sections given in polar form, we will follow the five step procedure outlined below.

> **Sketching the Graph of a Conic Section Given in Polar Form**
>
> **Step 1.** Write the equation in standard form if necessary.
>
> **Step 2.** Identify the conic section.
>
> **Step 3.** Plot the focus at the pole and sketch the directrix.
>
> **Step 4.** Plot the polar coordinates (r, θ) where θ is a quadrantal angle $\left(\theta = 0, \theta = \frac{\pi}{2}, \theta = \pi, \text{ and } \theta = \frac{3\pi}{2}\right)$ and determine which of these ordered pairs represents a vertex of the conic section.
>
> **Step 5.** Plot any additional ordered pairs if necessary then connect the ordered pairs with a smooth curve.

Example 3 Sketching the Graphs of Conic Sections Given in Polar Form

Each equation represents a conic section in polar form. Use the five step procedure for sketching the graphs of conic sections given in polar form.

a. $r = \frac{2}{3 - \cos\theta}$ b. $r + r\sin\theta = 3$ c. $r = \frac{12}{2 + 4\cos\theta}$

Solution

a. **Step 1.** First, multiply the numerator and denominator by $\frac{1}{3}$ to write the equation in standard form.

$$r = \frac{2}{3 - \cos\theta} \quad \text{Start with the original equation.}$$

$$r = \frac{2}{3 - \cos\theta} \cdot \frac{\frac{1}{3}}{\frac{1}{3}} \quad \text{Multiply the numerator and denominator by } \frac{1}{3}.$$

$$r = \frac{\frac{2}{3}}{1 - \frac{1}{3}\cos\theta} \quad \text{Simplify.}$$

The equation written in standard form is $r = \frac{\frac{2}{3}}{1 - \frac{1}{3}\cos\theta}$ which is of the form $r = \frac{ep}{1 - e\cos\theta}$.

Step 2. The equation $r = \frac{\frac{2}{3}}{1 - \frac{1}{3}\cos\theta}$ is of the form $r = \frac{ep}{1 - e\cos\theta}$ with $e = \frac{1}{3} < 1$. Therefore, the graph is an ellipse with a focus at the pole having a directrix **perpendicular** to the polar axis located p units to the **left** of the pole.

To solve for p, use the fact that $ep = \frac{2}{3}$. We can use this information to determine the value of p.

$$ep = \frac{2}{3} \quad \text{Start with the equation } ep = \frac{2}{3}.$$

$$\left(\frac{1}{3}\right)p = \frac{2}{3} \quad \text{Substitute } \frac{1}{3} \text{ for } e.$$

$$\cancel{3} \cdot \frac{1}{\cancel{3}}p = \frac{2}{\cancel{3}} \cdot \cancel{3} \quad \text{Multiply both sides by 3.}$$

$$p = 2 \quad \text{Simplify.}$$

Therefore, $p = 2$. Thus, the ellipse has a directrix **perpendicular** to the polar axis that is located $p = 2$ units to the **left** of the pole.

Step 3. We plot the focus at the pole and sketch the directrix that is located two units to the left of the pole. See Figure 30.

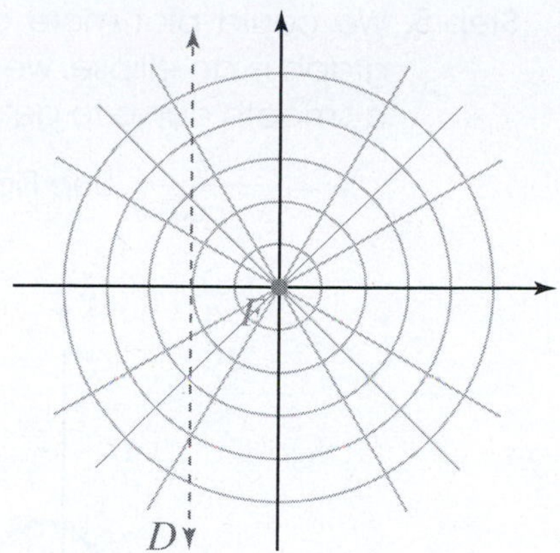

Figure 30

Step 4. We choose $\theta = 0, \theta = \frac{\pi}{2}, \theta = \pi$, and $\theta = \frac{3\pi}{2}$ then determine the corresponding values of r. See Table 1. Since the ellipse has a vertical directrix, we know that the vertices must lie along the polar axis or when $\theta = 0$ and $\theta = \pi$. We now plot the ordered pairs. See Figure 31.

Table 1

θ	$r = \frac{2}{3 - \cos\theta}$	(r, θ)
0	$r = \frac{2}{3 - \cos 0} = \frac{2}{3 - 1} = \frac{2}{2} = 1$	$(1, 0)$
$\frac{\pi}{2}$	$r = \frac{2}{3 - \cos\frac{\pi}{2}} = \frac{2}{3 - 0} = \frac{2}{3}$	$\left(\frac{2}{3}, \frac{\pi}{2}\right)$
π	$r = \frac{2}{3 - \cos\pi} = \frac{2}{3 - (-1)} = \frac{2}{4} = \frac{1}{2}$	$\left(\frac{1}{2}, \pi\right)$
$\frac{3\pi}{2}$	$r = \frac{2}{3 - \cos\frac{3\pi}{2}} = \frac{2}{3 - 0} = \frac{2}{3}$	$\left(\frac{2}{3}, \frac{3\pi}{2}\right)$

Vertices: $(1, 0)$ and $\left(\frac{1}{2}, \pi\right)$

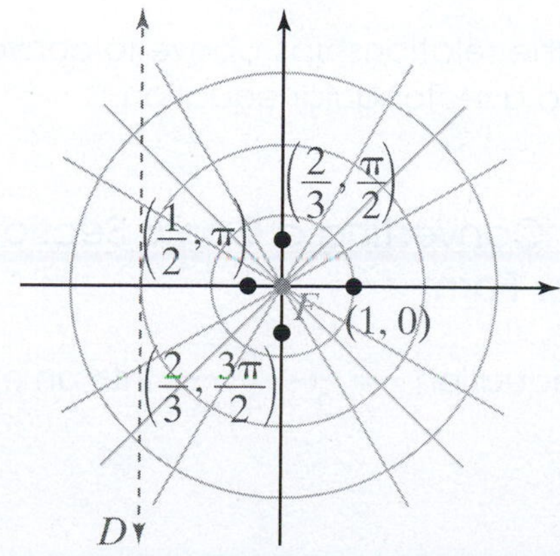

Figure 31 Points lying on the graph of $r = \frac{2}{3 - \cos\theta}$ whose angles are quadrantal angles

Step 5. We could plot more ordered pairs. However, since we know that the graph is an ellipse, we can connect the points plotted in Figure 31 with a smooth curve to get a good representation of the graph of $r = \dfrac{2}{3 - \cos\theta}$. See Figure 32.

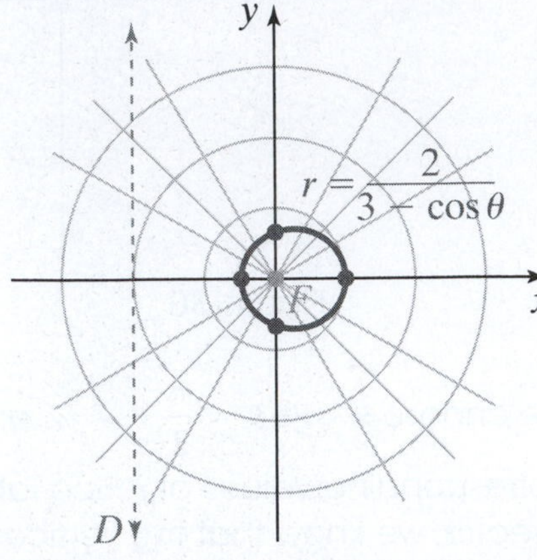

Figure 32 The graph of the ellipse $r = \dfrac{2}{3 - \cos\theta}$

Watch the **interactive video** to see how to sketch the graphs of b) $r + r\sin\theta = 3$ and c) $r = \dfrac{12}{2 + 4\cos\theta}$ or see the **graphs** of parts b) and c).

You Try It Work through this You Try It problem.

Work Exercises 13–21 in this textbook or in the MyLab Math Study Plan.

OBJECTIVE 3 CONVERTING AN EQUATION OF A CONIC SECTION FROM POLAR FORM TO RECTANGULAR FORM

In **Section 10.1** we converted equations from polar form to rectangular form using the following relationships.

$$x = r\cos\theta$$

$$y = r\sin\theta$$

$$r = \sqrt{x^2 + y^2} \text{ (or } r^2 = x^2 + y^2\text{)}$$

$$\tan\theta = \frac{y}{x}$$

We can use the relationships above to convert conic sections that are given in polar form into a rectangular equation.

Example 4 Converting a Conic Section Given in Polar Form into Rectangular Form

Convert the equation $r = \dfrac{6}{2 - 4\cos\theta}$ into an equation written in rectangular form.

Solution Start by multiplying both sides of the polar equation by $2 - 4\cos\theta$.

$r = \dfrac{6}{2 - 4\cos\theta}$ Start with the given equation in polar form.

$2r - 4r\cos\theta = 6$ Multiply both sides of the equation by $2 - 4\cos\theta$.

$2\sqrt{x^2 + y^2} - 4x = 6$ Substitute $\sqrt{x^2 + y^2}$ for r and x for $r\cos\theta$.

$2\sqrt{x^2 + y^2} = 6 + 4x$ Add $4x$ to both sides.

$\sqrt{x^2 + y^2} = 3 + 2x$ Divide both sides by 2.

$x^2 + y^2 = 4x^2 + 12x + 9$ Square both sides.

$3x^2 - y^2 + 12x + 9 = 0$ Combine like terms and set equal to 0.

Therefore, given the equation $r = \dfrac{6}{2 - 4\cos\theta}$ in polar form, we see that an equivalent equation in rectangular form is $3x^2 - y^2 + 12x + 9 = 0$.

View these steps to verify that the equation $3x^2 - y^2 + 12x + 9 = 0$ can be written as $\dfrac{(x+2)^2}{1} - \dfrac{y^2}{3} = 1$ which is the equation of a hyperbola written in standard from having a horizontal transverse axis. The center of the hyperbola is $(-2, 0)$. The coordinates of the two foci are $(0, 0)$ and $(-4, 0)$. See Figure 33.

Equation in Polar Form	Equation in Rectangular Form
$r = \dfrac{6}{2 - 4\cos\theta}$	$3x^2 - y^2 + 12x + 9 = 0$ or $\dfrac{(x+2)^2}{1} - \dfrac{y^2}{3} = 1$

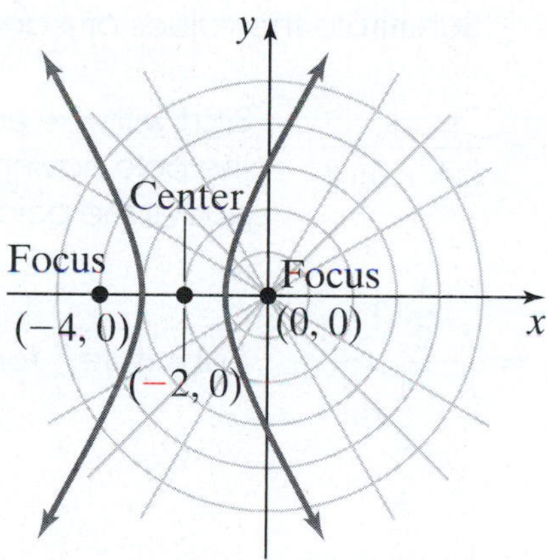

Figure 33

You Try It Work through this You Try It problem.

Work Exercises 22–27 in this textbook or in the MyLab Math Study Plan.

OBJECTIVE 4 DETERMINING THE EQUATION OF A CONIC SECTION IN POLAR FORM GIVEN THE ECCENTRICITY AND THE EQUATION OF THE DIRECTRIX

There are only two things that are required to determine the equation of a conic section in polar form:

1. The eccentricity
2. The equation of the directrix

If we are given the eccentricity and the equation of the directrix, then we can use that information to determine the equation in polar form.

Example 5 Determining the Equation of a Conic Section

Determine the polar equation in standard form of the conic section with a focus at the pole such that the eccentricity is $\frac{1}{2}$ and the equation of the directrix is $r = 3\csc\theta$.

Solution The eccentricity is $e = \frac{1}{2} < 1$, which means that the conic section is an ellipse. We can use a **reciprocal identity** and the relationship $y = r\sin\theta$ to convert the equation of the directrix into rectangular form.

$r = 3\csc\theta$ Write the original equation of the directrix.

$r = \dfrac{3}{\sin\theta}$ Use the reciprocal identity $\csc\theta = \dfrac{1}{\sin\theta}$.

$r\sin\theta = 3$ Multiply both sides by $\sin\theta$.

$y = 3$ Substitute y for $\sin\theta$.

Therefore, the equation of the directrix in rectangular form is $y = 3$. This is the equation of a horizontal line located three units above the polar axis. Thus, $p = 3$. Therefore, the equation of the conic must be of the form $r = \dfrac{ep}{1 + e\sin\theta}$ with $p = 3$ and $e = \frac{1}{2}$. Substitute the values of p and e to determine the equation.

$r = \dfrac{ep}{1 + e\sin\theta}$ Start with the equation of a conic with a focus at the pole having a horizontal directrix located p units above the polar axis.

$r = \dfrac{\frac{1}{2}\cdot 3}{1 + \frac{1}{2}\sin\theta}$ Substitute $\frac{1}{2}$ for e and 3 for p.

$r = \dfrac{\frac{3}{2}}{1 + \frac{1}{2}\sin\theta}$ Simplify.

Therefore, the equation of the conic section in standard form is $r = \dfrac{\frac{3}{2}}{1 + \frac{1}{2}\sin\theta}$.

You Try It Work through this You Try It problem.

Work Exercises 28–30 in this textbook or in the MyLab Math Study Plan.

11.4 Exercises

In Exercises 1–6, each equation represents a conic section in polar form. Identify the conic section, the position of the directrix, and the distance between the focus and directrix.

SbS 1. $r = \dfrac{3}{1 - \cos\theta}$ SbS 2. $r = \dfrac{2}{1 + \frac{1}{3}\sin\theta}$ SbS 3. $r = \dfrac{6}{1 + 2\cos\theta}$

SbS 4. $r = \dfrac{\frac{5}{4}}{1 - \sin\theta}$ SbS 5. $r = \dfrac{\frac{7}{3}}{1 + \frac{8}{9}\cos\theta}$ SbS 6. $r = \dfrac{\frac{6}{5}}{1 + \frac{3}{4}\cos\theta}$

In Exercises 7–12, each equation represents a conic section in polar form. Identify the conic section, the position of the directrix, and the distance between the focus and directrix.

SbS 7. $r = \dfrac{9}{5 - 5\cos\theta}$ SbS 8. $r = \dfrac{4}{5 + 3\sin\theta}$ SbS 9. $r = \dfrac{6}{3 - 7\sin\theta}$

SbS 10. $3r - 3r\cos\theta = 6$ SbS 11. $5r - 4r\sin\theta = 8$ SbS 12. $4r\sin\theta + 3r = 5$

In Exercises 13–21, each equation represents a conic section in polar form Use the five step procedure for sketching the graphs of conic sections given in polar form.

SbS 13. $r = \dfrac{3}{1 - \cos\theta}$ SbS 14. $r = \dfrac{4}{1 + \frac{1}{2}\sin\theta}$ SbS 15. $r = \dfrac{6}{1 + 2\cos\theta}$

SbS 16. $r = \dfrac{10}{5 - 5\cos\theta}$ SbS 17. $r = \dfrac{6}{5 - 3\sin\theta}$ SbS 18. $r = \dfrac{6}{2 + 4\cos\theta}$

SbS 19. $3r - 3r\cos\theta = 6$ SbS 20. $5r - 4r\sin\theta = 8$ SbS 21. $4r\sin\theta + 2r = 16$

In Exercises 22–27, convert each equation into a rectangular equation.

22. $r = \dfrac{3}{1 - \cos\theta}$ 23. $r = \dfrac{2}{1 + \frac{1}{3}\sin\theta}$ 24. $r = \dfrac{6}{1 + 2\cos\theta}$

25. $r = \dfrac{9}{5 - 5\cos\theta}$ 26. $r = \dfrac{4}{5 + 3\sin\theta}$ 27. $r = \dfrac{6}{3 - 7\sin\theta}$

28. Determine the polar equation in standard form of the conic section with a focus at the pole such that the eccentricity is 1 and the equation of the directrix is $r = 3\csc\theta$.

29. Determine the polar equation in standard form of the conic section with a focus at the pole such that the eccentricity is $\frac{3}{4}$ and the equation of the directrix is $r = -2\sec\theta$.

30. Determine the polar equation in standard form of the conic section with a focus at the pole such that the eccentricity is 2 and the equation of the directrix is $r = 4\sec\theta$.

Brief Exercises

In Exercises 31–39, sketch the graph of each conic section.

31. $r = \dfrac{3}{1 - \cos\theta}$

32. $r = \dfrac{4}{1 + \frac{1}{2}\sin\theta}$

33. $r = \dfrac{6}{1 + 2\cos\theta}$

34. $r = \dfrac{10}{5 - 5\cos\theta}$

35. $r = \dfrac{6}{5 - 3\sin\theta}$

36. $r = \dfrac{6}{2 + 4\cos\theta}$

37. $3r - 3r\cos\theta = 6$

38. $5r - 4r\sin\theta = 8$

39. $4r\sin\theta + 2r = 16$

11.5 Parametric Equations

THINGS TO KNOW

Before working through this section, be sure that you are familiar with the following concepts:

VIDEO | ANIMATION | INTERACTIVE

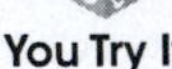
You Try It 1. Finding the Values of the Trigonometric Functions of Quadrantal Angles (Section 6.5)

You Try It 2. Evaluating Trigonometric Functions of Angles Belonging to the $\frac{\pi}{3}, \frac{\pi}{6}$, or $\frac{\pi}{4}$ Families (Section 6.5)

You Try It 3. Understanding the Unit Circle Definitions of the Trigonometric Functions (Section 6.6)

You Try It 4. Determining the Equation of a Parabola with a Horizontal Axis of Symmetry (Section 11.1)

You Try It 5. Sketching the Graph of an Ellipse (Section 11.2)

You Try It 6. Sketching the Graph of a Hyperbola (Section 11.3)

INTRODUCTION

Read this introduction before beginning Objective 1.

OBJECTIVES

1 Plotting Points to Sketch the Graph of a Plane Curve Defined Parametrically

2 Determining a Rectangular Representation for a Curve Defined Parametrically Using Substitution

3 Determining a Rectangular Representation for a Curve Defined Parametrically Using a Trigonometric Identity

Introduction to Section 11.5

In this section we will discuss **parametric equations**. A curve in the Cartesian plane, sometimes called a **plane curve**, is said to be parameterized when the set of all ordered pairs (x, y) on the curve are represented as functions of another variable. You have already seen an example of a set of parametric equations earlier in the text. Recall from **Section 6.6** that for any real number t, if (x, y) is a point on the unit circle corresponding to t, then the unit circle definitions of sine and cosine were defined as

$$x = \cos t, y = \sin t.$$

Figure 34 illustrates that every point that lies on the graph of the unit circle can be described by the ordered pair $(x, y) = (\cos t, \sin t)$ where t is the corresponding central angle or arc length from the point $(1, 0)$ to (x, y). The variable t is called a **parameter**. If $0 \le t < 2\pi$, then the parametric equation $x = \cos t, y = \sin t$ represent another way to describe the exact same set of points as the **rectangular equation** $x^2 + y^2 = 1$.

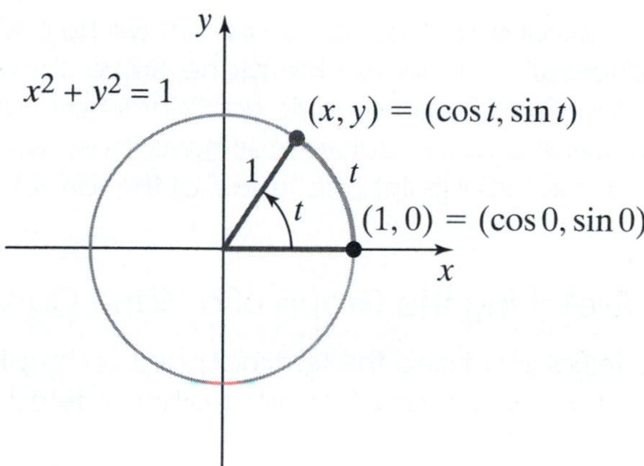

Figure 34 The unit circle can be described using the rectangular equation $x^2 + y^2 = 1$ or by using the parametric equations $x = \cos t, y = \sin t$ for $0 \le t < 2\pi$.

When a plane curve, such as the graph of the unit circle, does not pass the **vertical line test**, then it is impossible to describe the curve with a single function of the form $y = f(x)$. However, it is often be possible to describe the plane curve using parametric equations of the form $x = f(t)$ and $y = g(t)$. If you study calculus or physics, you will see that we need parametric equations to model many situations where one single function will not adequately suffice. See **Objective 5.**

> **Definition** Parametric Equations of a Plane Curve
>
> A **plane curve** is the set of all ordered pairs (x, y) such that $x = f(t)$ and $y = g(t)$, where t is in an interval I. The variable t is called a **parameter**, and the equations $x = f(t)$ and $y = g(t)$ are called **parametric equations** of the curve.

OBJECTIVE 1 PLOTTING POINTS TO SKETCH THE GRAPH OF A PLANE CURVE DEFINED PARAMETRICALLY

One way to sketch a plane curve that is defined parametrically is to plot ordered pairs using various values of the parameter. We then connect those ordered pairs with a smooth curve. If a parametric equation is defined on some predetermined interval such as $a \leq t \leq b$, then it is important to determine the **initial point** and the **terminal point**.

> **Definition** Initial Point and Terminal Point of a Plane Curve Defined Parametrically
>
> If a plane curve is defined parametrically as $x = f(t)$ and $y = g(t)$ on a closed interval $a \leq t \leq b$, then the ordered pair $(f(a), g(a))$ is called the **initial point** and the ordered pair $(f(b), g(b))$ is called the **terminal point**.

Determining the initial point and terminal point will help to understand the proper direction, or orientation, in which the plane curve should be sketched. Not all plane curves described by parametric equations have an initial point and terminal point. However, if a plane curve does have these two points, then the graph always "starts" at the initial point and "ends" at the terminal point.

Example 1 Sketching the Graph of a Plane Curve by Plotting Points

Determine the initial point and the terminal point of the plane curve defined by $x = 2t^2, y = 3t, -4 \leq t \leq 4$, then plot several other ordered pairs and sketch the plane curve.

Solution We obtain the initial point by evaluating $x = 2t^2$ and $y = 3t$ when $t = -4$ and we obtain the terminal point by evaluating $x = 2t^2$ and $y = 3t$ when $t = 4$.

Initial Point: $(2(-4)^2, 3(-4)) = (32, -12)$

Terminal Point: $(2(4)^2, 3(4)) = (32, 12)$

Now make a table of corresponding values of $t, x,$ and y to obtain several ordered pairs. See **Table 2**. We plot those ordered pairs and connect them with a smooth curve. Notice in **Figure 35** that the curve "starts" at the initial point, $(32, -12)$, and "ends" at the terminal point, $(32, 12)$. Watch this **animation** to see how to sketch this plane curve. The arrow heads represent the orientation or path of the curve as the value of the parameter t increases. The graph obtained in **Figure 35** appears to be a portion of a parabola. In **Example 3** we will show how to eliminate the parameter t to determine the **rectangular equation** of this parabola.

Table 2

t	$x = 2t^2$	$y = 3t$	(x, y)
-4	$x = 32$	$y = -12$	$(32, -12)$
-3	$x = 18$	$y = -9$	$(18, -9)$
-2	$x = 8$	$y = -6$	$(8, -6)$
-1	$x = 2$	$y = -3$	$(2, -3)$
0	$x = 0$	$y = 0$	$(0, 0)$
1	$x = 2$	$y = 3$	$(2, 3)$
2	$x = 8$	$y = 6$	$(8, 6)$
3	$x = 18$	$y = 9$	$(18, 9)$
4	$x = 32$	$y = 12$	$(32, 12)$

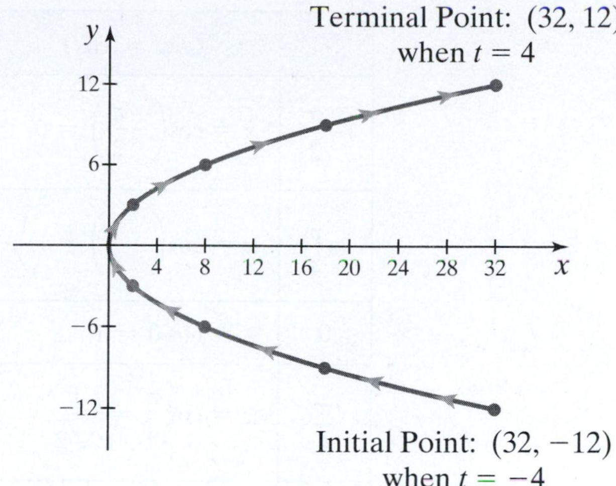

Figure 35 The plane curve described by the parametric equations $x = 2t^2, y = 3t, -4 \le t \le 4$

Example 2 Sketching the Graph of a Plane Curve by Plotting Points

Determine the initial point and the terminal point of the plane curve defined by $x = \cos t, y = \sin t, -\frac{\pi}{2} \le t \le \pi$, then plot several other ordered pairs and sketch the plane curve.

Solution We obtain the initial point by evaluating $x = \cos t$ and $y = \sin t$ when $t = -\frac{\pi}{2}$ and we obtain the terminal point by evaluating $x = \cos t$ and $y = \sin t$ when $t = \pi$.

Initial Point: $\left(\cos\left(-\frac{\pi}{2}\right), \sin\left(-\frac{\pi}{2}\right)\right) = (0, -1)$

Terminal Point: $(\cos(\pi), \sin(\pi)) = (-1, 0)$

Now make a table of corresponding values of t, x, and y to obtain several ordered pairs. See **Table 3**. We plot those ordered pairs and connect them with a smooth curve. As you can see in **Figure 36**, the plane curve is a portion of the graph of the **unit circle** that "starts" at the initial point, $(0, -1)$, and "ends" at the terminal point, $(-1, 0)$.

Table 3

t	$x = \cos t$	$y = \sin t$	(x, y)
$-\frac{\pi}{2}$	$x = \cos\left(-\frac{\pi}{2}\right) = 0$	$y = \sin\left(-\frac{\pi}{2}\right) = -1$	$(0, -1)$
$-\frac{\pi}{4}$	$x = \cos\left(-\frac{\pi}{4}\right) = \frac{1}{\sqrt{2}}$	$y = \sin\left(-\frac{\pi}{4}\right) = -\frac{1}{\sqrt{2}}$	$\left(\frac{1}{\sqrt{2}}, -\frac{1}{\sqrt{2}}\right)$
0	$x = \cos 0 = 1$	$y = \sin 0 = 0$	$(1, 0)$
$\frac{\pi}{4}$	$x = \cos\frac{\pi}{4} = \frac{1}{\sqrt{2}}$	$y = \sin\frac{\pi}{4} = \frac{1}{\sqrt{2}}$	$\left(\frac{1}{\sqrt{2}}, \frac{1}{\sqrt{2}}\right)$
$\frac{\pi}{2}$	$x = \cos\frac{\pi}{2} = 0$	$y = \sin\frac{\pi}{2} = 1$	$(0, 1)$
$\frac{3\pi}{4}$	$x = \cos\frac{3\pi}{4} = -\frac{1}{\sqrt{2}}$	$y = \sin\frac{3\pi}{4} = \frac{1}{\sqrt{2}}$	$\left(-\frac{1}{\sqrt{2}}, \frac{1}{\sqrt{2}}\right)$
π	$x = \cos \pi = -1$	$y = \sin \pi = 0$	$(-1, 0)$

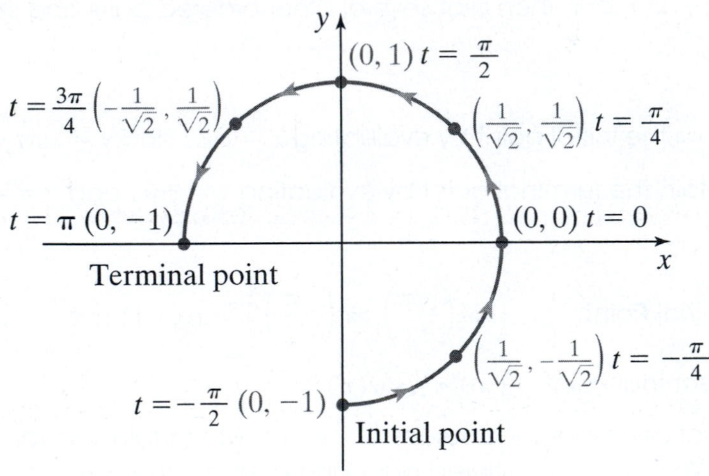

Figure 36 The plane curve described by the parametric equations $x = \cos t, y = \sin t, -\frac{\pi}{2} \leq t \leq \pi$

You Try It Work through this You Try It problem.

Work Exercises 1–8 in this textbook or in the MyLab Math Study Plan.

OBJECTIVE 2 DETERMINING A RECTANGULAR REPRESENTATION FOR A CURVE DEFINED PARAMETRICALLY USING SUBSTITUTION

It is sometimes possible to convert a given set of two parametric equations into an equivalent **rectangular equation**. We do this by eliminating the parameter in an attempt to write one equation in terms of x and y. One way to eliminate the parameter is to try to solve one of the parametric equations for the variable representing the parameter, then substitute that expression into the second equation. When using this method, carefully consider which of the two parametric equations is the easiest to solve for the parameter. The five-step method for determining a rectangular representation for a curve defined parametrically using substitution is stated below.

> **Determining a Rectangular Representation for a Curve Defined Parametrically Using Substitution**
>
> **Step 1.** Carefully observe the given parametric equations in an attempt to identify a relationship between the variables such that a substitution could be made to eliminate the parameter t.
>
> **Step 2.** Choose a parametric equation and solve for t or rewrite the parametric equation based on the observation made in step 1.
>
> **Step 3.** Substitute the expression from step 2 into the other parametric equation in place of the parameter to obtain the rectangular equation.
>
> **Step 4.** Write the rectangular equation in a familiar form whenever possible.
>
> **Step 5.** If necessary, restrict the domain and range of the rectangular equation to be consistent with any restrictions of the parameter or any restrictions of the parametric equations.

We start with the same set of parametric equations that were used in **Example 1.**

Example 3 Determining a Rectangular Representation for a Curve Defined Parametrically

Determine the rectangular equation of the curve whose parametric equations are given by $x = 2t^2, y = 3t, -4 \le t \le 4$, then sketch the plane curve.

Solution

Step 1. Observe that solving for t in the second equation $y = 3t$ would allow a substitution for t into the first equation $x = 2t^2$ that would eliminate the parameter t.

Step 2. Solve for t in the second equation $y = 3t$.

$y = 3t$ Start with the second parametric equation.

$\frac{y}{3} = t$ Divide both sides by 3 to solve for the parameter.

Step 3. Substitute the expression $\frac{y}{3}$ for t into the first equation.

$x = 2t^2$ Start with the first parametric equation.

$x = 2\left(\frac{y}{3}\right)^2$ Substitute $\frac{y}{3}$ for t.

$x = \frac{2y^2}{9}$ Simplify.

Step 4. The graph of the equation $x = \frac{2y^2}{9}$ is a parabola. The variable y is squared. This indicates that the graph is a parabola that has a **horizontal axis of symmetry**. To write this equation in **standard form** we multiply both sides of the equation by $\frac{9}{2}$.

$x = \frac{2y^2}{9}$ Start with the equation determined in **Step 3.**

$\frac{9}{2}x = \frac{9}{2} \cdot \frac{2y^2}{9}$ Multiply both sides by $\frac{9}{2}$.

$y^2 = \frac{9}{2}x$ Rewrite in **standard form.**

The equation $y^2 = \frac{9}{2}x$ is the equation of a parabola with the vertex at the origin with a **horizontal axis of symmetry.**

Step 5. The complete graph of the rectangular equation $y^2 = \frac{9}{2}x$ can be seen in **Figure 37**. Note that the domain of the graph seen in **Figure 37** is $0 \le x < \infty$. Although the graph in **Figure 37** represents a complete graph of the rectangular equation $y^2 = \frac{9}{2}x$, it is **not** an accurate representation of the graph of the original parametric equations because we were given restrictions on the parameter.

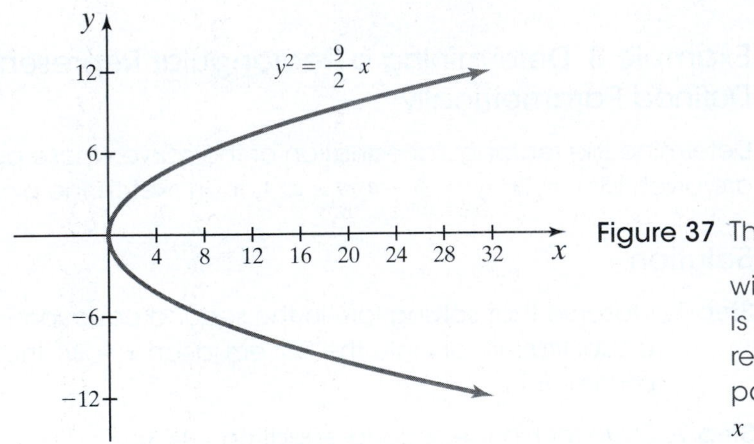

Figure 37 The graph of $y^2 = \frac{9}{2}x$ with domain $0 \le x < \infty$ is **not** an accurate representation of the parametric equations $x = 2t^2, y = 3t, -4 \le t \le 4$.

Recall that the parametric equations $x = 2t^2, y = 3t$ were defined only on the interval $-4 \le t \le 4$. Thus, there were restrictions imposed on the parameter. These restrictions must be considered when sketching the final graph of the parametric equations. Thus, given the restriction $-4 \le t \le 4$ we see that the domain of the rectangular equation $y^2 = \frac{9}{2}x$, which is $0 \le x < \infty$, must be modified. In **Example 1** we saw that when $t = -4$, the corresponding point that lies on the graph of the parametric equations is $(32, -12)$. Similarly, when $t = 4$, the corresponding point that lies on the graph of the parametric equations is $(32, 12)$. Therefore, the largest possible value of x is 32 and the rectangular equation representing the given parametric equations is $y^2 = \frac{9}{2}x, 0 \le x \le 32$. See **Figure 38**.

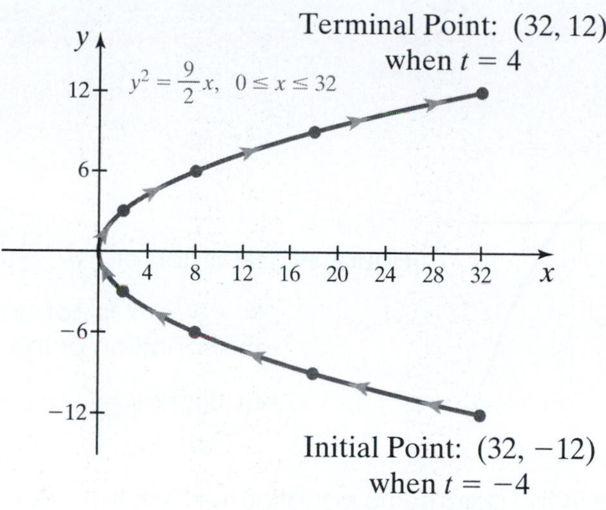

Figure 38 The plane curve described by the parametric equations $x = 2t^2, y = 3t, -4 \le t \le 4$, can also be described by the rectangular equation $y^2 = \frac{9}{2}x, 0 \le x \le 32$.

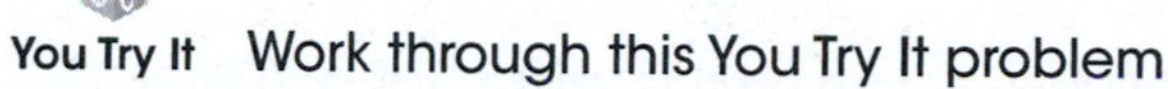

You Try It Work through this You Try It problem.

Work Exercises 9–12 in this textbook or in the MyLab Math Study Plan.

Example 4 Determining a Rectangular Representation for a Curve Defined Parametrically

Determine the rectangular equation of the curve whose parametric equations are given by $x = \sqrt{t}, y = -\frac{1}{2}t + 1$, then sketch the plane curve.

Solution

Step 1. Observe that isolating the parameter t by squaring both sides of the first equation $x = \sqrt{t}$ would allow a substitution for t into the second equation $y = -\frac{1}{2}t + 1$ that will eliminate the parameter t.

Step 2. Square both sides of the equation $x = \sqrt{t}$ to solve for t.

$x = \sqrt{t}$ Start with the first parametric equation.

$x^2 = t$ Square both sides to solve for t.

Step 3. Substitute the expression x^2 for t into the other equation.

$y = -\frac{1}{2}t + 1$ Start with the parametric equation describing y.

$y = -\frac{1}{2}(x^2) + 1$ Substitute x^2 for t.

Step 4. The equation $y = -\frac{1}{2}x^2 + 1$ is a **quadratic function** in standard form. The graph is a parabola that opens down having a vertex at the point $(0, 1)$.

Step 5. The complete graph of the rectangular equation $y = -\frac{1}{2}x^2 + 1$ having a domain of $-\infty < x < \infty$ can be seen in **Figure 39**. However, the graph in Figure 39 is **not** an accurate representation of the original parametric equations $x = \sqrt{t}, y = -\frac{1}{2}t + 1$.

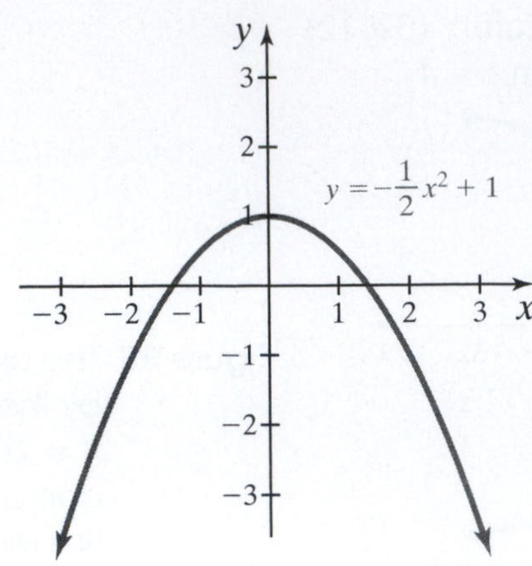

Figure 39 The graph of $y = -\frac{1}{2}x^2 + 1$ with domain $-\infty < x < \infty$ **is not** an accurate representation of the parametric equations $x = \sqrt{t}, y = -\frac{1}{2}t + 1$.

Note that in the parametric equation $x = \sqrt{t}$, the values of x will always be greater than or equal to 0. This forces a restriction on x in the rectangular equation $y = -\frac{1}{2}x^2 + 1$ to be $0 \le x < \infty$. Therefore, the rectangular equation representing the parametric equations $x = \sqrt{t}, y = -\frac{1}{2}t + 1$ is $y = -\frac{1}{2}x^2 + 1$, $0 \le x < \infty$. The plane curve that represents the parametric equations $x = \sqrt{t}, y = -\frac{1}{2}t + 1$ must be limited to **only** the right-hand portion of the graph of $y = -\frac{1}{2}x^2 + 1$. See Figure 40.

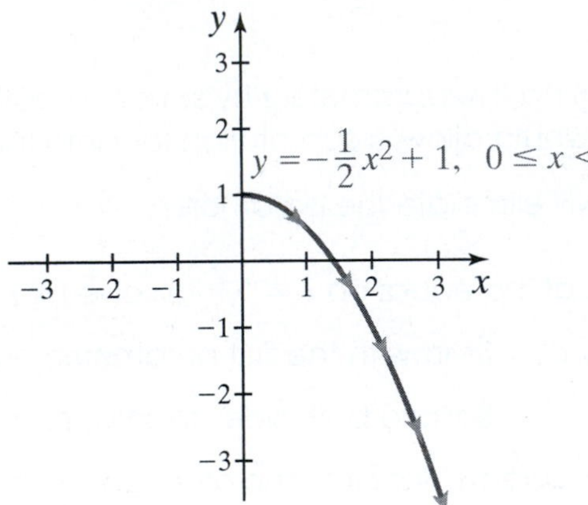

Figure 40 The plane curve described by the parametric equations $x = \sqrt{t}, y = -\frac{1}{2}t + 1$ can also be described by the rectangular equation $y = -\frac{1}{2}x^2 + 1, 0 \le x < \infty$.

You Try It Work through this You Try It problem.

Work Exercises 13–15 in this textbook or in the MyLab Math Study Plan.

In Examples 3 and 4, we chose one parametric equation and solved for the parameter, t. This allowed us to substitute the expression that represented t into the other equation, thus eliminating the parameter. Although this is often a convenient way to eliminate the parameter to determine the rectangular representation of parametric equations, it is not the only way. Sometimes the two given parametric equations are related to each other in some way. If we can recognize the relationship, then we might be able rewrite one of the parametric equations in terms of the other. This technique is especially useful when one of the parametric equations is a power of the other as in the following example.

Example 5 Determining a Rectangular Representation for a Curve Defined Parametrically

Determine the rectangular equation of the curve whose parametric equations are given by $x = t^6, y = t^2$, then sketch the plane curve.

Solution

Step 1. Observe that t^6 is equal to t^2 raised to the third power. This will allow us to make a substitution that will eliminate the parameter t in the first equation.

Step 2. Rewrite the parametric equation $x = t^6$ as $x = (t^2)^3$.

Step 3. Substitute y in place of t^2.

$x = (t^2)^3$ Start with the equation from Step 2.

$x = (y)^3$ Substitute y for t^2.

Step 4. We can solve the equation $x = y^3$ for y to find a more familiar form.

$x = y^3$ Start with the rectangular equation from Step 3.

$\sqrt[3]{x} = y$ Solve for y by taking the cube root of both sides.

Step 5. We see that the corresponding rectangular equation is $y = \sqrt[3]{x}$. The complete graph of the rectangular equation of $y = \sqrt[3]{x}$ having domain $-\infty < x < \infty$ is shown in **Figure 41**. However, the graph in Figure 41 does **not** accurately depict the graph of the plane curve given by the original parametric equations.

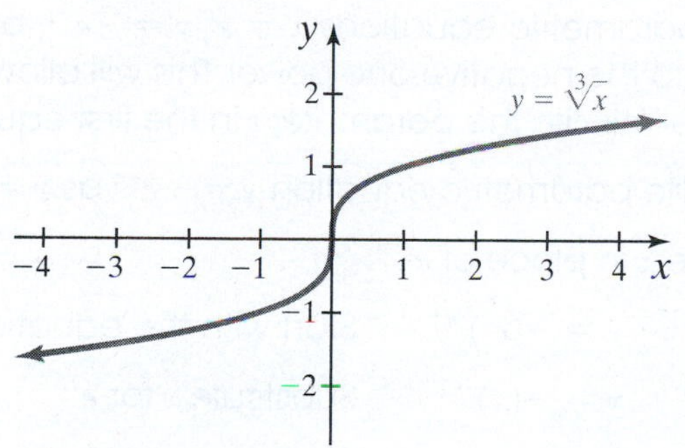

Figure 41 The graph of $y = \sqrt[3]{x}$ with domain $-\infty < x < \infty$ is **not** an accurate representation of the parametric equations $x = t^6, y = t^2$.

To determine which portion of the graph of $y = \sqrt[3]{x}$ will represent the plane curve described by the parametric equations, we examine the parametric equations. Note that for the parametric equation $x = t^6$, the values of x will always be greater than or equal to 0. This forces a restriction on x in the rectangular equation $y = \sqrt[3]{x}$ to be $0 \le x < \infty$. Therefore, the rectangular equation representing the parametric equations $x = t^6, y = t^2$ is $y = \sqrt[3]{x}, 0 \le x < \infty$.

Note that as the values of t approach negative infinity, the values of x and y both approach positive infinity. As t approaches 0, the graph approaches the origin and intersects the origin precisely at $t = 0$. The graph then traverses itself in the opposite direction as the values of t get larger. See Figure 42 or view this **table of values** to see some specific points that lie on the graph.

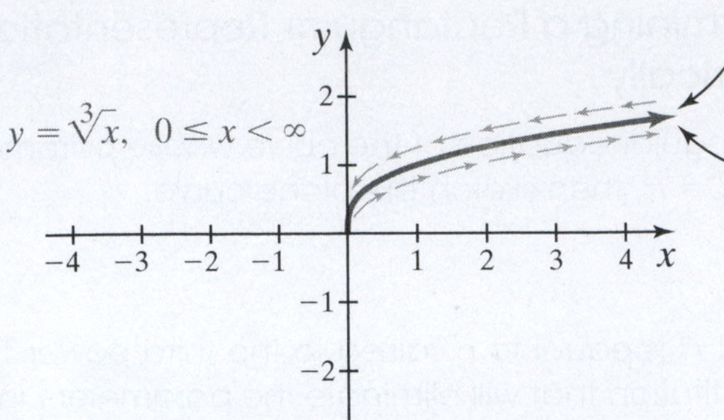
As t approaches negative infinity, the values of x and y approach positive infinity.
As t approaches positive infinity, the values of x and y approach positive infinity.
$y = \sqrt[3]{x},\ 0 \le x < \infty$
y
x

Figure 42 The plane curve described by the parametric equations $x = t^6, y = t^2$ can also be described by the rectangular equation $y = \sqrt[3]{x}, 0 \le x < \infty$.

You Try It Work through this You Try It problem.

Work Exercises 16–18 in this textbook or in the MyLab Math Study Plan.

Example 6 Determining a Rectangular Representation for a Curve Defined Parametrically

Determine the rectangular equation of the curve whose parametric equations are given by $x = e^t, y = -e^{-t}$, then sketch the plane curve.

Solution

Step 1. For the parametric equations $x = e^t, y = -e^{-t}$, observe that e^{-t} is equal to e^t raised to the negative one power. This will allow us to make a substitution that will eliminate the parameter t in the first equation.

Step 2. Rewrite the parametric equation $y = -e^{-t}$ as $y = -(e^t)^{-1}$.

Step 3. Substitute x in place of e^t.

$y = -(e^t)^{-1}$ Start with the equation from step 2.

$y = -(x)^{-1}$ Substitute x for e^t.

Step 4. Use an exponent property to write $y = -x^{-1}$ in a more familiar form.

$y = -x^{-1}$ Start with the rectangular equation from step 3.

$y = -\frac{1}{x}$ Write using the exponent property $b^{-m} = \frac{1}{b^m}$.

Step 5. We see that the corresponding rectangular equation is $y = -\frac{1}{x}$. The complete graph of $y = -\frac{1}{x}$ having domain $-\infty < x < 0$ or $0 < x < \infty$ is shown below in Figure 43. (Note that the graph of $y = -\frac{1}{x}$ is the graph of the reciprocal function, $y = \frac{1}{x}$, reflected about the x-axis.) Although the graph in Figure 43 is the complete graph of $y = -\frac{1}{x}$, it does **not** accurately depict the graph of the plane curve given by the original parametric equations.

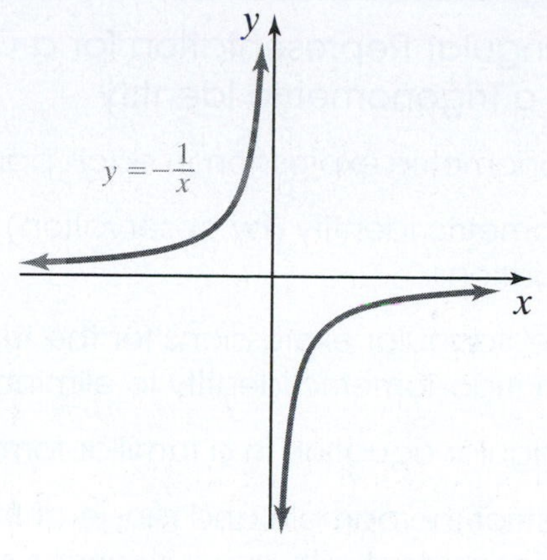

Figure 43 The graph of $y = -\frac{1}{x}$ with domain $-\infty < x < 0$ or $0 < x < \infty$ is **not** an accurate graph of the parametric equations $x = e^t, y = -e^{-t}$.

For the parametric equation $x = e^t$, the values of x are greater than zero. This forces a restriction on x in the rectangular equation $y = -\frac{1}{x}$ to be $0 < x < \infty$. Therefore, the rectangular equation representing the parametric equations $x = e^t, y = -e^{-t}$ is $y = -\frac{1}{x}, 0 < x < \infty$. Figure 44 shows the correct graph of the parametric equations $x = e^t, y = -e^{-t}$.

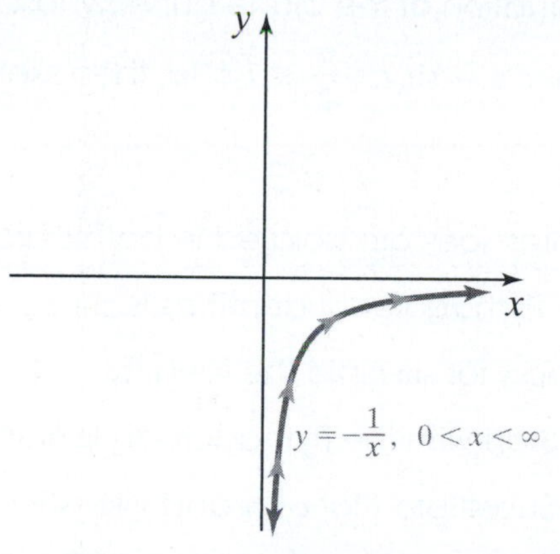

Figure 44 The plane curve described by the parametric equations $x = e^t, y = -e^{-t}$ can also be described by the rectangular equation $y = -\frac{1}{x}, 0 < x < \infty$.

You Try It Work through this You Try It problem.

Work Exercises 19–22 in this textbook or in the MyLab Math Study Plan.

OBJECTIVE 3 DETERMINING A RECTANGULAR REPRESENTATION FOR A CURVE DEFINED PARAMETRICALLY USING A TRIGONOMETRIC IDENTITY

In Examples 5 and 6 we eliminated the parameter by substituting an appropriate form of one of the equations into the other. When parametric equations are defined using trigonometric functions, this approach is not always possible. An alternate approach is to substitute both parametric equations into a known trigonometric identity to eliminate the parameter. It may be a good idea to review some of the **fundamental trigonometric identities** before proceeding. We now outline a five-step process for eliminating the parameter using trigonometric identities.

Determining a Rectangular Representation for a Curve Defined Parametrically Using a Trigonometric Identity

Step 1. Isolate the trigonometric expression in each parametric equation.

Step 2. Select a trigonometric identity (by observation) that relates the two parametric equations.

Step 3. Substitute the rectangular expressions for the trigonometric expressions into the chosen trigonometric identity to eliminate the parameter.

Step 4. Write the rectangular equation in a familiar form whenever possible.

Step 5. If necessary, restrict the domain and range of the rectangular equation to be consistent with any restrictions of the parameter or parametric equations.

We start with the same set of parametric equations that were used in **Example 2.**

Example 7 Determining a Rectangular Representation for a Curve Defined Parametrically Using a Trigonometric Identity

Determine the rectangular equation of the plane curve whose parametric equations are given by $x = \cos t, y = \sin t, -\frac{\pi}{2} \leq t \leq \pi$, then sketch the plane curve.

Solution

Step 1. The trigonometric expressions are isolated in both parametric equations.

Step 2. Recall that one of the **Pythagorean Identities** is $\sin^2 t + \cos^2 t = 1$.

Step 3. Substitute x for $\cos t$ and y for $\sin t$ into the identity.

$\sin^2 t + \cos^2 t = 1$ Start with the Pythagorean Identity $\sin^2 t + \cos^2 t = 1$.

$(y)^2 + (x)^2 = 1$ Substitute x for $\cos t$ and y for $\sin t$.

Step 4. The equation $(y)^2 + (x)^2 = 1$ can be rewritten as $x^2 + y^2 = 1$.

Step 5. The equation $x^2 + y^2 = 1$ is the equation of the unit circle. The complete graph of the unit circle is shown in Figure 45. However, the graph seen in Figure 45 is does **not** accurately depict the graph of the plane curve defined by the original parametric equations because we were given restrictions on the parameter.

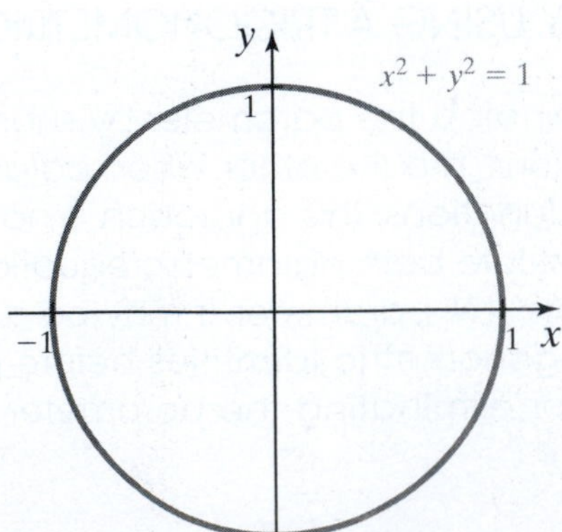

Figure 45 The complete graph of the unit circle is **not** an accurate graph of the parametric equations $x = \cos t, y = \sin t, -\frac{\pi}{2} \leq t \leq \pi$.

Recall that the parametric equations $x = \cos t, y = \sin t$ were defined only on the interval $-\frac{\pi}{2} \le t \le \pi$. Thus, there were restrictions imposed on the parameter which must be considered when sketching the final graph of the parametric equations. The graph the parametric equations will only be a portion of the graph of the unit circle of having an initial point at $\left(\cos\left(-\frac{\pi}{2}\right), \sin\left(-\frac{\pi}{2}\right)\right) = (0, -1)$ and a terminal point at $(\cos(\pi), \sin(\pi)) = (-1, 0)$. Therefore, the portion of the unit circle in Quadrant III must be excluded. See Figure 46.

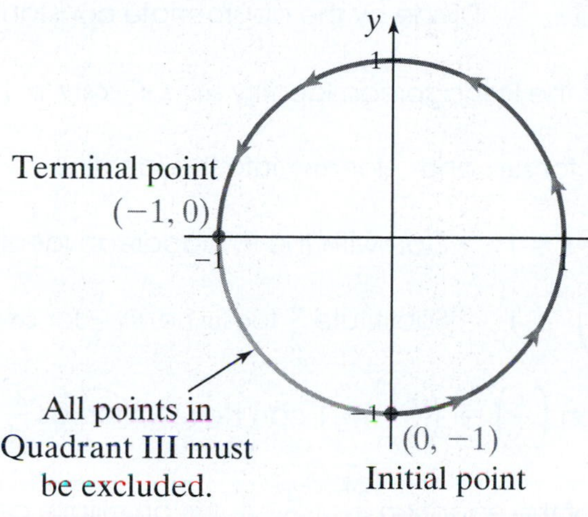

Figure 46 The plane curve described by the parametric equations $x = \cos t, y = \sin t, -\frac{\pi}{2} \le t \le \pi$ must exclude all points on the unit circle that are in Quadrant III.

The only portion of the unit circle in **Figure 46** that is not included is the set of points that lie in Quadrant III. These points have x- and y-coordinates that are both less than zero. Therefore, we must restrict the graph of $x^2 + y^2 = 1$ to include the cases where both x and y are greater than or equal to zero, where x is greater than or equal to zero and y is negative, and where y is greater than or equal to zero and x is negative. We represent this using the notation $x \ge 0$ or $y \ge 0$. Thus, the rectangular equation describing the parametric equations $x = \cos t, y = \sin t, -\frac{\pi}{2} \le t \le \pi$ is given by $x^2 + y^2 = 1, x \ge 0$ or $y \ge 0$. See Figure 47.

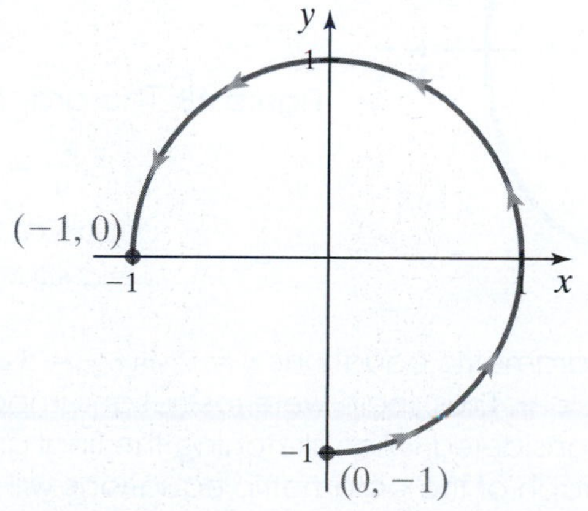

Figure 47 The plane curve described by the parametric equations $x = \cos t, y = \sin t, -\frac{\pi}{2} \le t \le \pi$ can also be described by the rectangular equation $x^2 + y^2 = 1, x \ge 0$ or $y \ge 0$.

Example 8 Finding a Rectangular Representation for a Curve Defined Parametrically Using a Trigonometric Identity

Determine the rectangular equation of the plane curve whose parametric equations are given by $x = 2\sin t, y = 3\cos t, 0 \le t \le \pi$, then sketch the plane curve.

Solution

Step 1. Isolate the trigonometric expressions in each parametric equation.

$x = 2\sin t$	Start with each parametric equation.	$y = 3\cos t$
$\frac{x}{2} = \sin t$	Divide by the appropriate constant.	$\frac{y}{3} = \cos t$

Step 2. We can use the Pythagorean Identity $\sin^2 t + \cos^2 t = 1$.

Step 3. Substitute $\frac{x}{2}$ for $\sin t$ and $\frac{y}{3}$ for $\cos t$ into the identity.

$\sin^2 t + \cos^2 t = 1$	Start with the Pythagorean Identity $\sin^2 t + \cos^2 t = 1$.
$\left(\frac{x}{2}\right)^2 + \left(\frac{y}{3}\right)^2 = 1$	Substitute $\frac{x}{2}$ for $\sin t$ and $\frac{y}{3}$ for $\cos t$.

Step 4. The equation $\left(\frac{x}{2}\right)^2 + \left(\frac{y}{3}\right)^2 = 1$ can be rewritten as $\frac{x^2}{4} + \frac{y^2}{9} = 1$.

Step 5. The graph of the equation $\frac{x^2}{4} + \frac{y^2}{9} = 1$ is an ellipse centered at the origin having a **vertical major axis.** The complete graph of this ellipse is shown in Figure 48. However, the graph seen in Figure 48 does **not** accurately depict the graph of the plane curve defined by the original parametric equations because we were given restrictions on the parameter.

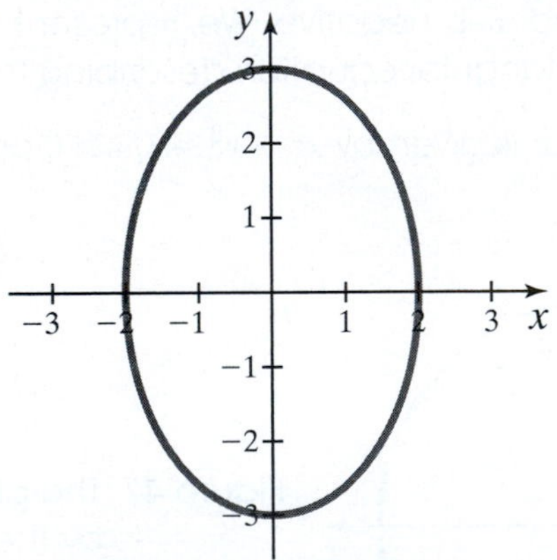

Figure 48 The graph of the ellipse $\frac{x^2}{4} + \frac{y^2}{9} = 1$ is not an accurate graph of the parametric equations $x = 2\sin t, y = 3\cos t, 0 \le t \le \pi$.

Recall that the parametric equations $x = 2\sin t, y = 3\cos t$ were defined only on the interval $0 \le t \le \pi$. Thus, there were restrictions imposed on the parameter which must be considered when sketching the final graph of the parametric equations. The graph of the parametric equations will only be a portion of the ellipse having an initial point at $(2\sin(0), 3\cos(0)) = (0, 3)$ and a terminal point at $(2\sin(\pi), 3\cos(\pi)) = (0, -3)$. See Figure 49. The rectangular equation describing the parametric equations $x = 2\sin t, y = 3\cos t, 0 \le t \le \pi$ is given by $\frac{x^2}{4} + \frac{y^2}{9} = 1, 0 \le x \le 2$.

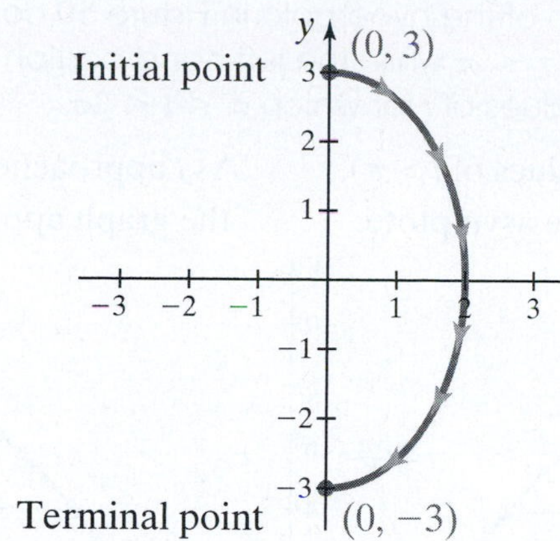

Figure 49 The plane curve described by the parametric equations $x = 2\sin t$, $y = 3\cos t$, $0 \le t \le \pi$ is a portion of the ellipse whose rectangular equation is given by $\frac{x^2}{4} + \frac{y^2}{9} = 1$, $0 \le x \le 2$. The initial point is at $(0, 3)$ and the terminal point is at $(0, -3)$.

Example 9 Finding a Rectangular Representation for a Curve Defined Parametrically Using a Trigonometric Identity

Determine the rectangular equation of the plane curve whose parametric equations are given by $x = 5\csc t$, $y = 4\cot t$, $0 < t < \pi$, $\pi < t < 2\pi$, then sketch the plane curve.

Solution To determine the rectangular equation that represents the parametric equations $x = 5\csc t$,$y = 4\cot t$, $0 < t < \pi$, $\pi < t < 2\pi$, we must use a trigonometric identity that relates the cosecant and cotangent functions. One identity that relates these two trigonometric functions is the Pythagorean identity $1 + \cot^2 t = \csc^2 t$. Now follow the five steps used for determining the rectangular equation for a curve defined parametrically using a trigonometric identity.

Step 1. Isolate the trigonometric expressions in each parametric equation.

$x = 5\csc t$	Start with each parametric equation.	$y = 4\cot t$
$\frac{x}{5} = \csc t$	Divide by the appropriate constant.	$\frac{y}{4} = \cot t$

Step 2. We can use the Pythagorean Identity $1 + \cot^2 t = \csc^2 t$.

Step 3. Substitute $\frac{x}{5}$ for $\csc t$ and $\frac{y}{4}$ for $\cot t$ into the identity.

$1 + \cot^2 t = \csc^2 t$	Start with the Pythagorean Identity $1 + \cot^2 t = \csc^2 t$.
$1 + \left(\frac{y}{4}\right)^2 = \left(\frac{x}{5}\right)^2$	Substitute for $\frac{x}{5}$ for $\csc t$ and $\frac{y}{4}$ for $\cot t$.

Step 4. The equation $1 + \left(\frac{y}{4}\right)^2 = \left(\frac{x}{5}\right)^2$ can be rewritten as $\frac{x^2}{25} - \frac{y^2}{16} = 1$.

Step 5. The rectangular equation $\frac{x^2}{25} - \frac{y^2}{16} = 1$ is a hyperbola centered at the origin having a **horizontal transverse axis**. The parameter t does not impose any restrictions on the variables x and y. The graph approaches asymptotes whose equations are $y = \frac{4}{5}x$ and$y = -\frac{4}{5}x$. See **Figure 50**. The graph will approach these asymptotes whenever the parametric equations $x = 5\csc t$ and $y = 4\cot t$ are undefined. These parametric equations are undefined for all values of t that are integer multiples of π. Several points are plotted in **Figure 50** along with the corresponding values of t to get a sense of the orientation of the plane curve. The right-hand portion of the hyperbola in **Figure 50** corresponds to values of t for which $0 < t < \pi$ while the left-hand portion of the hyperbola corresponds to values of t for which $\pi < t < 2\pi$.

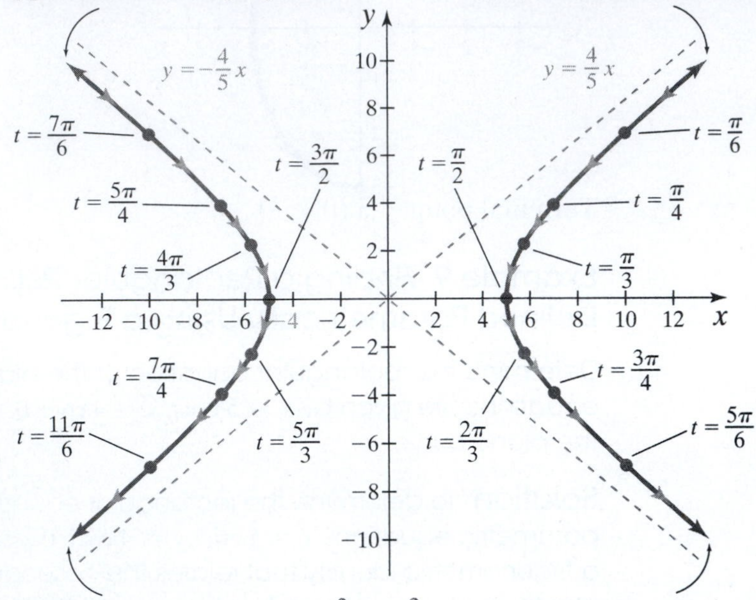

Figure 50 The hyperbola approaches an asymptote as the value of t approaches any integer multiple of π.

Thus, the rectangular equation describing the parametric equations $x = 5 \csc t, y = 4 \cot t, 0 < t < \pi, \pi < t < 2\pi$ is $\frac{x^2}{25} - \frac{y^2}{16} = 1$. See Figure 51.

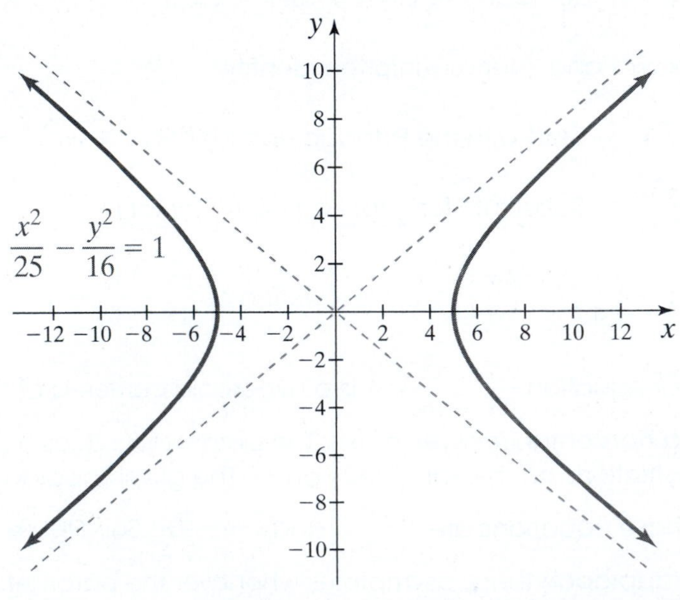

Figure 51 The plane curve described by the parametric equations $x = 5 \csc t$, $y = 4 \cot t, 0 < t < \pi, \pi < t < 2\pi$ can be described by the rectangular equation $\frac{x^2}{25} - \frac{y^2}{16} = 1$.

You Try It Work through this You Try It problem.

Work Exercises 23–28 in this textbook or in the MyLab Math Study Plan.

OBJECTIVE 4 DETERMINING A PARAMETRIC REPRESENTATION OF A PLANE CURVE THAT IS DEFINED BY A RECTANGULAR EQUATION

So far in this section we have found the rectangular equation of a plane curve defined parametrically, but how do we find a parametric representation for a curve that is defined using a rectangular equation? The fact is, there may be infinitely many pairs of parametric equations that describe the same plane curve. For example, consider the plane curve in Figure 52 having the rectangular equation $y = 1 - x^2, -1 \le x \le 1$. You should verify that three of the infinitely many pairs of parametric equations that can describe this plane curve are 1) $x = t, y = 1 - t^2, -1 \le t \le 1$; 2) $x = t^3, y = 1 - t^6, -1 \le t \le 1$; and 3) $x = \sin t$, $y = \cos^2 t$.

Rectangular Equation

$y = 1 - x^2, -1 \le x \le 1$

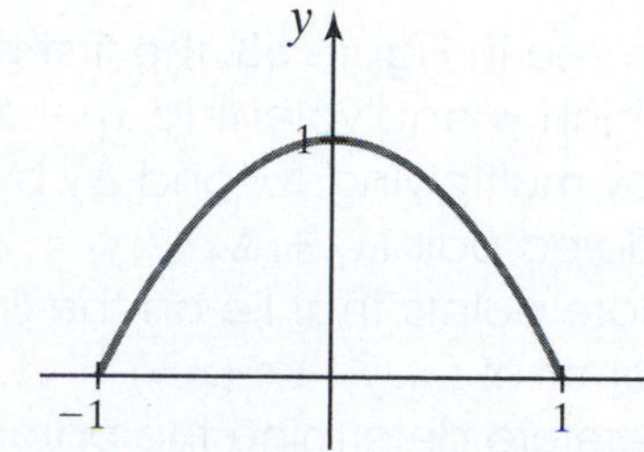

Parametric Equations

1. $x = t, y = 1 - t^2, -1 \le t \le 1$
2. $x = t^3, y = 1 - t^6, -1 \le t \le 1$
3. $x = \sin t, y = \cos^2 t$

Figure 52 The plane curve described by the rectangular equation $y = 1 - x^2, -1 \le x \le 1$ has infinitely parametric representations. Three such parametric representations of this plane curve are shown above.

As you can see from the plane curve in **Figure 52**, there are a variety of parametric representations of a curve that are defined by a rectangular equation. Therefore, there is no one method that can be used to determine parametric equations.

However, there are systematic ways that can be used to determine a parametric representation of many familiar plane curves such as lines, circles, and ellipses. We start by introducing a method for determining a parametric representation of a **linear equation in two variables.**

Suppose that a nonvertical line has slope $m = \dfrac{\text{change in } y}{\text{change in } x}$ and passes through a point (x_1, y_1) on the line. The slope $m = \dfrac{\text{change in } y}{\text{change in } x}$ can be rewritten using the notation $m = \dfrac{\Delta y}{\Delta x}$ where Δy means "change in y" and Δx means "change in x." (This notation is often used in calculus.)

If we start at the given point on the line (x_1, y_1), then we can determine the coordinates of another point on the line by traveling up (or down) Δy units then moving right (or left) Δx units. The coordinates of this point are $(x_1 + \Delta x \cdot 1, y_1 + \Delta y \cdot 1)$. We can continue this process again and again to find the coordinates of infinitely many more points on the line. Notice in **Figure 53** below that this process was used to determine the coordinates of four points that lie along a line with slope $m = \dfrac{\Delta y}{\Delta x}$ that passes through the given point (x_1, y_1).

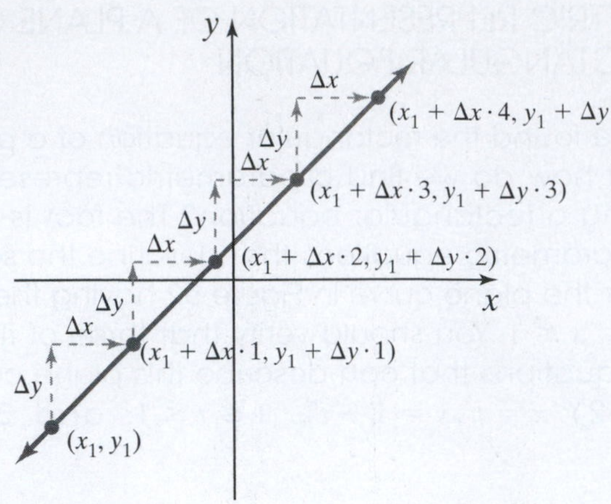

Figure 53 Infinitely many points on a nonvertical line can be determined using a given point (x_1, y_1) and the slope $m = \frac{\Delta y}{\Delta x}$.

As you can see in Figure 53, the first point was obtained by adding Δx to x_1 and Δy to y_1, which is equivalent to $x_1 + \Delta x \cdot 1$ and $y_1 + \Delta y \cdot 1$. The second point was obtained by multiplying Δx and Δy by 2 and adding those values to x_1 and y_1 to get the ordered pair $(x_1 + \Delta x \cdot 2, y_1 + \Delta y \cdot 2)$. We can continue this process to find infinitely more points that lie on the line. Thus, any point (x, y) on the line can be written in terms of (x_1, y_1) as $(x, y) = (x_1 + (\Delta x)t, y_1 + (\Delta y)t)$ for some real number t. We can therefore determine the parametric equations for any nonvertical line as follows.

> **A Parametric Representation of a Nonvertical Line**
>
> If a nonvertical line has slope $m = \frac{\Delta y}{\Delta x}$ and passes through the point (x_1, y_1), then a parametric representation of this line is
>
> $$x = x_1 + (\Delta x)t, y = y_1 + (\Delta y)t.$$

Note In the formula for the parametric equations above, any values of Δy and Δx can be used provided that the ratio $\frac{\Delta y}{\Delta x}$ is equal to the slope of the line. For example, if $m = -\frac{2}{3}$, then the equations below are all possible parametric representations for the line with slope $m = -\frac{2}{3}$ that passes through the point (x_1, y_1).

$$x = x_1 + 3t, y = y_1 - 2t \qquad \left(\frac{\Delta y}{\Delta x} = \frac{-2}{3} = -\frac{2}{3}\right)$$

$$x = x_1 - 3t, y = y_1 + 2t \qquad \left(\frac{\Delta y}{\Delta x} = \frac{2}{-3} = -\frac{2}{3}\right)$$

$$x = x_1 + 6t, y = y_1 - 4t \qquad \left(\frac{\Delta y}{\Delta x} = \frac{-4}{6} = -\frac{2}{3}\right)$$

$$x = x_1 - 9t, y = y_1 + 6t \qquad \left(\frac{\Delta y}{\Delta x} = \frac{6}{-9} = -\frac{2}{3}\right)$$

If a certain orientation of the line is required, then careful consideration must be given when choosing the values of Δy and Δx.

Example 10 Determining a Parametric Representation of a Line Defined by a Rectangular Equation

Determine a parametric representation of the line $y = -\frac{2}{3}x + 5$.

Solution The line $y = -\frac{2}{3}x + 5$ is in **slope-intercept form**. The slope is $m = \frac{\Delta y}{\Delta x} = -\frac{2}{3}$. We were not given a specific orientation for the graph. Thus, the choice of Δx and Δy does not matter provided that the ratio $\frac{\Delta y}{\Delta x} = -\frac{2}{3}$. For simplicity, we choose $\Delta y = -2$ and $\Delta x = 3$. The y-intercept is 5 so the line passes through the point $(0, 5)$. Let $(x_1, y_1) = (0, 5)$ where $x_1 = 0$ and $y_1 = 5$.

We use the equation $x = x_1 + (\Delta x)t$ with $x_1 = 0$ and $\Delta x = 3$ to get a parametric equation describing x.

$x = x_1 + (\Delta x)t$	Start with the equation describing a parametric equation for x.
$x = 0 + (3)t$	Substitute 0 for x_1 and 3 for Δx.
$x = 3t$	Simplify.

Now use the equation $y = y_1 + (\Delta y)t$ with $y_1 = 5$ and $\Delta y = -2$ to find a parametric equation describing y.

$y = y_1 + (\Delta y)t$	Start with the equation describing a parametric equation for y.
$y = 5 + (-2)t$	Substitute 5 for y_1 and -2 for Δy.
$y = 5 - 2t$	Simplify.

Therefore, a parametric representation of the line $y = -\frac{2}{3}x + 5$ is $x = 3t, y = 5 - 2t$.

Note this is only one of infinitely many parametric representations of this line. See Figure 54.

Rectangular Equation

$y = -\frac{2}{3}x + 5$

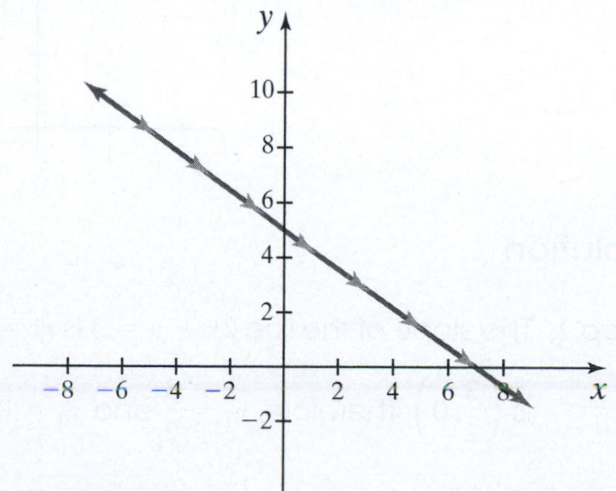

A Parametric Equations

$x = 3t, y = 5 - 2t$

Figure 54

When a line has an initial point (x_0, y_0) and a terminal point (x_T, y_T), we must impose a restriction on the parameter t. The simplest restriction is $0 \leq t \leq T$, where T is the value of the parameter corresponding to the terminal point.

We can follow the four steps outlined below to determine a parametric representation of a nonvertical line that has an initial point and a terminal point.

Determining a Parametric Representation of a Nonvertical Line that has an Initial Point and Terminal Point

Step 1. Determine the slope of the line $m = \frac{\Delta y}{\Delta x}$ and identify x_0 and y_0, where (x_0, y_0) is the initial point.

Step 2. Determine if Δy and Δx should be positive or negative depending of the orientation of line.

Step 3. Determine parametric equations of the form $x = x_0 + (\Delta x)t, y = y_0 + (\Delta y)t.$

Step 4. Restrict the parameter using the form $0 \le t \le T$, where T is the value of the parameter corresponding to the terminal point.

Example 11 Determining a Parametric Representation of a Line Defined by a Rectangular Equation

Determine a parametric representation for the linear equation $2x + y = 3, -1 \le x \le \frac{3}{2}$ such that initial point is $\left(\frac{3}{2}, 0\right)$ and the terminal point is $(-1, 5)$. See Figure 55.

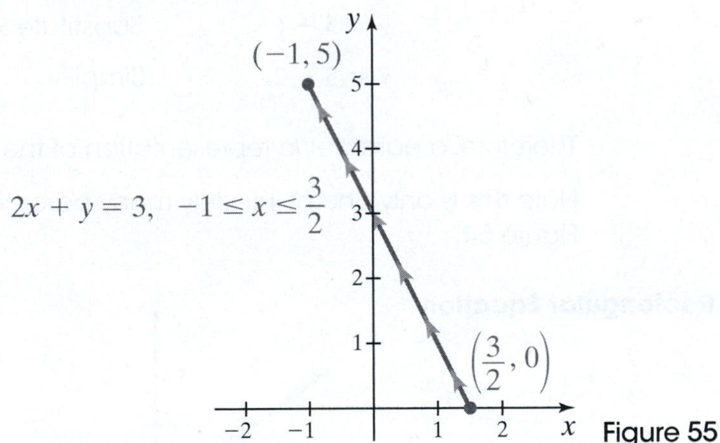

Figure 55

Solution

Step 1. The slope of the line $2x + y = 3$ is $m = -2 = \frac{\Delta y}{\Delta x} = \frac{-2}{1} = \frac{2}{-1}$. The initial point is $\left(\frac{3}{2}, 0\right)$. Therefore, $x_0 = \frac{3}{2}$ and $y_0 = 0.$

Step 2. The orientation is from right to left because the initial point is $\left(\frac{3}{2}, 0\right)$ and the terminal point is $(-1, 5)$. Thus, the values of x are decreasing and the values of y are increasing. Therefore we use the values of $\Delta x = -1$ and $\Delta y = 2$.

Step 3. We use the equation $x = x_0 + (\Delta x)t$ with $x_0 = \frac{3}{2}$ and $\Delta x = -1$ to get a parametric equation describing x.

$x = x_0 + (\Delta x)t$ Start with the equation describing a parametric equation for x.

$x = \frac{3}{2} + (-1)t$ Substitute $\frac{3}{2}$ for x_0 and -1 for Δx.

$x = \frac{3}{2} - t$ Simplify.

Now use the equation $y = y_0 + (\Delta y)t$ with $y_0 = 0$ and $\Delta y = 2$ to find a parametric equation describing y.

$y = y_0 + (\Delta y)t$ Start with the equation describing a parametric equation for y.

$y = 0 + (2)t$ Substitute 0 for y_0 and 2 for Δy.

$y = 2t$ Simplify.

Step 4. We now need to determine the value of T for the restriction $0 \le t \le T$. The value of T corresponds to the terminal point $(-1, 5)$. Use one of the parametric equations to solve for T. Here we choose $y = 2t$ because of its simplicity.

$y = 2t$ Start with one of the parametric equations.

$5 = 2T$ Substitute the y-coordinate of the terminal point for y and T for t.

$\frac{5}{2} = T$ Divide by 5 to solve for T.

Therefore, the terminal point corresponds to $T = \frac{5}{2}$ so the restriction of the parameter is $0 \le t \le \frac{5}{2}$. Thus, a parametric representation of the given line is $x = \frac{3}{2} - t, y = 2t, 0 \le t \le \frac{5}{2}$. See Figure 56.

Rectangular Equation

$2x + y = 3, -1 \le x \le \frac{3}{2}$

A Parametric Equations

$x = \frac{3}{2} - t, y = 2t, 0 \le t \le \frac{5}{2}$

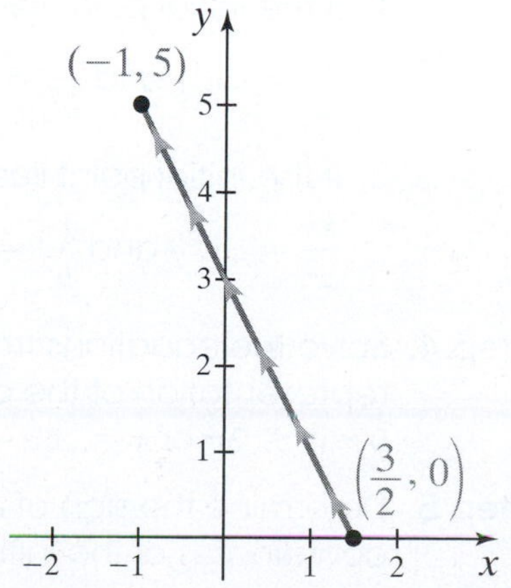

Figure 56

You Try It Work through this You Try It problem.

Work Exercises 29–32; 35 and 36 in this textbook or in the MyLab Math Study Plan.

We found a systematic way to determine a parametric representation of a non-vertical line. We can also systematically determine a parametric representation of other familiar plane curves such as circles and ellipses that are centered at the origin. To determine a parametric representation of these conic sections, we use the Pythagorean identity $\sin^2 t + \cos^2 t = 1$. This is because the rectangular equation of every circle or ellipse centered at the origin can be written as $\frac{x^2}{a^2} + \frac{y^2}{b^2} = 1$. If we can write the circle or ellipse in this form, then we can use the substitution $\frac{x^2}{a^2} = \sin^2 t$ and $\frac{y^2}{b^2} = \cos^2 t$ (or $\frac{x^2}{a^2} = \cos^2 t$ and $\frac{y^2}{b^2} = \sin^2 t$) to determine a parametric representation. The choice of this substitution depends on the location of the initial point. If the initial point lies along the y-axis, then we use the substitution $\frac{x^2}{a^2} = \sin^2 t$ and $\frac{y^2}{b^2} = \cos^2 t$ because the value of x must be 0 when $t = 0$. If the initial point lies along the x-axis, then we use the substitution $\frac{x^2}{a^2} = \cos^2 t$ and $\frac{y^2}{b^2} = \sin^2 t$ because the value of y must be 0 when $t = 0$. Note that if $a^2 = b^2$, then the equation is a circle.

We now outline a five-step procedure for determining a parametric representation of a circle or ellipse that is centered at the origin having an initial point that lies along an axis.

Determining a Parametric Representation of a Circle or Ellipse Centered at the Origin

Step 1. Sketch the graph of the circle or ellipse and plot the four points that lie along the x-axis and the y-axis. Also, indicate the proper given orientation of the plane curve.

Step 2. Write the equation of the circle or ellipse in the form $\frac{x^2}{a^2} + \frac{y^2}{b^2} = 1$. Note that if the equation is a circle, then $a^2 = b^2$.

Step 3. Use the Pythagorean Identity $\sin^2 t + \cos^2 t = 1$ to make one of the following substitutions:

1. If the initial point lies along the y-axis, then use the substitution $\frac{x^2}{a^2} = \sin^2 t$ and $\frac{y^2}{b^2} = \cos^2 t$.
2. If the initial point lies along the x-axis, then use the substitution $\frac{x^2}{a^2} = \cos^2 t$ and $\frac{y^2}{b^2} = \sin^2 t$.

Step 4. Solve the equations from **Step 3** for x and y to see that a parametric representation of the circle or ellipse is $x = \pm a \cos t, y = \pm b \sin t$, $0 \le t \le 2\pi$ or $x = \pm a \sin t, y = \pm b \cos t, 0 \le t \le 2\pi$.

Step 5. Determine the sign of a and b. The sign of a and b depend on the coordinates of the initial point and the orientation of the plane curve.

Example 12 Determining a Parametric Representation of a Circle Defined by a Rectangular Equation

Determine a parametric representation for the circle $x^2 + y^2 = 16$ such that the initial point is at $(0, -4)$ and the orientation is clockwise.

Solution

Step 1. The equation $x^2 + y^2 = 16$ represents a circle centered at the origin with a radius of 4 units. The graph of the circle is seen in **Figure 57**. Note that the initial point and terminal point is $(0, -4)$. We also plot the points $(-4, 0)$, $(0, 4)$, and $(4, 0)$. The orientation is clockwise.

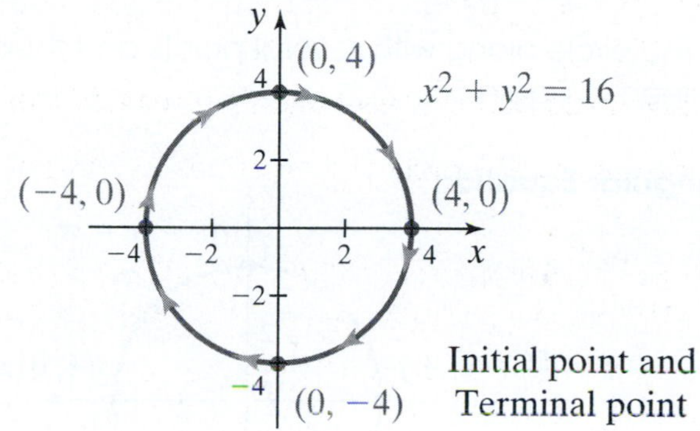

Figure 57

Step 2. We can divide both sides of the equation $x^2 + y^2 = 16$ by 16 to rewrite the equation as $\frac{x^2}{16} + \frac{y^2}{16} = 1$.

Step 3. The initial point, $(0, -4)$, lies along the y-axis and has a value of $x = 0$ when $t = 0$. Therefore, let $\frac{x^2}{16} = \sin^2 t$ and $\frac{y^2}{16} = \cos^2 t$. Using this substitution we see that the equation $\frac{x^2}{16} + \frac{y^2}{16} = 1$ is equivalent to the Pythagorean Identity $\sin^2 t + \cos^2 t = 1$.

Step 4. Solve the equations $\left(\frac{x^2}{16}\right) = \sin^2 t$ and $\left(\frac{y^2}{16}\right) = \cos^2 t$ for x and y, respectively.

$\left(\frac{x^2}{16}\right) = \sin^2 t$	Start with the two equations obtained from Step 3.	$\frac{y^2}{16} = \cos^2 t$
$x^2 = 16\sin^2 t$	Multiply both sides of both equations by 16.	$y^2 = 16\cos^2 t$
$x = \pm 4\sin t$	Solve by taking square roots of both sides.	$y = \pm 4\cos t$

We see that $x = \pm 4\sin t$ and $y = \pm 4\cos t$. Therefore, we get the following four possibilities for the parametric representation of the given circle.

1. $x = 4\sin t, y = 4\cos t, 0 \le t \le 2\pi$
2. $x = 4\sin t, y = -4\cos t, 0 \le t \le 2\pi$
3. $x = -4\sin t, y = 4\cos t, 0 \le t \le 2\pi$
4. $x = -4\sin t, y = -4\cos t, 0 \le t \le 2\pi$

Step 5. To determine which of the four parametric representations is correct, we must look at the initial point and we must consider the orientation.

First, the y-coordinate of the initial point is -4 which is a negative number. Therefore, the parametric equation representing y must be of the form $y = -4\cos t$ because when $t = 0$, $y = -4\cos 0 = -4(1) = -4$.

The orientation of the curve must be clockwise. The initial point has an x-coordinate of 0 and the circle has a clockwise orientation. This means that as the values of t start to increase, then the values of x must first start to be negative. Therefore, we must choose $x = -4\sin t$ as the parametric equation describing x. Thus, the parametric equations that describe the circle $x^2 + y^2 = 16$ having a clockwise orientation and an initial point at $(0, -4)$ are $x = -4\sin t$, $y = -4\cos t$, $0 \le t \le 2\pi$.

The circle along with several points and their corresponding values of t can be seen in Figure 58.

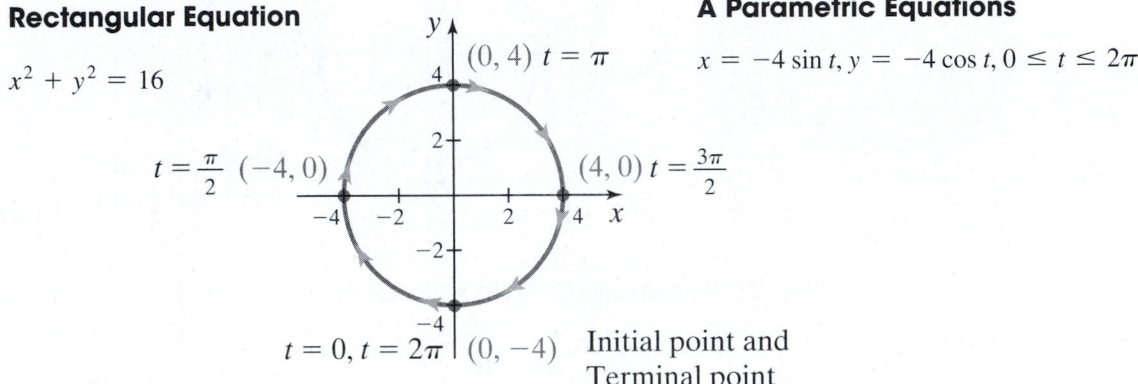

Figure 58 A parametric representation of the circle having the rectangular equation $x^2 + y^2 = 16$ is $x = -4\sin t$, $y = -4\cos t$, $0 \le t \le 2\pi$. This parametric representation has a clockwise orientation and an initial point at $(0, -4)$.

You Try It Work through this You Try It problem.

Work Exercises 33 and 34; 37–40 in this textbook or in the MyLab Math Study Plan.

OBJECTIVE 5 USING PARAMETRIC EQUATIONS TO MODEL THE MOTION OF AN OBJECT

Parametric equations are especially useful when describing the motion of an object. For example, suppose that you are on a pier that is h feet above the water and you decide to kick a ball into the water. Suppose that the ball is kicked with an initial velocity of v_0 at an angle θ with the horizontal. At any time t, the ball will have traveled some horizontal distance, $x = x(t)$, from the pier and will also have traveled a vertical distance, $y = y(t)$, from the top of the pier. See **Figure 59**.

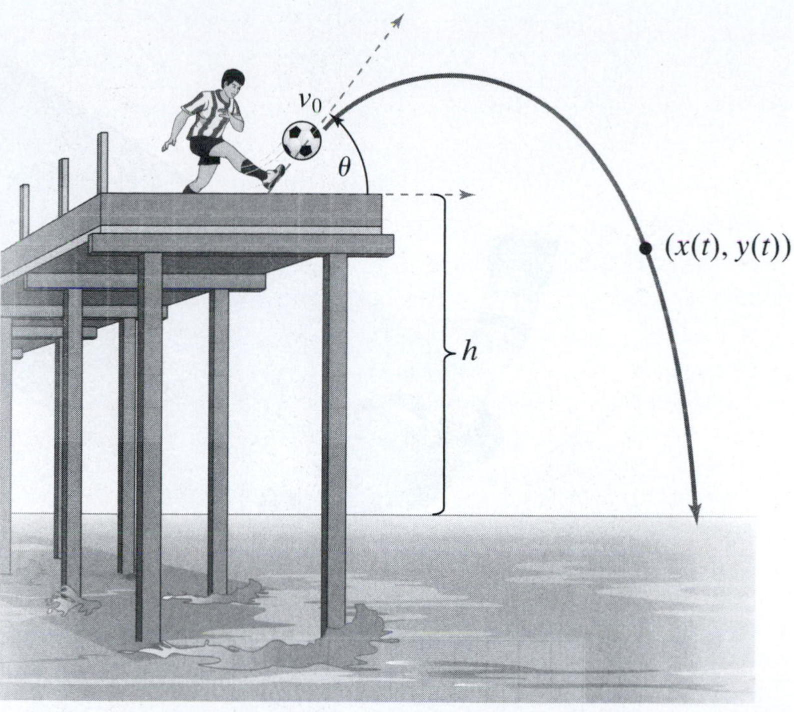

Figure 59

Assuming no outside factors (such as wind resistance) except the constant force of gravity, g, then it can be shown using calculus that we can describe the horizontal distance of the ball x and the vertical height of the ball y by the following parametric equations.

Parametric Equations Describing the Motion of an Object

If an object having an initial height, h, is propelled upward with an initial velocity of v_0 at an angle of θ with the horizontal, then

$$x = (v_0 \cos \theta)t \quad \text{and} \quad y = h + (v_0 \sin \theta)t - \frac{1}{2}gt^2,$$

where g is the constant acceleration due to gravity. Note that g is approximately 32 ft/sec/sec or 9.8 m/sec/sec.

Example 13 Using Parametric Equations to Model the Motion of an Object

A football player kicks a football with an initial velocity of 70 ft/sec at an angle of $\theta = 35°$ with the ground. Assuming no wind resistance, answer the following.

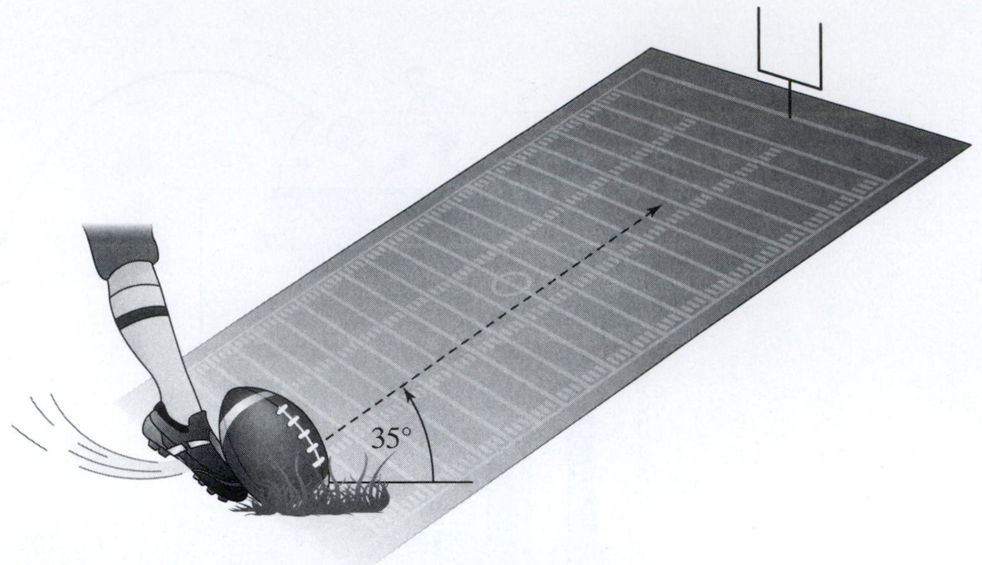

a. Determine the parametric equation describing the horizontal distance that the football travels from the point at which the ball was kicked.

b. Determine the parametric equation describing the vertical height of the football.

c. What is the horizontal distance that the ball has traveled at $t = 1.5$ sec? (Round to the nearest foot.)

d. What is the vertical height of the ball has traveled at $t = 1.5$ sec? (Round to the nearest foot.)

e. Assuming that the kicker accurately kicked the ball directly at the center of the 10 ft high crossbar between two goal posts that is 42 yards away, will the ball go over the crossbar?

Solution

a. Start with the parametric equation $x = (v_0 \cos \theta)t$ and substitute the appropriate values.

$x = (v_0 \cos \theta)t$	Start with the parametric equation describing the horizontal distance.
$x = (70 \cos 35°)t$	Substitute 70 for v_0 and 35° for θ.

The parametric equation describing the horizontal distance of the football is $x = (70 \cos 35°)t$.

b. Start with the parametric equation $y = h + (v_0 \sin \theta)t - \frac{1}{2}gt^2$ and substitute the appropriate values.

$y = h + (v_0 \sin \theta)t - \frac{1}{2}gt^2$	Start with the parametric equation describing the vertical height.
$y = (70 \sin 35°)t - \frac{1}{2}(32)t^2$	Substitute 70 for v_0, 35° for θ, and 32 for g.
$y = (70 \sin 35°)t - 16t^2$	Simplify

The parametric equation describing the vertical height of the football is $y = (70 \sin 35°)t - 16t^2$.

c. To determine the horizontal distance that the football has traveled when $t = 1.5$ sec, start with the parametric equation found in part (a) and substitute $t = 1.5$.

$x = (70\cos 35°)t$	Start with the parametric equation from part (a) that describes the horizontal distance.
$x = (70\cos 35°)(1.5)$	Substitute 1.5 for t.
$x \approx 86$ feet	Simplify and round to the nearest foot.

Therefore, the horizontal distance traveled by the football when $t = 1.5$ seconds is approximately 86 feet.

d. To determine the vertical height of the football when $t = 1.5$ sec, start with the parametric equation found in part (b) and substitute 1.5 for t.

$y = (70\sin 35°)t - 16t^2$	Start with the parametric equation from part (b) that describes the vertical distance.
$y = (70\sin 35°)(1.5) - 16(1.5)^2$	Substitute 1.5 for t.
$y \approx 24$ feet	Simplify and round to the nearest foot.

Therefore, the vertical height of the football when $t = 1.5$ seconds is approximately 24 feet.

e. To determine if the football goes over the 10-ft-high cross bar that is 42 yards away, first determine the time that it will take for the football to travel $42 \text{ yards} = 126$ feet.

$x = (70\cos 35°)t$	Start with the parametric equation describing the horizontal distance.
$126 = (70\cos 35°)t$	Set the horizontal distance equal to 126 feet (42 yards).
$\dfrac{126}{70\cos 35°} = t$	Solve for t.

At $t = \dfrac{126}{70\cos 35°} \approx 2.1974$ sec, the football will have traveled a horizontal distance of 42 yards. Therefore, the football has traveled the necessary horizontal distance to possibly go over the crossbar. Now we must determine if the height of the ball is more than 10 feet when $t = \dfrac{126}{70\cos 35°} \approx 2.1974$ sec.

Evaluate the parametric equation that describes the vertical height of the ball at $t = \dfrac{126}{70\cos 35°}$.

$y = (70\sin 35°)t - 16t^2$	Start with the parametric equation describing the vertical height.
$y = (70\sin 35°)\left(\dfrac{126}{70\cos 35°}\right) - 16\left(\dfrac{126}{70\cos 35°}\right)^2$	Evaluate y when $t = \dfrac{126}{70\cos 35°}$.
$y \approx 10.9693$ feet	Simplify.

We see that at the instant in time when the ball has traveled a horizontal distance of 42 yards, the height of the ball is greater than 10 feet. Therefore, the ball will go over the crossbar.

You Try It Work through this You Try It problem.

Work Exercises 41–43 in this textbook or in the MyLab Math Study Plan.

11.5 Exercises

In Exercises 1–8, determine the initial point and terminal point of the given parametric equations. Determine several other ordered pairs, then sketch the plane curve.

SbS 1. $x = -2 + 3t, y = -2t, 0 \le t \le 4$

SbS 2. $x = -\sin t, y = 3\sin t, -\frac{\pi}{2} \le t \le \frac{\pi}{2}$

SbS 3. $x = 1 + t^2, y = 2t, -3 \le t \le 2$

SbS 4. $x = 1 + 3t, y = 2 - t^2, -2 \le t \le 3$

SbS 5. $x = 2\cos t, y = 2\sin t, -\frac{\pi}{2} \le t \le \pi$

SbS 6. $x = 3\sin t, y = 3\cos t, 0 \le t \le \frac{3\pi}{2}$

SbS 7. $x = 3\cos t, y = 4\sin t, -\pi \le t \le \frac{\pi}{2}$

SbS 8. $x = 2\sin t, y = 5\cos t, -\frac{\pi}{4} \le t \le \pi$

In Exercises 9–28, determine the rectangular equation of the curve whose parametric equations are given, then sketch the plane curve.

SbS 9. $x = 5t, y = 2 - t, -1 \le t \le 1$

SbS 10. $x = 1 - 3t, y = 2t + 1, -1 \le t \le 2$

SbS 11. $x = 3t^2, y = -2t, -2 \le t \le 2$

SbS 12. $x = 4t, y = -32t^2, -3 \le t \le 3$

SbS 13. $x = \sqrt{t}, y = 1 - 2t$

SbS 14. $x = 2 - t, y = \sqrt{t},$

SbS 15. $x = \sqrt{t} + 1, y = \sqrt{t} - 3$

SbS 16. $x = -t^2, y = t^4$

SbS 17. $x = t^{\frac{3}{5}}, y = \sqrt[5]{t}$

SbS 18. $x = t^4, y = t^{-4}$

SbS 19. $x = e^t, y = -e^{2t}$

SbS 20. $x = -3e^{4t}, y = e^{4t}$

SbS 21. $x = -e^{-3t}, y = e^{3t}$

SbS 22. $x = \log_5 x^2, y = -\log_5 x$

SbS 23. $x = -\sin t, y = \csc t, 0 < t \le \frac{\pi}{2}$

SbS 24. $x = 3\sin t, y = 3\cos t, \pi \le t \le 2\pi$

SbS 25. $x = 2\sin t, y = 4\cos t, 0 \le t \le \pi$

SbS 26. $x = \sin t, y = 1 - \cos^2 t, 0 \le t \le \frac{\pi}{2}$

SbS 27. $x = \cos^2 t, y = \sin^2 t$

SbS 28. $x = 2\cot t, y = 3\csc t, 0 < t < \pi$

29. Determine a parametric representation of the line $y = 2x - 1$

30. Determine a parametric representation of the line $2x - 4y = 5$

SbS 31. Determine a parametric representation for the linear equation $2x + 5y = 5, -10 \leq x \leq 5$ such that initial point is $(-10, 5)$ the terminal point is $(5, -1)$.

SbS 32. Determine a parametric representation for the linear equation $3x - 2y = 4, -2 \leq x \leq 2$ such that initial point is $(2, 1)$ and the terminal point is $(-2, -5)$.

SbS 33. Determine a parametric representation for the circle $x^2 + y^2 = 25$ such that $(-5, 0)$ is the initial point and the orientation is counterclockwise.

SbS 34. Determine a parametric representation for the ellipse $\frac{x^2}{9} + \frac{y^2}{16} = 1$ such that $(0, 4)$ is the initial point and the orientation is clockwise.

For Exercises 35–40, determine a parametric representation for the given plane curve.

SbS 35.

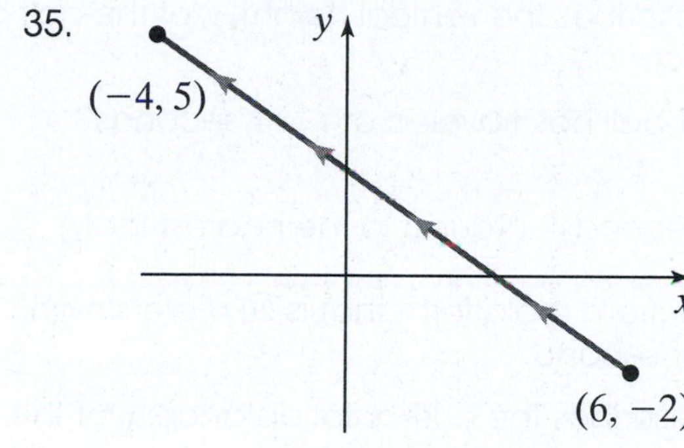

SbS 36.

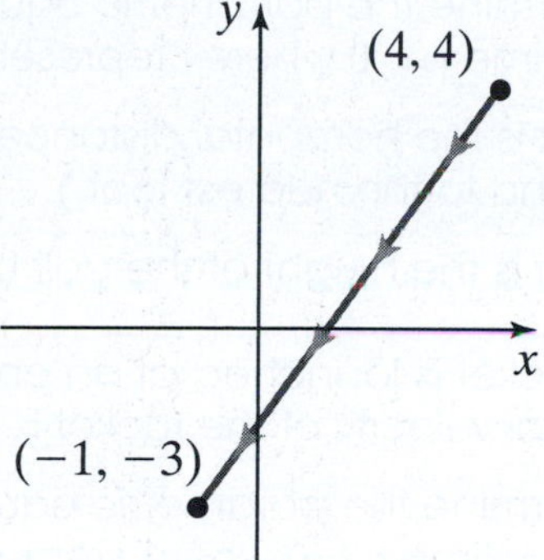

SbS 37.

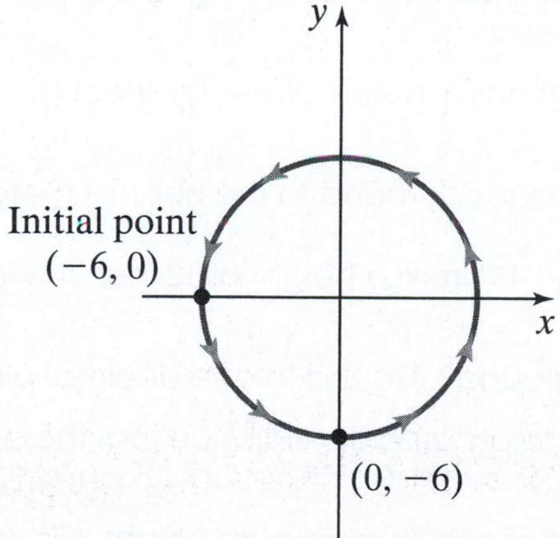

SbS 38.

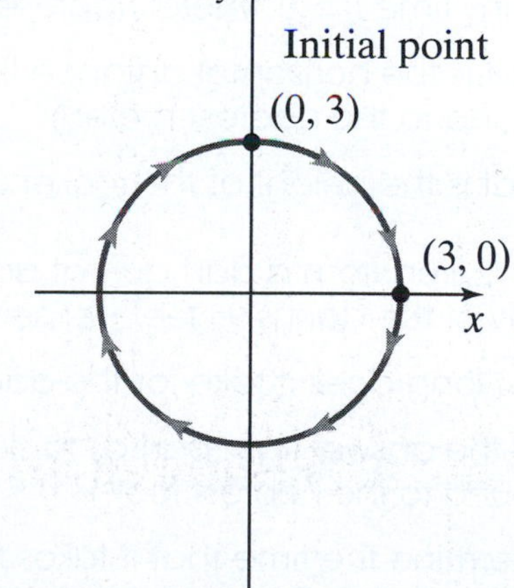

SbS 39.

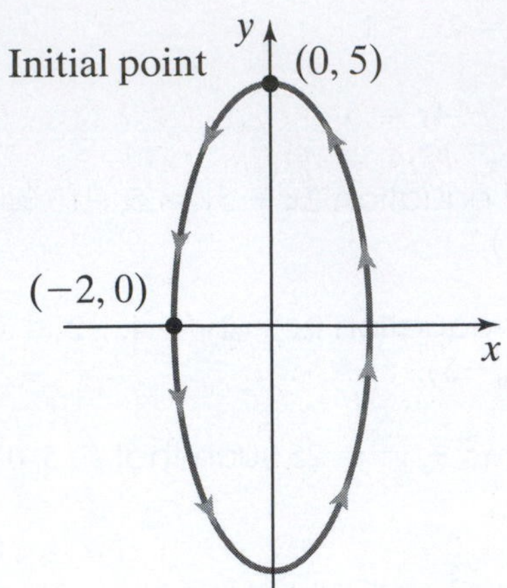

SbS 40.

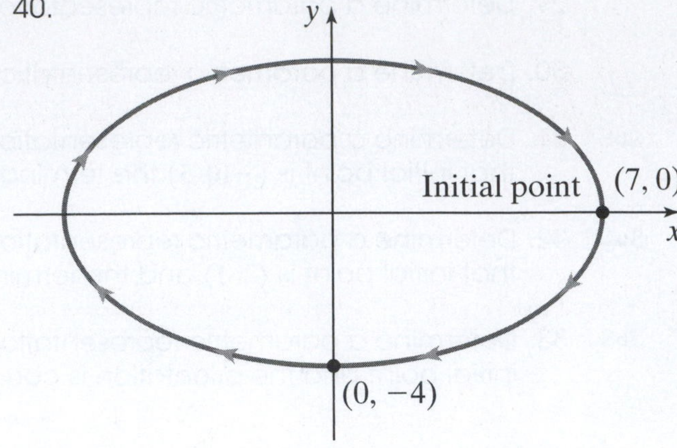

SbS 41. A golf ball is hit with an initial velocity of 120 feet per second at an angle of $\theta = 30°$.

a. Determine the parametric equation that describes the horizontal distance, x, of the golf ball at any time $t \geq 0$ where t represents time in seconds.

b. Determine the parametric equation that describes the vertical height, y, of the golf ball at any time $t \geq 0$ where t represents time in seconds.

c. What is the horizontal distance that the golf ball has traveled at $t = 2$ seconds? (Round to the nearest foot.)

d. What is the height of the golf ball at $t = 2$ seconds? (Round to the nearest foot.)

SbS 42. A toy rocket is launched at an angle of $\theta = 45°$ from a platform that is 20 meters high. The initial velocity of the rocket is 58 meters per second.

a. Determine the parametric equation that describes the horizontal distance, x, of the rocket at any time $t \geq 0$ where t represents time in seconds.

b. Determine the parametric equation that describes the vertical height, y, of the rocket at any time $t \geq 0$ where t represents time in seconds.

c. What is the horizontal distance that the rocket has traveled at $t = 1.5$ seconds? (Round to the nearest meter.)

d. What is the height of the rocket at $t = 1.5$ seconds? (Round to the nearest meter.)

SbS 43. A dart is fired from a dart gun at an angle of $\theta = 45°$ from a height of 12 feet. The initial velocity of the dart is 95 feet per second.

a. How long does it take for the dart to hit the ground? (Round to one decimal place.)

b. Use the answer from part a) to determine the total horizontal distance that the dart travels. (Round to the nearest foot.)

c. Determine the time that it takes for the dart to reach its maximum height. (Round to one decimal place.)

d. Use the answer from part c) to determine the maximum height of the dart. (Round to the nearest foot.)

Hint for part c) and part d): Recall that the coordinates of the vertex of the quadratic function $f(t) = at^2 + bt + c$ are $\left(\frac{-b}{2a}, f\left(\frac{-b}{2a}\right)\right)$.

Chapter 11 Summary

Key Concepts

11.1 The Parabola

The equation of a parabola in standard form with a **vertical axis of symmetry** is $(x-h)^2 = 4p(y-k)$ where the vertex is $V(h, k)$ and the distance from the vertex to the directrix is $|p|$.

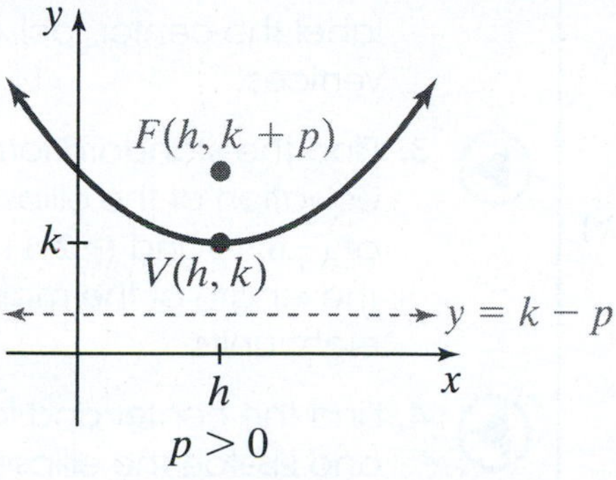

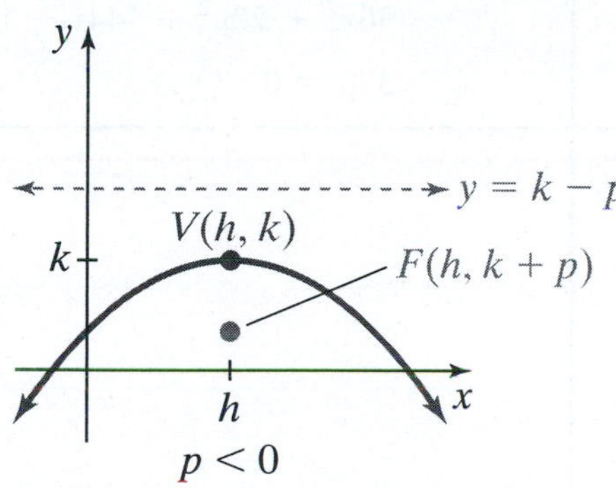

The equation of a parabola in standard form with a **horizontal axis of symmetry** is $(y-k)^2 = 4p(x-h)$ where the vertex is $V(h, k)$ and the distance from the vertex to the directrix is $|p|$.

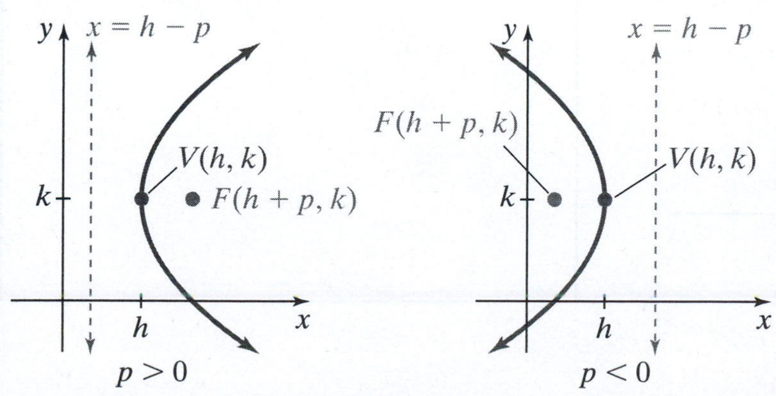

Examples/Videos

1. Find the vertex, focus, and directrix of the parabola $-(x+1)^2 = 4(y-3)$ and sketch its graph.
2. Find the vertex, focus, and directrix of the parabola $(y-3)^2 = 8(x+2)$ and sketch its graph.
3. Find the standard form of the equation of the parabola with focus $\left(-3, \frac{5}{2}\right)$ and directrix $y = \frac{11}{2}$.
4. Find the vertex, focus, and directrix and sketch the graph of the parabola $x^2 - 8x + 12y = -52$.

Key Concepts	Examples/Videos

11.2 The Ellipse

The equation of an ellipse in standard form with center (h, k) that has a **horizontal major axis** is $\frac{(x-h)^2}{a^2} + \frac{(y-k)^2}{b^2} = 1$, where $a > b > 0$ and $c^2 = a^2 - b^2$.

$(h, k+b)$ $F_2(h+c, k)$ $F_1(h-c, k)$ k $V_2(h+a, k)$ $V_1(h-a, k)$ $(h, k-b)$ h x y

1. Sketch the graph of the ellipse $\frac{x^2}{25} + \frac{y^2}{4} = 1$, and label the center, foci, and vertices.

2. Sketch the graph of the ellipse

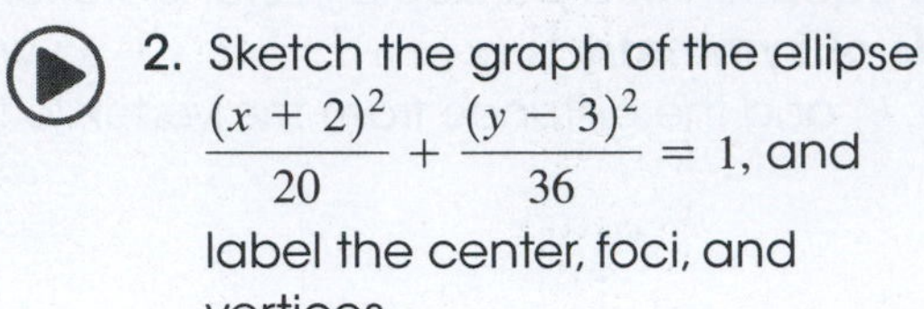

$\frac{(x+2)^2}{20} + \frac{(y-3)^2}{36} = 1$, and label the center, foci, and vertices.

3. Find the standard form of the equation of the ellipse with foci at $(-6, 1)$ and $(-2, 1)$ such that the length of the major axis is eight units.

4. Find the center and foci and sketch the ellipse $36x^2 + 20y^2 + 144x - 120y - 396 = 0$.

The equation of an ellipse in standard form with center (h, k) that has a **vertical major axis** is $\frac{(x-h)^2}{b^2} + \frac{(y-k)^2}{a^2} = 1$, where $a > b > 0$ and $c^2 = a^2 - b^2$.

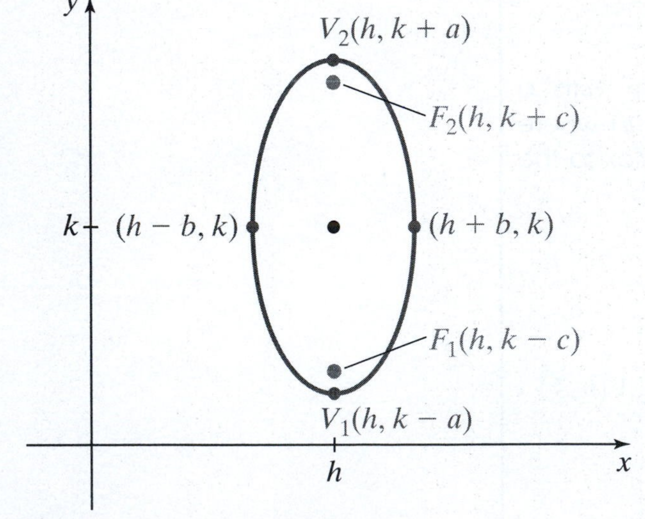

Key Concepts	Examples/Videos
11.3 The Hyperbola The equation of a hyperbola in standard form with center (h, k) that has a **horizontal transverse axis** is $\dfrac{(x-h)^2}{a^2} - \dfrac{(y-k)^2}{b^2} = 1$, where $b^2 = c^2 - a^2$. 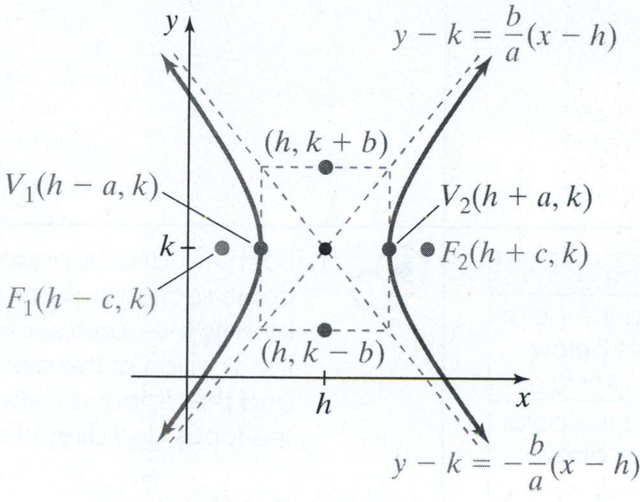	 **1.** Sketch the following hyperbolas. Determine the center, transverse axis, vertices, and foci and find the equations of the asymptotes. a. $\dfrac{(y-4)^2}{36} - \dfrac{(x+5)^2}{9} = 1$ b. $25x^2 - 16y^2 = 400$  **2.** Find the equation of the hyperbola with the center at $(-1, 0)$, a focus at $(-11, 0)$, and a vertex at $(5, 0)$.  **3.** Find the center, vertices, foci, and equations of asymptotes and sketch the hyperbola $12x^2 - 4y^2 - 72x - 16y + 140 = 0$.
The equation of a hyperbola in standard form with center (h, k) that has a **vertical transverse axis** is $\dfrac{(y-k)^2}{a^2} - \dfrac{(x-h)^2}{b^2} = 1$, where $b^2 = c^2 - a^2$. 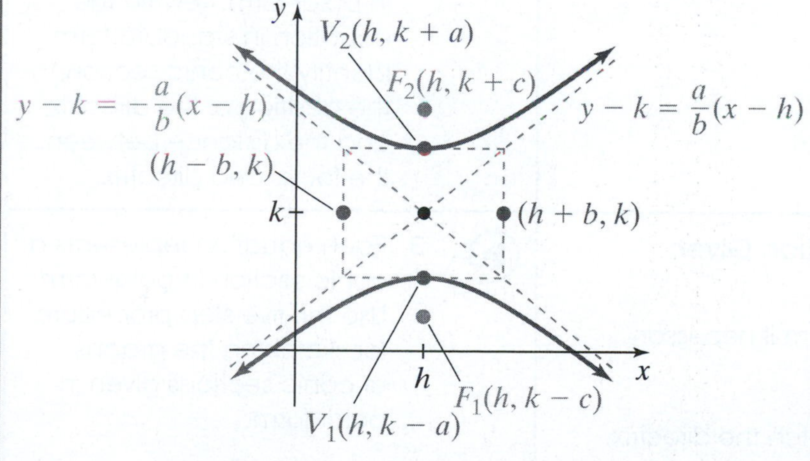	

Key Concepts	Examples/Videos

11.4 Conic Sections in Polar Form

Let D be a fixed line called the **directrix**. Let F be a fixed point called the **focus** located at the pole. Let $e > 0$ be a fixed constant called the **eccentricity**. A **conic section in polar form** is the set of all points P in the plane such that

$$\frac{d(P, F)}{d(P, D)} = e.$$

The conic section is a parabola if $e = 1$.

The conic section is an ellipse if $e < 1$.

The conic section is a hyperbola if $e > 1$.

Case	Equation	Description
1	$r = \dfrac{ep}{1 - e\sin\theta}$	Directrix **parallel** to the polar axis located p units **below** the polar axis.
2	$r = \dfrac{ep}{1 + e\sin\theta}$	Directrix **parallel** to the polar axis located p units **above** the polar axis.
3	$r = \dfrac{ep}{1 - e\cos\theta}$	Directrix **perpendicular** to the polar axis located p units **left** of the pole.
4	$r = \dfrac{ep}{1 + e\cos\theta}$	Directrix **perpendicular** to the polar axis located p units **right** the pole.

1. Each equation represents a conic section in polar form. Identify the conic section, the position of the directrix, and the distance between the focus and directrix.

a. $r = \dfrac{2}{1 - \cos\theta}$

b. $r = \dfrac{4}{1 + 3\sin\theta}$

2. The equation $3r\cos\theta + 4r = 5$ represents a conic section in polar form. Rewrite the equation in standard form. Identify the conic section, the position of the directrix, and the distance between the focus and directrix.

Sketching the Graphs of a Conic Section Given in Polar Form

Step 1. Write the equation in standard form if necessary.

Step 2. Identify the conic section.

Step 3. Plot the focus at the pole and sketch the directrix.

Step 4. Plot the polar coordinates (r, θ) where θ is a quadrantal angle $\left(\theta = 0, \theta = \frac{\pi}{2}, \theta = \pi, \text{and } \theta = \frac{3\pi}{2}\right)$ and determine which of these ordered pairs represents a vertex of the conic section.

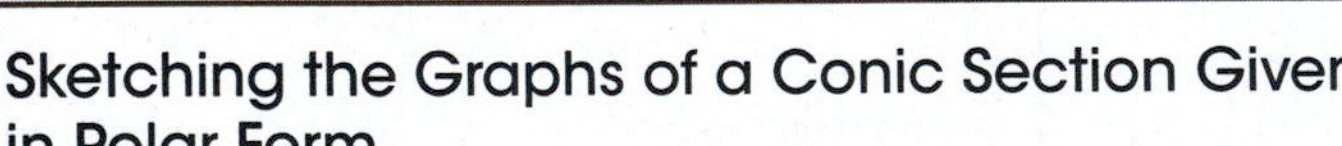

Step 5. Plot any additional ordered pairs if necessary then connect the ordered pairs with a smooth curve.

3. Each equation represents a conic section in polar form. Use the five step procedure for sketching the graphs of conic sections given in polar form.

a. $r = \dfrac{2}{3 - \cos\theta}$

b. $r + r\sin\theta = 3$

c. $r = \dfrac{12}{2 + 4\cos\theta}$

Key Concepts	Examples/Videos
Relationships Between Rectangular Form and Polar Form $x = r\cos\theta$ $y = r\sin\theta$ $r = \sqrt{x^2 + y^2}$ (or $r^2 = x^2 + y^2$) $\tan\theta = \frac{y}{x}$	**4.** Convert the equation $r = \frac{6}{2 - 4\cos\theta}$ into an equation written in rectangular form.
11.5 Parametric Equations A **plane curve** is the set of all ordered pairs (x, y) such that $x = f(t)$ and $y = g(t)$ where t is in an interval I. The variable t is called a **parameter**, and the equations $x = f(t)$ and $y = g(t)$ are called **parametric equations** of the curve. If a plane curve is defined parametrically as $x = f(t)$ and $y = g(t)$ on a closed interval $a \le t \le b$, then the ordered pair $(f(a), g(a))$ is called the **initial point** and the ordered pair $(f(b), g(b))$ is called the **terminal point.**	**1.** Determine the initial point and the terminal point of the plane curve defined by $x = 2t^2, y = 3t, -4 \le t \le 4$, then plot several other ordered pairs and sketch the plane curve. **2.** Determine the initial point and the terminal point of the plane curve defined by $x = \cos t, y = \sin t, -\frac{\pi}{2} \le t \le \pi$, then plot several other ordered pairs and sketch the plane curve.
Determining a Rectangular Representation for a Curve Defined Parametrically Using Substitution **Step 1.** Carefully observe the given parametric equations in an attempt to identify a relationship between the variables such that a substitution could be made to eliminate the parameter t. **Step 2.** Choose a parametric equation and solve for t or rewrite the parametric equation based on the observation made in Step 1. **Step 3.** Substitute the expression from Step 2 into the other parametric equation in place of the parameter to obtain the rectangular equation. **Step 4.** Write the rectangular equation in a familiar form whenever possible. **Step 5.** If necessary, restrict the domain and range of the rectangular equation to be consistent with any restrictions of the parameter or any restrictions of the parametric equations.	**3.** Determine the rectangular equation of the curve whose parametric equations are given by $x = 2t^2, y = 3t, -4 \le t \le 4$, then sketch the plane curve. **4.** Determine the rectangular equation of the curve whose parametric equations are given by $x = \sqrt{t}, y = -\frac{1}{2}t + 1$, then sketch the plane curve.

Key Concepts	Examples/Videos
Determining a Rectangular Representation for a Curve Defined Parametrically Using a Trigonometric Identity **Step 1.** Isolate the trigonometric expression in each parametric equation. **Step 2.** Select a trigonometric identity (by observation) that relates the two parametric equations. **Step 3.** Substitute the rectangular expressions for the trigonometric expressions into the chosen trigonometric identity to eliminate the parameter. **Step 4.** Write the rectangular equation in a familiar form whenever possible. **Step 5.** If necessary, restrict the domain and range of the rectangular equation to be consistent with any restrictions of the parameter or parametric equations.	 5. Determine the rectangular equation of the plane curve whose parametric equations are given by $x = 2\sin t, y = 3\cos t, 0 \le t \le \pi$, then sketch the plane curve.
Determining a Parametric Representation of a Circle or Ellipse Centered at the Origin 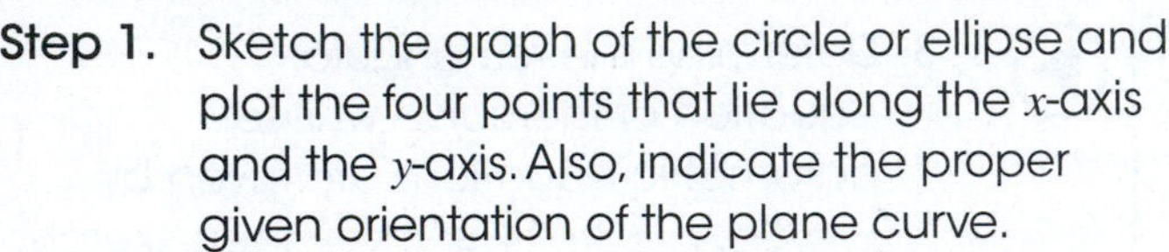 **Step 1.** Sketch the graph of the circle or ellipse and plot the four points that lie along the x-axis and the y-axis. Also, indicate the proper given orientation of the plane curve. **Step 2.** Write the equation of the circle or ellipse in the form $\frac{x^2}{a^2} + \frac{y^2}{b^2} = 1$. Note that if the equation is a circle, then $a^2 = b^2$. **Step 3.** Use the Pythagorean Identity $\sin^2 t + \cos^2 t = 1$ to make one of the following substitutions: 1. If the initial point lies along the y-axis, then use the substitution $\frac{x^2}{a^2} = \sin^2 t$ and $\frac{y^2}{b^2} = \cos^2 t$. 2. If the initial point lies along the x-axis, then use the substitution $\frac{x^2}{a^2} = \cos^2 t$ and $\frac{y^2}{b^2} = \sin^2 t$.	 6. Determine a parametric representation for the circle $x^2 + y^2 = 16$ such that the initial point is at $(0, -4)$ and the orientation is clockwise.

Key Concepts	Examples/Videos
Step 4. Solve the equations from **Step 3** for x and y to see that a parametric representation of the circle or ellipse is $x = \pm a\cos t, y = \pm b\sin t, 0 \le t \le 2\pi$ or $x = \pm a\sin t, y = \pm b\cos t, 0 \le t \le 2\pi$. **Step 5.** Determine the sign of a and b. The sign of a and b depend on the coordinates of the initial point and the orientation of the plane curve.	
Parametric Equations Describing the Motion of an Object If an object having an initial height, h, propelled upward with an initial velocity of v_0 at an angle of θ with the horizontal, then $x = (v_0\cos\theta)t$ and $y = h + (v_0\sin\theta)t - \frac{1}{2}gt^2$ where g is the constant acceleration due to gravity. Note that g is approximately 32 ft/sec/sec or 9.8 m/sec/sec.	7. A football player kicks a football with an initial velocity of 70 ft/sec at an angle of $\theta = 35°$ with the ground. Assuming no wind resistance, answer the following. a. Determine the parametric equation describing the horizontal distance that the football travels from the point at which the ball was kicked. b. Determine the parametric equation describing the vertical height of the football. c. What is the horizontal distance that the ball has traveled at $t = 1.5$ sec? (Round to the nearest foot.) d. What is the vertical height of the ball has traveled at $t = 1.5$ sec? (Round to the nearest foot.) e. Assuming that the kicker accurately kicked the ball directly at the center of the 10 ft high crossbar between two goal posts that is 42 yards away, will the ball go over the crossbar?

Chapter 11 Review Exercises

1. Determine the vertex, focus, and directrix of the parabola and sketch its graph.

$$(x - 1)^2 = -12(y - 4)$$

2. Find the equation of the parabola in standard form that has a focus at $(2, 4)$, and the equation of the directrix is $x = -4$.

3. Determine the vertex, focus, and directrix of the parabola and sketch its graph.

$$x^2 + 10x = 5y - 10$$

4. A parabolic eavesdropping device is used by a CIA agent to record terrorist conversations. The parabolic surface measures 120 centimeters in diameter at the top and is 90 centimeters deep at its center. How far from the vertex should the microphone be located?

5. Determine the center, foci, and vertices of the ellipse and sketch its graph.

$$\frac{(x-1)^2}{9} + \frac{(y+4)^2}{49} = 1$$

6. Determine the standard equation of the ellipse having foci at $(8, -1)$ and $(2, -1)$ such that the length of the major axis is twelve units.

7. Complete the square to write the equation $16x^2 + 20y^2 + 64x - 40y - 236 = 0$ in the form $\frac{(x-h)^2}{a^2} + \frac{(y-k)^2}{b^2} = 1$ or $\frac{(x-h)^2}{b^2} + \frac{(y-k)^2}{a^2} = 1$. Then, determine the center, foci, and vertices of the ellipse and sketch its graph.

8. An elliptical arch railroad tunnel 16 feet high at the center and 30 feet wide is cut through the side of a mountain. Find the equation of the ellipse if a vertex is represented by the point $(0, 16)$.

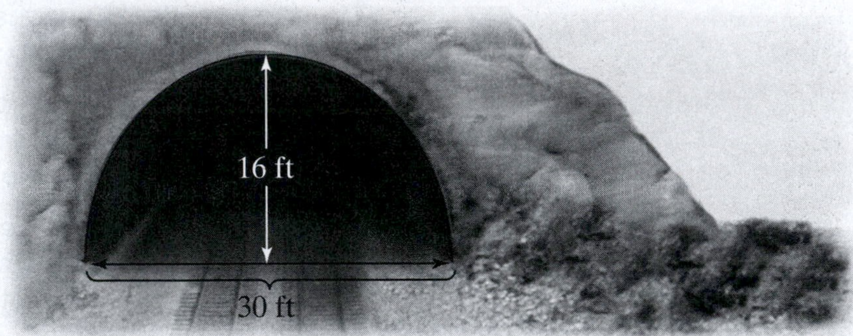

9. Determine the center, transverse axis, vertices, foci, and the equations of the asymptotes and sketch the hyperbola.

$$\frac{(y-4)^2}{25} - \frac{(x-2)^2}{36} = 1$$

10. Determine the standard equation of the hyperbola that has a center at $(-6, -1)$, a focus at $(-6, -9)$, and a vertex at $(-6, -5)$. Then sketch the graph.

11. Determine the center, vertices, foci, endpoints of the conjugate axis, and the equations of the asymptotes of the hyperbola, and sketch its graph.

$$y^2 - 9x^2 - 12y - 36x - 9 = 0$$

12. A light on a wall produces a shadow in the shape of a hyperbola. If the distance between the two vertices of the hyperbola is 14 inches and the distance between the two foci is 16 inches, find the equation of the hyperbola.

13. The equation $r = \dfrac{6}{1 + 2\cos\theta}$ represents a conic section in polar form. Identify the conic section, the position of the directrix, and the distance between the focus and directrix.

14. The equation $r = \dfrac{4}{1 + \frac{1}{2}\sin\theta}$ represents a conic section in polar form. Sketch the conic section.

15. Convert the equation $r = \dfrac{3}{1 - \cos\theta}$ into a rectangular equation.

16. Determine the polar equation in standard form of the conic section with a focus at the pole such that the eccentricity is 1 and the equation of the directrix is $r = 3\csc\theta$.

17. Determine the initial point and terminal point of the parametric equations $x = 1 + t^2$ and $y = 2t,\ -3 \leq t \leq 2$. Determine several other ordered pairs, then sketch the plane curve.

18. Determine the rectangular equation of the curve whose parametric equations are given by $x = 2\sin t$ and $y = 4\cos t, 0 \leq t \leq \pi$. Then sketch the plane curve.

19. Determine a parametric representation for the given plane curve.

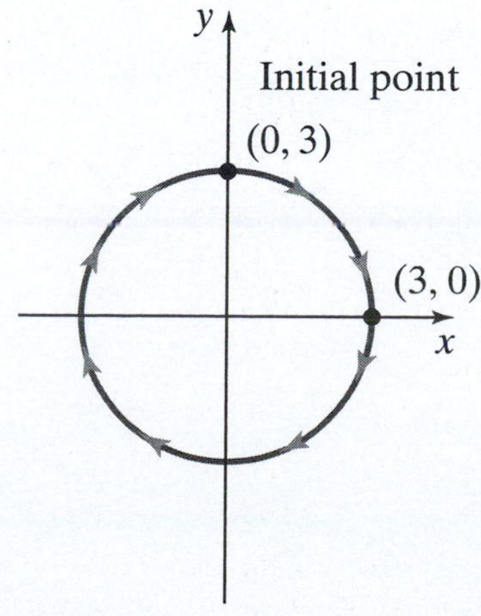

20. A golf ball is hit with an initial velocity of 120 feet per second at an angle $\theta = 30°$.

   a. Determine the parametric equation that describes the horizontal distance, x, of the golf ball at any time $t \geq 0$, where t represents time in seconds.

   b. Determine the parametric equation that describes the vertical height, y, of the golf ball at any time $t \geq 0$ where t represents time in seconds.

   c. What is the horizontal distance that the golf ball has traveled at $t = 2$ seconds? (Round to the nearest foot.)

   d. What is the height of the golf ball at $t = 2$ seconds? (Round to the nearest foot.)

CHAPTER TWELVE

Systems of Equations and Inequalities

CHAPTER TWELVE CONTENTS

12.1 Solving Systems of Linear Equations in Two Variables

THINGS TO KNOW

Before working through this section, be sure that you are familiar with the following concepts:

		VIDEO	ANIMATION	INTERACTIVE
 You Try It	1. Applications of Linear Equations (Section 1.2)			
 You Try It	2. Writing the Equation of a Line in Standard Form (Section 2.3)			
You Try It	3. Sketching Lines by Plotting Intercepts (Section 2.3)			
You Try It	4. Understanding the Definition of Parallel Lines (Section 2.4)			

OBJECTIVES

1 Verifying Solutions to a System of Linear Equations in Two Variables

2 Solving a System of Linear Equations Using the Substitution Method

OBJECTIVE 1 VERIFYING SOLUTIONS TO A SYSTEM OF LINEAR EQUATIONS IN TWO VARIABLES

In **Section 1.1**, a linear equation in one variable was defined as an equation that can be written in the form $ax + b = 0$, where a and b are real numbers with $a \neq 0$. This definition can be extended to more variables as follows.

> **Definition** Linear Equation in n Variables
>
> A linear equation in n variables is an equation that can be written in the form $a_1x_1 + a_2x_2 + \cdots + a_nx_n = b$ for variables $x_1, x_2, \ldots x_n$ and real numbers $a_1, a_2, \ldots a_n, b$, where at least one of $a_1, a_2, \ldots a_n$ is nonzero.

For example, $2x - 5y = -7$ is a linear equation in two variables, whereas $4x_1 + 9x_2 - 17x_3 = 11$ is a linear equation in three variables. The important thing to remember is that all variables of a linear equation have an exponent of 1. For much of this chapter, we learn how to solve **systems of linear equations**. We start our discussion of linear systems by considering linear systems in two variables.

> **Definition** System of Linear Equations in Two Variables
>
> A system of linear equations in two variables is the collection of two linear equations in two variables considered simultaneously. The solution to a system of equations in two variables is the set of all ordered pairs for which *both* equations are true.

Consider the following three systems of linear equations in two variables.

$$\begin{aligned} 3x - 2y &= -9 \\ x + y &= 2 \end{aligned} \qquad \begin{aligned} -3x_1 + 2x_2 &= 11 \\ 8x_1 - 9x_2 &= -2 \end{aligned} \qquad \begin{aligned} \sqrt{2}a + b &= \pi \\ -4a + \sqrt{3}b &= 1 \end{aligned}$$

In each system, different variables were used. It does not matter what we call the two variables as long as both equations are linear. According to the definition of a system of linear equations in two variables, the solution is the set of all ordered pairs for which both equations are true. We verify that the ordered pair $(-1, 3)$ is a solution to the system $\begin{aligned} 3x - 2y &= -9 \\ x + y &= 2 \end{aligned}$ in Example 1.

Example 1 Verify the Solution to a System of Linear Equations in Two Variables

Show that the ordered pair $(-1, 3)$ is a solution to the system $\begin{aligned} 3x - 2y &= -9 \\ x + y &= 2 \end{aligned}$.

Solution We can verify that the ordered pair $(-1, 3)$ is a solution to this system by substituting -1 for x and 3 for y in both equations.

First Equation		Second Equation
$3x - 2y = -9$		$x + y = 2$
$3(-1) - 2(3) \stackrel{?}{=} -9$	Substitute $x = -1$ and $y = 3$.	$(-1) + (3) \stackrel{?}{=} 2$
$-9 = -9$ ✓	Both equations are true.	$2 = 2$ ✓

Both equations yielding true statements indicate that the ordered pair $(-1, 3)$ is a solution to this system. ●

You Try It **Work through this You Try It problem.**

Work Exercises 1–4 in this textbook or in the MyLab Math Study Plan.

If a linear system has at least one solution, like the system shown in Example 1, it is said to be **consistent**, and the solution is the set of all ordered pairs that satisfy both equations. If the system does not have a solution, it is said to be **inconsistent**.

The solution to a system of two linear equations involving two variables can be viewed geometrically. Because the graph of each equation of the system is a line, we can sketch the two lines and geometrically view the solution. There are only three possibilities for the graph of a linear system of equations involving two variables:

1. The lines can have different slopes and thus intersect at a single point (**Figure 1a**). Systems that have at least one solution are said to be **consistent**. This consistent system has exactly one solution so it is said to be **consistent** and **independent**.
2. The two equations can be different representations of the same line and have the same graph (**Figure 1b**). In this case, there are infinitely many solutions. Because this system has at least one solution, it is said to be **consistent**. This consistent system has infinitely many solutions so it is called **consistent** and **dependent**.
3. The two lines can be parallel (have the same slope) and thus have no intersecting point (**Figure 1c**). This system is said to be **inconsistent** and has no solution.

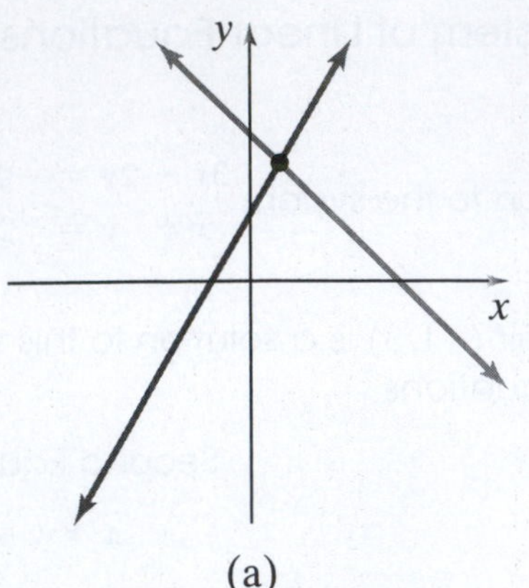

(a)
Consistent, Independent
One solution
Two lines are different, having one common point. The lines have different slopes.

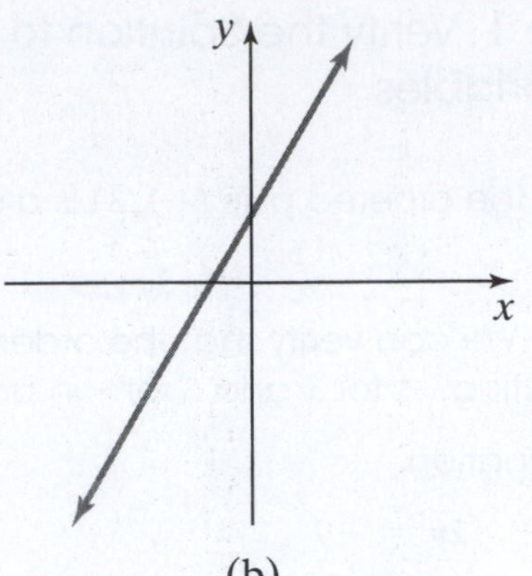

(b)
Consistent, Dependent
Infinitely many solutions
Two lines are the same, having infinitely many common points. The lines have the same slope and same y-intercepts.

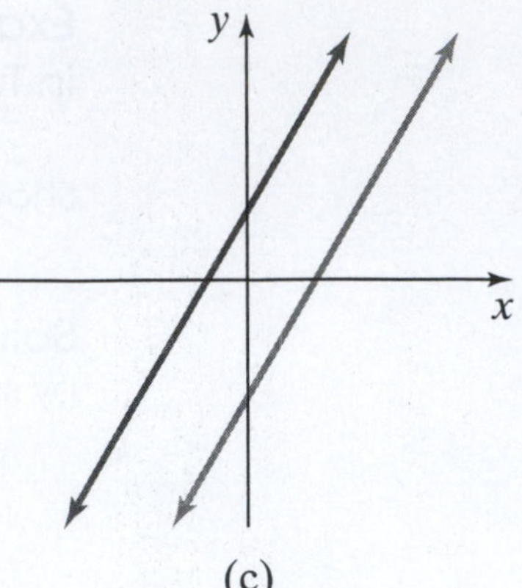

(c)
Inconsistent,
No solution
Two lines are different, having no common points. The lines have the same slope but different y-intercepts.

Figure 1

We now learn two algebraic methods used for solving linear systems involving two equations and two variables. These methods are known as the *substitution method* and the *elimination method.*

OBJECTIVE 2 SOLVING A SYSTEM OF LINEAR EQUATIONS USING THE SUBSTITUTION METHOD

The substitution method involves solving one of the equations for one variable in terms of the other, and then substituting that expression into the other equation. The substitution method can be summarized in the following four steps.

Solving a System of Equations by the Method of Substitution

Step 1. Choose an equation and solve for one variable in terms of the other variable.

Step 2. Substitute the expression from step 1 into the other equation.

Step 3. Solve the equation in one variable.

Step 4. Substitute the value found in step 3 into one of the original equations to find the value of the other variable.

The substitution method is used to solve the system in Example 2.

Example 2 Solve a System of Equations Using the Substitution Method

Solve the following system using the method of substitution: $\begin{aligned} 2x - 3y &= -5 \\ x + y &= 5 \end{aligned}$.

Solution

Step 1. If possible, choose an equation in which at least one **coefficient** is 1. In this example, both coefficients of the second equation are 1. We can choose to solve the second equation for either variable. In this case, we solve the second equation for y:

$x + y = 5$ Write the second equation.

$y = 5 - x$ Subtract x from both sides.

Step 2. Substitute $5 - x$ for y in the first equation:

$$2x - 3(5 - x) = -5$$

Step 3. Solve for x:

$2x - 3(5 - x) = -5$ Rewrite the equation from step 2.

$2x - 15 + 3x = -5$ Use the distributive property.

$5x - 15 = -5$ Combine like terms.

$5x = 10$ Add 15 to both sides.

$x = 2$ Divide both sides by 5.

Step 4. Because $x = 2$, we can solve for y by substituting $x = 2$ into one of the original equations. Here we use the simplified equation found in step 1.

$$y = 5 - (2) = 3$$

The solution of this system is the ordered pair $(2, 3)$. You can see in Figure 2 that this is the intersection of the two lines of this system.

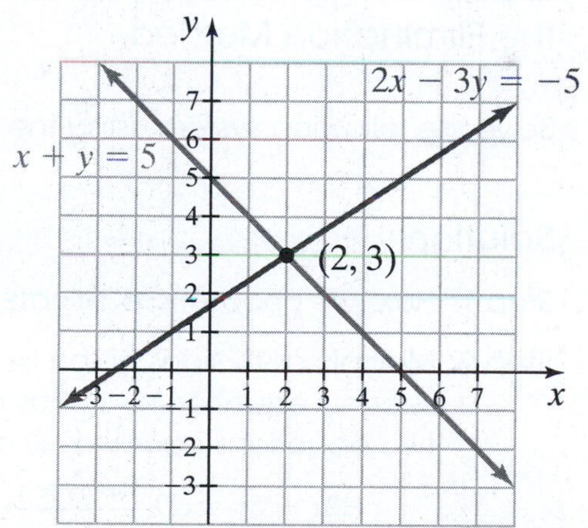

Figure 2 The solution to the system $\begin{aligned} 2x - 3y &= -5 \\ x + y &= 5 \end{aligned}$ is the ordered pair $(2, 3)$.

You Try It Work through this You Try It problem.

Work Exercises 5–9 in this textbook or in the MyLab Math Study Plan.

OBJECTIVE 3 SOLVING A SYSTEM OF LINEAR EQUATIONS USING THE ELIMINATION METHOD

Another method used to solve a system of two equations is called the elimination method. The elimination method involves adding the two equations together in an attempt to eliminate one of the variables. To accomplish this task, the coefficients of one of the variables must differ only in sign. This can be done by multiplying one or both of the equations by a suitable constant. The elimination method can be summarized in the following five steps.

> **Solving a System of Equations by the Method of Elimination**
>
> **Step 1.** Choose a variable to eliminate.
>
> **Step 2.** Multiply one or both equations by an appropriate nonzero constant so that the sum of the coefficients of one of the variables is zero.
>
> **Step 3.** Add the two equations together to obtain an equation in one variable.
>
> **Step 4.** Solve the equation in one variable.
>
> **Step 5.** Substitute the value obtained in step 4 into one of the original equations to solve for the other variable.

The elimination method is used to solve the system in Example 3.

Example 3 Solve a System of Equations Using the Elimination Method

Solve the following system using the method of elimination: $\begin{aligned} -2x + 5y &= 29 \\ 3x + 2y &= 4 \end{aligned}$.

Solution

Step 1. Here we choose to eliminate the variable x.

Step 2. Multiply both sides of the first equation by 3, and multiply both sides of second equation by 2, thus making the coefficient of the x-terms in the two equations equal to -6 and 6, respectively.

$$\begin{aligned} -2x + 5y &= 29 \xrightarrow{\text{Multiply by 3.}} 3(-2x + 5y) = 3(29) \longrightarrow -6x + 15y = 87 \\ 3x + 2y &= 4 \xrightarrow{\text{Multiply by 2.}} 2(3x + 2y) = 2(4) \longrightarrow 6x + 4y = 8 \end{aligned}$$

Step 3. We now add the two new equations together:

$$\begin{aligned} -6x + 15y &= 87 \\ \underline{6x + 4y} &\underline{= 8} \\ 19y &= 95 \end{aligned}$$

Step 4. Solve for y:

$$\begin{aligned} 19y &= 95 \\ y &= 5 \quad \text{Divide both sides by 19.} \end{aligned}$$

Step 5. To solve for x, we substitute $y = 5$ into one of the original equations. Here, we choose the first equation:

$-2x + 5y = 29$ Write the original first equation.

$-2x + 5(5) = 29$ Substitute 5 for y in the first equation.

$-2x + 25 = 29$ Simplify.

$-2x = 4$ Subtract 25 from both sides.

$x = -2$ Divide both sides by -2.

Therefore, the solution is the ordered pair $(-2, 5)$.

Using Technology

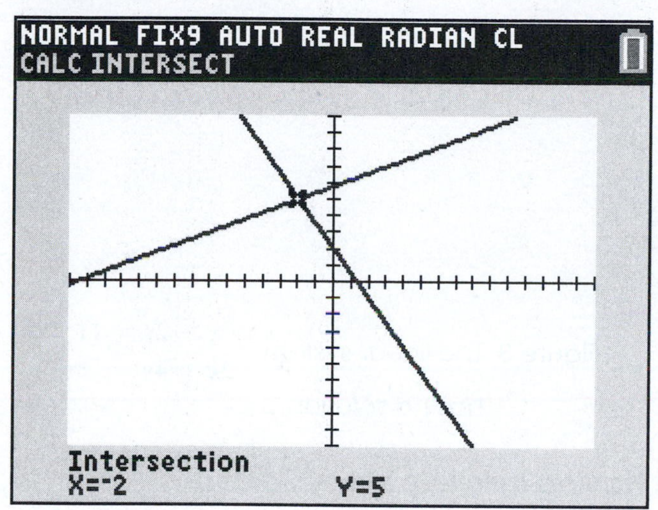

We can solve the system in Example 3 using a graphing utility by first solving both equations for y:

$$-2x + 5y = 29 \longrightarrow y_1 = \frac{2}{5}x + \frac{29}{5}$$

$$3x + 2y = 4 \longrightarrow y_2 = -\frac{3}{2}x + 2$$

Graph y_1 and y_2, and then use the INTERSECT feature to find that the point of intersection is $(-2, 5)$. ●

You Try It Work through this You Try It problem.

Work Exercises 10–14 in this textbook or in the MyLab Math Study Plan.

The systems in Examples 2 and 3 had one solution. We now look at an example of a system with no solution, and then solve a system that has infinitely many solutions.

Example 4 Solve a System with No Solution

Solve the system $\begin{aligned} x - 2y &= 11 \\ -2x + 4y &= 8 \end{aligned}$.

Solution We can solve this system using either the substitution method or the elimination method. In this case, we use the elimination method.

Step 1. We choose to eliminate the variable x because it has a **coefficient** of 1 in the first equation.

Step 2. Multiply both sides of the first equation by 2, thus making the coefficient of the x-terms in the two equations equal to 2 and -2, respectively.

$$x - 2y = 11 \xrightarrow{\text{Multiply by 2}} 2x - 4y = 22$$

$$-2x + 4y = 8 \xrightarrow{\text{Leave equation alone}} -2x + 4y = 8$$

Step 3. We now add the two new equations together:

$$\begin{aligned} 2x - 4y &= 22 \\ \underline{-2x + 4y} &\underline{= \ \ 8} \\ 0 &= 30 \end{aligned}$$

Notice how both variables were eliminated, leaving us with a **contradiction**. The number 0 is never equal to 30. This means that there is no solution to this system. This is an example of an **inconsistent** system. You can see in Figure 3 that the graphs of the two linear equations from this system are parallel. Thus, the two lines do not intersect, and the system has no solution.

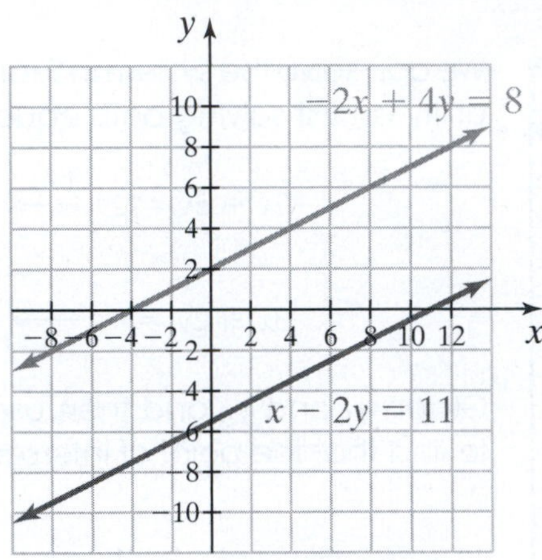

Figure 3 The linear system $\begin{aligned} x - 2y &= 11 \\ -2x + 4y &= 8 \end{aligned}$ has no solution.

Example 5 Solve a System Having Infinitely Many Solutions

Solve the system $\begin{aligned} -3x + 6y &= 9 \\ x - 2y &= -3 \end{aligned}$.

Solution We can solve this system using either the substitution method or the elimination method. Here, we choose the substitution method.

Step 1. We start by solving the second equation for x:

$x - 2y = -3$	Write the second equation.
$x = 2y - 3$	Add $2y$ to both sides.

Step 2. Substitute $2y - 3$ for x in the first equation:

$$-3(2y - 3) + 6y = 9$$

Step 3. Solve for y:

$-3(2y - 3) + 6y = 9$	Rewrite the equation from step 2.
$-6y + 9 + 6y = 9$	Use the distributive property.
$9 = 9$	Combine like terms.

The equation simplifies to the **identity** $9 = 9$, which is true for *any* value of y. Therefore, this system has infinitely many solutions. Note that we could have multiplied the second equation in the system by -3 to obtain the first equation. Because of this, both equations represent the same line and thus have the same graph. See Figure 4. The solution is the set of all ordered pairs that lie on the graph of either equation.

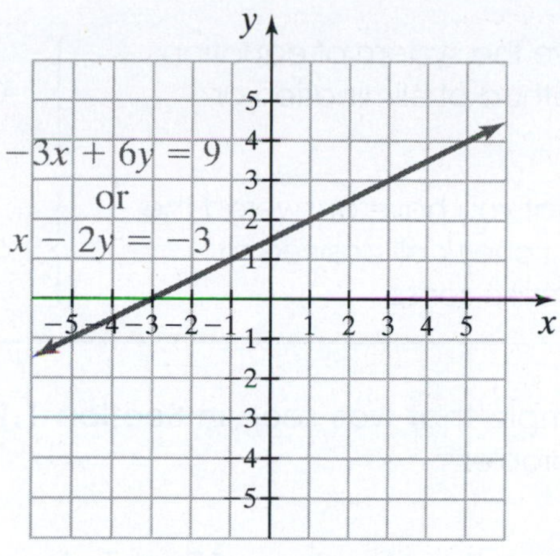

Figure 4 The system $\begin{aligned} -3x + 6y &= 9 \\ x - 2y &= -3 \end{aligned}$ has infinitely many solutions.

We can write the solution to this system in **ordered pair notation.** Recall in step 1 that we wrote x in terms of y as $x = 2y - 3$; thus, we can write the solution in ordered pair notation as $(2y - 3, y)$, where y is any real number. For example,

If $y = -2$, then $x = 2(-2) - 3 = -7$, so the ordered pair $(-7, -2)$ is a solution.

If $y = 0$, then $x = 2(0) - 3 = -3$, so the ordered pair $(-3, 0)$ is another solution.

Similarly, we could have solved either equation for y to obtain $y = \frac{1}{2}x + \frac{3}{2}$. So, an alternative way to describe the solution to this system in ordered pair notation is $\left(x, \frac{1}{2}x + \frac{3}{2}\right)$.

You Try It **Work through this You Try It problem.**

Work Exercises 15–20 in this textbook or in the MyLab Math Study Plan.

OBJECTIVE 4 SOLVING APPLIED PROBLEMS USING A SYSTEM OF LINEAR EQUATIONS

Applied problems can involve two or more unknown quantities. Sometimes we are able to use a single equation involving one variable to solve such problems as seen in Section 1.2. However, it is often easier to use two variables and create a system of two equations. The following steps are based on the four-step strategy for problem solving that was first introduced in Section 1.2.

Five-Step Strategy for Problem Solving Using Systems of Equations

Step 1. Read the problem several times until you have an understanding of what is being asked. If possible, create diagrams, charts, or tables to assist you.	Understand the problem.
Step 2. Choose variables that describe each unknown quantity. **Step 3.** Write a system of equations using the given information and the variables.	Devise a plan.

(Continued)

Step 4. Carefully solve the system of equations using the method of elimination or substitution.	Carry out the plan.
Step 5. Make sure that you have answered the question, and check all answers to ensure they make sense.	Look back.

We start with an example that was seen in **Section 1.2**. This time, we use two variables to solve the problem.

Example 6 Determine the Number of Touchdowns Thrown

Roger Staubach and Terry Bradshaw were both quarterbacks in the National Football League. In 1973, Staubach threw three touchdown passes more than twice the number of touchdown passes thrown by Bradshaw. If the total number of touchdown passes between Staubach and Bradshaw was 33, how many touchdown passes did each player throw?

Solution

Step 1. After carefully reading the problem, we see that we are trying to figure out how many touchdown passes were thrown by each player.

Step 2. Let

$$S = \text{ number of touchdowns thrown by Staubach}$$

$$B = \text{ number of touchdowns thrown by Bradshaw}$$

Step 3. The combined number of touchdown passes is 33, so the first equation is

$$\underbrace{\textit{Staubach's touchdown passes}}_{S} + \underbrace{\textit{Bradshaw's touchdown passes}}_{B} = \underbrace{\textit{Total touchdown passes}}_{33}$$

To find the second equation, we use the fact that Staubach threw three more than twice the number of touchdowns thrown by Bradshaw. Therefore, the second equation is $S = 2B + 3$. This equation can be rewritten as $S - 2B = 3$. Therefore, the system of equations is

$$\begin{aligned} S + B &= 33 \\ S - 2B &= 3 \end{aligned}$$

Step 4. Using the method of elimination and multiplying the first equation by 2, we obtain the system

$$\begin{aligned} 2S + 2B &= 66 \\ S - 2B &= 3 \end{aligned}$$

We can now add the two equations together to eliminate the variable B.

$$\begin{array}{r} 2S + 2B = 66 \\ \underline{S - 2B = 3} \\ 3S \qquad = 69 \end{array}$$

Dividing by 3 gives $S = 23$. To find B, replace S with 23 in the original first equation:

$$S + B = 33 \quad \text{Write the original first equation.}$$

$$23 + B = 33 \quad \text{Replace } S \text{ with 23.}$$

$$B = 10 \quad \text{Subtract 23 from both sides.}$$

Therefore, Staubach threw 23 touchdown passes, and Bradshaw threw 10 touchdown passes.

Step 5. To check, we see that the total number of touchdown passes is $23 + 10 = 33$ and the number of touchdown passes thrown by Staubach is exactly 3 more than twice the number of touchdown passes thrown by Bradshaw. ●

Example 7 Find the Number of Tickets Sold at a Jazz Festival

During one night at the jazz festival, 2,100 tickets were sold. Adult tickets sold for \$12, and child tickets sold for \$7. If the receipts totaled \$22,100, how many of each type of ticket were sold?

Solution

Step 1. We are asked to determine the number of adult tickets and the number of child tickets sold.

Step 2. Let

$$A = \text{number of adult tickets}$$

$$C = \text{number of child tickets}$$

Step 3. The number of adult tickets plus the number of child tickets totals 2,100, so we obtain our first equation

$$\underbrace{A}_{\textit{Number of adult tickets}} + \underbrace{C}_{\textit{Number of child tickets}} = \underbrace{2{,}100}_{\textit{Total number of tickets}}$$

To find the second equation, we use the fact that the value of an adult ticket is \$12 and the value of a child ticket is \$7; therefore,

$$\underbrace{12A}_{\textit{Total value of adult tickets}} + \underbrace{7C}_{\textit{Total value of child tickets}} = \underbrace{22{,}100}_{\textit{Total value of all tickets}}$$

Thus, the system of equations is $\begin{aligned} A + C &= 2{,}100 \\ 12A + 7C &= 22{,}100 \end{aligned}$.

Step 4. Using the method of elimination and multiplying the first equation by -7, we obtain the system

$$\begin{aligned} -7A - 7C &= -14{,}700 \\ 12A + 7C &= 22{,}100 \end{aligned}$$

Adding the equations together eliminates the variable C:

$$\begin{array}{r} -7A - 7C = -14{,}700 \\ \underline{12A + 7C = 22{,}100} \\ 5A = 7{,}400 \end{array}$$

Dividing by 5 gives $A = 1{,}480$. To find C, replace A with $1{,}480$ in the original first equation:

$A + C = 2{,}100$	Write the original first equation.
$1{,}480 + C = 2{,}100$	Replace A with $1{,}480$.
$C = 620$	Subtract $1{,}480$ from both sides.

Therefore, there were $1{,}480$ adult tickets sold and 620 child tickets sold.

Step 5. To check, we see that the total number of tickets is $1{,}480 + 620 = 2{,}100$ and the total value of all tickets sold is $(\$12)(1{,}480) + (\$7)(620) = \$22{,}100$.

Following are two more examples. Although the solutions to these examples are not worked out for you in the text, you may view them in detail by clicking on the appropriate video link.

Example 8 Mix Beans in a Food Plant

Twin City Foods, Inc., created a 10-lb bean mixture that sells for $5.75 by mixing lima beans and green beans. If lima beans sell for $.70 per pound and green beans sell for $.50 per pound, how many pounds of each bean went into the mixture?

Solution Watch the **video** to see that the mixture contains $3\frac{3}{4}$ pounds of lima beans and $6\frac{1}{4}$ pounds of green beans.

Example 9 Find the Speed of an Airplane

A small airplane flies from Seattle, Washington, to Portland, Oregon—a distance of 150 miles. Because the pilot encountered a strong headwind, the trip took 1 hour and 15 minutes. On the return flight, the wind is still blowing at the same speed. If the return trip took 45 minutes, what was the average speed of the airplane in still air? What was the speed of the wind?

Solution Watch the **video** to verify that the speed of the plane is 160 mph and the speed of the wind is 40 mph.

You Try It Work through this You Try It problem.

Work Exercises 21–30 in this textbook or in the MyLab Math Study Plan.

12.1 Exercises

In Exercises 1–4, two ordered pairs are given. Determine whether each ordered pair is a solution to the given system.

1. $\begin{aligned} -4x - 2y &= -2 \\ 5x + 7y &= 16 \end{aligned}$ Ordered pairs: $(-1, 3), (-3, 7)$

2. $\begin{aligned} \frac{1}{2}x_1 - \frac{1}{3}x_2 &= -\frac{7}{24} \\ -\frac{1}{4}x_1 + \frac{1}{2}x_2 &= \frac{5}{16} \end{aligned}$ Ordered pairs: $\left(-\frac{1}{2}, \frac{1}{4}\right), \left(-\frac{1}{4}, \frac{1}{2}\right)$

3. $\begin{aligned} 0.3a + 0.2b &= -0.12 \\ -0.7a - 0.1b &= -0.05 \end{aligned}$ Ordered pairs: $(0.2, -0.9), (-0.3, 0.4)$

4. $\begin{aligned} -2x + y &= -1 \\ \frac{2}{3}x - \frac{1}{3}y &= \frac{1}{3} \end{aligned}$ Ordered pairs: $(-2, 5), (3, 5)$

In Exercises 5–9, use the method of substitution to solve each system of linear equations.

5. $\begin{aligned} x &= 4 - y \\ 3x - 2y &= -3 \end{aligned}$

6. $\begin{aligned} x + 4y &= 5 \\ 2x - y &= -8 \end{aligned}$

7. $\begin{aligned} 2x - y &= 3 \\ -8x + 3y &= -8 \end{aligned}$

8. $\begin{aligned} 5x + 4y &= -6 \\ 8x - 4y &= -72 \end{aligned}$

9. $\begin{aligned} -\frac{2}{3}x + \frac{1}{2}y &= -\frac{1}{3} \\ -\frac{1}{2}x + \frac{3}{4}y &= -\frac{1}{2} \end{aligned}$

In Exercises 10–14, use the method of *elimination* to solve each system of linear equations

10. $\begin{aligned} x + 14y &= 20 \\ -x + 14y &= 15 \end{aligned}$

11. $\begin{aligned} 3x - 2y &= 20 \\ 2x + 6y &= -38 \end{aligned}$

12. $\begin{aligned} 8x - 6y &= -6 \\ 5x - 7y &= -33 \end{aligned}$

13. $\begin{aligned} 0.1x - 0.4y &= -2.4 \\ 0.7x - y &= -9.6 \end{aligned}$

14. $\begin{aligned} \frac{7}{16}x - \frac{3}{4}y &= -\frac{5}{2} \\ \frac{3}{4}x + \frac{5}{2}y &= 26 \end{aligned}$

In Exercises 15–20, use the substitution method or the elimination method to solve each system. If the system has infinitely many solutions, express the ordered pairs in terms of x or y as in Example 4.

15. $\begin{aligned} x + y &= 3 \\ -2x - 2y &= 1 \end{aligned}$

16. $\begin{aligned} 3x - 12y &= 6 \\ -2x + 8y &= -4 \end{aligned}$

17. $\begin{aligned} 8x - y &= -13 \\ y &= -8x \end{aligned}$

18. $\begin{aligned} -\frac{1}{2}x + \frac{2}{3}y &= -\frac{1}{3} \\ \frac{3}{2}x - 2y &= 1 \end{aligned}$

19. $\begin{aligned} 8x + 4y &= 5 \\ 2x + y &= -4 \end{aligned}$

20. $\begin{aligned} y &= \frac{5}{7}x - 9 \\ -15x + 21y &= -189 \end{aligned}$

In Exercises 21–30, use a system of equations to solve each problem.

21. Together, teammates Pedro and Ricky got 2,673 base hits last season. Pedro had 281 more hits than Ricky. How many hits did each player have?

22. Benjamin & Associates, a real estate developer, recently built 200 condominiums in McCall, Idaho. The condos were either two-bedroom units or three-bedroom units. If the total number of rooms in the entire complex is 507, how many two-bedroom units are there? How many three-bedroom units are there?

23. On a certain hot summer's day, 492 people used the public swimming pool. The daily prices are \$1.50 for children and \$2.50 for adults. The receipts for admission totaled \$929. How many children and how many adults swam at the public pool that day?

24. Scott invested a total of \$5,900 at two separate banks. One bank pays simple interest of 12% per year, whereas the other pays simple interest at a rate of 7% per year. If Scott earned \$568 in interest during a single year, how much did he have on deposit in each bank?

25. Thursday is ladies' night at the Slurp and Burp Bar and Grill. All adult beverages are \$2.50 for men and \$1.50 for women. A total of 956 adult beverages were sold last Thursday night. If the Slurp and Burp sold a total of \$1,996 in adult beverages last Thursday night, how many adult beverages were sold to women?

26. Farmer Brown planted corn and wheat on his 1,200 acres of land. The cost of planting and harvesting corn (which includes seed, planting, fertilizer, machinery, labor, and other costs) is \$285 per acre. The cost of planting and harvesting wheat is \$130 per acre. If Farmer Brown's total cost was \$245,900, how many acres of corn did he plant?

27. An airplane encountered a headwind during a flight between Joppetown and Jawsburgh which took 3 hours and 36 minutes. The return flight took 3 hours. If the distance from Joppetown to Jawsburgh is 1800 miles, find the airspeed of the plane (the speed of the plane in still air) and the speed of the wind, assuming both remain constant.

28. Gabby's piggy bank contains nickels and quarters worth \$9.10. If she has 46 coins in all, how many of each does she have?

29. Ned, the owner of Ned's Nut Shop, sells peanuts for \$3.25 per pound and cashews for \$6.00 per pound. Ned wants to create a 58.5-lb barrel of mixed nuts by mixing the peanuts and cashews together and sell it for \$4.00 per pound. How many pounds of each should Ned use?

30. Angela, the head of the office party planning committee, sent two coworkers, Kevin and Dwight, to the party store to purchase party hats and party whistles. Kevin purchased five packs of party hats and four packs of party whistles and paid a total of \$23.01. Dwight purchased four packs of party hats and five packs of party whistles, and paid a total of \$20.91. When they returned to the office, Angela informed them that they were supposed to buy a total of five packs of party hats and five packs of party whistles and sent them back to the store to straighten things out. How much should they be refunded?

12.2 Solving Systems of Linear Equations in Three Variables Using the Elimination Method

THINGS TO KNOW

Before working through this section, be sure that you are familiar with the following concepts:

VIDEO ANIMATION INTERACTIVE

 1. Solving a System of Linear Equations Using the Substitution Method (Section 12.1) 

 2. Solving a System of Linear Equations Using the Elimination Method (Section 12.1) 

OBJECTIVES

1 Verifying the Solution of a System of Linear Equations in Three Variables

2 Solving a System of Linear Equations Using the Elimination Method

3 Solving Consistent, Dependent Systems of Linear Equations in Three Variables

4 Solving Inconsistent Systems of Linear Equations in Three Variables

5 Solving Applied Problems Using a System of Linear Equations Involving Three Variables

SECTION 12.2 EXERCISES

OBJECTIVE 1 VERIFYING THE SOLUTION OF A SYSTEM OF LINEAR EQUATIONS IN THREE VARIABLES

In **Section 12.1** we solved systems of linear equations in two variables. In this section (and in **Section 12.3**) we will learn techniques that will enable us to solve systems of linear equations in three variables. An equation such as $2x + 3y + 4z = 12$ is called a **linear equation in three variables** because there are three variables and each variable term is **linear**.

Definition Linear Equation in Three Variables

A **linear equation in three variables** is an equation that can be written in the form $Ax + By + Cz = D$, where A, B, C, and D are real numbers, and A, B, and C are not all equal to 0.

When two or more linear equations in three variables are considered simultaneously, then this collection of linear equations forms a system of linear equations in three variables.

Definition System of Linear Equations in Three Variables

A **system of linear equations in three variables** is a collection of linear equations in three variables considered simultaneously. A solution to a system of linear equations in three variables is an **ordered triple** that satisfies all equations in the system.

Recall from Section 12.1 that if a system of two linear equations has one unique solution, then the solution can be geometrically represented by the intersection of two lines. The two lines intersect at exactly one point (or at a unique ordered pair.) Similarly, if a system of three linear equations in three variables has a unique solution, then the solution can be geometrically represented by the intersection of three planes. The three planes intersect at exactly one point (or at a unique **ordered triple**). See Figure 5.

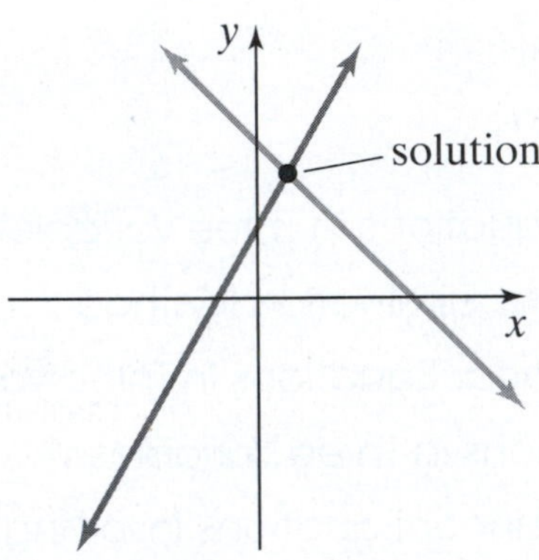

The figure above represents a system of two linear equations in two variables that has a unique solution. The solution can be represented by the point of intersection of two lines.

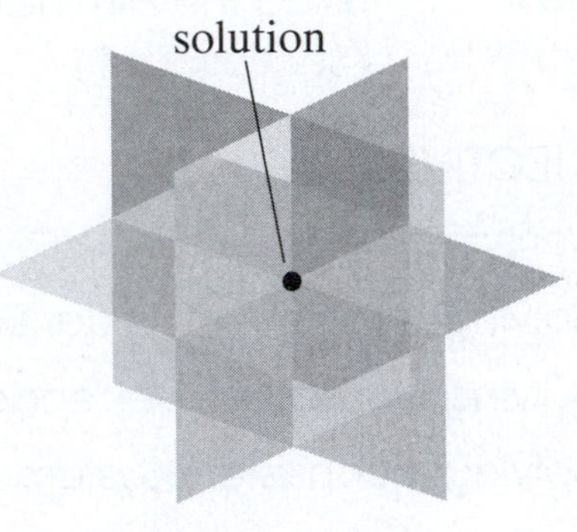

The figure above represents a system of three linear equations in three variables that has a unique solution. The solution can be represented by the point of intersection of three planes.

Figure 5

The following is an example of a system of three linear equations in three variables:

$$\begin{aligned} 2x + 3y + 4z &= 12 \\ x - 2y + 3z &= 0 \\ -x + y - 2z &= -1 \end{aligned}$$

The solution to this linear system is the set of all *ordered triples* (x, y, z) that satisfy all three equations. In Example 1, we show that the ordered triple $(1, 2, 1)$ is a solution to this linear system.

Example 1 Verify the Solution of a System of Linear Equations

Verify that the ordered triple $(1, 2, 1)$ is a solution to the following system of linear equations:

$$\begin{aligned} 2x + 3y + 4z &= 12 \\ x - 2y + 3z &= 0 \\ -x + y - 2z &= -1 \end{aligned}$$

Solution To verify that the ordered triple $(1, 2, 1)$ is a solution, we must substitute $x = 1, y = 2$, and $z = 1$ in each of the three equations and determine whether a true statement occurs.

$$2(1) + 3(2) + 4(1) \stackrel{?}{=} 12 \longrightarrow 2 + 6 + 4 \stackrel{?}{=} 12 \longrightarrow 12 = 12 \checkmark \text{ True}$$
$$(1) - 2(2) + 3(1) \stackrel{?}{=} 0 \longrightarrow 1 - 4 + 3 \stackrel{?}{=} 0 \longrightarrow 0 = 0 \checkmark \text{ True}$$
$$-(1) + (2) - 2(1) \stackrel{?}{=} -1 \longrightarrow -1 + 2 - 2 \stackrel{?}{=} -1 \longrightarrow -1 = -1 \checkmark \text{ True}$$

All three statements are true. This implies that the ordered triple $(1, 2, 1)$ is a solution to this system.

You Try It Work through this You Try It problem.

Work Exercises 1 and 2 in this textbook or in the MyLab Math Study Plan.

OBJECTIVE 2 SOLVING A SYSTEM OF LINEAR EQUATIONS USING THE ELIMINATION METHOD

A system of three linear equations in three variables can be geometrically represented by three planes in space. In Example 1, the system has one unique solution, the ordered triple $(1, 2, 1)$. If we were to graph the three planes represented by the three linear equations in Example 1, we would see that the three planes intersect at exactly one point. Systems of equations that have at least one solution are called **consistent systems.** A system that has exactly one solution is called a **consistent, independent** system. It is possible that a system of three linear equations in three variables has infinitely many solutions. These systems are called **consistent, dependent** systems. It is also possible for a system of three linear equations in three variables to have no solution at all. These systems are called **inconsistent systems. Figure 6** illustrates some possibilities of consistent systems and inconsistent systems.

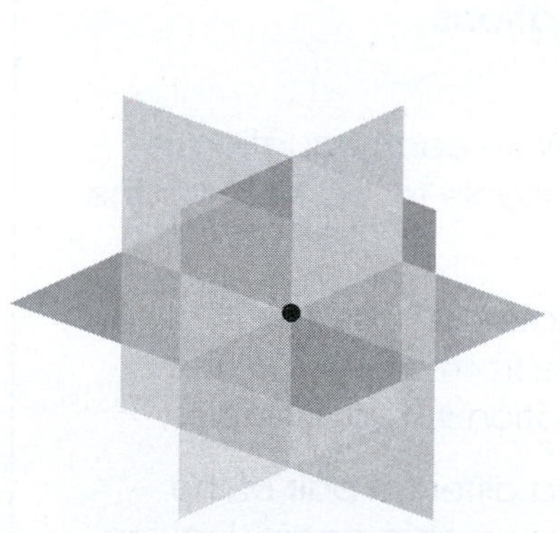

Consistent, Independent
One solution

Consistent, Dependent
Infinitely many solutions
(Planes intersect at a line.)

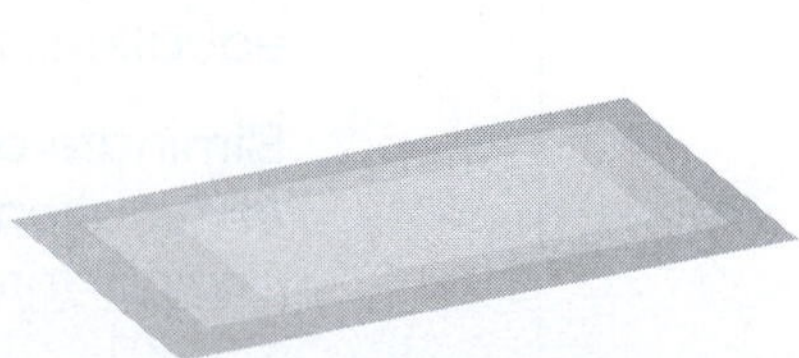

Consistent, Dependent
Infinitely many solutions
(Three equations describe the same plane.)

Figure 6 Continues

Inconsistent
No solution
(Three planes are parallel.)

Inconsistent
No solution
(Two planes are parallel.)

Inconsistent
No solution
(Planes intersect two at a time.)

Figure 6

To solve systems of linear equations in three variables by graphing would require us to graph **planes** in three dimensions, which is not a practical task. Instead, we will focus in this section on solving systems of linear equations in three variables by applying the elimination method. This method is similar to the elimination method used when solving systems of linear equations in two variables. See **Section 12.1** to review this method. (In **Section 12.3**, an alternative method will be introduced.)

The goal of the elimination method when solving a system of three linear equations in three variables is to reduce the system of three equations in three variables down to a system of two equations in two variables. At that point, we can reduce the two-equation system to a single equation in one variable and easily solve that equation as we did in **Section 12.1**. Then, using **back substitution**, we can find the values of the other two variables.

Before looking at an example, we present some guidelines for solving systems of linear equations in three variables by elimination.

Guidelines for Solving a System of Linear Equations in Three Variables by Elimination

Step 1. Write each equation In standard form. Write each equation in the form $Ax + By + Cz = D$ lining up the variable terms. Number the equations to keep track of them.

Step 2. Eliminate a variable from one pair of equations. Use the **elimination method** to eliminate a variable from any two of the original three equations, leaving one equation in two variables,

Step 3. Eliminate the same variable again. Use a different pair of the original equations and eliminate the same variable again, leaving one equation in two variables.

Step 4. Solve the system of linear equations in two variables. Use the resulting equations from steps 2 and 3 to create and solve the corresponding system of linear equations in two variables by **substitution** or **elimination**.

Step 5. Use back substitution to find the value of the third variable. Substitute the results from step 4 into any of the original equations to find the value of the remaining variable.

Step 6. Check the solution. Check the proposed solution in each equation of the system and write the solution set.

Example 2 Solving Systems of Linear Equations in Three Variables

Solve the following system:

$$\begin{aligned} 2x + 3y + 4z &= 12 \\ x - 2y + 3z &= 0 \\ -x + y - 2z &= -1 \end{aligned}$$

Solution We follow our **guidelines** for solving systems of linear equations in three variables. Follow the steps below, or watch this **video** for a detailed solution.

Step 1. The equations are already in **standard form** and all the variables are lined up. We rewrite the system and number each equation.

$$\begin{aligned} 2x + 3y + 4z &= 12 \quad &(1) \\ x - 2y + 3z &= 0 \quad &(2) \\ -x + y - 2z &= -1 \quad &(3) \end{aligned}$$

Step 2. We can eliminate any of the variables. For convenience, we will eliminate the variable x from equations **(1)** and **(2)**. We can do this by multiplying equation **(2)** by -2 and adding the equations together.

$$\begin{aligned} 2x + 3y + 4z &= 12 \quad &(1) \\ -2x + 4y - 6z &= 0 \quad &\text{Multiply (2) by } -2. \\ \hline 7y - 2z &= 12 \quad &\text{Add to produce new equation (4).} \end{aligned}$$

How do you know which variable to eliminate? Click **here** to find out.

Step 3. We need to eliminate the same variable, x, from a different pair of equations. We will use equations **(2)** and **(3)** for the second pair. Since the **coefficients** of x in these equations have the same **absolute value**, but opposite signs, we can eliminate the variable by simply adding the equations.

$$\begin{aligned} x - 2y + 3z &= 0 \quad &(2) \\ -x + y - 2z &= -1 \quad &(3) \\ \hline -y + z &= -1 \quad &\text{Add to produce new equation (5).} \end{aligned}$$

Step 4. Combining equations **(4)** and **(5)**, we form a system of linear equations in two variables.

$$\begin{aligned} 7y - 2z &= 12 \quad &(4) \\ -y + z &= -1 \quad &(5) \end{aligned}$$

To solve this system, we use the **elimination method** by multiplying equation **(5)** by 7 and adding the result to equation **(4)** to eliminate the variable y.

$$\begin{aligned} 7y - 2z &= 12 && (4) \\ \underline{-7y + 7z} &\underline{= -7} && \text{Multiply (5) by 7.} \\ 5z &= 5 && \text{Add.} \\ z &= 1 && \text{Divide by 5.} \end{aligned}$$

Since $z = 1$, we **back-substitute** this into **(5)** to solve for y.

$$\begin{aligned} -y + z &= -1 && (5) \\ -y + (1) &= -1 && \text{Substitute 1 for } z. \\ -y + 1 &= -1 && \text{Simplify.} \\ -y &= -2 && \text{Subtract 1.} \\ y &= 2 && \text{Divide by } -1. \end{aligned}$$

When **back-substituting**, we can use either of the equations, but often one equation is preferred over another. Can you see why we chose to use equation **(5)** instead of equation **(4)**? Click **here** for an explanation.

Step 5. Substitute 2 for y and 1 for z in any of the original equations, **(1)**, **(2)**, or **(3)**, and solve for x. We will back-substitute using equation **(2)**.

$$\begin{aligned} x - 2y + 3z &= 0 && (2) \\ x - 2(2) + 3(1) &= 0 && \text{Substitute 2 for } y \text{ and 1 for } z. \\ x - 4 + 3 &= 0 && \text{Multiply.} \\ x - 1 &= 0 && \text{Simplify.} \\ x &= 1 && \text{Add 1.} \end{aligned}$$

The solution to the system is the **ordered triple** $(1, 2, 1)$.

Step 6. Click **here** to check your answer.

You Try It **Work through this You Try It problem.**

Work Exercises 3–5 in this textbook or in the MyLab Math Study Plan.

Example 3 Solving Systems of Linear Equations in Three Variables Involving Fractions

Solve the following system:

$$\begin{aligned} \frac{1}{2}x + \quad y + \frac{2}{3}z &= 2 \\ \frac{3}{4}x + \frac{5}{2}y - 2z &= -7 \\ x + 4y + 2z &= 4 \end{aligned}$$

Solution If an equation in the system contains fractions, then it is often helpful to **clear the fractions** first. After doing this, follow the **guidelines** for solving a system of linear equations in three variables. Click **here** to check your answer, or watch this **video** for a detailed solution.

(eText Screens 12.2-1–12.2-36)

You Try It Work through this You Try It problem.

Work Exercises 6–8 in this textbook or in the MyLab Math Study Plan.

Example 4 Solving Systems of Linear Equations in Three Variables with Missing Terms

Solve the following system:

$$\begin{aligned} 2x + y &= 13 \\ 3x - 2y + z &= 8 \\ x + 2y - 3z &= 5 \end{aligned}$$

Solution We follow our **guidelines** for solving a system of linear equations in three variables.

Step 1. First, we rewrite and number each equation, lining up the variables. Notice that the first equation has a variable term missing, so we put a gap in its place.

$$\begin{aligned} 2x + y \quad\quad &= 13 \quad \textbf{(1)} \\ 3x - 2y + z &= 8 \quad \textbf{(2)} \\ x + 2y - 3z &= 5 \quad \textbf{(3)} \end{aligned}$$

Step 2. We can eliminate any of the variables, but we notice that one equation already has z eliminated. By selecting z as the variable to eliminate, we can move directly to **step 3.**

Step 3. Looking at equations **(2)** and **(3)**, we might be tempted to add these equations to eliminate y. However, in step 2, we selected z as the variable to eliminate, so we need to eliminate z again in this step.

Continue working through the solution. Click **here** to check your answer, or watch this **video** for a detailed solution.

You Try It Work through this You Try It problem.

Work Exercises 9–11 in this textbook or in the MyLab Math Study Plan.

OBJECTIVE 3 SOLVING CONSISTENT, DEPENDENT SYSTEMS OF LINEAR EQUATIONS IN THREE VARIABLES

Recall from **Section 12.1** that if a system of equations has at least one solution, then the system is consistent. If the system has infinitely many solutions, then the system is consistent and **dependent**. For the two-variable case, a consistent, dependent system can be geometrically represented by a pair of **coinciding lines**. For the three-variable case, consistent, dependent systems can be geometrically represented by three planes that intersect at a line or by three equations that all describe the same plane. See Figure 7.

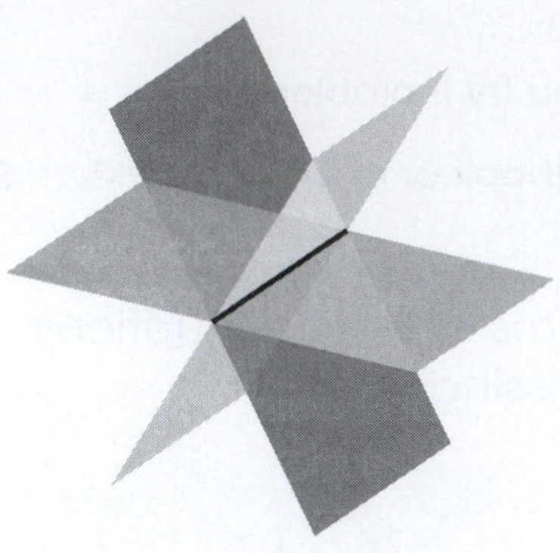

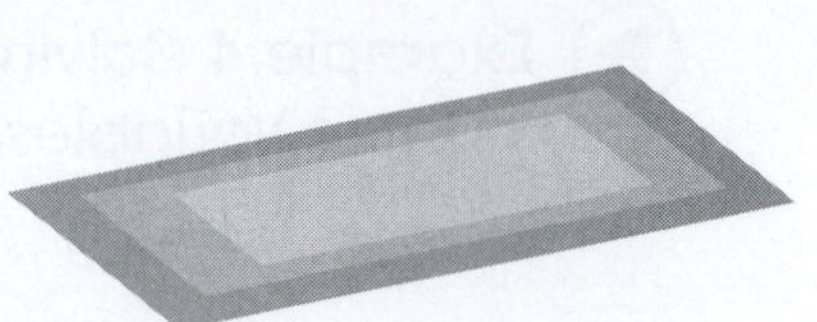

Consistent, Dependent
Infinitely many solutions
(Planes intersect at a line.)

Consistent, Dependent
Infinitely many solutions
(Three equations describe the same plane.)

Figure 7 A geometric representation of consistent, dependent systems in three variables.

When solving systems of linear equations in two variables, encountering an **identity** automatically meant that the system was **dependent** and had infinitely many ordered pair solutions. (See **Example 5** of Section 12.1.) However, unlike the two-variable case, identifying dependent systems in three variables takes a bit more work because we must consider more than one pairing of the equations in the system. Obtaining an identity with one pairing of equations is not sufficient to say that the system is dependent as we will see later in **Step 2 of Example 6**. However, during the solution process, if we encounter an **identity** after successfully obtaining a two-variable system of equations, then the system is dependent and the system has infinitely many solutions.

Example 5 Solving Systems of Dependent Linear Equations in Three Variables

Solve the following system.

$$x + 2y + 3z = 10 \qquad (1)$$

$$x + y + z = 7 \qquad (2)$$

$$3x + 2y + z = 18 \qquad (3)$$

Solution Eliminating the variable x using equations **(1)** and **(2)** results in the equation $-y - 2z = -3$. Eliminating the variable x again using equations **(1)** and **(3)** results in the equation $-4y - 8z = -12$. Combining these two equations gives the two-variable system:

$$-y - 2z = -3 \qquad (4)$$

$$-4y - 8z = -12 \qquad (5)$$

We can multiply equation **(4)** by -4 to get the new system:

$$4y + 8z = 12 \qquad (6)$$

$$-4y - 8z = -12 \qquad (5)$$

We now add equations **(6)** and **(5)**.

$$\begin{array}{rl} 4y + 8z = 12 & \textbf{(6)} \\ \underline{-4y - 8z = -12} & \textbf{(5)} \\ 0 = 0 & \end{array}$$

We see that adding equations **(6)** and **(5)** results in the **identity** $0 = 0$. Therefore, this two-variable system has infinitely many solutions and the system is said to be **dependent.** Solve either equation **(6)** or **(5)** for either variable. Solving for the variable y we get $y = 3 - 2z$.

Since $y = 3 - 2z$, we back-substitute this into **(1)** to solve for x.

$$x + 2(3 - 2z) + 3z = 10 \quad \text{Substitute } 3 - 2z \text{ for } y \text{ in equation } (1).$$

$$x + 6 - 4z + 3z = 10 \quad \text{Multiply.}$$

$$x + 6 - z = 10 \quad \text{Simplify.}$$

$$x = 4 + z \quad \text{Solve for } x.$$

Therefore, $x = 4 + z$ and $y = 3 - 2z$. Notice that the value of x and y depend on the choice of the variable z. The variable z is free to be any real number. Consequently, we say that z is a *free variable.*

Recall from **Example 5 of Section 12.1** that we can write the solutions to a **dependent** system of two equations in two variables in **ordered pair notation.** Similarly, the solution to a dependent system of three equations in three variables can be written in **ordered triple notation.** The solution to this **dependent** system can be written in ordered triple notation as $(4 + z, 3 - 2z, z)$ where z is free to be any real number.

Thus, this system has infinitely many solutions. For example, letting $z = 0, z = -1$, and $z = 3$, we get the following ordered triples as solutions to this system:

$$z = 0: \quad (4, 3, 0)$$

$$z = -1: \quad (3, 5, -1)$$

$$z = 3; \quad (7, -3, 3)$$

Try substituting these ordered triples into the original system to verify that all three are solutions.

Watch this **video** to see every step of the solution process to this system of three linear equations in three variables.

You Try It Work through this You Try It problem.

Work Exercises 12–15 in this textbook or in the MyLab Math Study Plan.

OBJECTIVE 4 SOLVING INCONSISTENT SYSTEMS OF LINEAR EQUATIONS IN THREE VARIABLES

Recall from **Section 12.1** that a system of linear equations in two variables could have no solution. Linear systems having no solutions are called **inconsistent systems.** For the two-variable case, this occurred if the two linear equations represented parallel lines because parallel lines have no points in common. For the three-variable case, a system is **inconsistent** if all three planes have no points in common. This happens if all three planes are parallel or if two of the planes are parallel. A third possibility of an inconsistent system is when the planes intersect two at a time. **Figure 8** geometrically illustrates three types of inconsistent systems in three variables.

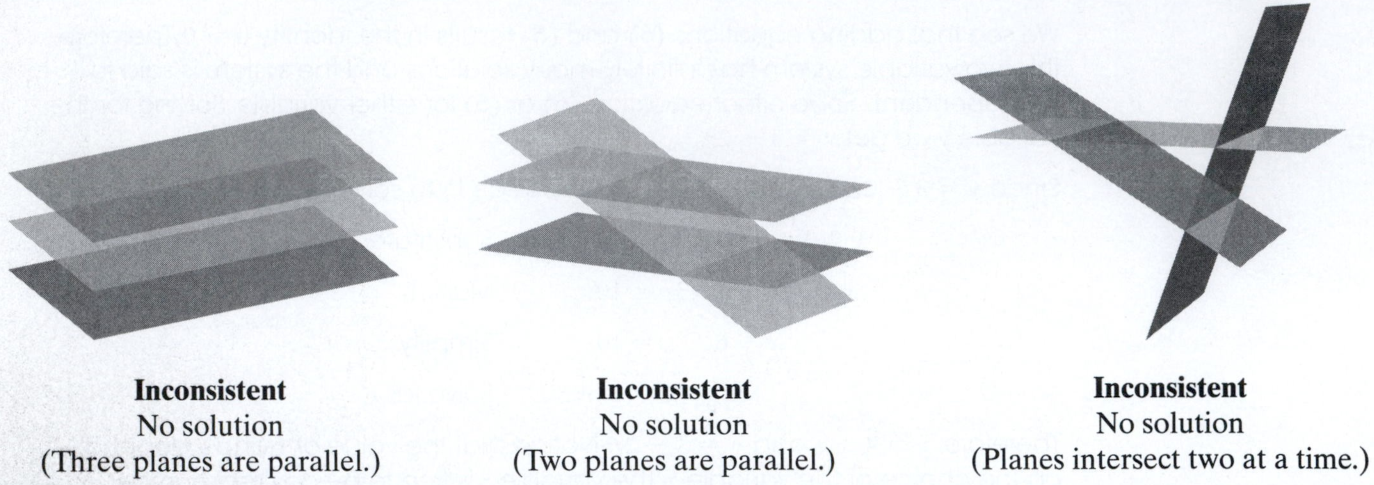

Inconsistent No solution (Three planes are parallel.)   **Inconsistent** No solution (Two planes are parallel.)   **Inconsistent** No solution (Planes intersect two at a time.)

Figure 8 A geometric representation of inconsistent systems in three variables.

Recall from **Example 4 of Section 12.1** that when solving an inconsistent system of two equations we will eventually obtain a **contradiction.** This is also true in the three-variable case. If we encounter a contradiction during the solution process, then the system is inconsistent and there is no solution. We illustrate this in Example 6.

Example 6 Solving Systems of Linear Equations in Three Variables with No Solution

Solve the following system:

$$\begin{aligned} x - y + 2z &= 5 \\ 3x - 3y + 6z &= 15 \\ -2x + 2y - 4z &= 7 \end{aligned}$$

Solution We follow our **guidelines** for solving a system of linear equations in three variables.

Step 1. The equations are already in standard form with all the variables lined up. We rewrite the system and number each equation.

$$\begin{aligned} x - y + 2z &= 5 && \textbf{(1)} \\ 3x - 3y + 6z &= 15 && \textbf{(2)} \\ -2x + 2y - 4z &= 7 && \textbf{(3)} \end{aligned}$$

Step 2. For convenience, we will eliminate the variable x from equations **(1)** and **(2)**. To do this, we multiply equation **(1)** by -3 and add the equations together.

$$\begin{aligned} -3x + 3y - 6z &= -15 && \text{Multiply (1) by } -3. \\ 3x - 3y + 6z &= 15 && \text{(2)} \\ \hline 0 &= 0 && \text{Add} \end{aligned}$$

The last line, $0 = 0$, is an **identity**. In the two-variable case, we would stop and say that the system had an infinite number of solutions. However, in systems of three variables, this is not necessarily the case.

Step 3. Let's continue our process and eliminate the variable x from the pairing of equations **(1)** and **(3)**. To do this, we multiply equation **(1)** by 2 and add the equations.

$$\begin{array}{rl} 2x - 2y + 4z = 10 & \text{Multiply (1) by 2.} \\ \underline{-2x + 2y - 4z = 7} & \text{(3)} \\ 0 = 17 & \text{Add.} \end{array}$$

The last line, $0 = 17$, is a **contradiction**, so the system is **inconsistent** and has no solution. Click **here** to find out why this occurs.

You Try It **Work through this You Try It problem.**

Work Exercises 12–17 in this textbook or in the MyLab Math Study Plan.

OBJECTIVE 5 SOLVING APPLIED PROBLEMS USING A SYSTEM OF LINEAR EQUATIONS IN THREE VARIABLES

Systems of linear equations can be used to solve a wide variety of applications. Following are two examples. We will solve each example using the elimination method.

Example 7 Real-Time Strategy Game

While playing a real-time strategy game, Joel created military units to defend his town: warriors, skirmishers, and archers. Warriors require 20 units of food and 50 units of gold. Skirmishers require 25 units of food and 35 units of wood. Archers require 32 units of wood and 32 units of gold. If Joel used 506 units of gold, 606 units of wood, and 350 units of food to create the units, how many of each type of military unit did he create?

Solution We want to find the number of each type of military unit created. There are three types of units: warriors, skirmishers, and archers. Each unit requires a certain amount of gold, wood, and food. We know how much of each resource is needed for each unit, and we know the total amount of each resource that is used. We summarize this information in the following table.

	Each Warrior	Each Skirmisher	Each Archer	Total
Units of Gold	50	0	32	506
Units of Wood	0	35	32	606
Units of Food	20	25	0	350

Let W, S, and A represent the number of warriors, skirmishers, and archers respectively. We can obtain the three equations based on the total units of gold, wood, and food used.

GOLD

We base our first equation on the total amount of gold used. Each warrior requires 50 units of gold, each skirmisher requires 0 units of gold, and each archer requires 32 units of gold. Joel used a total of 506 units of gold. Thus, we obtain the first equation

$$\underbrace{50W}_{\text{Units of gold for warriors}} + \underbrace{0S}_{\text{Units of gold for skirmishers}} + \underbrace{32A}_{\text{Units of gold for archers}} = \underbrace{506.}_{\text{Total units of gold}}$$

WOOD

The second equation is based on the total amount of wood used. Each warrior requires 0 units of wood, each skirmisher requires 35 units of wood, and each archer requires 32 units of wood. Joel used a total of 606 units of wood. Using this information, we obtain the second equation

$$\underbrace{0W}_{\text{Units of wood for warriors}} + \underbrace{35S}_{\text{Units of wood for skirmishers}} + \underbrace{32A}_{\text{Units of wood for archers}} = \underbrace{606.}_{\text{Total units of wood}}$$

FOOD

The third equation is based on the total amount of food used. Each warrior required 20 units of food, each skirmisher requires 25 units of food, and each archer requires 0 units of food. Joel used a total of 350 units of food. We now write the third equation

$$\underbrace{20W}_{\text{Units of food for warriors}} + \underbrace{25S}_{\text{Units of food for skirmishers}} + \underbrace{0A}_{\text{Units of food for archers}} = \underbrace{350.}_{\text{Total units of food}}$$

Using these three equations, we form the following system.

$$\begin{aligned} 50W + \ \ 0S + 32A &= 506 \\ 0W + 35S + 32A &= 606 \\ 20W + 25S + \ \ 0A &= 350 \end{aligned}$$

Now solve this system using the elimination method. Once you have solved this system, **check your answers**, or watch this **video** for a detailed solution.

Example 8 Buy Clothes Online

Wendy ordered 30 T-shirts online for her three children. The small T-shirts cost \$4 each, the medium T-shirts cost \$5 each, and the large T-shirts were \$6 each. She spent \$40 more purchasing the large T-shirts than the small T-shirts, Wendy's total bill was \$154, How many T-shirts of each size did she buy?

Solution Let S, M, and L represent the number of small, medium, and large T-shirts, respectively. Because a total of 30 T-shirts were purchased, we obtain the first equation

$$S + M + L = 30$$

Because small T-shirts cost \$4, we know that the total amount spent on small T-shirts is $4S$. Similarly, the total spent on medium T-shirts is $5M$, and the total spent on large T-shirts is $6L$. Therefore, because the total value of the T-shirts was $154, we obtain the second equation

$$4S + 5M + 6L = 154$$

Finally, Wendy spent $40 more buying large T-shirts than small T-shirts, so $6L = 4S + 40$, which is equivalent to the equation

$$4S - 6L = -40$$

Thus, we obtain the system
$$\begin{aligned} S + M + L &= 30 \\ 4S + 5M + 6L &= 154, \\ 4S \qquad - 6L &= -40 \end{aligned}$$

Watch this **video** to see how to use the elimination method to determine that Wendy purchased 8 small T-shirts, 10 medium T-shirts, and 12 large T-shirts.

You Try It Work through this You Try It problem.

Work Exercises 18–23 in this textbook or in the MyLab Math Study Plan.

12.2 Exercises

In Exercises 1 and 2, two ordered triples are given. Determine whether each ordered triple is a solution of the given system.

1. $(-1, 1, -2), (1, -1, 2)$
$$\begin{aligned} x + y + z &= -2 \\ -x - 3y - 2z &= 2 \\ 2x - 2y + 5z &= -14 \end{aligned}$$

2. $(2, -1, 4), (-2, 1, -4)$
$$\begin{aligned} \frac{1}{2}x + 3y - z &= 6 \\ -x + y + \frac{1}{4}z &= 2 \\ x - 4y + z &= -10 \end{aligned}$$

In Exercises 3–11, solve each system of linear equations using the elimination method.

3. $$\begin{aligned} x + y + z &= 4 \\ 2x - y - 2z &= -10 \\ -x - y + 3z &= 8 \end{aligned}$$

4. $$\begin{aligned} x - 2y + z &= 6 \\ 2x + y - 3z &= -3 \\ x - 3y + 3z &= 10 \end{aligned}$$

5. $$\begin{aligned} x - 2y + 2z &= 2 \\ 3x + 2y - 2z &= -1 \\ x - y - 2z &= 0 \end{aligned}$$

6. $$\begin{aligned} x - \frac{1}{2}y + \frac{1}{2}z &= -3 \\ x + y - z &= 0 \\ -3x - 3y + 4z &= 1 \end{aligned}$$

7. $$\begin{aligned} \frac{1}{3}x - \frac{2}{3}y + z &= 0 \\ \frac{1}{2}x - \frac{3}{4}y + z &= -\frac{1}{2} \\ -2x - y + z &= 7 \end{aligned}$$

8. $$\begin{aligned} x + y + 10z &= 3 \\ \frac{1}{2}x - y + z &= -\frac{5}{6} \\ -2x + 3y - 5z &= \frac{7}{3} \end{aligned}$$

9. $$\begin{aligned} -4x + 5y + 9z &= -9 \\ x - 2y + z &= 0 \\ 2y - 8z &= 8 \end{aligned}$$

10. $$\begin{aligned} 2x + 2y + z &= 9 \\ x + z &= 4 \\ 4y - 3z &= 17 \end{aligned}$$

11. $$\begin{aligned} x - y &= 7 \\ y - z &= 2 \\ x + z &= 1 \end{aligned}$$

In Exercises 12–17, solve each system of linear equations. If the system has infinitely many solutions, describe the solution with the equation of a plane or an ordered triple in terms of one variable.

12. $$\begin{aligned} 2x + 6y - 4z &= 8 \\ -x - 3y + 2z &= -4 \\ x + 3y - 2z &= 4 \end{aligned}$$

13. $$\begin{aligned} 2x - y + z &= -6 \\ x - \frac{1}{2}y + \frac{1}{2}z &= -3 \\ 4x - 2y + 2z &= -12 \end{aligned}$$

14. $$\begin{aligned} x + 2y - z &= 11 \\ x + 3y - 2z &= 14 \\ 3x + 7y - 4z &= 36 \end{aligned}$$

15. $$\begin{aligned} x - y + z &= 5 \\ 2x + 3y - 3z &= -5 \\ 3x + 2y - 2z &= 0 \end{aligned}$$

16. $$\begin{aligned} x - 4y + 2z &= 7 \\ \frac{1}{2}x - 2y + z &= 1 \\ -3x + y - 4z &= 9 \end{aligned}$$

17. $$\begin{aligned} 4x - y + z &= 8 \\ x + y + 3z &= 2 \\ 3x - 2y - 2z &= 5 \end{aligned}$$

In Exercises 18–23, Write a system of linear equations in three variables and then use the elimination method to solve the system.

18. While playing a real-time strategy game. Arvin created military units for a battle: long swordsmen, spearmen, and cross-bowmen. Long swordsmen require 60 units of food and 20 units of gold. Spearmen require 35 units of food and 25 units of wood, Crossbowmen require 25 units of wood and 45 units of gold. If Arvin used 1975 units of gold. 1375 units of wood, and 1900 units of food to create the units, how many of each type of military unit did he create?

19. The concession stand at a school basketball tournament sells hot dogs, hamburgers, and chicken sandwiches. During one game, the stand sold 16 hot dogs, 14 hamburgers, and 8 chicken sandwiches for a total of $89.00. During a second game, the stand sold 10 hot dogs. 13 hamburgers, and 5 chicken sandwiches for a total of $66.25. During a third game, the stand sold 4 hot dogs, 7 hamburgers, and 7 chicken sandwiches for a total of $49.75. Determine the price of each product.

20. Ben ordered 35 pizzas for an office party. He ordered three types: cheese, supreme, and pepperoni. Cheese pizza costs $9 each, pepperoni pizza costs $12 each, and supreme pizza costs $15 each. He spent exactly twice as much on the pepperoni pizzas as he did on the cheese pizzas. If Ben spent $420, how many pizzas of each type did he buy?

21. On opening night of the play *The Music Man,* 1010 tickets were sold for a total of $10,300. Adult tickets cost $12 each, children's tickets cost $10 each, and senior citizen tickets cost $7 each. If the total number of adult and children tickets sold exceeded twice the number of senior citizen tickets sold by 170 tickets, then how many tickets of each type were sold?

22. Tyler Hansbrough was the leading scorer of the 2009 NCAA Basketball champions, the North Carolina Tar Heels. Hansbrough scored a total of 722 points during the 2009 season. He made 26 more one-point free throws than two-point field goals, and his number of two-point field goals was two less than 25 times his number of three-point field goals. How many free throws, two-point field goals, and three-point field goals did Tyler Hansbrough make during the 2009 season? (*Source;* espn.com)

23. The number of new Facebook users, y (in millions), between September 2008 and March 2009 can be modeled by the equation $y = ax^2 + bx + c$, where x represents the age of the user. Using the ordered-pair solutions (15, 1), (35, 7), and (55, 3), create a system of linear equations in three variables for $a, b,$ and c. Do this by substituting each ordered-pair solution into the model, creating an equation in three variables. Solve the resulting system to find the coefficients of the model. Then use the model to predict the number of new Facebook users who were 25 years old. (*Source:* www.facebook.com)

12.3 Solving Systems of Linear Equations in Three Variables Using Gaussian Elimination and Gauss-Jordan Elimination

THINGS TO KNOW

Before working through this section, be sure that you are familiar with the following concepts:

VIDEO ANIMATION INTERACTIVE

You Try It
1. Solving a System of Linear Equations Using the Substitution Method (Section 12.1) 

You Try It
2. Solving a System of Linear Equations Using the Elimination Method (Section 12.1) 

You Try It
3. Verifying the Solution of a System of Linear Equations in Three Variables (Section 12.2)  

OBJECTIVES

1 Solving a System of Linear Equations Using Gaussian Elimination

2 Using an Augmented Matrix to Solve a System of Linear Equations

3 Solving Consistent, Dependent Systems of Linear Equations in Three Variables

4 Solving Inconsistent Systems of Linear Equations in Three Variables

5 Determining Whether a System Has No Solution or Infinitely Many Solutions

6 Solving Linear Systems Having Fewer Equations Than Variables

7 Solving Applied Problems Using a System of Linear Equations Involving Three Variables

SECTION 12.3 EXERCISES

OBJECTIVE 1 SOLVING A SYSTEM OF LINEAR EQUATIONS USING GAUSSIAN ELIMINATION

In Section 12.2, we used the method of elimination to solve systems of three equations involving three variables. In this section, will learn two more similar techniques that can be used to solve such systems. Before we learn these techniques, first consider the following system of equations.

$$\begin{aligned} x - 2y + 3z &= 0 \quad &(1)\\ -y + z &= -1 \quad &(2)\\ 5z &= 5 \quad &(3) \end{aligned}$$

Notice that equation **(3)** has only one variable, equation **(2)** contains two variables, and equation **(1)** has three variables. A system of this form is said to be in **triangular form.** Note that we can easily solve equation **(3)** by dividing both sides of the equation by 5 to get the solution $z = 1$. Using back substitution, we can solve for y and x.

Using the value of $z = 1$ in equation (2), we can solve for y:

$-y + z = -1$	Write equation (2).
$-y + (1) = -1$	Substitute 1 for z.
$-y = -2$	Subtract 1 from both sides.
$y = 2$	Divide both sides by -1.

Finally, using $z = 1$ and $y = 2$ in equation **(1)** gives

$x - 2y + 3z = 0$	Write equation (1).
$x - 2(2) + 3(1) = 0$	Substitute 2 for y and 1 for z.
$x - 1 = 0$	Simplify.
$x = 1$	Add 1 to both sides.

Therefore, the solution to this system of equations is the ordered triple $(1, 2, 1)$.

The process of writing a system of three linear equations in three variables into an **equivalent system** that is in **triangular form** and then using back substitution to solve for each variable is called **Gaussian elimination,** named after the famous German mathematician Carl Friedrich Gauss.

When using Gaussian elimination, there are three algebraic operations that we can use to reduce a system of linear equations into an equivalent system in triangular form. These algebraic operations are called the **elementary row operations** and are listed below.

ELEMENTARY ROW OPERATIONS

The following algebraic operations will result in an equivalent system of linear equations:

1. Interchange any two equations.
2. Multiply any equation by a nonzero constant.
3. Add a multiple of one equation to another equation.

Because Gaussian elimination can be a lengthy process, we use shorthand notation to document the elementary row operation performed at each step. This documentation is important because it will help you "retrace" your steps in case you make a mistake or want to study your work later. We use R_i to describe the i th equation of the system and R_j to describe the jth equation of the system. The shorthand notation is outlined as follows.

NOTATION USED TO DESCRIBE ELEMENTARY ROW OPERATIONS

Notation	**Meaning**
$R_i \Leftrightarrow R_j$	Interchange Rows i and j.
$kR_i \rightarrow \text{New } R_i$	k times Row i becomes New Row i.
$kR_i + R_j \rightarrow \text{New } R_j$	k times Row i plus Row j becomes New Row j.

Example 1 illustrates how we can use Gaussian elimination to solve a system of three linear equations. You should carefully work through Example 1 and take notes while you watch the video solution.

Example 1 Use Gaussian Elimination to Solve a System of Three Linear Equations

For the following system, use elementary row operations to find an equivalent system in triangular form and then use back substitution to solve the system:

$$\begin{aligned} 2x + 3y + 4z &= 12 \\ x - 2y + 3z &= 0 \\ -x + y - 2z &= -1 \end{aligned}$$

Solution Our goal is to use a series of elementary row operations to eliminate the variable x in one of the equations and eliminate the variables x and y in another equation. We typically start by trying to make the first coefficient of the first equation equal to 1. We can accomplish this by interchanging the first two equations. The other four elementary row operations are outlined as follows. You

can also watch the video to see each step worked out in detail.

$$\begin{aligned} 2x + 3y + 4x &= 12 \\ x - 2y + 3z &= 0 \\ -x + y - 2z &= -1 \end{aligned} \xrightarrow{R_1 \Leftrightarrow R_2} \begin{aligned} x - 2y + 3z &= 0 \\ 2x + 3y + 4z &= 12 \\ -x + y - 2z &= -1 \end{aligned} \xrightarrow{-2R_1 + R_2 \rightarrow \text{New } R_2} \begin{aligned} x - 2y + 3z &= 0 \\ 7y - 2z &= 12 \\ -x + y - 2z &= -1 \end{aligned}$$

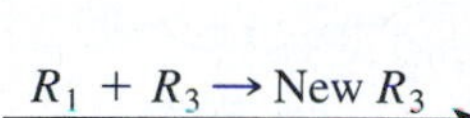

$$\xrightarrow{R_1 + R_3 \rightarrow \text{New } R_3} \begin{aligned} x - 2y + 3z &= 0 \\ 7y - 2z &= 12 \\ -y + z &= -1 \end{aligned} \xrightarrow{R_2 \Leftrightarrow R_3} \begin{aligned} x - 2y + 3z &= 0 \\ -y + z &= -1 \\ 7y - 2z &= 12 \end{aligned} \xrightarrow{7R_2 + R_3 \rightarrow \text{New } R_3} \begin{aligned} x - 2y + 3z &= 0 \\ -y + z &= -1 \\ 5z &= 5 \end{aligned}$$

Now that the linear system is written in triangular form, we can solve the third equation for z and then use back substitution to solve for the other two variables. You should verify that the solution to this system is the ordered triple $(1, 2, 1)$. ●

You Try It Work through this You Try It problem.

Work Exercises 1–4 in this textbook or in the MyLab Math Study Plan.

OBJECTIVE 2 USING AN AUGMENTED MATRIX TO SOLVE A SYSTEM OF LINEAR EQUATIONS

In Example 1, the variables x, y, and z were used. We actually could have used any three variables to describe this system. As it turns out, we do not need variables at all to solve a system of linear equations using Gaussian elimination! We need only the coefficients of each variable to perform the elementary row operations. We can simplify the Gaussian elimination process by simplifying the notation. Instead of writing three equations at each step, we only write the coefficients at each step. We accomplish this by writing the coefficients in a rectangular array called an **augmented matrix**. For example, the system from Example 1 can be written using an augmented matrix as follows:

System of Equations from Example 1

$$\begin{aligned} 2x + 3y + 4z &= 12 \\ x - 2y + 3z &= 0 \\ -x + y - 2z &= -1 \end{aligned}$$

Corresponding Augmented Matrix

$$\left[\begin{array}{rrr|r} 2 & 3 & 4 & 12 \\ 1 & -2 & 3 & 0 \\ -1 & 1 & -2 & -1 \end{array}\right]$$

Notice that each row of the corresponding augmented matrix represents the coefficients of each equation. The vertical bar is used to separate the coefficients of each variable from the constant coefficient that appears on the right side of each equation. We can perform the exact same elementary row operations as in Example 1 by writing an augmented matrix at each step. We now illustrate this process. For clarity, the corresponding system of equations is listed below each augmented matrix.

$$\left[\begin{array}{rrr|r} 2 & 3 & 4 & 12 \\ 1 & -2 & 3 & 0 \\ -1 & 1 & -2 & -1 \end{array}\right] \xrightarrow{R_1 \Leftrightarrow R_2} \left[\begin{array}{rrr|r} 1 & -2 & 3 & 0 \\ 2 & 3 & 4 & 12 \\ -1 & 1 & -2 & -1 \end{array}\right] \xrightarrow{-2R_1 + R_2 \rightarrow \text{New } R_2} \left[\begin{array}{rrr|r} 1 & -2 & 3 & 0 \\ 0 & 7 & -2 & 12 \\ -1 & 1 & -2 & -1 \end{array}\right] \xrightarrow{R_1 + R_3 \rightarrow \text{New } R_3}$$

$$\begin{aligned} 2x + 3y + 4z &= 12 \\ x - 2y + 3z &= 0 \\ -x + y - 2z &= -1 \end{aligned} \qquad \begin{aligned} x - 2y + 3z &= 0 \\ 2x + 3y + 4z &= 12 \\ -x + y - 2z &= -1 \end{aligned} \qquad \begin{aligned} x - 2y + 3z &= 0 \\ 7y - 2z &= 12 \\ -x + y - 2z &= -1 \end{aligned}$$

$$\left[\begin{array}{rrr|r} 1 & -2 & 3 & 0 \\ 0 & 7 & -2 & 12 \\ 0 & -1 & 1 & -1 \end{array}\right] \xrightarrow{R_2 \Leftrightarrow R_3} \left[\begin{array}{rrr|r} 1 & -2 & 3 & 0 \\ 0 & -1 & 1 & -1 \\ 0 & 7 & -2 & 12 \end{array}\right] \xrightarrow{7R_2 + R_3 \rightarrow \text{New } R_3} \left[\begin{array}{rrr|r} 1 & -2 & 3 & 0 \\ 0 & -1 & 1 & -1 \\ 0 & 0 & 5 & 5 \end{array}\right]$$

This matrix is now in triangular form. (All 0's below the diagonal.)

$$\begin{aligned} x - 2y + 3z &= 0 \\ 7y - 2z &= 12 \\ -y + z &= -1 \end{aligned} \qquad \begin{aligned} x - 2y + 3z &= 0 \\ -y + z &= -1 \\ 7y - 2z &= 12 \end{aligned} \qquad \begin{aligned} x - 2y + 3z &= 0 \\ -y + z &= -1 \\ 5z &= 5 \end{aligned}$$

Once we have a matrix written in triangular form, we can use the last row to solve for z and then use back substitution as before to solve for the remaining variables.

Just as in Example 1, the solution to this system is the ordered triple $(1, 2, 1)$.

Example 2 Solve a Linear System Using an Augmented Matrix (Triangular Form)

Create an augmented matrix and solve the following linear system using Gaussian elimination by writing an equivalent system in triangular form:

$$\begin{aligned} x + 2y - z &= 3 \\ x - 3y - 2z &= 11 \\ -x - 2y + 2z &= -6 \end{aligned}$$

Solution The augmented matrix that corresponds to the linear system is

$$\left[\begin{array}{rrr|r} 1 & 2 & -1 & 3 \\ 1 & -3 & -2 & 11 \\ -1 & -2 & 2 & -6 \end{array}\right]$$

We now proceed to use the following elementary row operations:

$$\left[\begin{array}{rrr|r} 1 & 2 & -1 & 3 \\ 1 & -3 & -2 & 11 \\ -1 & -2 & 2 & -6 \end{array}\right] \xrightarrow{(-1)R_1 + R_2 \rightarrow \text{New } R_2} \left[\begin{array}{rrr|r} 1 & 2 & -1 & 3 \\ 0 & -5 & -1 & 8 \\ -1 & -2 & 2 & -6 \end{array}\right] \xrightarrow{R_1 + R_3 \rightarrow \text{New } R_3} \left[\begin{array}{rrr|r} 1 & 2 & -1 & 3 \\ 0 & -5 & -1 & 8 \\ 0 & 0 & 1 & -3 \end{array}\right]$$

The augmented matrix is now in **triangular form**. Looking at the last row, we see that $z = -3$. The second row corresponds to the equation $-5y - z = 8$. Therefore,

$-5y - z = 8$	Use the augmented matrix to write equation (2).
$-5y - (-3) = 8$	Substitute -3 for z.
$-5y + 3 = 8$	Simplify.
$-5y = 5$	Subtract 3 from both sides.
$y = -1$	Divide both sides by -5.

Because $z = -3$ and $y = -1$, we can now use the first row to solve for x. You should verify that $x = 2$. Therefore, the solution to this linear system is the ordered triple $(2, -1 -3)$.

You Try It Work through this You Try It problem.

Work Exercises 5–10 in this textbook or in the MyLab Math Study Plan.

TRIANGULAR FORM, ROW-ECHELON FORM, AND REDUCED ROW-ECHELON FORM

When solving a system of three linear equations with three unknowns, only **triangular form** is needed to find the solution. However, it is often useful to further reduce the augmented matrix into one of two other simpler forms known as **row-echelon form** or **reduced row-echelon form**.

We see in Example 3 that the linear system can be reduced into the following triangular form:

Triangular Form

$$\left[\begin{array}{ccc|c} 1 & 2 & -1 & 3 \\ 0 & -5 & -1 & 8 \\ 0 & 0 & 1 & -3 \end{array}\right]$$

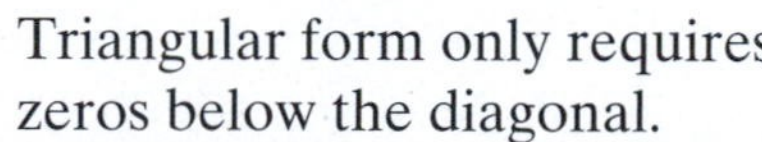

Triangular form only requires zeros below the diagonal.

To reduce this matrix into **row-echelon form** requires that all zeros remain below the diagonal and that all coefficients along the diagonal are 1. If we multiply the second row of the previous matrix by $-\frac{1}{5}$, we get the following equivalent matrix written in **row-echelon form**:

Row-Echelon Form

Row-echelon form requires zeros below the diagonal . . .

$$\left[\begin{array}{ccc|c} 1 & 2 & -1 & 3 \\ 0 & 1 & \frac{1}{5} & -\frac{8}{5} \\ 0 & 0 & 1 & -3 \end{array}\right]$$

. . . and 1's down the diagonal.

Reduced row-echelon form requires zeros below *and* above the diagonal and 1's down the diagonal. We can reduce the previous matrix into reduced row-echelon form by performing the following three elementary row operations.

$$\left[\begin{array}{ccc|c} 1 & 2 & -1 & 3 \\ 0 & 1 & \frac{1}{5} & -\frac{8}{5} \\ 0 & 0 & 1 & -3 \end{array}\right] \xrightarrow{(-2)R_2 + R_1 \to \text{New } R_1} \left[\begin{array}{ccc|c} 1 & 0 & -\frac{7}{5} & \frac{31}{5} \\ 0 & 1 & \frac{1}{5} & -\frac{8}{5} \\ 0 & 0 & 1 & -3 \end{array}\right] \xrightarrow{\left(-\frac{1}{5}\right)R_3 + R_2 \to \text{New } R_2}$$

$$\left[\begin{array}{ccc|c} 1 & 0 & -\frac{7}{5} & \frac{31}{5} \\ 0 & 1 & 0 & -1 \\ 0 & 0 & 1 & -3 \end{array}\right] \xrightarrow{\left(\frac{7}{5}\right)R_3 + R_1 \to \text{New } R_1} \left[\begin{array}{ccc|c} 1 & 0 & 0 & 2 \\ 0 & 1 & 0 & -1 \\ 0 & 0 & 1 & -3 \end{array}\right] \begin{array}{l} \Rightarrow x = 2 \\ \Rightarrow y = -1 \\ \Rightarrow z = -3 \end{array}$$

As you can see from the final matrix, we have now obtained all zeros above and below the diagonal. This final matrix is in **reduced row-echelon form**. The advantage to reducing the augmented matrix into reduced row-echelon form is that you can immediately read the solution by looking down the last column. In the previous reduced row-echelon matrix, you can clearly see that the solution is $x = 2, y = -1$, and $z = -3$. The process of reducing a system into **reduced row-echelon form** is called **Gauss-Jordan elimination**. We use Gauss-Jordan elimination from this point forward.

Example 3 Solve a Linear System Using Gauss-Jordan Elimination

Solve the following system using Gauss-Jordan elimination:

$$\begin{aligned} x_1 + x_2 + x_3 &= -1 \\ x_1 + 2x_2 + 4x_3 &= 3 \\ x_1 + 3x_2 + 9x_3 &= 3 \end{aligned}$$

Solution The augmented matrix that corresponds to the linear system is

$$\left[\begin{array}{ccc|c} 1 & 1 & 1 & -1 \\ 1 & 2 & 4 & 3 \\ 1 & 3 & 9 & 3 \end{array}\right]$$

We need to reduce this matrix into reduced row-echelon form. We must have 1s down the diagonal and zeros everywhere else to the left of the vertical bar. We can accomplish this by performing the following seven elementary row operations. Watch the **video** for a detailed step-by-step explanation.

$$\left[\begin{array}{ccc|c} 1 & 1 & 1 & -1 \\ 1 & 2 & 4 & 3 \\ 1 & 3 & 9 & 3 \end{array}\right] \xrightarrow{(-1)R_1 + R_2 \to \text{New } R_2} \left[\begin{array}{ccc|c} 1 & 1 & 1 & -1 \\ 0 & 1 & 3 & 4 \\ 1 & 3 & 9 & 3 \end{array}\right] \xrightarrow{(-1)R_1 + R_3 \to \text{New } R_3}$$

$$\left[\begin{array}{ccc|c} 1 & 1 & 1 & -1 \\ 0 & 1 & 3 & 4 \\ 0 & 2 & 8 & 4 \end{array}\right] \xrightarrow{(-1)R_2 + R_1 \to \text{New } R_1} \left[\begin{array}{ccc|c} 1 & 0 & -2 & -5 \\ 0 & 1 & 3 & 4 \\ 0 & 2 & 8 & 4 \end{array}\right] \xrightarrow{(-2)R_2 + R_3 \to \text{New } R_3}$$

$$\left[\begin{array}{ccc|c} 1 & 0 & -2 & -5 \\ 0 & 1 & 3 & 4 \\ 0 & 0 & 2 & -4 \end{array}\right] \xrightarrow{\left(\frac{1}{2}\right)R_3 \to \text{New } R_3} \left[\begin{array}{ccc|c} 1 & 0 & -2 & -5 \\ 0 & 1 & 3 & 4 \\ 0 & 0 & 1 & -2 \end{array}\right] \xrightarrow{(-3)R_3 + R_2 \to \text{New } R_2}$$

$$\left[\begin{array}{ccc|c} 1 & 0 & -2 & -5 \\ 0 & 1 & 0 & 10 \\ 0 & 0 & 1 & -2 \end{array}\right] \xrightarrow{2R_3 + R_1 \to \text{New } R_1} \left[\begin{array}{ccc|c} 1 & 0 & 0 & -9 \\ 0 & 1 & 0 & 10 \\ 0 & 0 & 1 & -2 \end{array}\right] \begin{array}{l} \Rightarrow x_1 = -9 \\ \Rightarrow x_2 = 10 \\ \Rightarrow x_3 = -2 \end{array}$$

You can see by looking at the last matrix that $x_1 = -9, x_2 = 10$, and $x_3 = -2$. Thus, the solution is the ordered triple $(-9, 10, -2)$.

You Try It Work through this You Try It problem.

Work Exercises 11–19 in this textbook or in the MyLab Math Study Plan.

OBJECTIVE 3 SOLVING CONSISTENT, DEPENDENT SYSTEMS OF LINEAR EQUATIONS IN THREE VARIABLES

So far in this section, every system of equations that we have encountered had one unique solution. Geometrically, this happens when the three planes described by each equation intersect at exactly one point. Systems of equations that have at least one solution are called **consistent systems.** When a system has *exactly one solution* then the system is called a **consistent, independent** system. It is possible for a system of three linear equations in three variables to have infinitely many solutions. If the system has infinitely many solutions, then the system is called a **consistent, dependent** system. For the two-variable case, a consistent, dependent system can be geometrically represented by a pair of **coinciding lines.** For the three-variable case, consistent, dependent systems can be geometrically represented by three planes that intersect at a line or by three equations that all describe the same plane. See **Figure 9**.

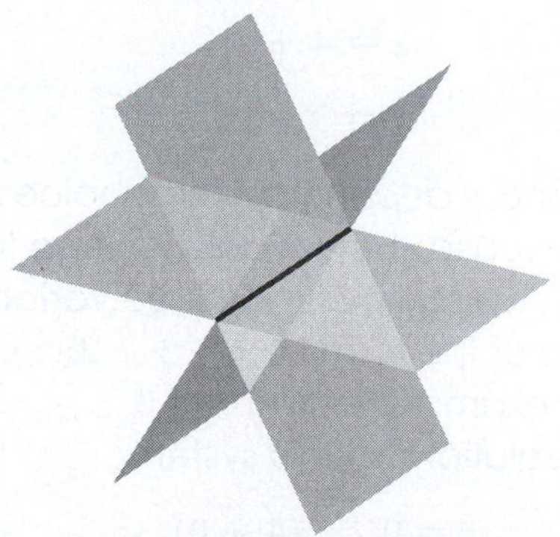

Consistent, Dependent
Infinitely many solutions
(Planes intersect at a line.)

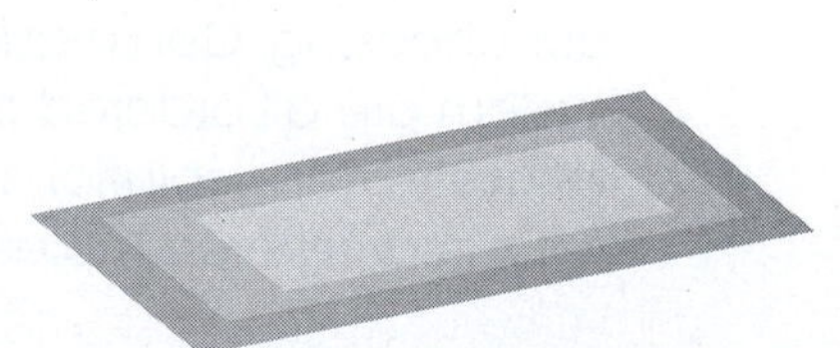

Consistent, Dependent
Infinitely many solutions
(Three equations describe the same plane.)

Figure 9 A geometric representation of consistent, dependent systems in three variables.

When using Gauss-Jordan elimination, we can determine that a system of three linear equations is a dependent system if one of the rows of the augmented matrix reduces to a row consisting of all zeros. We illustrate this in the following example.

Example 4 Solve System of Equations with Infinitely Many Solutions

Use Gauss-Jordan elimination to solve the system:

$$\begin{aligned} x + 2y + 3z &= 10 \\ x + y + z &= 7 \\ 3x + 2y + z &= 18 \end{aligned}$$

Solution Write the system using an augmented matrix, and use a series of **elementary row operations** to attempt to reduce this matrix into **reduced row-echelon form**. Watch the **video** to see each step worked out in detail.

$$\left[\begin{array}{ccc|c} 1 & 2 & 3 & 10 \\ 1 & 1 & 1 & 7 \\ 3 & 2 & 1 & 18 \end{array}\right] \xrightarrow{(-1)R_1 + R_2 \rightarrow \text{New } R_2} \left[\begin{array}{ccc|c} 1 & 2 & 3 & 10 \\ 0 & -1 & -2 & -3 \\ 3 & 2 & 1 & 18 \end{array}\right] \xrightarrow{(-3)R_1 + R_3 \rightarrow \text{New } R_3}$$

$$\left[\begin{array}{ccc|c} 1 & 2 & 3 & 10 \\ 0 & -1 & -2 & -3 \\ 0 & -4 & -8 & -12 \end{array}\right] \xrightarrow{(-1)R_2 \rightarrow \text{New } R_2} \left[\begin{array}{ccc|c} 1 & 2 & 3 & 10 \\ 0 & 1 & 2 & 3 \\ 0 & -4 & -8 & -12 \end{array}\right] \xrightarrow{(-2)R_2 + R_1 \rightarrow \text{New } R_1}$$

$$\left[\begin{array}{ccc|c} 1 & 0 & -1 & 4 \\ 0 & 1 & 2 & 3 \\ 0 & -4 & -8 & -12 \end{array}\right] \xrightarrow{4R_2 + R_3 \rightarrow \text{New } R_3} \left[\begin{array}{ccc|c} 1 & 0 & -1 & 4 \\ 0 & 1 & 2 & 3 \\ 0 & 0 & 0 & 0 \end{array}\right]$$

Notice that the last row of the final augmented matrix consists entirely of zeros. This corresponds to the **identity** $0 = 0$, which is true for every value of x, y, and z. The first two rows correspond to the equations

$$\begin{aligned} x \qquad - z &= 4 \\ y + 2z &= 3 \end{aligned}$$

We can solve the first equation for x and the second equation for y to obtain the equations

$$\begin{aligned} x &= 4 + z \\ y &= 3 - 2z \end{aligned}$$

Notice that the variables x and y *depend* on the choice of z. Thus, the system is said to be **dependent**. In this case, the variable z is free to be any real number of our choosing. Consequently, we say that z is a free variable. The solutions of this system are all ordered triples of the form $(4 + z, 3 - 2z, z)$. Thus, this system has infinitely many solutions. For example, letting $z = 0$, $z = -1$, and $z = 3$, we get the following ordered triples as solutions to this system:

$$\begin{aligned} z = 0&: \quad (4, 3, 0) \\ z = -1&: \quad (3, 5, -1) \\ z = 3&: \quad (7, -3, 3) \end{aligned}$$

Try substituting these ordered triples into the original system to verify that all three are solutions. ●

You Try It Work through this You Try It problem.

Work Exercises 20–23 in this textbook or in the MyLab Math Study Plan.

OBJECTIVE 4 SOLVING INCONSISTENT SYSTEMS OF LINEAR EQUATIONS IN THREE VARIABLES

Recall from **Section 12.1** that a system of linear equations in two variables could have no solution. Linear systems having no solutions are called **inconsistent systems.** For the two-variable case, this occurred if the two linear equations represented parallel lines because parallel lines have no points in common. For the three-variable case, a system is **inconsistent** if all three planes have no points in common. This happens if all three planes are parallel or if two of the planes are parallel. A third possibility of an inconsistent system occurs when the planes intersect two at a time. **Figure 10** geometrically illustrates three types of inconsistent systems in three variables.

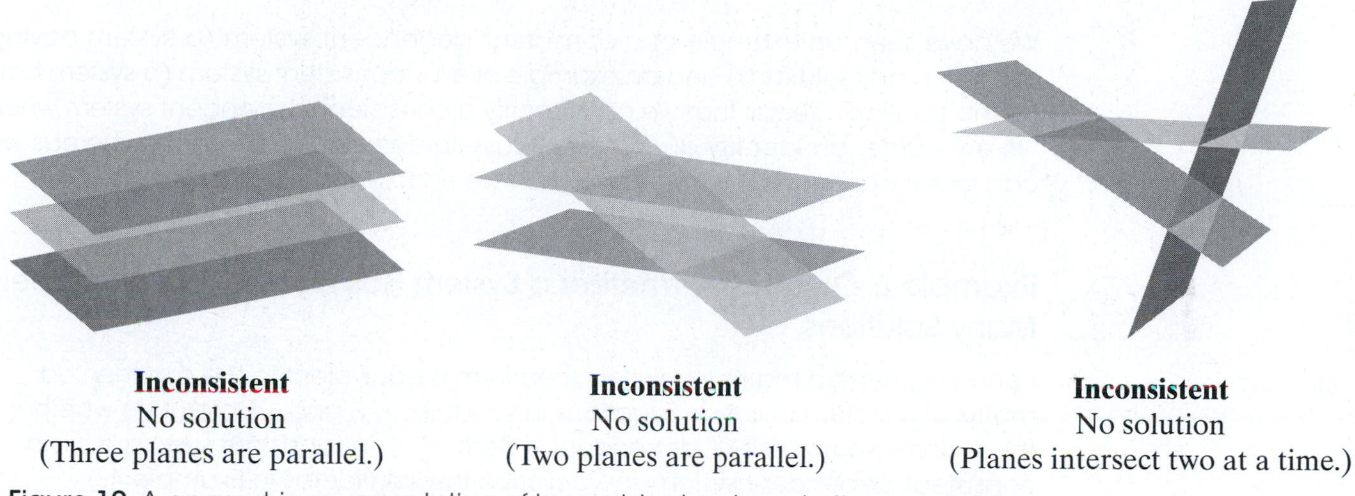

Figure 10 A geometric representation of inconsistent systems in three variables.

Recall from **Example 4 of Section 12.1** that when solving an inconsistent system of two equations we will eventually obtain a **contradiction**. This is also true in the three-variable case. When using Gauss-Jordan elimination, we can determine that a system of three linear equations is an inconsistent system if the three entries on the left-hand side of the augmented matrix reduces to all zeros while the entry on the right-hand side of the augmented matrix is a non-zero constant. We illustrate this in Example 5.

Example 5 Solve a System of Equations with No Solution

Use Gauss-Jordan elimination to solve the system:

$$\begin{aligned} x - y + 2z &= 4 \\ -x + 3y + z &= -6 \\ x + y + 5z &= 3 \end{aligned}$$

Solution Write the system using an augmented matrix and use a series of **elementary row operations** to attempt to reduce this matrix into **reduced row-echelon form**. Watch the **video** to see each step worked out in detail.

$$\left[\begin{array}{rrr|r} 1 & -1 & 2 & 4 \\ -1 & 3 & 1 & -6 \\ 1 & 1 & 5 & 3 \end{array}\right] \xrightarrow{R_1 + R_2 \rightarrow \text{New } R_2} \left[\begin{array}{rrr|r} 1 & -1 & 2 & 4 \\ 0 & 2 & 3 & -2 \\ 1 & 1 & 5 & 3 \end{array}\right] \xrightarrow{(-1)R_1 + R_3 \rightarrow \text{New } R_3}$$

$$\left[\begin{array}{rrr|r} 1 & -1 & 2 & 4 \\ 0 & 2 & 3 & -2 \\ 0 & 2 & 3 & -1 \end{array}\right] \xrightarrow{(-1)R_2 + R_3 \rightarrow \text{New } R_3} \left[\begin{array}{rrr|r} 1 & -1 & 2 & 4 \\ 0 & 2 & 3 & -2 \\ 0 & 0 & 0 & 1 \end{array}\right]$$

Looking at the final augmented matrix, we see that the last row represents the equation $0x + 0y + 0z = 1$ or, simply, $0 = 1$. Obviously, this is a **contradiction** (zero can never equal one). Therefore, we say that the system is inconsistent and has no solution. ●

You Try It Work through this You Try It problem.

Work Exercises 24–26 in this textbook or in the MyLab Math Study Plan.

OBJECTIVE 5 DETERMINING WHETHER A SYSTEM HAS NO SOLUTION OR INFINITELY MANY SOLUTIONS

We have seen an example of a consistent, dependent system (a system having infinitely many solutions) and an example of an inconsistent system (a system having no solution). Recall that we can identify a consistent, dependent system when we encounter an **identity** during the Gauss-Jordan solution process whereas we can identify an inconsistent system when we encounter a **contradiction.**

Example 6 Determine Whether a System Has No Solution or Infinitely Many Solutions

Each augmented matrix in row reduced form is equivalent to the augmented matrix of a system of linear equations in variables $x, y,$ and z. Determine whether the system is a consistent, dependent system or an inconsistent system. If it is a consistent, dependent system, the describe the solution as in Example 4.

a. $\left[\begin{array}{ccc|c} 1 & 0 & -2 & 5 \\ 0 & 1 & 3 & -2 \\ 0 & 0 & 0 & 0 \end{array}\right]$ **b.** $\left[\begin{array}{ccc|c} 1 & 0 & 0 & -4 \\ 0 & 1 & 0 & 6 \\ 0 & 0 & 0 & 10 \end{array}\right]$ **c.** $\left[\begin{array}{ccc|c} 1 & 0 & 0 & 3 \\ 0 & 1 & -2 & 4 \\ 0 & 0 & 0 & 0 \end{array}\right]$

Solution Watch the **interactive video** to verify that the augmented matrices in parts a and c represent consistent, dependent systems, whereas the augmented matrix in part b represents an inconsistent system. The infinite solutions to the system in part a can be described by all ordered triples of the form $(5 + 2z, -2 - 3z, z)$. The infinite solutions to the system in part c can be described by all ordered triples of the form $(3, 4 + 2z, z)$ or $\left(3, y, \frac{1}{2}y - 2\right)$. ●

You Try It Work through this You Try It problem.

Work Exercises 27–32 in this textbook or in the MyLab Math Study Plan.

OBJECTIVE 6 SOLVING LINEAR SYSTEMS HAVING FEWER EQUATIONS THAN VARIABLES

In every linear system that we have encountered thus far, the number of equations in the system was exactly the same as the number of variables. However, this does not always have to be the case. For example, suppose we encounter a system of linear equations in three variables that has only two equations.

We can geometrically illustrate such a system by sketching two planes. There are only three possible scenarios for such a system:

1. The two planes can be distinct parallel planes.
2. The two planes can intersect at a straight line.
3. The two planes can coincide; that is, the two equations describe the same plane. (See Figure 11).

Two planes are parallel.
No solution

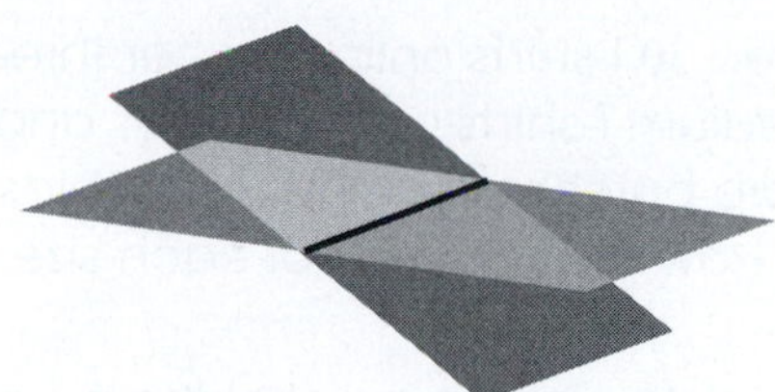

Two planes intersect at a line.
Infinitely many solutions

Two planes coincide.
Infinitely many solutions

Figure 11

Example 7 Solve a System Having Fewer Equations Than Variables

Solve the linear system using Gauss-Jordan elimination:

$$\begin{aligned} x + y + z &= 1 \\ 2x - 2y + 6z &= 10 \end{aligned}$$

Solution We write the system using an augmented matrix and obtain the following sequence of equivalent matrices.

$$\left[\begin{array}{ccc|c} 1 & 1 & 1 & 1 \\ 2 & -2 & 6 & 10 \end{array}\right] \xrightarrow{(-2)R_1 + R_2 \to \text{New } R_2} \left[\begin{array}{ccc|c} 1 & 1 & 1 & 1 \\ 0 & -4 & 4 & 8 \end{array}\right] \xrightarrow{\left(-\frac{1}{4}\right)R_2 \to \text{New } R_2}$$

$$\left[\begin{array}{ccc|c} 1 & 1 & 1 & 1 \\ 0 & 1 & -1 & -2 \end{array}\right] \xrightarrow{(-1)R_2 + R_1 \to \text{New } R_1} \left[\begin{array}{ccc|c} 1 & 0 & 2 & 3 \\ 0 & 1 & -1 & -2 \end{array}\right]$$

The last augmented matrix is in row reduced form. We see that the original system is equivalent to the system

$$\begin{aligned} x + 2z &= 3 \\ y - z &= -2 \end{aligned}$$

We can solve for x and y in terms of z. Thus, z is a free variable, and we find that

$$\begin{aligned} x &= 3 - 2z \\ y &= z - 2 \\ z &= Free \end{aligned}$$

The solution can be written in the form $(3 - 2z, z - 2, z)$.

You Try It Work through this You Try It problem.

Work Exercises 33–36 in this textbook or in the MyLab Math Study Plan.

OBJECTIVE 7 SOLVING APPLIED PROBLEMS USING A SYSTEM OF LINEAR EQUATIONS INVOLVING THREE VARIABLES

Systems of linear equations in three variables and Gauss-Jordan elimination can be used to solve a wide variety of applications. Following are two examples.

Example 8 Buy Clothes Online

Wendy ordered 30 T-shirts online for her three children. The small T-shirts cost \$4 each, the medium T-shirts cost \$5 each, and the large T-shirts were \$6 each. She spent \$40 more purchasing the large T-shirts than the small T-shirts, Wendy's total bill was \$154. How many T-shirts of each size did she buy?

Solution Let S, M, and L represent the number of small, medium, and large T-shirts, respectively. Because a total of 30 T-shirts were purchased, we obtain the first equation

$$S + M + L = 30$$

Because small T-shirts cost \$4, we know that the total amount spent on small T-shirts is $4S$. Similarly, the total spent on medium T-shirts is $5M$, and the total spent on large T-shirts is $6L$. Therefore, because the total value of the T-shirts was \$154, we obtain the second equation

$$4S + 5M + 6L = 154$$

Finally, Wendy spent \$40 more buying large T-shirts than small T-shirts, so $6L = 4S + 40$, which is equivalent to the equation

$$4S - 6L = -40$$

Thus, we obtain the system $\begin{aligned} S + M + L &= 30 \\ 4S + 5M + 6L &= 154. \\ 4S \qquad - 6L &= -40 \end{aligned}$

We now rewrite this system using the corresponding augmented matrix. Note that we must place a 0 in the middle column of the last row of the augmented matrix to represent the coefficient of the variable M of the third equation.

$$\left[\begin{array}{ccc|c} 1 & 1 & 1 & 30 \\ 4 & 5 & 6 & 154 \\ 4 & 0 & -6 & -40 \end{array}\right]$$

Watch the **video** to see how to use Gauss-Jordan elimination to rewrite the matrix in the following **reduced row-echelon form:**

$$\left[\begin{array}{ccc|c} 1 & 0 & 0 & 8 \\ 0 & 1 & 0 & 10 \\ 0 & 0 & 1 & 12 \end{array}\right]$$

Looking down the last column of this augmented matrix, we see that Wendy purchased 8 small T-shirts, 10 medium T-shirts, and 12 large T-shirts.

You Try It Work through this You Try It problem.

Work Exercises 37–41 in this textbook or in the MyLab Math Study Plan.

It takes two points to determine the equation of a line. Similarly, it takes three **noncollinear points** to determine the equation of the quadratic function whose graph passes through the points. Example 6 shows how to find the equation of a quadratic function given three points that lie on its graph. You may want to review quadratic functions from **Section 4.1**.

Example 9 Use Three Points to Determine a Quadratic Function

Determine the quadratic function whose graph passes through the three points $(1, -9)$, $(-1, -5)$, and $(-3, 7)$.

Solution A quadratic function has the form $f(x) = ax^2 + bx + c$. The points $(1, -9)$, $(-1, -5)$, and $(-3, 7)$ lie on the graph of f so

$$f(1) = a(1)^2 + b(1) + c = -9 \longrightarrow a + b + c = -9$$
$$f(-1) = a(-1)^2 + b(-1) + c = -5 \longrightarrow a - b + c = -5$$
$$f(3) = a(-3)^2 + b(-3) + c = 7 \longrightarrow 9a - 3b + c = 7$$

Therefore, we obtain the system and corresponding augmented matrix:

System of Equations **Corresponding Augmented Matrix**

$$\begin{aligned} a + b + c &= -9 \\ a - b + c &= -5 \\ 9a - 3b + c &= 7 \end{aligned} \qquad \left[\begin{array}{ccc|c} 1 & 1 & 1 & -9 \\ 1 & -1 & 1 & -5 \\ 9 & -3 & 1 & 7 \end{array}\right]$$

Using Gauss-Jordan elimination, we can rewrite this matrix in the following **reduced row-echelon form**. (View these **steps** to see the entire solution.)

$$\left[\begin{array}{ccc|c} 1 & 0 & 0 & 1 \\ 0 & 1 & 0 & -2 \\ 0 & 0 & 1 & -8 \end{array}\right]$$

We can see from the last column that $a = 1$, $b = -2$, and $c = -8$. Therefore, the quadratic function that passes through the points $(1, -9)$, $(-1, -5)$, and $(-3, 7)$ is $f(x) = x^2 - 2x - 8$.

You can see in **Figure 12** that the graph of $f(x) = x^2 - 2x - 8$ passes through the three given points.

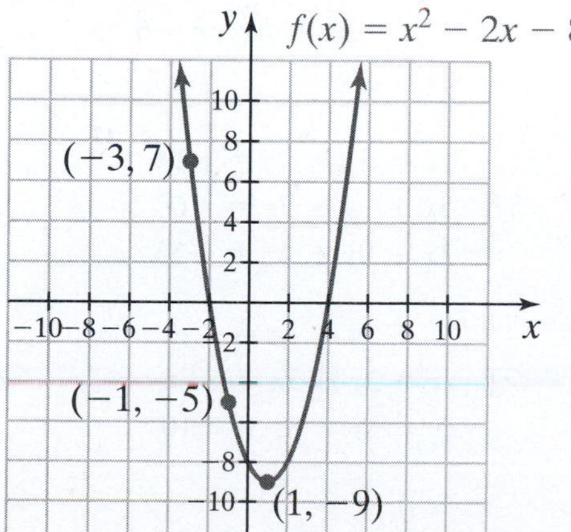

Figure 12 Graph of the quadratic function $f(x) = x^2 - 2x - 8$.

You Try It Work through this You Try It problem.

Work Exercises 42 and 43 in this textbook or in the MyLab Math Study Plan.

12.3 Exercises

In Exercises, 1–4 use Gaussian elimination to solve each linear system by finding an equivalent system in triangular form.

1. $\begin{aligned} x + y + z &= 4 \\ 2x - y - 2z &= -10 \\ -x - y + 3z &= 8 \end{aligned}$

2. $\begin{aligned} x - 2y + z &= 6 \\ 2x + y - 3z &= -3 \\ x - 3y + 3z &= 10 \end{aligned}$

3. $\begin{aligned} -4x_1 + 5x_2 + 9x_3 &= -9 \\ x_1 - 2x_2 + x_3 &= 0 \\ 2x_2 - 8x_3 &= 8 \end{aligned}$

4. $\begin{aligned} 2x_1 + 2x_2 + x_3 &= 9 \\ x_1 + x_3 &= 4 \\ 4x_2 - 3x_3 &= 17 \end{aligned}$

In Exercises 5–10, use Gaussian elimination and matrices to solve each system of linear equations. Write your final augmented matrix in triangular form and then solve for each variable using back substitution.

5. $\begin{aligned} x + y + z &= 0 \\ x - 2y + 3z &= -7 \\ -x - y + 4z &= 6 \end{aligned}$

6. $\begin{aligned} 2x - y + z &= -6 \\ x + y - z &= 0 \\ -3x - 3y + 4z &= 1 \end{aligned}$

7. $\begin{aligned} x_1 + 2x_2 + 3x_3 &= 11 \\ 3x_1 + 8x_2 + 5x_3 &= 27 \\ -x_1 + x_2 + 2x_3 &= 2 \end{aligned}$

8. $\begin{aligned} 2x_1 + 3x_3 &= 13 \\ -x_1 + 3x_2 &= -8 \\ 2x_1 - x_2 + 4x_3 &= 21 \end{aligned}$

9. $\begin{aligned} x_1 + x_2 + 10x_3 &= 3 \\ 3x_1 - 6x_2 + 6x_3 &= -5 \\ -2x_1 + 3x_2 + 5x_3 &= \frac{7}{3} \end{aligned}$

10. $\begin{aligned} x_1 - 2x_2 + 2x_3 &= 2 \\ 3x_1 + 2x_2 - 2x_3 &= -1 \\ x_1 - x_2 + 2x_3 &= 0 \end{aligned}$

In Exercises 11–26 use Gauss-Jordan elimination to solve each linear system and determine whether the system has a unique solution, no solution, or infinitely many solutions. If the system has infinitely many solutions, describe the solution as an ordered triple involving a free variable.

11. $\begin{aligned} x - y &= 7 \\ y - z &= 2 \\ x + z &= 1 \end{aligned}$

12. $\begin{aligned} x - 2y + 3z &= 0 \\ 4x - 6y + 8z &= -4 \\ -2x - y + z &= 7 \end{aligned}$

13. $\begin{aligned} -x_1 + 2x_2 - x_3 &= -15 \\ 3x_1 + x_2 + 4x_3 &= 34 \\ 2x_1 - 4x_2 + 3x_3 &= 40 \end{aligned}$

14. $\begin{aligned} -2x - y + 3z &= 0 \\ 4x + 3y + z &= -1 \\ x + y - 2z &= -\frac{1}{2} \end{aligned}$

15. $\begin{aligned} x_1 - x_3 &= 1 \\ -x_1 + 2x_2 - 3x_3 &= 12 \\ 2x_1 - 4x_2 &= -6 \end{aligned}$

16. $\begin{aligned} 2x_1 + x_2 + 6x_3 &= 1 \\ -x_1 - 3x_3 &= -2 \\ 5x_1 + 4x_2 &= -7 \end{aligned}$

17. $\begin{aligned} 2x - y + z &= -6 \\ x + y - z &= 0 \\ -3x - 3y + 4z &= 1 \end{aligned}$

18. $\begin{aligned} -x + 3y - z &= -14 \\ 5x + y + 7z &= 60 \\ 3x - 9y + 4z &= 53 \end{aligned}$

19. $\begin{aligned} x - 2y + 4z &= 2 \\ 4x - 6y + 8z &= -8 \\ -5x - y + z &= 32 \end{aligned}$

20. $\begin{aligned} x + y + z &= 3 \\ 2x + y - z &= 4 \\ 3x + y - 3z &= 5 \end{aligned}$

21. $\begin{aligned} x + y - z &= 2 \\ x + 2y - 3z &= 6 \\ 2x - 3y + 8z &= -16 \end{aligned}$

22. $\begin{aligned} x + y + z &= 2 \\ 4x + 2y + 3z &= 9 \\ 3x - y + z &= 8 \end{aligned}$

23. $\begin{aligned} x + 4y + 2z &= 5 \\ 3x + 7y + z &= 0 \\ 2x + 5y + z &= 1 \end{aligned}$

24. $\begin{aligned} x - 3y + 5z &= 6 \\ x - 4y + 13z &= 16 \\ x - 2y - 3z &= -5 \end{aligned}$

25. $\begin{aligned} x + y - z &= 3 \\ x + 2y + 2z &= 8 \\ 2x + y - 5z &= 2 \end{aligned}$

26. $\begin{aligned} x + y + z &= 2 \\ 3x + y - 2z &= 8 \\ 2x - y - 8z &= 5 \end{aligned}$

In Exercises 27–32, each augmented matrix in row reduced form is equivalent to the augmented matrix of a system of linear equations in variables x, y, and z. Determine whether the system is dependent or inconsistent. If the system is dependent, determine which variable is free and describe the solution as an ordered triple in terms of the free variable. See **Example 6.**

27. $\left[\begin{array}{ccc|c} 1 & 0 & 0 & 6 \\ 0 & 1 & 0 & 5 \\ 0 & 0 & 0 & -2 \end{array}\right]$

28. $\left[\begin{array}{ccc|c} 1 & 0 & -2 & -5 \\ 0 & 1 & 1 & 4 \\ 0 & 0 & 0 & 0 \end{array}\right]$

29. $\left[\begin{array}{ccc|c} 1 & 3 & 0 & 5 \\ 0 & 1 & 1 & 8 \\ 0 & 0 & 0 & 0 \end{array}\right]$

30. $\left[\begin{array}{ccc|c} 1 & 1 & 2 & 8 \\ 0 & 0 & 0 & 2 \\ 0 & 0 & 1 & 5 \end{array}\right]$

31. $\left[\begin{array}{ccc|c} 1 & 2 & 0 & 1 \\ 0 & 0 & 0 & 0 \\ 0 & 0 & 1 & -4 \end{array}\right]$

32. $\left[\begin{array}{ccc|c} 1 & 0 & 2 & -6 \\ 0 & 1 & -3 & 8 \\ 0 & 0 & 0 & 0 \end{array}\right]$

In Exercises 33–36, use Gauss-Jordan elimination to solve each linear system and determine whether the system has a unique solution, no solution, or infinitely many solutions. If the system has infinitely many solutions, describe the solution as an ordered triple involving a free variable.

33. $\begin{aligned} x + 2y - z &= 11 \\ x + 3y - 2z &= 14 \end{aligned}$

34. $\begin{aligned} x - 2y + z &= -2 \\ -3x + 6y + 3z &= -6 \end{aligned}$

35. $\begin{aligned} x + y + 5z &= 11 \\ 3x + y - z &= 7 \end{aligned}$

36. $\begin{aligned} x - y + z &= 5 \\ 3x + 2y - 2z &= 0 \end{aligned}$

In Exercises 37–43, write a system of linear equations in three variables and then use matrices to solve the system using Gaussian elimination or Gauss-Jordan elimination.

37. Ben was in charge of ordering 35 pizzas for the office party. He ordered three types of pizzas: cheese, supreme, and pepperoni. The cheese pizzas cost \$9 each, the pepperoni pizzas cost \$12 each, and the supreme pizzas cost \$15 each. He spent exactly twice as much on the pepperoni pizzas as he did on the cheese pizzas. If Ben spent a total of \$420 on pizza, how many pizzas of each type did he buy?

38. Nine hundred forty nine tickets were sold on opening night of the play "The Music Man" at the Lewiston Civic Theater. The total receipts for the performance were \$10,967. Adult tickets cost \$14 apiece, children tickets cost \$11 apiece, and senior citizen tickets cost \$8 apiece. If the combined number of adult and children tickets exceeded twice the number of senior citizen tickets by 151, determine how many tickets of each type were sold.

39. Gary and Larry's Tire Shop sells three types of tires: high performance, ultra performance, and extreme performance. High-performance tires sell for \$50/tire. Ultra performance tires sell for \$60/tire. Extreme performance tires sell for \$80/tire. Last week, the revenue generated by the sale of ultra performance tires exceeded the revenue from the sale of extreme performance tires by \$160. If the tire shop sold a total of 238 tires last week for a total revenue of \$14,440, how many of each type of tire were sold?

40. Amanda invested a total of \$2,200 into three separate accounts that pay 6%, 8%, and 9% annual interest. Amanda has three times as much invested in the account that pays 9% as she does in the account that pays 6%. If the total interest for the year is \$178, how much did Amanda invest at each rate?

41. Adam and Murph entered the inaugural Joppetown triathlon, which consisted of running, biking, and swimming. Adam averaged 8 mph during the running segment, 2 mph during the swimming segment, and 15 mph during the biking segment. Murph averaged 6 mph running, 3 mph swimming, and 20 mph biking. Adam's total time was 5 hours and 25 minutes. Murph's total time was 4 hours and 35 minutes. If the total of all three segments is 40 miles, how long was each segment?

42. Determine the quadratic function whose graph passes through the three points $(0, 0)$, $(-4, 36)$, and $(4, 60)$.

43. Determine the quadratic function whose graph passes through the three points $(-1, 2)$, $(3, 10)$, and $(-2, 15)$.

12.4 Partial Fraction Decomposition

THINGS TO KNOW

Before working through this section, be sure that you are familiar with the following concepts:

VIDEO ANIMATION INTERACTIVE

1. Solving a System of Linear Equations Using the Substitution Method (Section 12.1)

2. Solving a System of Linear Equations Using the Elimination Method (Section 12.1)

3. Solving a System of Linear Equations Using Gaussian Elimination (Section 12.3)

INTRODUCTION

Read this introduction before beginning Objective 1.

OBJECTIVES

1 Decomposing Rational Expressions of the Form $\frac{P(x)}{Q(x)}$, where $Q(x)$ has Only Distinct Linear Factors

2 Decomposing Rational Expressions of the Form $\frac{P(x)}{Q(x)}$, where $Q(x)$ has a Repeated Linear Factor

3 Decomposing Rational Expressions of the Form $\frac{P(x)}{Q(x)}$, where $Q(x)$ has a Distinct Prime Quadratic Factor

4 Decomposing Rational Expressions of the Form $\frac{P(x)}{Q(x)}$, where $Q(x)$ has a Repeated Prime Quadratic Factor

SECTION 12.4 EXERCISES

Introduction to Section 12.4

Recall that a rational expression in one variable is of the form $\frac{P(x)}{Q(x)}$, where $P(x)$ and $Q(x)$ are **polynomials** such that $Q(x) \neq 0$. When combining two or more rational expressions into a single rational expression, we must first determine the lowest common denominator (LCD). For example, suppose that we wish to subtract the rational expression $\frac{7}{x-1}$ from the rational expression $\frac{1}{x+3}$. The LCD is the product $(x-1)(x+3)$. The result is shown below.

$$\frac{1}{x+3} - \frac{7}{x-1} = \frac{1(x-1) - 7(x+3)}{(x+3)(x-1)} = \frac{x - 1 - 7x - 21}{(x+3)(x-1)} = \frac{-6x - 22}{x^2 + 2x - 3}$$

In this section, we will attempt to reverse this procedure. That is, we will start with a single rational expression and write it as the sum or difference of two or more simpler rational expressions. This process is called **partial fraction decomposition.** Using the example from above, we say that the partial fraction decomposition of the expression $\frac{-6x-22}{x^2+2x-3}$ is $\frac{1}{x+3} - \frac{7}{x-1}$. The two fractions $\frac{1}{x+3}$ and $\frac{-7}{x-1}$ are called **partial fractions.** Partial fraction decomposition is a procedure that is especially useful and often necessary in calculus.

The partial fraction decomposition of a rational expression of the form $\frac{P(x)}{Q(x)}$ can be performed when the following two criteria are satisfied:

1. The polynomials $P(x)$ and $Q(x)$ share no common factors.
2. The degree of $P(x)$ is less than the degree of $Q(x)$.

In all examples and exercises in this section, the two criteria above are always satisfied. We start by introducing rational expressions that involve distinct linear factors within the denominator.

OBJECTIVE 1 DECOMPOSING RATIONAL EXPRESSIONS OF THE FORM $\frac{P(x)}{Q(x)}$, WHERE $Q(x)$ HAS ONLY DISTINCT LINEAR FACTORS

Recall that a linear factor is an expression of the form $ax + b$ where $a \neq 0$. If the denominator of $\frac{P(x)}{Q(x)}$ is the product of n distinct linear factors, then the partial fraction decomposition of $\frac{P(x)}{Q(x)}$ will be of the form

$$\frac{P(x)}{Q(x)} = \frac{A_1}{a_1x + b_1} + \frac{A_2}{a_2x + b_2} + \cdots + \frac{A_n}{a_nx + b_n}$$

where $A_1, A_2, \ldots, A_n$ are constants to be determined.

We start with an example of a rational expression that contains two distinct linear factors in the denominator.

Example 1 Decomposing a Rational Expression in which the Denominator has Only Distinct Linear Factors

Determine the partial fraction decomposition of $\frac{x + 10}{2x^2 + 5x - 3}$.

Solution Start by factoring the denominator: $2x^2 + 5x - 3 = (2x - 1)(x + 3)$. We see that the denominator is the product of two distinct linear factors. Therefore, we introduce the two partial fractions $\frac{A}{2x - 1}$ and $\frac{B}{x + 3}$ to create the following equation:

$$\frac{x + 10}{(2x - 1)(x + 3)} = \frac{A}{2x - 1} + \frac{B}{x + 3}$$

Note that for convenience we use A and B to represent the unknown constants instead of A_1 and A_2.

To eliminate the denominators, multiply both sides of the rational equation by $(2x - 1)(x + 3)$.

$$(2x - 1)(x + 3)\frac{x + 10}{(2x - 1)(x + 3)} = \left(\frac{A}{2x - 1} + \frac{B}{x + 3}\right)(2x - 1)(x + 3)$$ Multiply both sides by $(2x - 1)(x + 3)$.

$$\cancel{(2x - 1)(x + 3)}\frac{x + 10}{\cancel{(2x - 1)(x + 3)}} = \frac{A}{\cancel{(2x - 1)}}\cancel{(2x - 1)}(x + 3) + \frac{B}{\cancel{x + 3}}(2x - 1)\cancel{(x + 3)}$$ Use the distributive property and cancel common factors.

$$x + 10 = A(x + 3) + B(2x - 1)$$ Write the resulting new equation.

We can now write the polynomial on the right-hand side in descending order by collecting like terms.

$x + 10 = Ax + 3A + 2Bx - B$ Use the distributive property.

$x + 10 = Ax + 2Bx + 3A - B$ Rearrange the terms.

$x + 10 = (A + 2B)x + (3A - B)$ Factor out an x. The polynomial on the right-hand side is now written in descending order.

Note that the polynomial on the left-hand side must be equivalent to the polynomial on the right-hand side. Thus, the corresponding coefficients must be equivalent. Therefore, the coefficient of the x-term on the left-hand side, 1, is equivalent to the coefficient of the x-term on the right-hand side, $A + 2B$.

$$1x + 10 = (A + 2B)x + (3A - B)$$

$$A + 2B = 1$$

Similarly, the coefficient of the constant term on the left-hand side, 10, is equivalent to the coefficient of the constant term on the right-hand side, $3A - B$.

$$x + 10 = (A + 2B)x + (3A - B)$$

$$3A - B = 10$$

The result of equating the coefficients of the left-hand side with the corresponding coefficients on the right-hand side leads to the following system of equations:

$$A + 2B = 1$$
$$3A - B = 10$$

We can solve the system of equations using the method of substitution or the method of elimination to obtain $A = 3$ and $B = -1$. Therefore,

$$\frac{x + 10}{(2x - 1)(x + 3)} = \frac{3}{2x - 1} - \frac{1}{x + 3}.$$

Note It is often possible to determine some or all of the constants of the partial fraction decomposition using a short-cut method that involves choosing appropriate values of x and substituting those values into the equation found after eliminating the denominators. We choose values of x that will make one of the linear factors equal to zero. This may result in the elimination of one or more of the variables that represent the undetermined constants. Using the equation $x + 10 = A(x + 3) + B(2x - 1)$ found in Example 1, we see that the linear factor $(x + 3)$ is equal to zero when $x = -3$. Thus, we can substitute $x = -3$ into the equation to eliminate the variable A to get the following:

$x + 10 = A(x + 3) + B(2x - 1)$	Start with the equation found after eliminating the denominators.
$-3 + 10 = A(-3 + 3) + B(2(-3) - 1)$	Substitute $x = -3$ to eliminate terms involving the variable A.
$7 = -7B$	Simplify.
$-1 = B$	Solve for B.

Similarly, we can substitute $x = \frac{1}{2}$ to get the following:

$x + 10 = A(x + 3) + B(2x - 1)$	Start with the equation found after eliminating the denominators.
$\frac{1}{2} + 10 = A\left(\frac{1}{2} + 3\right) + B\left(2\left(\frac{1}{2}\right) - 1\right)$	Substitute $x = \frac{1}{2}$ to eliminate terms involving the variable B.
$\frac{21}{2} = \frac{7}{2}A$	Simplify.
$3 = A$	Solve for A.

The values of $A = 3$ and $B = -1$ were precisely the values found in the solution to Example 1.

We now establish a seven-step procedure for determining the partial fraction decomposition of a rational expression of the form $\frac{P(x)}{Q(x)}$.

Steps for Determining the Partial Fraction Decomposition of $\frac{P(x)}{Q(x)}$

Step 1. Factor $Q(x)$.

Step 2. Set up an equation with $\frac{P(x)}{Q(x)}$ on the left-hand side and the correct partial fractions with the undetermined constants on the right-hand side.

Step 3. Multiply both sides of the equation created in step 2 by $Q(x)$ to eliminate all denominators.

Step 4*. If possible, choose appropriate values of x to readily solve for one or more undetermined constants. This may conclude the partial fraction decomposition procedure. If not, go on to step 5.

Step 5. Write the polynomial on the right-hand side of the equation found in step 3 in descending order.

Step 6. Equate the coefficients of $P(x)$ with the corresponding coefficients of the polynomial on the right-hand side of the equation, thus creating a system of equations.

Step 7. Solve the system of equations.

*Note that although step 4 usually simplifies the partial fraction decomposition process, it is not necessary.

Example 2 Decomposing a Rational Expression in which the Denominator has Only Distinct Linear Factors

Determine the partial fraction decomposition of $\frac{5x^2 - x + 1}{x^3 + 3x^2 - 4x}$.

Solution

Step 1. Factor the denominator: $x^3 + 3x^2 - 4x = x(x - 1)(x + 4)$. We see that the denominator is the product of three distinct linear factors.

Step 2. Introduce the three partial fractions $\frac{A}{x}$, $\frac{B}{x - 1}$, and $\frac{C}{x + 4}$ to create the following equation:

$$\frac{5x^2 - x + 1}{x^3 + 3x^2 - 4x} = \frac{A}{x} + \frac{B}{x - 1} + \frac{C}{x + 4}$$

Step 3. Multiply both sides of the equation from Step 2 by $x(x - 1)(x + 4)$ to get

$$5x^2 - x + 1 = A(x - 1)(x + 4) + Bx(x + 4) + Cx(x - 1)$$

Step 4. We can choose $x = 0, x = 1$, and $x = -4$ to solve for each constant.

$$x = 0\text{: } 5(0)^2 - (0) + 1 = A(0 - 1)(0 + 4) + B(0)(0 + 4) + C(0)(0 - 1)$$

$$1 = -4A$$

$$-\frac{1}{4} = A$$

(eText Screens 12.4-1–12.4-28)

$$x = 1: 5(1)^2 - (1) + 1 = A(1 - 1)(1 + 4) + B(1)(1 + 4) + C(1)(1 - 1)$$
$$5 = 5B$$
$$1 = B$$
$$x = -4: 5(-4)^2 - (-4) + 1 = A(-4 - 1)(-4 + 4) + B(-4)(-4 + 4) + C(-4)(-4 - 1)$$
$$85 = 20C$$
$$\frac{17}{4} = C$$

Therefore, $A = -\frac{1}{4}$, $B = 1$, and $C = \frac{17}{4}$. All constants have been determined so there is no need to go on to step 5. Thus, the partial fraction decomposition is

$$\frac{5x^2 - x + 1}{x^3 + 3x^2 - 4x} = \frac{-\frac{1}{4}}{x} + \frac{1}{x - 1} + \frac{\frac{17}{4}}{x + 4}.$$

 Watch this video to see every step of this solution.

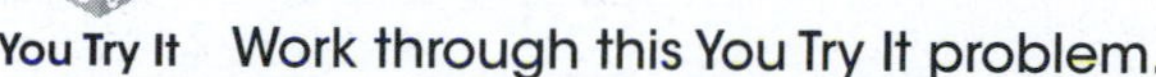

You Try It Work through this You Try It problem.

Work Exercises 1–14 in this textbook or in the MyLab Math Study Plan.

OBJECTIVE 2 DECOMPOSING RATIONAL EXPRESSIONS OF THE FORM $\frac{P(x)}{Q(x)}$, WHERE $Q(x)$ HAS A REPEATED LINEAR FACTOR

A repeated linear factor is of the form $(ax + b)^n$, where n is an integer such that $n \geq 2$. If the denominator of $\frac{P(x)}{Q(x)}$ has a repeated linear factor, then for every repeated linear factor of the form $(ax + b)^n$, we introduce n partial fractions of the form

$$\frac{A_1}{ax + b} + \frac{A_2}{(ax + b)^2} + \cdots + \frac{A_n}{(ax + b)^n}$$

Example 3 Setting up the Partial Fraction Decomposition of $\frac{P(x)}{Q(x)}$, where $Q(x)$ has Repeated Linear Factors

Set up the partial fraction decomposition for $\frac{x}{x^2(3x - 1)^3(x - 5)}$.

Do not solve for the constants.

Solution The denominator has two repeated linear factors, x^2 and $(3x - 1)^3$, and one distinct linear factor, $(x - 5)$. The correct partial fraction decomposition is of the form

$$\frac{x}{x^2(3x - 1)^3(x - 5)} = \frac{A}{x} + \frac{B}{x^2} + \frac{C}{3x - 1} + \frac{D}{(3x - 1)^2} + \frac{E}{(3x - 1)^3} + \frac{F}{x - 5},$$

where $A, B, C, D, E,$ and F are constants to be determined.

Notice that for convenience we use $A, B, C, \ldots$ instead of $A_1, A_2, A_3, \ldots$.

Example 4 illustrates how to follow the **steps for determining the partial fraction decomposition of** $\frac{P(x)}{Q(x)}$, where $Q(x)$ has a repeated linear factor.

Example 4 Decomposing a Rational Expression in which the Denominator Has a Repeated Linear Factor

Determine the partial fraction decomposition of $\frac{x-1}{x(x-2)^2}$.

Solution

Step 1. The denominator is already factored.

Step 2. Introduce the three partial fractions $\frac{A}{x}, \frac{B}{x-2}$, and $\frac{C}{(x-2)^2}$ to create the following equation:

$$\frac{x-1}{x(x-2)^2} = \frac{A}{x} + \frac{B}{x-2} + \frac{C}{(x-2)^2}$$

Step 3. Multiply both sides of the rational equation from step 2 by $x(x-2)^2$ to obtain the equation

$$x - 1 = A(x-2)^2 + Bx(x-2) + Cx$$

Step 4. We can solve for C by choosing $x = 2$.

$$x = 2\text{:}\quad 2 - 1 = A(2-2)^2 + B(2)(2-2) + C(2)$$

$$1 = 2C$$

$$\frac{1}{2} = C$$

Step 5. Write the polynomial on the right-hand side of the equation found in step 3 in descending order by collecting like terms. View **these steps** to see that we can write the equation found in step 3 as

$$x - 1 = (A + B)x^2 + (-4A - 2B + C)x + 4A$$

Step 6. The coefficient of the x^2-term on the left-hand side, 0, is equivalent to the coefficient of the x^2-term on the right-hand side, $A + B$.

$$0x^2 + x - 1 = (A + B)x^2 + (-4A - 2B + C)x + 4A$$

$$A + B = 0$$

The coefficient of the x-term on the left-hand side, 1, is equivalent to the coefficient of the x-term on the right-hand side, $-4A - 2B + C$.

$$1x - 1 = (A + B)x^2 + (-4A - 2B + C)x + 4A$$

$$-4A - 2B + C = 1$$

The coefficient of the constant term on the left-hand side, -1, is equivalent to the coefficient of the constant term on the right-hand side, $4A$.

$$x - 1 = (A + B)x^2 + (-4A - 2B + C)x + 4A$$

The result of equating the coefficients of the left-hand side with the corresponding coefficients on the right-hand side leads to the following system of equations:

$$\begin{aligned} A + B \quad &= 0 \\ -4A - 2B + C &= 1 \\ 4A \quad &= -1 \end{aligned}$$

Step 7. We know from step 4 that $C = \frac{1}{2}$. Solving the third equation for A give $A = -\frac{1}{4}$. Substituting $A = -\frac{1}{4}$ into the first equation, we get $B = \frac{1}{4}$. Therefore,

$$\frac{x - 1}{x(x - 2)^2} = \frac{-\frac{1}{4}}{x} + \frac{\frac{1}{4}}{x - 2} + \frac{\frac{1}{2}}{(x - 2)^2}.$$

Watch this **video** to see every step of this partial fraction decomposition process.

You Try It **Work through this You Try It problem.**

Work Exercises 15–26 in this textbook or in the MyLab Math Study Plan.

OBJECTIVE 3 DECOMPOSING RATIONAL EXPRESSIONS OF THE FORM $\frac{P(x)}{Q(x)}$, WHERE $Q(x)$ HAS A DISTINCT PRIME QUADRATIC FACTOR

A quadratic factor has the form $ax^2 + bx + c$, where $a \neq 0$. If the quadratic factor will not factor into the product of two linear factors using integer coefficients, then the quadratic factor is prime. If the denominator of $\frac{P(x)}{Q(x)}$ has a prime quadratic factor, then for every prime quadratic factor of the form $ax^2 + bx + c$, we introduce a partial fraction of the form

$$\frac{A_1x + B_1}{ax^2 + bx + c}.$$

Note For every linear factor of $Q(x)$, we introduced a partial fraction containing a constant over the linear factor, $\frac{A_1}{ax + b}$. For every prime quadratic factor of $Q(x)$, we now introduce a partial fraction containing a linear factor over the prime quadratic factor, $\frac{A_1x + B_1}{ax^2 + bx + c}$.

Example 5 Setting up the Partial Fraction Decomposition of $\frac{P(x)}{Q(x)}$, where $Q(x)$ has Distinct Prime Quadratic Factors

Set up the partial fraction decomposition for $\frac{5x^3 - 7x^2 - 8x + 1}{(x^2 + 1)(2x^2 + x + 7)(5x - 3)^2}$. Do not solve for the constants.

(eText Screens 12.4-1–12.4-28)

Solution The denominator has two prime quadratic factors, $x^2 + 1$ and $2x^2 + x + 7$, and one repeated linear factor of the form $(5x - 3)^2$. The correct partial fraction decomposition is of the form $\dfrac{5x^3 - 7x^2 - 8x + 1}{(x^2 + 1)(2x^2 + x + 7)(3x - 1)^2} =$ $\dfrac{Ax + B}{x^2 + 1} + \dfrac{Cx + D}{2x^2 + x + 7} + \dfrac{E}{5x - 3} + \dfrac{F}{(5x - 3)^2}$ where $A, B, C, D, E,$ and F are constants to be determined.

Notice that for convenience we use $A, B, C, D, \ldots$ instead of $A_1, B_1, A_2, B_2, \ldots$. ●

Example 6 Decomposing a Rational Expression in which the Denominator has a Prime Quadratic Factor

Determine the partial fraction decomposition of $\dfrac{8x^2 + 7}{x^3 - 1}$.

Solution

Step 1. The denominator is the **difference of two cubes** and factors as $x^3 - 1 = (x - 1)(x^2 + x + 1)$. The quadratic factor, $x^2 + x + 1$, does not factor with integer coefficients and is thus a prime quadratic factor.

Step 2. Introduce the two partial fractions $\dfrac{A}{x - 1}$ and $\dfrac{Bx + C}{x^2 + x + 1}$ to create the following equation:

$$\frac{8x^2 + 7}{x^3 - 1} = \frac{A}{x - 1} + \frac{Bx + C}{x^2 + x + 1}$$

Step 3. Multiply both sides of the rational equation from step 2 by $(x - 1)(x^2 + x + 1)$ to obtain the equation

$$8x^2 + 7 = A(x^2 + x + 1) + (Bx + C)(x - 1)$$

Step 4. We can solve for A by choosing $x = 1$.

$$x = 1\colon 8(1)^2 + 7 = A((1)^2 + 1 + 1) + (B(1) + C)(1 - 1)$$

$$15 = 3A$$

$$5 = A$$

Step 5. Write the polynomial on the right-hand side of the equation found in step 3 in descending order by collecting like terms. View **these steps** to see that we can write the equation found in step 3 as

$$8x^2 + 7 = (A + B)x^2 + (A - B + C)x + (A - C)$$

Step 6. Equating the coefficients of $8x^2 + 7$ with the corresponding coefficients of $(A + B)x^2 + (A - B + C)x + (A - C)$ leads to the following system of equations:

$$\begin{aligned} A + B \phantom{{}+C} &= 8 \\ A - B + C &= 0 \\ A - C &= 7 \end{aligned}$$

Step 7. Using the fact that $A = 5$, we can use the first equation to get $B = 3$. We can then use the second or third equation to get $C = -2$.

Thus, the partial fraction decomposition is $\dfrac{8x^2 + 7}{x^3 - 1} = \dfrac{5}{x - 1} + \dfrac{3x - 2}{x^2 + x + 1}$.

You Try It Work through this You Try It problem.

Work Exercises 27–35 in this textbook or in the MyLab Math Study Plan.

OBJECTIVE 4 DECOMPOSING RATIONAL EXPRESSIONS OF THE FORM $\dfrac{P(x)}{Q(x)}$, WHERE $Q(x)$ HAS A REPEATED PRIME QUADRATIC FACTOR

A repeated prime quadratic factor has the form $(ax^2 + bx + c)^n$, where $a \neq 0$ and where n is an integer such that $n \geq 2$. If the denominator of $\dfrac{P(x)}{Q(x)}$ has a repeated prime quadratic factor, then for every repeated prime quadratic factor of the form $(ax^2 + bx + c)^n$, we introduce n partial fractions of the form

$$\frac{A_1x + B_1}{ax^2 + bx + c} + \frac{A_2x + B_2}{(ax^2 + bx + c)^2} + \cdots + \frac{A_nx + B_n}{(ax^2 + bx + c)^n}$$

Example 7 Setting up the Partial Fraction Decomposition of $\dfrac{P(x)}{Q(x)}$, where $Q(x)$ has Repeated Prime Quadratic Factors

Set up the partial fraction decomposition for $\dfrac{x^4 + x^3 + x^2 + x + 1}{(x^2 + 4)^3(3x^2 + 10x + 1)^2}$.

Do not solve for the constants.

Solution The denominator has two repeated prime quadratic factors of the form $(x^2 + 4)^3$ and $(3x^2 + 10x + 1)^2$. The correct partial fraction decomposition is of the form $\dfrac{x^4 + x^3 + x^2 + x + 1}{(x^2 + 4)^3(3x^2 + 10x + 1)^2} = \dfrac{Ax + B}{x^2 + 4} + \dfrac{Cx + D}{(x^2 + 4)^2} + \dfrac{Ex + F}{(x^2 + 4)^3} + \dfrac{Gx + H}{3x^2 + 10x + 1} + \dfrac{Ix + J}{(3x^2 + 10x + 1)^2}$, where $A, B, C, D, E, F, F, H, I,$ and J are constants to be determined.

Notice that for convenience we use $A, B, C, D, \ldots$ instead of $A_1, B_1, A_2, B_2, \ldots$.

Example 8 Decomposing a Rational Expression in which the Denominator has a Repeated Prime Quadratic Factor

Determine the partial fraction decomposition of $\dfrac{3x^4 - 5x^3 + 7x^2 + x - 2}{(x - 1)(x^2 + 1)^2}$.

Solution

Step 1. The denominator is already factored.

Step 2. Introduce the three partial fractions

$$\frac{A}{x - 1}, \frac{Bx + C}{x^2 + 1}, \text{ and } \frac{Dx + E}{(x^2 + 1)^2}$$

to create the following equation:

$$\frac{3x^4 - 5x^3 + 7x^2 + x - 2}{(x-1)(x^2+1)^2} = \frac{A}{x-1} + \frac{Bx + C}{x^2+1} + \frac{Dx + E}{(x^2+1)^2}$$

Step 3. Multiply both sides of the rational equation from step 2 by $(x-1)(x^2+1)^2$ to obtain the equation

$$3x^4 - 5x^3 + 7x^2 + x - 2 = A(x^2+1)^2 + (Bx + C)(x-1)(x^2+1) + (Dx + E)(x-1).$$

Step 4. Choosing $x = 1$ will let us solve for A.

$$x = 1: 3(1)^4 - 5(1)^3 + 7(1)^2 + (1) - 2 = A((1)^2+1)^2 + (B(1) + C)(1-1)((1)^2+1) + (D(1) + E)(1-1)$$

$$4 = 4A$$

$$1 = A$$

Step 5. Write the polynomial on the right-hand side of the equation from step 3 in descending order by collecting like terms. View **these steps** to see that we can write the equation found in step 3 as

$$3x^4 - 5x^3 + 7x^2 + x - 2 = (A + B)x^4 + (-B + C)x^3 + (2A + B - C + D)x^2 + (-B + C - D + E)x + (A - C - E).$$

Step 6. Equating the coefficients of $3x^4 - 5x^3 + 7x^2 + x - 2$ with the corresponding coefficients of $(A + B)x^4 + (-B + C)x^3 + (2A + B - C + D)x^2 + (-B + C - D + E)x + (A - C - E)$ leads to the following system of equations:

$$A + B = 3$$
$$-B + C = -5$$
$$2A + B - C + D = 7$$
$$-B + C - D + E = 1$$
$$A - C - E = -2$$

Step 7. Because $A = 1$, we can use the first equation to get $B = 2$. Substituting $B = 2$ into the second equation yields $C = -3$. Substituting the known quantities into the third equation gives $D = 0$. It follows from the fourth or fifth equation that $E = 6$.

Therefore, the partial fraction decomposition is $\frac{3x^4 - 5x^3 + 7x^2 + x - 2}{(x-1)(x^2+1)^2} =$

$\frac{1}{x-1} + \frac{2x-3}{x^2+1} + \frac{6}{(x^2+1)^2}$. Watch this **video** to see each step of this solution.

You Try It Work through this You Try It problem.

Work Exercises 36–41 in this textbook or in the MyLab Math Study Plan.

12.4 Exercises

In Exercises 1–41, determine the partial fraction decomposition of each rational expression.

1. $\dfrac{3}{x(x-3)}$

2. $\dfrac{4x+30}{(x+9)(x+3)}$

3. $\dfrac{x}{(x-1)(x-2)}$

4. $\dfrac{x}{(x-2)(x+4)}$

5. $\dfrac{x-7}{(x-3)(x-5)}$

6. $\dfrac{6}{2x^2-11x+5}$

7. $\dfrac{x}{x^2+2x-35}$

8. $\dfrac{x}{x^2-3x+2}$

9. $\dfrac{8x^2-22x+20}{x(x+4)(x+5)}$

10. $\dfrac{x}{12x^2-17x-7}$

11. $\dfrac{7x^2-x-12}{x(x^2-1)}$

12. $\dfrac{7x^2-x-16}{x^3-x}$

13. $\dfrac{x^2}{(x^2-25)(x+1)}$

14. $\dfrac{2x+4}{x^3-x^2-42x}$

15. $\dfrac{-7x+11}{(x-5)^2}$

16. $\dfrac{x^2-6x+4}{(x-2)^3}$

17. $\dfrac{7x+197}{(x-4)^2(x+5)}$

18. $\dfrac{x^2}{(x-1)^2(x+4)}$

19. $\dfrac{x+3}{x^3-2x^2+x}$

20. $\dfrac{x^2-x+30}{x^3+6x^2}$

21. $\dfrac{4x^2}{(x-2)^2(x+2)^2}$

22. $\dfrac{x-4}{(x+3)(x+5)^2}$

23. $\dfrac{7x^3-2}{x^2(x+1)^3}$

24. $\dfrac{2}{(x-1)(x^2-1)}$

25. $\dfrac{6x+5}{x^4-25x^2}$

26. $\dfrac{2x^2+3x+6}{x^3-4x^2-16x+64}$

27. $\dfrac{6}{x(x^2+6)}$

28. $\dfrac{18x+3}{(x-1)(x^2+x+1)}$

29. $\dfrac{8x^2-38x+22}{(x-6)(x^2+5)}$

30. $\dfrac{-4}{x^3-27}$

31. $\dfrac{x^2-111}{x^4-x^2-72}$

32. $\dfrac{x+7}{x^2(x^2+9)}$

33. $\dfrac{-39x+52}{(x+3)^2(x^2+4)}$

34. $\dfrac{6x^2+11x+31}{x^3+3x^2+4x+12}$

35. $\dfrac{x^2+3x+9}{(x+3)(x^2-12x+6)}$

36. $\dfrac{x^3+x^2+5}{(x^2+4)^2}$

37. $\dfrac{5x^3+14x^2+22x-37}{(x^2+4x+8)^2}$

38. $\dfrac{4x^3+2x^2+15x+4}{(x^2+3)^3}$

39. $\dfrac{3x-1}{x(5x^2+3)^2}$

40. $\dfrac{x^4-2x^3+25x^2+5x+70}{(x+2)(x^2+4)^2}$

41. $\dfrac{3x^4+4x^2-2x+1}{x^6+2x^4+x^2}$

(eText Screens 12.5-1–12.5-34)

12.5 Systems of Nonlinear Equations

THINGS TO KNOW

Before working through this section, be sure that you are familiar with the following concepts:

		VIDEO	ANIMATION	INTERACTIVE
You Try It	1. Sketching the Graph of a Circle (Section 2.2)	▶	▶	
You Try It	2. Sketching Lines by Plotting Intercepts (Section 2.3)	▶		
You Try It	3. Graphing Quadratic Functions (Section 4.1)	▶		
You Try It	4. Use Transformations to Sketch the Graph of a Rational Function (Section 4.6)	▶		
You Try It	5. Sketching the Graph of a Logarithmic Function (Section 5.2)	▶		
You Try It	6. Solving Logarithmic Equations (Section 5.4)			▶
You Try It	7. Sketching the Graph of a Parabola (Section 11.1)	▶		
You Try It	8. Sketching the Graph of an Ellipse (Section 11.2)	▶		
You Try It	9. Sketching the Graph of a Hyperbola (Section 11.3)			▶
You Try It	10. Solving a System of Linear Equation Using the Method of Substitution (Section 12.1)	▶		
You Try It	11. Solving a System of Linear Equations Using the Method of Elimination (Section 12.1)	▶		

INTRODUCTION

Read this introduction before beginning Objective 1.

OBJECTIVES

1 Determining the Number of Solutions to a System of Nonlinear Equations

2 Solving a System of Nonlinear Equations Using the Substitution Method

3 Solving a System of Nonlinear Equations Using Substitution, Elimination, or Graphing

4 Solving Applied Problems Using a System of Nonlinear Equations

SECTION 12.5 EXERCISES

Introduction to Section 12.5

In Section 12.1 we solved systems of **linear equations in two variables**. The solution to a system of linear equations in two variables is represented by the set of all ordered pairs that satisfy both equations. Graphically, the solution can be represented by the point (or points) of intersection of the two lines.

In this section, we turn our attention to **nonlinear systems**. Systems of nonlinear equations in two variables contain at least one equation that is nonlinear. That is, at least one of the equations is *not* of the form $Ax + By = C$. Just like with linear systems, the real solutions to a nonlinear system can be graphically represented by all points of intersection between the graphs of both equations. In this section it is essential that you can sketch the graphs of nonlinear equations such as circles, parabolas, ellipses, and hyperbolas. Being able to sketch the equations of a nonlinear system will help you understand the nature of the solutions. Before we solve a system of nonlinear equations, take a moment to practice sketching the graphs of these familiar nonlinear equations. Click on a You Try It icon below to practice sketching the graphs of circles, parabolas, ellipses, and hyperbolas.

You Try It Practice sketching the graph of a circle that is in **standard form**.

You Try It Practice sketching the graph of a circle that is in **general form**.

You Try It Practice sketching the graph of a parabola that opens up or down (a quadratic function).

You Try It Practice sketching the graph of a parabola that opens left or right.

You Try It Practice sketching the graph of an ellipse that is in **standard form**.

You Try It Practice sketching the graph of an ellipse that is not in standard form.

You Try It Practice sketching the graph of a hyperbola that is in **standard form**.

You Try It Practice sketching the graph of a hyperbola that is not in standard form.

You Try It Practice sketching the graph of a rational function.

You Try It Practice sketching the graph of a logarithmic function.

OBJECTIVE 1 DETERMINING THE NUMBER OF SOLUTIONS TO A SYSTEM OF NONLINEAR EQUATIONS

As stated previously, the real solutions to a system of nonlinear equations can be represented by the points of intersection between the graphs of each equation. In Example 1, three different systems of nonlinear equations are shown. Without solving the systems, we can determine the number of solutions of each system by sketching the graph of each equation.

Example 1 Determining the Number of Solutions to a Nonlinear System

For each system of nonlinear equations, sketch the graph of each equation of the system and then determine the number of real solutions to each system. Do not solve the system.

a. $\begin{aligned} x^2 + y^2 &= 25 \\ x - y &= 1 \end{aligned}$ b. $\begin{aligned} x - y^2 &= 4 \\ x - y &= 6 \end{aligned}$ c. $\begin{aligned} x^2 + y^2 &= 9 \\ x^2 - y &= 3 \end{aligned}$

Solution

a. The graph of the equation $x^2 + y^2 = 25$ is a circle centered at the origin with a radius of 5 units. The graph of $x - y = 1$ is a line with a slope of 1 and a y-intercept of -1. The two graphs are shown in Figure 13.

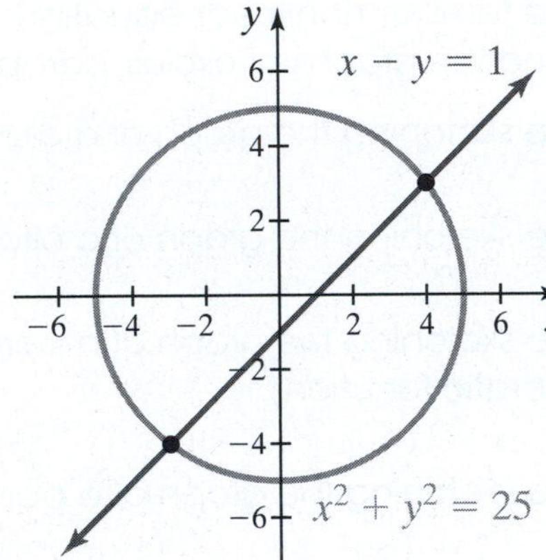

Figure 13

Because the graphs in Figure 13 intersect at two points, the system of nonlinear equations has two real solutions. Watch this **interactive video** to see how to sketch each equation.

b. The graph of the equation $x - y^2 = 4$ is a parabola with a **horizontal axis of symmetry**. The parabola has a vertex at the point $(4, 0)$ and opens to the right.

The graph of $x - y = 6$ is a line with a slope of 1 having a y-intercept of -6 and an x-intercept of 6. The two graphs are shown in Figure 14.

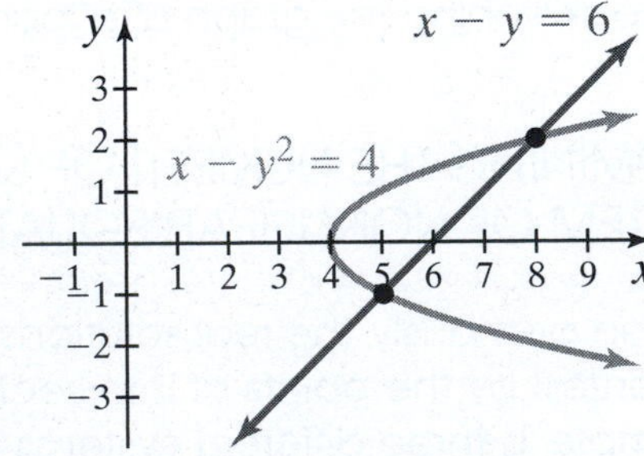

Figure 14

Because the graphs in Figure 14 intersect at two points, the system of nonlinear equations has two real solutions. Watch this **interactive video** to see how to sketch each equation.

c. The graph of the equation $x^2 + y^2 = 9$ is a circle centered at the origin with a radius of 3 units. The equation $x^2 - y = 3$ is equivalent to the equation $y = x^2 - 3$, which is a quadratic function whose graph is a parabola opening up having a vertex at the point $(0, -3)$. The two graphs are shown in Figure 15.

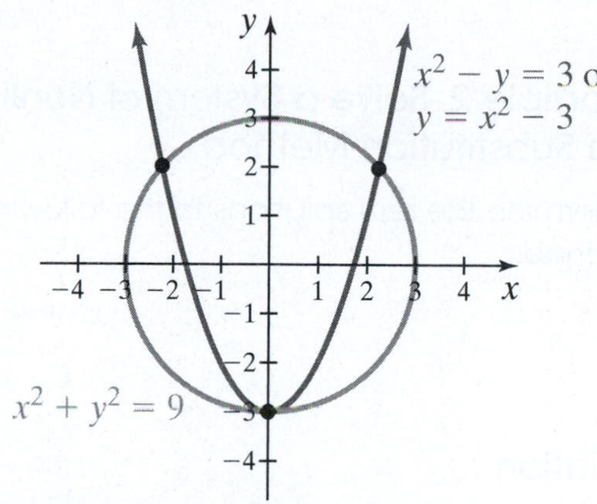

Figure 15

Because the graphs in Figure 15 intersect at three points, the system of nonlinear equations has three real solutions. Watch this **interactive video** to see how to sketch each equation.

You Try It Work through this You Try It problem.

Work Exercises 1–8 in this textbook or in the MyLab Math Study Plan.

OBJECTIVE 2 SOLVING A SYSTEM OF NONLINEAR EQUATIONS USING THE SUBSTITUTION METHOD

The substitution method for solving nonlinear systems of equations involves solving one of the equations for one variable in terms of the other and then substituting that expression into the other equation. This method is very similar to the substitution method used for solving a system of linear equations that was first introduced in **Section 12.1**. The substitution method for solving nonlinear systems of equations can be summarized in the following five steps.

Solving a System of Nonlinear Equations by the Substitution Method

Step 1. Choose an equation and solve for one variable (or expression) in terms of the other variable.

Step 2. Substitute the expression from step 1 into the other equation.

Step 3. Solve the equation in one variable.

Step 4. Substitute the value(s) found in step 3 into one of the original equations to find the value(s) of the other variable.

Step 5. Check each solution by substituting all proposed solutions into the other equation of the system.

(eText Screens 12.5-1–12.5-34)

Note Many nonlinear systems can be solved using more than one method. However, systems in which one equation involves all squared variable terms while the other equation is linear must be algebraically solved using the substitution method. The system shown in Example 2 is an example of a nonlinear system that must be solved algebraically using the method of substitution.

Example 2 Solve a System of Nonlinear Equations Using the Substitution Method

Determine the real solutions to the following system using the substitution method.

$$x^2 + y^2 = 25$$
$$x - y = 1$$

Solution

Step 1. If one of the equations of a nonlinear system is linear, then always solve the linear equation for one of the variables. Here we solve the linear equation for x.

$x - y = 1$ Write the second equation.

$x = y + 1$ Add y to both sides to solve for x.

Step 2. Substitute $y + 1$ for x in the first equation.

$$(y + 1)^2 + y^2 = 25$$

Step 3. Solve for y:

$(y + 1)^2 + y^2 = 25$ Rewrite the equation from step 2.

$y^2 + 2y + 1 + y^2 = 25$ Square: $(y + 1)^2 = y^2 + 2y + 1$.

$2y^2 + 2y - 24 = 0$ Combine like terms.

$y^2 + y - 12 = 0$ Divide both sides by 2.

$(y + 4)(y - 3) = 0$ Factor.

$y + 4 = 0$ or $y - 3 = 0$ Use the **zero product property**.

$y = -4$ or $y = 3$ Solve for y.

Step 4. We can solve for x by substituting $y = -4$ and $y = 3$ into either one of the original equations. Here, we choose the first equation $x^2 + y^2 = 25$.

$y = -4$: $x^2 + (-4)^2 = 25$ Substitute $y = -4$ into the first equation.

$x^2 + 16 = 25$ Square.

$x^2 = 9$ Subtract 16 from both sides.

$x = \pm 3$ Take the square root of both sides.

Thus, for $y = -4$, the possibilities for x are $x = 3$ and $x = -3$. This implies that two possible solutions to the nonlinear system are the ordered pairs $(3, -4)$ and $(-3, -4)$.

In a similar fashion, we can find the value(s) of x that correspond to $y = 3$. View **these steps** to verify that the values of x that correspond to

$y = 3$ are $x = \pm 4$. Therefore, two more possible solutions to the nonlinear system are the ordered pairs $(4, 3)$ and $(-4, 3)$.

Step 5. The four proposed solutions are the ordered pairs $(3, -4)$, $(-3, -4)$, $(4, 3)$, and $(-4, 3)$. We know that these four solutions check in the first equation, $x^2 + y^2 = 25$. Now we must check each proposed solution by substituting the appropriate values of x and y into the other equation, $x - y = 1$.

$(3, -4)$: $3 - (-4) \stackrel{?}{=} 1 \longrightarrow 3 + 4 \stackrel{?}{=} 1 \longrightarrow 7 \stackrel{?}{=} 1 \times \longrightarrow (3, -4)$ is **not** a solution.

$(-3, -4)$: $-3 - (-4) \stackrel{?}{=} 1 \longrightarrow -3 + 4 \stackrel{?}{=} 1 \longrightarrow 1 \stackrel{?}{=} 1 \checkmark \longrightarrow (-3, -4)$ is a solution.

$(4, 3)$: $4 - (3) \stackrel{?}{=} 1 \longrightarrow 4 - 3 \stackrel{?}{=} 1 \longrightarrow 1 \stackrel{?}{=} 1 \checkmark \longrightarrow (4, 3)$ is a solution.

$(-4, 3)$: $-4 - (3) \stackrel{?}{=} 1 \longrightarrow -4 - 3 \stackrel{?}{=} 1 \longrightarrow -7 \stackrel{?}{=} 1 \times \longrightarrow (-4, 3)$ is **not** a solution.

We see that the only two solutions to the nonlinear system are $(-3, -4)$ and $(4, 3)$. **Figure 16** shows the graphs of the two equations of the system. Note that the only two points of intersection are precisely the two points $(-3, -4)$ and $(4, 3)$. Watch this video to see every step of the solution to the given nonlinear system.

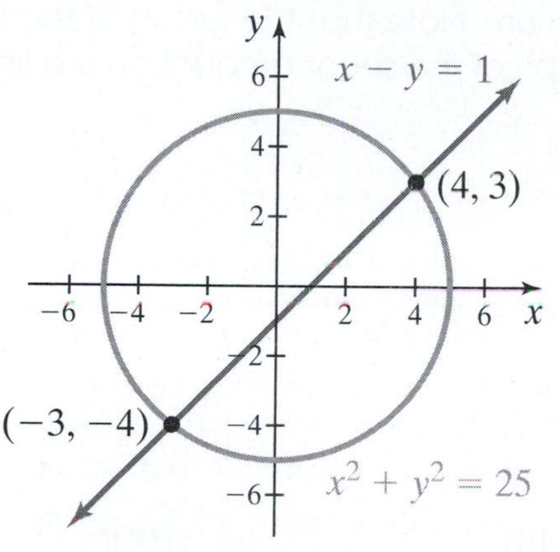

Figure 16 The solutions to the nonlinear system $\begin{aligned} x^2 + y^2 &= 25 \\ x - y &= 1 \end{aligned}$ are the ordered pairs $(-3, -4)$ and $(4, 3)$.

CAUTION It is absolutely critical to always check the solutions to a system of nonlinear equations by substituting each proposed solution into both equations of the system. In the previous example, there were four proposed solutions to the nonlinear system, but only two of those solutions checked in both equations.

Example 3 Solve a System of Nonlinear Equations Using the Substitution Method

Determine the real solutions to the following system using the substitution method.

$$\begin{aligned} 5x^2 - y^2 &= 25 \\ 2x + y &= 0 \end{aligned}$$

Solution

Step 1. The second equation is linear. Thus, solve the second equation for one of the variables. Here we solve the linear equation for y.

$2x + y = 0$ Write the second equation.

$y = -2x$ Subtract $2x$ from both sides to solve for y.

Step 2. Substitute $-2x$ for y in the first equation.

$$5x^2 - (-2x)^2 = 25$$

Step 3. Solve for x:

$5x^2 - (-2x)^2 = 25$	Rewrite the equation from step 2.
$5x^2 - 4x^2 = 25$	Square.
$x^2 = 25$	Combine like terms.
$x = \pm 5$	Take the square root of both sides.

Step 4. We can solve for x by substituting $x = 5$ or $x = -5$ into one of the original equations. Here, we choose the second equation of the form $y = -2x$.

$$x = 5\text{: } y = -2(5) = -10 \qquad x = -5\text{: } y = -2(-5) = 10$$

We see that the possible solutions are $(5, -10)$ and $(-5, 10)$.

Step 5. You should verify that $(5, -10)$ and $(-5, 10)$ are both solutions to the nonlinear system. View this **verification process**. Figure 17 shows the graphical solution to this system. Note that the graph of the first equation is a hyperbola and the graph of the second equation is a line.

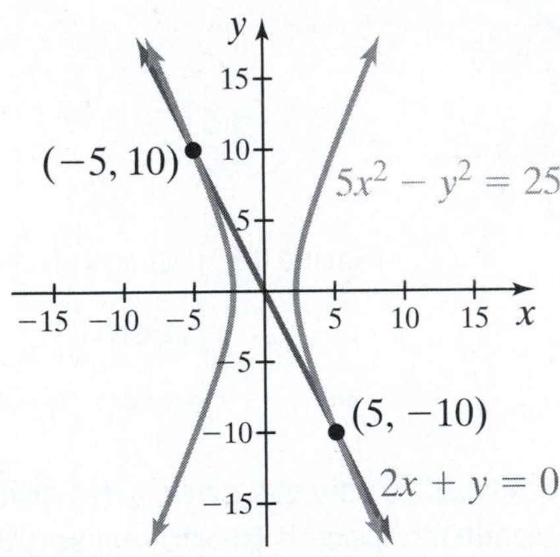

Figure 17 The solutions to the nonlinear system $\begin{matrix} 5x^2 - y^2 = 25 \\ 2x + y = 0 \end{matrix}$ are the ordered pairs $(5, -10)$ and $(-5, 10)$.

Example 4 Solve a System of Nonlinear Equations Using the Substitution Method

Determine the real solutions to the following system using the substitution method.

$$x^2 + 2y^2 = 18$$
$$xy = 4$$

Solution Try solving this nonlinear system on your own using the **five-step process for solving a nonlinear system by the method of substitution.**

When you think that you have accurately solved this system, **view the solution** or watch this **video** to see every step of the solution process.

You Try It **Work through this You Try It problem.**

Work Exercises 9–20 in this textbook or in the MyLab Math Study Plan.

OBJECTIVE 3 SOLVING A SYSTEM OF NONLINEAR EQUATIONS USING SUBSTITUTION, ELIMINATION, OR GRAPHING

We saw in Section 12.1 that systems of linear equations can be solved by adding a multiple of one equation to the other equation in attempt to eliminate one of the variables. This elimination method also works for certain nonlinear systems as well. Elimination is especially useful when both equation contain an x^2-term or a y^2-term. The elimination method can be summarized in the following six steps:

Solving a System of Nonlinear Equations by the Elimination Method

Step 1. Choose a variable to eliminate.

Step 2. Multiply one or both equations by an appropriate nonzero constant so that the sum of the coefficients of one of the terms of both equations is zero.

Step 3. Add the two equations together to obtain an equation in one variable.

Step 4. Solve the equation in one variable.

Step 5. Substitute the value(s) obtained from step 4 into one of the original equations to solve for the other variable.

Step 6. Check each solution by substituting all proposed solutions into the other equation of the system.

Example 5 Solve a System of Nonlinear Equations Using the Elimination Method

Determine the real solutions to the following system.

$$x^2 + y^2 = 9$$
$$x^2 - y = 3$$

Solution We can solve this system using either the substitution method or the elimination method. Here we choose the elimination method.

Step 1. Note that both equations contain an x^2-term. Thus, we can eliminate the variable x.

Step 2. Multiply both sides of the first equation by -1.

$$x^2 + y^2 = 9 \xrightarrow{\text{Multiply by } -1.} -x^2 - y^2 = -9$$
$$x^2 - y = 3 \xrightarrow{\text{Leave equation alone.}} x^2 - y = 3$$

Step 3. Now add the two new equations together:

$$\begin{array}{r} -x^2 - y^2 = -9 \\ \underline{x^2 - y = 3} \\ -y^2 - y = -6 \end{array}$$

Step 4. View these steps to verify that the solution to the quadratic equation $-y^2 - y = -6$ are $y = -3$ or $y = 2$.

Step 5. Solve for x by substituting $y = -3$ and $y = 2$ into one of the original equations. Here, we choose the second equation.

$y = -3$:		$y = 2$:
$x^2 - (-3) = 3$	Substitute the appropriate values into the second equation.	$x^2 - (2) = 3$
$x^2 + 3 = 3$	Simplify.	$x^2 - 2 = 3$
$x^2 = 0$	Isolate x^2.	$x^2 = 5$
$x = 0$	Take the square root of both sides.	$x = \pm\sqrt{5}$

We see that the three possible solutions are $(0, -3)$, $(\sqrt{5}, 2)$, and $(-\sqrt{5}, 2)$.

Step 6. You should take a moment to check that each of the three possible solutions also satisfies the first equation by substituting each ordered pair into the equation $x^2 + y^2 = 9$. Thus the solutions are the ordered pairs $(0, -3)$, $(\sqrt{5}, 2)$, and $(-\sqrt{5}, 2)$.

We could have solved this system using the substitution method. Watch this **interactive video** to see how to solve this system using the substitution method or the elimination method. Figure 18 shows the graphical solution to this nonlinear system.

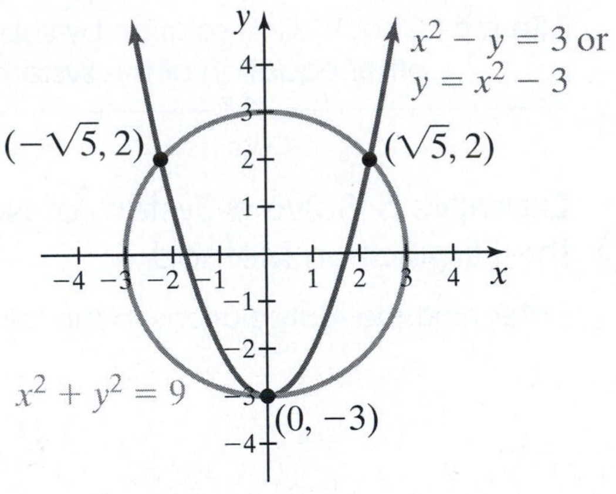

Figure 18 The solutions to the nonlinear system $\begin{aligned} x^2 + y^2 &= 9 \\ x^2 - y &= 3 \end{aligned}$ are the ordered pairs $(0, -3)$, $(\sqrt{5}, 2)$, and $(-\sqrt{5}, 2)$.

You Try It Work through this You Try It problem.

Work Exercises 21–29 in this textbook or in the MyLab Math Study Plan.

Example 6 Solve a System of Nonlinear Equations

Determine the real solutions to the following system.

$$y = \log(3x + 1) - 5$$
$$y = \log(x - 2) - 4$$

Solution We can solve this system using either the substitution method or the elimination method. Here we choose the substitution method.

Step 1. The variable y is isolated in both equations.

Step 2. Substitute $\log(3x+1)-5$ for y in the second equation.

$$\log(3x+1)-5=\log(x-2)-4$$

Step 3. Solve for x:

$\log(3x+1)-5=\log(x-2)-4$	Rewrite the equation from step 2.
$\log(3x+1)-\log(x-2)=1$	Subtract $\log(x-2)$ from both sides and add 5 to both sides.
$\log\left(\dfrac{3x+1}{x-2}\right)=1$	Use the **quotient rule for logarithms.**
$\dfrac{3x+1}{x-2}=10^1$	Rewrite in exponential form.
$3x+1=10(x-2)$	Multiply both sides by $x-2$.
$3x+1=10x-20$	Use the distributive property.
$21=7x$	Combine like terms.
$3=x$	Divide both sides by 7.

Step 4. We can solve for y by substituting $x=3$ into one of the original equations. Here, we choose the second equation of the form $y=\log(x-2)-4$.

$x=3$: $y=\log(3-2)-4$	Substitute $x=3$ into the second equation.
$y=\log(1)-4$	Simplify.
$y=-4$	Substitute 0 for $\log(1)$.

The only proposed solution is the ordered pair $(3,-4)$.

Step 5. Check the proposed solution by substituting the values of $x=3$ and $y=-4$ into the first equation, $y=\log(3x+1)-5$.

$$(3,-4):\ -4\stackrel{?}{=}\log(3(3)+1)-5\longrightarrow -4\stackrel{?}{=}\log(10)-5\longrightarrow -4\stackrel{?}{=}1-5\longrightarrow -4\stackrel{?}{=}-4\ ✓$$

The ordered pair $(3,-4)$ satisfies both equations and thus the solution is verified. Figure 19 shows the graphical solution to this nonlinear system.

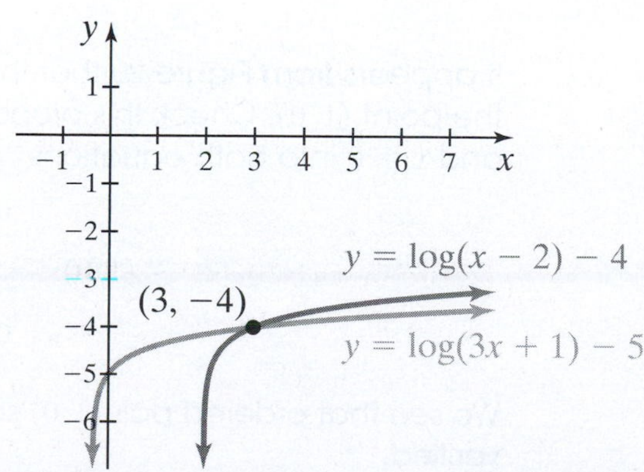

Figure 19 The solution to the nonlinear system $\begin{aligned} y&=\log(3x+1)-5\\ y&=\log(x-2)-4\end{aligned}$ is the ordered pair $(3,-4)$.

You Try It Work through this You Try It problem.

Work Exercises 30–31 in this textbook or in the MyLab Math Study Plan.

Example 7 Solve a System of Nonlinear Equations

Determine the real solutions to the following system.

$$y = \ln x$$
$$y = 1 - x$$

Solution First try using the substitution method by substituting the expression $1 - x$ for y in the first equation. This gives the following equation in one variable:

$$1 - x = \ln x$$

This equation cannot be solved using algebraic methods discussed in this eText. Solving by elimination yields a similar equation that cannot be solved using algebraic methods discussed in this eText.

Therefore, we can try to sketch the graphs of each equation and see if we can determine any points of intersection.

The graph of the natural logarithmic function, $y = \ln x$, and the graph of the linear function $y = 1 - x$, are shown below in Figure 20.

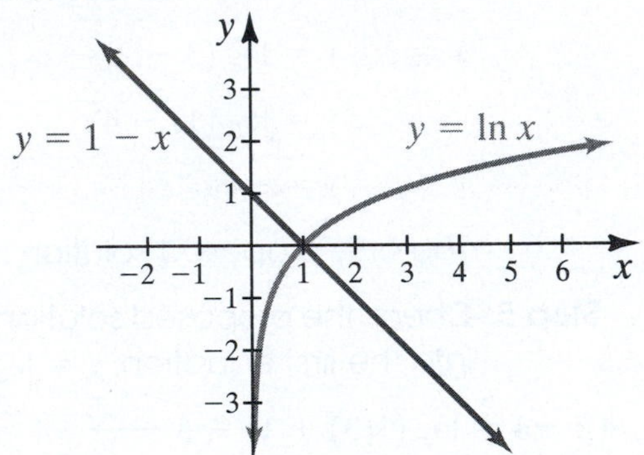

Figure 20 The graph of the nonlinear system
$y = \ln x$
$y = 1 - x$.

It appears from Figure 20 that the graphs of $y = \ln x$ and $y = 1 - x$ intersect at the point $(1, 0)$. Check this proposed solution by substituting the values of $x = 1$ and $y = 0$ into both equations.

$$(1,0): \quad \begin{array}{lcl} 0 \stackrel{?}{=} \ln(1) & & 0 \stackrel{?}{=} 0 \checkmark \\ & \longrightarrow & \\ 0 \stackrel{?}{=} 1 - 1 & & 0 \stackrel{?}{=} 0 \checkmark \end{array}$$

We see that ordered pair $(1, 0)$ satisfies both equations and thus the solution is verified.

Using Technology

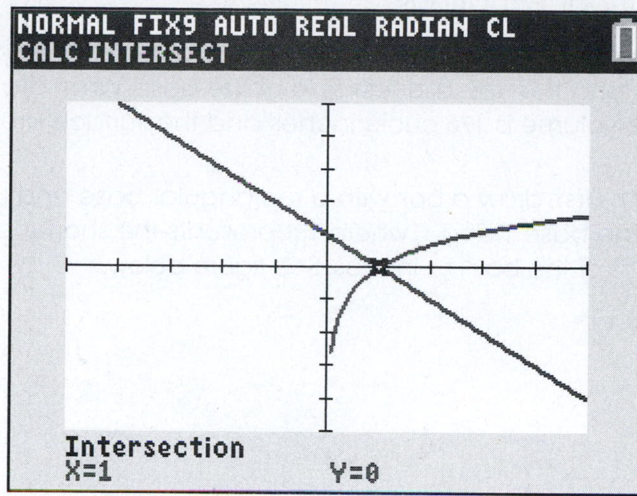

We can solve the nonlinear system from Example 7 using a graphing utility. Graph $y_1 = \ln x$ and $y_2 = 1 - x$, and then use the INTERSECT feature to find that the point of intersection is $(1, 0)$.

You Try It Work through this You Try It problem.

Work Exercises 32 in this textbook or in the MyLab Math Study Plan.

OBJECTIVE 4 SOLVING APPLIED PROBLEMS USING A SYSTEM OF NONLINEAR EQUATIONS

Many applications involving unknown quantities and many geometric applications can be solved using systems of nonlinear equations.

Example 8 Determining Two Unknown Numbers Using a System of Nonlinear Equations

Find two positive numbers such that the sum of the squares of the two numbers is 25 and the difference between the two numbers is 1.

Solution Let x and y represent the two numbers.

The sum of the squares of the two numbers is 25. Therefore, $x^2 + y^2 = 25$.
The difference between the two numbers is 1. Thus, $x - y = 1$.
We must therefore solve the following nonlinear system of equations:

$$\begin{aligned} x^2 + y^2 &= 25 \\ x - y &= 1 \end{aligned}$$

This is exactly the same system seen in Example 2. Refer to the **solution to Example 2** or watch the **video solution to Example 2** to see that the solutions to this system are the ordered pairs $(-3, -4)$ and $(4, 3)$. Therefore, the only positive numbers that have the given relationship are 3 and 4.

You Try It Work through this You Try It problem.

Work Exercises 33–37 in this textbook or in the MyLab Math Study Plan.

Example 9 Solve an Applied Problem Using a System of Nonlinear Equations

An open box (with no lid) has a rectangular base. The height of the box is equal in length to the shortest side of the base. What are the dimensions of the box if the volume is 176 cubic inches and the surface area is 164 square inches?

Solution First, draw a box with a rectangular base and label the lengths of the sides of the base x and y, where x represents the shorter of the two sides. Then, the height of the box is x inches. See figure below.

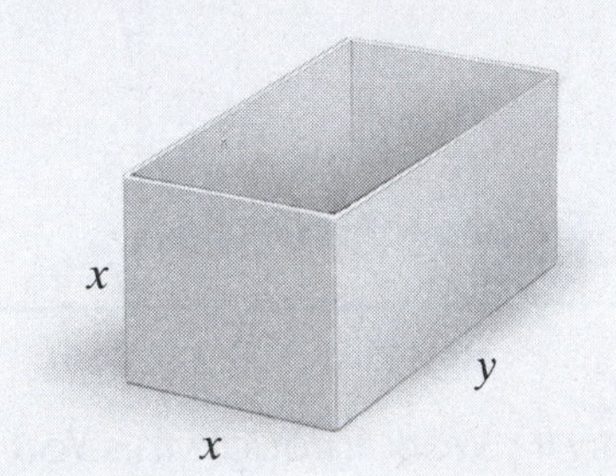

The surface area of the box is the equal to the sum of the areas of the five sides of the box, which is given to be 164 square inches. Therefore, $x^2 + x^2 + xy + xy + xy = 164$. This equation simplifies to $2x^2 + 3xy = 164$.

The volume of the box is 176 cubic inches, which is equal to the length times the width times the height. Thus, $x^2y = 176$. This gives the following system of nonlinear equations:

$$2x^2 + 3xy = 164$$
$$x^2y = 176$$

We can solve this system using the **five-step substitution method.**

Step 1. Solve the second equation for y.

$x^2y = 176$ Write the second equation.

$y = \dfrac{176}{x^2}$ Divide both sides by x^2.

Step 2. Substitute $y = \dfrac{176}{x^2}$ for y in the first equation.

$$2x^2 + 3x\left(\frac{176}{x^2}\right) = 164$$

Step 3. Solve for x:

$2x^2 + 3x\left(\dfrac{176}{x^2}\right) = 164$ Rewrite the equation from step 2.

$2x^2 + \dfrac{528}{x} = 164$ Multiply and cancel common factors.

$2x^3 + 528 = 164x$ Multiply both sides by x

$2x^3 - 164x + 528 = 0$ Subtract $164x$ from both sides.

$x^3 - 82x + 264 = 0$ Divide both sides by 2.

To solve the equation $x^3 - 82x + 264 = 0$, we can use the **rational zeros theorem** and synthetic division to verify that $x - 4$ is a factor of $x^3 - 82x + 264$.

$$\begin{array}{r|rrrr} 4 & 1 & 0 & -82 & 264 \\ & \downarrow & 4 & 16 & -264 \\ \hline & 1 & 4 & -66 & 0 \end{array}$$

The remainder is 0 when $x^3 - 82x + 264$ is divided by $x - 4$. Therefore, $x - 4$ is a factor of $x^3 - 82x + 264$.

Therefore, the polynomial $x^3 - 82x + 264$ factors as $(x - 4)(x^2 + 4x - 66)$. We can now solve the equation $x^3 - 82x + 264 = 0$.

$(x - 4)(x^2 + 4x - 66) = 0$ Write the left-hand side as the product of a linear factor and a quadratic factor.

$x - 4 = 0$ or $x^2 + 4x - 66 = 0$ Use the **zero product property**.

The solution to the equation $x - 4 = 0$ is $x = 4$. Solving the quadratic equation $x^2 + 4x - 66 = 0$ by completing the square gives $x = -2 \pm \sqrt{70}$. (View the **steps** to the completing the square process.) Because x represents the length of a side of the box, we can disregard the negative value $x = -2 - \sqrt{70}$. Therefore, the two possible values of x are $x = 4$ or $x = -2 + \sqrt{70}$.

Step 4. We can solve for x by substituting $x = 4$ and $x = -2 + \sqrt{70} \approx 6.37$ into one of the original equations. Here, we choose the second equation of the form $y = \dfrac{176}{x^2}$.

$x = 4$: $y = \dfrac{176}{(4)^2} = \dfrac{176}{16} = 11$ $\quad x = -2 + \sqrt{70} \approx 6.37$: $y = \dfrac{176}{(-2 + \sqrt{70})^2} \approx 4.34$

As stated in the original problem, the value of x had to be less than the value of y. Therefore, the only possible solution is $x = 4$ and $y = 11$.

Step 5. You should verify that the values of $x = 4$ and $y = 11$ also satisfy the equation $2x^2 + 3xy = 164$. Therefore, the dimensions of the box are 4 inches by 11 inches by 4 inches.

You Try It Work through this You Try It problem.

Work Exercises 38–42 in this textbook or in the MyLab Math Study Plan.

12.5 Exercises

For Exercises 1–8 a system of nonlinear equations is given. Sketch the graph of each equation of the system and then determine the number of real solutions to each system. Do not solve the system.

1. $\begin{aligned} x^2 - y &= -4 \\ 2x - y &= -5 \end{aligned}$

2. $\begin{aligned} x^2 + y^2 &= 16 \\ x + y &= 2 \end{aligned}$

3. $\begin{aligned} x - y^2 &= 0 \\ x - 2y &= -1 \end{aligned}$

4. $\begin{aligned} -x^2 + y^2 &= 16 \\ 8x - y &= 8 \end{aligned}$

5. $\begin{aligned} 2x^2 - y &= 3 \\ x^2 + 4y^2 &= 16 \end{aligned}$

6. $\begin{aligned} x^2 + y^2 &= 25 \\ 4x^2 - 16y^2 &= 100 \end{aligned}$

7. $x + y^2 = -2$
   $x^2 + y^2 = 4$

8. $4x^2 + y^2 = 36$
   $xy = -2$

For Exercises 9–20, determine the real solutions to each system of nonlinear equations using the substitution method.

9. $x + y = 5$
   $y = x^2 - 15$

10. $x^2 + y^2 = 100$
    $x - y = 2$

11. $y = x^2 - 2x + 7$
    $y = x^2 - 8x + 1$

12. $y^2 = x^2 - 64$
    $3y = x - 8$

13. $x + y = 2$
    $y = x^2 - 7x + 10$

14. $y = x + 2$
    $4x^2 + y^2 = 4$

15. $x + y = 2$
    $(x + 3)^2 + (y + 2)^2 = 25$

16. $xy = 30$
    $3x - y = -9$

17. $xy = 5$
    $x^2 + y^2 = 26$

18. $3x^2 + y^2 = 12$
    $xy = 3$

19. $x - y = 1$
    $x^2 - xy - y^2 = -19$

20. $7x^2 - 3xy + 9y^2 = 91$
    $x + 3y = 10$

For Exercises 21–32, determine the real solutions to each system of nonlinear equations.

21. $x^2 + y^2 = 80$
    $x^2 - y^2 = 48$

22. $4x^2 - 5y^2 = -16$
    $5x^2 + 2y^2 = 13$

23. $x^2 + y^2 = 9$
    $x^2 + (y - 1)^2 = 4$

24. $x^2 + y^2 = 16$
    $y^2 - 3x = 16$

25. $x^2 - 4y^2 + 60 = 0$
    $6x^2 + y^2 = 40$

26. $9x^2 - 5y^2 = -9$
    $10x^2 + 2y^2 = 7$

27. $x^2 + 4xy = 6$
    $2x^2 - xy = 3$

28. $2x^2 + y^2 = 1$
    $x^2 - 5y^2 = -15$

29. $x^3 + y = 0$
    $2x^2 + y = 0$

30. $y = \log_2(3x + 7) - 6$
    $y = \log_2(x + 5) - 5$

31. $y = \log_4(x + 22) + 3$
    $y = 5 - \log_4(x + 7)$

32. $y = \log_2 x$
    $y = 6 - x$

33. Find two numbers such that the sum of the two numbers is 22 and the product of the two numbers is 112.

34. Find two positive numbers such that the sum of the squares of the two numbers is 353 and the difference between the two numbers is 9.

35. Find two positive numbers such that the ratio of the two numbers is 6 to 5 and the product of the two numbers is 270.

36. The sum of the squares of the digits of a positive two-digit number is 97. The difference between the two digits is 5. Find this two-digit number.

37. The product of the three digits of a positive three-digit number is 56. The units digit is 2 less than the tens digit. If the difference between the hundreds digit and the tens digit is 3, then find this three-digit number.

38. The area of a rectangle is 78 square feet and the perimeter is 38 feet. Find the dimensions of the rectangle.

39. The area of a right triangle is 52 square meters. The length of the hypotenuse is $\sqrt{233}$ meters. Find the lengths of the legs of the right triangle.

40. The diagonal of a rectangular pen is 50 feet. The width of the pen is 7 feet less than the length. Find the length of the pen. (Round the length to one decimal place.)

41. An open box (with no lid) has a square base and four sides of equal height. What are the dimensions of the box if the volume is 972 cubic inches and the surface area is 513 square inches?

42. An open box (with no lid) has a rectangular base. The height of the box is equal in length to the shortest side of the base. What are the dimensions of the box if the volume is 468 cubic centimeters and the surface area is 306 square centimeters?

12.6 Systems of Inequalities

THINGS TO KNOW

Before working through this section, be sure that you are familiar with the following concept:

VIDEO ANIMATION INTERACTIVE

You Try It 1. Sketching the Graph of a Circle (Section 2.2)

You Try It 2. Sketching Lines by Plotting Intercepts (Section 2.3)

You Try It 3. Graphing Quadratic Functions (Section 4.1)

You Try It 4. Sketching the Graph of a Parabola (Section 11.1)

You Try It 5. Sketching the Graph of an Ellipse (Section 11.2)

You Try It 6. Sketching the Graph of a Hyperbola (Section 11.3)

You Try It 7. Solving a System of Nonlinear Equations Using the Method of Substitution (Section 12.5)

You Try It 8. Solving a System of Nonlinear Equations Using the Method of Elimination (Section 12.5)

OBJECTIVES

1 Determine If an Ordered Pair Is a Solution to an Inequality in Two Variables

2 Graphing a Linear Inequality in Two Variables

3 Graphing a Nonlinear Inequality in Two Variables

4 Determining If an Ordered Pair Is a Solution to a System of Inequalities in Two Variables

5 Graphing a System of Linear Inequalities in Two Variables

6 Graphing a System of Nonlinear Inequalities in Two Variables

SECTION 12.6 EXERCISES

OBJECTIVE 1 DETERMINE IF AN ORDERED PAIR IS A SOLUTION TO AN INEQUALITY IN TWO VARIABLES

In **Section 1.7** we solved **linear inequalities in one variable**. In **Section 1.9** we solved **polynomial inequalities** and **rational inequalities**. Polynomial and rational inequalities were examples of nonlinear inequalities in one variable. The solution set for inequalities in one variable is the set of all values of the variable that make the inequality true. We graph the solution set of inequalities in one variable on a number line.

The solution set to an inequality in two variables is the set of all ordered pairs that satisfy the inequality. Thus, given an inequality in the variables x and y, we can determine if the ordered pair (a, b) is a solution to the inequality by substituting the value of a in for x and b in for y. If a true statement results after this substitution, then (a, b) is a solution to the inequality.

Example 1 Determining If an Ordered Pair Is a Solution to an Inequality in Two Variables

Determine if the given ordered pair is a solution to the inequality $-2x + y^2 \leq 4$.

a. $(2, -1)$ b. $(-2, 0)$ c. $\left(-\frac{1}{2}, \frac{7}{2}\right)$

Solution

a.	$-2x + y^2 \leq 4.$	Write the original inequality.
	$-2(2) + (-1)^2 \overset{?}{\leq} 4$	Substitute 2 for x and -1 for y.
	$-3 \overset{?}{\leq} 4$ ✓	Simplify and evaluate. A true statement results.

The final statement is true, so $(2, -1)$ is a solution to the inequality. Work through parts (b) and (c) by yourself, then **check your answers**, or watch this **video** solution.

You Try It Work through this You Try It problem.

Work Exercises 1–4 in this textbook or in the MyLab Math Study Plan.

OBJECTIVE 2 GRAPHING A LINEAR INEQUALITY IN TWO VARIABLES

We now take our first look at graphing the solution set to an inequality in two variables. We start by graphing the solution to a **linear inequality in two variables**. Recall that a linear equation in two variables is of the form $Ax + By = C$, where A and B are not both equal to zero. Replacing the equal sign of this equation with an inequality symbol results in the following definition.

> **Definition** Linear Inequality in Two Variables
>
> A **linear inequality in two variables** is an inequality that can be written in the form $Ax + By < C$, where A, B, and C are real numbers, and A and B are not both equal to zero.
>
> **Note** The inequality symbol "$<$" can be replaced with $>$, $\leq$, or $\geq$.

Before sketching the solution to a linear inequality in two variables, we must first focus on the related linear equation. For example, to find the solutions to $x + y > 3$ or $x + y < 3$, we look at the related linear equation $x + y = 3$. For an ordered pair to satisfy this equation, the sum of its coordinates must be 3. Note that the ordered pairs $(-2, 5)$, $(1, 2)$, and $(3, 0)$ are all solutions to the equation $x + y = 3$ because the sum of the coordinates of each ordered pair is 3. Plotting and connecting these points gives the line shown in Figure 21.

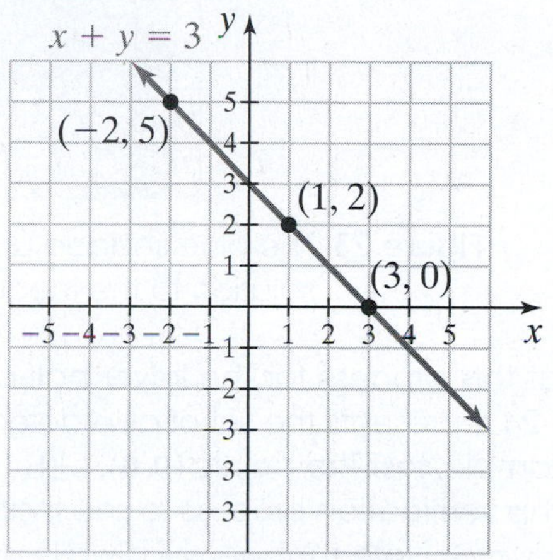

Figure 21

The line $x + y = 3$ divides the **coordinate plane** into two **half-planes**, an *upper half-plane* and a *lower half-plane*. In Figure 22, the upper half-plane is shaded blue and the lower half-plane is shaded red.

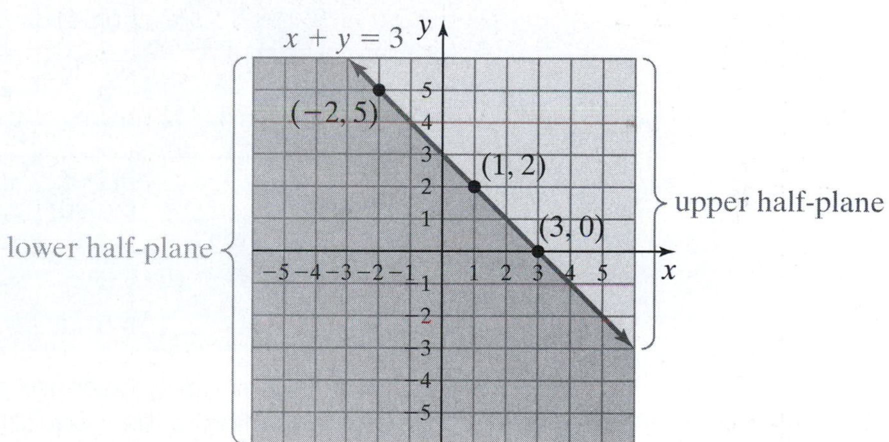

Figure 22 The line $x + y = 3$ divides the coordinate plane into two half-planes.

Let's now examine some points that lie in each half-plane to see which ones satisfy the inequalities $x + y > 3$ or $x + y < 3$.

First, choose any point that lies in the upper half-plane, say $(4, 2)$. The sum of its coordinates is $4 + 2 = 6$, which is larger than 3. Choose any other point in this region, such as $(0, 4)$. The sum of its coordinates is $0 + 4 = 4$, which is also greater than 3. For *any point* we choose in the upper half-plane, the sum of its coordinates will be greater than 3. This means that any ordered pair from the upper half-plane is a solution to the inequality $x + y > 3$. Thus, the area shaded blue in Figure 23 represents the set of all ordered pairs that are solutions to the inequality $x + y > 3$.

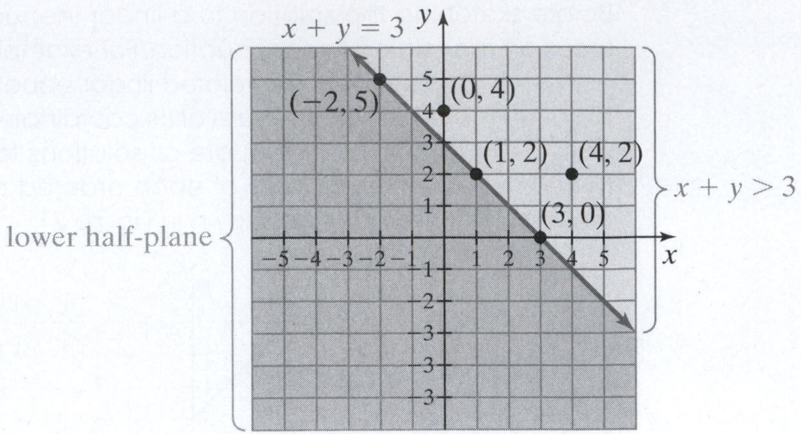

Figure 23 The blue shaded upper half-plane represents the solution to the inequality $x + y > 3$.

Repeat this process for the lower half-plane to see that the area shaded red in Figure 24 represents the set of all ordered-pair solutions to the inequality $x + y < 3$. For example, test the points $(0, 0)$, $(-1, 1)$, and $(-2, -2)$ from the lower half-plane. View this **verification process** to see that all three ordered pairs satisfy the inequality $x + y < 3$. Note that we will always use the point $(0, 0)$ as a test point unless it lies along the boundary line.

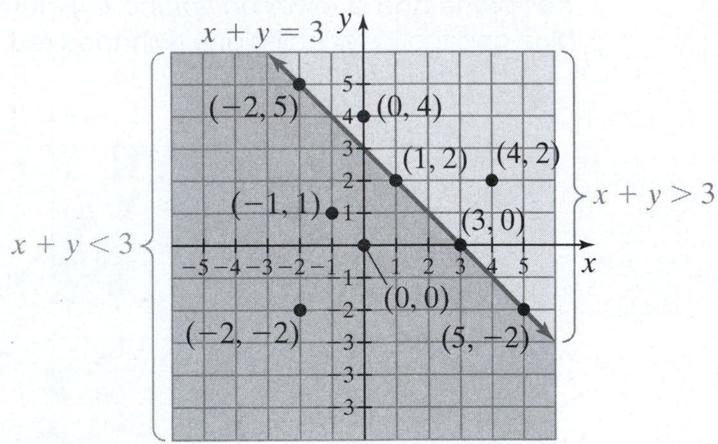

Figure 24 The red shaded lower half-plane represents the solution to the inequality $x + y < 3$.

The equation $x + y = 3$ acts as a **boundary line** that separates the solutions of the two inequalities $x + y > 3$ and $x + y < 3$. Based on this information, we can define a set of **steps for graphing linear inequalities in two variables**.

Steps for Graphing Linear Inequalities in Two Variables

Step 1. Find the **boundary line** for the inequality by replacing the inequality symbol with an equal sign and graphing the resulting equation. If the inequality is **strict**, graph the boundary line using a dashed line. If the inequality is **non-strict**, graph the boundary line using a solid line.

Step 2. Choose a **test point** that does not belong to the boundary line and determine if it is a solution to the inequality.

Step 3. If the test point is a solution to the inequality, then shade the half-plane that contains the test point. If the test point is not a solution to the inequality, then shade the half-plane that does not contain the test point. The shaded area represents the set of all ordered-pair solutions to the inequality.

Note When graphing a linear inequality in two variables, using a dashed line is similar to using an open circle when graphing a linear inequality in one variable on a number line. Likewise, using a solid line is similar to using a solid circle. Do you see **why**?

Example 2 Graphing a Linear Inequality in Two Variables

Graph each inequality.

a. $x - 2y \geq 4$ **b.** $3y < 2x$ **c.** $x < -2$

Solution

a. We follow the three-step process for **Graphing Linear Inequalities in Two Variables.**

Step 1. The **boundary line** is $x - 2y = 4$. We graph the line as a solid line because the inequality is **non-strict**. See **Figure 25a**.

Step 2. We choose the test point $(0, 0)$ and check to see if it satisfies the inequality.

$x - 2y \overset{?}{\geq} 4$	Write the original inequality.
$0 - 2(0) \overset{?}{\geq} 4$	Substitute 0 for x and 0 for y.
$0 \overset{?}{\geq} 4 \times$	Simplify and evaluate. A false statement results.

The point $(0, 0)$ is not a solution to the inequality.

Step 3. Because the test point is not a solution to the inequality, we shade the half-plane that does not contain $(0, 0)$. See Figure 25b. The shaded region, including the boundary line, represents all ordered-pair solutions to the inequality $x - 2y \geq 4$.

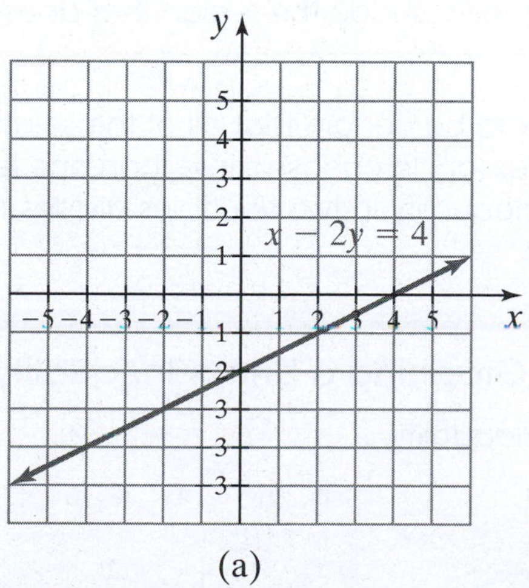

(a)

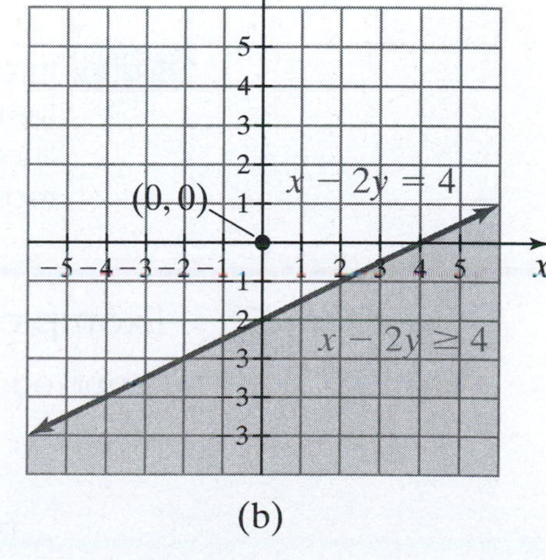

(b)

Figure 25

b. For the inequality $3y < 2x$, the boundary line is $3y = 2x$ (or $y = \frac{2}{3}x$.) Because the inequality is **strict**, we graph the boundary line using a dashed line. Choose a test point and complete the graph. Note here that we cannot use the point $(0, 0)$ as a test point because it lies along the boundary line. When you have completed the work, view the **graph of the solution** or watch this **interactive video** to see each step of the solution process.

c. For the inequality $x < -2$, the boundary line is the **vertical line** $x = -2$. Because the inequality is **strict**, we graph the boundary line using a dashed line. Choose a test point and complete the graph. When you have completed the work, view the **graph of the solution** or watch this **interactive video** to see each step of the solution process.

You Try It Work through this You Try It problem.

Work Exercises 5–14 in this textbook or in the MyLab Math Study Plan.

OBJECTIVE 3 GRAPHING A NONLINEAR INEQUALITY IN TWO VARIABLES

The procedure used for graphing a nonlinear inequality in two variables is nearly the same as the procedure used for sketching the graph of a linear inequality in two variables.

Steps for Graphing Nonlinear Inequalities in Two Variables

Step 1. Replace the inequality symbol with an equal sign and graph the resulting equation. If the inequality is **strict**, sketch the graph using dashes. If the inequality is **non-strict**, sketch the graph using a solid curve. This graph divides the coordinate plane into two or more regions.

Step 2. Choose one test point that does not belong to the graph of the equation from step 1 and determine if it is a solution to the inequality.

Step 3. If the test point is a solution to the inequality, shade the region that contains the test point. If the test point is not a solution to the inequality, shade the region that does not include the test point.

Note In order to be certain that all of the solution set has been identified, it may be necessary to choose more than one test point when the graph of a nonlinear inequality in two variables divides the coordinate plane into multiple regions.

Example 3 Graphing a Linear Inequality in Two Variables

Graph each inequality.

a. $x^2 + y^2 \geq 9$ b. $9y^2 - 4x^2 \leq 36$ c. $x - y^2 > 1$

Solution

a. We follow the steps for **Graphing Nonlinear Inequalities in Two Variables.**

Step 1. The inequality $x^2 + y^2 \geq 9$ is **non-strict**, so we graph the equation $x^2 + y^2 = 9$ (which is a circle centered at the origin having a radius of 3 units) using a continuous solid curve. See **Figure 26a**.

Step 2. We choose the test point $(0, 0)$, which lies inside the circle.

$x^2 + y^2 \overset{?}{\geq} 9$ Write the original inequality.

$0^2 + 0^2 \overset{?}{\geq} 9$ Substitute 0 for x and 0 for y.

$0 \overset{?}{\geq} 9 \times$ Simplify and evaluate. A false statement results.

The point $(0, 0)$ that lies inside the circle is not a solution to the inequality.

Step 3. Because the test point $(0, 0)$ is not a solution to the inequality, we shade the region that lies outside of the circle. See Figure 26b. The shaded region, including the graph of the circle, represents all ordered-pair solutions to the inequality $x^2 + y^2 \geq 9$.

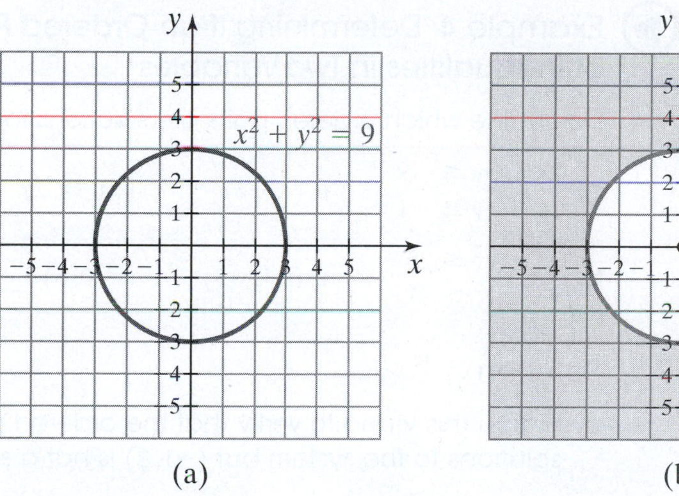

Figure 26

b. For the inequality $9y^2 - 4x^2 \leq 36$, the inequality is **non-strict**, so we graph the equation $9y^2 - 4x^2 = 36$ (which is a hyperbola centered at the origin having a vertical **transverse axis**) using a solid curve. Choose a test point and complete the graph. When you have completed the work, view the **graph of the solution** or watch this **interactive video** to see each step of the solution process.

c. For the inequality $x - y^2 > 1$, the inequality is **strict**, so we graph the equation $x - y^2 = 1$ (which is a parabola that opens to the right having a vertex at the point $(1, 0)$) using dashes. Choose a test point and complete the graph. When you have completed the work, view the **graph of the solution** or watch this **interactive video** to see each step of the solution process.

You Try It Work through this You Try It problem.

Work Exercises 15–24 in this textbook or in the MyLab Math Study Plan.

OBJECTIVE 4 DETERMINING IF AN ORDERED PAIR IS A SOLUTION TO A SYSTEM OF INEQUALITIES IN TWO VARIABLES

We now consider systems of inequalities in two variables.

> **Definition** System of Inequalities in Two Variables
>
> A **system of inequalities in two variables** is the collection of two or more inequalities in two variables considered together. If all inequalities are linear, then the system is a **system of linear inequalities in two variables**. If at least one inequality is nonlinear, then the system is a **system of nonlinear inequalities in two variables**.

Example 4 illustrates a system of linear inequalities in two variables and a system of nonlinear inequalities in two variables.

An ordered pair is a solution to a system of inequalities in two variables if the ordered pair satisfies all inequalities in the system.

Example 4 Determining If an Ordered Pair Is a Solution to a System of Inequalities in Two Variables

Determine which ordered pairs are solutions to the given system.

a. $\begin{aligned} 2x - 3y &\leq 9 \\ 2x - y &\geq -1 \end{aligned}$ i. $(1, -2)$ ii. $(-1, 2)$ iii. $(3, -1)$

b. $\begin{aligned} x^2 - y^2 &\leq 25 \\ x - y &> 1 \end{aligned}$ i. $(-1, -3)$ ii. $(0, 5)$ iii. $(-3, 4)$

Solution

a. Watch this **video** to verify that the ordered pairs $(1, -2)$ and $(3, -1)$ are solutions to the system but $(-1, 2)$ is not a solution.

b. Watch this **video** to verify $(-1, -3)$ is a solution whereas the ordered pairs $(0, 5)$ and $(-3, 4)$ are not solutions. ●

You Try It Work through this You Try It problem.

Work Exercises 25–30 in this textbook or in the MyLab Math Study Plan.

OBJECTIVE 5 GRAPHING A SYSTEM OF LINEAR INEQUALITIES IN TWO VARIABLES

The graph of a system of inequalities in two variables is the intersection of the graphs of each inequality in the system. We start by sketching **the graph of a system of linear inequalities in two variables**. Carefully watch this **animation** to see how to graph the following system of linear inequalities.

$$\begin{aligned} 2x - 3y &\leq 9 \\ 2x - y &> -1 \end{aligned}$$

We obtain the graph of a system of linear inequalities by graphing each inequality in the system and finding the region they have in common, if any. Thus, we follow the two-step process outlined below.

Steps for Graphing Systems of Linear Inequalities in Two Variables

Step 1. Use the **Steps for Graphing Linear Inequalities in Two Variables** to graph each inequality.

Step 2. Determine where the shaded regions overlap, if any. This overlapped (or shared) region represents the set of all solutions to the system of inequalities.

Example 5 Graphing a Linear Inequality in Two Variables

Graph each system of linear inequalities in two variables.

a. $\begin{aligned} x + y &> 2 \\ 2x - y &\le 6 \end{aligned}$

b. $\begin{aligned} x - 3y &> 6 \\ 2x - 6y &< -9 \end{aligned}$

c. $\begin{aligned} 4x &> y \\ x - 3y &< 9 \\ x + y &< 4 \end{aligned}$

Solution

a. To graph the system $\begin{aligned} x + y &> 2 \\ 2x - y &\le 6 \end{aligned}$, we follow the Steps for **Graphing Systems of Linear Inequalities in Two Variables**.

Step 1. **Graph $x + y > 2$.** The inequality $x + y > 2$ is **strict**, so we graph the **boundary line** $x + y = 2$ using a dashed line. The test point $(0, 0)$ does not satisfy the inequality $x + y > 2$ because $0 + 0 > 2$ is false, so we shade the half-plane that does not contain $(0, 0)$. See Figure 27a.

Graph $2x - y \le 6$. The inequality $2x - y \le 6$ is **non-strict**, so we graph the boundary line $2x - y = 6$ using a solid line. The test point $(0, 0)$ satisfies the inequality $2x - y \le 6$ because $2(0) - 0 \le 6$ is true, so we shade the half-plane that contains $(0, 0)$. See Figure 27b.

Step 2. We determine the region shared by both inequalities in the system. This is the purple-shaded region in **Figure 27c**. Any point in this region is a solution to this system of linear inequalities in two variables.

Watch this **interactive video** for a detailed solution.

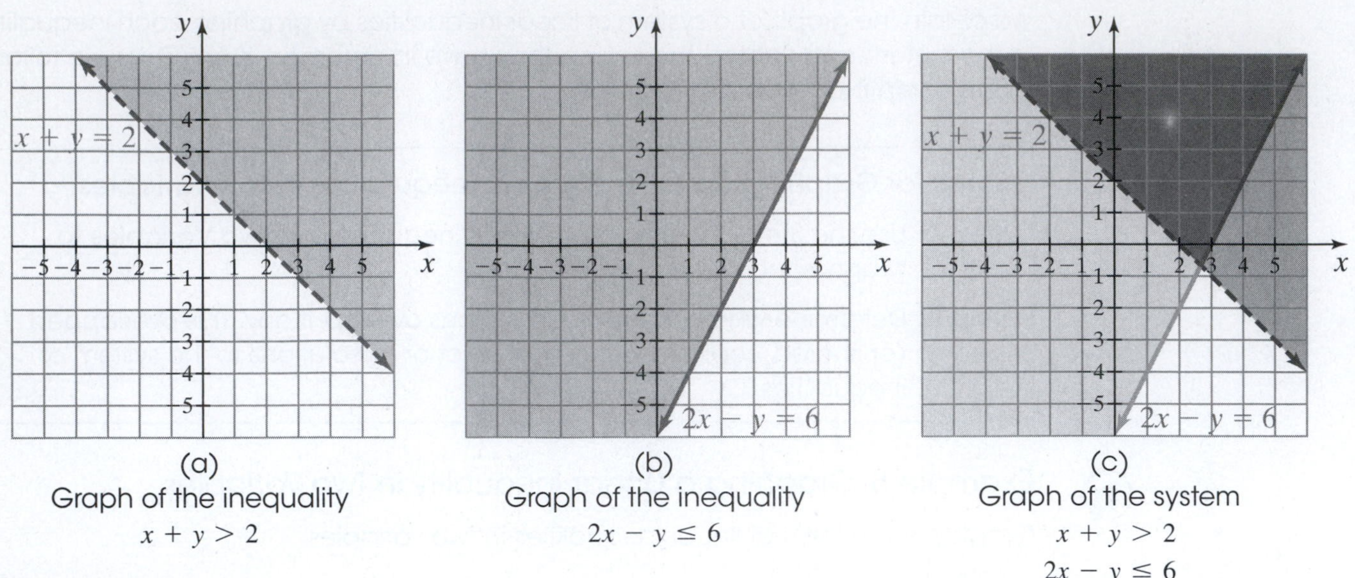

(a) Graph of the inequality $x + y > 2$

(b) Graph of the inequality $2x - y \le 6$

(c) Graph of the system $x + y > 2$, $2x - y \le 6$

Figure 27

CAUTION Only the ordered pairs in the purple-shaded region in Figure 27c, including the portion of the solid line $2x - y = 6$, are solutions to the system. Note that none of the points on the boundary line $x + y = 2$ are solutions to the system. Furthermore, the two boundary lines intersect at the point $\left(\frac{8}{3}, -\frac{2}{3}\right)$. View these steps to see how to determine this point of intersection. Because the inequality $x + y > 2$ is strict, this point is not part of the solution.

b. To graph the system $\begin{aligned} x - 3y &> 6 \\ 2x - 6y &< -9 \end{aligned}$, we once again follow the Steps for Graphing Systems of Linear Inequalities in Two Variables.

Step 1. **Graph $x - 3y > 6$.** The inequality $x - 3y > 6$ is **strict**, so we graph the **boundary line** $x - 3y = 6$ using a dashed line. The test point $(0, 0)$ does not satisfy the inequality $x - 3y > 6$, so we shade the half-plane that does not contain $(0, 0)$. See **Figure 28a**.

Graph $2x - 6y < -9$. The inequality $2x - 6y < -9$ is strict, so we graph the boundary line $2x - 6y = -9$ using a dashed line. The test point $(0, 0)$ does not satisfy the inequality $2x - 6y < -9$, so we shade the half-plane that does not contain $(0, 0)$. See **Figure 28b**.

Step 2. The boundary lines $x - 3y = 6$ and $2x - 6y = -9$ are parallel, and the shaded regions are on opposite sides of these parallel lines. Therefore, no region is shaded by both inequalities in the system. See **Figure 28c**. There are no ordered pairs that satisfy both inequalities, so there is no solution to the system.

Watch this **interactive video** for a detailed solution

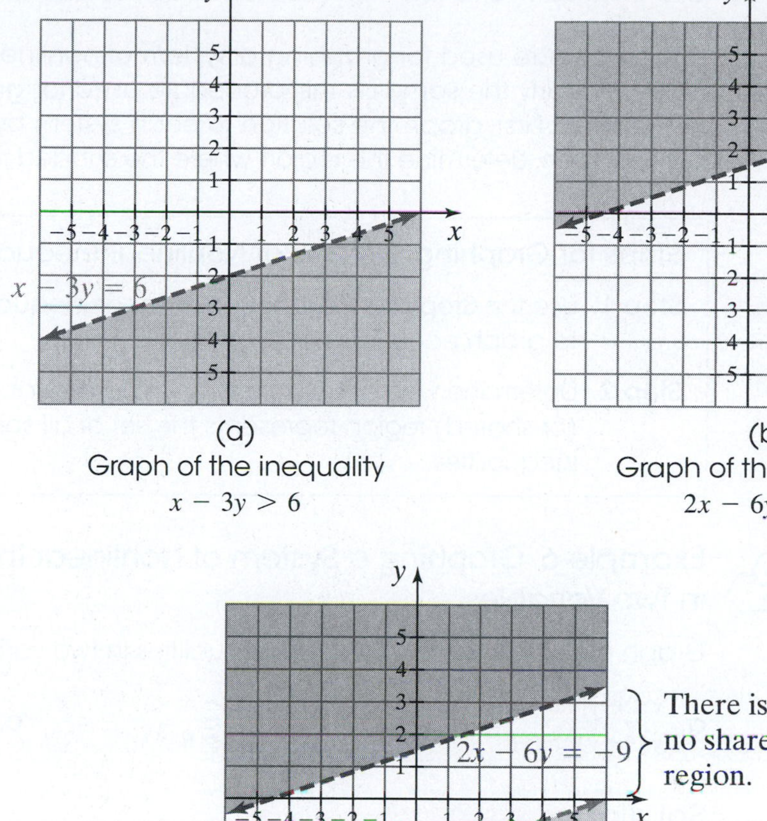

(a)
Graph of the inequality
$x - 3y > 6$

(b)
Graph of the inequality
$2x - 6y < -9$

2x − 6y = −9
There is no shared region.
x − 3y = 6

(c)
Graph of the system
$x - 3y > 6$
$2x - 6y < -9$

Figure 28

c. To graph the system

$$\begin{aligned} 4x &> y \\ x - 3y &< 9, \\ x + y &< 4 \end{aligned}$$

we once again follow the Steps for **Graphing Systems of Linear Inequalities in Two Variables.** Each inequality in the system is strict, so all of the boundary lines are dashed lines. Graph all three inequalities on the same coordinate plane and find the region shared by all three inequalities, if any. Once you are finished, you can **view the solution** or watch this **interactive video** to see a detailed solution. ●

You Try It Work through this You Try It problem.

Work Exercises 31–45 in this textbook or in the MyLab Math Study Plan.

OBJECTIVE 6 GRAPHING A SYSTEM OF NONLINEAR INEQUALITIES IN TWO VARIABLES

The procedure used for graphing a system of nonlinear inequalities in two variables is nearly the same as the procedure used for graphing a system of linear inequalities. First, graph the solution to each system by shading the appropriate regions. Then, determine the region where the shaded regions overlap.

Steps for Graphing Systems of Nonlinear Inequalities in Two Variables

Step 1. Use the **Steps for Graphing Nonlinear Inequalities in Two Variables** to graph each inequality.

Step 2. Determine where the shaded regions overlap, if any. This overlapped (or shared) region represents the set of all solutions to the system of inequalities.

Example 6 Graphing a System of Nonlinear Inequalities in Two Variables

Graph each system of nonlinear inequalities in two variables.

a. $\begin{aligned} x^2 + y^2 &\le 25 \\ x - y &> 1 \end{aligned}$ b. $\begin{aligned} x - y^2 &\ge 4 \\ x - y &\le 6 \end{aligned}$ c. $\begin{aligned} 4x^2 + 9y^2 &\le 36 \\ x^2 + y &\le -2 \end{aligned}$

Solution

a. To graph the system $\begin{aligned} x^2 + y^2 &\le 25 \\ x - y &> 1 \end{aligned}$, we follow the steps for **Graphing Systems of Nonlinear Inequalities in Two Variables**.

Step 1. **Graph $x^2 + y^2 \le 25$.** The inequality $x^2 + y^2 \le 25$ is **non-strict**, so we sketch the graph of $x^2 + y^2 = 25$ (which is a circle centered at the origin having a radius of 5 units) using a solid curve. The test point $(0, 0)$ satisfies the inequality $x^2 + y^2 \le 25$ because $0 + 0 \le 25$ is true, so we shade the region inside the circle containing $(0, 0)$. See **Figure 29a**.

Graph $x - y > 1$. The inequality $x - y > 1$ is **strict**, so we graph the boundary line $x - y = 1$ using a dashed line. The test point $(0, 0)$ does not satisfy the inequality $x - y > 1$ because $0 - 0 > 1$ is false, so we shade the half-plane that does not contain $(0, 0)$. See **Figure 29b**.

Step 2. We determine the region shared by both inequalities in the system. This is the purple-shaded region inside the circle and below the line. See Figure 29c. Any point in this region is a solution to this system of nonlinear inequalities in two variables. Watch this **interactive video** for a detailed solution. Note that the two graphs intersect at the points $(-3, -4)$ and $(4, 3)$. We can determine the coordinates of these points using the method of substitution. See **Example 2 from Section 12.5** or watch this video to see how to find these points. These points of intersection are not solutions to the system of inequalities because no points lying on the dashed boundary line $x - y = 1$ can be solutions. Thus, we represent these points using an open circle.

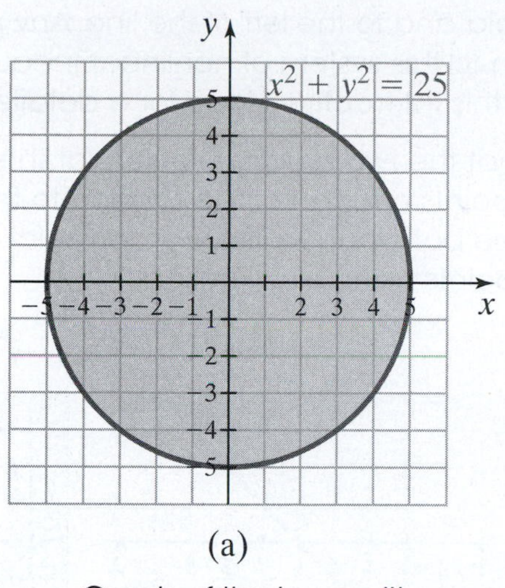

(a)

Graph of the inequality
$x^2 + y^2 \le 25$

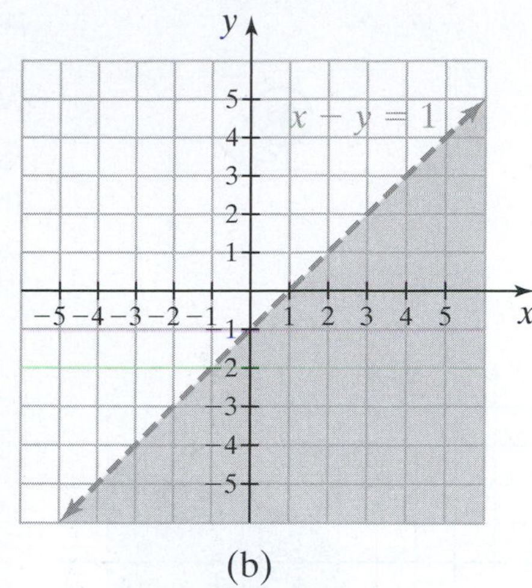

(b)

Graph of the inequality
$x - y > 1$

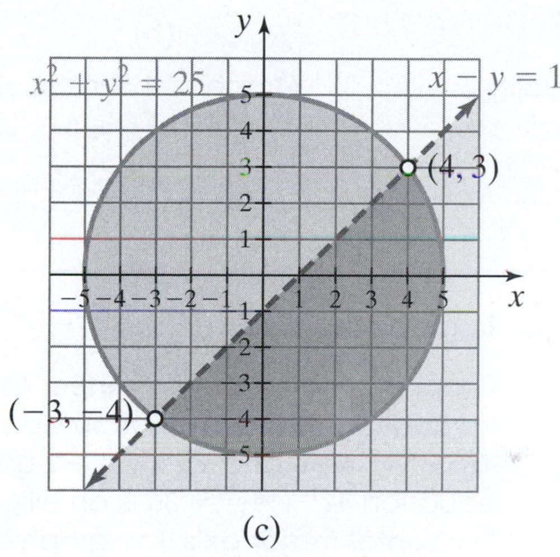

(c)

Graph of the system
$x^2 + y^2 \le 25$
$x - y > 1$

Figure 29

b. To graph the system $\begin{aligned} x - y^2 &\ge 4 \\ x - y &\le 6 \end{aligned}$, we follow the steps for **Graphing Systems of Nonlinear Inequalities in Two Variables**.

Step 1. **Graph $x - y^2 \ge 4$.** The inequality $x - y^2 \ge 4$ is **non-strict**, so we sketch the graph of $x - y^2 = 4$ (which is parabola that opens to the right having a vertex at the point $(4, 0)$) using a solid curve. The test point $(0, 0)$ does not satisfy the inequality $x - y^2 \ge 4$ because $0 - 0 \ge 4$ is false, so we shade the region inside the parabola. See **Figure 30a**.

Graph $x - y \le 6$. The inequality $x - y \le 6$ is non-strict, so we graph the boundary line $x - y = 6$ using a solid line. The test point $(0, 0)$ satisfies the inequality $x - y \le 6$ because $0 - 0 \le 6$ is true, so we shade the half-plane that contains $(0, 0)$. See **Figure 30b**.

Step 2. We determine the region shared by both inequalities in the system. This is the purple-shaded region seen in **Figure 30c** inside the parabola and to the left of the line. Any point in this region is a solution to this system of nonlinear inequalities in two variables. Watch this **interactive video** for a detailed solution.

Note that the two graphs intersect at the points $(5, -1)$ and $(8, 2)$. These points are part of the solution to the system of inequalities because both inequalities are non-strict. Therefore, we represent these points using a solid circle.

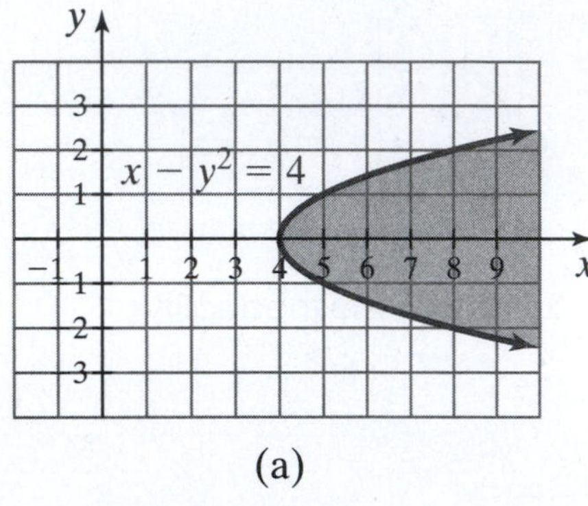

(a)

Graph of the inequality
$x - y^2 \geq 4$

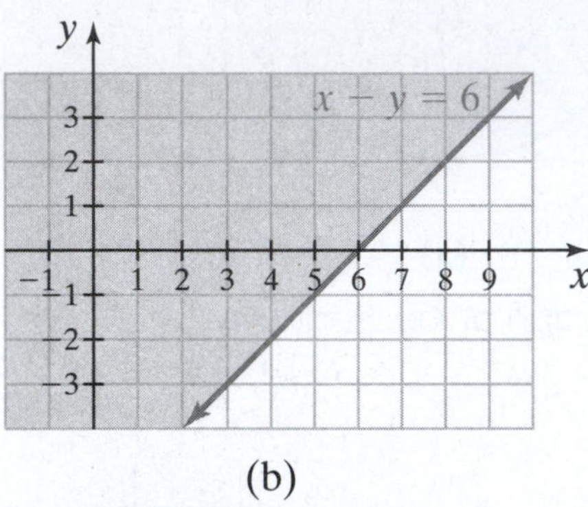

(b)

Graph of the inequality
$x - y \leq 6$

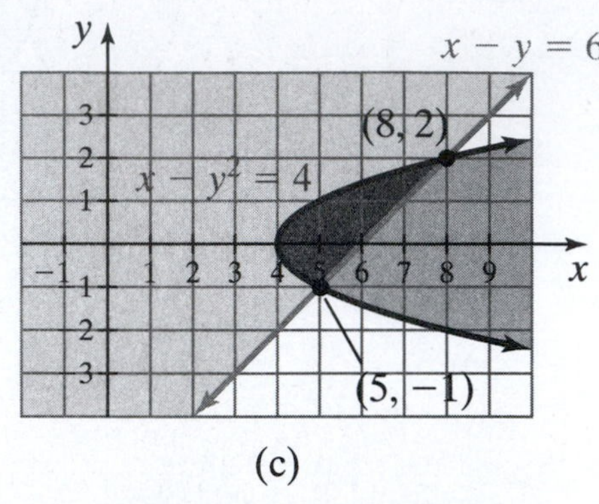

(c)

Graph of the system
$x - y^2 \geq 4$
$x - y \leq 6$

Figure 30

c. To graph the system $\begin{aligned} 4x^2 + 9y^2 &\leq 36 \\ x^2 + y &\leq -2 \end{aligned}$, we once again follow the **Steps for Graphing Systems of Nonlinear Inequalities in Two Variables**. Both inequalities are **non-strict**, so we sketch the graphs of the equations $4x^2 + 9y^2 = 36$ and $x^2 + y = -2$ using solid curves. Note that the graph of the equation $4x^2 + 9y^2 = 36$ is an ellipse centered at the origin having a **horizontal major axis**. The graph of the equation $x^2 + y = -2$ is a parabola opening down having a vertex at the point $(0, -2)$. Try graphing this system of nonlinear inequalities on your own. When you have completed the work, **view the graph** of this system or watch this **interactive video** to see that the two inequalities have exactly one point in common, the point $(0, -2)$. Therefore, the ordered pair $(0, -2)$ is the only solution to this system of nonlinear inequalities.

You Try It Work through this You Try It problem.

Work Exercises 46–58 in this textbook or in the MyLab Math Study Plan.

12.6 Exercises

In Exercises 1–4, determine if each ordered pair is a solution to the given inequality.

1. $2x + 5y > 10$
   a. $(-5, 4)$
   b. $(2, 3)$
   c. $\left(\frac{5}{2}, \frac{3}{5}\right)$

2. $6x - 5y \leq 30$
   a. $(0, 0)$
   b. $(2, -4)$
   c. $(2.5, -1.5)$

3. $x^2 - 2y^2 \geq 8$
   a. $(-4, -4)$
   b. $(3, -5)$
   c. $(6.5, 4.1)$

4. $x^2 - xy < -3$
   a. $(-1, -4)$
   b. $(2, 5)$
   c. $\left(-\frac{1}{2}, -\frac{9}{2}\right)$

In Exercises 5–14, graph each inequality.

5. $2x - y \geq -3$
6. $3x + 2y > 6$
7. $-x + 3y < -9$
8. $x + y \leq 0$
9. $y \leq \frac{5}{2}x - 1$
10. $4y < -3x$
11. $\frac{1}{2}x + \frac{2}{3}y > \frac{5}{6}$
12. $0.6x - 1.8y \leq 2.4$
13. $x \geq -1$
14. $y < 3$

In Exercises 15–24, graph each inequality.

15. $x^2 + y^2 \geq 49$
16. $x^2 + y^2 < 25$
17. $x^2 - y > 3$
18. $y - x^2 \leq 2$
19. $x^2 + 9y^2 < 81$
20. $4x^2 + y^2 \geq 4$
21. $x - y^2 \leq 2$
22. $y^2 > 4 - x$
23. $x^2 - 4y^2 \geq 16$
24. $y^2 - 9x^2 < 9$

In Exercises 25–30, determine if each ordered pair is a solution to the given system of inequalities in two variables.

25. $\begin{aligned} x - 2y &< 6 \\ 3x + y &> 2 \end{aligned}$
   a. $(1, 0)$
   b. $(4, -1)$
   c. $(3, -2)$

26. $\begin{aligned} 3x - 2y &\geq 6 \\ 4x + 3y &\leq 9 \end{aligned}$
   a. $(5, -3)$
   b. $(0, 0)$
   c. $(2, -2)$

27. $\begin{aligned} x - y &> -8 \\ 5x - 2y &\leq 10 \\ 4x + 3y &> 12 \end{aligned}$
   a. $(2, 4)$
   b. $(1, 1)$
   c. $(-1, 6)$

28. $\begin{aligned} x - y^2 &> 4 \\ x - y &\leq 6 \end{aligned}$
   a. $(4, 0)$
   b. $(5, -1)$
   c. $(6, 1)$

29. $\begin{aligned} 4x^2 + 9y^2 &< 36 \\ x^2 + y &\leq 2 \end{aligned}$
   a. $(-2, 0)$
   b. $(0, 2)$
   c. $(-1, 1)$

30. $\begin{aligned} x - y^2 &> 1 \\ x + y^2 &> -3 \\ y - 2x &\leq 3 \end{aligned}$
   a. $\left(\frac{3}{2}, 0\right)$
   b. $(2, 1)$
   c. $\left(\frac{5}{4}, \frac{1}{4}\right)$

In Exercises 31–45, graph each system of linear inequalities in two variables.

31. $\begin{aligned} x + y &\geq 4 \\ 2x - y &\leq 5 \end{aligned}$

32. $\begin{aligned} y &< -2x \\ y &> -3x \end{aligned}$

33. $\begin{aligned} 2x - y &< 6 \\ y &\geq \frac{3}{2}x + 3 \end{aligned}$

34. $\begin{aligned} x - 3y &\leq 6 \\ x &< 3 \end{aligned}$

35. $\begin{aligned} \frac{1}{3}x - \frac{1}{2}y &\leq 1 \\ y &\geq -2 \end{aligned}$

36. $\begin{aligned} x &> -4 \\ y &\geq 1 \end{aligned}$

37. $\begin{aligned} 3x - 5y &\geq 15 \\ -3x + 5y &\geq 15 \end{aligned}$

38. $\begin{aligned} x &> 3 \\ x &< -2 \end{aligned}$

39. $\begin{aligned} 6x - 8y &< 4 \\ -3x + 4y &\leq 12 \end{aligned}$

40. $\begin{aligned} y &\geq 2 \\ y &\leq 5 \end{aligned}$

41. $\begin{aligned} 2x - y &> 7 \\ y &< 2x + 5 \end{aligned}$

42. $\begin{aligned} 2x - 3y &\geq 12 \\ -x + 1.5y &< 6 \end{aligned}$

43. $\begin{aligned} 3x + 2y &\leq 4 \\ 3x - 2y &\geq -16 \\ x - 2y &\leq 4 \end{aligned}$

44. $\begin{aligned} 2x - y &> -3 \\ 5x + y &> 0 \\ x &> 1 \end{aligned}$

45. $\begin{aligned} 2x + y &\leq 4 \\ x + 3y &\leq 6 \\ x &\geq 0 \\ y &\geq 0 \end{aligned}$

In Exercises 46–58, graph each system of nonlinear inequalities in two variables.

46. $\begin{aligned} x^2 + y^2 &\leq 36 \\ x + y &< 4 \end{aligned}$

47. $\begin{aligned} x^2 + y^2 &> 4 \\ x^2 + y^2 &\leq 25 \end{aligned}$

48. $\begin{aligned} x + y &\leq 2 \\ y &\geq x^2 - 3 \end{aligned}$

49. $\begin{aligned} x - y^2 &\geq 0 \\ x - 3y &> -2 \end{aligned}$

50. $\begin{aligned} x^2 + y^2 &\leq 64 \\ x^2 - y &\leq 8 \end{aligned}$

51. $\begin{aligned} x^2 + y^2 &\leq 9 \\ y - x^2 &> 0 \end{aligned}$

52. $\begin{aligned} y - x^2 &> -7 \\ y - x &\leq -4 \end{aligned}$

53. $\begin{aligned} x^2 + y^2 &\geq 4 \\ x - y &\leq 2 \end{aligned}$

54. $\begin{aligned} y &\geq x^2 - 2x \\ y &< -2x^2 + 4x \end{aligned}$

55. $\begin{aligned} x + y^2 &\leq 2 \\ x^2 + 4x + y &\geq -4 \end{aligned}$

56. $\begin{aligned} x^2 + y^2 &\leq 4 \\ 2x - y &\geq 4 \end{aligned}$

57. $\begin{aligned} 9x^2 + 4y^2 &< 36 \\ x - y^2 &\geq 0 \end{aligned}$

58. $\begin{aligned} x^2 + 4y^2 &\leq 16 \\ x^2 - y^2 &> 4 \end{aligned}$

Chapter 12 Summary

Key Concepts	Examples/Videos
12.1 Solving Systems of Linear Equations in Two Variables **Solving a System of Equations by the Method of Substitution** **Step 1.** Choose an equation and solve for one variable in terms of the other variable. **Step 2.** Substitute the expression from step 1 into the other equation. **Step 3.** Solve the equation in one variable. **Step 4.** Substitute the value found in step 3 into one of the original equations to find the value of the other variable.	1. Solve the following system using the method of substitution. $2x - 3y = -5$ $x + y = 5$

Key Concepts	Examples/Videos
Solving a System of Equations by the Method of Elimination **Step 1.** Choose a variable to eliminate. **Step 2.** Multiply one or both equations by an appropriate nonzero constant so that the sum of the coefficients of one of the variables is zero. **Step 3.** Add the two equations together to obtain an equation in one variable. **Step 4.** Solve the equation in one variable. **Step 5.** Substitute the value found in step 4 into one of the original equations to find the value of the other variable.	2. Solve the following system using the method of elimination. $-2x + 5y = 29$ $3x + 2y = 4$
Five-Step Strategy for Problem Solving Using Systems of Equations **Step 1.** Read the problem several times until you have an understanding of what is being asked. If possible, create diagrams, charts, or tables to assist you. — Understand the problem. **Step 2.** Choose variables that describe each unknown quantity. **Step 3.** Write a system of equations using the given information and the variables. — Devise a plan. **Step 4.** Carefully solve the system of equations using the method of elimination or substitution. — Carry out the plan. **Step 5.** Make sure that you have answered the question, and check all answers to ensure they make sense. — Look back.	3. Twin City Foods, Inc., created a 10-lb bean mixture that sells for \$5.75 by mixing lima beans and green beans. If lima beans sell for \$.70 per pound and green beans sell for \$.50 per pound, how many pounds of each bean went into the mixture?
12.2 Solving Systems of Linear Equations in Three Variables Using the Elimination Method A **linear equation in three variables** is an equation that can be written in the form $Ax + By + Cz = D$, where A, B, C, and D are real numbers, and $A, B,$ and C are not all equal to 0. A **system of linear equations in three variables** is a collection of linear equations in three variables considered simultaneously. A solution to a system of linear equations in three variables is an **ordered triple** that satisfies all equations in the system.	1. Verify that the ordered triple $(1, 2, 1)$ is a solution to the following system of linear equations: $2x + 3y + 4z = 12$ $x - 2y + 3z = 0$ $-x + 2y - 2z = -1$

Key Concepts	Examples/Videos
Guidelines for Solving a System of Linear Equations in Three Variables by Elimination Step 1. **Write each equation in standard form.** Write each equation in the form $Ax + By + Cz = D$ lining up the variable terms. Number the equations to keep track of them. Step 2. **Eliminate a variable from one pair of equations.** Use the elimination method to eliminate a variable from any two of the original three equations, leaving one equation in two variables. Step 3. **Eliminate the same variable again.** Use a different pair of the original equations and eliminate the same variable again, leaving one equation in two variables. Step 4. **Solve the system of linear equations in two variables.** Use the resulting equations from steps 2 and 3 to create and solve the corresponding system of linear equations in two variables by substitution or elimination. Step 5. **Use back substitution to find the value of the third variable.** Substitute the results from step 4 into any of the original equations to find the value of the remaining variable. Step 6. **Check the solution.** Check the proposed solution in each equation of the system and write the solution set.	2. Solve the following system: $2x + 3y + 4z = 12$ $x - 2y + 3z = 0$ $-x + y - 2z = -1$ 3. Solve the following system: $2x + y = 13$ $3x - 2y + z = 18$ $3x + 2y - 3z = 15$
	4. Wendy ordered 30 T-shirts online for her three children. The small T-shirts cost \$4 each, the medium T-shirts cost \$5 each, and the large T-shirts were \$6 each. She spent \$40 more purchasing the large T-shirts than the small T-shirts. Wendy's total bill was \$154. How many T-shirts of each size did she buy?

Key Concepts	Examples/Videos

12.3 Solving Systems of Linear Equations in Three Variables Using Gaussian Elimination and Gauss-Jordan Elimination

Gaussian elimination is the process of writing a system of three linear equations in three variables into and **equivalent** system that is in **triangular form**, then using back substitution to solve for each variable.

Elementary Row Operations

1. Interchange any two equations.
2. Multiply any equation by a nonzero constant.
3. Add a multiple of one equation to another equation.

Notation Used to Describe Elementary Row Operations

Notation	Meaning
$R_i \Leftrightarrow R_j$	Interchange Rows i and j.
$kR_i \rightarrow \text{New } R_i$	k times Row i becomes New Row i.
$kR_i + R_j \rightarrow \text{New } R_j$	k times Row i plus Row j becomes New Row j.

1. For the following system, use elementary row operations to find an equivalent system in triangular form and then use back substitution to solve the system.

$$\begin{aligned} 2x + 3y + 4z &= 12 \\ x - 2y + 3z &= 0 \\ -x + 2y - 2z &= -1 \end{aligned}$$

A short-hand method to Gaussian elimination is to use an augmented matrix at each step of the process instead of writing out the entire set of equations at each step.

In Exercise 1 above, the corresponding augmented matrix for the given system is:

$$\left[\begin{array}{rrr|r} 2 & 3 & 4 & 12 \\ 1 & -2 & 3 & 0 \\ -1 & 1 & -2 & -1 \end{array}\right]$$

2. Create an augmented matrix and solve the following linear system using Gaussian elimination by writing an equivalent system in triangular form.

$$\begin{aligned} x + 2y - z &= 3 \\ x - 3y - 2z &= 11 \\ -x - 2y + 2z &= -6 \end{aligned}$$

When solving a system of three linear equations with three unknowns, only **triangular form** is needed to find the solution. However, it is often useful to further reduce the augmented matrix into one of two other simpler forms known as **row-echelon form** or **reduced row-echelon form**.

The process of reducing a system into reduced row-echelon form is called **Gauss-Jordan elimination.**

3. Solve the following system using Gauss-Jordan elimination.

$$\begin{aligned} x_1 + x_2 + x_3 &= -1 \\ x_1 + 2x_2 + 4x_3 &= 3 \\ x_1 + 3x_2 + 9x_3 &= 3 \end{aligned}$$

4. Use Gauss-Jordan elimination to solve the system.

$$\begin{aligned} x - y + 2z &= 4 \\ -x + 3y + z &= -6 \\ x + y + 5z &= 3 \end{aligned}$$

Key Concepts	Examples/Videos
12.4 Partial Fraction Decomposition The partial fraction decomposition of a rational expression of the form $\frac{P(x)}{Q(x)}$ can be performed when the following two criteria are satisfied: 1. The polynomials $P(x)$ and $Q(x)$ share no common factors. 2. The degree of $P(x)$ is less than the degree of $Q(x)$.	
If the denominator of $\frac{P(x)}{Q(x)}$ is the product of n distinct linear factors, then the partial fraction decomposition of $\frac{P(x)}{Q(x)}$ will be of the form $$\frac{P(x)}{Q(x)} = \frac{A_1}{a_1x + b_1} + \frac{A_2}{a_2x + b_2} + \cdots + \frac{A_n}{a_nx + b_n}$$ where $A_1, A_2, \ldots, A_n$ are constants to be determined.	1. Determine the partial fraction decomposition of $\frac{5x^2 - x + 1}{x^3 + 3x^2 - 4x}$.
A repeated linear factor is of the form $(ax + b)^n$, where n is an integer such that $n \geq 2$. If the denominator of $\frac{P(x)}{Q(x)}$ has a repeated linear factor, then for every repeated linear factor of the form $(ax + b)^n$, we introduce n partial fractions of the form $$\frac{A_1}{ax + b} + \frac{A_2}{(ax + b)^2} + \cdots + \frac{A_n}{(ax + b)^n}$$	2. Determine the partial fraction decomposition of $\frac{x - 1}{x(x - 2)^2}$.
A quadratic factor has the form $ax^2 + bx + c$, where $a \neq 0$. If the quadratic factor will not factor into the product of two linear factors using integer coefficients, then the quadratic factor is prime. If the denominator of $\frac{P(x)}{Q(x)}$ has a prime quadratic factor, then for every prime quadratic factor of the form $ax^2 + bx + c$, we introduce a partial fraction of the form $$\frac{A_1x + B_1}{ax^2 + bx + c}.$$	3. Determine the partial fraction decomposition of $\frac{8x^2 + 7}{x^3 - 1}$.

Key Concepts	Examples/Videos
A repeated prime quadratic factor has the form $(ax^2 + bx + c)^n$, where $a \neq 0$ and where n is an integer such that $n \geq 2$. If the denominator of $\frac{P(x)}{Q(x)}$ has a repeated prime quadratic factor, then for every repeated prime quadratic factor of the form $(ax^2 + bx + c)^n$, we introduce n partial fractions of the form $\frac{A_1x + B_1}{ax^2 + bx + c} + \frac{A_2x + B_2}{(ax^2 + bx + c)^2} + \cdots + \frac{A_nx + B_n}{(ax^2 + bx + c)^n}$.	**4.** Determine the partial fraction decomposition of $\frac{3x^4 - 5x^3 + 7x^2 + x - 2}{(x - 1)(x^2 + 1)^2}$.
12.5 Systems of Nonlinear Equations A **system of nonlinear equations** in two variables is a system of equations that contains at least one equation that is nonlinear. The solution to a system of nonlinear equations in two variables can be represented graphically by all points of intersection between the graphs of both equations.	**1.** For each system of nonlinear equations, sketch the graph of each equation of the system and then determine the number of real solutions to each system. Do not solve the system. a. $x^2 + y^2 = 25$ $x - y = 1$ b. $x - y^2 = 4$ $x - y = 6$ c. $x^2 + y^2 = 9$ $x^2 - y = 3$
Solving a System of Nonlinear Equations by the Substitution Method **Step 1.** Choose an equation and solve for one variable (or expression) in terms of the other variable. **Step 2.** Substitute the expression from step 1 into the other equation. **Step 3.** Solve the equation in one variable. **Step 4.** Substitute the value(s) found in step 3 into one of the original equations to find the value(s) of the other variable. **Step 5.** Check each solution by substituting all proposed solutions into the other equation of the system.	**2.** Determine the real solutions to the following system using the substitution method. $x^2 + y^2 = 25$ $x - y = 1$ **3.** Determine the real solutions to the following system using the substitution method. $5x^2 - y^2 = 25$ $2x + y = 0$

Key Concepts	Examples/Videos

Solving a System of Nonlinear Equations by the Elimination Method

Step 1. Choose a variable to eliminate.

Step 2. Multiply one or both equations by an appropriate nonzero constant so that the sum of the coefficients of one of the terms of both equations is zero.

Step 3. Add the two equations together to obtain an equation in one variable.

Step 4. Solve the equation in one variable.

Step 5. Substitute the value(s) obtained from step 4 into one of the original equations to solve for the other variable.

Step 6. Check each solution by substituting all proposed solutions into the other equation of the system.

4. Determine the real solutions to the following system.

$$x^2 + y^2 = 9$$
$$x^2 - y = 3$$

5. Determine the real solutions to the following system.

$$y = \log(3x + 1) - 5$$
$$y = \log(x - 2) - 4$$

12.6 Systems of Inequalities

A **linear inequality in two variables** is an inequality that can be written in the form $Ax + By < C$, where A, B, and C are real numbers, and A and B are not both equal to zero.

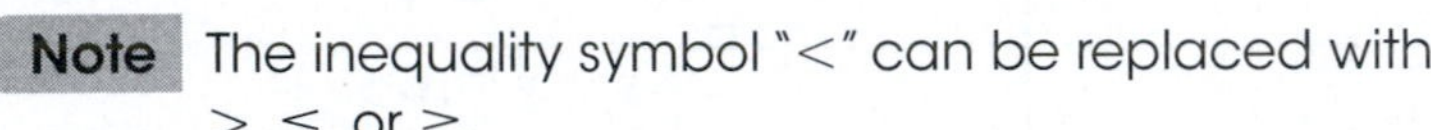

Note The inequality symbol "$<$" can be replaced with $>$, $\leq$, or $\geq$.

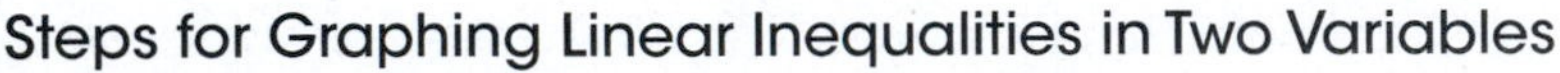

Steps for Graphing Linear Inequalities in Two Variables

Step 1. Find the **boundary line** for the inequality by replacing the inequality symbol with an equal sign and graphing the resulting equation. If the inequality is **strict**, graph the boundary line using a dashed line. If the inequality is **non-strict**, graph the boundary line using a solid line.

Step 2. Choose a **test point** that does not belong to the boundary line and determine if it is a solution to the inequality.

Step 3. If the test point is a solution to the inequality, then shade the half-plane that contains the test point. If the test point is not a solution to the inequality, then shade the half-plane that does not contain the test point. The shaded area represents the set of all ordered-pair solutions to the inequality.

1. Graph each inequality.

a. $x - 2y \geq 4$

b. $3y < 2x$

c. $x < -2$

Key Concepts	Examples/Videos
Steps for Graphing Nonlinear Inequalities in Two Variables **Step 1.** Replace the inequality symbol with an equal sign and graph the resulting equation. If the inequality is **strict**, sketch the graph using dashes. If the inequality is **non-strict**, sketch the graph using a solid curve. This graph divides the coordinate plane into two or more regions. **Step 2.** Choose one test point that does not belong to the graph of the equation from step 1 and determine if it is a solution to the inequality. **Step 3.** If the test point is a solution to the inequality, shade the region that contains the test point. If the test point is not a solution to the inequality, shade the region that does not include the test point.	2. Graph each inequality. a. $x^2 + y^2 \geq 9$ b. $9y^2 - 4x^2 \leq 36$ c. $x - y^2 > 1$
A **system of inequalities in two variables** is the collection of two or more inequalities in two variables considered together. If all inequalities are linear, then the system is a **system of linear inequalities in two variables**. If at least one inequality is nonlinear, then the system is a **system of nonlinear inequalities in two variables.**	3. Determine which ordered pairs are solutions to the given system. a. $2x - 3y \leq 9$ $2x - y \geq -1$ i. $(1, -2)$ ii. $(-1, 2)$ iii. $(3, -1)$ b. $x^2 - y^2 \leq 25$ $x - y > 1$ i. $(-1, -3)$ ii. $(0, 5)$ iii. $(-3, 4)$
Steps for Graphing Systems of Linear Inequalities in Two Variables **Step 1.** Use the **Steps for Graphing Linear Inequalities in Two Variables** to graph each inequality. **Step 2.** Determine where the shaded regions overlap, if any. This overlapped (or shared) region represents the set of all solutions to the system of inequalities.	4. Graph the system of linear inequalities in two variables. $2x - 3y \leq 9$ $2x - y \geq -1$ 5. Graph each system of linear inequalities in two variables. a. $x + y > 2$ $2x - y \leq 6$ b. $x - 3y > 6$ $2x - 6y < -9$ c. $4x > y$ $x - 3y < 9$ $x + y < 4$

Key Concepts	Examples/Videos
Steps for Graphing Systems of Nonlinear Inequalities in Two Variables **Step 1.** Use the **Steps for Graphing Nonlinear Inequalities in Two Variables** to graph each inequality. **Step 2.** Determine where the shaded regions overlap, if any. This overlapped (or shared) region represents the set of all solutions to the system of inequalities.	6. Graph each system of nonlinear inequalities in two variables. a. $\begin{aligned} x^2 + y^2 &\le 25 \\ x - y &> 1 \end{aligned}$ b. $\begin{aligned} x - y^2 &\ge 4 \\ x - y &\le 6 \end{aligned}$ c. $\begin{aligned} 4x^2 + 9y^2 &\le 36 \\ x^2 + y &\le -2 \end{aligned}$

Chapter 12 Review Exercises

1. Determine whether each ordered pair is a solution to the given system of equations.

$$\begin{aligned} -4x - 2y &= -2 \\ 5x + 7y &= 16 \end{aligned} \qquad \text{Ordered pairs: } (-1, 3), (-3, 7)$$

2. Use the method of substitution to solve the given system of linear equations.

$$\begin{aligned} x &= 4 - y \\ 3x - 2y &= -3 \end{aligned}$$

3. Use the method of elimination to solve the given system of linear equations.

$$\begin{aligned} 8x - 6y &= -6 \\ 5x - 7y &= -33 \end{aligned}$$

4. On a certain hot summer's day, 492 people used the public swimming pool. The daily prices are \$1.50 for children and \$2.50 for adults. The receipts for admission totaled \$929. How many children and how many adults swam at the public pool that day?

5. Determine whether each ordered triple is a solution to the given system of equations.

$$(-1, 1, -2), (1, -1, 2)$$

$$\begin{aligned} x + y + z &= -2 \\ -x - 3y - 2z &= 2 \\ 2x - 2y + 5z &= -14 \end{aligned}$$

6. Solve the given system using the elimination method.

$$\begin{aligned} x - 2y + 2z &= 2 \\ 3x + 2y - 2z &= -1 \\ x - y - 2z &= 0 \end{aligned}$$

7. Ben ordered 35 pizzas for an office party. He ordered three types: cheese, supreme, and pepperoni. Cheese pizza costs \$9 each, pepperoni pizza costs \$12 each, and supreme pizza costs \$15 each. He spent exactly twice as much on the pepperoni pizzas as he did on the cheese pizzas. If Ben spent \$420, how many pizzas of each type did he buy?

8. Use Gaussian elimination to solve the given linear system by finding an equivalent system in triangular form.

$$\begin{aligned} x - 2y + z &= 6 \\ 2x + y - 3z &= -3 \\ x - 3y + 3z &= 10 \end{aligned}$$

In Exercises 9 and 10, use Gaussian elimination and matrices to solve each system of linear equations. Write your final augmented matrix in triangular form and then solve for each variable using back substitution.

9. $$\begin{aligned} x + y + z &= 0 \\ x - 2y + 3z &= -7 \\ -x - y + 4z &= 6 \end{aligned}$$

10. $$\begin{aligned} x_1 + 2x_2 + 3x_3 &= 11 \\ 3x_1 + 8x_2 + 5x_3 &= 27 \\ -x_1 + x_2 + 2x_3 &= 2 \end{aligned}$$

In Exercises 11–13, determine the partial fraction decomposition of each rational expression.

11. $\dfrac{x}{(x-2)(x+4)}$

12. $\dfrac{x^2}{(x-1)^2(x+4)}$

13. $\dfrac{x+7}{x^2(x^2+9)}$

14. Sketch the graph of each equation in the given system and then determine the number of real solutions to each system. **Do not solve** the system.

$$\begin{aligned} x^2 + y^2 &= 16 \\ x + y &= 2 \end{aligned}$$

15. Determine the real solutions to the given system of nonlinear equations using the substitution method.

$$\begin{aligned} x + y &= 2 \\ y &= x^2 - 7x + 10 \end{aligned}$$

16. Find two positive numbers such that the sum of the squares of the two numbers is 353 and the difference between the two numbers is 9.

17. Determine if each ordered pair is a solution to the given inequality.

$$x^2 - 2y^2 \geq 8$$

a. $(-4, -4)$
b. $(3, -5)$
c. $(6.5, 4.1)$

18. Graph the inequality $2x - y \geq -3$.

19. Graph the inequality $x^2 + y^2 < 25$.

20. Graph the system of nonlinear inequalities in two variables.

$$\begin{aligned} 9x^2 + 4y^2 &< 36 \\ x - y^2 &\geq 0 \end{aligned}$$

CHAPTER THIRTEEN

Matrices

CHAPTER THIRTEEN CONTENTS

13.1 Matrix Operations

THINGS TO KNOW

Before working through this section, be sure that you are familiar with the following concepts:

VIDEO ANIMATION INTERACTIVE

You Try It

1. Solving a System of Three Equations in Three Variables Using Gauss-Jordan Elimination (Section 12.3) 

OBJECTIVES

1 Understanding the Definition of a Matrix

2 Adding and Subtracting Matrices

3 Scalar Multiplication of Matrices

4 Multiplication of Matrices

5 Applications of Matrix Multiplication

SECTION 13.1 EXERCISES

OBJECTIVE 1 UNDERSTANDING THE DEFINITION OF A MATRIX

In Section 12.3, we introduced matrices to display the coefficients of a system of linear equations in order to simplify the Gaussian elimination (or Gauss-Jordan elimination) process. For example, the linear system

$$\begin{aligned} 2x + 3y + 4z &= 12 \\ x - 2y + 3z &= 0 \\ -x + y - 2z &= -1 \end{aligned}$$

can be represented by the augmented matrix

$$\left[\begin{array}{rrr|r} 2 & 3 & 4 & 12 \\ 1 & -2 & 3 & 0 \\ -1 & 1 & -2 & -1 \end{array}\right].$$

The vertical bar was used to separate the left- and right-hand sides of the equations. We only used this vertical bar for convenience. We could have written this same matrix without the vertical bar as

$$\begin{bmatrix} 2 & 3 & 4 & 12 \\ 1 & -2 & 3 & 0 \\ -1 & 1 & -2 & -1 \end{bmatrix}.$$

This matrix has three rows and four columns. Therefore, we say that the size of this matrix is 3 by 4 (or 3×4). In general, a matrix is any rectangular array of numbers of any size.

Definition Matrix

A **matrix** is a rectangular array of numbers arranged in a fixed number, m, of rows and a fixed number, n, of columns. The size of a matrix is $m \times n$. The numbers in the array are called entries. Each entry is labeled a_{ij}, where i is the ith row of the entry and j is the jth column of the entry.

$$A = \begin{bmatrix} a_{11} & a_{12} & \cdots & a_{1n} \\ a_{21} & a_{22} & \cdots & a_{2n} \\ \vdots & \vdots & \vdots & \vdots \\ a_{m1} & a_{m2} & \cdots & a_{mn} \end{bmatrix}$$

Two matrices A and B are said to be equal $(A = B)$ provided both matrices are of the same size and the corresponding entries are the same.

We typically use capital letters to name a matrix. Following are some examples of matrices of various sizes:

$$A = \begin{bmatrix} 3 & -1 & 8 \\ 2 & -2 & 9 \\ 5 & 7 & -6 \end{bmatrix}, \quad B = [6 \ \ -5], \quad C = \begin{bmatrix} 2 & -1 \\ 3 & 1 \end{bmatrix}, \quad D = \begin{bmatrix} 3 & -1 \\ 9 & 6 \\ 2 & 7 \end{bmatrix},$$

3×3 Square matrix; 1×2 Row matrix; 2×2 Square matrix; 3×2 Matrix

$$E = \begin{bmatrix} 3 \\ 7 \\ 1 \\ -3 \end{bmatrix}, \quad F = \begin{bmatrix} 1 & 0 & 0 & 0 \\ 0 & 1 & 0 & 0 \\ 0 & 0 & 1 & 0 \\ 0 & 0 & 0 & 1 \end{bmatrix}$$

4×1 Column matrix; 4×4 Identity matrix

Matrices A, C, and F are **square matrices** because the number of rows equals the number of columns. (The size of matrix A is 3×3, the size of matrix C is 2×2, and the size of matrix F is 4×4.) Matrix B is called a **row matrix** because it only has one row. Similarly, matrix E is called a **column matrix** because it only has one column. Matrix F is an example of an **identity matrix**. An identity matrix is a square matrix with 1's down the diagonal (from left to right) and zeros everywhere else. (You will see why we call matrix F an identity matrix when we define matrix multiplication.) Each number in a matrix is called an entry.

We typically use lowercase letters and subscripts to denote each entry. For example, in matrix A, a_{23} refers to the entry of matrix A that is located in the second row and the third column. Therefore, $a_{23} = 9$. Most graphing calculators come equipped with matrix editors. Figure 1 shows matrix A with the entry $a_{23} = 9$ highlighted.

(eText Screens 13.1-1–13.1-24)

Using Technology

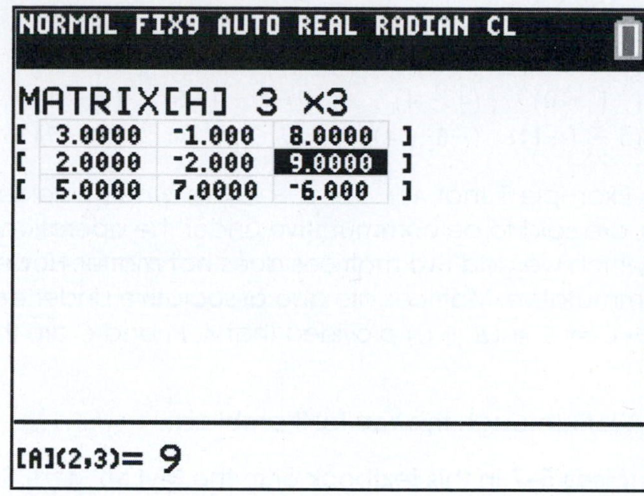

Figure 1 Using a graphing calculator to create a matrix.

You Try It Work through this You Try It problem.

Work Exercises 1–4 in this textbook or in the MyLab Math Study Plan.

OBJECTIVE 2 ADDING AND SUBTRACTING MATRICES

Just like with real numbers, we can add and subtract two matrices. To add or subtract matrices, they must be the *same size*. To add matrices, we simply add the corresponding entries. Similarly, to subtract matrices, we subtract the corresponding entries. Example 1 illustrates the addition and subtraction of two 3×4 matrices.

Example 1 Add and Subtract Matrices

$$\text{Let } A = \begin{bmatrix} 2 & -1 & 0 & 3 \\ -1 & 4 & -2 & 5 \\ 5 & -3 & 7 & 9 \end{bmatrix} \text{ and } B = \begin{bmatrix} -3 & -2 & 6 & 1 \\ 0 & 4 & -5 & 1 \\ -1 & 4 & 2 & 7 \end{bmatrix}.$$

a. Find matrix $A + B$.

b. Find matrix $B + A$.

c. Find matrix $A - B$.

Solution

a. Adding each corresponding entry of matrix A and matrix B, we obtain the matrix

$$A + B = \begin{bmatrix} (2 + (-3)) & (-1 + (-2)) & (0 + 6) & (3 + 1) \\ (-1 + 0) & (4 + 4) & (-2 + (-5)) & (5 + 1) \\ (5 + -1) & (-3 + 4) & (7 + 2) & (9 + 7) \end{bmatrix} = \begin{bmatrix} -1 & -3 & 6 & 4 \\ -1 & 8 & -7 & 6 \\ 4 & 1 & 9 & 16 \end{bmatrix}$$

b.

$$B + A = \begin{bmatrix} (-3 + 2) & (-2 + (-1)) & (6 + 0) & (1 + 3) \\ (0 + (-1)) & (4 + 4) & (-5 + (-2)) & (1 + 5) \\ (-1 + 5) & (4 + (-3)) & (2 + 7) & (7 + 9) \end{bmatrix} = \begin{bmatrix} -1 & -3 & 6 & 4 \\ -1 & 8 & -7 & 6 \\ 4 & 1 & 9 & 16 \end{bmatrix}$$

c. Subtracting each corresponding entry of matrix B from matrix A, we obtain the matrix

$$A - B = \begin{bmatrix} (2-(-3)) & (-1-(-2)) & (0-6) & (3-1) \\ (-1-0) & (4-4) & (-2-(-5)) & (5-1) \\ (5-(-1)) & (-3-4) & (7-2) & (9-7) \end{bmatrix} = \begin{bmatrix} 5 & 1 & -6 & 2 \\ -1 & 0 & 3 & 4 \\ 6 & -7 & 5 & 2 \end{bmatrix}$$ ●

Notice in Example 1 that $A + B = B + A$. This is no coincidence. Matrices, like real numbers, are said to be **commutative** under the operation of addition. That is, the order in which we add two matrices does not matter. However, matrix subtraction is *not* commutative. Matrices are also **associative** under addition, which means $(A + B) + C = A + (B + C)$ provided that A, B, and C are the same size.

You Try It Work through this You Try It problem.

Work Exercises 5–7 in this textbook or in the MyLab Math Study Plan.

OBJECTIVE 3 SCALAR MULTIPLICATION OF MATRICES

We can multiply any matrix by a real number. We call this real number a **scalar** and this operation is called scalar multiplication. Scalar multiplication is performed by multiplying every entry in the matrix by that scalar. For example, suppose matrix A is given by

$$A = \begin{bmatrix} a_{11} & a_{12} & \cdots & a_{1n} \\ a_{21} & a_{22} & \cdots & a_{2n} \\ \vdots & \vdots & \vdots & \vdots \\ a_{m1} & a_{m2} & \cdots & a_{mn} \end{bmatrix},$$

then if c is any real number, we define matrix cA as

$$cA = \begin{bmatrix} ca_{11} & ca_{12} & \cdots & ca_{1n} \\ ca_{21} & ca_{22} & \cdots & ca_{2n} \\ \vdots & \vdots & \vdots & \vdots \\ ca_{m1} & ca_{m2} & \cdots & ca_{mn} \end{bmatrix}.$$

Example 2 Scalar Multiplication and Addition of Matrices

If $A = \begin{bmatrix} -4 & 0 \\ 3 & 1 \end{bmatrix}$ and $B = \begin{bmatrix} 6 & 3 \\ -3 & -9 \end{bmatrix}$, find the following matrices: $-2A$, $\frac{1}{3}B$, and $-2A + \frac{1}{3}B$.

Solution Work through the **video** to verify that

$$-2A = \begin{bmatrix} 8 & 0 \\ -6 & -2 \end{bmatrix}, \quad \frac{1}{3}B = \begin{bmatrix} 2 & 1 \\ -1 & -3 \end{bmatrix}, \quad \text{and } -2A + \frac{1}{3}B = \begin{bmatrix} 10 & 1 \\ -7 & -5 \end{bmatrix}.$$

Figure 2 shows how a graphing utility can be used to perform scalar multiplication and matrix addition.

Using Technology 

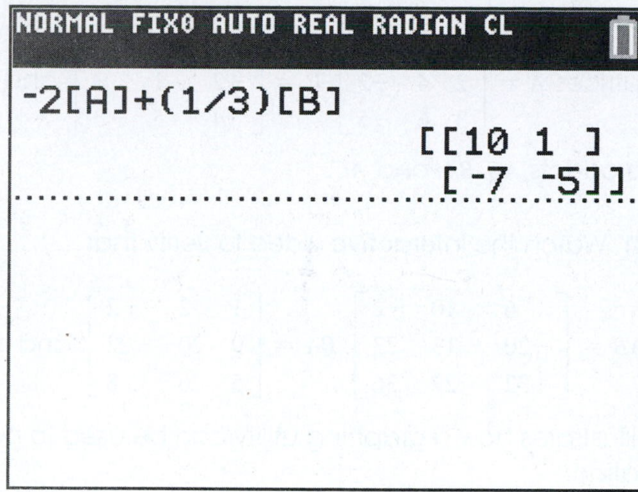

Figure 2 Using a graphing utility to compute matrix $-2A + \frac{1}{3}B$. ●

You Try It Work through this You Try It problem.

Work Exercises 8–10 in this textbook or in the MyLab Math Study Plan.

OBJECTIVE 4 MULTIPLICATION OF MATRICES

Unlike with real numbers, we cannot always multiply two matrices. The product of two matrices is defined only if the number of columns of the first matrix is equal to the number of rows of the second matrix. If the number of columns of the first matrix does not equal the number of rows of the second matrix, then we say that the product matrix is not defined.

MATRIX MULTIPLICATION

If A is an $m \times n$ matrix and B is an $n \times q$ matrix, then there exists an $m \times q$ product matrix AB. The element in the ith row and jth column of matrix AB is equal to the sum of the products of the corresponding entries from row i of matrix A and column j of matrix B.

We now illustrate matrix multiplication by multiplying the 2×3 and 3×2 matrices. Watch the **video** to see every step of the multiplication process.

$$\underset{2\times 3}{\overset{A}{\begin{bmatrix} 1 & 1 & 2 \\ 2 & 4 & -3 \end{bmatrix}}}\underset{3\times 2}{\overset{B}{\begin{bmatrix} 0 & -5 \\ 4 & -1 \\ 1 & 3 \end{bmatrix}}} = \begin{bmatrix} (1)(0)+(1)(4)+\ (2)(1) & (1)(-5)+(1)(-1)+\ (2)(3) \\ (2)(0)+(4)(4)+(-3)(1) & (2)(-5)+(4)(-1)+(-3)(3) \end{bmatrix} = \underset{2\times 2}{\overset{AB}{\begin{bmatrix} 6 & 0 \\ 13 & -23 \end{bmatrix}}}.$$

Same

Size of AB is 2×2

Notice that matrix AB has the same number of rows as matrix A and the same number of columns as matrix B. This will always be the case. In other words, if we multiply an $m \times n$ matrix by an $n \times q$ matrix, then the size of the resultant matrix will be $m \times q$.

Example 3 Matrix Multiplication

Given matrices $A = \begin{bmatrix} 1 & 1 & 2 \\ 2 & 4 & -3 \\ 3 & 6 & -5 \end{bmatrix}$, $B = \begin{bmatrix} 0 & -1 & 1 \\ -2 & 1 & 3 \\ 4 & 5 & -3 \end{bmatrix}$, and $C = \begin{bmatrix} 1 & 0 & 0 \\ 0 & 1 & 0 \\ 0 & 0 & 1 \end{bmatrix}$,

find the products AB, BA, and AC.

Solution Watch the **interactive video** to verify that

$$AB = \begin{bmatrix} 6 & 10 & -2 \\ -20 & -13 & 23 \\ -32 & -22 & 36 \end{bmatrix}, BA = \begin{bmatrix} 1 & 2 & -2 \\ 9 & 20 & -22 \\ 5 & 6 & 8 \end{bmatrix}, \text{ and } AC = \begin{bmatrix} 1 & 1 & 2 \\ 2 & 4 & -3 \\ 3 & 6 & -5 \end{bmatrix}.$$

Figure 3 illustrates how a graphing utility can be used to perform matrix multiplication.

Using Technology

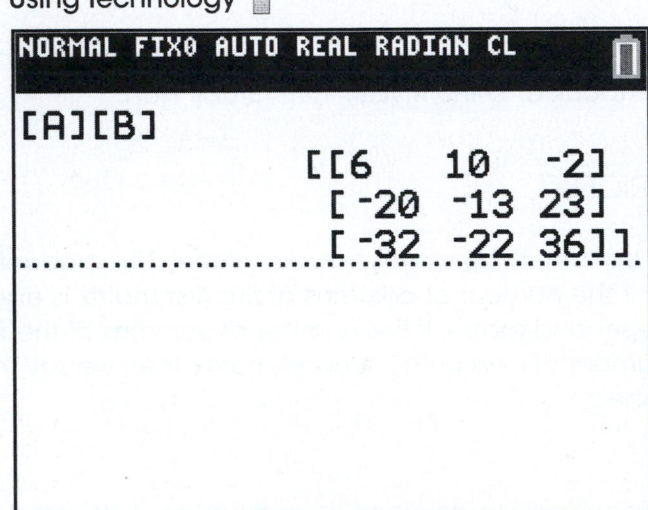

Figure 3 Using a graphing utility to compute matrices AB and BA.

You Try It Work through this You Try It problem.

Work Exercises 11–20 in this textbook or in the MyLab Math Study Plan.

In Example 3, notice that $AB \neq BA$. This implies that matrix multiplication is not commutative. Although matrix multiplication is not commutative in general, it is possible to find two matrices that commute under the operation of multiplication. For example, consider matrix A and matrix C from Example 3. We can easily verify that

$$AC = \begin{bmatrix} 1 & 1 & 2 \\ 2 & 4 & -3 \\ 3 & 6 & -5 \end{bmatrix}\begin{bmatrix} 1 & 0 & 0 \\ 0 & 1 & 0 \\ 0 & 0 & 1 \end{bmatrix} = \begin{bmatrix} 1 & 1 & 2 \\ 2 & 4 & -3 \\ 3 & 6 & -5 \end{bmatrix} = A \text{ and}$$

$$CA = \begin{bmatrix} 1 & 0 & 0 \\ 0 & 1 & 0 \\ 0 & 0 & 1 \end{bmatrix}\begin{bmatrix} 1 & 1 & 2 \\ 2 & 4 & -3 \\ 3 & 6 & -5 \end{bmatrix} = \begin{bmatrix} 1 & 1 & 2 \\ 2 & 4 & -3 \\ 3 & 6 & -5 \end{bmatrix} = A.$$

Notice that C is an **identity matrix**. We will typically use the notation I_n to denote an $n \times n$ identity matrix. For example,

$$I_2 = \begin{bmatrix} 1 & 0 \\ 0 & 1 \end{bmatrix}, I_3 = \begin{bmatrix} 1 & 0 & 0 \\ 0 & 1 & 0 \\ 0 & 0 & 1 \end{bmatrix}, I_4 = \begin{bmatrix} 1 & 0 & 0 & 0 \\ 0 & 1 & 0 & 0 \\ 0 & 0 & 1 & 0 \\ 0 & 0 & 0 & 1 \end{bmatrix}, \text{ and so on.}$$

Whenever an $m \times n$ matrix A is multiplied by the identity matrix I_n, the resultant matrix is always A. Thus, we have the following identity property for matrix multiplication. This property is extremely important to understand and is revisited in Section 13.2.

Identity Property for Matrix Multiplication

For any $m \times n$ matrix A, $I_m A = A$ and $AI_n = A$. If A is an $n \times n$ square matrix, then $AI_n = I_n A = A$.

Example 4 Multiplication of an Identity Matrix

If $A = \begin{bmatrix} 2 & -1 \\ 5 & 6 \\ 0 & -3 \end{bmatrix}$, find appropriate identity matrices I_m and I_n and verify that $I_m A = A$ and $AI_n = A$.

Solution A is a 3×2 matrix, so $m = 3$ and $n = 2$. Therefore,

$$I_m = I_3 = \begin{bmatrix} 1 & 0 & 0 \\ 0 & 1 & 0 \\ 0 & 0 & 1 \end{bmatrix} \text{ and } I_n = I_2 = \begin{bmatrix} 1 & 0 \\ 0 & 1 \end{bmatrix}. \text{ Thus,}$$

$$I_m A = I_3 A = \begin{bmatrix} 1 & 0 & 0 \\ 0 & 1 & 0 \\ 0 & 0 & 1 \end{bmatrix}\begin{bmatrix} 2 & -1 \\ 5 & 6 \\ 0 & -3 \end{bmatrix} = \begin{bmatrix} 2 & -1 \\ 5 & 6 \\ 0 & -3 \end{bmatrix} = A$$

$$\text{and } AI_n = AI_2 = \begin{bmatrix} 2 & -1 \\ 5 & 6 \\ 0 & -3 \end{bmatrix}\begin{bmatrix} 1 & 0 \\ 0 & 1 \end{bmatrix} = \begin{bmatrix} 2 & -1 \\ 5 & 6 \\ 0 & -3 \end{bmatrix} = A.$$

You Try It Work through this You Try It problem.

Work Exercises 21–23 in this textbook or in the MyLab Math Study Plan.

Summary of Matrix Operations

Assume matrix addition and matrix multiplication are defined for matrices $A, B,$ and C and let k be a scalar, then

1. $A + B = B + A$	Commutative Property for Matrix Addition
2. $(A + B) + C = A + (B + C)$	Associative Property for Matrix Addition
3. $k(A + B) = kA + kB$	Distributive Property of a Scalar with Matrix Addition
4. $(AB)C = A(BC)$	Associative Property for Matrix Multiplication
5. $A(B + C) = AB + AC$	Distributive Property
6. $(B + C)A = BA + CA$	Distributive Property
7. $k(AB) = (kA)B = A(kB)$	Associative Property for Scalar Multiplication
8. If A is an $m \times n$ matrix, then $I_mA = A$ and $AI_n = A$.	Identity Property

OBJECTIVE 5 APPLICATIONS OF MATRIX MULTIPLICATION

Matrices can be used as a convenient way to express data. For example, suppose that a certain sporting goods manufacturer has manufacturing plants in Atlanta, Boise, Columbus, and Detroit. The number of each type of ball produced (in thousands) during the month of June last year is represented by matrix A.

$$\begin{array}{l} \\ \text{Basketballs} \\ \text{Footballs} \\ \text{Golf balls} \end{array} \begin{array}{c} \begin{array}{cccc} \text{Atlanta} & \text{Boise} & \text{Columbus} & \text{Detroit} \end{array} \\ \begin{bmatrix} 8 & 5 & 4.5 & 10 \\ 7 & 3 & 2 & 8 \\ 10 & 0 & 8 & 15 \end{bmatrix} \end{array} = A$$

We can use this matrix to determine the number of each ball that was produced in each plant. For example, the Detroit plant produced 10,000 basketballs last June. We can deduce that the Boise plant does not produce golf balls (or perhaps the golf ball portion of the plant was shut down during June). Now, suppose that during the month of June, the selling price of a basketball was \$25, footballs sold for \$20, and golf balls sold for \$1.50. We can calculate the total revenue at each plant by multiplying matrix A on the left by the 1×3 row matrix $[25 \quad 20 \quad 1.5] = B$, which represents the selling price of each ball. Thus, we can calculate the revenue matrix BA:

$$BA = [25 \quad 20 \quad 1.5]\begin{bmatrix} 8 & 5 & 4.5 & 10 \\ 7 & 3 & 2 & 8 \\ 10 & 0 & 8 & 15 \end{bmatrix} = \begin{array}{cccc} \text{Atlanta} & \text{Boise} & \text{Columbus} & \text{Detroit} \\ [355 & 185 & 164.5 & 432.5] \end{array}$$

Therefore, we see that total revenue generated by each manufacturing plant in the month of June was

Atlanta:	\$355,000
Boise:	\$185,000
Columbus:	\$164,500
Detroit:	\$432,500

You Try It Work through this You Try It problem.

Work Exercises 24–26 in this textbook or in the MyLab Math Study Plan.

13.1 Exercises

1. $A = \begin{bmatrix} 2 & -5 & 7 & 9 \\ 1 & 3 & -4 & 11 \\ 8 & 6 & 1 & 0 \end{bmatrix}$

   a. What is the size of matrix A?

   b. Determine whether A is a square matrix, row matrix, column matrix, or none of these.

2. $A = \begin{bmatrix} 1 & 0 & 5 \\ 2 & 1 & 0 \\ 0 & -1 & 1 \end{bmatrix}$

   a. What is the size of matrix A?

   b. Determine whether A is a square matrix, row matrix, column matrix, or none of these.

3. $A = [4 \quad -2 \quad \pi \quad \sqrt{3}]$

   a. What is the size of matrix A?

   b. Determine whether A is a square matrix, row matrix, column matrix, or none of these.

4. $A = \begin{bmatrix} 3.9 \\ -1.2 \\ 5.3 \\ -2.7 \end{bmatrix}$

   a. What is the size of matrix A?

   b. Determine whether A is a square matrix, row matrix, column matrix, or none of these.

In Exercises 5–10, use matrices A and B to find the indicated matrix:

$$A = \begin{bmatrix} 1 & 0 & -2 \\ 3 & 1 & 4 \\ 5 & -1 & -5 \end{bmatrix}, \quad B = \begin{bmatrix} 3 & 4 & -1 \\ 5 & 0 & 2 \\ 3 & 4 & -5 \end{bmatrix}$$

5. $A + B$
6. $A - B$
7. $B - A$
8. $2A$
9. $\frac{1}{2}B$
10. $2A - \frac{1}{2}B$

In Exercises 11–14, let $(m \times n)$ denote a matrix of size $m \times n$. Find the size of the following products, or state that the product is not defined.

11. $(2 \times 3)(3 \times 2)$
12. $(3 \times 4)(4 \times 5)$
13. $(2 \times 3)(2 \times 3)$
14. $(1 \times 4)(4 \times 1)$

In Exercises 15–20, use matrices A, B, C, and D to find the indicated matrix:

$$A = \begin{bmatrix} -1 & 2 \\ 3 & 5 \end{bmatrix}, B = \begin{bmatrix} 1 & 3 & -1 \\ 4 & -2 & 0 \end{bmatrix}, C = \begin{bmatrix} 1 & 0 & -1 \\ 4 & 0 & 3 \\ 2 & -1 & -2 \end{bmatrix}, D = \begin{bmatrix} 1 & -2 \\ -3 & -4 \end{bmatrix}$$

15. AB
16. BC
17. $AB + BC$
18. A^2
19. ABC
20. $(A + D)B$

In Exercises 21–23, matrix A is given. Find appropriate identity matrices I_m and I_n such that $I_mA = A$ and $AI_n = A$.

21. $A = \begin{bmatrix} 5 & 1 & -1 \\ 3 & -2 & 4 \end{bmatrix}$

22. $A = \begin{bmatrix} 2 & -1 & 0 & 3 \\ -1 & 4 & -2 & 5 \\ 5 & -3 & 7 & 9 \end{bmatrix}$

23. $A = \begin{bmatrix} 3 & -1 \\ 9 & 6 \\ 2 & 7 \\ 0 & 4 \\ 1 & -3 \end{bmatrix}$

24. The youth from a local church are having a breakfast fund-raising event. They are planning on serving biscuits, pancakes, and waffles. The ingredients for one batch of each are as follows:

Biscuits: 3 cups baking mix, 1 egg, 1 cup milk, $\frac{1}{2}$ tablespoon cooking oil

Pancakes: 2 cups baking mix, 2 eggs, $\frac{2}{3}$ cup milk

Waffles: $2\frac{1}{2}$ cups baking mix, 1 egg, 1 cup milk, 2 tablespoons cooking oil

The ingredients can be summarized by matrix A.

$$\begin{array}{l} \\ \text{Biscuits} \\ \text{Pancakes} \\ \text{Waffles} \end{array} \begin{array}{c} \begin{array}{cccc} \text{Mix} & \text{Eggs} & \text{Milk} & \text{Oil} \end{array} \\ \begin{bmatrix} 3 & 1 & 1 & \frac{1}{2} \\ 2 & 2 & \frac{2}{3} & 0 \\ 2\frac{1}{2} & 1 & 1 & 2 \end{bmatrix} \end{array} = A$$

It was determined that they were going to need 16 batches of biscuits, 30 batches of pancakes, and 40 batches of waffles, which is summarized by matrix B.

$$\begin{array}{c} \\ \text{Batches} \end{array} \begin{array}{c} \begin{array}{ccc} \text{Biscuits} & \text{Pancakes} & \text{Waffles} \end{array} \\ \begin{bmatrix} 16 & 30 & 40 \end{bmatrix} \end{array} = B$$

a. Calculate matrix BA.

b. Interpret the meaning of each entry of matrix BA.

25. A lawn-mower manufacturer produces two types of lawn-mowers, a standard model and the self-propelled model. The production of each model requires assembly and quality assurance testing. The number of hours required for each model are given by the following matrix.

$$\begin{array}{l} \\ \text{Standard} \\ \text{Self-propelled} \end{array} \begin{array}{c} \begin{array}{cc} \text{Assembly} & \text{Testing} \end{array} \\ \begin{bmatrix} 2 & 0.5 \\ 2.5 & 1 \end{bmatrix} \end{array} = A$$

The company has three manufacturing plants located in New York, Los Angeles, and Beijing, China. The hourly labor rates (in dollars) for each plant is given by the following matrix.

$$\begin{array}{l} \\ \text{Assembly Testing} \end{array} \begin{array}{c} \begin{array}{ccc} \text{New York} & \text{Los Angeles} & \text{Beijing} \end{array} \\ \begin{bmatrix} \$12 & \$10 & \$4 \\ \$9 & \$8 & \$2 \end{bmatrix} \end{array} = B$$

a. Calculate matrix AB.
b. If ab_{ij} represents the entry of matrix AB in row i and column j, then interpret the meaning of the entries ab_{21}, ab_{12}, and ab_{23}.

26. A small clothing store has locations in Buffalo, Rochester, and Syracuse. The store sells only three items: sport coats, slacks, and shirts. During the first quarter, the Buffalo branch sold 100 sport coats, 120 pairs of slacks, and 250 shirts. The Rochester store sold 110 sport coats, 100 slacks, and 300 shirts. The Syracuse store sold 180 sport coats, 200 slacks, and 320 shirts. At each store, sport coats sell for \$100, slacks sell for \$50, and shirts sell for \$35. Use matrix multiplication to get a matrix that shows the total revenue for each store. What was the total revenue of all three stores combined?

13.2 Inverses of Matrices and Matrix Equations

THINGS TO KNOW

Before working through this section, be sure that you are familiar with the following concepts:

VIDEO ANIMATION INTERACTIVE

You Try It
1. Solving a System of Three Equations in Three Variables Using Gauss-Jordan Elimination (Section 12.3)

You Try It
2. Scalar Multiplication of Matrices (Section 13.1)

You Try It
3. Multiplication of Matrices (Section 13.1)

You Try It
4. Applications of Matrix Multiplication (Section 13.1)

OBJECTIVES

1 Understanding the Definition of an Inverse Matrix

2 Finding the Inverse of a 2×2 Matrix Using a Formula

3 Finding the Inverse of an Invertible Square Matrix

4 Solving Systems of Equations Using an Inverse Matrix

SECTION 13.2 EXERCISES

OBJECTIVE 1 UNDERSTANDING THE DEFINITION OF AN INVERSE MATRIX

In Section 13.1 we introduced the **identity matrix** and the **identity property for matrix multiplication**. Recall, if $A = \begin{bmatrix} 4 & 6 \\ -1 & -2 \end{bmatrix}$, then

$$AI_2 = \begin{bmatrix} 4 & 6 \\ -1 & -2 \end{bmatrix}\begin{bmatrix} 1 & 0 \\ 0 & 1 \end{bmatrix} = \begin{bmatrix} 4 & 6 \\ -1 & -2 \end{bmatrix} = A \quad \text{and}$$

$$I_2A = \begin{bmatrix} 1 & 0 \\ 0 & 1 \end{bmatrix}\begin{bmatrix} 4 & 6 \\ -1 & -2 \end{bmatrix} = \begin{bmatrix} 4 & 6 \\ -1 & -2 \end{bmatrix} = A.$$

Now, suppose there exists a 2×2 matrix B such that $AB = BA = I_2$. For example, if $A = \begin{bmatrix} 4 & 6 \\ -1 & -2 \end{bmatrix}$ and $B = \begin{bmatrix} 1 & 3 \\ -\frac{1}{2} & -2 \end{bmatrix}$, then

$$AB = \begin{bmatrix} 4 & 6 \\ -1 & -2 \end{bmatrix}\begin{bmatrix} 1 & 3 \\ -\frac{1}{2} & -2 \end{bmatrix} = \begin{bmatrix} 4-3 & 12-12 \\ -1+1 & -3+4 \end{bmatrix} = \begin{bmatrix} 1 & 0 \\ 0 & 1 \end{bmatrix} = I_2 \text{ and}$$

$$BA = \begin{bmatrix} 1 & 3 \\ -\frac{1}{2} & -2 \end{bmatrix}\begin{bmatrix} 4 & 6 \\ -1 & -2 \end{bmatrix} = \begin{bmatrix} 4-3 & 6-6 \\ -2+2 & -3+4 \end{bmatrix} = \begin{bmatrix} 1 & 0 \\ 0 & 1 \end{bmatrix} = I_2.$$

We say that matrix B is the **multiplicative inverse** of matrix A. We typically use the notation A^{-1} to denote the multiplicative inverse of matrix A.

Definition Multiplicative Inverse of a Matrix

Let A be an $n \times n$ square matrix. If there exists an $n \times n$ matrix A^{-1} such that $AA^{-1} = A^{-1}A = I_n$, then A^{-1} is called the **multiplicative inverse** of matrix A.

Note For the remainder of this section, the term *inverse* implies *multiplicative inverse*.

Example 1 Verify the Inverse of a Square Matrix

Verify that $A = \begin{bmatrix} 1 & 1 & 2 \\ 2 & 4 & -3 \\ 3 & 6 & -5 \end{bmatrix}$ and $B = \begin{bmatrix} 2 & -17 & 11 \\ -1 & 11 & -7 \\ 0 & 3 & -2 \end{bmatrix}$ are inverse matrices.

Solution To verify that the two matrices are inverses, we must show that $AB = BA = I_3$.

$$AB = \begin{bmatrix} 1 & 1 & 2 \\ 2 & 4 & -3 \\ 3 & 6 & -5 \end{bmatrix}\begin{bmatrix} 2 & -17 & 11 \\ -1 & 11 & -7 \\ 0 & 3 & -2 \end{bmatrix} = \begin{bmatrix} 1 & 0 & 0 \\ 0 & 1 & 0 \\ 0 & 0 & 1 \end{bmatrix} = I_3$$

$$BA = \begin{bmatrix} 2 & -17 & 11 \\ -1 & 11 & -7 \\ 0 & 3 & -2 \end{bmatrix}\begin{bmatrix} 1 & 1 & 2 \\ 2 & 4 & -3 \\ 3 & 6 & -5 \end{bmatrix} = \begin{bmatrix} 1 & 0 & 0 \\ 0 & 1 & 0 \\ 0 & 0 & 1 \end{bmatrix} = I_3$$

Because $AB = BA = I_3$, we have verified that matrix A is the inverse of matrix B and vice versa. Watch the **video** to see each step of the multiplication process.

You Try It Work through this You Try It problem.

Work Exercises 1–4 in this textbook or in the MyLab Math Study Plan.

OBJECTIVE 2 FINDING THE INVERSE OF A 2×2 MATRIX USING A FORMULA

The definition of the inverse of a matrix indicates that a matrix must be square in order for it to have an inverse. However, not every square matrix has an inverse. If a square matrix has an inverse, we say that the matrix is **invertible**. If a square matrix does not have an inverse, we say that the matrix is **singular**. There is an easy way to determine whether a 2×2 matrix is invertible or singular. Every square matrix has a number associated with it called the **determinant**. (We learn more about determinants in Section 13.3.) The formula for the determinant of a 2×2 matrix is as follows.

Determinant of a 2×2 Matrix

Let $A = \begin{bmatrix} a & b \\ c & d \end{bmatrix}$. The determinant of matrix A is denoted as $|A|$ or $\begin{vmatrix} a & b \\ c & d \end{vmatrix}$ and is defined by $|A| = \begin{vmatrix} a & b \\ c & d \end{vmatrix} = ad - cb$. If $|A| \neq 0$, then matrix A is invertible.

You can see that if $|A| \neq 0$, then matrix A has an inverse. For example, the matrix $A = \begin{bmatrix} 4 & 6 \\ -1 & -2 \end{bmatrix}$ is invertible because $a = 4, b = 6, c = -1, d = -2$, and $|A| = \begin{vmatrix} 4 & 6 \\ -1 & -2 \end{vmatrix} = (4)(-2) - (-1)(6) = -8 + 6 = -2 \neq 0$. Following is a formula for determining the inverse of a 2×2 matrix.

Formula for Determining the Inverse of a 2×2 Matrix

Let $A = \begin{bmatrix} a & b \\ c & d \end{bmatrix}$. If $|A| \neq 0$, then A is invertible and $A^{-1} = \dfrac{1}{|A|}\begin{bmatrix} d & -b \\ -c & a \end{bmatrix}$.

See the video proof.

Example 2 Find the Inverse of a 2×2 Matrix

Determine whether the following matrices are invertible or singular. If the matrix is invertible, find its inverse and then verify by using matrix multiplication.

a. $A = \begin{bmatrix} 2 & 3 \\ -4 & -6 \end{bmatrix}$ b. $B = \begin{bmatrix} 2 & -1 \\ 4 & 3 \end{bmatrix}$

Solution

a. $|A| = \begin{vmatrix} 2 & 3 \\ -4 & -6 \end{vmatrix} = (2)(-6) - (-4)(3) = -12 + 12 = 0.$

The determinant is zero, so matrix A is singular.

b. $|B| = \begin{vmatrix} 2 & -1 \\ 4 & 3 \end{vmatrix} = (2)(3) - (4)(-1) = 6 + 4 = 10.$

The determinant is nonzero, so matrix B is invertible.

$$B^{-1} = \frac{1}{10}\begin{bmatrix} 3 & 1 \\ -4 & 2 \end{bmatrix} = \begin{bmatrix} \frac{3}{10} & \frac{1}{10} \\ -\frac{2}{5} & \frac{1}{5} \end{bmatrix}$$

We can verify that this is the inverse matrix by showing that $BB^{-1} = B^{-1}B = I_2$. Watch the **video** to see the solution worked out in detail.

Using Technology

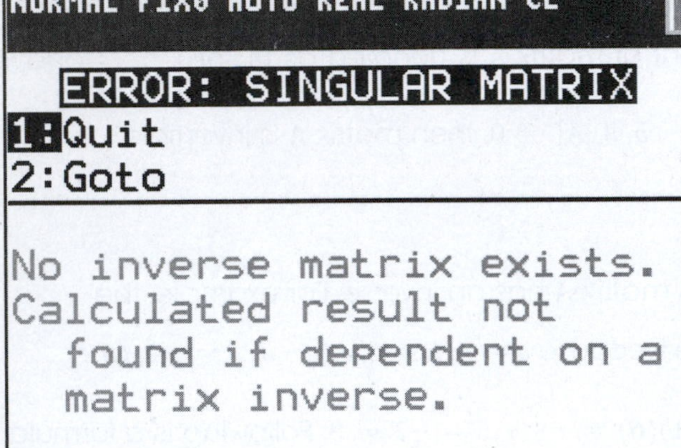

NORMAL FIX1 AUTO REAL RADIAN CL

[B]⁻¹

[[0.3 0.1]
[-0.4 0.2]]

Using the matrix editor feature of a graphing utility, we can attempt to find the inverse of the matrices from Example 2. You can see in the first screenshot that an error occurs because matrix A is singular.

You Try It Work through this You Try It problem.

Work Exercises 5–8 in this textbook or in the MyLab Math Study Plan.

OBJECTIVE 3 FINDING THE INVERSE OF AN INVERTIBLE SQUARE MATRIX

Before we learn how to find the inverse of any square invertible matrix, we start by showing an alternative way of finding the inverse of a 2×2 invertible matrix and then generalize this technique for any invertible $n \times n$ matrix. In Example 2, we saw that $B = \begin{bmatrix} 2 & -1 \\ 4 & 3 \end{bmatrix}$ was invertible.

Thus, there exists an inverse matrix B^{-1} such that $BB^{-1} = B^{-1}B = I_2$. If we let $B^{-1} = \begin{bmatrix} x & y \\ z & w \end{bmatrix}$, then

$$\underbrace{\begin{bmatrix} 2 & -1 \\ 4 & 3 \end{bmatrix}}_{B} \underbrace{\begin{bmatrix} x & y \\ z & w \end{bmatrix}}_{B^{-1}} = \underbrace{\begin{bmatrix} 1 & 0 \\ 0 & 1 \end{bmatrix}}_{I_2}.$$

Multiplying BB^{-1}, we get

$$\underbrace{\begin{bmatrix} 2x - z & 2y - w \\ 4x + 3z & 4y + 3w \end{bmatrix}}_{BB^{-1}} = \underbrace{\begin{bmatrix} 1 & 0 \\ 0 & 1 \end{bmatrix}}_{I_2}.$$

We can compare the corresponding entries of the previous two matrices to obtain the following two systems of equations:

$$\begin{aligned} 2x \quad - z &= 1 \\ 4x + 3z &= 0 \end{aligned} \quad \text{and} \quad \begin{aligned} 2y - \quad w &= 0 \\ 4y + 3w &= 1 \end{aligned}$$

These two systems correspond to the following two augmented matrices:

$$\left[\begin{array}{rr|r} 2 & -1 & 1 \\ 4 & 3 & 0 \end{array}\right] \quad \text{and} \quad \left[\begin{array}{rr|r} 2 & -1 & 0 \\ 4 & 3 & 1 \end{array}\right]$$

Because the coefficients of the left-hand side of the previous two matrices are exactly the same, we can combine the two augmented matrices into the following augmented matrix:

$$\left[\begin{array}{rr|r} 2 & -1 & 1 \\ 4 & 3 & 0 \end{array}\right] \quad \left[\begin{array}{rr|r} 2 & -1 & 0 \\ 4 & 3 & 1 \end{array}\right]$$

$$\downarrow$$

$$\left[\begin{array}{rr|rr} 2 & -1 & 1 & 0 \\ 4 & 3 & 0 & 1 \end{array}\right]$$

To solve for each variable, we use Gauss-Jordan elimination to write the left-hand side of the augmented matrix in **reduced row-echelon form.** View these **steps** to see how the previous augmented matrix can be reduced to the following matrix:

$$\left[\begin{array}{rr|rr} 1 & 0 & \frac{3}{10} & \frac{1}{10} \\ 0 & 1 & -\frac{2}{5} & \frac{1}{5} \end{array}\right]$$

Therefore, we see that $x = \frac{3}{10}$, $y = \frac{1}{10}$, $z = -\frac{2}{5}$, and $w = \frac{1}{5}$.

Because $B^{-1} = \begin{bmatrix} x & y \\ z & w \end{bmatrix}$, then $B^{-1} = \begin{bmatrix} \frac{3}{10} & \frac{1}{10} \\ -\frac{2}{5} & \frac{1}{5} \end{bmatrix}$.

You can see that this is precisely what we obtained for the inverse in Example 2. Note that we started with the augmented matrix

$$\left[\begin{array}{rr|rr} 2 & -1 & 1 & 0 \\ 4 & 3 & 0 & 1 \end{array}\right],$$

with B the left-hand block and I_2 the right-hand block.

then used Gauss-Jordan elimination to end up with the augmented matrix

$$\underbrace{\begin{bmatrix} 1 & 0 \\ 0 & 1 \end{bmatrix}}_{I_2}\Bigg|\underbrace{\begin{bmatrix} \frac{3}{10} & \frac{1}{10} \\ -\frac{2}{5} & \frac{1}{5} \end{bmatrix}}_{B^{-1}}.$$

This technique can be used to find the inverse of any invertible square matrix. The process is summarized as follows.

Steps for Finding the Multiplicative Inverse of an Invertible Square Matrix

If A is an $n \times n$ invertible matrix, then the multiplicative inverse of A can be obtained by following these steps:

Step 1. Form the augmented matrix $[A|I_n]$.

Step 2. Use Gauss-Jordan elimination to reduce A into the identity matrix I_n.

Step 3. The new augmented matrix is of the form $[I_n|A^{-1}]$, where A^{-1} is the multiplicative inverse of A.

Example 3 Find the Inverse of a 3×3 Matrix

If possible, determine the inverse of $A = \begin{bmatrix} 1 & 1 & 2 \\ 2 & 4 & -3 \\ 3 & 6 & -5 \end{bmatrix}$.

Solution To attempt to find the inverse of matrix A, we create the augmented matrix by attaching the identity matrix, I_3, to the right-hand side.

$$\left[\underbrace{\begin{matrix} 1 & 1 & 2 \\ 2 & 4 & -3 \\ 3 & 6 & -5 \end{matrix}}_{A}\;\middle|\;\underbrace{\begin{matrix} 1 & 0 & 0 \\ 0 & 1 & 0 \\ 0 & 0 & 1 \end{matrix}}_{I_3}\right]$$

We now use Gauss-Jordan elimination to attempt to write the matrix on the left-hand side as I_3. Watch this **video** to see each step of the Gauss-Jordan elimination process. The reduced augmented matrix is

$$\left[\underbrace{\begin{matrix} 1 & 0 & 0 \\ 0 & 1 & 0 \\ 0 & 0 & 1 \end{matrix}}_{I_3}\;\middle|\;\underbrace{\begin{matrix} 2 & -17 & 11 \\ -1 & 11 & -7 \\ 0 & 3 & -2 \end{matrix}}_{A^{-1}}\right].$$

Therefore, $A^{-1} = \begin{bmatrix} 2 & -17 & 11 \\ -1 & 11 & -7 \\ 0 & 3 & -2 \end{bmatrix}$. You should verify that $AA^{-1} = A^{-1}A = I_3$.

You Try It Work through this You Try It problem.

Example 4 Example of a 3 × 3 Singular Matrix

If possible, determine the inverse of $A = \begin{bmatrix} 5 & 0 & -1 \\ 1 & -3 & -2 \\ 0 & 5 & 3 \end{bmatrix}$.

Solution To attempt to find the inverse of matrix A, we create the augmented matrix by attaching the identity matrix, I_3, to the right-hand side.

$$\underbrace{\begin{bmatrix} 5 & 0 & -1 \\ 1 & -3 & -2 \\ 0 & 5 & 3 \end{bmatrix}}_{A}\!\!\left|\underbrace{\begin{matrix} 1 & 0 & 0 \\ 0 & 1 & 0 \\ 0 & 0 & 1 \end{matrix}}_{I_3}\right]$$

We now use a series of row operations to try to obtain I_3 on the left-hand side:

$$\left[\begin{array}{ccc|ccc} 5 & 0 & -1 & 1 & 0 & 0 \\ 1 & -3 & -2 & 0 & 1 & 0 \\ 0 & 5 & 3 & 0 & 0 & 1 \end{array}\right] \xrightarrow{R_1 \Leftrightarrow R_2} \left[\begin{array}{ccc|ccc} 1 & -3 & -2 & 0 & 1 & 0 \\ 5 & 0 & -1 & 1 & 0 & 0 \\ 0 & 5 & 3 & 0 & 0 & 1 \end{array}\right] \xrightarrow{-5R_1 + R_2 \rightarrow \text{New } R_2}$$

$$\left[\begin{array}{ccc|ccc} 1 & -3 & -2 & 0 & 1 & 0 \\ 0 & 15 & 9 & 1 & -5 & 0 \\ 0 & 5 & 3 & 0 & 0 & 1 \end{array}\right] \xrightarrow{\left(\frac{1}{15}\right) R_2 \rightarrow \text{New } R_2} \left[\begin{array}{ccc|ccc} 1 & -3 & -2 & 0 & 1 & 0 \\ 0 & 1 & \frac{3}{5} & \frac{1}{15} & -\frac{1}{3} & 0 \\ 0 & 5 & 3 & 0 & 0 & 1 \end{array}\right] \xrightarrow{(-5)R_2 + R_3 \rightarrow \text{New } R_3}$$

$$\left[\begin{array}{ccc|ccc} 1 & -3 & -2 & 0 & 1 & 0 \\ 0 & 1 & \frac{3}{5} & \frac{1}{15} & -\frac{1}{3} & 0 \\ 0 & 0 & 0 & -\frac{1}{3} & \frac{5}{3} & 1 \end{array}\right]$$

Because the last row of the matrix on the left-hand side is all zeros, there is no way to obtain the identity matrix. Therefore, matrix A does not have an inverse and thus is singular. In Section 13.3, we learn another way to determine whether a square matrix is invertible.

You Try It Work through this You Try It problem.

Work Exercises 9–14 in this textbook or in the MyLab Math Study Plan.

OBJECTIVE 4 SOLVING SYSTEMS OF EQUATIONS USING AN INVERSE MATRIX

Suppose we are given the linear equation $ax = b$, where $a \neq 0$. If $a^{-1} = \frac{1}{a}$ (the multiplicative inverse of a), then we can solve for x as follows:

$ax = b$	Write the original linear equation.
$a^{-1}ax = a^{-1}b$	Multiply both sides by the multiplicative inverse of a.
$\frac{1}{a}ax = a^{-1}b$	$a^{-1} = \frac{1}{a}$
$x = a^{-1}b$	$\frac{1}{a} \cdot a = 1$

(eText Screens 13.2-1–13.2-24)

We can use this same technique to solve systems of n linear equations with n variables as follows.

Suppose we are given the linear system
$$\begin{array}{ccccccccc} a_{11}x_1 &+& a_{12}x_2 &+& a_{13}x_3 &+& \cdots & a_{1n}x_n &= b_1 \\ a_{21}x_1 &+& a_{22}x_2 &+& a_{23}x_3 &+& \cdots & a_{2n}x_n &= b_2 \\ \vdots && \vdots && \vdots &&& \vdots & \vdots \\ a_{n1}x_1 &+& a_{n2}x_2 &+& a_{n3}x_3 &+& \cdots & a_{nn}x_n &= b_n \end{array}.$$

This system is equivalent to the matrix equation

$$\begin{bmatrix} a_{11} & a_{12} & a_{13} & \cdots & a_{1n} \\ a_{21} & a_{22} & a_{23} & \cdots & a_{2n} \\ \vdots & \vdots & \vdots & & \vdots \\ a_{n1} & a_{n2} & a_{n3} & & a_{nn} \end{bmatrix}\begin{bmatrix} x_1 \\ x_2 \\ \vdots \\ x_n \end{bmatrix} = \begin{bmatrix} b_1 \\ b_2 \\ \vdots \\ b_n \end{bmatrix}.$$

If we let $A = \underbrace{\begin{bmatrix} a_{11} & a_{12} & a_{13} & \cdots & a_{1n} \\ a_{21} & a_{22} & a_{23} & \cdots & a_{2n} \\ \vdots & \vdots & \vdots & & \vdots \\ a_{n1} & a_{n2} & a_{n3} & & a_{nn} \end{bmatrix}}_{\text{Coefficient Matrix}}$, $X = \underbrace{\begin{bmatrix} x_1 \\ x_2 \\ \vdots \\ x_n \end{bmatrix}}_{\text{Variable Matrix}}$, and $B = \underbrace{\begin{bmatrix} b_1 \\ b_2 \\ \vdots \\ b_n \end{bmatrix}}_{\text{Constant Matrix}}$, then the previous matrix equation can be written as $AX = B$. Matrix A is called the **coefficient matrix**, matrix X is called the **variable matrix**, and matrix B is called the **constant matrix**. If A is invertible, then we can determine the entries of matrix X by multiplying both sides of the matrix equation on the left by A^{-1}:

$AX = B$	Write the matrix equation.
$A^{-1}AX = A^{-1}B$	Multiply both sides of the equation on the left by A^{-1}.
$I_nX = A^{-1}B$	Definition of an inverse matrix ($A^{-1}A = I_n$)
$X = A^{-1}B$	Identity property for matrix multiplication ($I_nX = X$)

CAUTION Because matrix multiplication is not commutative, we must be careful when multiplying both sides of a matrix equation by another matrix. Notice that when we multiplied both sides of the previous equation by A^{-1}, we had to position A^{-1} on the left of each side. Note: If $AX = B$, then $A^{-1}AX \neq BA^{-1}$.

Example 5 Solve a System of Equations Using an Inverse Matrix

Solve the following system of linear equations using an inverse matrix:

$$-3x + 2y = 4$$
$$5x - 4y = 9$$

Solution Let $A = \begin{bmatrix} -3 & 2 \\ 5 & -4 \end{bmatrix}$, $X = \begin{bmatrix} x \\ y \end{bmatrix}$, and $B = \begin{bmatrix} 4 \\ 9 \end{bmatrix}$. The linear system is equivalent to the matrix equation $AX = B$:

$$\overbrace{\begin{bmatrix} -3 & 2 \\ 5 & -4 \end{bmatrix}}^{A}\overbrace{\begin{bmatrix} x \\ y \end{bmatrix}}^{X} = \overbrace{\begin{bmatrix} 4 \\ 9 \end{bmatrix}}^{B}$$

We now calculate A^{-1} using the formula for finding the inverse of a 2×2 invertible matrix:

$$|A| = \begin{vmatrix} -3 & 2 \\ 5 & -4 \end{vmatrix} = (-3)(-4) - (5)(2) = 2, \text{ so } A^{-1} = \frac{1}{2}\begin{bmatrix} -4 & -2 \\ -5 & -3 \end{bmatrix} = \begin{bmatrix} -2 & -1 \\ -\frac{5}{2} & -\frac{3}{2} \end{bmatrix}$$

To solve for X, we must multiply both sides of the matrix equation on the left by A^{-1}.

$$\underbrace{\begin{bmatrix} -3 & 2 \\ 5 & -4 \end{bmatrix}}_{A}\underbrace{\begin{bmatrix} x \\ y \end{bmatrix}}_{X} = \underbrace{\begin{bmatrix} 4 \\ 9 \end{bmatrix}}_{B}$$ Write the matrix equation $AX = B$.

$$\underbrace{\begin{bmatrix} -2 & -1 \\ -\frac{5}{2} & -\frac{3}{2} \end{bmatrix}}_{A^{-1}}\underbrace{\begin{bmatrix} -3 & 2 \\ 5 & -4 \end{bmatrix}}_{A}\underbrace{\begin{bmatrix} x \\ y \end{bmatrix}}_{X} = \underbrace{\begin{bmatrix} -2 & -1 \\ -\frac{5}{2} & -\frac{3}{2} \end{bmatrix}}_{A^{-1}}\underbrace{\begin{bmatrix} 4 \\ 9 \end{bmatrix}}_{B}$$ Multiply both sides of the equation on the left by A^{-1}.

$$\underbrace{\begin{bmatrix} 1 & 0 \\ 0 & 1 \end{bmatrix}}_{I_2}\begin{bmatrix} x \\ y \end{bmatrix} = \begin{bmatrix} -17 \\ -\frac{47}{2} \end{bmatrix}$$ Definition of an inverse matrix ($A^{-1}A = I_n$)

$$\begin{bmatrix} x \\ y \end{bmatrix} = \begin{bmatrix} -17 \\ -\frac{47}{2} \end{bmatrix}$$ Identity property for matrix multiplication ($I_nX = X$)

So, $x = -17$ and $y = -\frac{47}{2}$.

Example 6 Solve a System of Equations Using an Inverse Matrix

Solve the following system of linear equations using an inverse matrix:

$$\begin{aligned} 2x + 4y + z &= 3 \\ -x + y - z &= 6 \\ x + 4y \quad &= 7 \end{aligned}$$

Solution The linear system can be written as the following matrix equation:

$$\underbrace{\begin{bmatrix} 2 & 4 & 1 \\ -1 & 1 & -1 \\ 1 & 4 & 0 \end{bmatrix}}_{A}\underbrace{\begin{bmatrix} x \\ y \\ z \end{bmatrix}}_{X} = \underbrace{\begin{bmatrix} 3 \\ 6 \\ 7 \end{bmatrix}}_{B}$$

The solution can be obtained by multiplying both sides of this matrix equation by A^{-1} to obtain $X = A^{-1}B$. To find the inverse matrix, we create the augmented matrix by attaching the identity matrix, I_3, to the right-hand side.

$$\left[\underbrace{\begin{matrix} 2 & 4 & 1 \\ -1 & 1 & -1 \\ 1 & 4 & 0 \end{matrix}}_{A}\,\middle|\,\underbrace{\begin{matrix} 1 & 0 & 0 \\ 0 & 1 & 0 \\ 0 & 0 & 1 \end{matrix}}_{I_3}\right]$$

(eText Screens 13.2-1–13.2-24)

After a series of row operations, we obtain the following augmented matrix:

$$\underbrace{\left[\begin{array}{ccc|ccc} 1 & 0 & 0 & -4 & -4 & 5 \\ 0 & 1 & 0 & 1 & 1 & -1 \\ 0 & 0 & 1 & 5 & 4 & -6 \end{array}\right]}_{I_3 \qquad A^{-1}}$$

(View this **row reduction process**.)

Therefore, $A^{-1} = \begin{bmatrix} -4 & -4 & 5 \\ 1 & 1 & -1 \\ 5 & 4 & -6 \end{bmatrix}$.

The solution to the system of equations is

$$\begin{aligned} X &= A^{-1}B \\ &= \begin{bmatrix} -4 & -4 & 5 \\ 1 & 1 & -1 \\ 5 & 4 & -6 \end{bmatrix}\begin{bmatrix} 3 \\ 6 \\ 7 \end{bmatrix} \\ &= \begin{bmatrix} -1 \\ 2 \\ -3 \end{bmatrix}. \end{aligned}$$

Thus, $x = -1, y = 2$, and $z = -3$.

Using Technology

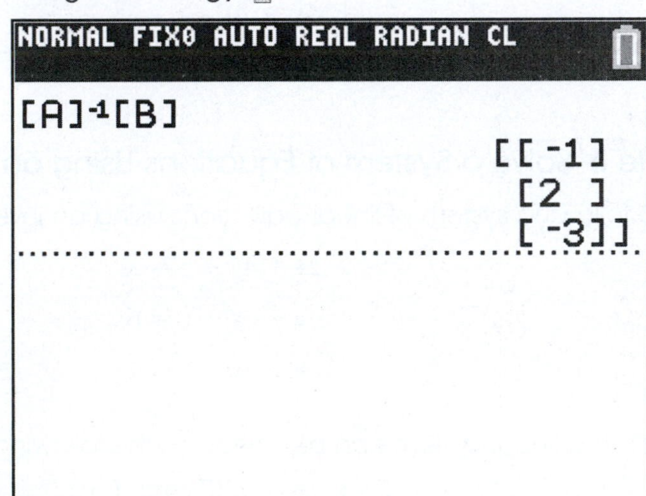

A graphing utility can be used to solve the linear system from Example 6. First, use the matrix editor feature to input the entries of the coefficient matrix A and the constant matrix B. Then, find the product $A^{-1}B$.

You Try It Work through this You Try It problem.

Work Exercises 15–21 in this textbook or in the MyLab Math Study Plan.

13.2 Exercises

In Exercises 1–4, determine whether matrix B is the inverse of matrix A by finding the products AB and BA.

1. $A = \begin{bmatrix} -3 & 1 \\ -5 & 2 \end{bmatrix}, B = \begin{bmatrix} -2 & 1 \\ -5 & 3 \end{bmatrix}$

2. $A = \begin{bmatrix} 2 & -4 \\ 1 & -3 \end{bmatrix}, B = \begin{bmatrix} \frac{3}{2} & -2 \\ \frac{1}{2} & -1 \end{bmatrix}$

3. $A = \begin{bmatrix} -28 & -13 & 3 \\ 2 & 1 & 0 \\ -7 & -3 & 1 \end{bmatrix}, B = \begin{bmatrix} 1 & 4 & -3 \\ -2 & -7 & 6 \\ 1 & 7 & -2 \end{bmatrix}$

4. $A = \begin{bmatrix} -5 & 7 & 8 \\ -5 & 8 & 9 \\ -5 & 9 & 8 \end{bmatrix}, B = \begin{bmatrix} -\frac{17}{10} & \frac{8}{5} & -\frac{1}{10} \\ -\frac{1}{2} & 0 & \frac{1}{2} \\ -\frac{1}{2} & 1 & -\frac{1}{2} \end{bmatrix}$

In Exercises 5–8, determine whether the given 2×2 matrix is invertible or singular. If it is invertible, find the inverse matrix.

5. $\begin{bmatrix} 2 & 1 \\ 5 & 3 \end{bmatrix}$

6. $\begin{bmatrix} -5 & 4 \\ -2 & 2 \end{bmatrix}$

7. $\begin{bmatrix} -3 & 6 \\ 4 & -8 \end{bmatrix}$

8. $\begin{bmatrix} \frac{1}{3} & \frac{5}{4} \\ \frac{6}{5} & \frac{9}{2} \end{bmatrix}$

In Exercises 9–14, find the inverse of each matrix if possible.

9. $\begin{bmatrix} 1 & -1 \\ 2 & 0 \end{bmatrix}$

10. $\begin{bmatrix} 1 & -6 & -2 \\ 0 & -1 & 0 \\ 2 & -11 & -3 \end{bmatrix}$

11. $\begin{bmatrix} 1 & 3 & 2 \\ 2 & -1 & 4 \\ 3 & 2 & 6 \end{bmatrix}$

12. $\begin{bmatrix} 1 & 4 & -3 \\ -2 & -7 & 6 \\ 1 & 7 & -2 \end{bmatrix}$

13. $\begin{bmatrix} 1 & 1 & 3 \\ -1 & -1 & -2 \\ -1 & 3 & 5 \end{bmatrix}$

14. $\begin{bmatrix} 1 & 0 & -2 & 0 \\ 0 & 1 & 0 & -5 \\ -4 & 0 & 9 & 0 \\ 0 & 2 & 1 & -9 \end{bmatrix}$

In Exercises 15–21, solve each linear system using an inverse matrix.

15. $\begin{aligned} -x + y &= -5 \\ -2x + 3y &= -13 \end{aligned}$

16. $\begin{aligned} -3x + 4y &= 3 \\ x - 2y &= -\frac{4}{3} \end{aligned}$

17. $\begin{aligned} -\frac{3}{10}x - \frac{1}{10}y &= -3 \\ \frac{2}{5}x - 1\frac{1}{5}y &= -7 \end{aligned}$

18. $\begin{aligned} x + y + 3z &= 11 \\ -x - y - 2z &= -7 \\ x - 3y - 5z &= -13 \end{aligned}$

19. $\begin{aligned} x + 4y - 3z &= -5 \\ -2x - 7y + 6z &= 11 \\ x + 7y - 2z &= 1 \end{aligned}$

20. $\begin{aligned} x + y - 2z &= -4 \\ x + 3y + 4z &= 8 \\ x + 2y - z &= 2 \end{aligned}$

21. $\begin{aligned} x - y + 2z + 3w &= 2 \\ 2x - y + 6z + 5w &= 6 \\ 3x - y + 9z + 6w &= 9 \\ 2x - 2y + 4z + 7w &= 5 \end{aligned}$

13.3 Determinants and Cramer's Rule

THINGS TO KNOW

Before working through this section, be sure that you are familiar with the following concept:

VIDEO ANIMATION INTERACTIVE

You Try It

1. Solving a System of Three Equations in Three Variables Using Gauss-Jordan Elimination (Section 12.3) 

You Try It

2. Finding the Inverse of a 2×2 Matrix Using a Formula (Section 13.2) 

You Try It

3. Solving Systems of Equations Using an Inverse Matrix (Section 13.2) 

OBJECTIVES

1 Using Cramer's Rule to Solve Linear Systems in Two Variables

2 Calculating the Determinant of an $n \times n$ Matrix by Expansion of Minors

3 Using Cramer's Rule to Solve Linear Systems in n Variables

SECTION 13.3 EXERCISES

OBJECTIVE 1 USING CRAMER'S RULE TO SOLVE LINEAR SYSTEMS IN TWO VARIABLES

In this section, we learn how determinants can be used to solve systems of linear equations. Let's start by considering the following system of linear equations:

$$ax + by = m$$
$$cx + dy = n$$

In Section 13.2, we saw that this linear system can be written in matrix form as

$$\underbrace{\begin{bmatrix} a & b \\ c & d \end{bmatrix}}_{A}\underbrace{\begin{bmatrix} x \\ y \end{bmatrix}}_{X} = \underbrace{\begin{bmatrix} m \\ n \end{bmatrix}}_{B}$$

We now define two new matrices, D_x and D_y. Matrix D_x is created by replacing the first column of the **coefficient matrix** by the entries of the **constant matrix**. So, $D_x = \begin{bmatrix} m & b \\ n & d \end{bmatrix}$. Similarly, matrix D_y is created by replacing the second column of the **coefficient matrix** by the entries of the **constant matrix**. Therefore, $D_y = \begin{bmatrix} a & m \\ c & n \end{bmatrix}$.

Recall that the determinant of the coefficient matrix is defined as $|A| = \begin{vmatrix} a & b \\ c & d \end{vmatrix} = ad - cb$. So, $|D_x| = \begin{vmatrix} m & b \\ n & d \end{vmatrix} = md - nb$ and $|D_y| = \begin{vmatrix} a & m \\ c & n \end{vmatrix} = an - cm$. If $|A| \neq 0$, then the linear system has a unique solution where x and y can be found by the following formula:

$$x = \frac{|D_x|}{|A|} \quad \text{and} \quad y = \frac{|D_y|}{|A|}$$

This technique of using determinants to solve a linear system is known as **Cramer's rule**.

> **Cramer's Rule for a System of Linear Equations in Two Variables**
>
> For the linear system of equations $\begin{matrix} ax + by = m \\ cx + dy = n \end{matrix}$, let $|A| = \begin{vmatrix} a & b \\ c & d \end{vmatrix}$, $|D_x| = \begin{vmatrix} m & b \\ n & d \end{vmatrix}$, and $|D_y| = \begin{vmatrix} a & m \\ c & n \end{vmatrix}$. If $|A| \neq 0$, then there is a unique solution to this system given by
>
> $$x = \frac{|D_x|}{|A|} \quad \text{and} \quad y = \frac{|D_y|}{|A|}$$

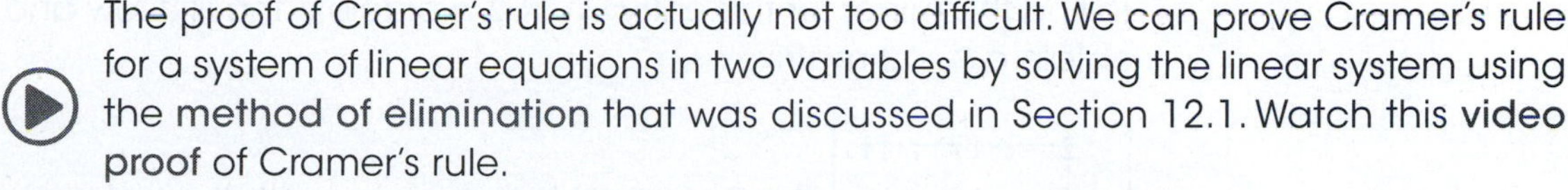

The proof of Cramer's rule is actually not too difficult. We can prove Cramer's rule for a system of linear equations in two variables by solving the linear system using the **method of elimination** that was discussed in Section 12.1. Watch this **video** proof of Cramer's rule.

Example 1 Use Cramer's Rule to Solve a Linear System in Two Variables

Use Cramer's rule to solve the following system:

$$\begin{aligned} 2x - 2y &= 4 \\ 3x + y &= -3 \end{aligned}$$

Solution For this particular system,

$$|A| = \begin{vmatrix} 2 & -2 \\ 3 & 1 \end{vmatrix} = (2)(1) - (3)(-2) = 8,$$

$$|D_x| = \begin{vmatrix} 4 & -2 \\ -3 & 1 \end{vmatrix} = (4)(1) - (-3)(-2) = -2, \quad \text{and}$$

$$|D_y| = \begin{vmatrix} 2 & 4 \\ 3 & -3 \end{vmatrix} = (2)(-3) - (3)(4) = -18.$$

Therefore, by Cramer's rule, the solution is

$$x = \frac{|D_x|}{|A|} = \frac{-2}{8} = -\frac{1}{4} \quad \text{and} \quad y = \frac{|D_y|}{|A|} = \frac{-18}{8} = -\frac{9}{4}.$$

You Try It Work through this You Try It problem.

Work Exercises 1–4 in this textbook or in the MyLab Math Study Plan.

Cramer's rule can actually be used to solve larger systems of n equations with n unknowns. Before we learn how to extend the use of Cramer's rule for linear systems of more than two equations, we must first learn how to find the determinant of larger square matrices.

(eText Screens 13.3-1–13.3-17)

OBJECTIVE 2 CALCULATING THE DETERMINANT OF AN $n \times n$ MATRIX BY EXPANSION OF MINORS

Before we learn how to calculate the determinant of any square matrix, we first learn how to calculate the determinant of a 3×3 matrix. For every entry of a 3×3 matrix, there is a 2×2 matrix associated with it. This 2×2 matrix is obtained by deleting the row and the column in which this entry appears. The determinants of these 2×2 matrices are called **minors**. In general, if a_{ij} is the entry in the ith row and jth column of a 3×3 matrix, then matrix M_{ij} is obtained by deleting the ith row and the jth column of this matrix. The determinant $|M_{ij}|$ is the minor corresponding to entry a_{ij}. For example, consider the following 3×3 matrix:

$$\begin{bmatrix} 2 & -17 & 11 \\ -1 & 11 & -7 \\ 0 & 3 & -2 \end{bmatrix}$$

To find the minor for the entry $a_{11} = 2$, we delete the first row and the first column to obtain a 2×2 matrix.

$\begin{bmatrix} 2 & -17 & 11 \\ -1 & 11 & -7 \\ 0 & 3 & -2 \end{bmatrix}$ The 2×2 matrix associated with the entry $a_{11} = 2$ is

$[M_{11}] = \begin{bmatrix} 11 & -7 \\ 3 & -2 \end{bmatrix}$.

Therefore, the minor for $a_{11} = 2$ is $|M_{11}| = \begin{vmatrix} 11 & -7 \\ 3 & -2 \end{vmatrix} = -1$. Similarly, the minor for the entry $a_{12} = -17$ is found by deleting the first row and second column.

$\begin{bmatrix} 2 & -17 & 11 \\ -1 & 11 & -7 \\ 0 & 3 & -2 \end{bmatrix}$ The 2×2 matrix associated with the entry $a_{12} = -17$ is

$[M_{12}] = \begin{bmatrix} -1 & -7 \\ 0 & -2 \end{bmatrix}$.

Thus, the minor for the entry $a_{12} = -17$ is $|M_{12}| = \begin{vmatrix} -1 & -7 \\ 0 & -2 \end{vmatrix} = (-1)(-2) - (0)(-7) = 2$.

To find the determinant of a 3×3 matrix, we use a technique called **expansion by minors.**

Determinant of a 3×3 Matrix Using Expansion by Minors

Let $A = \begin{bmatrix} a_{11} & a_{12} & a_{13} \\ a_{21} & a_{22} & a_{23} \\ a_{31} & a_{32} & a_{33} \end{bmatrix}$, then $|A| = a_{11}|M_{11}| - a_{12}|M_{12}| + a_{13}|M_{13}|$.

Example 2 Calculate the Determinant of a 3×3 Matrix

Calculate $|A|$ if $A = \begin{bmatrix} 2 & -17 & 11 \\ -1 & 11 & -7 \\ 0 & 3 & -2 \end{bmatrix}$.

Solution $|A| = a_{11}|M_{11}| - a_{12}|M_{12}| + a_{13}|M_{13}|$

$$= 2\begin{vmatrix} 11 & -7 \\ 3 & -2 \end{vmatrix} - (-17)\begin{vmatrix} -1 & -7 \\ 0 & -2 \end{vmatrix} + 11\begin{vmatrix} -1 & 11 \\ 0 & 3 \end{vmatrix}$$

$$= 2(-1) + 17(2) + 11(-3)$$

$$= -1$$

Using Technology

```
NORMAL FIX0 AUTO REAL RADIAN CL
[A]
                    [[2  -17 11]
                     [-1 11  -7]
                     [0  3   -2]]
det([A])
                              -1
```

We can use a graphing utility to find the determinant of the matrix from Example 2. ●

In Example 2, we calculated the determinant by expanding the minors across the first row. Actually, we can find the determinant using the expansions of minors across any row or down any column. We have to be careful to properly alternate the signs of the coefficients of the minors. The coefficient, a_{ij}, of each minor must be preceded by a "+" or "−", which can be determined by the formula $(-1)^{i+j}$. If the sum $i + j$ is odd, then the sign preceding the minor is $(-1)^{i+j} =$ "−". Likewise, if the sum $i + j$ is even, the sign preceding the coefficient is $(-1)^{i+j} =$ "+". The signs can be summarized in the following matrix:

$$\begin{bmatrix} + & - & + \\ - & + & - \\ + & - & + \end{bmatrix}$$

Example 3 Calculate the Determinant of a 3×3 Matrix by Expansion of Minors Down the First Column

Find $|A|$ using the matrix from Example 2 by expanding the minors down the first column.

Solution Expanding the minors down the first column, we get

$$|A| = \begin{bmatrix} 2 & -17 & 11 \\ -1 & 11 & -7 \\ 0 & 3 & -2 \end{bmatrix}$$

$$= 2\begin{vmatrix} 11 & -7 \\ 3 & -2 \end{vmatrix} - (-1)\begin{vmatrix} -17 & 11 \\ 3 & -2 \end{vmatrix} + 0\begin{vmatrix} -17 & 11 \\ 11 & -7 \end{vmatrix}$$

$$= 2(-1) + 1(1) + 0$$

$$= -1$$

Notice that we obtained the same value as we did in Example 2. When calculating determinants, it is typically easier to expand by minors across rows or down columns that contain zeros.

In **Example 4 of Section 13.2**, we found that the matrix $A = \begin{bmatrix} 5 & 0 & -1 \\ 1 & -3 & -2 \\ 0 & 5 & 3 \end{bmatrix}$ did not have an inverse (i.e., the matrix is **singular**). The determinant of this matrix can be found by expanding the minors down the first column:

$$|A| = \begin{bmatrix} 5 & 0 & -1 \\ 1 & -3 & -2 \\ 0 & 5 & 3 \end{bmatrix} = 5\begin{vmatrix} -3 & -2 \\ 5 & 3 \end{vmatrix} - 1\begin{vmatrix} 0 & -1 \\ 5 & 3 \end{vmatrix} = 5(1) - 1(5) = 0$$

The fact that the determinant of this matrix is zero is no coincidence, as stated in the following theorem.

Theorem

A square matrix A is invertible if and only if $|A| \neq 0$. (If $|A| = 0$, then A is a singular matrix.)

You Try It Work through this You Try It problem.

We can calculate the determinant of any $n \times n$ matrix in exactly the same way as with a 3×3 matrix by expansion of minors across any row or down any column. We have to make sure to alternate the signs of each coefficient properly. We define the determinant of any square matrix as follows.

Determinant of a Square Matrix

Let $A = \begin{bmatrix} a_{11} & a_{12} & \cdots & a_{1n} \\ a_{21} & a_{22} & \cdots & a_{2n} \\ \vdots & \vdots & \ddots & \vdots \\ a_{n1} & a_{n2} & \cdots & a_{nn} \end{bmatrix}$, then for each entry a_{ij} there is a corresponding minor $|M_{ij}|$, which is the determinant of a $(n - 1) \times (n - 1)$ matrix. The determinant $|A|$ is equal to the sum of the products of the entries of any row or column and the corresponding minors. (Each coefficient must be preceded by $(-1)^{i+j}$.)

Example 4 Calculate the Determinant of a 4 × 4 Matrix

Calculate $|A|$ if $A = \begin{bmatrix} 8 & 0 & 1 & 10 \\ -1 & 0 & 0 & -5 \\ 5 & 2 & -2 & 4 \\ -9 & 0 & 0 & -12 \end{bmatrix}$.

Solution Because the second column only has one nonzero entry, we calculate $|A|$ by expanding down the second column. The first entry of column 2 is $a_{12} = 0$. Note that $i = 1$ and $j = 2$, so $i + j = 1 + 2 = 3$, which is an odd number. Therefore, $(-1)^{i+j} = (-1)^{1+2} = (-1)^3 = -1$. The coefficient of the first term must be preceded by a negative sign. The signs then alternate.

$$\begin{aligned}|A| &= -0\cdot|M_{12}| + 0\cdot|M_{22}| - 2\cdot|M_{32}| + 0\cdot|M_{42}|\\ &= -2\cdot|M_{32}|\\ &= -2\cdot\begin{vmatrix} 8 & 1 & 10\\ -1 & 0 & -5\\ -9 & 0 & -12\end{vmatrix}\\ &= -2\cdot(-1)\left|\begin{vmatrix} -1 & -5\\ -9 & -12\end{vmatrix}\right.\\ &= -2\cdot(-1)(12-45)\\ &= -66\end{aligned}$$

Use expansion by minors down the second column.

Watch the video to see this solution worked out in detail.

You Try It Work through this You Try It problem.

Work Exercises 5–12 in this textbook or in the MyLab Math Study Plan.

OBJECTIVE 3 USING CRAMER'S RULE TO SOLVE LINEAR SYSTEMS IN n VARIABLES

Now that we know how to compute the determinant of an $n \times n$ matrix, we can extend the use of Cramer's rule to solve a system of linear equations in n variables.

Cramer's Rule for a System of Linear Equations in n Variables

Given the linear system

$$\begin{matrix} a_{11}x_1 + a_{12}x_2 + a_{13}x_3 + \cdots a_{1n}x_n = b_1\\ a_{21}x_1 + a_{22}x_2 + a_{23}x_3 + \cdots a_{2n}x_n = b_2\\ \vdots \quad\quad \vdots \quad\quad \vdots \quad\quad \vdots \quad\quad \vdots\\ a_{n1}x_1 + a_{n2}x_2 + a_{n3}x_3 + \cdots a_{nn}x_n = b_n\end{matrix},$$

or equivalently

$$\underbrace{\begin{bmatrix} a_{11} & a_{12} & a_{13} & \cdots & a_{1n}\\ a_{21} & a_{22} & a_{23} & \cdots & a_{2n}\\ \vdots & \vdots & \vdots & & \vdots\\ a_{n1} & a_{n2} & a_{n3} & & a_{nn}\end{bmatrix}}_{\substack{\text{Coefficient}\\ \text{Matrix } A}} \underbrace{\begin{bmatrix} x_1\\ x_2\\ \vdots\\ x_n\end{bmatrix}}_{\substack{\text{Variable}\\ \text{Matrix } X}} = \underbrace{\begin{bmatrix} b_1\\ b_2\\ \vdots\\ b_n\end{bmatrix}}_{\substack{\text{Constant}\\ \text{Matrix } B}},$$

if $|D_{x_i}| = \underbrace{\begin{bmatrix} a_{11} & a_{12} & \cdots & b_1 & \cdots & a_{1n}\\ a_{21} & a_{22} & \cdots & b_2 & \cdots & a_{2n}\\ \vdots & & \cdots & \vdots & & \vdots\\ a_{n1} & a_{n2} & \cdots & b_n & & a_{nn}\end{bmatrix}}_{\substack{\text{Replace the } ith \text{ column of } A \text{ with the}\\ \text{coefficients of the constant matrix.}}}$ and if $|A| \neq 0$,

then there is a unique solution to the linear system given by $x_i = \dfrac{|D_{xi}|}{|A|}$.

(eText Screens 13.3-1–13.3-17)

Example 5 Use Cramer's Rule to Solve a Linear System in Three Variables

Use Cramer's rule to solve the system $\begin{aligned} x + y + z &= 2 \\ x - 2y + z &= 5. \\ 2x + y - z &= -1 \end{aligned}$

Solution We must calculate the following four determinants:

$$|A| = \begin{vmatrix} 1 & 1 & 1 \\ 1 & -2 & 1 \\ 2 & 1 & -1 \end{vmatrix}, \ |D_x| = \begin{vmatrix} 2 & 1 & 1 \\ 5 & -2 & 1 \\ -1 & 1 & -1 \end{vmatrix},$$

$$|D_y| = \begin{vmatrix} 1 & 2 & 1 \\ 1 & 5 & 1 \\ 2 & -1 & -1 \end{vmatrix}, \text{ and } |D_z| = \begin{vmatrix} 1 & 1 & 2 \\ 1 & -2 & 5 \\ 2 & 1 & -1 \end{vmatrix}$$

Work through the **interactive video** to verify that $|A| = 9$, $|D_x| = 9$, $|D_y| = -9$, and $|D_z| = 18$. Therefore,

$$x = \frac{|D_x|}{|A|} = \frac{9}{9} = 1, \quad y = \frac{|D_y|}{|A|} = \frac{-9}{9} = -1, \text{ and } z = \frac{|D_z|}{|A|} = \frac{18}{9} = 2.$$

You Try It Work through this You Try It problem.

Work Exercises 13–18 in this textbook or in the MyLab Math Study Plan.

13.3 Exercises

In Exercises 1–4, use Cramer's rule to solve each linear system.

1. $\begin{aligned} x + y &= 3 \\ -2x + y &= 0 \end{aligned}$

2. $\begin{aligned} 2x - y &= 1 \\ -3x + 3y &= 0 \end{aligned}$

3. $\begin{aligned} 4x - y &= -4 \\ -3x + 3y &= -6 \end{aligned}$

4. $\begin{aligned} 2x - 3y &= 0 \\ 4x + 9y &= 5 \end{aligned}$

In Exercises 5–12, calculate $|A|$.

5. $A = \begin{bmatrix} 2 & 2 & 2 \\ -1 & 5 & -2 \\ 3 & 7 & 4 \end{bmatrix}$

6. $A = \begin{bmatrix} 7 & -3 & 8 \\ 1 & -5 & -2 \\ 3 & 7 & 4 \end{bmatrix}$

7. $A = \begin{bmatrix} 7 & 0 & 4 \\ -1 & 3 & -1 \\ 6 & 2 & 5 \end{bmatrix}$

8. $A = \begin{bmatrix} -1 & 0 & 2 \\ 4 & 0 & -1 \\ 3 & 0 & -4 \end{bmatrix}$

9. $A = \begin{bmatrix} 2 & 0 & 0 \\ 0 & 3 & 0 \\ 0 & 0 & 1 \end{bmatrix}$

10. $A = \begin{bmatrix} -5 & -1 & -2 \\ 8 & 10 & 0 \\ -7 & 5 & 0 \end{bmatrix}$

11. $A = \begin{bmatrix} 1 & 5 & 1 & 9 \\ 2 & -1 & 7 & 3 \\ 9 & 0 & 0 & 1 \\ 12 & -4 & -2 & 8 \end{bmatrix}$

12. $A = \begin{bmatrix} 0 & -1 & 6 & 6 \\ 7 & -3 & 0 & 2 \\ 1 & 0 & 2 & 3 \\ 3 & 0 & 4 & 8 \end{bmatrix}$

In Exercises 13–18, use Cramer's rule to solve each linear system.

13. $\begin{aligned} x + y - z &= 4 \\ 5x - y + 3z &= 6 \\ x + y - 5z &= 8 \end{aligned}$

14. $\begin{aligned} 3x + 3y + z &= 6 \\ 4x - y + z &= 8 \\ -x + y - 3z &= -2 \end{aligned}$

15. $\begin{aligned} -2x + 5y + 5z &= -4 \\ 5x + 6y + 5z &= -23 \\ -3x + 4y + 5z &= -3 \end{aligned}$

16. $\begin{aligned} x - z &= -1 \\ 2y - z &= 8 \\ 2x + 2y &= 0 \end{aligned}$

17. $\begin{aligned} 3x + 5z &= 0 \\ 2x + 4y &= -14 \\ -y + 2z &= -21 \end{aligned}$

18. $\begin{aligned} -2w + 6x - 9y + z &= 57 \\ -w + x - y + z &= 11 \\ w + x + y + z &= 1 \\ 3w + 2x + 2y + z &= -2 \end{aligned}$

Chapter 13 Summary

Key Concepts	Examples/Videos
13.1 Matrix Operations To add or subtract two matrices, they must first be the **same size.** To add two matrices of the same size, add the corresponding entries. Similarly, to subtract two matrices of the same size, subtract the corresponding entries.	1. Let $A = \begin{bmatrix} 2 & -1 & 0 & 3 \\ -1 & 4 & -2 & 5 \\ 5 & 3 & 7 & 9 \end{bmatrix}$ and $B = \begin{bmatrix} -3 & -2 & 6 & 6 \\ 0 & 4 & -5 & 1 \\ 1 & 4 & 2 & 7 \end{bmatrix}$. a. Find matrix $A + B$. b. Find matrix $B + A$. c. Find matrix $A - B$.
We can multiply any matrix by a real number. We call this real number a **scalar** and this operation is called **scalar multiplication**. Scalar multiplication is performed by multiplying every entry in the matrix by that scalar.	2. If $A = \begin{bmatrix} -4 & 0 \\ 3 & 1 \end{bmatrix}$ and $B = \begin{bmatrix} 6 & 3 \\ -3 & -9 \end{bmatrix}$, find the following matrices: $-2A$, $\frac{1}{3}B$, and $-2A + \frac{1}{3}B$.
Matrix Multiplication If A is an $m \times n$ matrix and B is an $n \times q$ matrix, then there exists an $m \times q$ product matrix AB. The element in the ith row and jth column of matrix AB is equal to the sum of the products of the corresponding entries from row i of matrix A and column j of matrix B.	3. Given matrices $A = \begin{bmatrix} 1 & 1 & 2 \\ 2 & 4 & -3 \\ 3 & 6 & -5 \end{bmatrix}$, $B = \begin{bmatrix} 0 & -1 & 1 \\ -2 & 1 & 3 \\ 4 & 5 & -3 \end{bmatrix}$, and $C = \begin{bmatrix} 1 & 0 & 0 \\ 0 & 1 & 0 \\ 0 & 0 & 1 \end{bmatrix}$, find the products AB, BA, and AC.

Key Concepts	Examples/Videos
We now illustrate matrix multiplication by multiplying the 2 × 3 and 3 × 2 matrices. $$\overset{A}{\begin{bmatrix} 1 & 1 & 2 \\ 2 & 4 & -3 \end{bmatrix}}\overset{B}{\begin{bmatrix} 0 & -5 \\ 4 & -1 \\ 1 & 3 \end{bmatrix}}$$ 2×3 3×2 Same Size of AB is 2×2 $$= \begin{bmatrix} (1)(0) + (1)(4) + (2)(1) & (1)(-5) + (1)(-1) + (2)(3) \\ (2)(0) + (4)(4) + (-3)(1) & (2)(-5) + (4)(-1) + (-3)(3) \end{bmatrix}$$ $$\overset{AB}{= \begin{bmatrix} 6 & 0 \\ 13 & -23 \end{bmatrix}}.$$ 2×2	
Summary of Matrix Operations Assume matrix addition and matrix multiplication are defined for matrices A, B, and C and let k be a scalar, then **1.** $A + B = B + A$ — **Commutative Property for Matrix Addition** **2.** $(A + B) + C = A + (B + C)$ — **Associative Property for Matrix Addition** **3.** $k(A + B) = kA + kB$ — **Distributive Property of a Scalar with Matrix Addition** **4.** $(AB)C = A(BC)$ — **Associative Property for Matrix Multiplication** **5.** $A(B + C) = AB + AC$ — **Distributive Property** **6.** $(B + C)A = BA + CA$ — **Distributive Property** **7.** $k(AB) = (kA)B = A(kB)$ — **Associative Property for Scalar Multiplication** **8.** If A is an $m \times n$ matrix, then $I_m A = A$ and $AI_n = A$. — **Identity Property**	

Key Concepts	Examples/Videos
13.2 Inverses of Matrices and Matrix Equations Let A be an $n \times n$ square matrix. If there exists an $n \times n$ matrix A^{-1} such that $AA^{-1} = A^{-1}A = I_n$, then A^{-1} is called the **multiplicative inverse** of matrix A.	1. Verify that $A = \begin{bmatrix} 1 & 1 & 2 \\ 2 & 4 & -3 \\ 3 & 6 & -5 \end{bmatrix}$ and $B = \begin{bmatrix} 2 & -17 & 11 \\ -1 & 11 & -7 \\ 0 & 3 & -2 \end{bmatrix}$ are inverse matrices.
Determinant of a 2×2 Matrix Let $A = \begin{bmatrix} a & b \\ c & d \end{bmatrix}$. The determinant of matrix A is denoted as $\lvert A \rvert$ or $\begin{vmatrix} a & b \\ c & d \end{vmatrix}$ and is defined by $\lvert A \rvert = \begin{vmatrix} a & b \\ c & d \end{vmatrix} = ad - cb$. If $\lvert A \rvert \neq 0$, then matrix A is invertible. **Formula for Determining the Inverse of a 2×2 Matrix** Let $A = \begin{bmatrix} a & b \\ c & d \end{bmatrix}$. If $\lvert A \rvert \neq 0$, then A is invertible and $A^{-1} = \frac{1}{\lvert A \rvert}\begin{bmatrix} d & -b \\ -c & a \end{bmatrix}$.	2. Determine whether the following matrices are invertible or singular. If the matrix is invertible find its inverse and then verify by using matrix multiplication. a. $A = \begin{vmatrix} 2 & 3 \\ -4 & -6 \end{vmatrix}$ b. $B = \begin{vmatrix} 2 & -1 \\ 4 & 3 \end{vmatrix}$
Steps for Finding the Multiplicative Inverse of an Invertible Square Matrix If A is an $n \times n$ invertible matrix, then the multiplicative inverse of A can be obtained by following these steps: **Step 1.** Form the augmented matrix $[A \vert I_n]$. **Step 2.** Use Gauss-Jordan elimination to reduce A into the identity matrix I_n. **Step 3.** The new auqmented matrix is of the form $[I_n/A^{-1}]$, where A^{-1} is the multiplicative inverse of A.	3. If possible, determine the inverse of $A = \begin{bmatrix} 1 & 1 & 2 \\ 2 & 4 & -3 \\ 3 & 6 & -5 \end{bmatrix}$.
Suppose that a system of n linear equations with n variables is given by: $a_{11}x_1 + a_{12}x_2 + a_{13}x_3 + \cdots a_{1n}x_n = b_1$ $a_{21}x_1 + a_{22}x_2 + a_{23}x_3 + \cdots a_{2n}x_n = b_2$ $\vdots \quad \vdots \quad \vdots \quad \vdots \quad \vdots$ $a_{n1}x_1 + a_{n2}x_2 + a_{n3}x_3 + \cdots a_{nn}x_n = b_n$	4. Solve the following system of linear equations using an inverse matrix: 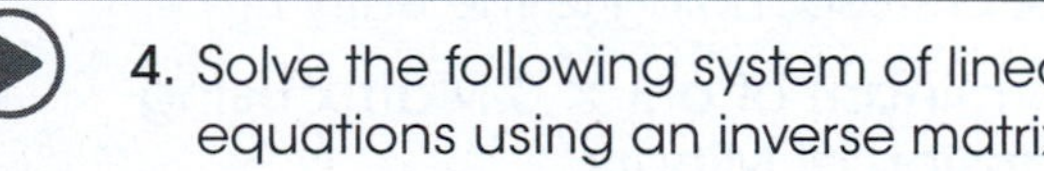 $-3x + 2y = 4$ $5x - 4y = 9$

(Summary)

Key Concepts	Examples/Videos

This system is equivalent to the matrix equation

$$\overset{A}{\begin{bmatrix} a_{11} & a_{12} & a_{13} & \cdots & a_{1n} \\ a_{21} & a_{22} & a_{23} & \cdots & a_{2n} \\ \vdots & \vdots & \vdots & & \vdots \\ a_{n1} & a_{n2} & a_{n3} & & a_{nn} \end{bmatrix}} \overset{X}{\begin{bmatrix} x_1 \\ x_2 \\ \vdots \\ x_n \end{bmatrix}} = \overset{B}{\begin{bmatrix} b_1 \\ b_2 \\ \vdots \\ b_n \end{bmatrix}}.$$

If A is invertible, then the entries of matrix X can be determined as follows:

$AX = B$	Write the matrix equation.
$A^{-1}AX = A^{-1}B$	Multiply both sides of the equation on the left by A^{-1}.
$I_nX = A^{-1}B$	Definition of an inverse matrix $(A^{-1}A = I_n)$
$X = A^{-1}B$	Identity property for matrix multiplication $(I_nX = X)$

13.3 Determinants and Cramer's Rule

Cramer's Rule for a System of Linear Equations in Two Variables

For the linear system of equations $\begin{aligned} ax + by &= m \\ cx + dy &= n \end{aligned}$, let

$$|A| = \begin{vmatrix} a & b \\ c & d \end{vmatrix}, |D_x| = \begin{vmatrix} m & b \\ n & d \end{vmatrix}, \text{ and } |D_y| = \begin{vmatrix} a & m \\ c & n \end{vmatrix}.$$

If $|A| \neq 0$, then there is a unique solution to this system given by

$$x = \frac{|D_x|}{|A|} \quad \text{and} \quad y = \frac{|D_y|}{|A|}$$

1. Use Cramer's rule to solve the following system.

$$\begin{aligned} 2x - 2y &= 4 \\ 3x + y &= -3 \end{aligned}$$

If a_{ij} is the entry of a 3×3 matrix that appears in the ith row and jth column, then the matrix M_{ij} is obtained by deleting the ith row and jth column of this matrix. The determinant $|M_{ij}|$ is called the **minor** corresponding to the entry a_{ij}.

Determinant of a 3×3 Matrix Using Expansion of Minors

Let $A = \begin{bmatrix} a_{11} & a_{12} & a_{13} \\ a_{21} & a_{22} & a_{23} \\ a_{31} & a_{32} & a_{33} \end{bmatrix}$, then

$$|A| = a_{11}|M_{11}| - a_{12}|M_{12}| + a_{13}|M_{13}|.$$

2. Calculate $|A|$ if $A = \begin{bmatrix} 2 & -17 & 11 \\ -1 & 11 & -7 \\ 0 & 3 & -2 \end{bmatrix}$.

3. Calculate $|A|$ by expanding the minors down the first column if

$$A = \begin{bmatrix} 2 & -17 & 11 \\ -1 & 11 & -7 \\ 0 & 3 & -2 \end{bmatrix}.$$

Key Concepts	Examples/Videos

The Determinant of a Square Matrix

Let $A = \begin{bmatrix} a_{11} & a_{12} & \cdots & a_{1n} \\ a_{21} & a_{22} & \cdots & a_{2n} \\ \vdots & \vdots & \ddots & \vdots \\ a_{n1} & a_{n2} & \cdots & a_{nn} \end{bmatrix}$, then for each entry a_{ij} there is a corresponding minor $|M_{ij}|$, which is the determinant of a $(n-1) \times (n-1)$ matrix. The determinant $|A|$ is equal to the sum of the products of the entries of any row or column and the corresponding minors.

(Each coefficient must be preceded by $(-1)^{i+j}$.)

4. Calculate $|A|$ if

$$A = \begin{bmatrix} 8 & 0 & 1 & 10 \\ -1 & 0 & 0 & -5 \\ 5 & 2 & -2 & 4 \\ -9 & 0 & 0 & -12 \end{bmatrix}.$$

Cramer's Rule for a System of Linear Equations in *n* Variables

Given the linear system

$$\begin{aligned} a_{11}x_1 + a_{12}x_2 + a_{13}x_3 + \cdots a_{1n}x_n &= b_1 \\ a_{21}x_1 + a_{22}x_2 + a_{23}x_3 + \cdots a_{2n}x_n &= b_2 \\ \vdots \quad\quad \vdots \quad\quad \vdots \quad\quad\quad \vdots \quad &\quad \vdots \\ a_{n1}x_1 + a_{n2}x_2 + a_{n3}x_3 + \cdots a_{nn}x_n &= b_n \end{aligned}$$

, or equivalently

$$\underbrace{\begin{bmatrix} a_{11} & a_{12} & a_{13} & \cdots & a_{1n} \\ a_{21} & a_{22} & a_{23} & \cdots & a_{2n} \\ \vdots & \vdots & \vdots & & \vdots \\ a_{n1} & a_{n2} & a_{n3} & & a_{nn} \end{bmatrix}}_{\text{Coefficient Matrix } A} \underbrace{\begin{bmatrix} x_1 \\ x_2 \\ \vdots \\ x_n \end{bmatrix}}_{\text{Variable Matrix } X} = \underbrace{\begin{bmatrix} b_1 \\ b_2 \\ \vdots \\ b_n \end{bmatrix}}_{\text{Constant Matrix } B},$$

If $|D_{x_i}| = \underbrace{\begin{bmatrix} a_{11} & a_{12} & \cdots & b_1 & \cdots & a_{1n} \\ a_{21} & a_{22} & \cdots & b_2 & \cdots & a_{2n} \\ \vdots & & \cdots & \vdots & & \vdots \\ a_{n1} & a_{n2} & \cdots & b_n & & a_{nn} \end{bmatrix}}_{\text{Replace the } ith \text{ column of } A \text{ with the coefficients of the constant matrix.}}$ and if $|A| \neq 0$,

then there is a unique solution to the linear system given by $x_i = \dfrac{|D_{xi}|}{|A|}$.

5. Use Cramer's rule to solve the system

$$\begin{aligned} x + y + z &= 2 \\ x - 2y + z &= 5 \\ 2x + y - z &= -1 \end{aligned}$$

Chapter 13 Review Exercises

For Exercises 1–3, use matrices A and B to find the indicated matrix.

$$A = \begin{bmatrix} 1 & 0 & -2 \\ 3 & 1 & 4 \\ 5 & -1 & -5 \end{bmatrix}, \quad B = \begin{bmatrix} 3 & 4 & -1 \\ 5 & 0 & 2 \\ 3 & 4 & -5 \end{bmatrix}$$

1. $A + B$
2. $A - B$
3. $2A - \frac{1}{2}B$
4. Let $(m \times n)$ denote a matrix of size $m \times n$. Find the size of the product matrix $(3 \times 4)(4 \times 5)$, or state that the product is not defined.
5. Find AB if $A = \begin{bmatrix} -1 & 2 \\ 3 & 5 \end{bmatrix}$ and $B = \begin{bmatrix} 1 & 3 & -1 \\ 4 & -2 & 0 \end{bmatrix}$.
6. The youth from a local church are having a breakfast fund-raising event. They are planning on serving biscuits, pancakes, and waffles. The ingredients for one batch of each are as follows:

   Biscuits: 3 cups baking mix, 1 egg, 1 cup milk, $\frac{1}{2}$ tablespoon cooking oil

   Pancakes: 2 cups baking mix, 2 eggs, $\frac{2}{3}$ cup milk

   Waffles: $2\frac{1}{2}$ cups baking mix, 1 egg, 1 cup milk, 2 tablespoons cooking oil

   The ingredients can be summarized by matrix A.

$$\begin{array}{l} \\ \text{Biscuits} \\ \text{Pancakes} \\ \text{Waffles} \end{array} \begin{array}{c} \begin{array}{cccc} \text{Mix} & \text{Eggs} & \text{Milk} & \text{Oil} \end{array} \\ \begin{bmatrix} 3 & 1 & 1 & \frac{1}{2} \\ 2 & 2 & \frac{2}{3} & 0 \\ 2\frac{1}{2} & 1 & 1 & 2 \end{bmatrix} \end{array} = A$$

   It was determined that they were going to need 16 batches of biscuits, 30 batches of pancakes, and 40 batches of waffles, which is summarized by matrix B.

$$\begin{array}{cc} & \begin{array}{ccc} \text{Biscuits} & \text{Pancakes} & \text{Waffles} \end{array} \\ \text{Batches} & \begin{bmatrix} 16 & 30 & 40 \end{bmatrix} \end{array} = B$$

   a. Calculate matrix BA.

   b. Interpret the meaning of each entry of matrix BA.

7. Determine whether matrix B is the inverse of matrix A by finding the products AB and BA.

$$A = \begin{bmatrix} -3 & 1 \\ -5 & 2 \end{bmatrix}, B = \begin{bmatrix} -2 & 1 \\ -5 & 3 \end{bmatrix}$$

For Exercises 8–10, find the inverse of each matrix if possible.

8. $\begin{bmatrix} -5 & 4 \\ -2 & 2 \end{bmatrix}$

9. $\begin{bmatrix} 1 & -6 & -2 \\ 0 & -1 & 0 \\ 2 & -11 & -3 \end{bmatrix}$

10. $\begin{bmatrix} 1 & 0 & -2 & 0 \\ 0 & 1 & 0 & -5 \\ -4 & 0 & 9 & 0 \\ 0 & 2 & 1 & -9 \end{bmatrix}$

11. Solve the system below using an inverse matrix.

$$\begin{aligned} x + y + 3z &= 11 \\ -x - y - 2z &= -7 \\ x - 3y - 5z &= -13 \end{aligned}$$

12. Use Cramer's rule to solve the following system of equations.

$$\begin{aligned} 4x - y &= -4 \\ -3x + 3y &= -6 \end{aligned}$$

13. Calculate $|A|$ if $A = \begin{bmatrix} 7 & -3 & 8 \\ 1 & -5 & -2 \\ 3 & 7 & 4 \end{bmatrix}$.

For Exercises 14 and 15, use Cramer's rule to solve each system of equations.

14. $$\begin{aligned} x + y - z &= 4 \\ 5x - y + 3z &= 6 \\ x + y - 5z &= 8 \end{aligned}$$

15. $$\begin{aligned} -2w + 6x - 9y + z &= 57 \\ -w + x - y + z &= 11 \\ w + x + y + z &= 1 \\ 3w + 2x + 2y + z &= -2 \end{aligned}$$

CHAPTER FOURTEEN

Sequences and Series; Counting and Probability

CHAPTER FOURTEEN CONTENTS

14.1 Introduction to Sequences and Series

THINGS TO KNOW

Before working through this section, be sure that you are familiar with the following concepts:

		VIDEO	ANIMATION	INTERACTIVE
 You Try It	1. Using the Vertical Line Test (Section 3.1)			

OBJECTIVES

1 Writing the Terms of a Sequence

2 Writing the Terms of a Recursive Sequence

3 Writing the General Term for a Given Sequence

4 Computing Partial Sums of a Series

5 Determining the Sum of a Finite Series Written in Summation Notation

6 Writing a Series Using Summation Notation

SECTION 14.1 EXERCISES

OBJECTIVE 1 WRITING THE TERMS OF A SEQUENCE

Consider the function $f(n) = 2n - 1$, where n is a **natural number**. The graph of this function consists of infinitely many ordered pairs of the form $(n, 2n - 1)$, where $n \geq 1$. Therefore, the ordered pairs that lie on the graph of this function are $(1, 1)$, $(2, 3)$, $(3, 5)$, $(4, 7)$, and so on. A portion of the graph of this function can be seen in Figure 1.

(eText Screens 14.1-1–14.1-22)

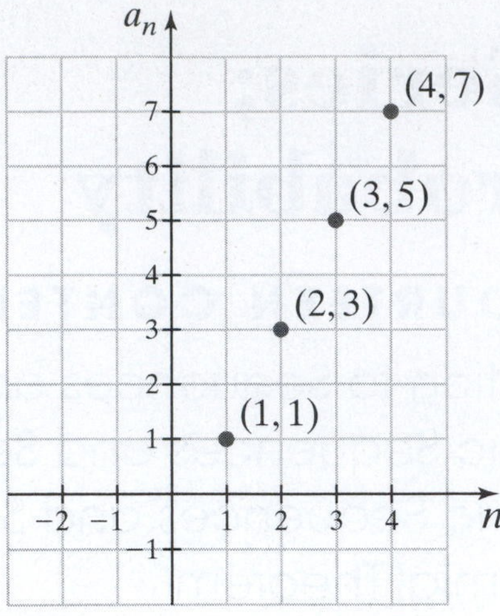

Figure 1 A portion of the graph of the function $f(n) = 2n - 1$, where n is a natural number.

Clearly, the graph seen in **Figure 1** is a function because the graph passes the **vertical line test**. Any function whose domain is the set of natural numbers is called an **infinite sequence**. Instead of using the conventional function notation $f(n)$ to name a sequence, we will use a subscript notation, such as a_n (read as "a sub n"), to name our sequences. We now formally define a sequence.

> **Definition** Sequence
>
> A **finite sequence** is a function whose domain is the finite set $\{1, 2, 3, \ldots, n\}$, where n is a **natural number**.
>
> An **infinite sequence** is a function whose domain is the set of all natural numbers.
>
> The range values of a sequence are called the **terms** of the sequence.

We now rename the sequence $f(n) = 2n - 1$ as $a_n = 2n - 1$. The first four terms of this sequence are $a_1 = 1, a_2 = 3, a_3 = 5,$ and $a_4 = 7$.

Using Technology

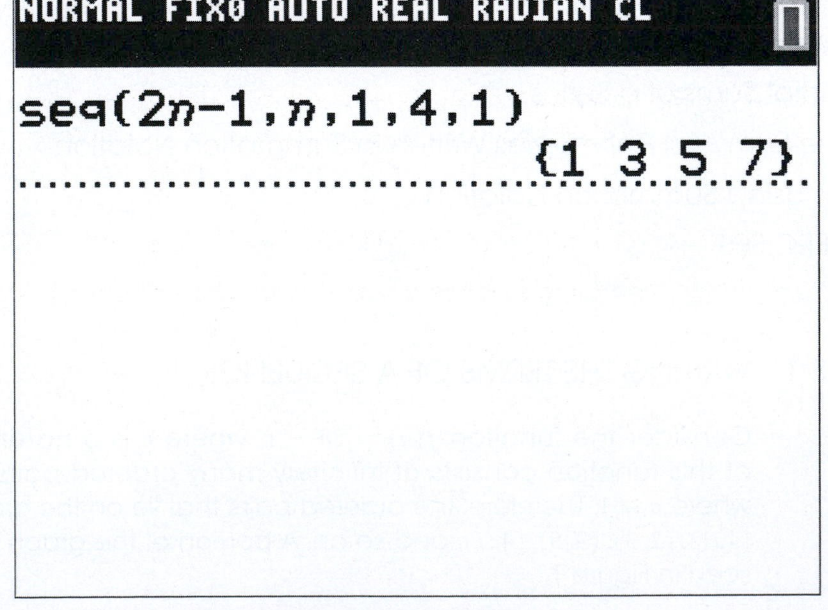

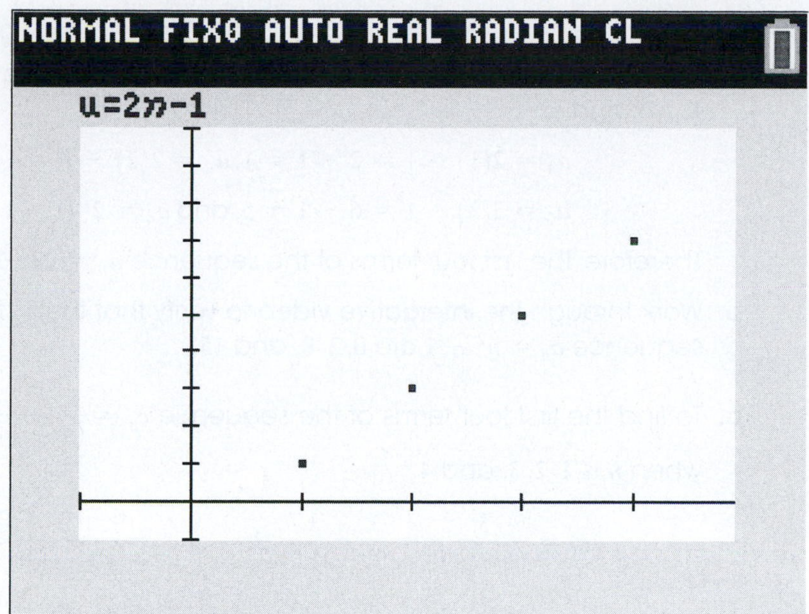

A graphing utility set to *sequence mode* can be used to sketch the graph of a sequence or determine the terms of a sequence. The figure on the left shows the first four terms of the sequence $a_n = 2n - 1$. The figure on the right shows the graph of the first four terms of the same sequence.

Before we start to find the terms of a sequence it is important to introduce factorial notation.

> **Definition** The Factorial of a Non-Negative Integer
>
> The factorial of a non-negative integer n, denoted as $n!$, is the product of all positive integers less than or equal to n. Thus, $n! = n(n - 1) \cdot \cdots \cdot 3 \cdot 2 \cdot 1$.

Note By definition, we say that zero factorial is equal to 1 or $0! = 1$.

Some examples of factorial notation are

$$5! = 5 \cdot 4 \cdot 3 \cdot 2 \cdot 1 = 120 \text{ and } 8! = 8 \cdot 7 \cdot 6 \cdot 5 \cdot 4 \cdot 3 \cdot 2 \cdot 1 = 40{,}320.$$

As you can see, the value of $n!$ gets large rather quickly. In fact, the value of $13!$ is over 6 billion, which is a rough estimate of the population of Earth! Example 1c illustrates how factorial notation can be used to define a sequence.

Example 1 Writing the Terms of a Sequence

Write the first four terms of each sequence whose nth term is given.

a. $a_n = 2n - 1$ **b.** $b_n = n^2 - 1$

c. $c_n = \dfrac{3^n}{(n - 1)!}$ **d.** $d_n = (-1)^n 2^{n-1}$

Solution

a. To find the first four terms of the sequence, we evaluate $a_n = 2n - 1$ when n is $1, 2, 3,$ and 4.

$$a_1 = 2(1) - 1 = 2 - 1 = 1, a_2 = 2(2) - 1 = 4 - 1 = 3,$$

$$a_3 = 2(3) - 1 = 6 - 1 = 5, \text{ and } a_4 = 2(4) - 1 = 8 - 1 = 7$$

Therefore, the first four terms of the sequence $a_n = 2n - 1$ are $1, 3, 5,$ and 7.

b. Work through the **interactive video** to verify that the first four terms of the sequence $b_n = n^2 - 1$ are $0, 3, 8,$ and 15.

c. To find the first four terms of the sequence $c_n = \dfrac{3^n}{(n-1)!}$, we evaluate c_n when n is $1, 2, 3,$ and 4.

$$c_1 = \frac{3^1}{(1-1)!} = \frac{3}{0!} = \frac{3}{1} = 3, \qquad c_2 = \frac{3^2}{(2-1)!} = \frac{9}{1!} = \frac{9}{1} = 9,$$

$$c_3 = \frac{3^3}{(3-1)!} = \frac{27}{2!} = \frac{27}{2}, \text{ and } c_4 = \frac{3^4}{(4-1)!} = \frac{81}{3!} = \frac{81}{6} = \frac{27}{2}$$

Therefore, the first four terms of the sequence $c_n = \dfrac{3^n}{(n-1)!}$ are

$$3, 9, \frac{27}{2}, \text{ and } \frac{27}{2}.$$

d. Work through the **interactive video** to verify that the first four terms of the sequence $d_n = (-1)^n 2^{n-1}$ are $-1, 2, -4,$ and 8.

The sequence $d_n = (-1)^n 2^{n-1}$ is an example of an alternating sequence because the successive terms alternate in sign.

You Try It Work through this You Try It problem.

Work Exercises 1–10 in this textbook or in the MyLab Math Study Plan.

OBJECTIVE 2 WRITING THE TERMS OF A RECURSIVE SEQUENCE

Some sequences are defined recursively. A **recursive sequence** is a sequence in which each term is defined using one or more of its previous terms. Typically, the first term of a recursive sequence is given, followed by the formula for the nth term of the sequence. The following example illustrates two recursive sequences.

Example 2 Writing the Terms of a Recursive Sequence

Write the first four terms of each of the following recursive sequences.

a. $a_1 = -3, a_n = 5a_{n-1} - 1$ for $n \geq 2$

b. $b_1 = 2, b_n = \dfrac{(-1)^{n-1} n}{b_{n-1}}$ for $n \geq 2$

Solution

a. The first four terms of this recursive sequence are $-3, -16, -81$, and -406. Work through this **interactive video** to verify.

b. The first four terms of this recursive sequence are $2, -1, -3$, and $\frac{4}{3}$. Work through this **interactive video** to verify.

Arguably the most famous recursively defined sequence is the Fibonacci sequence named after the 13th-century Italian mathematician **Leonardo of Pisa**, also known as Fibonacci. The Fibonacci sequence is defined in Example 3.

Example 3 Writing the Terms of the Fibonacci Sequence

The Fibonacci sequence is defined recursively by $a_n = a_{n-1} + a_{n-2}$, where $a_1 = 1$ and $a_2 = 1$. Write the first eight terms of the Fibonacci sequence.

Solution We are given that $a_1 = 1$ and $a_2 = 1$. We use the recursive formula $a_n = a_{n-1} + a_{n-2}$ to find the next six terms starting with $n = 3$.

$$a_3 = a_2 + a_1 = 1 + 1 = 2$$
$$a_4 = a_3 + a_2 = 2 + 1 = 3$$
$$a_5 = a_4 + a_3 = 3 + 2 = 5$$
$$a_6 = a_5 + a_4 = 5 + 3 = 8$$
$$a_7 = a_6 + a_5 = 8 + 5 = 13$$
$$a_8 = a_7 + a_6 = 13 + 8 = 21$$

You can see in Example 3 that each term of the Fibonacci sequence is the sum of the preceding two terms. We now write the first 12 terms of the Fibonacci sequence.

$$1, 1, 2, 3, 5, 8, 13, 21, 34, 55, 89, 144, \ldots$$

The numbers of this sequence are known as *Fibonacci numbers.* The Fibonacci sequence and Fibonacci numbers occur in many natural phenomena such as the spiral formation of seeds of various plants, the number of petals of a flower, and the formation of the branches of a tree. See **Exercise 44**.

You Try It Work through this You Try It problem.

Work Exercises 11–14 in this textbook or in the MyLab Math Study Plan.

OBJECTIVE 3 WRITING THE GENERAL TERM FOR A GIVEN SEQUENCE

Sometimes the first several terms of a sequence are given without listing the nth term. When this occurs, we must try to determine a pattern and use deductive reasoning to establish a rule that describes the general term, or nth term, of the sequence. Example 4 illustrates two such sequences.

Example 4 Finding the General Term of a Sequence

Write a formula for the nth term of each infinite sequence, then use this formula to find the 8th term of the sequence.

a. $\frac{1}{1}, \frac{1}{2}, \frac{1}{3}, \frac{1}{4}, \frac{1}{5}, \ldots$

b. $-\frac{2}{1}, \frac{4}{2}, -\frac{8}{6}, \frac{16}{24}, -\frac{32}{120}, \ldots$

Solution

a. The nth term of the sequence is $a_n = \frac{1}{n}$. Thus, the 8th term of this sequence is $a_8 = \frac{1}{8}$.

b. For this sequence, notice that the first term is negative and that terms alternate in sign. We can therefore represent the sign of each term as $(-1)^n$. Also, notice that the numerators are successive powers of 2. We now have the following pattern:

$$\begin{array}{ccccc} a_1 & a_2 & a_3 & a_4 & a_5 \\ \downarrow & \downarrow & \downarrow & \downarrow & \downarrow \\ -\frac{2}{1}, & \frac{4}{2}, & -\frac{8}{6}, & \frac{16}{24}, & -\frac{32}{120}, \ldots \\ \downarrow & \downarrow & \downarrow & \downarrow & \downarrow \\ \frac{(-1)^1 2^1}{1}, & \frac{(-1)^2 2^2}{2}, & \frac{(-1)^3 2^3}{6}, & \frac{(-1)^4 2^4}{24}, & \frac{(-1)^5 2^5}{120}, \ldots \end{array}$$

Finally, if we factor each successive denominator we get $1 = 1, 2 = 2 \cdot 1$, $6 = 3 \cdot 2 \cdot 1$, $24 = 4 \cdot 3 \cdot 2 \cdot 1$, and $120 = 5 \cdot 4 \cdot 3 \cdot 2 \cdot 1$. This suggests that the denominator of the nth term can be represented by $n!$ Therefore, the nth term of the sequence is $a_n = \frac{(-1)^n 2^n}{n!}$. The 8th term of this sequence is

$$a_8 = \frac{(-1)^8 2^8}{8!} = \frac{256}{40{,}320} = \frac{2}{315}.$$

If you would like to see this solution worked out in detail, watch this **video**. ●

You Try It Work through this You Try It problem.

Work Exercises 15–22 in this textbook or in the MyLab Math Study Plan.

OBJECTIVE 4 COMPUTING PARTIAL SUMS OF A SERIES

Suppose that we wanted to find the sum of the first four terms of the sequence $a_n = 2n - 1$. From Example 1 we saw that the first four terms of this sequence were $a_1 = 1$, $a_2 = 3$, $a_3 = 5$, and $a_4 = 7$. Therefore, the sum of the first four terms is $a_1 + a_2 + a_3 + a_4 = 1 + 3 + 5 + 7 = 16$. The expression $1 + 3 + 5 + 7$ is called a **series**.

> **Definition** Series
>
> Let $a_1, a_2, a_3, \ldots$ be a sequence. The expression of the form $a_1 + a_2 + a_3 + \cdots + a_n$ is called a **finite series.**
>
> The expression of the form $a_1 + a_2 + a_3 + \cdots + a_n + a_{n+1} + \cdots$ is called an **infinite series.**
>
> The sum of the first n terms of a series is called the nth **partial sum** of the series and is denoted as S_n.

For the series $1 + 3 + 5 + 7 + 9 + \cdots + 2n - 1$, the first five partial sums are as follows:

$$S_1 = 1$$
$$S_2 = 1 + 3 = 4$$
$$S_3 = 1 + 3 + 5 = 9$$
$$S_4 = 1 + 3 + 5 + 7 = 16$$
$$S_5 = 1 + 3 + 5 + 7 + 9 = 25$$

It appears that $S_n = n^2$. In fact, it can be shown that for any positive integer n, the sum of the series $1 + 3 + 5 + 7 + 9 + \cdots + 2n - 1$ is equal to n^2. We will be able to prove this assertion in Section 14.5 using a method called *mathematical induction*.

Example 5 Computing Partial Sums of a Series

Given the general term of each sequence, find the indicated partial sum.

a. $a_n = \frac{1}{n}$, find S_3. **b.** $b_n = (-1)^n 2^{n-1}$, find S_5.

Solution

a. The first three terms are $a_1 = 1, a_2 = \frac{1}{2}$, and $a_3 = \frac{1}{3}$. Therefore, the partial sum, S_3, is $S_3 = 1 + \frac{1}{2} + \frac{1}{3} = \frac{11}{6}$.

b. The first five terms are $b_1 = -1, b_2 = 2, b_3 = -4, b_4 = 8$, and $b_5 = -16$. Therefore, the partial sum, S_5, is $S_5 = -1 + 2 + (-4) + 8 + (-16) = -11$.

You Try It Work through this You Try It problem.

Work Exercises 23–28 in this textbook or in the MyLab Math Study Plan.

OBJECTIVE 5 DETERMINING THE SUM OF A FINITE SERIES WRITTEN IN SUMMATION NOTATION

Writing out an entire finite series of the form $a_1 + a_2 + a_3 + \cdots + a_n$ can be quite tedious, especially if n is fairly large. Fortunately, there is a convenient way to express a finite series using a short-hand notation called *summation notation* (also called *sigma notation*). This notation involves the use of the uppercase Greek letter sigma, which is written as Σ.

Definition Summation Notation

If $a_1, a_2, a_3, \ldots$ is a sequence, then the finite series $a_1 + a_2 + a_3 + \cdots + a_n$ can be written in **summation notation** as $\sum_{i=1}^{n} a_i$. The infinite series $a_1 + a_2 + \cdots + a_n + a_{n+1} + \cdots$ can be written as $\sum_{i=1}^{\infty} a_i$.

The variable i is called the **index of summation**. The number 1 is the **lower limit of summation** and n is the **upper limit of summation**.

The lower limit of summation, $i = 1$, below the sigma tells us which term to start with. The upper limit of summation, n, that appears above the sigma tells us which term of the sequence will be the last term to add. There is nothing special about the letter i that is used to represent the index of summation. We will often use different letters such as j or k. Also, it is not necessary for the lower limit of summation to start at 1. In Examples 6b and 6c, the lower limits of summation are 2 and 0, respectively.

Example 6 Determining the Sum of a Series Written in Summation Notation

Find the sum of each finite series.

a. $\sum_{i=1}^{5} i^2$ b. $\sum_{j=2}^{5} \frac{j-1}{j+1}$ c. $\sum_{k=0}^{6} \frac{1}{k!}$

(Round the sum to three decimal places.)

Solution

a. $\sum_{i=1}^{5} i^2 = 1^2 + 2^2 + 3^2 + 4^2 + 5^2 = 1 + 4 + 9 + 16 + 25 = 55$

b. $$\sum_{j=2}^{5} \frac{j-1}{j+1} = \frac{2-1}{2+1} + \frac{3-1}{3+1} + \frac{4-1}{4+1} + \frac{5-1}{5+1}$$

$$= \frac{1}{3} + \frac{2}{4} + \frac{3}{5} + \frac{4}{6} = \frac{21}{10}$$

c. $$\sum_{k=0}^{6} \frac{1}{k!} = \frac{1}{0!} + \frac{1}{1!} + \frac{1}{2!} + \frac{1}{3!} + \frac{1}{4!} + \frac{1}{5!} + \frac{1}{6!}$$

$$= 1 + 1 + \frac{1}{2} + \frac{1}{6} + \frac{1}{24} + \frac{1}{120} + \frac{1}{720} = \frac{1957}{720} \approx 2.718$$

You can work through this **interactive video** to see this solution worked out in detail. ●

(eText Screens 14.1-1–14.1-22)

Using Technology

```
NORMAL FIX9 AUTO REAL RADIAN CL
sum(seq(1/n!,n,0,6))
                        2.718055556
```

Using a TI-84 Plus, we can calculate the sum obtained in Example 6c. Notice that this number is a good approximation of the **number** e. In fact, it can be shown using calculus that the exact value of e is $e = \sum_{n=0}^{\infty} \frac{1}{n!}$. Other irrational numbers, such as π, can be represented by a series. Again, using calculus, it can be shown that the exact value of π is $\pi = \sum_{n=0}^{\infty} \frac{4 \cdot (-1)^n}{2n + 1}$.

You Try It Work through this You Try It problem.

Work Exercises 29–35 in this textbook or in the MyLab Math Study Plan.

OBJECTIVE 6 WRITING A SERIES USING SUMMATION NOTATION

Given the first several terms of a series, it is important to be able to rewrite the series using summation notation as in Example 7.

Example 7 Writing a Series Using Summation Notation

Rewrite each series using summation notation. Use 1 as the lower limit of summation.

a. $2 + 4 + 6 + 8 + 10 + 12$ **b.** $1 + 2 + 6 + 24 + 120 + 720 + \cdots + 3{,}628{,}800$

Solution

a. This series is the sum of six terms. Therefore, the lower limit of summation is 1 and the upper limit of summation is 6. Each term is a successive multiple of 2. So, one possible series is $\sum_{i=1}^{6} 2i$.

b. Notice that $1 = 1!, 2 = 2!, 6 = 3!, 24 = 4!, 120 = 5!, 720 = 6!,$ and $3{,}628{,}800 = 10!$

Thus, a possible series is $\sum_{n=1}^{10} n!$ ●

You Try It Work through this You Try It problem.

Work Exercises 36–43 in this textbook or in the MyLab Math Study Plan.

14.1 Exercises

In Exercises 1–10, write the first four terms of each sequence.

1. $a_n = 3n + 1$
2. $a_n = 4^n$
3. $a_n = \dfrac{4n}{n + 3}$
4. $a_n = (-4)^n$
5. $a_n = 5(n + 2)!$
6. $a_n = \dfrac{n^3}{(n + 1)!}$
7. $a_n = (-1)^n(5n)$
8. $a_n = \dfrac{3^n}{(-1)^{n+1} + 5}$
9. $a_n = \dfrac{(-1)^n}{(n + 5)(n + 6)}$
10. $a_n = \dfrac{(-1)^n(3)^{2n+1}}{(2n + 1)!}$

In Exercises 11–14, write the first four terms of each recursive sequence.

11. $a_1 = 7, a_n = 3 + a_{n-1}$ for $n \geq 2$
12. $a_1 = -1, a_n = n - a_{n-1}$ for $n \geq 2$
13. $a_1 = 6, a_n = \dfrac{a_{n-1}}{n^2}$ for $n \geq 2$
14. $a_1 = -4, a_n = 1 - \dfrac{1}{a_{n-1}}$ for $n \geq 2$

In Exercises 15–22, write a formula for the general term, or nth term, for the given sequence. Then find the indicated term.

15. $-1, 1, 3, 5, 7, \ldots; a_{11}$.
16. $\frac{1}{5}, \frac{2}{6}, \frac{3}{7}, \frac{4}{8}, \frac{5}{9}, \ldots; a_8$.
17. $1 \cdot 6, 2 \cdot 7, 3 \cdot 8, 4 \cdot 9, \ldots; a_7$.
18. $-2, 4, -8, 16, \ldots; a_7$.
19. $\frac{2}{5}, \frac{2}{25}, \frac{2}{125}, \frac{2}{625}, \ldots; a_6$.
20. $-6, 12, -24, 48, -96, \ldots; a_9$.
21. $-6, 24, -120, 720, \ldots; a_5$.
22. $\frac{3}{2}, \frac{9}{6}, \frac{27}{24}, \frac{81}{120}, \ldots; a_5$.

In Exercises 23–25, the first several terms of a sequence are given. Find the indicated partial sum.

23. $2, 4, 6, 8, 10, \ldots; S_4$
24. $3, -6, 9, -12, 15, -18, \ldots; S_9$
25. $\frac{1}{2}, -\frac{1}{4}, \frac{1}{8}, -\frac{1}{16}, \ldots; S_5$

In Exercises 26–28, the general term of a sequence is given. Find the indicated partial sum.

26. $a_n = 3n + 8; S_6$
27. $a_n = (-1)^n \cdot (4n); S_6$
28. $a_1 = 4, a_n = a_{n-1} - 8$ for $n \geq 2; S_8$

In Exercises 29–35, find the sum of each series.

29. $\sum_{i=1}^{9} i$
30. $\sum_{i=1}^{6} (4i + 3)$
31. $\sum_{i=1}^{7} i(i + 2)$
32. $\sum_{i=1}^{21} 7$
33. $\sum_{j=0}^{5} (j + 4)^2$
34. $\sum_{k=2}^{7} \dfrac{k!}{(k - 2)!}$
35. $\sum_{j=0}^{5} (j - 2)^3$

In Exercises 36–43, rewrite each series using summation notation. Use 1 as the lower limit of summation.

36. $1 + 2 + 3 + \cdots + 29$

37. $5 + 10 + 15 + \cdots + 50$

38. $1^2 + 2^2 + 3^2 + \cdots + 11^2$

39. $\frac{4}{5} + \frac{5}{6} + \frac{6}{7} + \cdots + \frac{12}{13}$

40. $2 + (-4) + 8 + (-16) + \cdots + (-256)$

41. $-\frac{1}{9} + \frac{1}{18} - \frac{1}{27} + \cdots + \frac{1}{54}$

42. $5 + \frac{5^2}{2} + \frac{5^3}{3} + \cdots + \frac{5^n}{n}$

43. $1 + 7 + \frac{7^2}{2!} + \frac{7^3}{3!} + \frac{7^4}{4!} + \cdots + \frac{7^n}{n!}$

44. The figure below shows the progression of the branching of a tree during each stage of development. Notice that the number of branches formed during a given stage is a Fibonacci number. Assuming that this branching pattern continues, how many branches will form during the 10th stage of development?

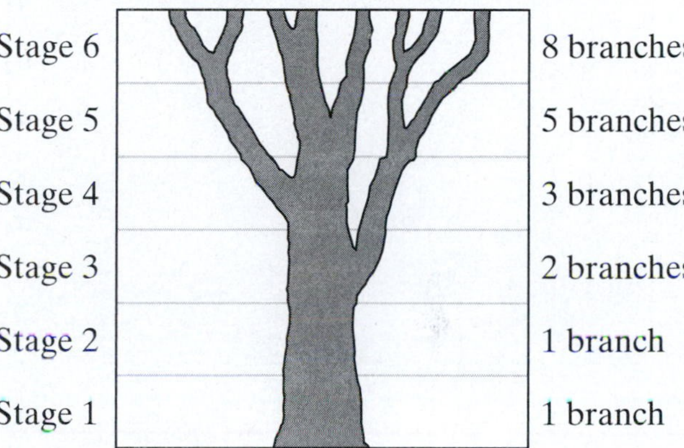

14.2 Arithmetic Sequences and Series

THINGS TO KNOW

Before working through this section, be sure that you are familiar with the following concepts:

VIDEO ANIMATION INTERACTIVE

You Try It

1. Solving a System of Linear Equations Using the Substitution Method (Section 12.1) 

You Try It

2. Solving a System of Linear Equations Using the Elimination Method (Section 12.1) 

You Try It

3. Determining the Sum of a Finite Series Written in Summation Notation (Section 14.1)

OBJECTIVES

1 Determining If a Sequence Is Arithmetic

2 Finding the General Term or a Specific Term of an Arithmetic Sequence

3 Computing the nth Partial Sum of an Arithmetic Series

4 Applications of Arithmetic Sequences and Series

SECTION 14.2 EXERCISES

(eText Screens 14.2-1–14.2-19)

OBJECTIVE 1 DETERMINING IF A SEQUENCE IS ARITHMETIC

In this section, we will work exclusively with a specific type of sequence known as an **arithmetic sequence**. A sequence is arithmetic if the difference in any two successive terms is constant. For example, the sequence

$$5, 9, 13, 17, \ldots$$

is arithmetic because the difference of any two successive terms is 4. The first term of this sequence is $a_1 = 5$ and the common difference is $d = 4$. Notice that we can rewrite the terms of this sequence as $5, 5 + 4, 5 + 2(4), 5 + 3(4), \ldots$

In general, given an arithmetic sequence with a first term of a_1 and a common difference of d, the first n terms of the sequence are as follows:

$$a_1$$

$$a_2 = a_1 + d$$

$$a_3 = a_2 + d = \underbrace{(a_1 + d)}_{a_2} + d = a_1 + 2d$$

$$a_4 = a_3 + d = \underbrace{(a_1 + 2d)}_{a_3} + d = a_1 + 3d$$

$$\vdots$$

$$a_n = a_1 + (n - 1)d$$

> **Definition** Arithmetic Sequence
>
> An **arithmetic sequence** is a sequence of the form $a_1, a_1 + d, a_1 + 2d, a_1 + 3d, a_1 + 4d, \ldots$, where a_1 is the first term of the sequence and d is the common difference. The general term, or nth term, of an arithmetic sequence has the form $a_n = a_1 + (n - 1)d$.

Example 1 Determining If a Sequence Is Arithmetic

For each of the following sequences, determine if it is arithmetic. If the sequence is arithmetic, find the common difference.

a. $1, 4, 7, 10, 13, \ldots$

b. $b_n = n^2 - n$

c. $a_n = -2n + 7$

d. $a_1 = 14, a_n = 3 + a_{n-1}$

Solution Watch this **interactive video** to verify that the sequences in parts a), c), and d) are arithmetic. The sequence in part b) is not arithmetic. Notice that the arithmetic sequence in part d) is a **recursive sequence**.

It is worth noting that every arithmetic sequence is a linear function whose domain is the natural numbers. A portion of the graphs of the arithmetic sequences from Example 1a and Example 1c are seen in **Figure 2**. Notice that the ordered pairs of each sequence are **collinear**.

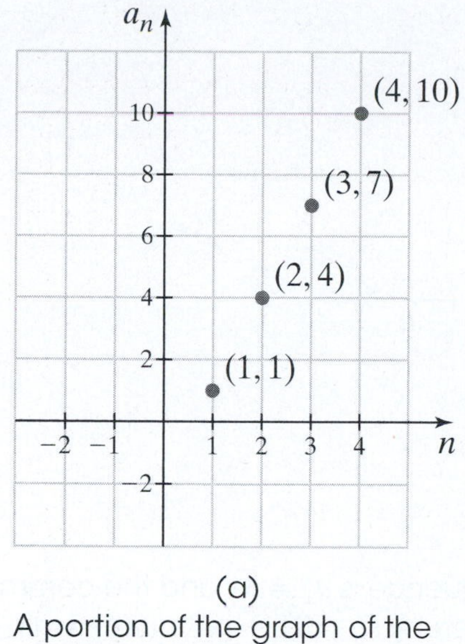

(a)
A portion of the graph of the sequence 1, 4, 7, 10, 13, . . .

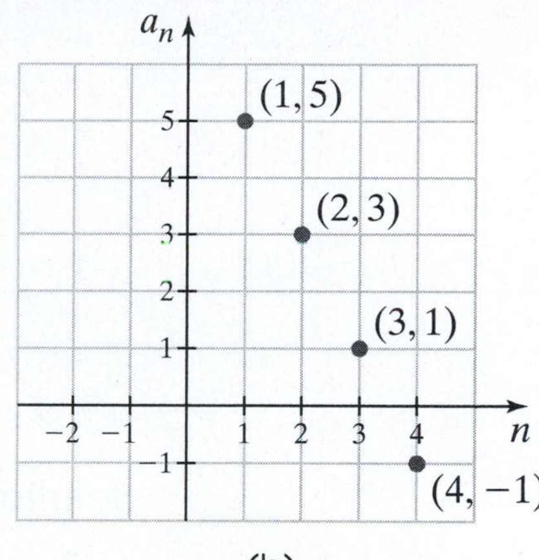

(b)
A portion of the graph of the sequence $a_n = -2n + 7$

Figure 2 The graph of every arithmetic sequence is represented by a set of ordered pairs that lies on a straight line.

TIP When the common difference of an arithmetic sequence is positive, the terms of the sequence *increase* and the graph is represented by a set of ordered pairs that lie along a line with positive slope. When the common difference of an arithmetic sequence is negative, the terms of the sequence *decrease* and the graph is represented by a set of ordered pairs that lies along a line with negative slope.

You Try It Work through this You Try It problem.

Work Exercises 1–6 in this textbook or in the MyLab Math Study Plan.

OBJECTIVE 2 FINDING THE GENERAL TERM OR A SPECIFIC TERM OF AN ARITHMETIC SEQUENCE

By the definition of an arithmetic sequence, the general term of an arithmetic sequence has the form $a_n = a_1 + (n - 1)d$. We can use this formula to find any term of an arithmetic sequence.

Example 2 Finding the General Term of an Arithmetic Sequence

Find the general term of each arithmetic sequence, then find the indicated term of the sequence. (In part c, only a portion of the graph is given. Assume that the domain of this sequence is all **natural numbers.**)

a. $11, 17, 23, 29, 35, \ldots; a_{50}$

b. $2, 0, -2, -4, -6, \ldots; a_{90}$

c. Find a_{31}.

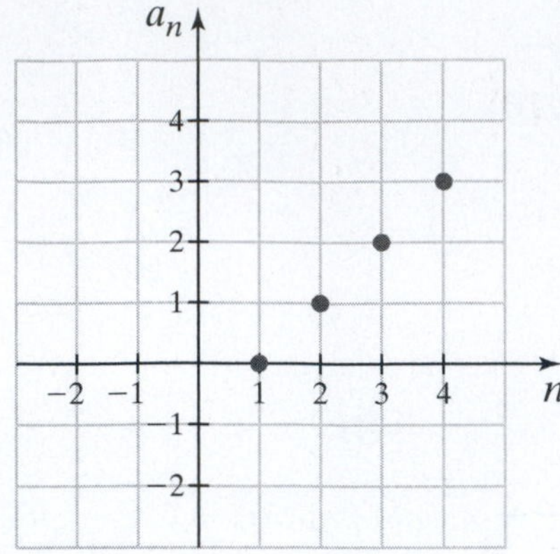

Solution

a. The first term of the sequence is $a_1 = 11$ and the common difference is $d = 6$. The general term is given by $a_n = 11 + (n - 1)(6) = 11 + 6n - 6 = 6n + 5$. Therefore, $a_{50} = 6(50) + 5 = 305$.

b. The first term of the sequence is $a_1 = 2$ and the common difference is $d = -2$. The general term is given by $a_n = 2 + (n - 1)(-2) = 2 - 2n + 2 = 4 - 2n$. Therefore, $a_{90} = 4 - 2(90) = -176$.

c. The first four terms of this sequence are $a_1 = 0, a_2 = 1, a_3 = 2$, and $a_4 = 3$. Thus, $a_1 = 0$ and the common difference is $d = 1$. The general term is given by $a_n = 0 + (n - 1)(1) = n - 1$. Thus, $a_{31} = 31 - 1 = 30$.

You may also watch this **interactive video** to see each of these solutions worked out in detail.

You Try It Work through this You Try It problem.

Work Exercises 7–12 in this textbook or in the MyLab Math Study Plan.

Example 3 Finding a Specific Term of an Arithmetic Sequence

a. Given an arithmetic sequence with $d = -4$ and $a_3 = 14$, find a_{50}.

b. Given an arithmetic sequence with $a_4 = 12$ and $a_{15} = -10$, find a_{41}.

Solution

a. We are given that $d = -4$ and $a_3 = 14$. We can use this information to solve for a_1.

$a_n = a_1 + (n - 1)d$	Use the formula for the general term of an arithmetic sequence.
$a_3 = a_1 + (3 - 1)(-4)$	Substitute $n = 3$ and $d = -4$.

Simplifying, we get $a_3 = a_1 - 8$. We can now substitute $a_3 = 14$ to solve for a_1.

$a_3 = a_1 - 8$	Start with the formula for a_3.
$14 = a_1 - 8$	Substitute $a_3 = 14$.
$22 = a_1$	Add 8 to both sides.

Using the formula $a_n = a_1 + (n-1)d$ with $a_1 = 22$ and $d = -4$, we can simplify to get $a_n = 26 - 4n$. Therefore, $a_{50} = 26 - 4(50) = -174$. Watch this **interactive video** to see every step of this solution.

b. Using the fact that $a_n = a_1 + (n-1)d$, we get $a_4 = a_1 + (4-1)d = 12$ and $a_{15} = a_1 + (15-1)d = -10$. This gives us the following system of linear equations:

$$a_1 + 3d = 12$$

$$a_1 + 14d = -10$$

Using the method of substitution or the method of elimination that were discussed in **Section 12.1** to solve this system, we get $a_1 = 18$ and $d = -2$. (Watch this **interactive video** to see how to solve this system using either method.) Using the formula $a_n = a_1 + (n-1)d$ with $a_1 = 18$ and $d = -2$, we can find the general term.

$a_n = a_1 + (n-1)d$	Use the formula for the general term of an arithmetic sequence.
$= 18 + (n-1)(-2)$	Substitute $a_1 = 18$ and $d = -2$.
$= 18 - 2n + 2$	Use the **distributive property**.
$= 20 - 2n$	Simplify.

The general term is $a_n = 20 - 2n$. Therefore, $a_{41} = 20 - 2(41) = -62$. Watch this **interactive video** to see this entire solution worked out in detail.

You Try It Work through this You Try It problem.

Work Exercises 13–18 in this textbook or in the MyLab Math Study Plan.

OBJECTIVE 3 COMPUTING THE nTH PARTIAL SUM OF AN ARITHMETIC SERIES

If $a_1, a_2, a_3, \ldots$ is an arithmetic sequence, then the expression $a_1 + a_2 + a_3 + \cdots + a_n + a_{n+1} + \cdots$ is called an **infinite arithmetic series** and can be written using summation notation as $\sum_{i=1}^{\infty} a_i$. Recall that the sum of the first n terms of a series is called the **nth partial sum** of the series and is given by $S_n = a_1 + a_2 + a_3 + \cdots + a_n$. We can also represent the nth partial sum using summation notation as $S_n = \sum_{i=1}^{n} a_i$. As you can see, the nth partial sum is simply the sum of a finite arithmetic series. Fortunately, there is a convenient formula for computing the nth partial sum of an arithmetic series.

> **Formula for the nth Partial Sum of an Arithmetic Series**
>
> The sum of the first n terms of an arithmetic series is called the **nth partial sum** of the series and is given by $S_n = \sum_{i=1}^{n} a_i = a_1 + a_2 + a_3 + \cdots + a_n$. This sum can be computed using the formula $S_n = \dfrac{n(a_1 + a_n)}{2}$.

(eText Screens 14.2-1–14.2-19)

Watch this **video** to see the derivation of this formula. (We will prove this formula again using mathematical induction in **Section 14.5**.)

The nth partial sum of an arithmetic series is simply the sum of a finite arithmetic series. An arithmetic series *must* be *finite* in order to compute the sum. This is not true for some other types of series. You will see how to find the sum of a special type of infinite series in Section 14.3.

Example 4 Finding the Sum of an Arithmetic Series

Find the sum of each arithmetic series.

a. $\sum_{i=1}^{20}(2i - 11)$

b. $-5 + (-1) + 3 + 7 + \cdots + 39$

Solution

a. We can use the formula $S_{20} = \dfrac{20(a_1 + a_{20})}{2}$ to compute the sum of the first 20 terms of this series.

$a_1 = 2(1) - 11 = -9$ Substitute $i = 1$ in the formula $2i - 11$ to find a_1.

$a_{20} = 2(20) - 11 = 29$ Substitute $i = 20$ in the formula $2i - 11$ to find a_{20}.

We now substitute $a_1 = -9$ and $a_{20} = 29$ into the formula $S_{20} = \dfrac{20(a_1 + a_{20})}{2}$.

$S_{20} = \dfrac{20(a_1 + a_{20})}{2}$ Use the formula for 20th partial sum of an arithmetic series.

$= \dfrac{20(-9 + 29)}{2}$ Substitute $a_1 = -9$ and $a_{20} = 29$.

$= 200$ Simplify.

Therefore, $\sum_{i=1}^{20}(2i - 11) = 200$. You may also watch this **interactive video** to see this solution worked out in detail.

b. Work through the **interactive video** to verify that the sum of this arithmetic series is 204.

You Try It Work through this You Try It problem.

Work Exercises 19–27 in this textbook or in the MyLab Math Study Plan.

OBJECTIVE 4 APPLICATIONS OF ARITHMETIC SEQUENCES AND SERIES

Example 5 Selling Newspaper Subscriptions

A local newspaper has hired teenagers to go door-to-door to try to solicit new subscribers. The teenagers receive \$2 for selling the first subscription. For each additional subscription sold, the newspaper will pay the teenagers 10 cents more than what was paid for the previous subscription. How much will the teenagers get paid for selling the 100th subscription? How much money will the teenagers earn by selling 100 subscriptions?

Solution The amount of money earned by selling one newspaper subscription can be represented by $a_1 = 2$. The money earned by selling the second subscription is $a_2 = 2.10$. The money earned by selling the third subscription is $a_3 = 2.20$. We see that the amount of money earned by selling n newspaper subscriptions is an arithmetic sequence with $a_1 = 2$ and $d = 0.10$. This sequence is defined by $a_n = 2 + (n - 1)(0.10) = 2 + (0.10)n - 0.10 = 0.10n + 1.90$.

The cash earned by selling the 100th subscription is the 100th term of this sequence, or $a_{100} = 0.10(100) + 1.90 = 11.90$. Therefore, the teenagers are paid \$11.90 for selling the 100th subscription.

To find the total amount earned by selling 100 subscriptions, we must find the sum of the series $\sum_{i=1}^{100}[(0.10)i + 1.90]$.

Using the formula $S_n = \dfrac{n(a_1 + a_n)}{2}$ with $n = 100$, $a_1 = 2$, and $a_{100} = 11.90$, we get

$$S_{100} = \frac{100(a_1 + a_{100})}{2} = \frac{100(2 + 11.90)}{2} = 50(13.90) = 695.$$

Thus, the teenagers will be paid \$695 if they sell 100 subscriptions.

Example 6 Seats in a Theater

A large multiplex movie house has many theaters. The smallest theater has only 12 rows. There are six seats in the first row. Each row has two seats more than the previous row. How many total seats are there in this theater?

Solution Try solving this word problem on your own. When you are done, watch this video to see if you are correct, then work through the following "You Try It" problem.

You Try It Work through this You Try It problem.

Work Exercises 28–35 in this textbook or in the MyLab Math Study Plan.

14.2 Exercises

In Exercises 1–6, determine if the sequence is arithmetic. If the sequence is arithmetic, find the common difference.

1. 8, 14, 20, 26, 32, . . .
2. 8, 11, 13, 16, 18, . . .
3. $a_n = \dfrac{3n + 1}{2}$
4. $a_n = n(n + 1)$
5. $a_1 = 8, a_n = 2 + a_{n-1}$
6. $a_1 = 5, a_n = 3a_{n-1} + 1$

In Exercises 7–12, find the general term of each arithmetic sequence then find the indicated term of the sequence. If the sequence is represented by a graph, assume that the domain of the sequence is all natural numbers.

7. $2, 7, 12, 17, \ldots;\ a_{10}$

8. $5, 1, -3, -7, \ldots;\ a_{31}$

9. $\frac{3}{2}, 3, \frac{9}{2}, 6, \frac{15}{2}, \ldots;\ a_{50}$

10. $5.0, 3.8, 2.6, 1.4, \ldots;\ a_{29}$

11. Find a_{17}.

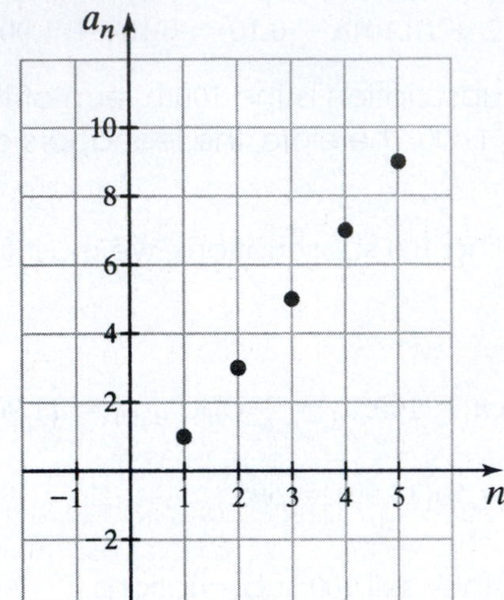

12. Find a_{11}.

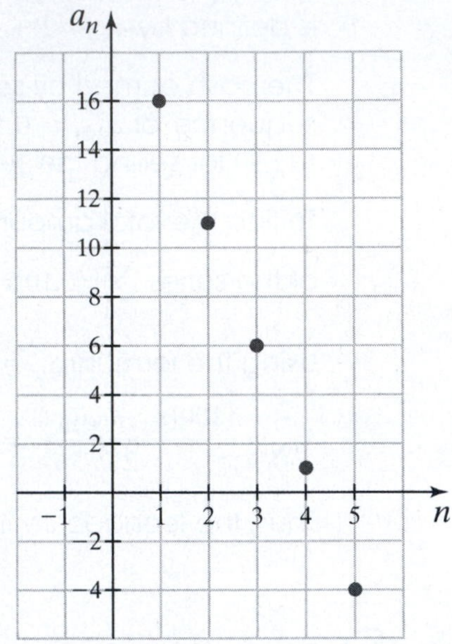

13. Given an arithmetic sequence with $d = 3$ and $a_8 = 5$, find a_{30}.

14. Given an arithmetic sequence with $d = -5$ and $a_7 = 11$, find a_{22}.

15. Given an arithmetic sequence with $a_5 = 4$ and $a_{22} = 55$, find a_{36}.

16. Given an arithmetic sequence with $a_6 = 4$ and $a_{20} = -52$, find a_{33}.

17. Given an arithmetic sequence with $a_{16} = 30$ and $a_{30} = 65$, find a_9.

18. Given an arithmetic sequence with $a_8 = -6$ and $a_{19} = -\frac{45}{2}$, find a_{34}.

In Exercises 19–27, find the indicated sum.

19. $\sum_{i=1}^{80} i$

20. $\sum_{j=1}^{10} (3j + 7)$

21. $7 + 10 + 13 + 16 + \cdots + 118$

22. $-14 + (-9) + (-4) + 1 + \cdots + 101$

23. $\sum_{i=3}^{14} (-7 - 9i)$

24. $1 + 11 + 21 + 31 + \cdots + a_{102}$

25. $6 + 14 + 22 + 30 + \cdots + (8n - 2)$

26. Find the sum of the first 100 odd integers.

27. Find the sum of the first 100 even positive integers.

28. A large multiplex movie house has many theaters. The largest theater has 40 rows. There are 12 seats in the first row. Each row has four seats more than the previous row. How many total seats are there in this theater?

29. A stack of logs has 47 logs on the bottom layer. Each subsequent layer has nine fewer logs than the previous layer. If the top layer has two logs, how many total logs are there in the pile?

30. A middle school mathematics teacher accepts a teaching position that pays \$31,000 per year. Each year, the expected raise is \$1,100. How much total money will this teacher earn teaching middle school mathematics over the first 12 years?

31. Suppose that you plan on taking a summer job selling magazine subscriptions. The magazine company will pay you \$1 for selling the first subscription. For each additional subscription sold, the magazine company will pay you 15 cents more than what was paid for the previous subscription. How much will you earn by selling 200 magazine subscriptions?

32. Two companies have offered you a job. Alpha Company has offered you \$35,000 per year with an annual raise of \$2,000. Beta Company has offered you a \$46,000 annual salary with an annual raise of \$800 per year. Which company will pay you more over the first 10 years?

33. A city fund-raiser raffle is raffling off 25 cash prizes. First prize is \$5,000. Each successive prize is \$200 less than the preceding prize. What is the value of the 25th prize? What is the total amount of cash given out by this raffle?

34. Larry's Luxury Rental Car Company rents luxury cars for up to 18 days. The price is \$300 for the first day, with the rental fee decreasing \$7 for each additional day. How much will it cost to rent a luxury car for 18 days?

35. A ball thrown straight up in the air travels 48 inches in the first tenth of a second. In the next tenth of a second, the ball travels 44 inches. After each additional tenth of a second, the ball travels 4 inches less than it did during the preceding tenth of a second. How long will it take before the ball starts coming back down? What is the total distance that the ball has traveled when it has reached its maximum height?

14.3 Geometric Sequences and Series

THINGS TO KNOW

Before working through this section, be sure that you are familiar with the following concepts:

VIDEO | ANIMATION | INTERACTIVE

You Try It 1. Using the Periodic Compound Interest Formula (Section 5.1)

You Try It 2. Solving a System of Linear Equations Using the Substitution Method (Section 12.1)

You Try It 3. Finding the General Term or a Specific Term of an Arithmetic Sequence (Section 14.2)

You Try It 4. Computing the nth Partial Sum of an Arithmetic Series (Section 14.2)

OBJECTIVES

OBJECTIVE 1 WRITING THE TERMS OF A GEOMETRIC SEQUENCE

Suppose that you have agreed to work for a successful business woman on a particular job for 21 days. She gives you two choices of payment. You can be paid \$100 for the first day and an additional \$50 per day for each subsequent day. Or, you can choose to be paid 1 penny for the first day with your pay doubling each subsequent day. Which method of payment would you choose? (We will revisit this question later in this section. See **Example 7**.) Notice that each payment method can be represented by a sequence:

Payment Method 1: $100, 150, 200, 250, 300, \ldots$

Payment Method 2: $0.01, 0.02, 0.04, 0.08, 0.16, \ldots$

The first method of payment is an **arithmetic sequence** with $a_1 = 100$ and $d = 50$. The second method of payment is an example of a **geometric sequence**. Each term of this geometric sequence can be obtained by multiplying the previous term by 2. The number 2 in this case is called the **common ratio**. We can obtain this common ratio by dividing any term of the sequence (except the first term) by the previous term. That is, $r = \frac{a_2}{a_1} = \frac{a_3}{a_2} = \cdots = \frac{a_{n+1}}{a_n}$. The first term of the payment method 2 sequence is $a_1 = 0.01$ and the common ratio is $r = 2$.

Notice that we can rewrite the terms of this sequence as 0.01, $(0.01)(2)$, $(0.01)(2^2)$, $(0.01)(2^3)$, ... In general, given a geometric sequence with a first term of a_1 and a common ratio of r, the first n terms of the sequence are

$$a_1$$

$$a_2 = a_1 r$$

$$a_3 = a_2 r = \underbrace{(a_1 r^2)}_{a_2} r = a_1 r^2$$

$$a_4 = a_3 r = \underbrace{(a_1 r^2)}_{a_3} r = a_1 r^3$$

$$\vdots$$

$$a_n = a_1 r^{n-1}$$

Definition Geometric Sequence

A **geometric sequence** is a sequence of the form $a_1, a_1r, a_1r^2, a_1r^3, a_1r^4, \ldots,$ where a_1 is the first term of the sequence and r is the common ratio such that $r = \frac{a_2}{a_1} = \frac{a_3}{a_2} = \cdots = \frac{a_{n+1}}{a_n}$ for all $n \geq 1$. The general term, or nth term, of a geometric sequence has the form $a_n = a_1r^{n-1}$.

A portion of the two sequences representing the two payment methods are sketched in Figure 3. The sequence representing payment method 1 (Figure 3a) is arithmetic. The ordered pairs of this sequence are **collinear**. The sequence representing payment method 2 (Figure 3b) is geometric. Notice that the ordered pairs of this sequence do not lie along a common line but rather lie on an **exponential curve**.

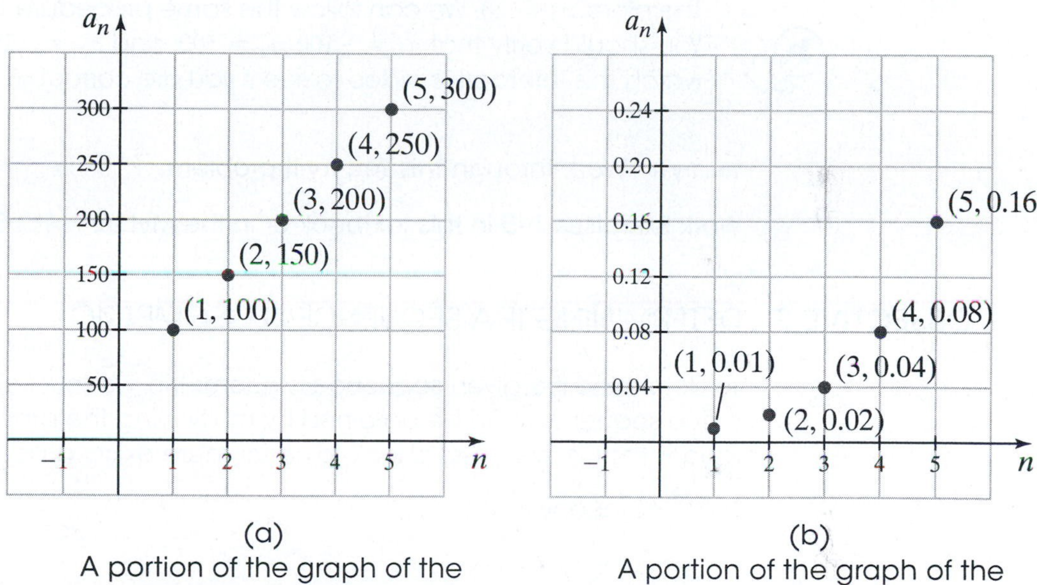

(a)
A portion of the graph of the sequence 100, 150, 200, 250, 300, . . .

(b)
A portion of the graph of the sequence 0.01, 0.02, 0.04, 0.08, 0.16, . . .

Figure 3 The graph of every arithmetic sequence is represented by a set of ordered pairs that lies on a straight line. The graph of a geometric sequence with $r > 0$ is represented by a set of ordered pairs that lie on an exponential curve.

If the first term and the common ratio of a geometric sequence are known, then we can determine the second term by multiplying the first term by the common ratio. The third term can be found by multiplying the second term by the common ratio. We continue this process to find the subsequent terms of the sequence.

Example 1 Writing a Geometric Sequence

a. Write the first five terms of the geometric sequence having a first term of 2 and a common ratio of 3.

b. Write the first five terms of the geometric sequence such that $a_1 = -4$ and $a_n = -5a_{n-1}$ for $n \geq 2$.

Solution

a. We are given that the first term is $a_1 = 2$ and the common ratio is $r = 3$. The second term is $a_2 = 2 \cdot 3 = 6$. The third term is $a_3 = 6 \cdot 3 = 18$. The fourth term is $a_4 = 18 \cdot 3 = 54$. The fifth term is $a_5 = 54 \cdot 3 = 162$. Therefore, the first five terms of this sequence are 2, 6, 18, 54, and 162.

b. This sequence is defined **recursively**. The first term is $a_1 = -4$. To find a_2, substitute the value of 2 for n in the formula $a_n = -5a_{n-1}$ and simplify:

$$\begin{aligned} a_n &= -5a_{n-1} && \text{Given recursive formula} \\ a_2 &= -5a_{2-1} && \text{Substitute.} \\ &= -5a_1 && \text{Simplify.} \\ &= -5(-4) && \text{Substitute } a_1 = -4. \\ &= 20 && \text{Multiply.} \end{aligned}$$

Therefore, $a_2 = 20$. We can follow this same procedure to find a_3, a_4, and a_5. You should verify that $a_3 = -100$, $a_4 = 500$, and $a_5 = -2500$ on your own, then watch this **interactive video** to see if you are correct.

You Try It Work through this You Try It problem.

Work Exercises 1–5 in this textbook or in the MyLab Math Study Plan.

OBJECTIVE 2 DETERMINING IF A SEQUENCE IS GEOMETRIC

To determine if a given sequence is geometric, we must check to see if each term of the sequence can be obtained by multiplying the previous term by a common ratio r. That is, we must check to see if there exists a constant value r such that $r = \dfrac{a_{n+1}}{a_n}$ for any $n \geq 1$.

Example 2 Determining If a Sequence Is Geometric

For each of the following sequences, determine if it is geometric. If the sequence is geometric, find the common ratio.

a. $2, 4, 6, 8, 10, \ldots$

b. $\dfrac{2}{3}, \dfrac{4}{9}, \dfrac{8}{27}, \dfrac{16}{81}, \dfrac{32}{243}, \ldots$

c. $12, -6, 3, -\dfrac{3}{2}, \dfrac{3}{4}, \ldots$

Solution

a. For this sequence, $\dfrac{a_2}{a_1} = \dfrac{4}{2} = 2$ and $\dfrac{a_3}{a_2} = \dfrac{6}{4} = \dfrac{3}{2}$. Since $\dfrac{a_2}{a_1} \neq \dfrac{a_3}{a_2}$, there does not exist a common ratio. Hence, this sequence is not geometric. (Note that this sequence is an **arithmetic sequence**.)

b. For this sequence, $\dfrac{a_2}{a_1} = \dfrac{\frac{4}{9}}{\frac{2}{3}} = \dfrac{4}{9} \cdot \dfrac{3}{2} = \dfrac{2}{3}$. Note that each term of this sequence (other than the first term) can be obtained by multiplying the previous term by $\dfrac{2}{3}$. Therefore, this sequence is geometric with a common ratio of $\dfrac{2}{3}$.

c. Try to determine if this sequence is geometric. Work through the interactive video to see if you are correct.

You Try It Work through this You Try It problem.

Work Exercises 6–10 in this textbook or in the MyLab Math Study Plan.

OBJECTIVE 3 FINDING THE GENERAL TERM OR A SPECIFIC TERM OF A GEOMETRIC SEQUENCE

By the definition of a geometric sequence, the general term, or nth term, of a geometric sequence has the form $a_n = a_1 r^{n-1}$. We can use this formula to find any term of a given geometric sequence.

Example 3 Finding the General Term of a Geometric Sequence

Find the general term of each geometric sequence.

a. $12, -6, 3, -\frac{3}{2}, \frac{3}{4}, \ldots$

b. $\frac{2}{3}, \frac{2}{9}, \frac{2}{27}, \frac{2}{81}, \frac{2}{243}, \ldots$

Solution

a. The first term of the sequence is $a_1 = 12$ and the common ratio is $r = \frac{a_2}{a_1} = \frac{-6}{12} = -\frac{1}{2}$. Therefore, $a_n = 12\left(-\frac{1}{2}\right)^{n-1}$.

b. The first term of the sequence is $a_1 = \frac{2}{3}$ and the common ratio is

$$r = \frac{a_2}{a_1} = \frac{\frac{2}{9}}{\frac{2}{3}} = \frac{1}{3}.\text{ Therefore, } a_n = \left(\frac{2}{3}\right)\left(\frac{1}{3}\right)^{n-1}.$$

Example 4 Finding a Specific Term of a Geometric Sequence

a. Find the 7th term of the geometric sequence whose first term is 2 and whose common ratio is -3.

b. Given a geometric sequence such that $a_6 = 16$ and $a_9 = 2$, find a_{13}.

Solution

a. We can use the formula $a_n = a_1 r^{n-1}$ with $a_1 = 2$ and $r = -3$ to find the general term. The general term is $a_n = 2(-3)^{n-1}$. Therefore, $a_7 = 2(-3)^{7-1} = 2(-3)^6 = 2(729) = 1{,}458$.

b. Since $a_6 = 16$, we can substitute $n = 6$ into the formula $a_n = a_1 r^{n-1}$ to get $a_6 = a_1 r^5 = 16$. Similarly, we can substitute $n = 9$ into the formula $a_n = a_1 r^{n-1}$ to get $a_9 = a_1 r^8 = 2$. This gives the following two equations.

$$(1)\quad a_1 r^5 = 16$$

$$(2)\quad a_1 r^8 = 2$$

Divide both sides of equation (1) by r^5 to get $a_1 = \frac{16}{r^5}$. Now, substitute $a_1 = \frac{16}{r^5}$ into equation (2) and solve for a_1.

$$a_1r^8 = 2 \quad \text{Start with equation (2).}$$

$$\left(\frac{16}{r^5}\right)r^8 = 2 \quad \text{Substitute } a_1 = \frac{16}{r^5}.$$

$$16r^3 = 2 \quad \frac{r^8}{r^5} = r^3$$

$$r^3 = \frac{1}{8} \quad \text{Divide both sides by 16.}$$

$$r = \frac{1}{2} \quad \text{Take the cube root of both sides.}$$

Now substitute $r = \frac{1}{2}$ into equation (1) to solve for a_1.

$$a_1r^5 = 16 \quad \text{Start with equation (1).}$$

$$a_1\left(\frac{1}{2}\right)^5 = 16 \quad \text{Substitute } r = \frac{1}{2}.$$

$$a_1\left(\frac{1}{32}\right) = 16 \quad \left(\frac{1}{2}\right)^5 = \frac{1}{32}$$

$$a_1 = 512 \quad \text{Multiply both sides by 32.}$$

We can now use the formula $a_n = a_1r^{n-1}$ with $a_1 = 512$ and $r = \frac{1}{2}$ to find the general term. The general term is $a_n = 512\left(\frac{1}{2}\right)^{n-1}$. Therefore, $a_{13} = 512\left(\frac{1}{2}\right)^{13-1} = 512\left(\frac{1}{2}\right)^{12} = \frac{512}{2^{12}} = \frac{512}{4{,}096} = \frac{1}{8}$.

You may wish to work through this **interactive video** to see these solutions worked out in detail.

You Try It **Work through this You Try It problem.**

Work Exercises 11–18 in this textbook or in the MyLab Math Study Plan.

OBJECTIVE 4 COMPUTING THE nTH PARTIAL SUM OF A GEOMETRIC SERIES

If $a_1, a_1r, a_1r^2, a_1r^3, \ldots$ is a geometric sequence, then the expression $a_1 + a_1r + a_1r^2 + a_1r^3 + \cdots + a_1r^{n-1} + \cdots$ is called an **infinite geometric series** and can be written in summation notation as $\sum_{i=1}^{\infty} a_1r^{i-1}$. Recall that the sum of the first n terms of a series is called the **nth partial sum** of the series and is given by $S_n = a_1 + a_1r + a_1r^2 + a_1r^3 + \cdots + a_1r^{n-1}$. We can also represent the nth partial sum using summation notation as $S_n = \sum_{i=1}^{n} a_1r^{i-1}$. As you can see, this nth partial sum is simply the sum of a finite geometric series. Fortunately, there is a convenient formula for computing the nth partial sum of a geometric series.

Formula for the nth Partial Sum of a Geometric Series

The sum of the first n terms of a geometric series is called the **nth partial sum** of the series and is given by $S_n = \sum_{i=1}^{n} a_1 r^{i-1} = a_1 + a_1 r + a_1 r^2 + a_1 r^3 + \cdots + a_1 r^{n-1}$.

This sum can be computed using the formula $S_n = \dfrac{a_1(1 - r^n)}{1 - r}$ for $r \neq 1$.

Watch this **video** to see the derivation of this formula. (We will also prove this formula using mathematical induction in **Section 14.5**.)

Example 5 Computing the nth Partial Sum of a Geometric Series

a. Find the sum of the series $\sum_{i=1}^{15} 5(-2)^{i-1}$.

b. Find the 7th partial sum of the geometric series $8 + 6 + \dfrac{9}{2} + \dfrac{27}{8} + \cdots$.

Solution

a. Using the formula $S_n = \dfrac{a_1(1 - r^n)}{1 - r}$ with $n = 15, a_1 = 5$, and $r = -2$, we get

$$S_{15} = \frac{5(1 - (-2)^{15})}{1 - (-2)} = \frac{5(1 - (-32{,}768))}{3} = \frac{5(32{,}769)}{3} = \frac{163{,}845}{3} = 54{,}615.$$

You may also watch this **interactive video** to see this solution worked out in detail.

b. Work through the **interactive video** to verify that the 7th partial sum of this series is $S_7 = \dfrac{14{,}197}{512}$. ●

You Try It Work through this You Try It problem.

Work Exercises 19–23 in this textbook or in the MyLab Math Study Plan.

OBJECTIVE 5 DETERMINING IF AN INFINITE GEOMETRIC SERIES CONVERGES OR DIVERGES

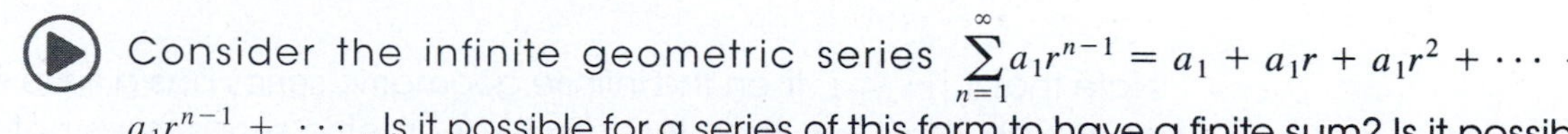

Consider the infinite geometric series $\sum_{n=1}^{\infty} a_1 r^{n-1} = a_1 + a_1 r + a_1 r^2 + \cdots + a_1 r^{n-1} + \cdots$. Is it possible for a series of this form to have a finite sum? Is it possible to add infinitely many terms and get a finite sum? The answer is **yes**, it is possible, but it depends on the value of r. Before we determine the value(s) of r for which an infinite geometric series has a finite sum, let's look at an example. Consider the following geometric series

$$\frac{1}{2} + \frac{1}{3} + \frac{2}{9} + \frac{4}{27} + \frac{8}{81} + \cdots$$

Note that $a_1 = \frac{1}{2}$ and $r = \frac{2}{3}$. We can use the formula $S_n = \frac{a_1(1 - r^n)}{1 - r}$ to find the nth partial sum for any value of n of our choosing. The nth partial sums for $n = 5, 10, 20,$ and 40 are given in **Table 1** as well as the value of r^n.

Table 1

n	$S_n = \frac{a_1(1 - r^n)}{1 - r} = \frac{\left(\frac{1}{2}\right)\left(1 - \left(\frac{2}{3}\right)^n\right)}{1 - \frac{2}{3}}$	$r^n = \left(\frac{2}{3}\right)^n$
5	1.3024691	0.1316872
10	1.4739877	0.0173415
20	1.4995489	0.0003007
40	1.4999999	0.0000001

Looking at Table 1, it appears that as n increases, the value of S_n is getting closer to $1.5 = \frac{3}{2}$. Also notice that as n increases, the value of $r^n = \left(\frac{2}{3}\right)^n$ is getting closer to 0. In fact, for any value of r between -1 and 1, the value of r^n will always approach 0 as n approaches infinity. We say, "For values of r between -1 and 1, r^n approaches zero as n approaches infinity" and write: For $|r| < 1$, $r^n \to 0$ as $n \to \infty$. Thus, if $|r| < 1$, $S_n = \frac{a_1(1 - r^n)}{1 - r} \approx \frac{a_1(1 - 0)}{1 - r} = \frac{a_1}{1 - r}$ for large values of n. Therefore, given an infinite geometric series with $|r| < 1$, the sum of the series is given by $S = \frac{a_1}{1 - r}$.

Note A formal proof of this formula requires calculus.

Formula for the Sum of an Infinite Geometric Series

Let $\sum_{n=1}^{\infty} a_1 r^{n-1} = a_1 + a_1 r + a_1 r^2 + a_1 r^3 + \cdots + a_1 r^{n-1} + \cdots$ be an infinite geometric series. If $|r| < 1$, then the sum of the series is given by $S = \frac{a_1}{1 - r}$.

Note that if $|r| < 1$, then the infinite geometric series has a finite sum and is said to **converge**. If $|r| \geq 1$, then the infinite geometric series does not have a finite sum and the series is said to **diverge**.

Example 6 Determining If an Infinite Geometric Series Converges or Diverges

Determine whether each of the following series converges or diverges. If the series converges, find the sum.

a. $\sum_{n=1}^{\infty} \frac{1}{2}\left(\frac{2}{3}\right)^{n-1}$ b. $3 - \frac{6}{5} + \frac{12}{25} - \frac{24}{125} + \cdots$

c. $12 + 18 + 27 + \frac{81}{2} + \frac{243}{4} + \cdots$

Solution

a. This is an infinite geometric series with $|r| = \left|\frac{2}{3}\right| < 1$. Since $|r| < 1$, the infinite series must converge and, thus, must have a finite sum. The sum of the series is

$$S = \frac{a_1}{1-r} = \frac{\frac{1}{2}}{1 - \frac{2}{3}} = \frac{\frac{1}{2}}{\frac{1}{3}} = \frac{1}{2} \cdot \frac{3}{1} = \frac{3}{2}.$$

b. For this infinite geometric series, the common ratio is $r = \frac{-\frac{6}{5}}{3} = -\frac{6}{5} \cdot \frac{1}{3} = -\frac{2}{5}$.

Since $|r| = \left|-\frac{2}{5}\right| = \frac{2}{5} < 1$, the infinite series converges. The sum is

$$S = \frac{a_1}{1-r} = \frac{3}{1 - \left(-\frac{2}{5}\right)} = \frac{3}{1 + \frac{2}{5}} = \frac{3}{\frac{7}{5}} = 3 \cdot \frac{5}{7} = \frac{15}{7}.$$

c. This infinite series diverges. Do you know why? Work through the **interactive video** to see why this series diverges.

You Try It Work through this You Try It problem.

Work Exercises 24–28 in this textbook or in the MyLab Math Study Plan.

OBJECTIVE 6 APPLICATIONS OF GEOMETRIC SEQUENCES AND SERIES

In Example 7, we revisit the question that was presented at the beginning of this section.

Example 7 Choosing a Payment Method

Suppose that you have agreed to work for a successful business woman on a particular job for 21 days. She gives you two choices of payment. You can be paid \$100 for the first day and an additional \$50 per day for each subsequent day. Or, you can choose to be paid 1 penny for the first day with your pay doubling each subsequent day. Which method of payment yields the most income?

Solution Each payment method can be represented by a sequence.

Payment method 1: 100, 150, 200, 250, . . .

Payment method 2: 0.01, 0.02, 0.04, 0.08, . . .

To find out which method of payment will yield the greatest income, we must find the sum of the first 21 terms of each sequence.

Payment method 1 is an **arithmetic sequence** with $a_1 = 100$ and $d = 50$. Using the **formula for the general term of an arithmetic sequence**, we get $a_n = 100 + (n - 1)50 = 100 + 50n - 50 = 50n + 50$. Note that $a_{21} = 50(21) + 50 = 1{,}100$. Using the **formula for the nth partial sum of an arithmetic series** with $n = 21$, we get $S_{21} = \frac{n(a_1 + a_{21})}{2} = \frac{21(100 + 1{,}100)}{2} = \$12{,}600$.

Payment method 2 is a geometric sequence with $a_1 = 0.01$ and $r = 2$. Using the **formula for the nth partial sum of a geometric series** with $n = 21$, we get $S_{21} = \frac{a_1(1 - r^{21})}{1 - r} = \frac{(0.01)(1 - 2^{21})}{1 - 2} \approx \$20{,}971.51$. Clearly, payment method 2 is the better choice. ●

Example 8 Total Amount Given to a Local Charity

A local charity received \$8,500 in charitable contributions during the month of January. Because of a struggling economy, it is projected that contributions will decline each month to 95% of the previous month's contributions. What are the expected contributions for the month of October? What is the total expected contributions that this charity can expect at the end of the year?

Solution The monthly contributions can be represented by a geometric sequence with $a_1 = 8{,}500$ and $r = 0.95$. Thus, the contributions for the nth month is given by $a_n = 8{,}500(0.95)^{n-1}$. The expected contributions for October, or when $n = 10$, are $a_{10} = 8{,}500(0.95)^{10-1} \approx 5{,}357.12$. Thus, the contributions for the month of October are expected to be about \$5,357.12.

The total contributions for the year can be written as the following finite geometric series:

$$8{,}500 + (8{,}500)(0.95) + (8{,}500)(0.95)^2 + \cdots + (8{,}500)(0.95)^{11} = \sum_{i=1}^{12} 8{,}500(0.95)^{i-1}$$

Using the **formula for the nth partial sum of a geometric series** with $n = 12$, we get $S_{12} = \frac{a_1(1 - r^{12})}{1 - r} = \frac{8{,}500(1 - 0.95^{12})}{1 - 0.95} \approx 78{,}138.79$.

Therefore, the charity can expect about \$78,138.79 in donations for the year. ●

You Try It Work through this You Try It problem.

Work Exercises 29–33 in this textbook or in the MyLab Math Study Plan.

Example 9 Expressing a Repeating Decimal as a Ratio of Two Integers

Every repeating decimal number is a **rational number** and can therefore be represented by the quotient of two integers. Write each of the following repeating decimal numbers as the quotient of two integers.

a. $0.\overline{4}$ **b.** $0.2\overline{13}$

Solution

a. We can rewrite $0.\overline{4}$ as $0.44444\ldots = \frac{4}{10} + \frac{4}{100} + \frac{4}{1{,}000} + \frac{4}{10{,}000} + \frac{4}{100{,}000} + \cdots$.

This is an infinite geometric series with $a_1 = \frac{4}{10}$ and $r = \frac{1}{10}$. Because $|r| = \left|\frac{1}{10}\right| < 1$, we know that the series converges. Using the formula $S = \frac{a_1}{1 - r}$

we see that $0.\overline{4} = \dfrac{\frac{4}{10}}{1 - \frac{1}{10}} = \dfrac{\frac{4}{10}}{\frac{9}{10}} = \dfrac{4}{9}$.

b. Carefully work through the **interactive video** to see that $0.2\overline{13} = \frac{211}{990}$.

You Try It **Work through this You Try It problem.**

Work Exercises 34–35 in this textbook or in the MyLab Math Study Plan.

ANNUITIES

In **Section 5.1** we derived a formula for **periodic compound interest**. This formula is used to determine the future value of a *one-time* investment. (Click on the You Try It icon to see a periodic compound interest practice exercise.) Suppose that instead of investing one lump sum, you wish to invest equal amounts of money at steady intervals. An investment of equal amounts deposited at equal time intervals is called an **annuity**. If these equal deposits are made at the end of a compound period, the annuity is called an **ordinary annuity**.

For example, suppose that you want to invest $\$P$ at the end of each payment period at an annual rate r, in decimal form. Then the interest rate per payment period is $i = \frac{r}{\text{number of payment periods per year}}$. We now summarize the total amount of the ordinary annuity after the first k payment periods.

End of 1st payment period: $\underbrace{P}_{\text{1st payment}}$

End of 2nd payment period: $\underbrace{P}_{\text{1st payment}} + \underbrace{Pi}_{\text{Interest earned on 1st payment}} + \underbrace{P}_{\text{2nd payment}} = \underbrace{P(1 + i)}_{\text{Total amount of 1st payment}} + \underbrace{P}_{\text{2nd payment}}$

End of 3rd payment period:

$$\underbrace{P(1 + i)}_{\text{Amount of 1st payment}} + \underbrace{P(1 + i)i}_{\text{Interest earned on amount of 1st payment}} + \underbrace{P}_{\text{2nd payment}} + \underbrace{Pi}_{\text{Interest earned on 2nd payment}} + \underbrace{P}_{\text{3rd payment}} = \underbrace{P(1 + i)^2}_{\text{Total amount of 1st payment}} + \underbrace{P(1 + i)}_{\text{Total amount of 2nd payment}} + \underbrace{P}_{\text{3rd payment}}$$

End of kth payment period: $\underbrace{P(1+i)^{k-1}}_{\text{Total amount of 1st payment}} + \underbrace{P(1+i)^{k-2}}_{\text{Total amount of 2nd payment}} + \cdots + \underbrace{P(1+i)}_{\text{Total amount of }(k-1)\text{st payment}} + \underbrace{P}_{k\text{th payment}}$

The total amount of the ordinary annuity after k payment periods is $A = P + P(1 + i) \cdots + P(1 + i)^{k-2} + P(1 + i)^{k-1}$.

This is a finite geometric series with $a_1 = P$ and a common ratio of $(1 + i)$. Thus, the amount of the annuity after the kth payment is

$$A = \frac{P(1 - (1 + i)^k)}{1 - (1 + i)} = \frac{P(1 - (1 + i)^k)}{-i} = \frac{P((1 + i)^k - 1)}{i}.$$

Amount of an Ordinary Annuity after the kth Payment

The total amount of an ordinary annuity after the kth payment is given by the formula

$$A = \frac{P((1 + i)^k - 1)}{i}$$

where

A = Total amount of annuity after k payments

P = Deposit amount at the end of each payment period

i = Interest rate per payment period

Example 10 Finding the Amount of an Ordinary Annuity

Chie and Ben decided to save for their newborn son Jack's college education. They decided to invest \$200 every 3 months in an investment earning 8% interest compounded quarterly. How much is this investment worth after 18 years?

Solution This is an ordinary annuity with $P = \$200$ and $i = \frac{0.08}{4} = 0.02$. What is k?

See if you can determine k and use the formula $A = \frac{P((1 + i)^k - 1)}{i}$ to determine the total amount of this annuity. When you are done, watch this **video** to see if you are correct.

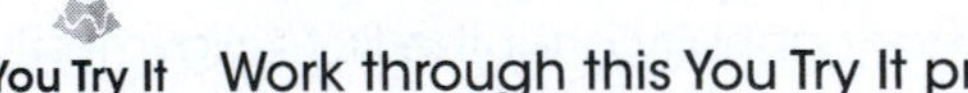

You Try It Work through this You Try It problem.

Work Exercises 36–38 in this textbook or in the MyLab Math Study Plan.

14.3 Exercises

In Exercises 1–5, write the first five terms of the geometric sequence with the given information.

1. The first term is 8 and the common ratio is 2.

2. The first term is 162 and the common ratio is $\frac{1}{3}$.

3. The first term is 25 and the common ratio is $-\frac{1}{5}$.

4. $a_n = 7a_{n-1}; a_1 = 3$

5. $a_n = -2a_{n-1}; a_1 = -4$

In Exercises 6–10, determine if the sequence is geometric. If the sequence is geometric, find the common ratio.

6. $4, 24, 144, 864, \ldots$

7. $-2, 2, -2, 2, \ldots$

8. $-3, 1, -1, -3, \ldots$

9. $2, -\frac{10}{3}, \frac{50}{9}, -\frac{250}{27}, \ldots$

10. $7.236, -5.7888, 4.63104, -3.704832, \ldots$

11. Determine the general term of the sequence $3, 6, 12, 24, \ldots$.

12. Determine the general term of the sequence $\frac{1}{2}, \frac{1}{8}, \frac{1}{32}, \frac{1}{128}, \ldots$.

13. Determine the general term of the sequence $\frac{1}{5}, -\frac{2}{15}, \frac{4}{45}, -\frac{8}{135}, \ldots$.

14. Find the 7th term of the geometric sequence whose first term is 5 and whose common ratio is 4.

15. Find the 6th term of the geometric sequence whose first term is 3,804 and whose common ratio is $-\frac{1}{4}$.

16. Find the 11th term of the geometric sequence $\$5,000, \$5,050, \$5,100.50, \ldots$.

17. Given a geometric sequence such that $a_4 = 108$ and $a_7 = 2916$, find a_{10}.

18. Given a geometric sequence such that $a_3 = 16$ and $a_8 = -\frac{1}{2}$, find a_{11}.

In Exercises 19–23, find the sum of each geometric series.

19. $\sum_{i=1}^{8} 7(-4)^{i-1}$

20. $\sum_{i=1}^{13} 2\left(\frac{3}{5}\right)^{i-1}$

21. $\sum_{i=1}^{10} 4(1.05)^{i-1}$

22. $2 + \frac{2}{3} + \frac{2}{9} + \frac{2}{27} + \cdots + \frac{2}{729}$

23. $1 - \frac{1}{2} + \frac{1}{4} - \frac{1}{8} + \cdots - \frac{1}{128}$

In Exercises 24–28, determine if each infinite geometric series converges or diverges. If the series converges, find the sum.

24. $-1 + \frac{1}{10} - \frac{1}{100} + \frac{1}{1000} - \cdots$

25. $343 + 49 + 7 + 1 + \cdots$

26. $\sum_{i=1}^{\infty} 7\left(\frac{1}{4}\right)^{i-1}$

27. $\sum_{i=1}^{\infty} \frac{1}{5}(3)^{i-1}$

28. $0.5 - 0.05 + 0.005 - 0.0005 + 0.00005 - \cdots$

29. Warren wanted to save money to purchase a new car. He started by saving $1 on the first of January. On the first of February, he saved $3. On the first of March he saved $9. So, on the first day of each month, he wanted to save three times as much as he did on the first day of the previous month. If Warren continues his savings pattern, how much will he need to save on the first day of September?

30. Suppose that you have accepted a job for 2 weeks that will pay $.07 for the first day, $.14 for the second day, $.28 for the third day, and so on. What will your total earnings be after 2 weeks?

31. Mary has accepted a teaching job that pays $25,000 for the first year. According to the Teacher's Union, Mary will get guaranteed salary increases of 4% per year. If Mary plans to teach for 30 years, what will be her total salary earnings?

32. A child is given an initial push on a rope swing. On the first swing, the rope swings through an arc of 12 feet. On each successive swing, the length of the arc is 80% of the previous length. After 10 swings, what is the total length the rope will have swung? When the child stops swinging, what is the total length the rope will have swung?

33. Randy dropped a rubber ball from his apartment window from a height of 50 feet. The ball always bounces $\frac{3}{5}$ of the distance fallen. How far does the ball travel once it is done bouncing?

34. Rewrite the number $0.\overline{7}$ as the quotient of two integers.

35. Rewrite the number $0.3\overline{25}$ as the quotient of two integers.

36. Kip contributes $200 every month to his 401(k). What will the value of Kip's 401(k) be in 10 years if the yearly rate of return is assumed to be 12% compounded monthly?

37. Mark and Lisa decide to invest $500 every 3 months in an education IRA to save for their son Beau's college education. What will the value of the IRA be after 10 years if the yearly assumed rate of return is 8% compounded quarterly?

38. Marv and Cindy decide to build a new home in 10 years. They will need $80,000 to purchase the lot of their dreams. How much should they save each month in an account that has an assumed yearly rate of return of 7% compounded monthly?

14.4 The Binomial Theorem

THINGS TO KNOW

Before working through this section, be sure that you are familiar with the following concepts:

VIDEO ANIMATION INTERACTIVE

You Try It 1. Multiplying Polynomials (Section R.4)

OBJECTIVES

1 Expanding Binomials Raised to a Power Using Pascal's Triangle

2 Evaluating Binomial Coefficients

3 Expanding Binomials Raised to a Power Using the Binomial Theorem

4 Finding a Particular Term or a Particular Coefficient of a Binomial Expansion

SECTION 14.4 EXERCISES

OBJECTIVE 1 EXPANDING BINOMIALS RAISED TO A POWER USING PASCAL'S TRIANGLE

In this section, we will focus on expanding algebraic expressions of the form $(a + b)^n$, where n is an integer greater than or equal to zero. Because $(a + b)$ is a binomial, we call the expansion of $(a + b)^n$ a *binomial expansion*. Consider the expansion of $(a + b)^4$.

$$(a + b)^4 = \underbrace{(a + b)(a + b)} \cdot \underbrace{(a + b)(a + b)}$$
$$= (a^2 + 2ab + b^2)(a^2 + 2ab + b^2)$$
$$= a^4 + 2a^3b + a^2b^2 + 2a^3b + 4a^2b^2 + 2ab^3 + a^2b^2 + 2ab^3 + b^4$$
$$= a^4 + 4a^3b + 6a^2b^2 + 4ab^3 + b^4$$

Although the expansion of $(a + b)^4$ using the method above is not too complicated, it would not be desirable to use this method to expand $(a + b)^n$ for large values of n.

The goal in this section is to try to develop a method for expanding expressions of the form $(a + b)^n$ without actually performing all of the multiplication. We start by studying the expanded forms of $(a + b)^n$ for $n = 0, 1, 2, 3, 4,$ and 5.

$$n = 0: (a + b)^0 = 1$$
$$n = 1: (a + b)^1 = 1a + 1b$$
$$n = 2: (a + b)^2 = 1a^2 + 2ab + 1b^2$$
$$n = 3: (a + b)^3 = 1a^3 + 3a^2b + 3ab^2 + 1b^3$$
$$n = 4: (a + b)^4 = 1a^4 + 4a^3b + 6a^2b^2 + 4ab^3 + 1b^4$$
$$n = 5: (a + b)^5 = 1a^5 + 5a^4b + 10a^3b^2 + 10a^2b^3 + 5ab^4 + 1b^5$$

The coefficients of each expansion are highlighted in red. These coefficients are known as **binomial coefficients**. Before we determine a pattern for these coefficients, let's first observe the exponent pattern. Notice in each expansion of

$(a + b)^n$, there are always $n + 1$ terms. The sum of the exponents of each term is always equal to n. Also note that the first term is always a^n (or $a^n b^0$) and the last term is always b^n (or $a^0 b^n$). As we look at the terms of each expansion from left to right, the exponent of the first variable decreases by 1 and the exponent of the second variable increases by 1. Thus, the exponent pattern of the variables of each expansion is $a^n b^0, a^{n-1}b^1, a^{n-2}b^2, a^{n-3}b^3, \ldots, a^1 b^{n-1}, a^0 b^n$. The pattern for the binomial coefficients is less obvious. To see the pattern for the coefficients, we start by rewriting the six expansions of $(a + b)^n$ again, this time we only write the coefficients. See **Figure 4**.

$n = 0$:	1
$n = 1$:	1 1
$n = 2$:	1 2 1
$n = 3$:	1 3 3 1
$n = 4$:	1 4 6 4 1
$n = 5$:	1 5 10 10 5 1

Figure 4 The coefficients of the expansions of $(a + b)^n$, also called Pascal's triangle.

Notice that the coefficients in Figure 4 form a "triangle." This triangle is known as Pascal's triangle, named after the French mathematician, **Blaise Pascal**. The first and last number of each row of Pascal's triangle is 1. Every other number is equal to the sum of the two numbers directly above it. We can now write the next row of Pascal's triangle, which is the row corresponding to $n = 6$.

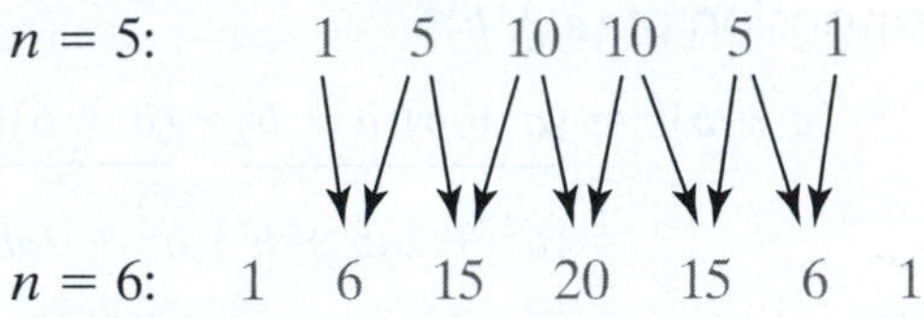

This new row of Pascal's triangle represents the coefficients of the expansion of $(a+b)^6$. Therefore,

$$\begin{aligned}(a + b)^6 &= a^6b^0 + 6a^5b^1 + 15a^4b^2 + 20a^3b^3 + 15a^2b^4 + 6a^1b^5 + a^0b^6 \\ &= a^6 + 6a^5b + 15a^4b^2 + 20a^3b^3 + 15a^2b^4 + 6ab^5 + b^6.\end{aligned}$$

Notice the pattern of the exponents of each variable. The exponent of variable a starts with 6 and decreases by 1 for each successive term until it equals 0. The exponent of variable b starts with 0 and increases by 1 each term until it equals 6.

See if you can create Pascal's triangle for values of n up to 10. When you are done, view this version of **Pascal's Triangle for $n = 0, 1, 2, \ldots 10$.**

Example 1 Using Pascal's Triangle to Expand a Binomial Raised to a Power

Use Pascal's triangle to expand each binomial.

a. $(x + 2)^4$ b. $(x - 3)^5$ c. $(2x - 3y)^3$

Solution

a. We start by looking at the row of **Pascal's triangle** corresponding with $n = 4$. We see that this row is $1\ 4\ 6\ 4\ 1$. Using these coefficients and the exponent pattern we get

$$\begin{aligned}(x + 2)^4 &= 1(x^4 \cdot 2^0) + 4(x^3 \cdot 2^1) + 6(x^2 \cdot 2^2) + 4(x \cdot 2^3) + 1(x^0 \cdot 2^4)\\ &= x^4 + 8x^3 + 24x^2 + 32x + 16\end{aligned}$$

b. The row of **Pascal's triangle** corresponding with $n = 5$ is $1\ 5\ 10\ 10\ 5\ 1$. Using these coefficients and the exponent pattern we get

$$\begin{aligned}(x - 3)^5 &= 1(x^5 \cdot (-3)^0) + 5(x^4 \cdot (-3)^1) + 10(x^3 \cdot (-3)^2)\\ &\quad +10(x^2 \cdot (-3)^3) + 5(x \cdot (-3)^4) + 1(x^0 \cdot (-3)^5)\\ &= x^5 - 15x^4 + 90x^3 - 270x^2 + 405x - 243\end{aligned}$$

c. See if you can use **Pascal's triangle** to show that $(2x - 3y)^3 = 8x^3 - 36x^2y + 54xy^2 - 27y^3$. Work through the **interactive video** to see the solution.

TIP The terms of the expansion of the form $(a - b)^n$ will *always* alternate in sign with the sign of the first term being positive.

You Try It Work through this You Try It problem.

Work Exercises 1–6 in this textbook or in the MyLab Math Study Plan.

OBJECTIVE 2 EVALUATING BINOMIAL COEFFICIENTS

Although Pascal's triangle is useful for determining the binomial coefficients of $(a + b)^n$ for fairly small values of n, it is not that useful for large values of n. For example, to find the binomial coefficients of the expansion of $(a + b)^{50}$ using Pascal's triangle, we would need to write the first 51 rows of the triangle to determine the coefficients. Fortunately, there is a convenient formula for the binomial coefficients. This formula requires the use of factorials. Recall, $n! = n \cdot (n-1) \cdot (n-2) \cdot \cdots \cdot 3 \cdot 2 \cdot 1$ and $0! = 1$. To establish a formula for the binomial coefficients, let's take another look at the expansion for $(a + b)^6$.

$$(a + b)^6 = 1a^6b^0 + 6a^5b^1 + 15a^4b^2 + 20a^3b^3 + 15a^2b^4 + 6a^1b^5 + 1a^0b^6$$

Table 2 shows the relationship between the variable parts of the expansion of the form $a^{n-r}b^r$ and the corresponding coefficients. Click on any of the coefficients in Table 2 to verify that the formula using factorial notation is true.

Table 2

Variables	Coefficient	Variables	Coefficient
a^6b^0	$1 = \dfrac{6!}{0!\cdot 6!}$	a^2b^4	$15 = \dfrac{6!}{4!\cdot 2!}$
a^5b^1	$6 = \dfrac{6!}{1!\cdot 5!}$	a^1b^5	$6 = \dfrac{6!}{5!\cdot 1!}$
a^4b^2	$15 = \dfrac{6!}{2!\cdot 4!}$	a^0b^6	$1 = \dfrac{6!}{6!\cdot 0!}$
a^3b^3	$20 = \dfrac{6!}{3!\cdot 3!}$		

You can see in Table 2 that for each pair of variables of the form $a^{n-r}b^r$, the corresponding binomial coefficient is of the form $\dfrac{n!}{r!\cdot(n-r)!}$. We will use the shorthand notation $\binom{n}{r}$, read as "n choose r", to denote a binomial coefficient.

Formula for a Binomial Coefficient

For nonnegative integers n and r with $n \geq r$, the coefficient of the expansion of $(a+b)^n$ whose variable part is $a^{n-r}b^r$ is given by

$$\binom{n}{r} = \frac{n!}{r!\cdot(n-r)!}.$$

Example 2 Evaluating Binomial Coefficients

Evaluate each of the following binomial coefficients.

a. $\binom{5}{3}$ b. $\binom{4}{1}$ c. $\binom{12}{8}$

Solution

a. $\binom{5}{3} = \dfrac{5!}{3!(5-3)!} = \dfrac{5!}{3!\cdot 2!} = \dfrac{5\cdot 4\cdot \cancel{3!}}{\cancel{3!}\cdot 2\cdot 1} = \dfrac{20}{2} = 10$

b. $\binom{4}{1} = \dfrac{4!}{1!(4-1)!} = \dfrac{4!}{1!\cdot 3!} = \dfrac{4\cdot \cancel{3!}}{1\cdot \cancel{3!}} = \dfrac{4}{1} = 4$

c. $\binom{12}{8} = \dfrac{12!}{8!(12-8)!} = \dfrac{12!}{8!\cdot 4!} = \dfrac{12\cdot 11\cdot 10\cdot 9\cdot \cancel{8!}}{\cancel{8!}\cdot 4\cdot 3\cdot 2\cdot 1} = \dfrac{11,880}{24} = 495$

Using Technology

```
NORMAL FIX0 AUTO REAL RADIAN CL
12 nCr 8
                              495
```

A calculator can be used to compute binomial coefficients. Typically, the key $\boxed{nCr}$ is used. The figure on the left shows the computation of $\binom{12}{8}$ using a graphing utility.

You Try It Work through this You Try It problem.

Work Exercises 7–10 in this textbook or in the MyLab Math Study Plan.

OBJECTIVE 3 EXPANDING BINOMIALS RAISED TO A POWER USING THE BINOMIAL THEOREM

Now that we know how to compute a binomial coefficient, we can state the Binomial Theorem.

Binomial Theorem

If n is a positive integer then,

$$(a+b)^n = \binom{n}{0}a^n + \binom{n}{1}a^{n-1}b + \binom{n}{2}a^{n-2}b^2 + \cdots + \binom{n}{n}b^n$$

$$= \sum_{i=0}^{n}\binom{n}{i}a^{n-i}b^i.$$

Example 3 Using the Binomial Theorem to Expand a Binomial Raised to a Power

Use the Binomial Theorem to expand each binomial.

a. $(x-1)^8$ b. $(\sqrt{x}+y^2)^5$

Solution

a. Work through the interactive video to verify that $(x-1)^8 = x^8 - 8x^7 + 28x^6 - 56x^4 + 70x^4 - 56x^3 + 28x^2 - 8x + 1$ using the Binomial Theorem.

b. The expansion of $(\sqrt{x} + y^2)^5$ is as follows.

$$(\sqrt{x} + y^2)^5 = \binom{5}{0}(\sqrt{x})^5 + \binom{5}{1}(\sqrt{x})^4(y^2) + \binom{5}{2}(\sqrt{x})^3(y^2)^2 + \binom{5}{3}(\sqrt{x})^2(y^2)^3 + \binom{5}{4}(\sqrt{x})(y^2)^4 + \binom{5}{5}(y^2)^5$$

$$= 1\cdot(\sqrt{x})^5 + 5\cdot(\sqrt{x})^4(y^2) + 10\cdot(\sqrt{x})^3(y^2)^2 + 10\cdot(\sqrt{x})^2(y^2)^3 + 5\cdot(\sqrt{x})(y^2)^4 + 1\cdot(y^2)^5$$

$$= x^2\sqrt{x} + 5x^2y^2 + 10x\sqrt{x}y^4 + 10xy^6 + 5\sqrt{x}y^8 + y^{10}$$

You Try It Work through this You Try It problem.

Work Exercises 11-17 in this textbook or in the MyLab Math Study Plan.

OBJECTIVE 4 FINDING A PARTICULAR TERM OR A PARTICULAR COEFFICIENT OF A BINOMIAL EXPANSION

We may want to find a particular term of a binomial expansion. Fortunately, we can use the Binomial Theorem to develop a formula for a particular term. We start by writing out the first several terms of $(a + b)^n$ using the Binomial Theorem.

$$(a + b)^n = \binom{n}{0}a^n + \binom{n}{1}a^{n-1}b + \binom{n}{2}a^{n-2}b^2 + \binom{n}{3}a^{n-3}b^3 + \cdots + \binom{n}{n}b^n$$

The first term is $\binom{n}{0}a^n$. The second term is $\binom{n}{1}a^{n-1}b$. The third term is $\binom{n}{2}a^{n-2}b^2$. Following this pattern, we can see that the formula for the $(r + 1)$st term (for $r \geq 0$) is given by $\binom{n}{r}a^{n-r}b^r$.

Formula for the $(r + 1)$st Term of a Binomial Expansion

If n is a positive integer and if $r \geq 0$, then the $(r + 1)$st term of the expansion of $(a + b)^n$ is given by

$$\binom{n}{r}a^{n-r}b^r = \frac{n!}{r!\cdot(n - r)!}a^{n-r}b^r.$$

Example 4 Finding a Particular Term of a Binomial Expansion

Find the third term of the expansion of $(2x - 3)^{10}$.

Solution Since we want to find the third term of this expansion, we will use the formula for the $(r + 1)$st term, which is equal to $\binom{n}{r}a^{n-r}b^r$ for $r = 2, n = 10, a = 2x$, and $b = -3$. Therefore, the third term is

$$\binom{10}{2}(2x)^{10-2}(-3)^2 = 45(256x^8)(9) = 103{,}680x^8$$

Watch this **video** to see every step of this solution.

Example 5 Finding a Particular Coefficient of a Binomial Expansion

Find the coefficient of x^7 in the expansion of $(x + 4)^{11}$.

Solution The formula for the $(r + 1)$st term of the expansion of $(x + 4)^{11}$ is given by the formula $\binom{11}{r}x^{11-r}4^r$.

The term containing x^7 occurs when $11 - r = 7$. Solving this equation for r we get $r = 4$. Therefore, the term involving x^7 is $\binom{11}{4}x^7 4^4$. Simplifying this expression we get $84{,}480x^7$. Thus, the coefficient of x^7 is $84{,}480$. Watch this **video** to see this solution worked out in detail.

You Try It Work through this You Try It problem.

Work Exercises 18–25 in this textbook or in the MyLab Math Study Plan.

14.4 Exercises

In Exercises 1–6, use Pascal's triangle to expand each binomial.

1. $(m + n)^6$
2. $(x + 5)^4$
3. $(x - y)^7$
4. $(x - 3)^5$
5. $(2x + 3y)^6$
6. $(3x^2 - 4y^3)^4$

In Exercises 7–10, evaluate each binomial coefficient.

7. $\binom{7}{1}$
8. $\binom{10}{4}$
9. $\binom{7}{7}$
10. $\binom{23}{3}$

In Exercises 11–17, use the Binomial Theorem to expand each binomial.

11. $(x + 2)^7$
12. $(x - 3)^6$
13. $(4x + 1)^5$
14. $(x + 3y)^4$
15. $(5x - 3y)^5$
16. $(x^4 + y^5)^6$
17. $(\sqrt{x} - \sqrt{2})^4$
18. Find the sixth term of the expansion of $(x + 4)^9$.
19. Find the fifth term of the expansion of $(a - b)^8$.
20. Find the third term of the expansion of $(3c - d)^7$.
21. Find the seventh term of the expansion of $(3x + 2)^{10}$.
22. Find the coefficient of x^5 in the expansion of $(x - 3)^{11}$.
23. Find the coefficient of x^4 in the expansion of $(4x + 1)^{12}$.
24. Find the coefficient of x^0 in the expansion of $\left(x^2 + \frac{1}{x}\right)^{12}$.
25. Find the coefficient of x^{10} in the expansion of $\left(x - \frac{3}{\sqrt{x}}\right)^{19}$.

14.5 Mathematical Induction

THINGS TO KNOW

Before working through this section, be sure that you are familiar with the following concepts:

VIDEO ANIMATION INTERACTIVE

You Try It
1. Computing the nth Partial Sum of an Arithmetic Series (Section 14.2) 

You Try It
2. Computing the nth Partial Sum of a Geometric Series (Section 14.3) 

OBJECTIVES

1 Writing Mathematical Statements

2 Using the Principle of Mathematical Induction to Prove Statements

SECTION 14.5 EXERCISES

OBJECTIVE 1 WRITING MATHEMATICAL STATEMENTS

In this section we will learn a technique that is often used in mathematics to prove certain mathematical statements that are declared to be true for all **natural numbers**. We will use the notation S_n to denote a mathematical statement. As a motivating example, consider the following mathematical statement

$$S_n: 1 + 2 + 3 + \cdots + n = \frac{n(n + 1)}{2} \text{ for all natural numbers } n$$

This statement suggests that the sum of the first n consecutive natural numbers is equal to the formula $\frac{n(n + 1)}{2}$ for *all* natural numbers. We can verify that this statement is true for the first natural number, $n = 1$, by adding the "first one terms" on the left side and by substituting 1 for n on the right side.

$$S_1: \underbrace{\quad 1 \quad}_{\substack{\text{The sum of the} \\ \text{"first one terms"}}} = \underbrace{\frac{1(1 + 1)}{2}}_{\text{Substitute } n = 1}$$

The statement S_1 is true because the left- and right-hand sides of the equal sign are both equal to 1. In Example 1, we show that the statement $S_n: 1 + 2 + 3 + \cdots + n = \frac{n(n + 1)}{2}$ is also true for the next four natural numbers.

Example 1 Verifying a Mathematical Statement for Specified Natural Numbers

Given the mathematical statement $S_n: 1 + 2 + 3 + \cdots + n = \frac{n(n + 1)}{2}$, write the statements $S_2, S_3, S_4,$ and S_5, and verify that each statement is true.

Solution

$S_2: \underbrace{1 + 2}_{3} = \underbrace{\frac{2(2+1)}{2}}_{3};$ $\quad S_3: \underbrace{1 + 2 + 3}_{6} = \underbrace{\frac{3(3+1)}{2}}_{6};$

$S_4: \underbrace{1 + 2 + 3 + 4}_{10} = \underbrace{\frac{4(4+1)}{2}}_{10};$ $\quad S_5: \underbrace{1 + 2 + 3 + 4 + 5}_{15} = \underbrace{\frac{5(5+1)}{2}}_{15}$

Because the left- and right-hand sides of the equal sign of each statement are equivalent, the statements S_2, S_3, S_4, and S_5 are true.

You Try It Work through this You Try It problem.

Work Exercises 1–4 in this textbook or in the MyLab Math Study Plan.

In Example 1, we wrote mathematical statements for *specific* natural numbers. It is important that we can write mathematical statements for *arbitrary* natural numbers as in Example 2.

Example 2 Writing a Mathematical Statement for Arbitrary Natural Numbers k and $k + 1$

Given the statement S_n: $1 + 2 + 3 + \cdots + n = \frac{n(n+1)}{2}$, write the statements S_k and S_{k+1} for natural numbers k and $k + 1$.

Solution To find S_k, we replace n with k to get S_k: $1 + 2 + 3 + \cdots + k = \frac{k(k+1)}{2}$. Similarly, to find S_{k+1} we replace n with $k + 1$ to get S_{k+1}: $1 + 2 + 3 + \cdots + k + (k + 1) = \frac{(k+1)((k+1)+1)}{2} = \frac{(k+1)(k+2)}{2}$. Watch this video to see each step of this solution.

You Try It Work through this You Try It problem.

Work Exercises 5–8 in this textbook or in the MyLab Math Study Plan.

OBJECTIVE 2 USING THE PRINCIPLE OF MATHEMATICAL INDUCTION TO PROVE STATEMENTS

So far in this section we have verified that the statement S_n: $1 + 2 + 3 + \cdots + n = \frac{n(n+1)}{2}$ is true for the first five natural numbers, but how do we verify that this statement is true for *all* natural numbers? The answer is that we must use a technique called *mathematical induction*. This is how mathematical induction works: To prove that a mathematical statement S_n is true for *all* natural numbers, we must first show that the statement S_1 is true. Next, is the **inductive step**.

We must prove that if S_k is true for any natural number k, then S_{k+1} is also true. Once we verify that S_{k+1} is true, it follows by this inductive step that S_{k+2} is true, S_{k+3} is true, and so on. Hence the statement is true for *all* natural numbers. We summarize this process by stating the principle of mathematical induction.

The Principle of Mathematical Induction

Let S_n be a mathematical statement involving the natural number n. The statement S_n is true for *all* natural numbers if the following two conditions are satisfied.

1. S_1 is true.
2. If, for any natural number k, S_k is true, then S_{k+1} is true.

Note The second condition is called the **inductive step**. We assume that S_k is true for some natural number k (this is called the **induction assumption**), then we must show that S_{k+1} is true.

Mathematical induction is like knocking down a sequence of dominoes. In order to knock down *all* of the dominoes, the first domino, S_1, must be knocked down. If we know that the kth domino, S_k, is knocked down and can verify that the $(k + 1)$st domino, S_{k+1}, will be knocked down, then we can conclude that *all* dominos will be knocked down. Note that both conditions must be true in order for all of the dominos to fall. Without knocking over the first domino, it is impossible to knock down *all* of the dominos. Similarly, if there is a domino that does not knock down a subsequent domino, then it is impossible to knock down *all* of the dominos. See Figure 5.

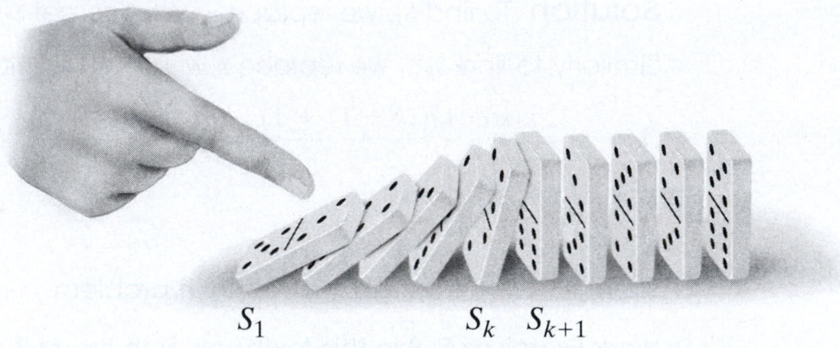

Figure 5 In order to knock down all dominos, the first domino, S_1, must fall. If it is assumed that domino S_k is knocked over and we can show that domino S_{k+1} will be knocked over, then we can conclude that *all* dominos will be knocked over.

We can now use the principle of mathematical induction to prove that the statement S_n: $1 + 2 + 3 + \cdots + n = \frac{n(n + 1)}{2}$ is true for *all* natural numbers.

Example 3 Using the Principle of Mathematical Induction to Verify a Formula

Prove that the mathematical statement S_n: $1 + 2 + 3 + \cdots + n = \frac{n(n+1)}{2}$ is true for all natural numbers n.

Solution

1. We need to show that S_1 is true.

   When we substitute $n = 1$ into the left- and right-hand side of this statement, we get $1 = \underbrace{\frac{1(1+1)}{2}}_{1}$.

   Thus, S_1 is true.

2. We assume that S_k is true for some natural number k. Thus, $1 + 2 + 3 + \cdots + k = \frac{k(k+1)}{2}$. This is our induction assumption. Our goal is to show that S_{k+1} is true.

   Therefore, our goal is to verify that the following mathematical statement is true.

$$S_{k+1}: 1 + 2 + 3 + \cdots + k + (k+1) = \frac{(k+1)((k+1)+1)}{2} = \frac{(k+1)(k+2)}{2}$$

Start with the left-hand side of this equation and show that it is equal to the right-hand side.

$1 + 2 + 3 + \cdots + (k+1)$	Start with the left-hand side of the statement S_{k+1}.
$= \underbrace{1 + 2 + 3 + \cdots + k}+ (k+1)$	The term preceding $k + 1$ is k.
$= \frac{k(k+1)}{2} + (k+1)$	Use the induction assumption.
$= \frac{k(k+1) + 2(k+1)}{2}$	Combine terms using a common denominator.
$= \frac{(k+1)(k+2)}{2}$	Factor.

Therefore, the statement S_{k+1}: $1 + 2 + 3 + \cdots + (k+1) = \frac{(k+1)(k+2)}{2}$ is true.

We have shown that S_1 is true. We have also shown that if the statement S_k is true, then the statement S_{k+1} is true. Therefore, by the principle of mathematical induction, S_n is true for all natural numbers. You may want to watch this **video** to see every step of this induction proof. ●

Example 4 Using the Principle of Mathematical Induction to Verify a Formula

Prove that the mathematical statement S_n: $1^3 + 2^3 + 3^3 + \cdots + n^3 = \frac{n^2(n+1)^2}{4}$ is true for all natural numbers n.

Solution Try following the same procedure as outlined in Example 3. When you are finished, watch this video to see if you are correct.

In Section 14.2 we studied arithmetic sequences and arithmetic series. Recall that if S_n is the nth partial sum of an arithmetic series, then this sum can be computed using the formula $S_n = \dfrac{n(a_1 + a_n)}{2}$. In Section 14.3 we studied geometric sequences and geometric series. If S_n is the nth partial sum of a geometric series, then this sum can be computed using the formula $S_n = \dfrac{a_1(1 - r^n)}{1 - r}$, where r is the common ratio. Watch this interactive video to see how to prove both of these formulas using the principle of mathematical induction.

So far, all of our examples involved verifying sum formulas. Mathematical induction is not limited to verifying only sum formulas. We can use the principle of mathematical induction to prove a variety of mathematical statements, as shown in Example 5.

Example 5 Using the Principle of Mathematical Induction to Verify a Mathematical Statement

Prove that the mathematical statement S_n: $2^n \geq 2n$ is true for all natural numbers n.

$$S_n: 2^n \geq 2n$$

Solution

1. We need to show that S_1 is true.

$$S_1: \underbrace{2^1}_{2} \geq \underbrace{2(1)}_{2}$$

   Since 2 is greater than or equal to 2, we have verified that S_1 is true.

2. We assume that S_k is true for some natural number k. Thus, $2^k \geq 2k$ is our induction assumption. Our goal is to show that S_{k+1} is true. Therefore, our goal is to verify that the inequality $2^{k+1} \geq 2(k + 1)$ is true. To do this, start with the left-hand side of this inequality and verify that it is greater than or equal to the right-hand side.

2^{k+1}	Start with the left-hand side of the statement S_{k+1}.
$= 2^k \cdot 2$	Use the product rule of exponents.
$= 2^k + 2^k$	$2^k + 2^k = 2 \cdot 2^k$
$\geq 2k + 2k$	$2^k \geq 2k$ by the induction assumption.
$\geq 2k + 2$	$2k \geq 2$
$= 2(k + 1)$	Factor.

Thus, $2^{k+1} \geq 2(k + 1)$ and we have verified that S_{k+1} is true. Therefore, by the principle of mathematical induction, the statement S_n: $2^n \geq 2n$ is true for all natural numbers.

You Try It Work through this You Try It problem.

Work Exercises 9–16 in this textbook or in the MyLab Math Study Plan.

14.5 Exercises

In Exercises 1–4, write the statements S_1, S_2, and S_3, and determine if each statement is true.

1. S_n: $3 + 9 + 15 + \cdots + (6n - 3) = 3n^2$
2. S_n: $1^2 + 2^2 + 3^2 + \cdots + n^2 = \dfrac{n(n+1)(2n+1)}{6}$
3. S_n: $2^n > n^2$
4. S_n: 3 is a factor of $n(n^2 - 1)$.

In Exercises 5–8, write the statements S_k and S_{k+1}.

5. S_n: $6 + 12 + 18 + \cdots + 6n = 3n(n + 1)$
6. S_n: $5 + 11 + 17 + \cdots + (6n - 1) = n(3n + 2)$
7. S_n: $1^2 + 2^2 + 3^2 + \cdots + n^2 = \dfrac{n(n+1)(2n+1)}{6}$
8. S_n: $\dfrac{1}{1\cdot 2} + \dfrac{1}{2\cdot 3} + \dfrac{1}{3\cdot 4} + \cdots + \dfrac{1}{n(n+1)} = \dfrac{n}{n+1}$

In Exercises 9–16, use the principle of mathematical induction to show that each mathematical statement is true for all natural numbers n.

9. S_n: $6 + 12 + 18 + \cdots + 6n = 3n(n + 1)$
10. S_n: $2 + 3 + 4 + \cdots + (n + 1) = \dfrac{1}{2}n(n + 3)$
11. S_n: $3 + 7 + 11 + \cdots + (4n - 1) = n(2n + 1)$
12. S_n: $1^2 + 2^2 + 3^2 + \cdots + n^2 = \dfrac{n(n+1)(2n+1)}{6}$
13. S_n: $3^0 + 3^1 + 3^2 + \cdots + 3^{n-1} = \dfrac{1}{2}(3^n - 1)$
14. S_n: $\dfrac{1}{1\cdot 2} + \dfrac{1}{2\cdot 3} + \dfrac{1}{3\cdot 4} + \cdots + \dfrac{1}{n(n+1)} = \dfrac{n}{n+1}$
15. S_n: $n(n + 1) + 4$ is divisible by 2.
16. S_n: $3^n \geq 3n$

14.6 The Theory of Counting

INTRODUCTION

Read this introduction before beginning Objective 1.

OBJECTIVES

1 Using the Fundamental Counting Principle

2 Using Permutations

3 Using Combinations

SECTION 14.6 EXERCISES

Introduction to Section 14.6

Counting is a task we learn at an early age. It usually starts with the simple, "How old are you?" and we hold up two fingers. But counting gets more complicated as we mature and our lives get more complex. For example, when the Louisiana lottery began, people were asking, the question, "How many possible sets of numbers are there if you can choose six from the numbers 1 to 44?" To answer this question, it is necessary to learn simple counting skills. The answer is 7,059,052. See **Example 12**.

In addition to practical purposes, counting skills are a great exercise to improve analytical reasoning skills, which can be used for all sorts of problem-solving and decision-making situations encountered in many walks of life. We start by introducing the **fundamental counting principle.**

OBJECTIVE 1 USING THE FUNDAMENTAL COUNTING PRINCIPLE

An **event** is a specific part of an experiment or task that has zero or more countable outcomes. For example, suppose that you are getting dressed for a day at the beach and you have two pairs of shorts and three shirts that all coordinate. The task is to get dressed. Getting dressed consists of two events: choosing your shorts to wear and choosing your shirt to wear. Choosing your shorts is an event that has two possible outcomes. Choosing a shirt is an event that has three possible outcomes. To determine the total number of possible ways to dress for the beach, we use the fundamental counting principle.

Definition The Fundamental Counting Principle

Suppose that an experiment or task is made up of two or more events. If the first event can occur in n_1 ways, the second event can occur in n_2 ways, and the third event can occur in n_3 ways and so on, then the number of possible outcomes of the experiment or task is $N = n_1 n_2 \ldots n_k$, where k is the total number of events.

The fundamental counting principle is sometimes referred to as the multiplication principle of counting because we simply multiply the number of possible outcomes of each event to determine the total number of possible outcomes of the experiment or task. In the example of getting dressed for the beach, the total number of possible outcomes is $N = n_1 n_2 = 2 \cdot 3 = 6$.

Example 1 Using the Fundamental Counting Principle

A woman wants to paint the exterior of her house using green for the shutters, peach for the siding, and off-white for the trim. She selects two shades of green, three shades of peach, and four shades of off-white that she likes. How many possible color schemes does this give her to choose from?

Solution We first recognize that the task of selecting a color scheme consists of three events.

Event 1: Choosing paint for the shutters; $n_1 = 2$ possible outcomes
Event 2: Choosing paint for the siding; $n_2 = 3$ possible outcomes
Event 3: Choosing paint for the trim; $n_3 = 4$ possible outcomes

By the fundamental counting principle, the total possible number of color schemes is $N = n_1 n_2 n_3 = 2 \cdot 3 \cdot 4 = 24$.

You Try It Work through this You Try It problem.

Work Exercises 1–3 in this textbook or in the MyLab Math Study Plan.

Example 2 Using the Fundamental Counting Principle

A student is to create a password for his **MyLab** Math homework account. He is informed that the password must consist of seven characters. How many possible ways can he create the password if the first three characters are letters followed by two digits and the last two characters can be letters or digits?

(Repeated letters are allowed and letters are not case sensitive. Also, repeated digits are allowed.)

Solution The letters are not case sensitive (there is no difference between a lowercase letter or an uppercase letter) and repetition of letters is allowed. Therefore, there are 26 possible letters to choose from for each character that can be a letter.

Because repetition of digits is allowed, there are 10 possible digits to choose from each time a digit can be chosen. We now consider the following seven events.

Event 1: Choose the first character, which must be a letter;	$n_1 = 26$ possible letters
Event 2: Choose the second character, which must be a letter;	$n_2 = 26$ possible letters
Event 3: Choose the third character, which must be a letter;	$n_3 = 26$ possible letters
Event 4: Choose the fourth character, which must be a digit;	$n_4 = 10$ possible digits
Event 5: Choose the fifth character, which must be a digit;	$n_5 = 10$ possible digits
Event 6: Choose the sixth character, which can be a letter or a digit;	$n_6 = 26 + 10 = 36$ possible characters
Event 7: Choose the seventh character, which can be a letter or a digit;	$n_7 = 26 + 10 = 36$ possible characters

By the fundamental counting principle, there are

$$N = n_1 \cdot n_2 \cdot n_3 \cdot n_4 \cdot n_5 \cdot n_6 \cdot n_7$$
$$= 26 \cdot 26 \cdot 26 \cdot 10 \cdot 10 \cdot 36 \cdot 36$$
$$= 2{,}277{,}849{,}600 \text{ possible ways to create the password.}$$

You Try It **Work through this You Try It problem.**

Work Exercises 4–6 in this textbook or in the MyLab Math Study Plan.

Sometimes the number of choices of a specific event is affected by the choice of a previous event. In this situation there will be more than one sequence of events to consider. Each sequence of events must be separated and counted. We call each sequence of events a **case,** and the technique is called **counting cases.** We illustrate this technique in Example 3 by slightly changing the password example seen previously.

Example 3 Using the Fundamental Counting Principle

A student is to create a password for his **MyLab** Math homework account. He is informed that the password must consist of seven characters. How many possible ways can he create the password if the first three characters are letters followed by two digits and the last two characters can be letters or digits?

(Repeated letters are not allowed and the letters are case sensitive. Also, repeated digits are allowed.)

Solution In this situation, repetition of letters is *not* allowed and the letters are case sensitive. The first three characters are letters so there are $26 \cdot 2 = 52$ distinct choices for the first character. There are 51 choices for the second character and there are 50 choices for the third character. The fourth and fifth characters are digits which can be repeated. Therefore there are 10 possible digits that can be used for the fourth and fifth characters. Note that the sixth and seventh characters can be either a letter or a digit. Since repetition of letters is not allowed, the number of possible outcomes for the seventh character is affected by the choice of the sixth character. So, we must consider two cases. The first case is to choose a letter as the sixth character. The second case is to choose a digit as the sixth character.

Case 1: The sixth character chosen is a letter:
In Case 1 we consider choosing a letter for the sixth character. There are only 49 possibilities because repetition of letters is not allowed. Then, there are 48 letters and 10 digits to choose from for the seventh character. Thus, the number of possibilities for the seventh character is $48 + 10 = 58$. Below is the number of outcomes for each of the seven events for this case.

$\frac{52}{n_1}$	$\cdot$	$\frac{51}{n_2}$	$\cdot$	$\frac{50}{n_3}$	$\cdot$	$\frac{10}{n_4}$	$\cdot$	$\frac{10}{n_5}$	$\cdot$	$\frac{49}{n_6}$	$\cdot$	$\frac{58}{n_7}$
choose a letter		*choose a letter*		*choose a letter*		*choose a digit*		*choose a digit*		*choose a letter*		*choose a letter or digit*

Using the Fundamental Counting Principle, we get

$$\begin{aligned} &n_1 \cdot n_2 \cdot n_3 \cdot n_4 \cdot n_5 \cdot n_6 \cdot n_7 \\ &= 52 \cdot 51 \cdot 50 \cdot 10 \cdot 10 \cdot 49 \cdot 58 \\ &= 37{,}684{,}920{,}000 \text{ possible passwords for Case 1.} \end{aligned}$$

Case 2: The sixth character chosen is a digit:
For Case 2 we consider the possibility of choosing a digit for the sixth character. Note that the number of choices for the first five characters is the same as in Case 1. For the sixth character, since repetition of digits is allowed, there are 10 possibilities. There are 49 letters and 10 digits to choose from for the seventh character. Therefore, there are $49 + 10 = 59$ possibilities for the seventh character.

$\frac{52}{n_1}$	$\cdot$	$\frac{51}{n_2}$	$\cdot$	$\frac{50}{n_3}$	$\cdot$	$\frac{10}{n_4}$	$\cdot$	$\frac{10}{n_5}$	$\cdot$	$\frac{10}{n_6}$	$\cdot$	$\frac{59}{n_7}$
choose a letter		*choose a letter*		*choose a letter*		*choose a digit*		*choose a digit*		*choose a digit*		*choose a letter or digit*

Using the Fundamental Counting Principle, we get

$$\begin{aligned} &n_1 \cdot n_2 \cdot n_3 \cdot n_4 \cdot n_5 \cdot n_6 \cdot n_7 \\ &= 52 \cdot 51 \cdot 50 \cdot 10 \cdot 10 \cdot 10 \cdot 59 \\ &= 7{,}823{,}\ 400{,}000 \text{ possible passwords for Case 2.} \end{aligned}$$

Counting these two cases we get a total of $N = 37{,}684{,}920{,}000 + 7{,}823{,}400{,}000 = 45{,}508{,}320{,}000$ possible passwords.

In the previous example we added the possibilities from both cases. If there are many cases to consider, it may be easier to count all cases together and then subtract the cases that do not apply. This technique is called **all minus the exclusion**. In Example 4, we show two different methods for counting the number of outfits a tourist can choose for an upcoming weekend trip.

Example 4 Using the Fundamental Counting Principle

Linda has five t-shirts of different colors, three pairs of shorts of different patterns, and four pairs of shoes of different colors. She is planning a weekend trip to Pensacola Beach, Florida. In how many ways can she combine these items if

a. the orange t-shirt clashes with the red shoes?

b. the red t-shirt clashes with the striped shorts?

Solution

a. Method I: Counting Cases

There are two non-clash cases for Linda. She can wear all t-shirts with all shorts and with the three non-red pairs of shoes. She can also wear her four non-orange t-shirts with all of her shorts and her red shoes.

Non-clash case 1: $\underline{\text{5 t-shirts}} \cdot \underline{\text{3 pairs of shorts}} \cdot \underline{\text{3 non-red pairs of shoes}} = 45$

Non-clash case 2: $\underline{\text{4 non-orange t-shirts}} \cdot \underline{\text{3 pairs of shorts}} \cdot \underline{\text{1 red pair of shoes}} = 12$

Therefore, we can count the cases to get $N = 45 + 12 = 57$ total possible non-clashing outfits.

Method II: All Minus the Exclusion

Using this method, we first determine the total number of possible ways for Linda to dress for the beach (ignoring clashes) and then subtract the number of outfits that clash and must be excluded.

Total possible outfits: $\underline{\text{5 t-shirts}} \cdot \underline{\text{3 pairs of shorts}} \cdot \underline{\text{4 pairs of shoes}} = 60$

Outfits that clash: $\underline{\text{1 orange t-shirt}} \cdot \underline{\text{3 pairs of shorts}} \cdot \underline{\text{1 red pairs of shoes}} = 3$

Now, subtract the outfits to exclude from the total possible outfits to get $N = 60 - 3 = 57$ total possible non-clashing outfits.

b. Method I: Counting Cases

Once again, there are two non-clash cases for Linda in this situation. She can wear all t-shirts with her two non-striped pairs of shorts and all pairs of shoes. She can also wear her four non-red t-shirts with the striped shorts and all pairs of shoes.

Non-clash case 1: $\underline{\text{5 t-shirts}} \cdot \underline{\text{2 non-striped pairs of shorts}} \cdot \underline{\text{4 pairs of shoes}} = 40$

Non-clash case 2: $\underline{\text{4 non-red t-shirts}} \cdot \underline{\text{1 striped pair of shorts}} \cdot \underline{\text{4 pairs of shoes}} = 16$

Counting these two cases, we get $N = 40 + 16 = 56$ total possible non-clashing outfits.

Method II: All Minus the Exclusion

From part (a) we know that there are 60 total possible outfits. The outfits that clash are the ones with one red shirt, one pair of striped shorts, and any of the four pairs of shoes.

Total possible outfits:	5 t-shirts	· 3 pairs of shorts	· 4 pairs of shoes	= 60
Outfits that clash:	1 red t-shirt	· 1 pair of striped shorts	· 4 pairs of shoes	= 4

Now, subtract the outfits to exclude from the total possible outfits to get $N = 60 - 4 = 56$ total possible non-clashing outfits. Watch this **video** to see the solution to this counting example worked out in detail.

You Try It Work through this You Try It problem.

Work Exercises 7–12 in this textbook or in the MyLab Math Study Plan.

Often in life, a certain number of objects need to be chosen from a specific group or class. The objects in the class have various attributes. Determining the total possible outcomes of an event will depend on the desired attributes. To practice this skill, work through Examples 5 and 6 after considering the following demographic information collected from a teacher of mathematics at the local community college.

Total number of students in the class:	50
Number of boys:	35
Number of girls:	15
Number of freshmen:	22
Number of sophomores:	13
Number of juniors:	9
Number of seniors:	6
Number of education majors:	7
Number of non-education majors:	43

Example 5 Using the Fundamental Counting Principle

The instructor chooses a student to dim the lights and a student to shut the door (to use the overhead projector). In how many ways can she do this if a student may do both jobs and

a. both are girls?

b. one is a boy and one is a girl?

c. at least one is a boy?

d. the door closer is a boy?

e. neither are seniors?

f. neither are education majors?

Solution There are two events (or jobs) to consider: the number of possible students who can dim the lights and the number of possible students to shut the door. Once we determine the number of students who can complete each job, we use the fundamental counting principle to determine the total possible outcomes.

a. Because both jobs can be completed by the same girl, we know that there are 15 girls that can do each job.

Case: Girl/Girl $\underline{\text{15 girls can dim the lights}} \cdot \underline{\text{15 girls can shut the door}} = 225$

Therefore, there are $N = 225$ possible ways for a girl to complete both jobs.

b. There are two cases for this situation.

Case 1: Boy/Girl $\underline{\text{35 boys can dim the lights}} \cdot \underline{\text{15 girls can shut the door}} = 525$

Case 2: Girl/Boy $\underline{\text{15 girls can dim the lights}} \cdot \underline{\text{35 boys can shut the door}} = 525$

Adding the total outcomes from each case, we see that there are $N = 525 + 525 = 1050$ ways for a boy and a girl to complete the two jobs.

c. There are two methods that can be used to count how many outcomes there are if at least one boy completes a job.

Method I: Counting Cases

Count each case in which at least one boy does one of the jobs.

Case 1: Boy/Girl $\underline{\text{35 boys can dim the lights}} \cdot \underline{\text{15 girls can shut the door}} = 525$

Case 2: Girl/Boy $\underline{\text{15 girls can dim the lights}} \cdot \underline{\text{35 boys can shut the door}} = 525$

Case 3: Boy/Boy $\underline{\text{35 boys can dim the lights}} \cdot \underline{\text{35 boys can shut the door}} = 1225$

Counting these three cases, we get $N = 525 + 525 + 1225 = 2275$ total possible ways for at least one boy to complete a job.

Method II: All Minus the Exclusion

The only case that we have to exclude is the case in which both jobs are completed by a girl. Therefore, we determine the total number of possible outcomes, then subtract the case in which both jobs are completed by a girl.

Total Possible Outcomes $\underline{\text{50 students can dim the lights}} \cdot \underline{\text{50 students can shut the door}} = 2500$

Case: Girl/Girl $\underline{\text{15 girls can dim the lights}} \cdot \underline{\text{15 girls can shut the door}} = 225$

Therefore, we subtract 225 from 2500 to get $N = 2500 - 225 = 2275$ total possible ways for at least one boy to complete a job.

d. In this situation, any student can dim the lights but only a boy can shut the door.

Case: Any Student/Boy $\underline{\text{50 students can dim the lights}} \cdot \underline{\text{35 boys can shut the door}} = 1750$

Thus, there are $N = 1750$ total possible ways for any student to dim the lights and a boy to shut the door.

e. There are 6 seniors in the class, so there are 44 non-seniors.

Case:Non-senior/Non-senior $\underline{\text{44 non-seniors can dim the lights}} \cdot \underline{\text{44 non-seniors can shut the door}} = 1936$

Thus, there are $N = 1936$ ways that a non-senior can do both tasks.

f. There are 43 non-education majors.

Case: Non-ed major/Non-ed major $\underline{\text{43 non-ed majors can dim the lights}} \cdot \underline{\text{43 non-ed majors can shut the door}} = 1849$ ●

Thus, there are $N = 1849$ ways that a non-education major can do both tasks.

Example 6 Using the Fundamental Counting Principle

Using the same **demographic information in Example 5**, if the instructor chooses one student to dim the lights and one to shut the door, in how many ways can she do this if a student may **not** do both jobs and

a. both are girls?
b. one is a boy and one is a girl?
c. at least one is a boy?
d. the door closer is a boy?
e. neither are seniors?
f. neither are education majors?

Solution Try to determine the number of outcomes for (a)–(f) on your own; then **check your answers** or watch this **interactive video** to see a complete solution.

You Try It Work through this You Try It problem.

Work Exercises 13–17 in this textbook or in the MyLab Math Study Plan.

OBJECTIVE 2 USING PERMUTATIONS

Sometimes, we are interested in counting the number of ways that objects can be arranged or ordered. For example, suppose that you purchased a new combination lock for your new mountain bike. The lock is a four-digit lock in which each digit is chosen from 0, 1, 2, 3, 4, 5, 6, 7, 8, and 9. Obviously, the precise order of the digits matters. This is an example of a **permutation**. (Perhaps we should call this lock a permutation lock.)

> **Definition** Permutation
>
> A **permutation** is an ordered arrangement of n objects taken r at a time.

We will consider three counting techniques that involve permutations.

PERMUTATIONS INVOLVING DISTINCT OBJECTS WITH REPETITION

The first involves situations in which repetition is allowed.

> **Theorem: Permutations Involving Distinct Objects with Repetition Allowed**
>
> The number N of different arrangements of n objects using $r \le n$ of them, in which
>
> **1.** order matters,
>
> **2.** an object can be repeated, and
>
> **3.** the n objects are distinct,
>
> is given by the formula $N = n^r$.

We have already seen a permutation of this type in **Example 5a**, where the selected students were distinct, a student was allowed to do both jobs, and the order of selection mattered because it determined the specific job. Now let's look at another example.

Example 7 Determining the Number of Permutations If Repetition Is Allowed

How many four-digit arrangements are possible for a bicycle permutation lock if repetition of digits is allowed?

Solution In any counting situation involving an arrangement, the order matters. Repetition of digits is allowed and there are ten distinct objects to choose from (the digits 0–9) Therefore, we use the permutation formula $N = n^r$ with $n = 10$ and $r = 4$. Thus, the number of total possibilities is $N = n^r = 10^4 = 10{,}000$. ●

You Try It Work through this You Try It problem.

Work Exercises 18–20 in this textbook or in the MyLab Math Study Plan.

PERMUTATIONS INVOLVING DISTINCT OBJECTS WITHOUT REPETITION

In the previous example, repetition was allowed. Many counting problems involve situations where repetition is not allowed. Before we introduce the permutation formula for these situations, consider the permutation lock problem from **Example 6**, but this time suppose that each digit of the combination must be unique. In other words, the repetition of digits is not allowed. In this situation, there would be 10 possibilities for the first digit, 9 possibilities for the second digit, and so on. Thus, the total number of possibilities would be

$$\underline{10 \text{ possible digits}} \cdot \underline{9 \text{ possible digits}} \cdot \underline{8 \text{ possible digits}} \cdot \underline{7 \text{ possible digits}} = 5040$$

To illustrate how to produce the general formula for permutations of this nature, let's rewrite the left-hand side of the equation above. Before doing so, you may need to recall the definition of the **factorial of a non-negative number** that was introduced in Section 14.1.

$10\cdot 9\cdot 8\cdot 7$	Start with the left-hand side of the equation above.
$= 10\cdot 9\cdot 8\cdot 7\cdot \dfrac{6\cdot 5\cdot 4\cdot 3\cdot 2\cdot 1}{6\cdot 5\cdot 4\cdot 3\cdot 2\cdot 1}$	Multiply by 1 in the form $1 = \dfrac{6\cdot 5\cdot 4\cdot 3\cdot 2\cdot 1}{6\cdot 5\cdot 4\cdot 3\cdot 2\cdot 1}$.
$= \dfrac{10!}{6!}$	Rewrite the numerator and denominator using **factorial notation**.
$= \dfrac{10!}{(10-4)!}$	Rewrite the denominator.

As you can see, the number of permutations of 10 digits taken 4 at a time without repetition is given by the formula $\dfrac{10!}{(10-4)!}$. We can generalize this permutation formula as $\dfrac{n!}{(n-r)!}$, where n is the number of distinct objects taken r at a time without repetition. This formula is restated in the following theorem.

Theorem: Permutations Involving Distinct Objects with Repetition Not Allowed

The number N of different arrangements of n objects using $r \le n$ of them, in which

1. order matters,

2. an object cannot be repeated, and

3. the n objects are distinct,

is given by the formula $N = {}_nP_r = \dfrac{n!}{(n-r)!}$.

The notation ${}_nP_r$ represents a permutation of n distinct objects taken r at a time where repetition is not allowed and order matters.

Many counting examples (and probability examples seen in Section 14.7) involve using a standard deck of playing cards. If you are unfamiliar with cards, read this **explanation of a standard deck of 52 playing cards**.

Example 8 Counting Cards Using Permutations

Four cards are drawn one at a time without replacement from a standard deck of 52 cards. In how many ways can they be drawn

a. when there are no restrictions as to what the cards are?

b. when all of the cards are face cards?

Solution

a. The cards are drawn one at a time, and therefore order matters. Thus, we count using a permutation. The cards are not replaced between draws, so repetition is not allowed. We are choosing from a deck of 52 distinct cards. Therefore, we use the formula $N = {}_nP_r = \frac{n!}{(n-r)!}$ with $n = 52$ and $r = 4$.

$$N = {}_{52}P_4 = \frac{52!}{(52-4)!} = \frac{52!}{48!} = \frac{52 \cdot 51 \cdot 50 \cdot 49 \cdot 48!}{48!} = \frac{52 \cdot 51 \cdot 50 \cdot 49 \cdot \cancel{48!}}{\cancel{48!}} = 6{,}497{,}400$$

Therefore, there are $N = 6{,}497{,}400$ ways to draw four cards at random from a standard card deck.

b. As in the solution to part (a), the order matters, repetition is not allowed, and the cards are distinct. Thus we use the formula $N = {}_nP_r = \frac{n!}{(n-r)!}$. Again we are choosing $r = 4$ cards. However, in this case, $n = 12$ because there are exactly 12 face cards in a standard deck. (View this **explanation of a standard deck of 52 playing cards** if you need to see why.) Thus, we use the formula $N = {}_nP_r = \frac{n!}{(n-r)!}$ with $n = 12$ and $r = 4$.

$$N = {}_{12}P_4 = \frac{12!}{(12-4)!} = \frac{12!}{8!} = \frac{12 \cdot 11 \cdot 10 \cdot 9 \cdot 8!}{8!} = \frac{12 \cdot 11 \cdot 10 \cdot 9 \cdot \cancel{8!}}{\cancel{8!}} = 11{,}880$$

Therefore, there are $N = 11{,}880$ ways to draw four face cards. ●

Example 9 Counting the Number of Competitors in a Race

There are nine competitors in a hurdle race.

a. In how many different ways can the nine competitors finish first, second, and third?

b. In how many different ways can all nine competitors finish the race?

Solution

a. The order in which the competitors finish the race matters, so we count using a permutation. Repetition is not allowed because a runner cannot finish in two places at once. Also, the nine competitors are all distinct. Thus, we use the formula $N = {}_nP_r = \frac{n!}{(n-r)!}$ with $n = 9$ and $r = 3$.

$$N = {}_9P_3 = \frac{9!}{(9-3)!} = \frac{9!}{6!} = \frac{9 \cdot 8 \cdot 7 \cdot 6!}{6!} = \frac{9 \cdot 8 \cdot 7 \cdot \cancel{6!}}{\cancel{6!}} = 504$$

There are $N = 504$ ways to arrange the nine competitors in first, second, and third place.

b. Again we use the permutation formula $N = {}_nP_r = \dfrac{n!}{(n-r)!}$, but this time we use $n = 9$ and $r = 9$ because we are interested in how the nine runners can finish in all nine places.

$$N = {}_9P_9 = \frac{9!}{(9-9)!} = \frac{9!}{0!} = \frac{9!}{1} = 362{,}880$$

There are $N = 362{,}880$ ways to arrange the nine competitors in all nine places.

Note Example 9b is a special case of the formula $N = {}_nP_r = \dfrac{n!}{(n-r)!}$, where $n = r$. When permuting n distinct objects n at a time, we get $N = {}_nP_n = \dfrac{n!}{(n-n)!} = \dfrac{n!}{0!} = \dfrac{n!}{1} = n!$. For example, there are $17!$ possible ways that 17 kindergarteners can be lined up to go to lunch, with 17 choices for who goes first, 16 choices for who goes next, and so on. ●

You Try It Work through this You Try It problem.

Work Exercises 21–27 in this textbook or in the MyLab Math Study Plan.

PERMUTATIONS INVOLVING SOME NON-DISTINCT OBJECTS

A third type of permutation involves a situation where some of the objects in a set of size n are not distinct.

> **Theorem: Permutations Involving Some Objects That Are Not Distinct**
>
> The number N of different arrangements of n objects taken n at a time in which
>
> **1.** order matters,
>
> **2.** n_1 are of one kind, n_2 are of a second kind, . . . , and n_k are of a kth kind, and
>
> **3.** some of the objects are not distinct,
>
> is given by the formula $N = \dfrac{n!}{n_1!n_2!\ldots n_k!}$, where $n_1 + n_2 + \ldots + n_k = n$.

We illustrate the use of this theorem in Example 10.

Example 10 Arranging Letters in a Word That Has Repeated Letters

How many distinguishable letter codes can be formed from the word SUCCESSFUL if every letter is used?

Solution We are arranging the letters so order matters, so we count using a permutation. There are ten letters in the word "SUCCESSFUL" but there are only six different kinds of letters. There are $n_1 = 3$ "S's," $n_2 = 2$ "U's," $n_3 = 2$ "C's," $n_4 = 1$ "E," $n_5 = 1$ "F," and $n_6 = 1$ "L". Some of theletters are obviously not distinct. Therefore, we use the formula $N = \dfrac{n!}{n_1!n_2! \ldots n_k!}$ with $n = 10, n_1 = 3, n_2 = 2, n_3 = 2, n_4 = 1, n_5 = 1,$ and $n_6 = 1.$

Therefore, the number of distinguishable arrangements is

$$N = \frac{n!}{n_1!n_2!n_3!n_4!n_5!n_6!} = \frac{10!}{3!2!2!1!1!1!} = 151{,}200.$$

You Try It Work through this You Try It problem.

Work Exercises 28–31 in this textbook or in the MyLab Math Study Plan.

OBJECTIVE 3 USING COMBINATIONS

Combinations are different from permutations. The difference is that order does not matter and there is never repetition of objects.

> **Definition** Combination
>
> A **combination** is an unordered arrangement of n distinguishable objects taken r at a time without repetition.

Before we establish a formula for the number of combinations of n distinguishable objects taken r at a time, consider the following example, which illustrates the difference between a permutation and a combination.

Example 11 Choosing Books at the Library

Suppose that you went to the library and you chose the following four American novels off the shelf:

Tom Sawyer, Little Women, The Grapes of Wrath, and *The Great Gatsby*

a. Suppose that you place two of these books back on the shelf. How many arrangements are there?

b. How many groups of two books chosen from these four books can you check out to take home?

Solution

a. When arranging books on the shelf, the order matters. We could use the permutation formula $N = {}_nP_r = \dfrac{n!}{(n-r)!}$ with $n = 4$ and $r = 2$ because the objects are distinct and no repetition is allowed.

However, to illustrate a point, we list each possibility separately.

Possibilities of Arranging Two of Four Novels on a Shelf

1. *Tom Sawyer, Little Women*	**5.** *Little Women, The Grapes of Wrath*
2. *Tom Sawyer, The Grapes of Wrath*	**6.** *Little Women, The Great Gatsby*
3. *Tom Sawyer, The Great Gatsby*	**7.** *The Grapes of Wrath, Tom Sawyer*
4. *Little Women, Tom Sawyer*	**8.** *The Grapes of Wrath, Little Women*

9. *The Grapes of Wrath, The Great Gatsby*
10. *The Great Gatsby, Tom Sawyer*
11. *The Great Gatsby, Little Women*
12. *The Great Gatsby, The Grapes of Wrath*

We see that there are 12 ways to arrange these four classic novels on a shelf two at a time.

b. When checking out the books to take home, the order in which the books are checked out does not matter. So checking out *Tom Sawyer* and *Little Women* is no different than checking out *Little Women* and *Tom Sawyer.* Below are the possible ways to check out two of these four novels.

Possibilities of Checking Out Two of Four Novels from the Library

1. *Tom Sawyer, Little Women*
2. *Tom Sawyer, The Grapes of Wrath*
3. *Tom Sawyer, The Great Gatsby*
4. *Little Women, The Grapes of Wrath*
5. *Little Women, The Great Gatsby*
6. *The Grapes of Wrath, The Great Gatsby*

As you can see, there are six ways that we can check out two out of the four great American novels.

Notice in Example 11a that there are 12 arrangements of the four novels taken two at a time whereas in Example 11b there are only six groups of four novels taken two at a time. This is because in Example 11b we do not need to count the same pair of books twice when checking the books out. The process of counting used in Example 11b is a combination.

In Example 11b, for each pair of books chosen, there are exactly $2 = 2!$ ways to arrange them. In fact, for any collection of r objects (where $r \geq 1$) there are always $r!$ ways to arrange them. Therefore, to produce the formula to describe the combination of n distinct objects taken r at a time, we simply start with the formula for a permutation of n distinct objects taken r at a time, then divide by $r!$.

Theorem: Combinations Involving Objects That Are Distinct

The number N of different arrangements of n objects using $r \leq n$ of them, in which

1. order does not matter,
2. an object cannot be repeated, and
3. the n objects are distinct,

is given by the formula $N = {}_nC_r = \dfrac{n!}{(n-r)!r!}$.

The notation ${}_nC_r$ represents a combination of n distinct objects taken r at a time where repetition is not allowed and order does not matter.

Example 12 Counting the Number of Possibilities in the Louisiana Lottery

The Louisiana lottery is a game in which six numbers are chosen from the numbers 1 to 44. Order does not matter and repetition of numbers is not allowed. How many possible sets of numbers are there in the Louisiana lottery?

Solution We use a combination to count the possible sets of numbers in the Louisiana lottery because the order does not matter. We must determine the number of combinations of the 44 numbers taken 6 at a time. Thus, we use the formula $N = {}_nC_r = \dfrac{n!}{(n-r)!r!}$, where $n = 44$ and $r = 6$.

$$N = {}_{44}C_6 = \frac{44!}{(44-6)!\cdot 6!} = \frac{44!}{38!\cdot 6!} = \frac{44\cdot 43\cdot 42\cdot 41\cdot 40\cdot 39\cdot \cancel{38!}}{\cancel{38!}\cdot 6!}$$

$$= \frac{44\cdot 43\cdot 42\cdot 41\cdot 40\cdot 39}{6\cdot 5\cdot 4\cdot 3\cdot 2\cdot 1} = 7{,}059{,}052$$

There are $N = 7{,}059{,}052$ possible sets of numbers in the Louisiana lottery.

You Try It Work through this You Try It problem.

Work Exercises 32–36 in this textbook or in the MyLab Math Study Plan.

Example 13 Counting the Number of Poker Hands

A poker hand consists of 5 cards that are dealt from a typical **standard deck of 52 cards.**

a. How many total possible poker hands are there?

b. How many possible poker hands are there that consist of exactly two aces and exactly two kings? (The fifth card is a non-ace and non-king.)

c. How many possible poker hands are there that consist of exactly two queens and exactly three hearts?

Solution

a. The order of the five cards dealt in a poker hand does not matter, so we use a combination to count the total number of poker hands. To determine the number of possible poker hands, we must determine the number of combinations of the 52 cards taken 5 at a time.

Thus, we use the formula $N = {}_nC_r = \dfrac{n!}{(n-r)!\cdot r!}$, where $n = 52$ and $r = 5$.

$$N = {}_{52}C_5 = \frac{52!}{(52-5)!\cdot 5!} = \frac{52!}{47!\cdot 5!} = \frac{52\cdot 51\cdot 50\cdot 49\cdot 48\cdot \cancel{47!}}{\cancel{47!}\cdot 5!}$$

$$= \frac{52\cdot 51\cdot 50\cdot 49\cdot 48}{5\cdot 4\cdot 3\cdot 2\cdot 1} = 2{,}598{,}960$$

Therefore, there are $N = 2{,}598{,}960$ possible poker hands.

b. To determine the number of possible poker hands that consist of exactly two aces and exactly two kings, there are three different events that must be considered.

Event 1: Choose the two aces.

There are four aces from which two must be chosen. The order does not matter, so we count using a combination. The number of possible ways to choose two aces is $_4C_2 = \mathbf{6}$.

Event 2: Choose the two kings.

There are four kings from which two must be chosen. Again, the order does not matter, so we count using a combination. The number of possible ways to choose two kings is $_4C_2 = \mathbf{6}$.

Event 3: Choose one card that is a non-ace and non-king.

There are 44 remaining non-ace and non-king cards from which 1 must be chosen. The order does not matter, so we once again count using a combination. The number of possible ways to choose the remaining card is $_{44}C_1 = \mathbf{44}$.

Using the fundamental counting principle, we multiply the number of possible outcomes of each event.

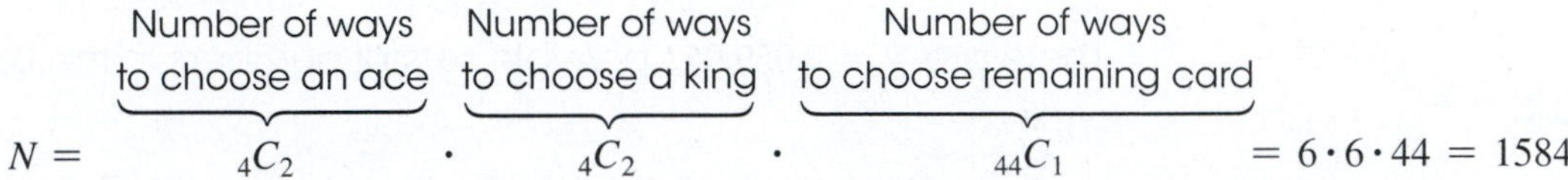

Thus, there are $N = 1584$ different poker hands that consist of exactly two aces and exactly two kings.

Note In the expression $_4C_2 \cdot {_4C_2} \cdot {_{44}C_1}$, the sum of the three values used for n $(4, 4, 44)$ is equal to 52. This can be used to check to see that you have considered all 52 cards in the deck. Also, the sum of the three values used for r $(2, 2, 1)$ is equal to 5. This can be used to check to see that you have chosen the correct number of cards.

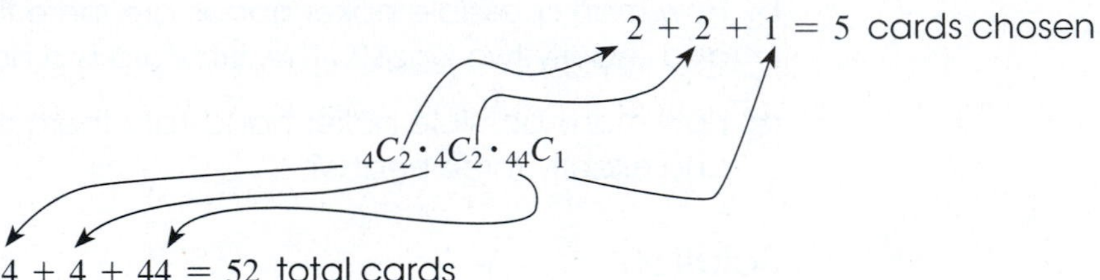

c. To determine the number of possible poker hands that consist of exactly two queens and exactly three hearts, there are two possible cases to consider. The first case is the case in which the poker hand contains the queen of hearts. The second case is the case in which the poker hand does not contain the queen of hearts.

Case 1: The hand contains the queen of hearts

For this case, there are four events to consider. Each event uses a combination because the order in which cards are chosen does not matter.

Event 1: Choose the queen of hearts.

There is one queen of hearts so there is $_1C_1 = \mathbf{1}$ way to choose the queen of hearts.

Event 2: Choose the second queen.

We must choose one out of the remaining three queens. Thus, there are $_3C_1 = 3$ ways to choose the remaining queen.

Event 3: Choose the remaining two hearts.

We must choose two out of the remaining 12 hearts. (One of the original 13 hearts, the queen of hearts, has already been accounted for in Event 1.) Thus, there are $_{12}C_2 = 66$ ways to choose the remaining two hearts.

Event 4: Choose the last card.

Out of the 36 non-queen and non-heart cards left, we must choose one of them to complete the five-card hand. Thus, there are $_{36}C_1 = 36$ ways to choose the last card.

Using the fundamental counting principle, we get the following:

$$\underbrace{{}_1C_1}_{\substack{\text{Number of ways}\\ \text{to choose the}\\ \text{queen of hearts}}} \cdot \underbrace{{}_3C_1}_{\substack{\text{Number of ways}\\ \text{to choose the}\\ \text{second queen}}} \cdot \underbrace{{}_{12}C_2}_{\substack{\text{Number of ways}\\ \text{to choose the}\\ \text{remaining two hearts}}} \cdot \underbrace{{}_{36}C_1}_{\substack{\text{Number of ways}\\ \text{to choose the}\\ \text{remaining card}}} = 1 \cdot 3 \cdot 66 \cdot 36 = 7128$$

Again, note that the sum of the four values used for n adds up to 52. $(1 + 3 + 12 + 36 = 52 \text{ total cards})$

The sum of the four values used for r adds up to 5. $(1 + 1 + 2 + 1 = 5 \text{ cards chosen})$

Case 2: The hand does not contain the queen of hearts

For this case, there are four events to consider. Each event uses a combination because the order in which cards are chosen does not matter.

Event 1: Consider the queen of hearts but do not choose it.

There is one queen of hearts so there is $_1C_0 = 1$ way not to choose the queen of hearts.

Event 2: Choose two queens out of the three queens that are not the queen of hearts. There are three queens that are not hearts in which two must be chosen, so there are $_3C_2 = 3$ ways to choose these two queens.

Event 3: Choose the three hearts.

We must choose three hearts out of the twelve non-queen hearts. Therefore, there are $_{12}C_3 = 220$ ways to choose the three hearts.

Event 4: Account for the cards remaining in the deck.

Out of the 36 non-queen and non-heart cards left, we do not choose any more cards because we already have a five-card hand. Thus, there are $_{36}C_0 = 1$ ways not to choose another card.

Using the fundamental counting principle, we get the following:

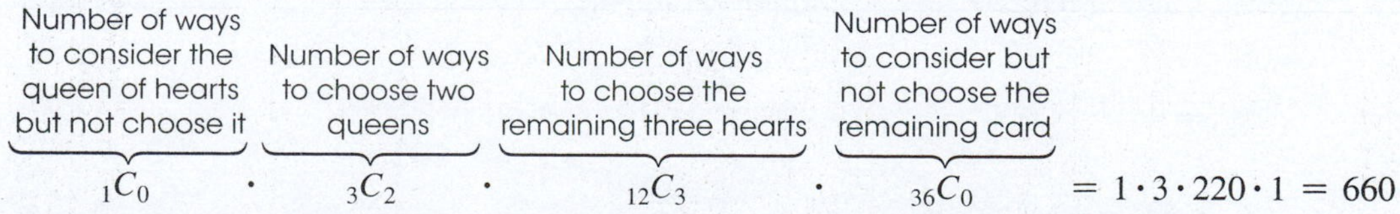

$$_1C_0 \cdot {}_3C_2 \cdot {}_{12}C_3 \cdot {}_{36}C_0 = 1 \cdot 3 \cdot 220 \cdot 1 = 660$$

Again, note that the sum of the four values used for n adds up to 52. $(1 + 3 + 12 + 36 = 52 \text{ total cards})$

The sum of the four values used for r adds up to 5. $(0 + 2 + 3 + 0 = 5 \text{ cards chosen})$

Adding the number of possible hands from each case, we get

$$N = \underbrace{{}_1C_1 \cdot {}_3C_1 \cdot {}_{12}C_2 \cdot {}_{36}C_1}_{7128} + \underbrace{{}_1C_0 \cdot {}_3C_2 \cdot {}_{12}C_3 \cdot {}_{36}C_0}_{660} = 7788$$

Therefore, there are $N = 7788$ possible hands that contain exactly two queens and exactly three hearts.

 Watch this **interactive video** for a detailed explanation of parts (a)–(c).

You Try It Work through this You Try It problem.

Work Exercises 37–45 in this textbook or in the MyLab Math Study Plan.

Given a counting problem, it is crucial that we first identify if it is a permutation or a combination. Always try to decide whether or not the order matters. If the order does not matter, then the problem involves a combination. If the order matters, then the problem involves a permutation and then the type of permutation must be determined. **Table 3** summarizes the fundamental counting principle and all of the counting techniques discussed in this section.

Table 3 The Fundamental Counting Principle and Counting Techniques

Fundamental Counting Principle	$N = n_1 n_2 \ldots n_k$	Suppose that an experiment or task is made up of two or more events. If the first event can occur in n_1 ways, the second event can occur in n_2 ways, and the third event can occur in n_3 ways and so on, then the number of possible outcomes of the experiment or task is $N = n_1 n_2 \ldots n_k$, where k is the total number of events.

Counting Techniques	Formula	Order Matters	Repetition Allowed	Distinct Objects
Permutations	$N = n^r$	✔	✔	✔
	$N = {}_nP_r = \dfrac{n!}{(n-r)!}$	✔	×	✔
	$N = \dfrac{n!}{n_1! \cdot n_2! \cdot \cdots \cdot n_k!}, n_1 + n_2 + \cdots + n_k = n$	✔	×	×
Combinations	$N = {}_nC_r = \dfrac{n!}{(n-r)! \cdot r!}$	×	×	✔

14.6 Exercises

1. A man has three pairs of slacks, four shirts, and five ties that all coordinate. How many different outfits of slacks, shirts, and ties can he put together?

2. The menu at Ben's Burgers is as follows:

Burgers	Sides	Drinks
Hamburger	Small fry	Water
Cheeseburger	Large fry	Coffee
Double cheeseburger	Small tater-tots	Tea
Double deluxe cheeseburger	Large tater-tots	Cola
	Salad	Root beer
		Sparkling cider

   How many ways can a person order one burger, one side, and one drink?

3. You have narrowed your choice of a new car to either a four-door sedan or an SUV. Each car comes in white, black, tan, or green. Each model has the option of a gas engine, diesel engine, or a hybrid engine. Each model also has the option of a manual transmission or an automatic transmission. How many different choices do you still have to consider?

4. All automobile license plates in a certain state consist of two capital letters followed by five digits. (Repeated letters are allowed but repeated digits are not allowed.) How many different license plates can there be for this state?

5. A student is taking a five-question quiz. The first two questions are True/False questions. The last three are multiple choice in which she can choose the answer A, B, C, or D. In how many ways can she answer the entire test?

6. A student is to create a password for her **MyLab Math** homework account. She is informed that the password must consist of six characters. How many possible ways can she create the password if the first two characters are digits followed by three letters and one digit? (Repeated letters are allowed and the letters are case sensitive. Also, repeated digits are not allowed.)

7. Amy is planning a week-end trip to Atlanta. She plans to take six t-shirts all of different colors, five pairs of shorts all of different patterns, and two pairs of shoes of different styles. How many outfits can she mix and match if

   a. all pieces of clothing match?

   b. her pink t-shirt clashes with her red shorts and her orange polka-dot shorts, but everything else coordinates?

In Exercises 8–12, refer to the following menu. In each exercise, assume that an order consists of one main item, one side dish, and one choice of bread.

Main	Side Dishes	Bread
fried chicken	carrots	cornbread
BBQ steak	french fries	french bread
baked ham	potato salad	biscuits
	baked potato	
	dirty rice	

8. How many possible orders are there if you like everything on the menu?

9. How many possible orders are there if you don't like potato salad?

10. How many possible orders are there if you only like BBQ steak served with a baked potato and french bread, but you like all other combinations?

11. How many possible orders are there if you only like fried chicken, and you do not like baked potatoes or cornbread?

12. How many possible orders are there if you like everything except french fries and french bread?

In Exercises 13–17, refer to the following demographic information that has been collected by an instructor after surveying his class.

Total number of students in the class:	46
Number of boys:	19
Number of girls:	27
Number of freshmen:	7
Number of sophomores:	16
Number of juniors:	12
Number of seniors:	11
Number of science majors:	31
Number of non-science majors:	15

13. If the instructor chooses one student to pass out papers, one student to clean the chalkboard, and one student to empty the trash, then how many ways can he do this if he picks all girls and a student may do more than one job?

14. If the instructor chooses one student to pass out papers, one student to clean the chalkboard, and one student to empty the trash, then how many ways can he do this if at least one student is a girl and a student may not do more than one job?

15. If the instructor chooses one student to pass out papers, one student to clean the chalkboard, and one student to empty the trash, then how many ways can he do this if all three students are not the same sex and a student may not do more than one job?

16. If the instructor chooses one student to pass out papers, one student to clean the chalkboard, and one student to empty the trash, then how many ways can he do this if at least one student is a senior and a student may do more than one job?

17. If the instructor chooses one student to pass out papers, one student to clean the chalkboard, and one student to empty the trash, then how many ways can he do this if at most one student is a senior and a student may not do more than one job?

18. In how many ways can four cards be drawn from a standard deck if the card is replaced between draws?

19. How many six-digit passwords are possible if repetition of digits is allowed?

20. A football coach must decide who is going to receive the following awards at the end-of-season banquet:

    Most Valuable Player, Mr. "Hustle," Most Improved, Most Inspirational

    How many ways can the 56 players on the team be chosen to win these four awards if a player can win more than one award?

21. Four customers are heading to the checkout lines at a local grocery store. In how many ways can three of the four people arrive and stand in line at the express checkout line?

22. How many ways can the letters of the word ORBIT be arranged if all of the letters are used without repetition?

23. How many ways can eight DVDs be arranged on a shelf?

24. A typical game of SCRABBLE begins with each player randomly choosing seven letter tiles. How many ways can a player arrange five of the seven letter tiles assuming that each tile is a different letter?

25. Three students are chosen from a class of 35. One will get an A, one will get a C, and the other will get an F. In how many different ways can these three students be chosen?

26. Three cards are drawn one at a time without replacement from a standard deck of 52 cards. In how many ways can exactly three hearts be drawn?

27. A football coach must decide who is going to receive the following awards at the end of season banquet:

    Most Valuable Player, Mr. "Hustle," Most Improved, Most Inspirational

    How many ways can the 56 players on the team be chosen to win these awards if no player can win more than one award?

28. How many distinguishable letter codes can be formed from the word MISSISSIPPI if every letter is used?

29. A child has 5 red balls, 3 blue balls, and 2 green balls. All 10 balls are to be lined up on a shelf. How many total distinguishable arrangements of the 10 balls are there?

30. A binary number is a number that consists of only the digits 0 or 1. How many distinguishable binary numbers can be formed from the binary number 11011001?

31. In how many ways can the algebraic expression $x^5y^3z^2$ be written as a product without using any exponents?

SbS 32. How many five-member committees can be formed from a class of 30 students?

SbS 33. A kindergarten child is allowed to pick two toys from the goody box for demonstrating good classroom behavior. How many pairs of toys can she pick from a box of 25 toys?

SbS 34. A basketball conference consists of 12 teams. How many total in conference games must be played for each team to play each other exactly once?

SbS 35. A football coach must decide who are going to be the team captains for the last game of the season. There will be four team captains for the last game. How many distinct sets of team captains are possible if there are 46 players on the team?

SbS 36. How many ways can 7 cards be dealt from a standard deck of 52 cards?

SbS 37. A neighborhood contains 5 coffee shops, 3 pizza restaurants, and 2 sushi restaurants. Three of the businesses are to be chosen at random. How many possible outcomes are there?

SbS 38. A neighborhood contains 5 coffee shops, 3 pizza restaurants, and 2 sushi restaurants. Three of the businesses are to be chosen at random. Of the total number of possible outcomes, how many would consist only of coffee shops?

SbS 39. A neighborhood contains 5 coffee shops, 3 pizza restaurants, and 2 sushi restaurants. Three of the businesses are to be chosen at random. Of the total number of possible outcomes, how many would consist only of pizza restaurants?

SbS 40. Three cards are dealt without replacement from a standard deck of fifty-two. In how many ways can this happen if there are no restrictions?

SbS 41. Three cards are dealt without replacement from a standard deck of fifty-two. In how many ways can this happen if all three are hearts?

SbS 42. Three cards are dealt without replacement from a standard deck of fifty-two. In how many ways can this happen if two cards are hearts and one is a spade?

SbS 43. A gin rummy hand consists of 7 cards dealt without replacement from a standard deck of 52 cards. How many gin rummy hands are there that contain any 7 cards?

SbS 44. A gin rummy hand consists of 7 cards dealt without replacement from a standard deck of 52 cards. How many gin rummy hands are there that contain exactly two aces and exactly two tens?

SbS 45. A gin rummy hand consists of 7 cards dealt without replacement from a standard deck of 52 cards. How many gin rummy hands are there that contain exactly two jacks and exactly five hearts?

Brief Exercises

46. Four customers are heading to the checkout lines at a local grocery store. In how many ways can three of the four people arrive and stand in line at the express checkout line?

47. How many ways can the letters of the word ORBIT be arranged if all of the letters are used without repetition?

48. How many ways can eight DVDs be arranged on a shelf?

49. A typical game of SCRABBLE begins with each player randomly choosing seven letter tiles. How many ways can a player arrange five of the seven letter tiles assuming that each tile is a different letter?

50. Three students are chosen from a class of thirty-five. One will get an A, one will get a C, and the other will get an F. In how many different ways can these three students be chosen?

51. Three cards are drawn one at a time without replacement from a standard deck of fifty-two cards. In how many ways can exactly three hearts be drawn?

52. A football coach must decide who is going to receive the following awards at the end of season banquet:

    Most Valuable Player, Mr. "Hustle," Most Improved, Most Inspirational

    How many ways can the 56 players on the team be chosen to win these awards if no player can win more than one award?

53. How many five-member committees can be formed from a class of 30 students?

54. A kindergarten child is allowed to pick two toys from the goody box for demonstrating good classroom behavior. How many pairs of toys can she pick from a box of 25 toys?

55. A basketball conference consists of 12 teams. How many total in conference games must be played for each team to play each other exactly once?

56. A football coach must decide who are going to be the team captains for the last game of the season. There will be four team captains for the last game. How many distinct sets of team captains are possible if there are 46 players on the team?

57. How many ways can 7 cards be dealt from a standard deck of 52 cards?

58. A neighborhood contains 5 coffee shops, 3 pizza restaurants, and 2 sushi restaurants. Three of the businesses are to be chosen at random. How many possible outcomes are there?

59. A neighborhood contains 5 coffee shops, 3 pizza restaurants, and 2 sushi restaurants. Three of the businesses are to be chosen at random. Of the total number of possible outcomes, how many would consist only of coffee shops?

60. A neighborhood contains 5 coffee shops, 3 pizza restaurants, and 2 sushi restaurants. Three of the businesses are to be chosen at random. Of the total number of possible outcomes, how many would consist only of pizza restaurants?

61. Three cards are dealt without replacement from a standard deck of fifty-two. In how many ways can this happen if there are no restrictions?

62. Three cards are dealt without replacement from a standard deck of fifty-two. In how many ways can this happen if all three are hearts?

63. Three cards are dealt without replacement from a standard deck of fifty-two. In how many ways can this happen if two cards are hearts and one is a spade?

64. A gin rummy hand consists of 7 cards dealt without replacement from a standard deck of 52 cards. How many gin rummy hands are there that contain any 7 cards?

65. A gin rummy hand consists of 7 cards dealt without replacement from a standard deck of 52 cards. How many gin rummy hands are there that contain exactly two aces and exactly two tens?

66. A gin rummy hand consists of 7 cards dealt without replacement from a standard deck of 52 cards. How many gin rummy hands are there that contain exactly two jacks and exactly five hearts?

67. Four customers are heading to the checkout lines at a local grocery store. In how many ways can three of the four people arrive and stand in line at the express checkout line?

68. How many ways can the letters of the word ORBIT be arranged if all of the letters are used without repetition?

69. How many ways can eight DVDs be arranged on a shelf?

70. A typical game of SCRABBLE begins with each player randomly choosing seven letter tiles. How many ways can a player arrange five of the seven letter tiles assuming that each tile is a different letter?

71. Three students are chosen from a class of thirty-five. One will get an A, one will get a C, and the other will get an F. In how many different ways can these three students be chosen?

72. Three cards are drawn one at a time without replacement from a standard deck of fifty-two cards. In how many ways can exactly three hearts be drawn?

73. A football coach must decide who is going to receive the following awards at the end of season banquet:

    Most Valuable Player, Mr. "Hustle," Most Improved, Most Inspirational

    How many ways can the 56 players on the team be chosen to win these awards if no player can win more than one award?

74. How many five-member committees can be formed from a class of 30 students?

75. A kindergarten child is allowed to pick two toys from the goody box for demonstrating good classroom behavior. How many pairs of toys can she pick from a box of 25 toys?

76. A basketball conference consists of 12 teams. How many total in conference games must be played for each team to play each other exactly once?

77. A football coach must decide who are going to be the team captains for the last game of the season. There will be four team captains for the last game. How many distinct sets of team captains are possible if there are 46 players on the team?

78. How many ways can 7 cards be dealt from a standard deck of 52 cards?

79. A bag contains five red balls, three blue balls, and two green balls. Three balls are to be selected at random. How many possible outcomes are there?

80. A bag contains five red balls, three blue balls, and two green balls. Three balls are to be selected at random. Of the total number of possible outcomes, how many would only contain red balls?

81. A bag contains five red balls, three blue balls, and two green balls. Three balls are to be selected at random. Of the total number of possible outcomes, how many would only contain blue balls?

82. Three cards are dealt without replacement from a standard deck of fifty-two. In how many ways can this happen if there are no restrictions?

83. Three cards are dealt without replacement from a standard deck of fifty-two. In how many ways can this happen if all three are hearts?

84. Three cards are dealt without replacement from a standard deck of fifty-two. In how many ways can this happen if two cards are hearts and one is a spade?

85. A gin rummy hand consists of 7 cards dealt without replacement from a standard deck of 52 cards. How many gin rummy hands are there that contain any 7 cards?

86. A gin rummy hand consists of 7 cards dealt without replacement from a standard deck of 52 cards. How many gin rummy hands are there that contain exactly two aces and exactly two tens?

87. A gin rummy hand consists of 7 cards dealt without replacement from a standard deck of 52 cards. How many gin rummy hands are there that contain exactly two jacks and exactly five hearts?

14.7 An Introduction to Probability

THINGS TO KNOW

Before working through this section, be sure that you are familiar with the following concepts:

VIDEO ANIMATION INTERACTIVE

You Try It

1. Using the Fundamental Counting Principle (Section 14.6)

You Try It

2. Using Combinations (Section 14.6)

INTRODUCTION

Read this introduction before beginning Objective 1.

OBJECTIVES

1 Understanding Probability

2 Determining Probability Using the Additive Rule and Venn Diagrams

3 Using Combinations to Determine Probabilities

4 Using Conditional Probability

5 Understanding Odds

SECTION 14.7 EXERCISES

Introduction to Section 14.7

Probability has been of interest to man as far back as recorded history can determine. Games of chance were played by the Assyrians, the Sumerians, and the Egyptians. Though their games differed from the games we play today, the probability of certain outcomes was as predictable. The theory of probability was developed further during the 1600s by famous mathematicians such as **Blaise Pascal** and **Pierre Fermat**. During the late 1800s, Russian mathematicians became interested in probability, and this further enhanced the probability theory we use today.

But probability is not just for games. It has practical uses in any situation where a decision needs to be made in the face of uncertainty, provided that the uncertainties can be quantified. Life insurance companies rely heavily on probability to set accident and death rates for specific age groups and sexes. Also, probability arises in evaluating the data gathered by all of the different sciences—physics, chemistry, biology, psychology, sociology, and political science. Probability theory is a major branch of mathematics, with hundreds—if not thousands—of specialists. Research institutions often have groups of mathematicians called probabilists. Their work has applications in such diverse areas as data transmission, quantum theory, neurophysics, and the analysis of financial markets.

But whatever your reasons for studying probability, you will find that it is a fascinating topic. You will be surprised how many times you will notice probability or odds used after you have studied the subject.

OBJECTIVE 1 UNDERSTANDING PROBABILITY

Most of us have tossed a coin and asked a friend to call "heads or tails." If the coin is "fair," then the probability that the coin lands on heads is $\frac{1}{2} = 0.5 = 50\%$. We can therefore think of probability as the likelihood that an event will occur. In the experiment of tossing a coin, the likelihood of the coin landing on heads is $\frac{1}{2} = 0.5 = 50\%$.

Definition Probability

Probability is the mathematical estimate of the likelihood that an event or sequence of events will occur.

Probabilities are usually written as fractions or decimals, and sometimes they are written as percents, depending on what seems most appropriate for the situation. Probabilities range from 0 to 1, or from 0% to 100%.

Every experiment has a total number of possible outcomes. In the experiment of tossing a fair coin, there are two possible outcomes. The first outcome is the coin landing on heads and the second outcome is the coin landing on tails. The set of all possible outcomes of an experiment is known as the **sample space.**

Definition Sample Space

The **sample space,** S, for an experiment is the set of all possible outcomes for the experiment. The number of possible outcomes of the sample space is denoted as $n(S)$.

Example 1 Determining the Sample Space of an Experiment

State the number of elements in the sample space of the following experiments.

a. Of the 20,000 students on a college campus, one student is selected to be the mascot at the football game this weekend.

b. One card is to be selected at random from a standard deck of 52 playing cards that has all of the diamonds removed.

c. One marble is to be selected at random from a marble bag that contains two green marbles, nine red marbles, four white marbles, and five blue marbles.

d. Three marbles are to be selected at random from a marble bag that contains two green marbles, nine red marbles, four white marbles, and five blue marbles.

Solution

a. The number of elements of the sample space is $n(S) = 20{,}000$.

b. There are 13 diamonds in a standard deck of 52 playing cards. So there are $52 - 13 = 39$ cards from which exactly one is selected. Thus the size of the sample space of this experiment is $n(S) = {}_{39}C_1 = 39$.

c. There are a total of 20 marbles in the bag from which exactly one is selected. Therefore, the size of the sample space of this experiment is $n(S) = {}_{20}C_1 = 20$.

d. There are a total of 20 marbles in the bag in which three marbles are selected. Therefore, the size of the sample space of this experiment is $n(S) = {}_{20}C_3 = 1140$. ●

Note In the solution to part c, above, we may have initially thought about considering counting only four possible outcomes for the sample space:

$$\{\text{green, red, white, blue}\}$$

This would suggest that the size of the sample space was 4, meaning that each color marble is equally likely to get chosen from the bag. This, of course, is not the case since the number of marbles of each color in the bag is not the same. Since we are counting the sample space in preparation for determining probability, we will see that we must consider all 20 marbles as if they are all distinct from one another. In the solution to part d, we also determined the size of the sample space by assuming that the marbles were distinct.

Thus, for part c, the list of total possible outcomes is really {green1, green2, red1, red2, . . . red9, white1, . . . white4, blue1, . . . blue5}. This is a total of 20 or ${}_{20}C_1$. For part d, the total possible outcomes are much more difficult to list. But, if we tried to list every possible grouping of three marbles with the understanding that all marbles were distinct, then we would get a list that starts to look something like this:

$$\{\{green1, green2, red1\}, \{green1, green2, red2\}, \{green1, green2, red3\}, \ldots\}$$

If we continued listing all outcomes, we would get a total of ${}_{20}C_3 = 1140$ groups of three marbles.

You Try It Work through this You Try It problem.

Work Exercises 1–7 in this textbook or in the MyLab Math Study Plan.

To compute the probability that an event E will occur, we must know the total number of possible ways that the event can occur and we must also know the size of the sample space of the experiment. The probability that the event will occur is the quotient of these two numeric values.

The Probability of an Event

Suppose that an experiment has a sample space S. The probability of event E occurring is denoted as $P(E)$ and is given by

$$P(E) = \frac{\text{the number of ways event } E \text{ can occur}}{\text{the total number of possible outcomes}} = \frac{n(E)}{n(S)}.$$

When studying probability, there are several principles that we must be aware of before proceeding. Three such principles are outlined below.

Principles of Probability

1. The probability of a certain event (an event that is guaranteed to happen) is 1, written as $P(\text{certain } E) = 1$.
2. The probability of an impossible event is 0, written as $P(\text{impossible } E) = 0$.
3. The sum of the probabilities of all possible outcomes of an experiment is 1, written as $P(E_1) + P(E_2) + \ldots + P(E_n) = 1$, where $E_1, E_2, \ldots, E_n$ are the possible outcomes of the experiment.

We must be careful when considering principle 3. The sum of the probabilities of all possible outcomes of an experiment is equal to 1 only when each event of the experiment is **mutually exclusive**. If the events are not mutually exclusive, the sum will not be 1. See the note at the end of Example 3.

Example 2 Determining Probabilities

A paint-ball gun ball hopper contains 11 yellow balls, 6 green balls, and 3 red balls. A ball is fired from the gun. Find the probabilities of each of the following events written as a fraction in lowest terms.

R: The ball is red

G: The ball is green

Y: The ball is yellow

Solution The experiment is firing a ball from a gun. The size of the sample space is $n(S) = 11 + 6 + 3 = 20$. The probability that a certain ball is fired is the ratio of the number of balls of that color to the size of the sample space. Thus, we get the following probabilities:

$$P(R) = \frac{n(R)}{n(S)} = \frac{3}{20}, \quad P(G) = \frac{n(G)}{n(S)} = \frac{6}{20} = \frac{3}{10}, \quad P(Y) = \frac{n(Y)}{n(S)} = \frac{11}{20}$$

The probability of firing a red ball is $\frac{3}{20}$.

The probability of firing a green ball is $\frac{6}{20} = \frac{3}{10}$.

The probability of firing a yellow ball is $\frac{11}{20}$.

Note In the previous example, the experiment was made up of exactly three events. The fired ball could be red, green, or yellow. Note that

$$P(R) + P(G) + P(Y) = \frac{3}{20} + \frac{6}{20} + \frac{11}{20} = \frac{20}{20} = 1,$$

which is exactly what we would expect from **principle 3** because these events are **mutually exclusive**.

You Try It Work through this You Try It problem.

Work Exercises 8–12 in this textbook or in the MyLab Math Study Plan.

Examples 3 through 5 pertain to the following demographic information collected from a teacher of mathematics at the local community college.

Total number of students in the class:	50
Number of boys:	35
Number of girls:	15
Number of freshmen:	22
Number of sophomores:	13
Number of juniors:	9
Number of seniors:	6

Example 3 Determining Probabilities

The instructor chooses one student to dim the lights and the next one to shut the door (to use the overhead projector). If a student may do both jobs, determine the following probabilities rounded to the nearest tenth of a percent as needed.

a. P(both are girls)

b. P(one is a girl and one is a boy)

c. P(at least one is a boy)

d. P(the door closer is a boy)

e. P(neither are seniors)

Solution The experiment is choosing a student to dim the lights and a student to shut the door. Because a student can do both jobs, there are 50 students that can be chosen for each job. Thus, by the **fundamental counting principle**, the size of the sample space is $n(S) = 50 \cdot 50 = 2500$.

a. Because both jobs can be completed by the same girl, we know that there are 15 girls that can do each job. Therefore, the number of outcomes in which both students are girls is

$$n(\text{both are girls}) = \underline{\text{15 girls can dim the lights}} \cdot \underline{\text{15 girls can shut the door}} = 225$$

The probability that both jobs are done by a girl is

$$P(\text{both are girls}) = \frac{n(\text{both are girls})}{n(S)} = \frac{225}{2500} = \frac{9}{100} = 0.09.$$

Thus, there is a 9% probability that two girls will be chosen to do both jobs.

b. First determine n(one is a girl and one is a boy), the number of outcomes in which one student picked is a girl and one student picked is a boy. To do this, we must consider two cases.

Case 1: Boy-Girl $\underline{\text{35 boys can dim the lights}} \cdot \underline{\text{15 girls can shut the door}} = 525$

Case 2: Girl-Boy $\underline{\text{15 girls can dim the lights}} \cdot \underline{\text{35 boys can shut the door}} = 525$

Adding the total outcomes from each case we see that $n(\text{one is a girl and one is a boy}) = 525 + 525 = 1050$. Thus,

$$P(\text{one is a girl and one is a boy}) = \frac{n(\text{one is a girl and one is a boy})}{n(S)} = \frac{1050}{2500} = \frac{21}{50} = 0.42.$$

Therefore, there is a 42% probability that one girl and one boy will be chosen to do the jobs.

c. To determine the number of outcomes in which at least one boy is chosen for a job, we can count the three separate cases in which at least one boy does one of the jobs.

Case 1: Boy-Girl $\underline{\text{35 boys can dim the lights}} \cdot \underline{\text{15 girls can shut the door}} = 525$

Case 2: Girl-Boy $\underline{\text{15 girls can dim the lights}} \cdot \underline{\text{35 boys can shut the door}} = 525$

Case 3: Boy-Boy $\underline{\text{35 boys can dim the lights}} \cdot \underline{\text{35 boys can shut the door}} = 1225$

Adding the outcomes from each case we see that the total number of outcomes in which at least one boy is chosen for a job is $n(\text{at least one is a boy}) = 525 + 525 + 1225 = 2275$.

Therefore, $P(\text{at least one is a boy}) = \frac{n(\text{at least one is a boy})}{n(S)} = \frac{2275}{2500} = \frac{91}{100} = 0.91$. So there is a 91% probability that at least one boy will be chosen for a job.

See **Example 4** for an alternate way to compute this same probability.

d. There are 50 students (boys and girls) that can dim the lights and 35 boys that can shut the door. Therefore, the total number of outcomes in which a boy can shut the door is

$$n(\text{the door closer is a boy}) = \underline{\text{50 students can dim the lights}} \cdot \underline{\text{35 boys can shut the door}} = 1750$$

Thus, $P(\text{the door closer is a boy}) = \frac{n(\text{the door closer is a boy})}{n(S)} = \frac{1750}{2500} = \frac{7}{10} = 0.7$.

Threrefore, there is a 70% probability that the person chosen to close the door will be a boy.

e. There are 6 seniors in the class so there are 44 non-seniors. Therefore, the number of possible outcomes for which two non-seniors do both jobs is

$$n(\text{neither are seniors}) = \underline{\text{44 non-seniors can dim the lights}} \cdot \underline{\text{44 non-seniors can shut the door}} = 1936$$

Thus, $P(\text{neither are seniors}) = \dfrac{n(\text{neither are seniors})}{n(S)} = \dfrac{1936}{2500} = \dfrac{484}{625} = 0.7744.$

Therefore, the probability that two non-seniors will be chosen to do both jobs is approximately 77.4%.

Note The sum of the probabilities in parts (a)–(e) of Example 3 do not equal 1. This is because we did not determine the probabilities for all **mutually exclusive events**. For example, the girls who are counted in part (a) when determining $P(\text{both are girls})$ are counted again when determining $P(\text{one is a girl and one is a boy})$ in part (b).

It can sometimes be easier to determine the probability of the **complement** of an event rather than determining the probability of the event itself.

Definition Complement

The **complement** of an event E is the set of all outcomes in the sample space that are not included in the outcomes of event E. The notation used for the complement of E is E'.

Some examples of the complement of an event are given below:

If A is the event that a tossed coin lands on heads, then A' is the event that the tossed coin lands on tails.

If B is the event that at least one girl is chosen for a two person committee, then B' is the event that no girls are chosen for the committee.

If C is the event that a heart is drawn from a deck of 52 cards, then C' is the event that a club, spade, or diamond is drawn.

The Complement Rule

Given the probability of an event E written as a fraction or a decimal, the probability of its complement can be found by subtracting the given probability from 1. That is, $P(E') = 1 - P(E)$. This equation can also be written as $P(E) = 1 - P(E')$.

Given the probability of an event E written as a percent, the probability of its complement can be found by subtracting the given probability from 100%. That is, $P(E') = 100\% - P(E)$. This equation can also be written as $P(E) = 100\% - P(E')$.

To illustrate the complement rule, consider Example 4, which is exactly the same exercise as Example 3c.

Example 4 Using the Compliment Rule to Determine Probability

Using the same **demographic information given in Example 3**, suppose that the instructor chooses one student to dim the lights and one to shut the door. If a student may do both jobs, determine the probability that at least one is a boy using the **complement rule**. Write the probability rounded correct to the nearest tenth of a percent as needed.

Solution The complement of the event "at least one is a boy" is the event "neither are boys" or "both are girls." From **Example 3a**, we see that $P(\text{both are girls}) = 9\%$. Therefore, by the complement rule, P(at least one is a boy) $= 100\%$ $-$ P(both are girls) $= 100\% - 9\% = 91\%$. Thus, the probability that at least one boy is chosen for a job is 91%. This is exactly the same probability that was obtained in **Example 3c**. ●

Example 5 Using the same **demographic information given in Example 3**, suppose that the instructor chooses one student to dim the lights and one to shut the door. If a student may **not** do both jobs, determine the following probabilities rounded correct to the nearest tenth of a percent as needed.

a. $P(\text{both are girls})$

b. $P(\text{one is a girl and one is a boy})$

c. $P(\text{at least one is a boy})$

d. $P(\text{the door closer is a boy})$

e. $P(\text{neither are seniors})$

Solution Try to determine the probabilities for (a)–(e) on your own, then **check** your answers or watch this **interactive video** to see a complete solution. ●

You Try It Work through this You Try It problem.

Work Exercises 13–17 in this textbook or in the MyLab Math Study Plan.

OBJECTIVE 2 DETERMINING PROBABILITY USING THE ADDITIVE RULE AND VENN DIAGRAMS

Often we are interested in determining the probability of the union or intersection of two events. For example, suppose that a single card is chosen at random from a **standard deck of 52 playing cards**. What is the probability of choosing a heart or an ace? This is an example of the probability of the union of two events. For this example we use the notation $P(H \text{ or } A) = P(H \cup A)$, where H is the event of drawing a heart and A is the event of drawing an ace. The union symbol, $\cup$, can be used interchangeably with the word *or*. We can determine $P(H \cup A)$ as follows:

First, there are 13 hearts in a standard deck of 52 cards, so the probability of choosing a heart is $P(H) = \frac{13}{52}$.

Second, there are 4 aces in a standard deck of 52 cards, so the probability of choosing an ace is $P(A) = \frac{4}{52} = \frac{1}{13}$.

It might appear that we can simply add these two probabilities. However, note that we have accounted for the probability of choosing the ace of hearts *twice* in this process. Therefore, we must account for this "double counting" by subtracting the probability of choosing the ace of hearts, which is $P(H \text{ and } A) = P(H \cap A) = \frac{1}{52}$. Recall that the symbol $\cap$ can be used inter-changeably with the word *and* which means "intersection." The notation $P(H \cap A)$ means "the probability of an ace and a heart."

Thus, the probability of choosing a heart or an ace is

$P(H \cup A) = P(H) + P(A) - P(H \cap A) = \frac{13}{52} + \frac{4}{52} - \frac{1}{52} = \frac{16}{52} = \frac{4}{13}$. This formula is known as the **additive rule of probability**.

> **The Additive Rule of Probability**
>
> For any two events A and B, $P(A \cup B) = P(A) + P(B) - P(A \cap B)$.

The additive rule states that the probability of the union of two events is equal to the sum of the probabilities of the two events minus the probability of the intersection of the two events. Note that if events A and B are **mutually exclusive**, then $P(A \cup B) = P(A) + P(B)$.

Example 6 Using the Additive Rule of Probability

Suppose $P(A) = 0.6$, $P(B) = 0.7$ and $P(A \cup B) = 0.9$. Find the following probability written as a decimal rounded to one decimal place as needed.

a. $P(A \cap B)$

b. $P((A \cap B)')$

Solution

a. We use the **additive rule of probability** to determine $P(A \cap B)$.

$P(A \cup B) = P(A) + P(B) - P(A \cap B)$	Write the additive rule of probability.
$P(A \cap B) = P(A) + P(B) - P(A \cup B)$	Solve for $P(A \cap B)$.
$= 0.6 + 0.7 - 0.9$	Substitute the given information.
$= 0.4$	Simplify.

b. We can use the **complement rule** to determine $P((A \cap B)')$.

$P((A \cup B)') = 1 - P(A \cap B)$	Write the formula representing the complement rule.
$= 1 - 0.4$	Substitute the information obtained from part (a).
$= 0.6$	Simplify.

You Try It Work through this You Try It problem.

Work Exercises 18–23 in this textbook or in the MyLab Math Study Plan.

A Venn diagram, named after British logician John Venn, can sometimes be used to determine probabilities.

Definition Venn Diagram

A Venn diagram is a visual relationship between sets usually drawn using circles or other shapes. The Venn diagram below illustrates two sets A and B drawn within a sample space S. The shaded region in common to both sets is called the intersection of sets A and B and is denoted as $A \cap B$.

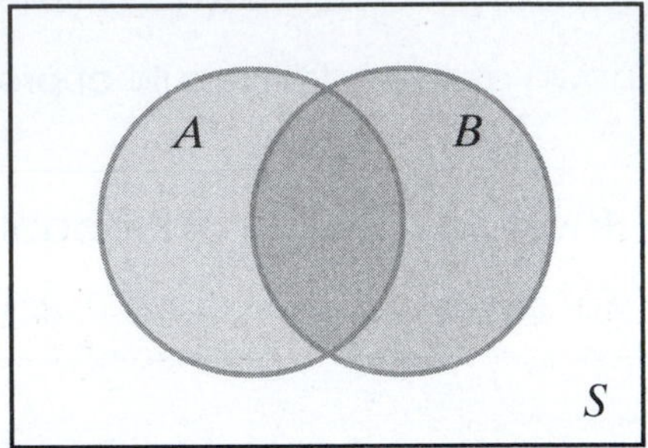

Example 7 Using a Venn Diagram to Determine Probabilities

In a building with 123 offices, 72 offices have computers in them. Forty-five offices have windows. Twenty-three offices have computers and windows. Construct a Venn diagram describing this information. If one office is chosen at random, determine the probability of the following events written as a fraction in lowest terms.

a. the office has a computer

b. the office has a window

c. the office has a computer and does not have a window

d. office does not have computer and does not have a window

Solution Watch this video to see how to construct the Venn diagram shown below.

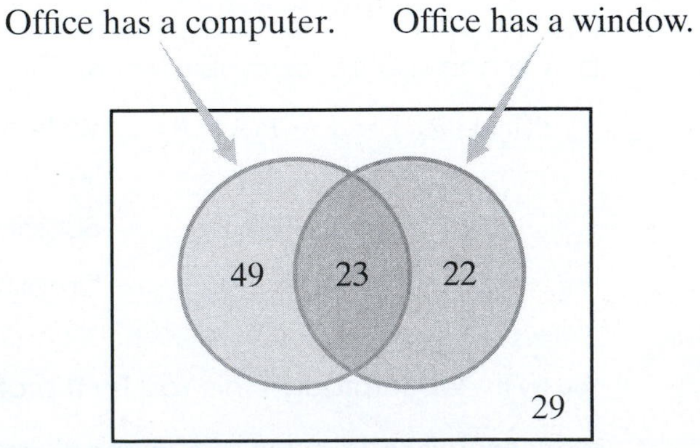

Office has a computer. Office has a window.

49 23 22

29

Using the **Venn diagram** and using the fact that the size of the sample space is 123, we can determine each probability.

a. P(the office has a computer) $= \frac{49 + 23}{123} = \frac{72}{123} = \frac{24}{41}$

b. P(the office has a window) $= \frac{22 + 23}{123} = \frac{45}{123} = \frac{15}{41}$

c. P(the office has a computer and does not have a window) $= \frac{49}{123}$

d. P(the office does not have computer and does not have a window) $= \frac{29}{123}$

You Try It Work through this You Try It problem.

Work Exercises 24–36 in this textbook or in the MyLab Math Study Plan.

OBJECTIVE 3 USING COMBINATIONS TO DETERMINE PROBABILITIES

In Section 14.6 we used the formula $N = {}_nC_r = \frac{n!}{(n-r)!r!}$ to count the number of different arrangements (combinations) of n objects taken r at a time when the following three criteria were satisfied:

1. order does not matter
2. an object cannot be repeated, and
3. the n objects are distinct

Combinations are often needed when determining certain probabilities.

Example 8 Using Combinations to Determine Probabilities

A club has 50 members, and you and your best friend are both members. Four people will be selected from this club to be in a promotional video. Determine each of the following probabilities rounded correct to the nearest tenth of a percent as needed.

a. P(you are chosen)

b. P(you and your best friend are chosen)

c. P(neither you nor your best friend are chosen)

d. P(at least one of you are chosen)

Solution First we must determine the size of the sample space. The order in which the four people are chosen does not matter. The four people cannot be repeated and the 50 members are distinct. Therefore, we can determine the size of the sample space using the formula $n(S) = {}_nC_r = \frac{n!}{(n-r)!r!}$, where $n = 50$ and $r = 4$.

(eText Screens 14.7-1–14.7-50)

$$n(S) = {}_{50}C_4 = \frac{50!}{(50-4)!4!} = \frac{50!}{46!4!} = \frac{50 \cdot 49 \cdot 48 \cdot 47 \cdot \cancel{46!}}{\cancel{46!} \cdot 4!} = \frac{50 \cdot 49 \cdot 48 \cdot 47}{4 \cdot 3 \cdot 2 \cdot 1} = 230{,}300$$

a. First determine the number of outcomes in which you can be chosen for the promotional video using the **fundamental counting principle.**

$$n(\text{you are chosen}) = \underbrace{{}_1C_1}_{\substack{\text{Number of ways}\\ \text{to choose you}}} \cdot \underbrace{{}_{49}C_3}_{\substack{\text{Number of ways to choose}\\ \text{the remaining three people}}}$$

Therefore, the probability that you will be chosen for the promotional video is

$$P(\text{you are chosen}) = \frac{n(\text{you are chosen})}{n(S)} = \frac{{}_1C_1 \cdot {}_{49}C_3}{{}_{50}C_4} = \frac{18{,}424}{230{,}300} = \frac{2}{25} = 0.08.$$

Thus, there is an 8% chance that you will be chosen for the promotional video.

b. The number of ways that you and your best friend can be chosen for the promotional video is

$$n(\text{you and your best friend are chosen}) = \underbrace{{}_1C_1}_{\substack{\text{Number of ways}\\ \text{to choose you}}} \cdot \underbrace{{}_1C_1}_{\substack{\text{Number of ways to}\\ \text{choose your best friend}}} \cdot \underbrace{{}_{48}C_2}_{\substack{\text{Number of ways to choose}\\ \text{the remaining two people}}}$$

Therefore, the probability that you and your best friend will be chosen for the promotional video is

$$P(\text{you and your best friend are chosen}) = \frac{n(\text{you and your best friend are chosen})}{n(S)}$$

$$= \frac{{}_1C_1 \cdot {}_1C_1 \cdot {}_{48}C_2}{{}_{50}C_4} = \frac{1{,}128}{230{,}300} = \frac{6}{1225} \approx 0.0049$$

Thus, there is approximately a 0.5% probability that you and your best friend will be chosen for the video.

c. The number of ways that neither you nor your friend are chosen for the promotional video is

$$n(\text{neither you nor your best friend are chosen}) = \underbrace{{}_1C_0}_{\substack{\text{Number of ways}\\ \text{to consider but}\\ \text{not choose you}}} \cdot \underbrace{{}_1C_0}_{\substack{\text{Number of ways to}\\ \text{consider but not choose}\\ \text{your best friend}}} \cdot \underbrace{{}_{48}C_4}_{\substack{\text{Number of ways to choose}\\ \text{4 people out of the}\\ \text{remaining 48 people}}}$$

Therefore, the probability that neither you nor your best friend will be chosen for the video is

$$P(\text{neither you nor your best friend are chosen}) = \frac{n(\text{neither you nor your best friend are chosen})}{n(S)}$$

$$= \frac{{}_1C_0 \cdot {}_1C_0 \cdot {}_{48}C_4}{{}_{50}C_4} = \frac{194{,}580}{230{,}300} = \frac{207}{245} \approx 0.845$$

There is approximately an 84.5% probability that neither you nor your best friend will be chosen for the video.

d. To determine the probability that at least one of you is chosen, we can use the **complement rule**. Note that the **complement** of "at least one of you is chosen" is "neither of you nor your best friend is chosen." In part (c), we determined that $P(\text{neither you nor your best friend are chosen}) = \frac{207}{245}$. Therefore, by the complement rule we get

$$P(\text{at least one of you are chosen}) = 1 - P(\text{neither you nor your best friend are chosen})$$
$$= 1 - \frac{207}{245} = \frac{38}{245} \approx 0.155$$

Thus, the probability that at least one of you is chosen for the video is approximately 15.5%.

You Try It Work through this You Try It problem.

Work Exercises 37–40 in this textbook or in the MyLab Math Study Plan.

Example 9 Using Combinations to Determine Probabilities

A box contains eight yellow, seven red, and six blue balls. Five balls are selected at random and the colors are noted. Determine each of the following probabilities written as a fraction in lowest terms.

a. $P(\text{all } 5 \text{ balls are yellow})$

b. $P(\text{exactly } 2 \text{ balls are red})$

c. $P(\text{at least one ball is blue})$

Solution Try to determine each probability on your own. Then **check your answers** or watch this **video** to see the entire worked out solution.

You Try It Work through this You Try It problem.

Work Exercises 41–46 in this textbook or in the MyLab Math Study Plan.

OBJECTIVE 4 USING CONDITIONAL PROBABILITY

Sometimes a situation arises where we want to determine the probability of an event B happening given that a separate event A has already happened. This is called **conditional probability** because we determine the probability of an event only after some condition has been met.

> **Definition** Conditional Probability
>
> The conditional probability of an event B in relation to an event A is the probability that event B will occur given that event A has already occurred. The notation used for conditional probability is $P(B|A)$, which is read as "the probability of B given A."

Conditional probability can be determined by two methods:

1. Reducing the sample space outright
2. Using a formula to guide the counting and reasoning

Example 10 Determining Conditional Probability by Reducing the Sample Space

The Venn diagram below shows that in a building with 123 offices, 72 offices have computers in them. Forty-five offices have windows to the outside of the building. Twenty-three offices have computers and windows. If one office is chosen at random, determine each of the following probabilities written as a fraction in lowest terms.

a. $P(C|W)$ b. $P(W|C)$

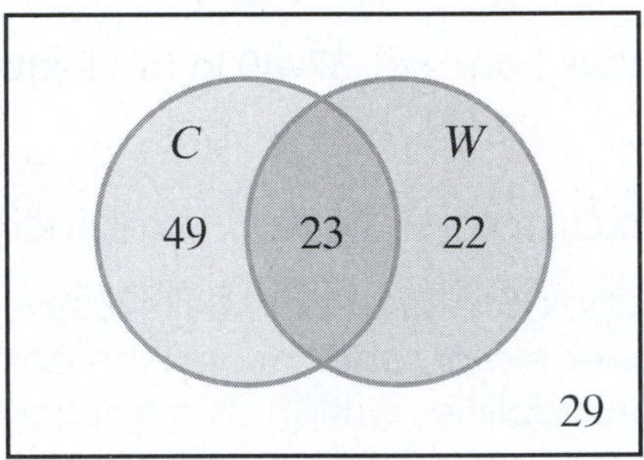

C: Offices that have a computer
W: Offices that have a window

Solution

a. To determine the probability that an office has a computer given that it has a window, we only consider the 45 offices that have a window. Thus, the sample space is reduced from 123 to 45. Of the offices that have a window, 23 of those offices have a computer. (This can be seen as the intersection of the two sets in the **Venn diagram**.) Thus, we get $P(C|W) = \frac{23}{45}$.

b. To determine the probability that an office has a window given that it has a computer, we only consider the 72 offices that have a computer. Thus, the sample space is reduced from 123 to 72. Of the offices that have a computer, 23 of those offices have a window. Thus, we get $P(W|C) = \frac{23}{72}$.

You Try It Work through this You Try It problem.

Work Exercises 47 and 48 in this textbook or in the MyLab Math Study Plan.

Note that in the previous example we determined the conditional probability by first determining the number of occurrences common to both events and then we divided that number by the number of occurrences of the given event. In the case of determining the conditional probability $P(C|W)$, we found that $P(C|W) = \dfrac{n(C \cap W)}{n(W)} = \dfrac{23}{45}$. Thus, we can establish a formula for determining conditional probability $P(B|A)$ as follows.

Multiply by 1 of the form $1 = \dfrac{\frac{1}{n(S)}}{\frac{1}{n(S)}}$

$$P(B|A) = \frac{n(A \cap B)}{n(A)} = \frac{n(A \cap B)}{n(A)} \cdot \frac{\frac{1}{n(S)}}{\frac{1}{n(S)}} = \frac{\frac{n(A \cap B)}{n(S)}}{\frac{n(A)}{n(S)}} = \frac{P(A \cap B)}{P(A)}$$

Conditional Probability Formula

The probability of an event B occurring given that event A has already occurred is given by the formula

$$P(B|A) = \frac{n(A \cap B)}{n(A)} = \frac{P(A \cap B)}{P(A)}.$$

Example 11 Determining Conditional Probability Using the Conditional Probability Formula

Suppose $P(A) = 0.6$, $P(B) = 0.7$, and $P(A \cup B) = 0.9$. Determine the following probabilities correct to two decimal places as needed.

a. $P(B|A)$　　　　b. $P(A|B)$

Solution First, determine $P(A \cap B)$ using the **additive rule of probability**.

$P(A \cup B) = P(A) + P(B) - P(A \cap B)$　　Write the additive rule of probability.

$P(A \cap B) = P(A) + P(B) - P(A \cup B)$　　Solve for $P(A \cap B)$.

$= 0.6 + 0.7 - 0.9$　　Substitute the given information.

$= 0.4$　　Simplify.

We can now use the conditional probability formula to determine each conditional probability.

a. $P(B|A) = \dfrac{P(A \cap B)}{P(A)}$　　Write the conditional probability formula.

$= \dfrac{0.4}{0.6}$　　Substitute 0.4 for $P(A \cap B)$ and substitute 0.6 for $P(A)$.

≈ 0.67　　Simplify.

b. $P(A|B) = \dfrac{P(A \cap B)}{P(B)}$ Write the conditional probability formula.

$= \dfrac{0.4}{0.7}$ Substitute 0.4 for $P(A \cap B)$ and substitute 0.7 for $P(B)$.

≈ 0.57 Simplify.

Using the conditional probability formula is not always the easiest way to determine conditional probability especially when the problem involves combinations. When combinations are involved, it is typically more straightforward to reduce the sample space as we see in the following example.

Example 12 Determining Conditional Probability

A box contains eight yellow, seven red, and six blue balls. Five balls are selected at random and the colors noted. Determine each of the following conditional probabilities written as a fraction reduced to lowest terms.

a. P(all yellow balls | all the same color) b. P(exactly 2 red balls | no blue balls)

c. P(at least 1 blue ball | no red balls)

Solution

a. First reduce the size of the sample space. That is, determine the number of possible outcomes of choosing five balls of the same color. We can determine the size of this reduced sample space by adding the number of outcomes of the three possibilities "choosing 5 yellow balls," "choosing 5 red balls," and "choosing 5 blue balls."

$$n(\text{all the same color}) = \underbrace{{}_8C_5}_{\substack{\text{Number of ways}\\ \text{to choose 5}\\ \text{yellow balls}}} + \underbrace{{}_7C_5}_{\substack{\text{Number of ways}\\ \text{to choose 5}\\ \text{red balls}}} + \underbrace{{}_6C_5}_{\substack{\text{Number of ways}\\ \text{to choose 5}\\ \text{blue balls}}}$$

The number of ways to choose all yellow balls is n(all yellow balls) $= {}_8C_5$. Therefore, the probability of choosing all yellow balls given that the balls are all of the same color is

$$P(\text{all yellow balls} \mid \text{all the same color}) = \frac{n(\text{all yellow balls})}{n(\text{all the same color})} = \frac{{}_8C_5}{{}_8C_5 + {}_7C_5 + {}_6C_5} = \frac{56}{56 + 21 + 6} = \frac{56}{83}$$

Try to determine the conditional probabilities for parts b and c on your own. Then **check your answers** or watch this **video** to see the entire worked out solution.

You Try It Work through this You Try It problem.

Work Exercises 49–54 in this textbook or in the MyLab Math Study Plan.

OBJECTIVE 5 UNDERSTANDING ODDS

Sometimes probabilities are stated in terms of **odds**. For example, the probability of a tossed coin landing on heads is $P(\text{heads}) = \frac{1}{2}$. But the odds of a tossed coin landing on heads are 1:1, read as "one to one." We interpret the notation 1:1 to mean that there is 1 way for a head to occur to one way that a tail can occur. We treat the odds notation $a{:}b$ in much the same way as we treat a fraction in the sense that we always want to write the odds notation in lowest terms. For example, the odds notation 6:4 should be rewritten as 3:2.

> **Definition** Odds
>
> Let E be an event, E' be the complement of E, and S be the sample space. The **odds of event E happening** is the ratio of the number of ways that event E can occur to the number of ways that event E cannot occur. The odds of an event E happening can be calculated using one of the following methods:
>
> $n(E){:}\,n(E')$, where $n(E') = n(S) - n(E)$
>
> $P(E){:}\,P(E')$, where $P(E') = 1 - P(E)$ and when $0 < P(E) < 1$
>
> $P(E){:}\,P(E')$, where $P(E') = 100\% - P(E)$ and when $0\% < P(E) < 100\%$

Example 13 Calculating Odds

A private school sold 2100 raffle tickets and you purchased 20. What are the odds of winning the raffle?

Solution Let E be the event of winning the raffle. Then the number of ways to win is $n(E) = 20$. The number of ways to not win is $n(E') = 2100 - 20 = 2080$. Therefore, the odds of winning the raffle are $n(E) : n(E') = 20 : 2080 = 1 : 104$.

Example 14 Calculating Odds

A paint-ball gun ball hopper contains 10 yellow balls, 6 green balls, and 3 red balls. A ball is fired from the gun. Find:

a. odds of firing a red ball

b. odds of firing a green ball

c. odds against firing a yellow ball

Solution

a. Let R be the event of firing a red ball. Then $n(R) = 3$ and $n(R') = 16$. Thus, the odds of firing a red ball are $n(R) : n(R') = 3 : 16$.

b. Let G be the event of firing a green ball. Then $n(G) = 6$ and $n(G') = 13$. Thus, the odds of firing a green ball are $n(G) : n(G') = 6 : 13$.

c. Let Y be the event of firing a yellow ball. Then $n(Y) = 10$ and $n(Y') = 9$. Thus, the odds against firing a yellow ball are $n(Y') : n(Y) = 9 : 10$.

You Try It Work through this You Try It problem.

Work Exercises 55–62 in this textbook or in the MyLab Math Study Plan.

14.7 Exercises

In Exercises 1–7, determine the size of the sample space for each experiment described.

1. A man has three pairs of slacks, four shirts, and five ties that all coordinate. He wants to choose a pair of slacks, a shirt, and a tie to wear to work.

2. A group of 20,000 math students were selected to participate in a study to determine how many hours each week that they spend working on mathematics.

3. Four coins are tossed and the results of "Heads or Tails" from each coin is recorded.

4. A standard deck of 52 cards has all of the face cards removed. One card is drawn at random.

5. Three cards are selected without replacement from a standard deck of 52 cards.

6. A marble bag contains 7 orange marbles, 4 red marbles, 9 green marbles, and 12 white marbles. One marble is selected at random.

7. A marble bag contains 7 orange marbles, 4 red marbles, 9 green marbles, and 12 white marbles. Four marbles are selected at random without replacement.

8. A paint-ball gun ball hopper contains 9 yellow balls, 12 green balls, and 7 red balls. A ball is fired from the gun. What is the probability that the ball fired from the gun is red?

9. There are five cars in a driveway: one tan van, one green sedan, one blue four-wheel-drive Bronco, one red Mustang convertible, and one green VW Super Beetle. All of the keys were hanging on a rack on the wall. The rack fell and the keys landed in a pile on the floor. If you randomly grab one set of keys, determine the following probabilities:

   a. P(keys are for the Mustang)   b. P(keys are for a green car)

Roulette is a game played in many casinos in which a wheel containing numbers of various colors is spun. A ball is dropped and eventually lands on a number. Use the roulette wheel shown below to answer Exercises 10–12.

10. What is the probability that the ball lands on red?

11. What is the probability that the ball lands on an odd number?

12. What is the probability that the ball lands on green?

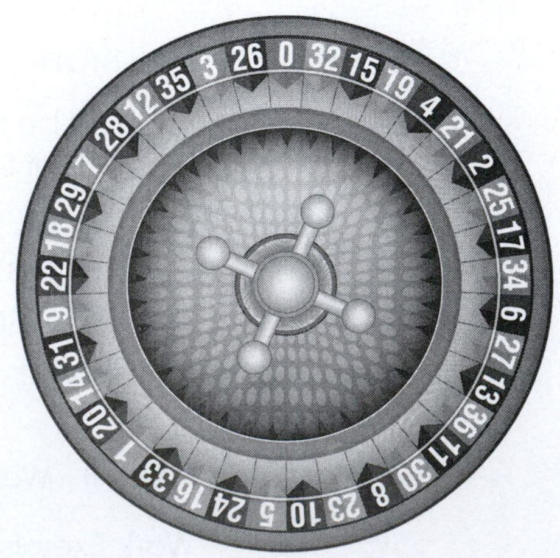

In Exercises 13–17, refer to the following demographic information that has been collected by an instructor after surveying her class.

Total number of students in the class:	46
Number of boys:	19
Number of girls:	27
Number of freshmen:	7
Number of sophomores:	16
Number of juniors:	12
Number of seniors:	11

13. The instructor randomly chooses one student to pass out papers, one student to clean the chalkboard, and one student to empty the trash. If a student may do more than one job, then what is the probability that the instructor will choose all girls to do the jobs?

14. The instructor randomly chooses one student to pass out papers, one student to clean the chalkboard, and one student to empty the trash. If a student may do more than one job, then what is the probability that the instructor will choose at least one girl to do the jobs?

15. The instructor randomly chooses one student to pass out papers, one student to clean the chalkboard, and one student to empty the trash. If a student may not do more than one job, then what is the probability that the instructor will choose all students of the same sex?

16. The instructor randomly chooses one student to pass out papers, one student to clean the chalkboard, and one student to empty the trash. If a student may do more than one job, then what is the probability that the instructor will choose at least one senior to do the jobs?

17. The instructor randomly chooses one student to pass out papers, one student to clean the chalkboard, and one student to empty the trash. If a student may not do more than one job, then what is the probability that the instructor will choose at most one senior to do the jobs?

For Exercises 18–21, suppose that $P(A) = 0.5$, $P(B) = 0.8$, and $P(A \cup B) = 0.9$.

18. Find $P(A \cap B)$.

19. Find $P(B')$.

20. Find $P(A')$.

21. Find $P((A \cap B)')$.

22. A single card is drawn from a standard deck of 52 cards. What is the probability that the card is black or a two?

23. A single card is drawn from a standard deck of 52 cards. What is the probability that the card is a face card or a club?

For Exercises 24–31, use the Venn diagram below to determine each probability.

24. $P(A)$

25. $P(B)$

26. $P(A \cap B)$

27. $P(A \cup B)$

28. $P(A')$

29. $P(B')$

30. $P(A' \cap B')$

31. $P(A' \cup B')$

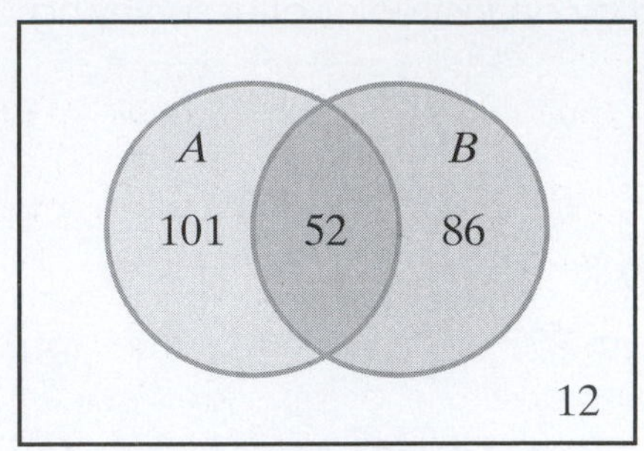

For Exercises 32–36, refer to the following information.

Two hundred sports fans were surveyed. It was found that 139 people were fans of the NFL (The National Football League), 64 were fans of MLB (Major League Baseball), and 35 were fans of both the NFL and MLB. One fan is selected at random. Create a Venn diagram and determine the indicated probabilities in Exercises 32–36.

32. What is the probability that the selected sports fan is a fan of the NFL?

33. What is the probability that the selected sports fan is a fan of MLB?

34. What is the probability that the selected sports fan is not a fan of the NFL?

35. What is the probability that the selected sports fan is a fan of MLB but not a fan of the NFL?

36. What is the probability that the selected sports fan is neither a fan of the NFL nor MLB?

SbS 37. You and your best friend are part of a crowd of 60 people attending a magic show. The magician will choose five people to assist him during the show. What is the probability that you are chosen?

SbS 38. You and your best friend are part of a crowd of 60 people attending a magic show. The magician will choose five people to assist him during the show. What is the probability that you and your best friend are chosen?

SbS 39. You and your best friend are part of a crowd of 60 people attending a magic show. The magician will choose five people to assist him during the show. What is the probability that neither you nor your best friend is chosen?

SbS 40. You and your best friend are part of a crowd of 60 people attending a magic show. The magician will choose five people to assist him during the show. What is the probability that at least one of you is chosen?

SbS 41. A box contains 10 yellow, 5 red, and 7 blue balls. Four balls are selected at random and the colors are noted. Determine the probability that all four balls are yellow.

SbS 42. A box contains 10 yellow, 5 red, and 7 blue balls. Four balls are selected at random and the colors are noted. Determine the probability that exactly three balls are blue.

SbS 43. A box contains 10 yellow, 5 red, and 7 blue balls. Four balls are selected at random and the colors are noted. Determine the probability that none of the balls are yellow.

SbS 44. A box contains 10 yellow, 5 red, and 7 blue balls. Four balls are selected at random and the colors are noted. Determine the probability that at least one ball is red.

SbS 45. Seven cards are dealt from a standard deck of fifty-two cards. What is the probability that exactly one of the cards is a face card?

SbS 46. Five cards are dealt from a standard deck of fifty-two cards. What is the probability that at least one of the cards is a face card?

The Venn diagram below shows that out of 1350 students surveyed who use an integrated eText in their College Algebra course, 994 students always watch a video before working a homework exercise. Eight hundred forty-five students always read the eText before working a homework exercise, and 720 students always watch a video and always read the eText before working a homework exercise. If one of these students is selected at random, determine the probabilities in Exercises 47 and 48.

47. $P(V|T)$

48. $P(T|V)$

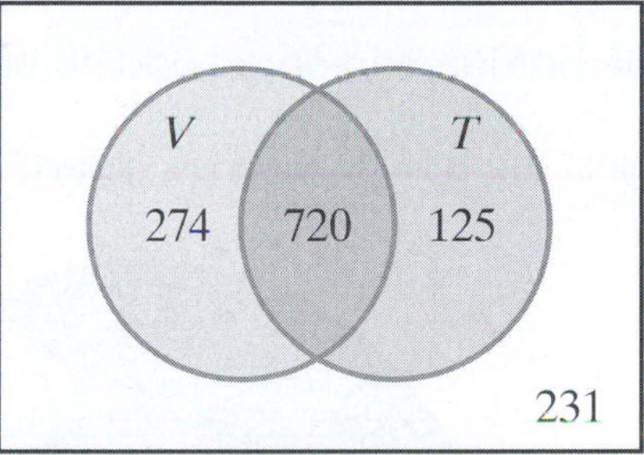

V: Always watch video before attempting homework
T: Always read eText before attempting homework

49. Suppose $P(A) = 0.4$, $P(B) = 0.6$, and $P(A \cup B) = 0.8$. Determine $P(B|A)$.

50. Suppose $P(A) = 0.4$, $P(B) = 0.6$, and $P(A \cup B) = 0.8$. Determine $P(A|B)$.

51. Two cards are drawn from a standard deck of fifty-two cards without replacement. What is the probability that the second card drawn is a diamond given that the first card drawn was a diamond?

52. A box contains 10 yellow, 5 red, and 7 blue balls. Five balls are selected at random and the colors are noted. Determine P(all yellow balls | all the same color).

53. A box contains 10 yellow, 5 red, and 7 blue balls. Six balls are selected at random and the colors are noted. Determine P(exactly 2 red balls | no blue balls).

54. A box contains 10 yellow, 5 red, and 7 blue balls. Five balls are selected at random and the colors are noted. Determine P(exactly 2 red balls | no blue balls).

55. A private school sold 1650 raffle tickets and you purchased 24. What are the odds of winning the raffle?

56. A marble bag contains 12 clear marbles, 8 red marbles, and 3 green marbles. One marble is chosen at random. Find the odds of choosing a clear marble.

57. A marble bag contains 12 clear marbles, 8 red marbles, and 3 green marbles. One marble is chosen at random. Find the odds of not choosing a clear marble.

58. Ten slips of paper are numbered consecutively 1–10 and placed in a hat. Two more slips of paper are numbered with a 5 and three more are numbered with a 6. Find the odds of drawing the number 5 from the hat.

59. Ten slips of paper are numbered consecutively 1–10 and placed in a hat. Two more slips of paper are numbered with a 5 and three more are numbered with a 4. Find the odds of drawing a number greater than 4.

Roulette is a game played in many casinos in which a wheel containing numbers of various colors is spun. A ball is dropped and eventually lands on a number. Use the roulette wheel shown below to answer Exercises 56–58.

60. What are the odds of the ball landing on red?

61. What are the odds of the ball landing on an odd number?

62. What are the odds against the ball landing on green?

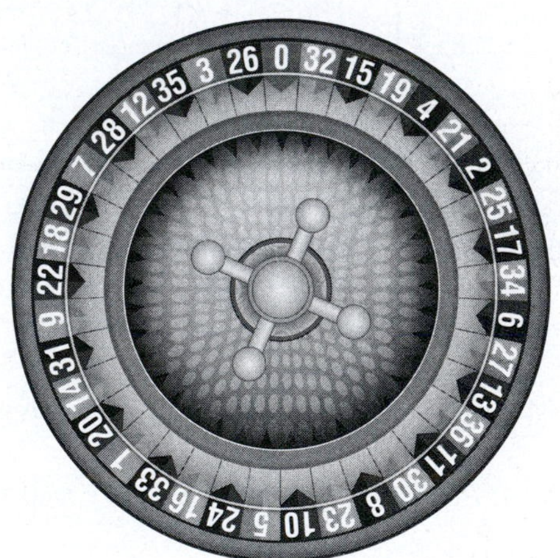

Brief Exercises

63. You and your best friend are part of a crowd of 60 people attending a magic show. The magician will choose five people to assist him during the show. What is the probability that you are chosen?

64. You and your best friend are part of a crowd of 60 people attending a magic show. The magician will choose five people to assist him during the show. What is the probability that you and your best friend are chosen?

65. You and your best friend are part of a crowd of 60 people attending a magic show. The magician will choose five people to assist him during the show. What is the probability that neither you nor your best friend is chosen?

66. You and your best friend are part of a crowd of 60 people attending a magic show. The magician will choose five people to assist him during the show. What is the probability that at least one of you is chosen?

67. A box contains 10 yellow, 5 red, and 7 blue balls. Four balls are selected at random and the colors are noted. Determine the probability that all four balls are yellow.

68. A box contains 10 yellow, 5 red, and 7 blue balls. Four balls are selected at random and the colors are noted. Determine the probability that exactly three balls are blue.

69. A box contains 10 yellow, 5 red, and 7 blue balls. Four balls are selected at random and the colors are noted. Determine the probability that none of the balls are yellow.

70. A box contains 10 yellow, 5 red, and 7 blue balls. Four balls are selected at random and the colors are noted. Determine the probability that at least one ball is red.

71. Seven cards are dealt from a standard deck of fifty-two cards. What is the probability that exactly one of the cards is a face card?

72. Five cards are dealt from a standard deck of fifty-two cards. What is the probability that at least one of the cards is a face card?

73. A box contains 10 yellow, 5 red, and 7 blue balls. Five balls are selected at random and the colors are noted. Determine $P(\text{all yellow balls} \mid \text{all the same color})$.

74. A box contains 10 yellow, 5 red, and 7 blue balls. Six balls are selected at random and the colors are noted. Determine $P(\text{exactly 2 red balls} \mid \text{no blue balls})$.

75. A box contains 10 yellow, 5 red, and 7 blue balls. Five balls are selected at random and the colors are noted. Determine $P(\text{exactly 2 red balls} \mid \text{no blue balls})$.

76. You and your best friend are part of a crowd of 60 people attending a magic show. The magician will choose five people to assist him during the show. What is the probability that you are chosen?

77. You and your best friend are part of a crowd of 60 people attending a magic show. The magician will choose five people to assist him during the show. What is the probability that you and your best friend are chosen?

78. You and your best friend are part of a crowd of 60 people attending a magic show. The magician will choose five people to assist him during the show. What is the probability that neither you nor your best friend is chosen?

79. You and your best friend are part of a crowd of 60 people attending a magic show. The magician will choose five people to assist him during the show. What is the probability that at least one of you is chosen?

80. A box contains 10 yellow, 5 red, and 7 blue balls. Four balls are selected at random and the colors are noted. Determine the probability that all four balls are yellow.

81. A box contains 10 yellow, 5 red, and 7 blue balls. Four balls are selected at random and the colors are noted. Determine the probability that exactly three balls are blue.

82. A box contains 10 yellow, 5 red, and 7 blue balls. Four balls are selected at random and the colors are noted. Determine the probability that none of the balls are yellow.

83. A box contains 10 yellow, 5 red, and 7 blue balls. Four balls are selected at random and the colors are noted. Determine the probability that at least one ball is red.

84. Seven cards are dealt from a standard deck of fifty-two cards. What is the probability that exactly one of the cards is a face card?

85. Five cards are dealt from a standard deck of fifty-two cards. What is the probability that at least one of the cards is a face card?

86. A box contains 10 yellow, 5 red, and 7 blue balls. Five balls are selected at random and the colors are noted. Determine P(all yellow balls | all the same color).

87. A box contains 10 yellow, 5 red, and 7 blue balls. Six balls are selected at random and the colors are noted. Determine P(exactly 2 red balls | no blue balls).

88. A box contains 10 yellow, 5 red, and 7 blue balls. Five balls are selected at random and the colors are noted. Determine P(exactly 2 red balls | no blue balls).

Chapter 14 Summary

Key Concepts	Examples/Videos
14.1 Introduction to Sequences and Series	
A **finite sequence** is a function whose domain is the finite set $\{1, 2, 3, \ldots, n\}$, where n is a natural number. An **infinite sequence** is a function whose domain is the set of all natural numbers. The range values of a sequence are called the **terms** of the sequence.	1. Write the first four terms of each sequence whose nth term is given. a. $a_n = 2n - 1$ b. $b_n = n^2 - 1$ c. $c_n = \dfrac{3^n}{(n-1)!}$ d. $d_n = (-1)^n 2^{n-1}$
A **recursive sequence** is a sequence in which each term is defined using one or more of its previous terms.	2. Write the first four terms of each of the following recursive sequences. a. $a_1 = -3, a_n = 5a_{n-1} - 1$ for $n \geq 2$ b. $b_1 = 2, b_n = \dfrac{(-1)^{n-1}n}{b_{n-1}}$ for $n \geq 2$
Let $a_1, a_2, a_3, \ldots$ be a sequence. The expression of the form $a_1 + a_2 + a_3 + \cdots + a_n$ is called a **finite series**. The expression of the form $a_1 + a_2 + a_3 + \cdots + a_n + a_{n+1} + \cdots$ is called an **infinite series**.	3. Find the sum of each finite series. a. $\sum_{i=1}^{5} i^2$ b. $\sum_{j=2}^{5} \dfrac{j-1}{j+1}$ c. $\sum_{k=0}^{6} \dfrac{1}{k!}$ (Round the sum to three decimal places.)

Key Concepts	Examples/Videos
The sum of the first n terms of a series is called the nth **partial sum** of the series and is denoted as S_n. If $a_1, a_2, a_3, \ldots$ is a sequence, then the finite series $a_1 + a_2 + a_3 + \cdots + a_n$ can be written in **summation notation** as $\sum_{i=1}^{n} a_i$. The infinite series $a_1 + a_2 + \cdots + a_n + a_{n+1} + \cdots$ can be written as $\sum_{i=1}^{\infty} a_i$. The variable i is called the **index of summation**. The number 1 is the **lower limit of summation** and n is the **upper limit of summation**.	
14.2 Arithmetic Sequences and Series An **arithmetic sequence** is a sequence of the form $a_1, a_1 + d, a_1 + 2d, a_1 + 3d, a_1 + 4d, \ldots,$ where a_1 is the first term of the sequence and d is the common difference. The general term, or nth term, of an arithmetic sequence has the form $a_n = a_1 + (n - 1)d$.	 1. For each of the following sequences, determine if it is arithmetic. If the sequence is arithmetic, find the common difference. a. $1, 4, 7, 10, 13, \ldots$ b. $b_n = n^2 - n$ c. $a_n = -2n + 7$ d. $a_1 = 14, a_n = 3 + a_{n-1}$  2. Find the general term of each arithmetic sequence, then find the indicated term of the sequence. (In part c, only a portion of the graph is given. Assume that the domain of this sequence is all **natural numbers**.) a. $11, 17, 23, 29, 35, \ldots; a_{50}$ b. $2, 0, -2, -4, -6, \ldots; a_{90}$ c. Find a_{31}. 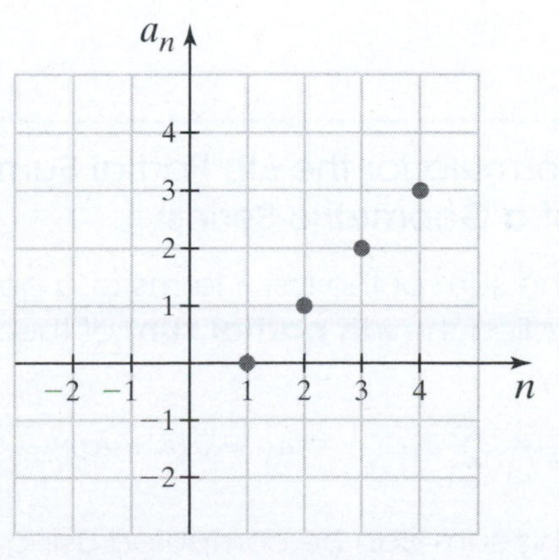

Key Concepts	Examples/Videos
Formula for the nth Partial Sum of an Arithmetic Series The sum of the first n terms of an arithmetic series is called the ***n*th partial sum** of the series and is given by $S_n = \sum_{i=1}^{n} a_i = a_1 + a_2 + a_3 + \cdots + a_n$. This sum can be computed using the formula $S_n = \frac{n(a_1 + a_n)}{2}$.	 3. Find the sum of each arithmetic series. a. $\sum_{i=1}^{20} (2i - 11)$ b. $-5 + (-1) + 3 + 7 + \cdots + 39$  4. A large multiplex movie house has many theaters. The smallest theater has only 12 rows. There are six seats in the first row. Each row has two seats more than the previous row. How many total seats are there in this theater?
14.3 Geometric Sequences and Series A **geometric sequence** is a sequence of the form $a_1, a_1r, a_1r^2, a_1r^3, a_1r^4, \ldots$, where a_1 is the first term of the sequence and r is the common ratio such that $r = \frac{a_2}{a_1} = \frac{a_3}{a_2} = \cdots = \frac{a_{n+1}}{a_n}$ for all $n \geq 1$. The general term, or nth term, of a geometric sequence has the form $a_n = a_1r^{n-1}$.	 1. a. Write the first five terms of the geometric sequence having a first term of 2 and a common ratio of 3. b. Write the first five terms of the geometric sequence such that $a_1 = -4$ and $a_n = -5a_{n-1}$ for $n \geq 2$.  2. For each of the following sequences, determine if it is geometric. If the sequence is geometric, find the common ratio. a. $2, 4, 6, 8, 10, \ldots$ b. $\frac{2}{3}, \frac{4}{9}, \frac{8}{27}, \frac{16}{81}, \frac{32}{243}, \ldots$ c. $12, -6, 3, -\frac{3}{2}, \frac{3}{4}, \ldots$  3. a. Find the 7th term of the geometric sequence whose first term is 2 and whose common ratio is -3. b. Given a geometric sequence such that $a_6 = 16$ and $a_9 = 2$, find a_{13}.
Formula for the nth Partial Sum of a Geometric Series The sum of the first n terms of a geometric series is called the ***n*th partial sum** of the series and is given by $S_n = \sum_{i=1}^{n} a_1r^{i-1} = a_1 + a_1r + a_1r^2 + a_1r^3 + \cdots + a_1r^{n-1}$. This sum can be computed using the formula $S_n = \frac{a_1(1 - r^n)}{1 - r}$ for $r \neq 1$.	 4. a. Find the sum of the series $\sum_{i=1}^{15} 5(-2)^{i-1}$. b. Find the 7th partial sum of the geometric series $8 + 6 + \frac{9}{2} + \frac{27}{8} + \cdots$.

Key Concepts	Examples/Videos
Formula for the Sum of an Infinite Geometric Series Let $\sum_{n=1}^{\infty} a_1 r^{n-1} = a_1 + a_1 r + a_1 r^2 + a_1 r^3 + \cdots + a_1 r^{n-1} + \cdots$ be an infinite geometric series. If $\lvert r \rvert < 1$, then the sum of the series is given by $S = \frac{a_1}{1 - r}$.	 5. Determine whether each of the following series converges or diverges. It the series converges, find the sum. a. $\sum_{n=1}^{\infty} \frac{1}{2}\left(\frac{2}{3}\right)^{n-1}$ b. $3 - \frac{6}{5} + \frac{12}{25} - \frac{24}{125} + \cdots$ c. $12 + 18 + 27 + \frac{81}{2} + \frac{243}{4} + \cdots$
14.4 The Binomial Theorem  The Expansion of $(a + b)^n$ for $n = 0, 1, 2, 3, 4$ and 5: $n = 0: (a + b)^0 = 1$ $n = 1: (a + b)^1 = 1a + 1b$ $n = 2: (a + b)^2 = 1a^2 + 2ab + 1b^2$ $n = 3: (a + b)^3 = 1a^3 + 3a^2b + 3ab^2 + 1b^3$ $n = 4: (a + b)^4 = 1a^4 + 4a^3b + 6a^2b^2 + 4ab^3 + 1b^4$ $n = 5: (a + b)^5 = 1a^5 + 5a^4b + 10a^3b^2 + 10a^2b^3 + 5ab^4 + 1b^5$ The coefficients of the expansion of $(a + b)^n$ is called **Pascal's Triangle** $n = 0$: 1 $n = 1$: 1 1 $n = 2$: 1 2 1 $n = 3$: 1 3 3 1 $n = 4$: 1 4 6 4 1 $n = 5$: 1 5 10 10 5 1 $n = 6$: 1 6 15 20 15 6 1	 1. Use Pascal's triangle to expand each binomial. a. $(x + 2)^4$ b. $(x - 3)^5$ c. $(2x - 3y)^3$

Key Concepts	Examples/Videos
Formula for a Binomial Coefficient For nonnegative integers n and r with $n \geq r$, the coefficient of the expansion of $(a + b)^n$ whose variable part is $a^{n-r}b^r$ is given by $\binom{n}{r} = \dfrac{n!}{r! \cdot (n - r)!}$. **Binomial Theorem** If n is a positive integer then, $(a + b)^n = \binom{n}{0}a^n + \binom{n}{1}a^{n-1}b + \binom{n}{2}a^{n-2}b^2 + \cdots + \binom{n}{n}b^n$ $= \sum_{i=0}^{n}\binom{n}{i}a^{n-i}b^i$.	2. Use the Binomial Theorem to expand each binomial. a. $(x - 1)^8$ b. $(\sqrt{x} + y^2)^5$
Formula for the $(r + 1)st$ Term of a Binomial Expansion If n is a positive integer and if $r \geq 0$, then the $(r + 1)$st term of the expansion of $(a + b)^n$ is given by $\binom{n}{r}a^{n-r}b^r = \dfrac{n!}{r! \cdot (n - r)!}a^{n-r}b^r$.	 3. Find the third term of the expansion of $(2x - 3)^{10}$.  4. Find the coefficient of x^7 in the expansion of $(x + 4)^{11}$.
14.5 Mathematical Induction **The Principle of Mathematical Induction** Let S_n be a mathematical statement involving the natural number n. The statement S_n is true for *all* natural numbers if the following two conditions are satisfied. **1.** S_1 is true. **2.** If, for any natural number k, S_k is true, then S_{k+1} is true.	 1. Prove that the mathematical statement S_n: $1 + 2 + 3 + \cdots + n = \dfrac{n(n + 1)}{2}$ is true for all natural numbers n.  2. Prove that the mathematical statement S_n: $1^3 + 2^3 + 3^3 + \cdots + n^3 = \dfrac{n^2(n + 1)^2}{4}$ is true for all natural numbers n.
14.6 The Theory of Counting **The Fundamental Counting Principle** Suppose that an experiment or task is made up of two or more events. If the first event can occur in n_1 ways, the second event can occur in n_2 ways, and the third event can occur in n_3 ways and so on, then the number of possible outcomes of the experiment or task is $N = n_1 n_2 \ldots n_k$, where k is the total number of events.	 1. A student is to create a password for his MyLab Math homework account. He is informed that the password must consist of seven characters. How many possible ways can he create the password if the first three characters are letters followed by two digits and the last two characters can be letters or digits? (Repeated letters are allowed and letters are not case sensitive. Also, repeated digits are allowed.)

<table>
<tr><th>Key Concepts</th><th>Examples/Videos</th></tr>
<tr><td></td><td>2. Linda has five t-shirts of different colors, three pairs of shorts of different patterns, and four pairs of shoes of different colors. She is planning a weekend trip to Pensacola Beach, Florida. In how many ways can she combine these items if
a. the orange t-shirt clashes with the red shoes?
b. the red t-shirt clashes with the striped shorts?</td></tr>
<tr><td>A permutation is an ordered arrangement of n objects taken r at a time.
Theorem: Permutations Involving Distinct Objects with Repetition Allowed
The number N of different arrangements of n objects using $r \le n$ of them, in which
1. order matters,
2. an object can be repeated, and
3. the n objects are distinct,
is given by the formula $N = n^r$.</td><td>3. How many four-digit arrangements are possible for a bicycle permutation lock if repetition of digits is allowed?</td></tr>
<tr><td>Theorem: Permutations Involving Distinct Objects with Repetition Not Allowed
The number N of different arrangements of n objects using $r \le n$ of them, in which
1. order matters,
2. an object cannot be repeated, and
3. the n objects are distinct, is given by the formula $N = {}_nP_r = \frac{n!}{(n-r)!}$.</td><td>4. There are nine competitors in a hurdle race.
a. In how many different ways can the nine competitors finish first, second, and third?
b. In how many different ways can all nine competitors finish the race?</td></tr>
</table>

Key Concepts	Examples/Videos
Theorem: Permutations Involving Some Objects That Are Not Distinct The number N of different arrangements of n objects taken n at a time in which 1. order matters, 2. n_1 are of one kind, n_2 are of a second kind, ..., and n_k are of a kth kind, and 3. some of the objects are not distinct, is given by the formula $N = \frac{n!}{n_1!n_2! \ldots n_k!}$, where $n_1 + n_2 + \cdots + n_k = n$.	5. How many distinguishable letter codes can be formed from the word SUCCESSFUL if every letter is used?
A **combination** is an unordered arrangement of n distinguishable objects taken r at a time. **Theorem: Combinations Involving Objects That Are Distinct** The number N of different arrangements of n objects using $r \leq n$ of them, in which 1. order does not matter, 2. an object cannot be repeated, and 3. the n objects are distinct, is given by the formula $N = {}_nC_r = \frac{n!}{(n-r)!r!}$.	6. The Louisiana lottery is a game in which six numbers are chosen from the numbers 1 to 44. Order does not matter and repetition of numbers is not allowed. How many possible sets of numbers are there in the Louisiana lottery?
14.7 An Introduction to Probability **Probability** is the mathematical estimate of the likelihood that an event or sequence of events will occur. The **sample space,** S, for an experiment is the set of all possible outcomes for the experiment. The number of possible outcomes of the sample space is denoted as $n(S)$.	1. State the number of elements in the sample space of the following experiments. a. Of the 20,000 students on a college campus, one student is selected to be the mascot at the football game this weekend. b. One card is to be selected at random from a **standard deck of 52 playing cards** that has all of the diamonds removed. c. One marble is to be selected at random from a marble bag that contains two green marbles, nine red marbles, four white marbles, and five blue marbles.

Key Concepts	Examples/Videos
	d. Three marbles are to be selected at random from a marble bag that contains two green marbles, nine red marbles, four white marbles, and five blue marbles.
The Probability of an Event Suppose that an experiment has a sample space S. The probability of event E occurring is denoted as $P(E)$ and is given by $P(E) = \frac{\text{the number of ways event } E \text{ can occur}}{\text{the total number of possible outcomes}} = \frac{n(E)}{n(S)}.$ **Principles of Probability** 1. The probability of a certain event (an event that is guaranteed to happen) is 1, written as $P(\text{certain } E) = 1$. 2. The probability of an impossible event is 0, written as $P(\text{impossible } E) = 0$. 3. The sum of the probabilities of all possible outcomes of an experiment is 1, written as $P(E_1) + P(E_2) + \cdots + P(E_n) = 1$, where $E_1, E_2, \ldots, E_n$ are the possible outcomes of the experiment.	2. A paint-ball gun ball hopper contains 11 yellow balls, 6 green balls, and 3 red balls. A ball is fired from the gun. Find the probabilities of each of the following events written as a fraction in lowest terms. R: The ball is red G: The ball is green Y: The ball is yellow
The **complement** of an event E is the set of all outcomes in the sample space that are not included in the outcomes of event E. The notation used for the complement of E is E'. **The Complement Rule** Given the probability of an event E written as a fraction or a decimal, the probability of its complement can be found by subtracting the given probability from 1. That is, $P(E') = 1 - P(E)$. This equation can also be written as $P(E) = 1 - P(E')$. Given the probability of an event E written as a percent, the probability of its complement can be found by subtracting the given probability from 100%. That is, $P(E') = 100\% - P(E)$. This equation can also be written as $P(E) = 100\% - P(E')$.	Total number of students in the class: 50 Number of boys: 35 Number of girls: 15 Number of freshmen: 22 Number of sophomores: 13 Number of juniors: 9 Number of seniors: 6

Key Concepts	Examples/Videos
	3. The demographic information above describes the students that attend a particular mathematics class at the local community college. Suppose that the instructor chooses one student to dim the lights and one to shut the door. If a student may **not** do both jobs, determine the following probabilities rounded correct to the nearest tenth of a percent as needed. a. P(both are girls) b. P(one is a girl and one is a boy) c. P(at least one is a boy) d. P(the door closer is a boy) e. P(neither are seniors)
The Additive Rule of Probability For any two events A and B, $P(A \cup B) = P(A) + P(B) - P(A \cap B)$.	4. Suppose $P(A) = 0.6$, $P(B) = 0.7$ and $P(A \cup B) = 0.9$. Find the following probability written as a decimal rounded to one decimal place as needed. a. $P(A \cap B)$ b. $P((A \cap B)')$
A **Venn diagram** is a visual relationship between sets usually drawn using circles or shapes. The Venn diagram below illustrates two sets A and B drawn within a sample space S. The shaded region in common to both sets is called the intersection of sets A and B and is denoted as $A \cap B$. A B S	5. In a building with 123 offices, 72 offices have computers in them. Forty-five offices have windows. Twenty-three offices have computers and windows. Construct a Venn diagram describing this information. If one office is chosen at random, determine the probability of the following events written as a fraction in lowest terms. a. the office has a computer b. the office has a window c. the office has a computer and does not have a window d. office does not have computer and does not have a window

Key Concepts

The **conditional probability** of an event B in relation to an event A is the probability that event B will occur given that event A has already occurred. The notation used for conditional probability is $P(B|A)$, which reads as "the probability of B given A."

Examples/Videos

6. The Venn diagram below shows that in a building with 123 offices, 72 offices have computers in them. Forty-five offices have windows to the outside of the building. Twenty-three offices have computers and windows. If one office is chosen at random, determine each of the following probabilities written as a fraction in lowest terms.

   a. $P(C|W)$ b. $P(W|C)$

C W
49 23 22
29

C: Offices that have a computer
W: Offices that have a window

Conditional Probability Formula

The probability of an event B occurring given that event A has already occurred is given by the formula

$$P(B|A) = \frac{n(A \cap B)}{n(A)} = \frac{P(A \cap B)}{P(A)}.$$

7. Suppose $P(A) = 0.6$, $P(B) = 0.7$, and $P(A \cup B) = 0.9$.

   Determine the following probabilities correct to two decimal places as needed.

   a. $P(B|A)$ b. $P(A|B)$

Let E be an event, E' be the complement of E, and S be the sample space. The **odds of event E happening** is the ratio of the number of ways that event E can occur to the number of ways that event E cannot occur. The odds of an event E happening can be calculated using one of the following methods:

$n(E): n(E')$, where $n(E') = n(S) - n(E)$

$P(E): P(E')$, where $P(E') = 1 - P(E)$ and when $0 < P(E) < 1$

$P(E): P(E')$, where $P(E') = 100\% - P(E)$ and when $0\% < P(E) < 100\%$

8. A paint-ball gun ball hopper contains 10 yellow balls, 6 green balls, and 3 red balls. A ball is fired from the gun. Find:

   a. odds of firing a red ball

   b. odds of firing a green ball

   c. odds against firing a yellow ball

Chapter 14 Review Exercises

1. Write the first four terms of the sequence $a_n = 5(n + 2)!$.

2. Write a formula for the general term, or nth term, for the sequence $-2, 4, -8, 16, \ldots$. Then find a_7.

3. Find the partial sum, S_5, of the sequence $\frac{1}{2}, -\frac{1}{4}, \frac{1}{8}, -\frac{1}{16}, \ldots$

4. Rewrite the series $2 + (-4) + 8 + (-16) + \cdots + (-256)$ using summation notation. Use 1 as the lower limit of the summation.

5. Find the general term of the arithmetic sequence $2, 7, 12, 17, \ldots$. Then find a_{30}.

6. Given an arithmetic sequence with $d = -5$ and $a_7 = 11$, find a_{22}.

7. Find the sum: $\sum_{j=1}^{10} (3j + 7)$

8. A large multiplex movie house has many theaters. The largest theater has 40 rows. There are 12 seats in the first row. Each row has four more seats than the previous row. How many total seats are there in this theater?

9. Write the first five terms of the geometric sequence whose first term of 162 and whose common ratio of $\frac{1}{3}$.

10. Find the seventh term of the geometric sequence whose first term is 5 and whose common ratio is 4.

11. Find the sum of the geometric series $\sum_{i=1}^{13} 2\left(\frac{3}{5}\right)^{i-1}$.

12. Determine if the infinite geometric series $\sum_{i=1}^{\infty} 7\left(\frac{1}{4}\right)^{i-1}$ converges or diverges. If the series converges, find the sum.

13. Evaluate the binomial coefficient $\binom{10}{4}$.

14. Use the Binomial Theorem to expand the binomial $(5x - 3y)^5$.

15. Find the seventh term of the expansion of $(3x + 2)^{10}$.

16. Find the coefficient of x^4 in the expansion $(4x + 1)^{12}$.

17. Use the principle of mathematical induction to show that the mathematical statement is true for all natural numbers n.

$$S_n\!: 2 + 3 + 4 + \cdots + (n + 1) = \frac{1}{2}n(n + 3)$$

18. A man has three pairs of slacks, four shirts, and five ties that all coordinate. How many different outfits of slacks, shirts, and ties can he put together?

19. How many five-member committees can be formed from a class of 30 students?

20. A neighborhood contains 5 coffee shops, 3 pizza restaurants, and 2 sushi restaurants. Three of the businesses are to be chosen at random. Of the total number of possible outcomes, how many would consist only of coffee shops?

21. Three cards are dealt without replacement from a standard deck of 52 cards. In how many ways can this happen if two cards are hearts and one is a spade?

22. A paint-ball gun ball hopper contains 9 yellow balls, 12 green balls, and 7 red balls. A ball is fired from the gun. What is the probability that the ball fired from the gun is red?

23. The demographic information of a classroom at a local high school is given below. The instructor randomly chooses one student to pass out papers, one student to clean the chalkboard, and one student to empty the trash. If a student may do more than one job, then what is the probability that the instructor will choose all girls to do the jobs?

Total number of students in the class:	46
Number of boys:	19
Number of girls:	27
Number of freshmen:	7
Number of sophomores:	16
Number of juniors:	12
Number of seniors:	11

24. Suppose that $P(A) = 0.5$, $P(B) = 0.8$, and $P(A \cup B) = 0.9$, find $P(A \cap B)$.

25. A box contains 10 yellow, 5 red, and 7 blue balls. Four balls are selected at random and the colors are noted. Determine the probability that at least one ball is red.

APPENDIX A

Degree, Minute, Second Form and Degree Decimal Form

Not every angle has an integer measurement. An angle can have any real number measurement. For example, we could have an angle that measures $41.23°$. This angle is an example of an angle written in **degree decimal form**, abbreviated DD. The size of this angle is 41 full degrees plus two tenths of one degree plus three one-hundredths of one degree.

Another way to describe the size of an angle uses a method that subdivides a degree into 60 equal parts called **minutes** and subdivides each minute into 60 equal parts called **seconds**. Just like we use the symbol ° to denote degrees, we use the symbol ′ to denote minutes and use the symbol ″ to denote seconds. For example, we could have an angle that measures $75°16'9''$. This angle is 75 degrees, 16 minutes, and 9 seconds.

This method of describing the measurement of an angle is called **degree, minute, second form**, abbreviated as DMS and is summarized below.

Degree, Minute, Second Form (DMS)

$360° = 1$	counterclockwise revolution
$1° = 60'$	(One degree is equal to 60 minutes.)
$1' = 60''$	(One minute is equal to 60 seconds.)
$1° = 3600''$	(One degree is equal to 3600 seconds.)

Example 1 Converting between Degree, Minute, Second and Degree Decimal Forms

a. Convert $41.23°$ to DMS form. Round to the nearest second.

b. Convert the angle $75°16'9''$ into DD form. Round to two decimal places.

Solution

a. $41.23° = 41° + 0.23°$ — Separate the whole part of the angle and the decimal part.

$= 41° + 0.23°\left(\frac{60'}{1°}\right)$ — There are 60 minutes in 1 degree.

$= 41° + 13.8'$ — Multiply.

$= 41° + 13' + 0.8'$ — Separate the whole part of the minutes and the decimal part.

$= 41° + 13' + 0.8'\left(\frac{60''}{1'}\right)$ — There are 60 seconds in 1 minute.

$= 41° + 13' + 48''$ — Multiply.

$= 41°13'48''$ — Rewrite in DMS form.

(eText Screens A.1-1–A.1-5)

b. $75°16'9'' = 75° + 16' + 9''$ — Separate the parts.

$= 75° + 16'\left(\frac{1°}{60'}\right) + 9''\left(\frac{1°}{3600''}\right)$ — There are 60 minutes in 1 degree and there are 3600 seconds in 1 degree.

$= 75° + 0.2\overline{6}° + 0.0025°$ — Multiply.

$\approx 75.27°$ — Rewrite in DD form and round to two decimal places.

Work through this **interactive video** to see these solutions worked out in detail.

Most graphing calculators are equipped with buttons that allow you to convert between the two forms. See Figure 1.

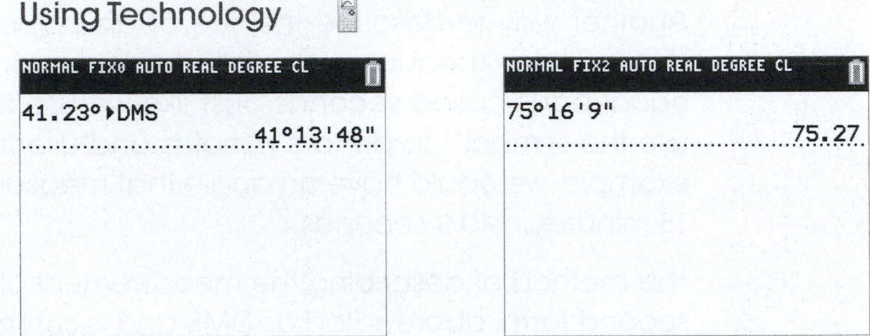

Figure 1 A graphing utility can be used to convert between degree, minute, second and degree decimal forms. When using this feature, it is important that the calculator is set to degree mode.

You Try It Work through this You Try It problem.

Work Exercises 1–6 in this textbook or in the **MyLab** Math Study Plan.

A.1 Exercises

In Exercises 1–3, convert each angle into degree, minute, second form. Round your answers to the nearest second.

1. $40.52°$
2. $73.345°$
3. $-13.273°$

In Exercises 4–6, convert each angle into degree decimal form. Round your answers to two decimal places.

4. $4°3'8''$
5. $59°31'22''$
6. $-57°21'58''$

APPENDIX B

Conic Section Proofs

APPENDIX B CONTENTS

B.1 Parabola Proofs

OBJECTIVES

1 Derivation of the Equation of a Parabola in Standard Form with a Vertical Axis of Symmetry

2 Derivation of the Equation of a Parabola in Standard Form with a Horizontal Axis of Symmetry

OBJECTIVE 1 DERIVATION OF THE EQUATION OF A PARABOLA IN STANDARD FORM WITH A VERTICAL AXIS OF SYMMETRY

To derive the equation of a parabola with a vertical axis of symmetry, start with the graph of a parabola with vertex $V(h, k)$, focus F, directrix D, which passes through the point $P(x, y)$ as shown in Figure 1.

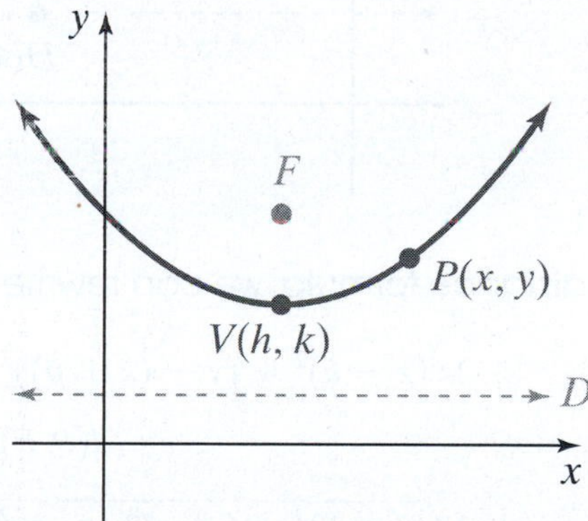

Figure 1

By the geometric definition of the parabola, the distance from any point that lies on the parabola to the focus must be equal to the distance from that point to the directrix. Therefore, the following equation must be true:

$$d(P, F) = d(P, D)$$

Because the vertex lies on the graph of the parabola, it implies that the distance from the vertex to the focus must be equal to the distance from the vertex to the directrix. Let $|p|$ be the distance from the vertex to the focus. Without loss of generality, suppose $p > 0$. Therefore, the coordinates of the focus must be $(h, k + p)$. Similarly, the equation of the directrix must be $y = k - p$. See Figure 2.

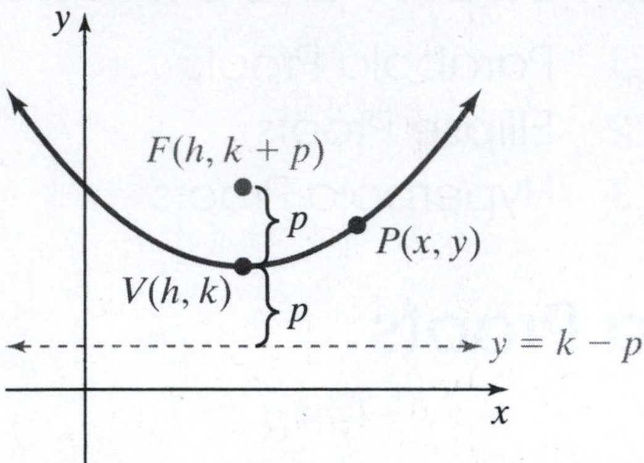

Figure 2

Consider the point $D(x, k - p)$ that lies on the directrix. By the geometric definition of the parabola, the distance from P to F must be the same as the distance from P to D. See Figure 3.

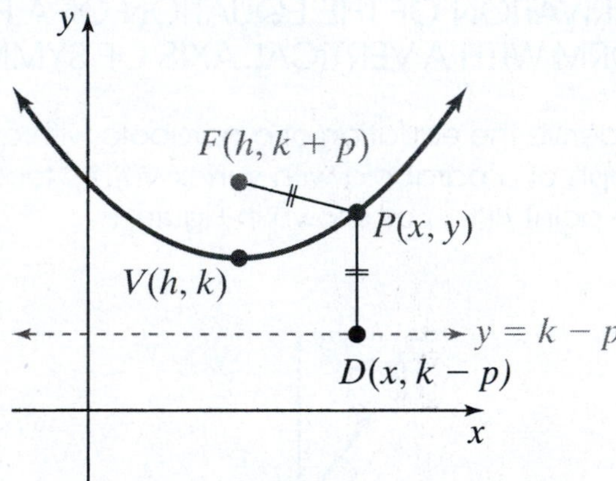

Figure 3

By the distance formula, we can rewrite the equation $d(P, F) = d(P, D)$ as

$$\sqrt{(x - h)^2 + (y - (k + p))^2} = \sqrt{(x - x)^2 + (y - (k - p))^2}$$

$$d(P, F) = d(P, D)$$

$$\sqrt{(x - h)^2 + (y - (k + p))^2} = \sqrt{(x - x)^2 + (y - (k - p))^2}$$

Squaring both sides of this equation and simplifying we get:

$$(x - h)^2 + (y - (k + p))^2 = (0)^2 + (y - (k - p))^2$$

$$(x - h)^2 + (y - k - p)(y - k - p) = (y - k + p)(y - k + p)$$

$$(x - h)^2 + y^2 - 2ky - 2py + 2pk + k^2 + p^2 = y^2 - 2ky + 2py - 2pk + k^2 + p^2$$

$$(x - h)^2 - 2py + 2pk = 2py - 2pk$$

$$(x - h)^2 = 4py - 4pk$$

$$(x - h)^2 = 4p(y - k)$$

Therefore, the equation of a parabola with a vertical axis of symmetry (and a horizontal directrix) is $(x - h)^2 = 4p(y - k)$. If p is positive, the graph opens upward. If p is negative, the graph opens downward.

Equation of a Parabola in Standard Form with a Vertical Axis of Symmetry

The equation of a parabola with a vertical axis of symmetry is $(x - h)^2 = 4p(y - k)$,

where

the vertex is $V(h, k)$,
$|p|$ = distance from the vertex to the focus = distance from the vertex to the directrix,
the focus is $F(h, k + p)$, and
the equation of the directrix is $y = k - p$.

The parabola opens *upward* if $p > 0$ or *downward* if $p < 0$.

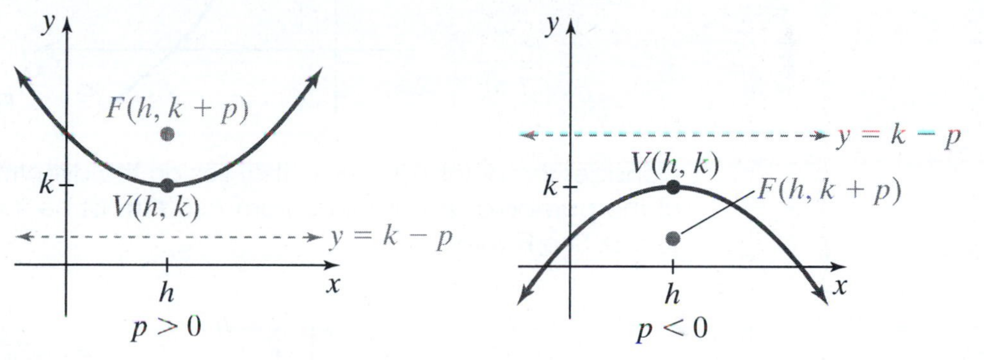

OBJECTIVE 2 DERIVATION OF THE EQUATION OF A PARABOLA IN STANDARD FORM WITH A HORIZONTAL AXIS OF SYMMETRY

To derive the equation of a parabola with a horizontal axis of symmetry, start with the graph of a parabola with vertex $V(h, k)$, focus F, directrix D, which passes through the point $P(x, y)$ as shown in Figure 4.

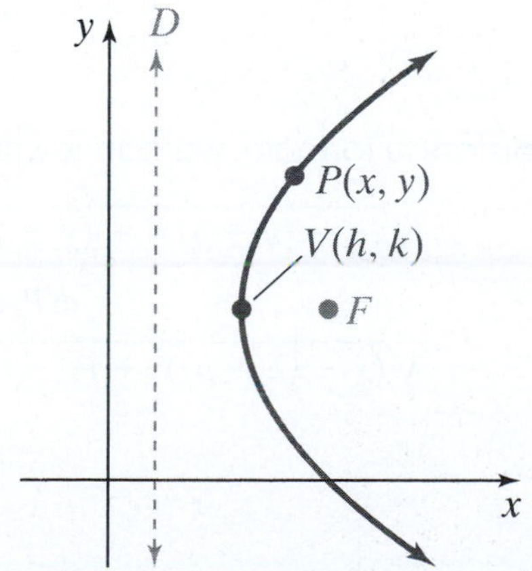

Figure 4

By the geometric definition of the parabola, the distance from any point that lies on the parabola to the focus must be equal to the distance from that point to the directrix. Therefore, the following equation must be true:

$$d(P, F) = d(P, D)$$

Because the vertex lies on the graph of the parabola, it implies that the distance from the vertex to the focus must be equal to the distance from the vertex to the directrix. Let $|p|$ be the distance from the vertex to the focus. Without loss of generality, suppose $p > 0$. Therefore, the coordinates of the focus must be $(h + p, k)$. Similarly, the equation of the directrix must be $x = h - p$. See Figure 5.

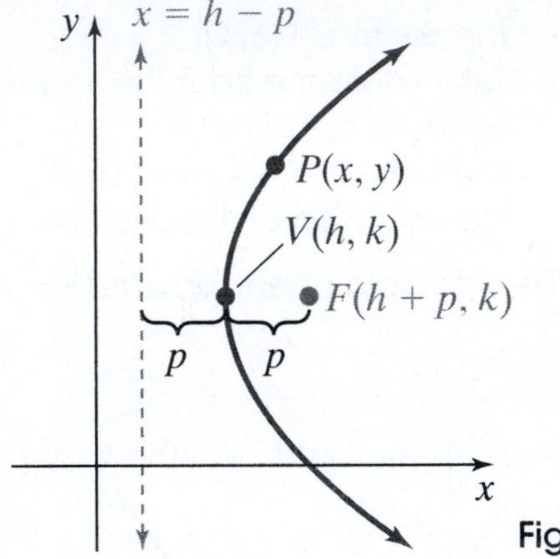

Figure 5

Consider the point $D(h - p, y)$ that lies on the directrix. By the geometric definition of the parabola, the distance from P to F must be the same as the distance from P to D. See Figure 6.

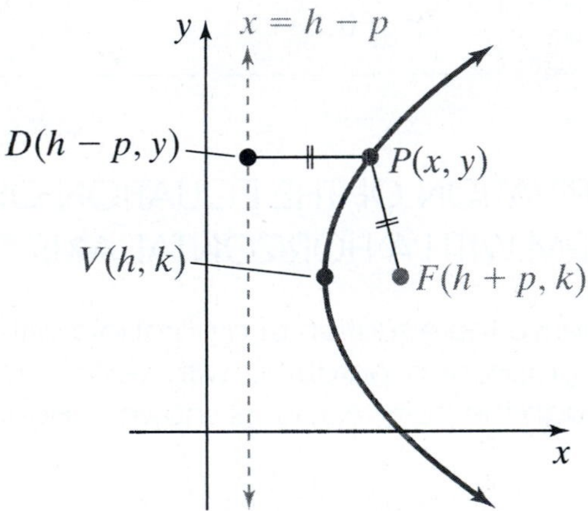

Figure 6

By the distance formula, we can rewrite the equation $d(P, F) = d(P, D)$ as

$$\sqrt{(x - (h + p))^2 + (y - k)^2} = \sqrt{(x - (h - p))^2 + (y - y)^2}$$

$$d(P, F) = d(P, D)$$

$$\sqrt{(x - (h + p))^2 + (y - k)^2} = \sqrt{(x - (h - p))^2 + (y - y)^2}$$

Squaring both sides of this equation and simplifying we get:

$$(x - (h + p))^2 + (y - k)^2 = (x - (h - p))^2 + (0)^2$$

$$(y - k)^2 + (x - h - p)(x - h - p) = (x - h + p)(x - h + p)$$

$$(y - k)^2 + x^2 - 2hx - 2px + 2ph + h^2 + p^2 = x^2 - 2hx + 2px - 2ph + h^2 + p^2$$

$$(y - k)^2 - 2px + 2ph = 2px - 2ph$$

$$(y - k)^2 = 4px - 4ph$$

$$(y - k)^2 = 4p(x - h)$$

Therefore, the equation of a parabola with a horizontal axis of symmetry (and a vertical directrix) is $(y - k)^2 = 4p(x - h)$. If p is positive, the graph opens to the right. If p is negative, the graph opens to the left.

Equation of a Parabola in Standard Form with a Horizontal Axis of Symmetry

The equation of a parabola with a horizontal axis of symmetry is
$(y - k)^2 = 4p(x - h)$,

where

the vertex is $V(h, k)$,
$|p|$ = distance from the vertex to the focus = distance from the vertex to the directrix,
the focus is $F(h + p, k)$, and
the equation of the directrix is $x = h - p$.

The parabola opens *right* if $p > 0$ or *left* if $p < 0$.

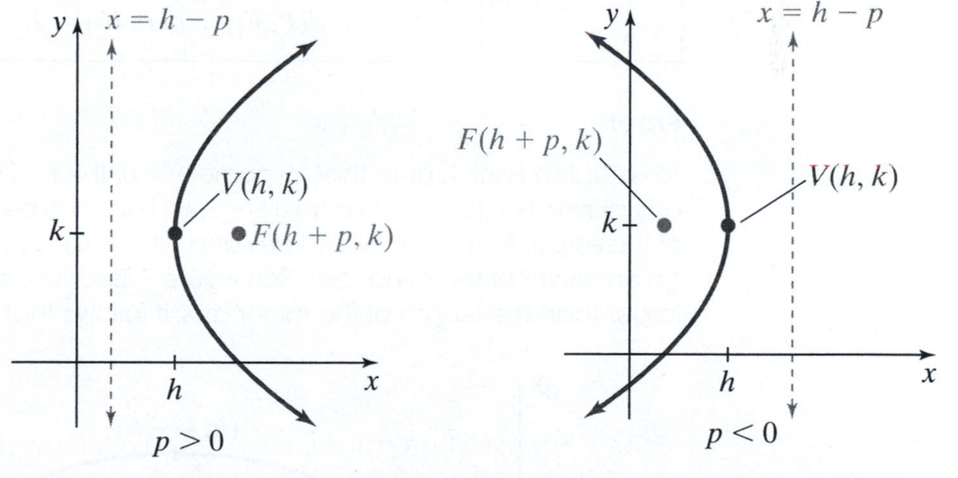

B.2 Ellipse Proofs

OBJECTIVES

1 Establishing Fact 1 for Ellipses

2 Establishing Fact 2 for Ellipses

3 Derivation of the Equation of an Ellipse in Standard Form with a Horizontal Major Axis

OBJECTIVE 1 ESTABLISHING FACT 1 FOR ELLIPSES

Fact 1 for Ellipses

Given the foci of an ellipse F_1 and F_2 and any point P that lies on the graph of the ellipse, the sum of the distances between P and the foci is equal to the length of the major axis or $d(F_1, P) + d(F_2, P) = 2a$.

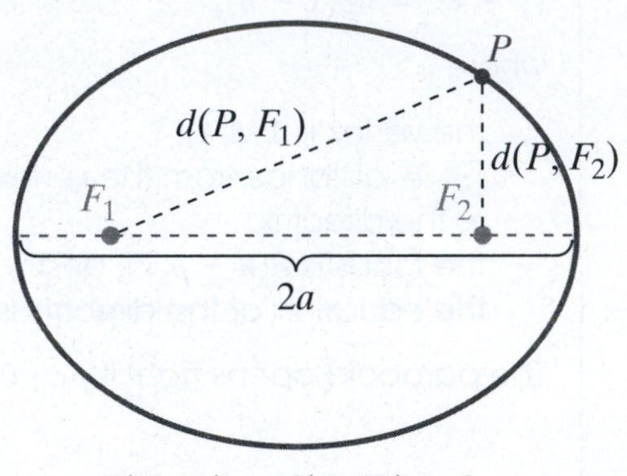

$$d(P, F_1) + d(P, F_2) = 2a$$

Proof

To establish Fact 1, note that for some $c > 0$, the foci are located along the major axis at points $F_1(h - c, k)$ and $F_2(h + c, k)$. Let $a > 0$ be the distance from the center of the ellipse to one of the vertices and let $b > 0$ be the distance from the center to an endpoint of the minor axis. See Figure 7. Because the length of the major axis is larger than the length of the minor axis, it follows that $a > b$

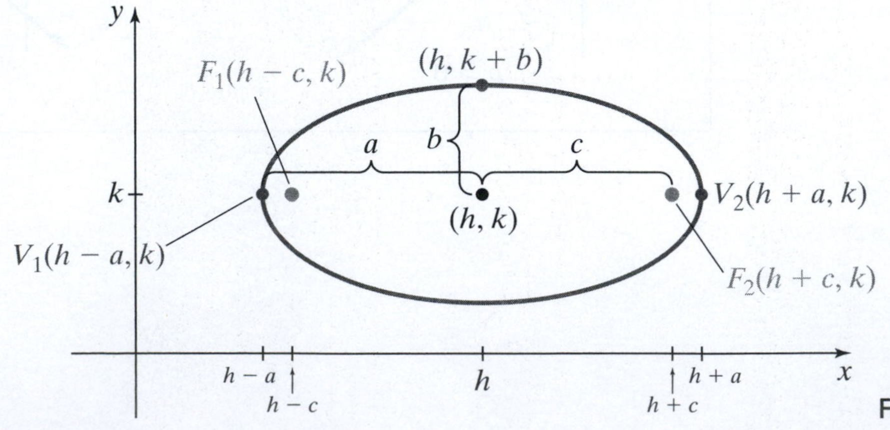

Figure 7

Consider the vertex $V_2(h + a, k)$. The distance from V_2 to F_2 is $(h + a) - (h + c) = a - c$ and the distance from V_2 to F_1 is $(h + a) - (h - c) = a + c$. See Figure 8.

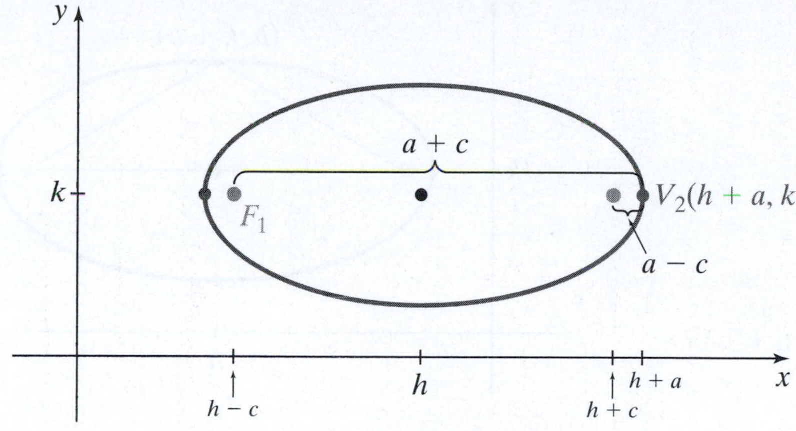

Figure 8

Thus, $d(V_2, F_1) + d(V_2, F_2) = (a + c) + (a - c) = 2a$. By the geometric definition of the ellipse, the sum of the distances from **any** point on the ellipse to the foci is a positive constant. Therefore, this positive constant must be equal to $2a$. This establishes Fact 1.

OBJECTIVE 2 ESTABLISHING FACT 2 FOR ELLIPSES

Fact 2 for Ellipses

Suppose $a > 0$ is the length from the center of an ellipse to a vertex, $b > 0$ is the length from the center to an endpoint of the minor axis, and $c > 0$ is the length from the center to a focus, then $b^2 = a^2 - c^2$ or equivalently, $c^2 = a^2 - b^2$.

Proof

Suppose that we label the coordinates of one of the endpoints of the minor axis as $(h, k + b)$. See Figure 9.

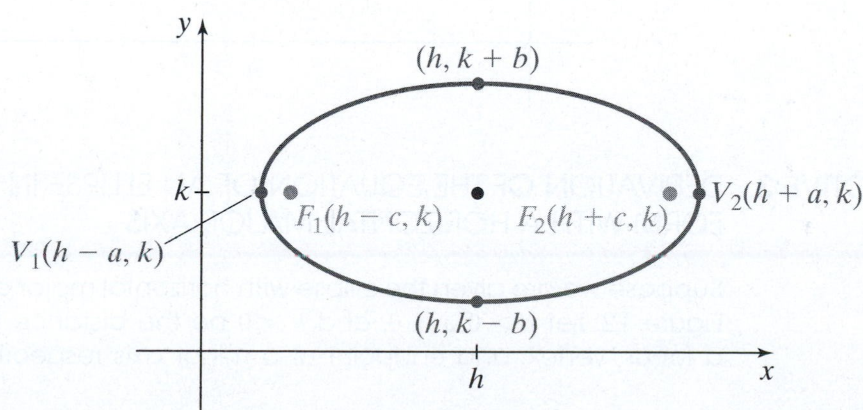

Figure 9

We can construct a triangle by connecting the point $(h, k + b)$ with the two foci as shown in Figure 10.

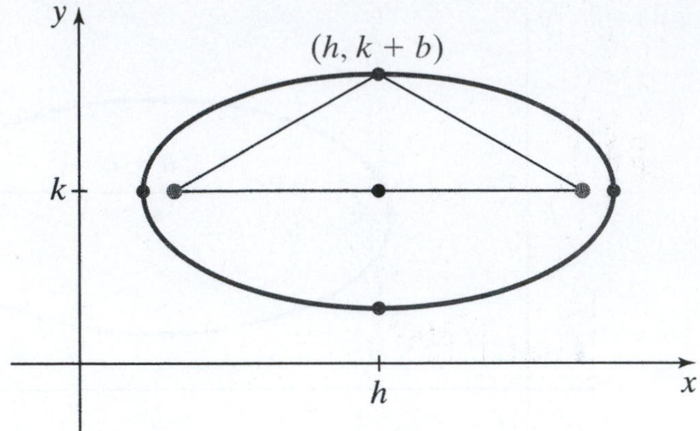

Figure 10

If we draw a line segment connecting the point $(h, k + b)$ with the center of the ellipse, we see that we have constructed two congruent right triangles. See **Figure 11**. The base of each right triangle has length c and the height of each right triangle has length b.

We know from Fact 1 that the sum of the distances between *any* point on the ellipse and the two foci is $2a$. Therefore, the sum of the distances between the point $(h, k + b)$ and the two foci is also equal to $2a$. Thus, the hypotenuse of each of the right triangles must be equal to a. Therefore, by the Pythagorean theorem we can establish the relationship $b^2 + c^2 = a^2$ or $b^2 = a^2 - c^2$. Thus, we have established Fact 2.

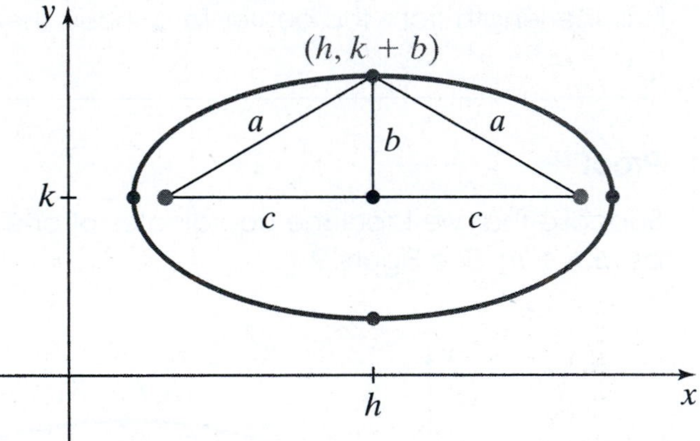

Figure 11

OBJECTIVE 3 DERIVATION OF THE EQUATION OF AN ELLIPSE IN STANDARD FORM WITH A HORIZONTAL MAJOR AXIS

Suppose we are given the ellipse with horizontal major axis centered at (h, k). See Figure 12. Let $c > 0, a > 0,$ and $b > 0$ be the distance between the center and a focus, vertex, and endpoint of a minor axis respectively. Because the ellipse has a horizontal major axis, we know that $a > b$. By the definition of the ellipse and by Fact 1, we know that for any point P that lies on the graph of the ellipse, $d(P, F_1) + d(P, F_2) = 2a$.

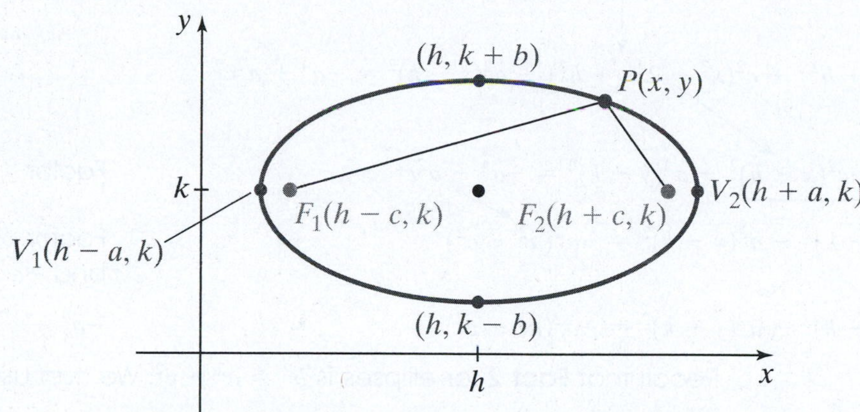

Figure 12

Using the distance formula, we can rewrite the equation $d(P, F_1) + d(P, F_2) = 2a$ as

$$\sqrt{(x-(h-c))^2+(y-k)^2}+\sqrt{(x-(h+c))^2+(y-k)^2}=2a$$

Subtract a radical from both sides.

$$\sqrt{(x-(h-c))^2+(y-k)^2}=2a-\sqrt{(x-(h+c))^2+(y-k)^2}$$

We now square both sides of the equation.

$$\left[\sqrt{(x-(h-c))^2+(y-k)^2}\right]^2=\left[2a-\sqrt{(x-(h+c))^2+(y-k)^2}\right]^2$$

$$(x-(h-c))^2+(y-k)^2=4a^2-4a\sqrt{(x-(h+c))^2+(y-k)^2}+(x-(h+c))^2+(y-k)^2$$

Subtract $(y-k)^2$ from both sides and simplify.

$$(x-(h-c))^2=4a^2-4a\sqrt{(x-(h+c))^2+(y-k)^2}+(x-(h+c))^2$$

$$x^2-2xh+2cx-2ch+h^2+c^2=4a^2-4a\sqrt{(x-(h+c))^2+(y-k)^2}+x^2-2xh-2cx+2ch+h^2+c^2$$

Combine like terms and subtract $4a^2$ from both sides.

$$4cx-4ch-4a^2=-4a\sqrt{(x-(h+c))^2+(y-k)^2}$$

Now divide both sides by -4.

$$-cx+ch+a^2=a\sqrt{(x-(h+c))^2+(y-k)^2}$$

We now square both sides and simplify.

$$\left[-cx+ch+a^2\right]^2=\left[a\sqrt{(x-(h+c))^2+(y-k)^2}\right]^2$$

$$(-cx+ch+a^2)(-cx+ch+a^2)=a^2[(x-(h+c))^2+(y-k)^2]$$

$$c^2x^2-2c^2xh-2a^2cx+2a^2ch+c^2h^2+a^4=a^2[x^2-2xh-2cx+2ch+h^2+c^2+(y-k)^2]$$

$$c^2x^2-2c^2xh-2a^2cx+2a^2ch+c^2h^2+a^4=a^2x^2-2a^2xh-2a^2cx+2a^2ch+a^2h^2+a^2c^2+a^2(y-k)^2$$

Collect the x and y terms on the left side.

$$-a^2x^2 + 2a^2xh - a^2h^2 + c^2x^2 - 2c^2xh + c^2h^2 - a^2(y-k)^2 = -a^4 + a^2c^2$$

$$\underbrace{-a^2x^2 + 2a^2xh - a^2h^2} + \underbrace{c^2x^2 - 2c^2xh + c^2h^2} - a^2(y-k)^2 = -a^4 + a^2c^2$$

Factor $-a^2$ out of the first three terms and factor c^2 out of the next three terms.

$$-a^2(x^2 - 2xh + h^2) + c^2(x^2 - 2xh + h^2) - a^2(y-k)^2 = -a^4 + a^2c^2$$

$$\underbrace{-a^2(x-h)^2 + c^2(x-h)^2} - a^2(y-k)^2 = -a^4 + a^2c^2$$

Factor.

$$(-a^2 + c^2)(x-h)^2 - a^2(y-k)^2 = -a^2(a^2 - c^2)$$

Factor out $(x-h)^2$ and $-a^2$.

$$-(a^2 - c^2)(x-h)^2 - a^2(y-k)^2 = -a^2(a^2 - c^2)$$

$-a^2 + c^2 = -(a^2 - c^2)$.

Recall that **Fact 2** for ellipses is $b^2 = a^2 - c^2$. We can use this fact to substitute b^2 for $a^2 - c^2$.

$$-(a^2 - c^2)(x-h)^2 - a^2(y-k)^2 = -a^2(a^2 - c^2)$$ Rewrite the equation from previous page.

$$-b^2(x-h)^2 - a^2(y-k)^2 = -a^2b^2$$ Substitute $b^2 = a^2 - c^2$.

$$\frac{-b^2(x-h)^2}{-a^2b^2} - \frac{a^2(y-k)^2}{-a^2b^2} = \frac{-a^2b^2}{-a^2b^2}$$ Divide both sides by $-a^2b^2$.

$$\frac{(x-h)^2}{a^2} + \frac{(y-k)^2}{b^2} = 1.$$

Therefore, the standard equation of an ellipse centered at (h, k) with a horizontal major axis is $\frac{(x-h)^2}{a^2} + \frac{(y-k)^2}{b^2} = 1$, where $a > b$ and $b^2 = a^2 - c^2$. We can derive the standard equation of an ellipse centered at (h, k) with a vertical major axis in a similar fashion to obtain the equation $\frac{(x-h)^2}{b^2} + \frac{(y-k)^2}{a^2} = 1$, where $a > b$ and $b^2 = a^2 - c^2$.

Equation of an Ellipse in Standard Form with Center (h, k)

Horizontal Major Axis

$$\frac{(x-h)^2}{a^2} + \frac{(y-k)^2}{b^2} = 1$$

- $a > b > 0$
- Foci: $F_1(h - c, k)$ and $F_2(h + c, k)$
- Vertices: $V_1(h - a, k)$ and $V_2(h + a, k)$
- Endpoints of minor axis: $(h, k - b)$ and $(h, k + b)$
- $c^2 = a^2 - b^2$

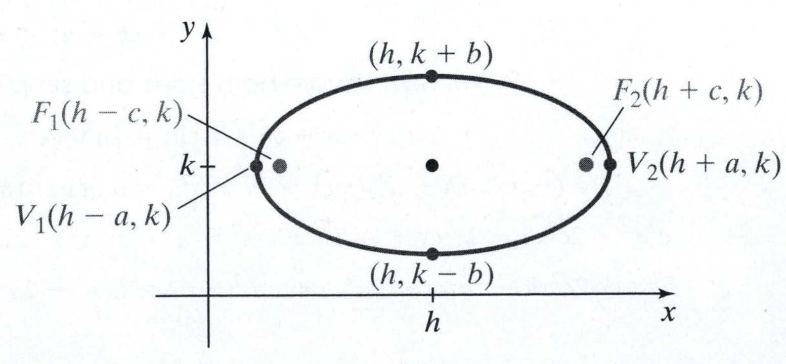

Equation of an Ellipse in Standard Form with Center (h, k)

Vertical Major Axis

$$\frac{(x-h)^2}{b^2} + \frac{(y-k)^2}{a^2} = 1$$

- $a > b > 0$
- Foci: $F_1(h, k - c)$ and $F_2(h, k + c)$
- Vertices: $V_1(h, k - a)$ and $V_2(h, k + a)$
- Endpoints of minor axis: $(h - b, k)$ and $(h + b, k)$
- $c^2 = a^2 - b^2$

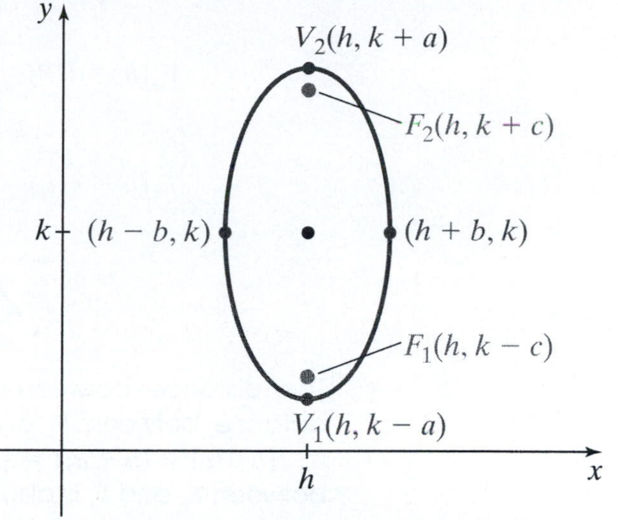

B.3 Hyperbola Proofs

OBJECTIVES

1 Establishing Fact 1 for Hyperbolas

2 Derivation of the Equation of a Hyperbola in Standard Form with a Horizontal Transverse Axis

OBJECTIVE 1 ESTABLISHING FACT 1 FOR HYPERBOLAS

Fact 1 for Hyperbolas

Given the foci of a hyperbola F_1 and F_2 and any point P that lies on the graph of the hyperbola, the difference of the distances between P and the foci is equal to $2a$. In other words, $|d(P, F_1) - d(P, F_2)| = 2a$. The constant $2a$ represents the length of the transverse axis.

Proof

Consider the hyperbola centered at (h, k) with vertices $V_1(h - a, k)$ and $V_2(h + a, k)$ and foci $F_1(h - c, k)$ and $F_2(h + c, k)$ as shown in Figure 13.

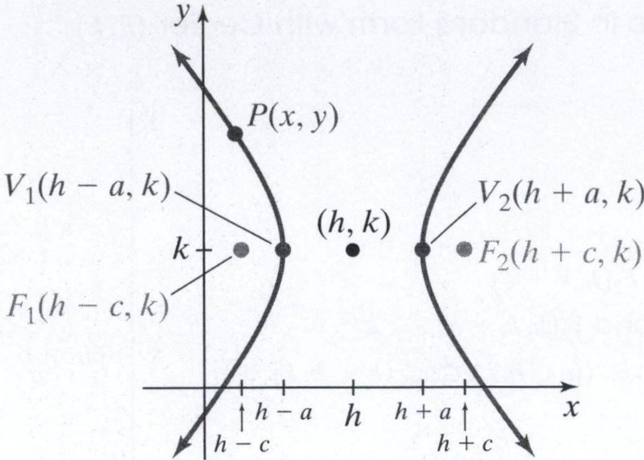

Figure 13

The distance between the center and either vertex is equal to a, so the distance between V_1 and V_2 is equal to $2a$. The distance between F_2 and V_2 is $|(h + c) - (h + a)| = |c - a| = c - a$ because $c > a$. Likewise, the distance between F_1 and V_1 is also equal to $c - a$. See Figure 14.

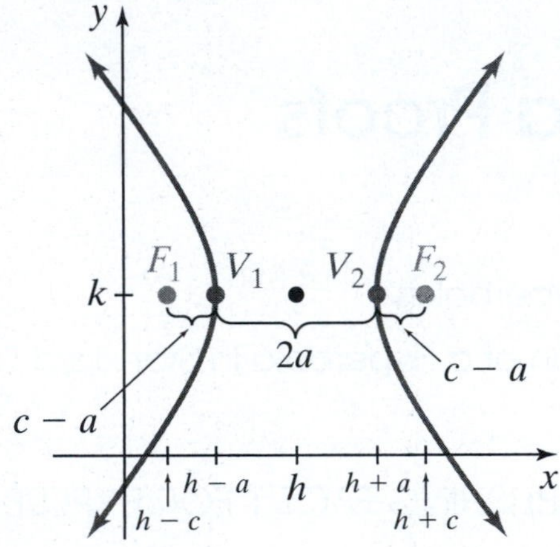

Figure 14

The distance between V_1 and F_2 is $d(V_1, F_2) = 2a + (c - a) = a + c$. The distance between V_1 and F_1 is $d(V_1, F_1) = c - a$. Therefore, $|d(V_1, F_2) - d(V_1, F_1)| = |a + c - (c - a)| = |a + c - c + a| = 2a$. Thus, by the geometric definition of the hyperbola, the difference between the distances from **any** point P on the hyperbola to the two foci must also be equal to $2a$. That is, $|d(P, F_1) - d(P, F_2)| = 2a$.

OBJECTIVE 2 DERIVATION OF THE EQUATION OF A HYPERBOLA IN STANDARD FORM WITH A HORIZONTAL TRANSVERSE AXIS

To derive the equation of the hyperbola with a horizontal transverse axis with center (h, k), use Figure 15. Without loss of generality, suppose that $d(P, F_2) > d(P, F_1)$. We will use Fact 1 for hyperbolas which states that $d(P, F_2) - d(P, F_1) = 2a$ to derive the equation.

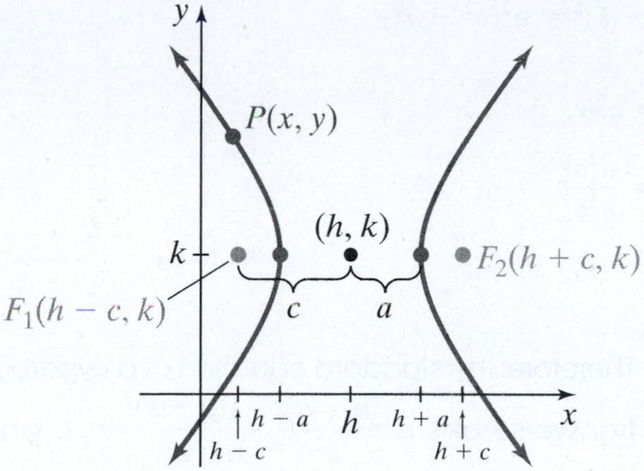

Figure 15

Using Fact 1 for hyperbolas and the distance formula, we get:

$$d(P, F_2) - d(P, F_1) = 2a$$

$$\sqrt{(x-(h+c))^2+(y-k)^2} - \sqrt{(x-(h-c))^2+(y-k)^2} = 2a$$

Now add a radical to both sides then square both sides of the equation.

$$\sqrt{(x-(h+c))^2+(y-k)^2} = 2a + \sqrt{(x-(h-c))^2+(y-k)^2}$$

$$\left[\sqrt{(x-(h+c))^2+(y-k)^2}\right]^2 = \left[2a\sqrt{(x-(h-c))^2+(y-k)^2}\right]^2$$

$$(x-(h+c))^2+(y-k)^2 = 4a^2 + 4a\sqrt{(x-(h-c))^2+(y-k)^2} + (x-(h-c))^2+(y-k)^2$$

Subtract $(y-k)^2$ from both sides and simplify.

$$(x-(h+c))^2 = 4a^2 + 4a\sqrt{(x-(h-c))^2+(y-k)^2} + (x-(h-c))^2$$

$$x^2 - 2hx - 2cx + 2ch + h^2 + c^2 = 4a^2 + 4a\sqrt{(x-(h-c))^2+(y-k)^2} + x^2 - 2hx + 2cx - 2ch + h^2 + c^2$$

$$-4a^2 - 4cx + 4ch = 4a\sqrt{(x-(h-c))^2+(y-k)^2}$$ Combine like terms and isolate the radical.

$$a^2 + cx - ch = -a\sqrt{(x-(h-c))^2+(y-k)^2}$$ Divide both sides by −4.

$$a^2 + cx - ch = -a\sqrt{(x-(h-c))^2+(y-k)^2}$$

Now square both sides and simplify.

$$[a^2 + cx - ch]^2 = \left[-a\sqrt{(x-(h-c))^2+(y-k)^2}\right]^2$$

$$(a^2 + cx - ch)(a^2 + cx - ch) = a^2[(x-(h-c))^2+(y-k)^2]$$

$$a^4 + 2a^2cx - 2a^2ch - 2c^2hx + c^2x^2 + c^2h^2 = a^2x^2 - 2a^2hx + 2a^2cx - 2a^2ch + a^2h^2 + a^2c^2 + a^2(y-k)^2$$

$$c^2x^2 - 2c^2hx + c^2h^2 - a^2x^2 + 2a^2hx - a^2h^2 - a^2(y-k)^2 = a^2c^2 - a^4$$

$$\underbrace{c^2x^2 - 2c^2hx + c^2h^2}\ -\ \underbrace{a^2x^2 + 2a^2hx - a^2h^2} - a^2(y-k)^2 = a^2c^2 - a^4$$ Factor c^2 out of the first three terms and factor $-a^2$ out of the next three terms.

$$c^2(x^2 - 2xh + h^2) - a^2(x^2 - 2xh + h^2) - a^2(y-k)^2 = a^2c^2 - a^4$$

$$c^2(x-h)^2 - a^2(x-h)^2 - a^2(y-k)^2 = a^2c^2 - a^4$$ $x^2 - 2xh + h^2 = (x-h)^2$

$(c^2 - a^2)(x - h)^2 - a^2(y - k)^2 = a^2(c^2 - a^2)$ Factor $(x - h)^2$ out of the first two terms.

$b^2(x - h)^2 - a^2(y - k)^2 = a^2b^2$ Let $b^2 = c^2 - a^2$.

$\frac{b^2(x - h)^2}{a^2b^2} - \frac{a^2(y - k)^2}{a^2b^2} = \frac{a^2b^2}{a^2b^2}$ Divide both sides by a^2b^2.

$\frac{(x - h)^2}{a^2} - \frac{(y - k)^2}{b^2} = 1$ Simplify.

Therefore, the standard equation of a hyperbola centered at (h, k) with a horizontal transverse axis is $\frac{(x - h)^2}{a^2} - \frac{(y - k)^2}{b^2} = 1$, where $b^2 = c^2 - a^2$. We can derive the standard equation of a hyperbola centered at (h, k) with a vertical transverse axis in a similar fashion to obtain the equation $\frac{(y - k)^2}{a^2} - \frac{(x - h)^2}{b^2} = 1$, where $b^2 = c^2 - a^2$.

Equation of a Hyperbola in Standard Form with Center (h, k)

Horizontal Transverse Axis

$$\frac{(x - h)^2}{a^2} - \frac{(y - k)^2}{b^2} = 1$$

- Foci: $F_1(h - c, k)$ and $F_2(h + c, k)$
- Vertices: $V_1(h - a, k)$ and $V_2(h + a, k)$
- Endpoints of conjugate axis: $(h, k - b)$ and $(h, k + b)$
- $b^2 = c^2 - a^2$
- Asymptotes: $y - k = \pm \frac{b}{a}(x - h)$

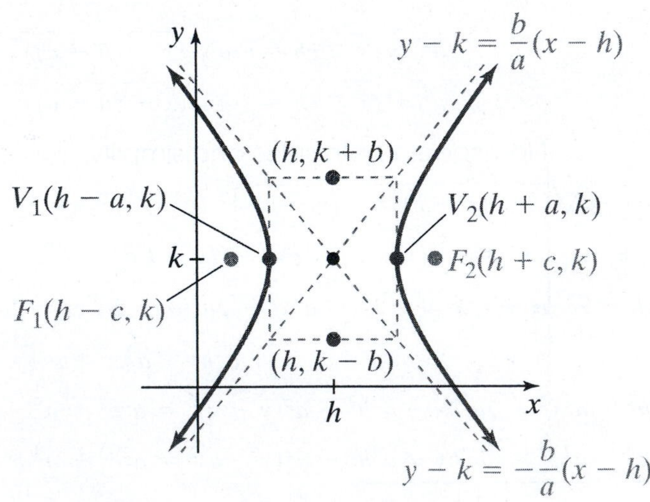

Equation of a Hyperbola in Standard Form with Center (h, k)

Vertical Transverse Axis

$$\frac{(y-k)^2}{a^2} - \frac{(x-h)^2}{b^2} = 1$$

- Foci: $F_1(h, k - c)$ and $F_2(h, k + c)$
- Vertices: $V_1(h, k - a)$ and $V_2(h, k + a)$
- Endpoints of conjugate axis: $(h - b, k)$ and $(h + b, k)$
- $b^2 = c^2 - a^2$
- Asymptotes: $y - k = \pm \frac{a}{b}(x - h)$

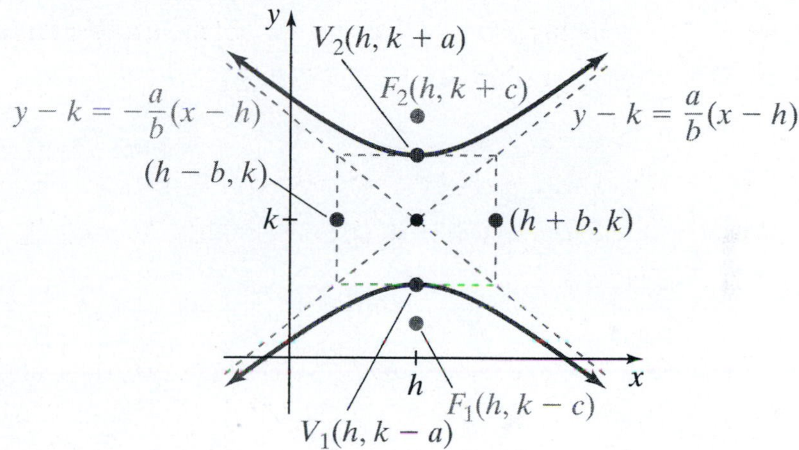

Answers

CHAPTER R

R.1 Exercises

1. natural, whole, integer, rational, real **2.** integer, rational, real **3.** rational, real **4.** irrational, real **5. a.** 1 **b.** 0, 1 **c.** $-17, 0, 1$ **d.** $-17, -\frac{25}{19}, 0, 0.331, 1$ **e.** $-\sqrt{5}, \frac{\pi}{2}$ **6. a.** none **b.** none **c.** -11 **d.** $-11, -3.\overline{2135}, -\frac{3}{9}, 21.1$ **e.** $\frac{\sqrt{7}}{2}$ **7. a.** open interval **b.** $\{x|0 < x < 3\}$ **c.** $(0, 3)$ **8. a.** closed infinite interval **b.** $\left\{x \middle| x \le \frac{1}{4}\right\}$ **c.** $\left(-\infty, \frac{1}{4}\right]$ **9. a.** half-open interval **b.** $\left\{x \middle| -\frac{3}{2} \le x < 2\right\}$ **c.** $\left[-\frac{3}{2}, 2\right)$ **10. a.** open infinite interval **b.** $\{x|x > -\sqrt{2}\}$ **c.** $(-\sqrt{2}, \infty)$ **11.** $\left\{x \middle| -\frac{1}{2} \le x \le 5\right\}$, **12.** $\left\{x \middle| 0 < x < \frac{5}{2}\right\}$, **13.** $\{x|x \le 3\}$, **14.** $\{x|x > -1\}$, **15.** $\left[-\frac{5}{2}, 1\right]$, **16.** $[0, 3)$, **17.** $\left[\frac{3}{4}, \infty\right)$, **18.** $(-4, \infty)$, **19.** $\{-2, 0, 6\}$ **20.** $\varnothing$

21. $\{-6, -4, -2, 0, 1, 2, 4, 5, 6, 9, 15\}$ **22.** $\{-6, -4, -2, -1, 0, 2, 3, 4, 6, 7, 11, 20\}$ **23.** $\{x|2 \le x < 9\}$ **24.** $\{x|x > 5\}$ **25.** $\{x|x \le -2\}$ **26.** $\{x|0 < x \le 14\}$ **27.** $\{x|x \le -3 \text{ or } x \ge 5\}$ **28.** $\{x|x < 6.4\}$ **29.** $\{x|\text{any real number}\}$ **30.** $\{x|x \ge -7\}$ **31.** $(-\infty, 3)$ **32.** $(-8, 3)$ **33.** $[0, 4]$ **34.** $(-\infty, 1) \cup (1, \infty)$ **35.** $[3, 0) \cup (0, \infty)$ **36.** $(-5, 1]$ **37.** 286 **38.** 2.2 **39.** 6 **40.** -3 **41.** $\pi - 2 \approx 1.14$ **42.** 11 **43.** 12 **44.** 22.5

R.2 Exercises

1. $m + 13$ **2.** $c + ab$ **3.** $10t + 11w$ **4.** $z \cdot 3$ **5.** nm **6.** $(v + 4)6$ **7.** $(4 + a) + 11$ **8.** $d + (a + 30)$ **9.** $zw + (a + y)$ **10.** $(45 \cdot y) \cdot z$ **11.** $(8 \cdot 2)(y + z)$ **12.** $132a + 48$ **13.** $20x + 20y$ **14.** $15a - 150 + 90b$ **15.** $5h + 10v - 30$ **16.** $x(w + z)$ **17.** $b(a - c)$ **18.** $q(p - r + t)$ **19.** 4^3, base $= 4$, exponent $= 3$ **20.** w^4, base $= w$, exponent $= 4$ **21.** $(-6y)^5$, base $= -6y$, exponent $= 5$ **22.** base $= 8$, exponent $= 3, 512$ **23.** base $= 5$, exponent $= 4, -625$ **24.** base $= -4$, exponent $= 3, -64$ **25.** base $= 2z$, exponent $= 5, 32z^5$ **26.** 9 **27.** $-\frac{99}{7}$ **28.** 3,088 **29.** $-\frac{3}{28}$ **30.** 116 **31.** 1,223 **32.** $\frac{17}{8}$ **33.** -4 **34.** -8 **35.** -5 **36.** 9 **37.** -21 **38.** 126 **39.** 7 **40.** -23 **41.** $\frac{9}{2}$ **42.** $-17x$ **43.** $9z^2$ **44.** $-14y^5$ **45.** $-10a + 30$ **46.** $10k^2 + 10k$ **47.** $9t - 25$ **48.** $21r - 12$ **49.** $-11w + 59$ **50.** $52p^2 - 1$ **51.** $\frac{17x}{4} - \frac{1}{2}$

R.3 Exercises

1. 3^5 **2.** k^9 **3.** x^{26} **4.** $a^{12}b^6$ **5.** t^7 **6.** $\frac{1}{r^{35}}$ **7.** w^{48} **8.** n^8 **9.** m^3 **10.** 1 **11.** -1 **12.** $\frac{1}{4}$ **13.** $-\frac{1}{16}$ **14.** $\frac{1}{n^3}$ **15.** $\frac{5}{x^6}$ **16.** $-\frac{7}{y^2}$ **17.** t^2 **18.** $5y^9$ **19.** $\frac{b^4}{a^{11}}$ **20.** $-\frac{5q^4}{p^7w^5}$ **21.** b^7 **22.** $\frac{1}{a^{14}}$ **23.** $\frac{1}{27}$ **24.** z^{18} **25.** $25x^4y^{10}$ **26.** $\frac{3y^{12}}{4x^5}$ **27.** 1 **28.** $\frac{1}{2a^8b^4c^6}$ **29.** $\frac{1}{q^{12}p^{16}w^{20}}$ **30.** $3x^4y^{15}$ **31.** 13 **32.** $\frac{1}{5}$ **33.** 6 **34.** -7

35. The square root is not a real number **36.** $-\frac{1}{10}$ **37.** y **38.** 9 **39.** $|k|$ **40.** 8 **41.** 5 **42.** $\frac{1}{2}$ **43.** -3
44. 8 **45.** $\frac{1}{16}$ **46.** $\frac{1}{x^{1/3}}$ **47.** x **48.** $x^{1/12}$ **49.** xy^2 **50.** $x^{5/7}y^2$ **51.** $\frac{3x}{y^{5/4}}$ **52.** $\frac{x^2}{2y^{9/5}}$ **53.** $\frac{x^4}{y^2}$
54. $\sqrt[4]{35}$ **55.** 6 **56.** $5x$ **57.** $\sqrt[4]{75xy^3}$ **58.** $6\sqrt{2}$ **59.** $2\sqrt[3]{7}$ **60.** $-3\sqrt{7}$ **61.** $4x\sqrt{3x}$ **62.** $-5x^2$
63. $15y^4\sqrt{5y}$ **64.** $28xy\sqrt[3]{2xy^2}$ **65.** $a^7\sqrt{a}$ **66.** $xy^2z^2\sqrt[4]{xz}$ **67.** $x^3y^3\sqrt[4]{xy^3}$ **68.** $6\sqrt{5}$ **69.** $84x\sqrt{2}$
70. $6x^2y^2\sqrt{10y}$ **71.** $3x\sqrt[3]{2x}$ **72.** $30x^2y$ **73.** $2x\sqrt[4]{3x}$ **74.** $\frac{7}{9}$ **75.** $\frac{2x\sqrt{3x}}{5y}$ **76.** $\frac{x\sqrt[3]{11}}{2}$ **77.** $\frac{2x^3\sqrt[3]{2}}{5y^2}$
78. $\frac{x\sqrt[4]{21x}}{2}$ **79.** $\frac{x^2}{3}$ **80.** 3 **81.** $2x^2y\sqrt{2y}$ **82.** $2x^5\sqrt{10}$ **83.** 10 **84.** $-2m\sqrt[3]{3}$ **85.** $\frac{x^2}{4y}$

R.4 Exercises

1. yes, 4, 5 **2.** no **3.** yes, 12, -1 **4.** no **5.** yes, 0, 0 **6.** no **7.** yes, 0 **8.** yes, 3 **9.** no **10.** yes, 7
11. $17x^2y - 7xy$ **12.** $30p^2q + 11pq^2$ **13.** $3a^3 - 3a^2b + 10b^3$ **14.** $3x^3 - 3xy - 4$ **15.** $-2m^2 + 16mn + 6n$
16. $a^4b^2 + 15a^2b^2 - 6ab$ **17.** $-\frac{2}{5}x^2 + \frac{17}{18}xy - \frac{5}{8}y^2$ **18.** $\frac{3a}{5} + \frac{3b}{8}$ **19.** $-x + 14$ **20.** $x^2 + 14x - 6$
21. $9x^3 - 15x^2 + 16x - 6$ **22.** $4x^5 + 7x^4 + 6x^3 + 8x^2 + 9x$ **23.** $-9x - 1$ **24.** $\frac{1}{2}x^2 + \frac{1}{12}x - \frac{3}{10}$
25. $-\frac{5}{56}x^3 + \frac{29}{15}x^2 - \frac{23}{18}x - \frac{1}{6}$ **26.** $x^4 - x^2 + 7x$ **27.** $-35x^{10} + 10x^4$ **28.** $6x^7 - 4x^5 + 14x^3 - 8x^2$
29. $x^3 + 10x^2 + 15x - 18$ **30.** $-x^4 - 2x^3 - 3x^2$ **31.** $\frac{5}{3}x^6 + 2x^5 - \frac{2}{3}x^4 - 3x^3$ **32.** $x^2 + 5x + 6$
33. $x^2 + 5x - 36$ **34.** $x^2 - 12x + 35$ **35.** $3x^2 + 17x + 10$ **36.** $15x^2 - 2x - 8$ **37.** $10x^2 - 51x + 56$
38. $3x^2 + 31x + 56$ **39.** $x^2 - \frac{x}{6} - \frac{1}{6}$ **40.** $\frac{2}{5}x^2 + \frac{181}{180}x + \frac{5}{8}$ **41.** $x^2 - 100$ **42.** $36x^2 - 4$ **43.** $x^2 + 10x + 25$
44. $16x^2 - 72x + 81$ **45.** $64x^2 - y^2$ **46.** $x^3 + 12x^2 + 48x + 64$ **47.** $x^3 - 18x^2 + 108x - 216$
48. $49p^2 - 28pq + 4q^2$ **49.** $27x^3 + 54x^2 + 36x + 8$ **50.** $16y^4 - 81$ **51.** $\frac{4}{25}x^2y^2 + \frac{12}{5}xy + 9$
52. $x^4 - 32x^2 + 256$ **53.** quotient $= 7x^2 - 26x + 79$, remainder $= -233$ **54.** quotient $= 9x^2 - 47$, remainder $= 2x + 244$ **55.** quotient $= 3x^2$, remainder $= -2x^2 + 7x + 6$ **56.** quotient $= x^2 - x - 9$, remainder $= 18x + 88$

R.5 Exercises

1. $13\sqrt{7}$ **2.** $\frac{-7\sqrt[3]{4}}{3}$ **3.** $3\sqrt{2} + 5\sqrt{6}$ **4.** $5\sqrt{7} - 12\sqrt[3]{6}$ **5.** $-6\sqrt[3]{6x}$ **6.** $-5\sqrt[3]{4y}$
7. $-8\sqrt{3x} + 3\sqrt{x}$ **8.** $-7x\sqrt{11}$ **9.** $\frac{5}{8}\sqrt{x}$ **10.** $-\frac{53\sqrt{3x}}{90}$ **11.** $6\sqrt{5} + 6\sqrt{x}$ **12.** $6\sqrt{a} + 2\sqrt{b} - a\sqrt{3}$
13. $7xy\sqrt[3]{2} - 5x\sqrt[3]{2}$ **14.** $10\sqrt{2}$ **15.** $-\sqrt{5}$ **16.** $11\sqrt[3]{4}$ **17.** $12\sqrt{3} - 2\sqrt{6}$ **18.** $\frac{7\sqrt{3}}{4}$
19. $5\sqrt{3} + 6\sqrt{5}$ **20.** $-11\sqrt{7} - 24\sqrt{2}$ **21.** $\sqrt[3]{9}$ **22.** $10\sqrt[3]{2} + 4\sqrt{2}$ **23.** $11\sqrt[3]{2} + 6$ **24.** $\frac{16}{5}\sqrt[3]{3}$
25. $3x\sqrt[4]{x}$ **26.** $24x\sqrt{5x}$ **27.** $5t\sqrt{t} + 6t$ **28.** $-2y\sqrt[3]{y}$ **29.** $2a^2b\sqrt{ab}$ **30.** $-m\sqrt[3]{3n} + 9m\sqrt[3]{6n}$
31. $\frac{23\sqrt{2}}{4y}$ **32.** $\frac{\sqrt[3]{a}}{14}$ **33.** $-4a\sqrt[4]{a^3}$ **34.** $7x^2y\sqrt{2x}$ **35.** $2x\sqrt{5} + \sqrt{15}$ **36.** $2x\sqrt{3} - 5\sqrt{2x}$
37. $2m\sqrt{3m} - 6m\sqrt{15}$ **38.** $3x\sqrt[3]{2} - \sqrt[3]{x^2}$ **39.** $29 - 13\sqrt{3}$ **40.** $6 + 3\sqrt{7} + 2\sqrt{5} + \sqrt{35}$
41. $x + 10\sqrt{x} + 24$ **42.** $20 - 6\sqrt{3x} - 6x$ **43.** $12 - 3\sqrt{y} - 4\sqrt{x} + \sqrt{xy}$ **44.** $\sqrt[3]{y^2} - \sqrt[3]{y} - 2$
45. $7z - 9$ **46.** $5 + 6y\sqrt{5} + 9y^2$ **47.** $9 - x$ **48.** $3a - 2b$ **49.** $4x + 20\sqrt{x} + 25$
50. $4m - 12\sqrt{mn} + 9n$ **51.** $\frac{2\sqrt{6}}{3}$ **52.** $\frac{\sqrt{15}}{5}$ **53.** $\frac{3\sqrt{2x}}{2x}$ **54.** $\frac{5\sqrt{11x}}{11x}$ **55.** $\frac{\sqrt[3]{18x}}{2x}$ **56.** $\frac{b\sqrt[4]{75a^3b}}{5a}$
57. $\frac{4\sqrt{3x}}{3x^2y^2}$ **58.** $\frac{\sqrt[3]{12x^2y}}{6y^2}$ **59.** $10 + 5\sqrt{3}$ **60.** $\frac{-20x + 4x\sqrt{6}}{19}$ **61.** $\frac{-3\sqrt{x} + 3}{x - 1}$ **62.** $\frac{3a + \sqrt{ab}}{9a - b}$
63. $\frac{m + 3\sqrt{m} - 28}{m - 49}$ **64.** $\frac{22\sqrt{2} + 21}{31}$

R.6 Exercises

1. $4(10x + 1)$ **2.** $w(x^4 + 1)$ **3.** $3y^7(1 + 2xy)$ **4.** $4x^2(x^2 - 5x + 4)$ **5.** $3wy(13w^8y^8 - 4wy + 4 + 5w)$ **6.** $(11 + 3a)(x + 7)$ **7.** $(8x + 1)(z + 1)$ **8.** $(9y - 7)(x^2 + 11)$ **9.** $(x + 3)(y + 5)$ **10.** $(a - 2)(b + 7)$ **11.** $(3x + 2)(3y + 4)$ **12.** $(x - 4)(3y - 2)$ **13.** $(x + 3)(5x + 4y)$ **14.** $(3x - 2)(x + y)$ **15.** $(x^2 + 4)(x + 9)$ **16.** $(x^2 - 7)(x - 1)$ **17.** $(x + 10)(x + 3)$ **18.** $(y + 7)(y - 2)$ **19.** $(w - 9)(w - 8)$ **20.** $(x - 8)(x + 6)$ **21.** prime **22.** $(x - 36)(x + 3)$ **23.** $(x + 4y)(x + 2y)$ **24.** $-(x - 6)(x + 4)$ **25.** $5(x + 5)(x - 3)$ **26.** $4w(x + 3)(x + 2)$ **27.** $5(x^2 - 6x - 5)$ **28.** $(3x + 5)(3x + 1)$ **29.** $(3y + 5)(4y + 3)$ **30.** $(x + 5)(2x - 5)$ **31.** $(2x - 3)(4x + 3)$ **32.** $(2y - 3)(5y - 4)$ **33.** $2(2x + 3)(4x + 3)$ **34.** $-(4x - 3)(5x + 4)$ **35.** $(x - 9)(x + 9)$ **36.** $(z + 5)^2$ **37.** $(x + 4)(x^2 - 4x + 16)$ **38.** $2(m - 12)^2$ **39.** $(8x - 7)(8x + 7)$ **40.** $(x - 7)(x^2 + 7x + 49)$ **41.** $3(6t - 7)(6t + 7)$ **42.** $(x + 8y - 5)(x + 8y + 5)$ **43.** $(5x + 6)(25x^2 - 30x + 36)$ **44.** $(2x - 7)(4x^2 - 16x + 19)$ **45.** $(w - 5)(w + 15)$ **46.** $(4x + 5)^2$ **47.** $x^{11}(x - 1)(x^2 + x + 1)$ **48.** $-(5x - 1)(5x + 1)(x^2 + 6)$ **49.** $y^4(x - 3)(x^2 + 3x + 9)$ **50.** $-(x^2 - x - 7)(x^2 + x + 7)$

R.7 Exercises

1. $1 - 3x$ for $x \neq 0$ **2.** $x - 9$ for $x \neq -9$ **3.** $\frac{9}{5}$ for $a \neq 3$ **4.** $x - 2$ for $x \neq -3$ **5.** -1 for $x \neq 13$ **6.** $2 - x$ for $x \neq -2$ **7.** $\frac{x - 2}{x - 6}$ for $x \neq -\frac{4}{3}, x \neq 6$ **8.** $x^2 + 5x + 25$ for $x \neq 5$ **9.** $\frac{x - 2}{2x^2 + 1}$ for $x \neq -\frac{1}{4}$ **10.** $\frac{1}{6x + 7}$ for $x \neq -\frac{7}{6}$ **11.** $\frac{3}{2(x - 1)}$ **12.** $-\frac{8}{7}$ **13.** $-\frac{5a}{3a + 1}$ **14.** $\frac{5(x + 2)}{2(x + 4)}$ **15.** $\frac{3a}{2(a - b)}$ **16.** $\frac{1}{6}$ **17.** $\frac{(x + 2)(x + 1)}{9}$ **18.** $\frac{x}{2}$ **19.** $\frac{7a^2}{a - b}$ **20.** $\frac{8}{(x + 6)(x + 7)}$ **21.** $\frac{1}{3}$ **22.** -1 **23.** $\frac{(x - 2)(x + 5)}{(x - 7)(5x + 1)}$ **24.** $\frac{40}{7x^3y}$ **25.** $\frac{(y - 4)(3x - 5)}{(y + 1)(2x - 3)}$ **26.** $\frac{5(3a + 2)}{a}$ **27.** $\frac{16}{x}$ **28.** $\frac{6x + 1}{x - 4}$ **29.** $\frac{(x - 2)(x + 2)}{3x + 1}$ **30.** $\frac{6x + 35}{49x^2}$ **31.** $\frac{9 - x}{x - 3}$ **32.** $\frac{2(x + 35)}{(x - 1)(x + 8)}$ **33.** $\frac{3x^2 - 4x - 21}{(x + 3)(x - 3)}$ **34.** $\frac{-11x + 4}{(x + 4)(x - 4)}$ **35.** $\frac{5x^2 - 144}{x(x - 6)(x + 6)}$ **36.** $\frac{7 - 39x}{12x^2y}$ **37.** $\frac{20a}{(a + b)(a - b)}$ **38.** $\frac{6(4z - 1)}{(z - 9)(z + 6)}$ **39.** $\frac{2(4x + 3y)}{(x - y)(x + y)}$ **40.** $\frac{2x^2 - 11x - 57}{(x + 9)(x + 3)(x - 3)}$ **41.** $\frac{x(5x - 34)}{(x + 6)(x - 2)(x - 6)}$ **42.** $\frac{-23y + 31}{(2y + 7)(y + 1)(y - 5)}$ **43.** $\frac{y + 45}{y(y - 3)^2}$ **44.** $\frac{(2x + 3)(x - 15)}{(x + 6)^2(2x - 9)}$ **45.** $\frac{7x + 1}{7x - 1}$ **46.** $\frac{x}{1 - 3x}$ **47.** $\frac{1}{(x - 6)(x + 5)}$ **48.** $\frac{6y + 5x}{5y - 6x}$ **49.** $\frac{x - 4}{4x}$ **50.** $\frac{6}{x + 7}$ **51.** $\frac{3}{x - 5}$ **52.** $\frac{x - 5}{x + 5}$ **53.** $\frac{x - 1}{x - 7}$ **54.** $\frac{6}{x - y}$ **55. a.** $\frac{4x^2 + 14x}{(x + 2)(x + 5)}$ **b.** $\frac{11}{7}$ inches **56. a.** $x + 6$ feet **b.** 14 feet **57. a.** $2(x + 4)$ millimeters **b.** 32 millimeters **58. a.** $\frac{(5x + 8)(x - 3)}{2x}$ **b.** $\frac{7}{2}$ square inches **59. a.** $\frac{1430x - 7150}{x(x - 10)}$ **b.** 24 hours **60. a.** $\frac{2x + 4}{x(x + 4)}$ **b.** $\frac{5}{24}$

Chapter R Review Exercises

1. a. 1 **b.** 1 **c.** $-17, 1$ **d.** $-17, \frac{25}{19}, 0.0331, 1$ **e.** $-\sqrt{5}, \frac{\pi}{2}$ **2. a.** closed infinite interval **b.** $\left\{x \mid x \leq \frac{1}{4}\right\}$ **c.** $\left(-\infty, \frac{1}{4}\right]$ **3.** $\{x \mid x > -1\}$ 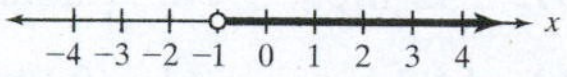

4. $[0, 3)$ 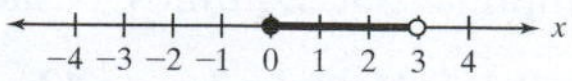

5. $[-3, 0) \cup (0, \infty)$ **6.** $q(p - r + t)$ **7.** $(-6y)^5$ The base is $-6y$, and the exponent is 5. **8.** 1223 **9.** -21 **10.** $52p^2 - 1$ **11.** $\frac{1}{2a^8b^4c^6}$ **12.** $x^{\frac{1}{12}}$ **13.** $15y^4\sqrt{5y}$ **14.** $6\sqrt{5}$ **15.** $\frac{2x\sqrt{3x}}{5y}$ **16.** $17x^2y - 7xy$ **17.** $6x^7 - 4x^5 + 14x^3 - 8x^2$ **18.** $15x^2 - 2x - 8$ **19.** $16x^2 - 72x + 81$

20. $7x^2 - 26x + 79$ with a remainder of -233 **21.** $-\sqrt{5}$ **22.** $24x\sqrt{5x}$ **23.** $6 + 3\sqrt{7} + 2\sqrt{5} + \sqrt{35}$

24. $\frac{3\sqrt{2x}}{2x}$ **25.** $\frac{3a + \sqrt{ab}}{9a - b}$ **26.** $4x^2(x - 4)(x - 1)$ **27.** $(5x + 4y)(x + 3)$ **28.** $(y + 7)(y - 2)$

29. $(2x - 5)(x + 5)$ **30.** $(x + 9)(x - 9)$ **31.** $\frac{2 - x}{2}$ **32.** $\frac{(y - 4)(3x - 5)}{(y + 1)(2x - 3)}$ **33.** $\frac{2x + 70}{(x - 1)(x + 8)}$

34. $\frac{5x^2 - 34x}{(x + 6)(x - 6)(x - 2)}$ **35.** $\frac{x - 4}{4x}$

CHAPTER 1

1.1 Exercises

SCE-1. 18 **SCE-2.** 24 **SCE-3.** 30 **SCE-4.** $x^2 - 1$ **SCE-5.** $8x^2$ **SCE-6.** $2w^2 - 7w + 3$ **SCE-7.** $x^2 - 1$ **SCE-8.** $x^2 - x$

1. linear **2.** nonlinear **3.** linear **4.** nonlinear **5.** linear **6.** nonlinear **7.** linear **8.** $x = \frac{1}{3}$ **9.** $x = 0$

10. $x = -\frac{9}{10}$ **11.** $x = \frac{11}{10}$ **12.** $x = -\frac{17}{5}$ **13.** $x = 8$ **14.** $y = \frac{2}{29}$ **15.** $p = -\frac{19}{46}$ **16.** $x = -25$

17. $x = -6$ **18.** $a = -\frac{3}{4}$ **19.** $x = 2.925$ **20.** $k = 128.5$ **21.** $x = 1$ **22.** rational **23.** non-rational

24. rational **25.** non-rational **26.** rational **27.** $x = -4$ **28.** $\varnothing$ **29.** $x = \frac{1}{4}$ **30.** $w = -\frac{3}{4}$ **31.** $\{7\}$

32. $\left\{\frac{8}{5}\right\}$ **33.** { } **34.** { } **35.** $x = -4$ **36.** $x = -\frac{2}{7}$

1.2 Exercises

SCE-1. $24t$ **SCE-2.** $23(7 + x)$ **SCE-3.** $(r - 5)(r + 5)$ **SCE-4.** $x(x + 2)$

1. $2n + 10$ **2.** $3(n + 6) - 5$ **3.** $\frac{3}{4n}$ **4.** $2n = \frac{n}{4} - 3$ **5.** $\frac{4n}{7} - 3 = 2n - 8$ **6.** $3n + \frac{1}{n} = 5$ **7.** 7 and 19 **8.** 12 feet, 14 feet, 16 feet **9.** Steve: 81 tickets, Tom: 40 tickets **10.** Let a and b represent the two numbers. $a = 4b$ and $\frac{1}{a} + \frac{1}{b} = \frac{5}{12}$ The two numbers are 3 and 12. **11.** Let a represent the number. $\frac{5 + a}{7 + a} = \frac{17}{23}$ The number is $\frac{2}{3}$.

12. 5 ft by 11 ft **13.** 6600 square yards **14.** 162 square feet **15.** 15 feet, 20 feet, 25 feet **16.** $x = \frac{3}{2}$ The area is 14 square feet. The lenghts of the base are $\frac{8}{3}$ feet and 2 feet. **17.** 14 quarters and 19 dimes **18.** 35 dimes, 41 quarters, and 5 silver dollars **19.** soda: 60, popcorn: 32, candy: 18 **20.** stock A: 64 shares, stock B: 32 shares, stock C: 16 shares **21.** 11.25 pounds of \$9.00/lb coffee and 18.75 pounds of \$5.00/lb coffee **22.** 20 gallons of the 80% orange juice **23.** $8\frac{1}{3}$ L of pure water should be mixed with a 5-L solution of 80% acid **24.** 21.6% **25.** 14.4% nitrogen, 16.8% phosphate, 4.8% potassium **26.** drain 2.25 L of coolant and fill up with 2.25 L of pure antifreeze **27.** \$8,000 at 4% simple annual interest and \$20,000 at 5% simple annual interest **28.** product A: \$21,000, product B: \$61,000 **29.** \$2500 at 8.5% and \$1500 at 10% **30.** \$3750 **31.** freight train: 30 mph, passenger train: 50 mph **32.** 22 hours **33.** 4 hours **34.** 30 mph **35.** 24 miles **36.** 3 mph **37.** 245 mph **38.** $\frac{5}{4}$ hours **39.** 120 minutes **40.** Joan: 60 minutes, Jane: 20 minutes **41.** 6 hours **42.** 450 min **43.** 4.8 hours **44.** 40 hrs

1.3 Exercises

SCE-1. $6\sqrt{2} - 2$ **SCE-2.** $9\sqrt{3} + 4$ **SCE-3.** $2\sqrt{5}$ **SCE-4.** $6\sqrt{3}$ **SCE-5.** $27 - 10\sqrt{2}$ **SCE-6.** $14 + 6\sqrt{5}$

1. i **2.** -1 **3.** -1 **4.** i **5.** $-i$ **6.** $-4 + 7i$ **7.** $10 - 11i$ **8.** -1 **9.** $3\sqrt{2}$ **10.** $6 + 8i$ **11.** $-1 - i$

12. $-10 + 5i$ **13.** $32 - 24i$ **14.** $1 - 2\sqrt{2}i$ **15.** 29 **16.** 2 **17.** $\frac{37}{4}$ **18.** 6 **19.** 1 **20.** $\frac{2}{25} - \frac{11}{25}i$

21. $\frac{2}{5} + \frac{1}{5}i$ **22.** $\frac{3}{4} + \frac{3}{4}i$ **23.** $\frac{12}{13} + \frac{5}{13}i$ **24.** $-7 + 6i$ **25.** $3 - 7i$ **26.** -6 **27.** -8 **28.** 4 **29.** $-2 - \sqrt{5}i$

30. $-\frac{1}{2} - \frac{3}{2}i$ **31.** $1 + \frac{\sqrt{2}}{2}i$

1.4 Exercises

SCE-1. 10 **SCE-2.** $-6 + 2\sqrt{11}$ **SCE-3.** $4 + 14i$ **SCE-4.** $6 - 2i\sqrt{29}$ **SCE-5.** $\frac{5}{3}$ **SCE-6.** $-3 + \sqrt{11}$

SCE-7. $\frac{2 + 7i}{3}$ **SCE-8.** $\frac{3 - i\sqrt{29}}{2}$ **SCE-9.** $(x + 4)(x + 6)$ **SCE-10.** $(x - 9)(x + 2)$ **SCE-11.** $(x - 7)(x - 3)$

SCE-12. $(2x + 1)(x + 4)$ **SCE-13.** $(3x - 5)(2x + 3)$

1. $x = 0, 8$ **2.** $x = -12, 0$ **3.** $m = 0$ or $m = -6$ **4.** $x = 0$ or $x = -3$ **5.** $x = -2$ or $x = 3$

6. $x = -5$ or $x = -4$ **7.** $t = 4$ or $t = 6$ **8.** $v = -7$ or $v = -4$ **9.** $x = -3$ or $x = \frac{1}{3}$ **10.** $n = -\frac{5}{6}$ or $n = 7$

11. $x = -4$ or $x = -\frac{1}{3}$ **12.** $r = -4$ or $r = \frac{3}{8}$ **13.** $m = -\frac{3}{4}$ or $m = \frac{5}{2}$ **14.** $z = -\frac{2}{7}$ or $z = \frac{3}{4}$ **15.** $x = \pm 8$

16. $x = \pm 8i$ **17.** $x = \pm 2\sqrt{6}$ **18.** $x = 6\sqrt{3}, -6\sqrt{3}$ **19.** $x = 15\sqrt{2}, -15\sqrt{2}$ **20.** $x = -5$ or $x = 1$

21. $x = 5 + 9\sqrt{2}, 5 - 9\sqrt{2}$ **22.** $x = -\frac{1}{2} \pm i$ **23.** $x = \frac{-1 + 4\sqrt{3}}{2}, \frac{-1 - 4\sqrt{3}}{2}$ **24.** $x = \frac{1 + 2i\sqrt{5}}{3}, \frac{1 - 2i\sqrt{5}}{3}$

25. 16 **26.** 25 **27.** $\frac{49}{4}$ **28.** $\frac{9}{4}$ **29.** $\frac{1}{64}$ **30.** $\frac{25}{36}$ **31.** $x = -8, 2$ **32.** $x = 4 \pm 3\sqrt{2}$ **33.** $x = \frac{5 \pm \sqrt{37}}{2}$

34. $x = -5 \pm 3i$ **35.** $x = \frac{-7 \pm i\sqrt{7}}{2}$ **36.** $x = \frac{1}{3}, -3$ **37.** $x = \frac{-3 \pm i\sqrt{7}}{2}$ **38.** $x = \frac{-12 \pm \sqrt{165}}{3}$

39. $x = \frac{-5 \pm i\sqrt{119}}{6}$ **40.** $x = -13, 2$ **41.** $x = \frac{1}{3}, -3$ **42.** $x = 4 \pm 3\sqrt{2}$ **43.** $x = \frac{2 \pm \sqrt{7}}{3}$

44. $x = \frac{4 \pm \sqrt{26}}{2}$ **45.** $x = \frac{1}{3}$ **46.** $x = \frac{-3 \pm i\sqrt{11}}{10}$ **47.** $x = \frac{1 \pm i\sqrt{127}}{8}$ **48.** $D = 0$: exactly one real solution

49. $D = 0$: exactly one real solution **50.** $D = 41$: two real solutions **51.** $D = -36$: two nonreal solutions

52–60. Same answers as 31–39. **61–68.** Same answers as 40–47

1.5 Exercises

1. $-\frac{4}{3}$ **2.** 11 or -12 **3.** $-\frac{5}{2} \pm \frac{\sqrt{313}}{2}$ **4.** 1, 3, and 5 or 3, 5, and 7 **5.** The two numbers are $5 + \sqrt{13}$ and $5 - \sqrt{13}$

6. 4 seconds **7.** 2 seconds **8.** $\frac{10}{3}$ cm by 9 cm **9.** 8 in. by 15 in. **10.** $\frac{7}{2}$ ft or 3.5 ft **11.** 25 mph **12.** 5 mph

13. 35 mph **14.** 120 min **15.** approximately 26 hr 10 min **16.** approximately 10 hr 43 min

1.6 Exercises

SCE-1. $(x + 20)(x + 3)$ **SCE-2.** $(y + 7)(y - 2)$ **SCE-3.** $(w - 12)(w - 8)$ **SCE-4.** $(x - 10)(x + 4)$

SCE-5. $(3x + 1)(4x + 7)$ **SCE-6.** $(3x - 5)(x + 3)$ **SCE-7.** $(3y - 2)(2y - 5)$

SCE-8. $3(x + 11)(x - 3)$ **SCE-9.** $5w(x + 2)(x + 3)$ **SCE-10.** $2(2x + 5)(5x + 1)$ **SCE-11.** $(x + 11)(x - 11)$

SCE-12. $3(6t + 7)(6t - 7)$ **SCE-13.** $(x + 8)(x^2 + 4)$ **SCE-14.** $x^2(x^2 - 3)(x - 1)$ **SCE-15.** $2x + 4$

SCE-16 $18 - 7x$ **SCE-17.** $4x^2 - 4x + 1$ **SCE-18.** $128x + 144$ **SCE-19.** $y + 13 + 6\sqrt{y + 4}$
SCE-20. $4x + 9 - 4\sqrt{4x + 5}$

1. $\{0, i, -i\}$ **2.** $\{-5, 0, 3\}$ **3.** $\{-3, -1, 1, 3\}$ **4.** $\{-\sqrt{3}i, \sqrt{3}i, 1\}$ **5.** $\left\{-2, -\frac{3}{2}, 2\right\}$ **6.** $\{-1, 1\}$

7. $\{-2, 2, -\sqrt{2}, \sqrt{2}\}$ **8.** $\left\{0, \frac{4}{13}\right\}$ **9.** $\left\{\frac{1}{8}, 8\right\}$ **10.** $\{-1, 2\}$ **11.** $\{256\}$ **12.** $\left\{-2, \frac{3}{2}\right\}$ **13.** $\left\{-2, \frac{1}{2}\right\}$

14. $x = 5$ **15.** $x = 4$ **16.** $x = 4$ **17.** $x = -2$ **18.** $y = 3$ **19.** $x = -3$ **20.** $x = \frac{-7}{2}$ **21.** $x = 8, -2$

22. $\{-2, 2, -\sqrt{2}, \sqrt{2}\}$ **23.** $\left\{0, \frac{4}{13}\right\}$ **24.** $\left\{\frac{1}{8}, 8\right\}$ **25.** $\{-1, 2\}$ **26.** $\{256\}$ **27.** $\left\{-2, \frac{3}{2}\right\}$ **28.** $\left\{-2, \frac{1}{2}\right\}$

29. $\{5\}$ **30.** $\{4\}$ **31.** $\{4\}$ **32.** $\{-2\}$ **33.** $\{3\}$ **34.** $\{-3\}$ **35.** $\left\{-\frac{7}{2}\right\}$ **36.** $\{-2, 8\}$

1.7 Exercises

SCE-1. $\left\{x \middle| -\frac{1}{2} \le x \le 5\right\}$, 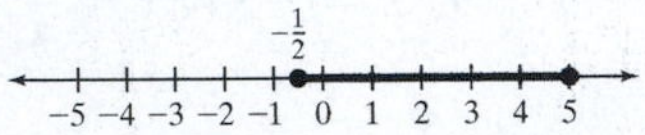 **SCE-2.** $\left\{x \middle| 0 < x < \frac{5}{2}\right\}$, 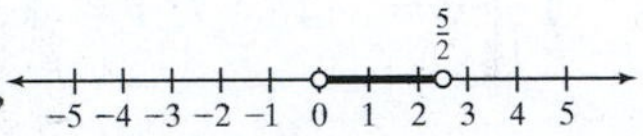

SCE-3. $\{x | x \le 3\}$, 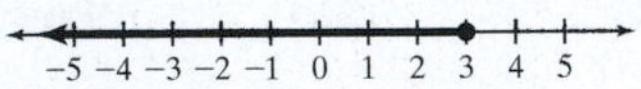 **SCE-4.** $\{x | x > -1\}$, 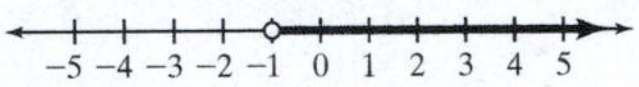

SCE-5. $\left[-\frac{5}{2}, 1\right]$, 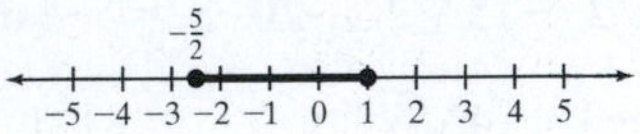 **SCE-6.** $[0, 3)$, 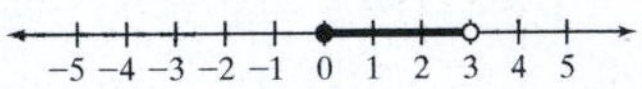

SCE-7. $\left[\frac{3}{4}, \infty\right)$, 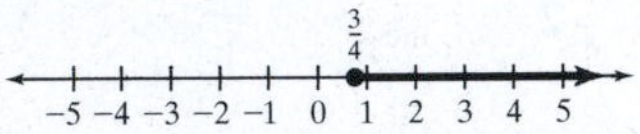 **SCE-8.** $(-4, \infty)$, 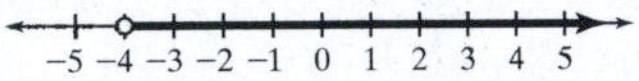

SCE-9. $\{x | 2 \le x < 9\}$ **SCE-10.** $\{x | x > 5\}$ **SCE-11.** $\{x | x \le -2\}$ **SCE-12.** $\{x | 0 < x \le 14\}$

SCE-13. $\{x | x \le -3 \text{ or } x \ge 5\}$ **SCE-14.** $\{x | x < 6.4\}$ **SCE-15.** $(-\infty, \infty)$ **SCE-16.** $\{x | x \ge -7\}$

SCE-17. $(-\infty, 3)$ **SCE-18.** $(-8, 3)$ **SCE-19.** $[0, 4]$ **SCE-20.** $(-\infty, 1) \cup (1, \infty)$ **SCE-21.** $[3, 0) \cup (0, \infty)$

SCE-22. $(-5, 1]$

1. $\left\{x \middle| x > -\frac{5}{2}\right\}$ **2.** $\left\{y \middle| y < \frac{1}{5}\right\}$ **3.** $\{a | a \ge -4\}$ **4.** $\left\{w \middle| w > \frac{30}{7}\right\}$ **5.** 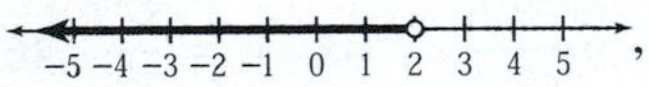, $(-\infty, 2)$

6. 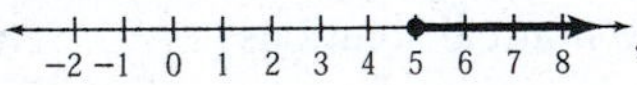, $[5, \infty)$ **7.** 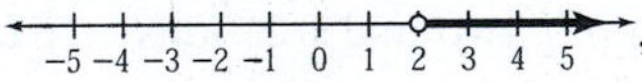, $(2, \infty)$

8. 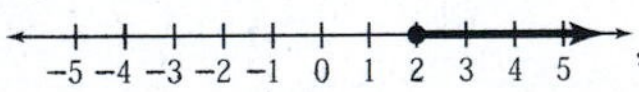, $[2, \infty)$ **9.** $[-1, \infty)$, 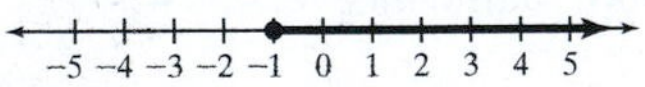

10. 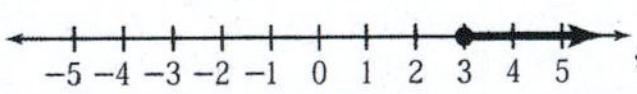, $[3, \infty)$ **11.** 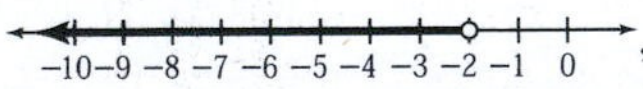, $(-\infty, -2)$

12. $(-\infty, -3]$, 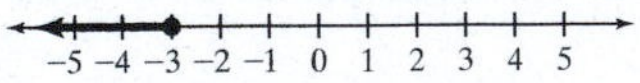 **13.** $\left(-\frac{32}{63}, \infty\right)$ **14.** $\left(\frac{13}{2}, \infty\right)$ **15.** $(-\infty, 74]$ **16.** $\left(-\infty, -\frac{3}{46}\right]$

17. $m < 5$, 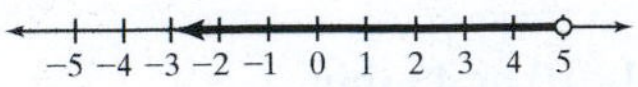 **18.** $x \ge \frac{37}{11}$, 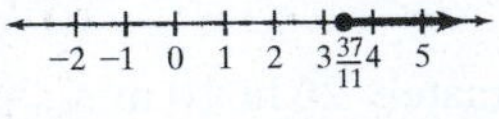

19. $p < -2$, 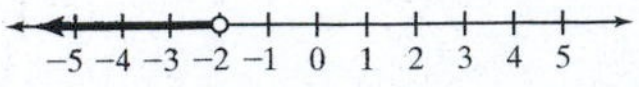 **20.** $k > \frac{7696}{449}$, 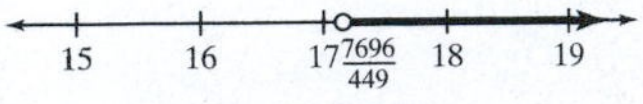 **21.** $(-1, 2]$

22. $\left[\frac{3}{2}, \frac{7}{2}\right]$, 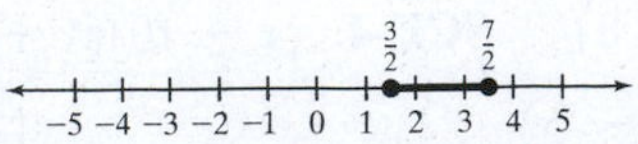 **23.** $\left[-\frac{11}{6}, -\frac{1}{2}\right)$, 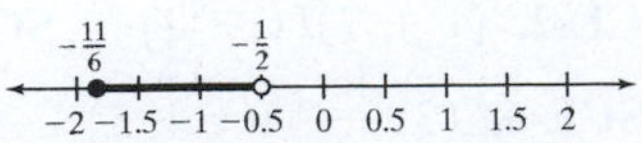

24. $\left(-\frac{3}{2}, \frac{11}{2}\right)$, 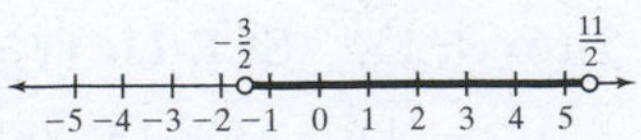 **25.** $[-10, -9]$, 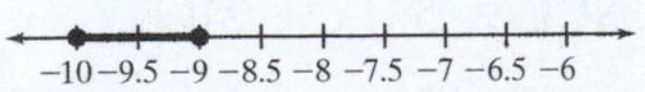

26. $(2, 11]$, 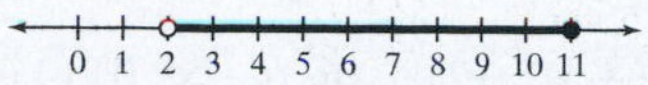

27. $(-7, 5)$, 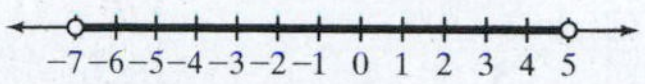

28. $\left(\frac{8}{5}, 4\right)$, 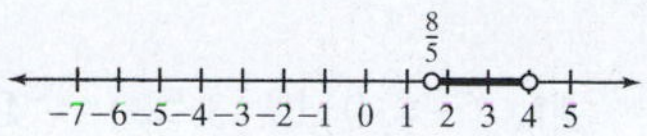

29. $[-4, -1)$ **30.** $[0, 5]$ **31.** $(-\infty, -8)$ **32.** $(-\infty, -3] \cup (4, \infty)$

33. $(1, 4]$ **34.** $\left(-\infty, \frac{1}{2}\right] \cup (7, \infty)$ **35.** no solution **36.** $(-3, \infty)$ **37.** $(-12, 2)$ **38.** $(-\infty, \infty)$ **39.** $[4, 5]$

40. $(-\infty, -5) \cup \left(-\frac{9}{5}, \infty\right)$ **41.** 5 hrs **42.** more than 320 miles **43.** 92 or higher **44.** 7′1′′ or taller

45. 28 feet to 68 feet **46.** 20 feet to 30 feet **47.** up to 9 gigabytes **48.** length $\leq$ 30 in. **49.** 105 guests
50. 92 plush toys **51.** more than 250 miles **52.** 159 bracelets

1.8 Exercises

SCE-1. $\left[-\frac{5}{2}, 1\right]$, 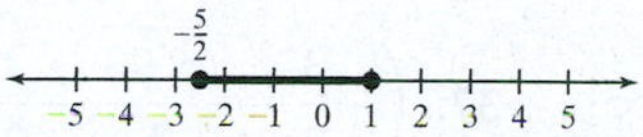

SCE-2 $[0, 3)$, 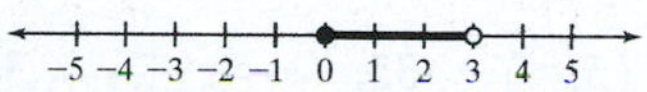

SCE-3. $\left[\frac{3}{4}, \infty\right)$, 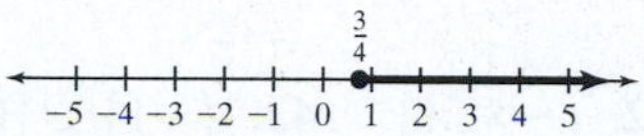

SCE-4. $(-4, \infty)$, 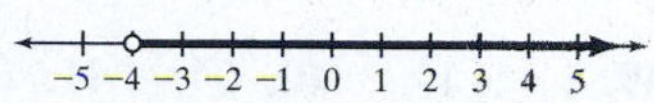

SCE-5. $[2, 9)$

SCE-6. $(-\infty, -3] \cup [5, \infty)$ **SCE-7.** $(-1, 2]$ **SCE-8** $\left[\frac{3}{2}, \frac{7}{2}\right]$, 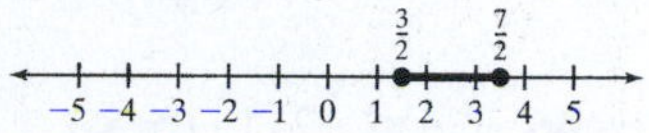

SCE-9. $[-4, -1)$

SCE-10. $(-\infty, -3] \cup (4, \infty)$ **SCE-11.** $(-\infty, \infty)$ **SCE-12.** $(-\infty, -5] \cup \left(-\frac{9}{5}, \infty\right)$

1. $\{-7, 7\}$ **2.** $\{-4, 6\}$ **3.** $\{0, -2\}$ **4.** $\left\{-\frac{2}{5}, -\frac{7}{5}\right\}$ **5.** $\left\{-3, \frac{2}{3}\right\}$ **6.** $\left\{0, \frac{16}{3}\right\}$ **7.** $\left\{-\frac{3}{7}\right\}$ **8.** $\varnothing$

9. $\{-5\}$ **10.** $\left\{-\frac{3}{7}, \frac{13}{7}\right\}$ **11.** $\{\}$ or $\varnothing$ **12.** $\{1, 6\}$ **13.** $\left\{-\frac{1}{10}, \frac{7}{10}\right\}$ **14.** $\left\{-2, -\frac{1}{4}\right\}$ **15.** $\left\{-1, -\frac{1}{3}\right\}$

16. $\left\{1 - \sqrt{2}, 1, 1 + \sqrt{2}\right\}$ **17.** $x = \frac{7}{4}, 3$ **18.** $\{-1, 0, 7\}$ **19.** $\left[-\frac{1}{3}, \frac{1}{3}\right]$ **20.** $(-3, 7)$ **21.** $\left(-2, \frac{7}{2}\right)$

22. $\left[-\frac{1}{5}, \frac{3}{5}\right]$ **23.** $\left\{\frac{7}{3}\right\}$ **24.** $\varnothing$ **25.** $\left[\frac{11}{5}, \frac{14}{5}\right]$ **26.** $\left(\frac{3}{8}, \frac{9}{8}\right)$ **27.** $(-\infty, -3) \cup (3, \infty)$

28. $(-\infty, -9) \cup (1, \infty)$ **29.** $\left(-\infty, \frac{1}{5}\right] \cup [1, \infty)$ **30.** $\left(-\infty, \frac{5}{7}\right) \cup \left(\frac{5}{7}, \infty\right)$ **31.** $(-\infty, \infty)$

32. $\left(-\infty, \frac{12}{7}\right] \cup [2, \infty)$ **33.** $\left(-\infty, \frac{9}{4}\right] \cup \left[\frac{15}{4}, \infty\right)$ **34.** $\left(-\frac{13}{3}, 3\right)$ **35.** $\left[\frac{1}{3}, \frac{7}{3}\right]$ **36.** $(-\infty, \infty)$

1.9 Exercises

SCE-1. $(x + 20)(x + 3)$ **SCE-2** $(y + 7)(y - 2)$ **SCE-3.** $(w - 12)(w - 8)$ **SCE-4.** $(3x - 5)(x + 3)$

SCE-5. $(4x + 7)(3x + 1)$ **SCE-6.** $(3y - 2)(2y - 5)$ **SCE-7.** $(x + 11)(x - 11)$ **SCE-8** $(x^2 + 4)(x + 8)$

SCE-9. $(x^2 - 3)(x - 1)$ **SCE-10.** $\dfrac{2(x + 35)}{(x - 1)(x + 8)}$ **SCE-11.** $\dfrac{3x^2 - 4x - 21}{(x + 3)(x - 3)}$ **SCE-12.** $\dfrac{6(4z - 1)}{(z - 9)(z + 6)}$

SCE-13. $\dfrac{2x^2 - 11x - 57}{(x + 9)(x + 3)(x - 3)}$ **SCE-14.** $\dfrac{x(5x - 34)}{(x + 6)(x - 2)(x - 6)}$ **SCE-15.** $\dfrac{(2x + 3)(x - 15)}{(x + 6)^2(2x - 9)}$

1. $(-\infty, -3] \cup [1, \infty)$ **2.** $\left[-\frac{2}{3}, 0\right]$ **3.** $[-4, 1] \cup [3, \infty)$ **4.** $(5, \infty)$ **5.** $(-\infty, -2]$ **6.** $(-\infty, 0) \cup (2, \infty)$

7. $\left(-\frac{3}{2} - \frac{\sqrt{93}}{2}, -\frac{3}{2} + \frac{\sqrt{93}}{2}\right)$ **8.** $(-\infty, 3] \cup [5, \infty)$ **9.** $[-1, 1]$ **10.** $\varnothing$ **11.** $(-4, 0) \cup (3, \infty)$ **12.** $(-\infty, 1]$

13. $\{-1\} \cup [0, \infty)$ **14.** $(1, 4)$ **15.** $[-3, 1)$ **16.** $(-\infty, -2) \cup (5, \infty)$ **17.** $(-3, 2]$ **18.** $(-\infty, 0) \cup (1, \infty)$

19. $(-\infty, -3] \cup (-2, 3]$ **20.** $(-3, 7) \cup (7, \infty)$ **21.** $(-\infty, -8) \cup [2, 3)$ **22.** $(-\infty, -2) \cup (-2, 3)$

23. $(-2,3) \cup (3,9)$ **24.** $(-\infty,-2) \cup (6,\infty)$ **25.** $(-1, 1]$ **26.** $\left(\frac{3}{2}, 8\right)$ **27.** $(-4, -2]$

28. $(-\infty, -2] \cup (0, 1] \cup (2, \infty)$ **29.** $\left(-3, -\frac{1}{2}\right] \cup (2, \infty)$ **30.** $\left(1, \frac{3}{2}\right]$ **31.** $(-\infty, -3] \cup [1, \infty)$ **32.** $\left[-\frac{2}{3}, 0\right]$

33. $[-4, 1] \cup [3, \infty)$ **34.** $(5, \infty)$ **35.** $(-\infty, -2]$ **36.** $(-\infty, 0) \cup (2, \infty)$ **37.** $\left(-\frac{3}{2} - \frac{\sqrt{93}}{2}, -\frac{3}{2} + \frac{\sqrt{93}}{2}\right)$

38. $(-\infty, 3] \cup [5, \infty)$ **39.** $[-1, 1]$ **40.** $\varnothing$ **41.** $(-4, 0) \cup (3, \infty)$ **42.** $(-\infty, 1]$ **43.** $\{-1\} \cup [0, \infty)$

44. $(1, 4)$ **45.** $[-3, 1)$ **46.** $(-\infty,-2) \cup (5,\infty)$ **47.** $(-3, 2]$ **48.** $(-\infty, 0) \cup (1, \infty)$ **49.** $(-\infty, -3] \cup (-2, 3]$

50. $(-3,7) \cup (7,\infty)$ **51.** $(-\infty, -8) \cup [2, 3)$ **52.** $(-\infty, -2) \cup (-2, 3)$ **53.** $(-2, 3) \cup (3, 9)$

54. $(-\infty, -2) \cup (6, \infty)$ **55.** $(-1, 1]$ **56.** $\left(\frac{3}{2}, 8\right)$ **57.** $(-4, -2]$ **58.** $(-\infty, -2] \cup (0, 1] \cup (2, \infty)$

59. $\left(-3, -\frac{1}{2}\right] \cup (2, \infty)$ **60.** $\left(1, \frac{3}{2}\right]$

Review Exercises

1. linear **2.** $x = \frac{11}{10}$ **3.** $x = -6$ **4.** $k = \frac{257}{2}$ **5.** non-rational **6.** $x = \frac{8}{5}$

7. Steve sold 81 raffle tickets, and Tim sold 40 raffle tickets. **8.** The garden has a width of 5 feet and a length of 11 feet.

9. 20 gallons **10.** 120 minutes (or 2 hours) **11.** $10 - 11i$ **12.** $-10 + 5i$ **13.** $\frac{2}{25} - \frac{11i}{25}$ **14.** $x = \frac{1}{3}, -3$

15. $x = -8, 2$ **16.** $x = \frac{1}{3}$ **17.** 2 seconds **18.** The rectangle has a width of 8 inches and a length of 15 inches.

19. 35 mph **20.** 120 minutes **21.** $x = 1, i\sqrt{3}, -i\sqrt{3}$ **22.** $x = -2, \frac{3}{2}$ **23.** $x = 4$ **24.** $y = 3$ **25.** $(-\infty,-3]$

26. $\left(\frac{13}{2}, \infty\right)$ **27.** $(-1, 2]$ **28.** No solution ($\varnothing$) **29.** $(-3, \infty)$ **30.** $x = -\frac{2}{5}, -\frac{7}{5}$ **31.** $\left[-\frac{1}{5}, \frac{3}{5}\right]$

32. $\left[\frac{1}{3}, \frac{7}{3}\right]$ **33.** $\left(-\frac{3}{2} - \frac{\sqrt{93}}{2}, -\frac{3}{2} + \frac{\sqrt{93}}{2}\right)$ **34.** $[-3, 1)$ **35.** $(-4, -2]$

CHAPTER 2

2.1 Exercises

SCE-1. $\frac{5}{2}$ **SCE-2.** $-\frac{5}{6}$ **SCE-3.** $\frac{1}{70}$ **SCE-4.** $6\sqrt{3}$ **SCE-5.** 25 **SCE-6.** $3\sqrt{10}$ **SCE-7.** 5

SCE-8. $x = -4$ **SCE-9.** $x = \frac{1}{4}$ **SCE-10.** $x = 5 \pm 9\sqrt{2}.$ **SCE-11.** $x = -\frac{1}{2} \pm i.$

SCE-12. $\{4\}$ **SCE-13.** $\{-2\}$

1. 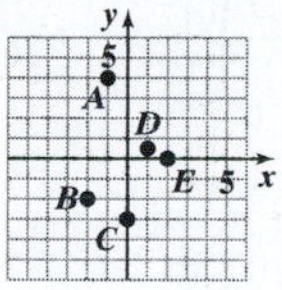 **a.** II **b.** III **c.** y-axis **d.** I **e.** x-axis **2.** 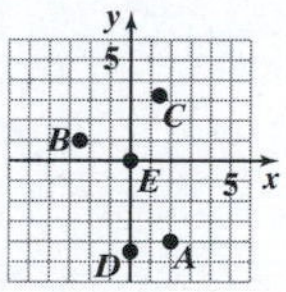 **a.** IV **b.** II **c.** I **d.** y-axis

e. origin **3.** 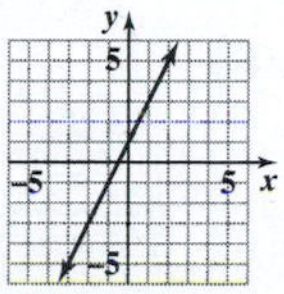 **4.** 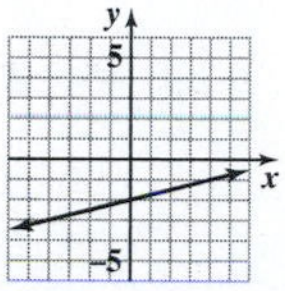 **5.** 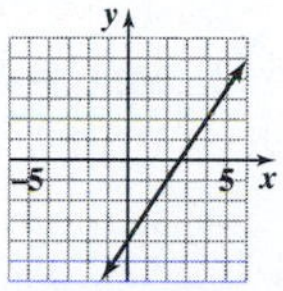 **6.** 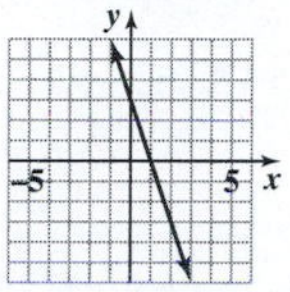 **7.** 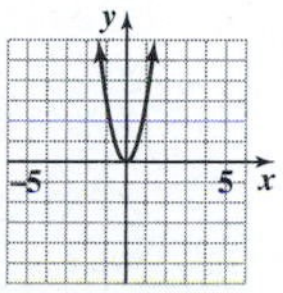 **8.**

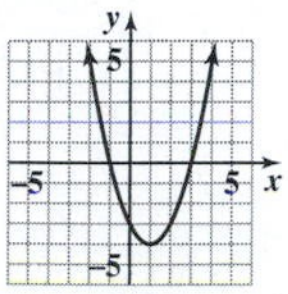

9. 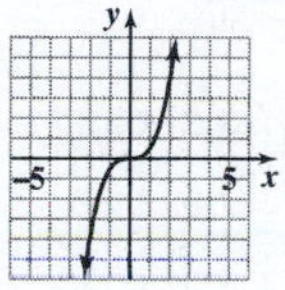 **10.** **a.** yes **b.** no **c.** yes **11.** **a.** yes **b.** no **c.** no **12.** **a.** no **b.** yes **c.** yes

13. x-intercept: $\frac{5}{3}$; y-intercept: -5 **14.** x-intercept: 4; y-intercept: $-\frac{8}{3}$ **15.** x-intercepts: -2, 4; y-intercept: -8

16. no x-intercepts; y-intercept: 6 **17.** x-intercept: -1; y-intercept: $-\frac{1}{3}$ **18.** x-intercepts: -2, 3; y-intercept: $-\frac{3}{2}$

19. x-intercept: 13; y-intercept $= \sqrt{3} - 4$ **20.** x-intercept: -1; y-intercepts: $2 \pm \sqrt{3}$ **21.** x-intercepts: -7, -1; no y-intercepts **22.** $\left(-\frac{1}{2}, -3\right)$ **23.** $(3, -2)$ **24.** $\left(-\frac{3}{2}, \frac{3}{4}\right)$ **25.** $\left(\frac{a+c}{2}, \frac{b+d}{2}\right)$ **26.** $\left(13, \frac{11}{2}\right)$ **27.** yes

28. no **29.** yes **30.** 5 **31.** $\sqrt{74} \approx 8.602$ **32.** $5\sqrt{5} \approx 11.180$ **33.** $\sqrt{5} \approx 2.236$ **34.** yes **35.** no

36. yes **37.** 0 or 6

2.2 Exercises

SCE-1. $25, (x+5)^2$ **SCE-2.** $\frac{25}{4}, \left(x - \frac{5}{2}\right)^2$ **SCE-3.** $16, (y-4)^2$ **SCE-4.** $\frac{81}{4}, \left(y - \frac{9}{2}\right)^2$ **SCE-5.** $\frac{25}{36}, \left(x + \frac{5}{6}\right)^2$

SCE-6. 25 and 16 **SCE-7.** $\frac{25}{4}$ and 9 **SCE-8.** $\frac{49}{4}$ and $\frac{121}{4}$ **SCE-9.** $\frac{1}{9}$ and $\frac{25}{196}$

1. $x^2 + y^2 = 1$ **2.** $(x+2)^2 + (y-3)^2 = 16$ **3.** $(x-1)^2 + (y+4)^2 = 9$ **4.** $\left(x + \frac{1}{4}\right)^2 + \left(y + \frac{1}{3}\right)^2 = 4$

5. $x^2 + (y-2)^2 = 25$ **6.** $(x+4)^2 + (y-7)^2 = 72$ **7.** $(x-3)^2 + (y+1)^2 = 13$

8. $(x-4)^2 + \left(y + \frac{5}{2}\right)^2 = \frac{65}{4}$ **9.** $(x-2)^2 + (y+3)^2 = 9$ **10.** $(x+4)^2 + (y-1)^2 = 16$

11. $C(0, 0), r = 1, x\text{-int} = \{\pm 1\}, y\text{-int} = \{\pm 1\}$ 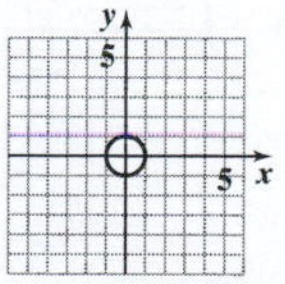 **12.** $C(0, 2), r = 2, x\text{-int} = \{0\}, y\text{-int} = \{0, 4\},$

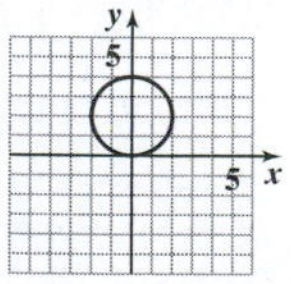

13. $C(1, -5)$, $r = 4$, no x-int, y-int $= \{-5 \pm \sqrt{15}\}$,

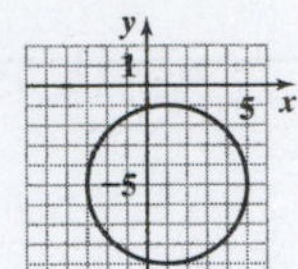

14. $C(-2, -4)$, $r = 6$, x-int $= \{-2 \pm 2\sqrt{5}\}$, y-int $= \{-4 \pm 4\sqrt{2}\}$,

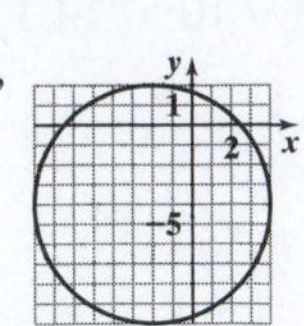

15. $C(4, -7)$, $r = 2\sqrt{3}$, no x-int, no y-int,

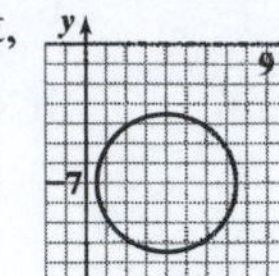

16. $C(-1, 3)$, $r = 2\sqrt{5}$, x-int $= \{-1 \pm \sqrt{11}\}$, y-int $= \{3 \pm \sqrt{19}\}$,

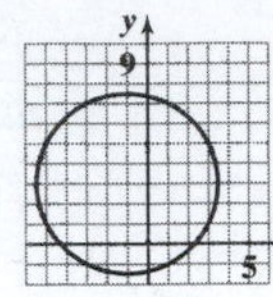

17. $C\left(\frac{1}{4}, -\frac{1}{2}\right)$, $r = 2$, x-int $= \left\{\frac{1}{4} \pm \frac{\sqrt{15}}{2}\right\}$, y-int $= \left\{-\frac{1}{2} \pm \frac{3\sqrt{7}}{4}\right\}$,

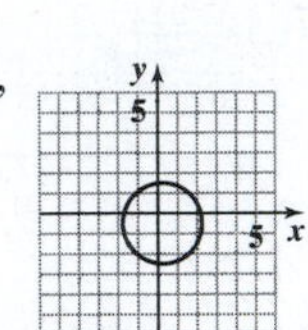

18. $C(-1, 2)$, $r = 2$, x-int $= \{-1\}$, y-int $= \{2 \pm \sqrt{3}\}$,

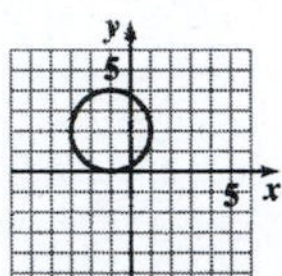

19. $C(5, -3)$, $r = 4$, x-int $= \{5 \pm \sqrt{7}\}$, no y-int,

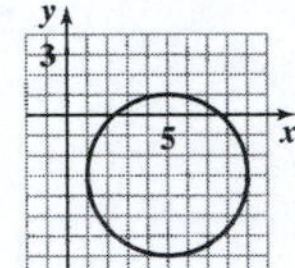

20. $C(2, 4)$, $r = 1$, no x-int, no y-int,

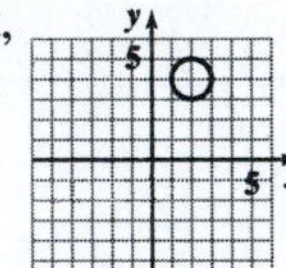

21. $C(0, -1)$, $r = 3$, x-int $= \{\pm 2\sqrt{2}\}$, y-int $= \{-4, 2\}$,

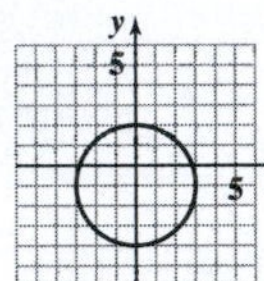

22. $C(3, 6)$, $r = 5\sqrt{2}$, x-int $= \{3 \pm \sqrt{14}\}$,

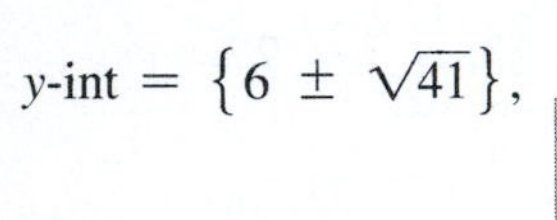
y-int $= \{6 \pm \sqrt{41}\}$,

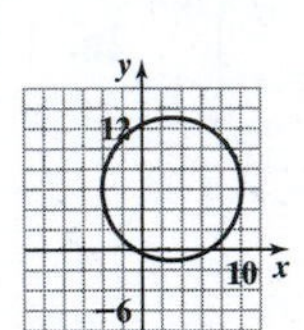

23. $C\left(\frac{3}{2}, \frac{1}{2}\right)$, $r = \sqrt{3}$, x-int $= \left\{\frac{3}{2} \pm \frac{\sqrt{11}}{2}\right\}$, y-int $= \left\{\frac{1}{2} \pm \frac{\sqrt{3}}{2}\right\}$,

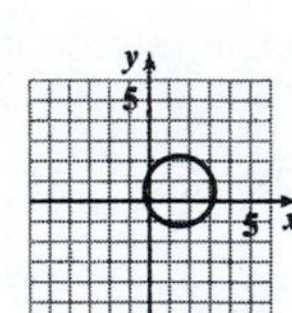

24. $C\left(-\frac{1}{3}, \frac{1}{4}\right)$, $r = \frac{3}{4}$, x-int $= \left\{-\frac{1}{3} \pm \frac{\sqrt{2}}{2}\right\}$, y-int $= \left\{\frac{1}{4} \pm \frac{\sqrt{65}}{12}\right\}$,

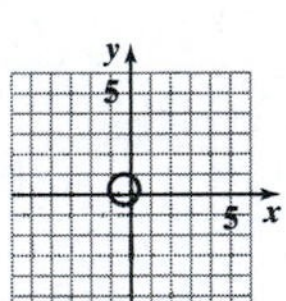

25. $C(1, -2)$, $r = 2$, x-int $= \{1\}$, y-int $= \{-2 \pm \sqrt{3}\}$,

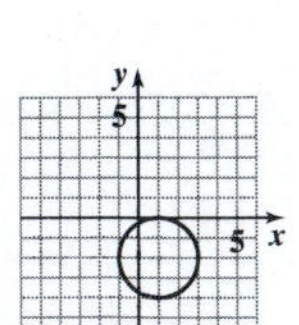

26. $C\left(\frac{1}{2}, -\frac{1}{4}\right)$, $r = 1$, x-int $= \left\{\frac{1}{2} \pm \frac{\sqrt{15}}{4}\right\}$, y-int $= \left\{-\frac{1}{4} \pm \frac{\sqrt{3}}{2}\right\}$,

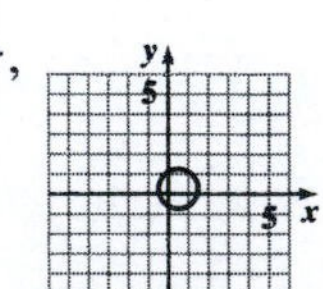

27. $C\left(\frac{1}{4}, \frac{1}{3}\right)$, $r = 2$, x-int $= \left\{\frac{1}{4} \pm \frac{\sqrt{35}}{3}\right\}$, y-int $= \left\{\frac{1}{3} \pm \frac{3\sqrt{7}}{4}\right\}$,

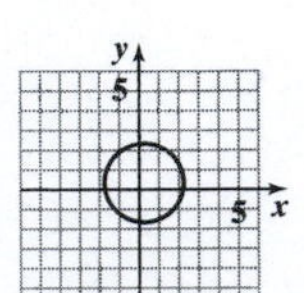

28. $C\left(-\frac{1}{6}, -1\right)$, $r = \sqrt{2}$,

$x\text{-int} = \left\{-\frac{7}{6}, \frac{5}{6}\right\}$, $y\text{-int} = \left\{-1 \pm \frac{\sqrt{71}}{6}\right\}$, **29.** $C(0, 0), r = 1$ **30.** $C(0, 2), r = 2$

31. $C(1, -5), r = 4$ **32.** $C(-2, -4), r = 6$ **33.** $C(4, -7), r = 2\sqrt{3}$ **34.** $C(-1, 3), r = 2\sqrt{5}$

35. $C\left(\frac{1}{4}, -\frac{1}{2}\right), r = 2$ **36.** $x\text{-int} = \{\pm 1\}, y\text{-int} = \{\pm 1\}$ **37.** $x\text{-int} = \{0\}, y\text{-int} = \{0, 4\}$

38. no x-int, $y\text{-int} = \{-5 \pm \sqrt{15}\}$ **39.** $x\text{-int} = \{-2 \pm 2\sqrt{5}\}, y\text{-int} = \{-4 \pm 4\sqrt{2}\}$ **40.** no x-int, no y-int

41. $x\text{-int} = \{-1 \pm \sqrt{11}\}, y\text{-int} = \{3 \pm \sqrt{19}\}$ **42.** $x\text{-int} = \left\{\frac{1}{4} \pm \frac{\sqrt{15}}{2}\right\}, y\text{-int} = \left\{-\frac{1}{2} \pm \frac{3\sqrt{7}}{4}\right\}$

43.
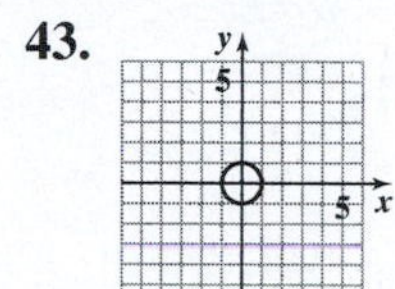

44.
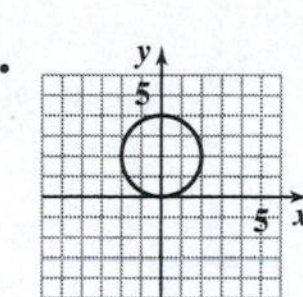

45.
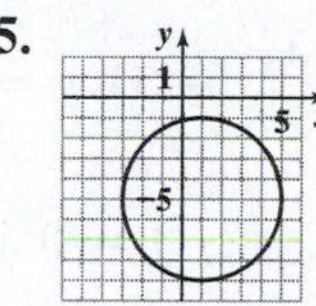

46.
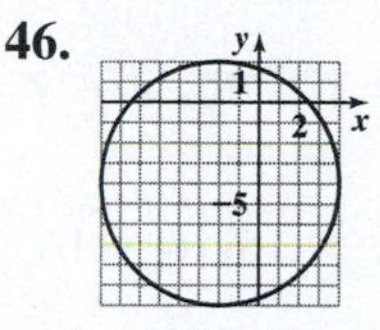

47.
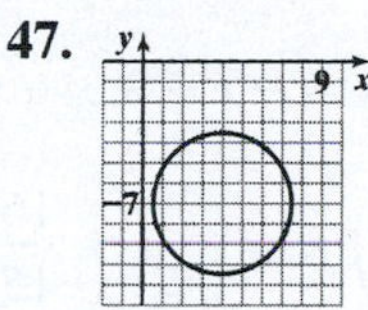

48.
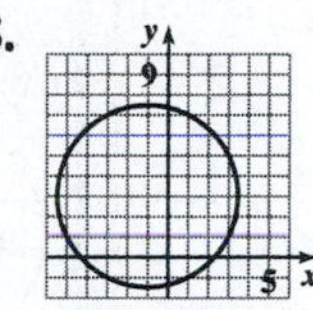

49.

50. $C(-1, 2), r = 2$ **51.** $C(5, -3), r = 4$ **52.** $C(2, 4), r = 1$ **53.** $C(0, -1), r = 3$

54. $C(3, 6), r = 5\sqrt{2}$ **55.** $C\left(\frac{3}{2}, \frac{1}{2}\right), r = \sqrt{3}$ **56.** $C\left(-\frac{1}{3}, \frac{1}{4}\right), r = \frac{3}{4}$ **57.** $C(1, -2), r = 2$ **58.** $C\left(\frac{1}{2}, -\frac{1}{4}\right), r = 1$

59. $C\left(\frac{1}{4}, \frac{1}{3}\right), r = 2$ **60.** $C\left(-\frac{1}{6}, -1\right), r = \sqrt{2}$ **61.** $x\text{-int} = \{-1\}, y\text{-int} = \{2 \pm \sqrt{3}\}$ **62.** $x\text{-int} = \{5 \pm \sqrt{7}\}$, no y-int **63.** no x-int, no y-int **64.** $x\text{-int} = \{\pm 2\sqrt{2}\}, y\text{-int} = \{-4, 2\}$ **65.** $x\text{-int} = \{3 \pm \sqrt{14}\}, y\text{-int} = \{6 \pm \sqrt{41}\}$

66. $x\text{-int} = \left\{\frac{3}{2} \pm \frac{\sqrt{11}}{2}\right\}, y\text{-int} = \left\{\frac{1}{2} \pm \frac{\sqrt{3}}{2}\right\}$ **67.** $x\text{-int} = \left\{-\frac{1}{3} \pm \frac{\sqrt{2}}{2}\right\}, y\text{-int} = \left\{\frac{1}{4} \pm \frac{\sqrt{65}}{12}\right\}$

68. $x\text{-int} = \{1\}, y\text{-int} = \{-2 \pm \sqrt{3}\}$ **69.** $x\text{-int} = \left\{\frac{1}{2} \pm \frac{\sqrt{15}}{4}\right\}, y\text{-int} = \left\{-\frac{1}{4} \pm \frac{\sqrt{3}}{2}\right\}$ **70.** $x\text{-int} = \left\{\frac{1}{4} \pm \frac{\sqrt{35}}{3}\right\}$, $y\text{-int} = \left\{\frac{1}{3} \pm \frac{3\sqrt{7}}{4}\right\}$ **71.** $x\text{-int} = \left\{-\frac{7}{6}, \frac{5}{6}\right\}, y\text{-int} = \left\{-1 \pm \frac{\sqrt{71}}{6}\right\}$

72.
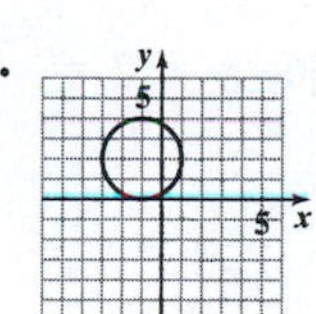

73.
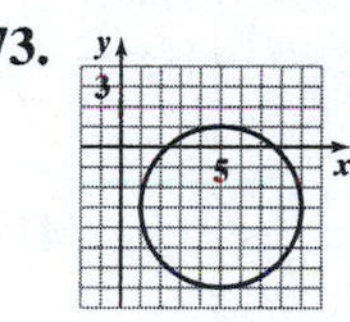

74. **75.** **76.** **77.** **78.** **79.**

80. **81.** **82.**

2.3 Exercises

SCE-1. $y = 2x - 5$ **SCE-2.** $y = -\frac{3}{2}x + \frac{7}{2}$ **SCE-3.** $y = 2x + 7$ **SCE-4.** $y = -5x + 7$ **SCE-5.** $y = \frac{2}{7}x + \frac{41}{7}$

SCE-6. $y = -\frac{3}{5}x - \frac{17}{5}$

1. $-\frac{8}{7}$ **2.** $\frac{1}{2}$ **3.** 0 **4.** The slope is undefined. **5.** $\frac{1}{2}$ **6.** $-\frac{1}{3}$ **7.** **8.**

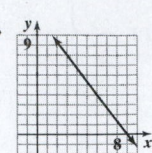

9. **10.** **11.** $y + 2 = \frac{1}{2}(x + 2)$ **12.** $y - 8 = -\frac{4}{3}(x - 3)$ **13.** $y - 3 = 3(x + 5)$

14. $y - 88 = \frac{5}{11}(x + 99)$ **15.** $y = x - 2$ **16.** $y = -\frac{1}{6}x + \frac{1}{2}$ **17.** $y = \frac{7}{9}x - \frac{8}{7}$ **18.** $y = 5$

19. $y - 2 = -1(x - 1), y = -x + 3, x + y = 3$ **20.** $y - 7 = -\frac{3}{2}(x + 5), y = -\frac{3}{2}x - \frac{1}{2}, 3x + 2y = -1$

21. $y - 1 = \frac{2}{15}\left(x + \frac{1}{2}\right), y = \frac{2}{15}x + \frac{16}{15}, 2x - 15y = -16$ **22.** $y - 4 = 0, y = 4, y = 4$

23. slope $= 4$, y-int $= -12$,

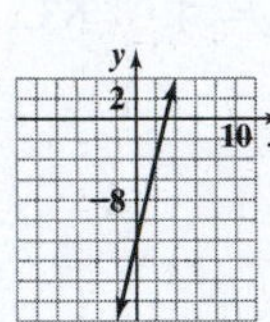

24. slope $= \frac{1}{3}$, y-int $= 3$,

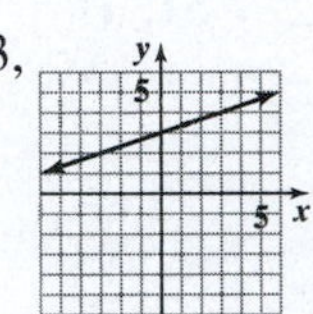

25. slope $= -\frac{5}{7}$, y-int $= \frac{12}{7}$,

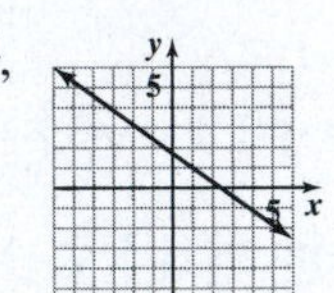

26. slope $= \frac{8}{11}$, y-int $= \frac{23}{11}$,

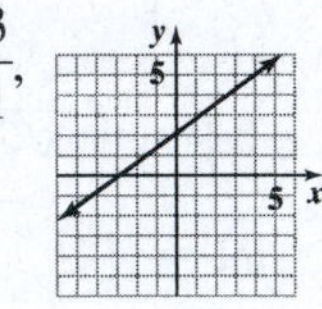

27.

28. **29.** **30.** **31.** $y = -2$ **32.** $x = 5$ **33.** $x = 4$ **34.** $y = -3$

35. $y - 2 = -1(x - 1)$ **36.** $y - 7 = -\frac{3}{2}(x + 5)$ **37.** $y - 1 = \frac{2}{15}\left(x + \frac{1}{2}\right)$ **38.** $y - 4 = 0$ **39.** $y = -x + 3$

40. $y = -\frac{3}{2}x - \frac{1}{2}$ **41.** $y = \frac{2}{15}x + \frac{16}{15}$ **42.** $y = 4$ **43.** $x + y = 3$ **44.** $3x + 2y = -1$ **45.** $2x - 15y = -16$

46. $y = 4$ **47.** slope $= 4$, y-int $= -12$ **48.** slope $= \frac{1}{3}$, y-int $= 3$ **49.** slope $= -\frac{5}{7}$, y-int $= \frac{12}{7}$

50. slope $= \frac{8}{11}$, y-int $= \frac{23}{11}$ **51.**

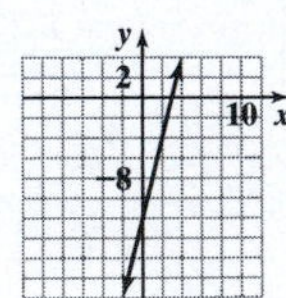

52.

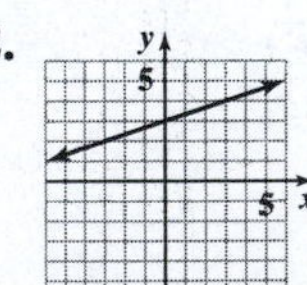

53.

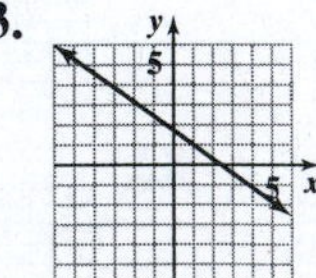

54.

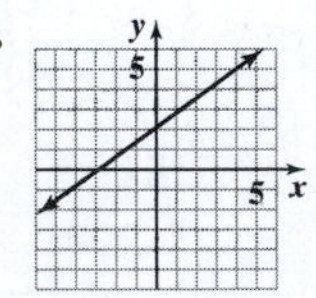

2.4 Exercises

SCE-1. $y = 2x - 5$ **SCE-2.** $y = -\frac{3}{2}x + \frac{7}{2}$ **SCE-3.** $y = 2x + 7$ **SCE-4.** $y = -5x + 7$ **SCE-5.** $y = \frac{2}{7}x + \frac{41}{7}$

SCE-6. $y = -\frac{3}{5}x - \frac{17}{5}$ **SCE-7.** slope $= 4$, y-int $= -12$, **SCE-8.** slope $= \frac{1}{3}$, y-int $= 3$,

SCE-9. slope $= -\frac{5}{7}$, y-int $= \frac{12}{7}$, **SCE-10.** slope $= \frac{8}{11}$, y-int $= \frac{23}{11}$,

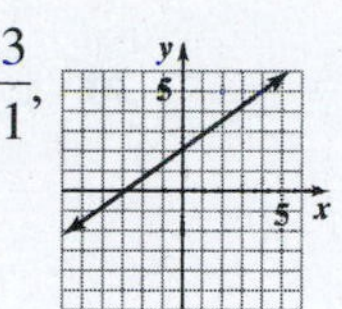

1. perpendicular **2.** parallel **3.** parallel **4.** perpendicular **5.** neither **6.** neither **7.** $y - 3 = \frac{1}{4}(x + 2)$, $y = \frac{1}{4}x + \frac{7}{2}$, $x - 4y = -14$ **8.** $y + 4 = \frac{3}{5}(x - 1)$, $y = \frac{3}{5}x - \frac{23}{5}$, $3x - 5y = 23$ **9.** $y + 5 = 3(x + 2)$, $y = 3x + 1$, $3x - y = -1$ **10.** $x = 11$ **11.** $y = -3$ **12.** $y - 3 = -4(x + 2)$, $y = -4x - 5$, $4x + y = -5$ **13.** $y + 4 = -\frac{5}{3}(x - 1)$, $y = -\frac{5}{3}x - \frac{7}{3}$, $5x + 3y = -7$ **14.** $y + 5 = -\frac{1}{3}(x + 2)$, $y = -\frac{1}{3}x - \frac{17}{3}$, $x + 3y = -17$ **15.** $y = -3$ **16.** $x = 11$ **17.** parallelogram and rhombus **18.** parallelogram **19.** neither **20.** parallelogram and rhombus **21.** $y - 3 = \frac{1}{4}(x + 2)$ **22.** $y + 4 = \frac{3}{5}(x - 1)$ **23.** $y + 5 = 3(x + 2)$ **24.** $y = \frac{1}{4}x + \frac{7}{2}$ **25.** $y = \frac{3}{5}x - \frac{23}{5}$ **26.** $y = 3x + 1$ **27.** $x - 4y = -14$ **28.** $3x - 5y = 23$ **29.** $3x - y = -1$ **30.** $y - 3 = -4(x + 2)$ **31.** $y + 4 = -\frac{5}{3}(x - 1)$ **32.** $y + 5 = -\frac{1}{3}(x + 2)$ **33.** $y = -4x - 5$ **34.** $y = -\frac{5}{3}x - \frac{7}{3}$ **35.** $y = -\frac{1}{3}x - \frac{17}{3}$ **36.** $4x + y = -5$ **37.** $5x + 3y = -7$ **38.** $x + 3y = -17$

Review Exercises

1. a. quadrant II **b.** quadrant III **c.** y-axis **d.** quadrant I **e.** x-axis

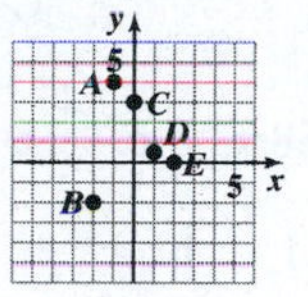

2.

3. a. yes b. no c. no **4.** x-intercepts: 3, −2; y-intercept: $-\frac{3}{2}$ **5.** $M\left(-\frac{3}{2}, 1\right)$ **6.** $d(A, B) = 5\sqrt{5}$ **7.** $(x + 2)^2 + (y - 3)^2 = 16$ **8.** $(x - 4)^2 + \left(y + \frac{5}{2}\right)^2 = \frac{65}{4}$ **9.** Center: $(4, -7)$, $r = 2\sqrt{3}$, no x-intercepts, no y-intercepts

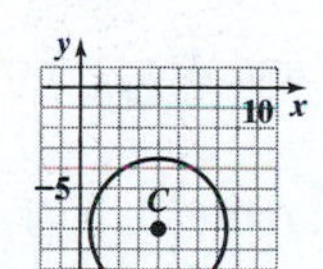

10. Center: $(5, -3)$, $r = 4$, x-intercepts: $5 + \sqrt{7}, 5 - \sqrt{7}$, no y-intercepts

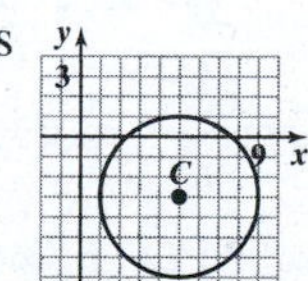

11. Center: $\left(\frac{1}{2}, -\frac{1}{4}\right)$, $r = 1$, x-intercepts:

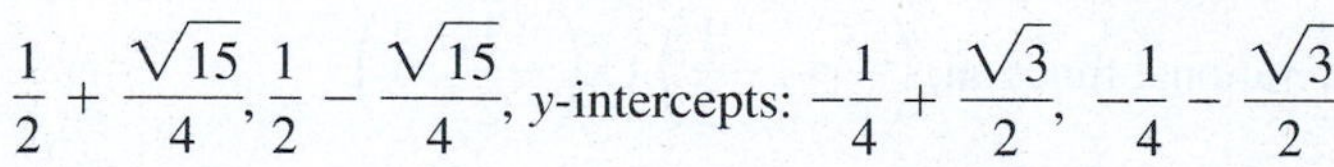

$\frac{1}{2} + \frac{\sqrt{15}}{4}, \frac{1}{2} - \frac{\sqrt{15}}{4}$, y-intercepts: $-\frac{1}{4} + \frac{\sqrt{3}}{2}, -\frac{1}{4} - \frac{\sqrt{3}}{2}$

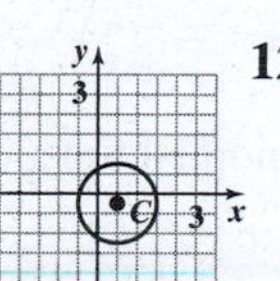

12. $m = \frac{1}{2}$ **13.**

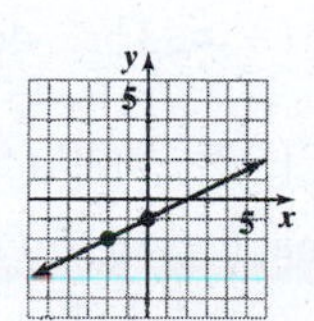

14. $y - 3 = 3(x + 5)$ **15.** $y = x - 2$ **16.** point-slope: $y + 5 = \frac{3}{2}(x - 3)$ OR $y - 7 = \frac{3}{2}(x + 5)$ Slope-intercept: $y = -\frac{3}{2}x - \frac{1}{2}$ Standard form: $3x + 2y = -1$ **17.** Slope is $\frac{1}{3}$. y-intercept: 3

18. **19.** $y = -2$ **20.** $x = 5$ **21.** Perpendicular **22.** $y + 4 = \frac{3}{5}(x - 1)$ OR $y = \frac{3}{5}x - \frac{23}{5}$ OR $3x - 5y = 23$ **23.** $x = 11$ **24.** $y + 4 = -\frac{5}{3}(x - 1)$ OR $y = -\frac{5}{3}x - \frac{7}{3}$ OR $5x + 3y = -7$ **25.** $x = 11$

CHAPTER 3

3.1 Exercises

SCE-1. $\left(\infty, \frac{2}{3}\right]$ **SCE-2.** $(\infty, -7] \cup [3, \infty)$ **SCE-3.** $[-4, 1] \cup [3, \infty)$ **SCE-4.** $(\infty, -3] \cup (1, \infty)$

SCE-5. $(-7, -3] \cup [3, \infty)$ **SCE-6.** $(-8, 2] \cup (3, \infty)$ **SCE-7.** $-18xh - 9h^2$ **SCE-8.** $\sqrt{x + h} - 2h - \sqrt{x}$

1. domain = {Tim, Audrie, Sheryl, Jason}, range = {Blue, Green, Brown}, function
2. domain = {Ronald, James, George}, range = {Sue, Laura, Holly, Nancy}, not a function
3. domain = $\{-1, 0, 1, 2\}$, range = $\{2, 3, 5, 7\}$, not a function **4.** domain = $\{-1, 0, 1, 2, 3\}$, range = $\{2, 3, 5, 7\}$, function
5. domain = $\{-3, -1, 1, 3\}$, range = $\{2\}$, function **6.** domain = $\{-3, -1, 1, 3\}$, range = $\{-2, 2\}$, not a function
7. Sample response: $(2, 5), (4, -1), (5, 0)$ **8.** function **9.** not a function **10.** function **11.** function
12. function **13. a.** -13 **b.** $-\frac{5}{2}$ **c.** $3x - 19$ **14. a.** 5 **b.** 5 **c.** $x^2 - 2x + 2$ **15. a.** 2 **b.** $\frac{7}{8}$
c. $\frac{a}{a + 1}$ **16. a.** 36π **b.** 288π **c.** $\frac{4}{3}\pi(r + 1)^3$ **17. a.** $2x - 3$ **b.** $2x + 2h - 5$ **18. a.** $22 - 7x$
b. $8 - 7x - 7h$ **19. a.** $x^2 - 1$ **b.** $x^2 + 2hx - 2x + h^2 - 2h$ **20. a.** $t^3 + 3t^2 + 2t$
b. $t^3 + 3t^2h + 3th^2 + h^3 - t - h$ **21. a.** -6 **b.** $\sqrt{x + h} - 2x - 2h$ **22.** 2 **23.** -7 **24.** $2x + h - 2$
25. $-18x - 9h$ **26.** $10x + 5h$ **27.** $3x^2 + 3xh + h^2 - 1$ **28.** $\frac{\sqrt{x + h} - 2h - \sqrt{x}}{h}$ **29.** function **30.** function
31. function **32.** not a function **33.** function **34.** function **35.** not a function **36.** not a function
37. function **38.** polynomial: $g(t) = 3t^2 + 5$ and $A(r) = \pi r^2$; rational: $f(x) = \frac{2x}{3x + 1}$ and $h(x) = -\frac{4}{x}$
39. polynomial: $m(t) = t(t + 2)$ and $H(x) = x$; rational: $h(t) = 3t^{-2}$; root: $f(x) = \sqrt[3]{1 + x}$
40. polynomial: $f(x) = -5$ and $h(t) = \frac{3t^2 + 5}{8}$; rational: $q(x) = \frac{8}{3x^2 + 5}$; root: $V(n) = \sqrt{\frac{n}{5}}$
41. polynomial: $f(x) = 8 - x^3$ and $H(y) = 4^{-1}$; rational: $T(t) = \frac{\sqrt[5]{7}}{t^5}$; root: $g(x) = \sqrt[5]{\frac{x^2 - 9}{2x}}$
42. polynomial function, $(-\infty, \infty)$ **43.** rational function, $(-\infty, 0) \cup (0, \infty)$ **44.** polynomial function, $(-\infty, \infty)$
45. root function, $(-\infty, 3]$ **46.** root function, $(-\infty, \infty)$ **47.** rational function, $\left(-\infty, -\frac{1}{2}\right) \cup \left(-\frac{1}{2}, \infty\right)$
48. root function, $(-\infty, 5) \cup (5, \infty)$ **49.** polynomial function, $(-\infty, \infty)$
50. rational function, $(-\infty, -7) \cup (-7, 6) \cup (6, \infty)$ **51.** polynomial function, $(-\infty, \infty)$ **52.** root function, $(-\infty, \infty)$
53. root function, $(-\infty, -1] \cup [1, \infty)$ **54.** rational function, $(-\infty, -1) \cup (-1, 0) \cup (0, 4) \cup (4, \infty)$
55. root function, $(-\infty, -5] \cup [3, \infty)$ **56.** polynomial function, $(-\infty, \infty)$ **57.** root function, $(-\infty, -3] \cup [3, \infty)$
58. root function, $(-\infty, 3) \cup (3, \infty)$ **59.** rational function, $(-\infty, \infty)$
60. root function, $(-4, 3] \cup (5, \infty)$ **61.** 2 **62.** -7 **63.** $2x + h - 2$ **64.** $10x + 5h$ **65.** $4x + 2h + 3$
66. $\frac{\sqrt{x + h} - 2h - \sqrt{x}}{h}$ **67.** $-18x - 9h$ **68.** $3x^2 + 3xh + h^2 - 1$

3.2 Exercises

SCE-1. $f(-x) = 3x^5 + 4x^4 - 3x^3 - 7x^2 - 21$ **SCE-2.** $f(-x) = x^4 - 7x^2 - 11$ **SCE-3.** $f(-x) = 3x^7 - 8x^3 + x$
SCE-4. $f(-x) = -\dfrac{2x^2 - 4}{x}$

1. $x\text{-int} = \frac{1}{2}$, $y\text{-int} = -1$ **2.** $x\text{-int} = \text{none}$, $y\text{-int} = 3$ **3.** $x\text{-int} = \{-2, 3\}$, $y\text{-int} = -6$
4. $x\text{-int} = \left\{\frac{7}{4} - \frac{\sqrt{73}}{4}, \frac{7}{4} + \frac{\sqrt{73}}{4}\right\}$, $y\text{-int} = -3$ **5.** $x\text{-int} = \{-2, -1, 1\}$, $y\text{-int} = -2$ **6.** $x\text{-int} = \left\{-6, 0, \frac{1}{2}\right\}$, $y\text{-int} = 0$
7. x-intercept: $\frac{1}{3}$; y-intercept: $-\frac{1}{2}$ **8.** x-intercept: 0; y-intercept: 0 **9.** x-intercepts: $\{-5, 5\}$; no y-intercept
10. x-intercept: 0; y-intercept: 0 **11.** x-intercepts: $-4, 3$; no y-intercept **12.** x-intercepts: $-\sqrt{7}, \sqrt{7}, 1$; y-intercept: 7
13. domain $= [-5, 3]$, range $= [0, 4]$ **14.** domain $= (-4, 5]$, range $= (-5, 5]$ **15.** domain $= \left(-\frac{\pi}{2}, \frac{\pi}{2}\right)$, range $= (-\infty, \infty)$ **16.** domain $= [-4, \infty)$, range $= (-\infty, 0]$ **17.** domain $= [-4, \infty)$, range $= (2, 5]$
18. domain $= (-\infty, -1) \cup (-1, 1) \cup (1, \infty)$, range $= (-\infty, 0] \cup (1, \infty)$ **19. a.** $(-\infty, -4) \cup (2, 5)$ **b.** $(-4, 2)$
c. none **20. a.** $(0, 2)$ **b.** $(-4, 0) \cup (2, 4)$ **c.** $(4, \infty)$ **21. a.** $(-3, 0)$ **b.** $(-\infty, -3) \cup (4, \infty)$ **c.** $(0, 4)$
22. a. none **b.** none **c.** $(-1, 0) \cup (0, 1) \cup (1, 2) \cup (2, 3)$ **23. a.** $x = 6$ **b.** $f(6) = 0$ **c.** $x = 2$ **d.** $f(2) = 6$
24. a. none **b.** none **c.** $x = 0$ **d.** $f(0) = 3$ **25. a.** $x = -2$ **b.** $f(-2) = -1$ **c.** $x = 2$ **d.** $f(2) = 2$
26. a. none **b.** none **c.** none **d.** none **27. a.** $x = 0$ **b.** $f(0) = 0$ **c.** none **d.** none **28.** odd
29. neither **30.** even **31.** odd **32.** even **33.** neither **34.** odd **35.** odd **36.** even **37. a.** $(-8, 9]$
b. $[-5, 3]$ **c.** increasing: $(-8, -6) \cup (-2, 1) \cup (4, 7)$, decreasing: $(-6, -2) \cup (7, 9)$, constant: $(1, 4)$
d. $x = -2$, $f(-2) = -5$ **e.** $x = -6$ and $x = 7$, $f(-6) = 3$ and $f(7) = 2$ **f.** $x = \left\{-\frac{15}{2}, -4, 5, 9\right\}$ **g.** -3
h. $\left(-8, -\frac{15}{2}\right] \cup [-4, 5] \cup \{9\}$ **i.** two **j.** -1 **38. a.** $(-\infty, \infty)$ **b.** $[-1, 7]$
c. increasing: $(-5, -4) \cup (0, 4) \cup (5, 6)$, decreasing: $(-6, -5) \cup (-4, 0) \cup (4, 5)$, constant $(-\infty, -6) \cup (6, \infty)$
d. $x = -5, x = 0$, and $x = 5$, $f(-5) = 5, f(0) = -1$, and $f(5) = 5$ **e.** $x = -4$ and $x = 4$, $f(-4) = 6$ and $f(4) = 6$
f. $x = \{-1, 1\}$ **g.** even **h.** $(-\infty, -1) \cup (1, \infty)$ **i.** true **j.** $(-\infty, -6] \cup [6, \infty)$ **39.** domain $= (-8, 9]$, range $= [-5, 3]$ **40.** increasing: $(-8, -6) \cup (-2, 1) \cup (4, 7)$, decreasing: $(-6, -2) \cup (7, 9)$, constant: $(1, 4)$
41. a. $x = -2$, $f(-2) = -5$ **b.** $x = -6$ and $x = 7, f(-6) = 3$ and $f(7) = 2$ **42.** -3
43. a. $\left(-8, -\frac{15}{2}\right] \cup [-4, 5] \cup \{9\}$ **44. a.** one **b.** -1 **45.** domain $= (-\infty, \infty)$, range $= [-1, 7]$
46. increasing: $(-5, -4) \cup (0, 4) \cup (5, 6)$, decreasing: $(-6, -5) \cup (-4, 0) \cup (4, 5)$, constant: $(-\infty, -6) \cup (6, \infty)$
47. a. $x = -5, x = 0$, and $x = 5, f(-5) = 5, f(0) = -1$, and $f(5) = 5$ **b.** $x = -4$ and $x = 4, f(-4) = 6$ and $f(4) = 6$
48. $x = \{-1, 1\}$ **49. a.** $(-\infty, -1) \cup (1, \infty)$ **b.** false **c.** $(-\infty, -6] \cup [6, \infty)$ **50.** even

3.3 Exercises

1. a, e, **2.** b, c, e, f, **3.** c, d, e, f, **4.** a, e, **5.** a, d, e,

6. b, c, d, e, f,

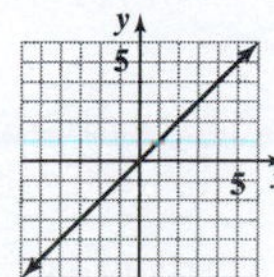

7. b, c, e, f,

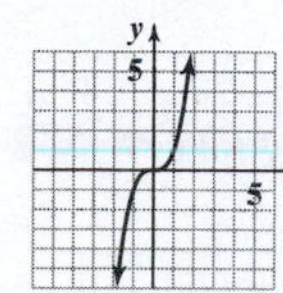

8. none, **9.** b, **10.** $f(x) = x$ for $-3 < x \le 1$

11. $f(x) = x^2$ for $x \le 1$

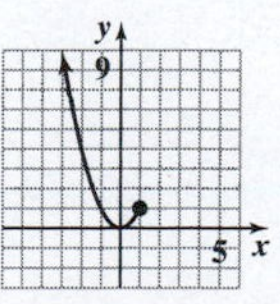

12. $f(x) = \sqrt{x}$ for $0 \le x < 4$

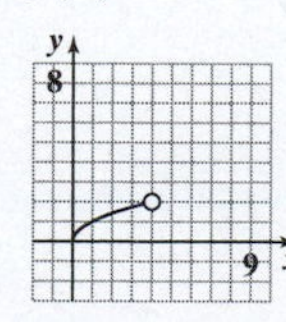

13. $f(x) = \frac{1}{x}$ for $x \ge -1$

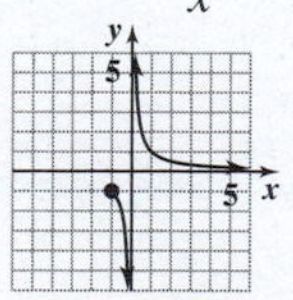

14. $f(x) = x^3$ for $-1 \le x \le 1$

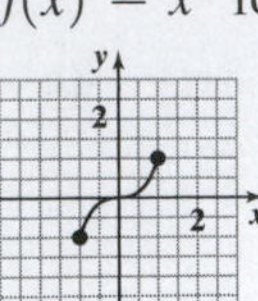

15. $f(x) = \sqrt[3]{x}$ for $x < 8$

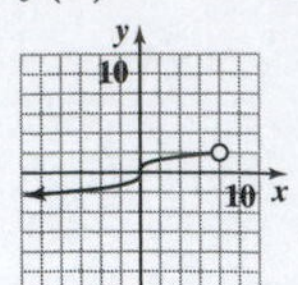

16. $f(x) = |x|$ for $-2 < x \le 1$

17. $f(x) = x$ for $x > -1$

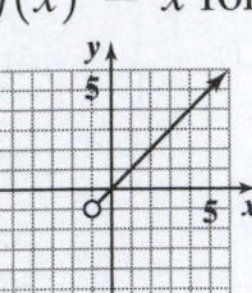

18. $f(x) = x^2$ for $-2 < x \le 3$

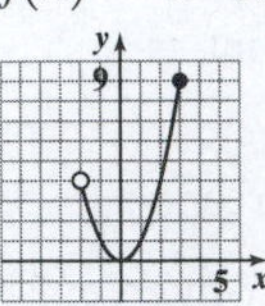

19. $f(x) = \sqrt{x}$ for $x > 1$

20. $f(x) = \dfrac{1}{x}$ for $-1 \le x < 2$

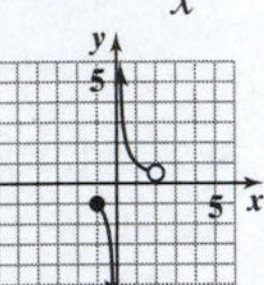

21. $f(x) = x^3$ for $x \ge -2$

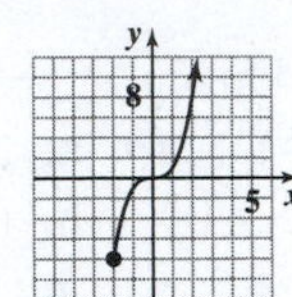

22. $f(x) = \sqrt[3]{x}$ for $-1 < x \le 8$

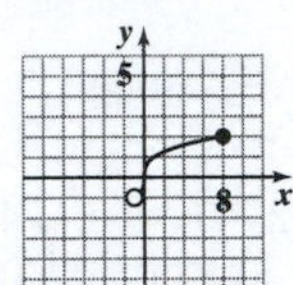

23. $f(x) = |x|$ for $x \le 3$

24. a. Domain: $(-1, \infty)$ **b.** $f(x) = x^2$ for $x > -1$

25. a. Domain: $[-1, 2)$ **b.** $f(x) = x^2$ for $-1 \le x < 2$

26. a. Domain: $(-\infty, 2)$ **b.** $f(x) = x$ for $x < 2$

27. a. Domain: $(-3, 1]$ **b.** $f(x) = x$ for $-3 < x \le 1$

28. a. Domain: $[-1, 4)$ **b.** $f(x) = |x|$ for $-1 \le x < 4$

29. a. Domain: $(-\infty, 2]$ **b.** $f(x) = |x|$ for $x \le 2$

30. a. Domain: $(-\infty, 1)$ **b.** $f(x) = x^3$ for $x < 1$

31. a. Domain: $(-1, 2]$ **b.** $f(x) = x^3$ for $-1 < x \le 2$

32. a. $f(-2) = 1, f(0) = 1, f(2) = -3$

b.

c. $(-\infty, \infty)$

d. $\{-3, 1\}$

33. a. $f(-2) = -5, f(0) = -1, f(2) = 4$

b.

c. $(-\infty, \infty)$

d. $(-\infty, 1] \cup (3, \infty)$

34. a. $f(-2) = -5,\ f(0) = -1,\ f(2) = 1$

b.

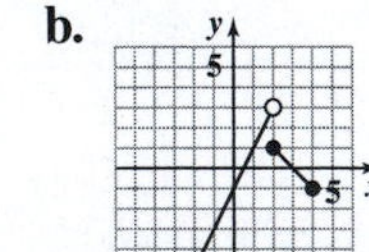

c. Domain: $[-2, 4]$

d. Range: $[-5, 3)$

35. a. $f(-2) = 4, f(0) = 0, f(2) = \dfrac{1}{2}$

b.

c. $(-\infty, \infty)$

d. $[0, \infty)$

36. a. $f(-2) = 3, f(0) = 0, f(2) = 1$

b.

c. $(-\infty, \infty)$

d. $[0, 2) \cup \{3\}$

37. a. $f(-2) = 4, f(0) = 2, f(2) = \sqrt{2}$

b.

c. $(-\infty, \infty)$

d. $(0, \infty)$

38. a. $f(-2) = -2, f(0) = 2, f(2) = 8$

b. 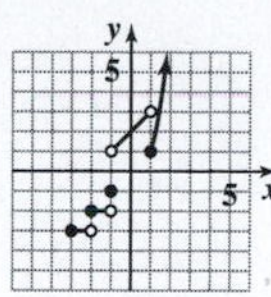

c. $[-3, \infty)$

d. $\{-3\} \cup \{-2\} \cup \{-1\} \cup [1, \infty)$

39. $f(x) = \begin{cases} 1 \text{ if } x < 0 \\ -1 \text{ if } x \geq 0 \end{cases}$

40. $f(x) = \begin{cases} x \text{ if } -4 \leq x \leq 2 \\ 3 \text{ if } x > 2 \end{cases}$

41. $f(x) = \begin{cases} |x| \text{ if } x < 1 \\ -x + 1 \text{ if } 1 \leq x \leq 4 \end{cases}$

42. $f(x) = \begin{cases} 2x + 3 \text{ if } -3 \leq x < -1 \\ x^2 \text{ if } -1 < x \leq 2 \end{cases}$

43. $f(x) = \begin{cases} \dfrac{1}{x} \text{ if } x \leq -\dfrac{1}{2} \\ \sqrt{x} \text{ if } 0 \leq x < 4 \end{cases}$

44. a. DIL 1036

b. $T(x) = \begin{cases} 0 \text{ if } 0 \leq x \leq 26,300 \\ 0.28(x - 26,300) \text{ if } x > 26,300 \end{cases}$

c. 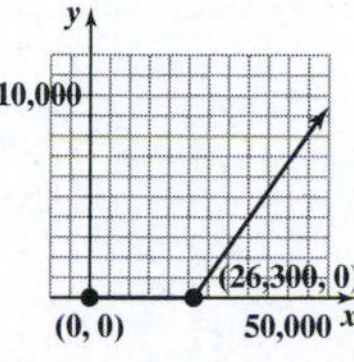

45. a. \$29.10

b. \$36.50

c. $C(x) = \begin{cases} 9.5 + 0.07x \text{ if } 0 \leq x \leq 350 \\ 34 + 0.05(x - 350) \text{ if } x > 350 \end{cases}$

d. 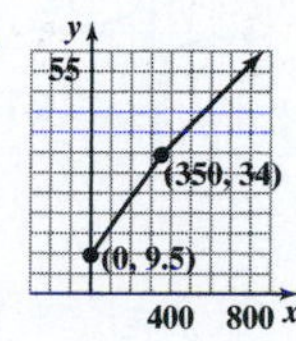

46. a. \$0.89

b. $P(x) = \begin{cases} 0.47 \text{ if } 0 < x \leq 1 \\ 0.68 \text{ if } 1 < x \leq 2 \\ 0.89 \text{ if } 2 < x \leq 3 \\ 1.10 \text{ if } 3 < x \leq 4 \\ 1.31 \text{ if } 4 < x \leq 5 \end{cases}$

c. 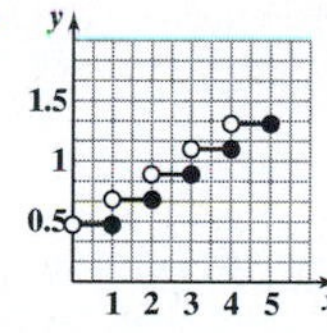

47. a. \$25.15

b. $P(x) = \begin{cases} 25 \text{ if } x \leq 500 \\ 25.05 \text{ if } 500 < x \leq 501 \\ 25.1 \text{ if } 501 < x \leq 502 \\ 25.15 \text{ if } 502 < x \leq 503 \\ 25.2 \text{ if } 503 < x \leq 504 \\ 25.25 \text{ if } 504 < x \leq 505 \end{cases}$

c. 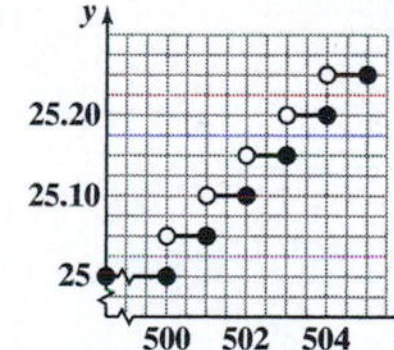

48. a. $f(-2) = 1, f(0) = 1, f(2) = -3$

b. 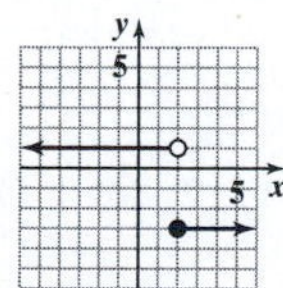

49. a. $f(-2) = -5, f(0) = -1, f(2) = 4$

b. 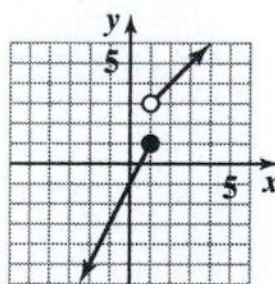

50. a. $f(-2) = -5, f(0) = -1, f(2) = 1$

b. 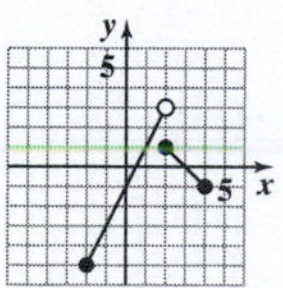

51. a. $f(-2) = 4, f(0) = 0, f(2) = \dfrac{1}{2}$

b. 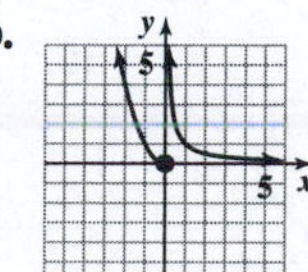

52. a. $f(-2) = 3, f(0) = 0, f(2) = 1$

b.

53. a. $f(-2) = 4, f(0) = 2, f(2) = \sqrt{2}$

b.

54. a. $f(-2) = -2, f(0) = 2, f(2) = 8$

b.

3.4 Exercises

1.

2.

3.

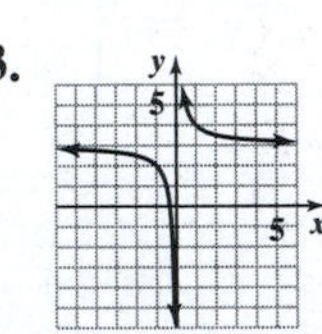

4.

5.

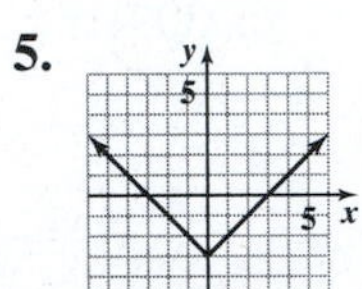

6.

7. $f(x) = x^2 - 1$ **8.** $f(x) = x^3 + 2$ **9.** $f(x) = \sqrt{x} - 3$ **10.** $f(x) = \frac{1}{x} + 2$ **11.** $f(x) = |x| + 4$

12. $f(x) = \sqrt[3]{x} - 3$

13.

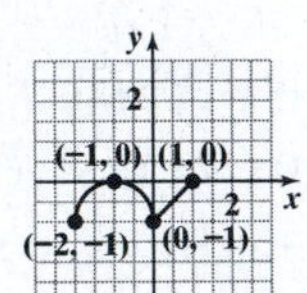

14.

15.

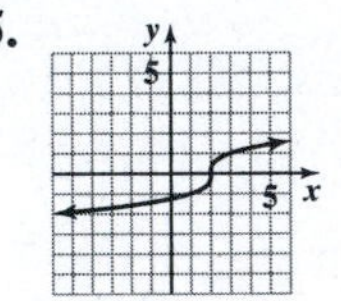

16.

17.

18.

19.

20.

21. $f(x) = (x + 2)^2$

22. $f(x) = (x - 1)^3$ **23.** $f(x) = \sqrt{x + 3}$ **24.** $f(x) = \frac{1}{x - 4}$ **25.** $f(x) = |x - 2|$ **26.** $f(x) = \sqrt[3]{x + 1}$

27.

28.

29.

30.

31.

32.

33.

34.

35. $f(x) = (x - 2)^2 + 1$ **36.** $f(x) = \sqrt{x + 1} - 3$ **37.** $f(x) = |x + 3| - 4$

38. $f(x) = (x - 1)^3 - 2$

39.

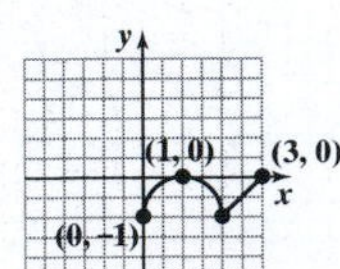

40.

41.

42.

43.

44.

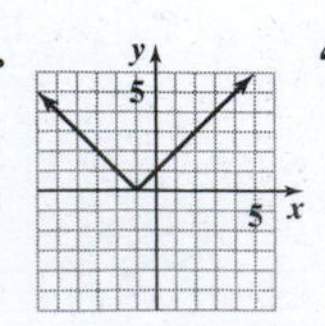

45.

46.

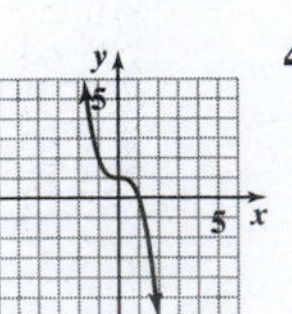

47.

48.

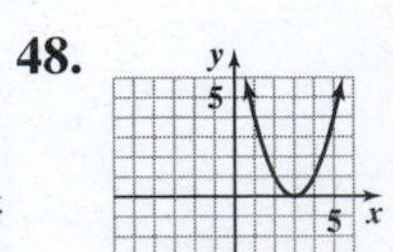

49.

50. 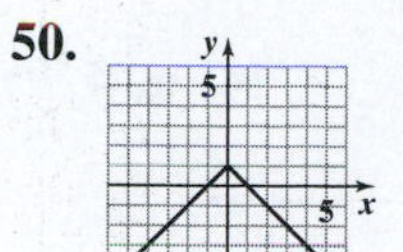

51. 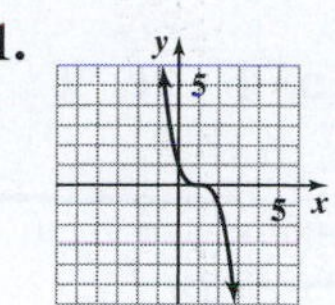

52. 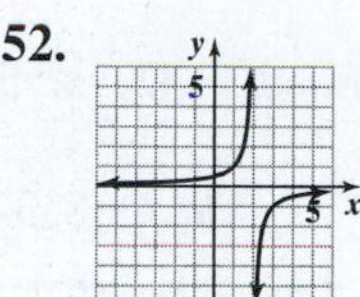

53.

54. 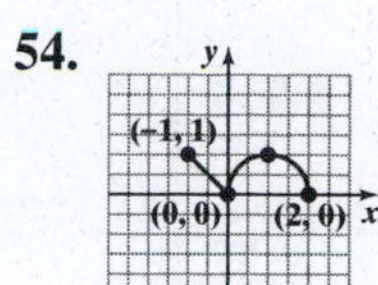

55. 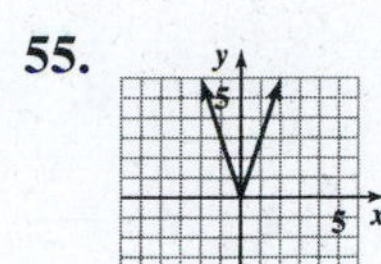

56. 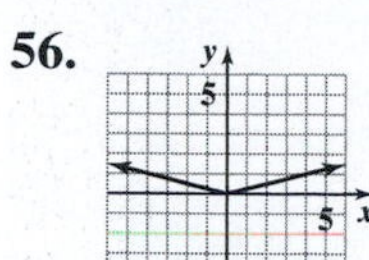

57. 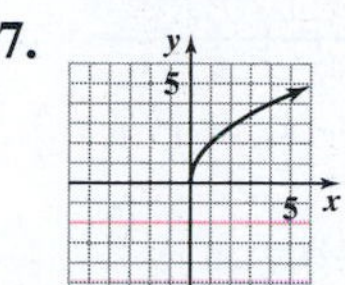

58.

59. 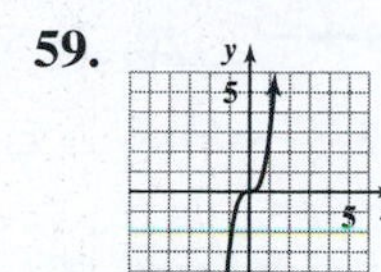

60. 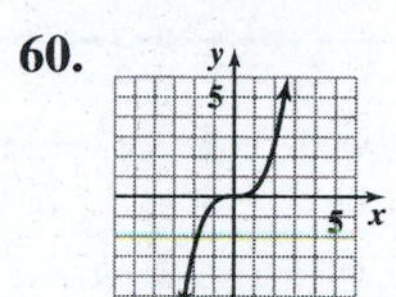

61. 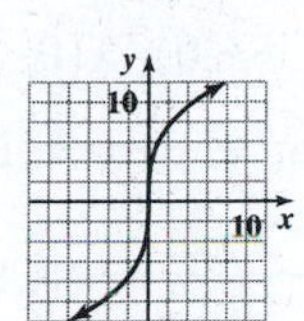

62. 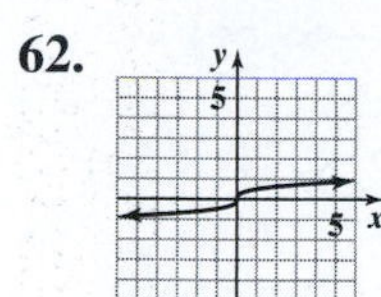

63. 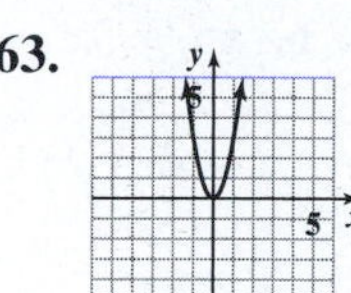

64. 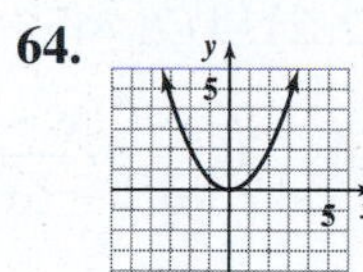

65. 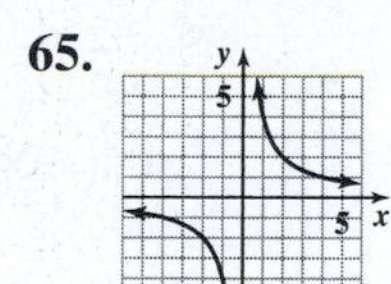

66. 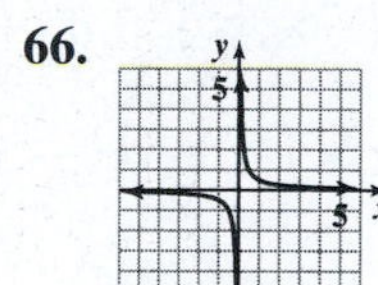

67. 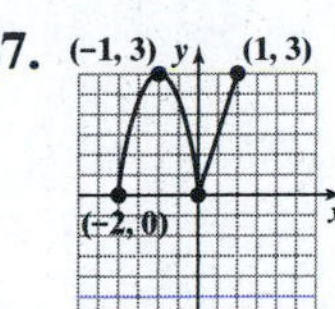

68. 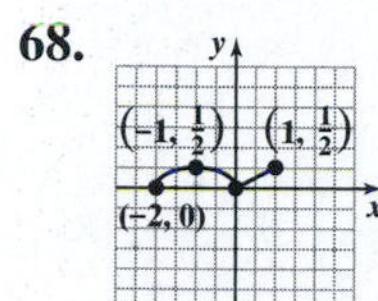

69. 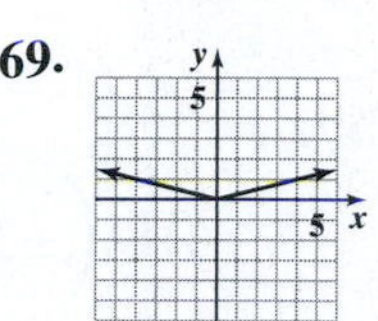

70. 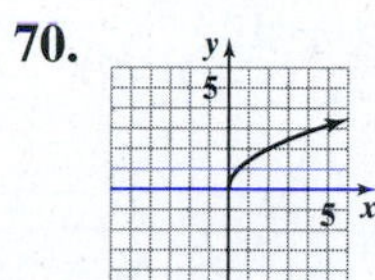

71. 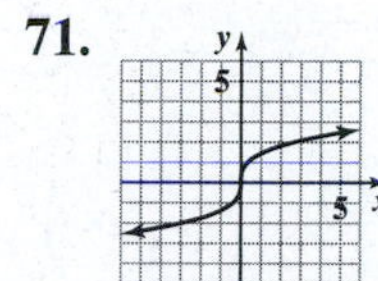

72. 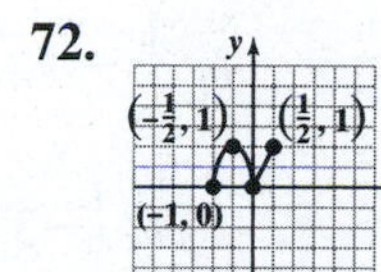

73. 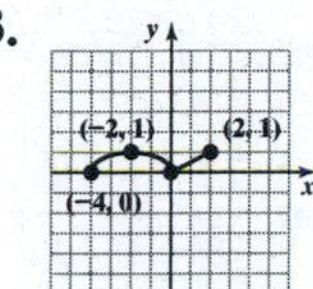

74. 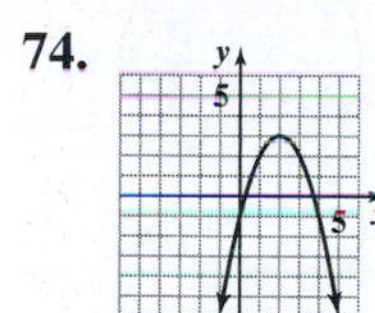

75. 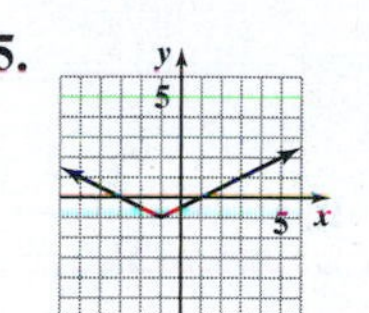

76. 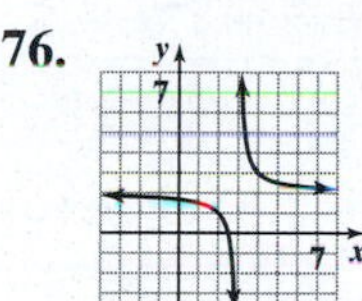

77. 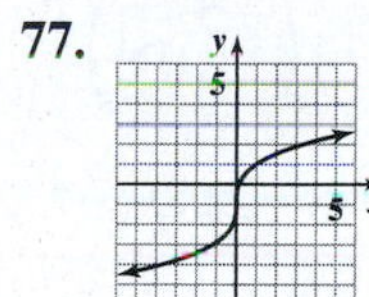

78. 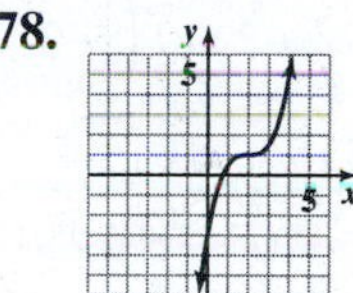

79. 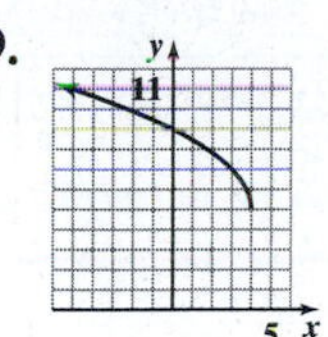

80. 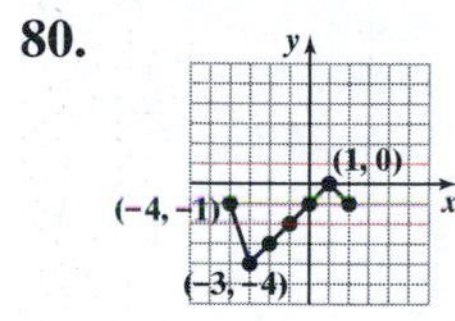

81. 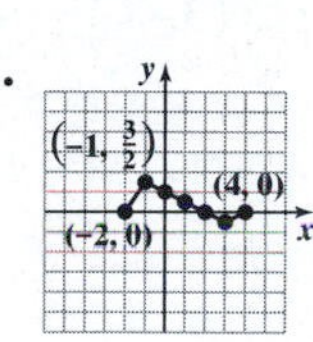

82. 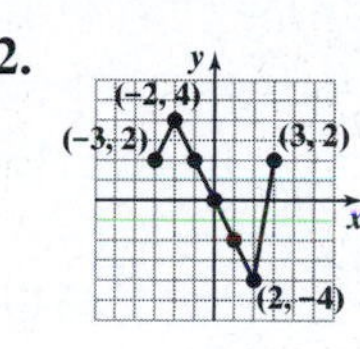

83. 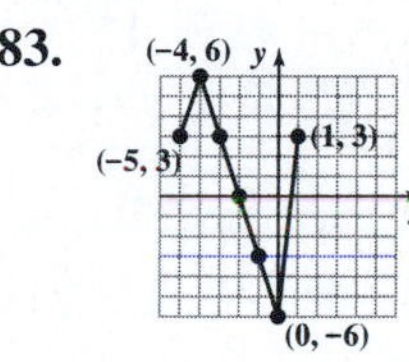

84. 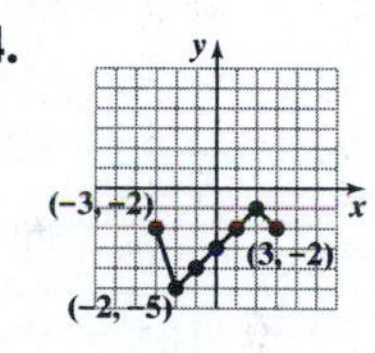

85. $f(x) = \begin{cases} |x - 2| & \text{if } x \le 2 \\ \sqrt{x - 2} & \text{if } x > 2 \end{cases}$

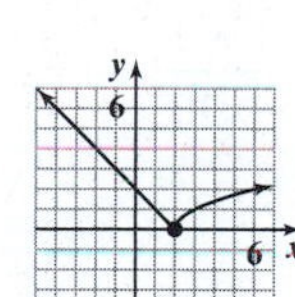

86. $f(x) = \begin{cases} -x^2 + 1 & \text{if } x < 0 \\ -\sqrt{x} - 3 & \text{if } x \ge 0 \end{cases}$

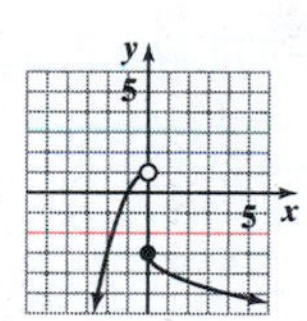

87. $f(x) = \begin{cases} (x + 1)^3 & \text{if } x \le 1 \\ |x - 2| - 2 & \text{if } x > 1 \end{cases}$

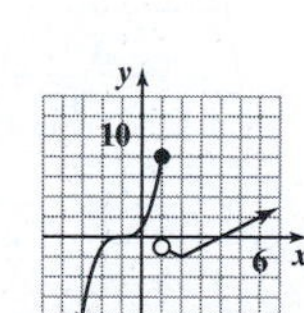

88. $f(x) = \begin{cases} \sqrt{-x} - 2 & \text{if } x \le 0 \\ (x - 1)^2 & \text{if } x > 0 \end{cases}$

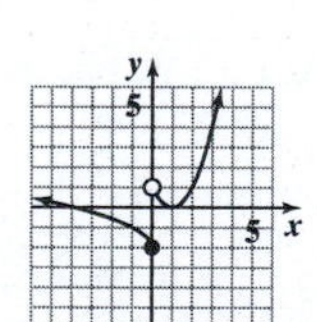

3.5 Exercises

SEC-1. $\dfrac{(2x + 1)(5x - 2)}{5x(x - 9)}$ **SEC-2.** $\dfrac{x(x - 5)}{(x - 2)(x + 2)}$ **SEC-3.** $\dfrac{(x - 7)^2(x + 7)}{(x - 3)(x + 2)(x + 8)}$ **SEC-4.** $\dfrac{7x + 29}{x + 3}$

SEC-5. $-\dfrac{2(3x + 4)}{x + 6}$ **SEC-6.** $\dfrac{3(x + 8)}{8x + 67}$ **SEC-7.** $\dfrac{12 + 3\sqrt{x + 5}}{\sqrt{x + 5} + 2}$

1. 9 **2.** −59 **3.** 1 **4.** $-\dfrac{1}{2}$ **5.** $\dfrac{4}{33}$ **6.** $-\dfrac{153}{11}$ **7.** undefined **8.** $7\sqrt{3}$ **9.** 186 **10.** $\dfrac{1}{36}$ **11. a.** 0

b. 8 **c.** −10 **d.** −1 **12. a.** −5 **b.** 0 **c.** 12 **d.** undefined **13. a.** 5 **b.** 1 **c.** 9 **d.** $\frac{2}{3}$ **14. a.** 4 **b.** 3 **c.** 4 **d.** $-\frac{1}{2}$ **15.** $(-\infty, 3)$, −5 −4 −3 −2 −1 0 1 2 3 4 5 **16.** $(-8, 3)$, −10 −8 −6 −4 −2 0 2 4 6 8 10 **17.** $[0, 4]$, −5 −4 −3 −2 −1 0 1 2 3 4 5 **18.** $(-\infty, 1) \cup (1, \infty)$, −5 −4 −3 −2 −1 0 1 2 3 4 5 **19.** $[-3, 0) \cup (0, \infty)$, −5 −4 −3 −2 −1 0 1 2 3 4 5 **20.** $(-5, 1]$, −5 −4 −3 −2 −1 0 1 2 3 4 5 **21. a.** $x^2 + x + 1, (-\infty, \infty)$ **b.** $x^2 - x + 3, (-\infty, \infty)$ **c.** $x^3 - x^2 + 2x - 2, (-\infty, \infty)$ **d.** $\frac{x^2 + 2}{x - 1}, (-\infty, 1) \cup (1, \infty)$ **22. a.** $x^2 + x - 9, (-\infty, \infty)$ **b.** $-x^2 + x - 1, (-\infty, \infty)$ **c.** $x^3 - 5x^2 - 4x + 20, (-\infty, \infty)$ **d.** $\frac{x - 5}{x^2 - 4}, (-\infty, -2) \cup (-2, 2) \cup (2, \infty)$ **23. a.** $x^2 - 5x - 11, (-\infty, \infty)$ **b.** $-x^2 + 7x + 21, (-\infty, \infty)$ **c.** $x^3 - x^2 - 46x - 80, (-\infty, \infty)$ **d.** $\frac{x + 5}{x^2 - 6x - 16}, (-\infty, -2) \cup (-2, 8) \cup (8, \infty)$ **24. a.** $\sqrt{x} + x + 6, [0, \infty)$ **b.** $\sqrt{x} - x - 6, [0, \infty)$ **c.** $x\sqrt{x} + 6x, [0, \infty)$ **d.** $\frac{\sqrt{x}}{x + 6}, [0, \infty)$ **25. a.** $\frac{1}{x} + \sqrt{x - 1}, [1, \infty)$ **b.** $\frac{1}{x} - \sqrt{x - 1}, [1, \infty)$ **c.** $\frac{\sqrt{x - 1}}{x}, [1, \infty)$ **d.** $\frac{1}{x\sqrt{x - 1}}, (1, \infty)$ **26. a.** $\frac{7x^2 + 18x + 11}{3x^2 + 14x + 15}, (-\infty, -3) \cup \left(-3, -\frac{5}{3}\right) \cup \left(-\frac{5}{3}, \infty\right)$ **b.** $\frac{5x^2 + 8x - 1}{3x^2 + 14x + 15}, (-\infty, -3) \cup \left(-3, -\frac{5}{3}\right) \cup \left(-\frac{5}{3}, \infty\right)$ **c.** $\frac{2x^2 + 5x + 2}{3x^2 + 14x + 15}, (-\infty, -3) \cup \left(-3, -\frac{5}{3}\right) \cup \left(-\frac{5}{3}, \infty\right)$ **d.** $\frac{6x^2 + 13x + 5}{x^2 + 5x + 6}, (-\infty, -3) \cup (-3, -2) \cup \left(-2, -\frac{5}{3}\right) \cup \left(-\frac{5}{3}, \infty\right)$ **27. a.** $\frac{2x^2 - 3x - 4}{x^3 - 3x^2 - 4x + 12}, (-\infty, -2) \cup (-2, 2) \cup (2, 3) \cup (3, \infty)$ **b.** $\frac{4 - 3x}{x^3 - 3x^2 - 4x + 12}, (-\infty, -2) \cup (-2, 2) \cup (2, 3) \cup (3, \infty)$ **c.** $\frac{x}{x^3 - 3x^2 - 4x + 12}, (-\infty, -2) \cup (-2, 2) \cup (2, 3) \cup (3, \infty)$ **d.** $\frac{x^2 - 3x}{x^2 - 4}, (-\infty, -2) \cup (-2, 2) \cup (2, 3) \cup (3, \infty)$ **28. a.** $\frac{2x^3 - 27x + 34}{x^4 - x^3 - 22x^2 + 16x + 96}, (-\infty, -4) \cup (-4, -2) \cup (-2, 3) \cup (3, 4) \cup (4, \infty)$ **b.** $\frac{-8x^2 - 5x + 94}{x^4 - x^3 - 22x^2 + 16x + 96}, (-\infty, -4) \cup (-4, -2) \cup (-2, 3) \cup (3, 4) \cup (4, \infty)$ **c.** $\frac{x + 5}{x^3 + 3x^2 - 10x - 24}$, $(-\infty, -4) \cup (-4, -2) \cup (-2, 3) \cup (3, 4) \cup (4, \infty)$ **d.** $\frac{x^3 - 4x^2 - 16x + 64}{x^3 + 4x^2 - 11x - 30}$, $(-\infty, -5) \cup (-5, -4) \cup (-4, -2) \cup (-2, 3) \cup (3, 4) \cup (4, \infty)$ **29. a.** $\sqrt{x + 2} + \sqrt{2 - x}, [-2, 2]$ **b.** $\sqrt{x + 2} - \sqrt{2 - x}, [-2, 2]$ **c.** $\sqrt{4 - x^2}, [-2, 2]$ **d.** $\frac{\sqrt{x + 2}}{\sqrt{2 - x}}, [-2, 2)$ **30. a.** $\sqrt{\frac{x + 1}{x + 2}} + \sqrt{x - 3}, [3, \infty)$ **b.** $\sqrt{\frac{x + 1}{x + 2}} - \sqrt{x - 3}, [3, \infty)$ **c.** $\sqrt{\frac{(x + 1)(x - 3)}{x + 2}}, [3, \infty)$ **d.** $\sqrt{\frac{x + 1}{(x + 2)(x - 3)}}, (3, \infty)$ **31. a.** $\sqrt[3]{x + 1} + \sqrt[3]{\frac{x - 1}{x + 2}}, (-\infty, -2) \cup (-2, \infty)$ **b.** $\sqrt[3]{x + 1} - \sqrt[3]{\frac{x^2 - 1}{x + 2}}, (-\infty, -2) \cup (-2, \infty)$ **c.** $\sqrt[3]{\frac{x^2 - 1}{x + 2}}, (-\infty, -2) \cup (-2, \infty)$ **d.** $\sqrt[3]{\frac{(x + 1)(x + 2)}{x - 1}}, (-\infty, -2) \cup (-2, 1) \cup (1, \infty)$ **32. a.** $\sqrt{7 - x} + \sqrt{x^2 - 2x - 15}, (-\infty, -3] \cup [5, 7]$ **b.** $\sqrt{7 - x} - \sqrt{x^2 - 2x - 15}, (-\infty, -3] \cup [5, 7]$ **c.** $\sqrt{-x^3 + 9x^2 + x - 105}, (-\infty, -3] \cup [5, 7]$ **d.** $\frac{\sqrt{7 - x}}{\sqrt{x^2 - 2x - 15}}, (-\infty, -3) \cup (5, 7]$ **33.** $\frac{6}{x + 1} + 1$ **34.** $\frac{2}{3x + 2}$ **35.** $3\sqrt{x + 3} + 1$ **36.** $\frac{2}{\sqrt{x + 3} + 1}$ **37.** $\sqrt{3x + 4}$ **38.** $\sqrt{\frac{3x + 5}{x + 1}}$ **39.** $9x + 4$ **40.** $\frac{2x + 2}{x + 3}$ **41.** $\sqrt{\sqrt{x + 3} + 3}$ **42.** $\frac{6}{\sqrt{x + 3} + 1} + 1$ or $\frac{7 + \sqrt{x + 3}}{\sqrt{x + 3} + 1}$ **43.** $\frac{2}{3\sqrt{x + 3} + 2}$ **44.** $\sqrt{\frac{4x + 10}{x + 1}}$ **45.** 7

46. 10 **47.** $\frac{2}{5}$ **48.** 1 **49.** 2 **50.** $\frac{\sqrt{14}}{2}$ **51.** -5 **52.** $\frac{10}{7}$ **53.** $\sqrt{5}$ **54.** 4 **55.** $\sqrt{6}$ **56.** $\frac{2}{11}$

57. a. -4 **b.** 0 **c.** 2 **d.** -1 **e.** -1 **f.** 0 **58. a.** 2 **b.** -2 **c.** 0 **d.** 1 **e.** 2 **f.** -1

59. $(-\infty,\infty),(-\infty,\infty)$ **60.** $(-\infty,\infty),(-\infty,\infty)$ **61.** $[0,\infty),(-\infty,\infty)$ **62.** $(-\infty,-2)\cup(-2,2)\cup(2,\infty)$, $(-\infty,0)\cup(0,\infty)$ **63.** $(-\infty,1)\cup(1,2)\cup(2,\infty),(-\infty,-1)\cup\left(-1,\frac{1}{2}\right)\cup\left(\frac{1}{2},\infty\right)$ **64.** $(-\infty,1)\cup(1,2)\cup(2,\infty)$, $(-\infty,-3)\cup(-3,3)\cup(3,\infty)$ **65.** $(-\infty,-3]\cup(2,\infty),[0,4)\cup(4,\infty)$ **66.** $(-\infty,0)\cup(0,4],(-\infty,2)\cup\left[\frac{9}{4},\infty\right)$

67. a. x^2+x+1 **b.** x^2-x+3 **c.** x^3-x^2+2x-2 **d.** $\frac{x^2+2}{x-1}$ **68. a.** x^2+x-9 **b.** $-x^2+x-1$ **c.** $x^3-5x^2-4x+20$ **d.** $\frac{x-5}{x^2-4}$ **69. a.** $x^2-5x-11$ **b.** $-x^2+7x+21$ **c.** $x^3-x^2-46x-80$ **d.** $\frac{x+5}{x^2-6x-16}$ **70. a.** $\sqrt{x}+x+6$ **b.** $\sqrt{x}-x-6$ **c.** $x\sqrt{x}+6x$ **d.** $\frac{\sqrt{x}}{x+6}$ **71. a.** $\frac{1}{x}+\sqrt{x-1}$ **b.** $\frac{1}{x}-\sqrt{x-1}$ **c.** $\frac{\sqrt{x-1}}{x}$ **d.** $\frac{1}{x\sqrt{x-1}}$ **72. a.** $\frac{7x^2+18x+11}{3x^2+14x+15}$ **b.** $\frac{5x^2+8x-1}{3x^2+14x+15}$ **c.** $\frac{2x^2+5x+2}{3x^2+14x+15}$ **d.** $\frac{6x^2+13x+5}{x^2+5x+6}$ **73. a.** $\frac{2x^2-3x-4}{x^3-3x^2-4x+12}$ **b.** $\frac{4-3x}{x^3-3x^2-4x+12}$ **c.** $\frac{x}{x^3-3x^2-4x+12}$ **d.** $\frac{x^2-3x}{x^2-4}$ **74. a.** $\frac{2x^3-27x+34}{x^4-3x^2-22x^2+16x+96}$ **b.** $\frac{-8x^2-5x+94}{x^4-x^3-22x^2+16x+96}$ **c.** $\frac{x+5}{x^3+3x^2-10x-24}$ **d.** $\frac{x^3-4x^2-16x+64}{x^3+4x^2-11x-30}$ **75. a.** $\sqrt{x+2}+\sqrt{2-x}$ **b.** $\sqrt{x+2}-\sqrt{2-x}$ **c.** $\sqrt{4-x^2}$ **d.** $\frac{\sqrt{x+2}}{\sqrt{2-x}}$

76. a. $\sqrt{\frac{x+1}{x+2}}+\sqrt{x-3}$ **b.** $\sqrt{\frac{x+1}{x+2}}-\sqrt{x-3}$ **c.** $\sqrt{\frac{(x+1)(x-3)}{x+2}}$ **d.** $\sqrt{\frac{x+1}{(x+2)(x-3)}}$

77. a. $\sqrt[3]{x+1}+\sqrt[3]{\frac{x-1}{x+2}}$ **b.** $\sqrt[3]{x+1}-\sqrt[3]{\frac{x-1}{x+2}}$ **c.** $\sqrt[3]{\frac{x^2-1}{x+2}}$ **d.** $\sqrt[3]{\frac{(x+1)(x+2)}{x-1}}$

78. a. $\sqrt{7-x}+\sqrt{x^2-2x-15}$ **b.** $\sqrt{7-x}-\sqrt{x^2-2x-15}$ **c.** $\sqrt{-x^3+9x^2+x-105}$ **d.** $\frac{\sqrt{7-x}}{\sqrt{x^2-2x-15}}$

3.6 Exercises

SEC-1. $\frac{-x+13}{x+5}$ **SEC-2.** $\frac{47x-29}{6x+40}$ **SEC-3.** $y=4x+20$ **SEC-4.** $y=\frac{7}{3}x-3$ **SEC-5.** $y=\frac{7x+1}{5x+8}$ **SEC-6.** $y=\frac{-4x-5}{8x-4}$ **SEC-7.** $y=(6-x)^3+3$ **SEC-8.** $y=-\sqrt{x-9}+4$

1. one-to-one **2.** not one-to-one **3.** one-to-one **4.** not one-to-one **5.** one-to-one **6.** one-to-one **7.** not one-to-one **8.** not one-to-one **9.** one-to-one **10.** one-to-one **11.** not one-to-one **12.** not one-to-one **13.** one-to-one **14.** one-to-one **15.** not one-to-one **16.** one-to-one **17.** one-to-one **18–24.** final result: $(f\circ g)(x)=(g\circ f)(x)=x$ **25.** , domain: $(-\infty,5]$, range: $[-1,\infty)$ **26.** , domain: $[2,\infty)$, range: $[0,\infty)$ **27.** , domain: $(-\infty,\infty)$, range: $(-\infty,\infty)$ **28.** , domain: $[-5,4]$, range: $[-4,2]$

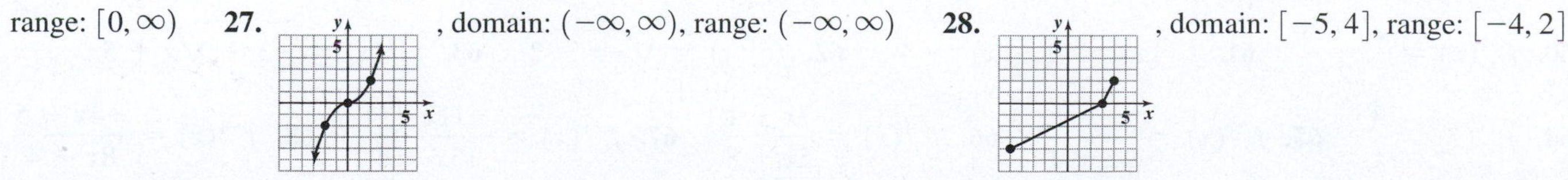

29. , domain: $[-4, 3]$, range: $[-4, 4]$ **30.** 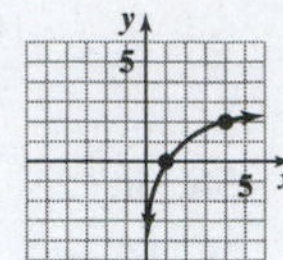 , domain: $(0, \infty)$, range: $(-\infty, \infty)$

31. a. $(-3, 4]$ **b.** $[-4, 3)$ **c.** -1 **d.** -1 **e.** -2 **f.** 0 **g.** -3 **h.** -4

32. $f^{-1}(x) = 3x + 15$, domain f = domain f^{-1} = range f = range $f^{-1} = (-\infty, \infty)$

33. $f^{-1}(x) = \dfrac{7x}{3} - 3$, domain f = domain f^{-1} = range f = range $f^{-1} = (-\infty, \infty)$ **34.** $f^{-1}(x) = \sqrt[3]{x - 5}$

Domain of f: $(-\infty, \infty)$ Domain of f^{-1}: $(-\infty, \infty)$ Range of f: $(-\infty, \infty)$ Range of f^{-1}: $(-\infty, \infty)$

35. $f^{-1}(x) = 4 - x^3$ Domain of f:$(-\infty, \infty)$ Domain of f^{-1}:$(-\infty, \infty)$ Range of f:$(-\infty, \infty)$ Range of f^{-1}:$(-\infty, \infty)$

36. $f^{-1}(x) = \dfrac{x^3 + 3}{2}$, domain f = domain f^{-1} = range f = range $f^{-1} = (-\infty, \infty)$

37. $f^{-1}(x) = (1 - x)^5 - 4$, domain f = domain f^{-1} = range f = range $f^{-1} = (-\infty, \infty)$

38. $f^{-1}(x) = \sqrt{-x - 2}$, domain f = range $f^{-1} = [0, \infty)$, domain f^{-1} = range $f = (-\infty, -2]$

39. $f^{-1}(x) = -3 - \sqrt{x + 5}$, domain f = range $f^{-1} = (-\infty, -3]$, domain f^{-1} = range $f = [-5, \infty)$

40. $f^{-1}(x) = \dfrac{3}{x}$, domain f = domain f^{-1} = range f = range $f^{-1} = (-\infty, 0) \cup (0, \infty)$

41. $f^{-1}(x) = \dfrac{1}{2x + 1}$, domain f = range $f^{-1} = (-\infty, 0) \cup (0, \infty)$, domain f^{-1} = range $f = \left(-\infty, -\dfrac{1}{2}\right) \cup \left(-\dfrac{1}{2}, \infty\right)$

42. $f^{-1}(x) = \dfrac{7x + 1}{5x + 8}$, domain f = range $f^{-1} = \left(-\infty, \dfrac{7}{5}\right) \cup \left(\dfrac{7}{5}, \infty\right)$, domain f^{-1} = range $f = \left(-\infty, -\dfrac{8}{5}\right) \cup \left(-\dfrac{8}{5}, \infty\right)$

43. $f^{-1}(x) = \dfrac{-11x + 1}{13x + 4}$, domain f = range $f^{-1} = \left(-\infty, -\dfrac{11}{13}\right) \cup \left(-\dfrac{11}{13}, \infty\right)$, domain f^{-1} = range f = $\left(-\infty, -\dfrac{4}{13}\right) \cup \left(-\dfrac{4}{13}, \infty\right)$ **44.** $f^{-1}(x) = \dfrac{-4x - 5}{8x - 4}$, domain f = range $f^{-1} = \left(-\infty, -\dfrac{1}{2}\right) \cup \left(-\dfrac{1}{2}, \infty\right)$, domain f^{-1} = range $f = \left(-\infty, \dfrac{1}{2}\right) \cup \left(\dfrac{1}{2}, \infty\right)$ **45.** $f^{-1}(x) = -2 + \sqrt{5 + x}$, domain f = range $f^{-1} = [-2, \infty)$, domain f^{-1} = range $f = [-5, \infty)$ **46.** $f^{-1}(x) = \dfrac{3}{2} - \dfrac{\sqrt{25 - 2x}}{2}$, domain f = range $f^{-1} = \left(-\infty, \dfrac{3}{2}\right]$, domain f^{-1} = range $f = \left(-\infty, \dfrac{25}{2}\right]$ **47.**

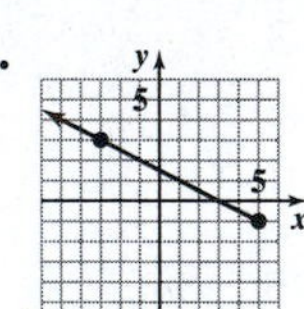

48.

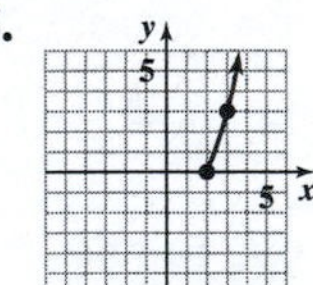

49.

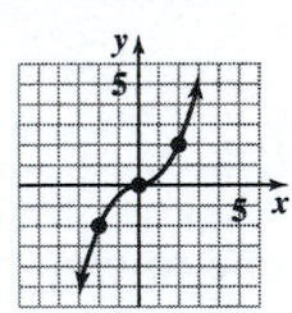

50.

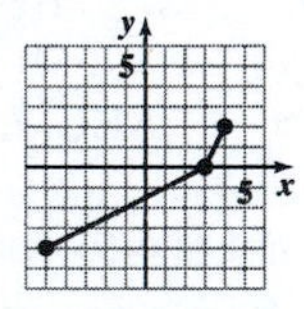

51.

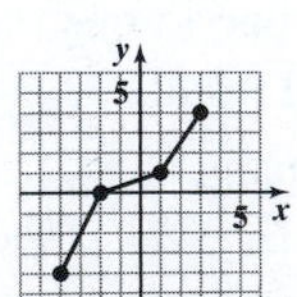

52.

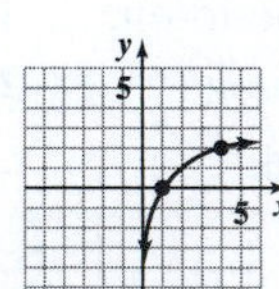

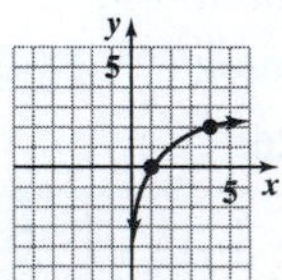

53. a. $(-3, 4]$ **b.** $[-4, 3)$ **54.** -1 **55. a.** -1 **b.** -2 **c.** 0 **d.** -3 **e.** -4 **56.** $f^{-1}(x) = 3x + 15$ **57.** $f^{-1}(x) = \dfrac{7x}{3} - 3$ **58.** $f^{-1}(x) = \sqrt[3]{x - 5}$ **59.** $f^{-1}(x) = 4 - x^3$

60. $f^{-1}(x) = \dfrac{x^3 + 3}{2}$ **61.** $f^{-1}(x) = (1 - x)^5 - 4$ **62.** $f^{-1}(x) = \sqrt{-x - 2}$ **63.** $f^{-1}(x) = -3 - \sqrt{x + 5}$

64. $f^{-1}(x) = \dfrac{3}{x}$ **65.** $f^{-1}(x) = \dfrac{1}{2x + 1}$ **66.** $f^{-1}(x) = \dfrac{7x + 1}{5x + 8}$ **67.** $f^{-1}(x) = \dfrac{-11x + 1}{13x + 4}$ **68.** $f^{-1}(x) = \dfrac{-4x - 5}{8x - 4}$

69. $f^{-1}(x) = -2 + \sqrt{5 + x}$ **70.** $f^{-1}(x) = \dfrac{3}{2} - \dfrac{\sqrt{25 - 2x}}{2}$

Review Exercises

1. sample answers: $(-1, 5), (2, 0), (4, -1)$ **2. a.** $f(-2) = 2$ **b.** $f(7) = \frac{7}{8}$ **c.** $f(a) = \frac{a}{a + 1}$

3. $2x + h - 2$ **4.** polynomial function; Domain: $(-\infty, \infty)$ **5.** rational functions; Domain: $(-\infty, -7) \cup (-7, 6) \cup (6, \infty)$ **6.** x-intercepts: $-4, 3$ no y-intercept **7.** D: $(-4, 5]$ R: $(-5, 5]$

8. increasing: $(-3, 0)$ constant: $(0, 4)$ decreasing: $(-\infty, -3)$ and $(4, \infty)$ **9.** odd **10. a.** Domain: $(-8, 9]$

b. Range: $[-5, 3]$ **c.** increasing: $(-8, -6)$ and $(-2, 1)$ and $(4, 7)$ decreasing: $(-6, -2)$ and $(7, 9)$ constant: $(1, 4)$

d. attains a relative minimum at an x-value of -2; relative minimum is -5 **e.** attains a relative maximum at x-values of -6 and 7; relative maxima are 3 and 2 **f.** real zeros of f are $-\frac{15}{2}, -4, 5, 9$ **g.** y-intercept: -3 **h.** $\left(-8, -\frac{15}{2}\right] \cup [-4, 5] \cup \{9\}$

11. $f(x) = x^3$ for $x \geq -2$ **12. a.** Domain: $(-1, \infty)$ **b.** $f(x) = x^2$ for $x > -1$

13. a. $f(-2) = 3, f(0) = 0, f(2) = 1$ **b.** **c.** Domain: $(-\infty, \infty)$ **d.** Range: $[0, 2) \cup \{3\}$

14. a. \$25.15

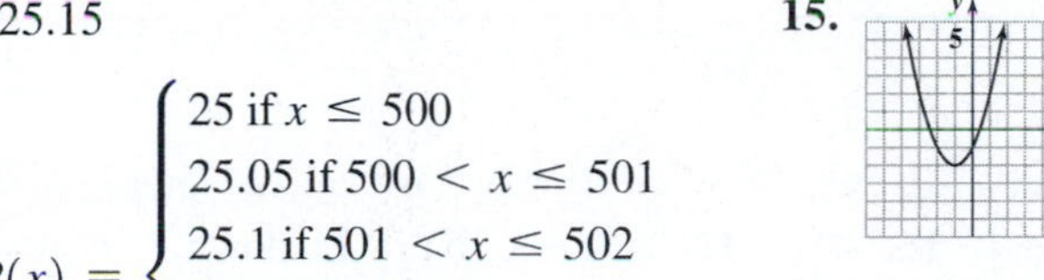

b. $P(x) = \begin{cases} 25 \text{ if } x \leq 500 \\ 25.05 \text{ if } 500 < x \leq 501 \\ 25.1 \text{ if } 501 < x \leq 502 \\ 25.15 \text{ if } 502 < x \leq 503 \\ 25.2 \text{ if } 503 < x \leq 504 \\ 25.25 \text{ if } 504 < x \leq 505 \end{cases}$

c.

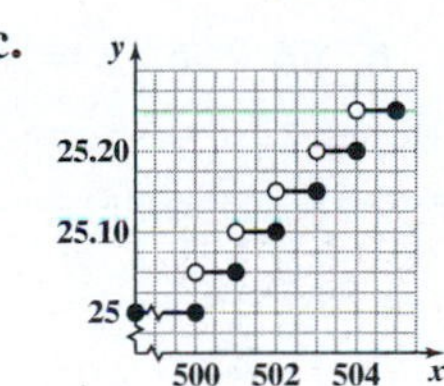

15. **16.** $f(x) = (x - 1)^3 - 2$ **17.** See 3.4 Exercise 82

18. See 3.4 Exercise 85 **19.** $(f + g)(3) = 9$ **20.** See 3.5 Exercise 27 **21.** 10 **22.** See 3.5 Exercise 66 **23.** Yes, it is a one-to-one function

24. **25.** $f^{-1}(x) = \frac{7x + 1}{5x + 8}$

CHAPTER 4

4.1 Exercises

SCE-1. 25 **SCE-2.** $\frac{25}{4}$ **SCE-3.** $\frac{25}{36}$

1. up **2.** down **3.** up **4.** down **5. a.** $(2, -4)$ **b.** up **c.** $x = 2$ **d.** 0, 4 **e.** 0 **f.**

g. domain: $(-\infty, \infty)$, range: $[-4, \infty)$ **6. a.** $(-1, -9)$ **b.** down **c.** $x = -1$ **d.** none **e.** -10 **f.**

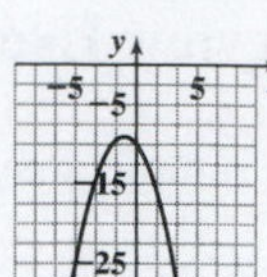

g. domain: $(-\infty, \infty)$, range: $(-\infty, -9]$ **7. a.** $(3, 2)$ **b.** down **c.** $x = 3$ **d.** 2, 4 **e.** -16 **f.**

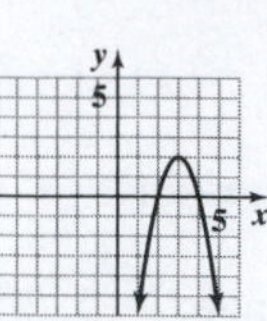

g. domain: $(-\infty, \infty)$, range: $(-\infty, 2]$ **8. a.** $(-2, 2)$ **b.** up **c.** $x = -2$ **d.** none **e.** 4 **f.**

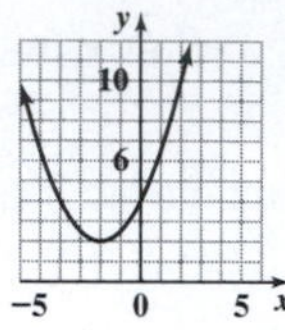

g. domain: $(-\infty, \infty)$, range: $[2, \infty)$ **9. a.** $(4, 2)$ **b.** down **c.** $x = 4$ **d.** $4 - 2\sqrt{2}, 4 + 2\sqrt{2}$ **e.** -2

f. **g.** domain: $(-\infty, \infty)$, range: $(-\infty, 2]$ **10. a.** $\left(-\dfrac{1}{3}, -4\right)$ **b.** up **c.** $x = -\dfrac{1}{3}$ **d.** $-\dfrac{1}{3} - \dfrac{2\sqrt{3}}{3}$,

$-\dfrac{1}{3} + \dfrac{2\sqrt{3}}{3}$ **e.** $-3\dfrac{2}{3}$ **f.** **g.** domain: $(-\infty, \infty)$, range: $[-4, \infty)$ **11. a.** $(4, -16)$ **b.** up **c.** $x = 4$

d. 0, 8 **e.** 0 **f.**

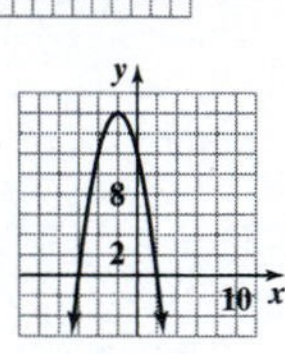

g. domain: $(-\infty, \infty)$, range: $[-16, \infty)$ **12. a.** $(-2, 16)$ **b.** down **c.** $x = -2$

d. $-6, 2$ **e.** 12 **f.** **g.** domain: $(-\infty, \infty)$, range: $(-\infty, 16]$ **13. a.** $(-1, -7)$ **b.** up **c.** $x = -1$

d. $-1 - \dfrac{\sqrt{21}}{3}, -1 + \dfrac{\sqrt{21}}{3}$ **e.** -4 **f.** **g.** domain: $(-\infty, \infty)$, range: $[-7, \infty)$ **14. a.** $\left(\dfrac{5}{4}, -\dfrac{49}{8}\right)$

b. up **c.** $x = \dfrac{5}{4}$ **d.** $-\dfrac{1}{2}, 3$ **e.** -3 **f.** **g.** domain: $(-\infty, \infty)$, range: $\left[-\dfrac{49}{8}, \infty\right)$

15. a. $(1, -5)$ **b.** down **c.** $x = 1$ **d.** none **e.** -6 **f.**

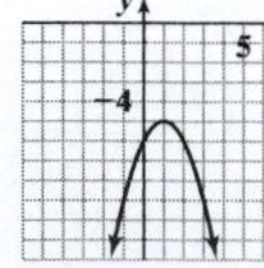

g. domain: $(-\infty, \infty)$, range: $(-\infty, -5]$

16. a. $(-6, -17)$ **b.** up **c.** $x = -6$ **d.** $-6 - \sqrt{34}, -6 + \sqrt{34}$ **e.** 1 **f.**

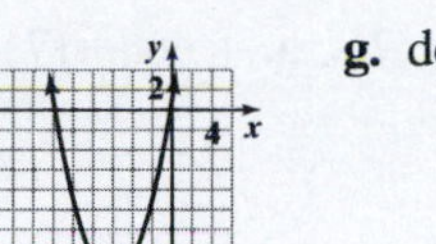

g. domain: $(-\infty, \infty)$, range: $[-17, \infty)$ **17. a.** $(-3, 8)$ **b.** down **c.** $x = -3$ **d.** $-3 - 2\sqrt{6}, -3 + 2\sqrt{6}$ **e.** 5 **f.**

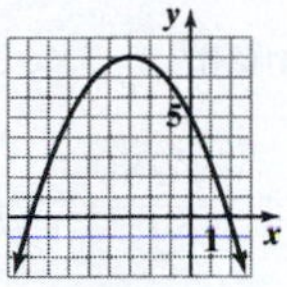

g. domain: $(-\infty, \infty)$, range: $(-\infty, 8]$ **18. a.** $\left(\frac{7}{6}, \frac{109}{12}\right)$ **b.** down **c.** $x = \frac{7}{6}$ **d.** $\frac{7}{6} - \frac{\sqrt{109}}{6}, \frac{7}{6} + \frac{\sqrt{109}}{6}$ **e.** 5 **f.**

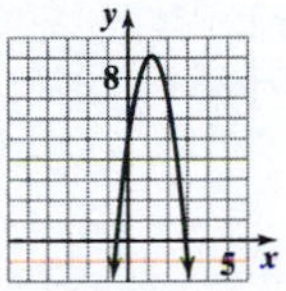

g. domain: $(-\infty, \infty)$, range: $\left(-\infty, \frac{109}{12}\right]$ **19. a.** $\left(-\frac{4}{3}, -\frac{25}{9}\right)$ **b.** up **c.** $x = -\frac{4}{3}$ **d.** $-3, \frac{1}{3}$

e. -1 **f.**

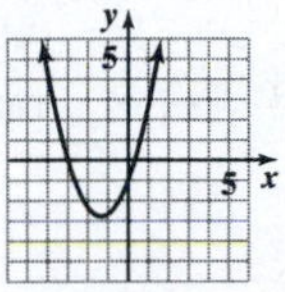

g. domain: $(-\infty, \infty)$, range: $\left[-\frac{25}{9}, \infty\right)$ **20. a.** $(12, 35)$ **b.** down **c.** $x = 12$

d. $12 - 2\sqrt{35}, 12 + 2\sqrt{35}$ **e.** -1 **f.**

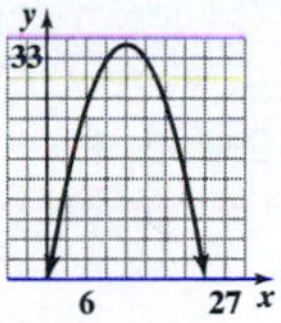

g. domain: $(-\infty, \infty)$, range: $(-\infty, 35]$ **21. a.** $(4, -16)$

b. up **c.** $x = 4$ **d.** 0, 8 **e.** 0 **f.**

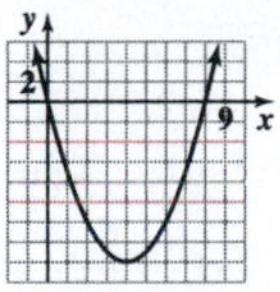

g. domain: $(-\infty, \infty)$, range: $[-16, \infty)$ **22. a.** $(-2, 12)$

b. down **c.** $x = -2$ **d.** $-2 - 2\sqrt{3}, -2 + 2\sqrt{3}$ **e.** 8 **f.**

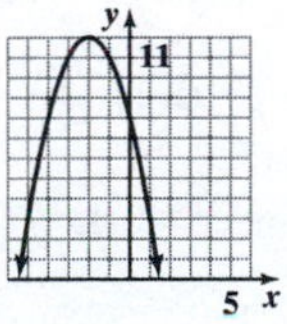

g. domain: $(-\infty, \infty)$, range: $(-\infty, 12]$

23. a. $(-1, -7)$ **b.** up **c.** $x = -1$ **d.** $-1 - \frac{\sqrt{21}}{3}, -1 + \frac{\sqrt{21}}{3}$ **e.** -4 **f.**

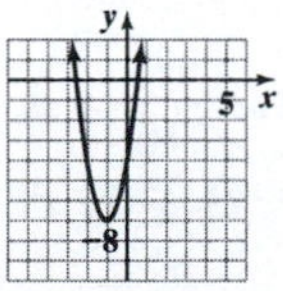

g. domain: $(-\infty, \infty)$, range: $[-7, \infty)$ **24. a.** $\left(\frac{5}{4}, -\frac{49}{8}\right)$ **b.** up **c.** $x = \frac{5}{4}$ **d.** $-\frac{1}{2}, 3$ **e.** -3 **f.**

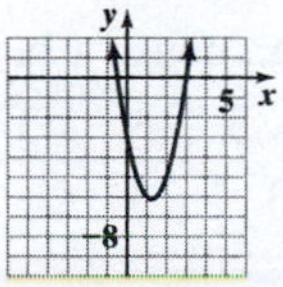

g. domain: $(-\infty, \infty)$, range: $\left[-\frac{49}{8}, \infty\right)$ **25. a.** $(1, -5)$ **b.** down **c.** $x = 1$ **d.** none **e.** -6 **f.**

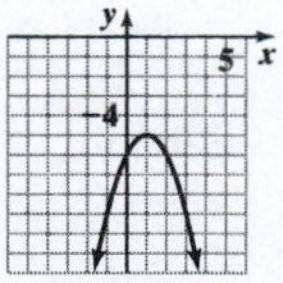

g. domain: $(-\infty, \infty)$,

range: $(-\infty, -5]$ **26. a.** $(-6, -17)$ **b.** up **c.** $x = -6$ **d.** $-6 - \sqrt{34}, -6 + \sqrt{34}$ **e.** 1 **f.**

g. domain: $(-\infty, \infty)$, range: $[-17, \infty)$ **27. a.** $\left(-\frac{27}{2}, \frac{263}{4}\right)$ **b.** down **c.** $x = -\frac{27}{2}$ **d.** $-\frac{27}{2} - \frac{\sqrt{789}}{2}$, $-\frac{27}{2} + \frac{\sqrt{789}}{2}$ **e.** 5 **f.**

g. domain: $(-\infty, \infty)$, range: $\left(-\infty, \frac{263}{4}\right]$ **28. a.** $\left(\frac{7}{6}, \frac{109}{12}\right)$ **b.** down

c. $x = \frac{7}{6}$ **d.** $\frac{7}{6} - \frac{\sqrt{109}}{6}, \frac{7}{6} + \frac{\sqrt{109}}{6}$ **e.** 5 **f.**

g. domain: $(-\infty, \infty)$, range: $\left(-\infty, \frac{109}{12}\right]$

29. a. $\left(-\frac{4}{3}, -\frac{25}{9}\right)$ **b.** up **c.** $x = -\frac{4}{3}$ **d.** $-3, \frac{1}{3}$ **e.** -1 **f.**

g. domain: $(-\infty, \infty)$, range: $\left[-\frac{25}{9}, \infty\right)$ **30. a.** $(12, 35)$ **b.** down **c.** $x = 12$ **d.** $12 - 2\sqrt{35}, 12 + 2\sqrt{35}$ **e.** -1 **f.**

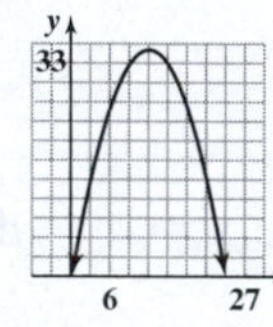

g. domain: $(-\infty, \infty)$, range: $(-\infty, 35]$ **31. a.** positive **b.** $(h, k) = (1, -2)$ **c.** 2 **d.** $f(x) = 2(x - 1)^2 - 2$
e. $f(x) = 2x^2 - 4x$ **32. a.** positive **b.** $(h, k) = (3, 0)$ **c.** $\frac{2}{9}$ **d.** $f(x) = \frac{2}{9}(x - 3)^2$ **e.** $f(x) = \frac{2}{9}x^2 - \frac{4}{3}x + 2$
33. a. positive **b.** $(h, k) = (2, 3)$ **c.** 3 **d.** $f(x) = 3(x - 2)^2 + 3$ **e.** $f(x) = 3x^2 - 12x + 15$
34. a. negative **b.** $(h, k) = (-2, 3)$ **c.** $-\frac{3}{4}$ **d.** $f(x) = -\frac{3}{4}(x + 2)^2 + 3$ **e.** $f(x) = -\frac{3}{4}x^2 - 3x$ **35. a.** negative
b. $(h, k) = (1, 1)$ **c.** $-\frac{1}{3}$ **d.** $f(x) = -\frac{1}{3}(x - 1)^2 + 1$ **e.** $f(x) = -\frac{1}{3}x^2 + \frac{2}{3}x + \frac{2}{3}$ **36. a.** negative
b. $(h, k) = (-1, -2)$ **c.** $-\frac{1}{16}$ **d.** $f(x) = -\frac{1}{16}(x + 1)^2 - 2$ **e.** $f(x) = -\frac{1}{16}x^2 - \frac{x}{8} - \frac{33}{16}$ **37. a.** $(2, -4)$ **b.** up
c. $x = 2$ **38. a.** $(-1, -9)$ **b.** down **c.** $x = -1$ **39. a.** $(3, 2)$ **b.** down **c.** $x = 3$ **40. a.** $(-2, 2)$
b. up **c.** $x = -2$ **41. a.** $(4, 2)$ **b.** down **c.** $x = 4$ **42. a.** $\left(-\frac{1}{3}, -4\right)$ **b.** up **c.** $x = -\frac{1}{3}$
43. a. 0, 4 **b.** 0 **44. a.** none **b.** -10 **45. a.** 2, 4 **b.** -16 **46. a.** none **b.** 4 **47. a.** $4 - 2\sqrt{2}, 4 + 2\sqrt{2}$
b. -2 **48. a.** $-\frac{1}{3} - \frac{2\sqrt{3}}{3}, -\frac{1}{3} + \frac{2\sqrt{3}}{3}$ **b.** $-3\frac{2}{3}$ **49. a.** **b.** domain: $(-\infty, \infty)$, range: $[-4, \infty)$

50. a. **b.** domain: $(-\infty, \infty)$, range: $(-\infty, -9]$ **51. a.** **b.** domain: $(-\infty, \infty)$, range: $(-\infty, -2]$ **52. a.**

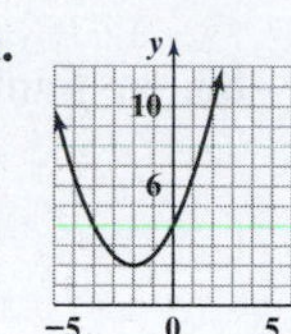

b. domain: $(-\infty, \infty)$, range: $[2, \infty)$ **53. a.**

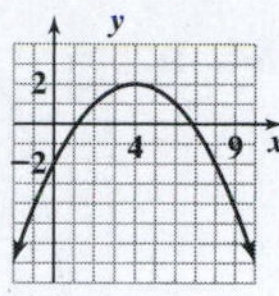

b. domain: $(-\infty, \infty)$, range: $(-\infty, 2]$ **54. a.**

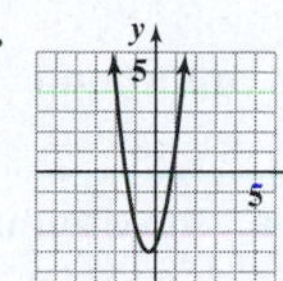

b. domain: $(-\infty, \infty)$, range: $[-4, \infty)$ **55. a.** $(-1, -7)$

b. up **c.** $x = -1$ **56. a.** $(1, -5)$ **b.** down **c.** $x = 1$ **57. a.** $(-6, -17)$ **b.** up **c.** $x = -6$ **58. a.** $\left(\frac{7}{6}, \frac{109}{12}\right)$

b. down **c.** $x = \frac{7}{6}$ **59. a.** $(12, 35)$ **b.** down **c.** $x = 12$ **60. a.** $-1 - \frac{\sqrt{21}}{3}, -1 + \frac{\sqrt{21}}{3}$ **b.** -4

61. a. none **b.** -6 **62. a.** $-6 - \sqrt{34}, -6 + \sqrt{34}$ **b.** 1 **63. a.** $\frac{7}{6} - \frac{\sqrt{109}}{6}, \frac{7}{6} + \frac{\sqrt{109}}{6}$ **b.** 5

64. a. $12 - 2\sqrt{35}, 12 + 2\sqrt{35}$ **b.** -1 **65. a.** **b.** domain: $(-\infty, \infty)$, range: $[-7, \infty)$

66. a.

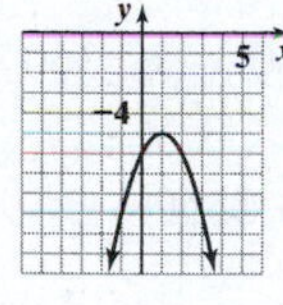

b. domain: $(-\infty, \infty)$, range: $(-\infty, -5]$ **67. a.** **b.** domain: $(-\infty, \infty)$,

range: $[-17, \infty)$ **68. a.** **b.** domain: $(-\infty, \infty)$, range: $\left(-\infty, \frac{109}{12}\right]$ **69. a.**

b. domain: $(-\infty, \infty)$, range: $(-\infty, 35]$ **70. a.** $(-1, -7)$ **b.** up **c.** $x = -1$ **71. a.** $(1, -5)$ **b.** down **c.** $x = 1$

72. a. $(-6, -17)$ **b.** up **c.** $x = -6$ **73. a.** $\left(\frac{7}{6}, \frac{109}{12}\right)$ **b.** down **c.** $x = \frac{7}{6}$ **74. a.** $(12, 35)$ **b.** down

c. $x = 12$ **75. a.** $-1 - \frac{\sqrt{21}}{3}, -1 + \frac{\sqrt{21}}{3}$ **b.** -4 **76. a.** none **b.** -6 **77. a.** $-6 - \sqrt{34}, -6 + \sqrt{34}$

b. 1 **78. a.** $\frac{7}{6} - \frac{\sqrt{109}}{6}, \frac{7}{6} + \frac{\sqrt{109}}{6}$ **b.** 5 **79. a.** $12 - 2\sqrt{35}, 12 + 2\sqrt{35}$ **b.** -1 **80. a.**

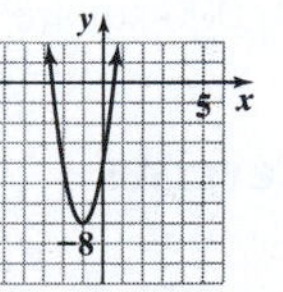

b. domain: $(-\infty, \infty)$, range: $[-7, \infty)$ **81. a.** **b.** domain: $(-\infty, \infty)$, range: $(-\infty, -5]$

82. a.

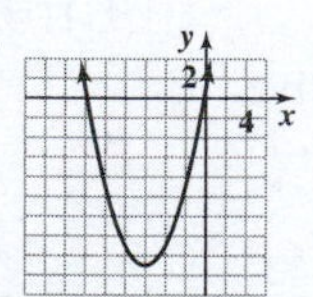

b. domain: $(-\infty, \infty)$, range: $[-17, \infty)$ **83. a.** **b.** domain: $(-\infty, \infty)$, range: $\left(-\infty, \frac{109}{12}\right]$ **84. a.** 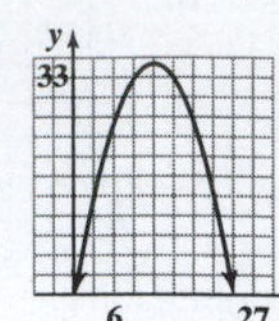 **b.** domain: $(-\infty, \infty)$, range: $(-\infty, 35]$ **85. a.** minimum **b.** -16

86. a. maximum **b.** 12 **87. a.** minimum **b.** -7 **88. a.** minimum **b.** $-\frac{49}{8}$ **89. a.** maximum **b.** -5

90. a. minimum **b.** -17 **91. a.** maximum **b.** $\frac{91}{2}$ **92. a.** maximum **b.** $\frac{109}{12}$ **93. a.** minimum

b. $-\frac{25}{9}$ **94. a.** maximum **b.** 35 **95.** $f(x) = 2(x-1)^2 - 2$ **96.** $f(x) = \frac{2}{9}(x-3)^2$ **97.** $f(x) = 3(x-2)^2 + 3$

98. $f(x) = -\frac{3}{4}(x+2)^2 + 3$ **99.** $f(x) = -\frac{1}{3}(x-1)^2 + 1$ **100.** $f(x) = -\frac{1}{16}(x+1)^2 - 2$ **101.** $f(x) = 2x^2 - 4x$

102. $f(x) = \frac{2}{9}x^2 - \frac{4}{3}x + 2$ **103.** $f(x) = 3x^2 - 12x + 15$ **104.** $f(x) = -\frac{3}{4}x^2 - 3x$ **105.** $f(x) = -\frac{1}{3}x^2 + \frac{2}{3}x + \frac{2}{3}$

106. $f(x) = -\frac{1}{16}x^2 - \frac{x}{8} - \frac{33}{16}$

4.2 Exercises

1. 3.5 sec, 61.025 m **2.** 3.75 sec, 78.9063 m **3.** 3.5 sec, 201 ft **4.** 87.8906 sec, 47.9453 ft **5.** 104.775 ft, 90.7256 ft
6. a. $\frac{w_0 v_0}{g}$ **b.** $\frac{v_0^2}{2g} + h_0$ **c.** 128, 65 **7. a.** \$292.50 **b.** 1,000, \$150,000 **c.** \$150 **8. a.** \$71.60 **b.** 360

c. \$36 **9. a.** $R(x) = \frac{-x^2}{20} + 1{,}000x$ **b.** $P(x) = \frac{-x^2}{20} + 900x - 5{,}000$ **c.** 9,000, \$4,045,000 **d.** \$550

10. a. $R(x) = \frac{-x^2}{40} + 8{,}000x$ **b.** $P(x) = \frac{-x^2}{40} + 4{,}000x - 20{,}000$ **c.** 80,000, \$159,980,000 **d.** \$6,000

11. 40 scooters, \$400, \$6,500 **12.** 1 pitcher, \$12, \$7 **13.** \$275, \$226,875 **14.** 1500 trees, \$112,500
15. \$172.50 **16.** \$60 **17.** 40 **18.** \$15, \$1125 **19.** 40 trees **20.** 5, 12,250
21. a. $A(x) = x(1800 - 2x)$ or $A(x) = 1800x - 2x^2$ **b.** 450 feet by 900 feet **c.** 405,000 square feet

22. a. $A(x) = x\left(1000 - \frac{1}{2}x\right)$ or $A(x) = 1000x - \frac{1}{2}x^2$ **b.** 500 feet by 1000 feet **c.** 500,000 square feet

23. 781,250 ft^2 **24. a.** $A(x) = x(4950 - 3x)$ or $A(x) = 4950x - 3x^2$ **b.** 825 feet **c.** 2475 feet
d. 2,041,875 square feet **25. a.** $A(x) = x(5200 - 4x)$ or $A(x) = 5200x - 4x^2$ **b.** 650 feet **c.** 2600 feet
d. 1,690,000 square feet **26.** $\frac{48}{17}$ in.

4.3 Exercises

1. polynomial, 7, $\frac{1}{3}$, -8 **2.** not polynomial **3.** not polynomial **4.** polynomial, 3, $-\frac{5}{11}$, 0 **5.** polynomial, 5, -9, 1
6. polynomial, 0, 12, 12 **7.** **8.** **9.** **10.** **11.**

12. **13.** **14.** **15.**

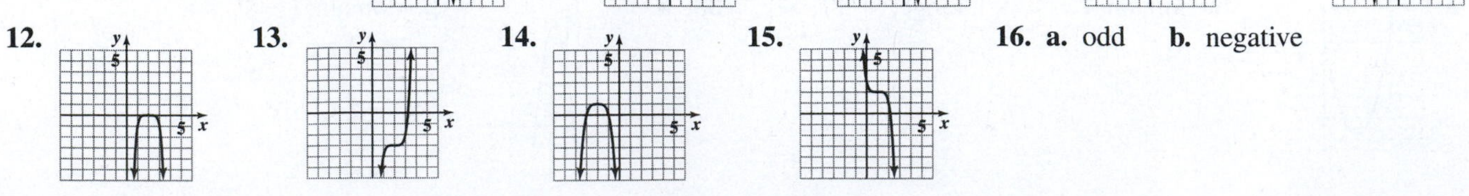

16. a. odd **b.** negative

17. a. even **b.** negative **18. a.** odd **b.** positive **19. a.** even **b.** positive **20. a.** odd **b.** negative **21. a.** even **b.** negative **22.** $-2, 5$ **23.** $-4, -3, 4$ **24.** $-6, \frac{1}{2}$ **25.** $-3, 2, 3$ **26.** $x = 2$, multiplicity 2, touches, $x = -1$, multiplicity 3, crosses **27.** $x = -1$, multiplicity 1, crosses, $x = -3$, multiplicity 2, touches **28.** $x = 0$ and $x = 3$, multiplicity 1, crosses, $x = 4$, multiplicity 4, touches **29.** $x = -\sqrt{2}$ and $x = \sqrt{2}$, multiplicity 2, touches **30.** $x = 2$, multiplicity 4, touches, $x = -2$, multiplicity 3, crosses **31.** $x = \frac{1 - \sqrt{13}}{2}$ and $x = \frac{1 + \sqrt{13}}{2}$, multiplicity 1, crosses, $x = 0$, multiplicity 2, touches **32.** $x = -\sqrt{3}$ and $x = \sqrt{3}$, multiplicity 1, crosses, $x = -1$, multiplicity 3, crosses **33.** $x = \sqrt{7}$, multiplicity 2, touches, $x = -\sqrt{7}$, multiplicity 5, crosses **34.**

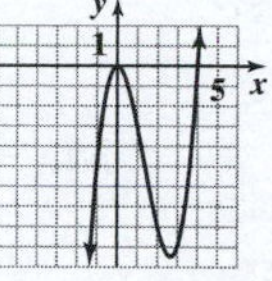

35. 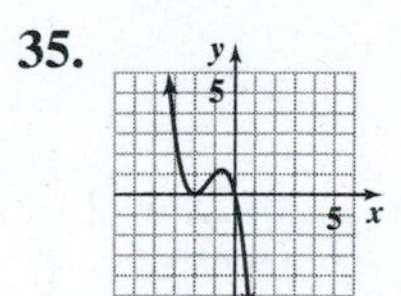 **36.** 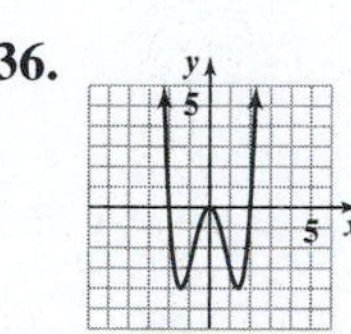 **37.** 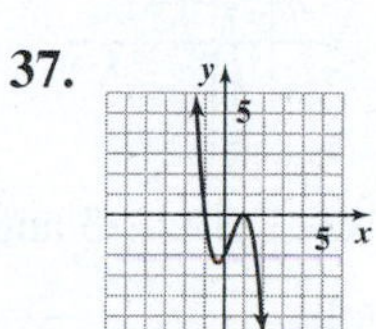 **38.** 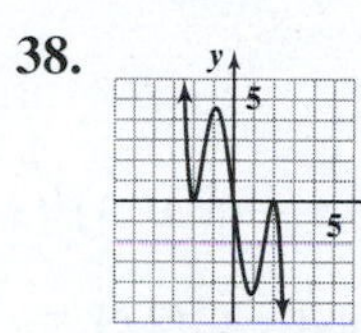 **39.** 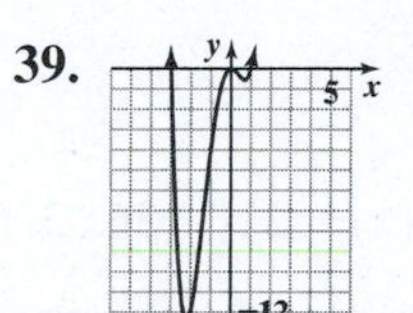 **40.** 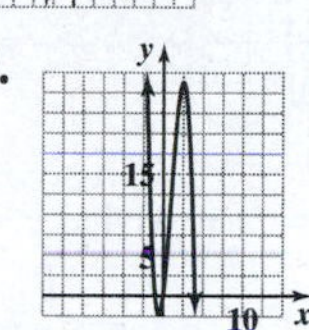

41. 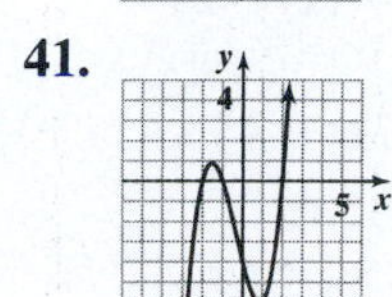 **42.** 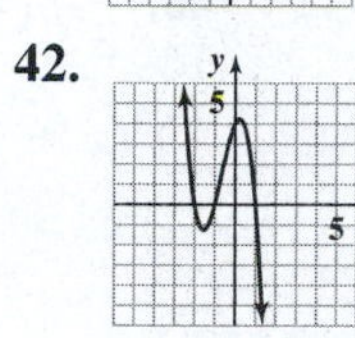 **43.** 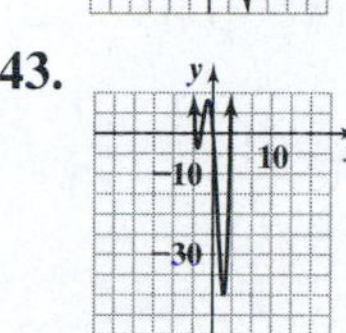 **44.** 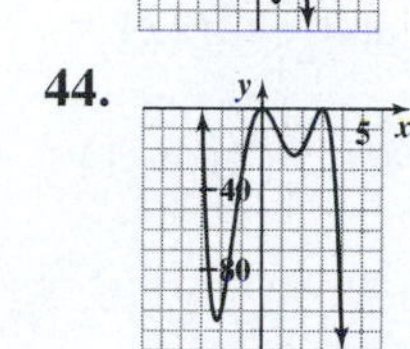 **45. a.** odd **b.** positive **c.** 0 **d.** $x = -2, x = 0$, and $x = 3$ all have odd multiplicity **e.** ii **46. a.** even **b.** negative **c.** 3 **d.** $x = -3$ has even multiplicity, and $x = -1$ and $x = 2$ have odd multiplicity **e.** iv **47. a.** even **b.** positive **c.** -3 **d.** $x = -3$ and $x = 1$ have even multiplicity, and $x = -1$ and $x = 3$ have odd multiplicity **e.** ii **48.** **49.**

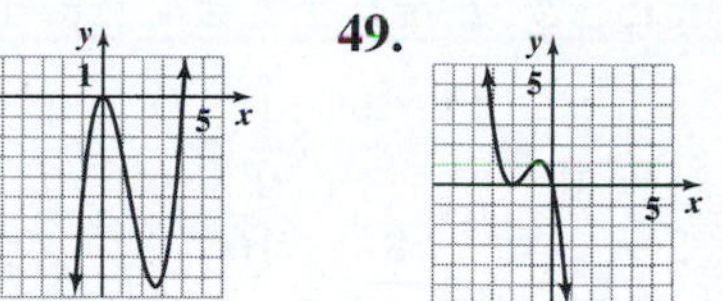

50. 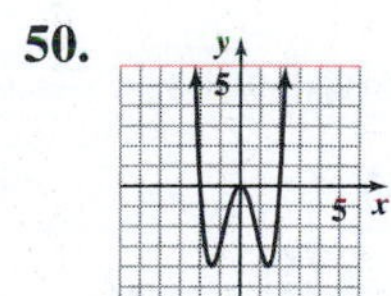 **51.** 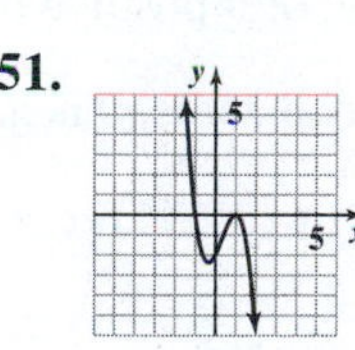 **52.** 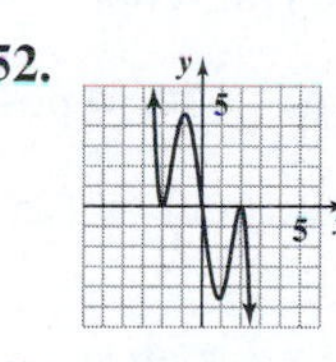 **53.** 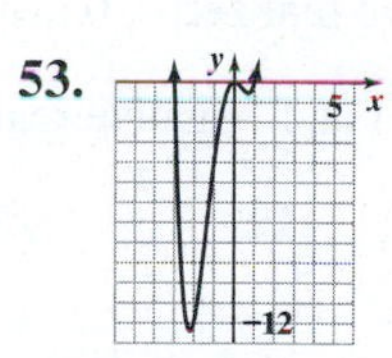 **54.** 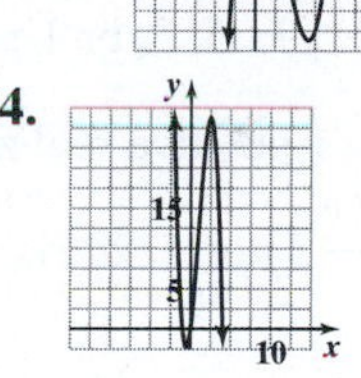 **55.** 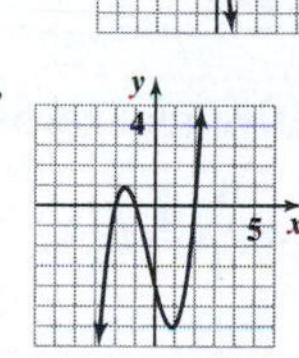

56. 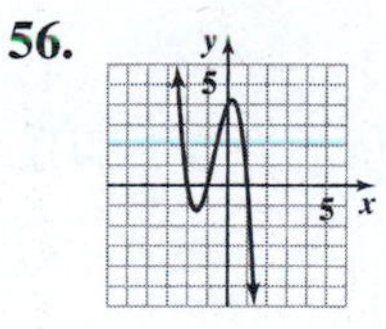 **57.** 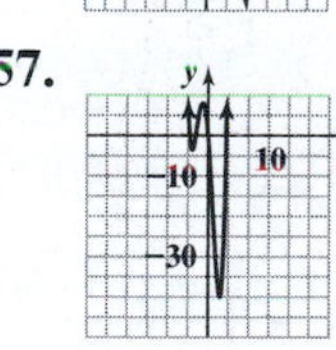 **58.** 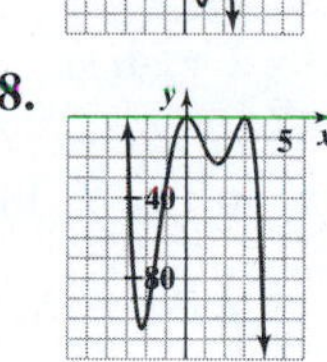 **59.** ii **60.** iv **61.** ii

4.4 Exercises

SCE-1. quotient: $9x^2 - 47$, remainder: $2x + 244$ **SCE-2.** quotient $= 3x^2$, remainder $= -2x^2 + 7x + 6$ **SCE-3.** quotient $= x^2 - x - 9$, remainder $= 18x + 88$

1. $f(x) = (x - 1)(x^2 - 2x - 1) - 6$ **2.** $f(x) = (x + 1)(x^2 + 4x - 7) + 6$ **3.** $f(x) = (x - 2)(6x^2 + 12x + 22) + 47$ **4.** $f(x) = (x + 3)(2x^2 - 7x + 21) - 59$ **5.** $f(x) = (x - 2)(x^3 + 3x^2 + 5x + 9) + 14$ **6.** $f(x) = (x + 1)(-2x^3 + 5x^2 - 6x + 13) - 11$ **7.** $f(x) = (x + 4)(-x^3 + 3x^2 - 12x + 43) - 172$ **8.** $f(x) = (x + 1)(3x^4 - 3x^3 + 3x^2 - 4x + 4) - 6$ **9.** $f(x) = (x - 1)(x^2 + x + 1)$ **10.** $f(x) = (x + 1)(x^4 - x^3 + x^2 - x + 1)$ **11.** -13 **12.** -7 **13.** 234 **14.** 101 **15.** -2 **16.** 0 **17.** factor **18.** not a factor **19.** factor **20.** factor **21.** not a factor **22.** factor **23.** factor **24.** factor **25.** factor **26.** not a factor **27.** -2 and -1, $f(x) = (x - 1)(x + 1)(x + 2)$,

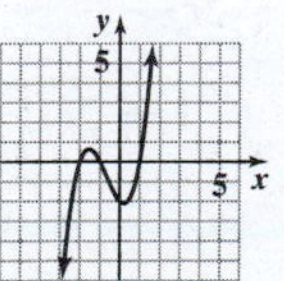

28. $3, f(x) = -(x - 3)^2(x + 1)$,

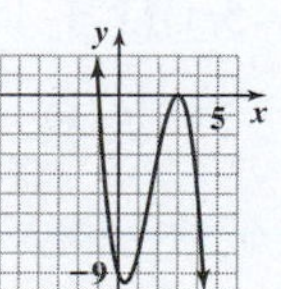

29. -3 and -1, $f(x) = (x + 1)(x + 3)(2x - 1)$,

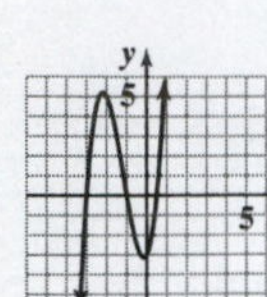

30. -1, $f(x) = \frac{1}{2}(x-2)(x+1)^2$,

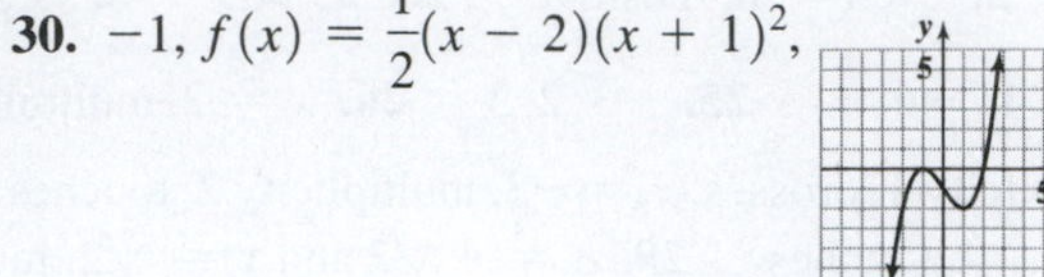

31. -2 and 1, $f(x) = -\frac{1}{3}(x-1)(x+2)(x+3)$,

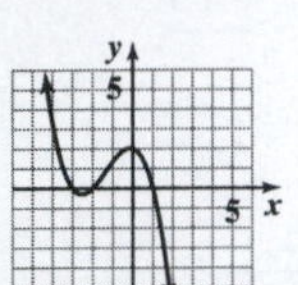

32. $\frac{1}{3}$, $f(x) = (x+2)^2(3x-1)$,

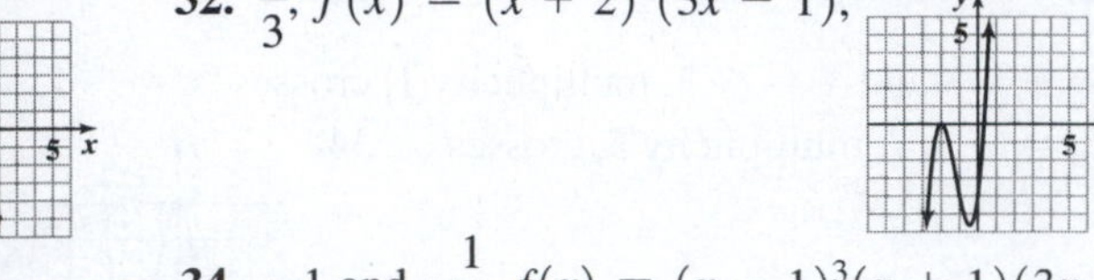

33. $-\frac{1}{4}$ and 1, $f(x) = (x-1)(x+1)^2(4x+1)$,

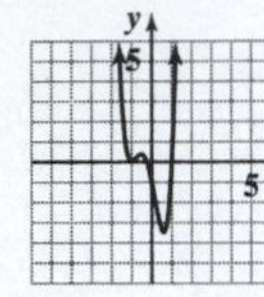

34. -1 and $-\frac{1}{3}$, $f(x) = (x-1)^3(x+1)(3x+1)$,

35. -2 and -1, $f(x) = (x-1)(x+1)(x+2)$ **36.** $3, f(x) = -(x-3)^2(x+1)$ **37.** -3 and $-1, f(x) = (x+1)(x+3)(2x-1)$ **38.** $-1, f(x) = \frac{1}{2}(x-2)(x+1)^2$ **39.** -2 and $-1, f(x) = -\frac{1}{3}(x-1)(x+2)(x+3)$ **40.** $\frac{1}{3}, f(x) = (x+2)^2(3x-1)$ **41.** $-\frac{1}{4}$ and $1, f(x) = (x-1)(x+1)^2(4x+1)$ **42.** -1 and $-\frac{1}{3}, f(x) = (x-1)^3(x+1)(3x+1)$

4.5 Exercises

SCE-1. $f(-x) = 3x^5 + 4x^4 - 3x^3 - 7x^2 - 21$ **SCE-2.** $f(-x) = x^4 - 7x^2 - 11$ **SCE-3.** $f(-x) = 3x^7 - 8x^3 + x$

1. $\pm1, \pm2$ **2.** $\pm1, \pm2, \pm4, \pm8, \pm16$ **3.** $\pm1, \pm2, \pm3, \pm4, \pm6, \pm12$ **4.** $\pm1, \pm2, \pm4, \pm8, \pm\frac{1}{3}, \pm\frac{2}{3}, \pm\frac{4}{3}, \pm\frac{8}{3}$ **5.** $\pm1, \pm2, \pm4, \pm\frac{1}{3}, \pm\frac{2}{3}, \pm\frac{4}{3}, \pm\frac{1}{9}, \pm\frac{2}{9}, \pm\frac{4}{9}$ **6.** $\pm1, \pm2, \pm4, \pm8, \pm16, \pm\frac{1}{3}, \pm\frac{2}{3}, \pm\frac{4}{3}, \pm\frac{8}{3}, \pm\frac{16}{3}, \pm\frac{1}{9}, \pm\frac{2}{9}, \pm\frac{4}{9}, \pm\frac{8}{9}, \pm\frac{16}{9}, \pm\frac{1}{27}, \pm\frac{2}{27}, \pm\frac{4}{27}, \pm\frac{8}{27}, \pm\frac{16}{27}$ **7.** 5, 3, or 1 positive real zeros, 0 negative real zeros **8.** 4, 2, or 0 positive real zeros, 0 negative real zeros **9.** 1 positive real zero, 3 or 1 negative real zeros **10.** 0 positive real zeros, 0 negative real zeros **11.** $-2, \frac{1}{4}, 3, f(x) = (x-3)(x+2)(4x-1)$ **12.** $-\frac{7}{2}, -\sqrt{3}, \sqrt{3}, f(x) = (2x+7)(x+\sqrt{3})(x-\sqrt{3})$ **13.** $\frac{1}{2}, -2+4i, -2-4i, f(x) = -(2x-1)(x-(-2+4i))(x-(-2-4i))$ **14.** $-\frac{1}{2}, 1, \frac{5}{3}$, $f(x) = -(x-1)^2(2x+1)(3x-5)$ **15.** $-\frac{1}{3}, 1, -1+5i, -1-5i, f(x) = (3x+1)(x-1)(x-(-1+5i))(x-(-1-5i))$ **16.** $-1, -\frac{5}{2}, \frac{-1-\sqrt{5}}{2}, \frac{-1+\sqrt{5}}{2}, f(x) = -(x+1)^2(2x+5)\left(x-\left(\frac{-1-\sqrt{5}}{2}\right)\right)\left(x-\left(\frac{-1+\sqrt{5}}{2}\right)\right)$ **17.** $\frac{1}{2}, -\frac{3}{2}, -2+2i, -2-2i, f(x) = (2x-1)^2(2x+3)(x-(-2+2i))(x-(-2-2i))$ **18.** $-\frac{1}{3}, \frac{1}{7}, \sqrt{2}, -\sqrt{2}$, $f(x) = (3x+1)^2(7x-1)(x-\sqrt{2})(x+\sqrt{2})$ **19.** $1, \frac{3}{5}, i, -i, f(x) = (x-1)^3(5x-3)(x-i)(x+i)$ **20.** $-\frac{1}{2}+\frac{\sqrt{3}}{2}i, -\frac{1}{2}-\frac{\sqrt{3}}{2}i, 1$ **21.** $i, -i, -1, 1$ **22.** $-1, 2, 3$ **23.** $\frac{1}{2}+\frac{\sqrt{7}}{2}i, \frac{1}{2}-\frac{\sqrt{7}}{2}i, -\frac{5}{2}, 1$ **24.** $-1+5i, -1-5i, -\frac{1}{3}, 1$ **25.** $-\frac{5}{2}-\frac{\sqrt{17}}{2}, -\frac{5}{2}+\frac{\sqrt{17}}{2}, -\frac{1}{6}, \frac{1}{2}$ **26.** answers may vary, $f(x) = (x-(5+i))(x-(5-i))(x-2)$ **27.** answers may vary, $f(x) = (x-1)^2(x-(3-2i))(x-(3+2i))$ **28.** $f(x) = -\frac{1}{78}(x-3i)(x+3i)(x-(1-5i))(x-(1+5i))(x+1)$ **29.** $f(x) = \frac{1}{4}(x-5)(x-2i)(x+2i)$ **30.** $f(x) = -(x+2)(x-1)^2(x-(1+i))(x-(1-i))$ **31.** $f(x) = 2(x+5)(x-5)(x-(-1-2i))(x-(-1+2i))$

32. $f(1) = -2$ and $f(2) = 7$ **33.** $f(0) = -1$ and $f(1) = 10$ **34.** $f(-1) = -6$ and $f(0) = 1$ **35.** 1.33

36. -1 and 1 **37.** $-.20$ **38.** **39.** **40.** 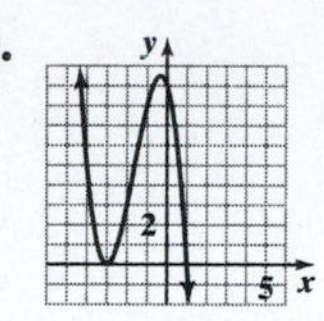 **41.** 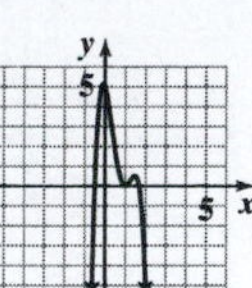

42. 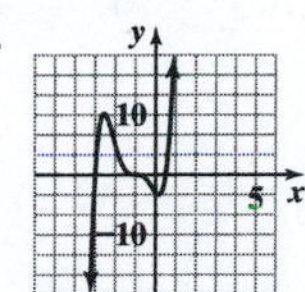 **43.** 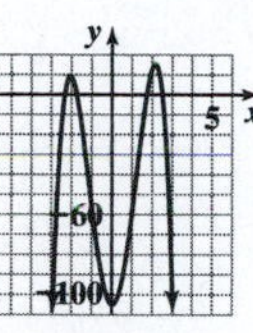 **44.** 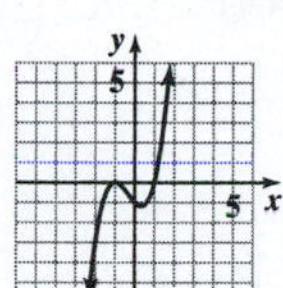 **45.** 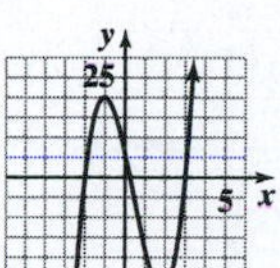 **46.** 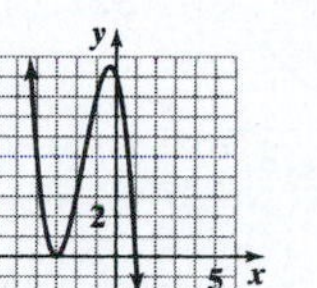 **47.** 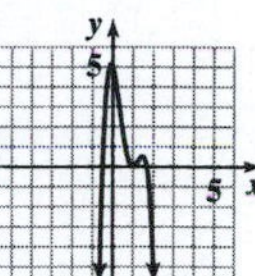

48. 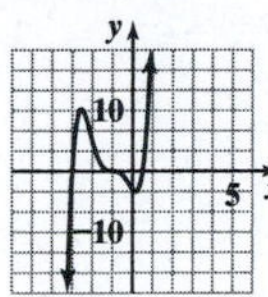 **49.** 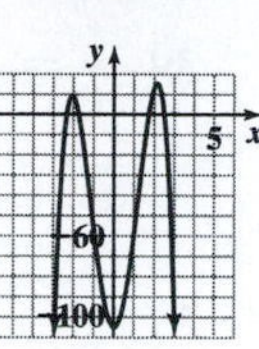

4.6 Exercises

1. a. $(-\infty, -1) \cup (-1, 1) \cup (1, \infty)$ **b.** 0 **c.** 0 **2. a.** $(-\infty, \infty)$ **b.** 0 **c.** 0 **3. a.** $(-\infty, 0) \cup (0, \infty)$ **b.** none **c.** $1, -1$ **4. a.** $(-\infty, -2) \cup (-2, \infty)$ **b.** $-\frac{1}{2}$ **c.** $\frac{1}{3}$ **5. a.** $(-\infty, -1) \cup (-1, 0) \cup (0, \infty)$ **b.** none **c.** $-4, 3$ **6. a.** $(-\infty, -1) \cup (-1, 5) \cup (5, \infty)$ **b.** $-\frac{6}{5}$ **c.** $-2, 1, 3$ **7.** $x = -1$, **8.** $x = 2$,

9. $x = -1$, 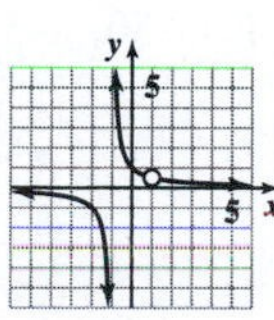 **10.** $x = 2$ and $x = 4$, 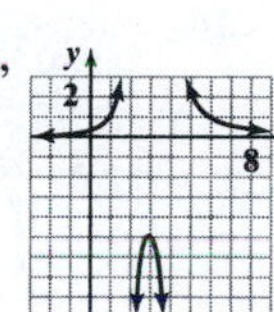 **11.** $x = -2$ and $x = 1$, 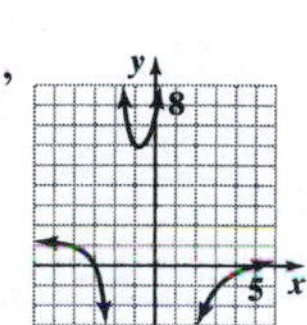

12. $x = -2$ and $x = 1$, 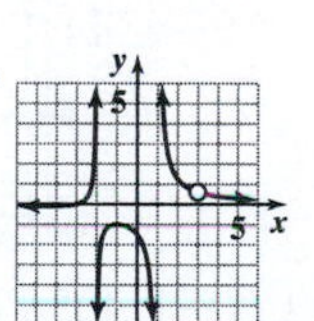 **13.** $y = 0$ **14.** none **15.** $y = \frac{3}{5}$ **16.** none **17.** $y = -\frac{4}{7}$ **18.** $y = -9$

19. $y = 0$ **20.** $y = \frac{3}{11}$ **21.** 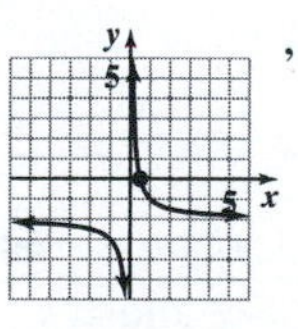 , $x = 0, y = -2$ **22.** 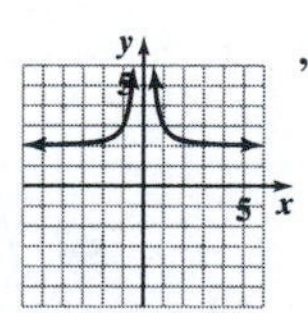 , $x = 0, y = 2$ **23.** , $x = 1$, $y = 0$ **24.** 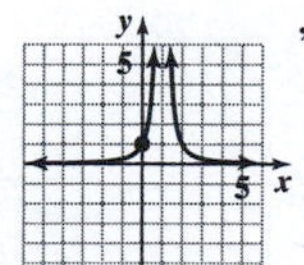 , $x = 1, y = 0$ **25.** 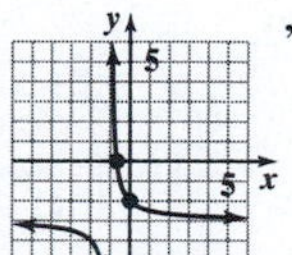 , $x = -1, y = -3$ **26.** 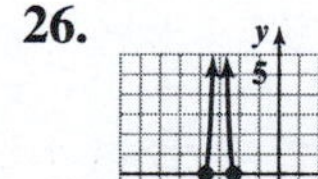 , $x = -3, y = -2$

27. 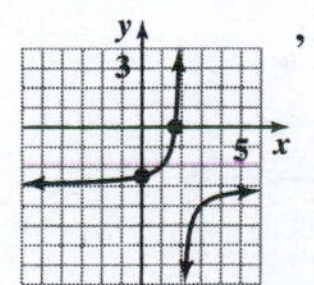 , $x = 2, y = -3$ **28.** 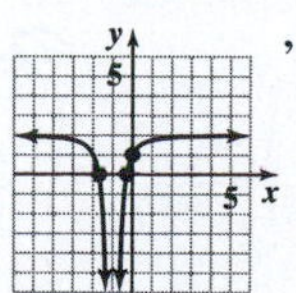 , $x = -1, y = 2$ **29.** 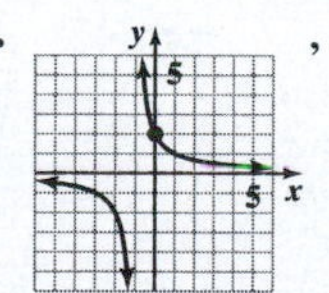 , $x = -1, y = 0$ **30.** 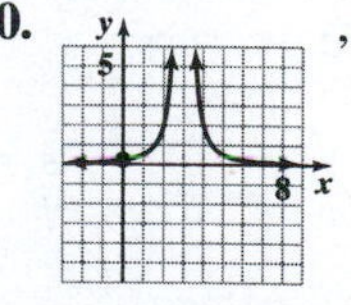 ,

$x = 3, y = 0$ **31.** , $x = -1, y = -1$ **32.** , $x = 1, y = 3$ **33.** $(3, 6)$,

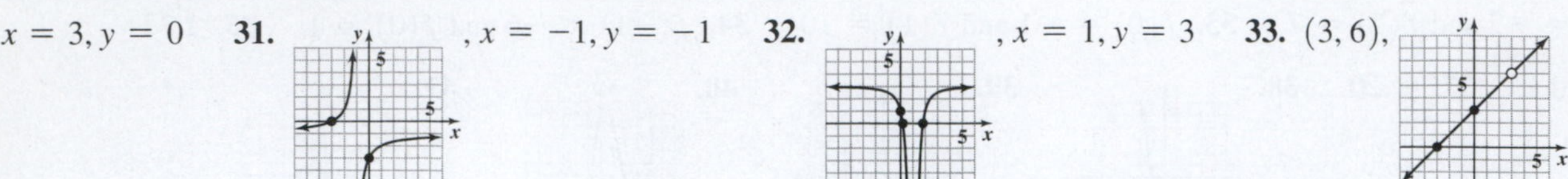

34. $(-4, -5)$, **35.** $(1, 1)$, , $x = 0, y = 0$ **36.** $\left(-2, -\frac{1}{5}\right)$, , $x = 3, y = 0$

37. $\left(-2, \frac{1}{9}\right)$, , $x = 1, y = 0$ **38.** $\left(1, \frac{1}{9}\right)$, , $x = -2, y = 0$ **39.** $y = 2x$ **40.** $y = x - 3$

41. $y = x - 1$ **42.** $y = 2x$ **43.** , $x = 1, y = 2$ **44.** , $x = 2$ and $x = -1, y = 0$

45. , $y = 1$ **46.** , $x = -2, y = 2x - 9$ **47.** , $x = -3$ and $x = 3, y = 0$

48. , $x = -3$ and $x = 3, y = 0$ **49.** , $x = -1, y = x + 1$ **50.** , $x = 3, y = 2$

51. , $x = -2, y = x - 2$ **52.** , $x = 4$ and $x = -1, y = 1$ **53.** , $x = -2, x = 1,$ and $x = 4, y = 2$ **54.** , $x = -3$ and $x = 3, y = 2x - 3$ **55.** $(-\infty, -1) \cup (-1, 1) \cup (1, \infty)$ **56.** $(-\infty, \infty)$

57. $(-\infty, 0) \cup (0, \infty)$ **58.** $(-\infty, -2) \cup (-2, \infty)$ **59.** $(-\infty, -1) \cup (-1, 0) \cup (0, \infty)$ **60.** $(-\infty, -1) \cup (-1, 5) \cup (5, \infty)$
61. $x = -1$ **62.** $x = 2$ **63.** $x = -1$ **64.** $x = 2$ and $x = 4$ **65.** $x = -2$ and $x = 1$ **66.** $x = -2$ and $x = 1$
67. **68.** **69.** **70.** **71.** **72.**

73. **74.** **75.** **76.** **77.** **78.**

79. **80.** **81.** **82.** **83.** 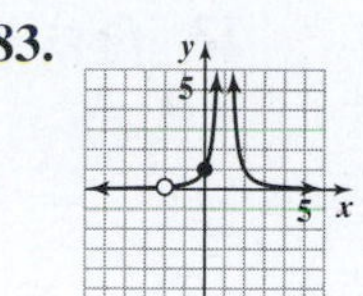 **84.** 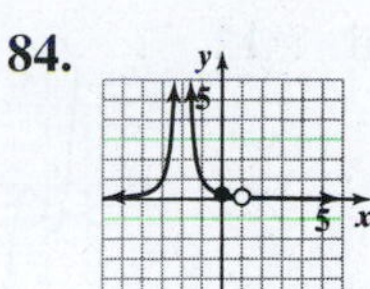

85. **86.** **87.** **88.** **89.** 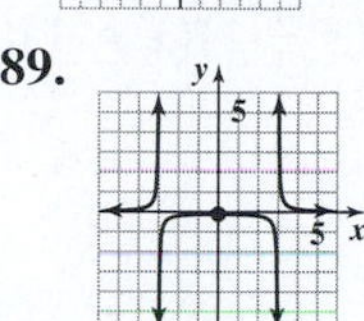 **90.** 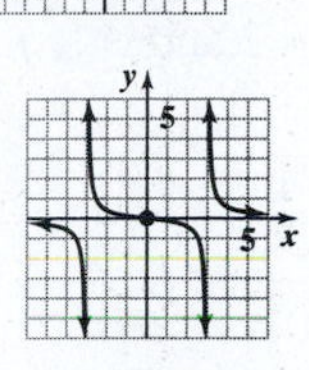

91. **92.** **93.** **94.** **95.** **96.** 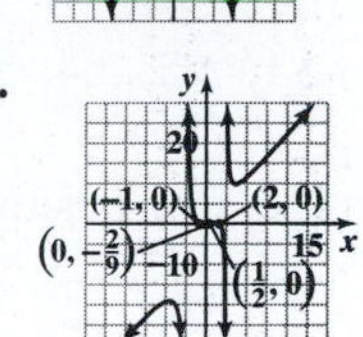

4.7 Exercises

1. a. $y = 3.5x$ **b.** $y = 24.5$ **2. a.** $S = 36t$ **b.** $S = 144$ **3. a.** $y = 24x^2$ **b.** $y = 600$ **4. a.** $M = 504r^2$ **b.** $M = 224$ **5. a.** $p = \frac{5}{8}n^3$ **b.** $p = 625$ **6. a.** $A = \frac{3}{4}m^3$ **b.** $A = 384$ **7.** 40 ft **8.** 16 m **9.** 900 gallons **10.** 396.9 m **11. a.** $y = \frac{600}{x}$ **b.** $y = 20$ **12. a.** $p = \frac{16.1}{q}$ **b.** $p = 11.5$ **13. a.** $n = \frac{518.4}{\text{m}^2}$ **b.** $n = 32.4$ **14. a.** $d = \frac{1200}{t^3}$ **b.** $d = 9.6$ **15.** 9 years old **16.** 3 ohms **17.** about 79 kg **18.** 18 lumens **19. a.** $y = \frac{(15x)}{z}$ **b.** $y = 20$ **20. a.** $A = 0.5bc$ **b.** $b = \frac{14}{3}$ **21. a.** $P = 4.8\text{nm}^2$ **b.** $P = 67.2$ **22. a.** $A = \frac{(12t^2)}{r^3}$ **b.** $A = 48$ **23.** \$5568 **24.** 5.2 horsepower **25.** 45 kilograms per square meter **26.** 47.75 revolutions per minute **27.** 18.75 ft or less **28.** 1638.4 watts

Review Exercises

1. a. $(3, 2)$ **b.** down **c.** $x = 3$ **d.** 2, 4 **e.** -16 **f.** **g.** domain: $(-\infty, \infty)$, range: $(-\infty, 2]$

2. a. $(-2, 12)$ **b.** down **c.** $x = -2$ **d.** $-2 - 2\sqrt{3}, -2 + 2\sqrt{3}$ **e.** 8 **f.**

g. domain: $(-\infty, \infty)$, range: $(-\infty, 12]$ **3. a.** negative **b.** $(h, k) = (-1, -2)$ **c.** $-\frac{1}{16}$ **d.** $f(x) = -\frac{1}{16}(x + 1)^2 - 2$ **e.** $f(x) = -\frac{1}{16}x^2 - \frac{x}{8} - \frac{33}{16}$ **4.** 3.5 sec, 61.025 m **5. a.** \$292.50 **b.** 1,000, \$150,000 **c.** \$150 **6. a.** $A(x) = x\left(1000 - \frac{1}{2}x\right)$ or $A(x) = 1000x - \frac{1}{2}x^2$ **b.** 500 feet by 1000 feet **c.** 500,000 square feet

7. polynomial, 5, −9, 1 **8. a.** odd **b.** negative **9.** −4, −3, 4 **10.** $x = 0$ and $x = 3$, multiplicity 1, crosses, $x = 4$, multiplicity 4, touches **11. a.** odd **b.** positive **c.** 0 **d.** $x = -2$, $x = 0$, and $x = 3$ all have odd multiplicity

e. ii **12.** 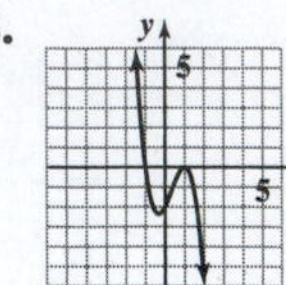 **13.** $f(x) = (x - 2)(6x^2 + 12x + 22) + 47$ **14.** 101 **15.** factor

16. -2 and -1, $f(x) = (x - 1)(x + 1)(x + 2)$, 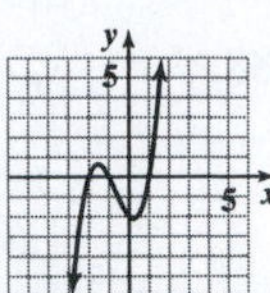

17. $\pm 1, \pm 2, \pm 4, \pm 8, \pm 16, \pm\frac{1}{3}, \pm\frac{2}{3}, \pm\frac{4}{3}, \pm\frac{8}{3}, \pm\frac{16}{3}, \pm\frac{1}{9}, \pm\frac{2}{9}, \pm\frac{4}{9}, \pm\frac{8}{9}, \pm\frac{16}{9}, \pm\frac{1}{27}, \pm\frac{2}{27}, \pm\frac{4}{27}, \pm\frac{8}{27}, \pm\frac{16}{27}$

18. 5, 3, or 1 positive real zeros, 0 negative real zeros **19.** $-2, \frac{1}{4}, 3, f(x) = (x - 3)(x + 2)(4x - 1)$

20. 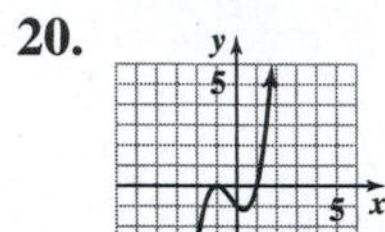 **21.** $x = 2$ and $x = 4$, 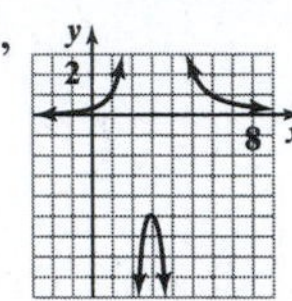 **22.** $y = \frac{3}{5}$ **23.** 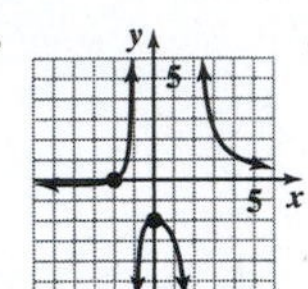, $x = 2$ and $x = -1, y = 0$

24. 40 ft **25.** about 79 kilograms

CHAPTER 5

5.1 Skill Check Exercises

SCE-1. 3^3 **SCE-2.** 2^{-3} **SCE-3.** $3^{1/4}$ **SCE-4.** 5^{-3x} **SCE-5.** $3^{x/3}$ **SCE-6.** 2^{1-x} **SCE-7.** $5^{7/4}$
SCE-8. $3^{4-x/5}$ **SCE-9.** 3^{x^2-2x} **SCE-10.** 2^{2x+x^2}

1. 0.303212 **2.** 5.805141 **3.** 20.085537 **4.** 0.818731 **5.** 1.395612 **6.** 88.426366 **7.** 32.725881

8. 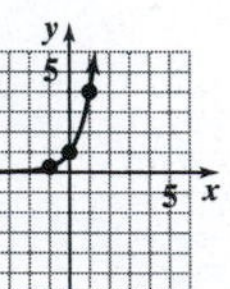 **9.** 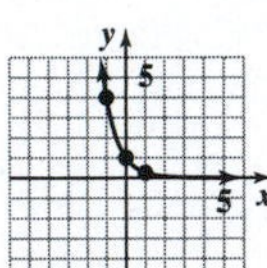 **10.** 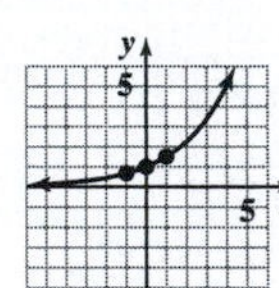 **11.** 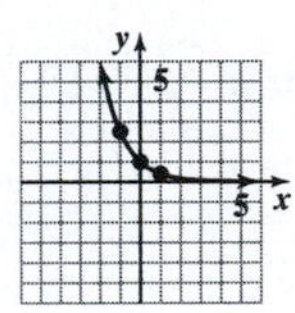 **12.** 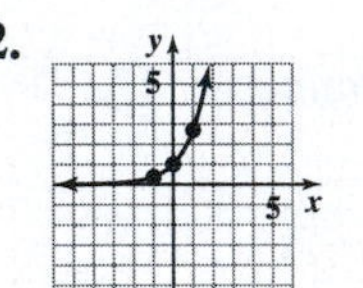 **13.** 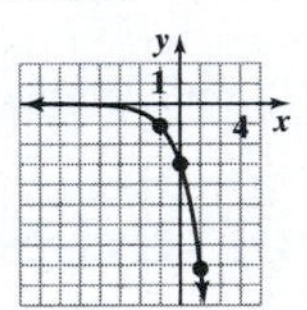

14. 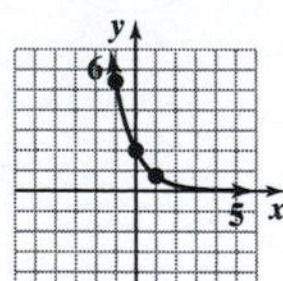 **15.** $f(x) = 10^x$ **16.** $f(x) = 2^x$ **17.** $f(x) = \left(\frac{2}{3}\right)^x$ **18.** $f(x) = \left(\frac{3}{4}\right)^x$ **19.** $f(x) = 8^x$

20. $f(x) = \left(\frac{1}{16}\right)^x$ **21.** $f(x) = 2.7^x$ **22.**, domain: $(-\infty, \infty)$, range: $(0, \infty)$, y-intercept: $\frac{1}{2}$, $y = 0$

23., domain: $(-\infty, \infty)$, range: $(-1, \infty)$, y-intercept: 0, $y = -1$ **24.**, domain: $(-\infty, \infty)$, range: $(-\infty, 0)$, y-intercept: -9, $y = 0$ **25.**, domain: $(-\infty, \infty)$, range: $(-\infty, -1)$, y-intercept: -3, $y = -1$

26. , domain: $(-\infty, \infty)$, range: $(0, \infty)$, y-intercept: $\frac{1}{2}$, $y = 0$ **27.** , domain: $(-\infty, \infty)$, range: $(-3, \infty)$, y-intercept: -2, $y = -3$ **28.** , domain: $(-\infty, \infty)$, range: $(1, \infty)$, y-intercept: 2, $y = 1$

29. , domain: $(-\infty, \infty)$, range: $(-2, \infty)$, y-intercept: 1, $y = -2$ **30.** 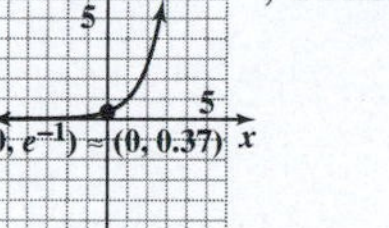

, domain: $(-\infty, \infty)$, range: $(0, \infty)$, $y = 0$ **31.** , domain: $(-\infty, \infty)$, range: $(-1, \infty)$, $y = -1$ **32.** , domain: $(-\infty, \infty)$, range: $(-\infty, 0)$, $y = 0$ **33.** 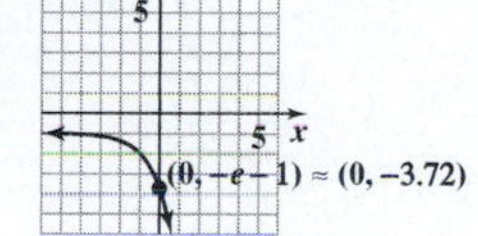

, domain: $(-\infty, \infty)$, range: $(-\infty, -1)$, $y = -1$

34. 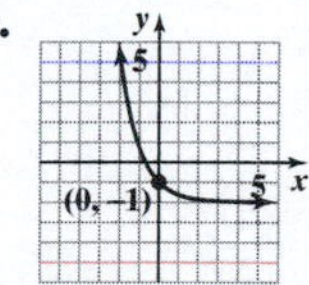

, domain: $(-\infty, \infty)$, range: $(-2, \infty)$, $y = -2$ **35.** 4 **36.** -1 **37.** $\frac{1}{4}$ **38.** 6 **39.** $-\frac{3}{5}$

40. $-\frac{7}{12}$ **41.** $\frac{1}{3}$ **42.** $-\frac{9}{5}$ **43.** $-1, 3$ **44.** $-1, 0, 2$ **45.** -2 **46.** $-\frac{1}{3}$ **47.** $\frac{1}{4}$ **48.** $-2, 2$ **49.** $-3, 4$ **50.** $-\frac{5}{2} - \frac{\sqrt{37}}{2}, -\frac{5}{2} + \frac{\sqrt{37}}{2}$ **51.** $-1, 1, 2$ **52.** $-1, 2, 3$ **53.** **a.** \$300,000 **b.** \$265,640 **c.** \$163,299 **54.** **a.** 130 lbs **b.** 201 lbs **c.** 235 lbs **55.** **a.** 2,560 bacteria **b.** 4,096 bacteria **c.** 8,589,934,592 bacteria **56.** **a.** 40 rabbits **b.** 62 rabbits **c.** 336 rabbits **d.** 1595 rabbits, 1599 rabbits, 1600 rabbits **57.** \$12,752.18 **58.** 7.35% compounded daily **59.** compounded quarterly at 4.5% **60.** $A(2) \approx$ \$6,798.89, $A(20) \approx$ \$20,942.06 **61.** \$55.23 **62.** \$332,570.82 **63.** \$5,862.47 **64.** \$469,068.04 **65.** \$2,000 **66.** \$10,202.05 **67.** a **68.** **a.** 2,000 people **b.** 3.5% **c.** 6,808 people **69.** **a.** $P(t) = 10e^{0.25t}$ **b.** 4,034 bacteria **70.** 5,488 people **71.** **a.** 8 rabbits **b.** 478,993 rabbits **72** , domain: $(-\infty, \infty)$, range: $(0, \infty)$, y-intercept: $\frac{1}{2}$, $y = 0$ **73.** , domain: $(-\infty, \infty)$, range: $(-1, \infty)$, y-intercept: 0, $y = -1$ **74.** , domain:

$(-\infty, \infty)$, range: $(-\infty, 0)$, y-intercept: -9, $y = 0$ **75.** 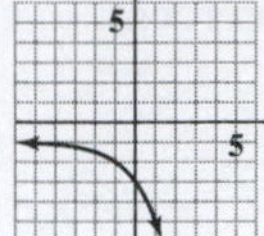 , domain: $(-\infty, \infty)$, range: $(-\infty, -1)$, y-intercept: -3, $y = -1$ **76.** 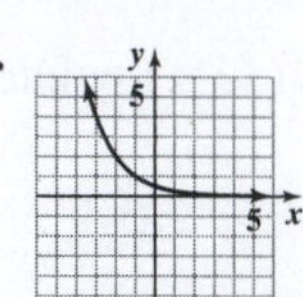 , domain: $(-\infty, \infty)$, range: $(0, \infty)$, y-intercept: $\frac{1}{2}$, $y = 0$ **77.** 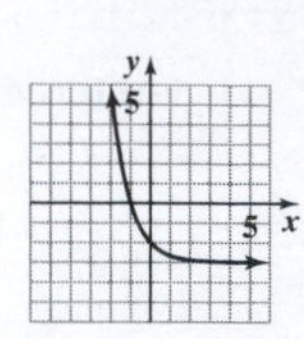 , domain: $(-\infty, \infty)$, range: $(-3, \infty)$, y-intercept: -2, $y = -3$ **78.** 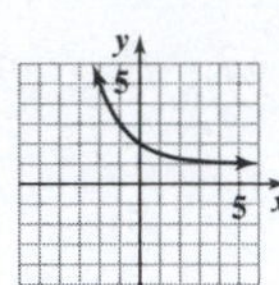 , domain: $(-\infty, \infty)$, range: $(1, \infty)$, y-intercept: 2, $y = 1$ **79.** 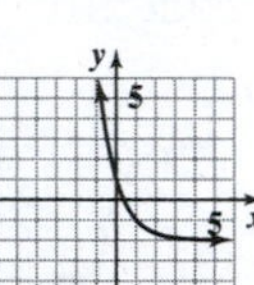 , domain: $(-\infty, \infty)$, range: $(-2, \infty)$, y-intercept: 1, $y = -2$ **80.** 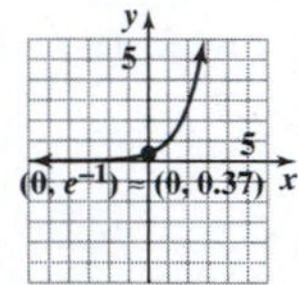

, domain: $(-\infty, \infty)$, range: $(0, \infty)$, $y = 0$ **81.** 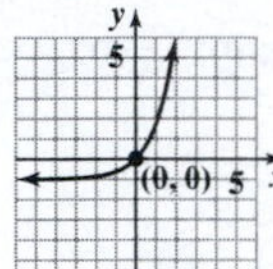

, domain: $(-\infty, \infty)$, range: $(-1, \infty)$, $y = -1$

82. (0, −e²) ≈ (0, −7.39), domain: $(-\infty, \infty)$, range: $(-\infty, 0)$, $y = 0$ **83.** 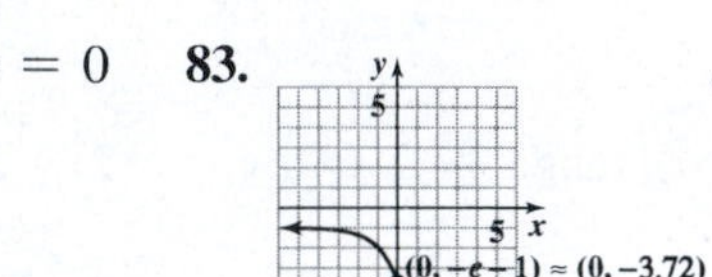

, domain: $(-\infty, \infty)$, range: $(-\infty, -1)$, $y = -1$ **84.** 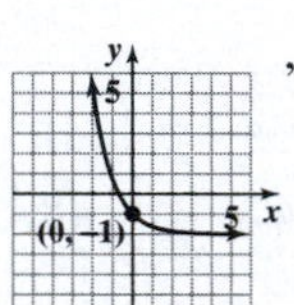

, domain: $(-\infty, \infty)$, range: $(-2, \infty)$, $y = -2$ **85.** 6,808 people

86. 4,034 bacteria **87.** 478,993 rabbits

5.2 Exercises

1. $\log_3 9 = 2$ **2.** $\log_{16} 4 = \frac{1}{2}$ **3.** $\log_2 \frac{1}{8} = -3$ **4.** $\log_{\sqrt{2}} W = \pi$ **5.** $\log_{1/3} 27 = t$ **6.** $\log_7 L = 5k$ **7.** $5^0 = 1$ **8.** $7^3 = 343$ **9.** $\sqrt{2}^6 = 8$ **10.** $4^L = K$ **11.** $a^3 = x - 1$ **12.** 3 **13.** $\frac{1}{2}$ **14.** -2 **15.** 4 **16.** $-\frac{3}{5}$ **17.** $-\frac{1}{3}$ **18.** -2 **19.** 11 **20.** 1 **21.** 0 **22.** -3 **23.** 1 **24.** M **25.** 0 **26.** 20 **27.** $\log 1000 = 3$ **28.** $\ln\left(\frac{1}{e}\right) = -1$ **29.** $\ln 2 = k$ **30.** $\log M = e$ **31.** $\ln Z = 10$ **32.** $e^0 = 1$ **33.** $10^6 = 1{,}000{,}000$ **34.** $10^L = K$ **35.** $e^4 = Z$ **36.** 4 **37.** -3 **38.** 0 **39.** $\frac{2}{3}$ **40.** e **41.** 49 **42.** 6 **43.** 4 **44.** (4, 1), (1, 0), $(0, \infty)$, $x = 0$ **45.** $(\frac{1}{3}, 1)$, (1, 0), $(0, \infty)$, $x = 0$ **46.** (4, 1), (2, 0), $(0, \infty)$, $x = 0$ **47.** (6, 1), (2, 0),

$(1, \infty)$, $x = 1$ **48.** , $(-1, \infty)$, $x = -1$ **49.** , $(0, \infty)$, $x = 0$ **50.** , $(2, \infty)$,

$x = 2$ **51.** , $(0, \infty)$, $x = 0$ **52.** , $(0, \infty)$, $x = 0$ **53.** , $(-1, \infty)$, $x = -1$

54.

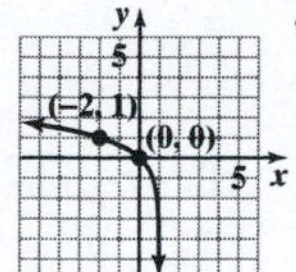

, $(-\infty, 1)$, $x = 1$ **55.**

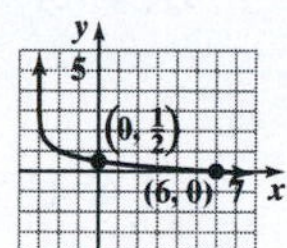

, $(-3, \infty)$, $x = -3$ **56.** $(-\infty, 0)$ **57.** $(-3, \infty)$

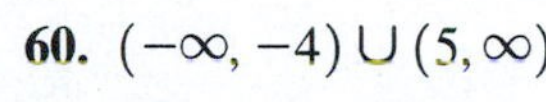

58. $\left(-\infty, \dfrac{1}{3}\right)$ **59.** $(-\infty, -3) \cup (3, \infty)$ **60.** $(-\infty, -4) \cup (5, \infty)$ **61.** $(-\infty, -5) \cup (8, \infty)$

62. $(-10, -2) \cup (3, \infty)$ **63.**

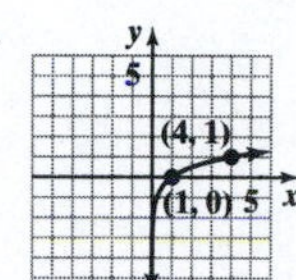

64. **65.** , $(0, \infty)$, $x = 0$ **66.** ,

$(1, \infty)$, $x = 1$ **67.**

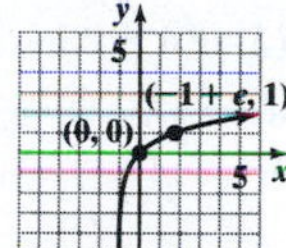

, $(-1, \infty)$, $x = -1$ **68.**

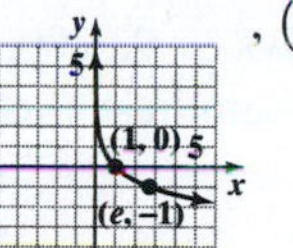

, $(0, \infty)$, $x = 0$ **69.** ,

$(2, \infty)$, $x = 2$ **70.**

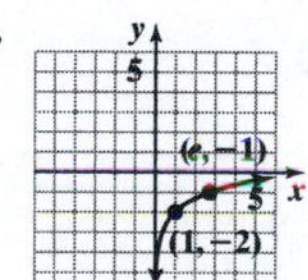

, $(0, \infty)$, $x = 0$ **71.**

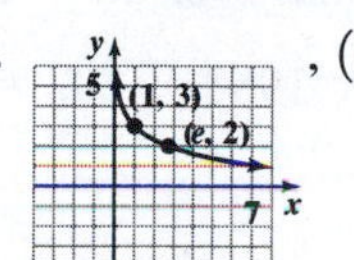

, $(0, \infty)$, $x = 0$ **72.**

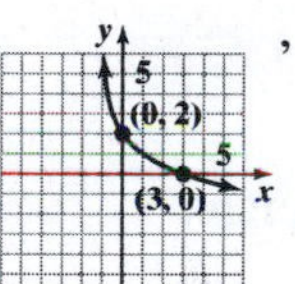

,

$(-1, \infty)$, $x = -1$ **73.**

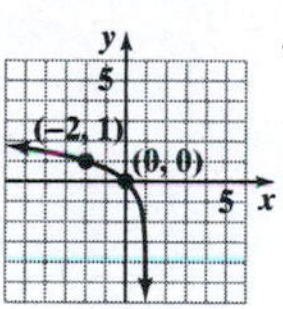

, $(-\infty, 1)$, $x = 1$ **74.**

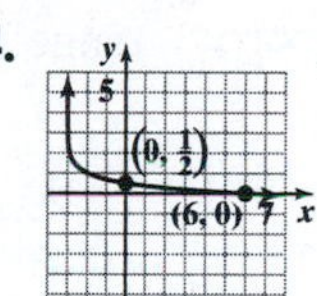

, $(-3, \infty)$, $x = -3$

5.3 Exercises

1. $\log_4 x + \log_4 y$ **2.** $\log 9 - \log t$ **3.** $3 \log_5 y$ **4.** $3 + \log_3 w$ **5.** $\ln 5 + 2$ **6.** $\dfrac{1}{4} \log_9 k$ **7.** $2 + \log P$

8. $5 - \ln r$ **9.** $6 + \log_{\sqrt{2}} x$ **10.** $\log_2 M - 5$ **11.** 40 **12.** $\dfrac{3}{11}$ **13.** 20 **14.** $\dfrac{3}{121}$ **15.** $2\log_7 x + 3\log_7 y$

16. $2\ln a + 3\ln b - 4\ln c$ **17.** $\dfrac{1}{2} \log x - 1 - 3 \log y$ **18.** $2 + \log_3 (x^2 - 25) = 2 + \log_3 (x - 5) + \log_3 (x + 5)$

19. $1 + \dfrac{1}{2} \log_2 x + \dfrac{1}{2} \log_2 y$ **20.** $-\dfrac{1}{6} + \dfrac{5}{2} \log_5 x - \dfrac{4}{3} \log_5 y$ **21.** $\dfrac{1}{5} + \dfrac{1}{5} \ln z - \dfrac{1}{2} \ln (x - 1)$

22. $\dfrac{1}{2} \log_3 x + \dfrac{5}{4} \log_3 y - \dfrac{1}{2}$ **23.** $-\dfrac{4}{3} \ln (x + 1) - 2\ln (x - 1)$ **24.** $3 + 3\log x - \dfrac{9}{2} \log (x - 4) - 5\log (x + 4)$

25. $\log_b (AC)$ **26.** $\log_4 \left(\dfrac{M}{N}\right)$ **27.** $\log_8 \left(x^2 \sqrt[3]{y}\right)$ **28.** 2 **29.** 4 **30.** $\ln \left(x^{\frac{5}{12}}\right)$ **31.** $\log_5 (x^2 - 4)$

32. $\log_3 \frac{x+1}{(x+4)\sqrt{6x}}$ **33.** $\ln \frac{x-1}{4x\sqrt{x+5}}$ **34.** $\ln[7x^2(x-3)^2]$ **35.** $\log_9(x-3)$ **36.** $\log\left(\frac{x+3}{x+1}\right)$

37. $\ln\left(\frac{1}{x-1}\right)$ **38.** 5 **39.** 36 **40.** 1 **41.** $\frac{7}{8}$ **42.** 2 **43.** 6 **44.** 2.8362 **45.** −0.1147 **46.** −2.6572

47. 4.7332 **48.** $\log_3(xw^2)$ **49.** $\log_5\left(\frac{1}{x^2}\right)$ **50.** $\log_2(xyw)$ **51.** $\ln(\sqrt{x^{15}})$ **52.** 5 **53.** $\frac{1}{20}$ **54.** 36 **55.** $\sqrt{5}$

5.4 Exercises

SCE-1. 1.2528 **SCE-2.** 5.1565 **SCE-3.** 1.6826 **SCE-4.** 1.8516 **SCE-5.** 738.1281 **SCE-6.** 6.0588
SCE-7. 11.5813 **SCE-8.** 72.5188

1. 1.4650 **2.** 12.7438 **3.** −1, 3 **4.** ∅ **5.** −13.3627 **6.** $\frac{1}{3}$ **7.** 4.4841 **8.** $-\frac{5}{6}$ **9.** 0.3334 **10.** 0.4833

11. 0.6931 **12.** 0.5988 **13.** −0.2055 **14.** 2.6889 **15.** −3.1192 **16.** $\frac{6}{5}$ **17.** 7 **18.** −3 **19.** 2, 3

20. $\frac{15}{4}$ **21.** $-\frac{99}{5}$ **22.** $\frac{23}{9}$ **23.** $\frac{1}{3}$ **24.** 8 **25.** 6 **26.** −8 **27.** $\frac{19}{6}$ **28.** 10 **29.** −5 **30.** $\frac{7}{5}$ **31.** 8

32. −1 and $\frac{1}{3}$ **33.** 6 **34.** ∅ **35.** 2.0933

5.5 Exercises

1. about 4 years 8 months **2.** about 7 years 9 months **3.** 13.73 years **4.** 17.33% **5.** Bank A: 4.5%, Bank B: 4.6%
6. 21,086 people **7.** 4,147 bacteria **8.** 252 days **9.** 200,119 people **10.** 53 days **11.** 64.84% **12. a.** 53.03 grams
b. 17 years **13. a.** 6,000 families **b.** 2,000 families **c.** 1988 **14. a.** 3,000 students **b.** $B = 374$
c. $K = \ln\left(\frac{29}{374}\right)$ **d.** ≈3.63 days **15. a.** 61,500 **b.** 40 **c.** $-\frac{\ln\left(\frac{43}{2}\right)}{6}$ **d.** 2009 **16.** just over 20 minutes
17. just before 11:52 P.M. **18.** about 58.6°F **19.** 1988 **20.** ≈3.63 days **21.** 2009

Review Exercises

1. $f(x) = \left(\frac{2}{3}\right)^x$ **2.** 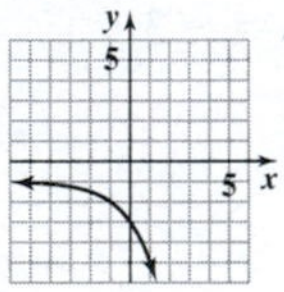 , domain: $(-\infty, \infty)$, range: $(-\infty, -1)$, y-intercept: −3, $y = -1$

3. −1 **4.** $-\frac{9}{5}$ **5.** $\log_7 L = 5k$ **6.** $4^L = K$ **7.** $-\frac{3}{5}$ **8.** 0 **9.** 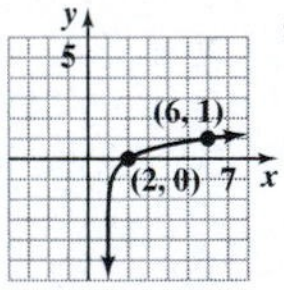 , $(1, \infty)$, $x = 1$

10. $3\log_5 y$ **11.** $3 + \log_3 w$ **12.** $-1 + \frac{1}{2}\log x - 3\log y$ **13.** $\frac{3}{11}$ **14.** $\log_8(x^2\sqrt[3]{y})$ **15.** 2 **16.** 2.8362

17. 1.4650 **18.** 4.4841 **19.** 0.5988 **20.** 2, 3 **21.** −8 **22.** about 7 years 9 months **23.** 4,147 bacteria

24. 53 days **25.** just over 20 minutes

CHAPTER 6

6.1 Exercises

SCE-1. $-\frac{7\pi}{5}$ **SCE-2.** 150 **SCE-3.** 4 **SCE-4.** 10 **SCE-5.** 16π **SCE-6.** -30π

1. 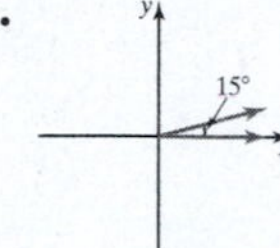

; Quadrant I **2.** ; y-axis **3.** ; Quadrant III **4.** ; Quadrant II

5. 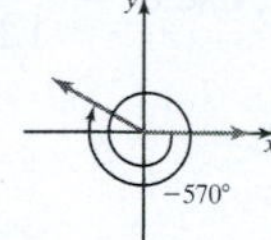

; Quadrant II **6.** 128° **7.** 98° **8.** 309°, −411° **9.** −990°, −270°, 90°, 450°

10. ; negative x-axis **11.** positive x-axis **12.** Quadrant IV 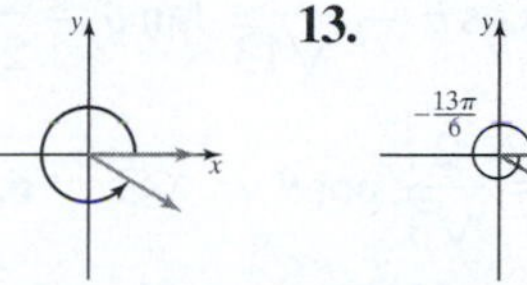 **13.** ; Quadrant IV

14. 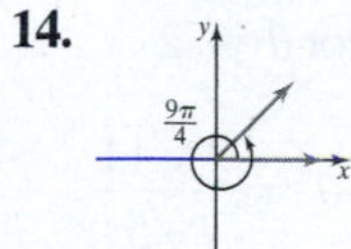

; Quadrant I **15.** Quadrant II **16.** 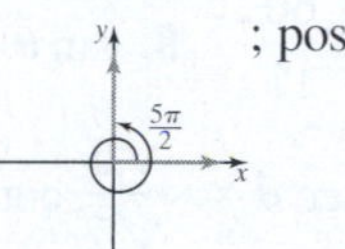

; positive y-axis **17.** positive y-axis 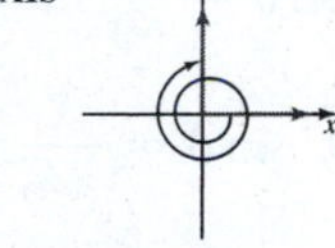

18. Quadrant III 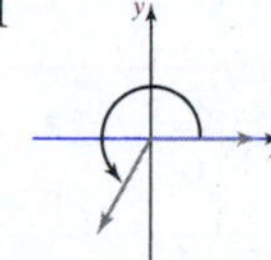 **19.** Quadrant II 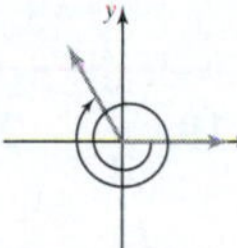 **20.** $\frac{\pi}{12}$ **21.** $\frac{5\pi}{6}$ **22.** -3π **23.** $\frac{8\pi}{15}$ **24.** $-\frac{119\pi}{60}$

25. 15° **26.** 108° **27.** −450° **28.** 205.71° **29.** −348° **30.** 171.89° **31.** $\frac{11\pi}{8}$ **32.** $\frac{5\pi}{6}$ **33.** $\frac{\pi}{3}$

34. $\frac{139\pi}{70}$ **35.** $\frac{13\pi}{12}, -\frac{35\pi}{12}$ **36.** $-\frac{11\pi}{2}, -\frac{3\pi}{2}, \frac{\pi}{2}, \frac{5\pi}{2}$

6.2 Exercises

1. 6 cm^2 **2.** ≈282.74 in.^{2} **3.** ≈33.51 m^2 **4.** ≈17.87 in.^{2} **5.** $\frac{60}{49}$ rad **6.** ≈70.03 ft **7.** ≈9.42 in. **8.** ≈226.19 cm

9. ≈81.16 m **10.** ≈212.93 km **11.** $\frac{8}{5}$ rad **12.** 1516 miles **13.** ≈11.67° **14.** 122° **15.** 12π rad/min

16. 100π rad/min **17.** ≈142.8 mph **18.** ≈7.85 m/min **19.** ≈64.71 mph **20.** ≈3.22 cm **21.** ≈85.68 mph

6.3 Exercises

1. obtuse, scalene **2.** right, scalene **3.** acute, scalene **4.** $\sqrt{193}$ **5.** 30 **6.** $\sqrt{137}$ **7.** 3 **8.** $\sqrt{7}$ **9.** ≈84.85 ft

10. ≈8.77 m **11.** 8 in. by 15 in. **12.** ≈5.74 ft **13.** $k = 2, RS = 7, QS = 10$ **14.** $k = \frac{5}{2}, HK = 10, QT = \frac{16}{5}$

15. $k = \frac{2}{\sqrt{3}}, LV = \frac{2\sqrt{5}}{\sqrt{3}}, LZ = 2\sqrt{3}$ **16.** $k = \frac{5}{\sqrt{2}}, DJ = 5\sqrt{2}, PT = \frac{12}{5}$ **17.** $AB = 25, XZ = 48, YZ = 14$

18. $RT = \frac{35}{2}, ST = \frac{37}{2}, KM = 35$ **19.** $DV = \frac{9}{\sqrt{5}}, LV = \frac{6}{\sqrt{5}}, BF = 4$ **20.** $HN = 3\sqrt{15}, BG = \frac{77}{3\sqrt{15}}, GX = \frac{112}{3\sqrt{15}}$

21. $10, 10\sqrt{2}$ **22.** $\sqrt{2}, 2\sqrt{2}$ **23.** $\frac{9}{2}, \frac{9\sqrt{3}}{2}$ **24.** $4\sqrt{2}, 4\sqrt{2}$ **25.** $2, 2\sqrt{3}$ **26.** $\frac{\sqrt{10}}{\sqrt{3}}, \frac{2\sqrt{10}}{\sqrt{3}}$ **27.** $\sqrt{2}, 2$

28. $\frac{\sqrt{5}}{\sqrt{3}}, \frac{2\sqrt{5}}{\sqrt{3}}$ **29.** $\frac{\sqrt{7}}{\sqrt{2}}, \frac{\sqrt{7}}{\sqrt{2}}$ **30.** $2\sqrt{15}, 4\sqrt{5}$ **31.** 202.5 ft **32.** ≈34.64 ft, 20 ft **33.** ≈2.67 ft **34.** ≈16,970.56 ft

35. length of bridge: 20.21 ft, distance between the two people: 40.41 ft **36.** 272 ft

6.4 Exercises

SCE-1. $\frac{\pi}{3}$ **SCE-2.** $\frac{\pi}{12}$

1. $\sin\theta = \frac{3}{5}, \cos\theta = \frac{4}{5}, \tan\theta = \frac{3}{4}, \csc\theta = \frac{5}{3}, \sec\theta = \frac{5}{4}, \cot\theta = \frac{4}{3}$ **2.** $\sin\theta = \frac{12}{13}, \cos\theta = \frac{5}{13}, \tan\theta = \frac{12}{5}, \csc\theta = \frac{13}{12},$ $\sec\theta = \frac{13}{5}, \cot\theta = \frac{5}{12}$ **3.** $\sin\theta = \frac{4}{\sqrt{137}}, \cos\theta = \frac{11}{\sqrt{137}}, \tan\theta = \frac{4}{11}, \csc\theta = \frac{\sqrt{137}}{4}, \sec\theta = \frac{\sqrt{137}}{11}, \cot\theta = \frac{11}{4}$

4. $\sin\theta = \frac{3}{\sqrt{13}}, \cos\theta = \frac{2}{\sqrt{13}}, \tan\theta = \frac{3}{2}, \csc\theta = \frac{\sqrt{13}}{3}, \sec\theta = \frac{\sqrt{13}}{2}, \cot\theta = \frac{2}{3}$ **5.** $\sin\theta = \frac{1}{2}, \cos\theta = \frac{\sqrt{3}}{2}, \tan\theta = \frac{1}{\sqrt{3}},$ $\csc\theta = 2, \sec\theta = \frac{2}{\sqrt{3}}, \cot\theta = \sqrt{3}$ **6.** $\cos\theta = \frac{12}{13}, \tan\theta = \frac{5}{12}, \csc\theta = \frac{13}{5}, \sec\theta = \frac{13}{12}, \cot\theta = \frac{12}{5}$ **7.** $\sin\theta = \frac{11}{61},$ $\tan\theta = \frac{11}{60}, \csc\theta = \frac{61}{11}, \sec\theta = \frac{61}{60}, \cot\theta = \frac{60}{11}$ **8.** $\sin\theta = \frac{1}{\sqrt{5}}, \cos\theta = \frac{2}{\sqrt{5}}, \csc\theta = \sqrt{5}, \sec\theta = \frac{\sqrt{5}}{2}, \cot\theta = 2$

9. $\sin\theta = \frac{1}{\sqrt{6}}, \cos\theta = \frac{\sqrt{5}}{\sqrt{6}}, \tan\theta = \frac{1}{\sqrt{5}}, \sec\theta = \frac{\sqrt{6}}{\sqrt{5}}, \cot\theta = \sqrt{5}$ **10.** $\sin\theta = \frac{2\sqrt{11}}{7}, \cos\theta = \frac{\sqrt{5}}{7}, \tan\theta = \frac{2\sqrt{11}}{\sqrt{5}},$ $\csc\theta = \frac{7}{2\sqrt{11}}, \cot\theta = \frac{\sqrt{5}}{2\sqrt{11}}$ **11.** $\sin\theta = \frac{1}{\sqrt{10}}, \cos\theta = \frac{3}{\sqrt{10}}, \tan\theta = \frac{1}{3}, \csc\theta = \sqrt{10}, \sec\theta = \frac{\sqrt{10}}{3}$ **12.** $\frac{1}{2}$ **13.** $\sqrt{2}$

14. $\frac{2}{\sqrt{3}}$ **15.** $\sqrt{2} - \sqrt{3}$ **16.** $2\sqrt{3} + 2$ **17.** $\frac{10}{3}$ **18.** $\frac{3}{4}$ **19.** $\frac{\pi}{4}$ **20.** $\frac{\pi}{6}$ **21.** $\frac{\pi}{6}$ **22.** $\frac{\pi}{6}$ **23.** $\frac{\pi}{6}$ **24.** $\frac{\pi}{4}$

25. $\tan\theta = \frac{7}{24}, \csc\theta = \frac{25}{7}, \sec\theta = \frac{25}{24}, \cot\theta = \frac{24}{7}$ **26.** $\tan\theta = \frac{3}{\sqrt{2}}, \csc\theta = \frac{\sqrt{11}}{3}, \sec\theta = \frac{\sqrt{11}}{\sqrt{2}}, \cot\theta = \frac{\sqrt{2}}{3}$

27. $\sin\theta = \frac{21}{29}, \cos\theta = \frac{20}{29}, \tan\theta = \frac{21}{20}, \cot\theta = \frac{20}{21}$ **28.** $\sin\theta = \frac{4}{\sqrt{23}}, \cos\theta = \frac{\sqrt{7}}{\sqrt{23}}, \tan\theta = \frac{4}{\sqrt{7}}, \cot\theta = \frac{\sqrt{7}}{4}$

29. 0 **30.** 1 **31.** 0 **32.** $\csc\theta$ **33.** $\sec\theta$ **34.** $\cos\left(-\frac{3\pi}{14}\right)$ **35.** $\tan 51°$ **36.** -1 **37.** $\tan\theta$

38. $\cot(105° - \theta)$ **39.** 0.7660 **40.** 1.0125 **41.** 1.7604 **42.** 0.8746

6.5 Exercises

SCE-1. $-8, 8$ **SCE-2.** $-\sqrt{22}, \sqrt{22}$ **SCE-3.** $-7, 7$ **SCE-4.** $-\sqrt{31}, \sqrt{31}$ **SCE-5.** $-10, 10$
SCE-6. $-\sqrt{11}, \sqrt{11}$ **SCE-7.** Quadrant IV **SCE-8.** x-axis

1. 0°, 90°, 180°, 270° **2.** $\frac{\pi}{6}, \frac{5\pi}{6}, \frac{7\pi}{6}, \frac{11\pi}{6}$ **3.** quadrantal family, 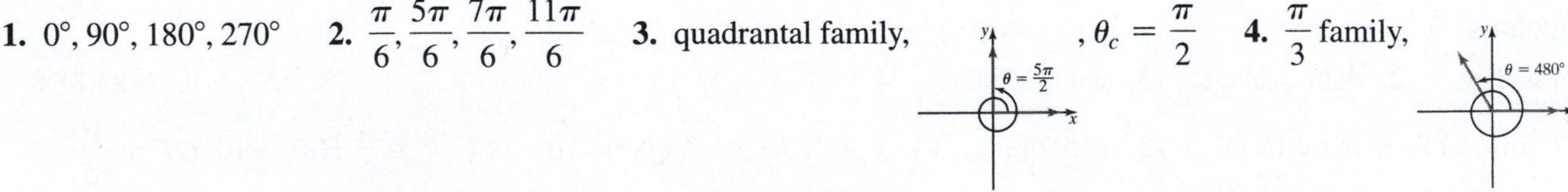

$, \theta_c = \frac{\pi}{2}$ **4.** $\frac{\pi}{3}$ family, , $\theta_c = \frac{2\pi}{3}$ **5.** $\frac{\pi}{3}$ family, 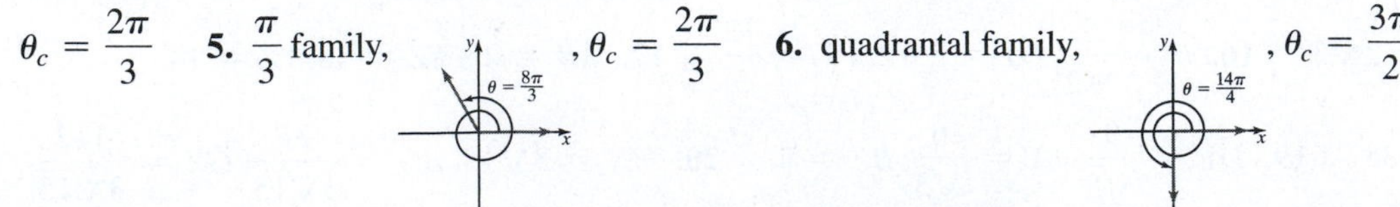

$, \theta_c = \frac{2\pi}{3}$ **6.** quadrantal family, , $\theta_c = \frac{3\pi}{2}$

7. $\frac{\pi}{4}$ family, 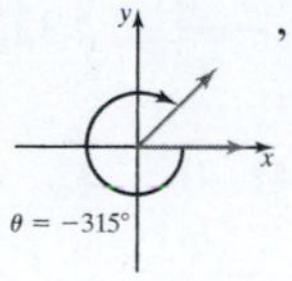 , $\theta_c = \frac{\pi}{4}$ **8.** quadrantal family, 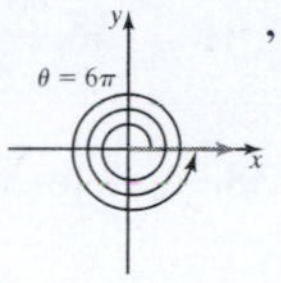 , $\theta_c = 0$ **9.** $\frac{\pi}{6}$ family, 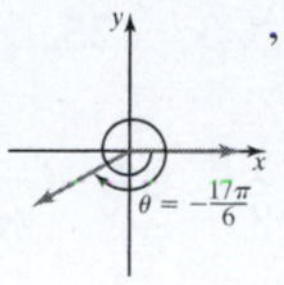 , $\theta_c = \frac{7\pi}{6}$

10. $\frac{\pi}{3}$ family, 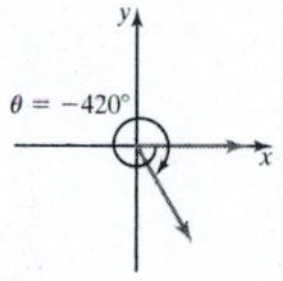 , $\theta_c = \frac{5\pi}{3}$ **11.** $\frac{\pi}{4}$ family, 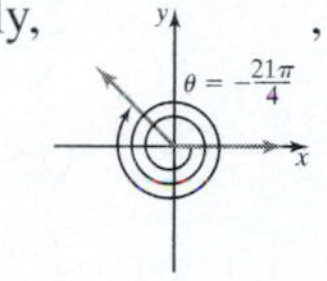 , $\theta_c = \frac{3\pi}{4}$ **12.** $\frac{\pi}{3}$ family, 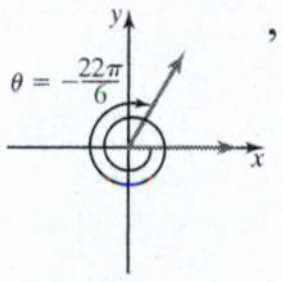 , $\theta_c = \frac{\pi}{3}$

13. $\sin\theta = \frac{3}{5}, \cos\theta = \frac{4}{5}, \tan\theta = \frac{3}{4}, \csc\theta = \frac{5}{3}, \sec\theta = \frac{5}{4}, \cot\theta = \frac{4}{3}$ **14.** $\sin\theta = \frac{12}{13}, \cos\theta = -\frac{5}{13}, \tan\theta = -\frac{12}{5}, \csc\theta = \frac{13}{12}, \sec\theta = -\frac{13}{5}, \cot\theta = -\frac{5}{12}$ **15.** $\sin\theta = -\frac{5}{\sqrt{29}}, \cos\theta = -\frac{2}{\sqrt{29}}, \tan\theta = \frac{5}{2}, \csc\theta = -\frac{\sqrt{29}}{5}, \sec\theta = -\frac{\sqrt{29}}{2}, \cot\theta = \frac{2}{5}$ **16.** $\sin\theta = -\frac{4}{\sqrt{17}}, \cos\theta = \frac{1}{\sqrt{17}}, \tan\theta = -4, \csc\theta = -\frac{\sqrt{17}}{4}, \sec\theta = \sqrt{17}, \cot\theta = -\frac{1}{4}$ **17.** $\sin\theta = \frac{\sqrt{3}}{2}, \cos\theta = -\frac{1}{2}, \tan\theta = -\sqrt{3}, \csc\theta = \frac{2}{\sqrt{3}}, \sec\theta = -2, \cot\theta = -\frac{1}{\sqrt{3}}$ **18.** $\sin\theta = -\frac{1}{\sqrt{2}}, \cos\theta = -\frac{1}{\sqrt{2}}, \tan\theta = 1, \csc\theta = -\sqrt{2}, \sec\theta = -\sqrt{2}, \cot\theta = 1$ **19.** 0 **20.** 0 **21.** undefined **22.** undefined **23.** -1 **24.** undefined **25.** 1 **26.** 0 **27.** undefined **28.** undefined **29.** 1 **30.** 0 **31.** Quadrants I and II **32.** Quadrants I and IV **33.** Quadrants II and IV **34.** Quadrants III and IV **35.** Quadrant IV **36.** Quadrant I **37.** Quadrant I **38.** $\frac{25}{24}$

39. $-\frac{1}{\sqrt{17}}$ **40.** $-\frac{5}{\sqrt{21}}$ **41.** $\frac{\pi}{3}$ **42.** $\frac{\pi}{4}$ **43.** $\frac{\pi}{3}$ **44.** $\frac{\pi}{4}$ **45.** $\frac{\pi}{3}$ **46.** $\frac{2\pi}{9}$ **47.** $\frac{3\pi}{7}$ **48.** 30° **49.** 45°

50. 19° **51.** $\frac{1}{2}$ **52.** $-\frac{1}{\sqrt{2}}$ **53.** $\frac{1}{\sqrt{3}}$ **54.** -2 **55.** $\sqrt{3}$ **56.** $-\frac{\sqrt{3}}{2}$ **57.** $\sqrt{2}$ **58.** -1 **59.** $\frac{1}{2}$ **60.** $-\sqrt{3}$

61. $\frac{2}{\sqrt{3}}$ **62.** $-\frac{2}{\sqrt{3}}$ **63.** $\frac{1}{\sqrt{2}}$ **64.** $-\frac{1}{2}$ **65.** $\frac{1}{\sqrt{3}}$ **66.** -2 **67.** 0 **68.** 0 **69.** undefined **70.** undefined

71. -1 **72.** undefined **73.** 1 **74.** 0 **75.** undefined **76.** undefined **77.** 1 **78.** 0 **79.** $\frac{1}{2}$ **80.** $-\frac{1}{\sqrt{2}}$

81. $\frac{1}{\sqrt{3}}$ **82.** -2 **83.** $\sqrt{3}$ **84.** $-\frac{\sqrt{3}}{2}$ **85.** $\sqrt{2}$ **86.** -1 **87.** $\frac{1}{2}$ **88.** $-\sqrt{3}$ **89.** $\frac{2}{\sqrt{3}}$ **90.** $-\frac{2}{\sqrt{3}}$

91. $\frac{1}{\sqrt{2}}$ **92.** $-\frac{1}{2}$ **93.** $\frac{1}{\sqrt{3}}$ **94.** -2

6.6 Exercises

SCE-1. $-\frac{3}{5}, \frac{3}{5}$ **SCE-2.** $-\frac{1}{2}, \frac{1}{2}$ **SCE-3.** $-\frac{2\sqrt{10}}{7}, \frac{2\sqrt{10}}{7}$ **SCE-4.** $-\frac{1}{\sqrt{2}}, \frac{1}{\sqrt{2}}$

1. $-\frac{\sqrt{5}}{3}$ **2.** $\frac{2\sqrt{6}}{5}$ **3.** $-\frac{4\sqrt{3}}{7}$ **4.** $-\frac{\sqrt{77}}{9}$ **5.** $-\frac{2\sqrt{11}}{3\sqrt{5}}$ **6.** $\left(\frac{1}{7}\right)^2 + \left(\frac{4\sqrt{3}}{7}\right)^2 = \frac{1}{49} + \frac{48}{49} = 1; \left(-\frac{1}{7}, \frac{4\sqrt{3}}{7}\right), \left(\frac{1}{7}, -\frac{4\sqrt{3}}{7}\right), \left(-\frac{1}{7}, -\frac{4\sqrt{3}}{7}\right)$ **7.** $\left(-\frac{1}{2}\right)^2 + \left(-\frac{\sqrt{3}}{2}\right)^2 = \frac{1}{4} + \frac{3}{4} = 1; \left(\frac{1}{2}, \frac{\sqrt{3}}{2}\right), \left(-\frac{1}{2}, \frac{\sqrt{3}}{2}\right), \left(\frac{1}{2}, -\frac{\sqrt{3}}{2}\right)$

8. $\left(-\frac{4}{\sqrt{17}}\right)^2 + \left(-\frac{1}{\sqrt{17}}\right)^2 = \frac{16}{17} + \frac{1}{17} = 1; \left(\frac{4}{\sqrt{17}}, \frac{1}{\sqrt{17}}\right), \left(-\frac{4}{\sqrt{17}}, \frac{1}{\sqrt{17}}\right), \left(\frac{4}{\sqrt{17}}, -\frac{1}{\sqrt{17}}\right)$ **9.** $\sin t = -\frac{4\sqrt{3}}{7}, \cos t = \frac{1}{7}, \tan t = -4\sqrt{3}, \csc t = -\frac{7}{4\sqrt{3}}, \sec t = 7, \cot t = -\frac{1}{4\sqrt{3}}$ **10.** $\sin t = -\frac{1}{\sqrt{17}}, \cos t = -\frac{4}{\sqrt{17}}, \tan t = \frac{1}{4}, \csc t = -\sqrt{17}, \sec t = -\frac{\sqrt{17}}{4}, \cot t = 4$ **11.** $\sin t = -\frac{\sqrt{3}}{2}, \cos t = -\frac{1}{2}, \tan t = \sqrt{3}, \csc t = -\frac{2}{\sqrt{3}}, \sec t = -2, \cot t = \frac{1}{\sqrt{3}}$

12. 1 **13.** 0 **14.** undefined **15.** -1 **16.** undefined **17.** 0 **18.** 0 **19.** -1 **20.** $\frac{1}{2}$ **21.** $-\frac{1}{\sqrt{2}}$ **22.** $\frac{1}{\sqrt{3}}$

23. $-\sqrt{2}$ **24.** $\sqrt{3}$ **25.** $-\frac{\sqrt{3}}{2}$ **26.** $\sqrt{2}$ **27.** -1 **28.** $\frac{1}{2}$ **29.** $-\sqrt{3}$ **30.** $\frac{2}{\sqrt{3}}$ **31.** $-\frac{2}{\sqrt{3}}$ **32.** $\frac{\sqrt{3}}{2}$

33. $-\frac{1}{2}$ **34.** $\frac{1}{\sqrt{3}}$ **35.** $-\frac{2}{\sqrt{3}}$ **36.** 1 **37.** 0 **38.** undefined **39.** -1 **40.** undefined **41.** 0 **42.** 0 **43.** -1

44. $\frac{1}{2}$ **45.** $-\frac{1}{\sqrt{2}}$ **46.** $\frac{1}{\sqrt{3}}$ **47.** $-\sqrt{2}$ **48.** $\sqrt{3}$ **49.** $-\frac{\sqrt{3}}{2}$ **50.** $\sqrt{2}$ **51.** -1 **52.** $\frac{1}{2}$ **53.** $-\sqrt{3}$

54. $\frac{2}{\sqrt{3}}$ **55.** $-\frac{2}{\sqrt{3}}$ **56.** $\frac{\sqrt{3}}{2}$ **57.** $-\frac{1}{2}$ **58.** $\frac{1}{\sqrt{3}}$ **59.** $-\frac{2}{\sqrt{3}}$

Review Exercises

1. y, 480°, x ; Quadrant II **2.** y, $-\frac{13\pi}{6}$, x ; Quadrant IV **3.** $\frac{\pi}{3}$ **4.** ≈ 282.74 in.^{2} **5.** ≈ 226.19 cm

6. 1516 miles **7.** ≈ 142.8 mph **8.** obtuse, scalene **9.** $\sqrt{7}$ **10.** ≈ 8.77 m **11.** $k = \frac{2}{\sqrt{3}}, LV = \frac{2\sqrt{5}}{\sqrt{3}}, LZ = 2\sqrt{3}$

12. $10, 10\sqrt{2}$ **13.** $\sqrt{2}, 2\sqrt{2}$ **14.** $\sin\theta = \frac{3}{\sqrt{13}}, \cos\theta = \frac{2}{\sqrt{13}}, \tan\theta = \frac{3}{2}, \csc\theta = \frac{\sqrt{13}}{3}, \sec\theta = \frac{\sqrt{13}}{2}, \cot\theta = \frac{2}{3}$

15. $\sin\theta = \frac{2\sqrt{11}}{7}, \cos\theta = \frac{\sqrt{5}}{7}, \tan\theta = \frac{2\sqrt{11}}{\sqrt{5}}, \csc\theta = \frac{7}{2\sqrt{11}}, \cot\theta = \frac{\sqrt{5}}{2\sqrt{11}}$ **16.** $\frac{\pi}{6}$ **17.** $\tan\theta = \frac{7}{24}, \csc\theta = \frac{25}{7},$

$\sec\theta = \frac{25}{24}, \cot\theta = \frac{24}{7}$ **18.** $\csc\theta$ **19.** $\cos\left(-\frac{3\pi}{14}\right)$ **20.** $\sin\theta = -\frac{4}{\sqrt{17}}, \cos\theta = \frac{1}{\sqrt{17}}, \tan\theta = -4, \csc\theta = -\frac{\sqrt{17}}{4},$

$\sec\theta = \sqrt{17}, \cot\theta = -\frac{1}{4}$ **21.** -1 **22.** Quadrants III and IV **23.** Quadrant IV **24.** $\frac{2\pi}{9}$ **25.** $-\frac{\sqrt{3}}{2}$

26. $-\frac{\sqrt{5}}{3}$ **27.** $\left(\frac{1}{7}\right)^2 + \left(\frac{4\sqrt{3}}{7}\right)^2 = \frac{1}{49} + \frac{48}{49} = 1; \left(-\frac{1}{7}, \frac{4\sqrt{3}}{7}\right), \left(\frac{1}{7}, -\frac{4\sqrt{3}}{7}\right), \left(-\frac{1}{7}, -\frac{4\sqrt{3}}{7}\right)$

28. $\sin t = -\frac{1}{\sqrt{17}}, \cos t = -\frac{4}{\sqrt{17}}, \tan t = \frac{1}{4},$ $\csc t = -\sqrt{17}, \sec t = -\frac{\sqrt{17}}{4}, \cot t = 4$ **29.** 0 **30.** $-\sqrt{3}$

CHAPTER 7

7.1 Exercises

SCE-1. $\frac{\pi}{10}$ **SCE-2.** 6

1. y, 1, -2π, 2π, x ; the following properties apply: The function is an odd function. The function is decreasing on the interval $\left(\pi, \frac{3\pi}{2}\right)$.

The domain is $(-\infty, \infty)$. The range is $[-1, 1]$. The y-intercept is 0. The zeros are of the form $n\pi$, where n is any integer. The function obtains a relative maximum at $x = \frac{\pi}{2} + 2n\pi$, where n is an integer. The function obtains a relative minimum at $x = \frac{3\pi}{2} + 2n\pi$, where n is an integer. **2.** $-\frac{11\pi}{6}, -\frac{7\pi}{6}, \frac{\pi}{6}, \frac{5\pi}{6}, \frac{13\pi}{6}$ **3.** $-\frac{7\pi}{4}, -\frac{5\pi}{4}, \frac{\pi}{4}$ **4.** $-\frac{8\pi}{3}, -\frac{7\pi}{3}, -\frac{2\pi}{3}, -\frac{\pi}{3}, \frac{4\pi}{3}$

5. iii **6.** i **7.** ii **8.** i **9.** 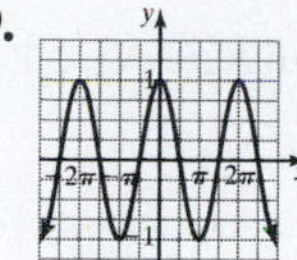 ; the following properties apply: The function is an even function. The function is increasing on the interval $\left(-\pi, -\frac{\pi}{2}\right)$. The domain is $(-\infty, \infty)$. The range is $[-1, 1]$. The y-intercept is 1. The zeros are of the form $\frac{(2n+1)\pi}{2}$, where n is any integer. The function obtains a relative maximum at $x = 2\pi n$, where n is an integer.

10. $-\frac{5\pi}{2}, -\frac{3\pi}{2}, -\frac{\pi}{2}, \frac{\pi}{2}$ **11.** $-\frac{9\pi}{4}, -\frac{7\pi}{4}, -\frac{\pi}{4}, \frac{\pi}{4}$ **12.** $-\frac{19\pi}{6}, -\frac{17\pi}{6}, -\frac{7\pi}{6}, -\frac{5\pi}{6}, \frac{5\pi}{6}, \frac{7\pi}{6}$ **13.** iii **14.** i **15.** ii

16. i **17.** amplitude: 3; range: $[-3, 3]$; 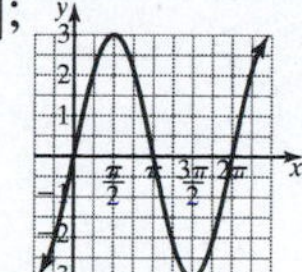 **18.** amplitude: 4; range: $[-4, 4]$; 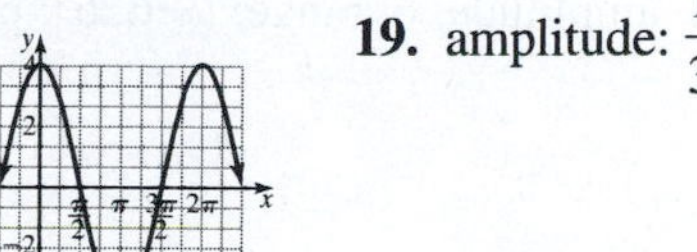 **19.** amplitude: $\frac{1}{3}$; range: $\left[-\frac{1}{3}, \frac{1}{3}\right]$; **20.** amplitude: $\frac{1}{4}$; range: $\left[-\frac{1}{4}, \frac{1}{4}\right]$; 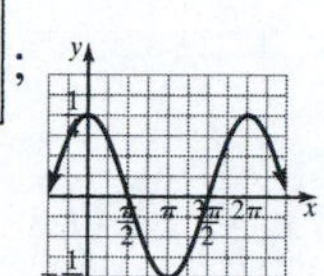 **21.** amplitude: $\frac{3}{2}$; range: $\left[-\frac{3}{2}, \frac{3}{2}\right]$; 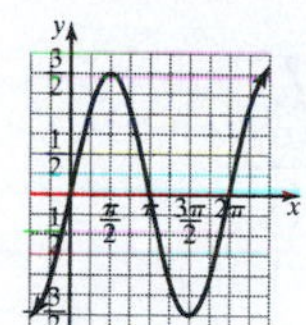

22. amplitude: 2; range: $[-2, 2]$; 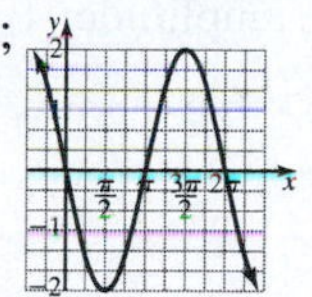 **23.** amplitude: 5; range: $[-5, 5]$; 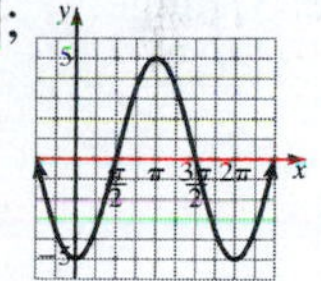

24. amplitude: $\frac{3}{4}$; range: $\left[-\frac{3}{4}, \frac{3}{4}\right]$; 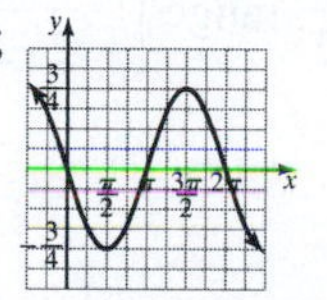 **25.** amplitude: $\frac{1}{2}$; range: $\left[-\frac{1}{2}, \frac{1}{2}\right]$; 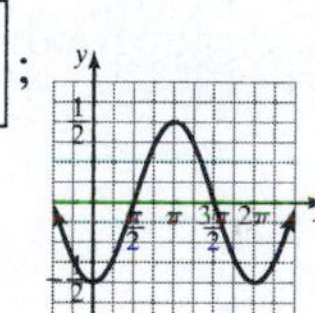 **26.** amplitude: $\frac{5}{4}$; range: $\left[-\frac{5}{4}, \frac{5}{4}\right]$; 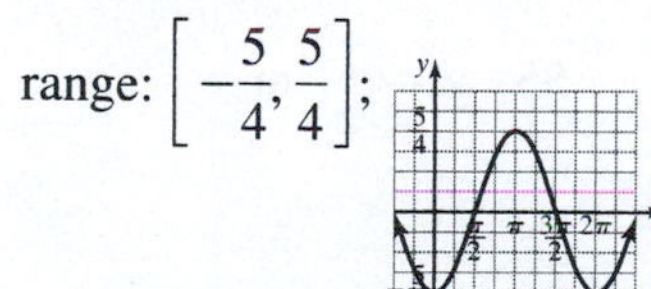

27. $\frac{2\pi}{3}$; 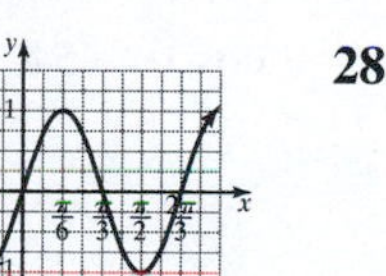 **28.** 6π; 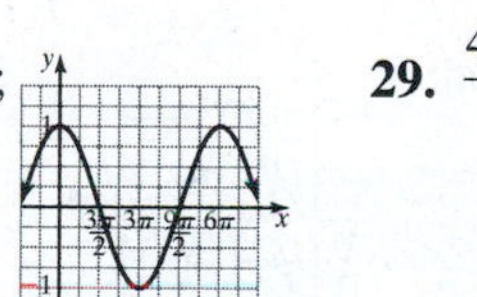 **29.** $\frac{4\pi}{3}$; 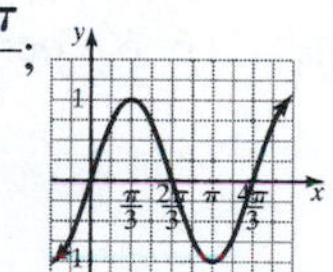

30. $\frac{2}{3}$; **31.** 6; **32.** $\frac{\pi}{2}$; **33.** 8π; 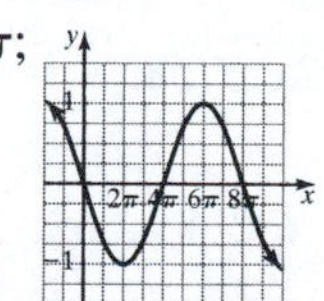 **34.** $\frac{4\pi}{5}$; 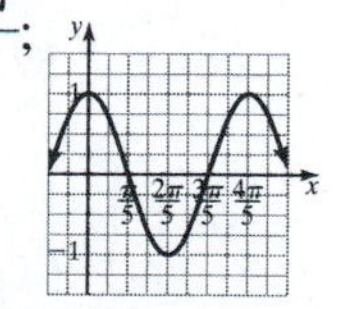

35. $\frac{2}{3}$; **36.** 4; **37.** amplitude: 2; range: $[-2, 2]$; period: $\frac{2\pi}{3}$; 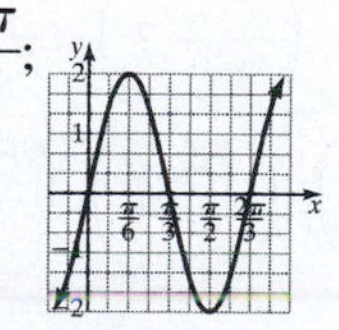

38. amplitude: $\frac{5}{3}$; range: $\left[-\frac{5}{3}, \frac{5}{3}\right]$; period: $\frac{\pi}{2}$; 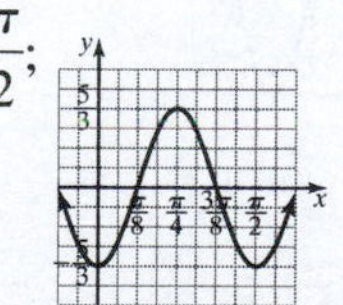 **39.** amplitude: $\frac{3}{4}$; range: $\left[-\frac{3}{4}, \frac{3}{4}\right]$; period: π; 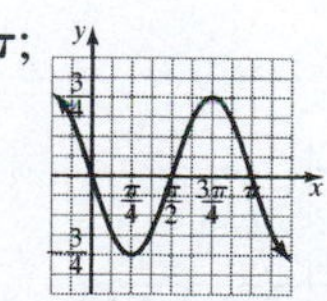

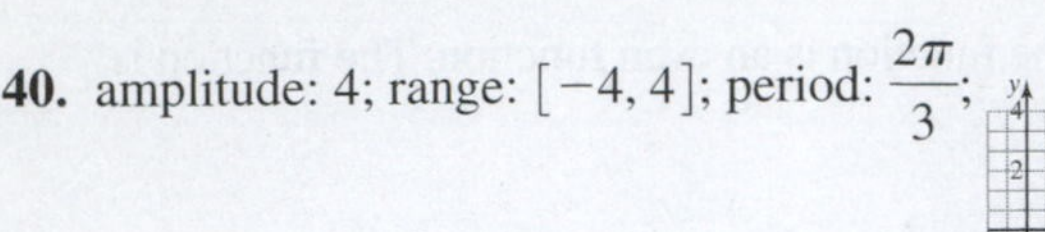

40. amplitude: 4; range: $[-4, 4]$; period: $\frac{2\pi}{3}$;

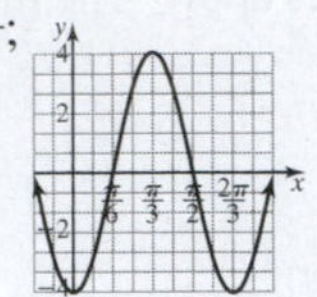

41. amplitude: 5; range: $[-5, 5]$; period: 4π;

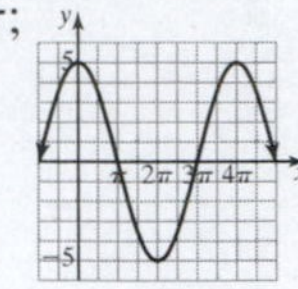

42. amplitude: $\frac{1}{2}$; range: $\left[-\frac{1}{2}, \frac{1}{2}\right]$; period: 8π;

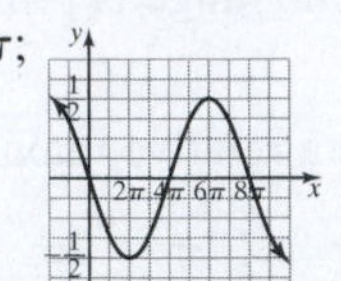

43. amplitude: 1; range: $[-1, 1]$; period: 10π;

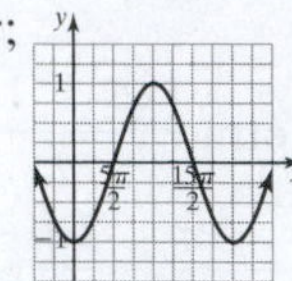

44. amplitude: 6; range: $[-6, 6]$; period: $\frac{4\pi}{3}$;

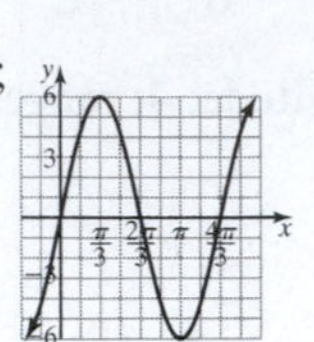

45. amplitude: 2; range: $[-2, 2]$; period: $\frac{3\pi}{2}$;

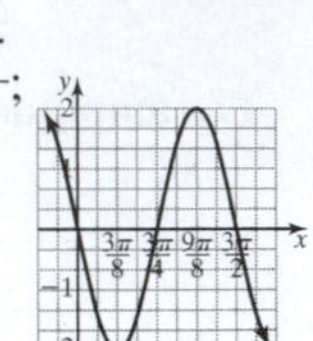

46. amplitude: 2; range: $[-2, 2]$; period: $\frac{8\pi}{5}$;

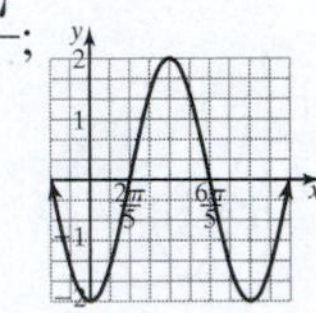

47. amplitude: 1; range: $[-1, 1]$; period: $\frac{4\pi}{5}$;

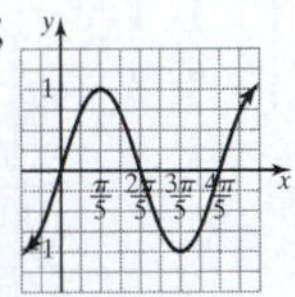

48. amplitude: 3; range: $[-3, 3]$; period: 4;

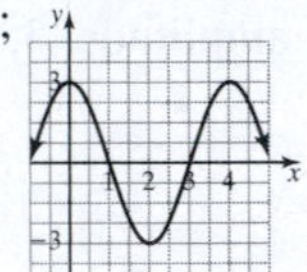

49. amplitude: 4; range: $[-4, 4]$; period: 6;

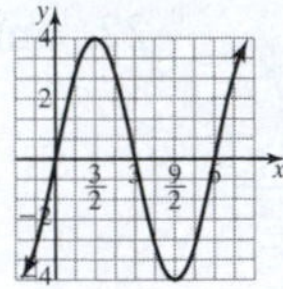

50. amplitude: $\frac{5}{4}$; range: $\left[-\frac{5}{4}, \frac{5}{4}\right]$; period: 2;

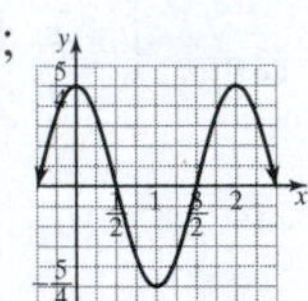

51. amplitude: $\frac{1}{2}$; range: $\left[-\frac{1}{2}, \frac{1}{2}\right]$; period: 2;

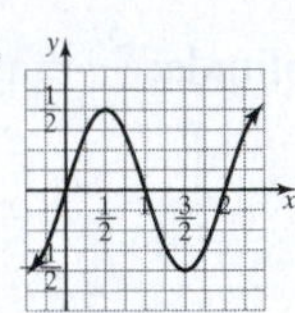

52. amplitude: 6; range: $[-6, 6]$; period: 3;

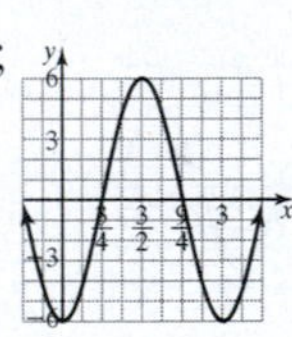

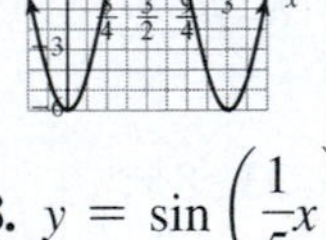

53. $y = 3\cos x$ **54.** $y = -4\sin x$ **55.** $y = \frac{1}{2}\sin x$

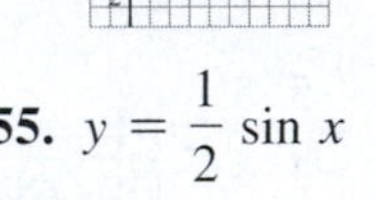

56. $y = -\frac{1}{3}\cos x$ **57.** $y = \cos 2x$ **58.** $y = \sin\left(\frac{1}{5}x\right)$ **59.** $y = 3\sin 4x$ **60.** $y = -5\cos 3x$ **61.** $y = 4\cos\left(\frac{1}{5}x\right)$

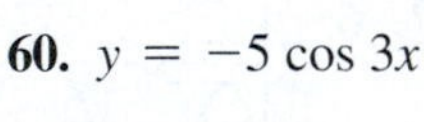

62. $y = -2\sin\left(\frac{1}{4}x\right)$ **63.** 3 **64.** 1 **65.** 2 **66.** $[-3, 3]$ **67.** $[-1, 1]$ **68.** $[-2, 2]$ **69.** 2π **70.** $\frac{2\pi}{3}$

71. $\frac{8\pi}{5}$ **72.** $(0, 0), \left(\frac{\pi}{2}, -3\right), (\pi, 0), \left(\frac{3\pi}{2}, 3\right), (2\pi, 0)$ **73.** $(0, 1), \left(\frac{\pi}{6}, 0\right), \left(\frac{\pi}{3}, -1\right), \left(\frac{\pi}{2}, 0\right), \left(\frac{2\pi}{3}, 1\right)$

74. $(0, -2), \left(\frac{2\pi}{5}, 0\right), \left(\frac{4\pi}{5}, 2\right), \left(\frac{6\pi}{5}, 0\right), \left(\frac{8\pi}{5}, -2\right)$

75.

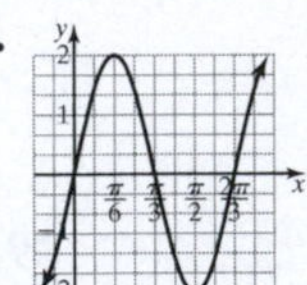

76. **77.**

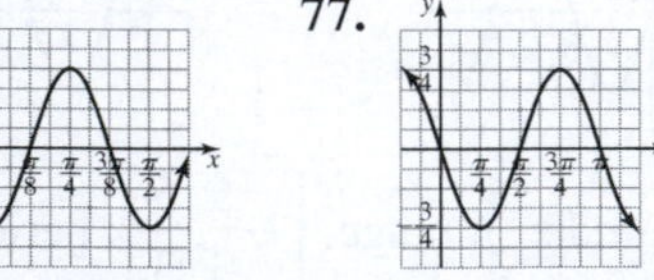

78.

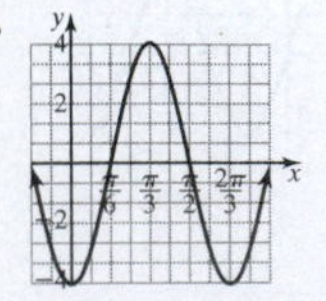

79.

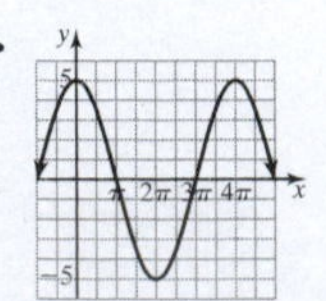

80.

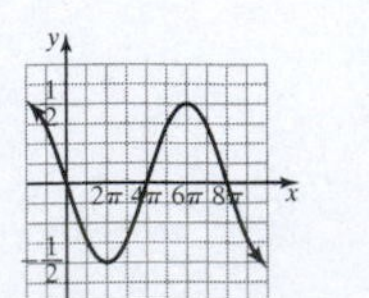

81.

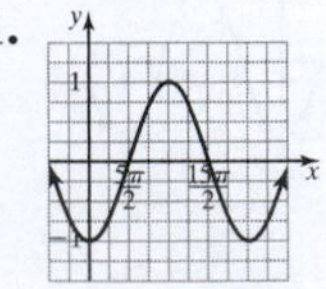

82.

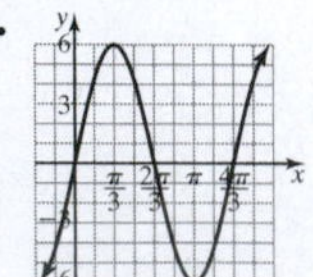

83.

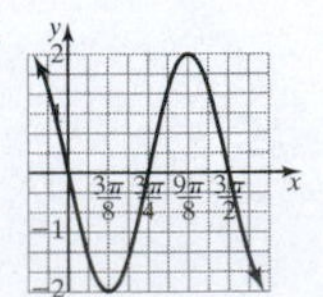

84. 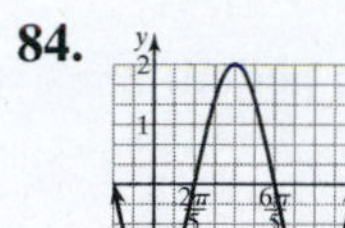 **85.** 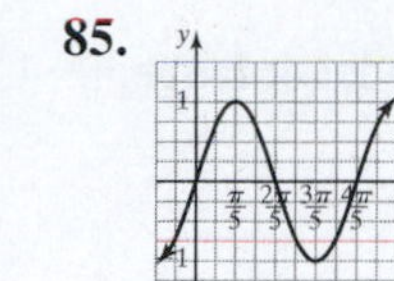 **86.** 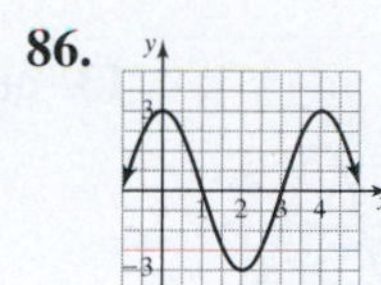 **87.** 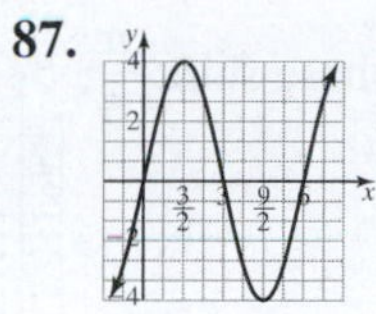 **88.** 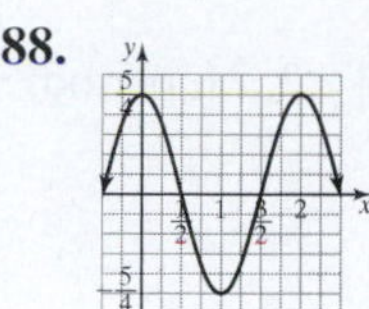 **89.** 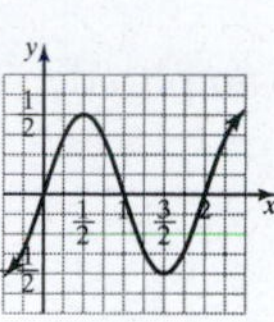

90. 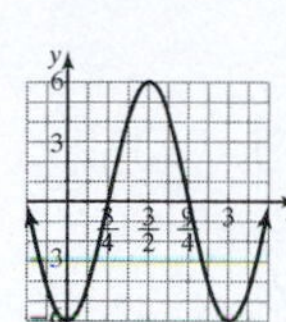 **91.** $y = 3\cos x$ **92.** $y = -4\sin x$ **93.** $y = \frac{1}{2}\sin x$ **94.** $y = -\frac{1}{3}\cos x$ **95.** $y = \cos 2x$

96. $y = \sin\left(\frac{1}{5}x\right)$ **97.** $y = 3\sin 4x$ **98.** $y = -5\cos 3x$ **99.** $y = 4\cos\left(\frac{1}{5}x\right)$ **100.** $y = -2\sin\left(\frac{1}{4}x\right)$

7.2 Exercises

SCE-1. $\left[\frac{\pi}{2}, \frac{3\pi}{2}\right]$ **SCE-2.** $\left[\frac{\pi}{18}, \frac{13\pi}{18}\right]$ **SCE-3.** $\left[\frac{3}{\pi}, 2 + \frac{3}{\pi}\right]$ **SCE-4.** $\frac{\pi}{4}$ **SCE-5.** $\frac{\pi}{8}$ **SCE-6.** $4\left(x - \frac{\pi}{4}\right)$

SCE-7. $5\left(x - \frac{\pi}{15}\right)$ **SCE-8.** $-3\left(x - \frac{\pi}{3}\right)$ **SCE-9.** $-4\left(x - \frac{\pi}{12}\right)$ **SCE-10.** $-2\left(x - \frac{\pi}{2}\right)$ **SCE-11.** $\pi\left(x - \frac{3}{\pi}\right)$

1. amplitude: 1; range: $[-1, 1]$; period: 2π; phase shift: π; **2.** amplitude: 1; range: $[-1, 1]$; period: 2π; phase shift: $-\frac{\pi}{2}$; 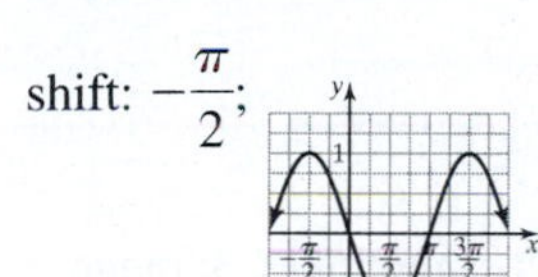 **3.** amplitude: 1; range: $[-1, 1]$; period: 2π; phase shift: $\frac{\pi}{6}$; 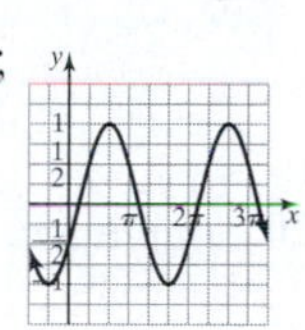 **4.** amplitude: 1; range: $[-1, 1]$; period: 2π; phase shift: $-\frac{2\pi}{3}$; 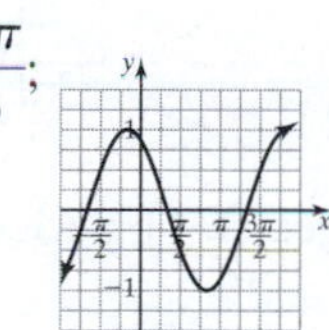 **5.** amplitude: 1; range: $[-1, 1]$; period: 2π; phase shift: 2; 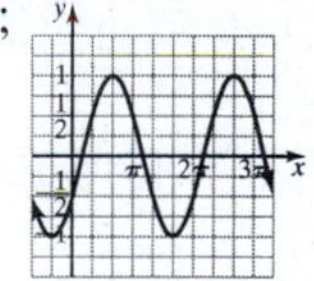

6. amplitude: 1; range: $[-1, 1]$; period: $\frac{2\pi}{3}$; phase shift: $-\frac{\pi}{3}$; 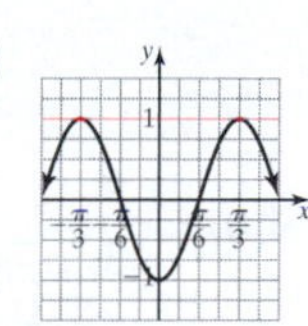 **7.** amplitude: 1; range: $[-1, 1]$; period: π; phase shift: $\frac{\pi}{2}$; **8.** amplitude: 3; range: $[-3, 3]$; period: $\frac{\pi}{2}$; phase shift: $\frac{\pi}{4}$; 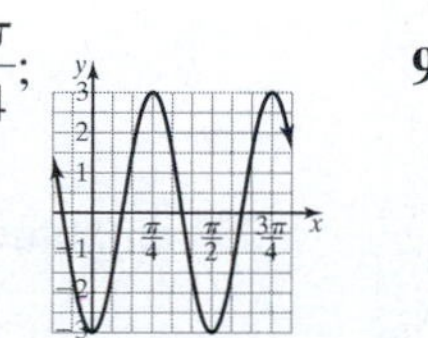 **9.** amplitude: 2; range: $[-2, 2]$; period: $\frac{\pi}{3}$; phase shift: $-\frac{\pi}{6}$; 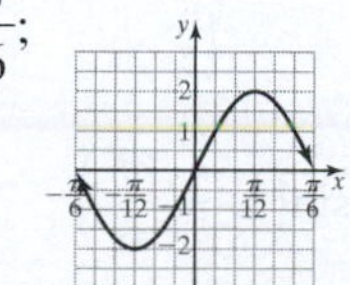 **10.** amplitude: 4; range: $[-4, 4]$; period: π; phase shift: $-\frac{\pi}{2}$; **11.** amplitude: 3; range: $[-3, 3]$; period: π; phase shift: $-\frac{\pi}{4}$; 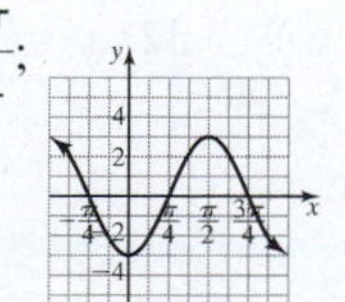

12. amplitude: 2; range: $[-2, 2]$; period: $\frac{2\pi}{3}$; phase shift: $\frac{\pi}{6}$; 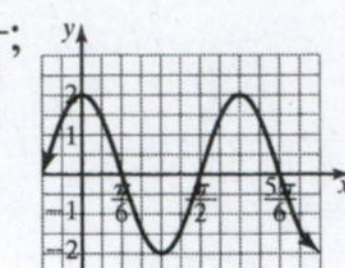 **13.** amplitude: 4; range: $[-4, 4]$; period: 2π; phase shift: $-\frac{\pi}{3}$; 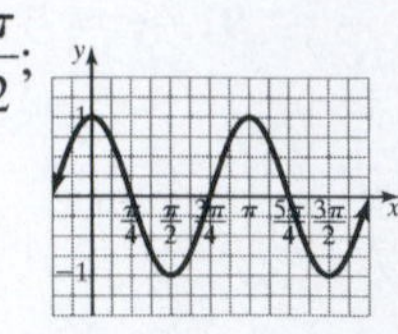 **14.** amplitude: 1; range: $[-1, 1]$; period: π; phase shift: $\frac{\pi}{2}$; 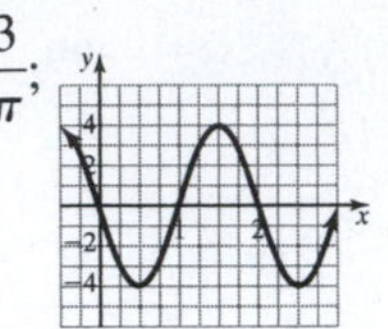

15. amplitude: 4; range: $[-4, 4]$; period: 2; phase shift: $\frac{3}{\pi}$; 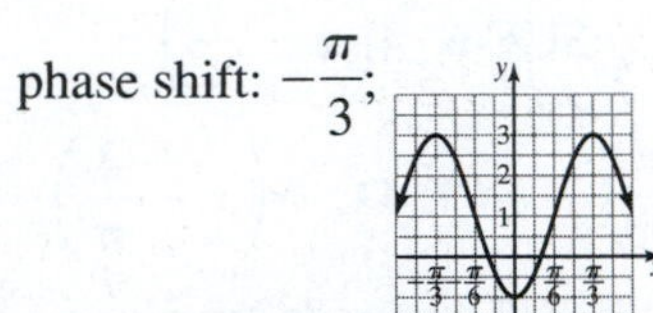 **16.** amplitude: 2; range: $[-1, 3]$; period: $\frac{2\pi}{3}$; phase shift: $-\frac{\pi}{3}$; 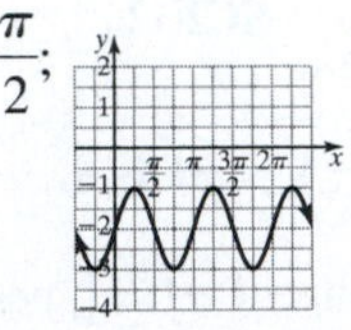 **17.** amplitude: 1; range: $[-3, -1]$; period: π; phase shift: $\frac{\pi}{2}$; 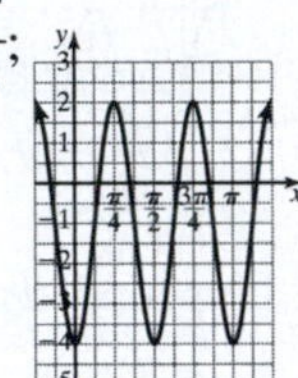

18. amplitude: 3; range: $[-4, 2]$; period: $\frac{\pi}{2}$; phase shift: $\frac{\pi}{4}$; 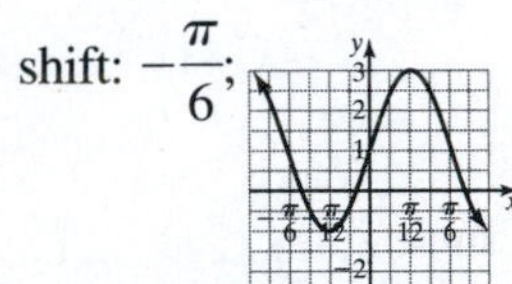 **19.** amplitude: 2; range: $[-1, 3]$; period: $\frac{\pi}{3}$; phase shift: $-\frac{\pi}{6}$; 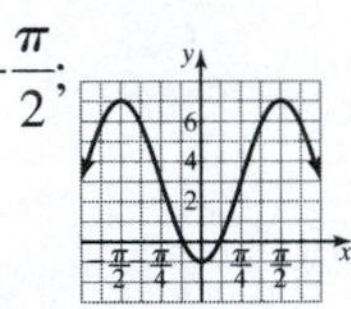 **20.** amplitude: 4; range: $[-1, 7]$; period: π; phase shift: $-\frac{\pi}{2}$; 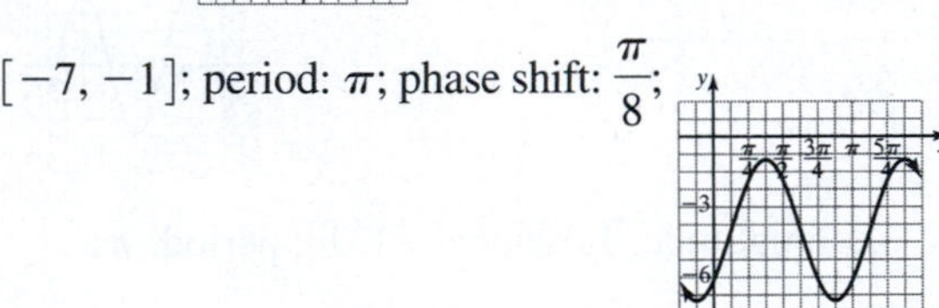 **21.** amplitude: 3; range: $[-7, -1]$; period: π; phase shift: $\frac{\pi}{8}$; 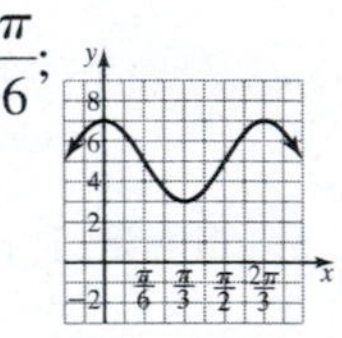

22. amplitude: 2; range: $[3, 7]$; period: $\frac{2\pi}{3}$; phase shift: $\frac{\pi}{6}$; 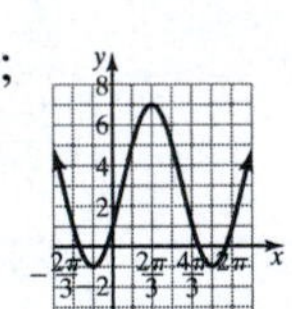

23. amplitude: 4; range: $[-1, 7]$; period: 2π; phase shift: $-\frac{\pi}{3}$; 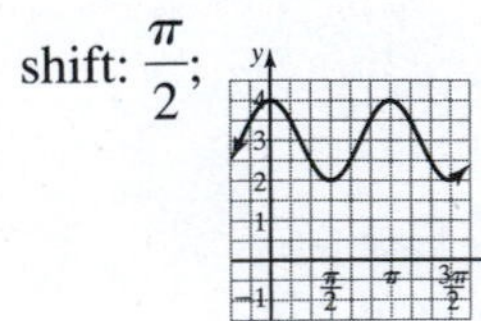 **24.** amplitude: 1; range: $[2, 4]$ period: π; phase shift: $\frac{\pi}{2}$; 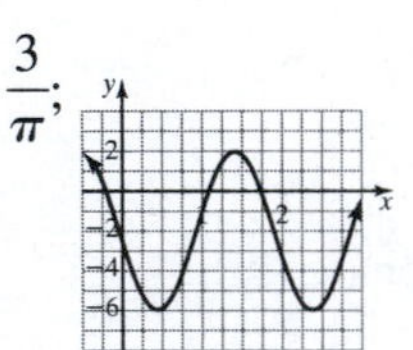

25. amplitude: 4; range: $[-6, 2]$; period: 2; phase shift: $\frac{3}{\pi}$; **26.** $y = \sin\left(x - \frac{\pi}{3}\right)$

27. $y = \cos\left(x + \frac{2\pi}{3}\right)$ **28.** $y = -3\cos(x + \pi)$ **29.** $y = -4\sin\left(x + \frac{\pi}{6}\right)$ **30.** $y = 2\sin\left(x - \frac{5\pi}{6}\right)$

31. $y = -5\cos(3x - \pi)$ **32.** $y = -3\cos\left(x - \frac{\pi}{4}\right) - 1$ **33.** $y = -2\sin\left(3x - \frac{\pi}{2}\right) + 4$ **34.** 1 **35.** 3 **36.** 2

37. $[-1, 1]$ **38.** $[-3, 3]$ **39.** $[-4, 0]$ **40.** 2π **41.** π **42.** 2π **43.** $\frac{\pi}{6}$ **44.** $-\frac{\pi}{2}$ **45.** $\frac{\pi}{3}$

46. $\left(\frac{\pi}{6}, 0\right), \left(\frac{2\pi}{3}, 1\right), \left(\frac{7\pi}{6}, 0\right), \left(\frac{5\pi}{3}, -1\right), \left(\frac{13\pi}{6}, 0\right)$ **47.** $\left(-\frac{\pi}{2}, -3\right), \left(-\frac{\pi}{4}, 0\right), (0, 3), \left(\frac{\pi}{4}, 0\right), \left(\frac{\pi}{2}, -3\right)$

48. $\left(\frac{\pi}{3}, -2\right), \left(\frac{5\pi}{6}, -4\right), \left(\frac{4\pi}{3}, -2\right), \left(\frac{11\pi}{6}, 0\right), \left(\frac{7\pi}{3}, -2\right)$ **49.** 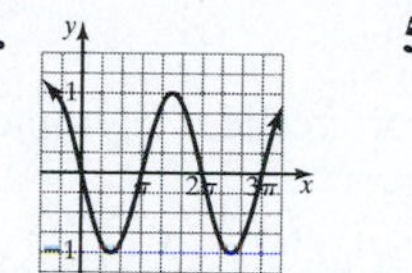 **50.** 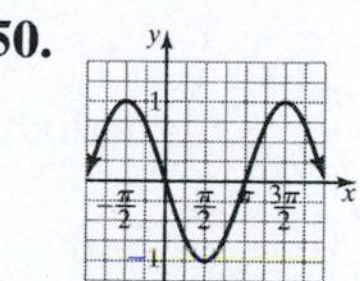 **51.** 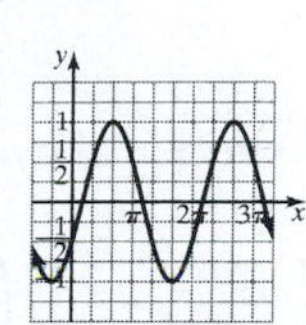

52. 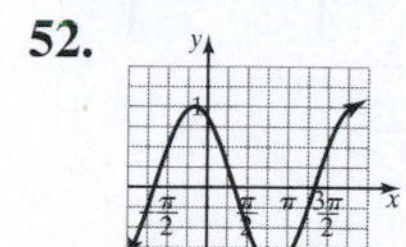 **53.** 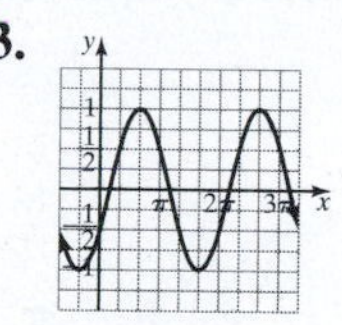 **54.** 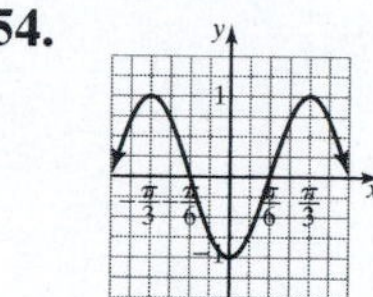 **55.** 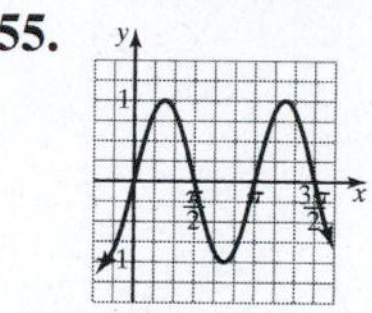 **56.** 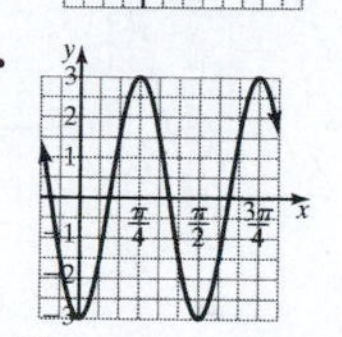 **57.** 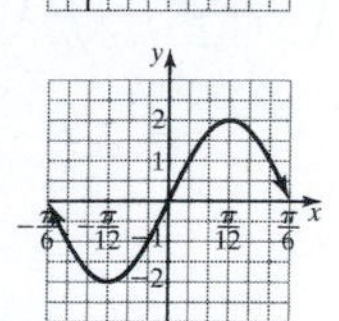

58. 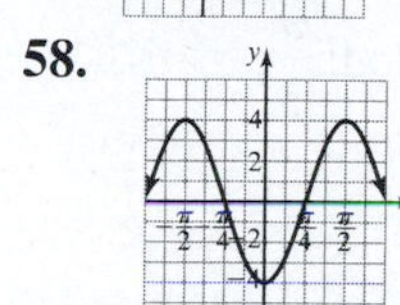 **59.** 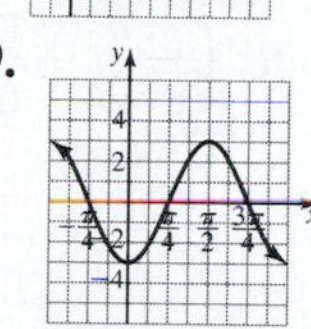 **60.** 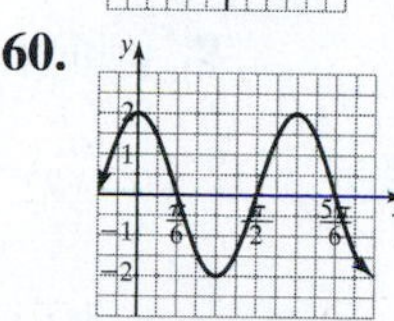 **61.** 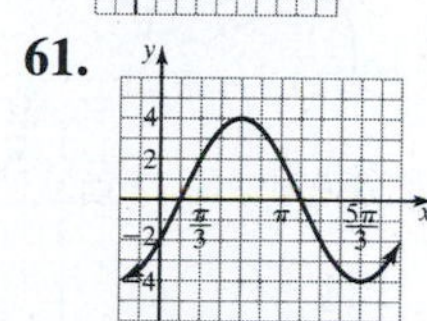 **62.** 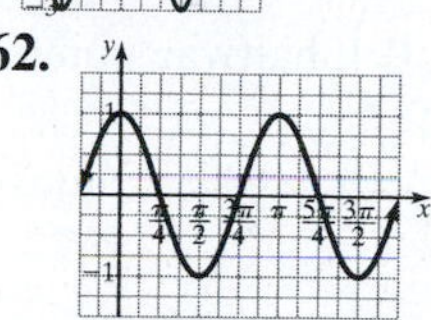

63. 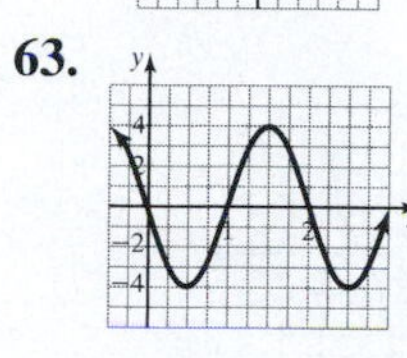 **64.** 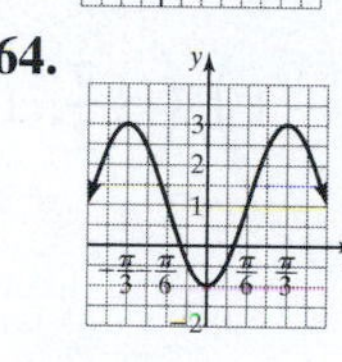 **65.** 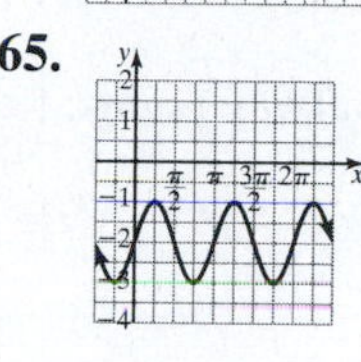 **66.** 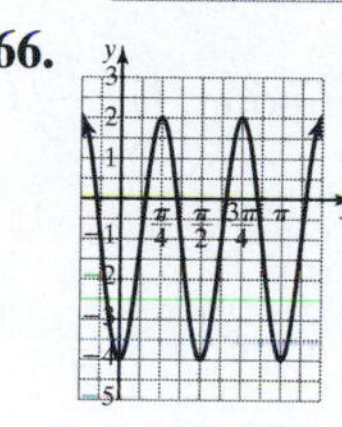 **67.** 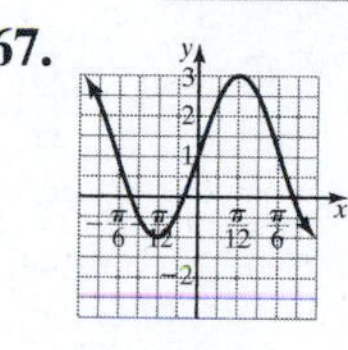 **68.** 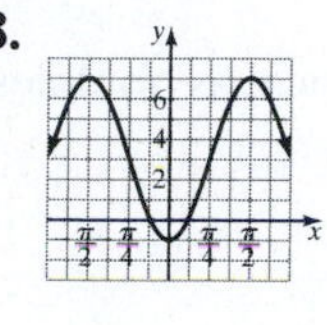

69. 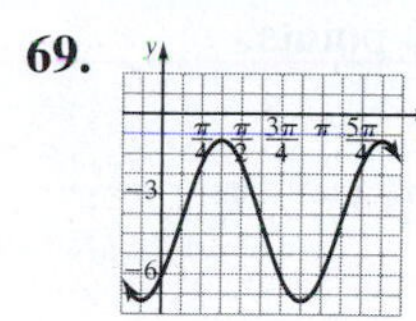 **70.** 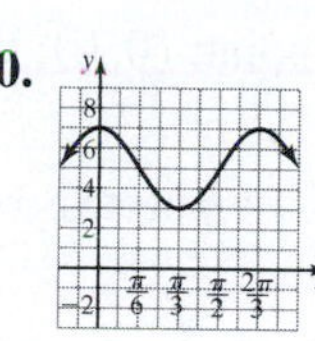 **71.** 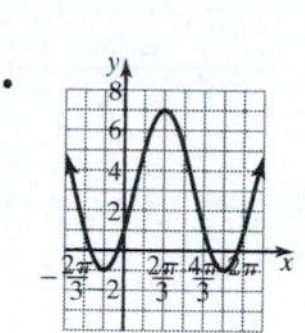 **72.** 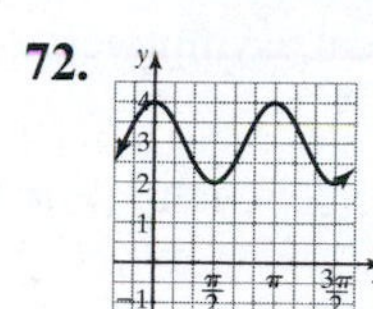 **73.** $y = \sin\left(x - \frac{\pi}{3}\right)$

74. $y = \cos\left(x + \frac{2\pi}{3}\right)$ **75.** $y = -3\cos(x + \pi)$ **76.** $y = -4\sin\left(x + \frac{\pi}{6}\right)$ **77.** $y = 2\sin\left(x - \frac{5\pi}{6}\right)$

78. $y = -5\cos(3x - \pi)$ **79.** $y = -3\cos\left(x - \frac{\pi}{4}\right) - 1$ **80.** $y = -2\sin\left(3x - \frac{\pi}{2}\right) + 4$

7.3 Exercises

SCE-1. $\left[\frac{\pi}{4}, \frac{3\pi}{4}\right]$ **SCE-2.** $\left[-\frac{\pi}{3}, 0\right]$ **SCE-3.** $\left[-\frac{3\pi}{2}, \frac{3\pi}{2}\right]$ **SCE-4.** $[0, 5\pi]$

1. ; the function is an odd function. The domain is all real numbers except odd integer multiples of $\frac{\pi}{2}$. The range is $(-\infty, \infty)$. Every halfway point has a y-coordinate of -1 or 1. The interval for the graph of the principal cycle is $\left(-\frac{\pi}{2}, \frac{\pi}{2}\right)$. The y-intercept is 0. The function has a period of $P = \pi$. The zeros are of the form $n\pi$, where n is an integer.

2. $\left(-\frac{5\pi}{3}, \sqrt{3}\right), \left(-\frac{2\pi}{3}, \sqrt{3}\right), \left(\frac{\pi}{3}, \sqrt{3}\right), \left(\frac{4\pi}{3}, \sqrt{3}\right), \left(\frac{7\pi}{3}, \sqrt{3}\right)$ **3.** $\left(-\frac{13\pi}{6}, -\frac{1}{\sqrt{3}}\right), \left(-\frac{7\pi}{6}, -\frac{1}{\sqrt{3}}\right), \left(-\frac{\pi}{6}, -\frac{1}{\sqrt{3}}\right)$

4. $\left(-\frac{7\pi}{4}, 1\right), \left(-\frac{5\pi}{4}, -1\right), \left(-\frac{3\pi}{4}, 1\right), \left(-\frac{\pi}{4}, -1\right), \left(\frac{\pi}{4}, 1\right), \left(\frac{3\pi}{4}, -1\right)$ **5.** iii **6.** principal cycle: $\left(-\frac{3\pi}{4}, \frac{\pi}{4}\right)$; period: π;

vertical asymptotes: $x = -\frac{3\pi}{4}, x = \frac{\pi}{4}$; center point: $\left(-\frac{\pi}{4}, 0\right)$; halfway points: $\left(-\frac{\pi}{2}, -1\right)$, $(0, 1)$;

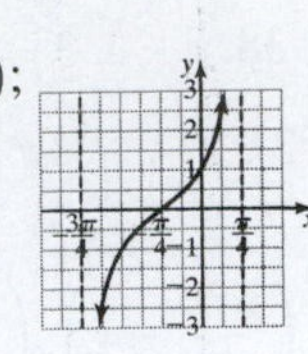

7. principal cycle: $\left(-\frac{\pi}{3}, \frac{2\pi}{3}\right)$; period: π; vertical asymptotes: $x = -\frac{\pi}{3}, x = \frac{2\pi}{3}$; center point: $\left(\frac{\pi}{6}, 0\right)$; halfway points: $\left(-\frac{\pi}{12}, -3\right), \left(\frac{5\pi}{12}, 3\right)$;

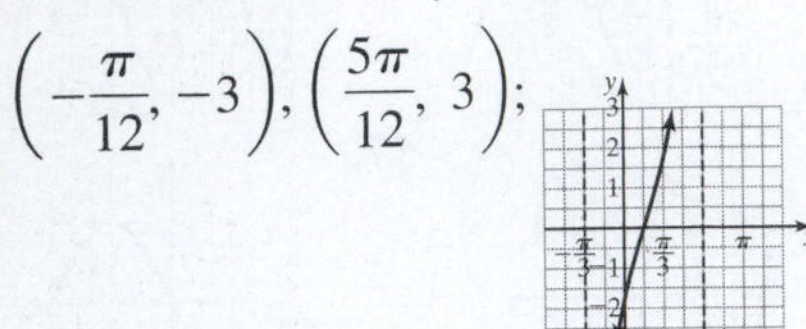

8. principal cycle: $\left(-\frac{\pi}{4}, \frac{3\pi}{4}\right)$; period: π; vertical asymptotes: $x = -\frac{\pi}{4}, x = \frac{3\pi}{4}$; center point: $\left(\frac{\pi}{4}, 0\right)$; halfway points: $(0, 2), \left(\frac{\pi}{2}, -2\right)$;

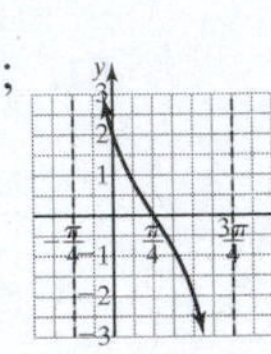

9. principal cycle: $\left(-\frac{\pi}{2}, -\frac{\pi}{6}\right)$; period: $\frac{\pi}{3}$; vertical asymptotes: $x = -\frac{\pi}{2}, x = -\frac{\pi}{6}$; center point: $\left(-\frac{\pi}{3}, 0\right)$; halfway points: $\left(-\frac{5\pi}{12}, -1\right), \left(-\frac{\pi}{4}, 1\right)$;

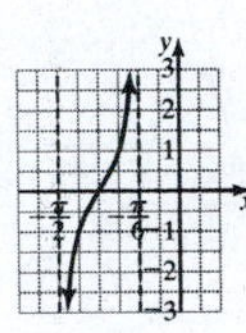

10. principal cycle: $\left(-\frac{\pi}{4}, \frac{\pi}{4}\right)$; period: $\frac{\pi}{2}$; vertical asymptotes: $x = -\frac{\pi}{4}, x = \frac{\pi}{4}$; center point: $(0, 0)$; halfway points: $\left(-\frac{\pi}{8}, -1\right), \left(\frac{\pi}{8}, 1\right)$;

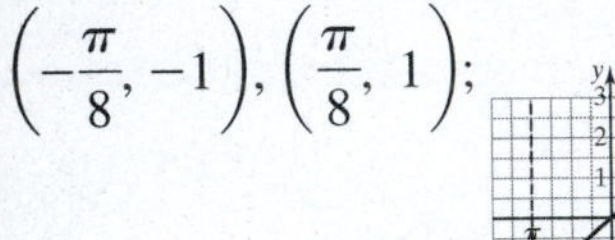

11. principal cycle: $(-\pi, \pi)$; period: 2π; vertical asymptotes: $x = -\pi, x = \pi$; center point: $(0, 0)$; halfway points: $\left(-\frac{\pi}{2}, -1\right), \left(\frac{\pi}{2}, 1\right)$;

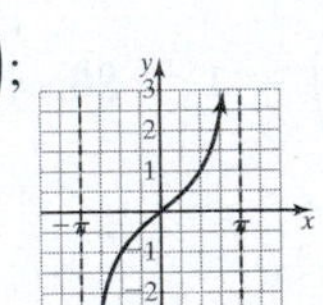

12. principal cycle: $\left(-\frac{\pi}{6}, \frac{\pi}{6}\right)$; period: $\frac{\pi}{3}$; vertical asymptotes: $x = -\frac{\pi}{6}, x = \frac{\pi}{6}$; center point: $(0, 0)$; halfway points: $\left(-\frac{\pi}{12}, 2\right), \left(\frac{\pi}{12}, -2\right)$;

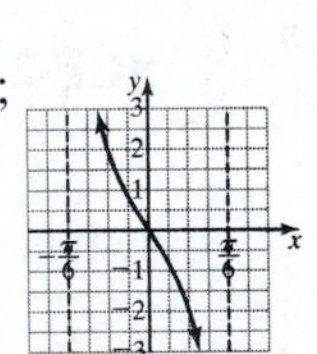

13. principal cycle: $(-3\pi, -\pi)$; period: 2π; vertical asymptotes: $x = -3\pi, x = -\pi$; center point: $(-2\pi, 0)$; halfway points: $\left(-\frac{5\pi}{2}, -1\right), \left(-\frac{3\pi}{2}, 1\right)$;

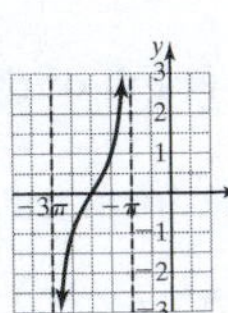

14. principal cycle: $\left(\frac{\pi}{4}, \frac{3\pi}{4}\right)$; period: $\frac{\pi}{2}$; vertical asymptotes: $x = \frac{\pi}{4}, x = \frac{3\pi}{4}$; center point: $\left(\frac{\pi}{2}, 0\right)$; halfway points: $\left(\frac{3\pi}{8}, 1\right), \left(\frac{5\pi}{8}, -1\right)$;

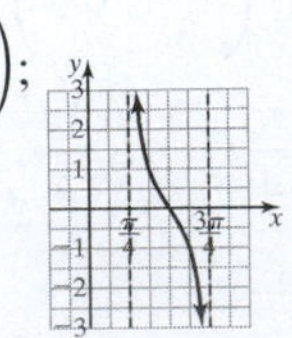

15. principal cycle: $\left(-\frac{3\pi}{4}, -\frac{\pi}{4}\right)$; period: $\frac{\pi}{2}$;

vertical asymptotes: $x = -\frac{3\pi}{4}, x = -\frac{\pi}{4}$; center point: $\left(-\frac{\pi}{2}, -1\right)$; halfway points: $\left(-\frac{5\pi}{8}, -4\right), \left(-\frac{3\pi}{8}, 2\right)$;

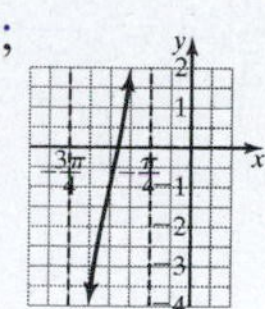

16. principal cycle: $\left(\frac{\pi}{4}, \frac{3\pi}{4}\right)$; period: $\frac{\pi}{2}$; vertical asymptotes: $x = \frac{\pi}{4}, x = \frac{3\pi}{4}$; center point: $\left(\frac{\pi}{2}, 1\right)$; halfway points: $\left(\frac{3\pi}{8}, 5\right), \left(\frac{5\pi}{8}, -3\right)$; **17.** principal cycle: $\left(-\frac{\pi}{6}, \frac{\pi}{6}\right)$; period: $\frac{\pi}{3}$; vertical asymptotes: $x = -\frac{\pi}{6}, x = \frac{\pi}{6}$; center point: $(0, 4)$; halfway points: $\left(-\frac{\pi}{12}, \frac{9}{2}\right), \left(\frac{\pi}{12}, \frac{7}{2}\right)$;

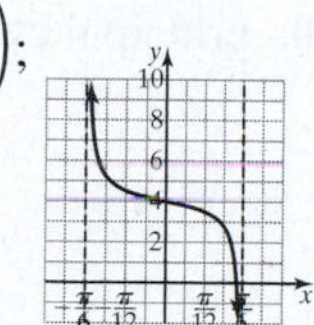

18. principal cycle: $(0, \pi)$; period: π; vertical asymptotes: $x = 0, x = \pi$; center point: $\left(\frac{\pi}{2}, 1\right)$; halfway points: $\left(\frac{\pi}{4}, 2\right), \left(\frac{3\pi}{4}, 0\right)$;

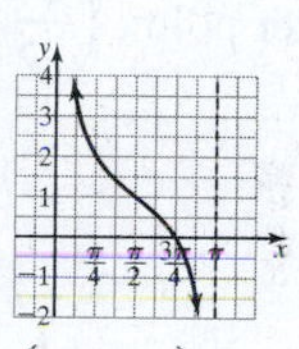

19. principal cycle: $\left(\frac{\pi}{4}, \frac{3\pi}{4}\right)$; period: $\frac{\pi}{2}$; vertical asymptotes: $x = \frac{\pi}{4}, x = \frac{3\pi}{4}$; center point: $\left(\frac{\pi}{2}, -1\right)$; halfway points: $\left(\frac{3\pi}{8}, -4\right), \left(\frac{5\pi}{8}, 2\right)$;

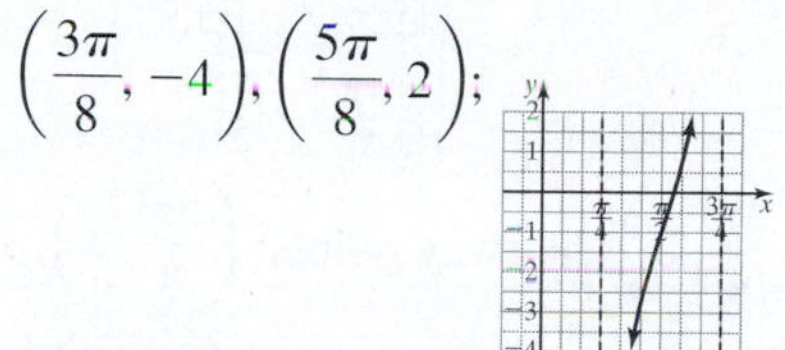

20. ; the function is an odd function. The range is $(-\infty, \infty)$. The zeros are of the form $\frac{n\pi}{2}$, where n is an odd integer. Every halfway point has a y-coordinate of -1 or 1. The domain is all real numbers except integer multiples of π. The function has a period of $P = \pi$. The interval for the graph of the principal cycle is $(0, \pi)$.

21. $\left(-\frac{11\pi}{6}, \sqrt{3}\right), \left(-\frac{5\pi}{6}, \sqrt{3}\right), \left(\frac{\pi}{6}, \sqrt{3}\right), \left(\frac{7\pi}{6}, \sqrt{3}\right), \left(\frac{13\pi}{6}, \sqrt{3}\right)$ **22.** $\left(-\frac{7\pi}{3}, -\frac{1}{\sqrt{3}}\right), \left(-\frac{4\pi}{3}, -\frac{1}{\sqrt{3}}\right), \left(-\frac{\pi}{3}, -\frac{1}{\sqrt{3}}\right)$ **23.** $\left(-\frac{7\pi}{4}, 1\right), \left(-\frac{5\pi}{4}, -1\right), \left(-\frac{3\pi}{4}, 1\right), \left(-\frac{\pi}{4}, -1\right), \left(\frac{\pi}{4}, 1\right), \left(\frac{3\pi}{4}, -1\right)$ **24.** i **25.** principal cycle: $\left(-\frac{\pi}{4}, \frac{3\pi}{4}\right)$; period: π; vertical asymptotes: $x = -\frac{\pi}{4}, x = \frac{3\pi}{4}$; center point: $\left(\frac{\pi}{4}, 0\right)$; halfway points: $(0, 1), \left(\frac{\pi}{2}, -1\right)$;

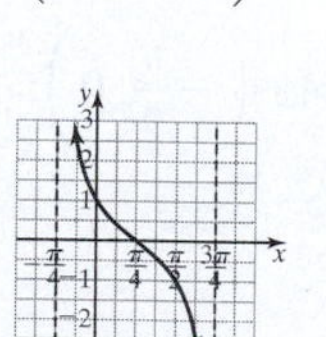

26. principal cycle: $\left(\frac{\pi}{2}, \frac{3\pi}{2}\right)$; period: π; vertical asymptotes: $x = \frac{\pi}{2}, x = \frac{3\pi}{2}$; center point: $(\pi, 0)$; halfway points: $\left(\frac{3\pi}{4}, 3\right), \left(\frac{5\pi}{4}, -3\right)$;

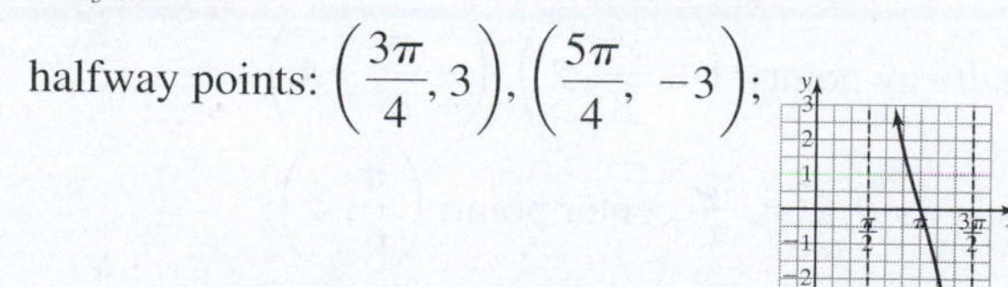

27. principal cycle: $\left(\frac{\pi}{3}, \frac{4\pi}{3}\right)$; period: π; vertical asymptotes:

$x = \frac{\pi}{3}, x = \frac{4\pi}{3}$; center point: $\left(\frac{5\pi}{6}, 0\right)$; halfway points: $\left(\frac{7\pi}{12}, -2\right), \left(\frac{13\pi}{12}, 2\right)$;

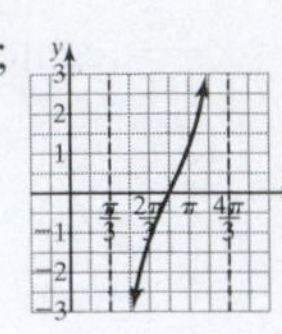

28. principal cycle: $\left(-\frac{\pi}{3}, 0\right)$; period: $\frac{\pi}{3}$; vertical asymptotes: $x = -\frac{\pi}{3}, x = 0$; center point: $\left(-\frac{\pi}{6}, 0\right)$; halfway points: $\left(-\frac{\pi}{4}, 1\right), \left(-\frac{\pi}{12}, -1\right)$;

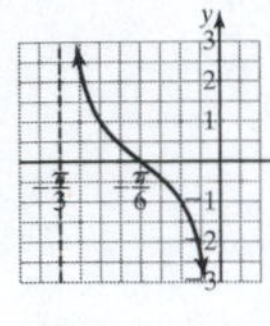

29. principal cycle: $\left(0, \frac{\pi}{2}\right)$; period: $\frac{\pi}{2}$; vertical asymptotes: $x = 0, x = \frac{\pi}{2}$; center point: $\left(\frac{\pi}{4}, 0\right)$; halfway points: $\left(\frac{\pi}{8}, 1\right), \left(\frac{3\pi}{8}, -1\right)$;

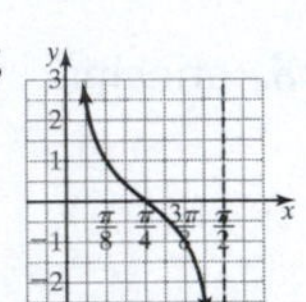

30. principal cycle: $(0, 3\pi)$; period: 3π; vertical asymptotes: $x = 0, x = 3\pi$; center point: $\left(\frac{3\pi}{2}, 0\right)$; halfway points: $\left(\frac{3\pi}{4}, 1\right), \left(\frac{9\pi}{4}, -1\right)$;

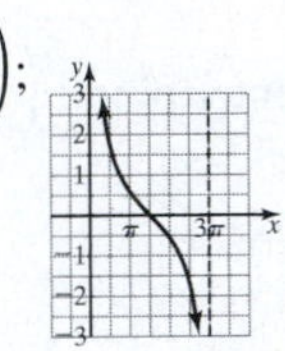

31. principal cycle: $\left(0, \frac{\pi}{3}\right)$; period: $\frac{\pi}{3}$; vertical asymptotes: $x = 0, x = \frac{\pi}{3}$; center point: $\left(\frac{\pi}{6}, 0\right)$; halfway points: $\left(\frac{\pi}{12}, -2\right), \left(\frac{\pi}{4}, 2\right)$;

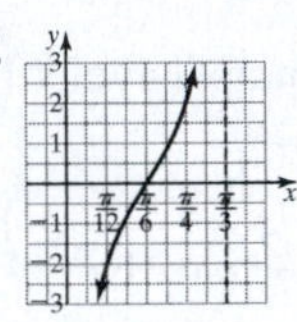

32. principal cycle: $\left(\frac{\pi}{2}, \pi\right)$; period: $\frac{\pi}{2}$; vertical asymptotes: $x = \frac{\pi}{2}, x = \pi$; center point: $\left(\frac{3\pi}{4}, 0\right)$; halfway points: $\left(\frac{5\pi}{8}, 3\right)$, $\left(\frac{7\pi}{8}, -3\right)$;

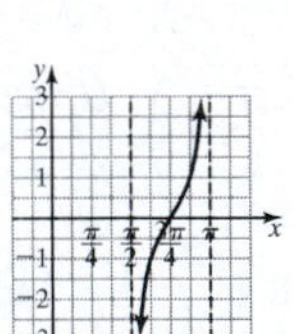

33. principal cycle: $\left(\frac{\pi}{2}, \pi\right)$; period: $\frac{\pi}{2}$; vertical asymptotes: $x = \frac{\pi}{2}, x = \pi$; center point: $\left(\frac{3\pi}{4}, 0\right)$; halfway points: $\left(\frac{5\pi}{8}, -1\right), \left(\frac{7\pi}{8}, 1\right)$;

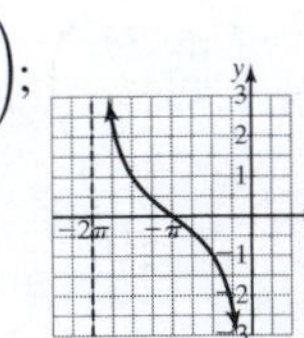

34. principal cycle: $(-2\pi, 0)$; period: 2π; vertical asymptotes: $x = -2\pi, x = 0$; center point: $(-\pi, 0)$; halfway points: $\left(-\frac{3\pi}{2}, 1\right), \left(-\frac{\pi}{2}, -1\right)$;

35. principal cycle: $\left(-\frac{\pi}{2}, 0\right)$; period: $\frac{\pi}{2}$; vertical asymptotes: $x = -\frac{\pi}{2}, x = 0$; center point: $\left(-\frac{\pi}{4}, -1\right)$; halfway points: $\left(-\frac{3\pi}{8}, 2\right), \left(-\frac{\pi}{8}, -4\right)$;

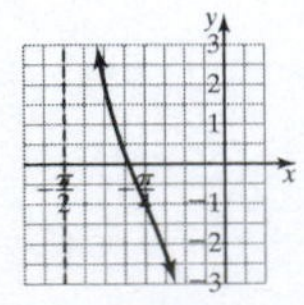

36. principal cycle: $\left(0, \frac{\pi}{3}\right)$; period: $\frac{\pi}{3}$; vertical asymptotes: $x = 0, x = \frac{\pi}{3}$; center point: $\left(\frac{\pi}{6}, 4\right)$;

halfway points: $\left(\frac{\pi}{12}, \frac{7}{2}\right), \left(\frac{\pi}{4}, \frac{9}{2}\right)$; 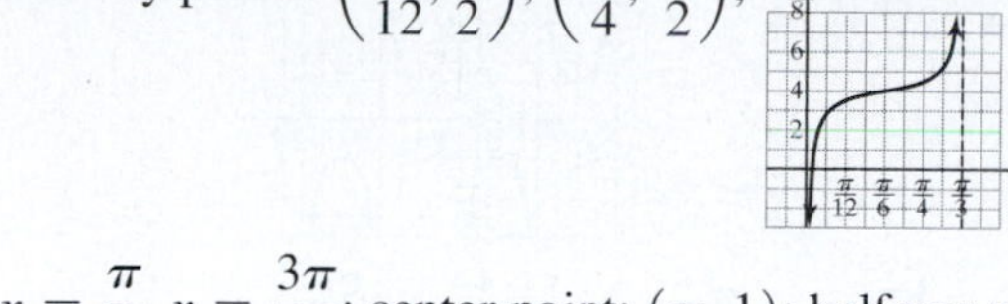

37. principal cycle: $\left(\frac{\pi}{2}, \frac{3\pi}{2}\right)$; period: π; vertical asymptotes: $x = \frac{\pi}{2}, x = \frac{3\pi}{2}$; center point: $(\pi, 1)$; halfway points: $\left(\frac{3\pi}{4}, 0\right), \left(\frac{5\pi}{4}, 2\right)$; 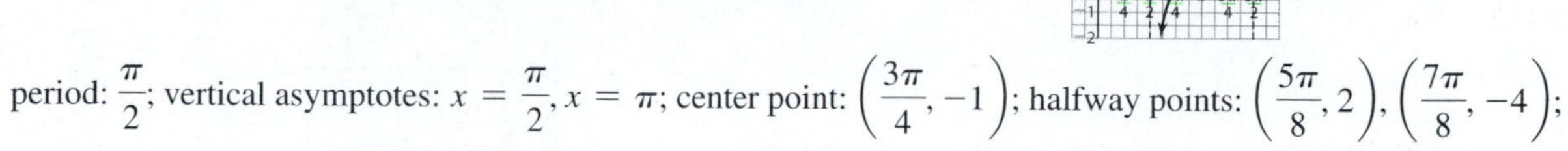

38. principal cycle: $\left(\frac{\pi}{2}, \pi\right)$; period: $\frac{\pi}{2}$; vertical asymptotes: $x = \frac{\pi}{2}, x = \pi$; center point: $\left(\frac{3\pi}{4}, -1\right)$; halfway points: $\left(\frac{5\pi}{8}, 2\right), \left(\frac{7\pi}{8}, -4\right)$; 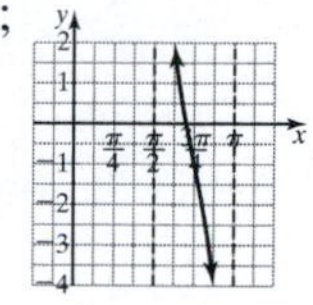

39. $y = 2\tan(x - \pi)$ or $y = -2\cot\left(x - \frac{\pi}{2}\right)$ **40.** $y = 3\tan\left(x - \frac{\pi}{6}\right)$ or $y = -3\cot\left(x + \frac{\pi}{3}\right)$

41. $y = -4\tan\left(x + \frac{\pi}{2}\right) + 1$ or $y = 4\cot(x + \pi) + 1$ **42.** $y = -5\tan\left(x - \frac{3\pi}{4}\right) + 2$ or $y = 5\cot\left(x - \frac{\pi}{4}\right) + 2$

43. $y = 2\tan\left(3x + \frac{\pi}{2}\right)$ or $y = -2\cot(3x + \pi)$ **44.** $y = 4\tan(2x + \pi) - 3$ or $y = -4\cot\left(2x + \frac{3\pi}{2}\right) - 3$

45. 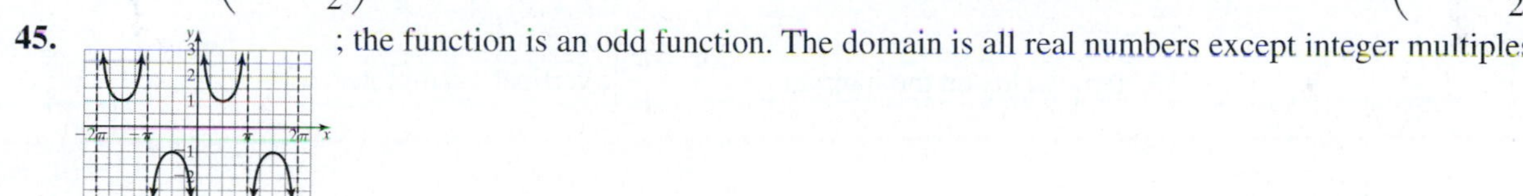 ; the function is an odd function. The domain is all real numbers except integer multiples of π. The function has a period of $P = 2\pi$. The vertical asymptotes are of the form $x = n\pi$, where n is an integer. The function obtains a relative maximum at $x = -\frac{\pi}{2} + 2\pi n$, where n is an integer. The range is $(-\infty, -1] \cup [1, \infty)$.

46. 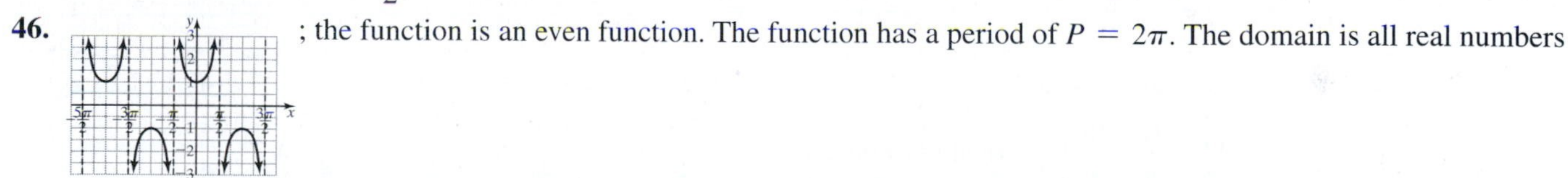 ; the function is an even function. The function has a period of $P = 2\pi$. The domain is all real numbers except odd integer multiples of $\frac{\pi}{2}$. The range is $(-\infty, -1] \cup [1, \infty)$. The function obtains a relative maximum at $x = n\pi$, where n is an odd integer. The vertical asymptotes are of the form $x = \frac{n\pi}{2}$, where n is an odd integer. **47.** two cycles on the interval $\left[-\frac{3\pi}{2}, \frac{5\pi}{2}\right]$; vertical asymptotes: $x = -\frac{3\pi}{2}, x = -\frac{\pi}{2}, x = \frac{\pi}{2}, x = \frac{3\pi}{2}, x = \frac{5\pi}{2}$; relative maximums: $(0, -2), (2\pi, -2)$; relative minimums: $(-\pi, 2), (\pi, 2)$; 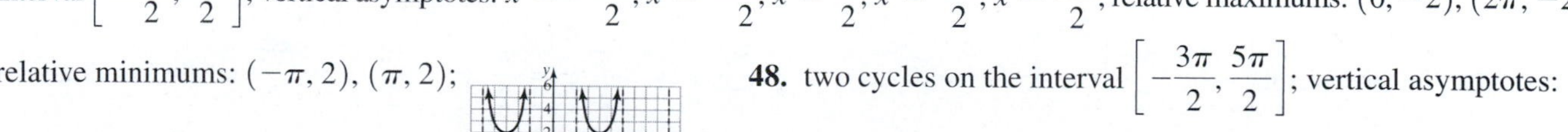

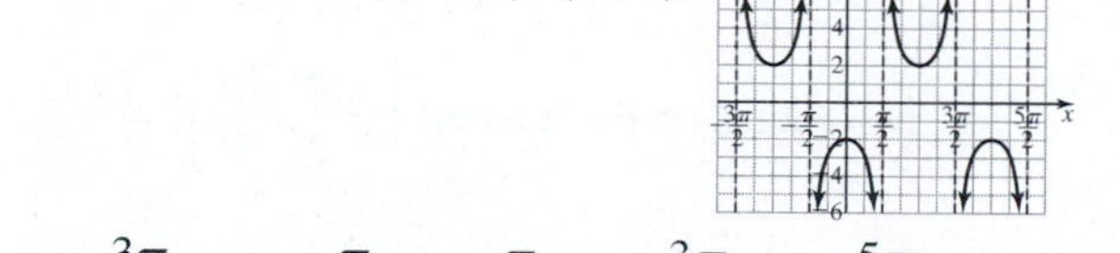

48. two cycles on the interval $\left[-\frac{3\pi}{2}, \frac{5\pi}{2}\right]$; vertical asymptotes: $x = -\frac{3\pi}{2}, x = -\frac{\pi}{2}, x = \frac{\pi}{2}, x = \frac{3\pi}{2}, x = \frac{5\pi}{2}$; relative maximums: $(-\pi, -2), (\pi, -2)$; relative minimums: $(0, 2), (2\pi, 2)$;

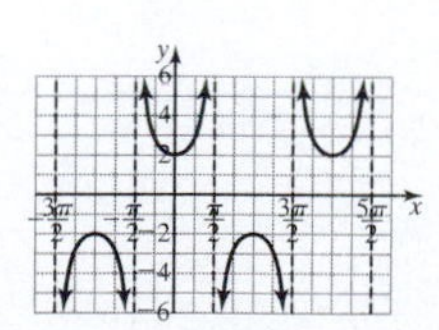

49. two cycles on the interval $\left[-\frac{7\pi}{4}, \frac{9\pi}{4}\right]$; vertical asymptotes: $x = -\frac{7\pi}{4}, x = -\frac{3\pi}{4}, x = \frac{\pi}{4}, x = \frac{5\pi}{4}$,

$x = \frac{9\pi}{4}$; relative maximums: $\left(-\frac{\pi}{4}, -1\right), \left(\frac{7\pi}{4}, -1\right)$; relative minimums: $\left(-\frac{5\pi}{4}, 1\right), \left(\frac{3\pi}{4}, 1\right)$;

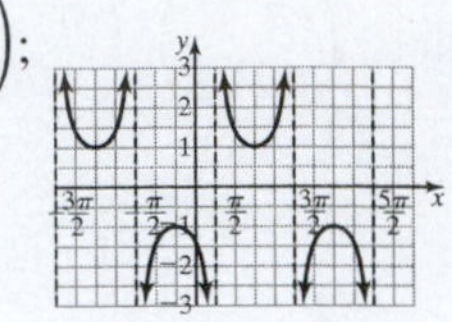

50. two cycles on the interval $\left[-\frac{7\pi}{4}, \frac{9\pi}{4}\right]$; vertical asymptotes: $x = -\frac{7\pi}{4}, x = -\frac{3\pi}{4}, x = \frac{\pi}{4}, x = \frac{5\pi}{4}, x = \frac{9\pi}{4}$; relative maximums: $\left(-\frac{5\pi}{4}, -1\right), \left(\frac{3\pi}{4}, -1\right)$; relative minimums: $\left(-\frac{\pi}{4}, 1\right), \left(\frac{7\pi}{4}, 1\right)$;

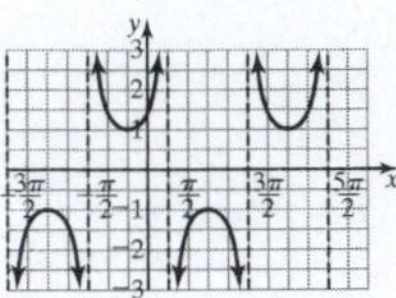

51. two cycles on the interval $\left[-\frac{\pi}{2}, \frac{5\pi}{6}\right]$; vertical asymptotes: $x = -\frac{\pi}{2}, x = -\frac{\pi}{6}, x = \frac{\pi}{6}, x = \frac{\pi}{2}, x = \frac{5\pi}{6}$; relative maximums: $(0, -2), \left(\frac{2\pi}{3}, -2\right)$; relative minimums: $\left(-\frac{\pi}{3}, 2\right), \left(\frac{\pi}{3}, 2\right)$;

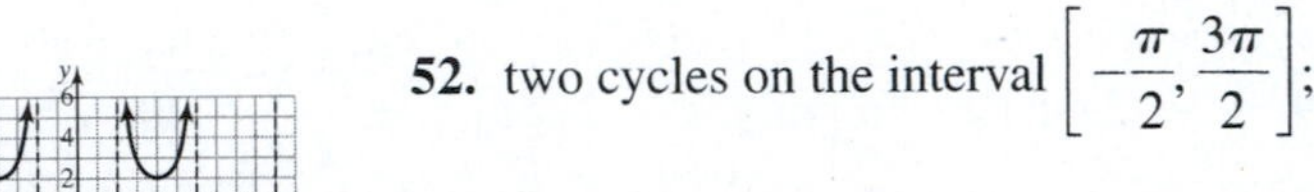

52. two cycles on the interval $\left[-\frac{\pi}{2}, \frac{3\pi}{2}\right]$; vertical asymptotes: $x = -\frac{\pi}{2}, x = 0, x = \frac{\pi}{2}, x = \pi, x = \frac{3\pi}{2}$; relative maximums: $\left(-\frac{\pi}{4}, -3\right), \left(\frac{3\pi}{4}, -3\right)$; relative minimums: $\left(\frac{\pi}{4}, 3\right), \left(\frac{5\pi}{4}, 3\right)$;

53. two cycles on the interval $\left[-\frac{3\pi}{4}, \frac{5\pi}{4}\right]$; vertical asymptotes: $x = -\frac{3\pi}{4}, x = -\frac{\pi}{4}, x = \frac{\pi}{4}, x = \frac{3\pi}{4}, x = \frac{5\pi}{4}$; relative maximums: $(0, -1), (\pi, -1)$; relative minimums: $\left(-\frac{\pi}{2}, 7\right), \left(\frac{\pi}{2}, 7\right)$;

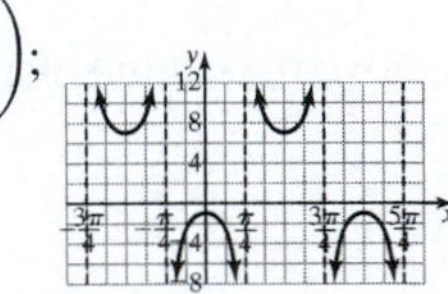

54. two cycles on the interval $\left[-\frac{7\pi}{4}, \frac{9\pi}{4}\right]$; vertical asymptotes: $x = -\frac{7\pi}{4}, x = -\frac{3\pi}{4}, x = \frac{\pi}{4}, x = \frac{5\pi}{4}, x = \frac{9\pi}{4}$; relative maximums: $\left(-\frac{5\pi}{4}, -7\right), \left(\frac{3\pi}{4}, -7\right)$; relative minimums: $\left(-\frac{\pi}{4}, -1\right), \left(\frac{7\pi}{4}, -1\right)$;

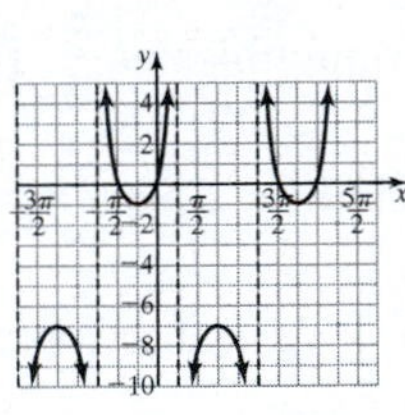

55. two cycles on the interval $\left[-\frac{3\pi}{2}, \frac{5\pi}{2}\right]$; vertical asymptotes: $x = -\frac{3\pi}{2}, x = -\frac{\pi}{2}, x = \frac{\pi}{2}, x = \frac{3\pi}{2}, x = \frac{5\pi}{2}$; relative maximums: $(-\pi, 3), (\pi, 3)$; relative minimums: $(0, 7), (2\pi, 7)$;

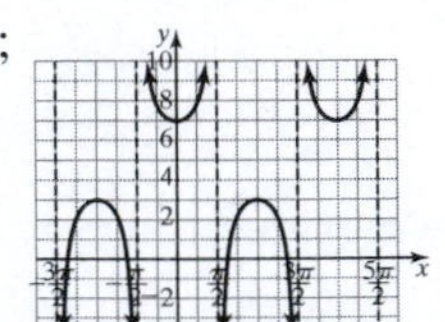

56. two cycles on the interval $\left[-\frac{\pi}{2}, \frac{3\pi}{2}\right]$;

vertical asymptotes: $x = -\frac{\pi}{2}, x = 0, x = \frac{\pi}{2}, x = \pi, x = \frac{3\pi}{2}$; relative maximums: $\left(-\frac{\pi}{4}, 2\right), \left(\frac{3\pi}{4}, 2\right)$; relative minimums: $\left(\frac{\pi}{4}, 4\right), \left(\frac{5\pi}{4}, 4\right)$;

57. $\left(-\frac{\pi}{3}, \frac{2\pi}{3}\right)$ **58.** $\left(0, \frac{\pi}{2}\right)$ **59.** $\left(-\frac{\pi}{2}, -\frac{\pi}{6}\right)$ **60.** $\left(\frac{\pi}{2}, \pi\right)$

61. $\left(\frac{\pi}{4}, \frac{3\pi}{4}\right)$ **62.** π **63.** $\frac{\pi}{2}$ **64.** $\frac{\pi}{3}$ **65.** $\frac{\pi}{2}$ **66.** $\frac{\pi}{2}$ **67.** $x = -\frac{\pi}{3}, x = \frac{2\pi}{3}$ **68.** $x = 0, x = \frac{\pi}{2}$

69. $x = -\frac{\pi}{2}, x = -\frac{\pi}{6}$ **70.** $x = \frac{\pi}{2}, x = \pi$ **71.** $x = \frac{\pi}{4}, x = \frac{3\pi}{4}$ **72.** $\left(\frac{\pi}{6}, 0\right)$ **73.** $\left(\frac{\pi}{4}, 0\right)$ **74.** $\left(-\frac{\pi}{3}, 0\right)$

75. $\left(\frac{3\pi}{4}, 0\right)$ **76.** $\left(\frac{\pi}{2}, 1\right)$ **77.** $\left(-\frac{\pi}{12}, -3\right), \left(\frac{5\pi}{12}, 3\right)$ **78.** $\left(\frac{\pi}{8}, 1\right), \left(\frac{3\pi}{8}, -1\right)$ **79.** $\left(-\frac{5\pi}{12}, -1\right), \left(-\frac{\pi}{4}, 1\right)$

80. $\left(\frac{5\pi}{8}, -1\right), \left(\frac{7\pi}{8}, 1\right)$ **81.** $\left(\frac{3\pi}{8}, -2\right), \left(\frac{5\pi}{8}, 4\right)$ **82.** $y = 2\tan(x - \pi)$ or $y = -2\cot\left(x - \frac{\pi}{2}\right)$

83. $y = 3\tan\left(x - \frac{\pi}{6}\right)$ or $y = -3\cot\left(x + \frac{\pi}{3}\right)$ **84.** $y = -4\tan\left(x + \frac{\pi}{2}\right) + 1$ or $y = 4\cot(x + \pi) + 1$

85. $y = -5\tan\left(x - \frac{3\pi}{4}\right) + 2$ or $y = 5\cot\left(x - \frac{\pi}{4}\right) + 2$ **86.** $y = 2\tan\left(3x + \frac{\pi}{2}\right)$ or $y = -2\cot(3x + \pi)$

87. $y = 4\tan(2x + \pi) - 3$ or $y = -4\cot\left(2x + \frac{3\pi}{2}\right) - 3$ **88.** $x = -\frac{\pi}{2}, x = \frac{\pi}{2}, x = \frac{3\pi}{2}, x = \frac{5\pi}{2}$

89. $x = -\frac{\pi}{2}, x = 0, x = \frac{\pi}{2}, x = \pi, x = \frac{3\pi}{2}$ **90.** $(-\pi, 2), (\pi, 2), (3\pi, 2)$ **91.** $\left(-\frac{\pi}{4}, 5\right), \left(\frac{3\pi}{4}, 5\right)$

92. $(0, -2), (2\pi, -2)$ **93.** $\left(\frac{\pi}{4}, -3\right), \left(\frac{5\pi}{4}, -3\right)$

94. **95.**

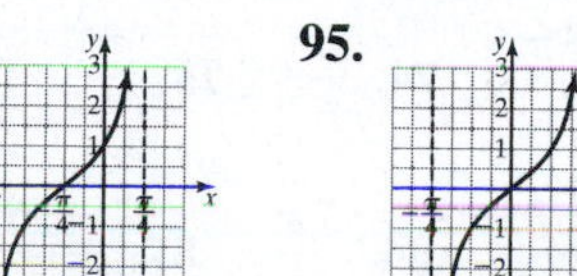

96.

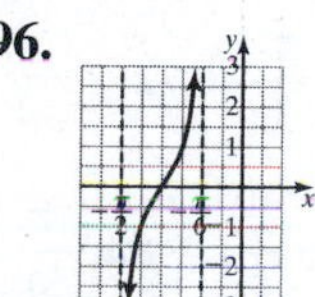

97.

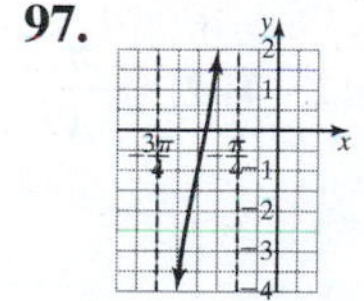

98.

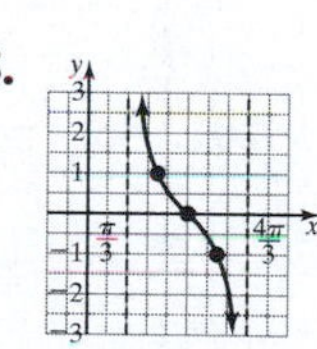

99.

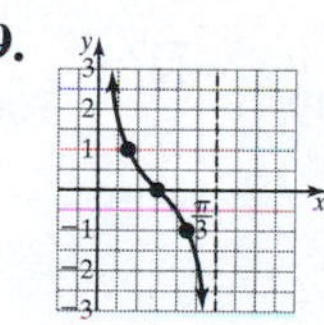

100.

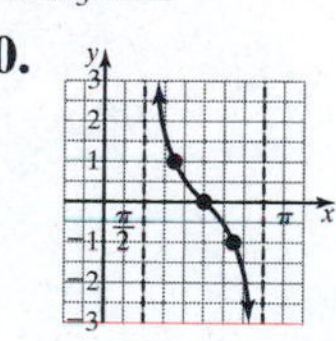

101.

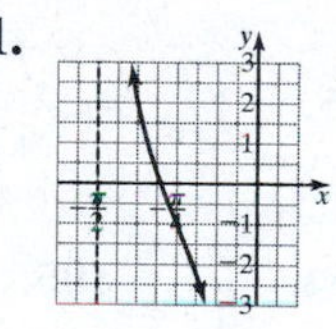

102.

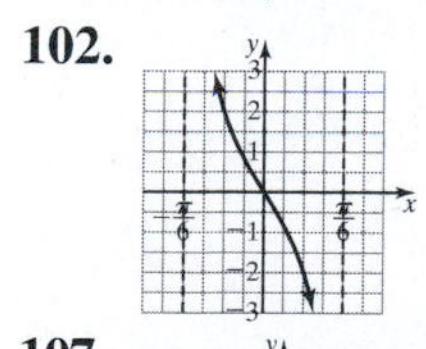

103.

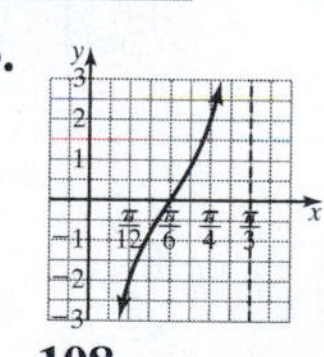

104.

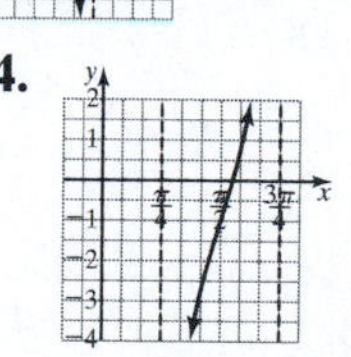

105.

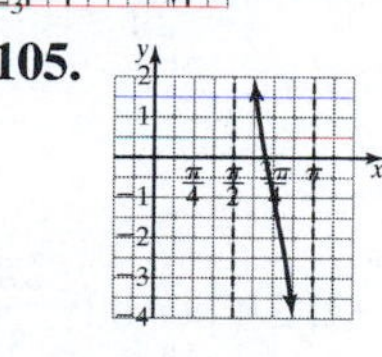

106.

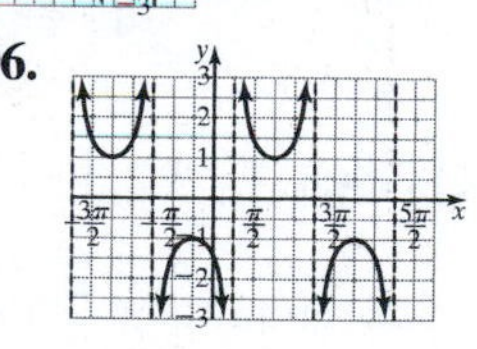

107.

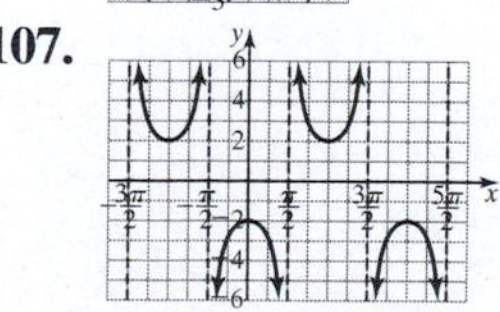

108.

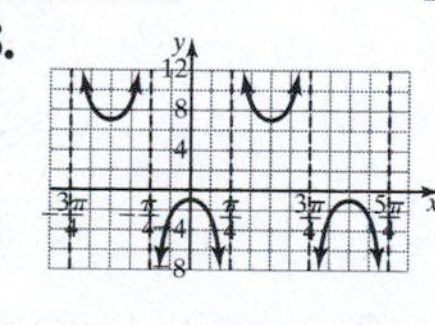

109.

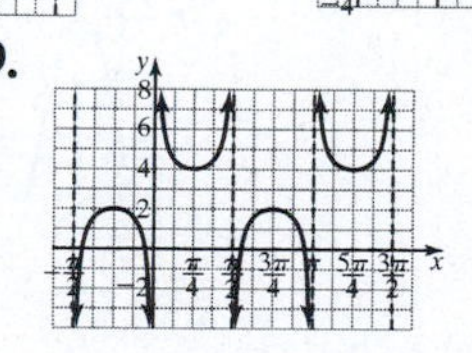

7.4 Exercises

1. $\frac{\pi}{3}$ **2.** $-\frac{\pi}{4}$ **3.** $\frac{\pi}{2}$ **4.** 0.3363 rad **5.** −1.0552 rad **6.** does not exist **7.** $\frac{\pi}{6}$ **8.** $\frac{3\pi}{4}$ **9.** 0 **10.** 1.2132 rad

11. 2.7389 rad **12.** does not exist **13.** $\frac{\pi}{6}$ **14.** $-\frac{\pi}{3}$ **15.** $\frac{\pi}{4}$ **16.** 0.6499 rad **17.** 1.3877 rad **18.** $\left[-\frac{\pi}{2}, \frac{\pi}{2}\right]$

19. $[0, \pi]$ **20.** $\left(-\frac{\pi}{2}, \frac{\pi}{2}\right)$ **21.** Quadrant I **22.** Quadrant IV **23.** y-axis **24.** Quadrant I **25.** Quadrant II

26. x-axis **27.** Quadrant I **28.** Quadrant IV **29.** Quadrant I **30.** $\frac{\pi}{3}$ **31.** $-\frac{\pi}{4}$ **32.** $\frac{\pi}{2}$ **33.** $\frac{\pi}{6}$ **34.** $\frac{3\pi}{4}$

35. 0 **36.** $\frac{\pi}{6}$ **37.** $-\frac{\pi}{3}$ **38.** $\frac{\pi}{4}$

7.5 Exercises

1. $\frac{1}{\sqrt{2}}$ **2.** $\frac{\sqrt{3}}{2}$ **3.** $\frac{1}{\sqrt{3}}$ **4.** $-\frac{\sqrt{3}}{2}$ **5.** $-\frac{1}{2}$ **6.** $-\sqrt{3}$ **7.** $\frac{7}{9}$ **8.** does not exist **9.** 35.4 **10.** $\frac{\pi}{7}$ **11.** $\frac{\pi}{5}$

12. $\frac{\pi}{3}$ **13.** $-\frac{\pi}{4}$ **14.** $\frac{3\pi}{4}$ **15.** $\frac{\pi}{6}$ **16.** $\frac{\pi}{3}$ **17.** $\frac{\pi}{4}$ **18.** $\frac{\pi}{2}$ **19.** 0 **20.** $-\frac{\pi}{3}$ **21.** $-\frac{\pi}{4}$ **22.** $\frac{5\pi}{6}$ **23.** $\frac{2\pi}{5}$

24. $\frac{\pi}{7}$ **25.** $\frac{\pi}{9}$ **26.** $-\frac{\pi}{12}$ **27.** $\frac{7\pi}{10}$ **28.** $-\frac{3\pi}{7}$ **29.** $\frac{\sqrt{3}}{2}$ **30.** does not exist **31.** $\frac{1}{2}$ **32.** $\frac{2}{\sqrt{3}}$ **33.** 2

34. $\frac{1}{\sqrt{2}}$ **35.** $\frac{\sqrt{3}}{2}$ **36.** -1 **37.** $\frac{7}{4}$ **38.** $\frac{\sqrt{74}}{7}$ **39.** $\frac{\sqrt{3}}{\sqrt{13}}$ **40.** $\frac{\sqrt{11}}{\sqrt{7}}$ **41.** $\frac{\pi}{6}$ **42.** does not exist **43.** $-\frac{\pi}{4}$

44. does not exist **45.** $\frac{3\pi}{4}$ **46.** 0 **47.** $\frac{9\pi}{22}$ **48.** $\frac{7\pi}{18}$ **49.** $\frac{5\pi}{6}$ **50.** $\frac{\pi}{6}$ **51.** $-\frac{\pi}{3}$ **52.** 0.0666 rad

53. does not exist **54.** 0.3285 rad **55.** $\sqrt{1-u^2}$ **56.** $\frac{\sqrt{1-4u^2}}{2u}$ **57.** $\frac{\sqrt{u^2-9}}{u}$ **58.** $\frac{\sqrt{11}}{u}$ **59.** $\frac{\sqrt{u^2+121}}{11}$

60. $\frac{1}{\sqrt{2}}$ **61.** $\frac{\sqrt{3}}{2}$ **62.** $\frac{1}{\sqrt{3}}$ **63.** $-\frac{\sqrt{3}}{2}$ **64.** $-\frac{1}{2}$ **65.** $-\sqrt{3}$ **66.** $\frac{7}{9}$ **67.** does not exist **68.** 35.4 **69.** $\frac{\pi}{7}$

70. $\frac{\pi}{5}$ **71.** $\frac{\pi}{3}$ **72.** $-\frac{\pi}{4}$ **73.** $\frac{3\pi}{4}$ **74.** $\frac{\pi}{6}$ **75.** $\frac{\pi}{3}$ **76.** $\frac{\pi}{4}$ **77.** $\frac{\pi}{2}$ **78.** 0 **79.** $-\frac{\pi}{3}$ **80.** $-\frac{\pi}{4}$ **81.** $\frac{5\pi}{6}$

82. $\frac{2\pi}{5}$ **83.** $\frac{\pi}{7}$ **84.** $\frac{\pi}{9}$ **85.** $-\frac{\pi}{12}$ **86.** $\frac{7\pi}{10}$ **87.** $-\frac{3\pi}{7}$ **88.** $\frac{\sqrt{3}}{2}$ **89.** does not exist **90.** $\frac{1}{2}$ **91.** $\frac{2}{\sqrt{3}}$ **92.** 2

93. $\frac{1}{\sqrt{2}}$ **94.** $\frac{\sqrt{3}}{2}$ **95.** -1 **96.** $\frac{7}{4}$ **97.** $\frac{\sqrt{74}}{7}$ **98.** $\frac{\sqrt{3}}{\sqrt{13}}$ **99.** $\frac{\sqrt{11}}{\sqrt{7}}$ **100.** $\frac{\pi}{6}$ **101.** does not exist **102.** $-\frac{\pi}{4}$

103. does not exist **104.** $\frac{3\pi}{4}$ **105.** 0 **106.** $\frac{9\pi}{22}$ **107.** $\frac{7\pi}{18}$

Review Exercises

1. 3 **2.** $\frac{\pi}{2}$ **3.** amplitude: 2; range: $[-2, 2]$; period: $\frac{2\pi}{3}$;

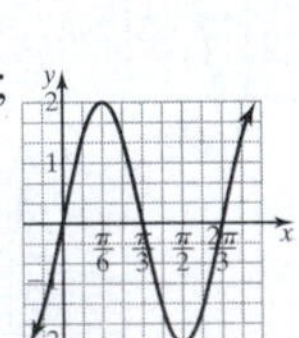

4. $y = 4\cos\left(\frac{1}{5}x\right)$ **5.** $-\frac{\pi}{2}$

6. amplitude: 1; range: $[-1, 1]$; period: 2π; phase shift: $-\frac{\pi}{2}$;

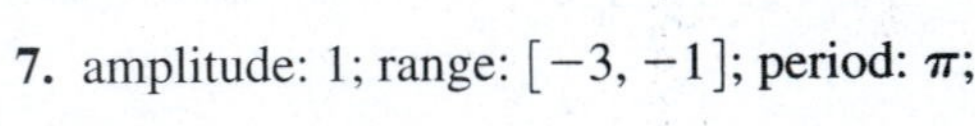

7. amplitude: 1; range: $[-3, -1]$; period: π;

phase shift: $\frac{\pi}{2}$;

8. $\left(0, \frac{\pi}{2}\right)$ **9.** $x = -\frac{\pi}{3}, x = \frac{2\pi}{3}$ **10.** principal cycle: $\left(-\frac{\pi}{2}, -\frac{\pi}{6}\right)$; period: $\frac{\pi}{3}$;

vertical asymptotes: $x = -\frac{\pi}{2}, x = -\frac{\pi}{6}$; center point: $\left(-\frac{\pi}{3}, 0\right)$; halfway points: $\left(-\frac{5\pi}{12}, -1\right), \left(-\frac{\pi}{4}, 1\right)$;

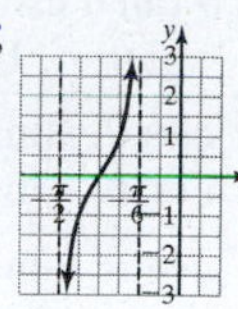

11. $y = -5\tan\left(x - \frac{3\pi}{4}\right) + 2$ or $y = 5\cot\left(x - \frac{\pi}{4}\right) + 2$ **12.** two cycles on the interval $\left[-\frac{3\pi}{2}, \frac{5\pi}{2}\right]$; vertical asymptotes: $x = -\frac{3\pi}{2}, x = -\frac{\pi}{2}, x = \frac{\pi}{2}, x = \frac{3\pi}{2}, x = \frac{5\pi}{2}$; relative maximums: $(0, -2), (2\pi, -2)$; relative minimums: $(-\pi, 2), (\pi, 2)$;

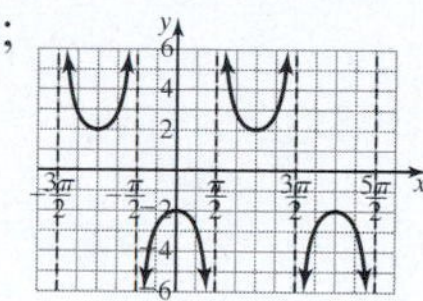

13. $\frac{\pi}{3}$ **14.** Quadrant IV **15.** $-\frac{\pi}{3}$ **16.** $\frac{1}{\sqrt{2}}$ **17.** $\frac{\pi}{4}$

18. 2 **19.** $\frac{7\pi}{18}$ **20.** $\frac{\sqrt{u^2 - 9}}{u}$

CHAPTER 8

8.1 Exercises

1. $\cos\theta$ **2.** $\cot\theta$ **3.** $\sin\theta$ **4.** $\cos\theta$ **5.** $\sin^2\theta$ **6.** $\tan^2\theta$; 1 **7.** $-\tan^2\theta$

8.
$$\begin{aligned} 1 + \cot^2(-\theta) &\stackrel{?}{=} \csc^2\theta \\ 1 + (\cot(-\theta))^2 &\stackrel{?}{=} \csc^2\theta \\ 1 + (-\cot\theta)^2 &\stackrel{?}{=} \csc^2\theta \\ 1 + \cot^2\theta &\stackrel{?}{=} \csc^2\theta \\ \csc^2\theta &= \csc^2\theta \end{aligned}$$

9.
$$\begin{aligned} \frac{\cot^2\beta + 1}{\csc\beta} &\stackrel{?}{=} \csc\beta \\ \frac{\csc^2\beta}{\csc\beta} &\stackrel{?}{=} \csc\beta \\ \csc\beta &= \csc\beta \end{aligned}$$

10.
$$\begin{aligned} \tan^2 4x + \csc^2 4x - \cot^2 4x &\stackrel{?}{=} \sec^2 4x \\ \tan^2 4x + 1 + \cot^2 4x - \cot^2 4x &\stackrel{?}{=} \sec^2 4x \\ \tan^2 4x + 1 &\stackrel{?}{=} \sec^2 4x \\ \sec^2 4x &= \sec^2 4x \end{aligned}$$

11.
$$\begin{aligned} (3\cos\theta - 4\sin\theta)^2 + (4\cos\theta + 3\sin\theta)^2 &\stackrel{?}{=} 25 \\ 9\cos^2\theta - 24\cos\theta\sin\theta + 16\sin^2\theta + 16\cos^2\theta + 24\cos\theta\sin\theta + 9\sin^2\theta &\stackrel{?}{=} 25 \\ 9\cos^2\theta + 16\sin^2\theta + 16\cos^2\theta + 9\sin^2\theta &\stackrel{?}{=} 25 \\ 9(\cos^2\theta + \sin^2\theta) + 16(\sin^2\theta + \cos^2\theta) &\stackrel{?}{=} 25 \\ 9(1) + 16(1) &\stackrel{?}{=} 25 \\ 25 &= 25 \end{aligned}$$

12.
$$\begin{aligned} 7\sin^2\theta + 4\cos^2\theta &\stackrel{?}{=} 4 + 3\sin^2\theta \\ 3\sin^2\theta + 4\sin^2\theta + 4\cos^2\theta &\stackrel{?}{=} 4 + 3\sin^2\theta \\ 3\sin^2\theta + 4(\sin^2\theta + \cos^2\theta) &\stackrel{?}{=} 4 + 3\sin^2\theta \\ 3\sin^2\theta + 4(1) &\stackrel{?}{=} 4 + 3\sin^2\theta \\ 4 + 3\sin^2\theta &= 4 + 3\sin^2\theta \end{aligned}$$

13.
$$\begin{aligned} \cos\theta\csc\theta &\stackrel{?}{=} \cot\theta \\ \cos\theta \cdot \frac{1}{\sin\theta} &\stackrel{?}{=} \cot\theta \\ \frac{\cos\theta}{\sin\theta} &\stackrel{?}{=} \cot\theta \\ \cot\theta &= \cot\theta \end{aligned}$$

14.
$$\begin{aligned} \tan(-x)\cos x &\stackrel{?}{=} -\sin x \\ -\tan x\cos x &\stackrel{?}{=} -\sin x \\ -\frac{\sin x}{\cos x} \cdot \cos x &\stackrel{?}{=} -\sin x \\ -\sin x &= -\sin x \end{aligned}$$

15.
$$\cos\theta\tan\theta\csc\theta \stackrel{?}{=} 1$$
$$\cos\theta\cdot\frac{\sin\theta}{\cos\theta}\cdot\frac{1}{\sin\theta} \stackrel{?}{=} 1$$
$$\frac{\cos\theta}{\cos\theta}\cdot\frac{\sin\theta}{\sin\theta} \stackrel{?}{=} 1$$
$$1 = 1$$

16.
$$\sec\theta - \cos\theta \stackrel{?}{=} \sin\theta\tan\theta$$
$$\frac{1}{\cos\theta} - \cos\theta \stackrel{?}{=} \sin\theta\tan\theta$$
$$\frac{1}{\cos\theta} - \cos\theta\cdot\frac{\cos\theta}{\cos\theta} \stackrel{?}{=} \sin\theta\tan\theta$$
$$\frac{1}{\cos\theta} - \frac{\cos^2\theta}{\cos\theta} \stackrel{?}{=} \sin\theta\tan\theta$$
$$\frac{1-\cos^2\theta}{\cos\theta} \stackrel{?}{=} \sin\theta\tan\theta$$
$$\frac{\sin^2\theta}{\cos\theta} \stackrel{?}{=} \sin\theta\tan\theta$$
$$\frac{\sin\theta\cdot\sin\theta}{\cos\theta} \stackrel{?}{=} \sin\theta\tan\theta$$
$$\sin\theta\cdot\frac{\sin\theta}{\cos\theta} \stackrel{?}{=} \sin\theta\tan\theta$$
$$\sin\theta\tan\theta = \sin\theta\tan\theta$$

17.
$$\csc t\sin t - \sin^2 t \stackrel{?}{=} \cos^2 t$$
$$\frac{1}{\sin t}\cdot\sin t - \sin^2 t \stackrel{?}{=} \cos^2 t$$
$$1 - \sin^2 t \stackrel{?}{=} \cos^2 t$$
$$\cos^2 t = \cos^2 t$$

18.
$$\frac{\csc\theta}{\sin\theta} - \cos(-\theta)\sec(-\theta) \stackrel{?}{=} \cot^2\theta$$
$$\frac{1}{\sin\theta}\cdot\csc\theta - \cos\theta\sec\theta \stackrel{?}{=} \cot^2\theta$$
$$\csc\theta\cdot\csc\theta - \cos\theta\cdot\frac{1}{\cos\theta} \stackrel{?}{=} \cot^2\theta$$
$$\csc^2\theta - 1 \stackrel{?}{=} \cot^2\theta$$
$$\cot^2\theta = \cot^2\theta$$

19.
$$\sec x + \tan^2 x\sec x \stackrel{?}{=} \sec^3 x$$
$$\sec x(1+\tan^2 x) \stackrel{?}{=} \sec^3 x$$
$$\sec x(\sec^2 x) \stackrel{?}{=} \sec^3 x$$
$$\sec^3 x = \sec^3 x$$

20.
$$\frac{\sin^2\beta - \cos^2\beta}{\sin\beta - \cos\beta} \stackrel{?}{=} \sin\beta + \cos\beta$$
$$\frac{(\sin\beta - \cos\beta)(\sin\beta + \cos\beta)}{\sin\beta - \cos\beta} \stackrel{?}{=} \sin\beta + \cos\beta$$
$$\sin\beta + \cos\beta = \sin\beta + \cos\beta$$

21.
$$\frac{\cot^3\theta + 1}{\cot\theta + 1} \stackrel{?}{=} \csc^2\theta - \cot\theta$$
$$\frac{(\cot\theta + 1)(\cot^2\theta - \cot\theta + 1)}{\cot\theta + 1} \stackrel{?}{=} \csc^2\theta - \cot\theta$$
$$\cot^2\theta - \cot\theta + 1 \stackrel{?}{=} \csc^2\theta - \cot\theta$$
$$\cot^2\theta + 1 - \cot\theta \stackrel{?}{=} \csc^2\theta - \cot\theta$$
$$\csc^2\theta - \cot\theta = \csc^2\theta - \cot\theta$$

22.
$$\frac{6\csc^2\theta - 7\csc\theta - 3}{1 + 3\csc\theta} \stackrel{?}{=} 2\csc\theta - 3$$
$$\frac{(2\csc\theta - 3)(3\csc\theta + 1)}{3\csc\theta + 1} \stackrel{?}{=} 2\csc\theta - 3$$
$$2\csc\theta - 3 = 2\csc\theta - 3$$

23.
$$\sin^4\theta - \cos^4\theta \stackrel{?}{=} 2\sin^2\theta - 1$$
$$(\sin^2\theta - \cos^2\theta)(\sin^2\theta + \cos^2\theta) \stackrel{?}{=} 2\sin^2\theta - 1$$
$$(\sin^2\theta - (1 - \sin^2\theta))(1) \stackrel{?}{=} 2\sin^2\theta - 1$$
$$2\sin^2\theta - 1 = 2\sin^2\theta - 1$$

24.
$$\frac{1 + 2\sec\theta}{\sec\theta} \stackrel{?}{=} 2 + \cos\theta$$
$$\frac{1}{\sec\theta} + \frac{2\sec\theta}{\sec\theta} \stackrel{?}{=} 2 + \cos\theta$$
$$\cos\theta + 2 \stackrel{?}{=} 2 + \cos\theta$$
$$2 + \cos\theta = 2 + \cos\theta$$

25.
$$\frac{\cot\theta + 1}{\csc\theta} \stackrel{?}{=} \sin\theta + \cos\theta$$
$$\frac{\cot\theta}{\csc\theta} + \frac{1}{\csc\theta} \stackrel{?}{=} \sin\theta + \cos\theta$$
$$\cot\theta\cdot\frac{1}{\csc\theta} + \sin\theta \stackrel{?}{=} \sin\theta + \cos\theta$$
$$\frac{\cos\theta}{\sin\theta}\cdot\sin\theta + \sin\theta \stackrel{?}{=} \sin\theta + \cos\theta$$
$$\cos\theta + \sin\theta \stackrel{?}{=} \sin\theta + \cos\theta$$
$$\sin\theta + \cos\theta = \sin\theta + \cos\theta$$

26.
$$\frac{\tan\alpha + \cot\alpha}{\tan\alpha} - \frac{\cot\alpha + \tan\alpha}{\cot\alpha} \stackrel{?}{=} \cot^2\alpha - \tan^2\alpha$$
$$\frac{\tan\alpha}{\tan\alpha} + \frac{\cot\alpha}{\tan\alpha} - \frac{\cot\alpha}{\cot\alpha} - \frac{\tan\alpha}{\cot\alpha} \stackrel{?}{=} \cot^2\alpha - \tan^2\alpha$$
$$1 + \cot\alpha \cdot \frac{1}{\tan\alpha} - 1 - \tan\alpha \cdot \frac{1}{\cot\alpha} \stackrel{?}{=} \cot^2\alpha - \tan^2\alpha$$
$$\cot\alpha \cdot \cot\alpha - \tan\alpha \cdot \tan\alpha \stackrel{?}{=} \cot^2\alpha - \tan^2\alpha$$
$$\cot^2\alpha - \tan^2\alpha = \cot^2\alpha - \tan^2\alpha$$

27.
$$\frac{\sin x + \tan x + 1}{\cos x} \stackrel{?}{=} \sec x + \tan x + \sin x \sec^2 x$$
$$\frac{\sin x}{\cos x} + \frac{\tan x}{\cos x} + \frac{1}{\cos x} \stackrel{?}{=} \sec x + \tan x + \sin x \sec^2 x$$
$$\tan x + \tan x \cdot \frac{1}{\cos x} + \sec x \stackrel{?}{=} \sec x + \tan x + \sin x \sec^2 x$$
$$\tan x + \frac{\sin x}{\cos x} \cdot \frac{1}{\cos x} + \sec x \stackrel{?}{=} \sec x + \tan x + \sin x \sec^2 x$$
$$\tan x + \sin x \cdot \frac{1}{\cos^2 x} + \sec x \stackrel{?}{=} \sec x + \tan x + \sin x \sec^2 x$$
$$\tan x + \sin x \sec^2 x + \sec x \stackrel{?}{=} \sec x + \tan x + \sin x \sec^2 x$$
$$\sec x + \tan x + \sin x \sec^2 x = \sec x + \tan x + \sin x \sec^2 x$$

28.
$$\frac{1 - \cos\theta}{\sin\theta} + \frac{\sin\theta}{1 - \cos\theta} \stackrel{?}{=} 2\csc\theta$$
$$\frac{1 - \cos\theta}{\sin\theta} \cdot \frac{1 - \cos\theta}{1 - \cos\theta} + \frac{\sin\theta}{1 - \cos\theta} \cdot \frac{\sin\theta}{\sin\theta} \stackrel{?}{=} 2\csc\theta$$
$$\frac{1 - 2\cos\theta + \cos^2\theta}{\sin\theta\,(1 - \cos\theta)} + \frac{\sin^2\theta}{\sin\theta\,(1 - \cos\theta)} \stackrel{?}{=} 2\csc\theta$$
$$\frac{1 - 2\cos\theta + \cos^2\theta + \sin^2\theta}{\sin\theta\,(1 - \cos\theta)} \stackrel{?}{=} 2\csc\theta$$
$$\frac{1 - 2\cos\theta + 1}{\sin\theta\,(1 - \cos\theta)} \stackrel{?}{=} 2\csc\theta$$
$$\frac{2 - 2\cos\theta}{\sin\theta\,(1 - \cos\theta)} \stackrel{?}{=} 2\csc\theta$$
$$\frac{2(1 - \cos\theta)}{\sin\theta\,(1 - \cos\theta)} \stackrel{?}{=} 2\csc\theta$$
$$\frac{2}{\sin\theta} \stackrel{?}{=} 2\csc\theta$$
$$2\csc\theta = 2\csc\theta$$

29.
$$\frac{\sin t}{1 + \cos t} + \cot t \stackrel{?}{=} \csc t$$
$$\frac{\sin t}{1 + \cos t} + \frac{\cos t}{\sin t} \stackrel{?}{=} \csc t$$
$$\frac{\sin t}{1 + \cos t} \cdot \frac{\sin t}{\sin t} + \frac{\cos t}{\sin t} \cdot \frac{1 + \cos t}{1 + \cos t} \stackrel{?}{=} \csc t$$
$$\frac{\sin^2 t}{(1 + \cos t)\sin t} + \frac{\cos t + \cos^2 t}{\sin t\,(1 + \cos t)} \stackrel{?}{=} \csc t$$
$$\frac{\sin^2 t + \cos t + \cos^2 t}{(1 + \cos t)\sin t} \stackrel{?}{=} \csc t$$
$$\frac{1 + \cos t}{(1 + \cos t)\sin t} \stackrel{?}{=} \csc t$$
$$\frac{1}{\sin t} \stackrel{?}{=} \csc t$$
$$\csc t = \csc t$$

30.
$$\frac{\sec\beta}{\sin\beta} - \frac{\sin\beta}{\sec\beta} \stackrel{?}{=} \frac{\tan^2\beta + \cos^2\beta}{\tan\beta}$$
$$\frac{\sec\beta}{\sin\beta} \cdot \frac{\sec\beta}{\sec\beta} - \frac{\sin\beta}{\sec\beta} \cdot \frac{\sin\beta}{\sin\beta} \stackrel{?}{=} \frac{\tan^2\beta + \cos^2\beta}{\tan\beta}$$
$$\frac{\sec^2\beta}{\sin\beta\sec\beta} - \frac{\sin^2\beta}{\sec\beta\sin\beta} \stackrel{?}{=} \frac{\tan^2\beta + \cos^2\beta}{\tan\beta}$$
$$\frac{\sec^2\beta - \sin^2\beta}{\sin\beta\sec\beta} \stackrel{?}{=} \frac{\tan^2\beta + \cos^2\beta}{\tan\beta}$$
$$\frac{1 + \tan^2\beta - (1 - \cos^2\beta)}{\sin\beta \cdot \dfrac{1}{\cos\beta}} \stackrel{?}{=} \frac{\tan^2\beta + \cos^2\beta}{\tan\beta}$$
$$\frac{1 + \tan^2\beta - 1 + \cos^2\beta}{\dfrac{\sin\beta}{\cos\beta}} \stackrel{?}{=} \frac{\tan^2\beta + \cos^2\beta}{\tan\beta}$$
$$\frac{\tan^2\beta + \cos^2\beta}{\tan\beta} = \frac{\tan^2\beta + \cos^2\beta}{\tan\beta}$$

31.
$$\frac{\sin\theta}{1 - \cos\theta} \stackrel{?}{=} \frac{1 + \cos\theta}{\sin\theta}$$
$$\frac{\sin\theta}{1 - \cos\theta} \cdot \frac{1 + \cos\theta}{1 + \cos\theta} \stackrel{?}{=} \frac{1 + \cos\theta}{\sin\theta}$$
$$\frac{\sin\theta\,(1 + \cos\theta)}{1 - \cos^2\theta} \stackrel{?}{=} \frac{1 + \cos\theta}{\sin\theta}$$
$$\frac{\sin\theta\,(1 + \cos\theta)}{\sin^2\theta} \stackrel{?}{=} \frac{1 + \cos\theta}{\sin\theta}$$
$$\frac{1 + \cos\theta}{\sin\theta} = \frac{1 + \cos\theta}{\sin\theta}$$

32. $$\frac{\sec t - 1}{\tan t} \stackrel{?}{=} \frac{\tan t}{\sec t + 1}$$
$$\frac{\sec t - 1}{\tan t} \cdot \frac{\sec t + 1}{\sec t + 1} \stackrel{?}{=} \frac{\tan t}{\sec t + 1}$$
$$\frac{\sec^2 t - 1}{\tan t\,(\sec t + 1)} \stackrel{?}{=} \frac{\tan t}{\sec t + 1}$$
$$\frac{\tan^2 t}{\tan t\,(\sec t + 1)} \stackrel{?}{=} \frac{\tan t}{\sec t + 1}$$
$$\frac{\tan t}{\sec t + 1} = \frac{\tan t}{\sec t + 1}$$

33. $$\cot^2 3x + \sec^2 3x - \tan^2 3x \stackrel{?}{=} \csc^2 3x$$
$$\cot^2 3x + 1 \stackrel{?}{=} \csc^2 3x$$
$$\csc^2 3x = \csc^2 3x$$

34. $$1 + \cot^2(-\theta) \stackrel{?}{=} \csc^2\theta$$
$$1 + (\cot(-\theta))^2 \stackrel{?}{=} \csc^2\theta$$
$$1 + (-\cot\theta)^2 \stackrel{?}{=} \csc^2\theta$$
$$1 + \cot^2\theta \stackrel{?}{=} \csc^2\theta$$
$$\csc^2\theta = \csc^2\theta$$

35. $$\tan(-x)\cos x \stackrel{?}{=} -\sin x$$
$$\tan(-x)\cos x \stackrel{?}{=} -\sin x$$
$$-\tan x \cos x \stackrel{?}{=} -\sin x$$
$$-\frac{\sin x}{\cos x} \cdot \cos x \stackrel{?}{=} -\sin x$$
$$-\sin x = -\sin x$$

36. $$\frac{\csc\theta + 1}{\cot\theta} \stackrel{?}{=} \sec\theta + \tan\theta$$
$$\frac{\csc\theta}{\cot\theta} + \frac{1}{\cot\theta} \stackrel{?}{=} \sec\theta + \tan\theta$$
$$\csc\theta \cdot \frac{1}{\cot\theta} + \tan\theta \stackrel{?}{=} \sec\theta + \tan\theta$$
$$\csc\theta \cdot \tan\theta + \tan\theta \stackrel{?}{=} \sec\theta + \tan\theta$$
$$\frac{1}{\sin\theta} \cdot \frac{\sin\theta}{\cos\theta} + \tan\theta \stackrel{?}{=} \sec\theta + \tan\theta$$
$$\frac{1}{\cos\theta} + \tan\theta \stackrel{?}{=} \sec\theta + \tan\theta$$
$$\sec\theta + \tan\theta \stackrel{?}{=} \sec\theta + \tan\theta$$

37. $$\frac{1 - \sec\theta}{\tan\theta} - \frac{\tan\theta}{1 - \sec\theta} \stackrel{?}{=} 2\cot\theta$$
$$\frac{1 - \sec\theta}{\tan\theta} \cdot \frac{1 - \sec\theta}{1 - \sec\theta} - \frac{\tan\theta}{1 - \sec\theta} \cdot \frac{\tan\theta}{\tan\theta} \stackrel{?}{=} 2\cot\theta$$
$$\frac{1 - 2\sec\theta + \sec^2\theta}{\tan\theta\,(1 - \sec\theta)} - \frac{\tan^2\theta}{(1 - \sec\theta)\tan\theta} \stackrel{?}{=} 2\cot\theta$$
$$\frac{1 - 2\sec\theta + \sec^2\theta - \tan^2\theta}{\tan\theta\,(1 - \sec\theta)} \stackrel{?}{=} 2\cot\theta$$
$$\frac{1 - 2\sec\theta + 1}{\tan\theta\,(1 - \sec\theta)} \stackrel{?}{=} 2\cot\theta$$
$$\frac{2 - 2\sec\theta}{\tan\theta\,(1 - \sec\theta)} \stackrel{?}{=} 2\cot\theta$$
$$\frac{2(1 - \sec\theta)}{\tan\theta\,(1 - \sec\theta)} \stackrel{?}{=} 2\cot\theta$$
$$\frac{2}{\tan\theta} \stackrel{?}{=} 2\cot\theta$$
$$2\cot\theta = 2\cot\theta$$

38. $$\frac{\csc\theta}{\sec\theta} + \frac{4\cos\theta}{\sin\theta} \stackrel{?}{=} 5\cot\theta$$
$$\csc\theta\cos\theta + 4\cot\theta \stackrel{?}{=} 5\cot\theta$$
$$\frac{1}{\sin\theta} \cdot \cos\theta + 4\cot\theta \stackrel{?}{=} 5\cot\theta$$
$$\cot\theta + 4\cot\theta \stackrel{?}{=} 5\cot\theta$$
$$5\cot\theta = 5\cot\theta$$

39. $$\frac{\cos^2 t + 3\cos t - 10}{\cos t + 5} \stackrel{?}{=} \frac{1 - 2\sec t}{\sec t}$$
$$\frac{(\cos t + 5)(\cos t - 2)}{\cos t + 5} \stackrel{?}{=} \frac{1 - 2\sec t}{\sec t}$$
$$\frac{\cos t - 2}{1} \stackrel{?}{=} \frac{1 - 2\sec t}{\sec t}$$
$$\frac{\cos t - 2}{1} \cdot \frac{\sec t}{\sec t} \stackrel{?}{=} \frac{1 - 2\sec t}{\sec t}$$
$$\frac{\cos t \cdot \sec t - 2\sec t}{\sec t} \stackrel{?}{=} \frac{1 - 2\sec t}{\sec t}$$
$$\frac{1 - 2\sec t}{\sec t} = \frac{1 - 2\sec t}{\sec t}$$

40. $$1 + \frac{1 - \cot^2 x}{1 + \cot^2 x} \stackrel{?}{=} 2\sin^2 x$$
$$\frac{1 + \cot^2 x}{1 + \cot^2 x} + \frac{1 - \cot^2 x}{1 + \cot^2 x} \stackrel{?}{=} 2\sin^2 x$$
$$\frac{1 + \cot^2 x + 1 - \cot^2 x}{1 + \cot^2 x} \stackrel{?}{=} 2\sin^2 x$$
$$\frac{2}{1 + \cot^2 x} \stackrel{?}{=} 2\sin^2 x$$
$$\frac{2}{\csc^2 x} \stackrel{?}{=} 2\sin^2 x$$
$$2\sin^2 x = 2\sin^2 x$$

41.
$$\begin{aligned}\sec^4\theta - \tan^4\theta &\stackrel{?}{=} 2\sec^2\theta - 1\\ (\sec^2\theta + \tan^2\theta)(\sec^2\theta - \tan^2\theta) &\stackrel{?}{=} 2\sec^2\theta - 1\\ (\sec^2\theta + \sec^2\theta - 1)(1) &\stackrel{?}{=} 2\sec^2\theta - 1\\ 2\sec^2\theta - 1 &= 2\sec^2\theta - 1\end{aligned}$$

42.
$$\begin{aligned}\frac{\tan(-\theta)}{\cot(-\theta)} - \sin\theta\csc(-\theta) &\stackrel{?}{=} \sec^2\theta\\ \frac{-\tan\theta}{-\cot\theta} - \sin\theta(-\csc\theta) &\stackrel{?}{=} \sec^2\theta\\ \frac{\tan\theta}{\cot\theta} + \sin\theta(\csc\theta) &\stackrel{?}{=} \sec^2\theta\\ \tan\theta\cdot\frac{1}{\cot\theta} + \sin\theta\cdot\frac{1}{\sin\theta} &\stackrel{?}{=} \sec^2\theta\\ \tan\theta\cdot\tan\theta + 1 &\stackrel{?}{=} \sec^2\theta\\ \tan^2\theta + 1 &\stackrel{?}{=} \sec^2\theta\\ \sec^2\theta &\stackrel{?}{=} \sec^2\theta\end{aligned}$$

43.
$$\begin{aligned}\frac{\cos^3\theta + \sin^3\theta}{\cos\theta + \sin\theta} &\stackrel{?}{=} 1 - \cos\theta\sin\theta\\ \frac{(\cos\theta + \sin\theta)(\cos^2\theta - \cos\theta\sin\theta + \sin^2\theta)}{\cos\theta + \sin\theta} &\stackrel{?}{=} 1 - \cos\theta\sin\theta\\ \cos^2\theta - \cos\theta\sin\theta + \sin^2\theta &\stackrel{?}{=} 1 - \cos\theta\sin\theta\\ \cos^2\theta + \sin^2\theta - \cos\theta\sin\theta &\stackrel{?}{=} 1 - \cos\theta\sin\theta\\ 1 - \cos\theta\sin\theta &= 1 - \cos\theta\sin\theta\end{aligned}$$

8.2 Exercises

SCE-1. $\frac{1+\sqrt{3}}{2\sqrt{2}}$ **SCE-2.** $-\frac{10}{\sqrt{2501}}$ **SCE-3.** $2+\sqrt{3}$ **SCE-4.** $\frac{-1-\sqrt{3}}{\sqrt{3}-1}$ **SCE-5.** Sample answer: $\frac{\pi}{4}+\frac{\pi}{6}$

SCE-6. Sample answer: $\frac{5\pi}{6}-\frac{\pi}{4}$

1. $\left(\frac{1}{2}\right)\left(\frac{1}{\sqrt{2}}\right)-\left(\frac{\sqrt{3}}{2}\right)\left(\frac{1}{\sqrt{2}}\right)$ **2.** $\left(\frac{\sqrt{3}}{2}\right)\left(\frac{1}{\sqrt{2}}\right)-\left(\frac{1}{2}\right)\left(\frac{1}{\sqrt{2}}\right)$ **3.** $\left(\frac{1}{\sqrt{2}}\right)\left(\frac{\sqrt{3}}{2}\right)+\left(\frac{1}{\sqrt{2}}\right)\left(\frac{1}{2}\right)$

4. $\left(\frac{1}{2}\right)\left(\frac{1}{\sqrt{2}}\right)+\left(\frac{\sqrt{3}}{2}\right)\left(\frac{1}{\sqrt{2}}\right)$ **5.** $\left(-\frac{\sqrt{3}}{2}\right)\left(-\frac{1}{\sqrt{2}}\right)-\left(-\frac{1}{2}\right)\left(-\frac{1}{\sqrt{2}}\right)$

6. $\left(-\frac{\sqrt{3}}{2}\right)\left(-\frac{1}{\sqrt{2}}\right)-\left(-\frac{1}{2}\right)\left(\frac{1}{\sqrt{2}}\right)$ **7.** $\left(-\frac{1}{\sqrt{2}}\right)\left(-\frac{\sqrt{3}}{2}\right)+\left(-\frac{1}{\sqrt{2}}\right)\left(-\frac{1}{2}\right)$

8. $\left(-\frac{1}{2}\right)\left(-\frac{1}{\sqrt{2}}\right)+\left(\frac{\sqrt{3}}{2}\right)\left(-\frac{1}{\sqrt{2}}\right)$ **9.** $\frac{1-\sqrt{3}}{2\sqrt{2}}$ **10.** $\frac{\sqrt{3}-1}{2\sqrt{2}}$ **11.** $\frac{\sqrt{3}+1}{2\sqrt{2}}$ **12.** $\frac{1+\sqrt{3}}{2\sqrt{2}}$ **13.** $\frac{\sqrt{3}-1}{2\sqrt{2}}$

14. $\frac{\sqrt{3}+1}{2\sqrt{2}}$ **15.** $\frac{\sqrt{3}+1}{2\sqrt{2}}$ **16.** $\frac{1-\sqrt{3}}{2\sqrt{2}}$ **17.** $\frac{1}{2}$ **18.** $\frac{\sqrt{3}}{2}$ **19.** $-\frac{1}{2}$ **20.** $-\frac{1}{\sqrt{2}}$

21. $\left(\frac{1}{\sqrt{2}}\right)\left(\frac{\sqrt{3}}{2}\right)-\left(\frac{1}{\sqrt{2}}\right)\left(\frac{1}{2}\right)$ **22.** $\left(\frac{1}{\sqrt{2}}\right)\left(\frac{\sqrt{3}}{2}\right)-\left(\frac{1}{\sqrt{2}}\right)\left(\frac{1}{2}\right)$ **23.** $\frac{1}{\left(\frac{1}{\sqrt{2}}\right)\left(-\frac{1}{2}\right)-\left(\frac{1}{\sqrt{2}}\right)\left(\frac{\sqrt{3}}{2}\right)}$

24. $\left(-\frac{\sqrt{3}}{2}\right)\left(\frac{1}{\sqrt{2}}\right)+\left(\frac{1}{2}\right)\left(\frac{1}{\sqrt{2}}\right)$ **25.** $\left(-\frac{\sqrt{3}}{2}\right)\left(\frac{1}{\sqrt{2}}\right)+\left(-\frac{1}{2}\right)\left(\frac{1}{\sqrt{2}}\right)$ **26.** $\frac{1}{\left(-\frac{1}{2}\right)\left(\frac{1}{\sqrt{2}}\right)+\left(-\frac{\sqrt{3}}{2}\right)\left(\frac{1}{\sqrt{2}}\right)}$

27. $\frac{\sqrt{3}-1}{2\sqrt{2}}$ **28.** $\frac{\sqrt{3}-1}{2\sqrt{2}}$ **29.** $\frac{2\sqrt{2}}{-1-\sqrt{3}}$ **30.** $\frac{-\sqrt{3}+1}{2\sqrt{2}}$ **31.** $\frac{-\sqrt{3}-1}{2\sqrt{2}}$ **32.** $\frac{2\sqrt{2}}{-1-\sqrt{3}}$

33. $\left(\frac{15}{17}\right)\left(\frac{3}{\sqrt{11}}\right)-\left(\frac{8}{17}\right)\left(-\frac{\sqrt{2}}{\sqrt{11}}\right)$ **34.** $\left(-\frac{15}{17}\right)\left(\frac{3}{4}\right)+\left(\frac{8}{17}\right)\left(\frac{\sqrt{7}}{4}\right)$ **35.** $\frac{45+8\sqrt{2}}{17\sqrt{11}}$ **36.** $\frac{-45+8\sqrt{7}}{68}$

37. $\left(\frac{1}{2}\right)\left(\frac{1}{\sqrt{2}}\right)+\left(\frac{\sqrt{3}}{2}\right)\left(\frac{1}{\sqrt{2}}\right)$ **38.** $\left(\frac{1}{2}\right)\left(\frac{1}{\sqrt{2}}\right)+\left(\frac{\sqrt{3}}{2}\right)\left(\frac{1}{\sqrt{2}}\right)$ **39.** $\left(\frac{1}{\sqrt{2}}\right)\left(\frac{\sqrt{3}}{2}\right)-\left(\frac{1}{\sqrt{2}}\right)\left(\frac{1}{2}\right)$

40. $\left(\frac{\sqrt{3}}{2}\right)\left(\frac{1}{\sqrt{2}}\right)-\left(\frac{1}{2}\right)\left(\frac{1}{\sqrt{2}}\right)$ **41.** $\left(-\frac{1}{2}\right)\left(-\frac{1}{\sqrt{2}}\right)+\left(-\frac{\sqrt{3}}{2}\right)\left(-\frac{1}{\sqrt{2}}\right)$ **42.** $\left(-\frac{1}{2}\right)\left(-\frac{1}{\sqrt{2}}\right)+\left(-\frac{\sqrt{3}}{2}\right)\left(\frac{1}{\sqrt{2}}\right)$

43. $\left(-\frac{1}{\sqrt{2}}\right)\left(-\frac{\sqrt{3}}{2}\right)-\left(-\frac{1}{\sqrt{2}}\right)\left(-\frac{1}{2}\right)$ **44.** $\left(\frac{\sqrt{3}}{2}\right)\left(-\frac{1}{\sqrt{2}}\right)-\left(-\frac{1}{2}\right)\left(-\frac{1}{\sqrt{2}}\right)$ **45.** $\frac{1+\sqrt{3}}{2\sqrt{2}}$ **46.** $\frac{1+\sqrt{3}}{2\sqrt{2}}$

47. $\frac{\sqrt{3}-1}{2\sqrt{2}}$ **48.** $\frac{\sqrt{3}-1}{2\sqrt{2}}$ **49.** $\frac{1+\sqrt{3}}{2\sqrt{2}}$ **50.** $\frac{1-\sqrt{3}}{2\sqrt{2}}$ **51.** $\frac{\sqrt{3}-1}{2\sqrt{2}}$ **52.** $\frac{-\sqrt{3}-1}{2\sqrt{2}}$ **53.** $\frac{1}{\sqrt{2}}$ **54.** $\frac{1}{2}$

55. $\frac{\sqrt{3}}{2}$ **56.** $-\frac{1}{2}$ **57.** $\left(\frac{1}{2}\right)\left(\frac{1}{\sqrt{2}}\right)+\left(\frac{\sqrt{3}}{2}\right)\left(\frac{1}{\sqrt{2}}\right)$ **58.** $\left(\frac{1}{2}\right)\left(\frac{1}{\sqrt{2}}\right)+\left(\frac{\sqrt{3}}{2}\right)\left(\frac{1}{\sqrt{2}}\right)$

59. $\frac{1}{\left(\frac{1}{\sqrt{2}}\right)\left(-\frac{1}{2}\right)+\left(\frac{1}{\sqrt{2}}\right)\left(\frac{\sqrt{3}}{2}\right)}$ **60.** $\left(\frac{1}{2}\right)\left(\frac{1}{\sqrt{2}}\right)-\left(-\frac{\sqrt{3}}{2}\right)\left(\frac{1}{\sqrt{2}}\right)$ **61.** $\left(-\frac{1}{2}\right)\left(\frac{1}{\sqrt{2}}\right)-\left(-\frac{\sqrt{3}}{2}\right)\left(\frac{1}{\sqrt{2}}\right)$

62. $\frac{1}{\left(-\frac{\sqrt{3}}{2}\right)\left(\frac{1}{\sqrt{2}}\right)-\left(-\frac{1}{2}\right)\left(\frac{1}{\sqrt{2}}\right)}$ **63.** $\frac{1+\sqrt{3}}{2\sqrt{2}}$ **64.** $\frac{1+\sqrt{3}}{2\sqrt{2}}$ **65.** $\frac{2\sqrt{2}}{-1+\sqrt{3}}$ **66.** $\frac{1+\sqrt{3}}{2\sqrt{2}}$

67. $\frac{-1+\sqrt{3}}{2\sqrt{2}}$ **68.** $\frac{2\sqrt{2}}{-\sqrt{3}+1}$ **69.** $\left(\frac{8}{17}\right)\left(\frac{3}{\sqrt{11}}\right)+\left(\frac{15}{17}\right)\left(-\frac{\sqrt{2}}{\sqrt{11}}\right)$

70. $\left(-\frac{4}{\sqrt{97}}\right)\left(\frac{2}{\sqrt{53}}\right)-\left(-\frac{9}{\sqrt{97}}\right)\left(-\frac{7}{\sqrt{53}}\right)$ **71.** $\frac{24-15\sqrt{2}}{17\sqrt{11}}$ **72.** $-\frac{71}{\sqrt{5141}}$ **73.** $\frac{\frac{1}{\sqrt{3}}+1}{1-\left(\frac{1}{\sqrt{3}}\right)(1)}$

74. $\frac{\sqrt{3}+1}{1-(\sqrt{3})(1)}$ **75.** $\frac{\sqrt{3}-1}{1+(\sqrt{3})(1)}$ **76.** $\frac{1-\sqrt{3}}{1+(1)(\sqrt{3})}$ **77.** $\frac{-\sqrt{3}-1}{1-(-\sqrt{3})(-1)}$ **78.** $\frac{\sqrt{3}-1}{1-(\sqrt{3})(-1)}$

79. $\frac{\frac{1}{\sqrt{3}}-1}{1+\left(\frac{1}{\sqrt{3}}\right)(1)}$ **80.** $\frac{1+\sqrt{3}}{1+(1)(-\sqrt{3})}$ **81.** $\frac{1+\sqrt{3}}{\sqrt{3}-1}$ **82.** $\frac{\sqrt{3}+1}{1-\sqrt{3}}$ **83.** $\frac{\sqrt{3}-1}{1+\sqrt{3}}$ **84.** $\frac{1-\sqrt{3}}{1+\sqrt{3}}$

85. $\frac{-\sqrt{3}-1}{1-\sqrt{3}}$ **86.** $\frac{\sqrt{3}-1}{1+\sqrt{3}}$ **87.** $\frac{1-\sqrt{3}}{\sqrt{3}+1}$ **88.** $\frac{1+\sqrt{3}}{1-\sqrt{3}}$ **89.** $\sqrt{3}$ **90.** $-\sqrt{3}$ **91.** $\frac{\frac{1}{\sqrt{3}}+1}{1-\left(\frac{1}{\sqrt{3}}\right)(1)}$

92. $\frac{\frac{1}{\sqrt{3}}+1}{1-\left(\frac{1}{\sqrt{3}}\right)(1)}$ **93.** $\frac{1-\left(\frac{1}{\sqrt{3}}\right)(-1)}{\frac{1}{\sqrt{3}}-1}$ **94.** $\frac{-\frac{1}{\sqrt{3}}-1}{1+\left(-\frac{1}{\sqrt{3}}\right)(1)}$ **95.** $\frac{\frac{1}{\sqrt{3}}-1}{1+\left(\frac{1}{\sqrt{3}}\right)(1)}$ **96.** $\frac{1+(-1)(\sqrt{3})}{-1-\sqrt{3}}$

97. $\frac{1+\sqrt{3}}{\sqrt{3}-1}$ **98.** $\frac{1+\sqrt{3}}{\sqrt{3}-1}$ **99.** $\frac{\sqrt{3}+1}{1-\sqrt{3}}$ **100.** $\frac{-1-\sqrt{3}}{\sqrt{3}-1}$ **101.** $\frac{1-\sqrt{3}}{\sqrt{3}+1}$ **102.** $\frac{1-\sqrt{3}}{-1-\sqrt{3}}$

103. $\frac{-\frac{1}{\sqrt{15}}+\frac{4}{3}}{1-\left(-\frac{1}{\sqrt{15}}\right)\left(\frac{4}{3}\right)}$ **104.** $\frac{-\frac{6}{5}-\frac{7}{\sqrt{42}}}{1+\left(-\frac{6}{5}\right)\left(\frac{7}{\sqrt{42}}\right)}$ **105.** $\frac{-3+4\sqrt{15}}{3\sqrt{15}+4}$ **106.** $\frac{-6\sqrt{42}-35}{5\sqrt{42}-42}$

107.
$$\sin\left(\frac{\pi}{2}+\theta\right)\stackrel{?}{=}\cos\theta$$
$$\sin\left(\frac{\pi}{2}\right)\cos\theta+\sin\theta\cos\left(\frac{\pi}{2}\right)\stackrel{?}{=}\cos\theta$$
$$(1)\cos\theta+\sin\theta\,(0)\stackrel{?}{=}\cos\theta$$
$$\cos\theta=\cos\theta$$

108.
$$\cos(\alpha+\beta)+\cos(\alpha-\beta)\stackrel{?}{=}2\cos\alpha\cos\beta$$
$$\cos\alpha\cos\beta-\sin\alpha\sin\beta+\cos\alpha\cos\beta+\sin\alpha\sin\beta\stackrel{?}{=}2\cos\alpha\cos\beta$$
$$2\cos\alpha\cos\beta=2\cos\alpha\cos\beta$$

109.
$$\frac{\cos(\alpha+\beta)}{\sin\alpha\cos\beta} \stackrel{?}{=} \cot\alpha - \tan\beta$$
$$\frac{\cos\alpha\cos\beta - \sin\alpha\sin\beta}{\sin\alpha\cos\beta} \stackrel{?}{=} \cot\alpha - \tan\beta$$
$$\frac{\cos\alpha\cos\beta}{\sin\alpha\cos\beta} - \frac{\sin\alpha\sin\beta}{\sin\alpha\cos\beta} \stackrel{?}{=} \cot\alpha - \tan\beta$$
$$\frac{\cos\alpha}{\sin\alpha} - \frac{\sin\beta}{\cos\beta} \stackrel{?}{=} \cot\alpha - \tan\beta$$
$$\cot\alpha - \tan\beta = \cot\alpha - \tan\beta$$

110.
$$\frac{\cos(\alpha-\beta)}{\cos(\alpha+\beta)} \stackrel{?}{=} \frac{1+\tan\alpha\tan\beta}{1-\tan\alpha\tan\beta}$$
$$\frac{\cos\alpha\cos\beta + \sin\alpha\sin\beta}{\cos\alpha\cos\beta - \sin\alpha\sin\beta} \stackrel{?}{=} \frac{1+\tan\alpha\tan\beta}{1-\tan\alpha\tan\beta}$$
$$\frac{\cos\alpha\cos\beta + \sin\alpha\sin\beta}{\cos\alpha\cos\beta - \sin\alpha\sin\beta} \cdot \frac{\frac{1}{\cos\alpha\cos\beta}}{\frac{1}{\cos\alpha\cos\beta}} \stackrel{?}{=} \frac{1+\tan\alpha\tan\beta}{1-\tan\alpha\tan\beta}$$
$$\frac{\frac{\cos\alpha\cos\beta + \sin\alpha\sin\beta}{\cos\alpha\cos\beta}}{\frac{\cos\alpha\cos\beta - \sin\alpha\sin\beta}{\cos\alpha\cos\beta}} \stackrel{?}{=} \frac{1+\tan\alpha\tan\beta}{1-\tan\alpha\tan\beta}$$
$$\frac{\frac{\cos\alpha\cos\beta}{\cos\alpha\cos\beta} + \frac{\sin\alpha\sin\beta}{\cos\alpha\cos\beta}}{\frac{\cos\alpha\cos\beta}{\cos\alpha\cos\beta} - \frac{\sin\alpha\sin\beta}{\cos\alpha\cos\beta}} \stackrel{?}{=} \frac{1+\tan\alpha\tan\beta}{1-\tan\alpha\tan\beta}$$
$$\frac{1+\tan\alpha\tan\beta}{1-\tan\alpha\tan\beta} = \frac{1+\tan\alpha\tan\beta}{1-\tan\alpha\tan\beta}$$

111.
$$\tan(\alpha-\beta) \stackrel{?}{=} \frac{\cot\beta - \cot\alpha}{\cot\alpha\cot\beta + 1}$$
$$\frac{\tan\alpha - \tan\beta}{1+\tan\alpha\tan\beta} \stackrel{?}{=} \frac{\cot\beta - \cot\alpha}{\cot\alpha\cot\beta + 1}$$
$$\frac{\tan\alpha - \tan\beta}{1+\tan\alpha\tan\beta} \cdot \frac{\cot\alpha\cot\beta}{\cot\alpha\cot\beta} \stackrel{?}{=} \frac{\cot\beta - \cot\alpha}{\cot\alpha\cot\beta + 1}$$
$$\frac{\tan\alpha\cot\alpha\cot\beta - \tan\beta\cot\alpha\cot\beta}{\cot\alpha\cot\beta + \tan\alpha\tan\beta\cot\alpha\cot\beta} \stackrel{?}{=} \frac{\cot\beta - \cot\alpha}{\cot\alpha\cot\beta + 1}$$
$$\frac{(\tan\alpha\cot\alpha)\cot\beta - (\tan\beta\cot\beta)\cot\alpha}{\cot\alpha\cot\beta + (\tan\alpha\cot\alpha)(\tan\beta\cot\beta)} \stackrel{?}{=} \frac{\cot\beta - \cot\alpha}{\cot\alpha\cot\beta + 1}$$
$$\frac{(1)\cot\beta - (1)\cot\alpha}{\cot\alpha\cot\beta + (1)(1)} \stackrel{?}{=} \frac{\cot\beta - \cot\alpha}{\cot\alpha\cot\beta + 1}$$
$$\frac{\cot\beta - \cot\alpha}{\cot\alpha\cot\beta + 1} = \frac{\cot\beta - \cot\alpha}{\cot\alpha\cot\beta + 1}$$

112.
$$\csc(\alpha-\beta) \stackrel{?}{=} \frac{\csc\alpha\csc\beta}{\cot\beta - \cot\alpha}$$
$$\frac{1}{\sin(\alpha-\beta)} \stackrel{?}{=} \frac{\csc\alpha\csc\beta}{\cot\beta - \cot\alpha}$$
$$\frac{1}{\sin\alpha\cos\beta - \sin\beta\cos\alpha} \stackrel{?}{=} \frac{\csc\alpha\csc\beta}{\cot\beta - \cot\alpha}$$
$$\frac{1}{\sin\alpha\cos\beta - \sin\beta\cos\alpha} \cdot \frac{\frac{1}{\sin\alpha\sin\beta}}{\frac{1}{\sin\alpha\sin\beta}} \stackrel{?}{=} \frac{\csc\alpha\csc\beta}{\cot\beta - \cot\alpha}$$
$$\frac{\frac{1}{\sin\alpha\sin\beta}}{\frac{\sin\alpha\cos\beta}{\sin\alpha\sin\beta} - \frac{\sin\beta\cos\alpha}{\sin\alpha\sin\beta}} \stackrel{?}{=} \frac{\csc\alpha\csc\beta}{\cot\beta - \cot\alpha}$$
$$\frac{\frac{1}{\sin\alpha} \cdot \frac{1}{\sin\beta}}{\frac{\cos\beta}{\sin\beta} - \frac{\cos\alpha}{\sin\alpha}} \stackrel{?}{=} \frac{\csc\alpha\csc\beta}{\cot\beta - \cot\alpha}$$
$$\frac{\csc\alpha\csc\beta}{\cot\beta - \cot\alpha} = \frac{\csc\alpha\csc\beta}{\cot\beta - \cot\alpha}$$

113. $\frac{1+\sqrt{3}}{2\sqrt{2}}$ **114.** -1 **115.** $\frac{-3-10\sqrt{14}}{42}$ **116.** $\sqrt{3}-2$ **117.** $\frac{1-\sqrt{3}}{2\sqrt{2}}$ **118.** $\frac{\sqrt{3}-1}{2\sqrt{2}}$ **119.** $\frac{\sqrt{3}+1}{2\sqrt{2}}$

120. $\frac{1+\sqrt{3}}{2\sqrt{2}}$ **121.** $\frac{\sqrt{3}-1}{2\sqrt{2}}$ **122.** $\frac{\sqrt{3}+1}{2\sqrt{2}}$ **123.** $\frac{\sqrt{3}+1}{2\sqrt{2}}$ **124.** $\frac{1-\sqrt{3}}{2\sqrt{2}}$ **125.** $\frac{1}{2}$ **126.** $\frac{\sqrt{3}}{2}$

127. $-\frac{1}{2}$ **128.** $-\frac{1}{\sqrt{2}}$ **129.** $\frac{\sqrt{3}-1}{2\sqrt{2}}$ **130.** $\frac{\sqrt{3}-1}{2\sqrt{2}}$ **131.** $\frac{2\sqrt{2}}{-1-\sqrt{3}}$ **132.** $\frac{-\sqrt{3}+1}{2\sqrt{2}}$ **133.** $\frac{-\sqrt{3}-1}{2\sqrt{2}}$

134. $\frac{2\sqrt{2}}{-1-\sqrt{3}}$ **135.** $\frac{1+\sqrt{3}}{2\sqrt{2}}$ **136.** $\frac{1+\sqrt{3}}{2\sqrt{2}}$ **137.** $\frac{\sqrt{3}-1}{2\sqrt{2}}$ **138.** $\frac{\sqrt{3}-1}{2\sqrt{2}}$ **139.** $\frac{1+\sqrt{3}}{2\sqrt{2}}$ **140.** $\frac{1-\sqrt{3}}{2\sqrt{2}}$

141. $\frac{\sqrt{3}-1}{2\sqrt{2}}$ **142.** $\frac{-\sqrt{3}-1}{2\sqrt{2}}$ **143.** $\frac{1}{\sqrt{2}}$ **144.** $\frac{1}{2}$ **145.** $\frac{\sqrt{3}}{2}$ **146.** $-\frac{1}{2}$ **147.** $\frac{1+\sqrt{3}}{2\sqrt{2}}$ **148.** $\frac{1+\sqrt{3}}{2\sqrt{2}}$

149. $\dfrac{2\sqrt{2}}{-1+\sqrt{3}}$ 150. $\dfrac{1+\sqrt{3}}{2\sqrt{2}}$ 151. $\dfrac{-1+\sqrt{3}}{2\sqrt{2}}$ 152. $\dfrac{2\sqrt{2}}{-\sqrt{3}+1}$ 153. $\dfrac{1+\sqrt{3}}{\sqrt{3}-1}$ 154. $\dfrac{\sqrt{3}+1}{1-\sqrt{3}}$

155. $\dfrac{\sqrt{3}-1}{1+\sqrt{3}}$ 156. $\dfrac{1-\sqrt{3}}{1+\sqrt{3}}$ 157. $\dfrac{-\sqrt{3}-1}{1-\sqrt{3}}$ 158. $\dfrac{\sqrt{3}-1}{1+\sqrt{3}}$ 159. $\dfrac{1-\sqrt{3}}{\sqrt{3}+1}$ 160. $\dfrac{1+\sqrt{3}}{1-\sqrt{3}}$

161. $\sqrt{3}$ 162. $-\sqrt{3}$ 163. $\dfrac{1+\sqrt{3}}{\sqrt{3}-1}$ 164. $\dfrac{1+\sqrt{3}}{\sqrt{3}-1}$ 165. $\dfrac{\sqrt{3}+1}{1-\sqrt{3}}$ 166. $\dfrac{-1-\sqrt{3}}{\sqrt{3}-1}$ 167. $\dfrac{1-\sqrt{3}}{\sqrt{3}+1}$

168. $\dfrac{1-\sqrt{3}}{-1-\sqrt{3}}$ 169. $\left(\dfrac{15}{17}\right)\left(\dfrac{3}{\sqrt{11}}\right)-\left(\dfrac{8}{17}\right)\left(-\dfrac{\sqrt{2}}{\sqrt{11}}\right)$ 170. $\left(-\dfrac{15}{17}\right)\left(\dfrac{3}{4}\right)+\left(\dfrac{8}{17}\right)\left(\dfrac{\sqrt{7}}{4}\right)$

171. $\left(\dfrac{8}{17}\right)\left(\dfrac{3}{\sqrt{11}}\right)+\left(\dfrac{15}{17}\right)\left(-\dfrac{\sqrt{2}}{\sqrt{11}}\right)$ 172. $\left(-\dfrac{4}{\sqrt{97}}\right)\left(\dfrac{2}{\sqrt{53}}\right)-\left(-\dfrac{9}{\sqrt{97}}\right)\left(-\dfrac{7}{\sqrt{53}}\right)$ 173. $\dfrac{-\dfrac{1}{\sqrt{15}}+\dfrac{4}{3}}{1-\left(-\dfrac{1}{\sqrt{15}}\right)\left(\dfrac{4}{3}\right)}$

174. $\dfrac{-\dfrac{6}{5}-\dfrac{7}{\sqrt{42}}}{1+\left(-\dfrac{6}{5}\right)\left(\dfrac{7}{\sqrt{42}}\right)}$ 175. $\dfrac{45+8\sqrt{2}}{17\sqrt{11}}$ 176. $\dfrac{-45+8\sqrt{7}}{68}$ 177. $\dfrac{24-15\sqrt{2}}{17\sqrt{11}}$ 178. $-\dfrac{71}{\sqrt{5141}}$

179. $\dfrac{-3+4\sqrt{15}}{3\sqrt{15}+4}$ 180. $\dfrac{-6\sqrt{42}-35}{5\sqrt{42}-42}$

8.3 Exercises

SCE-1. $\dfrac{\sqrt{2-\sqrt{2}}}{2}$ **SCE-2.** $\dfrac{\sqrt{2+\sqrt{2}}}{2}$ **SCE-3.** $-\dfrac{\sqrt{3}}{\sqrt{11}}$ **SCE-4.** $\dfrac{3}{\sqrt{14}}$ **SCE-5.** $\sqrt{3+2\sqrt{2}}$

1. $\sin\dfrac{\pi}{4}=\dfrac{1}{\sqrt{2}}$ 2. $\cos 150^\circ=-\dfrac{\sqrt{3}}{2}$ 3. $\tan\dfrac{7\pi}{4}=-1$ 4. $\cos\left(-\dfrac{5\pi}{6}\right)=-\dfrac{\sqrt{3}}{2}$ 5. $\sin 405^\circ=\dfrac{1}{\sqrt{2}}$

6. $\cos\left(-\dfrac{5\pi}{4}\right)=-\dfrac{1}{\sqrt{2}}$ 7. $\tan(-135^\circ)=1$ 8. $\cos 210^\circ=-\dfrac{\sqrt{3}}{2}$ 9. $\cos(-315^\circ)=\dfrac{1}{\sqrt{2}}$

10. $\cos\left(-\dfrac{9\pi}{4}\right)=\dfrac{1}{\sqrt{2}}$ 11. $2\left(\dfrac{2}{13}\right)\left(-\dfrac{\sqrt{165}}{13}\right)$ 12. $1-2\left(\dfrac{19}{20}\right)^2$ 13. $\dfrac{2\left(-\dfrac{7}{4\sqrt{2}}\right)}{1-\left(-\dfrac{7}{4\sqrt{2}}\right)^2}$

14. $2\left(-\dfrac{20}{29}\right)\left(\dfrac{21}{29}\right)$ 15. $2\left(\dfrac{8}{17}\right)^2-1$ 16. $\dfrac{2\left(-\dfrac{5}{12}\right)}{1-\left(-\dfrac{5}{12}\right)^2}$ 17. $2\left(-\dfrac{8}{17}\right)\left(-\dfrac{15}{17}\right)$ 18. $1-2\left(-\dfrac{12}{13}\right)^2$

19. $\dfrac{2\left(\dfrac{5}{12}\right)}{1-\left(\dfrac{5}{12}\right)^2}$ 20. $2\left(-\dfrac{10}{\sqrt{101}}\right)\left(-\dfrac{1}{\sqrt{101}}\right)$ 21. $\left(-\dfrac{8}{\sqrt{65}}\right)^2-\left(-\dfrac{1}{\sqrt{65}}\right)^2$ 22. $\dfrac{2\left(\dfrac{1}{15}\right)}{1-\left(\dfrac{1}{15}\right)^2}$ 23. $-\dfrac{4\sqrt{165}}{169}$

24. $-\dfrac{161}{200}$ 25. $\dfrac{112}{17\sqrt{2}}$ 26. $-\dfrac{840}{841}$ 27. $-\dfrac{161}{289}$ 28. $-\dfrac{120}{119}$ 29. $\dfrac{240}{289}$ 30. $-\dfrac{119}{169}$ 31. $\dfrac{120}{119}$ 32. $\dfrac{20}{101}$

33. $\dfrac{63}{65}$ 34. $\dfrac{15}{112}$ 35. $\sin\theta=\dfrac{5}{\sqrt{26}}$; $\cos\theta=\dfrac{1}{\sqrt{26}}$; $\tan\theta=5$ 36. $\sin\theta=\dfrac{1}{\sqrt{5}}$; $\cos\theta=-\dfrac{2}{\sqrt{5}}$; $\tan\theta=-\dfrac{1}{2}$

37. $\sin\theta=\dfrac{\sqrt{53+7\sqrt{53}}}{\sqrt{106}}$; $\cos\theta=-\dfrac{\sqrt{53-7\sqrt{53}}}{\sqrt{106}}$; $\tan\theta=-\dfrac{\sqrt{53}+7}{2}$ 38. $\sin\theta=\dfrac{\sqrt{7-2\sqrt{6}}}{\sqrt{14}}$;

$\cos\theta=-\dfrac{\sqrt{7+2\sqrt{6}}}{\sqrt{14}}$; $\tan\theta=\dfrac{2\sqrt{6}-7}{5}$ 39. $f(x)=-\dfrac{3}{2}+\dfrac{3}{2}\cos 2x$ 40. $f(x)=\dfrac{7}{2}+\dfrac{7}{2}\cos 2x$

41. $f(x) = \frac{3}{4} - \cos 2x + \frac{1}{4}\cos 4x$ **42.** $f(x) = -\frac{9}{8} - \frac{3}{2}\cos 2x - \frac{3}{8}\cos 4x$ **43.** $f(x) = -\frac{1}{4} + \frac{1}{4}\cos 4x$

44. $-\sqrt{\dfrac{1+\frac{1}{\sqrt{2}}}{2}}$ **45.** $\sqrt{\dfrac{1+\frac{1}{\sqrt{2}}}{2}}$ **46.** $\dfrac{-\frac{1}{\sqrt{2}}}{1+\frac{1}{\sqrt{2}}}$ **47.** $\sqrt{\dfrac{1+\frac{1}{\sqrt{2}}}{2}}$ **48.** $\sqrt{\dfrac{1+\frac{\sqrt{3}}{2}}{2}}$ **49.** $\dfrac{\frac{1}{2}}{1-\frac{\sqrt{3}}{2}}$

50. $-\dfrac{1}{\sqrt{\dfrac{1-\frac{1}{\sqrt{2}}}{2}}}$ **51.** $-\dfrac{1}{\sqrt{\dfrac{1+\frac{\sqrt{3}}{2}}{2}}}$ **52.** $-\dfrac{\sqrt{2+\sqrt{2}}}{2}$ **53.** $\dfrac{\sqrt{2+\sqrt{2}}}{2}$ **54.** $1-\sqrt{2}$ **55.** $\dfrac{\sqrt{2+\sqrt{2}}}{2}$

56. $\dfrac{\sqrt{2+\sqrt{3}}}{2}$ **57.** $2+\sqrt{3}$ **58.** $-\sqrt{4+2\sqrt{2}}$ **59.** $-2\sqrt{2-\sqrt{3}}$ **60.** $\sqrt{\dfrac{1-\left(-\frac{5}{13}\right)}{2}}$ **61.** $\sqrt{\dfrac{1+\left(-\frac{3}{5}\right)}{2}}$

62. $\dfrac{\frac{15}{17}}{1+\left(-\frac{8}{17}\right)}$ **63.** $\sqrt{\dfrac{1-\left(-\frac{21}{29}\right)}{2}}$ **64.** $\sqrt{\dfrac{1+\left(-\frac{5}{13}\right)}{2}}$ **65.** $\dfrac{\frac{8}{17}}{1+\left(-\frac{15}{17}\right)}$ **66.** $\sqrt{\dfrac{1-\left(-\frac{5}{13}\right)}{2}}$

67. $-\sqrt{\dfrac{1+\left(-\frac{8}{17}\right)}{2}}$ **68.** $\dfrac{1-\left(-\frac{3}{5}\right)}{-\frac{4}{5}}$ **69.** $\sqrt{\dfrac{1-\left(-\frac{4}{5}\right)}{2}}$ **70.** $-\sqrt{\dfrac{1+\left(-\frac{5}{13}\right)}{2}}$ **71.** $\dfrac{1-\left(-\frac{5}{17}\right)}{-\frac{2\sqrt{66}}{17}}$

72. $\sqrt{\dfrac{1-\left(-\frac{1}{\sqrt{10}}\right)}{2}}$ **73.** $-\sqrt{\dfrac{1+\left(-\frac{1}{\sqrt{290}}\right)}{2}}$ **74.** $\dfrac{-\frac{13}{\sqrt{170}}}{1+\left(-\frac{1}{\sqrt{170}}\right)}$ **75.** $\sqrt{\dfrac{1-\frac{5}{11}}{2}}$ **76.** $-\sqrt{\dfrac{1+\frac{7}{13}}{2}}$

77. $\sqrt{\dfrac{1-\frac{5}{19}}{1+\frac{5}{19}}}$ **78.** $\frac{3}{\sqrt{13}}$ **79.** $\frac{1}{\sqrt{5}}$ **80.** $\frac{5}{3}$ **81.** $\frac{5}{\sqrt{29}}$ **82.** $\frac{2}{\sqrt{13}}$ **83.** 4 **84.** $\frac{3}{\sqrt{13}}$ **85.** $-\frac{3}{\sqrt{34}}$

86. -2 **87.** $\frac{3}{\sqrt{10}}$ **88.** $-\frac{2}{\sqrt{13}}$ **89.** $-\frac{11}{\sqrt{66}}$ **90.** $\sqrt{\dfrac{\sqrt{10}+1}{2\sqrt{10}}}$ **91.** $-\sqrt{\dfrac{\sqrt{290}-1}{2\sqrt{290}}}$

92. $\dfrac{-13}{\sqrt{170}-1}$ **93.** $\sqrt{\frac{3}{11}}$ **94.** $-\sqrt{\frac{10}{13}}$ **95.** $\dfrac{\sqrt{7}}{2\sqrt{3}}$

96.

$$\frac{\tan^2\theta - 1}{1+\tan^2\theta} \stackrel{?}{=} -\cos 2\theta$$
$$\frac{\tan^2\theta - 1}{\sec^2\theta} \stackrel{?}{=} -\cos 2\theta$$
$$(\tan^2\theta - 1)\cdot\frac{1}{\sec^2\theta} \stackrel{?}{=} -\cos 2\theta$$
$$(\tan^2\theta - 1)\cdot\cos^2\theta \stackrel{?}{=} -\cos 2\theta$$
$$\tan^2\theta\cos^2\theta - \cos^2\theta \stackrel{?}{=} -\cos 2\theta$$
$$\sin^2\theta - \cos^2\theta \stackrel{?}{=} -\cos 2\theta$$
$$-(\cos^2\theta - \sin^2\theta) \stackrel{?}{=} -\cos 2\theta$$
$$-\cos 2\theta = -\cos 2\theta$$

97.

$$\cot 2\theta \stackrel{?}{=} \frac{1}{2}\sec\theta\csc\theta - \tan\theta$$
$$\frac{\cos 2\theta}{\sin 2\theta} \stackrel{?}{=} \frac{1}{2}\sec\theta\csc\theta - \tan\theta$$
$$\frac{1-2\sin^2\theta}{2\sin\theta\cos\theta} \stackrel{?}{=} \frac{1}{2}\sec\theta\csc\theta - \tan\theta$$
$$\frac{1}{2\sin\theta\cos\theta} - \frac{2\sin^2\theta}{2\sin\theta\cos\theta} \stackrel{?}{=} \frac{1}{2}\sec\theta\csc\theta - \tan\theta$$
$$\frac{1}{2}\cdot\frac{1}{\sin\theta}\cdot\frac{1}{\cos\theta} - \frac{\sin\theta}{\cos\theta} \stackrel{?}{=} \frac{1}{2}\sec\theta\csc\theta - \tan\theta$$
$$\frac{1}{2}\sec\theta\csc\theta - \tan\theta = \frac{1}{2}\sec\theta\csc\theta - \tan\theta$$

98. $\sec 2\theta \stackrel{?}{=} \dfrac{\csc^2\theta}{\csc^2\theta - 2}$

$\dfrac{1}{\cos 2\theta} \stackrel{?}{=} \dfrac{\csc^2\theta}{\csc^2\theta - 2}$

$\dfrac{1}{1 - 2\sin^2\theta} \stackrel{?}{=} \dfrac{\csc^2\theta}{\csc^2\theta - 2}$

$\dfrac{1}{1 - 2\sin^2\theta}\cdot\dfrac{\csc^2\theta}{\csc^2\theta} \stackrel{?}{=} \dfrac{\csc^2\theta}{\csc^2\theta - 2}$

$\dfrac{\csc^2\theta}{\csc^2\theta - 2\sin^2\theta\csc^2\theta} \stackrel{?}{=} \dfrac{\csc^2\theta}{\csc^2\theta - 2}$

$\dfrac{\csc^2\theta}{\csc^2\theta - 2} = \dfrac{\csc^2\theta}{\csc^2\theta - 2}$

99. $\tan\theta \stackrel{?}{=} \dfrac{1 - \cos 2\theta}{\sin 2\theta}$

$\tan\theta \stackrel{?}{=} \dfrac{1 - (1 - 2\sin^2\theta)}{2\sin\theta\cos\theta}$

$\tan\theta \stackrel{?}{=} \dfrac{2\sin^2\theta}{2\sin\theta\cos\theta}$

$\tan\theta \stackrel{?}{=} \dfrac{\sin\theta}{\cos\theta}$

$\tan\theta = \tan\theta$

100. $\sin^2\dfrac{\theta}{2} \stackrel{?}{=} \dfrac{\tan\theta - \sin\theta}{2\tan\theta}$

$\dfrac{1 - \cos\theta}{2} \stackrel{?}{=} \dfrac{\tan\theta - \sin\theta}{2\tan\theta}$

$\dfrac{1 - \cos\theta}{2}\cdot\dfrac{\tan\theta}{\tan\theta} \stackrel{?}{=} \dfrac{\tan\theta - \sin\theta}{2\tan\theta}$

$\dfrac{\tan\theta - \cos\theta\tan\theta}{2\tan\theta} \stackrel{?}{=} \dfrac{\tan\theta - \sin\theta}{2\tan\theta}$

$\dfrac{\tan\theta - \sin\theta}{2\tan\theta} = \dfrac{\tan\theta - \sin\theta}{2\tan\theta}$

101. $\tan\dfrac{\theta}{2} \stackrel{?}{=} \dfrac{1}{\csc\theta + \cot\theta}$

$\dfrac{\sin\theta}{1 + \cos\theta} \stackrel{?}{=} \dfrac{1}{\csc\theta + \cot\theta}$

$\dfrac{\sin\theta}{1 + \cos\theta}\cdot\dfrac{\frac{1}{\sin\theta}}{\frac{1}{\sin\theta}} \stackrel{?}{=} \dfrac{1}{\csc\theta + \cot\theta}$

$\dfrac{\frac{\sin\theta}{\sin\theta}}{\frac{1}{\sin\theta} + \frac{\cos\theta}{\sin\theta}} \stackrel{?}{=} \dfrac{1}{\csc\theta + \cot\theta}$

$\dfrac{1}{\csc\theta + \cot\theta} = \dfrac{1}{\csc\theta + \cot\theta}$

102. $\cot\dfrac{\theta}{2} - \cot\theta \stackrel{?}{=} \csc\theta$

$\dfrac{1}{\tan\frac{\theta}{2}} - \cot\theta \stackrel{?}{=} \csc\theta$

$\dfrac{1}{\tan\frac{\theta}{2}} - \cot\theta \stackrel{?}{=} \csc\theta$

$\dfrac{1 + \cos\theta}{\sin\theta} - \cot\theta \stackrel{?}{=} \csc\theta$

$\dfrac{1}{\sin\theta} + \dfrac{\cos\theta}{\sin\theta} - \cot\theta \stackrel{?}{=} \csc\theta$

$\csc\theta + \cot\theta - \cot\theta \stackrel{?}{=} \csc\theta$

$\csc\theta = \csc\theta$

103. $\dfrac{\sqrt{3}}{2}$ **104.** 0 **105.** $\dfrac{20}{29}$ **106.** $\dfrac{31}{81}$

107. $\sqrt{\dfrac{1 + \frac{1}{2}}{2}}$ **108.** $\sqrt{\dfrac{1 + \frac{1}{\sqrt{2}}}{2}}$ **109.** $\sqrt{\dfrac{1 + \frac{3}{7}}{2}}$ **110.** $\sqrt{\dfrac{1 + \frac{5}{11}}{2}}$ **111.** $-\sqrt{\dfrac{1 - \frac{1}{\sqrt{2}}}{2}}$ **112.** $\sqrt{\dfrac{1 + \frac{\sqrt{3}}{2}}{2}}$

113. $\sqrt{\dfrac{1 - \frac{7}{\sqrt{74}}}{2}}$ **114.** $\sqrt{\dfrac{1 + \frac{1}{\sqrt{26}}}{2}}$ **115.** $\dfrac{\sqrt{3}}{2}$ **116.** $\dfrac{\sqrt{2 + \sqrt{2}}}{2}$ **117.** $\dfrac{\sqrt{5}}{\sqrt{7}}$ **118.** $\dfrac{2\sqrt{2}}{\sqrt{11}}$

119. $-\dfrac{\sqrt{2 - \sqrt{2}}}{2}$ **120.** $\dfrac{\sqrt{2 + \sqrt{3}}}{2}$ **121.** $\dfrac{\sqrt{74 - 7\sqrt{74}}}{2\sqrt{37}}$ **122.** $\dfrac{\sqrt{26 + \sqrt{26}}}{2\sqrt{13}}$ **123.** $-\dfrac{4\sqrt{165}}{169}$

124. $-\dfrac{161}{200}$ **125.** $\dfrac{112}{17\sqrt{2}}$ **126.** $-\dfrac{840}{841}$ **127.** $-\dfrac{161}{289}$ **128.** $-\dfrac{120}{119}$ **129.** $\dfrac{240}{289}$ **130.** $-\dfrac{119}{169}$ **131.** $\dfrac{120}{119}$ **132.** $\dfrac{20}{101}$

133. $\dfrac{63}{65}$ **134.** $\dfrac{15}{112}$ **135.** $-\sqrt{\dfrac{1 + \frac{1}{\sqrt{2}}}{2}}$ **136.** $\sqrt{\dfrac{1 + \frac{1}{\sqrt{2}}}{2}}$ **137.** $\dfrac{-\frac{1}{\sqrt{2}}}{1 + \frac{1}{\sqrt{2}}}$ **138.** $\sqrt{\dfrac{1 + \frac{1}{\sqrt{2}}}{2}}$

139. $\sqrt{\dfrac{1 + \frac{\sqrt{3}}{2}}{2}}$ **140.** $\dfrac{\frac{1}{2}}{1 - \frac{\sqrt{3}}{2}}$ **141.** $-\dfrac{1}{\sqrt{\dfrac{1 - \frac{1}{\sqrt{2}}}{2}}}$ **142.** $-\dfrac{1}{\sqrt{\dfrac{1 + \frac{\sqrt{3}}{2}}{2}}}$ **143.** $-\dfrac{\sqrt{2 + \sqrt{2}}}{2}$

144. $\dfrac{\sqrt{2+\sqrt{2}}}{2}$ **145.** $1-\sqrt{2}$ **146.** $\dfrac{\sqrt{2+\sqrt{2}}}{2}$ **147.** $\dfrac{\sqrt{2+\sqrt{3}}}{2}$ **148.** $2+\sqrt{3}$ **149.** $-\sqrt{4+2\sqrt{2}}$

150. $-2\sqrt{2-\sqrt{3}}$ **151.** $\sqrt{\dfrac{1-\left(-\frac{5}{13}\right)}{2}}$ **152.** $\sqrt{\dfrac{1+\left(-\frac{3}{5}\right)}{2}}$ **153.** $\dfrac{\frac{15}{17}}{1+\left(-\frac{8}{17}\right)}$ **154.** $\sqrt{\dfrac{1-\left(-\frac{21}{29}\right)}{2}}$

155. $\sqrt{\dfrac{1+\left(-\frac{5}{13}\right)}{2}}$ **156.** $\dfrac{\frac{8}{17}}{1+\left(-\frac{15}{17}\right)}$ **157.** $\sqrt{\dfrac{1-\left(-\frac{5}{13}\right)}{2}}$ **158.** $-\sqrt{\dfrac{1+\left(-\frac{8}{17}\right)}{2}}$ **159.** $\dfrac{1-\left(-\frac{3}{5}\right)}{-\frac{4}{5}}$

160. $\sqrt{\dfrac{1-\left(-\frac{4}{5}\right)}{2}}$ **161.** $-\sqrt{\dfrac{1+\left(-\frac{5}{13}\right)}{2}}$ **162.** $\dfrac{1-\left(-\frac{5}{17}\right)}{-\frac{2\sqrt{66}}{17}}$ **163.** $\sqrt{\dfrac{1-\left(-\frac{1}{\sqrt{10}}\right)}{2}}$

164. $-\sqrt{\dfrac{1+\left(-\frac{1}{\sqrt{290}}\right)}{2}}$ **165.** $\dfrac{-\frac{13}{\sqrt{170}}}{1+\left(-\frac{1}{\sqrt{170}}\right)}$ **166.** $\sqrt{\dfrac{1-\frac{5}{11}}{2}}$ **167.** $-\sqrt{\dfrac{1+\frac{7}{13}}{2}}$ **168.** $\sqrt{\dfrac{1-\frac{5}{19}}{1+\frac{5}{19}}}$

169. $\dfrac{3}{\sqrt{13}}$ **170.** $\dfrac{1}{\sqrt{5}}$ **171.** $\dfrac{5}{3}$ **172.** $\dfrac{5}{\sqrt{29}}$ **173.** $\dfrac{2}{\sqrt{13}}$ **174.** 4 **175.** $\dfrac{3}{\sqrt{13}}$ **176.** $-\dfrac{3}{\sqrt{34}}$ **177.** -2

178. $\dfrac{3}{\sqrt{10}}$ **179.** $-\dfrac{2}{\sqrt{13}}$ **180.** $-\dfrac{11}{\sqrt{66}}$ **181.** $\sqrt{\dfrac{\sqrt{10}+1}{2\sqrt{10}}}$ **182.** $-\sqrt{\dfrac{\sqrt{290}-1}{2\sqrt{290}}}$ **183.** $\dfrac{-13}{\sqrt{170}-1}$

184. $\sqrt{\dfrac{3}{11}}$ **185.** $-\sqrt{\dfrac{10}{13}}$ **186.** $\dfrac{\sqrt{7}}{2\sqrt{3}}$ **187.** $\sqrt{\dfrac{1+\frac{1}{2}}{2}}$ **188.** $\sqrt{\dfrac{1+\frac{1}{\sqrt{2}}}{2}}$ **189.** $\sqrt{\dfrac{1+\frac{3}{7}}{2}}$

190. $\sqrt{\dfrac{1+\frac{5}{11}}{2}}$ **191.** $-\sqrt{\dfrac{1-\frac{1}{\sqrt{2}}}{2}}$ **192.** $\sqrt{\dfrac{1+\frac{\sqrt{3}}{2}}{2}}$ **193.** $\sqrt{\dfrac{1-\frac{7}{\sqrt{74}}}{2}}$ **194.** $\sqrt{\dfrac{1+\frac{1}{\sqrt{26}}}{2}}$ **195.** $\dfrac{\sqrt{3}}{2}$

196. $\dfrac{\sqrt{2+\sqrt{2}}}{2}$ **197.** $\dfrac{\sqrt{5}}{\sqrt{7}}$ **198.** $\dfrac{2\sqrt{2}}{\sqrt{11}}$ **199.** $-\dfrac{\sqrt{2-\sqrt{2}}}{2}$ **200.** $\dfrac{\sqrt{2+\sqrt{3}}}{2}$ **201.** $\dfrac{\sqrt{74-7\sqrt{74}}}{2\sqrt{37}}$

202. $\dfrac{\sqrt{26+\sqrt{26}}}{2\sqrt{13}}$

8.4 Exercises

1. $\frac{1}{2}\cos(4\theta)-\frac{1}{2}\cos(12\theta)$ **2.** $\frac{1}{2}\sin(10\theta)+\frac{1}{2}\sin(4\theta)$ **3.** $\frac{1}{2}\cos(4\theta)+\frac{1}{2}\cos(6\theta)$ **4.** $\frac{1}{2}\sin(3\theta)+\frac{1}{2}\sin(2\theta)$

5. $\dfrac{1+\sqrt{2}}{4}$ **6.** $\dfrac{1}{4}$ **7.** $\dfrac{1-\sqrt{2}}{2\sqrt{2}}$ **8.** $\dfrac{\sqrt{2}-1}{2\sqrt{2}}$ **9.** $2\sin\theta\cos(8\theta)$ **10.** $2\cos(4\theta)\cos\theta$ **11.** $2\sin(10\theta)\cos\theta$

12. $2\sin(4\theta)\sin\left(\frac{3\theta}{2}\right)$ **13.** $\frac{1}{\sqrt{2}}$ **14.** 0 **15.** $-\frac{1}{\sqrt{2}}$ **16.** $-\sqrt{2}$ **17.**

$$\frac{\sin\theta - \sin 3\theta}{\cos\theta + \cos 3\theta} \stackrel{?}{=} -\tan\theta$$
$$\frac{2\sin\left(\frac{\theta - 3\theta}{2}\right)\cos\left(\frac{\theta + 3\theta}{2}\right)}{2\cos\left(\frac{\theta + 3\theta}{2}\right)\cos\left(\frac{\theta - 3\theta}{2}\right)} \stackrel{?}{=} -\tan\theta$$
$$\frac{2\sin(-\theta)\cos 2\theta}{2\cos 2\theta\cos(-\theta)} \stackrel{?}{=} -\tan\theta$$
$$\frac{\sin(-\theta)}{\cos(-\theta)} \stackrel{?}{=} -\tan\theta$$
$$\frac{-\sin\theta}{\cos\theta} \stackrel{?}{=} -\tan\theta$$
$$-\tan\theta = -\tan\theta$$

18.

$$\frac{\cos\theta + \cos 3\theta}{\sin\theta + \sin 3\theta} \stackrel{?}{=} \cot 2\theta$$
$$\frac{2\cos\left(\frac{\theta + 3\theta}{2}\right)\cos\left(\frac{\theta - 3\theta}{2}\right)}{2\sin\left(\frac{\theta + 3\theta}{2}\right)\cos\left(\frac{\theta - 3\theta}{2}\right)} \stackrel{?}{=} \cot 2\theta$$
$$\frac{2\cos 2\theta\cos(-\theta)}{2\sin 2\theta\cos(-\theta)} \stackrel{?}{=} \cot 2\theta$$
$$\frac{\cos 2\theta}{\sin 2\theta} \stackrel{?}{=} \cot 2\theta$$
$$\cot 2\theta = \cot 2\theta$$

19.

$$\frac{\sin x + \sin y}{\cos x + \cos y} \stackrel{?}{=} \tan\frac{x+y}{2}$$
$$\frac{2\sin\frac{x+y}{2}\cos\frac{x-y}{2}}{2\cos\frac{x+y}{2}\cos\frac{x-y}{2}} \stackrel{?}{=} \tan\frac{x+y}{2}$$
$$\frac{\sin\frac{x+y}{2}}{\cos\frac{x+y}{2}} \stackrel{?}{=} \tan\frac{x+y}{2}$$
$$\tan\frac{x+y}{2} = \tan\frac{x+y}{2}$$

20.

$$\frac{\cos x + \cos y}{\cos x - \cos y} \stackrel{?}{=} -\cot\frac{x+y}{2}\cot\frac{x-y}{2}$$
$$\frac{2\cos\frac{x+y}{2}\cos\frac{x-y}{2}}{-2\sin\frac{x+y}{2}\sin\frac{x-y}{2}} \stackrel{?}{=} -\cot\frac{x+y}{2}\cot\frac{x-y}{2}$$
$$-\frac{\cos\frac{x+y}{2}}{\sin\frac{x+y}{2}}\cdot\frac{\cos\frac{x-y}{2}}{\sin\frac{x-y}{2}} \stackrel{?}{=} -\cot\frac{x+y}{2}\cot\frac{x-y}{2}$$
$$-\cot\frac{x+y}{2}\cot\frac{x-y}{2} = -\cot\frac{x+y}{2}\cot\frac{x-y}{2}$$

21.

$$\frac{\sin(6\theta) + \sin(8\theta)}{\sin(6\theta) - \sin(8\theta)} \stackrel{?}{=} -\frac{\tan(7\theta)}{\tan\theta}$$
$$\frac{2\sin\left(\frac{6\theta + 8\theta}{2}\right)\cos\left(\frac{6\theta - 8\theta}{2}\right)}{2\sin\left(\frac{6\theta - 8\theta}{2}\right)\cos\left(\frac{6\theta + 8\theta}{2}\right)} \stackrel{?}{=} -\frac{\tan(7\theta)}{\tan\theta}$$
$$\frac{2\sin(7\theta)\cos(-\theta)}{2\sin(-\theta)\cos(7\theta)} \stackrel{?}{=} -\frac{\tan(7\theta)}{\tan\theta}$$
$$\frac{\sin(7\theta)\cos(-\theta)}{\sin(-\theta)\cos(7\theta)} \stackrel{?}{=} -\frac{\tan(7\theta)}{\tan\theta}$$
$$\frac{\sin(7\theta)\cos\theta}{-\sin\theta\cos(7\theta)} \stackrel{?}{=} -\frac{\tan(7\theta)}{\tan\theta}$$
$$-\frac{\sin(7\theta)}{\cos(7\theta)}\cdot\frac{\cos\theta}{\sin\theta} \stackrel{?}{=} -\frac{\tan(7\theta)}{\tan\theta}$$
$$-\tan(7\theta)\cdot\cot\theta \stackrel{?}{=} -\frac{\tan(7\theta)}{\tan\theta}$$
$$-\tan(7\theta)\cdot\frac{1}{\tan\theta} \stackrel{?}{=} -\frac{\tan(7\theta)}{\tan\theta}$$
$$-\frac{\tan(7\theta)}{\tan\theta} = -\frac{\tan(7\theta)}{\tan\theta}$$

8.5 Exercises

SCE-1. $-\frac{7\pi}{4}, -\frac{13\pi}{12}, -\frac{5\pi}{12}, \frac{\pi}{4}, \frac{11\pi}{12}, \frac{19\pi}{12}$ **SCE-2.** $-1, -\frac{1}{2}$ **SCE-3.** $-1, 0$ **SCE-4.** $-1, -\frac{1}{2}$

1. $\theta = \frac{\pi}{3} + 2\pi k$ or $\theta = \frac{5\pi}{3} + 2\pi k$, where k is any integer; $\theta = \frac{\pi}{3}, \theta = \frac{5\pi}{3}$ **2.** $\theta = \frac{2\pi}{3} + \pi k$, where k is any integer; $\theta = \frac{2\pi}{3}, \theta = \frac{5\pi}{3}$ **3.** general formula: $\theta = \pi k$ where k is an integer; specific solutions: $\theta = 0, \pi$ **4.** $\theta = \frac{\pi}{2} + 2\pi k$, where k is any integer; $\theta = \frac{\pi}{2}$ **5.** $\theta = \frac{3\pi}{4} + \pi k$, where k is any integer; $\theta = \frac{3\pi}{4}, \theta = \frac{7\pi}{4}$ **6.** general formula: $\theta = \frac{\pi}{4} + \frac{\pi}{2}k$; Specific solutions: $\theta = \frac{\pi}{4}, \frac{3\pi}{4}, \frac{5\pi}{4}, \frac{7\pi}{4}$ **7.** $\theta = \frac{5\pi}{8} + \pi k$ or $\theta = \frac{7\pi}{8} + \pi k$, where k is any integer; $\theta = \frac{5\pi}{8}, \theta = \frac{7\pi}{8}, \theta = \frac{13\pi}{8}, \theta = \frac{15\pi}{8}$ **8.** $\theta = \frac{\pi}{4} + \frac{2\pi k}{3}$ or $\theta = \frac{5\pi}{12} + \frac{2\pi k}{3}$, where k is any integer; $\theta = \frac{\pi}{4}, \theta = \frac{5\pi}{12}, \theta = \frac{11\pi}{12}, \theta = \frac{13\pi}{12}, \theta = \frac{19\pi}{12}, \theta = \frac{7\pi}{4}$ **9.** $\theta = \frac{3\pi}{2} + 2\pi k$, where k is any integer; $\theta = \frac{3\pi}{2}$ **10.** $\theta = \frac{5\pi}{2} + 4\pi k$ or $\theta = \frac{7\pi}{2} + 4\pi k$, where k is any integer; no solution **11.** $\theta = \frac{2\pi}{3} + 4\pi k$ or $\theta = \frac{10\pi}{3} + 4\pi k$, where k is any integer; $\theta = \frac{2\pi}{3}$ **12.** $\theta = \frac{\pi}{3} + 2\pi k$, where k is any integer; $\theta = \frac{\pi}{3}$ **13.** $\theta = \frac{4\pi}{9} + \frac{4\pi k}{3}$ or $\theta = \frac{8\pi}{9} + \frac{4\pi k}{3}$, where k is any integer; $\theta = \frac{4\pi}{9}, \theta = \frac{8\pi}{9}, \theta = \frac{16\pi}{9}$ **14.** $\theta = \frac{\pi}{2} + \pi k, \theta = \frac{2\pi}{3} + 2\pi k$, or $\theta = \frac{4\pi}{3} + 2\pi k$, where k is any integer; $\theta = \frac{\pi}{2}, \theta = \frac{2\pi}{3}, \theta = \frac{4\pi}{3}, \theta = \frac{3\pi}{2}$ **15.** $\theta = \frac{\pi}{6} + \pi k$ or $\theta = \frac{5\pi}{6} + \pi k$, where k is any integer; $\theta = \frac{\pi}{6}, \theta = \frac{5\pi}{6}, \theta = \frac{7\pi}{6}, \theta = \frac{11\pi}{6}$ **16.** $\theta = \frac{\pi}{2} + 2\pi k, \theta = \frac{7\pi}{6} + 2\pi k$, or $\theta = \frac{11\pi}{6} + 2\pi k$, where k is any integer; $\theta = \frac{\pi}{2}, \theta = \frac{7\pi}{6}, \theta = \frac{11\pi}{6}$ **17.** $\theta = \frac{2\pi}{3} + 2\pi k$ or $\theta = \frac{4\pi}{3} + 2\pi k$, where k is any integer; $\theta = \frac{2\pi}{3}, \theta = \frac{4\pi}{3}$ **18.** $\theta = \pi k, \theta = \frac{\pi}{6} + \pi k$, or $\theta = \frac{5\pi}{6} + \pi k$, where k is any integer; $\theta = 0, \theta = \frac{\pi}{6}, \theta = \frac{5\pi}{6}, \theta = \pi, \theta = \frac{7\pi}{6}, \theta = \frac{11\pi}{6}$ **19.** $\theta = \frac{3\pi k}{2}$, where k is any integer; $\theta = 0, \theta = \frac{3\pi}{2}$ **20.** $\theta = \pi k, \theta = \frac{\pi}{6} + 2\pi k$, or $\theta = \frac{5\pi}{6} + 2\pi k$, where k is any integer $\theta = 0, \theta = \frac{\pi}{6}, \theta = \frac{5\pi}{6}, \theta = \pi$ **21.** $\theta = \frac{2\pi}{3} + 2\pi k, \theta = \pi + 2\pi k$, or $\theta = \frac{4\pi}{3} + 2\pi k$, where k is any integer; $\theta = \frac{2\pi}{3}, \theta = \pi, \theta = \frac{4\pi}{3}$ **22.** $\theta = \pi k$ or $\theta = \frac{\pi}{4} + \pi k$, where k is any integer; $\theta = 0, \theta = \frac{\pi}{4}, \theta = \pi, \theta = \frac{5\pi}{4}$ **23.** $\theta = \frac{\pi}{3} + 2\pi k$ or $\theta = \frac{5\pi}{3} + 2\pi k$, where k is any integer; $\theta = \frac{\pi}{3}, \theta = \frac{5\pi}{3}$ **24.** $\theta = \frac{\pi}{3} + \pi k$ or $\theta = \frac{2\pi}{3} + \pi k$, where k is any integer; $\theta = \frac{\pi}{3}, \theta = \frac{2\pi}{3}, \theta = \frac{4\pi}{3}, \theta = \frac{5\pi}{3}$ **25.** $\theta = \frac{\pi}{2} + \pi k$ or $\theta = \frac{\pi}{4} + \frac{\pi k}{2}$, where k is any integer; $\theta = \frac{\pi}{4}, \theta = \frac{\pi}{2}, \theta = \frac{3\pi}{4}, \theta = \frac{5\pi}{4}, \theta = \frac{3\pi}{2}, \theta = \frac{7\pi}{4}$ **26.** $\theta = \pi k, \theta = \frac{\pi}{3} + 2\pi k$, or $\theta = \frac{5\pi}{3} + 2\pi k$, where k is any integer; $\theta = 0, \theta = \frac{\pi}{3}, \theta = \pi, \theta = \frac{5\pi}{3}$ **27.** $\theta = \frac{\pi}{6} + \pi k$ or $\theta = \frac{5\pi}{6} + \pi k$, where k is any integer; $\theta = \frac{\pi}{6}, \theta = \frac{5\pi}{6}, \theta = \frac{7\pi}{6}, \theta = \frac{11\pi}{6}$ **28.** $\theta = \pi k$ or $\theta = \frac{\pi}{6} + \frac{\pi k}{3}$, where k is any integer; $\theta = 0, \theta = \frac{\pi}{6}, \theta = \frac{\pi}{2}, \theta = \frac{5\pi}{6}, \theta = \pi, \theta = \frac{7\pi}{6}, \theta = \frac{3\pi}{2}, \theta = \frac{11\pi}{6}$ **29.** $\theta = \frac{\pi k}{4}$, where k is any integer; $\theta = 0, \theta = \frac{\pi}{4}, \theta = \frac{\pi}{2}, \theta = \frac{3\pi}{4}, \theta = \pi, \theta = \frac{5\pi}{4}, \theta = \frac{3\pi}{2}, \theta = \frac{7\pi}{4}$ **30.** $\theta = \frac{3\pi}{4} + \pi k$, where k is any integer; $\theta = \frac{3\pi}{4}, \theta = \frac{7\pi}{4}$ **31.** $\theta = \frac{5\pi}{4} + 2\pi k$, where k is any integer; $\theta = \frac{5\pi}{4}$ **32.** $\theta = \frac{3\pi}{4} + \pi k$, where k is any integer; $\theta = \frac{3\pi}{4}, \theta = \frac{7\pi}{4}$ **33.** $\theta = \pi k$ or $\theta = \frac{2\pi}{3} + k\pi$, where k is any integer; $\theta = 0, \theta = \frac{2\pi}{3}, \theta = \pi, \theta = \frac{5\pi}{3}$ **34.** $\theta \approx 0.8038, \theta \approx 2.3378$

35. $\theta \approx 0.8892, \theta \approx 5.3939$ **36.** $\theta \approx 0.4636, \theta \approx 3.6052$ **37.** $\theta \approx 2.4981, \theta \approx 3.7851$ **38.** $\theta \approx 3.3430, \theta \approx 6.0818$ **39.** $\theta \approx 1.8158, \theta \approx 4.9574$ **40.** $\theta = \frac{\pi}{3}, \theta = \frac{5\pi}{3}$ **41.** $\theta = \frac{2\pi}{3}, \theta = \frac{5\pi}{3}$ **42.** $\theta = 0, \pi$ **43.** $\theta = \frac{\pi}{2}$ **44.** $\theta = \frac{3\pi}{4}, \theta = \frac{7\pi}{4}$ **45.** $\theta = \frac{\pi}{4}, \frac{3\pi}{4}, \frac{5\pi}{4}, \frac{7\pi}{4}$ **46.** $\theta = \frac{5\pi}{8}, \theta = \frac{7\pi}{8}, \theta = \frac{13\pi}{8}, \theta = \frac{15\pi}{8}$ **47.** $\theta = \frac{\pi}{4}, \theta = \frac{5\pi}{12}, \theta = \frac{11\pi}{12}, \theta = \frac{13\pi}{12}, \theta = \frac{19\pi}{12}, \theta = \frac{7\pi}{4}$ **48.** $\theta = \frac{3\pi}{2}$ **49.** no solution **50.** $\theta = \frac{2\pi}{3}$ **51.** $\theta = \frac{\pi}{3}$ **52.** $\theta = \frac{4\pi}{9}, \theta = \frac{8\pi}{9}, \theta = \frac{16\pi}{9}$ **53.** $\theta = \frac{\pi}{2}, \theta = \frac{2\pi}{3}, \theta = \frac{4\pi}{3}, \theta = \frac{3\pi}{2}$ **54.** $\theta = \frac{\pi}{6}, \theta = \frac{5\pi}{6}, \theta = \frac{7\pi}{6}, \theta = \frac{11\pi}{6}$ **55.** $\theta = \frac{\pi}{2}, \theta = \frac{7\pi}{6}, \theta = \frac{11\pi}{6}$ **56.** $\theta = \frac{2\pi}{3}, \theta = \frac{4\pi}{3}$ **57.** $\theta = 0, \theta = \frac{\pi}{6}, \theta = \frac{5\pi}{6}, \theta = \pi, \theta = \frac{7\pi}{6}, \theta = \frac{11\pi}{6}$ **58.** $\theta = 0, \theta = \frac{3\pi}{2}$ **59.** $\theta = 0, \theta = \frac{\pi}{6}, \theta = \frac{5\pi}{6}, \theta = \pi$ **60.** $\theta = \frac{2\pi}{3}, \theta = \pi, \theta = \frac{4\pi}{3}$ **61.** $\theta = 0, \theta = \frac{\pi}{4}, \theta = \pi, \theta = \frac{5\pi}{4}$ **62.** $\theta = \frac{\pi}{3}, \theta = \frac{5\pi}{3}$ **63.** $\theta = \frac{\pi}{3}, \theta = \frac{2\pi}{3}, \theta = \frac{4\pi}{3}, \theta = \frac{5\pi}{3}$ **64.** $\theta = \frac{\pi}{4}, \theta = \frac{\pi}{2}, \theta = \frac{3\pi}{4}, \theta = \frac{5\pi}{4}, \theta = \frac{3\pi}{2}, \theta = \frac{7\pi}{4}$ **65.** $\theta = 0, \theta = \frac{\pi}{3}, \theta = \pi, \theta = \frac{5\pi}{3}$ **66.** $\theta = \frac{\pi}{6}, \theta = \frac{5\pi}{6}, \theta = \frac{7\pi}{6}, \theta = \frac{11\pi}{6}$ **67.** $\theta = 0, \theta = \frac{\pi}{6}, \theta = \frac{\pi}{2}, \theta = \frac{5\pi}{6}, \theta = \pi, \theta = \frac{7\pi}{6}, \theta = \frac{3\pi}{2}, \theta = \frac{11\pi}{6}$ **68.** $\theta = 0, \theta = \frac{\pi}{4}, \theta = \frac{\pi}{2}, \theta = \frac{3\pi}{4}, \theta = \pi, \theta = \frac{5\pi}{4}, \theta = \frac{3\pi}{2}, \theta = \frac{7\pi}{4}$ **69.** $\theta = \frac{3\pi}{4}, \theta = \frac{7\pi}{4}$ **70.** $\theta = \frac{5\pi}{4}$ **71.** $\theta = \frac{3\pi}{4}, \theta = \frac{7\pi}{4}$ **72.** $\theta = 0, \theta = \frac{2\pi}{3}, \theta = \pi, \theta = \frac{5\pi}{3}$

Review Exercises

1.

$$\frac{\cot\theta + 1}{\csc\theta} \stackrel{?}{=} \sin\theta + \cos\theta$$
$$\frac{\cot\theta}{\csc\theta} + \frac{1}{\csc\theta} \stackrel{?}{=} \sin\theta + \cos\theta$$
$$\cot\theta \cdot \frac{1}{\csc\theta} + \sin\theta \stackrel{?}{=} \sin\theta + \cos\theta$$
$$\frac{\cos\theta}{\sin\theta} \cdot \sin\theta + \sin\theta \stackrel{?}{=} \sin\theta + \cos\theta$$
$$\cos\theta + \sin\theta \stackrel{?}{=} \sin\theta + \cos\theta$$
$$\sin\theta + \cos\theta = \sin\theta + \cos\theta$$

2.

$$\frac{\sec\beta}{\sin\beta} - \frac{\sin\beta}{\sec\beta} \stackrel{?}{=} \frac{\tan^2\beta + \cos^2\beta}{\tan\beta}$$
$$\frac{\sec\beta}{\sin\beta} \cdot \frac{\sec\beta}{\sec\beta} - \frac{\sin\beta}{\sec\beta} \cdot \frac{\sin\beta}{\sin\beta} \stackrel{?}{=} \frac{\tan^2\beta + \cos^2\beta}{\tan\beta}$$
$$\frac{\sec^2\beta}{\sin\beta\sec\beta} - \frac{\sin^2\beta}{\sec\beta\sin\beta} \stackrel{?}{=} \frac{\tan^2\beta + \cos^2\beta}{\tan\beta}$$
$$\frac{\sec^2\beta - \sin^2\beta}{\sin\beta\sec\beta} \stackrel{?}{=} \frac{\tan^2\beta + \cos^2\beta}{\tan\beta}$$
$$\frac{1 + \tan^2\beta - (1 - \cos^2\beta)}{\sin\beta \cdot \frac{1}{\cos\beta}} \stackrel{?}{=} \frac{\tan^2\beta + \cos^2\beta}{\tan\beta}$$
$$\frac{1 + \tan^2\beta - 1 + \cos^2\beta}{\frac{\sin\beta}{\cos\beta}} \stackrel{?}{=} \frac{\tan^2\beta + \cos^2\beta}{\tan\beta}$$
$$\frac{\tan^2\beta + \cos^2\beta}{\tan\beta} = \frac{\tan^2\beta + \cos^2\beta}{\tan\beta}$$

3. $$1 + \frac{1 - \cot^2 x}{1 + \cot^2 x} \stackrel{?}{=} 2\sin^2 x$$
$$\frac{1 + \cot^2 x}{1 + \cot^2 x} + \frac{1 - \cot^2 x}{1 + \cot^2 x} \stackrel{?}{=} 2\sin^2 x$$
$$\frac{1 + \cot^2 x + 1 - \cot^2 x}{1 + \cot^2 x} \stackrel{?}{=} 2\sin^2 x$$
$$\frac{2}{1 + \cot^2 x} \stackrel{?}{=} 2\sin^2 x$$
$$\frac{2}{\csc^2 x} \stackrel{?}{=} 2\sin^2 x$$
$$2\sin^2 x = 2\sin^2 x$$

4. $\frac{1}{2}$ **5.** $\frac{\sqrt{3} - 1}{2\sqrt{2}}$ **6.** $\frac{24 - 15\sqrt{2}}{17\sqrt{11}}$ **7.** $\frac{1 - \sqrt{3}}{\sqrt{3} + 1}$ **8.** -1

9. $\sin\frac{\pi}{4} = \frac{1}{\sqrt{2}}$ **10.** $\cos\left(-\frac{5\pi}{6}\right) = -\frac{\sqrt{3}}{2}$ **11.** $-\frac{4\sqrt{165}}{169}$ **12.** $-\frac{120}{119}$ **13.** $-\frac{\sqrt{2 + \sqrt{2}}}{2}$ **14.** $-\frac{3}{\sqrt{34}}$

15. $\frac{\sqrt{26 + \sqrt{26}}}{2\sqrt{13}}$ **16.** $\frac{1}{2}\cos(4\theta) - \frac{1}{2}\cos(12\theta)$ **17.** $-\frac{1}{\sqrt{2}}$ **18.** $\theta = \frac{2\pi}{3} + \pi k$, where k is any integer; $\theta = \frac{2\pi}{3}, \theta = \frac{5\pi}{3}$ **19.** $\theta = \frac{\pi}{2} + \pi k, \theta = \frac{2\pi}{3} + 2\pi k$, or $\theta = \frac{4\pi}{3} + 2\pi k$, where k is any integer; $\theta = \frac{\pi}{2}, \theta = \frac{2\pi}{3}, \theta = \frac{4\pi}{3}, \theta = \frac{3\pi}{2}$ **20.** $\theta = \frac{2\pi}{3} + 2\pi k, \theta = \pi + 2\pi k$, or $\theta = \frac{4\pi}{3} + 2\pi k$, where k is any integer; $\theta = \frac{2\pi}{3}, \theta = \pi, \theta = \frac{4\pi}{3}$

CHAPTER 9

9.1 Exercises

SCE-1. 27.58° **SCE-2.** 75.29° **SCE-3.** 42.41°

1. $c = 27.78$ in.; $A = 30.26°$; $B = 59.74°$ **2.** $a = 10.58$ cm; $A = 21.4°$; $B = 68.6°$ **3.** $b = 9.2$ m; $c = 12.19$ m; $B = 49°$ **4.** $a = 16.54$ cm; $c = 41.22$ cm; $B = 66.35°$ **5.** $a = 50.85$ m; $b = 70.27$ m; $B = 54.11°$ **6.** 6.36 in.; 12.47 in. **7.** 103.29 cm; 104.05 cm **8.** 101.69 ft **9.** 400.91 ft **10.** length of bridge: 11.1 m; distance between the two people: 36.72 m **11.** 8.31 ft **12.** 965,220 km **13.** 12.44 ft **14.** 8103 ft **15.** 2386 m

9.2 Exercises

SCE-1. 9.1 **SCE-2.** 32.3

1. yes **2.** no **3.** no **4.** yes **5.** yes **6.** no **7.** $B = 59°, b \approx 7.4, c \approx 8.4$ **8.** $A = 82°, b \approx 5.1, c \approx 5.6$ **9.** $C = 89°, b \approx 107.1, c \approx 151.5$ **10.** $C = 79°, a \approx 5.6, c \approx 7.4$ **11.** $B = 53°, b \approx 3.3, c \approx 1.5$ **12.** $A = 75°, a \approx 11.9, b \approx 10.4$ **13.** one triangle; $B \approx 35.4°, C \approx 104.6°, c \approx 12.2$ **14.** no triangle **15.** one triangle; $B = 90°, C = 60°, c \approx 6.4$ **16.** two triangles; $B_1 \approx 72.5°, C_1 \approx 52.5°, c_1 \approx 13.6; B_2 \approx 107.5°, C_2 \approx 17.5°, c_2 \approx 5.1$ **17.** two triangles; $B_1 \approx 83.7°, C_1 \approx 28.6°, c_1 \approx 7.7; B_2 \approx 96.3, C_2 \approx 16.0°, c_2 \approx 4.4$ **18.** no triangle **19.** two triangles; $B_1 \approx 59.2°, C_1 \approx 63.7°, c_1 \approx 14.2; B_2 \approx 120.8, C_2 \approx 2.1°, c_2 \approx 0.6$ **20.** 61.1 m **21.** 4.8 mi and 3.4 mi **22.** 20.5 mi and 12.4 mi **23.** 65.7 mi **24.** distance from Alpha: 56.6 mi, distance from Beta: 63.5 mi **25.** 493.9 ft **26.** one triangle **27.** no triangle **28.** one triangle **29.** two triangles **30.** two triangles **31.** no triangle **32.** two triangles

9.3 Exercises

SCE-1. 3.11 **SCE-2.** 34.3°

1. Law of Sines **2.** The triangle cannot be solved. **3.** Law of Cosines **4.** Law of Sines **5.** Law of Sines **6.** Law of Cosines **7.** $a \approx 3.0$ ft, $B \approx 58.4°, C \approx 96.6°$ **8.** $c \approx 9.7$ cm, $A \approx 54.1°, B \approx 23.9°$ **9.** $a \approx 3.4$ m, $B = 81°, C = 81°$ **10.** $c \approx 3.1, A \approx 41.0°, C \approx 119.0°$ **11.** $a \approx 10.4, B \approx 8.5°, C \approx 41.5°$ **12.** $A \approx 41.0°, B \approx 65.4°, C \approx 73.6°$ **13.** $A \approx 18.2°, B \approx 51.3°, C \approx 110.5°$ **14.** $A \approx 22.6°, B \approx 67.4°, C = 90°$ **15.** $A \approx 33.1°, B \approx 63.4°, C \approx 83.5°$ **16.** $A \approx 83.5°, B \approx 50.6°, C \approx 45.9°$ **17.** 1926.1 mi **18.** 1655.3 mi **19.** N 62.6° W **20.** 23.1° **21.** 16.3 cm **22.** 8.8 cm and 22.4 cm **23.** $a \approx 7.2, b = 9, c = 5, A \approx 53.1°, B \approx 93.2°, C \approx 33.7°$ **24.** 3106.9 ft **25.** 54.1°

9.4 Exercises

SCE-1. 22.19

1. 29.60 ft^2 **2.** 16.19 yd^2 **3.** 29.65 cm^2 **4.** 198.91 m^2 **5.** 16.73 ft^2 **6.** 258.13 km^2 **7.** 74.61 cm^2 **8.** 24.49 yd^2 **9.** 16.71 ft^2 **10.** 7.91 m^2 **11.** 405.50 km^2 **12.** 364.66 cm^2 **13.** 101.82 ft^2 **14.** 31.42 in.^{2} **15.** 31.55 m^2 **16.** 1892 ft^2 **17.** \$88,801 **18.** \$1831 **19.** 930 ft^2; yes **20.** 593.4 ft^2 **21.** \$77

Review Exercises

1. $c = 27.78$ in.; $A = 30.26°$; $B = 59.74°$ **2.** $b = 9.2$ m; $c = 12.19$ m; $B = 49°$ **3.** 103.29 cm; 104.05 cm **4.** 101.69 ft **5.** 12.44 ft **6.** $C = 79°, a \approx 5.6, c \approx 7.4$ **7.** one triangle; $B \approx 35.4°, C \approx 104.6°, c \approx 12.2$ **8.** two triangles; $B_1 \approx 72.5°, C_1 \approx 52.5°, c_1 \approx 13.6; B_2 \approx 107.5°, C_2 \approx 17.5°, c_2 \approx 5.1$ **9.** two triangles; $B_1 \approx 83.7°, C_1 \approx 28.6°, c_1 \approx 7.7; B_2 \approx 96.3, C_2 \approx 16.0°, c_2 \approx 4.4$ **10.** distance from Alpha: 56.6 mi, distance from Beta: 63.5 mi **11.** Law of Sines **12.** $A \approx 41.0°, B \approx 65.4°, C \approx 73.6°$ **13.** $A \approx 33.1°, B \approx 63.4°, C \approx 83.5°$ **14.** 3106.9 ft **15.** 54.1° **16.** 16.19 yd^2 **17.** 198.91 m^2 **18.** 16.71 ft^2 **19.** 31.42 in.^{2} **20.** \$88,801

CHAPTER 10

10.1 Exercises

1. 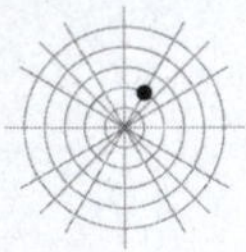 **2.** 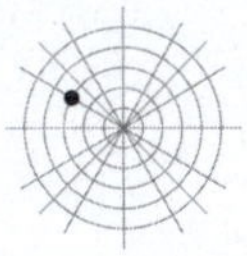 **3.** 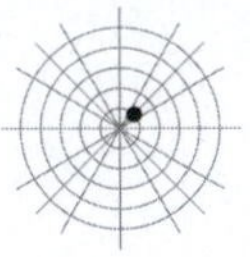 **4.** 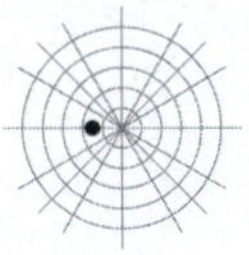 **5.** 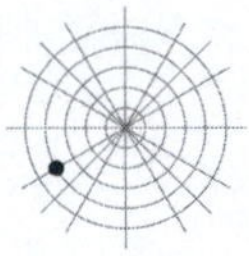 **6.**  **7.** 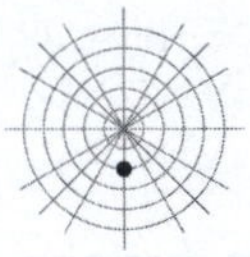 **8.**  **9.** 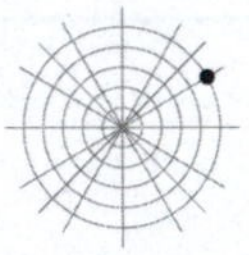 **10.** 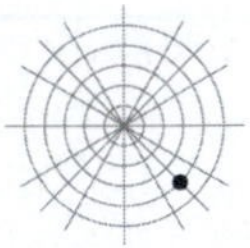 **11.** 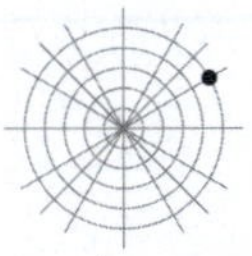 **a.** $P\left(5, -\frac{11\pi}{6}\right)$ **b.** $P\left(-5, \frac{7\pi}{6}\right)$ **c.** $P\left(5, \frac{13\pi}{6}\right)$ **12.**  **a.** $P\left(3, -\frac{2\pi}{3}\right)$ **b.** $P\left(-3, \frac{\pi}{3}\right)$ **c.** $P\left(3, \frac{10\pi}{3}\right)$ **13.** 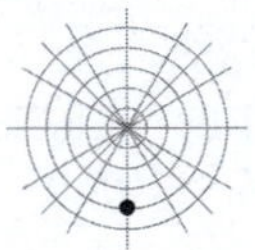 **a.** $P\left(4, -\frac{\pi}{2}\right)$ **b.** $P\left(-4, \frac{\pi}{2}\right)$ **c.** $P\left(4, \frac{7\pi}{2}\right)$ **14.** 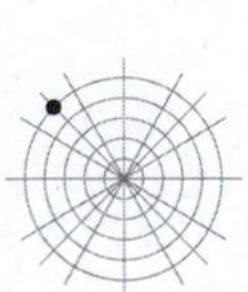 **a.** $P\left(5, -\frac{5\pi}{4}\right)$ **b.** $P\left(-5, \frac{7\pi}{4}\right)$ **c.** $P\left(5, \frac{11\pi}{4}\right)$ **15.** 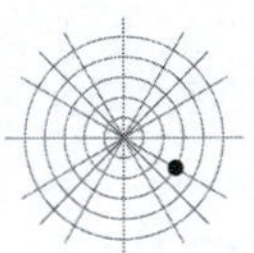 **a.** $P\left(3, -\frac{\pi}{6}\right)$ **b.** $P\left(-3, \frac{5\pi}{6}\right)$ **c.** $P\left(3, \frac{23\pi}{6}\right)$ **16.** $P\left(\frac{3}{\sqrt{2}}, \frac{3}{\sqrt{2}}\right)$ **17.** $P(-1, \sqrt{3})$ **18.** $P(0, -5)$ **19.** $P\left(-\frac{\sqrt{3}}{2}, -\frac{1}{2}\right)$ **20.** $P(2, -2\sqrt{3})$ **21.** $P(-\sqrt{3}, -1)$ **22.** $P\left(\frac{5}{\sqrt{2}}, -\frac{5}{\sqrt{2}}\right)$ **23.** $P(7, 0)$ **24.** $P\left(\frac{3}{2}, \frac{3\sqrt{3}}{2}\right)$ **25.** $P\left(-\frac{7\sqrt{3}}{2}, \frac{7}{2}\right)$ **26.** $P(3, 0)$ **27.** $P\left(\sqrt{2}, \frac{3\pi}{2}\right)$ **28.** $P(0.5, \pi)$ **29.** $P\left(1.5, \frac{\pi}{2}\right)$ **30.** $P\left(3\sqrt{2}, \frac{7\pi}{4}\right)$ **31.** $P\left(10, \frac{2\pi}{3}\right)$ **32.** $P\left(8, \frac{7\pi}{6}\right)$ **33.** $P\left(\sqrt{10}, \frac{\pi}{4}\right)$ **34.** $P\left(4, \frac{\pi}{3}\right)$ **35.** $P\left(12, \frac{5\pi}{6}\right)$ **36.** $P\left(\sqrt{53}, 2\pi - \tan^{-1}\frac{7}{2}\right)$ **37.** $P\left(\sqrt{3}, 2\pi - \tan^{-1}\frac{1}{\sqrt{2}}\right)$ **38.** $r = \frac{1}{7\cos\theta - 4\sin\theta}$ **39.** $r = 5$ **40.** $r = -3\sec\theta$ **41.** $r = 5\csc\theta$ **42.** $r = -4\sin\theta$ **43.** $r = 6\cos\theta$ **44.** $r = -2\cos\theta$ **45.** $r = 8\sin\theta$ **46.** $2x + 5y = 3$ **47.** $x^2 + y^2 = 3y$ **48.** $x^2 + y^2 = 16$ **49.** $x^2 + y^2 = -2x$ **50.** $y = \sqrt{3}x$ **51.** $x^2 + y^2 = 6y - 2x$ **52.** $x = 2$ **53.** $y = -3$ **54.** $P\left(2\sqrt{2}, \frac{\pi}{4}\right)$ **55.** $P\left(10, \frac{2\pi}{3}\right)$ **56.** $P\left(4, \frac{5\pi}{6}\right)$ **57.** $P\left(\sqrt{10}, \frac{3\pi}{4}\right)$ **58.** $P\left(2, \frac{4\pi}{3}\right)$ **59.** $P\left(12, \frac{7\pi}{6}\right)$

10.2 Exercises

SCE-1. $-\dfrac{7\pi}{4}, -\dfrac{13\pi}{12}, -\dfrac{5\pi}{12}, \dfrac{\pi}{4}, \dfrac{11\pi}{12}, \dfrac{19\pi}{12}$

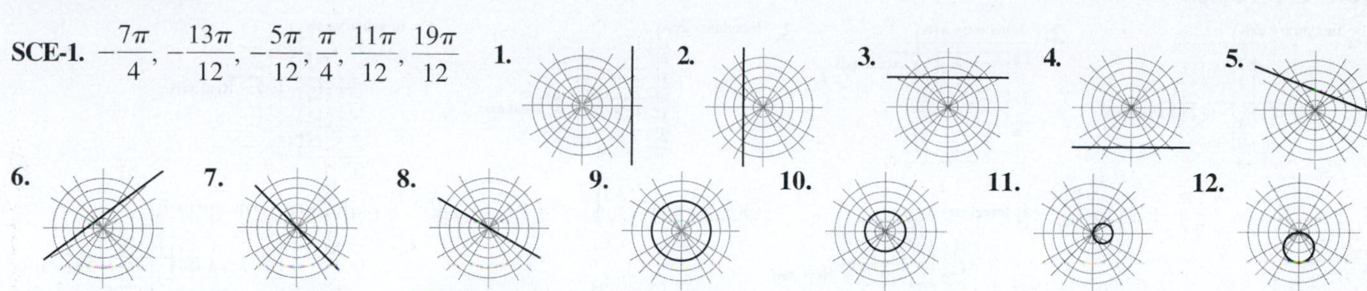

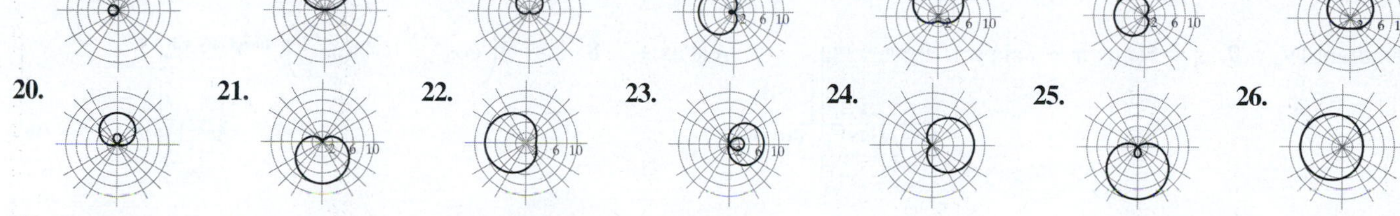

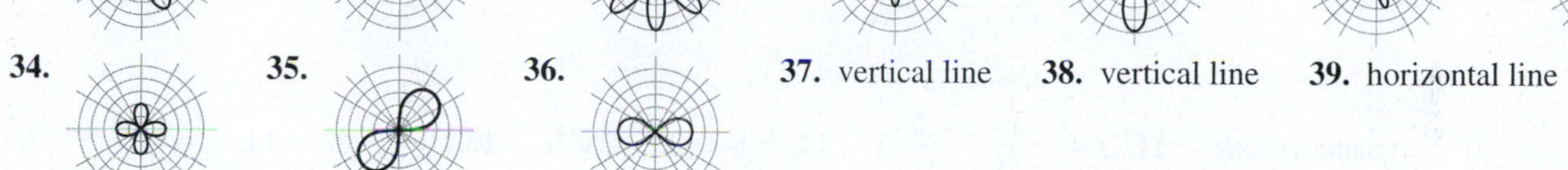

37. vertical line **38.** vertical line **39.** horizontal line **40.** horizontal line **41.** line **42.** line **43.** line through the pole **44.** line through the pole **45.** circle centered at the pole **46.** circle centered at the pole **47.** circle **48.** circle **49.** circle **50.** circle **51.** cardioid **52.** limacon with an inner loop **53.** limacon with a dimple **54.** cardioid **55.** limacon with no inner loop and no dimple **56.** limacon with an inner loop **57.** cardioid **58.** limacon with a dimple **59.** limacon with an inner loop **60.** cardioid **61.** limacon with an inner loop **62.** limacon with no inner loop and no dimple **63.** three-petal rose **64.** four-petal rose **65.** eight-petal rose **66.** five-petal rose **67.** three-petal rose **68.** five-petal rose **69.** eight-petal rose **70.** four-petal rose **71.** lemniscate **72.** lemniscate

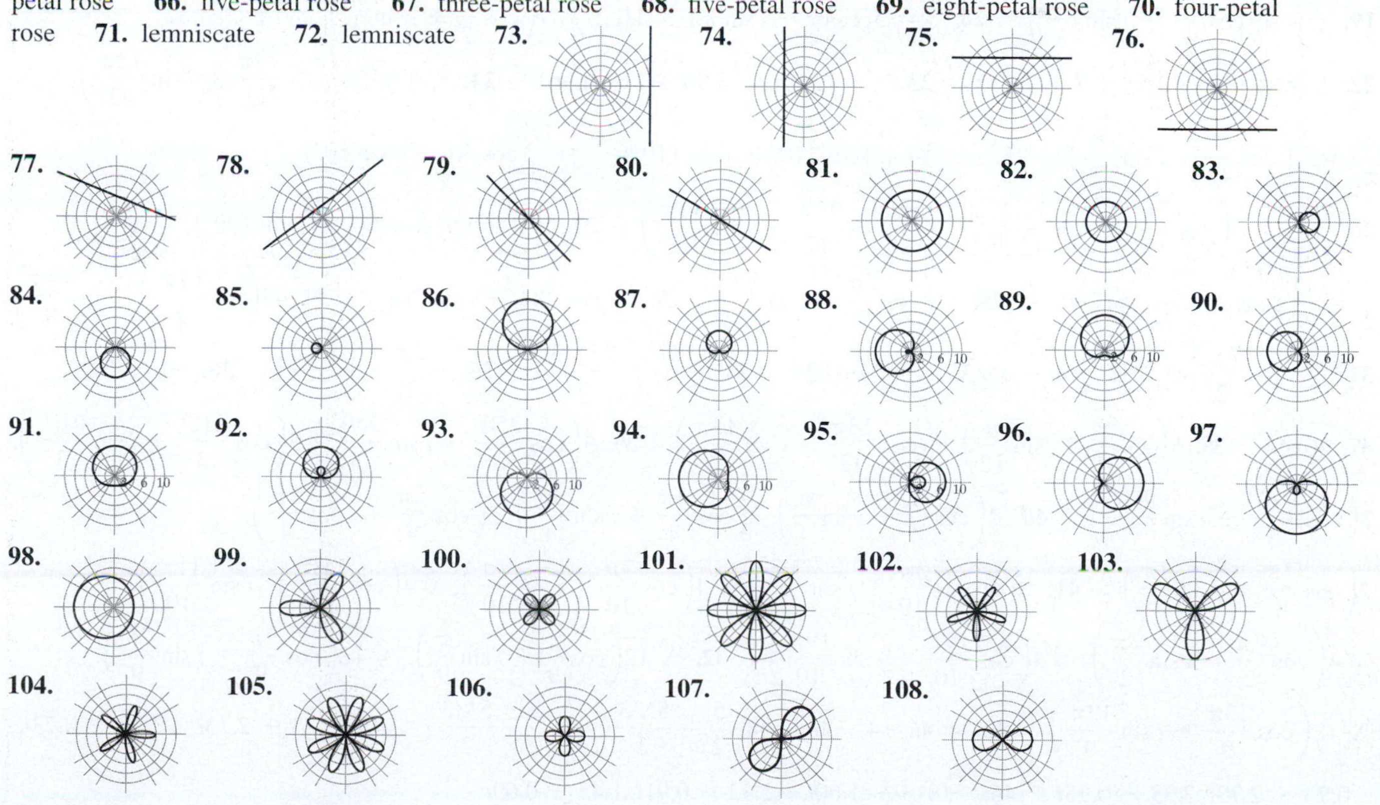

10.3 Exercises

1. 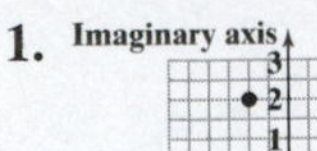

2. 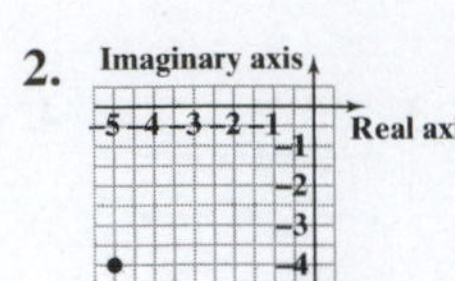

3. 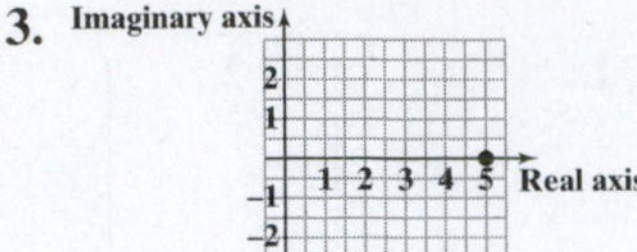

4.

5. $z = 3\left(\cos\frac{11\pi}{7} + i\sin\frac{11\pi}{7}\right)$; Quadrant IV **6.** $z = 2(\cos 280° + i\sin 280°)$; Quadrant IV **7.** $z = 5(\cos\pi + i\sin\pi)$; real axis **8.** $z = 3\left(\cos\frac{\pi}{3} + i\sin\frac{\pi}{3}\right)$; Quadrant I **9.** $z = 2(\cos 30° + i\sin 30°)$; Quadrant I **10.** $z = 5\left(\cos\frac{\pi}{2} + i\sin\frac{\pi}{2}\right)$; imaginary axis **11.** $z = \frac{5}{\sqrt{2}} - \frac{5}{\sqrt{2}}i$ **12.** $z = -2 - 2\sqrt{3}i$ **13.** $z = -8i$ **14.** $z = 5 - 5\sqrt{3}i$

15. $z = 9.51 + 3.09i$ **16.** $z = -15.89 - 8.45i$ **17.** $z = 4\sqrt{2}\left(\cos\frac{3\pi}{4} + i\sin\frac{3\pi}{4}\right)$ **18.** $z = 10\left(\cos\frac{11\pi}{6} + i\sin\frac{11\pi}{6}\right)$ **19.** $z = 4\left(\cos\frac{4\pi}{3} + i\sin\frac{4\pi}{3}\right)$ **20.** $z = 5(\cos\pi + i\sin\pi)$ **21.** $z = 7\left(\cos\frac{\pi}{2} + i\sin\frac{\pi}{2}\right)$ **22.** $z = \sqrt{13}(\cos 2.55 + i\sin 2.55)$ **23.** $z = \sqrt{17}(\cos 4.96 + i\sin 4.96)$ **24.** $z_1z_2 = 20\left(\cos\frac{13\pi}{42} + i\sin\frac{13\pi}{42}\right)$, $\frac{z_1}{z_2} = \frac{5}{4}\left(\cos\frac{\pi}{42} + i\sin\frac{\pi}{42}\right)$ **25.** $z_1z_2 = 42(\cos 110° + i\sin 110°)$, $\frac{z_1}{z_2} = \frac{7}{6}(\cos 50° + i\sin 50°)$ **26.** $z_1z_2 = 6\left(\cos\frac{7\pi}{12} + i\sin\frac{7\pi}{12}\right)$, $\frac{z_1}{z_2} = \frac{2}{3}\left(\cos\frac{11\pi}{12} + i\sin\frac{11\pi}{12}\right)$ **27.** $z_1z_2 = 99(\cos 100° + i\sin 100°)$, $\frac{z_1}{z_2} = \frac{11}{9}(\cos 200° + i\sin 200°)$ **28.** $32\left(\cos\frac{5\pi}{7} + i\sin\frac{5\pi}{7}\right)$ **29.** $81(\cos 284° + i\sin 284°)$ **30.** $49\left(\cos\frac{14\pi}{9} + i\sin\frac{14\pi}{9}\right)$ **31.** $\frac{27}{2} + \frac{27\sqrt{3}}{2}i$ **32.** $-4 - 4\sqrt{3}i$ **33.** -1024 **34.** $-\frac{27}{2} + \frac{27\sqrt{3}}{2}i$ **35.** $-\frac{25}{2} + \frac{25\sqrt{3}}{2}i$ **36.** $-8 + 8i$

37. 46,656 **38.** $6\left(\cos\frac{\pi}{12} + i\sin\frac{\pi}{12}\right), 6\left(\cos\frac{13\pi}{12} + i\sin\frac{13\pi}{12}\right)$ **39.** $3\left(\cos\frac{250°}{3} + i\sin\frac{250°}{3}\right), 3\left(\cos\frac{610°}{3} + i\sin\frac{610°}{3}\right), 3\left(\cos\frac{970°}{3} + i\sin\frac{970°}{3}\right)$ **40.** $2\left(\cos\frac{\pi}{4} + i\sin\frac{\pi}{4}\right), 2\left(\cos\frac{3\pi}{4} + i\sin\frac{3\pi}{4}\right), 2\left(\cos\frac{5\pi}{4} + i\sin\frac{5\pi}{4}\right), 2\left(\cos\frac{7\pi}{4} + i\sin\frac{7\pi}{4}\right)$ **41.** $\sqrt[5]{4}\left(\cos\frac{3\pi}{10} + i\sin\frac{3\pi}{10}\right), \sqrt[5]{4}\left(\cos\frac{7\pi}{10} + i\sin\frac{7\pi}{10}\right), \sqrt[5]{4}\left(\cos\frac{11\pi}{10} + i\sin\frac{11\pi}{10}\right), \sqrt[5]{4}\left(\cos\frac{3\pi}{2} + i\sin\frac{3\pi}{2}\right), \sqrt[5]{4}\left(\cos\frac{19\pi}{10} + i\sin\frac{19\pi}{10}\right)$ **42.** $\sqrt[3]{12}\left(\cos\frac{\pi}{9} + i\sin\frac{\pi}{9}\right), \sqrt[3]{12}\left(\cos\frac{7\pi}{9} + i\sin\frac{7\pi}{9}\right), \sqrt[3]{12}\left(\cos\frac{13\pi}{9} + i\sin\frac{13\pi}{9}\right)$ **43.** $4, 4i, -4, -4i$ **44.** $\frac{5}{2} + \frac{5\sqrt{3}}{2}i, -5, \frac{5}{2} - \frac{5\sqrt{3}}{2}i$ **45.** $0.73 + 2.73i, -2.73 + 0.73i, -0.73 - 2.73i, 2.73 - 0.73i$ **46.** $-0.13 + 1.60i, -1.32 - 0.91i, 1.45 - 0.69i$

10.4 Exercises

1. $\sqrt{61}$ **2.** 10 **3.** $\frac{\sqrt{1417}}{6}$ **4.** 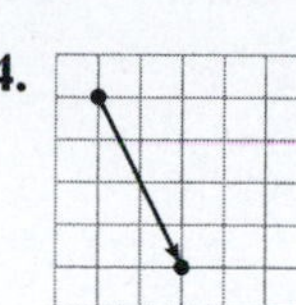 **5.** **6.** 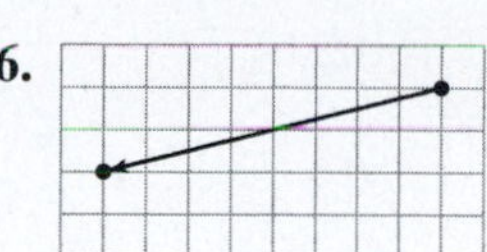 **7.** **8.** 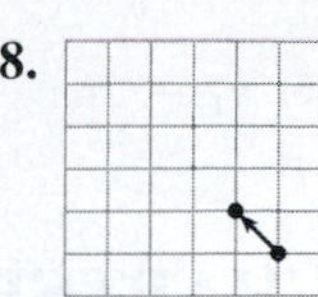

9. **10.** **11.** $<4, 3>$, 5 **12.** $<4, 3>$, 5 **13.** $<-6, 11>$, $\sqrt{157}$

14. $3\mathbf{i} + 5\mathbf{j}$ **15.** $-4\mathbf{i} + \mathbf{j}$ **16.** $-6\mathbf{i} - 7\mathbf{j}$ **17.** $-9\mathbf{i} + 21\mathbf{j}, 3\sqrt{58}$ **18.** $2\mathbf{i} + 6\mathbf{j}, 2\sqrt{10}$ **19.** $-8\mathbf{i} + 8\mathbf{j}, 8\sqrt{2}$

20. $3\mathbf{i} + 25\mathbf{j}, \sqrt{634}$ **21.** $\mathbf{i} - \frac{7}{3}\mathbf{j}, \frac{\sqrt{58}}{3}$ **22.** $-\frac{11}{4}\mathbf{i} + \frac{15}{4}\mathbf{j}, \frac{\sqrt{346}}{4}$ **23.** $-\mathbf{j}$ **24.** $\frac{4}{5}\mathbf{i} - \frac{3}{5}\mathbf{j}$ **25.** $\frac{5}{\sqrt{29}}\mathbf{i} - \frac{2}{\sqrt{29}}\mathbf{j}$

26. $\frac{1}{\sqrt{17}}\mathbf{i} - \frac{4}{\sqrt{17}}\mathbf{j}$ **27.** $\theta = 315°$ **28.** $\theta = 120°$ **29.** $\theta = 30°$ **30.** $\theta \approx 68.20°$ **31.** $\theta \approx 285.95°$

32. $-\frac{15\sqrt{3}}{2}\mathbf{i} + \frac{15}{2}\mathbf{j}$ **33.** $11\sqrt{2}\mathbf{i} - 11\sqrt{2}\mathbf{j}$ **34.** $\frac{7}{2}\mathbf{i} - \frac{7\sqrt{3}}{2}\mathbf{j}$ **35.** $-1.74\mathbf{i} + 9.85\mathbf{j}$ **36.** $-0.57\mathbf{i} - 0.35\mathbf{j}$

37. $50\sqrt{3}\mathbf{i} + 50\mathbf{j}$ **38.** $179.01\mathbf{i} - 18.82\mathbf{j}$ **39.** $20.68\mathbf{i} + 17.36\mathbf{j}$ **40.** $-13.68\mathbf{i} - 37.59\mathbf{j}$ **41.** 261.73 mph, N84.96°E

42. 314.48 mph, N83°W **43.** $-\mathbf{i} + 7\mathbf{j}$ **44.** $11\mathbf{i} + 16\mathbf{j}$ **45.** 519.37 pounds **46.** 1605.28 pounds, N73.78°W

47. 650.85 pounds, S52.90°E

10.5 Exercises

1. 28 **2.** -13 **3.** $140\sqrt{2}$ **4.** 37 **5.** -4 **6.** -63 **7.** 50 **8.** $\sqrt{1073}$ **9.** 54.16° **10.** 60.26° **11.** 135°

12. 161.57° **13.** 180° **14.** 2 **15.** $-\frac{21}{5}$ **16.** orthogonal **17.** neither **18.** parallel **19.** neither **20.** $\mathbf{v}_1 = 2\mathbf{i} - 2\mathbf{j}$, $\mathbf{v}_2 = -7\mathbf{i} - 7\mathbf{j}$ **21.** $\mathbf{v}_1 = -\frac{44}{53}\mathbf{i} + \frac{154}{53}\mathbf{j}, \mathbf{v}_2 = \frac{203}{53}\mathbf{i} + \frac{58}{53}\mathbf{j}$ **22.** $\mathbf{v}_1 = -\frac{15}{452}\mathbf{i} - \frac{1}{452}\mathbf{j}, \mathbf{v}_2 = \frac{241}{452}\mathbf{i} - \frac{3615}{452}\mathbf{j}$

23. 587.79 pounds **24.** 169.05 pounds, 362.52 pounds **25.** 1866.87 pounds **26.** 25.94° **27.** 200 foot-pounds

28. 27,189.23 foot-pounds **29.** 26,811.56 foot-pounds **30.** 48.2°

Review Exercises

1. **2.** **a.** $P\left(3, -\frac{2\pi}{3}\right)$ **b.** $P\left(-3, \frac{\pi}{3}\right)$ **c.** $P\left(3, \frac{10\pi}{3}\right)$ **3.** $P\left(\frac{5}{\sqrt{2}}, -\frac{5}{\sqrt{2}}\right)$

4. $P\left(8, \frac{7\pi}{6}\right)$ **5.** $r = -2\cos\theta$ **6.** **7.** **8.** **9.**

10. **11.** $z = 3\left(\cos\frac{\pi}{3} + i\sin\frac{\pi}{3}\right)$, ; Quadrant I **12.** $z = 5 - 5\sqrt{3}i$

13. $z = 10\left(\cos\frac{11\pi}{6} + i\sin\frac{11\pi}{6}\right)$ **14.** $z_1z_2 = 6\left(\cos\frac{7\pi}{12} + i\sin\frac{7\pi}{12}\right), \frac{z_1}{z_2} = \frac{2}{3}\left(\cos\frac{11\pi}{12} + i\sin\frac{11\pi}{12}\right)$

15. $\frac{27}{2} + \frac{27\sqrt{3}}{2}i$ **16.** $6\left(\cos\frac{\pi}{12} + i\sin\frac{\pi}{12}\right), 6\left(\cos\frac{13\pi}{12} + i\sin\frac{13\pi}{12}\right)$ **17.** 10 **18.**

19. $3\mathbf{i} + 25\mathbf{j}, \sqrt{634}$ **20.** $\theta = 30°$ **21.** $-\dfrac{15\sqrt{3}}{2}\mathbf{i} + \dfrac{15}{2}\mathbf{j}$ **22.** -4 **23.** $54.16°$ **24.** orthogonal

25. $\mathbf{v}_1 = 2\mathbf{i} - 2\mathbf{j}, \mathbf{v}_2 = -7\mathbf{i} - 7\mathbf{j}$

CHAPTER 11

11.1 Exercises

1. vertex: $(0, 0)$, focus: $(0, 4)$, directrix: $y = -4$, 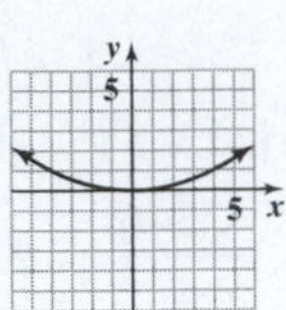 **2.** vertex: $(0, 0)$, focus: $(0, -2)$, directrix: $y = 2$,

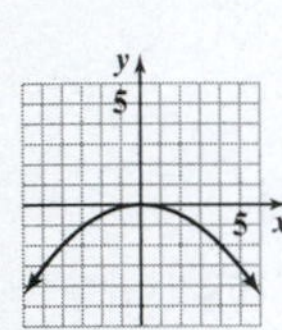 **3.** vertex: $(1, 4)$, focus: $(1, 1)$, directrix: $y = 7$, 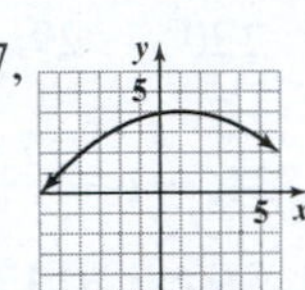 **4.** vertex: $(-3, 1)$, focus: $\left(-3, \dfrac{5}{2}\right)$,

directrix: $y = -\dfrac{1}{2}$, **5.** vertex: $(-2, -6)$, focus: $\left(-2, -\dfrac{19}{4}\right)$, directrix: $y = -\dfrac{29}{4}$, 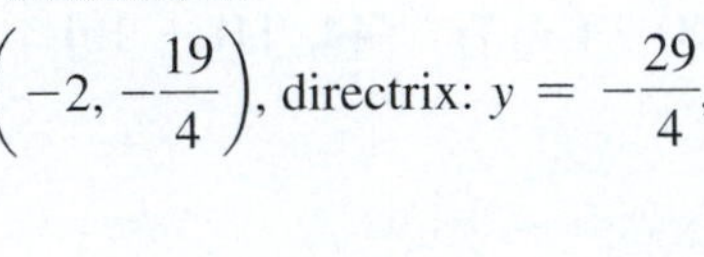 **6.** vertex:

$(0, 0)$, focus: $(1, 0)$, directrix: $x = -1$, 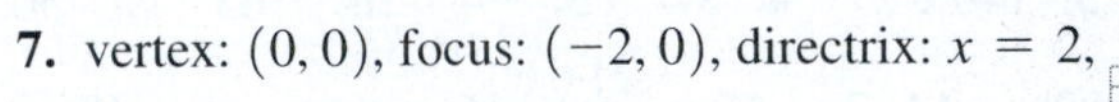 **7.** vertex: $(0, 0)$, focus: $(-2, 0)$, directrix: $x = 2$, 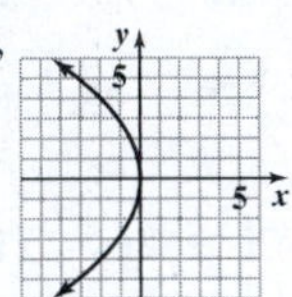

8. vertex: $(2, 5)$, focus: $(1, 5)$, directrix: $x = 3$, 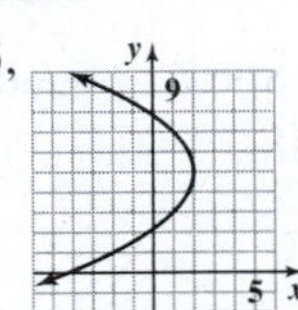 **9.** vertex: $(4, -3)$, focus: $(9, -3)$, directrix: $x = -1$,

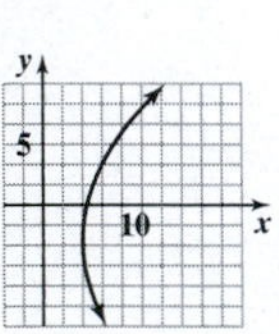 **10.** vertex: $(-3, -4)$, focus: $\left(-\dfrac{3}{4}, -4\right)$, directrix: $x = -\dfrac{21}{4}$, 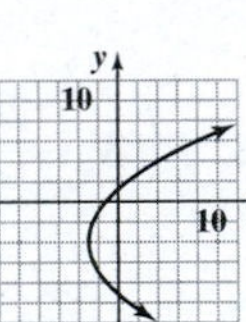 **11.** $y^2 = 8x$ **12.** $x^2 = -2y$

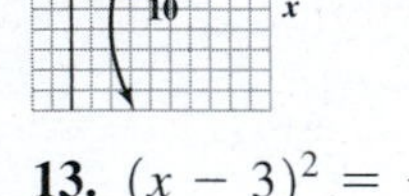

13. $(x - 3)^2 = -8(y + 3)$ **14.** $(y - 4)^2 = 12(x + 1)$ **15.** $(y + 2)^2 = -\left(x + \dfrac{11}{4}\right)$ **16.** $(x - 4)^2 = 2\left(y + \dfrac{1}{4}\right)$

17. $(y - 4)^2 = 16(x + 3)$ **18.** $8(x + 2) = (y - 4)^2$ and $-8(x - 2) = (y - 4)^2$ **19.** $-2(y - 2) = (x - 2)^2$ and

$8\left(y + \dfrac{1}{2}\right) = (x - 2)^2$ **20.** vertex: $(-5, -3)$, focus: $\left(-5, -\dfrac{7}{4}\right)$, directrix: $y = -\dfrac{17}{4}$, **21.** vertex: $(1, 6)$,

focus: $\left(-\dfrac{1}{2}, 6\right)$, directrix: $x = \dfrac{5}{2}$, **22.** vertex: $(1, 2)$, focus: $\left(1, \dfrac{7}{4}\right)$, directrix: $y = \dfrac{9}{4}$, 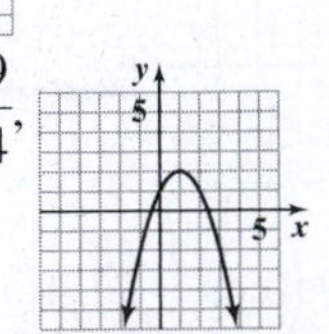

23. vertex: $(-1, -3)$, focus: $\left(-\dfrac{5}{4}, -3\right)$, directrix: $x = -\dfrac{3}{4}$, **24.** 10 cm **25.** $\dfrac{81}{32}$ in. **26.** $\dfrac{121}{16}$ ft

27. $-160(y - 40) = x^2$, no **28.** 14.4 m **29.** **30.** **31.**

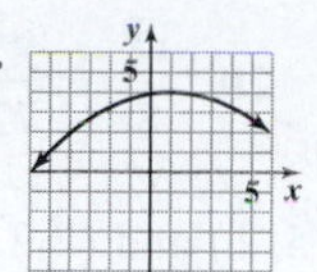

32.

33.

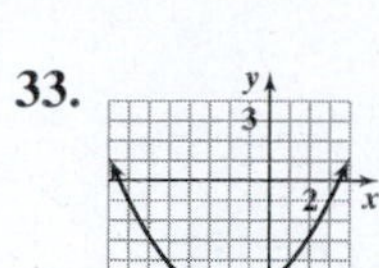

34. **35.** **36.** **37.**

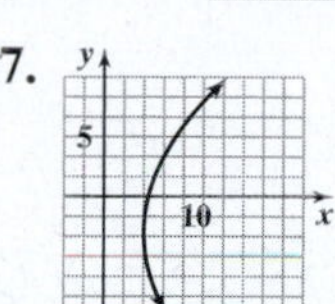

38.

39.

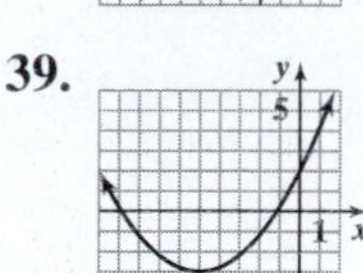

40. **41.** **42.**

11.2 Exercises

1. center: $(0, 0)$, foci: $(2\sqrt{3}, 0)$ and $(-2\sqrt{3}, 0)$, vertices: $(4, 0)$ and $(-4, 0)$,

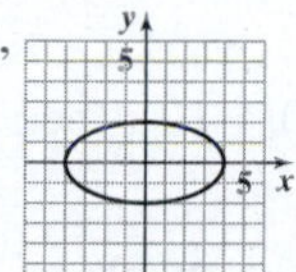

2. center: $(0, 0)$, foci: $(0, \sqrt{13})$ and $(0, -\sqrt{13})$, vertices: $(0, 5)$ and $(0, -5)$,

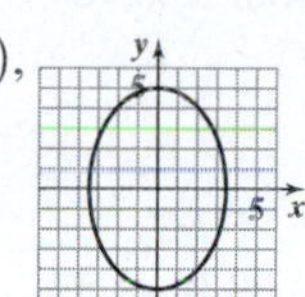

3. center: $(0, 0)$, foci: $(\sqrt{5}, 0)$ and $(-\sqrt{5}, 0)$, vertices: $(3, 0)$ and $(-3, 0)$,

4. center: $(0, 0)$, foci: $(0, \sqrt{15})$ and $(0, -\sqrt{15})$, vertices: $(0, 2\sqrt{5})$ and $(0, -2\sqrt{5})$,

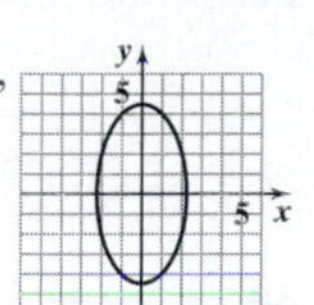

5. center: $(2, 4)$, foci: $(2 + \sqrt{11}, 4)$ and $(2 - \sqrt{11}, 4)$, vertices: $(8, 4)$ and $(-4, 4)$,

6. center: $(1, -4)$, foci: $(1, -4 + 2\sqrt{10})$ and $(1, -4 - 2\sqrt{10})$, vertices: $(1, 3)$ and $(1, -11)$,

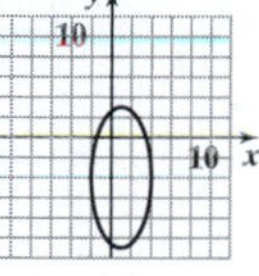

7. center: $(5, -1)$, foci: $(9, -1)$ and $(1, -1)$, vertices: $(0, -1)$ and $(10, -1)$,

8. center: $(-7, 2)$, foci: $(-7 - 2\sqrt{2}, 2)$ and $(-7 + 2\sqrt{2}, 2)$, vertices: $(-4, 2)$ and $(-10, 2)$,

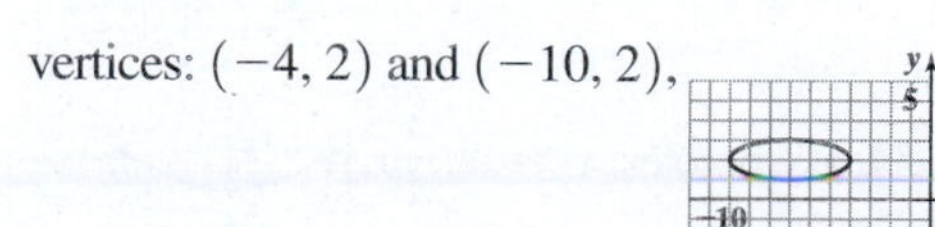

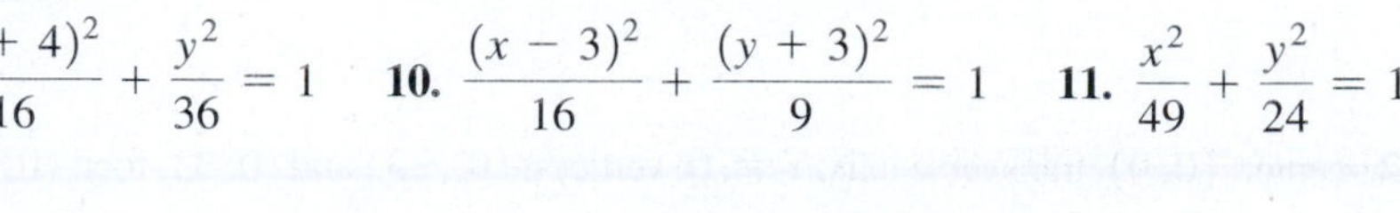

9. $\dfrac{(x + 4)^2}{16} + \dfrac{y^2}{36} = 1$ **10.** $\dfrac{(x - 3)^2}{16} + \dfrac{(y + 3)^2}{9} = 1$ **11.** $\dfrac{x^2}{49} + \dfrac{y^2}{24} = 1$

12. $\dfrac{(x - 3)^2}{36} + \dfrac{(y + 1)^2}{11} = 1$ **13.** $\dfrac{(x + 6)^2}{16} + \dfrac{(y - 5)^2}{25} = 1$ **14.** $\dfrac{(x - 4)^2}{100} + \dfrac{(y - 5)^2}{64} = 1$ **15.** $\dfrac{(x + 1)^2}{7} + \dfrac{(y - 4)^2}{16} = 1$

16. $\dfrac{(x - 2)^2}{20} + \dfrac{(y + 1)^2}{36} = 1$ **17.** $\dfrac{(x - 4)^2}{4} + \dfrac{(y - 1)^2}{53} = 1$ **18.** $\dfrac{(x - 1)^2}{1} + \dfrac{(y - 3)^2}{10} = 1$

19. $\dfrac{x^2}{16} + \dfrac{y^2}{12} = 1$ **20.** $\dfrac{x^2}{8} + \dfrac{y^2}{72} = 1$ **21.** $\dfrac{(x+2)^2}{20} + \dfrac{(y-1)^2}{16} = 1$, center: $(-2, 1)$, foci: $(-4, 1)$ and $(0, 1)$, vertices: $(-2 - 2\sqrt{5}, 1)$ and $(-2 + 2\sqrt{5}, 1)$, **22.** $\dfrac{(x+7)^2}{9} + \dfrac{(y+3)^2}{16} = 1$, center: $(-7, -3)$, foci: $(-7, -3 - \sqrt{7})$ and $(-7, -3 + \sqrt{7})$, vertices: $(-7, -7)$ and $(-7, 1)$,

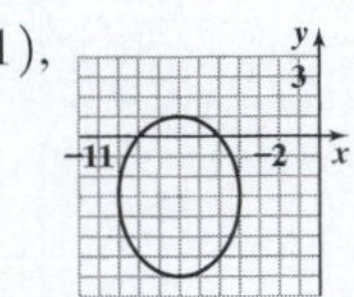

23. $\dfrac{(x+2)^2}{16} + \dfrac{(y-1)^2}{1} = 1$, center: $(-2, 1)$, foci: $(-2 - \sqrt{15}, 1)$ and $(-2 + \sqrt{15}, 1)$, vertices: $(-6, 1)$ and $(2, 1)$,

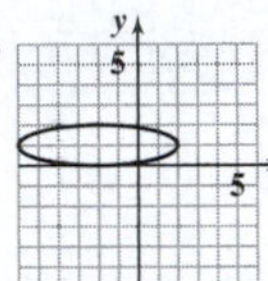

24. $\dfrac{x^2}{2} + \dfrac{(y+10)^2}{100} = 1$, center: $(0, -10)$, foci: $(0, -10 - 7\sqrt{2})$ and $(0, -10 + 7\sqrt{2})$, vertices: $(0, -20)$ and $(0, 0)$,

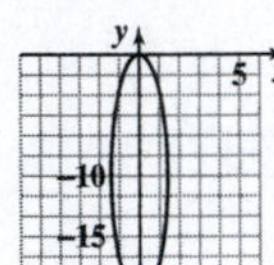

25. $\dfrac{x^2}{225} + \dfrac{y^2}{256} = 1$ **26.** $\dfrac{x^2}{81} + \dfrac{y^2}{441} = 1$ **27.** about 25.8 yds **28.** 74.83 cm to the left and right of center

29. 44.80 in. **30.** $\dfrac{x^2}{18{,}705{,}625} + \dfrac{y^2}{18{,}675{,}000} = 1$ **31.** about 39.69 ft **32.**

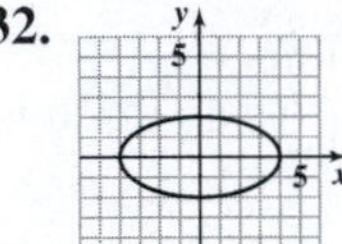

33.

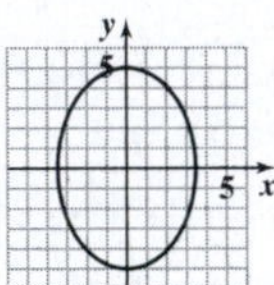

34.

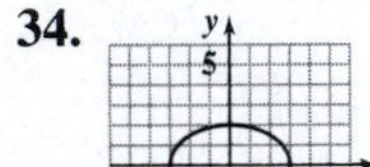

35.

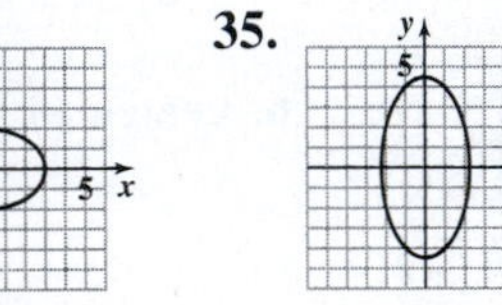

36.

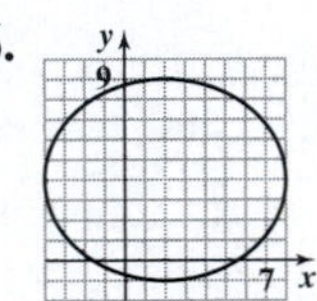

37.

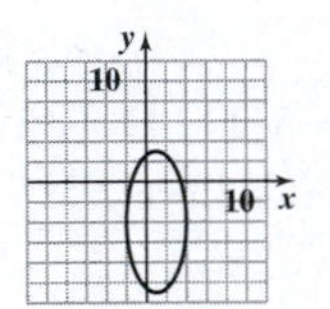

38.

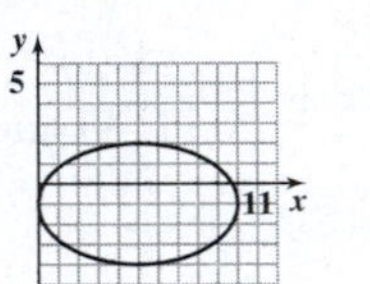

39.

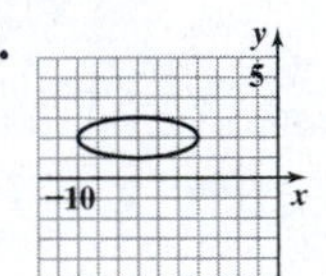

40.

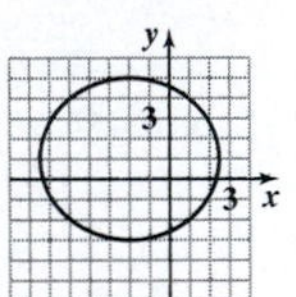

41.

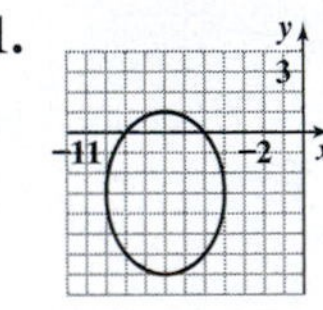

42.

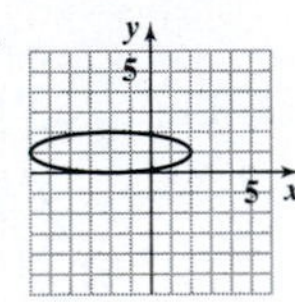

43.

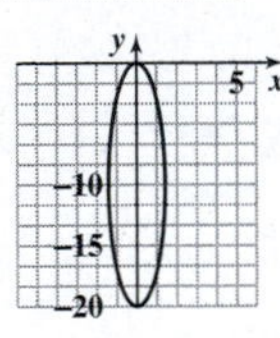

11.3 Exercises

1. center: $(0, 0)$, transverse axis: $y = 0$, vertices: $(-4, 0)$ and $(4, 0)$, foci: $(-5, 0)$ and $(5, 0)$, asymptotes: $y = \pm\dfrac{3}{4}x$,

2. center: $(0, 0)$, transverse axis: $x = 0$, vertices: $(0, -3)$ and $(0, 3)$, foci: $(0, -5)$ and $(0, 5)$, asymptotes: $y = \pm\dfrac{3}{4}x$,

3. center: $(2, 4)$, transverse axis: $x = 2$, vertices: $(2, -1)$ and $(2, 9)$, foci: $(2, 4 - \sqrt{61})$ and $(2, 4 + \sqrt{61})$, asymptotes:

$y - 4 = \pm\frac{5}{6}(x - 2)$, **4.** center: $(-1, -3)$, transverse axis: $y = -3$, vertices: $(-4, -3)$ and $(2, -3)$, foci: $(-1 - \sqrt{58}, -3)$ and $(-1 + \sqrt{58}, -3)$, asymptotes: $y + 3 = \pm\frac{7}{3}(x + 1)$, **5.** center: $(0, 0)$, transverse axis: $y = 0$, vertices: $(-\sqrt{5}, 0)$ and $(\sqrt{5}, 0)$, foci: $(-5, 0)$ and $(5, 0)$, asymptotes: $y = \pm 2x$,

6. center: $(1, 3)$, transverse axis: $x = 1$, vertices: $(1, 3 - 2\sqrt{5})$ and $(1, 3 + 2\sqrt{5})$, foci: $(1, -3)$ and $(1, 9)$, asymptotes: $y - 3 = \pm\frac{\sqrt{5}}{2}(x - 1)$,

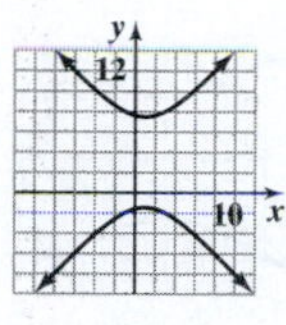

7. $\frac{x^2}{9} - \frac{y^2}{16} = 1$,

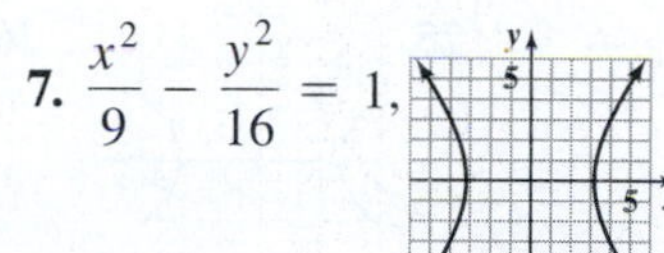

8. $\frac{y^2}{36} - \frac{x^2}{64} = 1$, **9.** $\frac{(x - 4)^2}{1} - \frac{(y + 4)^2}{3} = 1$,

10. $\frac{(y + 1)^2}{16} - \frac{(x + 6)^2}{48} = 1$, **11.** $\frac{(y - 6)^2}{4} - \frac{(x - 9)^2}{5} = 1$, **12.** $\frac{(x - 4)^2}{36} - \frac{(y + 3)^2}{49} = 1$,

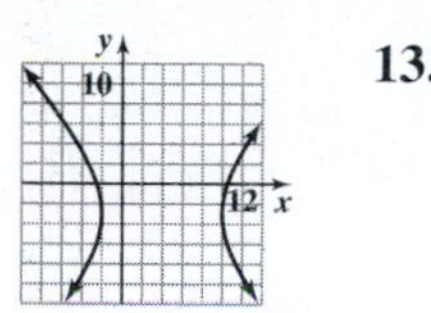

13. $\frac{y^2}{20} - \frac{x^2}{16} = 1$,

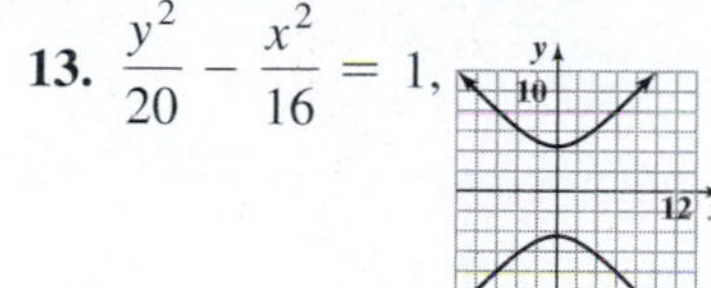

14. $\frac{(x - 2)^2}{16} - \frac{(y - 1)^2}{48} = 1$,

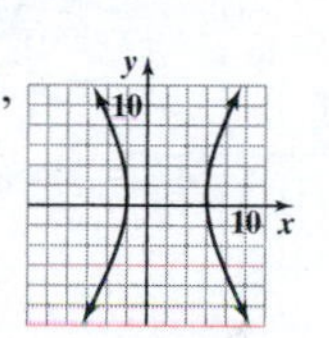

15. $\frac{(x - 2)^2}{4} - \frac{(y - 1)^2}{4} = 1$, center: $(2, 1)$, vertices: $(0, 1)$ and $(4, 1)$, foci: $(2 - 2\sqrt{2}, 1)$ and $(2 + 2\sqrt{2}, 1)$, endpoints of conjugate axis: $(2, -1)$ and $(2, 3)$, asymptotes: $y - 1 = \pm(x - 2)$, **16.** $\frac{(y + 3)^2}{2} - \frac{(x + 4)^2}{2} = 1$, center: $(-4, -3)$, vertices: $(-4, -3 - \sqrt{2})$ and $(-4, -3 + \sqrt{2})$, foci: $(-4, -5)$ and $(-4, -1)$, endpoints of conjugate axis: $(-4 - \sqrt{2}, -3)$ and $(-4 + \sqrt{2}, -3)$, asymptotes: $y + 3 = \pm(x + 4)$, **17.** $\frac{(y - 6)^2}{9} - (x + 2)^2 = 1$, center: $(-2, 6)$, vertices: $(-2, 3)$ and $(-2, 9)$, foci: $(-2, 6 - \sqrt{10})$ and $(-2, 6 + \sqrt{10})$, endpoints of conjugate axis: $(-3, 6)$ and $(-1, 6)$, asymptotes: $y - 6 = \pm 3(x + 2)$,

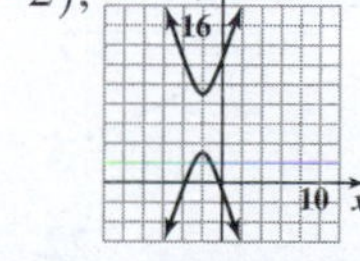

18. $\frac{(x + 5)^2}{16} - (y - 2)^2 = 1$, center: $(-5, 2)$, vertices: $(-9, 2)$ and $(-1, 2)$, foci: $(-5 - \sqrt{17}, 2)$ and $(-5 + \sqrt{17}, 2)$, endpoints of conjugate axis: $(-5, 1)$ and $(-5, 3)$, asymptotes:

$y - 2 = \pm\frac{1}{4}(x + 5)$, **19.** $\frac{(x-6)^2}{25} - \frac{(y+8)^2}{144} = 1$, center: $(6, -8)$, vertices: $(1, -8)$ and $(11, -8)$, foci: $(-7, -8)$ and $(19, -8)$, endpoints of conjugate axis: $(6, -20)$ and $(6, 4)$, asymptotes: $y + 8 = \pm\frac{12}{5}(x - 6)$, 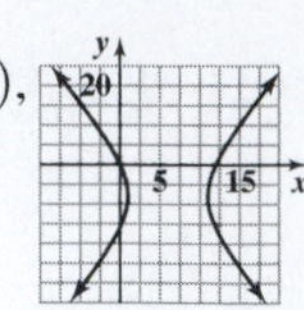

20. $\frac{(y-1)^2}{576} - \frac{(x+1)^2}{49} = 1$, center: $(-1, 1)$, vertices: $(-1, -23)$ and $(-1, 25)$, foci: $(-1, -24)$ and $(-1, 26)$, endpoints of conjugate axis: $(-8, 1)$ and $(6, 1)$, asymptotes: $y - 1 = \pm\frac{24}{7}(x + 1)$, **21.** $\frac{y^2}{49} - \frac{x^2}{15} = 1$

22. $\frac{x^2}{2500} - \frac{y^2}{7500} = 1$ **23. a.** $\frac{x^2}{900} - \frac{y^2}{2700} = 1$ **b.** 159 m **24.** $\frac{y^2}{324} - \frac{x^2}{1276} = 1$ **25.** 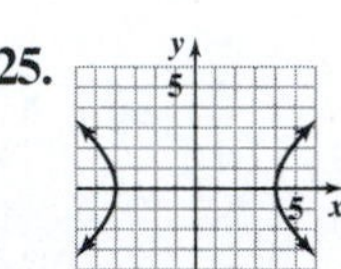 **26.** 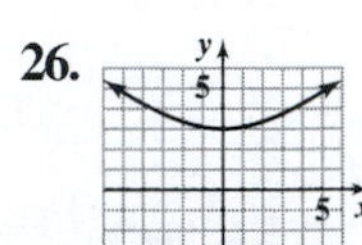

27. **28.** 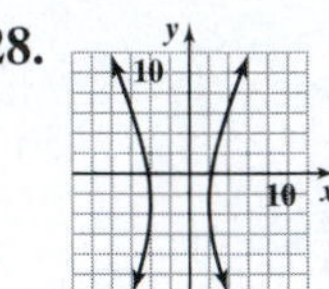 **29.** 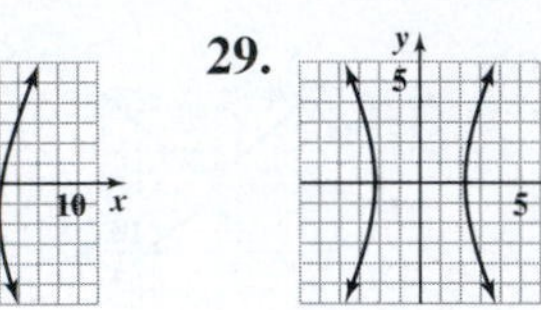 **30.** 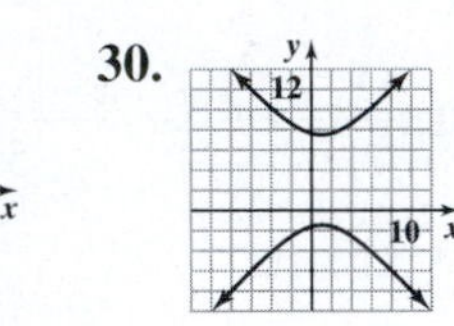 **31.** 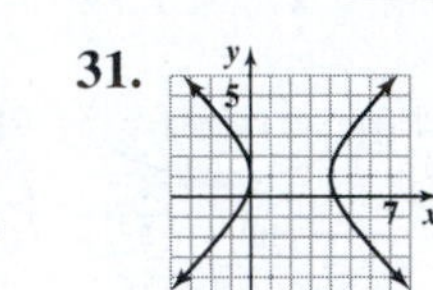 **32.** 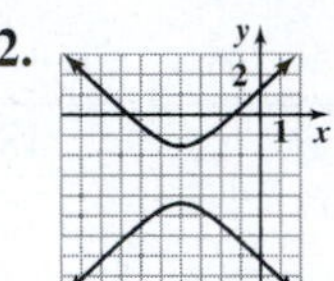

33. 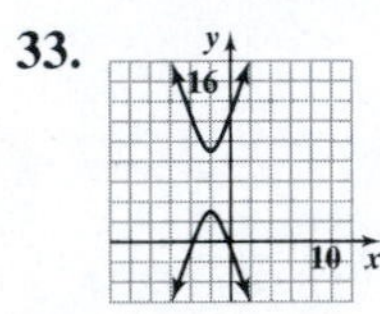 **34.** 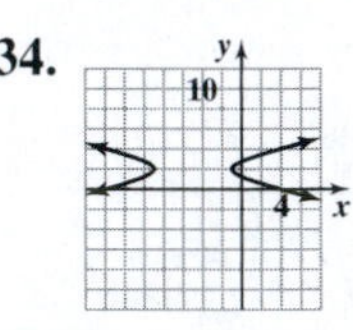 **35.** 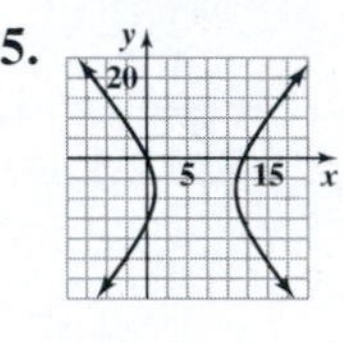 **36.** 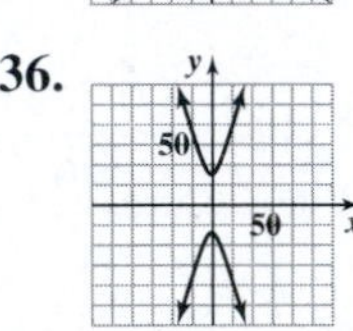

11.4 Exercises

1. parabola; directrix perpendicular to the polar axis located 3 units left of the pole **2.** ellipse; directrix parallel to the polar axis located 6 units above the pole **3.** hyperbola; directrix perpendicular to the polar axis located 3 units right of the pole **4.** parabola; directrix parallel to the polar axis located $\frac{5}{4}$ units below the pole **5.** ellipse; directrix perpendicular to the polar axis located $\frac{21}{8}$ units right of the pole **6.** ellipse; directrix perpendicular to the polar axis located $\frac{8}{5}$ units right of the pole **7.** parabola; directrix perpendicular to the polar axis located $\frac{9}{5}$ units left of the pole **8.** ellipse; directrix parallel to the polar axis located $\frac{4}{3}$ units above the pole **9.** hyperbola; directrix parallel to the polar axis located $\frac{6}{7}$ units below the pole **10.** parabola; directrix perpendicular to the polar axis located 2 units left of the pole **11.** ellipse; directrix parallel to the polar axis located 2 units below the pole **12.** hyperbola; directrix parallel to the polar axis located $\frac{5}{4}$ units above the pole

13. 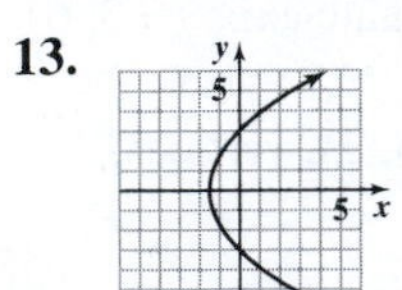 **14.** 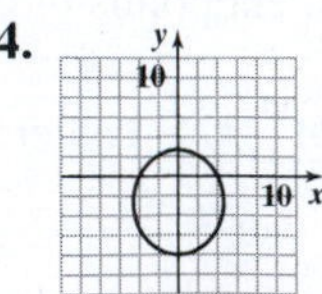 **15.** 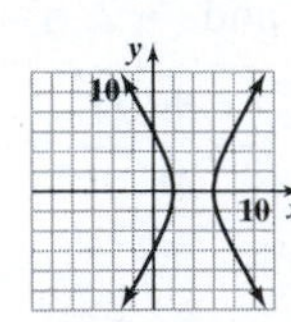 **16.** 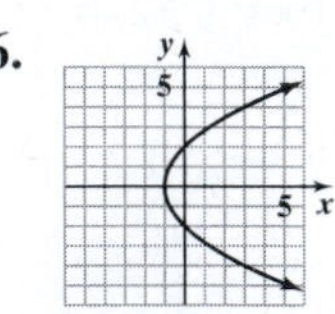 **17.** 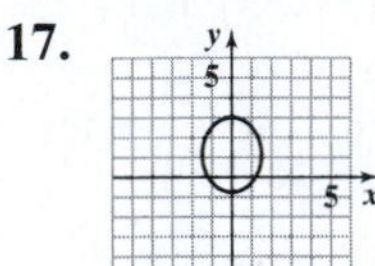

18. 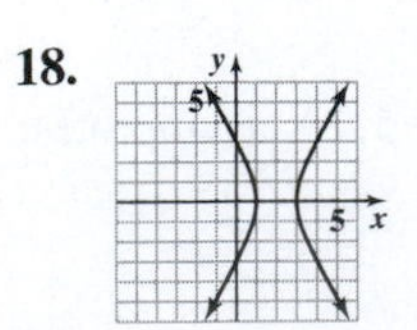 **19.** 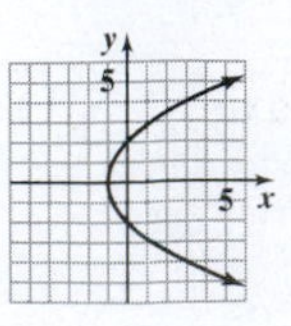 **20.** 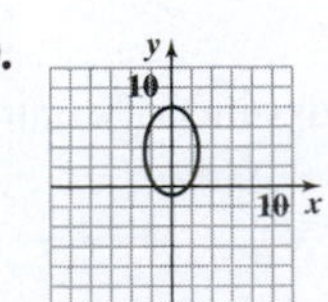 **21.** 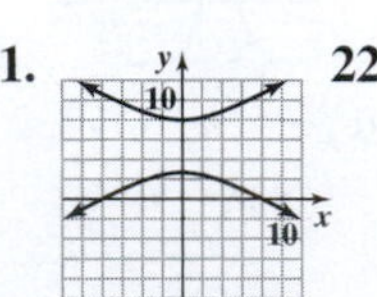 **22.** $y^2 - 6x - 9 = 0$

23. $9x^2 + 8y^2 + 12y - 36 = 0$ **24.** $-3x^2 + y^2 + 24x - 36 = 0$ **25.** $25y^2 - 90x - 81 = 0$

26. $25x^2 + 16y^2 + 24y - 16 = 0$ **27.** $9x^2 - 40y^2 - 84y - 36 = 0$ **28.** $r = \dfrac{3}{1 + \sin\theta}$ **29.** $r = \dfrac{\frac{3}{2}}{1 - \frac{3}{4}\cos\theta}$

30. $r = \dfrac{8}{1 + 2\cos\theta}$ **31.**

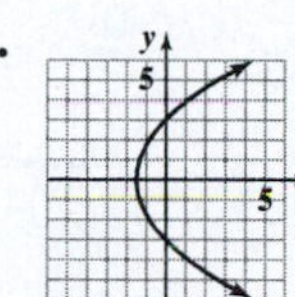

32. **33.**

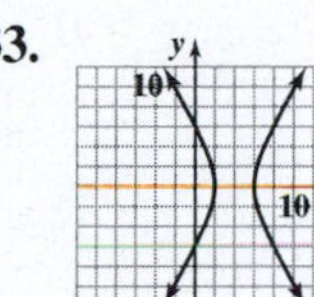

34.

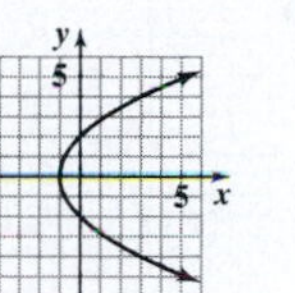

35.

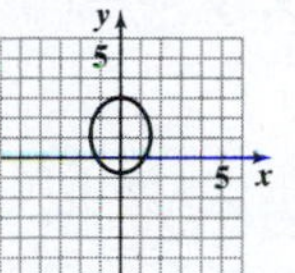

36. **37.**

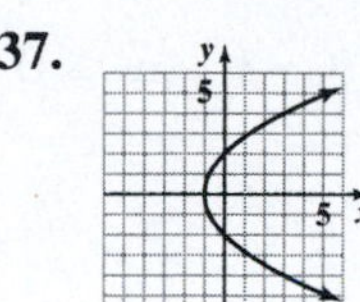

38.

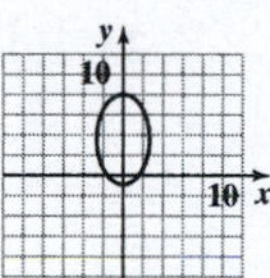

39.

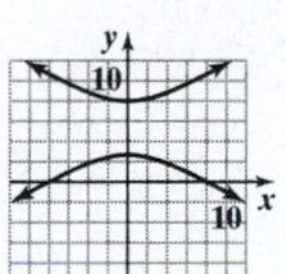

11.5 Exercises

1. initial point: $(-2, 0)$; terminal point: $(10, -8)$

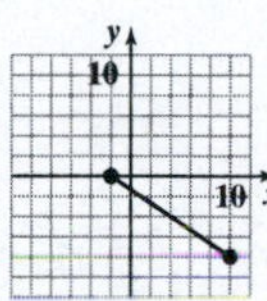

2. initial point: $(1, -3)$; terminal point: $(-1, 3)$

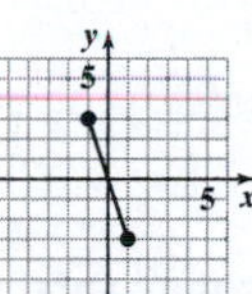

3. initial point: $(10, -6)$; terminal point: $(5, 4)$

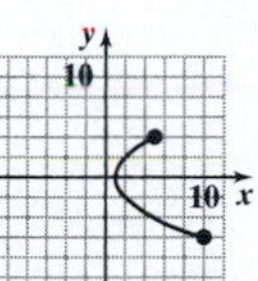

4. initial point: $(-5, -2)$; terminal point: $(10, -7)$

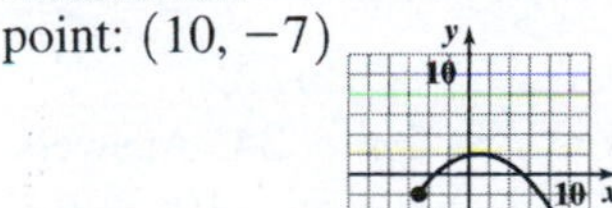

5. initial point: $(0, -2)$; terminal point: $(-2, 0)$

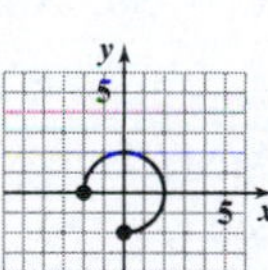

6. initial point: $(0, 3)$; terminal point: $(-3, 0)$

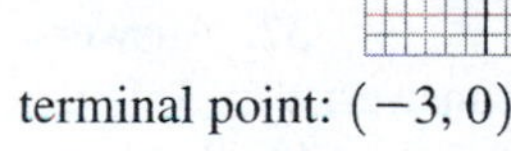

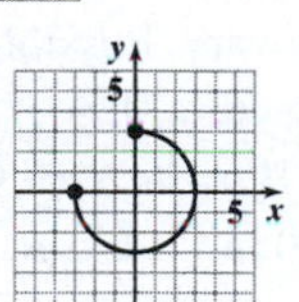

7. initial point: $(-3, 0)$; terminal point: $(0, 4)$

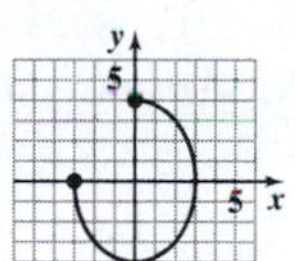

8. initial point: $\left(-\sqrt{2}, \dfrac{5\sqrt{2}}{2}\right)$; terminal point: $(0, -5)$

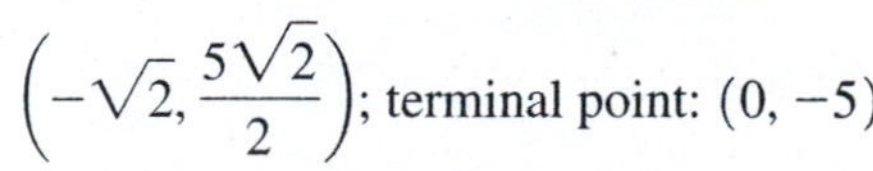

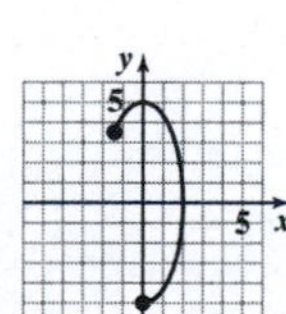

9. $y = -\dfrac{1}{5}x + 2, -5 \le x \le 5$

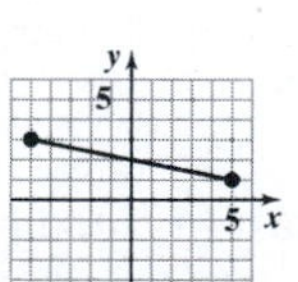

10. $y = -\dfrac{2}{3}x + \dfrac{5}{3}, -5 \le x \le 4$

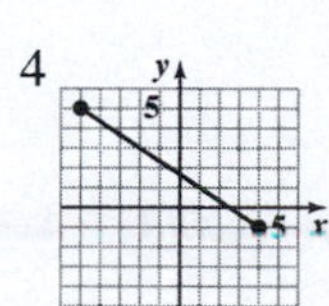

11. $x = \dfrac{3}{4}y^2, -4 \le y \le 4$

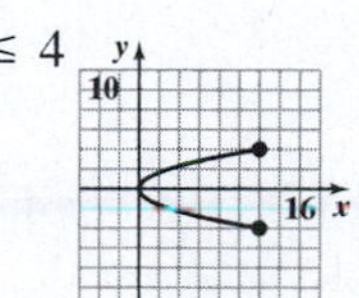

12. $y = -2x^2, -12 \le x \le 12$

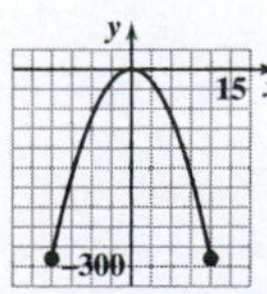

13. $y = 1 - 2x^2, x \ge 0$

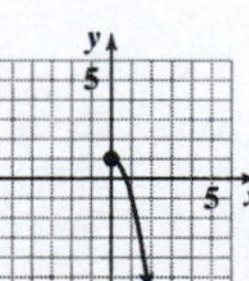

14. $x = 2 - y^2, y \ge 0$

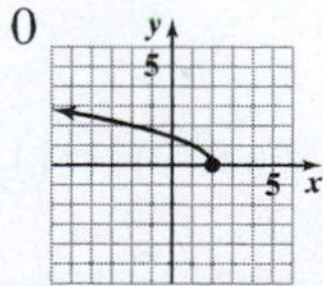

15. $y = x - 4, x \ge 1$

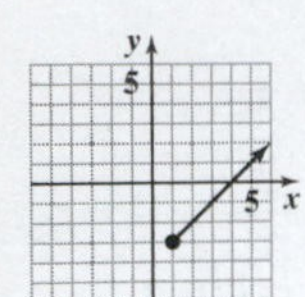

16. $y = x^2, x \le 0$

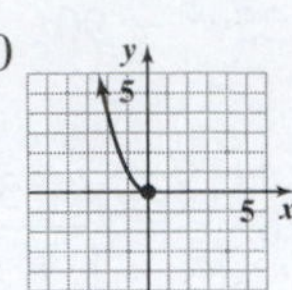

17. $y = \sqrt[3]{x}$

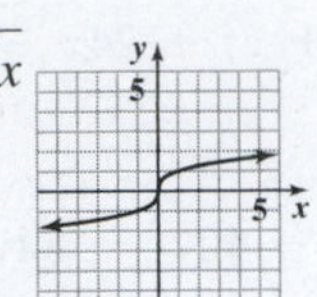

18. $y = \frac{1}{x}, x > 0$

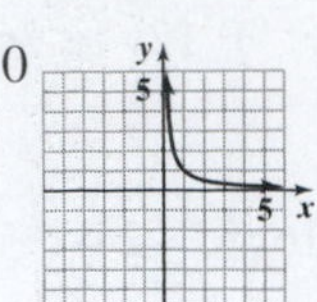

19. $y = -x^2, x > 0$

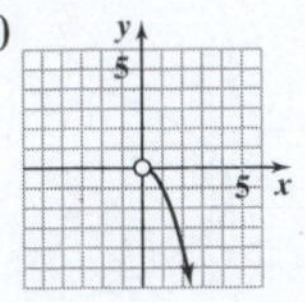

20. $y = -\frac{x}{3}, x > 0$

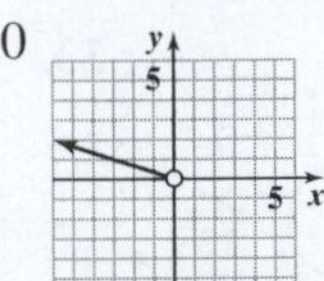

21. $y = -\frac{1}{x}, x < 0$

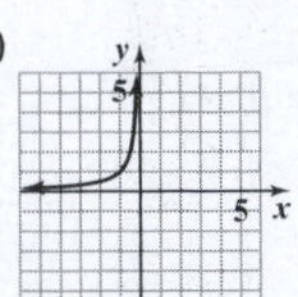

22. $y = -\frac{x}{2}$

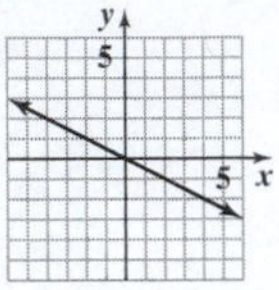

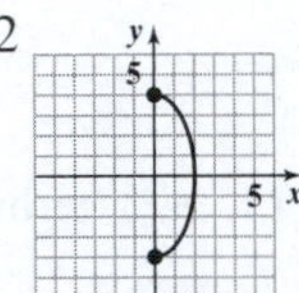

23. $y = -\frac{1}{x}, -1 \le x < 0$

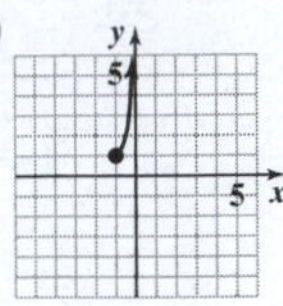

24. $x^2 + y^2 = 9, -3 \le x \le 0$

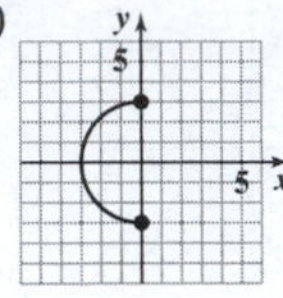

25. $4x^2 + y^2 = 16, 0 \le x \le 2$

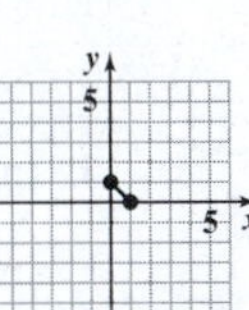

26. $y = x^2, 0 \le x \le 1, 0 \le y \le 1$

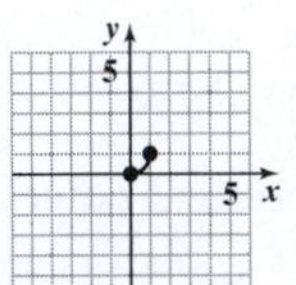

27. $x + y = 1, 0 \le x \le 1, 0 \le y \le 1$

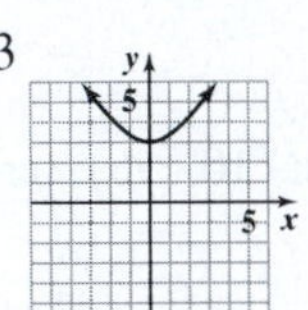

28. $\frac{y^2}{9} - \frac{x^2}{4} = 1, y \ge 3$

29. Answers may vary. Possible answer: $x = t, y = 2t - 1$. **30.** Answers may vary. Possible answer: $x = 2t, y = t - \frac{5}{4}$ **31.** Answers may vary. Possible answer: $x = 5t - 10, y = -2t + 5, 0 \le t \le 3$ **32.** Answers may vary. Possible answer: $x = -2t + 2$, $y = -3t + 1, 0 \le t \le 2$ **33.** Answers may vary. Possible answer: $x = -5 \cos t, y = -5 \sin t, 0 \le t \le 2\pi$ **34.** Answers may vary. Possible answer: $x = 3 \sin t, y = 4 \cos t, 0 \le t \le 2\pi$ **35.** Answers may vary. Possible answer: $x = -10t + 6$, $y = 7t - 2, 0 \le t \le 1$ **36.** Answers may vary. Possible answer: $x = -5t + 4, y = -7t + 4, 0 \le t \le 1$ **37.** Answers may vary. Possible answer: $x = -6 \cos t, y = -6 \sin t, 0 \le t \le 2\pi$ **38.** Answers may vary. Possible answer: $x = 3 \sin t$, $y = 3 \cos t, 0 \le t \le 2\pi$ **39.** Answers may vary. Possible answer: $x = -2 \sin t, y = 5 \cos t, 0 \le t \le 2\pi$ **40.** Answers may vary. Possible answer: $x = 7 \cos t, y = -4 \sin t, 0 \le t \le 2\pi$ **41. a.** $x = (60\sqrt{3})t$ **b.** $y = 60t - 16t^2$ **c.** 208 ft **d.** 56 ft **42. a.** $x = (29\sqrt{2})t$ **b.** $y = 20 + (29\sqrt{2})t - 4.9t^2$ **c.** 62 m **d.** 70 m **43. a.** 4.4 s **b.** 296 ft **c.** 2.1 s **d.** 83 ft

Review Exercises

1. vertex: $(1, 4)$, focus: $(1, 1)$, directrix: $y = 7$,

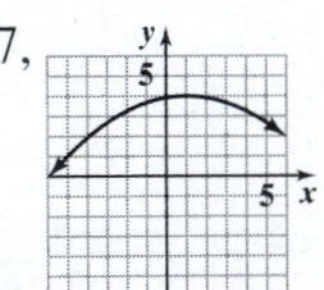

2. $(y - 4)^2 = 12(x + 1)$ **3.** vertex: $(-5, -3)$, focus: $\left(-5, -\frac{7}{4}\right)$, directrix: $y = -\frac{17}{4}$,

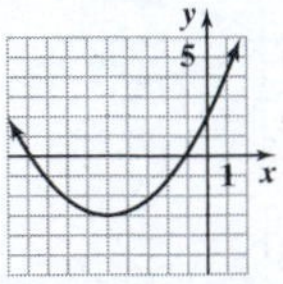

4. 10 cm **5.** center: $(1, -4)$, foci: $(1, -4 + 2\sqrt{10})$ and $(1, -4 - 2\sqrt{10})$, vertices: $(1, 3)$ and $(1, -11)$,

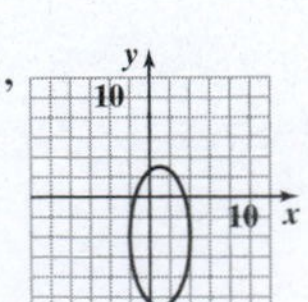

6. $\frac{(x-3)^2}{36} + \frac{(y+1)^2}{11} = 1$

7. $\dfrac{(x+2)^2}{20} + \dfrac{(y-1)^2}{16} = 1$, center: $(-2, 1)$, foci: $(-4, 1)$ and $(0, 1)$, vertices: $(-2 - 2\sqrt{5}, 1)$ and $(-2 + 2\sqrt{5}, 1)$,

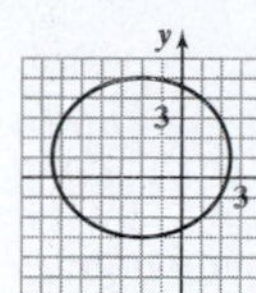

8. $\dfrac{x^2}{225} + \dfrac{y^2}{256} = 1$ **9.** center: $(2, 4)$, transverse axis: $x = 2$, vertices: $(2, -1)$ and $(2, 9)$, foci: $(2, 4 - \sqrt{61})$ and $(2, 4 + \sqrt{61})$, asymptotes: $y - 4 = \pm\dfrac{5}{6}(x - 2)$,

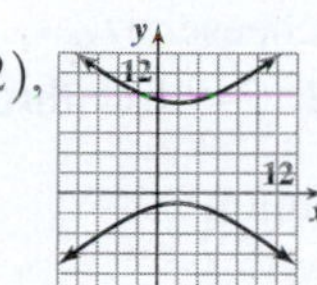

10. $\dfrac{(y+1)^2}{16} - \dfrac{(x+6)^2}{48} = 1$,

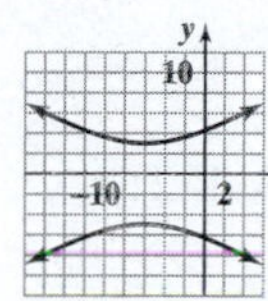

11. $\dfrac{(y-6)^2}{9} - (x+2)^2 = 1$, center: $(-2, 6)$, vertices: $(-2, 3)$ and $(-2, 9)$, foci: $(-2, 6 - \sqrt{10})$ and $(-2, 6 + \sqrt{10})$, endpoints of conjugate axis: $(-3, 6)$ and $(-1, 6)$, asymptotes: $y - 6 = \pm 3(x + 2)$,

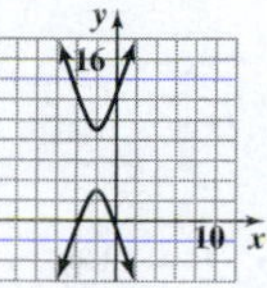

12. $\dfrac{y^2}{49} - \dfrac{x^2}{15} = 1$ **13.** hyperbola; directrix perpendicular to the polar axis located 3 units right of the pole

14.

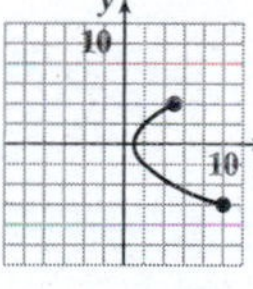

15. $y^2 - 6x - 9 = 0$ **16.** $r = \dfrac{3}{1 + \sin\theta}$ **17.** initial point: $(10, -6)$; terminal point: $(5, 4)$

18. $4x^2 + y^2 = 16, 0 \le x \le 2$ **19.** Answers may vary. Possible answer: $x = 3\sin t$, $y = 3\cos t, 0 \le t \le 2\pi$ **20. a.** $x = (60\sqrt{3})t$ **b.** $y = 60t - 16t^2$ **c.** 208 ft **d.** 56 ft

CHAPTER 12

12.1 Exercises

1. $(-1, 3)$ yes, $(-3, 7)$ no **2.** $\left(-\dfrac{1}{2}, \dfrac{1}{4}\right)$ no, $\left(-\dfrac{1}{4}, \dfrac{1}{2}\right)$ yes **3.** $(0.2, -0.9)$ yes, $(-0.3, 0.4)$ no **4.** $(-2, 5)$ no, $(3, 5)$ yes **5.** $(1, 3)$ **6.** $(-3, 2)$ **7.** $\left(-\dfrac{1}{2}, -4\right)$ **8.** $(-6, 6)$ **9.** $\left(0, -\dfrac{2}{3}\right)$ **10.** $\left(\dfrac{5}{2}, \dfrac{5}{4}\right)$ **11.** $(2, -7)$ **12.** $(6, 9)$ **13.** $(-8, 4)$ **14.** $(8, 8)$ **15.** no solution **16.** $(2 + 4y, y)$ or $\left(x, \dfrac{x-2}{4}\right)$ **17.** $\left(-\dfrac{13}{16}, \dfrac{13}{2}\right)$ **18.** $\left(x, \dfrac{3}{4}x - \dfrac{1}{2}\right)$ or $\left(\dfrac{2}{3} + \dfrac{4y}{3}, y\right)$ **19.** no solution **20.** $\left(x, \dfrac{5x}{7} - 9\right)$ or $\left(\dfrac{63}{5} + \dfrac{7y}{5}, y\right)$ **21.** Ricky: 1,196 hits, Pedro: 1,477 hits **22.** 93 two-bedroom and 107 three-bedroom **23.** 301 children and 191 adults. **24.** \$3,100 at 12% and \$2,800 at 7% **25.** 394 beverages **26.** 580 acres **27.** plane: 550 mph, wind: 50 mph **28.** 12 nickels and 34 quarters **29.** about 42.5 lbs of peanuts and about 16 lbs of cashews **30.** \$19.52

12.2 Exercises

1. $(-1, 1, -2)$ yes, $(1, -1, 2)$ no **2.** $(2, -1, 4)$ no, $(-2, 1, -4)$ yes **3.** $(-1, 2, 3)$ **4.** $(1, -2, 1)$ **5.** $\left(\frac{1}{4}, -\frac{1}{2}, \frac{3}{8}\right)$
6. $(-2, 3, 1)$ **7.** $(-3, 0, 1)$ **8.** $\left(-1, \frac{2}{3}, \frac{1}{3}\right)$ **9.** $(29, 16, 3)$ **10.** $(19, -7, -15)$ **11.** $(5, -2, -4)$
12. $x + 3y - 2z = 4$ **13.** $2x - y + z = -6$ **14.** $(5 - z, 3 + z, z)$ **15.** $(2, z - 3, z)$ **16.** no solution
17. no solution **18.** 20 long swordsmen, 20 spearmen, 35 crossbowmen **19.** \$1.50 for each hot dog, \$2.50 for each hamburger, \$3.75 for each chicken sandwich **20.** 10 cheese pizzas, 15 pepperoni pizzas, 10 supreme pizzas **21.** 520 adult tickets, 210 children tickets, 280 senior tickets **22.** 249 free throws, 223 two-point field goals, 9 three-point field goals
23. $a = -\frac{1}{80}, b = \frac{37}{40}, c = -\frac{161}{16}$; 5.25 million

12.3 Exercises

1. $(-1, 2, 3)$ **2.** $(1, -2, 1)$ **3.** $(29, 16, 3)$ **4.** $(19, -7, -15)$ **5.** $\left(-\frac{13}{3}, \frac{47}{15}, \frac{6}{5}\right)$ **6.** $(-2, 3, 1)$ **7.** $(3, 1, 2)$
8. $(-1, -3, 5)$ **9.** $\left(\frac{23}{21}, \frac{10}{7}, \frac{1}{21}\right)$ **10.** $\left(\frac{1}{4}, -2, -\frac{9}{8}\right)$ **11.** $(5, -2, -4)$ **12.** $(-3, 0, 1)$ **13.** $(-1, -3, 10)$
14. $\left(\frac{1}{2}, -1, 0\right)$ **15.** $\left(-2, \frac{1}{2}, -3\right)$ **16.** $\left(1, -3, \frac{1}{3}\right)$ **17.** $(-2, 3, 1)$ **18.** $(-3, -2, 11)$ **19.** $(-6, 0, 2)$
20. $(1 + 2z, 2 - 3z, z)$ **21.** $(-2 - z, 4 + 2z, z)$ **22.** $\left(\frac{5}{2} - \frac{z}{2}, -\frac{1}{2} - \frac{z}{2}, z\right)$ **23.** $(-7 + 2z, 3 - z, z)$ **24.** no solution
25. no solution **26.** $\left(\frac{21}{5}, -3, \frac{4}{5}\right)$ **27.** inconsistent **28.** $(-5 + 2z, 4 - z, z)$ **29.** $(5 - 3y, y, 8 - y)$ **30.** inconsistent
31. $(1 - 2y, y, -4)$ **32.** $(-6 - 2z, 8 + 3z, z)$ **33.** $(5 - z, 3 + z, z)$ **34.** $(2y, y, -2)$ **35.** $(-2 + 3z, 13 - 8z, z)$
36. $(2, -3 + z, z)$ **37.** 10 cheese, 15 pepperoni, 10 supreme **38.** 442 adults, 241 children, 266 senior citizens
39. 100 high performance, 80 ultra performance, 58 extreme performance **40.** \$200 at 6%, \$1,400 at 8%, \$600 at 9%
41. run: 10 miles, swim: 5 miles, bike: 25 miles **42.** $f(x) = 3x^2 + 3x$ **43.** $f(x) = 3x^2 - 4x - 5$

12.4 Exercises

1. $\frac{-1}{x} + \frac{1}{x - 3}$ **2.** $\frac{1}{x + 9} + \frac{3}{x + 3}$ **3.** $\frac{-1}{x - 1} + \frac{2}{x - 2}$ **4.** $\frac{\frac{1}{3}}{x - 2} + \frac{\frac{2}{3}}{x + 4}$ **5.** $\frac{2}{x - 3} - \frac{1}{x - 5}$

6. $\frac{\frac{2}{3}}{x - 5} - \frac{\frac{4}{3}}{2x - 1}$ **7.** $\frac{\frac{5}{12}}{x - 5} + \frac{\frac{7}{12}}{x + 7}$ **8.** $\frac{2}{x - 2} - \frac{1}{x - 1}$ **9.** $\frac{1}{x} - \frac{59}{x + 4} + \frac{66}{x + 5}$ **10.** $\frac{\frac{1}{25}}{3x + 1} + \frac{\frac{7}{25}}{4x - 7}$

11. $\frac{12}{x} - \frac{3}{x - 1} - \frac{2}{x + 1}$ **12.** $\frac{16}{x} - \frac{5}{x - 1} - \frac{4}{x + 1}$ **13.** $\frac{\frac{5}{12}}{x - 5} + \frac{\frac{5}{8}}{x + 5} - \frac{\frac{1}{24}}{x + 1}$ **14.** $\frac{-\frac{2}{21}}{x} + \frac{\frac{18}{91}}{x - 7} + \frac{-\frac{4}{39}}{x + 6}$

15. $\frac{-7}{x - 5} - \frac{24}{(x - 5)^2}$ **16.** $\frac{1}{x - 2} - \frac{2}{(x - 2)^2} - \frac{4}{(x - 2)^3}$ **17.** $\frac{-2}{x - 4} + \frac{25}{(x - 4)^2} + \frac{2}{x + 5}$

18. $\frac{\frac{9}{25}}{x - 1} + \frac{\frac{1}{5}}{(x - 1)^2} + \frac{\frac{16}{25}}{x + 4}$ **19.** $\frac{3}{x} - \frac{3}{x - 1} + \frac{4}{(x - 1)^2}$ **20.** $\frac{-1}{x} + \frac{5}{x^2} + \frac{2}{x + 6}$

21. $\frac{\frac{1}{2}}{x - 2} + \frac{1}{(x - 2)^2} - \frac{\frac{1}{2}}{x + 2} + \frac{1}{(x + 2)^2}$ **22.** $\frac{-\frac{7}{4}}{x + 3} + \frac{\frac{7}{4}}{x + 5} + \frac{\frac{9}{2}}{(x + 5)^2}$

23. $\frac{6}{x} - \frac{2}{x^2} - \frac{6}{x + 1} + \frac{3}{(x + 1)^2} - \frac{9}{(x + 1)^3}$ **24.** $\frac{\frac{1}{2}}{x + 1} - \frac{\frac{1}{2}}{x - 1} + \frac{1}{(x - 1)^2}$ **25.** $\frac{-\frac{6}{25}}{x} - \frac{\frac{1}{5}}{x^2} + \frac{\frac{7}{50}}{x - 5} + \frac{\frac{1}{10}}{x + 5}$

26. $\dfrac{\frac{13}{32}}{x+4}+\dfrac{\frac{51}{32}}{x-4}+\dfrac{\frac{25}{4}}{(x-4)^2}$ **27.** $\dfrac{1}{x}-\dfrac{x}{x^2+6}$ **28.** $\dfrac{7}{x-1}+\dfrac{-7x+4}{x^2+x+1}$ **29.** $\dfrac{2}{x-6}+\dfrac{6x-2}{x^2+5}$

30. $\dfrac{-\frac{4}{27}}{x-3}+\dfrac{\frac{4}{27}x+\frac{8}{9}}{x^2+3x+9}$ **31.** $\dfrac{1}{x+3}-\dfrac{1}{x-3}+\dfrac{7}{x^2+8}$ **32.** $\dfrac{\frac{1}{9}}{x}+\dfrac{\frac{7}{9}}{x^2}+\dfrac{-\frac{1}{9}x-\frac{7}{9}}{x^2+9}$

33. $\dfrac{3}{x+3}+\dfrac{13}{(x+3)^2}+\dfrac{-3x-4}{x^2+4}$ **34.** $\dfrac{4}{x+3}+\dfrac{2x+5}{x^2+4}$ **35.** $\dfrac{\frac{3}{17}}{x+3}+\dfrac{\frac{14}{17}x+\frac{45}{17}}{x^2-12x+6}$ **36.** $\dfrac{x+1}{x^2+4}-\dfrac{4x-1}{(x^2+4)^2}$

37. $\dfrac{5x-6}{x^2+4x+8}+\dfrac{6x+11}{(x^2+4x+8)^2}$ **38.** $\dfrac{4x+2}{(x^2+3)^2}+\dfrac{3x-2}{(x^2+3)^3}$ **39.** $\dfrac{-\frac{1}{9}}{x}+\dfrac{\frac{5}{9}x}{5x^2+3}+\dfrac{\frac{5}{3}x+3}{(5x^2+3)^2}$

40. $\dfrac{3}{x+2}+\dfrac{-2x+2}{x^2+4}+\dfrac{5x+3}{(x^2+4)^2}$ **41.** $\dfrac{-2}{x}+\dfrac{1}{x^2}+\dfrac{2x+2}{x^2+1}+\dfrac{2x}{(x^2+1)^2}$

12.5 Exercises

1. ; two real solutions **2.** ; two real solutions **3.** ; one real solution

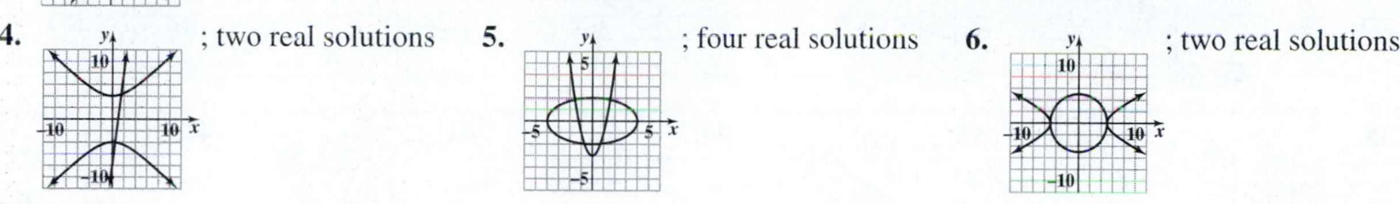

4. ; two real solutions **5.** ; four real solutions **6.** ; two real solutions

7. ; one real solution **8.** ; four real solutions **9.** $(-5, 10), (4, 1)$ **10.** $(-6, -8), (8, 6)$

11. $(-1, 10)$ **12.** $(-10, -6), (8, 0)$ **13.** $(2, 0), (4, -2)$ **14.** $\left(-\frac{4}{5}, \frac{6}{5}\right), (0, 2)$ **15.** $(0, 2), (1, 1)$
16. $(-5, -6), (2, 15)$ **17.** $(-5, -1), (-1, -5), (1, 5), (5, 1)$ **18.** $(-\sqrt{3}, -\sqrt{3}), (-1, -3), (1, 3), (\sqrt{3}, \sqrt{3})$
19. $(-3, -4), (6, 5)$ **20.** $\left(\frac{1}{3}, \frac{29}{9}\right), \left(3, \frac{7}{3}\right)$ **21.** $(-8, -4), (-8, 4), (8, -4), (8, 4)$
22. $(-1, -2), (-1, 2), (1, -2), (1, 2)$ **23.** $(0, 3)$ **24.** $(-3, -\sqrt{7}), (-3, \sqrt{7}), (0, -4), (0, 4)$
25. $(-2, -4), (-2, 4), (2, -4), (2, 4)$ **26.** $\left(-\frac{1}{2}, -\frac{3}{2}\right), \left(-\frac{1}{2}, \frac{3}{2}\right), \left(\frac{1}{2}, -\frac{3}{2}\right), \left(\frac{1}{2}, \frac{3}{2}\right)$ **27.** $\left(-\sqrt{2}, -\frac{1}{\sqrt{2}}\right), \left(\sqrt{2}, \frac{1}{\sqrt{2}}\right)$
28. no real solutions **29.** $(2, -8), (0, 0)$ **30.** $(3, -2)$ **31.** $(-6, 5)$ **32.** $(4, 2)$ **33.** 8 and 14 **34.** 17 and 8
35. 18 and 15 **36.** 94 or 49 **37.** 742 **38.** 6 ft by 13 ft, or 13 ft by 6 ft **39.** 8 m and 13 m **40.** 38.7 ft
41. 9 ft by 9 ft by 12 ft, or about 16.8 ft by 16.8 ft by 3.5 ft **42.** 6 ft by 13 ft by 6 ft

12.6 Exercises

1. a. no **b.** yes **c.** no **2. a.** yes **b.** no **c.** yes **3. a.** no **b.** no **c.** yes **4. a.** no **b.** yes **c.** no
5. **6.** **7.** **8.** **9.** **10.**

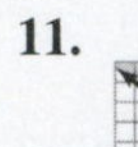

11. 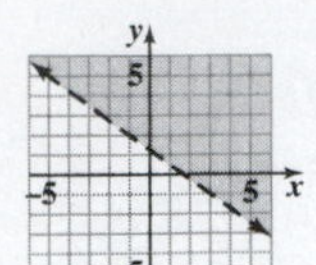 **12.** 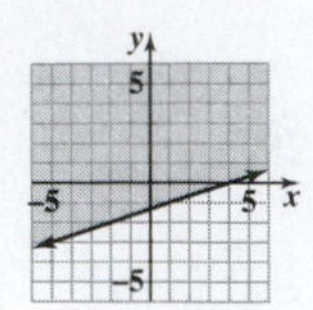 **13.** 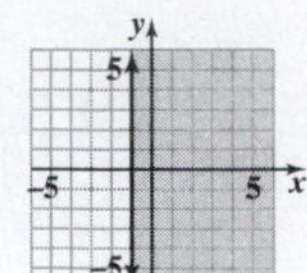 **14.** 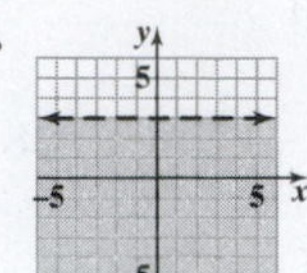 **15.** 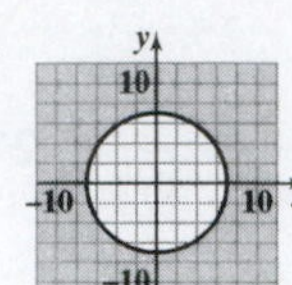 **16.** 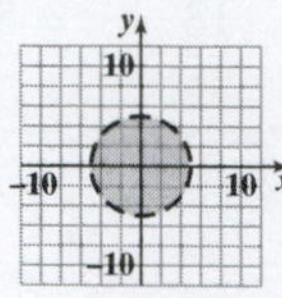

17. 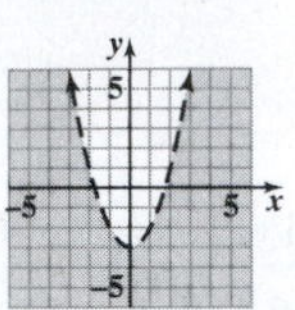 **18.** 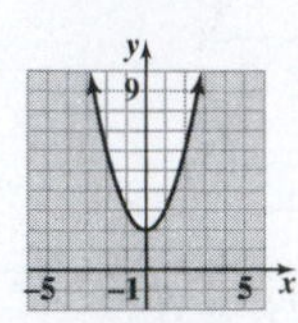 **19.** 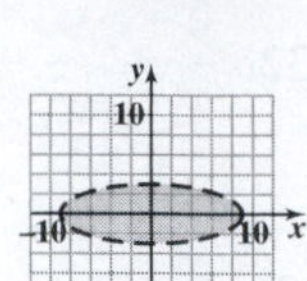 **20.** 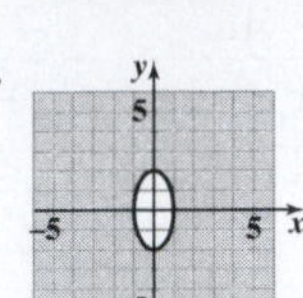 **21.** 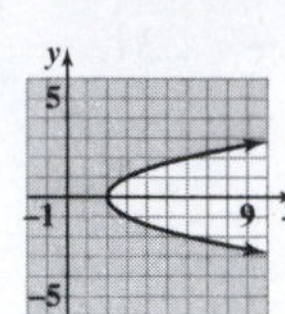 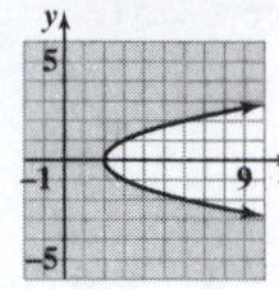 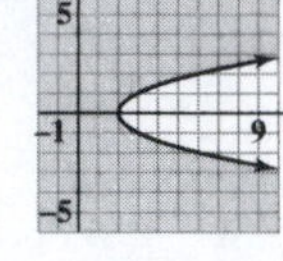 **22.** 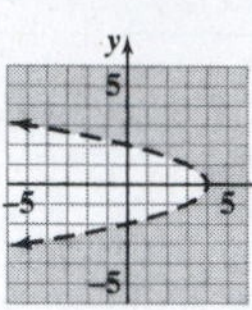 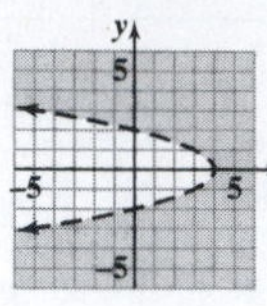 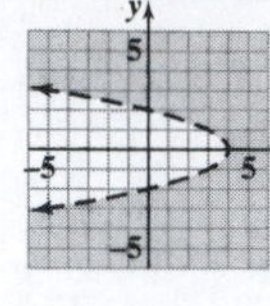

23. 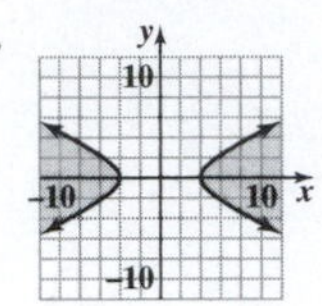 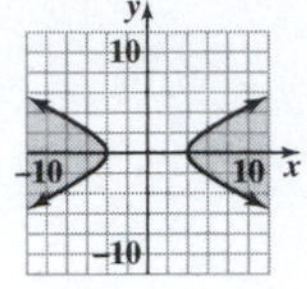 **24.** 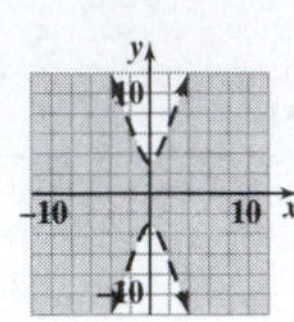 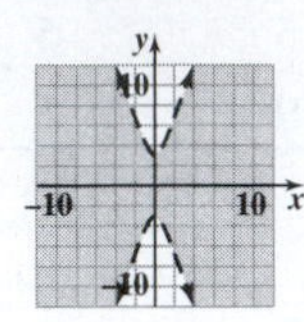 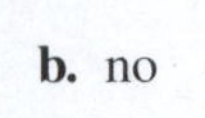 **25. a.** yes **b.** no **c.** no **26. a.** no **b.** no **c.** yes **27. a.** yes **b.** no **c.** yes **28. a.** no **b.** no **c.** yes **29. a.** no **b.** no **c.** yes **30. a.** yes **b.** no **c.** yes **31.** 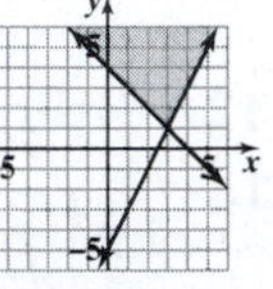

32. 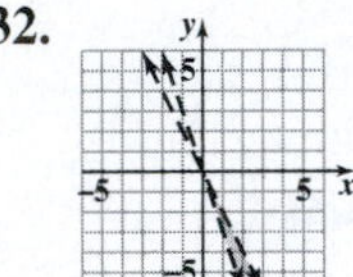 **33.** 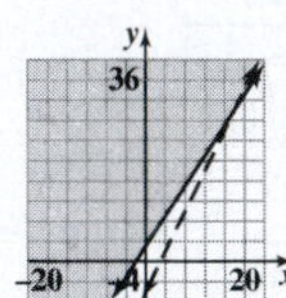 **34.** 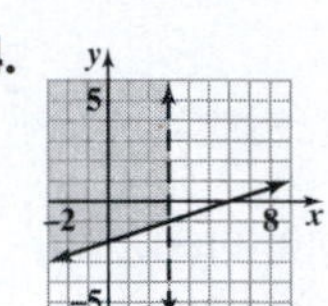 **35.** 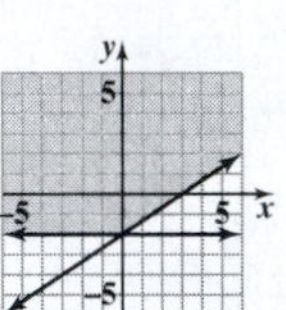 **36.** 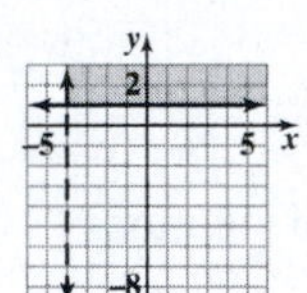 **37.** 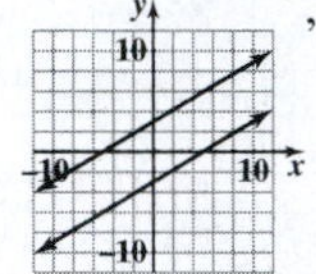, no solution **38.** 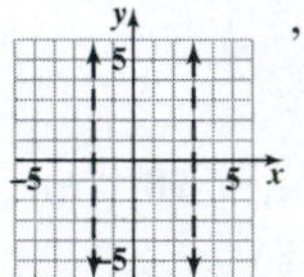, no solution **39.** 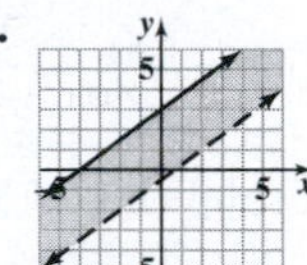 **40.** 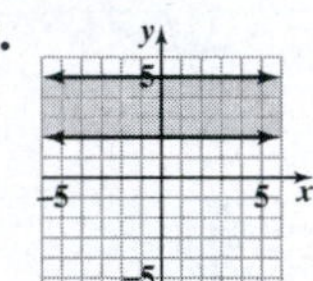 **41.** 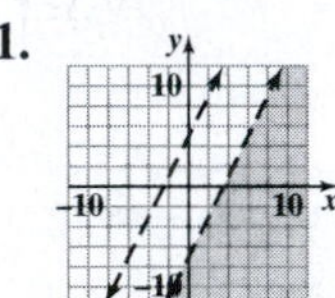 **42.** 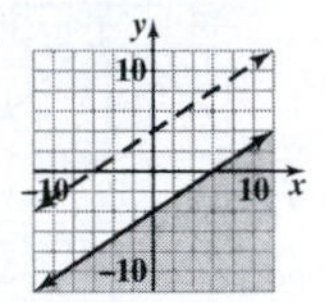

43. 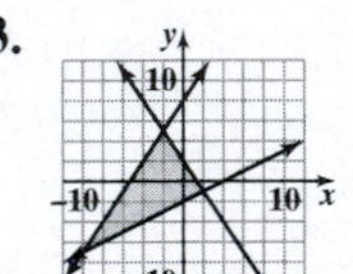 **44.** 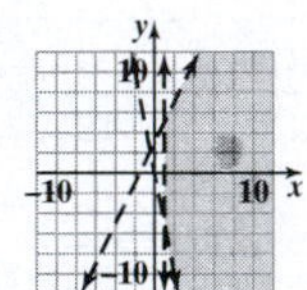 **45.** 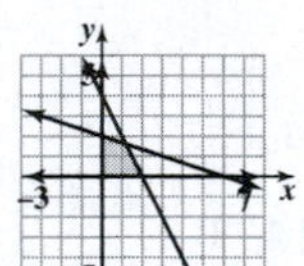 **46.** 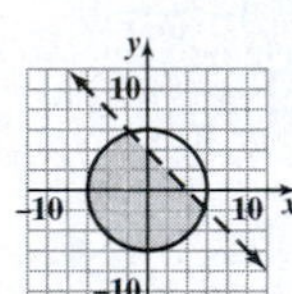 **47.** 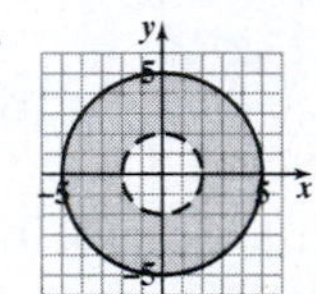 **48.** 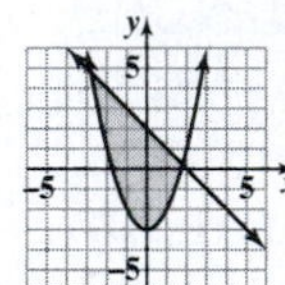

49. 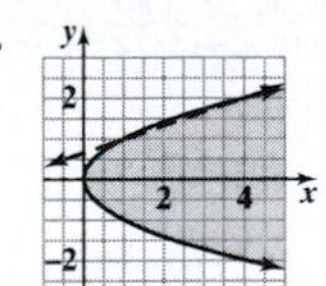 **50.** 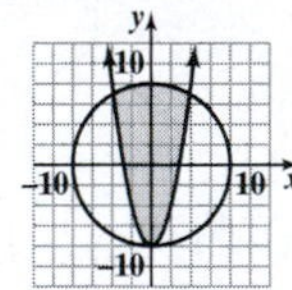 **51.** 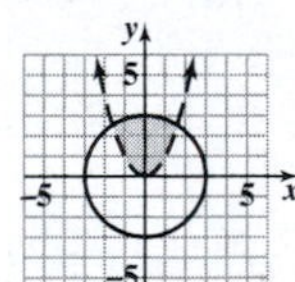 **52.** 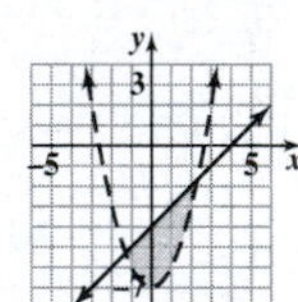 **53.** 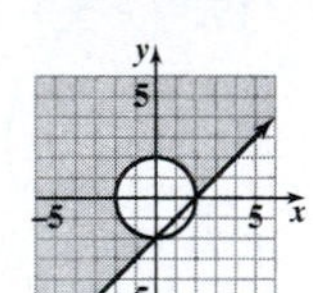 **54.** 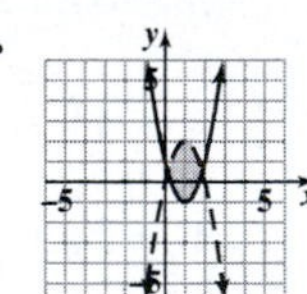

55. 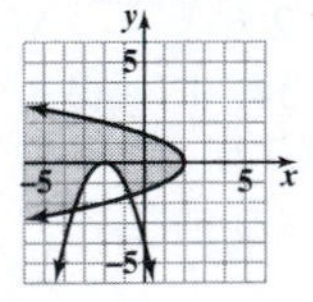 **56.** 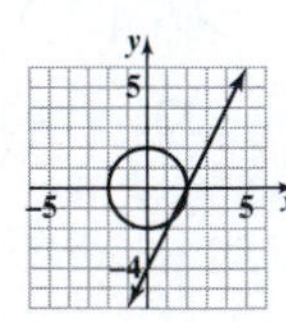 **57.** 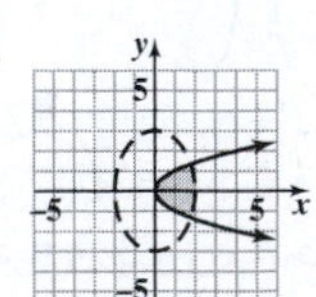 **58.** 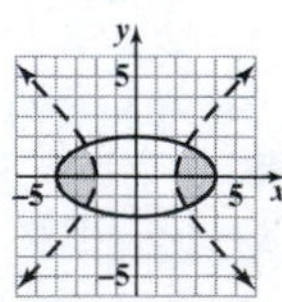

Review Exercises

1. $(-1, 3)$ yes, $(-3, 7)$ no **2.** $(1, 3)$ **3.** $(6, 9)$ **4.** 301 children and 191 adults. **5.** $(-1, 1, -2)$ yes, $(1, -1, 2)$ no **6.** $\left(\frac{1}{4}, -\frac{1}{2}, \frac{3}{8}\right)$ **7.** 10 cheese pizzas, 15 pepperoni pizzas, 10 supreme pizzas **8.** $(1, -2, 1)$ **9.** $\left(-\frac{13}{3}, \frac{47}{15}, \frac{6}{5}\right)$

10. (3, 1, 2) **11.** $\dfrac{\frac{1}{3}}{x-2} + \dfrac{\frac{2}{3}}{x+4}$ **12.** $\dfrac{\frac{9}{25}}{x-1} + \dfrac{\frac{1}{5}}{(x-1)^2} + \dfrac{\frac{16}{25}}{x+4}$ **13.** $\dfrac{\frac{1}{9}}{x} + \dfrac{\frac{7}{9}}{x^2} + \dfrac{-\frac{1}{9}x - \frac{7}{9}}{x^2+9}$

14. 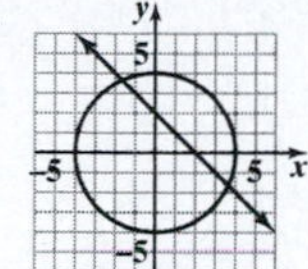

; two real solutions **15.** (2, 0), (4, −2) **16.** 17 and 8 **17. a.** no **b.** no **c.** yes

18. 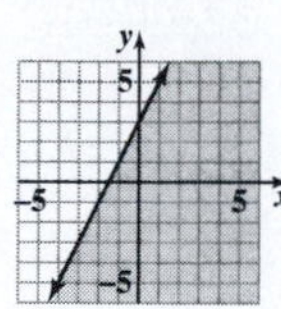

19. 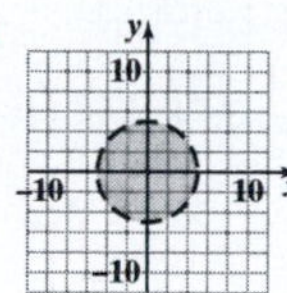

20. 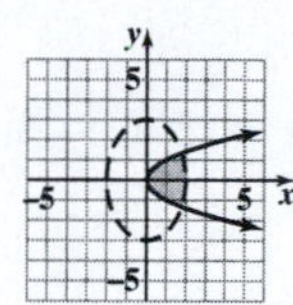

CHAPTER 13

13.1 Exercises

1. a. 3 × 4 **b.** none of these **2. a.** 3 × 3 **b.** square matrix **3. a.** 1 × 4 **b.** row matrix **4. a.** 4 × 1 **b.** column matrix **5.** $\begin{bmatrix} 4 & 4 & -3 \\ 8 & 1 & 6 \\ 8 & 3 & -10 \end{bmatrix}$ **6.** $\begin{bmatrix} -2 & -4 & -1 \\ -2 & 1 & 2 \\ 2 & -5 & 0 \end{bmatrix}$ **7.** $\begin{bmatrix} 2 & 4 & 1 \\ 2 & -1 & -2 \\ -2 & 5 & 0 \end{bmatrix}$ **8.** $\begin{bmatrix} 2 & 0 & -4 \\ 6 & 2 & 8 \\ 10 & -2 & -10 \end{bmatrix}$

9. $\begin{bmatrix} \frac{3}{2} & 2 & -\frac{1}{2} \\ \frac{5}{2} & 0 & 1 \\ \frac{3}{2} & 2 & -\frac{5}{2} \end{bmatrix}$ **10.** $\begin{bmatrix} \frac{1}{2} & -2 & -\frac{7}{2} \\ \frac{7}{2} & 2 & 7 \\ \frac{17}{2} & -4 & -\frac{15}{2} \end{bmatrix}$ **11.** 2 × 2 **12.** 3 × 5 **13.** not defined **14.** 1 × 1

15. $\begin{bmatrix} 7 & -7 & 1 \\ 23 & -1 & -3 \end{bmatrix}$ **16.** $\begin{bmatrix} 11 & 1 & 10 \\ -4 & 0 & -10 \end{bmatrix}$ **17.** $\begin{bmatrix} 18 & -6 & 11 \\ 19 & -1 & -13 \end{bmatrix}$ **18.** $\begin{bmatrix} 7 & 8 \\ 12 & 31 \end{bmatrix}$ **19.** $\begin{bmatrix} -19 & -1 & -30 \\ 13 & 3 & -20 \end{bmatrix}$

20. $\begin{bmatrix} 0 & 0 & 0 \\ 4 & -2 & 0 \end{bmatrix}$ **21.** $I_m = I_2 = \begin{bmatrix} 1 & 0 \\ 0 & 1 \end{bmatrix}$ and $I_n = I_3 = \begin{bmatrix} 1 & 0 & 0 \\ 0 & 1 & 0 \\ 0 & 0 & 1 \end{bmatrix}$ **22.** $I_m = I_3 = \begin{bmatrix} 1 & 0 & 0 \\ 0 & 1 & 0 \\ 0 & 0 & 1 \end{bmatrix}$ and

$I_n = I_4 = \begin{bmatrix} 1 & 0 & 0 & 0 \\ 0 & 1 & 0 & 0 \\ 0 & 0 & 1 & 0 \\ 0 & 0 & 0 & 1 \end{bmatrix}$ **23.** $I_m = I_5 = \begin{bmatrix} 1 & 0 & 0 & 0 & 0 \\ 0 & 1 & 0 & 0 & 0 \\ 0 & 0 & 1 & 0 & 0 \\ 0 & 0 & 0 & 1 & 0 \\ 0 & 0 & 0 & 0 & 1 \end{bmatrix}$ and $I_n = I_2 = \begin{bmatrix} 1 & 0 \\ 0 & 1 \end{bmatrix}$ **24. a.** $[208 \quad 116 \quad 76 \quad 88]$

b. 208 cups of baking mix, 116 eggs, 76 cups of milk, and 88 tablespoons of oil are needed to make the desired amount of food

25. a. $\begin{bmatrix} 28.5 & 24 & 9 \\ 39 & 33 & 12 \end{bmatrix}$ **b.** ab_{21} = it costs \$39 to produce a self-propelled mower in New York, ab_{12} = it costs \$24 to produce a standard mower in Los Angeles, ab_{23} = it costs \$12 to produce a self-propelled mower in Beijing

26. [Buffalo Rochester Syracuse] = [24, 750 26, 500 39, 200], \$90,450

13.2 Exercises

1. yes **2.** yes **3.** yes **4.** yes **5.** yes, $A^{-1} = \begin{bmatrix} 3 & -1 \\ -5 & 2 \end{bmatrix}$ **6.** yes, $A^{-1} = \begin{bmatrix} -1 & 2 \\ -1 & \frac{5}{2} \end{bmatrix}$ **7.** singular

8. singular **9.** $\begin{bmatrix} 0 & \frac{1}{2} \\ -1 & \frac{1}{2} \end{bmatrix}$ **10.** $\begin{bmatrix} -3 & -4 & 2 \\ 0 & -1 & 0 \\ -2 & 1 & 1 \end{bmatrix}$ **11.** singular matrix **12.** $\begin{bmatrix} -28 & -13 & 3 \\ 2 & 1 & 0 \\ -7 & -3 & 1 \end{bmatrix}$ **13.** $\begin{bmatrix} -\frac{1}{4} & -1 & -\frac{1}{4} \\ -\frac{7}{4} & -2 & \frac{1}{4} \\ 1 & 1 & 0 \end{bmatrix}$

14. $\begin{bmatrix} 9 & 0 & 2 & 0 \\ -20 & -9 & -5 & 5 \\ 4 & 0 & 1 & 0 \\ -4 & -2 & -1 & 1 \end{bmatrix}$ **15.** $x = 2, y = -3$ **16.** $x = -\frac{1}{3}, y = \frac{1}{2}$ **17.** $x = \frac{29}{4}, y = \frac{33}{4}$ **18.** $x = 1, y = -2, z = 4$

19. $x = 0, y = 1, z = 3$ **20.** $x = -10, y = 6, z = 0$ **21.** $x = 2, y = 3, z = 0, w = 1$

13.3 Exercises

1. $x = 1, y = 2$ **2.** $x = 1, y = 1$ **3.** $x = -2, y = -4$ **4.** $x = \frac{1}{2}, y = \frac{1}{3}$ **5.** 20 **6.** 164 **7.** 39 **8.** 0

9. 6 **10.** −220 **11.** 4,710 **12.** 42 **13.** $x = 2, y = 1, z = -1$ **14.** $x = 2, y = 0, z = 0$
15. $x = -3, y = 2, z = -4$ **16.** $x = -3, y = 3, z = -2$ **17.** $x = 35, y = -21, z = -21$
18. $w = -2, x = 4, y = -3, z = 2$

Review Exercises

1. $\begin{bmatrix} 4 & 4 & -3 \\ 8 & 1 & 6 \\ 8 & 3 & -10 \end{bmatrix}$ **2.** $\begin{bmatrix} -2 & -4 & -1 \\ -2 & 1 & 2 \\ 2 & -5 & 0 \end{bmatrix}$ **3.** $\begin{bmatrix} \frac{1}{2} & -2 & -\frac{7}{2} \\ \frac{7}{2} & 2 & 7 \\ \frac{17}{2} & -4 & -\frac{15}{2} \end{bmatrix}$ **4.** 3×5 **5.** $\begin{bmatrix} 7 & -7 & 1 \\ 23 & -1 & -3 \end{bmatrix}$

6. a. $[208 \quad 116 \quad 76 \quad 88]$ **b.** 208 cups of baking mix, 116 eggs, 76 cups of milk, and 88 tablespoons of oil are needed to make the desired amount of food **7.** yes **8.** yes, $A^{-1} = \begin{bmatrix} -1 & 2 \\ -1 & \frac{5}{2} \end{bmatrix}$ **9.** $\begin{bmatrix} -3 & -4 & 2 \\ 0 & -1 & 0 \\ -2 & 1 & 1 \end{bmatrix}$ **10.** $\begin{bmatrix} 9 & 0 & 2 & 0 \\ -20 & -9 & -5 & 5 \\ 4 & 0 & 1 & 0 \\ -4 & -2 & -1 & 1 \end{bmatrix}$

11. $x = 1, y = -2, z = 4$ **12.** $x = -2, y = -4$ **13.** 164 **14.** $x = 2, y = 1, z = -1$
15. $w = -2, x = 4, y = -3, z = 2$

CHAPTER 14

14.1 Exercises

1. 4, 7, 10, 13 **2.** 4, 16, 64, 256 **3.** $1, \frac{8}{5}, 2, \frac{16}{7}$ **4.** −4, 16, −64, 256 **5.** 30, 120, 600, 3600 **6.** $\frac{1}{2}, \frac{4}{3}, \frac{9}{8}, \frac{8}{15}$

7. −5, 10, −15, 20 **8.** $\frac{1}{2}, \frac{9}{4}, \frac{9}{2}, \frac{81}{4}$ **9.** $-\frac{1}{42}, \frac{1}{56}, -\frac{1}{72}, \frac{1}{90}$ **10.** $-\frac{9}{2}, \frac{81}{40}, -\frac{243}{560}, \frac{243}{4480}$ **11.** 7, 10, 13, 16 **12.** −1, 3, 0, 4

13. $6, \frac{3}{2}, \frac{1}{6}, \frac{1}{96}$ **14.** $-4, \frac{5}{4}, \frac{1}{5}, -4$ **15.** $a_n = 2n - 3, a_{11} = 19$ **16.** $a_n = \frac{n}{n+4}, a_8 = \frac{8}{12}$ **17.** $a_n = n(n+5), a_7 = 7 \cdot 12$

18. $a_n = (-1)^n \cdot 2^n, a_7 = -128$ **19.** $a_n = \frac{2}{5^n}, a_6 = \frac{2}{15{,}625}$ **20.** $a_n = (-1)^n \cdot 6 \cdot 2^{n-1}, a_9 = -1536$

21. $a_n = (-1)^n (n+2)!, a_5 = -5040$ **22.** $a_n = \frac{3^n}{(n+1)!}, a_5 = \frac{243}{720}$ **23.** 20 **24.** 15 **25.** $\frac{11}{32}$ **26.** 111

27. 12 **28.** −192 **29.** 45 **30.** 102 **31.** 196 **32.** 147 **33.** 271 **34.** 112 **35.** 27 **36.** $\sum_{i=1}^{29} i$ **37.** $\sum_{i=1}^{10} 5i$

38. $\sum_{i=1}^{11} i^2$ **39.** $\sum_{i=1}^{9} \frac{i+3}{i+4}$ **40.** $\sum_{i=1}^{8} (-1)^{i+1} \cdot 2^i$ **41.** $\sum_{i=1}^{6} \frac{(-1)^i}{9i}$ **42.** $\sum_{i=1}^{n} \frac{5^i}{i}$ **43.** $\sum_{i=1}^{n+1} \frac{7^{i-1}}{(i-1)!}$ **44.** 55

14.2 Exercises

1. yes, 6 **2.** no **3.** yes, $\frac{3}{2}$ **4.** no **5.** yes, 2 **6.** no **7.** $a_n = 5n - 3, a_{10} = 47$ **8.** $a_n = 9 - 4n, a_{31} = -115$

9. $a_n = \frac{3}{2}n, a_{50} = 75$ **10.** $a_n = 6.2 - 1.2n, a_{29} = -28.6$ **11.** $a_n = 2n - 1, a_{17} = 33$ **12.** $a_n = 21 - 5n, a_{11} = -34$

13. 71 **14.** −64 **15.** 97 **16.** −104 **17.** $\frac{25}{2}$ **18.** −45 **19.** 3240 **20.** 235 **21.** 2375 **22.** 1044 **23.** −1002 **24.** 51,612 **25.** $4n^2 + 2n$ **26.** 10,000 **27.** 10,100 **28.** 3600 **29.** 147 **30.** $444,600 **31.** $3185 **32.** Beta Company **33.** $200, $65,000 **34.** $4329 **35.** 1.3 sec, 312 in.

14.3 Exercises

1. 8, 16, 32, 64, 128 **2.** 162, 54, 18, 6, 2 **3.** $25, -5, 1, -\frac{1}{5}, \frac{1}{25}$ **4.** 3, 21, 147, 1029, 7203 **5.** −4, 8, −16, 32, −64

6. yes, 6 **7.** yes, −1 **8.** no **9.** yes, $-\frac{5}{3}$ **10.** yes, −.8 **11.** $a_n = 3 \cdot 2^{n-1}$ **12.** $a_n = \frac{1}{2}\left(\frac{1}{4}\right)^{n-1}$ **13.** $a_n = \frac{1}{5}\left(-\frac{2}{3}\right)^{n-1}$

14. 20,480 **15.** $-\frac{951}{256}$ **16.** ≈$5,523.11 **17.** 78,732 **18.** $\frac{1}{16}$ **19.** −91,749 **20.** ≈4.99 **21.** ≈50.31

22. $\frac{2186}{729}$ **23.** $\frac{85}{128}$ **24.** converges, $-\frac{10}{11}$ **25.** converges, $\frac{2401}{6}$ **26.** converges, $\frac{28}{3}$ **27.** diverges **28.** converges, $\frac{5}{11}$

29. $6561 **30.** $1146.81 **31.** ≈$1,402,123.44 **32.** ≈53.56 ft, 60 ft **33.** 200 ft **34.** $\frac{7}{9}$ **35.** $\frac{161}{495}$

36. ≈$46,007.74 **37.** ≈$30,200.99 **38.** ≈$462.20

14.4 Exercises

1. $m^6 + 6m^5n + 15m^4n^2 + 20m^3n^3 + 15m^2n^4 + 6mn^5 + n^6$ **2.** $x^4 + 20x^3 + 150x^2 + 500x + 625$ **3.** $x^7 - 7x^6y + 21x^5y^2 - 35x^4y^3 + 35x^3y^4 - 21x^2y^5 + 7xy^6 - y^7$ **4.** $x^5 - 15x^4 + 90x^3 - 270x^2 + 405x - 243$ **5.** $64x^6 + 576x^5y + 2160x^4y^2 + 4320x^3y^3 + 4860x^2y^4 + 2916xy^5 + 729y^6$ **6.** $81x^8 - 432x^6y^3 + 864x^4y^6 - 768x^2y^9 + 256y^{12}$ **7.** 7 **8.** 210 **9.** 1 **10.** 1771 **11.** $x^7 + 14x^6 + 84x^5 + 280x^4 + 560x^3 + 672x^2 + 448x + 128$ **12.** $x^6 - 18x^5 + 135x^4 - 540x^3 + 1215x^2 - 1458x + 729$ **13.** $1024x^5 + 1280x^4 + 640x^3 + 160x^2 + 20x + 1$ **14.** $x^4 + 12x^3y + 54x^2y^2 + 108xy^3 + 81y^4$ **15.** $3125x^5 - 9375x^4y + 11250x^3y^2 - 6750x^2y^3 + 2025xy^4 - 243y^5$ **16.** $x^{24} + 6x^{20}y^5 + 15x^{16}y^{10} + 20x^{12}y^{15} + 15x^8y^{20} + 6x^4y^{25} + y^{30}$ **17.** $x^2 - 4\sqrt{2}x^{\frac{3}{2}} + 12x - 8\sqrt{2}x^{\frac{1}{2}} + 4$ **18.** $129{,}024x^4$ **19.** $70a^4b^4$ **20.** $5103c^5d^2$ **21.** $1{,}088{,}640x^4$ **22.** 336,798 **23.** 126,720 **24.** 495 **25.** 19,779,228

14.5 Exercises

1. S_1: $3 = 3(1)^2$, S_2: $3 + 9 = 3(2)^2$, S_3: $3 + 9 + 15 = 3(3)^2$, yes **2.** S_1: $1^2 = \frac{1(2)(3)}{6}$, S_2: $1^2 + 2^2 = \frac{2(3)(5)}{6}$, S_3: $1^2 + 2^2 + 3^2 = \frac{3(4)(7)}{6}$, yes **3.** S_1: $2 > 1$, S_2: $4 > 4$, S_3: $8 > 9$, false **4.** S_1: 3 is not a factor of 0, S_2, 3 is a factor of 6, S_3: 3 is a factor of 24, false **5.** S_k: $6 + 12 + 18 + \cdots + 6k = 3k(k + 1)$, S_{k+1}: $6 + 12 + 18 + \cdots + 6k + 6(k + 1) = 3(k + 1)(k + 2)$ **6.** S_k: $5 + 11 + 17 + \cdots + (6k - 1) = k(3k + 2)$, S_{k+1}: $5 + 11 + 17 + \cdots + (6k - 1) + (6k + 5) = (k + 1)(3k + 5)$ **7.** S_k: $1^2 + 2^2 + 3^2 + \cdots + k^2 = \frac{k(k+1)(2k+1)}{6}$, S_{k+1}: $1^2 + 2^2 + 3^2 + \cdots + k^2 + (k+1)^2 = \frac{(k+1)(k+2)(2k+3)}{6}$ **8.** S_k: $\frac{1}{1 \cdot 2} + \frac{1}{2 \cdot 3} + \frac{1}{3 \cdot 4} + \cdots + \frac{1}{k(k+1)} = \frac{k}{k+1}$, S_{k+1}: $\frac{1}{1 \cdot 2} + \frac{1}{2 \cdot 3} + \frac{1}{3 \cdot 4} + \cdots + \frac{1}{k(k+1)} + \frac{1}{(k+1)(k+2)} = \frac{k+1}{k+2}$ **9.** Show true for S_1: $6 = 3 \cdot 1(1 + 1) = 6$ true. Assume S_k is true: S_k: $6 + 12 + 18 + \cdots + 6k = 3k(k + 1)$. We must show S_{k+1} is true. Start with the left-hand side of S_{k+1}. $6 + 12 + 18 + \cdots + 6k + 6(k + 1) = 3k(k + 1) + 6(k + 1) = 3k^2 + 3k + 6k + 6 = 3k^2 + 9k + 6 = 3(k + 1)(k + 2)$, so S_{k+1} is true.

10. Show true for S_1: $2 = \frac{1}{2}\cdot 1(1+3) = 2$ true. Assume S_k is true: S_k: $2+3+4+\cdots+(k+1) = \frac{1}{2}k(k+3)$. We must show S_{k+1} is true. Start with the left-hand side of S_{k+1}. $2+3+4+\cdots+(k+1)+(k+2) = \frac{1}{2}k(k+3)+(k+2) = \frac{1}{2}k^2+\frac{3}{2}k+k+2 = \frac{1}{2}k^2+\frac{5}{2}k+2 = \frac{1}{2}(k^2+5k+4) = \frac{1}{2}(k+1)(k+4)$, so S_{k+1} is true. **11.** Show true for S_1: $3 = 1(2\cdot 1+1) = 3$ true. Assume S_k is true: S_k: $3+7+11+\cdots+(4k-1) = k(2k+1)$. We must show S_{k+1} is true. Start with the left-hand side of S_{k+1}. $3+7+11+\cdots+(4k-1)+(4(k+1)-1) = k(2k+1)+(4(k+1)-1) = 2k^2+k+4k+4-1 = 2k^2+5k+3 = (k+1)(2k+3)$, so S_{k+1} is true. **12.** Show true for S_1: $1^2 = \frac{1(1+1)(2+1)}{6} = 1$ true. Assume S_k is true: S_k: $1^2+2^2+3^2+\cdots+k^2 = \frac{k(k+1)(2k+1)}{6}$. We must show S_{k+1} is true. Start with the left-hand side of S_{k+1}. $1^2+2^2+3^2+\cdots+k^2+(k+1)^2 = \frac{k(k+1)(2k+1)}{6}+(k+1)^2 = \frac{2k^3+3k^2+k+6k^2+12k+6}{6} = \frac{2k^3+9k^2+13k+6}{6} = \frac{(k+1)(k+2)(2k+3)}{6}$, so S_{k+1} is true. **13.** Show true for S_1: $3^0 = \frac{1}{2}(3^1-1) = \frac{1}{2}(2) = 1$ true. Assume S_k is true: S_k: $3^0+3^1+3^2+\cdots+3^{k-1} = \frac{1}{2}(3^k-1)$. We must show S_{k+1} is true. Start with the left-hand side of S_{k+1}. $3^0+3^1+3^2+\cdots+3^{k-1}+3^k = \frac{1}{2}(3^k-1)+3^k = \frac{(3^k-1)+2\cdot 3^k}{2} = \frac{3\cdot 3^k-1}{2} = \frac{3^{k+1}-1}{2} = \frac{1}{2}(3^{k+1}-1)$, so S_{k+1} is true. **14.** Show true for S_1: $\frac{1}{1\cdot 2} = \frac{1}{1+1} = \frac{1}{2}$ true. Assume S_k is true: S_k: $\frac{1}{1\cdot 2}+\frac{1}{2\cdot 3}+\frac{1}{3\cdot 4}+\cdots+\frac{1}{k(k+1)} = \frac{k}{k+1}$. We must show S_{k+1} is true. Start with the left-hand side of S_{k+1}. $\frac{1}{1\cdot 2}+\frac{1}{2\cdot 3}+\frac{1}{3\cdot 4}+\cdots+\frac{1}{k(k+1)}+\frac{1}{(k+1)(k+2)} = \frac{k}{k+1}+\frac{1}{(k+1)(k+2)} = \frac{k(k+2)}{(k+1)(k+2)}+\frac{1}{(k+1)(k+2)} = \frac{k^2+2k+1}{(k+1)(k+2)} = \frac{(k+1)^2}{(k+1)(k+2)} = \frac{k+1}{k+2}$, so S_{k+1} is true. **15.** Show true for S_1: $1(1+1)+4 = 6$, which is divisible by 2. Assume S_k: $k(k+1)+4$ is divisible by 2. We must show that S_{k+1} is true. That is, we must show that $(k+1)(k+2)+4$ is divisible by 2. $(k+1)(k+2)+4 = k(k+1)+2(k+1)+4 = k(k+1)+4+2(k+1)$ is divisible by 2 because $k(k+1)+4$ is divisible by 2 by the induction assumption, $2(k+1)$ is divisible by 2 since 2 is a factor, and the sum of two numbers divisible by 2 is also divisible by 2, so S_{k+1} is true. **16.** Show true for S_1: $3^1 \geq 3\cdot 1$ true. Assume S_k is true: S_k: $3^k \geq 3k$. We must show S_{k+1} is true. Start with the left-hand side of S_{k+1}. $3^{k+1} = 3\cdot 3^k = 3^k+3^k+3^k \geq 3k+3k+3k \geq 3k+2+1 = 3k+3 = 3(k+1)$, so S_{k+1} is true.

14.6 Exercises

1. 60 **2.** 120 **3.** 48 **4.** 20,442,240 **5.** 256 **6.** 101,237,760 **7. a.** 60 **b.** 56 **8.** 45 **9.** 36 **10.** 31 **11.** 8 **12.** 24 **13.** 19,683 **14.** 85,266 **15.** 67,716 **16.** 54,461 **17.** 78,540 **18.** 7,311,616 **19.** 1,000,000 **20.** 9,834,496 **21.** 24 **22.** 120 **23.** 40,320 **24.** 2520 **25.** 39,270 **26.** 1716 **27.** 8,814,960 **28.** 34,650 **29.** 2520 **30.** 35 **31.** 2520 **32.** 142,506 **33.** 300 **34.** 66 **35.** 163,185 **36.** 133,784,560 **37.** 120 **38.** 10 **39.** 1 **40.** 22,100 **41.** 286 **42.** 1014 **43.** 133,784,560 **44.** 476,784 **45.** 55,836 **46.** 24 **47.** 120 **48.** 40,320 **49.** 2520 **50.** 39,270 **51.** 1716 **52.** 8,814,960 **53.** 142,506 **54.** 300 **55.** 66 **56.** 163,185 **57.** 133,784,560 **58.** 120 **59.** 10 **60.** 1 **61.** 22,100 **62.** 286 **63.** 1014 **64.** 133,784,560 **65.** 476,784 **66.** 55,836 **67.** 24 **68.** 120 **69.** 40,320 **70.** 2520 **71.** 39,270 **72.** 1716 **73.** 8,814,960 **74.** 142,506 **75.** 300 **76.** 66 **77.** 163,185 **78.** 133,784,560 **79.** 120 **80.** 10 **81.** 1 **82.** 22,100 **83.** 286 **84.** 1014 **85.** 133,784,560 **86.** 476,784 **87.** 55,836

14.7 Exercises

1. 60 **2.** 20,000 **3.** 16 **4.** 40 **5.** 22,100 **6.** 32 **7.** 35,960 **8.** $\frac{1}{4}$ **9. a.** $\frac{1}{5}$ **b.** $\frac{2}{5}$ **10.** $\frac{18}{37}$ **11.** $\frac{18}{37}$ **12.** $\frac{1}{37}$ **13.** $\approx 20.2\%$ **14.** $\approx 93.0\%$ **15.** $\approx 25.7\%$ **16.** $\approx 56.0\%$ **17.** $\approx 86.2\%$ **18.** 0.4 **19.** 0.2

20. 0.5 **21.** 0.6 **22.** $\frac{7}{13}$ **23.** $\frac{11}{26}$ **24.** $\frac{153}{251}$ **25.** $\frac{138}{251}$ **26.** $\frac{52}{251}$ **27.** $\frac{239}{251}$ **28.** $\frac{98}{251}$ **29.** $\frac{113}{251}$ **30.** $\frac{12}{251}$ **31.** $\frac{199}{251}$ **32.** $\frac{139}{200}$ **33.** $\frac{8}{25}$ **34.** $\frac{61}{200}$ **35.** $\frac{29}{200}$ **36.** $\frac{4}{25}$ **37.** $\approx 8.3\%$ **38.** $\approx 0.6\%$ **39.** $\approx 83.9\%$ **40.** $\approx 16.1\%$ **41.** $\frac{6}{209}$ **42.** $\frac{15}{209}$ **43.** $\frac{9}{133}$ **44.** $\frac{141}{209}$ **45.** $\approx 34.4\%$ **46.** $\approx 74.7\%$ **47.** $\frac{144}{169}$ **48.** $\frac{360}{497}$ **49.** 0.5 **50.** ≈ 0.33 **51.** $\frac{12}{51}$ **52.** $\frac{126}{137}$ **53.** $\frac{60}{143}$ **54.** $\frac{400}{1001}$ **55.** 4 : 271 **56.** 12 : 11 **57.** 11 : 12 **58.** 1 : 4 **59.** 8 : 7 **60.** 18 : 19 **61.** 18 : 19 **62.** 36 : 1 **63.** $\approx 8.3\%$ **64.** $\approx 0.6\%$ **65.** $\approx 83.9\%$ **66.** $\approx 16.1\%$ **67.** $\frac{6}{209}$ **68.** $\frac{15}{209}$ **69.** $\frac{9}{133}$ **70.** $\frac{141}{209}$ **71.** $\approx 34.4\%$ **72.** $\approx 74.7\%$ **73.** $\frac{126}{137}$ **74.** $\frac{60}{143}$ **75.** $\frac{400}{1001}$ **76.** $\approx 8.3\%$ **77.** $\approx 0.6\%$ **78.** $\approx 83.9\%$ **79.** $\approx 16.1\%$ **80.** $\frac{6}{209}$ **81.** $\frac{15}{209}$ **82.** $\frac{9}{133}$ **83.** $\frac{141}{209}$ **84.** $\approx 34.4\%$ **85.** $\approx 74.7\%$ **86.** $\frac{126}{137}$ **87.** $\frac{60}{143}$ **88.** $\frac{400}{1001}$

Review Exercises

1. 30, 120, 600, 3600 **2.** $a_n = (-1)^n \cdot 2^n, a_7 = -128$ **3.** $\frac{11}{32}$ **4.** $\sum_{i=1}^{8} (-1)^{i+1} \cdot 2^i$ **5.** $a_n = 5n - 3, a_{10} = 47$ **6.** -64 **7.** 235 **8.** 3600 **9.** 162, 54, 18, 6, 2 **10.** 20,480 **11.** ≈ 4.99 **12.** converges, $\frac{28}{3}$ **13.** 210 **14.** $3125x^5 - 9375x^4y + 11250x^3y^2 - 6750x^2y^3 + 2025xy^4 - 243y^5$ **15.** $1{,}088{,}640x^4$ **16.** 126,720 **17.** Show true for S_1: $2 = \frac{1}{2} \cdot 1(1 + 3) = 2$ true. Assume S_k is true: S_k: $2 + 3 + 4 + \cdots + (k + 1) = \frac{1}{2}k(k + 3)$. We must show S_{k+1} is true. Start with the left-hand side of S_{k+1}. $2 + 3 + 4 + \cdots + (k + 1) + (k + 2) = \frac{1}{2}k(k + 3) + (k + 2) = \frac{1}{2}k^2 + \frac{3}{2}k + k + 2 = \frac{1}{2}k^2 + \frac{5}{2}k + 2 = \frac{1}{2}(k^2 + 5k + 4) = \frac{1}{2}(k + 1)(k + 4)$, so S_{k+1} is true. **18.** 60 **19.** 142,506 **20.** 10 **21.** 1014 **22.** $\frac{1}{4}$ **23.** $\approx 20.2\%$ **24.** 0.4 **25.** $\frac{141}{209}$

APPENDIX A

A.1 Exercises

1. 40°31′12″ **2.** 73°20′42″ **3.** −13°16′23″ **4.** 4.05° **5.** 59.52° **6.** −57.37°

Glossary

Acronym An acronym is a word (real or made up) formed by abbreviating the name of an organization or a multiword statement.

Acute Angles An acute angle is an angle with measure between 0 and 90 degrees.

Algebraic Expression An algebraic expression consists of one or more terms that may include variables, constants, and operating symbols such as + and −.

Alternate Form of the Dot Product of Two Vectors If $\mathbf{u}$ and $\mathbf{v}$ are non-zero vectors, then the alternate form of the dot product is $\mathbf{u} \cdot \mathbf{v} = \|\mathbf{u}\| \|\mathbf{v}\| \cos\theta$.

Alternate Form of the Law of Cosines If A, B, and C are the measures of the angles of any triangle and if a, b, and c are the lengths of the sides opposite the corresponding angles, then $\cos A = \frac{b^2 + c^2 - a^2}{2bc}$, $\cos B = \frac{a^2 + c^2 - b^2}{2ac}$, and $\cos C = \frac{a^2 + b^2 - c^2}{2ab}$.

Altitude An altitude of a triangle is a line segment from a vertex of a triangle perpendicular to the (possibly extended) opposite side. The length of the altitude is the distance between the vertex and the base.

Argument The argument of a function is the variable, term, or expression on which a function operates.

Arithmetic Sequence An arithmetic sequence is a sequence of the form $a_1, a_1 + d, a_1 + 2d, a_1 + 3d, a_1 + 4d, \ldots$, where a_1 is the first term of the sequence and d is the common difference. The general term, or nth term, of an arithmetic sequence has the form $a_n = a_1 + (n - 1)d$.

Arithmetic Series An arithmetic series is of the form $a_1 + (a_1 + d) + (a_1 + 2d) + \cdots + (a_1 + (n - 1)d) + \cdots$, where d is called the common difference.

Associative Property of Addition of Matrices If matrices A, B, and C have the same size, then $(A + B) + C = A + (B + C)$. In other words, because we can only add matrices two at a time, it does not matter which two we add together first.

Axis of Symmetry An axis of symmetry is an invisible line that divides the graph into two equal halves. An ellipse has a horizontal axis of symmetry and a vertical axis of symmetry.

Binomial A binomial is a polynomial expression consisting of two terms.

Bisect Two line segments bisect each other if the intersection of the segments divides each segment into two parts of equal length.

Boundary Line A boundary line is a line that separates the ordered pair solutions to a linear inequality in two variables from the ordered pairs that are not solutions to the inequality.

Boundary Point A boundary point of a polynomial inequality $p < 0$ is a number for which $p = 0$.

Carrying Capacity The carrying capacity is the maximum population size that can be regularly sustained by an environment.

Central Angle An angle whose vertex is at the center of a circle is called a central angle.

Closed Interval A closed interval is an interval of real numbers that include two endpoints.

Coefficient A coefficient is a constant of an algebraic expression or equation that stands alone or is multiplied by a variable.

Coincide Two or more planes are said to coincide if each plane is the same representation of the other planes. Planes that coincide, therefore, have infinitely many points in common.

Collinear Ordered pairs (or points) in the rectangular coordinate system are said to be collinear if they all lie on the same line.

Column Matrix A column matrix is a matrix consisting of m rows and one column.

Combined Variation When a variable is related to more than one other variable, we call this combined variation.

Common Logarithmic Function For $x > 0$, the common logarithmic function is defined by $y = \log x$ if and only if $x = 10^y$.

Commutative Property The commutative property is the algebraic property that states that a change in order produces the same result. Addition and multiplication are commutative because for any real numbers a and b, $a + b = b + a$ and $ab = ba$. Subtraction is *not* commutative.

Complement The complement of an event E is the set of all outcomes in the sample space that are not included in the outcomes of event E. The notation used for the complement of E is E'.

Complement Rule Given the probability of an event E written as a fraction or a decimal, the probability of its complement can be found by subtracting the given probability from 1.

Complementary Angles Two angles are called complementary angles if the sum of their measures is $90°$ (or $\frac{\pi}{2}$ radians).

Completely Factored Form The polynomial function $f(x) = a_n x^n + a_{n-1}x^{n-1} + a_{n-2}x^{n-2} + \cdots + a_1 x + a_0$, where $n \geq 1$ $a_n \neq 0$ is said to be in completely factored form if it is written in the form $f(x) = a_n(x - c_1)(x - c_2)(x - c_3) \cdots (x - c_n)$, where $c_1, c_2, c_3, \ldots c_n$ are complex numbers and not necessarily distinct. Note that in each factor of the form $(x - c)$, the coefficient of x is 1.

Complex Conjugate The complex conjugate of a complex number $a + bi$ is $a - bi$.

Complex Zero A complex zero is a zero of a polynomial $f(x)$ of the form $a + bi$, where a and b are real numbers.

Concentric Circles Concentric circles are circles that share a common center. When concentric circles intersect, their points of intersection form a hyperbola.

Constant A function f is constant on an interval (a, b) if, for any x_1 and x_2 chosen from the interval, $f(x_1) = f(x_2)$. The graph of a constant function is represented by a horizontal line segment.

Constant of Variation The constant of variation, or proportionality constant, is the constant ratio of two quantities that are directly related, or the constant product of two quantities that are inversely related.

Contradiction A contradiction (equation or inequality) is always false, regardless of the value of the variable, if any.

Coordinate Plane The coordinate plane is the plane represented by the rectangular coordinate system. It is also known as the Cartesian plane or the xy-plane.

Coterminal Angles Angles in standard position having the same terminal side are called coterminal angles.

Commutative Property of Matrices If matrices A and B have the same size, then $A + B = B + A$. In other words, the order in which we add matrices does not matter.

Decreasing A function f is decreasing on an interval (a, b) if, for any x_1 and x_2 chosen from the interval with $x_1 < x_2$, then $f(x_1) > f(x_2)$.

Dependent System A system of linear equations is dependent if all solutions of one equation are also solutions of the other equations. A dependent system has infinitely many solutions.

Dependent Variable A dependent variable is the variable whose value is determined by the value assumed by an independent variable.

Difference of Two Squares If x and y are any two algebraic expressions, then $x^2 - y^2$ is called a difference of two squares.

Element An element of a cone is a straight line segment joining the outer edge of the base of the cone with the vertex of the cone. The set of all elements creates the surface of a cone.

Ellipse An ellipse is the set of all points in a plane, the sum of whose distances from two fixed points is a positive constant. The two fixed points, $F1$ and $F2$, are called the *foci*.

Equation of a Circle Every equation of a circle can be written in general form. However, not every equation of the form $Ax^2 + By^2 + Cx + Dy + E = 0$ has a graph that is a circle.

Equation of a Circle in Standard Form Centered at the Origin A circle with radius r centered at the origin has equation $x^2 + y^2 = r^2$.

Equivalent System Two systems of linear equations are equivalent if and only if they have the same solution set.

Even A function $y = f(x)$ is even if for every x in the domain, $f(-x) = f(x)$. Even functions are symmetric about the y-axis.

Even Functions A function f is even if $f(-x) = f(x)$ for all x in the domain of f.

Exponential Function An exponential function is of the form $f(x) = b^x$, where $b > 0$ and $b \neq 1$. The domain is the set of all real numbers, whereas the range is the set of all positive real numbers. The constant, b, is called the base of the exponential function.

Extraneous Solution An extraneous solution is a solution obtained through algebraic manipulations that is *not* a solution to the original equation.

Fact 1 for Ellipses The sum of the distances from any point on the ellipse to the two foci is equal to $2a$, the length of the major axis.

Fact 1 for Hyperbolas Given the foci of a hyperbola F_1 and F_2 and any point P that lies on the graph of the hyperbola, the difference of the distances between P and the foci is equal to $2a$. In other words, $|d(p, F_1) - d(p, F_2)| = 2a$. The constant $2a$ represents the length of the transverse axis.

Factor A factor is one of two or more expressions that are multiplied together to form a product.

Factorial of a Non-Negative Integer The factorial of a non-negative integer n, denoted as $n!$, is the product of all positive integers less than or equal to n.

Four Quadrants The axes of a rectangular coordinate system divide the rectangular coordinate system into four quadrants.

Fundamental Counting Principle The Fundamental Counting Principle can be demonstrated in an experiment or task made up of two or more events. If the first event can occur in n_1 ways, the second event can occur in n_2 ways, and the third event can occur in n_3 ways and so on, then the number of possible outcomes of the experiment or task is $N = n_1 n_2 \ldots n_k$, where k is the total number of events.

Geometric Sequence A geometric sequence is a sequence of the form $a_1, a_1 r, a_1 r^2, a_1 r^3, a_1 r^4, \ldots$, where a_1 is the first term of the sequence and r is the common ratio such that $r = \frac{a_{n+1}}{a_n}$ for all $n \geq 1$. The general term, or nth term, of a geometric sequence has the form $a_n = a_1 r^{n-1}$.

Geometric Series If $a_1, a_1 r, a_1 r^2, a_1 r^3, \ldots$ is a geometric sequence, then $a_1 + a_1 r + a_1 r^2 + a_1 r^3 + \cdots + a_1 r^{n-1} + \cdots$ is called a geometric series.

Graph of an Equation The graph of an equation is the set of all ordered pairs (x, y) whose coordinates satisfy the equation.

Greatest Common Factor (GCF) The greatest common factor (GCF) is the largest number or algebraic expression that divides two or more algebraic expressions evenly.

Horizontal Asymptote A horizontal line $y = H$ is a horizontal asymptote of a function f if the values of $f(x)$ approach some fixed number H as the values of x approach ∞ or $-\infty$.

Horizontal Axis of Symmetry The horizontal axis of symmetry of a parabola is the invisible horizontal line that divides the parabola into two equal parts.

Horizontal Line Test Given the graph of a function f, if every horizontal line intersects the graph of f at most once, then f is a one-to-one function.

Horizontal Major Axis An ellipse has a horizontal major axis if the horizontal axis of symmetry passes through the two foci and is longer than the vertical axis of symmetry.

Identity An identity (equation or inequality) is always true, regardless of the value of the variable, if any.

Identity Matrix An identity matrix is a square matrix where all entries down the diagonal are 1's with zeros everywhere else.

Increasing A function f is increasing on an interval (a, b) if, for any x_1 and x_2 chosen from the interval with $x_1 < x_2$, then $f(x_1) < f(x_2)$.

Independent Variable An independent variable is the variable whose value determines the value of the other variables.

Index The index of a radical expression $\sqrt[n]{b}$ is the integer n.

Intercepts The intercepts of a graph are points where a graph crosses or touches a coordinate axis.

Inverse Cosine Function The inverse cosine function, denoted as $y = \cos^{-1} x$ is the inverse of $y = \cos x,\ 0 \le x \le \pi$.

Inverse Function Let f be a one-to-one function with domain A and range B. Then f^{-1} is the inverse function of f with domain B and range A.

Inverse Sine Function The inverse sine function, denoted as $y = \sin^{-1} x$ is the inverse of $y = \sin x,\ -\frac{\pi}{2} \le x \le \frac{\pi}{2}$.

Inverse Tangent Function The inverse tangent function, denoted as $y = \tan^{-1} x$ is the inverse of $y = \tan x, -\frac{\pi}{2} \le x \le \frac{\pi}{2}$.

Irrational Number An irrational number is a real number that *cannot* be expressed as the quotient of two integers $\frac{p}{q}$. Irrational numbers have an infinite, nonrepeating decimal representation.

Irrational Solution An irrational solution is a real number solution to an equation that *cannot* be expressed as the quotient of two integers.

Irrational Zero The value $x = r$ is an irrational zero of f if $f(r) = 0$ and if r *cannot* be written as the quotient of two integers.

Joint Variation A special case of combined variation occurs when a variable is directly proportional to the product of two or more other variables. This is called joint variation.

Law of Cosines If A, B, and C are the measures of the angles of any triangle and if a, b, and c are the lengths of the sides opposite the corresponding angles, then $a^2 = b^2 + c^2 - 2bc \cos A$, $b^2 = a^2 + c^2 - 2ac \cos B$, and $c^2 = a^2 + b^2 - 2ab \cos C$.

Law of Sines If A, B, and C are the measures of the angles of any triangle and if a, b, and c are the lengths of the sides opposite the corresponding angles, then $\frac{a}{\sin A} = \frac{b}{\sin B} = \frac{c}{\sin C}$ or $\frac{\sin A}{a} = \frac{\sin B}{b} = \frac{\sin C}{c}$.

Leading Coefficient The leading coefficient of a polynomial function is the constant that is multiplied by the variable of highest power.

Least Common Denominator (LCD) The least common denominator is the smallest multiple of all denominators in an algebraic expression or equation.

Linear Equation in n Variables A linear equation in n variables is an equation that can be written in the form $a_1x_1 + a_2x_2 + \cdots + a_nx_n = b$ for variables $x_1, x_2, \ldots x_n$ and real numbers $a_1a_2, \ldots a_n, b$, where at least one of $a_1, a_2, \ldots a_n$ is nonzero.

Linear Equation A linear equation in one variable is an equation that can be written in the form $ax + b = 0$, where a and b are real numbers and $a \ne 0$.

Linear Function A first-degree polynomial function of the form $f(x) = mx + b$ is called a linear function.

Linear Inequality in One Variable A linear inequality in one variable is an inequality that can be written in the form $ax + b < c$, where a, b, and c are real numbers and $a \ne 0$.

Logarithm Property of Equality The logarithm property of equality is if a logarithmic equation can be written as $\log_b u = \log_b v$, then $u = v$.

Magnitude of a Vector The magnitude of a vector of the form $\mathbf{v} = a\mathbf{i} + b\mathbf{j}$ is $\|\mathbf{v}\| = \sqrt{a^2 + b^2}$.

Method of Elimination The method of elimination is a method used to solve linear equations by adding a multiple of an equation to a multiple of another equation in the attempt to eliminate one of the variables.

Minor Axis The minor axis of an ellipse is the shorter of the two axes of symmetry.

Monomial Function A monomial function is a polynomial function consisting of a single term.

Mutually Exclusive Events Two or more events are mutually exclusive if they have no outcomes in common.

Natural Numbers The set of natural numbers, sometimes referred to as the *counting numbers,* is defined as $N = \{1, 2, 3, \ldots\}$.

Noncollinear Points Three points are said to be noncollinear if they do not all lie on the same line.

Non-Strict Inequality A non-strict inequality contains one or both of the following inequality symbols: $\le, \ge$. The possibility of equality is included in a non-strict inequality.

Number e The number e, called the *natural base,* is an irrational number. The number e rounded to 6 decimal places is $e \approx 2.718282$.

Obtuse Angles An obtuse angle is an angle with measure between 90 and 180 degrees.

Odd Functions A function f is odd if $f(-x) = -f(x)$ for all x in the domain of f.

One-to-One Function A function f is one-to-one if for any values $a \ne b$ in the domain of f, $f(a) \ne f(b)$.

Open Interval An open interval, denoted by the symbol (a, b), consists of all real numbers between a and b not including a and b.

Origin The origin is the point in the Cartesian plane where the x-axis and y-axis intersect. The origin has coordinates $(0, 0)$.

Parabola A parabola is the set of all points in a plane equidistant from a fixed point F and a fixed line D. The fixed point is called the focus, and the fixed line is called the directrix.

Parallelogram A parallelogram is a quadrilateral (four-sided polygon) in which opposite sides are parallel.

Perfect nth Power A perfect nth power is a rational number that can be written as the nth power of another rational number where n is a positive integer.

Periodic Compound Interest Formula Periodic compound interest can be calculated using the formula $A = P\left(1 + \frac{r}{n}\right)^{nt}$.

Periodic Function A function is said to be periodic if there is a positive number P such that $f(x + P) = f(x)$ for all x in the domain of f.

Point-Slope Form of the Equation of a Line Given the slope of a line m and a point on the line (x_1, y_1), the point-slope form of the equation of a line is given by $y - y_1 = m(x - x_1)$.

Polynomial Expression An algebraic expression in which the exponents of all variable factors are nonnegative integers is called a polynomial expression.

Polynomial Function The function $f(x) = a_nx^n + a_{n-1}x^{n-1} + a_{n-2}x^{n-2} + \cdots + a_1x + a_0$ is a polynomial function of degree n, where n is a nonnegative integer. The numbers $a_0, a_1, a_2, \ldots, a_n$ are called the coefficients of the polynomial function. The number a_n is called the leading coefficient, and a_0 is called the constant coefficient.

Polynomial Inequality A polynomial inequality is an inequality that can be written in the form $a_nx^n + a_{n-1}x^{n-1} + \cdots + a_1x + a_0 < 0$, where $a_0, a_1, \ldots, a_n$ are real numbers, and each exponent is an integer greater than or equal to 0.

Polynomial in Standard Form A polynomial in one variable is written in standard form when it is written in descending order with the term of highest power written first.

Present Value Formula Present value can be calculated using the formula $P = A\left(1 + \frac{r}{n}\right)^{-nt}$.

Prime Polynomial A prime polynomial is a term that indicates that a polynomial cannot be factored using *integer* coefficients. It is possible for a prime polynomial to be factored using rational or irrational coefficients.

Proportion A proportion is an equation involving two ratios in which the first ratio is equal to the second ratio.

Pythagorean Theorem Pythagorean Theorem states that given any right triangle, the sum of the squares of the lengths of the legs is equal to the square of the length of the hypotenuse.

Quadrantal Angle A quadrantal angle is an angle in standard position whose terminal side lies on an axis.

Quadratic Equation A quadratic equation is an equation that can be written in the form $ax^2 + bx + c = 0$, where a, b, and c are real numbers and $a \neq 0$.

Radical Conjugate The expression $\sqrt{5} - \sqrt{2}$ is called the radical conjugate of the expression $\sqrt{5} + \sqrt{2}$. The product of radical conjugates always yields an expression that does not contain a radical.

Radicand A radicand is the number, variable, or algebraic expression under the radical.

Range of a Parabola The range is the set of all values for the dependent variable. These are the second coordinates in the set of ordered pairs and are also known as *output values*.

Ratio A ratio is a comparison between two numbers.

Rational Equation A rational equation is an equation involving the quotient of two polynomial expressions where the polynomial in the denominator is non-constant.

Rational Expression A rational expression is the quotient of two polynomials.

Rational Function A rational function is of the form $f(x) = \frac{g(x)}{h(x)}$, where $g(x)$ and $h(x)$ are polynomial functions and $h(x) \neq 0$. The domain of a rational function is the set of all real numbers such that $h(x) \neq 0$.

Rational Inequality A rational inequality is an inequality that can be written in the form $\frac{p}{q} \geq 0$, where p and q are polynomials and $q \neq 0$.

Rational Number A rational number is a real number that can be expressed as the quotient of two integers $\frac{p}{q}$, where $q \neq 0$. All finite decimals and repeating decimals are rational numbers.

Rational Zero A rational zero is a real zero of a polynomial $f(x)$ of the form $\frac{p}{q}$, where p and q are integers and $q \neq 0$.

Rational Zeros Theorem Let f be a polynomial function of the form $f(x) = a_nx^n + a_{n-1}x^{n-1} + a_{n-2}x^{n-2} + \cdots + a_1x + a_0$ of degree $n \geq 1$, where each coefficient is an integer. If $\frac{p}{q}$ is a rational zero of f (where $\frac{p}{q}$ is written in lowest terms), then p must be a factor of the constant coefficient, a_0, and q must be a factor of the leading coefficient a_n.

Rationalizing a Denominator The process of removing radicals so that the denominator contains only a rational number is called rationalizing the denominator.

Ray A ray consists of an endpoint called the vertex and a half-line extending indefinitely in one direction.

Real Number A real number can be either a rational or irrational number. Rational numbers include integers, whole numbers, and natural numbers. Irrational numbers include decimals that neither repeat nor terminate.

Real Zero A real number $x = c$ is a real zero of a function f if $f(c) = 0$. Real zeros are also x-intercepts.

Reciprocal A reciprocal is a number or algebraic expression related to another number or algebraic expression in such a way that when multiplied together their product is 1.

Recursive Sequence A recursive sequence is a sequence in which each term is defined using one or more previous terms.

Reduced Row-Echelon Form An augmented matrix is said to be in reduced row-echelon form when all zeros appear above and below the diagonal *and* 1's appear down the diagonal.

Relative Maximum When a function $y = f(x)$ changes from increasing to decreasing at a point $(c, f(c))$, then f is said to have a relative maximum at $x = c$.

Relative Minimum When a function $y = f(x)$ changes from decreasing to increasing at a point $(c, f(c))$, then f is said to have a relative minimum at $x = c$.

Right Angle A right angle is an angle that has a measure of exactly $\frac{\pi}{2}$ radians (or $90°$).

Right Circular Cone A right circular cone is a cone that has a vertex directly above the center of a circular base.

Row Matrix A row matrix is a matrix consisting of one row and n columns.

Row-Echelon Form An augmented matrix is said to be in row-echelon form when all zeros appear under the diagonal *and* 1's appear down the diagonal.

Simple Interest The formula for simple interest is $I = Prt$, where $P =$ principal amount invested, $r =$ interest rate, and $t =$ time (in years).

Singular Matrix A square matrix is singular if it does not have an inverse matrix.

Slope-Intercept Form of a Line Given the slope of a line m and the y-intercept b, the slope-intercept form of the equation of a line is given by $y = mx + b$.

Square Function We can sketch the square function $f(x) = x^2$ by setting up a table of values and plotting points. The square function assigns to each number in the domain the square of that number.

Square Matrix A square matrix is a matrix that contains the same number of rows as columns.

Square Root Function We can sketch the square root function $f(x) = \sqrt{x}$ by setting up a table of values and plotting points. The square root function assigns to each number in the domain the square root of that number.

Square Root Property The solution to the quadratic equation $x^2 - c = 0$ or equivalently $x^2 = c$ is $x^2 = \pm\sqrt{c}$.

Standard Form A polynomial in one variable is written in standard form when it is written in descending order with the term of highest power written first.

Standard Form Equation of a Line The standard form of an equation of a line is given by $Ax + By = C$, where A, B, and C are real numbers such that and are both not zero.

Standard Position An angle is in standard position if the vertex is at the origin of a rectangular coordinate system and the initial side of the angle is along the positive x-axis.

Strict Inequality A strict inequality contains one or more of the following inequality symbols: $<, >, \neq$. There is no possibility of equality in a strict inequality.

Subset A subset is a set contained within a set. A subset can be the set itself. The subset of an interval must contain points within the interval.

Symmetry About the Origin The unit circle is symmetric about the origin. Thus, for any point (a, b) that lies on the graph, the point $(-a, -b)$ must also lie on the graph.

Symmetry About the x-Axis The unit circle is symmetric about the x-axis. Thus, for any point (a, b) that lies on the graph, the point $(a, -b)$ must also lie on the graph.

Symmetry About the y-Axis The unit circle is symmetric about the y-axis. Thus, for any point (a, b) that lies on the graph, the point $(-a, b)$ must also lie on the graph.

System of Linear Equations in Two Variables A system of linear equations in two variables is the collection of two linear equations considered simultaneously. The solution to a system of equations in two variables is the set of all ordered pairs for which *both* equations are true.

Test Value of a Polynomial or Rational Inequality After using the boundary points to divide the number line into intervals, a test value is a number chosen from each interval to test whether the inequality is positive or negative on the interval.

Test Value of a Polynomial Function When sketching the graph of a polynomial function, a test value is an x-value chosen in such a way as to help complete the graph of a function.

Transverse Axis The transverse axis of a hyperbola is the imaginary line segment whose endpoints are the vertices of the hyperbola.

Triangular Form A system of three linear equations is said to be in triangular form when the leading coefficient of the first variable in the second equation is zero and the coefficients of the first two variables of the third equation are zero.

Turning Points The turning points on the graph of a polynomial function are the points at which the graph changes from increasing to decreasing or from decreasing to increasing.

Unit Vector in the Same Direction of a Given Vector Given a nonzero vector $\mathbf{v} = a\mathbf{i} + b\mathbf{j}$, the unit vector $\mathbf{u}$ in the same direction as $\mathbf{v}$ is $\mathbf{u} = \frac{\mathbf{v}}{\|\mathbf{v}\|}$.

Vertex of a Parabola The vertex of a parabola is the lowest point (if the parabola opens *up*) or the highest point (if the parabola opens *down*) of the graph of the parabola.

Vertical Asymptote A vertical line $x = a$ is a vertical asymptote of a function f if the values of $f(x)$ approach ∞ or $-\infty$ as the values of x approach a.

Vertical Line Test A graph in the Cartesian plane is the graph of a function if and only if no vertical line intersects the graph more than once.

Vertical Major Axis An ellipse has a vertical major axis if the vertical axis of symmetry passes through the two foci and is longer than the horizontal axis of symmetry.

x-Intercept An x-intercept is the x-coordinate of a point where a graph crosses or touches the x-axis. (The y-coordinate is 0.)

y-Intercept A y-intercept is the y-coordinate of a point where a graph crosses or touches the y-axis. (The x-coordinate is 0.)

Zero A zero of a function is a value of x for which $f(x) = 0$.

Zero of a Function A zero of a function $f(x)$ is a value of x for which $f(x) = 0$.

Index

B

C

E

F

G

L

M

N

O

P

Q

R

S

U

V

Credits

Chapter R

R.1-1 Beer5020/Shutterstock

Chapter 1

1.1-1 supergenijalac/Shutterstock

Chapter 2

2.1-1 Luciano Mortula-LGM/Shutterstock

Chapter 3

3.1-1 Hugo Felix/123RF

Chapter 4

4.1-1 Andrii IURLOV/123RF

Chapter 5

5.1-1 alysta/Shutterstock

Chapter 6

6.1-1 ssuaphotos/Shutterstock

6.2-14 NASA

6.2-21 NASA

Chapter 7

7.1-1 Antony McAulay/Shutterstock

Chapter 8

8.1-1 Narathorn Kimprasoot/123RF

Chapter 9

9.1-1 iriana88w/123RF

Chapter 10

10.1-1 Eric Gevaert/123RF

Chapter 11

11.1-1 vlastas/Shutterstock

Chapter 12

12.1-1 sirtravelalot/Shutterstock

Chapter 13

13.1-1 Corina Rosu/123RF

Chapter 14

14.1-1 stuartburf/123RF